ISBN 978-0-265-97403-2
PIBN 10917921

CANADIAN HARDWARE AND METAL MERCHANT

The Weekly Organ of the Hardware, Metal, Heating, Plumbing and Contracting Trades in Canada.

VOL. XIII. MONTREAL AND TORONTO, JANUARY 5, 1901. NO. 1

The Demand for our

IMPERIAL OXFORD

kitchen range increases daily.

Every one sold causes other sales, so enthusiastic are householders over its improved features.

ITS EASE OF REGULATION
ECONOMY IN USING FUEL
DIFFUSIVE FLUE CONSTRUCTION
FRONT DRAW-OUT GRATE
OVEN THERMOMETER
DRAW-OUT OVEN RACK

and other conveniences, give it a superiority quickly appreciated.

You're sure of speedy sales when handling the Imperial Oxford. Write us for full information and price list if you have none in stock.

THE GURNEY FOUNDRY CO., Limited

TORONTO. WINNIPEG. VANCOUVER.

THE GURNEY-MASSEY CO., LIMITED, MONTREAL.

Steel Frame.

David Maxwell & Sons

ST. MARYS, ONT., CANADA

"Maxwell Favorite Churn"

PATENTED FEATURES: Improved Steel Stand, Roller Bearings, and Foot and Hand Lever Drive, and Detachable Driving Link Improved for season of 1901. Steel or Wood Frame as desired.

Wheelbarrows.

In Four different Sizes.

Lawn Mowers.

High and Low Wheels, from 12-in. to 20-in. widths. Cold Rolled Steel Shafting, Crucible Steel Knives and Cutting Plate.

If your Wholesale House does not offer you these articles

...SEND DIRECT TO US.

MAXWELL MOWER

8-inch Low Wheel.

"THE MAXWELL"

Lawn Mower

High Wheel 10 inches

OUR NEW YEAR WISHES.

We wish to convey to the hardware trade and every purchaser of our "**C**" brand horse shoe nails in Canada, our sincere appreciation of their orders during the past year.

Our aim shall be in the year 1901 and the century upon which we have now entered, to adhere to our high standard for quality and to spare no efforts on our part to maintain the position accorded to the "**C**" brand for the past 35 years, as being the best horse shoe nail in Canada and without a superior anywhere.

The quality is the best; the workmanship is the best; the patterns are the best, and they sell the best, and are the best for you to buy. The best resolution you can make for the year 1901 is to sell the best horse nails only, in which case your trade is ours.

CANADA HORSE NAIL COMPANY

MONTREAL.

The Annual Special Number of HARDWARE AND METAL will be issued March 9th this year, about two weeks earlier than corresponding issues in the past.

HARDWARE AND METAL

VOL. XIII. MONTREAL AND TORONTO, JANUARY 5, 1901. NO. 1.

President,
JOHN BAYNE MacLEAN,
Montreal.

THE MacLEAN PUBLISHING CO.
Limited.

Publishers of Trade Newspapers which circulate in the Provinces of British Columbia, North-West Territories, Manitoba, Ontario, Quebec, Nova Scotia, New Brunswick, P.E. Island and Newfoundland.

OFFICES

MONTREAL - - - - Board of Trade Building.
Telephone 1255.
TORONTO - - - - - - 10 Front Street East.
Telephone 2148.
LONDON, ENG. - - - - 109 Fleet Street, E.C.,
J. M. McKim,
MANCHESTER, ENG. - - - 18 St Ann Street.
H. S. Ashburner,
WINNIPEG - - - - Western Canada Block.
J. J. Roberts.
ST. JOHN, N. B. - - - No. 3 Market Wharf.
J. Hunter White,
NEW YORK. - - - - - - - 176 E. 88th Street,
W. J. Brandt.

Travelling Subscription Agents:
T. Donaghy. F. S. Millard.

Subscription, Canada and the United States, $2.00.
Great Britain and elsewhere - - - 12s.

Published every Saturday.

Cable Address { Adscript, London
Adscript, Canada.

THE ST. LAWRENCE LLOYDS.

THE fruits of the agitation that has been going on for some time for the lowering of marine insurance on vessels trading in the St. Lawrence channel, Canada's national waterway, are at last appearing.

Mr. Louis Boyer, of the legal firm of Dandurand, Brodeur & Boyer, Montreal, has given notice in the official Gazette of application for an Act to incorporate "The St. Lawrence Lloyds" for the purpose of carrying on an ocean and inland marine insurance business with the right to maintain and navigate ice-breaking and wreck-relieving steamers on the St. Lawrence River.

It is said that the idea of the formation of the new company arose out of the departure from Montreal, in the last days of November, of four vessels, belonging to the Algoma Central Railway Co., carrying steel rails, without any insurance. Mr. F. H. Clergue, of Sault Ste. Marie, came to Montreal at the time and found, to his extreme annoyance, that he could place no insurance on the boats. He immediately set to work and founded a "St. Lawrence Lloyds" by the aid of Montreal financiers. The capital of the company is set at $5,000,000. By this means the organizers hope to solve the difficulty of Canadian marine insurance rates.

The same company hope to keep navigation of the St. Lawrence open a few weeks longer each fall by the employment of a powerful ice-breaker, many plans of which have been before Montreal business men during the last few months.

Failure to accomplish all that was desired last year should excite, and not retard, the efforts in this, the New Year.

CONSOLIDATED COMPANIES.

"The Consolidations and Listed Stock Companies of the Iron and Allied Trades" is the title of a supplement to The Iron Age. It consists of 56 large pages, and gives the capital stock, the productive capacity, names of directors, etc., of the large combinations allied to the iron and steel trades which have been formed in the United States during the last few years. The supplement is a valuable contribution to the subject with which it deals.

A GOOD YEAR IN TORONTO.

THE past year has been a satisfactory one to all connected with building operations in Toronto. The report of the city commissioner for the year shows that permits have been taken out for the erection of buildings with a total value of $1,957,274, against $2,010,996 the previous year, showing a decrease of about $50,000. This, in view of the fact that 1899 was the best year since 1892, and that it was equal to 1896 and 1897 combined, shows a lot of building was done in Toronto last year.

But not only has the volume of trade done being larger, but it has been more remunerative than usual. The high range of prices for materials and the high wages paid had not been followed up by contractors as they might have been in previous years. Then the talk of a strike and the consequent discussion last spring as to the cost of materials spread abroad enough information to make householders expect high prices, and make it easier for contractors to get them.

A noticeable fact is that there is no falling off in the erection of buildings for industrial and commercial purposes. There has been a slight decrease in the number of residences erected, put permits to the value of $988,000 have been taken for warehouses, offices and factories, as compared with $724,000 the previous year. The extension of these works is all the more satisfactory in view of the fact that there will thus be given employment to more labor, and thus necessitate the erection of more dwelling houses in the future.

When you turn over a new leaf, there is nothing like will-power to keep it from being blown back.

HARDWAREMEN AND PRICES.

HARDWAREMEN should conform their prices to market conditions, just as a mariner sets the sails of his vessel to meet weather conditions.

Necessity compels him to do so if he is to make his business a success.

The man of business, like the yachtsman, must ever be on the alert to grasp opportunities to facilitate his progress. This means, paradoxical as it may seem, that he has got to conform his prices to the condition of the market, whether it be reducing them or advancing them.

Hardwaremen, as a rule, are not slow in following the market when it is declining. Competition compels them to do so. And it is wise that they should do so, for, if they do not, they will be handicapped in the race for trade.

But, the fact that they are compelled to follow the market when its tendency is downward emphasizes the necessity for their following it with equal readiness when the tendency is upward.

Of course, this is more easily said than done, but, just as competition facilitates the downward tendency of prices, so a little understanding among the merchants in a community would assist them in advancing their prices to a basis warranted by the condition of the general market.

PERSONALITY IN BUSINESS.

IT is not surprising that the phrase, "the man behind the gun" became popular throughout the Empire. There is so much of generality in our songs as well as in conversation that this definite reference to the personality responsible for the results produced won natural approval.

It is the same everywhere. Business men discuss this man's store; that man's windows; the advertisements of another. We even go to the extent of imitating what we consider he has done well.

But, after all, it is not the store in which a man starts his business or the amount of goods he is able to stock that is the best criterion of his probable success. It is rather the man himself. There are some men who would succeed in the face of seemingly insurmountable difficulties, and there are others who would fail where everything is most advantageous, and between these two classes are practically all the rest. A man's business, store, window, advertising, clerks, are pretty much what he makes of them.

Therefore, it would be well for business men generally to spend a few days during the opening weeks of the new century studying themselves to see wherein they lack in the qualities that go to make a man successful.

UNSALABLE GOODS.

THERE is in every store at the end of the year goods which have proved unsalable. The measure of them is largely in proportion to the ability of the merchant as a buyer. But, whatever the reason, the fact that there are goods on the shelves which should have been sold makes it evident that something should be done to dispose of them.

Whenever the wideawake merchant finds himself thus situated, he marks the price of such goods down until he has found customers. And not only does he mark them down, but he adopts ways and means to acquaint the public of the fact.

He will probably start a bargain-counter. He is also likely to make a display in his window, while advertising he, of course, does not overlook.

Aside altogether from the inconvenience of having goods in stock which cumber the ground, it is to be remembered that every dealer tied up in such a way means so much capital unrenumerative. There is no doubt about that. And it is well that every merchant who hesitates to mark a line of goods down to a figure which will probably make it attractive to buyers should remember this.

Every dollar that is not earning is helping to eat up the merchant's substance.

MORE REASON TO HOPE.

After an interview with the New Brunswick Government, Messrs. Chas. Burrill and B. F. Pearson, directors of The Dominion Iron and Steel Co., are reported to have said that within a year they would be building steel ships at St. John and Halifax. Five million dollars will be invested in two plants. Two thousand to three thousand hands will be employed.

The Government seems to look favorably upon the project and a subsidy will likely be granted.

PUTTY IS HIGHER.

A CHANGE is announced this week in the price of putty, quotations being 5c. per 100 lb. higher on bulk in barrels and 10 to 15c. on bladder in barrels.

The old and the new prices are now as follows:

	New prices.	Old prices.
Bulk in barrels	$2 00	$1 95
Bulk in less quantity	2 15	2 10
Bladder in barrels	2 20	2 10
Bladder in 100 or 200-lb. kegs or boxes, or loose	2 35	2 25
Bladder in 25-lb. tins in 100-lb. lots	2 45	2 35
Bladder in 12½-lb. tins in 100-lb. lots	2 75	2 60
Bladder, bulk or tins in less than 100-lb. lots	3 00	2 80

The above figures are f.o.b. Montreal, Ottawa, Toronto, Hamilton, London and Guelph. Freights may be equalized from Montreal, Ottawa and Toronto.

Quotations for the Maritime Provinces are 10c. per 100-lb. higher than those for the Provinces of Ontario and Quebec.

SHEETS AND PLATES.

THERE is nothing in the present condition of the market for plates and sheets that would indicate a lower range of values, at any rate for several months to come. We gather this from the perusal of the correspondence of a well-known metal house in Toronto.

Its correspondence is with several manufacturing firms in Great Britain and in the United States. For example, one of the largest manufacturing firms in Pittsburg says that, owing to the large number of orders on its books, it will be impossible to make delivery of sheets till June. A British concern reports that the mills are so loaded up with orders that only present stocks are available, while several firms report that they will be unable to make delivery of plates and sheets earlier than February and March.

Jobbers in this country are being somewhat inconvienced by this condition of affairs.

A WISE CHOICE FOR PRESIDENT.

THE St. John, N.B, Board of Trade showed good judgment in electing G. Wetmore Merritt to the presidency of that institution for the coming year. Mr. Merritt is not only a keen business man, but his interests are various and his connection so widespread that he should make a first-class president.

Mr. Merritt, who is a partner in the wholesale grocery firm of Merritt Bros. & Co., started his business life as a clerk with Turnbull & Co., wholesale grocers, St. John, in 1873. Eleven years later he was admitted into partnership, under the old style, which remained unchanged until 1893, when Mr. Turnbull retired from active business, and the present firm was formed, Mr. Turnbull continuing as a special partner till January 31, 1896.

In addition to his grocery business, Mr. Merritt is interested, to a large extent, in lumbering, gold and coal mining, and in the steamship carrying business. His investments in all these lines are reputed to be particularly successful.

In every way Mr. Merritt has proved himself deeply interested in the commercial development of St. John, most of his work in that respect being done through his connection with the board of trade of that city.

G. WETMORE MERRITT.

CHURCH VENTILATION.

Official inspection finds Chicago churches, says The Chronicle, uniformly defective in means of purifying the air. That germ diseases may be widely disseminated through places of assembly is a fact so well established that it is incumbent upon all ecclesiastical authorities to take note of it. In plague times prohibition of pilgrimages and dispersion of caravans has been the first indispensable to extirpation of a scourge. It is a pathetic commentary on the supposed advance of the building arts that in places of religious worship there should be uniform and complete lack of ventilating appliances other than doors and windows.

After a sexton has started furnace fires in a church he is expected, as a rule, to avoid waste of the heat thus expensively produced. Whether a congregation be rich or poor, or composed of various categories of good fortune, it contributes reluctantly to the support of its church. Pastors are under the general stress to save money by not spending it, and no class of men are subjected to more anxiety, as a rule, than they in finding the revenue wherewith to maintain property intrusted to their care. They naturally, but mistakenly, object to opening doors and windows of places of religious meeting during the cold weather. People are always to be found to complain if a church be cold, others if it be warm. Pastor and sexton alike try to meet an average standard by starting the furnaces none too soon and sealing up the edifice for the day after combustion begins to raise the temperature.

For these reasons, for which neither sexton nor pastor should be held blameworthy, the air in most churches is noxious during winter.

Air is especially liable to be exhausted or corrupted in the most popular churches where more than one service is held in a day. A thousand people gathered together within walls anywhere will throw off enough animal heat to raise temperature, and, in many cases, will throw off unavoidable morbid germs caught up in the swirl of a town or developed in unsanitary homes. Ventilation of churches and schools ought to be deemed of paramount importance in a great city, where of necessity all manner of people are compelled to have closest bodily contact with every other manner of people, juvenile, adult and senile, bathed and unbathed.

This necessity is aptly illustrated in an incident which happened in a public school since the arrival of frost this year. A kindergarten teacher undertook to loosen the jacket of a boy in order that he might more comfortably participate in gymnastic exercises. "Please, don't," gently protested the lad. "I'm sewed up for the winter."

A complete change of air ought to be effected by throwing open all the windows and doors at least before and after each church service. Ventilation of school rooms should be accomplished with corresponding frequency and thoroughness.

A BUSINESS MEN'S LEAGUE.

Notice is given in the Official Gazette that application will be made to the Quebec Legislature by Henry Miles, importer; Fred W. Evans, insurance manager; Charles Chaput, merchant; James W. Knox, merchant; George E. Drummond, manufacturer, all of Montreal, for a bill to incorporate them and others under the name of the "Montreal Business Men's League," with the following objects: To promote and develop tourist and sportsmen travel in Quebec Province; to encourage and facilitate the holding of conventions and other gatherings in the city of Montreal; to promote municipal improvements therein, more especially with regard to the cleanliness of streets and the embellishment of parks and squares, and other objects of a kindred character.

ANSWERS TO INQUIRERS.

A Hartford, Ont., subscriber writes: "We are in a position here to get what is known as cat tails. Can you inform me who are handlers of the same? I saw in some paper that furniture or upholstering firms used them for stuffing."

[Remarks: Can any of our readers supply the desired information?—The EDITOR.]

HOW TO SUCCEED IN BUSINESS.

SOME PERTINENT SUGGESTIONS FROM ONE WHO HAS SUCCEEDED.

JUST now, when prosperity is in everybody's mouth, and when hosts of new business enterprises are likely to be launched and hosts of old ones reorganized, it may be timely to consider the general subject of success in business.

Statisticians tell us that 98 per cent. of all persons who embark in business on their own account fail at one period or another in their lives.

Did it ever occur to the reader how great a loss to the whole community is involved in the large percentage of failures? asks J. E. S. in News-Tribune. It is suffered not alone in the great mass of bad debts, which after the failure occurs have to be borne by creditors, but every unprofitable business carried on at a loss, often a long time before the collapse comes, is a direct injury to every legitimate enterprise in the same line with which it comes into competition. It is an obvious wrong for one concern, which pays its employes regularly, and meets all its obligations honestly, to have to suffer the competition of another concern which is living on its creditors. Legitimately conducted business is unjustly assailed by rotten concerns, and every man who pays his debts indirectly has to bear a share of the losses entailed by those who do not pay.

The question arises, is failure in business avoidable? Would it be possible to reverse the figures and have but two per cent. of failures to 98 per cent. of successes? Would not the prosperity of the whole community be enhanced by such a condition? I believe all these may be answered in the affirmative.

How may it be done?

1. By the exercise of judgment in the choice of a field of business. The establishment of an art store or a jewelry store in a locality frequented only by a poor class of people one can readily see would be an act of folly. Such folly, only varying in degree, too often attends the establishment of a business enterprise. The question of adequacy of the field is not sufficiently considered. By choosing a comparatively unoccupied field capital is not wasted in driving another occupant from it. I think it is generally the safer and wiser policy to buy out a prior occupant than to waste capital and energy in driving him out. But, in any case, there must be a field adequate to the business expected to be done.

2. The scale of one's business must be adjusted both to the field and to the extent of his capital. The man who attempts to do a business which should command a capital of $50,000 with only $10,000 to work with, takes a great chance of making a failure of it. Nor must borrowing be relied upon. There should be sufficient capital in a business to create a condition of easiness, and where the capital is in excess of the needs it is an additional guarantee of success, provided the business is conducted with the same careful conservatism that it would be on a smaller capital. A successful business, I confidently assert, may be established on any amount of capital, no matter how limited, but the amount of business undertaken must be proportioned according to it.

3. I believe it will be found to be true, as a rule, that the most successful businesses have been begun in a small way and have grown up, while the large percentage of failures have occurred in cases where it has been set out to create a large business from the outset. It is the order of nature for things to grow up, and not come into existence full fledged, and I believe the rule applies preeminently in the realm of business. Many a legitimate business may be started upon a $500 capital, and may grow to become a great concern, but the projector must at the outset put himself in direct competition with rivals enjoying 20 times the capital, nor seek to grow too rapidly.

4. Do not borrow the capital to start upon, or any considerable portion of it. It is not safe. There are two important reasons for this: One is that the interest and repayment are a burden upon the business which it ought not to be subject to; the other is the fact that the sense of responsibility is much greater where a person is dealing absolutely with his own, and he is more likely to employ his means with wisdom and safety than where he is operating at another's risk.

5. Every business man should not only thoroughly understand bookkeeping, but should, to an extent, keep his own books. There is only one person in a concern who knows exactly all the time the true condition of the business, that is the bookkeeper. He may submit statements and trial balances and balance sheets to his principals, but in rare instances he can communicate by them that subtle knowledge which comes of daily and hourly familiarity with the accounts. The proprietor who does not keep his own books is always at a disadvantage, and he who does not understand the science is in absolute peril. I have no doubt that a very large share of the failures result from the ignorance of the true condition of one's business. The keeping of books is not an uninteresting occupation, especially where one has a personal interest in the results. I know a millionaire merchant who, until compelled by advancing age and infirmities to entrust the work to others, always kept his own ledger, his bookkeeper's work extending only to the journalizing and bill making. In large houses where this may be impracticable, a new system has come into vogue by which a secondary ledger is kept for the special use of the proprietor, in which the transactions are summed up in compact shape. If the entries in this could be made by the principal himself, the advantages of the system would be still increased.

6. Another frequent cause for business failure is indiscretion in giving credit. I imagine it is a very common thing for a man with a moderate capital to engage in the retail grocery business; in his zeal to gain business he gives credit to a host of irresponsible persons, and in the course of a few months discovers that his entire capital is absorbed in uncollectable accounts. Of course he has to give up, and another person comes in to repeat the experience. Wholesale houses do the same thing on a larger scale. For a small margin of profit they will risk the entire value of the goods sold. Every bankrupt house will be able to show you a long array of bad accounts, which, if they could have been collected, would have averted the failure.

And there is no use in catering to a dead-beat custom. Better far to keep the goods and lose the customer. A good plan is to have an understanding with the customer at the time of the sale as to when payment is to be made. A debtor will often pay at a time he has promised to, where he would allow the debt to run indefinitely where there is no date of payment fixed. Then, too, if there is an agreement to pay at a certain time and default is made, there is no reasonable obligation to further increase the credit, and the merchant may get off with a small loss, where by a looser system

it would be difficult to close the account, and a more serious loss would result.

If an absolutely cash business cannot be done, the next best thing is to have the terms of the sale distinctly understood. It is a common plan with the larger retail houses to render statements at the end of each month and expect payment on or before the tenth of the following month. Where customers are held rigidly to this rule, the losses by bad debts are probably very small, I have observed that the debtor always gives first attention to the payment of the bills of those who are most strict in insisting upon a prompt payment, and leaves to the last those bills rendered by easier, more careless or more good-natured dealers; and so, in case of inability to ultimately pay all, these easy ones have to suffer. It is never a cause for ill feeling on the part of an honest buyer that the seller insists upon the prompt payment for the goods sold. If one is disposed to resent it, he is just the customer that it will be profitable to lose.

Some people think that a great volume of business is a sign of success, and are tempted into risks which they ought not to take. A small, safe business is immeasurably preferable.

7. Another very important requisite to success is persistence, tenacity in sticking to a business through the to-be-expected period of discouragement. Too many persons embark in an enterprise, expend upon it a lot of intelligent thought, skill and industry, and then, when they see the returns are inadequate, throw it up or sell out at a sacrifice, forgetting that the efforts they have put into it are of the nature of capital, something essential to the enterprise, and which will one day bring returns. What we call the good-will of a business is simply the accummulation of all this brain and hand work not shown on the books, and often it is of even more value than the real money capital. Wherever an enterprise is abandoned, of course, all that has been done to advertise it, make friends and build up a trade is thrown away. It is really the same thing as throwing away cash, because if nursed and persistently stuck to it would in time have a cash value. Many a man, after a long period of discouragement, provided he has been working on correct lines, has suddenly found his business past the turning point, and as continuously prosperous as before it was the reverse. But this cannot be predicted upon a business improperly conducted, or where a false policy has been pursued.

A very important thing to be considered and built up is the good-will of a good class of customers. Poor customers there is nothing to be gained in cultivating. As Capt. E. B. Ward used to say: "Cultivate the best customers and let your rivals have the others." It it is possible to establish such relations with one's customers that no persuasion or inducements will tempt them to leave you. Great care to furnish entirely satisfactory goods, prompt readiness to rectify any errors or make good any imperfections, and zeal to see that no one is overcharged will always inspire a confidence which becomes a valuable asset in business. I have known merchants to consider only how they can swell their bills, forgetting that while they for the moment made a profit they endanger the permanent loss of

H. S. HOWLAND, SONS & CO.

WHOLESALE ONLY — 37-39 Front Street West, **Toronto.** — ONLY WHOLESALE

SPRING SKATES.

	LIST per pair.
No. 5—Ordinary Quality	$ .80
7—Best Quality	1.50
10— " " Plated	1.90
12— " " " Concaved Blades	2.80

All sizes to 12¼ inches.

LADIES' SKATES.

No. 415—Ordinary Quality, Plated	$1.40
1422—Best Quality, Plated	2.30
1424— " " " Con. Runners	2.70
447—Extra " " Light Runners	3.50

All sizes, 8½ to 10½ inches.

HOCKEY SKATES.

No. 510¼—Ordinary Quality, all sizes, 7 to 12 in.	$ .60
1531—Best Quality, all sizes, 8 to 12 in.	1.60

No. 631—Best Quality, all sizes, 9 to 12 in.	$1.80
632— " " Plated, all sizes, 9 to 12 in	2.50

No. 634—Best Quality, Plated, all sizes, 9 to 11½ in., Concave Blades	$2.90

No. 3697—Best Quality, Plated, all sizes, 10 to 12 in.	$4.00

Also the following lines of

Genuine "Starr" Halifax Skates.

		List per pair.
No. 90—"ACME" Spring, all sizes,	7 to 11 in.	$ .80
25— " " "	8 to 12 in.	1.80
7— " " "	9 to 11½ in.	3.00
20—Hockey, all sizes	9 to 11½ in.	1.60
20 PS— " "	9 to 11½ in.	1.80
25 PS— " "	9 to 12 in.	2.40

"Mic-Mac" Hockey Sticks. 1900 BRAND.

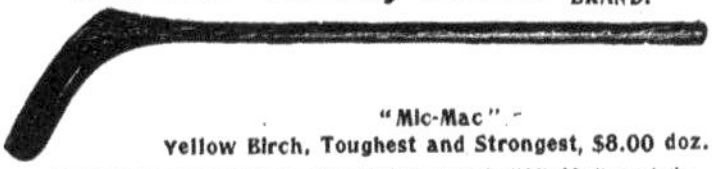

"Mic-Mac"

Yellow Birch, Toughest and Strongest, $8.00 doz.

Our stock this season has been carefully selected, and the "Mic-Mac" 1900 is the **BEST STICK** on the market.

The above are List Prices and subject to Liberal Trade Discount.

H. S. HOWLAND, SONS & CO., Toronto.

PROMPT SHIPMENTS — Graham Wire and Cut Nails are the Best. — OUR PRICES ARE RIGHT.

the customer. I once contracted with a dealer for a large quantity of carpets "made and laid." When the bills came in the thread, tacks and a lot of other things were charged extra. Of course it was my last transaction with that house.

The golden rule of business, and the one which in the long run is most profitable, is to aim at entire reciprocity of interest between buyer and seller. In other words, it should be as much the interest of the buyer to buy as it is of the seller to sell. Where both are equally advantaged, a basis of solid, profitable and permanent business is created, most valuable to all concerned.

8. Lastly, it is important to everyone embarking in business to establish for himself a bank credit. The concern which keeps no bank account may be set down as doing business in a very slouchy way. Perhaps the firm will tell you "we never have money enough on hand to make it worth while opening an account." One would not be far wrong in hazarding the reply that "you never will have." Although the amounts handled may be small, and the balance never great, it is still better to deposit in a bank all the receipts and pay at least all the larger bills by checks. Besides the element of safety in having a bank in charge of one's funds, there is a constant stimulus to increase the balance on hand; and a credit, which at any time may be valuable, is being built up. The concern which deposits regularly and never overdraws its account will imperceptibly come to have a standing at its bank which will do it good service when from any cause bank accommodation is needed.

Then by all means establish confidential relations with your banker. Let him know, of course in confidence, the exact condition of your business affairs. His advice may be valuable to you, and unless absolutely rotten, he will go a long way to help a customer through a pinch.

DID IT ON AN IVER JOHNSON.

Harry Elkes, the Iver Johnson flyer, made another good win when he defeated Jimmie Michaels in a special motor-paced race of 15 miles at Madison Square Garden, December 22. Not only did he defeat Michaels, but he also lowered the Garden track record held by that rider from 27.56 2-5 to 26.03 2 5.

Wm. Rodden & Co., founders, Montreal, have assigned to A. W. Stevenson, and a meeting of creditors will be held on January 10. The principal creditors are: Estate of Robert Hamilton, mortgage, $16,234.10; Crathern & Caverhill, $1,325; A. C. Leslie & Co., $1,650; S. E. L. Bricker, $8,000; George E. Douglas Trading Co., secured, $2,237.

CATALOGUES, BOOKLETS, ETC.

THE Canada Paint Co., Limited, are out with a handsome new "Century" catalogue, which has been printed in the firm's own premises and by their own staff of printers. Its neat style and high-class color-printing and decoration would lead one to think that it came forth from a specialty publishing house. It is strictly up to date, and can be relied upon as a most useful and handy price list. A number of new features are offered by this firm for 1901, full particulars of which will be found in the catalogue. Any reader of HARDWARE AND METAL who has not received this useful book by mail would do well to apply to the company's headquarters, Montreal.

AN ENAMELLED WARE CATALOGUE.

The Thos. Davidson Manufacturing Co., Limited, Montreal, Can., have undoubtedly reached a high standard of excellence in

their enamelled and embossing department. HARDWARE AND METAL has just received from them a calendar which has enamelled in six colors a face of an Indian chief in war dress. The design and coloring is so artistic, so true to life, that one wonders at the skill of the mechanical genius which has created the effect produced. The accompanying cut shows the design, but gives no idea of the brilliant effect that is produced by the contrasts and harmonies of the various striking colors used.

INTERIOR DECORATION IN METAL.

The hardware dealer who keep on hand literature which will help him to explain the advantages of any lines he wants to sell follows a wise custom. For this reason, a booklet which The Metallic Roofing Company of Canada, Limited, Toronto, have just issued should be of considerable value to everyone in the trade. It is an exquisitely-printed booklet, dealing with the use of metallic ceilings and walls. It is in no wise a catalogue, but is rather an educative pamphlet, gotten up with a view to convince readers that the artistic metal productions of this company are the proper material for interior decorations. And, from a perusal of the illustrations and reading matter of the booklet, it bids fair to convince many. Hardwaremen, therefore, should get one or more copies and interest probable customers in its contents.

BUSINESS CHANGES.

DIFFICULTIES, ASSIGNMENTS, COMPROMISES.

MAX COHEN, general merchant, Hawkesbury, Ont., has assigned to G. S. Bowie.

Alphonse Guimond, hardware dealer, 2628 Notre Dame street, Montreal, has assigned. Real Angers has been appointed provisional guardian. The chief creditors are Frothingham & Workman, $1,026; L. H. Hebert, $648; Nap. Sarrazin, $690; Dame Nap. Mathieu, $500; Caverhill, Learmont & Co., $714; A. Ramsay & Son, $400; Jas. Robertson & Co., $500. Total liabilities, $9.327. A meeting of the creditors has been called for January 10.

PARTNERSHIPS FORMED AND DISSOLVED.

Brett & Leighton, hardware dealers, Orangeville, Ont., have dissolved, M. J. Leighton retiring and S. Taylor being admitted under the style of Brett & Taylor.

John Lumsden & Co., manufacturers of tools, etc., Montreal, have dissolved, John Lumsden continuing.

Jack & Robertson, importers of metals, etc., Montreal, have dissolved, and Watson Jack has registered as proprietor under the style of Watson Jack & Co.

SALES MADE AND PENDING.

W. H. Hayes, blacksmith, Sussex, N.B., has sold out to H. H. Dryden.

E. H. Heaps & Co., sawmillers, etc., Vancouver, have removed to Cedar Grove, B. C.

C. W. Lurtey, machinst, Rossland, B.C., has sold ont to J. W. Pownall and J. L. Sanders.

The Dominion Cordage and Manufacturing Co., Limited, Peterboro', Ont., have been incorporated.

Stewart T. Patterson, dealer in agricultural implements, Rodney, Ont., has sold out to W. Livingstone.

A meeting has been held to organize The McFarlane-Neill Mfg. Co., Limited, manufacturers of hames, etc., St. Mary's Ferry, N. B.

Henry Yost, carriagemaker and blacksmith, Tavistock, Ont., has sold out to Kruspe Bros., Sebringville, who have moved to Tavistock.

FIRES.

McFadyen & McQuade, tinsmiths, Collingwood, Ont., have been burned out.

INDUSTRIAL GOSSIP.

Those having any items of news suitable for this column wi l confer a favor by forwarding them to this office addressed to the Editor.

SIR LOUIS DAVIES, Minister of Marine and Fisheries, in the Canadian Cabinet, has awarded to A. Wallace, Vancouver, a contract for a cruiser, 136 x 24 ft., for Pacific Coast cruising service, to cost between $60,000 and $70,000, and to The Albion Iron Works, Victoria, B.C., a contract for a 60 x 11-ft. cruiser, to be used at the mouth of Fraser River. The price of this cruiser is about $8,000.

The Dominion Rock Drill and Foundry Co., Napanee, Ont., are asking exemption from all taxes, except for school purposes. The ratepayers of Napanee will vote on the matter January 7.

The McArthur Export Co., lumbermen, Quebec, have applied for a charter.

The Owen Sound, Ont., Iron Works Co., Limited, are extending their premises.

The North-West Gold Dredging Co., Limited, Ottawa, have obtained a charter.

The Walkerville Wagon Co., Limited, Walkerville, Ont., have been incorporated.

The Hanover Portland Cement Co., Limited, Hanover, Ont., have been incorporated.

A. H. Raymond, of Philadelphia, has taken the management of The Perth Flax and Cordage Co., Stratford, Ont. He is also taking a financial interest in the concern, and new machinery sufficient to double its capacity is being installed.

During the holidays the McClary Manufacturing Co. have had a large staff at work making numerous changes to their factory at London. Two new boilers of the latest pattern and very large capacity have been placed in position, and the plant changed so as to economize labor and to make the work as systematic as possible. They have also made large additions to warehouse and factory, which will increase their capacity in both these departments. Work was resumed on Wednesday.

NEW RAILWAY FOR MANITOBA.

In reply to a deputation, Premier Roblin of Manitoba promised to have a railway built from Brandon towards Virden, reaching the boundary of the Province, and to have it ready for operation in time to haul out next season's grain. He gave no distinct pledge that the line would be built by the Government, although there was a possibility of the Government moving in that direction, neither did he give an answer as to what points would be touched by this railway.

MARKETS AND MARKET NOTES

QUEBEC MARKETS.

Montreal, January 4' 1901.

HARDWARE.

THE travellers are again on the road and orders are coming in. The volume of business being done is not large, as retailers are busy settling accounts and ending the year's business. The outlook, however, seems to be exceedingly bright. The sea of commerce is not disturbed by any strong wind of inflation or depression and with a fair pressure of steam good progress ought to be made. Movements from stocks are not large, but the making of spring contracts goes merrily on. Barb wire, lawn mowers, freezers, sprayers, rubber hose, poultry netting, green wire cloth are all demanding attention for spring delivery. Some fair amounts of nails have been moving this week at steady prices. Horseshoes are in light stock and still in fair request. Cordage is a little higher. Shelf goods are rather quiet. Wrenches have been put on a new price list.

BARB WIRE—From stock trade is very slow, but some orders are being placed for spring delivery at $3.20 f.o.b. Montreal in less than carlots.

GALVANIZED WIRE—Business is confined to ordering for spring. We quote : No. 5, $4.25 ; Nos. 6, 7 and 8 gauge $3.75 ; No. 9, $3.00 ; No. 10, $3.75 ; No. 11, $3.85 ; No. 12, $3.15 ; No. 13, $3.25 ; No. 14, $4.25 ; No. 15, $4.75 ; No. 16, $5.00.

SMOOTH STEEL WIRE—There are no new features to note. Business is slow at $2.80 per 100 lb.

FINE STEEL WIRE—There is nothing new to note, the discount remaining at 17½ per cent. off the list.

BRASS AND COPPER WIRE—The usual business is being done at last week's quotations. Discounts are 55 and 2½ per cent. on brass, and 50 and 2½ per cent. on copper.

FENCE STAPLES—This line is featureless. We quote : $3.25 for bright, and $3.75 for galvanized, per keg of 100 lb.

WIRE NAILS— Quite a number of small orders have come to hand this week. We quote: $2.85 for small lots and $2.75 for carlots, f.o.b. Montreal, Toronto, Hamilton, London, Gananoque, and St. John, N.B.

CUT NAILS — Cut nails are in moderate request. We quote as follows : $2.35 for small and $2.25 for carlots ; flour barrel nails, 25 per cent. discount; coopers' nails, 30 per cent. discount.

HORSE NAILS—Trade seems hardly so brisk as it was last week. The discounts are 50 per cent. on Standard and 50 and 10 per cent. on Acadia.

HORSESHOES — Stocks continue light and the demand brisk. We quote as follows : Iron shoes, light and medium pattern, No. 2 and larger, $3.50 ; No. 1 and smaller, $3.75 ; snow shoes, No. 2 and larger, $3.75 ; No. 1 and smaller, $4.00 ; X L steel shoes, all sizes, 1 to 5, No. 2 and larger, $3.60 ; No. 1 and smaller, $3.85 ; feather-weight, all sizes, $4.85; toe weight steel shoes, all sizes, $5.95 f.o.b. Montreal ; f.o.b. Hamilton, London and Guelph, 10c. extra.

POULTRY NETTING—Orders for spring delivery are being taken quite generally at

50 per cent. off. The situation has been unsettled lately, and we hear that some firms have been selling at 50 and 5 per cent. off.

GREEN WIRE CLOTH — The travellers report trade to be active on spring account at $1.50 per 100 sq. ft.

FREEZERS—Ice cream freezers are beginning to move, but the number ordered is not large, as yet.

SCREWS—The demand is only moderate this week at former quotations. Discounts are as follows: Flat head bright, 80 per cent. off list; round head bright, 75 per cent.; flat head brass, 75 per cent.; round head brass, 67½ per cent.

WRENCHES—New prices have been put on wrenches not long since. Agricultural wrenches are quoted at 60 and 10 per cent. off; "Acme," 50, 10 and 5 per cent. off; Whitman Barnes knife-handle, 6 to 12 in., 40 per cent. off; 15 in. up, 45 per cent. off.

BOLTS—A fair trade continues to be done in some lines, but the total amounts moving are not large. Discounts are: Carriage bolts, 65 per cent.; machine bolts, 65 per cent.; coach screws, 75 per cent.; sleigh shoe bolts, 75 per cent.; bolt ends, 65 per cent.; plough bolts, 50 per cent.; square nuts, 4½c. per lb. off list; hexagon nuts, 4¾c. per lb. off list; tire bolts, 67½ per cent.; stove bolts, 67½ per cent.

COTTERPINS — There is no change to note. We quote as follows: 55 per cent. off English list, or, according to American list, ¼-in. and under, 80 and 20 per cent., 5-16-in., 80 and 10 per cent., and ⅜-in., 70 and 10 per cent. off.

RIVETS—Continue quiet and unchanged. The discount on best iron rivets, section, carriage, and wagon box, black rivets, tinned do., coopers' rivets and tinned swedes rivets, 60 and 10 per cent.; swedes iron burrs are quoted at 55 per cent. off; copper rivets, 35 and 5 per cent. off; and coppered iron rivets and burrs, in 5-lb. carton boxes, are quoted at 60 and 10 per cent. off list.

CORDAGE — The tone of the cordage market is firm. Manila is now quoted at 13½c. per lb. for 7-16 and larger; sisal is worth 9½c. per lb. for 7-16 and larger, and lathyarn 9c. per lb.

SPADES AND SHOVELS — There is little being doing in this article just now. Discounts are 40 and 5 per cent.

TACKS—There is no change to note. We quote: Carpet tacks, in dozens and bulk, blued, 80 and 5 per cent. discount; tinned, 80 and 10 per cent.; cut tacks, blued, in dozens, 75 and 15 per cent. discount.

FIREBRICKS — Trade is dull at $18.50 to $26, as to brand.

CEMENT—Sales from stock do not amount to much. We quote: German, $2.50 to

$2.65; English, $2.40 to $2.50; Belgian, $1.90 to $2.15 per bbl.

METALS.

The metal market does not show as much strength as it did some time ago, but, on the whole, the situation is steady. Trade is quiet in most lines.

PIG IRON — There seems to be no predominant changing influence and the market appears steady. Summerlee is worth $24 to $25, and Canadian pig $19 to $20.

BAR IRON—Quite a fair amount of trading is being done at $1.65 to $1.70 per 100 lb.

BLACK SHEETS — This line is featureless at the present moment. The base price is $2.85 for 8 to 16 gauge.

GALVANIZED IRON—We hear that some import orders are being placed at very low prices. The travellers are making contracts for spring delivery. Out of stock there is not much moving. We quote: No. 28 Queen's Head, $4.65 to $4.85; Comet, No. 28, $4.30 to $4.55, and Appolo, 10¾ oz., $4 65 to $4.85.

INGOT COPPER—A fair amount of business is being done at 17½c.

INGOT TIN—The foreign markets are unsettled. Locally there is little trading being done. Lamb and Flag is selling at 33 to 34c.

LEAD—The market continues steady at $4.65.

LEAD PIPE—A fair amount of business is being done at 7c. for ordinary and 7½c. for composition waste, with 15 per cent. off.

IRON PIPE—A fair trade is being done. We quote as follows: Black pipe, ¼, $2.80 per 100 ft.; ⅜, $2.80; ½, $2.85; ¾, $3.05; 1-in., $4.35; 1¼, $5.95; 1½, $7.10; 2-in., $9.50. Galvanized, ½, $4.90; ¾, $5.40; 1 in., $7.35; 1¼, $9.75; 1½, $11.70; 2-in., $15.75.

TINPLATES — The market does not show much strength. We quote: Coke, $4.50; charcoal, $4.75.

CANADA PLATE—This article is not attracting much attention at the moment. Quotations are unchanged. We quote: 52's, $2.85; 60's, $3; 75's, $3.10; full polished, $3.75, and galvanized, $4.60.

TOOL STEEL—We quote: Black Diamond, 8c.; Jessop's 13c.

STEEL—We quote: Sleighshoe, $1.85; tire, $1.95; spring, $2.75; machinery, $2.75 and toe-calk, $2.50.

TERNE PLATES—A fair trade is still doing in terne plates at $8 25.

SWEDISH IRON—Unchanged at $4.25.

COIL CHAIN—There has been no change in local quotations. Some spring orders will likely be placed while the travellers are on their pesent trip. The tone of the market is firm. We quote as follows: No. 6, 11½c.; No. 5, 10c.; No. 4, 9½c.; No. 3, 9c.; ¼-inch, 7½c. per lb.; 5-16, $4.60; 5-16 exact, $5.10; ⅜, $4.20; 7-16, $4.00; ½, $3.75; 9-16, $3.65; ⅝, $3.35; ¾, $3.25; ⅞, $3.20; 1-in., $3.15.

SHEET ZINC — Values are steady at 6 to 6¼c.

ANTIMONY—Unchanged, at 10c.

GLASS.

The market for glass is without any new feature. We quote as follows: First break, $2; second, $2.10 for 50 feet; first break, 100 feet, $3.80; second, $4; third, $4.50; fourth, $4.75; fifth, $5.25; sixth, $5.75, and seventh, $6.25.

PAINTS AND OILS.

Owing to supplies coming in freely from Georgia aud South Carolina, turpentine has been reduced 1c. per gallon. Putty has been advanced 5c. per 100 lb. In dry colors, such as vermilion, permanent greens and oxides, a capital business has been done since our last report. In Canadian oxides, especially, some good orders have been received for shipment into the United States. Locally, a very fair inquiry exists for all classes of painting material, and inquiries are coming from all parts of the Dominion. As to price and prospects for paris green the market is not as yet formed and chemicals are unsettled, no quotations having as yet been available. Colors ground in oil and in japan have met with a fair measure of attention. White lead, as usual, at this time of year, is rather quiet, and quotations at the last meeting of the association were not changed. All lead products may be described as steady. Putty is moving freely at the advance. We quote :

A fair trade has been done this week, although orders were very scarce during the first few days. The feeling in linseed oil is a little stronger. We quote :

WHITE LEAD—Best brands, Government standard, $6.50 ; No. 1, $6.12½ ; No. 2, $5.75 ; No. 3, $5.37½, and No. 4, $5, all f.o.b. Montreal. Terms, 3 per cent. cash or four months.

DRY WHITE LEAD—$5.75 in casks ; kegs, $6.

RED LEAD—Casks, $5.50 ; in kegs, $5.75.

WHITE ZINC PAINT—Pure, dry, 8c.; No. 1, 6½c.; in oil, pure, 9c.; No. 1, 7½c.

PUTTY—We quote : Bulk, in barrels, $2 per 100 lb.; bulk, in less quantity, $2.15; bladders, in barrels, $2.20 ; bladders, in 100 or 200 lb. kegs or boxes, $2.35; in tins, $2.45 to $2.75 ; in less than 100-lb. lots, $3 f.o.b. Montreal, Ottawa, Toronto, Hamilton, London and Guelph. Maritime Provinces 10c. higher, f.o.b. St. John and Halifax.

LINSEED OIL—Raw, 80c.; boiled, 83c., in 5 to 9 bbls., 1c. less, 10 to 20-bbl. lots, open, net cash, plus 2c. for 4 months. Delivered anywhere in Ontario between Montreal and Oshawa at 2c. per gal. advance and freight allowed.

TURPENTINE—Single bbls., 59c.; 2 to 4 bbls., 58c.; 5 bbls. and over, open terms, the same terms as linseed oil.

MIXED PAINTS—$1.25 to $1.45 per gal.

CASTOR OIL—8¾ to 9¼c. in wholesale lots, and ½c. additional for small lots.

SEAL OIL—47½ to 49c.

COD OIL—32½ to 35c.

NAVAL STORES—We quote : Resins, $2.75 to $4.50, as to brand ; coal tar, $3.25 to $3.75 ; cotton waste, 4½ to 5½c. for colored, and 6 to 7½c. for white ; oakum, 5½ to 6½c., and cotton oakum, 10 to 11c.

SCRAP METALS.

There are no new features to notice this week. The demand seems rather quiet. Dealers are paying the following prices in the country : Heavy copper and wire, 13 to 13½c. per lb. ; light copper, 12c.; heavy brass, 12c.; heavy yellow, 8½ to 9c.; light brass, 6½ to 7c.; lead, 2¾ to 3c. per lb.; zinc, 2¼ to 2½c.; iron, No. 1 wrought, $13 to $14 per gross ton ; No. 1 cast, $13 to $14 ; stove plate, $8 to $9; light iron, No. 2, $4 a ton; malleable and steel, $4.

PETROLEUM.

The demand continues steadily brisk. We quote as follows : "Silver Star," 15 to 16c. ; "Imperial Acme," 16½ to 17½c. ; "S.C. Acme," 18 to 19c., and "Pratt's Astral," 19 to 20c.

HIDES.

The demand for hides continues slow, and there is little trading doing. We quote: Light hides, 8½c. for No. 1 ; 7½c. for No. 2, and 6½c. for No. 3. Calfskins, 8c. for No. 1 and 6c. for No. 2. Lambskins, 90c.

MONTREAL NOTES.

Rope is about 1c. per lb. higher.

Turpentine is 1c. per gal. lower.

The price list of wrenches has been recast.

Putty has been advanced in price 5 to 10c. per 100 lb.

Mixed paints are firm and, in some cases, higher, being now quoted at $1.25 to $1.45.

ONTARIO MARKETS.

TORONTO, January 4, 1901.

HARDWARE.

THE wholesale hardware trade is still enjoying its holidays' quiet feeling. The wholesale houses are still busy taking stock and the travellers remain in the warehouse lending their assistance to this task. On Monday next practically all the travellers will be again on the road and we may, therefore, look for an improvement in trade next week. The manufacturers are meeting this week in Toronto, but so far we hear of no change being made in any line appertaining to the iron trade. In fact, the only change worthy of note in any line is in putty, which is from 5 to 10c. per 100 lb. dearer. In all staple lines, such as nails and fencing wire, the volume of business is small indeed, but, of course, nothing else could be expected at this season. The letter order trade is fair for this time of the year.

BARB WIRE—Nothing is being done in the way of prompt shipment, but a few orders are being booked for spring delivery. We still quote f.o.b. Cleveland at $2.97½ in less than carlots, and $2.85 for carlots. From stock, Toronto, $3.10 per 100 lb.

GALVANIZED WIRE—Much the same remarks apply to this as to barb wire, the demand being only for spring shipment. We still quote No. 9 at $3.10, Toronto. The base price f.o.b. Cleveland is still $2.72½ per 100 lb.

SMOOTH STEEL WIRE—There are a few orders being taken for oiled and annealed wire for spring shipment, but nothing is being done in shipment from stock. There is a little movement in hay-baling wire from stock. The base price is $2.80 per 100 lb.

WIRE NAILS—Business in this line is almost at a standstill. The nail manufacturers are meeting in Toronto as we go to press, but so far no change in prices has been announced. We, therefore, still quote the base price at $2.85 per keg for less than carlots and $2.75 for carlots.

CUT NAILS—Dull and featureless with the base price unchanged at $2.35 per keg.

HORSESHOES—Very little doing, and prices are as before. We quote as follows f.o.b. Toronto : Iron shoes, No. 2 and larger, light, medium and heavy, $3.60 ; snow shoes, $3.85 ; light steel shoes, $3.70; featherweight (all sizes), $4.95 ; iron shoes, No. 1 and smaller, light, medium and heavy (all sizes), $3.85 ; snow shoes, $4 ; light steel shoes, $3.95 ; featherweight (all sizes), $4.95.

HORSE NAILS—Much the same remark applies to these as to horseshoes. Discount, 50 per cent. on standard oval head and 50 and 10 per cent. on Acadia.

SCREWS—Just a moderate trade is being done. Prices are as before. We quote wood screws as follows : Flat head bright, 80 per cent. off the list; round head bright, 75 per cent.; flat head brass, 75 per cent.; round head brass, 67½ per cent.; flat head bronze, 67½ per cent.; round head bronze, 62½ per cent.

BOLTS AND NUTS—These are quiet and unchanged. We quote: Carriage bolts (Norway), full square, 70 per cent.; carriage bolts, fulls quare, 70 per cent.; common carriage bolts, all sizes, 65 per cent.; machine bolts, all sizes, 65 per cent.; coach screws, 75 per cent.; sleighshoe bolts, 75 per cent.; blank bolts, 65 per cent.; bolt ends, 65 per cent.; nuts, square, 4½c. off; nuts, hexagon, 4¼c. off; tire bolts, 67½ per cent.; stove bolts, 67½; plough bolts, 60 per cent.; stove rods, 6 to 8c.

RIVETS AND BURRS — These are also quiet and without change. Discount, 60 and 10 per cent. on iron rivets; iron burrs, 55 per cent.; copper rivets and burrs, 35 and 5 per cent.

ROPE—Is dull and unchanged at last week's advance. Sisal, 9c. per lb. base, and manila 13c. Cotton rope is unchanged as follows: 3-16 in. and larger, 16½c.; 5-32 in., 21½c., and ¼ in., 22½c. per lb.

CUTLERY — There is practically nothing doing in this line as is to be expected after the holiday season.

BUILDING PAPER — There is a little movement in this line and stocks are now fairly complete. Ready roofing, 3-ply, $1.65 per square; ditto, 2-ply, $1.40 per square. Quotations are f. o. b. Toronto, Hamilton, London.

GREEN WIRE CLOTH—Very little is being done at the moment in the way of booking for future delivery. We still quote $1.50 per 100 sq. ft.

RULES AND PLANES—The Stanley Rule and Level Co. have issued a new list of prices on the articles they manufacture. A new list of prices has been issued, but the discounts are unchanged. The new list of prices, compared with those which went into force a year ago, is as follows:

BOXWOOD RULES.

	New list per doz.	Old list per doz.
No. 4	$11 00	$10 00
No. 12	16 00	14 00
No. 18	6 00	5 00
No. 26	11 00	9 00
No. 27	13 00	12 00
No. 29	4 50	3 50
No. 66	10 00	8 00
No. 66½	10 00	8 00
No. 67	6 00	5 00

STANLEY PLANES.

	New list each.	Old list each.
No. 100	$ 30	$ 25
No. 130	85	80

CEMENT—There is practically nothing doing. Prices are unaltered. We nominally quote in barrel lots: Canadian Portland, $2.80 to $3; Belgian, $2.75 to $3; English do., $3; Canadian hydraulic cements, $1.25 to $1.50; calcined plaster, $1.90; asbestos cement, $2.50 per bbl.

METALS.

This being the period of the year when most dealers are taking stock very little business is naturally being done in the metal trade. Although in the outside markets there have been some fluctuations in prices in some lines of metals, our figures are unchanged.

PIG IRON—Some good orders have been placed by large users of pig iron in Canada recently for delivery during the next six months and prices rule steady. For small lots of pig iron the Ontario furnaces are quoting $18, but for quantities this figure is shaded.

BAR IRON—The feeling in regard to bar iron is firm. A scarcity of scrap is reported, and it is asserted that the mills are selling bar iron close to the cost of production. The ruling base price is $1.70.

PIG TIN—There has been a great deal of fluctuation in prices in the outside markets during the past week, but the general tendency is toward lower prices. Locally, there is very little being done, but prices are unchanged at 33 to 34c. per lb.

TINPLATES—The demand for these is light with quotations as before.

TINNED SHEETS — Business is almost at a standstill in this line. We still quote 28 guage at 9 to 9½c. per lb.

TERNE PLATES—There is nothing doing in this line.

BLACK SHEETS — Like nearly all other metals very little is being done in black sheets. We quote $3.50 per 100 lb.

GALVANIZED SHEETS—The demand from stock has fallen off somewhat, but import orders are being booked with freedom. We quote English at $4.85 and American at $4.50 for ordinary quantities.

CANADA PLATES—There has been a little more movement in this line. We quote: All dull, $3.15; half and half, $3.25; and all bright, $3 85 to $4.

IRON PIPE—The demand keeps up fairly well, but jobbers' prices are rather low compared with the condition of the market. We quote as follows: Black pipe ¼ in., $3.00; ⅜ in., $3.00; ½ in., $3.00; ¾ in., $3.30; 1 in., $4.50; 1¼ in., $6.25; 1½ in., $7.75; 2 in., $10.40. Galvanized pipe is as follows: ½ in., $4.50; ¾ in., $5; 1 in., $7; 1¼ in., $9.50; 1½ in., $11.75; 2 in., $15.75.

SOLDER—A fairly good trade is being done. We quote half-and-half, 19 to 20c.; refined, 19c.

PIG LEAD — The demand is fair, with prices unchanged at 4¾ to 5c.

COPPER—There has been a small movement in ingot copper during the week, and trade is fair in sheet. We quote: Ingot,

19 to 20c.; bolt or bar, 23½ to 25c.; sheet, 23 to 23½c.

BRASS—Business continues quiet with the discount on rod and sheet unchanged at 15 per cent.

ZINC SPELTER — The demand has been rather good during the week. We quote 6 to 6½c. per lb.

SHEET ZINC—Business appears to have fallen off a little during the week, it now being characterized as quiet. We quote casks at $6.75 to $7, and part casks at $7 to $7.50 per 100 lb.

ANTIMONY — Trade is quiet and prices unchanged. We quote 11 to 11½c. per lb.

PAINTS AND OILS.

There is not much doing. Travellers have been out since Wednesday, but find little demand. Some orders for linseed oil for spring delivery are reported. Prices are easy, a decline of 2c. being noted. Turpentine is also 1c. lower, but since the reduction was made the market has strengthened, and the present quotations are not likely to be followed by lower ones. Putty has been advanced 5c. for bulk and 10c. for bladders. It is firm at the new prices. The manufacturers of white lead, who met late last week, decided not to make any change. An advance was expected by many. We quote:

WHITE LEAD—Ex Toronto, pure white lead, $6.62½; No. 1, $6.25; No. 2, $5.87½; No. 3, $5.50; No. 4, $4.75; dry white lead in casks, $6.

RED LEAD—Genuine, in casks of 560 lb., $5.50; ditto, in kegs of 100 lb., $5.75; No. 1, in casks of 560 lb., $5 to $5.25; ditto, kegs of 100 lb., $5.25 to $5.50.

LITHARGE—Genuine, 7 to 7½c.

ORANGE MINERAL—Genuine, 8 to 8½c.

WHITE ZINC—Genuine, French V.M., in casks, $7 to $7.25; Lehigh, in casks, $6.

PARIS WHITE—90c.

WHITING — 60c. per 100 lb.; Gilders' whiting, 75 to 80c.

GUM SHELLAC — In cases, 22c.; in less than cases, 25c.

PUTTY — Bladders, in bbls., $2.20; bladders, in 100 lb. kegs, $2.35; bulk in bbls., $2; bulk, less than bbls. and up to 100 lb., $2.15; bladders, bulk or tins, less than 100 lb., $3.

PLASTER PARIS—New Brunswick, $1.90 per bbl.

PUMICE STONE — Powdered, $2.50 per cwt. in bbls., and 4 to 5c. per lb. in less quantity; lump, 10c. in small lots, and 8c. in bbls.

LIQUID PAINTS—Pure, $1.20 to $1.30 per gal.; No. 1 quality, $1 per gal.

CASTOR OIL—East India, in cases, 10 to 10½c. per lb. and 10½ to 11c. for single tins.

LINSEED OIL—Raw, 1 to 4 barrels, 80c.; boiled, 84c.; 5 to 9 barrels, raw, 79c.; boiled, 83c., delivered. To Toronto, Hamilton, Guelph and London, 2c. less.

TURPENTINE—Single barrels, 57c.; 2 to 4 barrels, 56c., to all points in Ontario. For less quantities than barrels, 5c. per gallon extra will be added, and for 5-gallon packages, 50c., and 10 gallon packages, 80c. will be charged.

GLASS.

There is a strong disposition in favor of an advance, but as yet no change has been made. There is practically nothing doing. We still quote first break locally: Star, in 50-foot boxes, $2.10, and 100-foot boxes, $4; doublediamond under 26 united inches, $6, Toronto, Hamilton and London; terms 4 months or 3 per cent. 30 days.

OLD MATERIAL.

There is a fair delivery and a good demand. The feeling keeps steady. We quote jobbers' prices as follows: Agricultural scrap, 55c. per cwt.; machinery cast, 55c. per cwt.; stove cast, 40c.; No. 1 wrought 55c. per 100 lb.; new light scrap copper, 12c. per lb.; bottoms, 10½c.; heavy copper, 12½c.; coil wire scrap, 13c.; light brass, 7c.; heavy yellow brass, 10 to 10½c.; heavy red brass, 10½c.; scrap lead, 3c.; zinc, 2½c; scrap rubber, 7c.; good country mixed rags, 65 to 75c.; clean dry bones, 40 to 50c. per 100 lb.

SEEDS.

Though there is practically nothing doing, prices are nominally steady at $6 for the best values of alsike, and $5.50 to $6 for ordinary to the finest clover.

PETROLEUM.

A good movement continues. Prices keep firm as follows: Pratt's Astral, 16½ to 17c. in bulk (barrels, $1 extra); American water white, 16½ to 17c. in barrels; Photogene, 16 to 16½c.; Sarnia water white, 15½ to 16c. in barrels; Sarnia prime white, 14½ to 15c. in barrels.

COAL.

There is a large movement, but dealers state that they still have difficulty in getting cars. Prices are unchanged. We quote anthracite on cars Buffalo and bridges: Grate, $4.75 per gross ton and $4.24 per net ton; egg, stove and nut, $5 per gross ton and $4.46 per net ton.

MARKET NOTES.

Putty is 5 to 10c. per 100 lb. higher.

The Stanley Rule and Level Co., New Britain, Conn., has issued a new list of prices on the lines it manufactures. The discounts are unchanged, but the list is higher.

The stove and furnace manufacturers held a meeting in Toronto on Wednesday, but made no change in prices. And no change is anticipated, at any rate, for some time to come.

PAINT AND OIL AGENT WANTED.

A large manufacturing firm in England who manufacture paints and varnishes, and export oil, turpentine, harness oil, petroleum jelly, cycle oil, etc., is anxious to secure a thoroughly good firm or agent in Canada who will take up its line of goods. HARDWARE AND METAL will be pleased to put any reliable firm or individual in communication with the manufacturers in question. For particulars address, Agent, care of Advertising Department, HARDWARE AND METAL, Toronto.

A petition in insolvency was filed December 15 by Locke & Hodder, general merchants, Twillin Gate, Newfoundland.

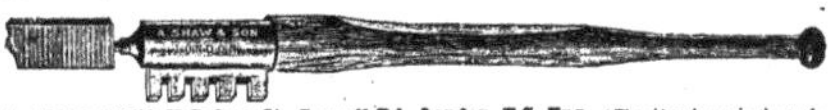

TRADE IN COUNTRIES OTHER THAN OUR OWN.

AMERICAN BARS IN ENGLAND.

THERE is no diminution in the importation of American steel bars for consumption by South Wales works. The imports up to November reached the enormous total of 41,000 tons, of which 25,000 were landed at Newport for the large sheet works in operation in that town. It is now reported, on a trustworthy authority, that an American shipping firm has contracted to deliver 20,000 tons of American bars at Newport for distribution among the various sheet and tinplate works in South Wales. There is, as will readily be seen, no improvement in the state of the local trade, and Welsh steel manufacturers may well be, as our South Wales correspondent puts it, at their "wit's end." The growth of the new American connection has badly disorganized both the steel and tinplate trades. It is worse than useless, however, for Welsh makers to stand idly by witnessing the disappearance of a valuable trade without making a real effort to retain it.—Hardwareman.

NEW YORK METAL MARKET.

The London metal market closed to-day at noon, but during the short business session the fluctuations in the price of pig tin were marked. At one time the market stood at £122 5s., and again at £121, but at the close the quotation for spot was £121 15s., or 17s. 6d. under Friday's closing. The trading in spot tin was light and in futures moderate in the English market, but here there was scarcely anything done. There was no call at the New York Metal Exchange, and speculative interest seemed to be entirely withheld. At the close spot tin was nominally quoted at 27c. Total imports of tin for the month of December amounted to 3,311 tons, leaving 1,445 tons afloat. The deliveries for consumption during December aggregate 2,400 tons, while the stock in store and loading is 3,041 tons.

COPPER—Continued firmness characterizes this market, although there was hardly a call for stock in any quantity to-day. Lake Superior is held at 17c., while for electrolytic and casting stock, 16⅝c. is quoted. In London this morning an advance of 2s. 6d. was recorded, the market closing firm.

PIG LEAD—There was virtually no market to-day, current wants of consumption being supplied and buyers deferring purchases for future needs until after the turn of the year. The close was dull at 4.37½c. for lots of 50 tons or over.

SPELTER—The market remained dull, with prices nominal and unchanged at 4.15c. There was no change in St. Louis or London.

ANTIMONY — Regulus remains steady though quiet at 9 to 10½c., as to brand and quantity.

IRON AND STEEL.—The last day of the year witnesses little buying or selling, as both buyers and sellers are occupied with other concerns which, for the time being, are of greater consequence. To-day was no exception to the general rule, and there was nothing new to be noted in any department. The record of the past two months, so far as the volume of business is concerned, was eminently satisfactory, and the trade is awaiting the opening of the new year and the dawning of a new century in the confident expectation of a continuance of the prosperous conditions of the immediate past.—New York Journal of Commerce, December 31, 1900.

PIG IRON IN ENGLAND.

The pig iron makers in the country have, in the majority of cases, obtained, within the last two or three weeks, such a marked abatement in the prices of raw materials that they are in a considerably better position than they were previously. The reduction in the price of coke has, generally speaking, been more considerable than the reduction in the price of pig iron. As a case in point, we may note the fact that in Cleveland warrants, as compared with this time last year, there is a fall of 11s. 9½d. per ton, but the fall in the cost of coke delivered at works from the highest point reached during the year has, in some cases, been 6d. to 1s. per ton more even than that. The imports of iron ores having fallen, too, during the past 11 months to the extent of something like 750,000 tons, we are justified in anticipating that the reduction in the total output of pig iron for the year will be at least 300,000 tons, and the fact that stocks are phenomenally low contributes to make the pig-iron situation a reasonably healthy one. The general expectation is that prices will fall somewhat lower still, but that this does not necessarily mean that pig iron will be in a worse position because the prices of raw material are likely to decline pari passu. The following is a statement of the public stocks of pig iron in tons :

		Decrease during 1900.
	Tons.	Tons.
Connal's at Glasgow	71,603	173,936
Connal's at Middlesbrough.	31,700 }	27,493
Railway Stores, " .	11,000 }	
Connal's at Middlesbrough, hematite................	555	8,948
Hematite, West Coast	22,863	175,384

—Iron and Coal Trades Review.

NEW OAR AND MAOHINE WORKS.

The new shops which the Ottawa and New York Railway built in Ottawa in consideration of a bonus $75,000 have been started.

The shops are located on Ann street, between King and Nicholas streets, and the main building is 125 ft. long, 60 ft. wide and two storeys in height. It is well equipped with machinery and tools necessary for the carrying on of the work. Adjoining is a blacksmith shop, 45 x 60 ft. in size, and a boiler house, 20 x 25 ft.

A boiler of 100 horse-power, along with a main engine of 75 horse-power, furnishes power for the shops, which have been in operation for five or six days, and at present 15 men are employed making repairs to cars.

A TRAVELLER'S ADVENTURES.

HARDWARE AND METAL had the pleasure a few days ago of meeting Mr. F. W. Franks, a representative of Aspinall's Enamel, Limited, of London, England. Mr. Franks is making a tour of the world, and he is more than an usually interesting man to talk with because of the incidents, many of them exciting, in which he has been directly and indirectly concerned.

It is about 15 months since he started out on his tour, which is chiefly a business one. During that time he has visited Gibraltar, Malta, Egypt, Ceylon, Straits Settlements, Dutch East Indies, Australia, New Zealand, Tasmania, the Philippine Islands, China and Japan. He is now in Canada and from here will go to New York and thence home.

In New Zealand he suffered the inconvenience of being quarantined on account of the bubonic plague. But that was only a small item in the list of his adventures. On his way to the Philippine Islands on the ss. Futami Maru, he was shipwrecked on Mindoro Island, the steamer striking a coral reef during a typhoon. On this Island, which was an uninhabited one, he lived for a week with 180 other persons, at the end of that time being fortunately taken off by a passing steamer, which the one boat they had saved from the wreck managed to intercept.

On the day he arrived at Hong Kong, the edict of the Dowager Empress of China was posted up urging the massacre of Europeans and Christians, and, strange to say, the day he reached Shanghai a similar edict was posted up there. In the latter city, an alarm was raised one night to the effect that the Boxers were marching on the city, and Mr. Franks, armed with a revolver which one of the volunteers loaned him, joined the ranks of those called out to defend the city. It was, fortunately, only an alarm.

Mr. Franks considers he has now only to be in a railway wreck to complete his list of adventures.

The visit of Mr. Franks to Canada is chiefly with a view to establishing distributing agencies in the chief commercial centres from the Atlantic to the Pacific. It is his first visit to Canada, and he is much surprised at the development which it has reached.

AN EXPENSIVE ERROR.

SOMEWHAT over a month ago the Canadian linseed oil market was disturbed and weakened by low quotations offered by a Montreal broker and merchant who was making contracts for February to August delivery at prices two to four cents below those of any other firm. At the time that these low prices were being scattered by telegraph over the country HARDWARE AND METAL made bold to say that it was difficult to understand how any firm could offer goods at such rates, and hinted that there must be a mistake somewhere.

Our suspicions have been only too correct, for the merchant who was revelling in such low prices now finds that his English house used a revised code of which he had not heard, and that he read the quotation 100s. a ton too low.

On the broker himself the misfortune entails a great loss, for he intends to fill all his contracts or pay damages. It has been prophesied in the trade that he would refuse to abide by his agreement, but we are glad to be able to say that he has no intention of ruining his reputation gained in 20 years of business experience by such a procedure. He will fill his contracts even at a loss of several thousand dollars unless his customers see fit to return the written agreement. Some firms have already voluntarily returned the papers, sympathizing with him in his misfortune. Others have refused to accede to his request to be allowed to withdraw and are claiming damages or the goods.

ICE TOOLS.

To the ice harvester it is of the utmost importance that he possesses the most perfect working tools it is possible to buy. None but the very best are economical to him. The cutting season is short, the weather is cold, the men are hurried, and an unnecessary delay on account of unmanagable or imperfect tools is both aggravating and expensive. W. T. Wood's ice tools have a world-wide reputation, and Rice, Lewis & Son, Limited, their Canadian agents, will be pleased to quote prices to all interested. Below we illustrate a few of their most useful tools:

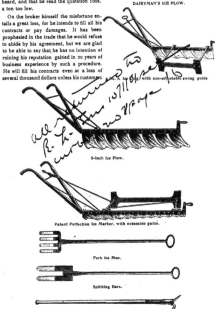

DAIRYMAN'S ICE PLOW.

No. A Ice Plow, with non-adjustable swing guide

8-Inch Ice Plow.

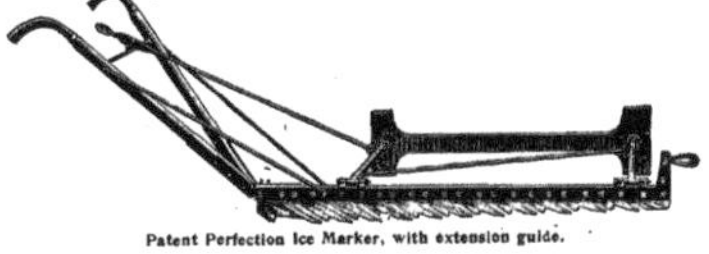

Patent Perfection Ice Marker, with extension guide.

Fork Ice Man.

Splitting Bars.

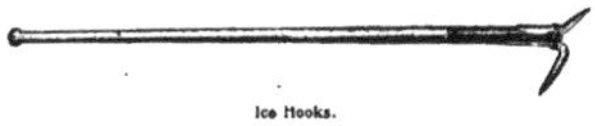

Ice Hooks.

John H. Birch, general merchant, etc., Dorchester Station, Ont., has been burned out; partially insured.

HEATING AND PLUMBING

GIVE US FRESH AIR.

THAT portion of civilized humanity which lives in large cities, says The Chicago Chronicle, is awaiting the coming of a benefactor to whom it will erect statues and sing pæans of praise. This benefactor will be the man who shall devise a cheap and effective method of ventilating flats and office buildings. No such method has yet been devised.

There is no ventilation of the class of buildings named. The unhappy tenant may choose between suffocation and pneumonia. He can either keep his windows down and stifle or he can raise them and creat a draft which shall be his undoing. He occupies a steam-heated box in which no provision has been made for fresh air. He breathes over and over again an atmosphere charged with carbonic acid gas. He may pay $5 per month, or he may pay $500, but he will get no ventilation.

The ordinary dwelling is better. It is usually heated by a furnace and that involves the introduction of fresh air into the house through the furnace flues. Usually, too, there is an open fire in the parlor or library or dining-room, and the open fire is an unsurpassed ventilator. Unless the householder be crazy enough to throw away his advantages and substitute steam heat for the furnace and fireplace, the average dwelling is fairly well ventilated.

But for the flat dweller and the tenant of the office building, there is no present hope. They are doomed to stifle or to shiver until some ingenious American—and it really doesn't seem as though he need be a marvel of ingenuity, either—shall devise a means of letting a little fresh air into steam-heated rooms without subjecting the tenants thereof to the peril of serious illness.

Is there not somewhere in this broad land a budding Edison who shall solve the problem and make his name blessed to countless generations yet unborn ?

NICKEL-PLATED LEAD PIPE.

Purdy, Mansell & Co. have successfully made an experiment which is likely to be of practical value. They put a piece of lead pipe through nickel plating apparatus. It was doubted whether the plating would hold to the soft lead and retain the clear, bright appearance that the plated brass pipe presents. The result of the experiment, however, was so satisfactory that the nickel-plated lead pipe could not be distinguished from plated brass pipe by the eye. The lead pipe retains its softness and pliability, but the plating is clear and firm. It has already been used in plumbing work by Purdy, Mansell & Co.

WILL MAKE STEEL PLATES.

On Saturday evening a meeting of the shareholders of The Dominion Iron and Steel Co., Limited, was held in Montreal, when the issue of $5,000,000 of preferred stock was ratified. A. J. Moxham, the general manager of the company, stated that the first iron furnace would be blown in during January and that many contracts for iron for European and American delivery had been secured. Work on the construction of the new steel-rolling plant will be commenced at once.

BUILDING PERMITS.

Building permits have been issued in Toronto to John Ewing, for three two-storey and attic residences at 135, 137 and 137½ North Beaconsfield avenue, to cost $5,000, and to G. B. Cameron, for a detached brick residence on Macpherson avenue, to cost $2,800.

SOME BUILDING NOTES.

Hanover, Ont., has offered The Knechtel Furniture Co. a bonus of $25,000 on condition that they will replace their building, which was destroyed by fire, with a better one. It is thought that the offer will be accepted.

Manager Abbott, of the Butte Hotel and concert hall, Phœnix, B.C., intends erecting, immediately in the rear of the hotel, an addition 60 x 70 ft. in size and two storeys in height. The ground floor will be fitted up as a concert hall, and the upper storey will be used as rooms for the hotel.

PLUMBING AND HEATING NOTES.

J. H. Midgley & Co., plumbers, Brandon, Man., have been succeeded by T. D. M. Osborne.

Owing to a falling off in the pressure from the Kingsville, Ont., natural gas wells and a fear that the flow may become very small, many householders in Windsor, Ont., have installed coal stoves to take the place of gas heaters. The plumbers, tinsmiths, gasfitters, coal dealers, etc., of Windsor, are consequently working night and day.

HOW A TRAP TEST SHOULD BE CONDUCTED.

THE following appeared in a recent issue of The Plumbers' Trade Journal : "The recent heralded test of traps in one of our eastern cities leads one to suppose that, after all was said and done, there was very little to demonstrate the true worth of a trap. An abnormal test of the sort represented does not demonstrate the actual sanitary worth of a trap, the one great fact being evident that, in order to stand a test of the sort reported, a trap must have mechanical features that of a necessity make it really insanitary.

"Built on lines for a perfect resistance of non-syphonic action proves a mechanical construction internally that is a danger in itself, permitting the deposit of grease, etc., that in time accumulate and make it a most insanitary trap.

"I have seen traps so constructed that are certainly a menace to the good health of those coming in contact with them in every-day usage.

"A normal test of traps on scientific principles, quantities of water used must be commensurate with that ordinarily used by the average person in basin or bath ; with the fall natural the test on lines of good plumbing from say three or four-storey construction, the fall from storey to storey gives the results of a good wholesome test.

"To exert a syphon on a trap and let it go just to see what it will do is not a proper test. A test of the sort I write will demonstrate the true worth of a trap as a sanitary fixture ; obviously, the best trap is the one that has the least internal construction, compartments, etc.; in fact, anything that is used for resistance is bad. A clear waterway made scientifically self-cleaning, pure and wholesome at all times under all conditions, those are the traps that count ; they are in the market and are being used every day ; have stood the test of time, and will always be recognized by the expert workman and sanitary engineer as being the nearest perfection, the best for all. There are a number of this class I write of, all good, all safe."

H. S. Skilson & Co., general merchants, Roland, Man., have been burned out. The loss is covered by insurance.

D. O. Bourdeau, of D. O. Bourdeau & Fils, general merchants, Victoriaville, Que., is dead.

Returned to Owner New York 4/1/01 See Cutback No 2 Page 3.

No. 219 Never Turn Screw Driver

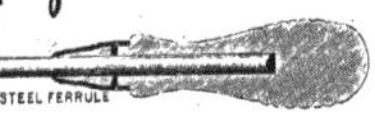

Fools never turn. Neither does No. 219. This turns screws and not in the handle.

GET OUR GREEN BOOK OF HARDWARE SPECIALIES.

SMITH & HEMENWAY CO., NEW YORK.

CANADIAN ADVERTISING is best done by THE E. DESBARATS ADVERTISING AGENCY MONTREAL.

ADVERTISING in WESTERN CANADA

will be Carefully, Efficiently, and Promptly attended to, by

The Roberts Advertising Agency,
WINNIPEG, CANADA.

KNOX HENRY

Heavy Hardware and Metal Broker
Room 220½ Board of Trade, MONTREAL.

TACKS, CUT NAILS, WIRE NAILS, HORSE SHOES, HORSE NAILS, SPIKES, BOLTS, NUTS, ETC.

SPECIALTIES – "C" Brand Horse Nails – Canada Horse Nail Co.

"BRASSITE" GOODS – Gunn Castor Co., Limited, Birmingham, Eng.

Manufacturers of

Heating Supplies

Pipe Fittings and Headers.
Large Manifolds made to Order.
Steam Traps and Appliances, etc.

The . . .
Jas. Morrison Brass Mfg. Co., Limited
TORONTO.

FIXING-UP TIME

follows stocktaking and is the best time to put in

BENNETT'S PATENT SHELF BOX
and the
KLONDIKE SAMPLE HOLDER.

J. S. BENNETT, 20 Sheridan Ave., Toronto.

N.B.–Don't forget we make boxes to fit your present shelving.

Standard Paint & Varnish Works.
Limited
Makers of High Grade
Varnishes, Japans,
Paints, Colors & Enamels.
Windsor, Ont.

Hardwood CHARCOAL in Bulk or Sacks.

WOOD ALCOHOL equalling Methylated Spirits as a solvent.

Manufactured only by..

THE STANDARD CHEMICAL CO., Limited

Factories { Fenelon Falls. Deseronto.

Gooderham Building, TORONTO

DIAMOND EXTENSION STOVE BANK

Patented, July 11th, 1898. Canadian Patent, June 14th, 1894.

They are easily adjusted and fitted to a stove by any one.

Please your customers by supplying them immediately with what they want.

EXTENDED.

Sold by Jobbers of . . .
Hardware
Tinware
and
Stoves.

Manufactured by THE ADAMS COMPANY, Dubuque, Iowa, U.S.A.
" " A. R. WOODYATT & CO., Guelph, Ontario.

LEADER CHURN

New Century Improvements.

FOUR DIFFERENT STYLES:
A—Steel Frame with double reversible Steel Lever.
B—Wood Frame with double reversible Steel Lever.
C—Steel Frame with Crank.
D—Wood Frame with Crank.

Styles A and B may be operated from a sitting or standing position.

Steel Frames and Hoops beautifully ALUMINIZED.

All LEADER CHURNS are equipped with BICYCLE BALL BEARINGS and PATENTED CREAM BREAKERS.

Stands are so constructed that they are particularly strong and rigid, and there is nothing to interfere with the placing of pail in the most convenient position for draining off buttermilk.

It Pays to Have the Best. None are Better Than the Leader.

THE
Dowswell Manufacturing Co.
Limited.

HAMILTON, ONT.

Eastern Agents: W. L. Haldimand & Son, Montreal, Que.

STANLEY RULE & LEVEL CO.,

NEW BRITAIN, CONN., U.S.A.

IMPROVED CARPENTERS' TOOLS

SOLD BY ALL HARDWARE DEALERS.

THE MAKING OF A RADIATOR.

THE Dominion Radiator Co., Limited, Toronto, always displays good taste in its advertising schemes. Another evidence of this is a New Year's card, or, rather, folder, which it has just issued. The first and second pages on the inside have illustrations showing two stages in the manufacture of the Safford radiator. The one shows the start, where the metal is being poured from the furnace. This illustration is herewith reproduced. The other shows the finished radiator. The whole folder is nicely designed and printed, and The Dominion Radiator Co., Limited, is to be congratulated.

The Start in Making a Radiator.

COMMERCIAL TRAVELLERS DINE.

THE twenty-eighth annual banquet of The Dominion Commercial Travellers' Association was held in the Place Viger Hotel, Montreal, on Saturday evening. There were about 200 guests present. The new president, T. L. Paton, occupied the chair.

The dinner proved a choice repast, but was equalled in point of quality by the excellence of the speeches which followed. After the toast of "The Queen," Mr. Paton referred to the growth of the association, which had increased from 251 members in 1875 to a membership of 3,485 in 1900. It now has an investment of $174,000. During the 25 years of its existence it had paid out no less than $274,819 in benefits.

The "Parliament of Canada" was responded to by Mayor Prefontaine, M.P., and F. D. Monk, M.P.; that of the "Legislature" by Hon. E. J. Flynn, leader of the Opposition in the Quebec Legislature. T. W. Burgess, president of the White Mountain Commercial Travellers' Association, and T. W. Barnard, jr., of Boston, brought fraternal greetings. Good speeches, good songs and good comradeship united to make a most enjoyable evening.

TORONTO TRAVELLERS' OFFICERS.

At the annual meeting of the Toronto City Travellers' Association, on Monday evening, the following officers were elected:

President—M. A. Muldrew, of Lumsden Bros.

First Vice-President—W. Anderson, of T. A. Lytle & Co.

Second Vice-President—W. A. Mitchell, of Todhunter, Mitchell & Co.

Chaplain—D. J. Ferguson, of Fairles Milling Co.

Guard—James Scott, of T. A. Lytle & Co.

Marshal—T. Holman, of The Christie, Brown Co., Limited.

Treasurer—J. Mortimer, of The Christie, Brown Co., Limited.

Secretary—W. F. Daniels, of Lyman, Knox & Co.

D. C. Thiesen, general merchant, Rosendoff, Man., has assigned to C. H. Newton, and a meeting of his creditors has been held.

INQUIRIES FOR CANADIAN PRODUCTS.

The following were among the recent inquiries relating to Canadian trade received at the High Commissioner's office, in London, England:

1. A north of England firm who are already engaged in the importation of eggs, cheese and butter, etc., are open to buy further supplies from Canada, and desire to be placed in communication with some large exporters in the Dominion.

2. The names of sound business firms in Canada who deal in mining materials are asked for by the manufacturers of steel wire screening for gold-mining.

3. Two applications have been received for names of asbestos mine owners in Canada.

4. The manufacturers of tinned, japanned, and enamelled hollow-ware, who have shipped several consignments of enamelled ware to Canada, are anxious to push the business and will be glad to hear from Canadian houses interested in it.

5. The names of manufacturers of the various kinds of wood pulp and oakum are asked for by a north of England firm.

[The names of the firms making the above inquiries, can be obtained on application to the editor of HARDWARE AND METAL. When asking for names, kindly give number of paragraph and date of issue.]

Mr. Harrison Watson, curator of the Canadian Section of the Imperial Institute, London, England, is in receipt of the following inquiries:

1. The proprietors of a patent water-feed filter and grease extractor would like to hear from Canadian firms prepared to introduce same in the Dominion.

2. A Liverpool firm desires information as to the production of corn oil in Canada and also the names of any manufacturers of the article.

3. A Birmingham firm asks for names of Canadian makers of dowels who can quote in good specification.

4. A French syndicate interested in wines, brandy, chocolate, preserves, etc., would be pleased to hear from a Canadian firm who would act as their representatives.

5. A South-African firm would like to secure the services of a reliable Canadian firm who could act as buying agent for timber and other lines in which they are interested. First-class references required.

MR. JEANDRON'S NEW POSITION.

W. J. Jeandron, representing J. C. McCarty & Co., manufacturers' agents, New York, for the past three years in Canada and Eastern States, leaves shortly to take up his residence in New York to act as American buyer for Rayer & Rangier Freres, hardware jobbers, Paris, France.

Mr. E. W. McCarty succeeds him in Canadian territory.

EMBARRASSMENT OR FAILURE?

By Hugo Kanzler.

IT seems to have become the custom of late to divide smashups into higher and lower classes, both of them having formerly been termed failures. But a marked distinction is now being made in the application of this term, depending largely upon the circumstances necessary to gloss over the true facts, or those not intended for publication. In the former case, the word "failure" is regarded as a harsh epithet, and the milder word "embarrassment" is substituted; while, if a small merchant meets with reverses, his misfortune is unfeelingly denominated a "failure," and he is not held in the same kind regard by the commercial community as in the other case. Of course, in order to justify the term "embarrassment" and administer the necessary sedatives to the creditors, delays are resorted to that are a mere mockery to the business world, more particularly so if the "embarrassed" firm has enjoyed unlimited credit for a number of years; and such artful pretexts are employed as "Books not posted," "Accounts to be investigated," "Expert bookkeepers to be employed," etc., in order to gain time to formulate plans while the storm is blowing over—to eventually induce creditors to accept a settlement satisfactory to the friends of the debtor.

The appointment of a receiver is unquestionably a most favorable modus operandi, as only in case of unreasonable delay on his part can the machinery of the courts be put into operation; and while the courts are ready to have their officers act promptly in administering the affairs of the defunct firm, it is nevertheless, with great hesitancy that the same rigid rule is adopted as to the receiver as under the assignment law or bankruptcy.

In the usual and ordinary course of business, the members of a firm look forward to the monthly trial balance on the first day of every month or shortly thereafter, and from this and the proper books kept by the confidential clerk a firm is well able to ascertain its true condition without having recourse to experts to do the work of commonsense bookkeepers.

Every firm or corporation, large or small, should see to it that a trial balance is handed down shortly after the close of the previous month; and it must now be regarded as a lax method of doing business if such statement is not rendered within at least ten days after the proper time. It seems incredible than any large mercantile concern at the present day should not know its true condition at least once every 30 days, more particularly so when the omission on the part of the small dealers to keep proper books is severely criticized in case of failure. And no substantial reason can be assigned why the same complaint ought not justly be made against the rich man's failure.

SUBSTITUTION CAUSES SUSPICION.

THE attempt by a dealer to sell his customer a substitute in place of the article the buyer calls for at once places that dealer under the ban of suspicion. The only reason why the buyer does not invariably realize the suspiciousness of substitution and promptly resent it, is probably because in many cases the money transaction involved is so small that it does not suggest the motive for fraud. Suppose a jeweler advertises a diamond at $50. He places it in his window. A would-be buyer enters the store and asks for this particular diamond. But the jeweler says: "I can give you that diamond if you want, but here's another that's just as good as the one advertised." The buyer's suspicions would be aroused at once. He would insist on the stone in the window and he'd keep an eye on it to see it wasn't changed. But in the case of a 50c. transaction it is different. The buyer is offered as "just as good" as a widely-advertised article, although substitution is just as suspicious in a 50c. transaction as in one involving $50.

Look at the question from another point of view. A sale of stock is advertised. There are horses with pedigrees and records to be sold. Farmer Brown attends the sale with the purpose of buying one of these good horses. But the seller says to him: "That horse you want is a good horse, of course, but I've got another here that is just as good which I'd like to sell you."

"Has he just as good a pedigree?"

"Well, no, he hasn't any pedigree to speak of."

"Has he any record?"

"Well, no, we never held a watch on him that I know of, but he's just as good as the horse you want."

Would Farmer Brown buy the "just as good" horse? The question answers itself. And yet, this same farmer will allow himself to be swindled time and again by accepting "just as good" articles in place of those called for. The article he called for has, so to speak, a pedigree and a record. It's a standard in the markets of the world. Yet, in place of this standard article, he will accept a substitute which nobody knows anything about—an untried, unproved article which has no record of value and no proof of origin.

Let the buyer who is offered a substitute bear in mind that substiution is suspicious, and that a substitute always carries the earmarks of a swindle.—Belleville Sun.

A CORNER FOR CLERKS.

ONE WAY TO KEEP TRACK OF A CLERK

JOHN BROWN, of Brownville, has the best store in that thrifty village. He has been bothered about to death by his clerks. There hasn't been anything they haven't done and won't do again if they have a chance. Their leading idea in life is to get ahead of him in some way and so far they have succeeded. For a long time he satisfied himself by discharging them when they proved not to his liking. He soon learned that he was teaching a

COMMERCIAL KINDERGARTEN

and that other storekeepers were ready to take his pupils by the time he was ready to graduate them. He found that his old idea of taking raw hands and moulding them into his pet forms, while it did carry out the idea of "clay in the hands of the potter," did as surely carry out the kindergarten thought, and of that he had had more than enough.

He made up his mind to

CHANGE HIS PLAN.

For some years now his showcase had furnished his youthful helpers shirt studs and sleeve buttons. His neckties went the same way. He supplied at less than cost the collars and cuffs which none of them had when they came to him, and while he had been willing to do this and so help the boys along in the world—at all events to get started in it—he began to find that it was only so much patience and generosity thrown away, and he made up his mind to have no more of it.

He noticed that the two clerks he now employed were pretty well fixed so far as goods were concerned which his stock could furnish, and he noticed, too, that both were showing those unmistakable signs which mean

AN EARLY GOOD BYE.

He sat down and made a little calculation—his books furnished him the needed data—and he found it would be money in his pocket to raise the boys' wages and save himself the trouble of breaking in another pair of clerks and the expense of supplying them with the usual outfit.

So far as he could judge, the trouble seemed to be in the fact that the boys began by being out nights and getting into the kind of mischief which ends in making them uneasy and discontented and good for nothing. Every case he could think of was traced directly or indirectly to that, and the problem, so far as he understood it, was how to prevent the young fellows from being up and out at night long after the time when they ought to be in bed.

To add to the difficulty, Brownville was at that stage of its existence when, like the meeting of the waters, it was neither rivulet nor river.

A BIG CLUMSY GAWK

of a place, it had spread itself over a large territory and had a frame like a giant, which the years in time might fill up, but there were no strong inducements for the boys to stay and grow up with it, and, the minute they were plumed for their flight, off to the city they went, and the Brownville which knew them once knew them no more forever. Like most places, as it grew it fought vigorously against the evils which attack the growing town. The saloon came and stayed. There were some billiard tables set up, and they thrived.

CARDS BEGAN TO BE PLAYED,

and almost before the people knew it the young folks began to be fast. The Sunday-school began to grow thin, and nobody but women went to church. In a word, while the town could not be said to be going down at the heel, it did seem to be a bad place for a boy who was inclined to fear being called a "wayback" or, what was far worse, "not up to date."

Mrs. Brown was in every sense of the word a helpmeet. She had no longings which took her away from her husband and his calling, and, while it had been years since she had given up her place behind the counter, she never cared to look beyond the horizon which shut in the Brownville store. When, therefore, the question was asked if she couldn't take the boys into the house, just as she did years ago, and she had been told the reason, like the devoted wife she was there was but one answer to be thought of and that was given promptly and heartily, and the childless woman made up her mind

TO TAKE THE BOYS

in and do for them and love them as if they were her own flesh and blood.

That night after closing the storekeeper had the boys stay for a while for a talk. "I've made up my mind," he began, "to raise your wages, boys. You've been doing good work and you've been faithful enough to please me, and, while I shan't give you much more, it's something, and it'll let you know anyway that I want to keep you. There are two conditions that I want to make and insist on if I raise your wages—one is that you live with me, and the other is that you are

AT HOME NIGHTS

by 9 o'clock; unless I know where you are and what you are doing. I'll give you good board and each of you shall have a good room; but I want you to be in at 9 o'clock and stay there. Think it over and tell me your decision to-morrow. I'll raise each of you 10 per cent. Good night."

The boys left the store on air and came back the next morning in the same frame of mind. Mrs. Brown came down during the morning to report that the rooms were ready and that afternoon saw the transfer of bag and baggage.

EVERYTHING WAS DONE

for the young men that could be thought of or asked for and Brown himself was forced to admit that he had hit on the only thing that could ever have worked with those fellows. They were honest to a dot. They were industrious to a fault. They meant well from first to last, and all they needed was just that little bit of restraint which John Brown had wit enough to insist upon; and on that and on every night, after the town clock struck nine and he knew both boys were in, he

LOCKED THE ONLY DOOR

they could get out of and put the key under his pillow; and every night Susan Brown heard him say to himself with infinite satisfaction, "There, darn ye! With the windows fastened on the outside, and the only key under my pillow, you can skin out and carouse all night if you can, and I'll never say a word!"

It was a good while before the boys found out that they were locked in from nine o'clock until morning. The first thought was rebellion; but when sober sense came to the front and they saw what an advantage the rest and the home had been to them they kept the matter to themselves, glad that "Uncle John," as they learned to call the storekeeper, had marked out the way and compelled them to walk in it; while Brown himself, to this day, affirms that "the only way to get along with clerks is to put 'em under lock and key and keep 'em there!"—Richard Malcolm Strong, in Michigan Tradesman.

Campbell & Cahill, dealers in agricultural implements, Rodney and West Lorne, Ont., whose premises in the latter place were burned the other day, have dissolved. Mr. Campbell continues in Rodney and Mr. Cahill in West Lorne.

HARDWARE SPECIALTIES.

THE Smith & Hemenway Co., one of the youngest and most enterprising houses in their line of business, herein illustrate a few of their new specialties:

219.

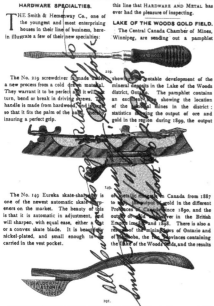

The No. 219 screwdriver is made under a new process from a cold drawn material. They warrant it to be perfect and it will not turn, bend or break in driving screws. The handle is made from hardwood, and is large so that it fits the palm of the hand, thereby insuring a perfect grip.

149.

The No. 149 Eureka skate-sharpener is one of the newest automatic skate-sharpeners on the market. The beauty of this is that it is automatic in adjustment, and will sharpen, with equal ease, either a flat or a convex skate blade. It is beautifully nickel-plated, and small enough to be carried in the vest pocket.

191.

The No. 191, Waldorf-Astoria can opener is a new departure in the can-opener line, being made from one "solid piece of steel" with a centre cut, which insures perfect gripping on the can. These can-openers are beautifully nickel-plated, and mounted on display cards to be placed on the counter.

In addition to the specialties herein illustrated, the Smith & Hemenway Co. are one of the largest manufacturers of this line of goods in the United States. They also manufacture a full line of vises, from amateur to the largest machine vise, and market the entire product of nippers and plyers of the Utica Drop Forge and Tool Company.

They state to us that they do nat publish a catalogue, but get out what they call the "Green Book of Hardware Specialties," which is one of the most unique things in this line that HARDWARE AND METAL has ever had the pleasure of inspecting.

LAKE OF THE WOODS GOLD FIELD.

The Central Canada Chamber of Mines, Winnipeg, are sending out a pamphlet showing the notable development of the mineral deposits in the Lake of the Woods district, Ontario. The pamphlet contains an excellent map showing the location of the principal mines in the district; statistics showing the output of ore and gold in the region during 1899, the output of metallic minerals in Canada from 1887 to 1899, the output of gold in the different Provinces of Canada since 1890, and the output of gold and silver in the British Empire in 1897 and 1898. There is also a resumé of the mining laws of Ontario and of Manitoba, the two Provinces containing the Lake of the Woods fields, and the results of 124 assay tests. As the purpose of this bureau is to disseminate reliable information and statistics through the medium of the press throughout the world, and as these essays are published under affidavit this pamphlet of the Central Canada Chamber of Mines should be of much value, especially to those immediately interested in mining affairs.

NEW CENTURY GREETINGS.

Upon another page will be found the new century greeting of that enterprising go-ahead company, the Canada Paint Company, of Montreal and Toronto. They are out for nineteen hundred and one, as the circular indicates, with a number of new features which the hardware trade will do well to study. This company have found it necessary to enlarge their manufacturing facilities very largely and energetic firms throughout Canada will, as heretofore, undoubtedly find it to their advantage to continue to push the manufactures of the Canada Paint Company.

CARLOADS OF MOLTEN IRON.

THE construction of a new bridge across the Monongahela, to be opened for service within the next few days, directs attention afresh to a striking feature of modern metallurgy. The usual way to make steel is to melt up cold pig iron, to which other materials are added, and then purify the mixture by burning out certain undesirable elements. Pig iron, however, is itself the product of a previous heating process, in which the ore is melted with carbonate of lime to remove the oxygen. It occurred to some ingenious Yankee a few years ago that, if the product of the blast furnace could be converted into steel before it had cooled sensibly, a great economy in fuel would be secured.

The new bridge just mentioned has been built for the Carnegie Company, and will be used to convey molten iron from the Carrie furnaces to the Homestead Steel Works, nearly a mile off. At the present time, Homestead obtains molten metal from Duquesne, about four and one-half miles away! The new route has been laid out so as to save time and distance, and, possibly, caloric, too. There has been for some time one "hot metal" bridge across the Monongahela, controlled by the Carnegie Company, and, besides the new one about to be opened, a third is in process of erection for the Jones & McLaughlin interest. It will thus be perceived that the practice has proved so successful that it is being rapidly extended.

One gets a vivid idea of this remarkable procedure when he reads about the precautions taken in the construction of the new bridge to prevent harm in case any of the melted metal leaks or slops over while in transit from the iron furnace to the steel works. The spaces between the ties are to be filled with sand so that no iron may fall to the decks of passing steamers. The ties will be of wood, but are to be protected by a covering of sand. On either side of the track their will be raised a screen of heavy metal plates, faced with firebrick and reaching to a height of four feet. An extension of thinner plates will bring the screen up six feet farther. The cars are ladle-shaped, and the molten metal runs directly into them when the furnaces are tapped. A locomotive then draws the train to the steel works at a moderate pace. The glowing freight, says The New York Tribune, is still in a fluid condition when it reaches the mixers there. If it were not, the cars would be ruined.

CURRENT MARKET QUOTATIONS.

January 4, 1901.

These prices are for such qualities and quantities as are usually ordered by retail dealers on the usual terms of credit, the lowest figures being for larger quantities and prompt pay. Large cash buyers can frequently make purchases at better prices. The Editor is anxious to be informed at once of any apparent errors in this list, as the desire is to make it perfectly accurate.

METALS.

Tin.

Lamb and Flag and Straits—56 and 28 lb. ingots, per lb.	0 33	0 34

Tinplates.

Charcoal Plates—Bright

M.L.S., equal to Bradley.		Per box
I.C., usual sizes	....	$7 00
I.X., "	....	8 50
I.X.X., "	....	10 00
Famous—		
I.C.	....	7 50
I.X.	....	8 50
I.X.X.	....	9 50
Raven & Vulture Grades—		
I.C., usual sizes	....	5 00
I.X., "	....	6 00
I.X.X., "	....	7 00
I.XXX., "	....	8 00
D.C., 12⅛x17	....	4 75
D.X.	....	5 50
D.X.X.	....	7 50

Coke Plates—Bright

Bessemer Steel—		
I.C., usual sizes	....	4 70
I.C., special sizes, base	....	4 50
20x28	....	8 75

Charcoal Plates—Terne

Dean or J. G. Grade—		
I.C., 20x28, 112 sheets	....	8 55
I.X., Terne Tin	....	10 75

Charcoal Tin Boiler Plates.

Cookley Grade—		Per lb.
X X., 14x56, 50 sheet bxs } " 14x60 " } " 14x65, " }	0 07	0 07½

Tinned Sheets.

72x30 up to 24 gauge	0 08	0 08½
" 26 "	0 08½	0 09
" 28 "	0 09	0 09½

Iron and Steel.

Common Bar, per 100 lbs	1 65	1 70
Refined " "	....	2 15
Horse Shoe Iron "	....	2 05
Hoop steel, 1½ to 3 in. base, extras for smaller sizes	....	3 10
Sleigh Shoe Steel " base	....	2 00
Tire Steel	....	2 00
Machinery iron finish	....	2 05
Cast Steel, per lb	0 00	0 00
Toe Calk Steel	....	2 39
T. Firth & Co's special cast steel, per lb.		0 13
Jessop's Tool Steel	....	0 13

Boiler Tubes.

1½-inch	....	0 21½
2 "	....	0 13½
2½ "	....	0 16
3 "	....	0 17½

Steel Boiler Plate.

¼ inch	....	2 25
3-16 inch	....	2 25
⅜ inch and thicker	....	2 25

Black Sheets.

18 gauge	....	3 10
20 gauge	....	3 10
22 to 24 "	....	3 20
26 "	....	3 30
28 "	....	3 40

Canada Plates.

All dull, 52 sheets	....	3 15
Half polished	....	3 25
All bright	3 85	4 00

Iron Pipe.

Black pipe—		
¼ inch	....	3 00
⅜ "	....	3 00
½ "	....	3 00
¾ "	....	3 31
1 "	....	4 50
1¼ "	....	6 25
1½ "	....	7 75
2 "	....	10 40
2½-6 inch, discount 60 p.c.		
Galvanized pipe—		
½ inch	....	4 50
¾ "	....	6 00
1 "	....	7 00
1¼ "	....	9 50
1½ "	....	11 75
2 "	....	15 75

Galvanized Sheets.

	G. C.	Comet.	Amer.	Queen's Head.
16 gauge		....	4 25	4 10
18 to 24 gauge	4 35	4 20	4 35	4 35
26 "	4 60	4 45	4 35	4 60
28 "	4 85	4 70	4 50	4 85

28 gauge American equals 26 gauge English.

Chain.

Proof Coil, 3-16 in., per 100 lb	....	....
" ¼ " "	8 00	8 50
" 5-16 " "	5 35	5
" ⅜ " "	4 35	4 85
" 7-16 " "	4 15	4 65
" ½ " "	4 35	4 50
" ⅝ " "	3 85	4 35
" ¾ " "	3 81	4 00
Halter, kennel and post chains,	40 and 50 p.c.	
Cow ties	40 p.c.	
Stall fixtures	35 p.c.	
Trace chain	25 p.c.	

Jack chain, iron, single and double, discount 35 p.c.

Jack chain, brass, single and double, discount 40 p.c.

Copper.

Ingot

English B. S., ton lots	0 19	0 20
Lake Superior	....	....

Bolt or Bar.

Cut lengths round, ½ to ⅞ in.	0 23½	0 25
" round and square 1 to 2 inches	0 23½	0 25

Sheet.

Untinned, 14 oz., and light, 16 oz., 14x48 and 14x60	0 23	0 23½
Untinned, 14 oz., and light, 16 oz., irregular sizes	0 23	0 23½

NOTE.—Extra for tinning, 2 cents per pound, and tinning and half planishing 3 cents per pound.

Tinned copper sheets	....	0 26
Planished	....	0 32

Braziers (In sheets.)

4x6ft. 25 to 30 lbs. ea., per lb.	....	0 25½
" 35 to 45 " "	....	0 24½
" 50-lb. and above, "	....	0 23½

Boiler and T. K. Pitts.

Plain Tinned, per lb	....	0 28
Spun, per lb	....	0 32

Brass.

Rod and Sheet, 14 to 30 gauge, 15 per cent.		
Sheets, hard-rolled, 2x4	0 24	0 25
Tubing, base, per lb	0 24	0 25

Zinc Spelter

Foreign, per lb	0 06	0 06½
Domestic "	....	....

Zinc Sheet.

5 cwt. casks	6 75	7 00
Part casks	7 00	7 50

Lead.

Imported Pig, per lb	0 04½	0 05
Bar, 1 lb	0 05½	0 05¾
Sheets, 2½ lbs. sq. ft., by roll	....	0 06½
Sheets, 3 to 6 lbs., "	....	0 06

NOTE.—Cut sheets ½ cent per lb. extra. Pipe, by the roll, usual weights per yard, lists at 7c. per lb. and 15 p.c. dis. f.o.b. Toronto.

NOTE.—Cut lengths, net price, waste pipe 8-ft. lengths lists at 7½ cents.

Shot.

Common, $6.50 per 100 lb.; chilled, $7.00 per 100 lb.; buck, seal and bal, $7.50. Discount, 7½ p.c. Prices are f.o.b. Toronto, Hamilton, Montreal, St. John and Halifax. Terms 3 per cent. cash, freights equalized on Montreal.

Soil Pipe and Fittings.

Discount, 60 and 10 per cent. on medium and extra heavy, and 60 per cent. on light.

Solder.

	Per lb.	Per lb.
Bar half-and-half	0 19	0 20
Refined	....	0 19
Wiping	....	0 18

NOTE.—Prices of this graded according to quantity. The prices of other qualities of solder in the market indicated by private brands vary according to composition.

Antimony.

Cookson's, per lb	0 11	0 11½

White Lead.

	Per cwt.
Pure	6 62½
No. 1 do	6 25
No. 2 do	5 87½
No. 3 do	5 50
No. 4 do	5 12½
Munro's Select Flake White	7 12½
Elephant and Decorators' Pure	6 87½

Red Lead.

Genuine, 560 lb. casks, per cwt.	....	$5 50
Genuine, 100 lb. kegs, per cwt.	....	5 75
No. 1, 560 lb. casks, per cwt.	....	5 25
No. 1, 100 lb. kegs, per cwt.	....	5 00

White Zinc Paint.

Elephant Snow White	0 08	0 09
Pure White Zinc	0 08	0 19
No. 1	0 06	0 07½
No. 2	0 05	0 6½

Dry White Lead.

Pure, casks	5 75
Pure, kegs	6 25
No. 1, casks	5 50
No. 1, kegs	00

Prepared Paints.

In ¼, ½ and 1 gallon tins.

Pure, per gallon	....	1 20
Second qualities, per gallon	....	1 00
Barn (in bbls.)	0 75	0 85
The Sherwin-Williams Paints	....	1 45
Canada Paint Co's Pure	....	1 20
Sanderson Pearcy's Pure	....	1 20
Stewart & Wood's Champion Pure.		1 20

Colors in Oil.

25 lb. tins, Standard Quality.

Venetian Red, per lb	0
Chrome Yellow	0
Golden Ochre	0
French "	0
Marine Black	0
" Green	0
Chrome "	0 0
French Imperial Green	0 [illegible]

Colors, Dry.

Yellow Ochre (J.C.) bbls	1 35	1 40
Yellow Ochre (J.F.L.S.), bbls	....	2 75
Yellow Ochre (Royal)	1 10	1 15
Brussels Ochre	....	2 00
Venetian Red (best), per cwt.	1 80	1 90
English Oxides, per cwt.	3 00	3 25
American Oxides, per cwt.	1 75	2 00
Canadian Oxides, per cwt.	1 75	2 00
Super Magnetic Oxides, 93 p.c.	2 00	2 25
Burnt Sienna, pure, per lb.	....	0 10
" Umber, " "	....	0 10
do Raw	....	0 09
Drop Black, pure	....	0 09
Chrome Yellows, pure	....	0 18
Chrome Greens, pure, per lb.	....	0 12
Golden Ochre	....	0 03½
Ultramarine Blue in 28-lb boxes, per lb.	0 08	0 24
Fire Proof Mineral, per 100 lb.	....	1 00
Genuine Eng. Litharge, per lb	....	0 07
Mortar Color, per 100 lb.	....	1 25
English Vermillion	....	0 80
Pure Indian Red, No. 45, lb.	....	0 80
Whiting, per 100 lb.	....	0 55

Blue Stone.

Casks, for spraying, per lb.	0 07
100-lb. lots, do. per lb.	0 08

Putty.

Bulk in bbls.	2 00
Bulk in less quantity	2 15
Bladders in bbls.	2 20
Bladders in kegs, boxes or loose	2 35
Bladders in 25-lb. tins	2 45
Bladders in 12½-lb. tins	2 75
Bladders in bulk or tins less than 100 lb	3 00

Varnishes.

(In 5-gal. lots.)	Per gal.	
Carriage, No. 1	2 90	3 30
" body	8 00	9 00
" rubbing	4 00	5 00
Gold Size, Japan	3 00	3 40
Brown Japan	2 40	2 80
Elastic Oak	2 90	3 30
Furniture, extra	2 40	2 80
" No. 1	1 60	2 00
Hard Oil Finish	2 70	3 10
Light Oil Finish	3 20	3 60
Demar	3 30	3 70
Shellac, white	4 40	4 80
" orange	4 00	4 40
Furniture Brown Japan	1 60	2 00
Black Japan	2 40	2 80
" " No. 1	1 60	2 00

The Imperial Varnish & Color Co's., Limited Elastilite Varnish 1 gal. can, each. $2 00.

Granatine Floor Finish, per gal. $2.00.

Maple Leaf Coach Enamels; Size 1, 60c.; Size 2, 35c.; Size 3, 20c. each.

Linseed Oil.

	Raw.	Boiled.
1 to 4 bbls delivered	$0 82	$0 85
5 to 9 bbls "	81	84

Toronto, Hamilton, London and Guelph 2c. less.

Turpentine.

Single barrel, freight allowed	....	0 64
2 to 4 barrels " "	....	0 63

Castor Oil.

East India, in cases, per lb.	0 10	0 10½
" " small lots	0 10½	0 11

Cod Oil, Etc.

Cod Oil, per gal	0 50	0 55
Pure Olive	....	1 20
" Neatsfoot	....	90

Glue.

Common	0 08½	0 09
French Medal	0 14	0 14½
Cabinet, sheet	0 12	0 13
White, extra	0 18	0 20
Gelatine	0 22	0 30
Strip	0 18	0 20
Coopers	0 19	0 20
Huttner	....	0 18

HARDWARE.

Ammunition.

Cartridges.
S. B. Caps. Dom. 50 and 5 per cent.
Rim Fire Pistol, dis. 40 p. c., Amer.
Rim Fire Cartridges, Dom., 50 and 5 p. c.
Central Fire Pistol and Rifle, 10 p.c. Amer.
Central Fire Cartridges, pistol sizes, Dom 30 per cent.
Central Fire Cartridges, Sporting and Military, Dom., 15 and 5 per cent.
Central Fire, Military and Sporting, Amer. add 5 p.c. to list. B.B. Caps, discount 40 per cent, Amer.
Loaded and empty Shells, "Trap" and "Dominion" grades, 25 per cent Rival and Nitro, net list.
Brass Shot Shells, 55 per cent.
Primers, Dom., 30 per cent.

Wads. per lb.
Best thick white felt wadding, in ¼-lb bags ... 1 00
Best thick brown or grey felt wads, in ¼-lb. bags ... 0 70
Best thick white card wads, in boxes of 500 each, 12 and smaller gauges 0 99
Best thick white card wads, in boxes of 500 each, 10 gauge ... 0 35
Best thick white card wads, in boxes of 500 each, 8 gauge ... 0 55
Thin card wads, in boxes of 1,000 each, 12 and smaller gauges ... 0 20
Thin card wads, in boxes of 1,000 each, 10 gauge ... 0 25
Thin card wads in boxes of 1,000 each, 8 gauge ...
Chemically prepared black edge grey cloth wads, in boxes of 250 each— Per M
11 and smaller gauge ... 0 60
9 and 10 gauges ... 0 70
7 and 8 gauges ... 0 90
5 and 6 gauges ... 1 10
Superior chemically prepared pink edge, best white cloth wads, in boxes of 250 each—
11 and smaller gauge ... 1 15
9 and 10 gauges ... 1 40
7 and 8 gauges ... 1 65
5 and 6 gauges ... 1 90

Adzes.

Discount, 20 per cent.

Anvils.

Per lb. ... 10 0 12½
Anvil and Vise combined 4 50
Wilkinson & Co.'s Anvils..lb. 0 09 0 09½
Wilkinson & Co.'s Vices..lb. 0 09¾ 0 10

Augers.

Gilmour's, discount 50 and 10 p.c. off list.

Axes.

Chopping Axes—
Single bit, per doz ... 6 50 10 00
Double bit, " ... 12 00 18 00
Bench Axes, 40 p.c.
Broad Axes, 33⅓ per cent.
Hunters' Axes ... 5 50 6 00
Boy's Axes ... 5 75 6 75
Splitting Axes ... 6 50 12 00
Handled Axes ... 7 00 10 00

Axle Grease.

Ordinary, per gross ... 5 75 6 00
Best quality ... 13 00 15 00

Bath Tubs.

Zinc 6 00
Copper, discount 15 p.c. off revised list

Baths.

Standard Enameled.
5½-inch rolled rim, 1st quality ... 30 00
" " " 2nd " ... 22 00

Anti-Friction Metal.

"Tandem" A ... per lb. 0 27
" B ... " 0 21
" C ... " 0 11½
Magnolia Anti-Friction Metal, per lb. 0 25

SYRACUSE SMELTING WORKS.
Aluminum, genuine ... 0 45
Dynamo ... 0 29
Special ... 0 25
Aluminum, 99 p.c. pure "Syracuse" .. 0 50

Bells.

Hand.
Brass, 60 per cent.
Nickel, 55 per cent.

Cow.
American make, discount 66⅔ per cent.
Canadian, discount 45 and 50 per cent.

Door.
Gongs, Sargant's ... 5 50 8 00
" Peterboro', discount 45 per cent.

Farm.
American, each ... 1 25 3 00

House.
American, per lb. ... 0 35 0 40

Bellows.

Hand, per doz. ... 3 35 4 75
Moulders', per doz. ... 7 50 10 00
Blacksmiths', discount 40 per cent.

Belting.

Extra, 50 and 10 per cent.
Standard, 60 per cent.
No. 1 Agricultural, 60 and 10 p.c.

Bits.

Auger.
Gilmour's, discount 50 and 10 per cent
Rockford, 50 and 10 per cent.
Jennings' Gen., net list.

Car.
Gilmour's, 47½ to 50 per cent.

Expansive.
Clark's, 40 per cent.

Gimlet.
Clark's, per doz. ... 0 65 0 90
Damond, Shell, per doz. ... 1 00 1 50
Nail and Spike, per gross ... 2 25 5 20

Blind and Bed Staples.

All sizes, per lb. ... 0 07½ 0 12

Bolts and Nuts. Per cent.

Carriage Bolts, full square, Norway ... 70
" " full square ... 70
Common Carriage Bolts, all sizes ... 65
Machine Bolts, all sizes ... 65
Coach Screws ... 75
Sleigh Shoe Bolts ... 75
Blank Bolts ... 65
Bolt Ends ... 65
Nuts, square ... 4½c. off
Nuts, hexagon ... 4¾c. off
Tire Bolts ... 67½
Stove Bolts ... 67½
Stove rods, per lb. ... 5½ to 6c.
Plough Bolts ... 60

Boot Calks.

Small and medium, ball, per M ... 4 25
Small heel, per M ... 4 50

Bright Wire Goods.

Discount ... 55 per cent.

Broilers.

Light, dis., 65 to 67½ per cent.
Reversible, dis., 65 to 67½ per cent.
Vegetable, per doz., dis. 37½ per cent.
Henis, No. 8, " 6 00
Henis, No. 9, " 7 00
Queen City " ... 7 50 0 00

Butchers' Cleavers.

German, per doz ... 6 00 11 00
American, per doz ... 12 00 20 00

Building Paper, Etc.

Plain building, per roll ... 0 30
Tarred lining, per roll ... 0 40
Tarred roofing, per 100 lb. ... 1 65
Coal Tar, per barrel ... 3 50
Pitch, per 100-lb. ... 0 85
Carpet felt, per ton ... 45 00

Bull Rings.

Copper, $2.00 for 2½ in. and $1.90 for 2 in.

Butts.

Wrought Brass, net revised list

Cast Iron.
Loose Pin, dis., 60 per cent.

Wrought Steel.
Fast Joint, dis. 60 and 10 per cent.
Loose Pin, dis. 60 and 10 per cent.
Berlin Bronzed, dis. 70, 70 and 5 per cent.
Gen. Bronzed, per pair ... 0 40 0 65

Carpet Stretchers.

American, per doz. ... 1 00 1 50
Bullard's, per doz. ... 6 50 ....

Castors.

Bed, new list, dis. 55 to 57½ per cent.
Plate, dis. 52½ to 57½ per cent.

Cattle Leaders.

Nos. 31 and 32, per gross ... 50 9 50

Cement.

Canadian Portland ... 2 80 3 00
English " 3 00
Belgian " ... 2 75 3 00
Canadian hydraulic ... 1 25 1 50

Chalk.

Carpenters, Colored, per gross 0 45 0 75
White lump, per cwt. ... 0 60 0 65
Red ... 0 05 0 06
Crayon, per gross ... 0 14 0 18

Chisels.

Socket, Framing and Firmer.
Broad's, dis. 70 per cent.
Warnock's, dis. 70 per cent.
P. S. & W. Extra 60 10 and 5 p.c.

Churns.

Revolving Churns, metal frames—No. 0, $8—No. 1, $8.50—No. 2, $9.00—No. 3, $10.00 No. 4, $12.00—No. 5, $16.00 each. Ditto, wood frames—20c. each less than above. Discounts: Delivered from factories, 58 p.c.; from stock in Montreal, 56 p.c. Terms, 4 months or 3 p.c. cash in 30 days

Clips.

Axle dis. 65 per cent.

Closets.

Plain Ontario Syphon Jet ... $8 00
Emb. Ontario Syphon Jet ... 8 50
Fittings ... 1 25
Plain Teutonic Syphon Washout ... 4 75
Emb. Teutonic Syphon Washout ... 5 25
Fittings ... 1 25
Plain Richelieu ... 3 75
Emb. Richelieu ... 4 00
Fittings ... 1 25
Closet connection ... 1 25
Basins, round, 14 in. ... 0 60
" oval, 17 x 14 in. ... 1 50
" " 19 x 15 in. ... 2 25

Compasses, Dividers, Etc.

American, dis. 62½ to 65 per cent.

Cradles, Grain.

Canadian, dis. 25 to 33⅓ per cent.

Crosscut Saw Handles.

S. & D., No. 3, per pair ... 17½
" " 5, " ... 22½
" " 6, " ... 15
Boynton pattern " ... 20

Door Springs.

Torrey's Rod, per doz. ... (15 p.c.) 2 00
Coil, per doz. ... 0 88 1 60
English, per doz. ... 2 00 4 00

Draw Knives.

Coach and Wagon, dis. 50 and 10 per cent.
Carpenters, dis. 70 per cent.

Drills.

Hand and Breast.
Millar's Falls, per doz. net list.

DRILL BITS.
Morse, is., 37½ to 40 per cent.
Standard dis. 50 and 5 to 55 per cent.

Faucets.

Common, cork-lined, dis. 35 per cent.

ELBOWS. (Stovepipe.)
No. 1, per doz. ... 1 80
No. 2, per doz. ... 1 60
Bright, 20c. per doz. extra.

ESCUTCHEONS.
Discount, 45 per cent.

ESCUTCHEON PINS.
Iron, discount 40 per cent.

FACTORY MILK CANS.
Discount off revised list, 40 per cent.

FILES.
Black Diamond, 50 and 10 to 60 per cent
Kearney & Foote, 60 and 10 p.c. to 50, 10, 10.
Nicholson File Co., 50 and 10 to 60 per cent.
Jowitt's, English list, 25 to 27½ per cent.

FORKS.
Hay, manure, etc., dis., 50 and 10 per cent, revised list.

GLASS—Window—Box Price.

Size United Inches.	Star Per 50 ft.	Star Per 100 ft	D. Diamond Per 50 ft.	D. Diamond Per 100 ft
Under 26	2 10	4 00	....	6 00
26 to 40	2 30	4 35	....	6 65
41 to 50	....	4 75	....	7 25
51 to 60	....	5 00	....	8 50
61 to 70	....	5 35	....	9 25
71 to 80	....	5 75	....	10 50
81 to 85	....	6 50	....	11 75
86 to 90	....	....	..	14 00
91 to 95	..	....	...	15 50
96 to 100	.. .	....	....	18 00

GAUGES.
Marking, Mortise, Etc.
Stanley's dis. 50 to 55 per cent.

Wire Gauges.
Winn's, Nos. 26 to 33, each ... 1 65 2 40

HALTERS.
Rope, ⅜ per gross ... .
" ½ " 9 00
" ⅝ to ¾ 14 00
Leather, 1 in., per doz. ... 3 87½ 4 00
" 1¼ in., " ... 5 15 5 20
Web, — per doz. ... 1 87 2 45

HAMMERS.
Nail
Maydole's, dis. 5 to 10 per cent. Can. dis. 25 to 27½ per cent.

Tack.
Magnetic, per doz. ... 1 10 1 20

Sledge.
Canadian, per lb. ... 0 07½ 0 08½

Ball Pean.
English and Can., per lb. ... 0 22 0 25

HANDLES.
Axe, per doz., net ... 1 50 2 00
Store door, per doz. ... 1 00 1 50

Fork.
C. & B., dis. 40 per cent. rev. list.

Hoe.
C. & B., dis. 40 per cent. rev. list.

Saw.
American, per doz. ... 1 00 1 25

Plane.
American, per gross ... 3 15 3 75

Hammer and Hatchet.
Canadian, 40 per cent.

Cross-Out Saws.
Canadian, per pair ... 0 13½

HANGERS. doz. pairs.
Steel barn door ... 5 85 6 00
Stearns, 4 inch 5 00
" 5 inch 6 50
Lane's covered—
No. 11, 5-ft. run 8 40
No. 11½, 10-ft. run 10 80
No. 12, 10-ft. run 12 60
No. 14, 15-ft. run 21 00
Lane's O.N.T. track, per foot. 4½

HARVEST TOOLS.
Discount, 50 and 10 per cent.

HATCHETS.
Canadian, dis. 40 to 42½ per cent.

HINGES.
Blind, Parker's, dis. 50 and 10 to 60 per cent
Heavy T and strap, 4-in., per lb. ... 0 06½
" " 5-in., " 0 06¼
" " 6-in., " 0 06
" " 8-in., " 0 05¾
" " 10-in., " 0 05½
Light T and strap, dis. 60 and 5 per cent.
Screw hook and hinge—
6 to 12 in., per 100 lbs. 4 50
14 in. up, per 100 lbs. 3 50
Per gro. pairs
Spring 12 00

HOES.
Garden, Mortar, etc., dis. 50 and 10 p.c.
Planter, per doz. ... 4 00 4 50

HOLLOW WARE
Discount, ... 45 and 5 per cent.

HOOKS.
Cast Iron.
Bird Cage, per doz. ... 0 50 1 10
Clothes Line, per doz. ... 0 27 0 63
Harness, per doz. ... 0 72 0 88
Hat and Coat, per gross ... 1 00 3 00
Chandelier, per doz. ... 0 50 1 00

Wrought Iron.
Wrought Hooks and Staples, Can., dis. 47½ per cent.

Wire.
Hat and Coat, discount 45 per cent.
Belt, per 1,000 ... 0 60
Screw, bright, dis. 55 per cent.

HORSE NAILS.
"C" brand 50 p.c. dis. } Oval head.
"M" brand 50 p.c. }
Acadian, 50 and 10 per cent.

HORSESHOES.

F.O.B. Montreal.

Iron Shoes.	No. 2 and larger.	No. 1. and smaller
Light, medium, and heavy...	3 50	3 75
Snow shoes................	3 75	4 00
Steel Shoes.		
Light....................	3 60	3 85
Featherweight (all sizes)......	4 85	4 85

F.O.B. Toronto, Hamilton, London and Guelph, 10c. per keg additional.

Toe weight steel shoes...... 6 70

JAPANNED WARE.

Discount, 45 and 5 per cent. off list, June 1899.

ICE PICKS.

Star per doz.................. 3 00 3 25

KETTLES.

Brass spun, 7½ p.c. dis. off new list.
Copper, per lb.............. 0 30 0 50
American, 60 and 10 to 65 and 5 p.c.

KEYS.

Lock, Can., dis., 45 p.c.
Cabinet, trunk, and padlock, Am. per gross.......... 60

KNOBS.

Door, japanned and N.P., per doz.................. 1 50 2 50
Bronze, Berlin, per doz...... 2 75 3 25
Bronze Genuine, per doz.... 6 00 9 00
Shutter, porcelain, F. & L. screw, per gross.......... 1 30 4 00
White door knobs—per doz. 1 25

HAY KNIVES.

Discount, 50 and 10 per cent.

LAMP WICKS.

Discount, 60 per cent.

LANTERNS.

Cold Blast, per doz.............. 7 50
No. 3 "Wright's"................. 8 50
Ordinary, with O burner.......... 4 25
Dashboard, cold blast 9 50
No. 0............................ 6 00
Japanning, 50c. per doz. extra.

LEMON SQUEEZERS.

per doz.

Porcelain lined,............ 2 20 5 60
Galvanized 1 87 3 85
King, wood................ 2 75 2 90
King, glass................ 4 00 4 50
All glass.................. 1 20 1 30

LINES.

Fish, per gross.............. 1 05 2 50
Chalk " 1 90 7 40

LOCKS

Canadian, dis. 45 p.c.
Russell & Erwin, per doz.... 3 00 3 25

Cabinet.

Eagle, dis. 30 p.c.

Padlock.

English and Am., per doz.... 50 6 00
Scandinavian, " 1 00 2 40
Eagle, dis. 20 to 25 p.c.

MACHINE SCREWS.

Iron and Brass.

Flat head, discount 25 p.c.
Round Head, discount 20 p.c.

MALLETS.

Tinsmiths', per doz.......... 1 25 1 50
Carpenters', hickory, per doz. 1 25 3 75
Lignum Vitae, per doz.. .... 3 85 5 00
Caulking, each............... 60 2 00

MATTOCKS

Canadian, per doz.......... 8 50 1 00

MEAT CUTTERS.

American, dis. 25 to 30 p.c.
German, 15 per cent.

MILK CAN TRIMMINGS

Discount, 25 per cent.

NAILS

Quotations are:	Cut.	Wire.
2d. and 3d	$3 35	$3 85
3d	3 00	3 52
4 and 5d..................	2 75	3 35
6 and 7d..................	2 65	3 20
8 and 9d..................	2 50	3 00
10 and 12d................	2 45	2 95
16 and 20d................	2 40	2 90
30, 40, 50 and 60d. (base)......	2 35	2 85

Galvanizing 2c. per lb. net extra.
Steel Cut Nails 10c. extra.
Miscellaneous wire nails, dis. 70 per cent.
Coopers' nails, dis. 30 per cent.
Flour barrel nails, dis. 25 per cent.

NAIL PULLERS.

German and American....... 1 85 3 50

NAIL SETS

Square, round, and octagon, per gross............... 3 38 4 00
Diamond 12 00 15 00

NETTING.

Poultry, 50 per cent. for McMullen's.

OAKUM. Per 100 lb

Navy 6 00
U.S. Navy.............. .... 7 25

OIL.

Water White (U.S.)......... .... 0 16½
Prime White (U.S.)......... .... 0 15½
Water White (Can.).......... .... 0 15
Prime White (Can.).......... .... 0 14

OILERS.

McClary's Model galvan. oil can, with pump, 5 gal., per doz.................. 0 00 10 00
Zinc and tin, dis. 50, 50 and 10.
Copper, per doz.............. 1 25 3 50
Brass, " 1 50 3 50
Malleable, dis. 25 per cent.

GALVANIZED PAILS.

Dufferin pattern pails, dis. 50 to 50 and 10 p.c.
Flaring pails, discount 45 per cent.
Galvanized washtubs, discount 45 per cent.

PIECED WARE.

Discount 40 per cent. off list, June, 1899.

PICKS.

Per doz..................... 6 00 9 00

PICTURE NAILS.

Porcelain head, per gross.... 1 75 3 00
Brass head, " 0 40 1 00

PICTURE WIRE.

Tin and gilt, discount 75 p.c.

PLANES.

Wood, bench, Canadian dis. 50 per cent American dis. 50.
Wood, fancy Canadian or American, 37½ to 40 per cent.

PLANE IRONS.

English, per doz.............. 2 00 5 00

PLIERS AND NIPPERS.

Button's Genuine per doz pairs, dis. 37½ 40 p.c.
Button's Imitation, per doz.. 5 00 9 00
German, per doz............ 0 60 2 60

PLUMBERS' BRASS GOODS.

Impression work, discount, 60 per cent.
Fuller's work, discount 65 per cent.
Rough stops and stop and waste cocks, discount, 60 per cent.
Jenkins' disk globe and angle valves, discount, 55 per cent.
Standard valves, discount, 60 per per cent.
Jenkins' radiator valves, discount 55 per cent.
" " " standard, dis., 60 p.c.
Quick opening valves, discount, 60 p.c.
No. 1 compression bath cock........ 2 00
No. 4 " " " 2 00
No. 7, Fuller's..................... 2 50
No. 4½, " 3 00

POWDER.

Velox Smokeless Shotgun Powder.

100 lb. or less..................... 0 85
1,000 lb. or more................... 0 80
Net 30 days.

PRESSED SPIKES.

Discount, 25 per cent.

PULLEYS.

Hothouse, per doz............ 0 55 1 00
Axle 0 22 0 33
Screw 0 27 1 00
Awning....................... 0 35 2 50

PUMPS

Canadian cistern............. 1 80 3 60
Canadian pitcher spout....... 1 40 2 10

PUNCHES.

Saddlers', per doz........... 1 00 1 85
Conductors', " 9 00 15 00
Tinners' solid, per set....... 0 00 0 72
" hollow, per inch.... 0 00 1 00

RANGE BOILERS

Galvanized, 30 gallons 6 50
" 35 " 7 50
" 40 " 8 50
Copper, 30 " 22 00
" 35 " 26 00
" 40 " 30 00
Discount off Copper Boilers 10 per cent.

RAKES.

Cast steel and malleable Canadian list 50 and 10 p.c. revised list.
Wood, 25 per cent.

RASPS AND HORSE RASPS.

New Nicholson horse rasp, discount 60 p.c.
Globe File Co.'s rasps, 60 and 10 to 70 p.c.
Heller's Horse rasps, 50 to 50 and 5 p.c.

RAZORS.

per doz.

Geo. Butler & Co.'s,.......... 8 00 18 00
Boker's 7 50 11 00
Wade & Butcher's.............. 3 60 10 00
Theile & Quack's 7 00 12 00
Elliot's 4 00 18 00

REAPING HOOKS.

Discount, 50 and 10 per cent.

REGISTERS.

Discount,........................ 40 per cent.

RIVETS AND BURRS.

Iron Rivets, discount 60 and 10 per cent.
Iron Burrs, discount 55 per cent.
Black and Tinned Rivets, 60 p.c.
Extras on Iron Rivets in 1-lb. cartons, ½c per lb.
Extras on Iron Rivets in ¼-lb. cartons, per lb.
Copper Rivets & Burrs, 35 and 5 p.c. dis. and cartons, 1c. per lb. extra, net.
Extras on Tinned or Coppered Rivets ½-lb. cartons, 1c. per lb.
Terms, 4 mos. or 3 per cent. cash 30 days.

RIVET SETS.

Canadian, dis. 35 37½ per cent.

ROPE, ETC.

	Sisal.	Manila.
7-16 in. and larger, per lb.	9	13
⅜ in......................	10	14
¼ and 5-16 in.		15
Cotton, 3-16 inch and larger		16½
" 5-32 inch..........		21½
" ⅛ inch..........		22½
Russia Deep Sea		15½
Jute		8
Lath Yarn		9¾
New Zealand Rope..................		10½

RULES.

Boxwood, dis. 75 and 10 p.c.
Ivory, dis. 37½ to 40 p.c.

SAD IRONS. per set

Mrs. Potts, No. 55, polished.......... 0 70
" No. 50, nickle-plated.... 75

SAND AND EMERY PAPER.

Dominion Flint Paper, 47½ per cent.
B & A. sand, 40 and 2½ per cent.
Emery, 40 per cent.

SAP SPOUTS.

Bronzed iron with hooks, per doz... 9 50

SAWS.

Hand, Disston's, dis. 12½ p.c.
S. & D., 40 per cent.
Crosscut, Disston's, per ft.... 0 35 0 55
S. & D., dis. 35 p.c. on Nos. 2 and 3.
Hack, complete, each........ 0 75 2 75
" frame only............ .... 0 75

SASH WEIGHTS.

Sectional, per 100 lbs......... 2 75 3 00
Solid, " 2 00 2 25

SASH CORD.

Per lb....................... 0 23 0 30

SAW SETS.

"Lincoln," per doz.......... .... 6 50

SCALES

B. S. & M. Scales, 45 p.c.
Champion, 65 per cent.
Fairbanks Standard, 35 p.c.
" Dominion, 55 p.c.
" Richelieu, 55 p.c.
Chatillon Spring Balances, 10 p.c.

SCREW DRIVERS.

Sargent's, per doz............ 0 65 1 00

SCREWS

Wood, F. H., iron, and steel, 80 p.
Wood R. H., " dis. 75 p.c.
" F. H., brass, dis. 75 p.c.
Wood, R. H., " dis. 67½ p.c.
" F.H., bronze, dis. 67½ p.c.
" R.H. " 62½ p.c.
Drive Screws, 80 per cent.
Bench, wood, per doz......... 3 25 4 00
" iron, " 4 25 5 75

SCYTHES.

Per doz, net................. 9 00

SCYTHE SNATHS.

Canadian, dis. 45 p.c.

SHEARS

Bailey Cutlery Co., full nickeled, dis. 60 p.c.
Seymour's, dis. 50 and 10 p.c.

SHOVELS AND SPADES.

Canadian, dis. 40 and 5 per cent.

SINKS.

Steel and galvanized, discount 45 per cent.

SNAPS.

Harness, German, dis. 25 p.c.
Lock, Andrews'.............. 4 50 11 50

SOLDERING IRONS.

1, 1½ lb., per lb....... .... 0 37
2 lb. or over, per lb. 0 34

SQUARES.

Iron, No. 493, per doz 2 40 2 55
No. 494, " 3 25 3 40
Steel, dis. 50 and 5 to 50 and 10 p.c., rev. list.
Try and bevel, dis. 50 to 52½ p.c.

STAMPED WARE.

Plain, dis., 75 and 12½ p.c. off revised list.
Retinned, dis., 75 p.c. off revised list.

STAPLES.

Galvanized 0 00 0 00
Plain 0 00 3 45
Coopers', discount 45 per cent.
Poultry netting staples, 40 per cent.

STOCKS AND DIES.

American dis. 25 p.c.

STONE. Per lb.

Washita........................	0 28	0 60
Hindostan......................	0 06	0 07
" slip, ...	0 09	0 09
Labrador.......................	....	0 13
" Axe..................	....	0 15
Turkey	....	0 50
Arkansas	0 00	1 50
Water-of-Ayr	0 00	0 10
Scythe, per gross	3 50	5 00
Grind, per ton	15 00	18 00

STOVE PIPES.

Nestable in crates of 25 lengths.

6 inch Per 100 lengths........ .... 8 00
7 inch " " 8 50

ENAMELINE STOVE POLISH.

No. 4—3 dozen in case, net cash $4 80
No. 6—3 dozen in case, " 8 40

TACKS BRADS, ETC.

Per cent.

Strawberry box tacks, bulk ......75 & 10
Cheese-box tacks, blued ....80 & 12½
Trunk tacks, black and tinned 85
Carpet tacks, blued ....80 & 5
" " tinned ....80 & 10
" " (in kegs) 40
Cut tacks, blued, in dozens only ..75 & 15
" ¼ weights...............60
Swedes, cut tacks, blued and tinned—
In bulk.........................80 & 10
In dozens.......................75
Swedes, upholsterers', bulk ....85 & 12½
" brush, blued & tinned, bulk..70
" gimp, blued, tinned and japanned.............75 & 12½
Zinc tacks 35
Leather carpet tacks 55
Copper tacks 50
Copper nails 52

Trunk nails, black		65 and 5
Trunk nails, tinned		65 and 10
Clout nails, blued and tinned		65 and 5
Chair nails		35
Cigar box nails		35
Patent brads		40
Fine finishing		40
Picture frame points		10
Lining tacks, in papers		10
" " in bulk		15
" " solid heads, in bulk		75
Saddle nails in papers		10
" " in bulk		15
Tufting buttons, 22 line, in dozens only		60
Tin capped trunk nails		15
Zinc glazier's points		5
Double pointed tacks, papers		90 and 10
" " " bulk		40

TAPE LINES.

English, ass skin, per doz	2 75	5 00
English, Patent Leather	5 50	9 75
Chesterman's each	0 90	2 85
" steel, each	0 80	8 00

THERMOMETERS

Tin case and dairy, dis. 75 to 75 and 10 p.c.

TRANSOM LIFTERS.

Payson's per doz ... 2 60

TRAPS. (Steel.)

Game, Newhouse, dis. 2½ p.c.
Game, H. & N., P. S. & W., 65 p.c.
Game, steel, 72½, 75 p.c.

TROWELS.

Disston's discount 10 per cent.

German, per doz	4 75	6 00

S. & D., discount 35 per cent.

TWINES.

Bag, Russian, per lb	....	0 27
Wrapping, cotton, per lb	0 22	0 26
Wrapping, mottled, per pack	0 50	0 60
Wrapping cotton, 3-ply		0 20
" " 4-ply	....	0 26
Mattress, per lb	0 33	0 45
Staging, "	0 27	0 35
Broom, "	0 30	0 55

VISES.

Hand, per doz	4 00	5 00
Bench, parallel, each	2 00	4 50
Coach, each	6 00	7 00
Peter Wright's, per lb	0 12	0 13
Pipe, each	5 50	9 00
Saw, per doz	6 50	13 00

ENAMELLED WARE.

White, Princess, Turquoise, Blue and White, discount 50 per cent.
Diamond, Famous, Premier, 50 and 10 p.c.
Granite or Pearl, Imperial, Crescent, 50, 10 and 10 per cent.

WIRE.

Brass wire, 50 to 50 and 2½ per cent. off the list.

Copper wire, 45 and 10 per cent. net cash 30 days, f.o.b. factory.

Smooth Steel Wire, base, $2.80 per 100 lb. List of extras: Nos. 2 to 5, advance 7c. per 100 lb.—Nos. 6 to 9, base—No. 10, advance 7c.—No. 11, 14c.—No. 12, 20c.—No. 13, 35c.—No. 14, 47c.—No. 15, 60c.—No. 16, 75c. Extras net per 100 lb.: Coppered wire, 60c.—tinned wire, $2—oiling, 10c.—special hay-bailing wire, 30c.—spring wire, $1—best steel wire, 75c.—bright soft drawn, 15c.—in 50 and 100-lb. bundles net, 10c.—in 25-lb. bundles net, 15c.—packed in casks or cases, 15c.—bagging or papering, 10c.

Fine Steel Wire, dis. 17½ per cent. List of extras: In 100-lb. lots: No. 17, $5—No. 18, $5.50—No. 19, $6—No. 20, $6.65—No. 21, $7—No. 22, $7.30—No. 23, $7.65—No. 24, $8—No. 25, $9—No. 26, $9.50—No. 27, $10—No. 28, $11—No 29, $12—No. 30, $13—No. 31, $14—No. 32, $15 No. 33, $16—No. 34, $17. Extras net—tinned wire, Nos. 17-25, $2—Nos. 26-31, $4—Nos. 32-34, $6. Coppered, 5c.—oiling, 10c.—in 25-lb. bundles, 15c.—in 5 and 10-lb. bundles, 25c.—in 1-lb. hanks, 50c.—in ½-lb. hanks, 75c.—in ¼-lb. hanks, $1—packed in casks or cases, 15c.—bagging or papering, 10c.

Galvanized Wire, per 100 lb.—Nos. 6, 7, 8, $3.85 No. 9, $3.10—No. 10, $4.00—No. 11, $4.05 No. 12, $3.25—No. 13, $3.35—No. 14, $4.40—No. 15, $4.90—No. 16, $5.15.

Clothes Line Wire, 19 gauge, per 1,000 feet 3 30

WIRE FENCING.

		F.O.B. Toronto
Galvanized 4 barb, 2¾ and 5 inches apart	....	3 10
Galvanized, 2 barb, 4 and 5 inches apart	....	3 10
Galvanized, plain twist	....	3 10

Galvanized barb, f.o.b. Cleveland, $2.97½ in less than carlots, and $2.85 in carlots. Terms, 60 days or 2 per cent. in 10 days.

Ross braid truss cable 4 50

WIRE CLOTH.

Painted Screen, per 100 sq. ft., net... 1 50
Terms, 4 months, May 1.; 3 p.c. off 30 days.

WRENCHES.

Acme, 35 to 37½ per cent.
Agricultural, 60 p.c.
Coe's Genuine, dis. 70 to 25 p.c.

Towers' Engineer, each	2 00	7 00
" S., per doz	5 80	6 00
G. & K.'s Pipe, per doz	....	3 40
Burrell's Pipe, each	....	3 00
Pocket, per doz	0 25	2 90

WRINGERS.

Leader	per doz.	$60 00
Royal Canadian	"	58 00
Royal American	"	50 00

Discount, 45 per cent.; terms 4 months, or p.c. 30 days.

WROUGHT IRON WASHERS

Canadian make, discount, 40 and 5 per cent.

PERSONS addressing advertisers will kindly mention having seen their advertisement in Canadian Hardware and Metal Merchant.

SEND for specimen copy of Phillips' Monthl Machinery Register, containing over 5,000 entries of new and second-hand machinery of every description. The oldest established and most successful medium in the world. Established 25 years for the purpose of introducing those who have machinery for sale, to those who wish to buy, has a circulation of about 50,000 copies per annum, all over the world, and is used for continual reference by a large number of firms. It is consequently a most valuable advertising medium for all engineers and manufacturers. Subscription, 6s. per annum, price per copy, 6d. Sole Proprietor, CHAS. D. PHILLIPS, M.I.M.E., Newport, Mon., England. Telegraphic address "Machinery, Newport, Mon.

IN BUYING——

LINSEED OIL

it is always well to get the purest and best—something you can recommend and guarantee to your customers.

Stewart Bros. & Spencer's

is the best. Name on every barrel. Special quotations for import.

J. WATTERSON & CO.

MONTREAL, Agents for Canada.

Lockerby & McComb

AGENTS IN CANADA

FOR THE

Celebrated P. & B.

Cold Storage Lining

AND

. . Ruberoid Roofing . .

P. S.--Prices on Application.

65 Shannon Street, MONTREAL.

BUSINESS NEWS

of any kind that is of value to business men supplied by our Bureau. We can give you market quotations from any town in Canada, reports from the city markets, stock quotations, etc. You can get commercial news from any Canadian paper through us.

Write us, giving us particulars of what you want and where you want it from, and we will quote you prices by return.

"Clippings from any Canadian paper on any subject."

CANADIAN PRESS CLIPPING BUREAU,

5C5 Board of Trade Bldg., MONTREAL, QUE.

Telephone Main 1255.

10 Front St. East, Toronto. Telephone 2148.

---

75 YEARS. *ESTABLISHED 1825.* *75 YEARS.*

CELEBRATED **HEINISCH** SHEARS.

Tailors' Shears,
Trimmers, Scissors,
Tinners' Snips, etc.

ACKNOWLEDGED THE BEST.

R. HEINISCH'S SONS CO. NEW YORK OFFICE, 90 Chambers St. NEWARK, N.J., U.S.A.

Not connected with any Shear Combination.

CHAS. F. CLARK, President. JARED CHITTENDEN, Treasurer.

...ESTABLISHED 1849...

BRADSTREET'S

Capital and Surplus, $1,500,000. Offices Throughout the Civilized World.

Executive Offices: Nos. 346 and 348 Broadway, New York City, U.S.A.

THE BRADSTREET COMPANY gathers information that reflects the financial condition and the controlling circumstances of every seeker of mercantile credit. Its business may be defined as of the merchants, by the merchants, for the merchants. In procuring, verifying and promulgating information no effort is spared, and no reasonable expense considered too great, that the results may justify its claim as an authority on all matters affecting commercial affairs and mercantile credit. Its offices and connections have been steadily extended, and it furnishes information concerning mercantile persons throughout the civilized world.

Subscriptions are based on the service furnished, and are available only by reputable wholesale, jobbing and manufacturing concerns, and by responsible and worthy financial, fiduciary and business corporations. Specific terms may be obtained by addressing the Company at any of its offices. **Correspondence Invited.**

—OFFICES IN CANADA—

HALIFAX, N.S.	HAMILTON, ONT.	LONDON, ONT.	MONTREAL, QUE.
OTTAWA, ONT.	QUEBEC, QUE.	ST. JOHN, N.B.	TORONTO, ONT.
VANCOUVER, B.C.	VICTORIA, B.C.	WINNIPEG, MAN.	

THOS. C. IRVING, Gen. Man. Western Canada, Toronto. JOHN A. FULTON, Gen. Man. Eastern Canada, Montreal.

Awarded a Gold Medal at **PARIS EXPOSITION** for superiority. That's proof enough of their quality, and clearly shows that they are the best.

The Bailey

Send for Catalogue and . . . . Price List.

Cutlery Co.

BRANTFORD, ONT.

Dominion Pattern

Cow Tie AND Stall Fixture

The special features of the tie and stall fixture are well shown in the illustration. As will be noticed the chain is very short, which prevents all danger of entanglement with the animal's foot. At the same time the form of the fixture is such that great freedom is allowed to the head. Because of the short chain this tie is much cheaper than the ordinary patterns.

The stall fixture is made from a tough quality of steel and is very strong. Also, owing to its circular cross-section, it is exceedingly rigid. Its simplicity, convenience, cheapness, and ease of attaching make it very popular with cow tie users.

This form of tie and stall fixture are sometimes called Niagara pattern.

American or Flat Link Chain,

for years the standard cow tie chain in "the States" is now rapidly coming in favor in Canada. Its short link, handsome appearance and smooth surface—which cannot injure the animal's neck—make it superior to all other styles of chain for cow ties.

For sale by all Jobbers; manufactured by

ONEIDA COMMUNITY, Limited, NIAGARA FALLS, ONT.

CANADIAN HARDWARE AND METAL MERCHANT

The Weekly Organ of the Hardware, Metal, Heating, Plumbing and Contracting Trades in Canada.

VOL. XIII. MONTREAL AND TORONTO, JANUARY 12, 1901. NO. 2

ALEXANDER GIBB

Manufacturers' Agent and Metal Broker,
13 St. John Street, Montreal.

Representing British and American Manufacturers. Correspondence invited from firms wishing to be represented in Canada.

McCaskill, Dougall & Co.
Fine Varnish & Japan
Manufacturers
Montreal
Price Lists on application

1901

Dundas Axes

are up-to-date in appearance and second to none in quality.

Dundas Axe Works

DUNDAS, CANADA.

MURALO

THE CENTURY'S WALL TINT

Wonderful strides were made in Wall Tints in the 19th century—Muralo, the greatest and best of all cold water wall tints, will lead in the 20th. Dealers should secure agencies quickly. It is well advertised and will be more so. The advertising methods will bring customers to your store.

It Sells. It's a Money Maker.

A. Ramsay & Son, Montreal

General Agents for Canada.

J. H. ASHDOWN, Winnipeg, Agent for Manitoba and N.W.T.

McLENNAN, McFEELY & CO., Limited, Vancouver, Agents for British Columbia.

VanTuyl & Fairbank

Petrolia, Ont.

Headquarters for . . .
Oil and Artesian Well Pumps, Casing, Tubing, Fittings, Drilling Tools, Cables, etc.

Ontario Nut Works, Paris

BROWN & CO.

Manufacturers of
All sizes of Hot Pressed Nuts, Square and Hexagon.

THE . . .

Waggoner Extension Ladder.

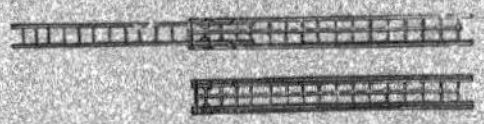

The strongest, lightest and most convenient ladder in the market. The only really satisfactory extension ladder made. Pulls up with a rope. Made in all lengths. Also extension and other step ladders, sawhorses, ironing boards, painters' trestles, etc. All first-class goods. Write for quotations to

The Waggoner Ladder Company, Limited, London, Ont.

Special list of low-priced Japanned and Regalvanized Wire Cloth.

24, 30, 36 in. wire, in 50 ft. rolls.

SAMPLES SENT WHEN DESIRED. WRITE FOR PRICES.

The B. GREENING WIRE CO., Limited

Hamilton, Ont., and Montreal, Que.

E. B. SALYERDS

Manufacturer of

Hockey Sticks

PRESTON,

Ontario, - Canada.

The Best Stick.
Made of Rock Elm.
Wholesale Trade Only Supplied.
Ask your Wholesale House for the Preston make of Stick.
Write for Prices.

"DAISY" CHURN

Has tempered steel cased bicycle ball bearings, strongest, neatest and most convenient frame. Only two bolts to adjust in setting up. Steel Bow Levers, suitable for either a standing or sitting posture. Has four wheels and adjustable feet to hold stand steady while churning. When churn is locked to stand the bow can be used as handles to move it about on the front wheels as handy as a baby carriage. Open on both sides to centre, giving free space for pail. Made with wood or steel stands, with Cranks only, or Bow Levers as desired.

Vollmar Perfect Washer

Has a most enviable record. A perfection of its kind—will wash more clothes in less time, do it better and easier, with less wear and tear, than any other machine.

THE

Wortman & Ward Mfg. Co., Limited

LONDON, ONT.

Eastern Branch, 60 McGill Street, Montreal, Que.

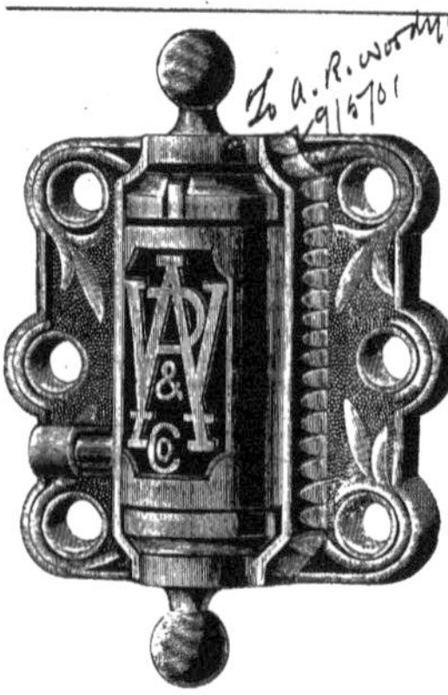

HOLD-BACK

SPRING HINGES.

NO. 20, JAPANNED.
NO. 120, ANTIQUE COPPER.

Have been sold for several years, and are acknowledged the most satisfactory style ever placed upon the market. Only the best material is used in its construction.

Sold by every Jobber in the Dominion.

Manufactured by— **A. R. WOODYATT & CO.,** GUELPH, ONT.

KEMP'S

Broad-Hoop Roll-Rim

Milk Can Bottoms

possess all the points which go to make perfection in Can Bottoms. They have been used by a criticizing public for two seasons, and their popularity is evidence of the satisfaction which they gave. The roll-rim has no sharp turns, which break the grain of the metal and lessen its wearing qualities. It has a broad wearing surface and will not damage floors. They do not cost more than inferior bottoms.

The Iron Clad Trimmings are made the same as the Broad Hoop, and differ from them only in having a narrower and thicker hoop, which does not require the roll-rim, and, therefore, can be sold cheaper.

For durability and finish, our Trimmings are unequalled.

Manufactured by

Kemp Manufacturing Co., Toronto, Canada.

HARDWARE AND METAL

VOL. XIII. MONTREAL AND TORONTO, JANUARY 12, 1901. NO. 2.

President,
JOHN BAYNE MacLEAN,
Montreal.

THE MacLEAN PUBLISHING CO.
Limited.

Publishers of Trade Newspapers which circulate in the Provinces of British Columbia, North-West Territories, Manitoba, Ontario, Quebec, Nova Scotia, New Brunswick, P.E. Island and Newfoundland.

OFFICES

MONTREAL . . . . Board of Trade Building. Telephone 1255.
TORONTO . . . . . . 10 Front Street East. Telephone 2148.
LONDON, ENG. . . . . 109 Fleet Street, E.C., J. M. McKim,
MANCHESTER, ENG. . . . 18 St Ann Street. H. S. Ashburner,
WINNIPEG . . . . Western Canada Block. J. J. Roberts.
ST. JOHN, N.B. . . . No. 3 Market Wharf. J. Hunter White,
NEW YORK. . . . . . . 176 E. 88th Street, W. J. Brandt.

Travelling Subscription Agents:
T. Donaghy. F. S. Millard.

Subscription, Canada and the United States, $2.00.
Great Britain and elsewhere . . . . 12s.

Published every Saturday.

Cable Address { Adscript, London / Adscript, Canada.

LINSEED OIL AND THE TARIFF.

INQUIRY for spring supplies of linseed oil is developing unusually early this season, supposed to be partly owing to the fact that merchants in Ontario had a good demand for this article during the fall months, on account of the unusually fine weather, so that their stocks are believed to be lower than ordinarily at this season.

During the past year a large quantity of linseed oil came into this market under the British preference clause in the Customs regulations, especially so after the month of July. Some interest is now being evoked about this supply for the coming year.

Linseed oil imported from England is produced from flax seed reaching there from four different sources, namely, East Indies, Russia, United States and Argentine Republic. The flax seed shipped from India usually comes from Calcutta, and is always quoted in the market prices in England at a slight advance over others. Oil produced from this seed may be entitled to the British preference, but, as all goods to come under this preference must have a substantial portion of British labor entering into their production, to the extent of not less than one-fourth of the value of such article in the condition in which it is exported, it is evident that linseed oil produced from seed of foreign growth cannot be imported into Canada under the preference clause of the tariff.

Oil produced from Calcutta seed usually is consumed in England in the months of December, January and February.

Flax is harvested in South America about the month of November, and seed exported in the month of January reaches England, freely, in February and March. Oil produced from this seed is then ready for shipment the last of March or the beginning of April, and, for the past seasons, it is oil from this seed that has usually reached Canada in the months of May and June.

As the exporter has to make a declaration, as the following, to enable the goods to be passed by the Customs here, and receive the British preference, it is believed that no producers of linseed oil in England, whose attention is called to this fact, will be prepared to sign the declaration for oils made from seed grown in the Argentine Republic:

Form of Certificate prescribed to be written, printed or stamped on the face or back of invoices of all articles, except raw and refined sugars, for entry under the British Preferential Tariff of Canada, when made and signed by an individual exporter personally.

I, (1) .... the exporter of the articles included in this invoice, have the means of knowing and do hereby certify that said invoice being from myself to (2) .... and amounting to (3) .... is true and correct; that all the articles included in the said invoice are bona fide the produce or manufacture of one or more of the following countries, viz.:—(4) .... and that a substantial portion of the labor of one or more of such countries has entered into the production of every manufactured article included in said invoice to the extent in each article of not less than one-fourth of the value of every such article in its present condition ready for export to Canada.

Signed........................

Dated at............................this

............................19....

LOWER PRICES ON SCREWS.

A REDUCTION has been made in the price of wood screws. The change was made a few hours after we went to press last week, and amounts to a reduction of about 20 per cent. The new discounts, together with those previously in force, are as follows:

	New discounts.	Old discounts.
Flat head, iron........................	85	80
Round head, iron	80	75
Flat head, brass	77½	75
Round head, brass	70	67½
Flat head, bronze	70	67½
Round head, bronze	65	62½

The reduction is the concomitant of a similar course which the United States market took a couple of weeks ago.

The market in the United States is rather demoralized owing to competition, and the association over there has found it necessary to dissolve the pool and adopt a less rigid agreement than that heretofore in existence.

A CALL FOR MORE STEAM.

SOME of the officials of the Intercolonial Railway should put on more steam and pull a little harder on the throttle or their system will stop altogether. This is the conclusion reached by many merchants who have had occasion to use the Intercolonial Railway service during the last few months. The reason is that it has been excessively, disappointingly and, in cases, expensively slow.

Accidents are bound to happen in the best regulated families, and some tardiness in delivery is excusable, but, when timely deliveries are the exception rather than the rule, there is need of an investigation into the principles on which the road is being run.

It is not only essential that freight should reach its destination, but also that it should reach it on time. If a railway company does not haul its freight in three times the length of time taken by an express company, then it is not providing reasonable accommodation. Judge the Intercolonial Railway freight service by that standard and one will not find it efficient.

The loudest complaints come from the Maritime Provinces where merchants have to rely on this one line entirely. The service this fall seems to have been execrable, and travellers who have toured through the country find that great dissatisfaction exists.

A merchant in Truro is reported on good authority to have entreated incessantly with the railway authorities to have a car, which had come into the station loaded with goods for him, moved into a position where the goods could be unloaded, but his prayers went unanswered for nine days and only on the tenth was the car shifted to where it could be approached.

We have heard of a case within the past month where it took more than a month to get a car of merchandise from Montreal to Moncton, N.B. We have been told that goods coming from Halifax to a Montreal agent were a month in transmission; this was just before Christmas, and, as part of the goods had to be sent to Vancouver, B.C., for the holiday trade, the agent was put to no small extra expense expressing stock that might have been freighted had the Intercolonial given even fairly good service.

A Montreal grocery house sent goods to Halifax for the Christmas trade, but, occupying 10 days in transmission, they were too late, and their value was discounted. There is no doubt that such instances could be multiplied, for the line has been in a congested condition for months.

On Halifax particularly this poor accommodation is having an injurious effect, for importers are sending orders to England to have all goods sent by St. John or Portland. Any way appears to be satisfactory rather than that via Halifax.

A leading Montreal dry goods importer remarked to a representative of this paper not long since: "We must have goods as soon as we can get them, else they are out of style before they arrive. According to my reckoning it takes about twice the time to bring goods from Liverpool by Halifax that it does by Portland. I won't have goods come by Halifax any more."

All the fault of this does not lie at the doors of the Intercolonial railway, for the steamers to Halifax are slower than those to Portland. We grant that. But how does it come that goods landed off the Allan Line steamer at Halifax, as she is leaving her mail, do not reach Montreal till a week after goods that came by the same boat have arrived via Portland? Surely this is to be accounted for only by slow train service.

Just as the Intercolonial by its slow service is driving the local freight into C.P.R. cars, so is it encouraging importers to have their goods brought in by Portland rather than by Halifax. This is serious. Steam is needed somewhere. Is it in the offices or on the road?

MR. JAMES PENNYCUICK DEAD.

Mr. James Pennycuick, inventor of the Luxfer prism, died at the Emergency Hospital, Toronto, on Wednesday, January 9. Mr. Pennycuick was a Scotchman by birth, and had lived in Newfoundland, Boston and Montreal, leaving the latter city nine years ago to take up his residence in Toronto. His wife is in Boston. The cause of death was epilepsy.

Deceased, who was 69 years of age at the time of his death, was a hardwareman early in life, and used to give interesting reminiscences of his experiences in that vocation.

A TOURIST ATTRACTION SCHEME.

THE energy of Mr. F. H. Clergue, of Sault Ste. Marie, appears to be as applicable to all requirements as it is tireless. Mr. Clergue has established various manufacturing industries in Sault Ste. Marie, he is developing iron mines in Northern Ontario, he is building a railway that will enormously facilitate the opening up of the country through which it will run, and now comes the announcement that next summer he will inaugurate two steamship lines in order to attract tourist travel to points on Lake Erie, Lake Huron, Lake Superior and the Georgian Bay. One route will be from Midland, via Parry Sound and Little Current, to Sault Ste. Marie, while the other will make its start at Toledo with the Sault as its terminus, calling en route at Detroit, Port Huron, Goderich, Kincardine, Southampton, Owen Sound, Collingwood and Parry Sound.

Indirectly, this all interests the merchants of the places to which these lines of steamships will bring passengers. Each year sees increased numbers of tourists attracted to points in Northern Ontario. Last summer, as we pointed out in a previous issue, the Grand Trunk Railway landed about 30,000 tourists at the Muskoka wharf. But, marked as has been the increase in the number of tourists attracted to that part of the country, the number would have been still greater had there been better hotel accommodation.

No doubt Mr. Clergue and those associated with him in his latest enterprise will take some steps to increase the accommodation for tourists. But the merchants at the points of attraction should interest themselves in the matter. Their shoulders are needed to the wheel. By, in season and out of season, urging the necessity of improved accommodation for travellers upon their fellow townsmen they can do a great deal toward bringing about the improvements desired. Self-interest, if nothing else, should actuate them, for there are none that reap greater benefit from tourist travel than the merchants.

NO STOVE CONSOLIDATION.

THE undertaking began early last year to consolidate the stove manufacturers in Ontario has failed.

The concerns which were to form the proposed consolidation were: The Buck Stove Co., Limited, Brantford; The McClary Mfg. Co., Limited, London; Stewart & Co., Woodstock; The Smart Manufacturing Co., Brockville; The Gurney-Tilden Co., Copp Bros., Burrow, Stewart & Milne and Bowes, Jamieson & Co., Hamilton. On the plant, etc., of all these firms an option was secured by the promoters, the basis of purchase being a cash one.

The amount of money wanted to carry out the idea was between $7,000,000 and $8,000,000. To interest that amount was one of the first difficulties experienced. And the manufacturers who had given options were asked to take part of the purchase price in stock. This, some of them, at anyrate, would not do. And now the time limit is expired and the deal is off, although a press despatch says that another option has been secured on the plant, etc., of some of the stove foundries in Hamilton.

A CANADIAN AGENT WANTED.

AT a recent meeting of the Bristol Chamber of Commerce, a letter was read from Mr. Savile Webb, urging that an effort be made to induce the Canadian Government to appoint an agent for that city. Mr. Webb, who is a member of the firm of Purnell, Webb & Co., Bristol, pointed out that the Council of the Chamber had, two years before, urged the Canadian authorities to make such appointment.

The Chamber concurred in the views of Mr. Webb and resolved to make another urgent request to the Canadian Government for the prompt appointment of an agent.

Trade between Canada and Bristol is growing, and a practical business man located there as the representative of Canada could do much towards still further developing it. A direct line of steamships has for some time been running between this country and that port.

The merchants of Bristol want the agent appointed, so that they can be kept informed as to tariff and other matters appertaining to Canada. But, useful as such an agent may be to the merchants of Bristol, he would be equally or more so to the businesss men in Canada interested in the export trade.

It is not necessary to dwell upon the importance of Bristol as a commercial centre. That is already well known. And that is all the more reason why there should be a speedy compliance, on the part of the Dominion Government, with the wishes of such an important body as the Bristol Chamber of Commerce.

THOROUGHNESS IN BUSINESS MATTERS.

THE importance of training for a commercial career is daily becoming more recognized. There was a time, and not very distant, either, when this was not so generally recognized as it is to-day.

Circumstances, we are told, alter cases. And the circumstances of to-day are such as to demand a higher state of efficiency in men intended for commercial careers than was the case even a decade ago.

To succeed in business men must know more than those whose places they are taking. "They must be up-to-date" as we commonly express it. Everywhere thinking men are preaching this doctrine.

In an address recently delivered in London, Eng., Sir Courtenay Boyle, K.C.B., permanent secretary of the Board of Trade, said that his experience was that in business there was as much technique to be learned, as much method to be acquired, as there was in any other of the spheres of life. There is no question as to the truthfulness of that statement. Of lawyers, of doctors, of skilled mechanics we demand knowledge in their respective vocations before we engage them. And if in such vocations, why not in that of business?

Of all vocations, there is none that calls for more intelligence, more practical knowledge, more executive ability than a commercial career. The business man must know when to buy, how to buy, and what to buy. He must be in a position to judge the quality of the goods he handles. He must be conversant with business customs and methods. In a word, he must be so well acquainted with the commercial machinery that he will know how to get the best results from it.

We have been passing through an age of drift. The most of us pick up a book to be entertained. To gather knowledge from it is often foreign to our thought. Our reading is consequently superficial. And it is the same, too often, in our daily vocation. We do not enter it with a determination to familiarize ourselves with it and thereby reach the highest rung in the ladder. We may desire to reach the topmost rung, but we do not care to climb there. We want to slide there, just as, when boys, we used to slide down the stair bannister. We forget that sliding takes us down and not up.

But we are realizing this. And born of this realization is a desire to be Thorough. It is time, too, for in every community there is a search for the Thorough man.

AUGERS AND AUGER BITS LOWER.

A change is announced in the price of Gilmour's augers and auger bits.

In the augers, both the list and discounts are new. The list price shows an advance of about 50c. per dozen all through, but the cost to the trade is about 20 per cent. less than before, on account of an increase in the discount.

No change has been made on auger bits, but the discount has been increased, so that the price is really from 10 to 12½ per cent. lower than before.

AN AGREEMENT COLLAPSES.

The agreement which has for some time existed among the manufacturers in regard to the price of stovepipe elbows has collapsed. The cause was the belief that some parties to the agreement were not keeping it.

Prices are now open and, as a result, lower. No. 1, which was formerly quoted at $1.80, is now being offered at $1.40, and No. 2 is quoted at $1.20, instead of $1.60.

It is asserted that purchases can be made at even lower figures than those named.

The agreement in regard to stovepipes has not been affected, although prices are this week $1 per 100 lengths lower. Five and six-inch pipes are now quoted at $7 per 100 lengths, and 7-in. at $7.50. It is a year since a change was last made in the price of stovepipes.

A TRAVELER'S CONTRACT.

WE have before us the copy of a contract made by a certain travelling man with a customer to whom he had just sold a bill of goods, upon which the traveller, while the ink was still wet, doubtless looked with admiration as expressing in the briefest possible form what he intended to say, but when it reached the credit man of his house, and particularly when the account became, as the credit man understood it, due, there did not seem to be so much reason for gratulation.

The clause of the contract relating to the time of payment read as follows:

The goods ordered herein to be settled by note due in two months from date of invoice; all on hand at end of two months to be credited on note given and new note given for like amount due in four months without interest.

Naturally enough, the purchaser, at the end of two months from the date of invoice, claimed four months more time on the goods remaining unsold; but this was not what the traveller had intended; he claimed that the note was to be due four months from the date of the invoice.

The question naturally arises, if that was what he meant, why did he not say so? The man who draws up a contract must cultivate the ability to detect, in forms of expressions which he is tempted to employ, other meanings than those which he intended to put into them. If he makes an agreement providing for "payment in four months," he certainly ought to know that unless he is careful to state the beginning of the period very clearly the other party to the contract will claim the interpretation which is most favorable to his own interests.

It may be said that this is a question for the schoolmaster rather than for the credit man; that what is required is the ability to write good, plain, unmistakable English. That is exactly the point. The travelling man must cultivate that ability. If he missed the training while a schoolboy, he must make it up by extra care now.

A young Kansas lawyer has convinced himself that the decrease of business for lawyers is due to the increase in general culture; that men carefully trained in the schools are less likely to find themselves in a position where the advice or assistance of a lawyer is necessary than one who has not availed himself of such advantages. We think he is right. Certainly, the travelling man whose "mind's eye" is keen to detect the various constructions that may be placed upon an ambiguous sentence will find that the contracts made by him involve his house in less differences of opinion with the customers than will he who is content to write what he thinks will express his intention, and leave the credit man to fight it out.—Credit Man.

INDUSTRIAL GOSSIP.

Those having any items of news suitable for this column will confer a favor by forwarding them to this office addressed to the Editor.

THE SEYBOLD & SONS CO., wholesale hardware dealers, Montreal, are seeking incorporation.

The Imperial Oil Co. propose establishing an oil depot to be situated at Fredericton, N.B.

The Ottawa and Hull Power and Manufacturing Co., Limited, Hull, Que., are seeking incorporation.

The Hamilton Bridge Works Co. has closed a contract to construct a big steel tow barge, about 200 feet long, for the Montreal Transportation Co.

On Monday, a by-law was submitted to the voters of Kingston, Ont., in favor of bonusing and exempting from taxes a smelter proposed for that city. It was carried by a vote of 1,296 to 166.

R. L. F. Strathy, formerly of Welland, Ont., has organized a company in Owen Sound, Ont., to manufacture a patent wire fence. He expects to have a 120 x 40 ft. two-storey factory running by spring. Fifty hands will be employed at the commencement, and in two years it is expected that 200 hands will be at work.

CANADIAN EXPORTS TO BRITIAN.

THE statistics for the year 1900, which have been compiled by the London, Eng., Board of Trade and which were issued this week, show that Canada's export trade with Great Britain continues to show steady development.

The three years preceding 1900 showed such a remarkable increase in exports of our food products that it was held by many that we would do well to maintain the volume of business done in 1899. Others were more optimistic, and expressed confidence in our ability to supply, and Britain's willingness to purchase, a still greater quantity of our produce.

The returns show that with the exception of butter and flour, which were not produced in as large quantity as in 1899, our exports show large increases.

Our exports of grains show increases of $2,025,000 in wheat; $2,045,000 in oats, and $180,000 in peas. This more than compensates for the decrease shown in our flour trade, $2,915,000. Our sales of cheese were $3,925,000 larger than in the previous year. This effectually offsets the loss in butter trade, which fell off $2,365,000 in the year. Our sales of bacon and hams were $2,290,000 larger than in 1899. Our exports of fish increased $1,415,000. Of eggs, we sent $275,000 more than in the previous year.

Apart from produce the only change of consequence was an increase in wood products. The sales of Canadian-sawn wood to Britain increased $2,110,000 during the year, while our sales of wood pulp were $655,000, and of hewn wood $350,000 greater than in 1899.

A FALLACIOUS POLICY.

SOME business houses which otherwise would have a fair rating are practically blacklisted because they pursue the highly fallacious policy of trying to take advantage of everyone with whom they do business, says an exchange. Some one connected with the house poses as a "kicker." He disputes every bill, every shipment of goods received, every order placed, and practically everything. By sedulously cultivating this unenviable trait for a few years he reaches the point where he tries to take advantage of people as a matter of course, and really thinks it is a trait of superior business attainments for him to be able to find fault as he does, and scale down bills, get allowances, etc.

No doubt a persistent faultfinder of this kind can succeed in getting many a concession. A business concern, in many instances, will sooner stand the loss than keep up a contention, although it may be positive that it is right. It prefers to stand that loss and either not do business with the faultfinder afterward, or watch him so closely that he cannot ply his favorite calling successfully. It doesn't pay to be small in regard to anything in this day and age of the world, and one small man can soon ruin a big business, if he is left to carry out his smallness. Faultfinding will speedily bear a crop of retribution, which has swamped more than one concern in the end. It not only pays to be honest for honesty's sake, but it is a splendid policy to pursue. To such an extent is this recognized by the leading mercantile agencies, that they invariably make a note in their reports whether a house deals fairly and above board, or whether it tries to take advantage of those with whom it does business.

If you discover that a house is in the habit of doing business of this kind, the thing to do is to shun it as you would the plague, because no matter how much you may endeavor to conduct trade with it satisfactorily, you are sure to suffer in the end. Frequently a small man is elevated to an important position in a large house, and he thinks he can earn his salary best by meanness of this kind. If he isn't pulled down, he will pull the house down, no matter what its reputation may have been in past years. That is why there is always an opportunity for honest, honorable and progressive young men in all lines of industry. Perhaps it is well that it is so, because otherwise the rising generations wouldn't have a chance.

COMPANY TO LIQUIDATE.

The Ossekeag Stamping Company, manufacturers of granite and tin ware at Hampton, N.B., has decided to go into liquidation. The industry, which is the largest in King's county, employing from 75 to 100 hands, was established by Charles Palmer and James E. Whittaker. On the death of Mr. Palmer it passed into the control of creditors, by whom the works were kept going. The Bank of Nova Scotia recently issued writs against the company to collect notes made by Fred S. Whittaker, of St. John, who is now in Dorchester penitentiary under a forgery charge. Though the company denies liability for the notes, it was decided, in view of the litigation, to go into liquidation.

The Kingston Locomotive Works are again showing signs of the activity of former years. Two hundred men are now employed and the men are all working full time. The force is being gradually increased.

CANADIAN TRADE WITH JAPAN.

SOME valuable suggestions regarding Canadian trade in Japan are contained in a letter from Malcolm C. Fenwick, of Kobe, formerly of this country, to George Anderson, who, it will be remembered, visited the land of the chrysanthemum as Canadian trade commissioner a couple of years ago. Mr. Fenwick has had considerable experience in Japan in the commission business and writes of what he knows. He says that during the past season he has sold goods in the following lines:

Foodstuffs — Canned goods (fruits and vegetables), packing-house products (hams, and bacon), butter in tins and wood, condensed milk (sweetened and unsweetened), cheese (small full cream, about 9 lb., most popular).

Dry Goods — Suspenders, furs, cotton fabrics. There is a large market in woollen cloths, woollen underclothing, woollen blankets, Mr. Fenwick says, which he has barely touched.

Sundries—Soaps, perfumes, cosmetiques, for which there is much and constant demand; iron, nails, watches, watch cases, jewellery, cutlery, bicycles, guns, sewing machines, in each of which there is an enormous trade.

Mr. Fenwick quotes the present through rate of the combined railways and steamship companies connecting with the east, and says he presumes the Canadian divisions of these lines will conform thereto. R. H. Countess, San Francisco, is the agent, and the following the present tariff per 100 pounds:

	Per car.	Less than car.
Canned goods............	$ 90	$1 50
Packing house products...	1 10	1 60
Piece goods..............	1 10	1 75
Machinery K. D. in pieces	1 00	2 50
Machinery K. D. in boxes.	1 00	2 00

The writer goes on to say that he receives a commission from the manufacturers on all goods, and usually gives the agency of a given product to a resident merchant, and then works up a trade through him by securing him orders. His idea of working up a trade for Canadian manufacturers is to secure a sample room temporarily in each port or large city, visited for a month or six weeks at a time periodically, and samples being displayed and advertisements published in the local papers, native and foreign. The Japanese are now making every effort to deal direct, and independent of the foreign commission merchant. Mr. Fenwick says if this were thought advisable, goods would have to be shipped against B.L. and freight paid by draft with order. This freight would also serve the second purpose of bargain money. He regards this as a popular scheme and one that would help to secure a footing against American, English, French and German goods already established. Mr. Fenwick concludes with an expression of opinion that our Canadian railways, especially the Canadian Pacific, should be prepared to do something better for Canadian trade than the regular through rates quoted above.—The Globe.

The owners of the Coldbrook Rolling Mills, St. John, N.B., have decided to rebuild the mills. It is estimated that the restoration will cost about $30,000.

H. S. HOWLAND, SONS & CO.

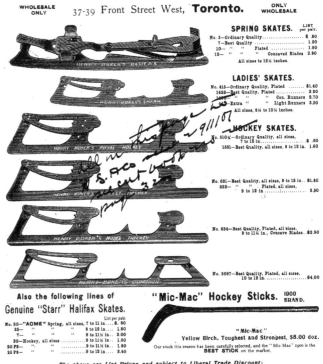

WHOLESALE ONLY — 37-39 Front Street West, **Toronto.** — ONLY WHOLESALE

SPRING SKATES.

	LIST per pair.
No. 5—Ordinary Quality	$ .80
7—Best Quality	1.50
10— " " Plated	1.90
12— " " " Concaved Blades	2.80

All sizes to 12½ inches.

LADIES' SKATES.

No. 415—Ordinary Quality, Plated	$1.40
1422—Best Quality, Plated	2.30
[illegible]— " " " Con. Runners	2.70
443—Extra " " Light Runners	3.50

All sizes, 8½ to 10½ inches.

HOCKEY SKATES.

No. 510½—Ordinary Quality, all sizes, 7 to 12 in.	$ .60
1531—Best Quality, all sizes, 8 to 12 in.	1.60

No. 631—Best Quality, all sizes, 9 to 12 in.	$1.80
632— " " Plated, all sizes, 9 to 12 in	2.50

No. 634—Best Quality, Plated, all sizes, 9 to 11½ in., Concave Blades	$2.90

No. 3697—Best Quality, Plated, all sizes, 10 to 12 in.	$4.00

Also the following lines of Genuine "Starr" Halifax Skates.

		List per pair.
No. 90—"ACME" Spring, all sizes,	7 to 11 in.	$. 80
25— " " "	8 to 12 in.	1.80
7— " " "	9 to 11½ in.	3.00
20—Hockey, all sizes	9 to 11½ in.	1.60
20 PS— " "	9 to 11½ in.	1.80
25 PS— " "	9 to 12 in.	2.40

"Mic-Mac" Hockey Sticks. 1900 BRAND.

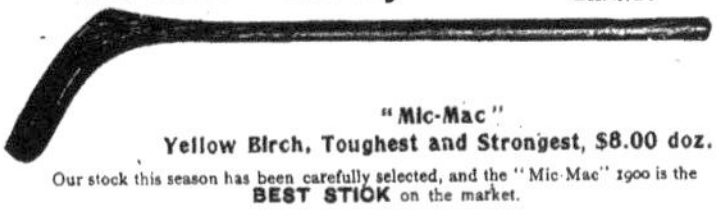

"Mic-Mac"

Yellow Birch, Toughest and Strongest, $8.00 doz.

Our stock this season has been carefully selected, and the "Mic-Mac" 1900 is the **BEST STICK** on the market.

The above are List Prices and subject to Liberal Trade Discount.

H. S. HOWLAND, SONS & CO., Toronto.

PROMPT SHIPMENTS — Graham Wire and Cut Nails are the Best. — OUR PRICES ARE RIGHT.

A COMBINATION OF NORTHERN AND SOUTHERN ENERGY.

ON page 27 will be found a combination advertisement of The Smith & Hemenway Co., Utica Drop Forge & Tool Co., and Thomson Bros. & Co., the latter manufacturers of the Seavey mitre box.

Mr. L. P. Smith, president of The Smith & Hemenway Co., is a Southern man, born and bred in the good old State of Tennessee, educated in the hardware business in Memphis and afterwards in St. Louis in one of the well-known jobbing houses in this branch of business.

Deciding to come East in 1895 in order to cast his lot with "the big fish," he formed the firm of Smith, Herlitz & Co., who were importers of hardware specialties, and continued under that name for something over a year.

His next venture was with Mr. J. B. Patterson, and the partnership under the name of Smith & Patterson was continued for two years, when early in 1898 an incorporated company with the name of The Smith & Hemenway Co. was formed on broader lines, and with Mr. J. F. Hemenway, a native of the "Empire State," as secretary and treasurer, whose earlier experience as manager and treasurer of the Empire Wringer Co., of Auburn, N.Y., and as assistant general manager and assistant treasurer of the American Wringer Co., of Providence, R.I., and New York City, had eminently fitted him to take an active part in pushing to the front the new enterprise. They combined forces with a view of becoming leaders in their special line of business, that is, manufacturing hardware and hardware specialties. For less than a year they continued at the old stand, No. 10 Warren street, New York City; but, finding the place entirely inadequate for the growing business, they moved to larger and more commodious quarters at No. 20 Warren street. After the expiration of another year they found these quarters entirely too small, and removed to No. 296 Broadway, in order to have room enough to conduct their growing business on the lines they desired.

The Smith & Hemenway Co. succeeded to the business of the following concerns: Smith & Patterson, Maltby-Henley Co., Bindley Automatic Wrench Co., Anderberg Importing Co., John Byrnes, glass-cutter manufacturer, all of New York City.

Soon after their organization, the Smith & Hemenway Co. associated themselves with the Utica Drop Forge and Tool Co., of Utica, New York. The Utica Drop Forge and Tool Co. are successors to the Interchangeable Tool Co. and the Russell Hardware and Implement Manufacturing Co. Their New York office is with the Smith & Hemenway Co., New York City, where all catalogues and quotations can be obtained.

Seavey Mitre Box.

At the organization both companies were small, but, both being composed of young blood, they forged their way forward until they have a line of hardware specialties and tools second to none in the world, showing conclusively that young blood and energy will assert itself under all conditions. The Smith & Hemenway Co.'s line comprises a large number of different articles in the hardware specialty line. The Utica Drop Forge and Tool Co. manufacture the largest line of nippers and pliers made by any one factory in the world.

The Smith & Hemenway Co. have recently organized The Schatz Hardware Manufacturing Co., Mount Carmel, Conn., for the manufacture of nail pullers and hardware specialties.

In the fall of 1899 The Smith & Hemenway Co. associated themselves with Thomson Bros. & Co., Lowell, Mass., manufacturers of the well known Seavy mitre box, and since that time have improved this article, until they have to-day what they might call perfection in this individual line.

FOR STORE AND WINDOW DISPLAYS.

A novelty to help store or window display is the Adjustable Display Stand which E. M. Marshall, of Strathroy, Ont., is offering the trade. It is made so that it can be hung on the wall, set down close on the base of the window, or used as an adjustable table. A more useful or practical piece of store furniture is hard to imagine, as it may be put to so many purposes, is so durable, and displays goods to perfection. A card to Mr. Marshall will bring prices and full description.

"The Marshall."

The other illustrations, which appear in the advertisement, will show to the reader how the display stand can be made to assume different shapes, thus adapting itself to the place and the purpose required. In recent novelties in store appliances it stands pre-eminent.

CREATING TRADE.

If a merchant were to close his store and suspend business every time trade lagged he would rightly be branded as a simpleton, remarks a contemporary. And yet in what essential would he differ from the advertiser who stops advertising for the same reason? One sells goods by means of spoken words, and the other by means of printed; their object is identical. It should be plain to the crudest understanding that the time to bid most aggressively for trade is when trade seems most elusive. The alert storekeeper, instead of waiting for something to turn up, turns up something. He changes his window display and show cards, offers particularly tempting values, and employs every device suggested by a nimble wit to turn dullness into activity. He is bold and persistent, and therefore in most instances wins his way.

Mr. W. S. Leslie, of Montreal, arrived in Toronto on Thursday morning.

Mr. S. H. Warnock, who has for some time, successfully represented Lewis Bros. & Co., Montreal, in Manitoba and the Northwest Territories, has severed his connection with this firm and engaged to travel for Lamplough & McNaughton, Montreal.

HOW SATIN WHITE IS MADE.

SATIN white is a pigment much used in the paper making and wall paper trades, for various purposes, on account of its being fairly cheap, being very light, its purity of color, and the fact that it has a strong lustre and takes a polish when rubbed. It can be made in several ways. The following are given by a writer in a London technical journal: 280 lb. of good, well-burnt quicklime are carefully slaked with water, and made into a thin milk. This is strained, so as to free it from grit and lumps of matter, and the liquor run into a tank. There is next added 90 lb. of soda crystals dissolved in 90 gal. of water, which is followed by a solution of 200 lb. of alumina sulphate in 200 gal. of water. The "white" precipitates out, is allowed to settle; the top liquor is run off, and fresh water run in to wash the white, which is then filtered off, pressed, and dried in the usual way.

A rather better quality is made by taking 3 cwt. of good quicklime, and straining, as before. To the milk of lime so made there is added a solution of 6 cwt. of sulphate of alumina in 600 gal. of water. The satin white precipitates out at once, and it is washed, filtered, pressed, and dried in the usual way.

The first of these two whites will consist of a mixture of sulphate of lime and alumina hydrate only. A preparation where magnesium carbonate replaces the lime is made in the following way: 100 lb. of magnesium chloride and 100 lb. alumina sulphate are dissolved in water, and 400 lb. of soda crystals are dissolved separately in water. The two solutions are mixed, and the "white," which is a mixture of carbonate of magnesia and alumina hydrate, is filtered off, washed, pressed, and dried in the usual manner. In making the white by any of these processes, the more dilute the liquids the finer will be the white which is produced.

CONDENSED OR "WANT" ADVERTISEMENTS.

Advertisements under this heading, 2c. a word each insertion; cash in advance. Letters, figures, and abbreviations each count as one word in estimating cost.

BRITISH BUSINESS CHANCES.

Firms desirous of getting into communication with British manufacturers or merchants, or who wish to buy British goods on the best possible terms, or who are willing to become agents for British manufacturers, are invited to send particulars of their requirements for

FREE INSERTION

in "Commercial Intelligence," to the Editor

"SELL'S COMMERCIAL INTELLIGENCE,"
168 Fleet Street, London, England.

"Commercial Intelligence" circulates all over the United Kingdom amongst the best firms. Firms communicating should give reference as to bona fides.

N.B.—A free specimen copy will be sent on receipt of a post card.

MARKETS AND MARKET NOTES

QUEBEC MARKETS.

Montreal, January 11, 1901.

HARDWARE.

As is to be expected at this time of year the hardware business is not brisk. The demand maintains fair proportions, however, and there is not that post-holiday sluggishness sometimes experienced in the first month of the year. Letter orders are quite frequent for sorting stocks of winter goods. Any real activity that is noticeable is to be found in connection with booking orders for spring delivery. Poultry netting is being sold now quite generally at 50 and 5 per cent. off, and green wire cloth at $1.35 per 100 sq. ft. The discounts on screws have been raised 5 per cent. all around. White lead has advanced 25c. per 100 lb. Some of the smaller sizes of galvanized wire have been lowered, and some of the larger raised in price. These comprise the changes in prices for the week.

BARB WIRE — A little future business continues to be done at $3.20 f.o.b. Montreal in less than carlots.

GALVANIZED WIRE — The price list has been altered slightly in the smaller sizes, some of the smaller sizes being reduced 20c., and some of the larger raised 10c. We quote: No. 5, $4.25; Nos. 6, 7 and 8 gauge $3.55; No. 9, $3.10; No. 10, $3.75; No. 11, $3.85; No. 12, $3.25; No. 13, $3.35; No. 14, $4.25; No. 15, $4.75; No. 16, $5.00.

SMOOTH STEEL WIRE—There is still little doing from stock, but a few lots of oiled and annealed wire for spring shipment are being booked. The price is $2.80 per 100 lb.

FINE STEEL WIRE—This line is featureless. The discount is 17½ per cent. off the list.

BRASS AND COPPER WIRE — Few inquiries are being received for these goods just now. Discounts are 55 and 2½ per cent. on brass, and 50 and 2½ per cent. on copper.

FENCE STAPLES—There is little doing in staples. We quote: $3.25 for bright, and $3.75 for galvanized, per keg of 100 lb.

WIRE NAILS—Prices remain unchanged. Trade continues in a quiet, steady way, at $2.85 for small lots and $2.75 for carlots, f.o.b. Montreal, Toronto, Hamilton, London, Gananoque, and St. John, N.B.

CUT NAILS — The existing prices of cut nails have been confirmed. A fair trade is passing. We quote as follows: $2.35 for small and $2.25 for carlots; flour barrel nails, 25 per cent. discount; coopers' nails, 30 per cent. discount.

HORSE NAILS—A small trade has been done this week with the discounts 50 per cent. on Standard and 50 and 10 per cent. on Acadia.

HORSESHOES — The demand keeps up remarkably well. We quote as follows: Iron shoes, light and medium pattern, No. 2 and larger, $3.50; No. 1 and smaller, $3.75; snow shoes, No. 2 and larger, $3.75; No. 1 and smaller, $4.00; X L steel shoes, all sizes, 1 to 5, No. 2 and larger, $3.60; No. 1 and smaller, $3.85; feather-weight, all sizes, $4.85; toe weight steel shoes, all sizes, $5.95 f.o.b. Montreal; f.o.b. Hamilton, London and Guelph, 10c. extra.

POULTRY NETTING—The ruling discount now is 50 and 5 per cent., where it was 50

a few weeks ago. Some future orders have been booked again this week.

GREEN WIRE CLOTH—The price of green wire cloth has been reduced to $1.35 per 100 sq. ft. Some spring business is being done.

FREEZERS—Ice cream freezers are meeting with more attention this week.

SCREWS—The discounts have been raised 5 per cent. on bright and 2½ per cent on brass screws. Discounts are as follows: Flat head bright, 85 per cent. off list; round head bright, 80 per cent.; flat head brass, 77½ per cent.; round head brass, 70 per cent.

BOLTS—There has been no change made in the prices of bolts, and a fair trade continues. Discounts are: Carriage bolts, 65 per cent.; machine bolts, 65 per cent.; coach screws, 75 per cent.; sleigh shoe bolts, 75 per cent.; bolt ends, 65 per cent.; plough bolts, 50 per cent.; square nuts, 4¾c. per lb. off list; hexagon nuts, 4¾c. per lb. off list; tire bolts, 67½ per cent.; stove bolts, 67½ per cent.

BUILDING PAPER — Spring orders in building paper are now being booked. We quote: Dry sheathing, 30c. per roll; cyclone dry do., 42c. per roll; straw do., 30c.; heavy straw do., $1.40 per 100 lb.; I.X.L., dry sheathing, 65c. per roll; cyclone, tarred do., 50c. per roll; tarred ordinary do., 40c. per roll; tarred felt, $1.60 per 100 lb.; ready roofing, 2-ply, 75c. per roll; 3-ply, $1 per roll.

RIVETS—There is nothing new to note. The discount on best iron rivets, section, carriage, and wagon box, black rivets, tinned do., coopers' rivets and tinned swedes rivets, 60 and 10 per cent.; swedes iron burrs are quoted at 55 per cent. off; copper rivets, 35 and 5 per cent. off; and coppered iron rivets and burrs, in 5-lb. carton boxes, are quoted at 60 and 10 per cent. off list.

CORDAGE — The tone of the market now seems to be steady. Manila is quoted at 13½c. per lb. for 7-16 and larger; sisal is worth 9½c. per lb. for 7-16 and larger, and lathyarn 9c. per lb.

SPADES AND SHOVELS—Business is of small proportions in this line. Discounts are 40 and 5 per cent.

TACKS—Prices remain as before. We quote: Carpet tacks, in dozens and bulk, blued, 80 and 5 per cent. discount; tinned, 80 and 10 per cent.; cut tacks, blued, in dozens, 75 and 15 per cent. discount.

FIREBRICKS—Very little trading is being done in this line. The price is $18.50 to $26, as to brand.

CEMENT—A small trade is passing. We quote: German, $2.50 to $2.65; English, $2.40 to $2.50; Belgian, $1.90 to $2.15 per bbl.

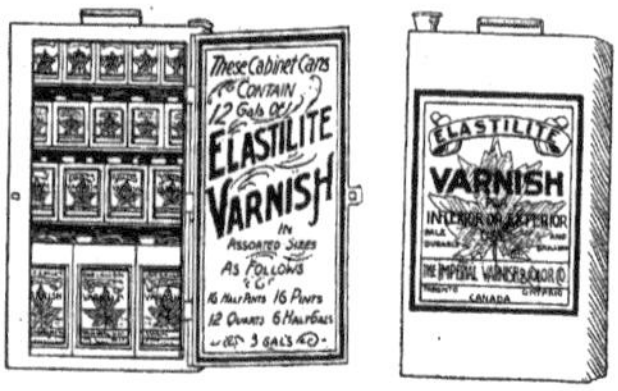

METALS

The metal market is steady, but there is little business being done as dealers are busy taking stock.

PIG IRON — Some sales have been made this week at unchanged figures. Canadian pig is worth $18 to $20, and Summerlee $24 to $25.

BAR IRON — The feeling is towards stationary prices. The ruling price is $1.65 to $1.70 per 100-lb.

BLACK SHEETS — There is but small inquiry for this article. The base price is $2.80 for 8 to 16 gauge.

GALVANIZED IRON—Quite a few additional spring orders have been booked in galvanized iron this week. We quote for immediate delivery: No. 28 Queen's Head, $4.70 to $5; Apollo, 10¾ oz., $4.70 to $5, and Comet, No. 28, $4.30 to $4.55.

INGOT COPPER—The price is unchanged at 17½c.

INGOT TIN—Foreign markets continue weak and Lamb and Flag is worth 33c. in the local market.

LEAD—Small lots are selling at $4.65.

LEAD PIPE—Trade is not active in this line. We quote: 7c. for ordinary and 7½c. for composition waste, with 15 per cent. off.

IRON PIPE—Trade continues of fair dimensions. We hear that the ruling prices on galvanized pipe are below our schedule. We quote as follows: Black pipe, ¼, $2.80 per 100 ft.; ⅜, $2.80; ½, $2.85; ¾, $3.05; 1-in., $4.35; 1¼, $5.95; 1½, $7.10; 2-in., $9.50. Galvanized, ½, $4.90; ¾, $5.40; 1-in., $7.35; 1¼, $9.75; 1½, $11.70; 2-in., $15.75.

TINPLATES—Inquiries are few with prices unchanged at $4.50 for coke and $4.75 for charcoal.

CANADA PLATE—Some movement is noticeable this week. We quote as follows: 52's, $2.90; 60's, $3; 75's, $3.10; full polished, $3.75, and galvanized, $4.60.

TOOL STEEL—We quote: Black Diamond, 8c.; Jessop's 12c.

STEEL—No change. We quote: Sleigh-shoe, $1.85; tire, $1.95; spring, $2.75; machinery, $2.75 and toe-calk, $2.50.

TERNE PLATES—Business is at a standstill in this line. We quote $8.25.

SWEDISH IRON—Steady at $4.25.

COIL CHAIN—The price remains unchanged. The spring business has not opened up well with the retailers yet. We quote: No. 6, 11½c.; No. 5, 10c.; No. 4, 9½c.; No. 3, 9c.; ¼-inch, 7¼c. per lb.; 5-16, $4.60; 5-16 exact, $5.10; ⅜, $4.20; 7-16, $4.00; ½, $3.75; 9-16, $3.65; ⅝, $3.35; ¾, $3.25; ⅞, $3.20; 1-in., $3.15.

SHEET ZINC — Values are steady at 6 to 6¼c.

ANTIMONY—Unchanged, at 10c.

GLASS.

The demand is very light. We quote: First break, $2; second, $2.10 for 50 feet; first break, 100 feet, $3.80; second, $4; third, $4.50; fourth, $4.75; fifth, $5.25; sixth, $5.75, and seventh, $6.25.

PAINTS AND OILS.

Turpentine is a little stronger in the Southern markets, but not sufficiently so to warrant an advance here this week. Should the firmness be maintained, in all probability there will be a slight appreciation in value of turpentine within the next few days. No change of moment is taking place in linseed oil, which continues in fair inquiry. General business has been tolerably brisk during the week, and the midwinter sluggishness generally experienced seems to be absent this season. The majority of groups of travellers are busily engaged preparing for their early spring trips. White lead is 25c. higher, and the quotations for Paris green are now published. We quote:

WHITE LEAD—Best brands, Government standard, $6.75; No. 1, $6.37½; No. 2, $6; No. 3, $5.62½, and No. 4, $5.25, all f.o.b. Montreal. Terms, 3 per cent. cash or four months.

DRY WHITE LEAD—$5.75 in casks; kegs, $6.

RED LEAD—Casks, $5.50; in kegs, $5.75.

WHITE ZINC PAINT—Pure, dry, 8c.; No. 1, 6½c.; in oil, pure, 9c.; No. 1, 7½c.

PUTTY—We quote: Bulk, in barrels, $2 per 100 lb.; bulk, in less quantity, $2.15; bladders, in barrels, $2.20; bladders, in 100 or 200 lb. kegs or boxes, $2.35; in tins, $2.45 to $2.75; in less than 100-lb. lots, $3 f.o.b. Montreal, Ottawa, Toronto, Hamilton, London and Guelph. Maritime Provinces 10c. higher, f.o.b. St. John and Halifax.

LINSEED OIL—Raw, 80c.; boiled, 83c., in 5 to 9 bbls., 1c. less, 10 to 20 bbl. lots, open, net cash, plus 2c. for 4 months. Delivered anywhere in Ontario between Montreal and Oshawa at 2c. per gal. advance and freight allowed.

TURPENTINE—Single bbls., 59c.; 2 to 4 bbls., 58c.; 5 bbls. and over, open terms, the same terms as linseed oil.

MIXED PAINTS—$1.25 to $1.45 per gal.

CASTOR OIL—8¾ to 9¼c. in wholesale lots, and ½c. additional for small lots.

SEAL OIL—47½ to 49c.

COD OIL—32½ to 35c.

NAVAL STORES—We quote: Resins, $2.75 to $4.50, as to brand; coal tar, $3.25 to $3.75; cotton waste, 4½ to 5½c. for colored, and 6 to 7½c. for white; oakum, 5½ to 6½c., and cotton oakum, 10 to 11c.

PARIS GREEN—Petroleum barrels, 16¾c. per lb.; arsenic kegs, 17c.; 50 and 100-lb. drums, 17½c.; 25-lb. drums, 18c.; 1-lb. packages, 18¾c.; ¼-lb. packages, 20¾c.; 1-lb. tins, 19½c.; ½-lb. tins, 21½c. f.o.b. Montreal; terms 3 per cent. 30 days, or four months from date of delivery.

SCRAP METALS.

The tone of the scrap metal market is steady with goods scarce. Dealers are paying the following prices in the country: Heavy copper and wire, 13 to 13½c. per lb.; light copper, 12c.; heavy brass, 12c.; heavy yellow, 8½ to 9c.; light brass, 6½ to 7c.; lead, 2¾ to 3c. per lb.; zinc, 2¼ to 2½c.; iron, No. 1 wrought, $13 to $14 per gross ton; No. 1 cast, $13 to $14; stove plate, $8 to $9; light iron, No. 2, $4 a ton; malleable and steel, $4.

PETROLEUM.

This midwinter necessity continues to go out freely. We quote: "Silver Star," 15 to 16c.; "Imperial Acme," 16½ to 17½c.; "S.C. Acme," 18 to 19c., and "Pratt's Astral," 19 to 20c.

HIDES.

Green hides are lower in sympathy with the decline in the United States. The demand for hides is improved. We quote: Light hides, 7½c. for No. 1; 6½c. for No. 2, and 5½c. for No. 3. Lambskins, 90c.

MONTREAL NOTES.

White lead is advanced 25c. per 100 lb.

A new price list is out on galvanized wire.

The discounts on screws have been raised.

Green wire cloth has been reduced 15c. per 100 sq. ft.

ONTARIO MARKETS.

TORONTO, January 11, 1901.

HARDWARE.

TRADE is gradually recovering from its holiday quietude. The travellers are again on the road, and the orders they have been sending in during the last few days appear to be, on the whole, rather better than expected. There has also been quite a nice complement of letter orders. No apparent improvement is shown in the demand for nails. Fence wires are still not wanted for immediate shipment, and but few orders are being booked for future delivery. Some orders are being booked n poultry netting for future shipment. The same is to be said in regard to green wire cloth. Quite a few orders have been received during the week for such small goods as tacks, shoe nails and shoe rivets. A fair trade is to be noted in rules and other lines of carpenters' tools. A nice trade is opening up in milk can trimmings. The week has witnessed a number of changes in prices, the most important of which is a reduction in wood screws. Stovepipes are $1 per 100 lengths lower, and elbows show a reduction of 40c. Lower prices also rule on Gilmour's augers and auger bits. White lead, on the other hand, is 25c. per 100 lb. higher.

BARB WIRE—Business is practically nil, for future as well as for present delivery. We quote $2.97½ f.o.b. Cleveland for less than carlots, and $2.85 in carlots. From stock, Toronto, $3.10 per 100 lb.

GALVANIZED WIRE—There is nothing scarcely doing. We still quote No. 9 at $3.10, Toronto. The base price f.o.b. Cleveland is still $2.72½ per 100 lb.

SMOOTH STEEL WIRE—A small business only is being done in both oiled and annealed and hay-baling wire. Base price is unchanged at $2.80 per 100 lb.

WIRE NAILS—Trade is still quiet and without any apparent improvement. The manufacturers, after a session of several days, decided to make no change in price. Base price is still, therefore, $2.85 per keg for less than carlots and $2.75 for carlots.

CUT NAILS—These are dull and featureless, with the base price still at $2.35 per keg.

HORSESHOES—Busines is moderate and without special feature. We quote as follows f.o.b. Toronto: Iron shoes, No. 2 and larger, light, medium and heavy, $3.60; snow shoes, $3.85; light steel shoes, $3.70; featherweight (all sizes), $4.95; iron shoes, No. 1 and smaller, light, medium and heavy (all sizes), $3.85; snow shoes, $4; light steel shoes, $3.95; featherweight (all sizes), $4.95.

HORSE NAILS—Quiet and unchanged. Discount, 50 per cent. on standard oval head and 50 and 10 per cent. on Acadia.

SCREWS—Prices are lower on wood screws by about 20 per cent., the manufacturers having decided upon a reduction shortly after we went to press last week. The reduction is in sympathy with a decline in the United States market, where the association has dissolved the pool owing to outside competition. We now quote: Flat head bright, 85 per cent. off the list; round head bright, 80 per cent.; flat head brass, 77½ per cent.; round head brass, 70 per cent.; flat head bronze, 70 per cent.; round head bronze, 65 per cent.

BOLTS AND NUTS—Trade is just fair and without any particular feature. We quote as follows: Carriage bolts (Norway), full square, 70 per cent.; carriage bolts, fulls quare, 70 per cent.; common carriage bolts, all sizes, 65 per cent.; machine bolts, all sizes, 65 per cent.; coach screws, 75 per cent.; sleighshoe bolts, 75 per cent.; blank bolts, 65 per cent.; bolt ends, 65 per cent.; nuts, square, 4½c. off; nuts, hexagon, 4¾c. off; tire bolts, 67½ per cent.; stove bolts, 67½; plough bolts, 60 per cent.; stove rods, 6 to 8c.

RIVETS AND BURRS — These are dull. Discount, 60 and 10 per cent. on iron rivets; iron burrs, 55 per cent.; copper rivets and burrs, 35 and 5 per cent.

ROPE—Business is still light. The hemp market is fairly firm, but the manufacturers in the United States are disinclined to pay present prices. We quote: Sisal, 9c. per lb. base, and manila, 13c. Cotton rope is unchanged as follows: 3-16 in. and larger, 16½c.; 5-32 in., 21½c., and ⅛ in., 22½c. per lb.

CUTLERY — The quietness which settled down in this line of trade after the holidays still obtains.

SPORTING GOODS—Only small quantities are going out.

BUILDING PAPER — Business is quiet. Ready roofing, 3-ply, $1.65 per square; ditto, 2-ply, $1.40 per square. Quotations are f.o.b. Toronto, Hamilton, London.

GILMOUR'S AUGERS AND AUGER BITS — These are lower in price. Fuller particulars will be found on our editorial pages.

GREEN WIRE CLOTH—Prices have been reduced 15c. per 100 lb., the quotation now being $1.35 per 100 sq. ft. A fairly good trade is being done for spring delivery.

SKATES—While trade is not active it is fair, and stocks in jobbers' hands are getting fairly well reduced.

HARVEST TOOLS—There is a great deal of hesitancy on the part of the retail trade to place orders for future delivery. Discounts 50, 10 and 5 per cent.

POULTRY NETTING—A fair number of orders are being booked for spring delivery. Discount off the Canadian list is still 50 per cent.

ENAMELLED WARE—Very little business is being done.

TINWARE—There is a little movement in tinware, particularly in milk can trimmings.

STOVEPIPES—Prices have been reduced $1, the quotations now being as follows: 5 and 6 in., $7 per 100 lengths, and 7 in., $7.50 per 100 lengths. It is about a year since the last change was made.

ELBOWS—The agreement in regard to the price of stovepipe elbows has been dissolved and lower prices rule. No. 1 are now quoted at $1.40, and No. 2 at $1.20 per dozen.

STOVES AND FURNACES—Very little is being done in stoves and practically nothing in furnaces.

CEMENT—The season is over. We nominally quote in barrel lots: Canadian Portland, $2.80 to $3; Belgian, $2.75 to $3; English do., $3; Canadian hydraulic cements, $1.25 to $1.50; calcined plaster, $1.90; asbestos cement, $2.50 per bbl.

METALS.

A slight improvement has taken place in the demand for metals, but the volume of business is still light.

PIG IRON—There is not much doing. For small quantities of pig iron the Ontario furnaces are quoting $18 per ton.

BAR IRON — The demand is fairly good with $1.70 as the ruling base price.

PIG TIN — The outside markets are quiet and lower, particularly in London. Locally the demand is fair for small quantities at 32 to 33c.

TINPLATES —While the movement is not large, it is fair for this time of the year.

TINNED IRON—Some fairly good shipments have been made dbring the past week on cheese-factory account.

TERNE PLATES—There has been a little better movement in this line during the past week.

BLACK SHEETS—The demand is fair. We quote $3.50 per 100 lb.

GALVANIZED SHEETS—While the demand is better than it was, trade is not yet active. We quote English at $4.85 and American at $4.50 for ordinary quantities.

CANADA PLATES—A little movement is still to be noted in them. We quote: All dull, $3.15; half and half, $3.25; and all bright, $3.85 to $4.

IRON PIPE—Trade keeps fair in this line. We quote as follows: Black pipe ¼ in., $3.00; ⅜ in., $3.00; ½ in., $3.00; ¾ in., $3.30; 1 in., $4.50; 1¼ in., $6.25; 1½ in., $7.75; 2 in., $10.40. Galvanized pipe is as follows: ½ in., $4.50; ¾ in., $5; 1 in., $7; 1¼ in., $9.50; 1½ in., $11.75; 2 in., $15.75.

SOLDER — The demand is good. We quote half-and-half, 19 to 20c.; refined, 19c.

LEAD—A moderate demand is to be noted. We quote 4¾ to 5c.

COPPER—Business is quiet in both sheet

and ingot copper. We quote: Ingot, 19 to 20c.; bolt or bar, 23½ to 25c.; sheet, 23 to 23½c.

BRASS—The demand is fair. Discount on rod and sheet 15 per cent.

ZINC SPELTER — Very little business is being done. We quote 16 to 16½c. per lb.

SHEET ZINC—In this line the demand has been fair during the past week. We quote casks at $6.75 to $7, and part casks at $7 to $7.50 per 100 lb.

ANTIMONY — A good trade is to be reported this week. We quote 11 to 11½c. per lb.

PAINTS AND OILS.

There is some improvement in the demand but the movement is still light. Orders for linseed oil for spring delivery are coming in well, as it is expected that a firm market will be found when the spring trade begins. Turpentine has advanced 2c. here, and is steady at the higher figure. White lead was advanced 25c. for all grades this week. This was unexpected, as at a meeting of the manufacturers a few days ago the matter was considered and no change was made. We quote:

WHITE LEAD—Ex Toronto, pure white lead, $6 87½; No. 1, $6.50; No. 2, $6.12½; No. 3, $5.75; No. 4 $5.37½; dry white lead in casks, $6.

RED LEAD—Genuine, in casks of 560 lb., $5.50; ditto, in kegs of 100 lb., $5.75; No. 1, in casks of 560 lb., $5 to $5.25; ditto, kegs of 100 lb.; $5.25 to $5.50.

LITHARGE—Genuine, 7 to 7½c.

ORANGE MINERAL—Genuine, 8 to 8½c.

WHITE ZINC—Genuine, French V.M., in casks, $7 to $7.25; Lehigh, in casks, $6.

PARIS WHITE—90c.

WHITING — 60c. per 100 lb.; Gilders' whiting, 75 to 80c.

GUM SHELLAC — In cases, 22c.; in less than cases, 25c.

PUTTY — Bladders, in bbls., $2.20; bladders, in 100 lb. kegs, $2.35; bulk in bbls., $2; bulk, less than bbls. and up to 100 lb., $2.15; bladders, bulk or tins, less than 100 lb., $3.

PLASTER PARIS—New Brunswick, $1.90 per bbl.

PUMICE STONE — Powdered, $2.50 per cwt. in bbls., and 4 to 5c. per lb. in less quantity; lump, 10c. in small lots, and 8c. in bbls.

LIQUID PAINTS—Pure, $1.20 to $1.30 per gal.; No. 1 quality, $1 per gal.

CASTOR OIL—East India, in cases, 10 to 10½c. per lb. and 10½ to 11c. for single tins.

LINSEED OIL—Raw, 1 to 4 barrels, 80c.; boiled, 84c.; 5 to 9 barrels, raw, 79c.; boiled, 83c., delivered. To Toronto, Hamilton, Guelph and London, 2c. less.

TURPENTINE—Single barrels, 59c.; 2 to 4 barrels, 58c., to all points in Ontario.

For less quantities than barrels, 5c. per gallon extra will be added, and for 5-gallon packages, 50c., and 10 gallon packages, 80c. will be charged.

GLASS.

There is not much doing, and there is little indication of an immediate advance. We still quote first break locally: Star, in 50-foot boxes, $2.10, and 100-foot boxes, $4; doublediamond under 26 united inches, $6. Toronto, Hamilton and London; terms 4 months or 3 per cent. 30 days.

OLD MATERIAL.

A strong feeling is manifested, but prices are unchanged. Delivery is moderate. We quote jobbers' prices as follows: Agricultural scrap, 55c. per cwt.; machinery cast, 55c. per cwt.; stove cast, 40c.; No. 1 wrought 55c. per 100 lb.; new light scrap copper, 12c. per lb.; bottoms, 10½c.; heavy copper, 12½c.; coil wire scrap, 13c.; light brass, 7c.; heavy yellow brass, 10 to 10½c.; heavy red brass, 10½c.; scrap lead, 3c.; zinc, 2½c.; scrap rubber, 7c.; good country mixed rags, 65 to 75c.; clean dry bones, 40 to 50c. per 100 lb.

HIDES, SKINS AND WOOL.

HIDES — Prices are steady with little doing. We quote as follows: Cowhides, No. 1, 7½c.; No. 2, 6½c.; No. 3, 5½c. Steer hides are worth 1c. more. Cured hides are quoted at 8½c.

SKINS — The market is dull with prices unchanged throughout. We quote as follows: No. 1 veal, 8-lb. and up, 8c. per lb.; No. 2, 7c.; dekins, from 40 to 60c. culls, 20 to 25c. Sheep are selling at 90 to 95c.

WOOL—A decline of 1c. is noted. The market is listless. We quote as follows: Combing fleece, 15 to 16c., and unwashed, 9½ to 10c.

PETROLEUM.

An advance of ½c. is noted throughout. There is a good movement. We quote as follows: Pratt's Astral, 17 to 17½c. in bulk (barrels, $1 extra); American water white, 17 to 17½c. in barrels; Photogene, 16½ to 17c.; Sarnia water white, 16 to 16½c. in barrels; Sarnia prime white, 15 to 15½c. in barrels.

COAL.

A good movement continues, but not as large as would be the case if more cars could be had. Prices are unchanged. We quote anthracite on cars Buffalo and bridges: Grate, $4.75 per gross ton and $4.24 per net ton; egg, stove and nut, $5 per gross ton and $4.46 per net ton.

MARKET NOTES.

Pig tin is 1c. per lb. lower.

Petroleum is ½c. per gal. dearer.

Turpentine has advanced 2c. per gal.

White lead is 25c. per 100 lb. higher.

Gilmour's augers and auger bits are cheaper.

Wood screws are quoted about 20 per cent. lower.

Green wire cloth has been reduced 15c. per 100 square feet.

Stovepipes are $1 per 100 lengths cheaper, and stovepipe elbows are quoted 40c. per doz. lower.

The outlook for export for pig iron, billets and all cruder forms is decidedly discouraging. It will take the foreigners some time to get over their scare, until they realize that the market here has changed. The financial situation is unfavorable, notably in Germany, and prices abroad have come down with a run. For some months deliveries on old orders will go forward, but then, unless there are new developments, we may expect a sharp decline.—Iron Age.

MANITOBA MARKETS.

WINNIPEG, January 7, 1900.

THE demand for all lines of shelf and heavy hardware, and paints and oils, is light indeed, practically nothing being done. The staffs of the various houses are chiefly employed in clearing the remnants of last year's business preparatory to stock-taking.

Price list for the week is as follows:

Barbed wire, 100 lb.	$3 75
Plain twist	3 75
Staples	4 25
Oiled annealed wire 10	3 95
" 11	4 00
" 12	4 05
" 13	4 20
" 14	4 35
" 15	4 45
Wire nails, 30 to 60 dy, keg	3 45
" 16 and 20	3 50
" 10	3 55
" 8	3 65
" 6	3 70
" 4	3 85
" 3	4 10
Cut nails, 30 to 60 dy.	3 00
" 20 to 40	3 05
" 10 to 16	3 10
" 8	3 15
" 6	3 20
" 4	3 30
" 3	3 65
Horsenails, 40 per cent. discount.	
Horseshoes, iron, No. 0 to No 1	4 90
No. 2 and larger	4 05
Snow shoes, No. 0 to No. 1	5 15
No. 2 and larger	4 90
Steel, No. 0 to No. 1	5 20
No. 2 and larger	4 95
Bar iron, $2.50 basis.	
Swedish iron, $4.50 basis.	
Sleigh shoe steel	3 00
Spring steel	3 25
Machinery steel	3 75
Tool steel, Black Diamond, 100 lb	8 50
Jessop	13 00
Sheet iron, black, 10 to 20 gauge, 100 lb.	3 50
20 to 26 gauge	3 75
28 gauge	4 00
Galvanized American, 16 gauge	2 54
18 to 22 gauge	4 50
24 gauge	4 75
26 gauge	5 00
28 gauge	5 25
Genuine Russian, lb	12
Imitation " "	8
Tinned, 24 gauge, 100 lb	7 55
26 gauge	8 80
28 gauge	8 00
Tinplate, IC charcoal, 20 x 28, box	10 75
" IX "	12 75
" IXX "	14 75
Ingot tin	35
Canada plate, 18 x 21 and 18 x 24	3 90
Sheet zinc, cask lots, 100 lb.	7 50
Broken lots	8 00
Pig lead, 100 lb.	6 00
Wrought pipe, black up to 2 inch	50 an 10 p.c.
" Over 2 inch	45 p.c.
Rope, sisal, 7-16 and larger	$10 00
" ⅜	10 50
" ¼ and 5-16	11 00
Manila, 7-16 and larger	13 50
" ⅜	14 00
" ¼ and 5-16	14 50
Solder	22
Cotton Rope, all sizes, lb.	17½
Axes, chopping	$7 50 to 12 00
" double bitts	12 00 to 18 00
Screws, flat head, iron, bright	75 and 10 p.c.
Round " "	70 p.c.
Flat " brass	70 p.c.
Round " "	60 and 5 p.c.
Coach	57½ p.c.
Bolts, carriage	42½ p.c.
Machine	45 p.c.
Tire	60 p.c.
Sleigh shoe	65 p.c.
Plough	40 p.c.
Rivets, iron	50 p.c.
Copper, No. 8	50c. lb.
Spades and shovels	40 p.c.
Harvest tools	50, and 10 p.c.
Axe handles, turned, s. g. hickory, doz.	$2 50
No. 1	1 50
No. 2	1 25
Octagon extra	1 75
No. 1	1 25
Files common	70, and 10 p.c.
Diamond	60
Ammunition, cartridges, Dominion R.F.	50 p.c.
Dominion, C.F., pistol	30 p.c.
" military	15 p.c.
American R.F.	30 p.c.
C.F. pistol	5 p.c.
C.F. military	10 p.c. advance.
Loaded shells:	
Eley's soft, 12 gauge	16 50
chilled, 12 guage	18 00
soft, 10 guage	21 00
chilled, 10 guage	23 00
American, M	16 25
Shot, Ordinary, per 100 lb	6 75
Chilled	7 50
Powder, F.F., keg	4 75
F.F.G.	5 00
Tinware, pressed, retinned	75 and 2½ p.c.
" " plain	70 and 15 p.c.
Graniteware, according to quality	50 p.c.

PETROLEUM.

Water white American	24½c.
Prime white American	23c.
Water white Canadian	21c.
Prime white Canadian	19c.

PAINTS, OILS AND GLASS.

Turpentine, pure, in barrels	$ 74
Less than barrel lots	79
Linseed oil, raw	87
Boiled	90
Lubricating oils, Eldorado castor	25½
Eldorado engine	24½
Atlantic red	27½
Renown engine	41
Black oil	23½ to 25
Cylinder oil (according to grade)	55 to 74
Harness oil	61
Neatsfoot oil	$ 1 00
Steam refined oil	85
Sperm oil	1 50
Castor oil per lb.	11½
Glass, single glass, first break, 16 to 25 united inches	2 25
26 to 40 per 50 ft.	2 50
41 to 50	5 50
51 to 60	6 00
61 to 70 per 100-ft. boxes	6 50
Putty, in bladders, barrel lots per lb.	2¼
kegs "	2½
White lead, pure per cwt.	7 25
No 1 "	7 00
Prepared paints, pure liquid colors, according to shade and color per gal.	$1.30 to $1.90

CATALOGUES, BOOKLETS, ETC.

A PRETTY CALENDAR.

E. K. Spinney, iron and hardware merchant, Yarmouth, N.S., has sent us one of the prettiest calendars of the season. It depicts a sailor-suited girl sitting on a rock on the wild sea shore, and she is beautiful enough to attract every eye within range to the notice that E. K. Spinney sells iron and hardware, is agent for several fire and marine insurance companies and for the White Star Line and Dominion Line S.S. Cos. The object of the calendar is thus accomplished. The calendar itself is large and clear, and has been printed with appropriate notices on each sheet.

A BANK'S CALENDAR.

The Royal Bank of Canada, Montreal, is distributing a useful and attractive calendar for the current year. This bank has developed rapidly in the last few years, and now has branches in all the principal centres in Canada. The calendar is one worth having, and no doubt any of our readers sending a post card with the request for one will get it.

DEMAND FOR BOOK AND BROOM WIRE.

The Peerless Wire Co., Hamilton, has recently received large orders for book and broom wire. Owing to the pressure of business the company is considering the advisability of running its mill night as well as day. This company, it will be remembered, makes a specialty of broom and book wire.

The sheriff's sale of the realty of C. Locke & Co., general merchants and fish dealers, Lockeport, N.S., is advertised for January 16.

HEATING AND PLUMBING

EXPLOSION OF GAS METERS.

THE following communication from J. A. Painchaud, Montreal, commenting on the article on "Explosion of Gas Meters," in The Metal Worker of December 22, appeared in that journal on January 5 :

"The statement of W. R. Park, relating to the particulars of gas meter explosion, clearly shows how exceedingly dangerous it is to have a mixture of air and illuminating gas, a fortiori acetylene, in a receptacle. It seems to me that the sooner acetylene gas generator manufacturers realize that only absolutely airless acetylene will afford absolute security and adopt some means of avoiding any possible admission of air into the system, the better it would be for the public and the acetylene industry. It would not take many reports of accidents similar to what happened at a ball in Aix-en Othe, France, where several lives were lost through an acetylene explosion, due to the meddling with the acetylene apparatus by a couple of ignorant guests, to cripple this promising industry for many years. As a true friend of acetylene, I believe that people should be made aware of the possible dangers of this illuminant. From a report made by an insurance inspector to me not long ago, of a generator manufacturer in the West, who tried to accelerate the emptying of acetylene from his apparatus in the inspector's presence by the same means as mentioned by Mr. Park, with the result of an explosion which fortunately did no more than frighten them, some generator manufacturers also need advice on this subject. A flame should never be applied to acetylene issuing from any opening except from proper burners. Plumbers, who are more liable than others to forget this in looking for leaks, should be made to understand it."

PROSPECTS ARE BRIGHT FOR WINNIPEG.

The outlook for the plumbing and heating trade in Winnipeg seems good, as several large contracts are being considered. The Bank of Hamilton intend making a 25 x 120 ft. addition, the same height as their present premises, to cost about $50,000. The new Y.M.C.A. building, to be two storeys high and to provide for four large stores, is to be started by W. F. Alloway and D. E. Sprague early next spring. John Leslie, furniture dealer, Main street, proposes expending $15,000 on alterations. The Lake of the Woods Milling Co. are having plans prepared for a new office building on McDermot street, to cost about $35,000. A syndicate of business men have decided to erect a large block for mercantile offices on the northwest corner of Arthur and McDermot streets. G. Olson, flour and feed dealer, King street, intends erecting a brick building near his present premises, to cost $20,000. Plans have also been prepared for many smaller business houses and residences.

PLUMBING AND HEATING NOTES.

Geo. A. Wooten & Co. have started as plumbers in Halifax.

Lindsay, Ont., voted in favor of municipal ownership of the electric light plant.

Thorold, Ont., carried, by a majority of 16, a by-law in favor of installing waterworks, on Monday.

Chas. H. Coursolles, who has been doing business in Ottawa as an electrician under the style of Cote & Coursolles, has started to do business under his own name.

On Monday, Parry Sound, Ont., voted on two by-laws, one for $29,500 for the purchase of an electric light plant and extension and improvements of waterworks system, and the other for $2,500 for the construction of a steel bridge across Seguin river. Both by-laws were carried.

The employes and friends of Purdy, Mansell & Co., plumbers, Toronto, intend holding their annual supper on Friday of next week at the Morris House, Lambton Mills. The party will leave Purdy, Mansell & Co.'s at 8 o'clock p.m., and proceed by carriage to the hotel. A big time is anticipated.

SOME BUILDING NOTES.

The council of St. Patrick's Home, Ottawa, propose erecting a new wing, to comprise a children's dormitory, lecture hall and chapel next spring.

Saxe & Archibald, architects, Montreal, will soon invite tenders for work on the Bellevue Building, St. Catherine street, Montreal, which M. S. Foley is erecting.

Permits have been issued in Toronto to L. J. Greenway for a two-storey and attic residence on Pearson street, near Roncesvalles avenue, to cost $1,800, and to the Macpherson estate (J. N. Townsend, architect) for a brick residence on the north side of Crescent road to cost $4,000.

A. F. Dunlop, architect, Montreal, is preparing plans for a front addition of marble 63 ft. and five additional storeys to the St. James street side of the Carsely departmental store, Montreal. A main feature of the new building will be one large entrance from St. James street, the vestibule to be very spacious, and the surroundings quite imposing in appearance. There will be two new rapid elevators, and a separate one from the basement to the kitchen, for restaurant work. The plans include a restaurant on the last storey.

PLUMBING AND HEATING CONTRACTS.

The John Ritchie Plumbing and Heating Co., Limited, Toronto, have the contract for alterations to the plumbing and heating of the addition to the Grand Union Hotel, Toronto.

PERSONAL MENTION.

Mr. Wm. Polson, president of the Polson Iron Works, shipbuilders and engine manufacturers, Toronto, died at his home, 102 Pembroke street, Toronto, on Tuesday.

Mr. Geo. F. Stephens, senior partner of the well-known jobbing firm, Messrs. G. F. Stephens & Co., Winnipeg, was in Montreal during the week visiting the different paint factories.

On Saturday morning of last week J. T. Peacock, of the James Morrison Brass Mfg. Co., Toronto, was presented with a handsome portmanteau by his fellow employes. Mr. Peacock is leaving the James Morrison Company to take an agency of the Canada Life Assurance Company at Port Hope, Ont.

It is understood that one of the largest locomotive manufacturing firms in the United States has decided to establish a branch at Sydney, Cape Breton.

20th Century Nail Pullers

WE MAKE 5 STYLES.
WE MAKE 5 GRADES.
THE GREEN BOOK TELLS THE STORY.

SMITH & HEMENWAY CO., Utica Drop Forge & Tool Co., **296 Broadway, New York City.**

CANADIAN ADVERTISING is best done by THE E. DESBARATS ADVERTISING AGENCY MONTREAL.

ADVERTISING in WESTERN CANADA

will be Carefully, Efficiently, and Promptly attended to, by

The Roberts Advertising Agency,
WINNIPEG, CANADA.

KNOX HENRY

Heavy Hardware and Metal Broker
Room 220½ Board of Trade, MONTREAL.

HORSE NAILS—"C" Brand Horse Nails—Canada Horse Nail Co.
"BRASSITE" GOODS—Gunn Castor Co., Limited, Birmingham, Eng.

Manufacturers of

Heating Supplies

Pipe Fittings and Headers.
Large Manifolds made to Order.
Steam Traps and Appliances, etc.

The . . .
Jas. Morrison Brass Mfg. Co., Limited
TORONTO.

FIXING-UP TIME

follows stocktaking and is the best time to put in

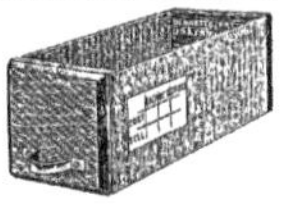

BENNETT'S PATENT SHELF BOX
and the
KLONDIKE SAMPLE HOLDER.

J. S. BENNETT, 20 Sheridan Ave., Toronto.

N.B.—**Don't forget** we make boxes to fit your present shelving.

Standard Paint & Varnish Works
Limited
Makers of High Grade
Varnishes, Japans,
Paints, Colors & Enamels
Windsor, Ont.

WIRE RODS!

Drawn to Decimal Sizes, Cut and Straightened, In Uniform Sizes. Prompt Shipment.

Chalcraft Screw Co., Limited, Brantford, Ont.

THE ADAMS STOVE PIPE REGISTER.

Design Patented. June 29, 1897.

Design Patented August 31, 1897.

Made by

The Adams Company

Dubuque, Iowa, U.S.A.

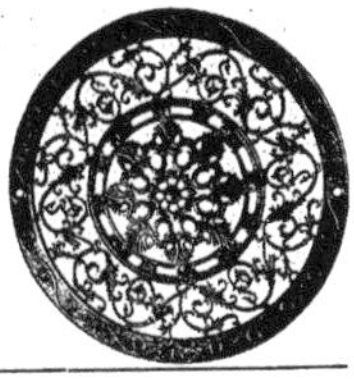

LEADER CHURN

New Century Improvements.

FOUR DIFFERENT STYLES:
A—Steel Frame with double reversible Steel Lever.
B—Wood Frame with double reversible Steel Lever.
C—Steel Frame with Crank.
D—Wood Frame with Crank.

Styles A and B may be operated from a sitting or standing position.

Steel Frames and Hoops beautifully ALUMINIZED.
All LEADER CHURNS are equipped with BICYCLE BALL BEARINGS and PATENTED CREAM BREAKERS.
Stands are so constructed that they are particularly strong and rigid, and there is nothing to interfere with the placing of pail in the most convenient position for draining off buttermilk.

It Pays to Have the Best. None are Better Than the Leader.

THE
Dowswell Manufacturing Co. Limited.
HAMILTON, ONT.
Eastern Agents: W. L. Haldimand & Son, Montreal, Que.

MATTRESS AND BROOM WIRE

HIGH GRADE, DOUBLE TINNED

Uniform Size and Temper Guaranteed.

Fine Annealed Brush and Market Wire,
TINNED WIRE OF ALL KINDS.

SAMPLES AND QUOTATIONS SENT ON APPLICATION.

The Peerless Wire Co., - Hamilton

TRADE IN COUNTRIES OTHER THAN OUR OWN.

MANUFACTURED IRON AND STEEL IN ENGLAND.

IN the finished branches of the trade an improvement was hardly to be expected at holiday time, and the depression which has now prevailed for some weeks past shows no diminution. Business has, of course, been on a very small scale during the past week, and there have been some further reductions. In South Staffordshire marked bars still command £10 10s., although that price is sometimes not obtained without difficulty, but common bars have been reduced to £8. A good deal is heard of American competition in South Wales and Scotland. Several cargoes from the United States are reported to be on their way over, and some orders for which British firms have been competing have apparently been placed across the Atlantic instead of in this country. At the same time a good many rumors are abroad as to the shutting down of iron works in this country, but many of them are unreliable, and the general situation is not so bad as these alarmist reports would make it seem to be.—Iron and Coal Trades Review.

PIG IRON IN GREAT BRITAIN.

The pig iron market has remained very quiet during the past week, and the Christmas holidays have reduced business to a very small compass. Although prices continue to show signs of weakness, no further reduction of any consequence has taken place. As we pointed out last week, the pig iron position is by no means so unsatisfactory as might be hastily assumed at first sight from the comparatively low rates now quoted. The fall in the price of coke, exceeding that of pig iron, the low stocks, and the better demand in the United States, all point to an improvement in the future, and certainly makers do not appear to be quite so despondent as they were a week or two ago, although they hardly expect anything but a quiet time over the winter. It is reported from the United States that fuel and ores are going down in price. This might prove an unfavorable factor leading to increased exports to Great Britain; but if the demand is equal to the production, as it seems not unlikely to be if present expectations are fulfilled in the New Year, British makers will have nothing to fear on this head.—Iron and Coal Trades Review.

NEW YORK METAL MARKET.

The promise of an upward movement in tin values held out by the recovery reported yesterday has now been fulfilled. At least there has been a break in the London market, and while not as decided as was yesterday's move in the other direction, has had a depressing effect. In the English market this morning there was a decline of 17s. 6d. in spot tin, but part of this loss was regained before the close, when the quotation stood at £120 10s., or 7s. 6d. under last night's figures. There was comparatively little trading there in spot and only a moderate business in futures. To the relatively better demand for the latter than for spot tin is probably due the fact that the decline in futures amounted to 5s. In the New York market the feeling was depressed, though the change in prices was not marked. The demand was light and the close dull, with sales of spot tin at 26.87½c. and buyers at 26.70c. January-February was nominally quoted at 26.50c. Two steamers were posted from London to-day with a total of 825 tons, making the stock afloat 2,440 tons. But 35 tons have arrived since the beginning of the month, 25 tons at Boston on Saturday last and 10 tons at this port to day.

COPPER—There was a slight decline in the London market, but no change in the situation here was to be noted, trade continuing light, while the firm tone of the market was retained. For Lake Superior, the quotation was 17c., while electrolytic and casting were held at 16½c.

PIG LEAD—Buyers are not anticipating requirements, and, as their current wants are small, there is little business doing at the moment. Prices are steady, however, on the basis of 4.37½c. in carload lots. The St. Louis market was reported by wire to be easy at 4.17½c.

SPELTER—The easy feeling previonsly noted still prevails, as a result of continued dullness. No further change in prices is reported, however, the qnoted range being 4 10 to 4.15c. St. Louis was easy, 3 90c. bid and 3.92½c. asked.

ANTIMONY—Regulus remains quiet, but steady, at the range of 9 to 10½c., as to brand and quantity.

TINPLATE—No change in the situation is noted. Deliveries on existing contracts make up the bulk of the current business, but an active demand is looked for later on, and the tone of the market is firm.

The improvement in trade appears to be making rather slow progress, but as sellers have not looked for any decided increase in the volume of business this month they are very well satisfied with the situation as it stands, finding ample encouragement in such inquiries as are received for the confidence in the future. Advices from Chicago state that business there is within moderate limits, but it was not thought that it will assume active proportions much before the close of the month, the disposition being to wait for the announcement of ore prices, which it is expected will be made in about two weeks.—New York Journal of Commerce, January 9.

BUSINESS CHANGES.

DIFFICULTIES, ASSIGNMENTS, COMPROMISES.

HUNTER & CO., (Morton E. Hunter) general merchants, Morewood, Ont., have assigned to Francis Elliot.

F. D. Ramsay & Co., general merchants, Chesley, Ont., have compromised.

Herbert Bond, harness dealer, Inwood, Ont., has assigned to Wm. T. Fuller.

Cyprien Primeau, general merchant, St. Urbain, Que., is offering 25c. on the dollar.

Romain Boursier, general merchant, Lefaivre, Ont., is offering 30c. on the dollar.

A. Leclair, general merchant, North Lancaster, Ont., is offering 40c. on the dollar.

The Stevens Mfg. Co., iron and brass founders, etc., London, Ont., have suspended.

The Hyde Trading Co., general merchants, Hyde, N.W.T., are asking for an extension.

H. Duchesneau, general merchant, Pointe Claire, Que., has compromised at 25c. on the dollar.

D. Ticker & Co., general merchants, St. Cyrille de Wendover, have assigned to V. E. Paradis.

Wm. Goldsmith, dealer in agricultural implements, Alexander, Man., is offering to compromise.

The estate of John Verret, general merchant, Becancour, Que., is offering 50c. on the dollar.

D. McLeod Vince has been appointed assignee of C. H. Taylor, general merchant, Hartland, N.B.

A statement of the affairs of L. A. Dion, general merchant, St. Eustache, Que., is being prepared.

Lamarche & Benoit have been appointed curators of A. D. Denis, general merchant, Farnham, Que.

A meeting of the creditors of Alex. J. McDonald, general merchant, Seaside, N.S., has been held.

Bilodeau & Chalifoux have been appointed curators of A. Guimond, hardware dealer, etc., Montreal.

J. Boydell & Co., general merchants, Robinson, Que., have assigned, and a meeting of their creditors has been held.

A meeting to appoint a liquidator for The British Columbia Iron Works Co., Limited, New-Westminister, B.C., has been called.

E. Christie, general merchant, South Mountain, Ont., has assigned to J. K. Allen, Kemptville, Ont., and a meeting of his creditors will be held on January 11.

PARTNERSHIPS FORMED AND DISSOLVED.

The John Tetrault Tool and Axe Works, Maissoneuve, Que., have dissolved.

H. St. Germain & Cie have registered as carriagemakers in St. Hyacinthe, Que.

Hooben & Wooten, wholesale and retail stove dealers, Halifax, have dissolved.

Dupont & Lacroix, have registered partnership as bicycle repairers, etc., Montreal.

Pinder & Kenzie, general merchants, Dutton, Ont., have dissolved, D. M. Kenzie continuing.

Morrow Bros., general merchants, Portage la Prairie, Man., have dissolved, Albert Morrow retiring.

Brett & Leighton, hardware dealers, etc., Orangeville, Ont., have dissolved. They are succeeded by Brett & Taylor.

SALES MADE AND PENDING.

S. A. Torrance, blacksmith, etc., Carleton Place, Ont., has sold out.

Lawther & Co., general merchants, Russel, Man., are selling out.

Charles Shaw, blacksmith, Caroll, Man., is advertising his business for sale.

Theodore R. Constantine, blacksmith, Elgin, N.B., is advertising his business for sale.

The stock of A. M. Wilson, general merchant, Barrington, N.S., has been sold by sheriff.

CHANGES.

George McKim, blacksmith, Omemee, Ont., has sold out to Richard Morton.

E. A. Baker & Co., flour and feed dealers, have been succeeded by Charles Gass.

M. A. Akesley, coal dealer, etc., Fredericton, N.B., has sold out to John S. Scott.

E. McCarthy & Co., general merchants, Condie, Man., have sold out to George H. Brown.

A. J. Ford & Co., general merchants, Woodham, Ont., have sold out to W. E. Doupe.

Anderson & Merrick, general merchants, Oakville, Man., have sold out to Alex. B. Dalzell.

Robinson & Co., general merchants, West Lorne, Ont., have sold out to P. J. Lindenman.

C. H. Clements & Co., general merchants, Aylesford, N.S., have sold out to Caldwell J. West.

Joseph Scott, dealer in agricultural implements, Souris, Man., has been succeeded by A. J. Hughes.

Dame Caroline Bergeron has registered as proprietress of Joseph Dion & Co., hardware dealers, Quebec.

R. J. Greenwood & Son, harness dealers, Shoal Lake, Man., have sold their Newdale, Man., branch to T. L. Grove.

H. W. Folkins, dealer in agricultural implements, etc., Sussex, N.B., has sold out to The Sussex Mercantile Co., Limited.

DEATHS.

Alex. Smith, tinsmith, etc., Stratford, Ont., is dead.

Marcus Oxner, of M. & H. Oxner, general merchants, Chester Basin, N.S., is dead.

Morfit & Raincock, late general merchants, Gladstone, Man., have assigned to C. H. Newton.

Weldon W. Melville, who bought out J. A. Phillips, general merchant, Bath, N.B., has leased the store occupied by Mr. Phillips and will continue the business. Mr. Phillips will devote all his time to buying and exporting farm implements.

HAS THE JOBBER ANY RIGHT TO DO A RETAIL BUSINESS?*

I WISH to call your attention at the outset to the fact that this question does not interest us alone in the State of Washington, but is agitating the minds of the retailers and jobbers all over the country, particularly in the South and Middle West and on this Coast, where so many of the jobbing houses do a retail business, while in the East the retail business is done almost exclusively by retail houses, the jobbers being distributors to the retail trade.

It is easy to account for this difference in policies, the South and West being newer sections of the country and the trade conditions not so thoroughly adjusted. Many of the jobbing houses located in the Southern and Western States started in business years ago as retail stores, and, as the country developed and their business grew got to doing a wholesale business, until to-day finds many of them immense, exclusive jobbers, while others are doing both wholesale and retail business in varying proportions.

Now these two systems of conducting a jobbing business cannot both be right. The question follows, which is right? From the standpoint of the retailer it is unfair for the jobber to load him up with all the goods he will buy, and then cut off his outlet for them by selling to his customers.

THE JOBBER WHO RETAILS

is unfair with the retail dealer when he claims the right to buy cheaper than he does, even if the retailer can use the same quantity of goods.

He is unfair with the manufacturer in trying to persuade him not to sell direct to the retail trade, when he is himself doing a retail business.

He is unfair with the legitimate jobber, who asks the manufacturer for a reasonable differential for distributing his goods, when it is shown on investigation that a large majority of the so-called jobbers on this coast are doing the principal retail business in their respective cities.

The retailer is not alone in his view of the matter, for he is backed by the jobbing houses which do a legitimate jobbing business.

I desire to call your attention to an address made by John Donnan at the Southern Hardware Jobbers' convention last June, when he discussed the question, "Can a manufacturer sell a jobber and a retail merchant in the same territory and conserve both interests?" If you have not read this article, I would refer you to The Iron Age of June 21, as I consider it one of the most comprehensive articles that has appeared in that magazine this year. In it he says: "I unhesitatingly state that I do not think a jobber has any right whatever to be a competitor of his customer."

I will also read a short letter from one of the Portland hardware jobbers, which I clipped from The Iron Age recently, as follows:

"We note with pleasure in the proceedings of a number of retail hardware associations the effort that is being made to

INDUCE THE JOBBER

to refrain from selling at retail. This effort during the year should in all cases develop into a demand, and, if not acceded to, should be taken past the jobber to the manufacturer, or trust, controlling lines of hardware, metals or other goods pertaining to the retail hardware business. We, as jobbers, ask and demand of the manufacturers that they refrain from selling to the retail trade, and if that point is not conceded, then that they grant a differential, which we are entitled to as distributers, relieving them of the expense and risk incurred in attempting to be manufacturers and jobbers.

THE RETAILER

is entitled to the same protection from the jobber that we ask from the manufacturer, and should not be forced to come into competition with him. In many cities, east and west, as well as on this coast, there are large firms that properly should come under the head of large retailers, rather than jobbers. True, they are on the jobbers' list and buy at bedrock, but that enables them to take an unfair advantage of a competitor confining himself to wholesale, while they have their retail profit to cut down their store expense if they hold to retail prices, and if not, as is often the case, their jobbing costs and carload rates of freight, to take undue advantage of retail competitor. We trust this issue will be fought to a finish."

When in San Francisco last June, I noticed one of the principal jobbing hardware houses had a prominent sign near the door which read, "We sell no hardware at retail," and I was told that none of the San Francisco jobbers do a retail business, all of which shows that the sentiment of a large percentage of the jobbers is with us on this point.

There are a number of important questions that will come up for our consideration in the association from time to time, but, as Mr. Bryan expresses it, I believe this to be

THE PARAMOUNT ISSUE,

for on it hinge most of the others.

We cannot be in favor of large freight differentials between carloads and less, nor of the manufacturer allowing the jobber who is selling his goods at retail much preference in the matter of price, if we have to sell in direct competition.

We must not overlook the fact that this system of doing business has been in force on this coast for years and a custom that has been practiced so long cannot be revolutionized at once.

All sorts of

RETALIATORY SCHEMES

have been suggested in the past for the purpose of "getting back at" the retail jobber. I tell you, gentlemen, this association is not organized for the purpose of antagonizing or "getting back at" any one. Two wrongs do not make a right. We are joined together for the purpose of drawing the interests of the hardware business closer, not for the purpose of fighting, and from the attitude that jobbers have taken toward us it will not be necessary to fight. They have intimated their willingness to give us a hearing, to discuss this and other matters with us, and grant everything in reason that we ask. They have come half way. It rests with us, by using wise counsel, prudent management and common sense, to have this and other differences adjusted.

*Paper read Before the Washington Hardware Association.

CANNOT TAX ENGINE AND BOILER.

On Monday Recorder Weir, of Montreal, gave judgment in a case of appeal from the assessor's decision in the case of Dame Jane Drummond, widow of the late John Redpath, and Francis Robert Redpath, in their quality of testamentary executors of the will of the late John Redpath, levying an assessment of $1,500 on a boiler and engine which does not belong to them as proprietors of premises containing it, but to the tenants leasing the premises.

In giving judgment, Recorder Weir refers to a section of a statute exempting from taxation machinery that is used for "motive power," which would clearly exempt the machinery in question. He then continues, "I, therefore, order that the assessment and valuation roll be amended by striking out the sum of $1,500 placed against the names of petitioners for machinery contained in their property, 45 St. Maurice street, each party to pay its own costs, as the presumption naturally was against the proprietors at the moment of valuation."

CURRENT MARKET QUOTATIONS

January 11, 1903.

These prices are for such qualities and quantities as are usually ordered by retail dealers on the usual terms of credit, the lowest figures being for larger quantities and prompt pay. Large cash buyers can frequently make purchases at better prices. The Editor is anxious to be informed at once of any apparent errors in this list as the desire is to make it perfectly accurate.

METALS.

Tin.

Lamb and Flag and Straits— 56 and 28 lb. ingots, per lb.	0 33	0 34

Tinplates.

Charcoal Plates—Bright

M.L.S., equal to Bradley.		Per box
I.C., usual sizes	....	$7 00
I.X., "	....	8 50
I.X.X., "	....	10 00
Famous—		
I.C.	....	7 50
I.X.	....	8 50
I.X.X.	....	9 50
Raven & Vulture Grades—		
I.C., usual sizes	....	5 00
I.X., "	....	6 00
I.X.X. "	....	7 00
I.X.X.X., "	....	8 00
D.C., 12½x17	....	4 75
D.X.	....	5 50
D.X.X	....	7 50

Coke Plates—Bright

Bessemer Steel—		
I.C., usual sizes	....	4 30
I.C., special sizes, base	....	4 50
20 x 28	....	8 75

Charcoal Plates—Terne

Dean or J. G. Grade—		
I.C., 20x28, 112 sheets	....	8 15
I.X., Terne Tin	....	10 75

Charcoal Tin Boiler Plates.

Cookley Grade—		Per lt.
X.X., 14x56, 50 sheet bxs / " 14x60 " / " 14x65, "	0 07	0 07½

Tinned Sheets.

72x30 up to 24 gauge	0 08	0 08½
" 26 "	0 08½	0 09
" 28 "	0 09	0 09½

Iron and Steel.

Common Bar, per 100 lbs.	1 65	1 70
Refined " "	....	2 15
Horse Shoe Iron "	....	2 05
Hoop steel, 1½ to 3 in. base, extras for smaller sizes	....	3 10
Sleigh Shoe Steel " base	....	2 00
Tire Steel	....	2 00
Machinery iron finish	....	2 05
Cast Steel, per lb	0 00	0 00
Toe Calk Steel	....	2 35
T. Firth & Co's special cast steel, per lb.		0 13
Jessop's Tool Steel	....	0 13

Boiler Tubes.

1½-inch	....	0 21½
2 "	....	0 13½
2½ "	....	0 16
3 "	....	0 17½

Steel Boiler Plate.

¼ inch	....	2 25
3-16 inch	....	2 25
⅜ inch and thicker	....	2 25

Black Sheets.

18 gauge	....	3 10
20 gauge	....	3 10
22 to 24 "	....	3 20
26 "	....	3 30
28 "	....	3 40

Canada Plates.

All dull, 52 sheets	....	3 15
Half polished	....	3 25
All bright	3 85	4 00

Iron Pipe.

Black pipe—		
¼ inch	....	3 00
⅜ "	....	3 00
½ "	....	3 10
¾ "	....	3 31
1 "	....	4 50
1¼ "	....	6 15
1½ "	....	7 75
2 "	....	10 40
2½-6 inch, discount 60 p.c.		
Galvanized pipe—		
½ inch	....	4 60
¾ "	....	5 60
1 "	....	7 80
1¼ "	....	9 50
1½ "	....	11 75
2 "	....	15 75

Galvanized Sheets.

	G. C. Comet.		Amer.	Queen's Head.
16 gauge	....		4 5	4 10
18 to 24 gauge	4 35	4 20	4 35	4 35
26 "	4 60	4 45	4 35	4 60
28 "	4 85	4 70	4 10	4 85

28 gauge American equals 26 gauge English.

Chain.

Proof Coil, 3-16 in., per 100 lb	....	....
" ¼ " "	8 00	8 50
" 5-16 " "	5 35	5
" ⅜ " "	4 35	4 85
" 7-16 " "	4 15	4 65
" ½ " "	4 35	4 50
" ⅝ " "	3 85	4 35
" ¾ " "	3 80	4 00

Halter, kennel and post chains, 40 and 50 p.c.
Cow ties .... 40 p.c.
Stall fixtures .... 35 p.c.
Trace chain .... 25 p.c.
Jack chain, iron, single and double, discount 35 p.c.
Jack chain, brass, single and double, discount 40 p.c.

Copper.

Ingot

English B. S., ton lots	0 19	0 20
Lake Superior	....	....

Bolt or Bar.

Cut lengths round, ½ to ⅞ in.	0 23½	0 25
" round and square 1 to 2 inches	0 23½	0 25

Sheet.

Untinned, 14 oz., and light, 16 oz., 14x48 and 14x60	0 23	0 23
Untinned, 14 oz., and light, 16 oz., irregular sizes	0 23	0 23½

NOTE.—Extra for tinning, 2 cents per pound, and tinning and half planishing 3 cents per pound.

Tinned copper sheets	....	0 26
Planished	....	0 32

Braziers (in sheets.)

4x6ft. 25 to 30 lbs. ea., per lb.	....	0 25½
" 35 to 45 " "	....	0 24½
" 50-lb. and above, "	....	0 23½

Boiler and T. K. Pitts

Plain Tinned, per lb	....	0 28
Spun, per lb	....	0 32

Brass.

Rod and Sheet, 14 to 30 gauge, 15 per cent.

Sheets, hard-rolled, 2x4	0 24	0 26
Tubing, base, per lb	0 24	0 25

Zinc Spelter

Foreign, per lb	0 06	0 06½
Domestic "	....	....

Zinc Sheet.

5 cwt. casks	6 75	7 00
Part casks	7 00	7 50

Lead.

Imported Pig, per lb	0 04¾	0 05
Bar, 1 lb	0 05½	0 05¾
Sheets, 2½ lbs. sq. ft., by	....	0 06¼
Sheets, 3 to 6 lbs., "	....	0 06

NOTE.—Cut sheets ½ cent per lb. extra.
Pipe, by the roll, usual weights per yard, lists at 7c. per lb. and 15 p.c. dis. f.o.b. Toronto.
NOTE.—Cut lengths, net price, waste pipe 8-ft. lengths lists at 7½ cents.

Shot.

Common, $6.50 per 100 lb.; chilled, $7.00 per 100 lb.; buck, seal and bal., $7.50. Discount, 7½ p.c. Prices are f.o.b. Toronto, Hamilton, Montreal, St. John and Halifax. Terms 3 per cent. cash, freights equalized on Montreal.

Soil Pipe and Fittings.

Discount, 60 and 10 per cent. on medium and extra heavy, and 60 per cent. on light.

Solder.

	Per lb.	Per lb.
Bar half-and-half	0 19	0 20
Refined	....	0 9
Wiping	....	0 18

NOTE.—Prices of this graded according to quantity. The prices of other qualities of solder in the market indicated by private brands vary according to composition.

Antimony.

Cookson's, per lb	0 11	0 11½

White Lead.

	Per cwt.
Pure	6 62½
No. 1 do	6 25
No. 2 do	5 87½
No. 3 do	5 50
No. 4 do	5 12½
Munro's Select Flake White	7 12½
Elephant and Decorators' Pure	6 87½

Red Lead.

Genuine, 560 lb. casks, per cwt.	....	$5 50
Genuine, 100 lb. kegs, per cwt.	....	5 75
No. 1, 560 lb. casks, per cwt.	....	5 25
No. 1, 100 lb. kegs, per cwt	....	5 50

White Zinc Paint.

Elephant Snow White	0 08	0
Pure White Zinc	0 08	0
No. 1	0 06	0 06½
No. 2	0 05	0 06½

Dry White Lead.

Pure, casks	75
Pure, kegs	25
No. 1, casks	50
No. 1, kegs	5 00

Prepared Paints.

In ¼, ½ and 1 gallon tins.

Pure, per gallon	....	1 25
Second qualities, per gallon	....	1 10
Barn (in bbls.)	0 75	0 85
The Sherwin-Williams Paints	....	1 45
Canada Paint Co's Pure	....	1 25
Toronto Lead & Color Co's Pure	....	1 25
Sanderson Pearcy's Pure	....	1 20
Stewart & Wood's Champion Pure		1 20

Colors in Oil.

25 lb. tins, Standard Quality.

Venetian Red, per lb	0 05
Chrome Yellow	0 11
Golden Ochre	0 06
French "	0 05
Marine Black	0 09
" Green	0 09
Chrome "	0 08
French Imperial Green	0 9

Colors, Dry.

Yellow Ochre (J.C.) bbls.	1 35	1 40
Yellow Ochre (J.F.L.S.) bbls	....	2 75
Yellow Ochre (Royal)	1 10	1 15
Brussels Ochre	....	2 00
Venetian Red (best), per cwt.	1 80	1 90
English Oxides, per cwt.	3 00	3 25
American Oxides, per cwt.	1 75	2 00
Canadian Oxides, per cwt.	1 75	2 00
Super Magnetic Oxides, 93p.c.	2 00	2 25
Burnt Sienna, pure, per lb.	....	0 10
" Umber, " "	....	0 10
do Raw	....	0 09
Drop Black, pure	....	0 09
Chrome Yellows, pure	....	0 18
Chrome Greens, pure, per lb.	....	0 12
Golden Ochre	....	0 03½

Ultramarine Blue in 28-lb. boxes, per lb	0 08	0 24
Fire Proof Mineral, per 100 lb	....	1 00
Genuine Eng. Litharge, per lb	....	0 07
Mortar Color, per 100 lb.	....	1 25
English Vermillion	....	0 80
Pure Indian Red, No. 45 lb.	....	0 80
Whiting, per 100 lb.	....	0 55

Blue Stone.

Casks, for spraying, per lb.	0 07
100-lb. lots, do. per lb	0 08

Putty.

Bulk in bbls.	2 00
Bulk in less quantity	2 15
Bladders in bbls.	2 20
Bladders in kegs, boxes or loose.	2 35
Bladders in 25-lb. tins	2 45
Bladders in 12½-lb. tins	2 75
Bladders in bulk or tins less than 100 lbs	3 00

Varnishes.

	In 5-gal. lots.	Per gal.
Carriage, No. 1	2 90	3 30
" body	8 00	9 00
" rubbing	4 00	5 00
Gold Size, Japan	3 00	3 40
Brown Japan	2 40	2 80
Elastic Oak	2 90	3 30
Furniture, extra	2 40	2 80
" No. 1	1 60	2 00
Hard Oil Finish	2 70	3 10
Light Oil Finish	3 20	3 60
Demar	3 30	3 70
Shellac, white	4 40	4 80
" orange	4 00	4 40
Furniture Brown Japan	1 60	2 00
Black Japan	2 40	2 80
" " No. 1	1 60	2 00

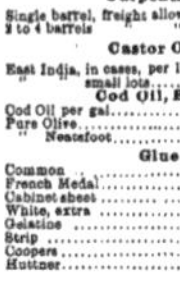

The Imperial Varnish & Color Co's., Limited Elastilite Varnish 1 gal. can, each, $2.00.

Granatine Floor Finish, per gal. $2.00.

Maple Leaf Coach Enamels; Size 1, 40c.; Size 2, 35c.; Size 3, 20c. each.

Linseed Oil.

	Raw.	Boiled.
1 to 4 bbls delivered	$0 81	$0 85
5 to 9 bbls "	81	84

Toronto, Hamilton, London and Guelph 2c. less.

Turpentine.

Single barrel, freight allowed	....	0 61
2 to 4 barrels "	....	0 63

Castor Oil.

East India, in cases, per lb.	0 10	0 10½
" " small lots	0 10½	0 11

Cod Oil, Etc.

Cod Oil per gal	0 50	0 55
Pure Olive	....	1 20
" Neatsfoot	....	90

Glue.

Common	0 08½	0 09
French Medal	0 14	0 14½
Cabinet sheet	0 12	0 13
White, extra	0 18	0 20
Gelatine	0 22	0 30
Strip	0 18	0 20
Coopers	0 19	0 20
Huttner	....	0 18

HARDWARE.

Ammunition.

Cartridges.

B. B. Caps. Dom. 50 and 5 per cent.
Rim Fire Pistol, dis. 40 p. c., Amer.
Rim Fire Cartridges, Dom., 50 and 5 p. c.
Central Fire Pistol and Rifle, 10 p.c. Amer.
Central Fire Cartridges, pistol sizes, Dom. 30 per cent.
Central Fire Cartridges, Sporting and Military, Dom., 15 and 5 per cent.
Central Fire, Military and Sporting, Amer, add 5 p.c. to list. B.B. Caps, discount 40 per cent, Amer.
Loaded and empty Shells, "Trap" and "Dominion" grades, 25 per cent Rival and Nitro, net list.
Brass shot Shells, 55 per cent
Primers, Dom., 30 per cent.

Wads — per lb

Best thick white felt wadding, in ½-lb bags ... 1 00
Best thick brown or grey felt wads, in ½-lb. bags ... 0 70
Best thick white card wads, in boxes of 500 each, 12 and smaller gauges 0 99
Best thick white card wads, in boxes of 500 each, 10 gauge ... 0 35
Best thick white card wads, in boxes of 500 each, 8 gauge ... 0 55
Thin card wads, in boxes of 1,000 each, 12 and smaller gauges ... 0 20
Thin card wads, in boxes of 1,000 each, 10 gauge ... 0 25
Thin card wads in boxes of 1,000 each, 8 gauge ...
Chemically prepared black edge grey cloth wads, in boxes of 250 each— Per M
11 and smaller gauge ... 0 60
9 and 10 gauges ... 0 70
7 and 8 gauges ... 0 90
5 and 6 gauges ... 1 10
Superior chemically prepared pink edge, best white cloth wads, in boxes of 250 each—
11 and smaller gauge ... 1 15
9 and 10 gauges ... 1 40
7 and 8 gauges ... 1 65
5 and 6 gauges ... 1 90

Adzes.

Discount, 20 per cent.

Anvils.

Per lb ... 10 0 12½
Anvil and Vise combined ... 4 50
Wilkinson & Co.'s Anvils..lb. 0 09 0 09½
Wilkinson & Co.'s Vices..lb. 0 09¾ 0 10

Augers.

Gilmour's, discount 65 and 5 p.c. off list.

Axes.

Chopping Axes—
Single bit, per doz ... 6 50 10 00
Double bit, " ... 12 00 18 00
Bench Axes, 40 p.c.
Broad Axes, 33⅓ per cent.
Hunters' Axes ... 5 50 6 00
Boy's Axes ... 5 75 6 75
Splitting Axes ... 6 50 12 00
Handled Axes ... 7 00 10 00

Axle Grease.

Ordinary, per gross ... 5 75 6 00
Best quality ... 13 00 15 00

Bath Tubs.

Zinc ... 6 00
Copper, discount 15 p.c. off revised list

Baths.

Standard Enameled.

5½-inch rolled rim, 1st quality ... 30 00
" " " 2nd " ... 22 00

Anti-Friction Metal.

"Tandem" A ... per lb. 0 27
" B ... " 0 21
" C ... " 0 11½
Magnolia Anti-Friction Metal, per lb. 0 25

SYRACUSE SMELTING WORKS.

Aluminum, genuine ... 0 45
Dynamo ... 0 29
Special ... 0 25
Aluminum, 99 p.c. pure "Syracuse" .. 0 50

Bells.

Hand.

Brass, 60 per cent.
Nickel, 55 per cent.

Cow.

American make, discount 66⅔ per cent.
Canadian, discount 45 and 50 per cent.

Door.

Gongs, Sargent's ... 5 50 8 00
" Peterboro', discount 45 per cent.

Farm.

American, each ... 1 25 3 00

House.

American, per lb ... 0 35 0 40

Bellows.

Hand, per doz ... 3 35 4 75
Moulders', per doz ... 7 50 10 00
Blacksmiths', discount 40 per cent.

Belting.

Extra, 50 and 10 per cent.
Standard, 60 per cent.
No. 1 Agricultural, 60 and 10 p.c.

Bits.

Auger.

Gilmour's, discount 60 and 5 per cent.
Rockford, 50 and 10 per cent.
Jennings' Gen., net list.

Car.

Gilmour's, 47½ to 50 per cent.

Expansive.

Clark's, 40 per cent.

Gimlet.

Clark's, per doz ... 0 65 0 90
Diamond, Shell, per doz ... 1 00 1 50
Nail and Spike, per gross ... 2 25 5 20

Blind and Bed Staples.

All sizes, per lb ... 0 07½ 0 12

Bolts and Nuts. Per cent.

Carriage Bolts, full square, Norway ... 70
" " full square ... 70
Common Carriage Bolts, all sizes ... 65
Machine Bolts, all sizes ... 65
Coach Screws ... 75
Sleigh Shoe Bolts ... 75
Blank Bolts ... 65
Bolt Ends ... 65
Nuts, square ... 4½c. off
Nuts, hexagon ... 4¾c. off
Tire Bolts ... 67½
Stove Bolts ... 67½
Stove rods, per lb ... 5½ to 8c.
Plough Bolts ... 60

Boot Calks.

Small and medium, ball, per M ... 4 25
Small heel, per M ... 4 50

Bright Wire Goods.

Discount ... 55 per cent.

Broilers.

Light, dis., 65 to 67½ per cent.
Reversible, dis., 65 to 67½ per cent.
Vegetable, per doz., dis. 37½ per cent.
Henis, No. 8, " ... 6 00
Henis, No. 9, " ... 7 00
Queen City " ... 7 50 0 00

Butchers' Cleavers.

German, per doz ... 6 00 11 00
American, per doz ... 12 00 20 00

Building Paper, Etc.

Plain building, per roll ... 0 30
Tarred lining, per roll ... 0 40
Tarred roofing, per 100 lb ... 1 65
Coal Tar, per barrel ... 3 50
Pitch, per 100-lb ... 0 85
Carpet felt, per ton ... 45 00

Bull Rings.

Copper, $2.00 for 2½ in. and $1.90 for 2 in.

Butts.

Wrought Brass, net revised list

Cast Iron.

Loose Pin, dis., 60 per cent.

Wrought Steel.

Fast Joint, dis. 60 and 10 per cent.
Loose Pin, dis. 60 and 10 per cent.
Berlin Bronzed, dis. 70, 70 and 5 per cent.
Gen. Bronzed, per pair ... 0 40 0 65

Carpet Stretchers.

American, per doz ... 1 00 1 50
Bullard's, per doz ... 6 50 ...

Castors.

Bed, new list, dis. 55 to 57½ per cent.
Plate, dis. 52½ to 57½ per cent.

Cattle Leaders.

Nos. 31 and 32, per gross ... 50 9 50

Cement.

Canadian Portland ... 2 80 3 00
English " ... 3 00
Belgian " ... 2 75 3 00
Canadian hydraulic ... 1 25 1 50

Chalk.

Carpenters, Colored, per gross 0 45 0 75
White lump, per cwt ... 0 60 0 65
Red ... 0 05 0 06
Crayon, per gross ... 0 14 0 18

Chisels.

Socket, Framing and Firmer.

Broad's, dis. 70 per cent.
Warnock's, dis. 70 per cent.
P. S. & W. Extra 60, 10 and 5 p.c.

Churns.

Revolving Churns, metal frames—No. 0, $8—No. 1, $8.50—No. 2, $9.00—No. 3, $10.00 No. 4, $12.00—No. 5, $16.00 each. Ditto wood frames—20c. each less than above. Discounts: Delivered from factories, 58 p.c.; from stock in Montreal, 56 p.c. Terms, 4 months or 3 p.c. cash in 30 days.

Clips.

Axle dis. 65 per cent.

Closets.

Plain Ontario Syphon Jet ... $8 00
Emb. Ontario Syphon Jet ... 8 50
Fittings ... 1 25
Plain Teutonic Syphon Washout ... 4 75
Emb. Teutonic Syphon Washout ... 5 25
Fittings ... 1 25
Low Down Teutonic, plain ... 14 50
" " " embossed ... 15 00
Plain Richelieu ... 3 75
Emb. Richelieu ... 4 00
Fittings ... 1 25
Low Down Ont. Syphon Jet, plain ... 20 00
" " " " " emb'd. 20 50
Closet connection ... 1 25
Basins, round, 14 in ... 0 80
" oval, 17 x 14 in ... 1 50
" " 19 x 15 in ... 2 25

Compasses, Dividers, Etc.

American, dis. 62½ to 65 per cent.

Cradles, Grain.

Canadian, dis. 25 to 33⅓ per cent.

Crosscut Saw Handles.

S. & D., No. 3, per pair ... 17½
" " 5, " ... 22½
" " 6, " ... 15
Boynton pattern " ... 20

Door Springs.

Torrey's Rod, per doz ... (15 p.c.) 2 00
Coil, per doz ... 0 88 1 60
English, per doz ... 2 00 4 00

Draw Knives.

Coach and Wagon, dis. 50 and 10 per cent.
Carpenters, dis. 70 per cent.

Drills.

Hand and Breast.

Millar's Falls, per doz. net list.

DRILL BITS.

Morse, dis., 37½ to 40 per cent.
Standard dis. 50 and 5 to 55 per cent.

Faucets.

Common, cork-lined, dis. 35 per cent.

ELBOWS. (Stovepipe.)

No. 1, per doz ... 1 4
No. 2, per doz ... 1 20
Bright, 20c. per doz. extra.

ESCUTCHEONS.

Discount, 45 per cent.

ESCUTCHEON PINS.

Iron, discount 40 per cent.

FACTORY MILK CANS.

Discount off revised list, 40 per cent.

FILES.

Black Diamond, 50 and 10 to 60 per cent.
Kearney & Foote, 60 and 10 p.c. to 60, 10, 10.
Nicholson File Co., 50 and 10 to 60 per cent.
Jowitt's, English list, 25 to 27½ per cent.

FORKS.

Hay, manure, etc., dis., 50 and 10 per cent. revised list.

GLASS—Window—Box Price.

Size United Inches.	Star Per 50 ft.	Star Per 100 ft.	D. Diamond Per 50 ft.	D. Diamond Per 100 ft.
Under 26	2 10	4 00	....	6 00
26 to 40	2 30	4 35	....	6 65
41 to 50	....	4 75	....	7 25
51 to 60	....	5 00	....	8 50
61 to 70	....	5 35	....	9 25
71 to 80	....	5 75	....	10 50
81 to 85	....	6 50	....	11 75
86 to 90	....	....	..	14 00
91 to 95	..	....	...	15 50
96 to 100	..	....	....	18 00

GAUGES

Marking, Mortise, Etc.

Stanley's dis. 50 to 55 per cent.

Wire Gauges.

Winn's, Nos. 26 to 33, each ... 1 65 2 40

HALTERS.

Rope, ⅜ per gross ... .
" ½ " 9 00
" ⅝ to ¾ 14 00
Leather, 1 in., per doz ... 3 87½ 4 00
" 1¼ in., " ... 5 15 5 20
Web, — per doz ... 1 87 2 45

HAMMERS.

Nail

Maydole's, dis. 5 to 10 per cent. Can. dis. 25 to 27½ per cent.

Tack.

Magnetic, per doz ... 1 10 1 20

Sledge.

Canadian, per lb ... 0 07½ 0 08½

Ball Pean.

English and Can., per lb ... 0 22 0 25

HANDLES.

Axe, per doz. net ... 1 50 2 00
Store door, per doz ... 1 00 1 50

Fork.

C. & B., dis. 40 per cent. rev. list.

Hoe.

C. & B., dis. 40 per cent. rev. list.

Saw.

American, per doz ... 1 00 1 25

Plane.

American, per gross ... 3 15 3 75

Hammer and Hatchet.

Canadian, 40 per cent.

Cross-Cut Saws.

Canadian, per pair ... 0 13½

HANGERS. doz. pairs.

Steel barn door ... 5 85 6 00
Stearns, 4 inch 5 00
" 5 inch 6 50
Lane's covered—
No. 11, 5-ft. run 8 40
No. 11½, 10-ft. run 10 80
No. 12, 10-ft. run 12 60
No. 14, 15-ft. run 21 00
Lane's O.N.T. track, per foot. 4½

HARVEST TOOLS.

Discount, 50 and 10 per cent.

HATCHETS.

Canadian, dis. 40 to 42½ per cent.

HINGES.

Blind, Parker's, dis. 50 and 10 to 60 per cent
Heavy T and strap, 4-in., per lb ... 0 06½
" " 5-in., " ... 0 06¼
" " 6-in., " ... 0 06
" " 8-in., " ... 0 05¾
" " 10-in., " ... 0 05½
Light T and strap, dis. 60 and 5 per cent.
Screw hook and hinge—
6 to 12 in., per 100 lbs 4 50
14 in. up, per 100 lbs 3 50
Per gro. pairs.
Spring 12 00

HOES.

Garden, Mortar, etc., dis. 50 and 10 p.c.
Planter, per doz ... 4 00 4 50

HOLLOW WARE.

Discount ... 45 and 5 per cent.

HOOKS.

Cast Iron.

Bird Cage, per doz ... 0 50 1 10
Clothes Line, per doz ... 0 27 0 63
Harness, per doz ... 0 72 0 88
Hat and Coat, per gross ... 1 00 3 00
Chandelier, per doz ... 0 50 1 00

Wrought Iron.

Wrought Hooks and Staples, Can., dis. 47½ per cent.

Wire.

Hat and Coat, discount 45 per cent.
Belt, per 1,000 ... 0 60
Screw, bright, dis. 55 per cent.

HORSE NAILS.

"C" brand 50 p.c. dis. } Oval head.
"M" brand 50 p.c. }
Acadian, 50 and 10 per cent.

HORSESHOES.

Iron Shoes.	F.O.B. Montreal No. 2 and larger.	No. 1. and smaller
Light medium, and heavy...	3 50	3 75
Snow shoes...	3 75	4 00
Steel Shoes.		
Light...	3 60	3 85
Featherweight (all sizes)...	4 85	4 85

F.O.B. Toronto, Hamilton, London and Guelph, 10c. per keg additional.

Toe weight steel shoes...... 6 70

JAPANNED WARE.

Discount, 45 and 5 per cent. off list, June 1899.

ICE PICKS.

Star per doz.... 3 00 3 25

KETTLES.

Brass spun, 7½ p.c. dis. off new list.

Copper, per lb.... 0 30 0 50

American, 60 and 10 to 65 and 5 p.c.

KEYS.

Lock, Can., dis., 45 p.c.

Cabinet, trunk, and padlock, Am. per gross.... 60

KNOBS.

Door, japanned and N.P., per doz...	1 50	2 50
Bronze, Berlin, per doz....	2 75	3 25
Bronze Genuine, per doz....	6 00	9 00
Shutter, porcelain, F. & L. screw, per gross...	1 30	4 00
White door knobs—per doz.		1 25

HAY KNIVES.

Discount, 50 and 10 per cent.

LAMP WICKS.

Discount, 60 per cent.

LANTERNS.

Cold Blast, per doz...	7 50
No. 3 "Wright's"...	8 50
Ordinary, with O burner...	4 25
Dashboard, cold blast...	9 50
No. 0...	6 00

Japanning, 50c. per doz. extra.

LEMON SQUEEZERS.

	per doz.	
Porcelain lined,...	2 20	5 60
Galvanized...	1 87	3 85
King, wood...	2 75	2 90
King, glass...	4 00	4 50
All glass...	1 20	1 30

LINES.

Fish, per gross...	1 05	2 50
Chalk " ...	1 90	7 40

LOCKS.

Canadian, dis. 45 p.c.

Russell & Erwin, per doz.... 3 00 3 25

Cabinet.

Eagle, dis. 30 p.c.

Padlock.

English and Am., per doz....	50	6 00
Scandinavian, "	1 00	2 40

Eagle, dis. 20 to 25 p.c.

MACHINE SCREWS.

Iron and Brass.

Flat head discount 25 p.c.

Round Head, discount 20 p.c.

MALLETS.

Tinsmiths', per doz...	1 25	1 50
Carpenters', hickory, per doz.	1 25	3 75
Lignum Vitae, per doz..	3 85	5 00
Caulking, each...	60	2 00

MATTOCKS.

Canadian, per doz... 8 50 1 00

MEAT CUTTERS.

American, dis. 25 to 30 p.c.

German, 15 per cent.

MILK CAN TRIMMINGS.

Discount, 25 per cent.

NAILS.

Quotations are:	Cut.	Wire.
2d. and 3d ...	$3 35	$3 85
3d ...	3 00	3 52
4 and 5d...	2 75	3 35
6 and 7d...	2 65	3 20
8 and 9d...	2 50	3 00
10 and 12d...	2 45	2 95
16 and 20d...	2 40	2 90
30, 40, 50 and 60d. (base)...	2 35	2 85

Galvanizing 2c. per lb. net extra.

Steel Cut Nails 10c. extra.

Miscellaneous wire nails, dis. 70 per cent.

Coopers' nails, dis. 30 per cent.

Flour barrel nails, dis. 25 per cent

NAIL PULLERS.

German and American...... 1 85 3 50

NAIL SETS

Square, round, and octagon, per gross...	3 38	4 00
Diamond...	12 00	15 00

NETTING.

Poultry, 50 per cent. for McMullen's.

OAKUM.

	Per 100 lb
Navy...	6 00
U. S. Navy...	7 25

OIL.

Water White (U.S.)...	0 16½
Prime White (U.S.)...	0 15½
Water White (Can.)...	0 15
Prime White (Can.)...	0 14

OILERS.

McClary's Model galvan. oil can, with pump, 5 gal., per doz... 0 00 10 00

Zinc and tin, dis. 50, 50 and 10.

Copper, per doz...	1 25	3 50
Brass, " ...	1 50	3 50

Malleable, dis. 25 per cent.

GALVANIZED PAILS.

Dufferin pattern pails, dis. 50 to 50 and 10 p.c.

Flaring pails, discount 45 per cent.

Galvanized washtubs, discount 45 per cent.

PIECED WARE.

Discount 40 per cent. off list, June, 1899.

PICKS.

Per doz... 6 00 9 00

PICTURE NAILS.

Porcelain head, per gross....	1 75	3 00
Brass head "	0 40	1 00

PICTURE WIRE.

Tin and gilt, discount 75 p.c.

PLANES.

Wood, bench, Canadian dis. 50 per cent American dis. 50.

Wood, fancy Canadian or American 7½ o 40 per cent.

PLANE IRONS.

English, per doz... 2 00 5 00

PLIERS AND NIPPERS.

Button's Genuine per doz pairs, dis. 37½ 40 p.c.

Button's Imitation, per doz..	5 00	9 00
German, per doz...	0 60	2 60

PLUMBERS' BRASS GOODS.

Impression work, discount, 60 per cent.

Fuller's work, discount 65 per cent.

Rough stops and stop and waste cocks, discount, 60 per cent.

Jenkins disk globe and angle valves, discount, 55 per cent.

Standard valves, discount, 60 per per cent.

Jenkins radiator valves discount 55 per cent.

" " " standard, dis., 60 p.c.

Quick opening valves discount, 60 p.c.

No. 1 compression bath cock...	2 00
No. 4 " " " ...	2 00
No. 7, Fuller's...	2 50
No 4½, " ...	3 00

POWDER.

VeloX Smokeless Shotgun Powder.

100 lb. or less...	0 85
1,000 lb. or more...	0 80

Net 30 days.

PRESSED SPIKES.

Discount 25 per cent.

PULLEYS.

Hothouse, per doz...	0 55	1 00
Axle...	0 22	0 33
Screw...	0 27	1 00
Awning...	0 35	2 50

PUMPS.

Canadian cistern...	1 80	3 00
Canadian pitcher spout....	1 40	2 10

PUNCHES.

Saddlers', per doz...	1 00	1 85
Conductors', " ...	9 00	15 00
Tinners' solid, per set...	0 00	0 72
" hollow, per inch....	0 00	1 00

RANGE BOILERS.

Galvanized, 30 gallons...	6 50
" 35 " ...	7 50
" 40 " ...	8 50
Copper, 30 " ...	22 00
" 35 " ...	26 00
" 40 " ...	30 00

Discount off Copper Boilers 10 per cent.

RAKES.

Cast steel and malleable Canadian list 50 and 10 p.c. revised list.

Wood, 25 per cent.

RASPS AND HORSE RASPS.

New Nicholson horse rasp, discount 60 p.c.

Globe File Co.'s rasps, 60 and 10 to 70 p.c.

Heller's Horse rasps, 50 to 50 and 5 p.c.

RAZORS.

	per doz.	
Geo. Butler & Co.'s,...	8 00	18 00
Boker's...	7 50	11 00
Wade & Butcher's...	3 60	10 00
Theile & Quack's...	7 00	12 00
Elliot's...	4 00	18 00

REAPING HOOKS.

Discount, 50 and 10 per cent.

REGISTERS.

Discount... 40 per cent.

RIVETS AND BURRS.

Iron Rivets, discount 60 and 10 per cent.

Iron Burrs, discount 55 per cent.

Black and Tinned Rivets, 60 p.c.

Extras on Iron Rivets in 1-lb. cartons, ½c per lb.

Extras on Iron Rivets in ½-lb. cartons, per lb.

Copper Rivets & Burrs, 35 and 5 p.c. dis. and cartons, 1c. per lb. extra, net.

Extras on Tinned or Coppered Rivets ½-lb. cartons, 1c. per lb.

Terms, 4 mos. or 3 per cent. cash 30 days.

RIVET SETS.

Canadian, dis. 35 37½ per cent.

ROPE ETC.

	Sisal.	Manila.
7-16 in. and larger, per lb.	9	13
⅜ in...	10	14
¼ and 5-16 in...		15
Cotton, 3-16 inch and larger		16½
" 5-32 inch...		21½
" ⅛ inch...		22½
Russia Deep Sea...		15½
Jute...		8
Lath Yarn...		9½
New Zealand Rope...		10½

RULES.

Boxwood, dis. 75 and 10 p.c.

Ivory, dis. 37½ to 40 p.c.

SAD IRONS.

	per set
Mrs. Potts, No. 55, polished...	0 70
No. 50, nickle-plated....	75

SAND AND EMERY PAPER.

Dominion Flint Paper, 47½ per cent.

B & A. sand, 40 and 2½ per cent.

Emery, 40 per cent.

SAP SPOUTS.

Bronzed iron with hooks, per doz... 9 50

SAWS.

Hand Disston's, dis. 12½ p.c.

S. & D., 40 per cent.

Crosscut, Disston's, per ft.... 0 35 0 55

S. & D., dis. 35 p.c. on Nos. 2 and 3.

Hack, complete, each...	0 75	2 75
" frame only...	....	0 75

SASH WEIGHTS.

Sectional, per 100 lbs...	2 75	3 00
Solid, " ...	2 00	2 25

SASH CORD.

Per lb... 0 23 0 30

SAW SETS.

"Lincoln," per doz... .... 6 50

SCALES.

B. S. & M. Scales, 45 p.c.

Champion, 65 per cent.

Fairbanks Standard, 35 p.c.

" Dominion, 55 p.c.

" Richelieu, 55 p.c.

Chatillon Spring Balances, 10 p.c.

SCREW DRIVERS.

Sargent's, per doz... 0 65 1 00

SCREWS

Wood, F. H., iron, and steel, 85 p.

Wood R. H., " dis. 81 p.c.

" F. H., brass, dis. 77½ p.c.

Wood, R. H., " dis. 70 p.c.

" F.H., bronze, dis. 70 p.c.

" R.H. " 65 p.c.

Drive Screws, 80 per cent.

Bench, wood, per doz...	3 25	4 00
" iron, " ...	4 25	5 75

SCYTHES.

Per doz, net... 9 00

SCYTHE SNATHS.

Canadian, dis. 45 p.c.

SHEARS.

Dailey Cutlery Co., full nickeled, dis. 60 p.c

Seymour's, dis. 50 and 10 p.c.

SHOVELS AND SPADES.

Canadian, dis. 40 and 5 per cent.

SINKS.

Steel and galvanized, discount 45 per cent.

SNAPS.

Harness, German, dis. 25 p.c.

Lock, Andrews'... 4 50 11 50

SOLDERING IRONS.

1, 1½ lb., per lb...	0 37
2 lb. or over, per lb...	0 34

SQUARES.

Iron, No. 493, per doz...	2 40	2 55
" No. 494, " ...	3 25	3 40

Steel, dis. 50 and 5 to 50 and 10 p.c., rev. list.

Try and bevel, dis. 50 to 52½ p.c.

STAMPED WARE.

Plain, dis., 75 and 12½ p.c. off revised list.

Retinned, dis., 75 p.c. off revised list.

STAPLES.

Galvanized...	0 00	0 00
Plain...	0 00	3 45

Coopers', discount 45 per cent.

Poultry netting staples, 40 per cent.

STOCKS AND DIES.

American dis. 25 p.c.

STONE.

		Per lb.
Washita...	0 28	0 60
Hindostan...	0 06	0 07
" slip,...	0 09	0 09
Labrador...	....	0 13
" Axe...	....	0 15
Turkey...	....	0 50
Arkansas...	0 00	1 50
Water-of-Ayr...	0 00	0 10
Scythe, per gross	3 50	5 00
Grind, per ton	15 00	18 00

STOVE PIPES.

Nestable in crates of 25 lengths.

5 and 6 inch Per 100 lengths	7 00
7 inch " "	7 50

ENAMELINE STOVE POLISH.

No. 4—3 dozen in case, net cash	$4 80
No. 6—3 dozen in case, "	8 40

TACKS BRADS, ETC.

	Per cent
Strawberry box tacks, bulk...	75 & 10
Cheese-box tacks, blued...	80 & 12½
Trunk tacks, black and tinned...	85
Carpet tacks, blued...	80 & 5
" " tinned...	80 & 10
" " (in kegs)...	40
Cut tacks, blued, in dozens only..	75 & 15
" ½ weights...	60
Swedes, cut tacks, blued and tinned—In bulk...	80 & 10
In dozens...	75
Swedes, upholsterers', bulk...	85 & 12½
" brush, blued & tinned, bulk..	70
" gimp, blued, tinned and japanned...	75 & 12½
Zinc tacks...	35
Leather carpet tacks...	55
Copper tacks...	50
Copper nails...	52

Trunk nails, black	65 and 5	
Trunk nails, tinned	65 and 10	
Clout nails, blued and tinned	65 and 5	
Chair nails	35	
Cigar box nails	35	
Patent brads	40	
Fine finishing	40	
Picture frame points	10	
Lining tacks, in papers	10	
" " in bulk	15	
" " solid heads, in bulk	75	
Saddle nails in papers	10	
" " in bulk	15	
Tufting buttons, 22 line, in dozens only	60	
Tin capped trunk nails	15	
Zinc glazier's points	5	
Double pointed tacks, papers	90 and 10	
" " " bulk	40	

TAPE LINES.

English, ass skin, per doz	2 75	5 00
English, Patent Leather	5 50	9 75
Chesterman's each	0 90	2 85
" steel, each	0 80	8 00

THERMOMETERS.

Tin case and dairy, dis. 75 to 75 and 10 p.c.

TRANSOM LIFTERS.

Payson's per doz 3 60

TRAPS. (Steel.)

Game, Newhouse, dis. 25 p.c.
Game, H. & N., P. S. & W., 65 p.c.
Game, steel, 72½, 75 p.c.

TROWELS.

Disston's discount 10 per cent.

German, per doz	4 75	6 00

S. & D., discount 35 per cent.

TWINES.

Bag, Russian, per lb	....	0 27
Wrapping, cotton, per lb	0 22	0 26
Wrapping, mottled, per pack	0 5)	0 (0
Wrapping, cotton, 3-ply		0 20
" " 4-ply	....	0 26
Mattress, per lb	0 33	0 45
Staging, "	0 27	0 35
Broom, "	0 30	0 55

VISES.

Hand, per doz	4 00	6 00
Bench, parallel, each	2 00	4 50
Coach, each	6 00	7 00
Peter Wright's, per lb	0 12	0 13
Pipe, each	5 50	9 00
Saw, per doz	6 50	12 00

ENAMELLED WARE.

White, Princess, Turquoise, Blue and White, discount 50 per cent.

Diamond, Famous, Premier, 50 and 10 p.c.

Granite or Pearl, Imperial, Crescent, 50, 10 and 10 per cent.

WIRE.

Brass wire, 50 to 50 and 2½ per cent. off the list.

Copper wire, 45 and 10 per cent. net cash 30 days, f.o.b. factory.

Smooth Steel Wire, base, $2.80 per 100 lb. List of extras: Nos. 2 to 5, advance 7c. per 100 lb.—Nos. 6 to 9, base—No. 10, advance 7c.—No. 11, 14c.—No. 12, 20c.—No. 13, 35c.—No. 14, 47c.—No. 15, 60c.—No. 16, 75c. Extras net per 100 lb.: Coppered wire, 60c.—tinned wire, $2—oiling, 10c.—special hay-bailing wire, 30c.—spring wire, $1—best steel wire, 75c.—bright soft drawn, 15c.—in 50 and 100-lb. bundles net, 10c.—in 25-lb. bundles net, 15c.—packed in casks or cases, 15c.—bagging or papering, 10c.

Fine Steel Wire, dis. 17½ per cent. List of extras: In 100-lb. lots: No. 17, $5—No. 18, $5.50—No. 19, $6—No. 20, $6.65—No. 21, $7—No. 22, $7.30—No. 23, $7.65—No. 24, $8—No. 25, $9—No. 26, $9.50—No. 27, $10—No. 28, $11—No. 29, $12—No. 30, $13—No. 31, $14—No. 32, $15 No. 33, $16—No. 34, $17. Extras net—tinned wire, Nos. 17-25, $2—Nos. 26-31, $4—Nos. 32-34, $6. Coppered, 5c.—oiling, 10c.—in 25-lb. bundles, 15c.—in 5 and 10-lb. bundles, 25c.—in 1-lb. hanks, 50c.—in ½-lb. hanks, 75c.—in ¼-lb. hanks, $1—packed in casks or cases, 15c.—bagging or papering, 10c.

Galvanized Wire, per 100 lb.—Nos. 6, 7, 8, $3.85 No. 9, $3.10—No. 10, $4.00—No. 11, $4.05 No. 12, $3.25—No. 13, $3.35—No. 14, $4.40—No. 15, $4.90—No. 16, $5.15.

Clothes Line Wire, 19 gauge, per 1,000 feet 3 30

WIRE FENCING. F.O.B. Toronto

Galvanized 4 barb, 2¼ and 5 inches apart	....	3 10
Galvanized, 2 barb, 4 and 5 inches apart	....	3 10
Galvanized, plain twist	....	3 10

Galvanized barb, f.o.b. Cleveland, $2.97½ in less than carlots, and $2.85 in carlots. Terms, 60 days or 2 per cent. in 10 days.

Ross braid truss cable 4 50

WIRE CLOTH.

Painted Screen, per 100 sq. ft., net .. 1 5)
Terms, 4 months, May 1. ; 3 p.c. off 30 days.

WRENCHES.

Acme, 35 to 37½ per cent.
Agricultural, 60 p.c.
Coe's Genuine, dis. 70 to 25 p.c.

Towers' Engineer, each	2 00	7 00
" S., per doz	5 80	6 00
G. & K.'s Pipe, per doz	....	3 40
Burrell's Pipe, each	....	3 00
Pocket, per doz	0 25	2 90

WRINGERS.

Leader	per doz.	$60 00
Royal Canadian	"	58 00
Royal American	"	50 00

Discount, 45 per cent.; terms 4 months, or 3 p.c. 30 days.

WROUGHT IRON WASHERS.

Canadian make, discount, 40 and 5 per cen

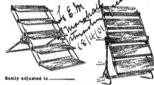

CANADIAN HARDWARE AND METAL MERCHANT

The Weekly Organ of the Hardware, Metal, Heating, Plumbing and Contracting Trades in Canada.

VOL. XIII. MONTREAL AND TORONTO, JANUARY 19, 1901. NO. 3

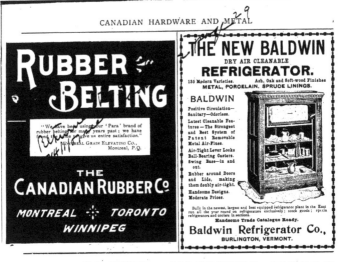

ALEXANDER GIBB

Manufacturers' Agent and Metal Broker,
13 St. John Street, **Montreal.**

Representing British and American Manufacturers. Correspondence invited from firms wishing to be represented in Canada.

McCaskill, Dougall & Co
Fine Varnish & Japan
Manufacturers.
Montreal
Price Lists on application

Dundas Axes

are ground sharp. If you are as sharp you will wait to see them before buying, or write for samples and prices to

Dundas Axe Works

DUNDAS, CANADA.

WANT TO MAKE MONEY?

Get a good

WALL TINT

GET

MURALO

The advertising methods of The Muralo Co. are up-to-date. The goods sell and stay sold. Repeat orders are certain and profits sure.

AGENTS.

A. RAMSAY & SON, - - - - - - MONTREAL.
J. H. ASHDOWN, - - - - - - - WINNIPEG.
McLENNAN, McFEELY & CO., - - - VANCOUVER.

VanTuyl & Fairbank

Petrolia, Ont.

Headquarters for . . .
Oil and Artesian Well Pumps, Casing, Tubing, Fittings, Drilling Tools, Cables, etc.

Ontario Nut Works, Paris

BROWN & CO.

Manufacturers of

All sizes of Hot Pressed Nuts, Square and Hexagon.

"DAISY" CHURN

Has tempered steel cased bicycle ball bearings, strongest, neatest and most convenient frame. Only two bolts to adjust in setting up. Steel Bow Levers, suitable for either a standing or sitting posture. Has four wheels and adjustable feet to hold stand steady while churning. When churn is locked to stand the bow can be used as handles to move it about on the front wheels as handy as a baby carriage. Open on both sides to centre, giving free space for pail. Made with wood or steel stands, with Cranks only, or Bow Levers as desired.

Vollmar Perfect Washer

Has a most enviable record. A perfection of its kind—will wash more clothes in less time, do it better and easier, with less wear and tear, than any other machine.

THE
Wortman & Ward Mfg. Co.,
Limited

LONDON, ONT.

Eastern Branch, 60 McGill Street, Montreal, Que.

IMPROVED STEEL WIRE TRACE CHAINS.

Every chain guaranteed. Most profitable and satisfactory chain to handle.

The B. GREENING WIRE CO., Limited

Hamilton, Ont., and Montreal, Que.

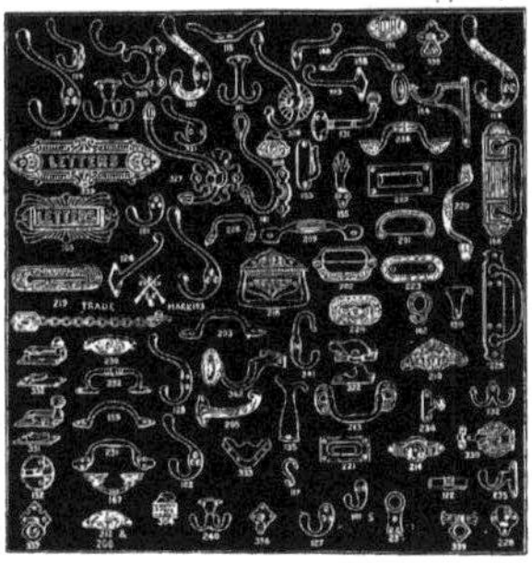

"BRASSITE"

REGISTERED TRADE MARK.

None genuine without the above "**Trade Mark.**"

"Gunn's"

Patent

"Brassite"

Goods.

Equal to **Solid Brass** in every particular. Cost less money—look and wear as well. Our sales are increasing all the time. Why not increase your sales?

THE GUNN CASTOR CO.

Limited.

KNOX HENRY, - Canadian Agent, 220½ Board of Trade, - **MONTREAL.**

Manufactured and Patented by

A. R. WOODYATT & CO., Guelph
CANADA.

THE WOODYATT

The Woodyatt Lawn Mower

has proved itself

THE BEST HIGH-GRADE MOWER

in the market.

Sizes—12, 14, 16, 18, 20 In. Cut.

THE STAR LAWN MOWER

is a Medium-Grade Mower of exceptional value.

Sizes—12, 14, 16 In. Cut.

THE STAR

Sold only through the Wholesale Trade.

KEMP'S

Broad-Hoop Roll-Rim

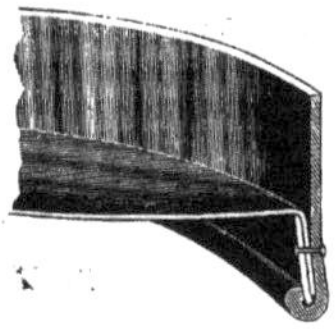

Milk Can Bottoms

possess all the points which go to make perfection in Can Bottoms. They have been used by a criticizing public for two seasons, and their popularity is evidence of the satisfaction which they gave. The roll-rim has no sharp turns, which break the grain of the metal and lessen its wearing qualities. It has a broad wearing surface and will not damage floors. They do not cost more than inferior bottoms.

The Iron Clad Trimmings are made the same as the Broad Hoop, and differ from them only in having a narrower and thicker hoop, which does not require the roll-rim, and, therefore, can be sold cheaper.

For durability and finish, our Trimmings are unequalled.

Manufactured by

Kemp Manufacturing Co., Toronto, Canada.

HARDWARE AND METAL

VOL. XIII. MONTREAL AND TORONTO, JANUARY 19, 1901. NO. 3.

President,
JOHN BAYNE MacLEAN,
Montreal.

THE MacLEAN PUBLISHING CO.
Limited.

Publishers of Trade Newspapers which circulate in the Provinces of British Columbia, North-West Territories, Manitoba, Ontario, Quebec, Nova Scotia, New Brunswick, P.E. Island and Newfoundland.

OFFICES

MONTREAL . . . . Board of Trade Building. Telephone 1255.
TORONTO . . . . . . 10 Front Street East. Telephone 2148.
LONDON, ENG. . . . . 109 Fleet Street, E.C., J. M. McKim,
MANCHESTER, ENG. . . . 18 St Ann Street. H. S. Ashburner,
WINNIPEG . . . . Western Canada Block. J. J. Roberts.
ST. JOHN, N. B. . . . No. 3 Market Wharf. J. Hunter White,
NEW YORK. . . . . . . 176 E. 88th Street, W. J. Brandt.

Travelling Subscription Agents :
T. Donaghy. F. S. Millard.

Subscription, Canada and the United States, $2.00.
Great Britain and elsewhere . . . 12s.

Published every Saturday.

Cable Address { Adscript, London. Adscript, Canada.

THE BICYCLES FOR 1901.

A MEMBER of HARDWARE AND METAL'S staff is in New York, at the Motor Exhibition, and writes : There are some surprises at the Sixth Annual Cycle Show which opened in Madison Square Garden this week. It is an exposition not only of bicycles of twentieth-century design, but of motor cycles and automobiles of improved types. As such it is more comprehensive and more generally attractive than any of its predecessors.

All available space in the Garden amphitheatre has been pressed into service for the show. Two great balconies were built over the arena boxes to accommodate the overflow. More than 100 individual exhibitors expose to the scrutiny of the public their respective contributions toward perfection in cycle and motor vehicle construction.

In the wheel models of 1901 on view are found improvements that will be welcomed by experienced riders. In their twentieth-century initial output manufacturers all seem to be animated by the same idea—that is, to provide for the comfort of the cyclist, even at the sacrifice of speed. A visit to the show demonstrates that the bicycle of 1901 is pre-eminently a comfortable bicycle.

In the new models seven features are conspicuous at the show. These all cater to ease of running and comfort. They are :

Cushion frames, of hygienic principles, on high-grade machines.
General use of chainless gear on better grades of machines.
General use of coaster brakes on same machines.
General use of handle bars that may be adjusted to any position without necessitating a dismount.
Use of improved spring saddles.
Slight lengthening in pedal cranks.
Small reduction in weight of better grades of machines.

With bicycle equipped with cushion frame and improved spring saddle the rider may pedal over cobblestones or other uneven pavement without the disagreeable jarring sensation experienced on a rigid machine. The use of adjustable handle-bars will obviate the cramping of the rider's wrists or arms because of being held constantly in one position. The advantages of the chainless gear and coaster brake already have been proved, while larger pedal cranks will give greater power in climbing grades.

These, however, are not the only changes in wheel construction that may be seen at the Garden. Scarcely a machine is exhibited that does not have some improvements in detail work. One model alone shows 11 alterations, most of which are in small details. In motor bicycles, tricycles and quadricycles, the display is probably the most complete ever made in this country. Many bicycle manufacturers have begun the building of small motor machines, and such as have vehicles ready for the market exhibit their models. The regular automobile exhibits, while necessarily curtailed, are interesting and instructive.

One established feature of the show is not lost sight of this year. That is the distribution of souvenirs. There are several voting contests, the prize in one of which, for the most popular schoolteacher, being a chainless bicycle. To the most popular public school scholar a chain wheel will be given ; while to the bicycle club member receiving the most votes a tandem will be awarded.

IN A NEW FORM.

The American Manufacturer comes to hand in a new form. Its pages have been reduced about one half in size, making them conform to the ordinary magazine. There has, however, been no diminution in the character of the articles. They are as interesting and as timely as under The Manufacturer's old style, which is, after all, the most important consideration in a trade or any other journal.

WATCH THE MARKETS.

The present year is not likely to witness the same sudden drop in values that 1900 did, but he is a wise merchant who keeps his stock prepared for eventualities. Keep your stock nicely sorted and watch the markets closely is the advice we would give members of the trade.

THE LATE MR. F. S. FOSTER.

THE death of Mr. Francis Stuart Foster, of the firm of The Watson, Foster Co., wall paper manufacturers, Montreal, which occurred at his late residence, 904 Dorchester street, Montreal, on Sunday, January 6, 1901, deprived Canada's business community of one of its most respected members and cut short one of the most promising business careers that could be prophesied for a young man.

Although he had acquired a leadership in his line of business, Mr. Foster had not had an exceedingly long experience, being born only somewhat over 41 years ago in Kingston. He received his education in the "Limestone City," attending the High School there, previous to the family's removal to Montreal.

It was in 1880 that Mr. Foster first became connected with the wall paper business, entering the firm of Watson & McArthur, which had just been formed, as bookkeeper. Four years later Mr. McArthur withdrew and the business was carried on by John C. Watson & Co. In 1891 Mr. Foster really entered into partnership in the firm, but not till 1894 did his name appear in the firm's style. In 1897 the business was formed into a joint stock company and the trading title has since been The Watson, Foster Co., Limited.

Mr. Foster's special duties belonged to the manufacturing part of the establishment, and he had acquired a thorough knowledge of the practical side of the business. In fact, to his ingenuity and enterprise is largely due the excellence which Canadian wall paper manufactories have learned to give to their products since they started to learn their business in 1880, on the adoption of the National Policy. Canadian merchants long found difficulty in selling the domestic-made wall decoration, but, thanks to the zeal and perseverance of such pioneers in the industry as Mr. Foster, we are now not only supplying our own trade, but entering into the export business as well.

Mr. Foster's influence extended into the office also, where his grasp of financial questions and his business ability of no mean order were valued very highly. Honesty and intergity were equally predominant with enterprise in his make-up. An example of his high principle, which he never would allow to be published, was shown one time, when, about eight months after he had made a settlement with an insurance company upon some losses the firm had sustained through fire, he found a mistake had been made in the valuation of some factory apparatus, and his firm sent the insurance company a cheque for $800. Acting upon such principles as actuated them in this case, he and his partners builded even better than they knew, and their business expanded to enormous proportions.

As a man, few business figures were held in respect equal to that enjoyed by Mr. Foster. Although he was very attentive to his private business, he had for some years been a member of the Montreal Board of Trade. He was a governor of the Montreal General Hospital and a warden of Christ Church Cathedral. His personality was affable, yet always impressive.

Mr. Foster had been away from business two years and eight months, seeking a recovery of health in different climes, but it was only during the last two months of his life he was seriously ill. He leaves a family of a widow and three children who, needless to say, have the warmest sympathy of his hosts of business friends.

The funeral service, rendered in the Cathedral in full chorus, was very impressive. The chief mourners were the two young sons of the deceased; Mr. W. Foster, brother; Mr. W. I. Gear, brother-in-law; Messrs. Hugh Watson and D. S. Boxer, partners of the deceased, and Wm. Cooper. Among the others present were: Sir M. W. Tait, Messrs. Alfred Griffin, George Creak, Capt. Riley, H. Adams, E. A. Barton, J. H. Hutchison, M. Fitzgibbon, David Smith, R. J. Notan, Lieut. Col. Butler, C. Richards, H. Ryan, George Howard, J. Fraser, C. P. Greaves, R. K. Howland, C. C. Howland, H. H. Howland. There was a large number of floral tributes sent by the immediate relatives of the deceased, and from Mr. and Mrs. W. B. Foster, Mr. and Mrs. Hugh Watson, G. Howland, Son & Co., Madame M. J. A. Prendergast, Cadieux & Derome, Colin McArthur, and Miss McArthur, the office and travelling staff.

HAD IT ONLY BEEN CANADA!

IT appears that, owing to the inability to maintain a comfortable temperature, it has been found necessary to close a number of schools in the United States. The cause, however, appears to be due more to lack of knowledge in regard to manipulation of furnaces and other heating apparatus than to the severity of the weather.

But we wonder what newspapers in the United States would have said had such a thing happened in Canada? They would certainly have said a great deal and emphasized it with headlines out of all proportion to the importance of the matter.

Canada has one of the best climates in the world, but one would imagine, from the statements made at times by sensational journals, that it was an annex of the North Pole.

A BUSINESS MAN'S ELECTION.

The election of Mr. George H. Gooderham as Public School Trustee in Ward 3, Toronto, is a matter for congratulation.

He is not only a young man of means and energy, but he is a business man and the descendant of a family which for more than half a century has exerted a great deal of influence upon the commercial career of the "Queen City."

It is to be hoped that more men of his stamp and ability will follow his example and allow themselves to be elected to positions of honor and trust in our various municipal institutions.

The ward-heeler and the professional politician have had their day. It is now time that practical business men like Mr. George H. Gooderham superseded them.

PRICE CUTTING IN DAWSON CITY.

A despatch from Ottawa states that, according to recent reports received there, the large mercantile houses of Dawson City are engaged in a price-cutting war on the smaller merchants. The despatch needlessly adds that the miners are making the most of their opportunity and are laying in large supplies of provisions at prices only a little higher than those charged in Seattle and Victoria.

NO INSOLVENCY LAW THIS SESSION.

MR. FORTIN, M.P. for Laval, will not introduce his insolvency bill in Parliament during the coming session. This is the report that comes from Ottawa, and it has been confirmed by the gentleman in question to a representative of this paper.

We learn that Mr. Fortin considers it useless to press his measure upon the House, thinking it foredestined to defeat. He says the banks are strongly opposed to it, the Maritime Provinces are giving it the cold shoulder, and the commercial organizations are only lukewarm in their support.

If these be the true considerations that weighed with Mr. Fortin when he was making his decision, it is truly unfortunate that he should not have been given more encouragement, for our business men and manufacturers are laboring under a veritable curse in being ill-provided with insolvency legislation. What with slow settlements, exorbitant legal charges, the custom of giving preferences and numerous subterfuges, a debt against an insolvent's estate is worth very little. And, in Ontario, a man cannot be compelled to assign. Nor would it be allowable in the other Provinces, if the law were tested, for the Federal Government is the only body constitutionally provided with the power to pass legislation for such a purpose. That is why we want a Dominion measure, and the boards of trade and other bodies interested should immediately agitate to have the matter discussed in the House during the coming session. It is a disgrace to Canada, to say nothing of the loss of trade, to have our insolvency laws advertised in the columns of English papers as they have been during the past few years. If we are so desirous of encouraging English trade as to adopt a preferential tariff, we should not be unmindful of the fact that we can mightily improve our business reputation in England by improved insolvency legislation.

The Montreal Chambre de Commerce is hitting the nail on the head when it approaches the Government on the matter, for it would appear that Mr. Fortin's ardor has received a severe dampening in Ottawa. The Government organs say that times have improved, that there is not the need for legislation now that there was two or three years ago, and that "eminent legal authorities are unable to see how the position of creditors in insolvency cases could be improved by a Dominion Insolvency Law." These are purely makeshift excuses.

In reply to the latter argument, all we have to say is that Mr. Fortin, M.P., a learned and practical Canadian jurist, has, in his insolvency bill, offered a remedy and it is idle talk to say a remedy cannot be found.

The argument that we stand in less need of such legislation than we did some years ago, carries as little weight. Happily, times have improved and failures are fewer, but insolvency legislation is not to prevent failures. Men still get into difficulties and will so long as business lasts. When they do, we want that creditors should get what they should out of the estate and that as speedily as possible. A case just came to our notice recently in which the settlement sheet of insolvent shows all the proceeds to have been gobbled up in winding-up expenses. Not even one cent was saved to pay on the rent account, and, of course, the ordinary claimants got nothing. And this occurs frequently.

Our business men are crying out for insolvency legislation and they must have it to save themselves and our national reputation.

PRICES ON PARIS GREEN.

The opening prices on paris green have been announced by the manufacturers. Compared with last year, they are slightly lower. The figures are as follows :

	Per lb.
Barrels	16¼c.
Kegs	17
50 and 100-lb. drums	17¼
25-lb. drums	18
1-lb. papers	18¼
1-lb. tins	19¼
½-lb. papers	20¼
½-lb. tins	21¼

Prices are guaranteed up to the time of shipment. On the strength of this, a few orders have been booked during the week.

TIME FOR STOCK-TAKING.

Now is a good time for careful stock-taking. Every merchant should take advantage of the comparative quiet which prevails at this season to find out just what stock he has on hand, in order that he may know just what is his financial standing. Laxity in this regard is a mistake.

DANGEROUS OUTSIDE VENTURES.

EVERY business man should think not only once or twice, but several times, before he branches out in some venture other than that in which he is immediately engaged. Even merchants with liberal capital cannot afford to ignore this principle, for, while their business may not be directly affected thereby, the old maxim in regard to too many irons being in the fire still holds good.

But, when merchants with but a moderate amount of capital take a part of the same to invest in something altogether aside from their regular business they are simply courting suicide. And then, this is a practice that is all too common. Not satisfied with the volume of business they are doing, or the profits they earn, they are induced to lend their time and money to some scheme or schemes which usually promise well but turn out bad.

If the business in which a man is engaged is not, for one or more reasons, to his liking, it is better that he should go out of it altogether than he should, while trying to retain it, devote a part of his time to another concern.

We have in mind at the moment a retail merchant who was doing a nice business and seemed prosperous. Being fond of horses he gradually drifted into speculation in them. Now his business is gone and he is in financial difficulties. This is only one of many similar instances that might be cited.

The successful man to-day is he who gives his undivided attention to the business in which he is engaged. There is no other alternative, for divided attention is like trying to travel simultaneously on two roads in order to reach a given point.

A WEAK TIN MARKET.

Pig tin has been the feature of the metal market for several days past.

Last week closed with prices much lower in both London and New York than they were at the opening. When the market opened this week the tendency was still downward, but at the time of writing the tone is steadier.

In Canada there have been some fluctuations, but quotations are much the same as they were a week ago.

SLOW DELIVERIES OF ENGLISH CUTLERY.

DURING the past year, the Canadian cutlery importers have not found the deliveries of English goods to improve, and the old grievance of slow delivery still exists. In many cases, it exists in an aggravated form.

So slow have the deliveries of some Sheffield manufacturers been during the last 12 months that agencies of 20 years' standing have been given up, and importers here who have a connection with the trade are concluding that the best firm to represent is that which gives the speediest deliveries. It has not been an uncommon thing lately for wholesalers to have to wait for 9 to 10 months to have orders filled. Naturally, this is an inconvenience, when it is taken into account that they do not wait nearly as long on any other class of importations.

The tendency to give the house which gives the quickest deliveries the preference is growing, and agents who have represented firms whose goods have given entire satisfaction when at last they arrived here have been compelled, for the sake of getting deliveries in a reasonable time, to transfer their allegiance to speedier cutlery lords.

Although the Sheffield manufacturers have been seriously handicapped in their productions this year by the South-African War, which called out a large number of their men, the entire cause of the slow delivery evil this year is not to be laid at the doors of the War Office.

If ever a remedy is to be applied to put this business on a proper footing, it must first be applied in Sheffield itself where the whole scheme of manufacturing must be changed. The Sheffield manufacturer must first learn—and learn from the simplest rudiments up—that speedy deliveries are not only desirable, but also necessary to stimulate trade. This he does not realize at the present moment. So long as he has enough orders on hand to keep his works running he does not appear to worry; a pile of unfilled orders does not cause him an anxious thought as it would an American manufacturer, who would immediately increase his capacity to keep his business up-to date. So long as he has orders on hand the Sheffield manufacturer rests as easy as do his employes, who invariably neglect their work to see a football or cricket match. They say it is not an uncommon thing to see the works closed down entirely for a half day for some more or less important sporting event.

Of course, a certain latitude must be allowed the English manufacturer, for his trade extends over an area measured only by the limits of the east and the west, and the number of patterns used the world over is so enormous that stocks cannot be kept to fill orders. But three or four months ought to be sufficient time to fill any order that is going to be filled, and when more than double this time is habitually taken, the complaints are well grounded.

The prices of English cuttlery have advanced during the past season, probably, from 5 to 10 per cent. all around. A Canadian agent, who has lately been in Sheffield two or three weeks, says that while he was there ebony advanced 75 per cent., bone over 100 per cent., pearl very materially; grinders and hafters were given a 10 per cent. increase in wages, and so scarce was labor that it was difficult to have cheap cutlery made at all. It is the general opinion that prices are up to stay, and that there will be no receding from the new values.

A GOOD ANNIVERSARY NUMBER.

The Age of Steel, St. Louis, Mo., is to be congratulated on the number which marks its 43rd anniversary. Few trade journals can boast a better advertising patronage, a more comprehensive and up-to-date news and editorial service, or a more attractive appearance than our St. Louis contemporary presents in this issue. The subjects discussed cover practically every branch of the steel industry, from the iron ore transportation problem to a comparison between the steel industry of America and Europe. Several of the articles are well illustrated.

E. W. GILLETT'S CALENDAR.

The calendar which E. W. Gillett is sending out this year is one of the striking productions of the season. The design is of a laughing negro boy, straddled over some packages of Gillett's goods, on one of which in large letters is the notice: "Gillett's Lye Eats Dirt." The calendar pad is big enough to be useful in any office or room.

HARDWARE SPECIALTIES HIS LINE.

Mr. James Burridge, Winnipeg, who is opening up an agency business in that city, has been in Toronto on a visit. Mr. Burridge has been a resident of Winnipeg for 20 years, and for some time was the manager of the branch business of The Gurney-Tilden Co., Hamilton. He severed his connection with that firm in May. He is starting up on his own account, and is prepared to accept the agency of manufacturers of hardware specialties.

BUSINESS CHANGES.

DIFFICULTIES, ASSIGNMENTS, COMPROMISES.

J. J. Brownley, stove and tinware dealer, Halifax, is reported to have left Halifax, and his landlord is in possession.

M. Forget, general merchant, St. Jerome, Que., has assigned to Gagnon & Caron.

Alfred Gibault, general merchant, St. Lucie de Doncaster, Que., has assigned.

J. G. Terryberry, general merchant, Burford, Ont., has assigned to Arthur G. Olive.

Cooper & Co., bicycle dealers, Brantford, Ont., have assigned to John P. Hemphill.

Jacob Lennis, hardware dealer, etc., Winnipeg, has assigned to C. H. Newton.

A. W. Stevenson has been appointed curator of Wm. Rodden & Co., founders, Montreal.

R. Bourcier, general merchant, Lefaivre, Ont., has compromised at 35c. on the dollar, cash.

Eugene Guay, general merchant, St. Jerome (Chicoutimi) Que., is offering 40c. on the dollar.

Assignment has been demaded of Alf. Mercier, general marchant, St. Angele (Rimouski), Que.

S. Renaud, general merchant, etc., St. Tite des Caps, Que., has compromised at 25c. on the dollar.

J. A. Andrews, tinsmith, etc., Kinburn, Ont., has assigned to W. A. Cole and a meeting of his creditors will be held on January 23.

A meeting of the creditors of J. A. Schneider, tinsmith, etc., Thornbury, Ont., who has assigned to Andrew Grier, has been called for January 21.

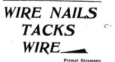

New copartnership has been registered by Joseph and W. J. Hunt under the style of Joseph Hunt & Son, general merchants, Mahon, Que.

Chisholm & Copeland, general merchants, Grenfell, Man., have dissolved, and R. A. Copeland continues under the style of R. A. Copeland & Co.

PARTNERSHIPS FORMED AND DISSOLVED.

Adams & Coate, hardware dealers, Kingsville, Ont., are about dissolving.

Partnership has been registered by Girard & Roy, general merchants, Ste. Flore, Que.

E. A. Walker, hardware dealer, Grenfell, Man., has admitted J. A. Walker under the style of E. A. Walker & Son.

Hay Bros., founders, Portage la Prairie, Man., have dissolved. Edward Hay continues.

The Rock Island Hardware Co., Rock Island, Que., have dissolved, and a new partnership has been registered.

James P. and Ralph B. Simmonds have registered partnership under the style of James Simmonds & Co., hardware dealers, Dartmouth, N.S.

SALES MADE AND PENDING.

The assets of T. Ross, general merchant, Aurque, Que., are advertised for sale.

The assets of E. Huared, general merchant, Bonfield, Ont., have been sold.

D. McCuig, dealer in agricultural implements, Treberne, Man., has sold out.

The business of John Wynn, blacksmith, Brussels, Ont., is advertised for sale.

The assets of Alphonse Guimond, hardware dealer, Montreal, are to be sold on January 22.

The assets of the estate of Hector Grenier, hardware dealer, Quebec, are to be sold on January 24.

The business of James Elsey, harness dealer, Mount Brydges, Ont., is advertised for sale.

The assets of C. Pearson & Co., general merchants, Cedar Hill, Que., are advertised for sale.

The assets of B. S. Chaiffer, general merchants, Magog, Que., are to be sold on January 21.

Edgar Scott, general merchant, Halifax, is advertising his stock for sale under warrant of distraint.

The business of the estate of Thos. McNeely, general merchant, Ladner, B.C., is advertised for sale.

F. I. Labrance, general merchant, Thetford Mines, Que., has sold out his business and is applying for a hotel license.

Schofield & Co., general merchants, Pincher Creek and McLeod, N.W.T., are advertising their McLeod business for sale.

CHANGES.

L. P. Venne is starting as hardware dealer in Montreal.

J. W. Saulnier, tinsmith, Weymouth Bridge, N.S., is giving up business.

D. A. Shindler has started as hardware and bicycle dealer in Atlin, B.C.

Pepper & Stonehouse are starting as blacksmiths, etc., in Forest, Ont.

Kersey & Kersey have succeeded Isaac Kersey as general merchant, Edy's Mills, Ont.

Charles Hartleib, hardware dealer, etc., Zurich, Ont., has sold out to Charles Greb.

John Roulson, general merchant, Garnet, Ont., has been succeeded by Chas. A. Walker.

H. S. Cook & Co. have started as manufacturers of stove polish, Yarmouth, N. S.

Alton, Beatty & Alton, dealers in agricultural implements, etc., Sidney, Man., have sold their harness business to James Tait.

FIRES.

P. A. Allen and J. M. Gibson, blacksmiths, and R. J. Nicholson, harness dealer, Brigden, Ont., have been burned out. All are insured.

DEATHS.

T. Rivard, of T. Rivard & Co., saddlers, Joliette, Que., is dead.

Thomas Gilbert, tinsmith, St. George East, Que., is dead.

P. Gagnon, saddler, Three Rivers, Que., is dead.

J. Milton O'Brien, blacksmith, Richibucto, N.B., is dead.

James Shea has started as harness dealer in Liverpool, N.S.

W. A. Maclauchlan has been appointed agent in St. John, N.B., for the cutlery firm of Askham & Co., Sheffield, Eng.

INQUIRIES FOR CANADIAN PRODUCTS.

The following was among the recent inquiries relating to Canadian trade received at the High Commissioner's office, in London, England:

1. A London firm ask to be placed in communication with Nova Scotian dealers in, and shippers of salted and dried codfish, packed in drums of 128 lb. each.

[The name of the firm making the above inquiry, can be obtained on application to the editor of HARDWARE AND METAL. When asking for names, kindly give number of paragraph and date of issue.]

Mr. Harrison Watson, curator of the Canadian Section of the Imperial Institute, London, England, is in receipt of the following inquiries:

1. A London house asks to be placed in correspondence with Canadian producers of lard oil.

2. A manufacturing company wishes to hear from Canadian producers of crude asbestos suitable for spinning purposes.

3. A Scotch cycle manufacturing company desires information as to prospects of securing trade in Canada, and invites correspondence from importers interested.

4. An old-established timber merchant contemplates adding a few lines of wood manufactures to his business, with which they could be advantageously worked. He would be pleased to hear from Canadian manufacturers equipped for export trade

"FAMOUS" TRAVELLERS MEET.

THURSDAY and Friday, January 3 and 4, at the head office of The McClary Mfg. Co., London, were spent in most instructive talks among all their travelling men. Representatives came from as far east as Nova Scotia and from the far West, making a gathering of 20 men interested in selling "Famous" goods in every

city and village in Canada. The meeting is an annual affair, to exchange views as to the best needs of the trade in different sections of the Dominion and to recommend such changes and add such new goods as the trade requires.

Needless to say that with such an enterprising company the interests and wants of their patrons are carefully considered and changes made accordingly.

As a result of this conference many new lines will be added this year and others still further improved, keeping the output of this company in the lead, as in the past. As a part of the proceedings some time was devoted to new goods, which will include lines of well-known refrigerators and oil stoves, besides other lines of new goods. Much useful and instructive information was given by members of the home office, which will be beneficial to many agents by aiding them with many of their problems, Mr. Foot (manager of sales department) speaking on general goods and new lines, Mr. Herrick (superintendent of stove department) on stove construction, and Mr. Irwin (furnace expert) on selling and setting furnaces.

On Thursday evening a banquet was given to the travellers and foremen of departments where some hours of social and instructive intercourse were spent.

The toast to "The Queen" called forth a reply with a true patriotic ring, including a description of the visit of the Canadian officers and men to Windsor Palace and their reception by the Queen.

The toast of "Our Guests" was replied to by representatives from Nova Scotia, Ontario and Manitoba, while other toasts, "The Office," "The Factory" and others were responded to by employes in those departments. "Reminiscences" was replied to by one who has been in the employ of the McClary Manufacturing Co. for nearly 40 years and his description spoke volumes for the enterprise of this company in having grown from a few small wooden buildings to the many acres of busy departments which it now covers.

The McClary Manufacturing Co. have many surprises in the shape of new lines to offer their agents and it will pay to hold orders for spring goods until their travellers call.

Immediately after the closing of the meeting on Friday evening the travellers took train for their various territories where they will be calling on their various customers with their "Famous" popular and useful lines.

The territory will be covered by the travellers named below:

With Headquarters at Montreal.—W. Owen, Nova Scotia, New Brunswick and Prince Edward Island; A. St. Arnaud, Northern Quebec; L. Tarleton, Ottawa District; H. LePage, Montreal City, and O. R. Anderson, Eastern Quebec.

With Headquarters at Toronto.—S. T. Smith, Eastern Ontario; W. E. Bulmer, Central Ontario; W. Jeffrey, Northern Ontario, and J. D. Laidlaw, Toronto City.

With Headquarters at London.—E. H. Grenfell, Southern Ontario; J. Chalmers, Northern Ontario Peninsula; M. F. Irwin, Western Ontario Peninsula, and D. G. Clark, London City.

With Headquarters at Winnipeg.—J. Brockest, Northwest Territories, and Mr. Anderson, Manitoba.

With Headquarters at Vancouver, B.C.—T. R. Ella, Interior British Columbia, and N. R. Turner, Coast cities.

H. S. HOWLAND, SONS & CO.

WHOLESALE ONLY — 37-39 Front Street West, **Toronto.** — ONLY WHOLESALE

HORSE SINGERS

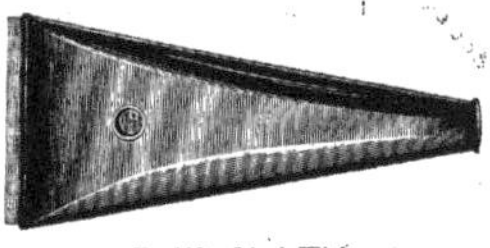

No. 162.—5-inch Wick.

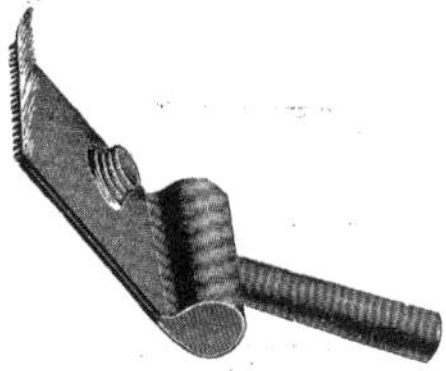

"ECLIPSE" 4-inch Wick.

No. 160.—5-inch Wick.

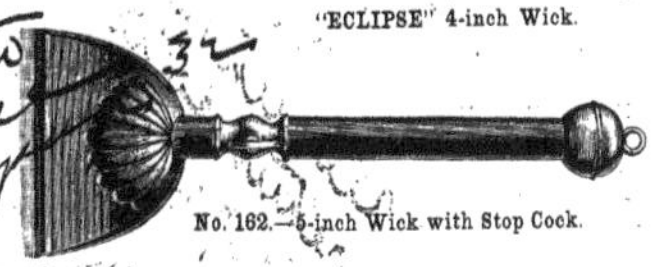

No. 162.—5-inch Wick with Stop Cock.

HORSE CLIPPERS.

"BOKER'S"
"Keen Cut."
"Perfection."
"Dandy."
1704, Ball-Bearing.

"NEWMARKET."

CHICAGO FLEXIBLE-SHAFT HORSE CLIPPERS.

No. 98.—To Swing with Rope.

No. 98.—Standard Machine.

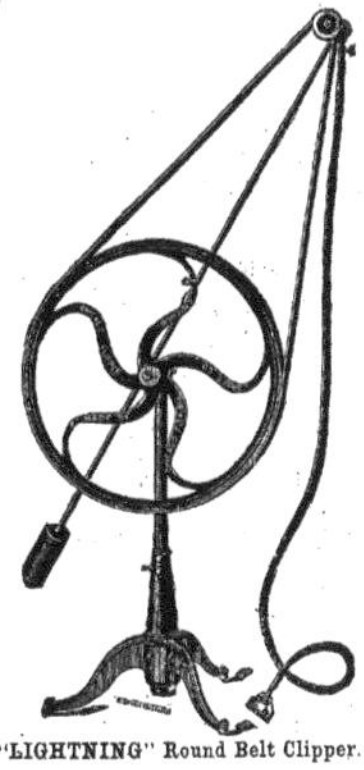

The "LIGHTNING" Round Belt Clipper.

H. S. HOWLAND, SONS & CO., Toronto.

OUR PRICES ARE RIGHT. — Graham Wire and Cut Nails are the Best. — WE SHIP PROMPTLY.

SEVERAL HARDWARE SPECIALTIES.

HARDWARE AND METAL last week gave a brief history of The Smith & Hemenway Co., 296 Broadway, New York. It is now proposed to deal with some of the specialties it makes.

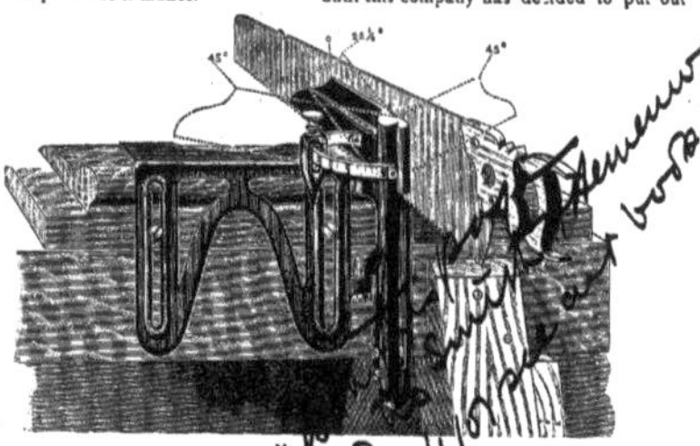

No. 159.

No. 159, Model No. 5, shows their 1901 pattern, which was ready for the market the first of this month. We mention a few of the points of superiority in this box: It is made entirely of iron and steel; can be folded up and carried in the pocket; no special saw is needed; any ordinary hand or cross-cut saw may be used; accurate and easily adjusted to any angle desired; all parts are interchangeable; it is the only mitre box that will saw moulding of any width or depth. It is made only in one size. One of the strong features is that a carpenter can carry it to the top of a ladder and saw moulding with as much ease as he can with the ordinary mitre box fixed on a permanent work bench. It retails at a popular price.

No. 1900 illustrates the celebrated Russell

No. 1900.

Staple Pulling Button Plyer. It is a combination of seven tools in one, being two staple pullers, two hammers, two wire cutters, a wire splicer, a small wrench and a pair of flat-nose pincers. This tool weighs 1⅛ lb., and no farmer or machinist can afford to be without one. It does its work perfectly, and is handy in the hands of any one.

No. 805, Improved Hall's Compound Nipper.—This is one of the latest patterns manufactured under the Hall's patent of the Utica Drop Forge and Tool Co. This tool is too well known to need any further comment.

No. 2000, Farmers' and Machinists' Universal Tool.—This is a combination of eight tools in one. The demand for a good farmers' and machinists' universal tool has steadily increased for the last two years, until this company has decided to put out one of the best that mechanical skill can produce. This is made from the finest quality of Brescian steel.

No. 888, Bent or Curved Nose Plyer.—

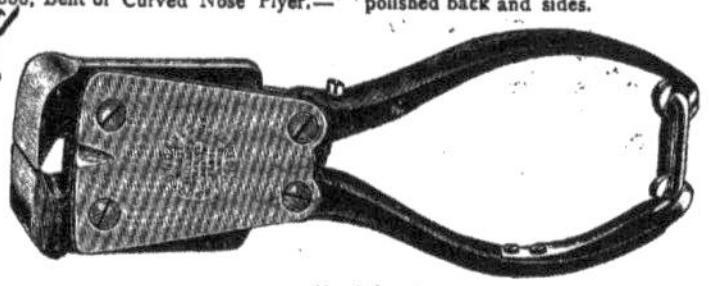

Np. 806.

This is especially adapted, so we are told, for electricians, machinists, textile mills, oculists and jewellers. The beauty of this tool can be seen at a glance.

No. 13, Diamond B.—The Smith & Hemenway Co. are owners of the genuine Giant Nail-Puller, and have, so we are informed, enjoyed the bulk of the nail-puller business in this country. Owing to the increased demand for a cheaper tool, they have put out what is known as their No. 13 Diamond B. In addition to this they also manufacture the Ajax, Eureka and Pipe Pullers. We are informed that they are the largest manufacturers of nail-pullers, varying in price from the cheapest to the best grade. They are at the present time equipping a factory in the vicinity of New York for the manufacture of nail-pullers and a few of their hardware specialties exclusively.

This company also puts out the No. 118 special linemen's or electricians' tool with insulated handles. The beauty of this can be seen at a glance.

The No. 427 is their special Ran-Tan-Ka-Rus Red Devil Razor, made from 60 small wires, the blank of which is herein illustrated. This is a new departure in the razor world. We are told that they have been experimenting upwards of four years to perfect this article. The value of this is that in hammering the wires together they make a perfect condensation of the metal, thereby insuring a Damascus effect. They positively guarantee that it is never necessary to hone this razor, and should it become worn sufficiently to require honing, it will be exchanged for a new one. It is beautifully Hamburg concaved with finely polished back and sides.

It would be well for all live, wide-awake hardware houses to keep their eye on this young firm. They are making their mark, and so deep that it will not be erased from the memory of the older hardware houses all over the country. The Smith & Hemenway Co. issue a unique catalogue, known to the trade as the "Green Book of Hardware Specialties," which will be sent gratis to anyone on application.

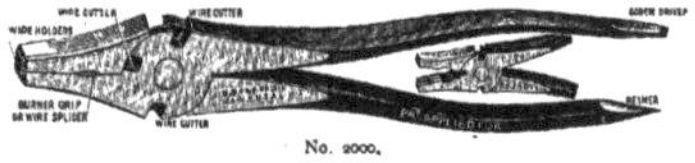

No. 2000.

ANSWERS TO INQUIRERS.

S. B. McC. & Co. writes: "Can you inform us where we can obtain Ladd's Discount Book, also price of same?"

[Remarks: Can any of our readers supply the desired information? — The EDITOR.]

CATALOGUES, BOOKLETS, ETC.

A GOOD ONE.

WITH the birth of each new year it is the custom of many wholesalers, manufacturers and retailers to send out booklets, calendars, etc., to their customers and probable customers. Some of these are very attractive, dainty, and useful. Money spent on the majority is wasted, for many are consigned to the

No. 868.

All cuts in this page returned to Smith & Hemenway 23/1/07 — cut book No. 1 page

waste basket. One of the tastiest and most novel sent out this year is the one gotten up by the Montreal Rolling Mills, Montreal. The cover is of white celluloid, on the front of which is printed a Union Jack in colors, and underneath the flag the words "Compliments Montreal Rolling Mills, Montreal." The effect is very pretty. The most useful part of the book lies in the calendars for the four years 1901, 1902, 1903, 1904, and its blank pages for memos. Those who have not received a book will get one by sending a request on a post card.

A NEW CENTURY CATALOGUE.

The Toronto Lead and Color Co., Limited, have just issued to their many patrons a new century catalogue and price list, a most complete, handsome and unique work, reflecting much credit upon the enterprise and management of the company. The paint, oil and varnish trade will, no doubt, obtain valuable information by keeping a copy before them. Any member who has not already received a copy will cheerfully be supplied with same upon application being made to the company's office, corner Leslie street and Eastern avenue, Toronto.

McCaskill, Dougall & Co., Montreal, have issued a beautiful, artistic calendar for 1901, advertising their famous productions. The illustration is a well-printed portrait of a pretty maiden. A copy may be had on application.

MR. IRVING TO LEAVE TORONTO.

Mr. John Irving, who for the past nine years has represented the Montreal Rolling Mills Co. in Toronto and Western Ontario, has accepted the position of sales agent with the Nova Scotia Steel Co., New Glasgow, N.S., and will shortly leave for New Glasgow, where he will in future reside.

Mr. A. H. Hough, who has represented the Montreal Rolling Mills Co. between

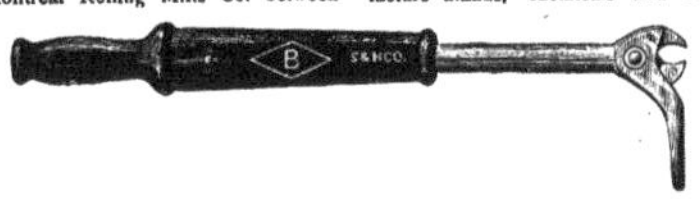

No. 13.

Montreal and Toronto, succeeds Mr. Irving in Toronto.

Mr. Irving has a great many friends in Toronto who will regret his departure from among them. HARDWARE AND METAL joins with them in wishing him success in his new sphere among the "Bluenoses."

No. 427.

AN IMPORTANT LAW DECISION.

On Friday Judge McDougall, of Toronto, gave a decision of importance to business men. Some time ago a Rossland, B.C., dealer named Gilmour sued Greville & Co., of Toronto, for an alleged debt amounting to $377.84. The case came up before the court of Kootenay county. The Toronto firm entered no defence, and a judgment was given in favour of the plaintiff with costs. Gilmour recently moved before the local County Court to have the judgment enforced.

In his judgment Judge McDougall declares that the judgment of the British Columbia court has no force in Ontario, and points out that it could not have, unless the defendant had gone to Kootenay in answer to the writ.

INDUSTRIAL GOSSIP.

Those having any items of news, suitable for this column will confer a favor by forwarding them to this office addressed to the Editor.

The Canadian Electric Chemical Co., Limited, have been incorporated in Sault Ste. Marie with $100,000 capital to manufacture alkalis, chemicals and chemical compounds, and all electrical, hydraulic, mechanical or automatic machinery. Provisional directors: Wm. V. Gibbs, Clayton E. Platt, F. H. Clergue, B. J. Clergue and Henry C. Hamilton.

A Vancouver despatch says that T. H. Davies & Co., iron manufacturers of Honolulu and Liverpool, have purchased Armstrong & Morrison's iron works in that city for $250,000.

BAD FIRE AT BRIGDEN, ONT.

On Thursday, last week, the business portion of Brigden, Ont., was practically wiped out by fire. The flames were first noticed in A. Harkness & Son's general store, a wooden building, which was speedily consumed, and from which the flames soon spread. Harkness & Son's loss is placed at $10,000; insurance $5,000.

Among the other losers were P. A. Allan and J. M. Gibson, blacksmiths; loss, $1,000; insurance, $400; and R. J. Nicholson, harnessmaker; loss, $1,500; fully insured.

No. 118.

MONTREAL RETAIL ASSOCIATION.

IT was a well-attended and enthusiastic meeting that the Retail Hardware and Paint Dealers' Association, of the city and district of Montreal, held in Monument National last Wednesday evening. President F. Martineau occupied the chair, and among those present were: Secretary Magnan, Treasurer A. Prudhomme, Messrs. Drysdale, Surveyer, Colleret, Denis, Dufault, Mailhot, Millen, Surveyer, U. Granger, Nap. Granger, A. A. Wilson, Huberdeau, G. Prudhomme, Belanger, Ponton, Jubinville, Young, Chausse, Tremblay, Couillard, Papineau, Shea, Beland, Marceau and others.

The meeting opened about 8.30 p.m., and until 10.45 p.m. the gathering indulged in a mild, sensible discussion of the present condition of Montreal hardware retailers.

After the minutes of the last meeting had been read and adopted, the president welcomed the new members and wished all present a happy and prosperous New Year. He said that the organization had already acquired a good deal of strength by presenting a long list of members. He outlined the objects of the association and gave its raison d'etre, stating that it had in immediate view the discontinuing of the wholesale houses selling retail, and particularly selling retail at wholesale prices. He instanced the case of a man going to a wholesale fur establishment on St. Paul street and asking for a fur coat; the first question put to him was: "Who are you; are you in business?" The reply coming in the negative, he was politely referred to a retailer. "Why should that not be done in the hardware business?" asked Mr. Martineau. He added that they did not expect to accomplish their objects in a day, but he felt sure that ways and means would be found of alleviating their grievances.

The secretary read a letter from Jenkins and Hardy, the secretaries of the Tarred Paper Manufacturers, stating that the request of the Retail Hardware Association had been laid before the manufacturers and that the latter wished to know, first, the names and addresses of the members of the retail association, and, second, what qualities of goods covered by the association were purchased in 12 months.

This gave rise to a lively discussion indulged in by Messrs. Belanger, Prudhomme, Millen, Wilson, Surveyer, Drysdale, Lariviere and Huberdeau. Finally Mr. Surveyer proposed, seconded by Mr. Belanger, that the request for a list of members be complied with and that the manufacturers be asked to estimate the amount of goods sold by the members of the association for themselves, as they are acquainted with the purchases of the Montreal retailers. This was adopted. The list of members now comprises 46 merchants, among whom are most of the important retailers of the city. Although this means more than half of all the hardware merchants in the district of Montreal, there are more names to be added. Here is the list of members as it reads at present:

Martineau, F., 1381 St. Catherine street.
Surveyer, E. J. A., 6 St. Lawrence street.
Prudhomme, A., 1940 Notre Dame street.
Prudhomme, G., 174 Centre street.
Belanger, E., 1213 Notre Dame street.
Drysdale, D., 648 Craig street.
Granger, N., 262 Lagauchetiere street.
Trudel, L., 1999 St. James street.
Denis, L. N., 313 St. Lawrence street.
Cauchon, O., 324 St. Lawrence street.
Denis, J. A., 235 St. Lawrence street.
Huberdeau, J., (H. Lamontagne & Co.) St. Lawrence street.
St. Amour, J. C., 927 Ontario street.
Leblanc, A., Rachel and Rivard streets.
Dufresne & Pratt, 135 St. Paul street.
Desforges & Geoffrion, 123 St. Paul street.
Provost & Baigne, 107-109 St. Paul street.
Granger, U., 1268 Ontario street.
John Millen & Sons 1325 St. Catherine street.
Tremblay, P., 2633 Notre Dame street.
Lavoie, F. U. (Pellascis Hardware Co.) 1901 Notre Dame street.
Amiot, Lecours & Lariviere, 593 St. Lawrence street.
Beland, C. J., 1379 Ontario street.
Leroux, A., 192 St. Antoine street.
Granger, S. J., 691 St Catherine street.
Dufault, A. (Ed. Cavanagh & Co.) 2453 Notre Dame street.
Duval, J. A., 1313 St. Catherine street.
Watts & Mailhot, 1031 Ontario street.
Noiseaux, L. N. and J. E., 2157 Notre Dame street.
Granger, W., 1192 Ontario street.
Marceau, U. A., 3595 Notre Dame street.
Sylvestre & Fils, H., 701 St. Lawrence street.
Shea, J. A., 991 St. Catherine srteet.
Magnan Freres, 306 St. Lawrence street.
Jubinville, P., 352 Rachel street.
Chausse, W., 947 St. Lawrence street.
Ponton, L., 325 Notre Dame street.
Couillard, L., 3174 Notre Dame street.
Papineau, E. J., 1123 St. Lawrence street.
Colleret, E. D., 25 St. Lawrence street.
Papineau, Z., 3293 Notre Dame street.
Wilson, A. A., 219 St. Paul street.
Young, Geerge, 1888 St. Catherine street.
L'Allemand, A., 2721 Notre Dame street.
James Walker Hardware Co., Limited, 234 St. James street.

James Robertson & Co. sent a letter to the secretary since the last meeting, saying they were in hearty sympathy with the association. This was read and received with appreciation. The reading of Messrs. Robertson's letter led to a discussion upon the association's relationship with the wholesalers and manufacturers. It was pointed out that the retailers wish the wholesalers nothing but good, but that they have some grievances which they would like the wholesalers to alleviate. The members of the association would like to meet the wholesalers and have a talk over the matter. It was decided that conferences were absolutely necessary and Mr. Lariviere's motion, seconded by Mr. Wilson, that the paper manufacturers be asked to attend the next meeting to discuss their common interests, was adopted.

Mr. Beland moved, seconded by Mr. Mailhot, that the different members of one firm should be considered as units, having only one vote per concern and paying only one fee, any employe of a house to have a right to represent it if he possess credentials to that effect.—Adopted.

Mr. Beland proposed and Mr. Gideon Prudhomme seconded a motion authorizing the secretary to send a list of members to all wholesale houses asking for such information.

Treasurer A. Prudhomme submitted a financial report to the meeting showing the finances to be in a highly-satisfactory condition, a balance of $109.10 remaining on hand.

Mr. Huberdeau brought forward a motion, which was seconded by Mr. Gideon Prudhomme, thanking Messrs. Martineau and Beland for the work done on behalf of the association in visiting the merchants of the city and authorizing the treasurer to pay their expenses.

"Not their champagne," interrupted Mr. Prudhomme, but, in spite of the objection, the motion passed with applause.

The next meeting will be held on Wednesday, February 6, when it is hoped that the tarred paper manufacturers will be present to confer.

THE STEEL INDUSTRY.

A lecture was given in the rotunda of the Toronto Board of Trade on Thursday evening by Mr. Walter Kennedy, mechanical engineer of the Cramp Ontario Steel Co., Limited, on "The Steel Industry of America." The lecture was under the auspices of the Canadian Manufacturers' Association, whose president, Mr. P. W. Ellis, was in the chair.

PERSONAL MENTION.

Mr. John Henderson, of Henderson & Potts, Halifax, was noticed in Montreal last week.

THE DEMAND EXCEEDED THE SUPPLY.

One of the prettiest calendars in the metal and hardware trades this year is that issued by M. & L. Samuel, Benjamin & Co. It represents an old-time coaching scene, and is done in natural colors. It is no wonder the demand has exceeded the supply.

HAD NO USE FOR ASSOCIATIONS.

I HAVE heard of a hardware dealer who joined an association and then resigned. He explained the situation in this way : "I became a member and attended the annual meeting. There were a lot of good fellows present and we had a pleasant time, especially at the dinner, though I couldn't see why they didn't put it on the table all at once instead of bringing on parts at a time and making you hungry before the next dish came along. Still, it was a good, square meal and my ticket didn't cost me anything, so I enjoyed and thought the whole show was a success. But I figured up that the trip cost me pretty near $18 and I haven't been able to see that I've sold any more goods just because I spent that money and joined the association. If you can show me that I'll get any of that money back by keeping on as a member, why, I'll do it, of course, but I don't see how I can make more money unless I sell more goods, and I'm plagued certain that the association hasn't helped me to sell a dollar's worth. Why, the other members are all dealers and do not have to buy my goods, so what's the use of my mixing up with people and spending money when they won't become my customers?"

There isn't a bit of use, Mr. Amos Bach ! You're the kind of a man who puts the brake on a wagon going up hill, for fear the horses will have too easy a job and won't earn their oats. Keep out of the association until you have had a surgical operation peformed on your head, so that a few modern and progressive ideas can get in and stay. The association doesn't want you, anyway, but it does want live men of advanced and advancing thought, and when I attend or hear from the meetings of the next two or three months I expect to learn that all these men are active members but that the dead ones have been buried. —Exchange.

CAT TAILS.

Editor HARDWARE AND METAL,—In answer to an inquiry of your correspondent, which appears in HARDWARE AND METAL of January 5, asking where he can find a market for cat tails, I beg to refer the inquirer to Messrs. Gum Choo & Wun Lung, Chinatown, British Columbia. They keep the "Yellow Dog" restaurant, a celestial feeding joint familiarly known locally as the "Stuffed Pup." Their patrons, it is said, will consume all the tails of the feline species which your Hartford correspondent can furnish. John Pigtail says : "Muchee plefer tailee catte to tailee oxee for soupee."

SKI HI.

P.S.—Manx cats are barred.

CONDENSED OR "WANT" ADVERTISEMENTS.

Advertisements under this heading, 2c. a word each insertion; cash in advance. Letters, figures, and abbreviations each count as one word in estimating cost.

FOR SALE.

TINSMITH AND PLUMBING BUSINESS for sale cheap. A splendid opening in a Nova Scotia town. Address, Jas. R. Gloster, Moncton, N. B. (5)

AGENT WANTED.

WE WANT TO APPOINT AN AGENT IN Canada to handle our well-known and reliable Glaziers' Diamonds. Good connection with the paint and oil branch of the hardware trade necessary. A Shaw & Son, 52 Rahere Street, E.C., London, Eng. (5)

SITUATION WANTED.

A YOUNG MAN WITH BEST REFERENCES, for seven years connected with the hardware business, having a thorough knowledge of book-keeping, would like to make a change. A week's notice solicited. Box 41, HARDWARE AND METAL, Toronto. (4)

J. H. Grout, of J. H. Grout & Co., manufacturers of agricultural implements, etc., Grimsby, Ont., is dead.

MARKETS AND MARKET NOTES

QUEBEC MARKETS.

Montreal, January 18, 1901.

HARDWARE.

THE wholesale houses are devoting more time to getting ready for spring business than they are to pushing sales for immediate shipment. In shipment from stock, trade is quiet, but prospects for the season's business are bright. Prices are now being settled for spring goods, and, while they will be somewhat lower than last year in many lines, the tendency at present seems steady. American travelers have been here this week quoting values for spring, and, while a few lines are lower, the general list stands unchanged, with some lines 5 to 10 per cent. higher. They report trade on the other side to be flourishing, and the tone of the market firm. Steel has been advanced 20c. during the last two weeks, and structural work is about $3 a ton higher. They also report that angles and crowbars have been advanced in price. Locally, prices are not much changed. Stovepipes and stovepipe elbows are lower. The list price of two numbers of Stanley wood rules has been lowered to $10 for No. 66 and $6 for No. 18. Agricultural wrenches have been changed to 60 and 10 and 10 per cent. for the Province of Quebec, and 60 and 10 per cent. for Ontario. "King Cutter" razors are now selling at $13.50 for white handle and $12 50 for black. Screen doors and windows are now quoted at values somewhat below those in vogue last year, the manufacturers not having succeeded in forming a combine. The discounts on augers and bits have been changed, and the new rates will be found in our schedule. Open prices now prevail on coil chain. Shelf goods are moving fairly well and heavy goods very slowly. The rebate of 7½c. per keg on nails is not now being allowed the Montreal retail merchants.

BARB WIRE — A few carlots are selling for spring, but business is not very active. The price is unchanged at $3.20 f.o.b. Montreal in less than carlots.

GALVANIZED WIRE—The same remarks apply to galvanized wire. The feeling is steady and business fair. We quote as follows: No. 5, $4.25; Nos. 6, 7 and 8 gauge $3.55; No. 9, $3.10; No. 10, $3.75; No. 11, $3.85; No. 12, $3.25; No. 13, $3.35; No. 14, $4.25; No. 15, $4.75; No. 16, $5.00.

SMOOTH STEEL WIRE—There is virtually nothing doing, 14 guage hay wire being the only line that is moving at all. The price is $2.80 per 100 lb.

FINE STEEL WIRE—A small trade is passing at 17½ per cent. off the list.

BRASS AND COPPER WIRE — There is always a little demand being experienced. Discounts are 55 and 2½ per cent. on brass, and 50 and 2½ per cent. on copper.

FENCE STAPLES—This line is featureless. We quote: $3.25 for bright, and $3.75 for galvanized, per keg of 100 lb.

WIRE NAILS—Business is quiet so far as carlots are concerned, but a fair number of small 25 and 50 keg shipments are being made at $2.85 for small lots and $2.75 for carlots, f.o.b. Montreal, Toronto, Hamilton, London, Gananoque, and St. John, N.B.

CUT NAILS—Business in this line is confined to small lots, which are moving as the consumers' demands compel pur-

chases. We quote as follows: $2.35 for small and $2.25 for carlots; flour barrel nails, 25 per cent. discount; coopers' nails, 30 per cent. discount.

HORSE NAILS—The demand is fair with the discount unchanged at 50 per cent. on standard, and 50 and 10 per cent. on Acadia.

HORSESHOES—Business is still flourishing and supplies are light. We quote as follows: Iron shoes, light and medium pattern, No. 2 and larger, $3.50; No. 1 and smaller, $3.75; snow shoes, No. 2 and larger, $3.75; No. 1 and smaller, $4.00; X L steel shoes, all sizes, 1 to 5, No. 2 and larger, $3.60; No. 1 and smaller, $3.85; feather-weight, all sizes, $4.85; toe weight steel shoes, all sizes, $5.95 f.o.b. Montreal; f.o.b. Hamilton, London and Guelph, 10c. extra.

POULTRY NETTING—Some business is being done at 50 and 5 per cent. for spring delivery. Travellers are meeting with fair success in this line.

GREEN WIRE CLOTH—The price is steady at $1.35 per 100 sq. ft. with contracts being freely made.

FREEZERS—Most houses are booking orders for ice cream freezers and quite a large number are being sold.

SCREEN DOORS AND WINDOWS—Open prices will prevail on screens during the coming season, manufacturers having failed to come to an agreement. We quote: Screen doors, plain cherry finish, $8.25 per doz.; do. fancy, $11.50 per doz.; windows, $2.25 to $3.50 per doz.

AUGERS AND BITS—The discounts on augers have been changed. The list adopted August 10, 1896, is still in use. Discounts are as follows: Augers, nut, short eye, long eye and boring machine, 60 and 5 per cent.; millwright's and rafting, 37½ per cent.; Thompson's, 32½ per cent., and ship, 12½ per cent. Auger bits, 60 and 5 per cent., and wood boxes, sets of 9, $1.90 net, and sets of 13, $2.50 net; car bits, 45 per cent., and dowel bits, 32½ per cent.

SCREWS—Prices are steady and a sorting up trade is being done. Discounts are as follows: Flat head bright, 85 per cent. off list; round head bright, 80 per cent.; flat head brass, 77½ per cent.; round head brass, 70 per cent.

BOLTS—Trade is confined to sorting-up proportions. Discounts are: Carriage bolts, 65 per cent.; machine bolts, 65 per cent.; coach screws, 75 per cent.; sleigh shoe bolts, 75 per cent.; bolt ends, 65 per cent.; plough bolts, 50 per cent.; square nuts, 4½c. per lb. off list; hexagon nuts, 4¼c. per lb. off list; tire bolts, 67½ per cent.; stove bolts, 67½ per cent.

BUILDING PAPER—Business on spring account is reported as being very successful.

We quote: Dry sheathing, 30c. per roll; cyclone dry do., 42c. per roll; straw do., 30c.; heavy straw do., $1.40 per 100 lb.; I.X.L., dry sheathing, 65c. per roll; cyclone, tarred do., 50c. per roll; tarred ordinary do., 40c. per roll; tarred felt, $1.60 per 100 lb.; ready roofing, 2-ply, 75c. per roll; 3-ply, $1 per roll.

RIVETS — A fair trade has been done in rivets this week. The discount on best iron rivets, section, carriage, and wagon box, black rivets, tinned do., coopers' rivets and tinned swedes rivets, 60 and 10 per cent.; swedes iron burrs are quoted at 55 per cent. off; copper rivets, 35 and 5 per cent. off; and coppered iron rivets and burrs, in 5-lb. carton boxes, are quoted at 60 and 10 per cent. off list.

STOVEPIPES — Quotations on stovepipes and elbows are lower. In stovepipes, 5 and 6 in. are selling at $7 per 100 lengths and 7 in., $7.50 per 100 lengths. We hear that some houses are cutting on elbows; perhaps $1.15 per doz. is a fair quotation.

CUTLERY—Only a small trade is being done in cutlery just now. "King Cutter" razors are now selling at $13.50 for white handles and $12.50 for black handles.

BUILDERS' HARDWARE—Some American travellers have been here this week and quotations are on the whole firm. Locally, agricultural wrenches have been changed to 60 and 10 and 10 per cent. for the Province of Quebec and 60 and 10 per cent. for Ontario. Stanley wood rules are changed in the list price of No. 66, which is now listed at $10, and No. 18, now listed at $6. Other sizes and discounts remain the same. Discount is 75 per cent.

CORDAGE—The cordage market is firm. Manila is quoted at 13½c. per lb. for 7-16 and larger; sisal is worth 9½c. per lb. for 7-16 and larger, and lathyarn 9½c. per lb.

SPADES AND SHOVELS—The prevailing tone of the market seems to be a certain firmness. Discounts are 40 and 5 per cent.

TACKS — Unchanged. A fair trade is passing. We quote: Carpet tacks, in dozens and bulk, blued, 80 and 5 per cent. discount; tinned, 80 and 10 per cent.; cut tacks, blued, in dozens, 75 and 15 per cent. discount.

FIREBRICKS—A winter quiet now prevails in this line. The price is $18.50 to $26, as to brand.

CEMENT—There is no demand being felt. We quote: German, $2.50 to $2.65; English, $2.40 to $2.50; Belgian, $1.90 to $2.15 per bbl.

METALS.

The metal market is quiet, but the undertone seems to be firm.

PIG IRON — The pig iron market is quiet and somewhat depressed at the moment. Canadian pig is worth $18 to $20, and Summerlee $24 to $25.

BAR IRON—The market is steady at the moment. The ruling price is $1.65 to $1.70.

BLACK SHEETS — The inquiry is small at $2.80 for 8 to 16 gauge.

GALVANIZED IRON—Some import orders for spring are being taken. We quote: No. 28 Queen's Head, $5 to $5.10; Apollo, 10¾ oz., $5 to $5.10t. Comet, No. 28, $4.50 with 25c. allowance in case lots.

INGOT COPPER—The ruling price is 17½c.

INGOT TIN—Values continue at the old level in foreign markets, and there is little business doing. Locally, Lamb and Flag is worth 33c.

LEAD — The market is steady with some transactions occurring at $4 65.

LEAD PIPE —There is nothing new to note. We quote: 7c. for ordinary and 7½c. for composition waste, with 15 per cent. off.

IRON PIPE—There is a little moving. The general trend of the market is strong. We quote as follows: Black pipe, ¼, $3 per 100 ft.; ⅜, $3; ½, $3; ¾, $3.15; 1-in., $4.50; 1¼, $6.10; 1½, $7.28; 2-in., $9.75. Galvanized, ½, $4.60; ¾, $5.25; 1-in., $7.50; 1¼, $9.80; 1½, $11.75; 2-in., $16.

TINPLATES—A small business is passing at $4.50 for coke and $4.75 for charcoal.

CANADA PLATE—Dealers seem to be eager to clear out stocks. We quote as follows: 52's, $2.90; 60's, $3; 75's, $3.10; full polished, $3.75, and galvanized, $4.60.

TOOL STEEL—We quote: Black Diamond, 8c.; Jessop's 12c.

STEEL—No change. We quote: Sleigh-shoe, $1.85; tire, $1.95; spring, $2.75; machinery, $2.75 and toe-calk, $2.50.

TERNE PLATES—Business is at a standstill in this line. We quote $8.25.

SWEDISH IRON—Steady at $4.25.

COIL CHAIN—A few small lots are moving. Open prices prevail now. We quote: No. 6, 11½c.; No. 5, 10c.; No. 4, 9½c.; No. 3, 9c.; ¼-inch, 7½c. per lb.; 5-16, $4.60; 5-16 exact, $5.10; ⅜, $4.20; 7-16, $4.00; ½, $3.75; 9-16, $3.65; ⅝, $3.35; ¾, $3.25; ⅞, $3.20; 1-in., $3.15. In carload lots an allowance of 10c. is made.

SHEET ZINC—There has been no change, the ruling price being 6 to 6¼c.

ANTIMONY—Unchanged, at 10c.

GLASS.

Some houses are preparing to take import orders for glass, while others do not intend to adopt this course. The market appears to be in a good condition to buy. We quote: First break, $2; second, $2.10 for 50 feet; first break, 100 feet, $3.80; second, $4; third, $4.50; fourth, $4.75; fifth, $5.25; sixth, $5.75, and seventh, $6.25.

PAINTS AND OILS.

Trade is opening up early and well, and paint manufacturers are extremely well pleased with the volume of business already done. All lines show an improved demand this week. We hear that values for summer delivery of oil are lower this week. As the market is falling, we cannot but advise our readers to defer the making of contracts fo summer supply. There are no changes in prices to report. We quote:—

WHITE LEAD—Best brands, Government standard, $6.75; No. 1, $6.37½; No. 2, $6; No. 3, $5.62½, and No. 4, $5.25, all f.o.b. Montreal. Terms, 3 per cent. cash

DRY WHITE LEAD—$5.75 in casks; kegs, $6.

RED LEAD—Casks, $5.50; in kegs, $5.75.

WHITE ZINC PAINT—Pure, dry, 8c.; No. 1, 6½c.; in oil, pure, 9c.; No. 1, 7½c.

PUTTY—We quote: Bulk, in barrels, $2 per 100 lb.; bulk, in less quantity, $2.15; bladders, in barrels, $2.20; bladders, in 100 or 200 lb. kegs or boxes, $2.35; in tins, $2.45 to $2.75; in less than 100-lb. lots, $3 f.o.b. Montreal, Ottawa, Toronto, Hamilton, London and Guelph. Maritime Provinces 10c. higher, f.o.b. St. John and Halifax.

LINSEED OIL—Raw, 80c.; boiled, 83c., in 5 to 9 bbls., 1c. less; 10 to 20-bbl. lots, open, net cash, plus 2c. for 4 months. Delivered anywhere in Ontario between Montreal and Oshawa at 2c. per gal. advance and freight allowed.

TURPENTINE—Single bbls., 59c.; 2 to 4 bbls., 58c.; 5 bbls. and over, open terms, the same terms as linseed oil.

MIXED PAINTS—$1.25 to $1.45 per gal.

CASTOR OIL—8¾ to 9¼c. in wholesale lots, and ½c. additional for small lots.

SEAL OIL—47½ to 49c.

COD OIL—32½ to 35c.

NAVAL STORES—We quote: Resins, $2.75 to $4.50, as to brand; coal tar, $3.25 to $3.75; cotton waste, 4½ to 5½c. for colored, and 6 to 7½c. for white; oakum, 5½ to 6½c., and cotton oakum, 10 to 11c.

PARIS GREEN—Petroleum barrels, 16¾c. per lb.; arsenic kegs, 17c.; 50 and 100-lb. drums, 17½c.; 25-lb. drums, 18c.; 1-lb. packages, 18½c.; ½-lb. packages, 20½c.; 1-lb. tins, 19½c.; ½-lb. tins, 21½c. f.o.b. Montreal; terms 3 per cent. 30 days, or four months from date of delivery.

SCRAP METALS.

The scrap metal market is quiet and steady. Dealers are paying the following prices in the country: Heavy copper and wire, 13 to 13½c. per lb.; light copper, 12c.; heavy brass, 12c.; heavy yellow, 8½ to 9c.; light brass, 6½ to 7c.; lead, 2¾ to 3c. per lb.; zinc, 2¼ to 2½c.; iron, No. 1 wrought, $13 to $14 per gross ton; No. 1 cast, $13 to $14; stove plate, $8 to $9; light iron, No. 2, $4 a ton; malleable and steel, $4.

PETROLEUM.

Trade has fallen off slightly. We quote: "Silver Star," 15 to 16c.; "Imperial Acme," 16½ to 17½c.; "S.C. Acme," 18 to 19c., and "Pratt's Astral," 19 to 20c.

HIDES.

Trade is in about the same position as last week. We quote: Light hides, 7½c. for No. 1; 6½c. for No. 2, and 5½c. for No. 3. Lambskins, 90c.

MONTREAL NOTES.

American building tools are firm.

Stovepipes and elbows are lower.

Nos. 66 and 18 Stanley wood rules are changed in regard to their listed prices.

It is rumored that the wholesale houses are trying to fix the prices of axes.

"King Cutter" razors are now selling at $13.50 for white handle and $12.50 for black.

Montreal retailers have been deprived of the rebate of 7½c. per keg that has recently been allowed them on cut and wire nails.

ONTARIO MARKETS.

TORONTO, January 18, 1901.

HARDWARE.

THE wholesale hardware trade is not of a particularly interesting character this week. Business can only be termed moderate, and the situation, as far as prices are concerned, is much about the same as a week ago. For prompt shipment the demand is light, most of the orders the travellers are sending being for future delivery. The orders for future delivery are not, however, large, as a rule. Practically all the business that is being done at the moment in fence wire is for future delivery. The feature of the nail trade is the booking of orders for spring delivery, a little business having developed in this particular during the past week. A few orders for future shipment are also being booked in harvest tools, spades and shovels, and churns. A little business has been done during the week in skates and sleigh bells. Payments are rather slow.

BARB WIRE—A few small lots have been booked for future delivery, but practically nothing is being done in the way of prompt shipment. We quote $2.97½ f.o.b. Cleveland for less than carlots, and $2.85 in carlots. From stock, Toronto, $3.10 per 100 lb.

GALVANIZED WIRE—Practically nothing is being done. We quote: Nos. 6, 7 and

8, $3.55; No. 9, $3.10; No., 10, $3.75; No. 11, $3.85; No. 12, $3.25; No. 13, $3.35; No. 14, $4.25; No. 15, $4.75, and No. 16, $5.

SMOOTH STEEL WIRE — A few small orders are being booked in oiled and annealed wire for future delivery, but nothing scarcely is being done in the way of prompt shipment. A little business has been done in hay-baling wire during the week. The base price is unchanged at $2.80 per 100 lb.

WIRE NAILS — Orders for future delivery have been taken during the past week, although not a great many. Business for immediate requirements is also light. Base price is unchanged at $2.85 per keg for less than carlots and $2.75 in carlots.

CUT NAILS—Immediate business is still small. For future delivery a little business is being done. The base price remains at $2.35 per keg.

HORSESHOES — Busines continues quiet at unchanged prices. We quote as follows f.o.b. Toronto: Iron shoes, No. 2 and larger, light, medium and heavy, $3.60; snow shoes, $3.85; light steel shoes, $3.70; featherweight (all sizes), $4.95; iron shoes, No. 1 and smaller, light, medium and heavy (all sizes), $3.85; snow shoes, $4; light steel shoes, $3.95; featherweight (all sizes), $4.95.

HORSE NAILS—Business is moderate. Discount, 50 per cent. on standard oval head and 50 and 10 per cent. on Acadia.

SCREWS—Business is fair at the recent reduction. We quote as follows: Flat head bright, 85 per cent. off the list; round head bright, 80 per cent.; flat head brass, 77½ per cent.; round head brass, 70 per cent.; flat head bronze, 70 per cent.; round head bronze, 65 per cent.

BOLTS AND NUTS—A fairly good trade is being done in bolts, and there is some talk of an attempt being made to raise prices. We quote: Carriage bolts (Norway), full square, 70 per cent.; carriage bolts, fulls quare, 70 per cent.; common carriage bolts, all sizes, 65 per cent.; machine bolts, all sizes, 65 per cent.; coach screws, 75 per cent.; sleighshoe bolts, 75 per cent.; blank bolts, 65 per cent.; bolt ends, 65 per cent.; nuts, square, 4½c. off; nuts, hexagon, 4¼c. off; tire bolts, 67½ per cent.; stove bolts, 67½; plough bolts, 60 per cent.; stove rods, 6 to 8c.

RIVETS AND BURRS—These remain quiet. Discount, 60 and 10 per cent. on iron rivets; iron burrs, 55 per cent.; copper rivets and burrs, 35 and 5 per cent.

ROPE—The demand for rope keeps quiet. The hemp markets rule steady to firm. We quote as follows: Sisal, 9c. per lb. base, and manila, 13c. Cotton rope is unchanged as follows: 3-16 in. and larger, 16½c.; 5-32 in., 21½c., and ⅛ in., 22½c. per lb.

CUTLERY — Being between the seasons, business is naturally light.

SPORTING GOODS—Very little is being done.

BUILDING PAPER — Business remains much about the same as it was a week ago. Ready roofing, 3-ply, $1.65 per square; ditto, 2-ply, $1.40 per square. Quotations are f.o.b. Toronto, Hamilton, London.

GREEN WIRE CLOTH—Some business is still being done on spring delivery account. We quote $1.35 per 100 sq. ft.

SKATES—A few of these have been going out during the past week.

CHURNS—Some orders are being booked for spring delivery.

SPADES AND SHOVELS—There is just the usual trade being done for immediate requirements, and a few orders are being booked for future delivery. Discount 40 and 5 per cent.

HARVEST TOOLS — Orders for spring delivery are being booked, and with rather more freedom than immediately preceding the holidays. Discount 50, 10 and 5 per cent.

POULTRY NETTING — Orders for future delivery are being booked with some freedom. Discount on Canadian make 50 and 5 per cent.

CEMENT—There is nothing doing. We nominally quote in barrel lots: Canadian Portland, $2.80 to $3; Belgian, $2.75 to $3; English do., $3; Canadian hydraulic cements, $1.25 to $1.50; calcined plaster, $1.90; asbestos cement, $2.50 per bbl.

METALS.

The metal markets are quiet, and, with the exception of lead, steady.

PIG IRON—Very little business is being done, but prices rule steady. The Canadian furnaces are quoting $17 for No. 2 in 100-ton lots.

BAR IRON—The demand continues fairly brisk at $1.65 to $1.70 per 100 lb.

PIG TIN—The outside markets are at the moment steady, but they have ruled weak, some sharp declines having taken place, particularly in London, Eng. Trade, locally, while not large, is fair for this time of the year. We quote 32 to 33c. as the ruling price.

TINPLATES — Coke plates are 15 to 25c. lower. The demand has improved. We now qnote bright coke plates as follows: I.C., usual sizes, $4.15; I.C., special sizes, base, $4.50; 20 x 28, $8.50.

TINNED SHEETS—The demand is more active than it was.

TERNE PLATES — Trade is quiet and featureless.

BLACK SHEETS—Business is rather quiet. We quote $3.50 per 100 lb.

GALVANIZED SHEETS—Orders are being booked for spring and summer delivery, and business is, on the whole, fairly good. We quote English at $4.75 and American at $4.50.

CANADA PLATE—Dealers are beginning to book for delivery next fall, and quite a few orders have been booked. We quote: All dull, $3.15; half and half, $3.25; and all bright, $3 85 to $4.

IRON PIPE—A fairly good trade is still to be noted. We quote: Black pipe ¼ in., $3.00; ⅜ in., $3 00; ½ in., $3.05; ¾ in., $3.20; 1 in., $4 60; 1¼ in., $6.35; 1½ in., $7.55; 2 in., $10.10. Galvanized pipe is as follows: ½ in., $4.65; ¾ in., $5.35; 1 in., $7.25; 1¼ in., $9.75; 1½ in., $11.25; 2 in., $15.50.

HOOP STEEL—Business is fair in this line.

COPPER — The demand for ingot copper has been active, and in sheet copper a fairly good trade is to be noted. We quote: Ingot, 19 to 20c.; bolt or bar, 23½ to 25c.; sheet, 23 to 23½c.

BRASS—Business is quiet. Discount on rod and sheet 15 per cent.

SOLDER — The demand is quiet. We quote: Half-and-half, guaranteed, 19c.; do., commercial, 18½c.; refined, 18½c., and wiping, 18c.

LEAD — Business is rather quiet. We quote 4¾ to 5c.

ZINC SPELTER—Not much doing. We quote 6 to 6½c. per lb.

ZINC SHEET—A fair business is to be noted. We quote casks at $6.75 to $7, and part casks at $7 to $7.50 per 100 lb.

ANTIMONY—Business is quiet at 11 to 11½c. per lb.

PAINTS AND OILS.

There is a considerable increase in the orders for immediate shipment, but the bulk of orders are, of course, for spring delivery. Jobbers are looking for a big spring trade, though price conditions are quite changed from a year ago. It will be remembered that at this time last year prices were all advancing, and dealers bought ahead freely. This year prices are high now, and some buyers are holding off for lower prices. There is no definite assurance of a decline, however, except in linseed oil, which, in any; case will not reach its low point till May. English oil is being sold for May shipment (which means arrival here about July 1) at 63c. Canadian crushers have placed contracts for seed which will make 67c. a possible price in May. At the moment, all lines are firm and there is little prospect of an immediate change. We quote:

WHITE LEAD—Ex Toronto, pure white lead, $6.87½; No. 1, $6.50; No. 2. $6.12½; No. 3, $5.75; No. 4. $5.37½; dry white lead in casks, $6.

RED LEAD—Genuine, in casks of 560 lb., $5.50; ditto, in kegs of 100 lb., $5.75; No. 1, in casks of 560 lb., $5 to $5.25; ditto, kegs of 100 lb.; $5.25 to $5.50.

LITHARGE—Genuine, 7 to 7½c.

ORANGE MINERAL—Genuine, 8 to 8½c.

WHITE ZINC—Genuine, French V.M., in casks, $7 to $7.25; Lehigh, in casks, $6.

PARIS WHITE—90c.

WHITING — 60c. per 100 lb.; Gilders' whiting, 75 to 80c.

GUM SHELLAC — In cases, 22c.; in less than cases, 25c.

PUTTY — Bladders, in bbls., $2.20; bladders, in 100 lb. kegs, $2.35; bulk in bbls., $2; bulk, less than bbls. and up to 100 lb., $2.15; bladders, bulk or tins, less than 100 lb., $3.

PLASTER PARIS—New Brunswick, $1.90 per bbl.

PUMICE STONE — Powdered, $2.50 per cwt. in bbls., and 4 to 5c. per lb. in less quantity; lump, 10c. in small lots, and 8c. in bbls.

LIQUID PAINTS—Pure, $1.20 to $1.30 per gal.; No. 1 quality, $1 per gal.

CASTOR OIL—East India, in cases, 10 to 10½c. per lb. and 10½ to 11c. for single tins.

LINSEED OIL—Raw, 1 to 4 barrels, 80c.; boiled, 83c.; 5 to 9 barrels, raw, 79c.; boiled, 82c., delivered. To Toronto, Hamilton, Guelph and London, 2c. less.

TURPENTINE—Single barrels, 59c.; 2 to 4 barrels, 58c., to all points in Ontario. For less quantities than barrels, 5c. per gallon extra will be added, and for 5-gallon packages, 50c., and 10 gallon packages, 80c. will be charged.

GLASS.

Trade is not active, though a fair number of orders for spring delivery are noted. We still quote first break locally: Star, in 50-foot boxes, $2.10, and 100-foot boxes, $4; doublediamond under 26 united inches, $6, Toronto, Hamilton and London; terms 4 months or 3 per cent. 30 days.

OLD MATERIAL.

The market is decidedly quiet. Stove cast and No. 1 wrought scrap have declined 5c. 100 lb. Scrap rubber is ½c. lower. We quote jobbers' prices as follows: Agricultural scrap, 55c. per cwt.; machinery cast, 55c. per cwt.; stove cast, 35c.; No. 1 wrought 50c. per 100 lb.; new light scrap copper, 12c. per lb.; bottoms, 10½c.; heavy copper, 12½c.; coil wire scrap, 13c.; light brass, 7c.; heavy yellow brass, 10 to 10½c.; heavy red brass, 10½c.; scrap lead, 3c.; zinc, 2½c; scrap rubber, 6½c.; good country mixed rags, 65 to 75c.; clean dry bones, 40 to 50c. per 100 lb.

PETROLEUM.

A good movement continues. Prices are steady since the advance of ½c. last week. We quote: Pratt's Astral, 17 to 17½c. in bulk (barrels, $1 extra); American water white, 17 to 17½c. in barrels; Photogene, 16½ to 17c.; Sarnia water white, 16 to 16½c. in barrels; Sarnia prime white, 15 to 15½c. in barrels.

COAL.

There is practically a famine in nut size. The delivery of other lines is reduced by the shortage of cars. Prices are unchanged. We quote anthracite on cars Buffalo and bridges: Grate, $4.75 per gross ton and $4.24 per net ton; egg, stove and nut, $5 per gross ton and $4 46 per net ton.

MARKET NOTES.

Scrap rubber has declined ½c. per lb. Stove cast, and No. 1 wrought scrap iron are 5c. per 100 lb. lower.

H. S. Howland, Sons & Co, agents for the "Micmac" hockey stick, report that, notwithstanding the large number they have sold this season, not one complaint has been received. They claim that the "Micmac" will stand more rough usage than any other stick.

The merchants in many of the Manitoba towns and villages have instituted early closing. Moosomin, Man., has recently added to the number. The stores there now close at 6.30 o'clock p.m. every evening except Saturdays and days before holidays.

MANITOBA MARKETS.

WINNIPEG, January 15, 1901.

THE hardware market, as far as Winnipeg is concerned, is remarkably quiet at the time of writing. All the houses are busy stock-taking and preparing for the spring business.

Price list for the week is as follows:

Barbed wire, 100 lb.	$3 45
Plain twist	3 45
Staples	3 95
Oiled annealed wire 10	3 95
" 11	4 00
" 12	4 05
" 13	4 20
" 14	4 35
" 15	4 45
Wire nails, 30 to 60 dy., keg	3 45
" 16 and 20	3 50
" 10	3 55
" 8	3 65
" 6	3 70
" 4	3 85
" 3	4 10
Cut nails, 30 to 60 dy.	3 00
" 20 to 40	3 05
" 10 to 16	3 10
" 8	3 15
" 6	3 20
" 4	3 30
" 3	3 65
Horsenails, 45 per cent. discount.	
Horseshoes, iron, No. 0 to No 1	4 65
No. 2 and larger	4 40
Snow shoes, No. 0 to No. 1	4 90
No. 2 and larger	4 40
Steel, No. 0 to No. 1	4 95
No. 2 and larger	4 70
Bar iron, $2.50 basis.	
Swedish iron, $4.50 basis.	
Sleigh shoe steel	3 00
Spring steel	3 25
Machinery steel	3 75
Tool steel, Black Diamond, 100 lb	8 50
Jessop	13 00
Sheet iron, black, 10 to 20 gauge, 100 lb.	3 50
20 to 26 gauge	3 75
28 gauge	4 00
Galvanized American, 16 gauge	2 54
18 to 22 gauge	4 50
24 gauge	4 75
26 gauge	5 00
28 gauge	5 25
Genuine Russian, lb	12
Imitation " "	8
Tinned, 24 gauge, 100 lb	7 55
26 gauge	8 80
28 gauge	8 00
Tinplate, IC charcoal, 20 x 28, box	10 75
" IX "	12 75
" IXX "	14 75
Ingot tin	35
Canada plate, 18 x 21 and 18 x 24	3 75
Sheet zinc, cask lots, 100 lb	7 50
Broken lots	8 00
Pig lead, 100 lb.	6 00
Wrought pipe, black up to 2 inch	50 an 10 p.c.
" Over 2 inch	50 p.c.
Rope, sisal, 7-16 and larger	$10 00
" ⅜	10 50
" ¼ and 5-16	11 00
Manila, 7-16 and larger	13 50
" ⅜	14 00
" ¼ and 5-16	14 50
Solder	21½
Cotton Rope, all sizes, lb.	16
Axes, chopping	$7 50 to 12 00
" double bitts	12 00 to 18 00
Screws, flat head, iron, bright	75 and 10 p.c.
Round " "	70 p.c.
Flat " brass	70 p.c.
Round " "	60 and 5 p.c.
Coach	57½ p.c.
Bolts, carriage	42½ p.c.
Machine	45 p.c.
Tire	60 p.c
Sleigh shoe	65 p.c.
Plough	40 p.c.
Rivets, iron	50 p.c.
Copper, No. 8	50c. lb.
Spades and shovels	40 p.c.
Harvest tools	50, and 10 p.c.
Axe handles, turned, s. g. hickory, doz.	$2 50
No. 1	1 50
No. 2	1 25
Octagon extra	1 75
No. 1	1 25
Files common	70, and 10 p.c.
Diamond	60
Ammunition, cartridges, Dominion R.F.	50 p.c.
Dominion, C.F., pistol	30 p.c.
" military	15 p.c.
American R.F.	30 p.c.
C.F. pistol	5 p.c.
C.F. military	10 p.c. advance.
Loaded shells:	
Eley's soft, 12 gauge	16 50
chilled, 12 guage	18 00
soft, 10 guage	21 00
chilled, 10 guage	23 00
American, M	16 25
Shot, Ordinary, per 100 lb	6 75
Chilled	7 50
Powder, F.F., keg	4 75
F.F.G.	5 00
Tinware, pressed, retinned	75 and 2½ p.c.
" " plain	70 and 15 p.c.
Graniteware, according to quality	50 p.c.

PETROLEUM.

Water white American	24½c.
Prime white American	23c.
Water white Canadian	21c.
Prime white Canadian	19c.

PAINTS, OILS AND GLASS.

Turpentine, pure, in barrels	$ 68
Less than barrel lots	73
Linseed oil, raw	87
Boiled	90
Lubricating oils, Eldorado castor	25½
Eldorado engine	24½
Atlantic red	27½
Renown engine	41
Black oil	23½ to 25
Cylinder oil (according to grade)	55 to 74
Harness oil	61
Neatsfoot oil	$1 00
Steam refined oil	85
Sperm oil	1 50
Castor oil, per lb.	11½
Glass, single glass, first break, 16 to 25 united inches	2 25
26 to 40, per 50 ft.	2 50
41 to 50	5 50
51 to 60	6 00
61 to 70, per 100-ft. boxes	6 50
Putty, in bladders, barrel lots, per lb.	2½
kegs "	2¾
White lead, pure, per cwt.	7 25
No 1 "	7 00
Prepared paints, pure liquid colors, according to shade and color, per gal.	$1.30 to $1.90

NOTES.

Mr. Falls, buyer for Geo. D. Wood & Co., the well-known hardware dealers, returned from the East last week.

THEIR 47TH ANNUAL REUNION.

The employes of Caverhill, Learmont & Co., Montreal, held their 47th annual dinner at Her Majesty's Cafe last Friday evening, and not only was it well attended, but the evening was a most enjoyable one throughout. Previous to the dinner they were present at the presentation of "Carmen" in the theatre, and shortly before midnight they assembled in the cafe, where an excellent dinner was served. The menu card was a characteristic hardware one.

After dinner the chairman, Mr. James Reid, rose and proposed "The Queen," which was most enthusiastically honored. The other toasts were: "Our Employers," proposed by Mr. J. W. Dowling, and responded to by Mr. Frank Ross Newman; "Our Guests," proposed by Mr. G. H. Cornell, and responded to by Messrs. Wm. Percival, R. W. Garth and Wm. Grose, and "The Ladies," proposed by Mr. J. R. Terrill. Mr. George McGowan proposed the health of the chairman, and Mr. Reid ably responded. The health of Mr. John Gouldthorpe, one of the oldest employes, and who has attended nearly every dinner, was also drunk with eclat. During the evening songs were given by Messrs. R. Platt, Alex. Bain, F. R. Newman, Jack Davidson, Fred Cockburn, Wm. Grose and Archie Macfarlane, and Mr. Dick Terrill gave a whistling solo. A vote of thanks was tendered to Mr. Wilfred Lawson for the able manner in which he had gotten up and carried out the dinner, and the proceedings terminated at an early hour with "God Save the Queen."

Ludger Doucet, sawmiller, Thetford Mines, Que., has been burned out.

HEATING AND PLUMBING

VIEWS ON THE PLUMBING BUSINESS.

By John E. Allen, of Des Moines.

WHAT are the conditions existing in our business to-day? Is there a future for it in the professions or are we on the downward road to obliteration? Are plumbing laws and inspectorship being a benefit to us or the public, and last, but not least, are we the only trade receiving the boasted protection granted by manufacturers and jobbers? Let us reason the matter together and see if our trade justifies us to devote the time and energy to it that is being done, and whether you or I would recommend our sons to continue in our footsteps by sticking to it through thick and thin.

Gentlemen, it's a fact that the business is in worse condition to-day than it has ever been and the future is not very bright. There is no longer any protection to the master plumber in any manner that it may be received. I could not purchase goods from a grocer, jeweller or boot and shoe manufacturer or jobber by still remaining in the plumbing business, having a card and letter heads printed, stating that I carried the goods mentioned above and deserved wholesale prices and purchased sufficient only to supply my family for their immediate wants. I would be refused point blank; this is protection that is protection, and it is right and proper that I should not be allowed to purchase at wholesale prices until there was no doubt in their minds that my intentions were honorable before quoting me prices, much less sell me the goods. Does this protection exist in the plumbing business? No sir, it does not! If an owner desires to furnish plumbing goods at wholesale prices, all he has to do is to have a bursted boiler painted, rent a store that has remained idle for some time with shelving already in, for 30 days purchase a number of cracked washnuts that can be had for the hauling of them away, have a card and letterhead printed and write the manufacturers and jobbers for prices, cash before goods are shipped, and he will not be disappointed in getting all he needs to finish his first and last job of plumbing to the injury and detriment of the legitimate trade. But no one can deny that under our present condition any manufacturer or jobber would call him a plumber and entitled to purchase goods. There is more of this work being done at the present time than has been known in the history of the trade.

Every implement store has in stock, sinks, lead pipe, lead traps, bibbs, check and wastes and iron pipe, bench vise stock and dies, and are allowed to sell to anyone whether in the trade or not, gradually undermining the jobbing part of the business. Under these conditions does it justify anyone to carry a large stock of goods when you can purchase in small quantities just as cheap, and save the interest on your investment, insurance, etc.? You can pick up daily, any newspaper in your large city and see advertisements calling your attention to cheap plumbing goods that your customer is posted on, and on account of the supposedly enormous profit in the business by some hook or crook they manage to get a catalogue and discount sheet. It is wonderful to see the number of plumbing and gas fixture catalogues that are carefully stowed away and treasured by your patrons.

It is a household word with owners of flats, "buy from so-and-so Wrecking Co." How long such methods will continue and then that they will be allowed to progress is the question. No one will doubt but that the situation is a trying one, and calls for immediate action on the part of all the trade before it is too late. I fear though that our business will gradually fall into other hands and before many years. The business will be a side line taken up by other trades with more capital, so that they can carry larger stocks and fill the long felt want of the manufacturer and jobber in our line. I have no desire to prevent any ambitious plumber from starting in the business free from restrictions. But I am strongly against usurpers being allowed to purchase goods at wholesale prices so easily; in fact, it is the easiest business on record to purchase material if you have the cash.

Some remedy certainly should be given us; why not adopt the one successfully carried out by the jobber? How many manufacturers will quote you prices and sell you direct, saving middleman's profit, and if they should, what would be the result?

The past year's experience certainly should prepare us all to settle this important matter once for all in a just and business-like manner, satisfying all and harmony and good fellowship exist forever.

True there are other improvements necessary which relate to ourselves individually, which have puzzled us for years and one is ruinous estimating, or I might truthfully say guessing. Anyone of us certainly would prefer to figure with a competitor who was conservative instead of the competitor who does not need to figure. We then stand a better chance of making a fair profit.—John E. Allen, in Plumbers' Trade Journal.

A CONTESTED CLAIM.

Lessard & Harris, plumbers, and L. Cohen & Son, coal dealers, both of Montreal, are contesting a claim for payment of certain goods which they bought from a clerk by the name of Marsily, in the employ of Mr. B. J. Coghlin. They gave the following account of the transaction: "Marsily left some samples of brass which he said had been consigned to a firm in Sherbrooke, who had refused them on account of quality, and that he was selling for account of the shippers, representing himself as a commercial agent. After 10 days of delay, it was agreed to purchase the goods at their regular market value for the purpose of remitting. The goods were delivered by a carter, who brought instructions from Marsily that he be paid $1.25. An invoice was presented and payment was made by cheque to Marsily's order. The cheque, which was endorsed by Marsily, bears the imprint of a rubber stamp, 'George Marsily, Agence Commerciale.' This closed the matter so far as Messrs. Cohen were concerned. They have since been informed, however, that Marsily told his employer he had sold the goods on time, and that he had obtained Mr. Coghlin's consent to do so. The goods were removed, as has been seen, by an outside carter whom Marsily represented to have been sent by the purchasers. Neither Cohen & Son nor Lessard & Harris were aware of Mr. Coghlin's interest in the goods, and only because so when they were asked for payment. Lessard & Harris have instituted a counter action against Mr. Coghlin."

BIG LIGHTING COMBINE PROPOSED.

According to a despatch from Montreal, Rodolphe Forget, president of the Royal Electric Co., of that city, is promoting a company to be known as the "Lighting and Power Co., of Montreal," with a capital of $25,000,000. It is proposed that this company will absorb Chambly Manufacturing Company, the Royal Electric Company, the Montreal Gas Company, and the Lachine Rapids Hydraulic and Land Company, and thus control the lighting and power of the

Nippers and Plyers

IS THE BEST GOOD ENOUGH FOR YOU?

ASK FOR THE GREEN BOOK.

Smith & Hemenway Co., UTICA DROP FORGE & TOOL CO.,
296 Broadway, New York City.

CANADIAN ADVERTISING is best done by THE E. DESBARATS ADVERTISING AGENCY MONTREAL.

ADVERTISING in WESTERN CANADA

will be Carefully, Efficiently, and Promptly attended to, by

The Roberts Advertising Agency,
WINNIPEG, CANADA.

KNOX HENRY

Heavy Hardware and Metal Broker
Room 220½ Board of Trade, MONTREAL.

HORSE NAILS—"C" Brand Horse Nails—Canada Horse Nail Co.
"BRASSITE" GOODS — Gunn Castor Co. Limited, Birmingham, Eng.

Manufacturers of

Heating Supplies

Pipe Fittings and Headers.
Large Manifolds made to Order.
Steam Traps and Appliances, etc.

The . . .
Jas. Morrison Brass Mfg. Co., Limited
TORONTO.

Standard Paint & Varnish Works
Limited
Makers of High Grade
Varnishes, Japans,
Paints, Colors & Enamels.
Windsor, Ont.

WIRE RODS!

Drawn to Decimal Sizes, Cut and Straightened, in Uniform Sizes. Prompt Shipment.

Chalcraft Screw Co., Limited, Brantford, Ont.

TRADE
NOBLES & HOARE NH LONDON
MARK

Nobles & Hoare.

CORNWALL ROAD STAMFORD STREET.
LONDON, ENG.

Manufacturers of
HIGH-CLASS VARNISHES ONLY

Which can be obtained direct from the works or from the principal Color Dealers in Canada.

LEADER CHURN

New Century Improvements.

FOUR DIFFERENT STYLES:
A—Steel Frame with double reversible Steel Lever.
B—Wood Frame with double reversible Steel Lever.
C—Steel Frame with Crank.
D—Wood Frame with Crank.

Styles A and B may be operated from a sitting or standing position.

Steel Frames and Hoops beautifully ALUMINIZED.
All LEADER CHURNS are equipped with BICYCLE BALL BEARINGS and PATENTED CREAM BREAKERS.
Stands are so constructed that they are particularly strong and rigid, and there is nothing to interfere with the placing of pail in the most convenient position for draining off buttermilk.

It Pays to Have the Best. None are Better Than the Leader.

THE
Dowswell Manufacturing Co.
Limited.
HAMILTON, ONT.
Eastern Agents: W. L. Haldimand & Son, Montreal, Que.

HOW TO SAVE GAS

..USE..

Peebles' Automatic Gas Governors

Gas, Fire and Stove Governor.

House Governor Burner.

Governor for Incandescents.

Mercurial Governor for Fixing at Meter

Sole Manufacturers, D. BRUCE PEEBLES & CO., Tay Works, Edinburgh, Scotland.

city. It is understood that several wealthy capitalists are behind Mr. Forget in his scheme.

MEETING OF BATHTUB MAKERS.

A MEETING of the manufacturers of steel-clad, copper-lined bathtubs, all-steel enameled tubs and solid-copper tubs was held in Buffalo, N.Y., on Thursday, January 10. A rumor was current during the closing part of last year to the effect that higher prices would be announced at this meeting. This, it is believed, will not be the case, owing to the fact that the manufacturers of enameled iron bathtubs have considerably reduced their prices, thus bringing the enameled tubs of the unguaranteed quality and the seconds of the first quality into close competition with steel-clad, copper-lined tubs. It would not surprise the trade if the price of steel-clad tubs was slightly reduced after this meeting.

There is very little profit in the steel-clad tub at the present price, but if the manufacturers intend to keep it in the market they will have to make some reduction, no matter how small, in order to keep the selling price at a safe distance from prices ruling on the lower grade enamelled iron tubs. That there will always be a market for a cheap tub is manifest to everybody, but it is not unlikely that in the near future this market will go to the all-steel painted tub. The present prices of the lower grade enamelled iron tubs and of full weight steel-clad tubs are so close together that the preference will go to the enamelled iron one, and, unless copper takes a drop in price, the steel-clad copper-lined tub may be put entirely out of the market. —Metal Worker, January 12.

SOME BUILDING NOTES.

The Amherst, N.S., council have appointed a committee to consider the advisability of erecting a new gaol.

The towns and villages in Lanark county, Ont., have carried a by-law to have a House of Industry erected in Perth, the county town.

The members of the St. John, N.B., Congregational church have decided to either build a new church or to make extensive repairs to the present one.

TORONTO BUILDING PERMITS.

Building permits have been issued in Toronto to The Wm. Davies Co., Limited, for a store building at the corner of College and Bathurst streets, to cost $3,600, and to W. S. Kellow, for a pair of two-storey and attic residences, near Albany avenue, on Wells street, to cost $5,000.

PLUMBING AND HEATING NOTES.

Joseph Lafrance & Co., plumbers, Montreal, have assigned.

Dupont & Leveille, plumbers, etc., Farnham, Que., have registered partnership.

The Frankford Electric Light Co., Limited, Frankford, Ont., have been incorporated.

Mrs. Felix Gaulin has registered as proprietress of F. Gaulin & Cie, plumbers, Granby, Que.

The Brome Lake Electric Power Co., Knowlton and Waterloo, Que., have been incorporated.

P. F. Moore and S. Walsh have registered partnership under the style of Moore & Co., plumbers, etc., St. John's Nfld.

Paquet & Godbout, St. Hyacinthe, Que., have secured the contract for an addition to the Convent of the Congregation St. Hyacinthe.

The Methodists of Wingham, Ont., have given S. Bennett the contract for building a new church to cost $11,400, exclusive of seats and furnace.

PLUMBING AND HEATING CONTRACTS.

The contracts for plumbing and kindred work in the G.T.R.'s new freight sheds at Montreal, which aroused such keen interest in the trade in the Montreal, Toronto and other centres, have at last been let. The Bennett & Wright Co., Limited, get the contracts for heating, ventilating and plumbing, while the conduit wiring and electric lighting has been given to the Western Electric Co., New York.

A CREDITABLE NEW YEAR NUMBER.

The new Year number of The Plumbing Trade Journal, New York, is one of the best of the kind that has been issued by its publishers. Its cover is of exceptionally artistic design, and is printed and embossed in red, green and gold. In addition to the usual amount of reading matter and illustrations, there are some excellent Christmas stories and poetry written especially for the plumbing trade.

HARDWARE CLERK GETS A COMMISSION.

W. R. H. Dann, one of the seven Canadians to whom commissions in the British army were granted recently upon the advice of the Earl of Minto, is employed in Geo. D. Wood & Co.'s hardware warehouse, Winnipeg. Mr. Dann, who is about 25 years of age, is a native of Ireland, but has resided for many years in Canada.

R. C. Cassady, dealer in agricultural implements, Boissevain, Man., has been succeeded by Owen Bell.

WHAT WE DO IN JANUARY.

GET ready for it first during the week intervening between Christmas' ending and the first day of the new year ; get ready for inventory by measuring rolls of belting, oilcloth, screen wire, poultry netting, rubber hose, leaving a memorandum slip with each line ; count the loose bolts, lag screws, twist drills, hand taps, cap and set screws, finished and semi-finished nuts ; count, weigh and mark and clean up everything that will permit of it and is not of a class constantly depleted by sales each day.

INVENTORY.

In this way it is wonderful how greatly the time of inventory is shortened. By the way, the inventory is never finished until every set of men in each department have submitted with their sheets a memorandum of all shortages for the want book (it is the ideal time of the year to find correctly your wants). White newspaper stock, cut to size, is splendid for purposes of stock-taking, using two men together, one to call off and one to set down ; then all to be handed in for copying in books Nos. 1, 2 and 3, as the case may be—leaving it in good shape for reference as an insurance record, and for the year's buying of new stocks as a guide. While copying, figuring and carrying out is being done in the office, all loose stocks and short goods, such as bolts, screws, nuts, washers, rivets, etc., can be filled up and rearranged for the year's beginning.

As the next best thing after finishing stock-taking would suggest getting at

NEGLECTED AND OVERDUE ACCOUNTS.

They are the bane of every merchant's existence, and at this time, as at no other of the year, very many of them can be collected. Many customers will settle then from whom it is hard to collect later in the year. The temptation is so great to depend for one's financing on those accounts that are easy to get and always certain to come in or to be had for the asking, while the slow ones are let run or neglected. It is surrpising what a little hard work will do at this time of year with a capable man back of the collections. Very many customers will not pay until asked to do so, and on the other hand only await the asking.

Many of us have a large country clientage—hard to get at and particularly hard when the account exists—yet a cleverly-worded letter stating your case and asking for funds in a way that will not antagonize will bring the majority of them on first trial. Threatening letters are no good—they are utterly without value. The old saying that you can coax where you cannot drive is essentially applicable in making collections. After starting the right man or men at this vital part of the month's work, it is well to put the best man in the house for the purpose at preparing the sale of

DEAD STOCK OR UNSALABLE GOODS.

Many of us insist that we do not have it—or but little of it. All of us have more or less, and nothing is lost just now by going over and bringing it out, giving it a table or counter of its own, marking any price on it that will sell it, and instructing every man in the house to make a special effort to see it disposed of. It occurs in different finishes of builders' hardware, locks, knobs, butts, etc., in old and old-style sash locks, sash lifts, escutcheons, cabinet trim, household goods, such as old-time coffee mills, clothes wringers, and various items scattered through the store, and easily found if hunted for.

Large sale cards in plain figures, at half their original value, with a little general effort, will sell every dollar's worth, and you are rid of it for all time.

LINES THAT HAVE NOT PAID

show up with every January invoice, and it becomes a good time to find out why—to either strengthen or change them, or drop them altogether. Perhaps the latter is the safest policy where good, strong efforts have been put forth on the goods without paying results. The same money put into another and better line may bring good profits.

It is a good time, too, to make a stronger effort, so that every month and each week of the coming year may be filled with goods that sell and that make the entire year a busy one—without the two old-time dull seasons supposed to belong to the hardware merchant. It is possible to add season goods, novelties and strong lines that will do this. More and more of it is being done by the merchants each year and the change in the business becomes a most agreeable one to all.

ECONOMIES FOR THE YEAR

may be better placed in January than any other month ; not the economies that come from cutting salaries of worthy employes—that is a false one—but the cutting off and shutting down on the little everyday leaks and expenses. The light and heat bills are nearly always excessive and can generally be improved on. The last few years have shown wonderful changes in methods of heating and lighting. Perhaps your drayage and delivery account (always a large one) can be lessened for the year—even with a growing business.

You may for the time being—for good reasons—have given up discounting your invoices. Make some arrangement to keep it up. No one thing creates as much revenue, no one thing is so abused and so greatly misunderstood. It is not only the fact that in an average store it will pay a good clerk's or bookkeeper's salary for the year, but it is a money investment not made or allowed in any bank. As stated in The Iron Age a number of times, 2 per cent. at 60 days equals an investment of nearly 16 per cent., and 2 per cent. at 30 days an investment of nearly 36 per cent. Everyone does not get the time to carefully go over and check invoices—it means dollars a great many times. There are other and many economies to be found.

THE MONTH OF SPRING CONTRACTS

is at hand in January. Your steel goods are to come in. It's a good time to make a new rack for them. Your bulk seed stock will need attention now, and no other one line pays so well if bought right and put in proper shape for sale. The paint stock is to be gone over and gotten ready, and new colors selected for the early spring months. Quotations are to be sought on and contracts made for lawn mowers, also ice cream freezers and refrigerators. A new and better line of hammocks are to be carried. Then there are lawn swings and lawn seats, screen doors and screen windows, poultry netting and screen wire, water coolers and filters—all to obtain quotations on and contract for. A very vital thing is to see that contracts or season goods get in early—very early—they nearly all have a spring dating, and in many cases you have largely sold and are reordering by the time your competitors' first shipment is in.

CULTIVATING THE TRADE,

old and new, should begin in January. Forms for attractive and well-worded personal letters should be gotten ready, the advertising of previous years improved upon, the reaching of trade that has never been in your house sought for. The manufacturers from whom you buy goods will aid you largely in your "spring cultivation" of trade with fresh cuts, new printed matter, catalogues, folders, vest pocket memoranda, metallic fence signs, food chopper and chafing receipts and books, tool catalogues, etc., which can be had for the asking.

January becomes the pivotal month of the year, and should have much attention. On it largely depends the results of the 11 months that follow—and the year's business. Instead of a month of dullness and rest, it should be a primal one—one full of good schemes for the year. Individual personal work will bring the year of 1901 up to what it should be.—"A Western Merchant," in Iron Age.

1813/ 1 see cur page 121

HENRY ROGERS, SONS & CO.,
SHEFFIELD, ENGLAND

Manufacturers of the celebrated

H.R.S.&Co

UNION JACK

CUTLERY

We make a specialty of

PLATED WARE, FRUIT KNIVES, ETC.

Our Canadian Representative carries a full line of samples.

Canadian Office:

6 St. Sacrament St., MONTREAL

F. A. YORK, Manager.

COOPER'S

ONE-PIECE ELBOWS.

Scheipe's Patent Stove Pipe

Best in the world.

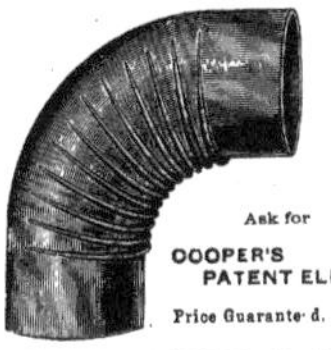

Ask for

COOPER'S PATENT ELBOW.

Price Guaranteed.

E. T. WRIGHT & CO.

Sole Owners. HAMILTON, ONT.

TRADE MARK J REGISTERED.

"JARDINE" HAND DRILLS

Five different sizes to suit all requirements. It pays to sell the best tools.

A. B. JARDINE & CO. HESPELER, ONT.

Valves and Plungers.

Only the very best leather and rubber are used in these goods, and all are carefully and evenly fitted, making them the best of their kind.

Berger Bros. Co.

PHILADELPHIA, U.S.A.

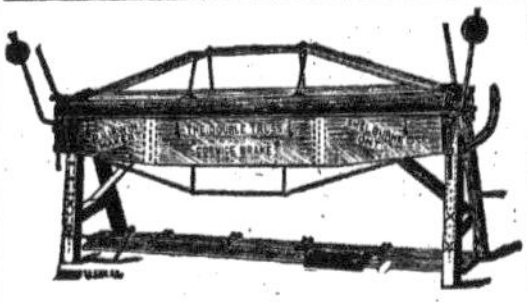

This eight-foot Brake bends 22-gauge iron and lighter, straight and true.

Price, $60

Very handy beader attachment, $15 extra if required.

Send for circulars and testimonials to

The Double Truss Cornice Brake Co. SHELBURNE, ONT.

The Latest and Best.

H. & R. Automatic Ejecting Single Gun.

Model 1900.

Steel and Twist Barrels in 30 and 32-inch.

12 Gauge.

Harrington & Richardson Arms Co.

Worcester, Mass., U.S.A.

Descriptive Catalogue on request.

STEVENS ...FINE TOOLS

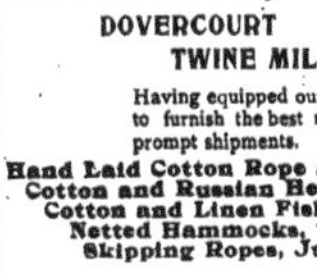

We make a perfect line of

CALIPERS and DIVIDERS

Also such tools as Surface Gauges, Tool Makers' Clamps, Center Punches, etc.

Write for our New Catalogue containing a description of our Tools. It is also a valuable hand-book of information for mechanics and people interested in such lines.

J. Stevens Arms & Tool Co.

P.O. Box 216, Chicopee Falls, Mass, U.S.A.

Carried by our representatives at Toronto and Montreal.

HUTCHISON, SHURLY & DERRETT

DOVERCOURT TWINE MILLS.

1078 BLOOR STREET WEST TORONTO.

Having equipped our Factory with entirely new machinery, we are prepared to furnish the best made goods in the market at closest prices and make prompt shipments.

Hand Laid Cotton Rope and Clothes Lines,
Cotton and Russian Hemp Plough Lines, plain and colored.
Cotton and Linen Fish Lines, laid and braided.
Netted Hammocks, white and colored, Tennis and Fly Nets.
Skipping Ropes, Jute, Hemp and Flax Twines.

THE TWENTIETH CENTURY MAN.

BY T. JAMES FERNLEY, SECY. TREAS. NATIONAL HARDWARE ASSOCIATION.

THE century is dead! Volumes by the thousand could be written in addition to the many thousands of volumes which have already been produced narrating the history of the past 100 years.

You have undoubtedly written to many gentlemen for an expression of views of the century which has passed, and you doubtless have many excellent articles which will be read by the favored subscribers of your organ. I leave it to others to write of the great events of the past century, and simply avail myself of the opportunity to say a few words to those who will be responsible for the making of the history of the twentieth century.

In our city, at the present time, we find many signs on the doors of leading institutions of industry, "Boy Wanted!" "Man Wanted!". We think we see hanging on the great door-knob of the twentieth century an immense placard bearing the words: "Man Wanted!"

We know the kind of men that have been wanted during the past century, but we are not quite sure that the same type of man is the one "wanted" for the twentieth century. Things are beginning to move very rapidly; stage coach and Conestoga wagon days are over; steam is giving out, electricity and compressed air seem to be the forces that will be used in the early days of the twentieth century. The man who is "wanted" may not be a wonderful genius. He certainly should not be a theorist, but must in every way be practical. It will not be so much a question of ability as of availability. By this we mean, the power to avail one's self of every opportunity to exert the talents which are inborn to the greatest extent and at the right time.

COMMON SENSE.

He must be a man endowed to the fullest extent with what is known as common sense. This quality with an education will be preferable, but if a man is not endowed with common sense, an education, in our opinion, will give emphasis to this lack to the detriment of the individual involved. We once heard of a lawyer who was cross-examining a witness. The lawyer had the reputation of not being particularly bright, although very well educated. A farmer who was noted for his common sense was under cross-examination by this lawyer, and, objecting to the way a certain question was put, the lawyer said to him: "Do you presume to object to my proposition? You, an ignorant farmer, while I graduated at two universities!"

The farmer replied: "That's nothing. I had a calf once that sucked from two cows, and the more it sucked, the greater calf he became."

QUICK DECISIONS NECESSARY.

Now, I would not have you understand that an education is not going to be necessary for the twentieth century man, but I feel that the man who is wanted is one who has common sense, one who knows what to do and when to do it. The man wanted in the twentieth century will be the one who is able to come to quick decisions. We know of many men who are willing, and who have ability, and would be available men were it not for the fact that they move too slowly. They do not come to quick decisions. It is dangerous for them to ride on a train that is run by compressed air—one of those trains that rushes up to the station, where the conductor has his hand on the rope connected with the air valve, even before the train comes to a stop, who calls out "All aboard!" and quickly shuts the gate until it reaches the next station. The twentieth century man, the man who is wanted, must be ready to step aboard as quickly as the train pulls up to the platform, and, if possible, before it comes to a stop.

ENTHUSIASM

The man wanted must be a man brim full of enthusiasm, not too dignified, only dignified to the extent of commanding proper respect, but not of that same dignified condition that he will be in after life leaves his body. We have seen many men in the latter days of the nineteenth century who have ability and would have been available men were it not for the fact that they are as cold as a corpse and as dignified.

The man who answers the call of the twentieth century—"Man Wanted!"—must be one full of enthusiasm; whatever he does must be done under high pressure; the gauge of the reservoir wherein the power is held must ever show a supply equal to any emergency. The twentieth century man must throw himself almost bodily into the work in which he is engaged.

MULES AS AMMUNITION.

We some time ago heard of an instance that happened before the Pacific railroads were built. A company of soldiers was crossing the mountains. Mule power was being used; a mountain howitzer was lashed to the mule's back. The company was attacked by Indians ambushed behind huge rocks. The attack was so sudden that the soldiers had not time to unlimber the guns and get into position. The captain was a young man full of enthusiasm. He whirled the mule around and fired the cannon from the mule's back. So great was the recoil of the cannon that it hurled cannon and mule end over end down the hill towards the rocks where the Indians were ambushed. They fled like sheep when they saw the strange shot coming.

The next day the chief was captured and brought into camp. The young captain asked him why he fled so yesterday when they had lost their gun and the whole party might have been scalped. The old chief straightened himself up and said: "Look at me! Me big Injun. Me no 'fraid little guns. Me no 'fraid big guns, but when white man fires whole mule at Injun me very much 'fraid."

The twentieth century man must be ever ready for emergencies such as confronted this little band of soldiers, and his very enthusiasm will lead him to victory.

BROADNESS.

Another quality which the man answering the call of the twentieth century must possess is that of broadness. This is a term which was very much used in the last days of the nineteenth century. Indeed, the writer has heard it used by some men who are extremely narrow. The broad-gauged man is one who constantly has before him the fact that "There are others." He must concede that the difference between the weight of his brain matter and that of his fellow is not very considerable and that by exchanging ideas with his fellowmen he can be made more efficient and more available for the development of work which the twentieth century demands.

COOPERATION THE KEYNOTE OF THE TWENTIETH CENTURY.

During the past six years I have had rare opportunity to study the characteristics of many men who were engaged in making the commercial history of that era, and I say, without any fear of successful contradiction, that the most successful of these were those who would rank as broad-gauged men. Cooperation will be the keynote of the twentieth century, therefore the man who is "wanted" is the one who will be willing to cooperate with his fellowmen in developing the best thought and plan of action.

It will not be the privilege of the present readers of this journal to review the history of the twentieth century, but we venture that those whose privilege it will be to read The American Artisan in the year 2001 will find that the men who will leave their imprint on the pages of history of the twentieth century will be those who have the qualities to which we have alluded.—American Artisan.

CURRENT MARKET QUOTATIONS

January 18 1901.

These prices are for such qualities and quantities as are usually ordered by retail dealers on the usual terms of credit, the lowest figures being for larger quantities and prompt pay. Large cash buyers can frequently make purchases at better prices The Editor is anxious to be informed at once of any apparent errors in this list as the desire is to make it perfectly accurate.

METALS.

Tin.

Lamb and Flag and Straits— 56 and 28 lb. ingots, per lb.	0 32	0 33

Tinplates.

Charcoal Plates—Bright

M.L.S., equal to Bradley.		Per box
I.C., usual sizes	....	$7 00
I.X., "	....	8 50
I.X.X., "	....	10 00
Famous—		
I.C.	....	7 50
I.X	....	8 50
I.X.X.	....	9 50
Raven & Vulture Grades—		
I.C., usual sizes	....	5 00
I.X., "	....	6 00
I.X.X "	....	7 00
I.X.X.X., "	....	8 00
D.C., 12½x17	....	4 75
D.X.	....	5 50
D.X.X	....	7 50

Coke Plates—Bright

Bessemer Steel—		
I.C., usual sizes	....	4 15
I.C., special sizes, base	....	4 50
20 x 28	....	8 50

Charcoal Plates—Terne

Dean or J. G. Grade—		
I.C., 20x28, 112 sheets	....	8 75
I.X., Terne Tin	....	10 75

Charcoal Tin Boiler Plates.

Cookley Grade—		Per lb.
X X., 14x56, 50 sheet bxs } " 14x60 " } " 14x65, " }	0 07	0 07½

Tinned Sheets.

72x30 up to 24 gauge	0 08	0 08½
" 26 "	0 08½	0 09
" 28 "	0 09	0 09½

Iron and Steel.

Common Bar, per 100 lbs.	1 65	1 70
Refined " "	....	2 15
Horse Shoe Iron "	....	2 05
Hoop steel, 1½ to 3 in. base, extras for smaller sizes	....	3 10
Sleigh Shoe Steel " base	....	2 00
Tire Steel	....	2 00
Machinery iron finish	....	2 05
Cast Steel, per lb	0 00	0 00
Toe Calk Steel	....	2 30
T. Firth & Co's special cast steel, per lb.		0 13
Jessop's Tool Steel	....	0 13

Boiler Tubes.

1½-inch	....	0 21½
2 "	....	0 13½
2½ "	....	0 16
3 "	....	0 17½

Steel Boiler Plate.

¼ inch	....	2 25
3-16 inch	....	2 25
⅜ inch and thicker	....	2 25

Black Sheets.

18 gauge	....	3 10
20 gauge	....	3 10
22 to 24 "	....	3 20
26 "	....	3 30
28 "	....	3 40

Canada Plates.

All dull, 52 sheets	....	3 15
Half polished	....	3 25
All bright	3 85	4 00

Iron Pipe.

Black pipe—		
¼ inch	....	3 00
⅜ "	....	3 00
½ "	....	3 05
¾ "	....	3 20
1 "	....	4 50
1¼ "	....	6 35
1½ "	....	7 55
2 "	....	10 10

2½-6 inch, discount 60 p c.

Galvanized pipe—		
½ inch	....	4 65
¾ "	....	6 35
1 "	....	7 95
1¼ "	....	9 75
1½ "	....	11 25
2 "	....	15 50

Galvanized Sheets.

	G C.	Comet.	Amer.	Queen's Head.,
16 gauge		....	4 15	4 00
18 to 24 gauge	4 25	3 85	4 35	4 25
26 "	4 50	4 10	4 35	4 50
28 "	4 75	4 35	4 60	4 75

28 gauge American equals 26 gauge English.

Chain.

Proof Coil, 3-16 in., per 100 lb	....	....
" ¼ " "	8 00	8 50
" 5-16 " "	5 35	5
" ⅜ " "	4 35	4 85
" 7-16 " "	4 15	4 65
" ½ " "	4 35	4 60
" ⅝ " "	3 85	4 35
" ¾ " "	3 80	4 00

Halter, kennel and post chains,	40 and 50 p.c.
Cow ties	40 p.c.
Stall fixtures	35 p.c.
Trace chain	25 p.c.

Jack chain, iron, single and double, discount 35 p c.

Jack chain, brass, single and double, discount 40 p.c.

Copper.

Ingot

English B. S., ton lots	0 19	0 20
Lake Superior	....	....

Bolt or Bar.

Cut lengths round, ⅜ to ⅞ in.	0 23½	0 25
" round and square 1 to 2 inches	0 23½	0 25

Sheet.

Untinned, 14 oz., and light, 16 oz., 14x48 and 14x60	0 23	0 23
Untinned, 14 oz., and light, 16 oz., irregular sizes	0 23	0 23½

NOTE.—Extra for tinning, 2 cents per pound, and tinning and half planishing 3 cents per pound.

Tinned copper sheets	....	0 26
Planished	....	0 32

Braziers (In sheets.)

4x6ft. 25 to 30 lbs. ea., per lb.	....	0 25½
" 35 to 45 " "	....	0 24½
" 50-lb. and above, "	....	0 23½

Boiler and T. K. Pitts

Plain Tinned, per lb	....	0 28
Spun, per lb	....	0 32

Brass.

Rod and Sheet, 14 to 30 gauge,	15 per cent.	
Sheets, hard-rolled, 2x4	0 24	0 25
Tubing, base, per lb	0 24	0 25

Zinc Spelter

Foreign, per lb	0 06	0 06½
Domestic "	....	....

Zinc Sheet.

5 cwt. casks	6 75	7 00
Part casks	7 00	7 50

Lead.

Imported Pig, per lb	0 04½	0 05
Bar, 1 lb.	0 05½	0 05¾
Sheets, 2½ lbs. sq. ft., by	....	0 05¾
Sheets, 3 to 6 lbs., "	....	0 06

NOTE.—Cut sheets ½ cent per lb. extra

Pipe, by the roll, usual weights per yard, lists at 7c. per lb. and 15 p.c. dis. f.o.b. Toronto.

NOTE.—Cut lengths, net price, waste pipe 8-ft. lengths lists at 7½ cents.

Shot.

Common, $6.50 per 100 lb.; chilled, $7.00 per 100 lb.; buck, seal and ball, $7.50. Discount, 7½ p.c. Prices are f o.b. Toronto, Hamilton, Montreal, St. John and Halifax. Terms 3 per cent. cash, freights equalized on Montreal.

Soil Pipe and Fittings.

Discount, 60 and 10 per cent. on medium and extra heavy, and 60 per cent. on light.

Solder.

	Per lb.	Per lb.
Bar half-and-half, guarant'd	....	0 19
Bar half-and-half, commer'l	....	0 18½
Refined	....	0 18½
Wiping	....	0 18

NOTE.—Prices of this graded according to quantity. The prices of other qualities of solder in the market indicated by private brands vary according to composition.

Antimony.

Cookson's, per lb.	0 11	0 11½

White Lead.

	Per 100 lb.
Pure	6 87½
No. 1 do	6 50
No. 2 do	6 12½
No. 3 do	5 75
No. 4 do	5 37½
Munro's Select Flake White	7 37½
Elephant and Decorators' Pure	7 12½
Brandram's B B. Genuine	8 00
" " Decorative	7 55
" " No. 1	6 85
" " No. 2	6 00

Red Lead.

Genuine, 560 lb. casks, per cwt.	....	$5 50
Genuine, 100 lb. kegs, per cwt.	....	5 75
No. 1, 560 lb. casks, per cwt.	....	5 25
No. 1, 100 lb. kegs, per cwt	....	5 00

White Zinc Paint.

Elephant Snow White	0 08	0 09
Pure White Zinc	0 08	0 09
No. 1	0 06	0 07½
No. 2	0 05	0 06½

Dry White Lead.

Pure, casks	5 75
Pure, kegs	6 25
No. 1, casks	5 50
No. 1, kegs	6 00

Prepared Paints.

In ¼, ½ and 1 gallon tins.

Pure, per gallon	....	1 25
Second qualities, per gallon	....	1 10
Barn (in bbls.)	0 75	0 85
The Sherwin-Williams Paints.	....	1 45
Canada Paint Co's Pure	....	1 25
Toronto Lead & Color Co's Pure	....	1 25
Sanderson Pearcy's Pure	....	1 20
Stewart & Wood's Champion Pure.		1 20

Colors in Oil.

25 lb. tins, Standard Quality.

Venetian Red, per lb.	0 05
Chrome Yellow	0 11
Golden Ochre	0 06
French "	0 05
Marine Black	0 09
" Green	0 09
Chrome "	0 08
French Imperial Green	0 09

Colors, Dry.

Yellow Ochre (J.C.) bbls.	1 35	1 40
Yellow Ochre (J.F.L.S.), bbls	....	2 75
Yellow Ochre (Royal)	1 10	1 15
Brussels Ochre	....	2 00
Venetian Red (best), per cwt.	1 80	1 90
English Oxides, per cwt.	3 00	3 25
American Oxides, per cwt.	1 75	2 00
Canadian Oxides, per cwt.	1 75	2 00
Super Magnetic Oxides, 93p.c.	2 00	2 25
Burnt Sienna, pure, per lb.	....	0 10
" Umber, " "	....	0 10
do Raw	....	0 09
Drop Black, pure	....	0 09
Chrome Yellows, pure	....	0 18
Chrome Greens, pure, per lb.	....	0 12
Golden Ochre	....	0 03½
Ultramarine Blue in 28-lb. boxes, per lb.	0 08	0 24
Fire Proof Mineral, per 100 lb	....	1 00
Genuine Eng. Litharge, per lb	....	0 07
Mortar Color, per 100 lb.	....	1 25
English Vermillion	....	0 80
Pure Indian Red, No. 45 lb.	....	0 80
Whiting, per 100 lb.	....	0 55

Blue Stone.

Casks, for spraying, per lb.	0 07
100-lb. lots, do. per lb.	0 08

Putty.

Bulk in bbls.,	2 00
Bulk in less quantity	2 15
Bladders in bbls.	2 20
Bladders in kegs, boxes or loose	2 35
Bladders in 25-lb. tins	2 45
Bladders in 12½-lb. tins	2 75
Bladders in bulk or tins less than 100lbs	3 00

Varnishes.

In 5-gal. lots.)		Per gal.
Carriage, No. 1	2 90	3 30
" body	8 00	9 00
" rubbing	4 00	5 00
Gold Size, Japan	3 00	3 40
Brown Japan	2 40	2 80
Elastic Oak	2 90	3 30
Furniture, extra	2 40	2 80
" No. 1	1 60	2 00
Hard Oil Finish	2 70	3 10
Light Oil Finish	3 20	3 60
Damar	3 30	3 70
Shellac, white	4 40	4 80
" orange	4 00	4 40
Furniture Brown Japan	1 60	2 00
Black Japan	2 40	2 80
" " No. 1	1 60	2 00

The Imperial Varnish & Color Co's., Limited Elastilite Varnish 1 gal. can, each. $2.00.

Granatine Floor Finish, per gal. $2.00.

Maple Leaf Coach Enamels: Size 1, 60c.; Size 2, 35c.; Size 3, 20c. each.

Linseed Oil.

	Raw.	Boiled.
1 to 4 bbls delivered	$0 82	$0 85
5 to 9 bbls "	81	84

Toronto, Hamilton, London and Guelph 2c. less.

Turpentine.

Single barrel, freight allowed	....	0 59
2 to 4 barrels " "	....	0 58

Castor Oil.

East India, in cases, per lb.	0 10	0 10½
" " small lots	0 10½	0 11

Cod Oil, Etc.

Cod Oil per gal.	0 50	0 55
Pure Olive	....	1 20
" Neatsfoot	....	90

Glue.

Common	0 08½	0 09
French Medal	0 14	0 14½
Cabinet sheet	0 12	0 13
White, extra	0 18	0 20
Gelatine	0 22	0 30
Strip	0 18	0 20
Coopers	0 19	0 20
Huttner	....	0 18

HARDWARE.

Ammunition.

Cartridges.

B. B. Caps. Dom. 50 and 5 per cent.
Rim Fire Pistol, dis. 40 p. c., Amer.
Rim Fire Cartridges, Dom., 50 and 5 p. c.
Central Fire Pistol and Rifle, 10 p.c. Amer.
Central Fire Cartridges, pistol sizes, Dom. 30 per cent.
Central Fire Cartridges, Sporting and Military, Dom., 15 and 5 per cent.
Central Fire, Military and Sporting, Amer. add 5 p.c. to list. B.B. Caps, discount 40 per cent. Amer.
Loaded and empty Shells, "Trap" and "Dominion" grades, 25 per cent Rival and Nitro, net list.
Brass shot Shells, 55 per cent
Primers, Dom., 30 per cent.

Wads — per lb
Best thick white felt wadding, in ¼-lb bags 1 00
Best thick brown or grey felt wads, in ¼-lb. bags 0 70
Best thick white card wads, in boxes of 500 each, 12 and smaller gauges 0 99
Best thick white card wads, in boxes of 500 each, 10 gauge 0 35
Best thick white card wads, in boxes of 500 each, 8 gauge 0 55
Thin card wads, in boxes of 1,000 each, 12 and smaller gauges 0 20
Thin card wads, in boxes of 1,000 each, 10 gauge 0 25
Thin card wads in boxes of 1,000 each, 8 gauge ....
Chemically prepared black edge grey cloth wads, in boxes of 250 each— Per M
11 and smaller gauge 0 60
9 and 10 gauges 0 70
7 and 8 gauges 0 90
5 and 6 gauges 1 10
Superior chemically prepared pink edge, best white cloth wads, in boxes of 250 each—
11 and smaller gauge 1 15
9 and 10 gauges 1 40
7 and 8 gauges 1 65
5 and 6 gauges 1 90

Adzes.

Discount, 20 per cent.

Anvils.

Per lb 10 0 12½
Anvil and Vise combined 4 50
Wilkinson & Co.'s Anvils..lb. 0 09 0 09½
Wilkinson & Co.'s Vices..lb. 0 09¾ 0 10

Augers.

Gilmour's, discount 60 and 5 p.c. off list.

Axes.

Chopping Axes—
Single bit, per doz 6 50 10 00
Double bit, " 12 00 18 00
Bench Axes, 40 p.c.
Broad Axes, 33⅓ per cent.
Hunters' Axes 5 50 6 00
Boy's Axes 5 75 6 75
Splitting Axes 6 50 12 00
Handled Axes 7 00 10 00

Axle Grease.

Ordinary, per gross 5 75 6 00
Best quality 13 00 15 00

Bath Tubs.

Zinc 6 00
Copper, discount 15 p.c. off revised list

Baths.

Standard Enameled.
5½-inch rolled rim, 1st quality 30 00
" " " 2nd " 24 00

Anti-Friction Metal.

"Tandem" A .... per lb. 0 27
" B .... " 0 21
" C .... " 0 11½
Magnolia Anti-Friction Metal, per lb. 0 25

SYRACUSE SMELTING WORKS.
Aluminum, genuine 0 45
Dynamo 0 29
Special 0 25
Aluminum, 99 p.c. pure "Syracuse" .. 0 50

Bells.

Hand.
Brass, 60 per cent.
Nickel, 55 per cent.

Cow.
American make, discount 66⅔ per cent.
Canadian, discount 45 and 50 per cent.

Door.
Gongs, Sargent's 5 50 8 00
" Peterboro', discount 45 per cent.

Farm.
American, each 1 25 3 00

House.
American, per lb 0 35 0 40

Bellows.

Hand, per doz 3 35 4 75
Moulders', per doz 7 50 10 00
Blacksmiths', discount 40 per cent.

Belting.

Extra, 50 and 10 per cent.
Standard, 60 per cent.
No. 1 Agricultural, 60 and 10 p.c.

Bits.

Auger.
Gilmour's, discount 60 and 5 per cent.
Rockford, 50 and 10 per cent.
Jennings' Gen., net list.

Car.
Gilmour's, 47½ to 50 per cent.

Expansive.
Clark's, 40 per cent.

Gimlet.
Clark's, per doz 0 65 0 90
Damond, Shell, per doz 1 00 1 50
Nail and Spike, per gross 2 25 5 20

Blind and Bed Staples.

All sizes, per lb 0 07½ 0 12

Bolts and Nuts. Per cent.

Carriage Bolts, full square, Norway 70
" " full square 70
Common Carriage Bolts, all sizes 65
Machine Bolts, all sizes 65
Coach Screws 75
Sleigh Shoe Bolts 75
Blank Bolts 65
Bolt Ends 65
Nuts, square .... 4½c. off
Nuts, hexagon .... 4¾c. off
Tire Bolts 67½
Stove Bolts 67½
Stove rods, per lb .... 5½ to 6c.
Plough Bolts 60

Boot Calks.

Small and medium, ball, per M 4 25
Small heel, per M 4 50

Bright Wire Goods.

Discount .... 55 per cent.

Broilers.

Light, dis., 65 to 67½ per cent.
Reversible, dis., 65 to 67½ per cent.
Vegetable, per doz., dis. 37½ per cent.
Henis, No. 8, " 6 00
Henis, No. 9, " 7 00
Queen City " 7 50 0 00

Butchers' Cleavers.

German, per doz 6 00 11 00
American, per doz 12 00 20 00

Building Paper, Etc.

Plain building, per roll 0 30
Tarred lining, per roll 0 40
Tarred roofing, per 100 lb 1 65
Coal Tar, per barrel 3 50
Pitch, per 100-lb 0 85
Carpet felt, per ton 45 00

Bull Rings.

Copper, $2.00 for 2½ in. and $1.90 for 2 in.

Butts.

Wrought Brass, net revised list

Cast Iron.
Loose Pin, dis., 60 per cent.

Wrought Steel.
Fast Joint, dis. 60 and 10 per cent.
Loose Pin, dis. 60 and 10 per cent.
Berlin Bronzed, dis. 70, 70 and 5 per cent.
Gen. Bronzed, per pair 0 40 0 65

Carpet Stretchers.

American, per doz 1 00 1 50
Bullard's, per doz .... 6 50 ....

Castors.

Bed, new list, dis. 55 to 57½ per cent.
Plate, dis. 52½ to 57½ per cent.

Cattle Leaders.

Nos. 31 and 32, per gross 50 9 50

Cement.

Canadian Portland 2 80 3 00
English " 3 00
Belgian " 2 75 3 00
Canadian hydraulic 1 25 1 50

Chalk.

Carpenters, Colored, per gross 0 45 0 75
White lump, per cwt 0 60 0 65
Red 0 05 0 06
Crayon, per gross 0 14 0 18

Chisels.

Socket, Framing and Firmer.
Broad's, dis. 70 per cent.
Warnock's, dis. 70 per cent.
P. S. & W. Extra 60, 10 and 5 p.c.

Churns.

Revolving Churns, metal frames—No. 0, $8—No. 1, $8.50—No. 2, $9.00—No. 3, $10.00 No. 4, $12.00—No. 5, $16.00 each. Ditto wood frames—20c. each less than above. Discounts: Delivered from factories, 58 p.c.; from stock in Montreal, 56 p.c. Terms, 4 months or 3 p.c. cash in 30 days.

Clips.

Axle dis. 65 per cent.

Closets.

Plain Ontario Syphon Jet 8 00
Emb. Ontario Syphon Jet 8 50
Fittings 1 25
Plain Teutonic Syphon Washout 4 75
Emb. Teutonic Syphon Washout 5 25
Fittings 1 25
Low Down Teutonic, plain 14 50
" " " embossed 15 00
Plain Richelieu 3 75
Emb. Richelieu 4 00
Fittings 1 25
Low Down Ont. Syphon Jet, plain .. 20 00
" " " " " emb'd. 20 50
Closet connection 1 25
Basins, round, 14 in 0 80
" oval, 17 x 14 in 1 50
" " 19 x 15 in 2 25

Compasses, Dividers, Etc.

American, dis. 62½ to 65 per cent.

Cradles, Grain.

Canadian, dis. 25 to 33⅓ per cent.

Crosscut Saw Handles.

S. & D., No. 3, per pair 17½
" " 5, " 22½
" " 6, " 15
Boynton pattern " 20

Door Springs.

Torrey's Rod, per doz .... (15 p.c.) 2 00
Coil, per doz 0 88 1 60
English, per doz 2 00 4 00

Draw Knives.

Coach and Wagon, dis. 50 and 10 per cent.
Carpenters, dis. 70 per cent.

Drills.

Hand and Breast.
Millar's Falls, per doz. net list.

DRILL BITS.
Morse, dis., 37½ to 40 per cent.
Standard dis. 50 and 5 to 55 per cent.

Faucets

Common, cork-lined, dis 35 per cent.

ELBOWS. (Stovepipe.)
No. 1, per doz 1 40
No. 2, per doz 1 20
Bright, 20c. per doz. extra.

ESCUTCHEONS.
Discount, 45 per cent.

ESCUTCHEON PINS.
Iron, discount 40 per cent.

FACTORY MILK CANS.
Discount off revised list, 40 per cent.

FILES.
Black Diamond, 50 and 10 to 60 per cent.
Kearney & Foote, 60 and 10 p.c. to 60, 10, 10.
Nicholson File Co., 50 and 10 to 60 per cent.
Jowitt's, English list, 25 to 27½ per cent.

FORKS.
Hay, manure, etc., dis., 50 and 10 per cent. revised list.

GLASS—Window—Box Price.

Size United Inches.	Star Per 50 ft.	Star Per 100 ft.	D. Diamond Per 50 ft.	D. Diamond Per 100 ft.
Under 26	2 10	4 00	....	6 00
26 to 40	2 30	4 35	....	6 65
41 to 50	....	4 75	....	7 25
51 to 60	....	5	....	8 50
61 to 70	....	5 35	....	9 75
71 to 80	....	5 75	....	10 50
81 to 85	....	6 50	....	11 75
86 to 90	....	....	..	14 00
91 to 95	..	....	...	15 50
96 to 100	..	....	....	18 00

GAUGES

Marking, Mortise, Etc.
Stanley's dis. 50 to 55 per cent.

Wire Gauges.
Winn's, Nos. 26 to 33, each 1 65 2 40

HALTERS.
Rope, ⅜ per gross ....
" ½ " 9 00
" ⅝ to ¾ 14 00
Leather, 1 in., per doz 3 87½ 4 00
" 1¼ in., " 5 15 5 20
Web, — per doz 1 87 2 45

HAMMERS.

Nail
Maydole's, dis. 5 to 10 per cent. Can. dis. 25 to 27½ per cent.

Tack.
Magnetic, per doz 1 10 1 20

Sledge.
Canadian, per lb 0 07½ 0 08½

Ball Pean.
English and Can., per lb 0 22 0 25

HANDLES.
Axe, per doz. net 1 50 2 00
Store door, per doz 1 00 1 50

Fork.
C. & B., dis. 40 per cent. rev. list.

Hoe.
C. & B., dis. 40 per cent. rev. list.

Saw.
American, per doz 1 00 1 25

Plane.
American, per gross 3 15 3 75

Hammer and Hatchet.
Canadian, 40 per cent.

Cross-Cut Saws.
Canadian, per pair 0 13½

HANGERS. doz. pairs.
Steel barn door 5 85 6 00
Stearns, 4 inch 5 00
" 5 inch 6 50
Lane's covered—
No. 11, 5-ft. run 8 40
No. 11½, 10-ft. run 10 80
No. 12, 10-ft. run 12 60
No. 14, 15-ft. run 21 00
Lane's O.N.T. track, per foot. 4½

HARVEST TOOLS.
Discount, 50 and 10 per cent.

HATCHETS.
Canadian, dis. 40 to 42½ per cent.

HINGES.
Blind, Parker's, dis. 50 and 10 to 60 per cent
Heavy T and strap, 4-in., per lb 0 06½
" " 5-in., " 0 06¼
" " 6-in., " 0 06
" " 8-in., " 0 05¾
" " 10-in., " 0 05½
Light T and strap, dis. 60 and 5 per cent.
Screw hook and hinge—
6 to 12 in., per 100 lbs 4 50
14 in. up, per 100 lbs 3 50
Per gro. pairs.
Spring 12 00

HOES.
Garden, Mortar, etc., dis. 50 and 10 p.c.
Planter, per doz 4 00 4 50

HOLLOW WARE.
Discount, ... 45 and 5 per cent

HOOKS.

Cast Iron.
Bird Cage, per doz 0 50 1 10
Clothes Line, per doz 0 27 0 63
Harness, per doz 0 72 0 88
Hat and Coat, per gross 1 00 3 00
Chandelier, per doz 0 50 1 00

Wrought Iron.
Wrought Hooks and Staples, Can., dis. 47½ per cent.

Wire.
Hat and Coat, discount 45 per cent.
Belt, per 1,000 .... 0 60
Screw, bright, dis. 55 per cent.

HORSE NAILS.
"C" brand 50 p.c. dis. } Oval head.
"M" brand 50 p.c. }
Acadian, 50 and 10 per cent.

HORSESHOES.

Iron Shoes. F.O.B. Montreal	No. 2 and larger.	No. 1. and smaller
Light, medium, and heavy...	3 50	3 75
Snow shoes...	3 75	4 00
Steel Shoes.		
Light...	3 60	3 85
Featherweight (all sizes)...	4 85	4 85

F.O.B Toronto, Hamilton, London and Guelph, 10c per keg additional.

Toe weight steel shoes...... 6 70

JAPANNED WARE.

Discount, 45 and 5 per cent. off list, June 1899.

ICE PICKS.

Star per doz... 3 00 3 25

KETTLES.

Brass spun, 7½ p.c. dis. off new list.
Copper, per lb... 0 30 0 50
American, 60 and 10 to 65 and 5 p.c.

KEYS.

Lock, Can., dis., 45 p.c.
Cabinet, trunk, and padlock, Am. per gross... 60

KNOBS.

Door japanned and N.P., per doz...	1 50	2 50
Bronze, Berlin, per doz...	2 75	3 25
Bronze Genuine, per doz...	6 00	9 00
Shutter, porcelain, F. & L. screw, per gross...	1 30	4 00
White door knobs—per doz.		1 25

HAY KNIVES.

Discount, 50 and 10 per cent.

LAMP WICKS.

Discount, 60 per cent.

LANTERNS.

Cold Blast, per doz...	7 50
No. 3 "Wright's"...	8 50
Ordinary, with O burner...	4 25
Dashboard, cold blast...	9 50
No. 0...	6 00

Japanning, 50c. per doz. extra.

LEMON SQUEEZERS.

	per doz.	
Porcelain lined...	2 20	5
Galvanized...	1 87	3
King, wood...	2 75	2
King, glass...	4 00	4 50
All glass...	1 20	1 30

LINES.

Fish, per gross...	1 05	2 50
Chalk " ...	1 90	7 40

LOCKS.

Canadian, dis. 45 p.c.
Russell & Erwin, per doz... 3 00 3 25

Cabinet.

Eagle, dis. 30 p.c.

Padlock.

English and Am., per doz...	50	6 00
Scandinavian, " ...	1 00	2 40

Eagle, dis. 20 to 25 p.c.

MACHINE SCREWS.

Iron and Brass.

Flat head discount 25 p.c.
Round Head, discount 20 p.c.

MALLETS.

Tinsmiths', per doz...	1 25	1 60
Carpenters', hickory, per doz.	1 25	3 75
Lignum Vitae, per doz...	3 85	5 00
Caulking, each...	60	2 00

MATTOCKS.

Canadian, per doz... 8 50 1 00

MEAT CUTTERS.

American, dis. 25 to 30 p.c.
German, 15 per cent.

MILK CAN TRIMMINGS.

Discount, 25 per cent.

NAILS.

Quotations are:	Cut.	Wire.
2d and 3d...	$3 35	$3 85
3d...	3 00	3 52
4 and 5d...	2 75	3 35
6 and 7d...	2 65	3 20
8 and 9d...	2 50	3 00
10 and 12d...	2 45	2 95
16 and 20d...	2 40	2 90
30, 40, 50 and 60d. (base)...	2 35	2 85

Galvanizing 2c. per lb. net extra.
Steel Cut Nails 10c. extra.
Miscellaneous wire nails, dis. 70 per cent.
Coopers' nails, dis. 30 per cent.
Flour barrel nails, dis. 25 per cent

NAIL PULLERS.

German and American... 1 85 3 50

NAIL SETS

Square, round, and octagon, per gross...	3 38	4 00
Diamond...	12 00	15 00

NETTING.

Poultry, 50 per cent. for McMullen's.

OAKUM.

	Per 100 lb
Navy...	6 00
U. S. Navy...	7 25

OIL.

Water White (U.S.)...	0 16½
Prime White (U.S.)...	0 15½
Water White (Can.)...	0 15
Prime White (Can.)...	0 14

OILERS.

McClary's Model galvan. oil can, with pump, 5 gal., per doz... 0 00 10 00
Zinc and tin, dis. 50, 50 and 10.

Copper, per doz...	1 25	3 50
Brass, " ...	1 50	3 50

Malleable, dis. 25 per cent.

GALVANIZED PAILS.

Dufferin pattern pails, dis. 50 to 50 and 10 p.c.
Flaring pails, discount 45 per cent.
Galvanized washtubs, discount 45 per cent.

PIECED WARE.

Discount 40 per cent. off list, June, 1899.

PICKS.

Per doz... 6 00 9 00

PICTURE NAILS.

Porcelain head, per gross...	1 75	3 00
Brass head " ...	0 40	1 00

PICTURE WIRE.

Tin and gilt, discount 75 p.c.

PLANES.

Wood, bench, Canadian dis. 50 per cent American dis. 50.
Wood, fancy Canadian or American 7½ o 40 per cent.

PLANE IRONS.

English, per doz... 2 00 5 00

PLIERS AND NIPPERS.

Button's Genuine per doz pairs, dis. 37½ 40 p.c.

Button's Imitation, per doz...	5 00	9 00
German, per doz...	0 60	2 60

PLUMBERS' BRASS GOODS.

Impression work, discount, 60 per cent.
Fuller's work, discount 65 per cent.
Rough stops and stop and waste cocks, discount, 60 per cent.
Jenkins disk globe and angle valves, discount, 55 per cent.
Standard valves, discount, 60 per per cent.
Jenkins radiator valves discount 55 per cent.
" " " standard, dis., 60 p.c.
Quick opening valves discount, 50 p.c.

No. 1 compression bath cock...	2 00
No. 4 " " " ...	2 00
No. 7, Fuller's...	2 50
No 4½, " ...	3 00

POWDER.

Velox Smokeless Shotgun Powder.

100 lb. or less...	0 85
1,000 lb. or more...	0 80

Net 30 days.

PRESSED SPIKES.

Discount 25 per cent.

PULLEYS.

Hothouse, per doz...	0 55	1 00
Axle...	0 22	0 33
Screw...	0 27	1 00
Awning...	0 35	2 50

PUMPS.

Canadian cistern...	1 80	3 60
Canadian pitcher spout...	1 40	2 10

PUNCHES.

Saddlers', per doz...	1 00	1 85
Conductors', " ...	9 00	15 00
Tinners' solid, per set...	0 00	0 72
" hollow, per inch...	0 00	1 00

RANGE BOILERS.

Galvanized, 30 gallons...		6 50
" 35 " ...		7 50
" 40 " ...		8 50
Copper, 30 " ...		22 00
" 35 " ...		26 00
" 40 " ...		30 00

Discount off Copper Boilers 10 per cent.

RAKES.

Cast steel and malleable Canadian list 50 and 10 p.c. revised list.
Wood, 25 per cent.

RASPS AND HORSE RASPS.

New Nicholson horse rasp, discount 60 p.c.
Globe File Co.'s rasps, 60 and 10 to 70 p.c.
Heller's Horse rasps, 50 to 50 and 5 p.c.

RAZORS.

	per doz.	
Geo. Butler & Co.'s...	8 00	18 00
Boker's...	7 50	11 00
Wade & Butcher's...	3 60	10 00
Theile & Quack's...	7 00	12 00
Elliot's...	4 00	18 00

REAPING HOOKS.

Discount, 50 and 10 per cent.

REGISTERS.

Discount... 40 per cent.

RIVETS AND BURRS.

Iron Rivets, discount 60 and 10 per cent.
Iron Burrs, discount 55 per cent.
Black and Tinned Rivets, 60 p.c.
Extras on Iron Rivets in 1-lb. cartons, ½c per lb.
Extras on Iron Rivets in ½-lb. cartons, per lb.
Copper Rivets & Burrs, 35 and 5 p.c. dis. and cartons, 1c. per lb. extra, net.
Extras on Tinned or Coppered Rivets ½-lb. cartons, 1c. per lb.
Terms, 4 mos. or 3 per cent. cash 30 days.

RIVET SETS.

Canadian, dis. 35 37½ per cent.

ROPE ETC.

	Sisal.	Manila.
7-16 in. and larger, per lb.	9	13
⅜ in...	10	14
¼ and 5-16 in...		15
Cotton, 3-16 inch and larger		16½
" 5-32 inch...		21½
" ⅛ inch...		22½
Russia Deep Sea...		15½
Jute...		8
Lath Yarn...		9½
New Zealand Rope...		10½

RULES.

Boxwood, dis. 75 and 10 p.c.
Ivory, dis. 37½ to 40 p.c.

SAD IRONS.

	per set
Mrs. Potts, No. 55, polished...	0 70
" No. 50, nickle-plated...	75

SAND AND EMERY PAPER.

Dominion Flint Paper, 47½ per cent.
B & A. sand, 40 and 2½ per cent.
Emery, 40 per cent.

SAP SPOUTS.

Bronzed iron with hooks, per doz... 9 50

SAWS.

Hand Disston's, dis. 12½ p.c.
S. & D., 40 per cent.
Crosscut, Disston's, per ft... 0 35 0 55
S. & D., dis. 35 p.c. on Nos. 2 and 3.
Hack, complete, each... 0 75 2 75
" frame only... .... 0 75

SASH WEIGHTS.

Sectional, per 100 lbs...	2 75	3 00
Solid, " ...	2 00	2 25

SASH CORD.

Per lb... 0 23 0 30

SAW SETS.

"Lincoln," per doz... .... 6 50

SCALES.

B. S. & M. Scales, 45 p.c.
Champion, 65 per cent.
Fairbanks Standard, 35 p.c.
" Dominion, 55 p.c.
" Richelieu, 55 p.c.
Chatillon Spring Balances, 10 p.c.

SCREW DRIVERS.

Sargent's, per doz... 0 65 1 00

SCREWS

Wood, F. H., iron, and steel, 85 p.
Wood R. H., " dis. 80 p.c.
" F. H., brass, dis. 77½ p.c.
Wood, R. H., " dis. 70 p.c.
" F.H., bronze, dis. 70 p.c.
" R.H. " 65 p.c.
Drive Screws, 80 per cent.

Bench, wood, per doz...	3 25	4 00
" iron, " ...	4 25	5 75

SCYTHES.

Per doz. net... 9 00

SCYTHE SNATHS.

Canadian, dis. 45 p.c.

SHEARS.

Bailey Cutlery Co., full nickeled, dis. 60 p.c.
Seymour's, dis. 50 and 10 p.c.

SHOVELS AND SPADES.

Canadian, dis. 40 and 5 per cent.

SINKS.

Steel and galvanized, discount 45 per cent.

SNAPS.

Harness, German, dis. 25 p.c.
Lock, Andrews... 4 50 11 50

SOLDERING IRONS.

1, 1½ lb., per lb...	0 37
2 lb. or over, per lb...	0 34

SQUARES.

Iron, No. 493, per doz...	2 40	2 55
" No. 494, " ...	3 25	3 40

Steel, dis. 50 and 5 to 50 and 10 p.c., rev. list.
Try and bevel, dis. 50 to 52½ p.c.

STAMPED WARE.

Plain, dis., 75 and 12½ p.c. off revised list.
Retinned, dis., 75 p.c. off revised list.

STAPLES.

Galvanized...	0 00	0 00
Plain...	0 00	3 45

Coopers', discount 45 per cent.
Poultry netting staples, 40 per cent.

STOCKS AND DIES.

American dis. 25 p.c.

STONE.

		Per lb.
Washita...	0 28	0 60
Hindostan...	0 06	0 07
" slip...	0 09	0 09
Labrador...	....	0 13
" Axe...	....	0 15
Turkey...	....	0 50
Arkansas...	0 00	1 50
Water-of-Ayr...	0 00	0 10
Scythe, per gross	3 50	5 00
Grind, per ton	15 00	18 00

STOVE PIPES.

Nestable in crates of 25 lengths.

5 and 6 inch Per 100 lengths	7 00
7 inch " "	7 50

ENAMELINE STOVE POLISH.

No. 4—3 dozen in case, net cash	$4 80
No. 6—3 dozen in case, "	8 40

TACKS BRADS, ETC.

	Per cent
Strawberry box tacks, bulk...	75 & 10
Cheese-box tacks, blued...	80 & 12½
Trunk tacks, black and tinned...	85
Carpet tacks, blued...	80 & 5
" " tinned...	80 & 10
" " (in kegs)...	40
Cut tacks, blued, in dozens only...	75 & 15
" ¼ weights...	60
Swedes, cut tacks, blued and tinned—	
In bulk...	80 & 10
In dozens...	75
Swedes, upholsterers', bulk...	85 & 12½
" brush, blued & tinned, bulk...	70
" gimp, blued, tinned and japanned...	75 & 12½
Zinc tacks...	35
Leather carpet tacks...	55
Copper tacks...	50
Copper nails...	52

Trunk nails, black.................65 and 5
Trunk nails, tinned.................65 and 10
Clout nails, blued and tinned....65 and 5
Chair nails 35
Cigar box nails................................ 35
Patent brads 40
Fine finishing................................ 40
Picture frame points 10
Lining tacks, in papers 10
" " in bulk.................. 15
" " solid heads, in bulk.... ... 75
Saddle nails in papers.................... 10
" " in bulk.................... 15
Tufting buttons, 22 line, in dozens only 60
Tin capped trunk nails.................... 15
Zinc glazier's points........................ 5
Double pointed tacks, papers.....90 and 10
" " " bulk 40

TAPE LINES.

English, ass skin, per doz....	2 75	5 00
English, Patent Leather....	5 50	9 75
Chesterman's each..........	0 90	2 85
" steel, each	0 80	8 00

THERMOMETERS.

Tin case and dairy, dis. 75 to 75 and 10 p.c.

TRANSOM LIFTERS.

Payson's per doz.......................... 2 50

TRAPS. (Steel.)

Game, Newhouse, dis. 25 p.c.
Game, H. & N., P. S. & W., 65 p.c.
Game, steel, 72½, 75 p.c.

TROWELS.

Disston's discount 10 per cent.
German, per doz............... 4 75 6 00
S. & D., discount 35 per cent.

TWINES.

Bag, Russian, per lb........	....	0 27
Wrapping, cotton, per lb....	0 22	0 25
Wrapping, mottled, per pack.	0 50	0 60
Wrapping, cotton, 3-ply......		0 20
" " 4-ply......	....	0 26
Mattress, per lb.............	0 33	0 45
Staging, "	0 27	0 35
Broom, "	0 30	0 55

VISES.

Hand, per doz...............	4 00	6 00
Bench, parallel, each........	2 00	4 50
Coach, each..................	6 00	7 00
Peter Wright's, per lb.......	0 12	0 13
Pipe, each....................	5 50	9 00
Saw, per doz	6 50	13 00

ENAMELLED WARE.

White, Princess, Turquoise, Blue and White, discount 50 per cent.
Diamond, Famous, Premier, 50 and 10 p.c.
Granite or Pearl, Imperial, Crescent, 50, 10 and 10 per cent.

WIRE.

Brass wire, 50 to 50 and 2½ per cent. off the list.
Copper wire, 45 and 10 per cent. net cash 30 days, f.o.b. factory.
Smooth Steel Wire, base, $2.80 per 100 lb. List of extras: Nos. 2 to 5, advance 7c. per 100 lb.—Nos. 6 to 9, base—No. 10, advance 7c.—No. 11, 14c.—No. 12, 20c.—No. 13, 35c.—No. 14, 47c.—No. 15, 60c.—No. 16, 75c. Extras net per 100 lb.: Coppered wire, 60c.—tinned wire, $2—oiling, 10c.—special hay-bailing wire, 30c.—spring wire, $1—best steel wire, 75c.—bright soft drawn, 15c.—in 50 and 100-lb. bundles net, 10c.—in 25-lb. bundles net, 15c.—packed in casks or cases, 15c.—bagging or papering, 10c.
Fine Steel Wire, dis. 17½ per cent. List of extras: In 100-lb. lots: No. 17, $5—No. 18, $5.50—No. 19, $6—No. 20, $6.65—No. 21, $7—No. 22, $7.30—No. 23, $7.65—No. 24, $8—No. 25, $9—No. 26, $9.50—No. 27, $10—No. 28, $11—No. 29, $12—No. 30, $13—No. 31, $14—No. 32, $15 No. 33, $16—No. 34, $17. Extras net—tinned wire, Nos. 17-25, $2—Nos. 26-31, $4—Nos. 32-34, $6. Coppered, 5c.—oiling, 10c.—in 25-lb. bundles, 15c.—in 5 and 10-lb. bundles, 25c.—in 1-lb. hanks, 50c.—in ½-lb. hanks, 75c.—in ¼-lb. hanks, $1—packed in casks or cases, 15c.—bagging or papering, 10c.
Galvanized Wire, per 100 lb.—Nos. 6, 7, 8, $3.55 No. 9, $3.10—No. 10, $3.75—No. 11, $3.85 No. 12, $3.25—No. 13, $3.35—No. 14, $4.25—No. 15, $4.75—No. 16, $5.00.
Clothes Line Wire, 19 gauge, per 1,000 feet.............. .... 3 30

WIRE FENCING. F.O.B. Toronto

Galvanized 4 barb, 3¼ and 5 inches apart.............. .... 3 10
Galvanized, 2 barb, 4 and 5 inches apart.............. .... 3 10
Galvanized, plain twist........ .... 3 10
Galvanized barb, f.o.b. Cleveland, $2.97½ in less than carlots, and $2.85 in carlots. Terms, 60 days or 2 per cent. in 10 days.
Ross braid truss cable........ .... 4 50

WIRE CLOTH.

Painted Screen, per 100 sq. ft., net.. 1 35
Terms, 4 months, May 1.; 3 p.c. off 30 days.

WRENCHES.

Acme, 35 to 37½ per cent.
Agricultural, 60 p.c.
Coe's Genuine, dis. 20 to 25 p.c.

Towers' Engineer, each......	2 00	7 00
" S., per doz..........	5 80	6 00
G. & K.'s Pipe, per doz......	....	3 40
Burrell's Pipe, each..........	....	3 00
Pocket, per doz..............	0 25	2 90

WRINGERS.

Leader.......................per doz. $60 00
Royal Canadian.......... " 58 00
Royal American.......... " 50 00
Discount, 45 per cent.; terms 4 months, or 3 p.c. 30 days.

WROUGHT IRON WASHERS.

Canadian make, discount, 40 and 5 per cen

PERSONS addressing advertisers will kindly mention having seen their advertisement in Canadian Hardware and Metal Merchant.

SEND for specimen copy of Phillips' Monthl Machinery Register, containing over 5,000 entries of new and second-hand machinery of every description. The oldest established and most successful medium in the world. Established 35 years for the purpose of introducing those who have machinery for sale, to those who wish to buy, has a circulation of about 50,000 copies per annum, all over the world, and is used for continual reference by a large number of firms. It is consequently a most valuable advertising medium for all engineers and manufacturers. Subscription, 5s. per annum, price per copy, 6d. Sole Proprietor, CHAS. D. PHILLIPS, M.I.M.E., Newport, Mon., England. Telegraphic address "Machinery, Newport, Mon."

IN BUYING——

LINSEED OIL

it is always well to get the purest and best—something you can recommend and guarantee to your customers.

Stewart Bros. & Spencer's

is the best. Name on every barrel. Special quotations for import.

J. WATTERSON & CO.

MONTREAL, Agents for Canada.

Lockerby & McComb

AGENTS IN CANADA

FOR THE

Celebrated P. & B.

Cold Storage Lining

AND

.. Ruberoid Roofing ..

P. S.—Prices on Application.

65 Shannon Street, MONTREAL.

BUSINESS NEWS

of any kind that is of value to business men supplied by our Bureau. We can give you market quotations from any town in Canada, reports from the city markets, stock quotations, etc. You can get commercial news from any Canadian paper through us.

Write us, giving us particulars of what you want and where you want it from, and we will quote you prices by return.

"Clippings from any Canadian paper on any subject."

CANADIAN PRESS CLIPPING BUREAU,

505 Board of Trade Bldg., MONTREAL, QUE.

Telephone Main 1255.

10 Front St. East, Toronto. Telephone 2148.

---

75 YEARS. *ESTABLISHED 1825.* *75 YEARS.*

CELEBRATED HEINISCH SHEARS.

Tailors' Shears,
Trimmers, Scissors,
Tinners' Snips, etc.

ACKNOWLEDGED THE BEST.

R. HEINISCH'S SONS CO.

NEW YORK OFFICE, 90 Chambers St.
NEWARK, N.J., U.S.A.

Not connected with any Shear Combination.

CHAS. F. CLARK, President. JARED CHITTENDEN, Treasurer.

...ESTABLISHED 1849...

BRADSTREET'S

Capital and Surplus, $1,500,000. Offices Throughout the Civilized World.

Executive Offices: Nos. 346 and 348 Broadway, New York City, U.S.A.

THE BRADSTREET COMPANY gathers information that reflects the financial condition and the controlling circumstances of every seeker of mercantile credit. Its business may be defined as of the merchants, by the merchants, for the merchants. In procuring, verifying and promulgating information no effort is spared, and no reasonable expense considered too great, that the results may justify its claim as an authority on all matters affecting commercial affairs and mercantile credit. Its offices and connections have been steadily extended, and it furnishes information concerning mercantile persons throughout the civilized world.

Subscriptions are based on the service furnished, and are available only by reputable wholesale, jobbing and manufacturing concerns, and by responsible and worthy financial, fiduciary and business corporations. Specific terms may be obtained by addressing the Company at any of its offices. **Correspondence Invited.**

—OFFICES IN CANADA—

HALIFAX, N.S.	HAMILTON, ONT.	LONDON, ONT.	MONTREAL, QUE.
OTTAWA, ONT.	QUEBEC, QUE.	ST. JOHN, N.B.	TORONTO, ONT.
VANCOUVER, B.C.	VICTORIA, B.C.	WINNIPEG, MAN.	

THOS. C. IRVING, Gen. Man. Western Canada, Toronto. JOHN A. FULTON, Gen. Man. Eastern Canada, Montreal.

Awarded a Gold Medal at **PARIS EXPOSITION** for superiority. That's proof enough of their quality, and clearly shows that they are the best.

Send for Catalogue and . . . . Price List.

The Bailey

Cutlery Co.

BRANTFORD, ONT.

Dominion Pattern

Cow Tie AND Stall Fixture

The special features of the tie and stall fixture are well shown in the illustration. As will be noticed the chain is very short, which prevents all danger of entanglement with the animal's foot. At the same time the form of the fixture is such that great freedom is allowed to the head. Because of the short chain this tie is much cheaper than the ordinary patterns.

The stall fixture is made from a tough quality of steel and is very strong. Also, owing to its circular cross-section, it is exceedingly rigid. Its simplicity, convenience, cheapness, and ease of attaching make it very popular with cow tie users.

This form of tie and stall fixture are sometimes called Niagara pattern.

American or Flat Link Chain,

for years the standard cow tie chain in "the States," is now rapidly coming in favor in Canada. Its short link, handsome appearance and smooth surface—which cannot injure the animal's neck—make it superior to all other styles of chain for cow ties.

For sale by all Jobbers; manufactured by

ONEIDA COMMUNITY, Limited, NIAGARA FALLS, ONT.

CANADIAN

HARDWARE

AND METAL MERCHANT

The Weekly Organ of the Hardware, Metal, Heating, Plumbing and Contracting Trades in Canada.

VOL. XIII. MONTREAL AND TORONTO, JANUARY 26, 1901. NO. 4

RUBBER & BELTING

"We are very much pleased indeed with the belts purchased, finding them everything that could be expected.
SAWYER & MASSEY CO.,
Hamilton, Ont.

THE
CANADIAN RUBBER Co.
MONTREAL · TORONTO
WINNIPEG

James Cartland & Son

Manufacturers of every description of Limited
CABINET, BUILDERS', FURNISHING AND NAVAL BRASSFOUNDRY
BIRMINGHAM, ENGLAND.

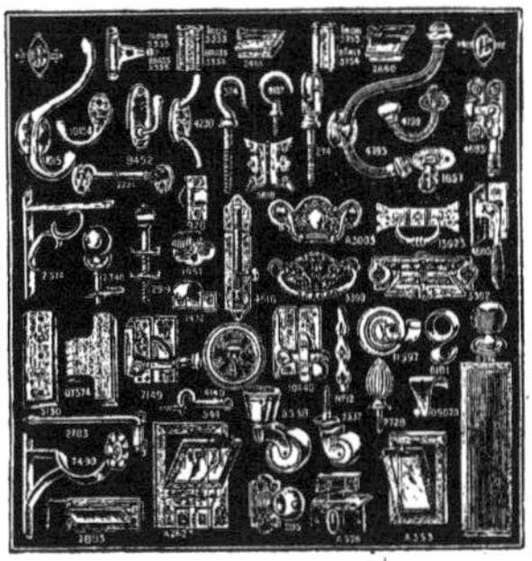

London Showrooms: 57 Holborn Viaduct, E.C.

SOME OF THE NEWER "YANKEE" TOOLS

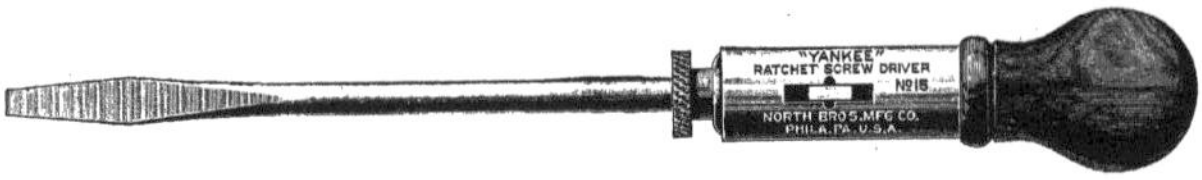

No. 15 "Yankee" Ratchet Screw Driver

RIGHT AND LEFT HAND, AND RIGID, WITH FINGER TURN ON BLADE—2, 3, 4 and 5-in. BLADES.

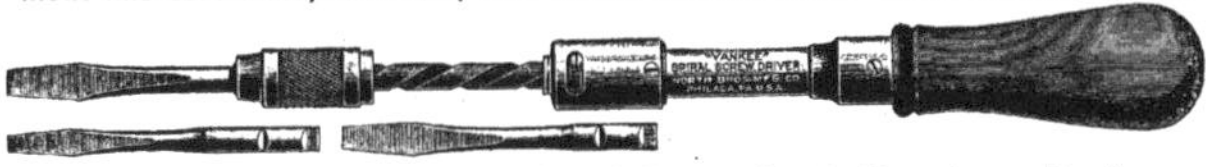

No. 20 "Yankee" Spiral-Ratchet Screw Driver

RIGHT HAND ONLY, AND RIGID. 3 SIZES, EXTREME LENGTH OPEN, INCLUDING BIT—14, 17 and 19-inches

Sold by Leading Jobbers throughout the Dominion.

NORTH BROS. MFG. CO.,

Philadelphia, Pa., U. S. A.

THE SUCCESS of our

IMPERIAL OXFORD

RANGE

ever since we put it on the market, has been enormous.

Its splendid construction and new patented features give it precedence over all others.

THE FRONT DRAW-OUT GRATE
DIFFUSIVE FLUE CONSTRUCTION
DRAW-OUT OVEN RACK

And other improvements, need only to be seen to be appreciated by your customers.

If you haven't them in stock, better write for full information and price list.

There's steady demand for them all over Canada.

THE GURNEY FOUNDRY CO., Limited

TORONTO. WINNIPEG. VANCOUVER.

THE GURNEY-MASSEY CO., LIMITED, MONTREAL.

Steel Frame.

David Maxwell & Sons

ST. MARYS, ONT., CANADA

"Maxwell Favorite Churn"

PATENTED FEATURES: Improved Steel Stand, Roller Bearings, and Foot and Hand Lever Drive, and Detachable Driving Link Improved for season of 1901. Steel or Wood Frame as desired.

Wheelbarrows.

In Four different Sizes.

Lawn Mowers.

High and Low Wheels, from 12-in. to 20-in. widths. Cold Rolled Steel Shafting, Crucible Steel Knives and Cutting Plate.

If your Wholesale House does not offer you these articles

...SEND DIRECT TO US.

MAXWELL MOWER

8-inch Low Wheel.

"THE MAXWELL" Lawn Mower

High Wheel 10 inches

ALEXANDER GIBB

Manufacturers' Agent and Metal Brok·r,
13 St. John Street, Montreal.

Representing British and American Manufacturers. Correspondence invited from firms wishing to be represented in Canada.

McCaskill, Dougall & Co
Fine-Varnish & Japan
Manufacturers.
Montreal
Price Lists on application

DUNDAS AXES

ARE

American in Sharpness,
Canadian in Reliability,

JUST WHAT YOU WANT TO SELL.

Dundas Axe Works

DUNDAS, ONT.

IN
BUSINESS
FOR
MONEY?

WHY DON'T YOU GET

MURALO

THE GREATEST OF ALL

WALL TINTS

It is a sure paying investment. The advertising methods bring you good business. It goes to every country in the world.

AGENTS.

A. RAMSAY & SON, - - - - - - - MONTREAL.
J. H. ASHDOWN, - - - - - - - - WINNIPEG.
McLENNAN, McFEELY & CO., - - VANCOUVER

VanTuyl & Fairbank

Petrolia, Ont.

Headquarters for . . .
Oil and Artesian Well Pumps, Casing, Tubing, Fittings. Drilling Tools, Cables, etc.

Ontario Nut Works, Paris

BROWN & CO.

Manufacturers of
All sizes of Hot Pressed Nuts, Square and Hexagon.

"DAISY" CHURN

Has tempered steel cased bicycle ball bearings, strongest, neatest and most convenient frame. Only two bolts to adjust in setting up. Steel Bow Levers, suitable for either a standing or sitting posture. Has four wheels and adjustable feet to hold stand steady while churning. When churn is locked to stand the bow can be used as handles to move it about on the front wheels as handy as a baby carriage. Open on both sides to centre, giving free space for pail. Made with wood or steel stands, with Cranks only, or Bow Levers as desired.

Vollmar Perfect Washer

Has a most enviable record. A perfection of its kind—will wash more clothes in less time, do it better and easier, with less wear and tear, than any other machine.

THE
Wortman & Ward Mfg. Co.,
Limited

LONDON, ONT.
Eastern Branch, 60 McGill Street, Montreal, Que.

BROWN'S PATENT STEEL WIRE CHAIN.

PATENT NO. 32840.

If you are interested in chains examine carefully the perfect mechanical construction of the Brown's. It is the most perfect chain made. We make it in 13 sizes. We use it exclusively in all our Halter, Dog, Tie-out, Cattle, Trace Chains, etc. You will make no mistake in handling our line exclusively.

THE B. GREENING WIRE CO., LIMITED

Hamilton and Montreal.

HARDWARE

Is not the subject about which we are talking now. The term here refers only to the medium through which we introduce

ALABASTINE

to the trade for the season of 1901. Up-to-date dealers are interested in handling goods that have both a record of value and proof of origin.

ALABASTINE is patented in Canada and other countries; is thoroughly advertised and time tested, and sells on its own merits. Dealers handling **ALABASTINE** run no risk whatever on account of infringement on the rights of others, or that they will have stocks left on their hands that they cannot sell.

Church's **ALABASTINE** is always in demand; is the only permanent and sanitary wall coating known to science.

If offered a substitute, or something "just as good," bear in mind that the party so offering may well be regarded with suspicion, and that a substitute always carries the earmarks of a swindle.

ALABASTINE is sold by leading dealers throughout Canada.

The attention of Contractors, Builders and Building Owners is called to our PARISTONE Wall Plaster and SHIELD BRAND Calcined Plaster. None Better.

Address, **THE ALABASTINE CO., LIMITED,** Paris, Ont.

The trade supplied by—
Sanderson Pearcy & Co., Toronto, Ont.
G. F. Stephens & Co., Winnipeg, Man.
Vancouver Hardware Co., Vancouver, B.C.
Wood, Vallance & Co., Hamilton, Ont.
William Hill, Montreal, P.Q.

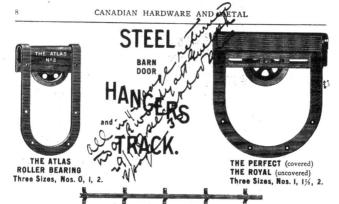

HARDWARE AND METAL

VOL. XIII. MONTREAL AND TORONTO, JANUARY 26, 1901. NO. 4.

President,
JOHN BAYNE MacLEAN,
Montreal.

THE MacLEAN PUBLISHING CO.
Limited.

Publishers of Trade Newspapers which circulate in the Provinces of British Columbia, North-West Territories, Manitoba, Ontario, Quebec, Nova Scotia, New Brunswick, P.E. Island and Newfoundland.

OFFICES

MONTREAL - - - - - - 232 McGill Street. Telephone 1255,
TORONTO - - - - - - 10 Front Street East. Telephone 2148,
LONDON, ENG. - - - - 109 Fleet Street, E.C., J. M. McKim,
MANCHESTER, ENG. - - - 18 St Ann Street, H. S. Ashburner,
WINNIPEG - - - - Western Canada Block, J. J. Roberts,
ST. JOHN, N.B. - - - No. 3 Market Wharf, J. Hunter White,
NEW YORK. - - - - - - 176 E. 88th Street,

Travelling Subscription Agents:
T. Donaghy. F. S. Millard.

Subscription, Canada and the United States, $2.00.
Great Britain and elsewhere - - - 12s.

Published every Saturday.

Cable Address { Adscript, London. Adscript, Canada.

HE WAS A GOOD PRESIDENT.

NO retiring president of the Toronto Board of Trade—for a great many years, at any rate—is more deserving of commendation than Mr. A. E. Kemp, M.P., the senior member of The Kemp Manufacturing Company.

When he assumed office, two years ago, it is true that the board was recovering from the ennui which had characterized it for some time. But Mr. Kemp, putting into action that same energy and business qualities that had proved so potent in the firm of which he is the head, soon caused a rapid multiplication in the membership of the board, and, what is more important still, made it a body whose recommendations and opinions were respected not only in the city of Toronto, but far and wide throughout the Dominion. All the credit is certainly not due to Mr. Kemp. He had good officers at his back and a membership that was beginning to wake up, but the standing of the board of trade in the community would not have been what it is to-day had there been a man in the presidential chair whose qualification for the office were less pronounced than those of Mr. Kemp.

CHANGES IN PRICES.

A FEW further changes in prices have taken place during the past week. While, of course, the general tendency of values in lines appertaining to hardware is downward, there are advances as well as reductions to be reported.

Mrs. Potts sad irons have dropped to 62½c. per set for polished and to 67½c. per set for nickel-plated. The previous quotations were 70 and 75c. respectively.

The discounts have been increased to 45 per cent. on 10-quart flaring sap buckets; 6, 10 and 14 quart I.C. flaring pails and creamer cans. The discount was formerly 40 per cent., at which figure it has ruled for some time.

Iron bench screws, grindstone fixtures and chest handles are all lower in price.

Wringers, as noted elsewhere, are also lower.

Ebony knives and forks are quoted higher by some of the English manufacturers owing to the scarcity of ebony. By this advance the prices of the lines mentioned have been advanced to figures about equal to those of knives and forks with white-bone handles.

We note as well a rise in the price of French mariners' compasses, magnifying glasses and goods of that description.

BUSINESS MEN IN THE SENATE.

BUSINESS men are beginning to multiply. Four men were appointed to the Upper House during the past week, and all are business men, three of whom, by their connection with large and important industries, are widely known in the commercial world.

The four gentlemen are Mr. A. T. Wood, of the wholesale hardware firm of Wood, Vallance & Co., Hamilton, and one of the founders of the Dominion Board of Trade, and one of the originators of the Ontario Cotton Co. Mr. Robert McKay, for many years a member of the firm of Joseph McKay & Bros., and a shareholder in the Montreal Rolling Mills Company, and in the Edwardsburg Starch Co., not to mention other commercial enterprises with which he was connected; Hon. Lyman Jones, general manager of the Massey-Harris Manufacturing Co., Toronto; Mr. George McHugh, an auctioneer in Lindsay, Ont. Messrs. Wood and McHugh have both sat in the House of Commons. Hon. Lyman Jones has been mayor of Winnipeg, and from 1888 to 1889 was Treasurer in the Greenway Administration, Manitoba. Mr. McKay was an unsuccessful aspirant for Parliamentary honors in 1896.

The appointment of business men like these to the Senate raises the ability and morale of that branch of Parliament, and makes less potent opposition to its existence.

The late Sir John A. Macdonald set the example in regard to business men for the Senate, when he appointed the late John Macdonald thereto. Neither Sir John nor his successors strictly adhered to the principle, but the principle is gradually asserting itself with more force, and for that let us be truly thankful.

A TUSSLE OVER FIXED PRICES.

THERE promises to be a lively tussle in Greater New York over the question of a minimum fixed price for patent medicines, which will, no doubt, excite the interest of all business men, whether or not they deal in the articles in question.

For some time a movement has been on foot to secure minimum prices on proprietary medicines, and within the last few days it was announced that 98 per cent. of the retail druggists of Greater New York had signed an agreement to that effect. It was further announced that this agreement was backed by the National Wholesale Druggists' Association and the Proprietary Association of America, both of which had covenanted not to supply retail dealers who refused to subscribe to the agreement.

The following is the schedule of prices as agreed upon by the 98 per cent. of the retail druggists: All 5, 10 and 15c. articles, full price; all 25c. articles, not less than 20c.; all 35c. articles, not less than 25c.; all 50c. articles, not less than 45c.; all 60c. articles, not less than 55c.; all 75c. articles, not less than 65c.; all $1 articles, not less than 85c.; all $1.25 articles, not less than $1.10; all $1.50 articles, not less than $1.25; all $2 articles, not less than $1.75. Infant food and beef extracts are not included in the list. It was decided that the agreement should be operative on Jan. 24.

A similar agreement is in operation in several cities of the United States and, it is claimed, with success.

But an obstacle to the success of the plan has arisen in Greater New York during the last few days. It is none other than the department stores which, through the Retail Dry Goods Association, have notified the promoters of the agreement that they do not propose to subscribe to it.

Nothing daunted, however, a joint committee, representing the manufacturers, the wholesalers and the retailers who are at the back of the agreement, decided on Friday last to put it in operation on January 24, as originally intended. A letter to that effect was sent to all retail druggists, grocers, dry goods, department stores and all handling patent medicines.

Our readers will possibly remember that a few months ago the courts in the United States held that manufacturers or wholesalers could refuse to supply goods to dealers who neglected to comply, in selling them, with the conditions stipulated by said manufacturers or wholesalers. If our memory serves us right it was in regard to the very matter of a fixed price on patent medicines.

Prices have for a long time been slaughtered on proprietary medicines and it is a pity that, through the perverseness of a few, the success of a scheme should be endangered which has for its object the discontinuance of a reprehensible practice. Agreements are distasteful to most people, but it is often necessary to do that which is distasteful in order that something which is more so may be circumvented.

THE DEATH OF THE QUEEN

The death of Her Majesty Queen Victoria and the accession of King Edward VII, are events that the commercial world cannot regard with indifference. The Queen's personality inspired much of the vigor and enthusiasm by means of which British trade and dominion have been extended since 1837. Her pure life and character are bright examples for all engaged in commercial pursuits. Her death is sincerely mourned by all her subjects, and the expressions of sympathy from Boards of Trade, Chambers of Commerce, etc., prove that the illustrious name of Victoria was a reality and a power in business life.

The new King we greet loyally and cordially. He has been a good son, attentive to all his public duties, a man of wide information, travel and knowledge of life. No better King could be found to preside over a great commercial Empire.

BEGAN AT THE WRONG END.

While the reaction against low-priced goods is gathering momentum, there are merchants who do not yet appear to be influenced by it. An instance of this came under our observation a few days ago.

A gentleman entered a retail store and asked for a certain article. He was a man in comfortable circumstances and well known by the clerk, yet, he was first shown the lowest-priced article of the kind in the store, and he had to ask no less than three times for the quality he wanted before it was produced. The clerk began at the wrong end, and he is not the only one who is daily doing the same thing.

A CHANCE FOR BUSINESS MEN.

ONE of the members of the Conservative party is out with a proposition to the effect that the policy of the Opposition during the session of Parliament should be mapped out and controlled by a committee of the party, rather than, as hitherto, the details being left to the leader.

The idea is a good one. Not that we are concerned in the welfare of the Conservative, any more than in the welfare of the Liberal, party. But we see in it an opportunity for the business men of one of the political parties to exercise a greater influence in Parliament than under the present system.

The leader of each of the political parties is usually, at one and the same time, a professional man and a professional politician. Consequently, he is not as well seized of the business requirements of the country as he who is a unit in the commercial world. Now then, with the programme or policy of one or both of the parties controlled by a committee, there is an opportunity for business men being placed on that committee. And the influence of a business man would obviously be greater on such a committee than could be possible under the conditions as they obtain to-day.

A NEW LIST ON WRINGERS.

A new list has been issued by manufacturers of wringers. The purport of them is a reduction in prices. The list is as follows:

Name.	Size. Inches.	Price. Per Doz.
Mogul	14 by 3½	$140
Ajax	14 by 2½	120
Ajax	12 by 2	96
Hamilton	12 by 2	120
Paragon	11 by 1¾	64
Cycle	11 by 1¾	62
Bayside	11 by 1¾	58
Colonial	11 by 1¾	58
Anchor	11 by 1¾	54
Improved Royal Canadian	11 by 1¾	64
Royal Canadian	11 by 1¾	50
Royal Dominion	11 by 1¾	50
Royal American	11 by 1¾	50
Premier	11 by 1¾	48
Novelty	10 by 1¾	44
Novelty	11 by 1¾	46
Novelty	12 by 1¾	48
Handy Bench	11 by 1¾	78
Universal	10 by 1¾	26
Universal	11 by 1¾	40
Crescent	10 by 1¾	32
Crescent	11 by 1¾	36
Star	11 by 1¾	50
New Eureka	11 by 1¾	5[illegible]
Rex	11 by 1¾	48
Eureka	10 by 1¾	4[illegible]
Eureka	12 by 1¾	48
Eureka	11 by 1¾	41
Magic	11 by 1⅝	36
Magic	11 by 1⅝	40
Magic	12 by 1⅝	41
Dux	10 by 1¾	3[illegible]
Dux	11 by 1¾	32
Dux	12 by 1¾	34

No change has been made in the discount, which is still 45 per cent., with the terms 3 per cent. 30 days, or 4 months.

TRADE COMMISSIONER FOR ENGLAND.

AT the last regular meeting of the Executive Committee of the Canadian Manufacturers' Association, Mr. Geo. H. Hees, of the large firm of Geo. H. Hees, Son & Co., Toronto, and chairman of the Tariff Committee of the Association, introduced for discussion a subject of great importance to the manufacturers of Canada; namely, the advisability of urging upon the Government the appointment of a Trade Commissioner for England.

In discussing this matter Mr. Hees spoke as follows:

"I desire to draw the attention of the Manufacturers' Association to a matter of great importance to every manufacturer and shipper in Canada; namely, the advisability of suggesting to the Government the appointment of a Trade Commissioner in England on the same lines as has been already done in Australia.

"Every exporter who has ever tried to find a market for his goods in England has felt the need of some such office as would be connected with a Trade Commissioner in order that he might be supplied with the information that is so necessary. At present he has to go single-handed and alone, groping for customers, and, after he has covered the ground as well as he can, is compelled to leave, feeling that he has left undone much that he might have done had proper facilities been at his disposal, such as would be afforded by a Trade Commissioner acting under the Dominion Government.

"We all know the splendid trade that has developed between Australia and Canada, and we can safely say that 75 per cent. of the business now being done between that country and Canada is due to the zeal and energy of our Trade Commissioner, Mr. Larke.

"An office fitted up in London, to be the headquarters of Canadian exporters, with all the information that is necessary to assist manufacturers and others in securing prospective customers, would undoubtedly meet with success greater in proportion to the much vaster population of the Mother Country."

Mr. Hees then outlined his proposition as follows:

"The appointment of a Trade Commissioner to Britain, with headquarters at London, would be a forward step in the direction of largely increasing the export trade of Canada, and would prove very popular with the manufacturers and producers of this country.

"The appointee should be a Canadian conversant with all sections from the Atlantic to the Pacific.

"To equip and furnish him for his work he should visit all the leading trade centres in the Dominion and meet the various boards of trade and merchants interested in the advancement of Canadian trade.

"An office should be opened in London, furnished entirely with Canadian furniture, carpets, etc., with a sufficient staff to answer all inquiries in regard to Canadian trade matters.

"London being the centre of the world's business, the Commissioner could easily ascertain the possibilities and probabilities of trade between other foreign countries and Canada.

"It would be the duty of the Commissioner to visit trade centres in Britain, such as Liverpool, Manchester, Birmingham, Leeds, Glasgow, Dundee, Aberdeen, Cardiff, Belfast, Dublin, etc., and to come into touch with the various Chambers of Commerce in these cities.

"It should be the object of the Commissioner to assist in bringing merchants in Britain and other foreign countries into close relations with the manufacturers and shippers of Canada, and for this purpose a well-equipped bureau of information should be maintained, the function of which would be to supply any needed information concerning foreign markets, the goods sold therein, the requirements of the markets and the names of the principal buyers in Britain and various parts of the world. Foreign merchants should be furnished with any desired information about goods that are manufactured or produced in Canada. A comprehensive directory of merchants in every part of the world should be kept, with full particulars about the lines of goods they handle, and with information as to whether they are interested in Canadian merchandise or not. These merchants would be brought in direct contact with the manufacturers and producers of Canada. Those who deal in Canadian goods and desire to increase the range of their business in this line, and who wish to be informed concerning Canadian goods which they could sell to advantage, should be invited to make their wants known to the Commissioner, with the assurance that their inquiries would receive prompt and careful attention.

"Samples of any merchandise wanted might be sent to the Commissioner. These samples could be placed in the hands of Canadian manufacturers who supply such goods, and would enable them to know exactly what is wanted by the buyer, and to submit prices and terms more intelligently.

"The Commissioner would be able to answer inquiries relative to shipping to any foreign countries either via Britain or direct from Canada.

"A trade index of those who manufacture goods suitable for export should be kept as follows:

"1st.—An alphabetical list of manufacturers and merchants, with a brief enumeration of the articles they manufacture and deal in, and other information helpful to the buyer.

"2nd.—The names of manufacturers and merchants grouped according to the articles manufactured and dealt in, an arrangement that will be of much assistance to the buyers who wish to find manufacturers and merchants in any particular line.

"3rd.—The registered cable addresses of those whose names are contained in the index."

GEO. H. HEES.

Mr. Hees drew attention to another important matter, of which he spoke as follows:

"I would also suggest that the Association ask the Government to recall Mr. Larke from Australia and post him on present conditions in Canada, as it is six years since he went out to Australia, and he has not since returned to Canada to take note of the great changes that have taken place in that period.

"Notwithstanding that Mr. Larke has been handicapped by his lack of intercourse with the manufacturers and exporters of Canada, he has succeeded in building up an enormous business. But how much more could he do if he returned and met the different exporters and manufacturers in all the various parts of Canada, and obtained from them up-to-date information as to the products which they are prepared to offer for sale. Could he then return to Australia, armed with this up-to-date information, he would have something new and original to present to prospective customers there, and the influence of such personal contact would at once be seen in the large trade which would result.

"We all know that great changes in the business world have taken place during the last six years, especially among manufacturers, and, unless a Commissioner meets with the manufacturers every year, or year and a half, and learns what is going on, he soon becomes obsolete, and from necessity talks ancient history.

"I would further recommend, if we succeed in inducing the Government to appoint a Trade Commissioner in England, that we should have him first become thoroughly posted as to the ability of Canadian firms at present to compete for foreign trade, and that once he has established his office he should return annually to confer with manufacturers and shippers in the various parts of Canada."

This important subject will be discussed by the Commercial Intelligence Committee, and finally dealt with at the next regular meeting of the Executive Committee, on Tuesday, February 12.

TORONTO BOARD OF TRADE.

The following officers have been elected by acclamation by the members of the Toronto Board of Trade:

President—A. E. Ames.
1st Vice-President—W. E. H. Massey.
Treasurer—J. L. Spink.
Harbor Board—W. A. Geddes and J. T. Matthews.

The election of the Council, the Board of Arbitration and the representatives to the Industrial Exhibition will take place on Tuesday week.

GEM JAR PACKAGES.

"THERE were more complaints about broken gem jars last year than in any previous year in our history," said Mr. Fred M. Watt, of the Toronto Glass Co., Limited, to HARDWARE AND METAL one day this week. "It is, therefore, easy to believe that there was more breakage."

"How do you account for the increase?" he was asked.

"I think it was largely, if not altogether due to a new regulation made by the railways at the beginning of the new shipping season last year. They then for the first time insisted that all glass should be sent over their lines 'at owner's risk.' We objected, but to no avail. They insisted on double freight charges unless the owner took the risk of shipment. Our customers, of course, did not want to pay double freight charges and took the risk."

"Why should this clause increase liability to damage?"

"For the simple reason that freight handlers are too human to take as much care with packages their company is not responsible for safe delivery of as if they were responsible. Every freight handler soon knew gem jars were shipped 'owner's risk' and treated them accordingly."

"Is this the only reason you assign for a large breakage?"

"No; not altogether. In sending out shipments we fill a car, when it is attached to a way freight. The jars for the nearest stations are placed in the centre of the car so that they can be easily reached. When half or more of the car has been emptied, there are at each end of the car two piles of boxes. If the car were to get an unusually bad jolt when being shunted about the whole front tier might give way. This may have lead to some of the most severe breakages reported. We will try to overcome that this year by having the cases so arranged that they will not rise from the floor in one abrupt tier at any time, but have them packed so that after every unloading the tiers will be graded in height from front to back."

"Do you not think you could improve the package?"

"I cannot conceive of any means of so doing. You can depend on it that we have spent much time and thought in trying to devise the most satisfactory package. It has been suggested that we increase the height and thickness of the cardboard. But that would be useless—come out to the works and I will show you."

We went out. Mr. Watt picked up the first case he came to in a great pile in the storeroom; then continued: "You see this cardboard comes slightly above where the jar begins to diminish in size. To bring it much above that point would be a sheer waste. There is enough there to keep each jar separate from all the others, and to keep them all from moving in the package. Thicker cardboard than we use would be practically no safer, but would add considerably to the cost of the jars to the merchant. We have tried everything we can conceive of to effect an improvement in our cases, and the present package is, in our opinion, the most practical that has been suggested."

"Do you anticipate as big a loss this year as was the case last season?"

"No. We have gone into the matter with the railways, and they have promised to do everything possible to prevent a recurrence of the troubles of last year. The most important change will be in the packing of goods in the car. If the railway companies give satisfactory service, there should be very little loss through breakage. The 'owner's risk' regulation will remain in force, however."

"Do many customers ask you to assume the risk?"

"A fair number do. But they forget that if we did that we would have to raise prices to meet the loss. If the losses amounted to 5 per cent., then prices would have to be advanced a like proportion. There are a number of customers who think we are, as sellers, bound to deliver our goods in sound condition. These must learn that this is not customary in any line of trade, unless delivery is specifically agreed to. When we deliver our goods to the railway and get our receipt, our responsibility ends. It is then a matter between the railway and the customer. We do our best to send out our goods in the best case to stand a trip in a freight train. We want to find a better package, but have failed, so far, and no suggestion that has been made has proven practicable."

Mr. James Kent, of Gowans, Kent & Co. was also seen by the representative of HARDWARE AND METAL. "While there may be objection to the proportion of gem jars that get broken," he said, "there is no reason for attributing this to the fault of the package. Many kinds have been tried. Up to a few years ago the jars were packed in straw. These packages caused such a storm of criticism because of the muss and dirt made in opening them that the glassmakers set about to construct a more satisfactory case. The present package is certainly the most practical that has yet been used. There is just enough cardboard between the jars for protection without adding any unnecessary cost to the jars.

"You can depend on it that, owing to the number of companies competing for business, if one of them could conceive of a more practical case, they would speedily use it—and the others would speedily follow suit.

"The real cause of the heavy breakage reported has undoubtedly been careless handling by trainmen. The package is small and light and could easily be tossed by one man to another. As some would be dropped, breakage would be the natural result.

"We always find it the case that where a package is small and light, the breakage is heavier than when a larger, heavier package is used, as trainmen are able to use it more carelessly."

DISASTROUS FIRE AT MONTREAL.

The worst fire that has ever visited Montreal took place on Wednesday evening. It started in M. Saxe & Co.'s clothing warehouse in the heart of the wholesale section of the city, and swiftly spread along St. Peter and St. Paul streets, destroying a score or more of important warehouses and the great Board of Trade building. Among the losers are: H. A. Nelson & Sons, fancy goods dealers; Choillan & Co., brokers; Seybold & Sons, wholesale hardware dealers; The St. Lawrence Fence Co.; W. H. De Courtney & Co., wholesale hardware dealers; The Thos. Davidson Manufacturing Co.

In the Board of Trade building, offices of the following were destroyed: Pillow, Hersey & Co.; Peck, Benny & Co.; The Dominion Commercial Travellers' Association; A. McKim & Co., advertising agents; G.T.R. and C.P.R. freight departments.

The office of HARDWARE AND METAL, situated in the Board of Trade building, was also destroyed. This will not, however, cause any interruption in business.

Herbert C. Coy has registered as proprietor of The H. Coy Novelty Co., manufacturers of metal goods, Montreal.

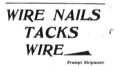

BUSINESS CHANGES.

DIFFICULTIES, ASSIGNMENTS, COMPROMISES.

NOE PAGE, general merchant, Crysler, Ont., has assigned to Daniel Davis, Cornwall, Ont.

C. Primeau, general merchant, St. Urbain, Que., has compromised.

J. A. Plamondon, general merchant, St. Raymond, Que., has assigned.

Hain & Co., general merchants, Midway, B.C., are asking an extension.

Alf. Mercier, general merchant, St. Andele (Rimouski), Que., has assigned.

L. B. Cormier, general merchant, Notre Dame, N.B., is offering to compromise.

E. A. Athinson, general merchant, L'Avenir, Que., has consented to assign.

H. Le Vasseur, general merchant, Fannystelle, Man., has been granted an extension.

J. L. Desilets, general merchant, St. Gertrude, Que., has assigned to Gagnon & Caron.

A meeting of the creditors of Edgar Scott, general merchant, Halifax, has been called.

Rosaire Bourbeau, general merchant, Victoriaville, Que., has assigned to J. McD. Hains.

Lalonde & Frere, general merchants, St. Benoit, Que., have assigned to Lamarche & Benoit.

Assignment has been demanded of F. A. Cantwell, general merchant, Franklin Centre, Que.

D. Licker & Co., general merchants, St. Cyrile de Wendover, Que., are offering 40c. on the dollar.

Masterson & Griffin, general merchants, Trout Lake, B.C., are reported to be asking an extension.

P. J. Stinson & Co., general merchants, Singhampton, Ont., have assigned to Thomas Brown.

The Standard Manufacturing Co., manufacturers of tin cans, etc., Toronto, have assigned to George Nicholson.

PARTNERSHIPS FORMED AND DISSOLVED.

S. Prevost & Co., hardware dealers, Montreal, have registered partnership.

Partnership has been registered by The Canadian Aluminum Works, Montreal.

Joseph Arthur, general merchant, Shanty Bay, Ont., is advertising his business for sale.

McGowan & Abraham, general merchants, Delhi, Ont., have dissolved. J. D. Abraham retires.

Capsey & Frary, general merchants, Frelighsburg, Que., have dissolved and Wells & Frary continue.

Reynolds & Reynolds, blacksmiths, carriagemakers, etc., Stroud, Ont., have dissolved. Sylvester Reynolds continues.

Alfred Doig, hardware dealer, Glenboro, Man., has taken in a partner under the style of Doig & Wilson.

D. W. Anderson & Co., general merchants, Harrow, Ont., have dissolved. D. W. Anderson continues alone.

SALES MADE AND PENDING.

A. Guimond, hardware dealer, Montreal, has sold his stock at 65c. on the dollar.

Walter Winning, hardware dealer, etc., Armstrong, B.C., is advertising his business for sale.

The assets of Mrs. Gariepy, general merchant, Lachine, Que., are to be sold on January 26.

The assets of Wright & Co., wholesale paper bag dealers, etc., Montreal, are to be sold on January 28.

The stock of The W. F. Horton Co., bicycle dealers, etc., London, Ont., will be sold by auction on January 29.

CHANGES.

Marchant Bros. are starting as paint dealers, etc., in Quebec.

John Graham, general merchant, Valetta, Ont., has sold out to Robertson Bros.

Louis Goldstein, general merchant, Rosenfieldt, Man., has retired from business.

C. W. Watson, general merchant, Ridgetown, Ont., has sold out to Joseph Baker.

Thomas Label, sawmiller, Fraserville, Que., has sold his mill to Price Bros. & Co.

J. M. Anguish, tinware dealer, Milton, Ont., has sold out to Suggett & Co.

Herbert J. Little, harness dealer, Lindsay, Ont., has sold out to The Rudd Harness Co.

H. D. Ashcroft, blacksmith, Nelson, B.C., has been succeeded by Reilly & Benoy.

Wm. Beattie & Co., general merchants, Ethel, Ont., have sold out to John McDonald.

Guilbault & Cote, hardware dealers, St. Boniface, Man., has sold out to Georgine Guilbault.

S. C. Cochrane, general merchant, Medicine Hat, N.W.T., has been succeeded by Cochrane & Sons.

F. W. Carscadden, dealer in agricultural implements, Strathcona, Man., has sold out to John C. Wainwright.

T. G. Hawthorn has registered as manager of the business of the American Axe Co., axe manufacturers, at Three Rivers, Que.

FIRES.

J. Corbett, general merchant, Brownsville, Ont., has been burned out; partially insured.

DEATHS.

Robert Millar, of Millar & Co., hardware dealers, Moosomin, Man., is dead.

W. McDonagh, of W. & J. McDonagh, potash manufacturers, Perth, Ont., is dead.

UNITED STATES COPPER PRODUCTION IN 1900.

THE regular monthly meeting of the Copper Producers' Association was held on Tuesday in New York, at which the figures of production and exports for December, compiled by Secretary Stanton, were given out as follows :

	Dec., 1900.	Nov., 1900.	Dec., 1899.
Production—			
By U. S. reporting mines	18,724	19,876	20,388
By outside sources (estimated)	3,400	3,400	3,400
By foreign reporting mines	8,483	7,752	7,360
Exports	11,223	9,508	15,550

All in tons of 2,240 lb.

The total production of the mines in the United States and the exports for the full calendar year 1900 are as follows :

	1900.	1899.	Incr'se.	Inc'se. p.c.
Production	268,787	262,206	6,581	2.9
Exports	159,602	119,812	39,790	33.2

The significance of these figures may be better appreciated when it is stated that producers allow for a 10 per cent. increase in production annually to meet increase in demand, whereas the increase in 1900 was less than 3 per cent. One reason for this probably is that the older mines devoted their attention to poorer grades of ore and rock which can be worked at a profit at current prices for copper, saving richer grades until later.

Production and exports for a series of years compare as follows :

	Production.	Exports.
1900	268,787	159,602
1899	262,000	119,812
1898	235,000	143,115
1897	216,000	129,210
1896	203,922	125,605

IMPROVEMENTS IN LANTERNS.

We would call the attention of the many readers of HARDWARE AND METAL to The Ontario Lantern Co.'s advertisement on page 17, where they show the new century "Banner" cold-blast lantern.

An embossed or corrugated lantern is an entirely new departure, which not only adds greatly to the appearance, but has the effect of strengthening the same very materially.

They have several new features combined in this lantern, all of which have been patented.

The company are now in a position to supply the jobbing trade with samples for their travelers.

The "Banner" will undoubtedly sell well during the coming season.

WILL EXECUTE ALL CONTRACTS

J. Watterson & Co., commission hardware and oil merchants, Montreal, have issued a circular to the trade stating that, owing to the fact that some of their competitors had asserted that their quotations for linseed oil sent out on November 28, 1900, were bogus, and that they would not be able to fill their contracts made on those quotations, they wish it thoroughly understood that the quotations were not bogus or issued with the object of depressing the market, and that they are ready to fill all orders booked by them on the basis of said quotations. They furthermore make the offer to forfeit $500 to any charitable object if any of their competitors can produce facts to show one instance where they have failed during the nine years they have been in business to fulfil any of their obligations.

A WOMAN'S IDEA.

A WOMAN, according to an exchange, thus gives her ideas regarding what a hardware store should be : "I like to see a look on everything in the store that seems to tell me I am welcome. I try to keep my own home in order and when I go out calling, either socially or on business, I want to find matters in such good shape that I can learn something about the everlasting fitness of things. If I see a store with dirty windows, I stay out. Things that are dirty on the outside are sure to be dirty within, and I don't want them. If the windows look all right but the goods in the store are not kept in order, I am quickly out on the street again. I don't like to trip over coal hods or bruise my knees against a stray keg of nails, but I do like to see all the goods nicely arranged, and in such a position that there is no trouble in either showing or looking at them. I like to be waited on by a courteous and accommodating clerk, not by one who acts as if he were performing an act of condescension in waiting on me at all. I especially object to the clerk who thinks that shirt-sleeves are good form behind the counter, whose finger-nails are in mourning, or who indicates other evidences of ill-breeding. It is a positive pleasure to go into some stores but positively disagreeable to go into others, and I keep away from these others as far as possible. I know it is said that women are a good deal of a nuisance about a store —sometimes—but I notice that they purchase a good many goods and I think that their peculiarities as buyers are worth catering to by the owners if they have any desire to increase their business. Besides, we women have a habit of talking, and if one of us doesn't like a store the others are likely to hear about it."

DOMINION INSOLVENCY LEGISLATION.

THE VIEWS OF BUSINESS MEN REGARDING ITS NECESSITY.

ALTHOUGH Mr. Fortin, M.P., who has been the particular champion in the House of Commons of a Dominion bankruptcy law during the last few years, has declared that he will not introduce a bill for that purpose at the ensuing session of Parliament, the question is by no means a dead one, as will be gathered from the following interviews with some of the leading merchants in Montreal :

ATTITUDE OF THE BANKS.

Mr. James Elliott, general-manager of the Molsons Bank, says the banks are not opposed to an insolvency law that would afford a means of giving a just, speedy and equitable division of estates among the creditors and that would prevent dishonest merchants from obtaining a clearance which they find no difficulty in getting at the present moment. He believed, too, that the law should be made general all over Canada, and thus remove the deficiencies in several of the Provincial Acts in operation now. But he is opposed to any measure that would disallow the banks having the estates of

TWO PERSONS AS SECURITY FOR THEIR PAPER.

This is one of the understandings upon which banks discount notes—they must have two names upon them. "Banks," said Mr. Elliott, "do not do business on ordinary business principles; they do not turn over goods at high profits like ordinary business concerns ; their rates of interest are low, consequently their returns must be sure. We cannot afford to incur any losses, so we must have two names on the paper we hold or we shall have to refuse to put money out on interest. So, you see, that any insolvency bill introduced into Parliament containing a provision to the effect that the banks cannot come upon two estates to recover the full value of the paper they hold will not meet with our approval so far as that part of the measure is concerned. Otherwise, I cannot see why the banks should be opposed to improved insolvency legislation. In fact, they ought to favor it."

A commercial man of the highest standing in Montreal, who does not wish his name disclosed, gives his opinion on insolvency matters as follows :

INSOLVENCY LAW BADLY NEEDED.

"We need a Dominion insolvency law very badly. Each of the Provinces has grievances that are decidedly vexatious, particularly to the wholesale merchants of Montreal. In Ontario, Manitoba and the Territories we have to meet chattel mortgages, which pop up here and there and everywhere, giving undue preference to certain creditors. We have a case on hand just now in Ontario where

A BANK'S CHATTEL MORTGAGE TAKES EVERYTHING.

"In the Maritime Provinces we meet with the same preferences."

"But are these not registered?"

"Yes, they are. But what good is that to us when merchants can give chattel mortgages after they have bought goods from us."

"Is the Quebec law satisfactory?"

THE QUEBEC LAW.

"The law as we have it in Quebec is the most satisfactory in the Dominion, but it is not an ideal law. It is too elastic. We can make a demand of assignment here, but it can be contested, and while it is being forced the debtor can dispose of all his goods. The law is not quick and practical enough."

EXCESSIVE CHARGES.

"What about excessive charges?"

"To-day they are certainly outrageous and should be regulated as suggested by Mr. Fortin in his draft of a bill. My opinion coincides with Police Magistrate Denison, of Toronto, who says that the lawyers are virtually robbing the community in the slick way they have of doing things. We have just had to deal with a case where an estate should easily have paid 75c. on the dollar, but after the lawyers had preyed to their full satisfaction it paid only ⅜c. on the dollar. These official notices of assignment, notices of meetings and all the rest of the tomfoolery form useless and needless expense. As Mr. Fortin suggests the assets of an insolvent should be put into the hands of the creditors who own it, not into the possession of any official receiving."

THE PRIVILEGES OF THE BANKS.

"Have the banks too much privilege?"

"Most decidedly they have. Why should they have a preferred claim over any other creditor? A bank can register its claim with the other creditors, but while the estate is being settled it goes on collecting on the paper it holds, and, then, when the time of giving dividends comes, it registers the balance of the claim which it has not collected. Thus it gets 100c. on the dollar for part of its debt which generally amounts to the whole. That is not right. The banks should hand over the paper they hold when an insolvent assigns."

"Do you think Mr. Fortin should introduce his measure this session?"

"Considering the needs of the country he should. Times are good now, and the present is the best time to put such a measure into operation. However, as there are so many lawyers in Parliament, who make their livelihood out of the existing state of affairs, I suppose Mr. Fortin runs a slim chance of having his bill passed if he should introduce it, and I don't blame him for not undertaking a task foredestined to bear no fruit."

VIEWS OF THE PRESIDENT OF THE CHAMBRE DE COMMERCE.

Mr. L. E. Geoffrion, manager of L. Chaput, Fils & Cie, Montreal, is an earnest advocate of a Dominion insolvency measure. He is president of the Chambre de Commerce, and he is striving to have a powerful deputation go to Ottawa to represent the commercial organizations of Montreal and urge the Government to remedy the existing evils. He is much opposed to the present methods of allowing preferences, and considers the privileges legislated to the banks extremely unjust.

"Why should the banks not be classed as ordinary creditors?" said he. "They make 6 per cent. on their money, while we make no more than 7 per cent. and have to handle goods. Why should these powerful corporations be allowed a preference? It is the height of injustice. In Ontario, the banks are allowed to seize the goods we ship our customers and hold them on their account, and then we have to whistle for the payment of the goods. The banks have no right to get a chattel mortgage on goods that are not paid for, and the sooner we have legislation to make such seizures criminal the better it will be for the commercial prosperity of this country.

"We want an insolvency law to cover the whole Dominion, and I want to see the Government take hold of the matter."

BUSINESS WILL GO ON AS USUAL.

The fire which destroyed the Montreal premises of H. A. Nelson & Sons is not likely to seriously interfere with the firm in supplying the grocery trade, for the broom factory is in Toronto, and the branch business there also carries in stock such lines as woodenware and cordage.

Arbuckle Bros. have started as dealers in agricultural implements, etc., in Carleton Place, Ont.

The Fredericton, N.B., Tourist Association have appointed the following officers : Chairman, C. F. Chestnut ; treasurer, F. B. Edgecombe, and secretary, R. P. Allen.

A BUSINESS OF ADVERTISING.

THAT manufacturers would reap far greater profits from their advertisements if they put the same business methods into advertising as they do in the other branches of their business is so often asserted, that, like many another self-evident fact, many allow it to pass unheeded.

A business man, a manufacturer of steam appliances in New York City, recently consolidated his business with a rival. The two men had carried the same-sized advertisements in the same papers. The first-mentioned manufacturer had made a study of catchy advertisements; made his own drawings, and originality stood out all over them. You could not turn the page where his advertisement was located without seeing and stoping to examine, perhaps to criticize his characteristic advertisement. The other manufacturer had used one plain, stereotyped, dignified, unchanged advertisement. After the consolidation the new concern took up the question of advertising, and, as might be expected, they each held divergent views as to its value. The result was a comparison of results, when the remarkable fact appeared that the man with the carefully-prepared advertisements had received some seven times as many responses to his advertisements as had the other.

Advertising is just as much a part, and an important part, of a manufacturer's business as is selling, since it is done as a means of selling. Why not, then, make it the most effective possible.

Business men read a man's characteristics in advertisements. A wide-awake manufacturer will do wide-awake advertising. His advertisements will reflect his business methods and genius. An up-to-date advertisement does not emanate from a dull or ultra-conservative manufacturer. A man's policy and genius are disclosed in his advertisements. The advertising pages of all leading papers in these later times are replete with genius and made beautiful with fine cuts. To such a point have these ideas been carried that the advertising columns of our standard industrial papers are read with as much interest as are the editorials. And often more can be learned from the advertising columns than from the reading pages. This fact enforces the necessity of attractive advertisements. — "Lapis," in Age of Steel.

The following officers have been elected by the Portage la Prairie, Man., Board of Trade: President, A. H. Dickins; vice-president, W. Bell; secretary-treasurer, H. W. B. Douglas; council — E. Brown, W. J. Cooper, Geo. Davidson, J. A. Marshall, C. R. Garland, C. S. Burley, W. J. May and Horace Ormond.

MARKETS AND MARKET NOTES

QUEBEC MARKETS.

Montreal, January 25, 1901.

HARDWARE.

TRADE shows a decided improvement this week. Retail merchants are now through stock-taking, have settled the year's accounts, and are looking through the stocks for shorts. Quite a number of letter orders have been received this week for seasonable goods. All shelf goods are moving freely, and some lines of heavy goods are in steady request. The travellers are meeting with good success in their search for spring orders, and good quantities of freezers, green wire cloth, netting, rubber hose, screen doors and windows, churns, building paper, etc., are being booked for spring shipment. In fact, the season's business is opening up satisfactorily in every way. The market is steady, and manufacturers seem to be so busy that they are stiffening in their attitude. Market changes have been few this week. The list on "Royal Dominion" and "Royal Canadian" wringers has been reduced from $58 to $50 per dozen, the discount remaining at 45 per cent., and the list price of "Ajax" has been reduced to $120 from $124; all other grades remain unaltered. Common sad irons have been slightly reduced, and are now selling at $3.55 to $3.60 per 100 lb. The discount on shoe nails has been slightly changed, and Swedes shoe nails, 14 gauge and heavier, are now selling at a discount of 65 per cent. Horseshoes are scarce.

BARB WIRE — A few lots are being booked for spring shipment, but business for immediate shipment is dead. The price is still $3.20 f.o.b. Montreal in less than carlots.

GALVANIZED WIRE—The market shows no change. Business is very quiet. We quote: No. 5, $4.25; Nos. 6, 7 and 8 gauge $3.55; No. 9, $3.10; No. 10, $3.75; No. 11, $3.85; No. 12, $3.25; No. 13, $3.35; No. 14, $4.25; No. 15, $4.75; No. 16, $5.00.

SMOOTH STEEL WIRE — The demand is limited, but a very few lots of hay-baling wire are moving. The price is $2.80 per 100 lb.

FINE STEEL WIRE—The usual inquiry has been experienced. The discount is 17½ per cent. off the list.

BRASS AND COPPER WIRE — There is nothing special to note. Discounts are 55 and 2½ per cent. on brass, and 50 and 2½ per cent. on copper.

FENCE STAPLES—A small trade is passing. We quote: $3.25 for bright, and $3.75 for galvanized, per keg of 100 lb.

WIRE NAILS—This is naturally a quiet season in nails, but business is quite up to the average for immediate shipment. There is little heavy business being done. We quote $2.85 for small lots and $2.75 for carlots, f.o.b. Montreal, Toronto, Hamilton, London, Gananoque, and St. John, N.B.

CUT NAILS—There have been no new developments in the nail business. Trade is confined almost entirely to immediate wants. We quote as follows: $2.35 for small and $2.25 for carlots; flour barrel nails, 25 per cent. discount; coopers' nails, 30 per cent. discount.

HORSE NAILS—A good trade is being

done at 50 per cent. on standard, and 50 and 10 per cent, on Acadia.

HORSESHOES—Jobbers say that the demand for horseshoes this year has been phenomenal. Supplies are light and the demand heavy. We quote: Iron shoes, light and medium pattern, No. 2 and larger, $3.50; No. 1 and smaller, $3.75; snow shoes, No. 2 and larger, $3.75; No. 1 and smaller, $4.00; X L steel shoes, all sizes, 1 to 5, No. 2 and larger, $3.60; No. 1 and smaller, $3.85; feather-weight, all sizes, $4.85; toe weight steel shoes, all sizes, $5.95 f.o.b. Montreal; f.o.b. Hamilton, London and Guelph, 10c. extra.

POULTRY NETTING — Business on spring account is still going on. The discount is 50 and 5 per cent.

GREEN WIRE CLOTH—The price remains as before at $1.35 per 100 sq. ft. Business is being freely done.

FREEZERS — Trade in freezers is in full swing this week. The demand has opened up well.

WRINGERS—Orders for spring shipment are now being booked. Within the past week the list price on Royal Dominion and Royal Canadian has been lowered from $58 to $50 per doz. The discount remains at 45 per cent. off. Ajax wringers have also been marked down from $124 list to $120. All other grades remain the same.

SAD IRONS — Common sad irons have been reduced about 20c. per 100 lb. by the manufacturers this week and are now quoted at $3.55 to $3.60 per 100 lb.

SCREEN DOORS AND WINDOWS — Trade has opened up well and fair orders for screens have been booked. We quote: Screen doors, plain cherry finish, $8.25 per doz.; do. fancy, $11.50 per doz.; windows, $2.25 to $3.50 per doz.

SCREWS—Like other shelf goods, screws are in good demand. Quite a number of shipments have been made this week in response to letter orders. Discounts are as follows: Flat head bright, 85 per cent. off list; round head bright, 80 per cent.; flat head brass, 77½ per cent.; round head brass, 70 per cent.

BOLTS — The demand for bolts is fair. Discounts are as follows: Carriage bolts, 65 per cent.; machine bolts, 65 per cent.; coach screws, 75 per cent.; sleigh shoe bolts, 75 per cent.; bolt ends, 65 per cent.; plough bolts, 50 per cent.; square nuts, 4½c. per lb. off list; hexagon nuts, 4¾c. per lb. off list; tire bolts, 67½ per cent.; stove bolts, 67½ per cent.

BUILDING PAPER—Shipments from stock are small, but a good deal of business on spring account is being done. We quote as follows: Dry sheathing, 30c. per roll; cyclone dry do., 42c. per roll; straw do., 30c.; heavy straw do., $1.40 per 100 lb.;

I.X L., dry sheathing, 65c. per roll; cyclone, tarred do., 50c. per roll; tarred ordinary do., 40c. per roll; tarred felt, $1.60 per 100 lb.; ready roofing, 2 ply, 75c. per roll; 3-ply, $1 per roll.

RIVETS — Trade continues unchanged. The discount on best iron rivets, section, carriage, and wagon box, black rivets, tinned do., coopers' rivets and tinned swedes rivets, 60 and 10 per cent.; swedes iron burrs are quoted at 55 per cent. off; copper rivets, 35 and 5 per cent. off; and coppered iron rivets and burrs, in 5-lb. carton boxes, are quoted at 60 and 10 per cent. off list.

CORDAGE—The situation is steady to firm. Manila is quoted at 13½c. per lb. for 7-16 and larger; sisal is worth 9½c. per lb. for 7-16 and larger, and lathyarn 9½c. per lb.

SPADES AND SHOVELS — Some business is being done at 40 and 5 per cent. discount.

HARVEST TOOLS—A few orders are being booked for spring shipment at 50, 10 and 5 per cent. discount.

TACKS—The feature this week is a change in the price of shoe nails, which are now quoted from 14 gauge and heavier, instead of 15 guage and heavier. Swedes shoe nails, soft steel nails and iron nails are now quoted at a discount of 65 per cent. Otherwise prices are unchanged. We quote: Carpet tacks, in dozens and bulk, blued, 80 and 5 per cent. discount; tinned, 80 and 10 per cent.; cut tacks, blued, in dozens, 75 and 15 per cent. discount.

FIREBRICKS—This line continues featureless. The price is $18.50 to $26, as to brand.

CEMENT—Inquiry is small. We quote: German, $2.50 to $2.65; English, $2.40 to $2.50; Belgian, $1.90 to $2.15 per bbl.

METALS.

The market is steady to strong in most lines. Pig iron is rather weak. Tin and copper seem to be rather firmer in primary markets.

PIG IRON—The market for Scotch pig on spot is firm at $24 to $25 for No. 1 Summerlee. No. 1 Hamilton is easier, and lower, and quoted at $20, although transactions have occurred below that figure.

BAR IRON—The market is steady at $1.65 to $1.70 in small lots.

BLACK SHEETS — The inquiry is small. Prices rule as before at $2.80 for 8 to 16 gauge.

GALVANIZED IRON—A few lots of galvanized iron are going out, but business seems pretty well confined to spring account. We quote: No. 28 Queen's Head, $5 to $5.10; Apollo, 10¾ oz., $5 to $5.10 Comet, No. 28, $4.50 with 25c. allowance in case lots.

INGOT COPPER— The ruling figure is steady at 17½c.

INGOT TIN—A small business is being done at the base 33c. for Lamb and Flag.

LEAD—Steady at $4 65.

LEAD PIPE—Business is confined to small proportions. We quote: 7c. for ordinary and 7½c. for composition waste, with 15 per cent. off.

IRON PIPE—A fair business is being done. We quote as follows: Black pipe, ¼, $3 per 100 ft.; ⅜, $3; ½, $3; ¾, $3.15; 1-in., $4.50; 1¼, $6.10; 1½, $7.28; 2-in., $9.75. Galvanized, ½, $4.60; ¾, $5.25; 1-in., $7.50; 1¼, $9.80; 1½, $11.75; 2-in., $16.

TINPLATES—There is nothing new to note. A seasonable trade is being done at $4.50 for coke and $4.75 for charcoal.

CANADA PLATE—Trade is not very brisk. We quote: 52's, $2.90; 60's, $3; 75's, $3.10; full polished, $3.75, and galvanized, $4.60.

TOOL STEEL—We quote: Black Diamond, 8c.; Jessop's 12c.

STEEL—No change. We quote: Sleighshoe, $1.85; tire, $1.95; spring, $2.75; machinery, $2.75 and toe calk, $2.50.

TERNE PLATES—Trade is quiet and featureless. We quote $8.25.

SWEDISH IRON—Steady at $4.25.

COIL CHAIN — Some business is being done. The tone of the market is firm. We quote: No. 6, 11½c.; No. 5, 10c.; No. 4, 9½c.; No. 3, 9c.; ¼-inch, 7½c. per lb.; 5-16, $4.60; 5-16 exact, $5.10; ⅜, $4.20; 7-16, $4.00; ½, $3.75; 9-16, $3.65; ⅝, $3.35; ¾, $3.25; ⅞, $3.20; 1-in., $3.15. In carload lots an allowance of 10c. is made.

SHEET ZINC—The ruling price is 6 to 6¼c.

ANTIMONY—Quiet, at 10c.

GLASS.

Business on spring account is beginning to be done by some houses. For immediate shipment, prices are unchanged. We quote: First break, $2; second, $2.10 for 50 feet; first break, 100 feet, $3.80; second, $4; third, $4.50; fourth, $4.75; fifth, $5.25; sixth, $5.75, and seventh, 6.25.

PAINTS AND OILS.

There are no special features this week. If anything, business has been on the quiet side. A number of the travelling salesmen have been laid up with the grippe and a good many firms throughout the country complain of being short-handed through the indisposition of members of the staff. Severe weather has also contributed to quiet the demand, except in paris green. This article has suddenly sprung into light, and the bookings have been pretty heavy. Dry colors do not show any great change. Kalsomine has been advanced 1c. per lb. Colors in oil are only in fair inquiry and house painting has been seemingly affected by the cold weather. Prices of liquid paints are well maintained. Varnishes are steady. Gold leaf is somewhat easier and is now quoted in single packs at $6 per pack. We quote:

WHITE LEAD—Best brands, Government standard, $6.75; No. 1, $6.37½; No. 2, $6; No. 3, $5.62½, and No. 4, $5.25, all f.o.b. Montreal. Terms, 3 per cent. cash or four months.

DRY WHITE LEAD—$5.75 in casks; kegs, $6.

RED LEAD—Casks, $5.50; in kegs, $5.75.

WHITE ZINC PAINT—Pure, dry, 8c.; No. 1, 6½c.; in oil, pure, 9c.; No. 1, 7½c.

PUTTY—We quote: Bulk, in barrels, $2 per 100 lb.; bulk, in less quantity, $2.15; bladders, in barrels, $2.20; bladders, in 100 or 200 lb. kegs or boxes, $2.35; in tins, $2.45 to $2.75; in less than 100-lb. lots, $3 f.o.b. Montreal, Ottawa, Toronto, Hamilton, London and Guelph. Maritime Provinces 10c. higher, f.o.b. St. John and Halifax.

LINSEED OIL—Raw, 80c.; boiled, 83c., in 5 to 9 bbls., 1c. less, 10 to 20 bbl. lots, open, net cash, plus 2c. for 4 months. Delivered anywhere in Ontario between Montreal and Oshawa at 2c. per gal. advance and freight allowed.

TURPENTINE—Single bbls., 59c.; 2 to 4 bbls., 58c.; 5 bbls. and over, open terms, the same terms as linseed oil.

MIXED PAINTS—$1.25 to $1.45 per gal.

CASTOR OIL—8¾ to 9¼c. in wholesale lots, and ½c. additional for small lots.

SEAL OIL—47½ to 49c.

COD OIL—32½ to 35c.

NAVAL STORES — We quote: Resins, $2.75 to $4.50, as to brand; coal tar, $3.25 to $3.75; cotton waste, 4½ to 5½c. for colored, and 6 to 7½c. for white; oakum, 5½ to 6½c., and cotton oakum, 10 to 11c.

PARIS GREEN—Petroleum barrels, 16¾c. per lb.; arsenic kegs, 17c.; 50 and 100-lb. drums, 17½c.; 25-lb. drums, 18c.; 1-lb. packages, 18½c.; ½-lb. packages, 20½c.; 1-lb. tins, 19½c.; ½-lb. tins, 21½c. f.o.b. Montreal; terms 3 per cent. 30 days, or four months from date of delivery.

SCRAP METALS.

The market is still quiet, with little business being transacted. Dealers are paying the following prices in the country: Heavy copper and wire, 13 to 13½c. per lb.; light copper, 12c.; heavy brass, 12c.; heavy yellow, 8½ to 9c.; light brass, 6½ to 7c.; lead, 2¾ to 3c. per lb.; zinc, 2¼ to 2½c.; iron, No. 1 wrought, $13 to $14 per gross ton; No. 1 cast, $13 to $14; stove plate, $8 to $9; light iron, No. 2, $4 a ton; malleable and steel, $4.

PETROLEUM.

A good trade continues to be done, but the demand is for smaller quantities. We quote: "Silver Star," 15 to 16c.; "Imperial Acme," 16½ to 17½c.; "S C. Acme," 18 to 19c., and "Pratt's Astral," 19 to 20c.

HIDES.

Tanners are not buying very freely, and business is rather slow. Dealers are paying 7½c. for No. 1 light, and tanners are asked to pay 8½c. for carlots. We quote: Light hides, 7½c. for No. 1; 6½c. for No. 2, and 5½c. for No. 3. Lambskins, 90c.

Mr. Stearnfield, representing The Markham Air Rifle Co., Plymouth, Mich., was in Toronto this week. He is said to have taken away some nice orders for "King" rifles.

ONTARIO MARKETS.

TORONTO, January 25, 1901.

HARDWARE.

A LITTLE better business is probably being done this week, but still the volume, generally speaking, is light. It is the general opinion that the volume of business is much smaller this month than it was for the corresponding period last year. Tinware is particularly quiet, although it is asserted that it is no worse than is usual at this season in previous years. Wire nails continue quiet, and in fence wires there is little or nothing being done outside of booking orders for future delivery, and, even in this particular, the volume of business is not large. In screws, bolts and rivets, there is the usual steady trade being done. Rope is quiet and featureless. In cuttlery, a slightly better trade is to be noted. A few orders are being booked for green wire cloth, poultry netting, harvest tools, and spades and shovels for future delivery. There have been a few changes in prices this week. French mariners' compasses and magnifying glasses are quoted higher. On flaring sap buckets and flaring pails and creamer cans the discount has been increased. A new list has been issued on wringers, which shows a decline in prices. Prices are also lower on iron bench screws, grindstone fixtures, chest handles and Mrs. Potts sad irons. Letter orders are fair and the same can be said of payments.

BARB WIRE—Nothing doing in shipments from stock and we hear of no orders being placed during the past few days for future delivery. We quote $2.97½ f.o.b. for less than carlots, and $2.85 in carlots. From stock, Toronto, $3.10 per 100 lb.

GALVANIZED WIRE—This is dull and featureless. We quote: Nos. 6, 7 and 8, $3.55; No. 9, $3.10; No., 10, $3.75; No. 11, $3.85; No. 12, $3.25; No. 13, $3.35; No. 14, $4.25; No. 15, $4.75, and No. 16, $5.

SMOOTH STEEL WIRE — A few orders for future delivery have been booked in oiled and annealed wire and a little has been done during the past week in hay-baling wire for prompt shipment. We quote the base price at $2.80 per 100 lb.

WIRE NAILS—A few orders are reported but they are only for small quantities. The base price is unchanged at $2.85 per keg for less than carlots and $2.75 in carlots.

CUT NAILS — These are still dull and featureless with the base price unchanged at $2.35 per keg.

HORSESHOES — Just a steady trade is to be reported. We quote as follows f.o.b. Toronto: Iron shoes, No. 2 and larger, light, medium and heavy, $3.60; snow shoes, $3.85; light steel shoes, $3.70 featherweight (all sizes), $4.95; iron shoes, No. 1 and smaller, light, medium and heavy (all sizes), $3.85; snow shoes, $4; light steel shoes, $3.95; featherweight (all sizes), $4.95.

HORSE NAILS—Business continues moderate. Discount, 50 per cent. on standard oval head and 50 and 10 per cent. on Acadia.

SCREWS—A fair sorting-up trade is being done at unchanged prices. We quote: Flat head bright, 85 per cent. off the list; round head bright, 80 per cent.; flat head brass, 77½ per cent.; round head brass, 70 per cent.; flat head bronze, 70 per cent.; round head bronze, 65 per cent.

BOLTS AND NUTS — There is the usual steady trade being done. We quote as follows: Carriage bolts (Norway), full square, 70 per cent.; carriage bolts, fulls quare, 70 per cent.; common carriage bolts, all sizes, 65 per cent.; machine bolts, all sizes, 65 per cent.; coach screws, 75 per cent.; sleighshoe bolts, 75 per cent.; blank bolts, 65 per cent.; bolt ends, 65 per cent.; nuts, square, 4½c. off; nuts, hexagon, 4¼c. off; tire bolts, 67½ per cent.; stove bolts, 67½; plough bolts, 60 per cent.; stove rods, 6 to 8c.

RIVETS AND BURRS—The demand is steady and prices unchanged. Discount, 60 and 10 per cent. on iron rivets; iron burrs, 55 per cent.; copper rivets and burrs, 35 and 5 per cent.

ROPE—Business is still quiet. We quote: Sisal, 9c. per lb. base, and manila, 13c. We quote cotton rope: 3-16 in. and larger, 16½c.; 5-32 in., 21½c., and ½ in., 22½c. per lb.

CUTLERY—There is little going out, and trade appears to be improving over that of the early part of the month.

SPORTING GOODS.— Some rifles, loaded shells and cartridges are wanted, but the demand is not brisk.

GREEN WIRE CLOTH — Orders are still being booked for future delivery at $1.35 per 100 sq. ft.

SCREEN WINDOWS—Shipments have gone forward this week to British Columbia. Difficulty was experienced in securing some of the sizes to complete orders.

WRINGERS—A new list has been issued by the Canadian manufacturers. It is lower than that which previously existed, while the discount is unchanged at 45 per cent. Terms 4 months or 3 per cent. 30 days. The list will be found on our editorial page.

MRS. POTTS SAD IRONS—The price of these has been reduced and we now quote No. 55, polished, 62½c.; No. 50, nickel-plated, 67½c.

HARVEST TOOLS — A few orders have been booked during the week for harvest tools for future delivery. Discount 50, 10 and 5 per cent.

SPADES AND SHOVELS—A little business has been done on future account during the past week. Discount 40 and 5 per cent.

POULTRY NETTING—No orders are being taken for present shipments, but a few are still being booked for future delivery. Discount on Canadian is 50 and 5 per cent. On English, prices are quoted at net figures.

TINWARE—There has been a reduction in the price of certain lines. On 10-quart flaring sap buckets; 6, 10, 14 quart I. C. flaring pails and creamer cans the discount is now 45 per cent. instead of 40 per cent. as formerly. The tinware trade is, as a rule, quiet, although quite a few milk can trimmings have gone out.

ENAMELLED WARE—Business in this line is still quiet.

ELBOWS—The reduction in prices has caused an improvement in the trade for delivery next fall. There is an idea that prices will be a little stiffer rather than lower, while the various manufacturers have been rather aggressive in securing orders.

CEMENT—There is nothing doing. We nominally quote in barrel lots: Canadian Portland, $2.80 to $3; Belgian, $2.75 to $3; English do., $3; Canadian hydraulic cements, $1.25 to $1.50; calcined plaster, $1.90; asbestos cement, $2.50 per bbl.

METALS.

While there is not an active trade doing in metals, as a rule, the demand is fair for this time of the year. Tin has shown a steadier tone.

PIG IRON—The market is quiet with the Canadian furnaces quoting $17 for No. 2 in 100-ton lots.

BAR IRON—A fair trade is still being done

at the base price of $1.65 to $1.70 per 100 lb.

PIG TIN—While the market is still somewhat irregular a better tone prevails than a week ago. Locally, however, business is quieter than it was with quotations unchanged at 32 to 33c., according to quantity.

TINPLATES — Quite a few tinplates are going, out showing that the tinsmiths in the country are making up stock in anticipation of their requirements.

TINNED SHEETS—The demand is more active, and there is already some scarcity of stocks on the local market. Spring stocks have not yet arrived, and wholesalers are finding that they will require them earlier than was anticipated.

TERNE PLATES — These are still quiet and unchanged in price.

BLACK SHEETS—Trade is fair with the base price unchanged at $3.50 per 100 lb.

GALVANIZED SHEETS—Trade has been fair, and a number of orders have been booked during the week for spring and summer delivery.

CANADA PLATES—A fairly-good trade is being done, the volume of business being larger than was expected at this time. Quite a few 5-box lots are still required. Import orders are being booked for next fall's delivery. From what can be gathered, local jobbers are not going to carry over as large stocks of Canada plates as was anticipated. For this they are quite gratified, in view of the probable future of the market as far as prices are concerned. We quote: All dull, $3; half and half, $3.15, and all bright, $3.65 to $3.75.

IRON PIPE—A fair business continues to be done in this line. We quote: Black pipe ¼ in., $3.00; ⅜ in., $3 00; ½ in., $3.50; ¾ in., $3.20; 1 in., $4.60; 1¼ in., $6.35; 1½ in., $7.55; 2 in., $10.10. Galvanized pipe is as follows: ½ in., $4.65; ¾ in., $5.35; 1 in., $7.25; 1¼ in., $9.75; 1½ in., $11.25; 2 in., $15 50.

HOOP STEEL—The demand during the past week has been fairly good with $3.10 as the base price.

COPPER — Ingot copper is rather quiet, but a good business is being done in bar and sheet. We quote: Ingot, 19 to 20c.; bolt or bar, 23½ to 25c.; sheet, 23 to 23½c.

BRASS—Trade continues quiet with the discount on rod and sheet still 15 per cent.

SOLDER—A fairly good business is reported in solder for small quantities.

LEAD — The demand is moderate at 4¾ to 5c.

ZINC SPELTER—Trade is rather quiet at 6 to 6½c. per lb.

ZINC SHEET—A fairly active trade is to be noted. We quote casks at $6.75 to $7, and part casks at $7 to $7.50.

PAINTS AND OILS.

There is no change. Flax seed has advanced in Chicago, but there is no indication that this will affect conditions here or in Great Britain. Turpentine fluctuates somewhat, but is practically at an unchanged basis. White lead is firm. There continues a good receipt of orders for spring delivery. We quote:

WHITE LEAD—Ex Toronto, pure white lead, $6.87½; No. 1, $6.50; No. 2, $6.12½; No. 3, $5.75; No. 4, $5.37½; dry white lead in casks, $6.

RED LEAD—Genuine, in casks of 560 lb., $5.50; ditto, in kegs of 100 lb., $5.75; No. 1, in casks of 560 lb., $5 to $5.25; ditto, kegs of 100 lb.; $5.25 to $5.50.

LITHARGE—Genuine, 7 to 7½c.

ORANGE MINERAL—Genuine, 8 to 8½c.

WHITE ZINC—Genuine, French V.M., in casks, $7 to $7.25; Lehigh, in casks, $6.

PARIS WHITE—90c.

WHITING — 60c. per 100 lb.; Gilders' whiting, 75 to 80c.

GUM SHELLAC — In cases, 22c.; in less than cases, 25c.

PUTTY — Bladders, in bbls., $2.20; bladders, in 100 lb. kegs, $2.35; bulk in bbls., $2; bulk, less than bbls. and up to 100 lb., $2.15; bladders, bulk or tins, less than 100 lb., $3.

PLASTER PARIS—New Brunswick, $1.90 per bbl.

PUMICE STONE — Powdered, $2.50 per cwt. in bbls., and 4 to 5c. per lb. in less quantity; lump, 10c. in small lots, and 8c. in bbls.

LIQUID PAINTS—Pure, $1.20 to $1.30 per gal.; No. 1 quality, $1 per gal.

CASTOR OIL—East India, in cases, 10 to 10½c. per lb. and 10½ to 11c. for single tins.

LINSEED OIL—Raw, 1 to 4 barrels, 80c.; boiled, 83c.; 5 to 9 barrels, raw, 79c.; boiled, 82c., delivered. To Toronto, Hamilton, Guelph and London, 2c. less.

TURPENTINE—Single barrels, 59c.; 2 to 4 barrels, 58c., to all points in Ontario. For less quantities than barrels, 5c. per gallon extra will be added, and for 5-gallon packages, 50c., and 10 gallon packages, 80c. will be charged.

GLASS.

A feeling is manifesting itself that prices for import will be reduced before long, but there is such an absence of definite information that there is no certainty of a decline. Stock prices are firm. We still quote first break locally: Star, in 50-foot boxes, $2.10, and 100-foot boxes, $4; double diamond under 26 united inches, $6, Toronto, Hamilton and London; terms 4 months or 3 per cent. 30 days.

OLD MATERIAL.

The market is easy, but prices are unaltered. We quote jobbers' prices as follows: Agricultural scrap, 55c. per cwt.; machinery cast, 55c. per cwt.; stove cast, 35c.; No. 1 wrought 50c. per 100 lb.; new light scrap copper, 12c. per lb.; bottoms, 10½c.; heavy copper, 12½c.; coil wire scrap, 13c.; light brass, 7c.; heavy yellow brass, 10 to 10½c.; heavy red brass, 10½c.; scrap lead, 3c.; zinc, 2½c; scrap rubber, 6½c.; good country mixed rags, 65 to 75c.; clean dry bones, 40 to 50c. per 100 lb.

PETROLEUM.

The demand keeps brisk. Prices are steady. We quote: Pratt's Astral, 17 to 17½c. in bulk (barrels, $1 extra); American water white, 17 to 17½c. in barrels; Photogene, 16½ to 17c.; Sarnia water white, 16 to 16½c. in barrels; Sarnia prime white, 15 to 15½c. in barrels.

COAL.

There is still a big shortage of nut size, and delivery of other lines is reduced by the shortage of cars. Prices are unchanged. We quote anthracite on cars Buffalo and bridges: Grate, $4.75 per gross ton and $4.24 per net ton; egg, stove and nut, $5 per gross ton and $4 46 per net ton.

MARKET NOTES.

French mariners' compasses and magnifying glasses are higher.

A new list of prices has been issued by the manufacturers of wringers.

The discount has been increased to 45 per cent on flaring sap buckets, flaring pails and creamery cans.

Mrs. Potts sad irons are quoted lower at 62½c. per set for polished, and 67½c. per set for nickel-plated.

H. S. Howland, Sons & Co. are in receipt of a shipment of German hardware and cutlery, mostly Boker's goods.

The hockey team of H. S. Howland, Sons & Co. on Wednesday night defeated the team of the Dominion Bank by 3 to 1. One of the staff declares that "Mic-Mac" hockey sticks helped to secure the victory.

MANITOBA MARKETS.

WINNIPEG, January 21, 1901.

TRAVELLERS are again on the road and report a fair amount of business. In the city, however, the wholesale houses have an air of profound calm, but time is not being wasted. Many houses are arranging matters in such a way as to handle with the greatest despatch the spring rush when it comes. There is every expectation of being an enormous amount of building and this will mean a corresponding trade in building hardware. No prices have been altered during the week.

Price list for the week is as follows :

Barbed wire, 100 lb.	$3 45
Plain twist	3 45
Staples	3 95
Oiled annealed wire 10	3 95
" 11	4 00
" 12	4 05
" 13	4 20
" 14	4 35
" 15	4 45
Wire nails, 30 to 60 dy, keg	3 45
" 16 and 20	3 50
" 10	3 55
" 8	3 65
" 6	3 70
" 4	3 85
" 3	4 10
Cut nails, 30 to 60 dy.	3 00
" 20 to 40	3 05
" 10 to 16	3 10
" 8	3 15
" 6	3 20
" 4	3 30
" 3	3 65
Horsenails, 45 per cent. discount.	
Horseshoes, iron, No. 0 to No 1	4 65
No. 2 and larger	4 40
Snow shoes, No. 0 to No. 1	4 90
No. 2 and larger	4 40
Steel, No. 0 to No. 1	4 95
No. 2 and larger	4 70
Bar iron, $2.50 basis.	
Swedish iron, $4.50 basis.	
Sleigh shoe steel	3 00
Spring steel	3 25
Machinery steel	3 75
Tool steel, Black Diamond, 100 lb	8 50
Jessop	13 00
Sheet iron, black, 10 to 20 gauge, 100 lb.	3 50
20 to 26 gauge	3 75
28 gauge	4 00
Galvanized American, 16 gauge	2 54
18 to 22 gauge	4 50
24 gauge	4 75
26 gauge	5 00
28 gauge	5 25
Genuine Russian, lb	12
Imitation " "	8
Tinned, 24 gauge, 100 lb	7 55
26 gauge	8 80
28 gauge	8 00
Tinplate, IC charcoal, 20 x 28, box	10 75
" IX "	12 75
" IXX "	14 75
Ingot tin	35
Canada plate, 18 x 21 and 18 x 24	3 75
Sheet zinc, cask lots, 100 lb	7 50
Broken lots	8 00
Pig lead, 100 lb.	6 00
Wrought pipe, black up to 2 inch	50 an 10 p.c.
" Over 2 inch	50 p.c.
Rope, sisal, 7-16 and larger	$10 00
" ⅜	10 50
" ¼ and 5-16	11 00
Manila, 7-16 and larger	13 50
" ⅜	14 00
" ¼ and 5-16	14 50
Solder	21½
Cotton Rope, all sizes, lb.	16
Axes, chopping	$7 50 to 12 00
" double bitts	12 00 to 18 00
Screws, flat head, iron, bright	75 and 10 p.c.
Round " "	70 p.c.
Flat " brass	70 p.c.
Round " "	60 and 5 p.c.
Coach	57½ p.c.
Bolts, carriage	42½ p.c.
Machine	45 p.c.
Tire	60 p.c
Sleigh shoe	65 p.c.
Plough	40 p.c.
Rivets, iron	50 p.c.
Copper, No. 8	50c. lb.
Spades and shovels	40 p.c.
Harvest tools	50, and 10 p.c.
Axe handles, turned, s. g. hickory, doz.	$2 50
No. 1	1 50
No. 2	1 25
Octagon extra	1 75
No. 1	1 25
Files common	70, and 10 p.c.
Diamond	60
Ammunition, cartridges, Dominion R.F.	50 p.c.
Dominion, C.F., pistol	30 p.c.
" military	15 p.c.
American R.F.	30 p.c.
C.F. pistol	5 p.c.
C.F. military	10 p.c. advance.
Loaded shells :	
Eley's soft, 12 gauge	16 50
chilled, 12 guage	18 00
soft, 10 guage	21 00
chilled, 10 guage	23 00
American, M	16 25
Shot, Ordinary, per 100 lb	6 75
Chilled	7 50
Powder, F.F., keg	4 75
F.F.G.	5 00
Tinware, pressed, retinned	75 and 2½ p.c.
" " plain	70 and 15 p.c.
Graniteware, according to quality	50 p.c.

PETROLEUM.

Water white American	24½c.
Prime white American	23c.
Water white Canadian	21c.
Prime white Canadian	19c.

PAINTS, OILS AND GLASS.

Turpentine, pure, in barrels	$ 68
Less than barrel lots	73
Linseed oil, raw	87
Boiled	90
Lubricating oils, Eldorado castor	25½
Eldorado engine	24½
Atlantic red	27½
Renown engine	41
Black oil	23½ to 25
Cylinder oil (according to grade)	55 to 74
Harness oil	61
Neatsfoot oil	$1 00
Steam refined oil	85
Sperm oil	1 50
Castor oil per lb.	11½
Glass, single glass, first break, 16 to 25 united inches	2 25
26 to 40 per 50 ft.	2 50
41 to 50	5 50
51 to 60	6 00
61 to 70 per 100-ft. boxes	6 50
Putty, in bladders, barrel lots per lb.	2½
kegs "	2¾
White lead, pure per cwt.	7 25
No 1 "	7 00
Prepared paints, pure liquid colors, according to shade and color per gal.	$1.30 to $1.90

CATALOGUES, BOOKLETS, ETC.

HENDERSON & POTTS' PRICE LIST.

Henderson & Potts, proprietors of The Nova Scotia Paint and Varnish Works, Halifax and Montreal, have issued their catalogue for 1901. In arranging this new catalogue, they have adopted a somewhat different plan from that used in former issues, namely, to have, as nearly as possible, one uniform discount for all lines of goods. Henderson & Potts are sole agents in Canada for Brandram Bros., which fact increases their facilities for putting first-class paints, white leads, varnishes, etc., on the market. Their catalogue is a neat, handy one, and is worth writing for.

MONTHLY STOCK LIST.

The Bourne-Fuller Co., Cleveland, O., are sending out the January number of their monthly stock list of heavy iron, galvanized iron sheets, steam pipe, gas pipe, water pipe, rivets, angles, fence channels, steel tire, spring steel, Norway iron, steel hoops and bands, machinery steel, bar iron, etc. These lists are a convenience which every cnstomer of this firm should appreciate.

A NEAT DESK CALENDAR.

A desk calendar for 1901 has been issued by the Canada Horse Nail Co., Montreal. It is neat and attractive. On one corner is the trade mark of the firm in gold, while the calendar itself is printed in red and brown. On the back of the calendar are a few reminders for the trade.

ARE REBUILDING THEIR PLANT.

The Dominion Snath Co., Waterville, Que., whose building and part of their plant was destroyed by fire on January 10, have issued a circular stating that they have already begun to rebuild, and with stock saved from the fire, and new stock coming in, they expect to fill all their orders promptly.

AN AUTOMATIC PISTOL.

THE invention of the automatic pistol illustrates an enormous improvement upon the ordinary revolver, which may be compared in the artillery to the invention of the Maxim over other quick-firing guns.

The "Browning" is the simplest, the most practical, the surest and strongest of all. Its most remarkable features are its accurate and quick-firing qualities. It represents really a perfect weapon.

Some revolvers have been called automatic because the ejection of the shells is made by one movement, the "breaking" of the weapon, but in the "Browning" pistol the ejection of the empty shells, the introduction of the cartridge into the chamber, and the cocking of the firing pin are entirely automatically effected by the action of the recoil.

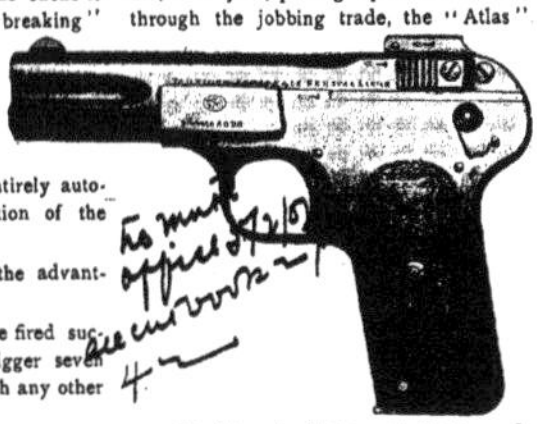

The "Browning" Automatic Pistol.

The following are a few of the advantages of this pistol:

Rapidity—Seven shots can be fired successively by pressing the trigger seven times, and in less time than with any other small arm.

Accuracy—With the revolver, on account of the recoil which throws the barrel upwards, the shooter must aim much lower than the object in view, and, if the handle is tightly gripped, the mouth of the barrel rises up when firing.

With the "Browning" pistol, the recoil acts firstly on the sliding mechanism, and does not affect the barrel until the bullet has left the latter, so that the accuracy of the shot is perfect, and is not influenced in any way by holding the handle more or less firmly.

Another point that unfavorably influences revolver shooting, is that the bullet must first of all go from the chamber of the revolving cylinder into the barrel, which is seldom in a mathematically precise line with the cylinder, whereas, in the "Browning" pistol, the bullet is placed in the same way as with a magazine rifle. Further, no loss of gas takes place from the breech as it is the case with any revolvers.

Rapidity and Accuracy—With this pistol, the recoil, being largely absorbed by the recoil-spring, does not jerk the arm out of the firing line; this, combined with the easiness of the trigger, allows of much quicker firing than a revolver.

Safety.—The pistol can be carried without the least danger. The safety-catch can easily be worked, but it is made in such a way that it cannot be moved accidentally. When the firing mechanism is locked, which can be seen at a glance, the pistol is absolutely harmless.

Construction.—The construction is symmetrical and of a most graceful shape. The weight is well apportioned; the hand being in its natural position renders accurate aim much easier, and, as there is no cylinder as in a revolver, the arm is absolutely flat and consequently very portable.

Weight.—The weight of a non-loaded pistol does not exceed 1¼ lb.

"ATLAS" BARN-DOOR HANGERS.

A. R. Woodyatt & Co., of Guelph, Ont., are, this year, placing upon the market, through the jobbing trade, the "Atlas" barn-door hangers with roller bearings, as shown on page 8, which they claim will equal, and in some respects, excel any similiar line made in the United States. As this firm have a reputation of turning out only the best quality of goods, we feel sure the trade will find it to their own advantage in future to carry this line in their stock.

COMPETING WITH THEIR CUSTOMERS.

Editor HARDWARE AND METAL,—I am sorry to see by your paper that the retail hardware merchants of Montreal are placed in the uncomfortable position of having to compete for trade with the wholesale houses from whom they buy. Fortunately, we hardware dealers of Toronto are better protected by the wholesale houses, and have very few complaints to make about their selling retail. The reading of the report of the meeting of the Retail Hardware Association in last week's paper reminded me of some trouble the Toronto retail grocers had with their wholesale houses in regard to the latter selling retail some years ago. The Retail Grocers' Association approached the wholesalers on the matter, and, after a conference had been held, the wholesalers one and all signed a pledge that they would cease selling retail entirely. This declaration was printed and a copy sent to each of the grocers of the city.

Thereafter, when any retailer detected a case of a wholesaler selling retail, he reported it to a committee of retailers, who waited on the wholesale house concerned in regard to the matter. If this proceeding was not efficacious for the purpose intended, the Grocers' Association boycotted the wholesaler who refused to abide by his pledge. I remember quite distinctly that one wholesaler was boycotted for some time, and so successfully that he was finally forced to come around and respect the wishes of the retail merchants.

I would suggest that the retailers and wholesalers hold a conference and discuss the matter frankly. By this means they ought to be able to arrive at an understanding.

TORONTO RETAILER.

HEATING AND PLUMBING

VERDIGRIS DUE TO COPPER BOILERS

IN his report to D. Stewart, acting warden of the St. Vincent de Paul penitentiary, Quebec, the medical officer of the institution, L. A. Fortier, drew attention to the danger of using copper-lined boilers and pipes for culinary purposes as follows:

"The said boilers being of pure copper with their long pipes for the escape of steam, I was compelled to examine them and inquire if there had been some verdigris observed inside of the said boilers and pipes. I received the assurance that no verdigris had been observed and that these pipes were kept perfectly clean by the steam running through the said copper pipes. My fears for a good while remained silent, but in presence of the large number of diseases in the alimentary canal of the convicts, I lately pushed an investigation so far as to discover numerous warts of verdigris growing inside of the said copper pipes.

"Copper does not change in dry air, but in moist air becomes covered with a green coat of carbonate, known as verdigris. In the present case the large and long surface exposed to the moisture is under the most favorable circumstances for the production of the said carbonate of copper, a poisonous salt of powerful action on animals. The soluble salts of copper are no more corrosive when largely diluted, but their action is but irritating or inflaming. The carbonate of copper is one of the soluble salts of the metallic copper.

"Experience teaches that copper becomes poisonous with time in small doses. In the present case, we are in presence of fractionally copper daily consumers. The elimination of copper is slow, being operated by the mucous membrane of the alimentary canal, the salivary glands, the liver and the kidneys.

"No wonder now that an immense surface of metallic copper being exposed to an easy production of verdigris, a soluble salt, becomes an active factor in the generation of a multitude of diseases in the alimentary canal of the convicts.

"But every convict is not equally apt to become sick by copper ingestion; it is of daily observation as for many other noxious substances. The five large boilers actually in use are in bad order, being now coated with tin. The present cooking system is a dangerous, manifestly a dangerous one.

"And I terminate my report with a remark taken from page 575 of Dispensatory of the United States: 'Vessels of copper which are not coated with tin should not be used in pharmaceutical or culniary operations; for, although the metal uncombined is inert, yet the risk is great that the vessels may be acted on and a poisonous salt formed.'"

SOME BUILDING NOTES.

WM. HODGSON, architect, Ottawa, is preparing plans for a new public school on Wellington street, Ottawa.

A new Presbyterian manse is being built at Lyn, Ont.

Work has been started on the Medical building for Queen's University, Kingston, Ont.

The Vancouver Times, in a recent editorial article, declared that a new high school is needed in that city.

Architect Ellis, Kingston, Ont., is preparing plans for a residence which he will build for John Marshall, M.A.] The new residence will be located on Union street.

Building permits have been granted in Toronto to Wm. Moss for a pair of semi-detached houses on Arthur street, to cost $3,800, and to Geo. Isaac, for a pair of semi-detached houses at 18 and 20 Dupont street, to cost $4,400.

A deputation representing the Alexandria school for girls, East Toronto, have asked the Ontario Government for a grant of $2,500, or half the cost, to erect a laundry and assembly hall. Premier Ross, in reply, stated that the Government would make the grant if possible.

Arthur Thompson, Ottawa, has had plans prepared by Mr. Baker, architect, New York, for an office building on the north-east corner of Sparks and Metcalfe streets. The building, which will be of stone and pressed brick, will be 66 x 99 ft. in size and six storeys high. The estimated cost is $140,000, but in Ottawa it is probable that the work can be done at a figure somewhat below the estimate. Modern conveniences will be installed throughout. It is proposed to start the building early in the spring and rush it through to completion.

FREE TRADE IN OLD LEAD.

It is a disgrace that plumbing worth $500 or $1,000 should be ripped out of the interior of a house and sold by thieves for what it will bring as old junk.

The question is raised as to whether the demands of the junk shop have not a tendency to create the supply which is furnished by the destroyers of plumbing in vacant houses.

The police might be assisted by a change in the law which would surround the sale of second-hand plumbing materials with some safeguards. If the vendor of old lead had to register his name and address, every transaction could be traced, and greater risks would attend the sale of the booty secured in these raids upon the plumbing of vacant houses.—Telegram, Toronto.

TORONTO PLUMBERS TO MEET.

For some time the Toronto Journeymen Plumbers' Union have been in negotiation with the Toronto Master Plumbers' Association, with a view to secure shorter hours and an advance in the rate per hour.

The minimum rate is now 27½c. per hour, the masters having agreed to a raise of 2½c. in August, 1898, previous to which time the minimum rate was 25c. per hour. The journeymen now want 31c. per hour. They further ask that the day's work be reduced from 9 hours per day to 8 hours.

It was reported by one of the daily papers on Wednesday that the masters had decided to forestall the possibility of a strike by declaring a lockout. There was no ground for the report, as the meeting to discuss the matter has not yet been held. It was reported also that a conference between the masters and the journeymen had been held on Wednesday evening. This was also erroneous, though a meeting is to be held this week.

PLUMBING AND HEATING NOTES.

Pelletier & Daniel have registered as plumbers in Montreal.

A. Desmarteau has been appointed curator of J. Lafrance & Co., plumbers, Montreal.

D. McKenzie & Co., plumbers, Winnipeg, Man., have dissolved. D. McKenzie continues.

Cornelius Brady has registered under the style of The United Incandescent Light Co., Montreal.

Manchester, Robertson & Allison, Limited, have been incorporated in St. John, N.B., with a capital of $800,000, to generate, sell and use electric power, light and heat in St. John.

QUICK WORK.

The Bennett & Wright Co., Limited, did some quick work on their contract for heating the new office building of the Grand Trunk Railway's headquarters in Montreal. In 10 days, they had the boilers and steam pipes in position and the fire started. And yet, the contract was one of the largest, if not the largest, ever awarded in Montreal. The statement made in last week's issue, that the contract was for the heating of the G. T. R. sheds, was hardly correct. It should have read "office building."

GLASS CUTTERS that cut glass.

We have 'em; You need 'em. Have you the GREEN BOOK? Ask for it. Contains good goods at right prices.

Smith & Hemenway Co., UTICA DROP FORGE & TOOL CO.,
296 Broadway, New York City.

CANADIAN ADVERTISING is best done by THE E. DESBARATS ADVERTISING AGENCY MONTREAL.

ADVERTISING in WESTERN CANADA

will be Carefully, Efficiently, and Promptly attended to, by

The Roberts Advertising Agency,
WINNIPEG, CANADA.

KNOX HENRY

Heavy Hardware and Metal Broker
Room 220½ Board of Trade, MONTREAL.

HORSE NAILS—"C" Brand Horse Nails—Canada Horse Nail Co.
"BRASSITE" GOODS—Gunn Castor Co., Limited, Birmingham, Eng.

Manufacturers of

Heating Supplies

Pipe Fittings and Headers.
Large Manifolds made to Order.
Steam Traps and Appliances, etc.

The . . .
Jas. Morrison Brass Mfg. Co., Limited
TORONTO.

Standard Paint & Varnish Works
Limited
Makers of High Grade
Varnishes, Japans,
Paints, Colors & Enamels.
Windsor, Ont.

THE . . .

Waggoner Extension Ladder.

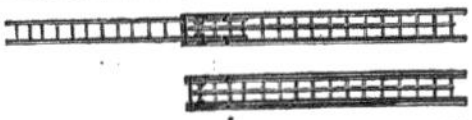

The strongest, lightest and most convenient ladder in the market. The only really satisfactory extension ladder made. Pulls up with a rope. Made in all lengths. Also extension and other step ladders, sawhorses, ironing boards, painters' trestles, etc. All first-class goods. Write for quotations to

The Waggoner Ladder Company, Limited, London, Ont.

E. B. SALYERDS
Manufacturer of

Hockey Sticks

PRESTON,
Ontario, - Canada.

The Best Stick.
Made of Rock Elm.
Wholesale Trade Only Supplied.
Ask your Wholesale House for the Preston make of Stick.
Write for Prices.

LEADER CHURN

New Century Improvements.

FOUR DIFFERENT STYLES:
A—Steel Frame with double reversible Steel Lever.
B—Wood Frame with double reversible Steel Lever.
C—Steel Frame with Crank.
D—Wood Frame with Crank.
Styles A and B may be operated from a sitting or standing position.

Steel Frames and Hoops beautifully ALUMINIZED.
All LEADER CHURNS are equipped with BICYCLE BALL BEARINGS and PATENTED CREAM BREAKERS.
Stands are so constructed that they are particularly strong and rigid, and there is nothing to interfere with the placing of pail in the most convenient position for draining off buttermilk.

It Pays to Have the Best. None are Better Than the Leader.

THE
Dowswell Manufacturing Co.
Limited.
HAMILTON, ONT.
Eastern Agents: W. L. Haldimand & Son, Montreal, Que.

Use Syracuse Babbitt Metal

IT IS THE BEST MADE.

For Paper and Pulp Mills, Saw and Wood Working Machinery, Cotton and Silk Mills, Dynamos, Marine Engines, and all kinds of Machinery Bearings.

Wire, Triangular and Bar Solder, Pig Tin, Lead, Ingot Copper, Ingot Brass, Antimony, Aluminum, Bismuth, Zinc Spelter, Phosphor Tin, Phosphor Bronze, Nickle, etc., always in stock.

Factories: { 332 William St., MONTREAL, QUE. and SYRACUSE, N.Y.

Syracuse Smelting Works

THE IRON AND STEEL INDUSTRY.

FOLLOWING is the balance of the address of Mr. Walter Kennedy, mechanical engineer of the Cramp Ontario Steel Co., delivered before the Canadian Manufacturers' Association, Toronto, on January 17, and unavoidably held over from our last issue:

Dealing with the growth of blast furnaces in the United States, Mr. Kennedy said he could not go back further than the old charcoal furnaces that were operated at one time, and had an average output of four or five tons a day. Twenty years ago 50 and 60 tons per day was considered a large output. He remembered when Lowthian Bell came from England to the United States to see the blast furnace near Youngstown, Ohio, which had an output of 60 tons per day. Ten years after that, 200 tons was considered a good day's work. Now there were at least 10 blast furnaces in the United States having a product of more than 600 tons a day; in fact, one had averaged 700 tons a day for a month. There were people who claimed that within the next few years there would be furnaces making 3,000 tons a day, though this seemed an extravagant estimate.

"I have been asked what is to be done at Collingwood," continued Mr. Kennedy. "Our idea is to equip a plant there suitable for the present needs of the local markets, and, if possible, something additional will be exported to foreign countries.

THE LATEST EQUIPMENTS.

"It will be equipped with machinery that is the latest and best of its kind for every department. The intention is to form an organization that will be capable of solving the different problems as they arise, just as Mr. Clergue and his associates have done at the Soo, and then encourage and branch out into any kind of business that seems most profitable and will build up other business in the Province of Ontario. (Applause.)

"Our plans for Collingwood contemplate the erection of four blast furnaces, one of which will be built immediately; 10 open-hearth furnaces, four of which will be built immediately; and a blooming mill, which breaks down large ingots and produces the billet that may be put into any kind of finished article. (Applause.) There has been some talk of starting a wire rod mill, as the wire fence business is becoming a very large one in the western part of this country. There are many lines that could be developed to advantage, but we do not propose trying to do everything at the start. I don't think it possible to try to cover at first so large a ground as The Carnegie Steel Company. Even the Carnegie Company began in a small way, and developed with proper management and skill.

A SAFE POLICY.

"Do I think Canada can compete with the United States in the manufacture of steel for export?

"This involves the question of coal," Mr. Kennedy answered. It would be highly important to put soft coal on the free list, so that it could be brought to the works and there converted into coke, saving a very large amount of fuel that was wasted at the coke ovens. In the United States they were not up to the Germans in the economic use of coal. Every good German firm had its plant for the saving of by-products, and if proper precautions were used, he thought it was possible in Canada to compete with the United States. (Applause.) The Carnegie Steel Company had lately purchased several thousand acres on the southern shore of Lake Erie, where they contemplated building steel works, at a cost of several million dollars, so as to have their new works on the shore of the great lakes. If they could operate successfully at that point there was certainly no disadvantage in coming a little further up the lake. There were many arguments now for assembling the raw material closer to the ore than the coal, because it required about only one-half as much coke now to make a ton of iron as it did some years ago, and the proportion now is about two tons of iron to one of coke.

THE CONSUMPTION OF FUEL

in the manufacture of iron has been cut down at least 15 or 20 per cent. in the last five or six years. The probability is that the reduction will continue, and if it does there will be another great argument on the point of cost. Duluth, I think, will become a great iron manufacturing centre. It may be some time, but it is very probable. All the locations on the lake shore in the Dominion of Canada will have a future very much of advancement equal to Duluth and equal to the United States. There are some kinds of manufactures, like the manufacture of fine wire and dental instruments—things of that kind—that can never be manufactured in any part of the world as cheaply as in Pittsburg, unless it is in China. The Chinese have facilities that should make them master of the situation when it comes to the struggle for cheapness and the survival of the fittest. That may yet be a long time, unless God in His providence should ordain otherwise.

NUMBER OF MEN TO BE EMPLOYED.

The number of men employed at Collingwood would probably not be less than 1,200, from that to 2,000 to commence work. But a great deal would have to depend on the grade of the finished material. It takes many more men to finish a ton of material in thin sheets than in heavy material such as bridges. A great deal also depends on how prosperous is the enterprise, how energetically it is pushed.

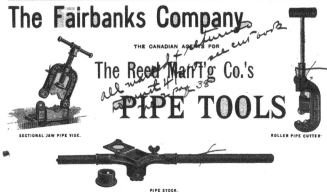

RELATION OF ADVERTISING TO THE COST OF GOODS.*

By Walter H. Cottingham.

One of our representatives a short time ago suggested that perhaps the recent change in price was made necessary by the large expenditure we are making in advertising and he made the suggestion that we give less advertising and a lower price.

Now, good advertising, as I have tried to explain during this session, does not make goods more expensive. Good advertising will lower the cost of doing business, and if it does not do this it is not good advertising. I want everybody to feel perfectly satisfied on this point.

We are often thus accused of making our prices high on account of our advertising.

The fact is, if we did not do so much advertising and do it so well, our prices would have to be higher.

To me, a proposition to increase advertising expenditures, or, let us say, to do better advertising in order to lower prices, would be more practical than a proposition to do less advertising to accomplish the same object.

Our advertising expenditure for the past year, while it was considerably larger in the aggregate, was materially lower in percentage to sales than it has been for many years past. Such results are what we aim to achieve in this department, and they are largely dependent upon the character of the advertising and the care with which it is put into effect.

Take as an illustration a man who is doing a business of $100,000 per annum, and let us suppose his expenses are $25,000, which is 25 per cent. to his sales. He wishes to increase this business and he decides to advertise. Let us suppose that he decides on an expenditure for this purpose of $7,500 per annum. His expenses are then increased to $32,500. By this expenditure, let us suppose, he is able to increase his sales to $130,000 and at this rate his expenses with advertising added would amount to same percentage as before, namely, 25 per cent. He has not increased the percentage of his expense and has sold $30,000 more goods, and if his net profit was 5 per cent. he has increased his net earnings by $1,500. The amount I have named for advertising such a business should bring even larger results.

What I want to make plain is that advertising well done does not increase expenses, but will lower them. That is the way we figure in our business.

We watch the results in a very thorough and careful manner. Each division and every department is charged with the amount of the advertising, and the amount spent in this way is constantly compared with the sales. If the sales warrant the expenditure it is all right, but if the results are not forthcoming, then there is something wrong.

An advertising report is furnished by each division quarterly. It shows the cost of each different line of advertising for each line of goods. It shows the amount of advertising used and the amount on hand, and it shows the total amount compared with the total sales. It takes a great deal of time and money to get up this report, but only in this way are we able to watch results and determine what is profitable and what is not.

This company are not going to throw away any money on advertising if they can help it.

Our advertising has been a great help in building up this large business and it has enabled us to increase our output and give us as low a cost as we can expect.

*Portion of an address delivered by Managing Director Cottingham at the recent convention of the representatives of the Sherwin-Williams Co. at Cleveland.

INQUIRIES REGARDING CANADIAN TRADE.

The following were among the recent inquiries relating to Canadian trade received at the High Commissioner's office, in London, England :

1. A German firm desirous of importing from Canada, carriages, sporting goods, boat motors, etc. ; all kinds of wood goods, office, and other furniture, toys, etc., will be glad to hear from Canadian exporters open to do business.

2. Canadian firms desiring a representative in Scotland can be furnished with the name of a gentleman in Glasgow who wishes to take up agencies.

[The names of the firms making the above inquiries, can be obtained on application to the editor of Hardware and Metal. When asking for names, kindly give number of paragraph and date of issue.]

Mr. Harrison Watson, curator of the Canadian Section of the Imperial Institute, London, England, is in receipt of the following inquiries :

1. A Glasgow house seeks supplies of Canadian oak staves for coopers' purposes and invite quotations.

2. A house possessing a considerable connection in Australia and New Zealand in boots and shoes would like to hear from Canadian manufacturers who are in a position to compete with American goods in the Australasian market.

3. A London timber house is prepared to undertake the agency of a first-class Canadian shipper of hardwoods. Old established connection.

4. A firm of Sheffield cutlery manufacturers would like to hear from a first-class Canadian house which could take up the sale of their goods for Canada.

QUEBEC BOARD OF TRADE.

THE sixtieth annual meeting of the Quebec Board of Trade, which was held on Tuesday afternoon, January 15, was attended by a large proportion of the prominent business men of the city, and proved to be one of the most successful meetings held for some time.

The report of the treasurer, James Brodie, showed the finances of the board to be in a healthful, satisfactory condition.

A communication was received from J. G. Scott, general-manager of the Canada Atlantic Railway, agreeing to place at the disposal of the Quebec Board of Trade a special train with sleepers, to leave Quebec and run through to Parry Sound at any date that may be convenient to the board after the opening of navigation in the spring.

On motion of J. B. Garneau, seconded by W. H. Wiggs, a resolution was passed condemning the rule recently inaugurated in Her Majesty's Customs in Quebec, viz.: That of insisting that all invoices (without exception) have to be checked before the entry can be passed. It was pointed out that other cities, including Montreal, were not "saddled" with this new rule. A motion was, therefore, passed requesting the Customs authorities of this port to rescind this rule, and that invoices be checked as heretofore, and refund entries be made in event of any errors, the collector to reserve the right to check the invoices of strangers.

A. B. Van Felson reported to the meeting that while freight from Montreal was delivered free in Three Rivers, that from Quebec was delivered at the expense of the receiver, notwithstanding the fact that the rate is the same and Montreal is five miles farther away than is Quebec.

The report of the council dealt with freight and passenger service to Quebec; the establishment of a Canadian Lloyds, with $5,000,000 capital, which will probably have the affect of prolonging navigation on the St. Lawrence; the establishment of abattoirs throughout the Province; the establishment of technical education in the Province; the laying of the corner stone of the Quebec bridge; the opening of the Great Northern Railway and its elevator, and the shipment of its first cargo of grain. Regarding bankruptcy, the council favors the enactment of a severe Act, uniform throughout the country, and has appointed a sub-committee which is now studying the matter in detail.

With respect to winter navigation the report says: "The most important question of the winter navigation of the St. Lawrence is now more than ever on the tapis, owing to the requirements of the Western grain trade, and it is evident that we are nearing a favorable solution of the problem. The principal requirement is the replacing of the lightships and gas buoys by permanent lights, an improvement which, we understand, has already been carried out on the lakes."

The following officers were elected for the ensuing year:

President—George Tanguay, M.P.P.
First Vice-President—John Ritchie.
Second Vice-President—P. J. Bazin.
Treasurer—D. J. Rattray.
Council—V. Chateauvert, J. G. Garneau, Joseph Winfield, G. E. Amyot, V. Lemieux, R. H. Smith, Napoleon Drouin, M. Joseph, D. Arcand, Nazaire Lavoie, P. B. Dumoulin and O. Poitras.

A vote of thanks was passed on motion of Messrs. Tanguay and Van Felson to the retiring president for his active interest in, and adequate efforts towards, the success of the Quebec Board of Trade during his term of office.

B.C.'S FIRST CONVERTER.

A DESPATCH from Greenwood, B.C., says: "The announcement that the British Columbia Copper Co. has let a contract to the E. P. Allis Co., Milwaukee, Wis., for the immediate construction of a converting plant, is one of great importance, as this will be the first converter in the Province. That the company should have placed the order before its smelter had turned out a pound of matte, leaves a little room to doubt the immense amount of confidence the directors place in the ability of Paul Johnson, the manager of the smelter department. He designed the local smelter, and most of the machinery is of his own invention. There now appears every probability that ere the installation of the bessemerising plant, the machinery for which is to be completed in three months, a second furnace will have been added to the reduction works, bringing the daily capacity up to 600 tons.

"The contract calls for a blowing engine, 40-ton electric crane, one stand of converters, crushing plant and accessories. Mr. H. V. Croll, manager of the Spokane branch of the E. P. Allis Co., stated that the cost would be in the neighborhood of $40,000 at the works, and by the time it is installed there an additional $15,000 will have been added. It is to be capable of treating a daily output of 600 tons of ore, producing roughly 40 tons of matte. This amount passing through the converter means about 20 tons of blister copper, averaging 98¼ to 99 per cent. The next step will be the making of electrolytic copper, and in time such a refining plant will be added. With the exception of the Butte smelters, none of the other Western reduction plants have yet installed converters. The Northport smelter has a calcining plant to reduce the sulphur excess in the ores."

Charles Pearson & Co., general merchants, Cedarhall, Que., have sold their stock at 75¼c. on the dollar to J. B. E. Bergevin, Matane, Que., and their book debts at 52c. on the dollar to P. Z. Dube, Amqui.

CURRENT MARKET QUOTATIONS

January 25, 1911.

These prices are for such qualities and quantities as are usually o dered by retail dealers on the usual te ms of c edit, the lowest figures being for larger quantities and prompt pay. Large cash buyers can frequently make purchases at better prices The Editor is anxious to be informed at once of any apparent errors in this list as the desire is to make it perfectly accurate.

METALS.

Tin.

Lamb and Flag and Straits—
56 and 28 lb. ingots, per lb. .... 0 32 .... 0 33

Tinplates.

Charcoal Plates—Bright

M.L.S., equal to Bradley. Per box
I.C., usual sizes $7 00
I.X., " 8 50
I.X.X., " 10 00

Famous—
I.C. 7 50
I.X 8 50
I.X.X 9 50

Raven & Vulture Grades—
I.C., usual sizes 5 00
I.X., " 6 00
I.X.X " 7 00
I.XXX., " 8 00
D.C., 12½x17 .... 4 75
D.X. 5 50
D.X.X 7 50

Coke Plates—Bright

Bessemer Steel—
I.C., usual sizes 4 15
I.C., special sizes, base 4 50
20 x 28 .... 8 50

Charcoal Plates—Terne

Dean or J. G. Grade—
I.C., 20x28, 112 sheets 8 75
I.X., Terne Tin 10 75

Charcoal Tin Boiler Plates.

Cookley Grade— Per lb.
X X., 14x56, 50 sheet bxs / " 14x60 " / " 14x65, " } 0 07 .... 0 07½

Tinned Sheets.

72x30 up to 24 gauge .... 0 08 .... 0 08½
" 26 " .... 0 08½ 0 09
" 28 " .... 0 09 .... 0 09½

Iron and Steel.

Common Bar, per 100 lbs. .... 1 65 .... 1 70
Refined " " 2 15
Horse Shoe Iron " 2 05
Hoop steel, 1½ to 3 in. base, extras for smaller sizes 3 10
Sleigh Shoe Steel " base 2 00
Tire Steel 2 00
Machinery iron finish 2 05
Cast Steel, per lb .... 0 00 .... 0 00
Toe Calk Steel 2 30
T. Firth & Co's special cast steel, per lb. 0 13
Jessop's Tool Steel 0 13

Boiler Tubes.

1½-inch 0 21½
2 " 0 13½
2½ " 0 16
3 " 0 17½

Steel Boiler Plate.

¼ inch 2 25
3-16 inch 2 25
⅜ inch and thicker 2 25

Black Sheets.

18 gauge 3 10
20 gauge 3 10
22 to 24 " 3 20
26 " 3 35
28 " 3 40

Canada Plates.

All dull, 52 sheets 3 00
Half polished 3 15
All bright .... 3 65 .... 3 75

Iron Pipe.

Black pipe—
¼ inch 3 00
⅜ " 3 00
½ " 3 15
¾ " 3 20
1 " 4 50
1¼ " 6 35
1½ " 7 55
2 " 10 10
2½-6 inch, discount 60 p c.

Galvanized pipe—
¼ inch 4 65
¾ " 5 35
1 " 7 05
1¼ " 9 75
1½ " 11 25
2 " 15 50

Galvanized Sheets.

	G. C.	Comet.	Amer.	Queen's Head.
16 gauge		....	4 5	4 00
18 to 24 gauge	4 25	3 85	4 35	4 25
26 "	4 50	4 10	4 35	4 50
28 "	4 75	4 35	4 50	4 75

28 gauge American equals 26 gauge English.

Chain.

Proof Coil, 3-16 in., per 100 lb ....
" ¼ " " .... 8 00 .... 8 50
" 5-16 " " 5 35 5
" ⅜ " " .... 4 35 .... 4 85
" 7-16 " " .... 4 15 .... 4 65
" ½ " " .... 4 35 .... 4 60
" ⅝ " " .... 3 85 .... 4 35
" ¾ " " .... 3 80 .... 4 00
Halter, kennel and post chains, 40 and 50 p.c.
Cow ties .... 40 p.c.
Stall fixtures .... 35 p.c.
Trace chain .... 25 p.c.
Jack chain, iron, single and double, discount 35 p c.
Jack chain, brass, single and double, discount 40 p.c.

Copper.

Ingot
English B. S., ton lots .... 0 19 .... 0 20
Lake Superior ....

Bolt or Bar.
Cut lengths round, ½ to ⅞ in. .... 0 23½ 0 25
" round and square 1 to 2 inches .... 0 23½ 0 25

Sheet.
Untinned, 14 oz., and light, 16 oz., 14x48 and 14x60 .... 0 23 .... 0 23
Untinned, 14 oz., and light, 16 oz., irregular sizes .... 0 23 .... 0 23½

NOTE.—Extra for tinning, 2 cents per pound, and tinning and half planishing 3 cents per pound.

Tinned copper sheets 0 26
Planished 0 32

Braziers (in sheets.)
4x6ft. 25 to 30 lbs. ea., per lb. 0 25½
" 35 to 45 " " 0 24½
" 50-lb. and above, " 0 23½

Boiler and T. K. Pitts
Plain Tinned, per lb 0 28
Spun, per lb 0 32

Brass.

Rod and Sheet, 14 to 30 gauge, 15 per cent.
Sheets, hard-rolled, 2x4 .... 0 24 .... 0 25
Tubing, base, per lb .... 0 24 .... 0 25

Zinc Spelter

Foreign, per lb .... 0 06 .... 0 06½
Domestic " ....

Zinc Sheet.

5 cwt. casks .... 6 75 .... 7 00
Part casks .... 7 00 .... 7 50

Lead.

Imported Pig, per lb .... 0 04¾ 0 05
Bar, 1 lb. .... 0 05½ 0 06¾
Sheets, 2½ lbs. sq. ft., by 0 06¾
Sheets, 3 to 6 lbs., " 0 06

NOTE.—Cut sheets ½ cent per lb. extra

Pipe, by the roll, usual weights per yard, lists at 7c. per lb. and 15 p.c. dis. f.o.b. Toronto.

NOTE.—Cut lengths, net price, waste pipe 8-ft. lengths lists at 7½ cents.

Shot.

Common, $6.50 per 100 lb.; chilled, $7.00 per 100 lb.; buck, seal and ball, $7.50. Discount, 7½ p.c. Prices are f.o.b. Toronto, Hamilton, Montreal, St. John and Halifax. Terms 3 per cent. cash, freights equalized on Montreal.

Soil Pipe and Fittings.

Discount, 60 and 10 per cent. on medium and extra heavy, and 60 per cent. on light.

Solder. Per lb. Per lb.

Bar half-and-half, guarant'd 0 19
Bar half-and-half, commer'l 0 18½
Refined 0 18½
Wiping 0 18

NOTE.—Prices of this graded according to quantity. The prices of other qualities of solder in the market indicated by private brands vary according to composition.

Antimony.

Cookson's, per lb. .... 0 11 .... 0 11½

White Lead. Per 100 lb.

Pure 6 87½
No. 1 do 6 50
No. 2 do 6 12½
No. 3 do 5 75
No. 4 do 5 37½
Munro's Select Flake White 7 37½
Elephant and Decorators' Pure 7 12½
Brandram's B B. Genuine 8 00
" " Decorative 7 55
" " No. 1 .... 6 85
" " No. 2 .... 6 00

Red Lead.

Genuine, 560 lb. casks, per cwt. $5 50
Genuine, 100 lb. kegs, per cwt. 5 75
No. 1, 560 lb. casks, per cwt. 5 25
No. 1, 100 lb. kegs, per cwt 5 00

White Zinc Paint.

Elephant Snow White .... 0 08 .... 0 09
Pure White Zinc .... 0 08 .... 0 09
No. 1 .... 0 06 .... 0 07½
No. 2 .... 0 05 .... 0 06½

Dry White Lead.

Pure, casks 5 75
Pure, kegs 6 25
No. 1, casks 5 50
No. 1, kegs 5 00

Prepared Paints.

In ¼, ½ and 1 gallon tins.
Pure, per gallon 1 25
Second qualities, per gallon 1 10
Barn (in bbls.) .... 0 75 .... 0 85
The Sherwin-Williams Paints 1 45
Canada Paint Co's Pure 1 25
Toronto Lead & Color Co's Pure 1 25
Sanderson Pearcy's Pure 1 20
Stewart & Wood's Champion Pure. 1 20

Colors in Oil.

25 lb. tins, Standard Quality.
Venetian Red, per lb. 0 05
Chrome Yellow 0 11
Golden Ochre 0 06
French " 0 05
Marine Black 0 09
" Green 0 09
Chrome " 0 08
French Imperial Green 0 09

Colors, Dry.

Yellow Ochre (J.C.) bbls .... 1 35 .... 1 40
Yellow Ochre (J.F.L.S.), bbls 2 75
Yellow Ochre (Royal) .... 1 10 .... 1 15
Brussels Ochre 2 00
Venetian Red (best), per cwt. .... 1 80 .... 1 90
English Oxides, per cwt. .... 3 00 .... 3 25
American Oxides, per cwt. .... 1 75 .... 2 00
Canadian Oxides, per cwt. .... 1 75 .... 2 00
Super Magnetic Oxides, 93p.c. .... 2 00 .... 2 25
Burnt Sienna, pure, per lb. 0 10
" Umber, " " 0 10
do Raw 0 09
Drop Black, pure 0 09
Chrome Yellows, pure 0 18
Chrome Greens, pure, per lb. 0 12
Golden Ochre 0 03½
Ultramarine Blue in 28-lb. boxes, per lb. .... 0 08 .... 0 24
Fire Proof Mineral, per 100 lb 1 00
Genuine Eng. Litharge, per lb 0 07
Mortar Color, per 100 lb. 1 25
English Vermillion 0 80
Pure Indian Red, No. 45 lb. 0 80
Whiting, per 100 lb. 0 55

Blue Stone.

Casks, for spraying, per lb. 0 07
100-lb. lots, do. per lb. 0 08

Putty.

Bulk in bbls. 2 00
Bulk in less quantity 2 15
Bladders in bbls. 2 20
Bladders in kegs, boxes or loose 2 35
Bladders in 25-lb. tins 2 45
Bladders in 12½-lb. tins 2 75
Bladders in bulk or tins less than 100lb 3 00

Varnishes.

In 5-gal. lots.). Per gal.
Carriage, No. 1 .... 2 90 .... 3 30
" body .... 8 00 .... 9 00
" rubbing .... 4 00 .... 5 00
Gold Size, Japan .... 3 00 .... 3 40
Brown Japan .... 2 40 .... 2 80
Elastic Oak .... 2 90 .... 3 30
Furniture, extra .... 2 40 .... 2 80
" No. 1 .... 1 60 .... 2 00
Hard Oil Finish .... 2 70 .... 3 10
Light Oil Finish .... 3 20 .... 3 60
Demar .... 3 30 .... 3 70
Shellac, white .... 4 40 .... 4 80
" orange .... 4 00 .... 4 40
Furniture Brown Japan .... 1 60 .... 2 00
Black Japan .... 2 40 .... 2 80
" " No. 1 .... 1 60 .... 2 00

The Imperial Varnish & Color Co's., Limited Elastilite Varnish 1 gal. can, each. $2.00.

Granatine Floor Finish, per gal. $2.00.

Maple Leaf Coach Enamels; Size 1, 60c; Size 2, 35c.; Size 3, 20c. each.

Linseed Oil.

Raw. Boiled.
1 to 4 bbls delivered .... $0 82 .... $0 85
5 to 9 bbls " 81 84

Toronto, Hamilton, London and Guelph 2c. less.

Turpentine.

Single barrel, freight allowed 0 59
2 to 4 barrels " " 0 58

Castor Oil.

East India, in cases, per lb. .... 0 10 .... 0 10½
" " small lots .... 0 10½ 0 11

Cod Oil, Etc.

Cod Oil per gal .... 0 50 .... 0 55
Pure Olive 1 20
" Neatsfoot 90

Glue.

Common .... 0 08½ 0 09
French Medal .... 0 14 .... 0 14½
Cabinet sheet .... 0 12 .... 0 13
White, extra .... 0 18 .... 0 20
Gelatine .... 0 22 .... 0 30
Strip .... 0 18 .... 0 20
Coopers .... 0 19 .... 0 20
Huttner 0 18

HARDWARE.

Ammunition.

Cartridges.

B. B. Caps, Dom. 50 and 5 per cent.
Rim Fire Pistol, dis. 40 p. c., Amer.
Rim Fire Cartridges, Dom., 50 and 5 p. c.
Central Fire Pistol and Rifle, 10 p. c. Amer.
Central Fire Cartridges, pistol sizes, Dom. 30 per cent.
Central Fire Cartridges, Sporting and Military, Dom., 15 and 5 per cent.
Central Fire, Military and Sporting, Amer. add 5 p.c. to list. B.B. Caps, discount 40 per cent, Amer.
Loaded and empty Shells, "Trap" and "Dominion" grades, 25 per cent. Rival and Nitro, net list.
Brass shot Shells, 55 per cent
Primers, Dom., 30 per cent.

Wads — per lb
Best thick white felt wadding, in ¼-lb bags ... 1 00
Best thick brown or grey felt wads, in ½-lb. bags ... 0 70
Best thick white card wads, in boxes of 500 each, 12 and smaller gauges 0 99
Best thick white card wads, in boxes of 500 each, 10 gauge ... 0 35
Best thick white card wads, in boxes of 500 each, 8 gauge ... 0 55
Thin card wads, in boxes of 1,000 each, 12 and smaller gauges ... 0 20
Thin card wads, in boxes of 1,000 each, 10 gauge ... 0 25
Thin card wads in boxes of 1,000 each, 8 gauge ...
Chemically prepared black edge grey cloth wads, in boxes of 250 each— Per M
11 and smaller gauge ... 0 60
9 and 10 gauges ... 0 70
7 and 8 gauges ... 0 90
5 and 6 gauges ... 1 10
Superior chemically prepared pink edge, best white cloth wads, in boxes of 250 each—
11 and smaller gauge ... 1 15
9 and 10 gauges ... 1 40
7 and 8 gauges ... 1 65
5 and 6 gauges ... 1 90

Adzes.

Discount, 20 per cent.

Anvils.

Per lb ... 10 0 12½
Anvil and Vise combined ... 4 50
Wilkinson & Co.'s Anvils..lb. 0 09 0 09½
Wilkinson & Co.'s Vices..lb. 0 09¾ 0 10

Augers.

Gilmour's, discount 65 and 5 p.c. off list.

Axes.

Chopping Axes—
Single bit, per doz ... 6 50 10 00
Double bit, " ... 12 00 18 00
Bench Axes, 40 p.c.
Broad Axes, 33⅓ per cent.
Hunters' Axes ... 5 50 6 00
Boy's Axes ... 5 75 6 75
Splitting Axes ... 6 50 12 00
Handled Axes ... 7 00 10 00

Axle Grease.

Ordinary, per gross ... 5 75 6 00
Best quality ... 13 00 15 00

Bath Tubs.

Zinc ... 6 00
Copper, discount 15 p.c. off revised list

Baths.

Standard Enameled.
5½-inch rolled rim, 1st quality ... 30 00
" " " 2nd " ... 22 00

Anti-Friction Metal.

"Tandem" A ... per lb. 0 27
" B ... " 0 21
" C ... " 0 11½
Magnolia Anti-Friction Metal, per lb. 0 25

SYRACUSE SMELTING WORKS.
Aluminum, genuine ... 0 45
Dynamo ... 0 29
Special ... 0 25
Aluminum, 99 p.c. pure "Syracuse" .. 0 50

Bells.

Hand.
Brass, 60 per cent.
Nickel, 55 per cent.

Cow.
American make, discount 66⅔ per cent.
Canadian, discount 45 and 50 per cent.

Door.
Gongs, Sargant's ... 5 50 8 00
" Peterboro', discount 45 per cent.

Farm.
American, each ... 1 25 3 00

House.
American, per lb ... 0 35 0 40

Bellows.

Hand, per doz ... 3 35 4 75
Moulders', per doz ... 7 50 10 00
Blacksmiths', discount 40 per cent.

Belting.

Extra, 50 and 10 per cent.
Standard, 60 per cent.
No. 1 Agricultural, 60 and 10 p.c.

Bits.

Auger.
Gilmour's, discount 60 and 5 per cent.
Rockford, 50 and 10 per cent.
Jennings' Gen., net list.

Car.
Gilmour's, 47½ to 50 per cent.

Expansive.
Clark's, 40 per cent.

Gimlet.
Clark's, per doz ... 0 65 0 90
Damond, Shell, per doz ... 1 00 1 50
Nail and Spike, per gross ... 2 25 5 20

Blind and Bed Staples.

All sizes, per lb ... 0 07½ 0 12

Bolts and Nuts. Per cent.

Carriage Bolts, full square, Norway ... 70
" " full square ... 70
Common Carriage Bolts, all sizes ... 65
Machine Bolts, all sizes ... 65
Coach Screws ... 75
Sleigh Shoe Bolts ... 75
Blank Bolts ... 65
Bolt Ends ... 65
Nuts, square ... 4½c. off
Nuts, hexagon ... 4¾c. off
Tire Bolts ... 67½
Stove Bolts ... 67½
Stove rods, per lb ... 5½ to 6c.
Plough Bolts ... 60

Boot Calks.

Small and medium, ball, per M ... 4 25
Small heel, per M ... 4 50

Bright Wire Goods.

Discount ... 55 per cent.

Broilers.

Light, dis., 65 to 67½ per cent.
Reversible, dis., 65 to 67½ per cent.
Vegetable, per doz., dis. 37½ per cent.
Henis, No. 8, " ... 6 00
Henis, No. 9, " ... 7 00
Queen City " ... 7 50 0 00

Butchers' Cleavers.

German, per doz ... 6 00 11 00
American, per doz ... 12 00 20 00

Building Paper, Etc.

Plain building, per roll ... 0 30
Tarred lining, per roll ... 0 40
Tarred roofing, per 100 lb ... 1 65
Coal Tar, per barrel ... 3 50
Pitch, per 100-lb ... 0 85
Carpet felt, per ton ... 45 00

Bull Rings.

Copper, $2.00 for 2½ in. and $1.90 for 2 in.

Butts.

Wrought Brass, net revised list

Cast Iron.
Loose Pin, dis., 60 per cent.

Wrought Steel.
Fast Joint, dis. 60 and 10 per cent.
Loose Pin, dis. 60 and 10 per cent.
Berlin Bronzed, dis. 70, 70 and 5 per cent.
Gen. Bronzed, per pair ... 0 40 0 65

Carpet Stretchers.

American, per doz ... 1 00 1 50
Bullard's, per doz ... 6 50 ....

Castors.

Bed, new list, dis. 55 to 57½ per cent.
Plate, dis. 52½ to 57½ per cent.

Cattle Leaders.

Nos. 31 and 32, per gross ... 50 9 50

Cement.

Canadian Portland ... 2 80 3 00
English " 3 00
Belgian " ... 2 75 3 00
Canadian hydraulic ... 1 25 1 50

Chalk.

Carpenters, Colored, per gross 0 45 0 75
White lump, per cwt ... 0 60 0 65
Red ... 0 05 0 06
Crayon, per gross ... 0 14 0 18

Chisels.

Socket, Framing and Firmer.
Broad's, dis. 70 per cent.
Warnock's, dis. 70 per cent.
P. S. & W. Extra 60, 10 and 5 p.c.

Churns.

Revolving Churns, metal frames—No. 0, $8—No. 1, $8.50—No. 2, $9.00—No. 3, $10.00 No. 4, $12.00—No. 5, $16.00 each. Ditto wood frames—20c. each less than above. Discounts: Delivered from factories, 58 p.c.; from stock in Montreal, 56 p.c. Terms, 4 months or 3 p.c. cash in 30 days.

Clips.

Axle dis. 65 per cent.

Closets.

Plain Ontario Syphon Jet ... $8 00
Emb. Ontario Syphon Jet ... 8 50
Fittings ... 1 25
Plain Teutonic Syphon Washout ... 4 75
Emb. Teutonic Syphon Washout ... 5 25
Fittings ... 1 25
Low Down Teutonic, plain ... 14 50
" " " embossed ... 15 00
Plain Richelieu ... 3 75
Emb. Richelieu ... 4 00
Fittings ... 1 25
Low Down Ont. Syphon Jet, plain ... 20 00
" " " " emb'd. 20 50
Closet connection ... 1 25
Basins, round, 14 in ... 0 80
" oval, 17 x 14 in ... 1 50
" " 19 x 15 in ... 2 25

Compasses, Dividers, Etc.

American, dis. 62½ to 65 per cent.

Cradles, Grain.

Canadian, dis. 25 to 33⅓ per cent.

Crosscut Saw Handles.

S. & D., No. 3, per pair ... 17½
" " 5, " ... 22½
" " 6, " ... 15
Boynton pattern " ... 20

Door Springs.

Torrey's Rod, per doz ... (15 p.c.) 2 00
Coil, per doz ... 0 88 1 60
English, per doz ... 2 00 4 00

Draw Knives.

Coach and Wagon, dis. 50 and 10 per cent.
Carpenters, dis. 70 per cent.

Drills.

Hand and Breast.
Millar's Falls, per doz. net list.

DRILL BITS.
Morse, dis., 37½ to 40 per cent.
Standard dis. 50 and 5 to 55 per cent.

Faucets.

Common, cork-lined, dis 35 per cent.

ELBOWS. (Stovepipe.)
No. 1, per doz ... 1 40
No. 2, per doz ... 1 20
Bright, 20c. per doz. extra.

ESCUTCHEONS.
Discount, 45 per cent.

ESCUTCHEON PINS.
Iron, discount 40 per cent.

FACTORY MILK CANS.
Discount off revised list, 40 per cent.

FILES.
Black Diamond, 50 and 10 to 60 per cent.
Kearney & Foote, 60 and 10 p.c. to 60, 10, 10.
Nicholson File Co., 50 and 10 to 60 per cent.
Jowitt's, English list, 25 to 27½ per cent.

FORKS.
Hay, manure, etc., dis., 50 and 10 per cent. revised list.

GLASS—Window—Box Price.

Size United Inches.	Star Per 50 ft.	Star Per 100 ft.	D. Diamond Per 50 ft.	D. Diamond Per 100 ft.
Under 26 ...	2;10	4 00	....	6 00
26 to 40 ...	2 30	4 35	....	6 65
41 to 50 ...	....	4 75	....	7 25
51 to 60 ...	....	5	....	8 50
61 to 70 ...	....	5f35	....	9 25
71 to 80 ...	....	5 75	....	10 50
81 to 85 ...	....	6 50	....	11 75
86 to 90 ...	....	....	..	14 00
91 to 95 ...	..	....	...	15 50
99 to 100 ...	..	....	....	18 00

GAUGES
Marking, Mortise, Etc.
Stanley's dis. 50 to 55 per cent.

Wire Gauges.
Winn's, Nos. 26 to 33, each ... 1 65 2 40

HALTERS.
Rope, ⅜ per gross ... .
" ½ " 9 00
" ⅝ to ¾ 14 00
Leather, 1 in., per doz ... 3 87½ 4 00
" 1¼ in., " ... 5 15 5 20
Web, — per doz ... 1 87 2 45

HAMMERS.
Nail
Maydole's, dis. 5 to 10 per cent. Can. dis. 25 to 27½ per cent.

Tack.
Magnetic, per doz ... 1 10 1 20

Sledge.
Canadian, per lb ... 0 07½ 0 08½

Ball Pean.
English and Can., per lb ... 0 22 0 25

HANDLES.
Axe, per doz. net ... 1 50 2 00
Store door, per doz ... 1 00 1 50

Fork.
C. & B., dis. 40 per cent. rev. list.

Hoe.
C. & B., dis. 40 per cent. rev. list.

Saw.
American, per doz ... 1 00 1 25

Plane.
American, per gross ... 3 15 3 75

Hammer and Hatchet.
Canadian, 40 per cent.

Cross-Cut Saws.
Canadian, per pair ... 0 13½

HANGERS. doz. pairs.
Steel barn door ... 5 85 6 00
Stearns, 4 inch 5 00
" 5 inch 6 50
Lane's covered—
No. 11, 5-ft. run 8 40
No. 11½, 10-ft. run 10 80
No. 12, 10-ft. run 12 60
No. 14, 15-ft. run 21 00
Lane's O.N.T. track, per foot. ... 4½

HARVEST TOOLS.
Discount, 50 and 10 per cent.

HATCHETS.
Canadian, dis. 40 to 42½ per cent.

HINGES.
Blind, Parker's, dis. 50 and 10 to 60 per cent
Heavy T and strap, 4-in., per lb ... 0 06½
" " 5-in., " ... 0 06¼
" " 6-in., " ... 0 06
" " 8-in., " ... 0 05¾
" " 10-in., " ... 0 05½
Light T and strap, dis. 60 and 5 per cent.
Screw hook and hinge—
6 to 12 in., per 100 lbs 4 50
14 in. up, per 100 lbs 3 50
Per gro. pairs.
Spring 12 00

HOES.
Garden, Mortar, etc., dis. 50 and 10 p.c.
Planter, per doz ... 4 00 4 50

HOLLOW WARE
Discount ... 45 and 5 per cent

HOOKS.
Cast Iron.
Bird Cage, per doz ... 0 50 1 10
Clothes Line, per doz ... 0 27 0 63
Harness, per doz ... 0 72 0 88
Hat and Coat, per gross ... 1 00 3 00
Chandelier, per doz ... 0 50 1 00

Wrought Iron.
Wrought Hooks and Staples, Can., dis. 47½ per cent.

Wire.
Hat and Coat, discount 45 per cent.
Belt, per 1,000 ... 0 60
Screw, bright, dis. 55 per cent.

HORSE NAILS.
"C" brand 50 p.c. dis. } Oval head.
"M" brand 50 p.c. }
Acadian, 50 and 10 per cent.

HORSESHOES.

Iron Shoes.	F.O.B. Montreal No. 2 and larger.	No. 1. and smaller
Light, medium, and heavy...	3 50	3 75
Snow shoes...	3.75	4 00
Steel Shoes.		
Light...	3 60	3 85
Featherweight (all sizes)...	4 85	4 85

F.O.B. Toronto, Hamilton, London and Guelph, 10c. per keg additional.

Toe weight steel shoes...... 6 70

JAPANNED WARE.

Discount, 45 and 5 per cent. off list, June 1899.

ICE PICKS.

Star per doz... 3 00 3 25

KETTLES.

Brass spun, 7½ p.c. dis. off new list.
Copper, per lb... 0 30 0 50
American, 60 and 10 to 65 and 5 p.c.

KEYS.

Lock, Can., dis., 45 p.c.
Cabinet, trunk, and padlock, Am. per gross... 60

KNOBS.

Door, japanned and N.P., per doz... 1 50 2 50
Bronze, Berlin, per doz... 2 75 3 25
Bronze Genuine, per doz... 6 00 9 00
Shutter, porcelain, F. & L. screw, per gross... 1 30 4 00
White door knobs—per doz. 1 25

HAY KNIVES.

Discount, 50 and 10 per cent.

LAMP WICKS.

Discount, 60 per cent.

LANTERNS.

Cold Blast, per doz... 7 50
No. 3 "Wright's"... 8 50
Ordinary, with O burner... 4 25
Dashboard, cold blast... 9 50
No. 0... 6 00
Japanning, 50c. per doz. extra.

LEMON SQUEEZERS.

per doz.
Porcelain lined,... 2 20 5 60
Galvanized... 1 87 3 85
King, wood... 2 75 2 90
King, glass... 4 00 4 50
All glass... 1 20 1 30

LINES.

Fish, per gross... 1 05 2 50
Chalk " ... 1 90 7 40

LOCKS.

Canadian, dis. 45 p.c.
Russell & Erwin, per doz.... 3 00 3 25

Cabinet.

Eagle, dis. 30 p.c.

Padlock.

English and Am., per doz.... 50 6 00
Scandinavian, " 1 00 2 40
Eagle, dis. 20 to 25 p.c.

MACHINE SCREWS.

Iron and Brass.
Flat head discount 25 p.c.
Round Head, discount 20 p.c.

MALLETS.

Tinsmiths', per doz... 1 25 1 50
Carpenters', hickory, per doz. 1 25 3 75
Lignum Vitae, per doz.. .... 3 85 5 00
Caulking, each... 60 2 00

MATTOCKS.

Canadian, per doz... 8 50 1 00

MEAT CUTTERS.

American, dis. 25 to 30 p.c.
German, 15 per cent.

MILK CAN TRIMMINGS.

Discount, 25 per cent.

NAILS.

Quotations are:	Cut.	Wire.
2d and 3d...	$3 35	$3 85
3d...	3 00	3 52
4 and 5d...	2 75	3 35
6 and 7d...	2 65	3 20
8 and 9d...	2 50	3 00
10 and 12d...	2 45	2 95
16 and 20d...	2 40	2 90
30, 40, 50 and 60d. (base)...	2 35	2 85

Galvanizing 2c. per lb. net extra.
Steel Cut Nails 10c. extra.
Miscellaneous wire nails, dis. 70 per cent.
Coopers' nails, dis. 30 per cent.
Flour barrel nails, dis. 25 per cent

NAIL PULLERS.

German and American... 1 85 3 50

NAIL SETS

Square, round, and octagon, per gross... 3 38 4 00
Diamond... 12 00 15 00

NETTING.

Poultry, 50 per cent. for McMullen's.

OAKUM.

Per 100 lb
Navy... 6 00
U. S. Navy... 7 25

OIL.

Water White (U.S.)... 0 16½
Prime White (U.S.)... 0 15½
Water White (Can.)... 0 15
Prime White (Can.)... 0 14

OILERS.

McClary's Model galvan. oil can, with pump, 5 gal., per doz... 0 00 10 00
Zinc and tin, dis. 50, 50 and 10.
Copper, per doz... 1 25 3 50
Brass, " ... 1 50 3 50
Malleable, dis. 25 per cent.

GALVANIZED PAILS.

Dufferin pattern pails, dis. 50 to 50 and 10 p.c.
Flaring pails, discount 45 per cent.
Galvanized washtubs, discount 45 per cent.

PIECED WARE.

Discount 40 per cent. off list, June, 1899.

PICKS.

Per doz... 6 00 9 00

PICTURE NAILS.

Porcelain head, per gross.... 1 75 3 00
Brass head " 0 40 1 00

PICTURE WIRE.

Tin and gilt, discount 75 p.c.

PLANES.

Wood, bench, Canadian dis. 50 per cent American dis. 50.
Wood, fancy Canadian or American 7½ to 40 per cent.

PLANE IRONS.

English, per doz... 2 00 5 00

PLIERS AND NIPPERS.

Button's Genuine per doz pairs, dis. 37½ 40 p.c.
Button's Imitation, per doz.. 5 00 9 00
German, per doz... 0 60 2 60

PLUMBERS' BRASS GOODS.

Impression work, discount, 60 per cent.
Fuller's work, discount 65 per cent.
Rough stops and stop and waste cocks, discount, 60 per cent.
Jenkins disk globe and angle valves, discount, 55 per cent.
Standard valves, discount, 60 per per cent.
Jenkins radiator valves discount 55 per cent.
" " " standard, dis., 60 p.c.
Quick opening valves discount, 60 p.c.
No. 1 compression bath cock... 2 00
No. 4 " " " ... 2 00
No. 7, Fuller's... 2 50
No 4½, " ... 3 00

POWDER.

Velox Smokeless Shotgun Powder.
100 lb. or less... 0 85
1,000 lb. or more... 0 80
Net 30 days.

PRESSED SPIKES.

Discount 25 per cent.

PULLEYS.

Hothouse, per doz... 0 55 1 00
Axle... 0 22 0 33
Screw... 0 27 1 00
Awning... 0 35 2 50

PUMPS.

Canadian cistern... 1 80 3 60
Canadian pitcher spout... 1 40 2 10

PUNCHES.

Saddlers', per doz... 1 00 1 85
Conductors', " ... 9 00 15 00
Tinners' solid, per set... 0 00 0 72
" hollow, per inch... 0 00 1 00

RANGE BOILERS.

Galvanized, 30 gallons... 6 50
" 35 " ... 7 50
" 40 " ... 8 50
Copper, 30 " ... 22 00
" 35 " ... 26 00
" 40 " ... 30 00
Discount off Copper Boilers 10 per cent.

RAKES.

Cast steel and malleable Canadian list 50 and 10 p.c. revised list.
Wood, 25 per cent.

RASPS AND HORSE RASPS.

New Nicholson horse rasp, discount 60 p.c.
Globe File Co.'s rasps, 60 and 10 to 70 p.c.
Heller's Horse rasps, 50 to 50 and 5 p.c.

RAZORS.

per doz.
Geo. Butler & Co.'s,... 8 00 18 00
Boker's... 7 50 11 00
Wade & Butcher's... 3 60 10 00
Theile & Quack's... 7 00 12 00
Elliot's... 4 00 18 00

REAPING HOOKS.

Discount, 50 and 10 per cent.

REGISTERS.

Discount... 40 per cent.

RIVETS AND BURRS.

Iron Rivets, discount 60 and 10 per cent.
Iron Burrs, discount 55 per cent.
Black and Tinned Rivets, 60 p.c.
Extras on Iron Rivets in 1-lb. cartons, ½c per lb.
Extras on Iron Rivets in ½-lb. cartons, per lb.
Copper Rivets & Burrs, 35 and 5 p.c. dis. and cartons, 1c. per lb. extra, net.
Extras on Tinned or Coppered Rivets ½-lb. cartons, 1c. per lb.
Terms, 4 mos. or 3 per cent. cash 30 days.

RIVET SETS.

Canadian, dis. 35 to 37½ per cent.

ROPE ETC.

	Sisal.	Manila.
7-16 in. and larger, per lb.	9	13
⅜ in...	10	14
¼ and 5-16 in. ...		15
Cotton, 3-16 inch and larger		16½
" 5-32 inch...		21½
" ⅛ inch...		22½
Russia Deep Sea...		15½
Jute...		8
Lath Yarn...		9¾
New Zealand Rope...		10½

RULES.

Boxwood, dis. 75 and 10 p.c.
Ivory, dis. 37½ to 40 p.c.

SAD IRONS.

per set.
Mrs. Potts, No. 55, polished... 0 62½
" No. 50, nickle-plated.... 0 67½

SAND AND EMERY PAPER.

Dominion Flint Paper, 47½ per cent.
B & A. sand, 40 and 2½ per cent.
Emery, 40 per cent.

SAP SPOUTS.

Bronzed iron with hooks, per doz... 9 50

SAWS.

Hand Disston's, dis. 12½ p.c.
S. & D., 40 per cent.
Crosscut, Disston's, per ft.... 0 35 0 55
S. & D., dis. 35 p.c. on Nos. 2 and 3.
Hack, complete, each... 0 75 2 75
" frame only... .... 0 75

SASH WEIGHTS.

Sectional, per 100 lbs... 2 75 3 00
Solid, " ... 2 00 2 25

SASH CORD.

Per lb... 0 23 0 30

SAW SETS.

"Lincoln," per doz... .... 6 50

SCALES.

B. S. & M. Scales, 45 p.c.
Champion, 65 per cent.
Fairbanks Standard, 35 p.c.
" Dominion, 55 p.c.
" Richelieu, 55 p.c.
Chatillon Spring Balances, 10 p.c.

SCREW DRIVERS.

Sargent's, per doz... 0 65 1 00

SCREWS

Wood, F. H., iron, and steel, 85 p.
Wood R. H., " dis. 80 p.c.
" F. H., brass, dis. 77½ p.c.
Wood, R. H., " dis. 70 p.c.
" F.H., bronze, dis. 70 p.c.
" R.H. " 65 p.c.
Drive Screws, 80 per cent.
Bench, wood, per doz... 3 25 4 00
" iron, " ... 4 25 5 75

SCYTHES.

Per doz. net... 9 00

SCYTHE SNATHS.

Canadian, dis. 45 p.c.

SHEARS.

Bailey Cutlery Co., full nickeled, dis. 60 p.c.
Seymour's, dis. 50 and 10 p.c.

SHOVELS AND SPADES.

Canadian, dis. 40 and 5 per cent.

SINKS.

Steel and galvanized, discount 45 per cent.

SNAPS.

Harness, German, dis. 25 p.c.
Lock, Andrews'... 4 50 11 50

SOLDERING IRONS.

1, 1½ lb., per lb... 0 37
2 lb. or over, per lb... 0 34

SQUARES.

Iron, No. 493, per doz... 2 40 2 55
" No. 494, " ... 3 25 3 40
Steel, dis. 50 and 5 to 50 and 10 p.c., rev. list.
Try and bevel, dis. 50 to 52½ p.c.

STAMPED WARE.

Plain, dis., 75 and 12½ p.c. off revised list.
Retinned, dis., 75 p.c. off revised list.

STAPLES.

Galvanized... 0 00 0 00
Plain... 0 00 3 45
Coopers', discount 45 per cent.
Poultry netting staples, 40 per cent.

STOCKS AND DIES.

American dis. 25 p.c.

STONE.

Per lb.
Washita... 0 28 0 60
Hindostan... 0 06 0 07
" slip, ... 0 09 0 09
Labrador... .... 0 13
" Axe... .... 0 15
Turkey... .... 0 50
Arkansas... 0 00 1 50
Water-of-Ayr... 0 00 0 10
Scythe, per gross 3 50 5 00
Grind, per ton 15 00 18 00

STOVE PIPES.

Nestable in crates of 25 lengths.
5 and 6 inch Per 100 lengths 7 00
7 inch " " 7 50

ENAMELINE STOVE POLISH.

No. 4—3 dozen in case, net cash $4 80
No. 6—3 dozen in case, " 8 40

TACKS BRADS, ETC.

Per cent
Strawberry box tacks, bulk... 75 & 10
Cheese-box tacks, blued... 80 & 12½
Trunk tacks, black and tinned... 85
Carpet tacks, blued... 80 & 5
" " tinned... 80 & 10
" " (in kegs)... 40
Cut tacks, blued, in dozens only... 75 & 15
½ weights... 60
Swedes, cut tacks, blued and tinned—
In bulk... 80 & 10
In dozens... 75
Swedes, upholsterers', bulk... 85 & 12½
" brush, blued & tinned, bulk.. 70
" gimp, blued, tinned and japanned... 75 & 12½
Zinc tacks... 35
Leather carpet tacks... 55
Copper tacks... 50
Copper nails... 52

Trunk nails, black	65 and 5
Trunk nails, tinned	65 and 10
Clout nails, blued and tinned	65 and 5
Chair nails	35
Cigar box nails	35
Patent brads	40
Fine finishing	40
Picture frame points	10
Lining tacks, in papers	10
" " in bulk	15
" " solid heads, in bulk	75
Saddle nails in papers	10
" " in bulk	15
Tufting buttons, 22 line, in dozens only	60
Tin capped trunk nails	15
Zinc glazier's points	5
Double pointed tacks, papers	90 and 10
" " " bulk	40

TAPE LINES.

English, ass skin, per doz	2 75	5 00
English, Patent Leather	5 50	9 75
Chesterman's each	0 90	2 85
" steel, each	0 80	8 00

THERMOMETERS.

Tin case and dairy, dis. 75 to 75 and 10 p.c.

TRANSOM LIFTERS.

Payson's per doz. 2 60

TRAPS. (Steel.)

Game, Newhouse, dis. 25 p.c.
Game, H. & N., P. S. & W., 65 p.c.
Game, steel, 72½, 75 p.c.

TROWELS.

Disston's discount 10 per cent.

German, per doz	4 75	6 00

S. & D., discount 35 per cent.

TWINES.

Bag, Russian, per lb	....	0 27
Wrapping, cotton, per lb	0 22	0 26
Wrapping, mottled, per pack	0 5J	0 60
Wrapping, cotton, 3-ply		0 20
" " 4-ply	....	0 25
Mattress, per lb	0 33	0 45
Staging, "	0 27	0 35
Broom, "	0 30	0 55

VISES.

Hand, per doz	4 00	6 00
Bench, parallel, each	2 00	4 50
Coach, each	6 00	7 00
Peter Wright's, per lb	0 12	0 13
Pipe, each	5 50	9 00
Saw, per doz	6 50	13 00

ENAMELLED WARE.

White, Princess, Turquoise, Blue and White, discount 50 per cent.

Diamond, Famous, Premier, 50 and 10 p.c.

Granite or Pearl, Imperial, Crescent, 50, 10 and 10 per cent.

WIRE.

Brass wire, 50 to 50 and 2½ per cent. off the list.

Copper wire, 45 and 10 per cent. net cash 30 days, f.o.b. factory.

Smooth Steel Wire, base, $2.80 per 100 lb. List of extras: Nos. 2 to 5, advance 7c. per 100 lb.—Nos. 6 to 9, base—No. 10, advance 7c.—No. 11, 14c.—No. 12, 20c.—No. 13, 35c.—No. 14, 47c.—No. 15, 60c.—No. 16, 75c. Extras net per 100 lb.: Coppered wire, 60c.—tinned wire, $2—oiling, 10c.—special hay-bailing wire, 30c.—spring wire, $1— best steel wire, 75c.—bright soft drawn, 15c.—in 50 and 100-lb. bundles net; 10c.—in 25-lb. bundles net, 15c.—packed in casks or cases, 15c.—bagging or papering, 10c.

Fine Steel Wire, dis. 17½ per cent. List of extras: In 100-lb. lots: No. 17, $5—No. 18, $5.50—No. 19, $6—No. 20, $6.65—No. 21, $7—No. 22, $7.30—No. 23, $7.65—No. 24, $8—No. 25, $9—No. 26, $9.50—No. 27, $10—No. 28 $11—No 29, $12—No. 30, $13—No. 31, $14—No 32, $15 No. 33, $16—No. 34, $17. Extras net—tinned wire, Nos. 17-25, $2—Nos. 26-31, $4—Nos. 32-34, $6. Coppered, 5c.—oiling, 10c.—in 25-lb. bundles, 15c.—in 5 and 10-lb. bundles, 25c.—in 1-lb. hanks, 50c.—in ½-lb. hanks, 75c.—in ¼-lb hanks, $1—packed in casks or cases, 15c.—bagging or papering, 10c.

Galvanized Wire, per 100 lb.—Nos. 6, 7, 8, $3.55 No. 9, $3.10—No. 10, $3.75—No. 11, $3.85 No. 12, $3.25—No. 13, $3.35—No. 14, $4.25—No. 15' $4.75—No. 16, $5.10.

Clothes Line Wire, 19 gauge, per 1,000 feet 3 30

WIRE FENCING. F.O.B. Toronto

Galvanized 4 barb, 2¼ and 5 inches apart	....	3 10
Galvanized, 2 barb, 4 and 6 inches apart	....	3 10
Galvanized, plain twist	....	3 10

Galvanized barb, f.o.b. Cleveland, $2.97½ in less than carlots, and $2.85 in carlots. Terms, 60 days or 2 per cent. in 10 days.

Ross braid truss cable 4 50

WIRE CLOTH.

Painted Screen, per 100 sq. ft., net .. 1 35
Terms, 4 months, May 1.; 3 p.c. off 30 days.

WRENCHES.

Acme, 35 to 37½ per cent.
Agricultural, 60 p.c.
Coe's Genuine, dis. 20 to 25 p.c.

Towers' Engineer, each	2 00	7 00
" S., per doz	5 80	6 00
G. & K.'s Pipe, per doz	....	3 40
Burrell's Pipe, each	....	3 00
Pocket, per doz	0 25	2 90

WRINGERS.

Leader	per doz.	$ ....
Royal Canadian	"	50 00
Royal American	"	50 00

Discount, 45 per cent.: terms 4 months, or 3 p.c. 30 days.

WROUGHT IRON WASHERS.

Canadian make, discount, 40 and 5 per cent

CANADIAN HARDWARE AND METAL MERCHANT

The Weekly Organ of the Hardware, Metal, Heating, Plumbing and Contracting Trades in Canada.

VOL. XIII. MONTREAL AND TORONTO, FEBRUARY 2, 1901. NO. 5

Garden Hose

Discounts for the new season now out.

WRITE FOR PRICES.

CANADIAN RUBBER CO.,

Montreal. Toronto. Winnipeg.

THE NEW BALDWIN

DRY AIR CLEANABLE

REFRIGERATOR.

135 Modern Varieties. Ash, Oak and Soft-wood Finishes

METAL, PORCELAIN, SPRUCE LININGS.

BALDWIN

Positive Circulation—Sanitary—Odorless.

Latest Cleanable Features—The Strongest and Best System of Patent Removable Metal Air-Flues.

Air-Tight Lever Locks Ball-Bearing Casters. Swing Base—in and out.

Rubber around Doors and Lids, making them doubly air-tight.

Handsome Designs. Moderate Prices.

Built in the newest, largest and best equipped refrigerator plant in the Eas run all the year round on refrigerators exclusively; stock goods; specia refrigerators and coolers in sections.

Handsome Trade Catalogue Ready.

Baldwin Refrigerator Co.,

BURLINGTON, VERMONT.

SOME OF THE NEWER "YANKEE" TOOLS

NO. 41 AUTOMATIC DRILL WITH DRILL POINTS IN HANDLE.

NO. 42 AUTOMATIC DRILL WITH DRILL POINTS IN BOX.

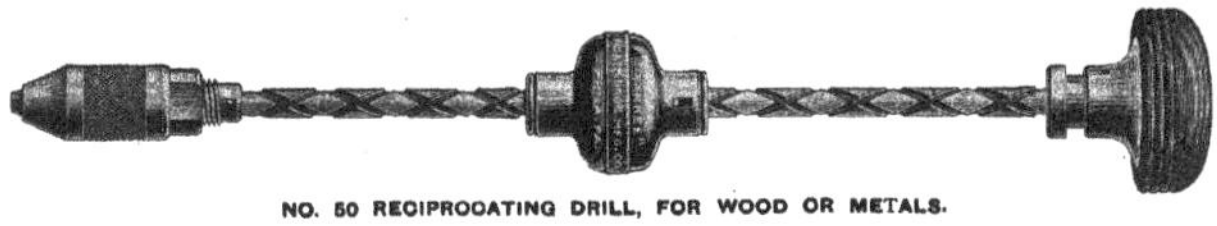

NO. 50 RECIPROCATING DRILL, FOR WOOD OR METALS.

Sold by Leading Jobbers throughout the Dominion.

NORTH BROS. MFG. CO.,

Philadelphia, Pa., U. S. A.

Our

IMPERIAL OXFORD

Represents the very highest development attained in range construction.

Its patented improved features give it precedence over all others—and the improvements need only to be seen to be appreciated.

ITS FRONT DRAW-OUT GRATE
DIFFUSIVE FLUE CONSTRUCTION
OVEN THERMOMETER
DRAW-OUT OVEN RACK

and handsome appearance are "talking points" that sell it everywhere.

Are you handling them?

If not, write for our catalogue and price list—they are leading favorites all over Canada.

THE GURNEY FOUNDRY CO., Limited

TORONTO. WINNIPEG. VANCOUVER.

THE GURNEY-MASSEY CO., LIMITED, MONTREAL.

Steel Frame.

David Maxwell & Sons

ST. MARYS, ONT., CANADA

"Maxwell Favorite Churn"

PATENTED FEATURES: Improved Steel Stand, Roller Bearings, and Foot and Hand Lever Drive, and Detachable Driving Link Improved for season of 1901. Steel or Wood Frame as desired

Wheelbarrows.

In Four different Sizes.

Lawn Mowers.

High and Low Wheels, from 12-in. to 20-in. widths. Cold Rolled Steel Shafting, Crucible Steel Knives and Cutting Plate.

If your Wholesale House does not offer you these articles

...SEND DIRECT TO US.

MAXWELL MOWER

8-inch Low Wheel.

"THE MAXWELL"

Lawn Mower

High Wheel 10 inches

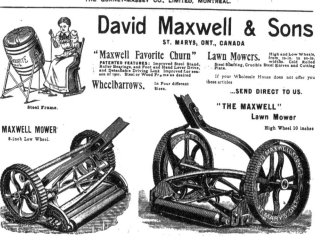

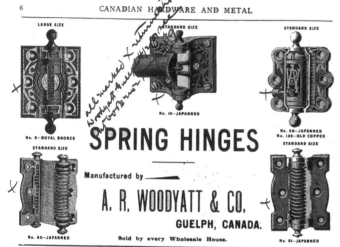

HARDWARE AND METAL

VOL. XIII. MONTREAL AND TORONTO, FEBRUARY 2, 1901. NO. 5.

President,
JOHN BAYNE MacLEAN,
Montreal.

THE MacLEAN PUBLISHING CO.
Limited.

Publishers of Trade Newspapers which circulate in the Provinces of British Columbia, North-West Territories, Manitoba, Ontario, Quebec, Nova Scotia, New Brunswick, P.E. Island and Newfoundland.

OFFICES

MONTREAL	232 McGill Street, Telephone 1255.
TORONTO	10 Front Street East, Telephone 2148.
LONDON, ENG.	109 Fleet Street, E.C., J. M. McKim.
MANCHESTER, ENG.	18 St Ann Street. H. S. Ashburner.
WINNIPEG	Western Canada Block, J. J. Roberts.
ST. JOHN, N.B.	No. 3 Market Wharf. J. Hunter White,
NEW YORK.	176 E. 88th Street,

Travelling Subscription Agents:
T. Donaghy. F. S. Millard.

Subscription, Canada and the United States, $2.00.
Great Britain and elsewhere 12s.

Published every Saturday.

Cable Address { Adscript, London. Adscript, Canada.

RETAIL STORE AMALGAMATIONS.

THE spirit of consolidation seems to be gradually reaching into and permeating every branch of industry and commercial activity.

The discussion of the "Trust" question—the question whether the influence of the great combinations of capital in recent years is likely to be good or bad, has waged long and waxed warm. The only apparent result has been to increase the tendency toward amalgamation.

Practically, the last department of trade to be affected by the movement is the retail business in the various lines. But evidences are accumulating that the retailers have begun to consider the question in relation to themselves.

Somewhat more than a year ago several of the retail merchants of Sussex, N.B., joined forces under the style of The Sussex Mercantile Co. That they have been satisfied with their experiment is evidenced by the fact that a few months ago they absorbed two other concerns in the place.

Several amalgamations have been reported from British Columbia lately. One of the most recent is the consolidation of the business of The Russell Hardware Co., W. M. Law & Co., general merchants, and Caulfield & Lamon, hardware dealers, Greenwood, B.C., under the style of The Russell-Law-Caulfield Co., Limited. The new concern has been incorporated with an authorized capital of $100,000.

There seems to be sound economy at the bottom of such consolidations as these. It is unquestionable that the business house which has sufficient capital to take all its discounts is at a big advantage over all competitors unable to do so. Not only in this respect, but in the saving of floor space, shelf room, bookkeeping, etc., affected by this means, there is an advantage which makes it reasonable to expect many such amalgamations as those mentioned above.

AN INFLUENCE ON PROFITS.

A merchant cannot afford to be on bad terms with his competitors any more than he can with his customers.

Customers are the only ones likely to gain from the flow of bad blood between merchants, and what the former gain the latter lose.

The better the terms on which merchants live the better the profits they earn.

AN OPPORTUNITY TO PLAN.

THE time of the year when trade is slack is the season that tests the quality of a merchant's or a clerk's energy or progressive spirit. The "dull days" show that altogether too many clerks and merchants are lacking in the true elements of progressiveness. The business man who is bound to make the best of the seasons of activity and rush is the one who finds in the slack season not a necessity to "kill time," but rather an opportunity to plan means of improving his stock, his store methods, the store itself, his advertising, and to study the business he is engaged in, the local conditions that have to be recognized in order to be most satisfactorily dealt with, and the characteristics of his help that he may be enabled to make the most of their services. There is no limit to the directions in which a merchant may direct his thoughts during a slack season. And thought, well directed, has a value equivalent to that of the hardest physical labor.

OPEN PRICES ON HORSESHOE NAILS.

The Canadian Horseshoe Nail Association was dissolved on January 1, and during this month open prices have prevailed on horseshoe nails.

As a result of the dissolution prices have been reduced this week, the discount being 50, 10 and 5 per cent. on oval head, and 50, 10 and 10 per cent. on countersunk head horse nails.

It seems that the break-away was made by concerns wishing a wider scope in which to do business.

A meeting is to be held in the first week in February, but reorganization is uncertain.

BUSINESS MEN IN PARLIAMENT SHOULD ORGANIZE.

THAT the Parliament of Canada should be conducted on business principles is now a recognized truism. It was not always so, for, by implication at any rate, it was generally held that the application of business principles to Parliamentary affairs was incompatible, and that those who held to the contrary were dreamers and faddists. Now, even the professional politician subscribes to the doctrine though, through ignorance or design, he seldom practices it.

But, while the belief in the soundness of the doctrine that business methods should be applied to Parliamentary practice is so general that no one probably would gainsay it, each session of the House forcibly reminds one that the leaven of business influence there is still very small.

The fact of the matter is that, while there are a good many business men in the House, and in theory business practice is a good thing and a necessary thing, the business men therein are practically without influence. And they are not lacking in influence because they are not numerous enough. There are over 80 business men in the House, or something like 37 per cent. of the total membership. The relatively small influence of the business men is not, therefore, due to lack of numbers. Nor is it due to the want of ability. The most useful members of the House are business men. It is due to lack of organization.

HARDWARE AND METAL is not an advocate of a third party, whether it be business men or any other class of men. Organization of the business men in Parliament does not mean obliteration of party lines any more than adherence to party principles means the renunciation of religious beliefs.

A man can be a Liberal or a Conservative and at the same time be a Roman Catholic or a Protestant.

The business men who are members of the House of Commons could in like manner have an organization of their own and yet at the same time still be associated with one or other of the two great political parties.

The representatives from the different Provinces hold their occasional conclaves. So sometimes do those of various religious beliefs. Why then should not business men? There is no reason why they should not. But there is every reason why they should.

As we have already intimated, the application of business methods to Parliamentary practice is essential to the successful conduct of the latter. No one will dispute that. It follows, therefore, that the more the business men in Parliament are working in unison on business questions the nearer is it possible to get to the ideal.

Supposing, for instance, the eighty-odd business men in the House were to get together and express themselves in favor of the much-desired insolvency law, does anyone for one moment imagine the Government would any longer defer introducing such a measure?

The Government is perfectly aware that the commercial exigencies of the country demand it, but it fears, as previous Administrations have feared, the political exigencies that the introduction of an insolvency bill might create.

Assured of the support of the business element, the Government would not be long in developing action in regard to this or any other question affecting the commercial interests of the country.

It should not be a difficult thing for the business men of the House of Commons to organize. There would be no tenets, either political or religious, to which they need subscribe. All that would be necessary would be to call a meeting, appoint a chairman and a secretary, and gather together again when it was necessary to consider, from a practical business standpoint, such Bills before the House as directly or indirectly affected the commercial interests of the country or to discuss measures of that character which it was thought wise should be brought forward.

Party exigencies would possibly prevent such an organization taking a united stand on every question of a commercial nature, but that is not an argument against its existence. On the contrary it is an argument for, rather than against, for it shows the necessity of controlling the party spirit when it conflicts with the commercial welfare of the country. And the longer such an organization existed the more potent would its influence become in regulating the action of Parliament.

Long before barb wire was used in actual warfare, it was the subject for warfare among manufacturers and dealers.

HEAVY PURCHASE OF PIG IRON.

A GREAT deal of interest is being taken in a large purchase of Bessemer pig iron by The Carnegie Steel Company. The quantity concerned is 150,000 tons.

The Carnegie Steel Company is already a large producer of pig iron. It has no less than 17 furnaces, and, with others that it is building, will have the facilities for making 2,700,000 tons of pig iron per annum.

But, in spite of its capacity for making pig iron, its capacity for making steel is greater; hence, in order to supply its steel plant, it is necessary to go into the open market. It is estimated that at present the Carnegie Company uses 900,000 tons of pig iron per annum in excess of what it produces.

Besides the purchases of Bessemer pig iron by the Carnegie interests, good quantities have been bought during the past month by The American Steel and Wire Company and other concerns.

As stocks of Bessemer pig iron have been decreasing for some time, the recent heavy purchases have given strength to the market, and some of the Valley furnaces have advanced their prices to $13, an appreciation of 50c. per ton. The price at which the sales were made to the Carnegie and other interests was $12.50 per ton. With ore costing the furnace $6.1c delivered, coke $1.75 to $2, and limestone and wages high, it is doubted by some authorities in the United States that the furnaces can make and sell pig iron at $12.50 per ton and earn a profit.

The United States Tinplate Trust has bought out another rival. It evidently has the "tin."

UNITED STATES PIG IRON PRODUCTION FOR 1900.

THE American Iron and Steel Association has received from the manufacturers complete statistics of the production of all kinds of pig iron in the United States, in 1900; also complete statistics of the stocks of pig iron which were on hand and for sale on December 31, 1900.

Production.—The total production of pig iron in 1900 was 13,789,242 gross tons, against 13,620,703 tons in 1899, 11,773,934 tons in 1898 and 9,652,680 tons in 1897. The production in 1900 was 168,539 tons greater than in 1899. The following table gives the half-yearly production of pig iron in the last four years.

Periods—	1897.	1898.	1899.	1900.
First half....	4,403,476	5,869,703	6,289,167	7,642,569
Second half..	5,249,204	5,904,231	7,331,536	6,146,673
Total ...	9,652,680	11,773,934	13,620,703	13 789,242

The production of pig iron in the second half of 1899 and the first half of 1900 aggregated 14,971,105 tons, or almost 15,000,000 tons.

It will be observed that there was a decline in production in the second half of 1900, as compared with the first half, of 1,495,896 tons.

The production of Bessemer pig iron in 1900 was 7,943,452 tons, against 8,202,778 tons in 1899.

The production of basic pig iron in 1900, all made with coke or mixed anthracite and coke, was 1,072,376 tons, against 985,033 tons in 1899.

The production of spiegeleisen and ferromanganese in 1900 was 255,977 tons, against 219,768 tons in 1899.

The production of charcoal pig iron in 1900 was 339,874 tons, against 284,766 tons in 1899.

Unsold Stocks.—Our statistics of unsold stocks do not include pig iron sold and not removed from the furnace bank, or pig iron in the hands of creditors, or pig iron manufactured by rolling mill owners for their own use, or pig iron in the hands of consumers. The stocks which were unsold in the hands of manufacturers or their agents on December 31, 1900, amounted to 442,370 tons, against 63,429 tons on December 31, 1899, and 338,053 tons on June 30, 1900.

Included in the stocks of unsold pig iron on hand on December 31, 1900, were 12,750 tons in the yards of The American Pig Iron Storage Warrant Co. which were yet under the control of the makers, the part in these yards not under their control amounting to 3,650 tons, which quantity, added to the 442,370 tons above mentioned, makes a total of 446,020 tons which were on the market at that date, against a similar total of 68,309 tons on December 31, 1899, and 342,907 tons on June 30, 1900. The total stocks in the above-named warrant yards on December 31, 1900, amounted to 16,400 tons, against 4,900 tons on December 31, 1899, and 5,800 tons on June 30, 1900.

Furnaces.—The whole number of furaaces in blast on December 31, 1900, was 232, against 289 on December 31, 1899, and 283 on June 30, 1900.

Production of all kinds of pig iron from 1897 to 1900 by States:

States—	Gross tons of 2,240 pounds 1897.	1898.	1899.	1900.
Massachusetts	3,284	3,661	2,476	3,310
Connecticut	8,336	6,336	10,129	10,733
New York	245,394	228,011	264,340	292,827
New Jersey	95,693	100,631	127,598	170,262
Pennsylvania	4,631,634	5,537,832	6,558,878	6,365,935
Maryland	193,702	190,974	284,477	291,073
Virginia	307,610	283,274	366,491	490,617
North Carolina and Georgia	17,092	13,762	17,835	28,084
Alabama	947,831	1,033,676	1,083,905	1,184,337
Texas	6,175	5,178	5,603	10,150
West Virginia	132,907	192,799	187,858	166,768
Kentucky	85,899	100,724	119,019	71,582
Tennessee	272,130	263,489	346,166	362,190
Ohio	1,372,889	1,986,356	2,378,212	2,470,911
Illinois	1,117,239	1,365,898	1,442,012	1,353,383
Michigan	132,578	147,640	184,443	163,712
Wisconsin and Minnesota	103,909	172,781	203,175	184,791
*Missouri and Colorado	30,465	141,010	178,880	150,204
Total	9,652,680	11,773,934	13,620,703	13,789,242

*Missouri, 23,883 tons; Colorado, 6,582 tons.

GERMANY'S PIG IRON PRODUCTION.

Germany's pig iron output for 1900 was 8,422,842 tons, an increase of 393,537 tons for the preceding year. The December output was 720,790 tons.

AUTOMOBILES IN THE POSTAL SERVICE.

An exchange states that the postal authorities in the United States are considering a project for making local mail collections by automobiles.

For some time automobiles have been in use in Berlin for this purpose, while in Toronto the local collections and the conveying the mails to and from the railway station have been done by the horseless carriage.

The United States is evidently a little behind Berlin and Toronto in the utilization of the automobile in the postal service.

GLASS DEARER IN THE UNITED STATES.

On Thursday last week contracts were signed by the National Window Glass Jobbers' Association of the United States for 700,000 boxes at an advance of 30 per cent. over former quotations. The cause of the advance is reported to be that, while in past years the American Window Glass Co. and the independent companies have been "at war" and prices were kept down by the keenness of competition, this year an understanding has been reached whereby prices have been raised and contracts divided in proportion to capacity. A further advance is anticipated.

This rise in prices in the United States will not affect the Canadian market.

ANOTHER DROP IN SCREWS.

For the second time since the beginning of the year, a decline is to be reported in the price of wood screws. The cause of these declines is American competition.

The discounts are now as follows: Flat head bright, 87½ and 10 per cent. off list; round head bright, 82½ and 10 per cent.; flat head brass, 80 and 10 per cent.; round head brass, 75 and 10 per cent.

FORT WILLIAM BOARD OF TRADE.

THE annual report of President Morton and the election of officers of the Fort William, Ont., Board of Trade, were the features of the annual meeting on Monday evening last week.

The report showed that since the organization of the board, ten years ago, Fort William has grown in population from 750 to about 5,000; that a town hall and town public schools have been erected at a cost of $50,000 which would be a credit to any city; a first-class system of waterworks and electric lighting have been installed, and that there has been a steady improvement in every respect.

During the past year the Standard Oil Co. have established in Fort William a branch from which they intend to supply the Canadian Northwest; the C.P.R. have materially increased their dockage and roundhouse facilities. Arpin, Scott & Finger have decided to locate large saw and planing mills in Fort William. It is likely the American Steel and Wire Co. will build iron ore docks and establish offices there.

The report concluded by suggesting that a strong effort be made to have a quarantine station established by the Dominion Government at Fort William.

The following officers were elected for 1901:

President—E. A. Morton, reelected.
Vice-President—C. W. Jarvis.
Secretary-Treasurer—E. R. Wayland.
Council—W. F. Hogarth, A. McDougall, E. S. Rutledge, S. C. Young, J. H. Perry, J. J. Wells, John King, Alex. Snelgrove, W. L. Morton, Don McKellar, W. H. Whalen, James Murphy.

PERSONAL MENTION.

Mr. John J. Drummond, managing-director of the Midland Iron Furnace was in Toronto last week.

Mr. A. O. Campbell, of the Vancouver Hardware Co., and Mr. Fred. Buscombe, crockery and fancy goods dealer, Vancouver, have left for a three months' tour of the manufacturing and trade centres of Eastern Canada, the United States, Great Britain and continental Europe.

CREDIT—WHO TO EXTEND IT TO AND FOR HOW LONG*

BY HERBERT F. HALEY.

"CREDIT—who to extend it to and for how long" is the bane of every merchant's successful existence; to extend credit to the worthy, who will show their appreciation of the favor by paying their bills promptly when they fall due, is a question which must be handled with every discretion and absolutely without sentiment.

THE BASIS OF CREDIT.

This worthiness and ability must be thoroughly determined as to time and amount before a single dollar's worth of goods are charged, and to allow the maxim "A credit well made is an account half collected" to always confront us.

In determining this credit, we must first consider the moral responsibility of our prospective customer, as well as his ability to pay; also how he has been paying our fellow-merchants.

DRAWING THE LINE.

No matter how good rating, this same customer has misfortunes, and to draw the lines at the proper time—good and strong—is the hardest proposition with which we have to deal.

Yet, if we do not draw them promptly, we invariably regret our inaction, and the result is a balance that is not only hard for us to carry out, but one which is too often left unpaid.

Human nature is sympathetic, and our former good customer expects us to share his misfortunes to the fullest.

AN INJURY TO BOTH.

In extending too much credit to a customer, we not only injure ourselves, but our customer as well. We lead him to extravagant living, buying goods that perhaps he would not have bought otherwise, soon becoming careless in his payments. Misfortunes of some kind overtake him, or, equally as bad, he decides to buy a home on the installment plan, uses our money to make the first payment, invests our money for us, but always in his wife's name.

The only way to avoid this is to

EXACT PROMPT PAYMENTS,

and in full, the day they are due; and, when extensions are given, make the time short, and see that the agreement is carried out to the letter.

The old saying, "That will be all right," is the most dangerous one that a merchant can use; it seems to mean at the time only a common courtesy, but later it means that if we have not the money coming from our "prompt payers," we will have to ask these same extensions from our creditors, which, if granted, are unpleasant, to say the least.

THINGS TO CONSIDER.

Often were we to consider in granting credit that we are risking, say, 80 per cent. of hard cash for a prospective 20 per cent. gain; and often, after our expenses are deducted from this 20 per cent., we have a net of, say, 8 per cent., or, in other words, we have risked 80 per cent. for a possible 8 per cent.—a ratio of 10 to 1, which certainly behooves us to make "caution" our ever watchword.

Credit and collections are so closely entwined that one cannot survive without the other. "Take care of the collections, and the credits will take care of themselves," is a very broad assertion, but one, which, if simmered down, contains a great deal of logic.

Many a good customer has been allowed to become careless in his payments on account of not being promptly and properly seen.

CREDIT AND BOOKKEEPING.

are also very closely linked. To successfully handle an account, you must know its standing at all times; to do this your books must be kept up to the minute, thus enabling you to quickly use tact and discretion in saying "Yes" or "No" at the proper time.

Keep track of your customer—as to all that pertains to him; of his successes or reverses; thus you will be in a position to increase or diminish the account, as the case may be.

BE CAREFULL.

We must be careful not to drive away a good customer whom it is safe to trust, and more careful not to extend credit to customers who either cannot or will not pay.

Fear and Friendly Hope and Envy watch the issue, while the lines "By which thou shalt be judged" are written down.

* Paper read before the Retail Grocers' Association, Chattanooga.

SALUTING THE QUEEN.

One of the best calendars and especially appropriate at the present time is "Soldiers of the Queen," which the Queen City Oil Co. are sending out. It represents a Highlander, a South Wales lancer and a Canadian mounted infantryman saluting and cheering a large portrait of the Queen. The drawing is excellent and there is no doubt that the whole calendar will be much appreciated by anyone fortunate enough to get one before the supply is exhausted. No doubt any of our readers may have one by writing them.

BUSINESS CHANGES.

DIFFICULTIES, ASSIGNMENTS, COMPROMISES.

A MEETING of the creditors of E. A. Atkinson, general merchant, L'Avenir, Que., has been called.

John W. Gordon, harness dealer, Embro, Ont., has assigned.

Alp. Boulanger, general merchant, St. Eugene, Que., has assigned.

Fremont & Co., general merchants, St. Felicite, Que., are offering 40c. on the dollar.

L. A. Dixon, general merchant, St. Eustache, Que., is offering 40c. cash on the dollar.

Leask & Rankin, merchants, Cranbrook, B.C., have assigned to Creighton R. Palmer.

A meeting of the creditors of Noe Page, general merchant, Crysler, Ont., has been called.

J. O. A. Dequire, general merchant, Glen Robertson, Ont., has assigned to Nap. Geneau.

G. Bremner & Son, general merchants, Cranbrook, B.C., have assigned to Robert E. Sherlock.

A meeting of the creditors of R. Bourbeau, general merchant, Victoriaville, Que., has been called.

John J. Wiens, general merchant, Low Farm, Man., has assigned to John Russell, and his stock has been sold.

Brown & Brown, general merchants, etc., Whitebourne, Newfoundland, have applied for declaration of insolvency.

Esdras Paradis, general merchant, Plessisville, Que., has assigned, and a meeting of his creditors has been called.

A meeting of the creditors of P. J. Stinson & Co., general merchants, Singhampton, Ont., will be held to-day (Friday).

Fanny Markson, general merchant, Glen Robertson, Ont., has assigned to A. Markson, Alexandria, Ont., and a meeting of her creditors will be held on February 4.

T. N. Gauthier, general merchant, Carillon, Que., has assigned to Kent & Turcotte. He is offering 50c. cash on the dollar.

PARTNERSHIPS FORMED AND DISSOLVED.

Moreau & Desjardins, blacksmiths, Montreal, have dissolved.

Roussin & Desjardins, blacksmiths, Montreal, have registered partnership.

Partnership has been registered by J. P. Luneau & Frere, blacksmiths, St. Helene, Que.

W. H. Otto & Co., general merchants, Elmira, Ont., have dissolved. W. H. Otto continues.

Van Blaricom & Currie, dealers in agricultural implements, Arden, Man., have dissolved.

Partnership has been registered by John McDougall & Co. as car wheel manufacturers in Montreal.

Partnership has been registered by Gregoire & Bourque, sawmillers, St. Germain, Que.

E. C. McLellan & Co., general merchants, Tatamagouche, N.S., have dissolved. The business will be continued by E. C. McLellan alone.

SALES MADE AND PENDING.

L. Robins, general merchant, Albuna, Ont., has sold out.

The assets of M. Forget, sawmiller, Quebec, have been sold.

T. W. Davis, hardware dealer, etc., Ripley, Ont., has sold out.

H. Davis, blacksmith, Kinnicott, Ont., is advertising his business for sale.

The assets of J. A. Andrews, tinsmith, Kinburn, Ont., are to be sold at auction.

The assets of J. E. I. Clavel, coal and wood merchant, Montreal, are to be sold.

T. G. Lewis & Co., hardware dealers, etc., Montreal, have retired from business.

Gaun Christie, general merchant, South Mountain, Ont., is advertising his business for sale by tender.

The stock of Lewin & Co., general merchants, Moosomin, Man., is advertised for sale by tender.

The stock of the estate of H. Grenier, hardware dealer, Quebec, has been sold at 75½c. on the dollar.

The assets of Eugene Guay, general merchant, St. Jerome (Chicoutimi), Que., are to be sold on Saturday.

The stock of the W. F. Horton Co., bicycle dealers, etc., London, Ont., is to be sold by auction on February 4.

E. C. Corbett, general merchant, Verschoyle and Mount Elgin, Ont., is advertising his business for sale.

The stock, etc., of the estate of E. J. Crawford, general merchant, Souris, Man., is advertised for sale by auction.

The stock, etc., of F. G. Terryberry, general merchant, Burford, Ont., is advertised for sale by auction to-day (Friday).

The stock of J. A. Plamondon, general merchant, St. Raymond, Que., has been sold at 53c. on the dollar to J. T. Marcotte, St. Bazile.

CHANGES.

Adams & Coate, hardware dealers, Kingsville, Ont., have sold out to Telfer & Oliver.

Lowther & Co., general merchants, etc., Russell, Man., have sold out to Smellie Bros. & Co.

Wellwood & Sales, carriagemakers, etc., Merlin, Ont., have been succeeded by Sales & Archer.

J. U. Charters, dealer in agricultural implements, Melita, Man., has been succeeded by James McCallum.

D. W. Mathewson & Co., general merchants, Lower Woodstock, N.B., have been succeeded by A. W. Hay.

The style of The Diamond Oil and Grease Co., Hamilton, Ont., has been changed to The Commercial Oil Co.

FIRES.

The stock of D. Irwin, general merchant, Elgin, Man., has been damaged by fire.

J. Fennell & Son, hardware and coal dealers, Berlin, Ont., have suffered loss by fire; insured.

Temple & McGuire, hardware dealers, and W. H. Bull, harness dealer, etc., Elgin, Man., have been burned out.

DEATHS.

C. J. Chisholm, importer of steel, etc., Montreal, is dead.

INQUIRIES REGARDING OANADIAN TRADE.

The following were among the recent inquiries relating to Canadian trade received at the High Commissioner's office, in London, England:

1. Inquiry has been received from an agent in London for names of Canadian firms desiring to be represented at the forthcoming Exhibition in Glasgow

2. A correspondent asks for information concerning the manufacture of soap, candles, starch, paper and turnery in Canada.

[The names of the firms making the above inquiries, can be obtained on application to the editor of HARDWARE AND METAL, Toronto. When asking for names, kindly give number of paragraph and date of issue.]

Mr. Harrison Watson, curator of the Canadian Section of the Imperial Institute, London, England, is in receipt of the following inquiries:

1. A Manchester firm of brokers would like to hear from Canadian shippers of tallow, paraffin wax, starch, resin, etc.

2. A Scotch firm asks for names of Canadian producers of excelsior.

3. An Irish firm desires to be placed in correspondence with Canadian makers of curled hair.

4. A firm manufacturing engineers' tools, turbines, fans, steam pumps, etc., would be prepared to appoint resident Canadian agent if an opening exists for the sale of above

PAINT VS. RUBBER.

A SWIFT, hot and spirited old-time game of hockey, was enjoyed by quite a crowd of spectators in the White Star rink, Montreal, on Tuesday evening last, when teams from The Canada Paint Co. and The Canadian Rubber Co., crossed sticks. The rubber men, while more elastic in their movements than the "Elephants," stormed The Canada Paint Co.'s goal without avail till near the close of the game. At each attack they harmlessly rebounded to the far end of the rink, while the puck in its various ramifications was shot into the goal of The Canadian Rubber Co. no less than four times, three goals being scored in the first half. Briefly, the match was won by The Canada Paint Co.'s team by 4 to 1. It is conceded on all sides that The Canadian Rubber Co.'s team played a brilliant game, and with a little more practice would give the paint men a hard rub. It was the team play of the paint men that carried the day, and Capt. Munro is to be congratulated, not only on his own brilliant work, but also on the effectiveness of his team. Mr. H. Brigger, the rubber captain, is also deserving of a good deal of commendation.

The feature of the match was a bold length-of-the-rink dash by Mr. Russell, of the paint staff, who is said to have limbs on him like those of a derrick, making him a tower of strength. The attacks of the catachouc men against him availed them little more than nothing.

This makes the second victory for the Canada Paint Co., they having defeated the Baylis Manufacturing Co. 2 goals to 1.

At the close of the match a French-Canadian spectator remarked: "B' gosh, zee puck, she go fro and too like zee chain lightning de grease. She remin' wan of my bob-tailed stallion Jeanette when she go two twenty on de (h)ice on ze grand course de trot."

A BRANCH HOUSE FOR WINNIPEG.

Caverhill, Learmont & Co. are opening a branch office and sample-room in the city of Winnipeg, which will be under the management of Mr. Thos. L. Waldon, well known by the hardware trade of Manitoba and the Northwest.

Leading lines of heavy hardware will be carried in stock in Winnipeg. Mr. Waldon will travel on the branch lines, giving a large share of his attention to the city of Winnipeg.

Mr. Norman J. Dinnen will cover the territory on the main line. Mr. John Burns, jr., and Mr. Colin C. Brown will retain their present districts in British Columbia.

CATALOGUES, BOOKLETS, ETC.

A TALK ABOUT RANGES.

THE Gurney Foundry Co., Limited, are sending out one of the most practical pamphlets HARDWARE AND METAL has read for some time. It purports to be a conversation between Mrs. Needarange and Mr. Active Hustler. Mrs. Needarange enters Mr. Hustler's store and asks to see a range. The manner in which the good points of the "Imperial Oxford" are described by the latter is well worth reading. Hardware merchants should read this carefully, and then pass it on to their clerks.

A DAINTY CALENDAR.

The Rollman Manufacturing Co., Mount Joy, Pa., have issued an attractive calendar, which should prove handy to hang up in an office, as, while it is of moderate size, the calendar pad is large enough to be easily read at some distance. Above the calendar pad is a half-tone drawing, showing a Rollman cherry seeder in position, with a couple of plates of luscious cherries near by. The calendar will be sent to any of the trade.

CLAYTON AIR COMPRESSORS.

The Clayton Air Compressor Works, New York, are sending out their catalogue No. 11. It includes descriptions of the entire machine and parts of compressors of every type, and for all pressures and for every purpose to which compressed air is applied. This firm make it their endeavor to avoid complicated innovations, adopting only such changes as have proven of practical value. Their claims for superiority are based upon the fundamental points of a good air compressor, such as simplicity of design, economy in consumption of power, efficiency in air compression, accessibility and durability of working parts and perfect automatic regulation.

On account of the reputation of this firm, their catalogue should be secured by all who use or handle air compressors. Their address is Havemeyer Building, Cortlandt street, New York.

MR. DACK IS PRESIDENT.

At the anual meeting of the Commercial Travellers' Mutual Benefit Society, on Saturday, the following were elected officers for the ensuing year:

President—W. B. Dack (acclamation).
Vice-President—Dan A. Rose.
Treasurer—John A. Ross.
Trustees for Toronto to fill vacancies on the board—John Orr, John Brasier, Geo. McQuillan, W. R. Madill and J. M. Woodland.
Trustees for Hamilton—John Hooper and E. A. Dalley.
Auditors—Henry Barber and H. J. M. Bryant.

President Dack, on assuming his office, made an appropriate and interesting address. Votes of thanks were tendered to the retiring members of the board, Messrs. W. J. Hopwood, R. L. Patterson, W. F. Smith, F. J. Zamaners and N. A. Cockburn.

ELECTRIC LIVERY IN TORONTO.

C. A. Ward, Arch. Fairgrieve, W. S. Jackson, A. M. Thompson, J. T. Smith and A. F. Dodge have been incorporated under the style of The Electric Cab Co., Limited, of Toronto, to operate automobiles in Toronto. A meeting of the company was held at the Arlington Hotel, and the following officers elected: President, C. A. Ward, manager of the Arlington Hotel; vice-president, A. M. Thompson, of The Canadian Motor Co., Limited; secretary-treasurer, W. S. Jackson, of James Robertson & Co.

The Cyclorama building, near the Union Station, has been rented by the company. Two electric tally-hos and a number of smaller vehicles are nearly ready, and business will begin in the course of a month or so.

WILL PAY 50C. ON THE DOLLAR.

A Kingston despatch says that the dividend statement of the defunct Kingston Locomotive Works Co., as prepared by A. F. Riddell and K. W. Blackweel, joint liquidators, and accepted by Judge Price, has been presented to the creditors. The liabilities are $339,494, on which the assets will pay 50c. on the dollar, or a total o $169,747.

The principal creditors are the Bank of Montreal, $172,533; F. Edgar, Montreal, $75,296; James W. Pyke & Co., Montreal, $40,185; The Canada Switch and Spring Co., Montreal, $16,351.

REDUCTION IN PRICE.

We learn from the Canadian agent, Mr. Knox Henry, 1 Place Royale, Montreal, that Brassite goods have been reduced about 20 per cent. all down the list. This will immensely increase the popularity of these goods in Canada, for, at these prices, their superiority of finish will give them a pre-eminent position on the market. Already inquiries are more numerous and orders heavier.

SYDNEY AS A PORT.

Over 17,000 vessels, including 700 ocean-going steamers, arrived in this harbor during the past season, their registered tonnage amounting to upwards of 1,000,000, and their crews numbering over 27,000 seamen. In point of arrivals, Sydney is thus to be classed among the great shipping ports of the world, and as regards ocean-going vessels, one of the greatest in Canada. —Sydney Record.

MONTREAL BOARD OF TRADE OFFICERS.

The following are the new officers of the Montreal Board of Trade:

President, Mr. Henry Miles; 1st Vice-president, Mr. F. W. Evans; 2nd Vice-president, R. W. MacDougall; Treasurer, A. J. Hodgson.

Members of Council—Geo. E. Drummond, W. I. Gear, A. E. Ellis, R. Wilson-Smith, Robt. Munn, Alex. McFee, Charles Chaput, Alex. McArthur, P. W. McLagan, A. B. Evans, W. H. Browne, J. C. Holden.

Board of Arbitration—James Crathern, E. B. Greenshields, John McKergow, Robert Archer, Charles F. Smith, Robert Bickerdike, Robert Reford, Edgar Judge, Robert McKay, David MacFarlane, Adam G. Thomson, Charles McLean.

RE-CUT FILES.

On account of the many inquiries from the large consumers of files throughout the Dominion during the past two or three years, The Globe File Manufacturing Co., of Port Hope, Ont., have been compelled to add a re-cutting department to their manufactory, and are now prepared to receive old files for re-cutting. Providing the blanks are shipped to them, sound, and free from rust, they claim that they can return the same goods re-cut, which will be in every respect equal to new files. They have issued a special net price list for this work, which may be had on application.

WALKERVILLE MATCH FACTORY BURNED.

The factory of The Walkerville, Ont., Match Co. was destroyed by fire on Friday, January 25, causing a loss estimated at $20,000 on stock and $5,000 on the building. The fire, in the opinion of Mr. Anderson, proprietor and manager of the company, was started by rats, and, owing to the inflammable nature of the contents, it spread rapidly. Two explosions were caused by carbide of potash becoming ignited. These blew the end and one side out of the building. A deplorable feature in regard to the fire was the death of two firemen, caused by the explosion.

OFF TO NEW GLASGOW.

Mr. John Irving, the Montreal rolling mills representative in Toronto and the West, whose appointment as sales agent of The Nova Scotia Steel Co. was noted a couple of weeks ago, leaves on Sunday night for New Glasgow, N.S., where he will in future reside.

A GRIPPE JOKE.

Mr. J. H. Lyons, of Sidney, Shepard & Co., Buffalo, is in Toronto on a brief visit. "Did you hear the latest grippe joke?" he said. And then by way of explanation added: "The traveller comes in, lays down his grip with the remark, 'I've got the grip, but (here he steps over it) I'm getting over it.'"

THE MONTREAL FIRE.

THE hardware trade suffered rather severely from the disastrous fire that raged in the wholesale section of Montreal a week ago last Wednesday night. The destruction of the magnificent Board of Trade building turned a large number of agents out of quarters for the time being, and this week they have been able to do nothing but look for suitable offices and fix them up. Withal the loss, the fire has show that there is sympathetic feeling down deep in the hearts of the business men towards their confreres. Many acts of kindness have been shown this week, and we do not need to say they will not be forgotten. Firms devoid of stock have been offered goods at special discounts by their contemporaries, desks have been freely offered and all sorts of invaluable offers have been made. Most of the fire sufferers have secured new quarters and are carrying on business as usual.

It is generally agreed that the Board of Trade building must be rebuilt. Action will probably be taken next week at the first meeting of council. The site will probably not be changed.

A. C. Leslie & Co., who had offices in the Board of Trade building, lost everything in their office, except the contents of the vault, which were saved in good condition. Rooms have been rented in the Merchants Bank building, St. James street, and business will be continued as usual.

Knox Henry, hardware agent, who had his office in the Board of Trade building, lost all his stock and office furniture. His vault preserved his papers, but his insurance did not come within $300 of covering his stock. He has taken temporary quarters at 1 Place Royale.

Alex. McArthur & Co., tarred and felt paper manufacturers, whose warehouse was situate next to Saxe & Co., lost their entire stock, which amounted to about $7,000. The loss is covered by insurance. All the firm's papers were saved. Temporary offices were taken at 87 St. Peter street, but a warehouse has been rented at 82 McGill street, and the offices have been removed to this address. Business is being carried on as usual.

View on Commissioners street, Montreal, showing where the big fire was stopped, two blocks away from where it originated. The tallest building is the ruin of Seybold, Sons & Co.'s big hardware establishment. The rear entrance was on Commissioners street, the front on St. Paul street.

The Thos. Davidson Manufacturing Co.'s salesroom at 474 St. Paul was completely gutted. The staff has been removed to their works where all their customers' wants will be attend to. A down town office will probably be fitted up in due time.

The Dominion Travellers' Association, which had its quarters in the Board of Trade building, lost everything, including some pictures and paintings of priceless value. New quarters have been fitted up at Room 9, Bank of Toronto Chambers.

H. A. Nelson & Sons' establishment was burned to the ground, forming one of the heaviest losses of the fire. The firm have decided to liquidate and go out of business so far as the Montreal end is concerned. The Toronto branch will be continued as heretofore. Temporary offices have been taken at 27 Common street.

Peck, Benny & Co. saved all their papers and documents that were kept in their three vaults, and business is being conducted as usual. The city offices are now situate in the Chesterfield Chambers, St. Alexis street.

Pillow & Hersey Manufacturing Co., Limited, were also sufferers. One vault, holding the heavy books, saved their contents, but the light copy books were destroyed. However, orders were filled on the following day without delay. The firm's city address is now 232 McGill street.

Seybold, Son & Co.'s large hardware establishment, facing on St. Paul street and running back to Commissioners street, was one of the last buildings attacked by the flames, but it was completely gutted, and only the rear wall on Commissioners street was left standing. Mr. Seybold and his office staff were in the office when the fire broke out, and were busy balancing the books for the year. The books and papers were all saved, but the entire stock of shelf hardware, valued at $82,000, was destroyed. Fortunately, the heavy hardware was stored elsewhere. The insurance amounts to $70,-500. On the morrow after the fire, Mr. Seybold rented a warehouse at 148 McGill street, but the structure was found to be defective, and the address has been changed to 18 and 20 St. Sacrament street, which affords an office connection with the heavy hardware warehouse in the rear. Here the firm will be located until a large new warehouse is built on the old spot. Mr. Seybold says he will erect quarters unexcelled in the city for convenience and safety; they will be especially for the hardware trade. Until he has laid in a stock of shelf goods, Mr. Seybold has made arrangements to fill all orders promptly through the courtesy of his confreres in the trade, and his customers need not fear that shipments will be slow or goods unsatisfactory.

The Imperial Oil Co., who were in the Board of Trade building, have changed their address to 71 St. James street. Other hardware losers by the fire were: The Northern Elevator Co.; The Copeland-Chatterson Co., (now in the Merchants Bank); J. B. Goode, hardware agent; James Hutton & Co., hardware agents; Magnolia Metal Co., American machinery; Beardmore Belting Co., and Gall-Schneider Oil Co., Limited. H. W. DeCourtenay Co., Commissioners street, lost a stock of heavy metals valued at $25,000.

A "PATENT LAW" DECISION.

THE Court of King's Bench at Montreal last week, gave an important decision on an action arising out of the Canadian patent law. The Asbestos and Asbestic Co., Montreal, appealed from a judgment which dismissed a demand for a perpetual injunction and quashed the interlocutary injunction issued. The Asbestos Co. alleged that they acquired from the Danville Asbestos and Slate Co. asbestos mines in the township of Shipton; that on July 5, 1899, the company transferred and assigned to them a trade mark obtained by the former and registered on February 3, 1896. This trade mark has the words, "Asbestic Wall Plaster" surmounting a trowel, on which is inscribed the letter A. They state that the William Sclater Co. have been and are still using the words of the trade mark, and selling what purports to be an asbestic wall plaster stamped and lablelled as such, and the public is led to believe that it is buying a product of the Asbestos Co. The appellants claim that they have extensively advertised their product and established a lucrative business in selling asbestic wall plaster and have acquired a right of property in the trade-mark words.

They asked for an injunction to restrain the Sclater Co. and their agents from further selling any goods or materials under the name of "asbestos wall plaster," and that the respondents be condemned to pay $1,000 damages.

The Sclater Co. pleaded to the effect that the Asbestos Co. could not by the alleged trade mark, obtain the uses of the words, "asbestic wall plaster," and the Government of Canada could not give them the sole right to use those words; also, that they had sold "asbestic wall plaster" long previous to February 3, 1896, and since, and have the right to make use of the words "asbestic wall plaster," the word "asbestic" being merely an indication and description of the article sold by them. Their plea was maintained and the action dismissed, the court being of opinion that the words "asbestic wall plaster" were descriptive of the materials of which the compound consisted. In appeal the judgment was held to be well founded, and it was confirmed.

WILL EXTEND THEIR WORKS.

At the annual meeting of The Brandon, Man., Machine Works Co., the following officers were elected: President, D. A. Hopper; vice-president, E. H. Johnson; manager, James Shirriff; secretary-treasurer, Fred. Adolph. It was decided to erect this year new works which will be larger and better fitted in every respect than the present premises. The company now employs over 30 men during the summer, and about 16 during the cold weather. With the extension of premises and largely increased plant now contemplated about 50 men will be employed all year round.

SELF-CHALKING CHALK LINE.

The Smith & Hemenway Co. are just putting on the market the automatic self-chalking chalk line. This is unique in its way, and is so constructed that it is impos-

Self-Chalking Chalk Line.

sible for the line to get bound in its operation. The flat spring keeps it wound perfectly tight both in drawing out and closing up. The accompanying cuts will

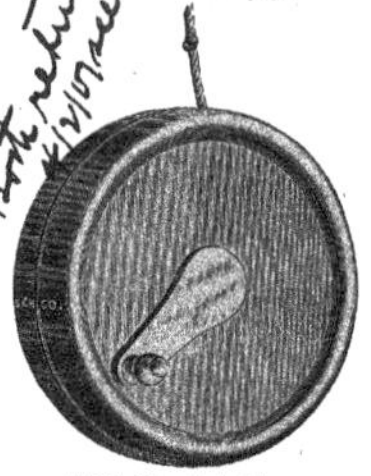

Self-Chalking Chalk Line.

give anyone an insight to the merit of the article.

The above firm publishes what is known as the "Green Book" of hardware specialties, which will be sent out to anyone in the hardware business on application. When writing for this kindly mention that you saw it in CANADIAN HARDWARE AND METAL MERCHANT.

P. Denis, general merchant, St. Cesaire, Que., has assigned to Lamarche & Benoit.

SIR FRANK SMITH'S ESTATE.

THE estate of the late Senator Sir Frank Smith, according to his will, is valued at over a million and a quarter. The estate is made up as follows: Real estate in Toronto, London and Ingersoll, Ont., $126,380; stock and bonds of Niagara Navigation Company and various bank stocks, $645,080; stock in gas companies, $257,077; bonds of various companies, $116,000; other stocks, $120,131; furniture, horses, carriages and sundry assets, $14,895; total, $1,279,564.

By the will, which is dated July 10, 1897, the Toronto General Trusts Corporation is appointed executor and trustee, and all the estate is devised to the trustee in trust. The trustee is authorized to sell any of the estate from time to time and to make investments on certain named securities, to give leases, also to change investments from time to time, and to retain, so long as the trustee thinks fit, any lands, property, assets of every kind. The succession duties are to be paid out of the capital of the estate. (These duties, amounting to about $65,000, go to charitable institutions of the Provincial Government.)

Sir Frank leaves to the House of Providence, Toronto, $1,000; St. Michael's Hospital, $1,000, and to the House of Industry, $1,000; to his niece, Mary Munro, $400 per annum during her life, and to four other nieces, daughters of his sister Margaret, $500 each. To his nephew, Andrew Munro, $500. In respect to his only surviving son he makes a provision of $4,000 per year. He gives $600 per annum out of income to each of his grandchildren so long as the parent of such grandchild is living. On the death of the parent of such grandchild the income of the grandchild is increased and such grandchild takes a share of the income of the estate in proportion to the number of grandchildren. One-third of the rest of the income is given to each of his daughters for her life. On the death of either of his two daughters now living the present husband of any such daughter is to receive $1,200 per annum.

At the expiration of 20 years from Sir Frank's death, or on the death of the last surviving of his children (whichever date or event shall last happen), the capital is to be divided between his grandchildren in equal shares.

The wish is expressed that John Foy and Robert H. McBride continue as directors of the Niagara Navigation Company, and directs that John Foy receive a power of attorney to represent his estate at meetings of shareholders of the Niagara Navigation Company and of the Home Savings and Loan Company. Any unexpected income of an infant, who may die before coming of age, falls into and forms part of the estate. He wishes his grandson, Frank A. Harrison (an orphan), to be brought up by one of his daughters, Mrs. Macdonald or Mrs. John Foy.

GLAZIERS' DIAMONDS.

THERE is a popular misconception that the diamonds used by glaziers are the rejected stones, of poor quality, that are worthless as gems, but such is not the case. It is incorrect, at least as far as it applies to such glaziers' diamonds as pass through the hands of A. Shaw & Sons. Mr. Robt. Shaw, the present head of this old-established house, in fact, the oldest house in the trade, recently gave us some information about the setting of these diamonds that is of interest. Mr. Shaw's great-grandfather founded this business over a century back, and invented the glaziers' diamonds now in universal use. During four generations, therefore, this house has been building up a reputation, with the result that now, not only do they supply their own regular customers, but they have a large and increasing trade with other manufacturers of glaziers' diamonds.

The diamonds used for glass cutting are imported chiefly from Brazil and South Africa. They come in the rough form and vary in size from 30 to the carat up to 10 to the carat. From the original parcels, but a small proportion is selected, the fine quality stones suitable for cutting glass being very scarce. Any diamond will scratch glass, but all will not cut, so as great care is required in selecting stones for glaziers' purposes as for jewelers'. If the stones are of the proper sort, of good shape, with good angles, etc., they do not require cutting or other treatment before being set.

The greatest care must be taken to have the cutting angles adjusted perfectly true, otherwise the diamond will not cut satisfactorily, but will soon become useless. A diamond-setter who knows his business can set a poor stone in such a way that it will do better work than, and outlast a good stone that has been set improperly. Things are not done by chance in the workrooms of A. Shaw & Son. Skilled workmen only are employed and the article in hand is put through numerous tests during its treatment, so by this means none but reliable cutters leave the factory.

In addition to the ordinary glaziers' diamonds, with and without racks, they produce folding diamonds, pocket knives with diamonds, and many devices in the form of circle arms, beam compasses, etc., for doing circular cutting and for cutting guage glass, glass shades etc. Mr. Shaw is prepared to reset diamonds for any readers of HARDWARE who may have stones that have become useless, the charge being very low as compared with the ultimate value of the reset stone.

R. K. McKenzie, general merchant, Middle River, N.S., is dead.

MARKETS AND MARKET NOTES

QUEBEC MARKETS.

Montreal, February 1, 1901.

HARDWARE.

ALTHOUGH trade continues to improve, the advance is slow and the volume of business being done just now is not great. Heavy goods are quiet, a few lots are being shipped from stock, but business is mostly confined to spring account. Nails remain in about the old position, inquiry being for small amounts. Wires are not attracting much attention. Shelf goods are the most active articles on the list and the demand for them continues in a sorting way. Poultry netting and green wire cloth are being booked for spring in fair-sized lots and ice cream freezers are now beginning to go. Glass is now being booked for import, although some houses refuse to take orders. Mrs. Potts sad irons have been slightly reduced during the week, and are now selling at 70c. for polished and 75c. for nickle-plated. Screen doors and windows are being ordered in good supply, while harvest tools and spades and shovels are also commencing to be booked.

BARB WIRE — Little or no business is doing, and the market is featureless. The price is still $3.20 f.o.b. Montreal in less than carlots.

GALVANIZED WIRE—No attention seems to be paid to this line, and very little business is being done in it. We quote: No. 5, $4.25; Nos. 6, 7 and 8 gauge $3.55; No. 9, $3.10; No. 10, $3.75; No. 11, $3.85; No. 12, $3.25; No. 13, $3.35; No. 14, $4.25; No. 15, $4.75; No. 16, $5.00.

SMOOTH STEEL WIRE — There is no new feature to note, and trade continues in the same lines as before. The price is $2.80 per 100.lb.

FINE STEEL WIRE — A small trade is passing The discount is 17½ per cent. off the list.

BRASS AND COPPER WIRE — The usual demand is being experienced. Discounts are 55 and 2½ per cent. on brass, and 50 and 2½ per cent. on copper.

FENCE STAPLES—Little business is being done as yet. We quote: $3.25 for bright, and $3.75 for galvanized, per keg of 100 lb.

WIRE NAILS—The demand for wire nails is not brisk, and only immediate needs create an inquiry. Quite a number of small lots have been shipped this week. We quote $2.85 for small lots and $2.75 for carlots, f.o.b. Montreal, Toronto, Hamilton, London, Gananoque, and St. John, N.B.

CUT NAILS — The demand is limited, only small lots are moving. We quote as follows: $2.35 for small and $2.25 for carlots; flour barrel nails, 25 per cent. discount; coopers' nails, 30 per cent. discount.

HORSE NAILS — The Manufacturers' Association has been dissolved and open prices now prevail. As yet values have not been changed. At their meeting this week, the manufacturers may reorganize. Discounts are still 50 per cent. on standard, and 50 and 10 per cent. on Acadia.

HORSESHOES — The demand continues

quite brisk. There has been no change in prices. We quote: Iron shoes, light and medium pattern, No. 2 and larger, $3.50; No. 1 and smaller, $3.75; snow shoes, No. 2 and larger, $3.75; No. 1 and smaller, $4.00; X L steel shoes, all sizes, 1 to 5, No. 2 and larger, $3.60; No. 1 and smaller, $3.85; feather-weight, all sizes, $4.85; toe weight steel shoes, all sizes, $5.95 f.o.b. Montreal; f.o.b. Hamilton, London and Guelph, 10c. extra.

POULTRY NETTING—The discount is s i'l 50 and 5 per cent. More spring business has been done this week.

GREEN WIRE CLOTH—Fair orders are being taken. The price remains at $1.35 per 100 sq. ft.

FREEZERS — A few orders are coming in, although some houses are not out with their quotations yet.

SAD IRONS—Mrs. Potts sad irons have been slightly reduced this week and are now selling at 70c. for full polished and 75c. for nickel-plated.

SCREEN DOORS AND WINDOWS — The travellers are booking fairly good orders. We quote: Screen doors, plain cherry finish, $8.25 per doz.; do. fancy, $11.50 per doz.; windows, $2.25 to $3.50 per doz.

SCREWS—A good sorting trade is being done and some letter orders have been received this week. Discounts are as follows: Flat head bright, 85 per cent. off list; round head bright, 80 per cent.; flat head brass, 77½ per cent.; round head brass, 70 per cent.

BOLTS—Inquiry has been moderate this week. Discounts are: Carriage bolts, 65 per cent.; machine bolts, 65 per cent.; coach screws, 75 per cent.; sleigh shoe bolts, 75 per cent.; bolt ends, 65 per cent.; plough bolts, 50 per cent.; square nuts, 4½c. per lb. off list; hexagon nuts, 4¾c. per lb. off list; tire bolts, 67½ per cent.; stove bolts, 67½ per cent.

BUILDING PAPER — Quite a number of orders for spring delivery have been placed this week, but trade is not at all brisk. We quote: Tarred felt, $1.70 per 100 lb.; 2-ply, ready roofing, 80c. per roll; 3-ply, $1.05 per roll; carpet felt, $2.25 per 100 lb.; dry sheathing, 30c. per roll; tar sheathing, 40c. per roll; dry fibre, 50c. per roll; tarred fibre, 60c. per roll; O. K. and I. X L, 65c. per roll; heavy straw sheathing, $28 per ton; slaters' felt, 50c. per roll.

RIVETS — A fair business is being done at unchanged prices. The discount on best iron rivets, section, carriage, and wagon box, black rivets, tinned do., coopers' rivets and tinned swedes rivets, 60 and 10 per cent.; swedes iron burrs are quoted at 55 per cent. off; copper rivets, 35 and 5 per cent. off; and coppered iron rivets and burrs, in 5-lb. carton boxes, are quoted at 60 and 10 per cent. off list.

CORDAGE — Prices are steady. Manila is quoted at 13c. per lb. for 7-16 and larger; sisal at 9c., and lathyarn 9c. per lb. In small lots ½c. per lb. higher is charged.

SPADES AND SHOVELS—A little business for spring delivery has been done this week. The discount is still 40 and 5 per cent. off the list.

HARVEST TOOLS—A few orders are being placed at 50, 10 and 5 per cent. discount.

TACKS—Prices are unchanged this week. We quote: Carpet tacks, in dozens and bulk, blued 80 and 5 per cent. discount; tinned, 80 and 10 per cent.; cut tacks, blued, in dozens, 75 and 15 per cent. discount.

CHURNS—Some spring business is being done at a discount of 56 per cent.

FIREBRICKS —A small jobbing trade is being done at $18.50 to $26, as to brand.

CEMENT — Few sales are taking place. We quote: German, $2.50 to $2.65; English, $2.40 to $2.50; Belgian, $1.90 to $2.15 per bbl.

METALS.

The market is generally steady. Canada plates are rather weak, cable quotations being somewhat lower this week. Tinplates, terne plates and galvanized iron are all beginning to move for next year. The rolling mills seem to be pretty busy, and are not disposed to make concessions.

PIG IRON—Summerlee continues at $24 to $25, with Canadian pig worth $19 to $20. It is said that the demand shows some slight improvement.

BAR IRON—The market is steady at $1.65 to $1.70 per 100 lb.

BLACK SHEETS—A little better inquiry is noticed this week. Prices rule at $2.80 for 8 to 16 gauge.

GALVANIZED IRON—Spring business is flourishing. Buying seems to be safe, as the market bears a firm tone. We quote: No. 28 Queen's Head, $5 to $5.10; Apollo, 10¾ oz., $5 to $5.10 Comet, No. 28, $4.50 with 25c. allowance in case lots.

INGOT COPPER—Some business is being done at 17½c.

INGOT TIN—Foreign markets appear to be very much unsettled, but yet steady. London quotes about £122, varying from day to day about this figure, which means 27c. laid down in New York. Prices here are 32 to 33c.

LEAD—The price of lead is steady at $4.65.

LEAD PIPE—A moderate demand is being experienced. We quote: 7c. for ordinary and 7½c. for composition waste, with 15 per cent. off.

IRON PIPE—Business is of fair volume. We quote as follows: Black pipe, ¼, $3 per 100 ft.; ⅜, $3; ½, $3; ¾, $3.15; 1-in., $4.50; 1¼, $6.10; 1½, $7.28; 2-in., $9.75. Galvanized, ½, $4.60; ¾, $5.25; 1-in., $7.50; 1¼, $9.80; 1½, $11.75; 2-in., $16.

TINPLATES—A fair demand continues to be experienced, and some business is doing on spring account. The ruling figures for immediate delivery are $4.50 for coke and $4.75 for charcoal.

CANADA PLATE—There is little buying going on just now, as cable advices indicate a falling market. Stocks in hand now are light, and bring full figures. We quote as follows: 52's, $2.90; 60's, $3; 75's, $3.10; full polished, $3.75, and galvanized, $4.60.

TOOL STEEL—We quote: Black Diamond, 8c.; Jessop's 12c.

STEEL—No change. We quote: Sleigh-shoe, $1.85; tire, $1.95; spring, $2.75; machinery, $2.75 and toe-calk, $2.50.

TERNE PLATES—Some little inquiry is coming in this week, and spring stocks are being arranged for. We quote $8.25.

SWEDISH IRON—Unchanged at $4.25.

COIL CHAIN — The coil chain market is quite firm, and an advance is anticipated. A good many orders are being booked. We quote: No. 6, 11½c.; No. 5, 10c.; No. 4, 9½c.; No. 3, 9c.; ¼-inch, 7½c. per lb.; 5-16, $4.60; 5-16 exact, $5.10; ⅜, $4.20; 7-16, $4.00; ½, $3.75; 9-16, $3.65; ⅝, $3.35; ¾, $3.25; ⅞, $3.20; 1-in., $3.15. In carload lots an allowance of 10c. is made.

SHEET ZINC—The ruling price is 6 to 6½c.

ANTIMONY—Quiet, at 10c.

GLASS.

Orders for spring delivery are being booked on a basis of $2.75 for first break, and $1.85 for second break for 50 feet. The primary markets are firm. Small amounts are being shipped from stock. We quote: First break, $2; second, $2.10 for 50 feet; first break, 100 feet, $3.80; second, $4; third, $4.50; fourth, $4.75; fifth, $5.25; sixth, $5.75, and seventh, 6.25.

PAINTS AND OILS.

Raw and burnt sienna ground in oil has been advanced 2c. per lb., and pure raw and burnt umber 1c. per lb. by the leading colormakers. A fair amount of business has been done since last week, and, as February is now at hand, prospects seem to be brightening. Turpentine is keeping low and steady, while linseed oil, for present delivery, is quite firm. A brisk inquiry has opened out for paris green, and all the jobbers are busily engaged placing orders. Spring sales seem to be larger than last year. Stocks in the country are light. There has been some inquiry heard for brick paint and oxides for spring shipment, and while at present the factories are not extra busy, it is thought that the trade will open out with vim in the first week of February. We quote:

WHITE LEAD—Best brands, Government standard, $6.75; No. 1, $6 37½; No. 2, $6; No. 3, $5.62½, and No. 4, $5.25, all f.o.b. Montreal. Terms, 3 per cent. cash or four months.

DRY WHITE LEAD—$5.75 in casks; kegs, $6.

RED LEAD—Casks, $5.50; in kegs, $5.75.

WHITE ZINC PAINT—Pure, dry, 8c.; No. 1, 6½c.; in oil, pure, 9c.; No. 1, 7½c.

PUTTY—We quote: Bulk, in barrels, $2 per 100 lb.; bulk, in less quantity, $2.15; bladders, in barrels, $2.20; bladders, in 100 or 200 lb. kegs or boxes, $2.35; in tins, $2.45 to $2.75; in less than 100-lb. lots, $3 f.o.b. Montreal, Ottawa, Toronto, Hamilton, London and Guelph. Maritime Provinces 10c. higher, f.o.b. St. John and Halifax.

LINSEED OIL—Raw, 80c.; boiled, 83c., in 5 to 9 bbls., 1c. less, 10 to 20 bbl. lots, open, net cash, plus 2c. for 4 months. Delivered anywhere in Ontario between Montreal and Oshawa at 2c. per gal. advance and freight allowed.

TURPENTINE—Single bbls., 59c.; 2 to 4 bbls., 58c.; 5 bbls. and over, open terms, the same terms as linseed oil.

MIXED PAINTS—$1.25 to $1.45 per gal.

CASTOR OIL—8¾ to 9½c. in wholesale lots, and ½c. additional for small lots.

SEAL OIL—47½ to 49c.

COD OIL—32½ to 35c.

NAVAL STORES—We quote: Resinsr $2.75 to $4.50, as to brand; coal tar, $3.25 to $3.75; cotton waste, 4½ to 5½c. fo, colored, and 6 to 7½c. for white; oakum, 5½ to 6½c., and cotton oakum, 10 to 11c.

PARIS GREEN—Petroleum barrels, 16½c. per lb.; arsenic kegs, 17c.; 50 and 100-lb. drums, 17½c.; 25-lb. drums, 18c.; 1-lb. packages, 18½c.; ½-lb. packages, 20½c.; 1-lb. tins, 19½c.; ½-lb. tins, 21½c. f.o.b. Montreal; terms 3 per cent. 30 days, or four months from date of delivery.

SCRAP METALS.

The market is quiet, and remains steady. Dealers are paying the following prices in the country: Heavy copper and wire, 13 to 13½c. per lb.; light copper, 12c.; heavy brass, 12c.; heavy yellow, 8½ to 9c.; light brass, 6½ to 7c.; lead, 2¾ to 3c. per lb.; zinc, 2¼ to 2½c.; iron, No. 1 wrought, $13 to $14 per gross ton; No. 1 cast, $13 to $14; stove plate, $8 to $9; light iron, No. 2, $4 a ton; malleable and steel, $4.

PETROLEUM.

A fair business is still kept up. We quote: "Silver Star," 15 to 16c.; "Imperial Acme," 16½ to 17½c.; "S.C. Acme," 18 to 19c., and "Pratt's Astral," 19 to 20c.

HIDES.

Trade is rather dull on account of poor demand from the tanners. Dealers are paying 7½c. for No. 1 light, and tanners are asked to pay 8½c. for carlots. We quote: Light hides, 7½c. for No. 1; 6½c. for No. 2, and 5½c. for No. 3 Lambskins, 90c.

ONTARIO MARKETS.

TORONTO, February 1, 1901.

HARDWARE.

THE wholesale hardware trade is lacking in activity, and the month which has just closed has experienced a much smaller business than January of 1900. It must be remembered, however, that in January, 1900, the demand was abnormally good on account of the anticipated advances in prices in many staple lines of hardware. This year the conditions, as far as prices are concerned, are the very opposite to what they were a year ago. Generally speaking, the demand is light both for present and for future shipment. Wire nails are still quiet and cut nails dull. Very few orders of any kind are reported in fencing wire. Horseshoes and horse nails are quiet, and on the latter prices are lower on account of the dissolution of the agreement among the manufacturers. Bolts and nuts are meeting with a fair demand. In screws, another decline, the second for the month, has taken place. In cutlery and sporting goods, business is still only light. In both tinware and enamelled ware trade is small. The feature of the trade this week is an improved business in screen doors and windows and

in green wire cloth for future delivery. Although prices are being quoted for binder twine for future delivery, very little business has so far been done.

BARB WIRE—There is no improvement, the demand being almost nil for importation as well as from stock. We quote $2.97 f.o.b. Cleveland for less than carlots, and $2.85 in carlots. From stock, Toronto, $3.10 per 100 lb.

GALVANIZED WIRE—On account of an idea that no lower prices will be seen, retailers are holding back their orders for future delivery. In fact, this may be said in regard to future business for all kinds of fence wire. We quote: Nos. 6, 7 and 8, $3.55; No. 9, $3.10; No. 10, $3.75; No. 11, $3.85; No. 12, $3.25; No. 13, $3.35; No. 14, $4.25; No. 15, $4.75, and No. 16, $5.

SMOOTH STEEL WIRE—There is a little being done in hay-baling wire, but, in oiled and annealed, business is practically at a standstill. The base price is unchanged at $2.80 per 100 lb.

WIRE NAILS—The chief business is confined to orders for future delivery, but, even in this particular, trade is small, indeed. The base price is unchanged at $2.85 per keg for less than carlots and $2.75 in carlots.

CUT NAILS—No improvement is to be noted in the demand, the volume of business still being very small. The base price is $2.35 per keg.

HORSESHOES—The season is getting pretty well over, and business, in consequence, is light. We quote as follows f.o.b. Toronto: Iron shoes, No. 2 and larger, light, medium and heavy, $3.60;

snow shoes, $3.85 ; light steel shoes, $3.70; featherweight (all sizes), $4.95 ; iron shoes, No. 1 and smaller, light, medium and heavy (all sizes), $3.85 ; snow shoes, $4 ; light steel shoes, $3.95 ; featherweight (all sizes), $4.95.

HORSE NAILS—The feature in this line is a decrease in prices consequent upon the dissolution of the Manufacturers' Association. The discounts are now : Oval head, 50, 10 and 5 per cent.; countersunk head, 50, 10 and 10 per cent.

SCREWS—On account of American competition still another decline, the second inside of three weeks, has taken place. The discounts are now as follows : Flat head bright, 87½ and 10 per cent.; round head bright, 82½ and 10 per cent.; flat head brass, 80 and 10 per cent.; round head brass, 75 and 10 per cent. Round head bronze is unchanged at 65 per cent., and flat head bronze at 70 per cent.. The screw trade is fair.

BOLTS AND NUTS—The steady trade noted in last week s issue continues. We quote as follows: Carriage bolts (Norway), full square, 70 per cent.; carriage bolts, fulls quare, 70 per cent.; common carriage bolts, all sizes, 65 per cent. ; machine bolts, all sizes, 65 per cent. ; coach screws, 75 per cent.; sleighshoe bolts, 75 per cent.; blank bolts, 65 per cent.; bolt ends, 65 per cent.; nuts, square, 4½c. off; nuts, hexagon, 4½c. off; tire bolts, 67½ per cent.; stove bolts, 67½ ; plough bolts, 60 per cent. ; stove rods, 6 to 8c.

RIVETS AND BURRS—Trade is steady with prices as before. Discount, 60 and 10 per cent. on iron rivets ; iron burrs, 55 per cent.; copper rivets and burrs, 35 and 5 per cent.

ROPE—Very little business is being done. We quote : Sisal, 9c. per lb. base, and manila, 13c.; cotton rope, 3-16 in. and larger, 16½c.; 5-32 in., 21½c., and ½ in., 22½c. per lb.

BINDER TWINE — Although prices are being quoted, very little business is being done. We quote : Pure manila, 10¾c. per lb.; mixed, 8½c. per lb.; sisal, 7¾c. per lb.

CUTLERY—The volume of business in this line is still only of a small character.

SPORTING GOODS — Business in this line continues quiet.

GREEN WIRE CLOTH—Business for future delivery has been a little more active during the past week. We quote $1.35 per 100 sq. ft.

SCREEN DOORS AND WINDOWS—There has also been an improvement in the number of orders in this line for spring delivery.

MRS. POTTS SAD IRONS—The market for sad irons is somewhat demoralized on account of the cutting in prices, and there is in consequence a wider range in quotations than is usual. Some are quoting No. 55, polished, at 62½c., while others are trying to get 65c. No. 50, nickel-plated, are quoted at from 67½ to 70c. An effort is being made to put a stop to the cutting that is being done.

ENAMELLED WARE—Dealers are only buying for immediate requirements and in small lots.

TINWARE—A fairly good shipment of milk-can trimmings are still being made. In other lines of tinware trade is dull.

LEATHER BELTING—This season has not yet opened up and trade in consequence is light. Discounts are 60 and 10 per cent. on standard, and 60 per cent. on extra.

HARVEST TOOLS — There is only an occasional order being booked for future delivery. Discount 50, 10 and 5 per cent.

SPADES AND SHOVELS — A light trade only is being done. Discount 40 and 5 per cent.

BUILDING PAPER—Very little business is being done and prices are unchanged.

POULTRY NETTING—A little is being done for future delivery. Discount on Canadian 50 and 5 per cent.

CEMENT—There is nothing doing. We nominally quote in barrel lots : Canadian Portland, $2.80 to $3 ; Belgian, $2.75 to $3; English do., $3 ; Canadian hydraulic cements, $1.25 to $1.50; calcined plaster, $1.90 ; asbestos cement, $2.50 per bbl.

METALS.

The tin and copper markets have been somewhat irregular during the week, but local quotations have not been changed. The metal trade, taken on the whole, is fairly good for this time of the year. Although the demand is not, perhaps, as brisk as it was last week.

PIG IRON—In consequence of some heavy purchases of Bessemer pig iron in the United States, a rather healthier tone has been imparted to the pig iron market generally. In Canada, trade is quiet and prices much as before. No. 2 Canadian iron is quoted at $17 in 100-ton lots.

BAR IRON —An active trade continues to be done in bar iron. The ruling base price is still $1.65 to $1.70 per 100 lb.

PIG TIN — The outside markets have fluctuated somewhat during the week with the tendency towards lower prices. Prices locally have not been quotably affected, the rulling figure still being 32c. per lb.

TINPLATES—The demand has been rather more active during the past week, and some orders have been booked for importation account. Quotations are about 25c. lower than they were a week ago.

TINNED SHEETS—Trade has also become

more active in this line, and it can now be termed fair. Prices are quoted as before.

BLACK SHEETS—These are quoted 10c. lower, the base price being $3.30 per 100 lb. Business is a little better than it was.

GALVANIZED SHEETS—Trade is moderate for shipment from stock, and orders are being booked freely for spring and summer delivery. Jobbers are making delivery from the United States for orders booked for February 1.

CANADA PLATES—Trade in this line continues fairly good. We quote: All dull, $3; half and half, $3.15, and all bright, $3.65 to $3.75.

IRON PIPE—The demand keeps good and prices unchanged. We quote: Black pipe ¼ in., $3.00; ⅜ in., $3 00; ½ in., $3.50; ¾ in., $3.20; 1 in., $4.60; 1¼ in., $6.35; 1½in., $7.55; 2 in., $10.10. Galvanized pipe is as follows: ½ in., $4.65; ¾ in., $5.35; 1 in., $7.25; 1¼ in., $9.75; 1½ in., $11.25; 2 in., $15 50.

HOOP STEEL—Trade keeps fair. The base price is unchanged at $3.10.

COPPER — There is rather more inquiry for ingot copper, but the volume of business is small. A fair trade is to be noted in sheet copper. We quote: Ingot, 19 to 20c.; bolt or bar, 23½ to 25c.; sheet, 23 to 23½c.

BRASS — Business is moderate. Discount on rod and sheet 15 per cent.

SOLDER — The demand is good, with prices as quoted last week.

LEAD—Trade is quiet at 4¾ to 5c.

ZINC SPELTER—This is still quiet with prices unchanged at 6 to 6½c. per lb.

ZINC SHEET—Only a few sheets are going out. We quote casks at $6.75 to $7, and part casks at $7 to $7.50.

ANTIMONY—There is more inquiry, but not a great deal of business is being done. We still quote 11 to 11½c. per lb.

PAINTS AND OILS.

The market is dull. The price of linseed oil has fluctuated at outside markets, but locally there is no change. Import shipments, which will be here about July 1, are offered at considerably reduced prices. Turpentine is stronger in the South, but unchanged here. Some orders for paris green are being booked, but the trade is not large. We quote:

WHITE LEAD—Ex Toronto, pure white lead, $6.87½; No. 1, $6.50; No. 2. $6.12½; No. 3, $5.75; No. 4. $5.37½; dry white lead in casks, $6.

RED LEAD—Genuine, in casks of 560 lb., $5.50; ditto, in kegs of 100 lb., $5.75; No. 1, in casks of 560 lb., $5 to $5.25; ditto, kegs of 100 lb.; $5.25 to $5.50.

LITHARGE—Genuine, 7 to 7½c.

ORANGE MINERAL—Genuine, 8 to 8½c.

WHITE ZINC—Genuine, French V.M., in casks, $7 to $7.25; Lehigh, in casks, $6.

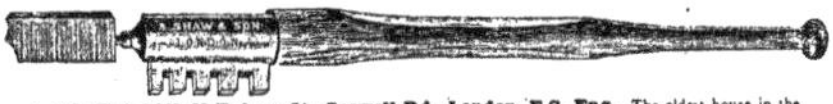

PARIS WHITE—90c.

WHITING — 60c. per 100 lb.; Gilders' whiting, 75 to 80c.

GUM SHELLAC — In cases, 22c.; in less than cases, 25c.

PARIS GREEN—Bbls., 16¾c.; kegs, 17c.; 50 and 100 lb. drums, 17½c.; 25-lb. drums, 18c.; 1-lb. papers, 18½c.; 1-lb. tins, 19½c.; ½-lb. papers, 20½c.; ½-lb. tins, 21½c.

PUTTY — Bladders, in bbls., $2.20; bladders, in 100 lb. kegs, $2.35; bulk in bbls., $2; bulk, less than bbls. and up to 100 lb., $2.15; bladders, bulk or tins, less than 100 lb., $3.

PLASTER PARIS—New Brunswick, $1.90 per bbl.

PUMICE STONE — Powdered, $2.50 per cwt. in bbls., and 4 to 5c. per lb. in less quantity; lump, 10c. in small lots, and 8c. in bbls.

LIQUID PAINTS—Pure, $1.20 to $1.30 per gal.; No. 1 quality, $1 per gal.

CASTOR OIL—East India, in cases, 10 to 10½c. per lb. and 10½ to 11c. for single tins.

LINSEED OIL—Raw, 1 to 4 barrels, 80c.; boiled, 83c.; 5 to 9 barrels, raw, 79c.; boiled, 82c., delivered. To Toronto, Hamilton, Guelph and London, 2c. less.

TURPENTINE—Single barrels, 59c.; 2 to 4 barrels, 58c., to all points in Ontario. For less quantities than barrels, 5c. per gallon extra will be added, and for 5-gallon packages, 50c., and 10 gallon packages, 80c. will be charged.

OLD MATERIAL.

Trade is not brisk, but prices are easy. We quote jobbers' prices as follows: Agricultural scrap, 55c. per cwt.; machinery cast, 55c. per cwt.; stove cast, 35c.; No. 1 wrought 50c. per 100 lb.; new light scrap copper, 12c. per lb.; bottoms, 10½c.; heavy copper, 12½c.; coil wire scrap, 13c.; light brass, 7c.; heavy yellow brass, 10 to 10½c.; heavy red brass, 10½c.; scrap lead, 3c.; zinc, 2½c; scrap rubber, 6½c.; good country mixed rags, 65 to 75c.; clean dry bones, 40 to 50c. per 100 lb.

GLASS.

There is still no information as to prices for import. Stock prices are firm. We still quote first break locally: Star, in 50 foot boxes, $2.10, and 100-foot boxes, $4; double diamond under 26 united inches, $6, Toronto, Hamilton and London; terms 4 months or 3 per cent. 30 days.

PETROLEUM.

The demand is falling off somewhat. Prices are steady. We quote: Pratt's Astral, 17 to 17½c. in bulk (barrels, $1 extra); American water white, 17 to 17½c. in barrels; Photogene, 16½ to 17c.; Sarnia water white, 16 to 16½c. in barrels; Sarnia prime white, 15 to 15½c. in barrels.

COAL.

Pea size has advanced 50c. per gross ton during the last few weeks. There is still a scarcity of nut sizes. Otherwise the market is easy, with a sufficient quantity offering. We quote anthracite on cars Buffalo and bridges: Grate, $4.75 per gross ton and $4.24 per net ton; egg, stove and nut, $5 per gross ton and $4 46 per net ton.

MARKET NOTES.

Quotations on tinplates are 25c. lower.

Wood screws have again been reduced in price.

A reduction of 10c. per 100 lb. is to be noted in black sheets.

New and higher discounts are this week being quoted on horse nails.

H. S. Howland, Sons & Co., are in receipt of a shipment of Samson's lumber crayons in blue black. The crayons are six inches long, by one inch in diameter.

MANITOBA MARKETS

WINNIPEG, January 28, 1901.

BUSINESS remains in practically the same condition as last week. Prices are firm, but demand in all lines is small. In paints and oils also there appears to be no change of price. Travellers for these lines, however, report better business than for hardware.

Price list for the week is as follows :

Barbed wire, 100 lb.	$3 45
Plain twist	3 45
Staples	3 95
Oiled annealed wire 10	3 95
" 11	4 00
" 12	4 05
" 13	4 20
" 14	4 35
" 15	4 45
Wire nails, 30 to 60 dy, keg	3 45
" 16 and 20	3 50
" 10	3 55
" 8	3 65
" 6	3 70
" 4	3 85
" 3	4 10
Cut nails, 30 to 60 dy.	3 00
" 20 to 40	3 05
" 10 to 16	3 10
" 8	3 15
" 6	3 20
4	3 30
" 3	3 65
Horsenails, 45 per cent. discount.	
Horseshoes, iron, No. 0 to No 1	4 65
No. 2 and larger	4 40
Snow shoes, No. 0 to No. 1	4 90
No. 2 and larger	4 40
Steel, No. 0 to No. 1	4 95
No. 2 and larger	4 70
Bar iron, $2.50 basis.	
Swedish iron, $4.50 basis.	
Sleigh shoe steel	3 00
Spring steel	3 25
Machinery steel	3 75
Tool steel, Black Diamond, 100 lb	8 50
Jessop	13 00
Sheet iron, black, 10 to 20 gauge, 100 lb.	3 50
20 to 26 gauge	3 75
28 gauge	4 00
Galvanized American, 16 gauge	2 54
18 to 22 gauge	4 50
24 gauge	4 75
26 gauge	5 00
28 gauge	5 25
Genuine Russian, lb	12
Imitation " "	8
Tinned, 24 gauge, 100 lb	7 55
26 gauge	8 80
28 gauge	8 00
Tinplate, IC charcoal, 20 x 28, box	10 75
" IX "	12 75
" IXX "	14 75
Ingot tin	35
Canada plate, 18 x 21 and 18 x 24	3 75
Sheet zinc, cask lots, 100 lb.	7 50
Broken lots	8 00
Pig lead, 100 lb.	6 00
Wrought pipe, black up to 2 inch	50 an 10 p.c.
" Over 2 inch	50 p.c.
Rope, sisal, 7-16 and larger	$10 00
" ⅜	10 50
" ¼ and 5-16	11 00
Manila, 7-16 and larger	13 50
" ⅜	14 00
" ¼ and 5-16	14 50
Solder	21½
Cotton Rope, all sizes, lb.	16
Axes, chopping	$ 7 50 to 12 00
" double bitts	12 00 to 18 00
Screws, flat head, iron, bright	75 and 10 p.c.
Round " "	70 p.c.
Flat " brass	70 p.c.
Round " "	60 and 5 p.c.
Coach	57½ p.c.
Bolts, carriage	42½ p.c.
Machine	45 p.c.
Tire	60 p.c
Sleigh shoe	65 p.c.
Plough	40 p.c.
Rivets, iron	50 p.c.
Copper, No. 8	50c. lb.
Spades and shovels	40 p.c.
Harvest tools	50, and 10 p.c.
Axe handles, turned, s. g. hickory, doz.	$2 50
No. 1	1 50
No. 2	1 25
Octagon extra	1 75
No. 1	1 25
Files common	70, and 10 p.c.
Diamond	60
Ammunition, cartridges, Dominion R.F.	50 p.c.
Dominion, C.F., pistol	30 p.c.
" military	15 p.c.
American R.F.	30 p.c.
C.F. pistol	5 p.c.
C.F. military	10 p.c. advance.
Loaded shells :	
Eley's soft, 12 gauge	16 50
chilled, 12 guage	18 00
soft, 10 guage	21 00
chilled, 10 guage	23 00
American, M	16 25
Shot, Ordinary, per 100 lb	6 75
Chilled	7 50
Powder, F.F., keg	4 75
F.F.G.	5 00
Tinware, pressed, retinned	75 and 2½ p.c.
" " plain	70 and 15 p.c.
Graniteware, according to quality	50 p.c.

PETROLEUM.

Water white American	24½c.
Prime white American	23c.
Water white Canadian	21c.
Prime white Canadian	19c.

PAINTS, OILS AND GLASS.

Turpentine, pure, in barrels	$ 68
Less than barrel lots	73
Linseed oil, raw	87
Boiled	90
Lubricating oils, Eldorado castor	25½
Eldorado engine	24½
Atlantic red	27½
Renown engine	41
Black oil	23½ to 25
Cylinder oil (according to grade)	55 to 74
Harness oil	61
Neatsfoot oil	$ 1 00
Steam refined oil	85
Sperm oil	1 50
Castor oil per lb.	11½
Glass, single glass, first break, 16 to 25 united inches	2 25
26 to 40 per 50 ft.	2 50
41 to 50	5 50
51 to 60	6 00
61 to 70 per 100-ft. boxes	6 50
Putty, in bladders, barrel lots per lb.	2½
kegs "	2¾
White lead, pure per cwt.	7 25
No 1 "	7 00
Prepared paints, pure liquid colors, according to shade and color..per gal.	$1.30 to $1.90

NOTES.

Over 450 commercial travellers' certificates have been issued in Manitoba during the present month.

T. Waldon, for the last 10 years Western traveller for Clare Bros., has severed his connection with that house and is taking a position in the East.

I. W. Martin, manager for the Gurney Co., has returned from a visit to the East, which included Hamilton, Toronto and Montreal in Canadian cities and many of the leading American cities.

THE GREENING CO.'S TRACE CHAIN

The B. Greening Wire Co., Hamilton, are placing, this year, on the market a trace chain, which their long experience in this business convinces them will be appreciated by their customers, enabling them to procure a chain combining the greatest amount of toughness with its already well-known strength. It is practically unbreakable with fair usage.

They have also decided to give their customers the advantage of the reduction in the price of wire, by giving them a cheaper chain, and expect to do a large business in this line during the present season.

The contract for a 1,500,000 bush. elevator to be erected at Port Arthur, Ont., has been let to J. A. Jamieson, Montreal, by Mackenzie & Mann. The elevator is to cost $350,000, and is to be completed next September, in time for next season's grain crop. The elevator is to be run in connection with the Canadian Northern Railway, now in course of construction.

THE INSOLVENCY QUESTION.

THIS week we give two additional opinions of Montreal business men, on the need of Dominion insolvency legislation :

MR. S. H. EWING'S VIEWS.

Mr. S. H. Ewing, of S. H. Ewing & Sons, Montreal, vice-president of the Molsons Bank and intimately connected with many other commercial concerns, is quite decided in affirming our need for an insolvency law that will cover all the Provinces, and is in favor of urging the Government to bring forward an insolvency measure.

"Do you consider the chattel mortgages of Ontario and the preferences of the Maritime Provinces to be unjust in their practical workings ?"

"I consider the chattel mortgages in the Provinces where they are legal can and have been very much abused by people taking a chattel mortgage and not registering till a few days before a failure, whereby the debtor is able to buy goods, as the seller is ignorant of the existence of a chattel mortgage. As to preferences in the Maritime Provinces, I consider them immoral and dishonest, and that they bring discredit on the trade of Canada generally, as it is well known that many foreign merchants do not care to place their property in the hands of parties who at any time can make a preference to any person they wish."

"Do you think the banks would or should object, to the adoption of an insolvency law ?"

"I do not see why the banks should object as their interests are the same as those of all other merchants."

"Do you find the winding-up charges excessive at present ?"

"Yes, very excessive in small estates. In many cases they eat up everything there is in the way of assets."

"Do you think they should be regulated by law ?"

"I think that certain charges should be taxed at a much less figure than they are at present. Referring to the law as it now stands, I think with small estates a curator should be appointed to wind up the affair for a nominal fee or percentage of the assets of the estate."

LOCKERBY BROS. WANT AN INSOLVENCY LAW.

Lockerby Bros., wholesale grocers, Montreal, are another firm that would like to see a Dominion insolvency law. They say that their losses under the old insolvency law were heavier than they are now because the assignees managed to absorb the bulk of the assets. But they believe that a law, such as that which Mr. Fortin suggests, would be practical and beneficial, for it would place the assets of an insolvent in the charge of the creditors who could wind up the estate as they pleased. They say they find it very inconvenient to be forced to keep track of different laws in all the Provinces, and they believe that a Dominion insolvency law should be passed. In their opinion the Government will not prove itself a business Government unless it takes hold of this matter.

The annual meeting of the Toronto Steel-Clad Bath and Metal Co., Limited, will be held on Tuesday next.

HEATING AND PLUMBING

DEFECTIVE FLUES.

OFTEN we are hustled out of an evening by the rushing and clanging of fire engines, hose reels, etc., and the street is turned into a pandemonium. For what? Why, there is a large fire raging, and in a few minutes at best a large business house or fine dwelling, together with its contents, is nothing but a smoldering pile of ashes and debris. The wise chief and long-headed inspector are called in to hold the inquest. The verdict—What is it? "A defective flue" is the finding in the majority of cases. Who is to blame? I answer—custom, old tradition, etc., and then the architect, the brick mason, or the tinner who constructs it. The architect reasons with himself that if he lines the flue with tile lining he will have a safe flue. That we will admit, if he stays on the ground and sees personally every section of the lining built in, which, unfortunately, he cannot do. The joints between these sections are often more unsafe than the naked brick flue would be, from the fact that the flue is obstructed and the products of combustion are forced between the outer surface of the lining and the inner surface of the brick covering, which is more loosely built by reason of the lining being present.

In other instances, the architect specifies that the flue must have an area of so many square inches and be well plastered on the inside. This is the poorest construction that could be used for a flue. In the course of time, the expansion and contraction, under the influence of heat, together with the rain beating down into the flue, loosens the plastering, and it falls off. If it fell off smoothly, there might still be a safe flue; but, unfortunately, this is not the case in a single instance. The plastering, in coming away from the brick, takes a long section of mortar from the joints between the brick. Thus, an opening is created through the wall of the flue, and the flue leaks and refuses to draw. Now is the stage where the wise man is called in to remedy the trouble. He decides at once that there must be a chimney pot put on, made of either terra cotta or sheet iron. He proceeds to measure the chimney top, and attaches his construction. Let us examine and analyze it.

The flue may be 9 x 9 in., with an area of 81 sq. in.; a chimney pot 8 in. in diameter has been attached, the area of which is 50 sq. in. If the pot is of terra cotta or of sheet iron, the same proportion holds good. What is the result? What is this wise man trying to do? Why, he is trying to perform something that cannot be done. He is trying to force the products of combustion from a flue having an area of 81 sq. in. through a chimney top having an area of 50 sq. in. In other words, he is trying to force a substance through an opening that is 30 per cent. smaller. Can it be done? I rather think not! Now this flue is in a fine condition for a fire in that building. The plaster has fallen off from the sides, making openings through the walls of the flue, and the stopping of the free outlet at the top crowds the heated gases back and they are forced through the walls of the flue, in contact with wood or what not, and, ergo, away she goes upward in smoke. Now if this supposed chimney had been pointed by a careful man, this supposed building could not have been burned in a hundred years from that cause.

The chimney pot, whether it be terra cotta or sheet iron, whether it be placed for ornament or utility, is the direct cause of a majority of fires. Millions in structures and goods have gone up in smoke through the direct medium of a chimney pot. To prove the theory, take a 3 or 4-in. hose attached to an engine in operation; punch a hole ¼ in. in diameter in the side of the hose. Then put a 1 or 2 in. nozzle on the hose and the water will be thrown 20 or 30 ft. through the ¼-in. hole. Now take off the nozzle and note the result. When the water can freely discharge from the full-sized opening of the hose very little will go through the ¼ in. hole. The action is the same in the case of a flue. If the top is left off, or one is attached of equal area with the opening in the flue, the air will draw inward instead of the gases being forced outward.

A flue should be built square, of good sound brick, laid in mortar made of two-thirds lime and one-third good cement. The brick should be bedded in the mortar and the joints struck smoothly, and the chimney carried above the highest point of the roof. Then there would not be any fires caused by defective flues. A flue should be at least 25 per cent. larger at the top than it is at the point where the smoke enters it. This increase should be carried gradually from the lowest point to the opening at the top; for example, say from 9 x 9 to 11 x 11 or 12 x 12 inches. A half brick might be knocked out of the side of a flue of this build and it would still be safe from fire.—By Caesar, in Metal Worker.

SOME BUILDING NOTES.

IT is probable that several fine blocks will be erected in Portage la Prairie, Man., this summer. The Agricultural Society propose erecting a $4,000 hall; R. P. Campbell intends building a business block; an addition will be built to the Merchants Bank; T. T. Bailey will build a business block; The Northern Pacific will probably build a depot there fitted with modern conveniences, and J. K. Hill contemplates erecting a drug store. It is also expected that several residences will be built.

P. Binder, Rodney, Ont., intends erecting a house on Main street.

P. Ready intends erecting a house at Kingscote, Ont., this summer.

W. T. Alexander, near Britton, Ont., intends erecting a new house next spring.

The Presbyterians of Moose Jaw, N.W.T., have decided to erect a church to cost $8,000.

Architect Witton is preparing plans for another storey to the New Royal Hotel, Hamilton.

Building permits have been issued in Toronto to Wm. Howland, for a residence at 20 Concord avenue, to cost $1,200; to Edward Drew, for a residence on Wilcox street, near Huron, to cost $3,500; and to P. Roach, for additions in the rear of 325 Queen street west, to cost $2,500. The total value of buildings for which permits have been taken during January is about $40,000, as compared with $109 000 last year and $31,800 in 1899.

PLUMBING AND HEATING CONTRACTS.

Lessard & Harris, Montreal, have secured the contract for the plumbing, heating and ventilating of Geoffrey Hales' Hospital, at Quebec; the plumbing and ventilating of St. Peter's Church, Sherbrooke, and the heating and ventilating of Sisters of St. Anne's Convent, St. Jacques, Que. This firm are busily engaged plumbing the new C.P.R. Telegraph building in Montreal.

THE CONFERENCE IN TORONTO.

The conference between committees representing the Toronto Master Plumbers' Association and the Toronto Journeymen Plumbers' Union has been more prolonged than was anticipated. Several meetings have been held, but the agreement has not yet been definitely settled. It is believed,

OUR AIM

To raise the standard of American Tools. We devote our time to manufacturing perfect goods. Ask for the **Green Book** of Hardware Specialties. It is full of good things.

SMITH & HEMENWAY CO.

296 Broadway, NEW YORK.

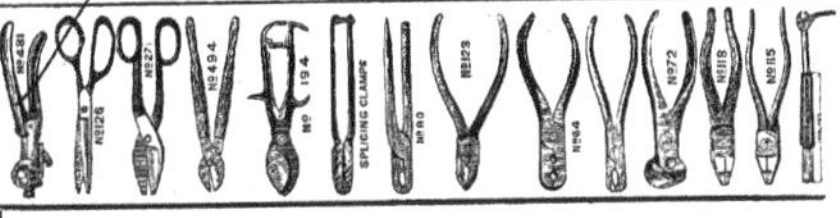

CANADIAN ADVERTISING is best done by THE E. DESBARATS ADVERTISING AGENCY MONTREAL.

ADVERTISING in WESTERN CANADA

will be Carefully, Efficiently, and Promptly attended to, by

The Roberts Advertising Agency,

WINNIPEG, CANADA.

KNOX HENRY

Heavy Hardware and Metal Broker
No. 1 Place Royale, MONTREAL.

HORSE NAILS—"C" Brand Horse Nails—Canada Horse Nail Co.

"BRASSITE" GOODS—Gunn Castor Co., Limited, Birmingham, Eng.

Manufacturers of

Heating Supplies

Pipe Fittings and Headers.
Large Manifolds made to Order.
Steam Traps and Appliances, etc.

The . . .
Jas. Morrison Brass Mfg. Co., Limited
TORONTO.

Standard Paint & Varnish Works
Limited
Makers of High Grade
Varnishes, Japans,
Paints, Colors & Enamels.
Windsor, Ont.

STANLEY RULE & LEVEL CO.,

NEW BRITAIN, CONN., U.S.A.

IMPROVED CARPENTERS' TOOLS

SOLD BY ALL HARDWARE DEALERS.

WIRE RODS!

Drawn to Decimal Sizes, Cut and Straightened, In Uniform Sizes. Prompt Shipment.

Chalcraft Screw Co., Limited, Brantford, Ont.

LEADER CHURN

New Century Improvements.

FOUR DIFFERENT STYLES:

A—Steel Frame with double reversible Steel Lever.
B—Wood Frame with double reversible Steel Lever.
C—Steel Frame with Crank.
D—Wood Frame with Crank.

Styles A and B may be operated from a sitting or standing position.

Steel Frames and Hoops beautifully ALUMINIZED.

All LEADER CHURNS are equipped with BICYCLE BALL BEARINGS and PATENTED CREAM BREAKERS.

Stands are so constructed that they are particularly strong and rigid, and there is nothing to interfere with the placing of pail in the most convenient position for draining off buttermilk.

It Pays to Have the Best. None are Better Than the Leader.

THE
Dowswell Manufacturing Co.
Limited.

HAMILTON, ONT.

Eastern Agents: W. L. Haldimand & Son, Montreal, Que.

HOW TO SAVE GAS

..USE..

Peebles' Automatic Gas Governors

Gas, Fire and Stove Governor.

House Governor Burner.

Governor for Incandescents.

Mercurial Governor for Fixing at Meter

Sole Manufacturers, D. BRUCE PEEBLES & CO., Tay Works, Edinburgh, Scotland.

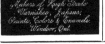

however, that the ultimate settlement will be on the basis that the minimum wage be 27½c. per hour; a day's work to consist of nine hours, and that the number of helpers be as hitherto.

PLUMBING AND HEATING NOTES.

The assets of J. Lafrance & Co., plumbers, etc., Montreal, are to be sold on February 7.

J. Williams, plumber, St. Thomas, Ont., has moved across the street to better premises.

ANOTHER BIG BLOCK FOR TORONTO.

William Davies, president of The Wm. Davies Co., Limited, who bought the property at the corner of Queen and James streets, Toronto, known as the Shaftesbury Hall, some time ago, has had plans for a five-storey business block prepared by Burke & Horwood, architects, Toronto. Tenders have been received from the various trades, but contracts will not be let for at least two weeks.

CALCIUM CARBIDE IN GERMANY.

THE United States Deputy Consul-General at Frankfort, writes: "On November 30 last, the convention of Swiss, Austrian, Swedish, Norwegian, and German manufacturers of calcium carbide, which was in session at Frankfort, combined in establishing price schedules and a mode of controlling the sale of their products. The Deutsche Gold und Silber Scheide-Anstalt, of Frankfort (which has branches in the United States and other countries), was appointed the sole agent for the sale of the syndicate's products. It is expected that by this combination the acetylene industry will be considerably strengthened. The members have adopted measures to avoid the fluctuating and ruinously low rates which, owing to heretofore existing sharp competition, have made the manufacture of their products unprofitable.

"The German acetylene industry is very important, there being at present in the Empire over 200,000 plants producing this gas. New patents for improved methods of production are constantly being issued. Thirty-two of the smaller towns in Germany are lighted by acetylene gas, and a number of other plants are in course of erection. The gas is also used by the railroads for lighting passenger cars.

"This year's production of calcium carbide in Germany is estimated at 20,000 metric tons, equivalent to 360,000 hectoliters (9,500,000 gallons) of petroleum."

The assets of Mrs. C. H. Gariepy, general merchant, Lachine, Que., have been sold.

THE GLASGOW EXHIBITION.

THE prospects that the Glasgow International Exhibition will be a thoroughly representative affair, are excellent. Official support has been secured from Russia, France, Austria, Japan, Denmark, India, Persia, Morocco, Australia and Canada. Though the United States will not be officially represented, manufacturers from that country have taken considerable space, especially in the machinery section. Some of the nationalities mentioned above are erecting special pavilions in addition to the space allotted to them in the main building. Russia, for example, is to have four, in order to fittingly display mining, timber, and other industries. One will be reserved for the display of the appurtenances of the Imperial estates, which are similar to the British Crown lands, and, by arrangement with the refreshment contractors, there will be a dining-room, in which dinners will be served in the Russian style, with wines, savouries, and other food products of the Empire which the Government are anxious to see introduced into other countries.

In the building to be occupied by the Japanese will be found a display of arts and manufactures, with native artisans at work illustrating some of the industries peculiar to that country. It will be surrounded by a Japanese garden, in itself no small attraction. Over 400 exhibitors are expected from France, whose section is being organized by a committee nominated by the French Government. Rhodesia's productions will include gold, industrial, and agricultural exhibits; Western Australia's display will include gold in various forms to the value of between £80,000 and £100,000; South Australia deals chiefly in wines; while the remainder will stage striking examples of their industries and resources.

In addition to 9,000 square feet in the main building, Canada is to have a special building, covering about 12,000 square feet, placed immediately at the main entrance to the grounds, wherein to exhibit minerals, manufactures, agricultural products, and fruit in season.

Free transportation will be given by the Dominion Government from the point of shipment. Exhibitors who do not care to have a special representative at the Exhibition, which will last from May to November, will have their exhibits cared for by the officials appointed by the Government. W. D. Scott, who represented Canada at the Paris World's Fair, is the Canadian Commissioner at Glasgow.

LIABILITY ON MINING SHARES.

OUR readers are well aware that we have always taken a firm stand against any attempts made by Canadian promoters of mining companies to sell shares to the investing public in this country by offering them at a considerable reduction below the face value. They always endeavor to assure their purchasers that these shares are fully paid and non-assessable, irrespective of any disparity between the price paid and the face value. This custom of selling shares at a discount has been so universally in vogue in Canada for some years past without the point of their non-assessability being called into question that one as come to regard the statement as one of fact. The recent judgment given at Rossland, by Mr. Justice Walkem, of the Supreme Court, has put quite another complexion on the matter, and shows that a number of holders of these twopenny ha'penny shares, which they thought they were picking at a bargain, have been living in a fool's paradise. The case which has resulted in this important decision, is one concerning the Kettle River Mines, Limited, capital $1,200,000 in shares of $1 each. The company issued 450,000 of these shares to the owners of the property, the promoters taking 495,000 shares for their trouble, and setting aside the balance for the development of the mine.

Regarding the 450,000 vendors' shares, these were given for value received, and, of course, are properly regarded as paid up and non-assessable, but the promoters sold 112,000 of their shares at a few cents each, the proceeds of which they apparently put in their own pockets, and when they saw that they could not sell the shares reserved for working capital, and that the shipwreck of the company was imminent, they endeavored to abandon the remainder of the promoters' shares which they held. This was owing to a meeting of the trustees of the company having made an assessment of 2c. per share on the 495,000 promoters' shares, which the promoters refused to pay and endeavored to divest themselves of their liability. In regard to the 112,000 shares which they had disposed of, the holders were sued in order to enforce payment of the assessment, but this, the judge decided, was not due from them, and he summed up the position in the following words:

"If you buy shares at 10c. each on certificates which represent them to be of a par value of $1 each, paid up, direct from the company, you must pay the difference between the 10c. and the par value, because you knew at the time you bought that you had not paid their face value. If, however, you have bought the same shares in the open market, on the same certificates, and at the same price, you are not responsible for the payment of the difference, as you are entitled to rely on the company's statement in the certificates that the shares are paid up and non assessable." The holders of shares in locally-registered Canadian companies will not have to sort out their certificates, and those which they bought direct from the company at a discount must be in future regarded as only partly paid up. If this decision should have the effect of making local mining companies postpone their operations until there is a reasonable amount of cash in their treasury it will have a good effect, for the number of abortive companies in Canada is enormous, and has done much to bring discredit upon mining undertakings. — British Columbia Review.

CORK FOUNDATIONS.

In Germany, cork is said to have been used with success in isolating the vibrations and consequent noise caused by machines installed in or near dwelling houses. A sheet made up of flat pieces of cork, in mosaic fashion, of corresponding size to the bed-plate of the machine, and held together by an iron frame, is laid under the machine. The source from which we gather this information also alludes to the isolation, by means of cork, of each bolt or connection between the bed-plate of the engine and its foundation, but does not describe in what manner such isolation is effected. It is, of course, apparent that the mere sandwiching of a sheet of cork between the bed-plate of a machine and its foundation would not necessarily absorb troublesome vibrations, since each anchor bolt is a medium whereby the vibrations of the engine may be transmitted to its foundation, thence to the floor and walls of the building unless effectively isolated from the same. How this effective isolation is obtained by the use of cork sheeting remains to be explained. For our part, we should be more inclined to rely upon a specially-prepared foundation of extra area and solidarity and to a careful balancing of the reciprocating parts of the machine.—Kuhlow.

DOES NOT INTERFERE WITH BUSINESS.

The Thos. Davidson Mfg. Co., Limited, beg to notify their customers that the fire in their local Montreal city office and sample-room does not interfere, in any way, with the prompt filling of orders. The staff of the local office are temporarily located at the general office, 187 Delisle street. Telephone "Main 3608."

Thos. Gebbie & Sons, general merchants, millers, etc., Howick, Que., have dissolved. Thos. Gebbie & Sons continue the milling business and Wm. Gebbie the general store.

BUSINESS MORALITY.

THERE are men in various positions who have few or no compunctions of conscience as to the manner of effecting their designs, though, for the sake of policy, they assume a disinterested purpose toward their intended victims, says John R. Ainslie, a writer in an exchange. Some thrive upon the misfortunes and necessities of others, and we are not disposed to doubt that many shrewd, calculating adepts in the practice of morality, from the respectable position they falsely occupy in the community, escape to quite an extent the observation of the public eye and profit in their deception.

I once read of an individual who had failed three times in business. The first time he was wholly unprepared for his misfortune; the second somewhat surprised him, but at the third he had become hardened, and remarked, with a peculiar expression of satisfaction, "I had them," meaning he had gotten the advantage over his creditors. This person kept up his respectability for a time, but it was evident to those who knew him that his ill-gotten gains gave him no peace, and he was not able to realize the joy and happiness of those who are governed by right principals. During the busy hours of the day his mind was occupied with the engrossing cares of business, but when the shades of darkness fell, he was of all men the most miserable. On retiring to his chamber he would walk the apartment for hours, lamenting his many misdeeds and the obligations he had violated.

It is to be regretted that the reputation of the dishonest man passes scrutiny for all business purposes, and that misrepresentation and fraud are allowed to pass unrebuked by public opinion. Most men characterize such proceedings as "clever" and "smart," and the men who are guilty of such questionable practices, instead of being shunned, are more likely to be regarded as desirable customers and companions, and it is not difficult to point out many individuals to-day who have gained wealth and position through unscrupulous methods and criminal carelessness as to the rights or welfare of others.

Not many years ago, I heard of a man who was in good standing in the community in which he lived, was known as a large dealer in merchandise, paid his bills with commendable promptness, securing the best discounts, but, invariably, from each remittance he would deduct an amount of shortages. Goods were doubly and trebly checked before being shipped, and every precaution adopted to insure correctness. Still the "shortages" continued. Then, a new plan was devised, and articles were enclosed without being charged—not only once, but two or three times. Of course, these "overs" were never reported. Then, convinced of his fraud and dishonesty, a bill was sent, covering all "overs" and "shortages," with an intimation that an immediate settlement was not only proper, but wise. Needless to say, payment was not long delayed. This man, to-day, is wealthy, has retired from business, and passes as a respectable member of society.

GROWING FLAX SEED.

The Commercial has received a letter from Northern Alberta, asking for information about flax seed, and stating that there is a movement on foot to grow flax seed in that section. The rich land of Northern Alberta should be particularly well adapted to the production of flax seed. The crop is one which is believed to be particularly well adapted to breaking in new land, and as a large area of new land will annually be prepared for crop in Northern Alberta for some years to come, no doubt considerable flax seed could be produced to advantage. Flax can be grown to advantage where wheat, oats and barley flourish. There is always a good market for flax seed. In fact, there is less liability of a depressed market for this commodity than for almost any other farm product. This being the case, it will undoubtedly be found a profitable crop in Northern Alberta, as well as in other grain sections of our western prairie country.—Commercial, Winnipeg.

ACETYLENE GAS VS. PETROLEUM.

U.S. Consul Hughes, of Coburg, under date of December 18, 1900, writes as follows:

"Up to the present time, Germany has imported each year from $25,000,000 to $30,000,000 worth of American petroleum. This industry, however, seems to be threatened somewhat by the introduction of acetylene as an illuminant, in a convenient and safe form, for house, store and other uses. This has resulted from the low price at which calcium carbide is being produced here, and also from the rise in the cost of petroleum in the German markets."

LOOKING FOR A SITE.

The Sherbrooke, Que., Examiner of a recent date notes that the Messrs. Croker, a firm of Fitchburg, Pa., paper machinery manufacturers, have had an interview with the council of the Sherbrooke Board of Trade with a view to their locating in that town. "They were satisfied," says The Examiner, "with what they saw of Sherbrooke, with its fine railway facilities and water power. It is not their intention to ask for a bonus, but look to the good-will of the people in that the price of any site they may agree upon would not be raised to an exorbitant price."

The Messrs. Croker are at present visiting other points in the Province of Quebec, but will again return to Sherbrooke.

R. C. Sharpe, harness dealer, Amherst, Ont., is dead.

CURRENT MARKET QUOTATIONS.

February 1, 1911.

These prices are for such qualities and quantities as are usually ordered by retail dealers on the usual terms of credit, the lowest figures being for larger quantities and prompt pay. Large cash buyers can frequently make purchases at better prices. The Editor is anxious to be informed at once of any apparent errors in this list as the desire is to make it perfectly accurate.

METALS.

Tin.

Lamb and Flag and Straits— 56 and 28 lb. ingots, per lb.	0 32	0 33

Tinplates.

Charcoal Plates—Bright

M.L.S., equal to Bradley.		Per box
I.C., usual sizes	....	$6 75
I.X., "	....	8 25
I.X.X., "	....	9 75
Famous—		
I.C.	....	6 75
I.X	....	8 25
I.X.X	....	9 50
Raven & Vulture Grades—		
I.C., usual sizes	....	5 00
I.X., "	....	6 00
I.X.X "	....	7 00
I.X.X.X., "	....	8 00
D.C., 12½x17	....	4 75
D.X.	....	5 50
D.X.X	....	7 50

Coke Plates—Bright

Bessemer Steel—		
I.C., usual sizes	....	4 15
I.C., special sizes, base	....	4 50
20 x 28	....	8 50

Charcoal Plates—Terne

Dean or J. G. Grade—		
I.C., 20x28, 112 sheets	....	8 75
I.X., Terne Tin	....	10 75

Charcoal Tin Boiler Plates.

Cookley Grade—		Per lb.
X.X., 14x56, 50 sheet bxs; " 14x60 "; " 14x65 "	0 07	0 07½

Tinned Sheets.

72x30 up to 24 gauge	0 08	0 08½
" 26 "	0 08½	0 09
" 28 "	0 09	0 09½

Iron and Steel.

Common Bar, per 100 lbs.	1 65	1 70
Refined " "	....	2 15
Horse Shoe Iron "	....	2 05
Hoop steel, 1½ to 3 in. base, extras for smaller sizes	....	3 10
Sleigh Shoe Steel " base	....	2 00
Tire Steel	....	2 00
Machinery iron finish	....	2 05
Cast Steel, per lb	0 00	0 00
Toe Calk Steel	....	2 30
T. Firth & Co's special cast steel, per lb.		0 13
Jessop's Tool Steel	....	0 13

Boiler Tubes.

1½-inch	....	0 12½
2 "	....	0 13
2½ "	....	0 15
3 "	....	0 16
3½ "	....	0 20
4 "	....	0 25

Steel Boiler Plate.

¼ inch	....	2 25
3-16 inch	....	2 25
⅜ inch and thicker	....	2 25

Black Sheets.

18 gauge	....	3 10
20 gauge	....	3 00
22 to 24 "	....	3 10
26 "	....	3 23
28 "	....	3 30

Canada Plates.

All dull, 52 sheets	....	3 00
Half polished	....	3 15
All bright	3 65	3 75

Iron Pipe.

Black pipe—		
¼ inch	....	3 00
⅜ "	....	3 00
½ "	....	3 15
¾ "	....	3 20
1 "	....	4 57
1¼ "	....	6 35
1½ "	....	7 56
2 "	....	10 10
2½-6 inch, discount 60 p.c.		
Galvanized pipe—		
½ inch	....	4 65
¾ "	....	5 35
1 "	....	7 75
1¼ "	....	9 75
1½ "	....	11 25
2 "	....	15 50

Galvanized Sheets.

	G. C.	Comet.	Amer.	Queen's Head.
16 gauge		....	4 5	4 00
18 to 24 gauge	4 25	3 85	4 35	4 25
26 "	4 50	4 10	4 35	4 50
28 "	4 75	4 35	4 50	4 75

28 gauge American equals 26 gauge English.

Chain.

Proof Coil, 3-16 in., per 100 lb	....	....
" ¼ " "	8 00	8 50
" 5-16 " "	5 35	5
" ⅜ " "	4 35	4 85
" 7-16 " "	4 15	4 65
" ½ " "	4 35	4 50
" ⅝ " "	3 85	4 35
" ¾ " "	3 80	4 40

Halter, kennel and post chains, 40 and 50 p.c.
Cow ties ... 40 p.c.
Stall fixtures ... 35 p.c.
Trace chain ... 25 p.c.
Jack chain, iron, single and double, discount 35 p.c.
Jack chain, brass, single and double, discount 40 p.c.

Copper.

Ingot

English B. S., ton lots	0 19	0 20
Lake Superior	....	....

Bolt or Bar.

Cut lengths round, ½ to ⅞ in.	0 23½	0 25
" round and square 1 to 2 inches	0 23½	0 25

Sheet.

Untinned, 14 oz., and light, 16 oz., 14x48 and 14x60	0 23	0 23
Untinned, 14 oz., and light, 16 oz., irregular sizes	0 23	0 23½

NOTE.—Extra for tinning, 2 cents per pound, and tinning and half planishing 3 cents per pound.

Tinned copper sheets	....	0 26
Planished	....	0 32

Braziers (In sheets.)

4x6ft. 25 to 30 lbs. ea., per lb.	....	0 25½
" 35 to 45 " "	....	0 24½
" 50-lb. and above, "	....	0 23½

Boiler and T. K. Pitts

Plain Tinned, per lb	....	0 28
Spun, per lb	....	0 32

Brass.

Rod and Sheet, 14 to 30 gauge, 15 per cent.

Sheets, hard-rolled, 2x4	0 24	0 25
Tubing, base, per lb	0 24	0 25

Zinc Spelter

Foreign, per lb	0 06	0 06½
Domestic "	....	....

Zinc Sheet.

5 cwt. casks	6 75	7 00
Part casks	7 00	7 50

Lead.

Imported Pig, per lb	0 04¾	0 05
Bar, 1 lb.	0 05½	0 05¾
Sheets, 2½ lbs. sq. ft., by	....	0 06½
Sheets, 3 to 6 lbs., "	....	0 06

NOTE.—Cut sheets ½ cent per lb. extra.

Pipe, by the roll, usual weights per yard, lists at 7c. per lb. and 15 p.c. dis. f.o.b. Toronto.

NOTE.—Cut lengths, net price, waste pipe 8-ft. lengths lists at 7½ cents.

Shot.

Common, $6.50 per 100 lb.; chilled, $7.00 per 100 lb.; buck, seal and bal., $7.50. Discount, 7½ p.c. Prices are f.o.b. Toronto, Hamilton, Montreal, St. John and Halifax. Terms 3 per cent. cash, freights equalized on Montreal.

Soil Pipe and Fittings.

Discount, 60 and 10 per cent. on medium and extra heavy, and 60 per cent. on light.

Solder.

	Per lb.	Per lb.
Bar half-and-half, guarant'd	....	0 19
Bar half-and-half, commer'l	....	0 18½
Refined	....	0 18½
Wiping	....	0 18

NOTE.—Prices of this graded according to quantity. The prices of other qualities of solder in the market indicated by private brands vary according to composition.

Antimony.

Cookson's, per lb.	0 11	0 11½

White Lead.

	Per 100 lb.
Pure	6 87½
No. 1 do	6 50
No. 2 do	6 12½
No. 3 do	5 75
No. 4 do	5 37½
Munro's Select Flake White	7 37½
Elephant and Decorators' Pure	7 12½
Brandram's B. B. Genuine	8 00
" " Decorative	7 55
" " No. 1	6 85
" " No. 2	6 00

Red Lead.

Genuine, 560 lb. casks, per cwt.	....	$5 50
Genuine, 100 lb. kegs, per cwt.	....	5 75
No. 1, 560 lb. casks, per cwt.	....	5 25
No. 1, 100 lb. kegs, per cwt.	....	5 00

White Zinc Paint.

Elephant Snow White	0 08	0 09
Pure White Zinc	0 08	0 09
No. 1	0 06	0 07½
No. 2	0 05	0 06½

Dry White Lead.

Pure, casks	5 75
Pure, kegs	6 25
No. 1, casks	5 50
No. 1, kegs	6 00

Prepared Paints.

In ¼, ½ and 1 gallon tins.

Pure, per gallon	....	1 25
Second qualities, per gallon	....	1 10
Barn (in bbls.)	0 75	0 85
The Sherwin-Williams Paints	....	1 45
Canada Paint Co's Pure	....	1 25
Toronto Lead & Color Co's Pure	....	1 25
Sanderson Pearcy's Pure	....	1 20
Stewart & Wood's Champion Pure		1 20

Colors in Oil.

25 lb. tins, Standard Quality.

Venetian Red, per lb	0 05
Chrome Yellow	0 11
Golden Ochre	0 06
French "	0 05
Marine Black	0 09
" Green	0 09
Chrome "	0 08
French Imperial Green	0 09

Colors, Dry.

Yellow Ochre (J.C.) bbls.	1 35	1 40
Yellow Ochre (J.F.L.S.), bbls	...	2 75
Yellow Ochre (Royal)	1 10	1 15
Brussels Ochre	....	2 00
Venetian Red (best), per cwt.	1 80	1 90
English Oxides, per cwt.	3 00	3 25
American Oxides, per cwt.	1 75	2 00
Canadian Oxides, per cwt.	1 75	2 00
Super Magnetic Oxides, 93p.c.	2 00	2 25
Burnt Sienna, pure, per lb.	....	0 10
" Umber, " "	....	0 10
do Raw,	....	0 09
Drop Black, pure	....	0 09
Chrome Yellows, pure	....	0 18
Chrome Greens, pure, per lb.	....	0 12
Golden Ochre	....	0 03½
Ultramarine Blue in 28-lb. boxes, per lb.	0 08	0
Fire Proof Mineral, per 100 lb	....	1
Genuine Eng. Litharge, per lb	....	0
Mortar Color, per 100 lb.	....	1
English Vermillion	....	0
Pure Indian Red, No. 45 lb.	....	0
Whiting, per 100 lb.	....	0 [illegible]

Blue Stone.

Casks, for spraying, per lb.	0 07
100-lb. lots, do. per lb.	0 08

Putty.

Bulk in bbls.	2 00
Bulk in less quantity	2 15
Bladders in bbls.	2 20
Bladders in kegs, boxes or loose	2 75
Bladders in 25-lb. tins	2 45
Bladders in 12½-lb. tins	2 75
Bladders in bulk or tins less than 100 lb.	3 00

Varnishes.

	In 5-gal. lots.	Per gal.
Carriage, No. 1	2 90	3 30
" body	8 00	9 00
" rubbing	4 00	5 00
Gold Size, Japan	3 00	3 40
Brown Japan	2 40	2 80
Elastic Oak	2 90	3 30
Furniture, extra	2 40	2 80
" No. 1	1 60	2 00
Hard Oil Finish	2 70	3 10
Light Oil Finish	3 20	3 60
Demar	3 30	3 70
Shellac, white	4 40	4 80
" orange	4 00	4 40
Furniture Brown Japan	1 60	2 00
Black Japan	2 40	2 80
" " No. 1	1 60	2 00

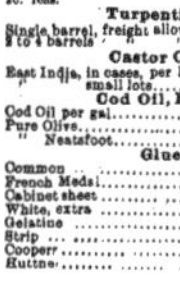

The Imperial Varnish & Color Co's., Limited Elastilite Varnish 1 gal. can, each. $2.00.

Granatine Floor Finish, per gal. $2.00.

Maple Leaf Coach Enamels: Size 1, 40c; Size 2, 35c.; Size 3, 20c. each.

Linseed Oil.

	Raw.	Boiled.
1 to 4 bbls delivered	$0 82	$0 85
5 to 9 bbls "	81	84

Toronto, Hamilton, London and Guelph 2c. less.

Turpentine.

Single barrel, freight allowed	....	0 59
2 to 4 barrels " "	....	0 58

Castor Oil.

East India, in cases, per lb.	0 10	0 10½
" small lots	0 10½	0 11

Cod Oil, Etc.

Cod Oil per gal.	0 50	0 55
Pure Olive	....	1 20
" Neatsfoot	....	90

Glue.

Common	0 08½	0 09
French Medal	0 14	0 14½
Cabinet sheet	0 12	0 13
White, extra	0 14	0 20
Gelatine	0 22	0 30
Strip	0 18	0 20
Cooper's	0 19	0 20
Huttner	....	0

HARDWARE.

Ammunition.

Cartridges.

B. B. Caps, Dom. 50 and 5 per cent.
Rim Fire Pistol, dis. 40 p. c., Amer.
Rim Fire Cartridges, Dom., 50 and 5 p. c.
Central Fire Pistol and Rifle, 10 p.c. Amer.
Central Fire Cartridges, pistol sizes, Dom. 30 per cent.
Central Fire Cartridges, Sporting and Military, Dom., 15 and 5 per cent.
Central Fire, Military and Sporting, Amer. add 5 p.c. to list. B.B. Caps, discount 40 per cent, Amer.
Loaded and empty Shells, "Trap" and "Dominion" grades, 25 per cent Rival and Nitro, net list.
Brass shot Shells, 55 per cent
Primers, Dom., 30 per cent.

Wads — per lb

Best thick white felt wadding, in ¼-lb bags.......... 1 00
Best thick brown or grey felt wads, in ½-lb. bags.......... 0 70
Best thick white card wads, in boxes of 500 each, 12 and smaller gauges 0 99
Best thick white card wads, in boxes of 500 each, 10 gauge.......... 0 35
Best thick white card wads, in boxes of 500 each, 8 gauge.......... 0 55
Thin card wads, in boxes of 1,000 each, 12 and smaller gauges.......... 0 20
Thin card wads, in boxes of 1,000 each, 10 gauge.......... 0 25
Thin card wads in boxes of 1,000 each, 8 gauge..........
Chemically prepared black edge grey cloth wads, in boxes of 250 each— Per M
11 and smaller gauge.......... 0 60
9 and 10 gauges.......... 0 70
7 and 8 gauges.......... 0 90
5 and 6 gauges.......... 1 10
Superior chemically prepared pink edge, best white cloth wads, in boxes of 250 each—
11 and smaller gauge.......... 1 15
9 and 10 gauges.......... 1 40
7 and 8 gauges.......... 1 65
5 and 6 gauges.......... 1 90

Adzes.

Discount, 20 per cent.

Anvils.

Per lb.......... 10 0 12½
Anvil and Vise combined 4 50
Wilkinson & Co.'s Anvils..lb. 0 09 0 09½
Wilkinson & Co.'s Vices..lb. 0 09¾ 0 10

Augers.

Gilmour's, discount 65 and 5 p.c. off list.

Axes.

Chopping Axes—
Single bit, per doz 6 50 10 00
Double bit, " 12 00 18 00
Bench Axes, 40 p.c.
Broad Axes, 33⅓ per cent.
Hunters' Axes 5 50 6 00
Boy's Axes.......... 5 75 6 75
Splitting Axes 6 50 12 00
Handled Axes 7 00 10 00

Axle Grease.

Ordinary, per gross.......... 5 75 6 00
Best quality.......... 13 00 15 00

Bath Tubs.

Zinc.......... .... 6 00
Copper, discount 15 p.c. off revised list

Baths.

Standard Enameled.

5½-inch rolled rim, 1st quality.......... 30 00
" " " 2nd " 22 00

Anti-Friction Metal.

"Tandem" A.......... per lb. 0 27
" B.......... " 0 21
" C.......... " 0 11½
Magnolia Anti-Friction Metal, per lb. 0 25

SYRACUSE SMELTING WORKS.

Aluminum, genuine.......... 0 45
Dynamo.......... 0 29
Special.......... 0 25
Aluminum, 99 p.c. pure "Syracuse".. 0 50

Bells.

Hand.

Brass, 60 per cent.
Nickel, 55 per cent.

Cow.

American make, discount 66⅔ per cent.
Canadian, discount 45 and 50 per cent.

Door.

Gongs, Sargant's.......... 5 50 8 00
" Peterboro', discount 45 per cent.

Farm.

American, each.......... 1 25 3 00

House.

American, per lb.......... 0 35 0 40

Bellows.

Hand, per doz.......... 3 35 4 75
Moulders', per doz.......... 7 50 10 00
Blacksmiths', discount 40 per cent.

Belting.

Extra, 60 and 10 per cent.
Standard, 60 per cent.
No. 1 Agricultural, 60 and 10 p.c.

Bits.

Auger.

Gilmour's, discount 60 and 5 per cent.
Rockford, 50 and 10 per cent.
Jennings' Gen., net list.

Car.

Gilmour's, 47½ to 50 per cent.

Expansive.

Clark's, 40 per cent.

Gimlet.

Clark's, per doz.......... 0 65 0 90
Diamond, Shell, per doz.......... 1 00 1 50
Nail and Spike, per gross.......... 2 25 5 20

Blind and Bed Staples.

All sizes, per lb.......... 0 07¾ 0 12

Bolts and Nuts. Per cent.

Carriage Bolts, full square, Norway.......... 70
" " full square.......... 70
Common Carriage Bolts, all sizes.......... 65
Machine Bolts, all sizes.......... 65
Coach Screws.......... 75
Sleigh Shoe Bolts.......... 75
Blank Bolts.......... 65
Bolt Ends.......... 65
Nuts, square..........4½c. off
Nuts, hexagon..........4⅜c. off
Tire Bolts.......... 67½
Stove Bolts.......... 67½
Stove rods, per lb..........5½ to 6c.
Plough Bolts.......... 60

Boot Calks.

Small and medium, ball, per M.......... 4 25
Small heel, per M.......... 4 50

Bright Wire Goods.

Discount .... 55 per cent.

Broilers.

Light, dis., 65 to 67½ per cent.
Reversible, dis., 65 to 67½ per cent.
Vegetable, per doz., dis. 37½ per cent.
Henis, No. 8, " 6 00
Henis, No. 9, " 7 00
Queen City " 7 50 0 00

Butchers' Cleavers.

German, per doz.......... 6 00 11 00
American, per doz.......... 12 00 20 00

Building Paper, Etc.

Plain building, per roll.......... 0 30
Tarred lining, per roll.......... 0 40
Tarred roofing, per 100 lb.......... 1 65
Coal Tar, per barrel.......... 3 50
Pitch, per 100-lb.......... 0 85
Carpet felt, per ton.......... 45 00

Bull Rings.

Copper, $2.00 for 2½ in. and $1.90 for 2 in.

Butts.

Wrought Brass, net revised list

Cast Iron.

Loose Pin, dis., 60 per cent.

Wrought Steel.

Fast Joint, dis. 60 and 10 per cent.
Loose Pin, dis. 60 and 10 per cent.
Berlin Bronzed, dis. 70, 70 and 5 per cent.
Gen. Bronzed, per pair 0 40 0 66

Carpet Stretchers.

American, per doz.......... 1 00 1 50
Bullard's, per doz.......... 6 50 ....

Castors.

Bed, new list, dis. 55 to 57½ per cent.
Plate, dis. 52½ to 57½ per cent.

Cattle Leaders.

Nos. 31 and 32, per gross.......... 50 9 50

Cement.

Canadian Portland.......... 2 80 3 00
English " 3 00
Belgian " 2 75 3 00
Canadian hydraulic.......... 1 25 1 50

Chalk.

Carpenters, Colored, per gross 0 45 0 75
White lump, per cwt.......... 0 60 0 65
Red.......... 0 05 0 06
Crayon, per gross.......... 0 14 0 18

Chisels.

Socket, Framing and Firmer.

Broad's, dis. 70 per cent.
Warnock's, dis. 70 per cent.
P. S. & W. Extra 60, 10 and 5 p.c.

Churns.

Revolving Churns, metal frames—No. 0, $8—No. 1, $8.50—No. 2, $9.00—No. 3, $10.00 No. 4, $12.00—No. 5, $16.00 each. Ditto wood frames—20c. each less than above. Discounts: Delivered from factories, 58 p.c.; from stock in Montreal, 56 p.c. Terms, 4 months or 3 p.c. cash in 30 days.

Clips.

Axle dis. 65 per cent.

Closets.

Plain Ontario Syphon Jet.......... $8 00
Emb. Ontario Syphon Jet.......... 8 50
Fittings.......... 1 75
Plain Teutonic Syphon Washout.......... 4 75
Emb. Teutonic Syphon Washout.......... 5 25
Fittings.......... 1 25
Low Down Teutonic, plain.......... 14 50
" " " embossed.......... 15 00
Plain Richelieu.......... 3 75
Emb. Richelieu.......... 4 00
Fittings.......... 1 25
Low Down Ont. Syphon Jet, plain.......... 20 00
" " " " " emb'd.......... 20 50
Closet connection.......... 1 25
Basins, round, 14 in.......... 0 80
" oval, 17 X 14 in.......... 1 50
" " 19 X 15 in.......... 2 25

Compasses, Dividers, Etc.

American, dis. 62½ to 65 per cent.

Cradles, Grain.

Canadian, dis. 25 to 33⅓ per cent.

Crosscut Saw Handles.

S. & D., No. 3, per pair.......... 17½
" " 5, " 22½
" " 6, " 15
Boynton pattern " 20

Door Springs.

Torrey's Rod, per doz.......... (15 p.c.) 2 00
Coil, per doz.......... 0 88 1 60
English, per doz.......... 2 00 4 00

Draw Knives.

Coach and Wagon, dis. 50 and 10 per cent.
Carpenters, dis. 70 per cent.

Drills.

Hand and Breast.

Millar's Falls, per doz. net list.

DRILL BITS.

Morse, dis., 37½ to 40 per cent.
Standard dis. 50 and 5 to 55 per cent.

Faucets.

Common, cork-lined, dis. 35 per cent.

ELBOWS. (Stovepipe.)

No. 1, per doz.......... 1 40
No. 2, per doz.......... 1 20
Bright, 20c. per doz. extra.

ESCUTCHEONS.

Discount, 45 per cent.

ESCUTCHEON PINS.

Iron, discount 40 per cent.

FACTORY MILK CANS.

Discount off revised list, 40 per cent.

FILES.

Black Diamond, 50 and 10 to 60 per cent.
Kearney & Foote, 60 and 10 p.c. to 60, 10, 10.
Nicholson File Co., 50 and 10 to 60 per cent.
Jowitt's, English list, 25 to 27½ per cent.

FORKS.

Hay, manure, etc., dis., 50 and 10 per cent, revised list.

GLASS—Window—Box Price.

Size United Inches.	Star Per 50 ft.	Star Per 100 ft.	D. Diamond Per 50 ft.	D. Diamond Per 100 ft.
Under 26..........	2 10	4 00	....	6 00
26 to 40..........	2 30	4 35	....	6 65
41 to 50..........	....	4 75	....	7 25
51 to 60..........	....	35	....	8 50
61 to 70..........	....	75	....	9 25
71 to 80..........	....	5	....	10 50
81 to 85..........	....	6 50	....	11 75
86 to 90..........	....	....	..	14 00
91 to 95..........	..	....	...	15 50
99 to 100..........	..	....	....	18 00

GAUGES

Marking, Mortise, Etc.

Stanley's dis. 50 to 55 per cent.

Wire Gauges.

Winn's, Nos. 26 to 33, each.......... 1 65 2 40

HALTERS.

Rope, ⅜ per gross.......... ....
" ½ " 9 00
" ⅝ to ¾ 14 00
Leather, 1 in., per doz.......... 3 87½ 4 00
" 1¼ in., " 5 15 5 20
Web, — per doz.......... 1 87 2 45

HAMMERS.

Nail

Maydole's, dis. 5 to 10 per cent. Can. dis. 25 to 27½ per cent.

Tack.

Magnetic, per doz.......... 1 10 1 20

Sledge.

Canadian, per lb.......... 0 07½ 0 08½

Ball Pean.

English and Can., per lb.......... 0 22 0 25

HANDLES.

Axe, per doz. net.......... 1 50 2 00
Store door, per doz.......... 1 00 1 50

Fork.

C. & B., dis. 40 per cent. rev. list.

Hoe.

C. & B., dis. 40 per cent. rev. list.

Saw.

American, per doz.......... 1 00 1 25

Plane.

American, per gross.......... 3 15 3 75

Hammer and Hatchet.

Canadian, 40 per cent.

Cross-Out Saws.

Canadian, per pair.......... 0 13½

HANGERS. doz. pairs.

Steel barn door.......... 5 85 6 00
Stearns, 4 inch.......... .... 5 00
" 5 inch.......... .... 6 50
Lane's covered—
No. 11, 5-ft. run.......... .... 8 40
No. 11½, 10-ft. run.......... .... 10 80
No. 12, 10-ft. run.......... .... 12 60
No. 14, 15-ft. run.......... .... 21 00
Lane's O.N.T. track, per foot. 4½

HARVEST TOOLS.

Discount, 50 and 10 per cent.

HATCHETS.

Canadian, dis. 40 to 42½ per cent.

HINGES.

Blind, Parker's, dis. 50 and 10 to 60 per cent
Heavy T and strap, 4-in., per lb.......... 0 06½
" " 5-in., " 0 06¼
" " 6-in., " 0 06
" " 8-in., " 0 05¾
" " 10-in., " 0 05⅝
Light T and strap, dis. 60 and 5 per cent.
Screw hook and hinge—
6 to 12 in., per 100 lbs.......... .... 4 50
14 in. up, per 100 lbs.......... .... 3 50
Per gro. pairs.
Spring.......... .... 12 00

HOES.

Garden, Mortar, etc., dis. 50 and 10 p.c.
Planter, per doz.......... 4 00 4 50

HOLLOW WARE

Discount.......... 45 and 5 per cent

HOOKS.

Cast Iron.

Bird Cage, per doz.......... 0 50 1 10
Clothes Line, per doz.......... 0 27 0 63
Harness, per doz.......... 0 72 0 88
Hat and Coat, per gross.......... 1 00 3 00
Chandelier, per doz.......... 0 50 1 00

Wrought Iron.

Wrought Hooks and Staples, Can., dis. 47½ per cent.

Wire.

Hat and Coat, discount 45 per cent.
Belt, per 1,000.......... 0 60
Screw, bright, dis. 55 per cent.

HORSE NAILS.

"C" brand 50, 10 and 5 p.c. } Oval head.
"M" brand 50, 10 and 5 p.c. }
Acadian, 50, 10 and 5 per cent.

HORSESHOES.

Iron Shoes.	F.O.B. Montreal No. 2 and larger.	No. 1. and smaller
Light, medium, and heavy...	3 50	3 75
Snow shoes	3 75	4 00
Steel Shoes.		
Light	3 60	3 85
Featherweight (all sizes)	4 85	4 85

F.O.B. Toronto, Hamilton, London and Guelph, 10c. per keg additional.
Toe weight steel shoes...... 6 70

JAPANNED WARE.
Discount, 45 and 5 per cent. off list, June 1899.

ICE PICKS.
Star per doz.................. 3 00 3 25

KETTLES.
Brass spun, 7½ p.c. dis. off new list.
Copper, per lb.............. 0 30 0 50
American, 60 and 10 to 65 and 5 p.c.

KEYS.
Lock, Can., dis., 45 p.c.
Cabinet, trunk, and padlock, Am. per gross............ 60

KNOBS.
Door, japanned and N.P., per doz.................... 1 50 2 50
Bronze, Berlin, per doz...... 2 75 3 25
Bronze Genuine, per doz.... 6 00 9 00
Shutter, porcelain, F. & L. screw, per gross........... 1 30 4 00
White door knobs—per doz. 1 25

HAY KNIVES.
Discount, 50 and 10 per cent.

LAMP WICKS.
Discount, 60 per cent.

LANTERNS.
Cold Blast, per doz............. 7 50
No. 3 "Wright's"............. 8 50
Ordinary, with O burner 4 25
Dashboard, cold blast 9 50
No. 0.......................... 6 00
Japanning, 50c. per doz. extra.

LEMON SQUEEZERS.

	per doz.	
Porcelain lined,	2 20	5 60
Galvanized	1 87	3 85
King, wood	2 75	2 90
King, glass	4 00	4 50
All glass	1 20	1 30

LINES.
Fish, per gross.............. 1 05 2 50
Chalk " 1 90 7 40

LOCKS.
Canadian, dis. 45 p.c.
Russell & Erwin, per doz.... 3 00 3 25

Cabinet.
Eagle, dis. 30 p.c.

Padlock.
English and Am., per doz.... 50 6 00
Scandinavian, " 1 00 2 40
Eagle, dis. 20 to 25 p.c.

MACHINE SCREWS.
Iron and Brass.
Flat head discount 25 p.c.
Round Head, discount 20 p.c.

MALLETS.
Tinsmiths', per doz........... 1 25 1 50
Carpenters', hickory, per doz. 1 25 3 75
Lignum Vitae, per doz.. .... 3 85 5 00
Caulking, each............. 60 2 00

MATTOCKS.
Canadian, per doz........... 8 50 1 00

MEAT CUTTERS.
American, dis. 25 to 30 p.c.
German, 15 per cent.

MILK CAN TRIMMINGS.
Discount, 25 per cent.

NAILS.

Quotations are :	Cut.	Wire.
2d and 3d	$3 35	$3 85
3d	3 00	3 52
4 and 5d	2 75	3 35
6 and 7d	2 65	3 20
8 and 9d	2 50	3 00
10 and 12d	2 45	2 95
16 and 20d	2 40	2 90
30, 40, 50 and 60d. (base)	2 35	2 85

Galvanizing 2c. per lb. net extra.
Steel Cut Nails 10c. extra.
Miscellaneous wire nails, dis. 70 per cent.
Coopers' nails, dis. 30 per cent.
Flour barrel nails, dis. 25 per cent

NAIL PULLERS.
German and American....... 1 85 3 50

NAIL SETS
Square, round, and octagon, per gross 3 38 4 00
Diamond 12 00 15 00

NETTING.
Poultry, 50 per cent. for McMullen's.

OAKUM. Per 100 lb
Navy 6 00
U. S. Navy...................... 7 25

OIL.
Water White (U.S.).......... .... 0 16½
Prime White (U.S.) 0 15½
Water White (Can.)........... .... 0 15
Prime White (Can.)............ .... 0 14

OILERS.
McClary's Model galvan. oil can, with pump, 5 gal., per doz.................. 0 00 10 00
Zinc and tin, dis. 50, 50 and 10.
Copper, per doz.............. 1 25 3 50
Brass, " 1 50 3 50
Malleable, dis. 25 per cent.

GALVANIZED PAILS.
Dufferin pattern pails, dis. 50 to 50 and 10 p.c.
Flaring pails, discount 45 per cent.
Galvanized washtubs, discount 45 per cent.

PIECED WARE.
Discount 40 per cent. off list, June, 1899.

PICKS.
Per doz 6 00 9 00

PICTURE NAILS.
Porcelain head, per gross.... 1 75 3 00
Brass head " 0 40 1 00

PICTURE WIRE.
Tin and gilt, discount 75 p.c.

PLANES.
Wood, bench, Canadian dis. 50 per cent American dis. 50.
Wood, fancy Canadian or American 7½ o 40 per cent.

PLANE IRONS.
English, per doz.............. 2 00 5 00

PLIERS AND NIPPERS.
Button's Genuine per doz pairs, dis. 37½ 40 p.c.
Button's Imitation, per doz.. 5 00 9 00
German, per doz............ 0 60 2 60

PLUMBERS' BRASS GOODS.
Impression work, discount, 60 per cent.
Fuller's work, discount 65 per cent.
Rough stops and stop and waste cocks, discount, 60 per cent.
Jenkins disk globe and angle valves, discount, 55 per cent.
Standard valves, discount, 60 per per cent.
Jenkins radiator valves discount 55 per cent.
" " " standard, dis., 60 p.c.
Quick opening valves discount, 60 p.c.
No. 1 compression bath cock........ 2 00
No. 4 " " " 2 00
No. 7, Fuller's..................... 2 50
No 4½, " 3 00

POWDER.
Velox Smokeless Shotgun Powder.
100 lb. or less 0 85
1,000 lb. or more........................ 0 80
Net 30 days.

PRESSED SPIKES.
Discount 25 per cent.

PULLEYS.
Hothouse, per doz............ 0 55 1 00
Axle 0 22 0 33
Screw 0 27 1 00
Awning........................ 0 35 2 50

PUMPS.
Canadian cistern.............. 1 80 3 60
Canadian pitcher spout...... 1 40 2 10

PUNCHES.
Saddlers', per doz........... 1 00 1 85
Conductors', " 9 00 15 00
Tinners' solid, per set........ 0 00 0 72
" hollow, per inch.... 0 00 1 00

RANGE BOILERS.
Galvanized, 30 gallons 6 50
" 35 " 7 50
" 40 " 8 50
Copper, 30 " 22 00
" 35 " 26 00
" 40 " 30 00
Discount off Copper Boilers 10 per cent.

RAKES.
Cast steel and malleable Canadian list 50 and 10 p.c. revised list.
Wood, 25 per cent.

RASPS AND HORSE RASPS.
New Nicholson horse rasp, discount 60 p.c.
Globe File Co.'s rasps, 60 and 10 to 70 p.c.
Heller's Horse rasps, 50 to 50 and 5 p.c.

RAZORS.

	per doz.	
Geo. Butler & Co.'s,	8 00	18 00
Boker's	7 50	11 00
Wade & Butcher's	3 60	10 00
Theile & Quack's	7 00	12 00
Elliot's	4 00	18 00

REAPING HOOKS.
Discount, 50 and 10 per cent.

REGISTERS.
Discount,......................... 40 per cent.

RIVETS AND BURRS.
Iron Rivets, discount 60 and 10 per cent.
Iron Burrs, discount 55 per cent.
Black and Tinned Rivets, 60 p.c.
Extras on Iron Rivets in 1-lb. cartons, ½c per lb.
Extras on Iron Rivets in ½-lb. cartons, per lb.
Copper Rivets & Burrs, 35 and 5 p.c. dis. and cartons, 1c. per lb. extra, net.
Extras on Tinned or Coppered Rivets ¼-lb. cartons, 1c. per lb.
Terms, 4 mos. or 3 per cent. cash 30 days.

RIVET SETS.
Canadian, dis. 35 to 37½ per cent.

ROPE ETC.

	Sisal.	Manila.
7-16 in. and larger, per lb.	9	13
⅜ in.	10	14
¼ and 5-16 in.		15
Cotton, 3-16 inch and larger		16¼
" 5-32 inch		21½
" ⅛ inch		22½
Russia Deep Sea		15½
Jute		8
Lath Yarn		9¾
New Zealand Rope		10½

RULES.
Boxwood, dis. 75 and 10 p.c.
Ivory, dis. 37½ to 40 p.c.

SAD IRONS. per set.
Mrs. Potts, No. 55, polished.......... 0 62½
" No. 50, nickle-plated.... 0 67½

SAND AND EMERY PAPER.
Dominion Flint Paper, 47½ per cent.
B & A. sand, 40 and 2½ per cent.
Emery, 40 per cent.

SAP SPOUTS.
Bronzed iron with hooks, per doz... 9 50

SAWS.
Hand Disston's, dis. 12½ p.c.
S. & D., 40 per cent.
Crosscut, Disston's, per ft.... 0 35 0 55
S. & D., dis. 35 p.c. on Nos. 2 and 3.
Hack, complete, each........ 0 75 2 75
" frame only............ .... 0 75

SASH WEIGHTS.
Sectional, per 100 lbs......... 2 75 3 00
Solid, " 2 00 2 25

SASH CORD.
Per lb........................ 0 23 0 30

SAW SETS.
"Lincoln," per doz.......... .... 6 50

SCALES.
B. S. & M. Scales, 45 p.c.
Champion, 65 per cent.
Fairbanks Standard, 35 p.c.
" Dominion, 55 p.c.
" Richelieu, 55 p.c.
Chatillon Spring Balances, 10 p.c.

SCREW DRIVERS.
Sargent's, per doz........... 0 65 1 00

SCREWS
Wood, F. H., iron, and steel, 85 p.
Wood R. H., " dis. 80 p.c.
" F. H., brass, dis. 77½ p.c.
Wood, R. H., " dis. 70 p.c.
" F.H., bronze, dis. 70 p.c.
" R.H. " 65 p.c.
Drive Screws, 80 per cent.
Bench, wood, per doz......... 3 25 4 00
" iron, " 4 25 5 75

SCYTHES.
Per doz, net 9 00

SCYTHE SNATHS.
Canadian, dis. 45 p.c.

SHEARS.
Bailey Cutlery Co., full nickeled, dis. 60 p.c.
Seymour's, dis. 50 and 10 p.c.

SHOVELS AND SPADES.
Canadian, dis. 40 and 5 per cent.

SINKS.
Steel and galvanized, discount 45 per cent.

SNAPS.
Harness, German, dis. 25 p.c.
Lock, Andrews'.............. 4 50 11 50

SOLDERING IRONS.
1, 1½ lb., per lb........ .... 0 37
2 lb. or over, per lb. 0 34

SQUARES.
Iron, No. 493, per doz 2 40 2 55
" No. 494, " 3 25 3 40
Steel, dis. 50 and 5 to 50 and 10 p.c., rev. list.
Try and bevel, dis. 50 to 52½ p.c.

STAMPED WARE.
Plain, dis., 75 and 12½ p.c. off revised list.
Retinned, dis., 75 p.c. off revised list.

STAPLES.
Galvanized 0 00 0 00
Plain 0 00 3 45
Coopers', discount 45 per cent.
Poultry netting staples, 40 per cent.

STOCKS AND DIES.
American dis. 25 p.c.

STONE.

	Per lb.	
Washita	0 28	0 60
Hindostan	0 06	0 07
" slip,	0 09	0 09
Labrador	....	0 13
" Axe	....	0 15
Turkey	....	0 50
Arkansas	0 00	1 50
Water-of-Ayr	0 00	0 10
Scythe, per gross	3 50	5 00
Grind, per ton	15 00	18 00

STOVE PIPES.
Nestable in crates of 25 lengths.
5 and 6 inch Per 100 lengths 7 00
7 inch " " 7 50

ENAMELINE STOVE POLISH.
No. 4—3 dozen in case, net cash $4 80
No. 6—3 dozen in case, " 8 40

TACKS BRADS, ETC.

	Per cent.
Strawberry box tacks, bulk	75 & 10
Cheese-box tacks, blued	80 & 12½
Trunk tacks, black and tinned	85
Carpet tacks, blued	80 & 5
" " tinned	80 & 10
" " (in kegs)	40
Cut tacks, blued, in dozens only	75 & 15
" ¼ weights	60
Swedes, cut tacks, blued and tinned—	
In bulk	80 & 10
In dozens	75
Swedes, upholsterers', bulk	85 & 12½
" brush, blued & tinned, bulk	70
" gimp, blued, tinned and japanned	75 & 12½
Zinc tacks	35
Leather carpet tacks	55
Copper tacks	50
Copper nails	52

Trunk nails, black	65 and 5
Trunk nails, tinned	65 and 10
Clout nails, blued and tinned	65 and 5
Chair nails	35
Cigar box nails	35
Patent brads	40
Fine finishing	40
Picture frame points	10
Lining tacks, in papers	10
" " in bulk	15
" " solid heads, in bulk	75
Saddle nails in papers	10
" " in bulk	15
Tufting buttons, 22 line, in dozens only	60
Tin capped trunk nails	15
Zinc glazier's points	5
Double pointed tacks, papers	90 and 10
" " " bulk	40

TAPE LINES.

English, ass skin, per doz	2 75	5 00
English, Patent Leather	5 50	9 75
Chesterman's each	0 90	2 85
" steel, each	0 80	8 00

THERMOMETERS.

Tin case and dairy, dis. 75 to 75 and 10 p.c.

TRANSOM LIFTERS.

Payson's per doz. 9 60

TRAPS. (Steel.)

Game, Newhouse, dis. 25 p.c.
Game, H. & N., P. S. & W., 65 p.c.
Game, steel, 72½, 75 p.c.

TROWELS.

Disston's discount 10 per cent.

German, per doz	4 75	6 00

S. & D., discount 35 per cent.

TWINES.

Bag, Russian, per lb	....	0 27
Wrapping, cotton, per lb	0 22	0 26
Wrapping, mottled, per pack	0 50	0 60
Wrapping, cotton, 3-ply		0 20
" " 4-ply	....	0 26
Mattress, per lb	0 33	0 45
Staging, "	0 27	0 35
Broom, "	0 30	0 55

VISES.

Hand, per doz	4 00	8 00
Bench, parallel, each	2 00	4 50
Coach, each	6 00	7 00
Peter Wright's, per lb	0 12	0 13
Pipe, each	5 50	9 00
Saw, per doz	6 50	13 00

ENAMELLED WARE.

White, Princess, Turquoise, Blue and White, discount 50 per cent.
Diamond, Famous, Premier, 50 and 10 p.c.
Granite or Pearl, Imperial, Crescent, 50, 10 and 10 per cent.

WIRE.

Brass wire, 50 to 50 and 2½ per cent. off the list.

Copper wire, 45 and 10 per cent. net cash 30 days, f.o.b. factory.

Smooth Steel Wire, base, $2.80 per 100 lb. List of eXtras: Nos. 2 to 5, advance 7c. per 100 lb.—Nos. 6 to 9, base—No. 10, advance 7c.—No. 11, 14c.—No. 12, 20c.—No. 13, 35c.—No. 14, 47c.—No. 15, 60c.—No. 16, 75c. EXtras net per 100 lb.: Coppered wire, 60c.—tinned wire, $2—oiling, 10c.—special hay-balling wire, 30c.—spring wire, $1—best steel wire, 75c.—bright soft drawn, 15c.—in 50 and 100-lb. bundles net, 10c.—in 25-lb. bundles net, 15c.—packed in casks or cases, 15c.—bagging or papering, 10c.

[F]ine Steel Wire, dis. 17½ per cent. List of eXtras: In 100-lb. lots: No. 17, $5—No. 18, $5.50—No. 19, $6—No. 20, $6.65—No. 21, $7—No. 22, $7.30—No. 23, $7.65—No. 24, $8—No. 25, $9—No. 26, $9.50—No. 27, $10—No. 28, $11—No. 29, $12—No. 30, $13—No. 31, $14—No. 32, $15—No. 33, $16—No. 34, $17. EXtras net—tinned wire, Nos. 17-25, $2—Nos. 26-31, $4—Nos. 32-34, $6. Coppered, 5c.—oiling, 10c.—in 25-lb. bundles, 15c.—in 5 and 10-lb. bundles, 25c.—in 1-lb. hanks, 50c.—in ½-lb. hanks, 75c.—in ¼-lb. hanks, $1—packed in casks or cases, 15c.—bagging or papering, 10c.

Galvanized Wire, per 100 lb.—Nos. 6, 7, 8, $3.55—No. 9, $3.10—No. 10, $3.75—No. 11, $3.85—No. 12, $3.25—No. 13, $3.35—No. 14, $4.25—No. 15, $4.75—No. 16, $5.00.

Clothes Line Wire, 19 gauge, per 1,000 feet 3 30

WIRE FENCING.	F.O.B. Toronto
Galvanized 4 barb, 2¼ and 5 inches apart	3 10
Galvanized, 2 barb, 4 and 6 inches apart	3 10
Galvanized, plain twist	3 10

Galvanized barb, f.o.b. Cleveland, $2.97½ in less than carlots, and $2.85 in carlots. Terms, 60 days or 2 per cent. in 10 days.

Ross braid truss cable 4 50

WIRE CLOTH.

Painted Screen, per 100 sq. ft., net .. 1 35
Terms, 4 months, May 1.; 3 p.c. off 30 days.

WRENCHES.

Acme, 35 to 37½ per cent.
Agricultural, 60 p.c.
Coe's Genuine, dis. 70 to 25 p.c.

Towers' Engineer, each	2 00	7 00
" S., per doz	5 80	6 00
G. & K.'s Pipe, per doz	....	3 40
Burrell's Pipe, each	....	3 00
Pocket, per doz	0 25	2 90

WRINGERS.

Leader	per doz.	$....
Royal Canadian	"	50 00
Royal American	"	50 00

Discount, 45 per cent.; terms 4 months, or 3 p.c. 30 days.

WROUGHT IRON WASHERS.

Canadian make, discount, 40 and 5 per cent

PERSONS addressing advertisers will kindly mention having seen their advertisement in Canadian Hardware and Metal Merchant.

SEND for specimen copy of Phillips' Monthl Machinery Register, containing over 5,000 entries of new and second-hand machinery of every description. The oldest established and most successful medium in the world. Established 25 years for the purpose of introducing those who have machinery for sale, to those who wish to buy, has a circulation of about 50,000 copies per annum, all over the world, and is used for continual reference by a large number of firms. It is consequently a most valuable advertising medium for all engineers and manufacturers. Subscription, 5s. per annum, price per copy, 6d. Sole Proprietor, CHAS. D. PHILLIPS, M.I.M.E., Newport, Mon., England. Telegraphic address "Machinery, Newport, Mon.

WESTERN

Incorporated 1851.

ASSURANCE COMPANY

Fire and Marine

Capital, subscribed	$2,000,000.00
Capital - - -	1,000,000.00
Assets, over - -	2,340,000.00
Annual Income -	2,290,000.00

Head Office: TORONTO, ONT.

Hon. Geo. A. Cox, President. J. J. Kenny, Vice-President. C. C. Foster, Secretary.

Lockerby & McComb

AGENTS IN CANADA

FOR THE

Celebrated P. & B.

Cold Storage Lining

AND

.. Ruberoid Roofing ..

P. S.—Prices on Application.

65 Shannon Street, MONTREAL.

BUSINESS NEWS

of any kind that is of value to business men supplied by our Bureau. We can give you market quotations from any town in Canada, reports from the city markets, stock quotations, etc. You can get commercial news from any Canadian paper through us.

Write us, giving us particulars of what you want and where you want it from, and we will quote you prices by return.

"Clippings from any Canadian paper on any subject."

CANADIAN PRESS CLIPPING BUREAU,

2.2 McGill Street, MONTREAL, QUE.

Telephone Main 1255.

10 Front St. East, Toronto. Telephone 2148.

---

75 YEARS. *ESTABLISHED 1825.* *75 YEARS.*

CELEBRATED **HEINISCH** SHEARS.

Tailors' Shears,
Trimmers, Scissors,
Tinners' Snips, etc.

ACKNOWLEDGED THE BEST.

R. HEINISCH'S SONS CO. NEW YORK OFFICE, 90 Chambers St. NEWARK, N.J., U.S.A.

Not connected with any Shear Combination.

CHAS. F. CLARK, President. JARED CHITTENDEN, Treasurer.

...ESTABLISHED 1849...

BRADSTREET'S

Capital and Surplus, $1,500,000. Offices Throughout the Civilized World.

Executive Offices: Nos. 346 and 348 Broadway, New York City, U.S.A.

THE BRADSTREET COMPANY gathers information that reflects the financial condition and the controlling circumstances of every seeker of mercantile credit. Its business may be defined as of the merchants, by the merchants, for the merchants. In procuring, verifying and promulgating information no effort is spared, and no reasonable expense considered too great, that the results may justify its claim as an authority on all matters affecting commercial affairs and mercantile credit. Its offices and connections have been steadily extended, and it furnishes information concerning mercantile persons throughout the civilized world.

Subscriptions are based on the service furnished, and are available only by reputable wholesale, jobbing and manufacturing concerns, and by responsible and worthy financial, fiduciary and business corporations. Specific terms may be obtained by addressing the Company at any of its offices. **Correspondence Invited.**

—OFFICES IN CANADA—

HALIFAX, N.S.	HAMILTON, ONT.	LONDON, ONT.	MONTREAL, QUE.
OTTAWA, ONT.	QUEBEC, QUE.	ST. JOHN, N.B.	TORONTO, ONT.
VANCOUVER, B.C.	VICTORIA, B.C.	WINNIPEG, MAN.	

THOS. C. IRVING, Gen. Man. Western Canada, Toronto. JOHN A. FULTON, Gen. Man. Eastern Canada, Montreal.

Awarded a Gold Medal at **PARIS EXPOSITION** for superiority. That's proof enough of their quality, and clearly shows that they are the best.

The Bailey

Send for Catalogue and . . . . Price List.

Cutlery Co.

BRANTFORD, ONT.

Dominion Pattern

Cow Tie AND Stall Fixture

The special features of the tie and stall fixture are well shown in the illustration. As will be noticed the chain is very short, which prevents all danger of entanglement with the animal's foot. At the same time the form of the fixture is such that great freedom is allowed to the head. Because of the short chain this tie is much cheaper than the ordinary patterns.

The stall fixture is made from a tough quality of steel and is very strong. Also, owing to its circular cross-section, it is exceedingly rigid. Its simplicity, convenience, cheapness, and ease of attaching make it very popular with cow tie users.

This form of tie and stall fixture are sometimes called Niagara pattern.

American or Flat Link Chain,

for years the standard cow tie chain in "the States, is now rapidly coming in favor in Canada. Its short link, handsome appearance and smooth surface—which cannot injure the animal's neck—make it superior to all other styles of chain for cow ties.

For sale by all Jobbers; manufactured by

ONEIDA COMMUNITY, Limited, NIAGARA FALLS, ONT.

CANADIAN

HARDWARE AND METAL MERCHANT

The Weekly Organ of the Hardware, Metal, Heating, Plumbing and Contracting Trades in Canada.

VOL. XIII. MONTREAL AND TORONTO, FEBRUARY 9, 1901. NO. 6

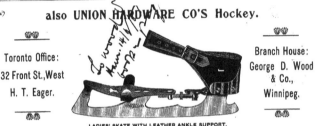

The Best Selling Range Ever Made.

Popular with dealers in every part of the country —because it is so enthusiastically praised by every buyer.

Our Imperial Oxford

has won its laurels—it is the favorite range of Canada—widely advertised, and everywhere appreciated for its practical superiority.

Are you familiar with its Diffusive Flue Construction Front Draw-Out Grate Draw-Out Oven Rack and other talking points ?

If there's any range business in your locality you'll get it by handling the Imperial Oxford. Fullest details if you write

THE GURNEY FOUNDRY CO., Limited

TORONTO. WINNIPEG. VANCOUVER.

THE GURNEY-MASSEY CO., LIMITED, MONTREAL.

David Maxwell & Sons

ST. MARYS, ONT., CANADA

Steel Frame.

"Maxwell Favorite Churn"

PATENTED FEATURES: Improved Steel Stand, Roller Bearings, and Foot and Hand Lever Drive, and Detachable Driving Link. Improved for season of 1901. Steel or Wood Frame as desired.

Wheelbarrows.

In Four different Sizes.

Lawn Mowers.

High and Low Wheels, from 12-in. to 20-in. widths. Cold Rolled Steel Shafting, Crucible Steel Knives and Cutting Plate.

If your Wholesale House does not offer you these articles

...SEND DIRECT TO US.

MAXWELL MOWER!

8-inch Low Wheel.

"THE MAXWELL" Lawn Mower

High Wheel 10 inches

EVERYTHING GOOD AND PROFITABLE IN WALL PAPER.

21-inch Ingrain Frieze, No. 2030.

21-inch Ingrain Frieze, No. 2043.

THE ABOVE ARE TWO OF MANY SUCCESSFUL FRIEZES IN OUR 1901 COLLECTION

IF OF INTEREST WE WILL MAIL SMALL BOOK OF INGRAIN SHADES WITH ILLUSTRATED MATCHED COMBINATIONS, OR SAMPLES OF ABOVE OR OTHERS IN ANY DESIRED SHADE

THE **WATSON, FOSTER CO.,** LIMITED

MONTREAL.

Feb. 1st.

ALEXANDER GIBB

Manufacturers' Agent and Metal Broker,
13 St. John Street, Montreal.

Representing British and American Manufacturers. Correspondence invited from firms wishing to be represented in Canada.

McCaskill, Dougall & Co
Fine Varnish & Japan
Manufacturers,
Montreal.
Price Lists on application

DUNDAS AXES

ARE——

American in Sharpness,
Canadian in Reliability,

JUST WHAT YOU WANT TO SELL.

Dundas Axe Works

DUNDAS, ONT.

MONEY
OR
SENTIMENT?

Will you stick to old fashioned

WALL TINTS

for sentiment, or will you go in for the latest and best for money?

MURALO

is the latest and best, because it is up-to-date and no old fashioned notions about it. Advertising plans are up-to-date, and dealers are backed up by one of the stronge.t financial institutions in America to assist in selling. Muralo goes to every country in the world.

AGENTS:

A. RAMSAY & SON, - - - - - - - MONTREAL.
J. H. ASHDOWN, - - - - - - - - WINNIPEG.
McLENNAN, McFEELY & CO., - - VANCOUVER.

MERCHANTS OF CANADA

Especially dealers in paints and **wall-coatings!** When you took account of stock at the dawn of the new century, did you find in store some goods that you had been induced to buy, and for which there is absolutely no demand? Possibly some that you carried over from 1899? Musingly, did you exclaim, "I got fooled when I bought these goods, but it was the fault of the man or firm who sold me; if I get fooled the second time, it will be my fault!"

The moral of it all is.—Buy goods that are manufactured by men with experience in their particular line, and in whom the people have confidence; who advertise their product, guaranteeing it as represented, and whose goods sell on their own merits.

Why put in stock, that in order to sell, you have to tell your customers is "the same thing" or "just as good" as something else? **CHURCH'S**

ALABASTINE

the permanent wall-coating, ready for use by mixing with cold water, is in demand everywhere. Is sanitary to the highest degree. Has a record of value, and proof of origin.

ALABASTINE for more than a quarter of a century has been prominently before the people. **Anyone can brush it on—no one can rub it off.**

ALABASTINE is the **acme of perfection** in wall-coating material, either for plain tinting, or the most elaborate work known to the decorator's art. Is adapted to all kinds of relief work, and is a special favorite with practical men.

ALABASTINE is manufactured in Canada. Patented in this and other countries. Do you sell it? If not, why not? Write for prices.

Address, **THE ALABASTINE CO., LIMITED,** Paris, Ont.

The trade supplied by—
Sanderson Pearcy & Co., Toronto, Ont.
G. F. Stephens & Co., Winnipeg, Man.
Vancouver Hardware Co., Vancouver, B.C.
Wood, Vallance & Co., Hamilton, Ont.
William Hill, Montreal, P.Q.

The attention of Contractors, Builders and Building Owners is called to our **PARISTONE** Wall Plaster, and **SHIELD BRAND** Calcined Plaster. None better.

VanTuyl & Fairbank

Petrolia, Ont.

Headquarters for . . .
Oil and Artesian Well Pumps, Casing, Tubing, Fittings, Drilling Tools, Cables, etc.

Ontario Nut Works, Paris

BROWN & CO.

Manufacturers of
All sizes of Hot Pressed Nuts, Square and Hexagon.

"DAISY" CHURN

Has tempered steel cased bicycle ball bearings, strongest, neatest and most convenient frame. Only two bolts to adjust in setting up. Steel Bow Levers, suitable for either a standing or sitting posture. Has four wheels and adjustable feet to hold stand steady while churning. When churn is locked to stand the bow can be used as handles to move it about on the front wheels as handy as a baby carriage. Open on both sides to centre, giving free space for pail. Made with wood or steel stands, with Cranks only, or Bow Levers as desired.

Vollmar Perfect Washer

Has a most enviable record. A perfection of its kind—will wash more clothes in less time, do it better and easier, with less wear and tear, than any other machine.

THE——
Wortman & Ward Mfg. Co.,
Limited
LONDON, ONT.
Eastern Branch, 60 McGill Street, Montreal, Que.

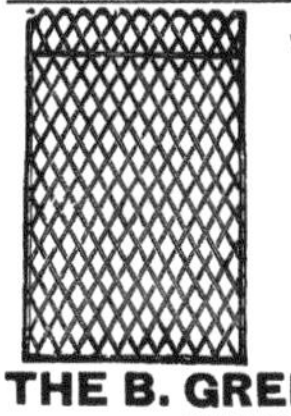

Wire Guards

FOR——

Store Fronts
Factory and Mill Windows
Basement Windows

THE B. GREENING WIRE CO., LIMITED

HAMILTON, ONT., AND MONTREAL, QUE.

CAST
Sap Spouts

These do not buckle and pull out of the trees if the wind blows.

THEY ARE SATISFACTORY UNDER ALL CONDITIONS.

THE "PERFECT" CLOTHES LINE HOOK.

Hold the line without a knot. The more weight hung on line, the tighter it binds. Clothes lines can be put up or taken down in an instant.

ORDER A SAMPLE FROM YOUR JOBBER.

Manufactured by **A. R. Woodyatt & Co., Guelph, Canada.**

SUGAR MAKERS' SUPPLIES

Sap Buckets.

Extra deep and straight. Three sizes. They possess many advantages over the Ordinary Flaring Buckets—being small in diameter they do not catch the rain or snow, and, as they are very deep, they hang perpendicularly, and, consequently, will not overflow until full. Covers supplied if required. They nest close for shipping or storing.

We can supply the Ordinary Flaring Sap Pails.

E. T. Sap Spouts.

Made of Retinned Steel. Strong and Durable. Only require a ⅜-in. hole in a tree. It does not cover the inside surface of the hole, therefore, a larger amount of Sap is obtained.

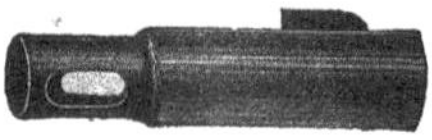

Maple Leaf Sap Spouts.

Made in Bronzed Steel. Require a ½-in. hole. Has a shoulder which prevents it being driven in too far. The hole in the tree is not exposed to wind and snow, consequently, sap will flow longer.

Sprup Cans.

Plain or Decorated. Made in ¼, ½ and 1 gallon sizes. Either Wine or Imperial Measure.

Kemp Manufacturing Co., Toronto, Canada.

HARDWARE AND METAL

VOL. XIII. MONTREAL AND TORONTO, FEBRUARY 9, 1901. NO. 6.

President,
JOHN BAYNE MacLEAN,
Montreal.

THE MacLEAN PUBLISHING CO.
Limited.

Publishers of Trade Newspapers which circulate in the Provinces of British Columbia, North-West Territories, Manitoba, Ontario, Quebec, Nova Scotia, New Brunswick, P.E. Island and Newfoundland.

OFFICES

MONTREAL - - - - - - 232 McGill Street, Telephone 1255.
TORONTO - - - - - - 10 Front Street East, Telephone 2148.
LONDON, ENG. - - - - 109 Fleet Street, E.C., J. M. McKim,
MANCHESTER, ENG. - - - 18 St Ann Street. H. S. Ashburner.
WINNIPEG - - - - Western Canada Block, J. J. Roberts.
ST. JOHN, N.B. - - - No. 3 Market Wharf, J. Hunter White,
NEW YORK. - - - - - - 176 E. 88th Street,

Subscription, Canada and the United States, $2.00.
Great Britain and elsewhere - - - 12s.

Published every Saturday.

Cable Address { Adscript, London. Adscript, Canada.

OUR LAWN MOWERS IN ENGLAND.

IN a quiet and unobtrusive way Canada is building up a nice export trade in lawn mowers of domestic manufacture.

Canadian-made lawn mowers are of a high standard and should sell well abroad, but, like everything else, they need to be "pushed" on the foreign market as well as on the lawn. This, fortunately, they have been, and the result is a growing export trade.

Unfortunately, lawn mowers are not classified in the trade returns. Consequently, the extent of the growth cannot be demonstrated by figures.

The chief market at present is Great Britain, but a trade is being developed with some of the continental countries of Europe. Only this week one Canadian manufacturer, A. R. Woodyatt & Co., Guelph, received a large order from Germany, and the same company is just shipping a carload of mowers to Edinburgh.

What the British market demands, as a rule, is goods of reliable quality. At present, Canadian products of many kinds stand high in public favor there. And our lawn mowers promise to eventually become so.

SOMETHING TO THINK ABOUT.

QUIET seasons in business should be times of active thinking on the part of hardware, as well as on the part of other merchants.

It is when there is little doing that the opportunity comes for preparing plans of campaign. In the height of business activity there is not much time for consecutive thought and careful planning, and when opportunity comes for doing so it should be grasped just as opportunities should be grasped for securing customers.

A subject which retail hardwaremen might discuss during the present quiet season is the formation of associations. There is now but one retail association in Canada, namely, that in Montreal, whereas there should be a dozen or two at the least.

Many of the associations across the border have been holding their annual conventions lately, and one cannot regret, when one reads the reports of the many interesting meetings, that there are not more of such educative institutions among the retail hardwaremen of Canada.

Think about these things.

He who would be successful in business should study not only the methods of others, but should experiment with ideas of his own.

TROUBLE OVER RAILWAY REBATES.

IN every country there are people who pride themselves that in this or that particular their own land is not like other lands. Not infrequently something happens which knocks their ideal to the ground.

For instance, it has been popularly supposed in Great Britain that that country was not as other countries in regard to railway freight rebates. Rebates might be given in the United States and other countries, but not in Great Britain.

If we are to judge from recent exposures, the Mother Country is by no means so free of the evil as it imagined. And if what some of the London papers say be true, the evil is of an aggravated, not of a mild type.

The Midland Railway Company, it appears, is the chief sinner. According to a secret rebate agreement between the railway and Rickett, Smith & Co., the latter were to be allowed 1¼ per cent. on all sums paid by them to the former for the carriage of coal.

The coal factors and merchants, who claim to have for some time suspected the existence of such an agreement, are up in arms, and the coal men, other than the firm which has been enjoying the special privileges, are demanding retrospective compensation.

The Society of Coal Merchants has resolved to take action against the railway, and as the stake involved is over $2,400,000 a good deal of interest is naturally excited.

The railway company pleads justification on the ground of the enormous through traffic which the Rickett Co. furnishes, but this is no solace for those who have been placed at a disadvantage on account of the rebate.

THE SITUATION IN MANITOBA.

WHILE one cannot ignore the fact that the partial failure of the wheat crop in Manitoba entailed a loss to the people in that Province, yet it is not sufficiently heavy to retard the steady development of that part of the Dominion. If it were it would be a reflection on the natural advantage possessed by the Province about which it has been our wont to boast.

It is only right and proper that business men should be guided by the conditions as they exist in Manitoba and cut their garments, financially speaking, according to their cloth. But care, on the other hand, is necessary in order that we do not fly to the other extreme. This we may possibly do if we overlook certain compensating factors. The wheat crop of Manitoba last year was undoubtedly much smaller than for several years. As the president of the Winnipeg Grain and Produce Exchange remarked in his recent annual report, in size it was closely pressed by that of 1887, when the yield was 12,350,000 bushels. But, while the crop was small, a good deal of it ranked high in quality, no less than 70 per cent. of the wheat inspected at Winnipeg being No. 1 hard, the highest percentage recorded. Furthermore, there is the testimony of millers who used last year's Manitoba wheat in their mills as to its excellent flour-producing qualities.

Then, it must be remembered that the dairy and live stock industries are expanding in a most substantial manner, so that the farmer in Manitoba is becoming less dependent upon grain producing as a source of revenue, although it will always probably be the chief source of money supply.

Although the partial failure of the Manitoba wheat crop last year no doubt meant a great loss to the farmer, it does not appear to have crippled him. This is evident from the reports of the loan companies doing business in that Province. One of them, the Canada Landed & National Investment Co., Limited, in its annual report, which was issued a few days ago, says that, notwithstanding the disappointing harvest in Manitoba, "payments by borrowers have been met very good indeed." And then it adds: "Manitoba is beyond any doubt a great and valuable Province, into which an industrious and frugal population is flowing steadily, and will become one of the greatest sources of the world's supply of wheat and flour, and dairy products as well."

In the annual report of the Winnipeg Grain and Produce Exchange, already referred to, Mr. William Martin, the president, estimates that this spring over 2,000,000 acres will be under wheat in Manitoba, and 500,000 acres in the Territories, "so that," he adds, "50,000,000 crop is no flight of the imagination."

Mr. Martin may be a little high in his estimates, but from what we can gather 1901 will see a much larger wheat acreage in Manitoba than last year. The acreage in Manitoba last year was 1,800,000, but, of course, the unusual drought prevented the Province from securing a crop that, under ordinary circumstances, would have approximated to it. In 1887 the acreage was only 432,134, and yet the yield was about as large as that of 1900.

ANOTHER BLAST FURNACE IN OPERATION.

ANOTHER step forward was taken in the iron industry of Canada on Saturday last when the charging of the first furnace of the Dominion Iron and Steel Co. at Sydney, N.S., took place. This furnace, which is 90 feet high and 18 feet in diameter, was charged with 400 tons of Belle Island ore, 225 tons of coke, and 125 tons of lime.

This is the first of four blast furnaces which the company is to shortly have in operation. The capacity of each furnace is 250 tons per day. The estimated cost of the blast furnaces is $2,500,000. Besides these a steel mill is to be constructed at a cost of $1,500,000 and coke ovens at a cost of about $1,250,000. It is expected that the full battery of furnaces will be in operation a couple of months hence.

There are now two blast furnaces in operation in Nova Scotia and six in the whole of the Dominion, there being, besides the two already mentioned, three in Ontario and two in Quebec. It is expected that Ontario will have a fourth by October next, when it is believed the furnace of the Cramp Company at Collingwood will be in blast.

THE TOURIST QUESTION IN NEW BRUNSWICK.

THE annual meeting of The New Brunswick Tourist Association, which was held in St. John, N.B. recently, shows that that body is not yet weary in well doing.

According to the report of the executive committee, two booklets were issued, one for distributing abroad, the other for visitors to St. John. Over 25,000 of the former were printed, and all but 600 disributed. The association has had illustrated articles, descriptive of New Brunswick, published in several newspapers, magazines and trade papers in Canada and in the United States. Over 22,000 picture post cards were issued by the association and sold to local stationers. Photographs for lantern slides and descriptive matter have been supplied lecturers in New England. These have already been shown in Boston, New York and Philadelphia. In several other ways efforts have been made to advertise the Province.

W. S. Fisher, the president of the association, sketched briefly the progress of the work in St. John since its inauguration, and told of the work of organization in other cities of Canada. He appealed for greater interest and more energetic work in bringing forward these advantages. A report was read showing that about 200 non-resident sportsmen took out licenses last year, bringing to the Government $6,000 for game and $3 000 for fishing licenses. These would spend in the Province about $200,000, and from ordinary tourists, numbering, say, 6,000 per week for 10 weeks, the receipts would probably be $2,400,000. An estimate of travel compiled from the different railways and steamship services showed that about 5,300 tourists per week visited New Brunswick last year.

The New Brunswick Tourist Association is to be congratulated on the success that has followed their efforts to advertise the great attractions of the Province to summer tourists. It should, moreover, encourage merchants and others in many sections of Canada to organize for the purpose of disseminating information regarding their respective localities.

TRADE IN COUNTRIES OTHER THAN OUR OWN.

THERE is some unevenness in the wire cloth market in the United States, as represented by the quotations of both manufacturers and jobbers. The low prices made by some makers are not, however, to be taken as representative of the market at large, in which the principal manufacturers are maintaining prices pretty steadily, some, indeed, anticipating that, with the advance of the season and the possible development of a scarcity, higher values will rule. The price to retail merchants who buy in ordinary small lots may be named as about 95c. to $1, a slightly lower figure being obtainable by those who buy in round lots from the manufacturers. —Iron Age.

SCOTCH IRON TRADE IN 1900.

The annual trade statement of the Scotch ironmasters shows that the production of iron in 1900 was 1,153,000 tons, a decrease of 13.000 tons.

The consumption in foundries was 295,000 tons, an increase of 106,000 tons. Malleable iron and steel, however, showed a decrease of no fewer than 109,000 tons.

The exports were 331,000 tons, or 16,000 tons more.

Stocks in Connal's stores are down 174,000 on the year, while iron in makers' yards is up to 31,000 tons.

IRON AND STEEL EXPORTS OF THE U.S.

We have opportnnity merely to refer in a general way to the exhibit of iron and steel exports for 1900, the advance sheets having just been received from the Treasery Department bureau. The $100.000,000 mark in this trade was passed in 1899, when the total was $105.689,645; but 1900, even with its slump in prices, goes well beyond this, showing the imposing total of $129,633 480 for all iron and steel sent abroad, including machinery. In 1897 the total was less than half this, or $62,737,250.

Pig iron exports last year amounted to 286.783 tons, as compared with 228,640 tons in 1899. The billet transactions of last summer were heavy, as is well known; for the year the shipments abroad were 107,476 tons, or more than four times the total of 25,605 tons in 1899. Steel rail exports were 356,245 gross tons in 1900, and of this amount 125,931 tons went to British North America. For 1899 the rail exports were only 171,272 to all countries. The exports of structural material were 67,714 gross tons, against 54,244 tons in 1899; steel bars, 81,366 gross tons; wire rods, 10,651 tons; wire, 78,014 tons; wire nails, 27,404 gross tons, against 37,500 tons in 1899 and 15 000 tons in 1898. The "pipes and fittings" sent abroad in 1900 were valued at $5,994,521, as against $6,763,396 in 1899.—Iron Trade Review.

WEAK METAL MARKETS IN ENGLAND.

S. W. Royse & Co., Manchester, Eng., in their monthly report on the metal trade, say: "Prices of pig iron show a heavy fall since the beginning of the year—about 8s. per ton in Scotch iron, and about 4s. 6d. per ton in Cleveland. Shipments from Glasgow and Middlesbrough during this year are only about half of what they were in the corresponding period of last year. These facts show a very bad state of trade, and, indeed, there has been little business doing at Glasgow latterly, and scarcely any at Middlesbrough during the last fortnight. The higher-priced metals also, have not commenced the year well. Copper has lost ground steadily, and is £2 5s. per ton cheaper than at the beginniug of the month. Tin, however, after a considerable drop early in the month, has advanced again and is practically unchanged. Spelter is practically unchanged. Lead is down 5s. per ton, and is easy."

THE IRON TRADE SITUATION.

Increased activity in finished materials, as compared with the earlier weeks of the year, is now apparent, and close students of conditions believe that buying will soon show itself that will give indications of the scale of operations in important consuming lines in 1901. Pig iron buying by foundries and mills is still postponed, and it may be some time before liberal contracting will be seen. There is no doubt of the better tone given to the market at large by the heavy purchases of Bessemer iron reported last week. These showed that in spite of the large increase in blast furnace capacity, the leading interests still require considerable outside iron. Two important consolidations are expected to be in the market later for Bessemer iron for shipment in the first half. The price has advanced 25c., as shown by sales of the past week, and it is now reported that Bessemer iron is not to be had below $13 at valley furnaces.—Iron Trade Review.

PROFITS IN THE BRITISH CYCLE TRADE.

Whatever other trades have languished in 1900 the wheels of the cycle manufacturing business, at any rate, would seem to have continued spinning briskly during all that time. In this connection, we have heard nothing of the complaints of bad trade and low prices with which aforetime we had become increasingly familiar. We are not sure if this has been the result of an increased demand on the part of the public for the means of enjoying this exhilarating method of locomotion. We should be inclined to say that to better organization on the part of manufacturers, rather than to any phenomenal increase in the number of sales, has success in this instance been due—success, too, which must be all the more gratifying, inasmuch as it has been attained despite a very sensible advance in the cost of materials and accessories.

The 27 leading cycle firms, representing together a total capital of £4,803,573, have earned in the twelvemonth last gone by a profit of £343.921 amongst them, which works out at an average of 2 9 per cent. That is not a very large amount, perhaps, but it is, at all events, something to go on with, and in these days small things are not to be despised. Very many businesses did not do so well as that even. In the same period the nine leading tyre manufacturing companies, who possess an aggregate capital of £5,344,212, earned a matter of £343.921, an average profit per company of 6.4 per cent. Here, again, the increment has been due chiefly to their improved method of carrying on business, by which the big fluctuations of the past have been entirely obviated.—Commerce, London.

HARDWARE TRADE IN THE STATES.

The important fact, so far as market values are concerned, is the announcement on Tuesday of an advance of $2 per ton in the products of the American Steel & Wire Company. This came upon the trade as a surprise, although it had been contemplated as a possibility.

There may be some difference of opinion as to the wisdom of the advance, as it will tend to stimulate competition, which, however, is not as yet at all a serious matter; but the effect on the market will probably be to give a more confident tone to values. There has of late been a good deal of activity in nails and wire, and the trade have been purchasing considerable quantities; but it is not thought that this was done in anticipation of the advance, as there was no public intimation that it was coming. In most lines of general and shelf hardware the market is held pretty steadily, with a very gradual drift toward somewhat lower values. The changes which are taking place are not, however, of sufficient importance to prevent the trade from buying in quantities to meet their requirements, and both wholesale and retail merchants are placing orders to keep their stocks up to a good working size. There are, indeed, some who think that with the passing of the dull season, which should now be nearly over, there may be a strengthening in values and some slight advance.—Iron Age, January 31.

THINGS I LEARNED WHILE STOCK-TAKING.*

FIRST, the necessity of thoroughness. Stock-taking being the time when all stock was handled, and probably parcels and articles examined in a way that time does not allow during ordinary business hours, it was essential that not one slip should be made, but everything turned out of hand should be perfect and ready for sale, or fit to be placed in the fixtures without requiring any additional attention.

2. To insure this, it was evident that overtime was necessary, as customers continually interrupted, making a chance of mistakes, preventing thorough examination of articles, and making stock-taking itself drag out for an indefinite period—a thing obviously undesirable.

3. Those firms whose assistants work till 9 o'clock each evening have their stock done almost as soon as those who work till 10 o'clock, and the workers do not have that tired, worn-out look with them all the time.

4. Overtime being paid for, and not only tea money given, as in some instances, put interest in the work for all the assistants, they endeavoring to use their energies and care to prove that they had earned it.

5. It is necessary to take every parcel out of the hole or fixture—a, to make sure that no odd line, as hooks, screws, broken lock, is left out of sight; b, so that the hole may be swept, and it was noticed in doing this to be best to sprinkle with damp sawdust to prevent the dust from flying to other fixtures, and then to sweep direct on to the dustpan held at the edge of the hole, so as to prevent the dirt from falling on the parcels below; c, that parcels may be properly sorted.

6. All broken parcels should be opened —a, that as many original ones as possible may be made; b, that the contents may be thoroughly examined for shortages, as locks without keys, or barrel bolts without staples, etc.; c, for sorting when lines, as cup hooks, screws, bolts, etc., are mixed; d, for taking out damaged articles not fit to sell at full price.

7. Careful parceling and dusting saves contents from damaging, so ragged parcels should have new paper and broken boxes be replaced by sound ones, and then be marked showing the nature and quantities of contents. If full quantities are in a parcel or box, the string should be tied in a knot, or the bow twisted under the fold of the paper or lid, to save time and labor when full parcels are required.

*A prize essay which appeared in Australian Ironmonger.

8. Fire and box irons, crosscut and circular saws, all bright steel goods, as well as guns, revolvers, etc., should be oiled to save from rust and kept in a dry place, and not downstairs where the damp gets to them. Wrought kettles, saucepans, etc., stained by straw, should be painted again with black, and if bruised they should be straightened before putting into stock, thus preventing delay when required. Ice chests, wood buckets, mirrors and woodenware should not be placed on the top storey where the sun plays on them, causing them to crack. Different sizes of lamp glasses should each have separate holes to prevent unnecessary handling and liability of breakage.

9. That ladders are the cause of a good deal of the untidiness of the stock, for if too long or short to reach the fixture they are leaned against the stock, pushing it out of place, and the assistant, grasping a corner of the parcel to pull it forward, tears it, or else leaves it pushed back in the holes.

10. When overstocks are made, tickets should always be placed over the stock hole to denote same; this prevents errors in double ordering, and often saves time in searching.

11. That samples should not be tied to stock parcels, but samples of cutlery, scissors, pen and pocket knives, razors, etc., should be kept in wrappers, and those of hinges, all classes of hooks, screws, locks, bolts, etc., be fastened on boards, and placed so as to attract customers' attention; this will save considerable time in serving, and will prevent the customers from handling the stock, thus saving from tarnished and often damaged articles, and the mixing of the keys in the case of locks.

12. When it is really necessary to show customers packages, sheep shears, shear stones, shovels, spades, handles, etc., to select from, only one parcel at a time should be opened, and not another till most of the contents of the first are sold, thus stopping an accumulation of lines well picked over which will need pushing later on.

13. That tins or boxes suitable to fixtures should be made to hold all classes of small lines constantly required, as hinges, screws, etc.; this will save time in serving, tying of parcels and keep the apperance of stock clean and attractive, as well as preventing two or three parcels of the one line being opened.

14. There should always be a counter for job or damaged articles, and the articles should be placed on this counter when noticed, and not allowed to wait till there is a large accumulation. All repairs when noticed should be put in the repair book; this will save keeping useless stock.

15. To prevent an accumulation of odd or obsolete lines, as special sizes and makes in lamp glasses, sheep shears, nails, varnishes, cutlery, new inventions, etc., only sufficient to supply orders in hand should be purchased, remembering in all stock ordering that, with the speedy means of transport we have compared with the earlier days of the colonies, it does not pay to stock heavily.

16. It is best to have a price book containing the selling price of all lines, thus saving labor in making or altering prices on parcels, preventing errors, as prices are constantly torn off parcels and tickets, or parcels are overlooked in remarking, then a customer may have his list price direct from the desk without going around to the various fixtures, and in invoicing errors and guesswork are dropped.

17. As the stock of a hole is taken, it should be entered direct in the stock book and a ticket branded T fastened on the hole or fixture. Assistants in entering up these lines should mark their entries with a T in the sales book, and at the finish all T sales lines can be added up and deducted from the gross amount.

GASOLINE LAMPS PROHIBITED.

A despatch from Toledo, Ohio, says: "State Oil Inspectors, Frank L. Baird, of Toledo and John R. Mallow, of Columbus, to-day issued a positive mandate that all manufacturers of gasoline lamps in the State must discontinue such manufacture and use at once. There are several large factories in Ohio and many thousand users. It is intended to serve notice on all at first and if the order is not obeyed in reasonable time, radical measures will be inaugurated. It is anticipated that the manufacturers will fight the matter through the courts. The statutes of Ohio are very positive on the subject, but have never been made effective by former State Oil Inspectors."

BUSINESS SUCCESS.

THE chief cry from all great institutions—railroad, big manufacturing establishments, trusts, insurance companies, publishing houses, banking and merchandising concerns—is for men of brains—clever, keen, enterprising men of executive ability—men who do things. For such men there is no practical limit to the salaries they can attain. Since the beginning of time there never was a period when genius, or even first-rate ability could command in the business world anything like the salary it commands to-day.

The fact is that capital alone is pitiably helpless. Brains mean more than capital the world over. Capital is much more dependent upon man than man is upon capital. The human being who thinks and works can do something without capital; capital can do nothing without human aid.

In business it is not so much a question of money as of brains. The strongest house with a weak management, I care not how old or how respectable its history, will go to the wall, while the weak house with a strong management will become big and powerful. This is inevitable. Man is king, not capital, and this will hold true throughout the ages, whether there be trusts or no trusts, combinations of capital or no combinations. Brains must at all times and under all conditions be reckoned with.

I am not so pessimistic as Mr. Croker about the future of the young man. This is a problem that the latter will work out for himself. There doubtless will be fewer individual business men, but it doesn't follow at all that there will be less successful men, and measured, too, by the dollar.

But what is success anyway? It cannot be measured alone by the accumulation of money. This would be a most imperfect and misleading measurement. Many things enter into the problem of working out a successful career. The very brief span of life allotted to man must be taken into consideration. If one sacrifices health, comfort, pleasure, family and friends merely to build up a name as the head of a business, gaining with all a fortune at middle life, has he lived wisely and well? Has his life been full and rich? Has he got all out of it that he was entitled to, has it meant to him what it should mean, according to his own estimate? With all his worries and strife—with all his business losses from failures and dishonesty—with meeting ruinous competition, and a thousand other annoying and trying conditions inevitable in the life of the business man who has carved out his own career—has he worked out the problem of living as well as the chum of his boyhood who has had all these 20 odd years a snug berth and salary?

The latter has had no serious cares, no worries, and no notes to pay. He has had time to be a good fellow—to be a good husband, and a good father, and to make friends—time to get pleasure out of each day and each week and each year as they went by—time to read and think, and grow broader, and sweeter, and wiser—time to keep health and youth. Possibly he is not worth as much in hard cash at 50 as his boyhood friend, and possibly he is worth a good deal more. At all events, he has sipped daily of the sweets of life, while the other has waited for success to crown his efforts before tasting these pleasures. But pleasures do not wait on any man. They must be taken as they pass by.—Munsey's

HARTNEY MERCHANTS MEET.

The business men of Hartney, Man., have formed an association for the purpose of advancing the interests of the town and the surrounding district. At their meeting on Wednesday evening, last week, it was decided to encourage the proposal to start a creamery at that place. The officers of the association are: President, James Innes: first vice-president, R. Shone; second vice-president, E. Chapin; secretary, T. D. Sutherland; treasurer, E. K. Strathy.

REVISED PRICE LIST ON "C" HORSE NAILS.

THE following is a copy of a circular which the Canada Horse Nail Company is sending out to the hardware trade under date of February 7:

We beg to submit to the hardware trade of Canada our revised trade list for "C" brand horse shoe nails; also a farrier's retail box price list which we have this day adopted for use in Ontario, Quebec and the Maritime Provinces.

The farrier's price list represents a price per box at which our nails may be sold to farriers in the Provinces above named, and which we shall require hereafter to be maintained as a minimum selling price by those who sell our "C" brand horse nails.

We desire, by the radical changes above announced, to secure for those who deal in our "C" brand horse nails, a legitimate business profit on their sale. This has not been the case, as you are aware, for some time past.

We are fully determined to spare no efforts on our part to assist those who sell our "C" brand horse nails, and who will maintain our quotations, to secure a fair profit on their sale.

We shall in future refuse to supply our nails to any firm against whom complaints are made and proved to be true of persistently underselling the quotations which we shall endeavor to establish.

We are the oldest and largest manufacturers of horse nails in Canada; our business being established in 1865. We manufacture only by the old reliable "hot forged" process from the best Swedish charcoal steel nail rods. This process and the material we use is the best known used by any maker.

Every nail is carefully examined and every box is warranted perfect. They may be returned at our expense if found otherwise.

All horse nails made and sold by us have our registered trade mark (the Gothic "C") and name in full on every box.

We ask the hardware trade for their loyal and generous support in making our efforts on their behalf successful.

YOUTH AS A FACTOR OF SUCCESS.

IN one of the articles recently published in The London Times concerning American supremacy, it is said :

"It has been asked what are the American manufacturer's advantages over his British confreres, and to what these advantages are due ? It is a bigger question than can be answered in a few words, but I will attempt to set down what appear to me to be the chief moving causes. Apart from physical resources—such as mineral wealth, etc., a subject already dealt with in former articles—perhaps the primary cause, if not the mainspring, of American enterprise is the consideration shown to youth. Mr. Lecky has said, in the introduction to 'Democracy and Liberty':

"'The respect for old age is one of the strongest English instincts, and is often carried so far that it will be found that men only attain their maximum of influence at a time when their faculties are manifestly declining.' The truth of this is strongly brought home to an Englishman on first visiting the United States. The great Carnegie Steel Works, which made a profit of between $40,000,000 and $42,000,000 in one year, afford an example ; one in many. Those who meet the founder of the company see a man full of vitality, but who has retired from the active management of the business at an age when many in this country look forward to years of control. The acting president is a young man, who was apparently not much above 30 years of age when he was appointed. There are three principal steel works owned by the company, each controlled by superintendents equally young.

"In the whole course of my last trip to the United States, when I made the matter one of close observation, I can remember only two instances of elderly men taking the leading part in the management of works, and in one of these the business, although of great reputation, did not give promise of further advancement.

"The Americans go on the principle that youth is the season of energy. As a man advances in life he has less to hope ; something has gone out of him. He ventures less and wins less. In this country we are overcautious, and, though our caution may avoid some mistakes, it loses more good chances.

"That the young men in the United States successfully fill positions for which we consider matured experience a first essential is due, no doubt, to a variety of causes, the first of which is to be found in the early treatment of children. I have sometimes been almost led to think there are no children in America, only some immature men and women."

WHITE LEAD IN THE STATES.

This market shows a moderate activity, with both contract deliveries and fresh orders, and prices are firmly held. Grinders are busy, and the outlet for dry lead is large, contract deliveries moving freely. The remaining supply is not large, so that orders in excess of contracts cannot command concessions in price. Distributing houses are receiving their stocks of lead in oil for spring division among their retail customers, and, upon the whole, the market appears brisk for a late January week. The foreign brands in oil show no change in price or other feature.—Paint, Oil and Drug Review, Chicago.

H. S. HOWLAND, SONS & CO.

WHOLESALE ONLY — 37-39 Front Street West, **Toronto.** — ONLY WHOLESALE

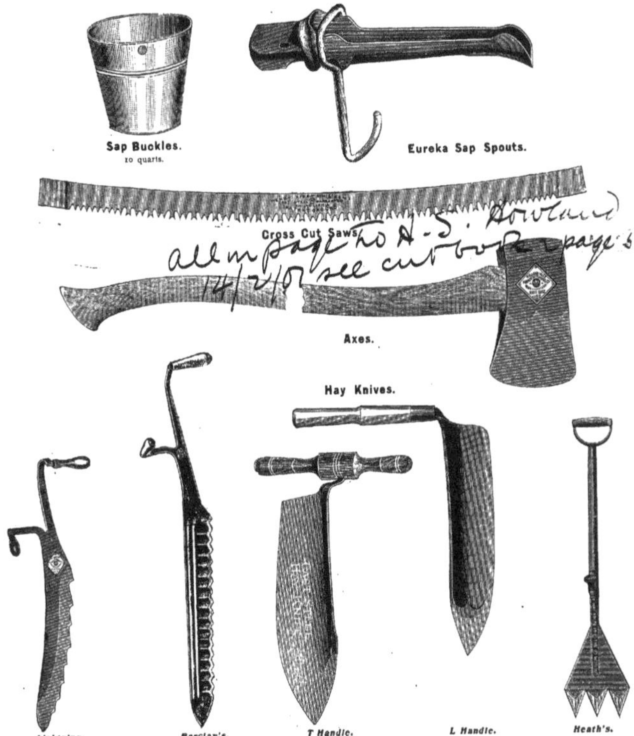

H. S. HOWLAND, SONS & CO., Toronto.

WE SHIP PROMPTLY. — Graham Wire and Cut Nails are the Best. — OUR PRICES ARE RIGHT.

HARDWARE SECTION OF THE TORONTO BOARD OF TRADE.

THE annual report of the Hardware and Metal Section of the Toronto Board of Trade, as presented by the chairman, Mr. Peleg Howland, was as follows:

An occasion for the calling together of this section has not arisen during the past year.

The volume of business, which promised at the beginning of the year to be very large, was checked in the late spring by the downward tendency of prices which set in at that time, and, as a consequence, purchases during the latter half of the year were comparatively light, an effort being made to reduce the rather heavy stocks existing. The short crops in Manitoba and the Northwest Territories also affected the sales of those doing business in that section. Notwithstanding these drawbacks, the year may be characterized as good.

Prospects may be said to be encouraging, particularly in this Province, where progress is being made in the development of the natural and industrial resources, and where the farmer is prosperous, the yield of agricultural products having been large and better than average prices having been realized. These conditions should lead to a good demand, with reasonable safety in granting credits, but do not warrant the elimination of caution, which is again recommended.

In spite of your partially successful efforts, freight discrimination against this city on the part of the railroad companies continues.

The difficulties of doing business in the Northwest country are intensified by the seeming determination of these companies to compel the distribution of all goods through Winnipeg by granting special traders' rates outward from that place.

Whether relief will come from the appointment of a railroad commission, which seems to be foreshadowed, will depend largely upon its composition and powers.

To be of any value, its members must be men absolutely incorruptible, of more than ordinary determination, and furnished with power to enforce their decisions. A judicial body, whose judgments must be referred to Government, will be practically useless.

HARDWARE VS. DRY GOODS.

A FAIR attendance was present at the hockey match last Monday evening at McQueen's rink, and once more the Drys were turned down, this time by the hardware players. After the game had been going only for a few minutes the "All-Wools" found they were up against something even harder than the tailors. Doc Stanton played in his usual position, cover point, and he was altogether too strong for his opponents, but whenever the rubber did pass him Charlie Boyd got it, and he would certainly lift the puck. Goalkeeper Crabb did good business, and made some splendid stops. Kendall and Broad, although spare men, are star forwards. The former apparently is an old-time shinney player. Terry, Whitehead, Sovereen and H. Pauline put up good games. The score stood 4-0 in the first half in favor of the hardware and the second 4-3 making the score 8-3 for the "Nail Handlers."

The teams lined up as follows:

ALL-WOOLS.		NAIL HANDLERS.
Crabb	Goal	Murdoch
Sovereen	Point	Boyd
Terry	Cover Point	Stanton
Pauline, V.	Forwards.	Kendall
Broad	Forwards.	Pauline, H.
Thompson	Forwards.	Austin
Wyckoff	Forwards.	Whitehead

Referee, Jack Cribb.

Timekeeper, W. E. Tisdale.

Goal umpires—W. Anderson and G. Winters.

—Reformer, Simcoe, January 31.

BUYERS AND SELLERS.

THIS is an age, probably more than any other when men are devising ways and means of facilitating business. The conditions under which business is done to-day necessitate it. And he who would keep in the van cannot afford to ignore the facilities thinking minds have provided.

One thing it is important that a man in business should know is the names and addresses of those from whom he can buy goods and of those to whom he can sell goods.

To secure this is usually a most difficult undertaking, and the wider his trade the more difficult it is. This has now been made easy for every business man, manufacturer, or wholesaler, no matter in what branch of trade he is engaged, consequent upon the appearance of a book which gives the name of every manufacturer in every branch of trade within the confines of the Dominion of Canada. "The Manufacturers' List Buyers' Guide of Canada" is the name of the book, and the publishers are The Manufacturers' List Co., 34 Victoria street, Toronto.

It tells where to obtain any article that a buyer may want. About 22,000 articles are indexed in the book, and the names of 7,800 manufacturers are alphabetically arranged for addressing purposes, giving the kind of factory of each. Not included in this, and also arranged alphabetically and classified, are 350 butter factories and creameries, 800 cheese factories, 250 fish, lobster and salmon packing houses, 150 electric light plants, 45 steam railway corporations, 500 shippers of grain, eggs, etc. Another valuable list is that of the classified manufacturers. In this 4,995 classes of goods are enumerated alphabetically, with the different makers of each below them. Altogether there are over 10,000 manufacturers named in the book.

To obtain the technical information contained in this book for classification it was necessary to visit each factory personally throughout the Dominion, as in no other way could it be gathered so completely and intelligibly. And no manufacturer has been omitted because he did not see fit to advertise in the book or subscribe for it.

It is well printed, and bound in a strong cloth cover, stamped in gold, contains 483 pages, 8 x 10, and is sent to any place in the Empire on receipt of $5.

The Department of Trade and Commerce has, through the King's printer, ordered a sufficient number of copies of this book for distribution among the Canadian commercial agents in Great Britain, Australia, South Africa, Norway and Sweden, Trinidad, Argentine Republic, etc.; also copies for the Glasgow Exhibition and Imperial Institute, London.

INQUIRIES REGARDING CANADIAN TRADE.

Mr. Harrison Watson, curator of the Canadian Section of the Imperial Institute, London, England, is in receipt of the following inquiries:

1. A company manufacturing crucible tool and mining drill steel, files, machine planing irons, etc., would be prepared to arrange for its agency with a first-class Canadian firm possessing the necessary connection.

2. A London house seeks the service of a good Canadian representative to introduce glues.

3. The manufacturer of a patent file-cutting machine wishes to appoint a Canadian agent.

4. A firm manufacturing carriage upholstery, etc., asks for names of Canadian shippers of sea grass.

5. A Leeds house wishes to secure the services of a responsible Canadian agent to attend to the purchase and shipment of apples on their behalf. First-class references required.

6. A London firm dealing in oils, wax, honey, minerals, drugs, gums, etc., would be pleased to hear from Canadian shippers of their lines.

7. A company using considerable quantities of asbestos is prepared to hear from Canadian owners of developed deposits of asbestos of good quality.

TWINE FACTORY IN CHATHAM.

The promoters of The Chatham Binder Twine Co. report that stock has been so readily subscribed that the factory is an assured fact. Mr. Cummings, one of the directors, has gone to Montreal and other points to make definite arrangements for the binder twine factory.

STEEL PLANT FOR OTTAWA.

It is understood that Canadian and United States capitalists will soon erect and operate in Ottawa a steel plant for the manufacture of tools and hardware.

CATALOGUES, BOOKLETS, ETC.

A GOOD CATALOGUE OF SPECIALTIES.

The catalogue of The Dowswell Manufacturing Co., Limited, for 1901 is one that should interest all Canadian hardware dealers. It comprises full illustrated descriptions of the clothes wringers, washing machines, mangles, revolving barrel churns, egg crates, lawn swings, garden hose reels, butter-workers, and repairs for these goods, which are manufactured by them.

The reputation this firm has obtained for producing up-to-date goods, and the great range in design, price and quality of their product is such that every salesman who handles these lines should make himself familiar with the contents of this book. A copy of it can be had from either the Hamilton office or from the eastern agents of the firm, W. L. Haldimand & Son, 32 and 34 St. Dizier street, Montreal.

AN ATTRACTIVE CATALOGUE.

We have just received a copy of Boeckh Bros. & Company's 1901 catalogue, which is more attractive this year than in former years. The cover is of an exceptionally neat design, and, while the coloring is not too bright, it is rich in its effect.

The late date of its appearance, we are informed, was due to the revising of the prices on all lines, and many reductions have been made, where it was possible, without cutting the quality of the goods.

It will be worth your while to write for this useful book. It will prove "a friend, indeed" when you are making up your orders.

ST. JOHN HARDWAREMEN MEET.

The seventh annual meeting of the St. John (N.B.) Iron and Hardware Association was held a few days ago in the board of trade rooms. The following officers were elected:

President—S. Hayward.
Vice-President—P. Carmichael.
Secretary-Treasurer—J. J. Barry.
Directors—R. B. Emerson, W. H. Thorne and Thomas McAvity.

The following resolution was unanimously adopted: "That the association lament the death of Her Most Gracious Majesty Queen Victoria, and desire to record our great sympathy for the irreparable loss that the British Empire has sustained, and at the same time express our devotion to her worthy successor, King Edward VII.

"Further resolved, that, in consequence of the death of Her Majesty Queen Victoria, the usual annual dinner be not held this year."

The Swansea Forging Co., Limited, Swansea, Ont., is applying for permission to surrender their charter of incorporation.

MARKETS AND MARKET NOTES

QUEBEC MARKETS.

Montreal, February 8, 1901.

HARDWARE.

ALTHOUGH trade is not brisk, it seems to be in a healthy condition. Business runs principally in shelf goods for immediate delivery, while for later shipment all lines are in demand. The travellers are booking nice orders for spring and summer goods, and in most lines there seems to be no hesitation in ordering good supplies. One of the healthiest signs in the hardware trade at the present moment is the fact that stocks throughout the country are light, and, if the consumptive demand comes on at all brisk, the wholesalers and manufacturers must soon feel it. Merchants have been buying on a falling market, and their stocks are now pretty well cleared out. This is proven by the number of letter and sorting orders that are being received. The feature of the week is the unsettled condition of the horse nail market. The general run of discounts now seems to be 50, 10 and 5 per cent. on oval head, and 50, 10 and 10 per cent. on countersunk head. The Canada Horse Nail Co. have decided to inaugurate a new price list altogether, mention of which will be found in another column. Their discounts will also be different. That the American wire market is improving is shown by the fact that The American Steel and Wire Co. have raised their prices of plain and barb wire and wire nails $2 per ton. This will not directly affect Canadian values, but it shows the trend of the market. Screws have taken a slump since our last report, due to a trade war being carried on by American manufacturers. All manufacturers' supplies, such as bolts, nuts and rivets, are in good demand. Payments are fair.

BARB WIRE—Trade is quiet, with few transactions occurring. The price is still $3.20 f.o.b. Montreal in less than carlots.

GALVANIZED WIRE—This line is featureless at present. The market is steady. We quote: No. 5, $4.25; Nos. 6, 7 and 8 gauge $3.55; No. 9, $3.10; No. 10, $3.75; No. 11, $3.85; No. 12, $3.25; No. 13, $3.35; No. 14, $4.25; No. 15, $4.75; No. 16, $5.00.

SMOOTH STEEL WIRE—The feature in this line is that the price has been advanced 10c. per 100 lb. across the border. The price here remains the same, $2.80 per 100 lb.

FINE STEEL WIRE—The usual demand continues. The discount is 17½ per cent. off the list.

BRASS AND COPPER WIRE—There is nothing new to note. A fair trade is being done. Discounts are 55 and 2½ per cent. on brass, and 50 and 2½ per cent. on copper.

FENCE STAPLES—Trade is of a limited character. We quote: $3.25 for bright, and $3.75 for galvanized, per keg of 100 lb.

WIRE NAILS—The advance on the part of the American Wire and Steel Company affects this product in the United States also. Trade here is quiet and consists of a small sorting nature. We quote $2.85 for small lots and $2.75 for carlots, f.o.b. Montreal, Toronto, Hamilton, London, Gananoque, and St. John, N.B.

CUT NAILS—Inquiry is very small, and,

as is to be expected at this time of year, trade is dull. Prices are steady. We quote as follows: $2.35 for small and $2.25 for carlots; flour barrel nails, 25 per cent. discount; coopers' nails, 30 per cent. discount.

HORSE NAILS—On account of the open prices that are now prevailing the condition of the horse nail market is unsettled. The ruling discounts are 50, 10 and 5 per cent. on oval head 50, 10 and 10 per cent. on countersunk head. "C" brand has been put on a new price list and has a discount of its own.

HORSESHOES — Trade keeps up remarkably well and the market is in a good condition. We quote: Iron shoes, light and medium pattern, No. 2 and larger, $3.50; No. 1 and smaller, $3.75; snow shoes, No. 2 and larger, $3.75; No. 1 and smaller, $4.00; X L steel shoes, all sizes, 1 to 5, No. 2 and larger, $3.60; No. 1 and smaller, $3.85; feather-weight, all sizes, $4.85; toe weight steel shoes, all sizes, $5.95 f.o.b. Montreal; f.o.b. Hamilton, London and Guelph, 10c. extra.

POULTRY NETTING — Business seems to have started very well, dealers buying freely at a discount of 50 and 5 per cent.

GREEN WIRE CLOTH—Values remain at $1.35 per 100 sq. ft. Business continues as before.

FREEZERS— New price lists have been issued and orders are being booked. "Peerless" is quoted as follows: "Two quarts, $1.85; 3 quarts, $2.10; 4 quarts, $2.50; 6 quarts, $3.20; 8 quarts, $4; 10 quarts, $5.25; 12 quarts, $6; 16 quarts, with fly wheel, $11; toy, 1 pint, $1.40.

SCREEN DOORS AND WINDOWS —Fair-sized orders are being booked at old figures. We quote: Screen doors, plain cherry finish, $8.25 per doz.; do. fancy, $11.50 per doz.; windows, $2.25 to $3.50 per doz.

SCREWS—Again, on account of the war being waged by the manufacturers of the United States, values have declined here. Discounts are now as follows: Flat head bright, 87½ and 10 per cent. off list; round head bright, 82½ and 10 per cent.; flat head brass, 80 and 10 per cent.; round head brass, 75 and 10 per cent.

BOLTS—The country manufacturers have been placing some good orders for immediate shipment this week. Discounts are: Carriage bolts, 65 per cent.; machine bolts, 65 per cent.; coach screws, 75 per cent.; sleigh shoe bolts, 75 per cent.; bolt ends, 65 per cent.; plough bolts, 50 per cent.; square nuts, 4½c. per lb. off list; hexagon nuts, 4¾c. per lb. off list; tire bolts, 67½ per cent.; stove bolts, 67½ per cent.

BUILDING PAPER — Spring business is being done as before. We quote as follows: Tarred felt, $1.70 per 100 lb.; 2-ply, ready roofing, 80c. per roll; 3-ply, $1.05 per roll; carpet felt, $2.25 per 100 lb.; dry sheathing, 30c. per roll; tar sheathing, 40c. per roll; dry fibre, 50c. per roll; tarred fibre, 60c. per roll; O.K. and I.X.L., 65c. per roll; heavy straw sheathing, $28 per ton; slaters' felt, 50c. per roll.

RIVETS — Some fair orders have been placed this week. The discount on best iron rivets, section, carriage, and wagon box, black rivets, tinned do., coopers' rivets and tinned swedes rivets, 60 and 10 per cent.; swedes iron burrs are quoted at 55 per cent. off; copper rivets, 35 and 5 per cent. off; and coppered iron rivets and burrs, in 5-lb. carton boxes, are quoted at 60 and 10 per cent. off list.

CORDAGE—A fair trade continues to be done at last week's quotations. Manila is quoted at 13c. per lb. for 7-16 and larger; sisal at 9c., and lathyarn 9c. per lb. In small lots ½c. per lb. higher is charged.

SPADES AND SHOVELS—Trade is rather small. The discount is still 40 and 5 per cent. off the list.

HARVEST TOOLS — A little business is being done on future account at 50, 10 and 5 per cent. discount.

TACKS — The demand is seasonable. We quote: Carpet tacks, in dozens and bulk, blued 80 and 5 per cent. discount; tinned, 80 and 10 per cent.; cut tacks, blued, in dozens, 75 and 15 per cent. discount.

CHURNS—Some spring business is being done at a discount of 56 per cent.

FIREBRICKS — Business is dull. The price is unchanged at $18.50 to $26, as to brand.

CEMENT — Supplies are not wanted. We quote: German, $2.50 to $2.65; English, $2.40 to $2.50; Belgian, $1.90 to $2.15 per bbl.

METALS.

The demand for metals on local and country account is not large for immediate shipment, but in all lines, except for Canada plate, spring business is being done.

PIG IRON—Canadian pig is worth about $19 on the Montreal market, Summerlee bringing $24 to $25.

BAR IRON—The feeling is steady. The rolling mills report the demand brisk and prices well maintained. Dealers are asking $1.65 to $1.70 per 100 lb.

BLACK SHEETS—A small seasonable demand has been felt this week. Prices rule at $2.80 for 8 to 16 gauge; $2.85 for 26 gauge, and $2.90 for 28 gauge.

GALVANIZED IRON — Some important orders have been placed this week at unchanged figures. We quote: No. 28 Queen's Head, $5 to $5.10; Apollo, 10¾ oz., $5 to $5.10; Comet, No. 28, $4.50, with 25c. allowance in case lots.

INGOT COPPER—The statistics for the end of January show on the half-month a decrease of 350 tons in American stocks, and an increase in afloats of 300 tons. The cable advices report a firmer market, and, although there is not much activity, prices are steady. Values here rule about 17½c.

INGOT TIN—The statistics for the end of the month are unfavorable, but the actual figures do not surprise the market. The visible supply for America increased 2,440 tons, and the total visible supply increased 2,131 tons. The market is quiet at 32 to 33c.

LEAD—The price of lead is steady at $4 65.

LEAD PIPE—A small demand only is being felt and the market is quiet. We quote: 7c. for ordinary and 7½c. for composition waste, with 15 per cent. off.

IRON PIPE—There is no change to note. We quote as follows: Black pipe, ¼, $3 per 100 ft.; ⅜, $3; ½, $3; ¾, $3.15; 1-in., $4.50; 1¼, $6.10; 1½, $7.28; 2-in., $9.75. Galvanized, ½, $4.60; ¾, $5.25; 1 in., $7.50; 1¼, $9.80; 1½, $11.75; 2-in., $16.

TINPLATES—Spring shipments are being

arranged for now. The ruling figures for immediate delivery are $4.50 for coke and $4.75 for charcoal.

CANADA PLATE—Business is rather slow on spring account. Some shipments for immediate requirements are being made. We quote: 52's, $2.90; 60's, $3; 75's, $3.10; full polished, $3.75, and galvanized, $4.60.

TOOL STEEL—We quote: Black Diamond, 8c.; Jessop's 13c.

STEEL—No change. We quote: Sleigh-shoe, $1.85; tire, $1.95; spring, $2.75; machinery, $2.75 and toe-calk, $2.50.

TERNE PLATES—Some few shipments are being made. We quote $8.25.

SWEDISH IRON—Unchanged at $4.25.

COIL CHAIN—A good business is being done in coil chain. American manufacturers report that they are very busy. We quote: No. 6, 11½c.; No. 5, 10c.; No. 4, 9½c.; No. 3, 9c.; ¼-inch, 7½c. per lb.; 5-16, $4.60; 5-16 exact, $5.10; ⅜, $4.20; 7-16, $4.00; ½, $3.75; 9-16, $3.65; ⅝, $3.35; ¾, $3.25; ⅞, $3.20; 1-in., $3.15. In carload lots an allowance of 10c. is made.

SHEET ZINC—The ruling price is 6 to 6¼c.

ANTIMONY—Quiet, at 10c.

GLASS.

Quite a large number of import orders have been booked this week on a basis of $1.75 for first break of 50 feet and $1.85 for second break. Small shipments are being made from stock. We quote as follows: First break, $2; second, $2.10 for 50 feet; first break, 100 feet, $3.80; second, $4; third, $4.50; fourth, $4.75; fifth, $5.25; sixth, $5.75, and seventh, $6.25.

PAINTS AND OILS.

The feature of the paint and oil market is the series of slumps in linseed oil for summer delivery on the English market; during the last two weeks it has declined 9c. per gal. This is due to the dullness prevailing in London, and low speculative offers have been made in anticipation of a huge River Platte crop. A reaction is expected. There has been no change locally as light stocks keep values firm, but linseed oil buyers are holding off, buying merely from hand-to-mouth. Last week in nearly all departments of the paint and oil business was decidedly flat. Saturday was universally observed as a holiday throughout the Dominion. It is thought that a brisk trade will not be done for a week or 10 days, as the weather is very severe and indoor painting and finishing has been seriously interfered with. White lead is looking brighter in the Old Country. Values in varnishes, dry colors and painters' supplies are generally unchanged. We quote:

WHITE LEAD—Best brands, Government standard, $6.75; No. 1, $6.37½; No. 2, $6; No. 3, $5.62½, and No. 4, $5.25, all f.o.b. Montreal. Terms, 3 per cent. cash or four months.

DRY WHITE LEAD—$5.75 in casks; kegs, $6.

RED LEAD—Casks, $5.50; in kegs, $5.75.

WHITE ZINC PAINT—Pure, dry, 8c.; No. 1, 6½c.; in oil, pure, 9c.; No. 1, 7½c.

PUTTY—We quote: Bulk, in barrels, $2 per 100 lb.; bulk, in less quantity, $2.15; bladders, in barrels, $2.20; bladders, in 100 or 200 lb. kegs or boxes, $2.35; in tins, $2.45 to $2.75; in less than 100-lb. lots, $3 f.o.b. Montreal, Ottawa, Toronto, Hamilton, London and Guelph. Maritime Provinces 10c. higher, f.o.b. St. John and Halifax.

LINSEED OIL—Raw, 80c.; boiled, 83c., in 5 to 9 bbls., 1c. less, 10 to 20 bbl. lots, open, net cash, plus 2c. for 4 months. Delivered anywhere in Ontario between Montreal and Oshawa at 2c. per gal. advance and freight allowed.

TURPENTINE—Single bbls., 59c.; 2 to 4 bbls., 58c.; 5 bbls. and over, open terms, the same terms as linseed oil.

MIXED PAINTS—$1.25 to $1.45 per gal.

CASTOR OIL—8¾ to 9¼c. in wholesale lots, and ½c. additional for small lots.

SEAL OIL—47½ to 49c.

COD OIL—32½ to 35c.

NAVAL STORES—We quote: Resinsr $2.75 to $4.50, as to brand; coal tar, $3.25 to $3.75; cotton waste, 4½ to 5½c. fo, colored, and 6 to 7½c. for white; oakum, 5½ to 6½c., and cotton oakum, 10 to 11c.

PARIS GREEN—Petroleum barrels, 16¾c. per lb.; arsenic kegs, 17c.; 50 and 100-lb. drums, 17½c.; 25-lb. drums, 18c.; 1-lb. packages, 18½c.; ½-lb. packages, 20½c.; 1-lb. tins, 19½c.; ½-lb. tins, 21½c. f.o.b. Montreal; terms 3 per cent. 30 days, or four months from date of delivery.

SCRAP METALS.

The market continues inactive, with values steady. Dealers are paying the following prices in the country: Heavy copper and wire, 13 to 13½c. per lb.; light copper, 12c.; heavy brass, 12c.; heavy yellow, 8½ to 9c.; light brass, 6½ to 7c.; lead, 2¾ to 3c. per lb.; zinc, 2¼ to 2½c.; iron, No. 1 wrought, $13 to $14 per gross ton; No. 1 cast, $13 to $14; stove plate, $8 to $9; light iron, No. 2, $4 a ton; malleable and steel, $4.

PETROLEUM.

A steadier business continues. We quote: "Silver Star," 15 to 16c.; "Imperial Acme," 16½ to 17½c.; "S C. Acme," 18 to 19c., and "Pratt's Astral," 19 to 20c.

HIDES.

Dealers still report trade slow, on account of poor demand from the tanners. Dealers are paying 7½c. for No. 1 light, and tanners are asked to pay 8½c. for carlots. We quote: Light hides, 7½c. for No. 1; 6½c. for No. 2, and 5½c. for No. 3 Lambskins, 90c.

ONTARIO MARKETS.

TORONTO, February 9, 1901.

HARDWARE.

ALTHOUGH trade is not active it is more so than it was a week ago. Nearly all orders for prompt shipment are small and of a sorting up nature. Quite a few orders are being booked for future delivery, but these also are small.

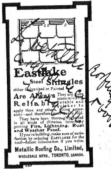

The orders that are being booked for future delivery are mostly poultry netting, screen doors and windows, green wire cloth, spades and shovels, harvest tools, etc. Barb wire has been advanced $2 per ton in the United States, but no change has been made in prices for the Canadian market. Locally, barb wire is dull. The same applies to galvanized wire. In smooth steel wire, business is light. Business in wire nails is still small, while cut nails are dull. Horse nails are moving a little better since the reduction in prices. Horseshoes are quiet. Cutlery and sporting goods continue quiet. Business, if anything, is a little better in screws. Very little is being done in enamelled ware and a moderate sorting-up trade is to be noted in tinware. Payments are on the whole fair.

BARB WIRE—Scarcely anything is doing in barb wire. Notwithstanding an advance of $2 per ton in the price in the United States, the combination over there has not so far made any change in prices for the Canadian market. We still, therefore, quote $2.97 f.o.b. Cleveland for less than carlots, and $2.85 in carlots. From stock, Toronto, $3.10 per 100 lb.

GALVANIZED WIRE—This is also higher in the United States, but unchanged here, with business almost nill, We quote: Nos. 6, 7 and 8, $3.55; No. 9, $3.10; No. 10, $3.75; No. 11, $3.85; No. 12, $3.25; No. 13, $3.35; No. 14, $4.25; No. 15, $4.75, and No. 16, $5.

SMOOTH STEEL WIRE—A few orders for oiled and annealed for future delivery have been booked during the past week, but nothing is doing in the way of immediate shipment in this particular line. A little, howevor, is being done in hay-baling wire for present shipment. The base price is $2.80 per 100 lb.

WIRE NAILS—Business is small indeed, and disappointing to most jobbers. The little that is being done is of a hand-to-mouth character. The base price is unchanged at $2.85 per keg for less than carlots and $2.75 for carlots. The advance of $2 per ton in the United States does not appear in any way to affect the Canadian market.

CUT NAILS—Business is almost nil in cut nails, and the proportion is even more un-unsatisfactory than that in wire nails. The base price is $2.35 per keg.

HORSESHOES — Business is only moderate, with prices unchanged We quote f.o.b. Toronto: Iron shoes, No. 2 and larger, light, medium and heavy, $3.60; snow shoes, $3.85; light steel shoes, $3.70; featherweight (all sizes), $4.95; iron shoes, No. 1 and smaller, light, medium and heavy (all sizes), $3.85; snow shoes, $4; light steel shoes, $3.95; featherweight (all sizes), $4.95.

HORSE NAILS—An error was made last week in quoting "C" brand at 50, 10 and 5 per cent, the price for that brand not having been fixed. A new list of prices and new discounts have been issued, and the latter is now 50 and 7½ per cent. On other brands of oval head horse nails the discount is 50, 10 and 5 per cent., as noted last week, and on countersunk head, 50, 10 and 10 per cent. Since the reduction in prices business has been a little better in horse nails.

SCREWS—Trade is fair with an indication that people are inclined to buy a little more freely since the reduction in prices were made. It is a fact worthy of note that the decline in wood screws is over 43 per cent. from the highest point. Discounts are: Flat head bright, 87½ and 10 per cent.; round head bright, 82½ and 10 per cent.; flat head brass, 80 and 10 per cent.; round head brass, 75 and 10 per cent. Round head bronze is unchanged at 65 per cent., and flat head bronze at 70 per cent.

BOLTS AND NUTS — Just a moderate business is being done. We quote as follows: Carriage bolts (Norway), full square, 70 per cent.; carriage bolts, full square, 70 per cent.; common carriage bolts, all sizes, 65 per cent.; machine bolts, all sizes, 65 per cent.; coach screws, 75 per cent.; sleighshoe bolts, 75 per cent.; blank bolts, 65 per cent.; bolt ends, 65 per cent.; nuts, square, 4½c. off; nuts, hexagon, 4¾c. off; tire bolts, 67½ per cent.; stove bolts, 67½; plough bolts, 60 per cent.; stove rods, 6 to 8c.

RIVETS AND BURRS — Rather a small trade is to be noted in this line. Discount, 60 and 10 per cent. on iron rivets, 55 per cent. on iron burrs, and 35 and 5 per cent. on copper rivets and burrs.

ROPE—Business continues quiet, with quotations as before. We quote as follows: Sisal, 9c. per lb. base, and manila, 13c.; cotton rope, 3-16 in. and larger, 16½c.; 5-32 in., 21½c., and ¼ in., 22½c. per lb.

BINDER TWINE — Very little is being done. We quote: Pure manila, 10¾c. per lb.; mixed, 8½c. per lb.; sisal, 7¾c. per lb.

CUTLERY—Taking it altogether, trade is not very brisk. The feature of business in this line is the demand on British Columbian account which the wholesalers are experiencing.

SPORTING GOODS — Business in this line is quiet and featureless.

GREEN WIRE CLOTH—A fairly good trade on future account is being done at $1.35 per 100 sq. ft.

SCREEN DOORS AND WINDOWS — The improvement noted last week in this line for future delivery has been maintained, quite a few orders having been received during the week.

ENAMELLED WARE—Business in this line is still quiet, being only of a small sorting-up nature.

TINWARE—There is still a good demand for milk can trimmings, and delivery is beginning to be made of sap buckets and sap spouts.

EAVETROUGH — An occasional order is being booked for spring delivery. Prices are without change.

HARVEST TOOLS — Some business for future delivery has been done during the past week. Discount 50, 10 and 5 per cent.

SPADES AND SHOVELS—A little business for spring delivery is also to be noted in this line. Discount 40 and 5 per cent.

POULTRY NETTING—A fairly good trade on future account is still being done. Discount on Canadian, 50 and 5 per cent.

CEMENT—There is nothing doing. We nominally quote in barrel lots: Canadian Portland, $2.80 to $3; Belgian, $2.75 to $3; English do., $3; Canadian hydraulic cements, $1.25 to $1.50; calcined plaster, $1.90; asbestos cement, $2.50 per bbl.

METALS.

Business in metals during the past week has been a little more active lately. The most active lines are Canada plates and galvanized sheets. Our quotations on terne plates are 25c. lower.

PIG IRON—Business is quiet and prices are rather easy. We quote $17 for Canadian iron in 100-ton lots.

BAR IRON—Business is fair, with the base price ruling at $1.65 to $1.70 per 100 lb.

PIG TIN—A fair movement is to be noted in this line, and prices are unchanged at 32 to 33c. per lb. Tin is lower in London, but steady in New York.

TINPLATES—Trade is fairly good and stocks, with at least some houses, are becoming depleted. Prices are unchanged.

TINNED SHEETS—A few shipments are being made for milk can purposes, but, generally speaking, trade is quiet.

TERNE PLATES—These are lower, at $8.50 to $10.50. Business is light.

BLACK SHEETS—Trade has been active during the week in both large and small lots. The base price is $3 30, as reduced last week.

GALVANIZED SHEETS — The demand is fair, and quite a few orders have been booked for future delivery. American are quoted at $4.50, and English at $4 75.

CANADA PLATES—A few lots are going out, but trade is, on the whole, quiet. A few orders have been booked for importation during the week. We quote: All dull, $3; half and half, $3.15, and all bright, $3.65 to $3.75.

IRON PIPE—Business is rather quiet this week with prices as before. We quote: Black pipe ¼ in., $3.00; ⅜ in., $3 00; ½ in., $3 05; ¾ in., $3.20; 1 in., $4 60; 1¼ in., $6.35; 1½in., $7.55; 2 in., $10.10. Galvanized pipe is as follows: ½ in., $4.65; ¾ in., $5.35; 1 in., $7 25; 1¼ in., $9.75; 1½ in., $11.25; 2 in., $15.50.

HOOP STEEL—Business in this line has improved a little during the past week. The base price is still $3.10.

COPPER—Ingot copper is quiet, but in sheet copper a fair trade has been experienced during the past week. We quote: Ingot, 19 to 20c.; bolt or bar, 23½ to 25c.; sheet, 23 to 23½c.

BRASS—A fair trade is to be noted. Discount on rod and sheet 15 per cent.

SOLDER—A fairly active trade has been experienced during the past week. We quote: Bar, half-and-half, guaranteed, 19c.; ditto, commercial, 18½c.; refined, 18½c.; wiping, 18c.

LEAD—No improvement is to be noted, the demand still being light at 4¾ to 5c.

ZINC SPELTER—Trade is quiet at 6 to 6½c. per lb.

ZINC SHEET—A fair trade is to be noted in this line. We quote casks at $6.75 to $7, and part casks at $7 to $7.50.

ANTIMONY—Business in this line is quiet, the improvement noted last week evidently not having been maintained. We quote 11 to 11½c.

PAINTS AND OILS.

The movement is light at the moment, but a large number of orders are coming in for spring delivery. The markets are steady. Linseed oil declined 4c. per gal. here last Saturday, and is now steady at the new prices. Turpentine and white lead are firm. We quote:

WHITE LEAD—Ex Toronto, pure white lead, $6 87½; No. 1, $6.50; No. 2, $6.12½; No. 3, $5.75; No. 4 $5.37½; dry white lead in casks, $6.

RED LEAD—Genuine, in casks of 560 lb., $5.50; ditto, in kegs of 100 lb., $5.75; No. 1, in casks of 560 lb., $5 to $5 25; ditto, kegs of 100 lb.; $5 25 to $5.50.

LITHARGE—Genuine, 7 to 7½c.

ORANGE MINERAL—Genuine, 8 to 8½c.

WHITE ZINC—Genuine, French V.M., in casks, $7 to $7.25; Lehigh, in casks, $6.

PARIS WHITE—90c.

WHITING — 60c. per 100 lb.; Gilders' whiting, 75 to 80c.

GUM SHELLAC — In cases, 22c.; in less than cases, 25c.

PARIS GREEN—Bbls., 16¾c.; kegs, 17c.; 50 and 100-lb. drums, 17½c.; 25-lb. drums, 18c.; 1-lb. papers, 18½c.; 1-lb. tins, 19½c.; ½ lb. papers, 20½c.; ½ lb. tins, 21½c.

PUTTY — Bladders, in bbls., $2.20; bladders, in 100 lb. kegs, $2.35; bulk in bbls., $2; bulk, less than bbls. and up to 100 lb., $2.15; bladders, bulk or tins, less than 100 lb., $3.

PLASTER PARIS—New Brunswick, $1.90 per bbl.

PUMICE STONE — Powdered, $2.50 per cwt. in bbls., and 4 to 5c. per lb. in less quantity; lump, 10c. in small lots, and 8c. in bbls.

LIQUID PAINTS—Pure, $1.20 to $1.30 per gal.; No. 1 quality, $1 per gal.

CASTOR OIL—East India, in cases, 10 to 10½c. per lb. and 10¾ to 11c. for single tins.

LINSEED OIL—Raw, 1 to 4 barrels, 76c.; boiled, 79c.; 5 to 9 barrels, raw, 75c.; boiled, 78c., delivered. To Toronto, Hamilton, Guelph and London, 2c. less.

TURPENTINE—Single barrels, 59c.; 2 to 4 barrels, 58c., to all points in Ontario. For less quantities than barrels, 5c. per gallon extra will be added, and for 5-gallon packages, 50c., and 10 gallon packages, 80c. will be charged.

GLASS.

There is not yet any information re prices on glass for import. The movement from stock is light at steady figures. We still quote first break locally: Star, in 50-foot boxes, $2.10, and 100-foot boxes, $4; double diamond under 26 united inches, $6, Toronto, Hamilton and London; terms 4 months or 3 per cent. 30 days.

OLD MATERIAL.

The movement is moderate though the demand is good. Prices are firm. We quote jobbers' prices as follows: Agricultural scrap, 55c. per cwt.; machinery cast, 55c. per cwt.; stove cast, 35c.; No. 1 wrought 50c. per 100 lb.; new light scrap copper, 12c. per lb.; bottoms, 10½c.; heavy copper, 12½c.; coil wire scrap, 13c.; light brass, 7c.; heavy yellow brass, 10 to 10½c.; heavy red brass, 10½c.; scrap lead, 3c.; zinc, 2½c; scrap rubber, 6½c.; good country mixed rags, 65 to 75c.; clean dry bones, 40 to 50c. per 100 lb.

PETROLEUM.

The demand is falling off somewhat. Prices are steady. We quote: Pratt's Astral, 17 to 17½c. in bulk (barrels, $1 extra); American water white, 17 to 17½c. in barrels; Photogene, 16½ to 17c.; Sarnia water white, 16 to 16½c. in barrels; Sarnia prime white, 15 to 15½c. in barrels.

COAL.

Pea size is firm at the high prices now quoted. Other sizes are steady, with a scarcity of nut. We quote anthracite on cars Buffalo and bridges: Grate, $4 75 per gross ton and $4.24 per net ton; egg, stove and nut, $5 per gross ton and $4 46 per net ton.

MARKET NOTES.

Linseed oil was reduced 4c. last Saturday.

MONTREAL RETAIL ASSOCIATION.

THE Montreal Retail Hardware Association held another well attended, enthusiastic and fruitful meeting at the Monument National last Wednesday evening.

President Martineau occupied the chair, and associated with him about the table were, besides the secretary, 1st Vice-President D. Drysdale and 2nd Vice-President L. J. A. Surveyer. Among others present were: A. Prudhomme, J. Prudhomme, U. Granger, W. Granger, E. D. Colleret, Leger (Lecroix & Leger), L. Prevost, Huberdeau, Chausse, Leblanc, Belanger, Jubinville, Watts, Mailhot, Leblanc and Mederic Martineau.

When the minutes had been read and approved, the president scanned the faces of the audience to see if he could seek out any of the tar-paper manufacturers who had been invited to the meeting to discuss the common interests of the manufacturers and retailers. Unsuccessful, he appealed to the secretary who read a letter from Mr. Patterson, stating that he would be unable to come on account of sickness. The absence of the other invited guests was not explained. However, a copy of a letter from the Canada Hardware Company to the Patterson Manufacturing Company was read, showing that at least one wholesale house has recognized the question of the retailers' pleas and is willing to aid them in the attainment of their objects. It read as follows;

The Patterson Manufacturing Co., Montreal.

DEAR SIRS,—Replying to yours of yesterday, we are pleased to say that we are of the opinion that the members of the Retail Hardware Association should be allowed 5 per cent. off the face of the invoice on building paper, coal tar, etc., as we know for a fact that the merchants of Montreal in general are selling these goods at a loss, as they have to compete with some of the smaller jobbers who are selling to consumers, and some large retailers who can buy the maximum quantity, so that the ordinary merchants are obliged to sell these goods at cost to keep their customers. We are also of the opinion that the jobbers should be allowed to give same rebates for same amount as the manufacturers, so that we can fill the orders of the larger buyers who can purchase $1,200 worth in a year.

Trusting this will have your consideration, we remain,

Yours truly,
The Canada Hardware Co.,
Per A. M. St. Arnaud, Mgr.

This letter aroused a keen discussion. Finally, Mr. Watts moved, seconded by Mr. Mailhot, that a delegation be appointed to wait on the manufacturers to explain to them the objects and aims of the association, the present difficulties of the retailers, and to ask them for a rebate of 7½ per cent. for the members of the association, to be paid every six months. This motion passed unanimously and the interviewing committee will be Messrs. Martineau, Prudhomme, Surveyer and Drysdale. A report upon these interviews may be expected next meeting, February 20.

A discussion then took place upon the fact of the wholesalers selling retail and the idea seemed to prevail that their entreaties had had no effect as yet. Some members threw out plans by which the retailers themselves might do something to have this practice discontinued or at least curtailed. One member said that a certain hardware merchant was getting his customers into the habit of going to the wholesale houses by sending them there for goods which he did not have in stock. An attempt will be made to have this stopped.

Mr. Belanger proposed, seconded by Mr. Huberdeau, that the committee appointed to wait on the paper manufacturers should also be charged with the duty of visiting the wholesalers and urging upon them the necessity of going out of the retail business. Adopted.

Mr. Mederic Martineau gave a notice of motion that Lacroix & Leger be admitted to the membership of the association at the next meeting.

Mr. Martineau also proposed a resolution of condolence to be tendered to Mr. E. Lecours, of Amoit, Lecours & Lariviere, who has lately passed through sad bereavement on account of the death of his wife. In the motion was included the idea that the letter of condolence be published in the daily newspapers and HARDWARE AND METAL. The president suggested that the latter paper be paid for the publication of this item, but a representative present informed him that this paper never published any paid reading-matter, and that the publishers would be pleased to publish an item of this character.

The gathering then adjourned to Wednesday, February 20.

HONORS FOR MR IRVING.

As Mr. John Irving, the newly-appointed agent for the Nova Scotia Steel Co., was taking leave of his firm, The Montreal Rolling Mills Co., last week, a family gathering was held in the manager's office, Montreal, and Mr. Irving was presented with a purse by the firm and a handsome cabinet of cut glassware by his fellow-employes. Mr. Irving leaves with his former associates' best wishes for his success in New Glasgow.

CRAMP STEEL WORKS STARTED.

The mayor of Collingwood, Mr. T. Silver, was a visitor at the Parliament buildings on Wednesday. Mr. Silver was there in connection with some additional water lots which the town proposes giving the Cramp Ontario Steel Co. The work on the company's ore docks will commence this week, and, it is said, that when completed they will be superior to the best ore docks at Cleveland. Mr. Silver says that the town is booming. Work is being carried on at the steel shipyard on the new steel ship for the Beatty Line. This ship will be one of the finest on the great lakes. The Cramp Ontario Steel Co. intend doing as much as possible in excavating the ground for their furnaces this winter. They expect to blow in the first furnace in October next.

ADVANCE IN WIRE AND WIRE NAILS.

Iron Age, January 31: An advance was made on January 20, by The American Steel and Wire Co. to take effect at once, of $2 per ton on wire nails and plain and barb wire. The advance came unexpectedly to the trade, as no intimation had been given of an advance at this time. It should be borne in mind, however, that shipments from Pittsburgh after February 1, to points affected by the reduction in freights will not net the full advance of 10c. The announcement of a reduction in freight rates has not had the effect of increasing orders for future delivery to any noticeable extent. Quotations at the advance are as follows, f.o.b. Pittsburgh, terms 60 days, or 2 per cent. discount for cash in 10 days:

To jobbers in carload lots	$2 30
To jobbers in less than carload lots	2 35
To retailers in carload lots	2 40
To retailers in less than carload lots	2 50

PERSONAL MENTION.

Mr. Duncan Gunn, hardware merchant, Manitou, Man., is in Toronto this week.

MAKES BUSINESS BETTER.

Editor HARDWARE AND METAL,—We have just bought over the stock of Adams & Coate, Kingsville, Ont., and, as we feel that the hardware business is far easier to run with such a valuable paper as yours, we would ask you to send it regularly to us, beginning February 2, 1901.

TELFER & OLIVER.

Kingsville, Ont., February 2, 1901.

MANITOBA MARKETS.

WINNIPEG, February 4. 1901.

BUSINESS continues exceedingly quiet, and no change of price is reported. Preparations for spring and the final winding up of stock-taking are the principal employments at the various wholesale houses.

Price list for the week is as follows:

Barbed wire, 100 lb.	$3 45
Plain twist	3 45
Staples	3 95
Oiled annealed wire 10	3 95
" 11	4 00
" 12	4 05
" 13	4 20
" 14	4 35
" 15	4 45
Wire nails, 30 to 60 dy, keg	3 45
" 16 and 20	3 50
" 10	3 55
" 8	3 65
" 6	3 70
" 4	3 85
" 3	4 10
Cut nails, 30 to 60 dy.	3 00
" 20 to 40	3 05
" 10 to 16	3 10
" 8	3 15
" 6	3 20
" 4	3 30
" 3	3 65
Horsenails, 45 per cent. discount.	
Horseshoes, iron, No. 0 to No 1	4 65
No. 2 and larger	4 40
Snow shoes, No. 0 to No. 1	4 90
No. 2 and larger	4 40
Steel, No. 0 to No. 1	4 95
No. 2 and larger	4 70
Bar iron, $2.50 basis.	
Swedish iron, $4.50 basis.	
Sleigh shoe steel	3 00
Spring steel	3 25
Machinery steel	3 75
Tool steel, Black Diamond, 100 lb	8 50
Jessop	13 00
Sheet iron, black, 10 to 20 gauge, 100 lb.	3 50
20 to 26 gauge	3 75
28 gauge	4 00
Galvanized American, 16 gauge	2 54
18 to 22 gauge	4 50
24 gauge	4 75
26 gauge	5 00
28 gauge	5 25
Genuine Russian, lb	12
Imitation " "	8
Tinned, 24 gauge, 100 lb	7 55
26 gauge	8 80
28 gauge	8 00
Tinplate, IC charcoal, 20 x 28, box	10 75
" IX "	12 75
" IXX "	14 75
Ingot tin	35
Canada plate, 18 x 21 and 18 x 24	3 75
Sheet zinc, cask lots, 100 lb	7 50
Broken lots	8 00
Pig lead, 100 lb.	6 00
Wrought pipe, black up to 2 inch	50 an 10 p.c.
" Over 2 inch	50 p.c.
Rope, sisal, 7-16 and larger	$10 00
" ⅜	10 50
" ¼ and 5-16	11 00
Manila, 7-16 and larger	13 50
" ⅜	14 00
" ¼ and 5-16	14 50
Solder	21½
Cotton Rope, all sizes, lb.	16
Axes, chopping	$ 7 50 to 12 00
" double bitts	12 00 to 18 00
Screws, flat head, iron, bright	75 and 10 p.c.
Round " "	70 p.c.
Flat " brass	70 p.c.
Round " "	60 and 5 p.c.
Coach	57½ p.c.
Bolts, carriage	42½ p.c.
Machine	45 p.c.
Tire	60 p.c
Sleigh shoe	65 p.c.
Plough	40 p.c.
Rivets, iron	50 p.c.
Copper, No. 8	50c. lb.
Spades and shovels	40 p.c.
Harvest tools	50, and 10 p.c.
Axe handles, turned, s. g. hickory, doz.	$2 50
No. 1	1 50
No. 2	1 25
Octagon extra	1 75
No. 1	1 25
Files common	70, and 10 p.c.
Diamond	60
Ammunition, cartridges, Dominion R.F.	50 p.c.
Dominion, C.F., pistol	30 p.c.
" military	15 p.c.
American R.F.	30 p.c.
C.F. pistol	5 p.c.
C.F. military	10 p.c. advance.
Loaded shells:	
Eley's soft, 12 gauge	16 50
chilled, 12 guage	18 00
soft, 10 guage	21 00
chilled, 10 guage	23 00
American, M	16 25
Shot, Ordinary, per 100 lb	6 75
Chilled	7 50
Powder, F.F., keg	4 75
F.F.G.	5 00
Tinware, pressed, retinned	75 and 2½ p.c.
" " plain	70 and 15 p.c.
Graniteware, according to quality	50 p.c.

PETROLEUM.

Water white American	24½c.
Prime white American	23c.
Water white Canadian	21c.
Prime white Canadian	19c.

PAINTS, OILS AND GLASS.

Turpentine, pure, in barrels	$ 68
Less than barrel lots	73
Linseed oil, raw	87
Boiled	90
Lubricating oils, Eldorado castor	25½
Eldorado engine	24½
Atlantic red	27½
Renown engine	41
Black oil	23½ to 25
Cylinder oil (according to grade)	55 to 74
Harness oil	61
Neatsfoot oil	$ 1 00
Steam refined oil	85
Sperm oil	1 50
Castor oil per lb.	11½
Glass, single glass, first break, 16 to 25 united inches	2 25
26 to 40 per 50 ft.	2 50
41 to 50	5 50
51 to 60	6 00
61 to 70 per 100-ft. boxes	6 50
Putty, in bladders, barrel lots per lb.	2½
kegs "	2¾
White lead, pure per cwt.	7 25
No 1 "	7 00
Prepared paints, pure liquid colors, according to shade and color per gal.	$1.30 to $1.90

NOTES.

H. B. Ashelman, who has been in charge of the National Cash Register Co.'s branch in this city for the past year, is leaving to rejoin his brother, B. F. Ashelman, in the Fargo agency of the company, where he was before coming to Winnipeg. Mr. Ashelman has made many friends in the Canadian West, who are sorry that he is returning to his old home. Mr. Whipple, of Wisconsin, succeeds Mr. Ashelman.

The partnership of Rosen & Duggan, Selkirk, has been dissolved, Jacob Duggan continuing the business.

The new Commercial Club of Winnipeg has purchased the old Ontario Bank for a club house, and will have the same handsomely fitted for the new club. It is a central location and admirably adapted for the purpose. The club is making rapid strides in membership.

ODD ADVERTISEMENTS.

An observer of the peculiarities of people copied, according to an exchange, the following from advertisements from various sources:

"Annual sale on. Don't go elsewhere to be cheated—come in here."

"A lady wants to sell her piano, as she is going abroad in a strong iron frame."

"Mr. Brown, furrier, begs to announce that he will make up gowns, capes, etc., for ladies of their own skins."

"Bulldog for sale; he will eat anything; very fond of children."

"Widow in comfortable circumstances wishes to marry two sons."

"To be disposed of, a small phæton, the property of a gentleman with a movable headpiece as good as new."

HEATING AND PLUMBING

SEWER VENTILATION.

IN a system of sewer ventilation now being introduced by J. Stone & Co., a Deptford firm, according to Hardwareman, water is employed as a means both to extract foul gas from the sewer and to purify it before it is discharged into the atmosphere. The apparatus consists of a small tank which is placed at any convenient position in the upper part of the sewer. In this tank, which is always filled with water up to a certain level, there is an ejector arrangement, laid horizontally, which is worked by a jet of water derived from the ordinary water main. The mere act of admitting the water from the main puts the appartus into action, and it so continues until the water is shut off. The suction produced by the jet is arranged to do three things: First, to draw in water from the tank (the ejector arrangement being submerged); secondly, to suck in air from the sewer; and, thirdly, to suck in fresh air from the outside.

Thus, the sewer gas is thoroughly churned up with clean water and fresh air by the action of the jet, and the result is to wash it of its offensive constituents, the water, after use, passing away down the sewer, while the purified air escapes to the atmosphere. Air taken from a sewer in Deptford, in which this apparatus has been fitted for about a year, was submitted to bacteriological examination. When an untreated sample was used to infect a gelatine culture, three days' incubation yielded 26 colonies of micro organisms visible to the naked eye, 19 being molds; but, with a treated sample used in the same way, only three colonies were to be discerned, and none of them were molds.

In another experiment the air in the sewer was artifically contaminated with sulphuretted hydrogen and ammonium sulphide so that when taken into the apparatus it contained 0.71 per cent. of the latter and 0.32 per cent. of the former; analysis of the issuing air showed that the sulphuretted hydrogen had been reduced to 0.04 per cent., while the amount of ammonium sulphide was inappreciable. The cost of water for treating 10,000 cubic feet of gas is put at about 1s. 6d., and it is considered that only intermittent employment of the apparatus, at times when the ordinary means of ventilation are particularly inefficient, would be required. One machine, it is stated, could deal satisfactorily with half a mile of sewer, and the cost of its installation would compare favorably with that of the tall ventilating shafts which are employed in some districts, even if they were erected only at such long intervals as 200 or 300 yards.

TORONTO'S BIG HOTEL.

The contract for the mammoth hotel, which is to be erected on Victoria, King and Colborne streets, Toronto, has at last been let. A meeting of the board of directors of the company was held on Wednesday to consider the tenders for demolishing the buildings now on the premises and construction of the hotel. The lowest tender, that of James Howard, jr., of Pittsburg, Penn., was accepted, and he will start as soon as the contracts are signed. His contract includes all the construction work that will be done. The plumbing, heating, etc., will probably be sublet.

SOME BUILDING NOTES.

The Baptists of Owen Sound, Ont., purpose building a new church this summer, which will probably cost from $10,000 to $12,000.

H. T. Godwin is preparing to build a house in Richmond, Ont.

John Weir is building a residence in Malvern, Ont.

McCormack & Hagarty, contractors, Brockville, Ont., are making an extension to the Canadian Oak Belting Co., Brockville. In the spring a large addition will be erected.

Thomas Hooper, architect, is calling for tenders for a two-storey brick residence and office at the corner of Douglas and Kane streets, Victoria. The building is to be modern in every respect, including electric bells, speaking tubes, and lighted by electricity.

WINNIPEG BUILDING PROSPEOTS.

The indications are that next summer will prove active in building operations in Winnipeg. It is understood that Architect Brown is preparing plans for a 700 room apartment house in the South End, to cost $50,000; S. Spence, of the Gault House, has secured a 75-ft. frontage on Portage avenue, where he intends to erect an up-to-date hotel; J. A. M. Aikins will erect a residence; The Toronto Type Foundry Co., Arthur Congdon and George Gregg have secured lots and, it is understood, will build this summer.

BRAZING BRASS TO COPPER.

THE best solder for uniting brass to copper is soft brass, which, as stated in a technical journal, will melt much easier than the brass which is to be joined to the copper, otherwise the work would melt at the same time as the solder.

The edge of the work must be carefully cleaned, and then the parts brought together in their proper place and secured with iron wire. The flux to be used is borax, rubbed up in water until it is like a fine cream. The solder, which may be in the form of beads, strips or wire, is next distributed along the joint. The amount of heat and the method of applying it depend entirely upon the size of the work to be done. If the work is small the blowpipe is by far the most convenient and safest, because if the heat is too great there is danger that the brass part of the work will be melted. The heat is to be applied until the solder melts. As soon as the solder melts or "flushes," the work should be struck so as to jar it just enough to make the solder flow into the joint. To find out whether the solder is soft enough for the work, a piece may be laid upon a bit of brass of the same kind as that of which the work is made and put it into the fire. If the solder melts considerably sooner than the brass it will be safe to use it for the work. If, on the other hand, they both melt about the same time, a softer solder will be needed. Spelter solder may also be used for the purpose.

PLUMBING AND HEATING NOTES.

J. A. Carslake, plumber, Stratford, Ont., has assigned to John B. Capatine.

Thomas Forest, plumber, Montreal, has assigned, and a meeting of his creditors will be held on February 12. His liabilities are about $6,500. The chief creditors are the Sun Life Insurance Company, mortgage, $3,450: the Jas. Robertson Co., $650; C. Rochon, mortgage, $600; J. A. Beaudoin, mortgage, $1,000; Theo. Theberge, $238.

A. L. Tanguay & Co. have registered as plumbers in St. Henri de Montreal, Que.

R. H. Smith, Tilbury, Ont., is advertising his electric light plant for sale.

During the month of January, Mr. Dore, sanitary engineer, Montreal, examined 27 building plans, visited 147 houses where plumbing required inspection, issued 16 notices about improper workmanship, and issued four plumbers' certificates. He complains that many plumbers are using lighter material than the by-law calls for.

LUMBER TRADE OF NEW BRUNSWICK.

THE spruce export of the year 1900 has been only a moderately paying one. Early sales for choice of dimensions were good, but not extraordinary when the enhanced cost of production is taken into consideration. Prices were not higher than first cost justified. Spruce deals in 9 and 11 by 3—particularly the latter size in long lengths—are scarce, and outside prices are demanded. Logs suitable for making these dimensions are obtainable only in limited and fast-diminishing quantities, a fact that consumers are slow to realize.

Advanced freight and high insurance rates had a curtailing effect on fall business, producers refusing to ship without seeing a prospect of realizing first cost; consequently, wintering stocks are slightly larger than last year, and shipments proportionately less.

Winter operations on the River Miramichi are on a reduced scale, assumed to be 20 to 25 per cent. less than last season's production. Shipments from the Miramichi are not able to keep pace with the cheaper productions from the ports of Nova Scotia and the Bay of Fundy, as the latter largely escape the high stumpage tax to which lumbermen in the northern portion of New Brunswick are subject; cheaper supplies and labor are also available.

The stock of merchantable spruce and pine wintering at Miramichi is about 33,000,000 superficial feet, against 32,000,000 last year and 40,000,000 in 1898; 44,500 superficial feet being pine, against 6,293,000 superficial feet last year.

The shipments from the principal ports, for the last two years, were as follows:

Port.	Year. 1899. Sup. feet.	Year. 1900. Sup. feet.
Miramichi (113,000,000 feet in 1898)	1[illegible],000,000	122,000,000
St. John (244,899,666 in 1897)	181,192,415	236,45[illegible],838
Moncton		41,509,444
Dalhousie		21,060,224
All other ports	112,807,585	64,970,494
Total for New Brunswick	426,000,000	489,000,000

The total shipments of deals from Nova Scotia last year were 146,294,110 feet, against 123,009,504 feet in the previous year.—Gustave Beutelspacher, U.S. Commercial Agent at Moncton, N.B.

THE COUNTRY GENERAL STORE.

If the books of the universe could be balanced in order to show the relative usefulness of the different branches of trade and industry, it is probable that the business of general storekeeping would rank high, remarks Merchants' Review, New York. It even might, like Ben Adhem's name, "lead all the rest." It has served as the training school of more great Americans than our colleges have done, and it fills a field in which hard work and anxiety are more commonly met with than the usual rewards of industry and perseverance.

The country storekeeper fills a position of great responsibility and ought to be better remunerated. He at least should receive more consideration when oleo legislation, pure food legislation and collection bills are on the tapis.

At the beginning of the new century this is a good time to reverse the traditional treatment of the country dealer and give him as much consideration as the pimply-faced bulbous-nosed saloon-keeper receives.

Success to the general stores of the country and their proprietors!

LIVING IN THE PAST.

SOME years ago, in Pennsylvania, says a writer in Stores and Hardware Reporter, I was intimately acquainted with a gentleman of the old school who had a bad habit of living in the past. He spent most of his time at home, because he was rich and could afford it, attending to his business through messengers, of whom I was frequently one. One day when I had returned from an errand on his account, and possibly wishing to avoid further trips, I said:

"Judge, why don't you put in a telephone? It doesn't cost much and it will be very much more convenient and save a good deal of time."

"Young man," he replied, "I have lived seventy five years on this earth and have got along and made money without needing telephones. I don't understand them and what I can't see through I won't use, so you needn't try to work off any new-fangled ideas on an old man who has made a success by following the ideas of his ancestors."

That settled me, of course, but I felt like telling him the story of an old fossil down in Alabama forty years ago who attended a town meeting for the purpose of considering a proposition to grant a franchise to a telegraph company. His speech illustrated the difficulties then encountered by the spirit of progress. It was long drawn out but he finally got down to the point of his argument.

"Fellow citizens," he concluded, "I don't see any use in this contraption, any way. They say it can carry letters and may be it can, but I've seen the wires and don't believe they're strong enough to carry cotton bales, and that's what we want."

Starr, Son & Franklin, and L. W. Sleep, hardware dealers, Wolfville, N.S., have decided to close at 6 p.m. on Mondays, Wednesdays and Saturdays during the winter.

1 to 25 quarts. With fly wheel 8 to 25 quarts.

1 to 25 quarts. With fly wheel 8 to 25 quarts.

Peerless Iceland Freezers

do the business. A complete line from one pint to forty quarts.

Get our Catalog.

DANA & CO.
Cincinnati.

New York Branch: 10 Warren St.
San Francisco Branch: 105 Front Street.

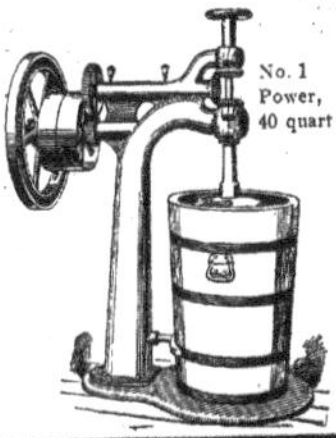

No. 1 Power, 40 quart

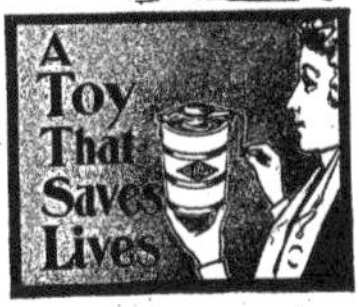

"ALWAYS RELIABLE."

The Fairbanks Standard Scales

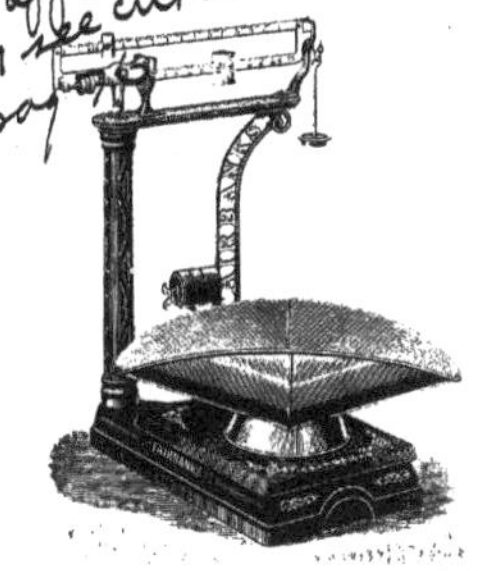

Our Counter Scales—
are most attractive and ornamental.

Watch this Space—
next week and see some of our other styles.

SEND FOR CATALOGUES AND PRICES.

The Fairbanks Company, 749 Craig Street Montreal, Que.

BUSINESS CHANGES.

DIFFICULTIES, ASSIGNMENTS, COMPROMISES.

ALEX. JARVIS, hardware dealer, Cornwall, Ont., has assigned to Jas. A. C. Cameron, and a meeting of his creditors will be held to-day (Saturday).

T. N. Gauthier, general merchant, Carillon, Que., has compromised.

F. X. Julien, general merchant, Lambton, Que., is offering to compromise.

Dugald Campbell, general merchant, Little Metis, Que., has assigned.

E. Ryerson, harness dealer, Hamilton, Ont., has assigned to E. A. Scott.

L. J. Desilets, general merchant, St. Gertrude, Que., is offering 50c. on the dollar, cash.

J. O. A. Deguire & Co., general merchants, Glen Robertson, Ont., have compromised.

J. O. Faubert & Co., general merchants, Barrington, Que., have assigned to Alex. Desmarteau.

J. McD. Hains has been appointed curator of Mrs. E. A. Atkinson, general merchant, L'Avenir, Que.

Arthur Hotte, general merchant, St. Cyrille de Wendover, Que., is offering 50c. on the dollar.

J. McD. Hains has been appointed curator of R. Bourbeau, general merchant, Victoriaville, Que.

V. E. Paradis has been appointed curator of Alph. Mercier, general merchant, St. Angele (Rimouski), Que.; also of F. Veilleux, grocer, etc., St. Francois, N.E., Que., and of Esdras Paradis, general merchant, Plessisville, Que.

PARTNERSHIPS FORMED AND DISSOLVED.

Bellefeuille & Corbeil, blacksmiths, Three Rivers, Que., have dissolved.

Partnership has been registered by The John Terreault Tool and Axe Works, Maissonneuve, Que.

Coppleman & Hartwell, general merchant, Wawanesa, Man., have dissolved. W. F. Hartwell continues.

W. W. Lewis & Co., general merchants, etc., Louisburg, N.S., have dissolved. W. W. Lewis continues under the old style.

SALES MADE AND PENDING.

The assets of J. E. I. Clavel, coal and wood dealer, Montreal, have been sold.

Morris Watts, dealer in agricultural implements, Cartwright, Man., has sold out.

The stock of the estate of E. J. Crawford, general merchant, Souris, Man., has been sold.

Charles Mason & Co., general merchants, Shelburne, Ont., are advertising their business for sale.

The stock of the estate of H. Bond, harness dealer, Inwood, Ont., is advertised to be sold by auction.

The assets of Lalonde & Frere, general merchants, St. Benoit, Que., have been sold.

Mrs. A. F. McDonald, harness dealer, Ingersoll, Ont., is advertising her business for sale.

Joseph Schnitzler, Mildmay, Ont., is offering his planing mill and electric light plant for sale.

CHANGES.

A. R. Smith, general merchant, Brussels, Ont., has removed to Stratford.

Alexander Cockburn, blacksmith, Cottam, Ont., is removing to Essex, Ont.

J. H. Partridge, blacksmith, Newdale, Man., has sold out to J. Livingston.

M. S. Houle, general merchant, Letellier, Man., is removing to St. Boniface, Man.

George Arnold, general merchant, Louisville, Ont., has sold out to F. H. Bedford.

J. G. Quarry, general merchant, Mount Carmel, Ont., has sold out to Hall & Glavin.

D. W. Marsh, general merchant, Calgary, N.W.T., has been succeeded by Wood & Greene.

Ann Carnahan, general merchant, etc., Wetaskiwin, N.W.T., has sold out to J. W. Herrick.

John Robinson, general merchant, Grand Valley, Ont., has been succeeded by Warren & Grayden.

P. P. Mailloux, wholesale saddler, hardware dealer, Montreal, has gone into voluntary assignment.

Mrs. Alphonse Guimond, has registered as proprietress of A. Guimond & Co., hardware dealers, Montreal.

Hunter & Moore, general merchants, Boissevain, Man., have been succeeded by Hunter, Moore & Aitkens.

FIRES.

On Wednesday night last week W. Mitchell's general store, Brownhill, Ont., was destroyed by fire. The loss is placed at $1,500, partly covered by insurance.

DEATHS.

P. Grandy, general merchant, Belle Oram, Nfld., is dead.

James Dixon, sash and door manufacturer, Lansdowne, Ont., is dead.

SHIPBUILDING AT HALIFAX.

The preposal to erect a shipbuilding plant in Halifax is obtaining hearty support from the merchants of that city. The Halifax Chronicle of last Saturday contains interviews with a score or so of the merchants, all of whom expressed warm approval of the scheme, and were in favor of the payment of a considerable bonus by the city for the establishment of such a plant.

Application is being made by The Canada Iron Furnace Co., Limited, authorizing the payment by Midland of a bonus of $50,000 to the company. Notice is given that it is intended to make provision in the Act for the employment of at least 200 men.

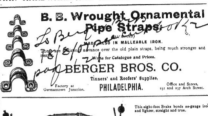

HOW PRICE-CUTTING BEGINS.

A JOBBER, according to an exchange, gives the following somewhat terse account of how price-cutting by jobbers generally commences :

"A travelling man starts out on his route and finds business exceedingly dull. A man sitting in the office of a wholesale house employing him drops him a note, asking him why no orders are received. The traveller writes back the condition of things, but does not send any memoranda with it. He works conscientiously and hard, but the roads are bad, farmers are not getting their produce to market and are unable to get to town to buy what little they need and are ready to purchase. The merchants on the traveller's route still refuse to place orders in advance of actual requirements, although he dilates nobly on the trade that soon must materialize.

"More letters come from the man paid to do the 'punching up' for the house. More explanations and the sole results of additional letters, now of an exceptionally severe tone. Then he gets desperate, walks into a store where he is well known and says: 'Mr. ——, my house, tells me I must sell goods; can't I take your order?' Something in the expression of the travelling man's face checks the refusal which was on the merchant's lips when he saw him coming, and a conference ensues, with the final result of an order for goods, it is true, but at prices which startle the 'house' when it is received. A very strong interrogation point comes in the next letter, and the poor traveller writes back that he had to do it to meet prices made by another house. That is a clincher for his employers, and they have to grin and bear it. Inside of a week the prices made by the salesman are known within a radius of a hundred miles, and are met by the representatives of other houses in the same line.

"I have told you what many of the jobbers really believe to be the genesis of most of the demoralization of jobbers' prices that you hear so much about, but I leave it to you to determine whether the travelling man is altogether to blame, and whether, if such an impossible state of affairs should exist as the transaction of business without the travellers, conditions would be difficult."

"OLD RUT" BUSINESS MEN.

It is queer how some business men of ample means, remarks Ad.-Writer, continue year after year in the old rut, doing business without advertising, and complaining of dull trade and too much competition, when they can look across the street and see some pushing young fellow who started on borrowed capital building a fine trade and hiring additional clerks each year, and succeeding admirably, by liberal advertising —letting the people know that he has goods to sell at the right prices, and goods that are fresher and better than "the other fellows keep," because he is constantly selling his and getting new stock. And still, these same merchants will ofttimes decline to give a drummer an order for certain articles because "they don't sell good any more—people call for articles of soap, coffee, hams, shoes, clothing, etc., that the factories or wholesale houses advertise." Thus, he sees all around him the righteous effect of advertising, but continues blind to the fact that it could be effectively applied to build up his business. The time is not far distant when this class of men will be compelled to retire from the field, which will be fully occupied by the more progressive business men—men who not only see, but profitably apply, the merits of advertising.

BAD DEBTS.

BAD debts! This is the reason for the wrecking of many a good man and of many a good business, says the National Provisioner:

When the store bookkeeper has finished adding up the accounts and has taken all credit for stock on hand and cash in the drawer, he takes up the commercial paper of the house and wades through these as "Bills Receivable" and "Bills Payable." When he has finished his work of auditing the accounts of his firm, he draws his credit balance and sees just where the business stands. His cold-blooded work has ended when he hands the result to his employer, who passes his eye over the footings of the various items and sees where the 12 months of 1900 have landed him. Maybe he has made both ends meet and maybe not. Perhaps he finds that he is heading for bankruptcy in the year 1901. At any rate, he finds a very annoying item, a big item, and more often than not the item which has caused all of his troubles. On a big stock of papers representing a long line of ghastly figures are the items over which appear the the words "Bad Debts," "Worthless Accounts," or some other accounting equivalent for goods sold upon which no payment has been made. The proprietor of that store lays his fevered cheek upon his nervous hand, gives a longing glance at them and then moans: "If I only had the money they represent, or even the half of it, I'd be all right and money to the good." But, alas! they represent what is gone. The storekeeper paid for it, but the purchaser from him got it for nothing.

Bad debts represent a species of over-trust and reckless business speculation which cannot be indulged in without accumulating the pile of accounts which represent the loss of profits and competence.

While one may not be able to collect these accounts or make them good, he can, at least, avoid the system of business which made them, so that the next New Year will find more money in the cash box and less of these "dead-beat" souvenirs in the account books.

CURRENT MARKET QUOTATIONS.

February 8, 1901.

These prices are for such qualities and quantities as are usually ordered by retail dealers on the usual terms of credit, the lowest figures being for larger quantities and prompt pay. Large cash buyers can frequently make purchases at better prices. The Editor is anxious to be informed at once of any apparent errors in this list as the desire is to make it perfectly accurate.

METALS.

Tin.

Lamb and Flag and Straits— 56 and 28 lb. ingots, per lb.	0 32	0 33

Tinplates.

Charcoal Plates—Bright

M.L.S., equal to Bradley.		Per box
I.C., usual sizes	....	$6 75
I.X. "	....	8 25
I.X.X. "	....	9 75
Famous—		
I.C.	....	6 75
I.X	....	8 25
I.X.X	....	9 50
Raven & Vulture Grades—		
I.C., usual sizes	....	5 00
I.X. "	....	6 00
I.X.X "	....	7 00
I.X.X.X. "	....	8 00
D.C., 12½x17	....	4 75
D.X.	....	5 50
D.X.X	....	7 50

Coke Plates—Bright

Bessemer Steel—		
I.C., usual sizes	....	4 15
I.C., special sizes, base	....	4 50
20 x 28	....	8 50

Charcoal Plates—Terne

Dean or J. G. Grade—		
I.C., 20x28, 112 sheets	....	8 75
I.X., Terne Tin	....	10 75

Charcoal Tin Boiler Plates.

Cookley Grade—		Per lb.
X.X., 14x56, 50 sheet bxs; " 14x60 "; " 14x65 "	0 07	0 07½

Tinned Sheets.

72x30 up to 24 gauge	0 08	0 08½
" 26 "	0 08½	0 09
" 28 "	0 09	0 09½

Iron and Steel.

Common Bar, per 100 lbs.	1 65	1 70
Refined " "	....	2 15
Horse Shoe Iron	....	2 05
Hoop steel, 1½ to 3 in. base, extras for smaller sizes	....	3 10
Sleigh Shoe Steel " base	....	2 00
Tire Steel	....	2 00
Machinery iron finish	....	2 05
Cast Steel, per lb	0 00	0 00
Toe Calk Steel	....	2 31
T. Firth & Co's special cast steel, per lb.		0 13
Jessop's Tool Steel	....	0 14

Boiler Tubes.

1½ inch	....	0 12½
2 "	....	0 13
2½ "	....	0 15
3 "	....	0 16
3½ "	....	0 22
4 "	....	0 25

Steel Boiler Plate.

¼ inch	....	2 25
3-16 inch	....	2 25
⅜ inch and thicker	....	2 25

Black Sheets.

18 gauge	....	3 00
20 gauge	....	3 00
22 to 24 "	....	3 10
26 "	....	3 20
28 "	....	3 30

Canada Plates.

All dull, 52 sheets	....	3 00
Half polished	....	3 15
All bright	3 65	3 75

Iron Pipe.

Black pipe—		
¼ inch	....	3 00
⅜ "	....	3 00
½ "	....	3 15
¾ "	....	3 20
1 "	....	4 50
1¼ "	....	6 35
1½ "	....	7 55
2 "	....	10 10
2½-6 inch, discount 60 p.c.		
Galvanized pipe—		
½ inch	....	4 65
¾ "	....	5 35
1 "	....	7 55
1¼ "	....	9 75
1½ "	....	11 25
2 "	....	15 50

Galvanized Sheets.

	G. C. Comet.	Amer.	Queen's Head.
16 gauge	....	4 05	4 00
18 to 24 gauge	4 25 / 3 85	4 35	4 25
26 "	4 50 / 4 10	4 35	4 50
28 "	4 75 / 4 35	4 50	4 75

28 gauge American equals 26 gauge English.

Chain.

Proof Coil, 3-16 in., per 100 lb	....	....
" ¼ " "	8 00	8 50
" 5-16 " "	5 35	5
" ⅜ " "	4 35	4 85
" 7-16 " "	4 15	4 65
" ½ " "	4 35	4 50
" ⅝ " "	3 85	4 35
" ¾ " "	3 80	4 00

Halter, kennel and post chains,	40 and 50 p.c.
Cow ties	40 p.c.
Stall fixtures	35 p.c.
Trace chain	25 p.c.

Jack chain, iron, single and double, discount 35 p.c.
Jack chain, brass, single and double, discount 40 p.c.

Copper.

Ingot

English B. S., ton lots	0 19	0 20
Lake Superior	....	....

Bolt or Bar.

Cut lengths round, ¼ to ⅞ in.	0 23½	0 25
" round and square 1 to 2 inches	0 23½	0 25

Sheet.

Untinned, 14 oz., and light, 16 oz., 14x48 and 14x60	0 23	0 23
Untinned, 14 oz., and light, 16 oz., irregular sizes	0 23	0 23½

NOTE.—Extra for tinning, 2 cents per pound, and tinning and half planishing 3 cents per pound.

Tinned copper sheets	....	0 26
Planished	....	0 32

Braziers (In sheets.)

4x6ft. 25 to 30 lbs. ea., per lb.	....	0 25½
" 35 to 45 " "	....	0 24½
" 50-lb. and above, "	....	0 23½

Boiler and T. K. Pitts

Plain Tinned, per lb	....	0 28
Spun, per lb	....	0 32

Brass.

Rod and Sheet, 14 to 30 gauge, 15 per cent.

Sheets, hard-rolled, 2x4	0 24	0 26
Tubing, base, per lb	0 24	0 25

Zinc Spelter

Foreign, per lb	0 06	0 06½
Domestic "	....	....

Zinc Sheet.

5 cwt. casks	6 75	7 00
Part casks	7 00	7 50

Lead.

Imported Pig, per lb	0 04¾	0 05
Bar, 1 lb	0 05½	0 06¼
Sheets, 2½ lbs. sq. ft., by	....	0 06¼
Sheets, 3 to 6 lbs., "	....	0 06

NOTE.—Cut sheets ½ cent per lb. extra

Pipe, by the roll, usual weights per yard, lists at 7c. per lb. and 15 p.c. dis. f.o.b. Toronto.

NOTE.—Cut lengths, net price, waste pipe 8-ft. lengths lists at 7¾ cents.

Shot.

Common, $6.50 per 100 lb.; chilled, $7.00 per 100 lb.; buck, seal and ball, $7.50. Discount, 7½ p.c. Prices are f.o.b. Toronto, Hamilton, Montreal, St. John and Halifax. Terms 3 per cent. cash, freights equalized on Montreal.

Soil Pipe and Fittings.

Discount, 60 and 10 per cent. on medium and extra heavy, and 60 per cent. on light.

Solder.

	Per lb.	Per lb.
Bar half-and-half, guarant'd	....	0 19
Bar half-and-half, commer'l	....	0 18½
Refined	....	0 18½
Wiping	....	0 18

NOTE.—Prices of this graded according to quantity. The prices of other qualities of solder in the market indicated by private brands vary according to composition.

Antimony.

Cookson's, per lb.	0 11	0 11½

White Lead.

	Per 100 lb.
Pure	6 87½
No. 1 do	6 50
No. 2 do	6 12½
No. 3 do	5 75
No. 4 do	5 37½
Munro's Select Flake White	7 37½
Elephant and Decorators' Pure	7 12½
Brandram's B. B. Genuine	8 00
" " Decorative	7 55
" " No. 1	6 85
" " No. 2	6 00

Red Lead.

Genuine, 560 lb. casks, per cwt.	....	$5 50
Genuine, 100 lb. kegs, per cwt.	....	5 75
No. 1, 560 lb. casks, per cwt.	....	5 25
No. 1, 100 lb. kegs, per cwt	....	5 00

White Zinc Paint.

Elephant Snow White	0 08	0 09
Pure White Zinc	0 08	0 09
No. 1	0 06	0 07½
No. 2	0 05	0 06½

Dry White Lead.

Pure, casks	5 75
Pure, kegs	6 25
No. 1, casks	5 50
No. 1, kegs	5 00

Prepared Paints.

In ¼, ½ and 1 gallon tins.

Pure, per gallon	....	1 25
Second qualities, per gallon	....	1 10
Barn (in bbls.)	0 75	0 85
The Sherwin-Williams Paints.	....	1 45
Canada Paint Co's Pure	....	1 25
Toronto Lead & Color Co's Pure	....	1 25
Sanderson Pearcy's Pure	....	1 20
Stewart & Wood's Champion Pure.		1 20

Colors in Oil.

25 lb. tins, Standard Quality.

Venetian Red, per lb.	0 05
Chrome Yellow	0 11
Golden Ochre	0 06
French "	0 05
Marine Black	0 09
" Green	0 09
Chrome "	0 08
French Imperial Green	0 09

Colors, Dry.

Yellow Ochre (J.C.) bbls.	1 35	1 40
Yellow Ochre (J.F.L.S.), bbls	...	2 75
Yellow Ochre (Royal)	1 10	1 15
Brussels Ochre	....	2 00
Venetian Red (best), per cwt.	1 80	1 90
English Oxides, per cwt.	3 00	3 25
American Oxides, per cwt.	1 75	2 00
Canadian Oxides, per cwt.	1 75	2 00
Super Magnetic Oxides, 93 p.c.	2 00	2 25
Burnt Sienna, pure, per lb.	....	0 10
" Umber, " "	....	0 10
do Raw	....	0 09
Drop Black, pure	....	0 09
Chrome Yellows, pure	....	0 18
Chrome Greens, pure, per lb.	....	0 12
Golden Ochre	....	0 03½
Ultramarine Blue in 28-lb. boxes, per lb.	0 08	0 24
Fire Proof Mineral, per 100 lb	....	1 00
Genuine Eng. Litharge, per lb	....	0 07
Mortar Color, per 100 lb.	....	1 25
English Vermillion	....	0 80
Pure Indian Red, No. 45 lb.	....	0 80
Whiting, per 100 lb.	....	0 55

Blue Stone.

Casks, for spraying, per lb.	0 07
100-lb. lots, do. per lb.	0 08

Putty.

Bulk in bbls.	2 00
Bulk in less quantity	2 15
Bladders in bbls.	2 20
Bladders in kegs, boxes or loose	2 35
Bladders in 25-lb. tins	2 45
Bladders in 12½-lb. tins	2 75
Bladders in bulk or tins less than 100 lb.	3 00

Varnishes.

(In 5-gal. lots.) Per gal.

Carriage, No. 1	2 90	3 30
" body	8 00	9 00
" rubbing	4 00	5 00
Gold Size, Japan	3 00	3 40
Brown Japan	2 40	2 80
Elastic Oak	2 90	3 30
Furniture, extra	2 40	2 80
" No. 1	1 60	2 00
Hard Oil Finish	2 70	3 10
Light Oil Finish	3 20	3 60
Demar	3 30	3 70
Shellac, white	4 40	4 80
" orange	4 00	4 40
Furniture Brown Japan	1 60	2 00
Black Japan	2 40	2 80
" " No. 1	1 60	2 00

The Imperial Varnish & Color Co's, Limited Elastilite Varnish 1 gal. can, each. $2.00.

Granatine Floor Finish, per gal. $2.00.

Maple Leaf Coach Enamels; Size 1, 60c.; Size 2, 35c.; Size 3, 20c. each.

Linseed Oil.

	Raw.	Boiled.
1 to 4 bbls delivered	$0 82	$0 85
5 to 9 bbls "	81	84

Toronto, Hamilton, London and Guelph 2c. less.

Turpentine.

Single barrel, freight allowed	...	0 19
2 to 4 barrels "	....	0 68

Castor Oil

East India, in cases, per lb.	0 10	0 10½
" " small lots	0 10½	0 11

Cod Oil, Etc.

Cod Oil per gal	0 50	0 55
Pure Olive	....	1 20
" Neatsfoot	....	90

Glue.

Common	0 08½	0 12
French Medal	0 14	0 14½
Cabinet sheet	0 12	0 13
White, extra	0 18	0 20
Gelatine	0 22	0 30
Strip	0 18	0 20
Coopers	0 19	0 20
Huttner	....	0

HARDWARE.

Ammunition.

Cartridges.

B. B. Caps. Dom. 50 and 5 per cent.
Rim Fire Pistol, dis. 40 p. c., Amer.
Rim Fire Cartridges, Dom., 50 and 5 p. c.
Central Fire Pistol and Rifle, 10 p.c. Amer.
Central Fire Cartridges, pistol sizes, Dom. 30 per cent.
Central Fire Cartridges, Sporting and Military, Dom., 15 and 5 per cent.
Central Fire, Military and Sporting, Amer, add 5 p.c. to list. B.B. Caps, discount 40 per cent. Amer.
Loaded and empty Shells, "Trap" and "Dominion" grades, 25 per cent Rival and Nitro, net list.
Brass Shot Shells, 55 per cent
Primers, Dom., 30 per cent.

Wads — per lb

Best thick white felt wadding, in ¼-lb bags 1 00
Best thick brown or grey felt wads, in ½-lb. bags 0 70
Best thick white card wads, in boxes of 500 each, 12 and smaller gauges 0 99
Best thick white card wads, in boxes of 500 each, 10 gauge 0 35
Best thick white card wads, in boxes of 500 each, 8 gauge 0 55
Thin card wads, in boxes of 1,000 each, 12 and smaller gauges 0 20
Thin card wads, in boxes of 1,000 each, 10 gauge 0 25
Thin card wads in boxes of 1,000 each, 8 gauge ....

Chemically prepared black edge grey cloth wads, in boxes of 250 each— Per M
11 and smaller gauge 0 60
9 and 10 gauges 0 70
7 and 8 gauges 0 90
5 and 6 gauges 1 10

Superior chemically prepared pink edge best white cloth wads, in boxes of 250 each—
11 and smaller gauge 1 15
9 and 10 gauges 1 40
7 and 8 gauges 1 65
5 and 6 gauges 1 90

Adzes.

Discount, 20 per cent.

Anvils.

Per lb .... 10 — 0 12½
Anvil and Vise combined .... — 4 50
Wilkinson & Co.'s Anvils..lb. 0 09 — 0 09½
Wilkinson & Co.'s Vices..lb. 0 09¾ — 0 10

Augers.

Gilmour's, discount 65 and 5 p.c. off list.

Axes.

Chopping Axes—
Single bit, per doz .... 6 50 — 10 00
Double bit, " .... 12 00 — 18 00
Bench Axes, 40 p.c.
Broad Axes, 33⅓ per cent.
Hunters' Axes .... 5 50 — 6 00
Boy's Axes .... 5 75 — 6 75
Splitting Axes .... 6 50 — 12 00
Handled Axes .... 7 00 — 10 00

Axle Grease.

Ordinary, per gross .... 5 75 — 6 00
Best quality .... 13 00 — 15 00

Bath Tubs.

Zinc .... — 6 00
Copper, discount 15 p.c. off revised list

Baths.

Standard Enameled.

5½-inch rolled rim, 1st quality 30 00
" " " 2nd " 22 00

Anti-Friction Metal.

"Tandem" A .... per lb. 0 27
" B .... " 0 21
" C .... " 0 11½
Magnolia Anti-Friction Metal, per lb. 0 25

SYRACUSE SMELTING WORKS.

Aluminum, genuine 0 45
Dynamo 0 29
Special 0 25
Aluminum, 99 p.c. pure "Syracuse" .. 0 50

Bells.

Hand.

Brass, 60 per cent.
Nickel, 55 per cent.

Cow.

American make, discount 66⅔ per cent.
Canadian, discount 45 and 50 per cent.

Door.

Gongs, Sargant's .... 5 50 — 8 00
" Peterboro', discount 45 per cent.

Farm.

American, each .... 1 25 — 3 00

House.

American, per lb .... 0 35 — 0 40

Bellows.

Hand, per doz .... 3 35 — 4 75
Moulders', per doz .... 7 50 — 10 00
Blacksmiths', discount 40 per cent.

Belting.

Extra, 50 and 10 per cent.
Standard, 60 per cent.
No. 1 Agricultural, 60 and 10 p.c.

Bits.

Auger.

Gilmour's, discount 60 and 5 per cent.
Rockford, 50 and 10 per cent.
Jennings' Gen., net list.

Car.

Gilmour's, 47½ to 50 per cent.

Expansive.

Clark's, 40 per cent.

Gimlet.

Clark's, per doz .... 0 65 — 0 90
Damond, Shell, per doz .... 1 00 — 1 50
Nail and Spike, per gross .... 2 25 — 5 20

Blind and Bed Staples.

All sizes, per lb .... 0 07½ — 0 12

Bolts and Nuts. Per cent.

Carriage Bolts, full square, Norway 70
" " full square 70
Common Carriage Bolts, all sizes 65
Machine Bolts, all sizes 65
Coach Screws 75
Sleigh Shoe Bolts 75
Blank Bolts 65
Bolt Ends 65
Nuts, square .... 4½c. off
Nuts, hexagon .... 4¾c. off
Tire Bolts 67½
Stove Bolts 67½
Stove rods, per lb .... 5½ to 6c.
Plough Bolts 60

Boot Calks.

Small and medium, ball, per M 4 25
Small heel, per M 4 50

Bright Wire Goods.

Discount .... 55 per cent.

Broilers.

Light, dis., 65 to 67½ per cent.
Reversible, dis., 65 to 67½ per cent.
Vegetable, per doz., dis. 37½ per cent.
Henis, No. 8, " .... — 6 00
Henis, No. 9, " .... — 7 00
Queen City " .... 7 50 — 0 00

Butchers' Cleavers.

German, per doz .... 6 00 — 11 00
American, per doz .... 12 00 — 20 00

Building Paper, Etc.

Plain building, per roll 0 30
Tarred lining, per roll 0 40
Tarred roofing, per 100 lb 1 65
Coal Tar, per barrel 3 50
Pitch, per 100-lb 0 85
Carpet felt, per ton 45 00

Bull Rings.

Copper, $2.00 for 2½ in. and $1.90 for 2 in.

Butts.

Wrought Brass, net revised list

Cast Iron.

Loose Pin, dis., 60 per cent.

Wrought Steel.

Fast Joint, dis. 60 and 10 per cent.
Loose Pin, dis. 60 and 10 per cent.
Berlin Bronzed, dis. 70, 70 and 5 per cent.
Gen. Bronzed, per pair .... 0 40 — 0 65

Carpet Stretchers.

American, per doz .... 1 00 — 1 50
Bullard's, per doz .... 6 50 — ....

Castors.

Bed, new list, dis. 55 to 57½ per cent.
Plate, dis. 52½ to 57½ per cent.

Cattle Leaders.

Nos. 31 and 32, per gross .... 50 — 9 50

Cement.

Canadian Portland .... 2 80 — 3 00
English " .... — 3 00
Belgian " .... 2 75 — 3 00
Canadian hydraulic .... 1 25 — 1 50

Chalk.

Carpenters, Colored, per gross 0 45 — 0 75
White lump, per cwt .... 0 60 — 0 65
Red .... 0 05 — 0 06
Crayon, per gross .... 0 14 — 0 18

Chisels.

Socket, Framing and Firmer.

Broad's, dis. 70 per cent.
Warnock's, dis. 70 per cent.
P. S. & W. Extra 60, 10 and 5 p.c.

Churns.

Revolving Churns, metal frames—No. 0, $8—No. 1, $8.50—No. 2, $9.00—No. 3, $10.00 No. 4, $12.00—No. 5, $16.00 each. Ditto wood frames—20c. each less than above. Discounts: Delivered from factories, 58 p.c.; from stock in Montreal, 56 p.c. Terms, 4 months or 3 p.c. cash in 30 days.

Clips.

Axle dis. 65 per cent.

Closets.

Plain Ontario Syphon Jet $8 00
Emb. Ontario Syphon Jet 8 50
Fittings 1 25
Plain Teutonic Syphon Washout 4 75
Emb. Teutonic Syphon Washout 5 25
Fittings 1 25
Low Down Teutonic, plain 14 50
" " " embossed 15 00
Plain Richelieu 3 75
Emb. Richelieu 4 00
Fittings 1 25
Low Down Ont. Syphon Jet, plain .. 20 00
" " " " " emb'd. 20 50
Closet connection 1 25
Basins, round, 14 in 0 60
" oval, 17 x 14 in 1 50
" " 19 x 15 in 2 25

Compasses, Dividers, Etc.

American, dis. 62½ to 65 per cent.

Cradles, Grain.

Canadian, dis. 25 to 33⅓ per cent.

Crosscut Saw Handles.

S. & D., No. 3, per pair 17½
" " 5, " 22½
" " 6, " 15
Boynton pattern " 20

Door Springs.

Torrey's Rod, per doz .... (15 p.c.) 2 00
Coil, per doz .... 0 88 — 1 60
English, per doz .... 2 00 — 4 00

Draw Knives.

Coach and Wagon, dis. 50 and 10 per cent.
Carpenters, dis. 70 per cent.

Drills.

Hand and Breast.

Millar's Falls, per doz. net list.

DRILL BITS.

Morse, dis., 37½ to 40 per cent.
Standard dis. 50 and 5 to 55 per cent.

Faucets.

Common, cork-lined, dis. 35 per cent.

ELBOWS. (Stovepipe.)

No. 1, per doz 1 40
No. 2, per doz 1 20
Bright, 20c. per doz. extra.

ESCUTCHEONS.

Discount, 45 per cent.

ESCUTCHEON PINS.

Iron, discount 40 per cent.

FACTORY MILK CANS.

Discount off revised list, 40 per cent.

FILES.

Black Diamond, 50 and 10 to 60 per cent.
Kearney & Foote, 60 and 10 p.c. to 60, 10, 10.
Nicholson File Co., 50 and 10 to 60 per cent.
Jowitt's, English list, 25 to 27½ per cent.

FORKS.

Hay, manure, etc., dis., 50 and 10 per cent. revised list.

GLASS—Window—Box Price.

Size United Inches.	Star Per 50 ft.	Star Per 100 ft.	D. Diamond Per 50 ft.	D. Diamond Per 100 ft.
Under 26	2 10	4 00	....	6 00
26 to 40	2 30	4 35	....	6 65
41 to 50	....	4 75	....	7 25
51 to 60	....	35	....	8 50
61 to 70	....	75	....	9 25
71 to 80	....	5	....	10 50
81 to 85	....	6 50	....	11 75
86 to 90	....	....	..	14 00
91 to 95	..	....	....	15 50
96 to 100	..	....	....	18 00

GAUGES

Marking, Mortise, Etc.

Stanley's dis. 50 to 55 per cent.

Wire Gauges.

Winn's, Nos. 26 to 33, each .... 1 65 — 2 40

HALTERS.

Rope, ⅜ per gross .... — ....
" ½ " .... — 9 00
" ⅝ to ¾ .... — 14 00
Leather, 1 in., per doz .... 3 87½ — 4 00
" 1¼ in., " .... 5 15 — 5 20
Web, — per doz .... 1 87 — 2 45

HAMMERS.

Nail

Maydole's, dis. 5 to 10 per cent. Can. dis. 25 to 27½ per cent.

Tack.

Magnetic, per doz .... 1 10 — 1 20

Sledge.

Canadian, per lb .... 0 07½ — 0 08½

Ball Pean.

English and Can., per lb .... 0 22 — 0 25

HANDLES.

Axe, per doz. net .... 1 50 — 2 00
Store door, per doz .... 1 00 — 1 50

Fork.

C. & B., dis. 40 per cent. rev. list.

Hoe.

C. & B., dis. 40 per cent. rev. list.

Saw.

American, per doz .... 1 00 — 1 25

Plane.

American, per gross .... 3 15 — 3 75

Hammer and Hatchet.

Canadian, 40 per cent.

Cross-Cut Saws.

Canadian, per pair 0 13½

HANGERS. doz. pairs.

Steel barn door .... 5 85 — 6 00
Stearns, 4 inch .... — 5 00
" 5 inch .... — 6 50
Lane's covered—
No. 11, 5-ft. run .... — 8 40
No. 11½, 10-ft. run .... — 10 80
No. 12, 10-ft. run .... — 12 60
No. 14, 15-ft. run .... — 21 00
Lane's O.N.T. track, per foot .... — 4½

HARVEST TOOLS.

Discount, 50 and 10 per cent.

HATCHETS.

Canadian, dis. 40 to 42½ per cent.

HINGES.

Blind, Parker's, dis. 50 and 10 to 60 per cent
Heavy T and strap, 4-in., per lb 0 06½
" " 5-in., " 0 06½
" " 6-in., " 0 06
" " 8-in., " 0 05¾
" " 10-in., " 0 05¾
Light T and strap, dis. 60 and 5 per cent.
Screw hook and hinge—
6 to 12 in., per 100 lbs .... — 4 50
14 in. up, per 100 lbs .... — 3 50

Per gro. pairs.

Spring .... — 12 00

HOES.

Garden, Mortar, etc., dis. 50 and 10 p.c.
Planter, per doz .... 4 00 — 4 50

HOLLOW WARE

Discount .... 45 and 5 per cent

HOOKS.

Cast Iron.

Bird Cage, per doz .... 0 50 — 1 10
Clothes Line, per doz .... 0 27 — 0 63
Harness, per doz .... 0 72 — 0 88
Hat and Coat, per gross .... 1 00 — 3 00
Chandelier, per doz .... 0 50 — 1 00

Wrought Iron.

Wrought Hooks and Staples, Can., dis. 47½ per cent.

Wire.

Hat and Coat, discount 45 per cent.
Belt, per 1,000 .... 0 60
Screw, bright, dis. 55 per cent.

HORSE NAILS.

"C" brand 50 and 7½ p.c. off new list } Oval
"M" brand 50, 10 and 5 p.c. } head
Countersunk, 50, 10 and 5 per cent.

HORSESHOES.

F.O.B. Montreal — No. 2 and larger / No. 1 and smaller

Iron Shoes.
Light, medium, and heavy... 3 50 3 75
Snow shoes... 3 75 4 00
Steel Shoes.
Light... 3 60 3 85
Featherweight (all sizes)... 4 85 4 85
F.O.B. Toronto, Hamilton, London and Guelph, 10c. per keg additional.
Toe weight steel shoes... 6 70

JAPANNED WARE.
Discount, 45 and 5 per cent. off list, June 1899.

ICE PICKS.
Star per doz... 3 00 3 25

KETTLES.
Brass spun, 7½ p.c. dis. off new list.
Copper, per lb... 0 30 0 50
American, 60 and 10 to 65 and 5 p.c.

KEYS.
Lock, Can., dis., 45 p.c.
Cabinet, trunk, and padlock, Am. per gross... 60

KNOBS.
Door, japanned and N.P., per doz... 1 50 2 50
Bronze, Berlin, per doz... 2 75 3 25
Bronze Genuine, per doz... 6 00 9 00
Shutter, porcelain, F. & L. screw, per gross... 1 30 4 00
White door knobs—per doz. 1 25

HAY KNIVES.
Discount, 50 and 10 per cent.

LAMP WICKS.
Discount, 60 per cent.

LANTERNS.
Cold Blast, per doz... 7 50
No. 3 "Wright's"... 8 50
Ordinary, with O burner... 4 25
Dashboard, cold blast... 9 55
No. 0... 6 00
Japanning, 50c. per doz. extra.

LEMON SQUEEZERS.
per doz.
Porcelain lined... 2 20 5 60
Galvanized... 1 87 3 85
King, wood... 2 75 2 90
King, glass... 4 00 4 50
All glass... 1 20 1 30

LINES.
Fish, per gross... 1 05 2 50
Chalk... 1 90 7 40

LOCKS.
Canadian, dis. 45 p.c.
Russell & Erwin, per doz... 3 00 3 25
Cabinet.
Eagle, dis. 30 p.c.
Padlock.
English and Am., per doz... 50 6 00
Scandinavian, "... 1 00 2 40
Eagle, dis. 20 to 25 p.c.

MACHINE SCREWS.
Iron and Brass.
Flat head discount 25 p.c.
Round Head, discount 20 p.c.

MALLETS.
Tinsmiths', per doz... 1 25 1 50
Carpenters', hickory, per doz. 1 25 3 75
Lignum Vitae, per doz... 3 85 5 00
Caulking, each... 60 2 00

MATTOCKS.
Canadian, per doz... 8 50 1 00

MEAT CUTTERS.
American, dis. 25 to 30 p.c.
German, 15 per cent.

MILK CAN TRIMMINGS.
Discount, 25 per cent.

NAILS.
Quotations are: — Cut. / Wire.
2d and 3d... $3 35 $3 85
3d... 3 00 3 52
4 and 5d... 2 75 3 35
6 and 7d... 2 65 3 20
8 and 9d... 2 50 3 05
10 and 12d... 2 45 2 95
16 and 20d... 2 40 2 90
30, 40, 50 and 60d. (base)... 2 35 2 85
Galvanizing 2c. per lb. net extra.
Steel Cut Nails 10c. extra.
Miscellaneous wire nails, dis. 70 per cent.
Coopers' nails, dis. 30 per cent.
Flour barrel nails, dis. 25 per cent.

NAIL PULLERS.
German and American... 1 85 3 50

NAIL SETS.
Square, round, and octagon, per gross... 3 38 4 00
Diamond... 12 00 15 00

NETTING.
Poultry, 50 per cent. for McMullen's.

OAKUM. Per 100 lb.
Navy... 6 00
U. S. Navy... 7 25

OIL.
Water White (U.S.)... 0 16½
Prime White (U.S.)... 0 15½
Water White (Can.)... 0 15
Prime White (Can.)... 0 14

OILERS.
McClary's Model galvan. oil can, with pump, 5 gal., per doz... 0 00 10 00
Zinc and tin, dis. 50, 50 and 10.
Copper, per doz... 1 25 3 50
Brass, "... 1 50 3 50
Malleable, dis. 25 per cent.

GALVANIZED PAILS.
Dufferin pattern pails, dis. 50 and 10 p.c.
Flaring pattern, discount 45 per cent.
Galvanized washtubs, discount 45 per cent.

PIECED WARE.
Discount 40 per cent. off list, June, 1899.
10-qt. flaring sap buckets, dis. 45 p.c.
6, 10 and 14-qt. flaring pails, dis. 45 p.c.
Creamer cans, dis. 45 p.c.

PICKS.
Per doz... 6 00 9 00

PICTURE NAILS.
Porcelain head, per gross... 1 75 3 00
Brass head "... 0 40 1 00

PICTURE WIRE.
Tin and gilt, discount 75 p.c.

PLANES.
Wood, bench, Canadian dis. 50 per cent. American dis. 50.
Wood, fancy Canadian or American 7½ to 40 per cent.

PLANE IRONS.
English, per doz... 2 00 5 00

PLIERS AND NIPPERS.
Button's Genuine per doz pairs, dis. 37½–40 p.c.
Button's Imitation, per doz... 5 00 9 00
German, per doz... 0 60 2 60

PLUMBERS' BRASS GOODS.
Impression work, discount, 60 per cent.
Fuller's work, discount 65 per cent.
Rough stops and stop and waste cocks, discount, 60 per cent.
Jenkins disk globe and angle valves, discount, 55 per cent.
Standard valves, discount, 60 per per cent.
Jenkins radiator valves, discount 55 per cent.
" " " standard, dis., 60 p.c.
Quick opening valves discount, 60 p.c.
No. 1 compression bath cock... 2 00
No. 4 " " "... 3 00
No. 7, Fuller's... 2 50
No. 4½, "... 3 00

POWDER.
Velox Smokeless Shotgun Powder.
100 lb. or less... 0 85
1,000 lb. or more... 0 80
Net 30 days.

PRESSED SPIKES.
Discount 25 per cent.

PULLEYS.
Hothouse, per doz... 0 55 1 00
Axle... 0 22 0 33
Screw... 0 27 1 00
Awning... 0 35 2 50

PUMPS.
Canadian cistern... 1 80 3 80
Canadian pitcher spout... 1 40 2 10

PUNCHES.
Saddlers', per doz... 1 00 1 85
Conductors', "... 9 00 15 00
Tinners' solid, per set... 0 00 0 72
" hollow, per inch... 0 00 1 00

RANGE BOILERS.
Galvanized, 30 gallons... 6 50
" 35 "... 7 50
" 40 "... 8 50
Copper, 30 "... 22 00
" 35 "... 26 00
" 40 "... 30 00
Discount off Copper Boilers 10 per cent.

RAKES.
Cast steel and malleable Canadian list 50 and 10 p.c. revised list.
Wood, 25 per cent.

RASPS AND HORSE RASPS.
New Nicholson horse rasp, discount 60 p.c.
Globe File Co.'s rasps, 60 and 10 to 70 p.c.
Heller's Horse rasps, 50 to 50 and 5 p.c.

RAZORS.
per doz.
Geo. Butler & Co.'s... 8 00 18 00
Boker's... 7 50 11 00
Wade & Butcher's... 3 60 10 00
Theile & Quack's... 7 00 12 00
Elliot's... 4 00 18 00

REAPING HOOKS.
Discount, 50 and 10 per cent.

REGISTERS.
Discount... 40 per cent.

RIVETS AND BURRS.
Iron Rivets, discount 60 and 10 per cent.
Iron Burrs, discount 55 per cent.
Black and Tinned Rivets, 60 p.c.
Extras on Iron Rivets in 1-lb. cartons, ½c per lb.
Extras on Iron Rivets in ½-lb. cartons, per lb.
Copper Rivets & Burrs, 35 and 5 p.c. dis. and cartons, 1c. per lb. extra, net.
Extras on Tinned or Coppered Rivets ½-lb. cartons, 1c. per lb.
Terms, 4 mos. or 3 per cent. cash 30 days.

RIVET SETS.
Canadian, dis. 35 to 37½ per cent.

ROPE ETC. — Sisal. / Manila.
7-16 in. and larger, per lb. 9 13
⅜ in... 10 14
¼ and 5-16 in... 15
Cotton, 3-16 inch and larger 16½
" 5-32 inch... 21½
" ⅛ inch... 22½
Russia Deep Sea... 15½
Jute... 8
Lath Yarn... 9⅞
New Zealand Rope... 10½

RULES.
Boxwood, dis. 75 and 10 p.c.
Ivory, dis. 37½ to 40 p.c.

SAD IRONS. per set.
Mrs. Potts, No. 55, polished... 0 62½ 0 65
" No. 50, nickle plated 0 67½ 0 70

SAND AND EMERY PAPER.
Dominion Flint Paper, 47½ per cent.
B & A. sand, 40 and 2½ per cent.
Emery, 40 per cent.

SAP SPOUTS.
Bronzed iron with hooks, per doz... 9 50

SAWS.
Hand, Disston's, dis. 12½ p.c.
S. & D., 40 per cent.
Crosscut, Disston's, per ft... 35 0 55
S. & D., dis. 35 p.c. on Nos. 2 and 3.
Hack, complete, each... 0 75 2 75
" frame only... 0 75

SASH WEIGHTS.
Sectional, per 100 lbs... 2 75 3 00
Solid, "... 2 00 2 25

SASH CORD.
Per lb... 0 23 0 30

SAW SETS.
"Lincoln," per doz... 6 50

SCALES.
B. S. & M. Scales, 45 p.c.
Champion, 65 per cent.
Fairbanks Standard, 35 p.c.
" Dominion, 55 p.c.
" Richelieu, 55 p.c.
Chatillon Spring Balances, 10 p.c.

SCREW DRIVERS.
Sargent's, per doz... 0 65 1 00

SCREWS
Wood, F. H., bright and steel, 67½ and 10 p.c.
Wood R. H., " dis. 82½ and 10 p.c.
" F. H., brass, dis. 80 and 10 p.c.
Wood, R. H., " dis. 75 and 10p.c.
" F.H., bronze, dis. 70 p.c.
" R.H. " 65 p.c.
Drive Screws, 80 per cent.
Bench, wood, per doz... 3 25 4 00
" iron, "... 4 25 5 75

SCYTHES.
Per doz. net... 9 00

SCYTHE SNATHS.
Canadian, dis. 45 p.c.

SHEARS.
Bailey Cutlery Co., full nickeled, dis. 60 p.c.
Seymour's, dis. 50 and 10 p.c.

SHOVELS AND SPADES.
Canadian, dis. 40 and 5 per cent.

SINKS.
Steel and galvanized, discount 45 per cent.

SNAPS.
Harness, German, dis. 25 p.c.
Lock, Andrews'... 4 50 11 50

SOLDERING IRONS.
1, 1½ lb., per lb... 0 37
2 lb. or over, per lb... 0 34

SQUARES.
Iron, No. 493, per doz... 2 40 2 55
" No. 494, "... 3 25 3 40
Steel, dis. 60 and 5 to 50 and 10 p.c., rev. list.
Try and bevel, dis. 50 to 52½ p.c.

STAMPED WARE.
Plain, dis., 75 and 12½ p.c. off revised list.
Retinned, dis., 75 p.c. off revised list.

STAPLES.
Galvanized... 0 00 0 00
Plain... 0 00 2 45
Coopers', discount 45 per cent.
Poultry netting staples, 40 per cent.

STOCKS AND DIES.
American dis. 25 p.c.

STONE. Per lb.
Washita... 0 28 0 60
Hindostan... 0 06 0 07
" slip... 0 09 0 09
Labrador... 0 13
" Axe... 0 15
Turkey... 0 50
Arkansas... 0 00 1 50
Water-of-Ayr... 0 00 0 10
Scythe, per gross 3 50 5 00
Grind, per ton 15 00 18 00

STOVE PIPES.
Nestable in crates of 25 lengths.
5 and 6 inch Per 100 lengths... 7 00
7 inch " "... 7 50

ENAMELINE STOVE POLISH.
No. 4—3 dozen in case, net cash... $4 80
No. 6—3 dozen in case, "... 8 40

TACKS, BRADS, ETC.
Per cent.
Strawberry box tacks, bulk... 75 & 10
Cheese-box tacks, blued... 80 & 12½
Trunk tacks, black and tinned... 85
Carpet tacks, blued... 80 & 5
" " tinned... 80 & 10
" " (in kegs)... 40
Cut tacks, blued, in dozens only... 75 & 15
" ½ weights... 60
Swedes, cut tacks, blued and tinned—
In bulk... 80 & 10
In dozens... 75
Swedes, upholsterers', bulk... 85 & 12½
" brush, blued & tinned, bulk... 70
" gimp, blued, tinned and japanned... 75 & 12½
Zinc tacks... 35
Leather carpet tacks... 55
Copper tacks... 50
Copper nails... 52

Trunk nails, black	65 and 5
Trunk nails, tinned	65 and 10
Clout nails, blued and tinned	65 and 5
Chair nails	35
Cigar box nails	35
Patent brads	40
Fine finishing	40
Picture frame points	10
Lining tacks, in papers	10
" " in bulk	15
" " solid heads, in bulk	75
Saddle nails in papers	10
" " in bulk	15
Tufting buttons, 22 line, in dozens only	60
Tin capped trunk nails	15
Zinc glazier's points	5
Double pointed tacks, papers	90 and 10
" " " bulk	40

TAPE LINES.

English, ass skin, per doz	2 75	5 00
English, Patent Leather	5 50	9 75
Chesterman's each	0 90	2 85
" steel, each	0 80	8 00

THERMOMETERS.

Tin case and dairy, dis. 75 to 75 and 10 p.c.

TRANSOM LIFTERS.

Payson's per doz. 2 80

TRAPS. (Steel.)

Game, Newhouse, dis. 25 p.c.
Game, H. & N., P. S. & W., 65 p.c.
Game, steel, 72½, 75 p.c.

TROWELS.

Disston's discount 10 per cent.

German, per doz.	4 75	6 00

S. & D., discount 35 per cent.

TWINES.

Bag, Russian, per lb.	....	0 27
Wrapping, cotton, per lb.	0 22	0 25
Wrapping, mottled, per pack	0 50	0 60
Wrapping, cotton, 3-ply		0 20
" " 4-ply	....	0 25
Mattress, per lb.	0 33	0 45
Staging, "	0 27	0 35
Broom, "	0 30	0 55

VISES.

Hand, per doz.	4 00	6 00
Bench, parallel, each	2 00	4 50
Coach, each	6 00	7 00
Peter Wright's, per lb.	0 12	0 13
Pipe, each	5 50	9 00
Saw, per doz	6 50	13 00

ENAMELLED WARE.

White, Princess, Turquoise, Blue and White, discount 50 per cent.

Diamond, Famous, Premier, 50 and 10 p.c.

Granite or Pearl, Imperial, Crescent, 50, 10 and 10 per cent.

WIRE.

Brass wire, 50 to 50 and 2½ per cent. off the list.

Copper wire, 45 and 10 per cent. net cash 30 days, f.o.b. factory.

Smooth Steel Wire, base, $2.80 per 100 lb. List of eXtras: Nos. 2 to 5, advance 7c. per 100 lb.—Nos. 6 to 9, base—No. 10, advance 7c.—No. 11, 14c.—No. 12, 20c.—No. 13, 35c.—No. 14, 47c.—No. 15, 60c.—No. 16, 75c. EXtras net per 100 lb.: Coppered wire, 60c.—tinned wire, $2—oiling, 10c.—special hay-bailing wire, 30c.—spring wire, $1—best steel wire, 75c.—bright soft drawn, 15c.—in 50 and 100-lb. bundles net, 10c.—in 25-lb. bundles net, 15c.—packed in casks or cases, 15c.—bagging or papering, 10c.

ine Steel Wire, dis. 17½ per cent List of eXtras: In 100-lb. lots: No. 17, $5—No. 18, $5.50—No. 19, $6—No. 20, $6.65—No. 21, $7—No. 22, $7.30—No. 23, $7.65—No. 24, $8—No. 25, $9—No. 26, $9.50—No. 27, $10—No. 28, $11—No. 29, $12—No. 30, $13—No. 31, $14—No. 32, $15 No. 33, $16—No. 34, $17. EXtras net—tinned wire, Nos. 17-25, $2—Nos. 26-31 $4—Nos. 32-34, $6. Coppered, 5c.—oiling, 10c.—in 25-lb. bundles, 15c.—in 5 and 10-lb. bundles, 25c.—in 1-lb. hanks, 50c.—in ¾-lb. hanks, 75c.—in ½-lb. hanks, $1—packed in casks or cases, 15c.—bagging or papering, 10c.

Galvanized Wire, per 100 lb.—Nos. 6, 7, 8, $[illegible]—No. 9, $3.10—No. 10, $3.75—No. 11, $3.85 No. 12, $3.25—No. 13, $3.35—No. 14, $4.25—No. 15, $4.75—No. 16, $5.00.

Clothes Line Wire, 19 gauge, per 1,000 feet	....	3 90

WIRE FENCING. F.O.B. Toronto

Galvanized 4 barb, 3¼ and 5 inches apart	....	3 10
Galvanized, 2 barb, 4 and 6 inches apart	....	3 10
Galvanized, plain twist	....	3 10

Galvanized barb, f.o.b. Cleveland, $2 97½ in less than carlots, and $2.85 in carlots. Terms, 60 days or 2 per cent. in 10 days.

Ross braid truss cable	....	4 50

WIRE CLOTH.

Painted Screen, per 100 sq. ft., net. 1 35
Terms, 4 months, May 1; 3 p.c. off 30 days.

WRENCHES.

Acme, 35 to 37½ per cent.
Agricultural, 60 p.c.
Coe's Genuine, dis. 20 to 25 p.c.

Towers' Engineer, each	2 00	7 00
" S., per doz.	5 80	6 00
G. & K.'s Pipe, per doz.	....	3 40
Burrell's Pipe, each	....	3 00
Pocket, per doz	0 25	2 90

WRINGERS.

Leader	per doz.	3....
Royal Canadian	"	50 00
Royal American	"	50 00

Discount, 45 per cent.; terms 4 months, or 3 p.c. 30 days.

WROUGHT IRON WASHERS.

Canadian make, discount, 40 and 5 per cent

PERSONS addressing advertisers will kindly mention having seen their advertisement in Canadian Hardware and Metal Merchant.

SEND for specimen copy of Phillips' Monthl Machinery Register, containing over 5,000 entries of new and second-hand machinery of every description. The oldest established and most successful medium in the world. Established 25 years for the purpose of introducing those who have machinery for sale, to those who wish to buy, has a circulation of about 50,000 copies per annum, all over the world, and is used for continual reference by a large number of firms. It is consequently a most valuable advertising medium for all engineers and manufacturers. Subscription, per annum, price per copy, 6d. Sole Proprietor, CHAS. D. PHILLIPS, M.I.M.E., Newport, Mon., England. Telegraphic address "Machinery, Newport, Mon."

BRITISH BUSINESS CHANCES.

Firms desirous of getting into communication with British manufacturers or merchants, or who wish to buy British goods on the best possible terms, or who are willing to become agents for British manufacturers, are invited to send particulars of their requirements for

FREE INSERTION

in "Commercial Intelligence," to the Editor

"SELL'S COMMERCIAL INTELLIGENCE,"
168 Fleet Street, London, England.

"Commercial Intelligence" circulates all over the United Kingdom amongst the best firms. Firms communicating should give reference as to bona fides.

N.B.—A free specimen copy will be sent on receipt of a post card.

Buy the Best.

HERCULES

Sash Cord.

Star Brand Cotton Rope
Star Brand Cotton Clothes Lines
Star Brand Cotton Twine

For Sale by all Wholesale Dealers

BUSINESS NEWS

of any kind that is of value to business men supplied by our Bureau. We can give you market quotations from any town in Canada, reports from the city markets, stock quotations, etc. You can get commercial news from any Canadian paper through us.

Write us, giving us particulars of what you want and where you want it from, and we will quote you prices by return.

"Clippings from any Canadian paper on any subject."

CANADIAN PRESS CLIPPING BUREAU,

232 McGill Street, MONTREAL, QUE.
Telephone Main 1255.
10 Front St. East, Toronto. Telephone 2148.

75 YEARS. *ESTABLISHED 1825.* *75 YEARS.*

CELEBRATED **HEINISCH** SHEARS.

Tailors' Shears,
Trimmers, Scissors,
Tinners' Snips, etc.

ACKNOWLEDGED THE BEST.

R. HEINISCH'S SONS CO. NEWARK, N.J., U.S.A. NEW YORK OFFICE, 90 Chambers St.

Not connected with any Shear Combination.

CHAS. F. CLARK, President. JARED CHITTENDEN, Treasurer.

...ESTABLISHED 1849...

BRADSTREET'S

Capital and Surplus, $1,500,000. Offices Throughout the Civilized World.

Executive Offices: Nos. 346 and 348 Broadway, New York City, U.S.A.

THE BRADSTREET COMPANY gathers information that reflects the financial condition and the controlling circumstances of every seeker of mercantile credit. Its business may be defined as of the merchants, by the merchants, for the merchants. In procuring, verifying and promulgating information no effort is spared, and no reasonable expense considered too great, that the results may justify its claim as an authority on all matters affecting commercial affairs and mercantile credit. Its offices and connections have been steadily extended, and it furnishes information concerning mercantile persons throughout the civilized world.

Subscriptions are based on the service furnished, and are available only by reputable wholesale, jobbing and manufacturing concerns, and by responsible and worthy financial, fiduciary and business corporations. Specific terms may be obtained by addressing the Company at any of its offices. **Correspondence Invited.**

—OFFICES IN CANADA—

HALIFAX, N.S.	HAMILTON, ONT.	LONDON, ONT.	MONTREAL, QUE.
OTTAWA, ONT.	QUEBEC, QUE.	ST. JOHN, N.B.	TORONTO, ONT.
VANCOUVER, B.C.	VICTORIA, B.C.	WINNIPEG, MAN.	

THOS. C. IRVING, Gen. Man. Western Canada, Toronto. JOHN A. FULTON, Gen. Man. Eastern Canada, Montreal.

Awarded a Gold Medal at **PARIS EXPOSITION** for superiority. That's proof enough of their quality, and clearly shows that they are the best.

Send for Catalogue and . . . . Price List.

The Bailey

Cutlery Co.
BRANTFORD, ONT.

Dominion Pattern

Cow Tie and Stall Fixture

The special features of the tie and stall fixture are well shown in the illustration. As will be noticed the chain is very short, which prevents all danger of entanglement with the animal's foot. At the same time the form of the fixture is such that great freedom is allowed to the head. Because of the short chain this tie is much cheaper than the ordinary patterns.

The stall fixture is made from a tough quality of steel and is very strong. Also, owing to its circular cross-section, it is exceedingly rigid. Its simplicity, convenience, cheapness, and ease of attaching make it very popular with cow tie users.

This form of tie and stall fixture are sometimes called Niagara pattern.

American or Flat Link Chain,

for years the standard cow tie chain in "the States," is now rapidly coming in favor in Canada. Its short link, handsome appearance and smooth surface—which cannot injure the animal's neck—make it superior to all other styles of chain for cow ties.

For sale by all Jobbers; manufactured by

ONEIDA COMMUNITY, Limited, NIAGARA FALLS, ONT.

CANADIAN HARDWARE AND METAL MERCHANT

The Weekly Organ of the Hardware, Metal, Heating, Plumbing and Contracting Trades in Canada.

VOL. XIII. MONTREAL AND TORONTO, FEBRUARY 16, 1901. NO. 7

Garden Hose

Discounts for the new season now out.

WRITE FOR PRICES.

* * *

CANADIAN RUBBER CO.,

Montreal. Toronto. Winnipeg.

THE NEW BALDWIN

DRY AIR CLEANABLE

REFRIGERATOR.

135 Modern Varieties. Ash, Oak and Soft-wood Finishes

METAL, PORCELAIN, SPRUCE LININGS.

BALDWIN

Positive Circulation—Sanitary—Odorless.

Latest Cleanable Features—The Strongest and Best System of Patent Removable Metal Air-Flues.

Air-Tight Lever Locks

Ball-Bearing Casters.

Swing Base—in and out.

Rubber around Doors and Lids, making them doubly air-tight.

Handsome Designs.

Moderate Prices.

Built in the newest, largest and best equipped refrigerator plant in the East run all the year round on refrigerators exclusively; stock goods; special refrigerators and coolers in sections.

Handsome Trade Catalogue Ready.

Baldwin Refrigerator Co.,

BURLINGTON, VERMONT.

SOME OF THE NEWER "YANKEE" TOOLS

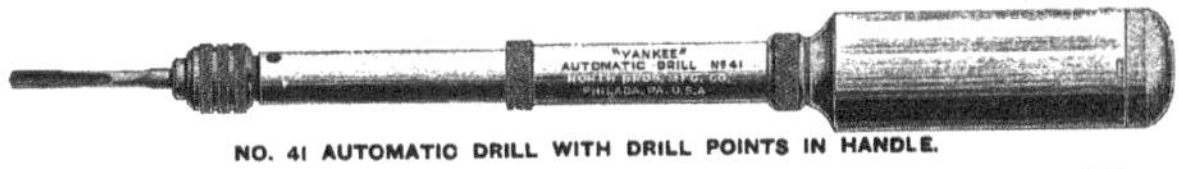

NO. 41 AUTOMATIC DRILL WITH DRILL POINTS IN HANDLE.

NO. 42 AUTOMATIC DRILL WITH DRILL POINTS IN BOX.

NO. 50 RECIPROCATING DRILL, FOR WOOD OR METALS.

Sold by Leading Jobbers throughout the Dominion.

NORTH BROS. MFG. CO.,

Philadelphia, Pa., U. S. A.

Woman is Responsible

for the sale of ice cream freezers.

Fifteen million women have considerable influence.

Ten thousand of them wrote to us in three months.

We are talking to them all the time—winter and summer.

They are all familiar with the Peerless Iceland Freezer.

Reports from them prove that we are right, and that one motion is enough.

We want your help in distributing Peerless Iceland Freezer.

DANA & CO.,
Cincinnati.

New York Branch, 10 Warren St.
San Francisco Branch, 105 Front St.

The Fairbanks Standard Scales

GIVE UNIVERSAL SATISFACTION.

We have Scales to suit all purposes.

"FAIRBANKS STANDARD"

TAKE NO SUBSTITUTE.

Write us for complete Catalogue and Prices.

THE FAIRBANKS COMPANY

749 Craig Street,

MONTREAL.

THE "WOODYATT" LAWN MOWER

Is still being made, to maintain the reputation it has gained during past seasons,

THOROUGHLY SATISFACTORY.

Only the best quality of material is used in its construction.

10½-Inch Drive Wheel. 12, 14, 16, 18, 20-Inch Cut.

Manufactured and Patented by . .

A. R. WOODYATT & CO., Guelph, Canada.

SUGAR MAKERS' SUPPLIES

Sap Buckets.

Extra deep and straight. Three sizes. They possess many advantages over the Ordinary Flaring Buckets—being small in diameter they do not catch the rain or snow, and, as they are very deep, they hang perpendicularly, and, consequently, will not overflow until full. Covers supplied if required. They nest close for shipping or storing.

We can supply the Ordinary Flaring Sap Pails.

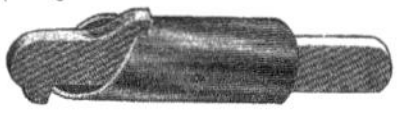

E. T. Sap Spouts.

Made of Retinned Steel. Strong and Durable. Only require a ⅜-in. hole in a tree. It does not cover the inside surface of the hole, therefore, a larger amount of Sap is obtained.

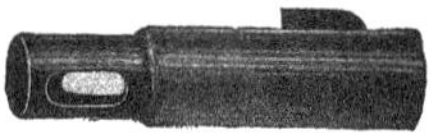

Maple Leaf Sap Spouts.

Made in Bronzed Steel. Require a ⅜-in. hole. Has a shoulder which prevents it being driven in too far. The hole in the tree is not exposed to wind and snow, consequently, sap will flow longer.

Syrup Cans.

Plain or Decorated. Made in ¼, ½ and 1 gallon sizes. Either Wine or Imperial Measure.

Kemp Manufacturing Co., Toronto, Canada.

HARDWARE AND METAL

VOL. XIII. MONTREAL AND TORONTO, FEBRUARY 16, 1901. NO. 7.

President,
JOHN BAYNE MacLEAN,
Montreal.

THE MacLEAN PUBLISHING CO.
Limited.

Publishers of Trade Newspapers which circulate in the Provinces of British Columbia, North-West Territories, Manitoba, Ontario, Quebec, Nova Scotia, New Brunswick, P.E. Island and Newfoundland.

OFFICES

MONTREAL - - - - - - 232 McGill Street, Telephone 1255.
TORONTO - - - - - - 10 Front Street East, Telephone 2148.
LONDON, ENG. - - - - 109 Fleet Street, E.C., J. M. McKim.
MANCHESTER, ENG. - - - 18 St Ann Street. H. S. Ashburner,
WINNIPEG - - - - - Western Canada Block. J. J. Roberts.
ST. JOHN, N.B. - - - No. 3 Market Wharf, I. Hunter White,
NEW YORK. - - - - - - 176 E. 88th Street.

Subscription, Canada and the United States, $2.00.
Great Britain and elsewhere 12s.

Published every Saturday.

Cable Address { Adscript, London. Adscript, Canada.

BRITISH AND UNITED STATES EXPORTS TO CANADA.

THE monthly report of the Canadian Department of Trade and Commerce gives the returns for the first five months regarding the exports to Canada from both the United States and Great Britain. The figures are taken from the official returns of Great Britain and the United States respectively.

The exports from Great Britain to Canada during the five months aggregated $9,345,856, and from the United States $38,105,096. This was an increase, compared with the same period in 1899, of nearly $6,000,000 in the figures of the latter country, although of less than $1,000,000 compared with 1898. The figures relating to the British exports show a decline of over $800,000, compared with the same period in 1899, and an increase of nearly $1,000,000, compared with 1898.

The figures of most interest to our readers are those relating to hardware and metals. As enumerated, for the five-month period ending November 30 for the past three years, they are as follows :

FIVE MONTHS' EXPORTS FROM GREAT BRITAIN.

	1898.	1899.	1900.
Hardware, unenumerated..........	$ 59,616	$ 48,884	$ 56,025
Cutlery..........	143,954	114,720	129,137
Pig iron..........	24,832	152,198	84,967
Bar, angle, bolt and rod..........	23,966	132,431	16,109
Railroad iron, all sorts..........	50,617	643,304	169,102
Hoops, sheets, boiler and armor plates..	217,798	384,141	218,762
Galvanized sheets...	200,019	188,776	156,926
Tinplates and sheets	416,102	770,761	892,824
Cast and wrought iron and all other manufactures.....	75,684	215,864	49,295
Old iron for remanufacture..........	2,428	29,520	2 025
Steel, unwrought....	93,701	534,398	115,020

FIVE MONTHS' EXPORTS FROM THE UNITED STATES.

	1898.	1899.	1900.
Builders' hardware, saws and tools........	$ 260,644	$ 261,266	$ 298,224
Steelbars or rails for railways ..	1,083,678	1,158,937	2,432,319

We regret that we cannot give the returns in greater detail, but we are reproducing them as they appear in the official returns before us and cannot well do more. Grouped as they are, however, it is evident that the trend of trade with Canada during the first five months of the present fiscal year has been more favorable to United States than to British products.

A DANGEROUS EXPERIMENT.

It is a fallacious idea to think you can all the time palm off on your customers a low-quality article at a low price and boast that its quality is high.

Part of the time you may be able to do it, but even that is dangerous, for it may prevent your being in business all the time.

HEAVY FIRE LOSSES.

FIRE losses in the United States and Canada during the month of January were unusually heavy.

According to the figures compiled by The Journal of Commerce, New York, the losses for the month aggregated $16,574,950, compared with $11,755,300 for January, 1900, and $10,718,000 for January, 1899.

Unfortunately, nearly 20 per cent. of the total loss was contributed by the three big fires in Montreal, the aggregate losses of which are placed at nearly $3,300,000.

The city with the next largest fire during January was New York, where the loss was $1,140,000.

With losses so heavy, one is led to wonder how much less they might have been had more precaution against fire been taken. One thing is certain, employes are often not as careful as they should be, and if the danger arising therefrom was more frequently impressed upon everyone in and about stores, warehouses and factories we believe fires would be less frequent than they now are.

EXPECT TO PROGRESS.

The merchant or clerk who believes he has reached the limit of the possibilities of his position is almost sure to find out that he has done so. On the other hand, the business man, whether he be employer or employe, who confidently looks forward to, and prepares for, growth or progress, finds new possibilities continually opening up before him. The habit of seeking about for opportunities for development is bound to bear fruit, sooner or later. It is essential, however, that preparation should be made to fill creditably the larger field one hopes to occupy.

STEEL SHIPBUILDING OUTLOOK IN CANADA.

CANADA once occupied an important place among the nations as a builder of vessels. That was in the days before wooden ships were supplanted by those of iron and steel. Now, the industry has shrunk out of all proportion to what it once was.

Compared with 27 years ago the number of vessels built in Canada is over 43 per cent. less, and in tonnage there is a decline of 88 per cent. Twenty-five years ago we sold to other countries 160 vessels, with an aggregate tonnage of 64,134 and a value of $2,189,270. In 1899 the number was only 14, the tonnage but 7,562 and the value $126,466. Here is a decline of 91 per cent. in the number of vessels, 88 per cent. in tonnage and 94 per cent. in value.

But most deplorable as has been the decline in the shipbuilding industry in Canada, we are not without hope for it in regard to the future.

The iron industry in this country is at a stage of development which scatters to the wind all doubt which may previously have existed as to its future. And concomitant with the development of the iron industry will be the growth of our shipbuilding industry.

During the last two or three years, the Bertram Company in Toronto have built a number of steel steamers and canal barges. Some of the former are among the finest on the lakes, and only a few days ago the keel was laid for a steel passenger steamer which will be larger than any yet constructed on Lake Ontario, she having a length of 340 ft. over all and will cost nearly $500,000. Then, on the upper lakes, there is the Cramp Company now being organized, which, in addition to its iron and steel plant to be erected at Collingwood, Ont., has already a large steel steamer in course of construction.

In the palmy days of shipbuilding in Canada, the centre of activity was in Nova Scotia and New Brunswick. And in the development of the steel shipbuilding industry which Canada now promises to experience, these Provinces are likely to be by no means unimportant factors.

Both Halifax, N.S., and St. John, N.B., have for some months been the centres of agitation for the establishment of steel shipbuilding yards. The Provincial Government of New Brunswick has already intimated its intention to assist the scheme as far as St. John is concerned, and only on Monday last there was a conference in St. John between the representatives of the City Council and members of the Provincial Cabinet with a view to arranging the nature of the assistance. The Dominion Steel Co. also appears to be interesting itself in the project. In Nova Scotia the agitation has so far advanced that application is to be made at the present session of the Dominion Parliament to incorporate the Dominion Shipbuilding Co.

The large iron and steel plant which is being erected at Sydney should greatly facilitate steel shipbuilding in the Maritime Provinces.

REFUSE DEFACED COINS.

A NUISANCE, which is steadily growing, and which, of late, has obtained much attention from business men, is the large number of mutilated coins that are in circulation. This increase is due to the fact that no organized attempt has been made to put a stop to it. Of course, it is illegal to deface coins in any way, but that does not deter the mischievously-inclined from doing that which the law declares they shall not do.

The street railway conductors in Hamilton have petitioned the Government in regard to the matter, and the retail grocers' association, of that city, has decided to back up that of the conductors with another petition and ask to have the mutilated coins called in.

In the House of Commons on Tuesday last the question came in for a little attention, and, in reply to a question, Hon. W. S. Fielding, Minister of Finance, said there was no provision in the Canadian law for the calling in of defaced or mutilated coins, and the Government had no authority at present to take such action. Persons who clipped or defaced coins or who altered such coins were liable to imprisonment. Defaced coins were not currency and not legal tender, and should be refused by everybody. Whether there should be an amendment in the Currency Act he said was a matter for consideration.

The position taken by the Minister of Finance clears the way for merchants to be decidedly firm in rejecting defaced coins. A coin with a hole caused by some silver having been bored out of it, or with such a hole plugged with tin or lead; a coin with its edge cut away or its features worn off by much handling is not currency, and so should not be accepted as money.

If all persons were firm in adhering to this ruling, those who bore holes in silver coins for the sake of the silver they thus secure would find their business a poor one, and those individuals who deface coins for the mere "fun of the thing" would have to themselves bear the loss which they occasion.

The subject is one which should be taken up by boards of trade and business men's organizations of various kinds throughout the country. This will put a stop to the practice of defacing of coin currency more quickly than ever legislative enactments can.

THE OFFICERS OF THE WINNIPEG BOARD OF TRADE.

MR. WILLIAM GEORGESON, who was, on February 5, unanimously, elected president of the Winnipeg Board of Trade, has been a resident of the "Prairie City" for over 20 years, and during all that time has been closely identified with the wholesale interests there.

In the early eighties he was manager for Thompson Codville Co.'s branch in Winnipeg, and, after the death of Mr. Thompson, sr., when the firm was reorganized as Codville & Co., Mr. Georgeson became a partner in the concern.

No man in the West is better posted on trade conditions, and, in addition to his actual knowledge, he has a charming, tactful manner and a pleasing personal appearance.

The vice-president, Mr. John Russell, is well and favorably known in Manitoba, while Mr. Andrew Strange, the treasurer, is one of Winnipeg's pioneers, whom not to know is to argue yourself unknown.

The office bearers of the board for the first year of the new century are men in every way calculated to add to the already established prestige of the Winnipeg Board of Trade.

TRADE IN COUNTRIES OTHER THAN OUR OWN.

A REMARKABLE demand is in progress for both black and galvanized sheets, not only from the mills, but also from local jobbers' stocks. The demand for black sheets is especially peculiar, as at this season black sheets are ordinarily quiet. The demand is very heavy for galvanized sheets, and stocks still continue inadequate to the demand. It might be expected, under the circumstances, that jobbers' prices on galvanized sheets would be advanced, as they say they have seldom sold on such a small margin.—Iron Age.

WIRE PICTURE CORD UNSETTLED.

The market for wire picture cord is in a very unsatisfactory condition. There appears to be no agreement among the manufacturers in regard to the list prices, some of them using the revised list of October 2, 1900 and others the old and lower list. Some of the prices reported are exceedingly low, and referred to as down to the cost of the wire. There is also a good deal of complaint in regard to the quality of the goods—the gauge of the wire, number of strands, shortage in measurement, quality of coating, etc. There is evidently an opportunity here for the trade to give careful attention to the matter of price and grade of goods purchased.—Iron Age.

PIG IRON IN GREAT BRITAIN.

In the pig iron market, business is exceedingly dull, and values have been falling away during the week. Warrants have gone down a great deal more, so that makers' iron, Cleveland, which we quoted at 48s. 7d. in our last, now stands at 47s.; while, in Glasgow, Scotch finish sells at 52s. 7½d., as against 56s. 8d. a week ago. In Middlesbrough, there seems to be a very apathetic feeling, and buying is confined to urgent requirements, because there is apparently no confidence in the maintenance of present prices, which are 47s. for No. 3 and 63s. for hematite. A similar feeling is noted elsewhere, and altogether the trade is in a very unsatisfactory state. At the present time, there should be a good deal of buying for spring delivery, but with a falling market this cannot, of course, be expected, and the outlook at the moment is certainly not very encouraging.—Iron and Coal Trades Review, February 1.

TINPLATE TRADE IN ENGLAND.

Sim & Coventry, Liverpool and London, in their report in regard to the tinplate market, say: "The year 1900 cannot be looked back upon with unmixed feelings of satisfaction so far as the tinplate trade is concerned. The year opened with a feeling of doubt as to whether the high prices then ruling could be sustained for a much longer period, the boom in the market having already lasted considerably longer than on some previous occasions. The prophecy of a well-known tinplate maker that before 1899 was out cokes would be up to 20s. had been lamentably falsified, and, as a matter of fact, the result showed that the opening year found the upward movement well nigh spent. The average price for cokes in January may be put down at 15s. 1½d., and although during February a sharp spurt took place until 15s. 9d. to 15s. 10½d. was paid, and this level was practically maintained for a month or six weeks, it proved to be the last dying struggle of the boom, and from that time dates a decline in values which has continued practically without break to the close of the year, the end of December finding the price of cokes at 12s. 10½d. f.o.b. Wales.

SCREWS LOWER IN THE UNITED STATES.

For some time the screw market has been irregular, with a decided drift toward lower prices. So much so has this been the case that the base discount of 87½ per cent. agreed upon by the manufacturers became largely nominal, in view of the series of extras which were more or less freely given. To meet this condition of things reduced prices were announced under date February 1 in which the base discount was made 90 per cent. The manufacturers are anticipating that this discouut will be pretty closely adhered to, so that the announced prices will represent the market as usual.—Iron Age.

MANUFACTURED IRON AND STEEL IN ENGLAND.

For finished material the market is dull at present, and, notwithstanding the recent reduction in prices and the general anticipation that bottom had been reached, some further declines are reported from various districts. In Middlesbro', for instance, plain iron columns have been put down 10s., while chairs have fallen to £3 17s. 6d. and £4 2s. 6d. Common bars are also 10s. lower, and best iron is correspondingly cheaper. These reductions, however, seem to be of little use in stimulating business. Lancashire makers are also feeling the prevailing depression, and although £5 5s. is quoted for steel billets, they have difficulty in realizing that price. The average selling price of Scotch iron during November and December is returned at £7 13s. 8d. per ton, which will have the effect of reducing wages north of Tweed. Mr. Waterhouse's statistics give the average selling price in the North of England during the same months as £8 5s. 2d., as compared with £8 5s. 11d. in September and October. This leaves wages unchanged.—Iron and Coal Trades Review.

NEW YORK METAL MARKET.

There is a fairly active demand for pig iron for prompt delivery, but with the current production closely concentrated there is not much stock to be picked up. There is little show of interest in futures as yet, and in some quarters there is increasing pressure to sell. This is particularly true of Southern foundry, which, in some instances, it is understood, has been offered at a concession from quoted prices. Pittsburg advices indicate a firm tone to the market for Western iron, but, reports from Philadelphia complain of dull trade, and intimate that the Carnegie affair has had a tendency to increase the feeling of depression, as it is not apparent how it will help to enlarge the consumption. There is something of a disposition on the part of buyers to await the outcome of the Carnegie-Morgan negotiations before taking further steps to cover their requirements, and business in the different lines is less active. There is a very firm tone to the market throughout, and in some quarters there is still talk of a probable early advance in the price of billets and rails at least.

TINPLATE—In a jobbing way business is of fair proportions, and this, together with regular deliveries, gives the market a moderately active appearance. The London market is slightly higher than a week ago.

PIG TIN—The London market had advanced 10s. up to noon on Tuesday, but after that a reaction set in and prices continued to decline until the close to-night, when they were 12s. 6d. below the last quotations of Monday and £1 2s. 6d. under yesterday's highest. Following the holiday of Tuesday there was a better feeling in the New York market, and despite the decline in London a much firmer tone was developed, buyers increasing their bids to 26.45c., which they offered for both spot February and March delivery, without bringing out sellers. The Kensington, sailing from this port to-day for Southampton, took 180 tons, part of the recent purchases made in this market for London.

COPPER—The market remains quiet, but the tone is steady and prices are maintained at 17c. for Lake Superior and 16⅝c. for electrolytic and casting. London has declined 10s. since Monday, the previous advance being thus lost.

PIG LEAD—The market remains dull, with prices somewhat nominal. The trust continues to quote on the basis of 4.37½c. for lots of 50 tons or over. In London, the price of soft Spanish has declined 2s. 6d. since Monday.

SPELTER—The market remains very dull, with the tendency in buyers' favor. Nominal quotations at the close were 3.90 to 3.95c. London shows a decline of 5s.

ANTIMONY—Regulus remains quiet, but steady, at the range of 8¾ to 10¼c., as to brand and quantity.

OLD METALS—The market is steady, though at the moment there is little demand.—New York Journal of Commerce, February 14.

TRADE CONDITIONS IN MANITOBA.

THE following is a part of an interesting address delivered by Mr. D. K. Elliott, the retiring president, at the recent annual meeting of the Winnipeg Board of Trade:

The business of the year past has in many respects fallen short of our expectations. The increased area placed under crop last spring inspired us with the reasonable hope of a corresponding increase of business, but the early drought, which affected every part of the country, retarded the growth of the grain, and this, followed by the heavy autumn rains, seriously

REDUCED THE YIELD OF ALL CEREALS.

The shortage of crop, and consequent decrease in business, has told heavily against the business interests of this city, which are so largely dependent upon the wheat crop of the West. When it is remembered that our total export of wheat for the crop of 1900 will not exceed 17,000,000 bush., and that principally in the lower grades, against 30,000,000 bush., largely of No. 1 hard, of the 1899 crop, it will be readily seen that the purchasing power of the farmer and his means for paying off existing liabilities have been very much curtailed.

It is gratifying, however, in spite of this to note that there is

NO TRACE OF PANIC,

nor any doubt in the minds of our people as to the future. All feel that the country is no longer on trial, but that stability and confidence are firmly established, and the past year may have its compensations in lessons of carefulness and frugality, taught to everyone, merchants and farmers alike. A reference to the last crop bulletin issued by the Provincial Government shows that farmers, too, are confident of the future, as is evidenced by the fact that there are already prepared for the crop of 1901, apart from the work of the coming spring, over one and a half millions of acres of land. I give the following extract from Dunn's Review of January 5, 1901, referring to failures for the past year: "Considering the severe loss to wheat-growers in Manitoba, that Province makes a splendid exhibit."

One of the best indications of the

FINANCIAL STRENGTH

of a country is the view taken of it as a field for investment by loan companies and capitalists, and I learn on good authority that money is being loaned on farm security as freely as ever, both in Manitoba and the Northwest Territories, and indeed that the West is looked upon as among the best and safest parts of the Dominion for such investments.

WINNIPEG BANK CLEARINGS

continue to show a satisfactory volume, and although the average decrease in Canada for the past year is nearly 4 per cent., that of Winnipeg is only about ¾ per cent. During the past year two of our leading banks, the Canadian Bank of Commerce and the Dominion Bank, have erected premises that would do honor to any city, and the Merchants Bank of Canada have now in the hands of their contractors a magnificent building of eight storeys, that will be, for many years to come, if one may judge from the plans, a prominent exhibition of the confidence felt in our country by the leading monetary institutions of Canada. In addition to these a number of fine warehouses and manufacturing blocks have been erected, and we believe the coming year will see a material addition to their number.

POSTAL MATTERS.

One important feature in the development of a new country is the necessity for a liberal and rapid expansion of the postal service, and it is a source of satisfaction that this Department of the Federal Government has shown a proper conception of the growth of the country and the requirements of its business interests. Daily mails have followed as quickly as possible daily train service, and the representations of your postal committee have been met by a ready response from the Department. A glance at statistics shows that there were, in Manitoba and the Territories, in the year 1880, 147 post offices; 10 years later, the number had increased to 523, and, in 1900, to the large number of 869.

IMMIGRATION.

The work of this Department of the Dominion Government is being prosecuted vigorously, and its splendid results are shown in the large number of 38,324 added to the population of the West during the year 1900. Of these, 14,000 were Canadians, and 5,136 came from the United States, and are principally well-to-do farmers, who have a thorough knowledge of the requirements of the country and can adapt themselves to the state of affairs existing here. They bring with them a large amount of money and personal effects, and, in most cases, either enter directly upon land previously selected by them or purchase improved farms and engage at once in the cultivation of their properties

It is reported by the Department that the foreigners who have come to our country are rapidly becoming self-sustaining, and exhibit a keen desire to remain upon and cultivate their farms, and it is evident that they will prove a valuable addition to our population.

There have been 8,827 land entries and sales during the year, aggregating nearly 2,000,000 acres, and, in addition to this, the C.P.R. company have made 2,283 contracts, containing 432,000 acres and amounting in cash to $1,377,715.48.

RAILROAD MATTERS.

While the year 1900 has not seen a marked activity in railway building, yet some important branches have extended in different parts of the country, opening up new districts and giving needed accommodation to others.

Commencing east of us, we find that the C.P.R. company have built a small spur east of Rat Portage, to one of the well-known mines in that district, and that the Lac du Bonnet branch, of about 22 miles, has opened up communication with that largely-timbered country drained by the English and Winnipeg rivers, both of which are tributary to this beautiful lake. Draining tile and superior brick for building and paving purposes are being manufactured, thus adding another to the large and varied number of our industrial concerns. Coming to the prairie country, a line is being extended from McGregor to a point on the Great North West Central, and the Pipestone branch has been carried forward 100 miles into the well-known, fertile and finely-situated Moose Mountain district. It has been found necessary by the C.P.R. to build a substantial steel bridge, supported on masonry piers, across the Red river, on account of the increasing traffic and larger engines used, and I have good reason to believe this will be followed soon by a large and commodious depot. It seems unfortunate that some reasonable arrangement could not have been arrived at between the city council and the C.P.R. and been ratified by the ratepayers for the building of a subway on Main street. The condition of things now existing is anything but satisfactory to either the railway company or the citizens, and, in addition to this, it has, I believe, caused the postponing for some time longer the erection of a splendid hotel in this city. The volume of traffic, both through and local, is increasing so rapidly that the C.P.R. has practically decided to establish, early next summer, two daily trans-continental trains, one a fast limited, making but few stops, and the other to accommodate local traffic.

Great progress, too, has been made by the Canadian Northern Railway Co., who, during the year 1900, constructed 224 miles of railway, and who expect before the end of the present year to have their line in operation from Port Arthur to Winnipeg, and from Winnipeg to a point within a very short distance of Prince Albert. They are now advertising for tenders for the construction of a magnificent steel bridge, with a draw span of 380 feet, to cross the Rainy river, and tenders are likely to be called for in a few days for a bridge across the Red river at Winnipeg. With the completion of these works, and the additional mileage to be constructed this year, this company expect to have not less than 1,100 miles of railway in operation, thus bringing them into the rank of the

THIRD RAILWAY IN CANADA.

It is pretty generally understood that a large portion of the fine timber, through which the road runs, will be brought into this city, and lumber mills established here for its manufacture, thus giving employment to a large number of men. Mining interests, too, should be greatly benefited, as the Canadian Northern runs largely through mining country tributary to the Seine river, and the improved transportation facilities should be of great value, both in shipping in machinery and exporting the products of the mines.

Thus it will be seen that very important railway works have been carried on during the year, and, if we are to believe current reports, even more will be accomplished before another crop is harvested.

CHAMBER OF MINES.

An institution which has come into considerable prominence during the past year is the Chamber of Mines, which was formed in this city early last spring, and is composed of thoroughly representative men, from all parts of Canada. The disinterested work of this voluntary association is meeting with favorable results; reliable information is being obtained and disseminated in the best possible form throughout the world, and this, doubtless, will be an important factor in inducing capitalists to invest in and develop the great mining region between this city and Lake Superior. The value of the work being accomplished has lately been recognized in the form of liberal grants and donations in support of the movement made to the chamber by the Dominion Government, the Government of Ontario, the Canadian Pacific Railway Co. and other influential corporations.

ELECTRIC CARS IN ENGLAND.

A London paper, calling attention to the fact that during the past year over 6,000 electric tramway cars were manufactured in America, says: "The magnitude of the business will be better understood by recalling the fact that there are probably not more than 6,000 tramway cars of all kinds in use in the United Kingdom to-day. Unfortunately, from the standpoint of British workmen, our manufacturers are not yet able, says The Sheffield Telegraph, to supply the increasing demand for cars, and from all parts of the country complaints come of great delay in filling orders. One large corporation was recently obliged to delay the opening of its electric tramways several months because cars could not be got." As to the iron trade, it says: "Short time is ruling in the steel and iron trades throughout the North and Midlands in consequence of the falling off in orders. A dozen forge works have closed down in the Midlands, and at many others the number of hands is considerably reduced. To ameliorate these conditions cheaper raw material—coal and pig iron—is regarded as imperative. With prices reduced to the 1896 7 level it is claimed that American manufacturers would be unable to compete with the British market."

BUSINESS CHANGES.

DIFFICULTIES, ASSIGNMENTS, COMPROMISES.

ARTHUR HOTTE, general merchant, St. Cyrille de Wendover, Que., has assigned, and is offering 50c. cash on the dollar.

Joseph Bernier, carriagemaker, L'Islet, Que., has assigned.

John D. Morrison, general merchant, Milan, Que., has assigned.

Chapleau & Leboeuf, contractors, Montreal, have consented to assign.

Mrs. E. G. E. McKee, general merchant, Orton, Ont., is offering to compromise.

Assignment has been demanded of F. X. Julien, general merchant, Lambton, Que.

Alf. Boulanger, general merchant, St. Eugene (L'Islet), Que., is offering 17c. on the dollar, cash.

A meeting of the creditors of G. Gibeault, general merchant, St. Lucie de Doncaster, Que., has been held.

PARTNERSHIPS FORMED AND DISSOLVED.

A. R. Pruneau & Co., coal dealers, Quebec, have dissolved.

Partnership has been registered by Roy L. Caron & Co., sawmillers, St. Julie, Que.

Hawkesworth & Springford, general merchants, Morris, Man., have dissolved.

Partnership has been registered by Lapointe & Beauchamp, general agents, etc., Montreal.

Partnership has been registered by Lemay & Marchand, general merchants, Shawenegan Falls, Que.

Boese & Unruh, general merchants, Rosthern, N.W.T., have dissolved; J. J. Boese continues.

Rosen & Dugan, general merchants, Selkirk West, Man., have dissolved; Jacob Rosen continues.

SALES MADE AND PENDING.

Aaron Lewis, tinsmith, Victoria, B.C., is selling out by auction.

The assets of L. J. Desilets, general merchant, St. Gertrude, Que., are to be sold.

The assets of X. Savard, general merchant, St. Felicien, Que., have been sold.

John Hiles, general merchant, Dungannon, Ont., is advertising his business for sale.

The assets of P. Denis, general merchant, St. Cesaire, Que., are to be sold on February 15.

The stock of the estate of J. W. Schoeman, hardware dealer, Virden, Man., is advertised for sale.

The Sackville Machine and Foundry Co., Sackville, N.B., are offering their business for sale by public auction.

Hugh McDonald, general merchant, Little Glace Bay, N.S., is advertising his property and plant for sale by auction March 4.

The stock, etc., of the estate of Johnson Bros., hardware dealers, Seaforth, Ont., is advertised for sale by auction on February 21.

CHANGES.

John Hisler, blacksmith, Okotoks, N.W.T., has sold out to Robert Graham.

Charlotte Ward, harness dealer, Seaforth, Ont., has sold out to Andrew Oke.

John Malcolm, hardware dealer, Rosebank, Man., has sold out to F. & E. Leggatt

Tocher & Klump, general store, Wapella, Man., have sold out to G. E. Nugent & Co.

D. C. Crosby, general merchant, Port Maitland, N.S., has sold out to James S. Gray.

Cattle & Porter, harness dealers, Ridgetown, Ont., have been succeeded by John Porter.

Miss A. Chalifoux, general merchant, Wendover, Ont., has been succeeded by W. J. Storey.

C. M. Sherwood, general merchant, Woodstock, N.B., is closing his branch store in that place.

A. J. McPherson, general merchant, Head of Millstream, N.B., has sold out to S. H. White & Co.

The Connors Bros., Limited, general merchants, etc., Black's Harbor, N.B., have applied for incorporation,

Thomas G. Holmes, founder, Clarksburg, Ont., has been succeeded by Ferguson & Rogers.

James E. Mattinson, sawmiller, Lower Stewiacke, N.S., has sold out to G. M. Mattinson.

M. W. McKim, dealer in agricultural implements, Elkhorn, Man., has been succeeded by McKim & Duxbury.

The assets of the estate of The W. F. Horton Co., bicycle dealers, etc., London, Ont., have been sold to S. V. Horton.

Armstrong & Morrison, manufacturers of steel pipes, etc., Vancouver, have sold out to The Vancouver Agency, Limited.

FIRES.

J. McLellan, harness dealer, Allenford, Ont., has been burned out.

B. Bolliver, sawmiller, Baker Settlement, N.S., has been burned out.

Jolly & Donaldson, general merchants, Allenford, Ont., have been burned out. The insurance is light.

DEATHS.

John Irvin, coal dealer, etc., Brampton, Ont., is dead.

Pearson, Covert & Pearson, Hamilton, have made application for the incorporation of the Dominion Shipbuilding Co., for the purpose of building, repairing, equipping, operating and maintaining ships.

BUSINESS FIRMS IN THE STATES.

THE following is an extract from the address of President Hanson, at the annual convention of the National Retail Grocers' Association of the United States, at Detroit: "Six or eight months ago, I received the following figures from Dun's Commercial Agency and they should be as near correct as any to be obtained. The number of concerns engaged in the trades in the United States is as follows:

Wholesale grocers	2,226
Retail grocers	109,145
General stores	121,558
Butchers and meat market men	38,900
Confectioners	7,917
Booksellers, stationers and newsdealers	5,101
Boots and shoes	22,280
Cigars and tobacco	22,817
Clothing	14,171
Drugs	37,146
Dry goods	14,539
Flour, grain and feed	15,143
Hardware	21,395
Harness and saddlery	15,143
Music and musical instruments	4,445
Milliners	18,268
Saloon and liquor dealers	94,094

Or a grand total of 572,178 in all lines.

"These figures show that the dealers in food products, such as groceries, meats, bakers' goods, confectionery, flour and feed, number 302,756, as against 269,394 in all other lines of trade, or 33,362 more in food products than in all others. This table also shows that the retail grocery business outranks all others in point of numbers, for all, or nearly all, general stores keep a line of groceries, and it is this large number that we must organize, and nearly every one of these lines of trade have national associations and hold national conventions, and get together and discuss methods for the betterment of their condition. In view of these stupendous figures and facts, I would recommend that the incoming officers do all they can to promote and push forward the work of organization as fast as possible until every State and Territory has a good, live State association and every State association is affiliated with the national association."

THE OSSEKEAG STAMPING CO.

The matter of the winding up of The Ossekeag Stamping Company, Limited, came up before Chief Justice Tuck in chambers on Tuesday, in St. John, N.B. Mr. Peter S. Archibald, who has been acting as provisional liquidator, was appointed permanently. Mr. H. A. Powell appeared for the petitioning creditor, and in response to inquiries from Mr. L. A. Currey and Mr. A. P. Barnhill, who appeared for creditors, stated that the raw material now on hand would be manufactured and sold. He doubted if the dividend to the creditors would be very large. The principal creditor is Senator Wood, who holds securities amounting to nearly $60,000. The question of disposing of the property would probably come up at a later period.

H. S. HOWLAND, SONS & CO.

WHOLESALE ONLY — 37-39 Front Street West, **Toronto.** — ONLY WHOLESALE

CLOTHES WRINGERS.

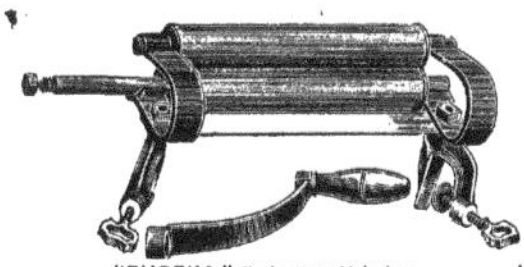

"EUREKA," Rolls 11 x 1¾ inches.

"CRESCENT," Rolls 10 x 1¾ inches, as cut.
"CRICKET," " 11 x 1¾ "

all to H. S. Howland 21/2/05 see cut book 2 page

"NEW MODEL" 1898, Rolls 11 x 1¾ inches.

"LIGHTNING," Rolls 11 x 1¾ inches.

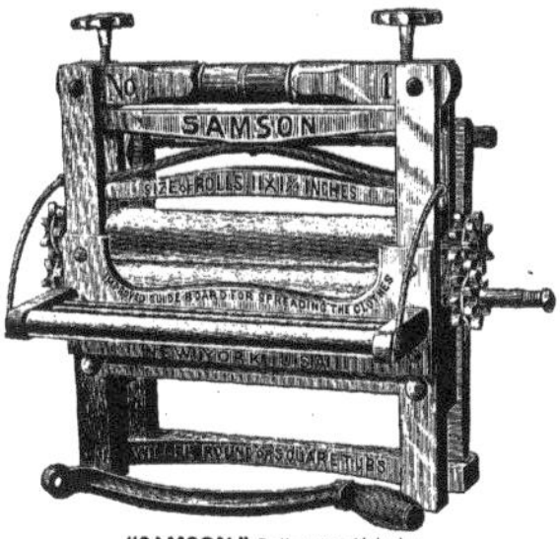

"SAMSON," Rolls 11 x 1¾ inches.

"THE RE-ACTING" Washing Machine.

H. S. HOWLAND, SONS & CO., Toronto.

WE SHIP PROMPTLY. — Graham Wire and Cut Nails are the Best. — OUR PRICES ARE RIGHT.

BRANTFORD BOARD OF TRADE.

THE annual meeting of the Brantford, Ont., Board of Trade, which was held on Tuesday of last week, was a satisfactory one to all members of the board.

President Major J. S. Hamilton occupied the chair. There was a good attendance. Eleven new members were admitted, while Wm. Grant and Alfred Watts were made life members.

The annual address of the president was a stirring one. During the year 115 new members were admitted, bringing the membership to 233.

The year's trade had been good. Two important industries have commenced operations in the city, viz.: The Malleable Iron Works of The Pratt & Letchworth Co., and The Farmers' Packing Co., Limited.

The Adams Wagon Co., Paris, Ont., have decided to remove to Brantford, and will shortly start the erection of their premises in the latter place. The old Consumers' Cordage Co.'s building in West Brantford, has been taken over by The Canada Farmers' Cordage Co., Limited, who expect to start operations in a few days. The number employed in the factories of the city during 1900 was 3,896, as compared with 3,515 in 1899. The wages paid amounted to $1,323,017 in 1900, and $1,234 888 in 1899.

The speaker commented on the high freight rates to the seaboard, which interfered with the export trade, and expressed the opinion that the Railway Committee of the Privy Council should see that the prices charged from Canadian points were not higher in proportion to those charged from points in the Western States. He also complained about the delay in payment of refunds of duty on exports. The secretary-treasurer's report showed the finances of the board to be in a healthy condition.

The following officers were elected :

President—Major J. S. Hamilton.
Vice-President—Lloyd Harris.
Secretary-Treasurer—George Hately.

The council and committee will be elected at the next regular meeting.

AN INCANDESCENT OIL LAMP.

Professor W. L. Emerson, Ottawa, has invented and interested a number of capitalists in an incandescent oil lamp. It is announced that the company, which has obtained a special charter from the Ontario Government, will shortly establish large works on Sussex street, Ottawa, where they will manufacture exclusively the new lighting and heating apparatus invented by Professor Emerson. It is claimed that an ordinary light will contain 2,000 candle power yet will be so small that it can be carried around in the coat pocket. A miniature table lamp will give 600 candle power and can be operated at one-tenth of a cent an hour. Professor Emerson says that his new light, which is generated entirely from coal oil, will be the cheapest and most powerful illumination extant, and will supersede, on account of the low cost and brilliancy, all other kinds. There is no danger of explosion and the power of an ordinary light at one tenth of a cent an hour is equal to that of the ordinary arc lamp which costs about 50c. an evening. In fact, it is said, the expense of operating the new light will not exceed the cost of an ordinary carbon in an arc electric lamp. Each light is self-contained being fed through a hollow wire, which is installed in the house very much the same as an ordinary dwelling is wired for electric lighting.

THE NEW LOCOMOTIVE WORKS.

On Monday The Canadian Locomotive Company, Limited, of Kingston, with a capital stock of $500,000, recently incorporated, had its inaugural meeting in Toronto, and elected the following officers : President, Hon. Wm. Harty; vice-president, M. J. Haney, Toronto ; managing-director, C. Birmingham, of Pittsburg ; treasurer, J. H. Birkett, Kingston, and superintendent, H. Tandy, Kingston. It will be remembered that Mr. Harty purchased the works from the liquidators of The Canadian Engine and Locomotive Company that had formerly carried them on in Kingston.

NEW COMPANIES INCORPORATED.

The following companies have been incorporated : The Strathy Wire Fence Co., Limited, Owen Sound, Ont.; the Lithographed Tin and Can Co., Limited, Toronto; the Canadian Oak Belting Co., Limited, Brockville, Ont.; the Petrolea Combination Rack Co., Limited, Petrolea, Ont.; the Northern Hardware Co., Limited, Sault Ste. Marie, Ont.; the Defiance Lantern and Stamping Co., Limited, Toronto.

GODERICH BOARD OF TRADE.

A strong Board of Trade was organized in Goderich, Ont., on Monday at a meeting attended by practically all the leading business men of the town. The following officers were elected :

President — R. S. Williams, of the Bank of Commerce.
Vice-President—J. H. Colborne, grain dealer.
Secretary — Jam s Mitchell, of The Goderich Star.
Treasurer—W. A. McKim, dry goods dealer.

A council composed of 12 business men was formed, and a platform for future work laid down.

The Wright Taper Roller Bearing Co., Limited, Montreal, has been incorporated.

KENTVILLE BOARD OF TRADE.

THE annual meeting of the Kentville, N.S., Board of Trade was held on Monday evening, January 21. President R. W. Eaton occupied the chair.

The president's address, which is a feature of the annual meeting, was a comprehensive one. He first reviewed the volume of trade and the conditions of business during the year. With the exception of the export apple trade, everything had been even more satisfactory than customary. The imports amounted to $12,435, an increase of almost $2,000. The export apple trade had been, however, exceedingly bad, the balance being nearly $100,000 on the wrong side. This he attributed to the inefficient and unsuitable class of vessels engaged in the ocean transport business during the season. These vessels received subsidies from the Government, but, as no provision had been made to secure suitable vessels, the subsidies served only to kill off competition and give employment to slow and unsuitable vessels. He was convinced that the Government should appoint a commission to inquire into the circumstances connected with the transportation and marketing of the apple crop of the Annapolis Valley with a view to remedying existing grievances.

In referring to the tourist business of the town, which is steadily increasing, the speaker stated that 20,000 copies of a booklet descriptive of the town had been issued.

The principal event of the year, as far as the board was concerned, was the annual meeting of the Maritime Board of Trade, held in Kentville on August 15, 16 and 17, last. The session, he considered, had been successful in every respect.

The industries of Kentville are in satisfactory condition. The Nova Scotia Carriage Co., which was organized during the year, has proven a valuable addition to the institutions of the town. All the other concerns have grown in size and strength during the year.

A vote of thanks was given to the president and secretary for the valuable service and reports.

The following officers were elected :

President—James Sealy.
Vice-President—W. P. Shaffner.
Secretary-Treasurer—G. E. Calkin.
Auditors—J. W. King, Dr. Saunders.

After the discussion it was unanimously resolved to endorse the action taken by the municipal council and the Kings County Board of Trade in reference to the appointment of a commission of inquiry in reference to apple transportion by the Federal Government.

A TAX PROPOSED ON NICKEL ORES.

A LENGTHY and comprehensive petition has been sent to the Ontario Government which is believed to have the solid support of all the iron manufacturers of the Dominion. It is officially signed by A. T. Wood, president; C. S Wilcox, general manager Hamilton Steel and Iron Co., Limited; The Nickel Steel Co. of Canada, John Patterson, secretary; The Canada Furnace Co., Limited, per Geo. E. Drummond, managing director and treasurer.

The petition sets forth the richness of the Ontario nickel mines and calls attention to the increased price of refined nickel in the markets of the world during the past 18 months, which gives Ontario a correspondingly less price for her share, the Province receiving now less than 28 per cent.

UNITED STATES LEGISLATION.

The legislation of the session to remedy this, the petitioners argue, needs supplementing. The United States prohibitory duty upon Canadian nickel and nickel alloys, while admitting the raw material free, is quoted as wholly destructive of the manufacturing of Ontario nickel.

USE ONTARIO ORE.

The petitioners say that the Ontario nickel ores are now, and for ten years have been, used in the construction of powerful navies and powerful guns by the United States and other foreign countries whose interests may at any time become adverse to the interests of Canada and the Empire.

WHAT IS ASKED.

The petition asks that the Government exercise the power granted in the Mines Act for the imposition of taxes upon nickel ores and upon nickel and copper ores and their partially treated products, whether the same be smelted and refined in Canada or not, and that the tax thus collected be paid as a bonus upon the manufacture of nickel-steel in Ontario, and they further ask, in order that the policy for the manufacture of nickel iron and nickel steel may be successfully carried on in Canada, and that Ontario may have such a monopoly of the raw material of this metal, that the Government convey no further lands or any title to or interest in any lands containing nickel ores to any individual or individuals, companies or corporations, who will not refine and use the nickel derived from such ores in the manufacture of this metal and its various alloys in Ontario.

The general store of George Dean, in Lobo, was destroyed by fire at an early hour Monday morning. Little was saved, and the insurance is light.

MARKETS AND MARKET NOTES

ONTARIO MARKETS.

TORONTO, February 15, 1901.

HARDWARE.

A RATHER better trade is to be noted this week, there being a better movement in hardware generally. Business, however, for immediate requirements is still of a hand-to-mouth character, and the greatest improvement is in the way of orders for future delivery, but even in this latter particular the buying is nothing like as free as it was a year ago. Of course, nothing else could be expected, for, while a year ago prices were tending upward, now the tendency is in the opposite direction. The outlook is for a fair spring and summer trade, as stocks generally throughout the country are not heavy. Payments are, on the whole, fair.

BARB WIRE—A little business has been done during the past week on future account, but only an occasional order is reported from stock. We quote: $2.97 f.o.b. Cleveland for less than carlots, and $2.85 in carlots. From stock, Toronto, $3.10 per 100 lb.

GALVANIZED WIRE — Business in this line is altogether confined to orders for future delivery, but even in this particular the orders are very small. We quote: No. 6, 7 and 8, $3.55; No. 9, $3.10; No. 10, $3.75; No. 11, $3.85; No. 12, $3.25; No. 13 $3.35; No. 14, $4.25; No. 15, $4.75, and No. 16, $5.

SMOOTH STEEL WIRE—Few odd ton lots are being booked for oiled and annealed for future delivery, and an occasional order is being received for hay-baling wire for prompt shipment. Base price is unchanged at $2.80 per 100 lb.

WIRE NAILS—Prompt business is very small indeed, but there is a little improvement in the orders for spring delivery, there evidently being a little more confidence in the market since the advance of 10c. per keg took place in the United States a few weeks ago. Locally, the price is steady and unchanged at $2.85 per keg in less than carlots, and $2.75 for carlots.

CUT NAILS—The reports from all quarters in regard to cut nails fail to show any improvement in business, the market still being decidedly dull. The base price is unchanged at $2.35 per keg.

HORSESHOES — These are still quiet and unchanged in price. We quote f.o.b. Toronto: Iron shoes, No. 2 and larger, light, medium and heavy, $3.60; snow shoes, $3.85; light steel shoes, $3.70; featherweight (all sizes), $4.95; iron shoes, No. 1 and smaller, light, medium and heavy (all sizes), $3.85; snow shoes, $4; light steel shoes, $3.95; featherweight (all sizes), $4.95.

HORSE NAILS—The demand is moderate and prices unchanged at recent decline. We quote oval head at 50 and 7½ per cent. discount on "C" brand, and 50, 10 and 5 per cent. on "M" brand. Both these nails, it will be remembered, are now sold from different lists. Countersunk head is still quoted at the discount of 50, 10 and 10 per cent.

SCREWS—Trade is good in this line with prices unchanged. Discounts are: Flat head bright, 87½ and 10 per cent.; round head bright, 82½ and 10 per cent.; flat head brass, 80 and 10 per cent.; round head

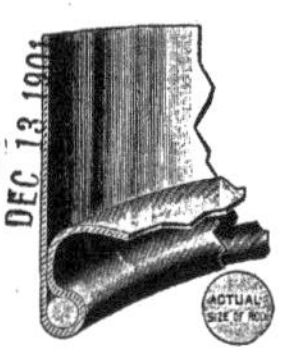

brass, 75 and 10 per cent. Round head bronze is unchanged at 65 per cent., and flat head bronze at 70 per cent.

BOLTS AND NUTS—Trade in this line is without any special feature, there being very little business doing, while prices are unchanged. We quote: Carriage bolts (Norway), full square, 70 per cent.; carriage bolts full square, 70 per cent.; common carriage bolts, all sizes, 65 per cent.; machine bolts, all sizes, 65 per cent.; coach screws, 75 per cent.; sleighshoe bolts, 75 per cent.; blank bolts, 65 per cent.; bolt ends, 65 per cent.; nuts, square, 4½c. off; nuts, hexagon, 4¾c. off; tire bolts, 67½ per cent.; stove bolts, 67½; plough bolts, 60 per cent.; stove rods, 6 to 8c.

RIVETS AND BURRS—There is very little being done in this line also. Discount, 60 and 10 per cent. on iron rivets, 55 per cent. on iron burrs, and 35 and 5 per cent. on copper rivets and burrs.

ROPE—The demand for rope continues small. We quote as follows: Sisal, 9c. per lb. base, and manila, 13c.; cotton rope, 3-16 in. and larger, 16½c.; 5-32 in., 21½c., and ⅛ in., 22½c. per lb.

BINDER TWINE—A good trade has been done so far this season, and it is estimated that fully 75 to 80 per cent. of the orders required have been placed with the manufacturers, and there is still a fair demand. Although our quotations are the same as a week ago, one large United States manufacturer, who does a large business in Canada, advanced prices ¼c. per lb. on February 15. We still quote pure manila, 10¾c. per lb.; mixed, 8½c. per lb.; sisal, 7¾c. per lb.

CUTLERY—General trade is still small, although quite a number of orders continue to arrive from the Coast.

SPORTING GOODS—Some loaded shells and cartridges are going out, but trade in this line is small for this time of the year.

CHURNS—A few orders are being received for immediate shipment, but business is mostly for later requirements.

GREEN WIRE CLOTH—Some business on future account is still being done at $1.35 per 100 sq. ft.

TINWARE—Milk can trimmings are still going out, and some delivery is being made of sap buckets and sap spouts.

HARVEST TOOLS—A few orders are coming in for future delivery. Discount, 50, 10 and 5 per cent.

SPADES AND SHOVELS—Orders for spring delivery, while a little more numerous than they were, are small individually. Discount, 40 and 5 per cent.

POULTRY NETTING—There is still a little being done for future delivery. Discount on Canadian, 50 and 5 per cent. English is quoted net.

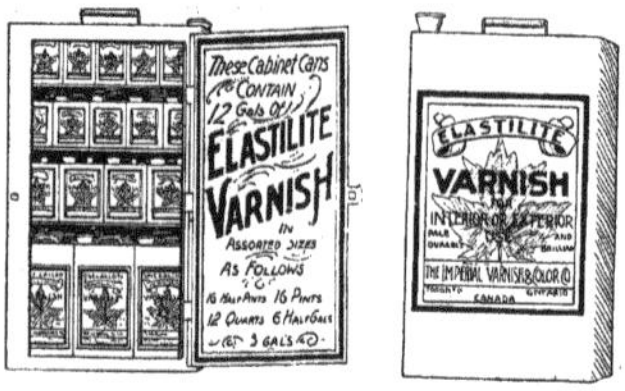

HOCKEY STICKS AND SKATES—There has been quite an active demand for hockey sticks and an occasional order is still being received for skates.

LAWN MOWERS—Orders are being taken for future delivery, and a few shipments have been made to the Coast during the past week.

CEMENT—There is nothing doing. We nominally quote in barrel lots: Canadian Portland, $2.80 to $3; Belgian, $2.75 to $3; English do., $3; Canadian hydraulic cements, $1.25 to $1.50; calcined plaster, $1.90; asbestos cement, $2.50 per bbl.

METALS.

The metal trade has been more active during the past week, and business may now be considered fairly good for this time of the year. The demand is principally for tin plates, black sheets, galvanized iron and copper.

PIG TIN—Up to three or four days ago the markets in London and New York were easy, but since then the tendency has been the other way, and at the moment prices are steady. Locally, a good business has been done during the past week both in large and small lots, and quotations are unchanged at 32 to 33c.

PIG IRON — There is not much business being done, but prices are steady. We quote $17 per ton for Canadian iron in 100 ton lots.

BAR IRON—Business is still fair, with the base price unchanged at $1.60 to $1.70 per 100 lb.

TINPLATES—Trade has been fair during the past week in small lots, and several orders have been taken for spring delivery. According to the latest cable advices prices are a little higher in the London, Eng., market than they were a week ago.

TINNED SHEETS — The demand during the past week has been fair with prices ranging from 9 to 9½c. for 23 gauge.

TERNE PLATES—Trade in this line is quiet and prices are unchanged at last week's edecline, our quotations still being $8.50 to $10.50.

BLACK SHEETS—The demand during the past week has been active, both in large and in small lots. The base price is $3.30.

GALVANIZED SHEETS—Business has improved during the past week and it is now fairly brisk. The primary markets are firm in price and ton lots for import are reported to be worth $4.65. We quote English at $4.75 for small lots and American at $4.50.

CANADA PLATES—Small lots are being shipped from stock and orders for fall delivery are being freely booked. We quote all dull, $3; half and half, $3.15, and all bright, $3.65 to $3.75.

IRON PIPE—Business in iron pipe is still rather quiet, with quotations much as before. It is understood, however, that an effort is being made among the jobbing trade to fix a uniform price on iron pipe, whether or not it will be successful, we cannot at the moment say. We quote: Black pipe ¼ in., $3.00; ⅜ in., $3 00; ½ in., $3.05; ¾ in., $3.20; 1 in., $4 60; 1¼ in., $6.35; 1½in., $7.55; 2 in., $10.10 Galvanized pipe is as follows: ½ in., $4 65; ¾ in., $5.35; 1 in., $7 25; 1¼ in., $9.75; 1½ in., $11.25; 2 in., $15 50.

HOOP STEEL—The demand is good, and the base price is still quoted at $3.10 for ordinary quantities.

COPPER—Ingot copper is quiet, but an active demand is being experienced for sheet copper. The outside markets are quiet but steady. We quote: Ingot, 19 to 20c.; bolt or bar, 23½ to 25c.; sheet, 23 to 23½c.

BRASS — The demand continues fairly good, with the discount on rod and sheet unchanged at 15 per cent.

SOLDER—Trade is fair, with prices unchanged. We quote: Bar, half-and-half, guaranteed, 19c.; ditto, commercial, 18½c.; refined, 18½c., and wiping, 18c.

LEAD—Locally, trade is quiet and the

outside markets are dull, with prices easier in London, Eng. Locally, we still quote 4¾ to 5c.

ZINC SPELTER — Trade continues quiet locally with the market dull and easy in both London and New York. We quote 6 to 6½c. per lb.

ZINC SHEET — Trade is dull at $6 75 to $7 for casks and $7 to $7.50 for part casks.

ANTIMONY—Business is still quiet at 11 to 11½c.

PAINTS AND OILS.

Orders for spring delivery continue to come in to jobbers in large numbers. As a rule they are large and well assorted. Quite a few orders for immediate delivery are being filled and sent out with invoices dated ahead. Linseed oil has again stiffened, an advance of 2c. being made last Saturday. We quote:

WHITE LEAD—Ex Toronto, pure white lead, $6.87½; No. 1, $6.50; No. 2, $6.12½; No. 3, $5.75; No. 4 $5.37½; dry white lead in casks, $6.

RED LEAD—Genuine, in casks of 560 lb., $5.50; ditto, in kegs of 100 lb., $5.75; No. 1, in casks of 560 lb., $5 to $5.25; ditto, kegs of 100 lb.; $5.25 to $5.50.

LITHARGE—Genuine, 7 to 7½c.

ORANGE MINERAL—Genuine, 8 to 8½c.

WHITE ZINC—Genuine, French V.M., in casks, $7 to $7.25; Lehigh, in casks, $6.

PARIS WHITE—90c.

WHITING — 60c. per 100 lb.; Gilders' whiting, 75 to 80c.

GUM SHELLAC — In cases, 22c.; in less than cases, 25c.

PARIS GREEN—Bbls., 16¾c.; kegs, 17c.; 50 and 100-lb. drums, 17½c.; 25-lb. drums, 18c.; 1-lb. papers, 18½c.; 1-lb. tins, 19½c.; ½-lb. papers, 20½c.; ½-lb. tins, 21½c.

PUTTY — Bladders, in bbls., $2.20; bladders, in 100 lb. kegs, $2.35; bulk in bbls., $2; bulk, less than bbls. and up to 100 lb., $2.15; bladders, bulk or tins, less than 100 lb., $3.

PLASTER PARIS—New Brunswick, $1.90 per bbl.

PUMICE STONE — Powdered, $2.50 per cwt. in bbls., and 4 to 5c. per lb. in less quantity; lump, 10c. in small lots, and 8c. in bbls.

LIQUID PAINTS—Pure, $1.20 to $1.30 per gal.; No. 1 quality, $1 per gal.

CASTOR OIL—East India, in cases, 10 to 10½c. per lb. and 10½ to 11c. for single tins.

LINSEED OIL—Raw, 1 to 4 barrels, 78c.; boiled, 81c.; 5 to 9 barrels, raw, 77c.; boiled, 80c., delivered. To Toronto, Hamilton, Guelph and London, 2c. more.

TURPENTINE—Single barrels, 59c.; 2 to 4 barrels, 58c., to all points in Ontario. For less quantities than barrels, 5c. per gallon extra will be added, and for 5-gallon packages, 50c., and 10 gallon packages, 80c. will be charged.

GLASS.

A large number of orders for shipment for stock in March and April are being received. Prices are steady. We still quote first break locally: Star, in 50-foot boxes, $2.10, and 100-foot boxes, $4; double diamond under 26 united inches, $6, Toronto, Hamilton and London; terms 4 months or 3 per cent. 30 days.

OLD MATERIAL.

There is no change. The demand keeps good, but deliveries are moderate. We quote jobbers' prices as follows: Agricultural scrap, 55c. per cwt.; machinery cast, 55c. per cwt.; stove cast, 35c.; No. 1 wrought 50c. per 100 lb.; new light scrap copper, 12c. per lb.; bottoms, 10½c.; heavy copper, 12½c.; coil wire scrap, 13c.; light brass, 7c.; heavy yellow brass, 10 to 10½c.; heavy red brass, 10½c.; scrap lead, 3c.; zinc, 2½c; scrap rubber, 6½c.; good country mixed rags, 65 to 75c.; clean dry bones, 40 to 50c. per 100 lb.

COAL.

There is a good movement, but nut is not yet as freely offered as might be desired. We quote anthracite on cars Buffalo and bridges: Grate, $4 75 per gross ton and $4.24 per net ton; egg, stove and nut, $5 per gross ton and $4 46 per net ton.

PETROLEUM.

There is a steady reduction in the volume of business, but for this time of year the movement is satisfactory. We quote: Pratt's Astral, 17 to 17½c. in bulk (barrels, $1 extra); American water white, 17 to 17½c. in barrels; Photogene, 16½ to 17c.; Sarnia water white, 16 to 16½c. in barrels; Sarnia prime white, 15 to 15½c. in barrels.

QUEBEO MARKETS.

Montreal, February 15, 1901.

HARDWARE.

TRADE has been quiet this week from every point of view. Travellers in the west still complain of a dull retail business, and while trade in the east is somewhat better, it is not at all brisk. Orders for immediate shipment are few and small while there seems to be no activity to make contracts for spring importation. Prices, however, are steady. Cut nails are dull and wire nails are not moving freely. Wire is as yet in small inquiry. Horse nails are in better demand since the decline, and horseshoes are still one of the most active articles on the market. Lawn mowers, garden hose, screens, green wire cloth, poultry netting, and freezers are all being considered for spring business. Screws are in good demand since the drop in prices, while such goods as bolts and other manufacturer's supplies continue to be inquired for in moderate quantities.

BARB WIRE—There is no new feature to note, prices remaining as before at $3.20 f.o.b. Montreal in less than carlots.

GALVANIZED WIRE—There is little on the market to interest the trade. Business is extremely quiet. We quote as follows: No. 5, $4.25; Nos. 6, 7 and 8 gauge $3.55; No. 9, $3.10; No. 10, $3.75; No. 11, $3.85; No. 12, $3.25; No. 13, $3.35; No. 14, $4.25; No. 15, $4.75; No. 16, $5.00.

SMOOTH STEEL WIRE—A fair demand has been noticed for this article throughout the week. Values are steady at $2.80 per 100 lb.

FINE STEEL WIRE—There is no change to note. The discount remains as before, 17½ per cent. off the list.

BRASS AND COPPER WIRE — A small trade is passing. Discounts are 55 and 2½ per cent. on brass, and 50 and 2½ per cent. on copper.

FENCE STAPLES—Only a few lots are being shipped. We quote: $3.25 for bright, and $3.75 for galvanized, per keg of 100 lb.

WIRE NAILS—A few small shipments have been made this week; but no one appears to be eager to lay in heavy stocks. We quote $2.85 for small lots and $2.75 for carlots, f.o.b. Montreal, Toronto, Hamilton, London, Gananoque, and St. John, N.B.

CUT NAILS -There is little doing, and prices are steady. We quote: $2.35 for small and $2.25 for carlots; flour barrel nails, 25 per cent. discount; coopers' nails, 30 per cent. discount.

HORSE NAILS—Since the reduction in prices, business has somewhat improved. Discounts remain as we gave them last week. The general discounts are 50, 10 and 5 per cent. on oval head and 50, 10 and 10 per cent. on countersunk head. "C" brand's new discount is 50 and 7½ per cent. on their own price list.

HORSESHOES—A brisk business continues to be done We quote: Iron shoes, light and medium pattern, No. 2 and larger, $3.50; No. 1 and smaller, $3.75; snow shoes, No. 2 and larger, $3.75; No. 1 and smaller, $4.00; X L steel shoes, all sizes, 1 to 5, No. 2 and larger, $3.60; No. 1 and smaller, $3.85; feather-weight, all sizes, $4.85; toe weight steel shoes, all sizes, $5.95 f.o.b. Montreal; f.o.b. Hamilton, London and Guelph, 10c. extra.

POULTRY NETTING — Some transactions have been made this week at former discount of 50 and 5 per cent.

GREEN WIRE CLOTH—A fairly good trade is being done on future account at $1.35 per 100 sq. ft.

FREEZERS—Business has progressed very favorably during the week. "Peerless" is quoted as follows: "Two quarts, $1.85; 3 quarts, $2.10; 4 quarts, $2 50; 6 quarts, $3.20; 8 quarts, $4; 10 quarts, $5.25; 12 quarts, $6; 16 quarts, with fly wheel, $11; toy, 1 pint, $1.40.

SCREEN DOORS AND WINDOWS—Additional orders on spring account have been booked this week. We quote: Screen doors, plain cherry finish, $8.25 per doz.; do. fancy, $11.50 per doz.; windows, $2.25 to $3 50 per doz.

SCREWS—A good trade is doing for immediate shipment. Discounts are: Flat head bright, 87½ and 10 per cent. off list; round head bright, 82½ and 10 per cent.; flat head brass, 80 and 10 per cent.; round head brass, 75 and 10 per cent.

BOLTS—The small manufacturers throughout the country have been purchasing freely of late. Discounts are as follows: Carriage bolts, 65 per cent.; machine bolts, 65 per cent.; coach screws, 75 per cent.; sleigh shoe bolts, 75 per cent.; bolt ends, 65 per cent.; plough bolts, 50 per cent.; square nuts, 4½c. per lb. off list; hexagon nuts, 4¾c. per lb. off list; tire bolts, 67½ per cent.; stove bolts, 67½ per cent.

BUILDING PAPER—Retailers are making contracts for their spring goods. The shipments being made at present are small. We quote: Tarred felt, $1.70 per 100 lb.; 2 ply, ready roofing, 80c. per roll; 3-ply, $1.05 per roll; carpet felt, $2.25 per 100 lb.; dry sheathing, 30c. per roll; tar sheathing, 40c. per roll; dry fibre, 50c. per roll; tarred fibre, 60c. per roll; O.K. and I.X L., 65c. per roll; heavy straw sheathing, $28 per ton; slaters' felt, 50c. per roll.

RIVETS—There has been no feature in this line. The discount on best iron rivets, section, carriage, and waggon box, black rivets, tinned do., coopers' riverts and tinned swedes rivets, 60 and 10 per cent.; swedes iron burrs are quoted at 55 per cent. off; copper rivets, 35 and 5 per cent. off; and coppered iron rivets and burrs, in 5-lb. carton boxes, are quoted at 60 and 10 per cent. off list.

CORDAGE—Trade is rather quiet, with prices unchanged. Manila is quoted at 13c. per lb. for 7-16 and larger; sisal at 9c., and lathyarn 9c. per lb. In small lots ½c. per lb. higher is charged.

SPADES AND SHOVELS—A small business for spring delivery is to be noticed at a discount of 40 and 5 per cent. off the list.

HARVEST TOOLS—For spring delivery there has also been some business done this week at a discount of 50, 10 and 5 per cent.

TACKS—A small business is doing at unchanged figures. We quote: Carpet tacks, in dozens and bulk, blued 80 and 5 per cent. discount; tinned, 80 and 10 per cent.; cut tacks, blued, in dozens, 75 and 15 per cent. discount.

CHURNS—Some spring business is being done at a discount of 56 per cent.

FIREBRICKS—Only a small amount of business is being done at $18.50 to $26, as to brand.

CEMENT—The winter demand is small. We quote: German, $2.50 to $2.65; English, $2.40 to $2.50; Belgian, $1.90 to $2.15 per bbl.

METALS.

The metal market is not, taken as a whole, in a very satisfactory condition. Iron is not very firm, and tin is unsettled. Copper is somewhat firmer. Coil chain is decidedly firm, and an advance is expected.

PIG IRON—Canadian pig iron is worth about $18 on the Montreal market, and No. 1 Summerlee about $22 to $23.

BAR IRON—The demand for bar iron is steady. Sales have been made at $1.60 to $1.65.

BLACK SHEETS—A fair trade has been done this week, some large lots having changed hands. Prices rule at $2.80 for 8 to 16 gauge; $2.85 for 26 gauge, and $2 90 for 28 gauge.

GALVANIZED IRON—The demand is fair and orders for import are being taken freely. We quote: No. 28 Queen's Head, $5 to $5.10; Apollo, 10¾ oz., $5 to $5.10; Comet, No. 28, $4.50, with 25c. allowance in case lots.

INGOT COPPER—Primary markets are a little stiffer, but values here remain at 17½c.

INGOT TIN—The London market has not gained during the week, and prices show no change. The price here remains at 33c.

LEAD—Unchanged at $4 65.

LEAD PIPE—A fair trade continues to be done. We quote: 7c. for ordinary and 7½c. for composition waste, with 15 per cent. off.

IRON PIPE—A good demand has been met with this week. We quote: Black pipe, ¼, $3 per 100 ft.; ⅜, $3; ½, $3; ¾, $3.15; 1-in., $4.50; 1¼, $6.10; 1½, $7.28; 2-in., $9 75. Galvanized, ½, $4.60; ¾, $5.25; 1 in., $7.50; 1¼, $9.80; 1½, $11.75; 2-in., $16.

TINPLATES—Stocks are low. Some shipments have been made this week, and future stocks are being contracted for. The ruling figures for immediate delivery are $4 50 for coke and $4.75 for charcoal.

CANADA PLATE—Trade is rather quiet, but some orders for spring importation are being taken. We quote: 52's, $2.90; 60's, $3; 75's, $3.10; full polished, $3.75, and galvanized, $4.60.

TOOL STEEL—We quote: Black Diamond, 8c.; Jessop's 13c.

STEEL—No change. We quote: Sleigh-

shoe, $1.85 ; tire, $1.95 ; spring, $2.75 ; machinery, $2.75 and toe-calk, $2.50.

TERNE PLATES—Trade is quiet at $8.25.

SWEDISH IRON—Unchanged at $4.25.

COIL CHAIN—The demand for coil chain is good, and a fair amount of trade has been done. Values are firm and an advance is expected. We quote as follows: No. 6, 11½c.; No. 5, 10c.; No. 4, 9½c.; No. 3, 9c.; ¼-inch, 7½c. per lb.; 5-16, $4.60; 5-16 exact, $5.10; ⅜, $4.20; 7-16, $4.00; ½, $3.75; 9-16, $3.65; ⅝, $3.35; ¾, $3.25; ⅞, $3.20; 1-in., $3.15. In carload lots an allowance of 10c. is made.

SHEET ZINC—The ruling price is 6 to 6¼c.

ANTIMONY—Quiet, at 10c.

GLASS.

Movements from stock are light. Some import orders are being taken on a basis of $1.75 for first break of 50 feet. We quote as follows: First break, $2 ; second, $2.10 for 50 feet ; first break, 100 feet, $3.80 ; second, $4 ; third, $4.50 ; fourth, $4.75; fifth, $5.25 ; sixth, $5.75, and seventh, $6.25.

PAINTS AND OILS.

After sinking so low that it could be shipped from England to the United States, pay an import duty of 20c. per wine gallon and undersell the American oil on its own market, linseed oil has taken a turn for the better on the English market. Sales have been made here for future delivery on the basis of 36c. first cost, meaning 52c. laid down here. What the outcome of this situation will be remains to be seen. The low prices for summer delivery have affected the spot market to the extent of 4c. per gallon. We quote :

WHITE LEAD—Best brands, Government standard, $6.75 ; No. 1, $6.37½ ; No. 2, $6 ; No. 3, $5.62½, and No. 4, $5.25, all f.o.b. Montreal. Terms, 3 per cent. cash or four months.

DRY WHITE LEAD—$5.75 in casks ; kegs, $6.

RED LEAD—Casks, $5.50 ; in kegs, $5.75.

WHITE ZINC PAINT—Pure, dry, 8c.; No. 1, 6½c.; in oil, pure, 9c.; No. 1, 7½c.

PUTTY—We quote : Bulk, in barrels, $2 per 100 lb.; bulk, in less quantity, $2.15; bladders, in barrels, $2.20 ; bladders, in 100 or 200 lb. kegs or boxes, $2.35; in tins, $2.45 to $2.75 ; in less than 100-lb. lots, $3 f.o.b. Montreal, Ottawa, Toronto, Hamilton, London and Guelph. Maritime Provinces 10c. higher, f.o.b. St. John and Halifax.

LINSEED OIL—Raw, 76c.; boiled, 79c., in 5 to 9 bbls., 1c. less, 10 to 20-bbl. lots, open, net cash, plus 2c. for 4 months. Delivered anywhere in Ontario between Montreal and Oshawa at 2c. per gal. advance and freight allowed.

TURPENTINE—Single bbls., 59c.; 2 to 4 bbls., 58c.; 5 bbls. and over, open terms, the same terms as linseed oil.

MIXED PAINTS—$1.25 to $1.45 per gal.

CASTOR OIL—8¾ to 9¼c. in wholesale lots, and ½c. additional for small lots.

SEAL OIL—47½ to 49c.

COD OIL—32½ to 35c.

NAVAL STORES — We quote : Resinsr $2.75 to $4.50, as to brand ; coal tar, $3.25 to $3.75 ; cotton waste, 4½ to 5½c. for colored, and 6 to 7½c. for white ; oakum, 5½ to 6½c., and cotton oakum, 10 to 11c.

PARIS GREEN—Petroleum barrels, 16¾c. per lb.; arsenic kegs, 17c.; 50 and 100-lb. drums, 17½c.; 25-lb. drums, 18c.; 1-lb. packages, 18½c.; ½-lb. packages, 20½c.; 1-lb. tins, 19½c.; ½-lb. tins, 21½c. f.o.b. Montreal; terms 3 per cent. 30 days, or four months from date of delivery.

SCRAP METALS.

The market is quiet and unchanged. Dealers are paying the following prices in the country : Heavy copper and wire, 13 to 13½c. per lb. ; light copper, 12c.; heavy brass, 12c. ; heavy yellow, 8½ to 9c. ; light brass, 6½ to 7c.; lead, 2¾ to 3c. per lb.; zinc, 2¼ to 2½c.; iron, No. 1 wrought, $13 to $14 per gross ton ; No. 1 cast, $13 to $14 ; stove plate, $8 to $9; light iron, No. 2, $4 a ton; malleable and steel, $4.

HIDES.

The market for green hides remain quiet but steady, dealers still paying 7½c. for No. 1 light, and tanners 8½c. for car lots. Quality is beginning to be complained of. We quote: Light hides, 7½c. for No. 1; 6½c. for No. 2, and 5½c. for No. 3. Lambskins, 90c.

PETROLEUM.

Business in petroleum has been rather quiet of late, the demand not being up to the average for the season. The tone of the market is steady. We quote : "Silver Star," 14½ to 15½c. ; "Imperial Acme," 16 to 17c. ; "S.C. Acme," 18 to 19c., and "Pratt's Astral," 18½ to 19½c.

MONTREAL NOTES.

H. W. DeCourtenay & Co., metal merchants, Montreal, who were burned out in the recent conflagration, have taken quarters at 86 and 88 McGill street, where they have a full stock of steel, including Thomas Firth & Sons' best tool steels.

HOCKEY IN MONTREAL.

In Montreal on Monday night the Colin McArthur Hockey team defeated the Canada Paint Company's crack aggregation by a score of 4 to 1.

The Canadian Rubber Company and the Baylis Manufacturing Company's teams played a very interesting game of hockey on the Ontario Rink, Montreal, on the 2nd, which was hotly contested, both teams being on their merits, but fortune favoured the Baylis', who won by the close score of 2 to 1.

THOMAS McVITTIE DEAD.

Thomas McVittie, who 50 or more years ago was engaged in the hardware business in Toronto, died at his home in Barrie, Ont., on Tuesday last week. For the past 30 years Mr. McVittie has resided in Barrie. For a time he was engaged in the hardware business, but in his declining years he went out of business and lived retired.

The moulding shop of the Kingston, Ont., Foundry was destroyed by fire on Wednesday night. The loss is placed at about $5,000; insurance $900.

MANITOBA MARKETS.

WINNIPEG, February 11, 1901.

TRADE is still quiet and collections slow. Wholesalers complain that orders for spring delivery are not up to last years' figures. The fact that there is not the stimulus of a rapidly rising market may have something to do with the apparent apathy of buyers. The following is the price list for the week.

Barbed wire, 100 lb.	$3 45
Plain twist	3 45
Staples	3 95
Oiled annealed wire 10	3 95
" 11	4 00
" 12	4 05
" 13	4 20
" 14	4 35
" 15	4 45
Wire nails, 30 to 60 dy, keg	3 45
" 16 and 20	3 50
" 10	3 55
" 8	3 65
" 6	3 70
" 4	3 85
" 3	4 10
Cut nails, 30 to 60 dy.	3 00
" 20 to 40	3 05
" 10 to 16	3 10
" 8	3 15
" 6	3 20
" 4	3 30
" 3	3 65
Horsenails, 45 per cent. discount.	
Horseshoes, iron, No. 0 to No 1	4 65
No. 2 and larger	4 40
Snow shoes, No. 0 to No. 1	4 90
No. 2 and larger	4 40
Steel, No. 0 to No. 1	4 95
No. 2 and larger	4 70
Bar iron, $2.50 basis.	
Swedish iron, $4.50 basis.	
Sleigh shoe steel	3 00
Spring steel	3 25
Machinery steel	3 75
Tool steel, Black Diamond, 100 lb	8 50
Jessop	13 00
Sheet iron, black, 10 to 20 gauge, 100 lb.	3 50
20 to 26 gauge	3 75
28 gauge	4 00
Galvanized American, 16 gauge	2 54
18 to 22 gauge	4 50
24 gauge	4 75
26 gauge	5 00
28 gauge	5 25
Genuine Russian, lb	12
Imitation " "	8
Tinned, 24 gauge, 100 lb	7 55
26 gauge	8 80
28 gauge	8 00
Tinplate, IC charcoal, 20 x 28, box	10 75
" IX "	12 75
" IXX "	14 75
Ingot tin	35
Canada plate, 18 x 21 and 18 x 24	3 75
Sheet zinc, cask lots, 100 lb.	7 50
Broken lots	8 00
Pig lead, 100 lb.	6 00
Wrought pipe, black up to 2 inch	50 an 10 p.c.
" Over 2 inch	50 p.c.
Rope, sisal, 7-16 and larger	$10 00
" ⅜	10 50
" ¼ and 5-16	11 00
Manila, 7-16 and larger	13 50
" ⅜	14 00
" ¼ and 5-16	14 50
Solder	21½
Cotton Rope, all sizes, lb.	16
Axes, chopping	$7 50 to 12 00
" double bitts	12 00 to 18 00
Screws, flat head, iron, bright	87½
Round " "	82½
Flat " brass	80
Round " "	75
Coach	57½ p.c.
Bolts, carriage	55 p.c.
Machine	55 p.c.
Tire	60 p.c
Sleigh shoe	65 p.c.
Plough	40 p.c.
Rivets, iron	50 p.c.
Copper, No. 8	35
Spades and shovels	40 p.c.
Harvest tools	50, and 10 p.c.
Axe handles, turned, s. g. hickory, doz.	$2 50
No. 1	1 50
No. 2	1 25
Octagon extra	1 75
No. 1	1 25
Files common	70, and 10 p.c.
Diamond	60
Ammunition, cartridges, Dominion R.F.	50 p.c.
Dominion, C.F., pistol	30 p.c.
" military	15 p.c.
American R.F.	30 p.c.
C.F. pistol	5 p.c.
C.F. military	10 p.c. advance.
Loaded shells:	
Eley's soft, 12 gauge black	16 50
chilled, 12 guage	18 00
soft, 10 guage	21 00
chilled, 10 guage	23 00
Shot, Ordinary, per 100 lb	6 75
Chilled	7 50
Powder, F.F., keg	4 75
F.F.G.	5 00
Tinware, pressed, retinned	75 and 2½ p.c.
" " plain	70 and 15 p.c.
Graniteware, according to quality	50 p.c.

PETROLEUM.

Water white American	25½c.
Prime white American	24c.
Water white Canadian	22c
Prime white Canadian	21c

PAINTS, OILS AND GLASS.

Turpentine, pure, in barrels	$ 68
Less than barrel lots	73
Linseed oil, raw	87
Boiled	90
Lubricating oils, Eldorado castor	25½
Eldorado engine	24½
Atlantic red	27½
Renown engine	41
Black oil	23½ to 25
Cylinder oil (according to grade)	55 to 74
Harness oil	61
Neatsfoot oil	$1 00
Steam refined oil	85
Sperm oil	1 50
Castor oil per lb.	11½
Glass, single glass, first break, 16 to 25 united inches	2 25
26 to 40 per 50 ft.	2 50
41 to 50 " 100 ft.	5 50
51 to 60 " " "	6 00
61 to 70 per 100-ft. boxes	6 50
Putty, in bladders, barrel lots per lb.	2½
kegs "	2¾
White lead, pure per cwt.	7 25
No 1 "	7 00
Prepared paints, pure liquid colors, according to shade and color. per gal.	$1.30 to $1.90

INDUSTRIAL GOSSIP.

Those having any items of news suitable for this column will confer a favor by forwarding them to this office addressed to the Editor.

A by-law granting a bonus of $10,000 to the Canadian Steam Carriage Co. was voted on by the property owners of Port Dover recently, and carried by 154 to 40.

The Cramp-Ontario Steel Co. have started work on their new docks at Collingwood. They expect to blow in their first furnace in October next.

B. C. Beach & Co., founders, Winchester, Ont,, intend erecting two new buildings. One of these will be 120 x 40, four storeys high, and will be used for offices, showroom, storage and mounting room. The other will be 60 x 75, and will be used as a moulding shop,

The keel of a new palace passenger steamer for the Richelieu and Ontario Navigation Co. is this week being laid at Bertram's shipyards. The steamer will cost between $450,000 to $475,000 to complete. Like the Kingston and Toronto, and other new vessels of the Richelieu Line, it has been designed by Mr. A. Angstrom, marine architect. She will be 340 ft. long over all. The Kingston is 290 ft. long.

A GOOD DEMAND FOR STOVES.

Owing to the great amount of natural gas sent across the Detroit river to Detroit city, the supply at Windsor, Ont., has recently been insufficient to keep the houses warm. consequently there is activity in the stove and furnace trade, as householders are taking out their gas fixtures and putting in stoves, etc.

SITUATION WANTED.

ABOUT March 15th next, well-posted young man will be open for position as pricer, confidential clerk or buyer. Twelve years' experience in wholesale trade. Best references. Address, Clerk, HARDWARE AND METAL, Montreal. (8)

For either new or old Buildings

OUR

Metallic Cornices

can't be equalled for practical service and artistic beauty.

They are light in weight, easily handled and economical in cost—giving fire-proof protection as well as a handsome effect.

Almost countless stock designs—or any particular size, shape or pattern to order. Write for full details.

METALLIC ROOFING CO., LIMITED

Wholesale Mfrs.

TORONTO, CANADA.

DO YOU HANDLE THE

AYLMER SPRAYERS?

IF NOT WHY NOT?

They are the most durable, easiest working and most simply constructed pump on the market, and give universal satisfaction. If our travellers have not yet called on you, write for catalogue. Liberal discounts to the trade.

AYLMER IRON WORKS, AYLMER, ONT

SPRAY PUMP

THE BEST IS THE CHEAPEST

Collins' Improved Spray Pump, the best pump made for spraying fruit trees, flowers, potato vines, vegetables of all kinds; also, for spraying cattle to protect them from the flies. Sprays equally as well in an upward position as downward. Made in heavy XX tin and galvanized iron. For sale by all Hardware Dealers. Price on application.

Manufactured by

The COLLINS MFG. CO.

34 West Adelaide, TORONT

DAVID PHILIP

MANUFACTURERS' AGENT

362½ Main St., WINNIPEG.

Correspondence invited from manufacturers of Staple or Heavy Hardware, Iron or Steel Bolts and Nuts, etc., either by carrying stock in Winnipeg or by selling direct from factory.

GOOD REFERENCES.

Book, Broom and Mattress Wire

High Grade, Double Tinned.

Fine Annealed Brush Wire

Soft Coppered Wire,

Tinned Wire of all kinds,

The Peerless Wire Co.

Hamilton, Ont.

SEALED TENDERS addressed to the undersigned, and endorsed "Tender for Elevator Wharf Extension at Depot Harbour, Ont." will be received at this office until Friday, 22nd February, 1901, for the construction of an additional length to the Elevator Wharf at Depot Harbour, Muskoka District, Ontario.

Plans and specifications can be seen at this Department; at the offices of Mr. H. A. Gray, resident engineer, Confederation Life Building, Toronto; The Resident Engineer, Room 411, Merchants Bank Building, St. James St., Montreal; Mr. Ph. Beland, Clerk of Works, Post Office, Quebec, and on application to the Postmaster, at Parry Sound, Ont. Forms of tender can also be obtained at the above mentioned places.

Persons tendering are notified that tenders will not be considered unless made on the printed forms supplied, and signed with their actual signatures.

The contractor will be required to conform to regulations to be made by the Governor-General-in-Council, respecting the accommodation, medical treatment and sanitary protection of the working men employed on the work.

Each tender must be accompanied by an accepted cheque on a chartered bank, made payable to the order of the Honourable the Minister of Public Works, for twenty-five thousand dollars ($25,000). The cheque will be forfeited if the party decline the contract or fail to complete the work contracted for. If the tender be not accepted, the cheque will be returned.

The Department does not bind itself to accept the lowest or any tender.

By order,
JOS. R. ROY,
Acting Secretary.

Department of Public Works of Canada.
Ottawa, January 22nd, 1901.

Newspapers inserting this advertisement without authority from the Department, will not be paid for it. (7)

MACKENZIE BROS.

HARDWARE
MANUFACTURERS' AGENTS,

Travellers covering Manitoba, Northwest Territories and British Columbia. | WINNIPEG, MAN.

CORRESPONDENCE SOLICITED.

What About **BRUSHES** for Spring Trade?

Our Revised Catalogue for 1901 now ready and our travellers are working your way. Don't buy till you get our prices. We guarantee good values and prompt shipment.

Meakins & Sons
Hamilton, Ont.

Meakins, Sons & Co.
Montreal, Que.

ROLLING MILLS MEN DINE.

THE first annual banquet of the staff and foremen of the Montreal Rolling Mills was held at the Queen's Hotel last Saturday evening, February 9. Covers were laid for about fifty, and everyone present spent a merry time in feasting and merriment. It was essentially a social function, and, as such, was an eminent success.

The printed menu was novel and appropriate, bearing a representation of a keg of "M" nails. There were sumptuous supplies called for by the menu, and they were done full justice to. Then the chairman, Mr. Wm. McMaster, proposed a toast to the King, and it was received with new-born enthusiasm. Letters of regret were read from Mr. A. F. Macpherson, the late secretary of the company; Mr. J. Irving, lately departed for Nova Scotia, and Mr. W. H. Dippel. There were still some guests present, however, and Mr. J. R. Kinghorn welcomed them in a few appropriate words, acknowledged by Mr. F. S. Hickey.

"Our Chief," was proposed by Mr. J. L. Waldie, secretary-treasurer, and the toast brought out a tribute to Mr. McMaster to which he neatly replied. "Our Works" was proposed by Mr. Wm. McMaster, and responded to by Mr. W. N. Fessenden. Neither were "The Ladies" neglected; benedict A. H. Huff proposed their health, and bachelor Thomas H. Moore ably responded on their behalf.

But there were more than speeches. Song and story played an important part in the evening's proceedings. Mr. F. L. Hickey sang several songs that were highly appreciated; Mr. H. Diplock told about the antics of "The Dude on the Street Car"; Mr. J. A. Gingras sang two comic songs entitled "What Cheer," and "The Old Guard," and Messrs. H. Diplock and C. J. Hempey gave a duet entitled "Larboard Watch"; Mr. W. E. Williams sang "The Tar's Farewell."

Then speeches were made by "one another." Mr. James Dunlop, the city traveller, made a happy hit when he proposed the functions be monthly instead of yearly. Mr. Morgan took down the house with a "cosmopolitan" speech, and Mr. Bradbury surprised the audience with the song "Then You'll Remember Me." The dinner committee, composed of Mr. A. McMaster, Geo. James and C. J. Hempy were also called upon for orations. Mr. G. Boyd materially increased his platform fame. Messrs. Myers and Hulick were other speech-makers. Mr. J. A. Gingras made the wittiest speech of the evening, vividly describing Mr. Miller's series of bath tubs in the new galvanizing plant.

THE INSOLVENCY BILL.

If the various commercial organizations intend to make further efforts to secure an insolvency law, it would appear that they ought to approach Mr. Fortin, M.P., on the matter. Mr. L. E. Geoffrion, president of the Montreal Chambre de Commerce, has received a leter from the Premier, saying that it would be useless for a deputation from the Board of Trade and Chambre de Commerce to wait upon the Government in regard to this matter, as "Mr. Fortin is already in possession of the facts." According to the tenor of this letter, Mr. Fortin has the bringing forward of the bill in his own hands. It would behoove those interested, then, to wait upon Mr. Fortin and assure him of support should he endeavor to legislate for this much-needed reform. At present Mr. Fortin thinks his bill has only lukewarm support from the commercial classes, and that the banks and some of the Provinces are actually opposed to the measure. Can he not be convinced that he has erred in judgment?

A. Sweet & Co., general merchants, Winchester, Ont., are having plans for an extensive addition to their new store prepared by an architect.

TRADE CHAT.

D. McKENZIE is starting as harness dealer and F. Lovenson as painter in Little Glace Bay, N.S.

Rigali & Rigali have started as painters in Quebec.

The Canadian Locomotive Co., Limited, has been incorporated.

Campbell & McBride are starting a hardware store in Renfrew, Ont.

Wallace McDonald has opened a general store in Little Glace Bay, N.B.

The Canada Linseed Oil Mills, Limited, Montreal, have been incorporated.

I. N. Waite is offering his stove and tinware business in Picton, Ont., for sale.

The Peat Fuel Co., of Canada, Limited, Fraserville, Que., have applied for incorporation.

At a meeting held the other evening a large number of the merchants east of the Don, Toronto, agreed to discontinue trading stamps.

Hawkins Bros., general merchants, Blind River and Spanish Station, Ont., have sold their store at Spanish Station to W. H. Graham.

Wm. Hope & Co., tinsmiths, etc., Perth, Ont., have sold out to A. T. McArthur, who has been in business in that town for many years.

The value of the products exported from St. John, N.B., up to February 1, this winter, was $2,813,695, as against $4,184,452 in the same period last year.

J. Lemmon and E. Lawrenson, for many years engaged in the heating and tinsmithing business, have opened up business on King street, near Princess, Kingston. They have purchased the stoves and tinsmithing stock of the late Squire Co., and will also add a new stock. Power & Son are preparing plans for the extension and improvement of the store. Mr. Lemmon was formerly foreman for the late Squire Co.

WILL ERECT A NEW WAREHOUSE.

C. A. Godson, mannager of James Robertson & Co.'s Vancouver agency, has secured premises on Hastings street, Vancouver, near the tram office, where a warehouse, to cost about $25 000, will be erected. The firm will move into the new premises as soon as completed, when the manufacture of white leads and ready-mixed paints will be started in Vancouver.

WINNIPEG RETAIL CLERKS.

The retail clerks held their regular meeting on Monday evening, President Bro. Trumble in the chair, who, in his opening address, said he was pleased to see such a well-attended meeting. Ex-President Calder was elected honorary president.—Winnipeg Free Press, February 6.

HORSE TRADE C MARK NAILS

Revised Hardware Trade Price List

Revised and Adopted. Montreal, Feb. 7th, 1901.

No.	14	12	11	10	9	8	7	6	5	4	
Length	3⅜	2⅞	2¾	2⅝	2½	2⅜	2¼	2⅛	2	1⅞	in.
List	$ .20	.20	.20	.20	.20	.22	.24	.28	.32	.48	cents.

In boxes of 25-lbs. each, (either loose, or in 5-lb. cardboard packages).
Extra for 1-lb. cardboard packages ½c. per lb. net.

RACE SHOE, OR PLATE NAILS.

EXTRA SELECTED.

No.	1	2	3	
Length	1½ inch.	1⅝ inch.	1¾ inch.	Short Oval and
List	$2.50	1.50	.75 per lb.	Short Oval.

In boxes of 5-lbs., 10-lbs and 25-lbs. each, 1-lb. packages.

Discount, 50 per cent. and 7½ per cent. (to Hardware Trade only).
Delivered free on board cars or boat at Montreal.
Terms Cash: Discount 3 per cent. for prompt settlement within 10 days.

PATTERNS AND SIZES.

Oval Head

Nos. 4 to 14.

Short Oval

Nos. 1 to 8.

Countersunk Head

Nos. 5 to 12.

Short Countersunk

Nos. 3 to 8.

FARRIERS' PRICE LIST.

Adopted Feb. 7th, 1901.

No.	14	12	11	10	9	8	7	6	5	4	
Length	3⅜	2⅞	2¾	2⅝	2½	2⅜	2¼	2⅛	2	1⅞	in.
Price	$2.50	2.50	2.50	2.50	2.50	2.75	3.00	3.50	4.00	6.00	per box.

In boxes of 25-lbs. each. Cash Discount 3 per cent.

SPECIAL NOTICE:

The above Farriers' list represents the minimum retail prices at which our "C" brand Horse Shoe Nails may be sold to Farriers in the Provinces of Ontario, Quebec, and Maritime Provinces.

Montreal, Feb. 7th, 1901. **CANADA HORSE NAIL COMPANY.**

HEATING AND PLUMBING

MONTREAL'S NEW BUILDING BY-LAW.

WITH a few amendments, the Montreal City Council have passed the proposed new building by-law, and it now only remains for the city fathers to set the date when the new regulations shall be enforced. Then the council will have done its duty to the contracting and plumbing trades of the city.

The general tendency of the by-law is to prevent the erection of shoddy buildings within the city limits, to make the erection of frame and cased buildings illegal and to fix much higher standards for fire protection, division walls, plumbing, etc. This will all prove additional protection for tenants against fires, diseases and collapse, but it will make building dearer, and this is exciting opposition among proprietors in those parts of the city where the cheaper class of houses are built—the outlying wards. As one alderman who represents one of the annexed wards remarked: "For the rents that the proprietors receive in the outlying wards they cannot afford to adopt all the expensive requirements of this new by-law, and those of us who are in the council would be simply cutting our throats if we voted for it."

However, in spite of such opposition, Aldermen Hart and Lamarche have pushed the measure clause by clause through the council, and will now strive to have it enforced at an early date.

The law cannot be enforced too soon, for the style of dwelling being erected is wanting in many respects. Even now Mr. Dore, the sanitary engineer, has ascertained that certain houses are being built in the city that are not having sewer pipes of the proper quality and weight put in them, and he is about to enter action against the owners of the houses.

Asked about this poor plumbing, an important plumber of the city remarked: "Oh, there is nothing new in that. It is disgraceful that the majority of houses being put up in the city are being furnished with such light sewer pipe installed in such a crude manner. It has gone on for years, but we hope this new law will stop it.

The Master Plumbers' Association intend to have extracts from the new by-law appertaining to the plumbing trade printed and published in the form of a pamphlet for the convenience and information of the master plumbers.

PLUMBING AND HEATING NOTES.

Laramee & Giroux, plumbers, Montreal, have dissolved.

Nicholas Connolly, of N. K. & M. Connolly, contractors, Quebec, is dead.

The assets of J. Lafrance & Co., plumbers, Montreal, have been sold.

Gagnon & Caron have been appointed curators of Thos. Forest, plumber, Montreal.

W. R. Walker, plumber and tinsmith, Welland, Ont., is advertising his business for sale.

James Robertson & Co. and Wm. Braid & Co. intend erecting two warehouses on Hastings street, Vancouver, to cost about $50,000.

Ritchie & Sharpe, contractors, Winnipeg, have the contract for Alloway's new block on Portage avenue, Winnipeg. The figure was about $30,000.

Plans have been prepared by M. C. Edey, architect, Ottawa, for a departmental store to be erected for T. Lindsay & Co. at the corner of Bank and Sparks streets, Ottawa. The store will extend the full block on Bank street, from Sparks to Wellington.

PLUMBERS ON STRIKE.

There was a hitch yesterday between the plumbers who are doing the plumbing and steamfitting work in connection with the new ventilating system now being installed in the city post office, and the Government engineer, Cowan. The reason why the plumbers have raised a contention is on account of non-union men being engaged to help them, while a large number of union men in the city are idle. The two plumbers quit work yesterday at 10 a.m., after their request for none but union men to assist them had been refused. The union men claim that the men taken on in their places are not mechanics, and say that they will not be responsible for work done with them.—Winnipeg Telegram, February 6.

AN IMPORTANT DECISION.

A decision of much importance to all contractors was given by Judge Morgan on Wednesday in the case of Walker & Craig vs. Ellis and Allridge. The plaintiffs are builders, Mr. Ellis an undertaker, and Mr. Allridge a foreman builder. On February 15, 1899, the defendants covenanted to pay $600 if one R. Smith failed to build four houses on the east side of Markham street in accordance with certain plans and specifications before November 10, 1899. Smith did not complete the buildings within the time, and the plaintiffs sued for the amount of the covenant. It appeared, however, that the plaintiffs had taken the contract away from Smith and given it back to him in an altered form without giving the defendants sufficient notice. His Honor held that although the defendants were liable under the original contract, its alteration had relieved them of all liability.

SOME BUILDING NOTES.

The Rodney Gas and Water Co., Limited, Rodney, Ont., have been incorporated.

James H. Quinn, contractor, Osnabruck, Ont., has assigned to John C. Milligan.

John Luney, Tecumseh avenue, London, Ont., has taken out a permit for a brick residence on James street, near Maitland, London.

Ex-Ald. S. J. Davis, Ottawa, is erecting nine houses on the old Metropolitan Athletic Grounds and intends starting to build two more shortly.

Improvements, to cost $15,000, are to be made to the Palmer House, Toronto. One of the changes will be the installation of an up-to date elevator.

Hudon, Hebert & Cie., wholesale hardware dealers, Montreal, are erecting a large warehouse on De Bresoles street, Montreal. It will contain five storeys 63 x 70 feet.

AN INSPECTOR FOR HAMILTON.

A. W. Harris, secretary of the Hamilton, Ont., Plumbers' Union, has written a letter to the Mayor directing his attention to the manner in which plumbing is done in the city, and requesting that a by-law governing the same be passed in the near future. Mr. Harris claims that sanitation should be a very great consideration in plumbing, but says this is not the case in Hamilton. Cheapness, Mr. Harris further says, is the only thing considered, and the tenant of a house is one who suffers, not the landlord. The appointment of an inspector is advocated in order that all plumbing work may be performed according to the by-law.

PLUMBING AND HEATING CONTRACTS.

Purdy, Mansell & Co. have secured the contract for the plumbing in the Queen City Yacht Club's new club house at the foot of York street.

PROSPECTS FOR A GOOD BUILDING SEASON.

PROSPECTS would indicate that there will be quite a large number of buildings erected in Montreal this coming season. Before the disastrous fire of January 23 occurred the architects were already busy at plans of projected erections, but now, when two blocks of business houses have been laid low, the outlook in the building trades becomes even much brighter.

It seems to be reasonably certain that a new Board of Trade building will be decided upon and begun immediately, and we can believe that it will be no less handsome than was its predecessor. Hutchison & Wood, architects, have already applied for permission to erect a new building for Silverman, Boulter & Co., but the city wishes to widen St. Paul street, and it is not decided that this fur and hat firm will rebuild on confined ground.

The road committee has given out that the new homologated line must be followed, and that if the proprietors build on the old line they will do so at their own risk, and get nothing if the city at any time were obliged to tear down the buildings. This move on the part of the road committee affects the other houses on St. Paul and St. Peter streets. Seybold, Son & Co. will build a large new and model hardware house.

Other new erections are on the tapis. The Bank of Montreal will put up a large handsome building for their main offices, and plans are about completed for the fine block which the Royal Insurance Co. will build this year on the west side of Place D'Armes. The material determined upon for this palatial edifice will be finely cut sandstone, the same blending with the company's six-storey building on the neighboring corner. There will be a grand entrance from Place D'Armes. The height of the new building from sidewalk to cornice will be 108 ft. The new portion will be seven storeys high, which will necessitate the addition of another storey to the corner block.

Mr. A. Raza has completed plans for a new block of stores for Mr. P. P. Martin, on St. Paul street, and five on Commissioners, the block extending through to that street. The frontage is 72 feet and depth 167 feet. The front will be of cut stone and the whole will present a solid and imposing appearance.

BUILDING PERMITS ISSUED.

The following building permits have been issued in Toronto: To Mrs. R. Almond, for alterations and additions to 634 and 636 Palmerston avenue, to cost $1,400; to Wm. Le Blanc, for three residences, 98-102 Dundas street, to cost $3 000; to George Gooderham, for alterations to 440 448 Front street, to cost $3,500, and The Canadian Hygiene Butter Co., Limited, for alterations to the building at the south-east corner of Berkeley and Queen streets, to cost $1,400.

ALUMINUM AS FUEL TO PRODUCE HEAT.

A mixture of powdered aluminum and oxide of iron has been patented throughout the world under the name of "Thermit," according to an exchange. It is said to be practically smokeless, and it emits when burning no noxious fumes; it is further free from liability to spontaneous combustion, but it only needs the light of a match to develop a temperature of 5,432 deg. F., which is far beyond the melting point of platinum (3 080 deg. F.)

This discovery of Dr. Hans Goldschmidt, of Essen, Ruhr, has found immediate applications of the most valuable kind in the industries, such as the welding of iron pipes, repairing castings, or making local additions of metal when necessary. For the repair of a 4-in. pipe all that is necessary is a clay-lined fire-clay crucible, a pair of tongs, a clamp, the necessary supply of thermit, and a box of matches, the whole weighing about 50 lb. After the weld, the pipes have been subjected to the several tests, even including drawing, without a sign of failure at the weld. The work is prepared in the following manner:

The pipe ends to be joined are cleaned by filing, and pressed together by clamps, a mould of sheet iron being fixed underneath and around the joint as a receptacle for the thermit; it is surrounded with mouldings and to prevent accident from leakage. The thermit is then ignited in a crucible of suitable size, and the white-hot fluid is poured into the sheet-iron mould until the joint is entirely surrounded by the fluid mass. Immediately on solidification a little sand is added to keep in the heat. It takes 1½ minutes to weld a 2½-in. pipe. The molten thermit does not adhere to the pipe, nor to the mould. Rails are joined in the same manner.

The extreme simplicity of this process when compared with electric welding or the use of a portable cupola will give thermit a preference in all sorts of work. The cost of jointing a pipe is said to be already cheaper than using either a coupling or flanges, and as aluminum is steadily decreasing in cost price, this handy process may be expected to gain popularity. On breakdowns of valuable machinery it should be of great service. We shall watch with much interest the progress of this invention.

John H. Wilson, wholesale and retail hardware dealer, Montreal, is dead.

THE FLAX SEED SITUATION.

ACCORDING to Beerbohm, the flax crop outlook for India is only moderate. The weather in India of late has been decidedly more favorable, and prospects in some districts are generally described as good. In the district supplying Calcutta quite an average crop is expected, but in Bombay, partly owing to the drought early in the season causing a decrease in the area sown, not much over half an average crop is expected; but even this would give a decidedly better result than last year, which is not saying much to encourage the bearishly inclined seed and oil man. The exports from India during the past seven years were in round numbers 5,600,000 bush. in 1900, 9,400,000 bush. in 1899, 9,600,000 bush. in 1898, 3,850,000 bush. in 1897, 7,080,000 bush. in 1896, 5,600,000 bush. in 1895, and 11,960,000 in 1894.

The crop in Argentine is always a Chinese puzzle, promising big and always turning out little. Does anyone in the oil trade remember of a single season where the outcome coresponded with the roseate Government crop report and much of the private information desseminated about this time in the year? It is always conservative, therefore, to reduce the figures thus paraded from 35 to 40 per cent. unless the history of the past be totally disregarded. The Government report mentioned by Snow in his cable is just about as expected. On wheat Snow says the crop has been estimated by the Government bureau at 106,000,000 bushels, about the same as last year. This would allow 72,000,000 bushels for export. Snow's opinion is that it is too high. The same report gives a flax crop of 650,000 tons, about 150,000 tons more than the highest of the private estimators, and 250,000 tons more than the lowest. This means a 23,214,000 bushel crop, or 5,357,000 bushels above the highest claim of private parties, and 8,928,000 bushels above the lowest claim. With expert (?) testimony tending to substantiate a crop ranging all the way from 17,286,000 to 23,214,000 bushels, it is not to be wondered at if the genuine seeker after the facts is inclined to take to the woods. The Argentine crop last year was a small one, approximately 22,000 tons, or 7,857,000 bushels, practically all of which was exported, and we do not expect to see this year more than a 400,000 ton crop, or 14,285,000 bushels, and will not be surprised if it falls below this figure.

Now, let us see what bearing the above may have on the general European situation—then what influence, if any, it will have on the American seed position. At the outside, not more than 8,000,000 bushels can be expected to reach Europe from India, and figuring the Argentine supply at 14,000,000 bushels, we have 22,000,000 bushels. Add 4,250,000 bushels from Russia, and there appears to be an available supply of 26,250,000 bushels, compared with total shipments in 1900 of 15,636,000 bushels, 19,696,000 bushels in 1899, and 18,836,000 bushels in 1898. It looks as though Europe would be entirely independent of American seed this year and abundantly able to help America out if our 1899 1900 crop should fail to furnish enough seed to go around. As it is not probable that this country will need much, if any, foreign seed this year, prices for the home grown article during the calendar year 1901 will most likely be governed by our own forces of supply and demand, coupled with the veto power which the price of foreign seed plus duty of 25c. per bushel and other shipping charges will exercise over domestic seed values.—Paint, Oil and Drug Review.

WHITBY BOARD OF TRADE.

The annual meeting of the Whitby, Ont., Board of Trade, was held on Friday last week. The following officers were elected:

President—J. B. Dow.
1st Vice-President—Dr. Adams.
2nd Vice-President—J. Ferguson.
Treasurer—J. B. Howden.
Secretary—F. H. Annes.
Directors—L. T. Barclay, Fred Hatch, Geo. Cormack, D. Galbraith, G. A. Ross, J. Thomson, A. T. Lawler, C. King, R. L. Huggard, Jas. Rutledge, J. H. Long, J. A. Watson, John Burns, J. Shaw, Col. Farewell and H. S. Newton; ex-officio, Mayor Ross and Dr. McGillivray, Chairman Board of Education.

President Dow, in his inaugural address, dealt with several important problems, chief among them being the installation of fire wards, the municipalization of street lighting, the development of a lakeside summer resort, and the building of a trolley line connecting the town with the water front.

A strong resolution in favor of the Government granting a bounty for the beet-sugar industry in order to establish it in Ontario was unanimously adopted.

INQUIRIES REGARDING CANADIAN TRADE.

The following were among the recent inquiries relating to Canadian trade received at the High Commissioner's office, in London, England:

1. The proprietors of a horse mart, with excellent facilities for the sale of imported animals, are desirous of getting into touch with Canadian exporters of horses.

2. The names of the principal paper and wood pulp makers in Canada are asked for by a North of England firm.

3. The proprietors of a saddle soap, for cleaning saddles, harness, military accoutrements, and brown leather goods generally, desire to place their Canadian agency in the hands of a responsible firm willing to take up the article.

4. A Staffordshire firm of sanitary pottery manufacturers make inquiry respecting the opening in Canada for such goods as they turn out—porcelain baisins, lavatories, wash-up sinks, enamelled fire-clay baths, fire-clay sinks for hospitals, etc.

5. A stationery firm manufacturing albums, scrap books, and fancy leather goods, inquire as to the prospect of doing business in Canada, and are open to appoint agents to represent them.

6. A London firm, who have a branch in Sydney, N.S.W., are anxious to get into touch with Canadian manufacturers of boots and shoes, rubber goods, etc., with a view to representing them in Australia.

[The names of the firms making the above inquiries, can be obtained on application to the editor of HARDWARE AND METAL, Toronto. When asking for names, kindly give number of paragraph and date of issue.]

CURRENT MARKET QUOTATIONS.

February 15, 1911.

These prices are for such qualities and quantities as are usually ordered by retail dealers on the usual terms of credit, the lowest figures being for larger quantities and prompt pay. Large cash buyers can frequently make purchases at better prices. The Editor is anxious to be informed at once of any apparent errors in this list as the desire is to make it perfectly accurate.

METALS.

Tin.

Lamb and Flag and Straits—56 and 28 lb. ingots, per lb.	0 32	0 33

Tinplates.

Charcoal Plates—Bright

M.L.S., equal to Bradley.		Per box
I.C., usual sizes	....	$6 75
I.X., "	....	8 25
I.X.X., "	....	9 75
Famous—		
I.C.	....	6 75
I.X.	....	8 25
I.X.X.	....	9 50
Raven & Vulture Grades—		
I.C., usual sizes	....	5 00
I.X., "	....	6 00
I.X.X. "	....	7 00
I.X.X.X. "	....	8 00
D.C., 12⅛x17	....	4 75
D.X.	....	5 50
D.X.X.	....	7 50

Coke Plates—Bright

Bessemer Steel—		
I.C., usual sizes	....	4 15
I.C., special sizes, base	....	4 50
20 x 28	....	8 50

Charcoal Plates—Terne

Dean or J. G. Grade—		
I.C., 20x28, 112 sheets	....	8 50
I.X., Terne Tin	....	10 50

Charcoal Tin Boiler Plates.

Cookley Grade—		Per lb.
X X., 14x56, 50 sheet bxs; " 14x60 "; " 14x65 "	0 07	0 07½

Tinned Sheets.

72x30 up to 24 gauge	0 08	0 08½
" 26 "	0 08½	0 09
" 28 "	0 09	0 09½

Iron and Steel.

Common Bar, per 100 lbs.	1 65	1 70
Refined " "	....	2 15
Horse Shoe Iron "	....	2 05
Hoop steel, 1½ to 3 in. base, extras for smaller sizes	....	3 10
Sleigh Shoe Steel " base	....	2 00
Tire Steel	....	2 00
Machinery iron finish	....	2 05
Cast Steel, per lb	0 90	0 00
Toe Calk Steel	....	2 30
T. Firth & Co's special cast steel, per lb.		0 13
Jessop's Tool Steel	....	0 13

Boiler Tubes.

1½-inch	....	0 12½
2 "	....	0 13
2½ "	....	0 15
3 "	....	0 16
3½ "	....	0 20
4 "	....	0 25

Steel Boiler Plate.

¼ inch	....	2 25
3-16 inch	....	2 25
⅜ inch and thicker	....	2 25

Black Sheets.

18 gauge	....	3 00
20 gauge	....	3 00
22 to 24 "	....	3 10
26 "	....	3 20
28 "	....	3 30

Canada Plates.

All dull, 52 sheets	....	3 00
Half polished	....	3 15
All bright	3 65	3 75

Iron Pipe.

Black pipe—		
¼ inch	....	3 00
⅜ "	....	3 00
½ "	....	3 15
¾ "	....	3 90
1 "	....	4 50
1¼ "	....	6 35
1½ "	....	7 55
2 "	....	10 10
2½-6 inch, discount 60 p c.		
Galvanized pipe—		
½ inch	....	4 65
¾ "	....	5 35
1 "	....	7 75
1¼ "	....	9 75
1½ "	....	11 25
2 "	....	15 50

Galvanized Sheets.

	G C. Comet.	Amer.	Queen's Head.
16 gauge		4 5	4 00
18 to 24 gauge 4 25	3 85	4 35	4 25
26 " 4 50	4 10	4 35	4 50
28 " 4 75	4 35	4 50	4 75

28 gauge American equals 26 gauge English.

Chain.

Proof Coil, 3-16 in., per 100 lb	....	....
" ¼ " "	8 00	8 50
" 5-16 " "	5 35	5
" ⅜ " "	4 35	4 85
" 7-16 " "	4 15	4 65
" ½ " "	4 35	4 50
" ⅝ " "	3 85	4 35
" ¾ " "	3 80	4 00

Halter, kennel and post chains,	40 and 50 p.c.
Cow ties	40 p.c.
Stall fixtures	35 p.c.
Trace chain	25 p.c.

Jack chain, iron, single and double, discount 35 p c.

Jack chain, brass, single and double, discount 40 p.c.

Copper.

Ingot

English B. S., ton lots	0 19	0 20
Lake Superior	....	....

Bolt or Bar.

Cut lengths round, ½ to ⅞ in.	0 23½	0 25
" round and square 1 to 2 inches	0 23½	0 25

Sheet.

Untinned, 14 oz., and light, 16 oz., 14x48 and 14x60	0 23	0 23
Untinned, 14 oz., and light, 16 oz., irregular sizes	0 23	0 23½

NOTE.—Extra for tinning, 2 cents per pound, and tinning and half planishing 3 cents per pound.

Tinned copper sheets	....	0 26
Planished	....	0 32

Braziers (In sheets.)

4x6ft. 25 to 30 lbs. ea., per lb.	....	0 25½
" 35 to 45 " "	....	0 24½
" 50-lb. and above, "	....	0 23½

Boiler and T. K. Pitts

Plain Tinned, per lb	....	0 28
Spun, per lb	....	0 32

Brass.

Rod and Sheet, 14 to 30 gauge, 15 per cent.

Sheets, hard-rolled, 2x4	0 24	0 25
Tubing, base, per lb	0 24	0 25

Zinc Spelter

Foreign, per lb	0 06	0 06½
Domestic "	....	....

Zinc Sheet.

5 cwt. casks	6 75	7 00
Part casks	7 00	7 50

Lead.

Imported Pig, per lb	0 04¾	0 05
Bar, 1 lb.	0 05½	0 05¾
Sheets, 2½ lbs. sq. ft., by	....	0 06¼
Sheets, 3 to 6 lbs., "	....	0 06

NOTE.—Cut sheets ½ cent per lb. extra.

Pipe, by the roll, usual weights per yard, lists at 7c. per lb. and 15 p.c. dis. f.o.b. Toronto.

NOTE.—Cut lengths, net price, waste pipe 8-ft. lengths lists at 7½ cents.

Shot.

Common, $6.50 per 100 lb.; chilled, $7.00 per 100 lb.; buck, seal and ball, $7.50. Discount, 7½ p.c. Prices are f.o.b. Toronto, Hamilton, Montreal, St. John and Halifax. Terms 3 per cent., cash, freights equalized on Montreal.

Soil Pipe and Fittings.

Discount, 60 and 10 per cent. on medium and extra heavy, and 60 per cent. on light.

Solder.

	Per lb.	Per lb.
Bar half-and-half, guarant'd	....	0 19
Bar half-and-half, commer'l	....	0 18½
Refined	....	0 18½
Wiping	....	0 18

NOTE.—Prices of this graded according to quantity. The prices of other qualities of solder in the market indicated by private brands vary according to composition.

Antimony.

Cookson's, per lb.	0 11	0 11½

White Lead.

	Per 100 lb.
Pure	6 87½
No. 1 do	6 50
No. 2 do	6 12½
No. 3 do	5 75
No. 4 do	5 37½
Munro's Select Flake White	7 37½
Elephant and Decorators' Pure	7 12½
Brandram's B B. Genuine	8 00
" " Decorative	7 55
" " No. 1	6 85
" " No. 2	6 00

Red Lead.

Genuine, 560 lb. casks, per cwt.	....	$5 50
Genuine, 100 lb. kegs, per cwt.	....	5 75
No. 1, 560 lb. casks, per cwt.	....	5 25
No. 1, 100 lb. kegs, per cwt.	....	5 00

White Zinc Paint.

Elephant Snow White	0 08	0 09
Pure White Zinc	0 08	0 09
No. 1	0 06	0 07½
No. 2	0 05	0 06½

Dry White Lead.

Pure, casks	5 75
Pure, kegs	6 25
No. 1, casks	5 50
No. 1, kegs	6 00

Prepared Paints.

In ¼, ½ and 1 gallon tins.

Pure, per gallon	....	1 26
Second qualities, per gallon	....	1 10
Barn (in bbls.)	0 75	0 85
The Sherwin-Williams Paints	....	1 45
Canada Paint Co's Pure	....	1 25
Toronto Lead & Color Co's Pure	....	1 25
Sanderson Pearcy's Pure	....	1 20
Stewart & Wood's Champion Pure.		1 20

Colors in Oil.

25 lb. tins, Standard Quality.

Venetian Red, per lb.	0
Chrome Yellow	0
Golden Ochre	0
French "	0
Marine Black	0
" Green	0
Chrome "	0 05
French Imperial Green	0 [illegible]

Colors, Dry.

Yellow Ochre (J.C.) bbls.	1 35	1
Yellow Ochre (J.F.L.S.), bbls	...	2
Yellow Ochre (Royal)	1 10	1
Brussels Ochre	....	2
Venetian Red (best), per cwt.	1 80	1
English Oxides, per cwt.	3 00	3
American Oxides, per cwt.	1 75	2
Canadian Oxides, per cwt.	1 75	2
Super Magnetic Oxides, 93p.c.	2 00	2
Burnt Sienna, pure, per lb.	....	0
" Umber, " "	....	0
do Raw	....	0 40
Drop Black, pure	....	0 [illegible]

Chrome Yellows, pure	....	0 18
Chrome Greens, pure, per lb.	....	0 12
Golden Ochre	....	0 03½
Ultramarine Blue in 28-lb. boxes, per lb.	0 08	0 24
Fire Proof Mineral, per 100 lb	....	1 00
Genuine Eng. Litharge, per lb	....	0 07
Mortar Color, per 100 lb.	....	1 25
English Vermillion	....	0 80
Pure Indian Red, No. 45 lb.	....	0 80
Whiting, per 100 lb.	....	0 55

Blue Stone.

Casks, for spraying, per lb.	0 07
100-lb. lots, do. per lb.	0 08

Putty.

Bulk in bbls.,	2 00
Bulk in less quantity	2 15
Bladders in bbls.	2 20
Bladders in kegs, boxes or loose	2 35
Bladders in 25-lb. tins	2 45
Bladders in 12½-lb. tins	2 75
Bladders in bulk or tins less than 100 lb	3 00

Varnishes.

In 5-gal. lots.)		Per gal.
Carriage, No. 1	2 90	3 30
" body	8 00	9 00
" rubbing	4 00	5 00
Gold Size, Japan	3 00	3 40
Brown Japan	2 40	2 80
Elastic Oak	2 90	3 30
Furniture, extra	2 40	2 80
" No. 1	1 60	2 00
Hard Oil Finish	2 70	3 10
Light Oil Finish	3 20	3 60
Demar	3 30	3 70
Shellac, white	4 40	4 80
" orange	4 00	4 40
Furniture Brown Japan	1 60	2 00
Black Japan	2 40	2 80
" " No. 1	1 60	2 00

The Imperial Varnish & Color Co's., Limited Elastilite Varnish 1 gal. can, each. $2.00.

Granatine Floor Finish, per gal. $2.00.

Maple Leaf Coach Enamels; Size 1, 60c.; Size 2, 35c.; Size 3, 20c. each.

Linseed Oil.

	Raw.	Boiled.
1 to 4 bbls delivered	$0 78	$0 81
5 to 9 bbls "	77	80

Toronto, Hamilton, London and Guelph 2c. less.

Turpentine.

Single barrel, freight allowed	...	0 59
2 to 4 barrels " "	....	0 58

Castor Oil.

East India, in cases, per lb.	0 10	0 10½
" small lots	0 10½	0 11

Cod Oil, Etc.

Cod Oil per gal.	0 50	0 55
Pure Olive	....	1 20
" Neatsfoot	....	90

Glue.

Common	0 08½	0 09
French Medal	0 14	0 14½
Cabinet sheet	0 12	0 13
White, extra	0 18	0 20
Gelatine	0 22	0 30
Strip	0 18	0 20
Coopers	0 19	0 20
Huttner	....	0

HARDWARE.

Ammunition.

Cartridges.

B. B. Caps. Dom. 50 and 5 per cent.
Rim Fire Pistol, dis. 40 p. c., Amer.
Rim Fire Cartridges, Dom., 50 and 5 p. c.
Central Fire Pistol and Rifle, 10 p.c. Amer.
Central Fire Cartridges, pistol sizes, Dom. 30 per cent.
Central Fire Cartridges, Sporting and Military, Dom., 15 and 5 per cent.
Central Fire, Military and Sporting, Amer. add 5 p.c. to list. B.B. Caps, discount 40 per cent. Amer.
Loaded and empty Shells, "Trap" and "Dominion" grades, 25 per cent Rival and Nitro, net list.
Brass Shot Shells, 55 per cent
Primers, Dom., 30 per cent.

Wads — per lb

Best thick white felt wadding, in ½-lb bags 1 00
Best thick brown or grey felt wads, in ½-lb. bags 0 70
Best thick white card wads, in boxes of 500 each, 12 and smaller gauges 0 99
Best thick white card wads, in boxes of 500 each, 10 gauge 0 35
Best thick white card wads, in boxes of 500 each, 8 gauge 0 55
Thin card wads, in boxes of 1,000 each, 12 and smaller gauges 0 20
Thin card wads, in boxes of 1,000 each, 10 gauge 0 25
Thin card wads in boxes of 1,000 each, 8 gauge ....

Chemically prepared black edge grey cloth wads, in boxes of 250 each— Per M
11 and smaller gauge 0 60
9 and 10 gauges 0 70
7 and 8 gauges 0 90
5 and 6 gauges 1 10

Superior chemically prepared pink edge, best white cloth wads, in boxes of 250 each—
11 and smaller gauge 1 15
9 and 10 gauges 1 40
7 and 8 gauges 1 65
5 and 6 gauges 1 90

Adzes.

Discount, 20 per cent.

Anvils.

Per lb. 10 0 12½
Anvil and Vise combined 4 50
Wilkinson & Co.'s Anvils..lb. 0 09 0 09½
Wilkinson & Co.'s Vices..lb. 0 09¾ 0 10

Augers.

Gilmour's, discount 65 and 5 p.c. off list.

Axes.

Chopping Axes—
Single bit, per doz 6 50 10 00
Double bit, " 12 00 18 00
Bench Axes, 40 p.c.
Broad Axes, 33⅓ per cent.
Hunters' Axes 5 50 6 00
Boy's Axes 5 75 6 75
Splitting Axes 6 50 12 00
Handled Axes 7 00 10 00

Axle Grease.

Ordinary, per gross 5 75 6 00
Best quality 13 00 15 00

Bath Tubs.

Zinc 6 00
Copper, discount 15 p.c. off revised list

Baths.

Standard Enameled.

5½-inch rolled rim, 1st quality 30 00
" " " 2nd " 22 00

Anti-Friction Metal.

"Tandem" A .... per lb. 0 27
" B .... " 0 21
" C .... " 0 11½
Magnolia Anti-Friction Metal, per lb. 0 25

SYRACUSE SMELTING WORKS.

Aluminum, genuine 0 45
Dynamo 0 29
Special 0 25
Aluminum, 99 p.c. pure "Syracuse" .. 0 50

Bells.

Hand.

Brass, 60 per cent.
Nickel, 55 per cent.

Cow.

American make, discount 66⅔ per cent.
Canadian, discount 45 and 50 per cent.

Door.

Gongs, Sargant's 5 50 8 00
" Peterboro', discount 45 per cent.

Farm.

American, each 1 25 3 00

House.

American, per lb. 0 35 0 40

Bellows.

Hand, per doz 3 35 4 75
Moulders', per doz 7 50 10 00
Blacksmiths', discount 40 per cent.

Belting.

Extra, 60 per cent.
Standard, 60 and 10 per cent.

Bits.

Auger.

Gilmour's, discount 60 and 5 per cent.
Rockford, 50 and 10 per cent.
Jennings' Gen., net list.

Car.

Gilmour's, 47½ to 50 per cent.

Expansive.

Clark's, 40 per cent.

Gimlet.

Clark's, per doz 0 65 0 90
Damond, Shell, per doz 1 00 1 50
Nail and Spike, per gross 2 25 5 20

Blind and Bed Staples.

All sizes, per lb. 0 07½ 0 12

Bolts and Nuts. Per cent.

Carriage Bolts, full square, Norway 70
" " full square 70
Common Carriage Bolts, all sizes 65
Machine Bolts, all sizes 65
Coach Screws 75
Sleigh Shoe Bolts 75
Blank Bolts 65
Bolt Ends 65
Nuts, square .... 4½c. off
Nuts, hexagon .... 4¾c. off
Tire Bolts 67½
Stove Bolts 67½
Stove rods, per lb. .... 5½ to 6c.
Plough Bolts 60

Boot Calks.

Small and medium, ball, per M 4 25
Small heel, per M 4 50

Bright Wire Goods.

Discount .... 55 per cent.

Broilers.

Light, dis., 65 to 67½ per cent.
Reversible, dis., 65 to 67½ per cent.
Vegetable, per doz., dis. 37½ per cent.
Henis, No. 8, " 6 00
Henis, No. 9, " 7 00
Queen City " 7 50 0 00

Butchers' Cleavers.

German, per doz 6 00 11 00
American, per doz 12 00 20 00

Building Paper, Etc.

Plain building, per roll 0 30
Tarred lining, per roll 0 40
Tarred roofing, per 100 lb. 1 65
Coal Tar, per barrel 3 50
Pitch, per 100-lb. 0 85
Carpet felt, per ton 45 10

Bull Rings.

Copper, $2.00 for 2½ in., and $1.90 for 2 in.

Butts.

Wrought Brass, net revised list

Cast Iron.

Loose Pin, dis., 60 per cent.

Wrought Steel.

Fast Joint, dis. 60 and 10 per cent.
Loose Pin, dis. 60 and 10 per cent.
Berlin Bronzed, dis. 70, 70 and 5 per cent.
Gen. Bronzed, per pair 0 40 0 65

Carpet Stretchers.

American, per doz 1 00 1 50
Bullard's, per doz .... 6 50 ....

Castors.

Bed, new list, dis. 55 to 57½ per cent.
Plate, dis. 52½ to 57½ per cent.

Cattle Leaders.

Nos. 31 and 32, per gross 50 9 50

Cement.

Canadian Portland 2 80 3 00
English " 3 00
Belgian " 2 75 3 00
Canadian hydraulic 1 25 1 50

Chalk.

Carpenters, Colored, per gross 0 45 0 75
White lump, per cwt 0 60 0 65
Red 0 05 0 06
Crayon, per gross 0 14 0 18

Chisels.

Socket, Framing and Firmer.

Broad's, dis. 70 per cent.
Warnock's, dis. 70 per cent.
P. S. & W. Extra 60, 10 and 5 p.c.

Churns.

Revolving Churns, metal frames—No. 0, $8—No. 1, $8.50—No. 2, $9.00—No. 3, $10.00—No. 4, $12.00—No. 5, $16.00 each. Ditto wood frames—20c. each less than above. Discounts: Delivered from factories, 58 p.c.; from stock in Montreal, 56 p.c. Terms, 4 months or 3 p.c. cash in 30 days.

Clips.

Axle dis. 65 per cent.

Closets.

Plain Ontario Syphon Jet 28 00
Emb. Ontario Syphon Jet 8 50
Fittings 1 25
Plain Teutonic Syphon Washout 4 75
Emb. Teutonic Syphon Washout 5 25
Fittings 1 25
Low Down Teutonic, plain 14 50
" " " embossed 15 00
Plain Richelieu 3 75
Emb. Richelieu 4 00
Fittings 1 25
Low Down Ont. Syphon Jet, plain .. 20 00
" " " " " emb'd. 20 50
Closet connection 1 25
Basins, round, 14 in. 0 60
" oval, 17 x 14 in. 1 50
" " 19 x 15 in. 2 25

Compasses, Dividers, Etc.

American, dis. 62½ to 65 per cent.

Cradles, Grain.

Canadian, dis. 25 to 33⅓ per cent.

Crosscut Saw Handles.

S. & D., No. 3, per pair 17½
" " 5, " 22½
" " 6, " 15
Boynton pattern " 20

Door Springs.

Torrey's Rod, per doz .... (15 p.c.) 2 00
Coil, per doz 0 88 1 60
English, per doz 2 00 4 00

Draw Knives.

Coach and Wagon, dis. 50 and 10 per cent.
Carpenters, dis. 70 per cent.

Drills.

Hand and Breast.

Millar's Falls, per doz. net list.

DRILL BITS.

Morse, dis., 37½ to 40 per cent.
Standard dis. 50 and 5 to 55 per cent.

Faucets

Common, cork-lined, dis. 35 per cent.

ELBOWS. (Stovepipe.)

No. 1, per doz 1 40
No. 2, per doz 1 20
Bright, 20c. per doz. extra.

ESCUTCHEONS.

Discount, 45 per cent.

ESCUTCHEON PINS.

Iron, discount 40 per cent.

FACTORY MILK CANS.

Discount off revised list, 40 per cent.

FILES.

Black Diamond, 50 and 10 to 60 per cent.
Kearney & Foote, 60 and 10 p.c. to 60, 10, 10.
Nicholson File Co., 50 and 10 to 60 per cent.
Jowitt's, English list, 25 to 27½ per cent.

FORKS.

Hay, manure, etc., dis., 50 and 10 per cent. revised list.

GLASS—Window—Box Price.

Size United Inches.	Star Per 50 ft.	Star Per 100 ft	D. Diamond Per 50 ft.	D. Diamond Per 100 ft.
Under 26	2 10	4 00	....	6 00
26 to 40	2 30	4 35	....	6 65
41 to 50	....	4 75	....	7 25
51 to 60	....	35	....	8 50
61 to 70	....	75	....	9 25
71 to 80	....	5	....	10 50
81 to 85	....	6 50	....	11 75
86 to 90	....	....	..	14 00
91 to 95	..	....	...	15 50
96 to 100	...	....	....	18 00

GAUGES

Marking, Mortise, Etc.

Stanley's dis. 50 to 55 per cent.

Wire Gauges.

Winn's, Nos. 26 to 33, each 1 65 2 40

HALTERS.

Rope, ⅜ per gross ....
" ½ " 9 00
" ⅝ to ¾ 14 00
Leather, 1 in., per doz 3 87½ 4 00
" 1¼ in., " 5 15 5 20
Web, — per doz 1 87 2 45

HAMMERS.

Nail

Maydole's, dis. 5 to 10 per cent. Can. dis. 25 to 27½ per cent.

Tack.

Magnetic, per doz 1 10 1 20

Sledge.

Canadian, per lb. 0 07½ 0 08½

Ball Pean.

English and Can., per lb. 0 22 0 25

HANDLES.

Axe, per doz. net 1 50 2 00
Store door, per doz 1 00 1 50

Fork.

C. & B., dis. 40 per cent. rev. list.

Hoe.

C. & B., dis. 40 per cent. rev. list.

Saw.

American, per doz 1 00 1 25

Plane.

American, per gross 3 15 3 75

Hammer and Hatchet.

Canadian, 40 per cent.

Cross-Cut Saws.

Canadian, per pair 0 13½

HANGERS. doz. pairs.

Steel barn door 5 85 6 00
Stearns, 4 inch 5 00
" 5 inch 6 50
Lane's covered—
No. 11, 5-ft. run 8 40
No. 11½, 10-ft. run 10 80
No. 12, 10-ft. run 12 60
No. 14, 15-ft. run 21 00
Lane's O.N.T. track, per foot. 4½

HARVEST TOOLS.

Discount, 50 and 10 per cent.

HATCHETS.

Canadian, dis. 40 to 42½ per cent.

HINGES.

Blind, Parker's, dis. 50 and 10 to 60 per cent
Heavy T and strap, 4-in., per lb. 0 06½
" " 5-in., " 0 06¼
" " 6-in., " 0 06
" " 8-in., " 0 05¾
" " 10-in., " 0 05¾
Light T and strap, dis. 60 and 5 per cent.
Screw hook and hinge—
6 to 12 in., per 100 lbs 4 95
14 in. up, per 100 lbs 3 50

Per gro. pairs.

Spring 12 00

HOES.

Garden, Mortar, etc., dis. 50 and 10 p.c.
Planter, per doz 4 00 4 50

HOLLOW WARE

Discount, ... 45 and 5 per cent

HOOKS.

Cast Iron.

Bird Cage, per doz 0 50 1 10
Clothes Line, per doz 0 27 0 63
Harness, per doz 0 72 0 88
Hat and Coat, per gross 1 00 3 00
Chandelier, per doz 0 50 1 00

Wrought Iron.

Wrought Hooks and Staples, Can., dis. 47½ per cent.

Wire.

Hat and Coat, discount 45 per cent.
Belt, per 1,000 .... 0 60
Screw, bright, dis. 55 per cent.

HORSE NAILS.

"C" brand 50 and 7½ p.c. off new list } Oval
"M" brand 50, 10 and 5 p.c. } head.
Countersunk, 50, 10 and 5 per cent.

HORSESHOES.

Iron Shoes.	F.O.B. Montreal No. 2 and larger.	No. 1. and smaller
Light, medium, and heavy.	3 50	3 75
Snow shoes	3 75	4 00
Steel Shoes.		
Light	3 60	3 85
Featherweight (all sizes)	4 85	4 85

F.O.B. Toronto, Hamilton, London and Guelph, 10c. per keg additional.

Toe weight steel shoes 6 70

JAPANNED WARE.

Discount, 45 and 5 per cent. off list, June 1899.

ICE PICKS.

Star per doz 3 00 3 25

KETTLES.

Brass spun, 7½ p.c. dis. off new list.
Copper, per lb 0 30 0 50
American, 60 and 10 to 65 and 5 p.c.

KEYS.

Lock, Can., dis., 45 p.c.
Cabinet, trunk, and padlock, Am. per gross 60

KNOBS.

Door, japanned and N.P., per doz	1 50	2 50
Bronze, Berlin, per doz	2 75	3 25
Bronze Genuine, per doz	6 00	9 00
Shutter, porcelain, F. & L. screw, per gross	1 30	4 00
White door knobs—per doz.		1 25

HAY KNIVES.

Discount, 50 and 10 per cent.

LAMP WICKS.

Discount, 60 per cent.

LANTERNS.

Cold Blast, per doz	7 50
No. 3 "Wright's"	8 50
Ordinary, with O burner	4 25
Dashboard, cold blast	9 50
No. 0	6 00

Japanning, 50c. per doz. extra.

LEMON SQUEEZERS.

	per doz.	
Porcelain lined,	2 20	5 60
Galvanized	1 87	3 85
King, wood	2 75	2 90
King, glass	4 00	4 50
All glass	1 20	1 30

LINES.

Fish, per gross	1 05	2 50
Chalk "	1 90	7 40

LOCKS.

Canadian, dis. 45 p.c.
Russell & Erwin, per doz 3 00 3 25

Cabinet.

Eagle, dis. 30 p.c.

Padlock.

English and Am., per doz 50 6 00
Scandinavian, " 1 00 2 40
Eagle, dis. 20 to 25 p.c.

MACHINE SCREWS.

Iron and Brass.

Flat head discount 25 p.c.
Round Head, discount 20 p.c.

MALLETS.

Tinsmiths', per doz	1 25	1 50
Carpenters', hickory, per doz.	1 25	3 75
Lignum Vitae, per doz	3 85	5 00
Caulking, each	60	2 00

MATTOCKS.

Canadian, per doz 8 50 1 00

MEAT CUTTERS.

American, dis. 25 to 30 p.c.
German, 15 per cent.

MILK CAN TRIMMINGS.

Discount, 25 per cent.

NAILS.

Quotations are:	Cut.	Wire.
2d and 3d	$3 35	$3 85
3d	3 00	3 52
4 and 5d	2 75	3 35
6 and 7d	2 65	3 20
8 and 9d	2 50	3 00
10 and 12d	2 45	2 95
16 and 20d	2 40	2 90
30, 40, 50 and 60d. (base)	2 35	2 85

Galvanizing 2c. per lb. net extra.
Steel Cut Nails 10c. extra.
Miscellaneous wire nails, dis. 70 per cent.
Coopers' nails, dis. 30 per cent.
Flour barrel nails, dis. 25 per cent

NAIL PULLERS.

German and American 1 85 3 50

NAIL SETS

Square, round, and octagon, per gross 3 38 4 00
Diamond 12 00 15 00

NETTING.

Poultry, 50 and 5 per cent. for McMullen's.

OAKUM. Per 100 lb.

Navy 6 00
U. S. Navy 7 25

OIL.

Water White (U.S.)	0 16½
Prime White (U.S.)	0 15½
Water White (Can.)	0 15
Prime White (Can.)	0 14

OILERS.

McClary's Model galvan. oil can, with pump, 5 gal., per doz 0 00 10 00
Zinc and tin, dis. 50, 50 and 10.
Copper, per doz 1 25 3 50
Brass, " 1 50 3 50
Malleable, dis. 25 per cent.

GALVANIZED PAILS.

Dufferin pattern pails, dis. 50 and 10 p.c.
Flaring pattern, discount 45 per cent.
Galvanized washtubs, discount 45 per cent.

PIECED WARE.

Discount 40 per cent. off list, June, 1899.
10-qt flaring san buckets. dis. 45 p.c.
6, 10 and 14-qt flaring pails, dis. 45 p.c.
Creamer cans, dis 45 p.c.

PICKS.

Per doz 6 00 9 00

PICTURE NAILS.

Porcelain head, per gross 1 75 3 00
Brass head " 0 40 1 00

PICTURE WIRE.

Tin and gilt, discount 75 p c.

PLANES.

Wood, bench, Canadian dis. 50 per cent American dis. 50.
Wood, fancy Canadian or American 7½ o 40 per cent.

PLANE IRONS.

English, per doz 2 00 5 00

PLIERS AND NIPPERS.

Button's Genuine per doz pairs, dis. 37½ 40 p.c.
Button's Imitation, per doz 5 00 9 00
German, per doz 0 60 2 60

PLUMBERS' BRASS GOODS.

Impression work, discount, 60 per cent.
Fuller's work, discount 65 per cent.
Rough stops and stop and waste cocks, discount, 60 per cent.
Jenkins disk globe and angle valves, discount, 55 per cent.
Standard valves, discount, 60 per per cent.
Jenkins radiator valves discount 55 per cent.
" " " standard, dis., 60 p.c.
Quick opening valves discount, 60 p.c.

No. 1 compression bath cock	2 00
No. 4 " " "	2 00
No. 7, Fuller's	2 50
No 4½, "	3 00

POWDER.

Velox Smokeless Shotgun Powder.

100 lb. or less 0 85
1,000 lb. or more 0 80
Net 30 days.

PRESSED SPIKES.

Discount 25 per cent.

PULLEYS.

Hothouse, per doz	0 55	1 00
Axle	0 22	0 33
Screw	0 27	1 00
Awning	0 35	2 50

PUMPS.

Canadian cistern 1 80 3 60
Canadian pitcher spout 1 40 2 10

PUNCHES.

Saddlers', per doz	1 00	1 85
Conductors', "	9 00	15 00
Tinners' solid, per set	0 00	0 72
" hollow, per inch	0 00	1 00

RANGE BOILERS.

Galvanized, 30 gallons		6 50
" 35 "		7 50
" 40 "		8 50
Copper, 30 "		22 00
" 35 "		26 00
" 40 "		30 00

Discount off Copper Boilers 10 per cent.

RAKES.

Cast steel and malleable Canadian list 50 and 10 p.c. revised list.
Wood, 25 per cent.

RASPS AND HORSE RASPS.

New Nicholson horse rasp, discount 60 p.c.
Globe File Co.'s rasps, 60 and 10 to 70 p.c.
Heller's Horse rasps, 50 to 50 and 5 p.c.

RAZORS.

	per doz.	
Geo. Butler & Co.'s,	8 00	18 00
Boker's	7 50	11 00
Wade & Butcher's	3 60	10 00
Theile & Quack's	7 00	12 00
Elliot's	4 00	18 00

REAPING HOOKS.

Discount, 50 and 10 per cent.

REGISTERS.

Discount ...... 40 per cent.

RIVETS AND BURRS.

Iron Rivets, discount 60 and 10 per cent.
Iron Burrs, discount 55 per cent.
Black and Tinned Rivets, 60 p.c.
Extras on Iron Rivets in 1-lb. cartons, ½c per lb.
Extras on Iron Rivets in ½-lb. cartons, per lb.
Copper Rivets & Burrs, 35 and 5 p.c. dis. and cartons, 1c. per lb. extra, net.
Extras on Tinned or Coppered Rivets ½-lb. cartons, 1c. per lb.
Terms, 4 mos. or 3 per cent. cash 30 days.

RIVET SETS.

Canadian, dis. 35 to 37½ per cent.

ROPE ETC.

	Sisal.	Manila.
7-16 in. and larger, per lb.	9	13
⅜ in	10	14
¼ and 5-16 in.		15
Cotton, 3-16 inch and larger		16½
" 5-32 inch		21½
" ⅛ inch		22½
Russia Deep Sea		15½
Jute		8
Lath Yarn		9¾
New Zealand Rope		10½

RULES.

Boxwood, dis. 75 and 10 p.c.
Ivory, dis. 37½ to 40 p.c.

SAD IRONS. per set.

Mrs. Potts, No. 55, polished 0 62½ 0 65
" No. 50, nickle-plated 0 67½ 0 70

SAND AND EMERY PAPER.

Dominion Flint Paper, 47½ per cent.
B & A. sand, 40 and 2½ per cent.
Emery, 40 per cent.

SAP SPOUTS.

Bronzed iron with hooks, per doz 9 50

SAWS.

Hand Disston's, dis. 12½ p.c.
S. & D., 40 per cent.
Crosscut, Disston's, per ft 35 0 55
S. & D., dis. 35 p.c. on Nos. 2 and 3.
Hack, complete, each 0 75 2 75
" frame only 0 75

SASH WEIGHTS.

Sectional, per 100 lbs 2 75 3 00
Solid, " 2 00 2 25

SASH CORD.

Per lb 0 23 0 30

SAW SETS.

"Lincoln," per doz 6 50

SCALES.

B. S. & M. Scales, 45 p.c.
Champion, 65 per cent.
Fairbanks Standard, 35 p.c.
" Dominion, 55 p.c.
" Richelieu, 55 p.c.
Chatillon Spring Balances, 10 p.c.

SCREW DRIVERS.

Sargent's, per doz 0 65 1 00

SCREWS

Wood, F. H., bright and steel, 87½ and 10 p.c.
Wood R. H., " dis. 82½ and 10 p.c.
" F. H., brass, dis. 80 and 10 p.c.
Wood, R. H., " dis. 75 and 10p.c.
" F.H., bronze, dis. 70 p.c.
" R.H. " 65 p.c.
Drive Screws, 80 per cent.
Bench, wood, per doz 3 25 4 00
" iron, " 4 25 5 75

SCYTHES.

Per doz, net 9 00

SCYTHE SNATHS.

Canadian, dis. 45 p.c.

SHEARS.

Bailey Cutlery Co., full nickeled, dis. 60 p.c.
Seymour's, dis. 50 and 10 p.c.

SHOVELS AND SPADES.

Canadian, dis. 40 and 5 per cent.

SINKS.

Steel and galvanized, discount 45 per cent.

SNAPS.

Harness, German, dis. 25 p.c.
Lock, Andrews' 4 50 11 50

SOLDERING IRONS.

1, 1½ lb., per lb 0 37
2 lb. or over, per lb. 0 34

SQUARES.

Iron, No. 493, per doz 2 40 2 55
" No. 494, " 3 25 3 40
Steel, dis. 50 and 5 to 50 and 10 p.c., rev. list.
Try and bevel, dis. 50 to 52½ p.c.

STAMPED WARE.

Plain, dis., 75 and 12½ p.c. off revised list.
Retinned, dis., 75 p.c. off revised list.

STAPLES.

Galvanized 0 00 0 00
Plain 0 00 3 45
Coopers', discount 45 per cent.
Poultry netting staples, 40 per cent.

STOCKS AND DIES.

American dis. 25 p.c.

STONE.	Per lb.	
Washita	0 28	0 60
Hindostan	0 06	0 07
" slip,	0 09	0 09
Labrador	....	0 13
" Axe	....	0 15
Turkey	....	0 50
Arkansas	0 00	1 50
Water-of-Ayr	0 00	0 10
Scythe, per gross	3 50	5 00
Grind, per ton	15 00	18 00

STOVE PIPES.

Nestable in crates of 25 lengths.

5 and 6 inch Per 100 lengths 7 00
7 inch " " 7 50

ENAMELINE STOVE POLISH.

No. 4—3 dozen in case, net cash $4 80
No. 6—3 dozen in case, " 8 40

TACKS BRADS, ETC.

	Per cent.
Strawberry box tacks, bulk	75 & 10
Cheese-box tacks, blued	80 & 12½
Trunk tacks, black and tinned	85
Carpet tacks, blued	80 & 5
" " tinned	80 & 10
" " (in kegs)	40
Cut tacks, blued, in dozens only	75 & 15
" ½ weights	60
Swedes, cut tacks, blued and tinned— In bulk	80 & 10
In dozens	75
Swedes, upholsterers', bulk	85 & 12½
" brush, blued & tinned, bulk	70
" gimp, blued, tinned and japanned	75 & 12½
Zinc tacks	35
Leather carpet tacks	55
Copper tacks	50
Copper nails	52

Trunk nails, black	65 and 5
Trunk nails, tinned	65 and 10
Clout nails, blued and tinned	65 and 5
Chair nails	35
Cigar box nails	35
Patent brads	40
Fine finishing	40
Picture frame points	10
Lining tacks, in papers	10
" " in bulk	15
" " solid heads, in bulk	75
Saddle nails in papers	10
" " in bulk	15
Tufting buttons, 22 line, in dozens only	60
Tin capped trunk nails	15
Zinc glazier's points	5
Double pointed tacks, papers	90 and 10
" " " bulk	40

TAPE LINES.

English, ass skin, per doz.	2 75	5 00
English, Patent Leather	5 50	9 75
Chesterman's each	0 90	2 85
" steel, each	0 80	8 00

THERMOMETERS.

Tin case and dairy, dis. 75 to 75 and 10 p.c.

TRANSOM LIFTERS.

Payson's per doz. ... 2 60

TRAPS. (Steel.)

Game, Newhouse, dis. 25 p.c.
Game, H. & N., P. S. & W., 65 p.c.
Game, steel, 72½, 75 p.c.

TROWELS.

Disston's discount 10 per cent.
German, per doz. ... 4 75 ... 6 00
S. & D., discount 35 per cent.

TWINES.

Bag, Russian, per lb.	....	0 27
Wrapping, cotton, per lb.	0 22	0 26
Wrapping, mottled, per pack.	0 50	0 60
Wrapping, cotton, 3-ply		0 20
" " 4-ply	....	0 26
Mattress, per lb.	0 33	0 45
Staging, "	0 27	0 35
Broom, "	0 30	0 55

VISES.

Hand, per doz.	4 00	6 00
Bench, parallel, each	2 00	4 50
Coach, each	6 00	7 00
Peter Wright's, per lb.	0 12	0 13
Pipe, each	5 50	9 00
Saw, per doz	6 50	13 00

ENAMELLED WARE.

White, Princess, Turquoise, Blue and White, discount 50 per cent.
Diamond, Famous, Premier, 50 and 10 p.c.
Granite or Pearl, Imperial, Crescent, 50, 10 and 10 per cent.

WIRE.

Brass wire, 50 to 50 and 2½ per cent. off the list.
Copper wire, 45 and 10 per cent. net cash 30 days, f.o.b. factory.
Smooth Steel Wire, base, $2.80 per 100 lb. List of extras: Nos. 2 to 5, advance 7c. per 100 lb.—Nos. 6 to 9, base—No. 10, advance 7c.—No. 11, 14c.—No. 12, 20c.—No. 13, 35c.—No. 14, 47c.—No. 15, 60c.—No. 16, 75c. Extras net per 100 lb.: Coppered wire, 60c.—tinned wire, $2—oiling, 10c.—special hay-bailing wire, 30c.—spring wire, $1—best steel wire, 75c.—bright soft drawn, 15c.—in 50 and 100-lb. bundles net, 10c.—in 25-lb. bundles net, 15c.—packed in casks or cases, 15c.—bagging or papering, 10c.
Fine Steel Wire, dis. 17½ per cent List of extras: In 100-lb. lots: No. 17, $5—No. 18, $5.50—No. 19, $6—No. 20, $6.65—No. 21, $7—No. 22, $7.30—No. 23, $7.65—No. 24, $8—No. 25, $9—No. 26, $9.50—No. 27, $10—No. 28, $11—No. 29, $12—No. 30, $13—No. 31, $14—No. 32, $15 No. 33, $16—No. 34, $17. Extras net—tinned wire, Nos. 17-25, $2—Nos. 26-31 $4—Nos. 32-34, $6. Coppered, 5c.—oiling, 10c.—in 25-lb. bundles, 15c.—in 5 and 10-lb. bundles, 25c.—in 1-lb. hanks, 50c.—in ½-lb. hanks, 75c.—in ¼-lb. hanks, $1—packed in casks or cases, 15c.—bagging or papering, 10c.
Galvanized Wire, per 100 lb.—Nos. 6, 7, 8, $3.55 No. 9, $3.10—No. 10, $3.75—No. 11, $3.85 No. 12, $3.25—No. 13, $3.35—No. 14, $4.25—No. 15, $4.75—No. 16, $5.00.
Clothes Line Wire, 19 gauge, per 1,000 feet ... 3 30

WIRE FENCING. F.O.B. Toronto

Galvanized 4 barb, 2¾ and 5 inches apart	3 10
Galvanized, 2 barb, 4 and 6 inches apart	3 10
Galvanized, plain twist	3 10

Galvanized barb, f.o.b. Cleveland, $2.97½ in less than carlots, and $2.85 in carlots. Terms, 60 days or 2 per cent. in 10 days.
Ross braid truss cable ... 4 50

WIRE CLOTH.

Painted Screen, per 100 sq. ft., net. ... 1 35
Terms, 4 months, May 1.; 3 p.c. off 30 days.

WRENCHES.

Acme, 35 to 37½ per cent.
Agricultural, 60 p.c.
Coe's Genuine, dis. 20 to 25 p.c.

Towers' Engineer, each	2 00	7 00
" S., per doz.	5 80	6 00
G. & K.'s Pipe, per doz.	....	3 40
Burrell's Pipe, each	....	3 00
Pocket, per doz.	0 25	2 90

WRINGERS.

Leader	per doz.	$.....
Royal Canadian	"	50 00
Royal American	"	50 00

Discount, 45 per cent.; terms 4 months, or 3 p.c. 30 days.

WROUGHT IRON WASHERS.

Canadian make, discount, 40 and 5 per cent

DISPLAY

Your stock on the front of

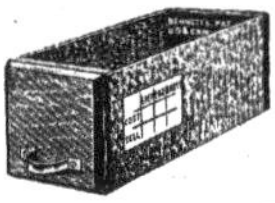

BENNETT'S PATENT SHELF BOX

and your store will be a great attraction to customers.

Prices and full particulars from

J. S. BENNETT, 20 Sheridan Ave., Toronto.

N.B.—**Don't forget** we make boxes to fit your present shelving.

PERSONS addressing advertisers will kindly mention having seen their advertisement in Canadian Hardware and Metal Merchant.

SEND for specimen copy of Phillips' Monthl Machinery Register, containing over 5,000 entries of new and second-hand machinery of every description. The oldest established and most successful medium in the world. Established 25 years for the purpose of introducing those who have machinery for sale, to those who wish to buy, has a circulation of about 50,000 copies per annum, all over the world, and is used for continual reference by a large number of firms. It is consequently a most valuable advertising medium for all engineers and manufacturers. Subscription, 6s. per annum, price per copy, 6d. Sole Proprietor, CHAS. D. PHILLIPS, M.I.M.E., Newport, Mon., England. Telegraphic address "Machinery, Newport, Mon."

Lockerby & McComb

AGENTS IN CANADA

FOR THE

Celebrated P. & B.

Cold Storage Lining

AND

. . Ruberoid Roofing . .

P. S.—Prices on Application.

65 Shannon Street, MONTREAL.

BUSINESS NEWS

of any kind that is of value to business men supplied by our Bureau. We can give you market quotations from any town in Canada, reports from the city markets, stock quotations, etc. You can get commercial news from any Canadian paper through us.

Write us, giving us particulars of what you want and where you want it from, and we will quote you prices by return.

"Clippings from any Canadian paper on any subject."

CANADIAN PRESS CLIPPING BUREAU,

232 McGill Street, MONTREAL, QUE.

Telephone Main 1255.

10 Front St. East, Toronto. Telephone 2148.

---

75 YEARS. *ESTABLISHED 1825.* *75 YEARS.*

CELEBRATED HEINISCH SHEARS.

Tailors' Shears,
Trimmers, Scissors,
Tinners' Snips, etc.

ACKNOWLEDGED THE BEST.

R. HEINISCH'S SONS CO. NEW YORK OFFICE, 90 Chambers St. NEWARK, N.J., U.S.A.

Not connected with any Shear Combination.

CHAS. F. CLARK, President. JARED CHITTENDEN, Treasurer.

...ESTABLISHED 1849...

BRADSTREET'S

Capital and Surplus, $1,500,000. **Offices Throughout the Civilized World.**

Executive Offices: Nos. 346 and 348 Broadway, New York City, U.S.A.

THE BRADSTREET COMPANY gathers information that reflects the financial condition and the controlling circumstances of every seeker of mercantile credit. Its business may be defined as of the merchants, by the merchants, for the merchants. In procuring, verifying and promulgating information no effort is spared, and no reasonable expense considered too great, that the results may justify its claim as an authority on all matters affecting commercial affairs and mercantile credit. Its offices and connections have been steadily extended, and it furnishes information concerning mercantile persons throughout the civilized world.

Subscriptions are based on the service furnished, and are available only by reputable wholesale, jobbing and manufacturing concerns, and by responsible and worthy financial, fiduciary and business corporations. Specific terms may be obtained by addressing the Company at any of its offices. **Correspondence Invited.**

—OFFICES IN CANADA—

HALIFAX, N.S.	HAMILTON, ONT.	LONDON, ONT.	MONTREAL, QUE.
OTTAWA, ONT.	QUEBEC, QUE.	ST. JOHN, N.B.	TORONTO, ONT.
VANCOUVER, B.C.	VICTORIA, B.C.	WINNIPEG, MAN.	

THOS. C. IRVING, Gen. Man. Western Canada, Toronto. JOHN A. FULTON, Gen. Man. Eastern Canada, Montreal.

Awarded a Gold Medal at **PARIS EXPOSITION** for superiority. That's proof enough of their quality, and clearly shows that they are the best.

Send for Catalogue and . . . . Price List.

The Bailey

Cutlery Co.

BRANTFORD, ONT.

Dominion Pattern

Cow Tie AND Stall Fixture

The special features of the tie and stall fixture are well shown in the illustration. As will be noticed the chain is very short, which prevents all danger of entanglement with the animal's foot. At the same time the form of the fixture is such that great freedom is allowed to the head. Because of the short chain this tie is much cheaper than the ordinary patterns.

The stall fixture is made from a tough quality of steel and is very strong. Also, owing to its circular cross-section, it is exceedingly rigid. Its simplicity, convenience, cheapness, and ease of attaching make it very popular with cow tie users.

This form of tie and stall fixture are sometimes called Niagara pattern.

American or Flat Link Chain,

for years the standard cow tie chain in "the States," is now rapidly coming in favor in Canada. Its short link, handsome appearance and smooth surface—which cannot injure the animal's neck—make it superior to all other styles of chain for cow ties.

For sale by all Jobbers; manufactured by

ONEIDA COMMUNITY, Limited, NIAGARA FALLS, ONT.

CANADIAN HARDWARE AND METAL MERCHANT

The Weekly Organ of the Hardware, Metal, Heating, Plumbing and Contracting Trades in Canada.

VOL. XIII. MONTREAL AND TORONTO, FEBRUARY 23, 1901. NO. 8

Wood, Vallance & Co.,

Iron Merchants.

Importers of British and Foreign

HARDWARE

HAMILTON, ONT.,
Canada.

Branch House: George D. Wood & Co., Winnipeg, Man.
Toronto Office: 32 Front Street West—H. T. Eager.
British Columbia Representative: W. G. Mackenzie, P.O. Box 460, Vancouver.

Wood, Vallance & Co., Hamilton, Ont.

GEO. D. WOOD & CO.,

Iron Merchants

Importers of British and Foreign

HARDWARE.

WINNIPEG, CANADA.

"PLYMOUTH" TWINE

Sells the best,
Works the best,
and makes the best money for the dealer.

"Plymouth" doesn't "fool the farmer." It gives him honest satisfaction, as it has qualities peculiar to itself, which cause it to work perfectly where other brands fail.

"Plymouth" Prolongs Profits.

Plymouth Binder Twine Agency — **McKinnon Bldg., Melinda St., Toronto, Can.**

The Demand for our

IMPERIAL OXFORD

kitchen range increases daily.

Every one sold causes other sales, so enthusiastic are householders over its improved features.

ITS EASE OF REGULATION
ECONOMY IN USING FUEL
ITS DIFFUSIVE FLUE CONSTRUCTION
FRONT DRAW-OUT GRATE
OVEN THERMOMETER
DRAW-OUT OVEN RACK

and other conveniences, give it a superiority quickly appreciated.

You're sure of speedy sales when handling the Imperial Oxford. Write us for full information and price list if you have none in stock.

THE GURNEY FOUNDRY CO., Limited

TORONTO. WINNIPEG. VANCOUVER.

THE GURNEY-MASSEY CO., LIMITED, MONTREAL.

Steel Frame Churn.

David Maxwell & Sons

ST. MARYS, ONT., CANADA

"Maxwell - Favorite Churn" PATENTED FEATURES: Improved Steel Stand, Roller Bearings, and Foot and Hand Lever Drive, and Detachable Driving Link Improved for season of 1901. Steel or Wood Frame as desired.

Wheelbarrows. In Four different Sizes.

Lawn Mowers. High and Low Wheels, from 12-in. to 20-in. widths. Cold Rolled Steel Shafting, Crucible Steel Knives and Cutting Plate.

If your Wholesale House does not offer you these articles

...SEND DIRECT TO US.

MAXWELL MOWER

8-inch Low Wheel.

"THE MAXWELL"
Lawn Mower

High Wheel 10 inches.

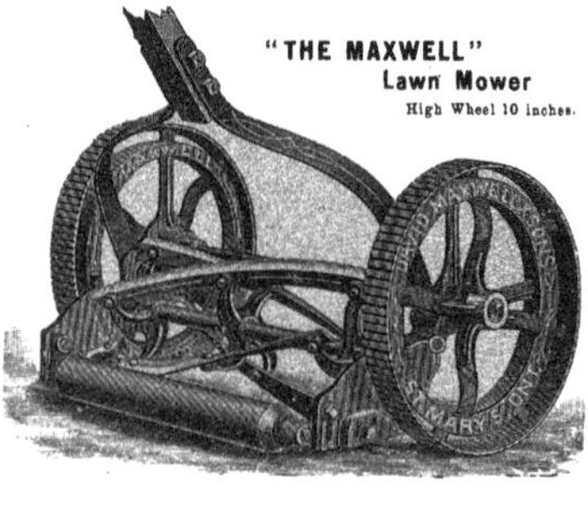

Woman is Responsible

for the sale of ice cream freezers.

Fifteen million women have considerable influence.

Ten thousand of them wrote to us in three months.

We are talking to them all the time—winter and summer.

They are all familiar with the Peerless Iceland Freezer.

Reports from them prove that we are right, and that one motion is enough.

We want your help in distributing Peerless Iceland Freezer.

DANA & CO.,
Cincinnati.

New York Branch, 10 Warren St.
San Francisco Branch, 105 Front St.

SECOND-HAND

Pipe Threading and Cutting Machine.

A BARGAIN.

We have in stock one No. 2 Armstrong Machine, complete with Countershaft and right hand Dies, cutting from 1-in. to 4-in. inclusive.

This machine has been used on one instalation only, and is in first-class condition and will be sold at a bargain.

Can be seen at our warerooms.

For prices and further information apply to

The Fairbanks Company

749 Craig Street,

MONTREAL.

ALEXANDER GIBB

Manufacturers' Agent and Metal Broker,
13 St. John Street, Montreal.

Representing British and American Manufacturers. Correspondence invited from firms wishing to be represented in Canada.

McCaskill, Dougall & Co
Fine Varnish & Japan
Manufacturers.
Montreal
Price Lists on application

DUNDAS AXES

ARE

American in Sharpness,
Canadian in Reliability,

JUST WHAT YOU WANT TO SELL.

Dundas Axe Works

DUNDAS, ONT.

MONEY IN WALL TINT.

MURALO is the latest and best up-to-date cold water wall tint in the world. Muralo challenges any old cold water compound or patented process of any make whatever. **MURALO** makes money for the dealer, because it is advertised properly, and once sold it speaks for itself. **MURALO** has an extensive foreign trade—all the world makes money out of it.

AGENTS:

A. RAMSAY & SON,	MONTREAL.
J. H. ASHDOWN,	WINNIPEG.
McLENNAN, McFEELY & CO.,	VANCOUVER

MERCHANTS OF CANADA

Especially dealers in paints and **wall-coatings!** When you took account of stock at the dawn of the new century, did you find in store some goods that you had been induced to buy, and for which there is absolutely no demand? Possibly some that you carried over from 1899? Musingly, did you exclaim, "I got fooled when I bought these goods, but it was the fault of the man or firm who sold me; if I get fooled the second time, it will be my fault?"

The moral of it all is:—Buy goods that are manufactured by men with experience in their particular line, and in whom the people have confidence; who advertise their product, guaranteeing it as represented, and whose goods sell on their own merits.

Why put in stock, that in order to sell, you have to tell your customers is "the same thing" or "just as good" as something else? **CHURCH'S**

ALABASTINE

the permanent wall-coating, ready for use by mixing with cold water, is in demand everywhere. Is sanitary to the highest degree. Has a record of value, and proof of origin.

ALABASTINE for more than a quarter of a century has been prominently before the people. **Anyone can brush it on—no one can rub it off.**

ALABASTINE is the **acme of perfection** in wall-coating material, either for plain tinting, or the most elaborate work known to the decorator's art. Is adapted to all kinds of relief work, and is a special favorite with practical men.

ALABASTINE is manufactured in Canada. Patented in this and other countries. Do you sell it? If not, why not? Write for prices.

Address, **THE ALABASTINE CO., LIMITED,** Paris, Ont.

The trade supplied by—
Sanderson Pearcy & Co., Toronto, Ont.
G. F. Stephens & Co., Winnipeg, Man.
Vancouver Hardware Co., Vancouver, B.C.
Wood, Vallance & Co., Hamilton, Ont.
William Hill, Montreal, P.Q.

The attention of Contractors, Builders and Building Owners is called to our **PARISTONE** Wall Plaster, and **SHIELD BRAND** Calcined Plaster. None better.

VanTuyl & Fairbank

Petrolia, Ont.

Headquarters for . . .
Oil and Artesian Well Pumps, Casing, Tubing, Fittings, Drilling Tools, Cables, etc.

Ontario Nut Works, Paris

BROWN & CO.

Manufacturers of

All sizes of Hot Pressed Nuts, Square and Hexagon.

"DAISY" CHURN

Has tempered steel cased bicycle ball bearings, strongest, neatest and most convenient frame. Only two bolts to adjust in setting up. Steel Bow Levers, suitable for either a standing or sitting posture. Has four wheels and adjustable feet to hold stand steady while churning. When churn is locked to stand the bow can be used as handles to move it about on the front wheels as handy as a baby carriage. Open on both sides to centre, giving free space for pail. Made with wood or steel stands, with Cranks only, or Bow Levers as desired.

Vollmar Perfect Washer

Has a most enviable record. A perfection of its kind—will wash more clothes in less time, do it better and easier, with less wear and tear, than any other machine.

THE

Wortman & Ward Mfg. Co.,

Limited

LONDON, ONT.

Eastern Branch, 60 McGill Street, Montreal, Que.

IMPROVED STEEL WIRE TRACE CHAINS.

Every chain guaranteed. Most profitable and satisfactory chain to handle.

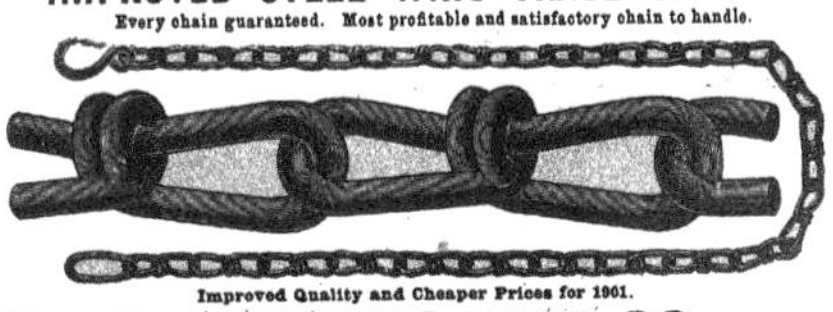

Improved Quality and Cheaper Prices for 1901.

The B. GREENING WIRE CO., Limited.

Hamilton, Ont., and Montreal, Que.

THE "WOODYATT" LAWN MOWER

Is still being made, to maintain the reputation it has gained during past seasons,

THOROUGHLY SATISFACTORY.

Only the best quality of material is used in its construction.

10½-Inch Drive Wheel. 12, 14, 16, 18, 20-Inch Cut.

Manufactured and Patented by . .

A. R. WOODYATT & CO., Guelph, Canada.

SUGAR MAKERS' SUPPLIES

Sap Buckets.

Extra deep and straight. Three sizes. They possess many advantages over the Ordinary Flaring Buckets—being small in diameter they do not catch the rain or snow, and, as they are very deep, they hang perpendicularly, and, consequently, will not overflow until full. Covers supplied if required. They nest close for shipping or storing.

We can supply the Ordinary Flaring Sap Pails.

E. T. Sap Spouts.

Made of Retinned Steel. Strong and Durable. Only require a ⅜-in. hole in a tree. It does not cover the inside surface of the hole, therefore, a larger amount of Sap is obtained.

Maple Leaf Sap Spouts.

Made in Bronzed Steel. Require a ⅜-in. hole. Has a shoulder which prevents it being driven in too far. The hole in the tree is not exposed to wind and snow, consequently, sap will flow longer.

Syrup Cans.

Plain or Decorated. Made in ¼, ½ and 1 gallon sizes. Either Wine or Imperial Measure.

Kemp Manufacturing Co., Toronto, Canada.

HARDWARE AND METAL

VOL. XIII. MONTREAL AND TORONTO, FEBRUARY 23, 1901. NO. 8.

President,
JOHN BAYNE MacLEAN,
Montreal.

THE MacLEAN PUBLISHING CO.
Limited.

Publishers of Trade Newspapers which circulate in the Provinces of British Columbia, North-West Territories, Manitoba, Ontario, Quebec, Nova Scotia, New Brunswick, P.E. Island and Newfoundland.

OFFICES

MONTREAL - - - - - - 232 McGill Street, Telephone 1255.
TORONTO - - - - - - 10 Front Street East, Telephone 2148.
LONDON, ENG. - - - - 109 Fleet Street, E.C., J. M. McKim.
MANCHESTER, ENG. - - - 18 St Ann Street. H. S. Ashburner,
WINNIPEG - - - - Western Canada Block. J. J. Roberts.
ST. JOHN, N.B. - - - No. 3 Market Wharf, J. Hunter White,
NEW YORK. - - - - - - 176 E. 88th Street.

Subscription, Canada and the United States, $2.00.
Great Britain and elsewhere - - - 12s.

Published every Saturday.

Cable Address { Adscript, London. Adscript, Canada.

LIGHT SPRING IMPORTATIONS FROM ENGLAND.

JUDGING from present indications, the orders to be placed by Canadian importers for May and June delivery of lead products, Canada plates and tinned and terne plates, as well as black sheets, will be exceedingly light this year. This is because the English market in these lines has been extremely easy since the close of navigation, and particularly since the first of the year, when spring importations begin to be considered.

Canadian dealers who imported Canada plate last year at prices ranging about £11 per box are losing money, for there has been a stiff decline to about £9. Tinned plates have suffered to the extent of about 13 or 14s. (some importers say more) since the close of navigation. Terne plates occupy a parallel position. And the markets are still easy.

There are those who look for further reductions in values, arguing that an inflation is invariably followed by an equally phenomenal depression. It would appear, however, that the English makers are adopting ways and means of preventing further declines, for, in a letter to their Canadian agents, A. C. Leslie & Co., Montreal, Richard Thomas & Co., one of the largest manufacturers of tinned and Canada plates, say: "That out of 495 mills in South Wales, 215 are idle and further stoppages are expected. We have had to close our works, owing to unremunerative prices, and there is no prospect of our reopening unless we can obtain at least the equivalent of what you paid some time ago or unless there is a fall in raw material." What the outcome of the struggle of the manufacturers against the depression will be remains to be determined. It will likely depend upon the tendency of the American market which at present is firm.

Black sheets form another line that is declining rapidly, and in this there is little business being done on import account. Since January 1 the cheaper grades of black sheets have declined fully £2 per box, and the better grades about 10s. since September.

Pig lead is another line to feel the tendency of the hour, it having been weakening for a couple of months and declining heavily for some weeks. Further declines are looked for, and, certainly, in the face of these conditions, lead products will not be imported in large shipments into Canada this spring.

These heavy and severe declines have immediately affected the spot market to some, although not to a great, extent. Terne plates are somewhat lower, and we hear that holders of Canada plates and tinned plates are sacrificing to clear; but this they need not do, for spot stocks of these latter goods are extremely light. White lead has declined ½c. per lb. this week, and dry white lead and red lead are ¼c. per lb. lower.

SHARP DECLINE IN WHITE LEAD.

WHITE lead has taken a rather peculiar turn as far as the Canadian market is concerned.

Only about a month ago an advance of 25c. per 100 lb. took place, and, although the outside markets are steady, there is in this country a sudden decline of 37½c. per 100 lb.

Quotations in 25-lb. packages are now as follows f.o.b. Toronto, Hamilton, London, Windsor, Ont., St. John, N.B., and Halifax, N.S.:

	Present price.	Old price.
Pure	$6 50	$6 87½
No. 1	6 12½	6 50
No. 2	5 75	6 12½
No. 3	5 37½	5 75
No. 4	5 00	5 37½

Prices f.o.b. Montreal and Ottawa are 12½c. per 100 lb. less. There is a freight allowance of 10c. per 100 lb. on shipments to Winnipeg, and shipments to other points in Manitoba and to the Northwest Territories to be equalized on Montreal.

Competition among the manufacturers in Canada is the cause of the reduction.

A lull in business should be an opportunity for creating a breeze with new ideas.

THE LINSEED OIL SITUATION.

LINSEED oil is in a most unsatisfactory condition, due to the weakness of prices to day and the uncertainly as to the future.

In Great Britain during the last three or four weeks there has been a decline of fully 10c. per gallon in price. And, as a result of this slump, British oil could be exported to the United States, in spite of the duty of 20c. per gallon, and sold there at something like 7c. per gallon below the price of the home product.

In Canada, quotations have been sent out by some of the brokers quoting linseed oil for shipment from Great Britain in May and June. If shipped in the former month it means that delivery will not be made in this country until June, and perhaps even July in some cases. Shipment of Canadian linseed oil, however, is made in April, May and June, the time of the year when linseed oil is most in demand in this country. It will be thus seen that the Canadian oil will be in the hands of the dealers in this country some time before the British oil is due to arrive.

The quotations on British linseed oil which have been submitted here have, as the trade is aware, much unsettled the Canadian market and made merchants reluctant to buy.

With a view to stimulating business the wholesale dealers in Canada, and particularly those in Toronto, have made a cut in prices that is unusually heavy for its sharpness.

At a meeting of the Toronto men on Wednesday the price was reduced 9c. per gallon, the same to go into effect on Monday, while in Montreal the decline is only 4c. per gallon, although that is, of course, a pretty sharp cut.

The price in Toronto, Hamilton and London is now as follows :

	Raw per gal. Cents.	Boiled per gal. Cents.
1 to 4 bbls.	68	71
5 to 9 "	67	70
10 to 19 "	66	60
20 to 59 "	65	68
Carload	64	67

The price to outside western points is as before, 1c. per gallon more freight being allowed.

It is claimed that, at the figures to which prices have been reduced in Toronto, at least some of the wholesalers are actually losing money, and it is the general opinion that there will be no further decline. On the contrary, there is more likely to be an advance than a decline on the local market.

It is, of course, unsafe to predict anything in regard to the future of linseed oil in the primary market. At present, there is a steadier feeling in both Great Britain and the United States. Regarding the situation in the United States as it at the moment stands, the following from The Paint, Oil and Drug Review, of February 16, may prove of interest :

The market has held steady during the past week on the 65c. per gallon (8⅝c. per lb.) basis, with some business in car load lots and smaller quantities doing at the current figures. Owing to the scarcity of available seed, much of the current receipts being shipped in on contracts and the bulk of the visible supply being under the control of two or three interests, the production of oil is believed to be far below the average for this time in the year. There are probably more idle mills in the country at the present time than at any recent corresponding date, making the situation somewhat anomalous in presenting a firm oil market and a weak seed market, with the former considerably above a parity with the latter. Considerable oil is moving on contracts, while the new business is of a hand-to-mouth character, buyers waiting for developments. The situation is pregnant with interesting possibilities, and it would be extremely hazardous to venture an opinion as to the outcome. The distribution of oil in small lots is enlarging as the busy season approaches, but as yet trade cannot be called active.

Much of the unsettled feeling which has predominated the linseed oil market is, as has been previously pointed out in this journal, due to the uncertainty in regard to the estimated large crop of linseed in the Argentine Republic. We have before us a report of the United States consul in Argentine. It is dated Buenos Ayres, December 13, 1900, and a part of it deals with the linseed crop. That particular part reads as follows :

The extraordinary rise in prices of linseed, due to the failure of the crops in India and other parts, encouraged larger sowings, estimated at 20 per cent. Fortunately, in nearly every district the crop has been a success, and, although the prospects of heavy shipments have brought about a fall in prices, the values still obtainable will give farmers splendid returns.

Some estimate that the quantity available for export will reach 550,000 tons. I fail to see where this tremendous quantity is to come from. Allowing for increased sowing, and for a larger average yield, I consider that if exports reach 400,000 tons, the figure will be an outside one.

To criticize a competitor in the presence of a customer is to give him a free "ad."

THE RAGE FOR BINDER TWINE FACTORIES.

CANADA undoubtedly has a great future before her as a manufacturing country, but there are branches of industry which zealous promoters are endeavoring to enlarge at a ratio far in excess of our requirements.

The binder twine industry is one of these.

The chief stock arguments being used are (1) that Canada imports 60 per cent. of her binder twine from the United States and (2) that, as the Farmers' Binder Twine Co., Brantford, paid a 60 per cent. dividend in 1898, 100 per cent. in 1899, and 90 per cent. in 1900, enormous profits are regular occurrences in the binder twine industry.

What is one man's meat, we are often told, is another man's poison. And certainly there was at least one binder twine concern that, during some of the years, at least, when the Farmers' Company was earning big dividends, was operated at a loss rather than at a profit, and that when it had the cheapest kind of labor. We refer to the factory at the Central Prison.

The explanation is that the Brantford factory was fortunate to buy its hemp when the market was low, while the Central Prison concern bought when the opposite conditions obtained. And another manufacturer who had bought hemp when the market was low found it more profitable, when the market was high, to sell his hemp instead of turning it into twine. The cause of the sharp advance in hemp was, of course, the Spanish American War, which led to the closing of the ports in Manila. The advance in the price of Manila hemp in 1898 was over 150 per cent. from the lowest points.

And then, if binder twine manufacturing is such a profitable industry in Canada, why did the Consumers Cordage Company close two of its factories even in the days when the industry was protected by the tariff, whereas now there is no protection ?

The fact of the matter is that binder twine making in Canada, like nearly all other industries, has to be carefully and intelligently managed to earn a fair profit. This, the business men of the country should instil into the minds of the farmers, for it is they whom the promoters are most actively working among.

AID FOR PROFESSIONAL POLITICIANS.

ANOTHER effort is being made to increase the sessional indemnity of the members of the Dominion Parliament. The indemnity is $1,000, and the proposal is to make it $1,500.

We ought not to be unjust to our representatives in Parliament; but it has yet to be demonstrated that the amount of sessional indemnity they now draw is unjust. To be frank, we do not think it is. True, as the advocates of an increased indemnity point out, the representatives in the Parliament of the new Australian Commonwealth are to receive $2,000. But that does not prove anything.

Tom Brown and Tom Smith may be filling similiar positions; and, while the latter may be receiving a salary of $1,000 less than the former, it does not follow that he is being underpaid. He may, in point of ability, be worth that much less, or, even should they be equal in ability, Tom Brown may be overpaid to the extent of $1,000.

But, then, if comparisons are to be made, why confine them to countries which pay a higher indemnity than Canada does? Why not, for instance, make comparison with the British Parliament? We, with pardonable pride boast that our Dominion Parliament is fashioned after that of the British Parliament. Surely, then, if we are to be guided by comparisons, we cannot afford to ignore the mother of modern Parliamentary institutions.

We do not, however, for one moment advocate that we should, at any rate at this stage of our history, follow the practice of the British Parliament in this particular; but it would be just as reasonable for us to do so as it is for the advocates of an increased indemnity to rest their case on the practice in countries which grant their representatives larger sums per session than we do in Canada.

The case should rest, not on comparisons, but on its own inherent merits.

Judged from a business standpoint the present indemnity is quite sufficient, particularly when we take into consideration the fact that each member is allowed mileage to and from his place of residence, whether or not he carries a pass in his pocket; and it is well known that nearly every member does possess such a document. Some of the members get over $500 in mileage allowance and numbers $100 and $200.

It was never intended that membership in Parliament should, like an ordinary vacation, be a source of revenue for supplying the requirements of daily life. Doubtless, there are certain men in Parliament to whom the sessional indemnity is the chief source of revenue, but that does not alter the original intention in regard to the indemnity.

The moment we recognize anything to the contrary that moment do we by implication acknowledge that, in electing a man to represent us in Parliament, we are providing him with a situation. This is a fact, and there is no getting away from it.

We fancy that the newspapers that are championing the increased indemnity cause would not for one moment favor a man for membership in Parliament to whom the $1,000 for attendance during the session was the attraction. But they are unconsciously allowing themselves to be used as the tools of those who are championing the cause of the professional politician. It is certainly not championing the cause of such men as Mr. Kemp, Mr. Brock, Mr. Bickerdike, Mr. Hyman and many other business men in the House. These men are not in Parliament for what "there is in it" for themselves. They are actuated by higher motives.

The larger the sessional indemnity is made the more attractive will a seat in Parliament become, not to the type of man most desired, but to the professional politician who is already a factor far too influential in the Legislatures of this country.

Push for payment those who owe you money or your creditors will push you out of business.

A HIGHER WHITING MARKET.

Whiting is one of the few articles that promises to rule higher than last year. At present the prices quoted by the manufacturers in Great Britain are about 10c. per 100 lb. higher than last year.

Two factors have contributed to this. And both are combinations. One has been ormed by manufacturers, all the large firms in England being members thereof. The other is among the steamship companies, who have increased freight rates by 1s. per ton.

KEEP IN MIND THE GENERAL GOOD.

THERE is an old saying to the effect that when thieves fall out honest men get their due. It may be true. But it is certain that when members of business men's organizations fall out that the influence of such organizations is weakened and the interests of the trade immediately concerned suffer.

Business men's associations can only be influential in bringing about reforms and preventing abuses when they are united and strong. An army that is characterized by internecine quarrels is never feared by the enemy. How then can an association, board of trade, or whatsoever name by which a business man's organization may be known, be respected or feared by those who should respect and fear it?

Members of business men's associations need to be patient, long suffering and not easily provoked, for, in organizations of that kind, as well as in all others, there are men small, jealous and mean, who are enough to try the patience of a saint let alone the average merchant whose own business matters give him worry enough. If a merchant cannot bear with this class of men he had better not identify himself with a business men's organization. But he should at the same time remember that, while such an organization is weakened by his refusing to associate himself with it, his own interests indirectly suffer as well.

Sink individual differences in order that the general good may float.

VIRDEN BOARD OF TRADE.

The following officers were elected at the annual meeting of the Virden, Man., Board of Trade the other day.

President—J. W. Higginbotham.

Vice-President—H. C. Simpson.

Secretary-Treasurer—Geo. H. Healey.

Council—B. Meek, H. J. Pugh, J. F. Fram, W. J. Wilcox, R. E. Trumball, D. McDonald, W. J. Kennedy, J. H. Agnew, R. Adamson, W. D. Craig, F. R. McLellan, J. T. Norsworthy.

Auditor—R. Adamson.

A feature of the annual report was the number of inquiries that are received by the secretary asking for information about the formation of boards of trade, and for copies of by-laws. Five such applications had been received during the year from various towns throughout Manitoba.

SUPPLIES FOR THE AFRICAN WAR AND CANADA'S CONTRIBUTION.

The War Office's Purchases in this Country Almost Nil—A Matter Which Should Receive the Attention of the Canadian Government, Its Commercial Agents and the Exporters in this Country.

OFFICE OF HARDWARE AND METAL.
London, England, February 4, 1901.

I HAVE long taken an interest in Canada and Canadian affairs, but this interest has been greatly increased since Canada came so heroically to the aid of the Motherland in her trouble in South Africa. We are large consumers of produce such as the Dominion of Canada raises, and, while I am not an advocate of paying Canada a better price than we would pay to any other country, yet, I believe all things being equal we should give the Dominion the preference. In view of the aid Canada gave us in the way of men for South Africa, I think it is only proper that we should buy all the produce we can for South Africa, particularly from your country. I have become so much interested of late that I have been led to investigate as to what our War Office was getting from Canada in the shape of supplies for man and beast in the South-African field. Through an officer in the War Office I have been able to secure information which I have no doubt is absolutely reliable. I must say that I am simply astonished, not at what supplies are being purchased from Canada, but what are not being purchased.

I might here say that there are in South Africa 210,000 of our soldiers, to say nothing of 16,000 Boer prisoners, a destitute population in many districts, refugees, all of whom have to be fed by the British Government. The number of men to be fed is not likely to decrease for some time to come, for Kitchener has asked the War Office for 150,000 more men and 100,000 horses. The papers here have stated that the number of men he had asked for was 50,000, but this is a mistake. Perhaps the papers in Canada have been better informed in this particular, although I suspect they have not. Besides all the men to be fed there are the beasts of burden, such as horses, mules and oxen. Then, of course, you must remember that we occasionally lose a convoy, so that at times we are really helping to feed a Boer army. From the officials at the War Office, to whom I have already referred, I have obtained some interesting data as to the monthly requirements of different kinds of produce in South Africa. Here they are :

Meat, ¼-lb. tins	25,000,000
Biscuits, in lb.	25,000,000
Canned and condensed vegetables	20,000,000
Hay and oats, tons	100,000
Bran, tons	5,000
Jam, lb.	3,000,000
Cheese, lb.	2,000,000
Bacon and hams, lb	1,000,000
Flour, tons	10,000
Salt, lb.	500,000

With such an immense quantity of produce required monthly in South Africa, one would imagine that Canada would get a goodly share, particularly as in many of these lines she is a large producer. But what are the actual facts? Canada is sending practically nothing, while Australia is doing an immense trade, and in just such articles as Canada is well qualified to export.

I notice that you get an occasional order in Canada for a couple of thousand tons of hay, or a couple of cars of supplies, and that you seem quite happy over the same. In view of the figures I have just given, I do not see much cause for satisfaction on the part of Canadians. There is no reason that I know of why Canadian producers do not participate in these orders except it be neglect on the part of the Canadian Government, its representatives in this country, or the exporters in Canada. Possibly all three are to blame. I have mentioned the Canadian officials on this side, but let me say by way of explanation, that Lord Strathcona is doing a good work here for Canada in a certain direction, and is most zealous, but, of course, he has not the time to look after matters of that kind, neither has Mr. Colmer, who is, no doubt, trying to do his best for Canada. What is really wanted is a practical business man as commercial agent in London. And then, of course, it is evident that your own exporters need to be awakened.

I might say that the list of articles I have given by no means represents all that are going to South Africa, and of which Canada could supply a part. For instance, thousands of tons of potatoes are going, so I am informed, there from Australia. Then, wooden huts are being made here in thousands, and are being sent to the army in South Africa. These houses are made in pieces, to be put together on the field, and surely this is a line in which Canada can do something as well as in food products. Wagons, too, as well as bicycles, are going out in large quantities. Only recently, in one order received here in England there were 400 wagons and 300 bicycles. Of horses, about 9,000 are required monthly. I understand that some orders for clothing are going to Canada, and one of the firms to which these orders are going is, I think, Sanford, or some name like that. The war is at present costing about £6,000,000 weekly, so you may gather by that what an immense quantity of supplies of different kinds are required.

What I would suggest is that you stir up your Government and its commercial agents in England. I understand that Canada has also a commercial agent in South Africa. It strikes me that he too might be a little more useful than he appears to be. Perhaps, however, I am judging him harshly. The exporters in Canada should also stir themselves up and send their agents to London and make contracts direct with the War Office here. If a sufficient supply was contracted for in Canada, ships could be chartered direct to South Africa from Canada, say, twice a month.

I feel that I shall not have written in vain

if this letter results in Canada getting a larger share of a trade to which she is most certainly entitled.

I might add that large supplies will be required monthly for probably a year, as from what I can gather from reliable sources the country in Africa, which has been the scene of the war, is much devastated, and the Boers as well as the army will have to be fed by the Government for goodness knows how long.

S.

A GOOD OPINION OF BRITISH COLUMBIA.

NEARLY every resident of British Columbia who visits Eastern Canada speaks of the Pacific Province with eulogistic appreciation of her present and enthusiastic optimism as to her future.

One of the calmest, yet thoroughly hopeful, opinions regarding the Province that HARDWARE AND METAL has heard was expressed this week by C. L. Lightfoot, Vancouver representative of The Gurney Foundry Co., Limited, in that district.

"Last spring, British Columbia was put to the test," said Mr. Lightfoot. "The Provincial Government was unstable; there was difficulty between the miners and the employers. The result was that almost every section of the Province was affected more or less by dullness of trade that caused a widespread feeling of depression. If the country had not contained within itself the essentials of permanent progress, dullness and depression would not have been the worst that would have to be reported.

"But, as everyone knows who is intimately acquainted with the various mining regions, those are such that as to insure a permanent as well as profitable development. In the coal mining regions both on Vancouver Island and in the Crow's Nest Pass district, the progress is steady. In the regions where the gold, silver, lead, etc., are mined, there is less steadiness. The questions of transportation, wages, etc., play such an important part that conditions are easily affected by a strike, a change in rates or any such occurrence. It was this district that caused the strain last spring. But the earth contains the minerals in such quantities that confidence in the country could not be badly shaken by a temporary depression.

"The result is that last fall when I again went over the ground confidence was restored in almost every town. Work is more plentiful, and still greater activity is looked for."

"You spoke of the importance of transportation to the mining country. How do the people there look upon the proposal to run a line into the mining region from the south?" asked HARDWARE AND METAL.

"My experience has been that every business man wants to see every person who desires to build a railroad in any direction build it. A short line from Vancouver to the Kootenay is what is greatly wanted now, and this is going to be built before long either by the C.P.R. or the Victoria, Vancouver and Eastern."

"Is there much agricultural land in the Province?"

"On Vancouver Island and along the Coast there is much excellent land. Besides, there are several extremely fertile valleys. In one of these various fruits are easily grown. These are sold locally and through the Northwest. There is some fine pasture land. I was talking to an old Dutch rancher last fall. He told me he had lost $40,000 last summer. I was amazed, as he did not look like a man who ever owned that much, and I knew the grazing had been specially good. I knew he was honest, yet, I must admit, there was a tone of doubt in my voice as I asked him to explain. 'Well,' said he, 'there was such good grass that I could have easily fed enough cattle there to make that much. And it nearly all was wasted.' If we looked at British Columbia from the old Dutchman's viewpoint, the world is losing a fortune there every year."

INQUIRIES REGARDING CANADIAN TRADE.

Mr. Harrison Watson, curator of the Canadian Section of the Imperial Institute, is in receipt of the following inquiries regarding Canadian trade:

1. A Copenhagen firm wishes to hear from Canadian manufacturers of pulp boards, desiring to establish a Danish connection.

2. An important London importer of poultry asks to be placed in communication with Canadian shippers.

3. Another Danish house seeks supplies of Canadian mica, and invites communications from producers.

4. A Midlands manufacturer, doing a large trade in steel and other metals with engineering firms in the United Kingdom, is prepared to arrange with experienced Canadian firms for the Canadian agency.

THE SYDNEY FURNACE.

While Mr. H. M. Whitney, president of the Dominion Iron and Steel Co., was in Montreal on Tuesday, he and his fellow-directors were much pleased to receive the following highly encouraging despatch from the vice-president and general manager of the steel works:

Sydney, C.B., February 12.

H. M. Whitney, Montreal:

Turned out splendidly; promise of a large product is now a certainty. The sulphur, which was our doubtful point, is under absolute control to-day, and this with unwashed coal. There is no longer room for doubt as to the quality or the good design of the blast furnace plant.

(Signed) A. J. Moxham.

TRADE IN COUNTRIES OTHER THAN OUR OWN.

A SLIGHT change has been made in the price of horseshoes, modifying the former arrangement by which there was a fixed difference between the base price of carload and less than carload lots. The carload price—namely, $3.50 on iron shoes and $3.25 on steel shoes, f.o.b. Pittsburg—is now given on orders of 10 kegs or more, but protection is given to large buyers, as the carload rate of freight is added on carload lots, and the less than carload rate on less than carload lots. Rebates on large quantities are unchanged.—Iron Age.

SNATHS AND CRADLES IN THE STATES.

There has been no change in the price of snaths and cradles since last season. The volume of business is reported by the manufacturers as not as great as up to the same date last year. This is in part accounted for by the fact that two years ago the demand was in excess of the supply, and, as a result, orders last season were not only placed quite early, but were unusually liberal. The hay crop was light in some sections last year, resulting in the carrying over by dealers of considerable quantities of goods, and it is quite possible that the aggregate sales of this season will fall somewhat below those of last year. Orders, however, are reported to be coming in satisfactorily, and the season promises to be up to the average.—Iron Age.

IRON TRADE IN SHEFFIELD.

Last week's pig-iron quotations have been maintained, and the apparent steadiness of the market seems to have stimulated business somewhat, as there have been more transactions in pig-iron, Bessemer billets, and steel sheets and rods than for some time past. Several crucible-steel makers also report a slight improvement in business, both on home and foreign account. Speaking generally, however, the iron and steel branches are in a most unsatisfactory condition, and the slight improvement reported may, after all, be due only to purchases by consumers who had depleted their stocks to a point which involved some risk. A significant feature of the situation is the fierce competition for orders for steel sheets, rods, etc., which obtains among proprietors of rolling-mills, forges, and tilts, and it is said that work is being taken at below cost in order to keep men and machinery employed. Price-lists have been abandoned, and the manufacturers are quoting specially for each inquiry. The rolling mills are working only three or four days a week. The diminished trade for finished goods deprives makers of the benefits they would otherwise have derived from the drop in prices of various classes of materials. The inquiry for shovels and spades is falling off.—Ironmonger.

INDUSTRIAL GOSSIP.

Those having any items of news suitable for this column will confer a favor by forwarding them to this office addressed to the Editor.

THE annual meeting of The St. John, N. B., Iron Works was held on Wednesday, last week. The reports showed the year to have been a profitable one. The old board of directors was elected as follows: John E. Moore (president), H. D. Troop, James Bender, W. H. Murray, Dr. W. W. White, Charles Miller and Charles McDonald (secretary-treasurer).

The Polson Iron Works, of Toronto, will furnish the engines and boilers for the Dominion steamers to be built on the Pacific Coast for the fishery protective service.

The Recinate Fire-Proofing Co., manufacturers of fireproof paints and preparations which render wood and fabrics safe from the flames, are looking for a location in Canada.

A despatch from Ottawa says that the Government has decided to have a new dredge of the latest design built at a cost of $250,000. The contract will likely go to a Toronto firm.

The repairs on the steamer City of Topeka, which are being made in Vancouver, will cost about $50,000, according to the statement of Supt. Miller, of The Pacific Coast Steamship Co. The damage to the hull consists of a large hole in the bow and two smaller ones further aft. In addition to these, there are a number of loose plates, and the general overhauling that will have to be given the vessel will complete the full cost of the $50,000.

An agreement has been signed by the municipal authorities of Port Hope, Ont., agreeing to exempt the Canada Radiator Co. from taxation for 10 years if that company will install an $11,000 plant and continuously employ at least 25 men.

The council of St. Henri, Que., is considering a proposal to exempt from taxes for 10 years the John Terrault Tool and Malleable Iron Works, the promoters of which propose to establish a new industry in St. Henri for the manufacture of axes and other tools from malleable steel under a new process.

A Duluth despatch says that Capt. A. B. Wolvin, of that city, and President James Wallace, of the American Shipbuilding Co., have made arrangements to erect a shipbuilding plant for their company in Halifax, N.S. The municipality has agreed to a bonus of $2 per ton on every ship built for 10 years and $1 for the succeeding 10. This will be the first yard owned by the American Shipbuilding Co. in Canada.

TAKING A TRIP TO CALIFORNIA.

Mr. H. Sapery, manager of the Montreal branch of the Syracuse Smelting Works, has taken a trip to California. On his way back he will visit the principal commercial centres in British Columbia and Manitoba, where the firm does an extensive trade. This firm's general business has opened up so well this year that it has been found necessary to increase the number of employes.

H. S. HOWLAND, SONS & CO.

ONLY WHOLESALE 37-39 Front Street West, **Toronto.** WHOLESALE ONLY

"Mic-Mac" Hockey Sticks

This Season's Make. This Season's Make.

Yellow Birch.

Toughest and Strongest.

The "Mic-Mac" Hockey Sticks are the best.

This season they are the favorite stick.

In a recent match in Toronto between two leading Hockey teams, all the players used "Mic-Mac" Sticks, **and only one was broken.**

THIS IS A GREAT RECORD.

Express orders filled promptly.

H. S. HOWLAND, SONS & CO., Toronto.

Graham Wire and Cut Nails are the Best.

M'CLARY'S EMPLOYES DANCE.

THE first annual ball and concert of the Employes' Benefit Society of The McClary Manufacturing Co., London, Ont., was held on Monday night. The affair was an immense success. Twelve hundred people, or close upon that great figure, participated in the event, and by their presence made it memorable. It is not easy to imagine so many pleasure-seekers gathered under one roof. It is no light task to picture the scene which such a gathering presented. Hundreds here; hundreds there. In the splendid new addition to the mammoth works of the great firm, which had been turned into ball and concert and card rooms for the occasion, there was little room to spare, go where you might. The employes of the company of themselves are a small army. Their friends are legion, and it became necessary to limit the number who might attend. This was at first placed at 500. But it was soon seen that the number must be doubled. And doubled it was.

And great as was the gathering, in point of dimensions, it was greater in point of enjoyment. The arrangements for the entertainment of the huge company were wonderfully complete. Despite the throng, there was absolutely no confusion. One might adopt any of the forms of amusement provided that appealed to the taste at almost any period of the evening. And it is to the credit of the committee that they should not only have accomplished so much and done it with so splendid regard to detail, but that the whole affair should have been mapped out and put into effect within three days. Connected with a firm of magnificent enterprise, it could be but expected of the employes that they would do something out of the ordinary. They did the extraordinary.

The McClary Employes' Benefit Society is no new organization, notwithstanding that this was their first attempt at a festive gathering of the kind. The society has accomplished a vast deal of good. It is still doing so. And there are the best reasons for believing that the good work will go on, for the organization is upon a firm basis. There are few, if any, of the employes who do not take a deep interest in the welfare of the society. Not only are the results in a financial sense satisfactory, but from the important standpoint of mutual sympathy and sociability the results are in the highest degree pleasing. Then the organization has the countenance and active aid of the members of the firm. Lieut.-Col. Gartshore was president of the general committee last night. Mr. McClary was honorary president.

Last night, as might be guessed, where there were so many young people assembled, the ballroom had the first call. The dance hall was upon the fourth floor. It was admirably suited to the purpose. Filled with a multitude of happy folk, with the strains of sweet music and with the laughter of the merry-making assemblage, it was difficult to imagine that within a few short weeks this great room would too be filled with the hum of industry. For the McClary Co. are making great strides in the world of commerce. The newly-finished building, with its extensive dimensions and its four floors, is but one of the numerous additions that have been made from time to time, and especially of late, to the works that are fast occupying a whole block in the very heart of the city. But the fact troubled no one. When another year has rolled around, the days of prosperity may have added one more to the buildings that comprise the whole, and have again provided a fit place for the holding of the second annual ball and concert.

The orchestra furnishing music for the merry dance was that of the employes, under the leadership of Mr. James Cresswell. It is probably superfluous to add anything as to the excellent quality of the music. The ball room was gaily decorated with flags and bunting. Arranged about the entire length and breadth of the hall were comfortable seats. There was an abundance of light. And the ventilation was better than could have been hoped for under the circumstances. The ladies were present in large numbers in honor of the event, and many of the costumes worn were exceedingly pretty. The dance was continued until a late hour. The advent of Lent was overcome by turning the clock back a couple of hours. And then in the wee, sma' hours of the coming day the festivities were brought to a close with the waltz, "Home, Sweet Home."

The providing of supper for the company of guests was a task that might have staggered a committee possessed of any but unwonted energy. Yet it was accomplished with comparative ease, and to the general satisfaction. On the second floor were long rows of neatly spread tables, and upon these, as the evening advanced, were furnished excellent repasts for all desiring to replenish the inner man.

The concert was, of course, given in the early part of the evening, so that those who delighted in an entertainment of the kind might later also have opportunity to take part in the dance. Lieut.-Col. Gartshore was chairman, and the programme introduced gave evident pleasure to the large crowd constantly present to be entertained. The chairman made a few timely remarks. He warned the company that this was the last time for a while upon which they would be able to enjoy themselves. The Lenten season began on the morrow, and they would then have to begin a fast of 40 days and 40 nights. He hoped that they would make the very best of the present, accordingly, and banish dull care.

Owing to the length of the programme, it was announced that encores were barred. The programme was given by the employes and their friends, and was of much general excellence. The stranger could not but have been struck with the fact that the McClary employes are a talented lot. The numbers given included the following: Piano selection, Mr. F. Gruber; song, Mr. J. Barrett; recitation, Miss Richardson; violin solo, Mr. B. Miles; duet, Messrs. J. Head and Donavan; instrumental duet, the Misses Herrick; recitation, Miss Nora Nicholson; song, Mr. I. Burke; musical selection, Mr. W. Watson; song, Mr. B. Joyce; duet, Misses Fitzwalter and Brighton; recitation, Miss Shoebottom; song, Mr. G. Burr; duet, Penny Brothers; piano selection, Mr. F. Gruber; song, Mr. B. Joyce; duet, the Misses Herrick; song, Mr. I. Burke; violin solo, Mr. B. Miles; duet, Messrs. J. Head and Donavan; song, Mr. Simpson; duet, Misses Fitzwalter and Brighton; song, Mr. J. Barrett; musical selection, Mr. W. Watson; song, Mr. G. Burr, concluding with the singing of "God Save the King."

The card-room was upon the third floor, between the concert and dance halls. Here were provided an abundance of card tables, and these were fully taken advantage of.

Some idea of the excellence of the arrangements made for the reception of guests may be gained from the fact that separate cloak-rooms were provided for ladies and gentlemen, each in charge of competent persons. The decorations in all parts of the building were very tasteful. To the committee all praise is due. The general committee was comprised as follows: John McClary, honorary president; W. M. Gartshore, president; Joseph Nicholson, first vice-president; James Pirie, second vice-president; H. Woodman, F. Bailey, John Head, Frank Couke, John Barned, M. G. Delaney, William Yelland, Charles Manning, E. Wingett, D. Wilson, S. Milliken, J. Bailey, Miss Porter, Miss Graham, Miss Ramsay, C. Donavan, Joseph Walcott, W. Lehman, R. Spencer, Miss Colter, Miss Selkirk and John Kenneally.

The executive committee consisted of ex-Ald. John Barned, chairman; Wm. Lehman, William Yelland, John Head, Ernest Wingett, Fred. Bailey and Misses Colter and Graham. To Mr. John Kenneally fell the onerous duties of secretary and treasurer. Neat programmes were printed, giving all details and bearing a representation of the magnitude of the McClary concern in the local works and the numerous important branches.

STEEL INDUSTRY IN CANADA AND UNITED STATES.

A Masterly Address Thereon by Mr. A. J. Moxham, General Manager of The Dominion Iron and Steel Co., Sydney, N.S.

AN address was delivered in the rotunda of the Board of Trade, Toronto, on Friday evening, February 15, by Mr. A. J. Moxham, general manager of The Dominion Iron and Steel Co., Sydney, N.S., on the iron and steel industry of Canada and the United States. The Canadian Manufacturers' Association, under whose auspices the lecture was delivered, is to be congratulated upon its forethought in bringing Mr. Moxham to the Queen City.

The address was, without doubt, the most masterly exposition of the iron and steel industry, particularly in as far as it relates to Canada, that any audience has ever listened to in this country. It was exhaustive, to the point, and gave information which was entirely new. The rotunda was crowded and rapt attention was given to Mr. Moxham while he was speaking and the heartiest kind of applause greeted him as he sat down. Mr. P. W. Ellis, president of the Canadian Manufacturers' Association occupied the chair.

Mr. Moxham spoke as follows:

Reduced to its final analysis, steel is a product resulting from the application of man's labor to three raw materials, viz., ore, coal and limestone.

These three ingredients by means of a blast furnace are converted into pig metal, and this in turn by means of the open-hearth furnace into steel.

Science to-day has acquired such control of the process that within wide limits it is capable of dealing with almost every quality of pig metal. The question of the percentage of phosphorous, silica, sulphur and other ingredients in the pig is to-day of far more importance in the light of the cost sheet than it is as any matter of necessity so far as the subsequent open-heaath practice is concerned. The day of trade secrets has gone by, and the treatment from the same quality of pig metal down to the finished steel is largely identical in all the different plants.

COST OF PIG METAL.

It is, therefore, a fact, broadly speaking, that the cost of the finished steel varies with that of the pig metal, and we need deal with this alone. Dividing the cost of pig metal between material and labor, the former is the variable, the latter the constant. It is true that between every individual plant and between every district there are some differences in the matter of labor, but it is also true that there need not be, that it is within the control of man's volition and can be eliminated if deemed economically desirable. Moreover, every improvement in the art and each development of machinery reduces these labor differences. With material it is different. There are no two great steel districts in all the world wherein the proximity of the raw material and the distance from the market is the same. No power of man's will, no good fairy with a golden wand can eliminate the space between. When the lake district was brought into prominent notice as the most economical point in the central west for making pig metal, Mr. Carnegie built his own railroad, known as the Pittsburk Bessemer Road, 155 miles long, connecting Lake Erie with Pittsburg, and it was announced that this had made Pittsburg a lake port. It always seemed to me that the 155 miles were still there.

A STUDY OF DISTANCES.

It is, therefore, to a study of distances between materials that we must turn when investigating the advantages of any given location. In a nutshell, it is entirely a matter of freight. At the very outset we are brought face to face with one great lesson, viz., that nowhere has nature grouped in one spot the three raw materials in the proper economic quantity and of the proper economic quality. It is as though the Great Master had said, "My gifts I freely give you, but they are worth your coming for them." I know this broad statement will be a surprise to many. We have all heard of places where the three elements were together. In fact, they existed one on top of the other, ready placed in proper proportions, and always in a hill of the proper height, provided with a nice level valley just fitted in size and location for a modern blast furnace—all that was needed was for the hill to topple down into the mouth of the furnace awaiting it below. We will all of us hear of these favored localities again. When they are brought to you and you are asked to believe in them remember Punch's good advice and "don't."

VARIETIES OF ORES.

You will ask me about the Black Band districts of Scotland and of Central Kentucky and the iron district of Birmingham, Ala. You will ask me whether ore and coal do not lie together in Nova Scotia and in Pennsylvania, and I have to answer yes, they do, and in many other places I could add to the list. Now, as I cannot afford to have you lose faith in me at the very beginning of our talk, let me explain. There is ore and ore. There are beds of very poor ore, carrying only from 10 per cent. to 20 per cent. of iron—which is the thing we want—and carrying enough silica to give us a respectable sand beach all round Nova Scotia, and this is the thing we don't want. These beds are generally of indisputable extent, and they also generally lay quite close to coal. There are also good ores carrying from 50 per cent. to even 65 per per cent. of iron with little or no silica, and low in phosphorous and sulphur, but these we find in Spain, Lake Superior or Ontario, and they do not lay close to the coal. Or if we do find them close to the coal, as in the Guysborough district of Nova Scotia, we find them pocketty and the beds of small extent. There is also coal and coal. There is coal like that of Central Kentucky and Illinois that will not coke well and is high in sulphur and ash, and there are coal like the Durham field of England, the Connelsville of Pennsylvania, that are the standards of excellence for steel-making. And, to these will be added before two years go by as an equally good standard for steel-making the coal fields of Cape Breton. The poorer coals of Kentucky are near very good ores; the good coals of Durham are near the low-grade ores of the Middlesboro' district.

BEST ARE FURTHEST APART.

It seems a law that the best of both are the furthest apart. Perhaps the district in which the three materials are more nearly grouped together is that of Birmingham, Ala. At the first start the ore and limestone were actually taken out of the same hill and the coal was obtained within a distance of from six to ten miles. But there, again, it was soon found that a little further off the ore was higher in iron, and in another direction was a coal that gave a better coke, until to-day the freight cost of assembling is a tangible part of the cost of pig iron. Nevertheless, the distances are not great, but, to offset this, neither is the percentage of iron in the ore. The whole district will, perhaps, not average over 40 per cent. ore. In the Middlesboro' district in England, which is close to the Durham coalfields, the native ore is taken out of the hills overtopping the works, but this ore will not run much over 25 to 30 per cent. before calcining, and does not exceed from 40 to 42 per cent. afterwards. I speak of the native ore because a large part of the ore used at Middlesboro' is imported ore, principally from Spain. So, when these matters are brought to your notice, do not condemn a steel venture because the raw materials are not grouped together and do not put all of your money into one which has everything inside the mill fence.

THE PERMISSIBLE COST.

Our next step is the inquiry, "What is the permissible cost of these materials?" The answer is short. The ore and the limestone delivered on cars or boat (as the case may be), at the mine must be "cheap as dirt," and the coal must not be far from the same. This is to be taken literally, not figuratively. In the Messabi, the Birmingham, the Middlesboro', the Luxembourg and at the Belle Island mines the actual price of mining and putting the ores on cars is less than the traditional contractor's price for removal of earth. It costs more per cubic yard to do the shipping for the ore in the Messabi range than it does to mine the ore. The limestone should not cost over 25c. a ton at the quarries. While there is greater latitude in coal, even this is within narrow limits. In the Connelsville district coal has been mined and put on cars under 40c. a ton. Now, as to the gist of our inquiry: "What are the limits within which these materials must lie?" We can only reach a conclusion by comparison. When we have made this I am hoping that I can convince you that it is 402 miles, or exactly the distance between our Belle Island ore mines and our Sydney coal.

CHEAPEST STEEL CENTRES TO-DAY.

The cheapest steel centres of the world to-day are the following: In England the Middlesboro' district, in Germany the Luxembourg district, in the United States the Central West and Alabama districts. Of these districts the one which overshadows the others is that of the central west, of which Pittsburg costs may be taken as the exponent. We will base our comparison on this, and so doing will be on safe ground. So, what is the freight cost of assembling the raw materials in Pittsburg? Please note we purpose taking actual costs, not market rates. It must be remembered that the haul is partly water, partly rail. It is well to find an equation between the two, and preferable to do so in terms of the rail freight. The actual cost of the lake haul of about 1,000 miles is in the neighborhood of 50c. per ton, or, say, five one-hundredths of a cent a mile. The actual cost of the railroad haul can be safely taken at four-tenths of a cent per ton per mile. We will, therefore, treat 1,000 miles of lake water carriage as equal to 125 miles of railroad carriage. Of necessity this is only closely approximate, for water freight varies largely with distance. The heaviest item in water freight is the lay time of the steamers when loading and unloading. This will be realized when I make the statement that during the open or operating season a lake ore boat is in port about 50 per cent. of her time while operating on an average route 1,000 miles long. And this in a district noted for having every improvement for quick loading and unloading. If the route be a short one the percentage of lay time, and consequently the cost of the freight per ton mile, must largely increase. With these deductions Pittsburg pays the following freight costs in the assemblage of raw material:

COST AT PITTSBURG.

Ore—From mines to upper lake port, 80 miles railroad freight. From upper to lower lake port, 1,000 miles water freight, equal to 125 miles railroad freight. From lower lake port to works, 155 miles railroad freight. Making a total on one ton of ore of 360 railroad miles. It takes 1.70 tons of 60 per cent. ore to make a ton of pig metal. Therefore, the latter calls for a total of 612 railroad ton miles.

Coal—From Connelsville district to Pittsburg, say 80 railroad miles. Taking the same quantity of coal, which is the amouut that is used, 1.70 tons, gives 136 railroad ton miles.

Limestone — From the Tyrone district, 130 miles. Say one-half ton limestone per ton of pig equals 65 railroad ton miles. Adding these together we have: Ore, 612 ton miles; coal, 136 ton miles; limestone, 65 ton miles; total, 813 ton miles. At four-tenths of a cent gives a freight cost of $3.25 per ton of pig iron made. Remember, again, this is freight cost, not the freight charged.

CANADA'S POSITION.

Now, where does Canada stand in this comparison as a steel manufacturer? If you will take a map and draw a line from British Columbia, on the west, to St. John's, Nfld., on the east, that line will run through four well-developed beds of ore of large quantity and of excellent quality, and all of them capable of being mined as cheap as dirt. They are related to the coal as follow:

1. In British Columbia coal beds in the American district not far from Seattle are contiguous, and those of the Crow's Nest Pass on the Canadian side. An assemblage of material is here possible inside of the standard we have taken.

2. On the northern shores of Lake Superior are large and pure beds. Part of these have found their resting place in the hands of our new industrial captain, Mr. Clergue. My expression is awkward. I doubt whether they will rest long either in their own beds or in his hands. This district is full of promise. It has only to bring its coal from the Connelsville or Pittsburg district, a distance of, say, 210 miles by rail to Lorain or Cleveland. Both ports are equipped with modern coal-loading plants. A short water haul equivalent to, say, 80 miles of rail haul will connect with the mines. On this basis 1.70 tons of coal at 290 miles equals 493 miles, which at four-tenths of a cent per mile gives a freight cost of $1.97, or $1.28 lower than the Pittsburg standard.

3. In the Ontario district exists ore of great promise, within reasonable distance of the coal of either the Connelsville or Punxatawny districts. Independently of the local supply is the Canadian lake ore to draw from. This and coal can be assembled within the Pittsburg margin. It is unfortunate for this district that there exists an imaginary line called a boundary, the community on either side of which have listened to that song of the devil called the tariff. Our own Government is to be congratulated, however, in that they are not quite as bad as their neighbors. They do let the coke in free, while the States do not let the ore in free. As a result it is cheaper to assemble and make the pig on the Canadian side.

ADVANTAGES OF BELLE ISLE.

4. Is Belle Island in Newfoundland? Here exists the now well-known Wabana ore. The economical point of its manufacture is at Sydney, directly on the cove beds. The cost of assembly is as follows:

Coal, nothing. Ore 402 miles sea freight. Owing to the relatively short distance the lake basis will not apply, as lay time in proportion to sea time will be the heavier. The actual cost will be 40c. per ton. Limestone 15c. per ton. We then have 1.8 tons of 54 per cent. ore at 40c., 72c.; one-half ton limestone, 7½c.; coal, nothing; total, 79½c.

Let me say while passing that this is the lowest assemblage cost in the world for the tonnage under consideration. As against the Pittsburg cost it represents a saving of $2.45½ per ton.

I have dealt with only four points as typical of what may follow. So far as present indications go, other large supplies are promised, but why go further?

From extreme west to extreme east, point

by point, is Canada favored by every natural condition, and the amount in sight is so great that it will last many times our day. We have so far proceeded only to the manufacture. What of the market?

The British Columbian district would command the eastern export markets. These are to-day somewhat limited in tonnage, and of a great assortment in kind, and this condition would have to be catered to. In addition to this a certain zone of contiguous territory which is rapidly growing up should form a good home market. It may perhaps be that this opportunity might "bide a wee" without loss, but whether it will or not is doubtful. Already it is throbbing in the hands of the promoters, and may be near fulfilment.

ONTARIO'S POSITION.

The central district, including Ontario, has a home market at its feet, and this means much. Already Canada is entering on her industrial renaissance, and her home market, now modest, is destined to grow by leaps and bounds. Even now the silent mighty tide of immigration is turning our way, bringing in its restless current the energy, pluck and never tiring pesistency of our American cousins across the way, bringing the farmer who will call for more ploughs, the blacksmith, the carriage builder, the wagon factory, the need of wire fences; in short, the demand for steel. It is well to realize that all the bountiful gifts of this grand heritage of ours are just now shared by only 5,000,000 people. The use of steel grows at an increasing ratio with growth of population, and is thus "twice blessed." Only a generation ago the consumption of iron per capita in the United States was 34 lb. per annum, Last year it was 150 lb. per capita per annum, and this on the greatly increased population. But even now there is a market waiting for supply, one that stretches from the north pole to the south, from the east to the west, and it is one where tariffs do not go. I mean the ocean. The central district can take her share of this as represented by the great lakes. She can put a large part of her steel into shipplates, boilerplates and channels, and her pig into triple expansion condensing engines, and so develop her shipbuilding and take her share of a trade even now knocking at the door. As to the Cape Breton district, situated on the seaboard, the whole wide world is her market, all of it, and in our modesty we do not ask for more.

TIDE WATER ADVANTAGES.

And at this point we must again revert to our cost comparison. We based this on the price at Pittsburg. To compete for the export business Pittsburg must get to tide water. She is now 500 miles from this, and it will cost her $2 to get there. In dealing with the finished steel it must also be remembered that it takes about 1 1-10 tons of pig to make the steel. So, taking 1 1-10 tons at $3 25, we have $3.57, to which add $2 freight on steel to seaboard, and we have $5.57, from which deduct Sydney's assemblage cost, 79½c., and there is left $4.78 as the net advantage in the cost basis. In actual practice this means more. We should further note that Sydney's tide water will average about 1,000 miles nearer to the world's market than that of Pittsburg. In all conservatism it would be safe to call the commercial difference all of $6 per ton. In a word, Canada's position as a steel-maker is something more than strong. It is simply invulnerable.

I will be pardoned for a few more words on the great promise of the Cape Breton district. So great is this promise that it has ceased to be local. It has ceased to be sectional. It has become international in its influence. To-day Sydney is a familiar name to every steel-maker in England, and before another year rolls around the leaders in this industry will have been with us to see for themselves the new centre. At this moment Sydney is debated with doubt and misgiving by the large German syndicates, and in the United States the strength of her position is conceded by every expert. Canada alone at this moment does not realize how splendid an opportunity is within her control.

PROPHETS OF EVIL ANSWERED.

As is usual in every new venture, Sydney has had her share of the prophets of evil. We are now in operation, and are making an excellent quality of pig metal out of nothing execept our own Wabana ore and Dominion coal. With the fact of what is doing, permit me for a moment to weigh some of the prominent predictions of likely trouble, if not failure. First, we were told that, although Cape Breton coals had coked at Everett, the coke was unfit for blast furnace use. It would not stand the burden. As answer, I would state that the third day after our start it was carrying the full burden; it has done so ever since, and will continue to do so indefinitely. We were further told that everyone knew that the coals were too high in sulphur. We did not heed much what everyone knew. We carefully analyzed the coal. We found it higher than Connelsville, but only slightly so, and we further found that the excess could be economically washed out, and so erected a washing plant. The coal-washer not being ready, we determined to start without it, and we did so, making our coke out of plain, every-day, unwashed Cape Breton coal. Our flux we knew to be unusually pure, and we felt this was worth something in controlling the sulphur. Our second cast from the furnace put us in control of the sulphur. The fifth cast brought it down to nineteen one-thousandths of 1 per cent. For most purposes it is too low, not too high. In every other ingredient Cape Breton coal equals, if it does not exceed, Connelsville. In ash the unwashed coal gives us a coke with from 6½ per cent. to 7½ per cent. ash, as against 10½ per cent. to 11 per cent. in the Connelsville; in fixed carbon, from 90¾ per cent. to 91 per cent., against 89½ per cent. in the Connelsville. So much for the coal. Now for the ore. The exposed surface ore gives us an average of about 52 per cent. iron, and as high as 11 per cent. silica. Our friends sympathized with us having so much silica to deal with. But the same pure flux which controlled the sulphur took a whack at the silica. The slag is so thoroughly basic that it holds down the silica, and we would not object to having even a little more left in the iron. The phosphorous has been spoken of as too high for basic pig. A few years ago, before the division of open-hearth practice into the primary and the finishing furnaces, this might have been true. It is not so to-day. In England and Germany iron much higher in phosphorous is used, and it is found that the basic slag which results is a valuable by-product, and commands a high price as a fertilizer. It is questionable whether we have quite enough phosphorous in our pig to give us this slag; but, if not, it can be cheaply added. For some very special brands of foundry iron the market demands are extremely low phosphorous. This we must obtain by the admixture of a neutral ore with the Belle Island ore. This we have found almost at our doors—not in large beds, but, as little of it will be needed, enough is at our disposal. At Belle Island doubts were developed as to whether we had as much ore as expected. Every test-hole put down, every heading driven, has proved the continuity of the ore and its improved quality when under cover.

CANADA'S STEEL FUTURE.

The existence of the submarine bed is proved and certain. On an extremely calm, smooth day its borders can be traced to a large extent by the naked eye. We have perhaps no right to put up our hopes too high, but we should also remember that there is as yet nothing to indicate small limits to this deposit. So, gentlemen, commencing at the west and overlapping, like Belle Island ore, into the sea in the east, do I find promise. West, central and east each has its proper sphere. There will be profitable room for many more steel works than those which are now projected. The raw material is favorably placed, and another year will convince Canadian capital that the time for doubt has gone. We need no more halting steps, but a stronger, firmer stride; we need no more speculation or hesitation, but a bolder reaching out for the harvest that lays ripening at our feet. It is there. It is ours. We have only to gather it.

A FEW WORDS FROM PRESIDENT WHITNEY.

Among those who occupied a seat at the chairman's table was Mr. H. M. Whitney, the president of the Dominion Iron & Steel Co., Limited. It was a strange coincidence that Mr. Whitney had arrived in Toronto not knowing that his general manager, Mr. Moxham, was there, or that he was to deliver an address that day. When Mr. Moxham concluded his address it was the most natural thing in the world for the chairman to ask Mr. Whitney to say a few words. During his remarks Mr. Whitney said he was impressed three years ago with the fundamental facts that coal, lime and ore were availiable for the enterprise, and knew it must be successful. He believed it would, as a basic industry, advantageously affect every business in the Dominion. He hoped they would have the pleasure of welcoming many Toronto business men on a visit to Sydney during the coming summer.

MAKING THE MACHINES IN CANADA.

In reply to a vote of thanks which was heartily tendered him, Mr. Moxham observed that he need not say how much they, on their side, reciprocated every hope of mutual advantage to be derived from the development of the steel interests of the country. It was true that they would export the great bulk of their product, but he made it evident that the company expects to do a large and profitable trade with Canadian manufacturers. As an instance of the development which might be expected in Canada, he stated that the company are building five very large blowing machines. No bigger machines of the sort exist in the world; some of equal size are in use in the United States. The U.S. manufacturers proved rather slow, and in a moment of wisdom, or of folly—he did not yet know which—he got permission to have them built in Canada. It was being done at the moment, and he thought they were doing it successfully. It was the prelude to more business which would come naturally, and things like that would put into the minds of their foundrymen and machinists the conviction that they could do the big things that other people did.

James Sparling's sash and door factory, Meaford, Ont., suffered about $6,000 damage by fire on Sunday. There is $2,700 insurance.

FEBRUARY IN THE HARDWARE STORE.

By H. C. W.

FEBRUARY is a hard month in which to lay down lines and rules to follow systematically. There is a little of everything to do, and that little runs into a great deal of importance before the month is ended. It's an aftermath, a lapping over of nearly all the work laid out for January, with a little entirely new added. The going into new sets of books is done with; the small and short accounts are, or should be, largely weeded out and collected. The leaks of the last year have been at least partially found and are on the way to being stopped or lessened. The dead stock sale is being pushed forward and the general cleaning up is well done with. Do we ever realize that we are at this season of the year like the drygoodsman, full of stocks for

REMNANTS AND REMNANT SALES?

There is the belting stock—the short ends and pieces, from 2 to 12 to 15 feet. We had one of the boys carefully measure them up, marking each piece and entering same on a memorandum. There were all sent back to the factory last February (and they will be accepted at their value direct or added to new rolls of stock), bringing us a credit slip from the manufacturer of $53 and odd cents.

There are remnants of floor oil cloth, table oil cloth, ducking, linoleum, screen wire, poultry netting, hardware cloth and remnants in many other lines, all to be measured and marked, or if already done for purposes of inventory, to be brought forward, kept marked and pushed out of the front door; and February is the month of the year for such work.

BINS, RACKS, ETC.,

for all sorts of stock, suggest repairing of the old and the making of many new ones. There cannot be too many bins, racks and shelves in any hardware store. There are axes to be mounted and axles to be matched up and tied in sets. The bar iron needs attention, and with it the tool steel rack. Brass and copper wire wants to be racked and numbered on the wall. Cast fittings, carriage and machine bolts and lag screws must all be taken care of for another year. Circular saws, cross cut saws and clothes wringers must have a stand or they will grow worse with handling. Dry colors and loose emery must have tin or wooden cases. Hinges of all kinds should have their various compartments, and so on through the line—steel goods, scythes, shovels, etc., and everything that stands up, falls down, or mixes. Every betterment of stock in this way tends closer toward the silent salesman and the disposal of goods with the least effort.

KEEPING AT COLLECTIONS

through the month of February, and until the work is as nearly done as ever it is, becomes a matter of necessity, in the light of the fact that without the most persistent and systematic attention throughout the year accounts will bunch and pile up—neglected! After all statements for January—covering every account on the books—are sent out, and the usual responses up to the 10th of the month are in, it is a capital plan to take long statement sheets, and from the balance sheet draw off every unpaid account, with ledger page and amount. Then start the best collector in the house out for those he can reach. He will collect money. Those who turn him down must set a date for payment, and that date must be religiously observed. If set ahead again, follow it up and keep on. In the meantime, those he cannot reach you can work on with stenographer and personal, strong but persuasive and well-worded letters, using city and country directory. It is wonderful the results that come from work of this kind, pursued faithfully through January and February.

THERE IS MONEY IN RENOVATING

and cleaning up. February is a good month in which to find it. How many linseed oil barrels have you in the cellar at 85c. each? How many turpentine barrels which cost you from 90c. to $1 and lubricating oil barrels from 50 to 60c.? The common shipping barrels, such as lantern globes come in, bring 10c. each. Your packing boxes all bring from 5 to 10c. each. Your wrought scrap is worth at least 75c. to $1 per cwt. Brass and copper scrap on hand is valuable. There is a pile of "warranted goods"—axes, hammers, hatchets, saws, etc.—not charged back or shipped to headquarters. They all take up room, they are of value to the other man. Get them ready, ship and sell everything named above. We expect to realize $150 to $175 during the month of February in these odds and ends, and count it nearly all clear gain.

RESAMPLING OF GOODS

is a splendid occupation for this month, and a clever young boy can do the work when laid out for him. You have changed many lines during the year—old samples are partly knocked off and disarranged. In many cases goods in the boxes are not what they are on the outside, and the resampling should mean, too, where necessary, remarking. Table cutlery should have attention and the razor and pocket knife samples;

so, too, the scissor stock, the tool stock, and any other bright, polished or plated goods in the house.

FEBRUARY FOR EARLY SHIPMENTS,

particularly season goods and those with a summer dating, means lots of dollars to the good, ahead of your competitors. The day is at hand when refrigerators and ice cream freezers sell nearly all the year around, and they can not get in too early. So also with the bulk seed stock (it takes a long time to arrange it properly), and much hard work is connected therewith. And the complete line of steel goods should be on hand before March 1. The first premature warm weather catches you otherwise short of hoes, rakes, spades, floral sets, etc., barring the fact, too, that it takes time to mark and arrange the stock for display and for sale.

KILL THE DULL SEASON

this year by using a part of February to decide on and select new lines of goods that will keep every man in the house doing something to make profit every day in the year. It is comparatively easy to have every month a busy one, and the added investment to secure it is not a heavy one if these lines are added by degrees. I would suggest small stocks of tinware, woodenware, plated silverware, cheap watches, box paints, small lines of tools not usually carried but often asked for, specialties in gold and silver paints, furniture and metal polishes, liquid glues and cements, and a hundred good novelties that leave a clear 100 per cent. profit every time they go out of the house. They pay rent and clerk hire; they are cash sales. A 20c. sale is better than a $2.50 keg of nails.—Iron Age.

AN ALUMINUM TRANSMISSION LINE

The Niagara Falls Power Co. has about completed its second power transmission line between Niagara Falls and Buffalo. The new line parallels the old line as far as Tonawanda, where it diverges and runs over a new right-of-way to Buffalo. It possesses special interest because of the fact that the new cables are made of aluminum. The three-phase current is transmitted by three cables, each composed of 37 strands. The old line consists of six copper cables, each of which has 19 strands. One advantage gained in the use of aluminum is that the cables being so much lighter, the span between poles, which in the old line is about 75 ft., averages 112½ ft. in the new line. On the completion of the aluminum line, the voltage of the current that is transmitted will be raised from 11,000 to 22,000 volts. When the line was first built, the electrical plant was designed with a view to this doubling of the voltage whenever the time was ripe to carry it out, and hence no material changes will be necessary.—Scientific American.

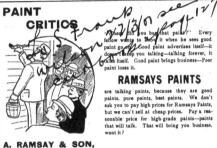

MARKETS AND MARKET NOTES

QUEBEC MARKETS.

Montreal, February 22, 1901.

HARDWARE.

BUSINESS continues to be rather quiet, particularly in heavy goods. There seems to be no circumstance to impel heavy purchases, and consequently dealers are buying from hand-to-mouth. Even spring bookings in such lines as poultry netting, green wire cloth, freezers, screens, lawn mowers, hose and churns, are rather light. Wires are moving a little more freely this week, particularly in smooth gauges. Horseshoes, shot, and horseshoe nails are all in good request. Nails are still very slow. The downward trend of English heavy metals has injuriously affected the demand for all grade of goods, and spring purchases are being deferred. Paris green and glass are important articles just now. Bright goods have been marked down this week, the discount now being 62½ per cent. in place of 55 per cent. Sisal rope has advanced ¼c. per lb. White lead is ¼c. per lb. lower, while dry white lead and red lead have declined ¼c. per lb. The Montreal trade is refusing to sell Mrs. Potts sad irons at cost, and quote polished at 67½c., nickel plated at 72½c.

BARB WIRE—There is nothing new to offer in the barb wire market. A few spring orders are being booked at $3.20 f.o.b. Montreal in less than carlots.

GALVANIZED WIRE—A small business is passing and some transactions in futures are also taking place. We quote as follows: No. 5, $4.25; Nos. 6, 7 and 8 gauge $3.55; No. 9, $3.10; No. 10, $3.75; No. 11, $3.85; No. 12, $3.25; No. 13, $3.35; No. 14, $4.25; No. 15, $4.75; No. 16, $5.00.

SMOOTH STEEL WIRE—Quite a few shipments have been made this week and business has been fair. Values are still at $2.80 per 100 lb.

FINE STEEL WIRE—There has been quite a call for fine wire this week, business taking place at 17½ per cent. off the list.

BRASS AND COPPER WIRE—Copper wire has been in fair demand this week. Discounts are still 55 and 2½ per cent. on brass, and 50 and 2½ per cent. on copper.

FENCE STAPLES—Trade is quiet. We quote: $3.25 for bright, and $3.75 for galvanized, per keg of 100 lb.

WIRE NAILS—Business is of small proportions, dealers buying from hand-to-mouth. We quote $2.85 for small lots and $2.75 for carlots, f.o.b. Montreal, Toronto, Hamilton, London, Gananoque, and St. John, N.B.

CUT NAILS—There is not much business doing in this line and trade is quiet. We quote: $2.35 for small and $2.25 for carlots; flour barrel nails, 25 per cent. discount; coopers' nails, 30 per cent. discount.

HORSE NAILS—A good business is doing. The general discounts are 50, 10 and 5 per cent. on oval head and 50, 10 and 10 per cent. on countersunk head. "C" brand's discount is 50 and 7½ per cent. on their own price list.

HORSESHOES—The demand continues quite brisk and prices are steady. We quote as follows: Iron shoes, light and medium pattern, No. 2 and larger, $3.50; No. 1 and smaller, $3.75; snow shoes, No. 2 and larger, $3.75; No. 1 and smaller, $4.00; X L steel shoes, all sizes, 1 to 5,

No. 2 and larger, $3.60; No. 1 and smaller, $3.85; feather-weight, all sizes, $4.85; toe weight steel shoes, all sizes, $5.95 f.o.b. Montreal; f.o.b. Hamilton, London and Guelph, 10c. extra.

POULTRY NETTING — A few orders are being booked on spring account at a discount of 50 and 5 per cent.

GREEN WIRE CLOTH—The same remark applies to green wire cloth; business is none too brisk. The price is still $1.35 per 100 sq. ft.

BRIGHT GOODS—The discount on bright goods, including screw eyes, screw hooks and gate hooks, has been raised from 50 to 62½ per cent.

FREEZERS—Summer supplies are being contracted for. "Peerless" is quoted as follows: Two quarts, $1.85; 3 quarts, $2.10; 4 quarts, $2.50 6 quarts, $3.20; 8 quarts, $4; 10 quarts, $5.25; 12 quarts, $6; 16 quarts, with fly wheel, $11; toy, 1 pint, $1.40.

SCREEN DOORS AND WINDOWS—Travellers report spring business as fair. We quote: Screen doors, plain cherry finish, $8.25 per doz.; do. fancy, $11.50 per doz.; windows, $2.25 to $3.50 per doz.

SCREWS — A fair trade is passing in screws. Discounts are as follows: Flat head bright, 87½ and 10 per cent. off list; round head bright, 82½ and 10 per cent.; flat head brass, 80 and 10 per cent.; round head brass, 75 and 10 per cent.

BOLTS—The demand seems to be very brisk, and some good sales have been made this week. Discounts are as follows: Carriage bolts, 65 per cent.; machine bolts, 65 per cent.; coach screws, 75 per cent.; sleigh shoe bolts, 75 per cent.; bolt ends, 65 per cent.; plough bolts, 50 per cent.; square nuts, 4½c. per lb. off list; hexagon nuts, 4¾c. per lb. off list; tire bolts, 67½ per cent.; stove bolts, 67½ per cent.

BUILDING PAPER — Some fair sales have been made on future account this week at unchanged prices. We quote as follows: Tarred felt, $1.70 per 100 lb.; 2-ply, ready roofing, 80c. per roll; 3-ply, $1.05 per roll; carpet felt, $2.25 per 100 lb.; dry sheathing, 30c. per roll; tar sheathing, 40c. per roll; dry fibre, 50c. per roll; tarred fibre, 60c. per roll; O.K. and I.X.L., 65c. per roll; heavy straw sheathing, $28 per ton; slaters' felt, 50c. per roll.

RIVETS — The aggregate of business is not large. The discount on best iron rivets, section, carriage, and waggon box, black rivets, tinned do., coopers' riverts and tinned swedes rivets, 60 and 10 per cent.; swedes iron burrs are quoted at 55 per cent. off; copper rivets, 35 and 5 per cent. off; and coppered iron rivets and burrs, in 5-lb. carton boxes, are quoted at 60 and 10 per cent. off list.

CORDAGE—There has been quite a brisk demand for twine this week. Sisal has been advanced ½c. per lb. Manila is quoted at 13c. per lb. for 7-16 and larger; sisal at 9c., and lathyarn 9c. per lb.

SPADES AND SHOVELS—Orders for spring delivery are small. The discount is still 40 and 5 per cent. off the list.

HARVEST TOOLS—Individually the orders are not large yet they are fairly numerous. The discount is 50, 10 and 5 per cent.

TACKS—This line fs featureless. We quote: Carpet tacks, in dozens and bulk, blued 80 and 5 per cent. discount; tinned, 80 and 10 per cent.; cut tacks, blued, in dozens, 75 and 15 per cent. discount.

CHURNS—Business continues to be done at a discount of 56 per cent.

FIREBRICKS—Only a small amount of business is being done at $18.50 to $26, as to brand.

CEMENT—The winter demand is small. We quote: German, $2.50 to $2.65; English, $2.40 to $2.50; Belgian, $1.90 to $2.15 per bbl.

METALS.

The English market in sheets and plates is rather demoralized, particularly in Canada plates and tinned plates. Black sheets are also affected and have dropped about £2 since the first of the year and are now within £1 of what they were at the beginning of 1899. Galvanized iron remains steady and a good trade is being done in pipe.

PIG IRON—The pig iron market is not very steady and lower values seem to prevail. Canadian pig is worth about $18 on the Montreal market, and No. 1 Summerlee about $23.

BAR IRON—The feeling in bar iron is stiff, and dealers are refusing offers of $1.60. We quote $1.65 to $1.70.

BLACK SHEETS—Little business is being done in black sheets, and the few shipments being made are small. The English market is easy and has declined about £2 since January 1 on lower grades and about 10s. since September last. This has not affected the spot market, but we may look for lower values when navigation opens. Hardly any import orders are being placed and there seems to be no tendency to transact business. Prices rule at $2.80 for 8 to 16 gauge; $2.85 for 26 gauge, and $2.90 for 28 gauge.

GALVANIZED IRON—The galvanized iron market is firm, although it is made rather quiet by the dull condition of the other metal markets. We quote: No. 28 Queen's Head, $5 to $5.10; Apollo, 10¾ oz., $5 to $5.10; Comet, No. 28, $4.50, with 25c. allowance in case lots.

INGOT COPPER—The foreign statistics for the middle of February, show, as compared with the figures of January 31, a decrease in the stocks of 350 tons, while the afloats remain unchanged. The market is firm locally at 17½c.

INGOT TIN—The feeling in tin seems to be slightly stronger than last week. A fair business is being done. The price here is 33 to 34c.

LEAD—The English market is weak. Locally the price is unchanged at $4.65.

LEAD PIPE—The demand continues to be well maintained. We quote: 7c. for ordinary and 7½c. for composition waste, with 15 per cent. off.

IRON PIPE—The pipe trade is one of the active lines of the metal market. Prices are steady. We quote: Black pipe, ¼, $3 per 100 ft.; ⅜, $3; ½, $3; ¾, $3.15; 1-in., $4.50; 1¼, $6.10; 1½, $7.28; 2-in., $9.75. Galvanized, ½, $4.60; ¾, $5.25; 1 in., $7.50; 1¼, $9.80; 1½, $11.75; 2-in., $16.

TINPLATES—Trade is slow on account of the demoralized condition of the English market. Because of the inability to secure paying prices, 215 out of 495 mills making tinned plates and Canada plates have shut down to wait for higher values or decrease in the cost of material. At present there seems to be no improvement in sight, but yet the mills seem to have adopted an artificial and sure course to improve the condition of the market. Whether values will recover depends upon the course of the American market. Of course, this appertains solely to spring deliveries of goods, not to spot values, which are quite steady. Stocks in hand are light. The ruling figures for immediate delivery are $4.50 for coke and $4.75 for charcoal.

CANADA PLATE—The remarks made about tinned plates apply as well to Canada plates which have declined about £2 per box since the close of navigation. Business is dull and only small import orders are being placed. We quote: 52's, $2.90; 60's, $3; 75's, $3.10; full polished, $3.75, and galvanized, $4.60.

TOOL STEEL—There is a fair demand for steel. We quote: Black Diamond, 8c.; Jessop's 13c.

STEEL—Unchanged. We quote: Sleighshoe, $1.85; tire, $1.95; spring, $2.75; machinery, $2.75 and toe-calk, $2.50.

TERNE PLATES—The remarks made in regard to tinned plates apply also to terne plates. We hear that some houses have reduced their prices on spot stocks to $8, being a reduction of 25c., which other houses are still trying to get.

COIL CHAIN—We are informed that American firms have raised their quotations on some gauges of chain. A good business is being done. We quote as follows: No. 6, 11½c.; No. 5, 10c.; No. 4, 9½c.; No. 3, 9c.; ¼-inch, 7½c. per lb.; 5-16, $4.65; 5-16 exact, $5.10; ⅜, $4.20; 7-16, $4.00; ½, $3.75; 9-16, $3.65; ⅝, $3.35; ¾, $3.25; ⅞, $3.20; 1-in., $3.15. In carload lots an allowance of 10c. is made.

SHEET ZINC—The ruling price is 6 to 6¼c.

ANTIMONY—Quiet, at 10c.

GLASS.

We understand that the business being done on import account is improving, orders being taken on a basis of $1.75 or $1.80 for the first break of 50 feet. We quote as follows: First break, $2; second, $2.10 for 50 feet; first break, 100 feet, $3.80; second, $4; third, $4.50; fourth, $4.75; fifth, $5.25; sixth, $5.75, and seventh, $6.25.

PAINTS AND OILS.

The linseed oil market shows little improvement, and there are few orders yet being placed for summer delivery. There will be small May and June arrivals. The lead market has taken a weak turn, and local prices are down ⅜c. per lb. for white lead and ¼c. per lb. for dry white lead and red lead. Turpentine is firm. A considerable amount of paris green is selling for spring stock, and putty is also going out freely. A further decline is not unlooked for in lead within the next year, but for some time the prices we quote will likely stand. We quote:

WHITE LEAD—Best brands, Government standard, $6.37½; No. 1, $6 00; No. 2, $5.62½; No. 3, $5.25, and No. 4, $4.87½ all f.o.b. Montreal. Terms, 3 per cent. cash or four months.

DRY WHITE LEAD—$5.50 in casks; kegs, $5.75.

RED LEAD—Casks, $5.25; in kegs, $5.50.

WHITE ZINC PAINT—Pure, dry, 8c.; No. 1, 6½c.; in oil, pure, 9c.; No. 1, 7½c.

PUTTY—We quote: Bulk, in barrels, $2 per 100 lb.; bulk, in less quantity, $2.15; bladders, in barrels, $2.20; bladders, in 100 or 200 lb. kegs or boxes, $2.35; in tins, $2.45 to $2.75; in less than 100-lb. lots, $3 f.o.b. Montreal, Ottawa, Toronto, Hamilton, London and Guelph. Maritime Provinces 10c. higher, f.o.b. St. John and Halifax.

LINSEED OIL—Raw, 76c.; boiled, 79c., in 5 to 9 bbls., 1c. less, 10 to 20-bbl. lots, open, net cash, plus 2c. for 4 months. Delivered anywhere in Ontario between Montreal and Oshawa at 2c. per gal. advance and freight allowed.

TURPENTINE—Single bbls., 59c.; 2 to 4 bbls., 58c.; 5 bbls. and over, open terms, the same terms as linseed oil.

MIXED PAINTS—$1.25 to $1.45 per gal.

CASTOR OIL—8¼ to 9¼c. in wholesale lots, and ½c. additional for small lots.

SEAL OIL—47½ to 49c.

COD OIL—32½ to 35c.

NAVAL STORES—We quote: Resinsr $2.75 to $4 50, as to brand; coal tar, $3 25 to $3.75; cotton waste, 4½ to 5½c. for colored, and 6 to 7½c. for white; oakum, 5½ to 6½c., and cotton oakum, 10 to 11c.

PARIS GREEN—Petroleum barrels, 16¾c. per lb.; arsenic kegs, 17c.; 50 and 100-lb. drums, 17½c.; 25-lb. drums, 18c.; 1-lb. packages, 18½c.; ½-lb. packages, 20½c.; 1-lb. tins, 19½c.; ½-lb. tins, 21½c. f.o.b. Montreal; terms 3 per cent. 30 days, or four months from date of delivery.

SCRAP METALS.

The metal market is still extremely quiet and prices are almost nominal. Holders are firm in their views. Dealers are paying the following prices in the country: Heavy copper and wire, 13 to 13½c. per lb.; light copper, 12c.; heavy brass, 12c.; heavy yellow, 8½ to 9c.; light brass, 6½ to 7c.; lead, 2¾ to 3c. per lb.; zinc, 2¼ to 2½c.; iron, No. 1 wrought, $13 to $14 per gross ton; No. 1 cast, $13 to $14; stove plate, $8 to $9; light iron, No. 2, $4 a ton; malleable and steel, $4.

HIDES.

The demand for hides is fair, and dealers are still paying 7½c. for No. 1 light, and tanners 8½c. for carlots. We quote: Light hides, 7½c. for No. 1; 6½c. for No. 2, and 5½c. for No. 3. Lambskins, 90c.

PETROLEUM.

Business is reduced to a spring volume. We quote: "Silver Star," 14½ to 15½c.; "Imperial Acme," 16 to 17c.; "S.C. Acme," 18 to 19c., and "Pratt's Astral," 18½ to 19½c.

MONTREAL NOTES.

Terne plates are lower.

White lead has been reduced ⅜c. per lb.

Coil chain is firm and advancing.

Sisal rope is quoted ½c. per lb. higher.

Red lead and dry white lead are ¼c. per lb. lower.

The discount on bright goods has been raised from 55 per cent. to 62½ per cent.

Small import orders are being placed for spring delivery of linseed oil, Canada or tinned plates, terne plates, or black sheets on account of the demoralized condition of the English market.

ONTARIO MARKETS.

TORONTO, February 22, 1901.

HARDWARE.

BUSINESS continues to improve a little. At the same time, however, there is not a large turnover. For prompt shipment the demand is only of a sorting-up character, but for spring delivery orders are numerous, although not as large as last year at this time. In general hardware prices are much the same as they were a week ago. Sisal rope, however, is 1½c. per lb. dearer, being now quoted at 9½c. base. Wholesale dealers in plumbers' supplies have come to an agreement in regard to prices, and as a result there is an advance in the figures on such lines as enamelled baths, closets, and range boilers. In most of these lines prices have been cut a great deal lately, and, at the figures now ruling, the jobbers claim they are only making a fair profit. A new list has been issued on wrought iron spikes. There has also been an advance of 10 to 15 per cent. in the price of screen doors and windows.

BARB WIRE—There is a little more of this in demand for immediate shipment, although the quantities wanted are very small. Little or nothing is being done in the way of future delivery. Prices are unchanged. We quote: $2.97 f.o.b. Cleveland for less than carlots, and $2.85 for carlots. From stock, Toronto, $3.10 per 100 lb.

GALVANIZED WIRE—There have been a few more inquiries this week, principally from manufacturers of wire fences. So far, however, these inquiries have not led to much business. We quote: No. 6, 7 and 8, $3.55; No. 9, $3.10; No. 10, $3.75; No. 11, $3.85; No. 12, $3.25; No. 13, $3.35; No. 14, $4.25; No. 15, $4.75, and No. 16, $5.

SMOOTH STEEL WIRE—There has been a little booking of wires in oiled and annealed for future delivery, but scarcely anything is being done in shipment from stock. A little hay-baling wire is going out for immediate shipment. The base price is still $2.80 per 100 lb.

WIRE NAILS—Business for immediate delivery is still decidedly small. Most of the business that is being done is for shipment next month. We still quote $2.85

per keg in less than carlots, and $2.75 for carlots.

CUT NAILS—These are as flat as ever, and there is no sign of any improvement in the demand. The base price is unchanged at $2.35 per keg.

HORSESHOES—There is still some demand for these, but trade is not as active as it was. Prices are without change. We quote f.o.b. Toronto: Iron shoes, No. 2 and larger, light, medium and heavy, $3.60; snow shoes, $3.85; light steel shoes, $3.70; featherweight (all sizes), $4.95; iron shoes, No. 1 and smaller, light, medium and heavy (all sizes), $3.85; snow shoes, $4; light steel shoes, $3.95; featherweight (all sizes), $4.95.

HORSE NAILS—A fair trade is being done in this line, and quotations rule as before. We quote "C" brand oval head at 50 and 7½ per cent. discount, and "M" brand at 50, 10 and 5 per cent., on their respective lists; countersunk head, 50, 10 and 10 per cent.

SCREWS—A moderate trade is to be noted in this line with prices as before. Discounts are: Flat head bright, 87½ and 10 per cent.; round head bright, 82½ and 10 per cent.; flat head brass, 80 and 10 per cent.; round head brass, 75 and 10 per cent. Round head bronze is unchanged at 65 per cent., and flat head bronze at 70 per cent.

BOLTS AND NUTS—Trade is moderate for this time of the year, and devoid of any special feature. We quote: Carriage bolts (Norway), full square, 70 per cent.; carriage bolts full square, 70 per cent.; common carriage bolts, all sizes, 65 per cent.; machine bolts, all sizes, 65 per cent.; coach screws, 75 per cent.; sleighshoe bolts, 75 per cent.; blank bolts, 65 per cent.; bolt ends, 65 per cent.; nuts, square, 4½c. off; nuts, hexagon, 4¾c. off; tire bolts, 67½ per cent.; stove bolts, 67½; plough bolts, 60 per cent.; stove rods, 6 to 8c.

RIVETS AND BURRS — There is the usual steady trade being done. Discount, 60 and 10 per cent. on iron rivets, 55 per cent. on iron burrs, and 35 and 5 per cent. on copper rivets and burrs.

ROPE—The feature of the rope trade this week is an advance of ½c. per lb. in sisal rope, the base being now 9½c. Some orders are being taken in rope for future delivery and trade in this line is fair for this time of the year. The base price of manila rope is unchanged at 13c.; cotton rope, 3-16 in. and larger, 16½c.; 5-32 in., 21½c., and ⅛ in., 22½c. per lb.

BINDER TWINE—Trade is moderate with prices unchanged. We quote pure manila, 10¾c. per lb.; mixed, 8½c. per lb.; sisal, 7¾c. per lb.

CUTLERY—Some shipments are still going forward to British Columbia, but otherwise trade continues quiet.

SPORTING GOODS—A few rifles are going out, and there is a little demand for loaded shells and cartridges.

CHURNS AND WRINGERS—Some churns are being booked and a few orders are being taken for immediate shipment. In wringers the trade is somewhat demoralized on account of the collapse of the association, which formerly controlled prices.

GREEN WIRE CLOTH—A few orders are still being booked on future account at $1.35 per 100 sq. ft.

POULTRY NETTING — Most of the orders for future delivery appear to have been taken, as trade is beginning to fall off. The discount on Canadian is 50 and 5 per cent., and English netting is quoted at net figures.

HARVEST TOOLS — A few orders are still being booked for future delivery, although trade is not active. Discount, 50, 10 and 5 per cent.

SPADES AND SHOVELS—A few orders are being taken for spades and shovels, but most of them are for future delivery. Business on future account is not as heavy as it was last year.

BUILDERS' SUPPLIES—Business is fair, quite a few lots going out.

BUILDING PAPER.—A fairly good trade is reported in building paper, with prices as before.

SCREEN DOORS AND WINDOWS—There has been an advance in prices of about 10 to 15 per cent. by the manufacturers. But, so far, the jobbers do not appear to have followed suit.

CEMENT—The indications point towards a heavy production this summer. Two companies were organized last fall, who intend to produce large quantities this summer. This week brings a report of the incorporation of another million dollar company, which has secured control over marl banks in Grey county, Ont., and which expects to start operations about May. At present there is nothing doing. We nominally quote in barrel lots: Canadian Portland, $2.80 to $3; Belgian, $2.75 to $3; English do., $3; Canadian hydraulic cements, $1.25 to $1.50; calcined plaster, $1.90; asbestos cement, $2.50 per bbl.

RANGE BOILERS—A fixed price has been made on galvanized boilers. We now quote as follows: 30 gallons, $7; 35 gallons, $8.25; 40 gallons, $9.50.

BOLTS — (Standard enamelled) First quality, $25; second quality, $21. These are the new prices which have been fixed by the jobbers.

CLOSETS—A fixed price has been made upon closets, some being quoted net figures and others by discounts. Reference to our "prices current" will show this.

METALS.

The metal trade is on the whole fairly good for this time of the year. Compared with last week business seems to be a little more active. In the United States the demand for steel is most active and premiums are being paid in order to get prompt delivery. Local quotations on metals are much the same as a week ago.

PIG IRON—Trade is rather quiet, most of the large foundrymen in Canada having placed their orders for supplies for the first half of the present year. The price for Canadian pig is still about $17.

BAR IRON—Trade continues fairly active at $1.65 to $1.70 per 100 lb. base.

PIG TIN—A fair amount of business is being done this week, both in large and in small lots. The outside markets have ruled firm until within the last couple of days, when they took an easier turn. Local quotations are unchanged, they still ruling at 32 to 33c.

TINPLATES — There is a moderate demand for the various grades. Some impor.

tant orders are being booked for spring and summer delivery. Prices are low in England, but reports received here state that it is believed that the bottom has been touched. Quotations locally are without change.

TINNED SHEETS—The demand is good and a still further improvement is looked for. Prices are still quoted at from 9 to 9½c. for 28 gauge.

TERNE PLATES—Inquiries have been a little more numerous during the past week, but the actual business is still light. Quotations range from $8.50 to $10.50, according to guage.

BLACK SHEETS—Trade is not as active as it was, but prices are unchanged, the base figure still being $3.30.

GALVANIZED SHEETS—Trade is good in this line, both from stock and on importation account. The ruling price for English is $4.75, and for American $4.50.

CANADA PLATES—Shipments from stock continue small, and import orders continue to be booked for next fall's delivery. We quote all dull, $3; half and half, $3.15, and all bright, $3.65 to $3.75.

IRON PIPE—Although the jobbers held a meeting, no arrangement has been made in regard to prices, and quotations remain as before. We quote as follows: Black pipe ¼ in., $3.00; ⅜ in., $3.00; ½ in., $3.05; ¾ in., $3.20; 1 in., $4.60; 1¼ in., $6.35; 1½ in., $7.55; 2 in., $10.10 Galvanized pipe is as follows: ½ in., $4.65; ¾ in., $5.35; 1 in., $7.25; 1¼ in., $9.75; 1½ in., $11.25; 2 in., $15.50.

HOOP, STEEL—The demand during the past week has been most active, with $3.10 as the ruling base price.

COPPER—The demand for ingot copper has improved during the past week, and trade continues good in sheet copper. We quote: Ingot, 19 to 20c.; bolt or bar, 23½ to 25c.; sheet, 23 to 23½c. per lb.

BRASS—The demand has fallen off during the week, trade being quiet. Discount on rod and sheet is unchanged at 15 per cent.

SOLDER—There is a fair demand, and prices are as before. We quote: Half-and-half, guaranteed, 19c.; ditto, commercial, 18½c.; refined, 18½c., and wiping, 18c.

LEAD—The demand in this line is good, with prices unchanged at 4¾ to 5c.

ZINC SPELTER—There is very little doing, and the ruling quotation is still 6 to 6½c. per lb.

ZINC SHEET—The demand in this line is still small. We quote casks at $6.75 to $7, and part casks at $7 to $7.50.

ANTIMONY—Little or nothing is being done in this line, and we still quote at 11 to 11½c. per lb.

PAINTS AND OILS.

There have been several important changes this week. The most important is a decline of 37½c. in white lead. The quotations on linseed oil which were issued on January 8 for delivery in April or later have been put into effect for immediate shipment. This means a decline in present quotations of 9c. for delivery to outside points, and 10c. in Toronto, Hamilton and London, yet, as practically no trade was likely to be done at the old prices, the decline does not mean cheaper oil to the retailer. Turpentine is stiff in the South and may be advanced here. Paris white and whiting have been raised 10c. per 100 lb. Other lines are unchanged. There is little doing. We quote:

WHITE LEAD—Ex Toronto, pure white lead, $6.50; No. 1, $6.12½; No. 2, $5.75; No. 3, $5.37½; No. 4, $5; dry white lead in casks, $6.

RED LEAD—Genuine, in casks of 560 lb., $5.50; ditto, in kegs of 100 lb., $5.75; No. 1, in casks of 560 lb., $5 to $5.25; ditto, kegs of 100 lb.; $5.25 to $5.50.

LITHARGE—Genuine, 7 to 7½c.

ORANGE MINERAL—Genuine, 8 to 8½c.

WHITE ZINC—Genuine, French V.M., in casks, $7 to $7.25; Lehigh, in casks, $6.

PARIS WHITE—90c. to $1 per 100 lb.

WHITING—70c. per 100 lb.; Gilders' whiting, 80c.

GUM SHELLAC—In cases, 22c.; in less than cases, 25c.

PARIS GREEN—Bbls., 16¾c.; kegs, 17c.; 50 and 100 lb. drums, 17½c.; 25-lb. drums, 18c.; 1-lb. papers, 18½c.; 1-lb. tins, 19½c.; ½ lb. papers, 20½c.; ½ lb. tins, 21½c.

PUTTY—Bladders, in bbls., $2.20; bladders, in 100 lb. kegs, $2.35; bulk in bbls., $2; bulk, less than bbls. and up to 100 lb., $2.15; bladders, bulk or tins, less than 100 lb., $3.

PLASTER PARIS—New Brunswick, $1.90 per bbl.

PUMICE STONE—Powdered, $2.50 per cwt. in bbls., and 4 to 5c. per lb. in less quantity; lump, 10c. in small lots, and 8c. in bbls.

LIQUID PAINTS—Pure, $1.20 to $1.30 per gal.; No. 1 quality, $1 per gal.

CASTOR OIL—East India, in cases, 10 to 10½c. per lb. and 10½ to 11c. for single tins.

LINSEED OIL—Raw, 1 to 4 barrels, 69c.; boiled, 72c.; 5 to 9 barrels, raw, 68c.; boiled, 71c., delivered. To Toronto, Hamilton, Guelph and London, 1c. less.

TURPENTINE—Single barrels, 59c.; 2 to 4 barrels, 58c., to all points in Ontario. For less quantities than barrels, 5c. per gallon extra will be added, and for 5-gallon packages, 50c., and 10 gallon packages, 80c. will be charged.

GLASS.

Orders for immediate shipments are light. The import business is unsatisfactory to jobbers, as orders are being taken at a much smaller margin to the jobber than usual. Advices from Belgium state that though glass there is about six points higher than a year ago there is little chance of a decline before next winter. We still quote first break locally as follows: Star, in 50-foot boxes, $2.10, and 100-foot boxes, $4; double diamond under 26 united inches, $6; Toronto, Hamilton and London; terms 4 months or 3 per cent. 30 days.

MARKET NOTES.

A new list has been issued on wrought iron spikes.

Sisal rope is ½c. per lb. higher, the base figure now being 9½c.

Galvanized range boilers, and closets are slightly higher. New prices have also been fixed on enamelled baths.

Jas. Hutton & Co., Montreal, who had their offices destroyed by the fire, have changed their address to 6 St. Sacrament street.

The "Mic-Mac" hockey stick this season is meeting with great success. The best teams and players all over Canada prefer the "Mic-Mac" stick.

M. & L. Samuel, Benjamin & Co. report a good trade in "Gordon Crown" galvanized sheets. "This brand," remarked a representative of the firm, "has been on the market for a number of years, and its reputation is being well sustained."

Bennett's Patent Shelf Boxes are in demand. The following merchants are equipping their stores with them: C. V. Macintosh, Liverpool, N.S.; W. Emery, Toronto; J. W. Franks, Woodbridge, Ont.; W. Hughes, Marmora; A. Childs & Son, Gravenhurst; M. S. Sutton, Andover, N.B.; J. Sheard, Fenelon Falls.

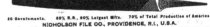

BUSINESS CHANGES.

DIFFICULTIES, ASSIGNMENTS, COMPROMISES.

LEMAY & MARCHAND, general merchants, Shawenegan Falls, Que., have assigned to Kent & Turcotte.

W. S. Grout, general merchant, Minto, Man., has assigned to C. H. Newton.

Henry Marquis, general merchant, etc. Edmundston, N.B., has suspended payment.

J. G. Fairbanks, general merchant, etc., Spruce Grove, N.W.T., has assigned to C. W. Cross.

A. Gibeault, general merchant, St. Lucie de Doncaster, Que., is offering 25c. cash on the dollar.

Antoine Paiement, tinsmith, etc., St. Anne de Prescott, Ont., has assigned to Louis J. Labrosse.

A meeting to appoint a curator for F. X. Julien, general merchant, Lambton, Que., has been called for February 28.

The sheriff is in possession of the business of Benor, Taylor & Co., general merchants, Alliston, Ont., and their stock has been sold.

P. S. Archibald has been appointed liquidator of The Ossekeag Stamping Co., Limited, manufacturers of tinware, Hampton, N.B.

PARTNERSHIPS FORMED AND DISSOLVED.

Darling Bros., machinists, Montreal, have dissolved.

Marceau & Juteau, manufacturers of files, Levis, Que., have registered dissolution.

Partnership has been registered by Hudon & Ouellett, general merchants, Black Lake, Que.

Hilborn & McTavish, general merchants, Paris, Ont., have dissolved; each continuing alone.

A. & T. Dell, general merchants, Niagara Falls South, Ont., have dissolved. A. E. Dell, continues.

Young & Paulin, hardware dealers, etc., Wingham, Ont., have dissolved. Alex. Young contiues.

Stovel & Strang, hardware dealers, Edmonton, N.W.T., have dissolved. James A. Stovel continues.

W. H. Luke, blacksmith, etc., Bothwell, Ont., has admitted Jas. Lindsay under the style of Luke & Lindsay.

Campbell & Hays and J. A. & H. A. McArthur, dealers in agricultural implements, Sussex, N.B., have amalgamated under the style of J. A. McArthur & Co.

SALES MADE AND PENDING.

G. W. Ray, general merchant, Newdale, Man., has sold out.

A. F. Elliott, general merchant, Alexander, Man., has sold out.

The assets of P. Denis, general merchant, St. Cesaire, Que., have been sold.

The assets of Alf. Mercier, general merchant, St. Angele, Que., are to be sold on February 22.

J. H. Lee, hardware dealer, Arnprior, Ont., has sold out.

R. McIvor, hardware dealer, Elkhorn, Man., is trying to sell out.

The assets of W. Rodden & Co., founders, Montreal, are to be sold on March 6.

The assets of L. J. Desilets, general merchant, St. Gertrude, Que., have been sold.

The assets of J. O. Faubert & Co., general merchants, Barrington, Que., have been sold.

The assets of R. Bourbeau, general merchant, Victoriaville, Que., are to be sold on February 22.

The assets of Mrs. E. A. Atkinson, general merchant, L'Avenir, Que., are to be sold on February 22.

The assets of A. Gibeault, general merchant, St. Lucie de Doncaster, Que., are to be sold on February 27.

The stock of the estate of P. J. Stinson & Co., general merchants, Singhampton, Ont., is advertised for sale by tender.

CHANGES.

W. J. Burgess, general merchant, Woodville, N.S., has sold out to W. B. Burgess.

J. A. Warner, general merchant, Fletwode, N.W.T., has sold out to Hourde & Warner.

Christian Karch, general merchant, Hespeler, Ont., has sold out to D. E. Morlock.

Wade & Johnson, general merchants, Fordwich, Ont., have sold out to G. E. McKee & Co.

Harriet McLennan, general store, Dalkeith, Ont., has been succeeded by Norman F. McLennan.

C. D. Lee, blacksmith, Glanworth, Ont., has sold out to J. N. Smith.

Bingham Bros., hardware dealers, Grand View, Man., are adding implements.

S. A. Thompson is closing up business as stove dealer and tinsmith in London, Ont.,

The stock of the estate of E. Ryerson, harness dealer, Hamilton, Ont., has been sold.

John Malcolm, hardware dealer, Rosebank, Man., has been succeeded by Leggett Bros.

The Canadian Chrome Iron Co., Limited, Sherbrooke, Que., have applied for incorporation.

Archibald L. Tanner, blacksmith, St. Thomas, Ont., has sold out to Simmington & Waite.

A. W. Littleproud, dealer in agricultural implements, Watford, Ont., have removed to London.

Wm. Crispin, painter and wall paper dealer, Stratford, Ont., has sold out to A. O. Neff.

A. S. Stewart, implement dealer, Prince Albert, N.W.T., has opened a branch at Stony Creek, N.W.T.

D. C. Peverett, dealer in agricultural implements, Rounthwaite, Man., has been succeeded by Peverett & McNab.

FIRES.

Harry Kellar, harness dealer, Deseronto, Ont., has been burned out.

The premises of Chas. Selby & Co., founders, Kingston, Ont., have been damaged by fire; insured.

DEATHS.

Charles Frank, glue manufacturer, Hamilton, Ont., is dead.

PERSONAL MENTION.

J. H. Ashdown, wholesale and retail hardware dealer, Winnipeg, is in Nelson, B.C., on business.

F. J. Marshall, hardware dealer, Orangeville, Ont., and his brother W. B. Marshall, Medicine Hat, N.W.T., were in Toronto this week.

Peter MacArthur, of MacArthur & Tracey, plumbers, tinsmiths, etc., and Thos. Cotter, of Cotter Bros., plumbers, Winnipeg, were in Toronto this week.

A. P. Love, of the geological staff of the Dominion, has resigned to take a position in the Dominion Steel Works at Sydney, Cape Breton, at $6,000 per year.

Wm. Paget, of Paget & Hay, hardware dealers, Unionville, Ont., was in Toronto on business this week. Mr. Paget reports that trade is in a most satisfactory condition in Unionville.

HEATING AND PLUMBING

MONTREAL'S NEW PLUMBING BY-LAW.

MONTREAL is at last to have a plumbing by-law. The aldermen are now in the by-law passing humor, and Alderman Lamarche, who has the measure in hand, is making use of his opportunities, and expects to have the new rules and regulations in force in two or three weeks. They are now engaging the attention of the council, and some of the sections have passed muster.

If the projected measure becomes law, it will be the most important law that Montreal plumbers have ever had to deal with. It aims to keep the science of plumbing up to a higher standard both by excluding inexperienced and incompetent master and journeymen plumbers from the business and by laying down rules to govern master plumbers in the performance of their work. One important section has already been decided upon, and that is, that in future each house must have a separate drain of its own leading to the sewer pipe. This has been a subject of keen discussion among the plumbers in the past, and they are to be congratulated upon the succees of their agitation for such a rule.

Another proposal incorporated in the bill is to have all master or journeymen plumbers licensed. To obtain the license they must pass an examination, to be set by an examining board, consisting of the sanitary inspector, city engineer, and a master plumber of at least 10 years' experience, to be appointed by the council. The object is to have a means by which incompetent plumbers may be kept out of the business, so that they will not endanger the public health or lower the standard of the profession.

It is expected that the bill will go through the council as drafted, with the exception of one clause, which aims to allow only iron pipe to be installed within a dwelling, the use of tile pipe to be prohibited entirely. It is not likely that this will go through, although Alderman Lamarche has not yet said die.

The by-law will likely read that both master and journeymen plumbers will be brought up for an examination about May 1.

AN INSPECTOR FOR ST. HENRI.

At the meeting of the St. Henri, Que., Council on Wednesday evening last week the report of the local Board of Health was adopted recommending the enactment of a by-law to enforce approved sanitary methods in plumber's work, under the supervision of an inspector to be appointed and paid by the council. Dr. Bernard was elected president and Dr. Lachapelle vice-president of the board; Dr. J. Lanctot was reappointed medical officer; George Nicholson, secretary, and W. Brisette, inspector.

PLUMBING AND HEATING CONTRACTS.

The Bennett & Wright Co., Limited, Toronto, have secured contracts for plumbing in the factory of The Imperial Starch Co., Limited, Prescott, Ont., and for plumbing and heating the new office building which Wm. Davies intends erecting on the site of the old Shaftesbury Hall, Queen street east, Toronto.

SOME BUILDING NOTES.

THE directors of the Sydenham Glass Co., Wallaceburg, Ont., have decided to rebuild at once the main building of their factory which was destroyed by fire on Tuesday.

M. Pritchard intends erecting a new store in North Wakefield, Ont.

The Presbyterian church at Wallacetown, Ont., is to be remodelled.

H. Morin has the contract to erect a church at St. Felicien, Que.

The parish churches at Riviere Ouelle and St. Pacome, Que., are being repaired.

There is an agitation for a new building to replace the Elm street school, St. John, N.B.

Brantford, Ont., Presbyterians have decided to build a new church in the East ward.

Tenders are asked before March 4 for a brick residence for Rev. P. H. Hauck, Markdale.

A new summer hotel to have a river frontage of 100 ft. is to be built at Port Lambton, Ont.

The Presbyterian church at Georgetown, Ont., which was destroyed by fire the other day, will be rebuilt at once.

McCulloch & Hill, contractors, Brampton, Ont., are to erect a two-storey residence for Geo. Hutchinson, Cheltenham, Ont.

Mr. Fraser, proprietor of the Chateau Belair, at the Island of Orleans, near Quebec, is building an extension to his hotel, which will give him larger kitchen accommodation and about 15 extra rooms, including two bathrooms and other modern conveniences.

W. H. Newman, contractor, Andrewsville, Ont., has the contract for enlarging J. A McCabe's hotel in Merricksville, Ont.

The Port Arthur Public School Board is advertising for plans and estimates for extensions to the present building in that town.

The Clarified Milk Co., Kingston, Ont., intend erecting a 40 x 70 three-storey building at the corner of Brock and Bagot streets, Kingston.

Simoneau & Dion, contractors, Sherbrooke, Que., have the contract for erecting a new academy at Coaticook, Que. The price is $10,500.

The Christian Workers of Hamilton, Ont., intend erecting a church on the corner of Park and Merrick streets, large enough to hold 600, with a Sunday-school hall to seat 200.

The Ottawa Public School Board have accepted the plans of Wm. Hodgson, architect, Montreal, for a $40,000 school on Wellington street. Tenders are being called for.

The erection of the new building for The Ottawa Produce Co., Ottawa, is being pushed forward rapidly. The Linde British system of mechanical refrigeration is being installed and a large force of men is employed.

The plans of Symons & Ray, Toronto, for the new buildings for Queen's University, Kingston, have been accepted. They will cost about $70,000. J. M. Power & Son, Kingston, will be the supervising architects.

St. Mary's Catholic Literary and Athletic Association will erect a handsome $20,000 club house on their property at the corner of Bathurst street and Macdonell square, Toronto. The club house will be thoroughly fitted in every particular.

The Canadian Pacific Railway Company have awarded contracts for the construction of a new station house at McAdam Junction, N.B. The new building is to be 130 feet long and 30 feet wide, two storeys high, and will be constructed of granite.

Wing Sang & Co., Chinese merchants, Dupont street, Vancouver, are building a fine three-storey brick block in the place of their former structure. The new building

20th CENTURY TOOLS.

Economize in the end by purchasing high-grade tools, best material and finest work. Consult the Green Book.

UTICA DROP FORGE & TOOL CO.,
296 Broadway, NEW YORK.

No. 1550.

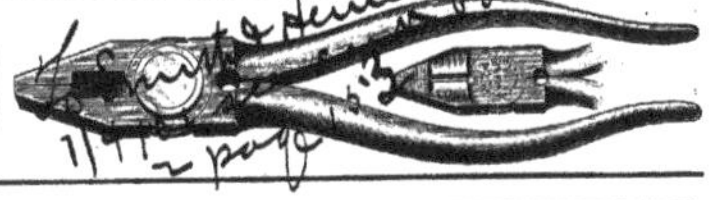

CANADIAN ADVERTISING is best done by THE E. DESBARATS ADVERTISING AGENCY MONTREAL.

ADVERTISING in WESTERN CANADA

will be Carefully, Efficiently, and Promptly attended to, by

The Roberts Advertising Agency,
WINNIPEG, CANADA.

KNOX HENRY

Heavy Hardware and Metal Broker
No. 1 Place Royale, MONTREAL.

HORSE NAILS—"C" Brand Horse Nails—Canada Horse Nail Co.
"BRASSITE" GOODS — Gunn Castor Co., Limited, Birmingham, Eng.

Manufacturers of

Heating Supplies

Pipe Fittings and Headers.
Large Manifolds made to Order.
Steam Traps and Appliances, etc.

The . . .
Jas. Morrison Brass Mfg. Co., Limited
TORONTO.

Standard Paint & Varnish Works, Limited
Makers of High Grade Varnishes, Japans, Paints, Colors & Enamels
Windsor, Ont.

THE . . .

Waggoner Extension Ladder.

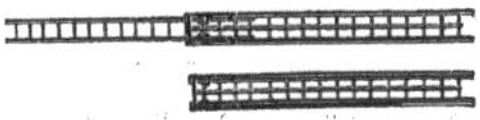

The strongest, lightest and most convenient ladder in the market. The only really satisfactory eXtension ladder made. Pulls up with a rope. Made in all lengths. Also eXtension and other step ladders, sawhorses, ironing boards, painters' trestles, etc. All first-class goods. Write for quotations to

The Waggoner Ladder Company, Limited, London, Ont.

WIRE RODS!

Drawn to Decimal Sizes, Cut and Straightened, In Uniform Sizes. Prompt Shipment.

Chalcraft Screw Co., Limited, Brantford, Ont.

LEADER CHURN

New Century Improvements.

FOUR DIFFERENT STYLES:
A—Steel Frame with double reversible Steel Lever.
B—Wood Frame with double reversible Steel Lever.
C—Steel Frame with Crank.
D—Wood Frame with Crank.

Styles A and B may be operated from a sitting or standing position.

Steel Frames and Hoops beautifully ALUMINIZED.

All LEADER CHURNS are equipped with BICYCLE BALL BEARINGS and PATENTED CREAM BREAKERS.

Stands are so constructed that they are particularly strong and rigid, and there is nothing to interfere with the placing of pail in the most convenient position for draining off buttermilk.

It Pays to Have the Best. None are Better Than the Leader.

THE
Dowswell Manufacturing Co.
Limited.
HAMILTON, ONT.
Eastern Agents: W. L. Haldimand & Son, Montreal, Que.

Use Syracuse Babbitt Metal

IT IS THE BEST MADE.

For Paper and Pulp Mills, Saw and Wood Working Machinery, Cotton and Silk Mills, Dynamos, Marine Engines, and all kinds of Machinery Bearings.

Wire, Triangular and Bar Solder, Pig Tin, Lead, Ingot Copper, Ingot Brass, Antimony, Aluminum, Bismuth, Zinc Spelter, Phosphor Tin, Phosphor Bronze, Nickle, etc., always in stock.

Factories: 332 William St., MONTREAL, QUE. and SYRACUSE, N.Y.

Syracuse Smelting Works

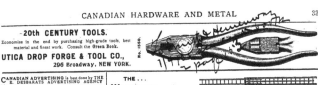

will cover three lots, while the old store was but one-third this width.

The Toronto Junction Public School Board have instructed Architect Ellis to prepare plans for a new four-roomed school, not to cost more than $6,000.

PLUMBING AND HEATING NOTES.

Higman & Co. have started as plumbers, etc., in Ottawa.

Chapleau & Leboeuf, contractors, Montreal, have assigned.

John O'Donnell, of O'Donnell Bros., plumbers, Toronto, is dead.

The Georgetown Electric Light Co., Limited, has been incorporated.

Boileau Freres have registered partnership as contractors at Isle Bisard, Que.

Arthur Lacosti has been appointed curator of A. Couvrette & Fils, contractors, Montreal.

Albert J. Smith has registered as proprietor of The Sun Light Gas Lamp Co., Lachine, Que.

The agreement between the Toronto Master Plumbers' Association and the journeymen's union of that city is not yet signed.

The Kingston Gas and Electric Light Co. have offered to sell their plant to the city for $373 000. The offer is being considered, but will not likely be accepted, as it is considered too high.

W. H. Meredith, president of the National Plumbers' Association, is running for a position on the Supreme Council of the Canadian Order of Foresters. The election takes place at the annual meeting of that body next week. Mr. Meredith's many friends in the plumbing trade will extend their earnest wish for success to his candidature.

BUILDING PERMITS ISSUED.

Building permits have been issued in Toronto to J. Wheeler for a pair of semi-detached brick dwellings on Smith street, near Broadview avenue, to cost $2,600; to Wm. Roaf, for alterations to a hotel at the corner of Bay and King streets, to cost $5,000.

The following permits have been issued in Ottawa: John Ball, frame dwelling, Eccles street, $800; Joseph Dupont, frame dwelling, Le Breton street, $600; Daniel Doherty, brick veneered dwelling, Besserer street, $1,300.

The following permits have been issued in Vancouver: A. P. Ingram, dwelling house, 309 Keefer street, cost $700; J. Magee, dwelling house, 789 Lansdowne street, cost $900; J. Magee, dwelling house, 79 Fifth avenue, cost $900.

THE PREFERENTIAL TARIFF IN CANADA.

IRONMONGER, London, Eng., February 2: "According to the latest returns which have reached us from the Department of Trade and Commerce of Canada, the total value of the imports into the Dominion during the year extending from October 1, 1899, to September 30, 1900, was $182,000,000, while the total value of the exports for the same period was $184,000,000, making the gross external trade of the country for the period specified $366,000,000, say, about £73,000,000. The Financial News, in its issue of January 28, issued an abstract of a report (which, in the absence of any specific statement, we presume deals with 1900), according to which the Canadian imports for that year reached the value of $189 000,000, while the exports were valued at $192,000,000, thus giving an aggregate trade for the year of $381,000,000, or over £73 000,000. The preferential tariff has evidently done a good deal to stimulate the trade of our great American colony, for the figures for the past few years show a remarkably steady increase. The total just referred to compares with $321,000,000 in 1899, $304,000,000 in 1898, $257 000,000 in 1897, and $239,000,000 in 1896.

Compared with the last year, prior to the advent of the present Administration, the total increase in the Dominion's foreign trade has reached the enormous sum of $142,000,000, equal to 60 per cent. How much of this increased turnover is due to the preferential tariff it is impossible to say, but it would seem that that measure has had a share in the growing prosperity of the country. The trade of the United Kingdom with Canada certainly shows a remarkable expansion. For years prior to the adoption of the preferential tariff our exports to the Dominion were falling off steadily, but the preference accorded to our goods had an immediate effect in arresting that decline, and an increase has been shown ever since. In 1897, the last fiscal year prior to the adoption of the new policy, the value of our exports to Canada stood at $29,000,000; in the first year of the new tariff they increased to $32,000,000, in the second to $37,000,000, and in the third to nearly $45,000,000.

John Ritter, hardware dealer, Newton, Ont., is advertising his business for sale.

The main building of the Sydenham glass factory at Wallaceburg was totally destroyed by fire on Tuesday. The loss will be over $10,000, but is fully covered by insurance. A lamp explosion in the engine room is believed to have caused the fire.

CORRESPONDENCE

INEXPERIENCED TRAVELLERS.

SIR,—It is surprising to see the number of inexperienced travellers sent out by some of our wholesale houses, with practically no training or personal knowledge of the goods they are handling. How do these firms expect them to do a successful trade battling against the old, tried, competent and experienced travellers. There is no doubt the financial end has a good deal to do with this state of affairs. But, take another view: If an experienced man with a connection can command a good trade and do double the business under the same expense as a greenhorn, he should get the preference; but this is not the case with a number of houses in Toronto and elsewhere. It is surprising that the merchants don't take more of their young men in warehouses and promote them as salesmen, for the reason that they are thoroughly conversant with all details of their business and lines they carry or manufacture. Some of these narrow-minded people will wake up and find the trade drifting into other channels, and some live, enterprising people growing head and shoulders over them. With the keenest competition of our American houses in nearly all lines of merchandise, travellers and managers are compelled to be on the move early and late. The sooner the Canadian manufacturers and wholesalers realize this the better for themselves, and place representatives out who know and understand their particular line of business.

A traveler can get acquainted with his territory in quarter the time he can get a knowledge of his line of goods and handle a customer to advantage.

These few lines may not meet the approval of some of our hard-hearted, money-grabbing managers, but it is a poor rule that don't work both ways.

Travellers would be repaid doubly for their time, if they could spare a few moments every week to read HARDWARE AND METAL and other trade papers. There are matters of importance, relating to the interests of themselves and their firms, such as changes in firms, new firms commencing, business in their territories, market values of raw materials, and many other items too numerous to mention. The writer always looks forward to secure a copy of your valuable publication each week of issue. Manufacturers and wholesalers and all classes of merchants should request their travellers and clerks to carefully read it through each week, as there is much information to be gained, which cannot be had from the daily press. TRAVELLER.

Toronto, February 16, 1901.

[Remarks: The subject touched upon by "Traveller" is an important one, and we would like to hear from more travellers on the question.—The Editor.]

SUMMER HOTEL FOR STURGEON POINT.

Most of the time at the annual meeting of the Lindsay, Ont., Board of Trade was devoted to consideration of a proposal to erect a summer hotel at Sturgeon Point, Ont. A communication was read from R. J. Matchett, who stated that he had consulted with a successful Toronto hotelman, and had received great encouragement. The G.T.R. had guaranteed to fill the hotel with guests if a proper building were erected. This should have about 200 bedrooms and be fitted up with all modern conveniences. It would cost about $50,000. If $10,000 were subscribed in Lindsay, Mr. Matchett stated that $15,000 or $25,000 could easily be raised in Toronto and the rest in New York, Rochester and other places.

A communication was also received from G. H. M. Baker, local manager for the Rathbun Co., asking the board to consider the advisability of having a blast furnace located at Lindsay. The cheapness of hardwood for fuel and the short rail haul of ore necessary made Lindsay, in Mr. Baker's opinion, a good point for the establishment of such an industry.

Action regarding both of these communications was deferred until next meeting.

HOW TO TREAT RAZORS.

If the purchaser returns the article to the ironmonger because it won't "go" properly, the latter is seldom sufficiently of an expert to point out to his customers where the trouble lies. A gentleman who has been connected with the razor trade for close on half a century states that very few people appreciate how tenderly a razor—especially one of the hollow-ground variety—should be treated. Nothing that is made in cutlery possesses such a thin, delicate edge, and consequently it is liable to injury unless treated with the greatest care. Improper stropping is a fruitful source of damage, few users performing the operation lightly enough. The finer the edge the lighter should it be stropped, the bare weight of the blade in some cases being too heavy. The razor should be stropped after using, and then carefully wiped, as the least speck of dirt, in conjunction with lather, will quickly eat into the edge and make it rough.

Perhaps if I put my friend's advice into rhyme, it may impress itself more indelibly upon the mind of the reader:

Take it up tenderly,
Strop it with care,
Wipe off the lather,
The blood, and the hair.
Put it back gingerly
Into its case,
And next time you use it
You'll not find a trace
Of stubble obscuring
Your beautiful face.

CURRENT MARKET QUOTATIONS.

February 2', 1901.

These prices are for such qualities and quantities as are usually ordered by retail dealers on the usual terms of credit, the lowest figures being for larger quantities and prompt pay. Large cash buyers can frequently make purchases at better prices. The Editor is anxious to be informed at once of any apparent errors in this list as the desire is to make it perfectly accurate.

METALS.

Tin.

Lamb and Flag and Straits— 56 and 28 lb. ingots, per lb.	0 32	0 33

Tinplates.

Charcoal Plates—Bright

		Per box
M.L.S., equal to Bradley.		
I.C., usual sizes	....	$6 75
I.X., "	....	8 25
I.X.X., "	....	9 75
Famous—		
I.C.	....	6 75
I.X	....	8 25
I.X.X.	....	9 50
Raven & Vulture Grades—		
I.C., usual sizes	....	5 00
I.X., "	....	6 00
I.X.X. "	....	7 00
I.XXX., "	....	8 00
D.C., 12½x17	....	4 75
D.X.	..	5 50
D.X.X	....	7 50

Coke Plates—Bright

Bessemer Steel—		
I.C., usual sizes	....	4 15
I.C., special sizes, base	....	4 50
20 x 28	....	8 50

Charcoal Plates—Terne

Dean or J. G. Grade—		
I.C., 20x28, 112 sheets	....	8 50
I.X., Terne Tin	....	10 50

Charcoal Tin Boiler Plates.

		Per lb.
Cookley Grade—		
X.X., 14x56, 50 sheet bxs / " 14x60 " / " 14x65 "	0 07	0 07½

Tinned Sheets

72x30 up to 24 gauge	0 08	0 08½
" 26 "	0 08½	0 09
" 28 "	0 09	0 09½

Iron and Steel.

Common Bar, per 100 lbs.	1 65	1 70
Refined " "	....	2 15
Horse Shoe Iron	....	2 05
Hoop steel, 1½ to 3 in. base, extras for smaller sizes	....	3 10
Sleigh Shoe Steel " base	....	2 00
Tire Steel	....	2 00
Machinery iron finish	....	2 05
Cast Steel, per lb	0 00	0 00
Toe Calk Steel	....	2 30
T. Firth & Co's special cast steel, per lb.		0 13
Jessop's Tool Steel	....	0 13

Boiler Tubes.

1½-inch	....	0 12½
2 "	....	0 13
2½ "	....	0 15
3 "	....	0 16
3½ "	....	0 20
4 "	....	0 25

Steel Boiler Plate.

¼ inch	....	2 25
3-16 inch	....	2 25
⅜ inch and thicker	....	2 25

Black Sheets.

18 gauge	....	3 10
20 gauge	....	3 00
22 to 24 "	....	3 10
26 "	....	3 20
28 "	....	3 30

Canada Plates.

All dull, 52 sheets	....	3 00
Half polished	....	3 15
All bright	3 65	3 75

Iron Pipe.

Black pipe—		
¼ inch	....	3 00
⅜ "	....	3 00
½ "	....	3 15
¾ "	....	3 20
1 "	....	4 50
1¼ "	....	6 35
1½ "	....	7 55
2 "	....	10 10
2½-6 inch, discount 61 p.c.		
Galvanized pipe—		
½ inch	....	4 65
¾ "	....	5 35
1 "	....	7 55
1¼ "	....	9 75
1½ "	....	11 25
2 "	....	15 50

Galvanized Sheets.

	G.C. Comet.		Amer.	Queen's Head.
16 gauge		....	4 5	4 60
18 to 24 gauge	4 25	3 85	4 35	4 25
26 "	4 50	4 10	4 35	4 50
28 "	4 75	4 25	4 60	4 75

28 gauge American equals 26 gauge English.

Chain.

Proof Coil, 3-16 in., per 100 lb	....	....
" ¼ " "	8 00	8 50
" 5-16 " "	5 35	5
" ⅜ " "	4 35	4 85
" 7-16 " "	4 15	4 65
" ½ " "	4 35	4 80
" ⅝ " "	3 85	4 35
" ¾ " "	3 80	4 00

Halter, kennel and post chains, 40 and 50 p.c.

Cow ties .... 40 p.c.

Stall fixtures .... 35 p.c.

Trace chain .... 25 p.c.

Jack chain, iron, single and double, discount 35 p.c.

Jack chain, brass, single and double, discount 40 p.c.

Copper.

Ingot

English B. S., ton lots	0 19	0 20
Lake Superior	....	....

Bolt or Bar.

Cut lengths round, ½ to ⅞ in.	0 23½	0 25
" round and square 1 to 2 inches	0 23½	0 25

Sheet.

Plain, 14 oz., and light, 16 oz., 14x48 and 14x60	0 23	0 23
Plain, 14 oz., and light, 16 oz., irregular sizes	0 23	0 23½
Tinned copper sheets	....	0 26
Planished	....	0 32

Braziers (In sheets.)

4x6ft. 25 to 30 lbs. ea., per lb.	....	0 25½
" 35 to 45 " "	....	0 24½
" 50-lb. and above, "	....	0 23½

Boiler and T. K. Pitts

Plain Tinned, per lb	....	0 28
Spun, per lb.	....	0 32

Brass.

Rod and Sheet, 14 to 30 gauge, 15 per cent.		
Sheets, hard-rolled, 2x4	0 24	0 25
Tubing, base, per lb	0 24	0 25

Zinc Spelter

Foreign, per lb	0 06	0 06½
Domestic "	....	...

Zinc Sheet.

5 cwt. casks	6 75	7 00
Part casks	7 00	7 50

Lead.

Imported Pig, per lb	0 04¾	0 05
Bar, 1 lb.	0 05½	0 05¾
Sheets, 2½ lbs. sq. ft., by	....	0 06¼
Sheets, 3 to 6 lbs., "	....	0 06

NOTE.—Cut sheets ½ cent per lb. extra.

Pipe, by the roll, usual weights per yard, list at 7c. per lb. and 15 p.c. dis. f.o.b. Toronto.

NOTE.—Cut lengths, net price, waste pipe 8-ft. lengths lists at 7½ cents.

Shot.

Common, $6.50 per 100 lb.; chilled, $7.00 per 100 lb.; buck, seal and ball, $7.50. Discount, 7½ p.c. Prices are f.o.b. Toronto, Hamilton, Montreal, St. John and Halifax. Terms 3 per cent. cash, freights equalized on Montreal.

Soil Pipe and Fittings.

Discount, 60 and 10 per cent. on medium and extra heavy, and 60 per cent. on light.

Solder.

	Per lb.	Per lb.
Bar half-and-half, guarant'd	....	0 19
Bar half-and-half, commer'l	....	0 18½
Refined	....	0 18½
Wiping	....	0 18

NOTE.—Prices of this graded according to quantity. The prices of other qualities of solder in the market indicated by private brands vary according to composition.

Antimony.

Cookson's, per lb.	0 11	0 11½

White Lead.

	Per 100 lb.
Pure	6 50
No. 1 do	6 12½
No. 2 do	5 75
No. 3 do	5 37½
No. 4 do	5 00
Munro's Select Flake White	7 37½
Elephant and Decorators' Pure	7 12½
Brandram's B B. Genuine	8 00
" " Decorative	7 55
" " No. 1	6 85
" " No. 2	6 00

Red Lead.

Genuine, 560 lb. casks, per cwt.	....	$5 50
Genuine, 100 lb. kegs, per cwt.	....	5 75
No. 1, 560 lb. casks, per cwt.	....	5 25
No. 1, 100 lb. kegs, per cwt.	....	5 00

White Zinc Paint.

Elephant Snow White	0 08	0 09
Pure White Zinc	0 08	0 09
No. 1	0 06	0 07½
No. 2	0 05	0 06½

Dry White Lead.

Pure, casks	5 75
Pure, kegs	6 25
No. 1, casks	5 50
No. 1, kegs	5 00

Prepared Paints.

In ¼, ½ and 1 gallon tins.

Pure, per gallon	....	1 25
Second qualities, per gallon	....	1 10
Barn (in bbls.)	0 75	0 85
The Sherwin-Williams Paints.	....	1 45
Canada Paint Co's Pure	....	1 25
Toronto Lead & Color Co's Pure	....	1 25
Sanderson Pearcy's Pure	....	1 20
Stewart & Wood's Champion Pure.		1 20

Colors in Oil.

25 lb. tins, Standard Quality.

Venetian Red, per lb.	0 05
Chrome Yellow	0 11
Golden Ochre	0 06
French "	0 05
Marine Black	0 09
" Green	0 09
Chrome "	0 08
French Imperial Green	0 09

Colors, Dry.

Yellow Ochre (J.C.) bbls.	1 35	1 40
Yellow Ochre (J.F.L.B.), bbls	...	2 75
Yellow Ochre (Royal)	1 10	1 15
Brussels Ochre	....	2 00
Venetian Red (best), per cwt.	1 80	1 90
English Oxides, per cwt.	3 00	3 25
American Oxides, per cwt.	1 75	2 00
Canadian Oxides, per cwt.	1 75	2 00
Super Magnetic Oxides, 93p.c.	2 00	2 25
Burnt Sienna, pure, per lb.	....	0 10
" Umber, " "	....	0 10
do Raw	....	0 09
Drop Black, pure	....	0 09
Chrome Yellows, pure	....	0 18
Chrome Greens, pure, per lb.	....	0 12
Golden Ochre	...	0 03¾
Ultramarine Blue in 28-lb. boxes, per lb.	0 08	0 24
Fire Proof Mineral, per 100 lb	....	1 00
Genuine Eng. Litharge, per lb	....	0 07
Mortar Color, per 100 lb.	....	1 25
English Vermillion	....	0 80
Pure Indian Red, No. 45 lb.	....	0 80
Whiting, per 100 lb.	....	0 55

Blue Stone.

Casks, for spraying, per lb.	0 07
100-lb. lots, do. per lb.	0 08

Putty.

Bulk in bbls.	2 00
Bulk in less quantity	2 15
Bladders in bbls.	2 20
Bladders in kegs, boxes or loose	2 35
Bladders in 25-lb. tins	2 45
Bladders in 12½-lb. tins	2 75
Bladders in bulk or tins less than 100 lb	3 00

Varnishes.

In 5-gal. lots.		Per gal.
Carriage, No. 1	2 90	3 30
" body	8 00	9 00
" rubbing	4 00	5 00
Gold Size, Japan	3 00	3 40
Brown Japan	2 40	2 80
Elastic Oak	2 90	3 30
Furniture, extra	2 40	2 80
" No. 1	1 60	2 00
Hard Oil Finish	2 70	3 10
Light Oil Finish	3 20	3 60
Demar	3 30	3 70
Shellac, white	4 40	4 80
" orange	4 00	4 40
Furniture Brown Japan	1 60	2 00
Black Japan	2 40	2 80
" " No. 1	1 60	2 00

The Imperial Varnish & Color Co's, Limited Elastilite Varnish 1 gal. can, each, $2.00.

Granatine Floor Finish, per gal. $2.00.

Maple Leaf Coach Enamels; Size 1, 60c.; Size 2, 35c.; Size 3, 20c. each.

Castor Oil.

East India, in cases, per lb.	0 10	0 10½
" " small lots	0 10½	0 11

Cod Oil, Etc.

Cod Oil per gal.	0 50	0 55
Pure Olive	....	1 20
" Neatsfoot	....	90

Glue.

Common	0 08½	0 09
French Medal	0 14	0 14½
Cabinet sheet	0 12	0 13
White, extra	0 18	0 20
Gelatine	0 22	0 30
Strip	0 18	0 20
Coopers	0 19	0 20
Huttner	....	0

HARDWARE.

Ammunition.

Cartridges.

B. B. Caps. Dom. 50 and 5 percent.
Rim Fire Pistol, dis. 40 p. c., Amer.
Rim Fire Cartridges, Dom., 50 and 5 p. c.
Central Fire Pistol and Rifle, 10 p.c. Amer.
Central Fire Cartridges, pistol sizes, Dom. 30 per cent.
Central Fire Cartridges, Sporting and Military, Dom., 15 and 5 per cent.
Central Fire, Military and Sporting, Amer, add 5 p.c. to list. B.B. Caps, discount 40 per cent, Amer.
Loaded and empty Shells, "Trap" and "Dominion" grades, 25 per cent Rival and Nitro, net list.
Brass shot Shells, 55 per cent
Primers, Dom., 30 per cent.

Wads — per lb
Best thick white felt wadding, in ¾-lb bags 1 00
Best thick brown or grey felt wads, in ¾-lb. bags 0 70
Best thick white card wads, in boxes of 500 each, 12 and smaller gauges 0 99
Best thick white card wads, in boxes of 500 each, 10 gauge 0 35
Best thick white card wads, in boxes of 500 each, 8 gauge 0 55
Thin card wads, in boxes of 1,000 each, 12 and smaller gauges 0 20
Thin card wads, in boxes of 1,000 each, 10 gauge 0 25
Thin card wads in boxes of 1,000 each, 8 gauge ....
Chemically prepared black edge grey cloth wads, in boxes of 250 each— Per M
11 and smaller gauge 0 60
9 and 10 gauges 0 70
7 and 8 gauges 0 90
5 and 6 gauges 1 10
Superior chemically prepared pink edge, best white cloth wads, in boxes of 250 each—
11 and smaller gauge 1 15
9 and 10 gauges 1 40
7 and 8 gauges 1 65
5 and 6 gauges 1 90

Adzes.

Discount, 20 per cent.

Anvils.

Per lb 10 0 12½
Anvil and Vise combined 4 50
Wilkinson & Co.'s Anvils..lb. 0 09 0 09½
Wilkinson & Co.'s Vices..lb. 0 09¾ 0 10

Augers.

Gilmour's, discount 65 and 5 p.c. off list.

Axes.

Chopping Axes—
Single bit, per doz 6 50 10 00
Double bit, " 12 00 18 00
Bench Axes, 40 p.c.
Broad Axes, 33⅓ per cent.
Hunters' Axes 5 50 6 00
Boy's Axes 5 75 6 75
Splitting Axes 6 50 12 00
Handled Axes 7 00 10 00

Axle Grease.

Ordinary, per gross 5 75 6 00
Best quality 13 00 15 00

Bath Tubs.

Zinc 6 00
Copper, discount 15 p.c. off revised list

Baths.

Standard Enameled.
5½-inch rolled rim, 1st quality 25 00
" " " 2nd " 21 00

Anti-Friction Metal.

"Tandem" A .... per lb. 0 27
" B .... " 0 21
" C .... " 0 11½
Magnolia Anti-Friction Metal, per lb. 0 25

SYRACUSE SMELTING WORKS.
Aluminum, genuine 0 45
Dynamo 0 29
Special 0 25
Aluminum, 99 p.c. pure "Syracuse" .. 0 50

Bells.

Hand.
Brass, 60 per cent.
Nickel, 55 per cent.

Cow.
American make, discount 66⅔ per cent.
Canadian, discount 45 and 50 per cent.

Door.
Gongs, Sargant's 5 50 8 00
" Peterboro', discount 45 per cent.

Farm.
American, each 1 25 3 00

House.
American, per lb 0 35 0 40

Bellows.

Hand, per doz 3 35 4 75
Moulders', per doz 7 50 10 00
Blacksmiths', discount 40 per cent.

Belting.

Extra, 60 per cent.
Standard, 60 and 10 per cent.

Bits.

Auger.
Gilmour's, discount 60 and 5 per cent.
Rockford, 50 and 10 per cent.
Jennings' Gen., net list.

Car.
Gilmour's, 47½ to 50 per cent.

Expansive.
Clark's, 40 per cent.

Gimlet.
Clark's, per doz 0 65 0 90
Damond, Shell, per doz 1 00 1 50
Nail and Spike, per gross 2 25 5 20

Blind and Bed Staples.

All sizes, per lb 0 07½ 0 12

Bolts and Nuts. Per cent

Carriage Bolts, full square, Norway 70
" " full square 70
Common Carriage Bolts, all sizes 65
Machine Bolts, all sizes 65
Coach Screws 75
Sleigh Shoe Bolts 75
Blank Bolts 65
Bolt Ends 65
Nuts, square .... 4½c. off
Nuts, hexagon .... 4¾c. off
Tire Bolts 67½
Stove Bolts 67½
Stove rods, per lb .... 5½ to 6c.
Plough Bolts 60

Boot Calks.

Small and medium, ball, per M 4 25
Small heel, per M 4 50

Bright Wire Goods.

Discount .... 55 per cent.

Broilers.

Light, dis., 65 to 67½ per cent.
Reversible, dis., 65 to 67½ per cent.
Vegetable, per doz., dis. 37½ per cent.
Henis, No. 8, " 6 00
Henis, No. 9, " 7 00
Queen City " 7 50 0 00

Butchers' Cleavers.

German, per doz 6 00 11 00
American, per doz 12 00 20 00

Building Paper, Etc.

Plain building, per roll 0 30
Tarred lining, per roll 0 40
Tarred roofing, per 100 lb 1 65
Coal Tar, per barrel 3 50
Pitch, per 100-lb 0 85
Carpet felt, per ton 45 00

Bull Rings.

Copper, $2.00 for 2½ in. and $1.90 for 2 in.

Butts.

Wrought Brass, net revised list

Cast Iron.
Loose Pin, dis., 60 per cent.

Wrought Steel.
Fast Joint, dis. 60 and 10 per cent.
Loose Pin, dis. 60 and 10 per cent.
Berlin Bronzed, dis. 70, 70 and 5 per cent.
Gen. Bronzed, per pair 0 40 0 65

Carpet Stretchers.

American, per doz 1 00 1 50
Bullard's, per doz .... 6 50 ....

Castors.

Bed, new list, dis. 55 to 57½ per cent.
Plate, dis. 52½ to 57½ per cent.

Cattle Leaders.

Nos. 31 and 32, per gross 50 9 50

Cement.

Canadian Portland 2 80 3 00
English " 3 00
Belgian " 2 75 3 00
Canadian hydraulic 1 25 1 50

Chalk.

Carpenters, Colored, per gross 0 45 0 75
White lump, per cwt 0 60 0 65
Red 0 05 0 06
Crayon, per gross 0 14 0 18

Chisels.

Socket, Framing and Firmer.
Broad's, dis. 70 per cent.
Warnock's, dis. 70 per cent.
P. S. & W. Extra 60, 10 and 5 p.c.

Churns.

Revolving Churns, metal frames—No. 0, $8—No. 1, $8.50—No. 2, $9.00—No. 3, $10.00 No. 4, $12.00—No. 5, $16.00 each. Ditto wood frames—20c. each less than above. Discounts: Delivered from factories, 58 p.c.; from stock in Montreal, 56 p.c. Terms, 4 months or 3 p.c. cash in 30 days.

Clips.

Axle dis. 65 per cent.

Closets.

Plain Ontario Syphon Jet $16 00
Emb. Ontario Syphon Jet 17 00
Fittings net 1 00
Plain Teutonic Syphon Washout 10 00
Emb. Teutonic Syphon Washout 11 00
Fittings net 1 25
Low Down Teutonic, plain 16 00
" " " embossed 17 00
Plain Richelieu net 3 75
Emb. Richelieu net 4 00
Fittings net 1 25
Low Down Ont. Sy. Jet, plain net .. 19 50
" " " " " emb'd. net 20 50
Closet connection net 1 25
Basins, round, 14 in 1 00
" oval, 17 x 14 in 2 50
" " 19 x 15 in 3 75
Discount 40 p.c., except on net figures.

Compasses, Dividers, Etc.

American, dis. 62½ to 65 per cent.

Cradles, Grain.

Canadian, dis. 25 to 33⅓ per cent.

Crosscut Saw Handles.

S. & D., No. 3, per pair 17½
" " 5, " 22½
" " 6, " 15
Boynton pattern " 20

Door Springs.

Torrey's Rod, per doz .... (15 p.c.) 2 00
Coil, per doz 0 88 1 60
English, per doz 2 00 4 00

Draw Knives.

Coach and Wagon, dis. 50 and 10 per cent.
Carpenters, dis. 70 per cent.

Drills.

Hand and Breast.
Millar's Falls, per doz. net list.

DRILL-BITS.
Morse, dis., 37½ to 40 per cent.
Standard dis. 50 and 5 to 55 per cent.

Faucets.

Common, cork-lined, dis. 35 per cent.

ELBOWS. (Stovepipe.)
No. 1, per doz 1 40
No. 2, per doz 1 20
Bright, 20c. per doz. extra.

ESCUTCHEONS.
Discount, 45 per cent.

ESCUTCHEON PINS.
Iron, discount 40 per cent.

FACTORY MILK CANS.
Discount off revised list, 40 per cent.

FILES.
Black Diamond, 50 and 10 to 60 per cent.
Kearney & Foote, 60 and 10 p.c. to 60, 10, 10.
Nicholson File Co., 50 and 10 to 60 per cent.
Jowitt's, English list, 25 to 27½ per cent.

FORKS.
Hay, manure, etc., dis., 50 and 10 per cent. revised list.

GLASS—Window—Box Price.

Size United Inches.	Star Per 50 ft.	Star Per 100 ft.	D. Diamond Per 50 ft.	D. Diamond Per 100 ft.
Under 26	2 10	4 00	....	6 00
26 to 40	2 30	4 35	....	6 65
41 to 50	....	4 75	....	7 25
51 to 60	....	35	....	8 50
61 to 70	....	75	....	9 25
71 to 80	....	5	....	10 50
81 to 85	....	6 50	....	11 75
86 to 90	....	....	..	14 00
91 to 95	..	....	....	15 50
96 to 100	..	....	....	18 00

GAUGES

Marking, Mortise, Etc.
Stanley's dis. 50 to 55 per cent.

Wire Gauges.
Winn's, Nos. 26 to 33, each 1 65 2 40

HALTERS.
Rope, ⅜ per gross ....
" ½ " 9 00
" ⅝ to ¾ 14 00
Leather, 1 in., per doz 3 87½ 4 00
" 1¼ in., " 5 15 5 20
Web, — per doz 1 87 2 45

HAMMERS.

Nail
Maydole's, dis. 5 to 10 per cent. Can. dis. 25 to 27½ per cent.

Tack.
Magnetic, per doz 1 10 1 20

Sledge.
Canadian, per lb 0 07½ 0 08½

Ball Pean.
English and Can., per lb 0 22 0 25

HANDLES.
Axe, per doz. net 1 50 2 00
Store door, per doz 1 00 1 50

Fork.
C. & B., dis. 40 per cent. rev. list.

Hoe.
C. & B., dis. 40 per cent. rev. list.

Saw.
American, per doz 1 00 1 25

Plane.
American, per gross 3 15 3 75

Hammer and Hatchet.
Canadian, 40 per cent.

Cross-Cut Saws.
Canadian, per pair 0 13½

HANGERS. doz. pairs.
Steel barn door 5 85 6 00
Stearns, 4 inch 5 00
" 5 inch 6 50
Lane's covered—
No. 11, 5-ft. run 8 40
No. 11½, 10-ft. run 10 80
No. 12, 10-ft. run 12 60
No. 14, 15-ft. run 21 00
Lane's O.N.T. track, per foot 4½

HARVEST TOOLS.
Discount, 50 and 10 per cent.

HATCHETS.
Canadian, dis. 40 to 42½ per cent.

HINGES.
Blind, Parker's, dis. 50 and 10 to 60 per cent
Heavy T and strap, 4-in., per lb 0 06½
" " 5-in., " 0 06¼
" " 6-in., " 0 06
" " 8-in., " 0 05¾
" " 10-in., " 0 05½
Light T and strap, dis. 60 and 5 per cent.
Screw hook and hinge—
6 to 12 in., per 100 lbs 4 50
14 in. up, per 100 lbs 3 50
Per gro. pairs.
Spring 12 00

HOES.
Garden, Mortar, etc., dis. 50 and 10 p.c.
Planter, per doz 4 00 4 50

HOLLOW WARE
Discount .... 45 and 5 per cent

HOOKS.

Cast Iron.
Bird Cage, per doz 0 50 1 10
Clothes Line, per doz 0 27 0 63
Harness, per doz 0 72 0 88
Hat and Coat, per gross 1 00 3 00
Chandelier, per doz 0 50 1 00

Wrought Iron.
Wrought Hooks and Staples, Can., dis. 47½ per cent.

Wire.
Hat and Coat, discount 45 per cent.
Belt, per 1,000 .... 0 60
Screw, bright, dis. 55 per cent.

HORSE NAILS.
"C" brand 50 and 7½ p.c. off new list } Oval
"M" brand 50, 10 and 5 p.c. } head.
Countersunk, 50, 10 and 10 per cent.

HORSESHOES.

F.O.B. Montreal

Iron Shoes.	No. 2 and larger.	No. 1. and smaller
Light, medium, and heavy. .	3 50	3 75
Snow shoes	3 75	4 00
Steel Shoes.		
Light	3 60	3 85
Featherweight (all sizes)	4 85	4 85

F.O.B. Toronto, Hamilton, London and Guelph, 10c. per keg additional.

Toe weight steel shoes...... 6 70

JAPANNED WARE.

Discount, 45 and 5 per cent. off list, June 1899.

ICE PICKS.

Star per doz.......... 3 00 3 25

KETTLES.

Brass spun, 7½ p.c. dis. off new list.

Copper, per lb.......... 0 30 0 50

American, 60 and 10 to 65 and 5 p.c.

KEYS.

Lock, Can., dis., 45 p.c.

Cabinet, trunk, and padlock, Am. per gross.......... 60

KNOBS.

Door, japanned and N.P., per doz	1 50	2 50
Bronze, Berlin, per doz	2 75	3 25
Bronze Genuine, per doz	6 00	9 00
Shutter, porcelain, F. & L. screw, per gross	1 30	4 00
White door knobs—per doz.		1 25

HAY KNIVES.

Discount, 50 and 10 per cent.

LAMP WICKS.

Discount, 60 per cent.

LANTERNS.

Cold Blast, per doz	7 50
No. 3 "Wright's"	8 50
Ordinary, with O burner	4 25
Dashboard, cold blast	9 50
No. 0	6 00

Japanning, 50c. per doz. extra.

LEMON SQUEEZERS.

per doz.

Porcelain lined,	2 20	5 60
Galvanized	1 87	3 85
King, wood	2 75	2 90
King, glass	4 00	4 50
All glass	1 20	1 30

LINES.

Fish, per gross	1 05	2 50
Chalk "	1 90	7 40

LOCKS.

Canadian, dis. 45 p.c.

Russell & Erwin, per doz.... 3 00 3 25

Cabinet.

Eagle, dis. 30 p.c.

Padlock.

English and Am., per doz	50	6 00
Scandinavian, "	1 00	2 40

Eagle, dis. 20 to 25 p.c.

MACHINE SCREWS.

Iron and Brass.

Flat head discount 25 p.c.

Round Head, discount 20 p.c.

MALLETS.

Tinsmiths', per doz	1 25	1 50
Carpenters', hickory, per doz.	1 25	3 75
Lignum Vitae, per doz	3 85	5 00
Caulking, each	60	2 00

MATTOCKS.

Canadian, per doz.......... 8 50 1 00

MEAT CUTTERS.

American, dis. 25 to 30 p.c.

German, 15 per cent.

MILK CAN TRIMMINGS.

Discount, 25 per cent.

NAILS.

Quotations are:	Cut.	Wire.
2d and 3d	$3 35	$3 85
3d	3 00	3 52
4 and 5d	2 75	3 35
6 and 7d	2 65	3 20
8 and 9d	2 50	3 00
10 and 12d	2 45	2 95
16 and 20d	2 40	2 90
30, 40, 50 and 60d. (base)	2 35	2 85

Galvanizing 2c. per lb. net extra.

Steel Cut Nails 10c. extra.

Miscellaneous wire nails, dis. 70 per cent.

Coopers' nails, dis. 30 per cent.

Flour barrel nails, dis. 25 per cent

NAIL PULLERS.

German and American....... 1 85 3 50

NAIL SETS

Square, round, and octagon, per gross	3 38	4 00
Diamond	12 00	15 00

NETTING.

Poultry, 50 and 5 per cent. for McMullen's.

OAKUM. Per 100 lb.

Navy	6 00
U. S. Navy	7 25

OIL.

Water White (U.S.)	0 16½
Prime White (U.S.)	0 15½
Water White (Can.)	0 15
Prime White (Can.)	0 14

OILERS.

McClary's Model galvan. oil can, with pump, 5 gal., per doz.......... 0 00 10 00

Zinc and tin, dis. 50, 50 and 10.

Copper, per doz	1 25	3 50
Brass, "	1 50	3 50

Malleable, dis. 25 per cent.

GALVANIZED PAILS.

Dufferin pattern pails, dis. 50 and 10 p.c.

Flaring pattern, discount 45 per cent.

Galvanized washtubs, discount 45 per cent.

PIECED WARE.

Discount 40 per cent. off list, June, 1899.

10-qt. flaring san buckets, dis. 45 p.c.

6, 10 and 14-qt. flaring pails, dis. 45 p.c.

Creamer cans, dis 45 p.c.

PICKS.

Per doz.......... 6 00 9 00

PICTURE NAILS.

Porcelain head, per gross	1 75	3 00
Brass head "	0 40	1 00

PICTURE WIRE.

Tin and gilt, discount 75 p.c.

PLANES.

Wood, bench, Canadian dis. 50 per cent American dis. 50.

Wood, fancy Canadian or American 7½ o 40 per cent.

PLANE IRONS.

English, per doz.......... 2 00 5 00

PLIERS AND NIPPERS.

Button's Genuine per doz pairs, dis. 37½ 40 p.c.

Button's Imitation, per doz.	5 00	9 00
German, per doz	0 60	2 60

PLUMBERS' BRASS GOODS.

Impression work, discount, 60 per cent.

Fuller's work, discount 65 per cent.

Rough stops and stop and waste cocks, discount, 60 per cent.

Jenkins disk globe and angle valves, discount, 55 per cent.

Standard valves, discount, 60 per per cent.

Jenkins radiator valves discount 55 per cent.

" " " standard, dis., 60 p.c.

Quick opening valves discount, 60 p.c.

No. 1 compression bath cock	2 00
No. 4 " " "	2 00
No. 7, Fuller's	2 50
No 4½, "	3 00

POWDER.

VeloX Smokeless Shotgun Powder.

100 lb. or less	0 85
1,000 lb. or more	0 80

Net 30 days.

PRESSED SPIKES.

Discount 25 pe cent.

PULLEYS.

Hothouse, per doz	0 55	1 00
Axle	0 22	0 33
Screw	0 27	1 00
Awning	0 35	2 50

PUMPS.

Canadian cistern	1 80	3 60
Canadian pitcher spout	1 40	2 10

PUNCHES.

Saddlers', per doz	1 00	1 85
Conductors', "	9 00	15 00
Tinners' solid, per set	0 00	0 72
" hollow, per inch	0 00	1 00

RANGE BOILERS.

Galvanized, 30 gallons		7 00
" 35 "		8 25
40 "		9 50
Copper, 30 "		22 00
" 35 "		26 00
" 40 "		30 00

Discount off Copper Boilers 10 per cent.

RAKES.

Cast steel and malleable Canadian list 50 and 10 p.c. revised list.

Wood, 25 per cent.

RASPS AND HORSE RASPS.

New Nicholson horse rasp, discount 60 p.c.

Globe File Co.'s rasps, 60 and 10 to 70 p.c.

Heller's Horse rasps, 50 to 50 and 5 p.c.

RAZORS.

per doz.

Geo. Butler & Co.'s,	8 00	18 00
Boker's	7 50	11 00
Wade & Butcher's	3 60	10 00
Theile & Quack's	7 00	12 00
Elliot's	4 00	18 00

REAPING HOOKS.

Discount, 50 and 10 per cent.

REGISTERS.

Discount.......... 40 per cent.

RIVETS AND BURRS.

Iron Rivets, discount 60 and 10 per cent.

Iron Burrs, discount 55 per cent.

Black and Tinned Rivets, 60 p.c.

Extras on Iron Rivets in 1-lb. cartons, ½c per lb.

Extras on Iron Rivets in ½-lb. cartons, per lb.

Copper Rivets & Burrs, 35 and 5 p.c. dis. and cartons, 1c. per lb. extra, net.

Extras on Tinned or Coppered Rivets ½-lb. cartons, 1c. per lb.

Terms, 4 mos. or 3 per cent. cash 30 days.

RIVET SETS.

Canadian, dis. 35 to 37½ per cent.

ROPE ETC.

	Sisal.	Manila.
7-16 in. and larger, per lb.	9½	13
⅜ in	10½	14
¼ and 5-16 in		15
Cotton, 3-16 inch and larger		16½
" 5-32 inch		21½
" ⅛ inch		22½
Russia Deep Sea		15½
Jute		8
Lath Yarn		9¾
New Zealand Rope		10½

RULES.

Boxwood, dis. 75 and 10 p.c.

Ivory, dis. 37½ to 40 p.c.

SAD IRONS. per set.

Mrs. Potts, No. 55, polished. ....

" No. 50, nickle-plated ....

SAND AND EMERY PAPER.

Dominion Flint Paper, 47½ per cent.

B & A. sand, 40 and 2½ per cent.

Emery, 40 per cent.

SAP SPOUTS.

Bronzed iron with hooks, per doz... 9 50

SAWS.

Hand Disston's, dis. 12½ p.c.

S. & D., 40 per cent.

Crosscut, Disston's, per ft.... 35 0 55

S. & D., dis. 35 p.c. on Nos. 2 and 3.

Hack, complete, each	0 75	2 75
" frame only	....	0 75

SASH WEIGHTS.

Sectional, per 100 lbs	2 75	3 00
Solid, "	2 00	2 25

SASH CORD.

Per lb.......... 0 23 0 30

SAW SETS.

"Lincoln," per doz.......... .... 6 50

SCALES.

B. S. & M. Scales, 45 p.c.

Champion, 65 per cent.

Fairbanks Standard, 35 p.c.

" Dominion, 55 p.c.

" Richelieu, 55 p.c.

Chatillon Spring Balances, 10 p.c.

SCREW DRIVERS.

Sargent's, per doz.......... 0 65 1 00

SCREWS

Wood, F. H., bright and steel, 87½ and 10 p.c.

Wood R. H., " dis. 82½ and 10 p.c.

" F. H., brass, dis. 80 and 10 p.c.

Wood, R. H., " dis. 75 and 10p.c.

" F.H., bronze, dis. 70 p.c.

" R.H. " 65 p.c.

Drive Screws, 80 per cent.

Bench, wood, per doz	3 25	4 00
" iron, "	4 25	5 75

SCYTHES.

Per doz. net.......... 9 00

SCYTHE SNATHS.

Canadian, dis. 45 p.c.

SHEARS.

Bailey Cutlery Co., full nickeled, dis. 60 p.c.

Seymour's, dis. 50 and 10 p.c.

SHOVELS AND SPADES

Canadian, dis. 40 and 5 per cent.

SINKS.

Steel and galvanized, discount 45 per cent.

SNAPS.

Harness, German, dis. 25 p.c.

Lock, Andrews'.......... 4 50 11 50

SOLDERING IRONS.

1, 1½ lb., per lb.	0 37
2 lb. or over, per lb.	0 34

SQUARES.

Iron, No. 493, per doz	2 40	2 55
" No. 494, "	3 25	3 40

Steel, dis. 50 and 5 to 50 and 10 p.c., rev. list.

Try and bevel, dis. 50 to 52½ p.c.

STAMPED WARE.

Plain, dis. 75 and 12½ p.c. off revised list.

Retinned, dis., 75 p.c. off revised list.

STAPLES.

Galvanized	0 00	0 00
Plain	0 00	3 45

Coopers', discount 45 per cent.

Poultry netting staples, 40 per cent.

STOCKS AND DIES.

American dis. 25 p.c.

STONE. Per lb.

Washita	0 28	0 60
Hindostan	0 06	0 07
" slip,	0 09	0 09
Labrador	....	0 13
" Axe	....	0 15
Turkey	....	0 50
Arkansas	0 00	1 50
Water-of-Ayr	0 00	0 10
Scythe, per gross	3 50	5 00
Grind, per ton	15 00	18 00

STOVE PIPES.

Nestable in crates of 25 lengths.

5 and 6 inch Per 100 lengths	7 00
7 inch " "	7 50

ENAMELINE STOVE POLISH.

No. 4—3 dozen in case, net cash	$4 80
No. 6—3 dozen in case, "	8 40

TACKS BRADS, ETC.

per cent.

Strawberry box tacks, bulk	75 & 10
Cheese-box tacks, blued	80 & 12½
Trunk tacks, black and tinned	85
Carpet tacks, blued	80 & 5
" " tinned	80 & 10
" " (in kegs)	40
Cut tacks, blued, in dozens only	75 & 15
¼ weights	60
Swedes, cut tacks, blued and tinned—	
In bulk	80 & 10
In dozens	75
Swedes, upholsterers', bulk	85 & 12½
" brush, blued & tinned, bulk	70
" gimp, blued, tinned and japanned	75 & 12½
Zinc tacks	35
Leather carpet tacks	55
Copper tacks	50
Copper nails	52

Trunk nails, black	65 and 5	
Trunk nails, tinned	65 and 10	
Clout nails, blued and tinned	65 and 5	
Chair nails	35	
Cigar box nails	35	
Patent brads	40	
Fine finishing	40	
Picture frame points	10	
Lining tacks, in papers	10	
" " in bulk	15	
" " solid heads, in bulk	75	
Saddle nails in papers	10	
" " in bulk	15	
Tufting buttons, 22 line, in dozens only	60	
Tin capped trunk nails	15	
Zinc glazier's points	5	
Double pointed tacks, papers	90 and 10	
" " " bulk	40	

TAPE LINES.

English, ass skin, per doz	2 75	5 00
English, Patent Leather	5 50	9 75
Chesterman's each	0 90	2 85
" steel, each	0 80	8 00

THERMOMETERS.

Tin case and dairy, dis. 75 to 75 and 10 p.c.

TRANSOM LIFTERS.

Payson's per doz. ... 2 60

TRAPS. (Steel.)

Game, Newhouse, dis. 25 p.c.
Game, H. & N., P. S. & W., 65 p.c.
Game, steel, 72½, 75 p.c.

TROWELS.

Disston's discount 10 per cent.

German, per doz	4 75	6 00

S. & D., discount 35 per cent.

TWINES.

Bag, Russian, per lb	....	0 27
Wrapping, cotton, per lb	0 22	0 26
Wrapping, mottled, per pack	0 50	0 60
Wrapping, cotton, 3-ply		0 20
" " 4-ply	....	0 26
Mattress, per lb	0 33	0 45
Staging, "	0 27	0 35
Broom, "	0 30	0 55

VISES.

Hand, per doz	4 00	6 00
Bench, parallel, each	2 00	4 50
Coach, each	6 00	7 00
Peter Wright's, per lb	0 12	0 13
Pipe, each	5 50	9 00
Saw, per doz	6 50	13 00

ENAMELLED WARE.

White, Princess, Turquoise, Blueand White, discount 50 per cent.
Diamond, Famous, Premier, 50 and 10 p.c.
Granite or Pearl, Imperial, Crescent, 50, 10 and 10 per cent.

WIRE.

Brass wire, 50 to 50 and 2½ per cent. off the list.

Copper wire, 45 and 10 per cent. net cash 30 days, f.o.b. factory.

Smooth Steel Wire, base, $2.80 per 100 lb. List of extras: Nos. 2 to 5, advance 7c. per 100 lb.—Nos. 6 to 9, base—No. 10, advance 7c.—No. 11, 14c.—No. 12, 20c.—No. 13, 35c.—No. 14, 47c.—No. 15, 60c.—No. 16, 75c. Extras net per 100 lb.: Coppered wire, 60c.—tinned wire, $2—oiling, 10c.—special hay-bailing wire, 30c.—spring wire, $1—best steel wire, 75c.—bright soft drawn, 15c.—in 50 and 100-lb. bundles net, 10c.—in 25-lb. bundles net, 15c.—packed in casks or cases, 15c.—bagging or papering, 10c.

Fine Steel Wire, dis. 17½ per cent List of extras: In 100-lb. lots: No. 17, $5—No. 18, $5.50—No. 19, $6—No. 20, $6.65—No. 21, $7—No. 22, $7.30—No. 23, $7.65—No. 24, $8—No. 25, $9—No. 26, $9.50—No. 27, $10—No. 28 $11—No 29, $12—No. 30, $13—No. 31, $14—No 32, $15 No. 33, $16—No. 34, $17. Extras net—tinned wire, Nos. 17-25, $2—Nos. 26-31 $4—Nos. 32-34, $6. Coppered, 5c.—oiling, 10c.—in 25-lb. bundles, 15c.—in 5 and 10-lb. bundles, 25c.—in 1-lb. hanks, 50c.—in ½-lb. hanks, 75c.—in ¼-lb. hanks, $1—packed in casks or cases, 15c.—bagging or papering, 10c.

Galvanized Wire, per 100 lb.—Nos. 6, 7, 8, $3.65 No. 9, $3.10—No. 10, $3.75—No. 11, $3.85 No. 12, $3.25—No. 13, $3.35—No. 14, $4.25—No. 15, $4.75—No. 16, $5.00.

Clothes Line Wire, 19 gauge, per 1,000 feet 3 30

WIRE FENCING. F.O.B. Toronto

Galvanized 4 barb, 2¾ and 5 inches apart	3 10
Galvanized, 2 barb, 4 and 6 inches apart	3 10
Galvanized, plain twist	3 10

Galvanized barb, f.o.b. Cleveland, $2.97½ in less than carlots, and $7.85 in carlots. Terms, 60 days or 2 per cent. in 10 days.

Ross braid truss cable 4 50

WIRE CLOTH.

Painted Screen, per 100 sq. ft., net ... 1 35
Terms, 4 months, May 1; 3 p.c. off 30 days.

WRENCHES.

Acme, 35 to 37½ per cent.
Agricultural, 60 p.c.
Coe's Genuine, dis. 20 to 25 p.c.

Towers' Engineer, each	2 00	7 00
" S., per doz	5 80	6 00
G. & K.'s Pipe, per doz	....	3 40
Burrell's Pipe, each	....	3 00
Pocket, per doz	0 25	2 90

WRINGERS.

Leader	per doz.	$....
Royal Canadian	"	50 00
Royal American	"	50 00

Discount, 45 per cent.; terms 4 months, or 3 p.c. 30 days.

WROUGHT IRON WASHERS.

Canadian make, discount, 40 and 5 per cent

COME TO STAY —BUILT TO LAST

and
every customer satisfied with

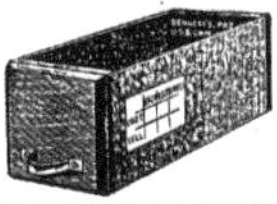

BENNETT'S PATENT SHELF BOX

Thousands now in use. Particulars from the patentee, who also makes the

KLONDIKE SAMPLE HOLDER.

J. S. BENNETT, 20 Sheridan Ave., Toronto.

N.B.—**Don't forget** we make boxes to fit your present shelving.

PERSONS addressing advertisers will kindly mention having seen their advertisement in Canadian Hardware and Metal Merchant.

SEND for specimen copy of Phillips' Monthl Machinery Register, containing over 5,000 entries of new and second-hand machinery of every description. The oldest established and most successful medium in the world. Established 25 years for the purpose of introducing those who have machinery for sale, to those who wish to buy, has a circulation of about 50,000 copies per annum, all over the world, and is used for continual reference by a large number of firms. It is consequently a most valuable advertising medium for all engineers and manufacturers. Subscription, 6s. per annum, price per copy, 6d. Sole Proprietor, CHAS. D. PHILLIPS, M.I.M.E., Newport, Mon., England. Telegraphic address "Machinery, Newport, Mon."

"BUILD TO-DAY THEN, STRONG AND SURE, WITH A FIRM AND AMPLE BASE." —Longfellow.

DO YOU?

WISH THUS TO BUILD an advertisement in the CONTRACT-RECORD, TORONTO will bring you tenders from the best contractors

---

BUSINESS NEWS

of any kind that is of value to business men supplied by our Bureau. We can give you market quotations from any town in Canada, reports from the city markets, stock quotations, etc. You can get commercial news from any Canadian paper through us.

Write us, giving us particulars of what you want and where you want it from, and we will quote you prices by return.

"Clippings from any Canadian paper on any subject."

CANADIAN PRESS CLIPPING BUREAU,

232 McGill Street, MONTREAL, QUE.

Telephone Main 1255.

10 Front St. East, Toronto. Telephone 2148.

---

75 YEARS. *ESTABLISHED 1825.* *75 YEARS.*

CELEBRATED **HEINISCH** SHEARS.

Tailors' Shears,
Trimmers, Scissors,
Tinners' Snips, etc.

ACKNOWLEDGED THE BEST.

R. HEINISCH'S SONS CO.

NEWARK, N.J., U.S.A. NEW YORK OFFICE, 90 Chambers St.

Not connected with any Shear Combination.

CHAS. F. CLARK, President. JARED CHITTENDEN, Treasurer.

...ESTABLISHED 1849...

BRADSTREET'S

Capital and Surplus, $1,500,000. **Offices Throughout the Civilized World.**

Executive Offices: Nos. 346 and 348 Broadway, New York City, U.S.A.

THE BRADSTREET COMPANY gathers information that reflects the financial condition and the controlling circumstances of every seeker of mercantile credit. Its business may be defined as of the merchants, by the merchants, for the merchants. In procuring, verifying and promulgating information no effort is spared, and no reasonable expense considered too great, that the results may justify its claim as an authority on all matters affecting commercial affairs and mercantile credit. Its offices and connections have been steadily extended, and it furnishes information concerning mercantile persons throughout the civilized world.

Subscriptions are based on the service furnished, and are available only by reputable wholesale, jobbing and manufacturing concerns, and by responsible and worthy financial, fiduciary and business corporations. Specific terms may be obtained by addressing the Company at any of its offices. **Correspondence Invited.**

—OFFICES IN CANADA—

HALIFAX, N.S. HAMILTON, ONT. LONDON, ONT. MONTREAL, QUE.
OTTAWA, ONT. QUEBEC, QUE. ST. JOHN, N.B. TORONTO, ONT.
VANCOUVER, B.C. VICTORIA, B.C. WINNIPEG, MAN.

THOS. C. IRVING, Gen. Man. Western Canada, Toronto. JOHN A. FULTON, Gen. Man. Eastern Canada, Montreal.

Awarded a Gold Medal at PARIS EXPOSITION for superiority. That's proof enough of their quality, and clearly shows that they are the best.

Send for Catalogue and . . . . Price List.

The Bailey

Cutlery Co.

BRANTFORD, ONT.

Dominion Pattern

Cow Tie AND Stall Fixture

The special features of the tie and stall fixture are well shown in the illustration. As will be noticed the chain is very short, which prevents all danger of entanglement with the animal's foot. At the same time the form of the fixture is such that great freedom is allowed to the head. Because of the short chain this tie is much cheaper than the ordinary patterns.

The stall fixture is made from a tough quality of steel and is very strong. Also, owing to its circular cross-section, it is exceedingly rigid. Its simplicity, convenience, cheapness, and ease of attaching make it very popular with cow tie users.

This form of tie and stall fixture are sometimes called Niagara pattern.

American or Flat Link Chain,

for years the standard cow tie chain in "the States," is now rapidly coming in favor in Canada. Its short link, handsome appearance and smooth surface—which cannot injure the animal's neck—make it superior to all other styles of chain for cow ties.

For sale by all Jobbers; manufactured by

ONEIDA COMMUNITY, Limited, NIAGARA FALLS, ONT.

CANADIAN HARDWARE AND METAL MERCHANT

The Weekly Organ of the Hardware, Metal, Heating, Plumbing and Contracting Trades in Canada.

VOL. XIII. | MONTREAL AND TORONTO, MARCH 2, 1901. | NO. 9

Garden Hose

Discounts for the new season now out.

WRITE FOR PRICES.

* * *

CANADIAN RUBBER CO.,

Montreal. Toronto. Winnipeg.

EXTENDED INSURANCE.

One of the many liberal features embodied in the **UNCONDITIONAL ACCUMULATIVE POLICY** issued by the

Confederation Life Association.

HEAD OFFICE--TORONTO.

is the provision for Extended Insurance. After three full annual premiums have been paid, the insured is entitled to Extended Insurance for the full amount of the policy for a term of years definitely stated therein. Paid-up and Cash Values also guaranteed.

Rates and full information sent on application to the Head Office, Toronto, or to any of the association's agents.

W. C. Macdonald, Actuary. **J. K. MACDONALD,** Managing Director

SOME OF THE NEWER "YANKEE" TOOLS

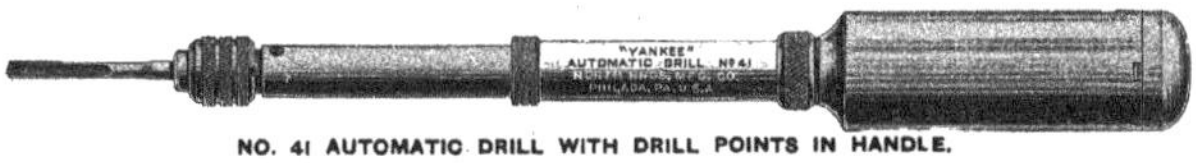

NO. 41 AUTOMATIC DRILL WITH DRILL POINTS IN HANDLE.

NO. 42 AUTOMATIC DRILL WITH DRILL POINTS IN BOX.

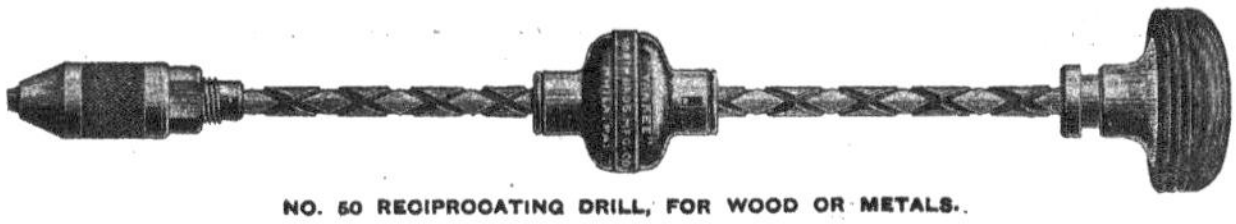

NO. 50 RECIPROCATING DRILL, FOR WOOD OR METALS.

Sold by Leading Jobbers throughout the Dominion.

NORTH BROS. MFG. CO.,

Philadelphia, Pa., U. S. A.

Established 1860. Incorporated 1895.

WHY YOU SHOULD BUY

OUR MILK CAN TRIMMINGS

"Davidson's" Patent Milk Can Bottom

All our Broad Hoop Milk Can Trimmings are made with our PATENT BOTTOM.

BECAUSE

our Patent **Broad Hoop** Bottom has all the advantages of three-piece bottom, but is practically one piece. The rim of patent bottom is turned in over edge of bottom proper, making smooth uniform edge that will not tear the floors of factories or waggons.

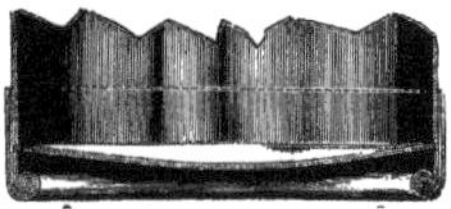

The above cut demonstrates more clearly than words, why everybody wants the

"Davidson" Patent Milk Can Bottom.

Hoops are all made 14 gauge and joined LONG WAY, thereby avoiding any possibility of splitting.

Sufficient space is left between bottom proper and rim to allow body of can to be inserted ¼ inch., making permanent joint.

Bottoms are thus sweated in with half the solder.

Bottoms are concave, draining to the centre, therefore, are easier to wash out. They will not corrode like those that drain to the side.

Top bands are SHOULDERED and have cut out at joint, making neater and cleaner job in half time than with old style hoop.

All bands have retinned edges.

For durability, finish and economy in making up our Trimmings are enequalled.

Patented February 9, 1894, and has been growing in popularity ever since its introduction.

The THOS. DAVIDSON MFG. CO., Limited, MONTREAL.

..TINPLATES..

COKE CHARCOAL

Standard sizes in stock.
Special sizes for import.

Inquiries Solicited. Please Write for Quotations.

SAMUEL, SONS & BENJAMIN, - - LONDON AND LIVERPOOL, ENGLAND.

M. & L. Samuel, Benjamin & Co.

General Importers and Exporters and Metal Merchants

27 Wellington Street West, - - TORONTO, ONT.

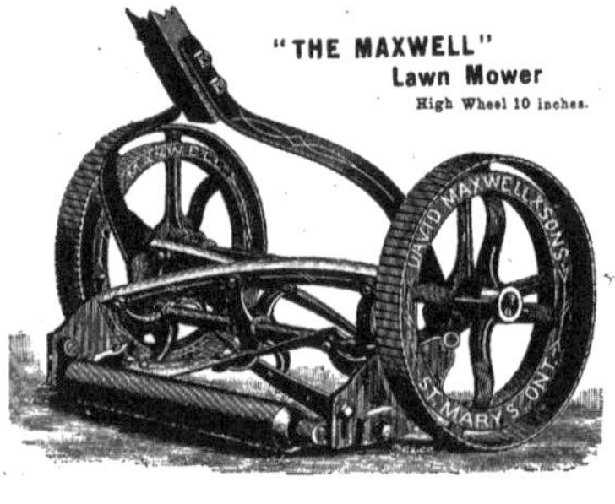
DAVID MAXWELL & SONS
ST. MARYS, ONT.

THE NEW BALDWIN

DRY AIR CLEANABLE

REFRIGERATOR.

135 Modern Varieties. Ash, Oak and Soft-wood Finishes

METAL, PORCELAIN, SPRUCE LININGS.

BALDWIN

Positive Circulation—Sanitary—Odorless.

Latest Cleanable Features—The Strongest and Best System of Patent Removable Metal Air-Flues.

Air-Tight Lever Locks

Ball-Bearing Casters.

Swing Base—in and out.

Rubber around Doors and Lids, making them doubly air-tight.

Handsome Designs.

Moderate Prices.

Built in the newest, largest and best equipped refrigerator plant in the East run all the year round on refrigerators exclusively; stock goods; special refrigerators and coolers in sections.

Handsome Trade Catalogue Ready.

Baldwin Refrigerator Co.,

BURLINGTON, VERMONT.

Intending Advertisers

should send in orders for space and copy for advertisements in the Special Number of

Hardware and Metal

without delay.

DAVID PHILIP

MANUFACTURERS' AGENT

362½ Main St., WINNIPEG.

Correspondence invited from manufacturers of Staple or Heavy Hardware, Iron or Steel Bolts and Nuts, etc., either by carrying stock in Winnipeg or by selling direct from factory.

GOOD REFERENCES.

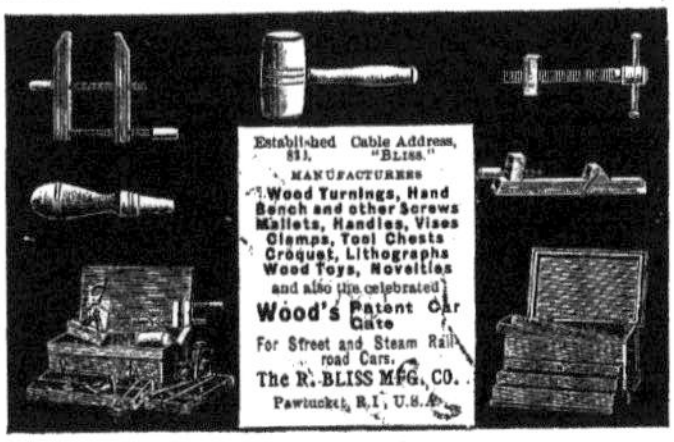

Broom and Mattress Wire

High Grade, Double Tinned.

Fine Annealed Brush Wire

Soft Coppered Wire

Tinned Wire of all kinds.

The Peerless Wire Co.

Hamilton, Ont.

What About **BRUSHES** for Spring Trade?

Our Revised Catalogue for 1901 now ready and our travellers are working your way. Don't buy till you get our prices. We guarantee good values and prompt shipment.

Meakins & Sons
Hamilton, Ont.

Meakins, Sons & Co.
Montreal, Que.

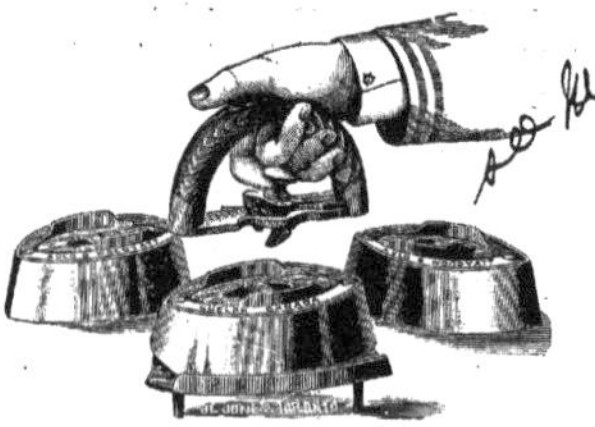

MRS. POTTS SAD IRONS.
No. 55, Plain Polished. No. 50, Nickel Plated.

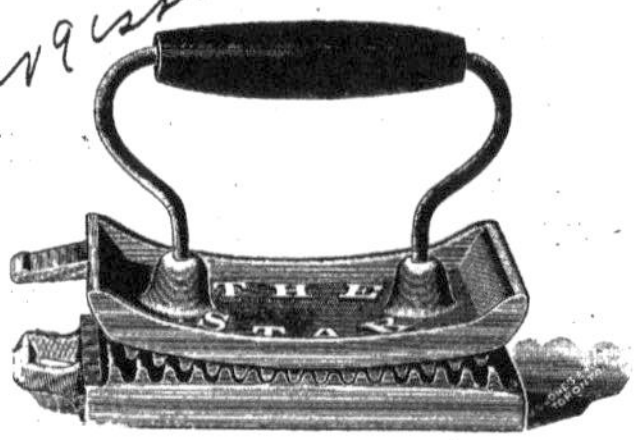

THE "STAR" FLUTING IRON.
Nickel Plated and Gold Bronze Finish.

ARE MADE FROM HIGH-GRADE IRON AND ARE WELL FINISHED.

Cost No More Than Inferior Makes, Give Far Better Satisfaction.

When Ordering From Your Jobber Ask For Our Make.

A. R. WOODYATT & CO., Guelph, Ont.

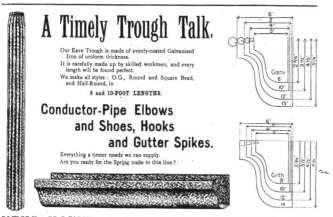

A Timely Trough Talk.

Our Eave Trough is made of evenly-coated Galvanized Iron of uniform thickness.
It is carefully made up by skilled workmen, and every length will be found perfect.
We make all styles: O.G., Round and Square Bead, and Half-Round, in

8 and 10-FOOT LENGTHS.

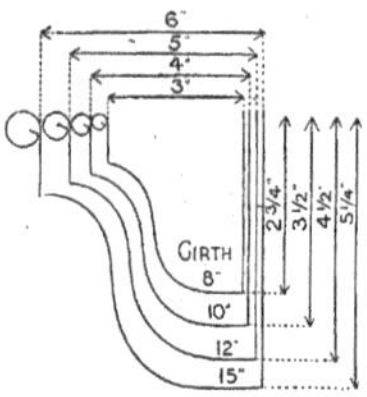

Conductor-Pipe Elbows and Shoes, Hooks and Gutter Spikes.

Everything a tinner needs we can supply.
Are you ready for the Spring trade in this line?

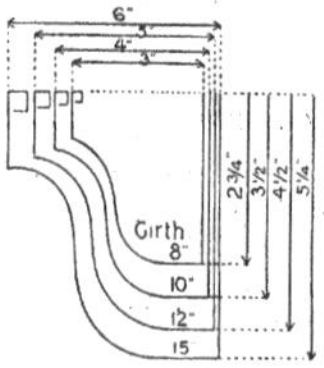

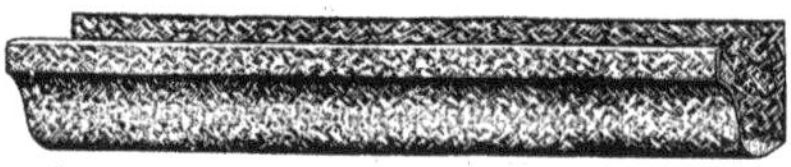

KEMP MANUFACTURING CO., TORONTO, ONT.

HARDWARE AND METAL

VOL. XIII. MONTREAL AND TORONTO, MARCH 2, 1901. NO. 9.

President,
JOHN BAYNE MacLEAN,
Montreal.

THE MacLEAN PUBLISHING CO.
Limited.

Publishers of Trade Newspapers which circulate in the Provinces of British Columbia, North-West Territories, Manitoba, Ontario, Quebec, Nova Scotia, New Brunswick, P.E. Island and Newfoundland.

OFFICES

MONTREAL	232 McGill Street, Telephone 1255.
TORONTO	10 Front Street East, Telephone 2148.
LONDON, ENG.	109 Fleet Street, E.C., J. M. McKim.
MANCHESTER, ENG.	18 St Ann Street. H. S. Ashburner,
WINNIPEG	Western Canada Block. J. J. Roberts.
ST. JOHN, N.B.	No. 3 Market Wharf, J. Hunter White,
NEW YORK.	176 E. 88th Street.

Subscription, Canada and the United States, $2.00.
Great Britain and elsewhere - - - 12s.

Published every Saturday.

Cable Address { Adscript, London. Adscript, Canada.

"EN REVOLTE OUVERTE."

LAST Friday's edition of La Presse, Montreal's leading French-Canadian organ, contained an article on its front page which, if it had not been so ridiculous, might be termed seriously misleading and harmful. It was written by a reporter who had conceived the idea that there was friction between the Dominion Wholesale Hardware Association and the Montreal Retail Hardware Association, and his attempts to find the cause for this friction were ludicrous in the extreme.

The article was headed "En Revolte Ouverte" (in open revolt), and asserted that the retail dealers were protesting against the action of the manufacturers and wholesale merchants. It claimed that the retailers thought that the wholesalers were making too much profit under the existing manufacturers' combine and its arrangements with the wholesalers, and that these latter people should share up their profit with the retailers. To obtain this, the retailers are reported as having organized a month ago. They had not yet been able to effect their object, and were seriously considering the advisability of appointing a buying committee for themselves and cease doing business with the jobbers.

Of course, such is all nonsense. The retailers hope not only to improve their own financial condition, but also that of the wholesalers. One grievance they have against the wholesalers, and that is that some of them sell retail. But this will cause no discord, for the wholesalers have nearly all testified to their unwillingness or refusal to do other than a wholesale trade.

They do not wish to curtail the wholesalers' profit. They care not what per cent. the wholesale merchants' make, be it 15, or 25 or 50 per cent. What they do want is a living profit for themselves, and they would have this profit come out of the pockets of their customers by the manufacturers setting retail prices, just as they dictate the values of the wholesale houses.

They would like the manufacturers to fix two scales of prices whereby both the wholesaler and retailer would make a living profit. Should this wish cause discord?

To us there seem to be no elements of unfriendliness in the relationship of the two associations. Indeed, several of the wholesale houses have written letters of hearty sympathy to the new-born organization, endorsing its formation.

That La Presse knew little of what is going on is proven by the fact that it said the association was only a month old and has held only two meetings. It was organized last September, and has since met twice a month.

LOWER PRICES ON FENCE WIRES.

THE price of barb wire, plain twist wire and galvanized wire has been reduced 15c. per 100 lb. by the manufacturers in the United States for the Canadian market.

Barb and plain twist wires are now quoted at $2.82½ per 100 lb. f.o.b. Cleveland in less than carload lots, and galvanized, Nos. 6 to 9 base, at $2.57½ per 100 lb. Carload lots of not less than 15 tons are quoted at 12½c. per 100 lb. less.

The competition of European wire seems to have been the cause of the reduction in price. The manufacturers in the United States are evidently determined to control the Canadian market. And the conditions are certainly in their favor.

It is worthy of note that retailers in Canada can buy, in Cleveland, barb wire and galvanized wire 27½c. per 100 lb. cheaper than can the retailers of the United States.

SCREEN DOORS AND WINDOWS.

The advance in the price of screen doors and windows noted in last week's issue is rather larger than then stated to be. It is from 10 to 20 per cent.

Door screens of the cheapest kind are now selling to the trade at $7.20 to $7.80 per dozen in 4-inch styles. Three-inch styles are 20c. per dozen less. Some at least of those in the trade have an idea that the 3-inch styles are too light.

Screen windows are quoted at $1.60 to $3.60 per dozen, according to size and extension.

It is thought that with the price of green wire cloth fixed, no further change will take place in in screen doors and windows.

CANADA AND THE BIG STEEL COMBINATION.

CANADA could not under ordinary circumstances be uninterested in the enormous iron and steel combination which is being consummated in the United States. The aspirations of this country itself in the direction of iron and steel development, and the juxtaposition of our neighbor to the south, to say nothing of Canada's position as a customer of the United States mills, could not well make it otherwise.

But the interest of the Dominion has been excited more than it otherwise would have been by the emphatic statements which have been made to the effect that the new iron and steel combination in question has obtained, or is to obtain, control of the iron and steel works of the Dominion Iron and Steel Co., Sydney, Nova Scotia, capitalized at $15,000,000. It is true that those connected with the Sydney company have denied the truthfulness of the statement. But the memory that similar schemes have been denied and afterwards consummated, makes people decidedly sceptical in regard to the denials of this kind made by the officials of large corporations. "Is it true?" is about all one hears in Canada at the moment in regard to the alleged deal.

The Carnegie-Morgan combination, whose ambitious undertaking and enormous capitalization are causing those interested in the iron and steel industry of two hemispheres to look on with astonishment, on Tuesday last filed a certificate of incorporation in Trenton, New Jersey. The name of the combination is The United States Steel Corporation, and its capital stock is $1,100,000,000, a sum that transcends all other corporated companies in existence. The stock is composed of $400,000,000 preferred, $400,000,000 common, and $300,000,000 bonds. It is provided in the articles of incorporation that a dividend of 7 per cent. shall be paid on the preferred stock.

The combinations which are to be swallowed up by this bigger trust are eight in all, and are: The Federal Steel Co., capital stock $99,745,200; American Steel and Wire Co., capital stock $90,000,000; National Tube Co., capital $80,000,000; American Tin Plate Co., capital $46,325,000; American Steel Hoop Co., capital $33,000,000; American Sheet Steel Co., capital $49,000,000, National Steel Co., capital $59,000,000, and The Carnegie Company, capital $200,000,000. In other words, eight concerns with an aggregate capital (preferred and common) of $647,702,200 become, when merged, one with a capitalization of $1,100,000,000, preferred and common.

The basis of the deal according to a statement made this week by J. P. Morgan & Co., the New York bankers who negotiated it, whereby the seven companies are to be taken over by the Carnegie-Morgan combination, is as follows:

	Present Stock.	Stock in new company to be exchanged for present stock of old companies.	
Federal Steel, preferred...	$53,260,000	$58,586,991	$............
" common	46,484,300	1,839,372	49,970,623
American Steel and Wire, preferred...	40,000,000	47,000,000	....
" common.....	50,000,000	....	51,250,000
National Tube, preferred...	40,000,000	50,000,000	....
" common....	40,000,000	3,250,000	50,000,000
National Steel, preferred...	27,000,000	33,750,000	....
" common ..	32,000,000	....	40,000,000
American Tin Plate, prefer'd	18,325,000	21,936,250	....
" common	28,000,000	5,600,000	35,000,000
American Steel Hoop, pre'd	14,000,000	14,000,000	....
" common	19,000,000	....	19,000,000
American Sheet Steel, pre'd	24,500,000	24,500,000	....
" common	24,500,000	....	24,500,000

This means that for their present aggregate outstanding preferred stock of $217,085,900 the seven companies are to be allotted $261,452,612 of the new company's preferred stock, and for their common of $239,984,300, common stock of the new company to the amount of $269,720,623. In other words, there will be an increase in the total capitalization of the seven companies of $74,103,035. As there was an over supply of water in their stock before, what must it be now?

As there has been a general feeling in the United States for some time that the iron and steel works at Sydney might prove a dangerous competitor for the Carnegie-Morgan combination, to make an effort to obtain possession of them is only what might be expected. Should the rumor be true, and the plant at Sydney become a part of the great combination in the United States, the question naturally arises, What will be its fate? Will the combination close the works or carry them on?

Candidly, we do not believe the combination will close the works. The capitalists who are at the back of the Carnegie-Morgan concern are not likely to put their millions into a plant merely for the purpose of closing it down. Mr. Moxham asserted, in his recent address in Toronto, that steel could be made at Sydney $6 per ton cheaper than at Pittsburg, and Mr. Moxham, until within the last couple of years, was actively connected with the steel trade of the United States. True, certain steel manufacturers in the United States have declared that at the best Sydney could not make steel at less than $2 per ton below the Pittsburg mills. But $2 is quite an item, particularly when competition is keen.

No, if the combination secures the Canadian plant it will work it, and regulate the price of its product by that obtaining in the mills it owns in the United States.

The only danger we apprehend is the influence of such a powerful combination, particularly in the politics of our country. We already have some idea what it means in the United States. Heretofore in Canada our experience of corporation influence in politics has been such as that which is exercised by railway companies, and that is bad enough.

A CONFERENCE IN MONTREAL.

The Montreal retail hardware merchants have invited the wholesale hardware dealers of the city to attend a meeting of their association, to be held in Monument National next Wednesday evening. Such a conference of the two bodies should be the occasion of fruitful discussion and the means of removing any misunderstandings that may exist. The retailers are to be congratulated on the step they have taken. We understand that most of the wholesale houses will be represented.

PREFERENTIAL TRADE WITH AUSTRALIA.

HARDWARE AND METAL has it upon good authority that it is the intention of the Dominion Government to send one of its members to Australia to negotiate a preferential tariff between that Commonwealth and the Dominion.

The Canadian Manufacturers' Association, led by Mr. George H. Hees, chairman of the tariff committee of that organization, has been an active champion of the movement for a preferential tariff between the two countries in question.

CANADA AND THE SOUTH-AFRICAN TRADE.

IN last week's issue of HARDWARE AND METAL appeared an article showing how little Canada was contributing in the way of supplies required by the army in South Africa. The attention of this journal has just been drawn to another matter which, at least, indirectly concerns Canada in South Africa.

It is the decision of the Government of New Zealand to call for tenders for the establishment of a subsidized line of steamers to ply monthly between that country and South Africa. As the absence of a direct service has been a long-felt want in New Zealand, the decision of the Government to supply that want has been hailed with a great deal of satisfaction by the business men, who have held meetings in various parts of the colony at which appreciative resolutions have been adopted.

According to a South-African paper there has long been a demand in that country for New Zealand oats and flour, but it has been difficult to meet it on account of the absence of proper transport facilities. The paper referred to is also of opinion that New Zealand might also do a large trade with South Africa in preserved and frozen meats, hams, bacon, butter and cheese.

The New Zealand Government, in taking the step it has, shows that it is ranging itself alongside those countries which are developing energy in regard to the South-African trade. The other countries are particularly the United States and Germany, whose consuls and commercial agents have for a year or more been trying to impress upon their respective Governments the importance of the South-African trade.

While other Governments and other people are getting wide awake to the importance of the trade, the Government of this country and the people of this country are, to say the least, not much concerned. The trouble in Canada is that we are too busy playing the game of politics to attend as we ought to matters commercial.

There is no country under the sun that possesses greater natural possibilities than the Dominion of Canada. Turn to her agricultural resources, her mineral resources, her forest resources, her fishery resources. Where can you find a country that possesses them all in such an abundance? Nowhere. But we leave it to foreigners to come in and lead in the development of our mining industry. And we leave it to foreigners to come in and lead in the development of our iron industry. Others lead, and we follow. It is not that we are deficient in capital. We have capital, and that in abundance. What we lack is enterprise and the faculty to initiate. And now, New Zealand, a country with an area of 104,471 square miles against our 3,519,000 square miles, and with a population of about 800,oco compared with our 6,000,000, has decided to put on a monthly line of steamships to run direct to South Africa, in order to overcome the disabilities experienced in sending merchandise thereto by way of Melbourne and Sydney.

That the trade of South Africa is worth reaching after is evident from the fact that the imports into Cape Colony alone are over $90,000,000 annually. In 1899 it was $93,476,738, and the value of the merchandise received from Canada was only $66,547. How Canada stood in comparison with other conutries may be gathered from the following table :

Great Britain	$63,986,111
Australasia	5,280,109
Canada	66,547
Other British possessions	4,621,309
Germany	3 566,172
United States	10,148,903
Other foreign countries	5,807,587

As to Canada's ability to contribute a larger share to the imports into Cape Colony is another question. However, there is not much doubt when we come to examine the class of goods that are imported.

PRINCIPAL IMPORTS INTO CAPE COLONY, 1899.

Agricultural implements and tools	$ 738,507
Ale and beer	503,958
Animals, living	251,835
Apparel, etc.	3,217,592
Bags	588,536
Boots and shoes	2,486,361
Butter and cheese	994,829
Carriages	682,015
Coal and coke	1,132,201
Cotton, and manufactures of	4,738,094
Drugs and chemicals	1,286,790
Haberdashery and millinery	5,651,723
Hardware and cutlery	3,978,500
Iron, sheet, corrugated, etc.	931,855
Leather	595,748
Machinery, agricultural and other	5,204,639
Meats, salted and preserved	1,186,021
Provisions, not elsewhere specified	2,599,438
Railway materials	1,239,486
Spirits and wines	1,228,055
Stationery, printing paper, etc	1,294,460
Tobacco	855,983
Wheat	2,451,948
Furniture	1,351,006
Wood and manufactures of, except furniture	2,182,359
Woollen manufactures	1,398,417

We do not for one moment claim that it is the duty of the Government to sell in foreign markets the products of our farms and factories, but we do hold that it is its duty to ascertain the requirements and peculiarities of foreign markets, and, through its agents, acquaint the people of this country with the result. And, furthermore, to encourage the establishment of transportation facilities. As far at least as South Africa is concerned these things are not being done.

NEW LIST ON PRESSED SPIKES.

Pressed spikes are now sold under a new system from that previously in vogue, they now being listed at the following standard sizes :

¼ inch, standard sizes, 4, 4½, 5 and 6 inches ; list, $4 75.

5-16 inch, standard sizes, 5, 6 and 7 inches ; list, $4 50.

⅜ inch, standard sizes, 6, 7 and 8 inches ; list, $4 25.

7 16 inch, standard sizes, 7 and 8 inches; list, $4.10.

½ inch, standard sizes, 8, 9, 10 and 12 inches ; list, $3 90.

⅝ inch, standard sizes, 10 and 12 inches; list, $3.90.

Discount on the above is 20 to 25 per cent. On lengths other than those specified 25c. per 100 lb. net extra will be charged.

The object sought by the manufacturers in reducing the numbers is to confine the trade to the above standard prices.

He who does not try to do his best will never occupy the best position in business or in anything else.

THINK MORE AND TALK LESS.

Members of Parliament at Ottawa are indignant because Hansard, containing the speeches of the previous day, is not in their hands till late in the afternoon. They have the remedy in their own hands. Let them think more and talk less.

A horse-meat canning factory started some time ago in Oregon has failed. The horse evidently would not "go" after it was put in the can.

THE ASSOCIATIVE MOVEMENT SPREADING.

IN forming a Retail Hardware and Paint Dealers' Association, the Montreal hardware merchants have wrought much good to themselves. But the benefits of their organization are extending beyond the bounds of the metropolis, for their example is as a star shedding its beams upon every community in Canada where hardwaremen are in business. In many other towns and cities there is talk of the forming of similar organizations to attain the same objects that the founders of the Montreal association have had in view. The following correspondence between Mr. Fred F. Quinn, of Chatham, and the president and secretary of the Montreal Retail Association explains itself :

CHATHAM, Ont., February 14, 1901.

Francois Martineau, Esq., President Montreal Retail Hardware Association, Montreal :

DEAR SIR,—I have seen the report of the Retail Hardware and Paint Dealers' Association in THE CANADIAN HARDWARE AND METAL MERCHANT, and am prompted to write and ask if you will give me any information that might help us at a meeting called for the same purpose here. I herewith mail to you a notice that I have sent out for a meeting to be held here next Thursday, and would ask that you give me a copy of your by-laws, rules, etc., and also an opinion of the benefits derived from such an association. Kindly give me this before Thursday next and oblige,

FRED. F. QUINN.

The reply was as follows :

MONTREAL, February 17, 1901.

Mr. F. F. Quinn, Chatham, Ont.

DEAR SIR,—At the request of Mr. Martineau, I answer yours of the 15th inst. It is a great pleasure to me to enclose a page of THE HARDWARE AND METAL in which is printed the constitution and by-laws of our association. I take this opportunity to congratulate you on such a good move, and I have no doubt but that you will find an association of this kind will be of great benefit to all its members and to all dealers of your district. With the best wishes for the success of your future association, I remain,

Yours, very respectfully,

A. MAGNAN, Secretary.

The circular letter sent out by Mr. Quinn reads as follows :

CHATHAM, ONT., February 13, 1901.

Dear Sir,—Some time ago I wrote you in reference to the hardware dealers of our county and adjacent counties forming themselves into an association for their own protection. This is not a question affecting Chatham alone, but should be as much to your benefit as to any other dealer elsewhere in the county. I have received favorable replies and have been encouraged to call a meeting for the organization of such an association. This meeting will take place in Chatham on Thursday, February 21, at the Garner House, say, at 3 o'clock. We can then thoroughly discuss the subject and I hope bring about the desired results. The hardware dealers of Chatham do not wish in any way to dictate to those on the outside and the meeting will be conducted in the fairest possible manner, and if there is no satisfactory result arrived at it will not be our fault. You will understand that there are a number of such associations in existence, both in Canada and the United States, which are gradually extending themselves and which have been found to work very satisfactorily to those interested. Do not think that by you staying away it will make no difference. Your presence is earnestly required, for every dealer can add to the importance of the occasion by his presence and advice, so be sure to make it a point to be there as this will be our only chance, for you can readily understand that if this meeting should fail it will be a hard matter to again get the dealers together. There are some towns and villages who probably think that they can protect themselves and consequently do not require to enter into any such an association, but you will bear in mind that this is not alone as to the matter of prices, but to protect the retail dealers generally both as against the jobber, manufacturer and those unprincipled purchasers, who, for a cent or two reduction, will go from one dealer to the other and deliberately misrepresent facts. Please give me a prompt reply to this, and oblige,

Yours, etc.,

FRED F. QUINN.

For the sake of the hardware merchants throughout the country who are watching with interest the progress of this association, we might say that the officers and members are day by day growing more enthusiastic about its usefulness and becoming more and more confident that it will accomplish reforms in the way of removing long-standing grievances. True, it has not accomplished much in a tangible way as yet, but its intangible results have been invaluable. Men have met together and become acquainted and have learned to admire those whom unknown they had long regarded with a jealous eye. This acquaintance and mutual confidence cannot fail to be of material usefulness in the near future when the association feels strong enough to strike out and adopt measures for the economic benefit of its members.

Meanwhile, the most tangible object entertained by the association is to arrive at a better understanding with the wholesalers in regard to their selling retail. This they stand in a fair way of accomplishing by a process we are not at liberty to discuss. When they have done this they will have passed their first milestone in the reaching of which they have learned to walk and then they will strike out with a stronger, longer and surer stride to attain greater achievements. In short, the formation of the association has been highly successful and the initiation of the force of union of retail interests into our commercial life has taken place under very fair auspices.

BETTER SERVICE TO AUSTRALIA.

A Vancouver despatch says that news has been received in that city from Australia to the effect that the Union Steamship Company of New Zealand has acquired a half interest in the business and vessels of the Canadian-Australian steamship line. As this company is the foremost steamship concern in Australasia this news will be received with pleasure by all Canadian concerns interested in Australian trade. It is understood that a better line than that running between San Francisco and Australia will be put on the Canadian-Australian line, though these changes will not likely be made this year.

LAUGHABLE TELEGRAPHIC MISTAKES.

The Leisure Hour for February gives an amusing instance of the mistakes which may arise when inexperienced persons try their hands at the compilation of telegraphic codes. The story goes that an Australian firm who had prepared such a code handed in a cablegram for transmission to England consisting of one word, which on arrival in the Black Country was read as "thanking," the translation whereof was "send 300 dozen rabbit-traps." The order created no little stir in the breasts of the recipients of the message, but the feelings of jubilation changed to mortification when a little later it was discovered that the word which should have been cabled was "banking," meaning "send three tons of cotton-waste."

This, writes "Vulcan" in Ironmonger, calls to mind another funny blunder of the Post Office, for the truth of which a colleague is sponsor. A commercial traveller, doing business at a distance from home, found that he could not conveniently spend the week-end with his family, and advised them accordingly. Next day he received a telegram from his wife urging him to return, as she was not at all well. He inquired by wire what was the matter, and received the startling reply, "I have got a child." Such an event being totally unexpected, the poor man was terribly upset, and at once rushed off home. Imagine what a mixture of sensations he must have experienced when he found that it was a "chill" and not a "child" that his wife had got. In both cases the mistakes arose through a misreading by the telegraphist of the Morse code, the signs for "th" differing only by one dot from the sign for "b," and the letter "l" by a dot from the letter "d."

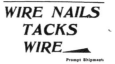

BUSINESS CHANGES.

DIFFICULTIES, ASSIGNMENTS, COMPROMISES.

W. H. AULT, general merchant, Finch, Ont., has assigned to R. E. Burns, Kingston, and a meeting of his creditors was held on February 25.

John Saulner, tinsmith, Bear River, N.S., has left that place.

I. J. Dupont, tinsmith, Farnham, Que., has assigned to Kent & Turcotte.

Taylor, Breen & Fraser, general merchants, Beulah, Man., are offering 65c. on the dollar.

Elizabeth Marshall, tinsmith, etc., Dunnville, Ont., has assigned to W. D. Swayze, Dunnville.

Premont & Co., general merchants, St. Felecite, Que., have compromised at 40c. on the dollar.

Arthur Hotte, general merchant, St. Cyrille de Wendover, Que., has compromised at 50c. on the dollar.

Morrison & Co., general merchants, Boissevain, Man., have assigned, and a meeting of their creditors has been held.

Mowat & Co., hardware dealers and G. V. Orser, carriagemaker, Trenton, Ont., have assigned to Geo. F. Hope, Belleville, Ont.

G. C. Ives & Bro., planing millers, etc., Colborne, Ont., have assigned to C. J. McCallum, and are asking an extension of time.

Oswald Smith has been appointed provisional liquidator under winding-up order of The David Inglis Co., Limited, general store and shingle mill, Flatlands, N.B.

PARTNERSHIPS FORMED AND DISSOLVED.

Decelles & Cordeau, general merchants, Farnham, Que., have dissolved.

The John Tetrault Tool and Axe Works, Maissonneuve, Que., have dissolved.

Trevethick & Kers, general merchants, Brinsley, Ont., have dissolved; J. Trevethick continues.

Humphreys & Teakles, general merchants, Sussex, N.S., have dissolved. Mr. Humphrey continues.

Martin & Sanderson, coal and wood dealers, Hamilton, Ont., have dissolved. John Martin continues.

Schwartz & Braun, dealers in agricultural implements, Altona, Man., have dissolved. Jacob Schwartz continues.

Gray & McKinnon, hardware and implement dealers, Pipestone, Man., have dissolved, Mr. Gray continuing.

M. J. Macleod, general merchant, Lacombe, N.W.T., has admitted S. Macleod under the style of M. J. & S. Macleod.

SALES MADE AND PENDING.

T. E. Risk, general merchant, Shetland, Ont., is advertising his business for sale.

The stock of Alf. Mercier, general merchant, St. Angele, Que., has been sold at 70c. on the dollar.

John Hall, general merchant, Dashwood, Ont., has sold out.

Walkon & Chapple, general merchants, Kirkton, Ont., have sold out.

W. H. Lanning, hardware dealer, Montreal, is selling out at auction.

Wm. Henderson, blacksmith, Mono Mills, Ont., is advertising his business for sale.

The assets of Alf Boulanger, general merchant, L'Islet, Que., have been sold.

Henry George, general merchant, Ninga, Man., is advertising his business for sale.

The assets of Mrs. E. A. Atkinson, general merchant, L'Avenir, Que., have been sold.

The assets of Fanny Markson, general merchant, Glen Robertson, Ont., are to be sold by tender.

The business of the estate of Robert Lewis, paint dealer, etc., London, Ont., is advertised for sale.

The stock of the estate of Johnston Bros., hardware dealers, Seaforth, Ont., has been sold at 62c. on the dollar to Allen Bros.

CHANGES.

D. Ross, general store, Ladysmith, Que., has sold out to J. F. Gronan.

J. P. Brown, dealer in agricultural implements, Goderich, Ont., has been succeeded by H. Parsons.

W. A. Cummings is starting as blacksmith in Digby, N.S.

C. Allen, coal dealer, etc., Simcoe, Ont., has sold out to C. Brookfield.

R. R. Argue, general merchant, Wilfrid, Ont., has sold out to L. W. Soper.

J. T. Cairns, general merchant, Varna, Ont., has sold out to J. E. Harnivel.

Roland G. Gordon, general merchant, Wingham, Ont., is closing out business.

The Remrose Co., general merchants, Lefroy, Ont., have sold out to E. B. Hill.

J. Clegg & Co., hardware dealers, Wingham, Ont., have sold out to James D. Burns.

G. W. Dunlap & Co., carriage hardware, trimmings, etc., have been succeeded by F. J. Henderson.

George A. Newton, harness dealer, Wingham, Ont., has sold out to George C. Manners, of Teeswater, Ont., who intends removing to Wingham.

FIRES.

E. C. Gates, general merchant, Middleton, Ont., has been burned out; insured.

The general store of Chas. A. Quick, Kingsville, Ont., has been destroyed by fire; insured.

DEATHS.

Joseph Woodruff, saw and grist miller, Sydenham, Ont., is dead.

Henri Croteau, agent for agricultural implements, D'Israeli, Que., is dead.

ROLLING MILLS TO LIQUIDATE.

ON Wednesday, Mr. Justice Meredith granted an order for the winding-up of The Abbott-Mitchell Rolling Mills Co., of Belleville. The order became operative on Thursday.

This was a result of a petition from R.W. Chisholm & Co., Buffalo; F. H. Stephens, Detroit; J. R. Walker, Montreal, and S. J. McCrudden, Belleville, asking the court to grant them an order by which their claims would not be prejudiced, should the Bank of Montreal sell the $50,000 worth of goods which that institution seized on January 25 in satisfaction of a debt of $32,000. The petitioners feared that a forced sale of the assets might not realize more than the $65,000 claim of the bondholders.

The sale of the goods seized by the Bank of Montreal was held Monday. There were present: Mr. A. J. Baxton, Worcester, Mass., a shareholder; Messrs. James Peck and Sinclair, of Peck, Benny & Co., Montreal; Mr. White, of Frost & Wood, Smith's Falls; Mr. Freeman, Ontario rolling mills, Hamilton; Mr. Gillies, Toronto; Messrs. Higgs and Sessesvain, Montreal Rolling Mills; Mr. Kloepfer, Guelph Rolling Mills; Mr. Near, of Pillow, Hersey & Co., Montreal; Mr. R. Hobson, Hamilton; Mr. R. Sylvester, Lindsay; Mr. Thomas Birkett, M.P., Ottawa; Mr. T. Lewis, Montreal; Mr. Birkett, Kingston; Mr. L. C. Marsh, Belleville, and many others.

At first the stock was offered en block, but as the top bid was only $3,500, which was lower than the reserve bid, it was put up in parcels and sold as follows: Bar iron, flat and round, 508 tons, $24 per ton, Mr. Kloepfer, of Guelph; manufactured steel, 57 tons, Sylvester Bros., Lindsay, at $23 50 per ton; railway spikes, 24 tons, at $30 per ton, to Peck, Benny & Co.; pressed spikes, 11¾ tons, at $43 40 per ton, to Peck, Benny & Co; pressed nails to the amount of 4 700 lb. were sold to W. Alford for $1.75 per cwt.; the washers, 3,250 lb., went to Dalton & Strange at $2 50 per cwt.; 55 tons of bolts went to Peck, Benny & Co. at $24 per ton; the scrap iron, $1,600 tons, was sold to the Montreal Rolling Mills Co. at $11.50 per ton; Peck, Benny & Co. secured 200 tons of steel billets at $20 per ton; the City Water Commissioners were lucky in securing between 800 and 1,000 tons of coal at $2.90 per ton; John Lewis & Co., Belleville, bought 876 kegs of cut nails at $1.75 per keg. The total amount realized was $54,271.88, considerably more than was anticipated.

PERSONAL MENTION.

Mr. John McLeod, general merchant, Prince Albert, N.W.T., is in Winnipeg on business.

Mr. Jesse Armstrong, of the Rome Brass and Copper Co., Rome, New York, was in Toronto this week on business.

Mr. J. Samuel, of Samuel, Sons & Benjamin, London and Liverpool, Eng., and of L. Samuel, Benjamin & Co., Toronto, is in Toronto on his annual visit.

Anderson & Langstaff, general merchants, Kemptville, Ont., are having a new office built in their store.

Mary L. Hannah, general merchant, Wingham, Ont., has admitted George Hannah, H. E. Jeffrey and W. H. Wightman under the style of Hannah & Co.

H. S. HOWLAND, SONS & CO.

WHOLESALE ONLY — 37-39 Front Street West, **Toronto.** — ONLY WHOLESALE

GARDEN SYRINGES

No. 600—18 x 1½ inch diameter, Brass, 2 Nipples.

INSECT EXTERMINATORS

Made in Tin Lacquered

or All Brass, Cannot Rust

"Cataract" Insect Exterminator is well made and very durable, equally effective on Fruit Trees, Plants and Flowers, in the Potato Patch, Poultry Yard, and on Cattle.

BRASS SPRAY PUMPS

WITH PATENT AGITATORS

No. 50

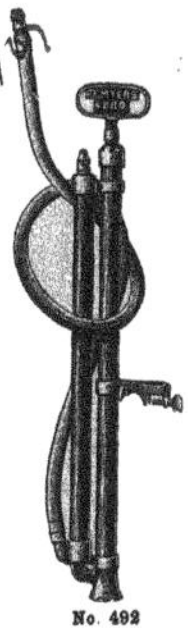

No. 492

No. 598

No. 632

H. S. HOWLAND, SONS & CO., Toronto.

WE SHIP PROMPTLY. GRAHAM WIRE AND CUT NAILS ARE THE BEST. OUR PRICES ARE RIGHT.

RETAIL BUSINESS VS. BANKING.*

By A. T. Nelson.

IN writing to you on this subject I do not wish to confine myself to any one particular line or branch of the business, neither to any one class of trade or locality. My subject allows me a wide range, therefore I will attempt to talk to you in a general way from the standpoint of a business man and a brother in the retail hardware trade.

Gentlemen, we are all bankers, or, in other words, dealers in merchandise, which represents dollars. The dollar is the motive power behind all business schemes. We, as bankers in merchandise, have our losses, our competition and our

DEAD STOCK,

the latter of which will compare well with the banker's bad paper, and, if not disposed of in time, may become a total loss. Why do we have this dead stock? Firstly, I think, because we sometimes overbuy, and again by changes of fashion, improved wares, etc. How much easier it is to sell the new article than the out-of-date and shelf-worn! Place the damaged, the out-of-style, the slow-seller for any cause in handy reach; push it out. If you can't get cost, get less.

Again, do you always take the position of a banker when making

A CREDIT SALE?

A loan of money—just so many dollars tied up in a stove, with your interest, cost of doing business and collecting all tied up in your profit. An illustration: You pay $20 for a stove; you want 25 per cent. gross profit, or $5. Say it costs you 5 per cent. to do business, leaving you a net profit of 10 per cent. As a hardware banker, how long can you have this account standing out?

If doing a little business of $10,000 per year, and you want, say, $1,000 per year as your salary,

YOU MUST HAVE 10 PER CENT.

on your sales. Therefore, you cannot afford to wait ten minutes on the sale I have described. Suppose you get 30 per cent. gross profit, you then have 5 per cent. upon which to carry your customer one year; but your $20 would have earned you this 5 per cent. loaned, without polishing, lifting or carting, guaranteeing, quarreling or repairing. Now, let us invest $20 in two ways and suppose that we are going to turn our stock over three times during the year. First cash sale, $20 invested, $25 sale; second sale, $25 invested, $31.25 sale; third sale, $31.25 invested, $39 sale; and at the end of the year your money in the bank, with an increase of $19—very nearly 100 per cent. Now the other investment: Twenty dollars invested, $25 sale on one year's time; a statement, a kick that the goods do not look as well as when sold and only $5 gross earnings on the investment for one year. This is just one of the secrets by which the catalogue house and cash store are able to make it hard for you to hold your job.

Please bear this in mind, gentlemen, and you will better understand the remarks I am about to make. I do not claim that this time trade can be successfully done away with in a day, but I do claim that the hardware banker and all other retail bankers in merchandise should use their influence to the fullest extent in reducing this unbusiness-like practice of buying trade with long-time inducements.

THE MAIL ORDER STORE.

Now I may touch a vital spot in the heart of some of our members, but, pardon me, I do not reflect on any one. Just a gentle slap at you all. Montgomery Ward & Co.! Bugbear! White Elephant! I am not in harmony with lines upon which so many of the associations are operating. They are spending too much time whining about the woes and ills that have beset them from department stores, and not enough time upon the ways of successfully combating these interests. I am afraid to get rid of these houses entirely, lest we become a trust ourselves.

In the first place, who are Montgomery Ward & Co., Sears, Roebuck & Co.? Perfect gentlemen, business men and financiers, so far as I am able to learn, out in the world for trade. There is nothing dishonest about it, nor underhanded in it. We don't own any trade, and the worst drawback to the business house that carries a large stock, employs help, pays taxes and all this, is the little store-box merchant who does his own draying in a wheelbarrow, lives above his store, slides down the banister at every click of the latch and knocks off a plum every now and then while you and your clerks are throwing stones at Montgomery Ward & Co.

THE CROSS ROAD MERCHANT.

The cross-road merchant that gets the number of a stove you have sold and sells a neighbor one like it, using your sale to show up the goods, then asks $1 over cost, the customer taking the stove from the station as shipped, without blackening—these are the fellows that bite. Nothing dishonest about it. How often the case, when besieged with dogs, while you are battling with the mastiff, the little cur slips in and carries off a branch of your trousers!

RETAILERS ADVERTISE THEM.

It would be just as reasonable for Montgomery Ward & Co. to appeal to Congress to prohibit us local hardware dealers from nipping at their trade as for us to try to force them off of the market by law. At the same time we are paying for their advertising by sifting their names broadcast among the country buyers, who will naturally think that we are being pinched and the department store is really a good place to buy, after all. Now the remedy: This is something that no one man can name for all localities, where conditions differ, competition differs and individual dispositions differ.

But I would suggest that every retail hardware dealer make a banker of himself.

TEACH FRIENDS AND CUSTOMERS,

as well as the customer of the department store, some lessons in finance. Teach them this prayer: "Oh, Mr. Hardwareman! I know that the secret of individual wealth lies in getting more than I spend; that my trade in your hands adds to your wealth; that with more wealth you are able to carry a better stock for me to select from, that you can furnish better help to wait upon me, buy more of my beef, drive a good team of my raising, and sell me goods cheaper than if I send my money to build up granite buildings in far off cities, where trusts and monopolies are reared."

After he has this well learned, then take up the following

FOR YOUR OWN GUIDANCE:

"Oh, Mr. Self: I begin to believe that the hardwareman is his own worst enemy; that the farmer is right in saving all the money he can; that a wealthy community means a healthy trade; that to make a wealthy community all merchants must themselves buy their family supplies at home; that a wealthy town must attract more money than it sends out; that (all things being equal) I will buy in my own State; that which my own State don't make or supply, some other State in my Union does. And I will handle goods made in the U.S.A. in preference to any others on earth." The principles I have set forth I believe will win. The details will have to be worked out by the parties interested.

Bear in mind: You can't do a healthy business in a town of paupers; that your personal interests are your first duty; that the joint interests of the town and community in which you do business form a part of your own and can best be served by unity of efforts and buying at home.

* Paper read before The Iowa Retail Hardware Dealers Association.

INQUIRIES REGARDING CANADIAN TRADE.

The following were among the recent inquiries relating to Canadian trade received at the High Commissioner's office, in London, England :

1. Inquiry is made for the names of one or two reliable firms in Canada who are in a position to ship cut wood for fruit crates to the Canary Islands.

2. The name of a Canadian firm interested in builders' materials is asked for by a London house having the monopoly of the export of some important lines of special wall tiles, both earthenware and opal-glass.

3. A manufacturers' agent in South Africa is desirous of taking up the representation of Canadian hardware, furniture, lumber, rubber and other firms.

4. The names of large exporters of fruit from Canada are asked for by a party who can sell on commission on the Liverpool market.

[The names of the firms making the above inquiries, can be obtained on application to the editor of HARDWARE AND METAL, Toronto. When asking for names, kindly give number of paragraph and date of issue.]

Mr. Harrison Watson, curator of the Canadian Section of the Imperial Institute, London, England, is in receipt of the following inquiries regarding Canadian trade:

1. A Nottingham produce and provision company desires to take up the handling of Canadian poultry and invites correspondence from Canadian shippers.

2. A Belfast house desires names of Canadian shippers of tallow.

3. A London firm of manufacturers is prepared to receive and report upon samples of infusorial earth, must be very white in color, light in gravity and fine in texture for their purposes.

4. A Liverpool firm manufacturing wheels and spokes desires quotations for oak pieces $1\frac{5}{8} \times 1\frac{5}{8}$ in. 7-inches long suitable for turning ; quantity, 500,000 pieces ; terms, c.i.f. Liverpool.

ANOTHER CONTRACT GONE.

An English paper just to hand says :

"Yet another British engineering contract has gone to America. The American Bridge Co., U.S.A., have succeeded in gaining the tender to supply certain bridges for the Uganda Railway.

"Ten British and three American firms applied for this contract, and, while the above company undertook to erect the work in 46 weeks, the only British firm who would specify a period gave 130 weeks as the time wanted.

"The contract price per ton is £18 erected, or £10 6s. free on board at any British port. The prices, unerected, of British manufacturers varied from £13 1s. 3d. to £19 17s. 6d. per ton."

McDougall & Co., general merchants, Renfrew and Dacre, Ont., have sold their branch at Dacre to Benj. Hunter.

MARKETS AND MARKET NOTES

QUEBEC MARKETS.

Montreal, March 1, 1901.

HARDWARE.

THE different wholesale houses throughout the city appear to be much busier this week than last, and they say that a fair volume of business for this season of the year is being done. Of course, there is no heavy buying being done, dealers not feeling any compulsion to buy heavily. But the sorting orders are numerous, and they show that trade is in a healthy condition. The orders for spring are coming in much faster than they were, and some houses assert their spring bookings total more than for any previous season. The demand from the Northwest is good, as also is that from the east. The formation of the big wire and steel trust in the United States has had a stiffening effect upon the market here, particularly in the different kinds of wire. There had been a supposition that barb wire might decline, but such an idea has been dispelled. The demand has somewhat improved for wires this week, while wire nails are also in better request. Cut nails are rather inactive. Shelf goods are in good request, and spring and summer goods are selling in all lines. A new list is out on pressed spikes. One of the most encouraging features of the market at the present moment is the satisfactory way in which paper is being met.

BARB WIRE—There has been a slightly better inquiry for wire this week, owing probably to the formation of the wire and steel trust in the United States. For spring delivery the general quotation is $3.20 f.o.b. Montreal, but we understand that this week some houses have reduced their values 15c.

GALVANIZED WIRE—A little more interest attaches to galvanized wire this week. Prices are unchanged. We quote as follows: No. 5, $4.25; Nos. 6, 7 and 8 gauge, $3.55; No. 9, $3.10; No. 10, $3.75; No. 11, $3.85; No. 12, $3.25; No. 13, $3.35; No. 14, $4.25; No. 15, $4.75; No. 16, $5.00.

SMOOTH STEEL WIRE—A good number of shipments have been made from stock this week, and spring bookings are being made a little more freely than heretofore. Values are still at $2.80 per 100 lb.

FINE STEEL WIRE—A steady trade is being done in all grades. The discount continues at 17½ per cent. off the list.

BRASS AND COPPER WIRE—A moderate amount of business is being done. Discounts are still 55 and 2½ per cent. on brass, and 50 and 2½ per cent. on copper.

FENCE STAPLES—Some fair amounts are moving. We quote: $3.25 for bright, and $3.75 for galvanized, per keg of 100 lb.

WIRE NAILS—The demand is improving. An advance is talked of in the United States. We quote $2.85 for small lots and $2.75 for carlots, f.o.b. Montreal, Toronto, Hamilton, London, Gananoque, and St. John, N.B.

CUT NAILS — There seems to be only a small inquiry for cut nails. We quote: $2.35 for small and $2.25 for carlots; flour barrel nails, 25 per cent. discount; coopers' nails, 30 per cent. discount.

HORSE NAILS — The volume of business continues to be large. The general discounts are 50, 10 and 5 per cent. on oval head and 50, 10 and 10 per cent. on countersunk head. "C" brand's discount is 50 and 7½ per cent. on their own price list.

HORSESHOES — Trade is hardly as brisk as it was, still a good business is being done. Prices are unchanged. We quote: Iron shoes, light and medium pattern, No. 2 and larger, $3.50; No. 1 and smaller, $3.75; snow shoes, No. 2 and larger, $3.75; No. 1 and smaller, $4.00; X L steel shoes, all sizes, 1 to 5, No. 2 and larger, $3.60; No. 1 and smaller, $3.85; feather-weight, all sizes, $4.85; toe weight steel shoes, all sizes, $5.95 f.o.b. Montreal; f.o.b. Hamilton, London and Guelph, 10c. extra.

POULTRY NETTING — Business on spring account is reported brisk this week. The discounts are 50 and 5 per cent.

GREEN WIRE CLOTH—Fair lots of goods have been booked for next season's trade. The price is still $1.35 per 100 sq. ft.

FREEZERS — This is one of the articles that are moving at the moment. Prices are unchanged. "Peerless" is quoted as follows: Two quarts, $1.85; 3 quarts, $2.10; 4 quarts, $2.50 6 quarts, $3.20; 8 quarts, $4; 10 quarts, $5.25; 12 quarts, $6; 16 quarts, with fly wheel, $11; toy, 1 pint, $1.40.

SCREEN DOORS AND WINDOWS — Prices are firm and a good business is being done. We quote: Screen doors, plain cherry finish, $8.25 per doz.; do. fancy, $11.50 per doz.; walnut, $7.40 per doz., and yellow, $7.45; windows, $2.25 to $3.50 per doz.

SCREWS — A moderate business is being done at old prices. Discounts are: Flat head bright, 87½ and 10 per cent. off list; round head bright, 82½ and 10 per cent.; flat head brass, 80 and 10 per cent.; round head brass, 75 and 10 per cent.

BOLTS — The demand continues to be fairly good for small lots. Discounts are: Carriage bolts, 65 per cent.; machine bolts, 65 per cent.; coach screws, 75 per cent.; sleigh shoe bolts, 75 per cent.; bolt ends, 65 per cent.; plough bolts, 50 per cent.; square nuts, 4¾c. per lb. off list; hexagon nuts, 4¾c. per lb. off list; tire bolts, 67½ per cent.; stove bolts, 67½ per cent.

SPIKES—A new list has been issued on pressed spikes. It now stands as follows: ¼ in., $4.75; 5-16, $4.50; ⅜, $4.25; 7-16, $4.10; ½, $3.90; ⅝, $3.90. The trade discount is 25 per cent.

BUILDING PAPER — Trade is opening out fairly well. We quote as follows: Tarred felt, $1.70 per 100 lb.; 2-ply, ready roofing, 80c. per roll; 3-ply, $1.05 per roll; carpet felt, $2.25 per 100 lb.; dry sheathing, 30c. per roll; tar sheathing, 40c. per roll; dry fibre, 50c. per roll; tarred fibre, 60c. per roll; O.K. and I.X.L., 65c. per roll; heavy straw sheathing, $28 per ton; slaters' felt, 50c. per roll.

RIVETS — There is nothing new to note. The discount on best iron rivets, section, carriage, and waggon box, black rivets, tinned do., coopers' riverts and tinned swedes rivets, 60 and 10 per cent.; swedes iron burrs are quoted at 55 per cent. off; copper rivets, 35 and 5 per cent. off; and coppered iron rivets and burrs, in 5-lb. carton boxes, are quoted at 60 and 10 per cent. off list.

CORDAGE—The movements in all sorts of cordage are quite large. Manila is worth 13c. per lb. for 7-16 and larger; sisal is selling at 9½c., and lathyarn 9½c. per lb.

SPADES AND SHOVELS—In the rural districts there is some inquiry for these goods. The discount is still 40 and 5 per cent. off the list.

HARVEST TOOLS—For this season of the year the inquiry seems to be fully up to the average. The discount is 50, 10 and 5 per cent.

TACKS—There has been no change. We quote: Carpet tacks, in dozens and bulk, blued 80 and 5 per cent. discount; tinned, 80 and 10 per cent.; cut tacks, blued, in dozens, 75 and 15 per cent. discount.

CHURNS—Trade in next season's goods seems to be rather brisk. The discount is 56 per cent.

FIREBRICKS—A fair winter trade is being done in firebricks. Prices are steady at $18 to $24 per 1,000, as to brand, ex store.

CEMENT — The demand for cement, as usual at this season of the year, is slow. In consequence, the market is dull, but the prospects for spring are encouraging, as the indications are that there will be considerable building done. The stock carried over from last summer was not excessive, and prices have ruled steady. We quote: German, $2.45 to $2.55; English, $2.30 to $2.40; Belgian, $1.95 to $2.05.

METALS.

Compared with last week, business is a little more active, but yet there are no large orders being placed. Steel products and iron pipe may probably be affected by the consolidation of many interests in the United States.

PIG IRON — There is quite a good deal of trading being done in pig iron. Canadian is worth about $18 on the Montreal market, and No. 1 Summerlee $22 to $23.

BAR IRON—The demand for bar iron is steady, and prices range from $1.65 to $1.70.

BLACK SHEETS — The depression noticed last week in black sheets continues, and there is little business being done for import. There are some spot transactions occurring at $2.80 for 8 to 16 gauge; $2.85 for 26 gauge, and $2.90 for 28 gauge.

GALVANIZED IRON — There is a firmer feeling in galvanized iron, and trading has been more freely entered into. We quote: No. 28 Queen's Head, $5 to $5.10; Apollo, 10¾ oz., $5 to $5.10; Comet, No. 28, $4.50, with 25c. allowance in case lots.

INGOT COPPER—The market is firm and unchanged, round lots selling at 17½c.

INGOT TIN — The foreign markets are eomewhat weaker again and buyers seem to be scarce. Lamb and Flag on the Montreal market is worth 33 to 34c.

LEAD—Is unchanged at $4.65.

LEAD PIPE—There have been some good movements in lead pipe, buyers operating freely. We quote: 7c. for ordinary and 7½c. for composition waste, with 15 per cent. off.

IRON PIPE—The feeling in pipe is firm, as the National Tube Co. are in the big consolidation. We quote: Black pipe, ¼, $3 per 100 ft.; ⅜, $3; ½, $3; ¾, $3.15; 1-in., $4.50; 1¼, $6.10; 1½, $7.28; 2-in., $9.75. Galvanized, ½, $4.60; ¾, $5.25; 1-in., $7.50; 1¼, $9.80; 1½, $11.75; 2-in., $16.

TINPLATES—For immediate delivery the demand for tinplates seems to be fairly brisk, but import orders are being made very lightly. From stock the ruling figures for immediate delivery are $4.50 for coke and $4.75 for charcoal.

CANADA PLATE—The feeling in regard to Canada plates does not appear to have improved. A few small orders are coming in for immediate delivery. We quote: 52's, $2.90; 60's, $3; 75's, $3.10; full polished, $3.75, and galvanized, $4.60.

TOOL STEEL—There is quite a good demand for tool steel. We quote: Black Diamond, 8c.; Jessop's 13c.

STEEL—Spring steel is particularly brisk and other varieties seem to be generally wanted. The feeling is firm. We quote: Sleighshoe, $1.85; tire, $1.95; spring, $2.75; machinery, $2.75 and toe-calk, $2.50.

TERNE PLATES—A fair business has been done in terne plates. Good brands are selling at $8.25; others at $8.

COIL CHAIN—The demand for coil chain is decidedly brisk and jobbers are placing big orders at the advanced prices. We quote: No. 6, 11½c.; No. 5, 10c.; No. 4, 9½c.; No. 3, 9c.; ¼-inch, 7½c. per lb.; 5-16, $4.65; 5-16 exact, $5.10; ⅜, $4.20; 7-16, $4.00; ½, $3.75; 9-16, $3.65; ⅝, $3.35; ¾, $3.25; ⅞, $3.20; 1-in., $3.15. In car-load lots an allowance of 10c. is made.

SHEET ZINC—The ruling price is 6 to 6¼c.

ANTIMONY—Quiet, at 10c.

GLASS.

Fair movements are noticeable in glass and the opening of navigation will likely see large importations. Import orders are still being taken on the basis of $1.75 to $1.80 for the first break of 50 feet. We quote as follows: First break, $2; second, $2.10 for 50 feet; first break, 100 feet, $3.80; second, $4; third, $4.50; fourth, $4.75; fifth, $5.25; sixth, $5.75, and seventh, $6.25.

PAINTS AND OILS.

Linseed oil has declined 4c. per gal., and turpentine is 3c. higher. The fall in oil in England appears to have been founded on statistical information, for it now turns out that there are 515,000 quarters of linseed afloat to England, against 272,000 quarters at this time last year. Business in all branches of the trade is good this week, and the manufacturers of mixed paints are working at night to keep up with orders. Lead and lead products are steady. We quote;

WHITE LEAD—Best brands, Government standard, $6.37½; No. 1, $6.00; No. 2, $5.62½; No. 3, $5.25, and No. 4, $4.87½ all f.o.b. Montreal. Terms, 3 per cent. cash or four months.

DRY WHITE LEAD—$5.50 in casks; kegs, $5.75.

RED LEAD—Casks, $5.25; in kegs, $5.50.

WHITE ZINC PAINT—Pure, dry, 8c.; No. 1, 6½c.; in oil, pure, 9c.; No. 1, 7½c.

PUTTY—We quote: Bulk, in barrels, $2 per 100 lb.; bulk, in less quantity, $2.15; bladders, in barrels, $2.20; bladders, in 100 or 200 lb. kegs or boxes, $2.35; in tins, $2 45 to $2.75; in less than 100-lb. lots, $3 f.o.b. Montreal, Ottawa, Toronto, Hamilton, London and Guelph. Maritime Provinces 10c. higher, f.o.b. St. John and Halifax.

LINSEED OIL—Raw, 72c.; boiled, 75c., in 5 to 9 bbls., 1c. less, 10 to 20 bbl. lots, open, net cash, plus 2c. for 4 months. Delivered anywhere in Ontario between Montreal and O,hawa at 2c. per gal. advance and freight allowed.

TURPENTINE—Single bbls., 62c.; 2 to 4 bbls., 61c.; 5 bbls. and over, open terms, the same terms as linseed oil.

MIXED PAINTS—$1.25 to $1.45 per gal.

CASTOR OIL—8¾ to 9¼c. in wholesale lots, and ½c. additional for small lots.

SEAL OIL—47½ to 49c.

COD OIL—32½ to 35c.

NAVAL STORES — We quote: Resinsr $2.75 to $4.50, as to brand; coal tar, $3 25 to $3.75; cotton waste, 4½ to 5½c. for colored, and 6 to 7½c. for white; oakum, 5½ to 6½c., and cotton oakum, 10 to 11c.

PARIS GREEN—Petroleum barrels, 16¾c. per lb.; arsenic kegs, 17c.; 50 and 100-lb. drums, 17½c.; 25-lb. drums, 18c.; 1-lb. packages, 18½c.; ½-lb. packages, 20½c.; 1-lb. tins, 19½c.; ½-lb. tins, 21½c. f.o.b. Montreal; terms 3 per cent. 30 days, or four months from date of delivery.

SCRAP METALS.

There is not a great deal of life in the scrap metal market, although the inquiry appears to be a little better this week. Dealers are paying the following prices in the country: Heavy copper and wire, 13 to 13½c. per lb.; light copper, 12c.; heavy brass, 12c.; heavy yellow, 8½ to 9c.; light brass, 6½ to 7c.; lead, 2¾ to 3c. per lb.; zinc, 2¼ to 2½c.; iron, No. 1 wrought, $13 to $14 per gross ton; No. 1 cast, $13 to $14; stove plate, $8 to $9; light iron, No. 2, $4 a ton; malleable and steel, $4.

HIDES.

The market appears to have improved on last week's condition. Dealers are paying 7½c. for No. 1 light, and tanners 8½c. for carlots. The demand seems brisker. We quote: Light hides, 7½c. for No. 1; 6½c. for No. 2, and 5½c. for No. 3. Lambskins, 90c.

PETROLEUM.

Business is being reduced to a spring volume. We quote: "Silver Star," 14½ to 15½c.; "Imperial Acme," 16 to 17c.; "S.C. Acme," 18 to 19c., and "Pratt's Astral," 18½ to 19½c.

MONTREAL NOTES.

James Hutton & Co. have moved into a commodious suite of offices in the Victoria Chambers, 232 McGill street, where they will do business hereafter.

ONTARIO MARKETS.

TORONTO, March 1, 1901.

HARDWARE.

QUITE an improvement can be reported this week in the wholesale hardware trade, and from this out an increasing trade is to be expected. In consequence of this improvement wholesale merchants and manufacturers are naturally expressing a good deal more confidence. The tendency of prices, however, in a good many lines is still downward. In staples the most important change this week is a reduction in the f.o.b. Cleveland price on barb wire, plain twist, and galvanized wire of 15c. per 100 lb. It is said, however, that this change is due to competition on the Canadian market of European wire. It should be stated that the price of these lines has not been reduced to the jobbers and retailers in the United States, being only for those in Canada. The manufacturers in Canada have reduced their prices on bright wire goods, the discount now being 60 per cent. Business is a little better on wire nails, but cut nails are as quiet as ever. Step-ladders are quoted lower, so is clothes line wire. Wringers continue to be demoralized in price. A fairly good trade is being done in such lumbermen's supplies as chain, rope, and cross-cut saws.

BARB WIRE—The feature in this line is the reduction of 15c. per 100 lb. in the f.o.b. Cleveland price for the Canadian market, and we now quote $2.82½ per 100 lb. for less than carlots, and $2.70 for carlots of not less than 15 tons. Quite a few orders have been booked for shipment for March 1, and with this lower price, of course, it is expected that quite a few shipments will be made. The cause of the reduction in price is understood to be the competition of European manufacturers on the Canadian market.

GALVANIZED WIRE—This, for the same reason as in barb wire, is 15c. per 100 lb. lower this week for shipment from Cleveland, and we now quote f.o.b. that point, $2.57½ for Nos. 6 to 9 base in less than carlots, 12½c. less for carlots of 15 tons. Some orders are being booked for spring delivery.

SMOOTH STEEL WIRE—The price of this is without change, the market being supplied by the home manufacturers. Not much business is being done either in oiled and annealed or hay-baling wire. Base price is $2.80 per 100 lb.

WIRE NAILS—The volume of business is a little larger, as spring approaches retailers evidently being more anxious to buy. At the same time, however, there is still room for improvement in this line. The base price is unchanged at $2.85 per keg in less than carlots, and $2.75 in carlots.

CUT NAILS—No improvement can yet be reported in cut nails. Prices are unchanged, the base figure still being $2.35 per keg.

HORSESHOES—A steady trade is being done at unchanged prices. We quote f.o.b. Toronto as follows: Iron shoes, No. 2 and larger, light, medium and heavy, $3.60; snow shoes, $3.85; light steel shoes, $3.70; featherweight (all sizes), $4.95; iron shoes, No. 1 and smaller, light, medium and heavy (all sizes), $3.85; snow shoes, $4;

light steel shoes, $3.95; featherweight (all sizes), $4 95.

HORSE NAILS — Trade is also steady in this line, with prices as before. We quote "C" brand oval head at 50 and 7½ per cent. discount off new list, and "M" brand at 60, 10 and 5 per cent. off old list; countersunk head, 50, 10 and 10 per cent.

SCREWS—Business is improving, and has been fairly good during the past week. We quote discounts as follows: Flat head bright, 87½ and 10 per cent.; round head bright, 82½ and 10 per cent.; flat head brass, 80 and 10 per cent.; round head brass, 75 and 10 per cent.; round head bronze, 65 per cent., and flat head bronze at 70 per cent.

BOLTS AND NUTS—There have been a good many inquiries during the past week, and quite a few orders have been received. A feature of the trade in this line is the demand which has been experienced during the past week for special lots. We quote as follows: Carriage bolts (Norway), full square, 70 per cent.; carriage bolts full square, 70 per cent.; common carriage bolts, all sizes, 65 per cent.; machine bolts, all sizes, 65 per cent.; coach screws, 75 per cent.; sleighshoe bolts, 75 per cent.; blank bolts, 65 per cent.; bolt ends, 65 per cent.; nuts, square, 4½c. off; nuts, hexagon, 4¾c. off; tire bolts, 67½ per cent.; stove bolts, 67½; plough bolts, 60 per cent.; stove rods, 6 to 8c.

RIVETS AND BURRS — A fairly good trade is also reported this week in rivets and burrs. Prices are unchanged, the discount being 60 and 10 per cent. on iron rivets, 55 per cent. on iron burrs, and 35 and 5 per cent. on copper rivets and burrs.

ROPE—There is quite a little rope going out for the lumber camps, and prices are unchanged. We quote sisal at 9½c. per lb. base; manila rope 13c. base; cotton rope, 3-16 in. and larger, 16½c.; 5 32 in., 21½c., and ⅛ in., 22½c. per lb.

BINDER TWINE—A fair trade is to be reported. We quote pure manila at 10¾c. per lb.; mixed, 8½c. per lb.; sisal, 7¾c. per lb.

CUTLERY — There is a steady and fair movement for this time of the year.

SPORTING GOODS—Trade in this line is quiet and steady.

WRINGERS—As noticed last week trade in this line is demoralized and it is asserted that some manufacturers are selling below cost. The best wringers it seems can now be purchased at $26 to $30 per dozen, net figures, according to quality and size of rollers.

GREEN WIRE CLOTH—A few orders are still coming to hand for future delivery and shipments are now being made. The price is steady and unchanged at $1.35 per 100 sq. ft.

CHAIN—There has been quite a good movement during the past week in chain for the lumber camps.

SAWS—The demand for crosscut saws has been much better this season than for some time. A fairly good demand is also reported for handsaws.

SCREEN DOORS AND WINDOWS—According to the new prices which the manufacturers have arranged, the cheapest screen doors now offering are quoted at $7.20 to $7.80 per dozen in 4 in. styles, with 3-in. styles, 20c. per dozen less. Screen windows are quoted at $1.60 to $3.60 per dozen, according to size and extensions. It is not expected that there will be any further changes in prices, and a steady trade is looked for.

PRESSED SPIKES — According to the change in price noted in a recent issue the following are the standard sizes and prices of pressed spikes: ¼ in., standard sizes are 4, 4½, 5 and 6 in., list $4.75; 5-16 in. are 5, 6 and 7 in., list $4.50; ⅜ in. are 6, 7 and 8 in., list $4.25; 7-16 in. are 7 and 8 in., list $4.10; ½ in. are 8, 9, 10 and 12 in., list $3 90; ⅝ in. are 10 and 12 in., list $3 90. The discount on the above is 20 to 25 per cent. off the list. In lengths other than those specified the charge will be 25c. per 100 lb. net extra.

BRIGHT WIRE GOODS—The manufacturers of bright wire goods in Canada have reduced their prices, the discount now being 60 per cent. instead of 55 per cent. Business is steady in this line.

STEP LADDERS—Prices in this line have also been reduced, and hardwood ladders are now being sold at 7 to 10c. per foot, according to quality. Business is beginning to open up in this line.

CLOTHESLINE WIRE—A reduction is to be noted this week in clothesline wire, and the base figures for 19 gauge are now $2.75 to $3 per 1,000 ft.

TINWARE — Shipments are still being made of milk can trimmings, and will continue to be made, of course, for some time yet. In tinware, generally, only a moderate business is being done.

SAP BUCKETS—A good many orders have been booked for shipment on March 1. Some firms have already made quite a few shipments.

POULTRY NETTING — Orders continue to be booked for spring shipment. Discount on Canadian is 50 and 5 per cent., and on English net figures are quoted.

HARVEST TOOLS—Some orders are being booked, particularly for manure forks. Discount, 50, 10 and 5 per cent.

SPADES AND SHOVELS—Quite a number of orders are being booked this week for spring shipment. Discount, 40 and 5 per cent.

BUILDING PAPER—A nice steady trade is being done at quotations.

RANGE BOILERS—A steady trade is being done at the new prices referred to last week. We quote: 30 gallons, $7; 35 gallons, $8.25; 40 gallons, $9.50.

METALS.

Business in metals on the local market has improved during the past week, and is now better than it has been for some months. Stocks in nearly all lines are light, and in some lines they are practically exhausted, particularly in tinplates and tinned iron. Supplies, however, are arriving. Local quotations are without quotable change.

PIG IRON—The market is beginning to feel the competition of Cape Breton iron, and it is possible that a slight decline may be noted. But, as prices are considered to be close to the cost of the production, it is expected that any reduction will be followed by an even greater advance. "A good sign," said a prominent iron man the other day, "is the determination of the Ontario smelters to hold firmly their present grip on this market. A representative of the Cape Breton concern was here the other day trying to get in here. Hamilton iron was out to meet his prices, and is to day 75c. to $1 below the lowest United States quotations, which in turn are as low, if not lower, than at any time this year."

BAR IRON—Business is fairly good and the outlook is for steady prices, particularly as the failure of the Belleville concern will for the time being, at least, mean one competitor less. The ruling base price is still $1.65 to $1.70 per 100 lb.

PIG TIN—The outside markets have been rather firmer this week. In London, early in the week, there was an advance of over 12s. per ton. Quotations, locally, are unchanged at 32 to 33c.

TINPLATES—Trade has been good during the past week, and orders are still being booked for importation. Business, generally speaking, is better.

TINNED SHEETS—Stocks are light, but the demand has been good during the past week. We quote 9 to 9½c. for 28 gauge.

BLACK SHEETS—Business in this line appears to have again improved a little, for the report this week is that the demand is good. The base figure is still $3 30.

GALVANIZED SHEETS—Trade from stock is fair. Although the dealers have pretty well placed their orders for importation, there are still a few orders being booked. Some of the jobbers are this week making delivery of American sheets ordered for future shipment. The ruling price for small quantities of English is $4 75, and for American $4 50.

CANADA PLATES—Business from stock is quiet, but there are a good many orders being booked on importation account. We quote all dull, $3; half and half, $3 15, and all bright, $3 65 to $3.75.

IRON PIPE—Prices are firmer and the American manufacturers have withdrawn their quotations. We quote: Black pipe ¼ in., $2.95; ⅜ in., $3 00; ½ in., $3.05; ¾ in., $3.30; 1 in., $4 70; 1¼ in., $6.40; 1½in., $7.70; 2 in., $10.25. Galvanized pipe is as follows: ½ in., $4.85; ¾ in., $5.25; 1 in., $7.25; 1¼ in., $9.75; 1½ in., $11.50; 2 in., $15 50.

STEEL—The demand for steel is active and jobbers report a good deal of difficulty in securing delivery from the manufacturers.

COPPER—In ingot copper business has improved during the past week and is now reported to be good. A good trade is also being done in sheet copper. We quote: Ingot, 19 to 20c.; bolt or bar 23½ to 25c.; sheet, 23 to 23½c. per lb.

BRASS—Business in this line has improved, and is this week termed fairly good. Discount, 15 per cent. on rod and sheet.

SOLDER—A good trade is reported in solder, with prices as before. We quote: Half-and-half, guaranteed, 19c.; ditto, commercial, 18½c.; refined, 18½c., and wiping, 18c.

LEAD—Business has been more active in this line during the past week. Prices are unchanged at 4¾ to 5c.

ZINC SPELTER—Trade is quiet with prices unchanged at 6 to 6½c. per lb.

ZINC SHEET—We quote casks at $6 75 to $7, and part casks at $7 to $7.50.

ANTIMONY—Trade is fair at 11 to 11½c. per lb.

PAINTS AND OILS.

Turpentine has advanced 3c. per gallon. White lead and linseed oil are firm since the reductions recently made. These reductions have considerably increased the movement of these lines and, to some extent, all lines. Sundries are firm throughout. We quote:

WHITE LEAD—Ex Toronto, pure white lead, $6 50; No. 1, $6 12½; No. 2, $5 75; No. 3, $5.37½; No. 4, $5; dry white lead in casks, $6.

RED LEAD—Genuine, in casks of 560 lb, $5 50; ditto, in kegs of 100 lb., $5.75; No. 1, in casks of 560 lb., $5 to $5 25; ditto, kegs of 100 lb.; $5 25 to $5 50.

LITHARGE—Genuine, 7 to 7½c.

ORANGE MINERAL—Genuine, 8 to 8½c.

WHITE ZINC—Genuine, French V.M., in casks, $7 to $7.25; Lehigh, in casks, $6.

PARIS WHITE—90c. to $1 per 100 lb.

WHITING—70c. per 100 lb.; Gilders' whiting, 80c.

GUM SHELLAC—In cases, 22c.; in less than cases, 25c.

PARIS GREEN—Bbls., 16¾c.; kegs, 17c.; 50 and 100 lb. drums, 17½c.; 25-lb. drums, 18c.; 1-lb. papers, 18½c.; 1-lb. tins, 19½c.; ½-lb. papers, 20½c.; ½ lb. tins, 21½c.

PUTTY—Bladders, in bbls., $2.20; bladders, in 100 lb. kegs, $2.35; bulk in bbls., $2; bulk, less than bbls. and up to 100 lb., $2.15; bladders, bulk or tins, less than 100 lb., $3.

PLASTER PARIS—New Brunswick, $1.90 per bbl.

PUMICE STONE—Powdered, $2.50 per cwt. in bbls., and 4 to 5c. per lb. in less quantity; lump, 10c. in small lots, and 8c. in bbls.

LIQUID PAINTS—Pure, $1.20 to $1.30 per gal.; No. 1 quality, $1 per gal.

CASTOR OIL—East India, in cases, 10 to 10½c. per lb. and 10½ to 11c. for single tins.

LINSEED OIL—Raw, 1 to 4 barrels, 69c.; boiled, 72c.; 5 to 9 barrels, raw, 68c.; boiled, 71c., delivered. To Toronto, Hamilton, Guelph and London, 1c. less.

TURPENTINE—Single barrels, 62c.; 2 to 4 barrels, 61c., to all points in Ontario. For less quantities than barrels, 5c. per gallon extra will be added, and for 5 gallon packages, 50c., and 10 gallon packages, 80c. will be charged.

GLASS.

A large number of orders for import are being taken, but at a close margin. The Belgian market keeps stiff. Stock prices are firm. We still quote first break locally as follows: Star, in 50-foot boxes, $2.10, and 100-foot boxes, $4; double diamond under 26 united inches, $6. Toronto, Hamilton and London; terms 4 months or 3 per cent. 30 days.

MARKET NOTES.

Turpentine is 3c. per gal. higher.

Iron pipe is quoted a little higher.

Stepladders are quoted lower.

The price of clothesline wire has been reduced, now being $2.75 to $3.

Barb wire, plain twist wire and galvanized wire are quoted 15c. per 100 lb. lower f.o.b. Cleveland.

The discount on bright wire goods has been increased to 60 per cent.

H. S. Howland, Sons & Co. have taken into stock press-button pocket knives. They have them in celluloid and stag horn handles, and in hunting as well as pocket knives. The blades are sprung open by merely pressing a button.

MANITOBA MARKETS.

WINNIPEG, February 23, 1901.

THE same conditions of trade prevail, and there are no changes of price to report. Merchants state that they do not anticipate any movement for two months at least. There has been a great gathering of implement men this week, and the indications seem good for all lines this spring.

The paint, oil and glass market is as quiet as hardware. With the single exception of a drop of 4c. on linseed oil there has not been a move all week.

The following is the price list for the week :

Barbed wire, 100 lb.	$3 45
Plain twist	3 45
Staples	3 95
Oiled annealed wire 10	3 95
" 11	4 00
" 12	4 05
" 13	4 20
" 14	4 35
" 15	4 45
Wire nails, 30 to 60 dy, keg	3 45
" 16 and 20	3 50
" 10	3 55
" 8	3 65
" 6	3 70
" 4	3 85
" 3	4 10
Cut nails, 30 to 60 dy.	3 00
" 20 to 40	3 05
" 10 to 16	3 10
" 8	3 15
" 6	3 20
" 4	3 30
" 3	3 65
Horsenails, 45 per cent. discount.	
Horseshoes, iron, No. 0 to No 1	4 65
No. 2 and larger	4 40
Snow shoes, No. 0 to No. 1	4 90
No. 2 and larger	4 40
Steel, No. 0 to No. 1	4 95
No. 2 and larger	4 70
Bar iron, $2.50 basis.	
Swedish iron, $4.50 basis.	
Sleigh shoe steel	3 00
Spring steel	3 25
Machinery steel	3 75
Tool steel, Black Diamond, 100 lb	8 50
Jessop	13 00
Sheet iron, black, 10 to 20 gauge, 100 lb.	3 50
20 to 26 gauge	3 75
28 gauge	4 00
Galvanized American, 16 gauge	2 54
18 to 22 gauge	4 50
24 gauge	4 75
26 gauge	5 00
28 gauge	5 25
Genuine Russian, lb	12
Imitation " "	8
Tinned, 24 gauge, 100 lb	7 55
26 gauge	8 80
28 gauge	8 00
Tinplate, IC charcoal, 20 x 28, box	10 75
" IX "	12 75
" IXX "	14 75
Ingot tin	35
Canada plate, 18 x 21 and 18 x 24	3 75
Sheet zinc, cask lots, 100 lb	7 50
Broken lots	8 00
Pig lead, 100 lb.	6 00
Wrought pipe, black up to 2 inch	50 an 10 p.c.
" Over 2 inch	50 p.c.
Rope, sisal, 7-16 and larger	$10 00
" ⅜	10 50
" ¼ and 5-16	11 00
Manila, 7-16 and larger	13 50
" ⅜	14 00
" ¼ and 5-16	14 50
Solder	21½
Cotton Rope, all sizes, lb.	16
Axes, chopping	$ 7 50 to 12 00
" double bitts	12 00 to 18 00
Screws, flat head, iron, bright	87½
Round " "	82½
Flat " brass	80
Round " "	75
Coach	57½ p.c.
Bolts, carriage	55 p.c.
Machine	55 p.c.
Tire	60 p.c
Sleigh shoe	65 p.c.
Plough	40 p.c.
Rivets, iron	50 p.c.
Copper, No. 8	35
Spades and shovels	40 p.c.
Harvest tools	50, and 10 p.c.
Axe handles, turned, s. g. hickory, doz.	$2 50
No. 1	1 50
No. 2	1 25
Octagon extra	1 75
No. 1	1 25
Files common	70, and 10 p.c.
Diamond	60
Ammunition, cartridges, Dominion R.F.	50 p.c.
Dominion, C.F., pistol	30 p.c.
" military	15 p.c.
American R.F.	30 p.c.
C.F. pistol	5 p.c.
C.F. military	10 p.c. advance.
Loaded shells :	
Eley's soft, 12 gauge black	16 50
chilled, 12 guage	18 00
soft, 10 guage	21 00
chilled, 10 guage	23 00
Shot, Ordinary, per 100 lb	6 75
Chilled	7 50
Powder, F.F., keg	4 75
F.F.G.	5 00
Tinware, pressed, retinned	75 and 2½ p.c.
" " plain	70 and 15 p.c.
Graniteware, according to quality	50 p.c.

PETROLEUM.

Water white American	25½c.
Prime white American	24c.
Water white Canadian	22c.
Prime white Canadian	21c.

PAINTS, OILS AND GLASS.

Turpentine, pure, in barrels	$ 68
Less than barrel lots	73
Linseed oil, raw	83
Boiled	86
Lubricating oils, Eldorado castor	25½
Eldorado engine	24½
Atlantic red	27½
Renown engine	41
Black oil	23½ to 25
Cylinder oil (according to grade)	55 to 74
Harness oil	61
Neatsfoot oil	$ 1 00
Steam refined oil	85
Sperm oil	1 50
Castor oil per lb.	11½
Glass, single glass, first break, 16 to 25 united inches	2 25
26 to 40 per 50 ft.	2 50
41 to 50 " 100 ft.	5 50
51 to 60 " " "	6 00
61 to 70 per 100-ft. boxes	6 50
Putty, in bladders, barrel lots per lb.	2½
kegs "	2¾
White lead, pure per cwt.	7 25
No 1 "	7 00
Prepared paints, pure liquid colors, according to shade and color..per gal.	$1.30 to $1.90

SARNIA BOARD OF TRADE.

The annual meeting of the Sarnia, Ont., Board of Trade was held on Wednesday last week, the president, Randal Kenny, in the chair. The principal matters considered were proposals to have the exhibition grounds of the Agricultural Society moved nearer the business portion of the town and to secure the erection of a 500,000 bushel elevator. Committees were appointed to consider both matters.

The following officers were elected for 1901 :

President—A. D. McLean.
1st Vice-President—Thos. Symington.
2nd Vice-President—David Milne.
Secretary—Col. C. S. Ellis.
Treasurer—Fred J. Winlow.
Council—Thos. Kenny, D. McCart, Jas. Watson, W. F. Lawrence, Dr. Poussette, W. J. Wiggins, John Cowan, Randal Kenny and Mayor Logie.

Chas. A. Quick's general store at Kingsville, Ont., was destroyed by fire Friday last week. The loss on the building is $6,000 ; insured for $4,000; loss on stock, $16,000 ; insured for $10,000. The post office and contents in rear of the store were also destroyed. The loss on plate glass in opposite stores will reach $1,000, partly insured. The fire probably originated from the heavy pressure of natural gas in the furnace.

"OWASSO"

Screen Doors and Windows

Are all carefully made from best materials.

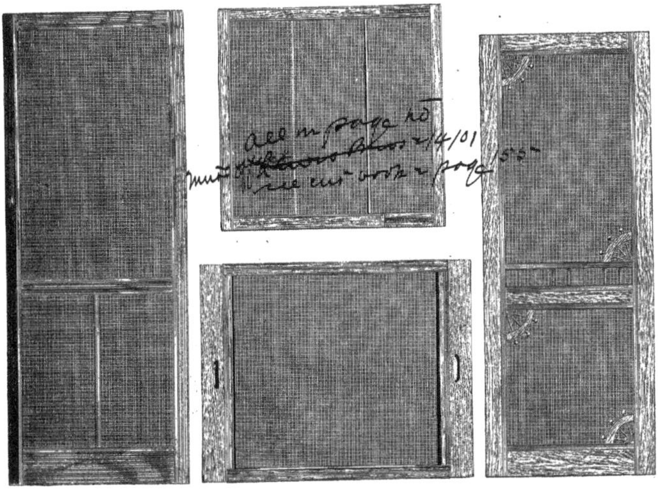

—ALSO A FULL LINE OF—

Hinges and Catches for Screen Doors.

LEWIS BROS. & CO.

WRITE FOR PRICES. Montreal, Que.

TRADE IN COUNTRIES OTHER THAN OUR OWN.

THE monthly pig iron statistics presented by The Iron Age show a heavy increase in the active capacity of blast furnaces in the United States during January, caused partly by the starting of new furnaces and by the blowing-in of stacks in the Central West. The number of furnaces in blast on February 1 was 271, as compared with 233 on January 1, and their weekly capacity was 278,258 gross tons, an increase for January of about 28,000 tons weekly, or 110,000 tons monthly. This brings the current production of pig iron nearly up to the capacity shown on July 1 of last year, and makes it equivalent to a rate of about 14,000,000 tons per year. A significant fact, however, is that along with this heavy increase in the output of pig iron the furnace stocks on February 1 show a falling-off of about 6,000 tons, as compared with January 1.

NEW YORK METAL MARKETS.

TIN — The London tin market opened weak and in the morning a further decline of 10s. was recorded, but under more liberal buying the market recovered, advancing to £121 10s. by noon and subsequently going 7s. 6d. higher, the close being 12s. 6d. above last night's quotation. Notwithstanding the stronger London cables, the New York market was easy. There was little or no speculative interest manifested and brokers complained of a lack of orders, as the consuming demand is very poor. The market closed dull at a decline of about 25 points, 26.50c. being bid and 26.80c. asked for spot and February, while there were sellers of May at 26.20c.

COPPER—There was a decided drop in London quotations to-day, spot showing a decline of 7s. 6d., while futures were 8s. 9d. lower. The close, however, was steady. In New York trade was very slow and prices somewhat nominal at 17c. for Lake Superior and 16⅝c. for electrolytic and casting. The consuming branches are reported to be extremely dull. One of the largest sales of sheet copper made in some time was completed yesterday and consisted of 40,000 lb. The Pennsylvania railroad was said to be the buyer, and the price paid was understood to be 19½c.

PIG LEAD—There was no change in the situation here or in St. Louis. London cables report a further decline of 2s. 6d. to the low price of £13 17s. 6d., which is equivalent to 3c. c. and f.

SPELTER—The tendency of the local market is still downward, and prices were 2⅝ points lower, at 3.95 to 4c. St. Louis was quiet, with 3.82½c. nominally quoted. A further decline of 2s. 6d. occurred in London.

ANTIMONY—For regulus there is a moderate demand, which is supplied at 8¼ to 10¼c., as to brand.

OLD METALS—Heavy cut copper and wire are firmer, as is also heavy brass, but in other lines there is no change, and we hear of little business.

IRON AND STEEL—Pending the settlement of the final details of the organization now known as the United States Steel Corporation, the market for all steel products is very firm, and it is confidently expected that a higher range of prices will be established very soon. Activity in all branches is reported. Billets are difficult to obtain for immediate shipment and still command a premium of $1 to $1.50 per ton. Pittsburg reports an increased demand and higher prices for iron and steel scrap and also an active business in finished material. Advices from Chicago are to the effect that pig iron is firmer there mainly on account of the more cheerful news from other trade centres. In finished material also a fairly-active demand has replaced the hesitating policy of buyers in that market. Cable advices show a continued decline in the English markets, pig iron warrants being down to 53s. 10d. in Glasgow, while No. 3 foundry in Middlesboro is now quoted at 46s.

TIN PLATE—The market is steady and unchanged under a good consuming demand.—New York Commercial Advertiser, February 27, 1901.

EARLY CLOSING ON SATURDAYS.

The Lord's Day Alliance, at the request of a number of Ottawa clerks, are circulating petitions to have the shops close earlier Saturday nights than at present. It is proposed to have the stores close at 9 or 9.30 o'clock. It is said that at present the stores are kept open too late and that the clerks are completely fatigued by the time closing hour arrives. It is held by the Alliance that if the stores were to close at 9 o'clock they would be the gainers. The bulk of the shopping is done before or up to that time, and the majority of those doing shopping later are those who go for a walk first and leave their shopping until going home.

REGINA BOARD OF TRADE.

At the annual meeting of the Regina, N.W.T., Board of Trade, the report showed that the principal work of the board last year had been the operation of a well borer. This had proved beyond all question that there is much water in the district around Regina, which knowledge had lead to the incoming of many settlers. Over 1,000 homestead entries had been made during the year, the largest number made at any point of entry in Manitoba or the Northwest. Delegates to the district had been entertained and leaflets distributed to possible settlers.

The following officers were elected :

President—J. W. Smith.
Vice-President—G. Michaelis.
Secretary—Wm. Trant.
Council—B. Spring-Rice, R. H. Williams, J. K. McInnis, J. M. Young, R. Martin, W. Molland, H. Armour, F. N. Darke and J. F. Bole.

GRINDING COFFEE BY ELECTRICITY

THE progressive and up-to-date grocer is constantly on the alert for fixtures that will lessen the labor of himself and his clerks, and at the same time add to the appearance of his store, which in the new century is essential.

The "Enterprise" electrically connected rapid-grinding and pulverizing mill was designed to meet the demand for a mill that would either granulate or pulverize and do so at a minimum cost. Twenty-five lb. of coffee can be ground ordinarily fine for one

cent. They are made in a variety of sizes and styles, and the grinding capacity is from 3 to 10 lb. per minute, according to size.

The machines are fitted with motors for either direct or alternating current and to suit the conditions existing in any locality. Those for direct current are so constructed as to enable the operator to instantly change the speed from fast to slow or vice versa. The fast speed is intended to be used only when coffee is to be granulated and the slow speed for pulverizing. The motor and mill have a direct connection, thus reducing loss of power to a minimum, which is a considerable saving as compared with mills run by motor having a belt connection.

Hardwaremen should try and induce grocers to use these machines.

GOOD THINGS ARE IMITATED.

ANY new inventions put on the market in these progressive times soon find many imitators, should they meet with the public approval. About two years ago, a leading Toronto manufacturer placed an adjustable show and display table on the market, made in various styles and sizes. They took so well with the general trade and were so highly commended by all who used them that half-a-dozen imitations were placed on the market in the short space of six months' time.

A conversation was overheard, a few days ago, in a large eastern city, regarding the merits and durability of Boeckh's adjustable table. The merchant called the attention of a group of travellers who were in his store to a table sent him on trial, remarking: "I have only had this in the store three weeks, and it has been to the repair shop once to have the legs repaired, and now has to go again to get shelves fixed, which are giving away under the weight of ordinary goods. Now, look at this one (pointing to Boeckh's), which I have had for two years and is in better shape after all kinds of rough usage than all the so-called improved ones I have seen on the market."

This is, no doubt, a good recommendation for Boeckh Bros. & Company's show-stand over other makes, but this is simply given in order to illustrate the result of buying inferior imitations, as it pays to buy the best in any line of goods, and they are bound to be the cheapest in the end.

SPRING AND SUMMER GOODS.

Lewis Bros. & Co., Montreal, have just issued Catalogue and Price List No. 27, which deals with spring and summer goods for 1901. The appearance of the book is neat and tasty, but its most remarkable feature is its compactness, all the information that a merchant could want about these goods being crowded into 38 pages. Coated poper has been employed in the make up of the publication, and the cuts are brought out distinctly and effectively. The index on the back page makes the work handy for reference purposes. Dealers can secure copies on application. In a short time Mr. Tamlin expects to have a catalogue of guns and ammunition before the firm's customers, and he has begun work on a large reference catalogue, which he intends shall startle the trade.

SUDBURY BOARD OF TRADE.

The following officers were elected at the annual meeting of the Sudbury, Ont., Board of Trade:

President—D. Baikie, reelected.
Vice-President—Jas. A. Orr.
Secretary—A. Fournier.
Treasurer—J. Purvis.
Council—T. J Ryan, Jas. Purvis, S. E. Wright, J. S. Gill, J. F. Black, F. Cochrane, R. Martin.

Lalonde & Lalonde are opening a general store at Plantagenet Springs, Ont.

HEATING AND PLUMBING

VAPOR PANS IN FURNACES.

THE question of vapor pans in furnaces is always an interesting one. No apology, therefore, is offered for reproducing the following from the pen of a correspondent of The Metal Worker :

"In Philadelphia a few years ago it was a common custom to make the best furnaces with an octagon shaped cast iron base. In one of the sides of this base, at the front, it was the common practice to place the water pans at the side of the ash pit. This made it convenient for filling and placing it where it could always be seen. Now, when the cold air supply was taken from out of doors, which 10 or 15 years ago was the exception and not the usual method by any means in Philadelphia, it was not uncommon to have the water pan freeze until it burst while the furnace was heating the house satisfactorily to the owner. Some people who had

SIMILAR EXPERIENCE,

and found that the water pans located at this point did not get warm enough to evaporate much water, resorted to an expedient which proved a benefit. They would get a piece of cast or wrought iron 1½ or 2 inches in diameter, or square, and put one end in the water pan and the other end against the fire pot of the furnace. The end against the fire pot naturally got hot and the heat was conducted down into the vapor pail, as it was called, warming the water so that a vapor would rise and moisten the air passing through the furnace. Just as soon as some expert furnace workman was called upon to make repairs to a furnace so equipped he would wonder what chump put that piece of iron there, and would remove it, as it had no business there, in his opinion.

TO OVERCOME FREEZING.

"The next step to overcome the freezing and to provide more efficient moistening was the manufacture of a long narrow water pan put into the top of the furnace, resting on the radiator of the furnace, with the end coming through the front of the furnace so that it could be readily filled. The common experience in filling was to pour the water in until it overflowed on the hot furnace. In some instances radiators have been cracked by the overflow of the water and gas let into the building, to say nothing about the dust stirred and steam generated while the catastrophe was in full bloom. A little care would prevent this calamity, but a similar disaster resulted from the vapor pan getting empty and becoming excessively hot and some member of the family, from a sense of duty, would fill the water pan, and instead of pouring the water in a little at the time, would pour it in copiously, when the pan would crack and the furnace receive a bath. Some people, in order to prevent this trouble, had a water pan inside of the furnace, and one outside the furnace, connected by means of pipes. In the outer pan or tank they would place a ball cock connected with the regular water service of the building. By this means the furnace was always supplied with water to be vaporized according to the strength of the fire. The advantage of the long furnace pan running well across the furnace casing was that all of the hot air pipes leading from the furnace stood a better show of having the air which flowed from them vaporized than when a comparatively small vapor pan was located on one side of the furnace.

"A few years ago, when natural gas was used in Western Pennsylvania, it was looked upon, when burnt in furnaces, as furnishing a dry heat, and a demand was developed for some method of

MOISTENING THE AIR

which flowed through each pipe. This resulted in the invention by a Western Pennsylvania furnaceman of what he called a vapor pan ring. This ring rested on the top of the outer casing of the furnace, and increased the diameter of the furnace body 3 inches on each side. This pan was 3 or 4 inches deep, and was located where the air in the cellar prevented its being overheated, and yet it was sufficiently warm to evaporate freely and furnish the additional moisture which is necessary to all the air flowing from each pipe of a hot-air furnace working properly. This vapor pan ring had another advantage—that of increasing the circumference of the furnace top, so that when a number of pipes were taken from the furnace there was more room to place them so that they would work advantageously without the danger of robbing each other, as sometimes happens when pipes are closely connected. Having given this history, so to speak, of the practice in the use of vapor pans, I hope that old furnacemen who know anything that has escaped my attention will bring it to the attention of 'T. W. F.'

A SUMMING UP.

"Having discussed the different phases of the question, as the lawyers say, I will 'sum up in an opinion.' It is my opinion that the proper place for locating the vapor pan in a furnace depends somewhat upon the construction of a furnace. If it is one of the type which works on the slow combustion theory, without excessively hot surfaces and without overheating the air, the long narrow pan reaching from one side of the furnace to the other at the top has its advantages, but in order to avoid the danger of its overflowing and the furnace being cracked when it is filled, it should by all means be connected with an outside tank supplied with water automatically by means of a ball cock. If the furnace is one of the red hot type it should be replaced by one having more surface of the indirect draft construction working at a lower temperature, which all give the vapor pan a proper chance. If moisture is not supplied artificially the dry air will extract it from furniture and the membranes of the people living in it. Frosty or wet windows will not please the eye, but the conditions which produce them will not cause personal discomfort to the body. Just a little less moisture in the air than this is the ideal amount."

HAD TO PAY PLUMBER'S BILL.

On Thursday, a suit brought by J. H. Parkes, plumber, against Wm. Milne, was settled in Toronto. The suit was for $53.68, and was brought by Mr. Parkes for plumbing work done and not paid for. The defence claimed that the work had been dragged along over the renting season, thus entailing a loss of about $40. This the plaintiff denied, claiming that the house, No. 33 Stewart street, was in an unfinished condition, which greatly interfered with the progress of the work.

A counter claim of $62.50, put in by the defendants, was disallowed, and judgment for $53 68 with costs was given the plaintiff by Judge Morson.

BUILDING PERMITS ISSUED.

The following building permits have been issued in Quebec: Paul Simard, for construction of dwelling on St. Valier street, estimated cost of which is $800 ; Mrs. Gunner, for repairing of property on Mountain Hill to extent of $2,500 ; J. I. Laroche, for construction of two-storey brick dwelling, the estimated cost of which is $2,500 ; Luc Parent, for repairing of property on Crown street, to extent of $2,000 ; Sir C. A. P. Pelletier, for repairing of property on Buade

street, to extent of $100; O. Belanger, for repairing of dwelling on St. Therese street, to extent of $200, and to D. McGee, for construction of a three storey brick house on the ramparts, the estimated cost of which is $10 000.

Building permits have been issued in Toronto to Mrs. M. Colwell, for a residence on Ossington avenue, to cost $1,200; to Mrs. Gillespie, for a residence at 48 Brooklyn avenue, to cost $1,400, and to Dr. Sheard, for an addition to his residence at Hanlan's Point, to cost $1 900.

SOME BUILDING NOTES.

THE land for a new theatre has been bought on St. Catherine street east, Montreal, and the building will likely be erected in the spring.

A new school is to be erected at Oakland, Man.

Mrs. McLeod, Red Deer, N.W.T., is adding a new wing to her house.

The trustees of S.S. No 9, Conroy, Ont., intend erecting a school this summer.

John Carroll, Middlemiss, Ont., is making preparations for building a brick dwelling house.

A company has been formed to erect a five-storey summer hotel at Chester, N.S., to cost $40,000.

J. & E. Brown have purchased the Lafferty block, Portage la Prairie, Man., and are now renovating it.

The Peter McSweeney Co., Moncton, N.B., intend erecting a three-storey block in that town early this summer.

W. Doherty, chairman of the building committee of the Methodist church, Clinton, Ont., is calling for tenders for a new church.

Improvements are being made to the stores of Anderson & Langstaff, Fraser Bros., J. R. Wallace and Miss Courtney, Kemptville, Ont.

A $2,000 addition is to be built at the rear of Murphy & Co.'s store on Sparks street, Ottawa. John O'Connor, Ottawa, has the contract.

Tenders have been received for a new hospital at Lindsay, Ont., but contracts have not yet been let. Taylor & Gordon, Montreal, are the architects.

The D. F. Brown Paper Co. have received tenders for the erection of a three-storey brick warehouse and factory on Canterbury street, St. John, N.B.

The ratepayers of Aurora, Ont., on Monday passed a bonus by-law of $10,000 and tax exemption for 10 years to Underhill & Sisman, boot manufacturers, Markham, Ont. A new factory will consequently be erected at once.

PLUMBING AND HEATING CONTRACTS.

The Bennett & Wright Co., Limited, have the contract for plumbing, heating, gas-fitting and electric wiring in a warehouse on Colborne street for R. Northcote.

J. Lamarche, Montreal, has been awarded the contract for roofing, gas and electric lighting, plumbing, heating and ventilating of St. Eusebe Catholic School. Mr. Lamarche intends to install an improved system of ventilation, and its trial will be watched with interest.

PLUMBING AND HEATING NOTES.

Alex. Newlands, contractor, Kingston, Ont., is dead.

Labelle & Labelle, have registered as plumbers in Montreal.

Milton S. Ruhland, contractor, Halifax, will open a branch in Sydney.

Alex. Desmarteau has been appointed curator of Chapleau & Leboeuf, contractors, Montreal.

I. H. Breck, dealer in electrical supplies, Kingston, Ont., has been succeeded by Breck & Halliday.

HEIGHT OF LIQUID IN A CASK.

In order to ascertain how far the liquid reaches in a keg, says Deutsche Destillateuren Zeitung, the following simple method may be employed:

"Take a glass tube, bent at right angles, whose long leg is equal to the height between the bunghole and the upper floor, while the shorter one need only be a few inches in length. The shorter end is now connected with the bung by a piece of rubber hose; the longer one is placed in a vertical position and the bung is opened. According to the law of communicating vessels, the liquid will rise in the tube to exactly the same height as in the cask, so that the level of the fluid can be ascertained with great accuracy."

THE BINDER TWINE TRADE.

Editor HARDWARE AND METAL,—In your issue of this date I noticed an article entitled "The Rage for Binder Twine Factories."

The reason why "Canada imports 60 per cent. of the twine she uses" is because the American twine is finer, stronger, and holds out the length per lb. that it is guaranteed to do, and the Canadian farmer being in most instances an up-to-date, progressive man knows a good thing when he uses it! Let the Canadian binder twine manufacturer make as good as article as his American competitor and he can hold the Canadian market; otherwise it will get away from him, "it may be for years, it may be forever." WANDERER.

MONTREAL, February 23, 1901.

HENRY ROGERS, SONS & CO.,

SHEFFIELD, ENGLAND

Manufacturers of the celebrated

H R S & Co

UNION JACK CUTLERY

We make a specialty of

PLATED WARE, FRUIT KNIVES, ETC.

Our Canadian Representative carries a full line of samples.

Canadian Office:

6 St. Sacrament St., MONTREAL

F. A. YORK, Manager.

If you want the BEST

Cold Blast Lantern

in the world, get WRIGHT'S.

—GET THE ORIGINAL—

We lead, others imitate.

E. T. WRIGHT & CO.

Manufacturers. HAMILTON, ONT.

TRADE MARK J REGISTERED.

"JARDINE" HAND DRILLS

Five different sizes to suit all requirements. It pays to sell the best tools.

A. B. JARDINE & CO.

HESPELER, ONT

BERGER'S PATENT PIPE FASTENER.

7 STYLES. . . .

No. 0, for brick, 3 inch.
No. 1, for brick, 4½ inch.
No. 2, for stone, 6 inch.
No. 3, for stone, 9 inch.
No. 4, for wood, 3 inch.
No. 5, for wood, 5 inch.
No. 6, to nail on wood.
No. 7, to screw in wood.

BEST ON EARTH. . . .

Pipe any distance desired from wall.
Concealed from view.
Handy to put up.
Solid as one piece.
No close measuring.
Send for sample FREE.

SEND FOR CATALOGUE OF TINNERS' AND ROOFERS' SUPPLIES.

BERGER BROS. & CO., 231 and 237 Arch St., PHILADELPHIA.

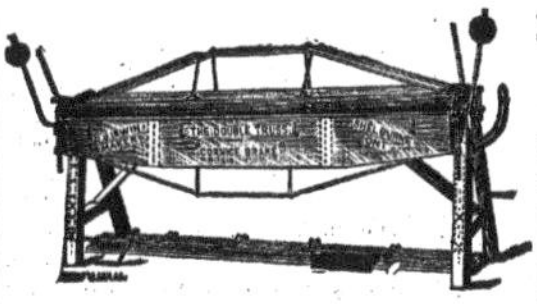

This eight-foot Brake bends 22-gauge iron and lighter, straight and true.

Price, $60

Very handy beader attachment, $15 extra if required.

Send for circulars and testimonials to

The Double Truss Cornice Brake Co. SHELBURNE, ONT.

The Latest and Best.

H. & R. Automatic Ejecting Single Gun.

Steel and Twist Barrels in 30 and 32-inch. 12 Gauge.

Model 1900.

Harrington & Richardson Arms Co.

Worcester, Mass., U.S.A.

Descriptive Catalogue on request.

STEVENS ARMS

MAKE AN EXCEPTIONALLY FINE LINE FOR THE HARDWARE TRADE

All Carefully Boxed—Shelf Goods in fact.
All Fully Guaranteed—By a company of 37 years' experience and reputation.
All Fine Sellers—Not a line which costs money to carry and moves slowly.

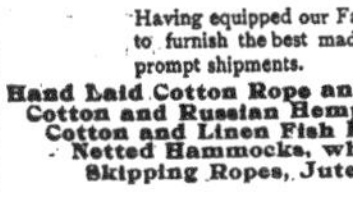

The Single Shot Pistol.

Full Nickel Plated, extremely accurate, 3½-Inch, Blued Barrel.

No. 1—22 Short R. F., List Price $2.50.

Send for Complete Catalogue to

J STEVENS ARMS & TOOL CO., P.O. Box 217, Chicopee Falls, Mass., U.S.A.

New York Office, 318 Broadway.

We desire to correspond with the trade everywhere

HUTCHISON, SHURLY & DERRETT

DOVERCOURT TWINE MILLS.

1078 BLOOR STREET WEST TORONTO.

Having equipped our Factory with entirely new machinery, we are prepared to furnish the best made goods in the market at closest prices and make prompt shipments.

Hand Laid Cotton Rope and Clothes Lines,
Cotton and Russian Hemp Plough Lines, plain and colored.
Cotton and Linen Fish Lines, laid and braided.
Netted Hammocks, white and colored, Tennis and Fly Nets.
Skipping Ropes, Jute, Hemp and Flax Twines.

CONVENTION WEEK IN WINNIPEG.

YEAR by year bonspiel time, with its cheap rates, has come to be looked upon as convention week, and between Monday, February 18, and Saturday the 23rd, some nine conventions met and transacted business. Many of these were the annual meetings of fraternal orders, and, in addition, the Manitoba Dairy Association, Sheep and Swine Breeders; Pure Bred Cattle Breeders' Association; Horse Breeders; Veterinary Association, Western Retail Lumber Dealers' Association, and others.

Among the speakers of note at the dairy and stock conventions was Prof. James W. Robertson, Dominion Agricultural and Dairy Commissioner; Prof. Day, of Guelph Agricultural College, and Alex. Galbraith, secretary of American Clydesdale Association. The meetings were all of a practical and helpful character and were extremely well attended.

The Winnipeg Industrial Exhibition Association held its annual meeting on Thursday, February 21, when Mr. F. W. Thompson, general manager in the West of the Ogilvie Milling Company, was elected president.

Mr. P. G. Van Vleet, proprietor and publisher of The Canadian Implement Trade, was a visitor during the week and attended a gathering of implement men.

Mr. Allan, of the Oneida Community Company, was in the city during the week looking after business interests here. Mr. Allan was on his way to the Coast and California. He reported trade in his line of goods fairly satisfactory.

A CHANGE OF PARTNERSHIP.

The well-known firm of Geo. Stephens & Co., Chatham, Ont., have made a change in their business and have added another partner to the firm, which will now be styled Geo. Stephens, Quinn & Douglas.

Mr. D. H. Douglas was a member of the firm of Tait & Douglas, hardware merchants of Campbellford, Ont., and brings to the concern to which he has been admitted as partner an experience of over 15 years in the hardware business. Mr. Douglas will have charge of the office and financial department of the business.

Mr. Stephens, the senior member of the firm, who is the originator and founder of this well-known business will still retain his interest in the company, and when not occupied with his Parliamentary duties, will continue to devote his time and energy to the management of the business.

Mr. Fred F. Quinn, the other member of the firm, who has had a long experience in the hardware business, and who is a practical tinsmith, sheet iron and metal worker, will have charge of the general management of the business during the absence of Mr. Stephens, and will, in every legitimate way, try to keep up the very enviable reputation of the firm, which Mr. Stephens, by his perseverance, integrity and constant attention to business has established.

It will be their aim in the future, as in the past, to keep only such goods as have gained a reputation for "best quality only," and in no case will they sacrifice quality for price. This has been the secret of their success in the past, and they will strictly adhere to this principle in the future.

THE DODGE CALCULATOR.

HARDWARE AND METAL is in receipt of a ingenious device called the "Dodge Calculator," issued by the Dodge Manufacturing Co., Toronto.

This device is not only a novelty but is also an instrument of considerable value and assistance to all mechanics, foremen, superintendents, etc., whose duty it is to figure up speeds of pulleys, gears, etc. We are informed that the Calculator is one illustration of the many uses to which the slide rule principle may be applied.

The Dodge Manufacturing Co. will be pleased to mail free, for the asking, the "Dodge Calculator."

The Wm. Hunter Co., Limited, general merchants, Silverton, B.C., expect to open a branch establishment at Phœnix, B.C., this week.

CATALOGUES, BOOKLETS, ETC.

A USEFUL BOOK.

About six months ago Mr. Godfrey S. Pelton, Montreal, made a gratuitous distribution among his customers of a "Ready Reckoner" he had published on his own account. Experience has proven it to be one of the handiest guides the Canadian importers can turn to to-day. It figures out on a scale of first cost prices just what it would cost to lay down chain, Canada plate, black sheets, hoop iron, galvanized iron, tinplates, sheet zinc, sheet lead, lead pipe, linseed oil and white lead ground in oil, on the Montreal market. To figure out from an f.o.b. cost in England, add on primage, insurance, freight, wharfage and duty charges is a task of no small concern, and Mr. Pelton is to be congratulated upon engratiating himself in the hearts of the importers by presenting them with his neat, compact and ready "Reckoner." It has been tried and has not been found wanting.

A PRETTY LANDSCAPE.

The Bourne-Fuller Co., manufacturers of iron, steel and pig iron, Cleveland, are this year sending out a very pretty calendar. It is somewhat different from the usual run of calendars that firms send out to their customers each year. It has a fine reproduction of a painting of a landscape, which measures 10 in. wide by 7 in. deep. Underneath this is pasted the calendar. There is really more picture about it than calendar, and it is worthy of a prominent position in any merchant's store.

CURRENT MARKET QUOTATIONS.

March 1, 1901.

These prices are for such qualities and quantities as are usually ordered by retail dealers on the usual terms of credit, the lowest figures being for larger quantities and prompt pay. Large cash buyers can frequently make purchases at better prices. The Editor is anxious to be informed at once of any apparent errors in this list as the desire is to make it perfectly accurate.

METALS.

Tin.

Lamb and Flag and Straits—
56 and 28 lb. ingots, per lb. 0 32 0 33

Tinplates.

Charcoal Plates—Bright

M.L.S., equal to Bradley. Per box
I.C., usual sizes $6 75
I.X., " 8 25
I.X.X., " 9 75

Famous—
I.C. 6 75
I.X 8 25
I.X.X 9 50

Raven & Vulture Grades—
I.C., usual sizes 5 00
I.X., " 6 00
I.X.X " 7 00
I.X.X.X., " 8 00
D.C., 12½x17 .... 4 75
D.X. 5 50
D.X.X 7 50

Coke Plates—Bright

Bessemer Steel—
I.C., usual sizes 4 15
I.C., special sizes, base 4 50
20 x 28 .... 8 50

Charcoal Plates—Terne

Dean or J. G. Grade—
I.C., 20x28, 112 sheets 8 50
I.X., Terne Tin 10 50

Charcoal Tin Boiler Plates.

Cookley Grade— Per lb.
X.X., 14x56, 50 sheet bxs }
" 14x60 " } 0 07 0 07½
" 14x65 " }

Tinned Sheets

72x30 up to 24 gauge 0 08 0 08½
" 26 " 0 08½ 0 09
" 28 " 0 09 0 09½

Iron and Steel.

Common Bar, per 100 lbs. 1 65 1 70
Refined " " 2 15
Horse Shoe Iron " 2 05
Hoop steel, 1½ to 3 in. base, extras for smaller sizes 3 10
Sleigh Shoe Steel " base 2 00
Tire Steel 2 00
Machinery iron finish 2 05
Cast Steel, per lb 0 00 0 00
Toe Calk Steel 2 37
T. Firth & Co's special cast steel, per lb. 0 13
Jessop's Tool Steel 0 13

Boiler Tubes.

1½-inch 0 12½
2 " 0 13
2½ " 0 15
3 " 0 16
3½ " 0 20
4 " 0 25

Steel Boiler Plate.

¼ inch 2 25
3-16 inch 2 25
⅜ inch and thicker 2 25

Black Sheets.

8 gauge 3 00
20 gauge 3 00
22 to 24 " 3 10
26 " 3 20
28 " 3 33

Canada Plates.

All dull, 52 sheets 3 00
Half polished 3 15
All bright 3 65 3 75

Iron Pipe.

Black pipe—
¼ inch 3 00
⅜ " 3 00
½ " 3 15
¾ " 3 20
1 " 4 30
1¼ " 6 35
1½ " 7 55
2 " 10 10
2½-6 inch, discount 60 p.c.

Galvanized pipe—
½ inch 4 65
¾ " 5 35
1 " 7 75
1¼ " 9 75
1½ " 11 25
2 " 15 50

Galvanized Sheets.

	G. C.	Comet.	Amer.	Queen's Head.
16 gauge		....	4 15	4 00
18 to 24 gauge	4 25	3 85	4 35	4 25
26 "	4 50	4 10	4 35	4 50
28 "	4 75	4 35	4 50	4 75

28 gauge American equals 26 gauge English.

Chain.

Proof Coil, 3-16 in., per 100 lb ....
" ¼ " " 8 00 8 50
" 5-16 " " 5 35 5
" ⅜ " " 4 35 4 85
" 7-16 " " 4 15 4 65
" ½ " " 4 35 4 50
" ⅝ " " 3 85 4 35
" ¾ " " 3 85 4 00

Halter, kennel and post chains, 40 and 50 p.c.
Cow ties .... 40 p.c.
Stall fixtures .... 35 p.c.
Trace chain .... 25 p.c.
Jack chain, iron, single and double, discount 35 p.c.
Jack chain, brass, single and double, discount 40 p.c.

Copper.

Ingot

English B. S., ton lots 0 19 0 20
Lake Superior ....

Bolt or Bar.

Cut lengths round, ½ to ⅞ in. 0 23½ 0 25
" round and square 1 to 2 inches 0 23½ 0 25

Sheet.

Plain, 14 oz., and light, 16 oz., 14x48 and 14x60 .... 0 23 0 23
Plain, 14 oz., and light, 16 oz., irregular sizes 0 23 0 23½
Tinned copper sheets 0 26
Planished 0 32

Braziers (In sheets.)

4x6ft. 25 to 30 lbs. ea., per lb. 0 25½
" 35 to 45 " " 0 24½
" 50-lb. and above, " 0 23½

Boiler and T. K. Pitts

Plain Tinned, per lb 0 28
Spun, per lb. 0 32

Brass.

Rod and Sheet, 14 to 30 gauge, 15 per cent.
Sheets, hard-rolled, 2x4 .... 0 24 0 25
Tubing, base, per lb 0 24 0 25

Zinc Spelter

Foreign, per lb 0 06 0 06½
Domestic " ....

Zinc Sheet.

5 cwt. casks 6 75 7 00
Part casks 7 00 7 50

Lead.

Imported Pig, per lb 0 04¾ 0 05
Bar, 1 lb. 0 05½ 0 06¼
Sheets, 2½ lbs. sq. ft., by 0 06¼
Sheets, 3 to 6 lbs., " 0 06

NOTE.—Cut sheets ½ cent per lb. extra.

Pipe, by the roll, usual weights per yard, list at 7c. per lb. and 15 p.c. dis. f.o.b. Toronto.

NOTE.—Cut lengths, net price, waste pipe 8-ft. lengths lists at 7½ cents.

Shot.

Common, $6.50 per 100 lb.; chilled, $7.00 per 100 lb.; buck, seal and bal, $7.50. Discount, 7½ p.c. Prices are f.o.b. Toronto, Hamilton, Montreal, St. John and Halifax. Terms 3 per cent. cash, freights equalized on Montreal.

Soil Pipe and Fittings.

Discount, 60 and 10 per cent. on medium and extra heavy, and 60 per cent. on light.

Solder. Per lb. Per lb.

Bar half-and-half, guarant'd 0 19
Bar half-and-half, commer'l 0 18¼
Refined 0 18½
Wiping 0 18

NOTE.—Prices of this graded according to quantity. The prices of other qualities of solder in the market indicated by private brands vary according to composition.

Antimony.

Cookson's, per lb. 0 11 0 11¾

White Lead. Per 100 lb.

Pure 6 50
No. 1 do 6 12½
No. 2 do 5 75
No. 3 do 5 37½
No. 4 do 5 00
Munro's Select Flake White 7 37½
Elephant and Decorators' Pure 7 12½
Brandram's B B. Genuine 8 00
" " Decorative 7 55
" " No. 1 .... 6 85
" " No. 2 .... 6 00

Red Lead.

Genuine, 560 lb. casks, per cwt. $5 50
Genuine, 100 lb. kegs, per cwt 5 75
No. 1, 560 lb. casks, per cwt 5 25
No. 1, 100 lb. kegs, per cwt 5 00

White Zinc Paint.

Elephant Snow White 0 08 0 09
Pure White Zinc 0 08 0 09
No. 1 .... 0 06 0 07½
No. 2 .... 0 05 0 06½

Dry White Lead.

Pure, casks 5 75
Pure, kegs 6 25
No. 1, casks 5 50
No. 1, kegs 5 00

Prepared Paints.

In ¼, ½ and 1 gallon tins.

Pure, per gallon 1 25
Second qualities, per gallon 1 10
Barn (in bbls.) 0 75 0 85
The Sherwin-Williams Paints 1 45
Canada Paint Co's Pure 1 25
Toronto Lead & Color Co's Pure 1 25
Sanderson Pearcy's Pure 1 20
Stewart & Wood's Champion Pure. 1 20

Colors in Oil.

25 lb. tins, Standard Quality.

Venetian Red, per lb 0 05
Chrome Yellow 0 11
Golden Ochre 0 08
French " 0 05
Marine Black 0 09
" Green 0 09
Chrome " 0 08
French Imperial Green 0 03

Colors, Dry.

Yellow Ochre (J.C.) bbls 1 35 1 40
Yellow Ochre (J.F.L.S.), bbls 2 75
Yellow Ochre (Royal) 1 10 1 15
Brussels Ochre 2 00
Venetian Red (best), per cwt. 1 80 1 90
English Oxides, per cwt 3 00 3 25
American Oxides, per cwt 1 75 2 00
Canadian Oxides, per cwt 1 75 2 00
Super Magnetic Oxides, 93 p.c. 2 00 2 25

Burnt Sienna, pure, per lb 0 01
" Umber, " " 0 10
do Raw 0 09
Drop Black, pure 0 09
Chrome Yellows, pure 0 18
Chrome Greens, pure, per lb. 0 12
Golden Ochre 0 03¾
Ultramarine Blue in 28-lb. boxes, per lb. 0 08 0 24
Fire Proof Mineral, per 100 lb 1 00
Genuine Eng. Litharge, per lb 0 07
Mortar Color, per 100 lb. 1 25
English Vermillion 0 80
Pure Indian Red, No. 45 lb. 0 80
Whiting, per 100 lb. 0 55

Blue Stone.

Casks, for spraying, per lb. 0 07
100-lb. lots, do. per lb 0 08

Putty.

Bulk in bbls. 2 00
Bulk in less quantity 2 15
Bladders in bbls. 2 20
Bladders in kegs, boxes or loose 2 35
Bladders in 25-lb. tins 2 45
Bladders in 12½-lb. tins 2 75
Bladders in bulk or tins less than 100 lb 3 00

Varnishes.

In 5-gal. lots. Per gal.

Carriage, No. 1 .... 2 3 30
" body 8 9 00
" rubbing 4 5 00
Gold Size, Japan 3 3 40
Brown Japan 2 2 80
Elastic Oak 2 3 30
Furniture, extra 2 2 80
" No. 1 .... 1 2 00
Hard Oil Finish 2 3 10
Light Oil Finish 3 3 60
Demar 3 3 70
Shellac, white 4 4 80
" orange 4 4 40
Furniture Brown Japan 1 2 00
Black Japan 2 90 2 80
" " No. 1 .... 1 60 2 00

The Imperial Varnish & Color Co's., Limited Elastilite Varnish 1 gal. can, each. $2.00.

Granatine Floor Finish, per gal. $2.00.

Maple Leaf Coach Enamels; Size 1, 60c.; Size 2, 35c.; Size 3, 20c. each.

Castor Oil.

East India, in cases, per lb. 0 10 0 10½
" " small lots 0 10½ 0 11

Cod Oil, Etc.

Cod Oil per gal 0 50 0 55
Pure Olive 1 20
" Neatsfoot 90

Glue.

Common 0 08½ 0 09
French Medal 0 14 0 14½
Cabinet sheet 0 12 0 13
White, extra 0 18 0 20
Gelatine 0 22 0 30
Strip 0 18 0 20
Coopers 0 19 0 20
Huttner 0 18

HARDWARE.

Ammunition.

Cartridges.

B. B. Caps. Dom. 50 and 5 per cent.
Rim Fire Pistol, dis. 40 p. c., Amer.
Rim Fire Cartridges, Dom., 50 and 5 p. c.
Central Fire Pistol and Rifle, 10 p.c. Amer.
Central Fire Cartridges, pistol sizes, Dom. 30 Per cent.
Central Fire Cartridges, Sporting and Military, Dom., 15 and 5 per cent.
Central Fire, Military and Sporting, Amer. add 5 p.c. to list. B.B. Caps, discount 40 per cent, Amer.
Loaded and empty Shells, "Trap" and "Dominion" grades, 25 per cent. Rival and Nitro, net list.
Brass Shot Shells, 55 per cent
Primers, Dom., 30 per cent.

Wads — per lb

Best thick white felt wadding, in ¼-lb bags 1 00
Best thick brown or grey felt wads, in ¼-lb. bags 0 70
Best thick white card wads, in boxes of 500 each, 12 and smaller gauges 0 99
Best thick white card wads, in boxes of 500 each, 10 gauge 0 35
Best thick white card wads, in boxes of 500 each, 8 gauge 0 55
Thin card wads, in boxes of 1,000 each, 12 and smaller gauges 0 20
Thin card wads, in boxes of 1,000 each, 10 gauge 0 25
Thin card wads in boxes of 1,000 each, 8 gauge ....
Chemically prepared black edge grey cloth wads, in boxes of 250 each— Per M
11 and smaller gauge 0 60
9 and 10 gauges 0 70
7 and 8 gauges 0 90
5 and 6 gauges 1 10
Superior chemically prepared pink edge, best white cloth wads, in boxes of 250 each—
11 and smaller gauge 1 15
9 and 10 gauges 1 40
7 and 8 gauges 1 65
5 and 6 gauges 1 90

Adzes.

Discount, 20 per cent.

Anvils.

Per lb 0 10 0 12½
Anvil and Vise combined 4 50
Wilkinson & Co.'s Anvils..lb. 0 (9 0 09½
Wilkinson & Co.'s Vices..lb. 0 09¾ 0 10

Augers.

Gilmour's, discount 65 and 5 p.c. off list.

Axes.

Chopping Axes—
Single bit, per doz 5 50 10 00
Double bit, " 11 00 18 00
Bench Axes, 40 p.c.
Broad Axes, 33⅓ per cent.
Hunters' Axes 5 50 6 00
Boy's Axes 5 75 6 75
Splitting Axes 6 50 12 00
Handled Axes 7 00 10 00

Axle Grease.

Ordinary, per gross 5 75 6 00
Best quality 13 00 15 00

Bath Tubs.

Zinc 6 00
Copper, discount 15 p.c. off revised list

Baths.

Standard Enameled.

5½-inch rolled rim, 1st quality 25 00
" " " 2nd " 21 00

Anti-Friction Metal.

"Tandem" A .... per lb. 0 27
" B .... " 0 21
" C .... " 0 11½
Magnolia Anti-Friction Metal, per lb. 0 25

SYRACUSE SMELTING WORKS.

Aluminum, genuine 0 45
Dynamo 0 20
Special 0 25
Aluminum, 99 p.c. pure "Syracuse" .. 0 50

Bells.

Hand.

Brass, 60 per cent.
Nickel, 55 per cent.

Cow.

American make, discount 66⅔ per cent.
Canadian, discount 45 and 50 per cent.

Door.

Gongs, Sargent's 5 50 8 00
" Peterboro', discount 45 per cent.

Farm.

American, each 1 25 3 00

House.

American, per lb 0 35 0 40

Bellows.

Hand, per doz 3 35 4 75
Moulders', per doz 7 50 10 00
Blacksmiths', discount 40 per cent.

Belting.

Extra, 60 per cent.
Standard, 60 and 10 per cent.

Bits.

Auger.

Gilmour's, discount 60 and 5 per cent.
Rockford, 50 and 10 per cent.
Jennings' Gen., net list.

Car.

Gilmour's, 47½ to 50 per cent.

Expansive.

Clark's, 40 per cent.

Gimlet.

Clark's, per doz 0 65 0 90
Diamond, Shell, per doz 1 00 1 50
Nail and Spike, per gross 2 25 5 20

Blind and Bed Staples.

All sizes, per lb 0 07½ 0 12

Bolts and Nuts.

Per cent

Carriage Bolts, full square, Norway 70
" " full square 70
Common Carriage Bolts, all sizes 65
Machine Bolts, all sizes 65
Coach Screws 75
Sleigh Shoe Bolts 75
Blank Bolts 65
Bolt Ends 65
Nuts, square .... 4½c. off
Nuts, hexagon .... 4¾c. off
Tire Bolts 67½
Stove Bolts 67½
Stove rods, per lb .... 5½ to 6c.
Plough Bolts 60

Boot Calks.

Small and medium, ball, per M 4 25
Small heel, per M 4 50

Bright Wire Goods.

Discount .... 60 per cent.

Broilers.

Light, dis., 65 to 67½ per cent.
Reversible, dis., 65 to 67½ per cent.
Vegetable, per doz., dis. 37½ per cent.
Henis, No. 8, " 6 00
Henis, No. 9, " 7 00
Queen City " 7 50 0 00

Butchers' Cleavers.

German, per doz 6 00 11 00
American, per doz 12 00 20 00

Building Paper, Etc.

Plain building, per roll 0 30
Tarred lining, per roll 0 40
Tarred roofing, per 100 lb 1 65
Coal Tar, per barrel 3 50
Pitch, per 100-lb 0 85
Carpet felt, per ton 45 00

Bull Rings.

Copper, $2.00 for 2½ in. and $1.90 for 2 in.

Butts.

Wrought Brass, net revised list

Cast Iron.

Loose Pin, dis., 60 per cent.

Wrought Steel.

Fast Joint, dis. 60 and 10 per cent.
Loose Pin, dis. 60 and 10 per cent.
Berlin Bronzed, dis. 70, 70 and 5 per cent.
Gen. Bronzed, per pair 0 40 0 65

Carpet Stretchers.

American, per doz 1 00 1 50
Bullard's, per doz .... 6 50 ....

Castors.

Bed, new list, dis. 55 to 57½ per cent.
Plate, dis. 52½ to 57½ per cent.

Cattle Leaders.

Nos. 31 and 32, per gross 50 9 50

Cement.

Canadian Portland 2 80 3 00
English " 3 00
Belgian " 2 75 3 00
Canadian hydraulic 1 25 1 50

Chalk.

Carpenters, Colored, per gross 0 45 0 75
White lump, per cwt 0 60 0 65
Red 0 05 0 06
Crayon, per gross 0 14 0 18

Chisels.

Socket, Framing and Firmer.

Broad's, dis. 70 per cent.
Warnock's, dis. 70 per cent.
P. S. & W. Extra 60, 10 and 5 p.c.

Churns.

Revolving Churns, metal frames—No. 0, $8—No. 1, $8.50—No. 2, $9.00—No. 3, $10.00 No. 4, $12.00—No. 5, $16.00 each. Ditto wood frames—20c. each less than above. Discounts: Delivered from factories, 58 p.c.; from stock in Montreal, 56 p.c. Terms, 4 months or 3 p.c. cash in 30 days.

Clips.

Axle dis. 65 per cent.

Closets.

Plain Ontario Syphon Jet $16 00
Emb. Ontario Syphon Jet 17 00
Fittings net 1 00
Plain Teutonic Syphon Washout 10 00
Emb. Teutonic Syphon Washout 11 00
Fittings net 1 25
Low Down Teutonic, plain 16 70
" " " embossed 17 00
Plain Richelieu net 3 75
Emb. Richelieu net 4 00
Fittings net 1 25
Low Down Ont. Sy. Jet, plain net .. 19 70
" " " " " emb'd. net 20 50
Closet connection net 1 25
Basins, round, 14 in 1 00
" oval, 17 x 14 in 2 50
" " 19 x 15 in 3 75
Discount 40 p.c., except on net figures.

Compasses, Dividers, Etc.

American, dis. 62½ to 65 per cent.

Cradles, Grain.

Canadian, dis. 25 to 33⅓ per cent.

Crosscut Saw Handles.

S. & D., No. 3, per pair 17½
" " 5, " 22½
" " 6, " 15
Boynton pattern " 20

Door Springs.

Torrey's Rod, per doz .... (15 p.c.) 2 00
Coil, per doz 0 88 1 60
English, per doz 2 00 4 00

Draw Knives.

Coach and Wagon, dis. 50 and 10 per cent.
Carpenters, dis. 70 per cent.

Drills.

Hand and Breast.

Millar's Falls, per doz. net list.

DRILL BITS.

Morse, dis., 37½ to 40 per cent.
Standard dis. 50 and 5 to 55 per cent.

Faucets.

Common, cork-lined, dis. 35 per cent.

ELBOWS. (Stovepipe.)

No. 1, per doz 1 40
No. 2, per doz 1 20
Bright, 20c. per doz. extra.

ESCUTCHEONS.

Discount, 45 per cent.

ESCUTCHEON PINS.

Iron, discount 40 per cent.

FACTORY MILK CANS.

Discount off revised list, 40 per cent.

FILES.

Black Diamond, 70 and 5 per cent.
Kearney & Foote, 60 and 10 to 70 per cent.
Nicholson File Co., 60 to 60 and 10 per cent.
Jowitt's, English list, 25 to 27½ per cent.

FORKS.

Hay, manure, etc., dis., 50 and 10 per cent. revised list.

GLASS—Window—Box Price.

Size United Inches.	Star Per 50 ft.	Star Per 100 ft	D. Diamond Per 50 ft.	D. Diamond Per 100 ft.
Under 26	2 10	4 00	....	6 00
26 to 40	2 30	4 35	....	6 65
41 to 50	....	4 75	....	7 25
51 to 60	....	35	....	8 50
61 to 70	....	75	....	9 85
71 to 80	....	5	....	10 50
81 to 85	....	6 50	....	11 75
86 to 90	....	....	..	14 00
91 to 95	..	....	....	15 50
96 to 100	....	....	....	18 00

GAUGES

Marking, Mortise, Etc.

Stanley's dis. 50 to 55 per cent.

Wire Gauges.

Winn's, Nos. 26 to 33, each 1 65 2 40

HALTERS.

Rope, ⅜ per gross ....
" ½ " 9 00
" ⅝ to ¾ 14 00
Leather, 1 in., per doz 3 87½ 4 00
" 1¼ in., " 5 15 5 20
Web, — per doz 1 87 2 45

HAMMERS.

Nail

Maydole's, dis. 5 to 10 per cent. Can. dis. 25 to 27½ per cent.

Tack.

Magnetic, per doz 1 10 1 20

Sledge.

Canadian, per lb 0 07½ 0 08½

Ball Pean.

English and Can., per lb 0 22 0 25

HANDLES.

Axe, per doz. net 1 50 2 00
Store door, per doz 1 00 1 50

Fork.

C. & B., dis. 40 per cent. rev. list.

Hoe.

C. & B., dis. 40 per cent. rev. list.

Saw.

American, per doz 1 00 1 25

Plane.

American, per gross 3 15 3 75

Hammer and Hatchet.

Canadian, 40 per cent.

Cross-Cut Saws.

Canadian, per pair 0 13½

HANGERS.

doz. pairs.

Steel barn door 5 85 6 00
Stearns, 4 inch 5 00
" 5 inch 6 50
Lane's covered—
No. 11, 5-ft. run 8 40
No. 11½, 10-ft. run 10 80
No. 12, 10-ft. run 12 60
No. 14, 15-ft. run 21 00
Lane's O.N.T. track, per foot. 4½

HARVEST TOOLS.

Discount, 50 and 10 per cent.

HATCHETS.

Canadian, dis. 40 to 42½ per cent.

HINGES.

Blind, Parker's, dis. 50 and 10 to 60 per cent
Heavy T and strap, 4-in., per lb 0 06½
" " 5-in., " 0 06¼
" " 6-in., " 0 06
" " 8-in., " 0 05¾
" " 10-in., " 0 05¾
Light T and strap, dis. 60 and 5 per cent.
Screw hook and hinge—
6 to 12 in., per 100 lbs 4 50
14 in. up, per 100 lbs 3 50

Per gro. pairs.

Spring 12 00

HOES.

Garden, Mortar, etc., dis. 50 and 10 p.c.
Planter, per doz 4 00 4 50

HOLLOW WARE

Discount .... 45 and 5 per cent

HOOKS.

Cast Iron.

Bird Cage, per doz 0 50 1 10
Clothes Line, per doz 0 27 0 63
Harness, per doz 0 72 0 88
Hat and Coat, per gross 1 00 3 00
Chandelier, per doz 0 50 1 00

Wrought Iron.

Wrought Hooks and Staples, Can., dis. 47½ per cent.

Wire.

Hat and Coat, discount 45 per cent.
Belt, per 1,000 .... 0 60
Screw, bright, dis. 55 per cent.

HORSE NAILS.

"C" brand 50 and 7½ p.c. off new list } Oval head.
"M" brand 50, 10 and 5 p.c. }
Countersunk: 50, 10 and 10 per cent.

HORSESHOES

Iron Shoes.	F.O.B. Montreal No. 2 and larger.	No. 1. and smaller
Light, medium, and heavy...	3 50	3 75
Snow shoes..................	3 75	4 00
Steel Shoes.		
Light........................	3 60	3 85
Featherweight (all sizes)......	4 85	4 85

F.O.B. Toronto, Hamilton, London and Guelph, 10c. per keg additional.

Toe weight steel shoes...... 6 70

JAPANNED WARE.

Discount, 45 and 5 per cent. off list, June 1899.

ICE PICKS.

Star per doz.................. 3 00 3 25

KETTLES.

Brass spun, 7½ p.c. dis. off new list.

Copper, per lb.............. 0 30 0 50

American, 60 and 10 to 65 and 5 p.c.

KEYS.

Lock, Can., dis., 45 p.c.

Cabinet, trunk, and padlock, Am. per gross............ 60

KNOBS.

Door, japanned and N.P., per doz........................	1 50	2 50
Bronze, Berlin, per doz......	2 75	3 25
Bronze Genuine, per doz....	6 00	9 00
Shutter, porcelain, F. & L. screw, per gross...........	1 30	4 00
White door knobs—per doz.		1 25

HAY KNIVES.

Discount, 50 and 10 per cent.

LAMP WICKS.

Discount, 60 per cent.

LANTERNS.

Cold Blast, per doz..................	7 50
No. 3 "Wright's"....................	8 50
Ordinary, with O burner	4 25
Dashboard, cold blast	9 50
No. 0..................................	6 00

Japanning, 50c. per doz. extra.

LEMON SQUEEZERS.

	per doz.	
Porcelain lined,..............	2 20	5 60
Galvanized	1 87	3 85
King, wood....................	2 75	2 90
King, glass....................	4 00	4 50
All glass.......................	1 20	1 30

LINES.

Fish, per gross................	1 05	2 50
Chalk "	1 90	7 40

LOCKS.

Canadian, dis. 45 p.c.

Russell & Erwin, per doz.... 3 00 3 25

Cabinet.

Eagle, dis. 30 p.c.

Padlock.

English and Am., per doz....	50	6 00
Scandinavian, "	1 00	2 40

Eagle, dis. 20 to 25 p.c.

MACHINE SCREWS.

Iron and Brass.

Flat head discount 25 p.c.

Round Head, discount 20 p.c.

MALLETS.

Tinsmiths', per doz...........	1 25	1 50
Carpenters', hickory, per doz.	1 25	3 75
Lignum Vitae, per doz..	3 85	5 00
Caulking, each................	60	2 00

MATTOCKS.

Canadian, per doz........... 5 50 6 50

MEAT CUTTERS.

American, dis. 25 to 30 p.c.

German, 15 per cent.

MILK CAN TRIMMINGS.

Discount, 25 per cent.

NAILS.

Quotations are:	Cut.	Wire.
2d and 3d	$3 35	$3 85
3d	3 00	3 52
4 and 5d..................	2 75	3 35
6 and 7d..................	2 65	3 20
8 and 9d..................	2 50	3 00
10 and 12d................	2 45	2 95
16 and 20d................	2 40	2 90
30, 40, 50 and 60d. (base)......	2 35	2 85

Galvanizing 2c. per lb. net extra.

Steel Cut Nails 10c. extra.

Miscellaneous wire nails, dis. 70 per cent.

Coopers' nails, dis. 30 per cent.

Flour barrel nails, dis. 25 per cent

NAIL PULLERS.

German and American....... 1 85 3 50

NAIL SETS

Square, round, and octagon, per gross	3 38	4 00
Diamond	12 00	15 00

NETTING.

Poultry, 50 and 5 per cent. for McMullen's.

OAKUM.	Per 100 lb.
Navy	6 00
U. S. Navy..........................	7 25

OIL.

Water White (U.S.)................	0 16½
Prime White (U.S.)	0 15½
Water White (Can.)................	0 15
Prime White (Can.)................	0 14

OILERS.

McClary's Model galvan. oil can, with pump, 5 gal., per doz.....................	0 00	10 00
Zinc and tin, dis. 50, 50 and 10.		
Copper, per doz..............	1 25	3 50
Brass, "	1 50	3 50

Malleable, dis. 25 per cent.

GALVANIZED PAILS.

Dufferin pattern pails, dis. 50 and 10 p.c.

Flaring pattern, discount 45 per cent.

Galvanized washtubs, discount 45 per cent.

PIECED WARE.

Discount 40 per cent. off list, June, 1899.

10-qt. flaring sap buckets, dis. 45 p.c.

6, 10 and 14-qt. flaring pails, dis. 45 p.c.

Creamer cans, dis 45 p.c.

PICKS.

Per doz 6 00 9 00

PICTURE NAILS.

Porcelain head, per gross....	1 75	3 00
Brass head "	0 40	1 00

PICTURE WIRE.

Tin and gilt, discount 75 p.c.

PLANES.

Wood, bench, Canadian dis. 50 per cent American dis. 50.

Wood, fancy Canadian or American 7½ to 40 per cent.

PLANE IRONS.

English, per doz.............. 2 00 5 00

PLIERS AND NIPPERS.

Button's Genuine per doz pairs, dis. 37½ 40 p.c.

Button's Imitation, per doz..	5 00	9 00
German, per doz............	0 60	2 60

PLUMBERS' BRASS GOODS.

Compression work, discount, 60 per cent.

Fuller's work, discount 65 per cent.

Rough stops and stop and waste cocks, discount, 60 per cent.

Jenkins disk globe and angle valves, discount, 55 per cent.

Standard valves, discount, 60 per per cent.

Jenkins radiator valves discount 55 per cent.

" " " standard, dis., 60 p.c

Quick opening valves discount, 60 p.c.

No. 1 compression bath cock........	2 00
No. 4 " " "	2 00
No. 7, Fuller's..........................	2 50
No 4½, "	3 00

POWDER.

VeloX Smokeless Shotgun Powder.

100 lb. or less	0 85
1,000 lb. or more........................	0 80

Net 30 days.

PRESSED SPIKES.

Discount 20 to 25 per cent.

PULLEYS.

Hothouse, per doz...........	0 55	1 00
Axle	0 22	0 33
Screw	0 27	1 00
Awning.....................	0 35	2 50

PUMPS.

Canadian cistern	1 80	3 60
Canadian pitcher spout....	1 40	2 10

PUNCHES.

Saddlers', per doz...........	1 00	1 85
Conductors', "	9 00	15 00
Tinners' solid, per set........	0 00	0 72
" hollow, per inch....	0 00	1 00

RANGE BOILERS.

Galvanized, 30 gallons	7 00
" 35 "	8 25
40 "	9 60
Copper, 30 "	22 00
" 35 "	26 00
" 40 "	30 00

Discount off Copper Boilers 10 per cent.

RAKES.

Cast steel and malleable Canadian list 50 and 10 p.c. revised list.

Wood, 25 per cent.

RASPS AND HORSE RASPS.

New Nicholson horse rasp, discount 60 to 60 and 10 p.c.

Globe File Co.'s rasps, 60 and 10 to 70 p.c.

Heller's Horse rasps, 50 to 50 and 5 p.c

RAZORS.

	per doz.	
Geo. Butler & Co.'s,..........	8 00	18 00
Boker's	7 50	11 00
" King Cut	12 50	50 00
Wade & Butcher's..............	3 60	10 00
Theile & Quack's	7 00	12 00
Elliot's	4 00	18 00

REAPING HOOKS.

Discount, 50 and 10 per cent.

REGISTERS.

Discount,.......................... 40 per cent

RIVETS AND BURRS.

Iron Rivets, black and tinned, discount 60 and 10 per cent.

Iron Burrs, discount 55 per cent.

Extras on Iron Rivets in 1-lb. cartons, ½c per lb.

Extras on Iron Rivets in ½-lb. cartons, per lb.

Copper Rivets & Burrs, 35 and 5 p.c. dis. and cartons, 1c. per lb. extra, net.

Extras on Tinned or Coppered Rivets ½-lb. cartons, 1c. per lb.

Terms, 4 mos. or 3 per cent. cash 30 days.

RIVET SETS.

Canadian, dis. 35 to 37½ per cent.

ROPE ETC.

	Sisal.	Manila.
7-16 in. and larger, per lb.	9½	13
⅜ in..........................	10½	14
¼ and 5-16 in.		15
Cotton, 3-16 inch and larger		16½
" 5-32 inch....		21½
" ⅛ inch..........		22½
Russia Deep Sea		15½
Jute		8
Lath Yarn		9½
New Zealand Rope...............		10½

RULES.

Boxwood, dis. 75 and 10 p.c.

Ivory, dis. 37½ to 40 p.c.

SAD IRONS.	per set.
Mrs. Potts, No. 55, polished..	....
" No. 50, nickle-plated	....

SAND AND EMERY PAPER.

Dominion Flint Paper, 47½ per cent.

B & A. sand, 40 and 5 per cent.

Emery, 40 per cent.

Garnet (Rurton's), 5 to 10 p.c. advance on list.

SAP SPOUTS.

Bronzed iron with hooks, per doz... 9 50

SAWS.

Hand Disston's, dis. 12½ p.c.

S. & D., 40 per cent.

Crosscut, Disston's, per ft.... 35 0 55

S. & D., dis. 35 p.c. on Nos. 2 and 3.

Hack, complete, each........	0 75	2 75
" frame only..............	....	0 75

SASH WEIGHTS.

Sectional, per 100 lbs........	2 75	3 00
Solid, "	2 00	2 25

SASH CORD.

Per lb......................... 0 23 0 30

SAW SETS.

"Lincoln," per doz............ .... 6 50

SCALES.

B. S. & M. Scales, 45 p.c.

Champion, 65 per cent.

Fairbanks Standard, 35 p.c.

" Dominion, 55 p.c.

" Richelieu, 55 p.c.

Chatillon Spring Balances, 10 p.c.

SCREW DRIVERS.

Sargent's per doz............ 0 65 1 00

SCREWS

Wood, F. H., bright and steel, 67½ and 10 p.c.

Wood R. H., " dis. 62½ and 10 p.c.

" F. H., brass, dis. 80 and 10 p.c.

Wood, R. H., " dis. 75 and 10 p.c.

" F.H., bronze, dis. 70 p.c.

" R.H. " 65 p.c.

Drive Screws, 80 per cent.

Bench, wood, per doz.........	3 25	4 00
" iron. "	4 25	5 75

Set, Case hardened, 60 per cent.

Square Cap, 50 and 5 per cent.

Hexagon Cap, 45 per cent.

SCYTHES.

Per doz. net 9 00

SCYTHE SNATHS.

Canadian, dis. 45 p.c.

SHEARS.

Bailey Cutlery Co., full nickeled, dis. 60 p.c.

Seymour's, dis. 50 and 10 p.c.

SHOVELS AND SPADES

Canadian, dis. 40 and 5 per cent.

SINKS.

Steel and galvanized, discount 45 per cent.

SNAPS.

Harness, German, dis. 25 p.c.

Lock, Andrews'............... 4 50 11 50

SOLDERING IRONS.

1, 1½ lb., per lb.......	0 37
2 lb. or over, per lb.	0 34

SQUARES.

Iron, No 493, per doz	2 40	2 65
" No. 494, "	3 25	3 40

Steel, dis. 50 and 5 to 50 and 10 p.c., rev. list.

Try and bevel, dis. 50 to 52½ p.c.

STAMPED WARE.

Plain, dis., 75 and 12½ p.c. off revised list.

Retinned, dis., 75 p.c. off revised list.

STAPLES.

Galvanized	3 75	4 00
Plain	3 5	3 50

Coopers', discount 45 per cent.

Poultry netting staples, 40 per cent.

STOCKS AND DIES.

American dis. 25 p.c.

STONE.		Per lb.
Washita........................	0 28	0 60
Hindostan..	0 06	0 07
" slip, ...	0 09	0 09
Labrador	....	0 13
" Axe...................	....	0 15
Turkey	....	0 50
Arkansas	0 00	1 50
Water-of-Ayr	0 00	0 10
Scythe, per gross	3 50	5 00
Grind, per ton	15 00	18 00

STOVE PIPES.

Nestable in crates of 25 lengths.

5 and 6 inch Per 100 lengths	7 00
7 inch " "	7 50

ENAMELINE STOVE POLISH.

No. 4—3 dozen in case, net case	$4 80
No. 6—3 dozen in case, "	8 40

TACKS BRADS, ETC.

	Per cent.
Strawberry box tacks, bulk	75 & 10
Cheese-box tacks, blued	80 & 12½
Trunk tacks, black and tinned	85
Carpet tacks, blued	80 & 5
" " tinned	80 & 10
" " (in kegs)	40
Cut tacks, blued, in dozens only ..	75 & 15
" ½ weights................	60
Swedes, cut tacks, blued and tinned—	
In bulk............................	80 & 10
In dozens.........................	75
Swedes, upholsterers', bulk	85 & 12½
" brush, blued & tinned, bulk..	70
" gimp, blued, tinned and japanned	75 & 12½
Zinc tacks	35
Leather carpet tacks	55
Copper tacks	50
Copper nails	52

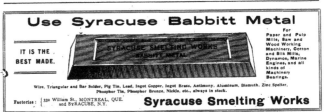

Trunk nails, black	65 and 5
Trunk nails, tinned	65 and 10
Clout nails, blued and tinned	65 and 5
Chair nails	35
Cigar box nails	35
Patent brads	40
Fine finishing	40
Picture frame points	10
Lining tacks, in papers	10
" " in bulk	15
" " solid heads, in bulk	75
Saddle nails in papers	10
" " in bulk	15
Tufting buttons, 22 line, in dozens only	60
Tin capped trunk nails	15
Zinc glazier's points	5
Double pointed tacks, papers	90 and 10
" " " bulk	40

TAPE LINES.

English, ass skin, per doz	2 75	5 00
English, Patent Leather	5 50	9 75
Chesterman's each	0 90	2 85
" steel, each	0 80	8 00

THERMOMETERS.

Tin case and dairy, dis. 75 to 75 and 10 p.c.

TRANSOM LIFTERS.

Payson's per doz. 2 60

TRAPS. (Steel.)

Game, Newhouse, dis. 25 p.c.
Game, H. & N., P. S. & W., 65 p.c.
Game, steel, 72½, 75 p.c.

TROWELS.

Disston's discount 10 per cent.

German, per doz	4 75	6 00

S. & D., discount 35 per cent.

TWINES.

Bag, Russian, per lb	....	0 27
Wrapping, cotton, per lb	0 22	0 26
Wrapping, mottled, per pack	0 50	0 60
Wrapping, cotton, 3-ply		0 30
" " 4-ply	....	0 26
Mattress, per lb	0 33	0 45
Staging, "	0 27	0 35
Broom, "	0 30	0 55

VISES.

Hand, per doz	4 00	6 00
Bench, parallel, each	2 00	4 50
Coach, each	6 00	7 00
Peter Wright's, per lb	0 12	0 13
Pipe, each	5 50	9 00
Saw, per doz	6 50	13 00

ENAMELLED WARE.

White, Princess, Turquoise, Blue and White, discount 50 per cent.
Diamond, Famous, Premier, 50 and 10 p.c.
Granite or Pearl, Imperial, Crescent, 50, 10 and 10 per cent.

WIRE.

Brass wire, 50 to 50 and 2½ per cent. off the list.

Copper wire, 45 and 10 per cent. net cash 30 days, f.o.b. factory.

Smooth Steel Wire, base, $2.80 per 100 lb. List of extras: Nos. 2 to 5, advance 7c. per 100 lb.—Nos. 6 to 9, base—No. 10, advance 7c.—No. 11, 14c.—No. 12, 20c.—No. 13, 35c.—No. 14, 47c.—No. 15, 60c.—No. 16, 75c. Extras net per 100 lb.: Coppered wire, 60c.—tinned wire, $2—oiling, 10c.—special hay-bailing wire, 30c.—spring wire, $1—best steel wire, 75c.—bright soft drawn, 15c.—in 50 and 100-lb. bundles net, 10c.—in 25-lb. bundles net, 15c.—packed in casks or cases, 15c.—bagging or papering, 10c.

Fine Steel Wire, dis. 17½ per cent. List of extras: In 100-lb. lots: No. 17, $5—No. 18, $5.50—No. 19, $6—No. 20, $6.65—No. 21, $7—No. 22, $7.30—No. 23, $7.65—No. 24, $8—No. 25, $9—No. 26, $9.50—No. 27, $10—No. 28, $11—No. 29, $12—No. 30, $13—No. 31, $14—No. 32, $15—No. 33, $16—No. 34, $17. Extras net—tinned wire, Nos. 17-25, $2—Nos. 26-31, $4—Nos. 32-34, $6. Coppered, 5c.—oiling, 10c.—in 25-lb. bundles, 15c.—in 5 and 10-lb. bundles, 25c.—in 1-lb. hanks, 50c.—in ½-lb. hanks, 75c.—in ¼-lb. hanks, $1—packed in casks or cases, 15c.—bagging or papering, 10c.

Galvanized Wire, per 100 lb.—Nos. 6, 7, 8, $3.55—No. 9, $3.10—No. 10, $3.75—No. 11, $3.85—No. 12, $3.25—No. 13, $3.35—No. 14, $4.25—No. 15, $4.75—No. 16, $[illegible]. Base sizes, Nos. 6 to 9, $2.57½ f.o.b. Cleveland.

Clothes Line Wire, 19 gauge, per 1,000 feet	2 75	3 00

WIRE FENCING. F.O.B. Toronto

Galvanized 4 barb, 2¾ and 5 inches apart	3 10
Galvanized, 2 barb, 4 and 6 inches apart	3 10
Galvanized, plain twist	3 10

Galvanized barb, f.o.b. Cleveland, $2.82½ in less than carlots, and $2.7? in carlots.

WIRE CLOTH.

Painted Screen, per 100 sq. ft., net 1 35
Terms, 4 months, May 1; 3 p.c. off 30 days.

WASTE COTTON. per lb.

Colored	4½ to 5
White, according to quality	6½ to 7¼

500-lb bale lots shaded.

WRENCHES.

Acme, 35 to 37½ per cent.
Agricultural, 60 p.c.
Coe's Genuine, dis. 20 to 25 p.c.

Towers' Engineer, each	2 00	7 00
" S., per doz	5 80	6 00
G. & K.'s Pipe, per doz	....	3 40
Burrell's Pipe, each	....	3 00
Pocket, per doz	0 25	2 90

WRINGERS.

Leader	per doz.	$30 00	33 00
Royal Canadian	"	26 00	18 00
Royal American	"	26 00	28 00
Sampson	"	....	30 00

Terms 4 months, or 3 p.c. 30 days.

WROUGHT IRON WASHERS.

Canadian make, discount, 40 and 5 per cent.

PERSONS addressing advertisers will kindly mention having seen their advertisement in Canadian Hardware and Metal Merchant.

CANADIAN
HARDWARE
AND METAL MERCHANT

The Weekly Organ of the Hardware, Metal, Heating, Plumbing and Contracting Trades in Canada.

VOL. XIII. MONTREAL AND TORONTO, MARCH 9, 1901. NO. 10

ALEXANDER GIBB

Manufacturers' Agent and Metal Broker,
13 St. John Street, Montreal.

Representing British and American Manufacturers. Correspondence invited from firms wishing to be represented in Canada.

McCaskill, Dougall & Co
Fine Varnish & Japan
Manufacturers.
Montreal
Price Lists on application

With______

DUNDAS AXES

in stock you may justly ask a clerk for an explanation if he misses a sale.

Without **DUNDAS AXES,** is it fair to blame the clerk when you have not bought the best value?

All First-Class Jobbers Sell Them.

IT SELLS!

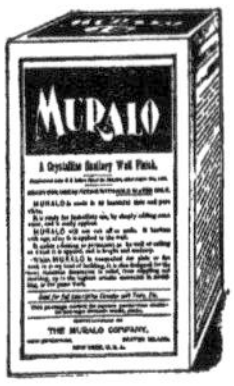

BECAUSE it's the best made.

BECAUSE it is the healthiest Wall Tint known.

BECAUSE it's the best advertised Wall Tint in the world.

BECAUSE once sold it recommends itself and sells again.

BECAUSE it is nicely put up, and there are many other "becauses" why

MURALO

is the best selling Wall Tint in the world.

AGENTS.

A. RAMSAY & SON, ... MONTREAL.
J. H. ASHDOWN, ... WINNIPEG.
McLENNAN, McFEELY & CO., . . . VANCOUVER.

DOES IT PAY TO EXPERIMENT

With untried preparations for wall-coating when you can get the well-known **CHURCH'S**

ALABASTINE

that sells readily on **its own merit**?

ALABASTINE is made in Canada, advertised extensively. No trouble to sell it, because you do not have to argue as to its comparative merit with something else.

ALABASTINE is the **acme of perfection** in wall-coating material, the recognized **standard the world over,** and is adapted to any and all classes of work, from a plain tint to the most elaborate decorating in relief.

ALABASTINE can be recoated from time to time as long as the wall stands. Sanitarians endorse it, and the people call for it. Manufactured in White and 20 beautiful tints.

Prices right; terms easy.

Is your stock complete? Order now, and be in line with goods and advertising matter for the early spring trade.

Address, **THE ALABASTINE CO., LIMITED,** Paris, Ont.

The trade supplied by—
Sanderson Pearcy & Co., Toronto, Ont.
G. F. Stephens & Co., Winnipeg, Man.
Vancouver Hardware Co., Vancouver, B.C.
Wood, Vallance & Co., Hamilton, Ont.
William Hill, Montreal, P.Q.

VanTuyl & Fairbank

Petrolia, Ont.

Headquarters for . . .
Oil and Artesian Well Pumps, Casing, Tubing, Fittings, Drilling Tools, Cables, etc.

Ontario Nut Works, Paris

BROWN & CO.

Manufacturers of
All sizes of Hot Pressed Nuts, Square and Hexagon.

"DAISY" CHURN

Has tempered steel cased bicycle ball bearings, strongest, neatest and most convenient frame. Only two bolts to adjust in setting up. Steel Bow Levers, suitable for either a standing or sitting posture. Has four wheels and adjustable feet to hold stand steady while churning. When churn is locked to stand the bow can be used as handles to move it about on the front wheels as handy as a baby carriage. Open on both sides to centre, giving free space for pail. Made with wood or steel stands, with Cranks only, or Bow Levers as desired.

Vollmar Perfect Washer

Has a most enviable record. A perfection of its kind—will wash more clothes in less time, do it better and easier, with less wear and tear, than any other machine.

THE______
Wortman & Ward Mfg. Co., Limited
LONDON, ONT.
Eastern Branch, 60 McGill Street, Montreal, Que.

THE B. GREENING WIRE CO., LIMITED

HAMILTON, ONT., AND MONTREAL, QUE.

MANUFACTURERS OF

GREENING Wire Rope

of every description and for all purposes.

Lang's Patent Wire Rope for Colliery and Mining Use.

GREENING

MRS. POTTS SAD IRONS.
No. 55, Plain Polished. No. 50, Nickel Plated.

THE "STAR" FLUTING IRON.
Nickel Plated and Gold Bronze Finish.

ARE MADE FROM HIGH-GRADE IRON AND ARE WELL FINISHED.

Cost No More Than Inferior Makes, Give Far Better Satisfaction.

When Ordering From Your Jobber Ask For Our Make.

A. R. WOODYATT & CO., Guelph, Ont.

A Timely Trough Talk.

OUR EAVE TROUGH

is made of evenly-coated Galvanized Iron of uniform thickness.

It is carefully made up by skilled workmen, and every length will be found perfect. We make all styles :

O.G., Round and Square Bead, and Half-Round
in 8 and 10-FOOT LENGTHS.

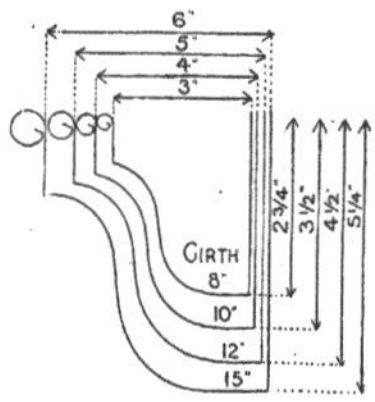

Conductor-Pipe, Elbows and Shoes, Hooks and Gutter Spikes.

Everything a tinner needs we can supply.
Are you ready for the Spring trade in this line ?

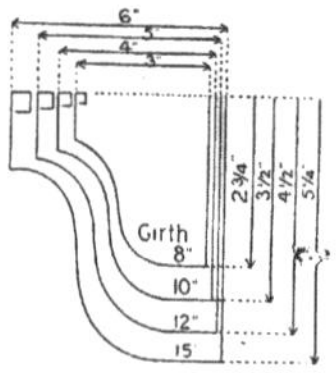

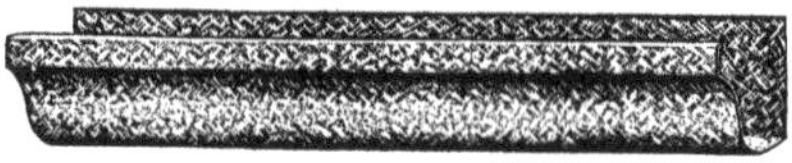

KEMP MANUFACTURING CO., TORONTO, ONT.

HARDWARE AND METAL

VOL. XIII. MONTREAL AND TORONTO, MARCH 9, 1901. NO. 10.

President,
JOHN BAYNE MacLEAN,
Montreal.

THE MacLEAN PUBLISHING CO.
Limited.

Publishers of Trade Newspapers which circulate in the Provinces of British Columbia, North-West Territories, Manitoba, Ontario, Quebec, Nova Scotia, New Brunswick, P.E. Island and Newfoundland.

OFFICES

MONTREAL - - - - - - - 232 McGill Street, Telephone 1255.
TORONTO - - - - - - - 10 Front Street East, Telephone 2148.
LONDON, ENG. - - - - - 109 Fleet Street, E.C., J. M. McKim.
MANCHESTER, ENG. - - - - 18 St Ann Street. H. S. Ashburner,
WINNIPEG - - - - - Western Canada Block. J. J. Roberts.
ST. JOHN, N.B. - - - - No. 3 Market Wharf, J. Hunter White,
NEW YORK. - - - - - - - 176 E. 88th Street.

Subscription, Canada and the United States, $2.00.
Great Britain and elsewhere - - - - 12s.

Published every Saturday.

Cable Address { Adscript, London.
Adscript, Canada.

A BAD PRACTICE.

WHEN a buyer places an order for goods he should be prepared to stand by his contract, unless, of course, the seller proves unfaithful to his part of the agreement or at the time of sale misrepresented the article.

When an order is booked with a wholesale house, whether it be for hardware, glass, paint or oil, that jobbing house has, without a doubt, purchased the goods which it is offering for sale, and, if they are not already in stock, the firm is bound to take delivery of them as per contract when the time comes, whether the market goes up or down. But there are some merchants, strange to say, who take a reckless view of the matter with reference to what they have purchased. If the market keeps firm or shows a good advance they congratulate themselves, but if prices decline they seem to think that they have a perfect right to cancel their order, or to demand a reduction in price.

This sort of thing we are pleased to say is not carried on by all the trade. There are good hardware firms—and lots of them—throughout Ontario who would not stoop to such unbusinesslike methods, and these firms are well known and have the respect of the trade.

What would the hardware merchant think, supposing there was a big advance in, say, window glass, if the wholesaler with whom he had placed his import order were to inform him later on that he had not sufficiently covered himself in his purchases in January, and, consequently, would not be able to fill his order unless he paid him the advanced price? He would simply be compelled to fill it, no matter what the loss might be.

When an order is once booked a drop in price does not warrant its being cancelled. This is business, whether the purchaser be wholesaler or retailer.

OILED AND ANNEALED WIRE LOWER.

A REDUCTION of 10 to 15c. per 100 lb. has taken place in the price of oiled and annealed wire. The net selling prices are now as follows :

	Per 100 lb.
Nos. 6 to 8 gauge	$2 90
No. 9	2 80
No. 10	2 87
No. 11	2 90
No. 12	2 95
No. 13	3 15
No. 14	3 37
No. 15	3 50
No. 16	3 65

The reduction on Nos. 9, 10 and 13 gauge is 10c. per 100 lb.; on No. 11 gauge 14c., and on No. 12 gauge 15c. Sizes other than these five and also other varieties of plain wire remain unchanged at the $2.80 base, with extras as before. The prices quoted for Nos. 9 to 13 include the charge of 10c. per 100 lb. for oiling.

Foreign competition is the cause of the changes. Barb wire and plain galvanized wire were, it will be remembered, reduced in price last week.

FREE TRADE IN RAILWAYS.

IN view of the fact that the application of certain railways for charters privileging them to construct lines in British Columbia is likely to meet with strong opposition from those interested in existing lines, it is advisable to note the view of the matter taken by the merchants and other business men in the district affected.

In another column of this issue is given a report of a meeting of the Associated Boards of Trade of Eastern British Columbia, at which every board of trade in the eastern mining district of that Province was represented. After a warm discussion of the matter, a resolution was almost unanimously passed, declaring as a general principle in favor of "Free trade in railways," that every bona fide railway company desirous of building railways in the Province should be allowed to do so, and petitioning the Dominion Parliament and the Provincial Legislature to grant the charter applied for by the Crow's Nest Pass Coal Company, and those applied for by other companies, who ask for nothing beyond the privilege of constructing railways in that Province.

With such a decided expression of opinion from the business men of British Columbia, the duty of the Parliament of the Dominion is clear.

THE MARITIME PROVINCES AND INSOLVENCY LEGISLATION.

IT has been stated that one of the reasons for an insolvency bill not being introduced at the present session of the Dominion Parliament is the opposition of the Maritime Provinces.

With a view to ascertaining what ground there was for this HARDWARE AND METAL wrote to several leading business men in New Brunswick, Nova Scotia and Prince Edward Island, reciting the charge and asking for their opinions as to its truthfulness. The views these gentlemen held on the question of insolvency were unknown to this paper, so that it cannot be said they were selected because of any opinions, favorable or otherwise, they might be expected to express. From the gentlemen to whom we wrote some replies have been received.

Mr. R. Innes, Coldbrook, N.S., president of the Kings County Board of Trade and largely interested in the apple export trade, writes: "I know of no opposition in this section of the country to a Dominion insolvency law, but the reverse. Our local law is incomplete and unsatisfactory. We would welcome a law that would force a man into insolvency at the proper time and give him a discharge."

Mr. J. A. Chipman, Halifax, writes: "I have noticed some opposition to a Bankruptcy Act in our Board of Trade and elsewhere, but I think it proceeds from selfish motives or a fear of working results, in view of past experience. That we should have a simple, equitable bankruptcy law is beyond question. The dignity and honor of our country demands it."

Mr. W. M. Jarvis, St. John, N.B., the well-known insurance man and an ex-president of the Maritime Board of Trade, says: "I believe there is very little, if any, feeling of opposition at St. John to a carefully-prepared Canadian bankruptcy law. We have considered the measures proposed from time to time in committee of the board of trade, and, I think, at meetings of the full board as well. Objections have been made to certain details, but not, as far as I recollect, to the principle itself. It is felt that any such measure should be carefully considered and well guarded to prevent needless delay and expense."

Mr. Frank C. Simson, of the wholesale drug firm of Simson Bros. & Co., Halifax, and an active board of trade man, declares: "I am surprised to note what you say about the people of the Maritime Provinces being opposed to a bankruptcy bill, as we are most decidedly in favor of it. While we have one in force here now, it has many weak points, and I was one of a committee to suggest improvements in the same."

The secretary of the St. John, N. B., Board of Trade, Mr. F. O. Allison, writes: "I am instructed to reply to your letter addressed to Mr. G. Wetmore Merritt, president of the Board of Trade, in reference to a Dominion bankruptcy law. In reply, I can speak for the members of the board of trade that the general feeling expressed at meetings held to consider this matter some little time ago seemed to be that in favor of the principle of the Dominion bankruptcy law. There is a difference of opinion in regard to details, especially those in connection with the delay and expense of winding up estates."

Mr. M. G. DeWolfe, Kentville, N.S., last year's president of the Maritime Board of Trade and a retired retail merchant, writes: "As far as Nova Scotia is concerned I doubt if the people are opposed to it. Personally, I think we should have a Dominion insolvency law, and do not see how any progressive, honest merchant, could be opposed to it."

Mr. Horace Haszard, Charlottetown, P.E.I., a well known business man and an ex-M.P. writes: "I am not aware of any more opposition to such a law in the Maritime Provinces than I think can be found amongst the merchants in the west. After our experience of a bankruptcy law we had in force some years ago, I confess I think we are better off as we are than as we were in the days when the official assignee generally managed to use up nearly all the assets of a moderate estate in commissions and other expenses. The Act, passed some time ago, preventing preferential assignments being made within 60 days of suspension of business seems to be fairly satisfactory. But, of course, a satisfactory bankruptcy law, if such could be framed, would be an improvement on the present state of affairs, and would, no doubt, inspire foreign creditors with more confidence in Canadian trade generally. We have not had many failures here of any account for some time past, and deserving men, even if hard pressed at times, can usually depend on the leniency of creditors, whilst the undeserving ones should not be provided with too much machinery for compromising with their creditors at 25 or 50c. on the dollar."

In the light of these letters from business men in different parts of the Maritime Provinces it is evident that the opposition in that part of the Dominion is light indeed. Of the gentlemen above quoted, Mr. Haszard is the only one that at all hesitates about the necessity of a Dominion insolvency law, and he clearly hesitates because of the fear of a repetition of the conditions as they obtained under the law 20 years ago. An efficient law, he emphatically declares, would be an improvement on the present state of affairs and would inspire confidence in us among foreign business men.

It is evident that the responsibility for the non-introduction of an insolvency bill does not lie with the Maritime Provinces. The strongest opponents of such a measure are, without question, the bankers, and their opposition is based on the refusal of business men to consent to their ranking for double liability, not on the general principle of an insolvency law.

THE PRICE OF SAD IRONS.

For some weeks, as the readers of HARDWARE AND METAL are aware, there has been a good deal of irregularity in the price of Mrs. Potts sad irons, some of the jobbers, in order to meet competition, having reduced their prices, while others endeavored to maintain the old figures.

Although no agreement exists, the prices are again uniform; those who were holding out for the old figures having reduced their quotations to the point at which others had already made theirs, namely, 62½c. per set for No. 55, polished, and 67½c. per set for No. 50, nickle-plated.

Whether these figures will remain [illegible] they are is a question. And for two reasons. In the first place, they are too close to cost, and, in the second place, an effort is being made to establish a fixed price, which, if successful, will probably mean a higher price.

LACK OF CAPITAL AS A CAUSE OF FAILURE.

OUT of a total number of 1,337 failures in Canada, Newfoundland, St. Pierre and Miquelon, 916 were due to lack of capital. This is 68.5 per cent. of the total failures of all kinds, compared with 74.1 per cent. in 1899, 69.1 per cent. in 1898, and 70.3 per cent. in 1897. In the United States the percentages due to lack of capital compared with the total number of failures were: 1900, 32 per cent.; 1899, 34.6; 1898, 34.1; 1897, 31.4.

Of all the causes of failure, according to a report just issued by Bradstreet's commercial agency, none are so prolific as lack of capital. And, judging from the figures we have given, the necessity of capital as a factor in business is not sufficiently recognized as to have exerted an influence for marked good upon men in mercantile life.

It is a laudable thing for young men to aspire to become merchants. Any young man behind the counter or the desk who has no such aspirations does not amount to a great deal. It is evident he has no ambition. And a man without ambition, like a locomotive without steam, cannot make headway.

But, just as capital cannot take the place of ambition so the latter cannot take the place of the former. The two are helpmeets to each other. And united they produce success.

Before even a moderate capital can be obtained the average man has got to travel a highway often long and usually beset with obstacles and discouragements without number. Those who have friends to help them are the few, not the many. He who has not the patience to plod along this highway may succeed in business by embarking in it with little or no capital, but the chances are against him. Of the total failures last year, as already pointed out, 68.5 per cent. were due to lack of capital.

Just the amount of capital a man should have in starting into business is not for us to say. A great deal depends upon the character of the business, the locality in which it is situated, the system in regard to cash or credit employed, and the experience and courage possessed by the one going into business.

A good many men have, even in late years, succeeded in business with only a small cash capital to start with. We have one wholesaler in mind at the present moment whose place of business is in Philadelphia. "I had," he once said to the writer, "when I began business as a wholesaler, a capital of $105,000. Of this amount $5,000 was cash I obtained from my father, and the balance was courage and confidence." This is an exceptional case, however. It is the rule we are dealing with.

But, capital is after all not wholly represented by dollars, or even by courage and confidence thrown in. What is just as essential as dollars is character. A man may be as rich as Crœsus and as courageous as a lion, and yet be badly handicapped for business if he is poor in character.

Money may be the foundation of business; but character is the corner stone.

Get money before you branch into business; but get character also.

A STRONG STEEL MARKET.

The strong position of steel is the feature of the metal market. In the United States the demand for steel is in excess of the ability of the manufacturers to supply. And in order to secure billets they have been paying premiums on even the prices of the pool.

Naturally the briskness of the demand for steel is stimulating business in pig iron, the price of which is firmer.

At its next meeting the pool is expected to advance the price of steel plates $2 a ton.

In Canada the wholesale houses report that they are being inconvenienced by their inability to get delivery of steel, and prices in this country are higher this week.

PRICES WITHDRAWN.

A firmer feeling has developed in the iron trade in Canada during the past week. The bar iron makers have withdrawn their quotations, and will only name prices on application. The makers of wrought iron pipe have, as their confreres in the United States did last week, withdrawn their prices.

In consequence of the higher prices on iron there is a firmer feeling on some of the staple manufactures thereof, particularly bolts, although no advances have taken place.

A COUNTER ATTRACTION TO DEPARTMENT STORES.

MANY of the evils that exist in the retail trade to day would be greatly minimized were merchants to adopt counteracting methods.

There are the department stores, for example, of the large cities. Their influence is felt by nearly all classes of merchants within a radius of 1,500 miles or more. A good many merchants seek relief through Legislative enactments. Whether relief can be obtained in this way is doubtful. But, at any rate, the Legislatures are slow to move, while the department stores, in the meantime, display a sleepless activity.

The most effective method, and the method that can be made to operate the quickest, is for merchants, both individually and collectively, to act for themselves. In other words, they should employ ways and means to attract the people in their respective communities from the department stores to their stores.

Down in Wolfville, N.S., where the influence of the Toronto department stores is felt, the merchants have for three seasons had what they term "a merchants' day." The day is usually a week before Christmas, and is selected at a meeting of the merchants of the town. But this is not all they do in unison, for, at the meeting in question, it is decided what line of goods each merchant shall put forward as a leader on "merchants' day." Then a large space is taken in the local papers by the combined merchants where each merchant advertises his bargains. Each merchant has, therefore, a bargain to offer separate and distinct from his fellow-merchants.

Such a scheme as this will do more to counteract the influence of the departmental store than all the laws which this country can enact in a century, and the merchants of Wolfville are to be congratulated upon their enterprise.

As will be seen from the letter of Mr. J. D. Chambers, printed on another page, all classes of business men participate in "merchants' day."

It is the worst thing a man can do for himself when he neglects to do the best he can for his employer.

CATALOGUES, BOOKLETS, ETC.

MARLIN FIRE ARMS CATALOGUE.

ONE of the most complete catalogues issued to the Canadian trade is the dainty book which the Marlin Fire Arms Co., New Haven, Conn., send out in description of their repeating rifles, carbines, muskets and shotguns and their shells, tools, sights and firearm sundries. This work is divided into three parts. The first is designed for the quick reference of dealers and consumers who desire briefly the details of the Marlin arms. The second is intended for the consumer who wishes more complete information in regard to the arms and ammunition used in them. By careful reading of this part a novice can reach a correct decision as to which arm will best suit his individual wants. The third part gives some hints that should interest, and possibly help, shooters in general. Every hardwareman should get this catalogue and read it carefully.

LUFKIN RULES, TAPES, ETC.

A catalogue containing illustrations and full descriptions of the standard and many special lines of measuring appliances, such as rules, tapes, etc., is one booklet that every hardwareman should possess. The Lufkin Rule Co., Saginaw, Mich., are just sending out their 1901 catalogue. This contains a full list of steel, metallic linen and leather measuring tapes, folding and straight steel pocket rules, yard, counter, pattern, board and log rules, log calipers, marking sticks for lumber crayons, lumber guages, boot calks, calk sets and punches, glass boards and rules. The pattern rule is accompanied by a pattern chart, with descriptions as to methods of using. The catalogue also contains a comparison of log tables, which is useful in ordering rules.

IVER JOHNSON ARMS AND CYCLES.

The Iver Johnson Arms and Cycle Works, Fitchburg, Mass., have earned an enviable reputation in two respects, the thorough quality of their product, both firearms and bicycles, and their artistic, effective manner of advertising. Their 1901 bicycle catalogue is in full keeping with their reputation in this regard, and is worth getting for the cover alone. But, of course, hardwaremen will want it for the information contained re Iver Johnson bicycles, where they are described in detail. It can be had on application.

THE ERICSSON SERIES.

The Ericsson Telephone Co., 296 Broadway, New York, have just issued the 5th number of their series of booklets on telephone matters. This number contains much technical information, yet is written in language and style for general reading. The leading article is devoted to "The Transposition of Wires," a timely subject. There is also a specially interesting article on the telephone and telegraph service in Switzerland. This series will be sent to the trade on application.

HAPPY THOUGHT GENERATORS.

The Guelph Acetylene Gas Generator Co., Guelph, Ont., are sending out their 1901 catalogue. It is devoted to a detailed description of their Happy Thought generators. This should prove a valuable booklet for hardwaremen to possess, as during the past year there have been great strides in the acetylene gas industry. The manufacture of calcium carbide has greatly extended in every country, especially Germany, France and the United States. In Canada, besides the extension of the Wilson Carbide Works in Merritton, two factories have been started in Ottawa. The generator made by this company has been accepted by the Canadian Fire Underwriters' Association as meeting all the requirements, hence there is no extra charge for insurance where it is installed. This booklet discusses acetylene gas questions so comprehensively that it is well worth writing for.

NEW SHIPYARD FOR MONTREAL.

A shipyard is to be established on the St. Lawrence, near Montreal, by The American Shipbuilding Company, of Cleveland. This company is a large builder of ships for lake service. It has found the inland waters too small for its ambitions, and is going to build vessels for the Atlantic and Pacific. The company has two ocean steamers in the course of construction at Cleveland. They will be completed in October, and will be cut in two and taken through the Canadian canals to the Atlantic in two sections. It is the intention of the company to join the two sections in the St. Lawrence, hence the shipyard. As a large amount of capital will have to be invested to provide machinery for joining the sections, the company has determined to go a little farther and invest enough capital to fit the St. Lawrence yard for shipbuilding. There will be no capital lying idle in the Canadian investment, for, when the St. Lawrence yard is not being used for joining sections of lake-built ships, it will be used for the construction of big freighters. It is figured that two will be built each year.

The two ships now building at Cleveland are 430 feet on the hull, 450 feet over all, 43 feet beam and 35 feet from deck to keel.

Two representatives of the firm were recently in Montreal, looking for a site for the shipyard.

THE PEERLESS ICELAND FREEZER.

THE Gurney Foundry Co., Toronto, Winnipeg and Vancouver, have secured the Canadian agency for the Peerless Iceland Freezer, manufactured by Dana & Co., Cincinnati, Ohio. The features claimed for this freezer are that it has the fewest parts and only one motion—all that is necessary for the effective freezing of

smooth, delicious ice cream. The method of working is that when the crank is turned the can revolves around the stationary dasher.

By this method a delicious ice cream is produced in three minutes. The dasher has two rows of leaves, screw-like in form, on opposite sides of the stem. Set at right angles to them are four arms carrying two hardwood scrapers, which scrape the cream from the sides of the revolving can before it becomes frozen hard, throwing it to the centre against the leaves, which lift it up and force it out again to the sides with a rolling motion, giving it four separate motions with each revolution of the crank. All this motion keeps the cream thoroughly mixed, and leaves no "soft spots" in the centre, or lumps on the outside.

There are no cogs on the can top. The tub top is so built that no salt or oil can get into the can. All metal parts of the freezer which come in contact with the cream are heavily plated with pure block tin. All outside parts are galvanized.

This freezer is made in all sizes, from 1 to 25 quarts. The sizes above 8 quarts are furnished if requested with fly wheels, similar to that shown in the accompanying cut.

SPLASHERS BECOME DASHERS.

CANADA Paints came into contact with Sherwin-Williams' Paints in Montreal one evening last week and there was quite a splash. The result was a very high color with a few scattered spots of black and blue. Both paints were found to be remarkable for their lasting and wearing qualities. The painting was done on ice by representative teams of the two establishments. People outside of the business call the game hockey.

The play is reported to have been exceptionally fast, both teams working tooth and nail to uphold their supremacy, but the team play of the Sherwin-Williams Company was too much for the individual efforts of the Canada Paintmen, who were defeated by four goals to one. The teams were as follows:

Canada Paint Company—Stubbs, goal; Miller, point; Speers, cover-point; Russell, Brown, Lamont and Munro (captain) forwards.

The Sherwin Williams Company—White, goal; Budge, point; McBreaty, cover-point; McGerrigie, McLaren, Bann and McMaster (captain) forwards.

The goals were scored by Munro for the Canada Paint Company and by Bann, McGerrigie and McLaren for Sherwin-Williams Company. It is now an understood thing that the Sherwin-Williams team will challenge for the Stanley Cup. The Canada Paint team have already gone into training to retrieve themselves next winter.

NOTHING DEFINITE YET.

It has been reported for some time, chiefly in Cape Breton papers, that the Montreal Rolling Mills had decided to erect a plant in Sydney, C.B. So far as HARDWARE AND METAL can learn nothing definite has as yet been decided upon, although the directors have had the matter under consideration for upwards of two years.

WILL GIVE EMPLOYES AN INTEREST.

R. Chestnut & Sons, founders, etc., Fredericton, N.B., have adopted the co-operative principle. Each man employed at the factory will be given his weekly wages as in the past. The proprietors will take a stated percentage as interest on the capital invested. At the end of the year whatever profits remain over and above the small percentage which goes to the proprietors, will be divided among the proprietors and workmen, the proprietors taking one-half and the remaining amount being divided among the men pro rata according to the wages which they receive.

It is believed that this will not only benefit the men by adding to their salary, but, by providing an incentive to better work, should increase their interest in, and value to, the business.

INDUSTRIAL GOSSIP.

Those having any items of news suitable for this column will confer a favor by forwarding them to this office addressed to the Editor.

THE Montreal Coal Co. have been incorporated in Montreal with $100,000 capital stock.

M. J. Killam is starting as general merchant and sawmiller, Liscomb Mills, N.S.

A 48 x 60 furniture factory is to be erected at Neustadt, Ont., this spring.

Morrisey & English are starting as dealers in mill supplies, St. John's, Newfoundland.

The Ontario Farmers' Cordage Co., Limited, of Brantford, with a capital stock of $100,000 have been incorporated.

The Argentine Peat Syndicate, Limited, Guelph, Ont., has been incorporated with $60,000 capital stock.

H. M. Pearl and S. K. Hamilton, Boston, Mass., are seeking incorporation as the North American Coal Co. with a capital of $1,000,000. Their coal areas are in Cape Breton.

Two roasters from the Vulcan Iron Works, Wilkesbarre, Pa., are being installed at the Grey and Bruce Cement Works, near Owen Sound, Ont. They will weigh over 100,000 lb., and will cost $12,000.

The Strathy Wire Fence Co., Limited, who are starting wire-fence works in Owen Sound, Ont., have elected J. R. Brown president, and R. L. F. Strathy secretary and manager of the company.

The National Iron Works, Limited, are erecting in Wingham, Ont., works which are to cost about $13,000. This company, of which John Galt is president and W. C. Bullock, managing director, will be the Canadian manufacturing agents for Glenfield Kennedy Co., Kilmarnock, Scotland.

McFarlane & Douglas intend erecting a 116 x 24 ft. foundry for the manufacture of galvanized iron, sheet metal, veneer works and similar lines of iron ware on Slater street, Ottawa. The building is to cost about $4,000, and it will be fitted with modern machinery, costing $3,500. Employment will be given to about 30 men at the start. Work will be rushed on the new structure and business will be started early in the summer.

A CASE OF FLOAT OR SINK.

SMITH: There should be no difficulty in floating the new steel combination. There will be lots of water in its stock.

BROWN: Yes; but too much water may sink it.

FILE WORKS CHANGE HANDS.

THE Arcade File Works, Anderson, Indiana, have been sold to the Nicholson File Works, Providence, R.I. The Allerton-Clark Co., 97 Chambers St., New York, selling agents of the Arcade files, have sent out the following circular in regard to the change:

Our connection with The Arcade File Works as the selling agents for its products having ceased, we extend to all our customers who have favored us with their file trade our thanks for their patronage, and our best wishes for their future success.

The well-known reputation of the Nicholson File Company furnishes sufficient guarantee that all business placed in their hands will receive prompt and business-like attention, and we trust that all who have handled Arcade files will continue to do so, feeling assured that their file interests could not be in better hands.

The following has been issued to the trade by The Nicholson File Company:

We herewith announce that we have purchased the Arcade File Works, of Anderson, Ind., and shall continue the manufacture of this brand of files at the works of the company.

The product of this factory has heretofore been handled by The Allerton-Clark Company, at their New York and Chicago stores, but, under the present ownership, this arrangement will not be continued.

All orders and inquiries for prices, and all correspondence in connection with the business of the Arcade File Works should be addressed to The Nicholson File Company, at Providence, R.I.

We shall endeavor to merit the continued patronage of all who have handled Arcade files, by our prompt, careful and courteous attention to whatever inquiries may be addressed to us, and to all business placed in our hands.

"QUICK-MEAL" STOVES.

The Gurney Foundry Co., Limited, Toronto, Winnipeg and Vancouver, will again act this summer as agents for "Quick-Meal" wickless oil and gasoline stoves, manufactured by The Ringen Stove Co., St. Louis, Mo. The stoves are steadily gaining ground because of the qualities which their name implies and their construction guarantees.

A HARDWARE AMALGAMATION.

Mr. R. E. Bingham, of Stayner, was in Toronto last week. Mr. Bingham was head of the firm of Bingham & Co., retail hardwaremen, who recently sold out to J. E. Doner. Since then Mr. Doner has amalgamated with T. W. Gibson, and the stock of the latter has been removed into the premises of the former.

GREENWOOD BOARD OF TRADE.

The annual convention of the Eastern British Columbia Board of Trade was held in Greenwood, B.C., on Saturday last week. Every Board of Trade in eastern British Columbia was represented. Mr. C. A. Galloway, of Greenwood, was elected president. Principal interest centred in the discussion of railway construction, which discussion resulted in the passage of a resolution endorsing the appointment of a Railway Commission, and declaring that the mining towns of British Columbia desire every bona fide company desirous of building railways in the Province should be allowed to do so.

CONFERENCE POSTPONED.

On account of the illness of President Martineau, the Retail Hardware Association of Montreal did not send out the invitations to the wholesale houses for a conference on Wednesday evening as was originally intended. We understand they will ask the wholesalers to attend the meeting to be held on the evening of March 20.

A GOOD PULLEY COVERING.

Machinists generally will appreciate the introduction of any covering which will effectually prevent belts from slipping. For this reason the Vacuum cement and pulley covering, manufactured by the Vacuum Cement and Pulley Co., New York, should have a ready sale in Canada. It is claimed that the use of this covering on pulleys will double the life of any leather belt.

NOVA SCOTIA STEEL CO.

The annual meeting of the Nova Scotia Steel Co. was held in New Glasgow, N.S., on Wednesday last week. The profits on the year were $665,272. Dividends of 4 per cent. on preferred stock and 10 per cent. on ordinary were declared.

The following directors were elected: Graham Fraser, G. F. McGay, J. D. McGregor, Thos. Cantley, Simon A. Fraser, New Glasgow; J. F. Stairs, J. W. Allison, Halifax, and Frank Ross, Quebec city.

The company's assets are now estimated at $5,208,331.55.

INQUIRIES AND ANSWERS.

"A.B." writes: "Could you give me the address of somebody who could sell me about 1,500 lb. of paraffin? This will oblige me very much."

[Remarks: The Queen City Oil Co. and McColl Bros. & Co., Toronto; B. S. Van Tuyl, Petrolea, Ont. These are all the names that come to mind at the moment. Possibly some of our readers may supply other names.—The Editor.]

"Quick" writes: "I desire to add window blinds to my business. Kindly insert some names of manufacturers of window blinds and window blind fixtures of all kinds."

[Remarks: Geo. H. Hees, Son & Co., 71 Bay street, Toronto, make a full line of both blinds and fixtures.—The Editor.]

PERSONAL MENTION.

H. G. McNaughton, Guelph, Ont., formerly representative of the Doherty Manufacturing Co., Sarnia, Ont., is again on the road, and meeting old friends in the trade in the capacity of representative of the Gurney Foundry Co., on the ground in Western Ontario lately covered by W. H. Smith.

AGENCIES WANTED.

A firm in British Columbia is open for an agency for an Old Country fire brick and fire clay company. British merchants desirous of appointing an agent would be good enough to address the Advertising Manager, The MacLean Publishing Company, Limited, Toronto.

G. W. Whitehead has started a general store at Carghill, Ont. He was formerly in business with his brother at Walkerton.

H. S. HOWLAND, SONS & CO.

ONLY WHOLESALE — 37-39 Front Street West, **Toronto.** — WHOLESALE ONLY

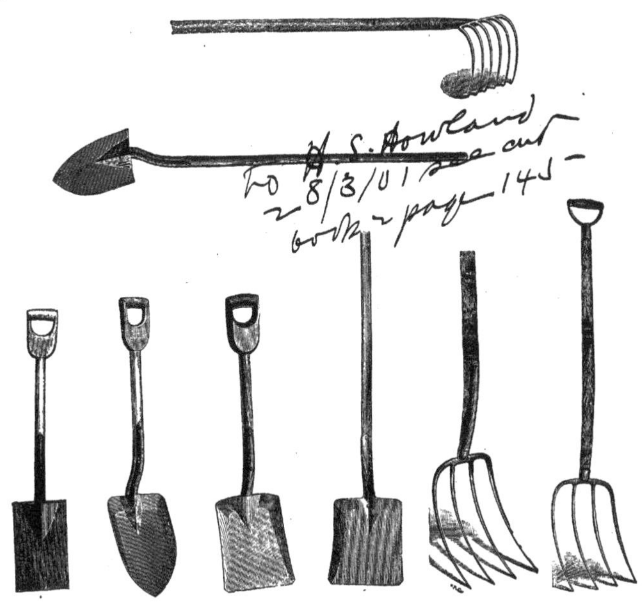

Spades, Shovels, Manure Forks and Drags.

H. S. HOWLAND, SONS & CO., Toronto.

WE SHIP PROMPTLY. OUR PRICES ARE RIGHT.

The Aylmer Iron Works Co., Limited, Aylmer, Ont., have been incorporated with a capital stock of $30,000.

Intending Advertisers

should send in orders for space and copy for advertisements in the Special Number of

Hardware and Metal

without delay.

LETTERS OF INTRODUCTION.

"I DISAPPROVE of letters of introduction," said an elderly New Orleans business man. "I won't give one under any circumstances. They are bad form, and border close on downright impertinence. What right have I, for example, to thrust a perfect stranger on my friend, John Smith, of Memphis or Chattanooga, without having at least asked Mr. Smith's permission or ascertained whether the introduction would be mutually agreeable. Then, again, such letters always mean either too little or too much. Most of us give good advice, without the least idea of incurring any responsibility—yet a letter of introduction is, or ought to be, an absolute indorsement of the bearer, and the recipient would be justified in holding the writer strictly accountable for any abuse of his hospitality. I believe this view is unassailable, but I must confess that I stopped writing letters of introduction myself on account of a little contretemps that had nothing to do with the proprieties of the question. It happened in this way :

"A certain friend asked me to give a letter to a young Englishman, introducing him to a former business partner of mine now living in Louisville. I didn't want to do it, but lacked moral courage to refuse, so I wrote up two letters—one, the introduction requested, and the other a brief note to the Louisville man explaining the circumstances and saying I didn't really know whether the Englishman was a gentleman or a horse thief. Two days later I got a telegram from my partner, saying that he had received a letter of introduction by mail, and was at a loss to know what to make of it. I had put the two inclosures in the wrong envelopes, and had given the Englishman the private note of repudiation."

"I suppose he read it, of course," remarked someone in the group of listeners.

"That's just what has been troubling me ever since," replied the old merchant. "I don't know whether he did or not. He presented it without turning a hair, and if he knew the contents he certainly made no sign. At least that is the report of my friend, who was so surprised when he ran his eye over the epistle that he nearly fell out of his chair. All this happened four years ago, and I haven't written a letter of introduction since. I wouldn't meet that Englishman again for a $1,000 bill, because if I did I wouldn't know whether to shake hands or get ready to fight."—New Orleans Times-Democrat.

John Henderson, city clerk, Ottawa, is asking for tenders for the erection of a new contagious disease hospital in that city.

MARKETS AND MARKET NOTES

QUEBEC MARKETS.

Montreal, March 8, 1901.

HARDWARE.

BUSINESS is brighter than it has been for many a day. Of the lighter lines of goods needed for summer trade dealers through the country are ordering freely for shipment after April 1, when the freight rates will be lower. This freedom in buying on future account has given business a good stimulus. For immediate delivery trade is also more active. Retailers report a better consumptive demand. Wire nails are moving out in larger quantities, while such staple lines as screws, bolts and rivets are in good request. In heavy metals the feeling does not seem to be improved, and little business is doing. The result of the unsettled feeling that has prevailed for some weeks in regard to the prices of Mrs. Potts sad irons has been that all dealers are now quoting 62½c. for No. 55 polished, and 67½c. for No. 50 nickle plated. This puts an end to the cutting. The discount on poultry netting has been raised to 50 and 10 per cent. Barb wire and galvanized wire have been reduced 15 cents per 100 lb. by the American manufacturers. We understand that the 10 cents extra charged for oiling annealed wire has been taken off certain sizes. Payments have been first class.

BARB WIRE—Barb wire is now selling at $2.82½ per 100 lb. f.o.b. Cleveland and Pittsburgh for less than carlots, and $2 70 for carlots of not less than 15 tons. Orders are being taken freely as the market is generally firm. The quotation f.o.b. Montreal is $3.05 per 100 lb.

GALVANIZED WIRE — A fair trade has been done in galvanized wire at the reduced prices. We quote: No. 5, $4.25; Nos. 6, 7 and 8 gauge, $3.55; No. 9 $3.10; No. 10, $3.75; No. 11, $3.85; No. 12, $3.25; No. 13, $3.35; No. 14, $4.25; No. 15, $4.75; No. 16, $5.00.

SMOOTH STEEL WIRE—Instead of a base price both oiled and annealed wire have been put on a net price this week for the following sizes: No. 9, $2.80; No. 10, $2.87; No. 11, $2.90; No. 12, $2.95; No. 13, $3.15 per 100 lb. f.o.b. Montreal, Toronto, Hamilton, London, St. John and Halifax. In other sizes oiled wire is 10c. per 100 lb. extra.

FINE STEEL WIRE—Prices are unchanged. A moderate business is passing at 17½ per cent. off the list.

BRASS AND COPPER WIRE — There has been no change in these articles. Discounts are still 55 and 2½ per cent. on brass, and 50 and 2½ per cent. on copper.

FENCE STAPLES — This line is without special feature. A quiet trade is being done. We quote: $3.25 for bright, and $3.75 for galvanized, per keg of 100 lb.

WIRE NAILS — The demand for nails is increasing, but the nature of the trade continues unchanged, small orders being the rule. The improvement is due simply to the increased call for consumptive purposes. We quote $2.85 for small lots and $2.75 for carlots, f.o.b. Montreal, London, Toronto and Hamilton.

CUT NAILS—The demand for cut nails is a little better this week, but it is not yet by any means active. We quote as follows: $2.35 for small and $2.25 for carlots; flour barrel nails, 25 per cent. discount; coopers' nails, 30 per cent. discount.

HORSE NAILS—The demand is moderate and prices are unchanged. "C" brand's discount is 50 and 7½ per cent. on their own list and "M" brand's discount is 50, 10 and 5 per cent. on oval head and 50, 10 and 10 per cent. on countersunk head.

HORSESHOES—A fair demand continues to be experienced and prices are unchanged. We quote as follows: Iron shoes, light and medium pattern, No. 2 and larger, $3.50; No. 1 and smaller, $3.75; snow shoes, No. 2 and larger, $3.75; No. 1 and smaller, $4.00; X L steel shoes, all sizes, 1 to 5, No. 2 and larger, $3.60; No. 1 and smaller, $3.85; feather-weight, all sizes, $4.85; toe weight steel shoes, all sizes, $5.95 f.o.b. Montreal; f.o.b. Hamilton, London and Guelph, 10c. extra.

POULTRY NETTING—An open market is now prevailing and we hear that the ruling discount is 50 and 10 per cent.

GREEN WIRE CLOTH—Orders are coming in freely at $1.35 per 100 sq. ft.

FREEZERS—Ice cream freezers are moving nicely. "Peerless" is quoted as follows: Two quarts, $1.85; 3 quarts, $2.10; 4 quarts, $2.50; 6 quarts, $3.20; 8 quarts, $4; 10 quarts, $5.25; 12 quarts, $6; 16 quarts, with fly wheel, $11; toy, 1 pint, $1.40.

SCREEN DOORS AND WINDOWS—Although the manufacturers have entered into a two years' agreement and have advanced prices, wholesalers continue to quote the same figures. A fair business is being done. We quote: Screen doors, plain cherry finish, $8.25 per doz.; do. fancy, $11.50 per doz.; walnut, $7.40 per doz., and yellow, $7 45; windows, $2.25 to $3.50 per doz.

SCREWS — An active business has been done this week. Discounts are: Flat head bright, 87½ and 10 per cent. off list; round head bright, 82½ and 10 per cent.; flat head brass, 80 and 10 per cent.; round head brass, 75 and 10 per cent.

BOLTS—There still seems to be a brisk demand for bolts, and good shipments are being made. Discounts are as follows: Carriage bolts, 65 per cent.; machine bolts, 65 per cent.; coach screws, 75 per cent.; sleigh shoe bolts, 75 per cent.; bolt ends, 65 per cent.; plough bolts, 50 per cent.; square nuts, 4½c. per lb. off list; hexagon nuts, 4¾c. per lb. off list; tire bolts, 67½ per cent.; stove bolts, 67½ per cent.

BUILDING PAPER — The inquiry for building paper is steady. We quote: Tarred felt, $1.70 per 100 lb.; 2 ply, ready roofing, 80c. per roll; 3-ply, $1.05 per roll; carpet felt, $2.25 per 100 lb.; dry sheathing, 30c. per roll; tar sheathing, 40c. per roll; dry fibre, 50c. per roll;

tarred fibre, 60c. per roll ; O.K. and I.X L., 65c. per roll ; heavy straw sheathing, $28 per ton ; slaters' felt, 50c. per roll.

RIVETS — A fair trade is reported on the whole. The discount on best iron rivets, section, carriage, and waggon box, black rivets, tinned do., coopers' rivets and tinned swedes rivets, 60 and 10 per cent ; swedes iron burrs are quoted at 55 per cent. off; copper rivets, 35 and 5 per cent. off; and coppered iron rivets and burrs, in 5-lb. carton boxes, are quoted at 60 and 10 per cent. off list.

CORDAGE—Some good shipments of rope have been made this week, while larger orders are being booked. Manila is worth 13c. per lb. for 7-16 and larger; sisal is selling at 9½c., and lathyarn 9½c. per lb.

SPADES AND SHOVELS—These form an active line, and the trade seems to be growing. The discount is still 40 and 5 per cent. off the list.

HARVEST TOOLS — Harvest tools are moving as well as can be expected at this time of year. The discount is 50, 10 and 5 per cent.

TACKS—No change. We quote as follows : Carpet tacks, in dozens and bulk, blued 80 and 5 per cent. discount ; tinned, 80 and 10 per cent ; cut tacks, blued, in dozens, 75 and 15 per cent. discount.

LAWN MOWERS—Lawn mowers are moving out freely. We quote : High wheel, 50 and 5 per cent. f.o.b. Montreal ; low wheel, in all sizes, $2.75 each net ; high wheel, 11-inch, 30 per cent. off.

FIREBRICKS—Business continues to be of a limited nature. Prices are steady at $18 to $24 per 1,000 as to brand, ex store.

CEMENT — The opening of the season bids fair to be bright. At present sales are few. We quote : German, $2.45 to $2.55 ; English, $2.30 to $2.40 ; Belgian, $1.95 to $2.05.

METALS.

There is little business being done in any of the metals. The immediate requirements are not heavy, and the unsettled condition of the markets precludes heavy buying for import. On the Canadian market, black sheets, Canada plate and terne plates are very scarce, and, although prices were easy some time ago, full values are obtainable to-day.

PIG IRON — Prices are none too firmly held. On the Montreal market Canadian pig is worth about $17 to $18. Summerlee is quoted at $23 to $24.

BAR IRON — The bar iron market is steady, at $1.65 to $1.70 per 100 lb.

BLACK SHEETS — The English market continues to be quite unsettled and dealers are afraid to operate. The Canadian market is rather bare of goods. For immediate delivery we quote : $2.80 for 8 to 16 gauge ; $2.85 for 26 gauge, and $2 90 for 28 gauge.

GALVANIZED IRON — The volume of business is not large. We quote : No. 28 Queen's Head, $5 to $5.10 ; Apollo, 10¾ oz., $5 to $5.10; Comet, No. 28, $4.50, with 25c. allowance in case lots.

INGOT COPPER—The market continues firm, the ruling price being 18c.

INGOT TIN — Primary markets are easier again this week and the outlook is unfavorable. Lamb and Flag is worth 33 to 34c.

LEAD—Rather weak at $4.65.

LEAD PIPE—A fair business continues to be done in lead pipe. We quote : 7c. for ordinary and 7½c. for composition waste, with 15 per cent. off.

IRON PIPE—The market is rather active and prices are firm. We quote: Black pipe, ¼, $3 per 100 ft.; ⅜, $3; ½, $3; ¾, $3.15; 1-in., $4.50; 1¼, $6.10; 1½, $7.28; 2-in., $9.75. Galvanized, ½, $4.60 ; ¾, $5.25 ; 1-in., $7.50 ; 1¼, $9.80 ; 1½, $11.75 ; 2-in., $16.

TINPLATES — Business is limited and values are by no means steady. For immediate delivery goods are worth $4 50 for coke and $4.75 for charcoal.

CANADA PLATE—No one is eager to buy heavily. Sales are very light. We quote : 52's, $2.90 ; 60's, $3 ; 75's, $3.10 ; full polished, $3.75, and galvanized, $4.60.

TOOL STEEL—There is quite a good demand for tool steel. We quote: Black Diamond, 8c.; Jessop's 13c.

STEEL—A fair inquiry has been experienced this week. We quote: Sleighshoe, $1.85; tire, $1.95; spring, $2.75; machinery, $2.75 and toe-calk, $2.50.

TERNE PLATES—The terne plate market is not in good condition for the early opening of the season's trade. The price varies from $8 to $8.25.

COIL CHAIN — This line is firm and active. We quote as follows: No. 6, 11½c.; No. 5, 10c.; No. 4, 9½c.; No. 3, 9c.; ¼-inch, 7½c. per lb.; 5-16 $4.65; 5-16 exact, $5.10; ⅜, $4.20; 7-16, $4.00; ½, $3.75; 9-16, $3.65; ⅝, $3.35; ¾, $3.25; ⅞, $3.20; 1-in., $3.15. In carload lots an allowance of 10c. is made.

SHEET ZINC—The ruling price is 6 to 6¼c.

ANTIMONY—Quiet, at 10c.

GLASS.

Import orders are being placed freely at close prices. The Belgian market is quite firm. We quote: First break, $2; second, $2.10 for 50 feet; first break, 100 feet, $3.80; second, $4; third, $4.50; fourth, $4.75; fifth, $5.25; sixth, $5.75, and seventh, $6.25.

PAINTS AND OILS.

For several weeks the trade in paints, colors and varnishes has been extremely active, not only for booking orders but also for actual shipment. The month of February was an agreeable surprise and the present month points to a heavy volume of trade all through the different lines. To particularize, liquid paints are perhaps the most active branch, but still there is a good healthy demand from all parts for coach colors and varnishes. Notwithstanding the erratic course of white lead, a fair amount of business is being done both for immediate and forward shipment. Sales of oil have been somewhat checked by the instability of the market. Turpentine is moving steadily. Paris green is experiencing a good healthy inquiry and large sales are reported this week. Dealers say that stocks are light. We quote:

WHITE LEAD—Best brands, Government standard, $6.37½; No. 1, $6.00; No. 2, $5.62½; No. 3, $5.25, and No. 4, $4.87½ all f.o.b. Montreal. Terms, 3 per cent. cash or four months.

DRY WHITE LEAD—$5.50 in casks; kegs, $5.75.

RED LEAD—Casks, $5.25; in kegs, $5.50.

WHITE ZINC PAINT—Pure, dry, 8c.; No. 1, 6½c.; in oil, pure, 9c.; No. 1, 7½c.

PUTTY—We quote: Bulk, in barrels, $2 per 100 lb.; bulk, in less quantity, $2.15; bladders, in barrels, $2.20; bladders, in 100 or 200 lb. kegs or boxes, $2.35; in tins, $2.45 to $2.75; in less than 100-lb. lots, $3 f.o.b. Montreal, Ottawa, Toronto, Hamilton, London and Guelph. Maritime Provinces 10c. higher, f.o.b. St. John and Halifax.

LINSEED OIL—Raw, 72c.; boiled, 75c., in 5 to 9 bbls., 1c. less, 10 to 20 bbl. lots, open, net cash, plus 2c. for 4 months. Delivered anywhere in Ontario between Montreal and Oshawa at 2c. per gal. advance and freight allowed.

TURPENTINE—Single bbls., 62c.; 2 to 4 bbls., 61c.; 5 bbls. and over, open terms, the same terms as linseed oil.

MIXED PAINTS—$1.25 to $1.45 per gal.

CASTOR OIL—8¾ to 9¼c. in wholesale lots, and ½c. additional for small lots.

SEAL OIL—47½ to 49c.

COD OIL—32½ to 35c.

NAVAL STORES—We quote: Resinsr $2.75 to $4.50, as to brand; coal tar, $3.25 to $3.75; cotton waste, 4½ to 5½c. for colored, and 6 to 7½c. for white; oakum, 5½ to 6½c., and cotton oakum, 10 to 11c.

PARIS GREEN—Petroleum barrels, 16¾c. per lb.; arsenic kegs, 17c.; 50 and 100-lb. drums, 17½c.; 25-lb. drums, 18c.; 1-lb. packages, 18½c.; ½-lb. packages, 20½c.; 1-lb. tins, 19½c.; ½-lb. tins, 21½c. f.o.b. Montreal; terms 3 per cent. 30 days, or four months from date of delivery.

SCRAP METALS.

The market remains in its morbid condition. Dealers are paying the following prices in the country: Heavy copper and wire, 13 to 13½c. per lb.; light copper, 12c.; heavy brass, 12c.; heavy yellow, 8½ to 9c.; light brass, 6½ to 7c.; lead, 2¾ to 3c. per lb.; zinc, 2¼ to 2½c.; iron, No. 1 wrought, $13 to $14 per gross ton; No. 1 cast, $13 to $14; stove plate, $8 to $9; light iron, No. 2, $4 a ton; malleable and steel, $4.

HIDES.

The demand for hides is fairly good and prices are being maintained at their present level. Dealers are paying 7½c. for No. 1 light, and tanners 8½c. for carlots. We quote: Light hides, 7½c. for No. 1; 6½c. for No. 2, and 5½c. for No. 3. Lambskins, 90c.

PETROLEUM.

The demand is falling off as the season advances. We quote: "Silver Star," 14½ to 15½c.; "Imperial Acme," 16 to 17c.; "S.C. Acme," 18 to 19c., and "Pratt's Astral," 18½ to 19½c.

MONTREAL NOTES.

Mrs. Potts sad irons have been reduced to 62½ and 67½c.

Barb and galvanized wires are 15 cents per 100 lb. lower.

Oiled and annealed wire on certain sizes have been set at an equal net price.

The discount on poultry netting has been raised from 50 and 5 per cent. to 50 and 10 per cent.

ONTARIO MARKETS.

TORONTO, March 8, 1901.

HARDWARE.

THE wholesale hardware trade continues to increase, and the outloook is for a thoroughly good spring trade. One of the satisfactory features of the situation is the requests that are coming forward for shipment of goods ordered some time ago for delivery in the spring. In fence wires, the feature is a change in the figures on oiled and annealed wire, this line now being sold at net prices. The figures quoted are 10 to 15c. per 100 lb. lower than a week ago. Business is a little better in fence wires. Although the demand for wire nails is not brisk, it is a little better than it was. The same cannot, however, be said in regard to cut nails. Trade is fairly good in such lines as screws, bolts, hinges, barn door hinges, and barn door traps. A little is being done in spades and shovels and harvest tools, and an occasional order is being received for poultry netting. Quite a few orders are being booked for loaded shells for future delivery. Trade in cutlery is small. Enamelled ware is quiet, and there is not a great deal being done in tinware. A fair sorting-up demand is reported in rope. Some business is being done in oil and gas stoves on future account. Prices have been withdrawn by the Canadian makers of iron pipe, and the feeling is stronger in bar iron.

BARB WIRE—Some business is being done on future account, but practically litt e is being done for immediate shipment. The price of barb wire from stock has been reduced to $3.05 per 100 lb. The f.o.b. price Cleveland was, it will be remembered, reduced last week, the figures being $2.82½ per 100 lb. for less than carlots and $2.70 for carlots of not less than 15 tons.

GALVANIZED WIRE—Local jobbers are also quoting this lower from stock in sympathy with the decline in the f.o.b. Cleveland price. The prices from stock are as follows: Nos. 6, 7 and 8, $3.50 to $3.85 per 100 lb., according to quantity; No. 9, $2.85 to $3.15; No. 10, $3.60 to $3.95; No. 11, $3.70 to $4.10; No. 12, $3 to $3.30; No. 13, $3.10 to $3.40; No. 14, $4.10 to $4.50; No. 15 $4.60 to $5.05; No. 16, $4.85 to $5.35. Nos. 6 to 9 base f.o.b. Cleveland are quoted at $2.57½ in less than carlots and 12c. less for carlots of 15 tons. There is a little better movement in galvanized wire this week.

SMOOTH STEEL WIRE—Orders are beginning to come in a little more freely for oiled and annealed wire. The method of quoting oiled and annealed wire has been changed, resulting in a reduction of from 10 to 15c. per 100 lb. The net selling prices are now as follows: Nos. 6 to 8, $2.90; 9, $2.80; 10, $2.87; 11, $2.90; 12, $2.95; 13, $3.15; 14, $3.37; 15, $3.50; 16, $3.65. Delivery points Toronto, Hamilton, London and Montreal, with freight equalized on these points.

WIRE NAILS—A fair business is reported this week for small lots for immediate delivery, and some orders are being booked for delivery at the end of the month. The deliveries of wire nails are much less than they were the same time a year ago. We still quote the base price at $2.85 per keg in less than carlots, and $2.75 in carlots.

CUT NAILS—No improvement can be noted in this line, trade still being flat. Belleville, as a delivery and freight equalization point, has been abolished. The delivery points are now Toronto, Hamilton, London, Montreal and St. John. The base price is still $2.35 per keg.

HORSESHOES—Trade continues fairly steady at unchanged prices. We quote f.o.b. Toronto as follows: Iron shoes, No. 2 and larger, light, medium and heavy, $3.60; snow shoes, $3.85; light steel shoes, $3.70; featherweight (all sizes), $4.95; iron shoes, No. 1 and smaller, light, medium and heavy (all sizes), $3.85; snow shoes, $4; light steel shoes, $3.95; featherweight (all sizes), $4.95.

HORSE NAILS—A steady trade is also to be noted in this line. We quote as follows: "C" brand oval head at 50 and 7½ per cent. discount off new list, and "M" brand at 60, 10 and 5 per cent. off old list; countersunk head, 50, 10 and 10 per cent.

SCREWS—Some business is being done for immediate shipment, and a few orders are being booked for delivery at the end of the month. The situation in screws is a little stronger, the manufacturers in the United States being less inclined to quote prices than they were a short time ago. We quote discounts as follows: Flat head bright, 87½ and 10 per cent.; round head bright, 82½ and 10 per cent.; flat head brass, 80 and 10 per cent.; round head brass, 75 and 10 per cent.; round head bronze, 65 per cent., and flat head bronze at 70 per cent.

BOLTS AND NUTS—A fairly good trade is being done in bolts, particularly in the carriage description. We quote as follows: Carriage bolts (Norway), full square, 70 per cent.; carriage bolts full square, 70 per cent.; common carriage bolts, all sizes, 65 per cent.; machine bolts, all sizes, 65 per cent.; coach screws, 75 per cent.; sleighshoe bolts, 75 per cent.; blank bolts, 65 per cent.; bolt ends, 65 per cent.; nuts, square, 4½c. off; nuts, hexagon, 4¼c. off; tire bolts, 67½ per cent.; stove bolts, 67½; plough bolts, 60 per cent.; stove rods, 6 to 8c.

RIVETS AND BURRS—There is just the usual steady trade to be noted in this line. We quote iron rivets, 60 and 10 per cent.; iron burrs, 55 per cent.; copper rivets and burrs, 35 and 5 per cent.

ROPE—A fair sorting-up trade is being done, and prices are unchanged. We quote as follows: Sisal at 9½c. per lb. base; manila rope 13c. base; cotton rope, 3-16 in. and larger, 16½c.; 5-32 in., 21½c., and ⅛ in., 22½c. per lb.

BINDER TWINE—We quote pure manila at 10¾c. per lb.; mixed, 8½c. per lb.; sisal, 7¾c. per lb.

CUTLERY—A little trade is being done in a sorting-up way.

SPORTING GOODS—Quite a few orders are being taken for loaded shells for future delivery. A few rifles are going out, and an odd gun or so is wanted.

GREEN WIRE CLOTH—A little business is being done at $1.35 per 100 sq. ft.

SCREEN DOORS AND WINDOWS—There is a little trade being done for future delivery. We quote 4-in. styles in windows at $7.20 to $7.80 per doz., and 3 in. styles 20c. per doz. less. Screen windows are quoted at $1.60 to $3.60 per doz., according to size and extension.

PRESSED SPIKES—There is the usual steady business being done and prices are unchanged as quoted last week.

TINWARE—There is a fairly good movement in sap buckets and in milk can trimmings, but, generally speaking, trade in tinware is only moderate.

ENAMELLED WARE—Trade continues quiet and without any particular feature.

POULTRY NETTING—Business on future account is fairly good. Discount on Canadian is now 50 and 10 per cent., and on English figures are net.

HARVEST TOOLS—An occasional order only is being received for harvest tools. Discount on these is unchanged at 50, 10 and 5 per cent.

SPADES AND SHOVELS—There is a moderate trade only being done. Discount, 40 and 5 per cent.

LEATHER CEMENT—The price of leather cements, which are used largely throughout the country for bicycle and shoemakers' purposes, have advanced considerably in sympathy with the higher price of crude rubber.

BUILDING PAPER—The demand is fairly good and prices are unchanged.

RANGE BOILERS—Trade is moderate and prices unchanged. We quote: 30 gallons, $7; 35 gallons, $8.25; 40 gallons, $9.50.

CEMENT—The movement has not yet started, but a big trade is looked for this spring. We quote Canadian portland $2 40 to $3; German, $3 to $3 15; English, $2.85 to $3; Belgian, $2.50 to $2.75; calcined plaster, $2.

GAS AND OIL STOVES—Some orders are being booked, and the manufacturers are making preparations for the spring trade.

METALS.

Business in metals is fairly good and the outlook appears to be quite as good as a year ago. The condition for buyers is more favorable than it was then, for, while prices are now considered to be at the bottom, a year ago the prospect was for a lower market.

PIG IRON—There is not a great deal of buying just now as the users of pig iron in

Canada have largely bought for their future requirements. The market is steady as to price, and the outlook is for a continuance of this condition. Shipments of pig iron from the new furnace at Sydney are on their way to the Toronto market.

BAR IRON—Prices have been withdrawn this week by the rolling mills and the latter will only now quote on application. The wholesalers are, however, still quoting to the retail trade at from $1.65 to $2.70 per 100 lb. base.

IRON PIPE — Although prices have been withdrawn by the manufacturers, jobbers are still quoting as before. The demand is fair. We quote as follows: Black pipe ¼ in., $2.95; ⅜ in., $3.00; ½ in., $3.05; ¾ in., $3.30; 1 in., $4.70; 1¼ in., $6.40; 1½ in., $7.70; 2 in., $10.25. Galvanized pipe is as follows: ½ in., $4.85; ¾ in., $5.25; 1 in., $7.25; 1¼ in., $9.75; 1½ in., $11.50; 2 in., $15.50.

STEEL — The steel market is firmer, and some of the manufacturers in Canada have this week withdrawn their prices. The situation this week is still very strong in the United States, where premiums are being paid on pool prices for steel billets.

PIG TIN—There have been some sharp declines in both New York and London during the past week, but no change has been made in figures of jobbers' locally. Trade, however, is dull, and the feeling easy in sympathy with the outside markets. We quote 32 to 33c.

TINPLATES — Business is a little more active than it was. This is partly due to the fact that depleted stocks have been replenished during the past week, and the jobbers are in consequence in a better position to do business.

TINNED SHEETS—These are now going out nicely, being required for making up into tin cans and cheese vats. We quote 9 to 9½c. for 28 gauge.

BLACK SHEETS—Trade is fair. Prices are being firmly held, buyers having been compelled to pay slightly higher prices than a short time ago.

GALVANIZED SHEETS—Trade has not yet properly opened up and will not until there has been an improvement in the weather. Consequently only a moderate business is being done. We quote: English at $4.75 in small lots, and American at $4.50, but these prices can be shaded for quantities.

CANADA PLATES—Local shipments are light, but import orders are being placed with a little more freedom. We quote all dull, $3. half-and-half, $3.15, and all bright, $3.65 to $3.75.

COPPER — Trade is much the same as it was a week ago, both in ingot and sheet. In the outside markets copper is fairly steady. We quote: Ingot, 19 to 20c.; bolt or bar, 23½ to 25c.; sheet, 23 to 23½c. per lb.

BRASS — Business is fair, and the discount on rod and sheet unchanged at 15 per cent.

SOLDER—A reasonable amount of business is being done. We quote: Half-and-half, guaranteed, 19c.; ditto, commercial, 18½c.; refined, 18½c., and wiping, 18c.

LEAD—Trade is fair. The outside markets are weak, prices having declined 15s. in London on Tuesday last, although it recovered 5s. before the close. Local quotations are unchanged at 4¾c. to 5c. per lb.

ZINC SPELTER — The market locally is dull and prices outside are weak. We still quote 6 to 6½c. per lb.

ZINC SHEET—We quote casks at 6¾ to 7c., and part casks at 7 to 7½c. per lb.

ANTIMONY—Trade is fair at 11 to 11½c. per lb.

PAINTS AND OILS.

As usual, during the first week of March, there has been a big movement of goods. Linseed oil is selling in great quantities. White lead is in brisk demand, Quite a lot of turpentine is selling, notwithstanding the fact that prices are lower in Savannah and a decline is looked for here. The demand for paris green is fair. Prepared paints, varnishes and sundries are in excellent request We quote:

WHITE LEAD—Ex Toronto, pure white lead, $6.50; No. 1, $6.12½; No. 2. $5.75; No. 3. $5.37½; No. 4. $5; dry white lead in casks, $6.

RED LEAD—Genuine, in casks of 560 lb., $5.50; ditto, in kegs of 100 lb., $5.75; No. 1, in casks of 560 lb., $5 to $5.25; ditto, kegs of 100 lb.; $5.25 to $5.50.

LITHARGE—Genuine, 7 to 7½c.

ORANGE MINERAL—Genuine, 8 to 8½c.

WHITE ZINC—Genuine, French V.M., in casks, $7 to $7.25; Lehigh, in casks, $6.

PARIS WHITE—90c. to $1 per 100 lb.

WHITING — 70c. per 100 lb.; Gilders' whiting, 80c.

GUM SHELLAC — In cases, 22c.; in less than cases, 25c.

PARIS GREEN—Bbls., 16¾c.; kegs, 17c.; 50 and 100-lb. drums, 17½c.; 25-lb. drums, 18c.; 1-lb. papers, 18½c.; 1-lb. tins, 19½c.; ½-lb. papers, 20½c.; ½-lb. tins, 21½c.

PUTTY — Bladders, in bbls., $2.20; bladders, in 100 lb. kegs, $2.35; bulk in bbls., $2; bulk, less than bbls. and up to 100 lb., $2.15; bladders, bulk or tins, less than 100 lb., $3.

PLASTER PARIS—New Brunswick, $1.90 per bbl.

PUMICE STONE — Powdered, $2.50 per cwt. in bbls., and 4 to 5c. per lb. in less quantity; lump, 10c. in small lots, and 8c. in bbls.

LIQUID PAINTS—Pure, $1.20 to $1.30 per gal.; No. 1 quality, $1 per gal.

CASTOR OIL—East India, in cases, 10 to 10½c. per lb. and 10½ to 11c. for single tins.

LINSEED OIL—Raw, 1 to 4 barrels, 69c.; boiled, 72c.; 5 to 9 barrels, raw, 68c.; boiled, 71c., delivered. To Toronto, Hamilton, Guelph and London, 1c. less.

TURPENTINE—Single barrels, 62c.; 2 to 4 barrels, 61c., to all points in Ontario. For less quantities than barrels, 5c. per gallon extra will be added, and for 5-gallon packages, 50c., and 10-gallon packages, 80c. will be charged.

GLASS.

Import orders continue large, but the margin keeps small as there is no indication of lower prices in Belgium. Stock quotations are firm, with a good movement recorded. We still quote first break locally as follows: Star, in 50-foot boxes, $2.10, and 100-foot boxes, $4 double diamond under 26 united inches, $6 Toronto, Hamilton and London; terms 4 months or 3 per cent. 30 days.

MARKET NOTES.

Oiled and annealed wire is sold under a new list.

The manufacturers of bar iron and iron pipe have withdrawn their prices.

Quotations on galvanized wire and barbed wire from stock are lower in sympathy with the decline in the f.o.b. Cleveland figures.

The Toronto branch of the McClary Manufacturing Co. is this week in receipt of a carload of "Leonard Cleanable" refrigerators.

D. W. Ross, general merchant, Parry Sound, Ont., has sold out to W. Adair & Co., who will continue in the present stand.

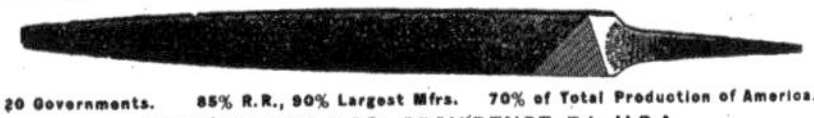

MANITOBA MARKETS.

WINNIPEG, March 4, 1901.

THE market continues very quiet and business dull, although some slight improvement is noticeable along certain lines.

Prices for the week remain unchanged, as follows :

Barbed wire, 100 lb.	$3 45
Plain twist	3 45
Staples	3 95
Oiled annealed wire 10	3 95
" 11	4 00
" 12	4 05
" 13	4 20
" 14	4 35
" 15	4 45
Wire nails, 30 to 60 dy, keg	3 45
" 16 and 20	3 50
" 10	3 55
" 8	3 65
" 6	3 70
" 4	3 85
" 3	4 10
Cut nails, 30 to 60 dy.	3 00
" 20 to 40	3 05
" 10 to 16	3 10
" 8	3 15
" 6	3 20
" 4	3 30
" 3	3 65
Horsenails, 45 per cent. discount.	
Horseshoes, iron, No. 0 to No 1	4 65
No. 2 and larger	4 40
Snow shoes, No. 0 to No. 1	4 90
No. 2 and larger	4 40
Steel, No. 0 to No. 1	4 95
No. 2 and larger	4 70
Bar iron, $2.50 basis.	
Swedish iron, $4.50 basis.	
Sleigh shoe steel	3 00
Spring steel	3 25
Machinery steel	3 75
Tool steel, Black Diamond, 100 lb	8 50
Jessop	13 00
Sheet iron, black, 10 to 20 gauge, 100 lb.	3 50
20 to 26 gauge	3 75
28 gauge	4 00
Galvanized American, 16 gauge	2 54
18 to 22 gauge	4 50
24 gauge	4 75
26 gauge	5 00
28 gauge	5 25
Genuine Russian, lb	12
Imitation " "	8
Tinned, 24 gauge, 100 lb	7 55
26 gauge	8 80
28 gauge	8 00
Tinplate, IC charcoal, 20 x 28, box	10 75
" IX "	12 75
" IXX "	14 75
Ingot tin	35
Canada plate, 18 x 21 and 18 x 24	3 75
Sheet zinc, cask lots, 100 lb	7 50
Broken lots	8 00
Pig lead, 100 lb.	6 00
Wrought pipe, black up to 2 inch	50 an 10 p.c.
" Over 2 inch	50 p.c.
Rope, sisal, 7-16 and larger	$10 00
" ⅜	10 50
" ¼ and 5-16	11 00
Manila, 7-16 and larger	13 50
" ⅜	14 00
" ¼ and 5-16	14 50
Solder	21½
Cotton Rope, all sizes, lb.	16
Axes, chopping	$7 50 to 12 00
" double bitts	12 00 to 18 00
Screws, flat head, iron, bright	87½
Round " "	82½
Flat " brass	80
Round " "	75
Coach	57½ p.c.
Bolts, carriage	55 p.c.
Machine	55 p.c.
Tire	60 p.c
Sleigh shoe	65 p.c.
Plough	40 p.c.
Rivets, iron	50 p.c.
Copper, No. 8	35
Spades and shovels	40 p.c.
Harvest tools	50, and 10 p.c.
Axe handles, turned, s. g. hickory, doz.	$2 50
No. 1	1 50
No. 2	1 25
Octagon extra	1 75
No. 1	1 25
Files common	70, and 10 p.c.
Diamond	60
Ammunition, cartridges, Dominion R.F.	50 p.c.
Dominion, C.F., pistol	30 p.c.
" military	15 p.c.
American R.F.	30 p.c.
C.F. pistol	5 p.c.
C.F. military	10 p.c. advance.
Loaded shells :	
Eley's soft, 12 gauge black	16 50
chilled, 12 guage	18 00
soft, 10 guage	21 00
chilled, 10 guage	23 00
Shot, Ordinary, per 100 lb	6 75
Chilled	7 50
Powder, F.F., keg	4 75
F.F.G.	5 00
Tinware, pressed, retinned	75 and 2½ p.c.
" " plain	70 and 15 p.c.
Graniteware, according to quality	50 p.c.

PETROLEUM.

Water white American	25½c,
Prime white American	24c.
Water white Canadian	22c.
Prime white Canadian	21c.

PAINTS, OILS AND GLASS

Turpentine, pure, in barrels	$ 68
Less than barrel lots	73
Linseed oil, raw	84
Boiled	87
Lubricating oils, Eldorado castor	25½
Eldorado engine	24½
Atlantic red	27½
Renown engine	41
Black oil	23½ to 25
Cylinder oil (according to grade)	55 to 74
Harness oil	61
Neatsfoot oil	$ 1 00
Steam refined oil	85
Sperm oil	1 50
Castor oil per lb.	11½
Glass, single glass, first break, 16 to 25 united inches	2 25
26 to 40 per 50 ft.	2 50
41 to 50 " 100 ft.	5 50
51 to 60 " " "	6 00
61 to 70 per 100-ft. boxes	6 50
Putty, in bladders, barrel lots per lb.	2½
kegs "	2¾
White lead, pure per cwt.	7 25
No 1 "	7 00
Prepared paints, pure liquid colors, according to shade and color per gal.	$1.30 to $1.90

The Ingersoll Metallic Manufacturing Co., Limited, Ingersoll, Ont., have been incorporated with $20,000 capital stock.

PETERBORO' BOARD OF TRADE.

A special meeting of the Peterboro' Board of Trade was held on Thursday night, last week, to discusss the establishment of a beet sugar industry in that city. President T. E. Bradburn occupied the chair.

The meeting was addressed at some length by Wm. Collins and Dawson Kennedy. Mr. Kennedy pointed out that the Ontario Government had expressed its willingness to assist in the project. There was no better soil in the world for the industry than in Ontario. An association has been formed in Ontario to foster the industry, which was almost sure to reach immense proportions. He thought that there would be at least ten factories established in Ontario, and there is no other business that a farmer can engage in that will be more remunerative for as little capital. As to the capital, it will cost about $25 per acre to cultivate the sugar beet, whilst the output will be worth about $50.

The following committee were appointed to look into the matter : G. W. Fitzgerald, warden of the county ; E. M. Elliott, president of the Farmers' Institute ; Dawson Kennedy ; William Collins, secretary of Farmers' Institute ; T. E. Bradburn, president of the Board of Trade ; Peter Campbell, McF. Wilson, H. LeBrun, Mayor Denne, Jos. Batten, Reeve Adams and J. H. Burnham. At a subsequent meeting of the committee, Mr. Kennedy was appointed chairman and Mr. Collins, secretary of the committee.

The North American Coal Co., Halifax, are applying for incorporation.

TRADE IN COUNTRIES OTHER THAN OUR OWN.

IN view of the advance in the price of raw material, the manufacturers of poultry netting in the United States are withdrawing outstanding quotations and naming slightly higher prices. The great bulk of the netting has been already purchased for the coming season, but this action will have a good effect on the market at large.

WIRE CLOTH HIGHER.

The increased cost of wire, together with the condition of the market and the moder-rate supply on hand, has resulted in an advance in the United States by the leading manufacturers of wire cloth, who are now quoting $1 per 100 square feet. It is thought not unlikely that, as the season advances, with the present prospect of a large business, that something of a scarcity may be developed.

THE BRITISH IRON MARKET.

S. W. Payse & Co., Manchester, Eng., in their monthly report say: "The iron trade continues in an unsatisfactory condition. Cleveland iron is about 6d. cheaper during this month, but business doing has only been small; home buyers are very cautious, and trade with the continent is hindered by the severe weather prevailing there. Many furnaces have recently been stopped in the Cleveland district, and the number now in operation is smaller than for many years past. Prices. of Scotch iron are some 2s. higher during the month, but the volume of business is very moderate. consumers only buying for pressing requirements."

LINSEED OIL IN CHICAGO.

The market is unsettled and lower. Demand is light and some apparent anxiety on the part of a few of the crushers to realize on some part of their holdings is in evidence. Deliveries on contracts are going forward quite freely, and most of the grinding trade will be provided for in this manner for some weeks to come. Exaggerated reports were sent out from Chicago during the week, misrepresenting the conditions here; this influence has not added to the tranquility of trade thought in the West, to say the least. The market is far from being demoralized and no 58c. oil is in sight. The market may be quoted at 60c. per 7½ lb. (8c. per lb.) as an inside figure, with but one mill naming that price. Others are holding at 1 to 2c. above that basis. Early in the week some second-hand oil sold at 61c.; this started the decline. In a jobbing way trade is very quiet, small manufacturing consumers buying 3 to 5 bbl. lots as they need it, but the retail dealers are holding off for mild weather.—Paint, Oil and Drug Review, March 6.

PLUMBERS' BRASS WORK IN THE STATES.

The manufacturers of Plumbers' brass work, at their meeting held in New York city on February 15, adopted a price list, which showed an advance in cost to the jobbers of an average of five per cent. over the previous quotations. The prices adopted were understood to be the extreme prices at which any manufacturer would sell his wares. No discount sheets were issued to the jobbing trade, unless upon application by the jobber. In no instance were the manufacturers to send discount sheets indiscriminately through the mail. It was the opinion of the meeting that considerable business was at hand. The only element which prevented the placing of these orders was the fact that the jobbing trade did not know which way the market was going to turn or what the manufacturers intended to do. It was contended by several of the members present at the meeting that if the manufacturers would only stand together and maintain their prices and show the jobbing trade that it was their intention to maintain prices and also to keep the market in a stable and unbroken condition during the spring months the jobbers would have no hesitation whatever in placing their orders for their needs. The market during the past few months has been the most remarkable one known for years. The jobbers' stocks of brass work are now smaller than they have been in any previous year at this season. This is owing to the fact that we have experienced an extraordinarily mild winter; a winter which enabled the plumbers to continue their work out of doors, and building has been exceptionally brisk up to the present time, thus keeping up the continued and even demand for all classes of goods entering into the plumbing supply lines. As we are now within a month of the busiest season of the year, and as the jobbing trade is sorely in need of new stock, and the manufacturers, owing to the demands made upon them from time to time to keep the jobber supplied in a hand-to-mouth way, have very little stock at the factories, the moment orders begin to pour into the various manufacturers, and the market shows a healthy and steady appearance, the prices are likely to go up in a material degree and stay there. The advance of 5 per cent., which was recently made, may shortly materialize into a 10 per cent. advance. This may not be done by joint action on the part of the manufacturers, but will simply be the effect of an increased demand on the part of the jobbers.—Metal Worker.

NEW YORK METAL MARKETS.

TIN—Under the unfavorable conditions previously noted, the market for pig tin continues to decline. In the New York market a further break of 20 points was recorded early in the day; 5 tons spot sold at 26.25c., but, although there were further offerings at that figure, there were no more sales, as buyers did not bid over 26c. The close here was dull and weak. March could have been bought at 26.15c., while 26c. was bid. There were no bids on later months, asking prices being 26c. on April, 25.50c. on May, 25.25c. on June and 25c. on July. The morning cables from London report a weak market, with a decline of 17s. 6d. on spot and of 10s. on futures, but after noon the feeling was better and the market closed steady without further change.

COPPER—The London market was slightly higher but closed quiet. There was nothing new in the situation locally, the demand being light, and prices somewhat nominal at 17c. for Lake Superior and 16⅝c. for electrolytic and casting.

PIG LEAD—There was another sensational break in the London market to day. This morning the price of soft Spanish declined 15s. to £13 12s. 6d., but after noon recovered somewhat, closing 5s. above the low point. Nothing new was presented in the New York market, which remained quiet with prices unchanged on the basis of 4.37½c. in lots of 50 tons or over. St. Louis telegrams reported that the market was firm at 4.22½ to 4.25c.

SPELTER—The market remains dull and weak. Prices, however, were without further change, the closing quotations being $3.92½ to 3.97½c. for spot and 3.90 to 3.95c. for March to May delivery. In St. Louis the market was still unsettled at 3.75 to 3.82½c.

ANTIMONY — There was a moderate jobbing demand for regulus at the range of 8¾ to 10¼c., as to quantity and brand.

OLD METALS — The market remains quiet, with prices unchanged.

IRON AND STEEL—Conditions in these markets are virtually unchanged. Locally more interest is being manifested in pig iron, and the market is hardening in sympathy with advices from other centres, but there has as yet been no actual change in quotations, and no very important transactions have come under our notice. Pittsburg continues to report an active market with a decided upward tendency to prices. It is reported there that talk is heard that an advance of $2 per ton in the price of steel plates will probably be made at the next meeting of the pool, and an advance [illegible] billets is also expected. The pig iron furnaces in the Pittsburg district are reported to be largely sold up until July.

TINPLATE — Continued activity is reported, but no fresh features were presented in this market to-day.—New York Journal of Commerce, March 6.

HEATING AND PLUMBING

FROZEN CONDUCTOR PIPES.

THE heavy snows and extreme cold weather during the month of February have already made a considerable business for roofers and tinsmiths in thawing out and repairing conductor pipes, and when the weather is milder it is probable that many will have to be replaced. The snow on the roof will thaw from the effects of the sun, even when the temperature is below the freezing point; as a result the water which flows down the conductor pipe freezes until the pipe becomes solid with ice, after which icicles will form on the outside of the conductor pipe until the weight is sufficient in some instances to tear the pipe away from its fastening to the building.

In other instances these huge icicles have partially melted away, so that they are supported by only a slender thread of ice, and in case of a little extra sunshine, or a heavy wind, icicles weighing more than 100 lb. fall on the sidewalk, endangering the lives of foot travellers, and in some instances it has been found necessary to put railings along the street to keep foot travellers outside of the danger line. These troubles have all been experienced where the conductor pipe runs, as it is customary in the majority of instances, on the outside of the building. The more modern practice of running the conductor pipe inside the building keeps the pipe warm enough so that it never becomes solid with ice, and no matter what the outside conditions may be, of freezing or thawing, any water resulting from a thaw finds escape through the open pipe. It is pointed out by those who have observed the icicles that invariably the greatest trouble is from conductor pipes disposing of water from the south side of roofs which are exposed to the rays of the sun.—Metal Worker.

TORONTO BUILDING PERMITS.

Building permits have been issued in Toronto to The Toronto Plate Glass Co. for a glass-bending factory at 209-213 Victoria street, to cost $1,200; to John C. Palmer, for an additional storey to the Palmer House, to cost $15,000 (Langley & Langley, architects); to W. Robinson, for a dwelling near College on Concord avenue, to cost $3,000; to J. J. Walsh, for three pairs semi-detached houses on Beatty avenue, near King street, to cost $15,000; Reid & Brown, for a foundry with galvanized iron roof, on Esplanade street, near West Market, to cost $14,000; to Joanna Duff for a pair houses at 20 and 22 Columbus avenue, to cost $1,000; to Mrs. Julia Grandier, for two dwellings on Clinton street, near College, to cost $1,200; to Caleb Evans, for a residence on Spadina road, near Prince Arthur avenue, to cost $5,000; to H. M. Death, for two residences at 765 and 767 Euclid avenue, to cost $2,500; to R. Score, for a corrugated iron roof to 77 King street west to cost $1,600; to Richard Dinnis, for two residences at 152 and 154 William street, to cost $2,600. The total for the week is $50,800. This improves the prospects, as the value of permits issued during January and February was only $82,510, as compared with $167,125 in 1900.

PLUMBING AND HEATING NOTES.

ARCHITECT MATTHEWS, London, Ont., is preparing plans for a $5,000 residence for D. B. Colbeck, Woodstock, and a $4,000 Methodist church at Brownsville, Ont.

John Greenway is erecting a residence near Oak River, Man.

A new brick manse will be erected at Largie, Ont., this season.

The Hornerite church will be erected in Cobden, Ont., during the summer.

Barracks for North West Mounted Police will be erected at Gleichen, N.W.T.

R. Westwood, Philadelphia, Pa., intends erecting a carpet factory in Cornwall.

Architect Wideman's plans for the Fenwick Presbyterian church, Guelph, Ont., have been accepted.

The freeholders of the new parish of St. Leo, Westmount, Montreal, have decided to erect a church at a cost of $30,000 and a presbytery to cost $10,000. Sherbrooke street and Argyle avenue is the proposed location.

Walter Suckling, Winnipeg, has purchased, on behalf of a client, the property on the northwest corner of McDermott and Albert streets, Winnipeg, and will, next spring, commence with the erection of a three-storey block with 135 feet frontage on McDermott avenue and 100 on Albert street. The building, it is estimated, will cost $65,000.

BUILDING PROSPECTS IN OTTAWA.

The outlook for Ottawa building trades is unusually bright this spring. Since November last, permits for buildings to the value of nearly $100,000 were issued, so builders are looking for the busiest season for many years.

THE APPRENTICESHIP QUESTION.

AT the recent meeting of the Master Plumbers' Association of Missouri, the committee on apprenticeship, according to a report in The Plumbers' Trade Journal, presented the following:

The National apprenticeship committee submitted an exhaustive report to the Baltimore convention of 1900, and same was referred to the executive board of 1901 for joint consideration with the committee—the intention being that an endorsement or a disapproval of the plans outlined would be submitted to the Kansas City convention by the executive board. Since that time the National executive board has held its mid-winter meeting in New York City and given the matter considerable attention. They concluded to recommend restrictions for local associations along the line mapped out by the apprenticeship committee, but do not consider it practical to adopt any uniforn National law at this time. This will, therefore, be the tenor of their report to the next convention. I submit the system that the apprenticeship committee desired adopted (in its general form at least). This system has been approved by the Ohio State Association, from the office of which the plan was first submitted:

"1. That the apprentices be governed by uniform system under the control of the National, State and local associations as follows: Certificates of apprenticeship to be issued by the National to the State association and from the State to the local, the locals to have an apprenticeship committee, consisting of three or more members, for the purpose of examining all applicants before apprenticeship.

"2. Applicants for apprenticeship must have the following qualifications: First, must be a boy of at least 16 years of age; second, must be sound physically; third, must have a good common school education; fourth, must be fairly bright and intelligent; fifth, must serve at least five years in apprenticeship.

"3. The apprentices are to be allotted as follows: Each and every plumbing shop having two men shall be allowed one apprentice, and for each and every three additional men he shall be allowed one apprentice. This to go into effect and to be in force for a term of five years."

An endorsement of such a system would help it on the floor of the next convention, but if the State convention here assembled does not favor a National law, then it would be the next best thing to concur in the conclusions of the National executive board and recommend the matter to local consideration.

OWN THEIR OWN LIGHTING PLANT.

The town of Woodstock, Ont., on Friday last took over the electric light plant and will in future run it themselves. Over two months ago the property owners voted on the purchase and it carried by a large majority. Since then, however, there has been considerable trouble over missing bonds, but up to date there are only $2,000 worth missing, and the town took over the plant, paying the purchase price, less the $2,000. When these bonds are found the other $2,000 will be paid the company. The purchase price is $14,000, $12,000 of which has been paid.

Wm. Rea, secretary of the Ottawa Public School Board, is asking bulk tenders for a new school building on Wellington street, Ottawa.

BUSINESS CHANGES.

DIFFICULTIES, ASSIGNMENTS, COMPROMISES.

A MEETING of the creditors of J. G. Fairbanks, general merchant, Spruce Grove, N.W.T., is to be held on Saturday.

Wm. H. Slight, builder, Cookstown, Ont., has assigned to John Elliott.

Joseph L. Brodeur, general merchant, St. Hyacinthe, Que, has assigned.

David Hacht, general merchant, Tichborne, Ont., is offering 30c. on the dollar, cash.

Walter Wardrop, general merchant, Lac des Bonnett, Man., has assigned to C. H. Newton.

Leonide Sicotte, general merchant, Boucherville, Que., has assigned to Kent & Turcotte.

A meeting of the creditors of Elizabeth Marshall, dealer in tinware, etc., Dunnville, Ont., was held this week.

B. J. Pattenier, machinist, Montreal, has assigned, and a meeting of his creditors has been called for February 14.

J. E. Hutton, general merchant, Thornbury, Ont., has assigned to Alfred Wood, and is offering to compromise.

Joseph Bernier, carriagemaker, L'Islet, Que., is offering 25c. on time, or 20c. cash, on the dollar. D. Arcand has been appointed curator of his estate.

PARTNERSHIPS FORMED AND DISSOLVED.

Leitch & Liphardt, hardware dealers, etc., Waterloo, Ont., have dissolved.

George Stephens & Co., hardware dealers, etc., Chatham, Ont., have admitted D. H. Douglas.

J. Sweeny has succeeded F. A. Empey as partner with Geo. A. Rendell as proprietors of the Eholt Trading Co., general merchants, Eholt, B.C.

Copartnership has been registered by J. J. Hughes and P. E. McFarlane under the style of Hughes & McFarlane, general merchants, Souris, P.E.I.

SALES MADE AND PENDING.

The assets of I. J. Dupont, hardware dealer, Farnham, Que., are to be sold.

The assets of R. Bourbeau, general merchant, Victoriaville, Que., have been sold.

Krotz & Walter, general merchants, Listowel, Ont., have sold out to Walter Bros.

The stock of the estate of W. S. Grout, general merchant, Minto, Man., has been sold.

The stock, etc., of the general store of W. J. Brompton, Moorefield, Ont., is advertised for sale by auction.

The stock of Alf. Boulanger, general merchant, St. Eugene (L'Islet), Que., has been sold at 64¼c.

The assets of The Ossekeag Stamping Co., Limited, manufacturers of tinware, Hampton, N.B., are advertised for sale by tender up to April 1.

The business of W. H. Michener, blacksmith, etc., Dunnville, Ont., is advertised for sale.

The stock, etc., of Mowat & Co., hardware dealers, Trenton, Ont., is advertised for sale on March 11.

The assets of A. Gibault, general merchant, St. Lucie de Doncaster, Que., are to be sold on the 12th inst.

The stock, etc., of the estate of Morrison & Co., general merchants, Boissevain, Man., is offered for sale by auction.

The business of John H. Grant & Co., manufacturers of agricultural implements, baskets, etc., Grimsby, Ont., is advertised for sale.

C. & V. B. Fullerton, general merchants, Parrsboro', N.S., have sold the Holmes hardware stock, which they recently purchased, to John W. Cameron.

CHANGES.

Duphin & Frere have registered as sawmillers in Stoke, Que.

Peladeau & David have registered as painters in Montreal.

Lacroix & Leger have registered as hardware dealers, etc., in Montreal.

Wells & Frary have registered as general merchants at Freligh̃sburg, Que.

W. J. Miller, carriagemaker, Carman, Man., has sold out to J. J. Levarton.

The business of J. W. Carwin, harness dealer, Elkhorn, Man., is to be wound up.

Wm. S. Brown has registered as proprietor of the Hub, general store, Cowansville, Que.

The stock of the estate of H. L. Moore, crockery dealer, St. John, N.B., has been sold to E. F. Copp.

Mrs. Nap. Bray has registered as proprietress of N. Bray & Co., carriagemakers, Coteau Station, Que.

FIRES.

J. F. Norton, general merchant, Cardigan, P.E.I., has been burned out; insured for $7,000.

Jeffrey Bros., manufacturers of carriages and implements, Petitcote, Que., have been burned out.

The stock of H. L. & J. T. McGowan, painters, etc., St. John, N.B., has been damaged by fire; insured.

DEATHS.

Peter Armstrong Sr., carriagemaker, Hamilton, is dead.

Francois Bouffard, general merchant, St. Pierre (Montmagny), is dead.

C. J. Marchildon, general merchant, St. Pierre les Becquets, Que., is dead.

Thomas Courtice, of Courtice & Jeffrey, wholesale and retail harness dealers, Port Perry, Ont., is dead.

Henry Rogers, Sons & Co.

Wolverhampton, England.

Manufacturers of

"Union Jack" Galvanized Sheets
Canada and Tin Plates
Black Sheets
Sleigh Shoes and Tyre Steel
Coil Chain, Hoop Iron
Sheet and Pig Lead
Sheet Zinc

Quotations can be had from
Canadian Office:
6 St. Sacrament St., MONTREAL

F. A. YORK, Manager.

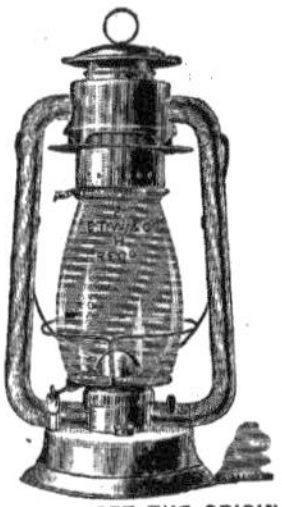

If you want the BEST
Cold Blast Lantern
in the world, get WRIGHT'S.

—GET THE ORIGINAL—
We lead, others imitate.

E. T. WRIGHT & CO.
Manufacturers. *HAMILTON, ONT.*

TRADE MARK J REGISTERED

"JARDINE" HAND DRILLS

Five different sizes to suit all requirements. It pays to sell the best tools.

A. B. JARDINE & CO.
HESPELER, ONT.

DIAMOND EXTENSION FRONT GRATE.

Ends Slide in Dovetails similar to Diamond Stove Back.

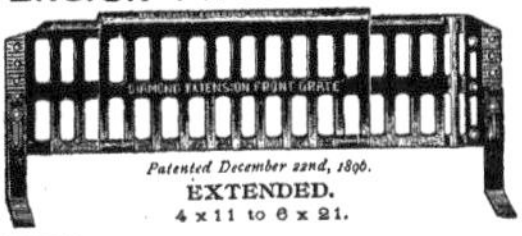

Patented December 22nd, 1896.
EXTENDED.
4 x 11 to 6 x 21.

Diamond Adjustable Cook Stove Damper

Patented March 14th, 1893.

For Sale by Jobbers of Hardware

Manufactured by THE ADAMS COMPANY, Dubuque, Iowa, U.S.A.
" A. R. WOODYATT & CO., Guelph, Ontario.

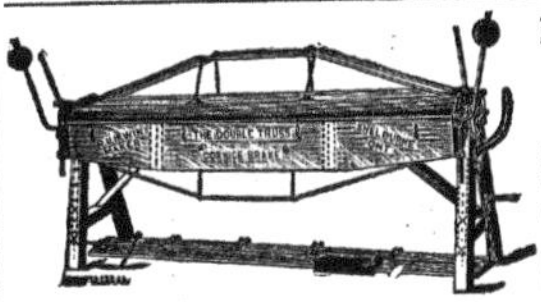

This eight-foot Brake bends 22-gauge iron and lighter, straight and true.

Price, $60

Very handy beader attachment, $15 extra if required.
Send for circulars and testimonials to

The Double Truss Cornice Brake Co. SHELBURNE, ONT.

The Latest and Best.

H. & R. Automatic Ejecting Single Gun.

Model 1900.

Steel and Twist Barrels in 30 and 32-inch.
12 Gauge.

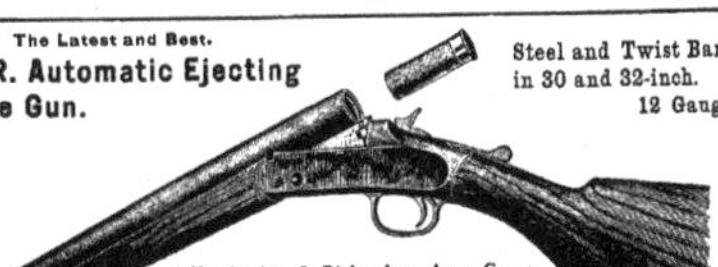

Harrington & Richardson Arms Co.
Worcester, Mass., U.S.A.
Descriptive Catalogue on request.

STEVENS IDEAL, NO. 44

This is as reliable and accurate a rifle as can be constructed. Placed at a moderate price to meet the demand for such a rifle. It is recommended without qualification and fully guaranteed. Made in the following styles:

.22 Long-Rifle R. F., 25 Stevens R. F., and .32 Long R. F. Standard length of barrel for rim-fire cartridges, 24 inches. Weight 7½ pounds.

.25-20 Stevens C. F., .32-40 C. F., .38-55 C. F., and .44-40 (.44 W. C. F) Standard length of barrel for center-fire cartridges, 26 inches. Weight, 7¾ pounds.

Half-octagon barrel, oiled walnut stock and fore-arm, rifle butt, case-hardened receiver, sporting rear and Rocky Mountain front sight.

Price, with standard length of barrel, $13.00.

Can be obtained of any of the leading jobbers in Canada at liberal discount from this price.

Send for complete catalogue of our full line of Rifles, Pistols and Machinists' Tools.

J. Stevens Arms & Tool Co., P. O. Box 217, Chicopee Falls, Mass., U.S.A.

HUTCHISON, SHURLY & DERRETT

DOVERCOURT TWINE MILLS. **1078 BLOOR STREET WEST TORONTO.**

Having equipped our Factory with entirely new machinery, we are prepared to furnish the best made goods in the market at closest prices and make prompt shipments.

Hand Laid Cotton Rope and Clothes Lines,
Cotton and Russian Hemp Plough Lines, plain and colored.
Cotton and Linen Fish Lines, laid and braided.
Netted Hammocks, white and colored, Tennis and Fly Nets.
Skipping Ropes, Jute, Hemp and Flax Twines.

THE TOWN OF GODERICH.

AT the recent annual meeting of the Goderich Board of Trade, the following interesting reference to the industries of the town appeared in the address of President Williams:

"Now, I want to say a few words with special reference to the town. Whilst we cannot at the moment boast that our population is the largest on record, it can I think be safely said that in no time of the town's history was it in better shape, nor its future brighter. We can well afford to be proud of it both as a town to live in, and as a manufacturing place, with its rail and lake facilities. It has a natural beauty. Its fine situation, its elevation, etc., etc., make it an ideal spot for a town to have been built. Of late years the place has vastly improved, many new buildings have gone up, both business places and houses, the latter mostly of an artistic nature, giving place to the old style frame structure, thereby adding to the beauty of the surroundings.

"The day of the dingy office and ill-lighted workroom in Canada is almost a thing of the past. In business matters we have vastly improved. We have our furniture factory, organ factory, knitting factory, bicycle and engine factory, saw mill, planing mills, an elevator second to none, and other industries employing a larger number of wage-earners than we ever had before; and a warm word of thanks is due to the few men amongst us who risked their money in establishing these industries, adding so materially to the prosperity of the town. May they succeed beyond their anticipation and more than double their money, is my wish. It is gratifying also to know that the "Big Mill," which has been idle so long, will soon be in full running order again, and may good luck be with its energetic manager! Further, we have perhaps more telephone services, more miles of electric wiring, more miles of sewerage and better drainage, more miles of water mains (though it must be admitted the water supplied us should be improved upon) than any other town of our population in the country. The soundness of the town may be illustrated by the fact that we have had but few failures in recent years. Chief amongst the reasons which have brought about at least some of these, may be noted the throwing upon the market of bankrupt stocks, a condition of affairs which I find exists all over the country, interfering with legitimate trade, demoralizing business generally in some lines, and being manifestly unjust to the hundred cents on the dollar man. Legislation is equired to correct this evil, and this might be brought about by the efforts of Boards of Trade." The officers elected were:

President—Mr. R. S. Williams.
Vice-President—Mr. Colborne.
Secretary—Mr. James Mitchell.
Treasurer—Mr. W. A. McKim.
New Council—F. W. Doty, S. A. McGaw, Robert McLean, Wm. Campbell, N. B. Smith, W. C. Goode, Alex. Saunders, G. F. Emerson, George Porter, G. M. Elliott, Joseph Beck and George Acheson.

MERCHANTS' DAY IN WOLFVILLE.

MERCHANTS' DAY, as advertised by the merchants of Wolfville, N.S., has now been in existence for three seasons. There is a difference of opinion among us as to its advantages. Some are in favor of it, and others are sceptical, but we think on the whole it has been beneficial in keeping at home trade that seemed to be drifting to other towns and especially to the departmental stores of Toronto that do a large business in this county.

Our method is to combine our advertising in the local papers by taking a large space and offering all the special bargains we can beside a general discount on all stock of 10 per cent. for that day. This reduces the cost to a very small figure for each one.

We agree at a meeting of all the merchants on the special goods that each will sell on that day in order that there may be no clashing, and the total list certainly offers great inducements to the customers. As our business is classified here, customers can make their list of dry goods, groceries, boots and shoes, hardware and drugs, etc., and by taking in each store get a lot of useful goods at very low figures.

We are careful to do exactly as we advertise; make our list of inducements—goods that the buying public need. Of course we drop a lot of profit but our sales are large and in the rush and excitement of a big sale lots of unseasonable and slow selling lines are worked off.

We have chosen the holiday season—generally a week before Christmas—to allow dealers in fancy goods the same opportunity of participating. The evil attending it is that so many wait from the date of the announcement of sale that trade is generally dull for a few days preceding the "merchants' day," but we seem to have our regular Christmas trade after it.

We think it has established more confidence among our customers that the local dealer is trying to do as well for them in the way of low prices as the city stores and prices quoted in catalogues. Personally, I think such a sale would be of great advantage also in the Spring, say, April or May.

If any of the readers of HARDWARE AND METAL wish any further details we should be pleased to give all the information we can.

J. D. CHAMBERS.

Wolfville, N.S., Feb. 22, 1901.

The assets of Lemay & Marchand, general merchants, Shawenegan Falls, Que., are offered for sale.

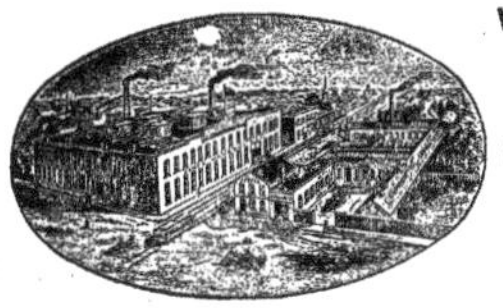

CURRENT MARKET QUOTATIONS.

March 8, 1901.

These prices are for such qualities and quantities as are usually ordered by retail dealers on the usual terms of credit, the lowest figures being for larger quantities and prompt pay. Large cash buyers can frequently make purchases at better prices. The Editor is anxious to be informed at once of any apparent errors in this list as the desire is to make it perfectly accurate.

METALS.

Tin.

Lamb and Flag and Straits—56 and 28 lb. ingots, per lb.	0 32	0 33

Tinplates.

Charcoal Plates—Bright

	Pe box
M.L.S., equal to Bradley.	
I.C., usual sizes	$6 75
I.X., "	8 25
I.X.X., "	9 75
Famous—	
I.C.	6 75
I.X	8 25
I.X.X	9 50
Raven & Vulture Grades—	
I.C., usual sizes	5 00
I.X., "	6 00
I.X.X "	7 0)
I.XXX., "	8 00
D.C., 12½x17	4 75
D.X.	5 50
D.X.X	7 50

Coke Plates—Bright

Bessemer Steel—	
I.C., usual sizes	4 15
I.C., special sizes, base	4 50
20 x 28	8 50

Charcoal Plates—Terne

Dean or J. G. Grade—	
I.C., 20x28, 112 sheets	8 50
I.X., Terne Tin	10 50

Charcoal Tin Boiler Plates.

		Per lb.
Cookley Grade—		
X.X., 14x56, 50 sheet bxs; " 14x60 "; " 14x65 "	0 07	0 07½

Tinned Sheets

72x30 up to 24 gauge	0 08	0 08½
" 26 "	0 08½	0 09
" 28 "	0 09	0 09½

Iron and Steel.

Common Bar, per 100 lbs.	1 65	1 70
Refined " "	....	2 15
Horse Shoe Iron "	....	2 05
Hoop steel, 1½ to 3 in. base, extras for smaller sizes	....	3 10
Sleigh Shoe Steel " base	....	2 00
Tire Steel	....	2 00
Machinery iron finish	....	2 05
Cast Steel, per lb	0 00	0 00
Toe Calk Steel	....	2 30
T. Firth & Co's special cast steel, per lb.		0 13
Jessop's Tool Steel	....	0 13

Boiler Tubes.

1½-inch	0 12½
2 "	0 13
2½ "	0 15
3 "	0 16
3½ "	0 20
4 "	0 25

Steel Boiler Plate.

[illegible]	2 25
[illegible] inch	2 25
½ inch and thicker	2 25

Black Sheets.

gauge	3 10
20 gauge	3 00
22 to 24 "	3 10
26 "	3 20
28 "	3 30

Canada Plates.

All dull, 52 sheets	....	3 00
Half polished	....	3 15
All bright	3 65	3 75

Iron Pipe.

Black pipe—	
¼ inch	3 00
⅜ "	3 00
½ "	3 15
¾ "	3 90
1 "	4 50
1¼ "	6 35
1½ "	7 55
2 "	10 10
2½-6 inch, discount 60 p.c.	
Galvanized pipe—	
½ inch	4 65
¾ "	5 35
1 "	7 15
1¼ "	9 75
1½ "	11 25
2 "	15 50

Galvanized Sheets.

	G. C. Comet.	Amer.	Queen's Head.
16 gauge	....	4 5	4 10
18 to 24 gauge	4 25 / 3 85	4 35	4 25
26 "	4 50 / 4 10	4 35	4 50
28 "	4 75 / 4 35	4 10	4 75

28 gauge American equals 26 gauge English.

Chain.

Proof Coil, 3-16 in., per 100 lb	....	....
" ¼ " "	8 00	8 50
" 5-16 " "	5 35	5
" ⅜ " "	4 35	4 85
" 7-16 " "	4 15	4 65
" ½ " "	4 35	4 50
" ⅝ " "	3 85	4 35
" ¾ " "	3 80	4 00

Halter, kennel and post chains, 40 and 50 p.c.
Cow ties ... 40 p.c.
Stall fixtures ... 35 p.c.
Trace chain ... 25 p.c.
Jack chain, iron, single and double, discount 35 p.c.
Jack chain, brass, single and double, discount 40 p.c.

Copper.

Ingot

English B. S., ton lots	0 19	0 20
Lake Superior	....	....

Bolt or Bar.

Cut lengths round, ½ to ⅞ in.	0 23½	0 25
" round and square 1 to 2 inches	0 23½	0 25

Sheet.

Plain, 14 oz., and light, 16 oz., 14x48 and 14x60	0 23	0 23
Plain, 14 oz., and light, 16 oz., irregular sizes	0 23	0 23½
Tinned copper sheets	....	0 26
Planished	....	0 32

Braziers (In sheets.)

4x6ft. 25 to 30 lbs. ea., per lb.	....	0 25½
" 35 to 45 " "	....	0 24½
" 50-lb. and above, "	....	0 23½

Boiler and T. K. Pitts

Plain Tinned, per lb	....	0 28
Spun, per lb	....	0 32

Brass.

Rod and Sheet, 14 to 30 gauge, 15 per cent.

Sheets, hard-rolled, 2x4	0 24	0 25
Tubing, base, per lb	0 24	0 25

Zinc Spelter

Foreign, per lb	0 06	0 06½
Domestic "	....	....

Zinc Sheet.

5 cwt. casks	6 75	7 00
Part casks	7 00	7 50

Lead.

Imported Pig, per lb	0 04½	0 05
Bar, 1 lb.	0 05½	0 05½
Sheets, 2½ lbs. sq. ft., by	....	0 06½
Sheets, 3 to 6 lbs., "	....	0 06

NOTE.—Cut sheets ½ cent per lb. extra.

Pipe, by the roll, usual weights per yard, lists at 7c. per lb. and 15 p.c. dis. f.o.b. Toronto.
NOTE.—Cut lengths, net price, waste pipe 8-ft. lengths lists at 7½ cents.

Shot.

Common, $6.50 per 100 lb.; chilled, $7.00 per 100 lb.; buck, seal and bal, $7.50. Discount, 7½ p.c. Prices are f.o.b. Toronto, Hamilton, Montreal, St. John and Halifax. Terms 3 per cent. cash, freights equalized on Montreal.

Soil Pipe and Fittings.

Discount, 60 and 10 per cent. on medium and extra heavy, and 60 per cent. on light.

Solder.

	Per lb.	Per lb.
Bar half-and-half, guarant'd	....	0 19
Bar half-and-half, commer'l	....	0 18½
Refined	....	0 18½
Wiping	....	0 18

NOTE.—Prices of this graded according to quantity. The prices of other qualities of solder in the market indicated by private brands vary according to composition.

Antimony.

Cookson's, per lb.	0 11	0 11½

White Lead.

	Per 100 lb.
Pure	6 50
No. 1 do	6 12½
No. 2 do	5 75
No. 3 do	5 37½
No. 4 do	5 00
Munro's Select Flake White	7 37½
Elephant and Decorators' Pure	7 12½
Brandram's B. B. Genuine	8 00
" " Decorative	7 55
" " No. 1	6 85
" " No. 2	6 00

Red Lead.

Genuine, 560 lb. casks, per cwt.	$5 50
Genuine, 100 lb. kegs, per cwt.	5 75
No. 1, 560 lb. casks, per cwt.	5 25
No. 1, 100 lb. kegs, per cwt.	5 00

White Zinc Paint.

Elephant Snow White	0 08	0 09
Pure White Zinc	0 08	0 09
No. 1	0 06	0 07½
No. 2	0 05	0 06½

Dry White Lead.

Pure, casks	5 75
Pure, kegs	6 25
No. 1, casks	5 50
No. 1, kegs	5 00

Prepared Paints.

In ¼, ½ and 1 gallon tins.

Pure, per gallon	....	1 25
Second qualities, per gallon	....	1 10
Barn (in bbls.)	0 75	0 85
The Sherwin-Williams Paints.	....	1 45
Canada Paint Co's Pure	....	1 25
Toronto Lead & Color Co's Pure	....	1 25
Sanderson Pearcy's Pure	....	1 20
Stewart & Wood's Champion Pure.		1 20

Colors in Oil.

25 lb. tins, Standard Quality.

Venetian Red, per lb.	0 05
Chrome Yellow	0 11
Golden Ochre	0 06
French "	0 05
Marine Black	0 09
" Green	0 09
Chrome "	0 08
French Imperial Green	0 09

Colors, Dry.

Yellow Ochre (J.C.) bbls.	1 35	1 40
Yellow Ochre (J.F.L.S.), bbls	....	2 75
Yellow Ochre (Royal)	1 10	1 15
Brussels Ochre	....	2 00
Venetian Red (best), per cwt.	1 80	1 90
English Oxides, per cwt.	3 00	3 25
American Oxides, per cwt.	1 75	2 00
Canadian Oxides, per cwt.	1 75	2 00
Super Magnetic Oxides, 93p.c.	2 00	2 25
Burnt Sienna, pure, per lb.	....	0 01
" Umber, " "	....	0 10
do Raw	....	0 09
Drop Black, pure	....	0 09
Chrome Yellows, pure	....	0 18
Chrome Greens, pure, per lb.	....	0 12
Golden Ochre	....	0 03½
Ultramarine Blue in 28-lb. boxes, per lb.	0 08	0 24
Fire Proof Mineral, per 100 lb.	....	1 00
Genuine Eng. Litharge, per lb	....	0 07
Mortar Color, per 100 lb.	....	1 25
English Vermillion	....	0 80
Pure Indian Red, No. 45 lb.	....	0 80
Whiting, per 100 lb.	....	0 55

Blue Stone.

Casks, for spraying, per lb.	0 07
100-lb. lots, do. per lb.	0 08

Putty.

Bulk in bbls.	2 00
Bulk in less quantity	2 15
Bladders in bbls.	2 20
Bladders in kegs, boxes or loose	2 35
Bladders in 25-lb. tins	2 45
Bladders in 12½-lb. tins	2 75
Bladders in bulk or tins less than 100 lbs	3 00

Varnishes.

	In 5-gal. lots.)	Per gal.
Carriage, No. 1	2 90	3 30
" body	8 00	9 00
" rubbing	4 00	5 00
Gold Size, Japan	3 00	3 40
Brown Japan	2 40	2 80
Elastic Oak	2 90	3 30
Furniture, extra	2 40	2 80
" No. 1	1 60	2 00
Hard Oil Finish	2 70	3 10
Light Oil Finish	3 20	3 60
Demar	3 30	3 70
Shellac, white	4 40	4 80
" orange	4 00	4 40
Furniture Brown Japan	1 60	2 00
Black Japan	2 40	2 80
" " No. 1	1 60	2 00

The Imperial Varnish & Color Co's., Limited Elastilite Varnish 1 gal. can, each. $2.00.

Granatine Floor Finish, per gal. $2.00.

Maple Leaf Coach Enamels; Size 1, 60c; Size 2, 35c.; Size 3, 20c. each.

Castor Oil.

East India, in cases, per lb.	0 10	0 10½
" " small lots	0 10½	0 11

Cod Oil, Etc.

Cod Oil per gal.	0 50	0 55
Pure Olive	....	1 20
" Neatsfoot	....	90

Glue.

Common	0 08½	0 09
French Medal	0 14	0 14½
Cabinet sheet	0 12	0 13
White, extra	0 18	0 20
Gelatine	0 22	0 30
Strip	0 18	0 20
Coopers	0 19	0 20
Huttner	....	0 18

HARDWARE.

Ammunition.

Cartridges.

S. B. Caps Dom. 50 and 5 per cent.
Rim Fire Pistol, dis. 40 p. c., Amer.
Rim Fire Cartridges, Dom., 50 and 5 p.c.
Central Fire Pistol and Rifle, 10 p.c. Amer.
Central Fire Cartridges, pistol sizes, Dom. 30 per cent.
Central Fire Cartridges, Sporting and Military, Dom., 15 and 5 per cent.
Central Fire, Military and Sporting, Amer. add 5 p.c. to list. B.B. Caps. discount 40 per cent, Amer.
Loaded and empty Shells, "Trap" and "Dominion" grades, 25 per cent Rival and Nitro, net list.
Brass shot Shells, 55 per cent
Primers, Dom., 30 per cent.

Wads	per lb
Best thick white felt wadding, in ½-lb bags	1 00
Best thick brown or grey felt wads, in ½-lb. bags	0 70
Best thick white card wads, in boxes of 500 each, 12 and smaller gauges	0 99
Best thick white card wads, in boxes of 500 each, 10 gauge	0 35
Best thick white card wads, in boxes of 500 each, 8 gauge	0 55
Thin card wads, in boxes of 1,000 each, 12 and smaller gauges	0 20
Thin card wads, in boxes of 1,000 each, 10 gauge	0 25
Thin card wads in boxes of 1,000 each, 8 gauge	
Chemically prepared black edge grey cloth wads, in boxes of 250 each—	Per M
11 and smaller gauge	0 60
9 and 10 gauges	0 70
7 and 8 gauges	0 90
5 and 6 gauges	1 10
Superior chemically prepared pink edge, best white cloth wads, in boxes of 250 each—	
11 and smaller gauge	1 15
9 and 10 gauges	1 40
7 and 8 gauges	1 65
5 and 6 gauges	1 90

Adzes.

Discount, 20 per cent.

Anvils.

Per lb.	0 10	0 12½
Anvil and Vise combined	....	4 50
Wilkinson & Co.'s Anvils..lb.	0 09	0 09½
Wilkinson & Co.'s Vices..lb.	0 09¾	0 10

Augers.

Gilmour's, discount 65 and 5 p.c. off list.

Axes.

Chopping Axes—		
Single bit, per doz	5 50	10 00
Double bit, "	11 00	18 00
Bench Axes, 40 p.c.		
Broad Axes, 33⅓ per cent.		
Hunters' Axes	5 50	6 00
Boy's Axes	5 75	6 75
Splitting Axes	6 50	12 00
Handled Axes	7 00	10 00

Axle Grease.

Ordinary, per gross	5 75	6 00
Best quality	13 00	15 00

Bath Tubs.

Zinc 6 00
Copper, discount 15 p.c. off revised list

Baths.

Standard Enameled.

5½-inch rolled rim, 1st quality	25 00
" " " 2nd "	21 00

Anti-Friction Metal.

"Tandem" A	per lb.	0 27
" B	"	0 21
" C	"	0 11½
Magnolia Anti-Friction Metal, per lb.		0 25

SYRACUSE SMELTING WORKS.

Aluminum, genuine	0 45
Dynamo	0 29
Special	0 25
Aluminum, 99 p.c. pure "Syracuse"	0 50

Bells.

Hand.

Brass, 60 per cent.
Nickel, 55 per cent.

Cow.

American make, discount 66⅔ per cent.
Canadian, discount 45 and 50 per cent.

Door.

Gongs, Sargant's 5 50 8 00
" Peterboro', discount 45 per cent.

Farm.

American, each 1 25 3 00

House.

American, per lb. 0 35 0 40

Bellows.

Hand, per doz.	3 35	4 75
Moulders', per doz.	7 50	10 00
Blacksmiths', discount 40 per cent.		

Belting.

Extra, 60 per cent.
Standard, 60 and 10 per cent.

Bits.

Auger.

Gilmour's, discount 60 and 5 per cent.
Rockford, 50 and 10 per cent.
Jennings' Gen., net list.

Car.

Gilmour's, 47½ to 50 per cent.

Expansive.

Clark's, 40 per cent.

Gimlet.

Clark's, per doz.	0 65	0 90
Damond, Shell, per doz.	1 00	1 50
Nail and Spike, per gross	2 25	5 20

Blind and Bed Staples.

All sizes, per lb. 0 07½ 0 12

Bolts and Nuts.

	Per cent
Carriage Bolts, full square, Norway	70
" " full square	70
Common Carriage Bolts, all sizes	65
Machine Bolts, all sizes	65
Coach Screws	75
Sleigh Shoe Bolts	75
Blank Bolts	65
Bolt Ends	65
Nuts, square	4½c. off
Nuts, hexagon	4¾c. off
Tire Bolts	67½
Stove Bolts	67½
Stove rods, per lb.	5½ to 6c.
Plough Bolts	60

Boot Calks.

Small and medium, ball, per M	4 25
Small heel, per M	4 50

Bright Wire Goods.

Discount .... 60 per cent.

Broilers.

Light, dis., 65 to 67½ per cent.
Reversible, dis., 65 to 67½ per cent.
Vegetable, per doz., dis. 37½ per cent.

Henis, No. 8, "	....	6 00
Henis, No. 9, "	....	7 00
Queen City "	7 50	0 00

Butchers' Cleavers.

German, per doz.	6 00	11 00
American, per doz.	12 00	20 00

Building Paper, Etc.

Plain building, per roll	0 30
Tarred lining, per roll	0 40
Tarred roofing, per 100 lb.	1 65
Coal Tar, per barrel	3 50
Pitch, per 100-lb.	0 85
Carpet felt, per ton	45 00

Bull Rings.

Copper, $2.00 for 2½ in. and $1.90 for 2 in.

Butts.

Wrought Brass, net revised list

Cast Iron.

Loose Pin, dis., 60 per cent.

Wrought Steel.

Fast Joint, dis. 60 and 10 per cent.
Loose Pin, dis. 60 and 10 per cent.
Berlin Bronzed, dis. 70, 70 and 5 per cent.
Gen. Bronzed, per pair 0 40 0 65

Carpet Stretchers.

American, per doz.	1 00	1 50
Bullard's, per doz.	6 50	....

Castors.

Bed, new list, dis. 55 to 57½ per cent.
Plate, dis. 52½ to 57½ per cent.

Cattle Leaders.

Nos. 31 and 32, per gross 50 9 50

Cement.

Canadian Portland	2 80	3 00
English "	....	3 00
Belgian "	2 75	3 00
Canadian hydraulic	1 25	1 50

Chalk.

Carpenters, Colored, per gross	0 45	0 75
White lump, per cwt.	0 60	0 65
Red	0 05	0 06
Crayon, per gross	0 14	0 18

Chisels.

Socket, Framing and Firmer.

Broad's, dis. 70 per cent.
Warnock's, dis. 70 per cent.
P. S. & W. Extra 60, 10 and 5 p.c.

Churns.

Revolving Churns, metal frames—No. 0, $8—No. 1, $8.50—No. 2, $9.00—No. 3, $10.00 No. 4, $12.00—No. 5, $16.00 each. Ditto wood frames—20c. each less than above. Discounts: Delivered from factories, 58 p.c.; from stock in Montreal, 56 p.c. Terms, 4 months or 3 p.c. cash in 30 days.

Clips.

Axle dis. 65 per cent.

Closets.

Plain Ontario Syphon Jet	$16 00
Emb. Ontario Syphon Jet	17 00
Fittings net	1 00
Plain Teutonic Syphon Washout	10 00
Emb. Teutonic Syphon Washout	11 00
Fittings net	1 25
Low Down Teutonic, plain	16 00
" " " embossed	17 00
Plain Richelieu net	3 75
Emb. Richelieu net	4 00
Fittings net	1 25
Low Down Ont. Sy. Jet, plain net	19 50
" " " " " emb'd. net	20 50
Closet connection net	1 25
Basins, round, 14 in.	1 00
" oval, 17 x 14 in.	2 50
" " 19 x 15 in.	3 75

Discount 40 p.c., except on net figures.

Compasses, Dividers, Etc.

American, dis. 62½ to 65 per cent.

Cradles, Grain.

Canadian, dis. 25 to 33⅓ per cent.

Crosscut Saw Handles.

S. & D., No. 3, per pair	17½
" " 5, "	22½
" " 6, "	15
Boynton pattern "	20

Door Springs.

Torrey's Rod, per doz. (15 p.c.)		2 00
Coil, per doz.	0 88	1 60
English, per doz.	2 00	4 00

Draw Knives.

Coach and Wagon, dis. 50 and 10 per cent.
Carpenters, dis. 70 per cent.

Drills.

Hand and Breast.

Millar's Falls, per doz. net list.

DRILL BITS.

Morse, dis., 37½ to 40 per cent.
Standard dis. 50 and 5 to 55 per cent.

Faucets.

Common, cork-lined, dis. 35 per cent.

ELBOWS. (Stovepipe.)

No. 1, per doz.	1 40
No. 2, per doz.	1 20
Bright, 20c. per doz. extra.	

ESCUTCHEONS.

Discount, 45 per cent.

ESCUTCHEON PINS.

Iron, discount 40 per cent.

FACTORY MILK CANS.

Discount off revised list, 40 per cent.

FILES.

Black Diamond, 70 and 5 per cent.
Kearney & Foote, 60 and 10 to 70 per cent.
Nicholson File Co., 60 to 60 and 10 per cent.
Jowitt's, English list, 25 to 27½ per cent.

FORKS.

Hay, manure, etc., dis., 50 and 10 per cent. revised list.

GLASS—Window—Box Price.

Size United Inches.	Star Per 50 ft.	Star Per 100 ft.	D. Diamond Per 50 ft.	D. Diamond Per 100 ft.
Under 26	2 10	4 00	....	6 00
26 to 40	2 30	4 35	....	6 65
41 to 50	....	4 75	....	7 25
51 to 60	....	....	....	8 50
61 to 70	....	5 25	....	9 25
71 to 80	....	5 75	....	10 50
81 to 85	....	6 50	....	11 75
86 to 90	....	....	..	14 00
91 to 95	..	....	....	15 50
96 to 100	..	....	....	18 00

GAUGES

Marking, Mortise, Etc.

Stanley's dis. 50 to 55 per cent.

Wire Gauges.

Winn's, Nos. 26 to 33, each 1 65 2 40

HALTERS.

Rope, ⅜ per gross	....	....
" ½ "	....	9 00
" ⅝ to ¾	....	14 00
Leather, 1 in., per doz.	3 87½	4 00
" 1¼ in., "	5 15	5 20
Web, — per doz.	1 87	2 45

HAMMERS.

Nail

Maydole's, dis. 5 to 10 per cent. Can. dis. 25 to 27½ per cent.

Tack.

Magnetic, per doz. 1 10 1 20

Sledge.

Canadian, per lb. 0 07½ 0 08½

Ball Pean.

English and Can., per lb. 0 22 0 25

HANDLES.

Axe, per doz. net	1 50	2 00
Store door, per doz.	1 00	1 50

Fork.

C. & B., dis. 40 per cent. rev. list.

Hoe.

C. & B., dis. 40 per cent. rev. list.

Saw.

American, per doz. 1 00 1 25

Plane.

American, per gross 3 15 3 75

Hammer and Hatchet.

Canadian, 40 per cent.

Cross-Cut Saws.

Canadian, per pair 0 13½

HANGERS.		doz. pairs.
Steel barn door	5 85	6 00
Stearns, 4 inch	....	5 00
" 5 inch	....	6 50
Lane's covered—		
No. 11, 5-ft. run	....	8 40
No. 11½, 10-ft. run	....	10 80
No. 12, 10-ft. run	....	12 60
No. 14, 15-ft. run	....	21 00
Lane's O.N.T. track, per foot	...	4½

HARVEST TOOLS.

Discount, 50 and 10 per cent.

HATCHETS.

Canadian, dis. 40 to 42½ per cent.

HINGES.

Blind, Parker's, dis. 50 and 10 to 60 per cent

Heavy T and strap, 4-in., per lb.	....	0 06½
" " 5-in., "	....	0 06¼
" " 6-in., "	....	0 06
" " 8-in., "	....	0 05¾
" " 10-in., "	....	0 05½
Light T and strap, dis. 60 and 5 per cent.		
Screw hook and hinge—		
6 to 12 in., per 100 lbs.	....	4 50
14 in. up, per 100 lbs.	....	3 50
		Per gro. pairs.
Spring	....	12 00

HOES.

Garden, Mortar, etc., dis. 50 and 10 p.c.
Planter, per doz. 4 00 4 50

HOLLOW WARE

Discount .... 45 and 5 per cent.

HOOKS.

Cast Iron.

Bird Cage, per doz.	0 50	1 10
Clothes Line, per doz.	0 27	0 63
Harness, per doz.	0 72	0 88
Hat and Coat, per gross	1 00	[illegible]
Chandelier, per doz.	0 50	[illegible]

Wrought Iron.

Wrought Hooks and Staples, Can., dis. 47½ per cent.

Wire.

Hat and Coat, discount 45 per cent.
Belt, per 1,000 .... 0 60
Screw, bright, dis. 55 per cent.

HORSE NAILS.

"C" brand 50 and 7½ p.c. off new list } Oval
"M" brand 50, 10 and 5 p.c. } head.
Countersunk, 50, 10 and 10 per cent.

HORSESHOES

F.O.B. Montreal

Iron Shoes.	No. 2 and larger.	No. 1. and smaller
Light, medium, and heavy.	3 50	3 75
Snow shoes	3 75	4 00
Steel Shoes.		
Light	3 60	3 85
Featherweight (all sizes)	4 85	4 85

F.O.B Toronto, Hamilton, London and Guelph, 10c. per keg additional.

Toe weight steel shoes...... 6 70

JAPANNED WARE.

Discount, 45 and 5 p.c. off list, June 1899.

ICE PICKS.

Star per doz.......... 3 00 3 25

KETTLES.

Brass spun, 7½ p.c. dis. off new list.

Copper, per lb........ 0 30 0 50

American, 60 and 10 to 65 and 5 p.c.

KEYS.

Lock, Can., dis., 45 p.c.

Cabinet, trunk, and padlock, Am. per gross.......... 60

KNOBS.

Door japanned and N.P., per doz............ 1 50 2 50

Bronze, Berlin, per doz...... 2 75 3 25

Bronze Genuine, per doz.... 6 00 9 00

Shutter, porcelain, F. & L. screw, per gross.......... 1 30 4 00

White door knobs—per doz. 1 25

HAY KNIVES.

Discount, 50 and 10 per cent.

LAMP WICKS.

Discount, 60 per cent.

LANTERNS.

Cold Blast per doz.......... 7 50

No. 3 "Wright's".......... 8 50

Ordinary, with O burner.......... 4 25

Dashboard, cold blast.......... 9 50

No. 0.......... 6 00

Japanning, 50c. per doz. extra.

LEMON SQUEEZERS.

Porcelain lined, ...per doz. 2 20 5 60

Galvanized.......... 1 87 3 85

King, wood.......... 2 75 2 90

King, glass.......... 4 00 4 50

All glass.......... 1 20 1 30

LINES.

Fish, per gross.......... 1 05 2 50

Chalk " 1 90 7 40

LOCKS.

Canadian, dis. 45 p.c.

Russell & Erwin, per doz.... 3 00 3 25

Cabinet.

Eagle, dis. 30 p.c.

Padlock.

English and Am., per doz.... 50 6 00

Scandinavian, " 1 00 2 40

Eagle, dis. 20 to 25 p.c.

MACHINE SCREWS.

Iron and Brass.

Flat head discount 25 p.c.

Round Head, discount 20 p.c.

MALLETS.

Tinsmiths', per doz.......... 1 25 1

Carpenters', hickory, per doz. 1 25 3

Lignum Vitae, per doz. 3 85 5 50

Caulking, each.......... 60 2 00

MATTOCKS.

Canadian, per doz.......... 5 50 6 50

MEAT CUTTERS.

American, dis. 25 to 30 p.c.

German, 15 per cent.

MILK CAN TRIMMINGS.

Discount, 25 per cent.

NAILS.

Quotations are:	Cut.	Wire.
2d and 3d	$3 35	$3 85
[illegible]	3 00	3 52
[illegible], 5d	2 75	3 35
6 and 7d	2 65	3 20
8 and 9d	2 50	3 00
10 and 12d	2 45	2 95
16 and 20d	2 40	2 90
30, 40, 50 and 60d. (base)	2 35	2 85

Galvanizing 2c. per lb. net extra.

Steel Cut Nails 10c. extra.

Miscellaneous wire nails, dis. 70 per cent.

Coopers' nails, dis. 30 per cent.

Flour barrel nails, dis 25 per cent

NAIL PULLERS.

German and American....... 1 85 3 50

NAIL SETS

Square, round, and octagon, per gross.......... 3 38 4 00

Diamond.......... 12 00 15 00

NETTING.

Poultry, 50 and 5 per cent. for McMullen's

OAKUM. Per 100 lb.

Navy.......... 6 00

U. S. Navy.......... 7 25

OIL.

Water White (U.S.).......... 0 16½

Prime White (U.S.).......... 0 15½

Water White (Can.).......... 0 15

Prime White (Can.).......... 0 14

OILERS.

McClary's Model galvan. oil can, with pump, 5 gal., per doz.......... 10 00

Zinc and tin, dis. 50, 50 and 10.

Copper, per doz.......... 1 25 3 50

Brass, " 1 50 3 50

Malleable, dis. 25 per cent.

GALVANIZED PAILS.

Dufferin pattern pails, dis. 45 p.c.

Flaring pattern, discount 45 per cent.

Galvanized washtubs, discount 45 per cent.

PIECED WARE.

Discount 40 per cent. off list, June, 1899.

10-qt. flaring pan buckets, dis. 45 p.c.

6, 10 and 14-qt. flaring pails, dis. 45 p.c.

Creamer cans, dis 45 p.c.

PICKS.

Per doz.......... 6 00 9 00

PICTURE NAILS.

Porcelain head, per gross.... 1 75 3 00

Brass head " 0 40 1 00

PICTURE WIRE.

Tin and gilt, discount 75 p.c.

PLANES.

Wood, bench, Canadian dis. 50 per cent American dis. 50.

Wood, fancy Canadian or American 7½ o 40 per cent.

PLANE IRONS.

English, per doz.......... 2 00 5 00

PLIERS AND NIPPERS.

Button's Genuine per doz pairs, dis. 37½ 40 p.c.

Button's Imitation, per doz.. 5 00 9 00

German, per doz.......... 0 60 2 60

PLUMBERS' BRASS GOODS.

Compression work, discount, 60 per cent.

Fuller's work, discount 65 per cent.

Rough stops and stop and waste cocks, discount, 60 per cent.

Jenkins disk globe and angle valves, discount, 55 per cent.

Standard valves, discount, 60 per per cent.

Jenkins radiator valves discount 55 per cent

" " " standard, dis., 60 p.c.

Quick opening valves discount, 60 p.c.

No. 1 compression bath cock.......... 2 00

No. 4 " " " 2 00

No. 7, Fuller's.......... 2 50

No 4½, " 3 00

POWDER.

Velox Smokeless Shotgun Powder.

100 lb. or less.......... 0 85

1,000 lb. or more.......... 0 80

Net 30 days.

PRESSED SPIKES.

Discount 20 to 25 per cent.

PULLEYS.

Hothouse, per doz.......... 0 55 1 00

Axle.......... 0 22 0 33

Screw.......... 0 27 1 00

Awning.......... 0 35 2 50

PUMPS.

Canadian cistern.......... 1 80 3 60

Canadian pitcher spout.... 1 40 2 10

PUNCHES.

Saddlers', per doz.......... 1 00 1 85

Conductors', ... 9 00 15 00

Tinners' solid, per set.......... 0 60 0 72

" hollow, per inch.... 0 00 1 00

RANGE BOILERS.

Galvanized, 30 gallons.......... 7 00

" 35 " 8 25

40 " 9 50

Copper, 30 " 22 00

" 35 " 26 00

" 40 " 30 00

Discount off Copper Boilers 10 per cent.

RAKES.

Cast steel and malleable, 50, 10 and 5 p.c.

Wood, 25 per cent.

RASPS AND HORSE RASPS.

New Nicholson horse rasp, discount 60 to 60 and 10 p.c.

Globe File Co.'s rasps, 60 and 10 to 70 p.c.

Heller's Horse rasps, 50 to 50 and 5 p.c.

RAZORS.

per doz.

Geo. Butler & Co.'s.......... 8 00 18 00

Boker's.......... 7 50 11 00

" King Cut.......... 12 50 50 00

Wade & Butcher's.......... 3 60 10 00

Theile & Quack's.......... 7 00 12 00

Elliot's.......... 4 00 18 00

REAPING HOOKS.

Discount, 50 and 10 per cent.

REGISTERS.

Discount,.......... 40 per cent.

RIVETS AND BURRS.

Iron Rivets, black and tinned, discount 60 and 10 per cent.

Iron Burrs, discount 55 per cent.

Extras on Iron Rivets in 1-lb. cartons, ½c per lb.

Extras on Iron Rivets in ½-lb. cartons, per lb.

Copper Rivets & Burrs, 35 and 5 p.c. dis. and cartons, 1c. per lb. extra, net.

Extras on Tinned or Coppered Rivets ½-lb. cartons, 1c. per lb.

RIVET SETS

Canadian, dis. 35 to 37½ per cent.

ROPE ETC.

	Sisal.	Manila.
7-16 in. and larger, per lb.	9½	13
⅜ in.	10½	14
¼ and 5-16 in.		15
Cotton, 3-16 inch and larger		16½
" 5-32 inch		21½
" ⅛ inch		22½
Russia Deep Sea		15½
Jute		8
Lath Yarn		9⅝
New Zealand Rope		10½

RULES.

Boxwood, dis. 75 and 10 p.c.

Ivory, dis. 37½ to 40 p.c.

SAD IRONS. per set

Mrs. Potts, No. 55, polished.. ...... 62½

" No. 50, nickle-plated 67½

SAND AND EMERY PAPER.

Dominion Flint Paper, 47½ per cent.

B & A. sand, 40 and 5 per cent.

Emery, 40 per cent.

Garnet (Rurton's), 5 to 10 p.c. advance on list.

SAP SPOUTS.

Bronzed iron with hooks, per doz... 9 50

SAWS.

Hand Disston's, dis. 12½ p.c.

S. & D., 40 per cent.

Crosscut, Disston's, per ft.... 35 0 55

S. & D., dis. 35 p.c. on Nos. 2 and 3.

Hack, complete, each.......... 0 75 2 75

" frame only.......... 0 75

SASH WEIGHTS.

Sectional, per 100 lbs.......... 2 75 3 00

Solid, " 2 00 2 25

SASH CORD.

Per lb.......... 0 23 0 30

SAW SETS.

"Lincoln," per doz.......... 6 50

SCALES.

B. S. & M. Scales, 45 p.c.

Champion, 65 per cent.

Fairbanks Standard, 35 p.c.

" Dominion, 55 p.c.

" Richelieu, 55 p.c.

Chatillon Spring Balances, 10 p.c.

SCREW DRIVERS.

Sargent's per doz.......... 0 65 1 00

SCREWS

Wood, F. H., bright and steel, 87½ and 10 p.c.

Wood R. H., " dis. 82½ and 10 p.c.

" F. H., brass, dis. 80 and 10 p.c.

Wood, R. H., " dis. 75 and 10 p.c.

" F.H., bronze, dis. 75 p.c.

" R.H. " 70 p.c.

Drive Screws, 87½ and 10 per cent.

Bench, wood, per doz.......... 3 25 4 00

" iron. " 4 25 5 75

Set. Case hardened, 60 per cent.

Square Cap, 50 and 5 per cent.

Hexagon Cap, 45 per cent.

SCYTHES.

Per doz. net.......... 9 00

SCYTHE SNATHS.

Canadian, dis. 45 p.c.

SHEARS.

Bailey Cutlery Co., full nickeled, dis. 60 p.c.

Seymour's, dis. 50 and 10 p.c.

SHOVELS AND SPADES.

Canadian, dis. 40 and 5 per cent.

SINKS.

Steel and galvanized, discount 45 per cent.

SNAPS.

Harness. German, dis. 25 p.c.

Lock, Andrews'.......... 4 50 11 50

SOLDERING IRONS.

1, 1½ lb., per lb.......... 0 37

2 lb. or over, per lb.......... 0 34

SQUARES.

Iron, No 493, per doz.......... 2 40 2 55

" No. 494, " 3 25 3 40

Steel, dis. 50 and 5 to 50 and 10 p.c., rev. list.

Try and bevel, dis. 50 to 52½ p.c.

STAMPED WARE.

Plain, dis. 75 and 12½ p.c. off revised list.

Retinned, dis., 75 p.c. off revised list.

STAPLES.

Galvanized.......... 3 75 4 00

Plain.......... 3 5 3 50

Coopers', discount 45 per cent.

Poultry netting staples, 40 per cent.

STOCKS AND DIES.

American dis. 25 p.c.

STONE. Per lb.

Washita		0 28	0 60
Hindostan		0 06	0 07
" slip,		0 09	0 09
Labrador		....	0 13
" Axe		....	0 15
Turkey		....	0 50
Arkansas		0 00	1 50
Water-of-Ayr		0 00	0 10
Scythe,	per gross	3 50	5 00
Grind,	per ton	15 00	18 00

STOVE PIPES.

5 and 6 inch Per 100 lengths 7 00

7 inch " " 7 50

ENAMELINE STOVE POLISH.

No. 4—3 dozen in case, net cash $4 80

No. 6—3 dozen in case, " 8 40

TACKS BRADS, ETC.

Per cent

Strawberry box tacks, bulk.......... 75 & 10

Cheese-box tacks, blued.......... 80 & 12½

Trunk tacks, black and tinned.......... 85

Carpet tacks, blued.......... 80 & 5

" " tinned.......... 80 & 10

" " (in kegs).......... 40

Cut tacks, blued, in dozens only.......... 75 & 15

" ½ weights.......... 60

Swedes, cut tacks, blued and tinned—

In bulk.......... 80 & 10

In dozens.......... 75

Swedes, upholsterers', bulk.......... 85 & 12½

" brush, blued & tinned, bulk.. 70

" gimp, blued, tinned and japanned.......... 75 & 12½

Zinc tacks.......... 35

Leather carpet tacks.......... 55

Copper tacks.......... 50

Copper nails.......... 52

Trunk nails, black.......... 65 and 5

Trunk nails, tinned.......... 65 and 10

Clout nails, blued and tinned.... 65 and 5

Chair nails.......... 35

Patent brads.......... 40

Fine finishing.......... 40

Picture frame points.......... 10

Lining tacks, in papers.......... 10

Lining tacks, in bulk		15
" " solid heads, in bulk		75
Saddle nails in papers		10
" " in bulk		15
Tufting buttons, 22 line, in dozens only		60
Tin capped trunk nails		15
Zinc glazier's points		5
Double pointed tacks, papers	90 and	10
" " " bulk		40

TAPE LINES.

English, ass skin, per doz.	2 75	5 00
English, Patent Leather	5 50	9 75
Chesterman's each	0 90	2 85
" steel, each	0 80	8 00

THERMOMETERS.

Tin case and dairy, dis. 75 to 75 and 10 p.c.

TRAPS. (Steel.)

Game, Newhouse, dis. 2½ p.c.
Game, H. & N., P. S. & W., 65 p.c.
Game, steel, 72½, 75 p.c.

TROWELS.

Disston's discount 10 per cent.

German, per doz.	4 75	5 00

S. & D., discount 35 per cent.

TWINES.

Bag, Russian, per lb.	....	0 27
Wrapping, cotton, per lb.	0 22	0 26
Wrapping, mottled, per pack.	0 50	0 60
Wrapping, cotton, 3-ply		0 20
" " 4-ply	....	0 26
Mattress, per lb.	0 33	0 45
Staging, "	0 27	0 35

VISES.

Hand, per doz.	4 00	6 00
Bench, parallel, each	2 00	4 50
Coach, each	6 00	7 00
Peter Wright's, per lb.	0 12	0 13
Pipe, each	5 50	9 00
Saw, per doz.	6 50	13 00

ENAMELLED WARE.

White, Princess, Turquoise, Blue and White, discount 50 per cent.
Diamond, Famous, Premier, 50 and 10 p.c.
Granite or Pearl, Imperial, Crescent, 50, 10 and 10 per cent.

WIRE.

Brass wire, 50 to 50 and 2½ per cent. off the list.

Copper wire, 45 and 10 per cent. net cash 30 days, f.o.b. factory.

Smooth Steel Wire, is quoted at the following net selling prices:

No. 6 to 8 gauge	$2 90
" 9 "	2 81
" 10 "	2 87
" 11 "	2 87
" 12 "	2 90
" 13 "	2 95
" 14 "	3 15
" 15 "	3 37
" 16 "	3 50
	3 65

Other sizes of plain wire outside of Nos. 9, 10, 11, 12 and 13 and other varieties of plain wire remain at $2.81 base with extras as before. The prices for Nos. 9 to 13 include the charge of 10c. for oiling. Extras net per 100 lb.: Coppered wire, 60c.—tinned wire, $2—oiling, 10c.—special hay-bailing wire, 30c.—spring wire, $1—best steel wire, 75c.—bright soft drawn, 15c.—in 50 and 100-lb. bundles net, 10c.—in 25-lb. bundles net, 15c.—packed in casks or cases, 15c.—bagging or papering, 10c.

Fine Steel Wire, dis. 17½ per cent. List of extras: In 100-lb. lots: No. 17, $5—No. 18, $5.50—No. 19, $6—No. 20, $6.65—No. 21, $7—No. 22, $7.30—No. 23, $7.65—No. 24, $8—No. 25, $9—No. 26, $9.50—No. 27, $10—No. 28, $11—No. 29, $12—No. 30, $13—No. 31, $14—No. 32, $15—No. 33, $16—No. 34, $17. Extras net—tinned wire, Nos. 17-25, $2—Nos. 26-31, $4—Nos. 32-34, $6. Coppered, 5c.—oiling, 10c.—in 25-lb. bundles, 15c.—in 5 and 10-lb. bundles, 25c.—in 1-lb. hanks, 50c.—in ½-lb. hanks, 75c.—in ¼-lb. hanks, $1—packed in casks or cases, 15c.—bagging or papering, 10c.

Galvanized Wire, per 100 lb.—Nos. 6, 7, 8, $3.50 to $3.85—No. 9, $2.83 to $3.15—No. 10, $3.60 to $3.95—No. 11, $3.70 to $4.10—No. 12, $3 to $3.30—No. 13, $3.10 to $3.40—No. 14, $4.10 to $4.50—No. 15, $4.60 to $5.05—No. 16, $4.85 to $5.35. Base sizes, Nos. 6 to 9, $2.57½ f.o.b. Cleveland.

Clothes Line Wire, 19 gauge, per 1000 feet	2 75	3 00

WIRE FENCING. F.O.B. Toronto

Galvanized barb	....	3 15
Galvanized, plain twist	....	3 15

Galvanized barb, f.o.b. Cleveland, $2.82½ in less than carlots, and $2.70 in carlots.

WIRE CLOTH.

Painted Screen, per 100 sq. ft., net	1 35

WASTE COTTON. per lb.

Colored	4½ to 5
White, according to quality	6½ to 7½

500-lb. bale lots shaded.

WRENCHES.

Acme, 35 to 37½ per cent.
Agricultural, 60 p.c.
Coe's Genuine, dis. 20 to 25 p.c.

Towers' Engineer, each	2 00	7 00
" S., per doz.	5 80	6 00
G. & K.'s Pipe, per doz.	....	3 40
Burrell's Pipe, each	....	3 00
Pocket, per doz.	0 25	2 90

WRINGERS.

Leader	per doz.	$30 00	31 00
Royal Canadian	"	26 00	28 00
Royal American	"	26 00	28 00
Sampson	"	....	30 00

Terms 4 months, or 3 p.c. 30 days.

WROUGHT IRON WASHERS.

Canadian make, discount, 40 and 5 per cent.

PERSONS addressing advertisers will kindly mention having seen their advertisement in Canadian Hardware and Metal Merchant.

SEND for specimen copy of Phillips' Monthly Machinery Register, containing over 5,000 entries of new and second-hand machinery of every description. The oldest established and most successful medium in the world. Established 25 years for the purpose of introducing those who have machinery for sale, to those who wish to buy, has a circulation of about 50,000 copies per annum, all over the world, and is used for continual reference by a large number of firms. It is consequently a most valuable advertising medium for all engineers and manufacturers. Subscription, 4s. per annum, price per copy, 6d. Sole Proprietor, CHAS. D. PHILLIPS, M.I.M.E., Newport, Mon., England. Telegraphic address: "Machinery, Newport, Mon."

BRITISH BUSINESS CHANCES.

Firms desirous of getting into communication with British manufacturers or merchants, or who wish to buy British goods on the best possible terms, or who are willing to become agents for British manufacturers, are invited to send particulars of their requirements for

FREE INSERTION

n "Commercial Intelligence," to the Editor

'SELL'S COMMERCIAL INTELLIGENCE,'
168 Fleet Street, London, England.

"Commercial Intelligence" circulates all over the United Kingdom amongst the best firms. Firms communicating should give reference as to bona fides.

N.B.—A free specimen copy will be sent on receipt of a post card.

Buy the Best.

HERCULES

Sash Cord.

Star Brand Cotton Rope
Star Brand Cotton Clothes Lines
Star Brand Cotton Twine

For Sale by all Wholesale Dealers

BUSINESS NEWS

of any kind that is of value to business men supplied by our Bureau. We can give you market quotations from any town in Canada, reports from the city markets, stock quotations, etc. You can get commercial news from any Canadian paper through us.

Write us, giving us particulars of what you want and where you want it from, and we will quote you prices by return.

"Clippings from any Canadian paper on any subject."

CANADIAN PRESS CLIPPING BUREAU,

232 McGill Street, MONTREAL, QUE.
Telephone Main 1255.
10 Front St. East, Toronto. Telephone 2148.

75 YEARS. ESTABLISHED 1825. 75 YEARS.

CELEBRATED HEINISCH SHEARS.

Tailors' Shears,
Trimmers, Scissors,
Tinners' Snips, etc.

ACKNOWLEDGED THE BEST.

R. HEINISCH'S SONS CO. NEW YORK OFFICE, 90 Chambers St. NEWARK, N.J., U.S.A.

Not connected with any Shear Combination.

CHAS. F. CLARK, President. JARED CHITTENDEN, Treasurer.

...ESTABLISHED 1849...

BRADSTREET'S

Capital and Surplus, $1,500,000. **Offices Throughout the Civilized World.**

Executive Offices: Nos. 346 and 348 Broadway, New York City, U.S.A.

THE BRADSTREET COMPANY gathers information that reflects the financial condition and the controlling circumstances of every seeker of mercantile credit. Its business may be defined as of the merchants, by the merchants, for the merchants. In procuring, verifying and promulgating information no effort is spared, and no reasonable expense considered too great, that the results may justify its claim as an authority on all matters affecting commercial affairs and mercantile credit. Its offices and connections have been steadily extended, and it furnishes information concerning mercantile persons throughout the civilized world.

Subscriptions are based on the service furnished, and are available only by reputable wholesale, jobbing and manufacturing concerns, and by responsible and worthy financial, fiduciary and business corporations. Specific terms may be obtained by addressing the Company at any of its offices. **Correspondence Invited.**

—OFFICES IN CANADA—

HALIFAX, N.S.	HAMILTON, ONT.	LONDON, ONT.	MONTREAL, QUE.
OTTAWA, ONT.	QUEBEC, QUE.	ST. JOHN, N.B.	TORONTO, ONT.
VANCOUVER, B.C.	VICTORIA, B.C.	WINNIPEG, MAN.	

THOS. C. IRVING, Gen. Man. Western Canada, Toronto. JOHN A. FULTON, Gen. Man. Eastern Canada, Montreal.

Awarded a Gold Medal at **PARIS EXPOSITION** for superiority. That's proof enough of their quality, and clearly shows that they are the best.

The Bailey

Cutlery Co.
BRANTFORD, ONT.

Send for Catalogue and . . . . Price List.

Dominion Pattern

Cow Tie and Stall Fixture

The special features of the tie and stall fixture are well shown in the illustration. As will be noticed the chain is very short, which prevents all danger of entanglement with the animal's foot. At the same time the form of the fixture is such that great freedom is allowed to the head. Because of the short chain this tie is much cheaper than the ordinary patterns.

The stall fixture is made from a tough quality of steel and is very strong. Also, owing to its circular cross-section, it is exceedingly rigid. Its simplicity, convenience, cheapness, and ease of attaching make it very popular with cow tie users.

This form of tie and stall fixture are sometimes called Niagara pattern.

American or Flat Link Chain,

for years the standard cow tie chain in "the States," is now rapidly coming in favor in Canada. Its short link, handsome appearance and smooth surface—which cannot injure the animal's neck—make it superior to all other styles of chain for cow ties.

For sale by all Jobbers; manufactured by

ONEIDA COMMUNITY, Limited, NIAGARA FALLS, ONT.

CANADIAN
HARDWARE
AND METAL MERCHANT

The Weekly Organ of the Hardware, Metal, Heating, Plumbing and Contracting Trades in Canada.

VOL. XIII. MONTREAL AND TORONTO, MARCH 16, 1901. NO. 11

Garden Hose

Discounts for the new season now out.

WRITE FOR PRICES.

* * *

CANADIAN RUBBER CO.,

Montreal. Toronto. Winnipeg.

THE NEW BALDWIN

DRY AIR CLEANABLE

REFRIGERATOR.

135 Modern Varieties. Ash, Oak and Soft-wood Finishes

METAL, PORCELAIN, SPRUCE LININGS.

BALDWIN

Positive Circulation—Sanitary—Odorless.

Latest Cleanable Features—The Strongest and Best System of Patent Removable Metal Air-Flues.

Air-Tight Lever Locks
Ball-Bearing Casters.
Swing Base—in and out.

Rubber around Doors and Lids, making them doubly air-tight.

Handsome Designs.
Moderate Prices.

Built in the newest, largest and best equipped refrigerator plant in the East run all the year round on refrigerators exclusively; stock goods; special refrigerators and coolers in sections.

Handsome Trade Catalogue Ready.

Baldwin Refrigerator Co.,

BURLINGTON, VERMONT.

SOME OF THE NEWER "YANKEE" TOOLS

NO. 41 AUTOMATIC DRILL WITH DRILL POINTS IN HANDLE.

NO. 42 AUTOMATIC DRILL WITH DRILL POINTS IN BOX.

NO. 50 RECIPROCATING DRILL, FOR WOOD OR METALS.

Sold by Leading Jobbers throughout the Dominion.

NORTH BROS. MFG. CO.,

Philadelphia, Pa., U. S. A.

"QUICK MEAL" SUMMER ...STOVES

Gasoline or Blue Flame Oil.

Used most and Valued Highest.

The popular favorites with Canadian housekeepers, because of their reliable trustworthiness, perfect adaptability for all kinds of kitchen work, and their simple ease of operation.

The "Quick Meal" Wickless Oil Stove is a marvel of perfect construction, made to please the popular demand, unique in its good points, an unfailing success.

Post yourself on the good points these fine stoves offer you. Send for our descriptive catalogue, and make up your mind to do a rousing Spring trade in these splendid quick-selling "Quick Meal" Stoves.

As in former years, we are sole Canadian Agents.

THE GURNEY FOUNDRY CO., LIMITED

TORONTO WINNIPEG VANCOUVER

THE GURNEY-MASSEY CO., Limited, - MONTREAL.

Steel Frame Churn.

David Maxwell & Sons

ST. MARYS, ONT., CANADA

"Maxwell Favorite Churn"

PATENTED FEATURES: Improved Steel Stand, Roller Bearings, and Foot and Hand Lever Drive, and Detachable Driving Link Improved for season of 1901. Steel or Wood Frame as desired.

Wheelbarrows.

In Four different Sizes.

Lawn Mowers.

High and Low Wheels, from 12-in. to 20-in. widths. Cold Rolled Steel Shafting, Crucible Steel Knives and Cutting Plate.

If your Wholesale House does not offer you these articles

...SEND DIRECT TO US.

MAXWELL MOWER

8-inch Low Wheel.

"THE MAXWELL" Lawn Mower

High Wheel 10 inches

How to Prevent a Belt from Slipping.

A BELT WILL LAST TWICE AS LONG

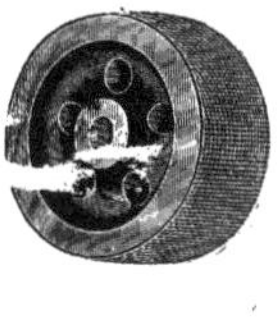

If our Pulley Covering is used, as thousands of manufacturers using leather belts for the transmission of power experience annoyance and trouble with slipping belts, from which arise many losses, such as time, power, etc.

No doubt you have tried many preparations, but the **"Vacuum Cement and Pulley Covering"** will prevent all belts from slipping, being free from all harmful ingredients. It will put your belts in better condition and keep them so, help them to do the proper work, as no other dressings or coverings ever invented have done before.

"The Vacuum Cement and Pulley Covering" is not the cheapest, but the best. It allows the use of a much slacker belt than usual; this slackness prevents belt strains, as well as strains on the machinery. The covering may be applied to any pulley, iron or wood, and is guaranteed for five years. We furnish the material, and the work can be done right in your own shop. We invite your careful investigation.

Write for full particulars.

The Vacuum Cement & Pulley Covering Co.,

Main Office, 1493 Third Ave., NEW YORK.

Meat & Food Chopper

No. 5, - $2.00

Rapid Grinding & Pulverizing Mills

No. 2½, - $4.50

Fruit, Wine & Jelly Press

No. 34, - $3.00

"ENTERPRISE"

PATENTED

HARDWARE SPECIALTIES

Comprising

Meat and Food Choppers, *35 sizes and styles for hand and power;* Rapid Grinding and Pulverizing Mills, *32 sizes and styles for hand and power;* Fruit, Wine and Jelly Presses; Meat Juice Extractors; Cherry Stoners; Raisin Seeders, *for hand and power;* Ice Shredders; Vegetable Slicers; Mrs. Potts' Cold Handle Irons; Sausage Stuffers and Lard Presses, etc., etc.

ARE THE BEST

Write for Descriptive Catalogue — *Sold by the Leading Jobbers of the Dominion*

The Enterprise Mg. Co. of Pa.

Philadelphia, Pa., U. S. A.

Raisin Seeder

No. 36, - .85

Meat Juice Extractor

No. 21, - $2.50

Meat & Food Chopper

No. 12, - $2.50

ALEXANDER GIBB

Manufacturers' Agent and Metal Broker,
13 St. John Street, Montreal.

Representing British and American Manufacturers. Correspondence invited from firms wishing to be represented in Canada.

McCaskill, Dougall & Co
Fine Varnish & Japan
Manufacturers.
Montreal
Price Lists on application

With

DUNDAS AXES

in stock you may justly ask a clerk for an explanation if he misses a sale.

Without **DUNDAS AXES**, is it fair to blame the clerk when you have not bought the best value?

All First-Class Jobbers Sell Them.

MURALO

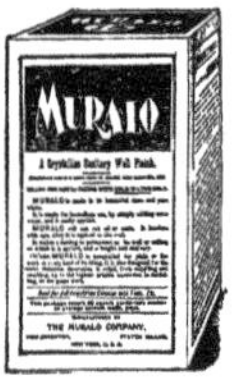

THE WALL TINT.

There has been more money made in Muralo in he past three years than all other wall tints put together. People were ready for something better than had been in wall tint, and, when they saw it, they flocked to it. When they tried it they wanted more; and now we see an enormous factory with numerous extension: , ıg to keep up with the demand in the domestic and foreign markets. Don't be deterred from trying Muralo because you have been disappointed with old-fashioned goods and methods.

AGENTS.

A. RAMSAY & SON, - - - - - - - MONTREAL.
J. H. ASHDOWN, - - - - - - - - WINNIPEG.
McLENNAN, McFEELY & CO., - - - VANCOUVER.

Church's ALABASTINE

The STANDARD WALL COATING.

Made in 20 Beautiful Tints

Anyone can brush it on.

Ready for use in Cold Water

No one can rub it off.

ALABASTINE IS A DURABLE COATING FOR WALLS AND CEILINGS. ENTIRELY DIFFERENT FROM ALL OTHER PREPARATIONS ON THE MARKET.

It is a cement that goes through a process of setting on the wall. Can be recoated from time to time as is necessary to renew, and grows harder with age. Kalsomines are temporary and rub and scale off, to the annoyance of all who use them.

Wall paper with its mouldy paste and poisonous colouring matter is unsanitary.

ALABASTINE is healthful and beautiful, has stood the test of time, is recommended by leading sanitarians throughout the country. There is nothing "just as good." Buy only the genuine.

Manufactured by **THE ALABASTINE CO., LIMITED, Paris, Ont.**

The trade supplied by—
Sanderson Pearcy & Co., Toronto, Ont.
G. F. Stephens & Co., Winnipeg, Man.
Vancouver Hardware Co., Vancouver, B.C.
Wood, Vallance & Co., Hamilton, Ont.
William Hill, Montreal, P.Q.

Ask your dealer for circular. Write us for Questions often asked, mentioning this paper.

"DAISY" CHURN

Has tempered steel cased bicycle ball bearings, strongest, neatest and most convenient frame. Only two bolts to adjust in setting up. Steel Bow Levers, suitable for either a standing or sitting posture. Has four wheels and adjustable feet to hold stand steady while churning. When churn is locked to stand the bow can be used as handles to move it about on the front wheels as handy as a baby carriage. Open on both sides to centre, giving free space for pail. Made with wood or steel stands, with Cranks only, or Bow Levers as desired.

Vollmar Perfect Washer

Has a most enviable record. A perfection of its kind—will wash more clothes in less time, do it better and easier, with less wear and tear, than any other machine.

THE
Wortman & Ward Mfg. Co.,
Limited
LONDON, ONT.
Eastern Branch, 60 McGill Street, Montreal, Que.

Wire Guards

FOR

Store Fronts
Factory and Mill Windows
Basement Windows

THE B. GREENING WIRE CO., LIMITED

HAMILTON, ONT., AND MONTREAL, QUE.

LEWIS BROS. & CO.,

MONTREAL.

PRUNING AND GARDEN SHEARS

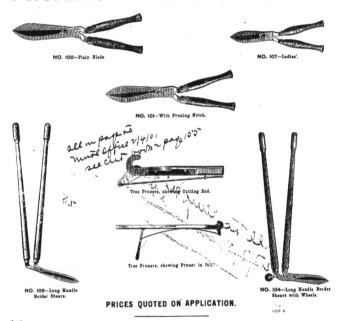

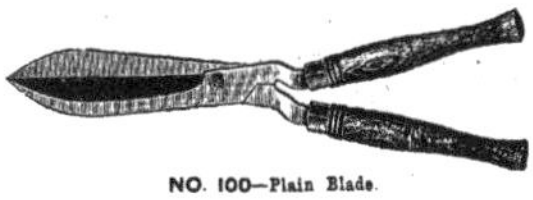

NO. 100—Plain Blade.

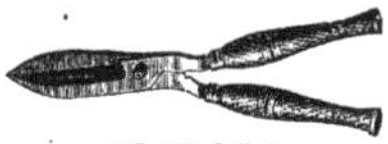

NO. 107—Ladies'.

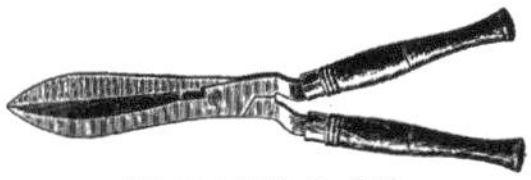

NO. 101—With Pruning Notch.

NO. 106—Long Handle Border Shears.

Tree Pruners, showing Cutting End.

Tree Pruners, showing Pruner in full.

NO. 104—Long Handle Border Shears with Wheels.

PRICES QUOTED ON APPLICATION.

LEWIS BROS. & CO.,

SPECIAL ATTENTION TO MAIL ORDERS.

Montreal

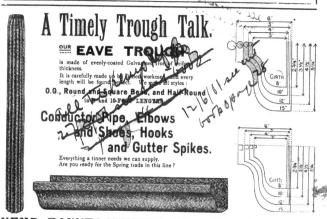

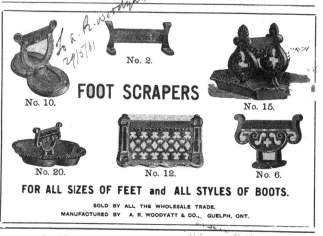

HARDWARE AND METAL

VOL. XIII. MONTREAL AND TORONTO, MARCH 16, 1901. NO. 11.

President,
JOHN BAYNE MacLEAN,
Montreal.

THE MacLEAN PUBLISHING CO.
Limited.

Publishers of Trade Newspapers which circulate in the Provinces of British Columbia, North-West Territories, Manitoba, Ontario, Quebec, Nova Scotia, New Brunswick, P.E. Island and Newfoundland.

OFFICES

MONTREAL - - - - - - 232 McGill Street, Telephone 1255.
TORONTO - - - - - - 10 Front Street East, Telephone 2148.
LONDON, ENG. - - - - 109 Fleet Street, E.C., J. M. McKim.
MANCHESTER, ENG. - - - 18 St Ann Street. H. S. Ashburner,
WINNIPEG - - - - Western Canada Block. J. J. Roberts.
ST. JOHN, N.B. - - - No. 3 Market Wharf, J. Hunter White,
NEW YORK. - - - - - - 176 E. 88th Street.

Subscription, Canada and the United States, $2.00.
Great Britain and elsewhere - - - 12s.

Published every Saturday.

Cable Address { Adscript, London. Adscript, Canada.

A LESSON IN TRADE ECONOMICS.

IT will probably be remembered that two or three years ago the bedstead manufacturers in Great Britain formed a combination and materially advanced prices. The Cuban correspondent of the British Trade Journal points out that, largely as a result of this, manufacturers in the United States are taking the trade away from the British manufacturer. Another contributory clause is the fact that the bedsteads made by the American concerns are lighter than those made in Great Britain, thus allowing them to be imported under a lower cost of duty. Bedsteads appraised for duty in Cuba are based on weight. The comparative light weight of the American bedsteads is due to the fact that the pillars are made of three-quarter inch and one-inch hollow castings, which enables the importer to save at least 50 per cent. in duties.

Aside altogether from the interest the report in question may have to the bedstead manufacturers of Great Britain, it contains an economic lesson, two-fold in its character, that has an application that is confined to no one class of goods or to no one country.

Nothing can reasonably be said against any class of business men, whether they be manufacturers or merchants, organizing for mutual benefit. Indeed, their doing so is to be highly commended, as long as the motive is a right one. If organization is sought for the purpose of improving methods of doing business, for the rectifying of evils or for insuring a fair profit, it is obvious the motive is praiseworthy rather than otherwise. If, on the other hand, the aim is big profits, or monopolistic hold upon particular lines of trade without consideration of others, then a great deal can be said against such organizations. The bedstead combination essayed to earn large profits, with what result we already have an illustration, as far as one country, at least, is concerned. And what occurred in one, no doubt, occurred in others.

Moderate profits tend to prevent competition; excessive, to multiply competitors.

Another lesson to be drawn is the necessity of continuity of effort to comply with the changing conditions of old markets and the requirements of the new. And the principle applies to individual customers as well as to national. In the instance under review, the United States manufacturer made a bedstead that would reduce the cost of duty, while the British manufacturer ignored this condition.

HIGHER PRICES ON ROPE.

FURTHER advances are to be noted this week in the price of rope, both manila and sisal being quoted ½c. per lb. higher. The base price for sisal is now 10c. per lb. and for manila, 13½c. per lb. New Zealand rope is quoted at 10c. and single tarred lathyarn at 9½c. per lb.

There is a demand, it is reported, for hemp for binder twine purposes, and, as the supply is rather light, the price is higher. This, in turn, has naturally affected the price of rope. At the higher figures rope is firm, and the manufacturers claim that a further appreciation is not improbable. As there is now an understanding among the manufacturers in regard to the price of rope this is made all the more possible.

He who does not want his business to give him the slip should keep his eye upon it.

QUOTATIONS ON POST CARDS.

Some discussion has been going on lately in the columns of trade papers in Great Britain over the practice of sending on post cards net quotations to customers.

We cannot say to what extent post cards are used for this purpose in Canada, but that they are at times used therefor we have no doubt. It is a practice not to be recommended, and for a two-fold reason. In the first place, there is always the risk of quotations thus given being seen by those whom it might be advisable should not see them. Then, altogether aside from the question of secrecy, a letter is better than a post card. It is more business-like, and creates a better impression.

A BUSINESSLIKE DEMAND FROM BUSINESS MEN.

THE Federal Government should have no hesitancy in establishing an assay office in Vancouver in accordance with the petition of the business men of that city.

A couple of years ago the Provincial Government of British Columbia established an assay office in Vancouver, with the result that a quantity of gold has been purchased by the banks in that city. But the extent of the Provincial affair is not great enough to attract to Vancouver the large quantity of gold which it is possible to attract there. "What is really wanted," to quote the 1890 report of the president of the Vancouver Board of Trade, "is action by the Dominion Government, and either the establishment of a mint or an assay office on a larger scale, where the market price would be paid for all gold offering."

The request of the Vancouver merchants is based on no visionary scheme. Its basis is business.

The news of the rich strikes of gold which had been made in the Klondike began to reach the outside world about the middle of 1897. By the time the output of 1898 was coming upon the market from the Klondike the Government of the United States had established an assay office at Seattle, Washington State, and thither Klondike gold went. To that city also went the great bulk of the business on Klondike account instead of to cities like Vancouver and Victoria in British Columbia; for where the gold is assayed there is the merchandise for the mining camps bought. And not only merchandise for the mining camps, but there is a great deal of money spent on articles for other purposes.

It is true the merchants of British Columbia are gradually receiving a larger share of the Klondike trade, and when the direct line of steamers from Vancouver to Skagway, which the C.P.R. is to put on when navigation opens, are running, they will receive a still larger share. But, as long as nearly all the gold from the Klondike goes to Seattle to be assayed, the business men in the British Columbian trade centres must be under disabilities which they should not be under and which it not only within the power but within the province of the Federal Government to remove.

The United States Government took action four years ago to assay Canadian gold in order to secure Canadian trade. And it has admirably succeeded in its efforts, for Seattle has developed enormously since Klondike gold came to its office to be assayed. After four years it is surely time for the Canadian Government to do what it should long ago have done. And it is to be hoped that the proposed legislation for the establishment of a mint at Vancouver will not be, like other legislation of a business nature often has been, shelved for a subsequent session.

ADVANCE IN IRON PIPE.

FOR some time the wholesale price of iron pipe in Toronto has been relatively lower than the figures quoted by the manufacturers warranted, due, as the trade well know, to the large stocks held by some of the jobbers. A sort of an understanding has at last been reached whereby figures will be more uniform. Prices, as a result, are slightly higher, but, notwithstanding this, they are lower than those quoted by the manufacturers. In view of this, it is possible we may see another advance in jobbers' figures before many days. Prices quoted by the jobbers are now as follows for black and galvanized pipe, respectively:

	Black.	Galvanized.
1/8 inch	$ 4 15	....
1/4 "	3 00	....
3/8 "	3 05	....
1/2 "	3 10	$ 4 75
3/4 "	3 35	5 25
1 "	4 70	7 25
1 1/4 "	6 40	9 50
1 1/2 "	7 70	11 50
2 "	10 25	15 50
2 1/2 "	21 85	....
3 "	28 69	....
3 1/2 "	36 10	....
4 "	41 04	....
5 "	55 10	....
6 "	71 44	....
7 "	89 30	....
8 "	107 16	....

PARIS GREEN INACTIVE.

The demand for paris green so far this season has not been up to the expectation of jobbers, many of whom looked for a rather active market during the first months this season on account of the small movement of this line.

It will be remembered that the season of 1899 was exceptionally dull, there being so little damage done to crops by bugs that a much smaller quantity of paris green than usual was absorbed. As retailers had stocked generously early that spring they were left with heavier stocks than they desired at the end of the season. This stock had to be cleared out before new orders were placed last year, hence the dull season.

This year, retailers are not, as a rule, overstocked, but their experience of two years ago has made them unwilling to buy largely before they need the goods.

THE SITUATION IN IRON AND STEEL.

IF the condition of the iron and steel market can be taken as any criterion, the outlook for the immediate future, at any rate, is assuring rather than otherwise.

The demand for pig iron in the United States is brisk, and there have been sales of Bessemer iron during the past week at an advance of $1 per ton over the figures of last week. The buying, we are told, is not for large lots, but it is known that many of the large consumers of iron have not yet fully covered their wants for the first half of the year, and the Valley furnacemen are not at all anxious sellers in consequence, especially in view of the firm condition of prices at the moment.

The price of iron ore has not yet been established for the year, but coke is 25c. per ton higher, and it is the opinion that there will be a further appreciation when the furnaces begin to make contracts for the second half of the year.

In Canada the large users of pig iron appear to have generally placed orders for their requirements of pig iron for the first half of the year.

Extraordinary strength still characterizes the steel market, and $2 to $5 premiums on the prices of the pool are being paid for billets, so anxious are buyers to have their orders filled. Manufacturers of plates are also, in some instances, receiving a premium of $2, but notwithstanding this no change was made in the scheduled price at the meeting which was held in New York last week. Hoops, bands and tire and toe-calk steel are all higher this week in Pittsburg. In tire and toe-calk steel the advance is $1 to $2 per ton.

CANADA AND THE SOUTH-AFRICAN MARKET.

CAPT. WYNNE, of H. R. Ives & Co., Montreal, was a transport officer in South Africa during the year the Canadian regiments were at the front, and consequently learned just how Great Britain victualled and cared for her army. Recognizing that his observations would be made from a commercial point of view, HARDWARE AND METAL has sought his opinion on the matter of Canada acquiring a share of the South-African trade.

He is firm in the conviction that if Canadian manufacturers and importers would study the market and adapt themselves to existing conditions, if the Canadian Government would see to it that we are placed in direct connection with the market, and that carriage charges are lowered, by the establishment of a line of steamers to ply between Canadian ports and Cape Town, and if there was a little more push and enthusiasm on the part of all concerned, Canada would develop an enormous trade in the South-African market.

"Did you see any quantities of Canadian goods while out there?"

"No, scarcely any. There was some jam, hay, butter and cheese. The packages of jam, I believe, were not filled, and the manufacturers seem to have tried to make as much profit as possible on the single transaction. They do not seem to have thought of building up a reputation and stimulating trade. I believe 300 tons of Canadian hay were thrown into the harbor at Cape Town. On the boat which took me over there was Canadian hay; some of it was good, but a good deal of it looked as if it had been taken off a marsh, and oftentimes we would find patches of thick, unpalatable clover sticks in the bales. The natural result is, the people of South Africa don't regard Canadian hay with favor. Of course, the Government bought it and distributed its purchases all over the country, whereas it should have confined itself to the best hay-producing districts. There was some Canadian butter there, but it went by way of England, where it was put into cans to make it portable. Our cheese, also, went via England, and the English middleman's big profit was tacked on to the price. I saw but little Canadian meat, biscuits, canned goods, bacon, flour, furniture, wagons, boots, or many other lines which we could supply as well as any other country trading in that market.

"Is it worth while taking steps to encourage this trade now?"

"Most certainly. If the war stopped to-day, South Africa would be importing huge quantities of foodstuffs and manufactured materials for five years at least. Both the army and the people must be victualled, clothed and housed and the country itself can produce but little."

"Can you suggest any way by which we could enter this market?"

"First of all, we must approach the market in a commercial spirit. When one is dealing with army contractors one is not doing business with a Government, as some Canadian manufacturers and exporters seem to have thought. The purchasers are men who want the best goods for the least cost, and they know when they get value. What we want in South Africa is a commercial agent—not a passive agent, but an active commercial agent. Mr. Moffatt who has represented us there is a first-class gentleman, but he serves no material purpose by hob-nobbing with Sir Alfred Milner, who does not buy a cent's worth of the vast quantity of goods going into the country. The Government should send a high-salaried man who will work and canvass for trade—a good commercial man, one who has proved himself worthy in business, not in politics. He should know this country and its manufacturers thoroughly and be able to recommend the proper sources of supplies. Moreover, he should make a study of South Africa's requirements and instruct the Canadian exporters in the art of putting up their goods in the proper style. For instance, we could make immense shipments of butter if our exporters would pack it in hermetically sealed tin cans as do the Englishmen with our butter or the Australians and Americans with their own. Our cold storage system is no good for this trade. Again, why do we not send potatoes as do the Americans who wash them and ship them in crates or boxes as they do lemons? There are many such pointers that could be thrown out by a competent man in South Africa.

"The best expedient I could suggest to work up the trade would be the running of a line of steamers from the supplying to the consuming market. Nearly all the goods we have sent have gone by England or New York. Sending them by England we incur an extra freight charge and a middleman's profit that would be wiped out if we traded directly. I am quite confident that we could profitably employ a line of steamers to ply at least monthly between Canadian and South-African ports. The people down there want our goods, and will give us the preference over American or German goods on every occasion, provided, of course, that they get even value. But we must do business in a business-like way—not a la politics. Let us send our cheese, butter, jam and flour direct and we will work up a wonderful trade in a short time. The number of lines of goods we could trade in is unlimited. The call for furniture is loud, and who can supply it more cheaply than we? A good wagon brings £75 in Cape Town, here it is worth $60.

"I could wish your call for a direct line of steamers might be a loud blast that would sound in the Government's ears. We don't want to send our goods by New York, for the Americans watch our shipments and immediately put their finger on the spot to which they are addressed, using us as advertising agents."

THE LEADERS IN BUSINESS.

It is much the same in business as in a foot race, the leaders are usually far in advance of the majority of their fellow competitors. It is the good training in the one as well as in the other instance that causes these things to be.

PERSONAL MENTION.

Mr. W. H. Wiggs, of The Mechanics Supply Company, Quebec, was a caller upon the Montreal office of HARDWARE AND METAL this week.

Mr. C. S. Archibald, representing The Maritime Nail Co., Limited, and The Portland Rolling Mills, Limited, St. John, N.B., was in Toronto this week.

IRON TRADE IN WOLVERHAMPTON.

There is more disposition to look upon the best side of things. Moreover, consumers and merchants are placing orders with greater freedom, so that there is some foundation for the more buoyant tone which is abroad. Buying is still confined for the most part to small lots for prompt delivery. Seeing, however, that stocks have been reduced to a minimum for months past, and that purchases have been regulated in rigid relation to actual commitments, the greater activity now apparent would seem to be the response to a definite expansion of demand. Albeit the brisker tendency, though it has improved the prevailing tone of the market, does not extend through all the ramifications of the trade, and there are many manufacturers whose situation is one of anxiety. The backward movement noted in other markets creates misgivings that the present rally may turn out nothing more than a "flash in the pan." Pig iron, though steadier than it has been latterly, is still seriously depressed by the absence of forward contracting. Buyers move with extreme caution, and the feeling of distrust thus engendered is communicated in a certain degree to the finished-iron trade.—Ironmonger.

TRADE IN COUNTRIES OTHER THAN OUR OWN.

THE new prices agreed upon by the manufacturers of malleable fittings have not yet been announced, but they are known to represent an advance of about 5 per cent.—Iron Age.

ENAMALED WARE IN THE UNITED STATES

The demand for enameled ware is extraordinarily heavy. Makers say they are receiving more business than they can take care of with reasonable promptness. Most of the manufacturers of these goods have raised prices sharply, the advance being relatively greatest on the lower grades. The advance, however, is not uniform. One of the largest manufacturing concerns in this line, however, announce that they have not advanced their prices and do not contemplate doing so in the near future.—Iron Age.

NEW YORK METAL MARKET.

TIN.—Although statistically no improvement could be found in the situation, the London market recovered partially to-day, closing 17s. 6d above last night's quotations, but still 10s. balow those of last Friday. The improvement in futures to-day was less pronounced than that in spot tin, the advance amounting to but 10s. In the New York market a firmer feeling prevailed, partly from sympathy with London and partly because the lower prices of yesterday brought out more inquiries from consumers. Spot showed an advance of 20 points for the day, closing at 25.80c. bid and 26c. asked. There were two arrivals to-day, aggregating 170 tons, and bringing the total for the month to date up to 795 tons. On March 27 about 25 tons of Banca tin will be sold at auction in Holland. The unsold stock of Banca in Holland is arge, amounting on February 28 to 6,907 tons, against 4,220 tons at the same date last year and 3,400 tons two years ago.

COPPER.—The London market showed a further decline of 3s. 9d. and closed quiet. In New York trade continues light, but prices are maintained at 17c. for Lake Superior and 16 5-8c. for electrolytic and casting.

PIG LEAD.—There was little business doing here and prices were unchanged on the basis of 4.37 1-2c. in lots of fifty tons or over. St. Louis was reported quiet but firm at 4.25c. In London soft Spanish declined 3s. 9d, reaching the lowest point for two years.

SPELTER.—The weakness continues and has resulted in a further decline, spot to May being now quoted at 3.90 to 3.95c. St. Louis was quiet, with 3.80c. bid. The London market showed a further decline of 5s., bringing the price down to £16 12s. 6d., which is equivalent to 3.60c. delivered in London.

ANTIMONY.—In a jobbing way a moderate business is being done in regulus at the range of 8 3-4 to 10 1-4c. as to brand.

OLD METALS.—Prices are maintained, though trade is light.

IRON AND STEEL.—Although the steel billet and plate pools have not seen fit to make the advance in prices which they have been expected to, almost from the beginning of the present activity second hands have placed a considerable premium on stock for immediate delivery and consumers pay it without protest. Activity in all branches of the steel trade continues unabated and the difficulty everywhere is to get stock as fast as wanted.

Western advices continue to report a large demand for pig iron, not only Bessemer, but foundry. Late Pittsburg advices state that heavy sales of the latter have been made there and that prices have advanced sharply. On the other hand, forge iron was reported to be a little weaker there. In New York and other eastern markets the absence of export demand is felt and the home trade in iron does not yet seem to have attained to very important dimensions.

TINPLATE.—The demand continues good and the market is firm at the quotations.—N. Y. Journal of Commerce, March 13.

CANCELLING ORDERS.

Editor HARDWARE AND METAL,—Your editorial headed "A Bad Practice" is timely and to the point. Among the failings of the average Ontario retailer one of the worst is his weakness for cancelling orders, and another the expectation of allowance for decline in market price. He never considers the propriety of offering to pay more for his goods in case of an advance and would be astonished if asked to do so, and yet, if prices happen to drop after he has placed an order, he looks for the lowest going, or, perhaps, "magnanimously" gives his wholesaler the option of meeting price or cancelling the order. This method is manifestly unfair and unbusinesslike; in fact, it is a case of "heads I win, tails you lose." Still, it goes on year after year, and, strange to say, the wholesalers, while they may murmur and complain, apparently make no effort to put a stop to it. Last year was a time of advancing values, and importers were filling orders during the greater part of the season at figures far below the price current at time of shipment, and their customers considered it only right. This year indications point the other way, and, already, I understand, the wholesale dealers are receiving letters, asking whether they are prepared to meet such and such a quotation. If the merchant should not agree, the order is cancelled, and, having already purchased goods to cover these orders, he is left with stock on his hands that may depreciate in value before he can again dispose of it. Of course, it is well known that such remarks do not apply to all our retail dealers. Probably one-half of those in the business are above such mean tactics, while a number who use them have no intention or desire to do anything that is not fair, but the practices complained of have become so common as to seem only regular methods of business. Their employers did it when they were clerks, perhaps, so that it is part of their business education.

To a great extent, I believe, commercial travellers—and even the suffering merchants themselves—have contributed to [illegible]cation. What is needed now is education the other way. Retailers must learn that buying ahead is a speculation, and that if they expect to profit by an advance in price they must also be prepared to stand a loss in the event of a drop. To impress this upon the trade there should be an ironclad agreement among wholesale dealers to make no rebates on account of falling markets and allow no cancellation of orders on that ground.

You, Mr. Editor, can also help to bring about the desired reform by keeping the matter before your readers with such articles as last week's editorial, and your assistance will be appreciated by

"ONE WHO HAS SUFFERED."

Toronto, March 14, 1901.

BINDER TWINE.

Editor HARDWARE AND METAL,—We wish to protest against the statement made by "Wanderer," in your last issue, regarding Canadian binder twine.

We have been mauufacturing binder twine for the past 20 years, and we claim that our twine has always been first-class in every respect, and will run more feet to the pound than we claim. We claim, and know, that the Lawson (Leeds, Eng.), system in use by us is superior to anything in the United States.

We would like to draw "Wanderer's" attention to the fact that all binder twine sold in this country must have the number of feet marked on each tag, and if it does not run the required number of feet the manufacturer is subject to a heavy fine.

"Wanderer" must have secured an American twine agency, or otherwise he is wandering into subjects that he does not know very much about.

CONSUMERS CORDAGE CO., LIMITED.

Montreal, March 9.

BUSINESS CHANGES.

DIFFICULTIES, ASSIGNMENTS, COMPROMISES.

M. J. LEITCH, general merchant, Michael's Bay, Ont., has assigned to A. E. Pavey, and a meeting of his creditors will be held to-day (Friday).

·lly, general merchant, Glenboro', Man., is asking an extension.

D. Levasseur, general merchant, Matane, Que., is offering 40c. on the dollar.

L. A. Corbiere, general merchant, Lacolle, Q. ., is offering 60c. on the dollar.

Chas. A. Quick, general merchant, Kingsville, Ont., is offering 85c. on the dollar.

David Hecht, general merchant, Tichborne, Ont., has assigned to John Carson.

Eliza Betts, general merchant, Mount Brydges, Ont., has compromised and retired.

Paul Bissonnette, general merchant, Casselman, Ont., is offering 55c. on the dollar.

L. G. Jourdoin, hardware dealer, Three Rivers, Que., has assigned to Gagnon & Caron.

J. A. Boyd & Son, tinsmiths, etc., St. Stephen, N.B., are offering 25c on the dollar.

E. Lepage & Co., departmental merchants, Montreal, are preparing a statement for creditors.

A meeting of the creditors of Leonide Sicotte, general merchant, Boucherville, Que., has been held.

M. Tisdall & Co., general merchants, Sturgeon Falls, Ont., have assigned to Richard Lee, Toronto.

Bilodeau & Chalifoux have been appointed curators of Louis Dore, coal and wood dealers, St. Henri de Montreal.

A meeting of the creditors of Roberge & Landry, general merchants, Thetford Mines, Que., has been called for March 25.

J. D. McLeod, general merchant, Prince Albert, N.W.T., has assigned to C. H. Newton, and a meeting of his creditors has been called.

The J. F. Brownscomb Co., general merchants, Uxbridge, Ont., have assigned to Henry Barber, Toronto. Their liabilities are placed at $33,958, with assets of $27,865.

PARTNERSHIPS FORMED AND DISSOLVED.

The Trail Hardware and Plumbing Co., Trail, B.C., have dissolved. M. A. Inglis continues.

I. Charbonneau & Co., sash and door manufacturers, St. Louis de Mile End, Que., have dissolved.

Smith & Arkell, hardware dealers, etc., Listowel, Ont., have dissolved; C. H. Smith continuing.

SALES MADE AND PENDING.

J. Seburger, painter, Listowel, Ont., is about to sell out.

Charles Girard, sawmiller, Beauce Junction, Que., has sold his mill.

John Ritter, hardware dealer, Newton, Ont., is offering his business for sale.

The stock of I. J. Dupont, hardware dealer, Farnham, Que., has been sold.

The assets of Wm. Rodden & Co., foundrymen, Montreal, have been sold.

The assets of A. J. McDonald, general merchant, Seaside, N.S., have been sold.

J. C. Price, general merchant, Ridgetown, Ont., is advertising his business for sale.

The assets of Prement & Co., general merchants, St. Felicite, Que., have been sold.

The stock of the estate of W. S. Grant, general merchant, Minto, Man., has been sold.

George E. Corbett, general merchant, Annapolis, N.S., is advertising his business for sale.

The stock of the estate of Morrison & Co., general merchants, Boissevain, Man., has been sold.

The business of John McRae, general merchant, etc., Quesnelle, B.C., is advertised for sale.

The business and property of L. B. Currie, general merchant, West Dublin, N.S., is advertised for sale.

The stock of Walter Wardrop, general merchant, Lac du Bonnett, Man., is advertised for sale by auction on March 14.

CHANGES.

Razz & Co. have registered as painters in Montreal.

John McKay, blacksmith, Tavistock, Ont., has sold out to H. Roedding.

Jobin, Chouinard & Co. have registered as machinists at St. Thecle, Que.

A. T. Wiley & Co. have registered as crockery dealers, etc., in Montreal.

Louis Baechelor, sawmiller, Milverton, Ont., has sold out to Peter McClelland.

M. McBean & Co., general merchants, Phœnix, B.C., have retired from business.

J. C. Parker, general merchant, Coldwater, Ont., has sold out to David Brown.

J. F. Ardill & Co., general merchants, Thornton, Ont., have sold out to Tomlinson Bros.

Thos. Lee & Co., harness dealers, Brandon, Man., have sold out to Samuel Nixon.

N. H. Turcotte, general merchant, Thetford, Que., is about to remove to St. Gentilly.

J. P. Porter, dealer in agricultural implements, Portage la Prairie, Man., has been succeeded by J. and E. Brown.

E. C. Crochu, general merchant, St. Agathe (Lotbiniere), Que., is about to open in Thetford.

F. X. Leduc, general merchant, St. Louis de Gonzaque, Que., is removing to Valleyfield.

Graber & Son, hardware and stove dealers, Stratford, Ont., have sold out to F. A. Campbell.

J. E. Gaudin, dealer in agricultural implements, Napinka, Man., has been succeeded by DeWitt & Gaudin.

The Maritime Glass Co., glass manufacturers, St. John, N.B., have been succeeded by the Maritime Art Glass Works.

FIRES.

White & Co., sawmillers, Inwood, Ont., have been burned out.

John Campbell & Son, manufacturers of carriages, London, Ont., were damaged to the extent of $40,000 by fire. The loss is largely covered by insurance.

DEATHS.

Eugene Boissonneau, general merchant, Newbois, Que., is dead.

THE TIN CAN DEAL.

A SPECIAL Chicago despatch to The Baltimore Sun upon the tin can consolidation, says: "The local subscriptions to the underwriting of the new American Tin Can Company were closed yesterday at the office of J. H. & W. H. Moore, and the papers were sent to their New York office, where the allotment will be made. Some trading in the underwriting privileges was reported at a premium; 101 bid and 102 and 103 asked was reported in the different brokerage houses.

"About 10 different concerns, situated all over the country, are absorbed. The principal ones besides the Chicago companies are located at Boston, New York, Philadelphia, Pittsburg, Cincinnati and St. Louis. Norton Brothers are the largest single company taken in. It is understood that they received $5,000,000 in stock. The Frank Diesel Co. got $350,000 and the Illinois Can Co. received $250,000. The other Chicago concerns taken in are much smaller. The Norton company is larger by five times than any of the other concerns included in the deal, and has been a leading factor in the tin can trade for many years. Oliver W. Norton, of Chicago, president of the Norton Brothers Co., will probably be elected president of the trust.

"The subscription allotted to Chicago was $3,000,000, which was distributed among brokers in the city. The stock has been on the Chicago market for the last four days. It is estimated that more than half of the stock will be taken up in New York. The allotment there was in the neighborhood of $6,000,000. Most of the companies took preferred stock in payment for their interests, but a few of the small concerns took cash offers. The trust will absorb about 60 per cent. of the output of the American tinplate companies.

"Economy in the distribution will effect one of the principal savings expected from the consolidation. Moore Brothers control the factories of The American Tinplate Co., which will insure favorable contracts for raw material to the Can Trust. The most valuable assets to the combination are the patents of the individual companies which will be controlled by the trust. This was a paramount issue in the deal, as it is claimed that it practically makes it impossible for outside concerns to exist."

H. S. HOWLAND, SONS & CO.

WHOLESALE ONLY 37-39 Front Street West, **Toronto.** ONLY WHOLESALE

BEST MADE.

CORN PLANTERS.

ALWAYS RIGHT.

"TRIUMPH."

"TRIUMPH," with Pumpkin Seed Attachment.

All to H.S. Howland cut book 2 28/3/01 page 15

BULL RINGS, Steel and Copper.

OX BALLS.

CATTLE LEADERS.

BULL SNAPS, with Chain, no Handle.

HOG RINGERS.

HOG TONGS.

H. S. HOWLAND, SONS & CO., Toronto.

WE SHIP PROMPTLY. Graham Wire and Cut Nails are the Best. OUR PRICES ARE RIGHT.

CREDITS IN THE HARDWARE STORE.*

IN order to meet your obligations you must also have a credit department. This department must consist solely of yourself. You, too, must look up the reputation of each one who asks for credit. You cannot be too careful in this matter. Don't be so anxious to sell goods on time that you will take every man's word as to his standing.

LAZY METHODS.

We are all optimists to an extent and in case of a bad credit try to make the best of it, hoping for better deals in the future. But to do as the fatalist, sit down and say it was meant to be so, is far from the successful path of a merchant's life. Often we find merchants too anxious to sell goods and see them go through the front door. You will find them ever watching their competitor, and if he sells one or two more stoves a resolve is made to catch up in number of sales. Then it is that the first man who comes in to buy a stove is sold one regardless of cost price and his standing. Often this first man is anything but a desirable customer. Being behind in sales, the merchant resolves to catch up and makes the sale, taking perhaps second or third mortgage on a cow as security. When payment time comes he finds the money is not forthcoming. He then endeavors to collect on the security given and finds that it is no good, which he could have found out when he made the sale, but the resolve to catch up in amount of sales and do business was so great that the standing of customer was forgotten.

THE LESSONS OF COLLECTION PERIODS.

At collection time he finds that selling goods so as to get paid for them is doing successful business and making money. Then, again, you can find merchants who are not satisfied without selling every one who comes in to buy, and at collection time they bump against a stump and discover that to sell every one who comes in on credit is a fatal business policy.

When a customer comes in to buy and the sale depends on a long credit and not on price given, such sales are unprofitable. To do so will injure your business and put you at a disadvantage, for will you not have to ask more credit of your jobber ?

ONE METHOD OF DETERMINING CREDITS.

Every merchant should see that in extending a line of credit, whoever gets it will be in a position to fulfill his promise to pay, or otherwise get good security, so that you, Mr. Merchant, can fulfill your promise to pay Mr. Jobber.

DONT'S.

Don't get a lot of long price notes, with a long extension of time for payment, in your safe, upon which nothing can be realized. If more merchants knew what an error it is to try and do all the business of their respective towns, fewer old notes and accounts would grace their assets.

QUICK SALES, SMALLER PROFITS.

The trouble with most merchants who fail lies in their inability to distinguish between the wisdom of making sure of the payment of an account and the folly of obtaining abnormally large profits on a long time sale. The thing to have in a business is something like what English bankers term "liquid assets," which, though returning small profits on short time sales, insure the payment of the account with the profit as well.

THE STORY OF A PUMP.

I therefore resolved to do little business and be sure of pay. On my first business day a farmer walked into the store and asked for a pump. I had a pump fitted up and loaded in the wagon for the farmer before payment was mentioned. The farmer walked into the store in a hurry and said : "You will have to charge this pump to me until I thrash." "Is that so," I said. "Well, I guess that is all right if you can satisfy me you will pay for it then." "Sure, I will pay for the pump. I always pay," was the prompt reply of the farmer. "Perhaps you do pay your accounts, but I don't know so. Therefore I wish your promise put in writing and security to show your good faith." Whereat the farmer got angry and let on that his feelings were more than hurt. Here I was with one of my first customers angry and we having trouble on the payment of a pump. I resolved, however, to carry my point, and went after the farmer something like this : "See here, I don't know as you pay your accounts, and you want me to trust you, which I am willing to do if you can satisfy me that you will pay the account when you thrash." I had him at once. To satisfy me he must make a showing of his worth or good intention to pay. His worth he could not show, so his good intention to pay was all that was left to get the pump on. I got out a note and filled it out, also an extra mortgage blank, and took security on the pump, and asked what security he was willing to give so as to satisfy me he would pay his note when due. At first the farmer thought the pump enough. But I showed him the pump was mine until paid for, and to take the pump only was to furnish my own security.

Then I got a cow as security which I took care to see was clear, and it was this cow that enabled me to get payment.

DON'T BE AFRAID TO QUESTION DOUBTFUL CREDITOR.

Don't be afraid to question a doubtful creditor. If he refuses to give you your asked-for information kindly refuse him credit. Often in the spring of the year customers come in to arrange for a credit through the summer ; then, Mr. Merchant, is your time to act. You can dictate terms upon which credit should be given, and if you fail to get good ones the fault is all your own. The credit given should carry with it no obligation of renewal at maturity, as too frequently these obligations are looked to by the customer. It is an important element in the merchant's resources in times of demand upon him that well selected and carefully inspected notes and accounts occupy a most important and responsible field. A danger arises when paper is floated, too easily and profits are made abnormally high so as to invite overselling and trading on the part of a merchant. But the dangers even here which intertwine are not beyond those which are liable to overtake a merchant in direct dealing with any customer who plans to practice dishonesty. It is in connection with all these that it is best when a credit is given that it be arranged at first so it must be paid without extension.

FIX THE LIMIT.

Don't be afraid to tell a customer that he can have a credit to the amount of so many dollars and no more. Give him to understand that when his limit is used up he can have no more credit, and I assure you your troubles at collection time will be limited to a very few.

*Paper read before the Dakota Retail Hardware Association, by M. O. Evenson.

CATALOGUES, BOOKLETS, ETC.

The Syracuse Smelting Works, manufacturers of babbitt metal, solder, etc., Syracuse, N.Y., have issued a calendar that is bound to attract attention. The central position of the calendar is a lithographed design showing two merry maidens coasting on a tandem ... a slide formed of Syracuse babbitt metal. At either side is a panel containing a full list of the lines handled by this firm. A calendar pad beneath the picture completes a good calendar.

CARE OF GAS ENGINES.

IN these days when gas engines are being discussed so extensively, an account of some troubles with gas engines may prove interesting, writes Albert Stritmatter in American Manufacturer. This type of engine too frequently goes into the hands of people who too often have the idea that all that is necessary for the successful running of the engine is to set it on the foundation bolts and open the gas or gasoline regulating valve.

In one foundry a gasoline engine was used to operate the cupola blast. During a heat the engine suddenly stopped. To prevent the iron from cooling in the cupola, the foreman allowed it to run out on the foundry floor. He at once telegraphed to the manufacturers for a man, who arrived the next day. On looking the engine over the machinist was unable to see anything wrong. The compression was good, and so was the spark. But on trying the gasoline pump to bring up the supply, he was unable to get gasoline. Further investigation developed the fact that the only trouble had been the

FAILURE OF THE ENGINEER

to fill up the tank and the gasoline had been used up during the heat.

Another party complained that he had been trying for two months to start his engine, but had been unable to run it for longer than a few minutes at a time. He said he had employed some expert machinist from a neighboring city. These people reported in a letter that they had been unable to locate the difficulty. They said their "expert" had been there without success. In the next paragraph they blandly asked whether it was necessary to use both the hot tube and the electric igniter, or whether only one was required at a time, thus disclosing the fact that while they might be experts at some things, they

DID NOT KNOW MUCH

about gas engines. Further investigation showed that the gas pressure was so strong that the gas bag was as hard as an inflated football, so that too much gas was fed to the engine.

In another instance, a printing office bought a 15 horse-power gas engine which had been installed by a local machinist, who was unable, however, to get it to run. A man was sent by the manufacturers, and on looking the plant over he noticed a couple of bricks on the gasometer. On inquiring the ...son for this, he was informed it was ...ep that 'dog-gone' thing down." On releasing the "thing," however, no further trouble was experienced.

A 35 h.p. gas engine was installed in an Ohio city with the pipe which supplied aid to the engine projecting into an alley, and the end of this pipe had been left open. A passing newsboy happened to notice the suction of air, without knowing the cause, and began throwing his cap upwards and watching its motion changed by the suction. Happening to throw it too near the pipe, he was greatly surprised and dismayed to see it disappear into the pipe, he knew not where. The owners of the engine, upon being informed of the mishap, were almost equally dismayed and puzzled as to how to get rid of the objectionable cap. They tried firing a charge through the pipe by holding the gas valve and igniting the charge, but this did not do any good. Finally, they went to work and took down the pipe, necessitating a shutting down of the plant for about three hours.

We all know that gas engines are like other machinery in that they must have intelligent care and attention in order to give the best results. But, as before stated, too many purchasers do not understand this, and as a result get into what are to them very puzzling difficulties, the remedies for which are really very simple.

MARKETS AND MARKET NOTES

QUEBEC MARKETS.

Montreal, March 15, 1901.

HARDWARE.

TRADE continues to improve and the outlook ahead is decidedly bright. The formation of the big American steel trust seems to have given business a good start, as dealers are ordering supplies quite freely. It is generally believed that screws and bolts will be advanced soon and we know that some wholesale houses have placed special orders this week in anticipation of such a move on the part of the manufacturers. All wire goods are firm and selling well. Wire nails are in better request, but principally for later shipment. Some brands of horse nails are lower in price. "M" brand, for instance, is selling at a discount of 60 per cent. for oval and city head, and 60 and 10 per cent. for new countersunk head on the old list. This is somewhat of a reduction. Cordage is ½c. higher. Spring goods, of all descriptions, are selling well. Another feature of the week is the strengthening of the metal market, which has taken a turn for the better in England. The retail trade, both locally and throughout the country, seems to be much brighter. The demand for stoves has set in somewhat earlier than usual. Wholesalers report payments firstclass.

BARB WIRE—Wire is moving freely at the lower quotations in view of the strong market on the other side of the line. There is a call for some immediate shipments, but the bulk of the orders are for forward delivery. The ruling quotation is $3.05 per 100 lb. f.o.b. Montreal.

GALVANIZED WIRE — Reports would indicate that a good trade is also being done in galvanized wire. There seems to be a great deal of confidence in the strength of the market. We quote: No. 5, $4.25; Nos. 6, 7 and 8 gauge, $3.55; No. 9, $3.10; No. 10, $3.75; No. 11, $3.85; No. 12, $3.25; No. 13, $3.35; No. 14, $4.25; No. 15, $4.75; No. 16, $5.00.

SMOOTH STEEL WIRE—A good trade is being done in all fence wires since the reduction in price. The oiled wire is competing very successfully with the imported galvanized. We quote both oiled and annealed as follows: No. 9, $2.80; No. 10, $2.87; No. 11, $2.90; No. 12, $2.95; No. 13, $3.15 per 100 lb. f.o.b. Montreal, Toronto, Hamilton, London, St. John and Halifax.

FINE STEEL WIRE—Business continues uninterrupted. The market is steady at a discount of 17½ per cent. off the list.

BRASS AND COPPER WIRE—These lines are featureless. Discounts are still 55 and 2½ per cent. on brass, and 50 and 2½ per cent. on copper.

FENCE STAPLES—The demand is increasing. We quote: $3.25 for bright, and $3.75 for galvanized, per keg of 100 lb.

WIRE NAILS—A little better inquiry has been experienced this week, but most orders coming in are for later shipment. We quote $2.85 for small lots and $2.75 for carlots, f.o.b. Montreal, London, Toronto and Hamilton.

CUT NAILS — The inquiry is limited, although the market is quite steady. We quote: $2.35 for small and $2.25 for carlots; flour barrel nails, 25 per cent. discount; coopers' nails, 30 per cent. discount.

HORSE NAILS—The market is unchanged. "C" brand is holding its price at a discount of 50 and 7½ per cent. off their own list. Other makes, including "M" brand, are quoting a discount of 60 per cent. on oval and city head, 60 and 10 per cent. on new countersunk head off the old list. We understand that the Monarch people are offering to make special brands for the wholesalers at very low figures. It is to be hoped that quality will be maintained.

HORSESHOES — The call for horseshoes has been somewhat subdued during the last few weeks. We quote as follows: Iron shoes, light and medium pattern, No. 2 and larger, $3.50; No. 1 and smaller, $3.75; snow shoes, No. 2 and larger, $3.75; No. 1 and smaller, $4.00; X L steel shoes, all sizes, 1 to 5, No. 2 and larger, $3.60; No. 1 and smaller, $3.85; feather-weight, all sizes, $4.85; toe weight steel shoes, all sizes, $5.95 f.o.b. Montreal; f.o.b. Hamilton, London and Guelph, 10c. extra.

POULTRY NETTING —Orders are coming in freely, the ruling discount being 50 and 10 per cent.

GREEN WIRE CLOTH — Retailers are ordering a good supply for immediate shipment. The price is $1.35 per 100 sq. ft.

FREEZERS — The firms pushing this line are doing a good business in it. "Peerless" is quoted as follows: Two quarts, $1.85; 3 quarts, $2; 4 quarts, $2.40; 6 quarts, $3.10; 8 quarts, $3 90; 10 quarts, $5.25; 12 quarts, $6; 16 quarts, with fly wheel, $11; toy, 1 pint, $1.40.

SCREEN DOORS AND WINDOWS — The spring demand is quite brisk. We quote: Screen doors, plain cherry finish, $8.25 per doz.; do. fancy, $11.50 per doz.; walnut, $7.40 per doz., and yellow, $7.45; windows, $2.25 to $3.50 per doz.

SCREWS—The tone is decidedly firm and there is a feeling toward a rise. Wholesalers have placed some large orders this week. Discounts are as follows: Flat head bright, 87½ and 10 per cent. off list; round head bright, 82½ and 10 per cent.; flat head brass, 80 and 10 per cent.; round head brass, 75 and 10 per cent.

BOLTS—A brisk business is being done in bolts and the feeling is firm. Discounts are: Carriage bolts, 65 per cent.; machine bolts, 65 per cent.; coach screws, 75 per cent.; sleigh shoe bolts, 75 per cent.; bolt ends, 65 per cent.; plough bolts, 50 per cent.; square nuts, 4½c. per lb. off list; hexagon nuts, 4¾c. per lb. off list; tire bolts, 67½ per cent.; stove bolts, 67½ per cent.

BUILDING PAPER—A fair business is being done at former prices. We quote: Tarred felt, $1.70 per 100 lb.; 2-ply, ready roofing, 80c. per roll; 3-ply, $1.05 per roll; carpet felt, $2.25 per 100 lb.; dry

sheathing, 30c. per roll; tar sheathing, 40c. per roll; dry fibre, 50c. per roll; tarred fibre, 60c. per roll; O.K. and I.X.L., 65c. per roll; heavy straw sheathing, $28 per ton; slaters' felt, 50c. per roll.

RIVETS — There is no change to note. The discount on best iron rivets, section, carriage, and waggon box, black rivets, tinned do., coopers' rivets and tinned swedes rivets, 60 and 10 per cent.; swedes iron burrs are quoted at 55 per cent. off; copper rivets, 35 and 5 per cent. off; and coppered iron rivets and burrs, in 5-lb. carton boxes, are quoted at 60 and 10 per cent. off list.

CORDAGE—The manufacturers have advanced prices ½c. per lb. The demand is rather brisk. Manila is now worth 13½c. per lb. for 7-16 and larger; sisal is selling at 10c., and lathyarn 10c.

SPADES AND SHOVELS—A spring business is being done. The discount is 40 and 5 per cent. off the list.

HARVEST TOOLS—Some orders for forward delivery have been placed this week. The discount is 50, 10 and 5 per cent.

TACKS—Unchanged. We quote as follows: Carpet tacks, in dozens and bulk, blued 80 and 5 per cent. discount; tinned, 80 and 10 per cent.; cut tacks, blued, in dozens, 75 and 15 per cent. discount.

LAWN MOWERS — Sales are aggregating a fair amount. We quote: High wheel, 50 and 5 per cent. f.o.b. Montreal; low wheel, in all sizes, $2.75 each net; high wheel, 11-inch, 30 per cent. off.

FIREBRICKS—As yet there is but little business being done. Prices are steady at $18 to $24 per 1,000 as to brand, ex store.

CEMENT — Preparations are now being made for the coming season's business. We quote: German, $2.45 to $2.55; English, $2.30 to $2.40; Belgian, $1.95 to $2.05.

METALS.

The feeling in metals has improved this week and although trade is not yet brisk it shows a decided increase. The English market has reacted a little. The trust having been formed in the United States where the Welsh manufacturers buy most of their billets, combined with the limitation of production by the closing down of so many mills has strengthened the market and apparently given it an upward turn. Canadian importers are importing a little more freely. In regard to Canadian manufactured goods, the trend of the market is also upward. The different rolling mills have withdrawn quotations on bar iron and prices are reported 5c. per 100 lb. higher. Steel is also somewhat firmer, as is iron pipe.

PIG IRON—The market here for pig iron is quiet, notwithstanding the advance in the United States. Sales of No. 1 Hamilton have been made at $18 to $18.50, and No. 1 Midland at about the same figures. No. 1 Summerlee is quoted at $20 to $20.50 for spring delivery. Sydney pig iron has been sold in the West at low figures.

BAR IRON—The rolling mills have withdrawn quotations for immediate delivery. A fair price from the wholesale houses now is $1.70 per 100 lb., although some are selling below that figure.

BLACK SHEETS — The English market has taken a better turn, and prices are now firm. Present quotations are: $2.80 for 8 to 16 gauge; $2.85 for 26 gauge, and $2.90 for 28 gauge.

GALVANIZED IRON—A fair trade is being done—better than last week. We quote: No. 28 Queen's Head, $5 to $5.10; Apollo, 10¾ oz., $5 to $5.10; Comet, No. 28, $4.50, with 25c. allowance in case lots.

INGOT COPPER—The market firm at 18c.

INGOT TIN—Primary markets are easy, and Lonnon has declined to £117. Quotations here are 31 to 32c.

LEAD—Unchanged at $4.65.

LEAD PIPE—The demand continues to be heavy. We quote as follows: 7c. for ordinary and 7½c. for composition waste, with 15 per cent. off.

IRON PIPE—The market is firm and the

manufacturers' prices somewhat higher. We quote as follows: Black pipe, ¼, $3 per 100 ft.; ⅜, $3; ½, $3; ¾, $3.15; 1-in., $4.50; 1¼, $6.10; 1½, $7.28; 2-in., $9.75. Galvanized, ½, $4.60; ¾, $5.25; 1-in., $7.50; 1¼, $9.80; 1½, $11.75; 2-in., $16.

TINPLATES — The tone of the market is somewhat improved in Wales and the brighter outlook is reflected here. For immediate delivery goods are worth $4.50 for coke and $4.75 for charcoal.

CANADA PLATE—The market has apparently reached bottom and is now in a process of reaction. We quote: 52's, $2.90; 60's, $3; 75's, $3.10; full polished, $3.75, and galvanized, $4.60.

TOOL STEEL—The tendency seems to be a higher level of values. Inquiries are more numerous. We quote: Black Diamond, 8c.; Jessop's 13c.

STEEL—There is every confidence in the present situation. We quote: Sleighshoe, $1.85; tire, $1.95; spring, $2.75; machinery, $2.75 and toe-calk, $2.50.

TERNE PLATES — There is a better inquiry, both for immediate and future delivery. The price is $8 to $8.25.

COIL CHAIN—Prices are firm. We quote: No. 6, 11½c.; No. 5, 10c.; No. 4, 9½c.; No. 3, 9c.; ¼-inch, 7½c. per lb.; 5-16, $4.65; 5-16 exact, $5.10; ⅜, $4.20; 7-16, $4.00; ½, $3.75; 9-16, $3.65; ⅝, $3.35; ¾, $3.25; ⅞, $3.20; 1-in., $3.15. In carload lots an allowance of 10c. is made.

SHEET ZINC—The ruling price is 6 to 6¼c.

ANTIMONY—Quiet, at 10c.

GLASS.

Most of the specifications sent in are for next month's delivery or open water. We quote: First break, $2; second, $2.10 for 50 feet; first break, 100 feet, $3.80; second, $4; third, $4.50; fourth, $4.75; fifth, $5.25; sixth, $5.75, and seventh, $6.25.

PAINTS AND OILS.

Linseed oil for immediate shipment has been reduced 3c. since our last report. A fair amount is being shipped. There has been no change in turpentine. Red lead is a little lower in England, but stocks are light in Canada, and there is but little tendency to lower prices just now. For spring importation, quotations are shaded 25c. per 100 lb. Trade in general goods is exceedingly bright, the tendency being to purchase all varnishes, stains and paints in tins in a handy form. Quotations on oxide of zinc have been lowered in Germany and England. There has been no change in the price of putty, but, in view of cheaper linseed oil, it is expected that prices may weaken somewhat when the cheaper supplies of oil come to hand. We quote:

WHITE LEAD—Best brands, Government standard, $6.37½; No. 1, $6.00; No. 2, $5.62½; No. 3, $5.25, and No. 4, $4.87½ all f.o.b. Montreal. Terms, 3 per cent. cash or four months.

DRY WHITE LEAD — $5.50 in casks; kegs, $5.75.

RED LEAD — Casks, $5.25; in kegs, $5.50.

WHITE ZINC PAINT—Pure, dry, 7c.; No. 1, 6c.; in oil, pure, 8c.; No. 1, 7c.; No. 2, 6c.

PUTTY—We quote: Bulk, in barrels, $2 per 100 lb.; bulk, in less quantity, $2.15; bladders, in barrels, $2 20; bladders, in 100 or 200 lb. kegs or boxes, $2.35; in tins, $2.45 to $2.75; in less than 100-lb. lots, $3 f.o.b. Montreal, Ottawa, Toronto, Hamilton, London and Guelph. Maritime Provinces 10c. higher, f.o.b. St. John and Halifax.

LINSEED OIL—Raw, 69c.; boiled, 73c.; in 5 to 9 bbls., 1c. less, 10 to 20 bbl. lots, open, net cash, plus 2c. for 4 months. Delivered anywhere in Ontario between Montreal and Oshawa at 2c. per gal. advance and freight allowed.

TURPENTINE—Single bbls., 62c.; 2 to 4 bbls., 61c.; 5 bbls. and over, open terms, the same terms as linseed oil.

MIXED PAINTS—$1.25 to $1.45 per gal.

CASTOR OIL—8¾ to 9¼c. in wholesale lots, and ½c. additional for small lots.

SEAL OIL—47½ to 49c.

COD OIL—32½ to 35c.

NAVAL STORES — We quote: Resins, $2.75 to $4 50, as to brand; coal tar, $3.25 to $3.75; cotton waste, 4½ to 5½c. for colored, and 6 to 7½c. for white; oakum, 5½ to 6½c., and cotton oakum, 10 to 11c.

PARIS GREEN—Petroleum barrels, 16¾c. per lb.; arsenic kegs, 17c.; 50 and 100-lb. drums, 17½c.; 25-lb. drums, 18c.; 1-lb. packages, 18½c.; ½-lb. packages, 20½c.; 1-lb. tins, 19½c.; ½-lb. tins, 21½c. f.o.b. Montreal; terms 3 per cent. 30 days, or four months from date of delivery.

SCRAP METALS.

The market is quiet and prices are

nominal. Dealers are paying the following prices in the country: Heavy copper and wire, 13 to 13½c. per lb.; light copper, 12c.; heavy brass, 12c.; heavy yellow, 8½ to 9c.; light brass, 6½ to 7c.; lead, 2¾ to 3c. per lb.; zinc, 2¼ to 2½c.; iron, No. 1 wrought, $13 to $14 per gross ton; No. 1 cast, $13 to $14; stove plate, $8 to $9; light iron, No. 2, $4 a ton; malleable and steel, $4.

HIDES.

The weaker feeling in the American market tends to keep prices down here. Dealers are still paying 7c. for No. 1 light. There is no activity in the demand. We quote: Light hides, 7c. for No. 1; 6c. for No. 2, and 5c. for No. 3. Lambskins, 90c.

PETROLEUM.

The demand is not so brisk. We quote as follows: "Silver Star," 14½ to 15½c.; "Imperial Acme," 16 to 17c.; "S C. Acme," 18 to 19c., and "Pratt's Astral," 18½ to 19½c.

MONTREAL NOTES.

Iron pipe is higher at the rolling mills.

Bar iron is 5c. higher at the rolling mills.

"M" brand horse nails have been reduced.

An advance in bolts and screws is talked of.

Cordage, both manila and sisal are ½c. per lb. higher.

ONTARIO MARKETS.

TORONTO, March 15, 1901.

HARDWARE.

CONTINUED improvement in business is to be noted in the wholesale hardware trade, and a more active trade from this out may be looked for. Prices on such staple lines as bar iron, iron pipe, and rope are stiff and advancing, and there is a general opinion that higher figures may be looked for in bolts. There is also a firm feeling in regard to screws. In barb wire forward orders are being shipped this week from Cleveland, and a few small orders are going out from stock. Much the same remarks apply to galvanized wire. There is very little being done in smooth steel wire of any kind. The movement in cutlery is fair for this time of the year. Sporting goods are quiet. Trade is fair in spades and shovels and in harvest tools. Further improvement is to be noted in the demand for wire nails. Cut nails are as dull as ever. Horseshoes are quiet. But a fair trade is to be noted in horse nails. Tinware is quiet, and the same may be said of enamelled ware. Inquiries are becoming more numerous for gas and oil stoves.

BARB WIRE—Forward orders are being shipped to their destination in Canada from Cleveland, and there has been little business done from stock. Business in that respect, however, is very light. There has been no change in prices and we quote f. o. b. Cleveland at $2.82½ per 100 lb. for less than carlots and $2 70 for carlots. The quotation from stock Toronto is $3.05 per 100 lb.

GALVANIZED WIRE — Forward orders from Cleveland are also being delivered this week, and a slight business is reported from stock. Prices are without change. We quote: Nos. 6, 7 and 8, $3.50 to $3.85 per 100 lb., according to quantity; No. 9, $2 85 to $3 15; No. 10, $3 60 to $3.95; No. 11, $3.70 to $4.10; No. 12, $3 to $3 30; No. 13, $3 10 to $3 40; No. 14, $4.10 to $4 50; No. 15 $4 60 to $5 05; No. 16, $4.85 to $5.35. Nos. 6 to 9 base f.o.b. Cleveland are quoted at $2.57½ in less than carlots and 12c. less for carlots of 15 tons. There is a little better movement in galvanized wire this week.

SMOOTH STEEL WIRE—There is little or nothing doing in either oiled or annealed or hay-baling wire. Net selling prices are as follows, as noted last week: Nos. 6 to 8, $2.90; 9, $2.80; 10, $2.87; 11, $2.90; 12, $2.95; 13, $3 15; 14, $3.37; 15, $3.50; 16, $3.65. Delivery points, Toronto, Hamilton, London and Montreal, with freights equalized on these points.

WIRE NAILS — The demand has improved, and a fair business is being done, although the orders are not individually large. The association meets early next month, and some are of the opinion that an advance is not improbable. The base price is still unchanged at $2.85 per keg in less than carlots, and $2.75 in carlots.

CUT NAILS—These are as dull as ever. The base price is $2.35 per keg. Delivery points, Toronto, Hamilton, London, Montreal and St. John, N.B.

HORSESHOES—The demand for horseshoes is rather light. We quote f. o. b. Toronto as follows: Iron shoes, No. 2 and larger, light, medium and heavy, $3 60; snow shoes, $3.85; light steel shoes, $3 70; featherweight (all sizes), $4.95; iron shoes, No. 1 and smaller, light, medium and heavy (all sizes), $3.85; snow shoes, $4; light steel shoes, $3.95; featherweight (all sizes), $4.95.

HORSE NAILS—Trade is fair in this line. Some of the manufacturers have again reorganized, and they are quoting oval head 50, 10 and 5 per cent. off old list, and countersunk head, 50, 10 and 10 per cent. The discount on "C" brand oval head is still 50 and 7½ per cent. off the new list.

STAPLES — These are a little lower in sympathy with the decline in wire. We quote: Galvanized, $3 50 to $4; plain, $3.25 to $3.75.

SCREWS—Trade continues fairly active

in screws, and the feeling in regard to prices shows increased confidence. We quote discounts as follows: Flat head bright, 87½ and 10 per cent.; round head bright, 82½ and 10 per cent.; flat head brass, 80 and 10 per cent.; round head brass, 75 and 10 per cent.; round head bronze, 65 per cent., and flat head bronze at 70 per cent.

BOLTS AND NUTS—Trade is fairly active in bolts of all kinds. The feeling noted last week in regard to the possibility of an advance in prices still obtains. We quote as follows: Carriage bolts (Norway), full square, 70 per cent.; carriage bolts full square, 70 per cent.; common carriage bolts, all sizes, 65 per cent.; machine bolts, all sizes, 65 per cent.; coach screws, 75 per cent.; sleighshoe bolts, 75 per cent.; blank bolts, 65 per cent.; bolt ends, 65 per cent.; nuts, square, 4½c. off; nuts, hexagon, 4¾c. off; tire bolts, 67½ per cent.; stove bolts, 67½; plough bolts, 60 per cent.; stove rods, 6 to 8c.

RIVETS AND BURRS—Business is steady and prices unchanged. We quote as follows: Iron rivets, 60 and 10 per cent.; iron burrs, 55 per cent.; copper rivets and burrs, 35 and 5 per cent.

ROPE—An advance of ½c. per lb. is to be noted in both sisal and manila rope. We now quote the base price as follows: Sisal 10c. per lb; manila, 13½c. per lb.; special manila, 11½c.; New Zealand, 10c. per lb.; single tarred lathyarn, 9½c. The demand is fairly good for rope, and prices are firm.

CUTLERY—There is a fair general sorting-up trade for this time of the year. Naturally, however, not a large trade is being done.

SPORTING GOODS—Orders are being taken for loaded shells, but business in sporting goods is on the whole quiet.

GREEN WIRE CLOTH—There is scarcely anything being done, and shipments have not yet been made of orders recently booked. Prices are unchanged at $1.35 per 100 sq. ft.

SCREEN DOORS AND WINDOWS—These are quiet. We quote 4 in. styles in doors at $7.20 to $7.80 per doz., and 3 in. styles 20c. per doz. less; screen windows, $1.60 to $3.60 per doz., according to size and extension.

TINWARE—Some disappointment is being expressed in regard to the trade in tinware, but the unfavorable winter weather is claimed to be the cause.

ENAMELLED WARE — Business is also quiet in enamelled ware, and there are no immediate signs of an improvement.

POULTRY NETTING — Shipments of imported wire are near at hand, and a little business is being done. The discount on Canadian is 50 and 10 per cent. and net figures are quoted on English.

HARVEST TOOLS—There is a little movement in some seasonable lines, but business does not amount to a great deal.

SPADES AND SHOVELS—Orders are being booked for forward delivery and quite a number of shipments are being made in spades and shovels. Some shipments are also being made in garden rakes and hoes.

RANGE BOILERS—Trade is still just moderate. We quote: 30 gallons, $7; 35 gallons, $8.25; 40 gallons, $9.50.

CEMENT — Trade has not yet opened up. We quote: Canadian portland $2.40 to $3; German, $3 to $3.15; English, $2.85 to $3; Belgian, $2.50 to $2.75; calcined plaster, $2.

COTTER PINS—Quite a radical change has taken place in the list of The Nettlefold Co's list of cotter pins, the American list having been adopted. These pins were formerly sold on the wire gauge system. By adopting the American list the sizes have been reduced by one half, they now being 8 instead of 16.

GAS AND OIL STOVES — Inquiries during the past week for gas and oil stoves have been more numerous, and trade in this line is expected to open up shortly.

METALS.

The weather has been unfavorable for the metal trade during the past week, and business is not as brisk as it was a week ago. The iron market is firm and steel is particularly strong.

PIG IRON — There is not a great deal being done, as far as can be learned, but the market is steady, and advices from the United States report an advance of $1 in Bessemer iron.

STEEL — The steel market continues decidedly strong, and in the United States premiums of $2 to $5 for billets are being paid. Some of the manufacturers in Canada have withdrawn their quotations during the past week.

BAR IRON—Prices are firmer and there is a fair demand. We quote the base price to retailers at $1.70 to $1.75 per 100 lb.

PIG TIN—There is a fair trade being done in some qualities. We quote Lamb and Flag at 32c. Prices are a little higher in New York on account of delayed arrivals. In England the market is steady as to price.

TINPLATES—There has been rather a nice movement in tinplates during the past week. Advices from Great Britain state that the manufacturers report that they cannot sell at figures lower than those now prevailing. Consequently, there is a steadier feeling as to prices.

TINNED SHEETS—A few of these are moving all the time, principally for dairy and cheese-making purposes. We quote 9 to 9½c. for 28 gauge.

BLACK SHEETS—Trade keeps fair, with 28 gauge quoted at $3.30.

GALVANIZED SHEETS—Business is only fair in this line, the stormy weather having interfered with business in this particular line. We quote English at $4.75 in small lots, and American at $4.50.

CANADA PLATES—Orders are still being booked for fall delivery, and some small orders are being received for shipment from stock. Prices are firm, and the difference in the manufacturers' price for all bright and ordinary is much greater than it has been for some years. We quote: All dull, $3, half-and half, $3.15, and all bright, $3.65 to $3.75.

COPPER—Ingot copper is in fair demand, and in sheet there is a good business being done. We quote: Ingot, 19 to 20c.; bolt or bar, 23½ to 25c.; sheet, 23 to 23½c. per lb. There was a sharp break in the London market on Wednesday, there being a decline of 12s. 6d. Prices in London are now lower than they have been since January 1900.

BRASS—Trade is improving and the discount on rod and sheet is unchanged at 15 per cent.

SOLDER—Trade is fair, and prices unchanged. We quote: Half-and-half, guaranteed, 18c.; ditto, commercial, 17½c.; refined, 17½c., and wiping, 17c.

LEAD—The demand is good and prices

in the outside markets are steady. We quote 4¾ to 5c.

LEAD PIPE—The price of lead pipe has been reduced, the discount now being 25 per cent. instead of 15 as formerly.

IRON PIPE—The jobbers in Toronto have agreed upon a uniform price for iron pipe, and the quotations are now as follows per 100 feet: Black pipe, ⅛ in., $4.15; ¼ in., $3; ⅜ in., $3.05; ½ in., $3.10; ¾ in., $3.35; 1 in., $4 70; 1¼ in., $6.40; 1½in., $7.70; 2 in., $10.25; 2½ in., $2.85; 3 in., $28 69; 3½ in, $36.10; 4 in., $41.04; 5 in., $55.10; 6 in., $71.44; 7 in., $89.30; 8 in., $107.16. Galvanized pipe, ½ in., $4.75; ¾ in., $5.25; 1 in., $7.25; 1¼ in., $9.50; 1½ in., $11.50; 2 in., $15 50.

ZINC SPELTER—Trade is quiet, locally, and the outside markets are dull and easy. We still quote 6 to 6½c. per lb.

ZINC SHEET—The demand is only moderate. We quote casks at 6¾ to 7c., and part casks at 7 to 7½c. per lb.

ANTIMONY—Business is more active and prices are unchanged at 11 to 11½c. per lb. In London, England, the market is easier.

PAINTS AND OILS.

There is a big movement. With the exception of paris green, which is rather slow, all lines are moving briskly. Turpentine fell about 1c. in Savannah some days ago, but is now steady. Linseed oil is unchanged, but the indications are that prices will be decidedly lower in June or July. White lead is firm. We quote:

WHITE LEAD—Ex Toronto, pure white lead, $6 50; No. 1, $6.12½; No. 2, $5.75; No. 3. $5.37½; No. 4. $5; dry white lead in casks, $6.

RED LEAD—Genuine, in casks of 560 lb., $5.50; ditto, in kegs of 100 lb., $5.75; No. 1, in casks of 560 lb., $5 to $5 25; ditto, kegs of 100 lb.; $5.25 to $5.50.

LITHARGE—Genuine, 7 to 7½c.

ORANGE MINERAL—Genuine, 8 to 8½c.

WHITE ZINC—Genuine, French V.M., in casks, $7 to $7.25; Lehigh, in casks, $6.

PARIS WHITE—90c. to $1 per 100 lb.

WHITING—70c. per 100 lb.; Gilders' whiting, 80c.

GUM SHELLAC—In cases, 22c.; in less than cases, 25c.

PARIS GREEN—Bbls., 16¾c.; kegs, 17c.; 50 and 100-lb. drums, 17½c.; 25-lb. drums, 18c.; 1-lb. papers, 18½c.; 1-lb. tins, 19½c.; ½-lb. papers, 20½c.; ½-lb. tins, 21½c.

PUTTY—Bladders, in bbls., $2.20; bladders, in 100 lb. kegs, $2.35; bulk in bbls., $2; bulk, less than bbls. and up to 100 lb., $2.15; bladders, bulk or tins, less than 100 lb., $3.

PLASTER PARIS—New Brunswick, $1.90 per bbl.

PUMICE STONE—Powdered, $2.50 per cwt. in bbls., and 4 to 5c. per lb. in less quantity; lump, 10c. in small lots, and 8c. in bbls.

LIQUID PAINTS—Pure, $1.20 to $1.30 per gal.; No. 1 quality, $1 per gal.

CASTOR OIL—East India, in cases, 10 to 10½c. per lb. and 10½ to 11c. for single tins.

LINSEED OIL—Raw, 1 to 4 barrels, 69c.; boiled, 72c.; 5 to 9 barrels, raw, 68c.; boiled, 71c., delivered. To Toronto, Hamilton, Guelph and London, 1c. less.

TURPENTINE—Single barrels, 62c.; 2 to 4 barrels, 61c., to all points in Ontario. For less quantities than barrels, 5c. per gallon extra will be added, and for 5-gallon packages, 50c., and 10 gallon packages, 80c. will be charged.

GLASS.

There is a fair movement from stock, with prices firm. A big import trade is being done, but at a very close margin, which jobbers claim is unsatisfactory to them. We still quote first break locally as follows: Star, in 50-foot boxes, $2.10, and 100-foot boxes, $4; double diamond under 26 united inches, $6. Toronto, Hamilton and London; terms 4 months or 3 per cent. 30 days.

HIDES, SKINS AND WOOL.

HIDES—The decline anticipated in these columns took place this week. Prices are ½ to ¾c. lower. We quote: Cowhides, No. 1, 7c.; No. 2, 6c.; No. 3, 5c. Steer hides are worth 1c. more. Cured hides are quoted at 7½c.

SKINS—Calfskins are 5c. higher, otherwise there is no change. The market is quiet. We quote: No. 1 veal, 8-lb. and up, 9c. per lb.; No. 2, 8c.; dekins, from 40 to 60c.; culls, 20 to 25c. Sheep skins, 90c. to $1.

WOOL—Unwashed has again been reduced ½c. We quote: Combing fleece, 14 to 15c., and unwashed, 8 to 9c.

COAL.

Owing to snow blockade on railroads running into Buffalo from the So[illegible] re is great difficulty in getting coal, es[illegible] bituminous, through that centre. [illegible] are unchanged. We quote anthra[illegible] cars Buffalo and bridges: Grate, [illegible] per gross ton and $4.24 per net ton; egg, stove and nut, $5 per gross ton and $4.46 per net ton.

PETROLEUM.

The season is about over and the market is easier. A decline of ½c. is noted. We quote: Pratt's Astral, 16½ to 17c. in bulk (barrels, $1 extra); American water white, 16½ to 17c. in barrels; Photogene, 16 to 16½c.; Sarnia water white, 15½ to 16c. in barrels; Sarnia prime white, 14½ to 15c. in barrels.

MARKET NOTES.

Staples are lower.

A new list has been issued on The Nettlefold Co.'s cotter pins.

The discount on lead pipe has been increased to 25 per cent.

An advance of ½c. per lb. has been made on sisal and manila rope.

H. S. Howland, Sons & Co. are in receipt of a shipment of wooden butter bowls in sizes from 13 to 21 metres in diameter.

Harland Bros., Clinton, Ont., have bought the hardware stock of the late David Johnson, Seaforth, and will open a store in the latter town.

MANITOBA MARKETS.

WINNIPEG, March 13, 1901.

THE general tone of business has distinctly improved during the week. The large amount of building being planned for the now opening season is increasing, the orders for builders' hardware and the amount of these orders are larger than was hoped for a month ago.

Since the formation of the Steel Trust there has been a firm feeling in iron and steel prices here, although no actual advance has as yet been noted.

Linseed oil has dropped 3c. per gallon since last writing and is now quoted at 80 and 83c. Dealers here seem to anticipate an advance in glass and prices are very firm. It is also expected that there will be a change in the price of white lead, as the market seems uncertain at the present writing and is rather week.

Quotations for the week are as follows :

Barbed wire, 100 lb.	$3 45
Plain twist	3 45
Staples	3 95
Oiled annealed wire 10	3 95
" 11	4 00
" 12	4 05
" 13	4 20
" 14	4 35
" 15	4 45
Wire nails, 30 to 60 dy, keg	3 45
" 16 and 20	3 50
" 10	3 55
" 8	3 65
" 6	3 70
" 4	3 85
" 3	4 10
Cut nails, 30 to 60 dy.	3 00
" 20 to 40	3 05
" 10 to 16	3 10
" 8	3 15
" 6	3 20
" 4	3 30
" 3	3 65
Horsenails, 45 per cent. discount.	
Horseshoes, iron, No. 0 to No 1	4 65
No. 2 and larger	4 40
Snow shoes, No. 0 to No. 1	4 90
No. 2 and larger	4 40
Steel, No. 0 to No. 1	4 95
No. 2 and larger	4 70
Bar iron, $2.50 basis.	
Swedish iron, $4.50 basis.	
Sleigh shoe steel	3 00
Spring steel	3 25
Machinery steel	3 75
Tool steel, Black Diamond, 100 lb	8 50
Jessop	13 00
Sheet iron, black, 10 to 20 gauge, 100 lb.	3 50
20 to 26 gauge	3 75
28 gauge	4 00
Galvanized American, 16 gauge	2 54
18 to 22 gauge	4 50
24 gauge	4 75
26 gauge	5 00
28 gauge	5 25
Genuine Russian, lb	12
Imitation " "	8
Tinned, 24 gauge, 100 lb	7 55
26 gauge	8 80
28 gauge	8 00
Tinplate, IC charcoal, 20 x 28, box	10 75
" IX "	12 75
" IXX "	14 75
Ingot tin	35
Canada plate, 18 x 21 and 18 x 24	3 75
Sheet zinc, cask lots, 100 lb	7 50
Broken lots	8 00
Pig lead, 100 lb.	6 00
Wrought pipe, black up to 2 inch ....50 an	10 p.c.
" Over 2 inch	50 p.c.
Rope, sisal, 7-16 and larger	$10 00
" ⅜	10 50
" ¼ and 5-16	11 00
Manila, 7-16 and larger	13 50
" ⅜	14 00
" ¼ and 5-16	14 50
Solder	21½
Cotton Rope, all sizes, lb.	16
Axes, chopping	$7 50 to 12 00
" double bitts	12 00 to 18 00
Screws, flat head, iron, bright	87½
Round " "	82½
Flat " brass	80
Round " "	75
Coach	57½ p.c.
Bolts, carriage	55 p.c.
Machine	55 p.c.
Tire	60 p.c
Sleigh shoe	65 p.c.
Plough	40 p.c.
Rivets, iron	50 p.c.
Copper, No. 8	35
Spades and shovels	40 p.c.
Harvest tools	50, and 10 p.c.
Axe handles, turned, s. g. hickory, doz.	$2 50
No. 1	1 50
No. 2	1 25
Octagon extra	1 75
No. 1	1 25
Files common	70, and 10 p.c.
Diamond	60
Ammunition, cartridges, Dominion R.F.	50 p.c.
Dominion, C.F., pistol	30 p.c.
" military	15 p.c.
American R.F.	30 p.c.
C.F. pistol	5 p.c.
C.F. military	10 p.c. advance.
Loaded shells :	
Eley's soft, 12 gauge black	16 50
chilled, 12 guage	18 60
soft, 10 guage	21 00
chilled, 10 guage	23 00
Shot, Ordinary, per 100 lb	6 75
Chilled	7 50
Powder, F.F., keg	4 75
F.F.G.	5 00
Tinware, pressed, retinned	75 and 2½ p.c.
" " plain	70 and 15 p.c.
Graniteware, according to quality	50 p.c.

PETROLEUM.

Water white American	25½c.
Prime white American	24c.
Water white Canadian	22c.
Prime white Canadian	21c.

PAINTS, OILS AND GLASS.

Turpentine, pure, in barrels	$ 68
Less than barrel lots	73
Linseed oil, raw	80
Boiled	83
Lubricating oils, Eldorado castor	25½
Eldorado engine	24½
Atlantic red	27½
Renown engine	41
Black oil	23½ to 25
Cylinder oil (according to grade)	55 to 74
Harness oil	61
Neatsfoot oil	$ 1 00
Steam refined oil	85
Sperm oil	1 50
Castor oil per lb.	11½
Glass, single glass, first break, 16 to 25 united inches	2 25
26 to 40 per 50 ft.	2 50
41 to 50 " 100 ft.	5 50
51 to 60 " " "	6 00
61 to 70 per 100-ft. boxes	6 50
Putty, in bladders, barrel lots per lb.	2½
kegs "	2¾
White lead, pure per cwt.	7 25
No 1 "	7 00
Prepared paints, pure liquid colors, according to shade and color..per gal.	$1.30 to $1.90

ANOTHER MAMMOTH ELEVATOR.

A Port Arthur, Ont., despatch says that the Canadian Northern Railway have let the first contracts for the improvements they intend making at that port. Plans are ready for a 2,000,000 bush. elevator. There are now two round houses, the last one fitted up with all modern machinery. Coal docks and other terminal foundations will be built on their three miles of terminal grounds at Port Arthur. It is expected that fully 5,000 men will be engaged in construction work there during the spring.

WHOLESALE TO PRIVATE BUYERS.

Bad times have undoubtedly fallen upon the cycle trade. People who have cycles to sell nowadays resort to as many dodges to get rid of them as a patent medicine agent. There is a cycle-making firm in Birmingham, for instance, who are carefully sending a circular letter headed "Wholesale to the Trade," and an illustrated price list marked "Private" and "Wholesale List," to all the likely and unlikely buyers to be found in the directories. It is a good old dodge to catch the private buyer. The idea is that the private buyer bursts himself in his endeavor to buy the high-grade machine offered at wholesale prices. But this idea is largely fallacious, because, as a rule, the private buyer does nothing of the kind. What he wants, generally, is a machine that has some reputation in its name, and this he is quite content to obtain through the regular agents. What usually happens, as a result of this kind of business-getting, is that the firm playing this game condemn themselves in the eyes of the cycle trade.—Hardwareman.

The stock, etc., of Benor, Taylor & Co., general merchants, Alliston, Ont., is advertised by sheriff to be sold to-day (Friday).

CONDENSED OR "WANT" ADVERTISEMENTS.

Advertisements under this heading, 2c. a word each insertion ; cash in advance. Letters, figures, and abbreviations each count as one word in estimating cost.

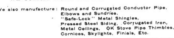

INDUSTRIAL GOSSIP.

Those having any items of news suitable for this column will confer a favor by forwarding them to this office addressed to the Editor.

MARCH & HENTHORN, boiler-makers, Belleville, Ont., have just completed for the steamer Alexandria one of the largest firebox boilers yet made in Canada, it being 19 feet long, 11½ feet in diameter and 20 feet high.

Dyment, Butterfield & Co., founders, Barrie, Ont., intend increasing their staff to 50 men.

Weir & Tizzard, of the Parry Sound Machine Works, have added a brass casting outfit to their plant.

The J. Still Manufacturing Co., Limited, St. Thomas, Ont., have been incorporated with $99,000 capital.

Hinde & Brownsword are putting about $50,000 new machinery in their terra cotta works at Anvil Island, B.C.

Shurly & Dietrich, Galt, Ont., have decided to start manufacturing harvest tools, particularly rakes, hoes, forks, etc.

The shipments of ore from Rossland, B.C., last week totalled 8,591 tons. This is a new record, the largest previous week's shipment being 7 814 tons.

The Peterborough Hydraulic Power Co., Limited, have been incorporated for the purpose of developing and selling or leasing power by means of water or otherwise. The share capital is $250,000, and the head office is at Peterborough, Ont.

Hall Bros., dealers in agricultural implements, Solsgirth, Man., have opened a branch at Birtle, Man.

A by-law has been passed in Tilsonburg, Ont., to raise $2,000 by debentures for building cement walks.

The Formosa Oil Co., Limited, have been incorporated, with headquarters in Formosa, Ont., and a share capital of $10,000.

By-laws to grant a $50,000 bonus and to exemp from taxation for 10 years a large smelter have been passed in Kaslo, B.C.

The Dominion Wrought Iron Wheel Co., Limited, have been incorporated with share capital of $20,000 and head office in Toronto.

A despatch from Sydney, N S., says the second blast furnace will be blown in about April 1. The output of the first furnace is now about 180 tons per day. The roof is being put on the blooming mill.

The Lincoln Electric Light and Power Co., Limited, have been incorporated with $400,000 share capital. The company intend to acquire and operate the property and franchise of the St. Catharines Electric Light and Power Company, Limited, St. Catharines, Ont.

AIDS TO BUSINESS.

A firm of ironmongers in the North of England have adopted a novel scheme for popularizing the instalment - purchase system. Recognizing the difficulty which a man's dependents usually experience in keeping up payments when the head of the family is suddenly taken away, the firm in question undertake in the event of the husband dying before a purchase is completed to hand over the goods to his widow without further charge. In addition to this all cash customers will in future receive a free insurance policy covering the amount of their purchases for 12 months.—Ironmonger.

MR. CHAYER'S NEW POSITION.

Mr. J. Chayer, formerly with H. A. Nelson & Sons, Co., Limited, has accepted a position with Lamplough & McNaughton, Montreal, and will take charge of the department covering such lines as are handled by the drug and fancy goods trade. Mr. Chayer has gone to Europe via New York, sailing on ss. Kaiser Wilhelm der Grosse. He will visit the various European centres and select samples suitable for the Canadian trade.

Glazer & Co. have registered as junk dealers in Montreal.

HEATING AND PLUMBING

SOME BUILDING NOTES.

G. M. Vance will erect a house in Shelburne, Ont., this summer.

Frank Nash is planning to erect a hotel in Regina, N.W.T., this summer.

The Oak Hall, St. John, N.B., is to be extended and new machinery installed.

Ed. Kichna intends erecting a brick residence in Sebringville, Ont., next summer.

A cottage hospital is to be established in Carberry, Man., by the ladies of that place.

The ratepayers of Chester, N.S., have voted to spend $6,000 on a new schoolhouse.

George Bryan is building a residence in Innisfail, N.W.T., for Thos. Evans, Dedham, Iowa.

J. J. McKillop and Dr. Teskey have decided to erect stores on adjoining lots in St. Thomas, Ont.

A. H. Pulford, Winnipeg, has decided to erect a $12,000 brick structure on Portage avenue, Winnipeg.

G. F. Stephens & Co., Winnipeg, have let contracts for a $2,000 block on Bannatyne avenue, in that city.

Chas. O'Reilly, proprietor of the Russell House, Smith's Falls, Ont., proposes to erect a new $20,000 hotel.

The Ottawa branch of the Salvation Army propose erecting a branch, to cost $7,000, on Cooper street, near Bank.

Edgar Ah Wing intends erecting a large brick building on Dupont street, Vancouver. M. O. Reete, contractor, Victoria, will construct the building.

W. D. McKillican has the contract to erect a new house at the corner of Douglas and Kane streets, Victoria, for Dr. Fraser. The contract price is $7,000.

Listowel, Ont., has passed a by-law granting the Hess Bros. Bent Chair Co. a loan of $10,000. The factory is expected to be running by the middle of July.

Architect A. W. Peene, Hamilton, is preparing plans for nine brick dwellings at the corner of Jackson and Spring streets for Dr. Simpson, of that city.

A. Hutchison, of Lavigueur & Hutchison, Quebec, is having a $5,000 residence built at the corner of St. Cyrille and De Salaberry streets, Quebec.

Plans for improvements on the Queen's Hotel, Barrie, Ont., to cost about $5,000 are being prepared by Gouinlock & Baker, architects, Toronto. The changes will include the removal of the lavatories and washroom, which are to be of marble to the basement, at an estimated cost of $2,000.

David Ouellet, architect, is asking for tenders for the erection of offices for the Metropolitan Life Insurance Co., Limited, St. John street, Quebec, before March 20.

Speagle Bros., Westport, Ont., have been awarded the contract for the erection of the new Catholic church in Lansdowne, Ont. The building will cost $6,000.

The D. F. Brown Paper Company have awarded the contract for their new brick warehouse and factory on Canterbury street, St. John, N.B., to Ald. Maxwell, of that city.

Tenders have been received by Stewart Mulvey, Secretary of the Winnipeg Public School Board, for additions to the school board offices, corner of Ellen and William streets, Winnipeg.

The Fire and Police Committee of the Vancouver Council have decided to advertise for competitive plans for two fire halls and a city jail. The cost of the buildings will aggregate nearly $60,000.

Building prospects in Portage la Prairie, Man., are bright. The Lafferty block is being repaired by J. E. Brown, at a cost of about $5,000; the Merchants Bank will erect an addition to their premises; J. K. Hill is advertising for tenders for a two-storey block 60 x 22 1-2 ft.; Campbell & Co. intend replacing their premises, which were destroyed by fire a short time ago, with a handsome brick block; the Northern Pacific are to erect a new station there and several other structures are to be erected.

MELTING SOLDERING COPPERS.

From W. S., Nova Scotia.—I have a large number of soldering coppers that are too small for use. Can I melt them over into larger coppers? If so, how is it done? I tried melting them in the crucible on a blacksmith's forge, but melted the crucible.

Answer.—Copper can be welded by making the surfaces clean and applying a mixture of 3 parts phosphate of soda and 1 part borax triturated together. Apply the powder when the metals are at a dull red heat by sprinkling it over the surface. Then heat to a cherry red and hammer as in welding iron; only carefully, as the copper becomes quite soft at the welding heat. The coppers may also be readily melted and cast in a warm iron mold. For this purpose a graphite or black lead crucible should be used in a fire box built of loose brick-work around a forge tuyere, a little larger than the crucible, so as to hold the heat around the crucible. In this way 4 or 5 pounds of copper may be melted.—Metal Worker.

BUILDING PERMITS ISSUED.

Building permits have been issued in Ottawa to Miss S. Sparks, for a $10,000 business block on Rideau street (Arnoldi & Ewart, architects); to J. Cuzner, brick hardware store, Queen street west, $2,000; John W. Hodgins, brick veneered dwelling, Concession street, $1,750; T. C. James, brick veneered dwelling, James street, $1,300; James Maloney, double brick veneered dwelling, Rideau street, $2,200; Mrs. C. A. Dorion, brick veneered dwelling, St. Patrick street, $1,400; J. T. Moxley, frame dwelling, Salisbury Place, $2,400; Robert Thackray, brick tenements, Bank street, $4,000.

The following permits have been taken out in Toronto: H. Smith, for a pair of dwellings at 68 and 70 Howland avenue, to cost $5,000; J. H. C. Durham, for a brick residence on Elm avenue, Rosedale, to cost $4,250; Albert Johnston, for a brick residence at 379 Berkeley street, costing $3,000; Macpherson estate, for two residences on Crescent road, near Yonge street, to cost $4,500; to Wm. Davies, for an office building, corner Queen and James streets, to cost $35,000.

PLUMBING AND HEATING CONTRACTS.

Quest & Co. have contracts for plumbing in several houses at Kew Beach, Toronto.

The Bennett & Wright Co., Toronto, have the contract for plumbing, heating, gas-fitting and electric wiring in a house for John Stark & Co., on Crescent road, Rosedale.

PLUMBING AND HEATING NOTES.

John F. Teed, contractor, Dorchester, N.B., is dead.

The stock, etc., of the estate of T. D. M. Osborne, plumber, etc., Brandon, Man., has been sold.

James A. Spence, proprietor of the electric light works in Colborne, Ont., has sold out to P. Coyne.

A Winnipeg despatch says that the premises of Cotter Bros., plumbers of that city, were damaged by fire on Tuesday.

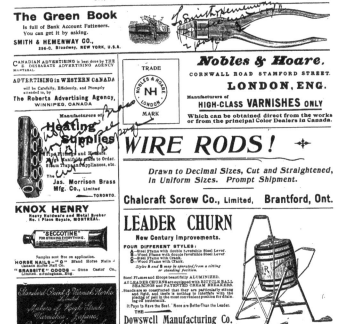

ANOTHER ALUMINUM SOLDER.

The greatest obstacle to the extension of the use of aluminum by sheet metal workers and other mechanics has been the absence of an absolutely efficient solder that will stand all tests and yet be easy of application by the ordinary workman with the common tools used in applying any solder. This want, according to an exchange, an Indiana firm claim to be able to fill in a new aluminum solder and flux which they are now ready to place on the market. By the use of this solder and flux, the inventors claim, aluminum can be soldered as quickly as any other metal with similar results to that given by the ordinary half and half solder. Any ordinary soldering iron, tinned with common solder, and using their solder and flux, will do the work.

The company give the following directions for the use of their preparation : " Care should be taken to scrape or file castings or sheet surfaces that are to be connected. Apply a light coating of flux and proceed as in the ordinary method with the iron or sweat solder. The aluminum solder will not flow like ordinary half and half solder, but will flow to some extent, care being taken to keep the iron hot. The solder and flux will not discolor the metal, but can be worked or polished the same as any other metal or alloy."

The company are making the solder and are prepared to furnish it in any quantity, large or small, full directions going with every stick. They will also be glad to furnish any information desired.

THIS PLUMBER LABORER ABUSED HIS WIFE.

A tall, rednosed Irishwoman, whose countenance showed marks of dissipation, approached the bench in one of the Greater New York police courts the day before Christmas and, in a voice loud enough to be heard on the Brooklyn bridge, exclaimed :

" Are ye the Jedge ? If yer is I want a warrunt quick for—"

" One minute, if you please, my good woman," interrupted his honor. " We are not deaf here, nor do we issue warrants in a hurry. Now what is your name and where do you reside ? "

" Me name is Bridget Purtell, sor, an' I lives in Eleventh sthrate, betune avenues C and B."

" I believe that locality is in our jurisdiction. For whom do you want the warrant, and for what ? "

" For me husband, to be shure. He's thritined to take the loife av me an' sphill ivery dhrop av me blud on the fleure."

" Did he strike you ? "

" Yis, av coorse he did. Phat else wud bring me here ? I'd be the noice dacent woman that I am to come here widout some rason ! "

" Why did he strike you ? "

" Faix an' I dunno, sor, except that about foive weeks—"

" Hold on, madam ; never mind what occurred weeks ago, but come down to last night."

" That's phat I'm shtriving to get at."

" Well, go ahead."

" Lasht noight Moike kem home from his wurruk, as a plumber's laborer, an' his supper was warum on the table. Faith, an' the furst thing I knows, he jumps up and says : ' You she-devil, for why don't yer have me supper lukewarum instead of boilin' hot.' Wud that he lets fly a hot potaty at my head. Says he, ' Get out of me soight or I'll brake some av yer bones.' ' I'm goin',' says I, an' it was smashed in smithereens on the dure. Yer 'Aner, then I got mad an' shtruck him shquare acrost the face wid a poker. There was the divil to pay for a minit, but I got the bhest of him, an' moind ye, made him go to bed widout a bite, an' it's fur him beatin' me I now want a warrunt."

" I see Bridget," said his Honor, " you are an asserter of woman's rights. Give a description of your better half and residence to that officer," pointing to Roundsman Reilly, " and he will see that your husband is well taken care of for a short time."

" Thank yer, Jedge," said mild-tempered Bridget, as she was leaving the court.

INQUIRIES REGARDING CANADIAN TRADE.

Mr. Harrison Watson, curator of the Canadian Section of the Imperial Institute, London, England, is in receipt of the following inquiries regarding Canadian trade :

1. Inquiry is made for the name of a responsible firm of Canadian manufacturers' agents wishing to undertake an agency in twines and cords.

2. A manufacturer of sauces seeks Canadian agents with good connection.

3. Another firm of importers desires names of Canadian shippers of poultry.

[The names of the firms making the above inquiries, can be obtained on application to the editor of HARDWARE AND METAL, Toronto. When asking for names, kindly give number of paragraph and date of issue.]

BUSINESS BETTER THAN LAST YEAR

Mr. James B. Campbell, of the Acme Canning Works, Montreal, has been in Toronto for some time. He attended the packers' convention last week. He reports business exceedingly brisk, the orders received being much ahead of those taken to date last year.

Jacques & Gray, coal and wood dealers, Toronto, have dissolved. D. B. Jacques & Co. continues.

A STRANGE CAUSE OF FIRE.

Fire may be caused by a bottle of water standing harmlessly on a table. A correspondent writes to Fire and Water, showing how this may be the case :

In my laboratory the other day I detected the odor of burning wood, and, seeking the cause, noticed a tiny wreath of smoke rising from the counter. Setting aside a flask of water that stood close by, I sponged over the burning spot with a damp cloth. Shortly after I again detected the odor of burning wood, when, to my surprise, I discovered another burning spot on the table close to the water flask. The flask was standing in the sunlight, thereby concentrating the rays to a focus on the top of the table, acting in this case as a burning glass. A handful of highly combustible material was thrown over the burning spot, catching fire almost immediately. I cite this instance merely as a warning to chemists and apothecaries who may not realize how easily a fire may be started in their storerooms by the sun shining through bottles, flasks and carboys of liquid, converting them for the time being into burning glasses of great power. I have in mind now the instance of a fire originating in a storeroom from this cause.

LIQUID-GLUE FROM BONE-GLUE.

A simple method of converting bone-glue into liquid-glue is much wanted. It is well-known that bone-glue when dissolved gelatinizes even as weak as 6 degrees B., and that it very soon goes mouldy when kept in solution. These facts, combined with its strong and disagreeable smell, prevent it from being kept for use in a liquid state. It is nevertheless strong glue, and hence the desideration mentioned in our first sentence. The process of conversion is as follows :

Having dissolved two kilos. of joiner's glue in 8 litres of hot water, the solution will be about 9 degrees B. In the meantime make a mixture of 40 grammes of per-oxide of barium, 60 grammes of water, and 20 grammes of strong sulphuric acid. Add this mixture to the glue solution, and keep the whole over the water bath at a temperature of about 80 degrees C. Sulphurous acid is given off, and the glue loses its power of gelatinization. It also acquires a pleasant smell, and will not mould even after months of exposure to the air. The mass should be evaporated down to about 2 litres. It is feebly acid and its sticking power is very great, and much greater than that of dextrine. Dried into sheets it looks like gum arabic. It costs to make about 3d. per kilo.

Very much improvement is effected in dextrine by the above treatment, as it becomes more tenacious and practically proof against mouldiness.

Ordinary glue can be purified by making it alkaline with ammonia and then keeping it at 80 degrees C. on the water bath for 24 hours. All mineral matters separate out, and the glue can be poured off from them. It can afterwards be bleached, if desired, or treated with peroxide of hydrogen (i.e., with peroxide of barium and sulphuric acid), as above directed, or it can be subjected to both processes.—Oesterr. Chemiker Zeitung.

A VISIT FROM A WESTERNER.

Mr. Merrick, of Merrick, Anderson & Co., manufacturers' agents, Winnipeg, arrived in Montreal on Wednesday and is spending some time visiting the various manufacturers whom his firm represents in Winnipeg. Mr. Merrick is a typical Westerner, smart and robust, and his bon-homme is particularly fetching in the East as in the West. He is full of good humor and has many interesting stories to tell of the boom times in Western Canada. Merrick, Anderson & Co. have a large growing trade in the Prairie and Mountain provinces, their warehouse in Winnipeg being quite a prominent feature of the place.

AS NECESSARY AS WIRE NAILS.

"I find it just as necessary," writes Mr. Fred. Haney, a Strathroy, Ont., hardware merchant, "to have your journal, Hardware and Metal, as it is to have wire nails. It keeps us posted on the markets."

THE BELLEVILLE ROLLING MILLS.

Belleville's hopes have again been raised in connection with the Steel Rolling Mills which went inta liquidation recently and which the city had bonused to the extent of $50,000. The plant has been closed down and will be sold at public auction within the coming month. However, it is rumored on good authority that the intending purchasers will continue to do business in Belleville.

AN IMPORTANT AMALGAMATION.

The hardware, stove and tin business of W. W. Chown & Co., Belleville, and the cheese and butter factory furnishing and stove and tin business of Mr. S. C. Chown, also of Belleville, have amalgamated. The firm is to be known in future as the W. W. Chown Company, Limited. Mr. W. W. Chown is elected president and managing director.

A BUSY NICKEL TOWN.

The splendid doings of Mr. Clergue at Sault Ste. Marie are finding a worthy counterpart in the enterprise of Dr. Ludwig Mond, of London, at Victoria Mines Station, a point on the Soo line, twenty-five miles north of Sudbury. Dr. Mond is a member of the great English chemical firm of Brunner, Mond & Co., which owns the patents of the Solvay process of separating copper from nickel, and experiments made in the Sudbury district convinced the doctor that he could reduce the cost of working the ore to such an extent that its commercial value would be doubled. A few months have served to transform Victoria Mines from a mere flag station into a busy town of considerable size. Half a million dollars will be spent by the English firm within the next twelve months, and their permanent investment will be over a million. It is the intention ultimately to carry on the work by means of a company with a capital of several millions of dollars.—Toronto Globe.

CURRENT MARKET QUOTATIONS.

March 15 1901.

These prices are for such qualities and quantities as are usually ordered by retail dealers on the usual terms of credit, the lowest figures being for larger quantities and prompt pay. Large cash buyers can frequently make purchases at better prices. The Editor is anxious to be informed at once of any apparent errors in this list, as the desire is to make it perfectly accurate.

METALS.

Tin.

Lamb and Flag and Straits— 56 and 28 lb. ingots, per lb.	0 32	0 33

Tinplates.

Charcoal Plates—Bright

		Per box
M.L.S., equal to Bradley.		
I.C., usual sizes	....	$6 75
I.X., "	....	8 25
I.X.X., "	....	9 75
Famous—		
I.C.	....	6 75
I.X.	....	8 25
I.X.X.	....	9 50
Raven & Vulture Grades—		
I.C., usual sizes	....	5 00
I.X., "	....	6 00
I.X.X. "	....	7 00
I.X.X.X., "	....	8 00
D.C., 12½x17	....	4 75
D.X.	....	5 50
D.X.X.	....	7 50

Coke Plates—Bright

Bessemer Steel—		
I.C., usual sizes	....	4 15
I.C., special sizes, base	....	4 50
20 x 28	....	8 50

Charcoal Plates—Terne

Dean or J. G. Grade—		
I.C., 20x28, 112 sheets	....	8 50
I.X., Terne Tin	....	10 50

Charcoal Tin Boiler Plates.

		Per lb.
Cookley Grade— X X., 14x56, 50 sheet bxs; " 14x60 "; " 14x65, "	0 07	0 07½

Tinned Sheets

72x30 up to 24 gauge	0 08	0 08½
" 26 "	0 08½	0 09
" 28 "	0 09	0 09½

Iron and Steel.

Common Bar, per 100 lbs.	1 65	1 70
Refined " "	....	2 15
Horse Shoe Iron "	....	2 05
Hoop steel, 1½ to 3 in. base, extras for smaller sizes	....	3 10
Sleigh Shoe Steel " base	....	2 00
Tire Steel	....	2 00
Machinery iron finish	....	2 05
Cast Steel, per lb	0 00	0 00
Toe Calk Steel	....	2 30
T. Firth & Co's special cast steel, per lb.		0 13
Jessop's Tool Steel	....	0 13

Boiler Tubes.

1½-inch	....	0 12½
2 "	....	0 13
2½ "	....	0 15
3 "	....	0 16
3½ "	....	0 20
4 "	....	0 25

Steel Boiler Plate.

¼ inch	....	2 25
3-16 inch	....	2 25
⅜ inch and thicker	....	2 25

Black Sheets.

gauge	....	3 00
20 gauge	....	3 00
22 to 24 "	....	3 10
26 "	....	3 20
28 "	....	3 30

Canada Plates.

All dull, 52 sheets	....	3 00
Half polished	....	3 15
All bright	3 65	3 75

Iron Pipe.

Black pipe—		
¼ inch	....	3 00
⅜ "	....	3 05
½ "	....	3 10
¾ "	....	3 35
1 "	....	4 70
1¼ "	....	6 47
1½ "	....	7 70
2 "	....	10 25
2½-6 inch, discount 60 p c.		
Galvanized pipe—		
½ inch	....	4 75
¾ "	....	5 25
1 "	....	7 25
1¼ "	....	9 50
1½ "	....	11 50
2 "	....	15 50

Galvanized Sheets.

	G. C.	Comet.	Amer.	Queen's Head.
16 gauge		....	4 5	4 00
18 to 24 gauge	4 25	3 85	4 35	4 25
26 "	4 50	4 10	4 35	4 50
28 "	4 75	4 35	4 50	4 75

28 gauge American equals 26 gauge English.

Chain.

Proof Coil, 3-16 in., per 100 lb	....	....
" ¼ " "	8 00	8 50
" 5-16 " "	5 35	5
" ⅜ " "	4 35	4 85
" 7-16 " "	4 15	4 65
" ½ " "	4 35	4 50
" ⅝ " "	3 85	4 35
" ¾ " "	3 80	4 00
Halter, kennel and post chains, 40 and 50 p.c.		
Cow ties	40 p.c.	
Stall fixtures	35 p.c.	
Trace chain	25 p.c.	
Jack chain, iron, single and double, discount 35 p.c.		
Jack chain, brass, single and double, discount 40 p.c.		

Copper.

Ingot

English B. S., ton lots	0 19	0 20
Lake Superior	....	....

Bolt or Bar.

Cut lengths round, ½ to ⅞ in.	0 23½	0 25
" round and square 1 to 2 inches	0 23½	0 25

Sheet.

Plain, 14 oz., and light, 16 oz., 14x48 and 14x60	0 23	0 23
Plain, 14 oz., and light, 16 oz., irregular sizes	0 23	0 23½
Tinned copper sheets	....	0 26
Planished	....	0 32

Braziers (In sheets.)

4x6ft. 25 to 30 lbs. ea., per lb.	....	0 25½
" 35 to 45 " "	....	0 24½
" 50-lb. and above, "	....	0 23½

Boiler and T. K. Pitts

Plain Tinned, per lb	....	0 28
Spun, per lb.	....	0 32

Brass.

Rod and Sheet, 14 to 30 gauge, 15 per cent.		
Sheets, hard-rolled, 2x4	0 24	0 25
Tubing, base, per lb.	0 24	0 25

Zinc Spelter

Foreign, per lb	0 06	0 06½
Domestic "	....	....

Zinc Sheet.

5 cwt. casks	6 75	7 00
Part casks	7 00	7 50

Lead.

Imported Pig, per lb	0 04¾	0 05
Bar, 1 lb.	0 05½	0 05¾
Sheets, 2½ lbs. sq. ft., by	....	0 06¼
Sheets, 3 to 6 lbs., "	....	0 06

NOTE.—Cut sheets ½ cent per lb extra.

Pipe, by the roll, usual weights per yard, lists at 7c. per lb. and 15 p.c. dis. f.o.b. Toronto.

NOTE.—Cut lengths, net price, waste pipe 8-ft. lengths lists at 7½ cents.

Shot.

Common, $6.50 per 100 lb.; chilled, $7.00 per 100 lb.; buck, seal and ball, $7.50. Discount, 7½ p.c. Prices are f.o.b. Toronto, Hamilton, Montreal, St. John and Halifax. Terms 3 per cent. cash, freights equalized on Montreal.

Soil Pipe and Fittings.

Discount, 60 and 10 per cent. on medium and extra heavy, and 60 per cent. on light.

Solder.

	Per lb.	Per lb.
Bar half-and-half, guarant'd	....	0 18
Bar half-and-half, commer'l	....	0 17½
Refined	....	0 17½
Wiping	....	0 17

NOTE.—Prices of this graded according to quantity. The prices of other qualities of solder in the market indicated by private brands vary according to composition.

Antimony.

Cookson's, per lb.	0 11	0 11½

White Lead.

	Per 100 lb.
Pure	6 50
No. 1 do	6 12½
No. 2 do	5 75
No. 3 do	5 37½
No. 4 do	5 00
Munro's Select Flake White	7 37½
Elephant and Decorators' Pure	7 12½
Brandram's B B. Genuine	8 00
" " Decorative	7 55
" " No. 1	6 85
" " No. 2	6 00

Red Lead.

Genuine, 560 lb. casks, per cwt.	....	$5 50
Genuine, 100 lb. kegs, per cwt.	....	5 75
No. 1, 560 lb. casks, per cwt.	....	5 25
No. 1, 100 lb. kegs, per cwt	....	5 00

White Zinc Paint.

Elephant Snow White	0 08	0 09
Pure White Zinc	0 08	0 09
No. 1	0 06	0 07½
No. 2	0 05	0 06½

Dry White Lead.

Pure, casks	5 75
Pure, kegs	6 25
No. 1, casks	5 50
No. 1, kegs	6 00

Prepared Paints.

In ¼, ½ and 1 gallon tins.

Pure, per gallon	....	1 25
Second qualities, per gallon	....	1 10
Barn (in bbls.)	0 75	0 85
The Sherwin-Williams Paints	....	1 45
Canada Paint Co's Pure	....	1 25
Toronto Lead & Color Co's Pure	....	1 25
Sanderson Pearcy's Pure	....	1 20
Stewart & Wood's Champion Pure.		1 20

Colors in Oil.

25 lb. tins, Standard Quality.

Venetian Red, per lb.	0 05
Chrome Yellow	0 11
Golden Ochre	0 06
French "	0 05
Marine Black	0 09
" Green	0 09
Chrome "	0 08
French Imperial Green	0 09

Colors, Dry.

Yellow Ochre (J.C.) bbls.	1 35	1 40
Yellow Ochre (J.F.L.S.), bbls	....	2 75
Yellow Ochre (Royal)	1 10	1 15
Brussels Ochre	....	2 00
Venetian Red (best), per cwt.	1 80	1 90
English Oxides, per cwt.	3 00	3 25
American Oxides, per cwt.	1 75	2 00
Canadian Oxides, per cwt.	1 75	2 00
Super Magnetic Oxides, 93p.c.	2 00	2 25
Burnt Sienna, pure, per lb.	....	0 01
" Umber, " "	....	0 10
do Raw	....	0 09
Drop Black, pure	....	0 09
Chrome Yellows, pure	....	0 18
Chrome Greens, pure, per lb.	....	0 12
Golden Ochre	....	0 03½
Ultramarine Blue in 28-lb. boxes, per lb.	0 08	0 24
Fire Proof Mineral, per 100 lb	....	1 00
Genuine Eng. Litharge, per lb	....	0 07
Mortar Color, per 100 lb.	....	1 25
English Vermillion	....	0 80
Pure Indian Red, No. 45 lb.	....	0 80
Whiting, per 100 lb.	....	0 55

Blue Stone.

Casks, for spraying, per lb.	0 07
100-lb. lots, do. per lb	0 08

Putty.

Bulk in bbls.	2 00
Bulk in less quantity	2 15
Bladders in bbls.	2 20
Bladders in kegs, boxes or loose	2 35
Bladders in 25-lb. tins	2 45
Bladders in 12½-lb. tins	2 75
Bladders in bulk or tins less than 100 lb.	3 00

Varnishes.

(In 5-gal. lots.)		Per gal.
Carriage, No. 1	2 90	3 30
" body	8 00	9 00
" rubbing	4 00	5 00
Gold Size, Japan	3 00	3 40
Brown Japan	2 40	2 80
Elastic Oak	2 90	3 30
Furniture, extra	2 40	2 80
" No. 1	1 60	2 00
Hard Oil Finish	2 70	3 10
Light Oil Finish	3 20	3 60
Demar	3 30	3 70
Shellac, white	4 40	4 80
" orange	4 00	4 40
Furniture Brown Japan	1 60	2 00
Black Japan	2 40	2 80
" " No. 1	1 60	2 00

The Imperial Varnish & Color Co's, Limited Elastilite Varnish 1 gal. can, each. $2.00.

Granatine Floor Finish, per gal. $2.00.

Maple Leaf Coach Enamels; Size 1, (0c; Size 2, 35c.; Size 3, 20c. each.

Castor Oil.

East India, in cases, per lb.	0 10	0 10½
" " small lots	0 10½	0 11

Cod Oil, Etc.

Cod Oil per gal.	0 50	0 55
Pure Olive	....	1 20
" Neatsfoot	....	90

Glue.

Common	0 08½	0 09
French Medal	0 14	0 14½
Cabinet sheet	0 12	0 13
White, extra	0 18	0 20
Gelatine	0 22	0 30
Strip	0 18	0 20
Coopers	0 19	0 20
Hutner	....	0 18

HARDWARE.

Ammunition.

Cartridges.

B. B. Caps. Dom. 50 and 5 per cent.
Rim Fire Pistol, dis. 40 p. c., Amer.
Rim Fire Cartridges, Dom., 50 and 5 p. c.
Central Fire Pistol and Rifle, 10 p.c. Amer.
Central Fire Cartridges, pistol sizes, Dom. 30 per cent.
Central Fire Cartridges, Sporting and Military, Dom., 15 and 5 per cent.
Central Fire, Military and Sporting, Amer. add 5 p.c. to list. B.B. Caps, discount 40 per cent. Amer.
Loaded and empty Shells, "Trap" and "Dominion" grades, 25 per cent Rival and Nitro, net list.
Brass Shot Shells, 55 per cent
Primers, Dom., 30 per cent.

Wads	per lb
Best thick white felt wadding, in ¼-lb bags	1 00
Best thick brown or grey felt wads, in ¼-lb. bags	0 70
Best thick white card wads, in boxes of 500 each, 12 and smaller gauges	0 99
Best thick white card wads, in boxes of 500 each, 10 gauge	0 35
Best thick white card wads, in boxes of 500 each, 8 gauge	0 55
Thin card wads, in boxes of 1,000 each, 12 and smaller gauges	0 20
Thin card wads, in boxes of 1,000 each, 10 gauge	0 25
Thin card wads in boxes of 1,000 each, 8 gauge	
Chemically prepared black edge grey cloth wads, in boxes of 250 each—	Per M
11 and smaller gauge	0 60
9 and 10 gauges	0 70
7 and 8 gauges	0 90
5 and 6 gauges	1 10
Superior chemically prepared pink edge, best white cloth wads, in boxes of 250 each—	
11 and smaller gauge	1 15
9 and 10 gauges	1 40
7 and 8 gauges	1 65
5 and 6 gauges	1 90

Adzes.

Discount, 20 per cent.

Anvils.

Per lb.	0 10	0 12½
Anvil and Vise combined	....	4 50
Wilkinson & Co.'s Anvils..lb.	0 09	0 09½
Wilkinson & Co.'s Vices..lb.	0 09¾	0 10

Augers.

Gilmour's, discount 65 and 5 p.c. off list.

Axes.

Chopping Axes		
Single bit, per doz	5 50	10 00
Double bit, "	11 00	18 00
Bench Axes, 40 p.c.		
Broad Axes, 33⅓ per cent.		
Hunters' Axes	5 50	6 00
Boy's Axes	5 75	6 75
Splitting Axes	6 50	12 00
Handled Axes	7 00	10 00

Axle Grease.

Ordinary, per gross	5 75	6 00
Best quality	13 00	15 00

Bath Tubs.

Zinc 6 00
Copper, discount 15 p.c. off revised list

Baths.

Standard Enameled.

5½-inch rolled rim, 1st quality	25 00
" " " 2nd "	21 00

Anti-Friction Metal.

"Tandem" A	per lb.	0 27
" B	"	0 21
" C	"	0 11½
Magnolia Anti-Friction Metal, per lb.		0 25

SYRACUSE SMELTING WORKS.

Aluminum, genuine	0 45
Dynamo	0 29
Special	0 25
Aluminum, 99 p.c. pure "Syracuse"	0 50

Bells.

Hand.

Brass, 60 per cent.
Nickel, 55 per cent.

Cow.

American make, discount 66⅔ per cent.
Canadian, discount 45 and 50 per cent.

Door.

Gongs, Sargent's	5 50	8 00
" Peterboro', discount 45 per cent.		

Farm.

American, each	1 25	3 00

House.

American, per lb.	0 35	0 40

Bellows.

Hand, per doz.	3 35	4 75
Moulders', per doz.	7 50	10 00
Blacksmiths', discount 40 per cent.		

Belting.

Extra, 60 per cent.
Standard, 60 and 10 per cent.

Bits.

Auger.

Gilmour's, discount 60 and 5 per cent.
Rockford, 50 and 10 per cent.
Jennings' Gen., net list.

Car.

Gilmour's, 47½ to 50 per cent.

Expansive.

Clark's, 40 per cent.

Gimlet.

Clark's, per doz	0 65	0 90
Damond, Shell, per doz.	1 00	1 50
Nail and Spike, per gross	2 25	5 20

Blind and Bed Staples.

All sizes, per lb	0 07½	0 12

Bolts and Nuts.

	Per cent
Carriage Bolts, full square, Norway	70
" " full square	70
Common Carriage Bolts, all sizes	65
Machine Bolts, all sizes	65
Coach Screws	75
Sleigh Shoe Bolts	75
Blank Bolts	65
Bolt Ends	65
Nuts, square	4½c. off
Nuts, hexagon	4⅝c. off
Tire Bolts	67½
Stove Bolts	67½
Stove rods, per lb.	5½ to 6c.
Plough Bolts	60

Boot Calks.

Small and medium, ball, per M	4 25
Small heel, per M	4 50

Bright Wire Goods.

Discount .... 60 per cent.

Broilers.

Light, dis., 65 to 67½ per cent.
Reversible, dis., 65 to 67½ per cent.
Vegetable, per doz., dis. 37½ per cent.

Henis, No. 8, "	....	6 00
Henis, No. 9, "	....	7 00
Queen City "	7 50	0 00

Butchers' Cleavers.

German, per doz.	6 00	11 00
American, per doz.	12 00	20 00

Building Paper, Etc.

Plain building, per roll	0 30
Tarred lining, per roll	0 40
Tarred roofing, per 100 lb.	1 65
Coal Tar, per barrel	3 50
Pitch, per 100-lb.	0 85
Carpet felt, per ton	45 00

Bull Rings.

Copper, $2.00 for 2½ in. and $1.90 for 2 in.

Butts.

Wrought Brass, net revised list

Cast Iron.

Loose Pin, dis., 60 per cent.

Wrought Steel.

Fast Joint, dis. 60 and 10 per cent.
Loose Pin, dis. 60 and 10 per cent.
Berlin Bronzed, dis. 70, 70 and 5 per cent.

Gen. Bronzed, per pair	0 40	0 65

Carpet Stretchers.

American, per doz.	1 00	1 50
Bullard's, per doz.	6 50	....

Castors.

Bed, new list, dis. 55 to 57½ per cent.
Plate, dis. 52½ to 57½ per cent.

Cattle Leaders.

Nos. 31 and 32, per gross	50	9 50

Cement.

Canadian Portland	2 80	3 00
English "	....	3 00
Belgian "	2 75	3 00
Canadian hydraulic	1 25	1 50

Chalk.

Carpenters Colored, per gross	0 45	0 75
White lump, per cwt.	0 60	0 65
Red	0 05	0 06
Crayon, per gross	0 14	0 18

Chisels.

Socket, Framing and Firmer.

Broad's, dis. 70 per cent.
Warnock's, dis. 70 per cent.
P. S. & W. Extra 60, 10 and 5 p.c.

Churns.

Revolving Churns, metal frames—No. 0, $8—No. 1, $8.50—No. 2, $9.00—No. 3, $10.00 No. 4, $12.00—No. 5, $16.00 each. Ditto wood frames—20c. each less than above. Discounts: Delivered from factories, 58 p.c.; from stock in Montreal, 56 p.c. Terms, 4 months or 3 p.c. cash in 30 days.

Clips.

Axle dis. 65 per cent.

Closets.

Plain Ontario Syphon Jet	$16 00
Emb. Ontario Syphon Jet	17 00
Fittings net	1 00
Plain Teutonic Syphon Washout	10 00
Emb. Teutonic Syphon Washout	11 00
Fittings net	1 25
Low Down Teutonic, plain	16 00
" " " embossed	17 00
Plain Richelieu net	3 75
Emb. Richelieu net	4 00
Fittings net	1 25
Low Down Ont. Sy. Jet, plain net	19 50
" " " " " emb'd. net	20 50
Closet connection net	1 25
Basins, round, 14 in.	1 00
" oval, 17 x 14 in.	2 50
" " 19 x 15 in.	3 75

Discount 40 p.c., except on net figures.

Compasses, Dividers, Etc.

American, dis. 62½ to 65 per cent.

Cradles, Grain.

Canadian, dis. 25 to 33⅓ per cent.

Crosscut Saw Handles.

S. & D., No. 3, per pair	17½
" " 5, "	22½
" " 6, "	15
Boynton pattern "	20

Door Springs.

Torrey's Rod, per doz. (15 p.c.)		2 00
Coil, per doz	0 88	1 60
English, per doz	2 00	4 00

Draw Knives.

Coach and Wagon, dis. 50 and 10 per cent.
Carpenters, dis. 70 per cent.

Drills.

Hand and Breast.

Millar's Falls, per doz. net list.

DRILL BITS.

Morse, dis., 37½ to 40 per cent.
Standard dis. 50 and 5 to 55 per cent.

Faucets.

Common, cork-lined, dis 35 per cent.

ELBOWS. (Stovepipe.)

No. 1, per doz	1 40
No. 2, per doz	1 20
Bright, 20c. per doz. extra.	

ESCUTCHEONS.

Discount, 45 per cent.

ESCUTCHEON PINS.

Iron, discount 40 per cent.

FACTORY MILK CANS.

Discount off revised list, 40 per cent.

FILES.

Black Diamond, 70 and 5 per cent.
Kearney & Foote, 60 and 10 to 70 per cent.
Nicholson File Co., 60 to 60 and 10 per cent.
Jowitt's, English list, 25 to 27½ per cent.

FORKS.

Hay, manure, etc., dis., 50 and 10 per cent. revised list.

GLASS—Window—Box Price.

Size United Inches.	Star Per 50 ft.	Star Per 100 ft.	D. Diamond Per 50 ft.	D. Diamond Per 100 ft.
Under 26	2 10	4 00	....	6 00
26 to 40	2 30	4	....	6 65
41 to 50	....	4 [illegible]	....	7 25
51 to 60	....	[illegible]	....	8 50
61 to 70	....	[illegible]	....	9 25
71 to 80	....	5 75	....	10 50
81 to 85	....	6 50	....	11 75
86 to 90	....	....	..	14 00
91 to 95	..	....	...	15 50
99 to 100	..	....	....	18 00

GAUGES

Marking, Mortise, Etc.

Stanley's dis. 50 to 55 per cent.

Wire Gauges.

Winn's, Nos. 26 to 33, each	1 65	2 40

HALTERS.

Rope, ⅜ per gross	....	....
" ½ "	....	9 00
" ⅝ to ¾	....	14 00
Leather, 1 in., per doz.	3 87½	4 00
" 1¼ in., "	5 15	5 20
Web, — per doz.	1 87	2 45

HAMMERS.

Nail

Maydole's, dis. 5 to 10 per cent. Can. dis. 25 to 27½ per cent.

Tack.

Magnetic, per doz.	1 10	1 20

Sledge.

Canadian, per lb.	0 07½	0 08½

Ball Pean.

English and Can., per lb.	0 22	0 25

HANDLES.

Axe, per doz. net	1 50	2 00
Store door, per doz.	1 00	1 50

Fork.

C. & B., dis. 40 per cent. rev. list.

Hoe.

C. & B., dis. 40 per cent. rev. list.

Saw.

American, per doz.	1 00	1 25

Plane.

American, per gross	3 15	3 75

Hammer and Hatchet.

Canadian, 40 per cent.

Cross-Out Saws.

Canadian, per pair	0 13½

HANGERS.

	doz. pairs.	
Steel barn door	5 85	6 00
Stearns, 4 inch	....	5 00
" 5 inch	....	6 50
Lane's covered—		
No. 11, 5-ft. run	....	8 40
No. 11½, 10-ft. run	....	10 80
No. 12, 10-ft. run	....	12 60
No. 14, 15-ft. run	....	21 00
Lane's O.N.T. track, per foot.	....	4½

HARVEST TOOLS.

Discount, 50 and 10 per cent.

HATCHETS.

Canadian, dis. 40 to 42½ per cent.

HINGES.

Blind, Parker's, dis. 50 and 10 to 60 per cent

Heavy T and strap, 4-in., per lb.	....	0 06½
" " 5-in., "	....	0 06¼
" " 6-in., "	....	0 06
" " 8-in., "	....	0 05¾
" " 10-in., "	....	0 05½
Light T and strap, dis. 60 and 5 per cent.		
Screw hook and hinge—		
6 to 12 in., per 100 lbs.	....	4 50
14 in. up, per 100 lbs.	....	3 50
		Per gro. pairs.
Spring	....	12 00

HOES.

Garden, Mortar, etc., dis. 50 and 10 p.c.		
Planter, per doz.	4 00	4 50

HOLLOW WARE

Discount .... 45 and 5 per cent.

HOOKS.

Cast Iron.

Bird Cage, per doz.	0 50	1 10
Clothes Line, per doz.	0 27	0 63
Harness, per doz.	0 72	0 88
Hat and Coat, per gross	1 00	3 00
Chandelier, per doz.	0 50	1 00

Wrought Iron.

Wrought Hooks and Staples, Can., dis. 47½ per cent.

Wire.

Hat and Coat, discount 45 per cent.
Belt, per 1,000 .... 0 60
Screw, bright, dis. 55 per cent.

HORSE NAILS.

"C" brand 50 and 7½ p.c off new list } Oval
"M" brand 60 per cent. } head
Countersunk, 60 and 10 per cent.

HORSESHOES

Iron Shoes.	F.O.B. Montreal; No. 2 and larger.	No. 1. and smaller
Light, medium, and heavy...	3 50	3 75
Snow shoes	3 75	4 00
Steel Shoes.		
Light	3 60	3 85
Featherweight (all sizes)	4 85	4 85

F.O.B. Toronto, Hamilton, London and Guelph, 10c. per keg additional.

Toe weight steel shoes...... 6 70

JAPANNED WARE.

Discount, 45 and 5 p.c. off list, June 1899.

ICE PICKS.

Star per doz.... 3 00 3 25

KETTLES.

Brass spun, 7½ p.c. dis. off new list.
Copper, per lb.... 0 30 0 50
American, 60 and 10 to 65 and 5 p.c.

KEYS.

Lock, Can., dis., 45 p.c.
Cabinet, trunk, and padlock, Am. per gross.... 60

KNOBS.

Door, japanned and N.P., per doz	1 50	2 50
Bronze, Berlin, per doz	2 75	3 25
Bronze Genuine, per doz	6 00	9 00
Shutter, porcelain, F. & L. screw, per gross	1 30	4 00
White door knobs—per doz.		1 25

HAY KNIVES.

Discount, 50 and 10 per cent.

LAMP WICKS.

Discount, 60 per cent.

LANTERNS.

Cold Blast per doz	7 50
No. 3 "Wright's"	8 50
Ordinary, with O burner	4 25
Dashboard, cold blast	9 50
No. 0	6 00

Japanning, 50c. per doz. extra.

LEMON SQUEEZERS.

Porcelain lined,......per doz.	2 20	5 60
Galvanized	1 87	3 85
King, wood	2 75	2 90
King, glass	4 00	4 50
All glass	1 20	1 30

LINES.

Fish, per gross	1 05	2 50
Chalk "	1 90	7 40

LOCKS.

Canadian, dis. 45 p.c.
Russell & Erwin, per doz.... 3 00 3 25

Cabinet.

Eagle, dis. 30 p.c.

Padlock.

English and Am., per doz....	50	6 00
Scandinavian, "	1 00	2 40

Eagle, dis. 20 to 25 p.c.

MACHINE SCREWS.

Iron and Brass.

Flat head discount 25 p.c.
Round Head, discount 20 p.c.

MALLETS.

Tinsmiths', per doz	1 25	1 50
Carpenters', hickory, per doz.	1 25	3 75
Lignum Vitae, per doz.	3 85	5 00
Caulking, each	60	2 00

MATTOCKS.

Canadian, per doz.... 5 50 6 50

MEAT CUTTERS.

American, dis. 25 to 30 p.c.
German, 15 per cent.

MILK CAN TRIMMINGS.

Discount, 25 per cent.

NAILS.

Quotations are:	Cut.	Wire.
2d and 3d	$3 35	$3 85
3d	3 00	[illegible] 52
4d and 5d	2 75	3 35
6d and 7d	2 65	3 20
8 and 9d	2 50	3 00
10 and 12d	2 45	2 95
16 and 20d	2 40	2 90
30, 40, 50 and 60d. (base)	2 35	2 85

Galvanizing 2c. per lb. net extra.
Steel Cut Nails 10c. extra.
Miscellaneous wire nails, dis. 70 per cent.
Coopers' nails, dis. 30 per cent.
Flour barrel nails, dis. 25 per cent

NAIL PULLERS.

German and American....... 1 85 3 50

NAIL SETS

Square, round, and octagon, per gross	3 38	4 00
iamond	12 00	15 00

NETTING.

Poultry, 50 and 5 per cent. for McMullen's

OAKUM.

	Per 100 lb.
Navy	6 00
U. S. Navy	7 25

OIL.

Water White (U.S.)	0 16½
Prime White (U.S.)	0 15½
Water White (Can.)	0 15
Prime White (Can.)	0 14

OILERS.

McClary's Model galvan. oil can, with pump, 5 gal., per doz.... 10 00
Zinc and tin, dis. 50, 50 and 10.

Copper, per doz	1 25	3 50
Brass, "	1 50	3 50

Malleable, dis. 25 per cent.

GALVANIZED PAILS.

Dufferin pattern pails, dis. 45 p.c.
Flaring pattern, discount 45 per cent.
Galvanized washtubs, discount 45 per cent.

PIECED WARE.

Discount 40 per cent. off list, June, 1899.
10-qt. flaring sap buckets, dis. 45 p.c.
6, 10 and 14-qt. flaring pails, dis. 45 p.c.
Creamer cans, dis. 45 p.c.

PICKS.

Per doz.... 6 00 9 00

PICTURE NAILS.

Porcelain head, per gross	1 75	3 00
Brass head "	0 40	1 00

PICTURE WIRE.

Tin and gilt, discount 75 p.c.

PLANES.

Wood, bench, Canadian dis. 50 per cent American dis. 50.
Wood, fancy Canadian or American 7½ o 40 per cent.

PLANE IRONS.

English, per doz.... 2 00 5 00

PLIERS AND NIPPERS.

Button's Genuine per doz pairs, dis. 37½ 40 p.c.

Button's Imitation, per doz..	5 00	9 00
German, per doz	0 60	2 60

PLUMBERS' BRASS GOODS.

Compression work, discount, 60 per cent.
Fuller's work, discount 65 per cent.
Rough stops and stop and waste cocks, discount, 60 per cent.
Jenkins disk globe and angle valves, discount, 55 per cent.
Standard valves, discount, 60 per per cent.
Jenkins radiator valves discount 55 per cent
" " " standard, dis., 60 p.c.
Quick opening valves discount, 60 p.c.

No. 1 compression bath cock	2 00
No. 4 " " "	2 00
No. 7, Fuller's	2 50
No 4½, "	3 00

POWDER.

Velox Smokeless Shotgun Powder.

100 lb. or less	0 85
1,000 lb. or more	0 80

Net 30 days.

PRESSED SPIKES.

Discount 20 to 25 per cent.

PULLEYS.

Hothouse, per doz	0 55	1 00
Axle	0 22	0 33
Screw	0 27	1 00
Awning	0 35	2 50

PUMPS.

Canadian cistern	1 80	3 00
Canadian pitcher spout	1 40	2 10

PUNCHES.

Saddlers', per doz	1 00	1 85
Conductors', "	9 00	15 00
Tinners' solid, per set	0 00	0 72
" hollow, per inch	0 00	1 00

RANGE BOILERS.

Galvanized, 30 gallons	7 00
" 35 "	8 25
" 40 "	9 50
Copper, 30 "	22 00
" 35 "	26 00
" 40 "	30 00

Discount off Copper Boilers 10 per cent.

RAKES.

Cast steel and malleable, 50, 10 and 5 p.c.
Wood, 25 per cent.

RASPS AND HORSE RASPS.

New Nicholson horse rasp, discount 60 to 60 and 10 p.c.
Globe File Co.'s rasps, 60 and 10 to 70 p.c.
Heller's Horse rasps, 50 to 50 and 5 p.c.

RAZORS.

	per doz.	
Geo. Butler & Co.'s	8 00	18 00
Boker's	7 50	11 00
" King Cutter	12 50	50 00
Wade & Butcher's	3 60	10 00
Theile & Quack's	7 00	12 00
Elliot's	4 00	18 00

REAPING HOOKS.

Discount, 50 and 10 per cent.

REGISTERS.

Discount.... 40 per cent.

RIVETS AND BURRS.

Iron Rivets, black and tinned, discount 60 and 10 per cent.
Iron Burrs, discount 55 per cent.
Extras on Iron Rivets in 1-lb. cartons, ½c per lb.
Extras on Iron Rivets in ½-lb. cartons, 1c. per lb.
Copper Rivets & Burrs, 35 and 5 p.c. dis. and cartons, 1c. per lb. extra, net.
Extras on Tinned or Coppered Rivets ½-lb. cartons, 1c. per lb.

RIVET SETS

Canadian, dis. 35 to 37½ per cent.

ROPE ETC.

	Sisal.	Manila.
7-16 in. and larger, per lb	10	13½
⅜ in.	11	14½
¼ and 5-16 in.		15½
Cotton, 3-16 inch and larger		16½
" 5-32 inch		21½
" ⅛ inch		22½
Russia Deep Sea		15½
Jute		8
Lath Yarn		9½
New Zealand Rope		10

RULES.

Boxwood, dis. 75 and 10 p.c.
Ivory, dis. 37½ to 40 p.c.

SAD IRONS.

	per set
Mrs. Potts, No. 55, polished..	62½
" No. 50, nickle-plated	67½

SAND AND EMERY PAPER.

Dominion Flint Paper, 47½ per cent.
B & A. sand, 40 and 5 per cent.
Emery, 40 per cent.
Garnet (Rurton's), 5 to 10 p.c. advance on list.

SAP SPOUTS.

Bronzed iron with hooks, per doz... 9 50

SAWS.

Hand Disston's, dis. 12½ p.c.
S. & D., 40 per cent.
Crosscut, Disston's, per ft.... 35 0 55
S. & D., dis. 35 p.c. on Nos. 2 and 3.

Hack, complete, each	0 75	2 75
" frame only		0 75

SASH WEIGHTS.

Sectional, per 100 lbs	2 75	3 00
Solid, "	2 00	2 25

SASH CORD.

Per lb.... 0 23 0 30

SAW SETS.

"Lincoln," per doz.... 6 50

SCALES.

B. S. & M. Scales, 45 p.c.
Champion, 65 per cent.
Fairbanks Standard, 35 p.c.
" Dominion, 55 p.c.
" Richelieu, 55 p.c.
Chatillon Spring Balances, 10 p.c.

SCREW DRIVERS.

Sargent's per doz.... 0 65 1 00

SCREWS

Wood, F. H., bright and steel, 87½ and 10 p.c.
Wood R. H., " dis. 82½ and 10 p.c.
" F. H., brass, dis. 80 and 10 p.c.
Wood, R. H., " dis. 75 and 10p.c.
" F.H., bronze, dis. 75 p.c.
" R.H. " 70 p.c.
Drive Screws, 87½ and 10 per cent.

Bench, wood, per doz	3 25	4 00
" iron. "	4 25	5 75

Set, Case hardened, 60 per cent.
Square Cap, 50 and 5 per cent.
Hexagon Cap, 45 per cent.

SCYTHES.

Per doz. net.... 9 00

SCYTHE SNATHS.

Canadian, dis. 45 p.c.

SHEARS.

Bailey Cutlery Co., full nickeled, dis. 60 p.c.
Seymour's, dis. 50 and 10 p.c.

SHOVELS AND SPADES.

Canadian, dis. 40 and 5 per cent.

SINKS.

Steel and galvanized, discount 45 per cent.

SNAPS.

Harness, German, dis. 25 p.c.
Lock, Andrews'.... 4 50 11 50

SOLDERING IRONS.

1, 1½ lb., per lb.	0 37
2 lb. or over, per lb.	0 34

SQUARES.

Iron, No. 493, per doz	2 40	2 55
" No. 494, "	3 25	3 40

Steel, dis. 50 and 5 to 50 and 10 p.c., rev. list.
Try and bevel, dis. 50 to 52½ p.c.

STAMPED WARE.

Plain, dis., 75 and 12½ p.c. off revised list.
Retinned, dis., 75 p.c. off revised list.

STAPLES.

Galvanized	3 50	4 00
Plain	3 15	3 75

Coopers', discount 45 per cent.
Poultry netting staples, 40 per cent.

STOCKS AND DIES.

American dis. 25 p.c.

STONE.

	Per lb.	
Washita	0 28	0 60
Hindostan	0 06	0 07
" slip,	0 09	0 09
Labrador	....	0 13
" Axe	....	0 15
Turkey	....	0 50
Arkansas	0 00	1 50
Water-of-Ayr	0 00	0 10
Scythe, per gross	3 50	5 00
Grind, per ton	15 00	18 00

STOVE PIPES.

5 and 6 inch Per 100 lengths	7 00
7 inch " "	7 50

ENAMELINE STOVE POLISH.

No. 4—3 dozen in case, net cash	$4 80
No. 6—3 dozen in case, "	8 40

TACKS BRADS, ETC.

	Per cent.
Strawberry box tacks, bulk	75 & 10
Cheese-box tacks, blued	80 & 12½
Trunk tacks, black and tinned	85
Carpet tacks, blued	80 & 5
" " tinned	80 & 10
" " (in kegs)	40
Cut tacks, blued, in dozens only	75 & 15
" ½ weights	60
Swedes, cut tacks, blued and tinned—	
In bulk	80 & 10
In dozens	75
Swedes, upholsterers', bulk	85 & 12½
" brush, blued & tinned, bulk	70
" gimp, blued, tinned and japanned	75 & 12½
Zinc tacks	35
Leather carpet tacks	55
Copper tacks	50
Copper nails	52
Trunk nails, black	65 and 5
Trunk nails, tinned	65 and 10
Clout nails, blued and tinned	65 and 5
Chair nails	35
Patent brads	40
Fine finishing	40
Picture frame points	10
Lining tacks, in papers	10

Lining tacks, in bulk		15
" " solid heads, in bulk		75
Saddle nails in papers		10
" " in bulk		15
Tufting buttons, 22 line, in dozens only		60
Tin capped trunk nails		15
Zinc glazier's points		5
Double pointed tacks, papers		90 and 10
" " " bulk		40

TAPE LINES.

English, ass skin, per doz	2 75	5 00
English, Patent Leather	5 50	9 75
Chesterman's each	0 90	2 85
" steel, each	0 80	8 00

THERMOMETERS.

Tin case and dairy, dis. 75 to 75 and 10 p.c.

TRAPS. (Steel.)

Game, Newhouse, dis. 25 p.c.
Game, H. & N., P. S. & W., 65 p.c.
Game, steel, 72½, 75 p.c.

TROWELS.

Disston's discount 10 per cent.

German, per doz	4 75	6 00

S. & D., discount 35 per cent.

TWINES.

Bag, Russian, per lb	....	0 27
Wrapping, cotton, per lb	0 22	0 26
Wrapping, mottled, per pack	0 53	0 60
Wrapping, cotton, 3-ply		0 20
" " 4-ply	....	0 26
Mattress, per lb	0 33	0 45
Staging, "	0 27	0 35

VISES.

Hand, per doz	4 00	6 00
Bench, parallel, each	2 00	4 50
Coach, each	6 00	7 00
Peter Wright's, per lb	0 12	0 13
Pipe, each	5 50	9 00
Saw, per doz	6 50	12 00

ENAMELLED WARE.

White, Princess, Turquoise, Blue and White, discount 50 per cent.
Diamond, Famous, Premier, 50 and 10 p.c.
Granite or Pearl, Imperial, Crescent, 50, 10 and 10 per cent.

WIRE.

Brass wire, 50 to 50 and 2½ per cent. off the list.
Copper wire, 45 and 10 per cent. net cash 30 days, f.o.b. factory.
Smooth Steel Wire, is quoted at the following net selling prices:

No. 6 to 8 gauge	$2 90
" 9 "	2 80
" 10 "	2 87
" 11 "	2 90
" 12 "	2 95
" 13 "	3 15
" 14 "	3 37
" 15 "	3 50
" 16 "	3 65

Other sizes of plain wire outside of Nos. 9, 10, 11, 12 and 13 and other varieties of plain wire remain at $2.80 base with extras as before The prices for Nos. 9 to 13 include the charge of 10c. for oiling. Extras net per 100 lb.: Coppered wire, 60c.—tinned wire, $2—oiling, 10c.—special hay-balling wire, 30c.—spring wire, $1—best steel wire, 75c.—bright soft drawn, 15c.—in 50 and 100-lb. bundles net, 10c.—in 25-lb. bundles net 15c.—packed in casks or cases, 15c.—bagging or papering, 10c.

Fine Steel Wire, dis. 17½ per cent List of extras: In 100-lb. lots: No. 17, $5—No. 18, $5.50—No. 19, $6—No. 20, $6.65—No. 21, $7—No. 22, $7.30—No. 23, $7.65—No. 24, $8—No. 25, $9—No. 26, $9.50—No. 27, $10—No. 28, $11—No. 29, $12—No. 30, $13—No. 31, $14—No. 32, $15—No. 33, $16—No. 34, $17. Extras net—tinned wire, Nos. 17-25, $2—Nos. 26-31 $4—Nos. 32-34, $6. Coppered, 5c.—oiling, 10c.—in 25-lb. bundles, 15c.—in 5 and 10-lb. bundles, 25c.—in 1-lb. hanks, 50c.—in ½-lb. hanks, 75c.—in ¼-lb. hanks $1—packed in casks or cases, 15c.—bagging or papering, 10c.

Galvanized Wire, per 100 lb.—Nos. 6,7,8, $3 50 to $3 8?—No. 9, $2.85 to $3.15—No. 10, $3.60 to $3.9?—No. 11, $3.70 to $4.10—No. 12, $3 to $3.30—No. 13, $3.10 to $3 4?—No. 14, $4.10 to $4.50—No. 15, $4.60 to $5.05—No. 16, $4.85 to $5.35. Base sizes, Nos. 6 to 9, $2.57½ f.o.b. Cleveland.

Clothes Line Wire, 19 gauge, per 1000 feet	2 75	3 00

WIRE FENCING.

	F.O.B. Toronto
Galvanized barb	3 15
Galvanized, plain twist	3 65

Galvanized barb, f.o.b. Cleveland, $2.82½ in less than carlots, and $2.7? in carlots.

WIRE CLOTH.

Painted Screen, per 100 sq. ft., net	1 35

WASTE COTTON.

	per lb.
Colored	4½ to 5
White, according to quality	6½ to 7½

500-lb bale lots shaded.

WRENCHES.

Acme, 35 to 37½ per cent.
Agricultural, 60 p.c.
Coe's Genuine, dis. 20 to 25 p.c.

Towers' Engineer, each	2 00	7 00
" S., per doz	5 80	6 00
G. & K.'s Pipe, per doz	....	3 40
Burrell's Pipe, each	....	3 00
Pocket, per doz	0 25	2 90

WRINGERS.

Leader	per doz.	$30 00	31 (0
Royal Canadian	"	26 00	18 00
Royal American	"	16 00	28 (0
Sampson	"	....	30 00

Terms 4 months, or 3 p.c. 30 days.

WROUGHT IRON WASHERS.

Canadian make, discount, 40 and 5 per cent.

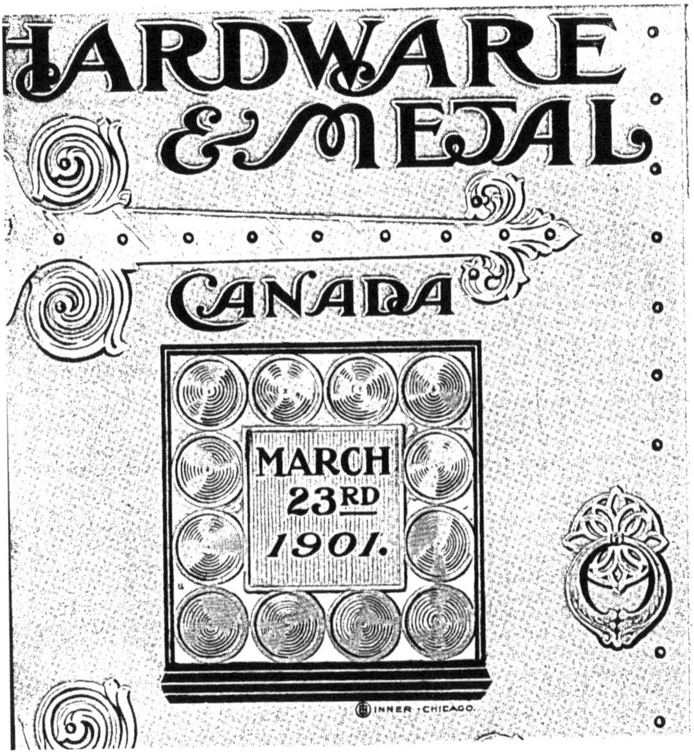
HARDWARE
& METAL
CANADA
MARCH
23RD
1901.

CANADIAN HARDWARE AND METAL MERCHANT

The Weekly Organ of the Hardware, Metal, Heating, Plumbing and Contracting Trades in Canada.

VOL. XIII. MONTREAL AND TORONTO, MARCH 23, 1901. NO. 12

ESTABLISHED 1864 INCORPORATED 1896

The Parmenter & Bulloch Co., Limited

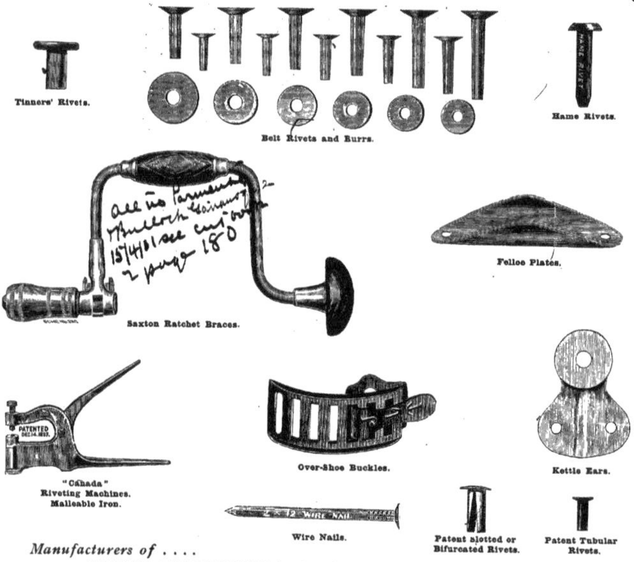

Manufacturers of . . . .

WIRE NAILS, RIVETS AND BURRS in all variety of material, form and finish, both regular and special.
COUNTERSUNK CLOUT, WAGON BOX AND HINGE NAILS.
ESCUTCHEON PINS in Brass and Steel.
CARPENTERS' BRACES of all kinds, Full Polished and Nickel Plated.
SHOE RIVETS in Steel and Brass.
PATENT BIFURCATED OR SLOTTED AND TUBULAR RIVETS, in White, Coppered and Japanned.
RIVETING MACHINES of every character. FELLOE PLATES, KETTLE EARS,
PATENT SHOE BUCKLES, HORSE BLANKET CLASPS AND ICE CREEPERS.

Head Office and Works:

GANANOQUE, ONT.

Branch Warehouses:

32 and 34 St. Dizier St.
Montreal.

29 Melinda St.
Toronto.

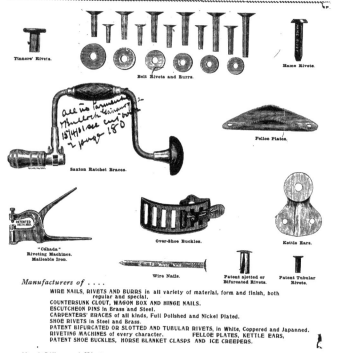

If you put

HENRY DISSTON & SON'S

Saws and Files

in the most prominent place in your store your customers will know at once that they are dealing with a first-class business house that handles the best and most satisfactory goods made.

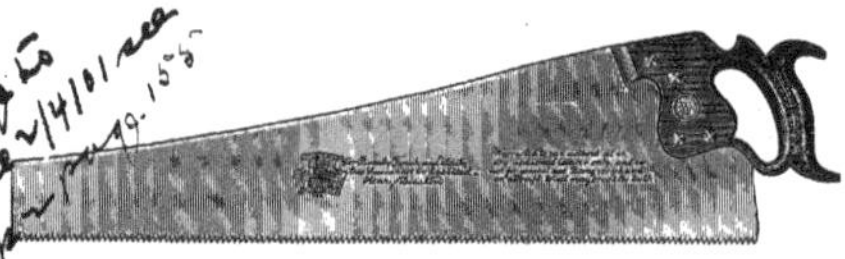

Ground perfectly by best of experts.

Will wear longer and not fill up like other files.

Henry Disston & Sons

Philadelphia, Pa., U.S.A.

AGENTS

Lewis Bros. & Co. - Montreal, Que.

Will be pleased to quote you best prices.

SMOKELESS POWDER

— IS BEST FOR —

Rifles, Shot Guns, Revolvers.

Clean

Safe

No Jar

Quick

Perfect Combus-tion

Long Range

NOT AFFECTED BY ANY CLIMATE.

— MANUFACTURED AT —

Barwick Works, Herts,

— BY —

The Smokeless Powder and Ammunition Company, LIMITED

LONDON, ENGLAND.

LEWIS BROS. & CO., MONTREAL,

Write for prices, it will pay you. Sole Agents for Canada.

DAVID MAXWELL & SONS

ST. MARYS, ONT., CANADA.

"Maxwell Favorite Churn"

PATENTED FEATURES—Improved Steel Stand, Roller Bearings, Foot and Hand Lever Drive, Detachable Links, Bolted Trunnions.

Steel Frame Churn.

Lawn Mowers

High and Low Wheels, from 12-in. to 20-in. widths, Cold Rolled Steel Shafting, Crucible Steel Knives and Cutting Plate.

Wood Frame Churn.

MAXWELL MOWER

8-inch Low Wheel.

"THE MAXWELL" Lawn Mower

High Wheel 10 inches.

WHEELBARROWS

Navvy,
General Purpose or
Garden Barrows.

All styles and sizes.

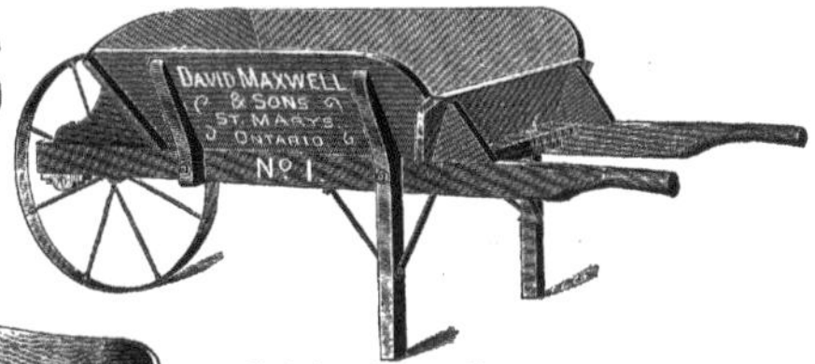

No. 1—General Purpose Barrow.

Navvy Barrow—Iron or Wood Wheel.

Our Lawn Mowers, Churns and Barrows are sold by the leading wholesale firms. If your wholesale house does not handle them write us direct.

DAVID MAXWELL & SONS

ST. MARYS, ONT., CANADA.

STERNS
4-in. and 5-in.

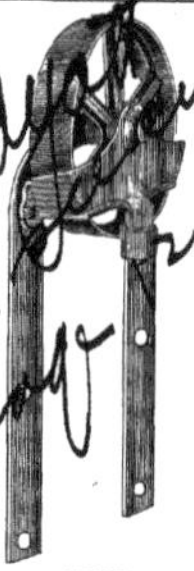

ACME
STEEL FRAME
4-in. and 5-in.

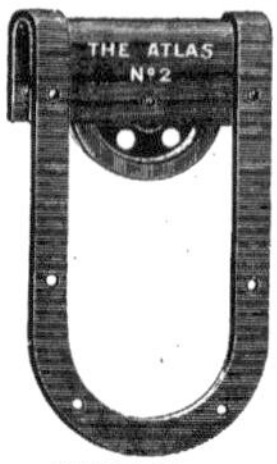

THE ATLAS
STEEL
ROLLER BEARING
No. 0 with 3-in. wheels
No. 1 with 3½-in. wheels
No. 2 with 4-in. wheels

SOLD ONLY THROUGH THE WHOLESALE TRADE

BARN DOOR HANGERS

MANUFACTURED BY . . . **A. R. Woodyatt & Co., Guelph, Canada.**

COWAN & BRITTON

GANANOQUE, ONT.

Established 1861.

Manufacturers . . .

- Wrought Steel Butts, of every description.
- Strap and Tee Hinges
- Screw Hook Hinges
- Washers
- Hasps and Staples
- Staples
- Hooks and Staples
- Nail Pullers
- Nails, of every description.
- Brass Yard Measures
- Little Giant Tack Pullers
- Ever Ready Electric Lights

Goods Unsurpassed and Prices the Lowest.

ORDERS SOLICITED

NONE BETTER

Beaver Liquid TRADE MARK **Pure Paints.**

As the beaver by his ingenuity captivates man, so does the quality of our Paints magnetize the trade. A Pure Linseed Oil Paint, in 49 most artistic shades. Every tin guaranteed.

Color Cards, Price Lists and Catalogues mailed upon application. . . .

The Toronto Lead & Color Co. LIMITED.

TORONTO.

HOW BUSINESS MAY BE DONE IN THE FUTURE.

The Tendency of Modern Trade is Toward Consolidation, Because the Administration of the Largest Mass is the Cheapest—Opportunities for Worthy Young Men Never So Great as at Present.

A VERY interesting article comes from Charles R. Flint, the capitalist and organizer of big corporations, on the direction in which trade is developing. He points out that there seems to be much confusion in the minds of the people as to the difference between a trust and an industrial company, due to the fact that those who talk most about them are not yet well informed, either as to their organization or operation. A trust was a syndicate of men who held stock certificates of several corporations and issued trust certificates therefor. Now, industrial interests are represented by shares of stock in regularly organized companies. Although strenuous efforts were made to develop the trust system, it was found to be imperfect. It was adopted when industrial combinations were in their infancy. I have always been opposed to the trust system of organization. They were not required to have any by-laws or keep any official minutes of their proceedings or to make any official reports. In general it might be said that they possessed great power without sufficient accountability.

The Supreme Court of the State of New York declared them illegal, and every lawyer who is informed in regard to industrial organizations will tell you that that decision has been accepted as final throughout the United States. But the word "trust" has since been applied to great industrial corporations, and as the word represents all that is best in human character, I see no reason why the word "trust" should not be adopted as a short name for industrial combinations, and may every officer and wage-earner in every "trust" realize that the shares of stock are widely distributed among widows, orphans and others dependent on its dividends for support, and live up to the true meaning of the word.

Tendency of Modern Trade is Toward Consolidation.

"The tendency of modern trade is toward consolidation, because the administration of the largest mass is the cheapest."

Centralized manufacture permits the highest development of special machinery and processes. The factory running full time on large volume, reduces the percentage of overhead charges. Direct sales on a large scale minimize the cost of distribution. Centralization of manufacture and distribution reduce aggregate stocks, and therefore save shop wear, storage, insurance and interest. Consolidated management results in fixing the standards of quality, the best standards being adopted; in avoiding waste and financial embarrassment through overproduction; in less loss by bad debts through comparisons of credit, and in securing the advantages of comparative accounting and comparative administration.

Industrial evolution, which is as inevitable and unalterable as the law of gravitation, has attained its, as yet, highest development here and in the United States, and we may anticipate a still higher development during the coming century. Every unprejudiced man must recognize its advantages, and that it is because of them that we are taking so important a position in the world's markets, increasing our national wealth, furthering the welfare and increasing the prosperity of our people. The great problems of the economics of production have been solved. What interests us most to-day is not so much the fact of our great industrial prosperity; it is, rather, the question whether the advantages of that prosperity are equitably divided among the contributors to it: 1, capital; 2, superintendence, and 3. labor.

1. The share to capital takes the form either of interest or dividends. Now, we find that the rate of interest paid to those furnishing money to industrial enterprises is decreasing. Fifty years ago the average rate throughout the United States was 8 per cent. per annum. Now it is less than 5 per cent. This general rule can be laid down: That the greater the confidence, the higher and more perfect the industrial organization, the lower the rate of interest.

Necessity for Intelligence Increasing.

2. Now, what is the position of the man of superior intelligence?—for superintendence stands midway between capital and labor. Highly-developed organizations, resulting in enormous volume of business, have increased, and will increase the necessity for intelligence, and as the supply of brains is not equal to the demand, the price of brains is high. The turning over of individual business to combinations has caused the retirement of old men to the advisory board for judgment and has made way for young men for action. You ask, "What chance have our young men?" While you are asking the question, those of ability and energy have already started on a career of successful industry. If the student will leave his books and the orator the stump and go to our factories, to our great farms, to our mines, to our lines of railway, they will find ten times as many men receiving over $3,000 per annum as there were 30 years ago.

Mr. Schwab, of Pittsburg, is a type. He started as a stake-driver of an engineering corps; to-day, though under 40 years of age, he is president of the largest iron company in the world, and I can point our 100 successful men to-day where you could not have named 10 under old conditions. But it is said they are dependent. Dependence upon each other is, however, the condition of civilization. The very word civilization implies community life, and community life means mutual dependence.

Wage-Earners Have Opportunities, Too.

3. Let us now consider the interests of the workingman in this economic evolution which has produced the perfect machinery and giant factories, supported by great aggregates of capital, represented by shares, which enable all to become investors. It is a fundamental fact that the man of superior ability cannot accumulate for himself without giving to the wage earners an opportunity to earn the larger share. The tendency is to day to a minimum of profits and to a maximum of wages. When profits become abnormal, they invite competition, and are immediately reduced, in which case the consumer is benefited solely. If they are not sufficiently abnormal to invite competition, then labor demands a larger share of the profit, in the form of increased wages, and it is either voluntarily or necessarily agreed to; in which case the body of wage-earners reap the advantage. And, inasmuch as the body of wage-earners is the great body of the community, it necessarily reaps the advantage in any case. Employes know almost as promptly as do the employers whether a mill is earning an extravagant profit. If it be, they at once demand their share, and the employer must

and inevitably does, succumb. It is thus that wages always tend to a maximum, and profits to a minimum.

Thus through cooperation and combination every interest is being benefited, but labor most of all. As wage earners become more intelligent, as they become overseers of machinery, they better understand these conditions. They have the intelligence to recognize that their greatest comfort and happiness is in furthering the industry of which they are a part. To-day one of the great advantages that the United States and Canada has over Europe is that their laborers are the more intelligent, and this advantage will become more and more effective. The European wage-earner, instead of welcoming labor-saving machinery as workingmen on this continent have done, has tried persistently to retard its general use, and the result has been that wages have been much lower in Europe. The American and Canadian workman has received more because he has produced more, and this is the great reason why notwithstanding our high wages, we are so, rapidly extending our trade with foreign markets. The best factory inevitably gets the most work. There is a continued struggle for existence between good factories and poor factories, and the good factory invariably wins.

Consolidation in Distribution.

The law of consolidation of capital and division of labor holds as good in the field of distribution as in that of production. It is inevitable and it is profitable. The department stores and the mail order stores sell for 10 per cent. instead of 30 per cent. profit, and the consumer saves 20 per cent. The profit obtained by the distributor of staples, on the way from the farmer to the consumer, is less than one-quarter what it was 30 years ago. The middleman will not disappear, but instead of being a distributor of every maker's goods he will represent only one maker of each article. This will be a great improvement on the present day of cut-throat competition. Dealers will have less worry, make more profits and have greater opportunities for advancement. They will, no doubt, be shareholders in the various concerns they represent.

Having reviewed the position of our great consolidated corporations as the result of an economic evolution, something should be said with regard to their capitalization. In general, there has been much greater conservatism in the capitalization of industrials than there was in the original capitalization of railroads. Our railroads were built principally for the amount of the bond issues, and the stock represented the capitalized hopes of the projectors. The issues of industrial bonds have been considerably below the actual value of the tangible assets, and industrial stock issues have generally been based on actual earning capacity. Still, it is undoubted that there has been more than one instance of marked over-capitalization of industrials, and no proper legislative measure to remedy this wrong or prevent its recurrence should be neglected. Fortunately, the evil caused by careless investing and unwise capitalization tends to correct itself by natural laws. Investors, confused by the few inflated industrials put out simultaneously with the sound ones, are afraid to buy, and the organizers, unable to sell their securities, now realize that sound capitalization is the best policy.

A Mining Party in the Klondyke.

In taking a comprehensive view of the organization and operation of industrial combinations, we cannot close our eyes to the fact that in the evolution of industrial progress, as in all human affairs, there are imperfections and abuses, for which it is our duty to try to find a remedy, but we should not permit those desirous of gaining political advantage to exaggerate these imperfections to such an extent as to bind us to the fact that it is through industrial combinations we are able to pay high wages and continue our prosperity. They would have us confine our vision to a dead tree in the landscape, instead of looking all around the horizon. It is such narrow-minded men who at times talk against our form of government, because some of its departmental machinery does not work to entire satisfaction. While believing in great organizations, while knowing that they are a necessity in order that this country should become a great power in the economic world and thereby continue the prosperity of the wage-earners of the land, I do not believe in large aggregations of wealth in the hands of individuals unfitted to wisely administer them.

The industrials to-day are owned by the many. While economic evolution is centralizing production in large corporations decentralization of ownership goes on simultaneously through the rapid distribution of shares. There are many hundred times more partners in manufacture, mining and railways than there were thirty years ago, and the number is rapidly increasing.

Inherited Control Made Impossible.

Under the old conditions of private ownership, the control of many of our industrial enterprises would have been inherited by one individual or family. Now, the control is subject to the same rule that prevails in the administration of our country, and that is the rule of the majority. It is seldom, and fortunately so, as preventing great aggregations of wealth in the hands of individuals or families, that the heirs of the industrial giants have the capacity to succeed to the direction of gigantic enterprises. Many inheritors of great fortunes, enervated by ease and luxury, prefer a life of indolence, or to chase the will-o'-the-wisps of society; others prefer to devote their time to literature or art; others, to enter upon scientific pursuits. Under the old conditions, they would have inherited the control of industries, but, under the present conditions of industrial consolidations, the majority of the stockholders—for, generally speaking, the numerical majority is also the majority in interest—elect as officers aspiring young men who, through years of application to a particular industry, have proved their ability and judgment to assume the responsibilities of leadership, and, owing to the higher evolution of our industrial organizations, these men are developing greater intelligence and superior ability to those who have preceded them. Thus, the fittest survive.

With our untold natural resources, with our inexhaustible supply of metals and coal, with our great forests, with every variety of soil and climate, with the most industrious, most intelligent and most contented of peoples working under the best conditions of modern methods, we in North America, Canadians and Americans, are destined to become the economic masters of the world.

There is but one good throw upon the dice, which is to throw them away.—Chatfield.

CASH AND CREDIT IN THE HARDWARE STORE.

A Diversity of Opinions in Regard Thereto by Retail Merchants.

HARDWARE stores in Toronto that do a purely cash business are few indeed. Every dealer says that his trade is as much a cash one as he can make it. But all, except a few, declare that it is next to impossible to give absolutely no credit, unless the business is limited to a small volume indeed. This I learned after interviewing a number of retail hardwaremen. The hardware sections of the large departmental stores, are of course, run on a strictly cash basis, but these are hardly to be classed with the ordinary store.

That it would be an excellent thing if buying and selling could be done by cash only, all dealers were willing to concede, but they differ on the question of the possibility of doing so.

In speaking to a number of hardwaremen I was struck by the apparent

FEELING OF ABSOLUTE CERTAINTY

with which most of them voiced their opinions, notwithstanding that many held exactly oppisite views. One dealer said that they always had done and were doing now a purely cash trade. Another declared that it wasn't possible to do it in Toronto, and that no man in the city was doing so. A third said that a cash trade could be done under certain conditions, but only under such conditions. The majority, however, were of the opinion that a business of any proportion would have to be conducted partly by giving credit.

A look into the business place of Russills at 159 King street east would almost lead one to believe that credit is given to everybody so numerous are the customers of that busy place. But the reverse is the case.

CREDIT IS GIVEN TO NOBODY.

Mr. Frank Russill, who, with his brother, carries on the business of their father, the late founder of the firm, declares that the business has been a cash one ever since it was opened. The late Mr. Russill began in the crockery business, doing a cash trade only, and afterwards added hardware, which is now the only line carried.

THE ADVANTAGE OF CASH.

"In the majority of cases," said Mr. Frank Russill, "we have found that when we give credit we get 'nipped.' The best of the trade won't ask for credit. When you sell your goods for cash you have money to put into more stock, and to enable you to take all the cash discounts that are going. We make all we can out of discounts and are able to give customers as good prices as they can get anywhere, leaving a fair margin of profit for ourselves. We are frequently approached by customers whom we know and asked to give them credit, and it is very difficult to refuse. But we always consider that there is a certain amount of risk with the best of customers, and we would rather have the goods than that kind of trade. Indeed, I think we have lost some trade in that way. In one instance we had a little difficulty with a customer, and his business went to another firm. He come back again, and we could easily have held his trade if we gave him credit, but our previous experience with him told us that it would be better to lose his trade altogether than to risk selling to him on credit. I believe that

THE ONLY WAY TO COMPETE

with the department stores is to adopt the same business principles as they do, that is to buy and sell for cash alone. If we sell $1,000 worth of goods on credit, there is usually about 25 per cent. lost on the whole, People frequently move away and it costs more to get the money than the money itself amounts to. People who ask for credit expect the same prices as those who pay cash, and nothing else will do them."

A LIMITED CREDIT TRADE.

Mr. Wm. Matheson, of 245 King street east, said that the cash business was certainly the most satisfactory, if it were possible to do one, but he doubted if any hardwareman in the city could work up much of a trade on a cash basis. "If a man starts in business and advertises his place as a strictly cash store, he might succeed and be able to keep his store going along those lines, but, when once a credit business has been done, it would be next to impossible to convert it into a cash trade altogether. We have cut off a number of customers who were always looking for credit, as we would rather do without their trade than have the bother of running after the accounts. The saving in time in bookings accounts and chasing them all around the place is one of the best arguments I know of against the credit system. What credit we do give is all at 30 days, and mostly with large manufacturing firms who run accounts here, so that our credit business is limited."

THE FARMERS' TRADE.

W. J. Whitten & Co. are located at 173 King street east, near the market, and the farmers' trade here is a considerable one. Mr. Whitten says that this trade is almost wholly a cash one. "The farmers will buy on credit in the country stores, but when they come to the city they are nearly always prepared to pay cash. Once in a while, a farmer who has been an old customer asks for credit, and generally gets it. It is hard to refuse a customer who has been trading with you for months and months. But our credit sales form a very small part of the business. We are as

NEAR TO BEING A CASH PLACE

as any other store, I think, but it is almost hopeless to expect to do a strictly cash business in the city. Many big firms run monthly accounts with us. They need some little thing now and then, and send a boy after it. They don't know the price, and are not going to give the boy the money. We are glad to oblige them by selling the goods at 30 days' credit, for the pay is sure, and, barring the fact that we are out of the money for that length of time, there is no particular advantage in selling to them for cash."

DROPPED CREDIT FOR CASH.

Wilkins & Co., at 168 King street east, have been in business for 20 years, and do both a wholesale and retail trade. They did not start in on a cash basis, but have made their business a cash one since. I found Mr. Wilkins just through with stocktaking, and he declared that they had had one of their most satisfactory years, business being done on a strictly cash basis. The goods were sold at close figures, and customers were given to understand it. "We have been frequently asked for credit," said Mr. Wilkins, "and by pretty good people too, but our answer to them is always that we must have the ready money so that we can buy the goods at the discount price and give the lowest prices to our customers. We impress upon them that every cash purchase of theirs enables us to sell them some other goods at a better price than if they got credit, for giving credit to one customer means, in time, giving it to a great many. It is easy to tell people that you can't give credit in such a way that you give no offence. In a credit business, there is generally from 10 to 25 per cent. of the sales lost, and we prefer the goods to the risk of losing both goods and money, or to running after accounts."

The Strathconas recently returned from South Africa—A snap shot at them as they were crossing an African river.

A number of those I spoke to were convinced that it was not possible to do altogether a cash business, though it was thought that it might be brought to nearly that, leaving as credit customers only the large business firms who would be inconvenienced by sending the money with every order for small amounts. One of these dealers said that when he started in business he did so with a resolve to conduct it on a cash basis.

HAD TO DISCONTINUE THE CASH TRADE.

"I knew nobody around here," he said, "and decided to give no credit anywhere. Things went all right for a while, but in time builders and big manufacturers in the neighborhood began to send in for small things they needed, which their regular trading places happened to be out of. I refused to sell even to them on credit at first, but soon realized that I was losing a good trade, a credit one, but one in which the money was sure, though I might have to wait a month for it. I commenced doing a partly credit business then and soon had some private accounts on my books. After being 'bitten' pretty badly, I weeded out all the undesirable creditors, and am now very careful to whom I give credit. Of course,

IF EVERYBODY DID A CASH BUSINESS

it would be better all around, but in the hardware business it is almost impossible. In groceries or dry goods the trade is all a household one, but the hardwareman has to supply the factories and the builders on the spur of the moment, and he cannot expect them to send the money with every order. Our buying is generally at 30 days. It is right to take all the cash discounts one can, but it isn't always convenient. We supply a number of households, too, who run monthly accounts with us. I never run after an account. I have found that painters and plumbers are the hardest men in this city to get anything out of. We have lost very little by bad accounts during the past year. I suppose we give a credit of about $150 a month, and only lost $10 of it in the year."

TRYING TO CONFORM TO CASH.

The Aikenhead Hardware Co., Adelaide street east, were of the same opinion as others in regard to doing a purely cash business. It couldn't be doue unless the business was confined wholly to private purchases. They always refused to open any new accounts even if the refusal lost them a customer, and were trying to make their business more and more a cash one.

A DEPARTMENT STORE MAN'S VIEWS.

"People don't expect credit when they come here," said the manager of Robt. Simpson's hardware department. "The aim of a department store is to supply goods at the lowest possible price, and usually their prices are lower than the small store. To do this they are compelled to buy and sell for cash. But, although people may not expect to get goods on credit, they do expect to get them cheaper than in any other store. I wouldn't like to give any opinions as to the advisability of small stores doing only a cash business, as I've never been in that business, but I believe that the great secret of success in small places that have to compete with the department store is to

GO IN FOR SPECIALIZING.

"Make a big thing out of some one line and keep pushing it. There has been a

The Strathconas recently returned from South Africa—A snap shot at them as they were crossing an African river.

Canadian Managers

JOHN LYSAGHT, LIMITED
BRISTOL

Makers of "QUEEN'S HEAD" Galvanized Iron.

AGENTS FOR

STANDARD CHAIN CO.

"LYDBROOK" AND "ALLAWAY'S" TINPLATES

A. C. Leslie & Co.,

Iron, Steel and Metal Merchants,

Montreal, Que.

AGENTS FOR

WHITE CROSS WIRE ROPE

CAMBRIA AGRICULTURAL STEELS

SANDERSON'S TOOL STEEL IN STOCK.

W. H. ANDERSON & SONS

Tool Manufacturers, Iron and Steel Workers

STONE CUTTERS' TOOLS, RAILROAD CONSTRUCTION TOOLS,
PAVING AND SEWER CONTRACTORS' TOOLS, ICE TOOLS,
ARTIFICIAL STONE AND CEMENT WORKERS' TOOLS,
MILL PICKS, COLD CHISELS, DRILLS, PUNCHES,
CAULKING TOOLS, HAMMERS, QUARRY TOOLS AND HOISTS,
DRAG AND WHEEL SCRAPERS, SHOVELS, WHEELBARROWS.

We quote Prices F.O.B. Windsor Ont., if desired.

W. H. ANDERSON & SONS

14 and 16 Macomb Street

Detroit, Mich.

CORRESPONDENCE SOLICITED.
ESTABLISHED 1871.

great deal said against the department stores, but most dealers will admit that they have forced them to adopt better and more up-to-date business methods. That is the reason of the success of department stores. They are always wide awake in taking advantage of anything new in conducting business, advertising, or anything else that will help them, and I think their example has done good to the small dealer. But as to doing a credit business, I cannot speak of it, never having had any experience in it. Our books are arranged for a cash system and nothing else; we'd have to change them if we were to give credit."

ANOTHER DEPARTMENT STORE MAN'S VIEWS.

The manager of T. Eaton's hardware section said practically the same. The only part of the business that was not strictly cash over the counter was done with some of the large institutions, like the hospitals, etc., of the city. These deposited a check with the firm and bought till the amount had been reached. He thought it would be difficult for the small stores to refuse credit to builders, manufacturing firms, etc., as they were the best customers. In their own trade they had none of these firms to sell to; the business was almost wholly a household trade, or tool sets, etc.

Mr. Thomas Meredith, of Thos. Meredith & Co., 156 King street east, was busy with his morning mail when I called, but found time to give his views on the subject. "We'd like to do a big cash trade very much," he said, "but it is an impossibility. There are advantages in both credit and cash systems, however.

GOOD CREDIT IS ALL RIGHT.

"Some men have made fortunes by it. If you do a safe credit business you have a better chance of keeping your trade. Your customers, when they have an account with you, will not go to the nearest store to get what they want, but come to you. People living in the west end of the city wait till they are coming down town to get what they want instead of going to the nearest store, as they would if they had no account here. When an old customer comes in and asks for credit we frequently take the risk, telling him, however, that we seldom do any credit business. We don't encourage private customers to look for credit. We say to them: 'Well, we never open any new accounts, but if you say you'll pay it the next time you come in, all right.' With

FACTORY TRADE

or selling to builders, etc., it is different. We'll do all the credit in that line that we can. I'd really rather do a credit business with them than a cash trade with private persons.

"In doing a cash trade, though, you don't have to run after accounts, and earn the money twice. It

COSTS AT LEAST 5 PER CENT. TO SELL ON CREDIT,

and you can afford to give that when selling for cash and still make as much money. Our location, near the market, enables us to do a big trade with the farmers, and this is all cash. The country trade is all cash and the city trade all credit, or nearly so. Our credit is never given for longer than 30 days if we can avoid it, though sometimes we are forced to extend it to a good deal longer. We try to do

AS MUCH CASH TRADE AS POSSIBLE,

and our business is gradually heading that way, but to make it a purely cash one would be very difficult. The only way to do it would be to advertise the place as a cash store, and give no credit to anybody or any firm, which would lose us some of our most valuable customers.

DECREASE IN THE CASH TRADE.

"It seems that the longer one is in business, the more credit one has to give. Ten years ago, only 75 per cent. of our trade was cash; now, only about 50 per cent. is cash. In 15 years' time," he concluded, laughing, "I suppose there'll be no cash."

DEPARTMENT STORE PRICES.

There seemed to be a feeling on the part of several dealers that the department stores got as good prices as any others on their hardware, notwithstanding that it was all cash trade. Some bargains could, no doubt, be had on special days, but, what was lost on one line was made up on another, so that, on the whole, their prices were no lower than those of the small store. One dealer said that, on some lines, he sold closer than the department stores. But, people, even after finding out that his prices were lower on some lines, would go to the department stores in preference. They generally come back to him in the end. "It's an old saying that people like to be humbugged," he said, "and they stand a better chance of it in a department store than in a small one."

A SHARP RETORT.

It is annoying to be approached by an absolute stranger who asks for credit on the mere strength of his name and address and the assurance that "it'll be all right," but there are such people, and the plain words of one dealer, although a little too much to the point for ordinary business purposes, at least served to keep him from being bothered more than once. A well-dressed woman entered the store of the dealer in question and wanted credit on some goods. He explained that they seldom gave credit, even to old customers.

"I'm quite sure it would be all right with you," he said, politely, "but it's against our business principles."

This was not enough, and she kept up her assurances of sure pay in three or four days. But continued refusals angered her, and she finally said with some heat that she could get credit in any other store in the city. "Why," she added, "I can go to Timothy Eaton's any day and get credit."

"Madam," he replied, "excuse me, but you lie! Timothy Eaton's own wife can't get credit there."

This is not exactly the retort courteous, but much can be forgiven the dealer whose time is taken up and whose patience is tried in such a way. H.

RETAIL CREDIT.

WHETHER or not a community is a desirable one in which to reside or do business is largely a matter of the cost of living therein, remarks a contemporary. That there is a noticeable difference between communities in this respect is not to be denied, neither is it strange that such is the case, when the policy of each is defined. It is not our purpose, in this article, to enter into a detailed analysis of methods, further than to illustrate briefly a universally admitted fact, namely, that loss due to credit is as important an item in determining this result as any other. It is a self-evident truth that that community which buys the nearest for cash, other things being equal, will create wealth and enjoy prosperity in a greater degree than where the reverse is the case.

Such being so, it becomes a more desirable place in which to locate, either in person or enterprise. Whether or not such conditions prevail is largely a question of methods by the retail merchants therein. If they dispense credit liberally, the loss therefrom will be large, and the price of goods consequently high. On the other hand, if credit is reduced to a minimum the opposite result will prevail. The class affected by these conditions directly is the public at large, then the merchant, and, lastly, the town as a whole.

The merchant through the decreased purchasing power of his customer, and the town by the undesirable conditions prevailing therein. Under a cash system those desirous of bettering their condition would accumulate faster, while those of the opposite class would be obliged to depend more upon their own exertions.

COMPETITION IN THE HARDWARE STORE.

By James H. Hamilton.

COMPETITION in the hardware trade is not confined simply to that between hardware dealers. In the world, there seems to be an absolute law of the survival of the fittest.

Referring to the competitors with whom the hardware dealer has to contend, he said :

On one hand, he has great monsters—the mail-order house, the departmental stores and the jobbers who will sell to a country boy as cheaply as to a dealer. On the other hand, he has

THE LEECH AND MICROBE KIND

—the grocery store trade ; the dollar, dime and nickel-racket stores ; the baking-powder and furniture-polish gift enterprises ; easy-payment furniture stores ; selling stoves on the installment plan ; second-hand stores ; lumber yards selling items of hardware ; range peddlers and farmer agents. Is it any wonder that it requires exertion to exist ? There is no business with which I am acquainted that has so many insidious competitors.

Yet there are some who think anything is fair that will down a competitor. Nails at 3c. when they cost 4c., loaded shells at 35c. when they cost 40c., barb wire at 3½c. when it costs more, and, if I cannot sell a stove at a profit, my competitor shall not. This is poor business policy, even should you win.

BE PLEASANT TO COMPETITORS.

How much more pleasant to greet your competitor and his family as friends and allies, meet them socially, get better acquainted, talk over business matters and trouble, agree upon fair and reasonable prices, blacklist deadbeats for each other, borrow and lend of each other, as necessity may require, be honorable and fair with each other, combine your forces to meet the common enemies, large and small ; to defeat them is to build yourselves up. By combined experience, exertion and capital, we ought to be able to buy and sell goods cheaper than our blood-leech and microbe competitors.

SUCCESS NOT GAINED BY PRICE CUTTING.

Price-cutting never leads to financial success. In your own mind just recall a few of the most prosperous dealers in different lines of trade, inquire into their manner of doing business. You will find that they are not price cutters. In a certain community there will be just about so many goods sold in a year. Because you are cutting the price of nails does not induce a laborer or merchant to build a new house, or because you cut $5 on a steel range does not induce a farmer to throw away a good cooking stove, in order to buy a new one from you. No, sir ; he should rather pay a range peddler $69 for one than to pay you $45.

UNDESIRABLE COMPETITORS.

Our meanest competitors at home are the easy-payment furniture house, selling stoves on the installment plan, and these so-called second-hand stores selling new, cheap goods, and the lumber yards that sell nails, roofing, cresting, etc., to their customers. For these blood-sucking leeches I know of no remedy but to treat them as serpents ; whenever you see a head, crush it if you can.

What the department stores have done for the cities, the

CATALOGUE HOUSES

are doing in the country. Our brothers in the larger cities find it an up-hill work to do a profitable, legitimate hardware business. I am personally acquainted with dealers in a large city, who occupied the whole three-storey buildings with a general stock of hardware and housefurnishings 20 years ago who to-day do not need one single room to carry such an assortment as they deem necessary to supply their demands. Year by year the growth in size and number of the great catalogue houses show that they are sapping the life of the country dealers. Some writers say to meet their prices. That is very good, if you had the opportunity, but nine times out of ten you do not get the chance. But if you did, where is your profit, as on most standard goods they sell them as cheap or cheaper than your jobber does to you ? I must confess that I do not know of any way to compete with him and live.

JOBBERS WHO SELL TO CONSUMERS.

Then there are some so-called jobbing houses, especially in sporting goods, heavy hardware and machinery supplies, who will send you a catalogue and price list, soliciting your trade, and will send the same lists to any country boy who will write for them, and sell him the goods, too, if he will send the money.

GIFT ENTERPRISES.

You are all acquainted with the gift enterprise of the baking powker, furniture polish and soap manufacturers. You have seen whole loads of enamelled ware, roasting pans, carving sets, butcher knives, toy wagons, etc., given away to induce people to pay a big price for an article otherwise not worth a penny.

SOME HONORABLE DEALERS.

But there are some honorable dealers in other lines of goods who so far forget their honor in business and the rights of others, that they will make such gifts, as the dry goods merchant giving away scales, the clothing dealer giving away sleds, boys' wagons, slates, etc., with every suit of boys' clothing. It seems that nothing is suitable for gifts excepting in the hardware line.

Then last, but not least, we have the range peddlers. I honor the range peddler, he is no cut price or gift enterprise competitor. He is indifferent to all competition. He works for his living. But, unlike my legitimate competitor, he wants good pay for his work, and he gets it just as easily as he could get half as much. You pay him his price, or he keeps his goods, and he makes money by it.

ELECTRIC LIGHT WORK THEIR SPECIALTY.

On page 7 of this issue will be found the advertisement of the Comstock Manufacturing Co., Comstock, Mich., manufacturers of the "Climax" automatic engine. About 80 per cent. of this firm's entire output goes for electric light work. They make these engines both for direct belting to generator and direct connected to generator on the shaft of the engine. For the last two years they have been behind with their orders, and are still far behind. Last year they added a number of new machines to their shops and extended their floor space by a new addition, and again this year they are adding several new special machines, and will add this spring 2,500 more square feet to their floor space. The export trade of the Comstock Company is extremely good—much larger than they had any anticipation it would be, and the number of foreign inquiries they are getting indicate they can still very much increase this branch of their business.

A WELL-EQUIPPED WHIP FACTORY.

The Hamilton Whip Co., Hamilton, have recently greatly enlarged their buildings, and they have now one of the largest and most complete plants of the kind in the world. They report a brisk trade, and have for months past been running their factory overtime. Having the latest and most improved automatic machinery, etc., they are in a position to manufacture reliable whips at the lowest possible prices consistent with best workmanship.

They have recently placed on the market a number of new leading lines of whips, which have been well received by the trade generally. They carry an enormous manufactured stock, and are in a position to complete all orders the day they are received.

For durability, quality, finish, etc., their goods have an enviable reputation throughout the whole Dominion.

The Largest Facilities.

The Best Quality of Work.

THE CANADIAN RUBBER CO.

of Montreal

MONTREAL. TORONTO. WINNIPEG.

Manufacturers of

Rubber Belting
Hose
Packing
Etc.

Seamless Tube

Write for Discounts.

THE ART OF WINDOW DRESSING.

A Series of Articles by Experts.

MANY hardware merchants do not realize what a great trade-bringing factor a good window trim is. They seem to think that window dressing is only intended for the dry goods and fancy goods stores, and are under the impression that such lines as hardware are not attractive-looking articles, and cannot be made so, owing to the nature of the goods and their colors, etc. This is a mistake, and the hardwareman who neglects to give attention to his window trims overlooks his best paying advertisement and one if properly arranged, ticketed, etc., will return him more dollars and cents for his time and trouble than any other scheme he can devise. Some of the most artistic and attractive trims I ever saw were made from a stock of hardware. 'Tis true that hardware articles are generally not very attractive in themselves, but when they are arranged in some artistic design, on a pretty background, or in some novel manner, they never fail to draw attention.

A CASE IN POINT.

The possibility of making goods talk for themselves was shown with much emphasis in one of the large hardware stores in Chicago recently. The proprietor or some of his assistants evinced great ingenuity in the arrangement of the goods. There was the usual display of bright steel wares in the showcases, but it was on the walls, where they could not fail to be seen, that the genius of the store trimmer showed itself. On each side of the store, near the front entrance, a space was divided off (about 10 ft. square), and covered over with jet black cloth, neatly bordered in yellow. Black cloth is much better than painting the space black, as the fabric possesses a softer surface and tends to show the polished steel tools to better advantage. In these spaces was an advertisement, fine and simple, in plain English, but the lettering was done with tools. A study of a hardware catalogue will show any bright clerk that, with chisels, bits, files, punches, cold chisels, and other straight tools of various lengths, any combination of letters can be made, even the crooked ones like S and R. In this sign the letters were of uniform size, but the line varied, giving the whole a very attractive appearance.

The following will convey an idea of how this novel ad. read :

> It is PLANE to be seen that our ADZE are strictly honest. This is on the LEVEL. Business with us is done on the SQUARE. You can CHAIN your faith to that. Low prices RULE here. If you don't see what you want AXE for it.

These ideas, if worked now and again in the windows, set people talking about your store and your novel ideas. What better advertisement can you have than to have people interested in your modes of attraction?

INDIVIDUAL DISPLAYS.

It is always preferable, when possible, to have individual displays, that is displays of one line of goods instead of a mixture of everything. These individual trims leave a better impression on the public. Trims composed of a little of everything only confuse the eye. People are attracted by a display of one particular line the same as the attention would be attracted to a big man or a big horse, etc. Let one trim be composed of a line of table cutlery, another time show shears, another time builders' materials, at another time all tinware, sporting goods, at house-cleaning time show all house-cleaning goods, in hot weather make a display of ice cream freezers, refrigerators, ice tongs, picks, lemonade squeezers, etc. The results from separate displays exceed, by far, those of a confusing mass of all lines.

WINDOW TICKETS.

Dressing windows without ticketing the goods shown is waste of time. I consider a hardware display robbed of its effectiveness as a sales-bringing trim, unless there is a good supply of price and descriptive cards in it. To assure yourself of this fact, just experiment some time on this. Put in a trim not ticketed and then one well price-ticketed and note the difference in results. Goods bearing a price card put in the window will generally sell themselves without the aid of a salesman, except to roll them up and take the money for them. As to the making of these cards, any ordinary salesman with a little practice will soon learn to make neat price tickets and large window cards. Water colors are easily made and are the cleanest for general use. They are made up by taking a package of "Diamond" dye (the color you desire to use). Put about quarter of the package in a tin cup. Add enough mucilage to make a thick paste and then add water. A lighter or darker shade can be had by increasing or diminishing the quantity of water. A sable brush, size 4 or 5, is the proper brush to use. White cardboard for the tickets can be bought at any stationer's for 4 or 5c. per sheet ; size 24 x 28 inches.

FRAMEWORK.

As hardware is a difficult line of goods from which to make bright, catchy trims,

An Attractive Miscellaneous Display.

Trade Mark.
"The Only Tools You'll Need."
DUNLOP TIRES
for 1901.
The advance couriers of trade will call upon you soon with the prettiest line of wheels you ever saw. When you order don't neglect the most important detail in the specifications—Dunlop Tires.
Short-sighted manufacturers may try to "ring in" cheap tires as substitues for "Dunlop's"—the tire that every buyer asks for.
We have bettered the Dunlop Tire this year with the thickened tread—more rubber on the wear surface—more resiliency—more value for the same money.
Dunlop Tires are simple—an ingenious application of the natural law of expansion forms the basis of their construction.
Endless wires that do not expand provide the rigidity of adherence to the rim. Without these wire protectors a few months' wear would render the outer cover liable to a "blow-off."
Bicycle dealers know the value of the Dunlop Tire. Twelvemonth guarantee.
The Thickened Tread
THE DUNLOP TIRE CO., Limited
TORONTO.
NOTE.—Perhaps you could make money handling the Dunlop Solid Rubber Carriage Tire. Write for the Catalogue and Price List.
Depots in Montreal, St. John, Winnipeg and Vancouver.

The Moffat Stove Co. Limited.
PEARL STOVES & RANGES
LOOKED AT FROM EVERY POINT OF VIEW THEY'RE ALL RIGHT
Weston, ONTARIO.

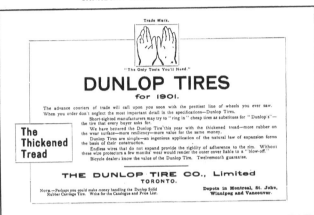

much of the brightness of it has to be made up in the backgrounds and frameworks. A good idea is to get a set of half-circles made of 1 x 4 in. lumber. About a dozen of these (in diameter about 3 ft.) is sufficient. They can be worked into circles, arches, and dozens of different designs, and arranged in the windows in artistic manners and goods displayed on them. They should be first covered with black or colored cambric or print. For nickel or cutlery, black looks best. These frames can be made for a trifle. They should be put together with screws or fine nails, so as when they are being taken down they will not be destroyed in trying to get them apart as they would be if large nails were used. The circles and designs made with these can be latticed in with white tape or crinkled paper. This makes a prettty effect. The illustrations

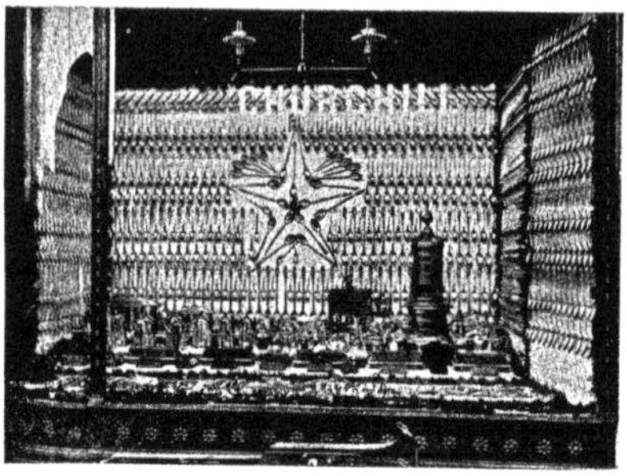

A Display of Cutlery, Oil Stoves, Etc.

will convey ideas that can be used in trims with these light frameworks.

DISPLAYS OF COAL STOVES.

In displays of coal stoves a good idea is to place an electric light in each stove, over this puff an orange sheet of tissue paper. Turn off all the window lights but those in the stoves and a beautiful effect will be had at night, the stoves all looking as if they had a good warm fire in them.

GARDENING TOOLS.

For a good display during the spring, a nice idea for showing tools used for gardening is made by sodding the bottom of the window, and, in the centre, make a flower-bed, in which place a few geraniums, etc., and on the grass could be placed a garden hose, watering can, rake, lawn clippers, lawn mower, etc. This would make a pretty, neat and attractive trim, and would be a big relief from the everyday trim. A card calling attention to gardening utensils would help it out.

HOUSE-CLEANING GOODS.

About the first week in April is a good time to get in a good display of house-cleaning goods. A big trade can be done at this time through good displays of this kind. Price and description tickets should be used freely, and a big card, appropriately worded, calling attention to the goods as being necessary for house-cleaning. This display should consist of everything that can be thought of throughout the stock, for instance: Brooms, whisks, tubs, pails, washboards, stepladders, whitewash brushes, whiting and calcimine, paints, enamels, varnishes, scrub brushes, dust pans, clothes-lines, stovepipe varnishes, picture hooks and wire, carpet sweepers and beaters, ironing boards and irons, clothes racks, mops, etc. A good display of this sort, well ticketed, is sure to interest every housewife.

CAMP SUPPLIES.

Another good, taking trim during the camping-out season is made by arranging a camp scene in the window. Throughout the scene arrange hardware articles useful in a camp. There are a hundred-and-one different articles that can be found to go with this display, such as saws, hammers, axes, hatchets, lanterns, camp stoves, cooking utensils, tinware, etc. Sporting goods and fishing tackle can also be shown in this manner. What is required is novelty—something out of the ordinary line of display as seen every day.

BIRD CAGES AND RAT TRAPS.

A display of bird cages and rat traps shown together is all right. A novel display that drew large crowds was recently seen in a hardware store. Rat traps (the cage kind) were piled up one on top of the other in a pyramid shape, and in each was seen a live rat. A card bore the following:

> RAT HYPNOTIZERS
> 15c. EACH.

In the hardware business, there are dozens of lines of goods that make good, attractive, business-bringing windows, and the trimmer can find varieties of ways for showing these wares, if he just takes time and thinks out novel, catchy ways for showing these goods. The illustrations give ideas that can be used in displays of cutlery, stoves, tinware, etc.

H. H.

WHAT MADE A DISPLAY EFFECTIVE.

"It is astonishing," said a hardware clerk to me a few days ago "what a difference is caused by little things in window display. This was demonstrated very forcibly in connection with a display I made a week or two ago.

"The goods shown were household smallwares, goods that most people like to have but don't really need. I had made a false bottom for the window, so that it rose in tiers about six inches deep from front to back. On each of these tiers I had a dozen or so small piles of goods. As I had paper of different colors for all the tiers, and the articles were not unattractive, the window was a good one. Many people looked at it, but few seemed to be persuaded by it to come in and get some of the goods shown.

"Feeling that something should be done to make it sell goods (a window is of no value unless it does that), I went outside to see if I could find out some way to improve it. While I was there two farmers came along and stopped to look at the window. One of them noticed some thermometers. He first asked me the price; then if they could be depended on. When I answered him, he laughed, and said that he had always thought good ones were very expensive. He guessed he would take one home.

"I was not slow to take the hint. My display lacked price tickets and information cards. Straightway I set to work to make them. I ticketed everything, and where I could say a good word for any of the articles I did so.

"The result was that while there did not seem to be more people looking at the display than before, the sales due to it were

much larger. Up to that time I thought tickets were good enough, but not at all necessary. Now I think many lines cannot properly be sold without them."

A MISTAKE OFTEN MADE.

"Could an inexperienced man make a good window display?" I asked a veteran trimmer some time ago.

A Stove Display.

"That depends," was the reply. "If he has the knack, his first window might be an excellent one."

"Are there any pitfalls that new men generally get into?" I again asked.

"Not many. The worst is the tendency to talk price, to emphasize cheapness. It is the same mistake as that which advertisers generally make. Everybody knows that an article can easily be sold at a low price. Consequently, a new man, in order to swell his sales, will be tempted to continually tell how cheap his goods are—what great bargains he offers. The result may be a big increase in the turnover, but when the net profit is accounted the results are often far from satisfactory. The able and successful merchant, however, is the one who gets a big turnover at a good margin. His energies are directed to this end, and, of course, his window trimming shows it. While he makes an excellent use of price tickets, the prices are accompanied by some reference to quality. I have found that since I have followed this principle in all advertising and window trimming I have earned the reputation of carrying goods of first-class quality. This enables me to get a fair profit."

PART OF A SYSTEM OF ADVERTISING.

"After experimenting with window trimming for some time," said a retailer the other day, "I have come to the conclusion that it is most forceful when treated as part of a system of advertising.

"For instance, if I want to push painters' supplies, I first advertise them in the local papers, devoting as much attention to quality as to price. Then I put them in my window. I make a background of the heavier goods, such as white lead kegs, paris green kegs and drums, putty bladders and tins, etc. I make brushes, putty-knives, etc., into designs, such as circles, stars, etc., at the back and sides. The centre space I fill up with all kinds of goods which painters use. At the front I arrange dry colors in small saucers. Besides having price tickets and hint cards, I put in some central place a card giving some reason why painters, amateur as well as professional, should secure their supplies right off and right here.

"Then, all my help have instructions to talk up our paint. We always aim to convince people that, while no startling bargains are offered, a sufficient reduction has been made in prices to make it a good time to buy. I find that when we join the forces of our newspaper advertising, window trimming and our own influence the results are excellent."

A SPORTING GOODS DISPLAY.

A window display in which sporting goods were the principal features was recently illustrated in an exchange. The top, back and sides of the window were enclosed. To the sides were neatly attached guns and rifles. At the back was described, by the aid of drapery, a large circle reaching almost from the top to the bottom of the back wall, with an oval on either side almost as deep. Within each of these ovals were hung two pairs of boxing gloves, and on either side of the central circle, only a little distance from the wall, was suspended a punching bag. Standing on the floor of the window and immediately in front of the large circle was an old-fashioned windmill. To the wings of this windmill were fastened penknives with the blades open at various degrees. To the well of the windmill facing the window were attached scissors of various sizes, some closed, others partly closed and others again wide open. Arranged on the floor of the window were lanterns, knives, revolvers, compasses and cutlery. Judging from the illustration the display must have been a particularly striking one, and from our brief description of it retail hardwaremen may be able to arrange a similar display.

A MOURNFUL BUT A STRIKING WINDOW TRIM.

In connection with the Queen's death a number of artistic windows were shown in Toronto. Rice Lewis & Son's corner window, on King and Victoria streets, was especially attractive, the simplicity and taste displayed in the design making it probably the best of all. In the front of the window, a few inches from the glass, a framework of light wood, about six feet square, formed the base of a pyramid, from which black goods were draped toward the back of the window, where the smaller end of the pyramid was about a foot and one half square. With this small end, covered with black goods, as a background a white marble bust of the Queen stood on a small projecting platform, the sides of the pyramid making, as it were, a deep frame for it. The contrast between the snow white of the bust and the black draping was very effective. The bust was made in Italy and only one other of the kind is believed to be in Canada. In value it is worth several hundred dollars. We herewith reproduce a rough sketch of the window display with

Rice Lewis & Son's Memorial Window.

the hope of not only giving our readers an idea of the effect of the display, but that it may suggest to them ideas whereby it might be utilized by themselves in window trims.

THE CANADA HARDWARE CO., Limited

Shelf and Heavy Hardware

ONLY WHOLESALE — MONTREAL. — WHOLESALE ONLY

THE GEM FOOD CHOPPER IS THE BEST.

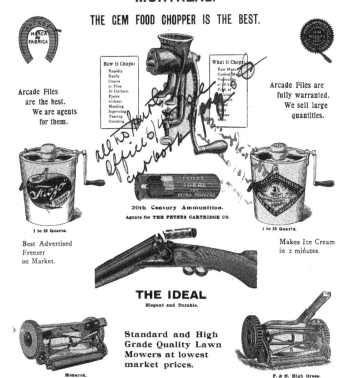

Arcade Files are the best. We are agents for them.

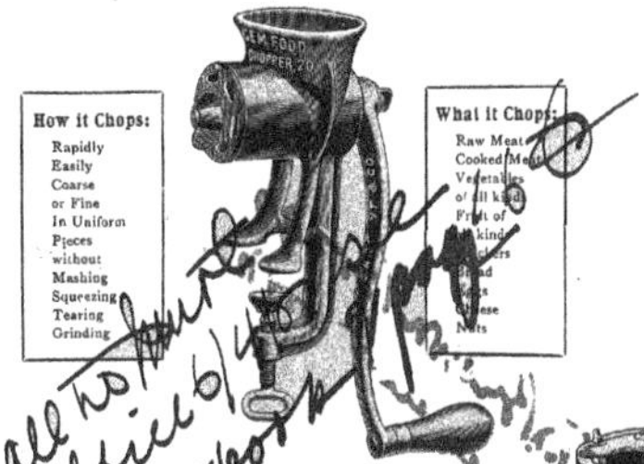

Arcade Files are fully warranted. We sell large quantities.

1 to 25 Quarts.

Best Advertised Freezer on Market.

20th Century Ammunition.

Agents for THE PETERS CARTRIDGE CO.

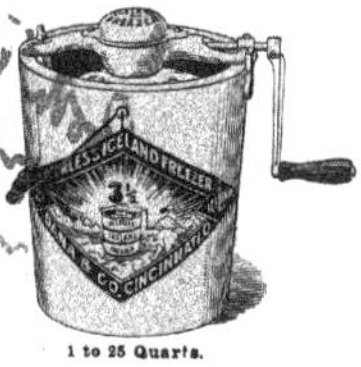

1 to 25 Quarts.

Makes Ice Cream in 2 minutes.

THE IDEAL

Elegant and Durable.

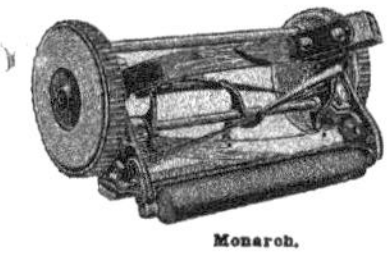

Monarch.

Standard and High Grade Quality Lawn Mowers at lowest market prices.

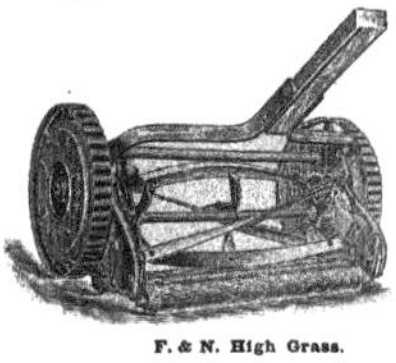

F. & N. High Grass.

Brown, Boggs & Co.

Wire Solder Machine.

MANUFACTURERS OF

- Can Making Machinery
- Canners' Processing Machinery
- Evaporating Machinery
- Tinsmiths' Tools and Machines
- Presses, Dies, Etc., and
- Sheet Metal Working Tools of any description
- A. C. Incandescent Arc Lamps
- Arc Lamp Riggings

Moore & Bristol Tomato Can Filler.

Hamilton, Canada.

INTERIOR ARRANGEMENT IN HARDWARE STORES.

Some of the Ideas Obtaining To-day.

WHAT changed conditions result in new methods is abundantly attested in the remarkable development in recent years in devices designed to economize space, in order to save time and money in the retail hardware store.

The advent of the big stores, with their extensive connection and their ever-increasing desire to spread the field of their activity, has caused such keenness of competition that new methods have been sought whereby the cost of doing business should be lessened and it made possible to reduce the margin of profit sufficiently to resist the encroachments of all competitors.

The effect has been to develop among hardwaremen and their assistants a desire for progress, the ability to design improvements and the courage to execute them. Every department of business is affected, from the bookkeeper's desk to the coal oil tank, and the time of every person about the store is economized, from the employer to the messenger boy. The change has been so widespread that to-day it would be difficult to determine what has been the most valuable improvement or what more generally adopted.

The introduction of plate glass windows is not of recent date, yet their use became general in this country within the memory of the majority of hardware dealers. Their advantage is to-day so generally recognized that little need be said in their favor now. In fact, the merchant of to-day is considering a step in advance of an altogether plate-glass front. The advantages of prismatic glass in giving light to the interior are such that, in many of the best stores now going up, there is a strip of prismatic glass from a foot to a yard in width placed above the plate front. Attention is now being devoted to the interior arrangement of windows so as to make them most effective. In many cases mirrors have been placed in the sides and back to double or triplicate the effect of the display. Artificial floors are prepared for use with the different classes of goods shown, so as to make the display more effective than would be the case were cutlery to be shown on the same level as is best for showing stoves, etc. The lighting of the windows and, in fact, the whole interior is also much considered.

ARRANGEMENT OF LIGHTS.

The most effective arrangement of lights in the window that I have seen was in a store erected about a year ago. There were 10 or 12 eight candle-power electric lights placed in the ceiling with reflectors arranged so as to throw the light directly onto the goods shown. Two more lights were placed in the side of the window with reflectors arranged with the same purpose.

DIVERSITY OF ARRANGEMENT.

There is great diversity in the appearance of the most effectively arranged hardware stores. The introduction of shelf boxes, saw and cutlery cabinets, has done much to make a hardware store look bright and attractive. But to expect, or to aim at the same neatness or even orderliness that are essential qualities of the up-to-date grocery store, would be sacrificing utility to appearance, and the former is the more important of the two.

Possibly the most noticeable improvement of recent years is the widespread introduction of either a

CASH CARRYING SYSTEM

or cash registers. Where a bookkeeper is employed the carrying system is undoubtedly the best. Where it is inconvenient to have a bookkeeper or a clerk to attend to change, the usefulness of a cash register is unquestioned. While the great majority of clerks are thoroughly honest, there are some who are weaker in this regard than the rest. It is, therefore, best that temptation should be removed as much as possible from all.

Though their use is not yet as general as that of the cash changing devices, wall cabinets, shelf boxes, drop shelves, etc., are rapidly becoming a part of hardware store fixtures. In the arrangement of these modern devices the hardwareman shows his thorough study of conditions and methods.

THE SAW CABINET

is so designed that all manner of carpenters', wood-workers' and butchers' saws are easily examined by a customer outside the counter. There are so placed, moreover, as to be opposite one of the divisions in the counter, so the customer can approach the case without going behind the counter.

The shelf-boxes and drop-shelves are now devoted to almost every kind of smallware in the store.

THE BEST BOXES

are a patent device, patented and manufactured by a former hardwareman, J. S. Bennett, Toronto, yet many dealers are satisfied with the home-made article, notwithstanding its greater bulkiness and less attractive appearance. The drop shelf has never, to my knowledge, been patented. The idea is to have a shelf attached by hinges at the bottom, and so connected by chains at the side that when open it makes a stand for holding goods to show customers. The best contrivance of this nature that I have seen is installed by Russill's-in-the-Market, Toronto.

On the front of these boxes or shelves, samples of the goods contained within are shown. Ingenuity is sometimes necessary to attach the goods firmly, but it is generally done well. In some cases—for instance, fine cutlery, such as good jack-knives, embroidery scissors, etc.—it has been found by more than one proprietor that attaching these goods to the outside of boxes or shelves depreciates their worth considerably. One dealer has, therefore, installed glass fronts to several shelves, and arranged these goods to show from within. Another has removed all his cutlery to his counter showcase.

The division of goods is also noteworthy. All

PAINTERS' SUPPLIES

are kept by the up-to-date merchant in one section of his store. One standing in front of this section will see a great variety of prepared paints on open shelves, dry colors in draw shelves or boxes, brushes, chamois, putty knives, etc., hung up or displayed on the counter, glass-cutting block, putty, oils, sundries, etc., all arranged to attract attention. Household smallwares are in another division, carpenters' tools in another, while garden tools, farmers' tools, etc., are in separate divisions of their own. If tinware, harness or carriage hardware be handled, they are in departments of the store distinct from others in the store. If stoves are sold, the centre of the store is generally devoted to them. Here, another improvement of recent years is noted. Whereas in past

years stoves were often shown on the floor without even being on their feet, now they are properly set up on stands from six inches to one foot high, so that a customer minutely examines before buying. It would seem that the hardwareman has learned that it

PAYS TO SELL STOVES

on their merits and to have a customer understand them before purchasing. This has been found true, in fact, about almost every article for sale in a hardware store.

The number of contrivances for saving time or space is steadily increasing. One of the best of these is the Bowser oil tank, which is practically a pump, which not only draws oil from the cellar, but measures the quantity with unfailing accuracy. This saves space in the store, only occupying about 5 x 3 ft., and also time, formerly spent going downstairs and in measuring.

A good device, which was described by HARDWARE AND METAL some time ago, is

A SHOT CASE

which one of the clerks in Russill's, Toronto, designed. This is made of wood, with a glass front and with compartments made of zinc. Each of these compartments is fed from the top with a funnel. When orders are being filled the shot or powder runs down a trough, the flow being controlled by a tin slide door. A pair of small scales are placed on top of this case, so that an order for shot or powder can be filled in remarkably quick time.

Racks for shovels, rakes, forks, etc., are now generally used. These are so placed that the goods can be easily removed by a customer waiting in the store. The arrangement is simply a number of small boards projecting from a long one, each of the small ones having several notches on which the heads of the forks, rakes, etc., easily grip. One rack I noticed the other day had, in addition, several iron rods suspended from the board and bent to sustain the shorter shovels, etc.

A RACK FOR AXES.

The favorite way among up-to-date men to hold axes seems to be to have a strong rack, in which holes are cut to contain iron rods. These rods are easily removed, so that when a customer desires to, he can have a more thorough examination of the goods. This style of rack is also used for holding twine. A good place to keep the twine is at one end of a counter. Vises are kept on a somewhat similar rack, except that instead of being placed on removable rods they are attached to firm boards. Whip racks are so common and simple as to be hardly worth noting here. Another simple device, which is not so common, however, is a

RACK FOR RAZOR STROPS.

This is merely a hoop with holders attached for the handles of the strops to grip. Axe handles are sometimes shown on a stand with notches into which the handles fit.

A SCHEME FOR HANDLING ROPE.

The custom of arranging rope in the cellar with an end running through holes bored in the floor to the store proper is steadily growing. Another arrangement is to have the rope hung on a large rack from which it is easily unwound.

A STAND FOR WRINGERS.

Some manufacturers have recognized the tendency towards making goods show well in small space. I have in mind a firm of wringer manufacturers who make a display stand on which their goods are shown effectively. But the effectiveness of this stand was increased by one Toronto dealer. The stand is sent out varnished about the same shade as the woodwork of the wringers. This dealer recognized that the effect would be better if there was a contrast between the stand and the goods, so the stand was painted a dark red. Another manufacturing firm sent out excellent stands for displaying carpet sweepers. The patented display stands which are now being sold so extensively have much to commend themselves as they can be adapted to so many purposes for either interior or window display.

THE COUNTER

of the up-to-date hardware store receives much attention. Instead of the bulky wooden arrangement of former years, at least one counter in every store to-day boasts a glass showcase on which are sliding boards, upon which goods are shown to customers. In these cases the more profitable lines are shown, such as cutlery, valuable tools, etc. Great care must be taken with these goods to have them always present a neat, attractive appearance. Underneath this case, in front of the counter, certain lines can be well kept. One of the best lines to keep in front of the counter is nails. These are most easily handled if the space for them is sufficiently open to permit a big handful being taken out. The canting bins are excellently adapted for this purpose.

RODS FOR THE CEILING.

Not content with making the most of his windows, floor, walls and counters, the modern hardwareman expends much ingenuity to make his ceiling contribute to swell his turnover. Rods are fastened, some close to the ceiling, some low enough to hang lanterns, etc. upon. These rods are devoted to displaying all the light, bulky articles.

An important feature of to-day's interior fixtures is the use of

PRICE TICKETS,

suggestion cards, etc. The policy of every up-to-date merchant is, of course, one price for all. So he tickets practically everything in his store so that a customer waiting to be served may practically do his or her own buying without any of the clerks' assistance. These cards are augmented by many placed throughout the store with seasonable suggestions. These are generally written by either the properietor himself or one of the clerks.

A significant fact of the development in store arrangement is that practically all the most up-to-date dealers encourage their clerks to make suggestions re improvements and, where advisable, allow the clerk to work out his ideas in his spare time. It is surprising how apt a number of clerks become in making improvements, writing cards, and in other similar work.

HOLDERS FOR GLASS GLOBES.

The accompanying illustrations show a couple of glass globe holders designed by an old English hardwareman. It will be noticed that one can be suspended and the

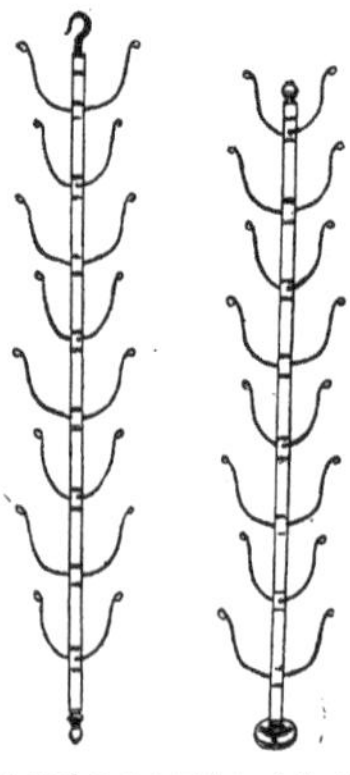

other stood on a counter or shelf. The inventor declares he saves 90 per cent. in breakage.

THE STORY OF A SUCCESSFUL HARDWARE BUSINESS

Founded on..... Fact...

FRED THOMPSON and Harry Morton had been warm friends for years. They had been chums at school, where the difference in their manners and appearance had excited much banter from their companions. Fred was early known as "Pudgy," because of his bulkiness and easy-going manner, while Harry was nicknamed "Pug," because of his early prowess with his fists and his readiness to use them. They had left school about the same time, both starting as messenger boys in stores in their home town, a place of about 4,000 population. Fred went into a dry goods, Harry into a hardware store.

The intimacy was maintained throughout the years during which they learned something of the tiresomeness of running messages, the difficulties in the pathway of the clerk, and some of the intricacies of business. To the astonishment of almost everyone, Fred seemed to make a much better clerk than his old chum. His manner with customers, while possibly not as brisk as some of the hurried ones would like, was so continuously genial, and his memory of names and faces so thorough, that he was a general favorite. Harry, while a favorite with his employers, did not "draw" trade because of his affability. He was constantly on the move, and when his customer was served he busied himself with some other duty. His knowledge of the goods he handled was, however, much superior to most clerks of his age.

One August evening, when they were both about 22 years of age, they were "up the river" together. Harry was rowing. Suddenly he stopped.

"Can you put up $500?" was his unexpected question.

"How far up?" came the rejoinder.

"I'm not fooling. Could you raise $500 without giving a mortgage on what you'd invest it in?"

"I've got more than that of my own," answered Fred.

"Well, I've been thinking for weeks that there is a fine opening for another hardware store in J———. I have $450, and can easily get the rest from my father. I'd like you to go shares with me."

"But I don't know the hardware trade."

"No; there would be a good deal for you to learn. But you get along with people so well that you'd soon be able to sell more in a day than I would."

Fred was cautious, and it took Harry months to convince him to make the venture. But he at last consented. A good stand was secured. This was stocked with the most up-to-date goods on the market. For two weeks before their store was opened to the public they devoted a "double-quarter-column" space to invitations to young and old to call in and examine their stock.

For the first year nothing but the regular hardware lines were kept. Then they accepted an agency for one of the most widely advertised bicycles. There was no bicycle repair shop in the town, and soon they were sending parts to Toronto to be either repaired or replaced. "This won't do," said Harry one day; we are losing a good chance to make a few dollars by not doing our own repairing."

"Well, I couldn't learn to do that, and you haven't time," answered Fred.

"No, but we must get someone who can do it."

"Would it pay?"

"If Frank (a younger brother of Harry's, who was acting as clerk) could do that in his spare time, it would pay well. I was thinking of advising him to take a year or two off and secure a position where he could learn the business. We could get a mechanic to take his place while he is away—one who could help around the store while he is not busy."

The matter was fully discussed, and it was finally agreed to try the experiment. Frank secured a position in Montreal. A capable man was hired to look after the repair shop. The installation of this department was well advertised, and it was put on a paying basis much sooner than anticipated.

This took up considerable space, and, before long, floor room in the store was congested. As the trade of the firm had steadily increased, and had been done on a fair margin, and as both of the partners had been content to take out of it only enough cash to pay necessary personal expenses, there was some profit in the first two years' business. This, and a portion of the original capital, was devoted to making an extension, which gave the necessary floor space. By judicious advertising, attention was directed to these changes in such a way as to emphasize the fact that the young firm was progressive.

While Fred devoted his time to convincing customers of the comparative values offered by Morton & Thompson, Harry devoted so much time to planning, scheming and studying that his partner often called him a "modern-method" crank. Many of his changes were slight, but some were entirely radical, necessitating a complete rearrangement of the goods in the store.

In two years Frank came back with a good general knowledge of mechanics. Instead of dismissing the man who had been secured to look after the repair work while Frank was away, the firm agreed that it would be far more advantageous to branch out a little more and to repair all kinds of small machinery, implements, etc. This was found profitable from the first, and the business done steadily increased until about two years ago a machine shop was opened, and first class lathes, punches, etc., installed, and the manufacture of hardware specialties started. Now a jobbing trade extending over several counties is done in many lines.

In the meantime the trade in the store has steadily increased. The town has grown materially during the past five years and several industries have been started. These have caused a demand for several lines hitherto not handled. These lines have been put into stock. In the case of one line which is neither made in Canada nor handled by the wholesale dealers here the name of the manufacturer in Europe was secured and the stock imported direct. In addition to their regular hardware lines they have put into stock a big range of both carriage hardware and harness. Last fall a second extension was necessary. When this was being made, shelf boxes, bicycle stepladders and other modern devices were installed.

The result is that while both partners are still young men they own a business which compares to advantage with any similar concern within a radius of 50 miles. The causes which contributed to these results were, in the writer's opinion, an ever-watchful desire for improvement and a careful study of conditions on the part of one member of the firm, a wise, genial treatment of customers on the part of the other, and the ability as a mechanic of the younger brother, who, by the way, has been admitted into partnership.

D. O. M.

ESTABLISHED 1874.

HENDERSON & POTTS

Halifax and Montreal.

"Anchor" White Lead and Zinc Paint
" Colored Paints in Oil
" Liquid House Paints
" Floor and Roofing Paints
" Varnishes for all purposes
" Lacquers, Blue and Gold for Canners
" Japans and Dryers
" Carriage Gloss Paints
" Pure Colors in Oil
" Superfine Colors in Japan
Marbleine Wall Tints
"Anchor" Straw Hat Enamel
Etc., Etc., Etc.

ANCHOR BRAND
REGISTERED
TRADE MARK

SOLE MANUFACTURERS IN CANADA OF

BRANDRAM BROS. & CO.
BRANDRAM'S
B.B.
WHITE LEAD
REGISTERED TRADE MARK
MANUFACTURERS, LONDON, ENGLAND

BRANDRAM BROS. & CO'S

London, England

B. B.

Genuine White Lead.

Brandram's Genuine B.B. is the *Best White Lead Made.* It is unequalled for whiteness, fineness and durability, and will cover more surface than any other White Lead Paint.

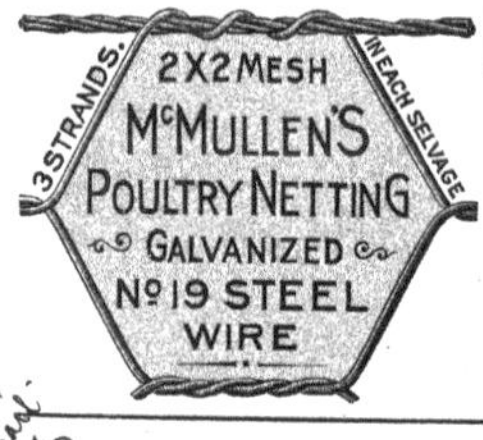

McMULLEN'S POULTRY NETTINGS

and LAWN FENCINGS are not surpassed in the world. Their WOVEN WIRE FENCINGS have stood 16 years of very successful testing.

TRADE DISCOUNT on Nettings:
1 TO 5 ROLLS, 55 AND 5 PER CENT. OFF LISTS.

For SPECIAL PRICES on large quantities write the manufacturers at PICTON, ONT.

THE ONTARIO WIRE FENCING CO., Limited

FOR SALE, also, by the Canadian jobbers.

THE B. GREENING WIRE CO., Limited — General Agents, HAMILTON and MONTREAL. and **JAMES COOPER,** MONTREAL, Ry. Fencing.

JAMES WRIGHT & CO.

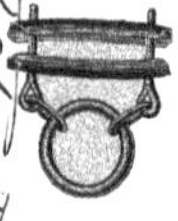

Carriage Hardware and Iron Merchants.

Carriage and Wagon Forgings.
Carriage Malleable Iron.
Carriage and Wagon Woodwork.

Bar Iron and Steel.
Wrought Iron Pipe.
Pipe Fittings, etc.

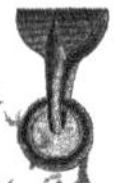

132 Dundas St., 133 Carling St., 443-445 Talbot St., LONDON, ONT.

We Can Quote Interesting Prices on

Wellington Bicycles

Lamps, Bells, Cyclometers, Toe Clips, Saddles, Cements, Oil and Lubricants, Wrenches, Pumps, Oilers, Stands, Etc. . . . .

1901 Catalogue now ready.

CAVERHILL, LEARMONT & CO., - MONTREAL.

BUSINESS AND ITS MANAGEMENT.

Pointed Articles Upon this Important Question.

AVERAGING ACCOUNTS.*

GEORGE W. STORCK.

AVERAGING accounts is a subject that few bookkeepers are familiar with, and without exaggeration I am free to confess that I do not believe that one bookkeeper in ten can average an account, and as one of our prominent members stated on one occasion that he knew of a bookkeeper who failed to obtain a good position "just because he didn't know the way." Gentlemen, this is not a joke. This has set me seriously to thinking about preparing an article on averaging accounts with a view of not having other positions lost on account of an average. I have carefully taken into consideration not only the theory but the practicability of each average, and shall fully illustrate same. Among the different methods which I shall illustrate will be a simple average, complex average, throwing a balance forward by means of part payments on account, etc. The first illustration will represent a single average. An easy method when there are numerous items is to average each month separately and then take the average dates of the different months together to get an average.

ILLUSTRATION NO. 1.

Purchases—January 1, $50.00. January 3, $10.00. January 10, $16.00. January 15, $31.00
" —February 3, 20.00. February 7. 40.00. February 10, 80.00. February 14, 60.00

January	1 x 50	equals	50
"	3 x 10	"	30
"	10 x 16	"	160
"	15 x 31	"	465

107 divided into 707 equals 7 days from January 0 is January 7, $107.

The average date is January 7.

February	3 x 20	equals	60
"	7 x 40	"	280
"	10 x 80	"	800
"	14 x 60	"	840

200 divided into 1980 equals 9 days from February 0 is February 9, $200.

Now take these averages of the two months together as follows:

Jan. 7—107 From Jan. 7 to Feb. 9.
Feb. 9—200 x 23 equals 6,600 equals 21 days added [to Jan. 7.
307

to January 28 the average date.

Rule—Multiply each amount by the figures of its posting date, divide the product thus obtained by the total amount of dollars, the result will be the number of days forward from the zero date of the month, which will be the day of the month on which the whole amount will be due. The average would be all right when the terms and dating were the same, but if the dating differed it would be figured somewhat differently, for examp'e:

June 10, $400 3 mo. Aug. 16, $500 2 mo.
Aug 28, $100 2 mo. Dec. 12, $300 1 mo.

ILLUSTRATION NO. 2.

June 10 3 mo. September 10 x 400 equals $400 equals 0
From September 10 to August 16, 2 mo., October 16, 36 x 500 equals 18 000
" " 10 to " 28, 2 mo., " 28, 48 x 100 equals 4,800
" " 10 to December 12, 1 mo., January 12, 124 x 300 equals 37,200

$1,300 divided into 60,000

equals 46 days forward from September 10 to October 26, or we might use the last date of maturity as a focal date to count back from. As follows:

From September 10 to January 12 is 124 days x $400 equals 49,600
" October 16 to " 12 is 88 " x 500 equals 44,000
" " 28 to " 12 is 76 " x 100 equals 7,600
" January 12 to " 12 x 300

$1,300 divided into 101,200

equals 78 days back from January 12 to October 26.

I do not approve of this method, for I believe that it is easier to add than to subtract, and, for that reason, I prefer the first method:

Rule—Illustration 2 is as follows:

Multiply each amount by the number of days from the maturity of the first item (as a focal date), divide the product thus obtained by the dollars of the account, the result will be the number of days to count forward from the focal date.

THE NEXT METHOD OF AVERAGING IS WHAT IS KNOWN AS THE INTEREST METHOD.

To get an average by interest when the items are all on one side. Take for the focal date the earliest one in the account (of maturity), ascertain the interest on each amount from the focal date to the due date of the item. Divide the total interest thus obtained by the interest (at the same rate) of the total dollars for one day. The result will be the number of days to count forward from the focal date, which will be the date of maturity.

ILLUSTRATION NO. 3.

Jan. 1....$200, 30 days, due Jan. 31
Feb. 10.... 100, 10 " " Feb. 20
Mar. 5.... 300, 60 " " May 4
Apr. 20.... 400, 30 " " May 20

Figured at 6 per cent. per annum.

From Jan. 31 to Jan. 31 (the earliest date of maturity).....	$200	
From Jan. 31 to Feb. 20, 20 days' interest at 6 per cent...	100	$0.33⅓
From Jan. 31 to May 4, 93 days' interest on at 6 per cent......	300	4.65
From Jan. 31 to May 20, 109 days' interest on at 6 per cent.	400	7.26⅔
	$1,000	$12.25

The interest on $2,000 for one day at 6 per cent. is 16⅔c.; this divided into the total amount of interest, $12.25=74 days forward from January 31, is April 15, the average date of the whole amount.

THROWING A BALANCE FORWARD.

This method of averaging accounts is seldom met with, and I believe there are very few who are familiar with same. The illustration which I shall show will cause the balance of the account to run forward several years.

We will say that we have made the following purchases of John Jones from January 1 to 12, amounting to $1,000, with a fall dating of November 1, 10 days (which is often the case with mills, glovers, shoes, etc.). On going throught our accounts on January 16, we make this amount of in-

*From a paper read before the New York Society of Accountants and Bookkeepers.

WELLS' SPECIALTIES.

Merchants will find it to their advantage to push the sale of these goods. Send for complete illustrated lists.

A. "THE WELLS LIGHT."
B. " "UNBREAKABLE" LAMPS AND OIL FEEDERS.
C. " "WASTE OIL" FILTERS.
D " "INDUSTRIAL" OIL LAMPS.
H " "LIGHTNING" PAINTING MACHINE.

A. C. WELLS & CO., 103 Midland Rd., St. Pancras. LONDON, ENG.

BRUSHES, BROOMS

..AND..

WOODENWARE

Mill, Cheese Factory and Tanners' Supplies.

Ordered Work a Specialty

JNO. ALLIN, - Oshawa.

WHEELER & BAIN

Manufacturers of

Galvanized Iron Eavetroughs, Corrugated Conductor Pipe, Corrugated Roofing and Siding, Cornices and Skylights, etc.

Write for Our Prices on GALVANIZED IRON SHEETS.

179 King St. East.
108 George St.,
TORONTO.

UNION MFG. CO.,

20 Breckenridge St.,
BUFFALO, N.Y.

Manufacturers of . . .

PATENTED

HARDWARE SPECIALTIES.

Complete line of
Foot Power

EMERY GRINDERS.

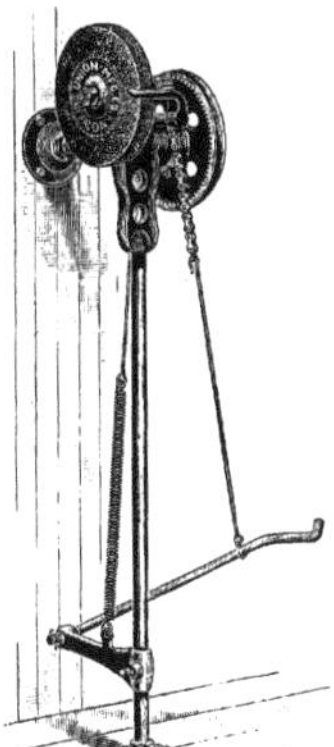

NO. 1, UNION TOOL GRINDER.

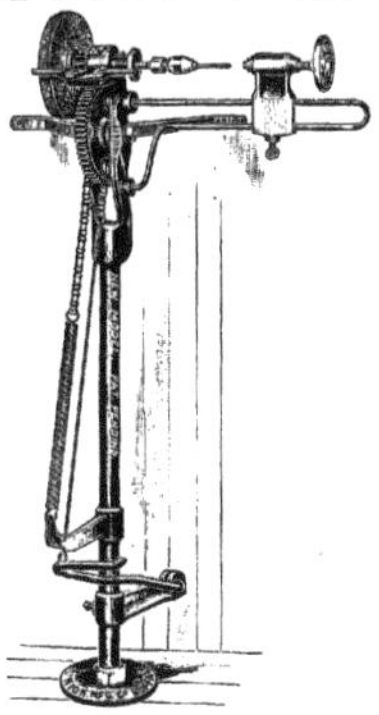

NEW MODEL.
NO. 2.

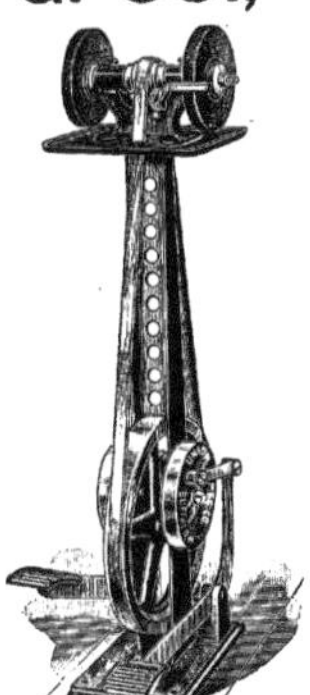

NO. 3.

All Steel Gear. With or Without Drilling Attachments.
Sold by all live Hardware Dealers throughout the World. Goods and Prices Guaranteed.

Union Mfg. Co.

BUFFALO, N.Y.
and
MT. FOREST,
Ontario.

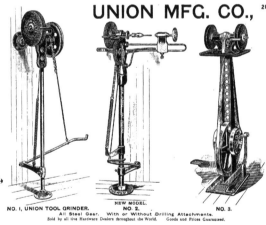

debtedness, and as we are somewhat flush we make them a part payment on account, $750, on this date, with the distinct understanding that purchases and payments are to be averaged and settlements made in accordance therewith.

ILLUSTRATION NO. 4.

Bills Jan. 1 to 12, due Nov. 1 to 11, $1,000.00.
Jan. 16, on account, $750.00.
1900.
Jan. 16—$ 750.00 from Jan. 16 to Nov. 11.
Nov. 11 —1,000.00 x 299 divided by 299,000 equals [1,196 days from.
Bal....... $250

Jan. 16, 1900, is 3 yrs., 3 mo. and 11 days from Jan. 16, 1900, or April 27, 1903.

Now, at a first glance, you will ask: Is this right?

Now I will put it in a more simple way.

John Jones has had the use of our $750 from January 16, 1900, to November 11, 1900—a period of 299 days. The interest on $750 for 299 days at 6 per cent. would amount to $37.37, which he would owe me for the use of my money on November 11, while the amount of the balance of $250—I have had the use of the same from November 11, 1900, to April 27, 1903, a period of 897 days. Now I would owe him interest at 6 per cent. per annum for 897 days on $250, which would amount to $37.37, so that you can see that the balance of $250 would have to run long enough to equal the amount of interest which I would be entitled to had I settled the account on November 11.

Another proof:

We will presume that this day is April 27, 1903, and we desire to make a settlement of the balance, $250.

ILLUSTRATION NO. 5.

Jan. 16, 1900 $750.00
Apr. 27, 1903 250.00 x 1,196
1,000).99,000(299 days.

299 days added to Jan. 16, 1900, would give us Nov. 11, the due date of our bills.

Let us look at this average again. Suppose, on January 16, 1901, we desire to settle the account, would we be entitled to any interest?

ILLUSTRATION NO. 6.

Bills due Nov. 11, $1,000. Paid on account, Jan. $750.00.
Jan. 16, 1900 $750.00
Jan. 16, 1901 250.00 x 365
1,000)91,250(91 days added.

to January 16, 1900—April 17, average date of our payment, and we would be entitled to interest on the full amount of $1,000 from April 17 to November 11, or about 6 months and 24 days at 6 per cent.—$34. Now, let us prove the correctness of this, and reason this out.

The bill of $1,000 was due on November 11, and we settle their account on January 16, 1901, or 66 days after the bills were past due, consequently he would be entitled to interest on $1,000 from November 11 to January 16, 1901, 66 days, at 6 per cent., or $11, while I would be entitled to interest on my payment of $750 from January 16, 1900, to January 16, 1901, one year, at 6 per cent., $45. Now, the difference between the $11 that he would be entitled to, and the interest $45, would be $34, the amount of interest we would be entitled to, thereby proving the correctness of the first method. In conclusion I wish to say a few words regarding this last average: You might come to the conclusion that if I was flush at the time I made this payment of $750 on account, why not anticipate the whole account instead of making a part payment and receiving less interest? In reply I would state that there are certain times in a year when we do not have so much money. If we anticipated, the most that we could obtain would be 6 per cent. p.a., and when the dull times come around, if the money market was "tight," we would perhaps have to go out and pay the same rate as we anticipated at, but applying the money on account in this way it tides you over the slack season, and should you then have more money lying in the bank than was necessary, you could always discount that particular account. Another thing that perhaps you have not noticed, as I did not speak of discounts, was that you were settling the account on a basis of deducting 1 per cent. a month and at the same time adding at the rate of 6 per cent. per annum.

RULE FOR CREDIT LIMIT.

Thos. P. Robbins.

One of the safeguards of the national banking system of the United States is the provision that no bank shall loan to any one individual or concern more than 10 per cent. of its own capital. This is a wise law, and, to the credit of our banks, it is observed closely.

If business generally was conducted with a view to allowing proper credit, restricted by a kindred rule applicable to the circumstances, there would be less loss and consequent derangement. It should be an axiom that large lines of credit ought to be extended only by concerns of large capital. We hear sometimes of houses, particularly in certain lines, meeting with losses of dimensions from 50 to 100 per cent. of their capital. It is not always the fact that the cause is ignorance, but usually the temptation to chance a credit with a hope of gain.

Some credit men may doubt that this state of affairs does exist, but if inquiry is made it will be clear that there is more or less unwise crediting of this sort. Let us suggest that if a concern has a capital of $25,000, not impared, a credit of $2,500 be allowed, and if their surplus is believed genuine, grant 10 per cent. on that. Do not even consider that if your own capital is $100,000 you can risk $10,000 in a particular case. Go the banks one better and make your rule apply on the credit asked and not on your own capacity. There is no desire to convey the impression that the banks have no consideration of the worth of the borrower. The fact is they watch this, and pretty closely, too.

Our sympathies naturally go out to the small concerns. Capital is small, the competition of the new combinations (we have no trusts) is severe, and it takes a long time for the enterprise to find the position that its principles desire for it. It is all the more necessary that the foundation should be well laid. Supposing a concern with $10,000 capital should find a profitable customer with a capital rating of $100,000, and purchasing $5,000 monthly, on 90 days' time. Should the small house undertake to extend the credit? No, not by any means. It is better to attempt to do small things first.

Some credit men cannot appreciate how it would feel to have to follow strictly a 10 per cent. limit, as their own experience has been in lines where the 10 per cent. has scarcely ever been expected, sales being large in number, well distributed, and not in large amounts. Their hearts go out, however, to houses who are known to have almost their whole capital tied up in single accounts gone wrong.

There are houses having not over 10 customers on their books, and still the credit men of such institutions have their troubles. Many deal with trade that buy in small amounts and are here to-day and there to-morrow. Some large concerns are reported in almost all of the failures in their line. To all of these and any others, let us suggest the application of a 10 per cent. plan where practical, and in the appropriate manner.

A REWARD FOR PROMPT PAYMENT.

A well known business man during a recent speech, said: "The sole and only object of giving a discount is a reward for prompt payment, and when prompt payment is not secured by this means, then the money is lost and the primary object of paying it is not obtained. The sum annually lost in this manner is an enormous one, and materially reduces the profits of the business. Doubtless much of this could be saved by concerted action if all would take hold of the matter with as much energy as we apply to our private business."

RATIO OF STORE EXPENSES TO SALES.

A bu-iness wi.h an income at its heels,
Furnishes always oil for its own wheels.

THE regulation of the expense account of a store is one of the most difficult problems which confronts the proprietor, and it is to the perusal of this account to which he is indebted for a majority of the wrinkles which time and worry bring to all of us. To the careless or thoughtless manner with which this account is suddenly reduced or increased, is due at least 90 per cent. of the business failures of our times. The finest business in the world can be seriously embarrassed, or entirely ruined, by improper management of its expenses, and an unusually prosperous year of sales be turned into a total loss, when the balance sheet is handed to the proprietor at the end of the year. It is not only against excessive expenses that the successful merchant must guard his business, but also he must not yield himself to the impulse to reduce his expenses too low, and while effecting a saving in cents, cause thereby a loss of dollars.

LOSSES FROM ECONOMY.

Business prudence is a very good quality, and most writers and speakers have extolled the virtue of prudence and economy until many business men, especially as they grow older, have come to believe that a saving in expense is a net gain equal to the amount by which they have reduced the account. While in some cases the nature of the business may demand a decided reduction in its expenses to meet a turn or change in trade, still I believe it will generally be found true that where a business has sustained certain necessary and proper expenses for a number of years at a reasonable profit, a reduction of the expense account will ultimately cause more loss than the temporary savings produced. This is shown very often in the matter of salaries. Take, for instance, a firm which has been employing a number of first-class salesmen, at a good salary and which has prospered and made money above its expenses. Now, suppose the business to have reached a firm and solid foundation, through the efforts of those salesmen, until it almost seems to "run itself." Then, suppose the owners in looking ever their expense account find the item of "salaries" quite large and prominent and decide to reduce their help or secure lower-priced men, with a view to pocketing the difference between what they were paying their old salesmen and the cheaper and less experienced substitutes. For a time there will be an apparent saving, the expense account will be less, and the momentum which the business has acquired will carry it along for a time, as before.

But soon customers begin to find fault; they are not properly or intelligently waited upon. Orders become mixed, and the stock confused, and, as a consequence, the customers become offended or disgusted, and bestow their patronage elsewhere.

RATIO OF EXPENSES.

Now, as to what should be the proper ratio of expenses to sales, of course this depends upon the nature and margin of profit which the business affords. It is safe to say, however, that legitimate expenses should be kept up to the point which will afford a reasonable living for the proprietor.

To illustrate this point, imagine a retail store which has, say, a cash capital, in stock, of $10,000. Suppose the sales to average $25,000, half cash and half 30 to 90 days' credit, and the per cent. of the gross profit 30 per cent. on sales. To support the proprietor it is necessary for him to make $1,500 clear, and he wishes to increase his sales for the future rather than reduce his store expenses and diminish his trade. Now let us glance at what his expenses would consist of, and their ratio:

Capital	$10,000
Sales per annum	25,000
Gross profit (30 p.c.)	7,500

EXPENSE ACCOUNT.

Items.	Amount.	Per cent. to sales.
Rent	$1,500	6
Clerks	2,250	9
Freight and cartage	300	1 1-5
Advertising	500	2
Telephone	50	1-5
Light	100	2-5
Fuel	50	1-5
Insurance	50	1-5
Postage	100	2-5
Taxes	150	3-5
Stationery	50	1-5
Bad bills, 2 p.c.	250	1
Interest on capital, 6 p.c.	600	2 2-5
Sundries	50	1-5
Totals	$6,000	24 p.c.

Under this set of expenses a net profit accrues to the proprietor of $1,500, which covers his personal expenses and 6 per cent. on his capital, which represents a gain of $600 to be added to his stock the following year. He is able to discount his bills, is paying fair wages to competent clerks and is advertising moderately. The result will be, other things being favorable, that the next year will show an increase of sales of, say, $2,000, and the gross profit will then be $8,100, and if he decides to help his increasing trade by more advertising or another clerk, his expenses will be limited to 24 per cent. of the sales ($27,000) and will amount to $6,480, an increase of expenses of $480. His net profit, aside from interest, under this set of expenses, will be $1,620, or a gain over the former year of $120 net. Now, as he has added $480 to his advertising bill or clerk hire, it is reasonable to suppose that his business will continue to increase in a few years to $30,000. The gross profit is now $9,000, and as he has continued to keep his store expenses up to the 24 per cent. standard, the expense account will figure up $7,200, leaving him a net profit of $1,800, which, with the 6 per cent. on original capital included in expense account, makes his total income $2,400 for the year.

AN ERROR SOME MAKE.

Just at this point a great many near-sighted merchants would fall into the error of supposing that a reduction of the expense account was the proper thing to increase their profits, and substitute cheaper help and curtail their advertising, these two being the items which are most readily reduced. The result would be that in a short time the trade would drop off much faster than it had grown, and, the rent and other "stationary" expenses being the same, the net income would fall below the living necessities of the proprietor.

On the other hand, had he still continued to keep up the ratio of expenses for clerks, advertising and the other items to a total of 24 per cent. of his sales, there would be no reason why, if the population was sufficient, his sales should not finally reach the amount of $50,000. The expenses would then be $12,000 and the gross profits $15,000, a net profit to the owner of $3,000 in addition to his interest on the capital, which would swell his total income to over $4,000. It is also apparent that as he sold so many more goods than formerly, he would be able to obtain better prices and rebates, which would cover the lower prices which he might be obliged at times to accept from his larger customers. This is the way his expense account would now appear:

Capital in stock	$10,000
Sales per annum	50,000
Gross profit (30 per cent.)	15,000

EXPENSE ACCOUNT.

Items.	Amount.	Per cent. to sales.
Rent	$3,000	6
Clerks	4,500	9
Freight and cartage	600	1 1-5
Advertising	1,000	2
Telephone	100	1-5
Light	200	2-5
Fuel	100	1-5
Insurance	100	1-5
Postage	200	2-5
Taxes	300	3 5
Stationery	100	1-5
Bad bills, 2 p.c.	500	1
Interest on capital, 6 p.c.	1,200	2 2-
Sundries	100	1-5
Total	$12,000	24 p.c.

The items which form this expense account are, of course, subject to the peculiarities of different businesses.

"If a man had £20 a year for his income, and spent £19 1s. 6d. he would be happy; but if he spent £20 0s. ½d. he would be miserable.—Ideas for Hardware Merchants.

BRANTFORD
Royal Blue House Pumps

FIG. 99
Lift Pump

Pitcher Spout
Closed Top

Nickle-Plated
Brass Cylinder

Sells Rapidly

Gives every
Satisfaction

Write for
Prices.

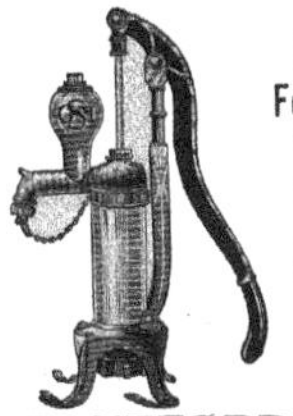

FIG. 100
Force and Lift
(See Engraving)

Syphon Spout

Nickle-Plated
Brass Cylinder

Many Sold

Handsome and
Efficient

BRANTFORD

Galvanized Steel Windmills and Steel Towers

Steel King Pumping Mill

Made in 8, 10, 12, 14 and 16 feet sizes.

For Power and Pumping.

Fitted with Patent Interchangeable Steel Roller and Ball Bearings.

Sprocket Drive
Gears Not Used
The Easiest Running
The Most Powerful
Angle Steel Frame

Will last a Lifetime.

SATISFACTION GUARANTEED.

GOOLD SHAPLEY & MUIR CO. LIMITED
BRANTFORD CAN.

James Cartland & Son
Manufacturers of every description of **Limited**

CABINET, BUILDERS', FURNISHING AND NAVAL BRASSFOUNDRY

BIRMINGHAM, ENGLAND.

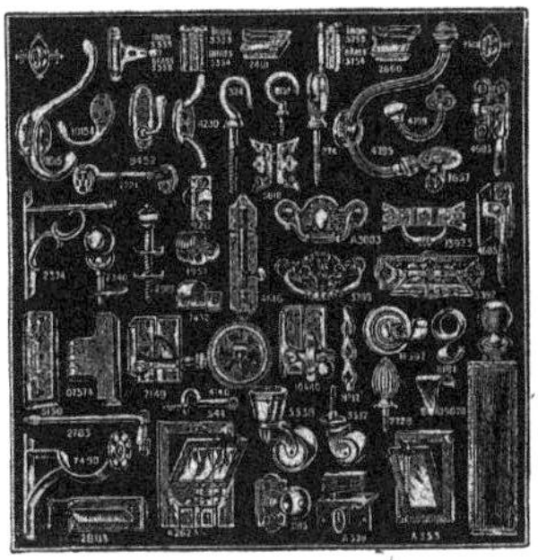

London Showrooms: 57 Holborn Viaduct, E.C.

THE CANADA SCREW COMPANY

BRANCHES:
69 Bay St., TORONTO.
446 and 448 St. Paul St., MONTREAL.

. . HAMILTON, ONTARIO

MANUFACTURERS OF

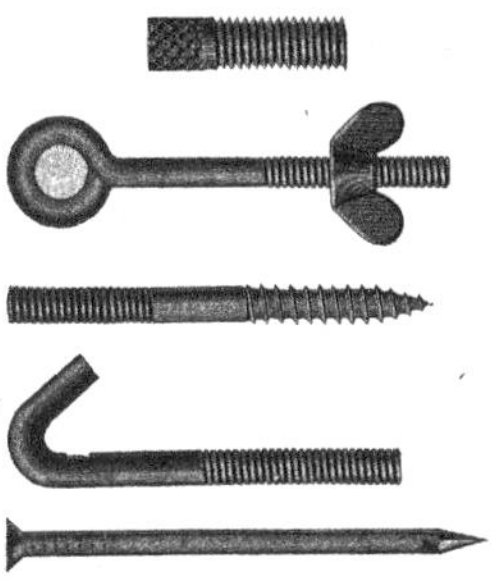

Wood Screws, Machine Screws, Screw Eyes and Hooks, Gate Hooks and Cornice Hooks, Tire Bolts, Stove Bolts, Stove Pins, Rivets, Elevator Bolts, Stove Rods, Nuts, Burrs, Wire Nails, and kindred articles of all varieties of material, form and finish.

Bright, Annealed, Oiled and Annealed, and Coppered Wire in coils and cut to length.

Special Threaded and Bent Wire Goods.

SPORTING GOODS IN THE HARDWARE STORE.

The Excellent Field there is in Canada for Sports of Various Kinds Should Induce Retail Hardwaremen to Make Greater Efforts to Push Sporting Goods.

Sporting goods are not a necessity of life, and hardware dealers should not handle them as they would a staple article like nails. There was a time in the history of this continent when man earned his daily bread by the rod and the gun, but now, by grace of our economic and social arrangements, he buys his meat at his very door. He lives on the products of the land and sea farms rather than on the fruits of the chase. When a man goes hunting or fishing in these days, he does it for pleasure, for sport, to rejuvenate his body or to recover his fallen spirits. It should be the policy of Canadian hardware merchants to increase this desire for pleasure. They can do so if they would only employ policy in trying.

It is said that there is no one who can sell more sporting goods than the sport himself. Now, all hardwaremen cannot afford the time to go hunting and fishing just to acquire the reputation of a "sport," but they can interest themselves in the sporting facilities of the neighborhood and in the success of those who are exploiting the game resources within a radius of their stores. By so doing the merchant brings himself into close touch with those who can be his customers. More than that, he educates himself and learns what goods he should lay in stock. He can study the demands of the sporting fraternity and as new discoveries are made he can lay them before his public without any trouble.

The "sport" is always looking for new ideas. This fancy is an ever-burning flame within his heart, a flame that the hardwaremen would do well to keep fanning if wind will do it. The continual employment of new goods and patents means extra outlay for the customer and increased profit for the merchant. Any information he may get, either through catalogue or traveller, he should lay before his customers for consideration. In this way he will be forwarding the interests of his patrons, but he will not be neglecting his own.

Moreover, he should use his window to push guns, ammunition, and fishing tackle. It is the best means of advertising he has. A window display moulds the young idea and turns his fancy towards sporting pleasures, while it is not without impression on the minds of the older "sports." There are hardwaremen, and their number is legion, who regard sporting goods as an unprofitable line, simply because the amount sold is not great in comparison with the amount of staples disposed of. But here lies their opportunity to work up a trade. It is useless to attempt to generate a demand for nails, people want only a certain amount and will buy no more than they want. With sporting goods it is different. People will buy these luxuries only as they are pressed to do so and the most enterprising and energetic merchant is the one who will get all the trade. Moreover, when the trade is once secured it is easily held, for a "sport" will go to that merchant for a gun who, he thinks, understands the business. One who takes no interest in rifles has a difficult task before him when he sets out to satisfy a customer. It is a line that requires knowledge and experience, so once gained, it is easily held.

Speaking to a retailer not long since, we learned that he had quit carrying fishing tackle because "my clerks stole more than I sold." We should counsel different remedies for such an evil. If a clerk is dishonest he should dismiss him ; if paid a decent wage an honest clerk will not be tempted to steal. There is no reason why fishing tackle should not be as profitable a line to carry as tacks. Indeed, it ought to be more so, for a little pushing will stimulate a trade in the former, while the only way to make a drive in tacks is with the hammer.

SPORT IN MANITOBA AND THE NORTHWEST.

BY C. HANBURY-WILLIAMS.

WHEN I have made all the money I deserve, and I am a multimillionaire, I am going to build a shooting lodge on the shores of one of our great lakes. I have the site in my mind's eye, and the plans all mapped out in my head ; all I am waiting for is my just deserts. There is a certain spot I wot of, within a very few hours of Winnipeg, sheltered on three sides by a curving belt of trees. For half a mile in front of it is a long stretch of swampy land, intersected by deep waterways that are fringed with tall reeds; beyond that again is a great inland sea. There are small creeks, or rather backwaters from the lakes, that run in and out of one another close by. There is

ONE PARTICULAR POINT

where I saw a man drop 70 mallards one morning, and bring them all to bag, too, for they fell on flat prairie or open water. At the right season of the year you can stand in front of my castle (in the air) and see black lines and triangles cut sharply out against the sky all round you, moving very swiftly, and you begin to wonder whether you have enough cartridges to hold out. You can hear the prairie chicken crowing like barn-door fowls ; and a little to the northeast is a bit of marshy ground, cattle-poached, and dappled with gleaming pools, where the snipe are nearly as thick as mosquitoes. A thin column of blue smoke curling up in the distance shows you where a few wandering Indians have pitched their camp, but there is no other indication of civilization in sight. Still, the neighborhood is well settled, and a short drive will bring you to a farm house where you can buy the finest butter and freshest eggs for uncivilized prices.

THE HUNTERS' PARADISE.

A very short railway journey will bring you to a country full of deer and the lordly wapiti, the king of the deer tribe the world over ; and down on the flat, boggy land by the lake shores the moose will stand knee-deep in water on the summer evenings, ready to lie down when the flies get bothering. All day you will breathe the wild free air of the prairie, and at night you are lulled to sleep by the surge and ripple and splash of the waves on the beach, broken now and then by the weird banshee cry of strange water-fowl. The other day I came

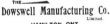

across some of the figures paid for grouse-moors and deer-forests in Scotland. They varied from $50,000 a year paid by Lord Farquhar for the rental of the forest of Mar to the Duke of Fife, down to the very modest $150 for which the Duke of Argyll lets a small shooting called St. Catherines. "You cannot get a decent grouse-moor nowadays in Scotland or England much under $6,000 a year," said the proprietor of one such moor, and part owner of another, to me a couple of months ago; "while for very much less than that amount these men could come out here, get all the shooting they want, if properly managed, and leave money behind. These men would be among the best advertising agents you could get, and their sevices would cost you nothing; in fact, they would be paying you. To talk of their wantonly slaughtering the game is ridiculous; all their training, taste and interest is against useless destruction. I intend to come out myself next year, and I hope to bring others with me."

I told him that I thought he was quite right. Personally I have shot in England, Scotland, Ireland, and Wales, and I say, quite deliberately, that when I have made my pile I shall still

PREFER SHOOTING IN MANITOBA

or the Northwest Territories. A man is his own boss here; he is permitted to use his own judgment in finding and killing game, and is not ordered about by a gamekeeper, and posted here, and bidden to walk there, as if he were an automatic shooting machine. He can work his own dogs; and, if he has to clean his own gun, so much the better for the gun and for himself. He may not return home to all the comforts that he would get in his own country, but, of course, I am going to fit up my shooting-lodge with a good bathroom, and keep a cook, and have a few books, and perhaps a billiard-room; and I find it rather difficult to think of much more that I shall want. All of this could be done for a good deal less money than the hiring of one middling good grouse-moor and here you can get your deer forest thrown in. You can get fox and wolf hunting too — with foxhounds bred from blood imported from England and Wales—over some of the finest hunting country in the world.

THE MONTREAL HUNT CLUB.

So far I have shown you one side of the shield; now I want to kick a little. I am informed on the best authority that the legislation with regard to game-shooting in this country is to receive early attention and alteration. There is no doubt that this is much needed. To-day a visitor to Manitoba from the Old Country, be he a British officer travelling to or from the East, or a London banker, cannot get a day's duck or chicken-shooting without being

MULCTED IN THE SUM OF $50,

a practically prohibitive tariff, for many of them may not be able to afford it, and those who can are so indignant at the imposition that they prefer to forego the sport altogether. With a very little management, excellent sport might be obtained here for a much smaller sum than is paid year after year for comparatively small shootings in England and Scotland, ever when all travelling expenses are taken into consideration. The lessees would have all the advantage of seeing a new country, and even if they did have to rough it a little, at first, through inexperience, still that element does not usually act as a deterrent to the genuine sportsman. The usual outcry from thoughtless settlers out here is that "we want all our shooting for ourselves." The answer is simple. In Manitoba and the Northwest Territories we have a good deal more land than in the whole of the British Isles, and our total population is very little larger than that of the single city of Bristol. We could get all the shooting we have now and still leave enough, and more than enough, for all the crowds that pour out of the great London railway stations in the first and second weeks of August.

DATES FOR THE CLOSE SEASON.

Another alteration that should be made this year is in the dates fixed for the close

Elastic Carbon Paint

A BIG THING.

LOOK INTO IT.

During the last year we have sold in Canada and the United States over five thousand (5,000) barrels, and have received from our patrons numerous voluntary testimonials, which are in many cases accompanied by repeat orders, and, owing to our success, many worthless, cheap imitations have been placed on the market, which are claimed by unscrupulous competitors to be as good as the genuine Elastic Carbon Paint, which is made only by ourselves. A trial of the goods will convince the most skeptical.

We are large importers of "Pure Spirits of Turpentine," in tank cars, enamelled white inside and used only for the transportation of this article, direct from the virgin Florida forests.

We also handle large quantities of pure Linseed Oil, bought from the most reputable manufacturers in England, both Turpentine and Linseed Oil being sold subject to chemical analysis.

Prices and Samples Cheerfully Submitted.

The Atlantic Refining Co.

TORONTO.

Corner Esplanade and Jarvis Sts.

Manufacturers and Importers of Illuminating and Lubricating Oils, Greases, etc.

season. Hitherto, these appear to have been settled on the broad principle that the less chance you give to sportsmen to shoot, the bigger will be your stock of game. At first sight, this reasoning appears sound, and it is strongly held by many who have given the subject little or no thought. Take the case of prairie-chicken shooting as an instance. As things are now, these birds may not be shot before October 1. By that time they are so wild and strong on the wing that dogs are practically useless, and big bags are difficult, if not impossible, to get. But, on the other hand, the coveys are not broken up; the old cocks are left to live until next spring and to drive off the younger birds, and the inevitable results of in-breeding, namely, disease, barren birds and small broods, are sure to follow. I wish here to quote words used last August by Sir Herbert Maxwell, in Blackwood's Magazine. Sir Herbert, I may say, was born and bred in a grouse-shooting country, and is a recognized authority in all matters concerning sport. He writes: "The more severely, but fairly, a moor is shot, the more surely will its grouse increase. Very different would the result be if grouse were left to the tender mercies of each other. Take any moor of high reputation you like—I care not whether in Yorkshire or Perthshire—put it in the hands of an extreme humanitarian, who objects on principles, which it is impossible not to respect, to the taking of animal life in sport, and see what the state of that moor will be at the end of three or four seasons. Instead of packs so numerous that they seem to darken the heavens as they fly over the boxes, there will be a very limited number of pairs of old birds, some barren, some with moderate families, each old cock jealously guarding his territory and furiously driving off every weaker individual of his own kind which presumes to crop his heather. These few strong old birds have asserted their rights, and, in virtue of their might, drive off all their younger relations. Now, this is no fancy picture drawn in favor of the sportsman. Any ornithologist or gamekeeper of experience will testify to its accuracy."

The classical authority on sport in all its branches in Great Britain is The Badminton Library. This is a compilation of volumes produced under the management of the late Duke of Beaufort and of Mr. Alfred Watson, editor of The Illustrated Sporting and Dramatic News. These gentlemen spared no pains in obtaining the services of writers whose names are of worldwide reputation on the particular subjects of which they treat, including every branch of athleticism. On page 103 of the volume by Lord Walsingham and Sir Ralph Payne Gallwey, in speaking of grouse being a nuisance on deer forests, as giving the alarm to stags (please remember that a deer forest in Scotland does not imply the presence of trees) occur these words:

"The simplest way of getting rid of them is not to shoot them. If left to themselves the old pair will take possession of a large hillside, establish fixity of tenure, with no squattings allowed for their own or for any other young birds. The result is the speedy diminution of their number."

Major-General R. T. Dashwood's name as a sportsman is known from the Himalayas to the Rocky Mountains. He is a constant contributor to The Field and other journals and magazines devoted to the interests of shooting, fishing and sports generally. He has

A Pack Train in the Klondyke.

KILLED BIG GAME

in at least three continents, but the country of his choice for the last 30 years has been Canada, which he has fished and hunted from the Atlantic to the Pacific. He is, perhaps, the only white man living to day who thoroughly understands "calling" a moose, and one of the finest trophies in the Naval and Military Club in London is the head of a moose which he slew by this means in Nova Scotia. In a letter to me, written last fall, he says:

"It is a mistake not to permit killing grouse (prairie chicken) before October 1, when the herds are packed and so wild as to render dogs useless. September 15 would be a much better date, in the interest not only of sportsmen, but of the stock of game. Nothing is so

INJURIOUS TO BREEDING

as leaving coveys unbroken at the end of the season. The law which makes it illegal to shoot ducks after sundown and so prevents flight shooting, when the birds come in to feed in certain places and are mostly rocketers and splendid shots, high over a man's head, is simply childish, and is, of course, ignored."

A similar alteration is required in the legislation which regulates the killing of big game. On this subject, I should like to quote the words of Mr. F. W. Stobart, who is a personal friend of mine, and probably has had more experience in killing moose and wapiti by "still hunting" than any white man in Manitoba:

"We open the season for killing moose, wapiti (or elk), etc., on October 15, just as the rutting season for these animals is about closed, when after nearly a month's running with the cows and fasting (for the bulls of the moose and wapiti will eat practically nothing at that time), the meat of the bulls is so rank and tough as to be practically unfit for human food. The cows alone are then in good condition, but, surely, if we wish to protect the game of this country, we should prevent the killing of cows altogether."

On this point General Dashwood says: "The heads of all deer are hard and stripped of the velvet by September 7, and the meat is in excellent condition for food. By far the most difficult period for hunting moose, elk, wapiti, etc., is when there is no snow on the ground, and the leaves are more or less on the trees; but it is much the pleasantest time to camp out in the woods, and, with regard to moose, is the season for "calling" that animal, an art hardly understood outside Nova Scotia, but an exceedingly exciting, though very uncertain kind of sport, as success depends not only on the caller, but on the weather, which must be absolutely still. Calling only lasts till about October 10. Yet, these two months, September and October, up to the 15th of the latter month in Manitoba, and to the 31st in the Northwest Territories, are included in the close time. On the other hand, later, when there is snow and the hunting is much easier, the time is open, which appears to me to be hardly logical or commonsense, having regard to the climate and habits of these animals. I should say that

A FAIR OPEN SEASON

for all kinds of deer would be from September 7 to December 7. This, if properly enforced, with a limit of two animals per species per gun, would not do any harm;

Henry Rogers, Sons & Co.

Wolverhampton, England.

Manufacturers of

"Union Jack" Galvanized Sheets
Canada and Tin Plates
Black Sheets
Sleigh Shoes and Tyre Steel
Coil Chain, Hoop Iron
Sheet and Pig Lead
Sheet Zinc

Quotations can be had from Canadian Office:
6 St. Sacrament St., MONTREAL

F. A. YORK, Manager.

KNOX HENRY

Heavy Hardware and Metal Broker
No. 1 Place Royale, MONTREAL.

Samples sent free on application.

HORSE NAILS—"C" Brand Horse Nails—Canada Horse Nail Co.
"BRASSITE" GOODS—Gunn Castor Co., Limited, Birmingham, Eng.

Manufacturers of

Heating Supplies

Pipe Fittings and Headers.
Large Manifolds made to Order.
Steam Traps and Appliances, etc.

The . . .
Jas. Morrison Brass Mfg. Co., Limited
TORONTO.

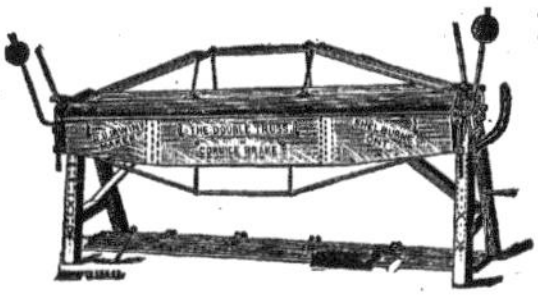

This eight-foot Brake bends 22-gauge Iron and lighter, straight and true.

Price, $60

Very handy beader attachment, $15 extra if required.
Send for circulars and testimonials to

The Double Truss Cornice Brake Co. SHELBURNE, ONT.

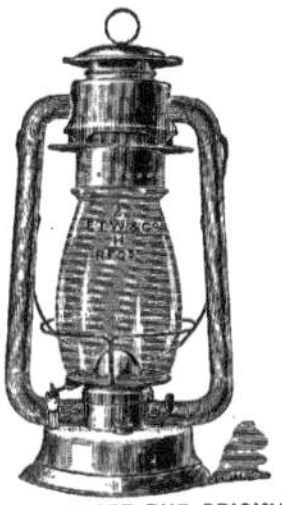

If you want the BEST
Cold Blast Lantern
in the world, get WRIGHTS.

—GET THE ORIGINAL—
We lead, others imitate.

E. T. WRIGHT & CO.
Manufacturers. HAMILTON, ONT.

The Latest and Best.

H. & R. Automatic Ejecting Single Gun.

Steel and Twist Barrels in 30 and 32-inch.
12 Gauge.

Model 1900.

Harrington & Richardson Arms Co.
Worcester, Mass., U.S.A.
Descriptive Catalogue on request.

STEVENS FINE TOOLS

We make a perfect line of

CALIPERS and DIVIDERS

Also such tools as Surface Gauges, Tool Makers' Clamps, Center Punches, etc.

Write for our New Catalogue containing a description of our Tools. It is also a valuable hand-book of information for mechanics and people interested in such lines.

J. Stevens Arms & Tool Co.
P.O. Box 216, Chicopee Falls, Mass., U.S.A.
Carried by our representatives at Toronto and Montreal.

TRADE MARK J REGISTERED.

"JARDINE" HAND DRILLS

Five different sizes to suit all requirements. It pays to sell the best tools.

A. B. JARDINE & CO.
HESPELER, ONT.

HUTCHISON, SHURLY & DERRETT

DOVERCOURT TWINE MILLS.

1078 BLOOR STREET WEST
TORONTO.

Having equipped our Factory with entirely new machinery, we are prepared to furnish the best made goods in the market at closest prices and make prompt shipments.

Hand Laid Cotton Rope and Clothes Lines,
Cotton and Russian Hemp Plough Lines, plain and colored.
Cotton and Linen Fish Lines, laid and braided.
Netted Hammocks, white and colored, Tennis and Fly Nets.
Skipping Ropes, Jute, Hemp and Flax Twines.

yet, if the time is considered too long, curtail it at the end, not at the beginning of the season.

"Again, the same reasoning with regard to breeding holds good in the case of the the deer tribe as well as that of grouse, for, as a matter of fact, with all animals that herd, it does good to kill off some of the big stags (which, alone, of course, a true sportsman cares to kill), as these big stags, in the case of cariboo and elk (so called wapiti), dominate herds and keep off younger, though for breeding purposes quite as efficient, animals, with the result that there are numerous barren does."

Another well-known sportsman, Major C. S. Cumberland—author of "Sport in the Pamirs," a constant contributor to The Field, who has penetrated into Central Asia where no other white man has ever been, and has probably forgotten more about the jungle than Rudyard Kipling (a personal friend of his) has ever learned—was travelling in this country last year. I have known him intimately all my life, and, again and again, in private conversation, he corroborated all I have quoted as insisted on by the authorities mentioned above. It must be remembered that these men are all

SPORTSMEN IN THE TRUE SENSE

of the word. They are "head-hunters," not hide-hunters or meat-hunters. They are fond not only of hunting, but of the quarry they hunt. Let me relate a little incident which occurred to one of them, whose name I shall not mention. His hunter, an Indian, after a long and arduous chase, brought him at last within sight of a couple of bull wapiti, which were fighting. These animals were so engaged that they allowed their pursuers to come close up to them and stand in full sight for some minutes before they took any notice of their presence. "Shoot, shoot!" whispered the Indian excitedly, but his companion never stirred. He wanted a five year-old, and neither of the combatants was more than four. He stood perfectly still, and watched the fight to a finish. Then he drew a deep breath of satisfaction, and said: "I've seen a good fight, and that was worth the hunt half a dozen times over." His guide to this day regards him as a kind of inspired lunatic, and is quite unable to understand the feeling that prevents a man from committing what seems to him to be wanton slaughter. These are the men who, I say, if

PROPERLY ENCOURAGED,

would come to this country, spend money here, and spend it freely; for to them hunting is the highest form of recreation. They would not interfere with our shooting; with more leisure, and more means than most of us, they would probably penetrate farther afield; and to them the idea of mere killing for the sake of killing would be utterly repugnant.

There are stories afloat of men coming to this country in private cars, and shooting chicken by the hundreds, leaving their bodies to rot on the ground. How far these stories are true, and how far exaggerated, I cannot tell, but I am sure that, at all events, they are of rare occurrence. Such men as these should be kept away at the point of the bayonet, if necessary. Limit the number of head of game to be shot by any one individual, if you will. The provision that "no person shall kill more than 100 chickens in one season, or more than 20 in one day," is a wise one, if properly enforced.

PUT A PROPORTIONATE LIMIT

on the deer tribe; but open the season earlier. In British Columbia the season for big game opens on August 1, in Wyoming on September 1, in Washington on the same day, and in Oregon on August 1. Let us have a chance of shooting chicken over dogs, instead of forcing us to begin when dogs are useless, and birds so wild as to be practically unapproachable, and so heavily feathered as to be almost unshootable. We have here one of the finest game countries in the world; let us make the most of it. I repeat that I am informed on the best authority that the legislation is to be altered this year. The sooner this is done, the better, and, above all, let the farmers and settlers understand the reason why it is done—that the change is made by the advice and with the support of men who not only have made a lifelong study of the preservation of game, but who also have the truest interests of the country at heart.

THE OFFICE HABIT.

WE are told that "the business man is naturally methodical, a born organizer," says Graphite. "His executive powers are out of proportion to his other gifts. Naturally, such a man takes pleasure in the detail of business. Instead of getting up at 10 o'clock and going out into the street hustling for business, his office holds him. He writes letters. He directs this agent and that. He sends messengers here, there, and everywhere to have everything done. This is a very laudable and effective kind of gilt, but if it holds a man in the office writing a latter or note when he could be out seeing his customers face to face, or dealing with men and getting new ideas while pushing his business, he is allowing the office habit to calmly fold him in its embrace; he is bringing about his own business ruin. The office should be worked for all it is worth, but there is no substitute for personal contact with other business men. Sitting at a desk the view grows narrow. It is bounded by the four walls of the office. It is useless for any one to expect to know what is going on in the world without going into that world. There is no substitute for going out and seeing with one's own eyes what other people are accomplishing and how they are doing it."

A TIN SHOP TIME CARD.

At the recent meeting of the Iowa Retail Hardware Association a time card, the following of which is a reproduction, was

TIME CARD. TIN SHOP.

J. F. RIS & BRO.,

Tinner............Dubuque, Iowa..........190

Time.

7

8

9

10

exhibited by Mr. J F. Ris, Dubuque. One of these is given to each man every morning. On it he enters the job on which he works, giving the time when he begins and when he gets through. Every night these cards are turned into the office and the proper entries are made from them. Thus the cost of every piece is known. His firm do a great deal of job work, but also make oil cans, wash boilers, air tights, gasoline ovens, etc., keeping four men busy. They get a better price for their wash boilers than for factory boilers, a good man making a boiler in an hour, to whom wages of 25c. per hour is paid.

The sincere, hard-working merchant, who gives scant attention to advertising, never gets but "so far." The advertiser who hasn't the right stuff behind his words—he never gets but "just so far," either. But the merchant with the right goods, who advertises them the right way—the bright way—he grows fast; and his name is sooner or later entered in the book of the world's successes.—Fame.

A. R. WOODYATT & CO.'S NEW FACTORY.

A. R. WOODYATT & Co., of Guelph, Ont., whose reputation for turning out reliable goods is so well known throughout the Dominion, commenced the 20th century in their new premises, an illustration of which is here shown.

This firm's business was started in a modest way in 1887, under the name of Auld & Woodyatt. In 1892, owing to a change in partnership, the firm name was altered to its present style. A. R. Woodyatt & Co.'s trade steadily increased, until, after building additions upon all the available space of the property then occupied by them, they eventually found themselves unable to properly handle their business, and were forced to look for new premises.

Although in receipt of some tempting offers to move to other places, they decided to stay in the city of their first venture, and purchased the property lately owned by The Guelph Woollen Mills Co., which included not only a large solid stone factory, but also about 8¼ acres of land, which would insure them from congestion for the future.

The factory has been entirely remodelled, one of the finest foundries it is possible to find has been built, and a private switch from the C.P.R. runs past the shipping room to the buildings especially erected for the storage of the iron, coke, coal and sand used.

In addition to equipping it with the newest and most improved machinery, under the immediate supervision of Mr. A. R. Woodyatt, who is a thoroughly practical mechanic, every appliance for the saving of labor has been secured, and, as every department has now got into smooth running order, Woodyatt & Co. claim they are now in a position to turn out their goods equal in every respect to any made by manufacturers of the same lines.

Besides manufacturing general cast hardware, they make a number of specialties, such as screen door hinges, steel barn door hangers and track, and also contract for special castings in grey and malleable iron and brass, and finish their goods not only in the ordinary japans, but also in the various plated finishes now coming so much into use.

But their leading line is lawn mowers. These are sold not only in the Canadian market, but also nearly all over the world, constant orders being received from far-off New Zealand and Australasia, and a large number are sold in Great Britain and other European countries by a representative of the firm who makes an annual trip over there.

HARDWARE AND METAL joins with their numerous other friends in wishing A. R. Woodyatt & Co. the same steady growth in the future as in the past.

THE SNOWSHOVEL.

MANY of us are old enough to remember when the only kind of snow-shovel to be had was made with a wooden blade. As in so many other lines, from battleships to mouse traps, the old wooden article is being replaced by its modern up-to-date successor manufactured from steel. This principle is illustrated in connection with the manufacture of snow-shovels which are now made almost exclusively of steel.

Some Canadians object to our country being called "Our Lady of the Snows," but the fact remains that for four or five months of the year all parts of our Dominion are visited with frequent and sometimes heavy falls of "the beautiful snow." The snow-shovel is therefore a necessity in this country wherever there are doorsteps, sidewalks or railroad tracks.

Last year the Eclipse Office Furniture Co., of Ottawa, began the manufacture of a superior line of steel blade snowshovels finished with a heavy coat of black enamel. Their shovels were so popular with the trade that they could not supply the demand.

They are preparing for a largely increased output for next season and will have a line of about a dozen different varieties, from the small child's snowshovel to the extra heavy line specially for railroad use, all with the finish which made their goods so attractive last season. The Eclipse Co.'s advertisement is on page 2.

DOOR PULLS.

No. 250—JAPANNED

No 300—ROYAL BRONZE
PACKED WITH SCREWS

SPRING HINGES.

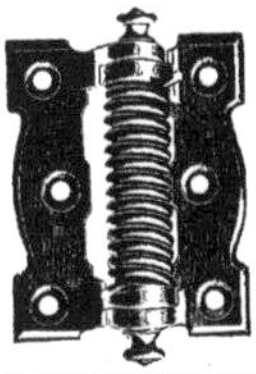

NON-HOLD-BACK PATTERN
No. 51—JAPANNED

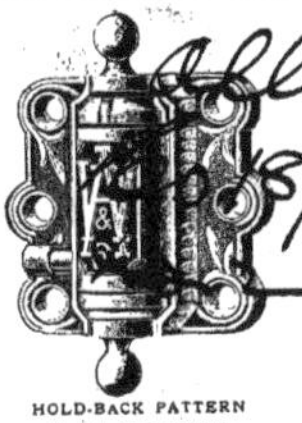

HOLD-BACK PATTERN
No. 20—JAPANNED
No. 120—OLD COPPER
No. 120—PACKED WITH SCREWS

HOLD-BACK PATTERN
No. 5—ROYAL BRONZE

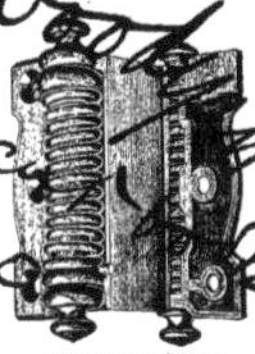

DOUBLE ACTING
No. 50—JAPANNED

LAWN MOWERS.

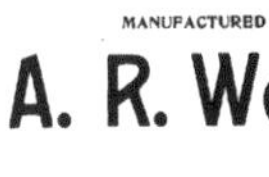

THE "STAR"
Sizes: 11-in., 14-in., 16-in. Cut.
9-in. Drive Wheel.

THE "PHILADELPHIA"
Sizes: 10-in., 12-in., 14-in., 16-in. Cut.
7½-in. Drive Wheel.

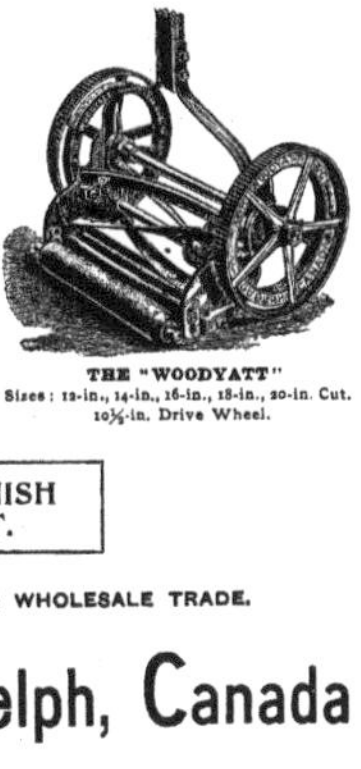

THE "WOODYATT"
Sizes: 12-in., 14-in., 16-in., 18-in., 20-in. Cut.
10½-in. Drive Wheel.

ALL OUR GOODS IN QUALITY AND FINISH
EQUAL THE BEST IN THE MARKET.

SOLD ONLY THROUGH THE WHOLESALE TRADE.

MANUFACTURED BY

A. R. Woodyatt & Co., Guelph, Canada

FRS. MARTINEAU, President Montreal Retail Hardware Association.

D. DRYSDALE, 1st Vice-President. A. MAGNAN, Secretary. A. PRUD'HOMME, Treasurer. L. J. A. SURVEYER, 2nd Vice-President.

THE RETAIL HARDWARE ASSOCIATION OF MONTREAL

A Sketch of Its History and of the Men who Guide It.

ONE of the most important moves made in the retail hardware trade during the past year has been the organization of an association by the retail hardware and paint dealers of Montreal. The birth of the association was purely indigenous; it stands without a relative in this country. It has a sheer origin and is wanting a pedigree. But from the first it has shown signs of strength and exceptional vitality, and it looks as if it will be as Adam and Eve have been to the human race, the first of a new species.

Many a hardware merchant throughout Canada is watching the progress of the organization and its accomplishments, wondering if such an association can be profitable to its members either financially or socially. So many of these trade organizations spring into a transient existence. Like the early spring flowers we pick in the woods, they blossom forth with a seeming superabundance of vitality to stand any ordinary hardship, but do not stand the tests of the seasons. The Montreal Hardware Association has blossomed forth, but, far from losing vitality, it goes on increasing its fund of energy. It is teaching a valuable lesson to Canadian retailers, showing them how they can improve their own condition and the condition of the trade by simply acting upon that old, trite, yet none the less truthful statement: "In union there is strength."

The head of the Hardware and Paint Dealers' Association of the city and district of Monteal is Mr. Francois Martineau. He was the real founder, and to-day no one denies him the honor. He is one of the largest hardware dealers in the city, and in his business career of 31 years he has learned how business should be conducted. During the last few years he has recognized that various grievances have been creeping into the trade, and the idea entered his mind that some at least of these could be removed by a union of the retail interests of the city. He had seen the wholesalers and the manufacturers combine successfully and he naturally entertained the idea that the retailers might likewise improve their status through the agency of an association. He waited on several of his confreres, and, although some of them threw him down, ridiculed his scheme as impracticable and asserted his wish to be father to his thought, he persevered for about two years before he felt confident enough to call a meeting of the trade. He then sent out the following card to all the retail hardware merchants of Montreal:

DEAR SIR,—I have the honor to invite you to be present at a meeting of the retail hardware and paint dealers, which is to be held in Monument National, room 4 (St. Lawrence street), Wednesday evening, September 19, at 8 o'clock.

The object of this meeting is to take counsel as to the best means to adopt to safeguard our common interests.

Hoping that you will consider it your duty to come to the meeting.

I have the honor to be, sir,
Your obedient servant,
FRS. MARTINEAU.

Montreal, Sept. 15, 1900.

Messrs. Lecours, Surveyer and Prud'homme were among the first to join in heartily with the movement. Assisted by these gentlemen, Mr. Martineau called upon nearly all the merchants, and set the case so strongly before them that they attended the first meeting in large numbers. About 60 were present, and the first thing they did was to accept Mr. Martineau's

proposition to form an association to be called the Retail Hardware and Paint Dealers' Association of the City and District of Montreal. The following officers were elected: President, Mr. Frs. Martineau; 1st vice-president, Mr. D. Drysdale; 2nd vice-president, Mr. L. J. A. Surveyer; treasurer, Mr. A. Prud'homme; secretary, Mr. A. Magnan.

A committee was appointed to draft a constitution, and the results of their labors appeared in the issue of HARDWARE AND METAL of November 3, 1900. The first evening fees were received 15 members paid up. The first year's fee is $6, and after that the annual amount of subscription is $5. At present there are 48 members on the roll, forming a thoroughly strong and deliberative organization. It is the intention to apply for Parliamentary incorporation. As nearly all the members are French, the proceedings generally take place in the French language.

The objects of the Association have been dealt with in various issues of this paper. The first and foremost aim is to create a spirit of fellowship and friendship among the members, to apply a lubricant to remove any friction existing between the different retailers in the city, and to help one another by the discussion of trade questions. That this aim has already been attained is a matter beyond dispute. Oftentimes have members of the association said to a representative of this paper that "It has been worth the fee to have these fellowship meetings, even if the association should accomplish no ulterior purpose. Before I met my confreres here, I did not know ten of them, and always regarded them with suspicion and jealousy. Now I can shake them by the hand in a spirit of trust and confidence."

Fellowship breeds financial profit. One retailer was heard to make these remarks one evening after a meeting: "I made my whole year's fees to-night at the meeting, for in conversation I find I have been paying too much for such and-such a line of goods." Then, with a flourish, he added: "Who says our association is a useless or needless organization?"

The next most important aim of the association is to induce the wholesalers to discontinue selling retail. They claim that their customers can go down to the wholesale establishments and buy goods as easily as they can themselves. There is no doubt they will stop this custom—in time; and it will be a success worthy of attainment.

On medium and large sales of Canadian manufactured goods the retailers claim that for one reason or another they make no profit. This they will try to remedy by having the manufacturers fix a price for them and allow them a rebate just as they do for the wholesalers. At present the small merchant makes nothing on the sale of a keg of nails, and they wish to fix it so he will make at least 25c.

The manufacturers of paper allow a man who buys $3,000 worth in a year, a rebate of 15 per cent.; this system they wish extended. This desideratum they will not have granted to them for some time, but the members are patient and persevering and they believe they can obtain it to the benefit of themselves, the wholesalers and the manufacturers.

The officers are to be congratulated upon the eminent success that has attended their efforts so far. They have proven themselves thoroughly capable of leading the association through dangerous paths. They are nearly all men of long experience. Mr. Martineau has been in the hardware business for more than 30 years, and is one of the most widely known men in his section of Montreal. He has grown up with that part of the city. Although he is a keen business man, one could not meet a more cordial or genial gentleman. For five years he was a member of the Quebec Legislature, where he adorned the Conservative side of the House. He was also a member of the city council for six years. In connection with his retail business, which he handles at 1381 St. Catherine street, he does quite a nice little jobbing trade with smaller retailers.

Mr. A. Prud'homme, the treasurer, is senior member of the firm of A. Prud'homme & Frere, trading at 1940 Notre Dame street. He has been in business somewhat over 20 years and has one of the largest retail businesses in the city. Mr. Prud'homme is a man of a deliberative turn of mind, yet is keen, and very popular in his social circles.

Mr. D. Drysdale, the 1st vice president, is the English member of the executive. He is a man of weight in the trade, and has worked up a large and profitable business at his store at 645 Craig street. As right-hand supporter of Mr. Martineau, Mr. Drysdale has filled his position with eminent success.

The 2nd vice-president is Mr. L. J. A. Surveyer, one of the active founders of the association. Mr. Surveyer has been in the hardware business in the city for 35 years, and has not failed to gain some rich experience in that time. He is a clear thinker, and his voice carries weight in the meetings of the association. He does a high-class trade in shelf goods at 6 St. Lawrence street.

The hard-working secretary is Mr. A. Magnan, of Magnan Freres, 306 8 St. Lawrence street. He is enthusiastic over the welfare of the society, and knows his duties from the ground up. Mr. Magnan learned his business with Jas. Walker & Co. and the Pallascio Hardware Co., and since he started his business three years ago he has proved himself a hustler, and, with his brother, has worked up a good trade.

IN THEIR THIRTY-SEVENTH YEAR.

POSSIBLY there is not another manufacturing concern in Canada better and more favorably known to the trade than The Parmenter & Bulloch Co., Limited, Gananoque, Ont. It is now 37 years since they commenced manufacturing in this busy town, and, due to their untiring zeal and progressive principles, it is long since that they have been recognized by the trade as among the leading makers of wire nails and rivets of every description. To this firm can be credited the introduction of wire nails in this country, and they have always maintained a high position in this industry. On page 6 of this number will be found their advertisement illustrating several lines of their manufacture, and particular attention is called to the "Canada" riveting machine which they have recently placed on the market with remarkable success. Attention is also called to the fact that they are now manufacturing the "Saxton" ratchet brace, which, owing to its neatness in construction, durability and simplicity, is sure to command a preference over other makes of ratchet braces. We would also remind the trade that this firm are the manufacturers of the patent bifurcated or slotted and tubular rivets. Hardwaremen will avoid trouble with their customers about quality when they buy Parmenter & Bulloch's goods.

HOW STOCK IS KEPT FRESH.

"The hardest thing about the retail shop business is to keep from buying more than you need," said a retailer. "Some years ago I awoke to the fact that I had considerable good money tied up in unsalable stock. I was buying carefully enough, as I thought, but the stuff would accumulate. The trouble was I could not resist a good bargain, or what I thought was a bargain and would please my trade. I would look at the shelves already pretty well stocked, but say to myself, 'Oh, well, it's a long time before I'll get them—three or four months—and by that time I can surely use them all right,' and down would go the order. Perhaps when the goods came I had 'others' and didn't need them nearly as much as I was going to; in short, I could have got along very nicely without them. So I made up my mind never to buy an article unless I absolutely had to have it—couldn't get along without it, and I ceased to speculate so far in the future, and the results have been a surprise to me."

W. W. GREENER'S CELEBRATED PIGEON GUNS

HOLD THE WORLD'S RECORDS AND WIN PRIZES EVERYWHERE.

THREE TIMES Winners of the GRAND PRIX DU CASINO and
TWICE of the TRIENNIAL CHAMPIONSHIP at Monte Carlo.

A Fortune Won With a Greener Gun.

In 1896-7 a well-known Pigeon shot won upwards of £3,500, including Cups, with his Greener Gun.

In 1894 Mr. J.A.R. Elliott made THE WORLD'S RECORD, killing 199 birds out of 200, in a two days' match with Dr. Carver, at 30 yards rise, 50 yards boundary.

From 1878 to 1901, over 120 PRIZES have been won at MONTE CARLO, worth upwards of £10,000, by shooters of all nationalities with Greener Guns

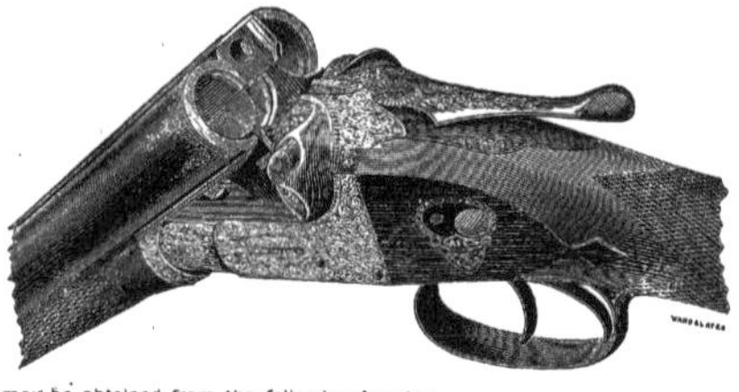

Greener's "Facile Princeps" **Hammerless Gun Price, £16.16.0.** (At Works, Birmingham, Eng.)

THE BEST VALUE FOR MONEY

IN THE WORLD.

It Is a Thoroughly Sound, Well Finished and Reliable Gun, and Its Shooting Qualities

Cannot be Surpassed.

Forester Hammer Gun Price, £13.13.0.
Dominion Hammer Gun Price, £9.9.0.

GUNS may be obtained from the following Agents:

MONTREAL T. W. BOYD & SON, NOTRE DAME STREET.
WINNIPEG HINGSTON-SMITH ARMS CO.
TORONTO RICE LEWIS & SON. AND OTHERS.

W.W. GREENER, ST. MARY'S SQUARE, BIRMINGHAM, ENGLAND

WM. CROFT & SONS.

ESTABLISHED 1855.

Manufacturers and Direct Importers of

Fishing Tackle

RODS
REELS
LINES
HOOKS

LARGEST and most COMPLETE STOCK in the Dominion.

FLIES
CASTS
BAITS
HAMMOCKS

DEALERS write for Illustrated Catalogue. Also direct importers of Smallwares and Notions.

53 Bay Street, Toronto, also Winnipeg and Quebec.

THE B. GREENING WIRE CO.
LIMITED

Office and Works:
HAMILTON, ONT.

Eastern Depot:
422 St. Paul Street
MONTREAL.

ESTABLISHED 1859

A Few... SEASONABLE LINES:

WIRE
Steel, Plain and Galvanized. For hay-baling, fencing, coppered spring, coppered soft, tinned, mattress and broom, galvanized hard coiled fence wire.

GALVANIZED NETTING
for all purposes. All meshes, widths and strengths in stock.

PAINTED SCREEN WIRE CLOTH
All widths, 18 to 48-inch, kept in stock.

WIRE CLOTH
Copper, Brass, Steel and Galvanized. All meshes and strengths, for all purposes.

TRACE CHAINS
Improved quality and cheaper price for 1901.

Brown's Patent Steel Wire Chains
Coil, tie-out, halter and dog chains, etc. Special chains made to order.

Greening's Patent Cattle Chains
and Stall Fixtures. Samples now ready for Fall trade.

WIRE ROPE
Standard and Lang's Lay. For derrick use, passenger and freight elevators, mining, tramways and other purposes.

Coppered Steel Furniture Springs
All standard sizes kept in stock.

STEEL WIRE DOOR MATS
Most sanitary and durable mat on the market. Made either plain or lettered.

PERFORATED SHEET METALS
Zinc, Copper, Brass, Steel, Galvanized Iron. For all purposes.

FOUNDRY SUPPLIES
Riddles, Steel Brushes, Bellows, Shovels, etc.

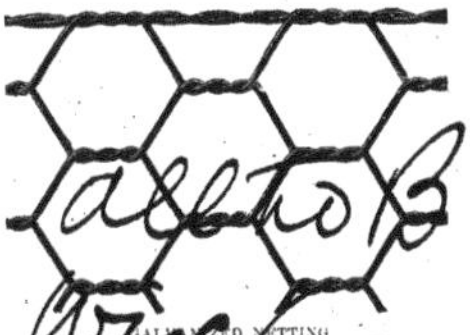

GALVANIZED NETTING

CHAINS—BROWN'S PATENT

CATTLE CHAINS

WIRE CLOTH

WIRE ROPE

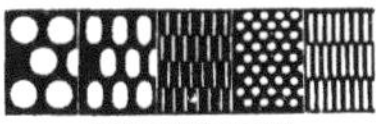

PERFORATED METALS

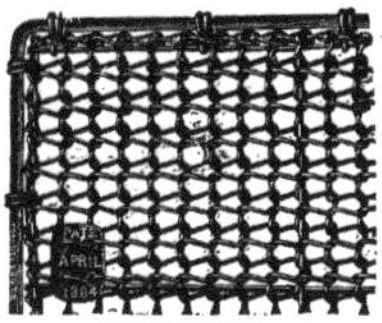

WIRE DOOR MATS

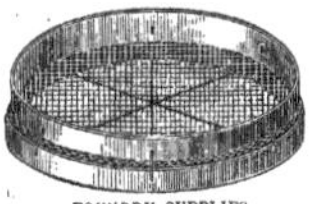

FOUNDRY SUPPLIES

Our last Catalogue is dated January 1900.
For Quality of Material and Workmanship, our goods are STANDARD IN THE MARKET.

LATEST FOOT POWER GRINDING MACHINERY.

WITH the ever-increasing demand for labor-saving implements in all branches of industry, the introduction of patented foot power grinding machinery by the Union Manufacturing Co., 20 Breckenridge street, Buffalo, N.Y. which has the distinction in this line of being the largest manufacturers on the continent, is not without interest.

The illustration here presented is known as their new model No. 2, with drilling attachment, a practical tool for all light grinding or drilling. The machine is furnished with two high-grade emery wheels, 6 x ¼, ready to grind. The pinion is cut from solid steel, engaging with the solid cast drive gear, operated with their ball clutch. The machine stands 42 in. high,

has 1 in. tubing for standard, cone bearing, speed is three to four thousand. The drilling attachment has for its bed a formed steel rod attached to socket inside of head, the movable tail stock sliding on bed fastened with thumb screw when adjusted to thickness of metal to be drilled. The chuck is a standard make, taking drills from 0 to ¼, and having a bushing that fits on the end of the wheel arbor. The drilling attachment can be procured any time and applied, as the machines are all made on the duplicate system.

The company manufacture a full line of these tool grinders, No. 1 being much improved, now furnished only with solid steel gear throughout.

No. 3 is a popular tool with all classes of mechanics. They also manufacture dentists' and jewellers' lathes, also foot and hand power punches, patented screw drivers, leather chisel handles, "Perfection" raisin seeders, can openers, etc.

The company occupies the large three-storey brick building at the above address, fitted throughout with the most modern machinery for the production of these goods. They also have a branch at Fort Erie, where Canadian orders are filled from.

The company will have an elaborate display of their lines at the coming Pan-American Exposition, and will be pleased to give any information required regarding these goods.

Their advertisement appears in another column of this issue, which will prove attractive to jobbers and the trade generally.

A PAINT FIRM'S GROWING TIME.

THE Standard Paint and Varnish Works Co., Limited, of Windsor, Ont., have made rapid strides since buying out the Canadian branch of the Acme White Lead and Color Works some years ago, being now well to the front in the paint and varnish business.

We all know the truth of the old proverb, "Honesty is the best policy," and evidently The Standard Paint and Varnish Co. are firm believers in its efficacy. Its management started out with the determination to give honest value for the money paid them, and the result is a rapidly-increasing list of customers, who know that they are fairly dealt with and that their interests are one with the interests of the firm with whom they are dealing.

The firm has lately increased its representation on the road, and they have now a branch office in Toronto. Additions have also been made to the factory and office buildings.

While all of their lines are giving splendid satisfaction, their "New Era" prepared house paint is a leader, and is in great demand. Its good wearing qualities and beautiful finish make it a favorite with all who desire beauty and durability in paint combined with ease of application. Their varnishes have also attained a first-class reputation throughout the Dominion, and we learn that this firm not only import large quantities of raw materials, but that they also can lay claim to a good share in the export trade of the past year.

In short, we can sum it all up by saying that their success in gaining customers has been satisfactory, and it shows that this firm is under the control of an enterprising and shrewd business management.

WIRE NAIL INDUSTRY FOUNDED BY A PRIEST.

IT was in Covington, Ky., that the first wire nails were made in America. In 1875 Father Goebel was pastor in charge of St. Augustine's Catholic Church in that city.

Before he came to the United States from Germany he had seen Frenchmen and Germans hammering nails out of wire. When he had established himself in the ministry at Covington he opened a forge in an old outbuilding standing in a brickyard. He started the making of wire nails first by hand, and gradually one improvement after another came to his mind and was carried out, until the nails made were more useful and could be made more cheaply. Soon after he began he improved upon the old nail by cutting barbs in its side, and by this they were made to hold more firmly. Then, to accelerate his work, he made a die, into which he slipped the wire that had been cut to proper lengths, and while resting on these dies the head was pounded on the nail. On an anvil he hammered on the point, and the barbs were cut in the sides by hand. It was the nail made to day.

It was about this time that the French introduced a machine that would do what Goebel was doing by hand, and as soon as the latter heard of it he imported one of these machines. The introduction of this machine was the real beginning of the wire nail industry into this country on a large scale. And this machine is now, according to The Chicago Record, in Chicago stacked in the attic at the large local plant of the American Steel and Wire Co. Covered with dust, as it is, and stored where it is never seen, it is, nevertheless, one of the epoch-markers of this industrial age, and from this comparative crude device sprung within a few years an industrial plant that is capitalized at $90,000,000, and that is making a good percentage on that large amount of stock. It was a queer machine when it was received, but the principle was right, and the great machines that to-day turn out hundreds of thousands of nails a day are constructed on identically the same plan. It was operated by hand, and the speed was 60 nails a minute. Goebel attached a flywheel, geared it to steam, and by other improvements increased the machine's speed to double this capacity, which was as many as 20 or 30 men working by hand could produce. This was the "single-header" machine, making one nail at each stroke, and this machine produces, with its present improvements, as high as 415 nails a minute, while the double-headers, producing two nails at a stroke, turn out from 550 to 600 a minute, or a total of 30,000 an hour.

The MacLean Publishing Co., Limited

President, JOHN BAYNE MACLEAN, Montreal.

Publishers of Trade Newspapers which circulate in the Provinces of British Columbia, Northwest Territories, Manitoba, Ontario, Quebec, Nova Scotia, New Brunswick, P.E. Island and Newfoundland.

OFFICES:

MONTREAL (Telephone 1255)	232 McGill Street.
TORONTO (Telephone 2148)	10 Front St. East.
LONDON, ENG. (J. Meredith McKim)	109 Fleet St. E.C.
MANCHESTER, ENG. (H. S. Ashburner)	18 St. Ann St.
WINNIPEG (J. J. Roberts)	Western Canada Block.
ST. JOHN, N.B. (J. Hunter White)	No 3 Market Wharf.
NEW YORK	176 East 88th Street.

Subscription, Canada and the United States, $2.00.
Great Britain and elsewhere - - - 12s.
Published every Saturday.
Cable Address; "Adscript," London; "Adscript," Canada.

BRITISH MANUFACTURERS AND THE CANADIAN MARKET.

BRITISH manufacturers and exporters of iron, steel and hardware, have, during the last few years, taken a greater and a more businesslike interest in the Canadian market than they previously did.

In the years gone by a good many of them conceived the idea that Canada was becoming Americanized. And they allowed a great deal of the trade with this country that once was theirs to go, as it were, by default, to the United States. Such events as the preferential tariff and the sending of the two contingents to South Africa caused them to realize that they were basing their calculations on a wrong premise.

But, although there has been a general awakening as to the wisdom of cultivating trade with Canada, the awakening as to methods for increasing the trade is not perhaps so general.

The trade of an individual can only be secured after a knowledge of his requirements has been obtained. It is the same with regard to the trade of a nation, for nations are only aggregations of individuals.

Realizing this a great many manufacturers and exporters have either paid Canada a visit or deputed representatives to do so. But there are a great many who have not.

Letters, circulars and catalogues they send in abundance. But these at best are poor substitutes for the live representative. They show what the manufacturer and the exporter have for sale, but they do not give the firms who send them a knowledge of the goods this country requires. The fact that these circulars and catalogues often refer to goods which are obsolete in Canada proves that.

The steady development of the United States trade with Canada in hardware and metals is largely due to the fact that many of the manufacturers and exporters in that country have representatives calling upon the merchants here as regularly as upon those living under the Stars and Stripes.

Nor is a visit paid to New York, Chicago or some other city in the United States sufficient to acquaint the British manufacturer with the requirements of the Canadian market. Yet, we occasionally hear of a case where the idea seems to obtain that it is sufficient.

Only a short time ago, for example, the head of a British manufacturing firm sailed for Canada via New York. He, however, got no nearer to Canada than New York, for when he arrived there a broker persuaded him to give him the agency for this country.

If the British manufacturer in question was anxious to develop trade with Canada—and he certainly was or he would not have crossed the Atlantic for the purpose of visiting this country—he was unwise in leaving the agency for this country with a New York broker. In the first place, the New York man could not keep in as close touch with the hardware trade in Canada as could a broker resident here. Another drawback, and one by no means insignificant, is the general dislike there is in this country to buy British goods through the medium of a foreign firm.

A great many firms in Great Britain, by the employment of proper methods, are largely increasing their trade with Canada. And many others can do so if they chose to go about it in the right way.

UNWISE INDUSTRIAL SCHEMES.

ZEAL for the welfare of the town in which they are situated often induces business men to support schemes for the starting of industries which are incompatible with the necessities of their respective localities.

Manufacturing industries are only helpful in so far as the towns in which they are started possess natural advantages for them in the way of adaptability of situation in regard to raw materials and convenience to the railway systems of the country. And yet, these matters are apparently frequently not given the slightest consideration. If they are, how is it that the ubiquitous promoter is so successful in inducing municipalities to give liberal bonuses and tax exemptions to industries that meet with such premature deaths?

We have in mind at the moment a town that has a name, but is dead. It became imbued some years ago with the idea that it was the ideal centre for manufacturing industries of various kinds. And the bonuses it gave were in keeping with its opinions. Practically all the industries thus established are dead and buried. The only thing that lives, unsavory odor and all that it has, is the heavy annual tax bill necessary to pay for the "horse" that is long since dead. Because of the high rate of taxation business men there are handicapped in their competition with their confreres in near-by towns. And what they hoped would become an advantage has become a disadvantage.

Just now the industrial promoter has taken a new lease of life in Canada. Iron works, cordage works and beet-sugar factories are his specialties. Scores of towns and villages are imbued with the idea that they are the ideal spots for one or all of these industries. Some of them are, no doubt, favorably situated for one or more of them. Others, again, are not.

Business men, whose influence is great, should exercise as much care and thought in the matter of assisting new industries as they would matters appertaining to their own business. This would greatly tend to minimize the dangers to which we have referred.

THE IRON INDUSTRY IN CANADA.

IN another part of this issue we print a special article dealing with the iron and steel industry of Canada. While, according to the figures there given, the output of pig iron last year was only 88,441 tons and that of steel rather less than 25,000 tons, it must be evident to even those who have considered the matter in a cursory way that the iron and steel industry of this country is on the eve of a development that five years ago the most sanguine could scarcely have anticipated.

There were in operation last year, either for the whole or part of that period, five blast furnaces and two steel plants. Within the last three months two more blast furnaces have been started. Before the year closes this number will be augmented, three more furnaces being under course of construction at Sydney, N.S., and one at Sault Ste. Marie. The Cramp Company, which is to establish iron and steel works in Collingwood, expects to be making pig iron by the beginning of 1902.

By the end of 1901 we may therefore confidently expect to see about double the number of furnaces in operation that there were at the close of 1900.

Then, besides the increase in the number of blast furnaces, there is the augmentation of the steel plants that will be experienced before the year closes. The Dominion Iron and Steel Co. is constructing ten 50-ton open-hearth furnaces at Sydney, and promises to be shortly producing steel blooms at the ratio of 60,000 tons per annum. At Sault Ste. Marie, Ont., Mr. Clergue's company has under construction a plant for the manufacture of nickel-steel, which will turn out 600 tons a day.

That we are, therefore, on the eve of a marked development in the iron and steel trade of the country is beyond speculation. It is an obvious fact.

The iron industry has been a long time coming to this stage, for it is over 160 years since, under the old regime, the first furnace was started in Canada. But patience is at last getting its own reward.

Although for generations it has been known that Canada was rich in iron ores, within the last few years discoveries have been made which have greatly increased the possibilities of this country as a producer of iron and steel. The Helen and other similar mines in the northern part of Ontario, discovered within the last few years, transcend in richness and in vastness anything that this country has yet possessed in the way of ore supplies. The enormous and easily accessible ores of Bell Island, Newfoundland, from which the furnaces in Nova Scotia are drawing their supplies, while unfortunately not in Canada are just about as conveniently situated as if they were.

In his address before the Manufacturers' Association in Toronto a few weeks ago, Mr. A. J. Moxham, the manager of The Dominion Iron and Steel Co., showed that whereas it costs $3.57 to assemble at Pittsburg the materials, such as ore, fuel and lime, necessary for making a ton of pig iron, at Sydney the cost is but 79½c., and at the furnace at Sault Ste. Marie it will be $1.97. Then, there is the matter of proximity to the European market. Sydney on the seaboard and Pittsburg 500 miles from it, the advantage of the former over the latter is undoubtedly material.

It is significant that the men who were the first of recent years to lead in the development of Canada's iron industry are citizens of the United States. If foreigners possess such confidence in the iron possibilities of this country, it would be an anomalous state of affairs indeed if the citizens of the Dominion did not.

COURTESY TO TRAVELLERS.

THE quality of courtesy should not be reserved to the home circle and for social acquaintance, but it should be incorporated into every field of a man's activity, into every relation of life. Courtesy should be as much a characteristic of the business man as of the social or political leader, for not only will it serve to cement friendship, but will add much to the enjoyment of life generally.

There are some churls who have been successful in business, some mean men who have accumulated large wealth. But they are in the minority and their success has been, as a rule, in spite of rather than due to their lack of courtesy.

The majority of merchants consider themselves courteous. They have become accustomed to greet their customers pleasantly, to serve them carefully and to receive any complaints or criticisms with due respect. But the true test of a merchant's courtesy is not his manner toward those to whom he sells, but his treatment of his help and of those he buys from.

One of the greatest difficulties that the commercial traveller has to contend with is the waste of his time, often occasioned by merchants compelling him to wait while they do innumerable things which might well be left for the short time necessary to talk to the traveller. This action is often due to carelessness on the part of the retailer; but, to whatever it may be due, it should not be.

The courteous merchant will make use of the first opportunity to find out what the traveller wants. If he can give an answer off-hand he will do so. This should be considered final by both parties, and if the merchant is busy the traveller should not wait longer. If an off-hand answer cannot be given, the courteous merchant will inform the traveller when he expects to be able to deal with him. If an appointment is made, honesty, as well as courtesy, demands that it be kept by both parties. In any event, the wise merchant will never forget that the traveller's time is precious as is his own, and will not cause any waste of it that he can by any means avoid.

There are undoubtedly some merchants who are mean in spirit and who purposely subject travellers to indignity as well as delay. Such men lose a great deal more than they gain. They soon are classified by travellers as mean, and never receive any of the advantages which "the knights of the grip" are frequently able to offer their best customers. And when adversity comes they realize that while they might have been making friends they have been hardening the hearts of men who are then not ready to concede anything, and to be satisfied with nothing less than cold justice. Courtesy, like charity, never fails.

DISCRIMINATION AGAINST BRITISH GOODS.

MONTREAL importers claim that the railways are charging exorbitant freight rates on certain lines of goods coming from England to Canada.

The conference rate on ingot tin from Liverpool to Montreal is 31s. 6d. in less than carload lots; on Canada plate, the rate is only 24s. Why this difference? Even on pig lead, much more difficult to handle, the freight charge is only 26s. 6d; on hoop iron, it is 26s. 6d.

Could it not be that the American trusts, who give the steamship companies much business, might induce them to formulate their charges according to their wishes? There is a grave suspicion in Montreal that such is the case.

Here is a glaring case of injustice that actually occurred a few days ago. On a certain line of manufactured goods, controlled largely by the American Steel and Wire Co. that was, the winter conference through rate from Liverpool to Montreal is given as 19s. 9d. for carloads, with 10 per cent. primage and 10 per cent. coal primage. Now, 13s. 4d. covers the rail rate from St. John to Montreal, leaving 6s. 5d. to pay for carriage by boat. Yet, the lowest rate quoted to St. John from Liverpool is 17s. 6d. Manchester and Liverpool liners are all the same. Why this extra 11s. 1d? Would it not appear that the conference steamship companies do not want business in this line of goods?

Furthermore, a conference line running from Liverpool to Boston absolutely refused to give a quotation on these goods coming from England and going to St. John via Boston. The St. John importer had thought he could get goods cheaper through Boston boats. In any ordinary case he could. But in this case he could not even get a quotation. We might say that he thereupon took a "tramp" steamer and got a comparatively low rate.

We Canadians are so eager to stimulate trade with Great Britain that we are willing to adopt a preferential tariff, and we should see to it that steamship companies, subsidized by Canadian and British Governments, are not allowed to offset that tariff many times over by discrimination in cartage charges.

CASH AND CREDIT IN THE HARDWARE STORE.

THERE are few questions that concern retail hardware merchants more than the cash and credit systems, respectively, of doing business.

Under the credit system loss of interest is as a rule entailed, and certainly loss of money through bad debts, which even the most careful of merchants occasionally experience. But, in spite of loss through bad debts, there is a fear on the part of many that, in adopting a purely cash system, losses still heavier might be entailed.

Naturally, every hardwareman desires to adopt that system which will entail the minimum of loss, whether it be through bad debts or loss of customers.

Obviously the nearer the retailer can get to the cash basis the better. But no hard and fast rule can be laid down. What is one man's meat is another man's poison is true in regard to this as to many other questions. Hence every man in business, or about to go into business, must largely determine for himself what is the best system to adopt.

If conditions were everywhere the same it would be a matter more easily to be determined, for then every merchant could be guided by the experience of his confreres. But they are not all the same.

In one locality the retailer has among his customers a number of contractors, or it may be manufacturers. To insist that they shall pay cash down for every article they buy would be most unbusiness-like. In another locality, however, the majority of the hardwareman's customers may be those whom to pay spot cash would be no inconvenience. There the field for the cultivation of the cash system is naturally inviting.

But one cannot plunge into the cash system with as much unconcern as an expert swimmer can plunge into a river. It must be done thoughtfully and carefully, and with a view to creating the minimum of friction.

When it is once decided to adopt the cash system it is well to give notice some time ahead. Advertise liberally in the papers and send a circular letter to each customer. And in both advertisement and letter set forth the advantages of the cash system to your customer as well as to yourself; for in all new departures the customer is more easily won over to it if it can be shown that he will share in the advantages which are likely to accrue therefrom.

THE TURPENTINE SITUATION.

TURPENTINE has declined 4c. per gal. in Toronto and west, and 2c. per gal. in Montreal and Eastern Ontario and Quebec points. The quotations on single barrels now are: Toronto and Western Ontario 58c.; Montreal and all points east of he 79th meridian, 60c.

This depreciation is a result of the situation in Savannah and other primary markets during the past few weeks. There has been great fluctuation during this period, but the net result is a decline of about 4c. in the last two weeks.

The weak feeling at primary points is naturally due to the heavy crop last season. The Savannah Naval Stores Review, of March 9, reports the crop and stocks on hand at primary points during the past four years as follows:

	Crop. Gals.	Stocks on hand. Gals.
1901	561,898	13.452
1900	529 144	7.172
1899	476.657	7.854
1898	......	19.934

It will be seen that each of the past three years has shown an increase in the total production. The crop of 1898, as will be remembered, was curtailed by heavy frosts in the turpentine belt, and prices went higher than had been the case for some time.

In the following year, the production was again small, and, notwithstanding the high prices, the demand was so brisk that stocks were, in March, 1900, even lower than in the previous year, while prices were still higher. Last season, prices started at a high level and gradually went up until the highest price in a decade was reached. It soon became manifest, however, that the crop was much larger than that of either of the previous seasons, and prices have experienced several sharp tumbles during the past four months.

The new crop will begin to come in during the next three or four weeks, and buyers do not seem anxious to stock up at present prices, which are, however, considerably lower than last year. The price of turpentine at Savannah about this time during the past six years has been as follows: 1896, 25c.; 1897, 27c.; 1898, 29c.; 1899, 38c.; 1900, 53½c.; 1901, 34c.

To prophesy as to whether present prices would be maintained or not would be folly as everything will depend on the crop, which no man can estimate beforehand.

The Hardware Store of The Twentieth Century.

An Expert Hardwareman Offers Suggestions as to How it Should be Fitted up.

WITH the old century there passed away forever the time when any kind of shack without fittings or fixtures would suffice as a place in which to carry on hardware or any other kind of business, and our object in penning these lines is to give our fellow-hardwaremen the benefit of a long experience in storefitting.

The subject is a large one, and in many points conflicting, as different objects must be worked for, which, in a measure, clash with each other, but we will endeavor to be pointed, plain and practical, hoping to help some and set all thinking in so important a matter. We urge using the best materials, and to utilize space in the display of stock so that your best salesmen will be your goods arranged in such a way as to persistently draw the attention of your customers.

It is not our intention to say anything about window dressing, a subject on which much has been written, and which will, in most cases, stand improvement. Windows generally indicate the appointments of the store behind them, though sometimes even well-dressed windows have the reverse of anything attractive to back them up, and are, like some peoples' good qualities, only skin deep. Such stores attract customers once, but drive them away permanently after the first acquaintance with them.

As we are in prosperous times and they bid fair to continue, many hardwaremen will be building stores for themselves, so we will briefly indicate what we consider to be the best materials with which to equip the store from the floor up, giving measurements where the same are a matter of importance.

THE FLOOR

should be double, the top being made of hardwood, oak preferred, not more than three inches wide, well tongued and nailed in the tongue. If it is laid in this width and consists of good kiln-dried lumber you will never be troubled with shrinkage.

THE WALLS

should be matched lined, also with narrow lumber well seasoned, and shrinkage will be reduced to a minimum. Bass, pine, birch or cedar are all suitable for this. The last two will be best finished in their natural colors. The first two should be stained before varnishing.

THE CEILING.

We would strongly advise a metallic ceiling as being the best, all points considered. It never allows the dust to come through from whatever may be above, and when nicely painted it adds greatly to the appearance of the store. It should always be painted in light colors. The most effective we have ever seen was painted white and relieved with blue and gold. The best height for a store is 12 feet from floor to ceiling.

THE WINDOWS

should start two feet from the pavement, and be carried up to the ceiling in clear glass. It is a great mistake to have anything along the top in the nature of colored lights. The purpose of a window is to let the light in, but these obstructions keep it out. When a store is less than 10 feet pitch and more than 40 feet long, light should be admitted from the rear; when 12 feet pitch it will generally be light enough without rear windows up to fifty feet in length. You cannot have too much light in your store. More light on your goods, more light in your office, and more light on every business transaction, is one of the imperative demands of the 20th century.

COUNTERS

should be 33 inches high, 24 to 30 inches wide, tops being made with hardwood and not varnished, but rubbed with oil only. We strongly recommend this, as varnished counters show scratches immediately on being used, and always wear badly, whereas oil-finished counters always look well when rubbed occasionally with a brush or coarse cloth and oil. Counters should be filled in with drawers and not bins, as these generally degenerate into muck holes, on account of the dirt and dust accumulating in them. Counters should always be kept clear, apart from two or three flat showcases containing pocket and table cutlery, razors and scissors.

SHELVING

should be uniform, both for the sake of appearance and general usefulness, the best measurements being 30 in. between standards, 12 in. wide, and ⅞ of an in. thick. Where the distance between standards is too great shelves will drop or sag in the centre, thoroughly spoiling the appearance of your store, and preventing shelf boxes fitting nicely. The base of shelving up to the same height as counters should be about 21 in. wide, the ledge being made of hardwood similar to your counter tops and finished in the same way. Under the ledge should be filled in with drawers varying from 18 to 30 in. wide, and from 4 to 8 in. deep. If filled in with plain divisions or lockers, they should be fitted with hinged covers or they will harbor dirt and have a very untidy appearance. We recommend drawers as being by far the best arrangement. The first four rows of shelving above ledge should be 4 in. between shelves, the next 6 in. between shelves, and then two rows about 8 in. Shelving should be carried to the ceiling, and any space above the 12 rows indicated divided into roomy places for stock. The old style of having a cornice and the space above it vacant looks unfinished, besides wasting valuable room and harboring dust. The 12 rows of shelving, of which we have given the width, should be filled with

SHELF BOXES

of different sizes, but all aliquot parts of 30 inches, that is 5, 6, 7½, 10, 12, 15 and 18 inches wide, so that different sized boxes may be put into one space and thus accommodate the different sizes in any line of hardware, which is a very important matter. The most serviceable shelf box is one made with wood front and back and metal body, as it takes up less room and is lighter than the cumbrous wooden shelf box of the last century. Formerly there were strong objections to the combination of metal and wood, on account of the necessarily rough connection between the two materials, but every objection has been overcome in a shelf box invented and patented by one of our Canadian hardwaremen, and is now on the market so completely finished in all sizes and at such prices that it will no longer pay hardwaremen to make their own shelf boxes, and no up-to-date merchant will fill his shelving with shelf boxes made entirely of wood after he has seen these, as they save a lot of shelf room.

SAMPLES

of the contents of shelf boxes should be fixed to the front, and these should be attached in such a manner as to be instantly removed and replaced. You thus incur no loss, as the samples can be changed before the goods become too shopworn to

H. S. HOWLAND, SONS & CO.

ONLY WHOLESALE

37-39 Front Street West, **Toronto.**

WHOLESALE ONLY

HANDSOME PRESENTS

— IN —

CUTLERY

with Solid Silver or Best Plated Flatware.

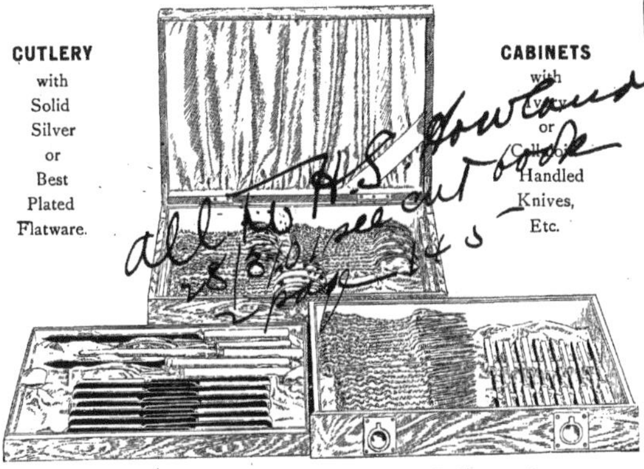

CABINETS

with Ivory or Celluloid Handled Knives, Etc.

Furnish us with list of pieces required and we will give you prices.

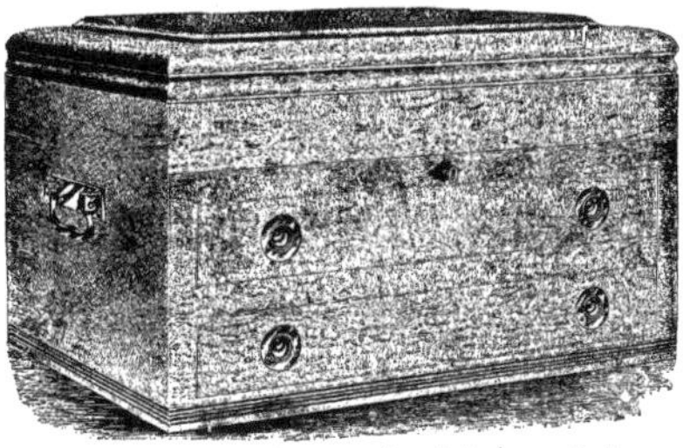

Cabinets made to hold any number of pieces required.

We carry a full line of Cased Goods, in Carvers, Dessert Sets, Fish Eaters, etc.

H. S. HOWLAND, SONS & CO., Toronto.

SEND FOR OUR CUTLERY CATALOGUE. PROMPT SHIPMENTS.

sell, and, in case of a cranky customer who has an idea that the sample is better than the bulk, you can give them the sample displayed. There are three spring sample holders on the market, but the simplest and best, also the cheapest, on account of its being made in Canada as well as the United States, has the cool name of the "Klondike" sample holder.

SHELVING FOR ENAMELLED WARE

should be 12 inches wide above the ledge and about 21 inches wide below the ledge, the space below the ledge being filled in with roomy drawers to carry many lines in housefurnishings. Above the ledge the first four shelves should be about 8 inches apart, the next two or three about 12 inches apart, and then one about 14 inches. This will give ample accommodation for all lines in enamel, japan and tinware. These shelves should be supported with small pillars turned out of 1⅛-inch lumber, and placed about 30 inches apart. This shelving also should be carried to the ceiling, when finished with a cornice, leaving an empty space above; it is often filled with rough parcels and looks untidy.

ROLLER STEPLADDERS

should be placed on all shelving. There are many kinds on the market. Our preference is for the one called the "Milbradt," on account of its simplicity. The "Iowa," too, is a good one.

STORING NAILS

is a vexed question with the hardwareman. They should be in bins, and, if possible, the bins should be covered individually, so as to keep out dirt and dust, also, that only the one in use is open, as too frequently a handful is returned to the wrong bin, entailing a loss of time in sorting out, or a great annoyance to your customer, whose feelings are as mixed as the nails he sometimes receives. We recently saw something new in nail bins, and for the benefit of the trade we trust they will be put on the market as they fill the bill completely.

CORDAGE.

It is a great mistake to have your sisal, manila and other cordage mounted on rollers in the store. It should be kept in the cellar, as it loses weight in the store through heat and draught. It should be passed through from the cellar at some convenient spot, so as to be handy, and, at the same time, not unsightly. You should have a drawer with short ends as samples, with descriptive tag attached, giving prices and full particulars.

RANGES, COOK STOVES AND HEATERS.

If you have room, stoves should be stood in a line so that customers can see all around them and your clerks can explain their flue arrangements, etc. If room is too scarce, make a double row of them, but always stand them on a platform raised about six inches from the floor. This will prove a great convenience in showing them, and will also prevent so much dirt and dust gathering under them.

SHELVING FOR PAINTS, VARNISHES, ENAMELS, ETC.

Goods of this class should be kept on narrow shelving accommodating not more than three rows from front to back. When shelving is deep and many tins are piled in front of each other, the different kinds are liable to get mixed, and you think you are out of some numbers, thus, customers are disappointed and sales lost. The stock of these goods is best kept in a dry cellar. Oils and varnishes in bulk should be kept in the cellar in a systematic manner with trays to catch the drippings. They should be provided with a wire gird raised about two inches on which to stand the measures, and these should be inverted after each using to drain them and keep them acurate.

COAL OIL TANKS.

This is another of the hardwareman's troubles, but we know of no better arrangement than tanks in the cellar and pumps in the store. The new self-registering pump, though expensive at first, is a sound investment, and will save its cost in a short time, as it does away with waste entirely and makes mistakes almost impossible.

GLASS RACK.

Glass should have a rack built to suit the different sizes. This is best at the back of the store, and starting, if possible, right in the corner with the front slanting, the narrow end being toward the front of the store and the broad end for your large sized glass right in the corner. In this way you economize space and also provide a place of safety for your more expensive sizes. The divisions in this rack should be just large enough to hold a box or case of glass. Handy to your glass-rack should be your glass board or table on which to cut glass. This should be placed good and solid on a firm counter. The base of this counter should be filled with drawers in which you can conveniently carry in the store a stock of dry colors, reserving a space clear to the floor for a bin for waste glass. For this purpose we would recommend a strong galvanized iron bin. The old barrel is decidedly out of place in a well fitted store. Make the bin out of 20-gauge iron, fitting it with a pair of strong handles, also a wooden bottom screwed on to the metal bottom to prevent your floor getting cut up,

HARVEST TOOLS AND SHOVELS.

These should be in a rack, a space being made for each kind carried. One of the cheapest and most convenient can be made out of half-inch iron pipe, attached to a solid platform, raised about six inches from the floor; this can be fastened on with flanges. The style and size of this rack largely depends upon the stock carried, and the ingenuity of the hardwareman.

SHOW TABLES.

In addition to the counters you need two or three show tables according to the size of the store. These should be about 28 in. high, 24 to 30 in. wide, and 6 to 8 ft. long. They should have a shelf raised about 6 in. from the floor and the full size of the table. If goods are nicely arranged on these and changed frequently, such tables will prove very effective silent salesmen. They do well for displaying special lines on which you are making a drive, such as the balance of anything toward the end of the season for it, and which you should never carry over if possible.

ADJUSTABLE TABLES

are excellent adjuncts in a store, and can be put to many uses in both store and window dressing; there are now two on the market and both very good.

In a short paper like this we have only been able to briefly touch the main points in connection with fitting the interior of the up-to-date or twentieth-century hardware store, leaving the enterprising and live hardwareman much to supply, but we trust our experience may be a help to some. Then the time we have devoted to it will not be lost, and we shall be satisfied in having helped our fellow hardwaremen to a better state of store equipment. BETA.

NEW LIGHT-PRODUCING GOODS.

The Ontario Lantern Co., of Hamilton, Ont., have just issued illustrations of the new goods which they are manufacturing for the coming season's trade, among which is the "New Century Banner" cold-blast lantern, with a great many improvements, which have been covered by patents. Then, they have recently put on the market the "Radiant" Shelby incandescent electric lamp, which they are now in a position to supply in the different candle powers and voltages. They claim that the "Radiant" Shelby has a great many advantages over the ordinary lamp, and, for efficiency and long life, stands at the head of the list.

They solicit sample orders from dealers in electrical supplies and from the large lighting stations. This company have also contracted with Mr. T. L. Willson, the "Carbide King of Canada," to manufacture for him a new acetylene gas lamp, which, for brilliancy of light and economy, will exceed anything that has ever been supplied in the past. They are also manufacturing a full line of lamps and lamp burners for kerosene oil, incandescent mantle burners, etc., and solicit orders and inquiries from the trade for the same.

THE CANCELLATION OF ORDERS.

ANOTHER LETTER ON THE QUESTION.

Editor HARDWARE AND METAL,—I am a subscriber to your journal, HARDWARE AND METAL, and welcome it as a weekly visitor to my desk.

I read your article of March 9. "A Bad Practice," also the letter from "One Who Has Suffered," in issue of March 16, commenting thereon, saying it was "timely and to the point"; also giving his views as to the manner in which retailers should do their business, notably with regard to booked orders.

"One Who Has Suffered" evidently does not feel that he yet has the retail trade sufficiently at his mercy, but would like to have them still more tightly tied up.

It makes all the difference in the world "whose ox is gored." Most of us feel our wounds pretty keenly, while not many of us feel very much over the troubles of others, and your correspondent is apparently one of this kind. One would think the manufacturers with their trusts, and the jobbing trade with their associations, meeting and fixing prices at which they will supply the retail trade, had things pretty well their own way; but this sufferer, possibly out-distanced in the race for business—appears to want an absolute "hold-up" on the retailers.

Manufacturers' agents and representatives of the jobbing houses start out in the beginning of the year looking for business. Retailers are besieged with this class of callers, and we extend the glad hand to many of them, wanting to book our orders oftentimes 12 months in advance of our requirements, and to get an order, they feel called upon to guarantee the price, which they do.

Now, what does a guaranteed price mean? It means that, in the event of an advance on the market, we will not have to pay any more than the price named, and that, should a decline take place up to the time of delivery, the price will be made on the basis of such decline.

"One Who Has Suffered" says he never considers the propriety of offering to pay more for his goods in the case of an advance, but magnanimously, in the event of a decline, gives his wholesaler the option of meeting the price or cancelling the order.

After fixing their price, wholesalers have their representatives call upon us, and they book our orders at guaranteed prices, and it matters not to the wholesaler how much prices may advance; he get his "rake-off" just the same, as he keeps himself covered—if he does not he should do so—as he goes along.

Now, Mr. Editor, if the manufacturers among themselves, or the jobbers between themselves, or, if in any case the one breaks faith with the other, resulting in a drop in price, would it be fair, I ask, that we, as retailers, should be compelled to pay—which was perhaps unwarrantably high—the price which these people started out to do business on? Certainly not! and no reputable business house would exact it.

I have cancelled orders on the lines complained of, and, while some retailers are exonerated from blame in this connection, as being "above such mean tactics," I hesitate to believe that there is one hardware merchant in Canada, of any standing, who has not at some time given "his wholesaler the option of meeting price or cancelling the order" for the reason complained of by "One Who Has Suffered."

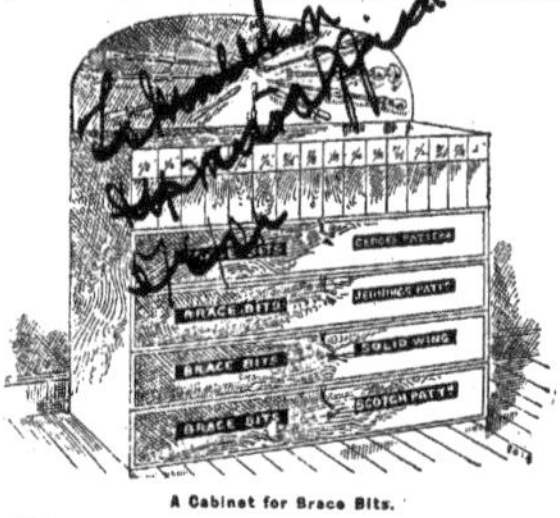

A Cabinet for Brace Bits.

I have been engaged in the hardware business for upwards of 20 years, and have in mind a particular wholesale house in Ontario with whom I have been doing business constantly during the whole of that time, and, after having become acquainted with them, I have not even asked for a copy of my order, well knowing I would be properly dealt with, and not infrequently, on receiving the invoice, I found the order filled at a less price than that at which the goods were booked, this too, without any reference to it having been made by either party. Again, I have had dealings with certain wholesale houses, the outcome of which was anything but satisfactory to me, and, was of such a nature that, if I again should attempt to do business with them, I would almost consider it necessary to call in my solicitors to look after my interests.

Retailers will require a good deal of "education the other way" to teach us that if we "expect to profit by an advance in price" we "must also be prepared to stand a loss in the event of a drop."

To do this it will be necessary for "One Who Has Suffered," as well as others whose cause he seeks to champion, to cease altogether the booking of orders for forward shipment, and, possibly, to call their travellers off the road entirely.

A good deal more may be said, Mr. Editor, along the line travelled by your correspondent, and I may return to the subject, but, as I have at present trespassed to a considerable extent upon your valuable space, I will conclude, trusting you will pardon me as this is a first offence.

"THE BEETON BELL."

Beeton, March 20, 1901.

A CABINET FOR BRACE BITS.

This is an illustration of a useful cabinet and stand for brace bits. It can be made in various sizes, but a handy size is 4 ft. deep and 12 inches thick. The top department in the accompanying sketch shows the position of the departments for the various sizes of bits, each one to have a narrow strip of wood nailed across the top, painted white and lettered black with the various sizes. The doors are secured with three hinges at bottom, so as to fall down, and fastened with Bale's patent catch at top. On the inside of the door should be the prices of the various sizes in the same department. The bits should all be placed in loose. Fastened to the back should be a piece of wood cut in half-circle, and a good assortment of bits fastened by means of staples should be shown."

MAY MOVE TO CHATHAM.

It is proposed to pass a by-law in Chatham, Ont., to loan Dowsley & Sons, steel buggy-spring manufacturers, Owen Sound, Ont., $20,000 for 20 years, and to exempt the company's factory from taxation, in consideration of the company operating a factory in Chatham and employing at least 40 hands the first year and 75 all years following. The loan is to be repaid in annual instalments of $1,000. The matter is to be brought up at the next meeting of the Chatham Council.

R. R. Neild, machinist, Stratford, Ont., has sold out to F. A. Leak.

WINNIPEG'S RETAIL HARDWARE STORES.

A Brief Sketch of their History and Business Methods.

SPECIAL numbers always call for a review of trade and a noting of the advances in supplies to meet the requirements of the market. The history of the retail hardware trade in Winnipeg shows a number of houses in the line who have done business in the city for many years—some reaching back to the early seventies—and have progressed steadily from small stocks of limited range in one-storey wooden buildings to fine modern stores with stocks of the latest and most improved goods in every line.

J. H. Ashdown's is the earliest hardware store of which your correspondent could get any trace. Mr. Ashdown opened here in a very small way, in 1870, carrying a limited stock of general hardware and stoves, and carrying on tinsmithing as well. To-day, the Ashdown retail hardware stores occupy a four-storey building at the corner of Main and Bannatyne streets, having a frontage of 60 feet by a depth of 225 feet, not to speak of the great wholesale house, of which a description was given in these pages some time ago. The stock carried is well assorted to the smallest detail, and the premises admirably arranged for transacting business swiftly and conveniently.

Next in point of age comes the house of R. Wyatt, which was opened in 1878. Mr. Wyatt coming here from London, Ont., where he had spent 10 years with the Clare Bros. Mr. Wyatt does a large and exclusively city trade, and, while carrying all lines, has made a specialty of housefurnishing hardware and furnaces. Mr. Wyatt finds that year by year he is growing more into a cash trade, which is the Mecca of all good business men; he is satisfied with the progress of his business so far, and is looking forward to a steady increase.

Next in point of time comes the firm of James Robertson & Company, formerly W. G. Pettigrew, which was opened about 1880. This firm has also a large wholesale business in the city.

C. A. Baskerville & Co. opened in 1882. Their store is near the C.P.R. depot on the west side of Main street, and, in addition to a large city, has a very considerable country trade.

Graham & Rolston is a house opened in 1889. At that time the firm was Skeed & Graham, Mr. Skeed having been for many years with J. H. Ashdown. Mr. Skeed only remained a short time with the new house, which was next known as C. W. Graham, and finally appeared under its present caption. From its first starting the house has been a popular one, and has steadily increased its business.

Anderson & Thomas in 1899 bought out the Campbell Bros., and since that time have worked up a very satisfactory business.

Templeton & Co. on Portage avenue are among the newer firms which, with tact and enterprise, are winning their way to a fair share of the constantly-growing trade of the city.

The retail hardware trade in Winnipeg is carried on very much on the same lines by all the houses. For instance, all carry all classes of shelf and heavy hardware, stoves, tin and granite ware, lamps, wooden ware, cutlery and silverware. With the single exception of Graham & Rolston, all deal in furnaces. Business is done on both cash and credit. No house does an exclusively cash trade, but all houses in this line are year by year doing more of a cash trade.

There is an excellent feeling of good fellowship among the men of the trade, and, in consequence, there is little cutting of prices. Trade is steady throughout the year with the customary increases in volume of business at certain seasons. There are no merchants in Winnipeg of which her citizens have a right to feel more justly proud than of her retail hardwaremen. They all stand high for integrity and public spirit, and their virtues are like their wares, of the enduring, but not obtrusive sort.

As the special number is also a spring number it will be of interest to readers to note something of the increasing business in the dairy supply trade. Owing, no doubt, to the fact that the failure of crop last year has pressed home to the farmer the desirability of having more than one string to his bow, and also that dairying is not attended with as great risk as grain raising, the inquiry for dairy supplies, such as butter-workers, churns and the like is several weeks earlier than usual and the inquiries are also more numerous. The old houses are all represented again this year. The Melotte, Alexandra, De Laval, are doing business at their old places and there is an air of activity unusual at this time in former seasons. Of importance among the new concerns is that of The Scott Dairy Supply Co., the formation of which was noted in these pages some time ago. As Mr. Scott, the head of the new firm, has had a long and honorab'e connection with dairy machinery and supplies in this country, when opening for himself he had the choice of many agencies particularly in the line of separators. After considerable deliberation the firm has undertaken the sale of the "Improved United States," manufactured by The Vermont Farm Machine Co., the largest manufacturers of dairy machinery in the world probably. This separator has been thoroughly tested in the Government Dairy School here, and in such capable hands is likely to give account of itself. The new firm have the best wishes of the trade generally for the success of their venture.

Another claimant for mention is the sample-room just being opened by T. L. Waldron, representative of Caverhill, Learmont & Co., Montreal. These premises are in the new Stovel block on the first floor, and being admirably lighted will show off the fine display of cutlery to the best advantage as well as the other lines of shelf hardware which will be carried. In addition to these rooms, Mr. Waldron has secured a large track warehouse for heavy goods.

E. C. H.

A WHOLESALE FIRM'S NEW WAREHOUSE.

The attention of the readers of HARDWARE AND METAL is called to the advertisement in another column of Thos. Birkett & Son Co., Limited, Ottawa.

The firm has been in business for the past 34 years, during which time they have worked up the largest wholesale and retail hardware trade in Eastern Ontario. Their aim has always been to give their customers good goods at an advance of a living profit.

Last fall the firm began to build a large brick wareroom on Canal street, next door to H. N. Bate & Sons, and near the Russell House. The building is large and central and well stocked with new goods from all the best manufacturers in England, Germany, United States and Canada.

The firm in their new premises will be in a position to give better attention and satisfaction to all their old customers, together with the many new ones whose intent it will be to give them a call and trade with them.

A large increase in the firm's business in their new premises, which for light, accommodation and convenience is second to none in the Dominion, is what one may confidently expect.

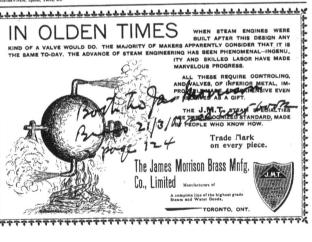

THE ADVERTISING ARENA.

THE FUNCTION OF THE ADVERTISING MANAGER.

EVERY man in business knows that to advertise properly requires time—time for management and time for free thought. For want of time there are legions of business men who neglect advertising. There are others who, for the same reason, do it half-heartedly. Both make serious mistakes.

Every manager of business is working on a plot of arable ground. It may be small; it may be large. If it is a hundred-acre farm it is divided into ten-acre lots. Each lot yields a profit through its own particular produce. In the country the space is divided up between wheat, oats, peas and pasture. Just so in business. The head of the firm may devote the largest lot to buying, and give it his special attention. Or the selling lot may occupy the preeminent place in his mind. Sometimes finance is the leader. Generally a man has his specialty. But whether or no he has one seed to which he looks for the largest profit, he must extend his watchful eye over the whole farm.

On the part of the man who has little time, advertising is often given only a small back lot, and even that does not receive due attention. Consequently, the yield in this field is discouraging, and the husbandman comes to the conclusion he has been laboring on stony ground. He forgets that the richest earth when neglected is not as profitable as sandy and stony soil, well cultivated, well watered and tenderly nursed.

A farmer does not pretend to be able to work a hundred-acre farm alone. Even with the aid of modern machinery he finds it an impossible task. He must bring in help. The "hired man" is given a team, a plough, a drill and a binder, and he goes out into that neglected back lot and he more than makes his wages.

Does that contain no lesson? The head of a firm in the city is also working a hundred-acre farm within his four walls. He has back lots that are neglected—and good rich soil they contain. Even with the aid of modern office machinery, even though all he has to do is sit on his high seat and drive, he cannot find time to supervise his whole business.

He should go out to the backwoods and take lessons from the agriculturist. He will come back and bring in a "hired man." On him he will bestow executive authority, and full power to act within his assigned sphere. And that man will develop his branch of the business and the profits of that firm will swell.

Business is becoming every day more specialized in its different departments. While there is still a place for the general supervisor, he is becoming more and more the two-hour-a-day man. Each department requires a man's special attention.

If a man wishes to have his advertisements bring him good results, he should place a man in charge of his advertising and catalogue department. It merits a person's whole attention. Ad.-writing has been reduced to a science, and a deep science it is. The making of a catalogue is a work of large dimensions. Combined, the tasks need one whole, entire brain. The number of firms who are adopting this policy is increasing. Thereby, they show consummate wisdom.

SMALL MERCHANT ADVERTISING.

It is a well-known fact, remarks an exchange, that a stranger can go into a new town and create more interest in himself and business in a given length of time than an old, well-known and respected citizen. The tendency seems to be to take up with strangers, partly, no doubt, on account of the novelty of the thing. The old-established business house once started down hill finds it much harder to put in paying business than a new business.

It is the same with advertising. A new business is much easier advertised than an old business. The merchant in the country [illegible] has the hardest problem in advertising; simply because he knows every man personally who is likely to read his advertising. They know his store, they have formed an opinion and learned the current prejudices, and in fact know the store better than he does. They know the kind of stock he carries and how his prices compare with the other stores, both locally and in other towns. They know the whole thing from start to finish.

It is a common expression that a prophet hath no honor in his own country. This idea seems to fit the advertising of the small country dealer. The man who can make advertising in the country pay, and very many of them do, is deserving of great credit.

Every advertiser should have a well-devised plan and have a system and explain that plan to the public. He should keep explaining his advantages to the public. He should do it regularly and at every opportunity.

FOR AND ABOUT PLACARDS.

NEVER attempt ornamentation on show cards unless you understand art, remarks American Advertiser. It is so very easy to over-do a card. A pleasing arrangement of a few words in black on white board will attract quicker and create a better impression than if done in gold and silver and red and blue, with birds and scroll work thrown in.

Study arrangement of words. Words can be piled to advantage. For example:

> **For Us**
> **Nothing**
> **is ever good**
> **enough**
> **if better**
> **exists.**

In this the words "good" and "better" are brought out by virtue of their positions.

Here is something for a clearance sale:

> **Take this home to-day**
> **for**
> **69c.**
> **You considered it a**
> **Bargain**
> **last week at a dollar.**

Any of the following phrases may be used on show cards, or supplementary to price tags:

The stamp of style.
It was.. It is....
New Century prices.
This is to satisfy you.
Bargains beckon you.
The price leans your way.
Worth makes it a bargain.
Bright things for dull days.
Chance of the new century.
Touched with temptingness.
Don't buy what you don't need.
Business suits at business prices.
The price helps to make it popular.
Ready-made, but custom goodness.
Your dollar is worth all it will buy.
Pretty goods—but more than pretty.
Don't economize at your own expense.
Fine furnishings for fastidious fellows
A price that gladdens the careful buyer.
A store where confidence dwells eternal.
We lose on these but we gain your favor.
Old Century goods at New Century prices.
Here are the things we know you will like.
A soft touch on the contents of your purse.
You can go farther but can't do any better.
He that serves quickly serves twice as well.
A margin of profit and a portion of pleasure.
Honest values stitched with truthful words.
This store is yours to the extent of your desires.

THE NOVA SCOTIA STEEL CO.
LIMITED

Manufacturers of

STEEL
Tire, Sleigh Shoe, Toe Caulk, Reeled Machinery, Spring, Angles, and all sizes of Merchant Bar, also special Agricultural and other sections.

POLISHED STEEL SHAFTING From $\frac{5}{8}$ to $3\frac{1}{2}$
GUARANTEED STRAIGHT AND TRUE TO 1-500 PART OF AN INCH.

SHEETS and PLATE STEEL
FROM 12 GAUGE TO $\frac{1}{2}$ INCH THICK AND UP TO 48 INCHES WIDE.

HEAVY STEEL FORGINGS and HAMMERED SHAFTING
NOTHING REQUIRED IN CANADA TOO LARGE FOR US.

STEAM and ELECTRIC CAR AXLES FISH PLATES
SPIKES, BOLTS, AND OTHER RAILROAD MATERIAL.

STEEL TEE RAILS, 12, 18 and 28 lb. per yard.

FERRONA PIG IRON for foundry use.

OWNERS OF THE SYDNEY MINES

(formerly the property of the General Mining Association of London).

Miners and Shippers of

OLD SYDNEY COAL

STEEL WORKS : New Glasgow. BLAST FURNACE : Ferrona.
COAL SHIPPING PIER : North Sydney.
HEAD OFFICE : New Glasgow, Nova Scotia, Canada.

THE Watson, Foster Co., LIMITED

MONTREAL.

WALL PAPERS

18 INCH INGRAIN FRIEZE NO. 1969.

A VERY DECORATIVE ROCOCO PATTERN WHICH MATCHES TWELVE LEADING SHADES. A SMALL SHADE BOOK OF INGRAINS, WITH ILLUSTRATED MATCH COMBINATIONS, OR SAMPLES, WILL BE SENT UPON REQUEST.

RE-ORDER NOW ANY PATTERNS YOU ARE LIKELY TO BE SHORT OF.

WORKS—ONTARIO STREET EAST, MONTREAL.

SAMPLE BOOKS OF ANY DESIRED GRADE OF WALL PAPERS SENT PREPAID TO THE TRADE.

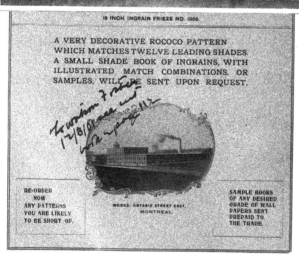

GREASY CANDLES

Smelly Oil Lamps,
Poor Electricity or Flickering Gas?

Isn't it about time to make a change and

GET more Light
For . . .
Less Money?

Get 100 Candle Power for 50c. a month.

GET the Light of Eight Oil Lamps for the cost of Two.

GET AN AUER GASOLINE LAMP.

Your money back if you don't like it.

Write for Catalogue. . .

AUER LIGHT CO.

E. Simpson & Co., Moose Jaw, Agents for the Territories. 1682 NOTRE DAME ST. MONTREAL

CORDAGE

ALL KINDS AND FOR ALL PURPOSES.

Manila Rope	Lathyarn
Sisal Rope	Shingleyarn
Jute Rope	Bale Rope
Russian Rope	Lariat Rope
Marline	Hemp Packing
Houseline	Italian Packing
Hambroline	Jute Packing
Clotheslines	Drilling Cables
Tarred Hemp Rope	Spunyarn
White Hemp Rope	Pulp Cord
Bolt Rope	Lobster Marlin
Hide Rope	Paper Cord
Halyards	Cheese Cord
Deep Sealine	Hay Rope
Ratline	Fish Cord
Plow Lines	Sand Lines

"RED THREAD" Transmission Rope from the finest quality Manila hemp obtainable, laid in tallow.

CONSUMERS CORDAGE COMPANY, Limited

Western Ontario Representative—
WM. B. STEWART,
Tel 94. 27 Front St. West, TORONTO.

MONTREAL, QUE.

HAMES SNAPS CURRY COMBS

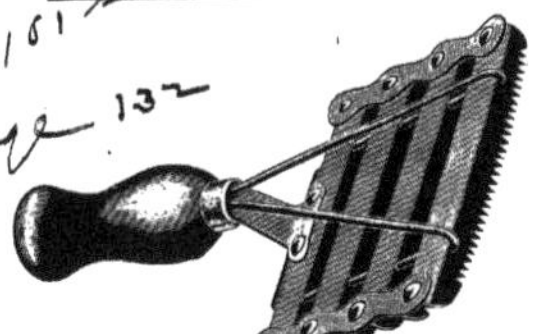

AND ALL KINDS OF.. **SADDLERY HARDWARE**

McKinnon Dash & Metal Works Co., Limited

NIAGARA FALLS and ST. CATHARINES, ONT.

A Hardware Manufacturer's Interesting Career

As Exemplified in the Life of T. Henry Asbury, President of The Enterprise Manufacturing Co.

ONE afternoon in September last a number of wholesale hardwaremen, a few manufacturers and others were standing on the deck of the Lord Stanley enjoying therefrom the magnificent view of the city of Quebec and its famous citadel. Among the group was a gentleman, medium in height, comfortable in build, and from whose pleasant face hung a pointed long white beard. The gentleman was T. Henry Asbury, the president of The Enterprise Manufacturing Co., of Philadelphia.

"I had intended only coming as far as Montreal," he said in reply to an inquiry I made shortly after I had had the privilege of being introduced to him. "But finding some of my customers were in Quebec attending the convention of the Wholesale Hardware Association, I decided I would come on here to see them."

"And right glad are we that you came," exclaimed some of those who were standing in the group.

Finally we began talking of trade matters, and I learned from Mr. Asbury that he took a great deal of interest in the Canadian trade and that his business with this country was steadily growing. During our conversation I was struck by the manner in which he spoke of the Canadian market. There was no bluster about it. There was nothing avaricious about it. While he did not say it in words, yet I felt that here was a man who wanted to do business with Canada because he liked her and her people. When some time afterwards I learned that Mr. Asbury was an Englishman by birth, having come to the United States when a young man, I began to put two and two together. And now I think I understand why he seemed to have a peculiar interest in the Canadian market.

Mr. Asbury was born over 62 years ago in that hive of iron and steel industry, Birmingham. His first occupation was errand boy to a silversmith in his native town at the magnificent wage of two shillings per week. He soon, however, found more congenial employment with a gunmaker. Next he went to a gunfinisher, and subsequent to that worked in a number of shops in the gun-making district of Birmingham, which at that time was the largest gun-making city in the world. Guns were made there for France, Germany, Russia, Turkey; in fact, for all civilized nations.

From the gun district and gun trade Mr. Asbury went to work with his brother, who was engaged in the manufacture of agricultural engines.

"I added largely to my stock of mechanical knowledge while with him," said Mr. Asbury. "Then I went back to Birmingham, and engaged with the Vulcan Iron Foundry, in order to attain full knowledge and experience, whereby to make me a thorough mechanic. I worked at both lathe and vise, and secured a fair knowledge of engine building. It was while engaged there that I determined to set my face toward America."

T. Henry Asbury.

"I was then 18 years of age. I went to Liverpool, and sailed away in the good old packet ship, Tonawanda, for Philadelphia. And that's all there is to it," said Mr. Asbury, with the air of one who had exhausted the subject.

"But your story has only commenced," expostulated the interviewer. "Please tell me about the start, and the fortunes that awaited you in Philadelphia."

"I can well remember some things about it," was the answer, "and we won't miss those that I have forgotten. The ship was supplied with what was called 'Government rations.' The law compelled the owners of the vessel to supply to each passenger so much flour, so much pork, so many beans, etc. Each of us went into the hold twice each week, and secured his rations. There was a cook who was supposed to prepare this for consumption, but there were so many of us that he soon took a back seat, and each passenger cooked for himself. We were seven weeks in coming, and so the most of us were sick, and we were all of us also sick of it, long before it was over. We had a rough voyage—I have been across 18 times since, and I never saw a rougher one.

"I landed in Philadelphia with just 62c. in my pocket, but a trunk full of clothes, packed by a thoughtful mother, and sufficient to last me for a year. I naturally lost no time in looking for work. For two weeks nothing came in sight. Then I saw an advertisement in The Ledger: 'Wanted—A young man, under instructions. Apply to Thomas Stewart, Drinker's Alley, Machinist.' I was there long before 7 o'clock in the morning waiting for Mr. Stewart. I asked him, 'Do you want a machinist?' 'I do not,' he said; 'but I do want a young man under instructions.' 'I am a machinst,' I answered, 'and I want the job. What wages will you pay?' He answered that he would give me $4.50 or $5 per week. I went to work at $5. At the end of the week he voluntarily raised me to $6, and at the end of the month he advanced me to $8.

"This was doing reasonably well, but I saw a chance to better myself, and took advantage of it. I went to Sharps' rifle factory, in West Philadelphia. They paid me $12, which I feel was as good as $18 now, when one takes into consideration the difference in rent and in the cost of living. I remained there until the panic of 1857 closed the establishment.

"I then engaged with Bement & Dougherty, and worked as a machinist until 1859. Dr. Gallagher had just come from Savannah with a patent rifle that he desired to make and market. He opened a small shop on the Frankford road, and I entered his employ, and made his first gun under his directions. I own that weapon to-day, as the doctor made me a present of it in 1896. You may believe that I prize it as a valued possession.

"A company was formed for the manufacture of this gun. They leased Slote's mill on the Fraser river; then built a factory on Twelfth and Thompson streets, and the same was run night and day at the commencement of the Civil War. Then they put up still another factory at Twenty-second street and Washington avenue, where

thousands of guns were made for our Government. At the close of the war they ceased to make this weapon, and entered upon the manufacture of the American button hole sewing machine.

"At the commencement of the war, Henry Disston took a Government contract for the manufacture of 30,000 cavalry bridle bits. Mr. Sharps, my old employer, had formerly made 400 of these, also for the Government. I had left Dr. Gallagher and gone back to Sharps, and had made the tools for the manufacture of these bits. Mr. Disston had learned of this, and engaged me to go with him and make the needed tools for the manufacture of the 30,000.

"In two months he decided to give me charge of his machine shop as foreman, which position I held until 1864, or until I believed that the time had come to go into business for myself."

"Please tell me about your start."

"There is not much to tell about that," Mr. Asbury responded. "It was a very humble beginning. All the recent years of labor had found me with but one purpose in view—the commencing of business for myself. I felt that I had passed through enough practical labor and experience to make me a master of the various branches of the machinist's trade. All I lacked was capital, but I had saved a little of that, and felt that the time for the venture had come.

"I commenced at the bottom of the ladder. This was in 1865. I transformed the third storey of my house into a workshop, and begun to make fair progress.

"One of my present partners in The Enterprise Manufacturing Co. is John G. Baker. At the time of which I am telling you he had invented and patented a machine for making glaziers' points—the little three-cornered metal bits that are used for holding the pane of glass in the window frame. It would cut out 5,000 in a minute. He soon found that it could also be used for saw-toothing, and sold it to Henry Disston for that purpose.

"It was when I was foreman in the shop of the latter. One day he came to me and asked if I did not want a machinist, as there was a man downstairs who wanted a job. I put him on, and it was Mr. Baker. After I had commenced for myself, he came to my house one Sunday and I told him I wanted him to come into business with me. 'But I have no money,' he said. 'Neither have I,' was my answer, 'and in that respect we will make a good combination.'

"He agreed to go in. We established ourselves in the third storey of No. 402 Library street, in a room 14 by 28 feet in size, and hung out a sign, 'T. Henry Asbury & Co.'

"We did a little jobbing work. We also commenced the manufacture of a measuring faucet for the drawing and measuring of molasses. After we had sold several hundred of these we were sued by Smith, Seltzer & Co. on the claim that we were infringing upon a patent owned by them. The litigation was finally ended by a compromise and a partnership. We were machinists and manufacturers; they were merchants. We could make the article, and they could sell it. It was so obviously to the advantage of all concerned, that we were not long in coming to an agreement. We put in our article for $20,000; they put theirs in at the same amount. We added $10,000 for working purposes, and The Enterprise Manufacturing Co. of Pennsylvania came into existence. The company was incorporated on September 8, 1866.

"There is not much more to tell," Mr. Asbury continued. "We moved into a small rented factory on Exchange place, Philadelphia, but soon outgrew it, and erected one of our own at Third and Dauphin streets. Both Mr. Baker and I worked at the bench, on various machines, along with other mechanics.

"Oh, let me tell you an incident about those measuring faucets. We once accumulated 400 dozen, and it gave us quite a load to carry them. So I set out on a tour, and visited 65 towns between Philadelphia and St. Joseph, Missouri. I had one or more dozen sent to each town ahead of me. I tried to sell them to the hardware dealers, but the reply occasionally was 'Young man, we have our cellar full of that kind of stuff. We don't want them.'

"I would then make arrangements with some one dealer to represent us. Then I would call on a dozen grocers and put the machine in practical operation. This was in the dead of winter. When the grocer saw the magic it performed upon his barrel of molasses, there was no trouble in persuading him that he needed one, and in getting him to pay $5 for it. This opened the eyes of the hardwareman, and he made up his mind that we had an article that would sell.

"After the faucet was well started, we began the manufacture of a bung-hole borer, then a coffee mill, and a tobacco cutter, and so, one by one, our great line of specialties came into existence."

Mr. Asbury was elected president of The Enterprise Co. in 1870. It now has an immense plant, devoted to the manufacture of patented hardware specialties and labor-saving machinery, its productions being known throughout the world. Mr. Asbury gives special attention to the affairs of the company. He has been largely interested in Florida, having served as president of the Kissimmey Land Co.; treasurer of the Florida Land and Improvement Co.; director in the Atlantic and Gulf Coast Canal and Okechocee Co. He is also a director in the Philadelphia Manufacturers' Mutual Fire Insurance Co. He also has large interests in real estate at Oak Lane, where he resides, together with his sons and daughters, who surround him in their delightful homes.

On October 3, 1857, he married Mary Elizabeth, daughter of Elias and Mary Rimmer Swann. They have three daughters and two sons, the latter being actively engaged with their father in The Enterprise Manufacturing Co.

"Mr. Asbury, I have heard with great interest of the labors you have performed—but what else have you managed to do or learn in all these years? What is about the best thing a man can have, after all?"

"Well," came the ready answer, "about the best thing I know of, is a contented mind. One should learn to do the best he can when at work, and lay aside all worry as to results, when that work is faithfully done.

"A man in my line of business soon learns another thing. That is, to first find a want, and then provide something to fill the same; and then to open your hand and show the people what you have—to advertise it. These three things have been our policy.

"I have also learned that it pays to make an article first-class in all respects. To hold the thing up in all lights and points of examination, and to ask: 'How can that be improved?' When you see where, go on and improve it. Spend money generously in labor-saving machinery and tools to produce it. Study the wants and comforts of your fellow mechanics, thus making sure of their good-will; trying to make life pleasant for them in the factory. I find that this pays. Some of our men have been here since 1865; a number of them for 30 years or more. They have never had a thought of going elsewhere. They have made themselves a part of the concern."

Besides his own visits to Great Britain, Mr. Asbury has from time to time sent representatives, with the result that The Enterprise Manufacturing Co. now do an extensive trade with that country. The company's trade with Canada has also made steady progress. The foreign trade of the company has become so important that it is now about 25 per cent. of its entire business.

THE ACME CAN WORKS.

As the country goes on producing more foodstuffs and as our manufacturing industries increase, the demand for tin enclosures grows apace and the can manufacturers must increase their capacity accordingly. This is just what the Acme Can Works of Montreal are doing. They have lately secured the handsome and commodious premises shown in the accompanying cut on Ontario street, and are in the process of removing their plant from St. Antoine street. Here they expect to turn out from 60,000 to 75,000 cans a day, employing a force of from 100 to 125 hands and running a well-balanced equipment of machinery. The firm are spending about $75,000 in getting ready for business and will have one of the best equipped plants on this continent.

The factory itself is built with solid brick walls and fireproof partitions. The main building contains two flats, each 200 feet long by 45 feet wide. At the front of first floor is the office, while the remaining space is devoted to general can-making. Next the wall on either side is a row of presses—about 30 in all; then come two lines of shears, one on each side, and then down the centre runs a row of 15 formers. This plant will allow of an immense output and permit of a great variety of patterns, ranging from a little 2-oz. fluid beef can to a 10 gallon square oil can.

At the rear is a series of pressure testers, after going through which the cans are dried and sent to the warehouses at the rear or to the storerooms on the upper flat. Upstairs, besides the two storerooms, are a general tin shop, a carpenter shop, a bake room, a paint shop, and a light, bright lunch-room for the workmen. The entire building is well ventilated and lighted.

The front extension on the right is the storeroom of tin and the raw products. It has been built to be fireproof and to keep contents perfectly dry. The ground floor is 40 x 50 feet, and is constructed to stand a pressure of 3,000 pounds to the square foot. The upper apartment wil be used as a storeroom. A covered entrance has been built to this addition, so that goods can be unloaded out of the wet on a rainy day.

The parallel projection at the rear is the machine shop, where all the dies are made. For this purpose there have been installed 9 lathes, 2 planers, 3 drills, a milling machine, a shaper, a special dye grinder and a few extra machines for special purposes. The machinery, on the whole, is thoroughly up-to-date and includes some patterns of the firm's own origination and patent. The motive power is electricity, the driving being done by two large motors.

The members of the Acme Can Works firm are men of enterprise. They are the only Canadian manufacturers of key-opening cans for meats, fruits and vegetables, and paint packages. They also have other special makes of their own. Within the next ten years they expect to see a wonderful improvement in the demand for tin cans, half of whose utility, they believe, has not yet been told. They are acting according to this idea, and have prepared a manufacturing establishment which will meet all the requirements of the Canadian trade. They intend to be ever ready to supply a demand, planning to keep in storehouse a stock of 3,000,000 to 4,000,000 cans.

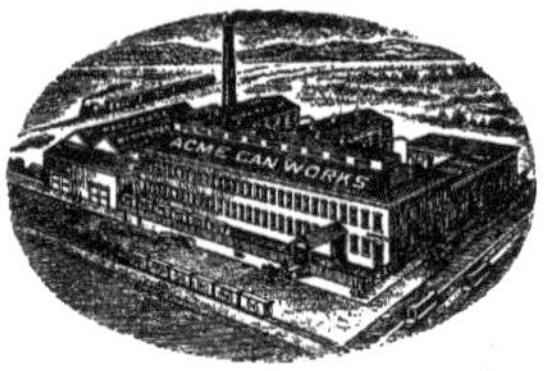

GOLD PAINT FORMULAE.

The formulæ of the various gold paints on the market are carefully guarded trade secrets. Essentially they consist of a bronze powder mixed with a varnish. The best bronze powder for the purpose is what is known in the trade as "French flake," a deep gold bronze. This bronze, as seen under the microscope, consists of tiny flakes or spangles of the bronze metal. As each minute flake forms a facet for the reflection of color, the paint made with it is much more brilliant than that prepared from finely powdered bronze.

For making gold paint like the so-called "washable gold enamel" that is sold by the manufacturers at the present time, it is necessary to mix a celluloid varnish with the French flake bronze powder. This varnish is made by dissolving transparent celluloid in amyl acetate in the proportion of about 5 per cent. of celluloid.

Transparent celluloid, finely shredded.. 1 oz.
Acetone, sufficient quantity.
Amyl acetate..................to make 20 oz.

Digest the celluloid in the acetone until dissolved and add the amyl acetate. From one to four ounces of flake bronze is to be mixed with this quantity of varnish. For silver paint or "aluminum enamel," flake aluminum bronze powder should be used in place of the gold. The celluloid varnish incloses the bronze particles in an impervious coating, air-tight and water-tight. As it contains nothing that will act upon the bronze, the latter retains its lustre for a long period, until the varnished surface becomes worn or abraded and the bronze thus exposed to atmospheric action.

All of the "gold" or, more properly, gilt furniture that is sold so cheaply by the furniture and department stores is gilded with a paint of this kind, and for that reason such furniture can be offered at a moderate price. The finish is surprisingly durable, and in color and lustre is a very close imitation of real gold leaf work. This paint is also used on picture frames of cheap and medium grades, taking the place of gold leaf or the lacquered silver leaf formerly used on articles of the better grades; it is also substituted for "Dutch metal," or imitation gold leaf, on the cheapest class of work.

A cheaper gold paint is made by using an inexpensive varnish composed of gutta-percha, gum damar, or some other varnish gum, dissolved in benzole, or in a mixture of benzole and benzine. The paint made with a celluloid amylacetate varnish give off a strong banana-like odor when applied, and may be readily recognized by this characteristic.

The impalpably powdered bronzes are called "lining" bronzes. They are chiefly used for striping or lining by carriage painters, in bronzing gas fixtures and metal work, in fresco and other interior decoration, and in printing; the use of a very fine powder in inks or paints admits of the drawing or printing of very delicate lines.

Lining bronze is also used on picture frames or other plastic ornamental work. Mixed with a thin, weak glue sizing it is applied over "burnishing clay," and, when dry, is polished with agate burnishers. The object thus treated, after receiving a finishing coat of a thin transparent varnish, imitates very closely in appearance a piece of finely cast antique bronze. To add still more to this effect the burnishing clay is colored the greenish-black that is seen in the deep parts of real antique bronzes, and the bronze powder, mixed with size, is applied only to the most prominent parts of "high lights" of the ornament.

Since the discovery of the celluloid amyl-acetate varnish, or bronze liquid, and its preservative properties on bronze powders, manufacturers have discontinued the use of liquids containing oils, turpentine, or gums, since their constituents corrode the bronze metal, causing their paint to finally turn black.

W. A. Dawson.

A VALUABLE AGENCY

THE best paint proposition today is The Sherwin-Williams Paint Agency. It means Good Goods, Good Methods and Good Advertising—the elements of business success.

It gets the business. It holds it. It helps all the trade of the store. It's profitable. It keeps growing. It means success in paint.

The struggle for business supremacy in all lines becomes keener every year. The pace gets faster and the fight hotter. Competition makes it so.

It's harder to succeed today than it used to be, but by employing the *right methods* you can succeed *quicker* and your success will be *greater* than that of others before you—quicker because methods have been improved, greater because the field is larger.

THE SHERWIN-WILLIAMS CO. helps every agent to improve his business, helps him to *lead* in his line—get there quicker—and provides him with the *right methods* to do it.

We bring to bear on the business of our agents the same methods that have made us successful, and properly applied they will never fail *anywhere* of the same results.

THE SHERWIN-WILLIAMS PAINTS are the best that can be made.

THE SHERWIN-WILLIAMS METHODS are the nearest perfect and most effective yet employed.

It will most assuredly pay you to secure our agency. It's an opportunity for you to share in one of the best and most complete systems that has ever been applied to a business. It's your opportunity to enlarge your business, and you ought not to let it slip.

SEND FOR THE "B-13" BOOKLET. A HANDSOME EXAMPLE OF PRINTING. IT EXPLAINS IN DETAIL THE PAINTS AND THE METHODS.

THE SHERWIN-WILLIAMS CO.

PAINT AND VARNISH MAKERS.

CLEVELAND.

CHICAGO, NEWARK, BOSTON, SAN FRANCISCO, NEW YORK, MONTREAL, TORONTO, KANSAS CITY.

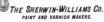

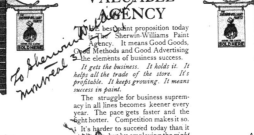

AN UP-TO-DATE GLASS IMPORTING HOUSE.

ONE of the best proofs of progressiveness is furnished in the business man who turns his losses to good account; who considers a reverse but as an opportunity for the greater progress. So The Toronto Plate Glass Importing Co., Hill & Rutherford, proprietors, have good claims to a reputation for progressiveness of the highest standard.

Last July their large warehouse on Victoria street, Toronto, was destroyed by fire. Temporary quarters were at once secured, and architects started to work on plans for a new warehouse on the site of the old and adjoining property. This warehouse is now almost completed.

Any person going over the premises, from the basement to the top floor, as a representative of HARDWARE AND METAL was shown by Mr. Hill the other day, would not fail to be impressed with the ingenious manner in which space has been economized and with the unique devices that have been introduced to insure the greatest saving in the time necessary for receiving, storing and shipping glass.

The exterior of the warehouse, a cut of which is shown, will be made exceedingly handsome by the use of white and colored glass or porcelain tiles, which this firm have used to great advantage in many buildings in recent years in both interior and exterior decoration. The tiles will be the shape and size of Roman bricks, and will give a substantial as well as attractive appearance to the front.

An unusual feature that has been introduced into the construction of this warehouse is the arrangement of the entrances, of which there are four. To the extreme left, as shown on the cut on this page, is the large goods-receiving entrance. Here a dray can be backed into the building and the goods removed by an overhead trolley crane. Next to this and included in the general entrance is the employes' doorway. In the centre is a large doorway leading to the offices. To the right is the entrance to the sales department. By this arrangement each department is separate from the others, yet, as all the dividing partitions are plate glass, of which over 1,000 sq. ft. are used, all goods are received and shipped immediately under the eye of the office.

The ground floor is devoted to the sales department and the storage of plate glass. In the sales department are hundreds of pockets containing single panes of window or sheet glass, thus making possible the delivery of assorted orders immediately after receipt of order. These pockets contain panes of the various sizes usually in demand, aggregating altogether over 100,000 ft. By a skilfully arranged system of steps and platforms the shipping clerks can reach any of the pockets, which are installed all the way from the floor to the ceiling, without using a ladder. The larger sizes are kept nearby in racks. Yet, notwithstanding the economy of labor effected by this arrangement, the movement of stock is so large and the necessity of speedy delivery so imperative that in addition to the men employed in filling orders, others are responsible for the daily replenishing of the pockets.

The remainder of the ground floor is devoted to plate glass of every conceivable size; also rolled and skylight glass, which are kept in racks. These racks are so firmly fastened to the floor that they could not be moved by any possible jarring or shaking. Half way to the ceiling is an elevated platform, on which mirror plate is stored in great quantities. This platform is reached by easy steps so that the workmen can easily take stock to or from it. Altogether, on this floor, 70,000 ft. of plate glass for windows and 35,000 ft. for mirrors will be kept in stock beside the big stock of rolled glasses.

The basement is devoted to window glass, single and double thick, under 24 x 30 in., in original packages. The boxes are received from the delivery wagons down a chute, and are piled in order according to size. Anyone walking through the many long corridors, neatly piled from the floor almost to the ceiling with unopened boxes, could not fail to be impressed with the magnitude of the stock carried by this firm.

The first floor contains window glass exceeding 24 x 30 in., in original packages. The back premises adjoining this floor are reserved for the manufacture and storage of stained, art leaded and fancy glass of all kinds. Here skilled artists and expert mechanics are engaged in designing and manufacturing the artistic productions for which this firm are noted.

The third floor is principally devoted to grinding glass, shipping and sand-cutting ornamental glass and making glass tiles. The process of shipping and sand-cutting is as difficult as it is interesting, but, owing to their large experience and the employment of expert mechanics in this as in other departments, this firm have steadily increased their output in this line. The use of glass tiles in Canada has increased greatly in recent years, and this firm have kept well in touch with every new development in the trade.

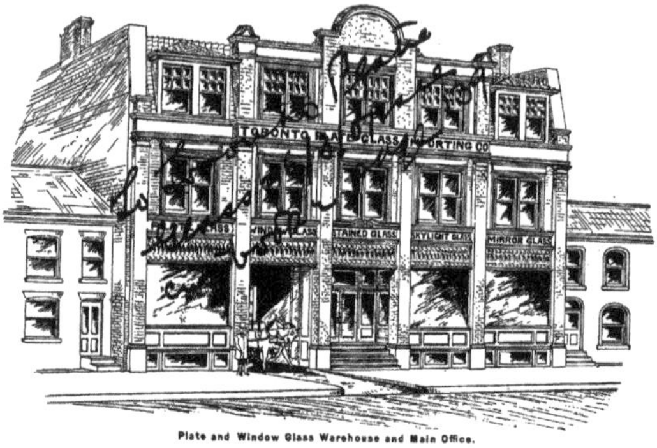

Plate and Window Glass Warehouse and Main Office.

On this floor is also the room devoted to the manufacture of the necessary distilled water used in the making of mirrors. By an ingenious contrivance enough distilled and warm water is secured in one operation to supply warm water to the entire factory and warehouse, as well as the bevelling, polishing and silvering plants, which are being installed on the second floor for the manufacture of both plain and bevelled mirror plate. The raw plates will be brought up by the hoist to this floor, and conveyed to the bevelling machinery. After being bevelled they are polished either by hand or machine. Then they are moved on to the silvering-room, where the most modern tables, heated by steampipes and covered by slate slabs, are installed. After the silvering operation is complete they are left at least 24 hours to dry, when they are packed and stored for shipment. The entire process of manufacture, from the receipt of the glass off the hoist to its shipment in packages, has been reduced to the basis of the greatest possible economy consistent with the large production necessary. Their stock of plain and bevelled German mirrors and shock mirror plates aggregates fully 100,000 sq. ft.

Another evidence of the firm's progressiveness is the fact that they are also erecting a bending plant, which is being constructed on the most modern lines. This building, a diagram of which is shown, is likely to be ready to start about the middle of May. The capacity of this plant will be in the neighborhood of 150,000 sq. ft. of bent glass annually. This amount is considerably greater than the demand now experienced for this kind of glass. But this firm have recognized that the sale of bent glass in this country has been limited by its high price and the time necessary to secure the filling of orders, and are confident that their enterprise in starting such a large plant, by which even large orders can be filled at a few days' notice and which will be operated at comparatively low cost, will effect such a material increase in the use of bent glass that their plant will be kept running at its full capacity.

The Mirror and Ornamental Glass Factory.

The Glass Bending Factory.

THOMAS FIRTH & SONS, Limited

SHEFFIELD, ENGLAND.

Best Tool, Drill, Chisel, Tap, Die, Punch and Sheet Steels.

THE STANDARD FOR PAST 50 YEARS IN CANADA AND UNITED STATES.

Full Stocks Carried.

Insist on having this Brand when ordering.

STEEL

H. W. DeCOURTENAY & Co.

86 and 88 McGill Street,
Montreal.

Best Tool Steels,
Rock and Mining Drill Steels,
Best Sheet Steels,

For
Lathe and Planer Tools,
Dies, Punches, Taps, Drills,
Blacksmith Tools, etc.

Cold Rolled Steel Shafting,
Bright Cold Rolled Steel Sheets,
Round, Square, Flat Bars, etc.

Steel for
Cant Hooks, Wedges, Shims,
and all Purposes.

The serious fire of January 23rd destroyed our **entire** stock of Steels, but are now receiving full **New Stocks**, and will be pleased to have your inquiry when in want.

ALEXANDER GIBB,

Manufacturers' Agent and Metal Broker,

Office and Sample Room,
22 St. John St.,

MONTREAL.

Galvanized Sheets.	Bar Copper.	Agricultural Implement Chain.
Tinplates.	Sheet Brass.	Canada Plates.
Terneplates.	Tobin Bronze.	Ingot Tin.
Tinned Taggers.	Norway Iron and Steel.	Tool Steel.
Sheet Steel.	Norway Nail Rods.	Circular Saw Plates.
Black Taggers.	Hoop and Band Steel.	Tyre Steel.
Imitation Russia Sheets.	Coil Chain.	Machinery Steel.
Sheet Iron.	B. B. and B. B. B. Crane Chain.	Zinc Sheets, Spelter and Oxide.
Genuine Russia Iron.	Dredge and Steam Shovel Chain.	Soldering Coppers.
Sheet Copper.	Trace Chain, etc.	Brass and Copper Tubing.

Wheelbarrows—all kinds.	Cutlery.
Store Trucks.	Seamless Steel Hollow-ware.
Washing Machines.	"Lava" Enameled Ware.
Handles (wood, all kinds).	"Lava" Enameled Filters.

White Lead (dry).	Orange Lead.	Paris and Milori Blue.
Red Lead.	Litharge.	Persian Red.

Also Earth, Mineral, and Chemical Colours.

INQUIRIES SOLICITED.

UNITED INSURANCE FOR HARDWAREMEN.

How it is Carried on in Minnesota and Ohio.

THE question of mutual insurance has come in for a great deal of consideration of late, particularly by retail organizations in the United States. In Canada the discussion so far has been largely confined to individual merchants.

At the recent session of the Ohio Hardware Association the question of mutual insurance was the subject of a special report by the committee on insurance. This report, which may give some light to those who are considering the question in Canada, is herewith reproduced. The report, as presented by the chairman, reads :

"You probably all know as much about mutual insurance as any one of our committee. What is not known is difficult to find out.

"I wrote the Commissioner of Insurance of this State and also the secretary of the Minneapolis Mutual Insurance Company, and received a reply in both cases, referring us to our State laws in the matter, and your committee put the questions to you for consideration as to whether we are in condition to assume insurance, whether it might be profitable for us, or might not, and whether we want to take it up, whether this is the time and place to pledge ourselves, and, if it be possible, to now secure a pledge to take enough insurance to comply with the State law. The limit of the Minneapolis policy is $3,000. The State limit is $5,000, beyond which you are not allowed to issue mutual insurance in this State ; and I think not in any other. If permissible, and the members would like it, I shall be very glad to read the law on the subject, viz. :

"Before incorporation, you must have—

"1. At least $500,000 of insurance subscribed.

"2. Not less than 200 separate risks.

"3. No risk shall exceed $5,000.

"4. A premium on insurance subscribed for one year must be paid in cash, aggregating not less than $10,000 in cash.

"5. Each subscriber must agree in writing to assume a liability to be named in the policy subject to call by the board of directors in a sum not less than three, nor more than five annual premiums.

"With these conditions met, you can organize a mutual fire insurance company under the regular insurance laws of Ohio, the plans and details of which we will work out for you just as soon as desired. You can limit the risks in your charter or in your by-laws to insurance carried on stocks of hardware, tools and implements, or can arrange to carry any of the lines of stock and trade that you deem advisable. Perhaps the best plan would be that no risk shall be taken except on hardware, tools and implements, unless the board of directors in regular meeting and after special examination authorize it.

"In a mutual company, the incorporation papers must be first submitted to the Attorney-General, and, if by him approved as being in accordance with the laws and demands of Ohio, shall be recorded by the Secretary of State, and a copy thereof deposited with the Superintendent of Insurance.

"The laws of Ohio provide :

"1. Manner in which the election of directors shall be held.

"2. How a company must invest its capital—that is, in United States bonds, Ohio State bonds, county, township and municipal bonds, bonds and mortgages on unencumbered real estate within Ohio worth 50 per cent. more than the sum loaned thereon, exclusive of the buildings.

"3. The stock of any national bank located in Ohio.

"4. First mortgage bonds of railroads within this State, etc.

"The statutes also provide the method and manner in which the company may invest accumulations ; provide for thorough examination by the Superintendent of Insurace; for the mode and manner of making assessments upon members, and enforcing the same ; for the filing of annual reports to the Superintendent of Insurance; for the cancellation of policies and the rates therefor; makes the premium notes given by the company not negotiable; provides for the examination of mutual companies by order of courts of common pleas in counties where the principal office is held ; in fact, throws around such an organization all safeguards in the way of directing, controlling and examining the company that are required of any standard company doing business in the State.

"There has been in the past in this State complaint that mutual companies have failed ; yet it is the best and cheapest insurance that the farmer has to-day, and there are no better safeguards provided in any of the States of our Union than are the Ohio insurance laws. Not a month goes by in which some mutual or other company is not investigated, and, if found wanting, they are debarred from doing business in Ohio. The examinations, I am told, are very thorough and very frequent. If I can answer any questions, please call upon me, although I think Secretary Gray, who has had some correspondence upon the subject, is probably better informed than I am."

In response to a question as to what has been the experience of the Minnesota organization, the chairman stated that it was very flattering, that they had made money and paid dividends. He called attention to copies of by-laws, rules and regulation of the Minnesota concern, which were for free distribution for those desiring them. The Minnesota organization had been in operation but one year, but had been authorized by the State Commissioner to issue an additional $500,000 insurance for the coming year.

To find how many were favorably disposed toward the organization of a mutual insurance company in the association, a straw vote was taken with but three votes in the negative.

Tellers were appointed to collect written slips from each member present stating amounts of insurance he would subscribe for under a satisfactory arrangement for a mutual insurance company. The tellers reported as the result of their account a total of $224,000 subscribed.

REVISED POCKET PRICE LIST.

The attention of the readers of HARDWARE AND METAL is drawn to the attractive advertisement of The Mechanic Supply Co., Quebec. This enterprising firm are able to handle any large or small orders, as they carry one of the largest and best assorted stocks in Canada.

They have recently issued a new and revised pocket price list which any steam-fitter or engineer may find convenient, a copy of which can be had on writing the firm and mentioning this paper.

HIGH-GRADE HICKORY HANDLES.

On page 90 will be found the advertisement of The Laing, Ritchie Co. of Essex, Limited. This firm manufactures a full line of handles, whiffletrees, neckyokes, extension ladders, bag trucks, wagon jacks, etc. The firm is up-to-date in every respect. It lately moved into new premises, which have been equipped in such a way that they take no second place for handle works in the Dominion. The firm's hickory axe handles cannot be surpassed in quality of timber, finish and grading. The factory is admirably located for this class of work. It will be to the interests of the hardware trade to correspond with this company.

YOU'LL BE LIKED BY YOUR CUSTOMERS AND DISLIKED BY THE ICE MAN

If you handle the

Leonard

CLEANABLE

Refrigerator

Eight Walls,
Complete Insulation,
Perfect Ventilation,

Make it the

Greatest Ice Saver

The Most Perfectly Constructed Refrigerator in the World.

THE

Dangler Wickless Oil Cooking Stove

NO VALVES to clog or get out of order.
Removable Oil Tank.
Sight Feed.

Visible Burner.
Quickly Taken Apart.
No Trouble to Keep Clean.

THE FUEL ECONOMIZER

We are the only Authorized Canadian Agents for the above. A pleasure to quote you.
Inquiries Solicited.

THE McCLARY MFG. CO.

London. Toronto. Montreal. Winnipeg. Vancouver.

NEW STEVENS RIFLES AND SHOTGUNS.

THE J. Stevens Arms and Tool Co., Chicopee Falls, Mass., are putting on the market a line of new shotguns, which, on account of the new features incorporated in them, and the reputation of their makers for sterling goods, should have a large sale.

The line consists of single-barrel shotguns in three numbers, 100, 110 and 120.

No. 100 is a non-ejector, but has an "electro steel" barrel, choke bored, pistol grip, rubber butt plate, case-hardened frame, bored for nitro powder.

No. 110 is the same as the No. 100, except that it is furnished with automatic shell ejector, and has checked capped pistol grip.

No. 120, which is known as the Fancy Ejector, has a patent fore end, an automatic

No. 100.

No. 120.

Maynard, Jr.

ejector, which is a special device whereby the operator can change it at will from an ejector to an extractor by two turns of a screw.

These shotguns are listed as follows: No. 100, $7.50; No. 110, $8.50; No. 120, $10. They are made in the usual weight and length of barrel, and in 12, 16 and 20 gauges. They are made exceedingly handsome by their fine finish.

About the middle of April this firm will also put on the market a new rifle, which will be called the Stevens-Maynard, Jr. This rifle has the same action as the old Maynard rifle, from which it was designed. Special attention has been paid to the quality, finish and accuracy of this rifle. It has an 18 in. half-octagon barrel, steel frame, walnut stock and steel butt plate. Some of its features are that it shoots 22-short only, and is accurate; all parts are machine-made, being true to gauge, and interchangeable, and that it weighs 2 pounds and 14 ounces. It is packed one in a box and lists at $3, with the usual discount to the trade.

THE CANADA PAINT COMPANY.

It is evidently the policy of The Canada Paint Company to keep every department well to the front. They are counted amongst the most astute and liberal advertisers in Canada. Last month, we are assured, was the busiest February for ready-mixed paints which they have ever experienced. Believing, also, that there will be a brisk demand this spring for a good grade of white lead, The Canada Paint Company have been advertising, with marked success, the "Painters Perfect White Lead"—a brand of white lead made solely by themselves, and said to possess extraordinary merit as regards its whiteness, permanence and covering properties. Shipments will be made either from Montreal or Toronto. Their advertisement will be found upon page 71.

A WINDOW DISPLAY IN OTTAWA.

Some of the Ottawa merchants make attractive window displays. One of particular mention this week is that of Blyth & Watt. On passing along Bank street one cannot help notice the handsome display of paints and varnishes, built in pyramid style, using small cans for the top and larger sized cans for the foundation, the whole being mounted on one of those adjustable display stands with metal ends. Then on top are different colored panels in a long row placed right above the stand and hung from the ceiling. For background they have some handsome banners issued by a Chicago paint firm.

In the corner window they have fitted up a bathroom, all the smallest details being attended to so as to make it a model up-to-date bathroom. Even the tooth brush and latest bath mats have not been overlooked.

Generally speaking, Ottawa merchants can hold their own for window display in any line of goods.

FURTHER FILE ABSORPTION.

A FEW weeks ago HARDWARE AND METAL announced that the Nicholson File Company, Providence, R.I., had purchased the business of the Arcade File Co. Now it has taken over the Kearney & Foot Co., of Paterson, N.J., as the following circulars which are being sent out to the trade announce:

The Nicholson File Company, of Providence, R.I., having purchased the property of the Kearney & Foot Company, of Paterson, N.J., the business of this company passes wholly into their hands. The works of the company at Paterson, N.J., and Kent, Ohio, will be under their control and management from this date, and all accounts due this company at this date must be paid to them.

The long and successful career of the Nicholson File Company has made the name a synonym for all that is of good repute, in quality of goods, and in business methods, and will secure to them the ready transfer of the trade which has been so kindly bestowed upon us, and for which we take this occasion to extend our most hearty thanks.

KEARNEY & FOOT COMPANY,
JAMES D. FOOT, President.

Having become sole owners of all the property of the Kearney & Foot Company, of Paterson, N.J., the factories of this company at Paterson, N.J., and Kent, Ohio, will be operated by us, in the continued manufacture of the well-known K. & F. brand of files.

All orders and inquiries for prices, all remittances, and all correspondence in connection with the business of the Kearney & Foot Company should be addressed to the Nicholson File Company at Providence, R.I.

A stock of the K. & F. files will continue to be carried at the store, Nos. 100 and 102 Reade street, in New York City, for the convenience of those doing business in that city, or in the adjacent cities, whose wants require the immediate filling of small orders.

We hope to retain the trade of all who have heretofore dealt in the K. & F. files, and it shall be our aim in the conduct of our business to merit their continued loyalty to this brand of files.

NICHOLSON FILE COMPANY,
SAMUEL M. NICHOLSON, President.

The following merchants are now putting in Bennett's patent shelf boxes: Fullerton & Zieman, Monkton, Ont.; A. W. Kelly, Collingwood, Ont.; Meyers Bros., Toronto, Ont.; W. Robinson, Selkirk, Man.

HEATING AND PLUMBING

ANNUAL MEETING OF THE MONTREAL ASSOCIATION.

AN adjourned annual meeting of the Montreal Master Plumbers' Association was held in St. Joseph's Hall, corner of St. Elizabeth and St. Catherine streets, Thursday evening, March 7. President Giroux presided over a good attendance. The yearly reports from the various committees were read. The auditors', financial secretary's and treasurer's reports were received and adopted on motion of Mr. John Watson, seconded by by Mr. P. C. Ogilvie. The secretary's report was also adopted on a motion of Mr. Sadler, seconded by Thos. Moll.

REPORT ON LEGISLATION AND APPRENTICESHIP.

A joint report was read by Mr. J. W. Hughes from the sanitary, legislation, arbitration and apprenticeship committees. This report read as follows :

JOINT REPORT OF COMMITTEES OF MASTER PLUMBERS' ASSOCIATION OF MONTREAL.

It being deemed desirable to present a joint report, the following is respectfully submitted :

The executive committee held three regular meetings principally in regard to the convention. The success of the convention is the best report your committee can make as to its work.

Sanitary Committee. — No regular or special meetings were held, there being no special work calling for same, but the members of committee were in regular attendance at the special meeting of the city aldermen in connection with the new by-laws issued in the city hall, and largely owing to their efforts several important features were added to the existing by-laws.

Arbitration.—Owing to the great harmony prevailing amongst our members, there was no work of a special nature for the care of the arbitration committee, but they were at their post at all times ready to give their services to any of our members who might have required the same.

Legislative Committee. — The principal work of the legislative committee was he watching of the legislation in connection with the city plumbing by-law. While the law is, in some respects, not all that your committe would desire, yet, it is in many points an improvement on the old one, and if our successors use their efforts and influence to secure its efficient enforcement, much good will result, not only to our citizens at large, but to the members of our association.

Apprenticeship committee.—Your apprenticeship committee regret to report that this important question is still in the unsatisfactory state that has prevailed for some time past. We fear that unless pressure is brought to bear from an outside source that matters will continue to drift. The plumbing class is still in existance, and at last reports was doing good work in training young men in methods not always possible to be given in the shop. As the new by-law calls for an examination, the young men of the future, either by shop methods or through the agency of the school, must receive some training if the standard of the trade is to be maintained.

The auditing committee will make a special report.

Chairman executive committee, J. A. Giroux, president ; chairman arbitration committee, F. J. Canult ; chairman legislative committee, John Watson ; chairman apprenticeship committee, J. Montpetit ; chairman sanitary committee, J. W. P. Cushes.

Mr. P. C. Ogilvie moved, and Mr. Thos. Moll seconded, that this report be received and adopted. Carried.

THE PRESIDENT'S ADDRESS.

The president, Mr. J. A. Giroux, then gave his report, which was received with applause. He spoke as follows :

GENTLEMEN,—I have the honor to submit to you, as president, the report of our deliberations for the past year. I regret to tell you that our meetings have not been as numerous as they should have been, on account of indifference on the part of members of our association. We ought to make it a matter of duty to attend our meetings, for they are held in our interest. We can, in this way, thoroughly discuss a large number of very important trade questions.

The officers of the association have been present at two meetings of the city board of health, and have discussed with the members of that commission several questions in connection with the construction—among other things—of tile drains. In spite of a strong opposition, we have succeeded in having several amendments introduced into the existing by-laws, and we hope that before long we shall have plumbing regulations that will be a credit to the city of Montreal.

You will remember that we were honored last June with a visit from the Confederation of Master Plumbers, who were in convention in our city. Eminent success, I do not fear to say, crowned the efforts of the reception committee, who performed their duties nobly. We showed our guests all our picturesque and interesting scenes which make our city such a favorite tourist resort, and we put a grand finish to our labors by a highly-successful banquet at the Windsor Hotel. I shall take advantage of this occasion to thank the corporation of the city of Montreal for the generous aid extended to us, and I thank especially our friend and confrere, Ald. Jos. Lamarche, for the aid he gave us on that occasion. I also cordially thank our supply men for the aid that they have given us and the honor they conferred on us by being present at our deliberations.

Gentlemen, this year this association will meet in Toronto. I am sure you will send there, as in the past, men who are worthy to represent us and who are capable of safeguarding our interests.

We have had no banquet this year, as yet. We have considered that, in view of the misfortune that has plunged the British Empire into grief, we should postpone that fraternal function.

I conclude, gentlemen, entreating you to take a more active interest in our deliberations. Our meetings are held on the first and third Thursdays of every month, and it is by meeting often and in large numbers that we can secure means towards the amelioration of our profession. Our association is powerful, and, if we wish, we can do much to safeguard the health of the citizens of our large and beautiful city.

The reports of the different committees will be submitted for your approbation.

The meeting then proceeded to the election of officers for the ensuing year. Mr. John Date was reelected honorary president, a position he has held for some time. He is the oldest plumber in Montreal where he has been plumbing houses for upwards of 50 years.

Mr. Carroll then moved that Mr. G. C. Denman be elected president. In amendment Mr. Denman moved that Mr. Hughes be elected president. Mr. Harris proposed an amendment to the amendment that Mr. Jas. A. Sadler be elected president. This was seconded by Mr. Hughes, who put forth Mr. Sadler's qualifications in an able manner, and carried unanimously.

Mr. Champagne was proposed for the vice-presidency by Mr. Moll, and Mr. Ogilvie's motion, that Mr. Moll be given the office, was carried. The second vice-presidency was assigned to Mr. P. C. Ogilvie, and the third vice-presidency to Mr. J. W. Harris.

Mr. E. C. Mount was reelected recording secretary on motion of Messrs. Watson and Harris. The new financial secretary will be Mr. P. J. Carroll. Mr. Giroux, the retiring president, was made the new treasurer. Mr. Hughes was made English corresponding secretary, while Mr. Lamarche will read and write the French correspondence.

The new president was then escorted to the chair by Messrs. Carroll and Denman and he assumed his position amid the applause of his confreres.

In a neat speech Mr. Hughes proposed a vote of thanks to the retiring president and officers for the work they had done during the past year. Mr. Harris, in seconding

MECHANICS W SUPPLY CO.
For Steam, Water or Gas
SUPPLIES
We are "right in it," and don't forget us when asking for prices.
We carry one of the largest and best assorted stocks in Canada.
Ask for our illustrated wall sheets and new revised pocket price list.
Mechanics Supply Co.
96 St. Peter Street
QUEBEC.
Long Distance Telephone 456.

the motion, also paid tribute to their efficiency. Carried.

CHAIRMEN OF COMMITTEES.

The chairmen of committees were then appointed as follows:

Sanitary Committee—Mr. Watson.
Arbitration Committee—Mr. Denman.
Auditing Committee—Mr. Champagne.
Legislation Committee—Mr. Thibeault.
Apprenticeship Committee—Mr. Jos. Brunet.

Mr. James Madden was received into membership. Mr. Watson gave notice of motion that the association hereafter meet only once a month instead of twice a month as now.

SOME BUILDING NOTES.

ARCHITECT STEWART McPHIE, Hamilton, has had plans accepted for two brick residences on Queen street south, three on West avenue, and a vestibule for the Macnab Street Presbyterian Church.

Mr. Cooke, Painswick, Ont., intends erecting a new house this summer.

Two new brick stores are being erected on Hastings street, Vancouver, adjoining the Mint.

The Brandon Ginger Ale Works, Brandon, N. W. T., have decided to erect larger premises.

R. Chestnut & Sons, Fredericton, N.B., have the contract for erecting a $4,000 residence for Dr. F. W. Barbour, of that place.

A military schoolhouse and a gymnasium are to be erected by the Halifax military authorities.

Mr. Smith, Ranton, Ont., intends erecting a new brick house in that place this spring.

The Portage and Lakeside Agricultural Society, Portage la Prairie, Man., intend erecting a $4,000 agricultural hall next summer.

A new schoolhouse is being erected at S.S. No. 9, Conroy, Ont.

J. R. Feick intends building a house in New Hamburg, Ont., this summer.

The congregation of the Swedish Lutheran Church, Winnipeg, have appointed Joan Mattson, O. Fagerberg, E. Anderson, O. Lindgren and J. Wallman a building committee to erect a new parsonage and to enlarge the church building on Henry street.

A brisk building season is expected in Midland, Ont., this year. R. Little and Mr. Grise intend erecting new store buildings. Thos. Chew expects to build several houses. Mr. Jeffrey is having plans prepared for a three-storey, 100 x 50 ft. building. Several others intend building during the summer.

THE MONTREAL ASSOCIATION'S NEW PRESIDENT.

THE newly-elected president of the Montreal Master Plumbers' Association has always been regarded as one of the hardest working members on the roll. He joined the association when it was first organized, indeed, perhaps before it was organized, and has always been one of those who supplied the backbone and staimina that gave it vitality and strength. No one has merited the honor more than he.

When the plumbers of Montreal were first organized, about 12 years ago, as a branch of the Montreal Contractors' Association, Mr. J. A. Sadler was one those who saw the spark flash fire and die. Then, when the plumbers formed their own corporate body, he was among those who still advocated organization and brought it about. Every year he has been an officer

President J. A. Sadler.

and his executive ability has been well employed. Twice he was financial secretary, which position he held last year; once he was auditor, and another year was vice-president. He though the hado ccupied an office long enough but, now, instead of retirement, which he wished, he is given the most responsible position—the presidency.

Mr. Sadler is an experienced plumber. He has been in the business for 35 years, 25 of which he has had his name on a sign above his own office door. For seven or eight years he has occupied his present stand at the corner of Bleury and St. Berthelet streets. As much by his personality, as by the enterprise and skill displayed in his profession, he has worked up an important business in his section of the city.

He enters upon his duties with the hope of infusing new life into the plumbers' association. He realizes that there is a good deal of work that the association can do, if the plumbers of the city would only join hands and do it.

PLUMBING AND HEATING NOTES.

D. P. Cane has opened a plumbing business in Vancouver.

The assets of Thomas Forest, plumber, Montreal, have been sold.

NATIONAL EXECUTIVE TO MEET.

"The sub-committee in Toronto of the executive committee of the National Association will probably be called to meet in a few days," said President W. H. Meredith to a representative of HARDWARE AND METAL this week. "In all probability I will call a meeting for next Monday. We will discuss the date of the next convention, which is to be held in Toronto this summer, and will set the per capita assessment for the year."

"When will the convention be held?"

"The executive must decide that. It was requested at the last convention by some members that the meeting be early in June, but I am afraid that it will be impossible to have everything ready before the end of the month, at which time we have met in former years."

"Have you decided on your programme yet?"

"No; but Secretary Mansell has sent, or is going to send, letters to the various associations asking them to state what matters they desire to have discussed. The answers we receive will largely determine the nature of the programme."

BUILDING PERMITS ISSUED.

Building permits have been issued in Hamilton, Ont., to M. Stow for a $1,000 residence on Duke street; and to E. B. Patterson for an $1,800 house on Fairleigh avenue for W. C. Toye.

The following building permits have been issued in Ottawa: Lieut.-Col. Macpherson, office and bakery, Somerset street, $1,200; James Walker, brick veneered house, Patterson avenue, $1,500; Thomas Clarey, four brick veneered cottages, Gilmour street, $8,000; Mrs. C. Trudeau, two solid brick dwellings, Stewart street, $6,000.

Building permits have been issued in Toronto to the Wind Engine & Pump Co., for a three-storey factory on Atlantic avenue, to cost $6,000; to Isaac Ritchie, for a residence at 284 Macpherson avenue,

to cost $1,750; to Thos. Maclean, for a pair of residences at 39-41 Cowan avenue, to cost $5,000; and to W. R. Membery, for additions to the Daly House Hotel, corner Front and Simcoe streets, to cost $5,000.

PLUMBING AND HEATING CONTRACTS.

Guest & Co., Toronto, have contracts for plumbing and gas-fitting in four houses on Queen street east for R. Borthwick; one on Queen street east for R. Gregory, and one on the corner of Ontario and Gerrard streets; and for alterations to plumbing, etc., in the Schiller House, Adelaide street east, Toronto.

The Bennett & Wright Co., Limited, Toronto, have contracts for heating a house on Indian road for Paul Syzeliske; for plumbing and gas fitting in two houses on Howard street for John Stark & Co., and in one house on Givens street for R. Langdon; in White, Allen & Co.'s factory on Ontario street, and for plumbing in the new factory of The T. Eaton Co., Limited.

EXPERIENCED BRUSHMAKERS.

The firm of Skedden & Co., brush manufacturers, Hamilton, was established in September, 1895 by the present proprietor, Edwin Skedden, who is a practical brushmaker, and understands the making of brushes of every description, from the heaviest machine brushes to the finest jewellers' and dentists' wheel brushes.

Their factory is well equipped with the latest machinery used in the brush-manufacturing business, and the work they produce is of a superior quality. The employes are all experienced brushmakers, and can be relied on for first-class work.

Mr. Skedden is well known throughout the Dominion as being one of the leading rifle shots of Canada, and has been a member of the Bisley Rifle Team on several occasions, where he has met with great success. The factory and warerooms are Nos. 4, 6, 8 and 10 Park street south, Hamilton, Ont.

THE DURHAM RUBBER CO., LTD.

The Durham Rubber Co., Limited, have moved their office from 60 Yonge street, Toronto, to their factories at Bowmanville, Ont. Experience has shown them that it is not in their best interests to have their office and factory separated.

In closer touch with their factory they are enabled to handle their fast-increasing business with more satisfaction to themselves as well as to their customers.

Business with them is "booming." Spring orders have been coming in so fast that for the past six weeks they have been steadily running overtime.

In Toronto they are represented by Mr. D. P. Sheerin, who continues to occupy their old warerooms at 60 Yonge street.

THREE USEFUL CONTRIVANCES.

THE Smith & Hemenway Co., of 296 Broadway, New York City, are just putting on the market the 1901 pattern "Russell" staple puller which we illustrate below. This shows the different operations to which the tool can be put and the improvement over the 1900 pattern will be appreciated at a glance by the user.

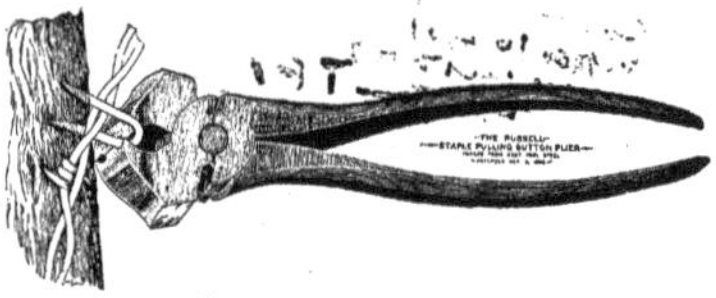

The 1901 "Russell" Staple Wire Puller.

Eight tools are combined in one, and this year's pattern has come into such favor that we are advised that they are from three to five weeks behind in their deliveries. It is needless to point out the superiority of this as it will recommend itself to the user. We are informed that the above company have a large export business on this tool to all the civilized countries on the globe.

No. 2000—Farmers' and Machinists' Universal Tool.

Another article which is meeting with great favor is their No. 2000, Farmers' and Machinists' Universal Tool, illustrated above. This is made from the finest quality of Brescian steel and is a combination of eight tools in one. As the demand for a good farmers' and machinists' universal tool has steadily increased for the last two years, this company decided to put out the best that mechanical skill can produce, and a trial will convince anyone of its superiority.

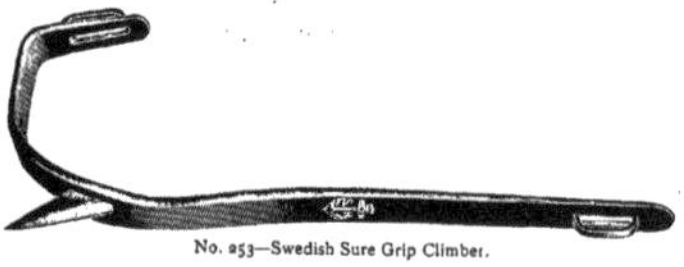

No. 253—Swedish Sure Grip Climber.

We also illustrate the Swedish Sure Grip Climber, which the above company have put on the market. The spur, in addition to being welded to the upright, is also riveted, thereby insuring a solid connection. We are informed that in a number of tests recently made, they have demonstrated beyond a question of a doubt, that the spur is there to stay and it is practically impossible to knock it out. They have been experimenting quite some time in the manufacture of climbers, and have now struck a standard and will hereafter warrant each and every climber, and should the spur loosen or come out, they will either refund the money to the purchaser, or replace the climber by a new pair.

This company publishes what is known as The Green Book of Hardware Specialities, and it would be well for anyone interested to write for a copy of same, as it gives information with reference to their goods. Their advertisement will be seen on page 101. When writing, please mention HARDWARE AND METAL.

LARGE IMPORTATION OF HOSE.

What is likely the largest import order of American rubber hose ever placed by a Canadian wholesale house was given a few days ago by Lewis Bros. & Co., Montreal. The shipment will include cotton, rubber-lined and rubber garden hose, and will run up to hundreds of thousands of feet. Buying in such quantity, the firm is able to secure special values. Fortunately for the concern also, values have advanced 20 per cent. since the order was placed. Quotations are now being offered to the retail trade on these goods, and we believe that they have already led to some large sales.

Radiant Shelby

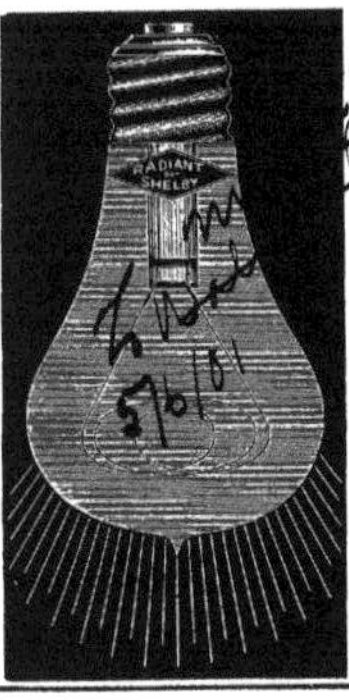

Incandescent Lamps

Patented in Canada.

HIGH EFFICIENCY. USEFUL LIGHT. LONG LIFE.

Manufactured by

The Ontario Lantern Co.,
HAMILTON, CAN.

We can supply the above Lamps any Base, Candle Power, or Voltage, and warrant our Lamps to be high grade in every respect.
Orders solicited from the large Lighting Companies and wholesale dealers.

WALTER GROSE, Montreal, Can.

SOLE SELLING AGENT.

NEW CENTURY BANNER

Cold Blast Tubular Lantern

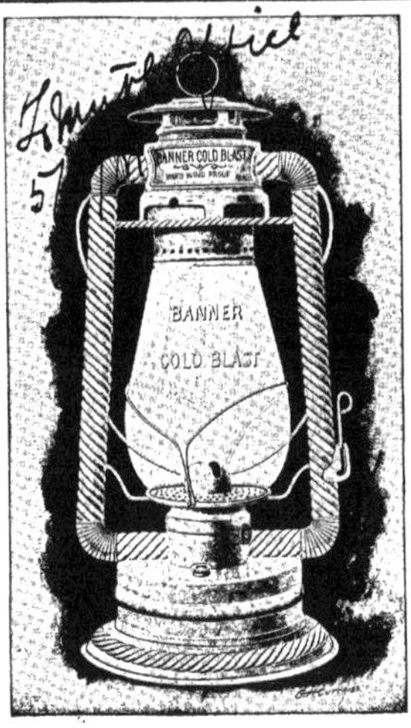

Warranted
HANDSOME IN DESIGN
PERFECT IN CONSTRUCTION

The best Lantern for outside light on land or sea.
Send for descriptive circulars.

Also Climax Safety Lanterns; "Little Bobs" Brass Lanterns; Gasoline and Acetylene Lamps.

Banner Lamp Burners; Gas Burners; and other Light Producing Goods.

Sold entirely by the Jobbing Trade throughout the Dominion.

Manufactured by

WALTER GROSE, Sole Selling Agent, Montreal, Can.

THE ONTARIO LANTERN CO. HAMILTON, CANADA.

AN IMPROVED RAILWAY MILK CAN.

THESE are days when a great deal of attention is being paid to the sanitary condition of the vessels in which milk is carried as well as to the sterilization of the milk itself. Among the firms that have been paying particular attention to the vessels in which milk is carried is the Kemp Manufacturing Co., of Toronto. One of the results of this particular attention is an improved railway milk can, an illustration of which is herewith given.

Kemp's Improved Railway Milk Can.

When designing the can the efforts of the company were centred in making a vessel that would, on the one hand, possess the minimum of possibility for the gathering of dust and other forms of dirt, and that would, on the other hand, possess the maximum of strength. In regard to both these points it appears to have been highly successful.

In ordinary railway milk cans the custom has heretofore been to make the neck, collar and breast in three pieces. In the improved can the Kemp Co. has just put upon the market, the collar and neck is in one piece, and this in turn is double-seamed into the breast. This is afterwards retinned, and is then not only without any crevices in which dirt can find a hiding place, but is as strong as if the neck, collar and breast were in one solid piece.

A new idea has also been developed in the joining of the breast and the body of the can. This is the insertion of a steel hoop or band between the breast and the body, thus greatly strengthening it, and furthermore prevents the cans rubbing together and becoming injured in transit. The handles are extra large and of great strength. Inside and underneath the can is a heavy steel hoop. This hoop is thoroughly sweated to the can, and leaves no place for dirt to congregate. The cover is made large, so as to extend all over the neck of the can and prevents dirt getting into the receiver, while the collar fits so close into the can as to prevent the milk slopping over the top.

The Kemp Manufacturing Co. are much gratified with their efforts, particularly in view of the fact that in competition with eight firms, Canadian and United States, they secured the order of the City Dairy Co., Limited, of Toronto, for 2,000 of these railway cans. One of the cuts we print is from a photograph of 1,000 of the cans which composed a part of the order. The order is believed to be the largest of the kind ever given in Canada. If placed one on top of the other these 2,000 cans would make a column of 4,000 feet, or over three quarters of a mile.

HONORING MR. FOOT.

THE members of the office staff of The McClary Manufacturing Co., London, assembled in the Palace Cafe, in that city, on Saturday evening last, and tendered a farewell dinner to Mr. John J. Foot, who has been promoted to the management of a branch for the Eastern Provinces just opened by the company at St. John, N.B. Upwards of 20 of the staff were present, and Mr. J. K. H. Pope was chairman. The dinner was all that could be desired, and, at its conclusion, Mr. Pope suitably introduced a short toast list, first calling on the company to drink lustily to the health of "The King." Col. Gartshore replied to "The McClary Co.," reviewing the growth of the business from the time he identified himself with the firm up to the present, when the office staff included 25 clerks. Col. Gartshore spoke in the highest terms of the business ability of Mr. Foot, who began 15 years ago in the lowest position in the office, and was now being rewarded for his diligence and strict attention to business by being placed in the highest position in the gift of the company. Reference was also made to the fact that Mr. Foot was the youngest of the branch managers.

In introducing "Our Guest," Mr. Pope also eulogized Mr. Foot and complimented him on his excellent advancement.

Mr. Foot made a suitable reply, thanking the previous speakers for the many kind expressions uttered by them, and expressing the regret he felt at leaving London, and especially the many warm friends he had made in the McClary office.

As the gathering was about to disperse Mr. Rowlands, the oldest employe in the London office, arose and after speaking of Mr. Foot's promotion, said that while the guest of the evening would long remember the farewell tendered him by the staff, it was the desire of the latter to give him greater occasion not to forget it. Mr. Rowlands thereupon presented Mr. Foot with a solid gold locket, on one side of which was his monogram, in raised letters, and on the other the following: "Presented to J. J. Foot by the office staff, McClary Mfg. Co., London, Ont., 1901."

Mr. E. H. Grenfell, the company's oldest traveller, also spoke of Mr. Foot and his steady advancement.

1,000 of Kemp's Improved Railway Milk Cans.

Dairy and Kitchen Utensils

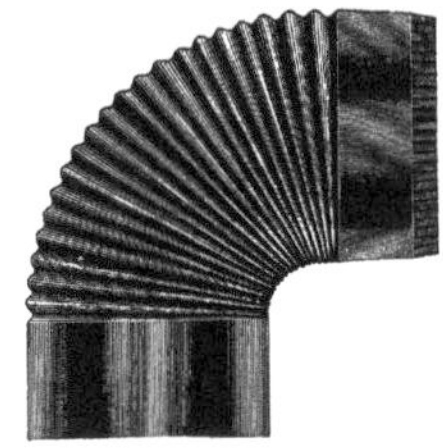

In Enameled Ware,
Tin, Sheet Steel and Copper Ware,
Galvanized Ware, Snow Shovels,
Coal Hods, Stove Pipe, Elbows,
Stove Boards, Oilers, Lanterns, Wire
Goods, Grocers' Cans and Lard Pails.

Metals, from stock or import.
A complete line of Tinners' Trimmings.

Prices on Application.

KEMP MANUFACTURING CO.

TORONTO
Canada

THE MAKING OF HIGH-CLASS VARNISHES.

IN conversation with Mr. James S. N. Dougall, of McCaskill, Dougall & Co., Montreal, we learn that the demand for this firm's varnishes is increasing, and that quite rapidly. This is a pleasing fact, inasmuch as McCaskill, Dougall & Co. have the only varnish factory in America built and run upon the English system. Best English varnishes are famous for their durability, American varnishes for their workability, McCaskill, Dougall & Co.'s varnishes for their durability and workability; thus there are three kinds of varnishes. In the Canadian and American markets, the last mentioned variety seems to be the favorite. It is used by the largest makers of railway cars, and carriage manufacturers. It has several times been put to a practical test and never came below first place. In a competition of nine makes in Boston last year, McCaskill, Dougall & Co.'s varnishes vanquished all; little wonder their export trade is increasing in America, Australia, West Indies and England. Some few notes on this well-known factory ought to be interesting to the trade. HARDWARE AND METAL has lately enjoyed the pleasure of an interesting visit to the place. In the most important part of the factory, we first met the manager, Mr. Caswell. Mr. Caswell is an Englishman, and the son of his father, who was once paid £1,000 in Brussels for his services in a varnish testing case. He is a practical man in the varnish trade and is well versed in chemistry; in addition, we might add, his manner is as pleasant as his position is important. The factory is in Point St. Charles, Montreal, situated in a block of land with about 60 yards frontage by 200 deep, bounded by Manufacturers, D'Argenson and St. Patrick streets. The size of the present buildings is shown by a cut in their advertisement. The first room behind the office and shipping department is the tank-storage room. Here there are 50 to 60 tanks, 500 to 600 gallons each, filled with ageing varnishes. The tanks are securely set on iron truss work.

The making of varnishes is not remarkable for its speed, for the product spends from 2 to 4 years in process of manufacture between the time it goes in and out of the establishment. The varnish is pumped into these tanks by an electric motor and machinery appliances, but the striking feature of the place is its cleanliness. Everything is kept as free from dust as possibly can be, the floor is cement, walls brick, with air space between and fireproof, the ceiling metallic and also fireproof. For safety the buttons are taken off the electric lamps to prevent a spark igniting any fumes that might gather. Behind this storeroom is the oil house. In this apartment are immense steel tanks containing oil in process of treatment, here also all turpentine is rectified, then behind this again is the thinning house, always clean as a new pin. In the thinning house is a motor used for pumping purposes built for the firm especially. There are no brushes in this pattern as these might generate sparks that would ignite any fumes liable to accumulate is such a place. In this room is installed a patent centric motion mill used in the manufacture of locomotive enamels, a business monopolized by this firm in Canada.

Through a fireproof door we came to the English system cooking house, a feature of the establishment and a pride of the firm. For a fire-place it is remarkably clean; the roof is high and the place well aired. There are 5 underground furnaces and the kettles sit on top of the furnace holes. The kettles are made in England and built so that one man can shift on a special lever truck a load of 1,000 lb. of liquids. Some of the thermometers hanging in this room are very expensive. On the other side of the building is the American cooking house where the cheaper varnishes are made. The fire-places here are more open and not so interesting. Then the rest of this right side of the establishment is a large brick and cemented floor shed for sorting some of the raw material—oil, gums, turpentine, etc. Turpentine is also stored in underground tanks. The shed is also utilized for the picking, and selecting of gums. The bulk of these goods are kept down town, brought up as needed, and charged up to the manufacturing account as they enter the works.

Then we step through a fireproof door into the office again. Partitioned off the shipping department is a canroom. All the cans used to hold their goods are kept absolutely free from dust in this room—everything possible is done to keep foreign matter out of the pure stuff. Above the office is Mr. Caswell's laboratory where he conducts his experiments. Here he analyzes everything that enters the building. He knows what goes into his varnish and does not wait till it is made before he tells its efficiency. Oftentimes a practical painter is brought in to test the value of finished goods from the users' standpoint to make sure that every can contains the standard value. Nothing is shipped before it is weighed and not found wanting. The appliances for doing this, as those for other purposes about the establishment, are practical, efficient, and of the most modern type. As at first mentioned, we purposed only in this article to give some particulars of this firm's manufacturing departments, so mention no details regarding their heavy stock of manufactured goods, barrelled, canned and cased ready for immediate transport.

THE PRESIDENT OF THE MARITIME ASSOCIATION.

MR. SAMUEL HAYWARD, the recently elected president of the Maritime Hardware Association, was born in Studholm, King's county, N.B., in 1840. He is of Loyalist stock, his great-grandfather having been an officer in the Revolutionary War, and, at its close, received a grant of land in Nova Scotia. Mr. Hayward started business on his own account at the early age of 16, but, afterward, owing to ill health, he went West, returning in 1870 to become a member of the wholesale hardware house of Warwick, Clark & Co. In 1874 Mr. Warwick retired, and the name of the firm was changed to Clark & Hayward. This was again changed to S. Hayward & Co. when, in 1877, Mr. Clark retired. In 1895 the business was put into a limited liability company and called The S. Hayward Co., with Mr. S. Hayward as president, a position which he still holds. Though of a retiring disposition, Mr. Hayward has had large business success, and The S. Hayward Co. is one of the largest hardware firms in the Lower Provinces.

Mr. Samuel Hayward.

SPURIOUS SAW SETS.

It seems that someone is imitating the Whiting saw sets. In order to put a stop to this, the manufacturer and patentee, R. Dillon, Oshawa, Ont., has decided to take legal proceedings against those in any way dealing in the spurious article. The genuine tool is stamped with the word "patented," and a printed circular is in each individual box, so that the hardwareman can readily ascertain whether or not he is getting the Whiting saw sets.

ADMITTED TO PARTNERSHIP.

THE accompanying cuts are fairly good photographs of the two eldest sons of Mr. John P. Seybold, Montreal, whom he has just admitted to partnership in his wholesale hardware business. The firm has been organized into a company, Mr. John P. Seybold being president, Mr. Gordon C. Seybold, vice-president, and Mr. Herbert B. Seybold, secretary. The style has been Seybold, Son & Co., but it now reads, The Seybold & Sons Company.

Although this change marks an era in the firm's history, it involves no radical change in the business. The head and controlling influence remains in the builder of the business, who, although he has been studying the hardware market since 1870, buying and selling for over 30 years, still retains all his vigor and business ability. The

Gordon C. Seybold.

sons are by no means unknown to the hardware trade. Mr. Gordon C. Seybold has been identified with his father's business for over 11 years. For five years he was in the office, while during the remainder of the time he has been on the road, confining himself more particularly to the Maritime Provinces. Yet he has traveled all over the country and is thoroughly capable of discharging his duty looking after the travellers and the outside work.

Mr. Herbert B. Seybold is two years his brother's elder, 28 being his sum total. He has been in the business for 13 years, and city traveler for the last eight years. The first five years of his business career he spent in the warehouse. Although only a young man he can boast of being the oldest hardware traveller, in point of experience, in Montreal. The new duties assigned him give him full charge of the city sales.

Herbert B. Seybold.

The infusion of this young blood into the firm ought to bring forth good results, for to say that the new partners are energetic and thorough business men, steady, reliable and capable, is to pay them no idle or undeserving compliment. They were both born in Montreal, educated at the high school, and talk French almost as fluently as English; and, moreover, have that flash of the eye and lightness of manner that is a necessary equipment in a business manager. Their tenacity to the business ought to be a source of gratitude to the father.

Mr. Seybold started business in a retail hardware store in 1870, at the corner of St. Joseph and Murray streets, but 10 years later he went into the wholesale business with his father. His father died in September, 1892, and Mr. Seybold carried on the business as before until his warehouse was destroyed by fire in the disastrous conflagration of January. As soon as possible handsome new quarters will be erected, and the firm ardently hope to be shipping goods out of new premises by December 1. Then there will be a continuation of that progress that has marked the name, Seybold, during the last 20 years—only in a more pronounced form.

EVOLUTION OF THE PAINT TRADE.

But who can paint
Like Nature!
Can imagination boast
Amid its gay creations
Hues like hers?
—Thomson.

A SHORT time ago a hardware traveller, in one of the western towns in Ontario, going into a store, addressed the proprietor as follows:

"Hallo, Mr. Cuttacks, I see you are right into the 'canned goods' business."

"Yes," replied Mr. C., "in the olden time I used to buy a cask of dry white lead, a barrel of yellow ochre, ultramarine blue by the box and varnish by the barrel, now I purchase almost every item suitable for my trade in small tins ready for use. Talking about varnish, last year a meek-looking man with a 'frontispiece' as gentle as Mary's proverbial lamb, came in and asked me for a pint of the best furniture varnish in his own can. 'Tom' had gone to dinner and 'Joe' was in the iron house. I went to the cellar to draw the varnish and when there heard footsteps overhead. Fearing that all was not right, I rushed up to find the individual had disappeared, and with him, all the money in the till and some Rodger jack knives. 'Again, in my haste,' Mr. Cuttacks sadly added, "I forgot to turn the tap in the varnish barrel; in the afternoon, I discovered that the cellar was covered with a coat of my beautiful furniture varnish. No more bulk goods for me!"

A visit to a first-class paint, color and varnish factory will show the most elaborate labor-saving machinery and a well-ordered system for putting up almost every article used in staining, painting, enamelling and varnishing. In a magazine article recently the query was propounded: "What becomes of all the pins?" One may well ask what becomes of all the empty tins? One firm in Montreal contracts for tins in immense quantities and they placed an order this spring for over a million and a quarter of labels which are used in the "canning" department upon tins ranging in size from a half pint to a gallon.

The transition from dry colors to colors in paste form, triple-strength tinting colors, and then to liquid consistency, with an occasional mixing in of what is known as "semi-paste," has been very rapid. The newest and up-to-date labor-saving machinery is employed in sealing, capping and labeling these handy articles, which are such a boon to the painting trade to day.

The chief articles which are put up in ready-mixed form, the utility and convenience of which are now unquestioned, are oil paints for every class of work, stains for wood, polish for straw hats and stove-pipes, varnishes and japans for all purposes, coach colors, enamels, colored varnishes, gloss paints, boat bottom composition, liquid filler for wood and iron, and other articles too numerous to mention. All this class and style of merchandise, it is conceded, has come to stay, and is still capable of expansion. At the same time, there is now a tendency amongst architects and practical painters to change from linseed oil paints and try a dry, sanitary, permanent and decorative paint, hard drying and free from all odors, with water only as the medium. It is said that to meet this demand a line of colors fulfilling these conditions are now being introduced by a prominent Canddian manufacturing company, and their reception will be watched with great interest.

WILLIAM H. EVANS.

AN EMBLEM OF HIS TRADE.

A verdant youth dropped into a jeweler's, and after gazing at some fraternity pins in the showcase, said to the proprietor:

"Them's mighty nice breastpins you got thar, mister."

"What kind of a pin would you like to look at?"

"How much is this one with a pair o' compasses and a square?" pointing to a Masonic pin.

"Five dollars."

"Five dollars, eh? You haven't got one with any handsaws upon it, have you? I'm just outer my time, and as I am goin' to set up as carpenter and jiner, I thought I'd like to have somethin' to wear so folks would know what I was doin'. Well, I'll take it, though I'd like one with a handsaw, but I guess mebbe that's plain enough. The compass is to mark out yer work, and the square is to measure it when marked out, and any durn fool knows that G stands for gimlet.

1 to 25 quarts. With fly wheel 8 to 25 quarts.

1 to 25 quarts. With fly wheel 8 to 25 quarts.

Peerless Iceland Freezers

do the business. A complete line from one pint to forty quarts.

Get our Catalog.

DANA & CO.
Cincinnati.

New York Branch: 10 Warren St.
San Francisco Branch: 105 Front Street.

No. 1 Power, 40 quart

THE Locomobile

is the standard of perfection in steam-driven carriages. It requires no expert to operate it. Its simple construction enables everyone to use it.

CATALOGUE ON APPLICATION.

BICYCLES are also manufactured by us. Our line comprises such well-known makes as the CRESCENT, E. & D., COLUMBIA, and STEARNS.

WRITE US FOR QUOTATIONS

National Cycle and Automobile Co'y, Limited

TORONTO, CAN.

MARKETS AND MARKET NOTES

QUEBEC MARKETS.

Montreal, March 22, 1901.

HARDWARE.

WINTER continues to drag along, preventing the spring trade from opening up. The trade generally look only for soft weather to set the season's business in full swing. Some houses report business a little quieter this week than last, but, on the whole, the volume seem to be fairly well maintained. Although there have been no changes in prices, the market is firm, and some large sales of screws, bolts and wires have been made. These transactions are occurring in anticipation of advances in values on the part of the American trust. Barb wire is selling rather freely, principally for delivery in two weeks. All other fence wires are in moderate request. Cut nails continue rather quiet; wire nails are a more active line. Poultry netting is reported to be somewhat firmer at the factories. Payments are fair.

BARB WIRE—There seems a moderate demand for barb wire, although purchases are not exceptionally heavy. The ruling quotation is $3.05 per 100 lb. f.o.b. Montreal.

GALVANIZED WIRE—Business in this line is moderate and prices are unchanged. We quote: No. 5, $4.25; Nos. 6, 7 and 8 gauge, $3.55; No. 9, $3.10; No. 10, $3.75; No. 11, $3.85; No. 12, $3.25; No. 13, $3.35; No. 14, $4.25; No. 15, $4.75; No. 16, $5.00.

SMOOTH STEEL WIRE—A fair trade is reported. We quote oiled and annealed as follows: No. 9, $2.80; No. 10, $2.87; No. 11, $2.90; No. 12, $2.95; No. 13, $3.15 per 100 lb. f.o.b. Montreal, Toronto, Hamilton, London, St. John and Halifax.

FINE STEEL WIRE — The demand is moderate and the market is steady at 17½ per cent. off the list.

BRASS AND COPPER WIRE — There is nothing new to note. Discounts are 55 and 2½ per cent. on brass, and 50 and 2½ per cent. on copper.

FENCE STAPLES — Trade shows no change. We quote: $3.25 for bright, and $3.75 for galvanized, per keg of 100 lb.

WIRE NAILS—A fair amount of business is being done in wire nails. We quote $2.85 for small lots and $2.75 for carlots, f.o.b. Montreal, London, Toronto and Hamilton.

CUT NAILS — The inquiry is limited, although the market is quite steady. We quote: $2.35 for small and $2.25 for carlots; flour barrel nails, 25 per cent. discount; coopers' nails, 30 per cent. discount.

HORSE NAILS—"C" brand is quoted at 50 and 7½ per cent. off its own list and other brands at 60 per cent. on oval and city head, and 60 and 10 per cent. on new countersunk head off the old list.

HORSESHOES — Business is only moderate. We quote as follows: Iron shoes, light and medium pattern, No. 2 and larger, $3.50; No. 1 and smaller, $3.75; snow shoes, No. 2 and larger, $3.75; No. 1 and smaller, $4.00; X L steel shoes, all sizes, 1 to 5, No. 2 and larger, $3.60; No. 1 and smaller, $3.85; feather-weight, all sizes, $4.85; toe weight steel shoes, all sizes, $5.95 f.o.b. Montreal; f.o.b. Hamilton, London and Guelph, 10c. extra.

POULTRY NETTING—The market is reported firmer, factories refusing to accept more business. The discount continues at 50 and 10 per cent.

GREEN WIRE CLOTH — There is no change in the situation. The price is still $1.35 per 100 sq. ft.

SCREEN DOORS AND WINDOWS — Trade is hardly as good as it was last week. We quote: Screen doors, plain cherry finish, $8.25 per doz.; do. fancy, $11.50 per doz.; walnut, $7.40 per doz., and yellow, $7.45; windows, $2.25 to $3.50 per doz.

SCREWS—The market is firm and demand brisk. Discounts are as follows: Flat head bright, 87½ and 10 per cent. off list; round head bright, 82½ and 10 per cent.; flat head brass, 80 and 10 per cent.; round head brass, 75 and 10 per cent.

BOLTS—The undertone of the market is quite firm. Discounts are as follows: Carriage bolts, 65 per cent.; machine bolts, 65 per cent.; coach screws, 75 per cent.; sleigh shoe bolts, 75 per cent.; bolt ends, 65 per cent.; plough bolts, 50 per cent.; square nuts, 4½c. per lb. off list; hexagon nuts, 4¾c. per lb. off list; tire bolts, 67½ per cent.; stove bolts, 67½ per cent.

BUILDING PAPER—Trade is moderate, and prices show no change. We quote: Tarred felt, $1.70 per 100 lb.; 2-ply, ready roofing, 80c. per roll; 3-ply, $1.05 per roll; carpet felt, $2.25 per 100 lb.; dry sheathing, 30c. per roll; tar sheathing, 40c. per roll; dry fibre, 50c. per roll; tarred fibre, 60c. per roll; O.K. and I.X.L., 65c. per roll; heavy straw sheathing, $28 per ton; slaters' felt, 50c. per roll.

RIVETS — A small trade is passing. The discount on best iron rivets, section, carriage, and waggon box, black rivets, tinned do., coopers' rivets and tinned swedes rivets, 60 and 10 per cent.; swedes iron burrs are quoted at 55 per cent. off; copper rivets, 35 and 5 per cent. off; and coppered iron rivets and burrs, in 5-lb. carton boxes, are quoted at 60 and 10 per cent. off list.

CORDAGE—The market is firm and the demand brisk. Manila is now worth 13½c. per lb. for 7-16 and larger; sisal is selling at 10c., and lathyarn 10c.

SPADES AND SHOVELS—The inquiry has been small this week, but some sales have been made at 40 and 5 per cent. off the list.

HARVEST TOOLS—There has been nothing new transpiring in the market this week. The discount is still 50, 10 and 5 per cent.

TACKS—Unchanged. We quote as follows: Carpet tacks, in dozens and bulk, blued 80 and 5 per cent. discount; tinned, 80 and 10 per cent.; cut tacks, blued, in dozens, 75 and 15 per cent. discount.

LAWN MOWERS—This line is not much

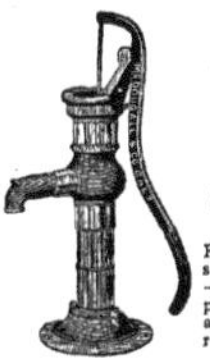

inquired for just yet. We quote: High wheel, 50 and 5 per cent. f.o.b. Montreal; low wheel, in all sizes, $2.75 each net; high wheel, 11-inch, 30 per cent. off.

FIREBRICKS—As yet there is but little business being done. Prices are steady at $18 to $24 per 1,000 as to brand, ex store.

CEMENT — Preparations are now being made for the coming season's business. We quote: German, $2.45 to $2.55; English, $2.30 to $2.40; Belgian, $1.95 to $2.05.

METALS.

The demand for metals continues to be rather quiet, chiefly owing to the backwardness of the season. The market seems to have righted itself, and most dealers feel that purchases are now safe in all lines. Ingot tin is easier, but this is the only break to a strong market.

PIG IRON—The pig iron market shows little change. No. 1 Hamilton is reported to be selling at $18 to $18.50, and No. 1 Summerlee is quoted at $20 to $20.50 for spring delivery.

BAR IRON—Dealers are quoting bar iron at $1.70 to $1.75 per 100 lb. Some large transactions have occurred below that figure.

BLACK SHEETS—A fair trade is doing. The tone of the market shows improvement. Present quotations are: $2.80 for 8 to 16 gauge; $2.85 for 26 gauge, and $2.90 for 28 gauge.

GALVANIZED IRON — The demand is reported to be only moderate. We quote: No. 28 Queen's Head, $5 to $5.10; Apollo, 10¾ oz., $5 to $5.10; Comet, No. 28, $4.50, with 25c. allowance in case lots.

INGOT COPPER—The market firm at 18c.

INGOT TIN—The London market shows a heavy break to £115 for spot goods. Quotations here are 31 to 32c.

LEAD—Unchanged at $4.65.

LEAD PIPE—The demand continues to be heavy. We quote as follows: 7c. for ordinary and 7½c. for composition waste, with 15 per cent. off.

IRON PIPE—The market is firm this week. We quote as follows: Black pipe, ¼, $3 per 100 ft.; ⅜, $3; ½, $3; ¾, $3.15; 1-in., $4.50; 1¼, $6 10; 1½, $7.28; 2-in., $9.75. Galvanized, ½, $4.60; ¾, $5.25; 1-in., $7.50; 1¼, $9.80; 1½, $11.75; 2-in., $16.

TINPLATES—Trade on spot is quiet, but the market is steady. We quote $4.50 for coke and $4 75 for charcoal.

CANADA PLATE—A few lots have been shipped this week, but the weather is against trade. We quote: 52's, $2.90; 60's, $3; 75's, $3.10; full polished, $3.75, and galvanized, $4.60.

TOOL STEEL—The tendency seems to be a higher level of values. Inquiries are more numerous. We quote: Black Diamond, 8c.; Jessop's 13c.

STEEL—The market is healthy. We quote: Sleighshoe, $1.85; tire, $1.95; bar, $1.85; spring, $2.75; machinery, $2.75 and toe-calk, $2.50.

TERNE PLATES—Some few sales are being made at $8 to $8.25.

COIL CHAIN—Prices are firm. We quote: No. 6, 11½c.; No. 5, 10c.; No. 4, 9½c.; No. 3, 9c.; ¼-inch, 7½c. per lb.; 5-16, $4.65; 5-16 exact, $5.10; ⅜, $4.20; 7-16, $4.00; ½, $3.75; 9-16, $3.65; ⅝, $3.35; ¾, $3.25; ⅞, $3.20; 1-in., $3.15. In carload lots an allowance of 10c. is made.

SHEET ZINC—The ruling price is 6 to 6¼c.

ANTIMONY—Quiet, at 10c.

GLASS.

Trade is not very active. The market is steady. We quote: First break, $2; second, $2.10 for 50 feet; first break, 100 feet, $3.80; second, $4; third, $4.50; fourth, $4.75; fifth, $5.25; sixth, $5.75, and seventh, $6.25.

PAINTS AND OILS.

Linseed oil and turpentine have each been reduced 2c. Trade is rather brisk, but when spring weather arrives a decided improvement is looked for. We quote:

WHITE LEAD—Best brands, Government standard, $6.37½; No. 1, $6.00; No. 2, $5.62½; No. 3, $5.25, and No. 4, $4.87½ all f.o.b. Montreal. Terms, 3 per cent. cash or four months.

DRY WHITE LEAD — $5.50 in casks; kegs, $5.75.

RED LEAD — Casks, $5.25; in kegs, $5.50.

WHITE ZINC PAINT—Pure, dry, 7c.; No. 1, 6c.; in oil, pure, 8c.; No. 1, 7c.; No. 2, 6c.

PUTTY—We quote: Bulk, in barrels, $2 per 100 lb.; bulk, in less quantity, $2.15; bladders, in barrels, $2.20; bladders, in 100 or 200 lb. kegs or boxes, $2.35; in tins, $2.45 to $2.75; in less than 100-lb. lots, $3 f.o.b. Montreal, Ottawa, Toronto, Hamilton, London and Guelph. Maritime Provinces 10c. higher, f.o.b. St. John and Halifax.

LINSEED OIL—Raw, 67c.; boiled, 70c., in 5 to 9 bbls., 1c. less, 10 to 20 bbl. lots, open, net cash, plus 2c. for 4 months. Delivered anywhere in Ontario between Montreal and Oshawa at 2c. per gal. advance and freight allowed.

TURPENTINE—Single bbls., 60c.; 2 to 4 bbls., 59c.; 5 bbls. and over, open terms, the same terms as linseed oil.

MIXED PAINTS—$1.25 to $1.45 per gal.

CASTOR OIL—8¾ to 9¼c. in wholesale lots, and ½c. additional for small lots.

SEAL OIL—47½ to 49c.

COD OIL—32½ to 35c.

NAVAL STORES — We quote: Resins, $2.75 to $4.50, as to brand; coal tar, $3.25 to $3.75; cotton waste, 4½ to 5½c. for colored, and 6 to 7½c. for white; oakum, 5½ to 6½c., and cotton oakum, 10 to 11c.

PARIS GREEN—Petroleum barrels, 16¾c. per lb.; arsenic kegs, 17c.; 50 and 100-lb. drums, 17½c.; 25-lb. drums, 18c.; 1-lb. packages, 18½c.; ½-lb. packages, 20½c.; 1-lb. tins, 19½c.; ½-lb. tins, 21½c. f.o.b. Montreal; terms 3 per cent. 30 days, or four months from date of delivery.

SCRAP METALS.

Buyers show more tendency to do business this week. Dealers are paying the following prices in the country: Heavy copper and wire, 13 to 13¾c. per lb.; light copper, 12c.; heavy brass, 12c.; heavy yellow, 8½ to 9c.; light brass, 6½ to 7c.; lead, 2¾ to 3c. per lb.; zinc, 2¾ to 2½c.; iron, No. 1 wrought, $13 to $14 per gross ton; No. 1 cast, $13 to $14; stove plate, $8 to $9; light iron, No. 2, $4 a ton; malleable and steel, $4.

HIDES.

Prices remain at the old level, and the market shows no improvement. We quote as follows: Light hides, 7c. for No. 1; 6c. for No. 2, and 5c. for No. 3. Lambskins, 90c.

PETROLEUM.

The demand is not brisk. We quote: "Silver Star," 14½ to 15½c.; "Imperial Acme," 16 to 17c.; "S.C. Acme," 18 to 19c., and "Pratt's Astral," 18½ to 19½c.

ONTARIO MARKETS.

TORONTO, March 22, 1901.

HARDWARE.

THE wholesale hardware trade continues to show increased activity and the prospects for spring trade continue to brighten. Among the changes which have taken place is an advance in rope halters of about 10 per cent., and an advance in wood, bench and fancy planes of American manufacture. Deliveries for spring trade are becoming more liberal, particularly in fence wires. Business is improving in wire nails, although it is not really active. Cut nails are as dull as ever, and in horse nails a fair business is to be noted. The same may be said of screws, bolts and nuts, and rivets and burrs. Spades and shovels and certain lines of harvest tools are in fair request, while trade in lumbermen's supplies is about over for the season. There is some demand for boot calks, and river drivers' supplies generally. A good business is being done in churns and in wringers. An improvement is to be noted this week in the demand for some lines of tinware, particularly sap buckets, and a better trade is reported for enamelled ware. Payments are, as a rule, fair.

BARB WIRE—Quite a demand has been experienced during the past week for small quantities of wire for shipment from stock. As far as can be learned, quite a little of this wire is for the use of settlers in the Northwest, who are able to carry it at low freight rates with the rest of their supplies. There is no change in price, and we quote f. o. b. Cleveland at $2.82½ per 100 lb. for less than carlots and $2.70 for carlots. From stock Toronto the figure is $3.05 per 100 lb.

GALVANIZED WIRE—There is not much doing in this line. Prices are without change. We quote: Nos. 6, 7 and 8, $3.50 to $3.85 per 100 lb., according to quantity; No. 9, $2.85 to $3.15; No. 10, $3.60 to $3.95; No. 11, $3.70 to $4.10; No. 12, $3 to $3.30; No. 13, $3.10 to $3.40; No. 14, $4.10 to $4.50; No. 15, $4.60 to $5.05; No. 16, $4.85 to $5.35. Nos. 6 to 9 base f.o.b. Cleveland are quoted at $2.57½ in less than carlots and 12c. less for carlots of 15 tons.

SMOOTH STEEL WIRE—Trade is not bad in oiled and annealed wire, there having been some improvement during the past week. But in hay-baling wire there is next to nothing deing done. Net selling prices are as follows, as noted last week: Nos. 6 to 8, $2.90; 9, $2.80; 10, $2.87; 11, $2.90; 12, $2.95; 13, $3.15; 14, $3.37; 15, $3.50; 16, $3.65. Delivery points, Toronto, Hamilton, London and Montreal, with freights equalized on these points.

WIRE NAILS—The demand for these has shown further improvement during the pas week, and quite a few small lots are going out. At the same time, however, trade cannot be called active. Prices are steady, and any changes which may be made at the meeting of the manufacturers a few weeks hence are more likely to be in the direction of an advance than a decline. The base price for less than carlots is still $2.85 per keg, and for carlots $2.75 per keg.

CUT NAILS—Trade in this line is still without any improvement, and the base price is unchanged at $2.35 per keg. Delivery points, Toronto, Hamilton, London, Montreal and St. John, N.B.

HORSESHOES—Business is just moderate and without any particular feature. We quote f.o.b. Toronto as follows: Iron shoes, No. 2 and larger, light, medium and heavy, $3.60; snow shoes, $3.85; light steel shoes, $3.70; featherweight (all sizes), $4.95; iron shoes, No. 1 and smaller, light, medium and heavy (all sizes), $3.85; snow shoes, $4; light steel shoes, $3.95; featherweight (all sizes), $4.95.

HORSE NAILS—Trade continues fair in horse nails. The discount on "C" brand oval head is unchanged at 50 and 7½ per cent. off the new list. The discount on other brands is 50, 10 and 5 per cent., and on countersunk head, 50, 10 and 10 per cent.

SCREWS—Trade continues fairly good, but without any new features as compared with a week ago. We quote discounts as follows: Flat head bright, 87½ and 10 per cent.; round head bright, 82½ and 10 per cent.; flat head brass, 80 and 10 per cent.; round head brass, 75 and 10 per cent.; round head bronze, 65 per cent., and flat head bronze at 70 per cent.

BOLTS AND NUTS—A fair trade is still being done in bolts and nuts and prices remain unchanged. We quote as follows: Carriage bolts (Norway), full square, 70 per cent.; carriage bolts full square, 70 per cent.; common carriage bolts, all sizes, 65 per cent.; machine bolts, all sizes, 65 per cent.; coach screws, 75 per cent.; sleighshoe bolts, 75 per cent.; blank bolts, 65 per cent.; bolt ends, 65 per cent.; nuts, square, 4½c. off; nuts, hexagon, 4¾c. off; tire bolts, 67½ per cent.; stove bolts, 67½; plough bolts, 60 per cent.; stove rods, 6 to 8c.

RIVETS AND BURRS—There is the usual steady trade being done. We quote as follows: Iron rivets, 60 and 10 per cent.; iron burrs, 55 per cent.; copper rivets and burrs, 35 and 5 per cent.

ROPE—Trade is rather quiet, and prices steady at last week's advance. We quote the base price as follows: Sisal 10c. per lb; manila, 13½c. per lb.; special manila, 11½c.; New Zealand, 10c. per lb.; single tarred lathyarn, 9½c.

CUTLERY — Trade has improved a little during the past week, although the volume of business is not large.

SPORTING GOODS — The demand is very light for all kinds of sporting goods.

GREEN WIRE CLOTH—There is very little being done. We quote $1.35 per 100 sq. ft.

SCREEN DOORS AND WINDOWS—Business is still light. We quote 4-in. styles in doors at $7.20 to $7.80 per doz., and 3 in. styles 20c. per doz. less; screen windows, $1.60 to $3.60 per doz., according to size and extension.

TINWARE—Business has improved during the past week, and shipments have been more active, particularly in sap buckets, as it is expected the maple syrup season will open with a rush, and dealers are anxious to be in a position to supply the necessary sap buckets. Most of the shipments for milk can trimmings have already been made.

ENAMELLED WARE—A little better demand is also to be noted in this line, although trade, generally speaking, is on the light side.

BUILDING PAPER—There is quite a good demand for building paper, roofing felt and carpet felt. Prices are without change.

POULTRY NETTING—There has been a good deal of poultry netting booked for future delivery, but there have not yet been many supplies come to hand. Discount on Canadian is 50 and 10 per cent. and net figures on English are about equal to a discount of 50, 10 and 5 per cent.

HARVEST TOOLS—Forward orders are being shipped, country customers being anxious to get delivery of such lines as rakes, hoes and manure forks. Discount, 50, 10 and 5.

SPADES AND SHOVELS—The demand for spades and shovels is fairly good the country over. The ruling discount is 40 and 5 per cent.

CHURNS—Trade has opened up in this line much earlier than usual, and a fair movement is reported. Discount, 58 per cent.

WRINGERS—A large trade is being done in this line, but, it is thought, without much profit either to the manufacturer or the wholesaler, there not now being any agreement as to prices.

ROPE HALTERS—There has been an advance of about 10 per cent. in prices. Small sisal halters are now quoted at $9.50 to $10 per gross, and large sisal at $12.50 to $13. The season is about over and prices are likely to remain firm for some time.

PLANES—There has been an advance by the United States manufacturers in the price of wood planes of about 20 per cent., and bench planes are also higher, the discount being 30 to 35 per cent. Fancy planes are quoted at from 15 to 20 per cent. off the list.

RANGE BOILERS—Just a moderate trade is being done. We quote : 30 gallons, $7 ; 35 gallons, $8.25 ; 40 gallons, $9.50.

GAS AND OIL STOVES — The demand for these has been a little better during the past week, and trade is fair for this time of the year.

CEMENT—The demand is beginning to open up. Notwithstanding the great increase in the production in Canada, prices are steady, as a big business is looked for. We quote : Canadian portland $2.40 to $3 ; German, $3 to $3.15 ; English, $2.85 to $3 ; Belgian, $2.50 to $2.75 ; calcined plaster, $2.

CHAIN—Trade is pretty well over for the season, but an occasional order is being received on account of the lumber camps.

METALS.

The metal trade, like the hardware trade, shows some improvement this week, although the demand is usually for small lots.

PIG IRON—The market keeps firm and trade fair.

BAR IRON—The market is strong and the demand active, with the mills, generally, filled with orders enough to keep them going for some time to come. The ruling prices are from $1.70 to $1.75, according to quantity.

PIG TIN—The outside markets have ruled weak, but there has been no change in local quotations, 32c. still being the ruling figure. There has been a fair amount of business done during the past week, although the quantities wanted are, as a rule, small. It should be noted that, according to the latest cable advices, the downward tendency of prices in London received a check on Wednesday.

TINPLATES—Trade has been better during the past week, both in coke and charcoal plates. Prices are as before.

TINNED SHEETS—Trade has been fair in this line and prices unchanged. We quote 9 to 9½c. for 28 gauge.

TERNE PLATES—A small business is being done, but, if anything, it is a little better than it was a week ago. Prices are easier.

GALVANIZED SHEETS—Some improvement is to be noted in the demand this week, and business may now be termed fair, although the outside markets are firm, particularly in the United States, where prices are 2c. higher than they were a short time ago. Local jobbers, in order to reduce present stocks, are quoting rather lower prices, as the shipments which will be shortly coming to hand are costing less money than the goods on hand. The ruling price for English galvanized iron is $4.60 and for American $4 50.

BLACK SHEETS—The demand is moderate, with 28 gauge still quoted at $3 30.

CANADA PLATES—Very little is being done in shipment from stock, but a fair number of orders are being booked for importation. The wholesalers are not, as a rule, however, courting orders for import. From stock we quote: All dull, $3. half-and half, $3.15, and all bright, $3.65 to $3.75.

COPPER—The demand is moderate, and a good trade is being done in sheet copper. We quote: Ingot, 19 to 20c.; bolt or bar, 23½ to 25c.; sheet, 23 to 23½c. per lb.

BRASS—Trade is quiet and discount unchanged at 15 per cent.

SOLDER—A fair business is being done. We quote as follows: Half-and-half, guaranteed, 18c.; ditto, commercial, 17½c.; refined, 17½c., and wiping, 17c.

IRON PIPE—The demand has continued brisk on account of the advancing markets. Another advance went into effect the latter end of this week. We quote as follows per 100 feet: Black pipe, ⅛ in., $4.35; ¼ in., $3.15; ⅜ in., $3 20; ½ in., $3 25; ¾ in., $3.52; 1 in., $4 93; 1¼ in., $6.72; 1½ in., $8.08; 2 in., $10.76; 2½ in., $22.94; 3 in., $36.12; 3½ in, $37.90; 4 in., $43.09; 5 in., $57.85; 6 in., $75.01; 7 in., $93 76; 8 in., $112 51. Galvanized pipe, ½ in., $5.52; ¾ in., $5 56; 1 in., $7 77; 1¼ in., $10.60; 1½ in., $12.70; 2 in., $16 90.

LEAD—Business is good. The outside markets are weak. Locally prices are unchanged at 4¾ to 5c. per lb.

LEAD PIPE—A moderate trade is to be reported. The discount is unchanged at 25 per cent.

ZINC SPELTER—In this line business is a little more active. We quote 6 to 6¼c. per lb.

ZINC SHEET—A good trade is being done at 6¾ to 7c. in cask lots and 7 to 7¼c. per lb. in part casks.

ANTIMONY—The improvement in the demand noted last week has not been maintained, trade this week being quiet. Quotations are unchanged at 11 to 11½c. per lb.

PAINTS AND OILS.

There is a fairly brisk spring movement. Jobbers state that orders have been rather smaller than usual, but the general opinion is that the spring and summer will witness a big movement as building operations are expected to be active. The feeling regarding linseed oil has been gloomy for some time on account of indications of a very low market in England this summer. A better feeling in now manifest, however, as cable reports have been received to the effect that an advance equivalent to 2½c. per gal. has taken place there. The turpentine market has weakened in Savannah owing to accumulation of stocks. Prices have declined 4c. here. This decline is considered to be temporary by some jobbers. Paris green is steady, but not moving briskly. The sale of sundries, liquid paints, etc., continues large. We quote as follows:

WHITE LEAD—Ex Toronto, pure white lead, $6 50; No. 1, $6.12½; No. 2. $5.75; No. 3. $5.37½; No. 4. $5; dry white lead in casks, $6.

RED LEAD—Genuine, in casks of 560 lb., $5.50; ditto, in kegs of 100 lb., $5.75; No. 1, in casks of 560 lb., $5 to $5 25; ditto, kegs of 100 lb.; $5 25 to $5.50.

LITHARGE—Genuine, 7 to 7½c.

ORANGE MINERAL—Genuine, 8 to 8½c.

WHITE ZINC—Genuine, French V.M., in casks, $7 to $7.25; Lehigh, in casks, $6.

PARIS WHITE—90c. to $1 per 100 lb.

WHITING—70c. per 100 lb.; Gilders' whiting, 80c.

GUM SHELLAC—In cases, 22c.; in less than cases, 25c.

PARIS GREEN—Bbls., 16¾c.; kegs, 17c.; 50 and 100 lb. drums, 17½c.; 25-lb. drums, 18c.; 1-lb. papers, 18½c.; 1-lb. tins, 19½c.; ½-lb. papers, 20½c.; ½-lb. tins, 21½c.

PUTTY—Bladders, in bbls., $2.20; bladders, in 100 lb. kegs, $2.35; bulk in bbls., $2; bulk, less than bbls. and up to 100 lb., $2.15; bladders, bulk or tins, less than 100 lb., $3.

PLASTER PARIS—New Brunswick, $1.90 per bbl.

PUMICE STONE—Powdered, $2.50 per cwt. in bbls., and 4 to 5c. per lb. in less quantity; lump, 10c. in small lots, and 8c. in bbls.

LIQUID PAINTS—Pure, $1.20 to $1.30 per gal.; No. 1 quality, $1 per gal.

CASTOR OIL—East India, in cases, 10 to 10½c. per lb. and 10½ to 11c. for single tins.

LINSEED OIL—Raw, 1 to 4 barrels, 69c.; boiled, 72c.; 5 to 9 barrels, raw, 68c.; boiled, 71c., delivered. To Toronto, Hamilton, Guelph and London, 1c. less.

TURPENTINE—Single barrels, 58c.; 2 to 4 barrels, 57c., to all points in Ontario. For less quantities than barrels, 5c. per gallon extra will be added, and for 5 gallon packages, 50c., and 10 gallon packages, 80c. will be charged.

GLASS.

The import business is about finished. The demand from stock is increasing and prices are steady. We quote Star in 100-foot boxes as follows: Under 26 in., $4; 26 to 40 in., $4.35; 41 to 50 in., $4.75; 51 to 60 in., $5; 61 to 70 in., $5.35; double diamond, under 26 in., $6; 26 to 40 in., $6.65; 41 to 50 in., $7.25; 51 to 60 in., $8.50; 61 to 70 in., $9.25, Toronto, Hamilton and London. Terms, 4 months or 3 per cent. 30 days.

COAL.

Owing to the blockade south of Buffalo last week, there was danger of a shortage of bituminous coal, but the deliveries have been heavy during the past week. Anthracite is in good supply. A steady trade is doing at unchanged prices. We quote anthracite on cars Buffalo and bridges: Grate, $4.75 per gross ton and $4.24 per net ton; egg, stove and nut, $5 per gross ton and $4.46 per net ton.

PETROLEUM.

The market is steady since last week's decline of ½c. A fair trade is doing. We quote: Pratt's Astral, 16½ to 17c. in bulk (barrels, $1 extra); American water white, 16½ to 17c. in barrels; Photogene, 16 to 16½c.; Sarnia water white, 15½ to 16c. in barrels; Sarnia prime white, 14½ to 15c. in barrels.

MARKET NOTES.

Wood, bench and fancy planes have been advanced in price by the United States manufacturers.

Rope halters are about 10 per cent. higher.

Iron pipe has again advanced in price.

The McClary Manufacturing Co., Limited, is in receipt of a large consignment of Dangler wickless blue flame oil stoves.

MANITOBA MARKETS.

WINNIPEG, March 18, 1901.

BUSINESS is showing some increased activity, but is not up to the standard of last year, at the same time. Prices have not changed for the week, and last list will stand for the coming week.

Quotations for the week are as follows :

Barbed wire, 100 lb	$3 45
Plain twist	3 45
Staples	3 95
Oiled annealed wire 10	3 95
" 11	4 00
" 12	4 05
" 13	4 20
" 14	4 35
" 15	4 45
Wire nails, 30 to 60 dy, keg	3 45
" 16 and 20	3 50
" 10	3 55
" 8	3 65
" 6	3 70
" 4	3 85
" 3	4 10
Cut nails, 30 to 60 dy.	3 00
" 20 to 40	3 05
" 10 to 16	3 10
" 8	3 15
" 6	3 20
" 4	3 30
" 3	3 65
Horsenails, 45 per cent. discount.	
Horseshoes, iron, No. 0 to No 1	4 65
No. 2 and larger	4 40
Snow shoes, No. 0 to No. 1	4 90
No. 2 and larger	4 40
Steel, No. 0 to No. 1	4 95
No. 2 and larger	4 70
Bar iron, $2.50 basis.	
Swedish iron, $4.50 basis.	
Sleigh shoe steel	3 00
Spring steel	3 25
Machinery steel	3 75
Tool steel, Black Diamond, 100 lb	8 50
Jessop	13 00
Sheet iron, black, 10 to 20 gauge, 100 lb.	3 50
20 to 26 gauge	3 75
28 gauge	4 00
Galvanized American, 16 gauge	2 54
18 to 22 gauge	4 50
24 gauge	4 75
26 gauge	5 00
28 gauge	5 25
Genuine Russian, lb	12
Imitation " "	8
Tinned, 24 gauge, 100 lb	7 55
26 gauge	8 80
28 gauge	8 00
Tinplate, IC charcoal, 20 x 28, box	10 75
" IX "	12 75
" IXX "	14 75
Ingot tin	35
Canada plate, 18 x 21 and 18 x 24	3 75
Sheet zinc, cask lots, 100 lb	7 50
Broken lots	8 00
Pig lead, 100 lb.	6 00
Wrought pipe, black up to 2 inch	50 an 10 p.c.
" Over 2 inch	50 p.c.
Rope, sisal, 7-16 and larger	$10 00
" ⅜	10 50
" ¼ and 5-16	11 00
Manila, 7-16 and larger	13 50
" ⅜	14 00
" ¼ and 5-16	14 50
Solder	21½
Cotton Rope, all sizes, lb.	16
Axes, chopping	$ 7 50 to 12 00
" double bitts	12 00 to 18 00
Screws, flat head, iron, bright	87½
Round " "	82½
Flat " brass	80
Round " "	75
Coach	57½ p.c.
Bolts, carriage	55 p.c.
Machine	55 p.c.
Tire	60 p.c.
Sleigh shoe	65 p.c.
Plough	40 p.c.
Rivets, iron	50 p.c.
Copper, No. 8	35
Spades and shovels	40 p.c.
Harvest tools	50, and 10 p.c.
Axe handles, turned, s. g. hickory, doz.	$2 50
No. 1	1 50
No. 2	1 25
Octagon extra	1 75
No. 1	1 25
Files common	70, and 10 p.c.
Diamond	60
Ammunition, cartridges, Dominion R.F.	50 p.c.
Dominion, C.F., pistol	30 p.c.
" military	15 p.c.
American R.F.	30 p.c.
C.F. pistol	5 p.c.
C.F. military	10 p.c. advance.
Loaded shells:	
Eley's soft, 12 gauge black	16 50
chilled, 12 guage	18 00
soft, 10 guage	21 00
chilled, 10 guage	23 00
Shot, Ordinary, per 100 lb	6 75
Chilled	7 50
Powder, F.F., keg	4 75
F.F.G.	5 00
Tinware, pressed, retinned	75 and 2½ p.c.
" " plain	70 and 15 p.c.
Graniteware, according to quality	50 p.c.

PETROLEUM.

Water white American	25½c.
Prime white American	24c.
Water white Canadian	22c.
Prime white Canadian	21c.

PAINTS, OILS AND GLASS

Turpentine, pure, in barrels	$ 68
Less than barrel lots	73
Linseed oil, raw	80
Boiled	83
Lubricating oils, Eldorado castor	25½
Eldorado engine	24½
Atlantic red	27½
Renown engine	41
Black oil	23½ to 25
Cylinder oil (according to grade)	55 to 74
Harness oil	61
Neatsfoot oil	$ 1 00
Steam refined oil	85
Sperm oil	1 50
Castor oil per lb.	11½
Glass, single glass, first break, 16 to 25 united inches	2 25
26 to 40 per 50 ft.	2 50
41 to 50 " 100 ft.	5 50
51 to 60 " " "	6 00
61 to 70 per 100-ft. boxes	6 50
Putty, in bladders, barrel lots per lb.	2½
kegs "	2¾
White lead, pure per cwt.	7 25
No 1 "	7 00
Prepared paints, pure liquid colors, according to shade and color. per gal.	$1.30 to $1.90

Mr. Aird, of the Canada Paint Company, is in town on his regular trip and reports finding trade very dull. Mr. Aird was unlucky enough to catch la grippe, but is now fully recovered.

Caverhill, Learmont & Co. are now well settled in their fine sample rooms in the Stovel block, McDermott avenue. Their western traveller has just returned from a trip through Alberta and reports trade in that section exceptionally good. Mr. T. L. Waldon has charge of the sample rooms here.

EUREKA REFRIGERATORS.

The Canadian hardware trade should find refrigerators a paying line to handle. The hardwareman is the natural distributor of these goods, the use of which in both the provision stores and the homes of the land is becoming so general, and if he posts himself thoroughly as to the essential qualities of a good refrigerator and has the right information for his customers concerning them he should make the line a profitable one. A good line of refrigerators is made by the Eureka Refrigerator Co., of Toronto. The refrigerator made by this company is noted for the dryness of air and the perfect circulation it affords. Both its lining and racks are of wood, so are odorless, non-corrosive, and as they contain no charcoal or other filling, never become foul by absorption or moisture. Full information regarding these refrigerators can be secured by writing to the company at 54 and 56 Noble street, Toronto.

THE IRON AND STEEL INDUSTRY OF CANADA.

What was Done Last Year, and What is Likely to be Done in the Near Future.

AFTER a struggle of over a century it can at last be said that the iron industry of Canada has an existence. Ten years ago we could not have said that without incurring the risk of being called to account. It is true, that for over a century pig iron has been made at the Radnor Forges, and that for many years blast furnaces have been producing iron in Nova Scotia. But this did not mean that Canada possessed, in its generally accepted sense, an iron industry, any more than that one balmy day makes a summer. We all hoped in those days that before our eyes grew dim we might see the desire of our heart and be satisfied. But, notwithstanding the possibilities we possessed, the years went by and the prophecies of the politicians were unfulfilled, and our hopes were unattained.

Everything, we are told, comes to him who waits. Whether this be true as a rule, is not a matter to be decided here. All we care to know just now is that those who have been waiting for the period when the iron industry could be said to have an existence in Canada have had their wish gratified, provided they have not within the last year or two gone hence.

A SIXTY YEAR PERIOD.

After the elapse of about 60 years between the failure of the last attempt to establish a blast furnace in Ontario and the inauguration of the present, a blast furnace was started at Hamilton, and it is in operation to-day. It is five years since this furnace was started. Two years ago the charcoal furnace at Deseronto was blown in, and three months ago the furnace which The Canada Iron Furnace Co., Limited, erected last summer at Midland, began the production of pig iron, thus making the third iron furnace in operation in Ontario, whereas, 10 years ago, there was not one even in embyro.

In Nova Scotia there is the furnace of The Dominion Iron and Steel Co., Limited, at Sydney, which was blown in for the first time the latter part of January last, and the furnace of The Nova Scotia Steel Co., New Glasgow. Quebec Province has the furnace of The Canada Iron Furance Co. at Radnor, and the furnace of The McDougal Co. at Drummondville, in intermittent operation.

In a few months hence we are promised three more of The Dominion Iron and Steel Co.'s furnaces in operation at Sydney, N.S., making four in all. At Sault Ste. Marie, Ont., Mr. Clergue is erecting a blast furnace, while by the beginning of next winter it is expected the furnace will be in operation which The Cramp Co. is to erect at Collingwood. A furnace is also projected at Kingston, although with but indifferent prospect of success at the moment.

FURNACES IN OPERATION.

By the end of 1901 we are therefore morally certain of having in active operation in Canada 10 or 11 blast furnaces. And, if their output is anywhere approximate to their capacity, we may expect to see them making pig iron at the rate of between 600,000 and 700,000 tons per annum. This would mean an increase of over 100 per cent. in the aggregate output of the seven furnaces now in operation in the Dominion if being run to their full capacity. The Sydney furnaces expect to be turning out pig iron at the rate of 300,000 tons per annum before the year closes.

With the exception of those at Radnor, Que., and Deseronto, Ont., all the pig iron furnaces in Canada use coke as fuel.

PRODUCTION OF PIG IRON IN 1900.

The output of pig iron in Canada last year was not as large by over 12,000 tons as it was in 1899, the total for the two years being 88,441 and 100,926 tons, respectively. The following table shows the production of pig iron in Canada during each of the three past calendar years :

PRODUCTION OF PIG IRON IN 1900, 1899 AND 1898.

	Output 1900.	Output 1899.	Output 1898.
Canada Iron Furnace Co. (7 mos.)	5,239	6,500	6,042
John McDougall & Co. (4 mos.)	800	......	.....
Deseronto Iron Co	10,244	11,616	....
Nova Scotia Steel Co	28,133	31,010	21,697
Hamilton Iron and Steel Co	43,925	51,800	48,253
	88,441	100,926	75,922

ORE USED.

The quantity of ore used in Canada last year was 165,829 tons, of which over 60 per cent. was foreign ; but, of course, classed in the latter are Newfoundland ores. The furnace at Midland, started about three months ago, is using Canadian ore, its supply coming from the now famous Helen mine in Northern Ontario. The other furnaces in Ontario are increasing the proportion of native ores which they use, but the drawback in the past has been inability to secure a steady enough supply to warrant their being more largely used. This difficulty, however, is being gradually remedied, work now being prosecuted more vigorously on several properties in the Province of Ontario. The Nova Scotia Steel Company last year used 19,000 tons of native ore and 35,000 tons of Newfoundland ore. The Dominion Iron and Steel Company will use Newfoundland ores exclusively in its furnaces. The Deseronto furnace used 17,636 tons of foreign ore and 793 tons of Canadian ore, while at the Hamilton furnace there were used 51,200 tons of foreign ore and 29,260 tons combined of Canadian ore and cinder from the company's own mill.

THE STEEL INDUSTRY.

There are at present in actual operation in Canada but two steel plants ; namely, that of The Nova Scotia Steel Co., Limited, and that of The Hamilton Steel and Iron Co., Limited. The steel plant of the latter company only began operations last year. There are two open-hearth furnaces in the plant, and each has a capacity of 40 to 45 tons per day. The quantity of steel produced during the time the plant was in operation last year was 2,900 tons. The output of The Nova Scotia Steel Co.'s plant was 22,000 tons, making a total of 24,900 tons in the country last year. In the very near future, however, the production of steel in Canada will be much increased. Down in Sydney The Dominion Iron and Steel Co. is erecting 10 open-hearth steel furnaces of a daily capacity of 50 tons, and the company expects to be shortly turning out steel blooms at the rate of 60,000 tons annually. Then, up at Sault Ste. Marie, Mr. Clergue's company has under construction a plant for the production of 600 tons of nickel-steel per day.

OUTLOOK FOR THE FUTURE.

With a population of not more than 6,000,000 at the outside, Canada, by the end of the first year in the twentieth century, will be fairly well supplied with blast furnaces. A good many think too well. Those who take the contrary view largely rest their case on the estimate that Canada annually consumes about 800,000 tons of

ALEXANDER GIBB

Manufacturers' Agent and Metal Broker,
13 St. John Street, Montreal.

Representing British and American Manufacturers. Correspondence invited from firms wishing to be represented in Canada.

McCaskill, Dougall & Co
Fine Varnish & Japan
Manufacturers.
Montreal
Price Lists on application

With

DUNDAS AXES

in stock you may justly ask a clerk for an explanation if he misses a sale.

Without **DUNDAS AXES,** is it fair to blame the clerk when you have not bought the best value?

All First-Class Jobbers Sell Them.

DOLLARS IN IT!

There are dollars in a good wall tint, because in the past a satisfactory wall tint could not be found. Old-fashioned wall tints are now useless and expensive alongside of

MURALO

which is the latest and best up-to-date, and best advertised wall tint in the world. Once sold it sells again. It is finer, better colored, more durable and easier applied than any other.

AGENTS.

A. RAMSAY & SON, - - - - - - - MONTREAL.
J. H. ASHDOWN, - - - - - - - - WINNIPEG.
McLENNAN, McFEELY & CO., - - - VANCOUVER.

"A WORD TO THE WISE IS SUFFICIENT."

Gold Dollars at 90c. each won't sell until you have convinced the customer that they are genuine. Have you the time to make the argument to do the convincing?

ALABASTINE is the standard wall-coating **"The World Over."** Would any dealer want to spend the time to convince a customer that a substitute, or something else, is "just as good" as **CHURCH'S**

ALABASTINE?

Besides, substitution is suspicious, and often bears earmarks of a swindle.

We do not claim to know all that can be learned about the manufacture of wall coatings, but twenty-five years of experience must count for something, and the fact that **ALABASTINE** is the goods that are in demand speaks volumes in its favor.

Never until **M. B. Church** introduced **ALABASTINE** was it found practicable to make wall-coating adapted to be mixed with cold water.

We own the patents on the cold water process, and this statement is our only argument at this time on that point.

ALABASTINE is healthful and beautiful, has stood the test of time; is manufactured in Canada, and recommended by leading sanitarians throughout the country. There is nothing "just as good." Buy only the genuine.

Manufactured by

THE ALABASTINE CO., LIMITED, Paris, Ont.

The trade supplied by—
Sanderson Pearcy & Co., Toronto, Ont.
G. F. Stephens & Co., Winnipeg, Man.
Vancouver Hardware Co., Vancouver, B.C.
Wood, Vallance & Co., Hamilton, Ont.
William Hill, Montreal, P.Q.

"DAISY" CHURN

Has tempered steel cased bicycle ball bearings, strongest, neatest and most convenient frame. Only two bolts to adjust in setting up. Steel Bow Levers, suitable for either a standing or sitting posture. Has four wheels and adjustable feet to hold stand steady while churning. When churn is locked to stand the bow can be used as handles to move it about on the front wheels as handy as a baby carriage. Open on both sides to centre, giving free space for pail. Made with wood or steel stands, with Cranks only, or Bow Levers as desired.

Vollmar Perfect Washer

Has a most enviable record. A perfection of its kind—will wash more clothes in less time, do it better and easier, with less wear and tear, than any other machine.

THE

Wortman & Ward Mfg. Co.,
Limited

LONDON, ONT.

Eastern Branch, 60 McGill Street, Montreal, Que.

Special list of low-priced Japanned and Regalvanized Wire Cloth.

24, 30, 36 in. wire, in 50 ft. rolls.

SAMPLES SENT WHEN DESIRED. WRITE FOR PRICES.

The B. GREENING WIRE CO., Limited

Hamilton, Ont., and Montreal, Que.

ONTARIO SILVER CO.,

Limited,

NIAGARA FALLS, CANADA.

Manufacturers of **FLATWARE, CUTLERY and ELECTRO PLATE.**

Ask for our Catalogue and Quotations.

Ontario Nut Works, Paris

BROWN & CO.

Manufacturers of

All sizes of Hot Pressed Nuts, Square and Hexagon.

iron. At present, taking into account the two new furnaces put into blast a few months ago, we are probably not producing one third of that quantity.

But, granting, with The Dominion Iron and Steel Co., at Sydney, running its full battery of four furnaces, and the projected furnaces at Collingwood and Kingston in operation, that the production of pig iron in Canada exceeds the consumptive capacity of the home market, does it follow that we should stay our hand ? Not at all. Such was not the policy of Great Britain. It was not the policy of Germany. It was not the policy of the United States. Why, therefore, should it be our policy ?

The agricultural, the lumbering, the mining and other of our natural resources are to no small extent dependent upon the foreign market for their development. And it will be much the same in regard to our iron industry.

The requirements of the home market in comparison with the vastness and richness of the iron resources of this country are as an ordinary hill is to the Himalayas. Of course our population is growing, and the larger it becomes the greater will be the iron consumptive requirements of the country. But we are ready to confess that we have not the patience to urge that the iron industry shall be developed only in the same ratio as the consumptive capacity of the country expands.

The New Blast Furnace at Midland.

ONE INDUSTRY BEGETS ANOTHER.

It is a law of the industrial world that one industry begets another. And of none is it truer than of that relating to iron. Where pig iron is made steel is made, and where steel is made there eventually develops the manufacture of such articles into the composition of which steel enters. We already witness that in our own country. A year or two ago we only had one concern making steel, but during the past year a steel plant has been built and put in operation alongside the blast furnace at Hamilton, Ont., and others are being constructed at Sault Ste. Marie, Ont., Collingwood, Ont., and Sydney, N.S. With that at New Glasgow there are now two steel plants in operation in Canada.

STEEL SHIPBUILDING.

Canada once occupied an important position among the nations as a builder of wooden ships. But the industry decreased as the iron and steel shipbuilding industry in Europe and in the United States increased. But during the past year, with the prominence which the iron industry is assuming, has come a decided revival in the shipbuilding industry. The establishment of yards for building steel ships is a live question in St. John, Halifax, Sydney and other places in the Maritime Provinces. In Toronto, steel steamers have been, and are being, constructed, which are equal to anything on the lakes. The same can be said of Collingwood, on the Georgian Bay. Out on the Pacific Coast they are apparently as ambitious as any other part of the Dominion in regard to steel shipbuilding, and in Vancouver an organized attempt is being made to secure the desideratum. So there is in Montreal.

The larger the source, the broader and deeper will be the stream it feeds. It is the same with the iron industry. The more iron we make in this country the broader and deeper will become the industrial stream whose source of supply is iron and its products.

To unduly force the development of the iron industry would be most unwise. But no one who is at all seized of the facts can well say that this is being done in Canada. It is true, we are going a great deal faster than we were 10 years ago. But for 10 years we were practically at a standstill.

CATALOGUES, BOOKLETS, ETC.

OXFORD GAS-RANGES.

THE Gurney Foundry Co., Toronto, Winnipeg and Vancouver, are sending out the 1901 spring catalogue of the gas cooking apparatus manufactured by them. There are several improvements in the Oxford ranges this year. They are now furnished with the oven thermometer which has been found so useful in cooking stoves. The ovens are asbestos lined. The doors are fitted with pedal foot openers. The most unique improvement is, however, a change in the construction of the burner by which the gas is conducted through a knife-edged slot, so that there is never enough gas in the slot at any one time to ignite the gas in the burner. This not only prevents an excess in the consumption of gas, but prevents that explosive noise which is so common in gas burners when being turned on or off. The catalogue gives full details as well as illustrations and prices of the gas ranges, stoves, heaters, etc., made by The Gurney Company, and can be had on application.

THE "KELSEY" WARM AIR GENERATOR.

The James Smart Manufacturing Co., Limited, Brockville, Ont., have secured the Canadian agency of the "Kelsey" Corrugated Warm Air Generator, which has come into great vogue in the principal cities of the Northern United States.

The "Kelsey" system of heating is unlike that of any other. It is claimed that by it the largest house can be heated by one fire without the use of radiators, steam pipes or coils. A daintily printed booklet, giving full details regarding this system, and illustrations of the plant and some of the fine residences that have been heated by it will be sent to any of the trade asking for it.

AN IMPROVED ROCK DRILL.

The Dominion Rock Drill and Foundry Co., Limited, of Napanee, are placing on the market a greatly improved new rock drill. It is the invention of the manager, Mr. Ed. J. Roy.

The Forsyth Acetylene Gas Generator Co., Limited, Stouffville, Ont., has been incorporated.

SAFE LOCK SHINGLE

CEILING PLATE, No. 557

IT WILL PAY YOU

to become acquainted with our line and method of doing business.

Our Interests are Identical

and we want you to interest yourself in us the same as we do in you and results will then be mutually satisfactory. Many of our lines are specialties, and help the dealer increase his profits. Why not send us **A Trial Order at Guaranteed Prices.**

We Manufacture:

Patent "Safe Lock" Metal Shingles,
Large and Small Sheet Siding,
Embossed Metal Ceilings and Side Walls,
Eave Trough, Conductor Pipe and Elbows,
Corrugated and Crimped Sheets,
O.K. Stove Pipe Thimbles,
Cornices, Finials, Skylights, Ventilators,
Ridgings, Window and Door Caps, etc.

PRESSED STEEL BRICK.

SKYLIGHTS & VENTILATORS

CORNICES AND FINIALS.

We are constantly adding money-making lines for the dealer.
Have you seen our Sheet Metal Pump and Galvanized Tubing?
DO YOU WANT AN AGENCY?

The Metal Shingle & Siding Co., Limited

PRESTON, ONT.

CORRESPONDENCE

THE BINDER TWINE QUESTION.

Editor HARDWARE AND METAL,—I am glad to see that my feeble remarks in re. binder twine have had the desired effect; viz., to set the Canadian manufacturers thinking. Yes, I am aware of the fact that the "Consumers Cordage Co." have been making binder twine for a number of years, and at times they have handicapped themselves, or, rather, the sale of their goods by giving certain implement manufacturers the control of their best lines, thus often driving good hardwaremen in various towns to look elsewhere for a good twine, and at the same time opening a door that the American manufacturer was quick to see and enter.

I am glad to know that there is a law against short measure and only wish it was more closely lived up to. No, I do not happen to have an American twine agency, but I know a man who has seen a man who has one, and the man told me that he would be foolish to buy Consumers Cordage Co. twine from me, when he had the price fully guaranteed on the American twine, settlement to be made any time after November 1, and unsold goods to be carried by the company—this I have seen.

It would seem from the wording of the article refered to above that the C. C. Co. stick to the old English ideas and let the Yankee sell all around them.

WANDERER.

Montreal, March 18.

DID NOT HANDLE HARDWARE.

Editor HARDWARE AND METAL—In your issue of February 23, under the heading "Business Changes," you say "R. McIvor, hardware dealer, Elkhorn, Man., is trying to sell out." I wish to correct this statement. R. McIvor never sold hardware, but was a harness dealer, he running in this place a branch of the Great West Saddlery Mfg. Co. I am the only hardware dealer in Elkhorn. In addition to hardware, I handle coal and wood, lumber and do undertaking.

G. SILVESTER.

Elkhorn, Man., March 11, 1901.

FISHING TACKLE FOR THE TRADE.

While hardwaremen have long been the natural distributors of fishing tackle in Canada, there has been a considerable increase of interest manifested in this line by the trade during the last few seasons. Because of the large margin, the advancing markets and the small deterioration in value of stocks fishing tackle has proven to be one of the most profitable lines the hardware dealer can handle.

One of the firms that have been affected most largely by the increased sale of this line is Wm. Croft & Son, Bay street, Toronto. This firm sell to the trade only, and, as they have an experience extending over nearly 50 years, they are in full touch with all demands of the trade, and their stock includes an assortment suitable for all localities from the Atlantic to the Pacific, and as they have sample-rooms in Quebec and Winnipeg, and seven travellers on the road, they are in close touch with the trade and are in an excellent position to fill all requirements. They issue a large catalogue containing fully illustrated descriptions and trade prices of the great range of fishing tackle handled by them. This catalogue can be had by any of the trade upon application.

TRADE CHAT.

Fred Anthony, who opened a hardware and tinsmithing business in Norval, Ont., some time ago, is putting in a stock of stoves.

The business of A. J. Jeffrey, hardware dealer, Stratford, Ont., which was established by Mr. Jeffrey's father, Wm. Jeffrey, sr., about 40 years ago, was sold last week to T. H. McCurdy, who will continue it in partnership with his brother, R. W. McCurdy, under the style of McCurdy Bros. The Messrs. McCurdy were formerly in the employ of The Hobbs Hardware Co., London., Ont. They will conduct a strictly cash business.

Gilpin Bros, hardware dealers, Orillia, Ont., have added about 40 ft. at the rear of their store, making it 120 ft. deep. They have also renovated the interior.

The Kaizer Brick Machine and Mfg. Co. has been formed in Winnipeg with the purpose of erecting a machine shop and foundry, and to manufacture brick machinery and bricks.

On Thursday last week a by-law was passed by St. Catharines, Ont., granting a percentage on the pay roll of the McKinnon Dash & Metal Works, to induce that firm to build a large factory there.

A peat factory will be erected in Beaverton, Ont., this spring, by Alex. Dobson.

CURRENT MARKET QUOTATIONS.

March 22, 1901.

These prices are for such qualities and quantities as are usually ordered by retail dealers on the usual terms of credit, the lowest figures being for larger quantities and prompt pay. Large cash buyers can frequently make purchases at better prices. The Editor is anxious to be informed at once of any apparent errors in this list as the desire is to make it perfectly accurate.

METALS.

Tin.

Lamb and Flag and Straits—56 and 28 lb. ingots, per lb.	0 32	0 33

Tinplates.

Charcoal Plates—Bright

		Per box
M.L.S., equal to Bradley.		
I.C., usual sizes	....	$6 75
I.X., "	....	8 25
I.X.X., "	....	9 75
Famous—		
I.C.	....	6 75
I.X.	....	8 25
I.X.X.	....	9 50
Raven & Vulture Grades—		
I.C., usual sizes	....	5 00
I.X., "	....	6 00
I.X.X "	....	7 00
I.XXX., "	....	8 00
D.C., 12½x17	....	4 75
D.X.	....	5 50
D.X.X	....	7 50

Coke Plates—Bright

Bessemer Steel—		
I.C., usual sizes	....	4 15
I.C., special sizes, base	....	4 50
20 x 28	....	8 50

Charcoal Plates—Terne

Dean or J. G. Grade—		
I.C., 20x28, 112 sheets	....	8 25
I.X., Terne Tin	....	10 25

Charcoal Tin Boiler Plates.

		Per lb.
Cookley Grade—		
X.X., 14x56, 50 sheet bxs / " 14x60 " / " 14x65, "	0 07	0 07½

Tinned Sheets

7 x30 up to 24 gauge	0 08	0 08½
" 26 "	0 08½	0 09
" 28 "	0 09	0 09½

Iron and Steel.

Common Bar, per 100 lbs.	1 70	1 75
Refined " "	....	2 15
Horse Shoe Iron "	....	2 05
Hoop steel, 1½ to 3 in. base, extras for smaller sizes	....	3 10
Sleigh Shoe Steel " base	....	2 00
Tire Steel	....	2 00
Machinery iron finish	....	2 05
Cast Steel, per lb	0 00	0 00
Toe Calk Steel	....	2 30
T. Firth & Co's special cast steel, per lb.		0 13
Jessop's Tool Steel	....	0 13

Boiler Tubes.

1½-inch	....	0 12½
2 "	....	0 13
2½ "	....	0 15
3 "	....	0 16
3½ "	....	0 20
4 "	....	0 25

Steel Boiler Plate.

¼ inch	....	2 25
3-16 inch	....	2 25
⅜ inch and thicker	....	2 25

Black Sheets.

gauge	....	3 00
20 gauge	....	3 00
22 to 24 "	....	3 10
26 "	....	3 20
28 "	....	3 30

Canada Plates.

All dull, 52 sheets	....	3 00
Half polished	....	3 15
All bright	3 65	3 75

Iron Pipe.

Black pipe—		
⅛ "	....	4 35
¼ inch	....	3 15
⅜ "	....	3 20
½ "	....	3 25
¾ "	....	3 52
1 "	....	4 93
1¼ "	....	6 74
1½ "	....	8 08
2 "	....	10 76
2½ "	....	22 91
3 "	....	31 12
3½ "	....	37 90
4 "	....	43 09
5 "	....	67 83
6 "	....	75 41
7 "	....	93 76
8 "	....	112 51
Galvanized pipe—		
½ inch	....	5 25
¾ "	....	5 56
1 "	....	7 77
1¼ "	....	10 60
1½ "	....	12 70
2 "	....	16 90

Galvanized Sheets.

	G. C. Comet.		Amer.	Queen's Head.
16 gauge		....	4 5	3 90
18 to 24 gauge	4 25	3 85	4 35	4 15
26 "	4 50	4 10	4 35	4 40
28 "	4 75	4 35	4 50	4 65

28 gauge American equals 26 gauge English.

Chain.

Proof Coil, 3-16 in., per 100 lb	....	....
" ¼ " "	7 50	8 00
" 5-16 " "	4 60	5 00
" ⅜ " "	4 00	4 50
" 7-16 " "	3 80	4 00
" ½ " "	3 60	3 75
" ⅝ " "	3 45	3 60
" ¾ " "	3 50	3 85

Halter, kennel and post chains, 40 and 50 p.c.
Cow ties 40 p.c.
Stall fixtures 35 p.c.
Trace chain 25 p.c.
Jack chain, iron, single and double, discount 35 p.c.
Jack chain, brass, single and double, discount 40 p.c.

Copper.

Ingot

English B. S., ton lots	0 19	0 20
Lake Superior	....	....

Bolt or Bar.

Cut lengths round, ½ to ⅞ in.	0 23½	0 25
" round and square 1 to 2 inches	0 23½	0 25

Sheet.

Plain, 14 oz., and light, 16 oz., 14x48 and 14x60	0 23	0 23
Plain, 14 oz., and light, 16 oz., irregular sizes	0 23	0 23½
Tinned copper sheets	....	0 26
Planished	....	0 32

Braziers (in sheets.)

4x6ft. 25 to 30 lbs. ea., per lb.	....	0 25½
" 35 to 45 " "	....	0 24½
" 50-lb. and above, "	....	0 23½

Boiler and T. K. Pitts

Plain Tinned, per lb	....	0 28
Spun, per lb	....	0 32

Brass.

Rod and Sheet, 14 to 30 gauge, 15 per cent.

Sheets, hard-rolled, 2x4	0 24	0 25
Tubing, base, per lb.	0 24	0 25

Zinc Spelter

Foreign, per lb	0 06	0 06½
Domestic "	....	....

Zinc Sheet.

5 cwt. casks	6 75	7 00
Part casks	7 00	7 50

Lead.

Imported Pig, per lb	0 04¾	0 05
Bar, 1 lb.	0 05½	0 05¾
Sheets, 2½ lbs. sq. ft., by	....	0 06¼
Sheets, 3 to 6 lbs., "	....	0 06

NOTE.—Cut sheets ½ cent per lb. extra.
Pipe, by the roll, usual weights per yard, lists at 7c. per lb. and 25 p.c. dis. f.o.b. Toronto.
NOTE.—Cut lengths, net price, waste pipe 8-ft lengths lists at 7½ cents.

Shot.

Common, $6.50 per 100 lb.; chilled, $7.00 per 100 lb.; buck, seal and ball, $7.50. Discount, 7½ p.c. Prices are f.o.b. Toronto, Hamilton, Montreal, St. John and Halifax. Terms 3 per cent. cash, freights equalized on Montreal.

Soil Pipe and Fittings.

Discount, 60 and 10 per cent. on medium and extra heavy, and 60 per cent. on light.

Solder.

	Per lb.	Per lb.
Bar half-and-half, guarant'd	....	0 18
Bar half-and-half, commer'l	....	0 17½
Refined	....	0 17½
Wiping	....	0 17

NOTE.—Prices of this graded according to quantity. The prices of other qualities of solder in the market indicated by private brands vary according to composition.

Antimony.

Cookson's, per lb.	0 11	0 11½

White Lead.

	Per 100 lb.
Pure	6 50
No. 1 do	6 12½
No. 2 do	5 75
No. 3 do	5 37½
No. 4 do	5 00
Munro's Select Flake White	7 37½
Elephant and Decorators' Pure	7 12½
Brandram's B B. Genuine	8 00
" " Decorative	7 55
" " No. 1	6 85
" " No. 2	6 00

Red Lead.

Genuine, 560 lb. casks, per cwt.	....	$5 50
Genuine, 100 lb. kegs, per cwt.	....	5 75
No. 1, 560 lb. casks, per cwt.	....	5 25
No. 1, 100 lb. kegs, per cwt.	....	5 00

White Zinc Paint.

Elephant Snow White	0 08	0 09
Pure White Zinc	0 08	0 09
No. 1	0 06	0 07½
No. 2	0 05	0 06½

Dry White Lead.

Pure, casks	5 75
Pure, kegs	6 25
No. 1, casks	5 50
No. 1, kegs	6 00

Prepared Paints.

In ¼, ½ and 1 gallon tins.

Pure, per gallon	....	1 25
Second qualities, per gallon	....	1 10
Barn (in bbls.)	0 75	0 85
The Sherwin-Williams Paints.	....	1 45
Canada Paint Co's Pure	....	1 25
Toronto Lead & Color Co's Pure	....	1 25
Sanderson Pearcy's Pure	....	1 20
Stewart & Wood's Champion Pure.		1 20

Colors in Oil.

25 lb. tins, Standard Quality.

Venetian Red, per lb.	0 05
Chrome Yellow	0 11
Golden Ochre	0 06
French "	0 05
Marine Black	0 09
" Green	0 09
Chrome "	0 08
French Imperial Green	0 09

Colors, Dry.

Yellow Ochre (J.C.) bbls.	1 35	1 40
Yellow Ochre (J.F.L.S.), bbls	....	2 75
Yellow Ochre (Royal)	1 10	1 15
Brussels Ochre	....	2 00
Venetian Red (best), per cwt.	1 80	1 90
English Oxides, per cwt.	3 00	3 25
American Oxides, per cwt.	1 75	2 00
Canadian Oxides, per cwt.	1 75	2 00
Super Magnetic Oxides, 93p.c.	2 00	2 25
Burnt Sienna, pure, per lb.	....	0 10
" Umber, " "	....	0 10
do Raw	....	0 09
Drop Black, pure	....	0 09
Chrome Yellows, pure	....	0 18
Chrome Greens, pure, per lb.	....	0 12
Golden Ochre	....	0 03½
Ultramarine Blue in 28-lb. boxes, per lb.	0 08	0 24
Fire Proof Mineral, per 100 lb	....	1 00
Genuine Eng. Litharge, per lb	....	0 07
Mortar Color, per 100 lb.	....	1 25
English Vermillion	....	0 80
Pure Indian Red, No. 45 lb.	....	0 80
Whiting, per 100 lb.	....	0 55

Blue Stone.

Casks, for spraying, per lb.	0 07
100-lb. lots, do. per lb.	0 08

Putty.

Bulk in bbls.	2 00
Bulk in less quantity	2 15
Bladders in bbls.	2 20
Bladders in kegs, boxes or loose	2 35
Bladders in 25-lb. tins	2 45
Bladders in 12½-lb. tins	2 75
Bladders in bulk or tins less than 100 lb	3 00

Varnishes.

(In 5-gal. lots.)		Per gal.
Carriage, No. 1	2	3
" body	8	9
" rubbing	4	5
Gold Size, Japan	3	3
Brown Japan	2	2
Elastic Oak	2	3
Furniture, extra	2	2
" No. 1	1	2
Hard Oil Finish	2	3
Light Oil Finish	3	3
Demar	3	3
Shellac, white	4	4
" orange	4	4
Furniture Brown Japan	1	2
Black Japan	2 90	2
" " No. 1	1 80	2 80

The Imperial Varnish & Color Co's, Limited Elastilite Varnish 1 gal. can, each, $2.00.

Granatine Floor Finish, per gal. $2.10.

Maple Leaf Coach Enamels; Size 1, 10c; Size 2, 35c.; Size 3, 20c. each.

Castor Oil.

East India, in cases, per lb.	0 10	0 10½
" " small lots	0 10½	0 11

Cod Oil, Etc.

Cod Oil per gal.	0 50	0 55
Pure Olive	....	1 20
" Neatsfoot	....	90

Glue.

Common	0 08½	0 09
French Medal	0 14	0 14½
Cabinet sheet	0 12	0 13
White, extra	0 18	0 20
Gelatine	0 22	0 30
Strip	0 18	0 20
Coopers	0 19	0 20
Huttner	....	0 18

HARDWARE.

Ammunition.

Cartridges.

B. B. Caps.Dom. 50 and 5 per cent.
Rim Fire Pistol, dis. 40 p. c., Amer.
Rim Fire Cartridges, Dom., 50 and 5 p.c.
Central Fire Pistol and Rifle, 10 p.c. Amer.
Central Fire Cartridges, pistol sizes, Dom. 30 per cent.
Central Fire Cartridges, Sporting and Military, Dom., 15 and 5 per cent.
Central Fire, Military and Sporting, Amer. add 5 p.c. to list. B.B. Caps, discount 40 per cent. Amer.
Loaded and empty Shells, "Trap" and "Dominion" grades, 25 per cent Rival and Nitro, net list.
Brass Shot Shells, 55 per cent
Primers, Dom., 30 per cent.

Wads — per lb

Best thick white felt wadding, in ½-lb bags 1 00
Best thick brown or grey felt wads, in ½-lb. bags 0 70
Best thick white card wads, in boxes of 500 each, 12 and smaller gauges 0 99
Best thick white card wads, in boxes of 500 each, 10 gauge 0 35
Best thick white card wads, in boxes of 500 each, 8 gauge 0 55
Thin card wads, in boxes of 1,000 each, 12 and smaller gauges 0 20
Thin card wads, in boxes of 1,000 each, 10 gauge 0 25
Thin card wads in boxes of 1,000 each, 8 gauge ....
Chemically prepared black edge grey cloth wads, in boxes of 250 each— Per M
11 and smaller gauge 0 60
9 and 10 gauges 0 70
7 and 8 gauges 0 90
5 and 6 gauges 1 10
Superior chemically prepared pink edge, best white cloth wads, in boxes of 250 each—
11 and smaller gauge 1 15
9 and 10 gauges 1 40
7 and 8 gauges 1 65
5 and 6 gauges 1 90

Adzes.

Discount, 20 per cent.

Anvils.

Per lb 0 10 0 12½
Anvil and Vise combined 4 50
Wilkinson & Co.'s Anvils..lb. 0 09 0 09½
Wilkinson & Co.'s Vices..lb. 0 09¾ 0 10

Augers.

Gilmour's, discount 65 and 5 p.c. off list.

Axes.

Chopping Axes
Single bit, per doz 5 50 10 00
Double bit, " 11 00 18 00
Bench Axes, 40 p.c.
Broad Axes, 33⅓ per cent.
Hunters' Axes 5 50 6 00
Boy's Axes 5 75 6 75
Splitting Axes 6 50 12 00
Handled Axes 7 00 10 00

Axle Grease.

Ordinary, per gross 5 75 6 00
Best quality 13 00 15 00

Bath Tubs.

Zinc 6 00
Copper, discount 15 p.c. off revised list

Baths.

Standard Enameled.

5½-inch rolled rim, 1st quality 25 00
" " " 2nd " 21 00

Anti-Friction Metal.

"Tandem" A .... per lb. 0 27
" B .... " 0 21
" C .... " 0 11½
Magnolia Anti-Friction Metal, per lb. 0 25

SYRACUSE SMELTING WORKS.

Aluminum, genuine 0 45
Dynamo 0 29
Special 0 25
Aluminum, 99 p.c. pure "Syracuse" .. 0 50

Bells.

Hand.

Brass, 60 per cent.
Nickel, 55 per cent.

Cow.

American make, discount 66⅔ per cent.
Canadian, discount 45 and 50 per cent.

Door.

Gongs, Sargant's 5 50 8 00
" Peterboro', discount 45 per cent.

Farm.

American, each 1 25 3 00

House.

American, per lb 0 35 0 40

Bellows.

Hand, per doz 3 35 4 75
Moulders', per doz 7 50 10 00
Blacksmiths', discount 40 per cent.

Belting.

Extra, 60 per cent.
Standard, 60 and 10 per cent.

Bits.

Auger.

Gilmour's, discount 60 and 5 per cent.
Rockford, 50 and 10 per cent.
Jennings' Gen., net list.

Car.

Gilmour's, 47½ to 50 per cent.

Expansive.

Clark's, 40 per cent.

Gimlet.

Clark's, per doz 0 65 0 90
Diamond, Shell, per doz 1 00 1 50
Nail and Spike, per gross 2 25 5 20

Blind and Bed Staples.

All sizes, per lb 0 07½ 0 12

Bolts and Nuts. Per cent

Carriage Bolts, full square, Norway 70
" " full square 70
Common Carriage Bolts, all sizes 65
Machine Bolts, all sizes 65
Coach Screws 75
Sleigh Shoe Bolts 75
Blank Bolts 65
Bolt Ends 65
Nuts, square .... 4½c. off
Nuts, hexagon .... 4¾c. off
Tire Bolts 67½
Stove Bolts 67½
Stove rods, per lb .... 5½ to 6c.
Plough Bolts 60

Boot Calks.

Small and medium, ball, per M 4 25
Small heel, per M 4 50

Bright Wire Goods.

Discount .... 60 per cent.

Broilers.

Light, dis., 65 to 67½ per cent.
Reversible, dis., 65 to 67½ per cent.
Vegetable, per doz., dis. 37½ per cent.
Henis, No. 8, " 6 00
Henis, No. 9, " 7 00
Queen City " 7 50 0 00

Butchers' Cleavers.

German, per doz 6 00 11 00
American, per doz 12 00 20 00

Building Paper, Etc.

Plain building, per roll 0 30
Tarred lining, per roll 0 40
Tarred roofing, per 100 lb 1 65
Coal Tar, per barrel 3 50
Pitch, per 100-lb 0 85
Carpet felt, per ton 45 00

Bull Rings.

Copper, $2.00 for 2½ in. and $1.90 for 2 in.

Butts.

Wrought Brass, net revised list

Cast Iron.

Loose Pin, dis., 60 per cent.

Wrought Steel.

Fast Joint, dis. 60 and 10 per cent.
Loose Pin, dis. 60 and 10 per cent.
Berlin Bronzed, dis. 70, 70 and 5 per cent.
Gen. Bronzed, per pair 0 40 0 65

Carpet Stretchers.

American, per doz 1 00 1 50
Bullard's, per doz .... 6 50 ....

Castors.

Bed, new list, dis. 55 to 57½ per cent.
Plate, dis. 52½ to 57½ per cent.

Cattle Leaders.

Nos. 31 and 32, per gross 50 9 50

Cement.

Canadian Portland 2 80 3 00
English " 3 00
Belgian " 2 75 3 00
Canadian hydraulic 1 25 1 50

Chalk.

Carpenters Colored, per gross 0 45 0 75
White lump, per cwt 0 60 0 65
Red 0 05 0 06
Crayon, per gross 0 14 0 18

Chisels.

Socket, Framing and Firmer.

Broad's, dis. 70 per cent.
Warnock's, dis. 70 per cent.
P. S. & W. Extra 60, 10 and 5 p.c.

Churns.

Revolving Churns, metal frames—No. 0, $8—No. 1, $8.50—No. 2, $9.00—No. 3, $10.00—No. 4, $12.00—No. 5, $16.00 each. Ditto wood frames—20c. each less than above. Discounts: Delivered from factories, 58 p.c.; from stock in Montreal, 56 p.c. Terms, 4 months or 3 p.c. cash in 30 days.

Clips.

Axle dis. 65 per cent.

Closets.

Plain Ontario Syphon Jet $16 00
Emb. Ontario Syphon Jet 17 00
Fittings net 1 00
Plain Teutonic Syphon Washout 10 00
Emb. Teutonic Syphon Washout 11 00
Fittings net 1 25
Low Down Teutonic, plain 16 00
" " " embossed 17 00
Plain Richelieu net 3 75
Emb. Richelieu net 4 00
Fittings net 1 25
Low Down Ont. Sy. Jet, plain net .. 19 00
" " " " " emb'd. net 20 50
Closet connection net 1 25
Basins, round, 14 in 1 00
" oval, 17 x 14 in 2 50
" " 19 x 15 in 3 75
Discount 40 p.c., except on net figures.

Compasses, Dividers, Etc.

American, dis. 62½ to 65 per cent.

Cradles, Grain.

Canadian, dis. 25 to 33⅓ per cent.

Crosscut Saw Handles.

S. & D., No. 3, per pair 17½
" " 5, " 22½
" " 6, " 15
Boynton pattern " 20

Door Springs.

Torrey's Rod, per doz .... (15 p.c.) 2 00
Coil, per doz 0 88 1 60
English, per doz 2 00 4 00

Draw Knives.

Coach and Wagon, dis. 50 and 10 per cent.
Carpenters, dis. 70 per cent.

Drills.

Hand and Breast.

Millar's Falls, per doz. net list.

DRILL BITS.

Morse, dis. 37½ to 40 per cent.
Standard dis. 50 and 5 to 55 per cent.

Faucets.

Common, cork-lined, dis 35 per cent.

ELBOWS. (Stovepipe.)

No. 1, per doz 1 40
No. 2, per doz 1 20
Bright, 20c. per doz. extra.

ESCUTCHEONS.

Discount, 45 per cent.

ESCUTCHEON PINS.

Iron, discount 40 per cent.

FACTORY MILK CANS.

Discount off revised list, 40 per cent.

FILES.

Black Diamond, 70 and 5 per cent.
Kearney & Foote, 60 and 10 to 70 per cent.
Nicholson File Co., 60 to 60 and 10 per cent.
Jowitt's, English list, 25 to 27½ per cent.

FORKS.

Hay, manure, etc., dis., 50 and 10 per cent. revised list.

GLASS—Window—Box Price.

Size United Inches.	Star Per 50 ft.	Star Per 100 ft	D. Diamond Per 50 ft.	D. Diamond Per 100 ft.
Under 26	2 10	4 00	....	6 00
26 to 40	2 30	4 35	....	6 65
41 to 50	....	4 75	....	7 25
51 to 60	....	5 30	....	8 50
61 to 70	...	5 35	....	9 75
71 to 80	....	5 75	....	10 50
81 to 85	....	6 50	....	11 75
86 to 90	....	....	..	14 00
91 to 95	..	....	...	15 50
99 to 100	...	....	....	18 00

GAUGES

Marking, Mortise, Etc.

Stanley's dis. 50 to 55 per cent.

Wire Gauges.

Winn's, Nos. 26 to 33, each 1 65 2 40

HALTERS.

Rope, ⅜ per gross ....
" ½ " 9 00
" ⅝ to ¾ 14 00
Leather, 1 in., per doz 3 87½ 4 00
" 1¼ in., " 5 15 5 20
Web, — per doz 1 87 2 45

HAMMERS.

Nail

Maydole's, dis. 5 to 10 per cent. Can. dis. 25 to 27½ per cent.

Tack.

Magnetic, per doz 1 10 1 20

Sledge.

Canadian, per lb 0 07½ 0 08½

Ball Pean.

English and Can., per lb 0 22 0 25

HANDLES.

Axe, per doz. net 1 50 2 00
Store door, per doz 1 00 1 50

Fork.

C. & B., dis. 40 per cent. rev. list.

Hoe.

C. & B., dis. 40 per cent. rev. list.

Saw.

American, per doz 1 00 1 25

Plane.

American, per gross 3 15 3 75

Hammer and Hatchet.

Canadian, 40 per cent.

Cross-Cut Saws.

Canadian, per pair 0 13½

HANGERS. doz. pairs.

Steel barn door 5 85 6 00
Stearns, 4 inch 5 00
" 5 inch 6 50
Lane's covered—
No. 11, 5-ft. run 8 40
No. 11½, 10-ft. run 10 80
No. 12, 10-ft. run 12 60
No. 14, 15-ft. run 21 00
Lane's O.N.T. track, per foot. 4½

HARVEST TOOLS.

Discount, 50 and 10 per cent.

HATCHETS.

Canadian, dis. 40 to 42½ per cent.

HINGES.

Blind, Parker's, dis. 50 and 10 to 60 per cent
Heavy T and strap, 4-in., per lb 0 06½
" " 5-in., " 0 06¼
" " 6-in., " 0 06
" " 8-in., " 0 05¾
" " 10-in., " 0 05½
Light T and strap, dis. 60 and 5 per cent.
Screw hook and hinge—
6 to 12 in., per 100 lbs 4 50
14 in. up, per 100 lbs 3 50

Per gro. pairs.

Spring 12 00

HOES.

Garden, Mortar, etc., dis. 50 and 10 p.c.
Planter, per doz 4 00 4 50

HOLLOW WARE

Discount, ... 45 and 5 per cent.

HOOKS.

Cast Iron.

Bird Cage, per doz 0 50 1 10
Clothes Line, per doz 0 27 0 63
Harness, per doz 0 72 0 88
Hat and Coat, per gross 1 00 3 00
Chandelier, per doz 0 50 1 00

Wrought Iron.

Wrought Hooks and Staples, Can., dis. 47½ per cent.

Wire.

Hat and Coat, discount 45 per cent.
Belt, per 1,000 .... 0 60
Screw, bright, dis. 55 per cent.

HORSE NAILS.

"C" brand 50 and 7½ p.c off new list } Oval
"M" brand 60 per cent. } head
Countersunk, 60 and 10 per cent.

HORSESHOES

Iron Shoes.	F.O.B. Montreal, No. 2 and larger.	No. 1. and smaller
Light, medium, and heavy	3 50	3 75
Snow shoes	3 75	4 00
Steel Shoes.		
Light	3 60	3 85
Featherweight (all sizes)	4 85	4 85
F.O.B Toronto, Hamilton, London and Guelph, 10c. per keg additional.		
Toe weight steel shoes		6 70

JAPANNED WARE.
Discount, 45 and 5 p.c. off list, June 1899.

ICE PICKS.
Star per doz. 3 00 3 25

KETTLES.
Brass spun, 7½ p.c. dis. off new list.
Copper, per lb. 0 30 0 50
American, 60 and 10 to 65 and 5 p.c.

KEYS.
Lock, Can., dis., 45 p.c.
Cabinet, trunk, and padlock, Am. per gross 60

KNOBS.

Door, japanned and N.P., per doz	1 50	2 50
Bronze, Berlin, per doz	2 75	3 25
Bronze Genuine, per doz	6 00	9 00
Shutter, porcelain, F. & L. screw, per gross	1 30	4 00
White door knobs—per doz.		1 25

HAY KNIVES.
Discount, 50 and 10 per cent.

LAMP WICKS.
Discount, 60 per cent.

LANTERNS.

Cold Blast per doz.	7 50
No. 3 "Wright's"	8 50
Ordinary, with O burner	4 25
Dashboard, cold blast	9 50
No. 0	6 00

Japanning, 50c. per doz. extra.

LEMON SQUEEZERS.

Porcelain lined, per doz.	2 20	5 60
Galvanized	1 87	3 85
King, wood	2 75	2 90
King, glass	4 00	4 50
All glass	1 20	1 30

LINES.

Fish, per gross	1 05	2 50
Chalk "	1 90	7 40

LOCKS.
Canadian, dis. 45 p.c.
Russell & Erwin, per doz. 3 00 3 25

Cabinet.
Eagle, dis. 30 p.c.

Padlock.
English and Am., per doz. 50 6 00
Scandinavian, " 1 00 2 40
Eagle, dis. 20 to 25 p.c.

MACHINE SCREWS.
Iron and Brass.
Flat head discount 25 p.c.
Round Head, discount 20 p.c.

MALLETS.

Tinsmiths', per doz.	1 25	1 50
Carpenters', hickory, per doz.	1 25	3 75
Lignum Vitae, per doz.	3 85	5 00
Caulking, each	60	2 00

MATTOCKS.
Canadian, per doz. 5 50 6 50

MEAT CUTTERS.
American, dis. 25 to 30 p.c.
German, 15 per cent.

MILK CAN TRIMMINGS.
Discount, 25 per cent.

NAILS.

Quotations are:	Cut.	Wire.
2d and 3d	$3 35	$3 85
3d	3 00	3 52
4 and 5d	2 75	3 35
6 and 7d	2 65	3 20
8 and 9d	2 50	3 00
10 and 12d	2 45	2 95
16 and 20d	2 40	2 90
30, 40, 50 and 60d. (base)	2 35	2 85

Galvanizing 2c. per lb. net extra.
Steel Cut Nails 10c. extra.
Miscellaneous wire nails, dis. 70 per cent.
Coopers' nails, dis. 30 per cent.
Flour barrel nails, dis. 25 per cent.

NAIL PULLERS.
German and American 1 85 3 50

NAIL SETS
Square, round, and octagon, per gross 3 38 4 00
Diamond 12 00 15 00

NETTING.
Poultry, 50 and 5 per cent. for McMullen's

OAKUM. Per 100 lb.

Navy	6 00
U. S. Navy	7 25

OIL.

Water White (U.S.)	0 16½
Prime White (U.S.)	0 15½
Water White (Can.)	0 15
Prime White (Can.)	0 14

OILERS.
McClary's Model galvan. oil can, with pump, 5 gal., per doz. 10 00
Zinc and tin, dis. 50, 50 and 10.
Copper, per doz. 1 25 3 50
Brass, " 1 50 3 50
Malleable, dis. 25 per cent.

GALVANIZED PAILS.
Dufferin pattern pails, dis. 45 p.c.
Flaring pattern, discount 45 per cent.
Galvanized washtubs, discount 45 per cent.

PIECED WARE.
Discount 40 per cent. off list, June, 1899.
10-qt. flaring sap buckets, dis. 45 p.c.
6, 10 and 14-qt. flaring pails, dis. 45 p.c.
Creamer cans, dis 45 p.c.

PICKS.
Per doz 6 00 9 00

PICTURE NAILS.
Porcelain head, per gross 1 75 3 00
Brass head " 0 40 1 00

PICTURE WIRE.
Tin and gilt, discount 75 p.c.

PLANES.
Wood, bench, Canadian dis. 50 per cent American dis. 50.
Wood, fancy Canadian or American 7½ to 40 per cent.

PLANE IRONS.
English, per doz. 2 00 5 00

PLIERS AND NIPPERS.
Button's Genuine per doz pairs, dis. 37½ 40 p.c.
Button's Imitation, per doz. 5 00 9 00
German, per doz. 0 60 2 60

PLUMBERS' BRASS GOODS.
Compression work, discount, 60 per cent.
Fuller's work, discount 65 per cent.
Rough stops and stop and waste cocks, discount, 60 per cent.
Jenkins disk globe and angle valves, discount, 55 per cent.
Standard valves, discount, 60 per per cent.
Jenkins radiator valves discount 55 per cent
" " " standard, dis., 60 p.c
Quick opening valves discount, 60 p.c.

No. 1 compression bath cock	2 00
No. 4 " " "	2 00
No. 7, Fuller's	2 50
No 4½, "	3 00

POWDER.
VeloX Smokeless Shotgun Powder.
100 lb. or less 0 85
1,000 lb. or more 0 80
Net 30 days.

PRESSED SPIKES.
Discount 20 to 25 per cent.

PULLEYS.

Hothouse, per doz.	0 55	1 00
Axle	0 22	0 33
Screw	0 27	1 00
Awning	0 35	2 50

PUMPS.
Canadian cistern 1 80 3 60
Canadian pitcher spout 1 40 2 10

PUNCHES.

Saddlers', per doz.	1 00	1 85
Conductors', "	9 00	15 00
Tinners' solid, per set	0 00	0 72
" hollow, per inch	0 00	1 00

RANGE BOILERS.

Galvanized, 30 gallons	7 00
" 35 "	8 25
" 40 "	9 50
Copper, 30 "	22 00
" 35 "	26 00
" 40 "	30 00

Discount off Copper Boilers 10 per cent.

RAKES.
Cast steel and malleable, 50, 10 and 5 p.c.
Wood, 25 per cent.

RASPS AND HORSE RASPS.
New Nicholson horse rasp, discount 60 to 60 and 10 p.c.
Globe File Co.'s rasps, 60 and 10 to 70 p.c.
Heller's Horse rasps, 50 to 50 and 5 p.c.

RAZORS.

	per doz.	
Geo. Butler & Co.'s	8 00	18 00
Boker's	7 50	11 00
" King Cutter	12 50	50 00
Wade & Butcher's	3 60	10 00
Theile & Quack's	7 00	12 00
Elliot's	4 00	18 00

REAPING HOOKS.
Discount, 50 and 10 per cent.

REGISTERS.
Discount 40 per cent.

RIVETS AND BURRS.
Iron Rivets, black and tinned, discount 60 and 10 per cent.
Iron Burrs, discount 55 per cent.
Extras on Iron Rivets in 1-lb. cartons, ½c per lb.
Extras on Iron Rivets in ½-lb. cartons, 1c. per lb.
Copper Rivets & Burrs, 35 and 5 p.c. dis. and cartons, 1c. per lb. extra, net.
Extras on Tinned or Coppered Rivets ½-lb. cartons, 1c. per lb.

RIVET SETS
Canadian, dis. 35 to 37½ per cent.

ROPE ETC.

	Sisal.	Manila.
7-16 in. and larger, per lb	10	13½
⅜ in.	11	14½
¼ and 5-16 in.		15½
Cotton, 3-16 inch and larger		16½
" 5-32 inch		21½
" ⅛ inch		22½
Russia Deep Sea		15½
Jute		8
Lath Yarn		9½
New Zealand Rope		10

RULES.
Boxwood, dis. 75 and 10 p.c.
Ivory, dis. 37½ to 40 p.c.

SAD IRONS. per set
Mrs. Potts, No. 55, polished 62½
" No. 50, nickle-plated 67½

SAND AND EMERY PAPER.
Dominion Flint Paper, 47½ per cent.
B & A. sand, 40 and 5 per cent.
Emery, 40 per cent.
Garnet (Rurton's), 5 to 10 p.c. advance on list.

SAP SPOUTS.
Bronzed iron with hooks, per doz. 9 50

SAWS.
Hand Disston's, dis. 12½ p.c.
S. & D., 40 per cent.
Crosscut, Disston's, per ft. 35 0 55
S. & D., dis. 35 p.c. on Nos. 2 and 3.
Hack, complete, each 0 75 2 75
" frame only 0 75

SASH WEIGHTS.
Sectional, per 100 lbs. 2 75 3 00
Solid, " 2 00 2 25

SASH CORD.
Per lb. 0 23 0 30

SAW SETS.
"Lincoln," per doz. 6 50

SCALES.
B. S. & M. Scales, 45 p.c.
Champion, 65 per cent.
Fairbanks Standard, 35 p.c.
" Dominion, 55 p.c.
" Richelieu, 55 p.c.
Chatillon Spring Balances, 10 p.c.

SCREW DRIVERS.
Sargent's per doz. 0 65 1 00

SCREWS
Wood, F. H., bright and steel. 87½ and 10 p.c.
Wood R. H., " dis. 82½ and 10 p.c.
" F. H., brass, dis. 80 and 10 p.c.
Wood, R. H., " dis. 75 and 10 p.c.
" F.H., bronze, dis. 75 p.c.
" R.H. " 70 p.c.
Drive Screws, 87½ and 10 per cent.
Bench, wood, per doz. 3 2[5]
" iron, " 4 2[5]
Set, Case hardened, 60 per cent.
Square Cap, 50 and 5 per cent.
Hexagon Cap, 45 per cent.

SCYTHES.
Per doz. net 9 00

SCYTHE SNATHS.
Canadian, dis. 45 p.c.

SHEARS.
Bailey Cutlery Co., full nickeled,
Seymour's, dis. 50 and 10 p.c.

SHOVELS AND SPADES
Canadian, dis. 40 and 5 per cent.

SINKS.
Steel and galvanized, discount 45

SNAPS.
Harness, German, dis. 25 p.c.
Lock, Andrews' 4 [illegible]

SOLDERING IRONS.
1, 1½ lb., per lb.
2 lb. or over, per lb.

SQUARES.
Iron, No. 493, per doz 2 40
" No. 494, " 3 25
Steel, dis. 50 and 5 to 50 and 10 p.c.,
Try and bevel, dis. 50 to 52½ p.c.

STAMPED WARE.
Plain, dis. 75 and 12½ p.c. off revised
Retinned, dis., 75 p.c. off revised list.

STAPLES.
Galvanized 3 50
Plain 3 2[illegible]
Coopers', discount 45 per cent.
Poultry netting staples, 40 per cent

STOCKS AND DIES.
American dis. 25 p.c.

STONE.

Washita	0 [illegible]	
Hindostan	0 06	
" slip,	0 06	
Labrador	....	
" Axe	....	0 15
Turkey	....	0 50
Arkansas	0 00	1 50
Water-of-Ayr	0 00	0 10
Scythe, per gross	3 50	5 00
Grind, per ton	15 00	18 00

STOVE PIPES.
5 and 6 inch Per 100 lengths 7 00
7 inch " " 7 50

ENAMELINE STOVE POLISH.
No. 4—3 dozen in case, net cash $4 80
No. 6—3 dozen in case, " 8 40

TACKS, BRADS, ETC.

	Per cent.
Strawberry box tacks, bulk	75 & 10
Cheese-box tacks, blued	80 & 12½
Trunk tacks, black and tinned	85
Carpet tacks, blued	80 & 5
" " tinned	80 & 10
" " (in kegs)	40
Cut tacks, blued, in dozens only	75 & 15
" ½ weights	60
Swedes, cut tacks, blued and tinned—	
In bulk	80 & 10
In dozens	75
Swedes, upholsterers', bulk	85 & 12½
" brush, blued & tinned, bulk	70
" gimp, blued, tinned and japanned	75 & 12½
Zinc tacks	35
Leather carpet tacks	55
Copper tacks	50
Copper nails	50
Trunk nails, black	65 and 5
Trunk nails, tinned	65 and 10
Clout nails, blued and tinned	65 and 5
Chair nails	35
Patent brads	40
Fine finishing	40
Picture frame points	10
Lining tacks, in papers	10

Lining tacks, in bulk		15
" " solid heads, in bulk		75
Saddle nails in papers		10
" " in bulk		15
Tufting buttons, 22 line, in dozens only		60
Tin capped trunk nails		15
Zinc glazier's points		5
Double pointed tacks, papers....90 and		10
" " " bulk		40

TAPE LINES.

English, ass skin, per doz	2 75	5 00
English, Patent Leather	5 50	9 75
Chesterman's each	0 90	2 85
" steel, each	0 80	8 00

THERMOMETERS.

Tin case and dairy, dis. 75 to 75 and 10 p.c.

TRAPS. (Steel.)

Game, Newhouse, dis. 25 p.c.
Game, H. & N., P. S. & W., 65 p.c.
Game, steel, 72½, 75 p.c.

TROWELS.

Disston's discount 10 per cent.

German, per doz	4 75	6 00

S. & D., discount 35 per cent.

TWINES.

Bag, Russian, per lb	....	0 27
Wrapping, cotton, per lb	0 22	0 26
Wrapping, mottled, per pack	0 50	0 60
Wrapping, cotton, 3-ply		0 20
" " 4-ply	....	0 26
Mattress, per lb	0 33	0 45
Staging, "	0 27	0 35

VISES.

Hand, per doz	4 00	6 00
Bench, parallel, each	2 00	4 50
Coach, each	6 00	7 00
Peter Wright's, per lb	0 12	0 13
Pipe, each	5 50	9 00
Saw, per doz	6 50	13 00

ENAMELLED WARE.

White, Princess, Turquoise, Blue and White, discount 50 per cent.
Diamond, Famous, Premier, 50 and 10 p.c.
Granite or Pearl, Imperial, Crescent, 50, 10 and 10 per cent.

WIRE.

Brass wire, 50 to 50 and 2½ per cent. off the list.

Copper wire, 45 and 10 per cent. net cash 30 days, f.o.b. factory.

Smooth Steel Wire, is quoted at the following net selling prices:

No. 6 to 8 gauge	$2 90
" 9 "	2 83
" 10 "	2 87
" 11 "	2 90
" 12 "	2 97
" 13 "	3 15
" 14 "	3 37
" 15 "	3 50
" 16 "	3 75

Other sizes of plain wire outside of Nos. 9, 10, 11, 12 and 13 and other varieties of plain wire remain at $2.81 base with extras as before. The prices for Nos. 9 to 13 include the charge of 1 c. for oiling. Extras net per 100 lb.: Coppered wire, 60c.—tinned wire, $2—oiling, 10c.—special hay-bailing wire, 30c.—spring wire, $1—best steel wire, 75c.—bright soft drawn, 15c.—in 50 and 100-lb. bundles net, 10c.—in 25-lb. bundles net 15c.—packed in casks or cases, 15c.—bagging or papering, 10c.

Fine Steel Wire, dis. 17½ per cent. List of extras: In 100-lb. lots: No 17, $5—No. 18, $5.50—No. 19, $6—No. 20 $6.65—No. 21, $7—No. 22, $7.30—No. 23, $7.65—No. 24, $8—No. 25, $9—No. 26, $9.50—No. 27, $10—No. 28, $11—No 29, $12—No. 30, $13—No. 31, $14—No. 32, $15 No. 33, $16—No. 34, $17. Extras net—tinned wire, Nos. 17-25, $2—Nos. 26-31 $4—Nos. 32-34, $6. Coppered, 5c.—oiling, 10c.—in 25-lb. bundles, 15c.—in 5 and 10-lb. bundles, 25c.—in 1-lb. hanks, 50c.—in ½-lb. hanks, 75c.—in ¼-lb. hanks, $1—packed in casks or cases, 15c.—bagging or papering, 10c.

Galvanized Wire, per 100 lb.—Nos. 6, 7, 8, $3 50 to $3 8 —No. 9, $2.85 to $3.15—No. 10, $3.60 to $3.95—No. 11, $3 70 to $4.10—No. 12, $3 to $3 30—No. 13, $3.10 to $3 4—No. 14, $4.10 to $4.50—No. 15, $4.60 to $5.05—No. 16, $4.85 to $5 35. Base sizes, Nos. 6 to 9, $2 57½ f.o.b. Cleveland.

Clothes Line Wire, solid 7 strand, No. 17, $4.25; No. 18, $2.65; No. 19, $2.35, f.o.b Hamilton, Toronto, Montreal.

WIRE FENCING.

	F.O.B. Toronto
Galvanized barb	3 15
Galvanized, plain twist	3 65

Galvanized barb, f.o.b. Cleveland, $2.82½ in less than carlots, and $2.7 in carlots.

WIRE CLOTH.

Painted Screen, per 100 sq. ft., net	1 35

WASTE COTTON.

	per lb.
Colored	4½ to 5
White, according to quality	6½ to 7¾

500-lb bale lots shaded.

WRENCHES.

Acme, 35 to 37½ per cent.
Agricultural, 60 p.c.
Coe's Genuine, dis. 20 to 25 p.c.

Towers' Engineer, each	2 00	7 00
" S., per doz	5 80	6 00
G. & K.'s Pipe, per doz	....	3 40
Burrell's Pipe, each	....	3 00
Pocket, per doz	0 25	2 90

WRINGERS.

Leader	per doz.	$37 00	31 00
Royal Canadian	"	36 00	38 00
Royal American	"	26 00	28 00
Sampson	"	....	30 00

Terms 4 months, or 3 p.c. 30 days.

WROUGHT IRON WASHERS.

Canadian make, discount, 40 and 5 per cent.

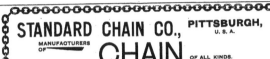

Index to Advertisements.

PERSONS addressing advertisers will kindly mention having seen their advertisement in Canadian Hardware and Metal Merchant.

SEND for specimen copy of Phillips' Monthly Machinery Register, containing over 5,000 entries of new and second-hand machinery of every description. The oldest established and most successful medium in the world. Established 25 years for the purpose of introducing those who have machinery for sale, to those who wish to buy, has a circulation of about 50,000 copies per annum, all over the world, and is used for continual reference by a large number of firms. It is consequently a most valuable advertising medium for all engineers and manufacturers. Subscription, 6s. per annum, price per copy, 6d. Sole Proprietor, CHAS. PHILLIPS, M.I.M.E., Newport, Mon., England. Telegraphic address: "Machinery, Newport, Mon."

BRITISH BUSINESS CHANCES.

Firms desirous of getting into communication with British manufacturers or merchants, or who wish to buy British goods on the best possible terms, or who are willing to become agents for British manufacturers, are invited to send particulars of their requirements for

FREE INSERTION

in "Commercial Intelligence," to the Editor

'SELL'S COMMERCIAL INTELLIGENCE,'
168 Fleet Street, London, England.

"Commercial Intelligence" circulates all over the United Kingdom amongst the best firms. Firms communicating should give reference as to bona fides.

N.B.—A free specimen copy will be sent on receipt of a post card.

Buy the Best.

HERCULES

Sash Cord.

Star Brand Cotton Rope
Star Brand Cotton Clothes Lines
Star Brand Cotton Twine

For Sale by all Wholesale Dealers

BUSINESS NEWS

of any kind that is of value to business men supplied by our Bureau. We can give you market quotations from any town in Canada, reports from the city markets, stock quotations, etc. You can get commercial news from any Canadian paper through us.

Write us, giving us particulars of what you want and where you want it from, and we will quote you prices by return.

"Clippings from any Canadian paper on any subject."

CANADIAN PRESS CLIPPING BUREAU,

232 McGill Street, MONTREAL, QUE.

Telephone Main 1255.

10 Front St. East, Toronto. Telephone 2148.

---

75 YEARS. *ESTABLISHED 1825.* *75 YEARS.*

CELEBRATED HEINISCH SHEARS.

Tailors' Shears,
Trimmers, Scissors,
Tinners' Snips, etc.

ACKNOWLEDGED THE BEST.

R. HEINISCH'S SONS CO. NEW YORK OFFICE, 90 Chambers St. NEWARK, N.J., U.S.A.

Not connected with any Shear Combination.

CHAS. F. CLARK, President. JARED CHITTENDEN, Treasurer.

...ESTABLISHED 1849...

BRADSTREET'S

Capital and Surplus, $1,500,000. **Offices Throughout the Civilized World.**

Executive Offices: Nos. 346 and 348 Broadway, New York City, U.S.A.

THE BRADSTREET COMPANY gathers information that reflects the financial condition and the controlling circumstances of every seeker of mercantile credit. Its business may be defined as of the merchants, by the merchants, for the merchants. In procuring, verifying and promulgating information no effort is spared, and no reasonable expense considered too great, that the results may justify its claim as an authority on all matters affecting commercial affairs and mercantile credit. Its offices and connections have been steadily extended, and it furnishes information concerning mercantile persons throughout the civilized world.

Subscriptions are based on the service furnished, and are available only by reputable wholesale, jobbing and manufacturing concerns, and by responsible and worthy financial, fiduciary and business corporations. Specific terms may be obtained by addressing the Company at any of its offices. **Correspondence Invited.**

—OFFICES IN CANADA—

HALIFAX, N.S.	HAMILTON, ONT.	LONDON, ONT.	MONTREAL, QUE.
OTTAWA, ONT.	QUEBEC, QUE.	ST. JOHN, N.B.	TORONTO, ONT.
VANCOUVER, B.C.	VICTORIA, B.C.	WINNIPEG, MAN.	

THOS. C. IRVING, Gen. Man. Western Canada, Toronto. JOHN A. FULTON, Gen. Man. Eastern Canada, Montreal.

Awarded a Gold Medal at **PARIS EXPOSITION** for superiority. That's proof enough of their quality, and clearly shows that they are the best.

Send for Catalogue and . . . . Price List.

The Bailey

Cutlery Co.

BRANTFORD, ONT.

Dominion Pattern

Cow Tie AND Stall Fixture

The special features of the tie and stall fixture are well shown in the illustration. As will be noticed the chain is very short, which prevents all danger of entanglement with the animal's foot. At the same time the form of the fixture is such that great freedom is allowed to the head. Because of the short chain this tie is much cheaper than the ordinary patterns.

The stall fixture is made from a tough quality of steel and is very strong. Also, owing to its circular cross-section, it is exceedingly rigid. Its simplicity, convenience, cheapness, and ease of attaching make it very popular with cow tie users.

This form of tie and stall fixture are sometimes called Niagara pattern.

American or Flat Link Chain,

for years the standard cow tie chain in "the States," is now rapidly coming in favor in Canada. Its short link, handsome appearance and smooth surface—which cannot injure the animal's neck—make it superior to all other styles of chain for cow ties.

For sale by all Jobbers; manufactured by

ONEIDA COMMUNITY, Limited, NIAGARA FALLS, ONT.

To Toronto Silver Plate Co 23/4/13, see cut book page 194

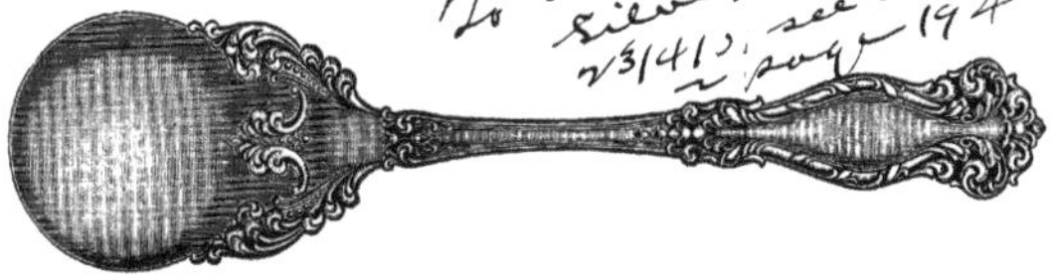

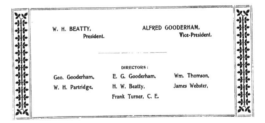

Particulars of the personnel, etc., etc., of a strictly **Canadian** Silverware Company, employing **Canadian** Capital. **Not** in the **Trust** or **Members** of any **Silverware Association** or **Combine.**

The Toronto Silver Plate Co.

LIMITED

Incorporated 1882.

Silversmiths and Manufacturers of Electro Plate.

CAPITAL, $100,000.00.

W. H. BEATTY, President.

ALFRED GOODERHAM, Vice-President.

DIRECTORS:

Geo. Gooderham, W. H. Partridge, E. G. Gooderham, H. W. Beatty, Frank Turner, C. E., Wm. Thomson, James Webster.

Factories and Salesrooms, King St. West, Toronto, Canada.

E. G. GOODERHAM, Managing Director and Sec.-Treas.

LONDON, ENG., Show Room, 23 Thavies Inn, Holburn Circus,

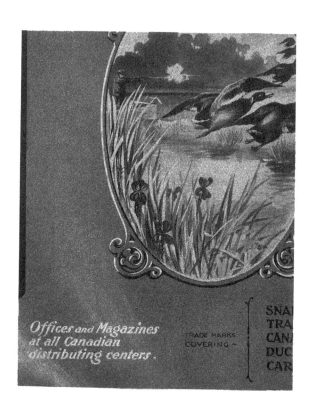

Offices and Magazines at all Canadian distributing centers.

TRADE MARKS COVERING –

SNA
TRA
CANA
DUC
CAR

CANADIAN HARDWARE AND METAL MERCHANT

The Weekly Organ of the Hardware, Metal, Heating, Plumbing and Contracting Trades in Canada.

VOL. XIII. MONTREAL AND TORONTO, MARCH 30, 1901. NO. 13

SUPPLIES FOR LUMBERMEN

SAWS

CANT HOOKS AND PEAVEYS
PIKE POLES

AXES

CROSS CUT SAWS

CHAIN AND BLOCKS
ROPE

SWEDES IRON

RINGS
HOOKS
CLEVISES

RICE LEWIS & SON

LIMITED

Cor. King and Victoria Sts., TORONTO.

Meat & Food Chopper

No. 5, - $2.00

Rapid Grinding & Pulverizing Mills

No. 2½, - $4.50

Fruit, Wine & Jelly Press

No. 34, - $3.00

"ENTERPRISE"

PATENTED

HARDWARE SPECIALTIES

Comprising

Meat and Food Choppers, *35 sizes aud styles for hand and power;* Rapid Grinding and Pulverizing Mills, *32 sizes and styles for hand and power;* Fruit, Wine and Jelly Presses; Meat Juice Extractors; Cherry Stoners; Raisin Seeders, *for hand and power;* Ice Shredders; Vegetable Slicers; Mrs. Potts' Cold Handle Irons; Sausage Stuffers and Lard Presses, etc., etc.

ARE THE BEST

Write for Descriptive Catalogue

Sold by the Leading Jobbers of the Dominion

The Enterprise Mg. Co. of Pa.

Philadelphia, Pa., U. S. A.

Raisin Seeder

No. 36, - .85

Meat Juice Extractor

No. 21, - $2.50

Meat & Food Chopper

No. 12, - $2.50

Garden Hose

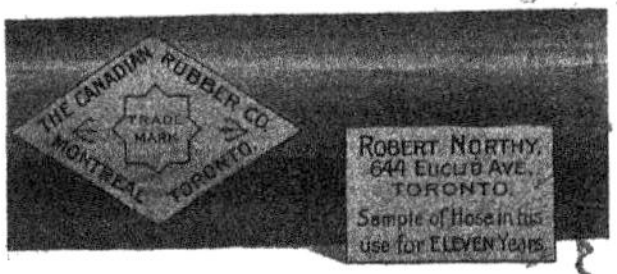

Discounts for the new season now out.

WRITE FOR PRICES.

* * *

CANADIAN RUBBER CO.,

Montreal. Toronto. Winnipeg.

THE NEW BALDWIN

DRY AIR CLEANABLE

REFRIGERATOR.

135 Modern Varieties. Ash, Oak and Soft-wood Finishes

METAL, PORCELAIN, SPRUCE LININGS.

BALDWIN

Positive Circulation—Sanitary—Odorless.

Latest Cleanable Features—The Strongest and Best System of Patent Removable Metal Air-Flues.

Air-Tight Lever Locks

Ball-Bearing Casters.

Swing Base—in and out.

Rubber around Doors and Lids, making them doubly air-tight.

Handsome Designs.

Moderate Prices.

Built in the newest, largest and best equipped refrigerator plant in the East run all the year round on refrigerators exclusively; stock goods; special refrigerators and coolers in sections.

Handsome Trade Catalogue Ready.

Baldwin Refrigerator Co.,

BURLINGTON, VERMONT.

SOME OF THE NEWER "YANKEE" TOOLS

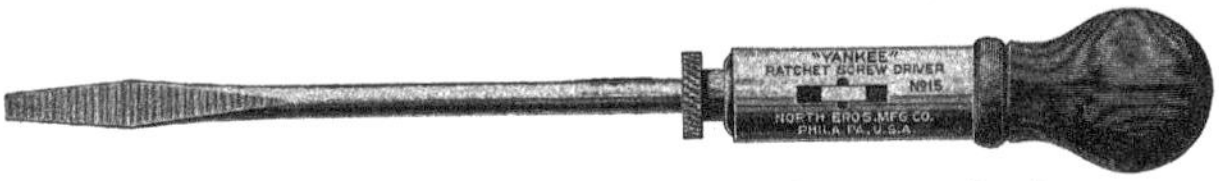

No. 15 "Yankee" Ratchet Screw Driver

RIGHT AND LEFT HAND, AND RIGID, WITH FINGER TURN ON BLADE—2, 3, 4 and 5-in. BLADES.

No. 20 "Yankee" Spiral-Ratchet Screw Driver

RIGHT HAND ONLY, AND RIGID. 3 SIZES, EXTREME LENGTH OPEN, INCLUDING BIT—14, 17 and 19-inches

Sold by Leading Jobbers throughout the Dominion.

NORTH BROS. MFG. CO.,

Philadelphia, Pa., U. S. A.

TRADE

C

MARK

Canada Horse Nail Company

Montreal

Established 1865.

Horse Nails

with the "**C**" brand and our name on the box, have our absolute guarantee that every nail is perfect.

They are hot forged from the best Swedish Charcoal Steel Nail Rods, combining the best process and the best material known for making a horse nail.

They will wear longest, as they are the toughest, and drive easier than any other, by reason of their faultless design and perfect hardened points.

Every dealer in horse nails will find it to his advantage to sell our "**C**" brand in preference to any other, as they are the best known to the Farrier trade, and conceded by them whenever used, to be the most satisfactory nail to use.

They are to be obtained from nearly all the leading wholesale dealers in Canada. If you cannot get our "**C**" brand on order from them, do not accept substitutes, but apply direct to us.

Woman is Responsible

for the sale of ice cream freezers.

Fifteen million women have considerable influence.

Ten thousand of them wrote to us in three months.

We are talking to them all the time—winter and summer.

They are all familiar with the Peerless Iceland Freezer.

Reports from them prove that we are right, and that one motion is enough.

We want your help in distributing Peerless Iceland Freezer.

DANA & CO.,

Cincinnati.

New York Branch, 10 Warren St.
San Francisco Branch, 105 Front St.

Lewis Bros. & Co.

Wholesale Hardware

MONTREAL.

FRICTIONLESS METAL

AN EXACT CUT OF FRICTIONLESS METAL BAR.

THE BEST for
- **High-Speed Machinery.**
- **Machinery with Heavy Bearings.**
- **Light Machinery.**

EVERY POUND OF FRICTIONLESS METAL FULLY GUARANTEED.

Put up in Boxes **25, 50 and 100 lbs.**

POSITIVELY THE BEST.

LEWIS BROS. & CO., MONTREAL

Sole Agents for Canada. Write for Prices.

No. 3

NOVELTY QUILT FRAME CLAMPS.

No. 4

JAPANNED or COPPERED.

FOR SPRING CLEANING, YOUR LADY CUSTOMERS WILL APPRECIATE THESE FOR THEIR CURTAIN FRAMES.

INSTEAD OF THE OLD WAY OF RESTING THE FRAMES ON CHAIRS, USE OLD BROOM HANDLES IN THE SOCKETS.

SAVES WORRY. **IMPROVES TEMPER.**

If you would like to see it, a sample will be mailed you if you will drop a post card to

A. R. WOODYATT & CO., GUELPH, ONT.

SOLD ONLY THROUGH THE WHOLESALE TRADE.

The VICTOR and DAISY Oil Stoves

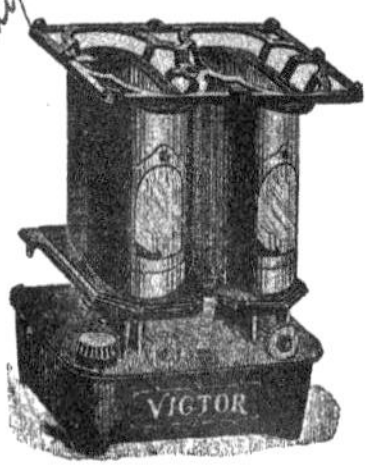

of to-day are the results of years of practical experience in oil stove making.

In them are embodied all the points which go to make up a perfect oil stove.

A stove which will not smoke
A stove which will not smell } but will give perfect satisfaction.

That's why our trade is increasing so rapidly.
Your stock would not be complete without these goods.
Made in one, two and three-burner.
We will be pleased to quote you.

KEMP MANUFACTURING CO.,

TORONTO, Canada

HARDWARE AND METAL

VOL. XIII. MONTREAL AND TORONTO, MARCH 30, 1901. NO. 13.

President,
JOHN BAYNE MacLEAN,
Montreal.

THE MacLEAN PUBLISHING CO.
Limited.

Publishers of Trade Newspapers which circulate in the Provinces of British Columbia, North-West Territories, Manitoba, Ontario, Quebec, Nova Scotia, New Brunswick, P.E. Island and Newfoundland.

OFFICES

MONTREAL - - - - - - - 232 McGill Street, Telephone 1255.
TORONTO - - - - - - - 10 Front Street East, Telephone 2148.
LONDON, ENG. - - - - 109 Fleet Street, E.C., J. M. McKim.
MANCHESTER, ENG. - - - 18 St Ann Street. H. S. Ashburner,
WINNIPEG - - - - Western Canada Block. J. J. Roberts.
ST. JOHN, N.B. - - - No. 3 Market Wharf, I. Hunter White,
NEW YORK. - - - - - - - 176 E. 88th Street.

Subscription, Canada and the United States, $2.00.
Great Britain and elsewhere - - - 12s.

Published every Saturday.

Cable Address { Adscript, London. Adscript, Canada.

APPRECIATIVE REMARKS.

HARDWARE AND METAL recently published part of an address of Mr. Walter H. Cottingham, general manager of The Sherwin-Williams Co., delivered at the annual convention of the representatives of that company. Mr. Harry Dwight Smith, the manager of the promoting department of The Sherwin-Williams Co., forwarded to Current Advertising, a New York periodical, a copy of HARDWARE AND METAL containing the address. The article was reproduced in Current Advertising, a copy of which has been sent to this office by Mr. A. C. Norris, secretary-treasurer of the Ontario Wind Engine and Pump Co., Limited, Toronto.

"In calling attention to this article," says Current Advertising, "Mr. Smith says: 'We also send the paper because it seems a very good example of a live, weekly trade paper. Nearly all the hardware papers in this country came to the writer's desk, but none of them is so interesting or so full of readable matter as HARDWARE AND METAL.'"

This leads Current Advertising to remark: "This is certainly high praise, and, coming from such a source, should make the paper to which it belongs, properly proud.

"Even those who are not engaged in the hardware business can readily understand from an examination of a copy of HARDWARE AND METAL, that it deserves much credit for the careful and exhaustive manner in which it covers the trade to which it is devoted."

The appreciative remarks of Mr. Smith and Current Advertising cannot be other than gratifying to us, and stimulate us to make HARDWARE AND METAL even more worthy of the esteem of our readers and contemporaries.

THE CLASSIFICATION OF STORES.

The early-closing by-law which is being considered in Victoria includes the following schedule of classifications of stores: Books and stationery, boot and shoe dealers, boot and shoe makers, bakers and confectioners, clothiers and merchant tailors, outfitters and gentlemen's furnishers, retail dry goods and milliners, fish and game dealers, furniture dealers (not second-hand), grocers, hardware, jewellers, pork and provision dealers, novelty stores, butchers.

In view of the fact that Toronto's early-closing by-law has become practically inoperative because of the difficulty as defining the term "grocery shop," it would be advisable to make the classification clearer by defining each of the divisions. It has been found difficult to decide what a grocer is in Toronto, and it would probably be even more difficult to make the distinction in law between other branches of trade.

FEDERAL BANKRUPTCY LAW.

A FEW weeks ago reference was made in these columns to a number of letters which had been received from business men in New Brunswick, Nova Scotia and Prince Edward Island, in reply to the charge that one of the reasons for the non-introduction of a Dominion bankruptcy Bill in the House of Commons was the opposition of the people in the Maritime Provinces. What little ground there was for this charge or excuse was evidenced by those letters.

We are now in receipt of another letter from Mr. W. S. Loggie, a well known business man in Chatham, N.B., and president of the Maritime Board of Trade, in which he says that as far as he can learn "there is no opposition to the principle of a federal bankruptcy law."

"I suppose," he adds, "it will be difficult to frame a law that will meet the views of bankers and merchants alike. The expenses of closing a small estate under the proposed law, I understand, would be excessive. I think, therefore, the details of such a law might be sent to boards of trade to consider and make suggestions."

Mr. Loggie's suggestion is a good one. It is not yet too late for the Government to draft a Bill, have it read once, submit it to the various boards of trade in the country, and let the House finally deal with it at the next session.

THE STEEL TRUST PROBLEM.

As far, at least, as mere organization is concerned, the enormous steel combination of the United States appears to be a success.

It will be remembered that the seven companies which were to be merged into one were to receive, in return for their stock, a certain number of shares in the new concern.

This proposition has been so favorably received by the shareholders of the various companies that up to the end of last week nearly all the stock had been deposited, the percentage being from 94 to 98 per cent. of the preferred and 92 to 99 of the common.

What will be the outcome of the consolidation in actual business experience remains to be seen. The interests it controls will naturally give it an enormous influence in the economic conditions of not only the United States, but of the world. Whether they be for good or for evil remains to be seen.

The American Steel and Wire Consolidation was swallowed up by this newer and greater organization before it had been in the world long enough to test the economic principles upon which it was based. The features it developed were not such, however, as will cause it to be kindly remembered. We have special reference to the action of Mr. Gates, the chairman of the board of directors, nearly a year ago, when the market was demoralized for some months by the mere ipse dixit of his company. Whether it was, as so many firmly believed, a stock-jobbing deal does not perhaps matter so much as the fact that it was within the power of one company to create a disturbance in the iron and steel trade that affected all commercial countries to a more or less extent. And The American Steel and Wire Co. was a much smaller concern than the new organization that has absorbed it and six others besides.

One effect of the new trust will certainly be to cause those engaged in the iron and steel industry in other countries to strengthen their own positions, either by combination, higher Customs tariffs, or improved machinery and business methods. The cable shows this. Only within the last few days the news is brought by cable that the Scotch and North of England steel manufacturers are combining with no other purpose than to meet the threatened competition of the Carnegie-Morgan aggregation. Even in the United States themselves, efforts are being made to rally forces to meet the new consolidation, which transcends all others in its magnitude.

The alarm that was first felt in Canada when it was rumored the Carnegie-Morgan company was negotiating to obtain control of the plant of The Dominion Iron and Steel Co. has subsided, but, if the latter proves to be able to make iron and steel at the low relative cost it is claimed it can, it is more likely than not that the big consolidation which is now being organized across the border will make an effort to make it a part and parcel of itself. This is, of course, largely speculation. But we are enabled to speculate this way because the absorption of the Canadian concern is a possibility.

One thing we may certainly expect is that the Carnegie-Morgan consolidation will not overlook the growing importance of the Canadian market, and will make an effort to keep out the products of the mills and factories of Europe similar to that which it produces.

We have already the example of The American Steel and Wire Co. to give ground for that belief. For some time the latter company has had a special price for the Canadian market on fence wires, with the object of keeping out those manufactured in Europe. At times this price has been as much as $11 below the figures the hardware trade in the United States were paying. The difference just now is about $3 per ton in favor of the Canadian dealer, there having been a recent reduction to meet the competition of European mills.

ANOTHER DROP IN WHITE LEAD.

A decline is again announced in the price of white lead. It is understood to be partly due to competition. The reduction is 12½c. per 100 lb. This makes the price for pure in Toronto $6.37½, and for No. 1 $6, and in Montreal $6.25 for pure and $5.87½ for No. 1.

SEWING MACHINES AS A SIDE LINE.

It is safe to say that the advent of the bicycle effected a radical change in hundreds of hardware stores throughout Canada. The hardwareman looked with indifference upon the old high-wheel bicycle as a line to handle. The safety won but slight attention during the first two or three seasons when high prices were paid for inferior wheels. Most of the large centres boasted bicyle emporiums which practically controlled the trade throughout the country. But, as their product became popular, bicycle manufacturers insisted that hardwaremen should handle them. The trade took hold of the wheel with diffidence, but it proved a most profitable line.

Then repairs became necessary, and soon hardwaremen were establishing repair shops. In many cases expert machinists were employed and repairs of all kinds made.

This revolution has paved the way for the hardwareman to take hold of several lines, which, until recently, he has not considered worth his attention. The sewing machine is one of these lines.

To-day, as has been true for years, this article, which is as much a staple in household affairs as either the wringer or the churn, is handled almost exclusively by agents or peddlers, who, for the very reason that their turnover is so small, have to do business on a much larger margin than the hardwareman would need. Many of them go to the expense of a horse and rig and endeavor to sell their machines from door to door. Some of them understand the machanism of their machine enough to make any necessary repairs. The aim of others is merely to get as many machines off their hands at as good a profit as possible.

The retail hardwareman who has installed a repair shop and who has a mechanic of even average ability around the place, should be able to easily compete with this class of machine dealers. He can do business at a closer margin and can at once make necessary repairs.

This line should be, moreover, a profitable one. The best machines are all sold with a good margin to the retailer, for the competition among sewing machine manufacturers is as keen to-day as was the competition between bicycle manufacturers five years ago.

TRADE IN COUNTRIES OTHER THAN OUR OWN.

AN advance of 15c. was made on black sheets on March 16. This again caught quite a number of manufacturing consumers who had not covered their needs, and probably caught some of the jobbers also. Considerable business has been placed at the full advance. The sheet mills are not able to run to their capacity, because of the scarcity of sheet bars.—Iron Age.

TINWARE HIGHER IN THE UNITED STATES.

Representatives of the leading eastern manufacturers of tinware met this week in New York and agreed upon a general advance in the price of stamped and pieced tinware and japanned ware, averaging 7½ to 10 per cent. Prices of tinware ruling for some time past, it is claimed, have left a very meagre margin of profit to the makers, in the case of the cheaper goods practically no profit at all. As the demand has been remarkably heavy, and promises to continue on a large scale, the element of competition has been practically obliterated, consequently no reason exists for cutting prices at the present time. With the strength which has been developed in all iron and steel products and a business which taxes the capacity of all, the tinware manufacturers saw no reason for continuing the quotation of unprofitable prices, hence the present advance.—Metal Worker, March 23.

ADVANCE IN PLUMBERS' BRASS WORK.

We understand that all of the manufacturers of plumbers' brass work who are in the Brass Manufacturers' Association have signed the prices agreed upon at the last meeting held in New York City on February 15. By this agreement the manufacturers have decided that they will maintain the prices then agreed upon, and in no instance will they meet or cut prices made by any manufacturer outside of the association. Ninety per cent. of the brass manufacturers in the United States are in the association, and as these manufacturers, representing 90 per cent. of the output of the United States, have signed the agreement and will maintain prices and sell at prices no lower than the association prices, it can readily be inferred that with the opening of the spring season prices will undergo a rapid advance. Within the next two or three weeks it will not be the extreme low prices which will be troubling the jobbers, but it will be the desire to learn where they can get their goods at any price. There are very few manufacturers who carry a fairly heavy stack of these goods on hand, as the continued low prices, the continued demand during the past winter and also the continued instability of the association prevented them from going very deeply into the manufacture or accumulation of large stocks.

WIRE NAILS IN THE UNITED STATES.

The volume of wire nails which is being distributed is large. Factories are having some difficulty in making shipments as promptly as purchasers desire. There is a feeling among the trade that an advance would not be surprising, but on the other hand expression is given to the idea by some that the consolidated interests under conservative counsels will be adverse to a radical advance in wire products.—Iron Age.

LINSEED SITUATION IN CHICAGO.

The nominal market is 60c. per 7½ lb. (8c. per lb.) but little or no business is doing and scarcely enough to test the market. We learn that 58c. oil has been sold by second hands, and crushers to move oil would be required to meet this price. There is also some talk about 57c. oil, but it is doubtful if that price will buy now. It is a matter of common report that the big trade are well covered ahead, and other factors are not interested at the moment. Deliveries on contracts are going forward freely, but on the score of new business the market remains practically lifeless. The advent of spring weather and the advance in seed latterly have improved the tone of the market and at the close it is not certain that any considerable quantity of oil could be picked up at more than 1 to 2c. below nominal figures. Single barrel quantities are bringing 61c. (8 2-15c.) for raw and 62c. (8 4-15c.) for boiled and better, and locally the demand has improved since the opening of the week. Stocks of oil held by mills being smaller than usual at the opening of the active consuming season does not argue for a weak market after demand becomes brisk, and the dilatoriness of buyers over taking hold now, causing heaviness and panicky feelings among dealers well stocked, can but add to the pressure of demand later on, which may put prices up beyond a basis warranted by the general situation. It is feared by some that the trade are so fixed that nearly all will want oil at once and under ordinary supply conditions this would be unfortunate. On the other hand, if spring should continue backward and painting operations be long retarded, lower prices may ensue.—Paint, Oil and Drug Review.

NEW YORK METAL MARKET.

TIN—The cable reported a substantial advance in the London tin market, which was said to be partly due to the buying on American account and partly from a desire to influence the auction of 2,500 tons Banca to be held in Holland to-morrow. Not all of the advance in London was retained, however, the market closing firm at £1 5s. above last night's quotation, but 5s. below the highest point of the day. The New York market, it is reported, was bid up by the bull interests and closed firm at 26.05c. bid and 26.50c. asked for spot, 26 to 26.37½c. for March, 25.62½ to 25.87½c. for April, and 25.25 to 25.60c. for May. The Maasdam added 60 tons to the spot stock, bringing the total receipts at Atlantic ports for the month to date up to 2,255 tons.

COPPER — Under an active demand the London market made a further advance of 3s. 9d. on spot and 2s. 6d. on future this morning, but the close was quiet. Nothing of fresh interest came to the surface in the New York market, which closed dull but steady, at 17c. for Lake Superior, and 16½c. for electrolytic and casting.

PIG LEAD—Trade continues on the hand-to-mouth order in the New York market, but prices are maintained on the basis of 4.37½c., in lots of 50 tons or more. St. Louis was reported steady by wire, with sales of 5 cars soft Missouri and 100 tons chemical at 4.22½c. There was no further change in the London market.

SPELTER—The movement continues very light, and the tone of the market is still weak. Spot quotations are unchanged at 3.85 and 3.90c. asked, but within the past few days shipments from the West have sold from 3.90 to 3.87½c., with more sellers at the lower figure. St. Louis was unchanged at 3.70c. London was steady.

ANTIMONY—Regulus is held at 8¾ to 10¼c., as to brand, and is in fair request.

OLD MATERIALS—Prices were steady, though the market is very quiet.

IRON AND STEEL—Advices from the West are to the effect that the activity in iron is beginning to abate, but that the volume of business is still heavy and above the average for the season. The combination of the furnaces in the Mahoming and Shenango valleys and the Pittsburg district, will, if completed, further strengthen the situation in Bessemer iron, but the impression prevails that the limit of the advance in prices has been about reached. In steel the market is very strong and demand continues active, with the tendency of prices on some descriptions still upward.

TINPLATE — Business continues good, orders for both prompt and forward delivery being liberal. Prices are firm and unchanged. — N. Y. Journal of Commerce, March 27.

WHOLESALE HARDWAREMEN AND THE CONSUMING TRADE.

A Conference in Montreal in Regard to the Question.

TO the wholesale and retail hardware merchants of Montreal, the importance of the conference that took place last Thursday night, March 21, in the Monument National, cannot be overestimated. The members of the retail association have succeeded in placing their grievances in their true light before the tribunal of wholesale merchants who have promised to do their utmost to mete out a full measure of justice. The giving of justice may lead to some important readjustments, profitable to wholesalers as well as retailers.

The conference took place, as our readers already know, at the instance of a series of invitations that were sent out by the Retail Hardware Association, written with the expressed intention of leading to a discussion for the common interests of the two branches of the trade. When President Martineau took the chair, he found grouped around him the representatives of five wholesale firms: Mr. G. C. Caverhill, of Caverhill, Learmont & Co.; Mr. W. Starke, of Howden, Starke & Co.; Mr. A. M. St. Arnaud, of The Canada Harness Co.; Mr. John P. Seybold, of The Seybold & Sons Co., and Mr. A. Jeannotte, of L. H. Hebert & Co. There were also present about 25 retailers.

THE QUESTION AT ISSUE.

The president opened the meeting in French. After welcoming the visitors, he gave a short outline of the history of the organization, explaining the object of its formation and its accomplishments. In the first place its raison d'etre is to provide a means by which the retail merchants will come to know one another, and by which they can meet to discuss matters relating to their trade. But there has always been present in the minds of the members of the association the idea that the organization should be a means of having the wholesalers discontinue selling to consumers. They felt they had a serious complaint to make on this matter, and it was for the purpose of having them hear this complaint that the wholesalers had been invited there that night. Their acceptance of the invitation shows that they were in sympathy with the association, and "I hope that this conference will do much good, not only for the retailers, but for the wholesalers as well. Neither of us can lose anything from understanding one another." He then called upon Vice-President Drysdale to repeat his words in English.

Mr. John Millen, of John Millen & Sons, also spoke and explained the desire of the association. "We often have customers come into our stores, who, when quoted a price, say that they have a friend in a wholesale hardware house who will get them those goods at wholesale prices. We don't believe that consumers should be able to purchase anything at a wholesale establishment, and would ask that the wholesale merchants present would suggest some means by which this will be stopped. We would like some assurance to that effect."

HE WISHED TO PROTECT THE RETAILER.

Mr. Caverhill was then called upon to address the meeting. He thanked the association for the honor of being invited, and congratulated the merchants upon their organization. He believed he was entitled to be called the "father" of the association; for 18 years he had advocated the formation of such an association at least 50 times a month. As a wholesaler, he wished to protect the retailers; without them, the wholesalers could not exist. But he believed the retail merchants were not adopting the best means to attain their desired objects. They should do more work among themselves and pay less attention to the wholesalers. He felt sure that all the wholesale houses represented there that night did not sell $5,000, perhaps not even $500, worth of goods to consumers in a year—certainly not enough for a retail association to bother about. The retailers should try among themselves to get a profit by setting prices and pledging the members to keep them. At first, the prices would not be maintained successfully, for there will always be some who are too dishonest to stick to an agreement, but, in time, these would be shamed into line. In this regard, he related some of the difficulties experienced by the wholesale association in keeping the prices uniform, but, he said, in spite of all such troubles, the wholesale association is growing stronger every day. They had shown the manufacturers that they could keep a uniform price, and more and more manufacturers were aiding them to make a reasonable profit. A trade association must grow in strength; it cannot attain to full vigor immediately upon its organization. He hoped that the members would persevere, and they must, in time, improve their status. (Applause.)

Mr. Belanger followed, and spoke from the retailer's point of view. He said, if he understood Mr. Caverhill aright, a competition existed between the wholesalers and retailers. Twenty years ago, when he started business, he used to hear the statement: "I can get this cheaper from Mr. Martineau, or Mr. Millen (retailers)"; now, my customers say: "I can get it cheaper from the wholesalers." He asserted that the wholesalers should not try to do two classes of trade. The retail association might try to control their own members, but they would not be able to set prices for the wholesalers, and the first step to take would be to have the wholesalers stop their retail trade.

Mr. Caverhill then explained that he could not enter into any argument in regard to the wholesalers selling retail. He knew nothing about this, for his firm sold no goods to consumers. Mr. Millen emphasized Mr. Belanger's point.

WILLING TO AID THE RETAILERS.

Mr. W. Starke was then called on for a speech, and he responded appropriately, thanking the association for the invitation and endorsing Mr. Caverhill's remarks. He assured the association that although the firm of which he is the vice-president did little city trade, they might soon enter the field, and that he was quite willing to aid the city retailers in alleviating their grievances. Like Mr. Caverhill, however, he thought they should commence among themselves. (Applause.)

Mr. St. Arnaud differed somewhat in opinion from Mr. Caverhill. As a wholesaler he sold no goods to the public, and he did not believe it was right for a wholesale house to do so. It made no difference what the amount was, the principle was wrong. He assured the association that if they struck the right spot they would get what they wanted and he promised to help them. (Applause.)

Mr. Seybold endorsed Mr. Caverhill's remarks. "The wholesalers have no time to sell retail," he said. "I turn customers out of my establishment every day. A neighbor whom I cannot turn away may get a keg of nails or some other trifling purchase."

Mr. Drysdale: "At wholesale price?"

Mr. Seybold: "No, at retail price. But if they were not neighbors, these people would not get anything. Outside of my neighbors, no consumers can buy from us." He then assured the association that if an explanation of what they wished was laid before the wholesale association, they would receive all possible help.

Mr. Jeannotte also addressed the meeting. He assured them that his house did not sell a cent's worth of goods to consumers. He asked the association to write all complaints to the wholesale association where they would be dealt with in an amicable spirit.

Mr. Millen: "The air is clearing. I believe we are going to get some satisfaction."

RETAILERS COULD HELP THEMSELVES.

Mr. Prudhomme, one of the leading retailers, spoke in English. He said that, although he was treasurer of the Retail Association (and had lots of money on that account), he was a "go-between" the jobbers and retailers, for he was in both businesses at once. Yet he did not think that any of his fellow-retailers could find any fault with his retail prices. (Applause.) He said that there was a great deal the retailers could do among themselves and they were doing it. But the subject of discussion to-night immediately concerned the jobbers. On Canadian-manufactured goods, he claimed, the ordinary city retailer made no profit.

He asked the wholesaler to set a price to consumers on nails, screws, carriage bolts, horseshoes and such lines whereby the smallest retailer would make a profit. He complained about wholesalers selling carriage bolts, horseshoes and horse nails to blacksmiths at same prices as to retailers. The wholesalers should give the small man a margin and then they can keep their own prices more easily. "You can do a lot to help us without taking a cent from your own pockets." he concluded.

RETAIL PRICE ON BLACKSMITHS' GOODS.

Mr. Starke explained that the Hardware Association had been for two years trying to fix a retail price on blacksmiths' goods. "In Ontario, the jobbers have been giving the retailers part of their profit. We have failed to do anything as yet. However, we hope to differentiate some day." He assured them it was a difficult matter to deal with. He also objected to the quantity rebate, because it left retailers on an unequal footing. It starts a lot of semi-jobbers in business and renders the situation complex in the extreme. "Some of you retailers buy rails at the same rate as we do. Is that fair to us or to the other retailers?" he said.

Mr. Beland came back to the point of jobbers selling consumers. He took up Mr. Caverhill's statement, endorsed by the other visitors, that not more than $500 worth of goods were sold this way in a year in Montreal. "Why, I know one house that sells more than that in a month. There are contractors whose trade I should have, who often give orders for $50 or $75 worth of goods to this certain house," he added. He asked why the rebate on nails had been taken off and thought it ought to be increased to 10c. per keg instead of 7½c. as it had been.

THE REBATE ON NAILS.

Mr. Jeannotte said the rebate on nails was taken off by the manufacturers, who said that some retailers were selling nails, under the protection of this rebate, at a price below the wholesalers' fixed quotation.

Mr. Caverhill also said that some retailers had abused their rebate. "I know five retailers, members of your association, who were selling one keg of nails as cheap as I was selling 150," he declared.

CUTTING IN THE PRICE OF NAILS.

Mr. Prudhomme, in reply, asserted that there were more than five members of the wholesale association that were cutting the prices of nails last year. In regard to the charge of the retailers cutting prices, he claimed that if the manufacturers would set the prices at which they should sell to consumers, it would be a great help, and, perhaps, a means of overcoming the difficulty.

SELLING TO CONTRACTORS.

Mr. Huberdeau urged upon the wholesalers the necessity of not selling to contractors, religious institutions, etc. He also expressed the opinion that the preferential list should be done away with in order to put all retailers on the same footing.

Mr. Belanger expressed his willingness to discontinue buying from the manufacturers, even if he has to pay 10 per cent. more than he now pays on any lines he buys direct, provided that the wholesalers will stop selling retail to contractors and other bodies.

THREE CLASSES OF HARDWAREMEN.

Mr. Magnan made one of the most forcible speeches of the evening. He suggested that there should be only three classes of hardware business men—the manufacturers, wholesalers and retailers. He thought it only proper that the manufacturers should sell only to the wholesale trade. He was also against the quantity basis. "As long as you have the quantity basis, you will always have the big people cutting prices," he concluded.

"That's right," said a wholesaler.

"But," said Mr. Magnan, "if the manufacturers should not sell to us, you should not sell to consumers, including convents, colleges, etc. You may be doing this as a charity, but we are in need of charity ourselves. Don't touch this trade, and the retailers will have a better chance and your own trade will be no less. The goods will still go through you, and your profit will be just as great. Favored thus, we will be able to pay 100c. on the dollar. There were 14 hardware failures in Montreal last year. What was the cause?"

Mr. Martineau then summed up the various points. He hoped now that the wholesalers understood the retailers' position and that they would help them. He pointed out that the wholesalers present did not sell retail, and said the association would have to find some way of reaching the guilty houses. He hoped that the visitors would favor the association with their presence again.

ASKED TO FORMULATE A LIST.

Mr. L. J. A. Surveyer asked Mr. Caverhill if he couldn't induce the manufacturers to fix prices for the retailers.

Mr. Caverhill replied that the retailers ought to formulate a list of wishes and have these laid before the wholesalers' and manufacturers' associations. He felt sure that the wholesalers would help them all they could. He could speak for Mr. Newman, his partner, who he knew was particularly anxious to have this desirable state of affairs brought about. He closed with some felicitous remarks. And everyone went away satisfied that the conference had served as the beginning of a movement of reform.

RESOLUTION OF CONDOLENCE.

Mr. Magnan then moved a resolution expressing the deep sorrow felt by the members of the association at the loss the trade has sustained through the death of Mr. Piche, the late hardware appraiser at Montreal. Mr. Drysdale seconded the motion, a copy of which will be sent to the family of deceased.

The meeting then adjourned to April 3.

F. E. GREENSHAW SOLD OUT.

F. E. Greenshaw, who has for the past 10 years conducted a hardware, implement, lumber and undertaking business in Shoal Lake, Man., under the name of C. H. Greenshaw, has sold out his hardware and implement business, but will continue as lumber dealer and undertaker.

Mr. Greensnaw's successors in the hardware and implement business are W. G. Eakins and W. J. Griffin, both of whom were formerly with The James Robertson Co., Limited, Winnipeg. Their firm style is Eakins & Griffin.

A North Sydney, N.S., despatch says that The Nova Scotia Steel Company will start the construction of an iron and steel plant in that place about May 1.

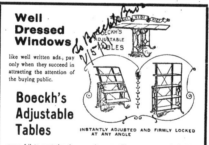

H. S. HOWLAND, SONS & CO.

WHOLESALE ONLY — 37-39 Front Street West, **Toronto.** — ONLY WHOLESALE

WASHING MACHINES.

DOWSWELL, SQUARE.

MOPS.

"COMBINATION."

DOWSWELL, ROUND RE-ACTING.

DASH CHURNS.

OAK DASH CHURN.

Plain Top.	Crib Top.
6 Gals.	6 Gals.
8 "	8 "
10 "	10 "

"LEADER" REVOLVING CHURNS.

STYLE B, WOOD FRAME.

STYLE A, STEEL FRAME.

No.	0	1	2	3	4	5
Capacity	7	9	15	20	25	35 Gallons.

H. S. HOWLAND, SONS & CO., Toronto.

WE SHIP PROMPTLY. — Graham Wire and Cut Nails are the Best. — OUR PRICES ARE RIGHT.

BUSINESS CHANGES.

DIFFICULTIES, ASSIGNMENTS, COMPROMISES.

A MEETING of the creditors of H. Cairns, general merchant, Sawyerville, Que., was held on Tuesday.

Frank Reardon, painter, etc., Halifax, is asking an extension.

H. V. Mooers, harness dealer, Woodstock, N.B., is offering 20c. on the dollar.

George A. Rollins, tinsmith, etc., Madoc, Ont., has assigned to Frederick Rollins.

George R. Garnett, general merchant, Murray Harbor South, P.E.I., has assigned.

Joseph Tays, general merchant, Port Moody, B.C., is offering 50c. on the dollar.

The bailiff is in possession of the business of Gustave Lichte, tinsmith, stove dealer, etc., Baden, Ont.

A meeting of the creditors of L. G. Jourdain, hardware dealer, Three Rivers, Que., has been held. Gagnon & Caron are curators of the estate.

PARTNERSHIPS FORMED AND DISSOLVED.

Fox Bros., general merchants, Swan River, Man., will dissolve April 1.

Brown & Yellowlees, hardware and lumber dealers, Ninette, Man., have dissolved. John Yellowlees continues.

SALES MADE AND PENDING.

The assets of B. J. Pettenar, machinist, Montreal, have been sold.

Mrs. C. J. Menard, general merchant, Lefaivre, Ont., has sold out.

C. Jacques, blacksmith, Tilbury, Ont., is advertising his business for sale.

The assets of Louis Dore, coal and wood dealer, Montreal, have been sold.

Albert Mickus, blacksmith, Wellesley, Ont., is advertising his business for sale.

Gibson Douglas, general merchant, Teviotdale, Ont., is advertising his business for sale.

The assets of F. X. Julien, general merchant, Lambton, Que., are to be sold on April 3.

The stock of J. D. McLeod, general merchant, Prince Albert, N.W.T., has been sold.

P. J. Lindeman, general merchant, etc., West Lorne, Ont., is advertising his hotel business for sale.

The assets of the estate of Walter Wardrop, general merchant, Whitemouth, Man., have been sold.

The assets of the estate of J. G. Fairbanks, general merchant, Spruce Grove, N.W.T., are advertised for sale by tender.

The stock of D. Campbell, general merchant, Little Metis, Que., has been sold at 58c. on the dollar to E. Hudon, St. Octave, Que.

CHANGES.

Henry George, general merchant, Ninga, Man., has given up business.

T. H. Easton, carriagemaker, Minnedosa, Man., is out of business.

Drouin & Plourde have registered as tinsmiths at Windsor Mills, Que.

R. Robertson, general merchant, Lanark, Ont., is retiring from business.

J. H. Black, general merchant, etc., Headingly, Man., is selling out.

Bell & Co., general merchants, Harrow, Ont., have sold out to John Stocker.

Keswick & Hammond, general merchant's St. Leonards, N.B., have closed out.

James Elsey, harness dealer, Mount Brydges, Ont., has sold out to S. J. Bond.

John Nasymth, general merchant, Lotus, Ont., has opened a branch in Lifford, Ont.

A. McDonald, general merchant, Caldwell's Mills, Ont., has started business at Lanark.

G. Drummond, general merchant, St. Aubert (L' Islet), Que., is about to remove to Rogersville, N.B.

W. F. Dibblee & Son, hardware dealers, Woodstock, N.B., are opening a branch at Centreville.

J. A. McArthur & Co., dealers in agricultural implements, etc., Sussex, Que., are opening a branch at Horton, N.B.

W. R. Wells, sawmiller, North Arm, B.C., has been succeeded by The South Vancouver Lumber Manufacturing Co., Limited.

The stock of the estate of Robert Lewis, painter and wall paper dealer, London, Ont., has been sold at 32c. on the dollar to Wm. Scarrow.

FIRES.

A. A. McCaull, general merchant, Elderslie, P.E.I., has been burned out.

W. H. Morrow, general merchant, Portage la Prairie, Man., has been burned out.

DEATHS.

J. Cecconi, general merchant, St. Pierre, N.S., is dead.

George P. Rodger, painter, Amherst, N.S., is dead.

PERMANENT AND DECORATIVE WATER PAINTS.

THE painter and decorator is frequently tempted to allow the enthusiasm of inventors to take possession of his mind and lead him to see things with their eyes, whereas the invariable practice should be to subject all claims and statements to the crucible of practical experiment.

Such experiment we now ask for the permanent and decorative water paints.

This paint rises as a phœnix from the ashes of the old line glue wall finishes, and becomes a peer in many respects to oil paints, while surpassing them in other respects, especially in artistic effects. There is no color limit to permanent water paints, any and every tint from white to black can be successfully used if they are properly mixed and judiciously applied.

They do not discolor with age, nor with impure gases in the air, dirt does not cling to them with the same tenacity as it does to oil paint, for the surface has no grip for the floating particles of carbon and other matter, but readily yields to washing down with sponge and cold water, eradicating all blemishes that would be fatal to the glue-size wall finish or to ingrain paper.

Permanent water paint is both an interior and exterior paint for decorative purposes. This paint is made in almost all special colors which are known to the trade, likewise in white and black, both of which can be used in toning down or neutralizing many of the colors.

Permanent water paint should be applied to new or clean walls; in no instance should it be coated over old wall finish or paper. It is much easier to apply than any glue-size mixture. The decorator can allay his fears, for should a second coat be necessary to give good results the under coat will not roll or rub up, as the binder is waterproof, hence, in applying the second coat, there can be but one result, that is, a good solid surface finish. This alone is invaluable to the painter, and will do more to bring the use of these paints into prominence than many of its other valuable features.

Permanent water paints can be applied to either inside or outside woodwork, and varnished over, giving it the stability of oil paint at a much less cost. As a brick paint it is permanent, much more so than oil colors. When applied on shingles it becomes a fireproof coating. As a sanitary paint it has no equal, for in its composition there is no animal or vegetable substance to soften by dampness, mildew, or rot with their unpleasant odors, which may not be so easily detected in bedrooms and living-rooms, which are probably daily aired or well ventilated by other means, but the rot is imperceptibly there, comtaminating the atmosphere. Much has already been done by scientists to remedy these conditions, and in the Canada Paint Company's sanitary water paints will be found another step in the direction of the ideal, for both sanitary and decorative effects.

HENRY CLUCAS.

London, Ont.

M. J. LEITCH HAS NOT FAILED.

In our issue of March 15, it was stated that M. J. Leitch, general merchant, Michael's Bay, Ont., had assigned. It should have read: "E. F. Leitch." We regret, exceedingly, the error, particularly as we understand that a Mrs. M. J. Leitch has recently started in business at Michael's Bay as a general merchant.

EDMONTON BOARD OF TRADE.

The annual meeting of the Edmonton, N.W.T., Board of Trade was attended by a large proportion of the business men of the town. Twelve new members were admitted.

President Gariepy's address showed that much good work has been done during the year. Roadways leading to the town have been improved. The train and mail service have been increased. The Government has made provision for the erection of a new court house. He suggested that during the next year an effort should be made to secure a new post office and a new immigration shed; and expressed the opinion that the headquarters of the North-West Mounted Police should be removed to Edmonton, and an experimental farm established in the district. He also advocated the installation of a good waterworks system.

The following officers were elected:

President—J. H. Gariepy, reelected.

Vice-President—A. Taylor.

Council—J. A. Stovel, F. J. Fisher, J. S. Willmott, H. Astley, W. Richardson, K. W. MacKensie, W. T. Henry, K. A. McLeod, J. A. Hallier and Robt. Lee.

IRON PIPE STILL ADVANCING.

For the third time within the last three weeks we announce an advance in the price of black iron pipe. The advance in the present instance, however, is in the sizes from ¼ to 2-inch inclusive. One-inch is now up to $5 per 100 feet, an advance of 7c. over last week and of 40c. over two weeks ago.

Prices are firmer in the United States, and trade both there and here is active.

AN ERROR IN DISCOUNT.

In the advertisement of the Ontario Wire Fencing Co., Limited, Picton, Ont., on page 36 of last week's edition of HARDWARE AND METAL, the trade discount on nettings was quoted: 1 to 5 rolls, 55 and 5 per cent. This was in error as the correct discount to the trade is 55 per cent.

A NEW RAIL SPIKE.

Mr. Eben Parkins, of the Portland Rolling Mills, Limited, St. John, N.B., has patented a new rail spike. This spike is a new departure, having cruciform, concave sides and Goldie point. Tests of this spike which have been made at McGill testing laboratories show that it possesses superior driving qualities, while its resistance to pressure is from 10 to 20 per cent. greater than standard rail spikes.

It is furthermore claimed that this shape of spike leaves the fibres of the wood in better condition, and that it combines strength with lightness, having 20 per cent. more spikes to 100 lb., a great advantage to the consumer.

MARKETS AND MARKET NOTES

QUEBEC MARKETS.

Montreal, March 29, 1901.

HARDWARE.

THE Dominion Wholesale Hardware Association has held its quarterly meetings this week, but there have been no interesting results given out. The manufacturers have also been in session all week. To date the only change is a decline in the price of shot. The discount has been raised from 7½ to 15 per cent. White lead declined ⅛c. per lb. this week. There has been a good call for wire, and some fair shipments have been made this week. The bulk of the goods, however, will be moving out in the first two weeks of April. The call for wire nails is well maintained and prices are firm. Cut nails are still slow. All lines of shelf goods are in fair request. The advent of mild weather has stimulated a trade for spring goods. Freezers are selling well, and poultry netting has had a good sale. The metals are weak, and little business has been done either for immediate requirements or for import. Stoves are selling exceedingly well this year, and utensils for making maple syrup and sugar have been sold freely.

BARB WIRE—The demand is good for barb wire, both for immediate and future requirements. Prices are steady at $3.05 per 100 lb. f.o.b. Montreal.

GALVANIZED WIRE — Some good lots have been sold this week, dealers ordering freely. We quote: No. 5, $4.25; Nos. 6, 7 and 8 gauge, $3.55; No. 9, $3.10; No. 10, $3.75; No. 11, $3.85; No. 12, $3.25; No. 13, $3.35; No. 14, $4.25; No. 15, $4.75; No. 16, $5.00.

SMOOTH STEEL WIRE—Fence wires are in fair request, while hay-baling is sometimes wanted. We quote oiled and annealed as follows: No. 9, $2.80; No. 10, $2.87; No. 11, $2.90; No. 12, $2.95; No. 13, $3.15 per 100 lb. f.o.b. Montreal, Toronto, Hamilton, London, St. John and Halifax.

FINE STEEL WIRE—The usual amounts are selling at 17½ per cent. off the list.

BRASS AND COPPER WIRE — There is nothing new in this line. Discounts are unchanged at 55 and 2½ per cent. on brass, and 50 and 2½ per cent. on copper.

FENCE STAPLES—There is a good trade doing in fence staples. We quote: $3.25 for bright, and $3.75 for galvanized, per keg of 100 lb.

WIRE NAILS—Thn wire nail manufacturers are in session this week, but as yet no change in price is announced. The demand from the retail trade is rather more brisk. We quote $2.85 for small lots and $2.75 for carlots, f.o.b. Montreal, London, Toronto and Hamilton.

CUT NAILS — The demand is rather slow. Prices are unchanged. We quote: $2.35 for small and $2.25 for carlots; flour barrel nails, 25 per cent. discount; coopers' nails, 30 per cent. discount.

HORSE NAILS—The demand has been rather slack this week. Prices are unchanged from last week. "C" brand is still quoted at 50 and 7½ per cent. off its own list and other brands at 60 per cent. on oval and city head, and 60 and 10 per cent. on new countersunk head off the old list.

HORSESHOES—Trade is rather quiet at the present moment, but it is expected to open up again soon. We quote as follows: Iron shoes, light and medium pattern, No. 2 and larger, $3.50; No. 1 and smaller, $3.75; snow shoes, No. 2 and larger, $3.75; No. 1 and smaller, $4.00; X L steel shoes, all sizes, 1 to 5, No. 2 and larger, $3.60; No. 1 and smaller, $3.85; feather-weight, all sizes, $4.85; toe weight steel shoes, all sizes, $5.95 f.o.b. Montreal; f.o.b. Hamilton, London and Guelph, 10c. extra.

POULTRY NETTING—Most retailers have placed their orders for spring goods. The ruling discount is still 50 and 10 per cent.

GREEN WIRE CLOTH—Business is going along as usual. The market is steady. The price is unchanged at $1.35 per 100 sq. ft.

SCREEN DOORS AND WINDOWS — There is no change to be noted. We quote as follows: Screen doors, plain cherry finish, $8.25 per doz.; do. fancy, $11.50 per doz.; walnut, $7.40 per doz., and yellow, $7.45; windows, $2.25 to $3.50 per doz.

SCREWS—There has been no change made in prices, but the market is firm and the demand brisk. Discounts are as follows: Flat head bright, 87½ and 10 per cent. off list; round head bright, 82½ and 10 per cent.; flat head brass, 80 and 10 per cent.; round head brass, 75 and 10 per cent.

BOLTS—The market is unchanged. The undertone is firm. Discounts are as follows: Carriage bolts, 65 per cent.; machine bolts, 65 per cent.; coach screws, 75 per cent.; sleigh shoe bolts, 75 per cent.; bolt ends, 65 per cent.; plough bolts, 50 per cent.; square nuts, 4½c. per lb. off list; hexagon nuts, 4¾c. per lb. off list; tire bolts, 67½ per cent.; stove bolts, 67½ per cent.

BUILDING PAPER—The spring demand continues. Prices are unchanged. We quote: Tarred felt, $1.70 per 100 lb.; 2-ply, ready roofing, 80c. per roll; 3-ply, $1.05 per roll; carpet felt, $2.25 per 100 lb.; dry sheathing, 30c. per roll; tar sheathing, 40c. per roll; dry fibre, 50c. per roll; tarred fibre, 60c. per roll; O.K. and I.X.L., 65c. per roll; heavy straw sheathing, $28 per ton; slaters' felt, 50c. per roll.

RIVETS — The market is without feature. The discount on best iron rivets, section, carriage, and waggon box, black rivets, tinned do., coopers' rivets and tinned swedes rivets, 60 and 10 per cent.; swedes iron burrs are quoted at 55 per cent. off; copper rivets, 35 and 5 per cent. off; and coppered iron rivets and burrs, in 5-lb. carton boxes, are quoted at 60 and 10 per cent. off list.

CORDAGE—There has been no change in the prices of cordage this week. The demand is moderate. Manila is now worth 13½c. per lb. for 7-16 and larger; sisal is selling at 10c., and lathyarn 10c.

SPADES AND SHOVELS — A better demand has been experienced for spades and shovels this week. The discount remains at 40 and 5 per cent. off the list.

HARVEST TOOLS — There is not much business doing in this line. The discount is still 50, 10 and 5 per cent.

TACKS—Unchanged. We quote as follows: Carpet tacks, in dozens and bulk, blued 80 and 5 per cent. discount; tinned, 80 and 10 per cent.; cut tacks, blued, in dozens, 75 and 15 per cent. discount.

LAWN MOWERS—Some sales have been made this week. We quote: High wheel, 50 and 5 per cent. f.o.b. Montreal; low wheel, in all sizes, $2.75 each net; high wheel, 11-inch, 30 per cent. off.

FIREBRICKS — Trade is very quiet at present. Prices are still fixed at $18 to $24 per 1,000 as to brand, ex store.

CEMENT—A new departure in the cement trade this season is the fact that American manufacturers are competing very keenly in this market against high-grade German cement, and, so far, they have been very successful, having secured all the big orders for spring delivery. Importers of German cement say they are not in a position to compete with America, owing to the fact that the latter take back the bags in which the cement is put up at the same price that buyers pay for them, while they have to pay the German manufacturers 30c. each for the barrels that contain their cement. We quote as follows: German, $2.45 to $2.55; English, $2.30 to $2.40; Belgian, $1.95 to $2.05.

METALS.

The market for metals, such as Canada plate, terne plates, tinplates, black sheets and galvanized sheets is weak, and there seems to be no improvement in sight, although hopes were expressed that the English market would react some days ago. There are only a few small orders being placed for early import. The discount on lead pipe is now 25 per cent.

PIG IRON—In spite of the increase in value of American pig iron, prices here show but little improvement. The English market continues to be demoralized, and No. 1 Summerlee is selling for import at $20 to $20.50. No. 1 Hamilton is worth $18 to $18.50.

BAR IRON — On the present mill price, values would be $1.70 to $1.75 per 100 lb. Transactions are, however, being made at lower figures.

BLACK SHEETS — The English market is easy and trade is quiet. On the Montreal market No. 28 gauge has been quoted as low as $2.50 in carlots. We quote $2.60 for 8 to 16 gauge; $2.65 for 26 gauge, and $2.75 for 28 gauge. Import orders are being taken at $2.40 to $2.45.

GALVANIZED IRON — There are few inclined to operate on the galvanized iron market, although it is in better condition than other metal markets. Import orders are being taken for No. 28 Queen's Head $4.25, the spot value is about $4.60; Apollo 10¾ oz., is worth $4.45 for immediate delivery and Comet, No. 28, $4.

INGOT COPPER—Values are steady at 15c.

INGOT TIN — The demand is small. Lamb and Flag is quoted generally at 31c.

LEAD—The market is weak at $4.

LEAD PIPE — A moderate trade is passing. We quote as follows: 7c. for ordinary and 7½c. for composition waste, with 25 per cent. off.

IRON PIPE — A good business is being done in iron pipe at unchanged prices. Our galvanized schednle may be slightly shaded. We quote as follows: Black pipe, ¼, $3 per 100 ft.; ⅜, $3; ½, $3; ¾, $3.15; 1-in., $4.50; 1¼, $6 10; 1½, $7.28; 2-in., $9.75. Galvanized, ½, $4.60; ¾, $5.25; 1-in., $7.50; 1¼, $9.80; 1½, $11.75; 2-in., $16.

TINPLATES — For import, coke tinplates are quoted at $4; cheap charcoal, at $4.25, and preferable grades, $4.50. For immediate delivery goods may be obtained at $4.15 to $4.25 for coke and $4.40 to $4.50 for charcoal. Business is dull.

Canada Plate—Some dealers are inquiring for goods to satisfy immediate wants. There is nothing doing on future account. The English market is weak. We quote: 52's, $2.60; 60's, $2.70; 75's, $2.80; full polished, $3.45, and galvanized, $4.30.

Tool Steel—The market is steady. We quote: Black Diamond, 8c.; Jessop's 13c.

Steel—A fair business is being done. We quote: Sleighshoe, $1.85; tire, $1.95; bar, $1.85; spring, $2.75; machinery, $2.75 and toe-calk, $2.50.

Terne Plates—For import, terne plates are quoted at $7. On spot they are selling for about $8. The market is not in good condition.

Coil Chain—Market is firm, but rather quiet just now. We quote as follows: No. 6, 11½c.; No. 5, 10c.; No. 4, 9½c.; No. 3, 9c.; ¼-inch, 7½c. per lb.; 5-16, $4.65; 5-16 exact, $5.10; ⅜, $4.20; 7-16, $4.00; ½, $3.75; 9-16, $3.65; ⅝, $3.35; ¾, $3.25; ⅞, $3.20; 1-in., $3.15. In car-load lots an allowance of 10c. is made.

Sheet Zinc—The ruling price is 6 to 6¼c.

Antimony—Quiet, at 10c.

Zinc Spelter—Is worth 5 to 5½c.

Solder—We quote: Bar solder, 18½c.; wire solder, 20c.

GLASS.

Import orders are still being taken on the basis of $1.70 or $1.75. The spot demand is fair. We quote: First break, $2; second, $2.10 for 50 feet; first break, 100 feet, $3.80; second, $4; third, $4.50; fourth, $4.75; fifth, $5.25; sixth, $5.75, and seventh, $6.25.

PAINTS AND OILS.

The feature of the week is the strengthening tendency of the English linseed oil market, which has reacted about 9c. At the present quotations it would cost 61c. to lay down oil for spring import. Import orders are being taken at 65 and 68c. in 10-bbl. lots. There is a scarcity in red lead. No supplies are expected until early in May. The market is bare. Holders are not making any concessions at all. Orange mineral and litharge are in fair supply, and stocks of dry white lead, while not abundant, are deemed to be quite sufficient to carry grinders till the opening of navigation. With milder weather there has been a much better demand for construction paints, such as oxide and graphite of various kinds. The carriage and implement paint trade is also fairly brisk and orders are liberal. Attention is now drawn to the new scale of ground white lead. A slight falling off is a surprise to some. It was made necessary on account of a little easiness in the English market. There is a good trade doing in all general painting material, and liquid paints are securing a veritable boom ahead of anything heretofore experienced. We quote:

White Lead—Best brands, Government standard, $6.25; No. 1, $5.87½; No. 2, $5.50; No. 3, $5.12½, and No. 4, $4.75 all f.o.b. Montreal. Terms, 3 per cent. cash or four months.

Dry White Lead—$5.50 in casks; kegs, $5.75.

Red Lead—Casks, $5.25; in kegs, $5.50.

White Zinc Paint—Pure, dry, 7c.; No. 1, 6c.; in oil, pure, 8c.; No. 1, 7c.; No. 2, 6c.

Putty—We quote: Bulk, in barrels, $2 per 100 lb.; bulk, in less quantity, $2.15; bladders, in barrels, $2.20; bladders, in 100 or 200 lb. kegs or boxes, $2.35; in tins, $2.45 to $2.75; in less than 100-lb. lots, $3 f.o.b. Montreal, Ottawa, Toronto, Hamilton, London and Guelph. Maritime Provinces 10c. higher, f.o.b. St. John and Halifax.

Linseed Oil—Raw, 70c.; boiled, 73c., in 5 to 9 bbls., 1c. less, 10 to 20 bbl. lots, open, net cash, plus 2c. for 4 months. Delivered anywhere in Ontario between Montreal and Oshawa at 2c. per gal. advance and freight allowed.

Turpentine—Single bbls., 60c.; 2 to 4 bbls., 59c.; 5 bbls. and over, open terms, the same terms as linseed oil.

Mixed Paints—$1.25 to $1.45 per gal.

Castor Oil—8¾ to 9¼c. in wholesale lots, and ½c. additional for small lots.

Seal Oil—47½ to 49c.

Cod Oil—32½ to 35c.

Naval Stores—We quote: Resins, $2.75 to $4.50, as to brand; coal tar, $3.25 to $3.75; cotton waste, 4½ to 5½c. for colored, and 6 to 7½c. for white; oakum, 5½ to 6½c., and cotton oakum, 10 to 11c.

Paris Green—Petroleum barrels, 16¾c. per lb.; arsenic kegs, 17c.; 50 and 100-lb. drums, 17½c.; 25-lb. drums, 18c.; 1-lb. packages, 18½c.; ½-lb. packages, 20½c.; 1-lb. tins, 19½c.; ½-lb. tins, 21½c. f.o.b. Montreal; terms 3 per cent. 30 days, or four months from date of delivery.

SCRAP METALS.

The demand for iron has materially improved during the week, and prices are 1c. per lb. higher. Canada will this year import her scrap from England instead of the United States. Dealers are paying the following prices in the country: Heavy copper and wire, 13 to 13½c. per lb.; light copper, 12c.; heavy brass, 12c.; heavy yellow, 8½ to 9c.; light brass, 6½ to 7c.; lead, 2¾ to 3c. per lb.; zinc, 2¼ to 2½c.; iron, No. 1 wrought, $14 to $15 per gross ton; No. 1 cast, $13 to $14; stove plate, $8 to $9; light iron, No. 2, $4 a ton; malleable and steel, $4.

HIDES.

The market is irregular, the tendency being towards an advance. We quote as follows: Light hides, 7c. for No. 1; 6c. for No. 2, and 5c. for No. 3. Lambskins, 90c.

PETROLEUM.

There has been no change in the market. We quote: "Silver Star," 14½ to 15½c.; "Imperial Acme," 16 to 17c.; "S.C. Acme," 18 to 19c., and "Pratt's Astral," 18½ to 19½c.

MONTREAL NOTES.

White lead is ⅛c. per lb. lower.

The discount on shot is raised from 7½ to 15 per cent.

The discount on lead pipe is now 25 per cent. instead of 15.

Mr. Charles Haldenby, of Messrs. Sanderson Pearcy & Co., Toronto, was in town on Monday, and spent some time visiting the leading paint factories. Mr. Haldenby is a favorite wherever he goes as he is of a most genial disposition. A number of his confreres who occasionally call upon him in Toronto and whom he always greets with a hearty welcome found his visit all too short.

ONTARIO MARKETS.

TORONTO, March 29, 1901.

HARDWARE.

AS spring approaches business expands. During the past week there has been a decided improvement in trade. The lots wanted are not as a rule large, but orders are numerous, and the wholesale houses are kept busy filling them. One assuring feature of the situation is the requests which are coming to hand asking for the delivery of forward orders earlier than the date for which they were originally placed. March is not usually a good month for letter orders, but a fair number of them are coming to hand. The manufacturers are holding their quarterly meetings in Montreal this week, but up to the time of writing the only changes they have made are reductions in the price of white lead and shot. Prices have again advanced on black iron pipe. Wholesalers report payments fair.

BARB WIRE—Trade is more active than it was on account of future delivery, but only a small business is being done. We quote f.o.b. Cleveland at $2.82½ per 100 lb. for less than carlots and $2 70 for carlots. From stock Toronto the figure is $3.05 per 100 lb.

GALVANIZED WIRE—Some orders are being booked, but not a great many. We quote: Nos. 6, 7 and 8, $3.50 to $3.85 per 100 lb., according to quantity; No. 9, $2.85 to $3.15; No. 10, $3.60 to $3.95; No. 11, $3.70 to $4.10; No. 12, $3 to $3.30; No. 13, $3 10 to $3 40; No. 14, $4.10 to $4 50; No. 15, $4.60 to $5.05; No. 16, $4.85 to $5.35. Nos. 6 to 9 base f.o.b. Cleveland are quoted at $2.57½ in less than carlots and 12c. less for carlots of 15 tons.

SMOOTH STEEL WIRE—A little is being done in oiled and annealed, and an occasional order is being received for hay-baling wire. Net selling prices for oiled and annealed are as noted last week: Nos. 6 to 8, $2 90; 9, $2.80; 10, $2.87; 11, $2.90; 12, $2.95; 13, $3 15; 14, $3.37; 15, $3.50; 16, $3.65. Delivery points, Toronto, Hamilton, London and Montreal, with freights equalized on these points.

WIRE NAILS—A fair demand has been experienced this week for immediate shipment. Inquiries are coming in more freely, and shipments during the early part of April are expected to be pretty heavy. The base price for less than carlots is still $2.85 per keg, and for carlots $2.75 per keg.

CUT NAILS—These are as dull and lifeless as ever. The base price is unchanged at $2.35 per keg. Delivery points are: Toronto, Hamilton, London, Montreal and St. John, N.B.

HORSESHOES—A moderate trade only is still to be noted. We quote f.o.b. Toronto as follows: Iron shoes, No. 2 and larger, light, medium and heavy, $3.60; snow shoes, $3.85; light steel shoes, $3.70; featherweight (all sizes), $4.95; iron shoes, No. 1 and smaller, light, medium and heavy (all sizes), $3.85; snow shoes, $4; light steel shoes, $3.95; featherweight (all sizes), $4 95.

HORSE NAILS—Trade is fair for this time of the year. Discount on "C" brand oval head is unchanged at 50 and 7½ per cent. off the new list and on other brands 50, 10 and 5 per cent., off the old list; countersunk head, 50, 10 and 10 per cent.

SCREWS—A fair trade is still being done. We quote discounts as follows: Flat head bright, 87½ and 10 per cent.; round head bright, 82½ and 10 per cent.; flat head brass, 80 and 10 per cent.; round head brass, 75 and 10 per cent.; round head bronze, 65 per cent., and flat head bronze at 70 per cent.

BOLTS AND NUTS—The demand for bolts is active, particularly with the makers who are well filled with orders. There have been some inquiries during the week for big lots. We quote: Carriage bolts (Norway), full square, 70 per cent.; carriage bolts full square, 70 per cent.; common carriage bolts, all sizes, 65 per cent.; machine bolts, all sizes, 65 per cent.; coach screws, 75 per cent.; sleighshoe bolts, 75 per cent.; blank bolts, 65 per cent.; bolt ends, 65 per cent.; nuts, square, 4½c. off; nuts, hexagon, 4¼c. off; tire bolts, 67½ per cent.; stove bolts, 67½; plough bolts, 60 per cent.; stove rods, 6 to 8c.

RIVETS AND BURRS—Trade is moderate. We quote as follows: Iron rivets, 60 and 10 per cent.; iron burrs, 55 per cent.; copper rivets and burrs, 35 and 5 per cent.

ROPE—A little business has been done from stock and a few orders have been booked for future delivery. We quote the base price as follows: Sisal 10c. per lb; manila, 13½c. per lb.; special manila, 11½c.; New Zealand, 10c. per lb.; single tarred lathyarn, 9½c.

CUTLERY—Some business has been done, but it does not amount to a great deal.

SPORTING GOODS—Quite a little gun and blasting powders are going out this week and a few guns and rifles. A good many orders are being booked for loaded shells for future delivery.

GREEN WIRE CLOTH—Not many orders are being taken just now. We quote $1.35 per 100 sq. ft.

SCREEN DOORS AND WINDOWS—Orders are still being taken for future delivery. We quote 4-in. styles in doors at $7.20 to $7.80 per doz., and 3 in. styles 20c. per

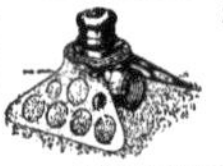

doz. less; screen windows, $1.60 to $3.60 per doz., according to size and extension.

TINWARE—Business is fairly good in sap buckets.

BUILDING PAPER—The demand keeps good with prices unchanged.

POULTRY NETTING—Supplies of English netting are being taken into stock, and retailers are getting their orders filled. Discount on Canadian 55 per cent.

HARVEST TOOLS —Quite a few orders have been booked, but not many deliveries have so far been made. Discount 50, 10 and 5 per cent.

SPADES AND SHOVELS — The feature of the trade in this line is the requests that are coming to hand from customers asking for the delivery of orders placed some time ago. Discount 40 and 5 per cent.

CHURNS — A fair number of orders have been booked, and quite a few deliveries are being made. Discount 58 per cent.

WRINGERS — There is some demand for these, the recent reduction in prices having stimulated the demand.

BRUSHES — The demand for brushes is active, and quite a little difficulty is being experienced in filling orders.

HORSE CLIPPERS — A good business is being experienced for horse clippers.

CEMENT—A fair movement is reported. Prices are steady. We quote: Canadian portland, $2.40 to $3; German, $3 to $3.15; English, $2.85 to $3; Belgian, $2.50 to $2.75; calcined plaster, $2.

METALS.

Orders are not as a rule individually large, but they are numerous, and even more so than a week ago. Boiler tubes and iron pipe are all higher. Shot is cheaper, the discount now being 15 per cent.

PIG TIN—The outside markets have ruled firm during the week. Local quotations are unchanged at 32 to 33c., and the demand has been fair.

PIG IRON — Prices continue firm with business moderate.

BAR IRON—There is a good inquiry and an active trade is being experienced. The mills are very busy. The ruling base price is $1.75 per 100 lb.

HOOP STEEL—Trade is active at $3.10 base.

STEEL BOILER PLATES—Prices are firm and we quote: ¼-inch, $2.40 to $2.50, and 3 16 inch, $2.60.

TINPLATES—A moderate business is being done with quotations as before.

TERNE SHEETS—A little more business is being done, but the volume is not large.

TINNED SHEETS—The demand is good for small lots. We quote 9 to 9½c. for small lots.

GALVANIZED SHEETS—A good trade is being done from stock, and orders are still being booked for importation. The ruling prices are $4.60 for English and $4.50 for American.

BLACK SHEETS—A good trade is reported this week. We still quote 28 gauge at $3.30.

CANADA PLATES—A few odd lots are moving out, but, in general, trade is quiet. From stock we quote: All dull, $3. half-and half, $3.15, and all bright, $3.65 to $3.75.

COPPER—Trade is more active in ingot copper than it was, and a good business is being done in sheet copper. Under the influence of an active demand in London, England, copper is higher in New York. Local prices are unchanged. We quote as follows: Ingot, 19 to 20c.; bolt or bar, 23½ to 25c.; sheet, 23 to 23½c. per lb.

BRASS—Trade has improved; discount unchanged at 15 per cent.

SOLDER—A fair demand is being experienced. We quote as follows: Half-and-half, guaranteed, 18c.; ditto, commercial, 17½c.; refined, 17½c., and wiping, 17c.

IRON PIPE—Prices on ½ to 2-inch have again been advanced, and inch pipe is now quoted at $5. We quote as follows per 100 feet: Black pipe, ⅛ in., $4.35; ¼ in., $3.25; ⅜ in., $3.30; ½ in., $3.35; ¾ in., $3.60; 1 in., $5.00; 1¼ in., $6.85; 1½ in., $8.25; 2 in., $11.00; 2½ in., $22.94; 3 in., $36.12; 3½ in., $37.90; 4 in., $43.09; 5 in., $57.85; 6 in., $75.01; 7 in., $93.76; 8 in., $112.51. Galvanized pipe, ½ in., $5.52; ¾ in., $5.56; 1 in., $7.77; 1¼ in., $10.60; 1½ in., $12.70; 2 in., $16.90.

BOILER TUBES—The manufacturers have advanced their prices about 10 per cent., and figures are firm at the advance.

LEAD—The demand continues good. The outside markets are quiet with prices steady. We still quote 4¾ to 5c. per lb.

LEAD PIPE—The discount is still 25 per cent., with the demand moderate.

ZINC SPELTER—A slightly better demand has been experienced during the past week. The tone of the outside markets is weak. Local quotations are unchanged at 6 to 6½c. per lb.

ZINC SHEET — A fairly active trade is being experienced. We quote 6¾ to 7c. for casks and 7 to 7½c. for part casks.

ANTIMONY — Trade is quiet and prices unchanged at 11 to 11½c. per lb.

PAINTS AND OILS.

The only change in prices this week is a decline of 25c. in white lead. Other lines are steady both locally and at primary market. There is a fair demand, but the demand is curtailed on account of the cool weather. A week or two of warm weather would cause a brisk movement of all staple lines. We quote as follows:

WHITE LEAD—Ex Toronto, pure white lead, $6.37½; No. 1, $6; No. 2. $5.67½; No. 3. $5.25; No. 4. $4.87½; dry white lead in casks, $6.

RED LEAD—Genuine, in casks of 560 lb., $5.50; ditto, in kegs of 100 lb., $5.75; No. 1, in casks of 560 lb., $5 to $5.25; ditto, kegs of 100 lb.; $5.25 to $5.50.

LITHARGE—Genuine, 7 to 7½c.

ORANGE MINERAL—Genuine, 8 to 8½c.

WHITE ZINC—Genuine, French V.M., in casks, $7 to $7.25; Lehigh, in casks, $6.

PARIS WHITE—90c. to $1 per 100 lb.

WHITING — 70c. per 100 lb.; Gilders' whiting, 80c.

GUM SHELLAC — In cases, 22c.; in less than cases, 25c.

PARIS GREEN—Bbls., 16¾c.; kegs, 17c.; 50 and 100 lb. drums, 17½c.; 25-lb. drums, 18c.; 1-lb. papers, 18½c.; 1-lb. tins, 19½c.; ½-lb. papers, 20½c.; ½-lb. tins, 21½c.

PUTTY —Bladders, in bbls., $2.20; bladders, in 100 lb. kegs, $2.35; bulk in bbls., $2; bulk, less than bbls. and up to 100 lb., $2.15; bladders, bulk or tins, less than 100 lb., $3.

PLASTER PARIS—New Brunswick, $1.90 per bbl.

PUMICE STONE — Powdered, $2.50 per cwt. in bbls., and 4 to 5c. per lb. in less quantity; lump, 10c. in small lots, and 8c. in bbls.

LIQUID PAINTS—Pure, $1.20 to $1.30 per gal.; No. 1 quality, $1 per gal.

CASTOR OIL—East India, in cases, 10 to 10½c. per lb. and 10½ to 11c. for single tins.

LINSEED OIL—Raw, 1 to 4 barrels, 69c.; boiled, 72c.; 5 to 9 barrels, raw, 68c.; boiled, 71c., delivered. To Toronto, Hamilton, Guelph and London, 1c. less.

TURPENTINE—Single barrels, 58c.; 2 to 4 barrels, 57c., to all points in Ontario. For less quantities than barrels, 5c. per gallon extra will be added, and for 5 gallon packages, 50c., and 10 gallon packages, 80c. will be charged.

GLASS.

There is practically nothing doing in import orders. The demand from stock is large and prices have decidedly stiffened on account of the steady rise in Belgium and England. One dealer has already advonced his quotations on star under 26 united inches to $4.25, an advance of 25c. There is a general movement for an advance, and though all the jobbers have not yet agreed to the change, it is probable that the following schedule may go into effect on Monday, April 1: Under 26 in., $4.15 26 to 40 in., $4.45; 41 to 50 in., $4.85; 51 to 60 in., $5.15; 61 to 70 in., $5.50; double diamond, under 26 in., $6; 26 to 40 in., $6.65; 41 to 50 in., $7.50; 51 to 60 in., $8.50; 61 to 70 in., $9.50. Toronto, Hamilton and London. Terms, 4 months or 3 per cent. 30 days.

PETROLEUM.

There is a moderate movement. Prices are steady. We quote: Pratt's Astral, 16½ to 17c. in bulk (barrels, $1 extra); American water white, 16½ to 17c. in barrels; Photogene, 16 to 16½c.; Sarnia water white, 15½ to 16c. in barrels; Sarnia prime white, 14½ to 15c. in barrels.

COAL.

There is a good movement for this time of year, as some orders had been delayed in delivery. Prices are steady. We quote anthracite on cars Buffalo and bridges: Grate, $4.75 per gross ton and $4.24 per net ton; egg, stove and nut, $5 per gross ton and $4.46 per net ton.

OLD MATERIAL.

The market is dull. There is a fair movement of machinery and stove cast iron, copper bottoms and heavy copper, heavy red brass, and prices have advanced on all these lines. The brass and copper is ½ to 1c. higher, and the cast iron is 5 to 15c. per cwt. dearer. We quote jobbers' prices as follows: Agricultural scrap, 55c. per cwt.; machinery cast, 60c. per cwt.; stove cast, 50c.; No. 1 wrought 50c. per 100 lb.; new light scrap copper, 12c. per lb.; bottoms, 11½c.; heavy copper, 13c.; coil wire scrap, 13c.; light brass, 7c.; heavy yellow brass, 10 to 10½c.; heavy red brass, 10½ to 11c.; scrap lead, 3c.; zinc, 2c; scrap rubber, 6½c.; good country mixed rags, 65 to 75c.; clean dry bones, 40 to 50c. per 100 lb.

MARKET NOTES.

Boiler tubes are firmer.

White lead is 12½c. per 100 lb. lower.

Machinery cast iron is 5c. and stove cast 15c. dearer.

The discount on shot has been increased to 15 per cent.

Scrap copper bottoms are 1c. and heavy copper ½c. higher.

Glass will probably be advanced 10 to 15c. per box on Monday.

The price of black iron has experienced another advance, the figure for 1-inch pipe now being $5 per 100 feet.

TRADE CHAT.

THE Chatham, Ont., Binder Twine Co., Limited, have been incorporated with $125,000 capital stock.

H. M. Borland, Coldwater, Ont., intends erecting a new store this spring.

Arrangements are being made to enlarge the Kingston, Ont., locomotive works.

Fire did about $1,000 damage to the Globe Paint Works, Toronto, on Monday night.

Aaron Child & Son, Gravenhurst, Ont., are fitting an up-to-date tinshop above their store.

The London, Ont., Fence Machine Co., Limited, have been incorporated with $40,000 capital stock.

The Mowat Hardware Co., of Trenton, Ont., Limited, have been incorporated with $10,000 capital stock.

The Imperial Corundum Co., Limited, have been incorporated with $1,000,000 capital stock and headquarters in Toronto.

The Manitoba Farmers' Hedge and Wire Fence Co., Limited, Brandon, Man., are seeking incorporation.

John McConnell, hardware and stove dealer, Harriston, Ont., has been succeeded by Wm. Beatty, of that town.

The Dominion Portland Cement Co., of Wiarton, Ont., Limited, have been incorporated with $250,000 capital stock.

The Gurney Scale Co., Hamilton, Ont., made their first shipment of scales to Cuba this week. The shipment, which includes 15 platform scales, weighing from 1,200 to 3,000 lb. each, was on an order received through sending a sample scale to that country a short time ago.

The officials of the Pan-American Exposition have decided to grant space for a warm-air heating device to but one house, the manufacturers of the Kelsey warm-air generators, although 60 different houses made application for the privilege. The Kelsey generators are made in Canada by The James Smart Mfg. Co., Limited, Brockville, Ont.

PERSONAL MENTION.

Mr. M. McConnell, hardware merchant, Cayuga, was in Toronto last week.

Mr. D. Bowyer, stove and tinware dealer, Listowel, Ont., was in Toronto on business last week.

Mr. Stephen King, stove dealer, etc., Ingersoll, Ont., was on a business trip to Toronto last week.

Mr. A. A. McMichael, manager for Jas. Robertson & Co., Toronto, was in Montreal on Monday.

The Mower

THAT WILL KILL ALL THE WEEDS IN YOUR LAWNS

If you keep the weeds cut so they do not go to seed, and cut your grass without breaking the small feeders of roots, the grass will become thick and weeds will disappear. **The Clipper will do it.**

CANADIAN PATENT FOR SALE.
SEND FOR CATALOGUE AND PRICES.

Clipper Lawn Mower Co.

NORRISTOWN, Pa.

CORDAGE

ALL KINDS AND FOR ALL PURPOSES.

Manila Rope	Lathyarn
Sisal Rope	Shingleyarn
Jute Rope	Bale Rope
Russian Rope	Lariat Rope
Marline	Hemp Packing
Houseline	Italian Packing
Hambroline	Jute Packing
Clotheslines	Drilling Cables
Tarred Hemp Rope	Spunyarn
White Hemp Rope	Pulp Cord
Bolt Rope	Lobster Marlin
Hide Rope	Paper Cord
Halyards	Cheese Cord
Deep Sealine	Hay Rope
Ratline	Fish Cord
Plow Lines	Sand Lines

"RED THREAD" Transmission Rope from the finest quality Manila hemp obtainable, laid in tallow.

CONSUMERS CORDAGE COMPANY,

Limited

Western Ontario Representative—
WM. B. STEWART,
Tel 94. 27 Front St. West, TORONTO.

MONTREAL, QUE.

What About TRUCKS?

We are prepared to quote you low prices on all styles of

Warehouse, Factory, Hotel, and Railway Trucks.

all to montreal office 3/4/01 see cut book page 160

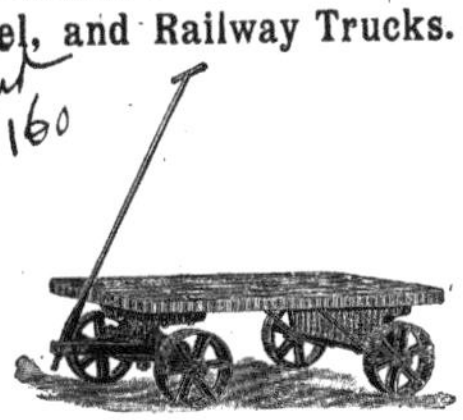

Rubber-Tired Wheels can be furnished if desired.

We have just received a large shipment and can ship promptly. Write for catalogues and prices.

THE FAIRBANKS COMPANY

749 Craig Street, MONTREAL.

CATALOGUES, BOOKLETS, ETC.

GUN AND AMMUNITION CATALOGUE.

Lewis Bros. & Co. have never issued a better all-round gun and ammunition catalogue than No. 28, which is just off the press, and which has been sent out to their customers this week. Every one of its 56 pages deserves the thorough study of the trade, for it contains much useful information and hints about a line of goods that hardwaremen might do well to push a little harder. It is replete with illustrations, showing clearly the style and mechanism of the various guns, rifles, pistols, revolvers and all necessary appurtenances, while the descriptions are terse and ample. Altogether it is an instructive work, and dealers in sporting goods should find it exceedingly useful for reference purposes. The prices are given on all goods, the quotations on ammunition being very ample. The make-up of the work is tasty. Colored ink has been employed with a pleasing and striking effect, and the cover and paper are of a quality that contribute a rich appearance. We are safe in saying that for a purely sporting goods catalogue this has never been excelled in Canada.

Any merchant who has not received a copy may obtain one upon application, mentioning this paper.

A DAINTY FREEZER CATALOGUE.

The North Bros. Manufacturing Co., Philadelphia, Pa., have issued an attractive catalogue, which, though small, contains full illustrations and descriptions of what they claim to be the largest variety of freezers made in any one factory. The features of North Bros.' freezers, which are known as the "Lightning," the "Gem," the "Blizzard," the "Crown," and the improved "Philadelphia Seaman," are their automatic twin scrapers, the construction of their cedar pails, which have electric welded wire hoops, and their drawn steel bottom cans, which will neither leak, break nor fall out of the body of the can. The firm furnish retailers who handle their freezers with circulars describing them for gratuitous distribution, also electros for advertising purposes free of charge.

WIRE GAUGE TABLE AND PRICE LIST.

W. F. Dennis & Co., 23 Billiter street, London, E.C., have prepared a second edition of their wire gauge table and price list. The first edition of this work, which was issued in January last year, was much sought for by users of wire in various countries. They have revised their prices to date, but have maintained their simple form of index for wires in general use, by which a buyer can, on opening the page at any gauge number he requires, not only ascertain the cost of the various descriptions of wire at the date of this issue, but also obtain correct information as to the Imperial standard size of each gauge in decimals of an inch and the approximate length of each size of wire in yards per cwt.

HAD A GOOD YEAR.

The regular annual meeting of The Robb Engineering Co., Limited, was held on Tuesday last week. A half-yearly dividend of 4 per cent. was declared, besides which ample amounts were set aside for depreciation, bad debts, interest and reserve fund.

The sales for the year 1900 have increased by about $50,000 over the previous year, and the opening of the present year has brought increasing business, the sales for the first two months of 1901 being about $8,000 more than for the same months of 1900 and orders are booked for fully three months' work in advance.

During the past year the company sold engines for electric tramways in England, Cuba, Brazil, Demerara and India, and have recently shipped a special compound engine for the municipal technical school, Manchester, England, to be used for electric lighting and as representative of the best British and foreign engineering practice.

The following directors were re elected: D. W. Robb, chairman and managing director; Rev. D. McGregor; W. B. Ross, K.C.; G. W. Cole, and A. G. Robb.

IT OUGHT TO BE CHEAP.

We would again draw the attention of the trade to the announcement made in our columns last week about a heavy order for hose placed by Lewis Bros & Co., of Montreal, some time ago. It is an important transaction, not only for the house mentioned, but also for the Canadian retail trade. Since the order was placed, and it was probably the largest ever given by a Canadian wholesale house, prices have advanced about 20 per cent., which, coupled with the quantity discount, enable Lewis Bros. & Co. to offer special value to their customers on this line. We wish to emphasize the importance of the opportunity.

IGNATIUS COCKSHUTT'S WILL.

The will of the late Ignatius Cockshutt, which has been admitted to probate upon application of the executors, Charles Cockshutt, Toronto, and Frank Cockshutt, Brantford, shows that, notwithstanding the large amount given away during his lifetime for charitable purposes, the deceased left over $400,000. With the exception of one charitable bequest of $5,000, the whole amount was left to the members of his family. Mr. Cockshutt had during his lifetime provided for the continuance of the many charitable institutions in which he was interested.

AN IMPROVED HORSE NAIL.

The Maritime Nail Co., Limited, St. John, N.B., have effected a great improvement in both the appearance and the quality of the "Monarch" horse nail, which they manufacture. As a result of nearly 800 conferences with horseshoers in Canada, important changes have been made in the construction of this nail, which now stands up well under the blows from the hammer, drives like a needle, holds well, and clinches excellently, as it is made 50 per cent. softer than formerly.

The material used is the highest grade of soft Swedish steel, exactly the same as used by all the best makers in Canada. While some processes reduce the tensile strength of this material, the "Monarch" process increases the same.

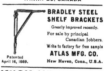

MUTUAL DEPENDENCE OF RETAILER AND JOBBER.

A paper read before the Iowa Retail Hardware Association, by H. E. Tredway.

I AM not certain but that my version of your committee's idea is given better expression by making my subject read "The Mutual Dependence of Retailer and Jobber." It is certainly easy to see that the jobber would soon look like "thirty cents." if the retailer withdrew his support; and I want to try and prove that the retailer would look likewise in time. I might concede him quite an extension in the time, but expire he must, if not drawing strength from the jobber's extended stock and thoroughly developed system of buying and selling behind him.

THE JOBBER'S ADVANTAGES.

Of course, I am going to talk on this subject as if it was applicable only to our mutual business, the hardware trade, and my arguments will be based on conditions in that trade which I know obtain; but they are applicable, I believe, to most other lines as well—some possibly less pronouncedly, but in many of them even more forcibly, because some one of the jobber's strong points exerts a greater influence in that particular line of business. That is, in these other lines, instant delivery may be of greatest importance; that is one of the ideal jobber's strong points. Maybe in another line great variety may be paramount; that is the jobber's strong point. In some lines which take first class or double first-class freight classifications a short haul by local freight is absolutely necessary to lowest delivered cost, and the jobber is so situated that he offers his customer a short haul. In perishable articles proximity is the surest prevention of loss en route, and the nearby jobber offers this. In so-called heavy commodities carload freight rates into the jobbing point and local freight rates out give the jobber so much of an advantage in his immediate local territory that not even the competitor in the next State or county can compete with him, let alone any remote mill or manufacturer; that is the jobber's strong point.

These are a few of the advantages which the existence of the jobber affords to the retail dealer, but as my paper is not a plea for the jobber but a study of the benefits accruing to each one of these factors through the existence of the other, let us then consider the reasons for the existence of these two factors.

1. Of course, the great and only reason why we all exist and work as business men is for the money there is in it. We are not after glory, but need the hard cash.

2. The retailer exists because every community demands a base of supply in its midst, and for that convenience are willing to pay a living profit on their purchases (if they cannot buy better outside). So, if you simmer it all down, there is just this one reason for the retailer's existence, but that one reason is all-powerful.

CAUSES FOR EXISTENCE OF THE JOBBER.

On the contrary, the jobber has three causes for his existence, none of them so great as the one for the retailer, but each of much moment in commercial economy.

1. The retailer seems willing to allow the jobber a living for the convenience of buying many of his goods at one time and in one place.

2. The average manufacturer is willing to concede the jobber a lower price for quantity purchases.

3. The railroads are willing to make the jobber lower rates on large quantities of freight in one shipment than they can or do make on small quantities in many shipments.

Primarily, as I before stated, in my business, either that of a retailer or jobber, there is but one vital point, net profit. If that is obtainable all else is secondary; but to the really successful business man there is the further consideration of increasing that net profit to a point where it affords more than a mere interest on the capital he invests and wages for the labor he puts in. Therefore, the wideawake dealer and jobber as well searches for the best methods of handling his business affairs, that the most favorable results may be recorded at the end of the year's business.

Now, this question of the mutual relation of retailer and jobber is influenced vitally by the consideration above stated, because, unless it is to the actual cheapening of the retailer's expense of doing business to have the jobber to assist him, there can be no plausible excuse for the existence of the jobber, and he becomes a useless encumberer of the earth.

But I believe there are many ways of showing the fact that in practically all lines of merchandise, and particularly hardware, the cost to the retailer, and consequently to the consumer, of any article is lessened through the offices of the jobber.

DISTRIBUTION OF GOODS.

As a general proposition, in the distribution of manufactured articles there are five methods or routes, each of which is more or less employed.

The first method is the primitive way from manufacturer to consumer without the interposition of any intermediate agent. In most commodities the least economical, because no manufacturer who produces on a large scale can find customers enough near him to buy his product.

The second method is from manufacturer through some transportation company to the consumer, and is possible in only a few lines, and in those probably not economical, because of the limited quantities a single consumer can buy, and the relatively high toll of the transportation company.

The third method involves the maker, transportation company, retailers and a consumer, and it is a method which I will dwell on later in my paper. I will only express my belief that this method is economical only where the manufacturer's line is large enough and weighty enough to make economical shipments, and where the line, too, is profitable enough to stand the extraordinary expense of specialty salesmen, for on strictly competitive staple goods it is a well proven fact that the ordinary retail dealers cannot under normal market conditions buy and ship sufficiently large quantities to satisfy both the manufacturer as to price and the transportation company as to the rate of freight.

The fourth method is the one which both the retailer and the jobber have to fear and to fight, and one which, owing to the extraordinary developments of the postal service and rapid transportation, has become a factor not only to be considered, but to be contested. This route is from manufacturer through a common carrier to a common enemy, the catalogue house, and from them through another common carrier to the consumer. The manufacturer here finds a customer who can buy ample quantities to make selling him economical. The first transportation company has all the weight necessary to lowest rates. The catalogue house has no expense for travelling men. There are only a few points at which this system is weak or vulnerable.

NECESSITY OF RETAILER.

There never has been, I believe, until recently any question of the need of the public for the ordinary retailer, because everybody conceded that the wants of the people were best served by him. There has arisen, however, that new factor in the distribution of merchandise, the result of the printer's art and of quickened mail and transportation facilities, which, though it can scarcely be said to threaten the life of the retailer proper, has grown to the proportions of a universally recognized competitor, and brings into consideration an entire new element as to the most economical method of the distribution of goods from the manufacturer to the consumer.

And with the advent of the catalogue house some question may have arisen as to the necessity for the old line retail dealer.

Against the jobber's right to exist, however, there have often been voiced insinuations, both by the manufacturer and retailer ; and to consider broadly the mutual relation, nay, vital dependence, of these two great factors of trade, the retailer and the jobber, upon each other it seems best to trace the development of the jobber and possibly thereby to prove his usefulness.

The first dealers in any new community are the general stores, whose stock must, of course, be so widely assorted as to cover not only the food requirements of the new settlement, but must furnish apparel to protect their bodies and material to erect their homes. Such dealers must draw their merchandise from a number of jobbers, and to them the jobber is an absolutely vital factor for existence ; without him their stock could never cover the diversified yet limited wants of a sparsely settled community. Think of what an impossible task it would be to attempt to buy even the hardware for such a store from the manufacturers direct, and how unprofitable such an attempt would be!

As a community grows and its wants become greater and more diversified, it outreaches the capacity of the general storeman, because he cannot meet all the wants, nor keep his stock up to their expectations, and retailers of special lines start in and offer better assortments of such goods. Why does this happen? Because Mr. General Storeman in himself or in his clerks has not and cannot have the capacity to so completely cover all the lines wanted by his community as to preclude the necessity for other stores carrying more complete independent lines, and what draws the competitor in, generally, is the knowledge that a well-stocked single line store will draw practically all the trade on that line away from a dealer who can give it at best only a part of his attention.

PURCHASING OF SUPPLIES FROM MANUFACTURER.

Now, from whom does this special line dealer draw his supplies? Let us suppose he is a hardwareman. Can he economically go to the manufacturers and buy the more complete assortment of hardware, which he must have in order to offer his little community any advantage over what they have already enjoyed? Suppose he should, he could obtain from one manufacturer, say, ten items of the thousand or two he needs, and he would have to see more manufacturers before he could possibly complete his lines, and with long distance freight charges to pay at local rates, the economy of his purchases, regardless almost of the prices he might obtain, would be more than doubtful, and the inconvenience and delay of it so great that apparent profis would wane before his stock was in hand.

We are supposing this dealer to be a hardwareman, and we are all more interested in hardware than anything else, and so, the other day, when I was thinking this talk over, and just to clinch my conclusions that an ordinary retail hardware stock could not within any reasonably comparative cost be bought exclusively or even in the main from the manfacturer direct, I ran casually through the index of our catalogue and found that buying from the same manufacturers we buy of and bunching his purchases with those who make the largest lines, he would have to open accounts with about 500 concerns. Thus he must have 1,000 shipments made him in a year, giving only two opportunities a year to replenish his assortment of each line.

Now, in any ordinary retail hardware business of, say, $8,000 (of course, you all do twice that, but I am speaking of the common, ordinary hardware dealer), split up into 1,000 shipments, you have a cost value of about $6 a shipment, and if he buys from the same shipping points we do his average freight charges on such shipments would be a prohibitive percentage of the cost value.

A lack of economy, therefore, in the general purchasing of goods from the manufacturer by the retailer being shown, and a necessity therefore proven for the existence of the jobber, for the convenience and profit of the retailer, a second question is brought up: Should any goods be bought of manufacturers? Can they be purchased to the retailer's ultimate profit? I say they cannot, and can bring forward some arguments to prove it. I admit first that sometimes you can and do buy certain articles a little cheaper by buying direct; I admit that you may occasionally in this way underbuy, and thereafter, unfortunately, undersell your competitor; but as this reduces his price and profit, and as the shoe is as likely to be on the other foot, it simply reduces the percentage of profit for both you and your neighbor in the long run.

Generally, however, you pay just about as much for these articles as if bought from the jobber, but in your judgment (and we jobbers always concede the judgment of the retailers good) the goods are a little more desirable for your trade, either because they are intrinsically better articles or because you have the exclusive sale of them; but, on the whole, you will generally admit that when the excess freight is paid the advantage even in apparent cost or value is not great, and the influential point to you is the exclusiveness of the goods. An excellent point, too, and one which no dealer or jobber can dispute.

ANOTHER VIEW OF THE QUESTION.

But let us look at this matter in another light. You have, let us say, practically hired these jobbers to obtain your supplies for you and to keep them on tap at points which will best serve you. You are paying them practically on a percentage basis for their buying, and the more buying they do for you collectively the lower this percentage basis can be, and is, made. You need not fully believe this, but reflect that the jobber's expenses are theoretically stationary while the sales fluctuate. For instance, the travelling expenses of a certain number of traveling men will be about so much each year; warehouse and office expenses are represented each year by a practically fixed amount, so that the variations come in largely on salaries to travellers and on gross sales. Salaries to salesmen are, it is true, based on what they do, but this is practically the only item which increases proportionotely with gross sales, and, as an increase in gross sales is the only way which the jobber can discover which will enable him to decrease the gross profit percentage which he must charge you for the work he assumes, the fact becomes apparent that the retailer's interests would be best served by restricting his purchases practically to the jobber, if he concedes in the first place that he must support and stand by the jobber.

In connection with the question of buying certain goods from the manufacturer, there is another argument to be presented. Conceded that the jobber does his business on a percentage basis, if you gave him this business to do he would do that on a percentage basis also. The jobber's salesmen have many times more goods to sell than the manufacturer's salesmen have, so the percentage of the salesmen's expenses in case the jobber sells the goods is reduced to a minimum by division of the expense with sales of other lines, while in the manufacturer's case it is increased to the maximum because of the limited sale.

As a consequence, you will find that almost without exception a manufacturer who canvasses and sells the dealer direct must keep his selling price well above mere cost to cover the cost of selling, while the jobber could, if permitted, sell the same goods at little, if any, increase in his fixed expenses. True, it might increase his net profits for the time being, but as experience has amply proven that there is competition enough among the jobbers to promptly eliminate any undue net profit, that consideration is not weighty.

JOBBER A NECESSITY TO THE RETAILER.

The foregoing argument simply goes to show to the retailer, who admits the jobber a necessity, that he (the retailer) should realize that the more manufacturers' salesmen he buys of the greater the sum of the travelling expenses he is loading upon himself and the retailers of the country; and if the manufacturer is given to understand that his goods offered through the regular channels of the jobbing trade would be more acceptable than if ordered direct, it would not take long to cut off the expense of the manufacturers' salesmen; nor will it take long for that manufacturer to discover that where it costs 2½ to 5 per cent. to sell his goods to the jobbing trade it has cost him 25 to 30 per cent. to sell to the retailer, and that he has cut off one large item of expense in doing business, and he can, when marketing through the jobber, materially reduce his price and still retain a manufacturer's margin.

But, to draw our final conclusions, the retail dealer who looks upon the jobber as his broker, who simply sees the necessary manufacturers, and by combining the business of many retailers is enabled to accomplish economies in many directions—the retailer who recognizes these facts easily sees the mutual relation of his interest to the jobber's.

Then the jobber who looks upon the prosperity of the retail dealer as his prosperity, and treats the enemy of the retailer as his enemy, is simply protecting the ground upon which he stands and without which he must fall.

MINERAL PRODUCTION IN BRITISH COLUMBIA.

FROM the returns of the mineral production of the Province of British Columbia for the year 1900, submitted to the local Legislature by the Provincial Minister of Mines, it is gathered that there was during the year an increase in the value of the output of the lode mines of the Province of $3,310,428 and a decrease in that of placer gold of $66,176, leaving a net increase in metallic minerals for the year of $3,244,252. There was also an increase of $1,013,238 in the value of coal and coke. No returns are yet available of the value of other non-metallic minerals so these cannot now be taken into account. The foregoing figures, though, exhibit practically last year's net increase, which was $4,257,490, as shown in detail in the following table:

	1899.	1900.	Increase	Decr'se.
Placer gold......	$1,344,900	$1,278,714		$66,176
Lode gold........	2,857,573	3,461,067	$ 6(3,494	
Silver...........	1,663,708	2 29.,(99	[illegible]	
Copper...........	1,352,4^3	1,615,289	263,836	
Lead.............	878,870	2 690,577	1,811 707	
Coal and coke....	4 053,651	5,066 899	1,013,238	
Total............	$12,100,155	$16,407,645	$4,323,666	$66,176

The mineral production of the Province for all years up to and including the year 1900 is as under:

Gold, placer........	$62,584,443
Gold, lode..........	12,820,546
Silver..............	13,635,708
Lead................	6,543,358
Copper..............	5,437,871
Coal and coke.......	49,426,700
Other minerals (approx.)...	2,000,000

The number of tons of metallic ore mined in the Province in 1899 was 287,343, and in 1900, 554,796, the increase for last year having been 267,453 tons. The respective increases for 1900 over 1898 were: In quantity of metallic ore mined, 338,852 tons, and in value of metallic minerals, coal and coke, $5,652,284, the latter being an increase of more than 50 per cent.—British Columbia Trade Budget.

A meeting of the citizens of Kingston, Ont., was called in the council chamber of that city, on March 19, for the purpose of considering the advisibility of establishing a branch of The Consolidated Phosphate Co., of London, Eng. F. W. Oates, of Toronto, the promotor of the company, who was present, addressed the meeting and explained that Brockville had offered greater inducements, but the company favored Kingston. A committee was formed to assist Mr. Oates in canvassing the city.

HEATING AND PLUMBING

SOME BUILDING NOTES.

IT has been decided to rebuild the college of St. Romuald, Quebec, which was destroyed by fire two years ago.

The Clarendon Hotel, Winnipeg, is to be enlarged this spring.

Tenders are asked for a residence for John Henderson, Cobourg, Ont.

Walter Miller intends building a brick house in Ridgetown, Ont., this spring.

Plans have been prepared for a new block of stores on Wellington street, Ottawa.

H. F. Teeter, Waterford, Ont., is preparing to build another storey to his hotel.

The London, Ont., Opera House, will be enlarged and made first class in every respect.

H. A. Manville is making preparations to erect a handsome brick block in Carberry, Man.

The congregation of Charles street Church, Halifax, have decided to build a new church.

Mrs. Wright has taken out a permit to erect a $2,000 residence on King street, London, Ont.

Power & Son, architects, Kingston, Ont., have prepared plans for a new collegiate institute in Cobourg, Ont.

A site has been purchased on the corner of Notre Dame and Nena streets, Winnipeg, for a new school building.

A 60 x 50 ft. two-storey summer hotel will be built at Whyteswold Beach, a resort near Winnipeg, this spring.

Suckling & Co., Winnipeg, intend erecting a new office building at the corner of McDermot and Albert streets.

The Cumberland Railway & Coal Co., have decided to build a new station at Parrsboro, N.S., during the coming summer.

The congregation of All Saints' Church, Winnipeg, have decided to build a rectory to cost about $4,000. It will adjoin the church property.

The new post office and court house which is to be erected in Rossland, B.C., will be started in a few days, but will not likely be completed this summer.

The Columbus Club, of Ottawa, Limited, composed of members of the Knights of Columbus, Ottawa, will erect a club hou e in that city to cost about $20,000.

It is believed that the vacant lot on St. James steet, Montreal, where The Montreal Star's temporary offices were located, will be the site of a new palatial building to be occupied by the consolidated electric lighting and gas companies. Those interested refuse to talk as yet.

W. E. Baker intends replacing the block on his site, Main street, Winnipeg, which was recently destroyed by fire, with a three-storey building which will have a 45 ft. frontage.

The Ottawa Building Co. will shortly commence the erection of a fine brick building on the vacant lot at the corner of Slater and Metcalfe streets. It will be 3½ storeys high and cost about $3,000.

Tenders are asked by to-day (Saturday) for the several works required in the erection of a boiler house, and also for the steam-heating and electric power plant of the proposed central heating system for Queen's University, Kingston, Ont. The plans are at the office of Arthur Ellis, architect, Kingston.

PLUMBING AND HEATING NOTES.

Louis Trudel, contractor, Montreal, is dead.

John A. Carslake, plumber, etc., Stratford, Ont., is offering to compromise.

Hugh McColl, electrical contractor, Ottawa, has assigned to E. A. Larmouth.

N. E. Broley, contractor, Rat Portage, Ont., has assigned to Fred. Armstrong, Rat Portage.

J. M. Marcotte has been appointed curator of Hamel & Bleau, contractors, Maisonneuve, Que.

S. S. Clarke, of the Bennett & Wright Plumbing and Heating Co., Limited, Toronto, has returned from a six weeks' trip to Southern California.

Bigg & McKenzie, tinsmiths, plumbers, etc., Picton, Ont., have dissolved. Mr. Bigg retires, but will continue in the employ of Mr. McKenzie. Arrangements are being made for a new firm, to be known as Johnston, Bigg & McKenzie, who will occupy the Hiram Welbanks block, now being fitted up for them.

BUILDING PROSPECTS IN HAMILTON

The indications point to a brisk building season at Hamilton this spring. The Christian Workers intend to erect a church at the corner of Merrick and Park streets, to cost about $8,000; the Church of the Ascension will build a $5,000 Sunday-school; the Aylmer Canning Co. will spend $3,000 in extending their works; a row of nine houses will be erected at the corner of Spring and Jackson streets; another row is being built at the corner of Stewart and Bay streets; it is probable that a new theatre will be erected; the American Hotel will have a fourth storey added to it, and the Royal Hotel is to have about $15 000 expended on it in improvements.

J. M. SHERLOCK NOT GUILTY.

On Monday the case against J. M. Sherlock, plumber, Toronto, who was charged with attempted blackmail against Joseph Wright, of the Bennett & Wright Co., Limited, Toronto, was tried before Judge McDougall. It will be remembered that Mr. Sherlock was an unsuccessful tenderer and the Bennett & Wright firm the successful tenderers for the plumbing work in the new Toronto City Hall.

The offence charged against Mr. Sherlock was that he visited Mr. Wright, and threatened that unless the latter gave him $3,000 he would make disclosures that would lead to trouble between Mr. Wright's company and the city authorities.

The trial showed that Mr. Sherlock did visit Mr. Wright and ask him for the money as charged, yet, as the latter had not the $3,000 demanded on his person, and as Mr. Sherlock did not threaten to cause injury to Mr. Wright, but rather manifested a desire to "get even" with Architect Lennox, who had made remarks which Mr. Sherlock considered to have done injury to himself, the charge of blackmail could not be sustained, and Mr. Sherlock was acquitted.

BUILDING PERMITS ISSUED.

Building permits have been issued in Ottawa to Thomas Cleary for brick cottages on Gilmour street to cost $8,000; to James Walker, for a $1,500 brick veneered house on Paterson avenue; to Col. Macpherson, for a $1,200 office and bakery on Somerset street; to George Gregory, for a $1,600 residence on Flora street; to James Kinmond, for a $2,000 house on Queen street; to D. P. Burke, for a $1,200 dwelling on Concession street; to James McVeity, for a $1,000 dwelling on Albert street and to McFarlane & Douglas for workshops on Slater street to cost $2,300.

Building permits have been issued in Toronto to The O'Keefe Brewery Co. for a five storey addition to the works on Victoria street, to cost $20,000; to James Dale, for a two-storey and attic residence on Withrow

avenue, to cost $2,500 ; to F. S. McCraney, for a residence on the east side Dovercourt road, near College street, to cost $3,250; to H. J. Clancy for a $1,900 residence at 219 Palmerston avenue ; to Wm. Booth, for two residences on Chicora avenue, to cost $4,-000 ; to Arthur Mitchell, for three two-storey houses on Harvard avenue, near Roncesvalles avenue, to cost $4,900 ; to R. Northcote for a $16,000 residence at the corner of Lowther avenue and Admiral road.

PLUMBING AND HEATING CONTRACTS.

Purdy, Mansell & Co., Toronto, have the contract for hot-water heating a store for P. Roach, Queen street west, Toronto.

The Bennett & Wright Co., Limited, Toronto, have contracts for plumbing, heating, electric wiring and ventilating additions to McMaster University ; for hot-water heating a large residence for E. F. Keating on Beverley street, Toronto, and for plumbing and electric wiring in the Bank of Hamilton branch at Wingham.

ORANGEVILLE BOARD OF TRADE.

The business men of Orangeville, Ont., have organized a board of trade. Some time ago, a committee was appointed to secure information regarding industries seeking suitable locations. Correspondence was presented from parties seeking inducements for the establishment of pork-packing, carpet, gas, biscuit, boot and shoe, anti-rust tinware and agricultural implement factories and a planing mill.

The following officers were elected :

President—E. Myers.

Vice-President—Mr. Gordon.

Secretary—E. Thompson.

Council—Messrs. Holland, Mann, Chapman, Turner, Still, Green, Claxton, T. Wright, H. Endacott, McIntyre, Brown and Stevenson.

GERMAN WIRE-NAIL TRUST.

United States Consul-General Guenther reports from Frankfort, February 19, 1901 : "The Frankfurther Zeitung states that the wire-nail trust, during the second half of 1900, sold 2,230,717.6 tons in Germany and 1,953,469.5 tons to foreign countries. While the sales to the latter were only about 300,000 tons less than those made in Germany, a profit of $280,270 was made on German sales, while the sales to foreign countries yielded a loss of $204,627. The price of wire nails in Germany is fixed by the trust at $2.70 per 100 lb.; for export, however, at only $1.51 per 100 lb. The German consumers have to pay an excessive price in order to enable the trust to sell its surplus to foreign countries at greatly reduced figures. And still the trust asks an increase of duties of 7 marks ($1.67) per 220 lb. The present tariff is 3 marks (71.4c.) per 220 lb."

MANITOBA MARKETS.

WINNIPEG, March 25, 1901.

TRADE is somewhat brighter than last week, but there are no changes in price to report. In paints and oils there has been another drop in linseed, which is now quoted at 77 and 80c. respectively.

The glass market shows great firmness here, although prices have not changed, but in view of the market abroad, and the fact that Winnipeg prices are already $1.25 per box lower than St. Paul, it would indicate at least an upward tendency here for some time to come.

Implement men are busy getting their goods to delivery points. Dairy supply men report a greatly increased demand for separators and dairy goods generally, indicating an increased production for the coming season.

Quotations for the week are as follows :

Barbed wire, 100 lb.	$3 45
Plain twist	3 45
Staples	3 95
Oiled annealed wire 10	3 95
" 11	4 00
" 12	4 05
" 13	4 20
" 14	4 35
" 15	4 45
Wire nails, 30 to 60 dy, keg	3 45
" 16 and 20	3 50
" 10	3 55
" 8	3 65
" 6	3 70
" 4	3 85
" 3	4 10
Cut nails, 30 to 60 dy.	3 00
" 20 to 40	3 05
" 10 to 16	3 10
" 8	3 15
" 6	3 20
" 4	3 30
" 3	3 65
Horsenails, 45 per cent. discount.	
Horseshoes, iron, No. 0 to No 1	4 65
No. 2 and larger	4 40
Snow shoes, No. 0 to No. 1	4 90
No. 2 and larger	4 40
Steel, No. 0 to No. 1	4 95
No. 2 and larger	4 70
Bar iron, $2.50 basis.	
Swedish iron, $4.50 basis.	
Sleigh shoe steel	3 00
Spring steel	3 25
Machinery steel	3 75
Tool steel, Black Diamond, 100 lb	8 50
Jessop	13 00
Sheet iron, black, 10 to 20 gauge, 100 lb.	3 50
20 to 26 gauge	3 75
28 gauge	4 00
Galvanized American, 16 gauge	2 54
18 to 22 gauge	4 50
24 gauge	4 75
26 gauge	5 00
28 gauge	5 25
Genuine Russian, lb	12
Imitation " "	8
Tinned, 24 gauge, 100 lb	7 55
26 gauge	8 80
28 gauge	8 00
Tinplate, IC charcoal, 20 x 28, box	10 75
" IX "	12 75
" IXX "	14 75
Ingot tin	35
Canada plate, 18 x 21 and 18 x 24	3 75
Sheet zinc, cask lots, 100 lb.	7 50
Broken lots	8 00
Pig lead, 100 lb.	6 00
Wrought pipe, black up to 2 inch	50 an 10 p.c.
" Over 2 inch	50 p.c.
Rope, sisal, 7-16 and larger	$10 00
" ⅜	10 50
" ¼ and 5-16	11 00
Manila, 7-16 and larger	13 50
" ⅜	14 00
" ¼ and 5-16	14 50
Solder	21½
Cotton Rope, all sizes, lb.	16
Axes, chopping	$ 7 50 to 12 00
" double bitts	12 00 to 18 00
Screws, flat head, iron, bright	87½
Round " "	82½
Flat " brass	80
Round " "	75
Coach	57½ p.c.
Bolts, carriage	55 p.c.
Machine	55 p.c.
Tire	60 p.c
Sleigh shoe	65 p.c.
Plough	40 p.c.
Rivets, iron	50 p.c.
Copper, No. 8	35
Spades and shovels	40 p.c.
Harvest tools	50, and 10 p.c.
Axe handles, turned, s. g. hickory, doz.	$2 50
No. 1	1 50
No. 2	1 25
Octagon extra	1 75
No. 1	1 25
Files common	70, and 10 p.c.
Diamond	60
Ammunition, cartridges, Dominion R.F.	50 p.c.
Dominion, C.F., pistol	30 p.c.
" military	15 p.c.
American R.F.	30 p.c.
C.F. pistol	5 p.c.
C.F. military	10 p.c. advance.
Loaded shells :	
Eley's soft, 12 gauge black	16 50
chilled, 12 guage	18 00
soft, 10 guage	21 00
chilled, 10 guage	23 00
Shot, Ordinary, per 100 lb	6 75
Chilled	7 50
Powder, F.F., keg	4 75
F.F.G.	5 00
Tinware, pressed, retinned	75 and 2½ p.c.
" " plain	70 and 15 p.c.
Graniteware, according to quality	50 p.c.

PETROLEUM.

Water white American	25½c.
Prime white American	24c.
Water white Canadian	22c.
Prime white Canadian	21c.

PAINTS, OILS AND GLASS.

Turpentine, pure, in barrels	$ 68
Less than barrel lots	73
Linseed oil, raw	77
Boiled	80
Lubricating oils, Eldorado castor	25½
Eldorado engine	24½
Atlantic red	27½
Renown engine	41
Black oil	23½ to 25
Cylinder oil (according to grade)	55 to 74
Harness oil	61
Neatsfoot oil	$ 1 00
Steam refined oil	85
Sperm oil	1 50
Castor oil per lb.	11½
Glass, single glass, first break, 16 to 25 united inches	2 25
26 to 40 per 50 ft.	2 50
41 to 50 " 100 ft.	5 50
51 to 60 " " "	6 00
61 to 70 per 100-ft. boxes	6 50
Putty, in bladders, barrel lots per lb.	2½
kegs "	2¾
White lead, pure per cwt.	7 25
No 1 "	7 00
Prepared paints, pure liquid colors, according to shade and color per gal.	$1.30 to $1.90

NOTES.

Mr. John Cane has purchased the hardware business of the late A. Schoenau at Virden, Man.

The fine steel bridge across the Assiniboine at Portage la Prairie is now open for traffic. It cost $20,000.

About $3 000 will be spent on the building and equipment of a creamery on the Carey road, near Victoria. Ald. Bryden, Victoria, will erect the plant.

A SUCCESSFUL HOCKEY SEASON.

THE annual report of the secretary-treasurer shows that the Canada Paint Co.'s hockey club, Montreal, has had a successful season. Several matches in which the team has been engaged have been reported in HARDWARE AND METAL, and it has been a feature that they have always shown themselves to be true and gentlemanly sports. The record on the ice has been as creditable as their record in the office or works on William street. The following is the annual report:

To the members of the Hockey Club:

It will be in your recollection that the Hockey Club was organized so recently as December last, and that the preliminary practices took place on the ice on the canal at the rear of the company's works. Subsequently, the committee decided to look elsewhere for more suitable ice, which was found at the White Star Rink. The company's employes kindly assisted to defray the expenses for the use of a proper rink for practice, as well as for contesting the merits of other clubs, and I have much pleasure in stating that the very liberal response enabled the club to be operated satisfactorily.

The first match was played on the White Star Rink with a team representing The Baylis Manufacturing Co. The annexed list shows the results of the various matches played during the season.

Considering the inexperience of several of the players (some of whom had not previously handled a hockey stick), the result is very satisfactory, indeed, and Capt. Allan Munro is to be congratulated upon the success attained.

The members desire to put on record their appreciation of the liberality of The Canada Paint Co. in generously assisting the club.

H. STUBBS,
Secretary-Treasurer.

Montreal, March 18, 1901.

The following is the schedule of games played:

Date.	Clubs.	Rink.	Won by.	Score.
Jan. 12	Canada Paint Co. vs. Baylis Mfg. Co.	White Star	C. P. Co	2 to 1
28	" " vs Canadian Rubber Co.	"	"	4 to 1
Feb. 1	" " vs. Colin McArthur Co.	"	C. McA.	4 to 1
4	" " vs. Heathers, of Westmount	"	C. P. Co	5 to 2
8	Ames-Holden Co. vs. Canada Paint Co.	Crystal	Draw	1 to 1
15	Canadian Rubber vs. " "	Ontario	C. P. Co	3 to 1
20	Montreal Gas Co. vs. " "	White Star	C. P. Co	4 to 1
22	Heathers vs. " "	Twin City	Draw	0 to 0
25	Pilkington Bros. vs. " "	White Star	Draw	1 to 1
27	Sherwin-Williams vs. " "	Crystal	S. W. Co	4 to 1
Mrch. 6	Canada Paint Co. vs. Montreal Gas Co	Minto	C. P. Co	2 to 0
8	Colin McArthurCo. vs. Canada Paint Co.	Minto	C. P. Co	3 to 0
15	Canada Paint Co. vs. Ames-Holden Co	Crystal	C. P. Co	4 to 0

Summary—Total games played, 13; Won, 8; lost, 2; drawn, 3.

The following players composed the team during the year: W. Lamont, D. Brown, H. Stubbs, A. J. Munro, D. Miller, A. Russell, W. Spiers, T. Lawlor, R. Campbell and J. P. Milloy.

INDICATIVE OF EXPANSION.

Mr. H. Roper, who has been for many years with Mr. G. S. Pelton, the Canadian representative of John Shaw & Sons, has taken a position with Mr. Alexander Gibb, manufacturers' agent, 13 St. John street, Montreal. This step has been forced upon Mr. Gibb by the expansion of his business which has outgrown his present capacity to handle it satisfactorily. It is his intention to put Mr. Roper on the road, and he will push for more business for the firms which Mr. Gibb represents. Mr. Roper is well and favorably known to the hardware trade, and, we think, should prove an acquisition to his new employer.

HARDWARE AND METAL is pleased to note that Mr. Gibb is meeting with so much success in connection with his agencies, necessitating the employment of a man of Mr. Roper's experience and connection.

AUTOMOBILE SLEIGH IN GERMANY.

United States Consul-General Guenther, of Frankfort, February 23, 1901, reports the appearance at Nuremburg of the first automobile sleigh. The vehicle glides along with great speed and a perfectly easy motion. It was constructed by the Nuremburg Motor Vehicle Factory Union.

NOVA SCOTIA'S COAL OUTPUT.

According to the annual report of the Department of Mines submitted to the Legislature of Nova Scotia a few days ago, the sale of coal of that Province to the United States, which in 1899 was 153,188 tons, increased to 624,273 tons in 1900. The total production was 3,238,245 tons, compared with 2,642,333 tons in the year 1899, showing the large increase of about 600,000 tons. As the revenue was calculated on sales, it is interesting to know that the total sales were 2,997.546 tons, as compared with 2,419,137 tons in 1899, the increase being 570,409 tons.

The output of the Dominion Coal is increasing rapidly, reaching, in February, 143,713 tons. For the fiscal year ending February 28, the output was 2,044,877 tons, as compared with 1,739.374 tons in 1900. The shipments for the same period were 1,885,605 tons.

Hayes Bros., general merchants, Head of Millstream, N.B., have been succeeded by A. J. McPherson.

In our issue of March 15, it was stated that M. J. Leitch, general merchant, Michael's Bay, Ont., had assigned to A. E. Pavey. The name should have been E. F. Leitch, not M. J. Leitch.

THE
CANADA
PAINT
COMPANY
LIMITED
Dry Color Makers.

ALEXANDER GIBB

Manufacturers' Agent and Metal Broker,
13 St. John Street, Montreal.

Representing British and American Manufacturers. Correspondence invited from firms wishing to be represented in Canada.

McCaskill, Dougall & Co.
Fine Varnish & Japan
Manufacturers,
Montreal
Price Lists on application

With——

DUNDAS AXES

in stock you may justly ask a clerk for an explanation if he misses a sale.

Without **DUNDAS AXES,** is it fair to blame the clerk when you have not bought the best value?

All First-Class Jobbers Sell Them.

MURALO

is a Wall Tint that

PAYS

and brings money to everyone who

HANDLES IT.

Muralo is a cold water wall tint, and there is nothing **just like it** in our world. A good many makers think they have something **just like it,** but they haven't, nor anything **quite** so good. The Muralo Co. is one of the largest and strongest financial manufacturing corporations in America. It protects every dealer of Muralo products from the least to the greatest. The Muralo Co. studies advertising methods. The Muralo Co. helps the dealer and invites customers to his store. These customers will buy Muralo and they will buy other things. Agents are wanted in every town and village in Canada. Do you want to hear about us?

—— Write to ——

A. RAMSAY & SON, - - - - - - - MONTREAL.
J. H. ASHDOWN, - - - - - - - - WINNIPEG.
McLENNAN, McFEELY & CO., - - - VANCOUVER.

Distributing Agents.

"A WORD TO THE WISE IS SUFFICIENT."

Gold Dollars at 90c. each won't sell until you have convinced the customer that they are genuine. Have you the time to make the argument to do the convincing?

ALABASTINE is the standard wall-coating **"The World Over."** Would any dealer want to spend the time to convince a customer that a substitute, or something else, is "just as good" as **CHURCH'S**

ALABASTINE?

Besides, substitution is suspicious, and often bears earmarks of a swindle.

We do not claim to know all that can be learned about the manufacture of wall coatings, but twenty-five years of experience must count for something, and the fact that **ALABASTINE** is the goods that are in demand speaks volumes in its favor.

Never until **M. B. Church** introduced **ALABASTINE** was it found practicable to make wall-coating adapted to be mixed with cold water.

We own the patents on the cold water process, and this statement is our only argument at this time on that point.

ALABASTINE is healthful and beautiful, has stood the test of time; is manufactured in Canada, and recommended by leading sanitarians throughout the country. There is nothing "just as good." Buy only the genuine.

Manufactured by——

THE ALABASTINE CO., LIMITED, Paris, Ont.

The trade supplied by—
Sanderson Pearcy & Co., Toronto, Ont.
G. F. Stephens & Co., Winnipeg, Man.
Vancouver Hardware Co., Vancouver, B.C.
Wood, Vallance & Co., Hamilton, Ont.
William Hill, Montreal, P.Q.

"DAISY" CHURN

Has tempered steel cased bicycle ball bearings, strongest, neatest and most convenient frame. Only two bolts to adjust in setting up. Steel Bow Levers, suitable for either a standing or sitting posture. Has four wheels and adjustable feet to hold stand steady while churning. When churn is locked to stand the bow can be used as handles to move it about on the front wheels as handy as a baby carriage. Open on both sides to centre, giving free space for pail. Made with wood or steel stands, with Cranks only, or Bow Levers as desired.

Vollmar Perfect Washer

Has a most enviable record. A perfection of its kind—will wash more clothes in less time, do it better and easier, with less wear and tear, than any other machine.

THE——
Wortman & Ward Mfg. Co.,
Limited
LONDON, ONT.
Eastern Branch, 60 McGill Street, Montreal, Que.

IMPROVED STEEL WIRE TRACE CHAINS.

Every chain guaranteed. Most profitable and satisfactory chain to handle.

Improved Quality and Cheaper Prices for 1901.

The B. GREENING WIRE CO., Limited,

Hamilton, Ont., and Montreal, Que.

ONTARIO SILVER CO.,

Limited,
NIAGARA FALLS, CANADA.
Manufacturers of FLATWARE, CUTLERY and ELECTRO PLATE. . . .
Ask for our Catalogue and Quotations.

Ontario Nut Works, Paris

BROWN & CO.

Manufacturers of
All sizes of Hot Pressed Nuts, Square and Hexagon.

"BRADLEY" STEEL SHELF BRACKET

THE Atlas Manufacturing Co., of New Haven, Conn., sole manufacturers of the "Bradley" steel shelf brackets, have had their goods on the market since 1893 and have met with success from the very start. At the present time about 1,000 miles of wire a year are being used in the

"Bradley" Steel Shelf Bracket.

PAT. APR. 5. 1892. NOV. 8, 1'99. APR. 4, 1899.

manufacture of these brackets. The idea of many that the brackets would not be sufficiently strong has been proven entirely unfounded, and their strength and durability are fully demonstrated. Readers will observe from the accompanying illustration that the brace is not curved, but is a straight line from one point of the bracket to another. The brace is made U shaped in cross section, which gives it great stiffness in all directions. A great many of the cheap brackets now made have a single flat piece of steel for the brace, and buckle very easily sidewise. This cannot happen where a brace is made U shaped, as in the "Bradley." It will also be noticed that the two principal screw holes on the upright arm are located near to where the bracket is angled, and that they are reinforced by a stiff plate, which is offset so as to give additional stiffness. This plate runs clear to the angle. This construction does away with all necessity for ribs running through the wall plates. The absence of these ribs is a great convenience in driving the screws. Now, it is clear that in order to break down this bracket either enough weight must be placed upon it to collapse the brace, which, as explained, is a straight line in all directions, or it must bend the wall plate at the point where the screws were put in, or else it must pull the wires apart, and this will require a weight very much greater than would appear at first thought and much greater than could be gotten onto a shelf. Recent tests show that a pair of 5 x 7 brackets will support a load of 400 to 700 lb., and in one case it required 925 lb. to break down the brackets.

IRON WORKS FOR TORONTO.

The Canadian General Electric Co., Peterboro' and Toronto, have secured 30 acres of land in Toronto Junction. Here, they first intend to erect a large general foundry to be followed by a pipe foundry, for the manufacture of gas and water pipes, hydrants, valves and general waterworks supplies, which the company are now making in the St. Lawrence Foundry, Toronto, and the facilities for the manufacture of which will be increased. Then, the company will erect machine shops, structural iron shops, blacksmith shop, pattern shop, power house, storehouse, stables, etc. Railway tracks will run through each shop, and electric cranes will be provided for handling work up to 50 tons. All the machinery in the shops will be operated by electric motors from current generated in the company's power house.

SHOP REGULATIONS IN ONTARIO.

In the Ontario Legislature on Tuesday, Hon. John Dryden introduced a bill to amend the Shops' Regulation Act. He explained that one clause was intended to place the onus of providing sanitary appliances on the owner rather than the tenant of the premises. Another clause was intended to make clearer the section providing that no employe in a bakeshop should be required to work more than 12 hours a day. At present a man might work eight hours on a Monday and eight hours on Tuesday, commencing at midnight, or 16 hours continuously, and yet, technically, he is said to have worked only eight hours on the same day. A further clause made it impossible for barbers to compel employes to work on Sunday.

THE BOOM AT SYDNEY.

Of the many indirect channels through which the Dominion and Steel Co. has benefitted the Canadian people, the Intercolonial Railway is probably the principal one. The increase in passenger and freight traffic to Sydney since the establishment of the works is enormous.

In 1899 there were 5,600 tons of freight landed at Sydney by the I.C.R., for which $42,800 was paid.

In 1900 there were 165,583 tons of freight landed at Sydney by the I.C.R., the freight charges on which amounted to $356,600.

In 1899 the I.C.R. carried 5,700 people from Sydney. The amount paid for tickets at the Sydney station was $9,050. In 1900 the I.C.R. carried 21,500 people from Sydney. The amount paid for tickets at the Sydney station was $71,340.

The above does not include the Sydney and Louisburg Railway. This road carried 4,000 passengers in 1899, and in 1900 it carried 14,000 passengers.

The cash remittances for the Cape Breton section of the I.C.R. in 1899 amounted to $150,695. In 1900 the cash remittances amounted to $539,800.

W. A. Bothwell, general merchant, Luton, Ont., has sold out to B. Tibbitts.

The Galt, Ont., Board of Trade was reorganized on Wednesday last week with F. H. Hayhurst, of James Warnock & Co. as president and A. G. Donaldson as secretary.

CURRENT MARKET QUOTATIONS.

March 29, 1901.

These prices are for such qualities and quantities as are usually ordered by retail dealers on the usual terms of credit, the lowest figures being for larger quantities and prompt pay. Large cash buyers can frequently make purchases at better prices. The Editor is anxious to be informed at once of any apparent errors in this list as the desire is to make it perfectly accurate.

METALS.

Tin.

Lamb and Flag and Straits— 56 and 28 lb. ingots, per lb.	0 32	0 33

Tinplates.

Charcoal Plates—Bright

M.L.S., equal to Bradley.		Per box
I.C., usual sizes	....	$6 75
I.X., "	....	8 25
I.X.X., "	....	9 75
Famous—		
I.C.	....	6 75
I.X.	....	8 25
I.X.X.	....	9 50
Raven & Vulture Grades—		
I.C., usual sizes	....	5 00
I.X., "	....	6 00
I.X.X. "	....	7 00
I.X.X.X., "	....	8 00
D.C., 12½x17	....	4 75
D.X.	....	5 50
D.X.X.	....	7 50

Coke Plates—Bright

Bessemer Steel—		
I.C., usual sizes	....	4 15
I.C., special sizes, base	....	4 50
20 x 28	....	8 50

Charcoal Plates—Terne

Dean or J. G. Grade—		
I.C., 20x28, 112 sheets	....	8 25
I.X., Terne Tin	....	10 25

Charcoal Tin Boiler Plates.

Cookley Grade—		Per lb.
X X., 14x56, 50 sheet bxs; " 14x60, "; " 14x65, "	0 07	0 07½

Tinned Sheets

72x30 up to 24 gauge	0 08	0 08½
" 26 "	0 08½	0 09
" 28 "	0 09	0 09½

Iron and Steel.

Common Bar, per 100 lbs.	1 70	1 75
Refined " "	....	2 15
Horse Shoe Iron "	....	2 05
Hoop steel, 1½ to 3 in. base, extras for smaller sizes	....	3 10
Sleigh Shoe Steel " base	....	2 00
Tire Steel	....	2 00
Machinery iron finish	....	2 05
Cast Steel, per lb	0 00	0 00
Toe Calk Steel	....	2 30
T. Firth & Co's special cast steel, per lb.		0 13
Jessop's Tool Steel	....	0 13

Boiler Tubes.

1½-inch	....	0 12½
2 "	....	0 13
2½ "	....	0 15
3 "	....	0 16
3½ "	....	0 20
4 "	....	0 25

Steel Boiler Plate.

¼ inch	....	2 25
3-16 inch	....	2 25
⅜ inch and thicker	....	2 25

Black Sheets.

gauge	....	3 00
20 gauge	....	3 00
22 to 24 "	....	3 10
26 "	....	3 20
28 "	....	3 30

Canada Plates.

All dull, 52 sheets	....	3 00
Half polished	....	3 15
All bright	3 65	3 75

Iron Pipe.

Black pipe—		
⅛ "	....	4 35
¼ inch	....	3 15
⅜ "	....	3 20
½ "	....	3 25
¾ "	....	3 52
1 "	....	4 93
1¼ "	....	6 72
1½ "	....	8 08
2 "	....	10 76
2½ "	....	22 91
3 "	....	31 12
3½ "	....	37 90
4 "	....	43 19
5 "	....	57 85
6 "	....	75 11
7 "	....	93 76
8 "	....	112 51
Galvanized pipe—		
¼ inch	....	5 25
¾ "	....	5 56
1 "	....	7 77
1¼ "	....	10 60
1½ "	....	12 71
2 "	....	16 90

Galvanized Sheets.

	G C.	Comet.	Amer.	Queen's Head.
16 gauge		....	4 5	3 90
18 to 24 gauge	4 25	3 85	4 35	4 15
26 "	4 50	4 10	4 35	4 40
28 "	4 75	4 35	4 60	4 65

28 gauge American equals 26 gauge English.

Chain.

Proof Coil, 3-16 in., per 100 lb	....	....
" ¼ " "	7 50	8 00
" 5-16 " "	4 60	5 00
" ⅜ " "	4 00	4 00
" 7-16 " "	3 80	4 00
" ½ " "	3 60	3 75
" ⅝ " "	3 45	3 60
" ¾ " "	3 50	3 35
Halter, kennel and post chains, 40 and 50 p.c.		
Cow ties	40 p.c.	
Stall fixtures	35 p.c.	
Trace chain	25 p.c.	

Jack chain, iron, single and double, discount 35 p.c.

Jack chain, brass, single and double, discount 40 p.c

Copper.

Ingot

English B. S., ton lots	0 19	0 20
Lake Superior	....	....

Bolt or Bar.

Cut lengths round, ½ to ⅞ in.	0 23½	0 25
" round and square 1 to 2 inches	0 23½	0 25

Sheet.

Plain, 14 oz., and light, 16 oz., 14x48 and 14x60	0 23	0 23
Plain, 14 oz., and light, 16 oz., irregular sizes	0 23	0 23½
Tinned copper sheets	....	0 26
Planished	....	0 32

Braziers (in sheets.)

4x6ft. 25 to 30 lbs. ea., per lb.	....	0 25½
" 35 to 45 " "	....	0 24½
" 50-lb. and above, "	....	0 23½

Boiler and T. K. Pitts

Plain Tinned, per lb	....	0 28
Spun, per lb	....	0 32

Brass.

Rod and Sheet, 14 to 30 gauge, 15 per cent.		
Sheets, hard-rolled, 2x4	0 24	0 25
Tubing, base, per lb	0 24	0 25

Zinc Spelter

Foreign, per lb	0 06	0 06½
Domestic "	....	..

Zinc Sheet.

5 cwt. casks	6 75	7 00
Part casks	7 00	7 50

Lead.

Imported Pig, per lb	0 04½	0 05
bar, 1 lb	0 05½	0 05¾
Sheets, 2½ lbs. sq. ft., by	....	0 06¼
Sheets, 3 to 6 lbs., "	....	0 06

NOTE.—Cut sheets ½ cent per lb. extra. Pipe, by the roll, usual weights per yard, lists at 7c. per lb. and 25 p.c. dis. f.o.b. Toronto.

NOTE.—Cut lengths, net price, waste pipe 8-ft lengths lists at 7½ cents.

Shot.

Common, $6.50 per 100 lb.; chilled, $7.00 per 100 lb.; buck, seal and ball, $7.50. Discount, 7½ p.c. Prices are f.o.b. Toronto, Hamilton, Montreal, St. John and Halifax. Terms 3 per cent. cash, freights equalized on Montreal.

Soil Pipe and Fittings.

Discount, 60 and 10 per cent. on medium and extra heavy, and 60 per cent. on light.

Solder.

	Per lb.	Per lb.
Bar half-and-half, guarant'd	....	0 18
Bar half-and-half, commer'l	....	0 17½
Refined	....	0 17½
Wiping	....	0 17

NOTE.—Prices of this graded according to quantity. The prices of other qualities of solder in the market indicated by private brands vary according to composition.

Antimony.

Cookson's, per lb	0 11	0 11½

White Lead.

	Per 100 lb.
Pure	6 50
No. 1 do	6 12½
No. 2 do	5 75
No. 3 do	5 37½
No. 4 do	5 00
Munro's Select Flake White	7 37½
Elephant and Decorators' Pure	7 12½
Brandram's B B. Genuine	8 00
" " Decorative	7 55
" " No. 1	6 85
" " No. 2	6 00

Red Lead.

Genuine, 560 lb. casks, per cwt.	....	$5 50
Genuine, 100 lb. kegs, per cwt.	....	5 75
No. 1, 560 lb. casks, per cwt.	....	5 25
No. 1, 100 lb. kegs, per cwt.	....	5 00

White Zinc Paint.

Elephant Snow White	0 08	0 09
Pure White Zinc	0 08	0 09
No. 1	0 06	0 07½
No. 2	0 05	0 06½

Dry White Lead.

Pure, casks	5 75
Pure, kegs	6 25
No. 1, casks	5 50
No. 1, kegs	5 00

Prepared Paints.

In ¼, ½ and 1 gallon tins.

Pure, per gallon	....	1 25
Second qualities, per gallon	....	1 10
Barn (in bbls.)	0 75	0 85
The Sherwin-Williams Paints.	....	1 45
Canada Paint Co's Pure	....	1 25
Toronto Lead & Color Co's Pure	....	1 25
Sanderson Pearcy's Pure	....	1 2
Stewart & Wood's Champion Pure.		1 20

Colors in Oil.

25 lb. tins, Standard Quality.

Venetian Red, per lb.	0 05
Chrome Yellow	0 11
Golden Ochre	0 06
French "	0 05
Marine Black	0 09
" Green	0 09
Chrome "	0 08
French Imperial Green	0 0[illegible]

Colors, Dry.

Yellow Ochre (J.C.) bbls.	1 35	1 40
Yellow Ochre (J.F.L.S.), bbls	...	2 75
Yellow Ochre (Royal)	1 10	1 15
Brussels Ochre	....	2 00
Venetian Red (best), per cwt.	1 80	1 90
English Oxides, per cwt.	3 00	3 25
American Oxides, per cwt.	1 75	2 00
Canadian Oxides, per cwt.	1 75	2 00
Super Magnetic Oxides, 93p.c.	2 00	2 25
Burnt Sienna, pure, per lb.	....	0 10
" Umber, " "	....	0 10
do Raw	....	0 09
Drop Black, pure	....	0 09
Chrome Yellows, pure	....	0 18
Chrome Greens, pure, per lb.	....	0 12
Golden Ochre	....	0 03½
Ultramarine Blue in 28-lb. boxes, per lb.	0 08	0 24
Fire Proof Mineral, per 100 lb	....	1 00
Genuine Eng. Litharge, per lb	....	0 07
Mortar Color, per 100 lb.	....	1 25
English Vermillion	....	0 80
Pure Indian Red, No. 45 lb.	....	0 80
Whiting, per 100 lb.	....	0 55

Blue Stone.

Casks, for spraying, per lb.	0 07
100-lb. lots, do. per lb.	0 08

Putty.

Bulk in bbls.	2 00
Bulk in less quantity	2 15
Bladders in bbls.	2 20
Bladders in kegs, boxes or loose	2 35
Bladders in 25-lb. tins	2 45
Bladders in 12½-lb. tins	2 75
Bladders in bulk or tins less than 100 lb	3 00

Varnishes.

(In 5-gal. lots.)		Per gal.
Carriage, No. 1	2 90	3 30
" body	8 00	9 00
" rubbing	4 00	5 00
Gold Size, Japan	3 00	3 40
Brown Japan	2 40	2 80
Elastic Oak	2 90	3 30
Furniture, extra	2 40	2 80
" No. 1	1 60	2 00
Hard Oil Finish	2 70	3 10
Light Oil Finish	3 20	3 60
Demar	3 30	3 70
Shellac, white	4 40	4 80
" orange	4 00	4 40
Furniture Brown Japan	1 60	2 00
Black Japan	2 40	2 80
" " No. 1	1 60	2 00

The Imperial Varnish & Color Co's, Limited Elastilite Varnish 1 gal. can, each. $2 00.

Granatine Floor Finish, per gal. $2.00.

Maple Leaf Coach Enamels; Size 1, 60c; Size 2, 35c.; Size 3, 20c. each.

Castor Oil

East India, in cases, per lb.	0 10	0 10½
" " small lots	0 10½	0 11

Cod Oil, Etc.

Cod Oil per gal.	0 50	0 55
Pure Olive	....	1 20
" Neatsfoot	....	90

Glue.

Common	0 08½	0 09
French Medal	0 14	0 14½
Cabinet sheet	0 12	0 13
White, extra	0 18	0 20
Gelatine	0 22	0 30
Strip	0 18	0 20
Coopers	0 19	0 20
Huttner	....	0 18

HARDWARE.

Ammunition.

Cartridges.

B. B. Cap Dom. 50 and 5 per cent.
Rim Fire Pistol, dis. 40 p. c., Amer.
Rim Fire Cartridges, Dom., 50 and 5 p. c.
Central Fire Pistol and Rifle, 10 p.c. Amer.
Central Fire Cartridges, pistol sizes, Dom. 30 per cent.
Central Fire Cartridges, Sporting and Military, Dom., 15 and 5 per cent.
Central Fire, Military and Sporting, Amer. add 5 p.c. to list. B.B. Caps, discount 40 per cent, Amer.
Loaded and empty Shells, "Trap" and "Dominion" grades, 25 per cent. Rival and Nitro, net list.
Brass shot Shells, 55 per cent.
Primers, Dom., 30 per cent.

Wads — per lb.

Best thick white felt wadding, in ½-lb bags 1 00
Best thick brown or grey felt wads, in ½-lb. bags 70
Best thick white card wads, in boxes of 500 each, 12 and smaller gauges 0 99
Best thick white card wads, in boxes of 500 each, 10 gauge 0 35
Best thick white card wads, in boxes of 500 each, 8 gauge 0 55
Thin card wads, in boxes of 1,000 each, 12 and smaller gauges 0 20
Thin card wads, in boxes of 1,000 each, 10 gauge 0 25
Thin card wads in boxes of 1,000 each, 8 gauge ....
Chemically prepared black edge grey cloth wads, in boxes of 250 each— Per M.
11 and smaller gauge 0 60
9 and 10 gauges 0 70
7 and 8 gauges 0 90
5 and 6 gauges 1 10
Superior chemically prepared pink edge, best white cloth wads, in boxes of 250 each—
11 and smaller gauge 1 15
9 and 10 gauges 1 40
7 and 8 gauges 1 65
5 and 6 gauges 1 90

Adzes.

Discount, 20 per cent.

Anvils.

Per lb. 0 10 0 12½
Anvil and Vise combined 4 50
Wilkinson & Co.'s Anvils. lb. 0 09 0 09½
Wilkinson & Co.'s Vices. lb. 0 09¾ 0 10

Augers.

Gilmour's, discount 65 and 5 p.c. off list.

Axes.

Chopping Axes
Single bit, per doz 5 50 10 00
Double bit, " 11 00 18 00
Bench Axes, 40 p.c.
Broad Axes, 33⅓ per cent.
Hunters' Axes 5 50 6 00
Boy's Axes 5 75 6 75
Splitting Axes 6 50 12 00
Handled Axes 7 00 10 00

Axle Grease.

Ordinary, per gross 5 75 6 00
Best quality 13 00 15 00

Bath Tubs.

Zinc 6 00
Copper, discount 15 p.c. off revised list

Baths.

Standard Enameled.

5½-inch rolled rim, 1st quality 25 00
" " " 2nd " 21 00

Anti-Friction Metal.

"Tandem" A .... per lb. 0 27
" B .... " 0 21
" C .... " 0 11½
Magnolia Anti-Friction Metal, per lb. 0 25

SYRACUSE SMELTING WORKS.

Aluminum, genuine 0 45
Dynamo 0 29
Special 0 25
Aluminum, 99 p.c. pure "Syracuse" .. 0 50

Bells.

Hand.

Brass, 60 per cent.
Nickel, 55 per cent.

Cow.

American make, discount 66⅔ per cent.
Canadian, discount 45 and 50 per cent.

Door.

Gongs, Sargant's 5 50 8 00
" Peterboro', discount 45 per cent.

Farm.

American, each 1 25 3 00

House.

American, per lb. 0 35 0 40

Bellows.

Hand, per doz 3 35 4 75
Moulders', per doz 7 50 10 00
Blacksmiths', discount 40 per cent.

Belting.

Extra, 60 per cent.
Standard, 60 and 10 per cent.

Bits.

Auger.

Gilmour's, discount 60 and 5 per cent.
Rockford, 50 and 10 per cent.
Jennings' Gen., net list.

Car.

Gilmour's, 47½ to 50 per cent.

Expansive.

Clark's, 40 per cent.

Gimlet.

Clark's, per doz 0 65 0 90
Diamond, Shell, per doz 1 00 1 50
Nail and Spike, per gross 2 25 5 20

Blind and Bed Staples.

All sizes, per lb 0 07½ 0 12

Bolts and Nuts. Per cent.

Carriage Bolts, full square, Norway 70
" " full square 70
Common Carriage Bolts, all sizes 65
Machine Bolts, all sizes 65
Coach Screws 75
Sleigh shoe Bolts 75
Blank Bolts 65
Bolt Ends 65
Nuts, square .... 4½c. off
Nuts, hexagon .... 4¾c. off
Tire Bolts 67½
Stove Bolts 67½
Stove rods, per lb. .... 5½ to 6c.
Plough Bolts 60

Boot Calks.

Small and medium, ball, per M 4 25
Small heel, per M 4 50

Bright Wire Goods.

Discount .... 60 per cent.

Broilers.

Light, dis., 65 to 67½ per cent.
Reversible, dis., 65 to 67½ per cent.
Vegetable, per doz., dis. 37½ per cent.
Henis, No. 8, " 6 00
Henis, No. 9, " 7 00
Queen City " 7 50 0 00

Butchers' Cleavers.

German, per doz 6 00 11 00
American, per doz 12 00 20 00

Building Paper, Etc.

Plain building, per roll 0 30
Tarred lining, per roll 0 40
Tarred roofing, per 100 lb. 1 65
Coal Tar, per barrel 3 50
Pitch, per 100-lb. 0 85
Carpet felt, per ton 45 00

Bull Rings.

Copper, $2.00 for 2½ in. and $1.90 for 2 in.

Butts.

Wrought Brass, net revised list

Cast Iron.

Loose Pin, dis., 60 per cent.

Wrought Steel.

Fast Joint, dis. 60 and 10 per cent.
Loose Pin, dis. 60 and 10 per cent.
Berlin Bronzed, dis. 70, 70 and 5 per cent.
Gen. Bronzed, per pair 0 40 0 65

Carpet Stretchers.

American, per doz 1 00 1 50
Bullard's, per doz .... 6 50 ....

Castors.

Bed, new list, dis. 55 to 57½ per cent.
Plate, dis. 52½ to 57½ per cent.

Cattle Leaders.

Nos. 31 and 32, per gross 50 9 50

Cement.

Canadian Portland 2 80 3 00
English " 3 00
Belgian " 2 75 3 00
Canadian hydraulic 1 25 1 50

Chalk.

Carpenters Colored, per gross 0 45 0 75
White lump, per cwt 0 60 0 65
Red 0 05 0 06
Crayon, per gross 0 14 0 18

Chisels.

Socket, Framing and Firmer.

Broad's, dis. 70 per cent.
Warnock's, dis. 70 per cent.
P. S. & W. Extra 60, 10 and 5 p.c.

Churns.

Revolving Churns, metal frames—No. 0, $8—No. 1, $8.50—No. 2, $9.00—No. 3, $10.00—No. 4, $12.00—No. 5, $16.00 each. Ditto, wood frames—20c. each less than above. Discounts: Delivered from factories, 58 p.c.; from stock in Montreal, 56 p.c. Terms, 4 months or 3 p.c. cash in 30 days.

Clips.

Axle dis. 65 per cent.

Closets.

Plain Ontario Syphon Jet $16 00
Emb. Ontario Syphon Jet 17 00
Fittings net 1 00
Plain Teutonic Syphon Washout 10 00
Emb. Teutonic Syphon Washout 11 00
Fittings net 1 25
Low Down Teutonic, plain 16 00
" " " embossed 17 00
Plain Richelieu net 3 75
Emb. Richelieu net 4 00
Fittings net 1 25
Low Down Ont. Sy. Jet, plain net .. 19 50
" " " " " emb'd. net 20 50
Closet connection net 1 25
Basins, round, 14 in. 1 00
" oval, 17 x 14 in. 2 50
" " 19 x 15 in. 3 75
Discount 40 p.c., except on net figures.

Compasses, Dividers, Etc.

American, dis. 62½ to 65 per cent.

Cradles, Grain.

Canadian, dis. 25 to 33⅓ per cent.

Crosscut Saw Handles.

S. & D., No. 3, per pair 17½
" " 5, " 22½
" " 6, " 15
Boynton pattern " 20

Door Springs.

Torrey's Rod, per doz ...... (15 p.c.) 2 00
Coil, per doz. 0 88 1 60
English, per doz. 2 00 4 00

Draw Knives.

Coach and Wagon, dis. 50 and 10 per cent.
Carpenters, dis. 70 per cent.

Drills.

Hand and Breast.

Millar's Falls, per doz. net list.

DRILL BITS.

Morse, dis., 37½ to 40 per cent.
Standard dis. 50 and 5 to 55 per cent

Faucets.

Common, cork-lined, dis 35 per cent.

ELBOWS. (Stovepipe.)

No. 1, per doz 1 40
No. 2, per doz 1 20
Bright, 20c. per doz. extra.

ESCUTCHEONS.

Discount, 45 per cent.

ESCUTCHEON PINS.

Iron, discount 40 per cent.

FACTORY MILK CANS.

Discount off revised list, 40 per cent.

FILES.

Black Diamond, 70 and 5 per cent.
Kearney & Foote, 60 and 10 to 70 per cent.
Nicholson File Co., 60 to 60 and 10 per cent.
Jowitt's, English list, 25 to 27½ per cent.

FORKS.

Hay, manure, etc., dis., 50 and 10 per cent. revised list.

GLASS—Window—Box Price.

Size United Inches.	Star Per 50 ft.	Star Per 100 ft.	D. Diamond Per 50 ft.	D. Diamond Per 100 ft.
Under 26	2 10	4 00	....	6 00
26 to 40	2 30	4 35	....	6 65
41 to 50	....	4 75	....	7 25
51 to 60	....	5 00	....	8 50
61 to 70	....	5 35	....	9 25
71 to 80	....	5 75	....	10 50
81 to 85	....	6 50	....	11 75
86 to 90	....	....	..	14 00
91 to 95	....	....	....	15 50
96 to 100	....	....	....	18 00

GAUGES

Marking, Mortise, Etc.

Stanley's dis. 50 to 55 per cent.

Wire Gauges.

Winn's, Nos. 26 to 33, each ... 1 65 2 40

HALTERS.

Rope, ⅜ per gross ....
" ½ " 9 00
" ⅝ to ¾ 14 00
Leather, 1 in., per doz 3 87½ 4 00
" 1¼ in., " 5 15 5 20
Web, — per doz 1 87 2 45

HAMMERS.

Nail

Maydole's, dis. 5 to 10 per cent. Can. dis. 25 to 27½ per cent.

Tack.

Magnetic, per doz 1 10 1 20

Sledge.

Canadian, per lb 0 07½ 0 08½

Ball Pean.

English and Can., per lb 0 22 0 25

HANDLES.

Axe, per doz. net 1 50 2 00
Store door, per doz 1 00 1 50

Fork.

C. & B., dis. 40 per cent. rev. list.

Hoe.

C. & B., dis. 40 per cent. rev. list.

Saw.

American, per doz 1 00 1 25

Plane.

American, per gross 3 15 3 75

Hammer and Hatchet.

Canadian, 40 per cent.

Cross-Cut Saws.

Canadian, per pair 0 13½

HANGERS. doz. pairs.

Steel barn door 5 85 6 00
Stearns, 4 inch 5 00
" 5 inch 6 50
Lane's covered—
No. 11, 5-ft. run 8 40
No. 11½, 10-ft. run 10 80
No. 12, 10-ft. run 12 60
No. 14, 15-ft. run 21 00
Lane's O.N.T. track, per foot. ... 4½

HARVEST TOOLS.

Discount, 50 and 10 per cent.

HATCHETS.

Canadian, dis. 40 to 42½ per cent.

HINGES.

Blind, Parker's, dis. 50 and 10 to 60 per cent.
Heavy T and strap, 4-in., per lb 0 06½
" " 5-in., " 0 06¼
" " 6-in., " 0 06
" " 8-in., " 0 05¾
" " 10-in., " 0 05½
Light T and strap, dis. 60 and 5 per cent.
Screw hook and hinge—
6 to 12 in., per 100 lbs 4 50
14 in. up, per 100 lbs 3 50
Per gro. pairs.
Spring 12 00

HOES.

Garden, Mortar, etc., dis. 50 and 10 p.c.
Planter, per doz 4 00 4 50

HOLLOW WARE

Discount, ... 45 and 5 per cent.

HOOKS.

Cast Iron.

Bird Cage, per doz 0 50 1 10
Clothes Line, per doz 0 27 0 63
Harness, per doz 0 72 0 88
Hat and Coat, per gross 1 00 3 00
Chandelier, per doz 0 50 1 00

Wrought Iron.

Wrought Hooks and Staples, Can., dis. 47½ per cent.

Wire.

Hat and Coat, discount 45 per cent.
Belt, per 1,000 .... 0 60
Screw, bright, dis. 55 per cent.

HORSE NAILS.

"C" brand 50 and 7½ p.c. off new list } Oval
"M" brand 60 per cent. } head.
Countersunk, 60 and 10 per cent.

HORSESHOES

F.O.B. Montreal. No. 2 and larger. / No. 1 and smaller.

Iron Shoes.
Light, medium, and heavy... 3 50 3 75
Snow shoes... 3 75 4 00
Steel Shoes.
Light... 3 60 3 85
Featherweight (all sizes)... 4 85 4 85
F.O.B. Toronto, Hamilton, London and Guelph, 10c. per keg additional.
Toe weight steel shoes... 6 70

JAPANNED WARE.
Discount, 45 and 5 p.c. off list, June 1899.

ICE PICKS.
Star per doz... 3 00 3 25

KETTLES.
Brass spun, 7½ p.c. dis. off new list.
Copper, per lb... 0 30 0 50
American, 60 and 10 to 65 and 5 p.c.

KEYS.
Lock, Can., dis., 45 p.c.
Cabinet, trunk, and padlock, Am. per gross... 60

KNOBS.
Door, japanned and N.P., per doz... 1 50 2 50
Bronze, Berlin, per doz... 2 75 3 25
Bronze Genuine, per doz... 6 00 9 00
Shutter, porcelain, F. & L. screw, per gross... 1 30 4 00
White door knobs—per doz. 1 25

HAY KNIVES.
Discount, 50 and 10 per cent.

LAMP WICKS.
Discount, 60 per cent.

LANTERNS.
Cold Blast, per doz... 7 50
No. 3 "Wright's"... 8 50
Ordinary, with O burner... 4 25
Dashboard, cold blast... 9 50
No. 0... 6 00
Japanning, 50c. per doz. extra.

LEMON SQUEEZERS.
Porcelain lined,... per doz. 2 20 5 60
Galvanized... 1 87 3 85
King, wood... 2 75 2 90
King, glass... 4 00 4 50
All glass... 1 30 1 30

LINES.
Fish, per gross... 1 05 2 50
Chalk "... 1 90 7 40

LOCKS.
Canadian, dis. 45 p.c.
Russell & Erwin, per doz... 3 00 3 25

Cabinet.
Eagle, dis. 30 p.c.

Padlock.
English and Am., per doz... 50 6 00
Scandinavian, "... 1 00 2 40
Eagle, dis. 20 to 25 p.c.

MACHINE SCREWS.
Iron and Brass.
Flat head discount 25 p.c.
Round Head, discount 20 p.c.

MALLETS.
Tinsmiths', per doz... 1 25 1 50
Carpenters', hickory, per doz. 1 25 3 75
Lignum Vitae, per doz... 3 85 5 00
Caulking, each... 60 2 00

MATTOCKS.
Canadian, per doz... 5 50 6 50

MEAT CUTTERS.
American, dis. 25 to 30 p.c.
German, 15 per cent.

MILK CAN TRIMMINGS.
Discount, 25 per cent.

NAILS.
Quotations are: Cut. Wire.
2d and 3d... $3 35 $3 85
3d... 3 00 3 52
4 and 5d... 2 75 3 35
6 and 7d... 2 65 3 20
8 and 9d... 2 50 3 00
10 and 12d... 2 45 2 95
16 and 20d... 2 40 2 90
30, 40, 50 and 60d. (base)... 2 35 2 85
Galvanizing 2c. per lb. net extra.
Steel Cut Nails 10c. extra.
Miscellaneous wire nails, dis. 70 per cent.
Coopers' nails, dis. 30 per cent.
Flour barrel nails, dis. 25 per cent

NAIL PULLERS.
German and American... 1 85 3 50

NAIL SETS
Square, round, and octagon, per gross... 3 38 4 00
Diamond... 12 00 15 00

NETTING.
Poultry, 50 and 5 per cent. for McMullen's

OAKUM.
Per 100 lb.
Navy... 6 00
U. S. Navy... 7 25

OIL.
Water White (U S)... 0 16½
Prime White (U.S)... 0 15⅔
Water White (Can.)... 0 15
Prime White (Can.)... 0 14

OILERS.
McClary's Model galvan. oil can, with pump, 5 gal., per doz... 10 00
Zinc and tin, dis. 50, 50 and 10.
Copper, per doz... 1 25 3 50
Brass, "... 1 50 3 50
Malleable, dis. 25 per cent.

GALVANIZED PAILS.
Dufferin pattern pails, dis. 45 p.c.
Flaring pattern, discount 45 per cent.
Galvanized washtubs, discount 45 per cent.

PIECED WARE.
Discount 40 per cent. off list, June, 1899.
10-qt. flaring sap buckets, dis. 45 p.c.
6, 10 and 14-qt. flaring pails, dis. 45 p.c.
Creamer cans, dis 45 p.c.

PICKS.
Per doz... 6 00 9 00

PICTURE NAILS.
Porcelain head, per gross... 1 75 3 00
Brass head "... 0 40 1 00

PICTURE WIRE.
Tin and gilt, discount 75 p.c.

PLANES.
Wood, bench, Canadian dis. 50 per cent. American dis. 50.
Wood, fancy Canadian or American 7½ to 40 per cent.

PLANE IRONS.
English, per doz... 2 00 5 00

PLIERS AND NIPPERS.
Button's Genuine per doz pairs, dis. 37½ 40 p.c.
Button's Imitation, per doz... 5 00 9 00
German, per doz... 0 60 2 60

PLUMBERS' BRASS GOODS.
Compression work, discount, 60 per cent.
Fuller's work, discount 65 per cent.
Rough stops and stop and waste cocks, discount, 60 per cent.
Jenkins disk globe and angle valves, discount, 55 per cent.
Standard valves, discount, 60 per per cent.
Jenkins' radiator valves discount 55 per cent.
" " " standard, dis., 60 p.c.
Quick opening valves discount, 60 p.c.
No. 1 compression bath cock... 2 00
No. 4 " " "... 2 00
No. 7, Fuller's... 2 50
No 4½, "... 3 00

POWDER.
VeloX Smokeless Shotgun Powder.
100 lb. or less... 0 85
1,000 lb. or more... 0 80
Net 30 days.

PRESSED SPIKES.
Discount 20 to 25 per cent.

PULLEYS.
Hothouse, per doz... 0 55 1 00
Axle... 0 22 0 33
Screw... 0 27 1 00
Awning... 0 35 2 50

PUMPS.
Canadian cistern... 1 80 3 60
Canadian pitcher spout... 1 40 2 10

PUNCHES.
Saddlers', per doz... 1 00 1 85
Conductors', "... 9 00 15 00
Tinners' solid, per set... 0 00 0 72
" hollow, per inch... 0 00 1 00

RANGE BOILERS.
Galvanized, 3 gallons... 7 00
" 35 "... 8 25
" 40 "... 9 50
Copper, 30 "... 22 00
" 35 "... 26 00
" 40 "... 30 00
Discount off Copper Boilers 10 per cent.

RAKES.
Cast steel and malleable, 50, 10 and 5 p.c.
Wood, 25 per cent.

RASPS AND HORSE RASPS.
New Nicholson horse rasp, discount 60 to 60 and 10 p.c.
Globe File Co.'s rasps, 60 and 10 to 70 p.c.
Heller's Horse rasps, 50 to 50 and 5 p.c.

RAZORS.
per doz.
Geo. Butler & Co.'s,... 8 00 18 00
Boker's... 7 50 11 00
" King Cutter... 12 50 50 00
Wade & Butcher's... 3 60 10 00
Theile & Quack's... 7 00 12 00
Elliot's... 4 00 18 00

REAPING HOOKS.
Discount, 50 and 10 per cent.

REGISTERS.
Discount,... 40 per cent.

RIVETS AND BURRS.
Iron Rivets, black and tinned, discount 60 and 10 per cent.
Iron Burrs, discount 55 per cent.
Extras on Iron Rivets in 1-lb. cartons, ½c. per lb.
Extras on Iron Rivets in ½-lb. cartons, 1c. per lb.
Copper Rivets & Burrs, 35 and 5 p.c. dis. and cartons, 1c. per lb. extra, net.
Extras on Tinned or Coppered Rivets ½-lb. cartons, 1c. per lb.

RIVET SETS
Canadian, dis. 35 to 37½ per cent.

ROPE ETC.
Sisal. Manila.
7-16 in. and larger, per lb. 10 13½
⅜ in... 11 14½
¼ and 5-16 in... 15½
Cotton, 3-16 inch and larger 16½
" 5-32 inch... 21½
" ⅛ inch... 22½
Russia Deep Sea... 15½
Jute... 8
Lath Yarn... 9½
New Zealand Rope... 10

RULES.
Boxwood, dis. 75 and 10 p.c.
Ivory, dis. 37½ to 40 p.c.

SAD IRONS.
per set.
Mrs. Potts, No. 55, polished... 62½
" No. 50, nickle-plated... 67½

SAND AND EMERY PAPER.
Dominion Flint Paper, 47½ per cent.
B & A. sand, 40 and 5 per cent.
Emery, 40 per cent.
Garr et (Ru t n r), 5 to 10 p.c. advance on list.

SAP SPOUTS.
Bronzed iron with hooks, per doz... 9 50

SAWS.
Hand Disston's, dis. 12½ p.c.
S. & D., 40 per cent.
Crosscut, Disston's, per ft... 35 0 55
S. & D., dis. 35 p.c. on Nos. 2 and 3.
Hack, complete, each... 0 75 2 75
" frame only... 0 75

SASH WEIGHTS.
Sectional, per 100 lbs... 2 75 3 00
Solid, "... 2 00 2 25

SASH CORD.
Per lb... 0 23 0 30

SAW SETS.
"Lincoln," per doz... 6 50

SCALES.
B. S. & M. Scales, 45 p.c.
Champion, 65 per cent.
Fairbanks Standard, 35 p.c.
" Dominion, 55 p.c.
" Richelieu, 55 p.c.
Chatillon Spring Balances, 10 p.c.

SCREW DRIVERS.
Sargent's per doz... 0 65 1 00

SCREWS
Wood, F. H., bright and steel, 87½ and 10 p.c.
Wood R. H., " dis. 82½ and 10 p.c.
" F. H., brass, dis. 80 and 10 p.c.
Wood, R. H., " dis. 75 and 10 p.c.
" F.H., bronze, dis. 75 p.c.
" R.H. " 70 p.c.
Drive Screws, 87½ and 10 per cent.
Bench, wood, per doz... 3 25 4 00
" iron, "... 4 25 5 75
Set, Case hardened, 60 per cent.
Square Cap, 50 and 5 per cent.
Hexagon Cap, 45 per cent.

SCYTHES.
Per doz, net... 9 00

SCYTHE SNATHS.
Canadian, dis. 45 p.c.

SHEARS.
Bailey Cutlery Co., full nickeled, dis. 60 p.c.
Seymour's, dis. 50 and 10 p.c.

SHOVELS AND SPADES.
Canadian, dis. 40 and 5 per cent.

SINKS.
Steel and galvanized, discount 45 per cent.

SNAPS.
Harness, German, dis. 25 p.c.
Lock, Andrews'... 4 50 11 50

SOLDERING IRONS.
1, 1½ lb., per lb... 0 37
2 lb. or over, per lb... 0 34

SQUARES.
Iron, No 493, per doz... 2 40 2 55
" No. 494, "... 3 25 3 40
Steel, dis. 50 and 5 to 50 and 10 p.c., rev. list.
Try and bevel, dis. 50 to 52½ p.c.

STAMPED WARE.
Plain, dis. 75 and 12½ p.c. off revised list.
Retinned, dis., 75 p.c. off revised list.

STAPLES.
Galvanized... 3 50 4 00
Plain... 3 15 3 75
Coopers', discount 45 per cent.
Poultry netting staples, 40 per cent.

STOCKS AND DIES.
American dis. 25 p.c.

STONE.
Per lb.
Washita... 0 28 0 60
Hindostan... 0 06 0 07
" slip,... 0 09 0 09
Labrador... 0 13
" Axe... 0 15
Turkey... 0 50
Arkansas... 0 00 1 50
Water-of-Ayr... 0 00 0 10
Scythe, per gross 3 50 5 00
Grind, per ton 15 00 18 00

STOVE PIPES.
5 and 6 inch Per 100 lengths... 7 00
7 inch " "... 7 50

ENAMELINE STOVE POLISH.
No. 4—3 dozen in case, net cash... $4 80
No. 6—3 dozen in case, "... 8 40

TACKS BRADS, ETC.
Per cent.
Strawberry box tacks, bulk... 75 & 10
Cheese-box tacks, blued... 80 & 12½
Trunk tacks, black and tinned... 85
Carpet tacks, blued... 80 & 5
" " tinned... 80 & 10
" " (in kegs)... 40
Cut tacks, blued, in dozens only... 75 & 15
" ½ weights... 60
Swedes, cut tacks, blued and tinned—
In bulk... 80 & 10
In dozens... 75
Swedes, upholsterers', bulk... 85 & 12½
" brush, blued & tinned, bulk... 70
" gimp, blued, tinned and japanned... 75 & 12½
Zinc tacks... 35
Leather carpet tacks... 55
Copper tacks... 50
Copper nails... 52
Trunk nails, black... 65 and 5
Trunk nails, tinned... 65 and 10
Clout nails, blued and tinned... 65 and 5
Chair nails... 35
Patent brads... 40
Fine finishing... 40
Picture frame points... 10
Lining tacks, in papers... 10

Lining tacks, in bulk	15
" " solid heads, in bulk	75
Saddle nails in papers	10
" " in bulk	15
Tufting buttons, 22 line, in dozens only	60
Tin capped trunk nails	15
Zinc glazier's points	5
Double pointed tacks, papers	90 and 10
" " " bulk	40

TAPE LINES.

English, ass skin, per doz	2 75	5 00
English, Patent Leather	5 50	9 75
Chesterman's each	0 90	2 85
" steel, each	0 80	8 00

THERMOMETERS.

Tin case and dairy, dis. 75 to 75 and 10 p.c.

TRAPS. (Steel.)

Game, Newhouse, dis. 25 p.c.
Game, H. & N., P. S. & W., 65 p.c.
Game, steel, 72½, 75 p.c.

TROWELS.

Disston's discount 10 per cent.

German, per doz	4 75	6 00

S. & D., discount 35 per cent.

TWINES.

Bag, Russian, per lb	....	0 27
Wrapping, cotton, per lb	0 22	0 26
Wrapping, mottled, per pack	0 50	0 60
Wrapping, cotton, 3-ply		0 20
" " 4-ply	....	0 26
Mattress, per lb	0 33	0 45
Staging, "	0 27	0 35

VISES.

Hand, per doz	4 00	6 00
Bench, parallel, each	2 00	4 50
Coach, each	6 00	7 00
Peter Wright's, per lb	0 12	0 13
Pipe, each	5 50	9 00
Saw, per doz	6 50	13 00

ENAMELLED WARE.

White, Princess, Turquoise, Blue and White, discount 50 per cent.
Diamond, Famous, Premier, 50 and 10 p.c.
Granite or Pearl, Imperial, Crescent, 50, 10 and 10 per cent.

WIRE.

Brass wire, 50 to 50 and 2½ per cent. off the list.

Copper wire, 45 and 10 per cent. net cash 30 days, f.o.b. factory.

Smooth Steel Wire, is quoted at the following net selling prices:

No. 6 to 8 gauge	$2 90
" 9 "	2 80
" 10 "	2 87
" 11 "	2 90
" 12 "	2 95
" 13 "	3 15
" 14 "	3 37
" 15 "	3 50
" 16 "	3 65

Other sizes of plain wire outside of Nos. 9, 10, 11, 12 and 13, and other varieties of plain wire remain at $2.80 base with extras as before. The prices for Nos. 9 to 13 include the charge of 1 c. for oiling. Extras net per 100 lb.: Coppered wire, 60c.—tinned wire, $2—oiling, 10c.—special hay-bailing wire, 30c.—spring wire, $1—best steel wire, 75c.—bright soft drawn, 15c.—in 50 and 100-lb. bundles net, 10c.—in 25-lb. bundles net 15c.—packed in casks or cases, 15c.—bagging or papering, 10c.

Fine Steel Wire, dis. 17½ per cent List of extras: In 100-lb. lots: No 17, $5—No. 18, $5.50—No. 19, $6—No. 20 $6.65—No. 21, $7—No. 22, $7.30—No. 23, $7.65—No. 24, $8—No. 25, $9—No. 26, $9.50—No. 27, $10—No. 28, $11—No. 29, $12—No. 30, $13—No. 31, $14—No. 32, $15 No. 33, $16—No. 34, $17. Extras net—tinned wire, Nos. 17-25, $2—Nos. 26-31 $4—Nos. 32-34, $6. Coppered, 5c.—oiling, 10c.—in 25-lb. bundles, 15c.—in 5 and 10-lb. bundles, 25c.—in 1-lb. hanks, 50c.—in ¼-lb. hanks, 75c.—in ⅛-lb. hanks, $1—packed in casks or cases, 15c.—bagging or papering, 10c.

Galvanized Wire, per 100 lb.—Nos. 6,7,8, $3 50 to $3 85—No. 9, $2.85 to $3.15—No. 10, $3.60 to $3.95—No. 11, $3 70 to $4.10—No. 12, $3 to $3 30—No. 13, $3.10 to $3 40—No. 14, $4.10 to $4.50—No. 15, $4.60 to $5.05—No. 16, $4.85 to $5 35. Base sizes, Nos. 6 to 9, $2.57½ f.o.b. Cleveland.

Clothes Line Wire, solid 7 strand, No. 17, $4.25; No. 18, $2.65; No. 19, $2.35, f.o.b Hamilton, Toronto, Montreal.

WIRE FENCING. F.O.B. Toronto

Galvanized barb	....	3 15
Galvanized, plain twist	....	3 05

Galvanized barb, f.o.b. Cleveland, $2.82½ in less than carlots, and $2.70 in carlots.

WIRE CLOTH.

Painted Screen, per 100 sq. ft., net	1 35

WASTE COTTON. per lb.

Colored	4½ to 5
White, according to quality	5½ to 7½

500-lb bale lots shaded.

WRENCHES.

Acme, 35 to 37½ per cent.
Agricultural, 60 p.c.
Coe's Genuine, dis. 70 to 25 p.c.

Towers' Engineer, each	2 00	7 00
" S., per doz	5 80	6 00
G. & K.'s Pipe, per doz	....	3 40
Burrell's Pipe, each	....	3 00
Pocket, per doz	0 25	2 90

WRINGERS.

Leader	per doz.	$30 00	31 00
Royal Canadian	"	36 00	38 00
Royal American	"	16 00	28 00
Sampson	"	....	30 00

Terms 4 months, or 3 p.c. 30 days.

WROUGHT IRON WASHERS.

Canadian make, discount, 40 and 5 per cent.

PERSONS addressing advertisers will kindly mention having seen their advertisement in Canadian Hardware and Metal Merchant.

"THE EMLYN" SAW BENCH

Made in 6 sizes. Best value obtainable. Specially designed for export. With or without "Emlyn" Patent Guard. Sole maker—

CHARLES D. PHILLIPS,

Cables— "Machinery," Newport.

Emlyn Engineering Works. NEWPORT, MON., ENGLAND.

DO YOU HANDLE THE

AYLMER SPRAYERS?

IF NOT WHY NOT?

They are the most durable, easiest working and most simply constructed pump on the market, and give universal satisfaction. If our travellers have not yet called on you, write for catalogue. Liberal discounts to the trade.

AYLMER IRON WORKS, AYLMER, ONT.

Lockerby & McComb

AGENTS IN CANADA

FOR THE

Celebrated P. & B.

Cold Storage Lining

AND

.. Ruberoid Roofing ..

P. S.—Prices on Application.

65 Shannon Street, MONTREAL.

BUSINESS NEWS

of any kind that is of value to business men supplied by our Bureau. We can give you market quotations from any town in Canada, reports from the city markets, stock quotations, etc. You can get commercial news from any Canadian paper through us. Write us, giving us particulars of what you want and where you want it from, and we will quote you prices by return.

"Clippings from any Canadian paper on any subject."

CANADIAN PRESS CLIPPING BUREAU,

232 McGill Street, MONTREAL, QUE.

Telephone Main 1255.

10 Front St. East, Toronto. Telephone 2148.

---

75 YEARS. *ESTABLISHED 1825.* *75 YEARS.*

CELEBRATED **HEINISCH** SHEARS.

Tailors' Shears,
Trimmers, Scissors,
Tinners' Snips, etc.

ACKNOWLEDGED THE BEST.

R. HEINISCH'S SONS CO. NEW YORK OFFICE, 90 Chambers St. NEWARK, N.J., U.S.A.

Not connected with any Shear Combination.

CHAS. F. CLARK, President. — JARED CHITTENDEN, Treasurer.

...ESTABLISHED 1849...

BRADSTREET'S

Capital and Surplus, $1,500,000. **Offices Throughout the Civilized World.**

Executive Offices: Nos. 346 and 348 Broadway, New York City, U.S.A.

THE BRADSTREET COMPANY gathers information that reflects the financial condition and the controlling circumstances of every seeker of mercantile credit. Its business may be defined as of the merchants, by the merchants, for the merchants. In procuring, verifying and promulgating information no effort is spared, and no reasonable expense considered too great, that the results may justify its claim as an authority on all matters affecting commercial affairs and mercantile credit. Its offices and connections have been steadily extended, and it furnishes information concerning mercantile persons throughout the civilized world.

Subscriptions are based on the service furnished, and are available only by reputable wholesale, jobbing and manufacturing concerns, and by responsible and worthy financial, fiduciary and business corporations. Specific terms may be obtained by addressing the Company at any of its offices. **Correspondence Invited.**

—OFFICES IN CANADA—

HALIFAX, N.S. OTTAWA, ONT. VANCOUVER, B.C. HAMILTON, ONT. QUEBEC, QUE. VICTORIA, B.C. LONDON, ONT. ST. JOHN, N.B. WINNIPEG, MAN. MONTREAL, QUE. TORONTO, ONT.

THOS. C. IRVING, Gen. Man. Western Canada, Toronto. JOHN A. FULTON, Gen. Man. Eastern Canada, Montreal.

Awarded a Gold Medal at PARIS EXPOSITION for superiority. That's proof enough of their quality, and clearly shows that they are the best.

The Bailey

Cutlery Co.

BRANTFORD, ONT.

Send for Catalogue and . . . . Price List.

Dominion Pattern

Cow Tie AND Stall Fixture

The special features of the tie and stall fixture are well shown in the illustration. As will be noticed the chain is very short, which prevents all danger of entanglement with the animal's foot. At the same time the form of the fixture is such that great freedom is allowed to the head. Because of the short chain this tie is much cheaper than the ordinary patterns.

The stall fixture is made from a tough quality of steel and is very strong. Also, owing to its circular cross-section, it is exceedingly rigid. Its simplicity, convenience, cheapness, and ease of attaching make it very popular with cow tie users.

This form of tie and stall fixture are sometimes called Niagara pattern.

American or Flat Link Chain,

for years the standard cow tie chain in "the States," is now rapidly coming in favor in Canada. Its short link, handsome appearance and smooth surface—which cannot injure the animal's neck—make it superior to all other styles of chain for cow ties.

For sale by all Jobbers; manufactured by

ONEIDA COMMUNITY, Limited, NIAGARA FALLS, ONT.

CANADIAN HARDWARE AND METAL MERCHANT

The Weekly Organ of the Hardware, Metal, Heating, Plumbing and Contracting Trades in Canada.

VOL. XIII. MONTREAL AND TORONTO, APRIL 6, 1901. NO. 14

Garden Hose

Discounts for the new season now out.

WRITE FOR PRICES.

* * *

CANADIAN RUBBER CO.,

Montreal. Toronto. Winnipeg.

James Cartland & Son

Manufacturers of every description of **Limited**

CABINET, BUILDERS', FURNISHING AND NAVAL BRASSFOUNDRY

BIRMINGHAM, ENGLAND.

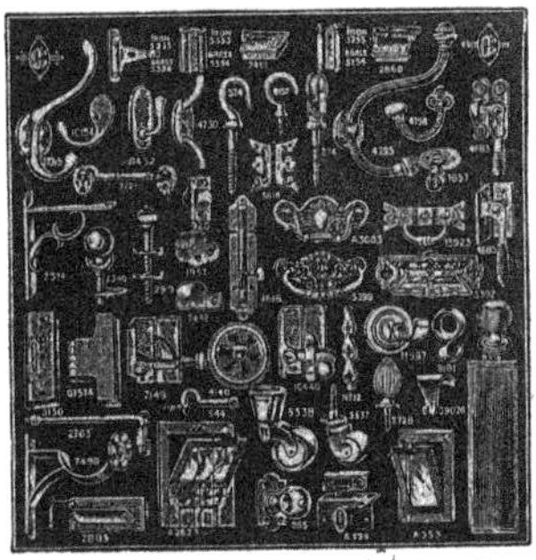

London Showrooms: 57 Holborn Viaduct, E.C.

SOME OF THE NEWER "YANKEE" TOOLS

NO. 41 AUTOMATIC DRILL WITH DRILL POINTS IN HANDLE.

NO. 42 AUTOMATIC DRILL WITH DRILL POINTS IN BOX.

NO. 50 RECIPROCATING DRILL, FOR WOOD OR METALS.

Sold by Leading Jobbers throughout the Dominion.

NORTH BROS. MFG. CO.,

Philadelphia, Pa., U. S. A.

"Quick Meal"

Gasoline or Blue Flame Oil. Summer Stoves

Used Most and Valued Highest.

The popular favorites with Canadian housekeepers, because of their reliable trustworthiness, perfect adaptability for all kinds of kitchen work, and their simple ease of operation.

The "Quick Meal" Wickless Oil Stove is a marvel of perfect construction, made to please the popular demand, unique in its good points, an unfailing success.

Post yourself on the good points these fine stoves offer you Send for our descriptive catalogue, and make up your mind to do a rousing Spring trade in these splendid quick-selling "Quick Meal" Stoves.

AS IN FORMER YEARS, WE ARE SOLE CANADIAN AGENTS.

The Gurney Foundry Co., Limited

THE GURNEY-MASSEY CO., LIMITED, MONTREAL

TORONTO, WINNIPEG, VANCOUVER

David Maxwell & Sons

ST. MARYS, ONT., CANADA

Steel Frame Churn.

"Maxwell Favorite Churn"

PATENTED FEATURES: Improved Steel Stand, Roller Bearings, and Foot and Hand Lever Drive, and Detachable Driving Link Improved for season of 1901. Steel or Wood Frame as desired.

Wheelbarrows.

In Four different Sizes.

Lawn Mowers.

High and Low Wheels, from 12-in. to 20-in. widths. Cold Rolled Steel Shafting, Crucible Steel Knives and Cutting Plate.

If your Wholesale House does not offer you these articles

...SEND DIRECT TO US.

MAXWELL MOWER

8-inch Low Wheel.

"THE MAXWELL" Lawn Mower

High Wheel 10 inches.

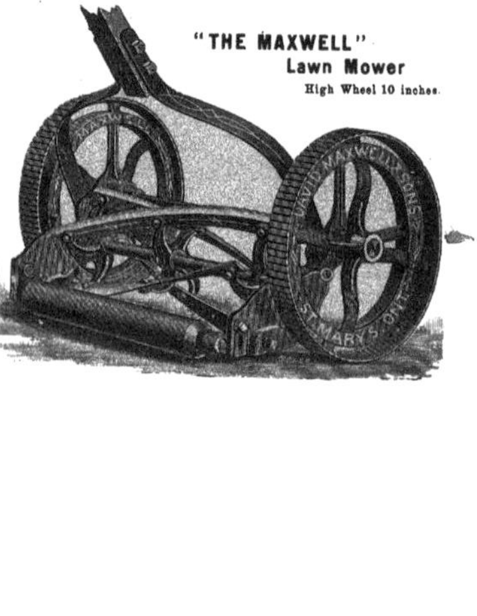

EVERYTHING GOOD AND PROFITABLE IN **WALL PAPER.**

21-inch Ingrain Frieze, No. 2030

21-inch Ingrain Frieze, No. 2043

THE ABOVE ARE TWO OF MANY SUCCESSFUL FRIEZES
IN OUR 1901 COLLECTION
IF OF INTEREST WE WILL MAIL SMALL BOOK OF
INGRAIN SHADES WITH ILLUSTRATED MATCHED
COMBINATIONS, OR SAMPLES OF ABOVE OR OTHERS
IN ANY DESIRED SHADE

THE **WATSON, FOSTER CO.,** LIMITED

Feb. 1st. MONTREAL.

LEWIS BROS. & CO., MONTREAL

Our Stock of

Lawn and Garden Hose

and Fixtures is most complete.

It will pay you to get our Prices on Hose.

Caldwell Hose Straps.

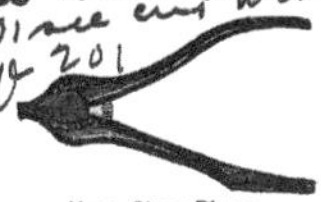

Hose Strap Plyers.

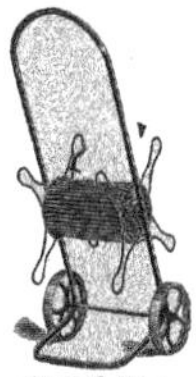

No. K. C. No. 1.

Cooper Hose Mender, ½, ¾, and 1-inch.

Brass Hose Coupling.

Hose Band, ½ and ¾-in.

Iron Hose Mender, ½, ¾ and 1-in.

Wood Hose Mender, ½ and ¾-in.

Gem Nozzle.

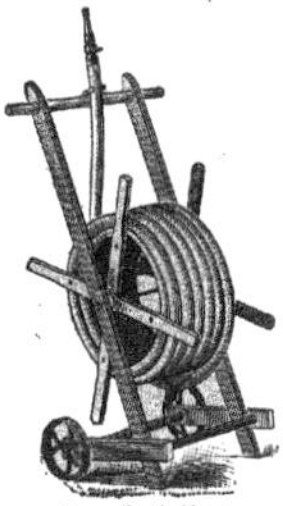

Hose Reel, No. 1.

Magic Nozzle.

Write for Prices.

LEWIS BROS. & CO.

Wholesale Hardware, MONTREAL.

NOVELTY QUILT FRAME CLAMPS.

No. 3

JAPANNED or COPPERED.

No. 4

FOR SPRING CLEANING, YOUR LADY CUSTOMERS WILL APPRECIATE THESE FOR THEIR CURTAIN FRAMES.

INSTEAD OF THE OLD WAY OF RESTING THE FRAMES ON CHAIRS, USE OLD BROOM HANDLES IN THE SOCKETS.

SAVES WORRY. **IMPROVES TEMPER.**

If you would like to see it, a sample will be mailed you if you will drop a post card to

A. R. WOODYATT & CO., GUELPH, ONT.

SOLD ONLY THROUGH THE WHOLESALE TRADE.

The VICTOR and DAISY Oil Stoves

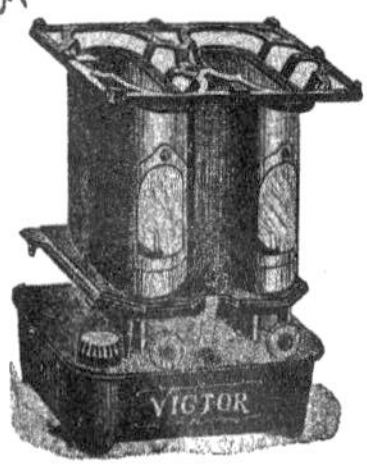

of to-day are the results of years of practical experience in oil stove making.

In them are embodied all the points which go to make up a perfect oil stove.

A stove which will not smoke } **but will give perfect**
A stove which will not smell } **satisfaction.**

That's why our trade is increasing so rapidly.
Your stock would not be complete without these goods.
Made in one, two and three-burner.
We will be pleased to quote you.

KEMP MANUFACTURING CO.,

TORONTO, Canada

HARDWARE AND METAL

VOL. XIII. MONTREAL AND TORONTO, APRIL 6, 1901. NO. 14.

President,
JOHN BAYNE MacLEAN,
Montreal.

THE MacLEAN PUBLISHING CO.
Limited.

Publishers of Trade Newspapers which circulate in the Provinces of British Columbia, North-West Territories, Manitoba, Ontario, Quebec, Nova Scotia, New Brunswick, P.E. Island and Newfoundland.

OFFICES

MONTREAL - - - - - - 232 McGill Street, Telephone 1255.
TORONTO - - - - - - 10 Front Street East, Telephone 2148.
LONDON, ENG. - - - - 109 Fleet Street, E.C., J. M. McKim.
MANCHESTER, ENG. - - - 18 St Ann Street. H. S. Ashburner,
WINNIPEG - - - - Western Canada Block. J. J. Roberts.
ST. JOHN, N.B. - - - No. 3 Market Wharf, J. Hunter White,
NEW YORK. - - - - - - 176 E. 88th Street.

Subscription, Canada and the United States, $2.00.
Great Britain and elsewhere - - - - 12s.

Published every Saturday.

Cable Address { Adscript, London. Adscript, Canada.

THE ADVANCE IN GLASS.

AS anticipated in last week's issue, a general advance was made in glass prices at the beginning of the week. The stiffness that led to this advance here was largely due to the growing firmness in Europe. The strike which started in Belgium last August is still on, and is curtailing the production there to a considerable extent.

The production in Great Britain has been larger than usual, but the increase has not been sufficient to meet the shortage in Belgium.

The natural consequence is that the Belgian exporters have had more business than they could handle, and have taken full advantage of the conditions. Orders sent at a basis ruling some time ago have been refused, and quotations have been sent out which are considerably higher than the figures of a few weeks ago.

The situation in Canada seems to be fully as strong as that in Belgium. The import order business has been smaller than last year. As stocks here are no heavier than they were a year ago, when a shortage was so feared, and as it is reasonable to expect a larger movement from stock than a year ago, the jobbers have followed the advance in Belgium. Further advances are predicted here.

HUMANITARIAN POLICY.

PERCY LANGMUIR, a writer on foundry economy, lays a great deal of stress on the treatment of employes as a feature of economical management. "There is little or no sentiment in the matter," he says. "It is simply better treatment, better results—a policy which will admit of untold expansion in the foundry world."

In every factory, in every warehouse, in every store, there are, undoubtedly, employes whom any amount of humanitarian treatment will not induce to put forth better effort on their employers' behalf. But all are not the same. The majority are not the same.

Whether it be the man in the factory or the man in the office, the humanitarian policy will, in the long run, be found to be the best policy. People do not expect to turn up "sixes" with every throw of the dice, and yet they go on throwing. In humanitarian treatment of employes, the chances of success are much greater.

A coercive policy may pay with slaves, but with the average free man, never.

A RETAILER'S MISTAKE.

THE representative of a well-known manufacturing concern was recently attracted by the low figure at which a retailer in Toronto had marked the selling price of the line of goods his firm produced. Without revealing his identity he stepped into the store and made a purchase of the goods, but finding the quantity each individual could buy was limited, he secured several boys to enter the store and secure additional quantities. As the figure at which he bought was two cents a package below the manufacturer's figure, a near-by departmental store readily took over the goods at the figure at which the representative had purchased them.

A few days later the representative again visited the retailer's store and found him still ticketing the goods in question at the same low price. After explaining who he was he asked the retailer how he could afford to sell the article at such a low figure.

"O, I'm not making anything, but I'm not losing; I'm selling at actual cost."

"No, you are not selling at cost," said the representative, "you are losing a clear two cents a package on the first cost."

And so he was, for it turned out that he thought there were a certain number of packages in a case when there were really 10 less. Naturally he was much confused when he learned his mistake. But he never ought to have based his selling price on what he thought was the number of packages there were in the case. He should have learned for a certainty what the actual number was.

The moral in this incident is obvious.

Stiff iron prices give backbone to trade.

BUSINESS VS. POLITICAL ENERGY.

HARDWARE AND METAL has repeatedly urged the business men of this country not to put their faith in legislators, but to develop energy themselves if they ever hoped to induce Federal or Provincial Parliaments to pass laws in conformity with the commercial necessities of the country.

Support of this principle has come from a quarter least expected; namely, from a member of one of the legislative bodies of the country.

A few evenings ago, the Toronto Board of Trade held a special meeting to consider the development of New Ontario. Among the gentlemen who spoke was Mr. W. H. Hoyle, the representative of North Ontario in the Ontario Legislature. Mr. Hoyle is a party man, and a pretty strong one, too; but he is also a practical and a successful business man. And during the three years he has occupied a seat in the Ontario Legislature he has doubtless learned a great deal about the ways of the average politician. At any rate, his own business instincts have not been destroyed, for he declared at the meeting of the Toronto Board of Trade referred to that if the business men present hoped to succeed in their efforts to develop New Ontario they must not "put their faith in legislators, but in their own energy and enterprise."

Coming from a man who is a legislator, this advice carries with it more weight than it could possibly have had were it given by a private citizen. But, judging from the past, it will require the preaching of a great deal of such gospel before business men universally shall be awakened to a sense of their sin of omission in this particular.

Business men, as represented by boards of trade and other commercial bodies, have, time and again, to use a slang phrase, been turned down by Governments, Federal and Provincial, in regard to commercial matters.

The power behind the throne is not the business man; it is the politician. If the latter wants any favors he stands at the door and knocks, and knocks hard, until it is opened and his wants are satisfied. He is enterprising; he is energetic, and he gets his reward.

The business man, on the other hand, when he wants anything for the welfare of the commercial interests gently taps at the door of the Cabinet or of the Parliament. If the door is not forthwith opened he ceases knocking and departs.

Party exigencies demand that the politician's demands shall be satisfied. He controls votes which he will not hesitate to use against his party, if, by so doing, he can further his own ends. There is no such danger in regard to the business man. In his demands for the improvement of laws or regulations of commercial significance he is not seeking the gratification of personal aspirations or ambitions. His thought is the welfare of the business community and, of course, indirectly, that of the country. But whether his demands are satisfied or not he falls into line on election day at the crack of the party whip. If he is Conservative he is Conservative still; if he is Liberal he is Liberal still.

THE DUTY AGAINST UNITED STATES SILVER COINS.

THE existence of laws, sometimes interesting ones, are often not known to us until revealed by accident.

Probably not one person in a hundred is aware that the Canadian Customs tariff imposes a duty on silver coins of the United States. Yet it does. A reader of HARDWARE AND METAL learned this the other day, to his great surprise, when he was taxed 20 per cent. on a small sum of United States silver coinage which came through the mails.

The clause of the tariff authorizing the imposition of a duty on the silver coinage of our neighbors is really in the free list. It is numbered 473 and in part reads: "Coins, cabinets of, collections of medals and of other antiquities including collections of postage stamps; gold and silver coins, except United States silver coins." Being thus denied the right of the free list, and not enumerated elsewhere, United States silver coin comes under the unenumerated goods clause and is subject therefor to a duty of 20 per cent.

The provision for taxing United States silver coinage has been in the Customs tariff for some years, notwithstanding that so few people appear to be aware of the fact. And it was created because of the large amount of United States silver coinage that was being brought into Canada by cattle-drovers and others and put into circulation here until it became, as a Customs official put it, "a nuisance."

In spite of the 20 per cent. duty, however, a good deal of United States silver coinage finds its way into Canada and circulates here, and at par, too, notwithstanding that Canadian silver coinage, although purer than the former, will not be taken at all in many parts of the United States.

The official trade returns do not show the amount of silver coins that are brought into Canada from the United States, but, of gold coins, we last year imported over $7,400,000 from that country, while our total imports of coin, silver and gold, all told, was only about $8,000,000.

It may be, perhaps, interesting in this connection to note that our total exports of gold coin last year were $6,903,562, of which all but $400 went to the United States, and of silver, $83,440, of which all but $650 went to the United States.

CHANGES IN WIRE NAILS AND TACKS.

THE manufacturers have concluded their meetings in Montreal, and while a few changes have been made, none of them are of great importance.

It was maintained by some of the wire nail manufacturers that they could not afford to maintain a difference of 10c. per keg between the price of carlots and less than carlots. It was furthermore urged that the difference of 10c. per keg, together with the lower freight rates which the purchaser of carlots of wire nails received, placed the small dealer at too great a disadvantage. As a result it was decided to increase the price on carlots 2½c. per 100 lb., and to allow the old quotation to stand on less quantities, thus reducing the difference between carlots and less than carlots to 7½c. per keg.

A reduction was made in the price of a few lines of tacks. It is principally in the descriptions used by manufacturers. This will be gathered from the following, showing the new and old discount:

	New Discount.	Old Discount.
Carpet tacks, blued	80 & 15	80 & 5
Carpet tacks, tinned	80 & 20	80 & 10
Upholsterers' tacks, bulk	85, 12½ & 12½	85 & 12½
Cooper nails	52½	52

The changes are due, it is understood, to increased competition among the manufacturers.

THE CANCELLATION OF ORDERS.

Editor HARDWARE AND METAL,—I have looked over the comments on your editorial of March 9 with some interest, and am pleased to see good points brought forward on the subject of business honor in soliciting orders for forward delivery of goods and as to how far a man is bound to accept goods as agreed upon, in case a decline in values takes place before the date for delivery. Your correspondent, "The Beeton Bell," presents his views at some length, bringing in some special phases that were not under consideration in your article. One point referred to by him is where an order is taken and a guarantee to meet any reduction is made by the house accepting it.

All will agree that where this is the case the party giving the order has all the advantage on his side, and the party soliciting orders under such one-sided conditions must take the consequences.

Another case supposed is where manufacturers combine and put up prices above market values, and, before orders are filled, the combine reduce the selling price and try to make their patrons pay the enhanced value. Notwithstanding they are obliged to make reductions to a legitimate basis before the delivery of the goods, the retail dealer who finds himself in such a fix will always have the sympathy of the business public.

But, let us consider the subject simply as first discussed. A wholesale merchant in many cases must arrange for the goods he is to supply to the retailer some months before they are to be shipped out. His traveller goes out on his route and it takes some weeks to cover his district. As his orders are taken he forwards them in to the office. Possibly, the house has a dozen men out. Their orders booked are condensed, and, in an ordinary business way, the firm must cover these by booking with the manufacturer, possibly, in Great Britain or elsewhere, to arrive in one, two or three months. And, while these goods are in transit, the raw material from which the goods are made comes down in price and the market weakens. Another set of travellers are out who began their routes where the first left off, and reach, at a late date in the season, the section where others began. The retail man finds the market is weakening and may come down still lower. He secures the lowest quotations possible, perhaps for double the quantity his first order is for, and at once writes the house who holds his first order, and says, "Unless you meet the figures I have to-day had given me I cancel my order," and calls himself an honest man.

Might I suggest this savors of a hold-up if "The Beeton Bell" will allow me to use his words.

I was at church last Sunday evening and heard a sermon on "Man's Duty to His Fellow." The preacher discussed a business transaction and said: "When making a bargain you should get a profit for yourself. It is your due. You must take care to get it, but not all the profit. You must let the other man have some. If you don't, do you know what you are? I will tell. You are a thief and don't forget it." I do not say what the retail man is who performs the hold-up by saying: "Cancel my order you booked and provided to supply or meet the price I am now quoting" no matter if it does take all your profit away, but I leave him with the preacher.

AN OLD TRAVELLER.

Toronto, April 2, 1901.

TRAVELLERS FROM THE WEST.

HARDWARE AND METAL has, during the past week, run across three travellers who have just returned from the Northwest and British Columbia. They were Messrs. A. T. Chambers, of H. S. Howland, Sons & Co., Toronto; H. G. Allen, of The Oneida Community, Limited, Niagara Falls; and J. T. Webb, of The Thos. Davidson Manufacturing Co., Limited, Montreal.

Their reports regarding the trade conditions corroborated each other. They all stated that, while trade was quiet in Manitoba and payments slow, the conditions in these respects were not as bad as anticipated, owing to the failure of the wheat crop last year. As the farmers had put in their spare time last fall in breaking new ground the acreage under wheat will be much larger. The outlook for the coming season is, therefore, good. They all reported trade fairly brisk in British Columbia.

Mr. J. T. Webb will be on his old territory in Ontario in the course of the next few days.

HE WAS SATISFIED.

A merchant who had his doubts about the results he was getting from his advertisement in the local papers, determined to test the efficacy of his mediums. He had two stores in nearby towns. In each he put a table of special bargains at greatly reduced prices. The goods on the table in each store had the prices plainly marked on them. The merchant inserted an advertisement in his customary medium, and in it he advertised the bargain counter in one store and said nothing about the like counter in the other. The results surprised him. The bargain counter he advertised emptied itself faster than he could fill it, and the clerks in that store kept busy. People drove straight through the town wherein were the bargains he had not advertised and came for miles to patronize the other store, when they might, had they known it, have bought the same goods for the same price almost at their doors. That merchant is now thoroughly converted to belief in the efficacy of the newspaper ad.

HOW TO DO BUSINESS.

RULES FOR BUYERS.

1. Buy to please your customers and not the manufacturer.
2. Give your customers the best values you can for your money.
3. Buy close enough to meet all competitors' prices.
4. Study prevailing styles—be first to show what you think will increase business.
5. Never overstock your store or department.
6. Keep posted on the different qualities and grades of your stock.
7. Be punctual in keeping business engagements.
8. Buy from those you know to be strictly just.
9. Do not trust to memory—keep a memorandum.
10. Never accept favors from drummers.
11. Never talk of employer's business to outsiders.
12. Keep posted on methods of up-to-date business houses.

RULES FOR EMPLOYES.

1. Get the confidence of your employer.
2. Be honest and accurate.
3. Be pleasant to rich and poor alike.
4. Never misrepresent goods to customers.
5. Be punctual as to business hours.
6. Meet all customers half way.
7. Study your stock—keep it clean and in place.
8. Always work for your employer's interest.
9. Make no engagements for business hours.
10. Always keep good company outside the store.
11. Do not insist on a customer buying what he or she does not want.
12. Do not talk too much to customers—answer their questions politely. — Sam Whitmire, in Advertising World.

T. R. Armstrong, Newbury, Ont., has opened a furniture and hardware store in Wardsville, Ont.

CORRESPONDENCE

A SCHEME TO MEET DEPARTMENT STORE COMPETITION.

Editor HARDWARE AND METAL,—I read with interest in HARDWARE AND METAL of March 9, of the united action the merchants of Wolfville, N.S., are taking to fight the large city departmental stores, which, I think, a step in the right direction, and is most commendable. But, in my opinion, they do not go far enough, as they have only one bargain day in the year, while the departmental stores are advertising a bargain day every week in the year.

The best way to fight departmental stores is with their own thunder, or, in other words, with their own weapons.

I would like to suggest a method to the readers of HARDWARE AND METAL, which, I think, would be most effectual in combating the influence of the big departmental stores in country places. The big departmental stores depend upon the city trade as their paying trade, and in the city, with the exception of a few bargain-day cut prices, which are mostly for out-of-season goods, remnants and other damaged or short-measure goods, such as wall paper, spools, tape, twine, yarn, hemp, which have not got the full complement of feet or yards as are usually made for and sold by the general trade, and which are made expressly for departmental store leaders, they keep up the price of their goods in the city, and, in some cases, especially with the wealthy class, they get fabulous profits. All their cut-throat business is done in the country, with the exception of goods carried by a few specialty merchants in the cities, such as druggists, grocers, confectioners and jewellers, whom they endeavor to wreck as speedily as possible by making leaders of their best-paying lines, when they know they cannot depend upon the steady and constant trade of the country people.

The method I would suggest is that all the first-class country towns, where they have no departmental stores, such as Woodstock, Stratford, St. Catharines, Galt, Paris, Berlin, Perth, Smith's Falls, Almonte, Pembroke, and all the small villages unite as one firm or company and get up a first-class catalogue, such as Eaton gets up, with cuts and prices of all the goods actually sold. In this catalogue I would place all kinds of goods suitable for all classes of city trade, the cost of catalogue to be borne by all the merchants of the town or by a dozen of towns combined. To put the price of all goods listed therein right down to cost, with only margin enough added to pay for cost of catalogues, freight on goods from Montreal, Toronto, etc., and other incidentals. Flood the cities with these catalogues; pursue exactly the same method in the cities that Eaton and others do in the country. Keep up the prices at home, and sell at cost or nearly at cost in the cities.

The goods could be imported direct to the town from where the catalogues are issued or you could send an agent to the city. Say, one to Montreal and one to Toronto and open a sample room and have a sample of all goods listed in your catalogue to choose from. He could also see faithfully to the distribution of the catalogues to every family in the city; and as the orders came in, which would only be filled for spot cash, they could be sent to the towns, or they could be sent in direct to the wholesale house or manufacturer in the city, saving the cost of express or freight on the goods.

If all the country merchants in the country were to combine and pursue some such method as above suggested, I am of the opinion that if it was faithfully carried out they could wreck every departmental store in the country inside of two years. As self preservation is the first law of nature, it behooves us to wreck them, as they are apparently going to wreck us or reduce our business to such an unprofitable condition as to deprive us of anything but the most meagre existence. SUBSCRIBER.

Cobden, Ont., March 25.

TERM "PENNY" AS APPLIED TO NAILS.

Editor HARDWARE AND METAL,—Is it not high time manufacturers discarded the obsolete term "dy" in marking nail kegs?

Country merchants never order nails in this old-fashioned way (for some have been known to get 4-inch for 4 dy.), and it's very mystifying to the rising generation who have enough to learn about hardware without anything superfluous.

This expression originated in England many years ago, when 3-inch nails were made 100 to the lb.; thus, a ten-penny nail being 1,000 nails to 10 lb., a six-penny nail being 1,000 nails to six lb., but to-day nails are much heavier, as you will find by weighing that there are 71 cut, or 67 wire nails in a pound of 3-inch, consequently, this expression is as useless to us now as pounds, shillings and pence.

I shall certainly push that factory's goods which is up-to-date enough to mark nails in inches.

Another good idea would be to use different colors for stencilling wire or cut nails, this would be a great advantage in distinguishing them when stored together, which is nearly always the case.

J. HARDIE,
Commercial Traveller.

Montreal, March 29, 1901.

RICE LEWIS & SON'S PAINT DEPARTMENT.

WHILE an old-established firm, Rice Lewis & Son, Limited, are always casting about for something new, like all progressive firms. They recently decided to create a ready-mixed paint department, and, although this department has only been in existence a short time, the firm is more than pleased with the results.

"It is," explained a member of the firm, "an entirely new departure with us, and the results have been gratifying beyond our expectations. I do not know that we ever created a department that took better. The repeat orders we have received clearly indicate that it is now beyond the experimental stage. In some instances we have duplicated and trebled our original specifications."

The brand under which the paint is put upon the market is the "Jewel." The paint is in tins and fractions of tins. The peculiar qualities claimed for "Jewel" paints are their economy, owing to the large space they will cover, and freedom from blistering.

MAKE YOURSELF INDISPENSABLE.

A man should always stand on his own feet, take advantage of opportunities, and be honest and diligent. To succeed, you must make yourself indispensable and not set a limit to the time of your working hours, but do your work to the best of your ability and let pleasure be of secondary importance. The right type of man finds pleasure in his work, and employers are looking for such. Men who compel recognition by their work cannot be restrained from forging ahead. It is not always the man who is smartest who makes the greatest advancement; it is he of bulldog tenacity, he who cannot be discouraged and never gives up.—C. M. Schwab.

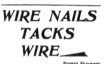

BUSINESS CHANGES.

DIFFICULTIES, ASSIGNMENTS, COMPROMISES.

MOHR & RYAN, general merchants and lumber dealers, Killaloe Station, Ont., have assigned to Alex. P. Mutchmor.

Alp. Letourneau, general merchant, Little Mechins, Que., has assigned.

Alex. Cameron, general merchant, North Lochaber, N.S., has assigned.

H. Cairns, general merchant, Sawyerville, Que., is offering 50c. on the dollar.

L. G. Jourdain, hardware dealer, Three Rivers, Que., is offering 30c. cash on the dollar.

A meeting of the creditors of Paul Bissonette, general merchant, Casselman, Ont., has been held.

A statement of the affairs of Flavien Paquette, general merchant, Paquetville, Que., is being prepared.

A meeting of the creditors of Messrs. George A. Rollins, tinware dealers, etc., Madoc, Ont., has been called for to-day (Friday).

PARTNERSHIPS FORMED AND DISSOLVED.

S. Prevost & Co., hardware dealers, Montreal, have dissolved.

Alex. Young, hardware dealer, Wingham, Ont., has admitted his sons into the business.

Hodgkinson & Fisher, blacksmiths, Kincardine, Ont., have dissolved. Joseph Fisher continues.

Hermeston & Nichol, hardware dealers, Walkerton, Ont., have dissolved. W. J. Hermeston continues.

The Arlington Bicycle Co., Montreal, have dissolved. Robt. A. Radmore has registered as proprietor.

Peter McKay, and David Johnston have registered partnership as general merchants under the style of McKay & Co., Thorburn, N.S.

SALES MADE AND PENDING.

Henry Living, hardware dealer, Ottawa, has sold out.

The stock of B. J. Petteres, machinist, Montreal, has been sold.

H. W. Murphy, general merchant, Auburn Station, N S., has sold out.

The assets of Hugh MacColl, electrical contractor, etc., Ottawa, are for sale.

The assets of Joseph Bernier, carriage-maker, L'Islet, Que., are to be sold April 16.

J. C. Price, general merchant, Ridgetown, Ont., is advertising his business for sale.

The stock of the estate of Ruth Perry, general merchant, Gad's Hill, Ont., is to be sold.

CHANGES.

Mary Mills, general merchant, Orwell, Ont., has sold out to Joseph Slade.

Wm. Brown, hardware dealer, Belmont, Ont., has sold out to McPherson Bros.

Eliza Betts, general merchant, Mount Brydges, Ont., has been succeeded by Chas. McGregor.

T. N. Wheeler, general merchant, Shedden, Ont., is about to remove to London.

Butchart & Somersall, hardware dealers, Carman, Man., have been succeeded by Butchart & Co.

J. & R. Young, general merchants, Tracadie, N.B., have been succeeded by Charles Robinson.

The stock of J. H. McMillan & Co., grocer, Nanaimo, B.C., has been sold to Hugh A. McMillan.

John Price, general merchant, etc., Port Stanley, Ont., has sold his general store business to W. A. Day.

Joseph Dore has registered as proprietor of Leduc & Dore, coal and wood dealers, St. Henri de Montreal.

The business of George R. Garrett, general merchant, Murray Harbor South, P.E.I., is to be sold out.

Isaac Hirtle, grocer, Lunenburg, N.S., has registered consent for his wife, Tamon Hirtle, to do business in her own name.

J. Ritter, hardware dealer, Newton, Ont., has sold out to Wm. Attig.

C. Lajole & Cie have registered as coal and wood dealers in Montreal.

F. L. Hamilton, blacksmith, Cromarty, Ont., has sold out to H. M. Lachlin.

Bain & Hunter, bicycle dealers, etc., Hamilton, Ont., have sold out to A. D. Wise.

John Steel, stove and tinware dealer, Dunnville, Ont., has been succeeded by Chas. Herring.

FIRES.

The Shuswap Milling Co., sawmillers, Kamloops, B.C., have been burned out.

DEATHS.

Wm. Pendrigh, brass founder, Yarmouth, N.S., is dead.

John Carruthers, president and manager of the Prescott Emery Wheel Co., is dead.

W. F. Dibblee & Son, hardware dealers, Woodstock, N.B., are opening a hardware store in St. John.

McCurdy Bros., who, as reported in HARDWARE AND METAL of March 23, succeeded A. J. Jeffrey, hardware dealer, Stratford, Ont., desire to extend through this paper thanks to the trade who were their customers while they were on the road for the Hobbs Hardware Co., London, Ont., for past courtesies.

A NOVEL SELF-HEATING SAD IRON.

ONE of the drawbacks to the more general introduction of the self-heating sad iron is the necessity of using as fuel a high-grade gasoline, which, in country villages, cannot always be readily obtained. The drawback has been simply enough avoided in a new sad iron invented by Mr. Iver Wickland, of West Superior, Wis. Besides overcoming the difficulty mentioned, the inventor has also devised a burner which completely consumes the vapor formed, and has provided a generator which maintains a constant pressure.

The oil reservoir is supported at one end of the iron. From the lower end of the reservoir a tube extends through the lower portion of the iron beneath a metal shield. The tube is connected with a retort in which the oil is received for the generation of gas. The retort communicates with a valve-casing provided with a needle valve and arranged to discharge the vapor in a burner tube located directly above the shield. Openings in the lower portion of the burner tube discharge the gas downward on each side of the shield. The forward portion of the iron constitutes a cup for oil. When ignited this oil will heat the retort and adjacent parts sufficiently to generate vapor. By opening the needle valve the vapor is mixed with air, forced into the burner tube and ignited.

The iron can be kept hot for more than 60 hours at a cost of 20c. The heat can be regulated as desired. Only one iron is required for the house laundry. No stove is needed to heat several irons. The combustion of the gas is so complete that no odor is perceptible.—Scientific American.

ELECTRIC GENERATORS BOUGHT IN THE UNITED STATES.

M. P. Davis, of Quebec, Canada, who has obtained the contract for the construction of the Shieks Island plant, which is to be utilized for operating the canal docks in that vicinity is now placing a number of important contracts for equipment, and so forth. The plant, which is to be a water-power one, will develop 4,000 units. The Bullock Electric Manufacturing Company, of Cincinnati, Ohio, has secured the contract for the building of the first of four three phase electric generators to be installed in the plant. It is estimated that fully $500,000 will be expended for equipment, etc.—New York Journal of Commerce.

Geo. W. Brown, Southampton, N.B., is promoting a company to build a steamer to carry freight and passengers between Fredericton and St. John, N.B. Over $5,000 of the necessary capital stock has been subscribed.

H. S. HOWLAND, SONS & CO.

WHOLESALE ONLY — 37-39 Front Street West, **Toronto.** — ONLY WHOLESALE

"NEWMARKET" Horse Clipper.

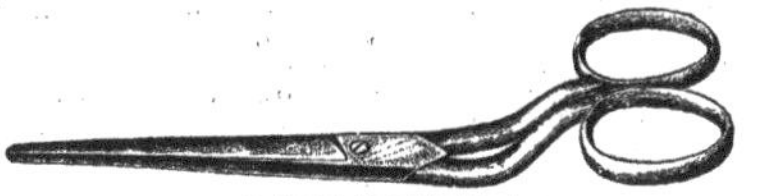

Cattle-Marking Scissors.

Boker's
"Dandy" Horse Clippers.
"Keen Cut" Horse Clippers.
"Perfection" Horse Clippers.
No. 1704. Ball-Bearing
with Improved Steel Ball-Bearing.

Also a full line of
Toilet Clippers.

Horse Clippers

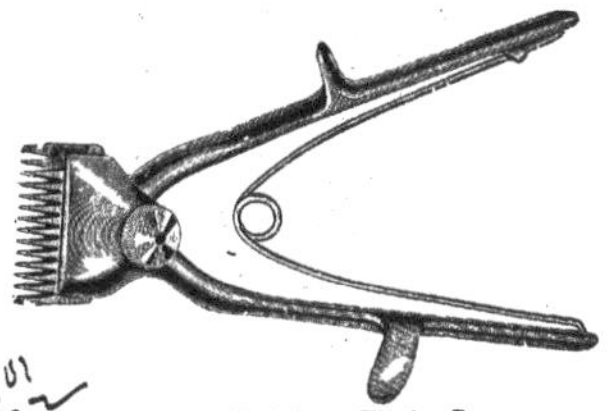

For Cutting Fetlocks or Clipping Dogs.
Coarse Teeth.

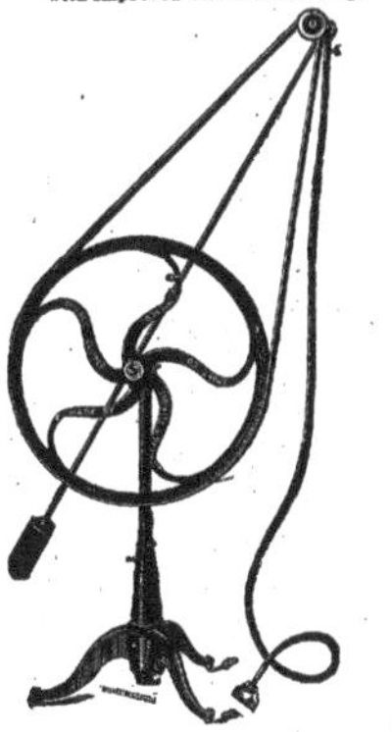

The "Lightning" Round Belt Clipper.
Has a 24-in. Balance Wheel and carries same Shaft and Knife as the 98 machine.

No. 98. Standard Machine.
Mounted on an Improved Collapsible Stand.

Chicago Flexible Horse Clippers.

Chicago Horse Clipper.
Knife and Handles.
Knife Only—Top.
Knife Only—Bottom.

H. S. HOWLAND, SONS & CO., Toronto.

WE SHIP PROMPTLY. — Graham Wire and Cut Nails are the Best. — OUR PRICES ARE RIGHT.

ROMANCE OF SHOP FITTING.

WHEN you look through a smart looking shop-window at the glittering goods inside, does it ever strike you that the window may cost more than the goods behind it? says an English paper.

There is a new boot-shop in Holborn of which the plates of glass are 16 ft. x 18 ft. These cost £50 apiece. There are five of them. They are framed in carved wood. Each frame cost £75. Behind each is an air-tight showcase. On each case £70 was spent. Each of these window-cases is decorated, and has shelves of glass five-eighths of an inch thick. Another £40. At night the whole glass front gleams with electric lamps. These cost £20 for each window. If you glance above the window you will see the name of the company in large letters on a front of stoneware. The letters are of teak, a very expensive Indian wood, and are bolted to the bed they rest on. Another £150 went into this one item.

If you saw one of those huge sheets of plate-glass leaning against a wall in Messrs. Sage's factory you would find that, by pressing your finger on the centre, the glass, thick as it is, was not rigid. There would be a play of three or four inches. Try the same experiment with the same glass in the window. It is stiff as a wall. But your eyes will be sharp if you can see the almost invisible glass tube which props the window from inside.

Jewellers' shops cost most to fit. There is in Tiffany's a slab of porphyry which had to be specially quarried for them in Norway. One of those revolving showcases which you see in jewellers' windows for the display of rings costs £80. To this must be added £16 for the motor which turns the cone, and several shillings apiece for the fifty-four electric lamps which light it.

Fancy shops run jewellers very close. Four thousand pounds was recently spent on a Bond street shop merely in alterations. There is a fancy goods shop, recently rebuilt, in St. Paul's Churchyard, of which the interior fitting of one room only cost £1,500.

Bending plate glass into segments of a circle trebles its value. Bending it into irregular shapes increases its cost six times. There is an example of this in the fancy goods shop last named. Two glass showcases, which are made to fit round and hide the steel pillars which support the roof, cost £100 apiece. All cases for jewellery and fancy goods are made absolutely air-tight, and that without the employment of rubber. Another thing which adds to the expense of showcases is that they must be made to stand all climates. Glass and wood both expand with heat and moisture, and contract when it is cold and dry. The wood used must be seasoned for five years at least before using, and allowance must be made for the expansion of the glass. You would not think there was room for a single bit of metal in the narrow wood moulding around a case. There is a screw every inch.

HAVE ENLARGED THEIR STORE.

Gilpin Bros., who have conducted a steadily growing hardware business in Orillia, Ont., for twelve years, have just finished enlarging and improving their store. The depth of the store has been increased to 120 feet. This great depth furnishes capacious shelf-room as the shelves run full length of both sides and along the rear to a height of eleven feet, bicycle ladders giving easy access to all. The ceilings, which are arched, are sixteen feet in height, and are finished in wood, painted white to give added light and brightness, and relieved by blue trimmings. New windows have been installed at the top and rear of the store, furnishing abundance of light to the entire premises. In addition to their hardware stock Gilpin Bros. carry plumbing supplies, graniteware and tinware, electrical goods, etc.

TO UTILIZE THE SCRAP.

The daily newspapers of Montreal published at the beginning of the week a somewhat startling story to the effect that The Thos. Davidson Manufacturing Co. would shortly establish large steel works for the manufacture of steel ingots, and hinted that in the course of time the production of this furnace would be large enough to supply the heaviest Canadian consumers of steel. There is some truth in the rumor, but it was exploded in the reports.

The fact is, The Thos. Davidson Manufacturing Co. have a good deal of waste scrap which they have decided to utilize by having it made into steel ingots for their own use. A little furnace of 10 tons a day capacity will be built in conjunction with their present works. At present nothing more comprehensive is contemplated.

OSSEKEAG STAMPING WORKS SOLD

On Tuesday the Ossekeag Stamp ng Works, Hampton, N.B., which were offered for sale under liquidation, were purchased jointly by A. E. Kemp, M P., of the Kemp Manufacturing Co., Limited, Toronto; J. D. Davidson, of the Thos. Davidson Manufacturing Co., Limited, Montreal, and Col. Gartshore, of the McClary Manufacturing Co., Limited, London. The price is reported to be about $80,000. It has not yet been decided by the purchasers whether they will continue to operate the works at Hampton or not.

THE DOOM OF THE SIXPENNY KNIFE

IT is said in Sheffield that the sixpenny pocket-knife, so dear to the heart of the British schoolboy, is doomed. With it will disappear one of the most familar lines handled by the ironmonger, and one which has yielded a satisfactory profit withal. But, owing to the increased cost of material and labor, the manufacturer, it is said, now finds it impossible to make a knife to retail at sixpence, and even the two-bladed shilling knife is threatened with extinction. Both, it is said, may cease to be made, in Sheffield at least, before the end of the year. It should be understood that we are now referring to knives that will cut, and are fitted with best forged blades. Hitherto the typical Sheffield sixpenny knife has been an article of utility, fit to point a lead-pencil or whittle a stick. If the makers will use flied blades and a commoner steel, they may still make a knife to retail at sixpence, but the ancient repute of the article will fall away, and ere long the youth of Britain will turn from the thing in disgust.

It is probable that, in the future, the whilom sixpenny and shilling knives will be advanced to 8d. and 1s. 3d. respectively, at which prices the maker will be able to pay higher rates to his workmen and reap a moderate profit for himself. The makers say that, so far as this important branch of their trade is concerned, they have for years been giving the public too much value for the money; "change for sixpence," in fact. Moreover, labor is scarce and machinery cannot be used in the production of such patterns of these knives as are bought in this country, while an enormous number of patterns must be stocked, and most the orders are for microscopic quantities of many kinds, causing much trouble and delay—few firms having every pattern ordered finished in stock. The threatened departure from such even prices as sixpence and a shilling will, we fear, prove inconvenient to the ironmonger, who will naturally view the change with disfavor. We suppose that knives to retail at these prices will continue to be asked for, and will somehow be supplied in future; but most buyers in this country will probably prefer to pay a little more for a sound article. It is satisfactory to know that German competition in these cheap goods is not nearly so dangerous a factor nowadays as it was a few years ago. The cost of production has proportionately increased more in Germany than in Sheffield.—Ironmonger.

GAVE THE WRONG FIRM NAME.

In the report of the conference of the Montreal wholesale and retail hardware dealers in last week's issue, reference was made to the presence of Mr. A. M. St. Arnaud at the conference. The paragraph referred to him as of "The Canada Harness Co.," whereas the firm Mr. St. Arnaud is a member of and represented at the conference is the Canada Hardware Co., Limited, Montreal.

The saw and grist mills of John B. Adam, general merchant, etc., Kilburn, N. B., have been burned; insured for $3,000.

PORTLAND CEMENTS

Best German, Belgian and English Brands.

Fire Bricks, Fire Clay, Flue Linings.
Flue Covers, Drain Pipes
Scotch and Canadian Tiles,
Calcined Plaster,
Granite Hard Wall Plaster,
Wheelbarrows,
Scotch Cranes and Derricks,
Cement Concrete Mixers, Garden Vases,
Ornamental Chimney Tops,
Mortar Stains.

A full stock of Builders' and Contractors' Supplies.
Write for Quotations.

W. McNALLY & CO., Montreal

JAS B CAMPBELL. WILLIAM PRATT.

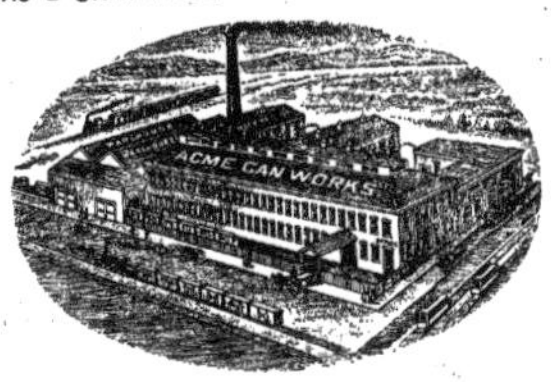

The Acme Can Works

Manufacturers of

Paint and Color Cans, round and square, Varnish and Oil Cans, Paint Irons, Paint Packages, Lye Tins,

and every description of Tin or can required by the trade. Write us for anything you need in our line. We will be pleased to quote you prices.

OFFICE AND FACTORY
Ontario St. and Jeanne D'Arc Avenue, - - MONTREAL.

THE MAN WHO PULLS THE CHAIN

WOULD APPRECIATE THE

Yale Chain Blocks.

Yale Differential Block.

Load always self sustained. One man can lift 800 lbs. four feet per minute. Three men pull 216 lbs. and handle 30 feet of chain.

The Differential are durable, smooth and easy working. For occasional use.

The Duplex for general use.

The Triplex, most efficient block made, twice the efficiency of the screw-gear type, and triple that of the Differential Type.

A STOCK CARRIED.

Yale Triplex Block.

Load always self sustained. One man can lift 2,000 lbs. four feet per minute. One man pulls 82 lbs. and handles 31 feet of chain.

THE FAIRBANKS COMPANY

749 Craig Street, MONTREAL, P.Q.

Markets and Market Notes

QUEBEC MARKETS.

Montreal, April 4, 1901.

HARDWARE.

THE season's trade has commenced in earnest this week when the summer freight rates went into force. These rates are not as low as expected, the extra 25 per cent. that was levied last year still being quoted. Yet there seems to be no hope of a reduction, as spring shipments are now being made. Business is reported to be good, although in metals the unsettled tone of the market dampers trade quite appreciably. Yet even in metals some goods must be had for immediate wants, and a considerable hand-to-mouth trade is passing. The firm tone that has been noted in wire nails during the last few weeks materialized at last week's meetings only in the raising of the price on carload lots, the difference being now only 7½c. off per keg instead of 10c. as formerly. Putty has been reduced 10c., and the discounts on tacks have been slightly raised. These changes, in addition to those mentioned last week on white lead, shot and lead pipe, constitute the alterations preparatory to the real opening of spring business The market in almost all lines is now firm. Screws and bolts were not advanced, as some anticipated, but such a move is still being mooted. All wires are selling freely, and summer goods, such as poultry netting, green wire cloth, screens, lawn mowers and hose, are moving out in good quantities. Wheelbarrows and churns are also being called for. On the whole, hardware is in a healthy condition.

BARB WIRE—Large quantities of barb wire are being shipped this week. Prices are unchanged at $3.05 per 100 lb. f.o.b. Montreal.

GALVANIZED WIRE—Fence wires are in good demand, and heavy shipments have been made this week. The market is steady. We quote: No. 5, $4.25; Nos. 6, 7 and 8 gauge, $3.55; No. 9, $3.10; No. 10, $3.75; No. 11, $3.85; No. 12, $3.25; No. 13, $3.35; No. 14, $4.25; No. 15, $4.75; No. 16, $5.00.

SMOOTH STEEL WIRE—Nos. 9, 10 and 11 are being ordered freely, and a good volume of business is the result. We quote oiled and annealed: No. 9, $2.80; No. 10, $2.87; No. 11, $2.90; No. 12, $2.95; No. 13, $3 15 per 100 lb. f.o.b. Montreal, Toronto, Hamilton, London, St. John and Halifax.

FINE STEEL WIRE—A small trade has been done this week. The market is featureless, and the discount is unchanged at 17½ per cent. off the list.

BRASS AND COPPER WIRE—These articles show no change. The discount on brass is 55 and 2½ per cent., and on copper 50 and 2½ per cent.

FENCE STAPLES—The spring demand has set in, and a good many shipments have been made this week. We quote: $3.25 for bright, and $3.75 for galvanized, per keg of 100 lb.

WIRE NAILS—The difference on carload lots of nails has been reduced from 10c. to 7½c. per keg. This denotes a strengthening market. Otherwise, prices are unchanged. The demand is not heavy, yet it is active. We quote $2.85 for small lots and $2.77½ for carlots, f.o.b. Montreal, London, Toronto and Hamilton.

CUT NAILS—There has been no change

in cut nails. Business shows a slight improvement. We quote: $2.35 for small and $2.25 for carlots; flour barrel nails, 25 per cent. discount; coopers' nails, 30 per cent. discount.

HORSE NAILS—There is not a great deal of business doing. Prices are unchanged. "C" brand is quoted at 50 and 7½ per cent. off its own list and other brands at 60 per cent. on oval and city head, and 60 and 10 per cent. on new countersunk head off the old list. It seems that the Acadian nails are not being made since the reduction in prices.

HORSESHOES—A better movement has set in this week. We quote as follows: Iron shoes, light and medium pattern, No. 2 and larger, $3.50; No. 1 and smaller, $3.75; snow shoes, No. 2 and larger, $3.75; No. 1 and smaller, $4.00; X L steel shoes, all sizes, 1 to 5, No. 2 and larger, $3.60; No. 1 and smaller, $3.85; feather-weight, all sizes, $4.85; toe weight steel shoes, all sizes, $5.95 f.o.b. Montreal; f.o.b. Hamilton, London and Guelph, 10c. extra.

POULTRY NETTING—The shipments being made his week would indicate that good orders have been booked during the last two months. The ruling discount is still 50 and 10 per cent.

GREEN WIRE CLOTH—The market is steady and business fair. The price is still $1.35 per 100 sq. ft.

SCREEN DOORS AND WINDOWS—The spring demand seems to be fully up to the average. We quote as follows: Screen doors, plain cherry finish, $8 25 per doz; do. fancy, $11.50 per doz; walnut, $7.40 per doz., and yellow, $7 45; windows, $2.25 to $3 50 per doz.

SCREWS—The tone of the market is firm and trading is free. Discounts are as follows: Flat head bright, 87½ and 10 per cent. off list; round head bright, 82½ and 10 per cent.; flat head brass, 80 and 10 per cent.; round head brass, 75 and 10 per cent.

BOLTS—The market is steady at previous quotations. Discounts are as follows: Carriage bolts, 65 per cent.; machine bolts, 65 per cent.; coach screws, 75 per cent.; sleigh shoe bolts, 75 per cent.; bolt ends, 65 per cent.; plough bolts, 50 per cent.; square nuts, 4½c. per lb. off list; hexagon nuts, 4¾c. per lb. off list; tire bolts, 67½ per cent.; stove bolts, 67½ per cent.

BUILDING PAPER—This week has seen some good shipments. Prices are steady and unchanged. We quote as follows: Tarred felt, $1.70 per 100 lb.; 2 ply, ready roofing, 80c. per roll; 3-ply, $1.05 per roll; carpet felt, $2.25 per 100 lb.; dry sheathing, 30c. per roll; tar sheathing, 40c. per roll; dry fibre, 50c. per roll; tarred fibre, 60c. per roll; O.K. and I.X.L.,

65c. per roll; heavy straw sheathing, $28 per ton; slaters' felt, 50c. per roll.

RIVETS — There is nothing new to note. The discount on best iron rivets, section, carriage, and waggon box, black rivets, tinned do., coopers' rivets and tinned swedes rivets, 60 and 10 per cent.; swedes iron burrs are quoted at 55 per cent. off; copper rivets, 35 and 5 per cent. off; and coppered iron rivets and burrs, in 5-lb. carton boxes, are quoted at 60 and 10 per cent. off list.

BINDER TWINE — A good, seasonable business is being done. We quote: Blue Ribbon, 11½c.; Red Cap, 9½c.; Tiger, 8½c.; Golden Crown, 8c.; Sisal, 8¼c.

CORDAGE—Rope is selling in moderate quantities only. Manila is now worth 13½c. per lb. for 7-16 and larger; sisal is selling at 10c., and lathyarn 10c.

SPADES AND SHOVELS — A few orders have come on this week. The discount is still 40 and 5 per cent. off the list.

HARVEST TOOLS — Some inquiries have been received this week. The discount is unchanged at 50, 10 and 5 per cent.

TACKS—The discount has been raised slightly this week. Carpet tacks, in dozens and bulk, are now sold at a discount of 80 and 15 per cent., instead of 80 and 5 per cent; tinned at 80 and 20 per cent., instead of 80 and 10 per cent.; and cut tacks, blued, in dozens, 80 per cent., instead of 75 and 15 per cent.

LAWN MOWERS—Some business is doing. We quote: High wheel, 50 and 5 per cent. f.o.b. Montreal; low wheel, in all sizes, $2 75 each net; high wheel, 11-inch, 30 per cent. off.

FIREBRICKS—As yet there have not been many inquiries received. Prices are fixed at $18 to $20 per 1,000 as to brand ex store.

CEMENT—American cement continues to hold favor for the better grade, and it would look as if German cement will be imported this year. We quote as follows: German, $2.45 to $2.55; English, $2.30 to $2.40; Belgian, $1.95 to $2.05. American in bags, $2.30 to $2.40 per 400 lb.

METALS.

The feeling on the other side of the water is reported to be better this week, and there seems to be a willingness on the part of dealers here to cooperate. Prices are unchanged. For immediate delivery there has been some better business done this week.

PIG IRON—Some of the product of the Sydney furnaces has been shown here, and is reported first class. Canadian iron is worth from $18 to $18.50, while Summerlee for early import is quoted at $20 to $20.50.

BAR IRON—It is said that some houses who stocked up at low prices have not advanced quotations to the present cost basis, which would make the price $1.70 to $1.75.

BLACK SHEETS — The English market is somewhat firmer this week, and the feeling here is a little better. For import retailers are quoted $2.40 to $2.45. For spot goods, we quote $2.60 for 8 to 16 gauge; $2.65 for 26 gauge, and $2.75 for 28 gauge.

GALVANIZED IRON — There is a better feeling dominating the galvanized iron market also. No. 28 Queen's Head, $4.60; Apollo, 10¾ oz., is worth $4.45 for immediate delivery, and Comet is quoted at $4.30 to $4.50.

INGOT COPPER—The market is steady at 18c.

INGOT TIN — The foreign markets are fluctuating about the same figure as last week. Locally, the demand is light at 31c. for Lamb and Flag.

LEAD—Some sales have occurred at $4.

LEAD PIPE — Prices are unchanged. We quote as follows: 7c. for ordinary and 7½c. for composition waste, with 25 per cent. off.

IRON PIPE — The market is firm at last week's quotations. A fair business has been done. We quote as follows: Black pipe, ¼, market is again reported to be rallying, yet, there is but little local activity. For import, coke tinplates are quoted at $4; cheap charcoal, at $4.25, and preferable grades, $4.50. For immediate delivery goods may be obtained at $4.15 to $4.25 for coke and $4.40 to $4.50 for charcoal.

CANADA PLATE—Prices of Canada plate seem to vary with the quantity of stock on hand. There is no one who wishes to invest heavily in this line at present, and there is a tendency to defer purchases. We quote: 52's, $2.60; 60's, $2.70; 75's, $2.80; full polished, $3.45, and galvanized, $4.30.

TOOL STEEL—A fair business is being done. We quote: Black Diamond, 8c.; Jessop's 13c.

STEEL—The market for steel is fairly good, the inquiries this week being numerous. We quote: Sleighshoe, $1.85; tire, $1.95; bar, $1.85; spring, $2.75; machinery, $2.75 and toe-calk, $2.50.

TERNE PLATES—There seems to be more willingness to operate, for import orders are being taken at $7. On spot they are worth $7.75 and some sales have been made below this figure.

COIL CHAIN—The market is steady under a rather smaller demand. We quote: No. 6, 11½c.; No. 5, 10c.; No. 4, 9½c.; No. 3, 9c.; ¼-inch, 7½c. per lb.; 5-16, $4.65; 5-16 exact, $5.10; ⅜, $4.20; 7-16, $4.00; ½, $3.75; 9-16, $3.65; ⅝, $3.35; ¾, $3.25; ⅞, $3.20; 1-in., $3.15. In carload lots an allowance of 10c. is made.

SHEET ZINC—The ruling price is 6 to 6¼c.

ANTIMONY—Quiet, at 10c.

ZINC SPELTER—Is worth 5 to 5½c.

SOLDER—We quote: Bar solder, 18½c.; wire solder, 20c.

GLASS.

Some have advanced quotations on glass, but the majority are clinging to the lower prices. The market is decidedly firm both for spot and import goods. We quote as follows: First break, $2; second, $2.10 for 50 feet; first break, 100 feet, $3.80; second, $4; third, $4.50; fourth, $4.75; fifth, $5.25; sixth, $5.75, and seventh, $6.25.

PAINTS AND OILS.

The English linseed-oil market has continued to advance, and it would now cost 64 and 67c. to lay oil down in Montreal, a rise of 12 to 13c. from the lowest point reached. So sharp is this turn that an advance on spot goods is being talked of. At present, they are still selling at 67 and 70c. Turpentine is steady. Putty has been reduced 10c. per 100 lb. White lead is rather quiet, but prices are steady. Shipments of all kinds of paints and oils have been somewhat larger this week, on account of the lower freight tariff that went into effect on the first of the month. We quote:

WHITE LEAD—Best brands, Government standard, $6.25; No. 1, $5.87½; No. 2, $5.50; No. 3, $5.12½, and No. 4, $4.75 all f.o.b. Montreal. Terms, 3 per cent. cash or four months.

URNED
18 1904

DRY WHITE LEAD—$5.50 in casks; kegs, $5.75.

RED LEAD—Casks, $5.25; in kegs, $5.50.

WHITE ZINC PAINT—Pure, dry, 7c.; No. 1, 6c.; in oil, pure, 8c.; No. 1, 7c.; No. 2, 6c.

PUTTY—We quote: Bulk, in barrels, $1 90 per 100 lb.; bulk, in less quantity, $2.05; bladders, in barrels, $2.10; bladders, in 100 or 200 lb. kegs or boxes, $2.25; in tins, $2.55 to $2 65; in less than 100-lb. lots, $3 f.o.b. Montreal, Ottawa, Toronto, Hamilton, London and Guelph. Maritime Provinces 10c. higher, f.o.b. St. John and Halifax.

LINSEED OIL—Raw, 67c.; boiled, 70c., in 5 to 9 bbls., 1c. less, 10 to 20 bbl. lots, open, net cash, plus 2c. for 4 months. Delivered anywhere in Ontario between Montreal and Oshawa at 2c. per gal. advance and freight allowed.

TURPENTINE—Single bbls., 60c.; 2 to 4 bbls., 59c.; 5 bbls. and over, open terms, the same terms as linseed oil.

MIXED PAINTS—$1.25 to $1.45 per gal.

CASTOR OIL—8¾ to 9¼c. in wholesale lots, and ½c. additional for small lots.

SEAL OIL—47½ to 49c.

COD OIL—32½ to 35c.

NAVAL STORES—We quote: Resins, $2.75 to $4.50, as to brand; coal tar, $3.25 to $3.75; cotton waste, 4½ to 5½c. for colored, and 6 to 7½c. for white; oakum, 5½ to 6½c., and cotton oakum, 10 to 11c.

PARIS GREEN—Petroleum barrels, 16¼c. per lb.; arsenic kegs, 17c.; 50 and 100-lb. drums, 17½c.; 25-lb. drums, 18c.; 1-lb. packages, 18½c.; ½-lb. packages, 20½c.; 1-lb. tins, 19½c.; ½-lb. tins, 21½c. f.o.b. Montreal; terms 3 per cent. 30 days, or four months from date of delivery.

SCRAP METALS.

Except in woollen rags, trade in all lines is in a healthy condition. The demoralization of the woollen industry has, however, caused a decline of about one-third in the price of woollen rags. The metal market is firm, except lead, which is somewhat lower. Dealers are paying the following prices in the country: Heavy copper and wire, 13 to 13½c. per lb.; light copper, 12c.; heavy brass, 12c.; heavy yellow, 8½ to 9c.; light brass, 6½ to 7c.; lead, 2½ to 2¾c. per lb.; zinc, 2¼ to 2½c.; iron, No. 1 wrought, $15 to $16 per gross ton f.o.c. Montreal; No. 1

cast, $13 to $14; stove plate, $8 to $9; light iron, No. 2, $4 a ton; malleable and steel, $4; rags, country, 70 to 80c. per 100 lb.; old rubbers, 6¼c. per lb.

HIDES.

There is quite a quantity of poor hides coming on the market, and they have a depreciating effect. We quote: Light hides, 7c. for No. 1; 6c. for No. 2, and 5c. for No. 3. Lambskins, 90c.

PETROLEUM.

The demand is falling off. Prices are unchanged. We quote: "Silver Star," 14½ to 15½c.; "Imperial Acme," 16 to 17c.; "S.C. Acme," 18 to 19c., and "Pratt's Astral," 18½ to 19½c.

MARKET NOTES.

Tacks are a little lower in price.

Scrap lead has been reduced ¼c.

The discount on nails in carload lots is now only 7½c. instead of 10c. per keg.

Putty has been marked down 10c. per 100 lb. in sympathy with the recent drop in linseed oil.

The English linseed oil market has reacted about 12 to 13c. from the lowest point, and spot values are stiffer.

Cleland Bros., Meaford, who are represented in Montreal by Mr. Alexander Gibb, are this year manufacturing quite a number of new lines of barrows and warehouse trucks. Catalogues may be had on application.

ONTARIO MARKETS.

TORONTO, April 4, 1901.

HARDWARE.

TRADE is good. Customers are not ordering in large quantities, but the demand is brisk for present requirements and for small lots of goods for future delivery. The manufacturers have concluded their meetings in Montreal, and the only additional changes are on carlots of wire nails, tacks and putty. In wire nails there is an advance of 2½c. per keg on carlots, in tacks a slight reduction, and in putty a reduction of 10c. per 100 lb. Wire nails, fence wires, screws, bolts, churns, wringers, spades and shovels, screen doors and windows, are all in fairly good request. A little better business seems to be doing in cutlery and sporting goods. An improvement has taken place in the demand for enamelled ware and tinware. Oil stoves are being inquired for quite freely.

BARB WIRE—A fair trade is being done. We quote f.o.b. Cleveland at $2.82½ per 100 lb. for less than carlots and $2.70 for carlots. From stock Toronto the figure is $3.05 per 100 lb.

GALVANIZED WIRE — Business in this line is only moderate. We quote as follows: Nos. 6, 7 and 8, $3.50 to $3.85 per 100 lb., according to quantity; No. 9, $2.85 to $3.15; No. 10, $3.60 to $3.95; No. 11, $3.70 to $4.10; No. 12, $3 to $3.30; No. 13, $3.10 to $3.40; No. 14, $4.10 to $4.50; No. 15, $4.60 to $5.05; No. 16, $4.85 to $5.35. Nos. 6 to 9 base f.o.b. Cleveland are quoted at $2.57½ in less than carlots and 12c. less for carlots of 15 tons.

SMOOTH STEEL WIRE—There is a little oiled and annealed wire moving, but the demand is not as good as it was this time last year. There is practically nothing doing in hay-baling wire. Net selling prices for oiled and annealed are as noted last week: Nos. 6 to 8, $2.90; 9, $2.80; 10, $2.87; 11, $2.90; 12, $2.95; 13, $3.15; 14, $3.37; 15, $3.50; 16, $3.65. Delivery points, Toronto, Hamilton, London and Montreal, with freights equalized on these points.

WIRE NAILS—The price of wire nails in carlots has been advanced 2½c. per keg. For less than carlots prices are unchanged. The difference between carlots and less than carlots is now 7½c. per keg, instead of 10c., as before. Delivery points: Toronto, Hamilton, London, Gananoque and Montreal.

CUT NAILS—No change has taken place either in prices or in business, the latter still being dull. The base price is $2.35 per keg. Delivery points are: Toronto, Hamilton, London, Montreal and St. John, N.B.

HORSE NAILS—Trade is only moderate. Discount on "C" brand oval head is unchanged at 50 and 7½ per cent. off the new list and on other brands 50, 10 and 5 per cent., off the old list; countersunk head, 50, 10 and 10 per cent.

HORSESHOES — Business is moderate. We quote f.o.b. Toronto as follows: Iron shoes, No. 2 and larger, light, medium and heavy, $3.60; snow shoes, $3.85; light steel shoes, $3.70; featherweight (all sizes), $4.95; iron shoes, No. 1 and smaller, light, medium and heavy (all sizes), $3.85; snow shoes, $4; light steel shoes, $3.95; featherweight (all sizes), $4.95.

SCREWS—A steady trade is being done. At the meeting of the manufacturers, in Montreal, no change was made in prices. We quote discounts as follows: Flat head bright, 87½ and 10 per cent.; round head bright, 82½ and 10 per cent.; flat head brass, 80 and 10 per cent.; round head brass, 75 and 10 per cent.; round head bronze, 65 per cent., and flat head bronze at 70 per cent.

BOLTS AND NUTS—The expected has not happened in regard to prices, the manufacturers having met and adjourned without making any change. The demand is active for all kinds of bolts. We quote as follows: Carriage bolts (Norway), full square, 70 per cent.; carriage bolts full square, 70 per cent.; common carriage bolts, all sizes, 65 per cent.; machine bolts, all sizes, 65 per cent. ; coach screws, 75 per cent.; sleighshoe bolts, 75 per cent.; blank bolts, 65 per cent.; bolt ends, 65 per cent.; nuts, square, 4½c. off; nuts, hexagon, 4¼c. off; tire bolts, 67½ per cent.; stove bolts, 67½ ; plough bolts, 60 per cent. ; stove rods, 6 to 8c.

RIVETS AND BURRS—A fair trade is being done We quote as follows : Iron rivets, 60 and 10 per cent. ; iron burrs, 55 per cent.; copper rivets and burrs, 35 and 5 per cent.

ROPE—A steady trade is being done this week in rope. We quote the base price as follows : Sisal 10c. per lb. ; manila, 13½c. per lb.; special manila, 11½c.; New Zealand, 10c. per lb.; single tarred lathyarn, 9½c.

CUTLERY—Some nice orders are being shipped to the Maritime Provinces, but the larger orders are going forward to the Northwest Territories and British Columbia.

SPORTING GOODS — Quite a few guns have been sold during the past week and a nice trade is being done in loaded shells. There is some demand for powder.

ENAMELLED WARE — An advance has taken place this week in the price of enamelled ware, the discount on "Granite," "Pearl," "Crescent" and "Imperial" wares now being 50, 10 and 10 per cent. An improved demand has been experienced the past week in enamelled ware.

GALVANIZED BUCKETS—A change took place in straight or improved pattern galvanized buckets this week, the discount now being 45 per cent.

GREEN WIRE CLOTH — Some forward orders were shipped during the past week. Price is unchanged at $1.35 per 100 sq. ft.

SCREEN DOORS AND WINDOWS—A few orders have been booked during the past week for later delivery. We quote 4-in. styles in doors at $7.20 to $7.80 per doz., and 3 in. styles 20c. per doz. less ; screen windows, $1.60 to $3.60 per doz., according to size and extension.

BUILDING PAPER—The demand is good, and there is now an ample supply.

POULTRY NETTING—The orders booked have been pretty well delivered. Those who bought Canadian netting obtained delivery some time ago, and those who bought British have obtained delivery within the last week or two. Discount, 55 per cent.

CHURNS, WRINGERS — Supplies in some makes of churns, washers and wringers are not equal to the demand, and some difficulty is being experienced in filling orders. Discount on churns, 58 per cent.

HARVEST TOOLS — Forward orders are being shipped. Retailers are asking for delivery of manure forks, hoes and rakes. Discount 50, 10 and 5 per cent.

SPADES AND SHOVELS — A fairly good trade is being done in spades and shovels. Discount 40 and 5 per cent.

TACKS—A reduction has taken place in the price of certain lines of tacks. The following are the lines in which the changes have been made together with the new discounts : Carpet tacks, blued, 80 and 15 per cent ; carpet tacks, tinned, 80 and 20 per cent ; cut tacks, blued, in dozens only, 80 per cent. ; Swedes, upholsterers', bulk, 85, 12½ and 12½ per cent.; copper nails, 52½ per cent.

BRIGHT WIRE GOODS—These are selling fairly well. Discount 62½ per cent.

BINDER TWINE—Prices are firm. We understand that pure manila quotations are mostly nominal, as manufacturers have experienced so much difficulty in obtaining hemp of good enough quality to make 650 ft. binder twine, that many of them, including some of the largest, have practically withdrawn prices for that grade. We quote : American—Sisal and standard, 8½c.; manila, 10½c.; pure manila, 11½ to 12c. per lb. Canadian—Sisal, 8½c.; pure manila, 11½c. per lb.

CEMENT—The demand is improving. Price of Canadian portland is 20c. lower. We quote : Canadian portland, $2.40 to $2.80; German, $3 to $3.15; English, $3 ; Belgian, $2.50 to $2.75 ; Canadian hydraulic, $1.25 to $1.50.

OIL AND GAS STOVES—A good demand has sprung up for oil stoves during the past week. A few gas stoves are being delivered in the city, but no shipments are yet being made to outside points.

REFRIGERATORS AND ICE CREAM FREEZERS—Delivery is now being asked for and shipments are being forwarded.

METALS.

The metal trade is not, perhaps, as brisk as it was a week ago. Pig tin and zinc sheets are quoted lower. Quotations on solder are a little firmer.

PIG IRON—Prices are strong and the demand is fair.

BAR IRON—The demand continues good, and the different mills are gradually working their prices up. We quote base price at $1.70 to $1.75, according to quality.

HOOP STEEL — Trade keeps good at $3.10 per 100 lb. base.

STEEL BOILER PLATES—There is just a moderate business. We quote : ¼-inch, $2.40 to $2.50, and 3 16 inch, $2.60.

PIG TIN — Prices are 1c. lower. The London market has ruled weaker, prices there having declined £1 1s. 3d. in four or five days. In New York, on the other hand, prices are steady. Locally, quotations are 1c. per lb. lower at 31 to 32c.

TINPLATES — Trade is only fair, and is about equally divided between coke and charcoal plates.

TINNED SHEETS—Not many shipments are now being made, customers having been pretty well supplied. We quote 9 to 9½c. for small lots from stock.

TERNE SHEETS — Very little is being done this line.

GALVANIZED SHEETS—Import orders are arriving, and the demand for shipment from stock has in consequence fallen off. Prompt delivery of United States sheets cannot be had, the mills over there being short of material. Where it usually takes two to three weeks to get delivery it is now something like six weeks. We quote English sheets from stock at $4.60, and United States sheets at $4.50.

BLACK SHEETS—Trade is fair and prices unchanged at $2.30 for 28 gauge.

CANADA PLATES—Business is still quiet in Canada plates. All dull, $3, half-and-half, $3.15, and all bright, $3.65 to $3.75.

COPPER—The demand for ingot copper is more active and a good trade is being experienced in sheet copper. Prices are easier, both at home and abroad. We quote as follows : Ingot, 19c.; bolt or bar, 23½ to 25c.; sheet, 23 to 23½c. per lb.

BRASS—A rather better demand is being experienced. Discount 15 per cent.

SOLDER—A good trade is being done in solder. We quote as follows : Half-and-half, guaranteed, 18½c.; ditto, commercial, 18c. ; refined, 18c., and wiping, 17½c.

IRON PIPE—Prices are unchanged at the recent advance. We quote as follows per 100 feet: Black pipe, ⅛ in., $4.35; ¼ in., $3 25; ⅜ in., $3.30; ½ in., $3.35; ¾ in., $3.60; 1 in., $5.00; 1¼ in., $6.85; 1½ in., $8.25; 2 in., $11.00; 2½ in., $22.94; 3 in., $36.12; 3½ in., $37.90; 4 in., $43.09; 5 in., $57 85; 6 in., $75 01; 7 in., $93.76; 8 in., $112 51. Galvanized pipe, ½ in., $5.52; ¾ in., $5.56; 1 in., $7.77; 1¼ in., $10.60; 1½ in., $12.70; 2 in., $16 90.

LEAD—The demand for pig lead is active. We quote 4¾ to 5c. per lb. Prices are cabled lower in London, but they are steady in New York.

ZINC SPELTER—Trade is quiet, with the outside markets steady as to price. We quote 6 to 6¼c. per lb.

ZINC SHEET—Prices are lower. We quote casks at 6½c. and part casks at 6¾c. per lb.

ANTIMONY—Trade is quiet and prices unchanged at 11 to 11½c. per lb.

PAINTS AND OILS.

Though the weather has not been condusive to brisk business, there has been a very fair movement. Prepared paints, varnishes and general sundries are moving nicely, but the demand for linseed oil, turpentine, white lead and paris green is moderate. Putty is 10c. lower. Red lead is 25c. cheaper. Other lines are unchanged here. There will likely be an advance in whiting. Linseed oil is stiffening in Great Britain. We quote as follows:

WHITE LEAD—Ex Toronto, pure white lead, $6 37½; No. 1, $6; No. 2, $5.67½; No. 3, $5.25; No. 4, $4 87½; genuine dry white lead in casks, $5.37½.

RED LEAD—Genuine, in casks of 560 lb., $5.25; ditto, in kegs of 100 lb., $5.50; No. 1, in casks of 560 lb., $5; ditto, kegs of 100 lb., $5 25.

LITHARGE—Genuine, 7 to 7½c.

ORANGE MINERAL—Genuine, 8 to 8½c.

WHITE ZINC—Genuine, French V.M., in casks, $7 to $7.25; Lehigh, in casks, $6.

PARIS WHITE—90c. to $1 per 100 lb.

WHITING—70c. per 100 lb.; Gilders' whiting, 80c.

GUM SHELLAC—In cases, 22c.; in less than cases, 25c.

PARIS GREEN—Bbls., 16¾c.; kegs, 17c.; 50 and 100 lb. drums, 17½c.; 25-lb. drums, 18c.; 1-lb. papers, 18½c.; 1-lb. tins, 19½c.; ½-lb. papers, 20½c.; ½-lb. tins, 21½c.

PUTTY—Bladders, in bbls., $2.10; bladders, in 100 lb. kegs, $2.25; bulk in bbls., $1.90; bulk, less than bbls. and up to 100 lb., $2.05; bladders, bulk or tins, less than 100 lb., $2.90.

PLASTER PARIS—New Brunswick, $1.90 per bbl.

PUMICE STONE—Powdered, $2.50 per cwt. in bbls., and 4 to 5c. per lb. in less quantity; lump, 10c. in small lots, and 8c. in bbls.

LIQUID PAINTS—Pure, $1.20 to $1.30 per gal.; No. 1 quality, $1 per gal.

CASTOR OIL—East India, in cases, 10 to 10½c. per lb. and 10½ to 11c. for single tins.

LINSEED OIL—Raw, 1 to 4 barrels, 69c.; boiled, 72c.; 5 to 9 barrels, raw, 68c.; boiled, 71c., delivered. To Toronto, Hamilton, Guelph and London, 1c. less.

TURPENTINE—Single barrels, 58c.; 2 to 4 barrels, 57c., to all points in Ontario. For less quantities than barrels, 5c. per gallon extra will be added, and for 5-gallon packages, 50c., and 10 gallon packages, 80c. will be charged.

GLASS.

The advance anticipated last week went into effect on Monday. Advices received from Belgium show that the market there is steadily advancing because of the heavy demand and the curtailment of production by the strike. We quote: Under 26 in., $4.15 26 to 40 in., $4 45; 41 to 50 in., $4.85; 51 to 60 in., $5 15; 61 to 70 in., $5.50; double diamond, under 26 in., $6; 26 to 40 in., $6.65; 41 to 50 in., $7.50; 51 to 60 in., $8.50; 61 to 70 in., $9.50, Toronto, Hamilton and London. Terms, 4 months or 3 per cent. 30 days.

OLD MATERIAL

The market is dull with prices unchanged. We quote jobbers' prices: Agricultural scrap, 55c. per cwt.; machinery cast, 60c. per cwt.; stove cast, 50c.; No. 1 wrought 50c. per 100 lb.; new light scrap copper, 12c. per lb.; bottoms, 11½c.; heavy copper, 13c.; coil wire scrap, 13c.; light brass, 7c.; heavy yellow brass, 10 to 10½c.; heavy red brass, 10½ to 11c.; scrap lead, 3c.; zinc, 2c; scrap rubber, 6½c.; good country mixed rags, 65 to 75c.; clean dry bones, 40 to 50c. per 100 lb.

COAL.

It is reported that prices are easier at the mines, but as no official notice of a reduction has been received here no change is yet noted. It is expected that opening prices will be given out shortly. We quote anthracite on cars Buffalo and bridges: Grate, $4.75 per gross ton and $4.24 per net ton; egg, stove and nut, $5 per gross ton and $4.46 per net ton.

MARKET NOTES.

Zinc sheet is quoted lower.

Pig tin is quoted 1c. per lb. lower.

Some descriptions of tacks have been reduced in price.

The price of wire nails has been increased to $2.77½ per keg.

H. S. Howland, Sons & Co. are in receipt of a shipment of "Star" nail pullers.

Putty is 10c. lower. Red lead is 25c. lower. Whiting is likely to advance shortly.

The discount on straight or improved-pattern galvanized buckets, Nos. 12, 14 and 16, has been fixed at 45 per cent.

The price of enamelled ware has been advanced, the discount on "Crescent," "Imperial," "Pearl" and "Granite" now being 50, 10 and 10 per cent.

H. S. Howland, Sons & Co. have taken into stock a shipment of Boker's ball-bearing horse clippers, which have been scarce. They have also taken into stock "Newmarket" horse clippers.

HAVE INSTALLED A FOUNDRY.

The Vancouver Engineering Co., who have succeeded Armstrong & Morrison, manufacturers of steel pipes, etc., Vancouver, are installing a complete brass and iron foundry equipment and extending the works generally. They have secured a staff of engineers and draughtsmen, and will undertake all sorts of design work, especially marine work.

TIN CANS ARE DEARER IN THE UNITED STATES.

As a result of the tin-can combination in the United States, the price of 2-lb. cans has been advanced to $1.80 per 100, and on 3-lb. tins to $2.40 per 100.

HEATING AND PLUMBING

PLUMBERS' WAGES AND HOURS OF LABOR.

TORONTO MASTERS AND MEN REACH AN AGREEMENT.

FOR several weeks a joint committee, composed of Messrs. J. H. Wilson, Wm. Mansell, W. J. McGuire, Kenneth J. Allison and Henry Hogarth, representing the Toronto Master Plumbers', Steam and Gas Fitters' Association, and Messrs. Arthur Davies, W. N. Brogbon, David Bell, Terry McCann and Herbert Johnson, representing the Toronto Journeymen Plumbers', Steam and Gas Fitters' Union, have been meeting in conference with the purpose of drawing up a new agreement between these two bodies regarding the hours of labor, wages and the relationship generally between employer and employe in the trade.

On Friday night, last week, the joint committee completed its labors and the new agreement was signed by all the members. The agreement is as follows:

Clause 1—That the hours of labor shall be from 8 a.m. to 5 p.m. with one hour for dinner; Saturdays from 8 a.m. to 12 noon.

Clause 2—That the minimum rate of wages shall be 27½c. per hour; the men now receiving 27½c. to get 30c.; those getting 30c. to get 32½c. per hour, which shall be the maximum rate.

Clause 3—All overtime to be paid at the following rate: Saturday afternoon, time and a half; from 5 p.m. to 8 a.m., time and a half; statute holidays and Sundays, double time.

Clause 4—Wages to be paid before 5.15 p.m. on Fridays or 12.15 noon on Saturdays on pay week, with a recommendation that every second Friday be pay day.

Clause 5—Time spent to and from out-of-town work during working hours only to be paid at usual rate of wages, and if travelling all night a sleeper berth to be provided by employer.

Clause 6—All expenses for board and fares on out-of-town work to be paid by employer.

Clause 7—In case of any grievance arising, said grievance shall be referred to a committee of five from each association; committees to meet within three days' notice thereof, and to have full power to settle disputes.

Clause 8—That no plumber, steam or gas fitter shall perform any labor pertaining to his trade or put in any material supplied by or for any other person than his employer, who shall be a bona-fide master plumber, and a member of the Master Plumbers', Steam and Gas Fitters' Association; but no enforcement of this clause to take place before being sanctioned by a joint committee of the grievance committee of both associations. Three months will be given from date of signing this agreement before it shall be enforced.

Clause 9—The members of the Master Plumbers', Steam and Gas Fitters' Association shall employ none but members of the Journeymen Plumbers', Steam and Gas Fitters' Union.

Clause 10—That no boy at either trade be allowed a kit of tools until he has served three years at the trade.

Clause 11—That the time for enforcement of this agreement takes effect from April 1, 1901, and to stay in force until January 1, 1904. In case either party to this agreement wishes to change, add to, or amend the above, they shall be given at least three months' notice in writing prior to the termination of this agreement.

There are many changes in this agreement from the old, which was drawn up on July 4, 1899. The hours of labor were formerly 9 hours per day from April to November inclusive. The minimum rate was 25c. per hour, but there was no maximum rate. Statute holiday work was paid for at the rate of time and a half. The provision that employers shall provide a sleeper berth for employes traveling to and from work by night is altogether new, as are also clause 9 and that part of clause 8 by which the journeymen of the union agree to work for no other employer than those belonging to the Master Plumbers' Steam and Gas Fitters' Association.

CAUGHT STEALING LEAD PIPE.

On Thursday last week Joseph Boucher and Raoul Routhier were caught while stealing plumbing from the house of C. P. Rouleau, 36 Stewart street, Ottawa. The police, on entering the house, found four boys in an upper room. Two of them escaped, but the two named above were caught.

The officers found a bag of plunder, which included lead pipe, brass taps and gas fixtures. The pipe weighed about 150 lb., and was cut into short pieces. The plumbing of the house, from cellar to attic, is ruined. The owner, Mr. Rouleau, estimates the damage at $200. He had just had the plumbing system repaired.

BUILDING PERMITS ISSUED.

Building permits have been issued in Hamilton, Ont., to J. H. Larkin for three brick dwellings on Stuart street between Bay and Caroline streets, to cost $3 900, to W. P. Witton, for a $5,000 wing to the city hospital. The total permits issued in Hamilton during March were rated at $29,455, against $10,000 last year.

C. A. Heinrich, who has conducted a tinware business in Waterloo, Ont., for several years, has sold his stock to Mr. Geiger, of New Hamburg, Ont., who will continue the business in Waterloo.

SOME BUILDING NOTES.

DR. CHAMBERS has let the contract for the building of a new brick store to Mr. McGrath, Souris, Man.

Dr. R. Jelly intends erecting a brick residence in Harrietsville, Ont.

G. Connors will erect a new house in Harlowe, Ont., this summer.

Paul Chapelle, Bradford, Ont., is building a general store in Brown Hill, Ont.

The Episcopalians of Elgin, Man., intend erecting a new church this summer.

A new Presbyterian church will be erected at Burnt Lake, N.W.T., this summer.

A public library building will probably be erected in Napanee, Ont., this summer.

Mr. Lyttle is building a store and dwelling house at Knee Hill Valley, N.W.T.

Thos. Farlow intends erecting new buildings on the McIntosh property, Milton, Ont.

The Murray street Baptist church, Peterboro', Ont., will erect a 22 x 46 ft. addition, to cost about $800.

John Letter, contractor, Waterloo, Ont., intends erecting a two storey, 100 x 24½ ft., business building in that place.

Rev. M. Addison, Surrey, N.B., is asking tenders for the erection of a church at Albert Mines, N.B., before April 20.

J. A. Neff, Ingersoll, Ont., is asking tenders for the erection of a two storey business block, 82½ x 26 feet, in Ingersoll, Ont.

Tenders are asked by to morrow (Saturday) for the erection of a large hotel in the Muskoka lake district by Beaumont Jarvis, architect, Toronto.

John McMullen, proprietor of the Revere House, Brockville, Ont., has decided on the expenditure of about $10,000 in renewing and improving his hotel.

At a meeting of the Catholic School Board, Montreal, the contract for St. Gabriel school was awarded to Raymond Chartrand, of that city, at $26,799.50.

Tenders are asked by April 18, by W. H. Shillinglaw, Brandon, Man., or Thomas Miller, Moose Jaw, N.W.T., for the erection of a Presbyterian church at Moose Jaw.

Contracts have been let for six houses to be erected in Waterloo, Ont., this summer. A. K. Roesch will build two; Mr. Huhn, of Heidelberg, Ont., two; John Wendell, one, and John Nichill, one.

Base Ball
Lawn Tennis
Lacrosse
Foot Balls
Boxing Gloves
Striking Bags
Croquet Sets
Hammocks
Air Rifles
Toy Pistols
Toy Garden Sets
Sand Pails
Sail Boats

Summer Catalogue Ready.

WRITE FOR IT.

Flags
Paper Lanterns
Splint Baskets
Velocipedes
Tricycles
Toy Carts
Barrows
Baby Wagons
Baby Carriers
Express Wagons
Doll Cabs
Doll Go-Carts
Baby Swings

NERLICH & CO.

146-148 Front Street West,
(Opposite Union Station)
TORONTO

ESTABLISHED
1858.

ELEVATORS

LEITCH & TURNBULL'S, HAMILTON, CANADA.

INCORPORATED
1900.

Our reputation comes from **Canada Elevator Works.** the satisfactory working of our

High-Class Passenger and Freight Elevators

found in the majority of the large Public Buildings, Factories, and Warehouses in the Dominion of Canada, and other countries.

Toronto Municipal Buildings.

The LEITCH & TURNBULL CO., Limited, Hamilton, Can.

MANITOBA MARKETS.

WINNIPEG, April 2, 1901.

THE volume of business has much improved since last writing, and the demand for shelf and heavy hardware is now active. As before noted, there is a special demand for building hardware.

In paints and oils, no change of price has occurred during the week, but there is a feeling that linseed oil will go lower. There has probably never been a more active demand for lumber than there is at present.

Hardware prices and paints and oils are as follows :

Barbed wire, 100 lb.	$3 45
Plain twist	3 45
Staples	3 95
Oiled annealed wire 10	3 95
" 11	4 00
" 12	4 05
" 13	4 20
" 14	4 35
" 15	4 45
Wire nails, 30 to 60 dy, keg	3 45
" 16 and 20	3 50
" 10	3 55
" 8	3 65
" 6	3 70
" 4	3 85
" 3	4 10
Cut nails, 30 to 60 dy.	3 00
" 20 to 40	3 05
" 10 to 16	3 10
" 8	3 15
" 6	3 20
" 4	3 30
" 3	3 65
Horsenails, 45 per cent. discount.	
Horseshoes, iron, No. 0 to No 1	4 65
No. 2 and larger	4 40
Snow shoes, No. 0 to No. 1	4 90
No. 2 and larger	4 40
Steel, No. 0 to No. 1	4 95
No. 2 and larger	4 70
Bar iron, $2.50 basis.	
Swedish iron, $4.50 basis.	
Sleigh shoe steel	3 00
Spring steel	3 25
Machinery steel	3 75
Tool steel, Black Diamond, 100 lb	8 50
Jessop	13 00
Sheet iron, black, 10 to 20 gauge, 100 lb.	3 50
20 to 26 gauge	3 75
28 gauge	4 00
Galvanized American, 16 gauge	2 54
18 to 22 gauge	4 50
24 gauge	4 75
26 gauge	5 00
28 gauge	5 25
Genuine Russian, lb	12
Imitation " "	8
Tinned, 24 gauge, 100 lb	7 55
26 gauge	8 80
28 gauge	8 00
Tinplate, IC charcoal, 20 x 28, box	10 75
" IX "	12 75
" IXX "	14 75
Ingot tin	35
Canada plate, 18 x 21 and 18 x 24	3 75
Sheet zinc, cask lots, 100 lb	7 50
Broken lots	8 00
Pig lead, 100 lb.	6 00
Wrought pipe, black up to 2 inch....50 an	10 p.c.
" Over 2 inch	50 p.c.
Rope, sisal, 7-16 and larger	$10 00
" ⅜	10 50
" ¼ and 5-16	11 00
Manila, 7-16 and larger	13 50
" ⅜	14 00
" ¼ and 5-16	14 50
Solder	21½
Cotton Rope, all sizes, lb.	16
Axes, chopping	$ 7 50 to 12 00
" double bitts	12 00 to 18 00
Screws, flat head, iron, bright	87½
Round " "	82½
Flat " brass	80
Round " "	75
Coach	57½ p.c.
Bolts, carriage	55 p.c.
Machine	55 p.c.
Tire	60 p.c
Sleigh shoe	65 p.c.
Plough	40 p.c.
Rivets, iron	50 p.c.
Copper, No. 8	35
Spades and shovels	40 p.c.
Harvest tools	50, and 10 p.c.
Axe handles, turned, s. g. hickory, doz.	$2 50
No. 1	1 50
No. 2	1 25
Octagon extra	1 75
No. 1	1 25
Files common	70, and 10 p.c.
Diamond	60
Ammunition, cartridges, Dominion R.F.	50 p.c.
Dominion, C.F., pistol	30 p.c.
" military	15 p.c.
American R.F.	30 p.c.
C.F. pistol	5 p.c.
C.F. military	10 p.c. advance.
Loaded shells:	
Eley's soft, 12 gauge black	16 50
chilled, 12 guage	18 00
soft, 10 guage	21 00
chilled, 10 guage	23 00
Shot, Ordinary, per 100 lb	6 75
Chilled	7 50
Powder, F.F., keg	4 75
F.F.G.	5 00
Tinware, pressed, retinned	75 and 2½ p.c.
" " plain	70 and 15 p.c.
Graniteware, according to quality	50 p.c.

PETROLEUM.

Water white American	25½c.
Prime white American	24c.
Water white Canadian	22c,
Prime white Canadian	21c.

PAINTS, OILS AND GLASS.

Turpentine, pure, in barrels	$ 68
Less than barrel lots	73
Linseed oil, raw	77
Boiled	80
Lubricating oils, Eldorado castor	25½
Eldorado engine	24½
Atlantic red	27½
Renown engine	41
Black oil	23½ to 25
Cylinder oil (according to grade)	55 to 74
Harness oil	61
Neatsfoot oil	$ 1 00
Steam refined oil	85
Sperm oil	1 50
Castor oil ... per lb.	11½
Glass, single glass, first break, 16 to 25 united inches	2 25
26 to 40 ... per 50 ft.	2 50
41 to 50 ... " 100 ft.	5 50
51 to 60 ... " " "	6 00
61 to 70 ... per 100-ft. boxes	6 50
Putty, in bladders, barrel lots ... per lb.	2½
kegs ... "	2¾
White lead, pure ... per cwt.	7 25
No 1 ... "	7 00
Prepared paints, pure liquid colors, according to shade and color .. per gal.	$1.30 to $1.90

NOTES.

Mr. Geo. Hunt, who represents The Cockshutt Plow Co. at Bagot, is erecting a handsome new warehouse.

Eakins & Griffin, who have for some years been in the employ of The James Robertson Co., this city, have purchased the hardware business of C. H. Greenshaw, at Shoal Lake.

THIS IS REALLY STARTLING.

"My friend," said the debtor to the blustering bill collector, "have you ever stopped to think that if all fellows like me paid our bills regularly you'd be out of a job?"—Syracuse Herald.

OBSERVATIONS—ON CREDIT.

CHAS. D. WETTACH.

THE old subject of credit is still being presented in so many shapes and forms that it will soon be a wonder if there will be anything left of it that will make it presentable in some other way.

Credit, so to speak, is simply the confidence reposed by one person in another to do a certain thing or things at a certain time—a contract pure and simple in which everyone who has anything to do in this department has his own method of construing it, and here are a few of mine :

1. On the application for credit ; what is to be considered in the granting of such credit ?

2. On the application for credit ; what is to be considered as the refusal of such credit.

3. The relation that exists between creditor and debtor.

(a) The grantor of credit.

(b) the acceptor of credit.

Following the lines laid down above, the first proposition presents itself in this light. As a rule, credit is asked for when ready cash is not at hand, and it is desirable to obtain merchandise on the strength of certain statements regarding capital, reputation, experience, ability, speculative chances for success. On this first point there is vast difference of opinion, and it will follow that, admitting that the mercantile reports are absolutely valuable, yet it is not wise to pass by the experience of those whose names have either been given or obtained otherwise as reference. The points to be considered in such information would naturally be the length of time, the amount of purchases, if in accordance with good judgment, and the promptness in meeting obligations. In determining the basis on agency reports, it is questionable whether capital is to be considered in preference to character and habits and the reputation borne in the community ; so that it is requisite that the one must depend largely upon the other, and, undoubtedly, capital being a convenient factor, yet we have a vast amount of evidence where character, reputation and integrity succeeded where capital failed. Therefore, these two factors must be closely scrutinized.

The next point would lead us to closely observe when refusing credit :

First.—Where the reputation and habits savor of dishonesty, incapability and lack of business principles, such as making statements that on the face are not absolutely true ; in other words, apparently excessive in proportion to the capital, amount of annual business and general demands of the community.

Second.—Insufficiency of capital must also enter into consideration, yet it is not wholly sufficient to refuse credit on this alone, as the capital may be amply sufficient, yet so tied up and unavailable that promptness in meeting obligations is an impossibility. Frequently, on this point, credit is refused intuitively ; and how often, when second thought is taken, in place of refusing, credit is granted ; and at the close of the year the very same account is charged off as doubtful, having been returned by attorneys as worthless.

Taking up the third and last point, the grantor of credit is entitled to receive from the prospective customer the fullest confidence, as it is frequently within the power of the grantor to point out weak places that need strengthening. How often when a prospective customer is asked to furnish a statement of his affairs he rudely, oftimes insultingly, declines, stating that his credit is unlimited elsewhere, and the order is countermanded. This at once shows a lack of the principles that justify credit. Happily it appears that this refusal to comply with the request of the seller is becoming less frequent, and those who are apylying for credit in many cases feel that those with whom they intend to deal have at heart the best wishes to their success.

INQUIRIES REGARDING CANADIAN TRADE.

The following were among the recent inquiries relating to Canadian trade received at the High Commissioner's office, in London, England :

1. An Aberdeen firm of merchants are desirous of being placed in communication with shippers of Canadian cereals and produce.

2. An English firm of stylographic and fountain penmakers make inquiry respecting the opening for such goods in Canada and the appointment of agents.

3. A Northampton firm of boot and shoe manufacturers, who do a small trade with Canada, are desirous of extending it, and wish to get into touch with boot and shoe makers in the Dominion.

4. Inquiry is made concerning the prospect of successfully introducing to the Canadian market a new coffee extract, by a firm who propose to appoint agents on commission.

Mr. Harrison Watson, curator of the Canadian Section of the Imperial Institute, London, England, is in receipt of the following inquiries :

5. A firm in Turin, Italy, wish to secure several first-class Canadian agencies in such lines as manufactures of wood, dried apples, tallow, lard, oil, mica, etc.

6. The proprietors of a patent bicycle brake wish to make arrangements for sale of same in Eastern Canada.

7. A London firm in touch with consumers of manganese ores invite correspondence from Canadian shippers.

[The names of the firms making the above inquiries, can be obtained on application to the editor of HARDWARE AND METAL, Toronto. When asking for names, kindly give number of paragraph and date of issue.]

CURRENT MARKET QUOTATIONS.

April 4, 1901.

These prices are for such qualities and quantities as are usually ordered by retail dealers on the usual terms of credit, the lowest figures being for larger quantities and prompt pay. Large cash buyers can frequently make purchases at better prices. The Editor is anxious to be informed at once of any apparent errors in this list as the desire is to make it perfectly accurate.

METALS.

Tin.

Lamb and Flag and Straits—56 and 28 lb. ingots, per lb.	0 31	0 32

Tinplates.

Charcoal Plates—Bright

		Per box
M.L.S., equal to Bradley.		
I.C., usual sizes	....	$6 75
I.X., "	....	8 25
I.X.X., "	....	9 75
Famous—		
I.C.	....	6 75
I.X	....	8 25
I.X.X	....	9 50
Raven & Vulture Grades—		
I.C., usual sizes	....	5 00
I.X., "	....	6 00
I.X.X "	....	7 00
I.X.X.X., "	....	8 00
D.C., 12½x17	....	4 75
D.X.	....	5 50
D.X.X	....	7 50

Coke Plates—Bright

Bessemer Steel—		
I.C., usual sizes	....	4 15
I.C., special sizes, base	....	4 50
20 x 28	....	8 50

Charcoal Plates—Terne

Dean or J. G. Grade—		
I.C., 20x28, 112 sheets	....	8 25
I.X., Terne Tin	....	10 25

Charcoal Tin Boiler Plates.

		Per lb.
Cookley Grade—		
X X., 14x56, 50 sheet bxs; " 14x60, "; " 14x65, "	0 07	0 07½

Tinned Sheets

72x30 up to 24 gauge	0 08	0 08½
" 26 "	0 08½	0 09
" 28 "	0 09	0 09½

Iron and Steel.

Common Bar, per 100 lbs.	1 70	1 75
Refined " "	....	2 20
Horse Shoe Iron	....	2 05
Hoop steel, 1½ to 3 in. base, extras for smaller sizes	....	3 10
Sleigh Shoe Steel " base	....	2 00
Tire Steel	....	2 00
Machinery iron finish	....	2 05
Cast Steel, per lb	0 00	0 00
Toe Calk Steel	....	2 30
T. Firth & Co's special cast steel, per lb.		0 13
Jessop's Tool Steel	....	0 13

Boiler Tubes.

1½-inch	....	0 12½
2 "	....	0 13
2½ "	....	0 15
3 "	....	0 16
3½ "	....	0 20
4 "	....	0 25

Steel Boiler Plate.

¼ inch	....	2 25
3-16 inch	....	2 25
⅜ inch and thicker	....	2 25

Black Sheets.

18 gauge	....	3 00
20 gauge	....	3 00
22 to 24 "	....	3 10
26 "	....	3 20
28 "	....	3 30

Canada Plates.

Al dull, 52 sheets	....	3 00
Half polished	....	3 15
All bright	3 65	3 75

Iron Pipe.

Black pipe—		
⅛ "	....	3 35
¼ inch	....	3 35
⅜ "	....	3 30
½ "	....	3 35
¾ "	....	3 60
1 "	....	5 00
1¼ "	....	6 85
1½ "	....	8 25
2 "	....	11 10
2½ "	....	22 91
3 "	....	30 12
3½ "	....	37 90
4 "	....	43 19
5 "	....	57 65
6 "	....	75 11
7 "	....	93 76
8 "	....	112 51
Galvanized pipe—		
½ inch	....	5 25
¾ "	....	5 56
1 "	....	7 77
1¼ "	....	10 60
1½ "	....	12 70
2 "	....	16 90

Galvanized Sheets.

	G C.	Comet.	Amer.	Queen's Head.
16 gauge		....	4 15	3 90
18 to 24 gauge	4 15	3 85	4 35	4 15
26 "	4 40	4 10	4 35	4 40
28 "	4 65	4 35	4 50	4 65

28 gauge American equals 26 gauge English.

Chain.

Proof Coil, 3-16 in., per 100 lb	....	....
" ¼ " "	8 00	8 50
" 5-16 " "	4 70	5 00
" ⅜ " "	4 05	4 70
" 7-16 " "	3 90	4 25
" ½ " "	3 70	4 10
" 9-16 " "	3 65	4 65
" ⅝ " "	3 35	3 90
" ¾ " "	3 60	4 10

Halter, kennel and post chains,	40 and 50 p.c.
Cow ties	40 p.c.
Stall fixtures	35 p.c.
Trace chain	25 p.c.

Jack chain, iron, single and double, discount 35 p.c.

Jack chain, brass, single and double, discount 40 p.c.

Copper.

Ingot

English B. S., ton lots	0 16	0 19
Lake Superior	....	....

Bolt or Bar.

Cut lengths round, ½ to ⅞ in.	0 23½	0 25
" round and square 1 to 2 inches	0 23½	0 25

Sheet.

Plain, 14 oz., and light, 16 oz., 14x48 and 14x60	0 23	0 23
Plain, 14 oz., and light, 16 oz., irregular sizes	0 23	0 23½
Tinned copper sheets	....	0 26
Planished	....	0 32

Braziers (in sheets.)

4x6ft. 25 to 30 lbs. ea., per lb.	....	0 25½
" 35 to 45 " "	....	0 24½
" 50-lb. and above, "	....	0 23½

Boiler and T. K. Pitts

Plain Tinned, per lb	....	0 28
Spun, per lb.	....	0 32

Brass.

Rod and Sheet, 14 to 30 gauge,		15 per cent.
Sheets, hard-rolled, 2x4	0 24	0 25
Tubing, base, per lb.	0 24	0 25

Zinc Spelter

Foreign, per lb	0 06	0 06½
Domestic "	....	..

Zinc Sheet.

5 cwt. casks	....	00 6¼
Part casks	....	00 6¾

Lead.

Imported Pig, per lb	0 04¾	0 05
Bar, 1 lb.	0 05½	0 05¾
Sheets, 2½ lbs. sq. ft., by	....	0 06¾
Sheets, 3 to 6 lbs., "	....	0 06

NOTE.—Cut sheets ½ cent per lb. extra. Pipe, by the roll, usual weights per yard, lists at 7c. per lb. and 25 p.c. dis. f.o.b. Toronto.

NOTE.—Cut lengths, net price, waste pipe 8-ft. lengths lists at 7¼ cents.

Shot.

Common, $6.50 per 100 lb.; chilled, $7.00 per 100 lb.; buck, seal and ball, $7.50. Discount, 15 p.c. Prices are f.o.b. Toronto, Hamilton, Montreal, St. John and Halifax. Terms 3 per cent. cash, freights equalized on Montreal.

Soil Pipe and Fittings.

Discount, 60 and 10 per cent. on medium and extra heavy, and 60 per cent. on light.

Solder.

	Per lb.	Per lb.
Bar half-and-half, guarant'd	....	0 18½
Bar half-and-half, commer'l	....	0 18
Refined	....	0 17½
Wiping	....	0 17

NOTE.—Prices of this graded according to quantity. The prices of other qualities of solder in the market indicated by private brands vary according to composition.

Antimony.

Cookson's, per lb.	0 11	0 11½

White Lead:

	Per 100 lb.
Pure	6 50
No. 1 do	6 12½
No. 2 do	5 75
No. 3 do	5 37½
No. 4 do	5 00
Munro's Select Flake White	7 37½
Elephant and Decorators' Pure	7 12½
Brandram's B B. Genuine	8 00
" " Decorative	7 55
" " No. 1	6 85
" " No. 2	6 00

Red Lead.

Genuine, 560 lb. casks, per cwt.	....	$5 50
Genuine, 100 lb. kegs, per cwt.	....	5 75
No. 1, 560 lb. casks, per cwt.	....	5 25
No. 1, 100 lb. kegs, per cwt	....	5 00

White Zinc Paint.

Elephant Snow White	0 08	0 09
Pure White Zinc	0 08	0 09
No. 1	0 06	0 07½
No. 2	0 05	0 06½

Dry White Lead.

Pure, casks	5 75
Pure, kegs	6 25
No. 1, casks	5 50
No. 1, kegs	5 00

Prepared Paints.

In ¼, ½ and 1 gallon tins.

Pure, per gallon	....	1 25
Second qualities, per gallon	....	1 10
Barn (in bbls.)	0 75	0 85
The Sherwin-Williams Paints.	....	1 45
Canada Paint Co's Pure	....	1 25
Toronto Lead & Color Co's Pure	....	1 25
Sanderson Pearcy's Pure	....	1
Stewart & Wood's Champion Pure.		1 20

Colors in Oil.

25 lb. tins, Standard Quality.

Venetian Red, per lb.	0 05
Chrome Yellow	0 11
Golden Ochre	0 06
French "	0 05
Marine Black	0 09
" Green	0 09
Chrome "	0 08
French Imperial Green	0 09

Colors, Dry.

Yellow Ochre (J.C.) bbls	1 35	1 40
Yellow Ochre (J.F.L.S.), bbls	...	2 75
Yellow Ochre (Royal)	1 10	1 15
Brussels Ochre	...	2 00
Venetian Red (best), per cwt.	1 80	1 90
English Oxides, per cwt.	3 00	3 25
American Oxides, per cwt.	1 75	2 00
Canadian Oxides, per cwt.	1 75	2 00
Super Magnetic Oxides, 93p.c.	2 00	2 25
Burnt Sienna, pure, per lb.	....	0 10
" Umber, " "	....	0 10
do Raw	....	0 09
Drop Black, pure	....	0 09
Chrome Yellows, pure	....	0 18
Chrome Greens, pure, per lb.	....	0 12
Golden Ochre	....	0 03½
Ultramarine Blue in 28-lb. boxes, per lb.	0 08	0 24
Fire Proof Mineral, per 100 lb	....	1 00
Genuine Eng. Litharge, per lb	....	0 07
Mortar Color, per 100 lb.	....	1 25
English Vermillion	....	80
Pure Indian Red, No. 45 lb.	....	80
Whiting, per 100 lb.	....	0 55

Blue Stone.

Casks, for spraying, per lb.	0 07
100-lb. lots, do. per lb	0 08

Putty.

Bulk in bbls.	1 90
Bulk in less quantity	2 05
Bladders in bbls.	2 10
Bladders in kegs, boxes or loose	2 25
Bladders in 25-lb. tins	2 35
Bladders in 12½-lb. tins	2 65
Bladders in bulk or tins less than 100lb	2 90

Varnishes.

	In 5-gal. lots.).	Per gal.
Carriage, No. 1	2 90	3 30
" body	8 00	9 00
" rubbing	4 00	5 00
Gold Size, Japan	3 00	3 40
Brown Japan	2 40	2 80
Elastic Oak	2 90	3 30
Furniture, extra	2 40	2 80
" No. 1	1 60	2 00
Hard Oil Finish	2 70	3 10
Light Oil Finish	3 20	3 60
Demar	3 30	3 70
Shellac, white	4 40	4 80
" orange	4 00	4 40
Furniture Brown Japan	1 60	2 00
Black Japan	2 40	2 80
" " No. 1	1 60	2 00

The Imperial Varnish & Color Co's, Limited Elastilite Varnish 1 gal. can, each. $3.00.

Granatine Floor Finish, per gal., $2 50.

Maple Leaf Coach Enamels; Size 1, 60c. Size 2, 35c.; Size 3, 20c. each.

Castor Oil.

East India, in cases, per lb.	0 10	0 10½
" " small lots	0 10½	0 11

Cod Oil, Etc.

Cod Oil per gal	0 50	0 55
Pure Olive	....	1 20
" Neatsfoot	....	90

Glue.

Common	0 08½	0 09
French Medal	0 14	0 14½
Cabinet sheet	0 12	0 13
White, extra	0 18	0 20
Gelatine	0 22	0 30
Strip	0 18	0 20
Coopers	0 19	0 20
Huttner	....	0 18

HARDWARE.

Ammunition.

Cartridges.

B. B Cap Dom. 50 and 5 per cent.
Rim Fire Pistol, dis. 40 p. c., Amer.
Rim Fire Cartridges, Dom., 50 and 5 p. c.
Central Fire Pistol and Rifle, 10 p.c. Amer.
Central Fire Cartridges, pistol sizes, Dom. 30 per cent.
Central Fire Cartr'dges, Sporting and Military, Dom., 15 and 5 per cent.
Central Fire, Military and Sporting, Amer, add 5 p.c. to list. B.B. Caps. discount 40 per cent, Amer.
Loaded and empty Shells, "Trap" and "Dominion" grades, 25 per cent. Rival and Nitro, net list.
Brass Shot Shells, 55 per cent.
Primers, Dom., 30 per cent.

Wads — per lb.

Best thick white felt wadding, in ¼-lb bags 1 00
Best thick brown or grey felt wads, in ¼-lb. bags 70
Best thick white card wads, in boxes of 500 each, 12 and smaller gauges 0 99
Best thick white card wads, in boxes of 500 each, 10 gauge 0 35
Best thick white card wads, in boxes of 500 each, 8 gauge 0 55
Thin card wads, in boxes of 1,000 each, 12 and smaller gauges 0 20
Thin card wads, in boxes of 1,000 each, 10 gauge 0 25
Thin card wads in boxes of 1,000 each, 8 gauge ....
Chemically prepared black edge grey cloth wads, in boxes of 250 each— Per M.
11 and smaller gauge 0 60
9 and 10 gauges 0 70
7 and 8 gauges 0 90
5 and 6 gauges 1 10
Superior chemically prepared pink edge, best white cloth wads, in boxes of 250 each—
11 and smaller gauge 1 15
9 and 10 gauges 1 40
7 and 8 gauges 1 65
5 and 6 gauges 1 90

Adzes.

Discount, 20 per cent.

Anvils.

Per lb 0 10 0 12½
Anvil and Vise combined 4 50
Wilkinson & Co.'s Anvils..lb. 0 09 0 09½
Wilkinson & Co.'s Vices..lb. 0 09¾ 0 10

Augers.

Gilmour's, discount 65 and 5 p.c. off list.

Axes.

Chopping Axes
Single bit, per doz 5 50 10 00
Double bit, " 11 00 18 00
Bench Axes, 40 p.c.
Broad Axes, 33⅓ per cent.
Hunters' Axes 5 50 6 00
Boy's Axes 5 75 6 75
Splitting Axes 6 50 12 00
Handled Axes 7 00 10 00

Axle Grease.

Ordinary, per gross 5 75 6 00
Best quality 13 00 15 00

Bath Tubs.

Zinc 6 00
Copper, discount 15 p.c. off revised list

Baths.

Standard Enameled.

5½-inch rolled rim, 1st quality 25 00
" " " 2nd " 21 00

Anti-Friction Metal.

"Tandem" A .... per lb. 0 27
" B .... " 0 21
" C .... " 0 11½
Magnolia Anti-Friction Metal, per lb. 0 25

SYRACUSE SMELTING WORKS.

Aluminum, genuine 0 45
Dynamo 0 29
Special 0 25
Aluminum, 99 p.c. pure "Syracuse" .. 0 50

Bells.

Hand.

Brass, 60 per cent.
Nickel, 55 per cent.

Cow.

American make, discount 66⅔ per cent.
Canadian, discount 45 and 50 per cent.

Door.

Gongs, Sargant's 5 50 8 00
" Peterboro', discount 45 per cent.

Farm.

American, each 1 25 3 00

House.

American, per lb 0 35 0 40

Bellows.

Hand, per doz 3 35 4 75
Moulders', per doz 7 50 10 00
Blacksmiths', discount 40 per cent.

Belting.

Extra, 60 per cent.
Standard, 60 and 10 per cent.

Bits.

Auger.

Gilmour's, discount 60 and 5 per cent.
Rockford, 50 and 10 per cent.
Jennings' Gen., net list.

Car.

Gilmour's, 47½ to 50 per cent.

Expansive.

Clark's, 40 per cent

Gimlet.

Clark's, per doz 0 65 0 90
Damond, Shell, per doz 1 00 1 50
Nail and Spike, per gross 2 25 5 20

Blind and Bed Staples.

All sizes, per lb 0 07½ 0 12

Bolts and Nuts. Per cent.

Carriage Bolts, full square, Norway 70
" " full square 70
Common Carriage Bolts, all sizes 65
Machine Bolts, all sizes 65
Coach Screws 75
Sleigh Shoe Bolts 75
Blank Bolts 65
Bolt Ends 65
Nuts, square .... 4½c. off
Nuts, hexagon .... 4¾c. off
Tire Bolts 67½
Stove Bolts 67½
Stove rods, per lb .... 5½ to 8c.
Plough Bolts 60

Boot Calks.

Small and medium, ball, per M 4 25
Small heel, per M 4 50

Bright Wire Goods.

Discount .... 60 per cent.

Broilers.

Light, dis., 65 to 67½ per cent.
Reversible, dis., 65 to 67½ per cent.
Vegetable, per doz., dis. 37½ per cent.
Henis, No. 8, " 6 00
Henis, No. 9, " 7 00
Queen City " 7 50 0 00

Butchers' Cleavers.

German, per doz 6 00 11 00
American, per doz 12 00 20 00

Building Paper, Etc.

Plain building, per roll 0 30
Tarred lining, per roll 0 40
Tarred roofing, per 100 lb 1 65
Coal Tar, per barrel 3 50
Pitch, per 100-lb 0 85
Carpet felt, per ton 45 00

Bull Rings.

Copper, $2.00 for 2½ in. and $1.90 for 2 in.

Butts.

Wrought Brass, net revised list

Cast Iron.

Loose Pin, dis., 60 per cent.

Wrought Steel.

Fast Joint, dis. 60 and 10 per cent.
Loose Pin, dis. 60 and 10 per cent.
Berlin Bronzed, dis. 70, 70 and 5 per cent.
Gen. Bronzed, per pair 0 40 0 65

Carpet Stretchers.

American, per doz 1 00 1 50
Bullard's, per doz .... 6 50 ....

Castors.

Bed, new list, dis. 55 to 57½ per cent.
Plate, dis. 52½ to 57½ per cent.

Cattle Leaders.

Nos. 31 and 32, per gross 50 9 50

Cement.

Canadian Portland 2 50 2 80
English " 3 00
Belgian " 2 10 2 15
Canadian hydraulic 1 25 1 50

Chalk.

Carpenters Colored, per gross 0 45 0 75
White lump, per cwt 0 60 0 65
Red 0 05 0 06
Crayon, per gross 0 14 0 18

Chisels.

Socket, Framing and Firmer.

Broad's, dis. 70 per cent.
Warnock's, dis. 70 per cent.
P. S. & W. Extra 60, 10 and 5 p.c.

Churns.

Revolving Churns, metal frames—No. 0, $8—No. 1, $8.50—No. 2, $9.00—No. 3, $10.00—No. 4, $12.00—No. 5, $16.00 each. Ditto, wood frames—20c. each less than above. Discounts: Delivered from factories, 58 p.c.; from stock in Montreal, 56 p.c. Terms, 4 months or 3 p.c. cash in 30 days.

Clips.

Axle dis. 65 per cent.

Closets.

Plain Ontario Syphon Jet $16 00
Emb. Ontario Syphon Jet 17 00
Fittings net 1 00
Plain Teutonic Syphon Washout 10 00
Emb. Teutonic Syphon Washout 11 00
Fittings net 1 25
Low Down Teutonic, plain 16 00
" " " embossed 17 00
Plain Richelieu net 3 75
Emb. Richelieu net 4 00
Fittings net 1 25
Low Down Ont. Sy. Jet, plain net .. 19 50
" " " " " emb'd. net 20 50
Closet connection net 1 25
Basins, round, 14 in 1 04
" oval, 17 x 14 in 2 50
" " 19 x 15 in 3 75
Discount 40 p.c., except on net figures.

Compasses, Dividers, Etc.

American, dis. 62½ to 65 per cent.

Cradles, Grain.

Canadian, dis. 25 to 33⅓ per cent.

Crosscut Saw Handles.

S. & D., No. 3, per pair 17½
" " 5, " 22½
" " 6, " 15
Boynton pattern " 20

Door Springs.

Torrey's Rod, per doz .... (15 p.c.) 2 00
Coil, per doz 0 88 1 60
English, per doz 2 00 4 00

Draw Knives.

Coach and Wagon, dis. 50 and 10 per cent.
Carpenters, dis. 70 per cent.

Drills.

Hand and Breast.

Millar's Falls, per doz. net list.

DRILL BITS.

Morse, dis., 37½ to 40 per cent.
Standard dis. 50 and 5 to 55 per cent

Faucets.

Common, cork-lined, dis. 35 per cent.

ELBOWS. (Stovepipe.)

No. 1, per doz 1 40
No. 2, per doz 1 20
Bright, 20c. per doz. extra.

ESCUTCHEONS.

Discount, 45 per cent.

ESCUTCHEON PINS.

Iron, discount 40 per cent.

FACTORY MILK CANS.

Discount off revised list, 40 per cent.

FILES.

Black Diamond, 70 and 5 per cent.
Kearney & Foote, 60 and 10 to 70 per cent.
Nicholson File Co., 60 to 60 and 10 per cent.
Jowitt's, English list, 25 to 27½ per cent.

FORKS.

Hay, manure, etc., dis., 50 and 10 per cent. revised list.

GLASS—Window—Box Price.

Size United Inches	Star Per 50 ft.	Star Per 100 ft.	D. Diamond Per 50 ft.	D. Diamond Per 100 ft.
Under 26	2 15	4 15	....	6 00
26 to 40	2 30	4 45	....	6 65
41 to 50	....	4 85	....	7 50
51 to 60	....	5 15	....	8 50
61 to 70	....	5 50	....	9 50
71 to 80	....	6 00	....	10 50
81 to 85	....	6 50	....	11 75
86 to 90	....	....	..	14 00
91 to 95	..	....	...	15 50
96 to 100	.. ..	....	....	18 00

GAUGES

Marking, Mortise, Etc.

Stanley's dis. 50 to 55 per cent.

Wire Gauges.

Winn's, Nos. 26 to 33, each... 1 65 2 40

HALTERS.

Rope, ⅜ per gross ....
" ½ " 9 00
" ⅝ to ¾ 14 00
Leather, 1 in., per doz 3 87½ 4 00
" 1¼ in., " 5 15 5 20
Web, — per doz 1 87 2 45

HAMMERS.

Nail

Maydole's, dis. 5 to 10 per cent. Can. dis. 25 to 27½ per cent.

Tack.

Magnetic, per doz 1 10 1 20

Sledge.

Canadian, per lb 0 07½ 0 08½

Ball Pean.

English and Can., per lb 0 22 0 25

HANDLES.

Axe, per doz. net 1 50 2 00
Store door, per doz 1 00 1 50

Fork.

C. & B., dis. 40 per cent. rev. list.

Hoe.

C. & B., dis. 40 per cent. rev. list.

Saw.

American, per doz 1 00 1 25

Plane.

American, per gross 3 15 3 75

Hammer and Hatchet.

Canadian, 40 per cent.

Cross-Cut Saws.

Canadian, per pair 0 13½

HANGERS. doz. pairs.

Steel barn door 5 85 6 00
Stearns, 4 inch 5 00
" 5 inch 6 50
Lane's covered—
No. 11, 5-ft. run 8 40
No. 11½, 10-ft. run 10 80
No. 12, 10-ft. run 12 60
No. 14, 15-ft. run 21 00
Lane's O.N.T. track, per foot. 4½

HARVEST TOOLS.

Discount, 50 and 10 per cent.

HATCHETS.

Canadian, dis. 40 to 42½ per cent.

HINGES.

Blind, Parker's, dis. 50 and 10 to 60 per cent.
Heavy T and strap, 4-in., per lb 0 06½
" " 5-in., " 0 06¼
" " 6-in., " 0 06
" " 8-in., " 0 05¾
" " 10-in., " 0 05½
Light T and strap, dis. 60 and 5 per cent.
Screw hook and hinge—
6 to 12 in., per 100 lbs 4 50
14 in. up, per 100 lbs 3 50
Per gro. pairs.
Spring 12 00

HOES.

Garden, Mortar, etc., dis. 50 and 10 p.c.
Planter, per doz 4 00 4 50

HOLLOW WARE

Discount, ... 45 and 5 per cent.

HOOKS.

Cast Iron.

Bird Cage, per doz 0 50 1 10
Clothes Line, per doz 0 27 0 63
Harness, per doz 0 72 0 88
Hat and Coat, per gross 1 00 3 00
Chandelier, per doz 0 50 1 00

Wrought Iron.

Wrought Hooks and Staples, Can., dis. 47½ per cent.

Wire.

Hat and Coat, discount 45 per cent.
Belt, per 1,000 .... 0 60
Screw, bright, dis. 55 per cent.

HORSE NAILS.

"C" brand 50 and 7½ p.c. off new list } Oval-
"M" brand 60 per cent. } head.
Countersunk, 60 and 10 per cent.

HORSESHOES

Iron Shoes.	F.O.B. Montreal. No. 2 and larger.	No. 1. and smaller.
Light, medium, and heavy	3 50	3 75
Snow shoes	3 75	4 00
Steel Shoes.		
Light	3 60	3 85
Featherweight (al lsizes)	4 85	4 85

F.O.B. Toronto, Hamilton, London and Guelph, 10c. per keg additional.

Toe weight steel shoes...... 6 70

JAPANNED WARE.

Discount, 45 and 5 p.c. off list, June 1899.

ICE PICKS.

Star per doz.......... 3 00 3 25

KETTLES.

Brass spun, 7½ p.c. dis. off new list.

Copper, per lb.......... 0 30 0 50

American, 60 and 10 to 65 and 5 p.c.

KEYS.

Lock, Can., dis., 45 p.c.

Cabinet, trunk, and padlock, Am. per gross.......... 60

KNOBS.

Door, japanned and N.P., per doz.	1 50	2 50
Bronze, Berlin, per doz.	2 75	3 25
Bronze Genuine, per doz.	6 00	9 00
Shutter, porcelain, F. & L. screw, per gross	1 30	4 00
White door knobs—per doz.		1 25

HAY KNIVES.

Discount, 50 and 10 per cent.

LAMP WICKS.

Discount, 60 per cent.

LANTERNS.

Cold Blast, per doz.	7 50
No. 3 "Wright's"	8 50
Ordinary, with O burner	4 25
Dashboard, cold blast	9 50
No. 0	6 00

Japanning, 50c. per doz. extra.

LEMON SQUEEZERS.

Porcelain lined, per doz.	2 20	5 60
Galvanized	1 87	3 85
King, wood	2 75	2 90
King, glass	4 00	4 50
All glass	1 20	1 30

LINES.

Fish, per gross	1 05	2 50
Chalk "	1 90	7 40

LOCKS.

Canadian, dis. 45 p.c.

Russell & Erwin, per doz.... 3 00 3 26

Cabinet.

Eagle, dis. 30 p.c.

Padlock.

English and Am., per doz.	50	6 00
Scandinavian, "	1 00	2 40

Eagle, dis. 20 to 25 p.c.

MACHINE SCREWS. Iron and Brass.

Flat head discount 25 p.c.

Round Head, discount 20 p.c.

MALLETS.

Tinsmiths', per doz.	1 25	1 50
Carpenters', hickory, per doz.	1 25	3 75
Lignum Vitae, per doz.	3 85	5 00
Caulking, each	60	2 00

MATTOCKS.

Canadian, per doz.......... 5 50 6 50

MEAT CUTTERS.

American, dis. 25 to 30 p.c.

German, 15 per cent.

MILK CAN TRIMMINGS.

Discount, 25 per cent.

NAILS.

Quotations are:	Cut.	Wire.
2d and 3d	$3 35	$3 85
3d	3 00	3 52
4 and 5d	2 75	3 35
6 and 7d	2 65	3 20
8 and 9d	2 50	3 00
10 and 12d	2 45	2 95
16 and 20d	2 40	2 90
30, 40, 50 and 60d. (base)	2 35	2 85

Wire nails in carlots are $2.77½

Galvanizing 2c. per lb. net extra.

Steel Cut Nails 10c. extra.

Miscellaneous wire nails, dis. 70 and 10 p.c.

Coopers' nails, dis. 30 per cent.

Flour barrel nails, dis. 25 per cent.

NAIL PULLERS.

German and American....... 1 85 3 50

NAIL SETS

Square, round, and octagon, per gross	3 38	4 00
Diamond	12 00	15 00

NETTING.

Poultry, 50 and 5 per cent. for McMullen's

OAKUM.

	Per 100 lb.
Navy	6 00
U. S. Navy	7 25

OIL.

Water White (U.S.)	0 16½
Prime White (U.S.)	0 15½
Water White (Can.)	0 15
Prime White (Can.)	0 14

OILERS.

McClary's Model galvan. oil can, with pump, 5 gal., per doz. 10 00

Zinc and tin, dis. 50, 50 and 10.

Copper, per doz.	1 25	3 50
Brass, "	1 50	3 50

Malleable, dis. 25 per cent.

GALVANIZED PAILS.

Dufferin pattern pails, dis. 45 p.c.

Flaring pattern, discount 45 per cent.

Galvanized washtubs, discount 45 per cent.

PIECED WARE.

Discount 40 per cent. off list, June, 1899.

10-qt. flaring sap buckets, dis. 45 p.c.

6, 10 and 14-qt. flaring pails, dis. 45 p.c.

Creamer cans, dis. 45 p.c.

PICKS.

Per doz.......... 6 00 9 00

PICTURE NAILS.

Porcelain head, per gross	1 75	3 00
Brass head "	0 40	1 00

PICTURE WIRE.

Tin and gilt, discount 75 p.c.

PLANES.

Wood, bench, Canadian dis. 50 per cent. American dis. 50.

Wood, fancy Canadian or American 7½ o 40 per cent.

PLANE IRONS.

English, per doz.......... 2 00 5 00

PLIERS AND NIPPERS.

Button's Genuine per doz pairs, dis. 37½ 40 p.c.

Button's Imitation, per doz.	5 00	9 00
German, per doz.	0 60	2 60

PLUMBERS' BRASS GOODS.

Compression work, discount, 60 per cent.

Fuller's work, discount 65 per cent.

Rough stops and stop and waste cocks, discount, 60 per cent.

Jenkins disk globe and angle valves, discount, 55 per cent.

Standard valves, discount, 60 per per cent.

Jenkins' radiator valves discount 55 per cent.

" " " standard, dis., 60 p.c.

Quick opening valves discount, 60 p.c.

No. 1 compression bath cock	2 00
No. 4 " " "	2 00
No. 7, Fuller's	2 50
No 4½, "	3 00

POWDER.

Velox Smokeless Shotgun Powder.

100 lb. or less	0 85
1,000 lb. or more	0 80

Net 30 days.

PRESSED SPIKES.

Discount 20 to 25 per cent.

PULLEYS.

Hothouse, per doz.	0 15	1 00
Axle	0 22	0 33
Screw	0 27	1 00
Awning	0 35	2 50

PUMPS.

Canadian cistern	1 80	3 60
Canadian pitcher spout	1 40	2 10

PUNCHES.

Saddlers', per doz.	1 00	1 85
Conductors', "	9 00	15 00
Tinners' solid, per set	0 00	0 72
" hollow, per inch	0 00	1 00

RANGE BOILERS.

Galvanized, 3 gallons	7 00
" 35 "	8 25
" 40 "	9 50
Copper, 30 "	22 00
" 35 "	26 00
" 40 "	30 00

Discount off Copper Boilers 10 per cent.

RAKES.

Cast steel and malleable, 50, 10 and 5 p.c.

Wood, 25 per cent.

RASPS AND HORSE RASPS.

New Nicholson horse rasp, discount 60 to 60 and 10 p.c.

Globe File Co.'s rasps, 60 and 10 to 70 p.c.

Heller's Horse rasps, 50 to 50 and 5 p.c.

RAZORS.

	per doz.	
Geo. Butler & Co.'s	8 00	18 00
Boker's	7 50	11 00
" King Cutter	12 50	50 00
Wade & Butcher's	3 60	10 00
Theile & Quack's	7 00	12 00
Elliot's	4 00	18 00

REAPING HOOKS.

Discount, 50 and 10 per cent.

REGISTERS.

Discount.......... 40 per cent.

RIVETS AND BURRS.

Iron Rivets, black and tinned, discount 60 and 10 per cent.

Iron Burrs, discount 55 per cent.

Extras on Iron Rivets in 1-lb. cartons, ½c. per lb.

Extras on Iron Rivets in ½-lb. cartons, 1c. per lb.

Copper Rivets & Burrs, 35 and 5 p.c. dis. and cartons, 1c. per lb. extra, net.

Extras on Tinned or Coppered Rivets ½-lb. cartons, 1c. per lb.

RIVET SETS

Canadian, dis. 35 to 37½ per cent.

ROPE ETC.

	Sisal.	Manila.
7-16 in. and larger, per lb	10	13½
⅜ in.	11	14½
¼ and 5-16 in.		15½
Cotton, 3-16 inch and larger		16½
" 5-32 inch		21½
" ⅛ inch		22½
Russia Deep Sea		15½
Jute		8
Lath Yarn		9½
New Zealand Rope		10

RULES.

Boxwood, dis. 75 and 10 p.c.

Ivory, dis. 37½ to 40 p.c.

SAD IRONS. per set.

Mrs. Potts, No. 55, polished	62½
" No. 50, nickle-plated	67½

SAND AND EMERY PAPER.

Dominion Flint Paper, 47½ per cent.

B & A. sand, 40 and 5 per cent.

Emery, 40 per cent.

Garnet (Rurton's), 5 to 10 p.c. advance on list.

SAP SPOUTS.

Bronzed iron with hooks, per doz... 9 50

SAWS.

Hand Disston's, dis. 12½ p.c.

S. & D., 40 per cent.

Crosscut, Disston's, per ft.... 35 0 55

S. & D., dis. 35 p.c. on Nos. 2 and 3.

Hack, complete, each	0 75	2 75
" frame only	....	0 75

SASH WEIGHTS.

Sectional, per 100 lbs.	2 75	3 00
Solid, "	2 00	2 25

SASH CORD.

Per lb.......... 0 23 0 30

SAW SETS.

"Lincoln," per doz.......... 6 50

SCALES.

B. S. & M. Scales, 45 p.c.

Champion, 65 per cent.

Fairbanks Standard, 35 p.c.

" Dominion, 55 p.c.

" Richelieu, 55 p.c.

Chatillon Spring Balances, 10 p.c.

SCREW DRIVERS.

Sargent's per doz.......... 0 65 1 00

SCREWS.

Wood, F. H., bright and steel, 87½ and 10 p.c.

Wood R. H., " dis. 82½ and 10 p.c.

" F. H., brass, dis. 80 and 10 p.c.

Wood, R. H., " dis. 75 and 10 p.c.

" F.H., bronze, dis. 75 p.c.

" R.H. " 70 p.c.

Drive Screws, 87½ and 10 per cent.

Bench, wood, per doz.	3 25	4 00
" iron. "	4 25	5 75

Set, Case hardened, 60 per cent.

Square Cap, 50 and 5 per cent.

Hexagon Cap, 45 per cent.

SCYTHES.

Per doz. net.......... 9 00

SCYTHE SNATHS.

Canadian, dis. 45 p.c.

SHEARS.

Bailey Cutlery Co. full nickeled, dis. 60 p.c.

Seymour's, dis. 50 and 10 p.c.

SHOVELS AND SPADES.

Canadian, dis. 40 and 5 per cent.

SINKS.

Steel and galvanized, discount 45 per cent.

SNAPS.

Harness, German, dis. 25 p.c.

Lock, Andrews'.......... 4 50 11 50

SOLDERING IRONS.

1, 1½ lb., per lb.	0 37
2 lb. or over, per lb.	0 34

SQUARES.

Iron, No. 493, per doz	2 40	2 55
" No. 494, "	3 25	3 40

Steel, dis. 50 and 5 to 50 and 10 p.c., rev. list.

Try and bevel, dis. 50 to 52½ p.c.

STAMPED WARE.

Plain, dis. 75 and 12½ p.c. off revised list.

Retinned, dis., 75 p.c. off revised list.

STAPLES.

Galvanized	3 50	4 00
Plain	3 15	3 75

Coopers', discount 45 per cent.

Poultry netting staples, 40 per cent.

STOCKS AND DIES.

American dis. 25 p.c.

STONE.

	Per lb.	
Washita	0 28	0 60
Hindostan	0 06	0 07
" slip,	0 09	0 09
Labrador	....	0 13
" Axe	....	0 15
Turkey	....	0 50
Arkansas	0 00	1 50
Water-of-Ayr	0 00	0 10
Scythe, per gross	3 50	5 00
Grind, per ton	15 00	18 00

STOVE PIPES.

5 and 6 inch Per 100 lengths	7 00
7 inch " "	7 50

ENAMELINE STOVE POLISH.

No. 4—3 dozen in case, net cash	$4 80
No. 6—3 dozen in case, "	8 40

TACKS BRADS, ETC

	Per cent.
Strawberry box tacks, bulk	75 & 10
Cheese-box tacks, blued	80 & 12½
Trunk tacks, black and tinned	85
Carpet tacks, blued	80 & 5
" " tinned	80 & 10
" " (in kegs)	40
Cut tacks, blued, in dozens only	75 & 15
" ½ weights	60
Swedes, cut tacks, blued and tinned—	
In bulk	80 & 10
In dozens	75
Swedes, upholsterers', bulk	85 & 12½
" brush, blued & tinned, bulk	70
" gimp, blued, tinned and japanned	75 & 12½
Zinc tacks	35
Leather carpet tacks	55
Copper tacks	50
Copper nails	52
Trunk nails, black	65 and 5
Trunk nails, tinned	65 and 10
Clout nails, blued and tinned	65 and 5
Chair nails	35
Patent brads	40
Fine finishing	40
Picture frame points	10
Lining tacks, in papers	10

Lining tacks, in bulk		15
" " solid heads, in bulk		75
Saddle nails in papers		10
" " in bulk		15
Tufting buttons, 22 line, in dozens only		60
Tin capped trunk nails		15
Zinc glazier's points		5
Double pointed tacks, papers ....90 and 10		
" " " bulk		40

TAPE LINES.

English, ass skin, per doz	2 75	5 00
English, Patent Leather	5 50	9 75
Chesterman's each	0 90	2 85
" steel, each	0 80	8 00

THERMOMETERS.

Tin case and dairy, dis. 75 to 75 and 10 p.c.

TRAPS. (Steel.)

Game, Newhouse, dis. 2) p.c.
Game, H. & N., P. S. & W., 65 p.c.
Game, steel, 72½, 75 p.c.

TROWELS.

Disston's discount 10 per cent.

German, per doz	4 75	6 00

S. & D., discount 35 per cent.

TWINES.

Bag, Russian, per lb	....	0 27
Wrapping, cotton, per lb	0 22	0 26
Wrapping, mottled, per pack	0 50	0 60
Wrapping, cotton, 3-ply		0 20
" " 4-ply	....	0 26
Mattress, per lb	0 33	0 45
Staging, "	0 27	0 35

VISES.

Hand, per doz	4 00	6 00
Bench, parallel, each	2 00	4 50
Coach, each	6 00	7 00
Peter Wright's, per lb	0 12	0 13
Pipe, each	5 50	9 00
Saw, per doz	6 50	13 00

ENAMELLED WARE.

White, Princess, Turquoise, Blue and White, discount 50 per cent.
Diamond, Famous, Premier, 50 and 10 p.c.
Granite or Pearl, Imperial, Crescent, 50, 10 and 10 per cent.

WIRE.

Brass wire, 50 to 50 and 2½ per cent. off the list.
Copper wire, 45 and 10 per cent. net cash 30 days, f.o.b. factory.

Smooth Steel Wire, is quoted at the following net selling prices:

No. 6 to 8 gauge		$2 90
" 9	"	2 8)
" 10	"	2 87
" 11	"	2 90
" 12	"	2 93
" 13	"	3 15
" 14	"	3 37
" 15	"	3 50
" 16	"	3 65

Other sizes of plain wire outside of Nos. 9, 10, 11, 12 and 13, and other varieties of plain wire remain at $2.8) base with extras as before. The prices for Nos. 9 to 13 include the charge of 1c. for oiling. Extras net per 100 lb.: Coppered wire, 60c.—tinned wire, $2—oiling, 10c.—special hay-bailing wire, 30c.—spring wire, $1—best steel wire, 75c.—bright soft drawn, 15c.—in 50 and 100-lb. bundles net, 10c.—in 25-lb. bundles net 15c.—packed in casks or cases, 15c.—bagging or papering, 10c.

Fine Steel Wire, dis. 17½ per cent. List of extras: In 100-lb. lots: No. 17, $5—No. 18, $5.50—No. 19, $6—No. 20 $6.65—No. 21, $7—No. 22, $7.30—No. 23 $7.65—No. 24, $8—No. 25, $9—No. 26, $9.50—No. 27, $10—No. 28, $11—No. 29, $12—No. 30, $13—No. 31, $14—No. 32, $15 No. 33, $16—No. 34, $17. Extras net—tinned wire, Nos. 17-25, $2—Nos. 26-31 $4—Nos. 32-34, $6. Coppered, 5c.—oiling, 10c.—in 25-lb. bundles, 15c.—in 5 and 10-lb. bundles, 25c.—in 1-lb. hanks, 50c.—in ¼-lb. hanks, 75c.—in ½-lb. hanks, $1—packed in casks or cases, 15c.—bagging or papering, 10c.

Galvanized Wire, per 100 lb.—Nos. 6, 7, 8, $3 5) to $3 8—No. 9, $2.85 to $3.15—No. 10, $3.60 to $3.9—No. 11, $3 70 to $4.10—No. 12, $3 to $3 30—No. 13, $3.10 to $3 [illegible]—No. 14, $4.10 to $4.50—No. 15, $4.60 to $5.05—No. 16, $4.85 to $5 35. Base sizes, Nos. 6 to 9, $2.57½ f.o.b. Cleveland.

Clothes Line Wire, solid 7 strand, No. 17, $4.25; No. 18, $2.65; No. 19, $2.35, f.o.b. Hamilton, Toronto, Montreal.

WIRE FENCING. F.O.B. Toronto

Galvanized barb	....	3 (5
Galvanized, plain twist	....	3 (5

Galvanized barb, f.o.b. Cleveland, $2.82½ in less than carlots, and $2.70 in carlots.

WIRE CLOTH.

Painted Screen, per 100 sq. ft., net.. 1 35

WASTE COTTON. per lb.

Colored	4½ to 5
White, according to quality	6½ to 7¾

50(-lb bale lots shaded.

WRENCHES.

Acme, 35 to 37½ per cent.
Agricultural, 60 p.c.
Coe's Genuine, dis. 70 to 25 p.c.

Towers' Engineer, each	2 00	7 00
" S., per doz	5 80	6 00
G. & K.'s Pipe, per doz	....	3 40
Burrell's Pipe, each	....	3 00
Pocket, per doz	0 25	2 90

WRINGERS.

Leader	per doz.	$30 00	33 (0
Royal Canadian	"	26 00	28 00
Royal American	"	16 00	28 (0
Sampson	"	....	30 00

Terms 4 months, or 3 p.c. 30 days.

WROUGHT IRON WASHERS.

Canadian make, discount, 40 and 5 per cent.

CANADIAN
HARDWARE
AND METAL MERCHANT

The Weekly Organ of the Hardware, Metal, Heating, Plumbing and Contracting Trades in Canada.

VOL. XIII. MONTREAL AND TORONTO APRIL 13, 1901. NO. 15

NEW CYCLE AND SPORTING GOODS FOR 1901

THE TRADE

will be pleased with our new assortment of goods for the coming season. Our travellers are now on the road with samples ; if they have not seen you, a card to us will bring them to you. It will pay to get our new prices.

WRITE FOR CATALOGUE AND PRICES.

RICE LEWIS & SON

LIMITED

Cor. King and Victoria Sts., TORONTO.

How to Prevent a Belt from Slipping.

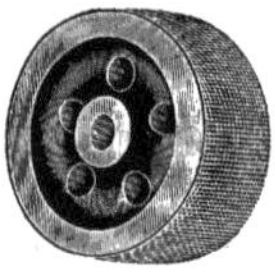

A BELT WILL LAST TWICE AS LONG

If our Pulley Covering is used, as thousands of manufacturers using leather belts for the transmission of power experience annoyance and trouble with slipping belts, from which arise many losses, such as time, power, etc.

No doubt you have tried many preparations, but the **"Vacuum Cement and Pulley Covering"** will prevent all belts from slipping, being free from all harmful ingredients. It will put your belts in better condition and keep them so, help them to do the proper work, as no other dressings or coverings ever invented have done before.

"The Vacuum Cement and Pulley Covering" is not the cheapest, but the best. It allows the use of a much slacker belt than usual ; this slackness prevents belt strains, as well as strains on the machinery. The covering may be applied to any pulley, iron or wood, and is guaranteed for five years. We furnish the material, and the work can be done right in your own shop. We invite your careful investigation.

Write for full particulars.

The Vacuum Cement & Pulley Covering Co.,

Main Office, 1493 Third Ave., NEW YORK.

THE EASE WITH WHICH THE

White Mountain Freezer

MAKES money for the dealer and makes ice cream for his customers, makes it the most attractive and satisfactory freezer to sell.

A White Mountain Freezer sold means a perfectly satisfied customer, a permanent advertisement and a liberal profit. Thousands of dealers selling thousands of **White Mountain Freezers** will testify that this states the case exactly as it is.

OUR AGENTS IN THE DOMINION ARE

The McClary Mfg. Co.

LONDON, TORONTO, MONTREAL, WINNIPEG AND VANCOUVER.

Write them for complete descriptive catalogue, special information or prices on our complete line.

The WHITE MOUNTAIN FREEZER CO.

Nashua, N. H.

HARDWARE AND METAL

VOL. XIII. MONTREAL AND TORONTO, APRIL 13, 1901. NO. 15.

President,
JOHN BAYNE MacLEAN,
Montreal.

THE MacLEAN PUBLISHING CO.
Limited.

Publishers of Trade Newspapers which circulate in the Provinces of British Columbia, North-West Territories, Manitoba, Ontario, Quebec, Nova Scotia, New Brunswick, P.E. Island and Newfoundland.

OFFICES

MONTREAL - - - - - - 232 McGill Street, Telephone 1255.
TORONTO - - - - - - 10 Front Street East, Telephone 2148.
LONDON, ENG. - - - - 109 Fleet Street, E.C., W. H. Miln.
MANCHESTER, ENG. - - - 18 St Ann Street. H. S. Ashburner,
WINNIPEG - - - - Western Canada Block. J. J. Roberts,
ST. JOHN, N.B. - - - No. 3 Market Wharf, J. Hunter White.
NEW YORK. - - - - - - 176 E. 88th Street.

Subscription, Canada and the United States, $2.00.
Great Britain and elsewhere - - - 12s.

Cable Address { Adscript, London. Adscript, Canada.

Published every Saturday.

STEEL SHIP SUBSIDIES IN NOVA SCOTIA.

A BILL has been passed by the Nova Scotia Legislature empowering municipalities within the Province to subsidize steel ship building companies to the extent of $100,000, provided the ratepayers approve.

The Bill was the subject of a great deal of discussion before it was passed, there being a strong opposition to a subsidy being granted upon anything other than results. In other words, it was contended that the subsidy should only be paid on vessels of iron or steel when completed. The promoters of the Bill were stronger, however, and it was finally passed without any conditions as to payment by results.

The principle on which the bonus on pig iron and steel is based under Federal authority is on results. It is the same with regard to the Beet Sugar Bounty Bill passed by the Ontario Legislature a few days ago. And it seems to us it would have been the better plan for the Nova Scotian Legislature to have adopted.

Against subsidies wisely given and properly safeguarded nothing can be said from a business standpoint. Otherwise, a great deal is to be said against them. There are municipalities in this country which are almost poverty stricken because of the subsidy mania which possessed them in the years gone by.

As HARDWARE AND METAL has pointed out in previous issues, the country is possessed with the same mania now. There are cities and towns throughout the Dominion which are possessed with the idea that they have the ideal spots for the location of iron industries, twine and rope factories, beet sugar factories or other industries, which the ubiquitous promoter is busily trying to establish. There is, undoubtedly, room for each and all of those industries in the country, but there is not room for them in every city or town.

With commendable zeal, Nova Scotia is desirous of occupying among the shipbuilding parts of the world a position analogous with that which she occupied in the days of wooden ships. Nearly every port of importance wants to possess a steel shipbuilding industry. And each appears to be afraid that the other will get ahead of it. This is all right and to be commended. But herein lies the very danger of a lax subsidy law, for, with the competition among the different ports keen, there is always a chance that too much may be paid for even a shipbuilding industry.

While the burden resultant from such possible mistakes would naturally largely rest on the municipality making it, it must not be forgotten that indirectly the country as a whole would suffer also.

IDEAS FOR SPRING TRADE.

WITH the opening up of spring hardwaremen are naturally looking for an increased trade. In this improved trade all will share to some extent, but the greater share will be enjoyed by those who put forth special efforts to secure it, just as the yachtsman who is alert for every favorable breeze makes better headway than he who is careless in this respect.

During the spring time as well as during other seasons of the year, scarcely too much thought and care can be given to the goods which are then seasonable. While one can often profit from the ideas of others, there is, after all, nothing which tends to more successfully attract trade than ideas and schemes that are original.

It is individuality stamped upon a business that makes it successful. He who is satisfied to be an atom, just like other atoms in the business world, of course attracts no attention.

Individuality comes of thinking, of devising and of acting. And in devising ways and means of profiting by the spring trade the merchant should not only set his own mental powers to work, but should endeavor to stimulate the mental powers of his clerks along the same lines as well.

There are the window displays, the advertising, the arrangement of goods in the interior of the store, all afford excellent opportunities for the development of original and trade-making ideas.

THE GOVERNMENT'S STEEL RAIL PURCHASE.

THE Hon. Mr. Blair, Minister of Railways and Canals, announced in the House of Commons on Tuesday that his Department had made a contract in October last with the Clergue company at Sault Ste. Marie, Ont., whereby the latter was to deliver, by a certain time next year, 25,000 tons of steel rails at the wharf, at either Montreal or Levis, the price to be $32.85 per ton. It was also agreed to take a like quantity during each of the four following years at the prices which would be ruling in England at the time of delivery.

The announcement has naturally created a great deal of interest. At present, no steel rails are being made in Canada. Indeed, steel rails never have been produced in Canada. Neither, strictly speaking, were even iron rails ever made in this country. Forty-one years ago a mill was started in Toronto to re-roll iron rails, but the advent of steel rails compelled it to go out of existence. In 1864, The Great Western Railway Company started a rolling mill in Hamilton, Ont., to patch and re-roll iron rails, but it ceased to exist eight years later. From that day to this we have not had in Canada what, even by courtesy, might be called a rail-making plant.

We are not likely, however, to long remain in that position. The fact that Mr. Clergue has contracted to, next year, supply the Government with 25,000 tons for use on the Intercolonial Railway makes that evident. But the fact that the Clergue company has already a plant in course of construction for the making of rails from nickle steel is still better proof, while down at Sydney there is being constructed the steel works of The Dominion Iron and Steel Co., from which rails will be turned out before a great while. Then there is also the recently completed steel works of The Hamilton Iron and Steel Co.

The action of the Government in placing the order in question has caused quite a little adverse criticism. The main points of the contention are: (1) That tenders should have been called for; (2) that the price paid was higher than that ruling, a decline having taken place a few days before the contract was signed.

Under ordinary conditions the objections would have been well taken; but they were not ordinary. Mr. Clergue, on behalf of the company that is making such an industrial revolution at Sault Ste. Marie, Ont., came to the Government with a proposition to establish a nickle-steel rail industry. What he wanted, of course, was some encouragement from the Government. And he got the encouragement in the shape of an order for 25,000 tons of rails at $32.85, delivered at the points named by the Government. It was certainly, with the iron and steel market in the condition it was last fall, not the way an ordinary business corporation would have placed a similar order. But the Government was desirous of giving some assistance to an industry which we have practically never had but which we have for a generation been trying to get. There is not an industry in this country that has not, either directly from the Dominion Treasury or through a protective tariff, received Governmental assistance. And the difference in the market price of steel rails when the 25,000 tons in question are delivered and the price which the Government contracted to pay will represent the measure of that assistance. If the Government is to be condemned on that score, then must the whole system of protection to home industries be condemned.

In 1879 a duty of 15 per cent. was placed upon iron rails, but steel rails have always been on the free list.

When Sir Charles Tupper brought down his famous iron tariff in May, 1887, he explained that it was the intention of the Government to make an exception of steel rails and allow them to remain on the free list, because of a desire not to retard the railway development of the country. He, however, said that in his judgement the iron policy which the Government had adopted would place Canada in the position where she would be able to provide her own rails within the next ten years. In his budget speech of 1888 he again referred to the subject, speaking as follows :

> I have been pressed, and strongly pressed, to take another step in that direction, for the purpose of having steel rails manufactured in our country. I mentioned to the House a year ago that Canada was the only country in the world possessing 12,000 miles of railway within her borders that did not manufacture its own steel rails, and I had the evidence presented to me that, by giving proper protection, such protection as we gave the other branches of the iron industry, we might succeed in establishing rolling mills for steel rails. But we had to take into consideration the fact of the enormous importance of railway development of a country like Canada, and under those circumstances we felt that we must postpone, at all events for this year, making such a change as would lead to the establishment of rolling mills in this country for the manufacture of our own rails.

The following year did not see a duty imposed for the encouragement of steel-rail making, neither has any year since. In the United States there is a duty on steel rails of 7.20c. per lb.

Speaking in round numbers, Canada's imports of steel rails during the last 15 years have ranged from 1,500,000 to 3,000,000 tons. Last year the quantity was 2,617,646 cwt. and the value $2,787,866.

We have more often condemned than commended the Government for its actions, but we do not see in the present instance how we can do any other than commend it.

INSOLVENCY LAW POSTPONED.

FOR the first time in some years, we have now a definite idea of how the Dominion Government feel on the question of insolvency legislation.

The report laid before the council of the Montreal Board of Trade by Mr. Robert Munro, from the insolvency committee, lets in some light on the Government's intentions. He had an interview with the Minister of Justice (Hon. David Mills), who said that the subject of insolvency was under consideration by the Ministers.

Apparently, the wish of the Federal authorities is that the Provincial laws shall be brought to embody all the reforms demanded by the commercial interests. The removal of preferences in Nova Scotia, for instance, which is one of the reforms demanded by the trade, will be sought by Provincial legislation first. If the Nova Scotian Legislature does not see its way to modify the present law, then the Dominion Government will consider the wisdom of enacting a Federal law, dealing, we infer, only with this phase of the question. But no promise is made that anything will be done this session of Parliament.

We are sorry the Government cannot carry out the wishes of the commercial interests in this matter during the present year. A general Dominion law is preferable to Provincial enactments. The time to pass one could easily enough be secured by squeezing out pro Boer resolutions and other political "tommy-rot" by which the time of our Parliament is so much wasted.

TRADE IN COUNTRIES OTHER THAN OUR OWN.

MANUFACTURED IRON AND STEEL IN GREAT BRITAIN.

THE finished branches of the trade have been somewhat quieter during the past week, but makers are fairly well booked ahead, taking the position all through, and do not look for any scarcity of work through the spring and summer at any rate. There is a disposition, too, to expect an improvement in the amount of new business coming forward after the holidays. The drop which has taken place in values lately is sufficiently apparent in Mr. Waterhouse's return, which gives the average selling price of manufactured iron in the north of England during January and February as £7 10s. 9d., as against £8 5s. 2d in the previous two months, a reduction of 14s. 5d. per ton. It is plainly apparent that there must be further reductions in subsequent returns as the year advances, for the ascertained prices are above present market values. This, of course, arises from orders having been in hand from last year, which were taken at higher rates. In the Midlands the ascertained selling price is £7 19s. There have been several reductions in prices in the Cleveland district this week. In iron castings, chairs and floor plates are down 2s. 6d., while iron ship plates and girder plates are cheaper to the same extent, and steel railway sleepers are reduced £1, ship plates 5s., hoops 10s., and boiler plates 2s. 6d. In Lancashire, iron bars are weaker, iron hoops are 2s. 6d. lower, and in steel billets and boiler plates a similar reduction has been made. In South Wales, steel rails are 5s. cheaper, and common bars in South Staffordshire are down to £7.—Iron and Coal Trades Review, March 29.

PIG IRON IN GREAT BRITAIN.

In the pig iron department of the trade the amount of business doing is confined within comparatively narrow limits, buyers being unwilling to make any extensive purchases while prices are as weak as they have now become. All are naturally anxious to place their orders at the lowest possible rates, and the slackening in business noticeable since our last, accompanied as it has been by further slight reductions in prices, has shaken their belief that bottom has been touched, and they are holding off again to secure better terms in the future. It is more than doubtful, however, whether their hopes will be realized, for the present reduced production is only about sufficient to meet current requirements, while in some kinds of pig, notably level and forge qualities, the output is not equal to the consumption, and in these circumstances prices can hardly continue to fall for any length of time. Makers are not expecting much further cheapening of coke apparently, for they are buying forward more freely than for some time past, and the only way in which they can hope for any relief from the present high cost of production, which at the present low rates leaves but the narrowest margin of profit, despite the reduction in fuel, is at the expense of the workmen, and the reduction in wages in the North of England, which will result from the ascertainment of the average selling price of iron, will be a step in this direction so far as the Cleveland district is concerned. Hematite is weaker than ordinary iron in both Barrow and Middlesbrough, mixed numbers in the latter centre having gone down to 56s. per ton —Iron and Coal Trades Review.

THE LEAD MARKET IN ENGLAND.

Lead has been a firm market, and the weakness which recently characterized it has, temporarily at least, passed away. During the early part of the week a large business is reported to have been done in Spanish at £12 15s. to £12 17s. 6d., amounting, it is said, to between 800 and 900 tons. On Thursday, when a slump is generally looked for, as much as £13 3s. 9d. was paid for distant delivery, and there is a fair amount of inquiry for arrival stuff by consumers. With any marked revival in the building trade, lead would not remain long at its present level.—Iron and Steel Trades' Journal.

NEW YORK METAL MARKET

TIN—The London pig tin market which has been closed since Thursday noon of last week opened on Thursday at an advance of 12s. 6d., which was possibly a reflection of the higher figures established here on Monday through the efforts of the chief holders of spot stock. English buyers evidently did not respond to the improvement in their market, which closed quiet at a decline of 2s. 6d. from the highest point of the day. This was a disappointment to the bulls on this side who had evidently looked for a substantial advance in London. The New York market was consequently dull and tame, closing at 26c. bid and 26.50c. asked for spot. For April, 25.80c. was bid, but later deliveries were not mentioned. To add to the depressing influences in this market a cable from London announced that the Mesaba due here from that port on March 15 will bring 970 tons, while the Polarstjernen now due from the Straits will land here 575 tons from Singapore and from 50 to 150 tons from Penang. The Manitou from London brought 183 tons, making the arrivals for the month to date 405 tons. Stocks afloat including that on the Mesaba amount to 4,034 tons.

COPPER—The downward movement of prices in London was renewed at the opening of that market to day after the interval covered by the Easter holiday. There was little business done in spot or futures, and the close was quiet at a decline of 2s. 6d. from last Thursday's quotations. Nothing new was presented in the New York market, which remained quiet but for deliveries on existing contracts. Prices, however, were steady and unchanged at 17c. for Lake Superior and 16⅝c. for electrolytic and casting.

PIG LEAD—The settlement of the Guggenheimer affair was not reflected in the market for the metal, which closed quiet and steady at 4.37½c. for 50-ton lots or more. St. Louis was reported quiet at 4.22½c. There was no change in London.

SPELTER—This market was somewhat firmer on increased buying for local consumption, and prices advanced to 3.92½ to 3.97½c. Advices by wire from St. Louis reported that market steady at 3.75 to 3.77½c. The London market continued to decline, and closed 2s. 6d. lower than on Thursday last.

OLD METALS—The market is very quiet, but prices are steady and unchanged.

PIG IRON—For pig iron the demand continues heavy, local dealers reporting more business than at any time for the past three months. Full prices are readily obtained, but the tendency toward a further advance is not so pronounced, and no quotable change has been made for several days. Eor finished materials including sheets and plates there is still a very active demand, and with most buyers it is more a question of supplies than of price, so difficult is it to get even small orders for prompt delivery filled.

TINPLATE—There is no change to be noted in the situation, which is characterized by such activity of demand as taxes the producing capacity to its utmost. Prices are firm and unchanged.

GALT BOARD OF TRADE.

There is a vigorous movement on foot to reorganize the Galt Board of Trade. A meeting for this purpose was held on Tuesday night, but after consideration it was determined to postpone the matter for two weeks and to increase the membership of the board in the meantime, so that a strong working committee may be appointed. Six new members were received, and many more are expected.

UTILIZING THE STORE SERVICE.

At the beginning of the busy retail season is a good time to think about getting all that is possible out of the store service—the firm as well as the sales force. From about the middle of March everybody is busy, so busy as to think nothing more could be done, but the work is like a crowd of people—always room for one more. Nor is it always so much a question of doing more work as it is doing better that which is already on hand.

The conduct of a store requires as much diplomacy as dogged persistence, and both must be forced to work together. Maybe the suggestion will seem a little strange, but it is a truth that the merchant must be more or less careful with the treatment of himself, and govern his own actions from the standpoint of how much he can get out of himself and how best to control and make most profitable his own work.

The man who is the best controller of his own actions, makes his personal machinery work the smoothest, longest and best and gets the best results from all his efforts is the man who will also have the most efficient, best controlled, most willing, most enthusiastic and most reliable lot of people in the work about him. He will always have a good store, because the spirit of doing things right and doing them well is as contagious as the spirit of lounging and the hope of doing just as little as possible.

INFECTIOUS SPIRIT.

Nothing in the world is more infectious than the attempt to make results better than the results of like attempts have been before. If the merchant will stop to think about it he will discover the spirit within his own work. He is not satisfied that certain things are done; he wants to do them better and is searching for the methods whereby the desired improvement can be accomplished. The merchant who has the don't-care spirit back of even the little details always finds his balance sheet considerably disappointing at the end of a season of selling which should show large returns.

The manner of doing things with a vim and energy and doing lots of them does not imply that the merchant has to personally look after all the details and flying ends of the store business. That is where many men make a mistake and worry their time into a lot of small matters when they might turn them over to others and give their time to the planning and mapping out of management.

DIPLOMATIC WORK.

Right in there is where a good bit of diplomatic work can be pushed through. Every merchant has ways of his own in managing the people in his charge, and usually thinks that he does it pretty well, but there is one method of procedure which is not used as much as it should be and which is really the easiest of all ways, both for the merchant and the employe. More flies are caught with sugar than with vinegar, and that means as many kinds of human flies as there are kinds of work to be done.

Everybody from merchant down to bundle boy knows that and knows when the sugar is being put out to catch, but he is not averse to it nor harmed by being frankly told that it is the method to be employed for his capture. We know that a clerk is very well aware when he is being led, when he is doing work for the reason that he wishes to please the firm by doing it right, as well as helping himself forward, but he is also very glad to do it that way. It is the pride of cooperation that comes as a result of being asked pleasantly, of having responsibility thrust upon them, and of feeling that much does really depend on them.

In the rush of spring business is a most excellent time to begin delegating to other people some of the work the merchant has been doing and lay out some of the work that nobody has yet done, but which everybody feels should be done. The successful general is the one who can lay out the campaign and lead his armies to victory, giving all the minor work to others, yet keeping track of the whole thing himself. The successful merchant is the same character of man. That we do not mean that the detail worker is not successful, but we mean that he expends an amount of energy on details that could be made to bring a vastly larger amount of returns if applied in other ways.

DON'T DO IT ALL

The point is something like this: You haven't got to superintend the unpacking of every case and watch every piece that comes from it, the unwrapping of new goods and the placing of them in the shelves doesn't have to be watched minutely by you; every sale doesn't have to be boosted along or assisted by some effort of yours; you need not mark all price tickets; you don't have to watch every entry on the books; you don't have to turn on the lights and wait until they are turned off; you can have a carefully selected stock without picking out every pattern yourself; other people can drive nails and turn screws; your millinery trade does not hinge on your personal selection of braids, flowers and feathers; the arrangement of goods in show cases, on ledges and in windows can be accomplished by other people than yourself.

YOUR EXPERIENCE.

It will not leave you idle nor shirmishing about something to occupy your mind and hands. It will give you an opportunity to become more completely the master of your business. When you were an employe you had great respect for your employer who had the actual knowledge of all the work which he expected you to do, but did not think the more of him because he hustled around and tried to do it all himself. And, more than that, we are of an opinion that has a pretty fair basis, that you thought him welcome to all those duties so long as he was willing to do them without calling on you. Now that you are the employer, cannot you see how the same plan will work with you?

It is not because the people about the store are particularly lazy, but they become indifferent when no responsibility rests upon them. The infectiousness of work comes not alone from seeing others hustle; it also comes from having something to hustle for. Don't study on how much more work can be placed on the shoulders of your store people, but study out a better system of attending to the whole store service, and you will find that there is plenty of time for everyone to do something more.

DO IT TOGETHER.

Talk with the boys who unpack the goods; tell them how it is best done; put them on their honor to do it right, and you'll find it to be done as well as though you did it yourself; if it isn't, discharge the boys and get others—there are lots of good ones left.

Talk with the people who have charge of the stocks. If they have a better way of doing things than you, let them do it that way; you don't have to turn in and do it yourself. If the dress goods, linens, underwear, or what not are kept well it isn't necessary to worry because it isn't done just your way. The result is what you are after, and when the result doesn't come there is plenty of time to take a hand.

Get your people started right and they will be as enthusiastic as you are in the betterment of the store. Be sincere and let them understand that you are as anxious for their improvement as for the increase of your sales. Make them to know that your present and their future are linked together.

Don't try to do it all yourself. No man was ever made who could, and your endeavor is sure to leave undone many things of more importance than some you do.—General Merchant.

EXPERIENCE IN CASH TRADE.

THE experience of merchants with the cash and credit systems of trading is interesting and instructive. In an interior Illinois town there is a storekeeper who is doing a cash business, and his sales run from $135 to $400 a day. He also keeps an account of the cost of each article and what it is sold for, and knows each night the result of the day's business. This merchant is doing a satisfactory business, and is getting on in the business world. It is a clean, neat way of operating a retail store.

Another Illinois firm, located at Manton, has been experimenting with cash and credit. Prior to three years ago they had used credit so extensively that they found they had tied up about all their capital, amounting to several thousand dollars, on their books. The firm also owed a few thousand dollars, and were in a state of perplexity.

To get out of the hole it was decided t stop credit, collect as closely as possible and sell for cash. In a year they had cash resources and were discounting their bills. Having thus got on easy street, it seems strange that the cash system should not have been maintained. The merchants, however, confess that they were weak enough to slide back to giving credit, and, before they really knew it, were in the dumps again.

Just as this fix was discovered the store had a fire and quite a loss resulted. This sort of discipline determined these people to readopt the cash system, and they say they will stick to it this time. As a result of having tried the two plans of selling goods these merchants express the following views :

"Under the credit system a year's failure of crops may wipe a fellow out but there is no fear of this with the cash plan. Then we think the cash plan enables the merchant to satisfy his customers who watch mail-order houses and buying. There is less dissatisfaction and discontent with our goods and prices.

"Also, on a cash basis there is little or no fear that some new merchant is going to come to town and knock you out. This fear is constant with the credit merchant. The worry from this and the possible loss of bad accounts keeps a credit man in a bad state."

It is such object lessons as these that should be advisory to all merchants who are worrying along with the credit system. They can benefit themselves by studying the experience of others and the proof that the cash storekeeper does not lose, but, on the other hand, gains, by adhering to that system.

The cash-dealing merchant does not have to vex himself to get garnishment laws passed by the Legislature. He does not have to chase those that owe him, and take his pay for the goods he has sold on the instalment plan.

If the merchant will stop and think it over carefully he will stop lending his money to Tom, Dick and Harry and allow them to pay as they please, or not at all. As a cash merchant he will do as much, very likely more, business, and he will have mercantile peace and prosperity and something to show for his labor.—Exchange.

AN IMPORTANT QUESTION.

Mr. H. L. Hjermstad, the author of a prize paper on the subject "What Should the Retailer Do to Secure New Customers, and How Can He Hold Them?" says : "I have never resorted to any schemes, premiums, cut prices, or soliciting of orders from house to house, as I do not believe in such means to obtain business, and do not for an instant believe that custom obtained in this manner will become permanent, as nothing but the best quality, best service, coupled with a reasonable price, will keep a customer permanently."

THE REWARD OF INDUSTRY.

A vagrant who had been sentenced to death begged to be taken before the King, that he might plead for his life. When he had been brought to the throne, the King looked down upon him and angrily said : "Thou worm, why comest thou adding to the troubles of thy monarch? Dost think, oh, thou crawling, cringing thing, that thy fate is worthy of the notice of a king? Begone, thou drone—out of my sight! Thou hast never done a thing in all thy worthless life. Thou art like a rotten shingle —useless. There is not one little reason why I should spare thee. Away with him!" "But, oh gracious King, hear me," the vagrant cried. "Thou sayest I never did anything in my life. Nay, thou wrongest me. Even now I am doing something." "What is it?" the King demanded. "Letting my whiskers grow." At this the King was so well pleased that he not only restored the man's liberty, but made him oil inspector at a salary of $12,000 per year, with a cheap boy to do the inspecting.—Chicago Times-Herald.

THE COAL OIL DUTY.

A SPECIAL meeting of the executive Committee of the Canadian Manufacturers' Association was held on Tuesday to consider the motion of Mr. E. R. Clarkson, Hamilton, asking the Government to remove the duty on coal oil. It was decided that not sufficient facts had been laid before the association to clearly show "that the price of oil has been raised to an exorbitant extent as a result of the duty," but a motion was adopted calling upon the Government "to appoint a commission to fully investigate the facts connected with the Canadian oil industry prior to adopting legislation affecting this industry."

There is scarcely any article of commerce in the Custom's tariff around which has centered so much discussion as coal oil, and yet most people are at "sixes and sevens" in regard to the actual merits of the case. This is, perhaps, largely due to the fact that party politics have been industriously mixed up with it. If the suggestion of the executive committee of the Manufacturers' Association is acted upon by the Government, it is probable we shall get more light upon the question than we have hitherto had. We hope, therefore, that the memorial of the Canadian Manufacturers' Association will find favor with the Government.

TRADING STAMP BILL PASSED.

On Thursday the Ontario Legislature went into committee of the whole to discuss the Trading Stamp Bill introduced by Geo. P. Graham, M.P.P., for Brockville. The bill, which empowers municipalities to prohibit the handling of trading stamps under the present system, leaving, however, to merchants or manufacturers the right to give out their own coupons, which they must redeem themselves, was reported favorably by the committee by a majority of 48 to 20.

NEW CEMENT COMPANIES.

The production of cement in Canada continues to rapidly increase. During the next few days the Manitoba Cement Company will start their works, their plant having all been installed and a good sample manufactured, but the warehouse is not yet ready to receive stocks The mines and works are about five miles from the thriving town of Miami, on the Morris and Brandon branch of the N. P. and M. Railway. The plant for the manufacturing of cement is complete, and when tested a few days ago worked highly satisfactorily. The plant, which includes the latest improvements for the manufacture of cement, gives a capacity of 250 bbls. per day.

Another company has been organized to manufacture cement in Ontario. This organization, known as the Dominion Portland Cement Co., has $250,000 capital stock, and is composed of Wiarton, Walkerton and Hepworth business men. Marl beds, which are declared inexhaustable, have been located between Wiarton and Oxenden. The Township of Keppel, of which Oxenden is part, has offered the company exemption from taxes to locate the works in the latter place. An agitation is on foot in Wiarton to offer inducements to the company to locate there.

H. S. HOWLAND, SONS & CO.

WHOLESALE ONLY — 37-39 Front Street West, **Toronto.** — ONLY WHOLESALE

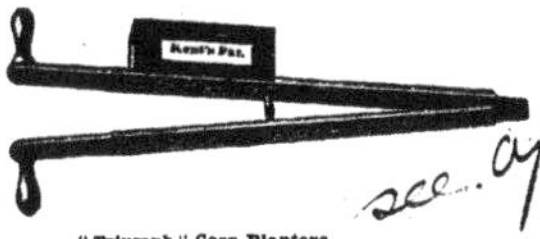

"Triumph" Corn Planters.
" With Pumpkin Seed Attachment.

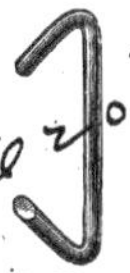

Hog Rings.
Nos. 1, 2, 3.

Hog Ringers, Cast.
" Malleable.

RAKES

Malleable Iron.

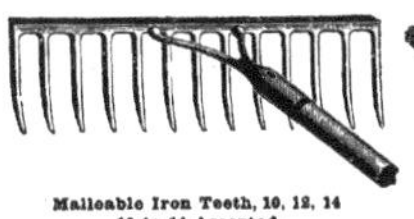

Malleable Iron Teeth, 10, 12, 14
10 to 14 Assorted.

Steel.

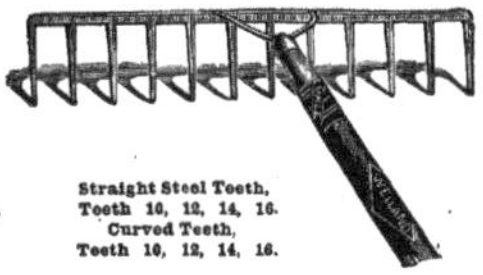

Straight Steel Teeth,
Teeth 10, 12, 14, 16.
Curved Teeth,
Teeth 10, 12, 14, 16.

Lawn.

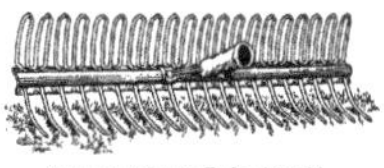

"Queen" Lawn Rake (as cut.)
24 Steel-Tinned Teeth.
"Gibb," Wire Teeth, Wood Back.

PRUNERS

Tree Pruners.

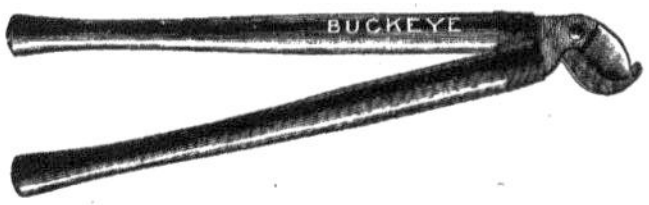

No. 35 Cast Steel Blades, Length 26 in.
No 38 Cast Steel Blades, Length 41 in.
will cut 1¼ in. Stick with ease

Pruning Shears.

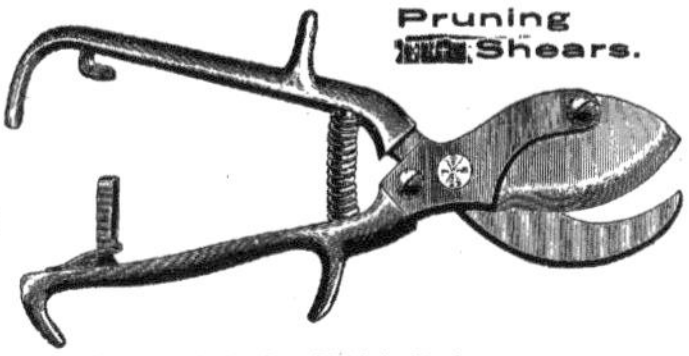

"Boker's" 1636, 8 in. long, Flat Spiral Spring.
"Boker's" 7191, 8½ in. long, Flat Spiral Spring, Bow Handles.
"American" 12, 8½ in, long, Flat Spiral Spring.
"American" 0, 9 in long, Brass Coil Spring.

Long Handle Tree Pruners.
Feet 6, 8, 10, 12.

H. S. HOWLAND, SONS & CO., Toronto.

WE SHIP PROMPTLY. — Graham Wire and Cut Nails are the Best. — OUR PRICES ARE RIGHT

THE MEETING OF DEPARTMENT STORE COMPETITION.

Editor HARDWARE AND METAL—In trying to counteract the evil effects of departmental stores, I think with your correspondent, "Subscriber," it is best to fight them with their own weapons, although not just in the manner suggested by him, as I think it is very doubtful if city buyers could be induced to purchase from country dealers.

No doubt the injury done to small towns and villages is very great, but "Subscriber" may rest assured that the ruin departmental-store methods has brought on both retailers and wholesalers in large cities is most serious.

Now, I suggest: First—Let all classes of business men in a city drop their petty jealousy, form themselves into a self-preservation society and subscribe to a common fund, to be used under properly appointed directors in fighting the stores. Second—Arrange that the business place of every member of the society shall be used in rotation as the leading bargain emporium for a particular day. Third—Watch the "ads." of the depatmentals and whatever goods they offer at cut rates, say, on Monday, sell the same class of goods at the society depot on Tuesday, but at prices below the departmental stores' offerings of the day before, and so on with every day of the week, following cut prices with a deeper cut on the next day. Of course, the special store and its one-day-price must be well advertised in the papers.

The public, and especially bargain-hunters, would soon learn that they could save money by waiting one day longer, and would gradually forsake the departmentals and patronize the cheapest place. Of course, the expenses would be considerable, but if every dealer in each line of business affected by the departmental stores would unite in one society the expense to each individual would be small.

I hope this matter may be thoroughly ventilated by your readers, and have no doubt that some method will ultimately be suggested whereby the evil effects of departmentals may be mitigated, if not overcome.

RETAIL.

Hamilton, Ont., April 8, 1901.

ENCOURAGING TRADE WITH GREAT BRITAIN.

There are a number of Canadians residing in Great Britain who are anxious to see the manufacturers of Canada do a larger business with the Mother Country. One of them is Mr. J. H. Moore, formerly a hardware merchant in Hamilton, Ont., but during the last three years a resident of London, Eng. And so much interested is he that he returned to Canada five or six weeks ago with the object largely of trying to induce manufacturers who had not yet done so to make an effort to cater for the British trade. In the little time he has had at his disposal he has interested quite a number, and some have already consigned goods to the metropolis. Mr. Moore, who sailed on Monday last for London, Eng., will be only too glad to correspond with any manufacturers or others who may desire information about the British market. His address is 67 Aldersgate, London, E.C.

TRADE-PAPER ADVERTISING.

THE first mission of any ad. is to be seen—not its greatest mission, but its first, says Charles A. Bates in Current Advertising. To be seen, the ad. must be conspicuous, and conspicuous in a different way from the ads. that are around it. This means display.

But after an ad. has been seen, its next and greatest mission is to convince. To do this it must say something about the goods advertised, and say that thing in a way which will carry conviction to the reader. More than this, each and every ad. which appears in any given trade journal should be a part of a carefully-prepared advertising plan. The effect of ads. which have gone before or which are to come after should be considered. Each ad. should occupy its space in this advertising story, and should do its part toward adding point to point in convincing the reader of the trade journal that your goods are the best for his purpose You wish to make him believe that your goods are the kind which will bring him more business and better business, which it will pay him to carry, and for certain specific reasons. In order to get his entire and undivided attention you should give these reasons.

If your goods, no matter what they may be, ought to be bought there are certainly some reasons why they ought to be bought. There are some reasons why your present customers prefer to buy their goods from you instead of from your competitors. What are those reasons? You must know what they are and you ought to print them in your trade paper ads. The facts which make you and your establishment preferred by your present customers would operate to bring new customers if you told what they are and kept on telling year in and year out. That is what your trade journal space is for.

If you sent a man on the road and he simply called on possible customers, presented a card having printed on it your name and your line of business, and then stood with his mouth shut to see what would happen, you would think that he was either a fool or crazy. You wouldn't expect him to sell goods. On the contrary, you would expect him to injure seriously the reputation and prestige of your house. You expect a salesman to do something more than call attention to your name and address. You expect him to wax eloquent over your facilities for rendering just the kind of service that the prospective customer desires. You expect him to explain in detail just how more money can be made and a more satisfactory business can be done by handling your goods than in any other way. That is what a travelling salesman is for. And that is precisely what a trade journal ad. is for. If you can make your trade journal ads. talk just the way a really good salesman does, you will come just as near having a perfectly written ad. as it is possible to come.

In still another respect this trade journal advertising is like the work of a traveller. When your traveller goes into a store and asks for an order and a merchant says he doesn't want anything in your line just now, the traveller goes away and comes back again when he is making another trip. He does not feel discouraged nor lose the hope of finally making that man a customer. The man was approached at the wrong time, and the drummer keeps coming back every time he is in town, in the hope that, sooner or later, his visit and the man's requirements for your line of goods will coincide, or that some time he will find his man in the kind of mood which will enable him to make a sale.

It is just the same way with trade journal ads. You cannot expect one advertisement to do the work in every case or in the majority of cases. You have no good reason for saying that because big results do not pour in, trade journal advertising is no good. You are simply making weekly or monthly calls upon a certain number of men who ought to become your customers. Through their trade journal you are approaching them at regular intervals and asking them for their trade. The fact that they do not immediately fall on your neck and embrace you does not go to prove that they are not impressed by your advertising or that sooner or later you will not be able to make them your customers. The thing to do is to keep right on with regular, systematic advertising campaign and wait, just as the traveller does, for your advertisement and their wants to coincide.

MARKETS AND MARKET NOTES

QUEBEC MARKETS.

Montreal, April 12, 1901.

HARDWARE.

THE wholesale houses are decidedly busy, and they will likely have more business than they can conveniently handle until the opening of navigation. Dealers at water points are placing good orders for shipment by the first boats. Shelf goods are selling well, particularly house-furnishing and builders' hardware. At the meeting of the Lantern Association, last week, the price of cold-blast lanterns was reduced from $7.50 to $7 per dozen, and that of ordinary from $4 25 to $4 per dozen. Hollow-ware is now selling at 40 per cent. discount, instead of 45 per cent. Horse nails are being shipped in good quantities, and we hear that some concessions have been given on countersunk head, but, as a general rule, dealers are adhering to our schedule. Acadian nails and other inferior grades are being sold at a discount of 63⅓ per cent. "Monarch" brand is quoted at a discount of 65 per cent. The tone of the market is stiffer in England, the United States and Germany. We understand that some low offers of German wire that had been sent here have since been withdrawn. The English market is still fluctuating, yet dealers have enough confidence to order goods for July shipment. The United States steel market is buoyant, and the markets here are firmer, in sympathy. A regrettable feature of business at the present moment is the poor reception of paper; we have heard some loud complaints about payments this week.

BARB WIRE—The market is in a healthy condition. Fair quantities of goods are being shipped, and more will be sent out in the course of the next few weeks. The price is still $3.05 per 100 lb. f.o.b. Montreal.

GALVANIZED WIRE—Trade is quite up to the average upon a steady market. We quote as follows: No. 5, $4.25; Nos. 6, 7 and 8 gauge, $3.55; No. 9, $3.10; No. 10, $3.75; No. 11, $3.85; No. 12, $3.25; No. 13, $3.35; No. 14, $4.25; No. 15, $4.75; No. 16, $5.00.

SMOOTH STEEL WIRE—All fence wires are active. The market is steady to firm, steel being buoyant at all outside points. We quote oiled and annealed: No. 9, $2.80; No. 10, $2.87; No. 11, $2.90; No. 12, $2.95; No. 13, $3 15 per 100 lb. f.o.b. Montreal, Toronto, Hamilton, London, St. John and Halifax.

FINE STEEL WIRE — The demand is moderate. The discount is still 17½ per cent. off the list.

BRASS AND COPPER WIRE—There have been no new features to note this week. The discount on brass is 55 and 2½ per cent., and on copper 50 and 2½ per cent.

FENCE STAPLES—A fairly brisk trade is doing in fence staples at previous quotations. We quote: $3.25 for bright, and $3.75 for galvanized, per keg of 100 lb.

WIRE NAILS—Wire nails are active under a brisk inquiry for this time of year. A better demand is looked for in two or three weeks. We quote $2.85 for small lots and $2.77½ for carlots, f.o.b. Montreal, London, Toronto, Hamilton and Gananoque.

CUT NAILS—A fair demand has been experienced this week. We quote: $2.35

for small and $2.25 for carlots; flour barrel nails, 25 per cent. discount; coopers' nails, 30 per cent. discount.

HORSE NAILS—Since the market was declared open the range of prices on horse nails has been irregular. There have been further concessions made during the past week on all brands, except "C," the makers of which are holding to their established discount of 50 and 7½ per cent. The general discount on other nails, including "W" brand, is 60 per cent. on oval and city head, and 60 and 10 per cent. on new countersunk head off the old list. Some Acadian nails are being cleared out at a discount of 66⅔ per cent., but the dealers seem to have no further use for the inferior grade. "Monarch" nails are quoted at a discount of 60 and 10 per cent., or a straight discount of 65 per cent. Business is good.

HORSESHOES—As yet trade is rather quiet, but an improvement is anticipated soon. We quote as follows: Iron shoes, light and medium pattern, No. 2 and larger, $3.50; No. 1 and smaller, $3.75; snow shoes, No. 2 and larger, $3.75; No. 1 and smaller, $4.00; X L steel shoes, all sizes, 1 to 5, No. 2 and larger, $3.60; No. 1 and smaller, $3.85; feather-weight, all sizes, $4.85; toe weight steel shoes, all sizes, $5.95 f.o.b. Montreal; f.o.b. Hamilton, London and Guelph, 10c. extra.

POULTRY NETTING—Good shipments are being made. An extra concession has been given in some cases lately, but we believe the cutting is through with. We quote the discount at 50 and 10 per cent.

GREEN WIRE CLOTH—There is no new feature to the market. The price is still $1.35 per 100 sq. ft.

SCREEN DOORS AND WINDOWS—Trade is fair and there is still room for improvement. We quote as follows: Screen doors, plain cherry finish, $8.25 per doz.; do. fancy, $11.50 per doz.; walnut, $7.40 per doz., and yellow, $7.45; windows, $2.25 to $3.50 per doz.

SCREWS — Trade is brisk. Prices are unchanged. Discounts are as follows: Flat head bright, 87½ and 10 per cent. off list; round head bright, 82½ and 10 per cent.; flat head brass, 80 and 10 per cent.; round head brass, 75 and 10 per cent.

BOLTS — Bolts form one of the most active lines on the market. The demand from manufacturers is extremely heavy. Discounts are as follows: Carriage bolts, 65 per cent.; machine bolts, 65 per cent.; coach screws, 75 per cent.; sleigh shoe bolts, 75 per cent.; bolt ends, 65 per cent.; plough bolts, 50 per cent.; square nuts, 4¼c. per lb. off list; hexagon nuts, 4¾c. per lb. off list; tire bolts, 67½ per cent.; stove bolts, 67½ per cent.

BUILDING PAPER—Business is quite good at unchanged prices. We quote as follows: Tarred felt, $1.70 per 100 lb.; 2 ply, ready roofing, 80c. per roll; 3-ply, $1.05 per roll; carpet felt, $2.25 per 100 lb.; dry sheathing, 30c. per roll; tar sheathing, 40c. per roll; dry fibre, 50c. per roll; tarred fibre, 60c. per roll; O.K. and I.X.L., 65c. per roll; heavy straw sheathing, $28 per ton; slaters' felt, 50c. per roll.

RIVETS — Fair sorting orders have been filled this week. The discount on best iron rivets, section, carriage, and waggon box, black rivets, tinned do., coopers' rivets and tinned swedes rivets, 60 and 10 per cent.; swedes iron burrs are quoted at 55 per cent. off; copper rivets, 35 and 5 per cent. off; and coppered iron rivets and burrs, in 5-lb. carton boxes, are quoted at 60 and 10 per cent. off list.

BINDER TWINE — The movement in this line has hardly begun yet. We quote: Blue Ribbon, 11½c.; Red Cap, 9¼c.; Tiger, 8¼c.; Golden Crown, 8c.; Sisal, 8¼c.

CORDAGE — Business continues quite brisk. Manila is now worth 13½c. per lb. for 7-16 and larger; sisal is selling at 10c., and lathyarn 10c.

SPADES AND SHOVELS — Some good orders have come in this week. The discount is still 40 and 5 per cent. off the list.

HARVEST TOOLS — Inquiries as yet are not numerous. The discount is unchanged at 50, 10 and 5 per cent.

TACKS—Business is fairly good at steady prices. We quote: Carpet tacks, in dozens and bulk, 80 and 15 per cent.; tinned, 80 and 20 per cent., and cut tacks, blued, in dozens, 80 per cent.

LAWN MOWERS—These articles are not selling as freely yet as they will be, but still quite a few inquiries have been received. We quote: High wheel, 50 and 5 per cent. f.o.b. Montreal; low wheel, in all sizes, $2.75 each net; high wheel, 11-inch, 30 per cent. off.

FIREBRICKS — There has been no noticeable improvement in the volume of business. The demand is still quiet. Prices are $18 to $20 per 1,000 as to brand ex store.

CEMENT—Trade is beginning to pick up, starting somewhat earlier than last year. We quote as follows: German, $2.45 to $2.55; English, $2.30 to $2.40; Belgian, $1.95 to $2.05. American in bags, $2.30 to $2.40 per 400 lb.

METALS.

The English market in sheet metals is still fluctuating, but the feeling is much healthier than it was, and large dealers here have placed orders as far ahead as for July shipment, showing increased confidence in the market. This drop in prices has left dealers here with some stocks which were greatly reduced earlier in the season, and now on account of a light supply prices are being held firmly a good deal above import figures.

PIG IRON—In spite of the advance in the United States, prices here show no great change, a fact due partly to the depression in England and partly to the competition between the three chief Canadian furnaces. Canadian iron is worth from $18 to $18.50, while Summerlee for early import is quoted at $20 to $21.

BAR IRON—A fairly good trade is doing. Carlots have been sold this week at $1.65, but the general figure is $1.70.

BLACK SHEETS — The English market is fairly steady. Stocks here are rather light, and it is not always that the required gauges can be secured. For spot goods, we quote $2.65 for 8 to 16 gauge; $2.70 for 26 gauge, and $2.75 for 28 gauge.

GALVANIZED IRON—In spite of the low quotations offered on import account, prices on spot goods are being fairly well maintained in most cases. No. 28 Queen's Head is worth $4.65; Apollo, 10¾ oz., $4.50, and Comet, $4.40 to $4.45, with a 15c. reduction for case lots.

INGOT COPPER—There is no change in the market. Sellers are firm at 18c.

INGOT TIN—The London market is somewhat firmer this week. Lamb and Flag is quoted at 31 to 32c.

LEAD—The ruling price is 4c. per lb.

LEAD PIPE—No change. We quote: 7c. for ordinary and 7 1/2c. for composition waste, with 25 per cent. off.

IRON PIPE—The situation in steel makes iron pipe strong. Business is moderately brisk. We quote as follows: Black pipe, 1/4, $3 per 100 ft.; 3/8, $3; 1/2, $3; 3/4, $3.15; 1-in., $4.50; 1 1/4, $6.10; 1 1/2, $7.28; 2-in., $9.75. Galvanized, 1/2, $4.60; 3/4, $5.25; 1-in., $7.50; 1 1/4, $9.80; 1 1/2, $11.75; 2, in., $16.

TINPLATES—Dealers are trying to maintain prices on present stocks, which are quite light, but, nevertheless, are being sold at a loss. For immediate delivery they are worth $4.05 to $4.15 for coke and $4.30 to $4.40 for charcoal.

CANADA PLATE—The sorting demand makes up a fair volume of business. We quote: 52's, $2.60; 60's, $2.70; 75's, $2.80; full polished, $3.45, and galvanized, $4.30.

SHEET STEEL—Fair quantities are selling. We quote: Nos. 22 and 24, $3, and Nos. 18 and 20, $2.85.

TOOL STEEL—Prices are unchanged at 8c. for Black Diamond, and 13c. for Jessop's.

STEEL—The market continues strong, but prices are unchanged. We quote as follows: Sleighshoe, $1.85; tire, $1.95; bar, $1.85; spring, $2.75; machinery, $2.75 and toe-calk, $2.50.

TERNE PLATES—Fair shipments of goods have been made this week. The ruling price is $7.75 to $8.

COIL CHAIN—As the boats are being fitted up for the season's trade, a good demand has been experienced during the last few days on both local and outside account. We quote as follows: No. 6, 11 1/2c.; No. 5, 10c.; No. 4, 9 1/2c.; No. 3, 9c.; 1/4-inch, 7 1/2c. per lb.; 5-16, $4.65; 5-16 exact, $5.10; 3/8, $4.20; 7-16, $4.00; 1/2, $3.75; 9-16, $3.65; 5/8, $3.35; 3/4, $3.25; 7/8, $3.20; 1-in., $3.15. In carload lots an allowance of 10c. is made.

SHEET ZINC—The ruling price is 6 to 6 1/4c.

ANTIMONY—Quiet, at 10c.

ZINC SPELTER—Is worth 5 to 5 1/2c.

SOLDER—We quote: Bar solder, 18 1/2c.; wire solder, 20c.

GLASS.

A good business is being done in glass, both for immediate and future delivery. We quote as follows: First break, $2; second, $2.10 for 50 feet; first break, 100 feet, $3.80; second, $4; third, $4.50; fourth, $4.75; fifth, $5.25; sixth, $5.75, and seventh, $6.25.

PAINTS AND OILS.

The orders continue to be numerous, but small. But signs would indicate that they must soon improve in size. Linseed oil is steady in England, but in sympathy with the recent advance there prices are 2c. higher. Turpentine has been reduced 3c. We quote:

WHITE LEAD—Best brands, Government standard, $6.25; No. 1, $5.87 1/2; No. 2, $5.50; No. 3, $5.12 1/2, and No. 4, $4.75 all f.o.b. Montreal. Terms, 3 per cent. cash or four months.

DRY WHITE LEAD—$5.50 in casks; kegs, $5.75.

RED LEAD—Casks, $5.25; in kegs, $5.50.

WHITE ZINC PAINT—Pure, dry, 7c.; No. 1, 6c.; in oil, pure, 8c.; No. 1, 7c.; No. 2, 6c.

PUTTY—We quote: Bulk, in barrels, $1 90 per 100 lb.; bulk, in less quantity, $2 05; bladders, in barrels, $2.10; bladders, in 100 or 200-lb. kegs or boxes, $2.25; in tins, $2.55 to $2.65; in less than 100-lb. lots, $3 f.o.b. Montreal, Ottawa, Toronto, Hamilton, London and Guelph. Maritime Provinces 10c. higher, f.o.b. St. John and Halifax.

LINSEED OIL—Raw, 69c.; boiled, 72c., in 5 to 9 bbls., 1c. less, 10 to 20 bbl. lots, open, net cash, plus 2c. for 4 months. Delivered anywhere in Ontario between Montreal and Oshawa at 2c. per gal. advance and freight allowed.

TURPENTINE—Single bbls., 57c.; 2 to 4 bbls., 56c.; 5 bbls. and over, open terms, the same terms as linseed oil.

MIXED PAINTS—$1.25 to $1.45 per gal.

CASTOR OIL—8 3/4 to 9 1/4c. in wholesale lots, and 1/2c. additional for small lots.

SEAL OIL—47 1/2 to 49c.

COD OIL—32 1/2 to 35c.

NAVAL STORES—We quote: Resins, $2.75 to $4.50, as to brand; coal tar, $3 25 to $3.75; cotton waste, 4 1/2 to 5 1/2c. for colored, and 6 to 7 1/2c. for white; oakum, 5 1/2 to 6 1/2c., and cotton oakum, 10 to 11c.

PARIS GREEN—Petroleum barrels, 16 3/4c. per lb.; arsenic kegs, 17c.; 50 and 100 lb. drums, 17 1/2c.; 25-lb. drums, 18c.; 1-lb. packages, 18 1/2c.; 1/2-lb. packages, 20 1/2c.; 1-lb. tins, 19 1/2c.; 1/2-lb. tins, 21 1/2c. f.o.b. Montreal; terms 3 per cent. 30 days, or four months from date of delivery.

SCRAP METALS.

The market is improving in activity. The tone is buoyant. Dealers are paying the following prices in the country: Heavy copper and wire, 13 to 13 1/2c. per lb.; light copper, 12c.; heavy brass, 12c.; heavy yellow, 8 1/2 to 9c.; light brass, 6 1/2 to 7c.; lead, 2 1/2 to 2 3/4c. per lb.; zinc, 2 1/4 to 2 1/2c.; iron, No. 1 wrought, $15 to $16 per gross ton f.o.c. Montreal; No. 1 cast, $13 to $14; stove plate, $8 to $9; light iron, No. 2, $4 a ton; malleable and steel,

$4; rags, country, 70 to 80c. per 100 lb.; old rubbers, 6½c. per lb.

HIDES.

The market for green hides is quiet and easier in sympathy with the West and Chicago, dealers paying 6½ to 7c. for No. 1 light. We quote: Light hides, 6½c. for No. 1; 5½c. for No. 2, and 4½c. for No. 3. Lambskins, 10c.; sheepskins, 90c.; calfskins, 8c. for No. 1 and 6c. for No. 2.

PETROLEUM.

Business is only moderate. We quote as follows: "Silver Star," 14½ to 15½c.; "Imperial Acme," 16 to 17c.; "S.C. Acme," 18 to 19c., and "Pratt's Astral," 18½ to 19½c.

ONTARIO MARKETS.

TORONTO, April 12, 1901.

HARDWARE.

WITH the improvement in the spring weather the wholesale hardware trade continues to expand, the wholesale houses all being decidedly busy. The orders are not, as a rule, for large quantities, but they are numerous, and the general tone of the market is healthy, and payments are fairly good. The only disagreeable feature of trade at the moment is the difficulty that is being experienced in getting delivery of goods from manufacturers. These remarks apply particularly to cask goods, which are made in both Canada and the United States. There have not been many changes in prices during the week. Probably the most important is a decline of 50c. per dozen in the price of lanterns. Delivery is being made of fence wires and some fresh orders are coming forward. There is a fairly good trade being done in wire nails, and, while cut nails are still dull, there is an improvement even in this line. Business is fairly brisk in both horse nails and horseshoes. A fairly good business is still being maintained in screws. Bolts are in good demand and a fair trade is being done in rivets and burrs. There has been a fair demand for rope. In powder, an increased movement is to be noted. In both enamelled ware and tinware an increased trade is being experienced this week. Some shipments of screen doors and windows are going out. A good trade is being done in poultry netting. Harvest tools and spades and shovels are meeting with a seasonable demand. A fair trade is being done in churns and wringers, and in ice cream freezers and refrigerators business appears to be even better than is usual at this time of the year.

BARB WIRE—Book orders are being delivered and people are asking for delivery earlier than originally intended. The mills are behind some two or three weeks with their orders. The price is unchanged. We quote f.o.b. Cleveland at $2.82½ per 100 lb. for less than carlots and $2.70 for carlots. From stock Toronto the figure is $3.05 per 100 lb.

GALVANIZED WIRE — There is just a moderate trade being done. We quote as follows: Nos. 6, 7 and 8, $3.50 to $3.85 per 100 lb., according to quantity; No. 9, $2.85 to $3.15; No. 10, $3.60 to $3.95; No. 11, $3.70 to $4.10; No. 12, $3 to $3.30; No. 13, $3.10 to $3.40; No. 14, $4.10 to $4.50; No. 15, $4.60 to $5.05; No. 16, $4.85 to $5.35. Nos. 6 to 9 base f.o.b. Cleveland are quoted at $2.57½ in less than carlots and 12c. less for carlots of 15 tons.

SMOOTH STEEL WIRE — Quite a little oiled and annealed wire is going out this week, but in hay-baling wire there is little or nothing doing. Net selling prices for oiled and annealed are as noted last week: Nos. 6 to 8, $2.90; 9, $2.80; 10, $2.87; 11, $2.90; 12, $2.95; 13, $3.15; 14, $3.37; 15, $3.50; 16, $3.65. Delivery points, Toronto, Hamilton, London and Montreal, with freights equalized on these points.

WIRE NAILS—Quite a quantity of wire nails are going out, but the individual orders coming in are not large, on account of the booked orders which are being filled. Prices are unchanged at last week's advance. We quote: Less than carlots at $2.85, and carlots at $2.77½ to the retail trade. Delivery points: Toronto, Hamilton, London, Gananoque and Montreal.

CUT NAILS—There are a few nails going out, and the demand is principally for 4d nails. At the same time, however, generally speaking, trade in cut nails is still decidedly dull. The base price is $2.35 per keg. Delivery points are: Toronto, Hamilton, London, Montreal and St. John, N.B.

HORSE NAILS—These are going out fairly well this week. Discount on "C" brand, oval head, 50 and 7½ per cent. off new list and on other brands 50, 10 and 5 per cent., off the old list; countersunk, 50, 10 and 10 per cent.

HORSESHOES — There is a good demand for horseshoes, although the orders are individually small. We quote f.o.b. Toronto: Iron shoes, No. 2 and larger, light, medium and heavy, $3.60; snow shoes, $3.85; light steel shoes, $3.70; featherweight (all sizes), $4.95; iron shoes, No. 1 and smaller, light, medium and heavy (all sizes), $3.85; snow shoes, $4; light steel shoes, $3.95; featherweight (all sizes), $4.95.

SCREWS—Business is fairly good, and some of the orders are for good-sized lots. We still quote discounts as follows: Flat head bright, 87½ and 10 per cent.; round head bright, 82½ and 10 per cent.; flat head brass, 80 and 10 per cent.; round head brass, 75 and 10 per

cent.; round head bronze, 65 per cent. and flat head bronze at 70 per cent.

BOLTS AND NUTS—Business in bolts and nuts continues fairly good at unchanged prices. We quote: Carriage bolts (Norway), full square, 70 per cent.; carriage bolts full square, 70 per cent.; common carriage bolts, all sizes, 65 per cent.; machine bolts, all sizes, 65 per cent.; coach screws, 75 per cent.; sleighshoe bolts, 75 per cent.; blank bolts, 65 per cent.; bolt ends, 65 per cent.; nuts, square, 4½c. off; nuts, hexagon, 4¼c. off; tire bolts, 67½ per cent.; stove bolts, 67½; plough bolts, 60 per cent.; stove rods, 6 to 8c.

RIVETS AND BURRS — Copper rivets are moving out freely this week, and there is a fair demand for the iron description. We quote as follows: Iron rivets, 60 and 10 per cent.; iron burrs, 55 per cent.; copper rivets and burrs, 35 and 5 per cent.

ROPE—The demand for rope during the past week has been fair, with prices ruling as before.

CUTLERY—Some business is being done, but it is not heavy. For this time of the year, it is probably about as good as usual.

SPORTING GOODS—An increased demand is to be noted this week for gunpowder. In loaded shells, a fair business is being done. Some arms are going out.

ENAMELLED WARE — There has been a rather better trade in enamelled ware, but the business seems to be more active in the lines which were not affected by the recent advance. We quote: "Granite," "Pearl," "Crescent" and "Imperial" wares at 50, 10 and 10 per cent.; white, "Princess," "Turquoise." blue and white, 50 per cent.; "Diamond," "Famous" and "Premier," 50 and 10 per cent.

TINWARE—An improved business is also to be noted in tinware.

LANTERNS—On account of competition the manufacturers have reduced their prices 50c. per dozen, and we now quote as follows: Plain cold blast, $7 per dozen; plain ordinary, with 0 burner, $4 per doz.; dashboard cold blast, $9; ditto No. 0, $5.75; japanning, 50c. extra. Coppered plated cold blast, $2 per dozen over prices of plain cold blast.

GREEN WIRE CLOTH — This is beginning to move out fairly well. We quote $1.35 per 100 sq. ft. as before.

SCREEN DOORS AND WINDOWS—Some shipments have been made during the past week, but it is rather early for free shipments to be made. We quote 4-in. styles in doors at $7.20 to $7.80 per doz., and 3 in. styles 20c. per doz. less; screen windows, $1.60 to $3.60 per doz., according to size and extension.

BUILDING PAPER — Business continues fairly good with prices unchanged at 30c. for plain building, 40c. for tarred lining, and $1.65 for tarred roofing.

POULTRY NETTING — There is a good demand for poultry netting all over the country. The discount on Canadian is 55 per cent.

CHURNS, WRINGERS—There is a good demand and quite a few shipments are being made.

ICE CREAM FREEZERS AND REFRIGERATORS. — The season has opened up particularly well for trade in these two lines, and jobbers report that they have sent out more than is usual at this time of the year.

HARVEST TOOLS — There is a good demand for manure forks, which are hard to get from the manufacturers; a few orders are being filled in other lines coming under the classification of harvest tools. Discount 50, 10 and 5 per cent.

SPADES AND SHOVELS — Quite a number of booked orders have been filled during the past week, but not a great many orders have come to hand. Discount 40 and 5 per cent.

TACKS—Trade is fair for this time of the year, at last week's reduction in prices. The following are the lines in which the changes have been made together with the new discounts: Carpet tacks, blued, 80 and 15 per cent; carpet tacks, tinned, 80 and 20 per cent; cut tacks, blued, in dozens only, 80 per cent.; Swedes, upholsterers', bulk, 85, 12½ and 12½ per cent.; copper nails, 52½ per cent.

BRIGHT WIRE GOODS—The demand continues fairly good for bright wire goods at the discount of 62½ per cent.

BINDER TWINE —There is just a moderate business being done, and prices are the same as given last week. We quote as follows: American—Sisal and sstandard, 8½c.; manila, 10¼c.; pure manila, 11½ to 12c. per lb. Canadian—Sisal, 8¼c.; pure manila, 11½c. per lb.

OIL AND GAS STOVES — The demand for oil stoves continues good, and quite a number have been shipped during the past week. Not a great deal is yet being done in gas stoves.

LAWN MOWERS—An active trade is being done in lawn mowers.

CEMENT—A good movement is reported. The indications point toward a big trade this season. Prices are steady at old quotations. We quote: Canadian portland $2.40 to $2.80; German, $3 to $3.15; English, $3; Belgian, $2.50 to $2.75; Canadian hydraulic, $1.25 to $1.50.

METALS.

The metal trade, generally, shows a little more activity than it did a week ago. The most active lines are galvanized sheets, black sheets and pig metal. Our quotations are a little lower on zinc spelter, pig lead and antimony.

PIG IRON—While there is not a great deal being done, prices are firmly maintained.

BAR IRON—A good business is being done and prices are firm, the ruling base price being $1.70 to $1.75 per 100 lb., according to quality.

STEEL—The tendency of the steel market is still upward, and a good trade is being done.

STEEL BOILER PLATES—Business continues moderate. We quote ¼ inch, $2.40 to $2.50, and 3 16 inch, $2.60. The demand has been a little better during the past week, being now fairly brisk. Prices are somewhat irregular in England and not so firm in New York as they were. Locally, quotations are unchanged at 31 to 32c.

TINPLATES—Trade has been fair during the past week, although the quantities going out are not large.

TINNED SHEETS—Several inquiries have been received during the past week, but actual business has been only moderate. We quote 9 to 9½c. for small lots from stock.

TERNE SHEETS —Trade in this line continues only light.

GALVANIZED SHEETS—The demand has improved although the quantities wanted are not large. Jobbers are this week making delivery of English galvanized sheets. The ruling prices from stock are $4.60 for English and $4.50 for United States sheets.

BLACK SHEETS—Business has increased during the past week, and it is now steady at $2.30 for 28 gauge.

CANADA PLATES —Trade is a little more active, although the quantities wanted from stock are small. Import orders are still being booked. We quote: All dull, $3, half-and-half, $3.15, and all bright, $3.65 to $3.75.

COPPER—An active trade has been done in ingot copper during the past week, and a fair amount of business is reported in sheet copper. Prices are fairly steady in the outside market. We quote : Ingot, 19c.; bolt or bar, 23½ to 25c.; sheet, 23 to 23½c. per lb.

BRASS — The improvement noted last week in the demand is being maintained. Discount 15 per cent. on rod and sheet.

SOLDER—Business continues fairly good with prices unchanged. We quote : Half-and-half, guaranteed, 18½c.; ditto, commercial, 18c. ; refined, 18c., and wiping, 17c.

IRON PIPE — The demand continues fairly good. We quote as follows per 100 feet : Black pipe, ⅛ in., $4.35 ; ¼ in., $3.25 ; ⅜ in., $3.30 ; ½ in., $3.35 ; ¾ in., $3.60; 1 in., $5.00; 1¼ in., $6.85; 1½ in., $8.25; 2 in., $11.00; 2½ in., $22.94; 3 in., $36.12; 3½ in , $37.90; 4 in., $43.09; 5 in., $57 85 ; 6 in., $75.01 ; 7 in., $93.76; 8 in., $112 51. Galvanized pipe, ½ in., $5.52; ¾ in., $5.56; 1 in., $7.77; 1¼ in., $10.60; 1½ in., $12.70; 2 in., $16 90.

LEAD—The price of lead has been reduced ¼c. per lb., being now quoted at 4½ to 4¾c. per lb. The outside markets have been a little steadier during the past few days. Business locally is fairly good.

ZINC SPELTER—Trade is quiet, and, although outside markets are a little steadier, prices locally are ½c. lower, the figures now being 5½ to 6c. per lb.

ZINC SHEET—Business is a little better, possibly owing to last week's reduction in prices. We quote 6½c. for casks and 6¾c. for part casks.

ANTIMONY—Prices in this line have been reduced in sympathy with the easier market abroad. This is the first time local quotations have been changed for a long while. The price is now 10½ to 11c.

CHAIN—There has been a fairly good demand for chain during the past week, and quotations rule as before.

PAINTS AND OILS.

There is a good spring trade doing. Prepared paints, varnishes, sundries, linseed oil and putty are in excellent request. Turpentine is moving fairly well. Paris green is slow. All lines are steady in value, except turpentine, in which there was a decline of 3c. last Saturday. Since that date prices have fallen 1 to 1½c. in Savannah, but it is reported steadier now at that centre. We quote as follows :

WHITE LEAD—Ex Toronto, pure white lead, $6.37½ ; No. 1, $6; No. 2. $5.67½ ; No. 3. $5.25 ; No. 4. $4 87½ ; genuine dry white lead in casks, $5.37½.

RED LEAD—Genuine, in casks of 560 lb., $5.25; ditto, in kegs of 100 lb., $5.50 ; No. 1, in casks of 560 lb., $5 ; ditto, kegs of 100 lb., $5.25.

LITHARGE—Genuine, 7 to 7½c.

ORANGE MINERAL—Genuine, 8 to 8½c.

WHITE ZINC—Genuine, French V.M., in casks, $7 to $7.25; Lehigh, in casks, $6.

PARIS WHITE—90c. to $1 per 100 lb.

WHITING — 70c. per 100 lb. ; Gilders' whiting, 80c.

GUM SHELLAC — In cases, 22c.; in less than cases, 25c.

PARIS GREEN—Bbls., 16¾c.; kegs, 17c.; 50 and 100-lb. drums, 17½c.; 25-lb. drums, 18c.; 1-lb. papers, 18½c.; 1-lb. tins, 19½c.; ½-lb. papers, 20½c.; ½-lb. tins, 21½c.

PUTTY —Bladders, in bbls., $2.10; bladders, in 100 lb. kegs, $2.25; bulk in bbls., $1.90 ; bulk, less than bbls. and up to 100 lb., $2.05 ; bladders, bulk or tins, less than 100 lb., $2.90.

PLASTER PARIS—New Brunswick, $1.90 per bbl.

PUMICE STONE — Powdered, $2.50 per cwt. in bbls., and 4 to 5c. per lb. in less quantity ; lump, 10c. in small lots, and 8c. in bbls.

LIQUID PAINTS—Pure, $1.20 to $1.30 per gal.; No. 1 quality, $1 per gal.

CASTOR OIL—East India, in cases, 10 to 10½c. per lb. and 10½ to 11c. for single tins.

LINSEED OIL—Raw, 1 to 4 barrels, 69c.; boiled, 72c.; 5 to 9 barrels, raw, 68c.; boiled, 71c., delivered. To Toronto, Hamilton, Guelph and London, 1c. less.

TURPENTINE—Single barrels, 56c.; 2 to 4 barrels, 55c., delivered. Toronto, Hamilton and London 1c. less. For less quantities than barrels, 5c. per gallon extra will be added, and for 5-gallon packages, 50c., and 10 gallon packages, 80c. will be charged.

GLASS.

Advices from Belgium report a continuance of firm prices, and the market here is stiff. There is a good movement from stock. We quote : Under 26 in., $4.15 26 to 40 in., $4.45 ; 41 to 50 in., $4 85; 51 to 60 in., $5.15 ; 61 to 70 in., $5.50; double diamond, under 26 in., $6 ; 26 to 40 in., $6.65 ; 41 to 50 in., $7.50 ; 51 to 60 in., $8.50; 61 to 70 in., $9.50, Toronto, Hamilton and London. Terms, 4 months or 3 per cent. 30 days.

HIDES, SKINS AND WOOL.

HIDES—The market is dull at unchanged figures. We quote: Cowhides, No. 1, 6½c. ; No. 2, 5½c.; No. 3, 4½c. Steer hides are worth 1c. more. Cured hides are quoted at 7 to 7½c.

SKINS—There is little doing. Prices are unchanged. We quote : No. 1 veal, 8-lb. and up, 8c. per lb.; No. 2, 7c.; dekins, from 40 to 60c.; culls, 20 to 25c. Sheepskins, 90c. to $1.

WOOL—There is nothing doing. We quote : Combing fleece, 14 to 15c., and unwashed, 8 to 9c.

COAL.

Prices have been reduced for April delivery to encourage buying. Jobbers say that there will be an advance for May shipment. We quote anthracite on cars Buffalo and bridges as follows: Grate, $4.25 per gross ton and $3.79 per net ton ; egg, stove and nut, $4.50 per gross ton and $4.01 per net ton, for April shipment.

PETROLEUM.

There is a moderate movement. Prices are steady. We quote: Pratt's Astral, 16½ to 17c. in bulk (barrels, $1 extra) ; American water white, 16½ to 17c. in barrels; Photogene, 16 to 16½c.; Sarnia water white, 15½ to 16c. in barrels; Sarnia prime white, 14½ to 15c. in barrels.

MARKET NOTES.

Turpentine is 3c. lower.

Coal for April delivery is 25c. per gross ton cheaper.

Zinc spelter and antimony are quoted ½c. per lb. lower.

A reduction of ¼c. per lb. has taken place in pig lead.

Lanterns have been reduced 50c. per dozen on account of competition.

H. S. Howland, Sons & Co. have purchased a large stock of curry combs which they are offering at a low figure.

F. X. Julien, general merchant, Lambton, Que., has sold his stock at 71 1-4c. on the dollar to V. Dion, St. George.

MANITOBA MARKETS.

WINNIPEG, April 8, 1901.

BUSINESS is steady and fairly active in all lines and prices have remained without change for the entire week.

The additional drop in linseed oil has not taken place, but is still anticipated as one of the possibilities of the near future.

Implement trade is active; in fact, more so than was actually looked for.

The demand for dairy supplies is also most encouraging.

Quotations for the week are as follows:

Barbed wire, 100 lb.	$3 45
Plain twist	3 45
Staples	3 95
Oiled annealed wire 10	3 95
" 11	4 00
" 12	4 05
" 13	4 20
" 14	4 35
" 15	4 45
Wire nails, 30 to 60 dy, keg	3 45
" 16 and 20	3 50
" 10	3 55
" 8	3 65
" 6	3 70
" 4	3 85
" 3	4 10
Cut nails, 30 to 60 dy.	3 00
" 20 to 40	3 05
" 10 to 16	3 10
" 8	3 15
" 6	3 20
" 4	3 30
" 3	3 65
Horsenails, 45 per cent. discount.	
Horseshoes, iron, No. 0 to No 1	4 65
No. 2 and larger	4 40
Snow shoes, No. 0 to No. 1	4 90
No. 2 and larger	4 40
Steel, No. 0 to No. 1	4 95
No. 2 and larger	4 70
Bar iron, $2.50 basis.	
Swedish iron, $4.50 basis.	
Sleigh shoe steel	3 00
Spring steel	3 25
Machinery steel	3 75
Tool steel, Black Diamond, 100 lb	8 50
Jessop	13 00
Sheet iron, black, 10 to 20 gauge, 100 lb.	3 50
20 to 26 gauge	3 75
28 gauge	4 00
Galvanized American, 16 gauge	2 54
18 to 22 gauge	4 50
24 gauge	4 75
26 gauge	5 00
28 gauge	5 25
Genuine Russian, lb	12
Imitation " "	8
Tinned, 24 gauge, 100 lb.	7 55
26 gauge	8 80
28 gauge	8 00
Tinplate, IC charcoal, 20 x 28, box	10 75
" IX "	12 75
" IXX "	14 75
Ingot tin	35
Canada plate, 18 x 21 and 18 x 24	3 75
Sheet zinc, cask lots, 100 lb.	7 50
Broken lots	8 00
Pig lead, 100 lb.	6 00
Wrought pipe, black up to 2 inch	50 an 10 p.c.
" Over 2 inch	50 p.c.
Rope, sisal, 7-16 and larger	$10 00
" ⅜	10 50
" ¼ and 5-16	11 00
Manila, 7-16 and larger	13 50
" ⅜	14 00
" ¼ and 5-16	14 50
Solder	21½
Cotton Rope, all sizes, lb.	16
Axes, chopping	$7 50 to 12 00
" double bitts	12 00 to 18 00
Screws, flat head, iron, bright	87½
Round " "	82½
Flat " brass	80
Round " "	75
Coach	57½ p.c.
Bolts, carriage	55 p.c.
Machine	55 p.c.
Tire	60 p.c
Sleigh shoe	65 p.c.
Plough	40 p.c.
Rivets, iron	50 p.c.
Copper, No. 8	35
Spades and shovels	40 p.c.
Harvest tools	50, and 10 p.c.
Axe handles, turned, s. g. hickory, doz.	$2 50
No. 1	1 50
No. 2	1 25
Octagon extra	1 75
No. 1	1 25
Files common	70, and 10 p.c.
Diamond	60
Ammunition, cartridges, Dominion R.F.	50 p.c.
Dominion, C.F., pistol	30 p.c.
" military	15 p.c.
American R.F.	30 p.c.
C.F. pistol	5 p.c.
C.F. military	10 p.c. advance.
Loaded shells:	
Eley's soft, 12 gauge black	16 50
chilled, 12 guage	18 00
soft, 10 guage	21 00
chilled, 10 guage	23 00
Shot, Ordinary, per 100 lb	6 75
Chilled	7 50
Powder, F.F., keg	4 75
F.F.G.	5 00
Tinware, pressed, retinned	75 and 2½ p.c.
" " plain	70 and 15 p.c.
Graniteware, according to quality	50 p.c.

PETROLEUM.

Water white American	25½c.
Prime white American	24c.
Water white Canadian	22c.
Prime white Canadian	21c.

PAINTS, OILS AND GLASS.

Turpentine, pure, in barrels	$ 68
Less than barrel lots	73
Linseed oil, raw	77
Boiled	80
Lubricating oils, Eldorado castor	25½
Eldorado engine	24½
Atlantic red	27½
Renown engine	41
Black oil	23½ to 25
Cylinder oil (according to grade)	55 to 74
Harness oil	61
Neatsfoot oil	$ 1 00
Steam refined oil	85
Sperm oil	1 50
Castor oil per lb.	11½
Glass, single glass, first break, 16 to 25 united inches	2 25
26 to 40 per 50 ft.	2 50
41 to 50 " 100 ft.	5 50
51 to 60 " " "	6 00
61 to 70 per 100-ft. boxes	6 50
Putty, in bladders, barrel lots per lb.	2½
kegs "	2½
White lead, pure per cwt.	7 25
No 1 "	7 00
Prepared paints, pure liquid colors, according to shade and color per gal.	$1.30 to $1.90

WILL INCREASE THEIR FACILITIES.

The Canada Horse Nail Co., Montreal, has been incorporated with $100,000 capital stock. The directors of the company are James Ferrier, the former proprietor; John Torrance, steamship agent; Margaret Watson Ferrier, wife of John Torrance; Robert Ferrier Macfarlane, passenger agent, and William Small, manager, all of Montreal.

The company has, during the past 35 years, manufactured horse nails exclusively, but, owing to the present war in prices on this article, it is proposed by the company to make provision to enlarge their manufacturing facilities, especially with a view to the putting in of an additional plant for the manufacture of horseshoes and other articles associated with the business.

PERSONAL MENTION.

Mr. W. H. Carrick, general manager of The Gurney Foundry Co., Limited, Toronto, left this week for a visit to Great Britain and Ireland.

INDUSTRIAL GOSSIP.

Those having any items of news suitable for this column will confer a favor by forwarding them to this office addressed the Editor.

J. D. McArthur, contractor, Winnipeg, has decided to erect a large, up-to-date sawmill at Lac du Bonnett, Que.

The Victoria Machinery Co., Victoria, B.C., have just completed the erection of a new boiler shop at their works, Rock Bay, and the latest improved pneumatic tools have been installed.

Mr. Oliver Richards, of Victoria, B.C., has gone to Great Britain to purchase a plant for the establishment of marine ways and shipbuilding yards. The company will employ from 60 to 70 workmen, and the plant will have a capacity for building vessels up to 2,000 tons. He expects to return to Victoria in August and start the erection of his new works.

The E. Long Mfg. Co., Orillia, Ont., have closed a deal this week with W. H. Croker, by which they become possessors of the old Orillia foundry, which has been lying idle for some years. A portion of the machinery will be installed in the Long company's works, and the foundry building may also be moved to the site of the latter, the producing capacity of which will be much increased by this purchase.

T. W. DAVIS, RIPLEY, DEAD.

T. W. Davis, formerly hardware dealer and tinsmith, Ripley, Ont., died at his home in that place on Monday after a brief illness. Mr. Davis had for some time been suffering from inflammatory rheumatism which affected his heart. About a week previous to his death he was attacked by pains in the region of the appendix which developed into appendicitis. Heart failure was the direct cause of death. The late Mr. Davis was for many years one of the most active and progressive citizens of Ripley, and had built up a large hardware and tinsmithing business from which he retired a short time ago.

A BIG CLAIM AGAINST TORONTO.

The Metallic Roofing Company have entered against the municipality of Toronto a claim for $21,150 for damages which they state have been caused by reason of the city not removing the Elliott & Neelon construction plant from a piece of land which the city sold to the company, and on which the latter decided to erect additions to their works but were unable to do so by the presence of the plant.

THE COPP BROS.' DIFFICULTY.

It is to be hoped that the financial difficulties of The Copp Bros., Limited, will be settled satisfactorily, as has been predicted by the president, W. J. Copp, and that the company will continue in business. The cause of the suspension was the issuance of a writ against the firm by the Merchants Bank for $37,500.

Morgan Bros., whip, saddlery and hardware manufacturers, Hamilton, Ont., were destroyed by fire on Thursday night. The loss is estimated at $35,000. Insurance to the extent of $30,000 was carried.

Elastic Carbon Paint

A
BIG
THING.

LOOK
INTO
IT.

During the last year we have sold in Canada and the United States over five thousand (5,000) barrels, and have received from our patrons numerous voluntary testimonials, which are in many cases accompanied by repeat orders, and, owing to our success, many worthless, cheap imitations have been placed on the market, which are claimed by unscrupulous competitors to be as good as the genuine Elastic Carbon Paint, which is made only by ourselves. A trial of the goods will convince the most skeptical.

We are large importers of "Pure Spirits of Turpentine," in tank cars, enamelled white inside and used only for the transportation of this article, direct from the virgin Florida forests.

We also handle large quantities of pure Linseed Oil, bought from the most reputable manufacturers in England, both Turpentine and Linseed Oil being sold subject to chemical analysis.

Prices and Samples Cheerfully Submitted.

The Atlantic Refining Co.
TORONTO.

Corner Esplanade
and Jarvis Sts.

Manufacturers and Importers of
Illuminating and Lubricating Oils, Greases, etc.

HEATING AND PLUMBING

ADVANCE IN RADIATION IN THE UNITED STATES.

THROUGHOUT the country, the heating contractors received, last Monday, notices to the effect that quotations on radiation had been withdrawn, that an advance approximating 5 per cent. had been made, and that new quotations would be furnished on application. Already sufficient time has elapsed since this announcement to allow the advance to be considered by the manufacturers of these goods, and the impression quite generally prevails that the advance was too slight, and that, even with it, radiation is selling at least 1c. per foot too low to bear a proper relation to the cost of iron and labor. While this feeling exists quite generally, it is improbable that radiation will continue through the season at the price now obtaining; in fact, some hold to the opinion that May 1 may be the date of a further advance, to be influenced more or less by the condition of the trade and the outlook in the iron market.

Some of the independent manufacturers of radiation, who had expected to keep in harmony with those who have cooperated in the advance, are disappointed that the advance was not 5 per cent. on the base price, instead of 5 per cent. in the auxiliary discounts. In consequence of the advance and the feeling existing in reference to it the heating contractors will do well to place their specified orders for goods for contracts closed and to be careful to withdraw bids based on the prices prevailing at the present time that have not been accepted. They should incorporate in their bids the statement that the bid is subject to any advance that may occur in any of the materials required. The mild advance made will certainly cause less hardship to those who will be called upon to carry out a contract based on the old figures. The price of boilers seems to be holding remarkably firm, while it is noted that fittings are both strong in price and scarce in the market, factors that are likely to cause an advance if the demands of the trade should suddenly assume any magnitude.—Metal Worker, April 6.

CAST-IRON SOIL PIPE AND FITTINGS

The market for cast-iron soil pipe and fittings was especially stiff this week. It is pretty certain that prices on this line of goods have not yet reached their highest level, and, judging from present indications, it seems likely that there will be a further advance. Consumers would not be far wrong in placing their orders for immediate wants at the present current prices, with the reservation of cancelling their orders if the advance does not take place.—Metal Worker, April 6.

SOME BUILDING NOTES.

THE MUSKOKA NAVIGATION CO. have almost completed plans for the erection of a big summer hotel of 225 rooms, on Big Island, in Lake Rosseau, Muskoka District, Ont., which is to be completed by July 1. The hotel will contain 64 bathrooms, and will be lighted by electricity, and will be steam-heated throughout. The cost will be about $60,000.

J. Erwin will build a house at Sharbot Lake this summer.

John Tate is preparing to erect a new hotel at Inwood, Ont.

A Methodist church will be built in Richmond, Que., this summer.

Walker Bros. & Button, Wingham, Ont., are erecting an addition to their store.

A. D. Cameron will erect an 80 x 55 ft. building in Buckingham, Que., this summer.

E. Z. Labrosse, Vankleek Hill, Ont., will erect one or two brick houses this summer.

Boyden, Harris & Campbell, furniture dealers, etc., Ottawa, intend enlarging their premises.

A $250,000 church will be erected by the Roman Catholics of Westmount, near Montreal.

Thomas Miller, Moose Jaw, N.W.T., is asking for tenders for a Presbyterian church to seat 600 persons.

W. H. Davis, contractor, Ottawa, will erect, it is understood, a $100,000 building on Rideau street, Ottawa.

Alex. C. Mitchell is erecting a residence on West Wellington street, Ottawa. Geo. E. Wilson is the architect.

Sproul & Burley, contractors, St. John, N.B., have contracts for remodelling the Congregational Church, St. John.

D. McFarlane, Parry Sound, Ont., is advertising for tenders for a four-roomed brick school house in Parry Sound.

Improvements to the amount of $10,000 are being made in the Albion Hotel, Montreal, by James Devlin, the proprietor.

Tenders for alterations and additions to the Bank of Hamilton, Winnipeg, are asked for by D. B. Dick, Toronto, before April 19.

H. Gillien, proprietor of the Midas Hotel, Ottawa, will erect a new four-storey hotel on Rideau street between Mosgrove and William, to cost $7,500.

G. W. Mitchell, Cobourg, Ont., is advertising for tenders for the erection of a collegiate institute at Cobourg. Power & Son, Kingston, Ont., are the architects.

A site has been chosen and plans are being prepared for a new library and alumni hall for the Guelph Agricultural College. The two buildings are to cost about $75,000.

Among the buildings to be built in Grand Valley, Ont., this summer will be a $5,000 Presbyterian church, a post office and store for Postmaster Taylor, and a block for Dr. Hopkins.

A. H. Pulford, Winnipeg, will build a three-storey business block with a 54-ft. frontage on Portage avenue. Messrs. Sprague, Alloway and Champion will erect a two-storey building with 69 ft. frontage between this building and the Y.M.C.A.

PLUMBING AND HEATING NOTES.

J. C. Thibault & Co., plumbers, Arthabaskaville, Que., has assigned.

John E. Fitzgerald, plumber, etc., St. John, N.B., has assigned to the sheriff.

Bigg & Mackenzie, plumbers, Picton, Ont., are removing into larger premises.

A. Paquette & Co., contractors, etc., St. Henri de Montreal, Que., have assigned.

T.D.M. Osborne, plumber, etc., Brandon, Man., has compromised at 50c. on the dollar.

J. Ed. McDonald has registered as proprietor of The Best Heat and Light Co., New Glasgow, N.S.

The Amherstburg Electric Light, Heat and Power Co., Limited, have suffered loss by fire; partially insured.

TORONTO'S PALACE HOTEL.

On Tuesday the contract for the construction of Toronto's new hotel was let to Issley & Horn, a new firm of general contractors, composed of William A. Issley, formerly of The George A. Fuller Co., Chicago, and Thomas W. Horn, president of The Luxfer Prism Co., Toronto.

The contractors are pushing the work of demolishing the old buildings with great vigor. Issley & Horn, who will sub-let

contracts to the various trades, have received the tenders, and it is expected that the contracts will be signed in a few days. E. J. Lennox is the architect in charge.

HARD SOLDERING CAST IRON.

A PROCESS for hard soldering cast iron has, according to an exchange, been brought out in Germany in which the cast iron surfaces are cleaned by means of an acid in the usual way, fixed together, and the soldering places covered or surrounded with a paste consisting of suboxide of copper and borax. This paste is prepared by mixing suboxide and borax, by boiling them together so intimately that the suboxide of copper is surrounded by a layer of borax-absorbing oxide, which excludes the action of the atmosphere upon the suboxide during the heating process required for soldering. For the borax, other suitable fluxes, such as glass or water-glass, etc., may be substituted. While hard soldering the cast iron the borax melts and protects, as is well known, the clean surface of the iron against oxidation, removes any oxide thereon, and also protects the suboxide of copper against the action of the oxygen in the atmosphere.

Consequently the suboxide of copper, likewise heated to a red heat, transfers its oxygen to the red hot cast iron surface, which oxygen combines, with the graphite contained in the cast iron surface to form carbon monoxide or dioxide, thus decarbonizing the surfaces, while the metallic copper becomes disassociated in a very finely divided condition. At the same time the hard solder is added, and as this solder, which is brought upon the surfaces to be soldered in the well-known manner, is likewise melted by the heat, it alloys itself with the incandescent particles of copper, and this new alloy immediately combines with the red hot decarbonizing soldering surfaces of the cast iron.

BUILDING PERMITS ISSUED.

Building permits have been issued in Hamilton to Thomas Lovejoy for two dwellings on William street, to cost $2,100; to Thos. H. Sellery for a dwelling at 451 York street, to cost $1,000; F. J. Rastrick & Son, for a $1,400 dwelling on Main street west for Mrs. E. Gore, and to Charles Mills for a $1,000 brick addition to the foundry of The D. Moore Co., Limited.

Building permits have been issued in Ottawa to A. C. Mitchell, dwelling, Wellington street, $2,500; F. W. Burman, dwelling, Preston street, $2,200; H. Gallien, solid brick hotel, Rideau street, $7,500; Mrs. Mary Baxter, dwelling, Rochester street, $2,000; A. Delorme, brick veneering and repairs to present hotel, Besserer street, $3,000; Patrick Hanrahan, two dwellings, Osgoode street, $3,100; Louis Jarvis, dwelling, Gilmour street, $1,000; E. Morel, dwelling, Cooper street, $2,500; A. Huckels & Co., factory, Slater street, $4,000; E. Cockburn, dwelling, Albert street, $1,800; J. Lavigne, Duke street, $4,500.

Building permits have been issued in Toronto to Geo. Gooderham for six dwellings on Worts avenue to cost $8,000; to R. M. Ogilvie for two houses on Albany avenue, near Bloor street, to cost $6,000; to E. A. Drummer for two $1,500 houses on St. Clarens avenue, north of College street; to Adams Bros. for a $2,500 factory at the corner of King and Frederick streets; to W. T. Atkinson, for a $2,000 dwelling at 207 Crawford street; to E. Jones, for a $4,500 factory at 32 Prince Arthur avenue; to the Dominion Bank, for a $15,000 branch building at the corner of Bloor and Bathurst streets; to G. F. Bromley, for a $1,500 dwelling on St. Vincent street, near St. Albans street; to Frank Arnoldi, for a $6,000 residence on Strickland avenue; to J. D. Lumus, for a $5,500 dwelling at 98 Bedford Road; to W. L. Lunn, for a $1,000 residence at 224 Palmerston avenue.

PLUMBING AND HEATING CONTRACTS.

The John Ritchie Plumbing and Heating Co., Limited, Toronto, have the contract for remodelling the plumbing of W. F. Thomson's hotel at Rose Point, Parry Sound, Ont.

A NEW HOTEL FOR OTTAWA.

On the Clemow property which has 155 ft. frontage on Rideau street; 200 ft. on McKenzie avenue and 200 ft. on Sussex street, Ottawa, a palatial, 10-storey hotel is to be erected. The ground floor will be devoted to the essential business features of a modern hotel, and will be occupied by a bank, safety deposit vault, bar, cafe and restaurant, tobacco and drug stores, ticket office, etc. The rest of the hotel will be fitted up as a centre for tourists, politicians, business men and for social functions, and will be thoroughly modern.

A NEW POWDER COMPANY.

The Ontario Powder Company has been incorporated with $100,000 capital stock to take over the business of manufacturing and selling explosives carried on by Daniel Smith and Colin Angus Macpherson of Kingston, under the name of the Ontario Powder Works. The directors are Angus Macpherson of Kingston, Hugh Macpherson of Nelson, B.C., Frederick Hall-Hooper of Brownsburg, Que., and Francis King of Kingston.

BUSINESS CHANGES.

DIFFICULTIES, ASSIGNMENTS, COMPROMISES.

A MEETING of the creditors of Mohs & Ryan, general merchants, etc., Killaloe Station, Ont., has been held.

Frank Reardon, wholesale paint dealer, Halifax, has assigned.

Joseph Pauze, hardware dealer, Montreal, is offering 35 cents cash on the dollar.

M. Weidman, hardware dealer, etc., Winnipeg, has assigned to C. H. Newton.

Wells & Frary, general merchants, Frelighsburg, Que., are seeking an extension.

Roberge & Landry, general merchants, Thetford Mines, Que., are offering 50 cents on the dollar.

Miles Birkett, hardware dealer, Ottawa, has assigned, and a meeting of creditors will be held on the 15th inst.

SALES MADE AND PENDING.

Henry W. Wright, blacksmith, Truro, N.S., has sold out.

T. F. Ruttan, general merchant, Strathcona, Ont., has sold out.

Sam. Charette, general merchant, The Brook, Ont., has sold out.

Henry Huth, blacksmith, Lakelet, Ont., is advertising his business for sale.

The stock, etc., of R. Perry, general merchant, Gad's Hill, Ont., has been sold by auction.

The assets of Paul Bissonette, general merchant, Casselman, Ont., are to be sold at auction.

The stock, etc., of the estate of N. Holmes, general merchant, Macgregor, Man., has been sold by auction.

PARTNERSHIPS FORMED AND DISSOLVED.

Anderson & Hamilton, machinists, Nanaimo, B.C., have dissolved.

Fliteroft & Strickland, carriagemakers, Hamilton, Ont., have dissolved.

Simard & Fils, coal and wood dealers, etc., St. Johns, Que., have dissolved.

E. H. Phelps & Co., manufacturers of carriage woodwork, Merritton, Ont., have dissolved. E. H. Phelps retires, but F. N. Hara and J. S. Hara continue under unchanged style.

CHANGES.

Frank Wade, blacksmith, Petrolea, Ont., has sold out to W. A. Simpson.

King Warden & Son, Limited, have registered as founders, in Montreal.

W. M. Butchart, hardware dealer, Huntsville, Ont., is giving up business there.

Harry Philbin has registered as proprietor of Philbin & Co., hardware dealers, Westmount, Que.

The stock of R. O. Hoffman, general merchant, Holstein, Ont., has been sold to John Waddell.

John McArthur has registered as proprietor of the John McArthur Co., painters, etc., Montreal.

C. Pratt, general merchant, Stony Plain, N.W.T., has sold out to Schwartz & Co., and has removed to Spruce Grove.

O. G. & M. J. Rutledge, dealers in agricultural implements, Killarney, Man., have been succeeded by George Winram.

FIRES.

The stock of Alex. Ramsay & Son, wholesale paint and oil dealers, Montreal, has been partially damaged by fire and water; insured.

CORRESPONDENCE

THE CANCELLING OF ORDERS.

Editor HARDWARE AND METAL, — In answer to "An Old Traveller," who, in last week's issue of your journal, claims that I departed from the question at issue, as laid down in your article of March 9, allow me to say that, to the ordinarily intelligent mind, I think I made it quite plain that I was not dealing alone with your editorial, but also with a letter from a correspondent which appeared on March 16, and that I adhered as closely to the text set me, and perhaps understood it fully as well, as did the preacher discoursing "Man's Duty to His Fellow."

I do not retract anything which I stated in my letter to you of March 23rd, and, as I take it that you have no space to devote to anything that is not of interest to the people whom you serve, I will not trespass on your space further than to simply inform "An Old Traveller" that I have no time to spare dealing with the idle ravings of a coward who is afraid to disclose his identity; or with one whose mind has, perhaps, become enfeebled with his age, but would exhort him to keep right on with his Sabbath-day devotions, and cease pouring out his idle vaporings upon a business public.

"THE BEETON BELL."

Beeton, Ont., April 10.

RAINY RIVER BOARD OF TRADE.

The following officers have been elected by the Board of Trade of the District of Rainy River, Ont., for 1901 1902:

President—C. E. Neads.
Vice-President—Jacob Hose.
Secretary—E. A. Chapman.
Council—George Drewry, Wm. Margach, W. A. Weir, Frank Gardner, D. C. Cameron, J E. Rice, M. Kyle and John Dean.

VALUE OF HIGH-PRICED STOVES.

THIS suggestion is one which we commend to the thoughtful consideration of those for whom it has interest. The stove retailer who handles a line of standard goods is often at a loss how to explain to a possible customer why a stove he offers is enough better than one which looks as well and seems to be as large, and which can be purchased somewhat cheaper at a furniture or department store. Whether the manufacturer can make the difference clear to the understanding of an average customer is a question, especially if he has to do it at second hand, so to speak, by first making it clear to the dealer and through him to the customer. This reminds us that there is a dearth of the kind of stove trade literature which the retailer would find most useful in dispos ing of his stock. Most of that which has come under our notice has for its object to induce the dealer to purchase. Of that which would help him to sell there is less than could be profitably used. The attempts which the manufacturers have made from time to time to supply the dealer with literature designed to assist him in making sales at retail seem to have been inspired by the idea that the average customer is impressed by nursery rhymes and amusing caricatures. Perhaps she may be, but the selling of a stove is not so easy a matter that the choice of a prudent buyer, between one make and another, is likely to be in any important degree influenced by such means.

In our judgment every line of standard stoves should have, for the education of the dealer and the information of his customers, a circular containing exactly the information the housekeeper is likely to find comprehensible and useful, and dealers should be supplied with such literature as freely as the requirements of their trade may demand. The writing of such circulars demands a high grade of talent, and very few in the business have it. If we are not mistaken, clever women would do this work much better than men, and no doubt it would pay every large manufacturer to have attached to his office staff a capable woman to give her whole time to the preparation of the kind of literature which will interest women and give them just the information they want about the stoves they are asked to buy. The points which interest the manufacturer, and which he is, as the rule, most desirous of communicating to the retailer, do not commonly interest the housekeeper. That which appeals to her and most often influences her judgment in selection the average dealer is apt to lose sight of if he knows it. He repeats what has been told him by the manufacturer to secure his orders, but of this very little appeals to the housekeeper. What he does not know and cannot tell is what the user of a stove for the practical purposes of heating and cooking would find most interesting and instructive.

It is safe generalization that the more efficient the co-operation of the manufacturer with the retailer the larger and more profitable his wholesale distribution. Unfortunately, his judgment of what the retailer wants in the way of trade literature is often at fault. This subject will repay careful consideration, and we are not sure it does not offer an opportunity for a new profession for capable women who combine good sense and domestic experience with the literary facility of which so many are giving evidence in practical journalism. In what is done in this direction it is well at the outset to assume that the average housekeeper, needing so expensive an article of furniture as a stove, will be more interested in a brief, lucid and intelligible explanation of the practical advantages of the stove she is considering than in doggerel or pictures of impossible housemaids in light opera costumes, doing impracticable things in ways which would insure a "calling down" from the head of the family. There is a place for this kind of thing, and lots of it to put there; but there is also a place for a kind of stove-trade literature, which, in the fewest possible words and with the clearest possible illustrations, will give the retailer the means of showing why a $30-stove which he handles is a better stove, and will give better satisfaction, than a $20-stove, which someone not in the stove business at all is handling as a side line.—Metal Worker.

ANTHRACITE COAL IN MANITOBA.

It has long been known that there is an abundance of lignite coal in Manitoba, but it comes as a welcome surprise to learn that anthracite coal has been found there.

The discovery was made last week by James Thompson, a contractor acting for the Anthracite Coal Co., of Winnipeg, while drilling near Souris. After passing through some small strata of iron-tinged clay and one or two seams of coal an inch or two thick, he finally struck a bed of solid clay, 40 feet thick. A seam of sandstone two feet thick was next passed. Under this was a seam of coal 26 inches thick. This coal was similar in grade to Canadian anthracite, a coal of superior quality. The total depth of the boring before the seam was struck was 108 feet.

Needless to say, considerable interest has been aroused by this discovery. It is claimed by some experts that when real development takes place in the Souris district, a great quantity of high-grade coal will be placed on the market. While this discovery does not give sufficient foundation for such hope as this, it must surely be encouraging, and all Canadians will hope that this find may prove but an earnest of what is to be done later.

THE PRICE OF CANS.

Regarding the increased cost of cans mail advices from Baltimore state: "Prices on No. 2 size have been advanced from $1.65 to $1.80 per 100, No. 3s from $2.15 to $2.40 per 100, and gallons from $5 to $5.50 per 100. This means an increase of about 2c. per dozen on No. 2 goods, 3c. per dozen on No. 3. and 6c. per dozen on gallon goods. Whether this is the first step in a definitely settled policy of gradually advancing prices, or whether it is a feeler, on the result of which future action is to be based, remains to be seen. With practical control of the can situation of the country, it seems quite probable that prices will be further advanced."

A DRUMMER'S LITTLE GIRL.

My papa is a travelling man,
Some people call him "drummer,"
He goes away in August and
Gets home again next summer.

I don't know my papa very well—
I wish I knew him better;
But every week I take my pen
And write a big, long letter.

And mamma says some day he'll come
(I thought I would have fainted)
And she will keep him in the house
Until we get acquainted.

Now, ain't that funny, don't you think?
It gives me lots of bother,
To think a great, big girl like me
Don't really know her father!

I don't really know how to act.
Of course, he'll have to "Miss" me;
But, goodness gracious! it won't do
To let a strange man kiss me.

And when he first comes in the house
I won't know how to greet him;
I guess I'll call him "Mr. Papa,"
And say I'm pleased to meet him.

Oh, papa! it's plaguey mean to have
One's papa for a drummer.
I wish he'd come in autumn and
Stay winter, spring and summer.

—John A. Condit, in Business.

SHORTER HOURS IN NANAIMO, B.C.

A despatch from Nanaimo, B.C., says: "The Merchants' Employes Association is again bestirring itself to secure the boon which its members enjoyed last year, namely, a weekly half-holiday. It is proposed also to urge the closing of all stores at six o'clock, except on Saturday evenings and the evenings preceding holidays. Another proposal is to do away with the half-holiday weekly, and instead allow each employe a week's holidays on full pay every summer, as is the custom in the Old Country. Nanaimo has the distinction of being the only city on the Coast where this matter is dealt with in an organized manner. The employers and the employes are on the best of terms, and any proposal which is practical and sound will always receive the kindly attention of the employers."

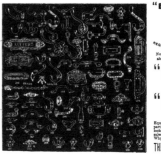

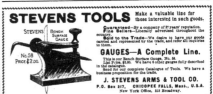

THE PROS AND CONS OF A BANKRUPTCY LAW.

THE United States and Canada have had somewhat similar experiences in the matter of national bankruptcy laws. Each country was led to make trial of such legislation; each, because of defects in the laws adopted and resulting abuses, was led to discard them; and in each, almost immediately, there arose a demand for a reenactment. In the case of the United States the demand for reenactment was acceeded to two years ago. For two years a national bankruptcy law has been in operation, and the result is discussed in an article in The April North American Review, by Mr. W. H. Hotchkiss, one of the referees under the law for New York.

Mr. Hotchkiss refers to the first bankruptcy law in England, that of Henry VIII., in 1542, which was simply a measure for the imprisonment of persons who did not pay their debts. It was not till 1705 that provision for a discharge became a feature of English legislation. It is now the chief feature of all bankruptcy laws, and also the cause of the chief objection to them.

The records of the United States courts during the period covered in Mr. Hotchkiss's review are hardly fair evidence of what in its ordinary operation the law would show. Of some 24,000 discharges granted, about 12,000 were in the cases of wage-earners, largely men who had been in business, and failing, and being without ability either to meet or escape from their obligations, were driven back to the ranks. The courts, in dealing with them, have been clearing off arrears, and not the actual casualties of the present time. This fact has not, however, prevented complaints against the law as one intended for the wholesale cancellation of old debts. It is pointed out that in actual practice it is all but impossible to prevent a discharge in undeserving cases, that there is no limit to the number of times a man may be discharged, that a fraudulent preference is not an objection to a discharge, while the penalties for flagrant frauds on creditors are few and the frauds are not easily proven. There is a demand for amendment. There is a call for a suspension of the law for four years, to be followed by a period of operation for one year, and then another period of suspension, and so on.

There is a fairly strong call for repeal, and among those who are raising it the bankers' associations are prominent. Heads of great mercantile establishments join in this demand, and some lawyers, thinking the law causes a decrease in litigation, say it should go. When Congress closed its work last month there were repeal as well as suspension Bills before it. The Act has not, therefore, been an unqualified success in the minds of the public. Neither has it been the failure that was feared. The previous bankruptcy law was repealed because its fees and delays permitted the looting of estates.

Under the law of 1898, Mr. Hotchkiss claims, delays are rendered impossible, while reports furnished to the Attorney-General at Washington show that, the country over, the average cost of administration during the past year was less than $40 in voluntary cases, and $210 in involuntary ones, the latter, as a rule, including the collection and distribution of assets. The American Bar Association, the National Association of Credit Men, the Commercial Law League of America and boards of trade and chambers of commerce in various parts of the country have pronounced in favor of retaining the law, as has, also, Mr. E. C. Brandenburg, the bankruptcy expert of the Department of Justice.

The referees in bankruptcy under the law have also given their ideas which are, naturally perhaps, in favor of an improved law. To make the law more perfect there are asked amendments making it more difficult to obtain discharges, and to shut the door on shady or chronic bankruptcy; to include corporations within its voluntary features; to meet cases of innocent preferences, such as alimony, etc.

The American Bar Association favors a provision that any debtor who substantially diminishes his property by gambling, speculation or reckless management may be adjudged a bankrupt at the instance of his creditors. These demands are generally familiar to those who remember the last Canadian bankruptcy law. The first especially suggests that the United States law has the faults of practically all laws which provide for the release of a debtor from obligations which his assets are not sufficient to meet. Still, with all its defects, inherent or incidental, commercial experience indicates that bankruptcy laws are a need of every country of extended trade. Rome had one; so had Genoa and Venice in the Middle Ages.

Holland, France and Great Britain have them, and Germany, as soon as it became an empire instead of a lot of small kingdoms, found one necessary. There is even talk of arranging international bankruptcy conventions. Mr. Hotchkiss's argument is fairly effective of its purpose of showing that more good than evil comes from fairly administered bankruptcy legislation; and in this it may not be without its value in Canada where opinion yet halts at the proposal to again put a bankrupt Act on the statute book.—Gazette, Montreal.

THE FUTURE OF THE KLONDIKE.

ON appreciative article on the Klondike gold fields appeared in a recent issue of The Financial News, London, England. In part it read as follows:

"As respects the actual yield of gold, as stated in the Government returns, this, so far as Klondike is concerned, may be taken to be very much under the mark; for the reason that the very heavy royalty of 10 per cent. on the gross output is looked upon by the miners as an unfair and extortionate demand, with the consequence that large quantities of gold are constantly leaving the country which have paid but a small toll to the royalty returns.

"The royalty further operates adversely to the introduction of proper plant and to such systematic development as the conditions of the gold field now begin to require. To explain why this should be so necessitates a brief description of the deposits carrying the gold. These are ably described in the report made by the Canadian Government geologist, Mr. K. G. McConnell, B.A., last year, who divides them into four descriptions of gravel, the most recently deposited being the gravels found in the present creek beds. Then come the terrace gravels and, higher again, the old river gravels, and, lastly, the white quartz drift—the oldest and most important series of all. These last deposits lie from 200 ft. to 300 ft. or more above the levels of the present creek-beds, and, considering their extent and richness, may be said to be untouched.

"One frequently hears it remarked that the Klondike is worked out; but this seems quite an absurdity to anyone who has a practical knowledge of the field. Some few creek claims are nearly worked out by individual effort, so far as the richer leads of pay-dirt are concerned; but the mass of the gravels still left in the claims will yet yield fortunes to those who rework them with suitable appliances. I am sure I am within the mark in stating that, despite the gold already taken out in 1897, 1898, 1899, and 1900, mostly from creek claims (at least, £12,500,000), more than 75 per cent. of the gold yet remains to be extracted from the creek deposits, not to mention the hill gravels and the white quartz drift.

"The latter is of enormous extent, and, so far as it had been worked—round the fringes of the deposits—has given exceptionally high returns. Probably over 130,000,000 cubic yards of the quartz lie in the Bonanza Eldorado Valley, 150,000,000 in Hunber Valley, and 50,000,000 to 60,000,000 in Quartz Creek Valley. The value of this gravel will run anywhere from 5s. to £30 per cubic yard. As for the old river gravels, their bulk is enormous, and practically they lie untouched; but, as a rule, their gold values are much below those of the White Quartz Drift, but would still be considered fair in ordinary hydraulic mining. Within the past twelve months very considerable discoveries are reported to have been made of quartz reefs and other veins, carrying gold and other minerals.

"With all this wealth in view, it is only natural for the question to be put: Why have so many London companies been failures? The answer is simple enough. Until the summer of 1900 it was scarcely possible, at any cost, to get in plant of such a character as to make it practicable to substitute mechanical for manual labor. Now that the White Pass and Yukon Railway can deliver goods below the White Horse Rapids, and that heavy freight can be readily brought in and conveyed down stream to Dawson, one of the principal causes of non-success is removed, and excessive working cost should become a thing of the past, though shareholders must be prepared to find the necessary cash working capital. A proper equipment for mechanically working ground on a payable scale cannot be provided much under £10,000, exclusive of wages, fuel, etc.

"But before working conditions can be such as to encourage sufficient capital it is quite essential that the royalty should be reduced to such a figure as will give the investor a fair chance, and that the Canadian Government should, without hesitation or delay, follow the enlightened examples afforded by Australian and New Zealand legislation, and give liberal assistance to all parties who are prepared to spend money on means of communication, upon the bettering of the water supply, and in bringing in large plants for treating the poorer deposits. The Klondike is admirably administered, the climate is excellent, and the people are sober and hard-working, whilst the extent, regularity, and richness of the deposits hold out the certainty of large returns for many a long year."

HOW TO BRAZE CAST IRON.

The reason that cast iron cannot be brazed with spelter, as wrought iron can, says a contemporary, is that the graphitic carbon in the former prevents the adhesion of the spelter, as a layer of dust prevents the adhesion of cement to stone or brick.

A process to remove this graphite has been patented in Germany, consisting essentially in applying to the surfaces to be united an oxide of copper and protecting them against the influence of the air with borax or silicate of soda. When the joint is heated the oxide of copper gives up its oxygen to the graphite, converting it into carbonic oxide gas, which escapes in bubbles, while particles of metallic copper are deposited on the iron.

Any oxide of iron which may be formed is dissolved by the borax, and the surfaces of the iron, thus freed from graphite, unite readily with the spelter, which is run into the joint before it cools, the copper already deposited on the iron assisting the process. The inventor claims that cast iron can in this way be readily brazed in an ordinary blacksmith's forge.

F. A. Cantwell, general merchant, Franklin Centre, Que., has assigned.

T. J. Sears, general merchant, Lochaber, N.S., has purchased the livery business of A. K. McDonald, Antigonish, N.S.

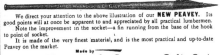

CURRENT MARKET QUOTATIONS.

April 12, 1901.

These prices are for such qualities and quantities as are usually ordered by retail dealers on the usual terms of credit, the lowest figures being for larger quantities and prompt pay. Large cash buyers can frequently make purchases at better prices. The Editor is anxious to be informed at once of any apparent errors in this list as the desire is to make it perfectly accurate.

METALS.

Tin.

Lamb and Flag and Straits—56 and 28 lb. ingots, per lb.	0 31	0 32

Tinplates.

Charcoal Plates—Bright

		Per box
M.L.S., equal to Bradley.		
I.C., usual sizes	....	$6 75
I.X., "	....	8 25
I.X.X., "	....	9 75
Famous—		
I.C.	....	6 75
I.X	....	8 25
I.X.X	....	9 50
Raven & Vulture Grades—		
I.C., usual sizes	....	5 00
I.X., "	....	6 00
I.X.X "	....	7 00
I.XXX., "	....	8 00
D.C., 12½x17	....	4 75
D.X.	....	5 50
D.X.X	....	7 50

Coke Plates—Bright

Bessemer Steel—		
I.C., usual sizes	....	4 15
I.C., special sizes, base	....	4 50
20 x 28	....	8 50

Charcoal Plates—Terne

Dean or J. G. Grade—		
I.C., 20x28, 112 sheets	....	8 25
I.X., Terne Tin	....	10 25

Charcoal Tin Boiler Plates.

		Per lb.
Cookley Grade—		
X X., 14x56, 50 sheet bxs; " 14x60, "; " 14x65, "	0 07	0 07½

Tinned Sheets

72x30 up to 24 gauge	0 08	0 08½
" 26 "	0 08½	0 09
" 28 "	0 09	0 09½

Iron and Steel.

Common Bar, per 100 lbs.	1 70	1 75
Refined " "	....	2 20
Horse Shoe Iron	....	2 05
Hoop steel, 1½ to 3 in. base, extras for smaller sizes	....	3 10
Sleigh Shoe Steel " base	....	2 00
Tire Steel	....	2 00
Machinery iron finish	....	2 05
Cast Steel, per lb	0 00	0 00
Toe Calk Steel	....	2 30
T. Firth & Co's special cast steel, per lb.		0 13
Jessop's Tool Steel	....	0 13

Boiler Tubes.

1½-inch	....	0 12½
2 "	....	0 13
2½ "	....	0 15
3 "	....	0 16
3½ "	....	0 20
4 "	....	0 25

Steel Boiler Plate.

¼ inch	....	2 25
3-16 inch	....	2 25
⅜ inch and thicker	....	2 25

Black Sheets.

18 gauge	....	3 00
20 gauge	....	3 00
22 to 24 "	....	3 10
26 "	....	3 20
28 "	....	3 30

Canada Plates.

Al dull, 52 sheets	....	3 00
Half polished	....	3 15
All bright	3 65	3 75

Iron Pipe.

Black pipe—		
⅛ "	....	3 35
¼ inch	....	3 35
⅜ "	....	3 30
½ "	....	3 35
¾ "	....	3 60
1 "	....	5 00
1¼ "	....	6 85
1½ "	....	8 25
2 "	....	11 (0
2½ "	....	22 91
3 "	....	30 12
3½ "	....	37 90
4 "	....	43 (9
5 "	....	57 85
6 "	....	75 (1
7 "	....	93 76
8 "	....	112 51
Galvanized pipe—		
½ inch	....	5 25
¾ "	....	5 56
1 "	....	7 77
1¼ "	....	10 60
1½ "	....	12 70
2 "	....	16 90

Galvanized Sheets.

	G O.	Comet.	Amer.	Queen's Head.
16 gauge		....	4 25	3 90
18 to 24 gauge	4 15	3 85	4 35	4 15
26 "	4 40	4 10	4 35	4 40
28 "	4 65	4 35	4 50	4 65

28 gauge American equals 26 gauge English.

Chain.

Proof Coil, 3-16 in., per 100 lb	....	....
" ¼ " "	8 30	8 50
" 5-16 " "	4 70	5 00
" ⅜ " "	4 05	4 C0
" 7-16 " "	3 9J	4 25
" ½ " "	3 10	4 10
" 9-16 " "	3 85	4 05
" ⅝ " "	3 35	3 90
" ¾ " "	3 60	4 10
Halter, kennel and post chains,	40 and 50 p.c.	
Cow ties	40 p.c.	
Stall fixtures	35 p.c.	
Trace chain	25 p.c.	

Jack chain, iron, single and double, discount 35 p.c.

Jack chain, brass, single and double, discount 40 p.c.

Copper

Ingot

English B. S., ton lots	....	0 19
Lake Superior	....	....

Bolt or Bar.

Cut lengths round, ½ to ⅞ in.	0 23½	0 25
" round and square 1 to 2 inches	0 23½	0 25

Sheet.

Plain, 14 oz., and light, 16 oz., 14x48 and 14x60	0 23	0 23
Plain, 14 oz., and light, 16 oz., irregular sizes	0 23	0 23½
Tinned copper sheets	....	0 26
Planished	....	0 32

Braziers (In sheets.)

4x6ft. 25 to 30 lbs. ea., per lb.	....	0 25½
" 35 to 45 " "	....	0 24½
" 50-lb. and above, "	....	0 23½

Boiler and T. K. Pitts

Plain Tinned, per lb	....	0 28
Spun, per lb	....	0 32

Brass.

Rod and Sheet, 14 to 30 gauge, 15 per cent.

Sheets, hard-rolled, 2x4	0 24	0 25
Tubing, base, per lb.	0 24	0 25

Zinc Spelter

Foreign, per lb	0 06	0 06½
Domestic "	....	..

Zinc Sheet.

5 cwt. casks	....	00 6½
Part casks	....	(0 6¾

Lead.

Imported Pig, per lb	0 04¾	0 05
Bar, 1 lb.	0 05½	0 05¾
Sheets, 2½ lbs. sq. ft., by	....	0 06¾
Sheets, 3 to 6 lbs., "	....	0 06

Note.—Cut sheets ½ cent per lb. extra. Pipe, by the roll, usual weights per yard, lists at 7c. per lb. and 25 p.c. dis. f.o.b. Toronto.

Note.—Cut lengths, net price, waste pipe 8-ft. lengths lists at 7½ cents.

Shot.

Common, $6.50 per 100 lb.; chilled, $7.00 per 100 lb.; buck, seal and ball, $7.50. Discount, 15 p.c. Prices are f.o.b. Toronto, Hamilton, Montreal, St. John and Halifax. Terms 3 per cent. cash, freights equalized on Montreal.

Soil Pipe and Fittings.

Discount, 60 and 10 per cent. on medium and extra heavy, and 60 per cent. on light.

Solder.

	Per lb.	Per lb.
Bar half-and-half, guarant'd	....	0 18½
Bar half-and-half, commer'l	....	0 18
Refined	....	0 17½
Wiping	....	0 17

Note.—Prices of this graded according to quantity. The prices of other qualities of solder in the market indicated by private brands vary according to composition.

Antimony.

Cookson's, per lb.	0 11	0 11½

White Lead.

	Per 100 lb.
Pure	6 50
No. 1 do	6 12½
No. 2 do	5 75
No. 3 do	5 37½
No. 4 do	5 00
Munro's Select Flake White	7 37½
Elephant and Decorators' Pure	7 12½
Brandram's B B. Genuine	9 00
" " Decorative	7 55
" " No. 1	6 85
" " No. 2	6 00

Red Lead.

Genuine, 560 lb. casks, per cwt.	....	$5 50
Genuine, 100 lb. kegs, per cwt.	....	5 75
No. 1, 560 lb. casks, per cwt.	....	5 25
No. 1, 100 lb. kegs, per cwt	....	5 00

White Zinc Paint.

Elephant Snow White	0 08	0 09
Pure White Zinc	0 08	0 09
No. 1	0 06	0 07½
No. 2	0 05	0 06½

Dry White Lead.

Pure, casks	5 75
Pure, kegs	6 25
No. 1, casks	5 50
No. 1, kegs	5 00

Prepared Paints.

In ¼, ½ and 1 gallon tins.

Pure, per gallon	....	1 25
Second qualities, per gallon	....	1 1C
Barn (in bbls.)	0 75	0 85
The Sherwin-Williams Paints.	....	1 45
Canada Paint Co's Pure	....	1 25
Toronto Lead & Color Co's Pure	....	1 25
Sanderson Pearcy's Pure	....	1
Stewart & Wood's Champion Pure.		1 20

Colors in Oil.

25 lb. tins, Standard Quality.

Venetian Red, per lb.	0 05
Chrome Yellow	0 11
Golden Ochre	0 06
French "	0 05
Marine Black	0 09
" Green	0 09
Chrome "	0 08
French Imperia Green	0 09

Colors, Dry.

Yellow Ochre (J.C.) bbls.	1 35	1 40
Yellow Ochre (J.F.L.S.), bbls	...	2 75
Yellow Ochre (Royal)	1 10	1 15
Brussels Ochre	....	2 00
Venetian Red (best), per cwt.	1 80	1 90
English Oxides, per cwt.	3 00	3 25
American Oxides, per cwt.	1 75	2 00
Canadian Oxides, per cwt.	1 75	2 00
Super Magnetic Oxides, 93 p.c.	2 00	2 25
Burnt Sienna, pure, per lb.	....	0 10
" Umber, " "	....	0 10
do Raw	....	0 09
Drop Black, pure	....	0 09
Chrome Yellows, pure	....	0 18
Chrome Greens, pure, per lb.	....	0 12
Golden Ochre	....	0 03½
Ultramarine Blue in 28-lb. boxes, per lb.	0 08	0 24
Fire Proof Mineral, per 100 lb	....	1 00
Genuine Eng. Litharge, per lb	....	0 07
Mortar Color, per 100 lb.	....	1 25
English Vermillion	....	0 80
Pure Indian Red, No. 45 lb.	....	0 80
Whiting, per 100 lb.	....	0 55

Blue Stone.

Casks, for spraying, per lb.	0 07
100-lb. lots, do. per lb	0 08

Putty.

Bulk in bbls.	1 90
Bulk in less quantity	2 05
Bladders in bbls.	2 10
Bladders in kegs, boxes or loose	2 25
Bladders in 25-lb. tins	2 35
Bladders in 12½-lb. tins	2 65
Bladders in bulk or tins less than 100 lb	2 90

Varnishes.

	In 5-gal. lots.	Per gal.
Carriage, No. 1	2 90	3 30
" body	8 00	9 00
" rubbing	4 00	5 00
Gold Size, Japan	3 00	3 40
Brown Japan	2 40	2 80
Elastic Oak	2 90	3 30
Furniture, extra	2 40	2 80
" No. 1	1 60	2 00
Hard Oil Finish	2 70	3 10
Light Oil Finish	3 20	3 60
Demar	3 30	3 70
Shellac, white	4 40	4 80
" orange	4 00	4 40
Furniture Brown Japan	1 60	2 00
Black Japan	2 40	2 80
" " No. 1	1 60	2 00

The Imperial Varnish & Color Co's., Limited Elastilite Varnish 1 gal. can, each. $3.00.

Granatine Floor Finish, per gal., $2 50.

Maple Leaf Coach Enamels; Size 1, 60c. Size 2, 30c.; Size 3, 20c. each.

Castor Oil.

East India, in cases, per lb.	0 10	0 10½
" " small lots	0 10½	0 11

Cod Oil, Etc.

Cod Oil per gal.	0 50	0 55
Pure Olive	....	1 20
" Neatsfoot	....	90

Glue.

Common	0 08½	0 09
French Medal	0 14	0 14½
Cabinet sheet	0 12	0 13
White, extra	0 18	0 20
Gelatine	0 22	0 30
Strip	0 18	0 20
Coopers	0 19	0 20
Huttner	....	0 18

HARDWARE.

Ammunition.

Cartridges.

B. B Cap Dom. 50 and 5 per cent.
Rim Fire Pistol, dis. 40 p. c., Amer.
Rim Fire Cartridges, Dom., 50 and 5 p. c.
Central Fire Pistol and Rifle, 10 p.c. Amer.
Central Fire Cartridges, pistol sizes, Dom. 30 per cent.
Central Fire Cartridges, Sporting and Military, Dom., 15 and 5 per cent.
Central Fire, Military and Sporting, Amer. add 5 p.c. to list. B.B. Caps, discount 40 per cent. Amer.
Loaded and empty Shells, "Trap" and "Dominion" grades, 25 per cent. Rival and Nitro, net list.
Brass Shot Shells, 55 per cent.
Primers, Dom., 30 per cent.

Wads — per lb.

Best thick white felt wadding, in ¼-lb bags ... 1 00
Best thick brown or grey felt wads, in ½-lb. bags ... 70
Best thick white card wads, in boxes of 500 each, 12 and smaller gauges ... 0 99
Best thick white card wads, in boxes of 500 each, 10 gauge ... 0 35
Best thick white card wads, in boxes of 500 each, 8 gauge ... 0 55
Thin card wads, in boxes of 1,000 each, 12 and smaller gauges ... 0 20
Thin card wads, in boxes of 1,000 each, 10 gauge ... 0 25
Thin card wads in boxes of 1,000 each, 8 gauge ...
Chemically prepared black edge grey cloth wads, in boxes of 250 each— Per M.
11 and smaller gauge ... 0 60
9 and 10 gauges ... 0 70
7 and 8 gauges ... 0 90
5 and 6 gauges ... 1 10
Superior chemically prepared pink edge, best white cloth wads, in boxes of 250 each—
11 and smaller gauge ... 1 15
9 and 10 gauges ... 1 40
7 and 8 gauges ... 1 65
5 and 6 gauges ... 1 90

Adzes.

Discount, 20 per cent.

Anvils.

Per lb ... 0 10 — 0 12½
Anvil and Vise combined ... 4 50
Wilkinson & Co.'s Anvils..lb. 0 09 — 0 09½
Wilkinson & Co.'s Vices..lb. 0 09¾ — 0 10

Augers.

Gilmour's, discount 65 and 5 p.c. off list.

Axes.

Chopping Axes
Single bit, per doz ... 5 50 — 10 00
Double bit, " ... 11 00 — 18 00
Bench Axes, 40 p.c.
Broad Axes, 33⅓ per cent.
Hunters' Axes ... 5 50 — 6 00
Boy's Axes ... 5 75 — 6 75
Splitting Axes ... 6 50 — 12 00
Handled Axes ... 7 00 — 10 00

Axle Grease.

Ordinary, per gross ... 5 75 — 6 00
Best quality ... 13 00 — 15 00

Bath Tubs.

Zinc ... 6 00
Copper, discount 15 p.c. off revised list

Baths.

Standard Enameled.

5½-inch rolled rim, 1st quality ... 25 00
" " " 2nd " ... 21 00

Anti-Friction Metal.

"Tandem" A ... per lb. 0 27
" B ... " 0 21
" C ... " 0 11½
Magnolia Anti-Friction Metal, per lb. 0 25

SYRACUSE SMELTING WORKS.

Aluminum, genuine ... 0 45
Dynamo ... 0 29
Special ... 0 25
Aluminum, 99 p.c. pure "Syracuse" ... 0 50

Bells.

Hand.

Brass, 60 per cent.
Nickel, 55 per cent.

Cow.

American make, discount 66⅔ per cent.
Canadian, discount 45 and 50 per cent.

Door.

Gongs, Sargant's ... 5 50 — 8 00
" Peterboro', discount 45 per cent.

Farm.

American, each ... 1 25 — 3 00

House.

American, per lb ... 0 35 — 0 40

Bellows.

Hand, per doz ... 3 35 — 4 75
Moulders', per doz ... 7 50 — 10 00
Blacksmiths', discount 40 per cent.

Belting.

Extra, 60 per cent.
Standard, 60 and 10 per cent.

Bits.

Auger.

Gilmour's, discount 60 and 5 per cent.
Rockford, 50 and 10 per cent.
Jennings' Gen., net list.

Car.

Gilmour's, 47½ to 50 per cent.

Expansive.

Clark's, 40 per cent

Gimlet.

Clark's, per doz ... 0 65 — 0 90
Diamond, Shell, per doz ... 1 00 — 1 50
Nail and Spike, per gross ... 2 25 — 5 20

Blind and Bed Staples.

All sizes, per lb ... 0 07½ — 0 12

Bolts and Nuts. Per cent.

Carriage Bolts, full square, Norway ... 70
" " full square ... 70
Common Carriage Bolts, all sizes ... 65
Machine Bolts, all sizes ... 65
Coach Screws ... 75
Sleigh Shoe Bolts ... 75
Blank Bolts ... 65
Bolt Ends ... 65
Nuts, square ... 4½c. off
Nuts, hexagon ... 4⅝c. off
Tire Bolts ... 67½
Stove Bolts ... 67½
Stove rods, per lb ... 5½ to 6c.
Plough Bolts ... 60

Boot Calks.

Small and medium, ball, per M ... 4 25
Small heel, per M ... 4 50

Bright Wire Goods.

Discount ... 62½ per cent.

Broilers.

Light, dis., 65 to 67½ per cent.
Reversible, dis., 65 to 67½ per cent.
Vegetable, per doz., dis. 37½ per cent.
Henis, No. 8, " ... 6 00
Henis, No. 9, " ... 7 00
Queen City " ... 7 50 — 0 00

Butchers' Cleavers.

German, per doz ... 6 00 — 11 00
American, per doz ... 12 00 — 20 00

Building Paper, Etc.

Plain building, per roll ... 0 30
Tarred lining, per roll ... 0 40
Tarred roofing, per 100 lb ... 1 65
Coal Tar, per barrel ... 3 50
Pitch, per 100-lb ... 0 85
Carpet felt, per ton ... 45 00

Bull Rings.

Copper, $2.00 for 2½ in. and $1.90 for 2 in.

Butts.

Wrought Brass, net revised list

Cast Iron.

Loose Pin, dis., 60 per cent.

Wrought Steel.

Fast Joint, dis. 60 and 10 per cent.
Loose Pin, dis. 60 and 10 per cent.
Berlin Bronzed, dis. 70, 70 and 5 per cent.
Gen. Bronzed, per pair ... 0 40 — 0 65

Carpet Stretchers.

American, per doz ... 1 00 — 1 50
Bullard's, per doz ... 6 50 — ....

Castors.

Bed, new list, dis. 55 to 57½ per cent.
Plate, dis. 52½ to 57½ per cent.

Cattle Leaders.

Nos. 31 and 32, per gross ... 50 — 9 50

Cement.

Canadian Portland ... 2 50 — 2 80
English " ... . — 3 00
Belgian " ... 2 50 — 2 75
Canadian hydraulic ... 1 25 — 1 50

Chalk.

Carpenters Colored, per gross 0 45 — 0 75
White lump, per cwt ... 0 60 — 0 65
Red ... 0 05 — 0 06
Crayon, per gross ... 0 14 — 0 18

Chisels.

Socket, Framing and Firmer.

Broad's, dis. 70 per cent.
Warnock's, dis. 70 per cent.
P. S. & W. Extra 60, 10 and 5 p.c.

Churns.

Revolving Churns, metal frames—No. 0, $8—No. 1, $8.50—No. 2, $9.00—No. 3, $10.00—No. 4, $12.00—No. 5, $16.00 each. Ditto, wood frames—20c. each less than above. Discounts: Delivered from factories, 58 p.c.; from stock in Montreal, 58 p.c. Terms, 4 months or 3 p.c. cash in 30 days.

Clips.

Axle dis. 65 per cent.

Closets.

Plain Ontario Syphon Jet ... $16 00
Emb. Ontario Syphon Jet ... 17 00
Fittings net ... 1 00
Plain Teutonic Syphon Washout ... 10 00
Emb. Teutonic Syphon Washout ... 11 00
Fittings net ... 1 25
Low Down Teutonic, plain ... 16 00
" " " embossed ... 17 00
Plain Richelieu net ... 3 75
Emb. Richelieu net ... 4 00
Fittings net ... 1 25
Low Down Ont. Sy. Jet, plain net ... 19 50
" " " " " emb'd. net 20 50
Closet connection net ... 1 25
Basins, round, 14 in ... 1 00
" oval, 17 x 14 in ... 2 50
" " 19 x 15 in ... 3 75
Discount 40 p.c., except on net figures.

Compasses, Dividers, Etc.

American, dis. 62½ to 65 per cent.

Cradles, Grain.

Canadian, dis. 25 to 33⅓ per cent.

Crosscut Saw Handles.

S. & D., No. 3, per pair ... 17½
" " 5, " ... 22½
" " 6, " ... 15
Boynton pattern " ... 20

Door Springs.

Torrey's Rod, per doz ... (15 p.c.) 2 00
Coil, per doz ... 0 88 — 1 60
English, per doz ... 2 00 — 4 00

Draw Knives.

Coach and Wagon, dis. 50 and 10 per cent.
Carpenters, dis. 70 per cent.

Drills.

Hand and Breast.

Millar's Falls, per doz. net list.

DRILL BITS.

Morse, dis., 37½ to 40 per cent.
Standard dis. 50 and 5 to 55 per cent

Faucets.

Common, cork-lined, dis. 35 per cent.

ELBOWS. (Stovepipe.)

No. 1, per doz ... 1 40
No. 2, per doz ... 1 20
Bright, 20c. per doz. extra.

ESCUTCHEONS.

Discount, 45 per cent.

ESCUTCHEON PINS.

Iron, discount 40 per cent.

FACTORY MILK CANS.

Discount off revised list, 40 per cent.

FILES.

Black Diamond, 70 and 5 per cent.
Kearney & Foote, 60 and 10 to 70 per cent.
Nicholson File Co., 60 to 60 and 10 per cent.
Jowitt's, English list, 25 to 27½ per cent.

FORKS.

Hay, manure, etc., dis., 50 and 10 per cent. revised list.

GLASS—Window—Box Price.

Size United Inches.	Star Per 50 ft.	Star Per 100 ft.	D. Diamond Per 50 ft.	D. Diamond Per 100 ft.
Under 26	2 15	4 15	....	6 00
26 to 40	2 30	4 45	....	6 65
41 to 50	....	4 85	....	7 50
51 to 60	....	5 15	....	8 50
61 to 70	....	5 50	....	9 50
71 to 80	....	6 00	....	10 50
81 to 85	....	6 50	....	11 75
86 to 90	....	....	....	14 00
91 to 95	....	....	....	15 50
96 to 100	....	....	....	18 00

GAUGES

Marking, Mortise, Etc.

Stanley's dis. 50 to 55 per cent.

Wire Gauges.

Winn's, Nos. 26 to 33, each ... 1 65 — 2 40

HALTERS.

Rope, ⅜ per gross ... . — ....
" ½ " ... . — 9 00
" ⅝ to ¾ ... . — 14 00
Leather, 1 in., per doz ... 3 87½ — 4 00
" 1¼ in., " ... 5 15 — 5 20
Web, — per doz ... 1 87 — 2 45

HAMMERS.

Nail

Maydole's, dis. 5 to 10 per cent. Can. dis. 25 to 27½ per cent.

Tack.

Magnetic, per doz ... 1 10 — 1 20

Sledge.

Canadian, per lb ... 0 07½ — 0 08½

Ball Pean.

English and Can., per lb ... 0 22 — 0 25

HANDLES.

Axe, per doz. net ... 1 50 — 2 00
Store door, per doz ... 1 00 — 1 50

Fork.

C. & B., dis. 40 per cent. rev. list.

Hoe.

C. & B., dis. 40 per cent. rev. list.

Saw.

American, per doz ... 1 00 — 1 25

Plane.

American, per gross ... 3 15 — 3 75

Hammer and Hatchet.

Canadian, 40 per cent.

Cross-Cut Saws.

Canadian, per pair ... 0 13½

HANGERS. doz. pairs.

Steel barn door ... 5 85 — 6 00
Stearns, 4 inch ... . — 5 00
" 5 inch ... . — 6 50
Lane's covered—
No. 11, 5-ft. run ... . — 8 40
No. 11½, 10-ft. run ... . — 10 80
No. 12, 10-ft. run ... . — 12 60
No. 14, 15-ft. run ... . — 21 00
Lane's O.N.T. track, per foot ... . — 4½

HARVEST TOOLS.

Discount, 50 and 10 per cent.

HATCHETS.

Canadian, dis. 40 to 42½ per cent.

HINGES.

Blind, Parker's, dis. 50 and 10 to 60 per cent.
Heavy T and strap, 4-in., per lb ... 0 06½
" " 5-in., " ... 0 06¼
" " 6-in., " ... 0 06
" " 8-in., " ... 0 05¾
" " 10-in., " ... 0 05½
Light T and strap, dis. 60 and 5 per cent.
Screw hook and hinge—
6 to 12 in., per 100 lbs ... 4 50
14 in. up, per 100 lbs ... 3 50
Per gro. pairs.
Spring 12 00

HOES.

Garden, Mortar, etc., dis. 50 and 10 p.c.
Planter, per doz ... 4 00 — 4 50

HOLLOW WARE

Discount ... 45 and 5 per cent.

HOOKS.

Cast Iron.

Bird Cage, per doz ... 0 50 — 1 10
Clothes Line, per doz ... 0 27 — 0 55
Harness, per doz ... 0 72 — 0 88
Hat and Coat, per gross ... 1 00 — 3 00
Chandelier, per doz ... 0 50 — 1 00

Wrought Iron.

Wrought Hooks and Staples, Can., dis. 47½ per cent.

Wire.

Hat and Coat, discount 45 per cent.
Belt, per 1,000 ... 0 60
Screw, bright, dis. 55 per cent.

HORSE NAILS.

"C" brand 50 and 7½ p.c. off new list } Oval-
"M" brand 60 per cent. } head
Countersunk, 60 and 10 per cent.

HORSESHOES

F.O.B. Montreal.

Iron Shoes. — No. 2 and larger. — No. 1. and smaller.

Light, medium, and heavy... 3 50 3 75
Snow shoes........ 3 75 4 00

Steel Shoes.

Light........ 3 60 3 85
Featherweight (al lsizes)..... 4 85 4 85
F.O.B. Toronto, Hamilton, London and Guelph, 10c. per keg additional.
Toe weight steel shoes...... 6 70

JAPANNED WARE.

Discount, 45 and 5 p.c. off list, June 1899

ICE PICKS.

Star per doz........ 3 00 3 25

KETTLES.

Brass spun, 7½ p.c. dis. off new list.
Copper, per lb........ 0 30 0 50
American, 60 and 10 to 65 and 5 p.c.

KEYS.

Lock, Can., dis., 45 p.c.
Cabinet, trunk, and padlock, Am. per gross........ 60

KNOBS.

Door, japanned and N.P., per doz........ 1 50 2 50
Bronze, Berlin, per doz...... 2 75 3 25
Bronze Genuine, per doz.... 6 00 9 00
Shutter, porcelain, F. & L. screw, per gross........ 1 30 4 00
White door knobs—per doz. 1 25

HAY KNIVES.

Discount, 50 and 10 per cent.

LAMP WICKS.

Discount, 60 per cent.

LANTERNS.

Cold Blast, per doz........ 7 00
No. 3 "Wright's"........ 8 50
Ordinary, with O burner........ 4 00
Dashboard, cold blast........ 9 00
No. 0........ 5 75
Japanning, 50c. per doz. extra.

LEMON SQUEEZERS.

Porcelain lined,......per doz. 2 20 5 60
Galvanized........ 1 87 3 85
King, wood........ 2 75 2 90
King, glass........ 4 00 4 50
All glass........ 1 20 1 30

LINES.

Fish, per gross........ 1 05 2 50
Chalk " 1 90 7 40

LOCKS.

Canadian, dis. 45 p.c.
Russell & Erwin, per doz.... 3 00 3 25

Cabinet.

Eagle, dis. 30 p.c.

Padlock.

English and Am., per doz.... 50 6 00
Scandinavian, " 1 00 2 40
Eagle, dis. 20 to 25 p.c.

MACHINE SCREWS. Iron and Brass.

Flat head discount 25 p.c.
Round Head, discount 20 p.c.

MALLETS.

Tinsmiths', per doz........ 1 25 1 50
Carpenters', hickory, per doz. 1 25 3 75
Lignum Vitae, per doz.. .... 3 85 5 00
Caulking, each........ 60 2 00

MATTOCKS.

Canadian, per doz........ 5 50 6 50

MEAT CUTTERS.

American, dis. 25 to 30 p.c.
German, 15 per cent.

MILK CAN TRIMMINGS.

Discount, 25 per cent.

NAILS.

Quotations are: — Cut. — Wire.

2d and 3d........ $3 85 $3 85
3d........ 3 00 3 52
4 and 5d........ 2 75 3 35
6 and 7d........ 2 65 3 20
8 and 9d........ 2 50 3 00
10 and 12d........ 2 45 2 95
16 and 20d........ 2 40 2 90
30, 40, 50 and 60d. (base)..... 2 35 2 85
Wire nails in carlots are $2.77½.
Galvanizing 2c. per lb. net extra.
Steel Cut Nails 10c. extra.
Miscellaneous wire nails, dis. 70 and 10 p.c.
Coopers' nails, dis. 30 per cent.
Flour barrel nails, dis. 25 per cent

NAIL PULLERS.

German and American...... 1 85 3 50

NAIL SETS

Square, round, and octagon, per gross........ 3 38 4 00
Diamond........ 12 00 15 00

NETTING.

Poultry, 50 and 5 per cent. for McMullen's

OAKUM. Per 100 lb.

Navy........ 6 00
U.S. Navy........ 7 25

OIL.

Water White (U.S.)........ 0 16½
Prime White (U.S.)........ 0 15½
Water White (Can.)........ 0 15
Prime White (Can.)........ 0 14

OILERS.

McClary's Model galvan. oil can, with pump, 5 gal., per doz........ 10 00
Zinc and tin, dis. 50, 50 and 10.
Copper, per doz........ 1 25 3 50
Brass, " 1 50 3 50
Malleable, dis. 25 per cent.

GALVANIZED PAILS.

Dufferin pattern pails, dis. 45 p.c.
Flaring pattern, discount 45 per cent.
Galvanized washtubs, discount 45 per cent.

PIECED WARE.

Discount 40 per cent. off list, June, 1899.
10-qt. flaring sap buckets, dis. 45 p.c.
6, 10 and 14-qt. flaring pails, dis. 45 p.c.
Creamer cans, dis. 45 p.c.

PICKS.

Per doz........ 6 00 9 00

PICTURE NAILS.

Porcelain head, per gross.... 1 75 3 00
Brass head " 0 40 1 00

PICTURE WIRE.

Tin and gilt, discount 75 p.c.

PLANES.

Wood, bench, Canadian dis. 50 per cent. American dis. 50.
Wood, fancy Canadian or American 7½ o 40 per cent.

PLANE IRONS.

English, per doz........ 2 00 5 00

PLIERS AND NIPPERS.

Button's Genuine per doz pairs, dis. 37½ 40 p.c.
Button's Imitation, per doz.. 5 00 9 00
German, per doz........ 0 60 2 60

PLUMBERS' BRASS GOODS.

Compression work, discount, 60 per cent.
Fuller's work, discount 65 per cent.
Rough stops and stop and waste cocks, discount, 60 per cent.
Jenkins disk globe and angle valves, discount, 55 per cent.
Standard valves, discount, 60 per per cent.
Jenkins' radiator valves discount 55 per cent.
" " " standard, dis., 60 p.c.
Quick opening valves discount, 60 p.c.
No. 1 compression bath cock........ 2 00
No. 4 " " " 2 00
No. 7, Fuller's........ 2 50
No 4½, " 3 00

POWDER.

Velox Smokeless Shotgun Powder.

100 lb. or less........ 0 85
1,000 lb. or more........ 0 80
Net 30 days.

PRESSED SPIKES.

Discount 20 to 25 per cent.

PULLEYS.

Hothouse, per doz........ 0 55 1 00
Axle........ 0 22 0 33
Screw........ 0 27 1 00
Awning........ 0 35 2 50

PUMPS.

Canadian cistern........ 1 80 3 60
Canadian pitcher spout.... 1 40 2 10

PUNCHES.

Saddlers', per doz........ 1 00 1 85
Conductors', " 9 00 15 00
Tinners' solid, per set........ 0 00 0 72
" hollow, per inch.... 0 00 1 00

RANGE BOILERS.

Galvanized, 3 gallons........ 7 60
" 35 " 8 25
" 40 " 9 50
Copper, 30 " 22 00
" 35 " 26 00
" 40 " 30 00
Discount off Copper Boilers 10 per cent.

RAKES.

Cast steel and malleable, 50, 10 and 5 p.c.
Wood, 25 per cent.

RASPS AND HORSE RASPS.

New Nicholson horse rasp, discount 60 to 60 and 10 p.c.
Globe File Co.'s rasps, 60 and 10 to 70 p.c.
Heller's Horse rasps, 50 to 50 and 5 p.c.

RAZORS.

per doz.

Geo. Butler & Co.'s,........ 8 00 18 00
Boker's........ 7 50 11 00
" King Cutter........ 12 50 50 00
Wade & Butcher's........ 3 60 10 00
Theile & Quack's........ 7 00 12 00
Elliot's........ 4 00 18 00

REAPING HOOKS.

Discount, 50 and 10 per cent.

REGISTERS.

Discount........ 40 per cent.

RIVETS AND BURRS.

Iron Rivets, black and tinned, discount 60 and 10 per cent.
Iron Burrs, discount 55 per cent.
Extras on Iron Rivets in 1-lb. cartons, ½c. per lb.
Extras on Iron Rivets in ½-lb. cartons, 1c. per lb.
Copper Rivets & Burrs, 35 and 5 p.c. dis. and cartons, 1c. per lb. extra, net.
Extras on Tinned or Coppered Rivets ½-lb. cartons, 1c. per lb.

RIVET SETS

Canadian, dis. 35 to 37½ per cent.

ROPE ETC.

— Sisal. — Manila.

7-16 in. and larger, per lb 10 13½
⅜ in........ 11 14½
¼ and 5-16 in........ 15½
Cotton, 3-16 inch and larger 16½
" 5-32 inch........ 21½
" ⅛ inch........ 22½
Russia Deep Sea........ 15½
Jute........ 8
Lath Yarn........ 9½
New Zealand Rope........ 10

RULES.

Boxwood, dis. 75 and 10 p.c.
Ivory, dis. 37½ to 40 p.c.

SAD IRONS. per set.

Mrs. Potts, No. 55, polished........ 62½
" No. 50, nickle-plated........ 67½

SAND AND EMERY PAPER.

Dominion Flint Paper, 47½ per cent.
B & A. sand, 40 and 5 per cent.
Emery, 40 per cent.
Garnet (Rurton's), 5 to 10 p.c. advance on list.

SAP SPOUTS.

Bronzed iron with hooks, per doz... 9 50

SAWS.

Hand Disston's, dis. 12½ p.c.
S. & D., 40 per cent.
Crosscut, Disston's, per ft.... 35 0 55
S. & D., dis. 35 p.c. on Nos. 2 and 3.
Hack, complete, each........ 0 75 2 75
" frame only........ 0 75

SASH WEIGHTS.

Sectional, per 100 lbs........ 2 75 3 00
Solid, " 2 00 2 25

SASH CORD.

Per lb........ 0 23 0 30

SAW SETS.

"Lincoln," per doz........ 6 50

SCALES.

B. S. & M. Scales, 45 p.c.
Champion, 65 per cent.
Fairbanks Standard, 35 p.c.
" Dominion, 55 p.c.
" Richelieu, 55 p.c.
Chatillon Spring Balances, 10 p.c.
Warren Champion 65 p.c.
" Standard 45 p.c.

SCREW DRIVERS.

Sargent's per doz........ 0 65 1 00

SCREWS

Wood, F. H., bright and steel, 87½ and 10 p.c.
Wood R. H., " dis. 82½ and 10 p.c.
" F. H., brass, dis. 80 and 10 p.c.
Wood, R. H., " dis. 75 and 10 p.c.
" F.H., bronze, dis. 75 p.c.
" R.H. " 70 p.c.
Drive Screws, 87½ and 10 per cent.
Bench, wood, per doz........ 3 25 4 00
" iron, " 4 25 5 75
Set, Case hardened, 60 per cent.
Square Cap, 50 and 5 per cent.
Hexagon Cap, 45 per cent.

SCYTHES.

Per doz, net........ 9 00

SCYTHE SNATHS.

Canadian, dis. 45 p.c.

SHEARS.

Bailey Cutlery Co., full nickeled, dis. 60 p.c.
Seymour's, dis. 50 and 10 p.c.

SHOVELS AND SPADES.

Canadian, dis. 40 and 5 per cent.

SINKS.

Steel and galvanized, discount 45 per cent.

SNAPS.

Harness, German, dis. 25 p.c.
Lock, Andrews'........ 4 50 11

SOLDERING IRONS.

1, 1½ lb., per lb........ 0 37
2 lb. or over, per lb........ 0 34

SQUARES.

Iron, No. 493, per doz........ 2 40 2 55
" No. 494, " 3 25 3 40
Steel, dis. 50 and 5 to 50 and 10 p.c., rev. list.
Try and bevel, dis. 50 to 52½ p.c.

STAMPED WARE.

Plain, dis., 75 and 12½ p.c. off revised list.
Retinned, dis., 75 p.c. off revised list.

STAPLES.

Galvanized........ 3 50 4 00
Plain........ 3 15 3 75
Coopers', discount 45 per cent.
Poultry netting staples, 40 per cent.

STOCKS AND DIES.

American dis. 25 p.c.

STONE. Per lb.

Washita........ 0 28 0 60
Hindostan........ 0 06 0 07
" slip,........ 0 09 0 09
Labrador........ .... 0 13
" Axe........ .... 0 15
Turkey........ .... 0 50
Arkansas........ 0 00 1 50
Water-of-Ayr........ 0 00 0 10
Scythe, per gross 3 50 5 00
Grind, per ton 15 00 18 00

STOVE PIPES.

5 and 6 inch Per 100 lengths 7 00
7 inch " " 7 50

ENAMELINE STOVE POLISH.

No. 4—3 dozen in case, net cash $4 80
No. 6—3 dozen in case, " 8 40

TACKS BRADS, ETC.

Per cent.

Strawberry box tacks, bulk ......75 & 10
Cheese-box tacks, blued ......80 & 12½
Trunk tacks, black and tinned 85
Carpet tacks, blued ....80 & 15
" " tinned ....80 & 20
" " (in kegs) 40
Cut tacks, blued, in dozens only . 80
" ¼ weights........60
Swedes, cut tacks, blued and tinned—
In bulk........80 & 10
In dozens........75
Swedes, upholsterers', bulk... 85, 12½ & 12½
" brush, blued & tinned, bulk..70
" gimp, blued, tinned and japanned........75 & 12½
Zinc tacks........35
Leather carpet tacks 55
Copper tacks........50
Copper nails........52½
Trunk nails, black........65 and 5
Trunk nails, tinned........65 and 10
Clout nails, blued........65 and 5
Chair nails........35
Patent brads........40
Fine finishing........40
Picture frame points........10
Lining tacks, in papers........10

STANDARD CHAIN CO., PITTSBURGH, U. S. A.

MANUFACTURERS OF

CHAIN OF ALL KINDS.

Proof Coil, B.B., B.B.B., Crane, Dredge Chain, Trace Chains, Cow Ties, etc.

ALEXANDER GIBB, Montreal. For Provinces of Ontario and Quebec. —Canadian Representatives— A. C. LESLIE & CO., Montreal. For other Provinces.

Lining tacks, in bulk		15
" " solid heads, in bulk		75
Saddle nails in papers		10
" " in bulk		15
Tufting buttons, 22 line, in dozens only		60
Tin capped trunk nails		15
Zinc glazier's points		5
Double pointed tacks, papers....90 and		10
" " " bulk		40

TAPE LINES.

English, ass skin, per doz.	2 75	5 00
English, Patent Leather	5 50	9 75
Chesterman's each	0 90	2 85
" steel, each	0 80	8 00

THERMOMETERS.

Tin case and dairy, dis. 75 to 75 and 10 p.c.

TRAPS. (Steel.)

Game, Newhouse, dis. 25 p.c.
Game, H. & N., P. S. & W., 65 p.c.
Game, steel, 72½, 75 p.c.

TROWELS.

Disston's discount 10 per cent.

German, per doz.	4 75	6 00

S. & D., discount 35 per cent.

TWINES.

Bag, Russian, per lb.	....	0 27
Wrapping, cotton, per lb.	0 22	0 26
Wrapping, mottled, per pack.	0 50	0 60
Wrapping, cotton, 3-ply		0 20
" " 4-ply	....	0 26
Mattress, per lb.	0 33	0 45
Staging, "	0 27	0 35

VISES.

Hand, per doz.	4 00	6 00
Bench, parallel, each	2 00	4 50
Coach, each	6 00	7 00
Peter Wright's, per lb.	0 12	0 13
Pipe, each	5 50	9 00
Saw, per doz	6 50	13 00

ENAMELLED WARE.

White, Princess, Turquoise, Blueand White, discoun: 50 per cent.
Diamond, Famous, Premier, 50 and 10 p.c.
Granite or Pearl, Imperial, Crescent, 50, 10 and 10 per cent.

WIRE.

Brass wire, 50 to 50 and 2½ per cent. off the list.

Copper wire, 45 and 10 per cent. net cash 30 days, f.o.b. factory.

Smooth Steel Wire, is quoted at the following net selling prices:

No. 6 to 8 gauge		$2 90
" 9 "		2 80
" 10 "		2 87
" 11 "		2 90
" 12 "		2 95
" 13 "		3 15
" 14 "		3 37
" 15 "		3 50
" 16 "		3 65

Other sizes of plain wire outside of Nos. 9, 10, 11, 12 and 13, and other varieties of plain wire remain at $2.80 base with extras as before. The prices for Nos. 9 to 13 include the charge of 10c. for oiling. Extras net per 100 lb.: Coppered wire, 60c.—tinned wire, $2—oiling, 10c.—special hay-bailing wire, 30c.—spring wire, $1—best steel wire, 75c.—bright soft drawn, 15c.—in 50 and 100-lb. bundles net, 10c.—in 25-lb. bundles net 15c.—packed in casks or cases, 15c.—bagging or papering, 10c.

Fine Steel Wire, dis. 17½ per cent List of extras: In 100-lb. lots: No 17, $5—No. 18, $5.50—No. 19, $6—No. 20 $6.65—No. 21, $7—No. 22, $7.30—No. 23 $7.65—No. 24, $8—No. 25, $9—No. 26, $9.50—No. 27, $10—No. 28, $11—No. 29, $12—No. 30, $13—No. 31, $14—No. 32, $15 No. 33, $16—No. 34, $17. Extras net—tinned wire, Nos. 17-25, $2—Nos. 26-31 $4—Nos. 32-34, $6. Coppered, 5c.—oiling, 10c.—in 25-lb. bundles, 15c.—in 5 and 10-lb. bundles, 25c.—in 1-lb. hanks, 50c.—in ½-lb. hanks, 75c.—in ¼-lb. hanks, $1—packed in casks or cases, 15c.—bagging or papering, 10c.

Galvanized Wire, per 100 lb.—Nos. 6, 7, 8, $3 50 to $3 8?—No. 9, $2.85 to $3.15—No. 10, $3.60 to $3.9?—No. 11, $3 70 to $4.10—No. 12, $3 to $3 30—No. 13, $3.10 to $3 40—No. 14 $4.10 to $4.50—No. 15, $4.60 to $5.05—No. 16, $4 85 to $5 35. Base sizes, Nos. 6 to 9, $2.57½ f.o.b. Cleveland.

Clothes Line Wire, solid 7 strand, No. 17, $4.25; No. 18, $2.65; No. 19, $2.35, f.o.b Hamilton, Toronto, Montreal.

WIRE FENCING. F.O.B. Toronto

Galvanized barb	3 05
Galvanized, plain twist	3 05

Galvanized barb, f.o.b. Cleveland, $2.82½ in less than carlots, and $2.70 in carlots.

WIRE CLOTH.

Painted Screen, per 100 sq. ft., net	1 35

WASTE COTTON. per lb.

Colored	4½ to 6
White, according to quality	6½ to 7½

500-lb. bale lots shaded.

WRENCHES.

Acme, 35 to 37½ per cent.
Agricultural, 60 p.c.
Coe's Genuine, dis. 20 to 25 p.c.

Towers' Engineer, each	2 00	7 00
" S., per doz.	5 80	6 00
G. & K.'s Pipe, per doz.	....	3 40
Burrell's Pipe, each	....	3 00
Pocket, per doz.	0 25	2 90

WRINGERS.

Leader	per doz.	$30 00	33 00
Royal Canadian	"	26 00	28 00
Royal American	"	26 00	28 00
Sampson	"	....	30 00

Terms 4 months, or 3 p.c. 30 days.

WROUGHT IRON WASHERS.

Canadian make, discount, 40 and 5 per cent

DAVID PHILIP

MANUFACTURERS' AGENT

362½ Main St., WINNIPEG.

Correspondence invited from manufacturers of Staple or Heavy Hardware, Iron or Steel Bolts and Nuts, etc., either by carrying stock in Winnipeg or by selling direct from factory.

GOOD REFERENCES.

KNOX HENRY

Heavy Hardware and Metal Broker
No. 1 Place Royale, MONTREAL.

Samples sent free on application.

HORSE NAILS—"C" Brand Horse Nails—Canada Horse Nail Co.

"BRASSITE" GOODS — Gunn Castor Co., Limited Birmingham, Eng.

Standard Paint & Varnish Works
Limited
Makers of High Grade
Varnishes, Japans,
Paints, Colors & Enamels
Windsor, Ont.

DIAMOND STOVE PIPE DAMPER.

U.S. Patent June 26th, 1895. Canadian Patent December 13th, 1891

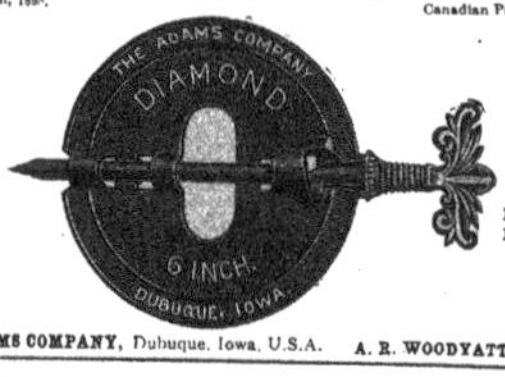

Nickle Handle.

Made by **THE ADAMS COMPANY**, Dubuque, Iowa, U.S.A. **A. R. WOODYATT & CO.**, Guelph, Ont.

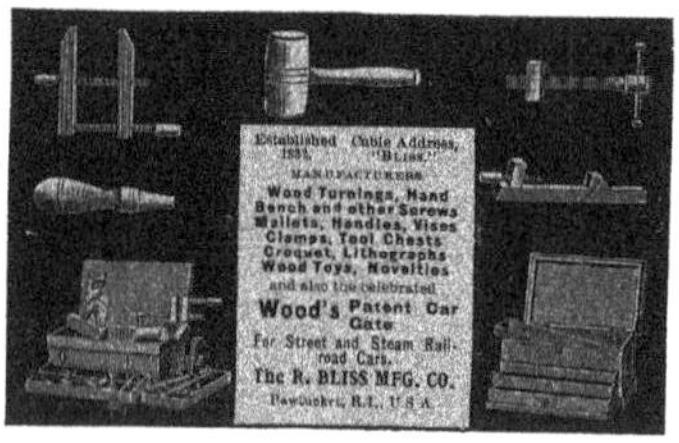

CANADIAN
HARDWARE
AND METAL MERCHANT

The Weekly Organ of the Hardware, Metal, Heating, Plumbing and Contracting Trades in Canada.

VOL. XIII. MONTREAL AND TORONTO APRIL 20, 1901. NO. 16

The Best Selling Range Ever Made.

Popular with dealers in every part of the country because it is so enthusiastically praised by every buyer.

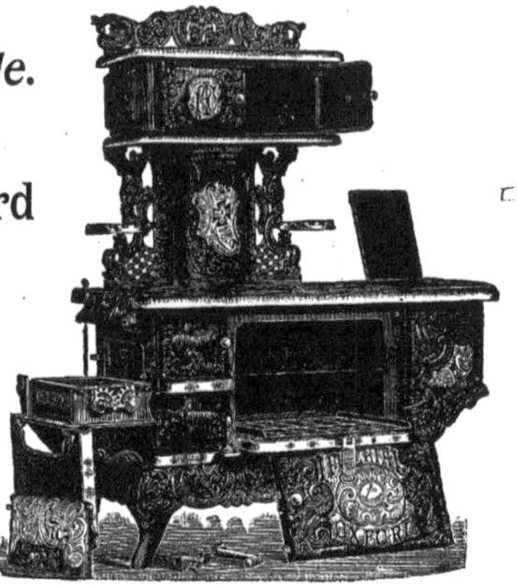

Our Imperial Oxford

has won its laurels—it is the favorite range of Canada—widely advertised, and everywhere appreciated for its practical superiority.

Are you familiar with its
Diffusive Flue Construction
Front Draw-Out Grate
Draw-Out Oven Rack
and other talking points. ?

If there's any range business in your locality you'll get it by handling the Imperial Oxford. Fullest details if you write

THE GURNEY FOUNDRY CO., Limited

TORONTO. WINNIPEG. VANCOUVER.

THE GURNEY-MASSEY CO., LIMITED, MONTREAL.

For the first season since 1897, we are in the market for the supply of

Galvanized Wire, Plain Twist AND Barb Wire Fencing

OF OUR OWN MANUFACTURE.

4 Barb, 6" apart "A" — 4 Barb, 4" apart "B"

2 Wires "C" — 3 Wires "D"

2 Barb, 5" apart—Long Barb "G" — 2 Barb, 5" apart—Short Barb "I"

When ordering Fencing do not forget

Bright and Galvanized Staples.

We solicit your orders also for

Wire of all kinds and for all purposes, of Steel, Brass or Copper.

Wire Nails and Wood Screws

Bright Wire Goods — Cotter Pins
Jack Chain — Door Pulls
and "Crescent" Coat and Hat Hooks
Bed, Blind, and Netting Staples.

Prices quoted upon application to

Dominion Wire Manufacturing Co.

MONTREAL and TORONTO.

BURMAN & SONS' CELEBRATED CLIPPERS for Horsemen and Barbers.

Established 1871. BIRMINGHAM, ENG.

NO. 3—POWER CLIPPER, with "Wrist Joint."
As supplied to The Czar of Russia. The King of Denmark. Earl Roberts, Etc., Etc.

NO. 297.

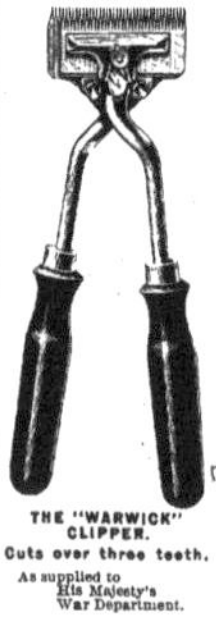

THE "WARWICK" CLIPPER.
Cuts over three teeth.
As supplied to His Majesty's War Department.

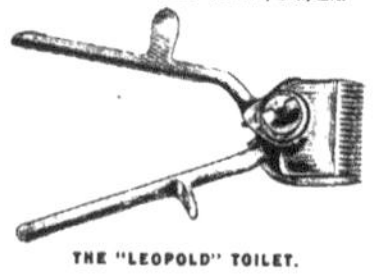

THE "LEOPOLD" TOILET.

SEND FOR PRICE LIST AND TERMS.

DELORME BROS., Agents, Debresoles Street, Montreal

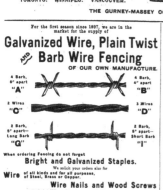

WOOD, VALLANCE & CO.,

Iron Merchants.

Importers of British and Foreign

HARDWARE

HAMILTON, ONT.,
Canada.

Branch House: George D. Wood & Co., Winnipeg, Man.
Toronto Office: 32 Front Street West—H. T. Eager.
British Columbia Representative: W. G. Mackenzie, P.O. Box 460, Vancouver.

Wood, Vallance & Co., Hamilton, Ont.

GEO. D. WOOD & CO.,

Iron Merchants

Importers of British and Foreign

HARDWARE.

WINNIPEG, CANADA.

THE NEW BALDWIN

DRY AIR CLEANABLE

REFRIGERATOR.

135 Modern Varieties. Ash, Oak and Soft-wood Finishes
METAL, PORCELAIN, SPRUCE LININGS.

BALDWIN

Positive Circulation—Sanitary—Odorless.

Latest Cleanable Features—The Strongest and Best System of Patent Removable Metal Air-Flues.

Air-Tight Lever Locks Ball-Bearing Casters, Swing Base—in and out.

Rubber around Doors and Lids, making them doubly air-tight.

Handsome Designs. Moderate Prices.

Built in the newest, largest and best equipped refrigerator plant in the East run all the year round on refrigerators exclusively; stock goods; special refrigerators and coolers in sections.

Handsome Trade Catalogue Ready.

Baldwin Refrigerator Co.,
BURLINGTON, VERMONT.

JAS. B. CAMPBELL. WILLIAM PRATT.

The Acme Can Works

Manufacturers of

Paint and Color Cans, round and square, Varnish and Oil Cans, Paint Irons, Paint Packages, Lye Tins,

and every description of Tin or can required by the trade. Write us for anything you need in our line. We will be pleased to quote you prices.

OFFICE AND FACTORY
Ontario St. and Jeanne D'Arc Avenue, - - MONTREAL.

THE WALKER UNIVERSAL TOOL AND CUTTER GRINDER

FLOOR
GRINDING MACHINES

Machine Tools,
Mill Supplies,
Pipe Fittings.

We are making a specialty of Machine Tools. Send for our catalogues.

The Fairbanks Company,
749 Craig Street,
MONTREAL.

1 to 25 quarts.
With fly wheel 8 to 25 quarts.

1 to 25 quarts.
With fly wheel 8 to 25 quarts.

Peerless Iceland Freezers

do the business. A complete line from one pint to forty quarts.

Get our Catalog.

DANA & CO.
Cincinnati.

New York Branch: 10 Warren St.
San Francisco Branch: 105 Front Street.

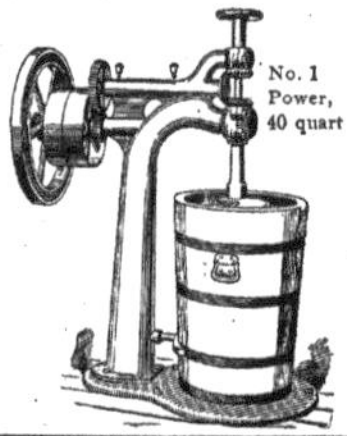

No. 1 Power, 40 quart

HENRY ROGERS, SONS & CO.,

SHEFFIELD, ENGLAND

Manufacturers of the celebrated

H R S & C°

UNION JACK CUTLERY

We make a specialty of

PLATED WARE, FRUIT KNIVES, ETC.

Our Canadian Representative carries a full line of samples.

Canadian Office: 6 St. Sacrament St., MONTREAL.

F. A. YORK, Manager.

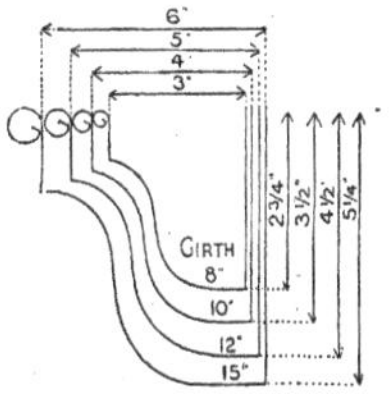

EAVETROUGH AND CONDUCTOR PIPE

Write for prices

E. T. WRIGHT & CO., Mfrs.
HAMILTON, CANADA.

TRADE MARK J REGISTERED

"JARDINE" HAND DRILLS

Five different sizes to suit all requirements. It pays to sell the best tools.

A. B. JARDINE & CO.
HESPELER, ONT.

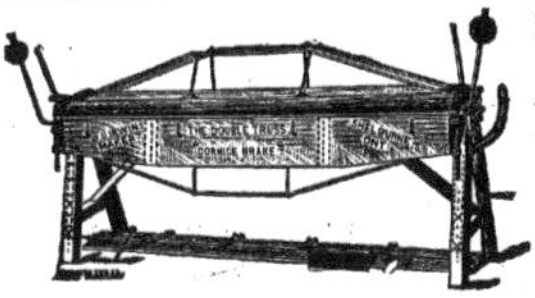

This eight-foot Brake bends 22-gauge iron and lighter, straight and true.

Price, $60

Very handy beader attachment, $15 extra if required.

Send for circulars and testimonials to

The Double Truss Cornice Brake Co. SHELBURNE, ONT.

The Latest and Best.

H. & R. Automatic Ejecting Single Gun.

Model 1900.

Steel and Twist Barrels in 30 and 32-inch.
12 Gauge.

Harrington & Richardson Arms Co.
Worcester, Mass., U.S.A.
Descriptive Catalogue on request.

STEVENS

Goods make an Up-to-date line for the retail merchant

Rifles, Shot-Guns, Pistols

FINE MECHANICAL TOOLS

Victor Bicycles

Handled by all the Leading Jobbing Houses.

J. Stevens Arms & Tool Co., P.O. Box 217 Chicopee Falls, Mass., U.S.A.

ASPINALL'S

O. White for Inside,
Indian White—Outside
for
Decorators' Use.
Imperial Gallons
and ½-Gallons.

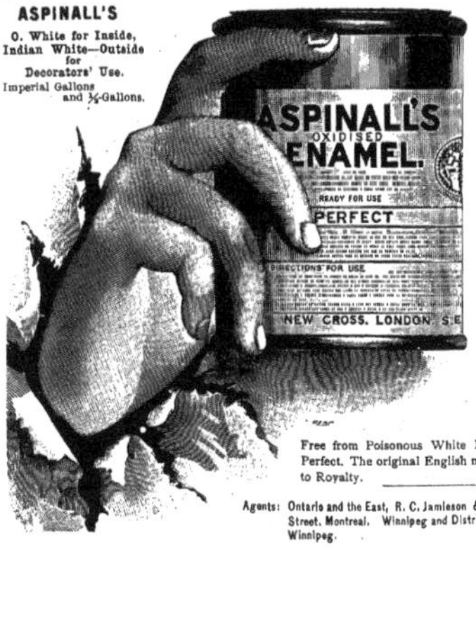

Free from Poisonous White Lead. Colours Perfect. The original English make as supplied to Royalty.

Agents: Ontario and the East, R. C. Jamieson & Co., 13 St. John Street, Montreal. Winnipeg and Distric J. H. Ashdown, Winnipeg.

HARDWARE AND METAL

VOL. XIII. MONTREAL AND TORONTO, APRIL 20, 1901. NO. 16.

President,
JOHN BAYNE MacLEAN,
Montreal.

THE MacLEAN PUBLISHING CO.
Limited.

Publishers of Trade Newspapers which circulate in the Provinces of British Columbia, North-West Territories, Manitoba, Ontario, Quebec, Nova Scotia, New Brunswick, P.E. Island and Newfoundland.

OFFICES

MONTREAL - - - - - - - 232 McGill Street, Telephone 1255.
TORONTO - - - - - - - 10 Front Street East, Telephone 2148.
LONDON, ENG. - - - - - 109 Fleet Street, E.C., W. H. Miln.
MANCHESTER, ENG. - - - 18 St Ann Street. H. S. Ashburner,
WINNIPEG - - - - - Western Canada Block. J. J. Roberts,
ST. JOHN, N.B. - - - - No. 3 Market Wharf, I. Hunter White.
NEW YORK. - - - - - - - 176 E. 88th Street.

Subscription, Canada and the United States, $2.00.
Great Britain and elsewhere - - - - 12s.

Cable Address { Adscript, London. Adscript, Canada.

Published every Saturday.

A PROFITABLE AGREEMENT.

AT the conference in Montreal some weeks ago, Mr. John Millen, of John Millen & Sons, told an interesting and instructive little story that has not yet appeared in print. Mr. Millen and Mr. Martineau, the president of the Montreal Retail Hardware Association, are two of the largest and best rated retailers in the city, and it so happens that they have long been business neighbors competing for east-end trade. Speaking of what confidence such as that bred by an association would do, Mr. Millen spoke of this little incident :

About 10 years ago (perhaps more) he and Martineau were in keen competition for patronage, and each would resort to price-cutting methods to secure trade. But they were wide enough awake to recognize that in heaping coals of fire on the "other fellow's" head they got burned themselves. So they came to an agreement about certain prices.

There was no written agreement, but as friends they relied on one another's word. They also agreed that if either had a suspicion that the other was cutting the established prices, he should test his neighbor by despatching a messenger to purchase the article in question. It seems that Mr. Millen sometimes did resort to this means of testing Mr. Martineau's fidelity to a bargain, but he never found him selling below the established price. Mr. Martineau never had occasion to put Mr. Millen to the test.

This affords an illustration of what mutual agreements among hardwaremen can do. We would not be so bold as to say that these agreements between Messrs. Martineau and Millen made them what they are, but we have their own words for it that they put a good deal of money in their pockets.

He who has an eye to the future will do his best for the business of to-day. That which a man does to-day is the foundation for the future.

RESTRICTIONS ON HORSE NAILS.

In view of the open market in horse nails between makers, The Canada Horse Nail Co. announce that they have, in response to their friends in the trade, removed the restrictions imposed by them as to the selling price of their goods, until the situation is different to that which exists at present.

SUMMER RESORT FOR MERCHANTS.

THE business men in the Maritime Provinces are each year taking increased interest in the subject of tourist travel. They would be a dull and unappreciative people, indeed, if they did not.

No part of this North American continent possesses greater natural attractions for summer tourists than the Maritime Provinces. The climate is delightful, being balmy and bracing, while, for fishing, hunting, bathing and boating, they are the ideal. The scenery abounds in variety and beauty. The smiling valleys and picturesque hills are all summer clothed in a rich greenness which to the people in Western Canada appears remarkable, while the rugged, rock-bound coasts never cause one to tire in gazing upon them.

But the Maritime Provinces have their interest historically as well as in regard to natural endowments. The very air seems laden with it. And the fact that their history contains so much that is romantic makes a visit to that part of the Dominion all the more interesting.

Some of the tourist associations in the Maritime Provinces are this spring making a special effort to induce business men in Western Canada to visit their part of the country during the coming summer. With this object in view negotiations are pending with the railway companies to arrange excursions.

To a great many business men and others in Western Ontario the Maritime Provinces are a terra incognita. It is to be hoped that many of those who have not yet spent a holiday in that most interesting part of the Dominion will do so this year.

CANADIAN IRON FOR THE BRITISH MARKET.

A STEAMER sailed from Sydney, N.S., on Thursday, with 2,400 tons of pig iron for the British market. This alone is about one-third of the total quantity of pig iron exported from Canada all last year.

But the significance of this is that it is the initiatory step in the development of a plan to secure a market in Great Britain for the iron and steel product of the Sydney works.

On the same day that the steamer in question sailed, Mr. James Ross, of Montreal, one of the directors of The Dominion Iron and Steel Co., returned from a visit to Great Britain, and, in an interview with a daily newspaper, he said that his company had secured an order in Great Britain for 150,000 tons of steel, valued at $3,000,000. This is even more interesting news than the announcement of the shipment of the first lot of iron from the Sydney furnaces to Great Britain.

The chief object the promoters of the Sydney works had in view from the very start was the development of the export trade in iron and steel. And for this purpose the works are exceptionally well situated, both geographically and in regard to raw material.

Sydney, which has one of the finest natural harbors in the world, has not only the advantage of being right on the seaboard, whereas Pittsburg is 500 miles inland, is 828 miles nearer Liverpool than New York and 878 miles nearer than Philadelphia. And not only is its geographical position favorable from a shipping standpoint, but it is even more so from the raw material assembling standpoint. The coal and the limestone are right at the door of the works, while the ore—except the small quantity that will be imported for mixing purposes—is but 400 miles away, whereas the Pittsburg furnaces are compelled to bring their ore from the north shore of Lake Superior, 1,000 miles away. Sydney has, therefore, it will be observed, an enormous advantage over Pittsburg, its great rival, in regard to the assembling of the necessary raw materials as well as in regard to juxtaposition to the British and other European markets.

As Mr. Moxham showed in his address before the Canadian Manufacturers' Association in Toronto a couple of months ago, the cost of assembling the raw materials at Sydney necessary to make one ton of pig metal was 79½c., while the cost for the same purpose at Pittsburg was $3.25, a saving in the assembling of the raw material alone of $2 45½ per ton in favor of the Canadian enterprise.

Now, in regard to steel. Taking into consideration the facts already pointed out, of advantages in cost of assembling raw material and the absence of any railroad charges for carriage to the seaboard, the Sydney plant has a net advantage in the cost basis of making steel of between $5 and $6 per ton over the plants at Pittsburg.

It is only about three months since the one of the four contemplated furnaces at Sydney was started up, but for a month or more its pig metal product has been coming into Toronto and other western Ontario cities. And what is still more noteworthy is the fact that shipments have also been made to Philadelphia.

At present the export trade in Canadian pig iron is but small. It is, however, growing. Last year it was by far the largest on record, as will be seen by the following table giving exports since 1892 :

EXPORTS OF PIG IRON.

	Tons.	Value.
1892	3	$ 95
1893	12	330
1895	259	6,202
1896	1,940	45,363
1897	2,677	65,555
1898	2,403	61,029
1899	2,188	50,767
1900	6,261	137,651

In 1894 no pig iron was exported. The countries to which our pig iron was exported last year were :

	Tons.	Value.
Great Britain	2,115	$ 36,647
Newfoundland	716	24,200
Holland	9	144
United States	3,421	76,660
Total	6,291	$137,651

With the Sydney furnace now in operation, and catering to a more or less extent to the foreign trade, we may confidently look forward to a large increase in our pig iron export trade within the next year or two.

ADVANCE IN THE PRICE OF BOLTS.

The anticipated advance in the price of bolts has at last taken place, the discounts having been reduced on Wednesday. The new discounts, compared with those previously ruling, are as follows :

	New Discount. Per Cent.	Old Discount. Per Cent.
Carriage bolts, Norway iron	65	70
" " full square	65	70
" " common	60	65
Machine bolts, all sizes	60	65
Coach screws	70	75
Sleighshoe bolts	72½	75
Blank bolts	60	65
Bolt ends	62½	65
Plough bolts	60	60
Nuts, square, all sizes, per lb. off	4c.	4½c.
Nuts, hexagon, per lb. off	4½c.	4¾c.

The advance is about 10 per cent., and is the result of the steady advance in the price of bar iron.

THE CHANGES IN IRON PIPE.

IRON pipe continues to advance in price, there having been a fair appreciation in quotations this week. In fact, on the Toronto market there were two changes in jobbers' quotations within that number of days.

In the first instance, the change was rather in the direction of an adjustment of the figures on black pipe from 2-inch upward, those from ⅛ to 2-inch inclusive being unchanged. There was some adjustment also in the quotations on galvanized pipe. On Thursday there was another change. By this black pipe was again advanced, 1-inch size being increased to $5.15 per 100 lb. There was also an advance in the price of galvanized pipe, but, as in the adjustment of Tuesday, there was a lowering of quotations ; the figures are still below those of last week. Subjoined we give the quotations per 100 feet now ruling together with those which appeared in our issue of April 13 :

BLACK PIPE.

Inch.	New Prices.	Old Prices.
⅛	$ 4.35	$ 4.35
¼	3.25	3.35
⅜	3.30	3.30
½	3.50	3.35
¾	3.65	3.60
1	5.15	5.00
1¼	7.00	6.85
1½	8 40	8.25
2	11 25	11.00
2½	20.95	22.94
3	24.55	30.12
3½	30.75	37.90
4	39 00	43.09
4½	41 80	.....
5	47.35	57.85
6	62.10	75.01

GALVANIZED PIPE.

½	4 90	5 25
¾	5.25	5.56
1	7.55	7.77
1¼	10.30	10.60
1½	12 35	12.70
2	16.50	16.90

TRADE CONDITIONS IN CANADA.

THERE is a great deal in the trade situation in Canada which is of an assuring character. In all staple lines of merchandise, spring business is opening up well. Individual orders are not, as a rule, large, but they are numerous; and not only are they numerous, but, in several seasonable lines which had been booked for future shipment, delivery is being urgently asked for before the date originally specified. This is a feature of the trade situation which is gratifying, showing, as it does, that stocks are not only not heavy, but that consumers are already beginning to inquire for the goods wanted at this time of the year. The wholesale houses in the different branches of trade are, in consequence, busily employed. The same can be said of the manufacturers, many of whom report that they are booked with orders enough to keep them going for some time to come.

In a country like Canada, where agriculture is the chief industry, it follows that our farmers must be in a prosperous condition if the country, commercially, is to be healthy.

The past six months has been a trying period to the farmers of Manitoba, owing to the partial failure of the wheat crop last year. This has naturally affected the general trade of that Province, but not to the extent it was a few months ago anticipated it would. This is the unanimous expression of the reports we have received from various reliable sources.

A well known essayist has told us that there is compensation in all things. And the compensation the farmers of Manitoba had for their short wheat crop was the additional time they had at their disposal, which they utilized in breaking new ground. The result is that this spring they have a much larger acreage ready for sowing wheat and other cereals than they had last year.

About the farming industry of Ontario there is not a particle of doubt. At no period in the history of the country was it probably ever in such a satisfactory condition. The cereal crops in that Province last year, both in quality and quantity, were excellent, while for butter, cheese and live stock there is a demand at good prices for all the farmer can produce. For horses and hogs the demand really exceeds the supply. Another evidence of the prosperity of the farmers in Ontario is the large number of new buildings that are being erected by them in nearly every part of the Province, to say nothing of the repairs that are being made.

Fall wheat has come through the past winter in Ontario in excellent condition.

In the Kootenay country, trade conditions are somewhat unsettled, but, taking British Columbia as a whole, business is not only fairly good at the moment, but the outlook is bright. The direct line of steamers which is to be put on between Vancouver and Skagway by the C.P.R. will divert to Victoria and Vancouver a great deal of the trade from the Klondike, that has hitherto gone to Seattle. Hitherto, there has been no regular line of steamships running out of British Columbian ports on Klondike trade account. Owing to this passengers and freight from points in Canada destined for the Klondike usually had to go via Seattle. The mint that is to be established at Vancouver by the Dominion Government will divert Klondike trade as well as Klondike gold from Seattle.

The year 1901 will be a red-letter year in the iron and steel industry of Canada. The starting up of the plants at Sydney, N.S.; Sault Ste. Marie, Ont.; Collingwood, Ont., and Midland, Ont., places that beyond all peradventure. Then there is the coal industry. On both coasts and in the Crow's Nest Pass the mines are experiencing an excellent trade, and the prospects for the future, with the additional demands there will be, particularly for the coal of the Nova Scotian and Crow's Nest Pass mines, are for even a better trade than that which is now being experienced.

The lumber industry is in a fairly healthy condition, although some engaged therein are of opinion that the turnover is perhaps not so far as large as it was anticipated it would be. The past winter has been favorable for operations in the woods, and a great deal of timber has been got out, but the quantity of dry lumber which the mills carried over from last year is small.

Perhaps one of the least satisfactory features of the situation is the commercial failures in Canada during the past quarter. This will be gathered from a glance at the accompanying table, compiled from Dun's report, giving the total and the number which occurred under the classifications, "manufacturing" and "trading."

The principal increase in liabilities was in Quebec Province, and one large failure swelled the figures for British Columbia, but Ontario reported a material improvement, both in number and amount. "The most pleasing feature," says Dun's Review, "of the quarter's statement is the lack of defaults in the banking and financial class."

It is generally conceded that the best of trade barometers are the railways and the bank clearing returns. Taking these for the past quarter and applying them to the trade conditions in Canada we think they corroborate what we have already said in regard to the trade conditions in Canada. The gross earnings of the two great Canadian railway systems during the three months, compared with the same period in 1900, were as follows:

FAILURES IN CANADA—FIRST QUARTER:

Provinces.	Total Commercial.			Manufacturing.		Trading.	
	No.	Assets.	Liabilities.	No.	Liabilities.	No.	Liabilities.
Ontario	139	$ 538,343	$ 661,580	33	$ 74,781	105	$ 585,399
Quebec	158	1,315,131	1 843,076	26	573,748	129	1,177,758
British Columbia	35	625,050	469,250	3	6,500	32	462,750
Nova Scotia	33	42,000	102,550	4	9 500	29	93,050
Manitoba	38	161,656	145 800	4	14,700	33	128 600
New Brunswick	17	50,969	100,466	3	39,000	13	44,466
Prince Edward Island	4	6,600	11,000	..	......	4	11,000
Total	424	$2,740,649	$3,333,722	73	$ 718,229	345	$2,503,023
" 1900	406	1 976,798	2,754,041	78	538,058	318	2,177,503
" 1899	363	3,163,116	4,241,411	92	2,433,155	267	1,803,906

GROSS RAILWAY EARNINGS FOR THREE MONTHS.

	1901.	1900.
Grand Trunk	$6,545,800	$6,213,304
Canadian Pacific	6,500,000	6,384,071

The Canadian Pacific has, it should be pointed out, 466 greater mileage than in the first quarter of 1900.

The returns in regard to the bank clearings are even more satisfactory than those in regard to the railways. For the three months they are as follows:

BANK CLEARINGS FOR THREE MONTHS.

	1901.	1900.
Montreal	$191,835,364	$171,985,788
Toronto	146,267,420	123,560,728
Winnipeg	24,621,434	23,930,215
Halifax	19,653,268	17,930,884
Vancouver	9,420,442	9,570,998
Hamilton	10,004,931	9 954,152
St. John, N.B.	8,695,523	7,886,551
Victoria	7,662,627	7,146,605
Total	$418,161,009	$371,965,921

Business in Canada is devoid of any speculative tendency, wholesalers and retailers alike buying for requirements. This is a good thing, tending, as it does, to keep trade in a healthy condition.

TRADE IN COUNTRIES OTHER THAN OUR OWN.

BAR IRON IN THE UNITED STATES.

A despatch from Pittsburg says: "The report that a meeting of the bar iron manufacturers had been held in New York this week, at which representatives were present from Jones & Laughlins, Carnegie Steel Company, Cambria Steel Company and other large steel concerns, is misleading. The facts are that none of these concerns makes a pound of iron, their entire product being of steel. There is no association among the bar iron mills, most of these in the Central West being controlled by the American Steel Hoop Company and Republic Iron & Steel Company. Price agreements on iron bars, or for that matter on any kind of iron and steel products, are really not necessary at this time, as the heavy demand not only keeps prices up, but keeps them steadily advancing. It is only a short while ago since iron bars were selling at about 1.20 cents at maker's mill; but the Republic Iron & Steel Company and American Steel Hoop Company are both holding firm now for 1.50 cents a pound at mill, and have all the business they can take care of at that price."

IRON TRADE IN THE STATES.

Labor cost to Valley furnacemen has been raised again, the demands of the men being granted in the week, resulting in a higher wage than was ever paid on the present level of pig iron prices. Foundry iron markets have been active, and orders for delivery in the second half of the year are now on furnace books in large volume. It is plain that consumption is now at the highest rate the country has known.

Foundries of nearly all descriptions are contributing to the increase, and even the stove manufacturers, who have found cause for thinking a wage reduction was necessary, have agreed to the old scale for another year. The most significant fact in connection with finished material is that mills which have not run for months and some of which were scarcely expected to be active again, have been started up. This is particularly true in bars, sheets and pipe. Rail mills are so well sold that foreign business is not being sought as in the closing months of last year. London agents representing other American products have been notified that foreign business is not to be gone after for delivery in the next few months. The structural demand is assuming prodigious proportions, and it can be seen that the rush of specifications later in the season may easily duplicate the conditions in 1899 in this trade.—Iron Trade Review, April 4.

THE BRITISH TINPLATE TRADE.

The inquiry for tinplates continues vigorous in all directions, and has resulted in some good lines being placed out, both for prompt and forward deliveries. The cheerful feature of this demand consists in the fact that a large portion of it comes from California, as well as the Eastern States of America. The demand has embraced "all sorts and conditions" of tinplates, from light plates of all weights down to special sizes, and to this extent no doubt the market is indebted to the "American boom" in the iron trade. While prices on the other side are advancing continuously, those on this side have remained stationary, so that the trade here has profited by this difference in the condition of the two markets. The Continental, Canadian, and home trades have all been full of life, and this happy "combination" has brought about a more favorable tone to the market. Owing to the retrogression in bars, makers have been able to meet this demand to a much larger extent, and many of the most important works have filled their order books over the present half.—Iron and Coal Trades Review, April 5.

THE HARDWARE TRADES IN ENGLAND.

With the end of the quarter, it cannot be said that the trade in the hardware manufacturing centres shows any improvement though the outlook for after the holidays is decidedly more encouraging. One of the chief obstacles to a revival of business in the Midlands was removed last week by the all-round reduction declared in coal and iron, and ironworkers' wages. Naturally, this movement should be followed up in a similar manner by the manufacturers of finished goods. There still remains room for improvement in the matter of prices especially is this observable in the brass and copper trades. Shipping orders for heavy lines are rather scarce owing to the keenness of American and German competition, but as prices are steadily rising in the States while they are falling in this country, no doubt our manufacturers will soon be able to recover lost ground. The continued dullness of the building trade is restricting business in many lines, including builders' castings, bath and lavatory work, stove grates, door and window fittings, etc. The unseasonable spell of wintry weather has almost paralyzed the demand for garden tools and outdoor goods, though fortunately for the ironmonger, he has made up this deficit by the extra demand for heating appliances!—Hardwareman.

PIG IRON IN GREAT BRITAIN.

The pig-iron market is quieter than last week, and prices are not so well supported. This keeps consumers from buying to any great extent, and in most districts they are again restricting their purchases to supplies urgently needed for immediate use. The current production in South Staffordshire, however, is being absorbed, makers finding little or no difficulty in disposing of it at the present low rates. In the Cleveland district the out-turn of forge qualities is still very restricted, and the price of the iron is rapidly approaching that of No. 3. There is now less difference between the two than there has been for a long time, and forge may in the future become dearer than No. 3, which is now quoted at 45s. 3d. In Barrow, hematite is quiet at unchanged prices. The steelmakers are the largest users. In Glasgow Scotch warrants have fallen since our last, being quoted at 52s. 9 1-2d. as against 53s. 8d. a week ago.—Iron and Coal Trades Review, April 5.

NEW YORK METAL MARKET

COPPER.—The upward movement of copper prices in the London market, which is attributed largely to the renewal of speculative interest with the return of a hopeful feeling in trade circles there, made further progress to-day, the markets closing at an advance of 10s. on spot and 11s. 3d. on futures. There was considerable activity, though the buying, especially of futures, was not as heavy as yesterday. The New York market does not respond to the better conditions in London. Prices here, while firm, are unchanged at 17c. for Lake Superior and 16 5-8c. for electrolytic and casting.

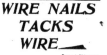

PIG TIN.—The London market, affected by the same influences, it is said, that have caused the rise in copper there, showed a further advance of £1 2s. 6d. on spot and £1 10s. on futures. In view of the anticipated increase in spot supplies the New York market is rather flat, but prices at the close were 15 to 25 points higher in sympathy with London, closing at 26c. bid and 26.50 asked for spot. The shipments from the Straits for the first half of the month were comparatively large, amounting to 2,210 tons, compared with 1,510 tons for the corresponding period last year. There is considerable criticism in the trade for the manner in which American tin statistics are twisted to suit the purposes of certain speculative operators in London, as shown by their monthly circular letters. An instance of this is found in a circular issued by one London firm and just received here, in which the American visible supply on April 1 is given as 3,408 tons on the spot and afloat, whereas, according to the statistics of The New York Metal Exchange, the quantity afloat on that day alone amounted to 3,419 tons. Another English firm gives the American visible supply on

April 1 as 900 tons on the spot and 2,400 tons afloat, although there were actually at that date 1,170 tons on the dock and discharging from vessels in port, not to speak of stocks in warehouse. According to these English statisticians the American factories during the past few months have been compelled to shut down frequently because of a lack of material, when, as a matter of fact, there has at no time been a scarcity of spot stook, although it is true that it has been largely under the control of one firm, which supplies the largest consumers. The official figures of the American visible supply on April 1 differ radically from those given in the circulars above mentioned, being as follows: In store and landing, 2,873 tons: afloat, 3,419 tons; total, 6,292 tons.

PIG LEAD.—The New York market was very quiet, but prices are maintained on the basis of $4.37 1-2c for lots of 50 tons or over. St. Louis was steady at $4.20 to $4.22 1-2. In London soft Spanish declined 3s. 9d.

SPELTER.—The market remains quiet but steady at 3.90 to 3.95c. St. Louis was reported firm with sales of 100 tons spot at $3.80 and 100 tons May at $3.82 1-2. There was a sharp rise in London, that market closing 7s. 6d. above last night's quotation.

ANTIMONY.—Regulus is steady and in moderate request at 8 1-2c. to 10 1-4c., as to brand.

OLD METALS.—Trade is light, but the market is steady at the quotations.

IRON.—Advices from Pittsburg refer to current reports of large sales of Bessemer pig iron for delivery over the last half of the year which, although not verified as to quantity or price, are believed to have been made. Mill iron also is reported to be active there, while foundry was in good demand. Locally a very good business is being done, but there are no very heavy transactions in foundry iron or other pig metal. There is no abatement of the demand for finished material, but there has been no further change in prices. Scotch warrants in Glasgow continue to show improvement, to-day's cable quotations being 4d. above those of yesterday.

TINPLATE.—The market is firm and active but without new feature.—N. Y. Journal of Commerce, April 17.

THE ROBERTSON-GODSON COMPANY.

The Vancouver branch of the James Robertson Co., Montreal and Toronto, has been reorganized. C. A. Godson, who has acted as manager of the Vancouver branch, has been admitted into full partnership and the firm will now be known as The Robertson-Godson Company. A large warehouse is being erected and the business of the branch will be extended, though most lines will still be made in the factories in Toronto and Montreal.

CANADA AND THE GLASGOW EXHIBITION.

HARDWAREMAN, Birmingham:—"Many of the Canadian engineers and manufacturers, judging from appearances, intend to extensively patronize the forthcoming Glasgow Exhibition in preference to the Pan-American Exhibition to be held about the same time at Buffalo. They have arrived at this conclusion, not from any sentimental 'preferential' spirit, but after mature consideration of the advantages to be derived from an exchange of ideas with the Motherland. In an eminently practical, if not convincing, way they argue that if the United States manufacturers can successfully invade the British market there is no reason why Canadians cannot do the same. Their optimism augurs well for the future.

"Certain it is that if they do not educate themselves sufficiently to successfully invade this country with their manufactures, there is no place in the whole of the world of more interest to engineers especially than Glasgow, where many of the problems in civic Government, and municipal engineering have been wholly or partially solved. A study of these problems alone, which have been the despair of workers in those spheres in other great cities of the world, will have its educational advantages, and be well worth the time and money spent in their study. Let us hope that the interest of the Canadian engineers and exhibitors alike will be stimulated, and that the exchange of ideas will be reciprocal and mutually advantageous."

ELDER BROWN.

A. A. Brown, the representative of the McClary Manufacturing Co., Montreal, who, it will be remembered, travelled for many years all over Ontario and counts in the hardware trade of that Province hosts of friends, has just been elected to the responsible position of elder of St. Paul's Presbyterian Church, Montreal, one of the leading Presbyterian churches of the city.

PERSONAL MENTION.

Mr. C. H. Whetworth, who is opening a new store in Smith's Falls, was in Montreal during the past week placing his orders for goods.

APPRECIATES HARDWARE AND METAL.

Editor HARDWARE AND METAL,—We appreciate very much your efforts in getting up such a newsy, up-to-date trade paper, and find that your articles and forecasts on price fluctuations are a great assistance to us in our business. In fact, we could hardly do business without your paper.

McDONALD & McCRARY.

Florence, Ont., April 9, 1901.

RETAILERS VS. DEPARTMENTAL STORES.

(By A Toronto Retailer.)

MANY solutions have been offered through your columns for the solving of the supposed unfair competition of the departmental stores. The fact remains, however, that the departmental store is in high favor with the buying public. This seems to indicate conclusively that their method is the popular one. Now, if, as retailers, we wish to secure favor with the public why not conduct our business on the popular plan? It will possibly require our making changes, considerable exertion, and must be backed up with a considerable amount of determination to run our business for the benefit of the profit and loss account regardless of sentiment.

1. Cut out the bad debts; sell for cash only. This will save lots of bad accounts, the loss of omitting to charge, and the expense of borrowing capital to carry on a credit business. Our goods cost us money and we have no more right to loan them out at 30, 60 or 90 days than we have to loan so much money. We did not enter the business as a loaning concern, yet we have adopted this extravagant method against the departmental store's spot cash. Some say this cannot be done. Our keenest competition, the departmental store, does it and they are popular. It is a saving and popular. What more then do we want?

2. Have one price and one price only. Let every customer feel he is getting the same treatment others are getting, no better, no worse. It will give him confidence in prices asked and he leaves the store under no obligation.

3. Systemize the delivery. Customers form their opinions from small things connected with the business. A slack, irregular delivery denotes a slack, irregular business generally. It carries weight, either for or against. Our competitors have aided themselves by it. We must not be behind. Establish a method of doing everything connected with your business. Stick right to it. It gives the impression that the concern is an up-to-date one and a going concern. People patronize a live business because it is a live business, and thus keep it alive.

4. To increase buying power, to more evenly compete with the departmental store, cut off some of the wholesalers selling the same lines. Get a good house, give them all the trade and make your trade worth somebody's time to do you a favor. The wholesaler will be interested in the success of his customer, and will be found more ready to aid him than when the account is scattered.

5. Advertise. Don't throw money away but let people know where you are and what you have to sell, and for how much. Publishing prices gives confidence in those prices.

6. Check both ways. Demand an invoice with your goods. See that you get and give exactly 16 ounces to the pound, 36 inches to the yard, and 12 only to the dozen.

There is an old law, "The survival of the fittest." It was here before the departmental store. We have both to live under that law, and if we liven up and get to be the "fittest," the departmental store will go, or vice versa.

NOVA SCOTIA COAL DEAR.

Speaking before the Miners' Arbitration Commission, in Halifax, last Saturday, H. D. McKenzie, coal dealer, Halifax, stated that he had just been offered a cargo of Scotch coal delivered at St. John at 17s. 9d., including cost, freight and insurance, which, after the 40 cents duty was paid, would mean $4.05. Mr. McKenzie believed that the scarcity of soft coal was largely due to the Massachusetts contract which the Dominion Coal Company had made, amounting to 800,000 tons. Prices in Nova Scotia were higher than in any other part of Canada, and he believed they were higher than in any other coal-producing country in the world.

A RECORD IN STEEL-MAKING.

Figures which have been collected and published by the Association of Iron and Steel Manufacturers of Germany show that the total production of steel in that country in 1900 was 6,645,869 tons, of which 6,223,417 tons were produced by the basic process, a system that was not applied until 1879, so that this remarkable advance has been achieved within twenty years. So recently as 1894, the total make of basic steel in Germany was only 3,241,000 tons, so that the output has practically doubled within the last six years. Of the total output of basic steel in 1900, 4,141,587 tons were produced by the Bessemer process, and 1,997,765 tons by the open-hearth process. It may be added that while in Great Britain four-fifths of our total steel output is by the older acid process, in Germany that process yields only one-sixteenth of the whole.—Iron and Coal Trades Review.

INDUSTRIAL GOSSIP.

Those having any items of news suitable for this column will confer a favor by forwarding them to this office addressed the Editor.

The Ontario Powder Co., Limited, Kingston, Ont., has been incorporated.

The Ladysmith, B.C., Iron Works Co. are clearing the ground for a foundry and iron works.

The Gurney-Tilden Co. have had plans prepared for a fourth storey on their warehouse building, Hamilton.

The D. Moore Co., Hamilton, have added a storey to their foundry buildings, and have secured the necessary land for further additions.

The rolling mills of Peck, Benny & Co., Montreal, have been closed down on account of lack of water power, the water having been taken out of the Lachine canal for repairs. About 500 men are laid off. The works will not be in full operation until May 1.

The Grey and Bruce Cement Works have placed a contract with Goldie & McCullough, Galt, Ont., for shaftings, pulleys, sidings, etc. When these are installed their new works will be ready for operations, which may begin early in June.

H. S. HOWLAND, SONS & CO.

WHOLESALE ONLY — 37-39 Front Street West, **Toronto.** — ONLY WHOLESALE

"Triumph" Corn Planters.
" With Pumpkin Seed Attachment.

Hog Rings.
Nos. 1, 2, 3.

Hog Ringers, Cast.
" Malleable.

RAKES

Malleable Iron.

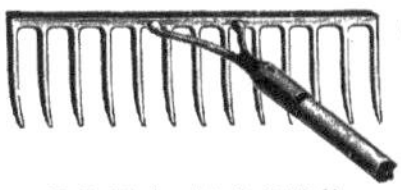

Malleable Iron Teeth, 10, 12, 14
10 to 14 Assorted.

Steel.

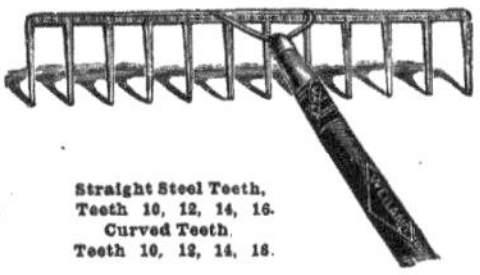

Straight Steel Teeth,
Teeth 10, 12, 14, 16.
Curved Teeth.
Teeth 10, 12, 14, 16.

Lawn.

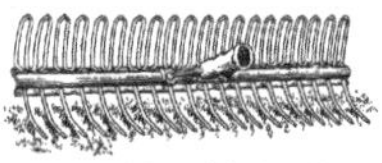

"Queen" Lawn Rake (as cut.)
24 Steel-Tinned Teeth.
"Gibb," Wire Teeth, Wood Back.

PRUNERS

Tree Pruners.

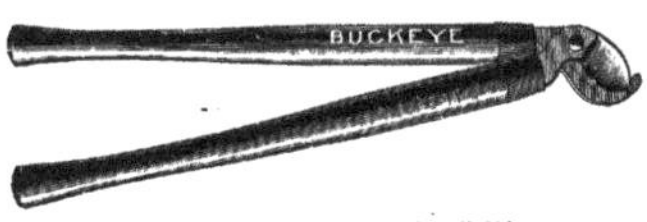

No. 35 Cast Steel Blades, Length 26 in.
No 38 Cast Steel Blades, Length 41 in.
will cut 1¼ in. Stick with ease

Pruning Shears.

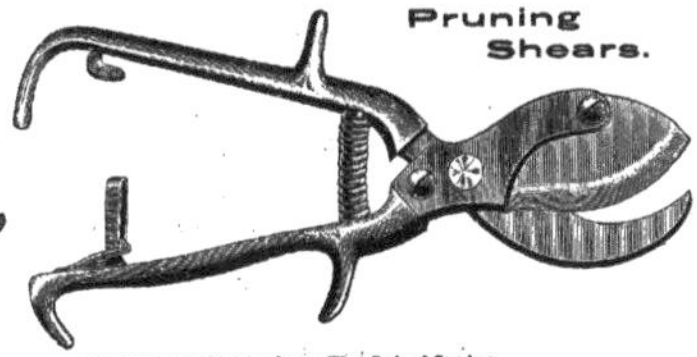

"Boker's" 1636, 8 in. long, Flat Spiral Spring.
"Boker's" 7191, 8½ in. long, Flat Spiral Spring, Bow Handles.
"American" 12, 8½ in. long, Flat Spiral Spring
"American" 9, 9 in. long, Brass Coil Spring.

Long Handle Tree Pruners,
Feet 6, 8, 10, 12.

H. S. HOWLAND, SONS & CO., Toronto.

WE SHIP PROMPTLY. — Graham Wire and Cut Nails are the Best. — OUR PRICES ARE RIGHT

CATALOGUES, BOOKLETS, ETC.

A VOLUME OF MAGNITUDE.

IF a firm's catalogue is any criterion by which to judge its business, Caverhill, Learmont & Co. have done much by the publication of a catalogue this week to individualize themselves before the Canadian trade and show themselves to be a leading wholesale house. We can safely say that their Catalogue No. 40, just off the press, is the best hardware catalogue ever published by a Canadian house. It is a book of dictionary size. There are over 1,000 pages, each 9 x 12 in., bound in a heavy, serviceable cloth, with leather corners. But, prententious as the book is on the outside, its magnitude and excellence within is still more striking. The index itself, although run in a concise, double-column form, occupies 20 pages, leading to a vast store of information. Every article that a retail hardwareman can need is catalogued in this work, and its completeness shows that it has long been in the course of preparation. The paper used is a fine variety, serving to bring out the thousands of cuts clearly. The half-tones are uniformly good, and their profusion is as surprising as it is desirable.

Everything capable of illustration seems to be described by the photograph, a fact which the retail trade will not be slow to appreciate. As a consequence, the remarks are, as a rule, terse and short. In most cases the prices are given, but it is explained in the introduction that when ordering the quotations should be confirmed. Tables of weights and measures, as well as of standard lists, have been included, and no doubt they will be found useful. In fact, no space or expense has been spared to give the customer of a wholesale hardware establishment all the information that he could wish for to aid him in making out a letter order. A touch of the pleasing is added by the reproduction of several cuts of the firm's establishments on the opening pages, and it is a feature difficult to escape notice that on each page there is a light impression of reading: "Caverhill, Learmont & Co., Wholesale Hardware Merchants, Montreal."

Any firm who is favored with a loan of this work should consider that they have been given a decided treat. It ought to do much to add additional importance and dignity to the firm's name.

Alexander Gibb, St. John street, Montreal, Canadian agent for The Springfield Machine Screw Co., has handed us a neat little catalogue that his firm have issued for the convenience of their customers. It contains clear, explanatory illustrations and price lists of the various goods turned out at their factory, including belt punches, calipers, crank keys for bicycles, hack-saw frames, nail sets, knurling tools, saddlers' punches, scratch awls and gauges, solid punches, tack pullers, wood-rim tools, etc. These are goods that hardwaremen often want on the spur of the moment, and this catalogue will be found handy for reference and ordering purposes. The products of this firm are well known for their invention and quality the world over, and they can safely be purchased through the mail. They are securing quite a hold on the Canadian market. Merchants can secure copies of this catalogue by applying to Mr. Gibb and mentioning this paper.

GLASS MAY ADVANCE.

Notwithstanding the high prices of window glass now ruling, the indications point towards an advance. The continuance of the strike in Belgium has caused such a shortage in the production in that country that the large houses there which have taken import orders in Canada have advised jobbers here that they will have great difficulty in filling orders given this season before the close of navigation. Last year, import orders were large, and were delivered in good time, but it will be remembered that, in 1899, owing to shortage, many orders were not delivered until after navigation closed, and it looks as if history would repeat itself in this regard. The English glass market is high too, and the agents here of English houses keep their prices stiff.

Stocks in jobbers' hands are fairly generous, but some report scarcity of certain lines, so the appearance seems to be that notwithstanding the likelihood of large stocks being held by some retailers who placed liberal import orders last season the heavy consumption will cause a shortage, which will cause an advance in stock prices, especially as the stock prices ruling here are lower than glass bought in Belgium at to-day's basis could be laid down here.

A HINT TO DEPARTMENT STORES.

It is sad to think that the big stores of Toronto do not keep mints, smelters and lead refineries for sale. If they did the express company could hardly handle the business it would do with B.C.—Ledge, New Denver, B.C.

The Acadia Copper Co., Limited, Halifax, have been incorporated.

Munroe & McIntyre are starting as hardware dealers in Alvinston, Ont.

The Sydney Hardware Co., Limited, Sydney, N.S., have been incorporated.

CARRYING TOO LARGE STOCKS.

AN observant traveller, recently returned cust, noticed that the stock carried by dry goods and general dealers in the Western Provinces, in the majority of cases, were much larger than is the rule in the Province of Ontario, and apparently larger than is warranted by the extent of the trade that is to be done on the capital employed. This was almost a necessity some years ago, when railroad and steamboat navigation in this newer country was so infrequent as to compel a merchant to lay in a goodly store at one time to carry until the next opportunity for replenishing. The custom cannot, however, be excused on such grounds now.

A trader can have the goods on his shelves in a few days after ordering. He can "sort up" as frequently as he pleases. Perhaps this undesirable feature of storekeeping by our western merchants is being perpetuated from sheer force of long habit. We know that the best governed wholesale firms discourage such a practice by Ontario merchants, and evidences are not wanting that their efforts are gradually bearing fruit. Buyers are beginning to realize that there is every advantage in purchasing frequently—as their needs demand. The stock is always fresher.

There is less likelihood of a surplus of unseasonable goods, and slaughter prices to get ready cash when there's an inopportune bill to meet. The general liability is always smaller, and the position of the man who buys in this way is consequently far more likely to be one of ease than if he tried to anticipate the wants of his customers for a whole season in advance.

A failure now and then gives point to this argument. The insolvent is caught with a large stock in dull times. His tempting inducements in the shape of big discounts at clearing sales only put off the evil day, and failure comes at last. But we are told that the tendency to overstock shows some really gratifying indications of being brought under proper control. It certainly is a question demanding the best consideration of wholesaler and retailer. It must be clear to all concerned that a practice fraught with so much menace to success in trade is one which should be discontinued without delay.—Nelson Economist.

AN EARLY-CLOSING DECISION.

An interesting point was made in a decision bearing on the early-closing by-law of Winnipeg given by Magistrate Dawson last week. M. Conway, who holds an auctioneer's license, was charged with violation of the early-closing by-law, came up for hearing. It was clearly proven that the defendant sold the goods after the closing hours named in the by-law. The defendant admitted selling the goods after 6 p.m., but took the ground that he was not liable to a fine owing to the wording of his auctioneer's license, which read to the effect that the licensee could sell goods by auction "at any time or place." The magistrate held that the by-law was good and fined Mr. Conway $5 without costs.

IRON PRODUCTION IN THE UNITED STATES.

THE Iron Age, in its usual monthly review of capacity of pig iron furnaces in blast, says: Owing to the blowing out of a considerable number of furnaces for repairs, the starting of some stacks of magnitude has not had the effect of increasing current capacity as it would otherwise have done. During the next few months, one by one, some large modern furnaces will make their first cast. The old works are nearly up to their full capacity.

The decline in furnace stock shown by our statement is heavy, amounting as it does to about 67,000 tons in one month. The weekly capacity of the furnaces in blast on April 1 compares as follows with that of the preceding periods :

	Furnaces in blast.	Capacity per Week. Gross Tons.
April 1, 1901	250	296,676
March 1	2[illegible]8	292,899
February 1	245	278,258
January 1	233	250 351
December 1, 1900	211	228,846
November 1	201	215 3[illegible]4
October 1	213	223,169
September 1	228	231,774
August 1	240	244,426
July 1	284	283,413
June 1	293	296 3[illegible]6
May 1	29[illegible]	293,850
April 1	291	289,482

FURNACE STOCKS.

The position of furnace stocks sold and unsold, as reported to us, was as below on April 1, the same furnaces being represented as in former months. This does not include the holdings of the steel works producing their own iron :

Stocks	Nov. 1.	Jan 1.	Mar. 1.	Apr. 1
Anthracite and coke	573,952	489,453	455,840	388,712
Charcoal	67,514	69,213	80,603	78,163
Total	641,466	558,663	536,443	466,875

WARRANT STOCKS.

The American Pig Iron Storage Warrant Co. report the following stocks :

Stocks.	Nov. 1	Jan. 1.	Mar. 1.	Apr. 1.
Coke and anthracite	18,500	15,000	12,600	12,000
Charcoal	1,500	1,400	1,400	1,400
Totals	20,000	16,400	14,000	13,400

IMPORTANCE OF FIRE INSURANCE.

THE series of destructive fires of recent occurrence in our large cities, entailing heavy losses and great inconvenience to business, at a season when jobbing houses are carrying heavy stocks of goods preparatory to an anticipated heavy spring output of merchandise, has aroused the attention of those who have large interests at stake, to the necessity of carrying a full insurance upon their stocks, writes L. R. Nolley in an exchange. Such prudence on the part of merchants handling large quantities of merchandise is so well recognized that no intimation is necessary to convince them of the desirability of availing themselves of this safeguard against loss. Experience teaches that it is the wise man who is careful at all times to entrench himself against danger. The money invested in fire insurance amply repays the merchant in the feeling of security it gives him.

More than a word of caution is needed to the retail merchant, particularly those in villages and larger country towns; the inflammable character of whose buildings and their environments, with their limited facilities for extinguishing fires, has advanced the cost of risk on the part of insurance companies to a point which makes it onerous for the merchants of limited resources, yet they cannot afford to be remiss in so important a factor. The jobber feels more confidence in the ability of the customer who keeps himself well insured against loss by fire. The cost of insurance should be regarded as one of the legitimate expenses of every well regulated establishment, and when this fact is known and recognized, it strengthens the credit of the country merchant.

A pertinent question upon our letter heads and bill sheets should be: "Are you fully covered by fire insurance?"

PRINCE EDWARD ISLAND AS A TOURIST RESORT.

The Island Farmer, Summerside, P.E.I., in its issue of April 10, had the following:

"We have in former issues directed the attention to the benefits which might reasonably be expected from the organization of a P. E. Island Tourists' Association. Year after year, tourists in increasing numbers visit our Province, and express themselves as delighted with our genial climate, bathing, boating, fishing and other attractions. The quiet of our country life and manner of living is an almost irresistible attraction to those whom business cares and the crowded, heated and unhealthy city drive from their homes during the summer months. Our garden island offers attractions equalled by few of the best summer and health resorts on the continent, and yet we have done practically nothing to encourage tourist travel. The comparatively few who visit us during the summer are either those who are repeating former visits or who come on the recommendation of those who had been here before them. . . Seventy-five per cent. of these would for their holiday, prefer camp life in the country, in the woods, or by the shore to hotel life under the most modern conditions. They are not as much concerned about the elegance of their accommodation as upon the distinctively rural features where outdoor amusements, surf bathing, fishing, boating, etc., can be enjoyed, combined with good, plain, wholesome food such as the majority of our farm houses can furnish.

There are hundreds of farm houses along our shores, and by our best fishing streams which would be looked upon as ideal resting places by many in search of a holiday in the country.

MARKETS AND MARKET NOTES

QUEBEC MARKETS

Montreal, April 19, 1901.

HARDWARE.

THE hardware market is in a decidedly healthy condition this week, and the volume of business is in accordance with the improved feeling. All the wholesale houses are exceedingly busy both on local and country account. Reports from various manufacturing concerns, both in Montreal and outside of the city, confirm previous opinion regarding the strong general situation. Furnaces and mills are booking orders for delivery some months ahead. There is no indication of an immediate advance in prices, yet there is confidence in the market as at present constituted. Bar iron, pipe and linseed oil are the strongest articles, all having been advanced this week. There has been a slight break on some lines of butts and hinges, wrought butts now being quoted at a discount of 65, 10 and 2½ per cent.; chest hinges, 65 and 5; hinge hasps, 65 and 5, and light T strap hinges, 65 and 5. These involve small reductions. All wires are in active demand, and The American Steel and Wire Co. are behind in their shipments. Wire nails are reported scarce, but there will probably be plenty of goods forthcoming when the mills recommence operations. Spring goods are all moving well. Payments are not first class.

BARB WIRE—The demand for barb wire is very heavy and the shipments from the other side are weeks behind. The price remains as before at $3.05 per 100 lb. f.o.b. Montreal. There is a little speculative demand, but most of the goods being ordered are going into consumption.

GALVANIZED WIRE—Brisk movements are reported in all fence wires. We quote as follows: No. 5, $4.25; Nos. 6, 7 and 8 gauge, $3.55; No. 9, $3.10; No. 10, $3.75; No. 11, $3.85; No. 12, $3.25; No. 13, $3.35; No. 14, $4.25; No. 15, $4.75; No. 16, $5 00.

SMOOTH STEEL WIRE—Trade continues to be very brisk in seasonable sizes. We quote oiled and annealed: No. 9, $2.80; No. 10, $2.87; No. 11, $2.90; No. 12, $2.95; No. 13, $3.15 per 100 lb. f.o.b. Montreal, Toronto, Hamilton, London, St. John and Halifax.

FINE STEEL WIRE—The demand is steady and the market featureless. The discount is still 17½ per cent. off the list.

BRASS AND COPPER WIRE—A fair business is being done. The discount on brass is 55 and 2½ per cent., and on copper 50 and 2½ per cent.

FENCE STAPLES—Some good shipments of staples have been made this week at former prices. We quote: $3.25 for bright, and $3.75 for galvanized, per keg of 100 lb.

WIRE NAILS—The inquiry for nails is quite brisk. At present stocks are low, due to the closing down of two of the mills at this season of the year because of repairs being made in the canal. We quote $2.85 for small lots and $2.77½ for carlots, f.o.b. Montreal, London, Toronto, Hamilton and Gananoque.

CUT NAILS—A little better demand for cut nails is reported this week. We quote: $2.35 for small and $2.25 for carlots; flour barrel nails, 25 per cent. discount; coopers' nails, 30 per cent. discount.

HORSE NAILS—A heavy demand is being experienced for horse nails since open prices have prevailed. "C" brand is selling at a discount of 50 and 7½ per cent. off their own list, and "M" brand at 60 per cent. off the old list on oval and city head and 65 per cent. off on countersunk head. It seems, however, that even better discounts are being offered on other brands. The market is strong, in spite of the higher discounts, which are due solely to competition. The demand is becoming somewhat speculative.

HORSESHOES—There are fair movements of horseshoes, although this is not the season for heavy shipments. We quote: Iron shoes, light and medium pattern, No. 2 and larger, $3.50; No. 1 and smaller, $3.75; snow shoes, No. 2 and larger, $3.75; No. 1 and smaller, $4.00; X L steel shoes, all sizes, 1 to 5, No. 2 and larger, $3.60; No. 1 and smaller, $3.85; feather-weight, all sizes, $4.85; toe weight steel shoes, all sizes, $5.95 f.o.b. Montreal; f.o.b. Hamilton, London and Guelph, 10c. extra.

POULTRY NETTING — Some business is being done, but new business is chiefly confined to local concerns just now. The discount remains at 50 and 10 per cent.

GREEN WIRE CLOTH—A regular spring business is being done at $1.35 per 100 sq. ft.

SCREEN DOORS AND WINDOWS — Inquiries for all kinds of screens are numerous, and a good business is passing. We quote as follows: Screen doors, plain cherry finish, $8.25 per doz.; do. fancy, $11.50 per doz.; walnut, $7.40 per doz., and yellow, $7.45; windows, $2.25 to $3.50 per doz.

SCREWS — The week shows a good business. Discounts are as follows: Flat head bright, 87½ and 10 per cent. off list; round head bright, 82½ and 10 per cent.; flat head brass, 80 and 10 per cent.; round head brass, 75 and 10 per cent.

BOLTS—The demand for bolts has fallen off somewhat this week, but a moderate trade is still being experienced in this line. Discounts are as follows: Carriage bolts, 65 per cent.; machine bolts, 65 per cent.; coach screws, 75 per cent.; sleigh shoe bolts, 75 per cent.; bolt ends, 65 per cent.; plough bolts, 50 per cent.; square nuts, 4¼c. per lb. off list; hexagon nuts, 4¾c. per lb. off list; tire bolts, 67½ per cent.; stove bolts, 67½ per cent.

BUILDING PAPER — Building paper is moving out in good quantities. We quote: Tarred felt, $1.70 per 100 lb.; 2-ply, ready roofing, 80c. per roll; 3-ply, $1.05 per roll; carpet felt, $2.25 per 100 lb.; dry sheathing, 30c. per roll; tar sheathing, 40c. per roll; dry fibre, 50c. per roll;

tarred fibre, 60c. per roll; O.K. and I.X.L., 65c. per roll; heavy straw sheathing, $28 per ton; slaters' felt, 50c. per roll.

RIVETS—A moderate business is passing on a featureless market. Discount on best iron rivets, section, carriage, and waggon box, black rivets, tinned do., coopers' rivets and tinned swedes rivets, 60 and 10 per cent.; swedes iron burrs are quoted at 55 per cent. off; copper rivets, 35 and 5 per cent. off; and coppered iron rivets and burrs, in 5-lb. carton boxes, are quoted at 60 and 10 per cent. off list.

BUTTS AND HINGES—The makers have slightly raised some of the discounts. Wrought butts are now quoted at a discount of 65, 10 and 2½ per cent., instead of 60 and 10 per cent. as formerly. Chest hinges are now selling at 65 and 5 off; hinge hasps, 65 and 5 per cent.; light T strap hinges, 65 and 5 per cent.

BINDER TWINE—A little more attention is being paid to binder twine this week. We quote: Blue Ribbon, 11½c.; Red Cap, 9¼c.; Tiger, 8¼c.; Golden Crown, 8c.; Sisal, 8¼c.

CORDAGE—A continued broad movement is reported in cordage. Manila is now worth 13½c. per lb. for 7-16 and larger; sisal is selling at 10c.; and lathyarn 10c.

SPADES AND SHOVELS—All garden tools are selling well, and spades and shovels are among the tools being called for. The discount is still 40 and 5 per cent. off the list.

HARVEST TOOLS—As yet the general line is not selling very well, but such goods as forks are moving out freely. The discount is still 50, 10 and 5 per cent.

TACKS—A nice business is being done. We quote: Carpet tacks, in dozens and bulk, 80 and 15 per cent.; tinned, 80 and 20 per cent., and cut tacks, blued, in dozens, 80 per cent.

LAWN MOWERS—Some orders are being taken and fair numbers are being shipped. We quote: High wheel, 50 and 5 per cent. f.o.b. Montreal; low wheel, in all sizes, $2.75 each net; high wheel, 11-inch, 30 per cent. off.

FIREBRICKS—As the opening of navigation is decided, improvement in trade is looked for. Prices are still $18 to $24 per 1,000 as to brand ex store.

CEMENT—Business is opening well, although there is not the same confidence manifested as there was last year. We quote as follows: German, $2.45 to $2.55; English, $2.30 to $2.40; Belgian, $1.95 to $2.05. American in bags, $2.30 to $2.40 per 400 lb.

METALS.

In iron and steel the markets are firm. Pig iron is unchanged on account of Canadian competition, but the manufacturers of bar iron have advanced their prices 5 cents per 100 lb. this week. Steel is also firm and we quote a slight advance. The English sheet metal market remains much as it has been. Lead is somewhat easier again.

PIG IRON—In spite of the improvement abroad, Canadian pig iron is selling at former quotations, $18 to $18.50. Summerlee for early import is worth $20 to $21.

BAR IRON—The iron market is quite firm; manufacturerers advanced prices 5 cents per 100 lb. this week. Merchants' bar is now worth $1.75 and horseshoe $2.

BLACK SHEETS—Business is somewhat more lively this week. The market is fairly steady at $2.65 for 8 to 16 gauge; $2.70 for 26 gauge, and $2 75 for 28 gauge.

GALVANIZED IRON—Orders are numerous but small. Business will not be brisk until navigation opens. We quote: No. 28 Queen's Head, $4 65; Apollo, 10¾ oz., $4.50, and Comet, $4.40 to $4 45, with a 15c. reduction for case lots.

INGOT COPPER—The markets, both here and outside, are firm. A good business is being done at 18 to 19c.

INGOT TIN—The tin market continues fairly steady, Strait tin being worth 26c. in New York. Lamb and Flag is quoted at 31c. here.

LEAD—Lead is cabled decidedly weak, and the feeling here is easier. We quote $3.80 to $3.90 per 100 lb.

LEAD PIPE—A fair business is being done at unchanged prices. We quote: 7c. for ordinary and 7½c. for composition waste, with 25 per cent. off.

IRON PIPE—The pipe market is strong, and we note a slight advance on black. We quote: Black pipe, ¼, $3 per 100 ft.; ⅜, $3; ½, $3.05; ¾, $3.30; 1-in., $4.70; 1¼, $6.40; 1½, $7.70; 2-in. $10.25. Galvanized, ½, $4.60; ¾, $5.25; 1-in., $7.50; 1¼, $9.80; 1½, $11.75; 2-in., $16.

TINPLATES—The English market shows no improvement on the week. Bessemer cokes are quoted at 11s. 10½d. to 12s. per box. Locally, business is only moderate at $3.80 to $4 for coke, and $4.15 to $4.35 for charcoal.

CANADA PLATE — Orders are fairly numerous, but uniformly small. We quote: 52's, $2.60; 60's, $2.70; 75's, $2.80; full polished, $3.45, and galvanized, $4.30.

STEEL — The steel market has strengthened somewhat, and we note an advance. We quote: Sleighshoe, $1.95; tire, $2; bar, $1.95; spring, $2.75; machinery, $2.75 and toe-calk, $2.50.

SHEET STEEL—The market is firm and unchanged. Business is fairly good. We quote: Nos. 22 and 24, $3, and Nos. 18 and 20, $2.85.

TOOL STEEL — We still quote 8c. for Black Diamond, and 13c. for Jessop's.

TERNE PLATES—A fair business is being done at $7.75 to $8, with some allowance for large lots.

COIL CHAIN — A fairly good trade is being done in coil chain at the former quotations. We quote as follows: No. 6, 11½c.; No. 5, 10c.; No. 4, 9½c.; No. 3, 9c.; ¼-inch, 7½c. per lb.; 5-16, $4.65; 5-16 exact, $5.10; ⅜, $4.20; 7-16, $4.00; ½, $3.75; 9-16, $3.65; ⅝, $3.35; ¾, $3.25; ⅞, $3.20; 1-in., $3.15. In carload lots an allowance of 10c. is made.

SHEET ZINC—The ruling price is 6 to 6¼c.

ANTIMONY—Quiet, at 10c.

ZINC SPELTER—Is worth 5c.

SOLDER—We quote: Bar solder, 18½c.; wire solder, 20c.

GLASS.

A fair business is being done in glass, but the bulk of the movement will not take place until the opening of navigation. We quote as follows: First break, $2; second, $2.10 for 50 feet; first break, 100 feet, $3.80; second, $4; third, $4.50; fourth, $4.75; fifth, $5.25; sixth, $5.75, and seventh, $6.25.

PAINTS AND OILS.

The great strength of linseed oil has given the market a powerful fillip and parties who are looking for lower prices for liquid paints and varnishes in May and June will have to seek consolation elsewhere because, in view of the advance in oil in England, there is no tendency to lower prices for first-class established brands. Some cutting is being done in liquid paints of the class which will not stand investigation, but for genuine first class handy paints prices are well maintained and customers are experiencing a great deal of satisfaction in selling a really A1 stock. Varnishes and japans are very firm. Dry colors are meeting with a very good inquiry. There has been no change in white lead and shipments are not excessive. They are, however, about up to the average. Paris green is in a lull at present. Orders taken early in the season have been distributed, and it is not thought that there will be very much stir in this department until the middle of May. Linseed oil is 3c. higher. We quote:

WHITE LEAD—Best brands, Government standard, $6.25; No. 1, $5 87½; No. 2, $5.50; No. 3. $5.12½, and No. 4. $4.75 all f.o.b. Montreal. Terms, 3 per cent. cash or four months.

DRY WHITE LEAD—$5.50 in casks; kegs, $5.75.

RED LEAD—Casks, $5.25; in kegs, $5.50.

WHITE ZINC PAINT—Pure, dry, 7c.; No. 1, 6c.; in oil, pure, 8c.; No. 1, 7c.; No. 2, 6c.

PUTTY—We quote: Bulk, in barrels, $1 90 per 100 lb.; bulk, in less quantity, $2.05; bladders, in barrels, $2.10; bladders, in 100 or 200 lb. kegs or boxes, $2.25; in tins, $2.55 to $2.65; in less than 100-lb. lots, $3 f.o.b. Montreal, Ottawa, Toronto, Hamilton, London and Guelph. Maritime Provinces 10c. higher, f.o.b. St. John and Halifax.

LINSEED OIL—Raw, 72c.; boiled, 75c. in 5 to 9 bbls., 1c. less, 10 to 20 bbl. lots, open, net cash, plus 2c. for 4 months. Delivered anywhere in Ontario between Montreal and Oshawa at 2c. per gal. advance and freight allowed.

TURPENTINE—Single bbls., 57c.; 2 to 4 bbls., 56c.; 5 bbls. and over, open terms, the same terms as linseed oil.

MIXED PAINTS—$1.25 to $1.45 per gal.

CASTOR OIL—8¾ to 9¼c. in wholesale lots, and ½c. additional for small lots.

SEAL OIL—47½ to 49c.

COD OIL—32½ to 35c.

NAVAL STORES — We quote: Resins, $2 75 to $4.50, as to brand; coal tar, $3.25 to $3 75; cotton waste, 4½ to 5½c. for colored, and 6 to 7½c. for white; oakum, 5½ to 6½c., and cotton oakum, 10 to 11c.

PARIS GREEN—Petroleum barrels, 16¾c. per lb.; arsenic kegs, 17c.; 50 and 100 lb. drums, 17¼c.; 25-lb. drums, 18c.; 1-lb. packages, 18½c.; ½-lb. packages, 20½c.; 1-lb. tins, 19½c.; ½-lb. tins, 21½c. f.o.b. Montreal; terms 3 per cent. 30 days, or four months from date of delivery.

SCRAP METALS.

The market is strong and active. Dealers are paying the following prices in the country: Heavy copper and wire, 13 to 13½c. per lb.; light copper, 12c.; heavy brass, 12c.; heavy yellow, 8½ to 9c.; light brass, 6½ to 7c.; lead, 2½ to 2¾c. per lb.; zinc, 2¼ to 2½c.; iron, No. 1 wrought, $15 to $16 per gross ton f.o.c. Montreal; No. 1 cast, $13 to $14; stove plate, $8 to $9; light iron, No. 2. $4 a ton; malleable and steel, $4; rags, country, 70 to 80c. per 100 lb.; old rubbers, 6¼c. per lb.

HIDES.

The hide market continues easy with the demand fairly good. We quote: Light hides, 6½c. for No. 1; 5½c. for No. 2, and 4½c. for No. 3. Lambskins, 10c.; sheepskins, 90c.; calfskins, 8c. for No. 1 and 6c. for No. 2.

PETROLEUM.

Business is only moderate. We quote as follows: "Silver Star," 14½ to 15½c.; "Imperial Acme," 16 to 17c.; "S.C. Acme," 18 to 19c., and "Pratt's Astral," 18½ to 19½c.

MONTREAL NOTES.

Linseed oil is 3c. lower.

Iron pipe is quoted higher again.

Sleighshoe steel is 10c. higher.

Bar iron is quoted 5c. per 100 lb. higher.

The discounts on wrought butts and some lines of hinges have been slightly raised.

Lewis Bros. & Co., Montreal, report that they have met with great success during the past year in handling the "Frictionless" metal. It is a perfectly efficient babbitt metal that is taking the place of high grade metals.

ONTARIO MARKETS.

TORONTO, April 19, 1901.

HARDWARE.

FURTHER improvement is to be noted this week in the wholesale hardware trade, and the shipping rooms of all the houses are piled up with goods ready to be sent out. One of the railway companies stated this week that its shipments from Toronto during the month of March were the largest for the same month of any previous year. While the demand for goods is so brisk, the orders are not, as a rule, for large quantities, but nearly all dealers appear to be urgently in need of seasonable goods. At any rate they are pressing for delivery. So great is the demand for seasonable goods that a great deal of difficulty is being experienced by the wholesale houses in getting their orders filled which they placed with the manufacturers. A good business is being done in fence wire, and an active demand is being experienced for wire nails. Cut nails are still quiet. Business is seasonably quiet in horseshoes and horse nails. An active trade is reported in screws, and a fair business is being done in bolts and nuts. Rope is going out in fairly good quantities. Increased activity is to be noted in both enamelled ware and tinware. Quite a few screen doors and windows are going out, and a little business is to be noted in green wire cloth. Building paper is in fair demand. There is quite an active movement in ice cream freezers and refrigerators. Shipments of harvest tools and spades and shovels are increasing. The same may be said in regard to oil stoves. A fairly good trade is being done in lawn mowers. Binder twine is quiet. There have been a few changes in prices during the past week, notably in bolts and nuts, iron pipe and bar iron. Payments are, on the whole, fair.

BARB WIRE—There is a large demand for barb wire, and some difficulty is being experienced in getting delivery from the United States manufacturers. So great has the difficulty become that jobbers have been compelled to order from the Canadian manufacturer and pay a higher price than that ruling in Cleveland. We quote f.o.b. Cleveland at $2.82½ per 100 lb. for less than carlots and $2.70 for car. lots. From stock Toronto the figure is $3.05 per 100 lb.

GALVANIZED WIRE—A good demand is also reported this week for plain galvanized wire, at unchanged prices. We quote as follows: Nos. 6, 7 and 8, $3.50 to $3.85 per 100 lb., according to quantity; No. 9, $2.85 to $3.15; No. 10, $3.60 to $3.95; No. 11, $3.70 to $4.10; No. 12, $3 to $3.30; No. 13, $3.10 to $3.40; No. 14,

$4.10 to $4.50; No. 15 $4.60 to $5.05; No. 16, $4.85 to $5.35. Nos. 6 to 9 base f.o.b. Cleveland are quoted at $2.57½ in less than carlots and 12c. less for carlots of 15 tons.

SMOOTH STEEL WIRE—A good demand is springing up for oiled and annealed wire, but hay-baling wire is dull. Net selling prices for oiled and annealed are as follows: Nos. 6 to 8, $2.90; 9, $2.80; 10, $2.87; 11, $2.90; 12, $2.95; 13, $3.15; 14, $3.37; 15, $3.50; 16, $3.65. Delivery points, Toronto, Hamilton, London and Montreal, with freights equalized on these points.

WIRE NAILS—The improvement noted last week has continued, and both manufacturers and wholesalers report an active demand this week. We quote: Less than carlots at $2.85, and carlots at $2.77½ to the retail trade. Delivery points: Toronto, Hamilton, London, Gananoque and Montreal.

CUT NAILS—There is no improvement in the demand for cut nails, the volume of business still being very small. We quote the base price at $2.35 per keg. Delivery points are: Toronto, Hamilton, London, Montreal and St. John, N.B.

HORSE NAILS—The demand for horse nails, as is usual at this time of the year, is light. Discount on "C" brand, oval head, 50 and 7½ per cent. off new list and on other brands 50, 10 and 5 per cent., off the old list; countersunk, 50, 10 and 10 per cent.

HORSESHOES—This line is also seasonably quiet. We quote f.o.b. Toronto: Iron shoes, No. 2 and larger, light, medium and heavy, $3.60; snow shoes, $3.85; light steel shoes, $3.70; featherweight (all sizes), $4.95; iron shoes, No. 1 and smaller, light, medium and heavy (all sizes), $3.85; snow shoes, $4; light steel shoes, $3.95; featherweight (all sizes), $4.95.

SCREWS—An active trade is being done in screws, and prices are steady and unchanged. We still quote discounts: Flat head bright, 87½ and 10 per cent.; round head bright, 82½ and 10 per cent.; flat head brass, 80 and 10 per cent.; round head brass, 75 and 10 per cent.; round head bronze, 65 per cent., and flat head bronze at 70 per cent.

BOLTS AND NUTS—The feature of trade in this line is an advance of about 10 per cent. in prices. The change took effect on Wednesday, and has for some time been anticipated. Dealers, in consequence, are fairly well supplied with stocks, and, naturally, stand to make a little money. We quote: Carriage bolts (Norway), full square, 65 per cent.; carriage bolts full square, 65 per cent.; common carriage bolts, all sizes, 60 per cent.; machine bolts, all sizes, 60 per cent.; coach screws, 70 per cent.; sleighshoe bolts, 72½ per cent.; blank bolts, 60 per cent.; bolt ends, 62½ per cent.; nuts, square, 4c. off; nuts, hexagon, 4¼c. off; tire bolts, 67½ per cent.; stove bolts, 67½; plough bolts, 60 per cent.; stove rods, 6 to 8c.

RIVETS AND BURRS—Business in rivets and burrs has been fairly good during the past week. Prices are steady and unchanged. We quote as follows: Iron rivets, 60 and 10 per cent.; iron burrs, 55 per cent.; copper rivets and burrs, 35 and 5 per cent.

ROPE—Trade is fairly active; the demand is principally from vessel owners. A feature of the trade is the demand for tow-lines. The base price is steady and unchanged at 10c. for sisal and 13½c. for manila per lb.

CUTLERY—Trade is fairly good for this time of the year. Of course, at this season, an active demand is not usually experienced.

SPORTING GOODS—There is quite a little ammunition going out, and there is a steady trade being done in loaded shells. In guns and rifles, there is not much being done.

ENAMELLED WARE—Business in enamelled ware has improved during the past week, and a fair trade is now reported. The same remarks would apply to tinware. The discounts on enamelled ware are as follows: We quote: "Granite," "Pearl," "Crescent" and "Imperial" wares at 50, 10 and 10 per cent.; white, "Princess,"

"Turquoise." blue and white, 50 per cent.; "Diamond," "Famous" and "Premier," 50 and 10 per cent.

GREEN WIRE CLOTH—While the demand is not heavy, some shipments are being made this week. The price is steady and unchanged at $1.35 per 100 sq. ft.

SCREEN DOORS AND WINDOWS — Large quantities of screen doors and windows are being shipped this week.

BUILDING PAPER — A large quantity of building paper and roofing felt is being sold this week. We quote: Plain building, 30c.; tarred lining, 40c., and tarred roofing, $1 65.

POULTRY NETTING—There has been a large trade in this line and orders are still coming forward. Discount is 55 per cent. on Canadian.

ICE CREAM FREEZERS AND REFRIGERATORS—The demand for both ice cream freezers and refrigerators is active, and during the past week the wholesale houses have received requests from a number of their customers asking that goods ordered for delivery May 1 be immediately shipped.

HARVEST TOOLS — Trade is fairly good in harvest tools. In such early lines as rakes, hoes and manure forks the demand is active. Discount 50, 10 and 5 per cent.

SPADES AND SHOVELS—Some nice orders are being put through this week, and, in general, business is good. Discount 40 and 5 per cent.

TACKS—A moderate business is being done at unchanged prices. We quote as follows: Carpet tacks, blued, 80 and 15 per cent; carpet tacks, tinned, 80 and 20 per cent; cut tacks, blued, in dozens only, 80 per cent.; Swedes, upholsterers', bulk, 85, 12½ and 12½ per cent.; copper nails, 52½ per cent.

BRIGHT WIRE GOODS—Business continues fairly good in bright wire goods, and the discount is unchanged at 62½ per cent.

BINDER TWINE — The volume of business in this line is small. We quote as follows: American—Sisal and standard, 8½c.; manila, 10½c.; pure manila, 11½ to 12c. per lb. Canadian—Sisal, 8½c.; pure manila, 11½c. per lb.

OIL AND GAS STOVES — A good deal of activity has developed in regard to this line during the past week, and a rather large business is now being done. Not a great deal of attention is yet being paid to gas stoves.

LAWN MOWERS—Business in lawn mowers continues fairly good.

EAVETROUGH AND CONDUCTOR PIPE—Eavetrough and conductor pipe are going out freely, the shipments being extra heavy. Prices are, however, demoralized on account of the competition among the manufacturers. As the figures which are now ruling are practically at cost, the dealers throughout the country are taking advantage of the situation to lay in good supplies before any advance takes place. We are informed on good authority that an effort is being made to induce the manufacturers to put prices on a more satisfactory basis.

SPRING HINGES—Quite a few of these are going out, although the busiuess is not yet what might be termed large. We quote $12 per gross pair.

MECHANICS' TOOLS—There is a good demand for mechanics' tools and for builders' supplies in general.

CHAIN—Trade is quiet and prices unchanged at quotations.

CEMENT—A big trade is doing. During the past week some large contracts have been made at close figures. We quote barrel lots: Canadian portland $2.25 to $2 80; German, $3 to $3.15; English, $3; Belgian, $2.50 to $2.75; Canadian hydraulic, $1.25 to $1.50.

METALS.

Business is more active than it was a week ago, and the demand is fairly well distributed, although the demand is chiefly for steel, galvanized sheets, black sheets and tinplates. The metal market generally appears to be firmer than it has been for some time.

PIG IRON—The market continues firm, and business is fairly good. The idea as to price for Canadian pig iron is $16.50 for No. 2 and $16 for No. 3.

BAR IRON—The bar iron market continues firm, and prices are about 5c. higher than they were a week ago. The base figure is now $1.75 to $1.80 per 100 lb.

STEEL—This market is decidedly active and prices are steadily advancing in a number of lines. The Canadian manufacturers are unusually busy, and one of the principal makers is several months behind with his orders.

PIG TIN—Business has been fair during the week, although, perhaps, not as brisk as it was a week ago. The outside markets have been very strong, and in London prices advanced £2 10s. in three days. Locally, no change has been made as the jobbers are still quoting 31 to 32c. per lb.

TINPLATES—There are few, if any, large quantities going out, although the movement, generally, is steady at unchanged prices.

TINNED SHEETS—A little improvement is to be noted in the demand and quite a few shipments have been made during the week. We quote 9 to 9½c.

GALVANIZED SHEETS—Dealers throughout the country, now that the mild weather is upon us, are in a hurry for the delivery of galvanized sheets. There is, in consequence, quite a little activity in this line. As noted in a previous issue, there is still quite a little difficulty being experienced in getting delivery of galvanized sheets from the manufacturers in the United States. On account of the threatened strike among the employes, the prospects for an immediate improvement in the delivery are not bright. The ruling prices from stock are $4 60 for English and $4.50 for United States sheets.

BLACK SHEETS — There has been a decided improvement during the week. The demand is now active for both small and large lots. We quote 28 gauge at $2.30.

CANADA PLATES — Trade in Canada plates continues quiet. We quote: All dull, $3; half-and-half, $3.15, and all bright, $3.65 to $3.75.

COPPER — The demand for ingot copper has eased off a little during the week, but the demand for sheet copper is good. We quote: Ingot, 19c.; bolt or bar, 23½ to 25c.; sheet, 23 to 23½c. per lb.

BRASS — The demand for brass has slightly improved during the past week. The discount on rod and sheet is unchanged at 15 per cent.

SOLDER—A fairly active trade is still being done in solder, and prices rule as before. We quote: Half-and-half, guaranteed, 18½c.; ditto, commercial, 18c.; refined, 18c., and wiping, 17c.

IRON PIPE — The iron pipe market is firmer, and higher prices are this week being quoted by manufacturers and jobbers on most sizes. These prices will be found in detail in our editorial pages.

LEAD—The demand is fair, and prices are unchanged at last week's reduction. We quote 4½ to 4¾c. per lb.

ZINC SPELTER — Trade is a little more active, and prices are unchanged at 5¾ to 6c. per lb.

ZINC SHEET—More business is still to be reported in this line, although the demand is not altogether active. We quote 6½c. for casks and 6¾c. for part casks.

ANTIMONY—Prices are the same as last week, namely, 10½ to 11c. per lb.

PAINTS AND OILS.

There is a good general business doing. Linseed oil, because of the manifest firmness of the market, is in excellent demand. White lead, turpentine, prepared paints and general sundries are all selling well. Paris green is rather quiet. Turpentine has declined 2c. The price is 18c. lower than at this time last year, 10c. lower than in 1899, and 7c. higher than in 1898, when prices were at such a low basis. The movement of prices will depend on the crop, which will be gathered during the next few months, but which cannot at the moment be estimated. Red led is 25c. dearer. We quote as follows:

WHITE LEAD—Ex Toronto, pure white lead, $6.37½; No. 1, $6; No. 2. $5.67½; No. 3. $5.25; No. 4. $4.87½; genuine dry white lead in casks, $5.37½.

RED LEAD—Genuine, in casks of 560 lb., $5.50; ditto, in kegs of 100 lb., $5.75; No. 1, in casks of 560 lb., $5; ditto, kegs of 100 lb., $5.25.

LITHARGE—Genuine, 7 to 7½c.

ORANGE MINERAL—Genuine, 8 to 8½c.

WHITE ZINC—Genuine, French V.M., in casks, $7 to $7.25; Lehigh, in casks, $6.

PARIS WHITE—90c. to $1 per 100 lb.

WHITING — 70c. per 100 lb.; Gilders' whiting, 80c.

GUM SHELLAC — In cases, 22c.; in less than cases, 25c.

PARIS GREEN—Bbls., 16¾c.; kegs, 17c.; 50 and 100-lb. drums, 17½c.; 25-lb. drums, 18c.; 1-lb. papers, 18½c.; 1-lb. tins, 19½c.; ½-lb. papers, 20½c.; ½-lb. tins, 21½c.

PUTTY — Bladders, in bbls., $2.10; bladders, in 100 lb. kegs, $2.25; bulk in bbls., $1.90; bulk, less than bbls. and up to 100 lb., $2 05; bladders, bulk or tins, less than 100 lb., $2.90.

PLASTER PARIS—New Brunswick, $1.90 per bbl.

PUMICE STONE — Powdered, $2.50 per cwt. in bbls., and 4 to 5c. per lb. in less quantity; lump, 10c. in small lots, and 8c. in bbls.

LIQUID PAINTS—Pure, $1.20 to $1.30 per gal.; No. 1 quality, $1 per gal.

CASTOR OIL—East India, in cases, 10 to 10½c. per lb. and 10½ to 11c. for single tins.

LINSEED OIL—Raw, 1 to 4 barrels, 71c.; boiled, 74c.; 5 to 9 barrels, raw, 70c.; boiled, 73c., delivered. To Toronto, Hamilton, Guelph and London, 1c. less.

TURPENTINE—Single barrels, 54c.; 2 to 4 barrels, 53c., delivered. Toronto, Hamilton and London 1c. less. For less quantities than barrels, 5c. per gallon extra will be added, and for 5-gallon packages, 50c., and 10 gallon packages, 80c. will be charged.

GLASS.

The indications point toward an advance. It is feared that import orders will not be to hand before the close of navigation and prices are stiffening in Belgium. We quote as follows: Under 26 in., $4.15 26 to 40 in., $4.45; 41 to 50 in., $4.85; 51 to 60 in., $5.15; 61 to 70 in., $5.50; double diamond, under 26 in., $6; 26 to 40 in., $6.65; 41 to 50 in., $7.50; 51 to 60 in., $8.50; 61 to 70 in., $9.50, Toronto, Hamilton and London. Terms, 4 months or 3 per cent. 30 days.

HIDES, SKINS AND WOOL.

HIDES—The market is dull at unchanged figures. We quote: Cowhides, No. 1, 6½c.; No. 2, 5½c.; No. 3, 4½c. Steer hides are worth 1c. more. Cured hides are quoted at 7 to 7½c.

SKINS—A fair trade is doing. Prices are unchanged. We quote: No. 1 veal, 8-lb. and up, 8c. per lb.; No. 2, 7c.; dekins, from 40 to 60c.; culls, 20 to 25c. Sheepskins, 90c. to $1.

WOOL—There is nothing doing. We quote: Combing fleece, 14 to 15c., and unwashed, 8 to 9c.

COAL.

The prices made last week still maintain. We quote April delivery anthracite on cars Buffalo and bridges as follows: Grate, $4.25 per gross ton and $3.79 per net ton; egg, stove and nut, $4.50 per gross ton and $4.01 per net ton, for April shipment.

MARKET NOTES.

Iron pipe is dearer.

Bolts and nuts have advanced 10 per cent.

H. S. Howland, Sons & Co. are in receipt of a shipment of Boker's fencing plyers, and are in a position to fill orders for all sizes in both patterns. They are also in receipt of a shipment of copper-plated oilers, which can be retailed at 10c. each.

A VISIT FROM MR. SMITH.

Mr. Harry Dwight Smith, manager of the promoting department of The Sherwin-Williams Co., Cleveland, O., and editor of the company's two bright publications, passed through Toronto on Saturday, en route home, after having paid a visit to the New York, Boston and Montreal branches.

He informed HARDWARE AND METAL, during a brief conversation, that he found business good at all the branches he visited, and, compared with last year, larger in volume.

Mr. Smith is one of the many bright young men that occupy responsible positions with a bright and enterprising firm.

HEATING AND PLUMBING

HOUSE PIPING FOR ACETYLENE GAS

THERE have been a good many inquiries of late from both prospective users of acetylene gas and from plumbers and fitters who are installing generators for house and store lighting about the sizes of piping required, the effect of acetylene on iron pipe, probability of leakage, etc. We are glad to notice this interest, especially on the part of the trade, as an indication of a desire and intention to require a practical knowledge on the subject, and to treat jobs of acetylene work with the same thoroughness and care as is given by the better class of plumbers and gas fitters to their old lines of work.

The first requisite in house-piping for acetylene, is to have ample pipe capacity for its distribution. Of course, the fact that only about one-tenth as much acetylene is required for a given amount of light as of ordinary city gas makes it not only possible, but wiser to use smaller pipes for acetylene, but it does not follow that the pipes can be only one-tenth as large for the following reason:

Acetylene is about twice as heavy as city gas and this makes it more sluggish in its movement. As pipes decrease in size the effect of the friction of the gas increases decidedly.

With the small sizes of pipe, the possibility of stoppage because of rust or scale, oily matter, thread cuttings, or the use of too much lead or shellac on the threads, must be considered.

It is of vital importance with any illuminating gas to carry the full pressure desired up to each burner, in order to obtain a uniform size of flame without fluctuation. If a part of the system is too small, a higher pressure at the generator is necessary to keep the burners on the small pipes up to the standard, and this gives too much pressure for the best results on the burners nearest the generator. Another bad result of small pipes is that the burning on or off of one or two burners on the system changes the rate of burning of all the other lights to a greater or less extent.

The expense for labor, which is a large part of the cost of piping, is no less for small sizes. The difference in the cost of the proper size pipe is so little more than the cost of a smaller size, that it need not be considered.

We advise, therefore, as a general rule, that nothing less than 1-4-inch pipe be used for a single burner and that the main riser and branches be made correspondingly large.

Fittings should be of malleable iron and should be examined carefully for sand holes, cracks and flaws. Care in selecting pipe and fittings often saves a lot of work and expense.

All joints should be made up in white lead or shellac. Reasonable care to avoid cross threading and to see that threads are perfect and that each joint is tightened up, will make a job that will test out right and stand as long as the building lasts.

When the piping is completed and before the fixtures are put up, all the openings should be capped and an air pressure of at least six inches mercury column pumped up on the system.

A good job should hold this six-inch pressure for twenty-four hours without appreciable loss.

House piping for acetylene gas should grade back to the generator, or a drip, so that if there should be condensation or other liquid in the pipes it would return by gravity to a point where it could be easily disposed of. This may seem unnecessary, as there is no likelihood of there being any condensation from acetylene gas, but it costs no more to run pipe to grade and the result is certainly not bad.

A few people seem to think that acetylene must leak more or less, and that even if a piping job is perfect to start with acetylene gas will start leaks which would not occur with city gas.

We wish to state emphatically that acetylene gas will not decompose iron, nor "eat it up," as one man puts it. If a job is tight—no sand holes, split pipe, or defective joints—it will carry acetylene gas without leakage indefinitely. Don't allow anyone to try to explain a leak by laying it to the quality of the gas. A leaky job means either poor material or poor work.

Acetylene gas is ten times richer than city gas for lighting, therefore, only about one-tenth as large a quantity is required, but it costs several times as much per cubic foot of gas, and, consequently, a leak of one cubic foot per hour means a loss in money of perhaps five or six times the amount which would result from leakage of one cubic foot of city gas. This makes a leak of acetylene gas just so much worse than a leak of any other illuminating gas.

So far as danger is concerned, any leak of gas, whether it is coal gas, water gas, sewer gas, or acetylene gas, is dangerous just in proportion to the quantity which escapes, and acetylene is no more or less dangerous than either of the others.

Acetylene has a strong and characteristic odor which makes known any escape, or leak, at once. In every respect it is as safe as the most cautious person could desire.

Briefly stated—If a job of piping and fixtures is put in properly by a man who knows how there will be no leaks. If there is poor work and it should leak, the odor will give instant warning.—Plumbers' Trade Journal.

SOME BUILDING NOTES.

A. Hackett is building a residence in Phoenix, B.C.

James B. Leighton is building a house at Savonas, B.C.

Dr. Charbonneau intends erecting a brick house and office in Lanark, Ont.

The Methodists of Clarksburg, Ont., intend erecting a new parsonage.

Tenders are asked before April 24, for a Baptist church in Guysboro', N.S.

The British American Hotel, Kingston, Ont., is being extensively repaired.

G. A. Cliff is building a business block for Miss Barrett, milliner, Napanee, Ont.

James McCue is preparing to build an elevator at Melancthon, Ont., next summer.

Ald. John McLeod will build a $3,000 residence for James Baker, Kingston, Ont.

A. Larose is erecting a hotel at the corner of Queen and Bridge streets, Ottawa.

The Presbyterian Church at Georgetown, Ont., which was seriously damaged by fire, is being repaired.

The congregation of Home Memorial Church, Stratford, Ont., are considering the erection of a new church.

Tenders are asked by C. W. Keeling up to to-day (Saturday), for the erection of a brick house and woollen mill at Cargill, Ont.

P. A. Lamonde, contractor, Quebec, intends erecting a 75 x 40 ft. freight shed and a three-storey 75 x 35 ft. warehouse in that city.

A new post office, a bank building, and stores for J. Carnie, J. S. Armitage, S. Appleby and J. S. Brown & Co., will be erected in Paris, Ont., this summer.

Presbyterian, Methodist and Anglican churches are to be erected in North Sydney, N.S., this summer. The Bank of Nova Scotia and the Union Bank will erect two bank buildings. Mrs. Angus Young intends building a three-storey business block, large enough for four stores on the ground floor.

Building operations are active in Winnipeg. John Leslie is spending $16,000 on enlarging the Hargrave block. Architect Russell is calling for tenders for a new residence on Smith street, and is preparing plans for a house on Ellen street for H. J. Lauderkin. Judge Bain is building a house on Rosslyn road which will cost about $15,000. John Way has started to put up a $10,000 house at Armstrong's Point, and F. H. Welfy is erecting a residence on Sherbrooke street.

PORTLAND CEMENTS

Best German, Belgian and English Brands.

Fire Bricks, Fire Clay,
Flue Linings,
Drain Pipes,
Calcined Plaster;
Granite Hard Wall Plaster,
Wheelbarrows,
Mortar Stains.

A full stock of Builders' and Contractors' Supplies. Write for Quotations.

W. McNally & Co.,
MONTREAL.

CANADIAN ADVERTISING is best done by THE E. DESBARATS ADVERTISING AGENCY, MONTREAL.

BROOM AND MATTRESS WIRE *High Grade, Double Tinned*

Fine Annealed Brush Wire. Soft Coppered Wire. Tinned Wire of all kinds.

THE PEERLESS WIRE CO.,
Hamilton, Ont.

THE
PATENT ASBESTOS STOVE PIPE THIMBLE
THE ONLY FIREPROOF THIMBLE MADE.

Adjusts itself to any thickness of floor, from 8 to 16 inches. Light, cool and strong. Cast iron on upper side which is made to receive our cast iron register, which can be inserted when pipe is taken down; also can be opened or closed at will, ventilate room if desired, leaving a solid, smooth, surface to carpet over. For sale by all hardware, tin and stove dealers. Send for catalogue of about 25 first-class hardware specialties.

Manufactured by

THE COLLINS MANUFACTURING CO.,
34 Adelaide Street West, TORONTO.

HIGH-GRADE STEAM GOODS
FOR ALL USES.

Relief Valve

Manufacturers and Importers.

The James Morrison Brass Mfg. Co., Limited
TORONTO.

PAINTS THAT SELL

are paints that work easy, that wear well, that give the dealer a confidence that everything is all right. Paints that are made to please a customer, to make him satisfied, that bring him back again and again for more.

RAMSAYS PAINTS

are J-st that kind. A dealer can sell them and retain his custom ; he can get his profit and does not need to bo her what his neighbor does with some other brand. Isn't that a business paint to handle? Won't it pay you best to have the best paint made--and have it at a price that pays you well? That's Ramsays Paint.

A. RAMSAY & SON,
PAINTMAKERS,
Est'd 1842. MONTREAL.

"I CAN'T MANAGE THESE OLD ELBOWS"
"WHY DON'T YOU GET ALLEN'S FLAT CRIMPED ELBOW NO LEAK"

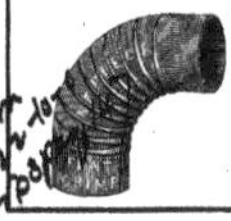

2, 3, and 4-in. Tin and Galvanized Conductor Elbows, for sale by all leading jobbing houses.

Made by GEO. ALLEN
BURLINGTON, ONT.

WIRE RODS!

Drawn to Decimal Sizes, Cut and Straightened, In Uniform Sizes. Prompt Shipment.

Chalcraft Screw Co., Limited, Brantford, Ont.

BUILDING PERMITS ISSUED.

The following permits have been issued in Ottawa: Holbrook and Sutherland, four tenements and two cottages, Gilmour street, to cost $10,000; Mrs. A. Cummings, brick veneered house, Somerset street, $900; Joseph Grant, warehouse, York street, $3,500; Rev. M. E. Harnois, addition to Sacred Heart juniorate, $1,000; Donald Skuce, brick veneered house, Maria street, $1,000; A. W. Fleck, brick house on Wilbrod street, Ottawa, to cost $15,000.

Building permits have been issued in Toronto as follows: Noble J. Craif, pair semi-detached dwellings at 610 and 612 Bathurst street, $6,000; Allan Maclean, dwelling, 96 Amelia street, $1,700; F. S. Duff, dwelling Shaw street, $1,200; Consumers' Gas Co., two-storey extension to retort house, Front street, $12,000; Robt. Mulholland, brick and stone residence, Huron street, near Lowther avenue, $6,000; Henry Crowther estate, brick palm room, 100 Yonge street, $3,000; Chas. Stark, residence, Howard street, near Parliament, $1,000; W. E. Willmott, dwelling on Crescent road, $3,500; Richard Tuthill, store and apartment dwelling, King and Cowan avenue, $8,000; Henry Cawthra estate, remodelling building, southwest corner of Yonge and Wellington streets, $8,500; J. W. Lee & Co., two-storey brick addition to a factory, Atlantic avenue, near Central Prison, $6,000; B. G. Austin, dwelling, Margueretta street, north of Bloor, $1,200.

GAS LIGHT.

A recent number of The Contemporary Review contains an exceedingly able article on "Gas Light," by an author who, in the plainest, yet most accurate of language, describes the revolution wrought in artificial illumination by the discovery of the incandescent mantle. In sober truth this story may be called a romance. For years highly successful attempts were made to improve upon the original "hole in a pipe," from which coal-gas was consumed; the fishtail, batswing, argand, and regenerative burners were invented, each being an improvement over its predecessors in the economy with which the carbon of the gas was rendered luminous. Now all these devices are superseded; the inherent illuminating power of the gas is no longer of primary importance, calorific value becomes the chief criterion, and this change has been mainly effected by the researches of one man—Dr. Carl Auer von Welsbach. Of course, Welsbach did not "invent" the incandescent system of lighting; nothing of great benefit to mankind is ever "invented"; like Topsy, it "grows." Drummond, Fahnehjelm, Clamond, were all workers in the same field; but the greatest of them all was Bunsen. Nevertheless, to Welsbach belongs the credit of devising the first incandescent burner of any real practical utility; but he was dependent upon the atmospheric flame. The mantle is to be seen in every town where coal-gas has penetrated, and it is found in most streets and most houses of the civilized cities throughout the world.

PLUMBING AND HEATING NOTES.

Phillips & Wilson, contractors, Hamilton, have dissolved.

The Montreal Gas Engine Co. has registered to do business in Montreal.

The Oxford Electric Co., Limited, Oxford, N.S., have been incorporated.

Weeks & Son, practical plumbers, have opened at 840 Pender street, Vancouver.

The creditors of John E. Fitzgerald, plumber, St. John, N.B., met on Tuesday.

The Mahone Bay, N. S. Electric Light & Power Co., Limited, have been incorporated.

Weeks & McIntyre, plumbers, Vancouver, B.C., have dissolved. S. J. McIntyre continues.

The Blockhouse Electric Light Power Co., Limited, Blockhouse, N.S., have been incorporated.

Mrs. Ernest Boisclair has registered as proprietress of Boisclair & Co., contractors, Montreal.

PLUMBING AND HEATING CONTRACTS.

Carroll Bros., Montreal, have been given the contract for the plumbing, heating and roofing of three houses in Maissonneuve for Mr. P. Rafferty. They also have the contract for the plumbing, heating and electric-lighting of a store and dwelling on St. Antoine street for Mr. N. Grant. Mr. W. A. Doran is making some alterations in the heating and plumbing of some houses on Ontario street and Lorne Crescent, Montreal; Carroll Bros. will do the work.

J. W. Hughes & Co., Montreal, are installing the heating apparatus in the new crematorium in Mount Royal Cemetery, Montreal.

Mr. Joseph Lamarche, Montreal, has secured the contract of plumbing, heating, lighting and ventilating of the new St. Gabriel street school being put up by the Catholic Commissioners; the structure will cost about $30,000. He also has the contract for the plumbing and heating of the new post office being erected in Hochelaga, as well as for three new roofs for the Ryan estate in Montreal.

PROPOSE TO BUILD TINWARE WORKS.

J. L. Board and W. S. Cumming, who were recently engaged manufacturing tinware and other articles in Chicago, but who sold out their business to the Steel Trust, are asking inducements from Sarnia for the establishment of works in that place. They offer to build works to employ about four hundred hands and to make machine tools, tinware, etc., for the export trade.

HAMMERS MADE OF RAWHIDE.

"The common idea of a hammer, no doubt," said a dealer in tools to an exchange, "would be that it was an implement made to pound with, and having a head of iron or steel. The pounding part of that would certainly be all right, but not all hammer heads are made of metal; there are some hammers, in fact, with head made of rawhide.

"Where the head would be on an ordinary hammer there is on the rawhide hammer seat, at right angles across the end of the handle, a short section of iron pipe. The rawhide that forms the hammer head is first cut into an oblong strip, which is then, beginning at one end, snugly rolled up. The roll, thus formed is put through an iron pipe, being made long enough so that it will project an inch or more at either end. The ends of the solid rawhide are trimmed off flat and true, like the face of any hammer, making this a two-faced hammer.

"The rawhide hammer is used for various purposes, largely in place of a mallet, for instance, for pounding on punches, and on chisel handles. It is used where pounding is to be done on polished metal surfaces; it serves the purpose without scratching the metal. Rawhide hammers are made in various sizes.

"Then there is a rawhide implement that is called a mallet, in which the head is formed in the same manner as the rawhide hammer head, but joined in the handle direct, without being held there in a holder. The rawhide mallet is also made in various sizes; it is a smaller and lighter tool than the hammer.

"Another rawhide pounding tool is the rawhide maul, heavier than the hammer, and made in various sizes. The head of the rawhide maul is made of disks of rawhide laid together to a sufficient thickness and held together by iron caps top and bottom, through which, as of course through the rawhide as well, the maul handle passes. The block of rawhide thus made is turned into the usual maul form. Built up as it is of compacted layers placed crosswise of the handle, the striking surface of the maul, as is the case with the hammer and the mallet in the manner in which they are made, presents the rawhide in a mass edgewise. The rawhide is used, for example, by artificial flower makers, pounding all day long on dies and punches, cutting out flowers and leaves.

"These rawhide hammers and mallets and mauls cost about three times as much as corresponding wooden mallets would cost—they last about ten times as long."

RADIANT SHELBY LAMPS.

The Ontario Lantern Company, represented by Walter Grose, are now in a position to supply all demands for Radiant Shelby Incandescent lamps. Wherever used since they have been put on the market, they have been found to give every satisfaction. The double spiral carbon set down well into the bulb has proven to be all that was expected of it.

BUSINESS CHANGES.

DIFFICULTIES, ASSIGNMENTS, COMPROMISES.

GIRARD & CO., general merchants, St. Liboire, Que., have offered to compromise.

Z. Paquet, general merchant, Roberval, Que., has assigned.

Benson & Borland, coal dealers, Quebec, are offering 25c. on the dollar.

E. H. Williams, hardware dealer, Sintaluta, Man., has assigned to J. W. Moody.

L. G. Jourdain, hardware dealer, Three Rivers, Que., has compromised at 35c. on the dollar.

Wilks & Michaud have been appointed curators of F. A. Cantwell, general merchant, Franklin Centre, Que.

A meeting of the creditors of M. Gillespie & Co., planing millers, Alvinston, Ont., has been called for April 20.

Application has been made for the appointment of a liquidator of the Copp Bros. Co., Limited, manufacturers of stoves, etc., Hamilton.

J. E. Tremblay, general merchant, St. Anne de Bellevue, Que., has assigned, and a meeting of his creditors will be held on April 19, (to-day.) The principal creditors are: Liddell, Lesperance & Co., $1,128; T. Toupin, $372; A. Robitaille & Co., $500; Letang Hardware Co., $500; Daoust & Lalonde, $263; A. Ramsay & Co., $250; D. R. Ronaldson, $350; Caverhill, Kissock & Co., $223; Victoria-Montreal Insurance Co., $400.

PARTNERSHIPS FORMED AND DISSOLVED.

Bell & Flett, stove dealers, etc., Vancouver, have dissolved.

Girouard Bros., general merchants, Somerset, Man., have dissolved.

A. G. Fox & Co., general merchants, Burnside, Man., have admitted W. R. Lee.

E. Prefontaine & Co., coal and wood dealers, Longueuil, Que., have dissolved.

Campbell & Farrell, blacksmiths, Glencoe, Ont., have dissolved. J. J. Campbell continues.

O'Brien & Allan, blacksmiths, Hamilton, Ont., have dissolved. Charles O'Brien continues.

T. F. Kirkham, stove and tinware dealer, Lethbridge, N.W.T., has admitted D. Stewart into partnership.

The Sherbrooke Iron and Metal Co., Sherbrooke, Que., have dissolved, and a new partnership has been registered.

Perez E. Lloyd and Geo. Caldwell have registered partnership under the style of the Kentville Coal and Lumber Co., Kentville, N.S.

SALES MADE AND PENDING.

The assets of Miles Birkett, hardware dealer, Ottawa, are to be sold.

David Plato, blacksmith, Fort Erie, Ont., is advertising his business for sale.

E. & C. Thompson, general merchants, Elmsdale, Ont., are offering their business for sale.

The assets of D. Levasseur, general merchant, Matane, Que., are to be sold to-day (Friday).

Oliver & McArthur, harness dealers, Dauphin, Man., are advertising their business for sale.

The real estate of the estate of T. Ross, general merchant, Amqui, Que., is to be sold on May 3.

The business of M. G. McEwan, hardware dealer, Qu'Appelle Station, N.W.T., is advertised for sale.

The stock of the estate of Ruth Perry, general merchant, Gad's Hill, Ont., has been sold to J. H. Birch at 67c. on the dollar.

The stock of Joseph Bernier, carriage maker, L'Islet, Que., has been sold at 43c. on the dollar to Letourneau & Anger, Victoriaville, Que.

CHANGES.

Rigali & Rigali have registered as painters in Quebec.

R. S. Hannah, hardware dealer, Mitchell, Ont., is about closing up.

Scott & Wallace, blacksmiths, Penobsquis, N.B., have discontinued business.

A. Barrisdale, blacksmith, Stratford, Ont., has sold out to Kalbfleisch & Kochel.

Robert Munn, blacksmith, Winthrop, Ont., has sold out to Robert Pethick.

Charles Gough, blacksmith, Elkhorn, Man., has sold out to Wm. MacLeod.

H. P. Read is giving up business as general merchant in Bear River, N.S.

John Chatham, general merchant, Massie, Ont., has sold out to W. D. Dunnington.

E. J. Boucher, general merchant, Boucherville, Que., has sold out to T. A. Boucher.

Wm. Burnard, general merchant, Britton, Ont., has sold out to Robert A. Thompson.

Alex. Boivin will continue as painter in Montreal, doing business in his wife's name.

Thomas E. Risk, general merchant, Shetland, Ont., has sold out to H. H. Mann.

R. McIvor, harness dealer, Elkhorn, Man., has been succeeded by F. A. McCullagh.

Findlay Chisholm, dealer in agricultural implements, Milton, Ont., has given up business.

William Stuckey, planing miller, etc., Grand Valley, Ont., has been succeeded by James Campbell.

James Sutherland & Son, blacksmiths, Grenfell, Man., have been succeeded by R. P. Sutherland.

W. F. Hartwell, general merchant, Wawanesa, Man., has opened a branch at Swan Lake.

OUTDOOR WORK FOR THE TINSMITH.

Always after the winter there are a good many repairs needed around the barns, residences and stores of any community, writes "O. D. S." in Metal Worker. Owing to the heavy snows of the winter and the freezing and thawing quite a number of eavetroughs, gutters and conductor pipes not only need leaks stopped, but they also need painting and better fastening to keep them in place. Roofs that were not painted last fall should not be left to go through the summer without painting now, and it is probable that if they are gone over and all the leaks soldered and made tight before the painting is done the property owner will derive a decided benefit.

Naturally, however, the property owner is now looking for an opportunity to spend money in this direction. The new strawberries of the greengrocer are more likely to catch his odd dollars than the tinsmith. It is quite probable that the tinsmith has more need of these stray dollars than the greengrocer, but he may look on the greengrocer as a successful competitor if he does not make display or do something to induce the property owner to spend his odd dollar with him. It has been the custom of some successful tinsmiths to send to the property owners of their community at this time, postal cards or circulars, calling attention to the fact that at this season they are better able to do outdoor work than they will be later on when the building season opens and new work occupies their time. This is a pertinent fact, and if it is supplemented with a suggestion that the conductor pipe, the eavetrough and the roof should all be looked after, and the tin roofs in particular will be better for a coat of paint, many a profitable job will be done and some new customers secured.

The masons will find no fault if the tinsmith suggests that the chimney should be reflashed and pointed up. This is the season for shops doing outdoor work to make a personal canvass, supplemented by a distribution of circulars and a suitable "ad." and reading notices in the local paper. Those who show their enterprise in these lines will certainly be repaid for their trouble.

BIRMINGHAM SMALL ARMS COMPANY.

Lewis Bros. & Co., Montreal, Canadian agents for the Birmingham Small Arms Co., Limited, expect to do a good trade in military rifles this summer. It is thought that the military spirit engendered by the South-African War, and the necessity for target practice disclosed by the same event, will lead to a more wide-spread use of the rifle in Canada. Dr. Borden's wise provision of further facilities for target practice is only one of the signs of this change that has come over the country.

Not only will the Government rifles be more generally used, but there will be a greater desire among young men to own a rifle of their own and to become familar with its use. They cannot buy a better gun than that turned out by the Birmingham Small Arms Co., Limited. They manufacture the rifle in use by all the forces of Greater Britain, an approbation that cannot be excelled.

Their rifles are all carefully tested for alignment of sights and grouping of shots and are guaranteed accurate. They are all submitted to Government inspection and stamped with the viewer's mark. The three principal guns are the Lee-Enfield Magazine Tatget Rifle, the Lee-Metford Magazine Target Rifle and the Lee-Enfield magazine Sporting Rifle. Besides these there are included in their lines the Martini-Metford, the Martini-Metford carbines and special sporting rifles.

These goods can be secured through any wholesale house in the Dominion. Catalogues will be ready for the post in a few days. Meanwhile write for one.

ABUSE OF DISCOUNTS.

One of the flagrant abuses in trade to-day, writes F. H. Woodward in an exchange, is the taking of discount on bills long after the time allowed for discount, one which when figured in dollars and cents would astonish manufacturers and dealers and open their eyes to one of the reasons for the lack of profit in the past few years. The manufacturer or dealer is in a great measure to blame for this abuse, as in his zeal to increase his sales he becomes lenient to a degree, and believing or fearing that his competitors permit the evil, relaxes his vigilance, and his customers finding no rebuff in their robbery, for robbery it is, grow bolder, and from a few days' overtime they go to such lengths that they demand the discount on bills when goods arrive, or claim to have certain days to draw checks, or give some other plausible excuse, resenting any protest from the vendor as unwarranted and uncalled for, claiming that the vendor's competitors allow it.

Should you go to a bank to have a note for $500 payable in four months discounted at 6 per cent., the interest or discount of $10 would be deducted and you would be given the balance and would not expect any different treatment. But if you sell $500 worth of merchandise to a customer, at four months, discount 2 per cent. ten days, and the customer takes twenty days to discount, he has robbed you of ten days' interest and you permit it. Figure up the interest you lose by this injustice and you will realize the robbery you are suffering. The remedy lies in your own hands. Insist that if bills are to be discounted in ten or fifteen days, or whatever time is customary in your line of business, those terms be acceded to. If customers refuse, show them in unmistakable terms the injury to yourself and their own loss of credit, for beyond doubt their credit is injured far more than they realize by their own acts and we believe the abuse will be rectified. All abuses are small at first and only become evils as they are permitted to grow. So reform may be slow at the start, but let manufacturers and dealers take a firm stand for the principle and the abuse will be ended.

CURRENT MARKET QUOTATIONS.

April 19, 1901.

These prices are for such qualities and quantities as are usually ordered by retail dealers on the usual terms of credit, the lowest figures being for larger quantities and prompt pay. Large cash buyers can frequently make purchases at better prices. The Editor is anxious to be informed at once of any apparent errors in this list as the desire is to make it perfectly accurate.

METALS.

Tin.

Lamb and Flag and Straits—56 and 28 lb. ingots, per lb.	0 31	0 32

Tinplates.

Charcoal Plates—Bright

		Per box
M.L.S., equal to Bradley.		
I.C., usual sizes	....	$6 75
I.X., "	....	8 25
I.X.X., "	....	9 75
Famous—		
I.C.	....	6 75
I.X	....	8 25
I.X.X	....	9 50
Raven & Vulture Grades—		
I.C., usual sizes	....	5 00
I.X., "	....	6 00
I.X.X "	....	7 00
I.XXX., "	....	8 00
D.C., 12⅛x17	....	4 75
D.X.	....	5 50
D.X.X	....	7 50

Coke Plates—Bright

Bessemer Steel—		
I.C., usual sizes	....	4 15
I.C., special sizes, base	....	4 50
20 x 28	....	8 50

Charcoal Plates—Terne

Dean or J. G. Grade—		
I.C., 20x28, 112 sheets	....	8 25
I.X., Terne Tin	....	10 25

Charcoal Tin Boiler Plates.

		Per lb.
Cookley Grade—		
X X., 14x56, 50 sheet bxs; " 14x60, "; " 14x65, "	0 07	0 07½

Tinned Sheets

72x30 up to 24 gauge	0 08	0 08½
" 26 "	0 08½	0 09
" 28 "	0 09	0 09½

Iron and Steel.

Common Bar, per 100 lbs	1 70	1 80
Refined " "	....	2 25
Horse Shoe Iron "	....	2 10
Hoop steel, 1½ to 3 in. base, extras for smaller sizes	....	3 10
Sleigh Shoe Steel " base	....	2 10
Tire Steel	....	2 15
Machinery iron finish	....	2 15
Cast Steel, per lb	0 00	0 00
Toe Calk Steel	....	2 40
T. Firth & Co's special cast steel, per lb.		0 13
Jessop's Tool Steel	....	0 13

Boiler Tubes.

1½-inch	....	0 12½
2 "	....	0 13
2½ "	....	0 15
3 "	....	0 16
3½ "	....	0 20
4 "	....	0 25

Steel Boiler Plate.

¼ inch	....	2 40
3-16 inch	....	2 50
⅜ inch and thicker	....	2 60

Black Sheets.

18 gauge	....	3 00
20 gauge	....	3 00
22 to 24 "	....	3 10
26 "	....	3 20
28 "	....	3 30

Canada Plates.

Al dull, 52 sheets	....	3 00
Half polished	....	3 15
All bright	3 65	3 75

Iron Pipe.

Black pipe—		
⅛ "	....	4 45
¼ inch	....	3 25
⅜ "	....	3 30
½ "	....	3 50
¾ "	....	3 65
1 "	....	5 15
1¼ "	....	7 10
1½ "	....	8 40
2 "	....	10 70
2½ "	....	20 25
3 "	....	24 55
3½ "	....	30 75
4 "	....	39 00
5 "	....	47 35
6 "	....	62 10
Galvanized pipe—		
½ inch	....	4 90
¾ "	....	5 25
1 "	....	7 55
1¼ "	....	10 30
1½ "	....	12 35
2 "	....	16 50

Galvanized Sheets.

	G C. Comet.	Amer.	Queen's Head.
16 gauge	....	4 5	3 90
18 to 24 gauge	4 15 3 85	4 35	4 15
26 "	4 41 4 10	4 35	4 40
28 "	4 65 4 35	4 50	4 65

28 gauge American equals 26 gauge English.

Chain.

Proof Coil, 3-16 in., per 100 lb	....	....
" ¼ " "	8 00	8 50
" 5-16 " "	4 70	5 00
" ⅜ " "	4 05	4 50
" 7-16 " "	3 90	4 25
" ½ " "	3 70	4 10
" 9-16 " "	3 65	4 05
" ⅝ " "	3 35	3 90
" ¾ " "	3 50	4 10

Halter, kennel and post chains, 40 and 50 p.c.
Cow ties ... 40 p.c.
Stall fixtures ... 35 p.c.
Trace chain ... 25 p.c.
Jack chain, iron, single and double, discount 35 p.c.
Jack chain, brass, single and double, discount 40 p.c.

Copper.

Ingot

English B. S., ton lots	....	0 19
Lake Superior	....	....

Bolt or Bar.

Cut lengths round, ⅜ to ⅞ in.	0 23½	0 25
" round and square 1 to 2 inches	0 23½	0 25

Sheet.

Plain, 14 oz., and light, 16 oz., 14x48 and 14x60	0 23	0 23
Plain, 14 oz., and light, 16 oz., irregular sizes	0 23	0 23½
Tinned copper sheets	....	0 26
Planished	....	0 32

Braziers (In sheets.)

4x6ft. 25 to 30 lbs. ea., per lb.	....	0 25½
" 35 to 45 " "	....	0 24½
" 50-lb. and above, "	....	0 23½

Boiler and T. K. Pitts

Plain Tinned, per lb	....	0 28
Spun, per lb	....	0 32

Brass.

Rod and Sheet, 14 to 30 gauge, 15 per cent.		
Sheets, hard-rolled, 2x4	0 24	0 25
Tubing, base, per lb	0 24	0 25

Zinc Spelter

Foreign, per lb	0 05½	0 06
Domestic "	....	...

Zinc Sheet.

5 cwt. casks	....	00 6½
Part casks	....	00 6¾

Lead.

Imported Pig, per lb	0 04½	0 04¾
Bar, 1 lb.	0 05½	0 06¼
Sheets, 2½ lbs. sq. ft., by	....	0 06¼
Sheets, 3 to 6 lbs., "	....	0 06

NOTE.—Cut sheets ½ cent per lb. extra. Pipe, by the roll, usual weights per yard, lists at 7c. per lb. and 25 p.c. dis. f.o.b. Toronto.

NOTE.—Cut lengths, net price, waste pipe 8-ft. lengths lists at 7½ cents.

Shot.

Common, $6.50 per 100 lb.; chilled, $7.00 per 100 lb.; buck, seal and ball, $7.50. Discount, 25 p.c. Prices are f.o.b. Toronto, Hamilton, Montreal, St. John and Halifax. Terms 3 per cent. cash, freights equalized on Montreal.

Soil Pipe and Fittings.

Discount, 60 and 10 per cent. on medium and extra heavy, and 60 per cent. on light.

Solder.

	Per lb.	Per lb.
Bar half-and-half, guarant'd	....	0 18½
Bar half-and-half, commer'l	....	0 18
Refined	....	0 17½
Wiping	....	0 17

NOTE.—Prices of this graded according to quantity. The prices of other qualities of solder in the market indicated by private brands vary according to composition.

Antimony.

Cookson's, per lb	0 10½	0 11

White Lead.

	Per 100 lb.
Pure	6 37
No. 1 do	6 00
No. 2 do	5 62½
No. 3 do	5 25
No. 4 do	4 87½
Munro's Select Flake White	7 37½
Elephant and Decorators' Pure	7 12½
Brandram's B. B. Genuine	8 00
" " Decorative	7 55
" " No. 1	6 85
" " No. 2	6 00

Red Lead.

Genuine, 560 lb. casks, per cwt.	....	$5 50
Genuine, 100 lb. kegs, per cwt.	....	5 75
No. 1, 560 lb. casks, per cwt.	....	5 25
No. 1, 100 lb. kegs, per cwt	....	5 00

White Zinc Paint.

Elephant Snow White	0 08	0 09
Pure White Zinc	0 08	0 09
No. 1	0 06	0 07½
No. 2	0 05	0 06½

Dry White Lead.

Pure, casks	5 75
Pure, kegs	6 25
No. 1, casks	5 50
No. 1, kegs	5 00

Prepared Paints.

In ¼, ½ and 1 gallon tins.

Pure, per gallon	....	1 25
Second qualities, per gallon	....	1 10
Barn (in bbls.)	0 75	0 85
The Sherwin-Williams Paints.	....	1 45
Canada Paint Co's Pure	....	1 25
Toronto Lead & Color Co's Pure	....	1 25
Sanderson Pearcy's Pure	....	1 20
Stewart & Wood's Champion Pure.		1 20

Colors in Oil.

25 lb. tins, Standard Quality.

Venetian Red, per lb.	0 05
Chrome Yellow	0 11
Golden Ochre	0 06
French "	0 05
Marine Black	0 09
" Green	0 09
Chrome "	0 08
French Imperial Green	0 09

Colors, Dry.

Yellow Ochre (J.C.) bbls.	1 35	1 40
Yellow Ochre (J.F.L.S.), bbls	....	2 75
Yellow Ochre (Royal)	1 10	1 15
Brussels Ochre	....	2 00
Venetian Red (best), per cwt.	1 80	1 90
English Oxides, per cwt.	3 00	3 25
American Oxides, per cwt.	1 75	2 00
Canadian Oxides, per cwt.	1 75	2 00
Super Magnetic Oxides, 93p.c.	2 00	2 25
Burnt Sienna, pure, per lb.	....	0 10
" Umber, " "	....	0 10
do Raw	....	0 09
Drop Black, pure	....	0 09
Chrome Yellows, pure	....	0 18
Chrome Greens, pure, per lb.	....	0 12
Golden Ochre	....	0 03½
Ultramarine Blue in 28-lb. boxes, per lb.	0 08	0 24
Fire Proof Mineral, per 100 lb	....	1 00
Genuine Eng. Litharge, per lb	....	0 07
Mortar Color, per 100 lb.	....	1 25
English Vermillion	....	0 80
Pure Indian Red, No. 45 lb.	....	0 80
Whiting, per 100 lb.	....	0 55

Blue Stone.

Casks, for spraying, per lb.	0 07
100-lb. lots, do. per lb	0 08

Putty.

Bulk in bbls.	1 90
Bulk in less quantity	2 05
Bladders in bbls.	2 10
Bladders in kegs, boxes or loose	2 25
Bladders in 25-lb. tins	2 35
Bladders in 12½-lb. tins	2 65
Bladders in bulk or tins less than 100 lb.	2 90

Varnishes.

	In 5-gal. lots.)	Per gal.
Carriage, No. 1	2 90	3 30
" body	8 00	9 00
" rubbing	4 00	5 00
Gold Size, Japan	3 00	3 40
Brown Japan	2 40	2 80
Elastic Oak	2 90	3 30
Furniture, extra	2 40	2 80
" No. 1	1 60	2 00
Hard Oil Finish	2 70	3 10
Light Oil Finish	3 20	3 60
Demar	3 30	3 70
Shellac, white	4 40	4 80
" orange	4 00	4 40
Furniture Brown Japan	1 60	2 00
Black Japan	2 40	2 80
" " No. 1	1 60	2 00

The Imperial Varnish & Color Co's., Limited Elastilite Varnish 1 gal. can, each. $3.00.

Granatine Floor Finish, per gal., $2.50.

Maple Leaf Coach Enamels; Size 1, 60c.; Size 2, 35c.; Size 3, 20c. each.

Castor Oil.

East India, in cases, per lb.	0 10	0 10½
" " small lots	0 10½	0 11

Cod Oil, Etc.

Cod Oil per gal.	0 50	0 55
Pure Olive	....	1 20
" Neatsfoot	....	90

Glue.

Common	0 08½	0 09
French Medal	0 14	0 14½
Cabinet sheet	0 12	0 13
White, extra	0 18	0 20
Gelatine	0 22	0 30
Strip	0 18	0 20
Coopers	0 19	0 20
Huttner	....	0 18

HARDWARE.

Ammunition.

Cartridges.

B. B Cap Dom. 50 and 5 per cent.
Rim Fire Pistol, dis. 40 p. c., Amer.
Rim Fire Cartridges, Dom., 50 and 5 p. c.
Central Fire Pistol and Rifle, 10 p.c. Amer.
Central Fire Cartridges, pistol sizes, Dom. 30 per cent.
Central Fire Cartridges, Sporting and Military, Dom., 15 and 5 per cent.
Central Fire, Military and Sporting, Amer. add 5 p.c. to list. B.B. Caps, discount 40 per cent. Amer.
Loaded and empty Shells, "Trap" and "Dominion" grades, 25 per cent. Rival and Nitro, net list.
Brass Shot Shells, 55 per cent.
Primers, Dom., 30 per cent.

Wads — per lb.
Best thick white felt wadding, in ¼-lb bags ... 1 00
Best thick brown or grey felt wads, in ½-lb. bags ... 70
Best thick white card wads, in boxes of 500 each, 12 and smaller gauges ... 0 99
Best thick white card wads, in boxes of 500 each, 10 gauge ... 0 35
Best thick white card wads, in boxes of 500 each, 8 gauge ... 0 55
Thin card wads, in boxes of 1,000 each, 12 and smaller gauges ... 0 20
Thin card wads, in boxes of 1,000 each, 10 gauge ... 0 25
Thin card wads in boxes of 1,000 each, 8 gauge ...
Chemically prepared black edge grey cloth wads, in boxes of 250 each— Per M.
11 and smaller gauge ... 0 60
9 and 10 gauges ... 0 70
7 and 8 gauges ... 0 90
5 and 6 gauges ... 1 10
Superior chemically prepared pink edge, best white cloth wads, in boxes of 250 each—
11 and smaller gauge ... 1 15
9 and 10 gauges ... 1 40
7 and 8 gauges ... 1 65
5 and 6 gauges ... 1 90

Adzes.

Discount, 20 per cent.

Anvils.

Per lb. ... 0 10 — 0 12½
Anvil and Vise combined ... 4 50
Wilkinson & Co.'s Anvils..lb. 0 09 — 0 09½
Wilkinson & Co.'s Vices..lb. 0 09½ — 0 10

Augers.

Gilmour's, discount 65 and 5 p.c. off list.

Axes.

Chopping Axes
Single bit, per doz ... 5 50 — 10 00
Double bit, " ... 11 00 — 18 00
Bench Axes, 40 p.c.
Broad Axes, 33⅓ per cent.
Hunters' Axes ... 5 50 — 6 00
Boy's Axes ... 5 75 — 6 75
Splitting Axes ... 6 50 — 12 00
Handled Axes ... 7 00 — 10 00

Axle Grease.

Ordinary, per gross ... 5 75 — 6 00
Best quality ... 13 00 — 15 00

Bath Tubs.

Zinc ... 6 00
Copper, discount 15 p.c. off revised list

Baths.

Standard Enameled.
5½-inch rolled rim, 1st quality ... 25 00
" " " 2nd " ... 21 00

Anti-Friction Metal.

"Tandem" A ... per lb. 0 27
" B ... " 0 21
" C ... " 0 11½
Magnolia Anti-Friction Metal, per lb. 0 25
Frictionless metal, per lb ... 0 23

SYRACUSE SMELTING WORKS.
Aluminum, genuine ... 0 45
Dynamo ... 0 29
Special ... 0 25
Aluminum, 99 p.c. pure "Syracuse" .. 0 50

Bells.

Hand.
Brass, 60 per cent.
Nickel, 55 per cent.

Cow.
American make, discount 66⅔ per cent.
Canadian, discount 45 and 50 per cent.

Door.
Gongs, Sargent's ... 5 50 — 8 00
" Peterboro', discount 45 per cent.

Farm.
American, each ... 1 25 — 3 00

House.
American, per lb. ... 0 35 — 0 40

Bellows.

Hand, per doz ... 3 35 — 4 75
Moulders', per doz ... 7 50 — 10 00
Blacksmiths', discount 40 per cent.

Belting.

Extra, 60 per cent.
Standard, 60 and 10 per cent.

Bits.

Auger.
Gilmour's, discount 60 and 5 per cent.
Rockford, 50 and 10 per cent.
Jennings' Gen., net list.

Car.
Gilmour's, 47½ to 50 per cent.

Expansive.
Clark's, 40 per cent

Gimlet.
Clark's, per doz ... 0 65 — 0 90
Damond, Shell, per doz ... 1 00 — 1 50
Nail and Spike, per gross ... 2 25 — 5 20

Blind and Bed Staples.

All sizes, per lb ... 0 07½ — 0 12

Bolts and Nuts. Per cent.

Carriage Bolts, full square, Norway ... 65
" " full square ... 65
Common Carriage Bolts, all sizes ... 60
Machine Bolts, all sizes ... 60
Coach Screws ... 70
Sleigh Shoe Bolts ... 72½
Blank Bolts ... 60
Bolt Ends ... 62½
Plough Bolts ... 60
Nuts, square ... 4 c. off
Nuts, hexagon ... 4½c. off
Tire Bolts ... 67½
Stove Bolts ... 67½
Stove rods, per lb ... 5½ to 6c.
Plough Bolts ... 60

Boot Calks.

Small and medium, ball, per M ... 4 25
Small heel, per M ... 4 50

Bright Wire Goods.

Discount ... 62½ per cent.

Broilers.

Light, dis., 65 to 67½ per cent.
Reversible, dis., 65 to 67½ per cent.
Vegetable, per doz., dis. 37½ per cent.
Henis, No. 8, " ... 6 00
Henis, No. 9, " ... 7 00
Queen City " ... 7 50 — 0 00

Butchers' Cleavers.

German, per doz ... 6 00 — 11 00
American, per doz ... 12 00 — 20 00

Building Paper, Etc.

Plain building, per roll ... 0 30
Tarred lining, per roll ... 0 40
Tarred roofing, per 100 lb ... 1 65
Coal Tar, per barrel ... 3 50
Pitch, per 100-lb ... 0 85
Carpet felt, per ton ... 45 00

Bull Rings.

Copper, $2.00 for 2½ in. and $1.90 for 2 in.

Butts.

Wrought Brass, net revised list

Cast Iron.
Loose Pin, dis., 60 per cent.

Wrought Steel.
Fast Joint, dis. 65, 10 and 2½ per cent.
Loose Pin, dis. 65, 10 and 2½ per cent.
Berlin Bronzed, dis. 70, 70 and 5 per cent.
Gen. Bronzed, per pair ... 0 40 — 0 65

Carpet Stretchers.

American, per doz ... 1 00 — 1 50
Bullard's, per doz ... 6 50 — ....

Castors.

Bed, new list, dis. 55 to 57½ per cent.
Plate, dis. 52½ to 57½ per cent.

Cattle Leaders.

Nos. 31 and 32, per gross ... 50 — 9 50

Cement.

Canadian Portland ... 2 50 — 2 80
English " ... . — 3 00
Belgian " ... 2 50 — 2 75
Canadian hydraulic ... 1 25 — 1 50

Chalk.

Carpenters Colored, per gross ... 0 45 — 0 75
White lump, per cwt ... 0 60 — 0 65
Red ... 0 05 — 0 06
Crayon, per gross ... 0 14 — 0 18

Chisels.

Socket, Framing and Firmer.
Broad's, dis. 70 per cent.
Warnock's, dis. 70 per cent.
P. S. & W. Extra 60, 10 and 5 p.c.

Churns.

Revolving Churns, metal frames—No. 0, $8—No. 1, $8.50—No. 2, $9.00—No. 3, $10.00—No. 4, $12.00—No. 5, $16.00 each. Ditto, wood frames—20c. each less than above. Discounts: Delivered from factories, 58 p.c.; from stock in Montreal, 58 p.c. Terms, 4 months or 3 p.c. cash in 30 days.

Clips.

Axle dis. 65 per cent.

Closets.

Plain Ontario Syphon Jet ... $16 00
Emb. Ontario Syphon Jet ... 17 00
Fittings net ... 1 00
Plain Teutonic Syphon Washout ... 10 90
Emb. Teutonic Syphon Washout ... 11 00
Fittings net ... 1 25
Low Down Teutonic, plain ... 16 00
" " " embossed ... 17 00
Plain Richelieu net ... 3 75
Emb. Richelieu net ... 4 00
Fittings net ... 1 25
Low Down Ont. Sy. Jet, plain net ... 19 50
" " " " " emb'd. net 20 50
Closet connection net ... 1 25
Basins, round, 14 in ... 1 00
" oval, 17 x 14 in ... 2 50
" " 19 x 15 in ... 3 75
Discount 40 p.c., except on net figures.

Compasses, Dividers, Etc.

American, dis. 62½ to 65 per cent.

Cradles, Grain.

Canadian, dis. 25 to 33⅓ per cent.

Crosscut Saw Handles.

S. & D., No. 3, per pair ... .17½
" " 5, " ... .22½
" " 6, " ... .15
Boynton pattern " ... .20

Door Springs.

Torrey's Rod, per doz ... (15 p.c.) — 2 00
Coil, per doz ... 0 88 — 1 60
English, per doz ... 2 00 — 4 00

Draw Knives.

Coach and Wagon, dis. 50 and 10 per cent.
Carpenters, dis. 70 per cent.

Drills.

Hand and Breast.
Millar's Falls, per doz. net list.

DRILL BITS.
Morse, dis., 37½ to 40 per cent.
Standard dis. 50 and 5 to 55 per cent

Faucets.

Common, cork-lined, dis. 35 per cent.

ELBOWS. (Stovepipe.)
No. 1, per doz ... 1 40
No. 2, per doz ... 1 20
Bright, 20c. per doz. extra.

ESCUTCHEONS.
Discount, 45 per cent.

ESCUTCHEON PINS.
Iron, discount 40 per cent.

FACTORY MILK CANS.
Discount off revised list, 40 per cent.

FILES.
Black Diamond, 70 and 5 per cent.
Kearney & Foote, 60 and 10 to 70 per cent.
Nicholson File Co., 60 to 60 and 10 per cent.
Jowitt's, English list, 25 to 27½ per cent.

FORKS.
Hay, manure, etc., dis., 50 and 10 per cent. revised list.

GLASS—Window—Box Price.

Size United Inches.	Star Per 50 ft.	Star Per 100 ft.	D. Diamond Per 50 ft.	D. Diamond Per 100 ft.
Under 26	2 15	4 15	....	6 00
26 to 40	2 30	4 45	....	6 65
41 to 50	....	4 85	....	7 50
51 to 60	....	5 15	....	8 50
61 to 70	....	5 50	....	9 50
71 to 80	....	6 00	....	10 50
81 to 85	....	6 50	....	11 75
86 to 90	....	....	..	14 00
91 to 95	..	....	...	15 50
99 to 100	.. .	....	....	18 00

GAUGES

Marking, Mortise, Etc.
Stanley's dis. 50 to 55 per cent.

Wire Gauges.
Winn's, Nos. 26 to 33, each ... 1 65 — 2 40

HALTERS.
Rope, ⅜ per gross ... . — ....
" ½ " ... . — 9 00
" ⅝ to ¾ ... . — 14 00
Leather, 1 in., per doz ... 3 87½ — 4 00
" 1¼ in., " ... 5 15 — 5 20
Web, — per doz ... 1 87 — 2 45

HAMMERS.

Nail
Maydole's, dis. 5 to 10 per cent. Can. dis. 25 to 27½ per cent.

Tack.
Magnetic, per doz ... 1 10 — 1 20

Sledge.
Canadian, per lb ... 0 07½ — 0 08½

Ball Pean.
English and Can., per lb ... 0 22 — 0 25

HANDLES.
Axe, per doz. net ... 1 50 — 2 00
Store door, per doz ... 1 00 — 1 50

Fork.
C. & B., dis. 40 per cent. rev. list.

Hoe.
C. & B., dis. 40 per cent. rev. list.

Saw.
American, per doz ... 1 00 — 1 25

Plane.
American, per gross ... 3 15 — 3 75

Hammer and Hatchet.
Canadian, 40 per cent.

Cross-Out Saws.
Canadian, per pair ... 0 13½

HANGERS. doz. pairs.
Steel barn door ... 5 85 — 6 00
Stearns, 4 inch ... . — 5 00
" 5 inch ... . — 6 50
Lane's covered—
No. 11, 5-ft. run ... . — 8 40
No. 11½, 10-ft. run ... . — 10 80
No. 12, 10-ft. run ... . — 12 60
No. 14, 15-ft. run ... . — 21 00
Lane's O.N.T. track, per foot ... . — 4½

HARVEST TOOLS.
Discount, 50 and 10 per cent.

HATCHETS.
Canadian, dis. 40 to 42½ per cent.

HINGES.
Blind, Parker's, dis. 50 and 10 to 60 per cent.
Heavy T and strap, 4-in., per lb ... 0 06½
" " 5-in., " ... 0 06¼
" " 6-in., " ... 0 06
" " 8-in., " ... 0 05¾
" " 10-in., " ... 0 05½
Light T and strap, dis. 65 and 5 per cent.
Screw hook and hinge—
6 to 12 in., per 100 lbs ... . — 4 50
14 in. up, per 100 lbs ... . — 3 50
Per gro. pairs.
Spring ... . — 12 00

HOES.
Garden, Mortar, etc., dis. 50 and 10 p.c.
Planter, per doz ... 4 00 — 4 50

HOLLOW WARE
Discount ... 45 and 5 per cent.

HOOKS.

Cast Iron.
Bird Cage, per doz ... 0 50 — 1 10
Clothes Line, per doz ... 0 27 — 0 63
Harness, per doz ... 0 72 — 0 88
Hat and Coat, per gross ... 1 00 — 3 00
Chandelier, per doz ... 0 50 — 1 00

Wrought Iron.
Wrought Hooks and Staples, Can., dis. 47½ per cent.

Wire.
Hat and Coat, discount 45 per cent.
Belt, per 1,000 ... 0 60
Screw, bright, dis. 55 per cent.

HORSE NAILS.
"C" brand 50 and 7½ p.c. off new list } Oval-head
"M" brand 60 per cent. }
Countersunk, 60 and 10 per cent.

HORSESHOES

Iron Shoes.	F.O.B. Montreal. No. 2 and larger.	No. 1. and smaller.
Light, medium, and heavy...	3 50	3 75
Snow shoes....	3 75	4 00
Steel Shoes.		
Light....	3 60	3 85
Featherweight (al lsizes)....	4 85	4 85

F.O.B. Toronto, Hamilton, London and Guelph, 10c. per keg additional.
Toe weight steel shoes...... 6 70

JAPANNED WARE.
Discount, 45 and 5 p.c. off list, June 1899

ICE PICKS.
Star per doz.... 3 00 3 25

KETTLES.
Brass spun, 7½ p.c. dis. off new list.
Copper, per lb.... 0 30 0 50
American, 60 and 10 to 65 and 5 p.c.

KEYS.
Lock, Can., dis., 45 p.c.
Cabinet, trunk, and padlock, Am. per gross.... 60

KNOBS.
Door, japanned and N.P., per doz.... 1 50 2 50
Bronze, Berlin, per doz...... 2 75 3 25
Bronze Genuine, per doz.... 6 00 9 00
Shutter, porcelain, F. & L. screw, per gross.... 1 30 4 00
White door knobs—per doz. 1 25

HAY KNIVES.
Discount, 50 and 10 per cent.

LAMP WICKS.
Discount, 60 per cent.

LANTERNS.
Cold Blast, per doz.... 7 00
No. 3 "Wright's".... 8 50
Ordinary, with O burner.... 4 00
Dashboard, cold blast.... 9 00
No. 0.... 5 75
Japanning, 50c. per doz. extra.

LEMON SQUEEZERS.
Porcelain lined,......per doz. 2 20 5 60
Galvanized.... 1 87 3 85
King, wood.... 2 75 2 90
King, glass.... 4 00 4 50
All glass.... 1 20 1 30

LINES.
Fish, per gross.... 1 05 2 50
Chalk " 1 90 7 40

LOCKS.
Canadian, dis. 45 p.c.
Russell & Erwin, per doz.... 3 00 3 25

Cabinet.
Eagle, dis. 30 p.c.

Padlock.
English and Am., per doz.... 50 6 00
Scandinavian, " 1 00 2 40
Eagle, dis. 20 to 25 p.c.

MACHINE SCREWS. Iron and Brass.
Flat head discount 25 p.c.
Round Head, discount 20 p.c.

MALLETS.
Tinsmiths', per doz.... 1 25 1 50
Carpenters', hickory, per doz. 1 25 3 75
Lignum Vitae, per doz.... 3 85 5 00
Caulking, each.... 60 2 00

MATTOCKS.
Canadian, per doz.... 5 50 6 50

MEAT CUTTERS.
American, dis. 25 to 30 p.c.
German, 15 per cent.

MILK CAN TRIMMINGS.
Discount, 25 per cent.

NAILS.

Quotations are:	Cut.	Wire.
2d and 3d....	$3 35	$3 85
3d....	3 00	3 52
4 and 5d....	2 75	3 35
6 and 7d....	2 65	3 20
8 and 9d....	2 50	3 00
10 and 12d....	2 45	2 95
16 and 20d....	2 40	2 90
30, 40, 50 and 60d. (base)....	2 35	2 85

Wire nails in carlots are $2.77½
Galvanizing 2c. per lb. net extra.
Steel Cut Nails 10c. extra.
Miscellaneous wire nails, dis. 70 and 10 p. c.
Coopers' nails, dis. 30 per cent.
Flour barrel nails, dis. 25 per cent.

NAIL PULLERS.
German and American...... 1 85 3 50

NAIL SETS
Square, round, and octagon, per gross.... 3 38 4 00
Diamond.... 12 00 15 00

NETTING.
Poultry, 50 and 5 per cent. for McMullen's

OAKUM. Per 100 lb.
Navy.... 6 00
U. S. Navy.... 7 25

OIL.
Water White (U.S.).... 0 16½
Prime White (U.S.).... 0 15½
Water White (Can.).... 0 15
Prime White (Can.).... 0 14

OILERS.
McClary's Model galvan. oil can, with pump, 5 gal., per doz.... 10 00
Zinc and tin, dis. 50, 50 and 10.
Copper, per doz.... 1 25 3 50
Brass, " 1 50 3 50
Malleable, dis. 25 per cent.

GALVANIZED PAILS.
Dufferin pattern pails, dis. 45 p.c.
Flaring pattern, discount 45 per cent.
Galvanized washtubs, discount 45 per cent.

PIECED WARE.
Discount 40 per cent. off list, June, 1899.
10-qt. flaring sap buckets, dis. 45 p.c.
6, 10 and 14-qt. flaring pails, dis. 45 p.c.
Creamer cans, dis. 45 p.c.

PICKS.
Per doz.... 6 00 9 00

PICTURE NAILS.
Porcelain head, per gross.... 1 75 3 00
Brass head " 0 40 1 00

PICTURE WIRE.
Tin and gilt, discount 75 p c.

PLANES.
Wood, bench, Canadian dis. 50 per cent. American dis. 50.
Wood, fancy Canadian or American 7½ to 40 per cent.

PLANE IRONS.
English, per doz.... 2 00 5 00

PLIERS AND NIPPERS.
Button's Genuine per doz pairs, dis. 37½ 40 p.c.
Button's Imitation, per doz.. 5 00 9 00
German, per doz.... 0 60 2 60

PLUMBERS' BRASS GOODS.
Compression work, discount, 60 per cent.
Fuller's work, discount 65 per cent.
Rough stops and stop and waste cocks, discount, 60 per cent.
Jenkins disk globe and angle valves, discount, 55 per cent.
Standard valves, discount, 60 per per cent.
Jenkins' radiator valves discount 55 per cent.
" " " standard, dis., 60 p.c.
Quick opening valves discount, 60 p.c.
No. 1 compression bath cock.... 2 00
No. 4 " " " 2 00
No. 7, Fuller's.... 2 50
No 4½, " 3 00

POWDER.
Velox Smokeless Shotgun Powder.
100 lb. or less.... 0 85
1,000 lb. or more.... 0 80
Net 30 days.

PRESSED SPIKES.
Discount 20 to 25 per cent.

PULLEYS.
Hothouse, per doz.... 0 55 1 00
Axle.... 0 22 0 33
Screw.... 0 27 1 00
Awning.... 0 35 2 50

PUMPS.
Canadian cistern.... 1 80 3 60
Canadian pitcher spout.... 1 40 2 10

PUNCHES.
Saddlers', per doz.... 1 00 1 85
Conductors', " 9 00 15 00
Tinners' solid, per set.... 0 00 0 72
" hollow, per inch.... 0 00 1 00

RANGE BOILERS.
Galvanized, 3 gallons.... 7 00
" 35 " 8 25
" 40 " 9 50
Copper, 30 " 22 00
" 35 " 26 00
" 40 " 30 00
Discount off Copper Boilers 10 per cent.

RAKES.
Cast steel and malleable, 50, 10 and 5 p.c.
Wood, 25 per cent.

RASPS AND HORSE RASPS.
New Nicholson horse rasp, discount 60 to 60 and 10 p.c.
Globe File Co.'s rasps, 60 and 10 to 70 p.c.
Heller's Horse rasps, 50 to 50 and 5 p.c.

RAZORS.

	per doz.	
Geo. Butler & Co.'s....	8 00	18 00
Boker's....	7 50	11 00
" King Cutter....	12 50	50 00
Wade & Butcher's....	3 60	10 00
Theile & Quack's....	7 00	12 00
Elliot's....	4 00	18 00

REAPING HOOKS.
Discount, 50 and 10 per cent.

REGISTERS.
Discount,.... 40 per cent.

RIVETS AND BURRS.
Iron Rivets, black and tinned, discount 60 and 10 per cent.
Iron Burrs, discount 55 per cent.
Extras on Iron Rivets in 1-lb. cartons, ½c. per lb.
Extras on Iron Rivets in ½-lb. cartons, 1c. per lb.
Copper Rivets & Burrs, 35 and 5 p.c. dis. and cartons, 1c. per lb. extra, net.
Extras on Tinned or Coppered Rivets ½-lb. cartons, 1c. per lb.

RIVET SETS
Canadian, dis. 35 to 37½ per cent.

ROPE ETC.

	Sisal.	Manila.
7-16 in. and larger, per lb	10	13½
⅜ in....	11	14½
¼ and 5-16 in....		15½
Cotton, 3-16 inch and larger		16½
" 5-32 inch....		21½
" ⅛ inch....		22½
Russia Deep Sea....		15½
Jute....		8
Lath Yarn....		9½
New Zealand Rope....		10

RULES.
Boxwood, dis. 75 and 10 p.c.
Ivory, dis. 37½ to 40 p.c.

SAD IRONS. per set.
Mrs. Potts, No. 55, polished.... 62½
" No. 50, nickle-plated.... 67½

SAND AND EMERY PAPER.
Dominion Flint Paper, 47½ per cent.
B & A. sand, 40 and 5 per cent.
Emery, 40 per cent.
Garnet (Rurton's), 5 to 10 p.c. advance on list.

SAP SPOUTS.
Bronzed iron with hooks, per doz... 9 50

SAWS.
Hand Disston's, dis. 12½ p.c.
S. & D., 40 per cent.
Crosscut, Disston's, per ft.... 35 0 55
S. & D., dis. 35 p.c. on Nos. 2 and 3.
Hack, complete, each.... 0 75 2 75
" frame only.... 0 75

SASH WEIGHTS.
Sectional, per 100 lbs.... 2 75 3 00
Solid, " 2 00 2 25

SASH CORD.
Per lb.... 0 23 0 30

SAW SETS.
"Lincoln," per doz.... 6 50

SCALES.
B. S. & M. Scales, 45 p.c.
Champion, 65 per cent.
Fairbanks Standard, 35 p.c.
" Dominion, 55 p.c.
" Richelieu, 55 p.c.
Chatillon Spring Balances, 10 p.c.
Warren Champion 65 p.c.
" Standard 45 p.c.

SCREW DRIVERS.
Sargent's per doz.... 0 65 1 00

SCREWS
Wood, F. H., bright and steel, 87½ and 10 p.c.
Wood R. H., " dis. 82½ and 10 p.c.
" F. H., brass, dis. 80 and 10 p.c.
Wood, R. H., " dis. 75 and 10 p.c.
" F.H., bronze, dis. 75 p.c.
" R.H. " 70 p.c.
Drive Screws, 87½ and 10 per cent.
Bench, wood, per doz.... 3 25 4 00
" iron, " 4 25 5 75
Set, Case hardened, 60 per cent.
Square Cap, 50 and 5 per cent.
Hexagon Cap, 45 per cent.

SCYTHES.
Per doz. net.... 9 00

SCYTHE SNATHS.
Canadian, dis. 45 p.c.

SHEARS.
Bailey Cutlery Co., full nickeled, dis. 60 p.c.
Seymour's, dis. 50 and 10 p.c.

SHOVELS AND SPADES.
Canadian, dis. 40 and 5 per cent.

SINKS.
Steel and galvanized, discount 45 per cent.

SNAPS.
Harness, German, dis. 25 p.c.
Lock, Andrews'.... 4 50 11

SOLDERING IRONS.
1, 1½ lb., per lb.... 0 37
2 lb. or over, per lb.... 0 34

SQUARES.
Iron, No. 493, per doz.... 2 40 2 55
" No. 494, " 3 25 3 40
Steel, dis. 50 and 5 to 50 and 10 p.c., rev. list.
Try and bevel, dis. 50 to 52½ p.c.

STAMPED WARE.
Plain, dis. 75 and 12½ p.c. off revised list.
Retinned, dis., 75 p.c. off revised list.

STAPLES.
Galvanized.... 3 50 4 00
Plain.... 3 25 3 75
Coopers', discount 45 per cent.
Poultry netting staples, 40 per cent.

STOCKS AND DIES.
American dis. 25 p.c.

STONE.

		Per lb.
Washita....	0 28	0 60
Hindostan....	0 06	0 07
" slip,....	0 09	0 09
Labrador....	....	0 13
" Axe....	....	0 15
Turkey....	....	0 50
Arkansas....	0 00	1 50
Water-of-Ayr....	0 00	0 10
Scythe, per gross	3 50	5 00
Grind, per ton	15 00	18 00

STOVE PIPES.
5 and 6 inch Per 100 lengths.... 7 00
7 inch " " 7 50

ENAMELINE STOVE POLISH.
No. 4—3 dozen in case, net cash.... $4 80
No. 6—3 dozen in case, " 8 40

TACKS BRADS, ETC.

	Per cent.
Strawberry box tacks, bulk....	75 & 10
Cheese-box tacks, blued....	80 & 12½
Trunk tacks, black and tinned....	85
Carpet tacks, blued....	80 & 15
" " tinned....	80 & 20
" " (in kegs)....	40
Cut tacks, blued, in dozens only....	80
" ½ weights....	60
Swedes, cut tacks, blued and tinned—	
In bulk....	80 & 10
In dozens....	75
Swedes, upholsterers', bulk....	85, 12½ & 12½
" brush, blued & tinned, bulk....	70
" gimp, blued, tinned and japanned....	75 & 12½
Zinc tacks....	35
Leather carpet tacks....	55
Copper tacks....	50
Copper nails....	52½
Trunk nails, black....	65 and 5
Trunk nails, tinned....	65 and 10
Clout nails, blued....	65 and 5
Chair nails....	35
Patent brads....	40
Fine finishing....	40
Picture frame points....	10
Lining tacks, in papers....	10

Lining tacks, in bulk		15
" " solid heads, in bulk		75
Saddle nails in papers		10
" " in bulk		15
Tufting buttons, 22 line, in dozens only		60
Tin capped trunk nails		15
Zinc glazier's points		5
Double pointed tacks, papers		90 and 10
" " " bulk		40

TAPE LINES.

English, ass skin, per doz	2 75	5 00
English, Patent Leather	5 50	9 75
Chesterman's each	0 90	2 85
" steel, each	0 80	8 00

THERMOMETERS.

Tin case and dairy, dis. 75 to 75 and 10 p.c.

TRAPS. (Steel.)

Game, Newhouse, dis. 25 p.c.
Game, H. & N., P. S. & W., 65 p.c.
Game, steel, 72½, 75 p.c.

TROWELS.

Disston's discount 10 per cent.

German, per doz	4 75	6 00

S. & D., discount 35 per cent.

TWINES.

Bag, Russian, per lb	....	0 27
Wrapping, cotton, per lb	0 22	0 26
Wrapping, mottled, per pack	0 50	0 60
Wrapping, cotton, 3-ply		0 20
" " 4-ply	....	0 26
Mattress, per lb	0 33	0 45
Staging, "	0 27	0 35

VISES.

Hand, per doz	4 00	5 00
Bench, parallel, each	2 00	4 50
Coach, each	6 00	7 00
Peter Wright's, per lb	0 12	0 13
Pipe, each	5 50	9 00
Saw, per doz	6 50	13 00

ENAMELLED WARE.

White, Princess, Turquoise, Blueand White, discount 50 per cent.
Diamond, Famous, Premier, 50 and 10 p.c.
Granite or Pearl, Imperial, Crescent, 50, 10 and 10 per cent.

WIRE.

Brass wire, 50 to 50 and 2½ per cent. off the list.

Copper wire, 45 and 10 per cent. net cash 30 days, f.o.b. factory.

Smooth Steel Wire, is quoted at the following net selling prices:

No. 6 to 8 gauge	$2 90
" 9 "	2 80
" 10 "	2 87
" 11 "	2 90
" 12 "	2 93
" 13 "	3 15
" 14 "	3 37
" 15 "	3 50
" 16 "	3 65

Other sizes of plain wire outside of Nos. 9, 10, 11, 12 and 13, and other varieties of plain wire remain at $2.80 base with extras as before. The prices for Nos. 9 to 13 include the charge of 10c. for oiling. Extras net per 100 lb.: Coppered wire, 60c.—tinned wire, $2—oiling, 10c.—special hay-bailing wire, 30c.—spring wire, $1— best steel wire, 75c.—bright soft drawn, 15c.—in 50 and 100-lb. bundles net, 10c.—in 25-lb. bundles net 15c.—packed in casks or cases, 15c.—bagging or papering, 10c.

Fine Steel Wire, dis. 17½ per cent. List of extras: In 100-lb. lots: No. 17, $5—No. 18, $5.50—No. 19, $6—No. 20, $6.65—No. 21, $7—No. 22, $7.30—No. 23, $7.65—No. 24, $8—No. 25, $9—No. 26, $9.50—No. 27, $10—No. 28, $11—No. 29, $12—No. 30, $13—No. 31, $14—No. 32, $15 No. 33, $16—No. 34, $17. Extras net—tinned wire, Nos. 17-25, $2—Nos. 26-31 $4—Nos. 32-34, $6. Coppered, 5c.—oiling, 10c.—in 25-lb. bundles, 15c.—in 5 and 10-lb. bundles. 25c.—in 1-lb. hanks, 50c.—in ½-lb. hanks, 75c.—in ¼-lb. hanks, $1—packed in casks orcases, 15c.—bagging or papering, 10c.

Galvanized Wire, per 100 lb.—Nos. 6,7,8, $3 50 to $3 85—No. 9, $2.85 to $3.15—No. 10, $3.60 to $3.95—No. 11, $3.70 to $4.10—No 12, $3 to $3 30—No. 13, $3.10 to $3.40—No. 14, $4.10 to $4.50—No. 15, $4.60 to $5.05—No. 16, $4.85 to $5 35. Base sizes, Nos. 6 to 9, $2.57½ f.o.b. Cleveland.

Clothes Line Wire, solid 7 strand, No. 17 $4.25; No. 18, $2.65; No. 19, $2.35, f.o.b Hamilton, Toronto, Montreal.

WIRE FENCING.

	F.O.B. Toronto
Galvanized barb	3 15
Galvanized, plain twist	3 15

Galvanized barb, f.o.b. Cleveland, $2.82½ in less than carlots, and $2.70 in carlots.

WIRE CLOTH.

Painted Screen, per 100 sq. ft., net .. 1 35

WASTE COTTON.

	per lb.
Colored	4½ to 5
White, according to quality	6½ to 7¾

500-lb bale lots shaded.

WRENCHES.

Acme, 35 to 37½ per cent.
Agricultural, 60 p.c.
Coe's Genuine, dis. 70 to 25 p.c.

Towers' Engineer, each	2 00	7 00
" S., per doz	5 80	6 00
G. & K.'s Pipe, per doz	....	3 40
Burrell's Pipe, each	....	3 00
Pocket, per doz	0 25	2 90

WRINGERS.

Leader	per doz.	$30 0	33 00
Royal Canadian	"	26 00	28 00
Royal American	"	26 00	28 00
Sampson	"	....	30 00

Terms 4 months, or 3 p.c. 30 days.

WROUGHT IRON WASHERS.

Canadian make, discount, 40 and 5 per cent.

CANADIAN HARDWARE AND METAL MERCHANT

The Weekly Organ of the Hardware, Metal, Heating, Plumbing and Contracting Trades in Canada.

VOL. XIII. MONTREAL AND TORONTO APRIL 27, 1901. NO. 17

FOLDING POCKET LUNCH BOX

OPEN.

"THE PERFECTION."

Covered outside with black cloth (waterproof).

CLOSED.

No complicated parts.
Easy to operate.
One move opens it.
One move closes it.
Can be washed.

THE THOS. DAVIDSON MFG. CO., Limited, MONTREAL, QUE.

Galvanized Sheets

From stock or for Importation.

SAMUEL, SONS & BENJAMIN, - - LONDON AND LIVERPOOL, ENGLAND

M. & L. Samuel, Benjamin & Co.

General Importers and Exporters and Metal Merchants

27 Wellington Street West, - - TORONTO, ONT.

LEWIS BROS. & CO.

Wholesale Hardware, MONTREAL.

LAWN MOWERS, LAWN RAKES, Etc.

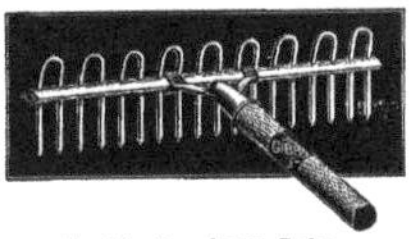

No. 14—Boss Lawn Rake.

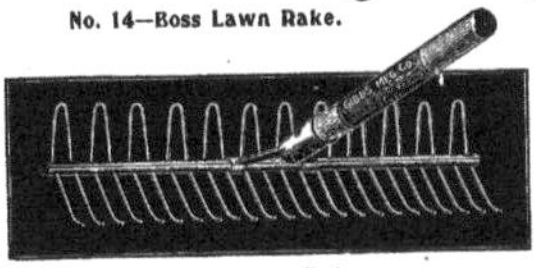

Gibbs' Lawn Rake.

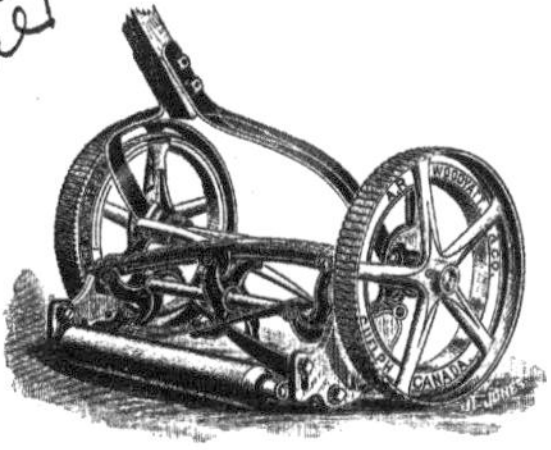

All Styles and Sizes of **LAWN MOWERS**

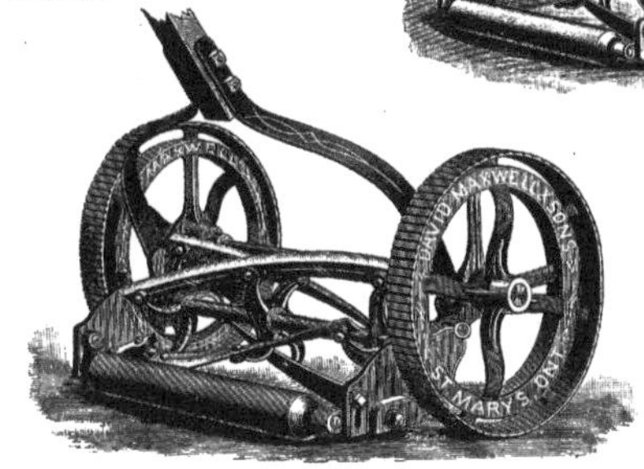

Write for Complete Catalogue of Lawn Requisites.

Lewis Bros. & Co.,

SPECIAL ATTENTION TO MAIL ORDERS.

MONTREAL.

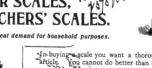

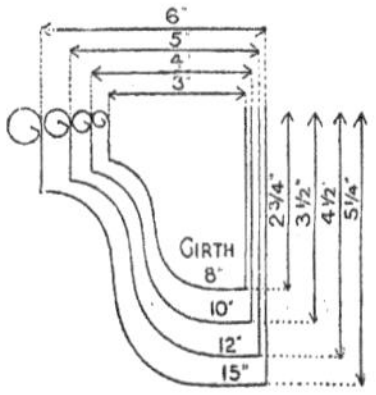

6"
5"
4"
3"
2¾"
3½"
4½"
5¼"
GIRTH
8"
10"
12"
15"

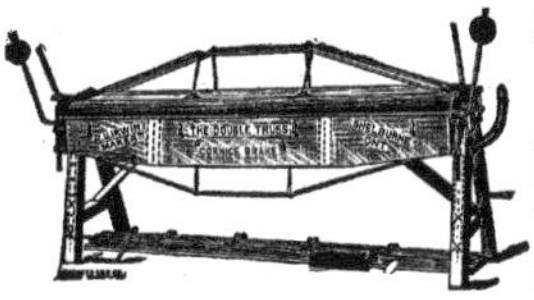

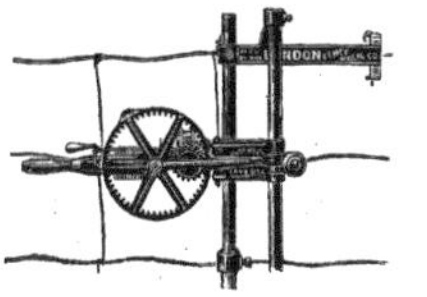

HARDWARE AND METAL

VOL. XIII. MONTREAL AND TORONTO, APRIL 27, 1901. NO. 17.

President,
JOHN BAYNE MacLEAN,
Montreal.

THE MacLEAN PUBLISHING CO.
Limited.

Publishers of Trade Newspapers which circulate in the Provinces of British Columbia, North-West Territories, Manitoba, Ontario, Quebec, Nova Scotia, New Brunswick, P.E. Island and Newfoundland.

OFFICES

MONTREAL - - - - - - 232 McGill Street, Telephone 1255.
TORONTO - - - - - - 10 Front Street East, Telephone 2148.
LONDON, ENG. - - - - 109 Fleet Street, E.C., W. H. Miln.
MANCHESTER, ENG. - - - 18 St Ann Street. H. S. Ashburner,
WINNIPEG - - - - Western Canada Block. J. J. Roberts,
ST. JOHN, N.B. - - - No. 3 Market Wharf, J. Hunter White.
NEW YORK. - - - - - - 176 E. 88th Street.

Subscription, Canada and the United States, $2.00.
Great Britain and elsewhere - - - - 12s.

Cable Address { Adscript, London. Adscript, Canada.

Published every Saturday.

OUR TRADE WITH RUSSIA.

A REPRESENTATIVE of the Canadian Pacific Railway, Mr. W. Whyte, is shortly to leave for Russia to take a trip over the trans-Siberian railway with a view to making arrangements to try and expand trade between Canada and Russia.

We shall await the result of the Canadian Pacific Railway Co.'s efforts with interest. Canada's trade with Russia at present is small, but it is increasing. And, as the United States is in bad odor with the Muscovite at present, the opportunity for increasing our trade with him should be all the better.

According to a newspaper report, one of the lines of Canadian manufactured goods which it is proposed to push in Russia is agricultural implements. These are already our chief article of export to that country, $35,599 worth being sent there in 1900. In all, the exports of iron and steel and manufactures thereof aggregated $57,033. As this bears a proportion of over 80 per cent. to the total exports of all kinds, it is obvious that what else we send to Russia is insignificant.

The next largest item to those already enumerated is ships, of which we sent $11,688 worth. In 1899 we exported over $10,000 worth of coal to Russia, but last year the amount was only $432. Of wood and manufactures of we sent $9,331 worth, but our exports in 1900 were nil. In fact, outside agricultural implements and other manufactures of iron and steel there is no line in which we do a steady export trade with Russia. In these particular lines our trade during the past three years was as follows:

	1898.	1899.	1900.
Agricultural implements...	$ 9,723	$11,362	$35,599
All other iron and steel and manufactures of....	8,041	16,872	21,434
Total	$17,764	$28,234	$57,033

Our aggregate trade with Russia last year was $95,217, of which $70,558 was exports and $24,659 imports. Our imports consist largely of furs, as will be gathered from the fact that the value of the furs brought in last year was $21,718.

Under the treaty of 1859 between Great Britain and Russia the reciprocal most favored nation stipulations are applicable to Canada. According to a report from the United States consuls at St. Petersburg and Odessa, the new rates imposed by the Russian Government on certain imports from the United States, notably, machinery, in retaliation for the imposition of the countervailing duty on sugar, show an increase of from 20 to 30 per cent.

A MATTER FOR GRATIFICATION.

ONE of the things which the press had to deplore a few years ago was the depreciation in the value of the farm lands of Ontario. The recent meeting of the Canada Company in London, Eng., not only brings this fact to our mind, but also reminds us that the condition of the farm lands in Ontario is more satisfactory to-day than it was five or six years ago.

At the meeting in question, the chairman, in referring to the land disposed of, said that there was not, in the whole quantity of land disposed of, a single instance of an acre being sold below the valuation of 1894. Of lots valued in 1894 below the average valuation of $9.32 an acre, there were disposed of 4,328¼ acres (valued in 1894 at $4.55 an acre) realizing $7 an acre, an increase of $2.45 an acre, or 53¼ per cent. Of lots valued in 1894 above the average valuation of $9.32 an acre, there were disposed of 2,403 acres (valued in 1894 at $19.62 an acre), realizing $24.54 an acre, an increase of $4.92 an acre, or 25 per cent. Lots redisposed of—3,148¼ acres, which reverted to the company at $13.73 an acre—realized $14.58 an acre, an increase of $0.85 an acre, or 6.15 per cent. As a general result, 9,880 acres were disposed of at $13.68 an acre, an increase over the 1894 valuation of $2.54 an acre, or not less than 22¾ per cent.

The shareholders of the company were naturally gratified, but there is food for gratification for those whose interests are in the country and not in the company.

It is because the number of gullible people is so great that the efforts of the ubiquitous promoter in floating visionary schemes are so successful.

THE PRICES OF A MONOPOLY.

MATTERS are continually cropping up, which show how the business men of this country are in the power of the Standard Oil Co. A year or more ago it was the power it exercised over the railways to the disadvantage of the Canadian manufacturers and merchants who tried to escape from the disabilities imposed upon them by the Oil Trust by importing oil from the independent concerns in the United States. Then, there was the question of the supply of crude oil for fuel purposes, the arbitrary action of the Standard Oil Co. compelling many of the manufacturers of this country who had put oil burning fixtures under their boilers to again revert, at great expense, to coal as fuel.

During the last few days another phase of the irksomeness of the Standard Oil Co.'s methods of dealing with those in Canada who are dependent upon it for their supplies has been drawn to our attention. We have reference to the arbitrary and exorbitant price of benzine.

As most people are aware, benzine is one of the articles of raw material used in several of the manufacturing industries in Canada. At one time there were in Canada five refineries from which benzine could be purchased. Now there is but one. That one is at Sarnia, and is the property of the Standard Oil Co. All the others were bought up by the latter and then demolished, their plant being sold for scrap. It thus practically controls the Canadian market; and, as is usual with monopolies, charges monopolistic prices.

In the United States, manufacturers can purchase benzine at less than 5½c. per gal. in large quantities; in Canada, manufacturers have got to pay 13½c. per gal. And it is a peculiar fact that the price at which Canadian refined benzine shall sell in Canada is determined by the officials of the Standard Oil Co. at the head office in New York.

The import duty on benzine is 5c. per gal. Most manufacturers, from necessity or otherwise, feel they are well off if they can add to the selling price of their products 25 per cent. of the protection they enjoy. But The Standard Oil Co. is not satisfied with the addition of even the whole of the duty. Its idea is the cost, plus the duty, plus what its monopolistic powers enable it to put on.

Wherein lies the remedy for this state of affairs is somewhat of a problem. One thing is certain, a remedy must in time be found somewhere. The removal of the duty would possibly give some relief; but such powerful concerns as the Oil Trust seem to be invulnerable to even the enactments of Parliaments and Congresses. And our faith has, in consequence, become so weak that we even fear that our representative institutions will be unable to come to the rescue. Were the Parliament of our country dominated less by concern for party exigencies and more by sound business commonsense, hope for a remedy from that quarter might be stronger.

When there is a monopoly it should be controlled by the country; if it is not, it will control the country. And that is what the Standard Oil Co. appears to be doing.

We understand the benzine question is to be brought before the Canadian Manufacturers' Association. It is to be hoped it will not dismiss it as it did the coal oil question a week or two ago.

CANADA'S EXPORT BARLEY TRADE.

WHEN the McKinley tariff of 1890 shut Canadian barley out of the United States market it was the general opinion that the barley-growing industry in this country had received its quietus, for practically all our exports of that particular cereal went to the breweries of the neighboring Republic. Take, for example, the exports during the year preceding the inauguration of the McKinley tariff. Their total to all countries was nearly 10,000,000 bushels, of which over 9,900,000 bushels alone went to the United States. Only 6,312 bushels went to Great Britain; and in 1888 the quantity exported to the latter country was but 1,687 bushels.

Some hope was entertained, after the advent of the McKinley tariff, that a trade in barley might be developed with the Mother Country. With this end in view, a good deal of two-rowed barley was sown. And by 1892 Canada was able to export 2,439,959 bushels of barley to Great Britain. But the results were not satisfactory, due to some extent to the dishonest practice of mixing other descriptions of barley with the two-rowed kind.

By 1896 the quantity exported to Great Britain was down to 45,769 bushels, while our total to all countries was only 840,725 bushels, against nearly 10,000,000 bushels up to the time the McKinley tariff came into existance. In 1899 the results were still worse, for the total exports were only 238,948 bushels. But again was exemplified the old saying, that the darkest hour is just before the dawn.

In 1900, Canadian barley began to get a better footing in the British market, and in that year we shipped there 1,753,135 bushels, against 116,131 bushels in 1899. The United States took about 40,000 bushels more than they had in previous years, and the exports to "other countries" jumped from 443 bushels in 1899 to 238,679 in 1900, while the sum total of our exports was 100,000 bushels larger than in any previous year since 1892.

The revival in the export barley trade last year has not forsaken us this year. The returns for the eight months show this, the total being 1,666,294 bushels. During the eight months, Great Britain has taken 1,336,448 bushels, the United States 182,022 bushels, Belgium 144,394 bushels and "other countries" 3,430 bushels. The demand for barley on British export account is still active.

We are still a long way from the export trade which existed up to 1890, but the outlook for the barley industry is certainly brighter than it has been at any time since the McKinley tariff so badly crippled it.

As Great Britain imports something like 360,000,000 bushels of barley per annum, it is evident we have an unlimited field in which to develop our export trade in this particular cereal.

The United States, it might be pointed out, exports over 4,000,000 bushels of barley to Great Britain annually.

MEETINGS POSTPONED.

As this is a busy season for the retail hardwareman, the executive of the Montreal Retail Hardware Association have decided to postpone all meetings till Wednesday, June 5. A progressive plan of campaign will then be mapped out and operated upon.

BRITISH BUDGET AND CANADA'S COAL TRADE.

NOT since 1846, when Sir Robert Peel abolished the corn laws, has a tariff of such importance and such wide interest been brought down in the British House of Commons as that brought down by Sir Michael Hicks-Beach on the 18th inst. The tariff of 1846 was the climax of the agitation for free trade which had been so vigorously carried on, led by Cobden, Bright, Villiers and others, in the years preceding. What the tariff brought down a few days ago is the precursor of, is a subject for speculation.

Although the import duties of 4s. 2d. per cwt. on refined sugar, of 2s. per cwt. on molasses and syrup, of 1s. 8d. per cwt. on glucose and an export duty of 1s. per ton on coal are born of the necessities of the revenue and not of the principles of protection to any of the industries concerned, yet one cannot ignore the fact that they nevertheless contain within themselves the seeds of protection which may in time produce a system labelled and known by that name.

"There is," says Justin McCarthy in his "History of Our Times," "no more chance of a reaction against free trade in England than there is of a reaction against the rule of three." Under free trade England has prospered enormously; and to change from it to protection might be an unwise thing, yet most people will consider that Mr. McCarthy's statement too positive. The sugar refineries, which only a few years ago supplied 80 per cent. of the home requirements, now only supply about 40 per cent. For some time the refiners have been pleading for protection against the bounty-fed sugars of Europe. And they have had quite a respectable support from commercial men and financial papers. It has not always been known by the term "protection"; "countervailing duties" has been the common expression, in favor of which the late Mr. Gladstone expressed himself in 1888. Then there is the iron and steel industry, in regard to which the increasing competition of the United States and Germany is creating quite a little alarm. Naturally, with the alarm has come a desire, lightly expressed so far, for tariff protection. And those who believe that in a Customs tariff is the panacea for foreign competition will certainly not be discouraged by the new tariff.

The protection contained in the new tariff may not be larger than a grain of mustard seed, but it is a seed, and we may depend upon it that there are those in Great Britain who will endeavor to cultivate it. As we can be just as certain that the orthodox free traders, who are in a large majority, will endeavor to destroy the seed, we may look for the reentering of the tariff into the political arena.

In the meantime, what most people will be interested in is the effect of the tariff in its commercial ramifications.

Of the articles which have come in for a change under the new tariff, none probably create a wider interest than coal, every manufacturer, as will as every householder, in Great Britain being interested to some extent in the article.

We have no access to figures giving the output of coal in Great Britain in 1900, but the quantity in 1899 was over 220,000,000 tons, the largest on record. But, notwithstanding the largeness of the output, the demand has exceeded it. Germany, Denmark, Holland, France, Spain and other countries all increased their demand last year for British coal, while a further tax upon the resources of the country's coal mines was the heavy purchases of the British Admiralty, on account of their different coaling stations. In consequence of the heavy foreign demand, the exports have sprung up to over 46,000,000 tons and prices appreciated 50 to 90 per cent. over 1899. For some time a powerful section of the trade and daily press has been advocating the imposition of an export duty, for the three-fold purpose of (1) raising revenue to assist in meeting the expenditure on war account; (2) for conserving the home supply; (3) for cheapening the cost to the home manufacturers and railways, which were heavily handicapped by the high price of coal.

"We do not think," said The Iron and Steel Trades Journal in an issue last September, "that a moderate duty on coal would very considerably reduce exports, though it would act as a certain check on the same, but it would put our manufacturers and the public at large in a better position in buying their coal than our foreign consumers are." The paper just quoted considered 2s. per ton would be a moderate duty. The Chancellor of the Exchequer has made it half that sum; but the coal-mine owners are anything but pleased. As some of them last year paid dividends of 33⅓ and 50 per cent. there is not much likelihood of their being ruined by the imposition of the export duty.

The export coal trade of Great Britain exceeds that of any other country. That of no other country even approximates it. Take the exports of the leading coal producing countries in Europe for 1899 as an example, and we find the following: Great Britain, 43,108,000 tons; Germany, 13,943,000 tons; France, 1,229,000 tons; Belgium, 4,563,000 tons.

Naturally, the tendency of the export tax will be to restrict the shipment of British coal to foreign customers. At present, with the demand as active as it is, it may not have any perceptible effect, but the less active the foreign demand is the more marked will be the influence of the tax.

None of the coal-producing countries of Europe are able to supply even their home trade. The United States has, within the last year or two, become the largest coal producing nation, and, owing to the extraordinary high prices ruling in Great Britain, has lately been exporting coal to Europe. Whether the high prices in Great Britain will be a greater advantage to the coal exporting trade of the United States than the British export tax remains to be seen. But of the exporting coal countries, the United States, nevertheless, appears to us to be likely to gain the most by the export tax on British coal.

Canada should reap some advantage, and more in the distant future than in the near future.

We have over 97,000 square miles of coal lands in Canada, not including the known but unexplored areas in the far north, and are producing over 5,000,000 tons a year, and yet the demand is so good that we are told that the mines in Nova Scotia cannot increase their export trade until the facilities for increasing the production have been secured.

The iron works at Sydney will demand a great deal of coal from the Nova Scotian mines, while New England, notwithstanding the duty, draws quite a supply from the same source. The output of the British Columbian coal mines goes largely to California.

Canada sends some coal to France, Germany and other European countries that depend largely upon Great Britain for their supplies, and in time she may increase that trade, but there is not much likelihood of it in the immediate future.

The quantity of Canadian coal exported last year was the largest on record, being 1,641,031 tons. Of that quantity, 23,097 tons went to Great Britain, 108,462 tons to Newfoundland, 34,829 tons to Hawaii, 1,430,437 tons to the United States, 5,982 tons to Australia, 200 tons to Denmark, 1,056 tons to France, 1,307 tons to Germany and 386 tons to Holland. In all, we exported to 19 different countries.

THE PRIMARY PRINCIPLES OF FOREIGN EXCHANGE.

HAVING in mind the primary principles of foreign exchange, we will consider their application to "sterling" exchange and how by their means the rates for buying and selling are derived.

In the first place, we will take the rate for demand exchange and must start from the foundation fact that the par value of £1 sterling is $4.86656, because the sovereign or pound sterling contains 113,001597 grains of pure gold, which in gold coin of the fineness of that of the United States is equal to $4.86656.

A QUESTION OF TRANSPORTATION.

If gold could be transported as easily and at no greater cost or risk than attends the sending of bills of exchange, the rates of exchange between countries on a gold basis would vary only as influenced by the rate of interest; but we know considerable expense attends the shipments of gold, and if the supply of mercantile exchange is not sufficient to cover the amount of sterling exchange issued by the bankers this expense must be met, for gold being the medium of settlement it must be procured and shipped.

To ship to London from New York £100,000 in gold would involve the following expenditure at least:

£100 000, at par	$486 656 66
Freight, say 1/4 of 1 p. c.	608 32
Insurance, say 1-16 of 1 p. c	304 16
Boxing, cartage, etc., say 1 32 of 1 p.c.	152 08
	$487,721 22

Thus, if the banker in New York received $488,000 for his demand draft of £100,000 on London, he is in a position to send either sovereigns or their equivalent in gold, to meet his obligation there.

It is evident, then, that if gold has to be shipped, $487,721.22 at least must be obtained for the demand draft; but as a profit has to be made and certainly one-sixteenth of one per cent., which would only just make it pay, we will add that to the above and we get $488,025.38 so then, when the rate for demand exchange is quoted by the banker as 4.88 gold is said to be at the "shipping point," because at this rate the banker can under ordinary and normal conditions, procure the gold and pay costs attending shipment of same to England and so provide the necessary funds to meet demand exchange sold at that rate.

As a rule, shipments of gold take place only when the supply of mercantile paper is short of the demand and the rate of exchange consequently rises to the shipping point; yet it does happen that shipments of gold may be remunerative when the rate for demand exchange is below shipping point.

GOLD RESERVES.

This condition might arise when, for some reason, it was found necessary for the English bankers to increase their gold reserves, and the Bank of England, on whom they rely for their gold, might have to offer a premium on gold to attract enough to meet the requirements, thus apparently enhancing the value of the metal.

Gold, in this case, is the requisite; and since there may be such a supply of commercial sterling exchange available in New York that demand bills could be readily obtained at say 4.86 3-4, yet this exchange, while perfectly serviceable to offset a demand draft on London, because it will supply the demand in gold there, is of no value for the purpose of adding to the gold reserves of that country, since it represents payment for merchandise and in fact tends to diminish the English gold supply; so the Bank of England will offer a premium to attract gold.

If we presume that such a premium was 3-8ths of one per cent. or seven shillings and six pence on the £100, then the equivalent of £100,000, with the premium added, would be, at par, $488,481.60. In other words, the rate obtained against the gold shipment would be nearly 4.88 1-2, while the rate for sterling exchange on demand stood at 4.86 3-4 in New York.

If this premium were offered for one month's use of the gold, and the banker sold his demand exchange against it in a month at $4.86 3-4, the transaction, aside from other contingencies, would only pay the exporting banker, if the rate of interest in the home market was under two per cent., for it would be investing $487,721.22 (the £100,000 exported, plus the cost as previously stated) at about two per cent. per annum, as follows:

£100,000, plus premium of £375 at 4.86 3/4, equals	$488,575 31
Deduct £100,000 and cost of exporting.	487,721 22
Margin of profit	$ 854 09
Interest obtainable in the home market on $48,772 22, for one month at 2 p.c.	$ 812 87

It is to be presumed that some other inducement would be required before a banker would ship gold under the above conditions, since it is evident that the premium on the gold will not pay him unless the rate obtained in the home market for money was less than two per cent.; we have, however, assumed that to be the prevailing rate.

OTHER MEANS OF ATTRACTING GOLD.

The other means of attracting gold, and that which most frequently operates in this way, is a sufficiently higher rate of interest in the foreign market but the higher rate must prevail for a sufficiently long period to make the transaction profitable; and in such case the gold shipment may be regarded as a loan and against this at some future date, the end of the long period to wit, the banker should sell his demand exchange at a rate sufficiently high for him to be recouped in New York without reducing the profit earned, or better still, make a further profit on the exchange sold against it.

EXCHANGE AGAINST GOLD SHIPMENTS.

In considering the subject of exchange against the gold shipment we will do so from various points of view that appeal to the banker, because profitable employment of money in his business and the gain from loan operations is frequently associated with his exchange transactions and each has a direct influence on the other.

Assuming, then, that in addition to the premium of 3-8th of one per cent. on the £100,000 gold exported a rate of interest one-half per cent. above that available in the home market be offered for a period of three months it becomes evident that we have a time loan and consequently we might operate "time exchange" against it but as we are at present considering demand exchange, the other aspect will be referred to in a later article.

We will suppose that the inducement to ship gold is an increased interest rate of the half of one per cent. and we will estimate the rate for time loans in New York as being three per cent. and for call money two per cent. per annum on the highest class of securities, these being about what would prevail in ordinary quiet state of the money markets in New York and London, although it has been the rule in the past for higher rates to prevail in the American money market.

The loan to the Bank of England will be at the rate of 3 1-2 per cent. per annum and will net the following, viz.:

Amount of loan	£100,000 0 0
Add premium 3/8 of 1 p. c.	375 0 0
Add interest at 3 1/2 p. c. for three months	875 0 0
Thus the sum exported will have grown in 3 months to	£101,250 0 0

against which the banker can then sell his demand exchange; and, if the rate is as good as when he shipped the gold, it would net on

£101,250 sold at 4.86 3/4	$492,834 37
At an outlay of 3 months, of	487,721 22
Showing a profit of	5,113 15

or over 4 1-5 per cent. per annum, although the rate for loans in the home market was only 3 per cent.

Now, it would not be wise to presume that the rate at the end of three months will be as good as when this transaction was entered on, for the fact of shipping gold tends to put up the rate of interest in the home market, in which case the rate of exchange will fall unless the rate of interest in the foreign market goes up also; and should the rate of exchange to the importing point, the profits would be reduced to 3.15 per cent. per annum, which would still be on the right side but scarcely enough to warrant the business, so there must be some more certain way of assuring a profit than waiting the risky eventualities of three months, and this is set forth in a subsequent chapter.

THE RISE AND FALL OF EXCHANGE.

To fully explain the above statement relating to exchange "falling to the import point," we will refer to the fact set forth that the exporting point is arrived at when the rate of exchange equals the par value plus the costs of exporting, which amount of about 7-32 of one per cent. on the amount exported.

As the same expenses attend importing gold, it follows that when the rate of exchange for demand drafts on London is at par minus 7-23, which is $4.85 1-2, then the importing point is touched; so if the banker sold his £101,250 at $4.85 1-2 it would net only $491,568.75, and after recouping his first outlay of $487,721.22 his profit would be $3,847.53, or 3.15 per cent. per annum on the capital employed, as previously stated. The exporting and importing rates are here given at a minimum of profit. As a rule, the banker would consider a half cent higher or lower than these rates respectively as desirable for actual operations in gold.

Having explained the circumstances which cause the rate of $4.88 for demand exchange to be called the "shipping point" and $4.85 1-2 the "importing point," and indicate the conditions which are conclusive in producing these rates, it is advisable to remark on the rates that are between these points.

Ordinarily, in either buying or selling, one desires to get par if the business be in exchange or anything else, but particularly so when dealing with moneys; hence, if purchasing a demand draft on London for £100, we would hope to get it for $486.65 and some small addition as commission for the banker's services. This addition is from 1-32 on large amounts to 1-4 or 1-2 of one per cent. on very small sums, thus on £100 it would probably be $1.22; but, as has been explained, it is not always possible for the banker to cover his drafts on London at par, hence an addition to par value of $1.22 might not pay him for his services, so it follows that what is usual when dealing with internal exchange, of purchasing a draft and paying a commission, although the principle is the same, must in foreign exchange give place to a system of exchange rates which not only includes the banker's commission but takes into account demand and supply, interest and the value of the metals as represented in the coin; these cause fluctuations between the "exporting" and "importing" rates, hence the above rate of $4.8665, with $1.22 added, is quoted at $4.88, which includes banker's commission and all other contingencies; so that the purchaser of exchange from the banker has full benefit of low rates when exchange is under par and of high rates of selling exchange when the rate is over par.

Now, if the exporter can get par for the goods he has exported he will, as a rule, sell his exchange; but if there be more of such exchange offering than there is demand it follows that the banker will grade his rate accordingly and if the supply be very abundant the rate will fall to the importing point; or the contrary, rise to the exporting rate if the demand be in excess of the supply.

So demand and supply are the factors in exchange rates between what we will call the minimum and maximum rates, while the shipment of gold, either way, has a tendency to produce the par or medium rate.

INTEREST AND RATE OF EXCHANGE.

The rate of interest, however, has much to do in fixing the rate of exchange, even when the supply is abundant or scarce. For instance, if sterling exchange is abundant so far as its supply is produced from the balance of trade being in favor of this country, but the exporters, who have money in England against which they can issue exchange, prefer to leave the money there rather than take a rate much under par, because they are able to put it out at as good or a better rate of interest than can be obtained in the home market; or rather than sell their exchange at low rates here they might purchase goods in England on which a profit could be derived by importing or by shipping to some foreign land; hence these things act to protect the seller of merchandise from being obliged to sell his exchange at a price below the intrinsic value.

On the other hand it may suit the sellers of exchange to accept a low rate so that they may obtain a higher rate of interest in the home market (and this is most usual when exchange is low), or the sellers might be induced to accept a low rate for their exchange and use the proceeds to purchase for cash and thus gain more on their discounts than has apparently been lost by accepting less than par for the exchange.

So then, while demand and supply govern rates of exchange, the rates of interest at home and abroad react on these to govern demand and supply, causing exchange to fluctuate from day to day, and at times even from hour to hour and can be counted on with but little more certainty than the price of stocks and shares which themselves influence and are influenced by interest rates and the rates of exchange, because, like foreign bills, they are a form of international exchange.

If the interest rate be the same both in London and New York and the supply of sterling exchange equal to the demand, the prevailing rate on demand exchange would be $4.86-6 plus the banker's charge of from 1-32nd to 1-2 on sales by the banker, according to the amount of the transaction; and the same minus that charge on purchases; and the variations from these rates will be governed by the disparity between the demand and supply, these being themselves subject to the influence of the prevailing rates of interest in the home and foreign markets; a difference of one per cent. in either market being, as a rule, enough to attract gold.

No hard and fast rule can be laid down as applicable at all times, but all matters that affect the exchange market must be considered in arriving at the rate that should prevail; and while interest is the prime factor it is, as shown, but one of several in producing results on the rates of foreign exchange.—The Bookkeeper.

A SIGN OF EXPANSION.

The London Fence Machine Company, whose advertisement may be found in this issue, have recently added to their plant an improved coiling machine which will be in operation in a few days. They have bought heavily with the American mills and their warehouse and siding in London commands direct entry from Cleveland and Conneaut Harbor. The G. T. R., C. P. R., L. E. & D. R., and M. C. R. lines and their connections from London give every facility for reshipment after being coiled. This firm will guage their prices to protect the dealers as far as possible.

Their "London" Fence Machine has won golden opinions from those who have handled them and has made an unprecedented success in this line as a practical machine.

AN EXPLANATION.

R. S. Hannah, hardware merchant, Mitchell, Ont., writes:

R. S. Hannah, hardware dealer, Mitchell, Ont., is about closing up.

Re above, which appeared in your issue of April 20. This is misleading. The facts are: I have bought out a hardware business in London, and as soon as I sell my stock here I am going to move to London.

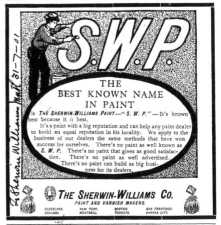

CANADIAN-ENGLISH SERVICE.

B. R. McAuley, of St. John, N.B., has interested British capitalists in a plan which means much if carried out. The proposal is to build and run a fleet of 10,000-ton steamers between Canada and England. To attract all British trade through Canadian ports, the plan is to ask the Canadian Government to increase the preference on British goods from 33⅓ to 35⅓ per cent., when such are imported through Canadian ports; also to grant a bonus of 1 per cent. on exports of Canadian meats, dairy products, etc., through Canadian ports.

CARD OF WEIGHTS.

The American Sheet Steel Company have just issued, for the convenience of their customers, a card of weights of "Apollo Best Bloom" galvanized sheets, showing the weight of sheets, number in a bundle and weight per bundle of galvanized sheets in all the standard sizes and gauges, from No. 10 to 30 inclusive. Copies of these handy cards may be obtained by the trade on application to the advertising department of The American Sheet Steel Co., Battery Park building, New York.

SERIOUS ILLNESS OF MR. RUTLEDGE.

The many friends of Mr. H. G. Rutledge, who was for over a quarter of a century buyer for M. & L. Samuels, Benjamin & Co., Toronto, will be sorry to learn that he has been stricken with paralysis. His left side has been attacked, his leg and arm being rendered powerless. He is in no immediate danger.

Mr. Rutledge was known from one end of Canada to the other and has a host of warm friends in every quarter.

ENLARGED STEEL PLANT AT THE SOO.

Mr. E. V. Clergue, promoter of the large Sault Ste. Marie enterprises, stated in Toronto on Monday that his company, in addition to completing the 600-ton steel plant, which should be ready by August or September, plans are being made for a second iron and steel plant with a capacity of 2,000 tons, which will be completed within 18 months.

ANSONIA BRASS AND COPPER CO.

Mr. Alexander Gibb, 13 St. John street, Montreal, who has for the past four years been representative of the Ansonia Brass and Copper Co. for Montreal and district, has been appointed their representative for the Dominion.

Some four years ago, when starting out as a manufacturers' agent, Mr. Gibb applied for the representation of the firm referred to, but they refused to entertain his application. Nothing daunted, he, however, by persistent application, succeeded in getting them to allow him to represent them for Montreal and district, and has, during the time he has acted for them, built up a large business for their products. It is a tribute to Mr. Gibb's ability as a salesman that they have now extended his territory. Mr. Gibb, or his representative, proposes to make frequent visits to the West, and it is to be hoped he will not only be able to retain the old customers of the Ansonia Brass and Copper Co., but be able to add to their numbers.

H. S. HOWLAND, SONS & CO.

WHOLESALE ONLY — 37-39 Front Street West, **Toronto.** — ONLY WHOLESALE

INSECT EXTERMINATORS.

"Cataract,"

Tin Lacquered.
All Brass, cannot rust.

No 11—All Tin.
No. 14—Tin Barrel, Brass Reservoir.
No. 16—All Brass, Lacquered

Grass Catchers.

Japanned Tin Grass Catchers. Inches, 16, 18, 20.
Grass Catchers will fit the "Woodyatt" Mowers only.

Grass Shears.

No. 100—7½ Inch, Plain Blade.
No. 101—8 " Notched Blade.
No. 101—9 " " "
No. 101—10 " " "

LAWN MOWERS.

"Star."

Length of Knives.	No. of Knives.	Diameter of Wheel.
12 Inches.	3	9 Inch.
14 "	3	9 "
16 "	3	9 "

"Woodyatt."

Length of Knives	No. of Knives.	Diameter of Wheel.
12 Inches.	4	10½ Inch.
14 "	4	10½ "
16 "	4	10½ "
18 "	4	10½ "
20 "	4	10½ "

H. S. HOWLAND, SONS & CO., Toronto.

OUR PRICES ARE RIGHT — Graham Wire and Cut Nails are the Best. — WE SHIP PROMPTLY.

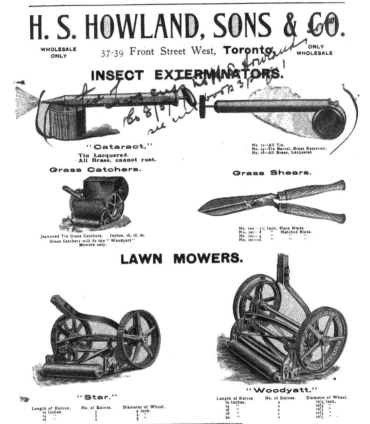

TRADE IN COUNTRIES OTHER THAN OUR OWN.

VIEWS ON THE BRITISH IRON TRADE.

Views of The Iron and Coal Trades Review: "The first three months of the year 1901 have now expired, and the opportunity is afforded by the publication of the official returns of exports of estimating the course of our foreign trade during that period. Unfortunately that course, like the rake's progress, is from bad to worse. The total quantity of iron and steel exported in the first three months of 1901 was 616,826 tons, against 962,457 tons in the first three months of 1900, so that there has been a decrease of 345,631 tons, or 35 per cent., compared with the corresponding quarter of 1900. When compared with the first three months of 1898, which was regarded as a prosperous period, although less so than 1900, the decrease is 13 per cent. The decrease of value in 1901, against 1900, has been £2,218,000, which is at the rate of £8,872,000 a year, but the decrease in value for the month of March alone, in respect of 1901, compared with 1900, has been £911,000, which is at the rate of nearly eleven millions sterling a year. This is one of the most serious falls that the British iron trade has ever experienced within so short a period.

"Unfortunately for those who are engaged in the production of iron and steel in the United Kingdom, this phenomenal reduction of our exports has been coincident with a very material advance in the volume of our imports, which were 207,178 tons for the quarter, against 132,838 tons for last year. The following are the details:—

	1900. Tons.	1901. Tons.
Pig-iron	20,949	37,539
Puddled iron	1,040	1,189
Bars, angles, etc	15,057	19,986
Unwrought steel	6,099	45,335
Girders, beams, etc.	25,200	25,519
Rails	8,404	15,221
Tyres and axles	575	885
Unenumerated	55,514	61,501
Totals	132,838	207,178

"Here we have an increase for the three months of over 74,000 tons, or 56 per cent."

Views of The Iron and Steel Trades' Journal: "Although the results of the trade returns for March are not wholly satisfactory, we still want tangible proof that, compared with normal times, our foreign trade is going to the wall. A cursory glance at the table which we annex certainly gives one a bad impression; but then we must bear in mind that the year 1900, or at least the first quarter thereof, witnessed a 'boom' almost without precedent in the item which shows the greatest falling-off over the quarter, and during the past month, and which naturally most interests our readers—metals and the articles manufactured therefrom. The decrease in our exports in this branch during March was £965,328, and for the quarter £2,280,466; but as we have already pointed out this compares with a period when the activity in the iron trade was chiefly responsible for an increase of £3,427,000. Without necessarily adopting a cum grano principle, we are inclined to look upon the falling-off merely as a result of the period of famine which, generally speaking, follows one of plenty, and through which we are now passing in this particular department. It cannot certainly be Yankee competition, for the prices which have been ruling in this country for some time past have been much below theirs: in fact, iron has been recently shipped from these shores to the States."

THE TINPLATE TRADE IN ENGLAND.

The Board of Trade returns for the first quarter of this year are distinctly disappointing, the decrease in the export of tin plates, amounting to about 220,000 boxes under the same period of last year; and added to this, there has been a falling-off of some 50 per cent. in the shipments of black plates. In the face of the recent large increase in the production, this combined decrease in the exports is far from encouraging, and precludes any hope of further advance in prices. Whatever help, therefore, that markets can look for must, of necessity be confined to reductions in wages and raw materials. It is consequently to be hoped that the men will face the situation at the end of June, when the rearrangement of the wages list is to be taken in hand, and thus do away with the disadvantages of any further labor troubles.

The holidays this week have interferred with business somewhat, but the inquiry keeps up and is fairly active in all directions, resulting in orders being placed out for prompt and forward deliveries to a considerable extent at prices showing little or no variation.

The impression gains ground that The American Steel Trust will not interfere, for the present, with the course of business on this side, but will confine their operations to money-making out of the "boom" in their own country.—Iron and Coal Trades' Review, April 12.

THE UNITED STATES IRON MARKET.

The iron market shows continued signs of strength in nearly every quarter. The pig-iron statistics for April 1 are instructive, indicating that production is now within a few hundred tons a week of high point, while consumption, if stock reports could be taken as indicating that the iron leaving furnace piles had been melted, is at the highest rate on record. There is a strong sentiment in the trade against further advances, and with plates and structural material at 1.60c. Pittsburg, and bars within $1 or $2 of that level, stress is laid on the danger of curtailing consumption by advances beyond the point of largest employment of iron and steel. The heavy demand upon malleable works, a good index of railroad and agricultural works buying, still keeps up, but capacity to produce is greater than ever, and the advances made by malleable works in 1899 will not be repeated. Just at the moment the buying of Bessemer pig-iron is light, steel works having provided for their requirements in the first half of the year. But inquiry is being made for metal for the second half and considerable sales may soon be announced. Furnaces are holding firmly to $16.25 in the Valley for second quarter iron and ask $16 for the second half. Some coke contracts for the third and fourth quarters have been made, but in most cases furnacemen are holding off, expecting better prices. Foundry iron is in good demand and some contracts have been made for the second half.—Iron Trade Review.

PIG IRON TRADE IN GREAT BRITAIN.

In this, as in other sections of the market, business has been limited during the past week; but some manufacturers of finished material are giving out orders for immediate execution, being short of stocks at works. Prices show little change, taking the trade all round, the limited production helping to strengthen the market and enabling quotations to be maintained. In the Cleveland district the number of furnaces in operation has declined to 51, as against 69 a year ago, and the output has decreased to 460,000 in the quarter just ended, 18 per cent. less than in the same period of 1900. This curtailment of the production has, of course, not been without its effect upon the market, and it is encouraging to note that the shipments from Middlesbrough during April have, up to the present time, shown a substantial improvement on those of last month, and trade with the Continent is apparently reviving. In Manchester there is a very moderate request for pig, and prices are weak and irregular, while in South Staffordshire there is a somewhat better tone, but very little doing.—Iron and Coal Trades' Review.

NEW YORK METAL MARKET

TIN.—The early cables from London reported an advance of 12s. 6d. in the price of spot tin and after noon a further rise of 2s. 6d. occurred. Futures went up 12s. 6d. in the morning, but lost 2s. 6d. later. The improvement in the English market was without influence here, spot and April being nominally quoted at 26c. The stock afloat, as posted to-day, has increased to 4,229 tons. Included in this are 515 tons from London, which will be due here next Monday and 950 tons which are expected to arrive here by direct steamer on May 7. This, with the recent considerable arrivals, appears to be too strong an argument against an advance to admit of the manifestation of much buying interest at the present time.

PIG LEAD. There was nothing new in the local market, trade being still of the hand-to-mouth order, with prices maintained on the basis of 4.37 1-2c. in lots of 50 tons or more. St. Louis was steady at 4.20 to 4.22 1-2c. as to brand. There was a slight reaction in London, that market closing at a decline of 2s. 6d.

SPELTER.—There were no fresh developments in this market, which closed steady at 3.95 to 4c. St. Louis and London were without change.

ANTIMONY.—The market for regulus was quiet but steady at 8 1-2 to 10 1-4c. as to brand.

OLD METALS.—There was no demand, but the steady tone of the market was retained.

IRON AND STEEL.—The only feature out of the ordinary in the iron market to-day was the announcement that the principal producers had determined upon an advance of $2 in the price of steel rails to be effective from May 1. The demand for rails from home and export buyers continues heavy and it is estimated that since last fall orders for fully 2,000,000 tons have been booked. The movement in pig iron is confined almost exclusively to deliveries on existing contracts, but there is no abatement of the firm feeling. In finished materials the rush seems to be over, for the present at least, but the mills are full of orders that will absorb their production for some months to come. Old material is still receiving a fair amount of attention and under small supplies is firmly held. The cable from Glasgow to-day reported an advance of 5s. in the price of Scotch pig iron warrants and prices on foundry iron at Middlesboro' were also firmer and a shade higher.

TINPLATE.—There was nothing new in the situation, the movement continuing on a liberal scale, while prices are maintained at the previous quotations.—N. Y. Journal of Commerce, April 24.

MARKETS AND MARKET NOTES

QUEBEC MARKETS

Montreal, April 26, 1901.

HARDWARE.

THE hardware business has almost fully righted itself after some months of depression, and now, except in sheet metals, where there is still some hesitation to operate, the market is in good shape, and steady enough to insure safe dealing. The most important feature of the week is the change in the prices of bolts and nuts, and the fact that the retail trade is in future to be allowed a preference of 5 per cent. on bolts and ½c. per lb. on nuts. The purchases of wire have been heavy this week, and retailers are clamoring for speedy shipment. In fact, all along the line of goods people are wanting supplies, and it would appear that the orders that were looked in vain for last February are just coming in now. Stocks in the country have been slow. Navigation is now open, and the first boats are leaving well loaded. Wire nails continue in good request, as also are horse nails. Horseshoes are moving somewhat more freely. Poultry netting and green wire cloth are being called for in fair quantities. Freezers, lawn mowers, churns, wheelbarrows, etc., are all in request. In fact, in some of these lines the manufacturers are weeks behind in their orders. Builders' hardware is one of the most active lines in the market.

BARB WIRE—There has been another heavy demand for barb wire this week and some heavy shipments have been made. Orders cannot be filled without some little delay. The price is still $3.05 per 100 lb. f.o.b. Montreal.

GALVANIZED WIRE—The demand has been exceedingly heavy this week for fence sizes. Shipments are a little behind. We quote as follows: No. 5, $4.25; Nos. 6, 7 and 8 gauge, $3.55; No. 9, $3.10; No. 10, $3.75; No. 11, $3.85; No. 12, $3.25; No. 13, $3.35; No. 14, $4.25; No. 15, $4.75; No. 16, $5.00.

SMOOTH STEEL WIRE—Trade in both oiled and annealed is good, particularly the former. Prices are steady. We quote oiled and annealed: No. 9, $2.80; No. 10, $2.87; No. 11, $2.90; No. 12, $2.95; No. 13, $3.15 per 100 lb. f.o.b. Montreal, Toronto, Hamilton, London, St. John and Halifax.

FINE STEEL WIRE—A moderate inquiry is being experienced. The discount is 17½ per cent. off the list.

BRASS AND COPPER WIRE—The usual business is passing. The discount on brass is 55 and 2½ per cent., and on copper 50 and 2½ per cent.

FENCE STAPLES—A brisk movement is to be reported in fence staples at former quotations. We quote: $3.25 for bright, and $3.75 for galvanized, per keg of 100 lb.

WIRE NAILS—The good demand that we have noted for nails during the past few weeks continues and nails form one of the most active lines on the market. We quote $2.85 for small lots and $2.77½ for carlots, f.o.b. Montreal, London, Toronto, Hamilton and Gananoque.

CUT NAILS—A fair trade is being done, and some fair orders have been filled during the week. Prices are unchanged. We quote: $2.35 for small and $2.25 for carlots; flour

barrel nails, 25 per cent. discount; coopers' nails, 30 per cent. discount.

HORSE NAILS—The market is decidedly irregular, and discounts of 66 2/3 and 60 and 10 and 10 per cent. on countersunk are sometimes spoken of in large lots. "C" brand is being held firmly at the former discount, 50 and 7½ per cent., and on this nail no concession is being made. "M" brand is selling at 60 per cent. off the old list on oval and city head and 65 per cent. off countersunk head.

HORSESHOES—A better trade is doing this week. Prices are unchanged. We quote: Iron shoes, light and medium pattern, No. 2 and larger, $3.50; No. 1 and smaller, $3.75; snow shoes, No. 2 and larger, $3.75; No. 1 and smaller, $4.00; X L steel shoes, all sizes, 1 to 5, No. 2 and larger, $3.60; No. 1 and smaller, $3.85; feather-weight, all sizes, $4.85; toe weight steel shoes, all sizes, $5.95 f.o.b. Montreal; f.o.b. Hamilton, London and Guelph, 10c. extra.

POULTRY NETTING — Shipments are being made freely. The discount is still 55 per cent.

GREEN WIRE CLOTH—Business is very satisfactory. Sales are being made at $1.35 per 100 sq. ft.

SCREEN DOORS AND WINDOWS—There is nothing new to note in this market. It is rather difficult to fill all orders immediately on receipt. We quote: Screen doors, plain cherry finish, $8.25 per doz.; do. fancy, $11.50 per doz.; walnut, $7.40 per doz., and yellow, $7.45; windows, $2.25 to $3.50 per doz.

SCREWS — Discounts remain the same. The demand for spring stocks has been very brisk. Discounts are as follows: Flat head bright, 87½ and 10 per cent. off list; round head bright, 82½ and 10 per cent.; flat head brass, 80 and 10 per cent.; round head brass, 75 and 10 per cent.

BOLTS — The discounts on bolts have been changed, some having been raised and some lowered. All retailers are now allowed an extra discount of 5 per cent. on bolts and ½c. per lb. on nuts. Discounts are as follows: Norway carriage bolts, 65 per cent.; common, 60 per cent.; machine bolts, 60 per cent.; coach screws, 70 per cent.; sleigh shoe bolts, 72½ per cent.; blank bolts, 70 per cent.; bolt ends, 62½ per cent.; plough bolts, 60 per cent.; tire bolts, 67½ per cent.; stove bolts, 67½ per cent. To any retailer an extra discount of 5 per cent. is allowed. Nuts, square, 4c. per lb. off list; hexagon nuts, 4½c. per lb. off list. To all retailers an extra discount of ½c. per lb. is allowed.

BUILDING PAPER—A good spring business is being done. We quote as follows:

Tarred felt, $1.70 per 100 lb.; 2-ply, ready roofing, 80c. per roll; 3-ply, $1.05 per roll; carpet felt, $2.25 per 100 lb.; dry sheathing, 30c. per roll; tar sheathing, 40c. per roll; dry fibre, 50c. per roll; tarred fibre, 60c. per roll; O.K. and I.X.L., 65c. per roll; heavy straw sheathing, $28 per ton; slaters' felt, 50c. per roll.

RIVETS—There is no feature to note. Discount on best iron rivets, section, carriage, and wagon box, black rivets, tinned do., coopers' rivets and tinned swedes rivets, 60 and 10 per cent.; swedes iron burrs are quoted at 55 per cent. off; copper rivets, 35 and 5 per cent. off; and coppered iron rivets and burrs, in 5-lb. carton boxes, are quoted at 60 and 10 per cent. off list.

BINDER TWINE — Movements this week are rather larger. We quote: Blue Ribbon, 11¼c.; Red Cap, 9¾c.; Tiger, 8¾c.; Golden Crown, 8c.; Sisal, 8¼c.

CORDAGE—The sales of cordage during the past few weeks have been quite large. Manila is now worth 13½c. per lb. for 7-16 and larger; sisal is selling at 10c., and lathyarn 10c.

SPADES AND SHOVELS—Spring goods are being called for freely. The discount is still 40 and 5 per cent. off the list.

HARVEST TOOLS — Harvest tools have been very active and heavy orders are being placed, despite the general impression that left-over stocks were heavy. The discount is still 50, 10 and 5 per cent.

TACKS—Good sorting orders are to hand this week. Prices are unchanged. We quote as follows: Carpet tacks, in dozens and bulk, 80 and 15 per cent.; tinned, 80 and 20 per cent., and cut tacks, blued, in dozens, 80 per cent.

LAWN MOWERS — Many inquiries are coming in for lawn mowers. We quote as follows: High wheel, 50 and 5 per cent. f.o.b. Montreal; low wheel, in all sizes, $2.75 each net; high wheel, 11-inch, 30 per cent. off.

FIREBRICKS—The spring prices for firebricks are $1.25 per 1,000 lower than they were a year ago, with Scotch quoted at $17.50 to $22 and English at $17 to $21 per 1,000 as to brand. The demand for spot stock is fair, and prices are unchanged at $18 to $24 per 1,000 as to brand ex store.

CEMENT—Since the fine spring weather has set in there has been a decided improvement in the demand for cement, both for prompt and future delivery, and some large sales have taken place. Belgian brands are 10c. per bbl. lower and German and English a little higher than a year ago. We quote as follows: German, $2.45 to $2.55; English, $2.30 to $2.40; Belgian, $1.95 to $2.05. American in bags, $2.30 to $2.40 per 400 lb.

METALS.

The iron and steel markets are firm, but the English plate market is still unsteady with a downward tendency. Shipments from stock are a little larger this week, but dealers are buying for immediate shipment only.

PIG IRON—Quotations on Canadian pig remain unchanged at $18 to $18.50. Summerlee is quoted at $20 to $21.

BAR IRON — The market remains firm under heavy trading. Merchants' bar is now worth $1.70 to $1.75 and horseshoe $2.

BLACK SHEETS —Inquiries are somewhat more numerous this week. Dealers are asking for spot goods $2.65 for 8 to 16 gauge; $2.70 for 26 gauge, and $2.75 for 28 gauge.

GALVANIZED IRON — Some trade has been done this week for immediate delivery only. We quote as follows: No. 28 Queen's Head, $4.65; Apollo, 10¾ oz., $4.50, and Comet, $4.40 to $4.45, with a 15c. reduction for case lots.

INGOT COPPER — Some fair transactiosn have been entered into this week, and the market, both here and abroad, is firm. On the Montreal market the ruling quotation is 18c.

INGOT TIN —Some better business has

been done during the past few days at 30c. per lb.

LEAD—The lead market has taken on a stiffer appearance. On this market the ruling quotation is $3 75 per 100 lb.

LEAD PIPE—Business is good. We quote: 7c. for ordinary and 7½c. for composition waste, with 25 per cent. off.

IRON PIPE—The rise that has taken place of late has not prevented active buying. We quote: Black pipe, ¼, $3 per 100 ft.; ⅜, $3; ½, $3 05; ¾, $3 30; 1-in., $4.70; 1¼, $6.40; 1½, $7.70; 2-in. $10 25. Galvanized, ½, $4.60; ¾, $5.25; 1-in., $7.50; 1¼, $9.80; 1½, $11.75; 2 in., $16.

TINPLATES—The English market is still unsteady, with a downward tendency. Reports say that stocks are increasing in Wales rather than diminishing. Locally, there is a good demand for spot goods, and importers are bringing in good stocks on early steamers. Coke plates are worth $3.80 to $4 and charcoal, $4.15 to $4.35.

CANADA PLATE — A fair distributing trade is being done, but business is not heavy. We quote: 52's, $2.60; 60's, $2.70; 75's, $2.80; full polished, $3.45, and galvanized, $4.30.

STEEL — The demand for steel is fairly brisk, and the market is firm to strong. We quote: Sleighshoe, $1.95; tire, $2; bar, $1.95; spring, $2.75; machinery, $2.75 and toe-calk, $2.50.

SHEET STEEL—There is a fair demand for sheet steel at the present moment. We quote: Nos. 22 and 24, $3. and Nos. 18 and 20, $2.85.

TOOL STEEL—Black Diamond is worth 8c. and Jessop's, 13c.

TERNE PLATES—The demand for terne plates is fair. The price rules about $7.75.

COIL CHAIN—The last few weeks have seen some good shipments of coil chain. We quote: No.6, 11½c.; No.5, 10c.; No.4, 9½c.; No. 3, 9c.; ¼-inch, 7½c. per lb.; 5-16, $4.65; 5-16 exact, $5.10; ⅜, $4.20; 7-16, $4.00; ½, $3.75; 9-16, $3.65; ⅝, $3.35; ¾, $3.25; ⅞, $3.20; 1-in., $3.15. In carload lots an allowance of 10c. is made.

SHEET ZINC—Orders are being taken for May delivery at 5c. On spot we quote 5¾ to 6c.

ANTIMONY—Quiet, at 10c.

ZINC SPELTER—Is worth 5c.

SOLDER—We quote: Bar solder, 18½c.; wire solder, 20c.

GLASS.

A good general business is being done at unchanged prices. We quote as follows: First break, $2; second, $2.10 for 50 feet; first break, 100 feet, $3.80; second, $4; third, $4.50; fourth, $4.75; fifth, $5.25; sixth, $5.75, and seventh, $6.25.

PAINTS AND OILS.

Owing to competition in the South and the arrival of spring goods, turpentine has a downward tendency. The firmer feeling in linseed oil is fully maintained and on the whole the oil market is in a healthy condition. White lead, after experiencing a long period of weakness, has suddenly developed a decided strong tone on the Continent. All lead products, including litharge, orange mineral, etc., are firmer in sympathy. Prices remain the same as last week. We quote:

WHITE LEAD—Best brands, Government standard, $6.25; No. 1, $5.87½; No. 2, $5.50; No. 3, $5.12½, and No. 4, $4.75 all f.o.b. Montreal. Terms, 3 per cent. cash or four months.

DRY WHITE LEAD—$5.50 in casks; kegs, $5.75.

RED LEAD—Casks, $5.25; in kegs, $5.50.

WHITE ZINC PAINT—Pure, dry, 7c.; No. 1, 6c.; in oil, pure, 8c.; No. 1, 7c.; No. 2, 6c.

PUTTY—We quote: Bulk, in barrels, $1 90 per 100 lb.; bulk, in less quantity, $2.05; bladders, in barrels, $2.10; bladders, in 100 or 200 lb. kegs or boxes, $2.25; in tins, $2.55 to $2 65; in less than 100-lb. lots, $3 f.o.b. Montreal, Ottawa, Toronto, Hamilton, London and Guelph. Maritime Provinces 10c. higher, f.o.b. St. John and Halifax.

LINSEED OIL—Raw, 72c.; boiled, 75c. in 5 to 9 bbls., 1c. less, 10 to 20 bbl. lots, open, net cash, plus 2c. for 4 months Delivered anywhere in Ontario between Montreal and Oshawa at 2c. per gal. advance and freight allowed.

TURPENTINE—Single bbls., 57c.; 2 to 4 bbls., 56c.; 5 bbls. and over, open terms, the same terms as linseed oil.

MIXED PAINTS—$1.25 to $1.45 per gal.

CASTOR OIL—8¾ to 9½c. in wholesale lots, and ½c. additional for small lots.

SEAL OIL—47½ to 49c.

COD OIL—32½ to 35c.

NAVAL STORES — We quote: Resins, $2.75 to $4.50, as to brand; coal tar, $3.25 to $3 75; cotton waste, 4½ to 5½c. for colored, and 6 to 7½c. for white; oakum, 5½ to 6½c., and cotton oakum, 10 to 11c.

PARIS GREEN—Petroleum barrels, 16¾c. per lb.; arsenic kegs, 17c.; 50 and 100. lb. drums, 17½c.; 25-lb. drums, 18c.; 1-lb. packages, 18½c.; ½-lb. packages, 20½c.; 1-lb. tins, 19½c.; ½-lb. tins, 21½c. f.o.b. Montreal; terms 3 per cent. 30 days, or four months from date of delivery.

SCRAP METALS.

Quite a spring activity has been imparted to the scrap metal market during the past few days. Scrap iron is especially firm. Dealers are paying the following prices in the country: Heavy copper and wire, 13 to 13½c. per lb.; light copper, 12c.; heavy brass, 12c.; heavy yellow, 8½ to 9c.; light brass, 6½ to 7c.; lead, 2½ to 2¾c. per lb.; zinc, 2¼ to 2½c.; iron, No. 1 wrought, $15 to $16 per gross ton f.o.c. Montreal; No. 1 cast, $13 to $14; stove plate, $8 to $9; light iron, No. 2, $4 a ton; malleable and steel, $4; rags, country, 70 to 80c. per 100 lb.; old rubbers, 6¼c. per lb.

HIDES.

The feeling has been somewhat steadier this week, consequent upon an improved demand. We quote as follows: Light hides, 6½c. for No. 1; 5½c. for No. 2, and 4½c. for No. 3. Lambskins, 10c.; sheepskins, 90c.; calfskins, 8c. for No. 1 and 6c. for No. 2.

PETROLEUM.

The demand is falling off. We quote as follows: "Silver Star," 14½ to 15½c.; "Imperial Acme," 16 to 17c.; "S.C. Acme," 18 to 19c., and "Pratt's Astral," 18½ to 19½c.

ONTARIO MARKETS.

TORONTO, April 27, 1901.

HARDWARE.

WHILE trade is perhaps not as active as it was a week ago there is still a good business being done, and the shipping rooms of the different wholesale houses are the scene of a great deal of activity. A good deal of merchandise is being got ready for shipment by the regular steamship lines plying on the upper lakes as soon as they begin running. Some shipments have already been made by what may be termed irregular lines. For the trade in Western Ontario where the season opens earlier quite a few harvest tools, wire, screen doors and windows, spades and shovels, spray pumps and syringes are being sent forward. Some demand is being experienced this week for corn planters. Wire nails are not going out as briskly as they were, which is to be expected in view of the recent brisk demand. Cut nails appear to be about as dull as ever. Fence wire is selling well, and the scarcity noted in barbed wire still exists. In bolts, an active trade is being done, and in screws, rivets and burrs, business is fair. A little better business is being done in both sporting goods and cutlery. Quite a few lawn mowers have been booked, but very few so far have gone forward. There has been quite an improvement during the week in the demand for enamelled ware and tinware. Ice cream freezers and refrigerators are in brisk demand. Quite a few oil stoves have gone out during the week. Payments are fair.

BARB WIRE—The scarcity of barb wire referred to last week has become even more pronounced, and most of the jobbers do not appear to have any at all in stock, while the manufacturers in the United States are four and five weeks behind with their orders. The price is unchanged at $3 05 per 100 lb. stock Toronto. We quote f.o.b. Cleveland as follows: $2.82½ per 100 lb. for less than carlots and $2 70 for carlots.

GALVANIZED WIRE—The demand for galvanized wire is fairly good, and the same may be said of the supply. We quote as follows: Nos. 6, 7 and 8, $3.50 to $3.85 per 100 lb., according to quantity; No. 9, $2.85 to $3.15; No. 10, $3.60 to $3.95; No. 11, $3.70 to $4 10; No. 12, $3 to $3 30; No. 13, $3 10 to $3 40; No. 14, $4 10 to $4 50; No. 15, $4.60 to $5.05: No. 16, $4.85 to $5.35. Nos. 6 to 9 base f.o.b. Cleveland are quoted at $2.57½ in less than carlots and 12c. less for carlots of 15 tons.

SMOOTH STEEL WIRE—Orders that were booked for May delivery are being asked for by the retail trade throughout the country. It is thought that the urgency for oiled and annealed wire is in some degree caused by the scarcity of barb wire. All the factories are reported to be in a good position to supply oiled annealed wire. Net selling prices for oiled and annealed are as follows: Nos. 6 to 8, $2 90; 9, $2.80; 10, $2.87; 11, $2.90; 12, $2.95; 13, $3.15; 14, $3.37; 15, $3 50; 16, $3 65. Delivery points, Toronto, Hamilton, London and Montreal, with freights equalized on these points.

WIRE NAILS—A good many small lots are going out, but the large buyers, having bought freely some time ago, are not yet in the market. "While the month will show a large shipment," said one well-known manufacturer, "there is not much business coming in this week. There are quite a few small orders, but no heavy ones." We quote as follows: Less than carlots at $2.85, and carlots at $2.77½ to the retail trade. Delivery points: Toronto, Hamilton, London, Gananoque and Montreal.

CUT NAILS—A few shingle nails are going out, but otherwise trade is dull. We quote the base price at $2.35 per keg for less than carlots, and $2.25 for carlots. Delivery points: Toronto, Hamilton, London, Montreal and St. John, N.B.

HORSE NAILS—A fairly good trade is being done in this line, for, although no large orders are being received, nearly every order sent in by the travellers contains an item for horse nails. Discount on "C" brand, oval head, 50 and 7½ per cent. off new list and on other brands 50, 10 and 5 per cent., off the old list; countersunk, 50, 10 and 10 per cent.

HORSESHOES—There is not a great deal of business being done, but for this time of the year the demand is fair. We quote as follows f.o.b. Toronto: Iron shoes, No. 2 and larger, light, medium and heavy, $3.60; snow shoes, $3.85; light steel shoes, $3.70; featherweight (all sizes), $4.95; iron shoes, No. 1 and smaller, light, medium and heavy (all sizes), $3.85; snow shoes, $4; light steel shoes, $3.95; featherweight (all sizes), $4 95.

SCREWS—There is the usual steady trade being done, and prices are unchanged. We still quote discounts as follows: Flat head bright, 87½ and 10 per cent.; round head bright, 82½ and 10 per cent.; flat head brass, 80 and 10 per cent.; round head brass, 75 and 10 per cent.; round head bronze, 65 per cent., and flat head bronze at 70 per cent.

BOLTS AND NUTS—There is a good demand for all kinds of bolts and nuts. The demand from the machinery manufacturers appears to be particularly good. We quote: Carriage bolts (Norway), full square, 65 per cent.; carriage bolts full square, 65 per cent.; common carriage bolts, all sizes, 60 per cent.; machine bolts, all sizes, 60 per cent.; coach screws, 70 per cent.; sleighshoe bolts, 72½ per cent.; blank bolts, 60 per cent.; bolt ends, 62½ per cent.; nuts, square, 4c. off; nuts, hexagon, 4½c. off; tire bolts, 67½ per cent.; stove bolts, 67½; plough bolts, 60 per cent.; stove rods, 6 to 8c.

RIVETS AND BURRS—There is quite a lot of these going out, but the demand is principally for copper rivets. We quote as follows: Iron rivets, 60 and 10 per cent.; iron burrs, 55 per cent.; copper rivets and burrs, 35 and 5 per cent.

ROPE—Although the orders are, as a rule, for small lots, trade is fairly active. The base price is steady and unchanged at 10c. for sisal and 13½c. for manila.

CUTLERY—There is a fair movement this week in general lines of cutlery. The line which is most in demand at the moment is pocket knives.

SPORTING GOODS—Trade is a little more active this week. Quite a little gunpowder is going out, and there is quite a little demand for rifles and guns.

ENAMELLED WARE—Trade is much more active in both enamelled ware and tinware than it was a week ago, and prices are steady and unchanged. The discounts on enamelled ware are as follows: "Granite," "Pearl," "Crescent" and "Imperial" wares at 50, 10 and 10 per cent.; white, "Princess," "Turquoise," blue and white, 50 per cent.; "Diamond," "Famous" and "Premier," 50 and 10 per cent.

GREEN WIRE CLOTH—The movement in green wire cloth is fairly good and prices are unchanged at $1.35 per 100 sq. ft.

SCREEN DOORS AND WINDOWS—Shipments in this line have been quite active during the week.

BUILDING PAPER—A great many orders are on file, and quite a few shipments will be made during the next few days. We quote: Plain building, 30c.; tarred lining, 40c., and tarred roofing, $1.65.

POULTRY NETTING—There has been a fair movement in poultry netting during the past week. Discount, 55 per cent.

ICE CREAM FREEZERS AND REFRIGERATORS—Trade in these lines appears to be on the whole even more active than it was a week ago.

CHURNS—An active business is being done in this line, and the same may be said of washing machines, the demand for which has been stimulated by the low prices.

HARVEST TOOLS—There is some business being done, but the demand is mostly, at the moment, from the smaller dealers. The line for which the demand is most active is manure forks and goods of that description. Discount 50, 10 and 5 per cent.

SPADES AND SHOVELS—Business is more active. Quite a few shipments have been made during the week. Discount 40 and 5 per cent.

TACKS—The usual steady trade is being done.

BRIGHT WIRE GOODS—Trade is steady, and the discount is unchanged at 62½ per cent.

BINDER TWINE—Trade is still seasonably quiet in this line, and prices are steady and unchanged. We quote as follows: American—Sisal and standard, 8½c.; manila, 10½c.; pure manila, 11½ to 12c. per lb. Canadian—Sisal, 8½c.; pure manila, 11½c. per lb.

OIL AND GAS STOVES—The season is now in full blast in regard to oil stoves, and quite an active trade is, therefore, being done. Some business is being done in gas stoves.

LAWN MOWERS—Quite a few orders have been booked, but not many shipments, so far, have been made.

EAVETROUGH—The movement in eavetrough and conductor pipe continues active, with prices still very low.

MECHANICS' TOOLS—The activity is still maintained in mechanics' tools, and the same may be said of builders' supplies generally.

CHAIN—While trade is not as brisk as it was, there are still a few orders coming in for boom chain.

CEMENT—There is a big movement. While prices are unchanged, a considerable reduction is made on large lots. We quote barrel lots: Canadian portland $2.25 to $2.80; German, $3 to $3.15; English, $3; Belgian, $2.50 to $2.75; Canadian hydraulic, $1.25 to $1.50.

METALS.

The metal trade, generally speaking, has been good during the past week and prices are, as a rule, steady and unchanged.

PIG IRON—The demand for pig iron is not active, but prices rule steady. No. 2 Canadian pig iron is worth $16.50 in large lots.

BAR IRON—The situation is even stronger than it was a week ago, while the demand exceeds the supply, some of the mills being quite a distance behind in their orders, and are unwilling to accept new business for large quantities, for nearby shipment, at any rate. The base price is now $1.85 to $1.90 from stock.

STEEL—The steel market continues very strong with the tendency of prices upwards. We now quote tire steel at $2 30 to $2.50; sleighshoe steel at $2.10 to $2 25; reeled machinery steel, $3. Hoop steel is unchanged at $3 10.

PIG TIN—There has been a fairly good trade in small lots. During the last few days there has been quite an advance in the price of tin in London, England, but New York has not responded. Locally, quotations are unchanged at 31 to 32c. per lb.

TINPLATES—No large shipments are being made, but still there is a fair demand for small lots.

TINNED SHEETS—There is quite an active demand, which is usual at this time of year, as most of the business will be done during the next few weeks. We quote 9 to 9½c. per lb.

TERNE PLATES—There is a little more activity in this line, although the movement is not large.

GALVANIZED SHEETS—The trade has been very brisk during the past week, both from stock, and on importation account. Customers, who ordered for May shipment are asking for earlier delivery, but wholesalers say that the request comes too late to have the shipments made earlier than orginally booked for. Dealers' stocks throughout the country are light. The ruling prices from stock are still $4.60 for English and $4.50 for United States galvanized sheets.

BLACK SHEETS—Trade has been good during the past week, both for large and small lots; the price of 28 gauge still rules at $2.30.

CANADA PLATES—There has been a little better movement in Canada plates than for some time past. Import orders are still being booked. We quote: All dull, $3; half-and half, $3.15, and all bright, $3.65 to $3.75.

COPPER — In ingot copper trade has been confined to small lots. There is good business being done in sheet copper. Copper has advanced during the last few days in London, England, but prices in New York remain much as before. We quote: Ingot, 19c.; bolt or bar, 23½ to 25c.; sheet, 23 to 23½c. per lb.

BRASS—Trade in brass is quiet, and the discount on rod and sheet is 15 per cent., as before.

SOLDER—The demand keeps fairly good, and prices are unchanged. We quote: Half-and-half, guaranteed, 18½c.; ditto, commercial, 18c.; refined, 18c., and wiping, 17c.

IRON PIPE—There has been no further change in prices, and the demand continues active.

LEAD—Business is not as brisk as it was. The market in London, Eng., is weak and lower. Locally no change has taken place, the ruling price still being 4½ to 4¾c. per lb.

ZINC SPELTER — Business in this line is more active than it was, and prices are firmer and higher in the outside markets. Locally, we still quote 5½ to 6c. per lb.

ZINC SHEET—Business is fairly good in this line. We quote 6½c. for casks and 6¾c. for part casks.

ANTIMONY — There is some business being done in small lots. We quote 10½ to 11c. per lb.

PAINTS AND OILS.

The trade is unusually good for March. Linseed oil is in good demand. The movement of turpentine is fair. Paris green is the slowest article on the list. The sale of all sundries has been excellent. "The past two months have been, said one house, "the busiest in our recollection." Another house stated: "The demand is, in fact, unprecedented. It is due, of course, to the price conditions this year. You know the market was high during January and February, and dealers postponed their buying. Now they all need supplies, and we are doing a good business." Prices are unchanged throughout. Turpentine is not moving as briskly at primary points as is usual at this season. Neither the offerings nor the demand are up to the standard. Prices are easy. Linseed oil is scarce, and is likely to continue so for some time, as the demand is larger than the present output of Canadian refineries, and British oil, which is advancing, cannot be here for some time yet. Other lines are steady in value. We quote as follows:

WHITE LEAD—Ex Toronto, pure white lead, $6.37½; No. 1, $6; No. 2, $5.67½; No. 3, $5.25; No. 4, $4.87½; genuine dry white lead in casks, $5.37½.

RED LEAD—Genuine, in casks of 560 lb., $5.50; ditto, in kegs of 100 lb., $5.75; No. 1, in casks of 560 lb., $5; ditto, kegs of 100 lb., $5.25.

LITHARGE—Genuine, 7 to 7½c.

ORANGE MINERAL—Genuine, 8 to 8½c.

WHITE ZINC—Genuine, French V.M., in casks, $7 to $7.25; Lehigh, in casks, $6.

PARIS WHITE—90c. to $1 per 100 lb.

WHITING — 70c. per 100 lb.; Gilders' whiting, 80c.

GUM SHELLAC — In cases, 22c.; in less than cases, 25c.

PARIS GREEN—Bbls., 16¾c.; kegs, 17c.; 50 and 100 lb. drums, 17½c.; 25-lb. drums, 18c.; 1-lb. papers, 18½c.; 1-lb. tins, 19½c.; ½-lb. papers, 20½c.; ½-lb. tins, 21½c.

PUTTY — Bladders, in bbls., $2.10; bladders, in 100 lb. kegs, $2.25; bulk in bbls., $1.90; bulk, less than bbls. and up to 100 lb., $2.05; bladders, bulk or tins, less than 100 lb., $2.90.

PLASTER PARIS—New Brunswick, $1.90 per bbl.

PUMICE STONE — Powdered, $2.50 per cwt. in bbls., and 4 to 5c. per lb. in less quantity; lump, 10c. in small lots, and 8c. in bbls.

LIQUID PAINTS—Pure, $1.20 to $1.30 per gal.; No. 1 quality, $1 per gal.

CASTOR OIL—East India, in cases, 10 to 10½c. per lb. and 10½ to 11c. for single tins.

LINSEED OIL—Raw, 1 to 4 barrels, 71c.; boiled, 74c.; 5 to 9 barrels, raw, 70c.; boiled, 73c., delivered. To Toronto, Hamilton, Guelph and London, 1c. less.

TURPENTINE—Single barrels, 54c.; 2 to 4 barrels, 53c., delivered. Toronto, Hamilton and London 1c. less. For less quantities than barrels, 5c. per gallon extra will be added, and for 5-gallon packages, 50c., and 10 gallon packages, 80c. will be charged.

GLASS.

There is a strong agitation among the jobbers for an advance. The constantly-rising European market, the probability of late delivery of import orders and the small stocks carried are given as sound reasons for an advance. A meeting of the jobbers is to be held to consider the matter, and a rise may take place next week. We quote as follows: Under 26 in., $4.15; 26 to 40 in., $4.45; 41 to 50 in., $4.85; 51 to 60 in., $5.15; 61 to 70 in., $5.50; double diamond, under 26 in., $6; 26 to 40 in., $6.65; 41 to 50 in., $7.50; 51 to 60 in., $8.50; 61 to 70 in., $9.50, Toronto, Hamilton and London. Terms, 4 months or 3 per cent. 30 days.

OLD MATERIAL.

The market is dull with prices unchanged. We quote jobbers' prices: Agricultural scrap, 55c. per cwt.; machinery cast, 60c. per cwt.; stove cast, 50c.; No. 1 wrought 50c. per 100 lb.; new light scrap copper, 12c. per lb.; bottoms, 11½c.; heavy copper, 13c.; coil wire scrap, 13c.; light brass, 7c.; heavy yellow brass, 10 to 10½c.; heavy red brass, 10½ to 11c.; scrap lead, 3c.; zinc, 2c; scrap rubber, 6½c.; good country mixed rags, 65 to 75c.; clean dry bones, 40 to 50c. per 100 lb.

PETROLEUM.

The demand is quiet. Prices are easy. We quote as follows: Pratt's Astral, 16½ to 17c. in bulk (barrels, $1 extra); American water white, 16½ to 17c. in barrels; Photogene, 16 to 16½c.; Sarnia water white, 15½ to 16c. in barrels; Sarnia prime white, 14½ to 15c. in barrels.

COAL.

There is no change for April delivery, and no information re May prices can be secured. We quote April delivery anthracite on cars Buffalo and bridges as follows: Grate, $4.25 per gross ton and $3.79 per net ton; egg, stove and nut, $4.50 per gross ton and $4.01 per net ton, for April shipment.

The Canadian Fire Engine Co., Limited, London, Ont., have been incorporated.

INDUSTRIAL GOSSIP.

Those having any items of news suitable for this column will confer a favor by forwarding them to this office addressed the Editor.

THE Federation Brand Salmon Canning Co. are having a Linde plant for freezing and cold storage installed in their works.

The Canadian Chrome Iron Co., Limited, have been incorporated.

The total output of the mines in the Rossland district up to the end of last week was 120,119 tons. Last week's output was 8,258 tons.

Schmidt & Co., brass founders, Winnipeg, intend erecting a three storey, 29 x 100 ft., foundry and warehouse, to cost about $40,000, near the present premises, Albert street, Winnipeg.

The Farmers' Binder Twine and Agricultural Implement Manufacturing Co. of Brantford, Ont., Limited, has been authorized to increase its share capital from $100,000 to $155,000.

It is reported that the five largest stationary engine manufacturing companies in the United States are to be consolidated with a capital of $15,000,000. The companies concerned are: The E. P. Allis Co., of Milwaukee; the Pennsylvania Iron Works Co., of Philadelphia; the Gates Iron Works Co., of Chicago; the Fraser & Chalmers Co., of Chicago, and the Dixon Manufacturing Co., of Scranton, Pa.

INQUIRIES AND ANSWERS.

PREMIUMS WANTED.

We are desirous of obtaining the names of a few firms who handle goods suitable for premium purposes, such as glassware, etc. Can you furnish us with the same. Thanking you in advance for a reply.

PREMIUMS.

Halifax, April 18, 1901.

PERSONAL MENTION.

Mr. W. A. Drummond, who for the past three years has had the management of the hardware department of The Robert Simpson Co., Limited, Toronto, has severed his connection with that firm.

THE ATTRACTIONS OF HALIFAX.

The members of the Nova Scotia Tourist Association are to be congratulated on the booklet "Halifax, Nova Scotia," which they are now sending out. The publications of this body have all been good, but in this work an unusually high standard of excellence has been reached, especially in the illustrations, which are superb, presenting in a most captivating style the attractions which Halifax presents the tourist or traveller.

It is not surprising that this association has been successful in its work of developing tourist travel. They have in a central part of Halifax an "Information Bureau," where visitors can obtain reports of hotel service, boarding house accommodation, shooting, fishing, etc., in almost every town and village in the Province. Last year 30,000 booklets, similar to the one now being sent out, were distributed. Altogether the association expended $2,300, but the general verdict was that it was a profitable investment and this year they are more energetic than ever in their useful work.

HEATING AND PLUMBING

SMOKE FLUES FOR FURNACES.

THE importance of the smoke flue in its relation to the success of any heating apparatus is not always given a sufficient consideration, writes M. L. Person in an exchange. Practically speaking the successful operation of the apparatus depends very largely on the construction of the smoke flue, and where this is defective the operation is far from satisfactory, with the result that the apparatus itself is blamed for a fault which does not belong to it.

In the case of a hot air furnace, the smoke flue should have a greater height than any nearby building, so that the draft may be steady and not be changed by air currents produced by wind sweeping around the corners. This flue should be used for the furnace draft alone and have no other openings, being continuous along its entire length and as perpendicular as possible. Rectangular flues should have dimensions not less than 8 x 12 inches, and the diameter of round flues ought not to be less than 10 inches, with a preference for 12 inches. Round flues are preferable to rectangular because of their area in proportion to the friction against the draft on the sides. A square stovepipe is never used and this is a practical illustration of the fact that the round form of smoke flue is better.

On the other hand a smoke flue may be too large. When its interior dimensions are in excess of its ability to provide an upward movement and prevent down drafts, the fire will get low and the percentage of consumed fuel to the heat given out is unnecessarily large. The space in the flue should be no greater than can be heated by the surplus heat from the smoke and gas coming from the smoke pipe and for which it provides both a draft and an exit. The space in the flue should average at least one-third more area than that in the smoke pipe.

The chimney itself should be built from the bottom of the cellar, with the smoke flue extending several feet below the collar, forming a pocket for the accumulation of soot. This pocket should have an opening through which the soot can be removed, but the door leading therefrom should always be kept closed when a draft on the furnace is desired. If the chimney is of brick, the walls should be double in order to prevent a waste of heat on the outside. The smoke pipe should be connected with the flue by a thimble of galvanized iron, not allowing the pipe to go so far into the chimney that the outlet from the pipe is cut off or reduced. The pipe should extend no farther than one-third into the chimney stack and even this distance may be reduced with safety if the pipe is properly secured.

Soft coal requires a larger flue and stronger draft than is necessary for hard coal if a steady heat is desired, but if the draft is too strong it can be controlled by inserting a damper in the smoke pipe between the check damper and the heater. Trouble often arises in a new building because the heating apparatus is put at work before the mortar in the chimney has thoroughly dried out. The gradual evaporation from the mortar will remove the difficulty, but in order to obtain the best results from the start the chimney should become thoroughly dry before the building is occupied. If this is not done, so much heat is required to evaporate the moisture in the mortar that the smoke is cooled and the draft is checked. So far as is possible the mortar between the bricks in a chimney should be scraped off so as to avoid unnecessary friction.

It is the practical experience of furnace builders as well as of furnace users, who have given careful thought to the subject, that the best apparatus ever made will not do good work unless it is given fair treatment. A furnace is like an animate being in that it can be given too much work for its capacity or be supplied so liberally with fuel and drafts that it becomes a glutton at the expense of the owner, who will find fault with it in either case. Yet the supposed problem of adjusting a furnace to its requirements is not a problem at all if the principles of construction and use are properly applied. A poor furnace of sufficient size, with a suitable smoke flue, one that is well piped and has registers in the right positions, will do better work than a thoroughly good furnace that is deprived of these essentials or has the further disadvantage of having an area to heat that is greater than its capacity. These points are to be considered by furnace contractors who will enter the furnace field for the first time this year, nor are they to be neglected by others who have been in the field for some time but are not always successful with their work.

PLUMBING AND HEATING NOTES.

Mrs. Henri Beliveau has registered as proprietress of the business of Henri Beliveau, electrician, Montreal.

The Provincial Light, Heat and Power Co., Three Rivers, Que. have applied for incorporation.

Alex. Desmarteau has been appointed curator of A. Paquette & Co., contractors, St. Henri de Montreal.

Albert Ward, plumber, etc., Stratford, Ont., has accepted a position with the McClary Manufacturing Co., as traveller, and has admitted Joseph Myers, who will manage the business while Mr. Ward is on the road.

SOME BUILDING NOTES.

The German Lutherans are erecting a church in Morden, Man.

The C. A. R. intend erecting a station at the end of Rochester street, Ottawa.

Wm. H. Davis, Sandy Hill, Ont., intends erecting a $100,000 hotel on the Howe property, Rideau street, Ottawa.

The Yale B. C. Power Company has been incorporated under the Water Clauses Consolidation Act, with a capital of $200,000.

Dr. McQueene is erecting a new residence in Cypress River, Man.

R. J. Devlin has started to build a new block on Wellington street, Ottawa.

Building operations in Wingham, Ont., are brisk this summer. In addition to the new Methodist church and the National Iron Works building, contracts have been made for new residences for J. G. Stewart, Wm. Nicholson and W. H. Rintoul.

PLUMBING AND HEATING CONTRACTS.

W. J. Green has the contract for plumbing, and Parkinson & Co. for heating and galvanized iron work, in the new Park street school, London, Ont.

The John Ritchie Plumbing and Heating Co., Limited, Toronto, have the contract for steam-heating the Benson House, Lindsay, Ont.

The Bennett & Wright Co., Limited, Toronto, have the contract for heating, plumbing and lighting the new palace hotel on King street, Toronto; for plumbing, heating and gas-fitting in a residence for A. Willmott, Crescent road, Rosedale; for heating and plumbing a house on Lowther avenue for Mrs. Geo. Scott; for heating, plumbing and gas-fitting an apartment house, corner of King street and Cowan avenue; for plumbing and heating two houses on Howard street and one at 126 Munro street for John Stark & Co., and for heating the Bank of Commerce branch building at the corner of Bathurst and Queen streets.

BUILDING PERMITS ISSUED.

The Winnipeg building inspector reports that permits have been issued by him for spring building operations to date to the value of $401,830, and that he expects three times that amount during the present season.

Building permits have been issued in Hamilton, Ont., to Stewart McPhie, for two dwellings at the corner of Main and Queen streets for Thomas Lees, to cost $3,600; to W. A. Peene, for a brick addition to F. W. Fearman's factory on Rebecca street, to cost $3,000. The latter company contemplate other extensive additions in the near future.

The following building permits have been issued in Toronto, to T. Grafton, for a $5,000 residence at the corner of First and Logan avenues; to Henry Arnold, for a $2,900 dwelling at the corner of Lippincott and Harbord streets; to W. D. Wilson, for a $4,000 residence at the corner of Walmer road and Bernard avenue; to A. J. Woodley, for four $2,000 residences at 1, 3, 5 and 7 Dupont street; to M. W. Haynes, for a $2,200 dwelling on Summerhill avenue; to T. Wright & Son, for a $3,200 residence on Roxborough avenue, east of Avenue road; to J. B. and A. F. Milligan, for a pair of residences, 63 and 65 John street, to cost $2,000; to Separate School Board, for a $2,800 school on Queen street, near Bolton avenue; to Geo. C. Watson, for a $3,000 dwelling on Dufferin street, near King; to W. J. Gage & Co., Limited, for a $3,400 addition to warehouse on Piper street.

BUSINESS CHANGES.

DIFFICULTIES, ASSIGNMENTS, COMPROMISES.

ROBERGE & Landry, general merchants, Thetford Mines, Que., have compromised at 50 cents on the dollar.

H. Cairns, general merchant, Sawyerville, Que., has compromised.

D. Gillanders, general merchant, Wellington, Ont., has compromised.

Alex. Manson has been appointed assignee of A. Cameron, general merchant, Lochaber, N.S.

The creditors of Alph. Letourneau, general merchant, Little Metis, Que., met this week.

A. A. McCaull, general merchant, etc., Ellerslie, P.E.I., is offering 50 cents on the dollar.

F. Paquette, general merchant, Paquetteville, Que., is offering 25 cents on the dollar.

Lamarche & Benoit have been appointed curators of J. E. Tremblay, general merchant, Ste. Anne de Bellevue, Que.

SALES MADE AND PENDING.

The assets of Joseph Brodeur, general merchant, St. Hyacinthe, Que., are to be sold.

C. S. Coggins, general merchant, etc., Penobsquis, N.B., is selling out.

The assets of Joseph Quinlan, general merchant, Manotick, Ont., are to be sold.

The stock of D. Levisseur, general merchant, Matane, Que., was sold at 60 1-2c. on the dollar.

The assets of Paul Bissonnette, general merchant, Casselman and South Indian, Ont., are to be sold.

The assets of Alex. Cameron, general merchant, North Lochaber, N.S., are advertised for sale by auction on May 3.

CHANGES.

W. D. Mackie, general merchant, Dunrea, Man., has sold out to J. H. Fawcett.

A. Dulmage, general merchant, Lakelet, Ont. has sold out to Wm. Burkfield.

D. Nesbitt, general merchant, Wellwood, Man., has been succeeded by W. Nesbitt.

W. N. Secord, general merchant, Winona, Ont., has been succeeded by Budge Bros.

Graves & McGuire, general merchants, Vienna, Ont., are removing to Wallaceburg.

Richard Common, general merchant, Winthrop, Ont., has removed to Newbridge, Ont.

J. R. Bellamy, general merchant, Black Bank, Ont., has been succeeded by Wm. Duffin.

T. W. Thompson, general merchant, Barwick, Ont., has sold out to C. R. Langstaff.

Margaret Baker, general merchant, St. Augustine, Ont., has been succeeded by Charles Moss.

DEATHS.

Henry R. Buzzell, of Buzzell Bros., general merchants, Cowansville, Que., is dead.

Walter McCormick, of Wm. McCormick & Son, general merchants, Annapolis, N.S., is dead.

A RETAILER'S CATALOGUE.

A. Sweet & Co., general merchants, Winchester, Ont., have adopted a method of meeting the competition of the large city departmental that is both unique and enterprising. They are "meeting the enemy with his own weapons" by issuing a semi-annual catalogue which is almost as comprehensive and up-to-date as those sent out by the large city concerns. The pages are replete with illustrations, price-lists, etc., of every line of goods handled by Sweet & Co. The effect of the work as an advertising medium is much increased by the argument in the preface, "Why customers should patronize this house," which reads in part as follows: "Our sales for 1900 were larger than for any previous year in the history of our business, and so far 1901 shows a substantial increase over the corresponding months of last year. . . . We pay cash and save the discounts. We are willing to work on a close margin. We mark all our goods in plain figures and are not afraid to publish our prices. . . . People who formerly sent away to the city stores tell us they prefer to buy here, because they can make their selections personally, can fit on garments and see just what they are getting before paying their money, and then they have the privilege of returning any article not found satisfactory."

TRADE CHAT.

N. Tremblay, sawmiller, Les Eboulements, Que., is dead.

Robert McLaren is opening a blacksmith shop at Cardigan, P.E I.

Marchand & Brateau, carriagemakers, Montreal, have dissolved.

H. White, harness dealer, Parrsboro', N.S., has given up business.

Thos. McKay, blacksmith, Tavistock, Ont., has sold out his business.

Geo. Wood, blacksmith, Carrville, Ont., is advertising his business for sale.

F. Reardon, wholesale paint dealer, etc., Halifax, is offering 30c. on the dollar.

J. B. Cane has bought out James A. Shoeman, hardware dealer, Virden, Man.

James R. Bower's planing mills at Shelburne, N S., have been destroyed by fire; no insurance.

E. E. Gauvin & Co., hardware dealers, and Charles Benault, saddler, Magog, Que., have been burned out.

The stock of the estate of E. H. Williams, tinsmith, Sintaluta, Man., is advertised for sale by tender to-day (Saturday).

John Finnegan & Son, carriage dealers, etc., and Peter Milne, coal dealer, Belleville, have suffered loss by fire.

It is understood that arrangements have been made for the establishment of an automobile factory in London, Ont., capitalized at $250,000.

J. Trelford, who opened a hardware store in Proton, Ont., some time ago, has been successful in securing a good trade. His business is steadily extending.

TORONTO TRAVELLERS TO MEET.

The regular monthly meeting of the City Travellers' Association of Toronto will be held in St. George's Hall on Friday evening this week. After the business is over a social evening will be spent, music and euchre being the attractions.

CATALOGUES, BOOKLETS, ETC.

BLISS MNFG. CO. CATALOGUES.

THE Bliss Mnfg. Co., Pawtucket, R.I., have issued two new catalogues. The smaller of these is devoted entirely to tool chests, of which this firm make a great variety from small boys' chests listing at $15 per gross to machinists' or jewelers' chests listing as high as $18 each. The larger of the two catalogues is more general, including descriptions of wood hardware and specialties, wood turnings of every descriptions, mill supplies, tool chests and field croquet sets. As this firm manufacture many excellent lines, and as full descriptions and list prices are given, these catalogues should be of considerable value to hardware dealers.

A HARDWARE RETAILER'S BOOKLET.

M. Weichel & Son, hardware dealers, Elmira and Waterloo, Ont., are about as ambitious as they are enterprising. About 20 years ago Mr. Weichel started business in Elmira in a small way and succeeded in building up a good, steadily-growing trade. Five years ago the firm opened a branch in Waterloo, ten miles distant, which has also proven a success. They have prepared and are now distributing among their customers a neat memorandum booklet showing why their business has grown and how they are now in a position to give their customers the best possible value. Half of the booklet is devoted to this end; the other half is ruled off for memoranda, which should make the booklet useful to customers as well as a good advertisement for M. Weichel & Son.

TINSMITH'S PATTERN MANUAL.

The American Artisan Press, 69 Dearborn street, Chicago, are issuing a second edition of the Tinsmith's Pattern Manual, the first edition of which was published eight years ago, and which proved so valuable that the edition was exhausted some time ago. The new edition has been thoroughly revised and brought up-to-date. This work is not only authoritative, but comprehensive, the first chapters preparing the beginner for the difficult technical operations which are described in the succeeding chapters. This work should be secured by all tinsmiths who have not a copy of the first edition.

Joseph McAdam, St. Thomas, Ont., has been appointed liquidator to wind up the Copp Bros. Co., stove manufacturers, etc., Hamilton, Ont., the firm being unable to meet the claims of the Merchants Bank. Copp Bros. have been in business for half a century, and employed 130 hands.

"WHY" IN ADVERTISING.

BUYERS of goods are not satisfied with being told that the goods are the best. They are not satisfied with bare assertions. They want to know "why," says C. A. Bates.

There is always some reason why the maker or seller of goods believes that his goods are more desirable than others in his line.

There is always some reason why each man in business believes he has a right to be in business. There is some reason why he thinks people should trade with him rather than with his competitors. Generalities won't do, he must come down to facts, or at least what he thinks are facts.

The question "why?" is in every one's mind.

The more definitely and completely it can be answered, the better for the advertiser.

If a man expects to sell goods nowadays, he must be able to answer convincingly.

The other day I talked to a hardware-man from Indiana. He has by far the best store in his place—the best store in his county or in several surrounding counties. He wanted more trade than he had—said he deserved more trade than he had—said there was in his county possible trade in his line double the amount that he is getting.

I looked over his ads. and told him that I could not see no reason under the sun why anybody should ever buy anything in his store; told him that I didn't believe he deserved trade.

That stirred him up and in five minutes he gave me more reasons "why" people should trade with him that he had given in his ads. for a year.

He hadn't told people "why."

When he started to write an ad. he was burdened with the idea that he was performing some sort of literary gymnastics, and that he must say something fine and star-spangled if he died in the attempt. As a matter of fact, he filled his ads. with words that meant absolutely nothing; they told nothing about his store, they gave no idea of his reasons for expecting trade. They would fit any other store under the sun just as well as they would fit his store.

He didn't realize that the thing to do was to write the news of his business.

Business news is as interesting to people as any other news. People really want to spend their money.

The man who receives his salary on Saturdays generally doesn't have any money left by the next Thursday.

People who receive quarterly incomes usually have to economize for a month before the income comes. They have spent the previous quarter's money long ago.

They are going to spend their money somewhere. They are going to spend it in the store that best answers their question "Why?"

BE CAUTIOUS AND FIRM.

"Once establish yourself and your mode of life as what they really are, and your foot is on solid ground, whether for the gradual step onward for the sudden spring over the precipice. From these maxims let me deduce another, once wrote Bulwer Lytton.

"Learn to say 'No' with decision; 'Yes' with caution. 'No' with decision whenever it meets a temptation. 'Yes' with caution whenever it implies a promise. A promise given is a bond inviolable. A man is already of consequence in the world when it is known we can implicitly rely on him. I have frequently seen such a man preferred to a long list of applicants for some important charge; he has been lifted at once into station and fortune merely because he has this reputation—that when he knows a thing, he knows; and when he says he will do a thing, he will do it."

STEPPING STONES TO SUCCESS.

ALL the world longs for and strives for success. It's the powerful magnet around which restless humanity whirls and struggles in its desperate efforts to gain admittance within its enchanting circle.

It's the world's "Grand Prix." It's the reward of merit, the laurel of superiority and the distinguishing badge of pre-eminence.

It's worth striving for, it's worth the great price which it commands. Every one has a theory for gaining it.

The writer's theory is simply this:

First resolve to succeed, next fix your aim, then work mightily for it.

Resolve earnestly, aim high, work unceasingly.

The winning of success is a serious matter—it requires a firm resolution at the start. Make up your mind first of all that you will succeed. Begin there.

Aim. The fixing of your aim is important. You must have something definite to work for, some shining light ahead to steer by, or you'll drift upon the rocks that lie hidden in numbers along the dangerous course.

Decide early exactly what your line of work is to be and let your decision be arrived at only after a thorough study of your ability, capacity and inclination. Find the work you are best adapted for. When you have discovered it fix your aim for achievement in it. Place the mark high and keep your eyes ever on it. Think about it, dream about it, expect it, work for it, reach it.

Work. Work mightily. The resolve and the aim will count for nothing without the work. Fortunate is the man who has a large and ready capacity for work. Difficult it will be for the one who finds work a hardship, a drudgery, an unpleasant task—something to be quickly done with. Such will require extraordinary ambition to keep alive within the necessary motive power of energy.

The trouble with most people comes from their desire of winning success by some lucky stroke. They are looking for some short cut, some royal road, by which they will be enabled quickly and easily to reach the lofty heights. They are always ready to take a long chance and have great faith in their "lucky star," and are constantly expecting to stumble into something some time that'll lead to fame and fortune. They are lured away by every quick and easy scheme that comes to their notice.

But it won't do. Precious time is wasted in such ways and the prize is further out of reach than ever. If you're going to win it you must train for it. The course is a hard one, the competition is the very keenest—it's tremendous, it's world-wide, and the one who dares hope to finish in the lead must be prepared for a mighty struggle. He must be swift with great endurance, or go down with the crowd.

Get in training.

Be a prize winner.—The Chameleon.

MANITOBA MARKETS.

WINNIPEG, April 22, 1901.

BUSINESS is steady in all lines of hardware, and prices are being well maintained. In iron and steel, the outlook of the early spring is not being fulfilled, as all classes of these goods are advancing in price on the outside markets. Many American mills have withdrawn all their old quotations from this market, and their mills are refusing further orders. As usual at this season, there is a good demand for bluestone for the treatment of seed grain. In paints, oils and glass, the demand is good. The surprise of the week was the 3c. advance of linseed oil, which was entirely unexpected. Just for the moment, there is an easier feeling in turpentine, and, to practically cash customers, figures are being made lower than actual market quotations, but it is understood that this is only temporary.

A good deal of interest is being exhibited in binder twine. The fact that the soil is in fine condition for seeding and the acreage prepared for seeding large has no doubt induced some dealers to place large orders early. On the other hand, a good many will not buy until there is at least some chance of estimating on the possible amount of the crop. A good many orders were placed when twine was at 7½ to 8c.

Quotations for the week are as follows:

Barbed wire, 100 lb	$3 45
Plain twist	3 45
Staples	3 95
Oiled annealed wire 10	3 95
" 11	4 00
" 12	4 05
" 13	4 20
" 14	4 35
" 15	4 45
Wire nails, 30 to 60 dy, keg	3 45
" 16 and 20	3 50
" 10	3 55
" 8	3 65
" 6	3 70
" 4	3 85
" 3	4 10
Cut nails, 30 to 60 dy.	3 00
" 20 to 40	3 05
" 10 to 16	3 10
" 8	3 15
" 6	3 20
" 4	3 30
" 3	3 65
Horsenails, 45 per cent. discount.	
Horseshoes, iron, No. 0 to No 1	4 65
No. 2 and larger	4 40
Snow shoes, No. 0 to No. 1	4 90
No. 2 and larger	4 40
Steel, No. 0 to No. 1	4 95
No. 2 and larger	4 70
Bar iron, $2.50 basis.	
Swedish iron, $4.50 basis.	
Sleigh shoe steel	3 00
Spring steel	3 25
Machinery steel	3 75
Tool steel, Black Diamond, 100 lb	8 50
Jessop	13 00
Sheet iron, black, 10 to 20 gauge, 100 lb.	3 50
20 to 26 gauge	3 75
28 gauge	4 00
Galvanized American, 16 gauge	2 54
18 to 22 gauge	4 50
24 gauge	4 75
26 gauge	5 00
28 gauge	5 25
Genuine Russian, lb	12
Imitation " "	8
Tinned, 24 gauge, 100 lb	7 55
26 gauge	8 80
28 gauge	8 00
Tinplate, IC charcoal, 20 x 28, box	10 75
" IX "	12 75
" IXX "	14 75
Ingot tin	35
Canada plate, 18 x 21 and 18 x 24	3 75
Sheet zinc, cask lots, 100 lb	7 50
Broken lots	8 00
Pig lead, 100 lb	6 00
Wrought pipe, black up to 2 inch	50 an 10 p.c.
" Over 2 inch	50 p.c.
Rope, sisal, 7-16 and larger	$10 00
" ⅜	10 50
" ¼ and 5-16	11 00
Manila, 7-16 and larger	13 50
" ⅜	14 00
" ¼ and 5-16	14 50
Solder	21½
Cotton Rope, all sizes, lb	16
Axes, chopping	$ 7 50 to 12 00
" double bitts	12 00 to 18 00
Screws, flat head, iron, bright	87½
Round " "	82½
Flat " brass	80
Round " "	75
Coach	57½ p.c.
Bolts, carriage	55 p.c.
Machine	55 p.c.
Tire	60 p.c
Sleigh shoe	65 p.c.
Plough	40 p.c.
Rivets, iron	50 p.c.
Copper, No. 8	35
Spades and shovels	40 p.c.
Harvest tools	50, and 10 p.c.
Axe handles, turned, s. g. hickory, doz.	$2 50
No. 1	1 50
No. 2	1 25
Octagon extra	1 75
No. 1	1 25
Files common	70, and 10 p.c.
Diamond	60
Ammunition, cartridges, Dominion R.F.	50 p.c.
Dominion, C.F., pistol	30 p.c.
" military	15 p.c.
American R.F.	30 p.c.
C.F. pistol	5 p.c.
C.F. military	10 p.c. advance.
Loaded shells:	
Eley's soft, 12 gauge black	16 50
chilled, 12 guage	18 00
soft, 10 guage	21 00
chilled, 10 guage	23 00
Shot, Ordinary, per 100 lb	6 75
Chilled	7 50
Powder, F.F., keg	4 75
F.F.G.	5 00
Tinware, pressed, retinned	75 and 2½ p.c.
" " plain	70 and 15 p.c.
Graniteware, according to quality	50 p.c.

PETROLEUM.

Water white American	25½c.
Prime white American	24c.
Water white Canadian	22c,
Prime white Canadian	21c.

PAINTS, OILS AND GLASS

Turpentine, pure, in barrels	$ 68
Less than barrel lots	73
Linseed oil, raw	80
Boiled	83
Lubricating oils, Eldorado castor	25½
Eldorado engine	24½
Atlantic red	27½
Renown engine	41
Black oil	23½ to 25
Cylinder oil (according to grade)	55 to 74
Harness oil	61
Neatsfoot oil	$ 1 00
Steam refined oil	85
Sperm oil	1 50
Castor oil per lb.	11½
Glass, single glass, first break, 16 to 25 united inches	2 25
26 to 40 per 50 ft.	2 50
41 to 50 " 100 ft.	5 50
51 to 60 " " "	6 00
61 to 70 per 100-ft. boxes	6 50
Putty, in bladders, barrel lots per lb.	2¼
kegs "	2½
White lead, pure per cwt.	7 25
No 1 "	7 00
Prepared paints, pure liquid colors, according to shade and color per gal.	$1.30 to $1.90

NOTES.

It is rumored that The Lac du Bonnet Brick Co. will open an office in Winnipeg this season.

J. T. Black of The McClary Manufacturing Co., is calling for tenders for the purchase of the hardware stock of E. H. Williams, Sintaluta.

CURRENT MARKET QUOTATIONS

April 26, 1901.

These prices are for such qualities and quantities as are usually ordered by retail dealers on the usual terms of credit, the lowest figures being for larger quantities and prompt pay. Large cash buyers can frequently make purchases at better prices. The Editor is anxious to be informed at once of any apparent errors in this list as the desire is to make it perfectly accurate.

METALS.

Tin.

Lamb and Flag and Straits—56 and 28 lb. ingots, per lb.	0 31	0 32

Tinplates.

Charcoal Plates—Bright

		Per box
M.L.S., equal to Bradley.		
I.C., usual sizes	....	$6 75
I.X., "	....	8 25
I.X.X., "	....	9 75
Famous—		
I.C.	....	6 75
I.X	....	8 25
I.X.X	....	9 50
Raven & Vulture Grades—		
I.C., usual sizes	....	5 00
I.X., "	....	6 00
I.X.X "	....	7 00
I.XXX., "	....	8 00
D.C., 12⅛x17	....	4 75
D.X.	....	5 50
D.X.X	....	7 50

Coke Plates—Bright

Bessemer Steel—		
I.C., usual sizes	....	4 15
I.C., special sizes, base	....	4 50
20 x 28	....	8 50

Charcoal Plates—Terne

Dean or J. G. Grade—		
I.C., 20x28, 112 sheets	....	8 25
I.X., Terne Tin	....	10 25

Charcoal Tin Boiler Plates.

		Per lb.
Cookley Grade—		
X X., 14x56, 50 sheet bxs; " 14x60, "; " 14x65, "	0 07	0 07½

Tinned Sheets

72x30 up to 24 gauge	0 08	0 08½
" 26 "	0 08½	0 09
" 28 "	0 09	0 09½

Iron and Steel.

Common Bar, per 100 lbs	1 85	1 90
Refined " "	....	2 35
Horse Shoe Iron "	....	2 25
Hoop steel, 1½ to 3 in. base, extras for smaller sizes	3 10	3 25
Sleigh Shoe Steel " base	....	2 10
Tire Steel	2 30	2 50
Reeled Machinery	....	3 00
ast Steel, per lb	0 00	0 00
Toe Calk Steel	....	2 40
T. Firth & Co's special cast steel, per lb.		0 13
Jessop's Tool Steel	....	0 13

Boiler Tubes.

1½-inch	....	0 12½
2 "	....	0 13
2½ "	....	0 15
3 "	....	0 16
3½ "	....	0 20
4 "	....	0 25

Steel Boiler Plate.

¼ inch	....	2 40
3-16 inch	....	2 50
⅜ inch and thicker	....	2 60

Black Sheets.

18 gauge	....	3 00
20 gauge	....	3 00
22 to 24 "	....	3 10
26 "	....	3 20
28 "	....	3 30

Canada Plates.

Al dull, 52 sheets	....	3 00
Half polished	....	3 15
All bright	3 65	3 75

Iron Pipe.

Black pipe—		
⅛ "	....	4 45
¼ inch	....	3 25
⅜ "	....	3 30
½ "	....	3 50
¾ "	....	3 66
1 "	....	5 15
1¼ "	....	7 10
1½ "	....	8 40
2 "	....	11 25
2½ "	....	20 95
3 "	....	24 55
3½ "	....	30 75
4 "	....	39 00
5 "	....	47 35
6 "	....	62 10
Galvanized pipe—		
½ inch	....	4 90
¾ "	....	5 25
1 "	....	7 55
1¼ "	....	10 30
1½ "	....	12 35
2 "	....	16 50

Galvanized Sheets.

	G C.	Comet.	Amer.	Queen's Head.
16 gauge		....	4 25	3 90
18 to 24 gauge	4 15	3 85	4 35	4 15
26 "	4 40	4 10	4 35	4 40
28 "	4 65	4 35	4 50	4 65

28 gauge American equals 26 gauge English.

Chain.

Proof Coil, 3-16 in., per 100 lb	....	....
" ¼ " "	8 00	8 50
" 5-16 " "	4 70	5 00
" ⅜ " "	4 05	4 50
" 7-16 " "	3 90	4 25
" ½ " "	3 10	4 10
" 9-16 " "	3 65	4 65
" ⅝ " "	3 35	3 90
" ¾ " "	3 60	4 40

Halter, kennel and post chains, 40 and 50 p.c.
Cow ties .... 40 p.c.
Stall fixtures .... 35 p.c.
Trace chain .... 25 p.c.
Jack chain, iron, single and double, discount 35 p.c.
Jack chain, brass, single and double, discount 40 p.c.

Copper.

Ingot

English B. S., ton lots	....	0 19
Lake Superior	....	....

Bolt or Bar.

Cut lengths round, ½ to ⅞ in.	0 23½	0 25
" round and square 1 to 2 inches	0 23½	0 25

Sheet.

Plain, 14 oz., and light, 16 oz., 14x48 and 14x60	0 23	0 23
Plain, 14 oz., and light, 16 oz., irregular sizes	0 23	0 23½
Tinned copper sheets	....	0 26
Planished	....	0 32

Braziers (In sheets.)

4x6ft. 25 to 30 lbs. ea., per lb.	....	0 25½
" 35 to 45 " "	....	0 24½
" 50-lb. and above, "	....	0 23½

Boiler and T. K. Pitts

Plain Tinned, per lb	....	0 28
Spun, per lb	....	0 32

Brass.

Rod and Sheet, 14 to 30 gauge, 15 per cent.

Sheets, hard-rolled, 2x4	0 24	0 25
Tubing, base, per [illegible]	0 24	0 25

Zinc Spelter

Foreign, per lb	0 05½	0 06
Domestic "	....	..

Zinc Sheet.

5 cwt. casks	....	00 6½
Part casks	....	00 6¾

Lead.

Imported Pig, per lb	0 04½	0 04¾
Bar, 1 lb	0 05½	0 06½
Sheets, 2½ lbs. sq. ft., by	....	0 06½
Sheets, 3 to 6 lbs., "	....	0 06

NOTE.—Cut sheets ½ cent per lb. extra. Pipe, by the roll, usual weights per yard, lists at 7c. per lb. and 25 p.c. dis. f.o.b. Toronto.

NOTE.—Cut lengths, net price, waste pipe 8-ft. lengths lists at 7½ cents.

Shot.

Common, $6.50 per 100 lb.; chilled, $7.00 per 100 lb.; buck, seal and ball, $7.50. Discount, 25 p.c. Prices are f o.b. Toronto, Hamilton, Montreal, St. John and Halifax. Terms 3 per cent. cash, freights equalized on Montreal.

Soil Pipe and Fittings.

Discount, 60 and 10 per cent. on medium and extra heavy, and 60 per cent. on light.

Solder.

	Per lb.	Per lb.
Bar half-and-half, guarant'd	....	0 18½
Bar half-and-half, commer'l	....	0 18
Refined	....	0 17½
Wiping	....	0 17

NOTE.—Prices of this graded according to quantity. The prices of other qualities of solder in the market indicated by private brands vary according to composition.

Antimony.

Cookson's, per lb	0 10½	0 11

White Lead.

	Per 100 lb.
Pure	6 37
No. 1 do	6 00
No. 2 do	5 62½
No. 3 do	5 55
No. 4 do	4 87½
Munro's Select Flake White	7 37½
Elephant and Decorators' Pure	7 12½
Brandram's B B. Genuine	8 00
" " Decorative	7 55
" " No. 1	6 85
" " No. 2	6 00

Red Lead.

Genuine, 560 lb. casks, per cwt.	....	$5 50
Genuine, 100 lb. kegs, per cwt.	....	5 75
No. 1, 560 lb. casks, per cwt.	....	5 25
No. 1, 100 lb. kegs, per cwt	....	5 00

White Zinc Paint.

Elephant Snow White	0 08	0 09
Pure White Zinc	0 08	0 09
No. 1	0 06	0 07½
No. 2	0 05	0 06½

Dry White Lead.

Pure, casks	5 75
Pure, kegs	6 25
No. 1, casks	5 50
No. 1, kegs	5 00

Prepared Paints.

In ¼, ½ and 1 gallon tins.

Pure, per gallon	....	1 25
Second qualities, per gallon	....	1 10
Barn (in bbls.)	0 75	0 85
The Sherwin-Williams Paints	....	1 45
Canada Paint Co's Pure	....	1 25
Toronto Lead & Color Co's Pure	....	1 25
Sanderson Pearcy's Pure	....	1 20
Stewart & Wood's Champion Pure.		1 20

Colors in Oil.

25 lb. tins, Standard Quality.

Venetian Red, per lb.	0 05
Chrome Yellow	0 11
Golden Ochre	0 06
French "	0 05
Marine Black	0 09
" Green	0 09
Chrome "	0 08
French Imperia Green	0 09

Colors, Dry.

Yellow Ochre (J.C.) bbls.	1 35	1 40
Yellow Ochre (J.F.L.S.), bbls	...	2 75
Yellow Ochre (Royal)	1 10	1 15
Brussels Ochre	....	2 00
Venetian Red (best), per cwt.	1 80	1 90
English Oxides, per cwt.	3 00	3 25
American Oxides, per cwt.	1 75	2 00
Canadian Oxides, per cwt.	1 75	2 00
Super Magnetic Oxides, 93p.c.	2 00	2 25
Burnt Sienna, pure, per lb.	....	0 10
" Umber, " "	....	0 10
do Raw	....	0 09
Drop Black, pure	....	0 09
Chrome Yellows, pure	....	0 18
Chrome Greens, pure, per lb.	....	0 12
Golden Ochre	....	0 03½
Ultramarine Blue in 28-lb. boxes, per lb.	0 08	0
Fire Proof Mineral, per 100 lb	....	1
Genuine Eng. Litharge, per lb	....	0
Mortar Color, per 100 lb.	....	1
English Vermillion	....	0
Pure Indian Red, No. 45 lb.	....	0 [illegible]
Whiting, per 100 lb.	....	0 [illegible]

Blue Stone.

Casks, for spraying, per lb.	0 07
100-lb. lots, do. per lb.	0 08

Putty.

Bulk in bbls.	1
Bulk in less quantity	2
Bladders in bbls.	2
Bladders in kegs, boxes or loose	2
Bladders in 25-lb. tins	2
Bladders in 12½-lb. tins	2 90
Bladders in bulk or tins less than 100 lb	2 [illegible]

Varnishes.

	In 5-gal. lots.).	Per gal.
Carriage, No. 1	2 90	3 30
" body	8 00	9 00
" rubbing	4 00	5 00
Gold Size, Japan	3 00	3 40
Brown Japan	2 40	2 80
Elastic Oak	2 90	3 30
Furniture, extra	2 40	2 80
" No. 1	1 60	2 00
Hard Oil Finish	2 70	3 10
Light Oil Finish	3 20	3 60
Demar	3 30	3 70
Shellac, white	4 40	4 80
" orange	4 00	4 40
Furniture Brown Japan	1 60	2 00
Black Japan	2 40	2 80
" " No. 1	1 60	2 00

The Imperial Varnish & Color Co's., Limited Elastilite Varnish 1 gal. can, each. $3.00.

ranatine Floor Finish, per gal., $2.50.

Maple Leaf Coach Enamels; Size 1, 60c.; Size 2, 35c.; Size 3, 20c. each.

Castor Oil.

East India, in cases, per lb.	0 10	0 10½
" " small lots	0 10½	0 11

Cod Oil, Etc.

Cod Oil per gal	0 50	0 55
Pure Olive	....	1 20
" Neatsfoot	....	90

Glue.

Common	0 08½	0 02
French Medal	0 14	0 14½
Cabinet sheet	0 12	0 13
White, extra	0 18	0 20
Gelatine	0 22	0 30
Strip	0 18	0 20
Coopers	0 19	0 20
Huttner	....	0 18

HARDWARE.

Ammunition.

Cartridges.

B. B. Cap Dom. 50 and 5 per cent.
Rim Fire Pistol, dis. 40 p. c., Amer.
Rim Fire Cartridges, Dom., 50 and 5 p. c.
Central Fire Pistol and Rifle, 10 p.c. Amer.
Central Fire Cartridges, pistol sizes, Dom. 30 per cent.
Central Fire Cartridges, Sporting and Military, Dom., 15 and 5 per cent.
Central Fire, Military and Sporting, Amer, add 5 p.c. to list. B.B. Caps, discount 40 per cent, Amer.
Loaded and empty Shells, "Trap" and "Dominion" grades, 25 per cent. Rival and Nitro, net list.
Brass Shot Shells, 55 per cent.
Primers, Dom., 30 per cent.

Wads — per lb.

Best thick white felt wadding, in ¼-lb bags ... 1 00
Best thick brown or grey felt wads, in ½-lb. bags ... 70
Best thick white card wads, in boxes of 500 each, 12 and smaller gauges ... 0 99
Best thick white card wads, in boxes of 500 each, 10 gauge ... 0 35
Best thick white card wads, in boxes of 500 each, 8 gauge ... 0 55
Thin card wads, in boxes of 1,000 each, 12 and smaller gauges ... 0 20
Thin card wads, in boxes of 1,000 each, 10 gauge ... 0 25
Thin card wads in boxes of 1,000 each, 8 gauge ...
Chemically prepared black edge grey cloth wads, in boxes of 250 each— Per M.
11 and smaller gauge ... 0 60
9 and 10 gauges ... 0 70
7 and 8 gauges ... 0 90
5 and 6 gauges ... 1 10
Superior chemically prepared pink edge, best white cloth wads, in boxes of 250 each—
11 and smaller gauge ... 1 15
9 and 10 gauges ... 1 40
7 and 8 gauges ... 1 65
5 and 6 gauges ... 1 90

Adzes.

Discount, 20 per cent.

Anvils.

Per lb ... 0 10 ... 0 12½
Anvil and Vise combined ... 4 50
Wilkinson & Co.'s Anvils..lb. 0 09 ... 0 09½
Wilkinson & Co.'s Vices..lb.. 0 09¾ ... 0 10

Augers.

Gilmour's, discount 65 and 5 p.c. off list.

Axes.

Chopping Axes
Single bit, per doz ... 5 50 ... 10 00
Double bit, " ... 11 00 ... 18 00
Bench Axes, 40 p.c.
Broad Axes, 33⅓ per cent.
Hunters' Axes ... 5 50 ... 6 00
Boy's Axes ... 5 75 ... 6 75
Splitting Axes ... 6 50 ... 12 00
Handled Axes ... 7 00 ... 10 00

Axle Grease.

Ordinary, per gross ... 5 75 ... 6 00
Best quality ... 13 00 ... 15 00

Bath Tubs.

Zinc ... 6 00
Copper, discount 15 p.c. off revised list

Baths.

Standard Enameled.

5½-inch rolled rim, 1st quality ... 25 00
" " " 2nd " ... 21 00

Anti-Friction Metal.

"Tandem" A ... per lb. 0 27
" B ... " 0 21
" C ... " 0 11½
Magnolia Anti-Friction Metal, per lb. 0 25
Frictionless metal, per lb ... 0 23

SYRACUSE SMELTING WORKS.

Aluminum, genuine ... 0 45
Dynamo ... 0 29
Special ... 0 25
Aluminum, 99 p.c. pure "Syracuse" .. 0 50

Bells.

Hand.

Brass, 60 per cent.
Nickel, 55 per cent.

Cow.

American make, discount 66⅔ per cent.
Canadian, discount 45 and 50 per cent.

Door.

Gongs, Sargent's ... 5 50 ... 8 00
" Peterboro', discount 45 per cent.

Farm.

American, each ... 1 25 ... 3 00

House.

American, per lb ... 0 35 ... 0 40

Bellows.

Hand, per doz ... 3 35 ... 4 75
Moulders', per doz ... 7 50 ... 10 00
Blacksmiths', discount 40 per cent.

Belting.

Extra, 60 per cent.
Standard, 60 and 10 per cent.

Bits.

Auger.

Gilmour's, discount 60 and 5 per cent.
Rockford, 50 and 10 per cent.
Jennings' Gen., net list.

Car.

Gilmour's, 47½ to 50 per cent.

Expansive.

Clark's, 40 per cent

Gimlet.

Clark's, per doz ... 0 65 ... 0 90
Diamond, Shell, per doz ... 1 00 ... 1 50
Nail and Spike, per gross ... 2 25 ... 5 20

Blind and Bed Staples.

All sizes, per lb ... 0 07½ ... 0 12

Bolts and Nuts. Per cent.

Carriage Bolts, full square, Norway ... 65
" " full square ... 65
Common Carriage Bolts, all sizes ... 60
Machine Bolts, all sizes ... 60
Coach Screws ... 70
Sleigh Shoe Bolts ... 72½
Blank Bolts ... 60
Bolt Ends ... 62½
Plough Bolts ... 60
Nuts, square ... 4 c. off
Nuts, hexagon ... 4¼c. off
Tire Bolts ... 67½
Stove Bolts ... 67½
Stove rods, per lb ... 5½ to 6c.

Boot Calks.

Small and medium, ball, per M ... 4 25
Small heel, per M ... 4 50

Bright Wire Goods.

Discount ... 62½ per cent.

Broilers.

Light, dis., 65 to 67½ per cent.
Reversible, dis., 65 to 67½ per cent.
Vegetable, per doz., dis. 37½ per cent.
Henis, No. 8, " ... 6 00
Henis, No. 9, " ... 7 00
Queen City " ... 7 50 ... 0 00

Butchers' Cleavers.

German, per doz ... 6 00 ... 11 00
American, per doz ... 12 00 ... 20 00

Building Paper, Etc.

Plain building, per roll ... 0 30
Tarred lining, per roll ... 0 40
Tarred roofing, per 100 lb ... 1 65
Coal Tar, per barrel ... 3 50
Pitch, per 100-lb ... 0 85
Carpet felt, per ton ... 45 00

Bull Rings.

Copper, $2.00 for 2½ in. and $1.90 for 2 in.

Butts.

Wrought Brass, net revised list

Cast Iron.

Loose Pin, dis., 60 per cent.

Wrought Steel.

Fast Joint, dis. 65, 10 and 2½ per cent.
Loose Pin, dis. 65, 10 and 2½ per cent.
Berlin Bronzed, dis. 70, 70 and 5 per cent.
Gen. Bronzed, per pair ... 0 40 ... 0 65

Carpet Stretchers.

American, per doz ... 1 00 ... 1 50
Bullard's, per doz ... 6 50 ... .

Castors.

Bed, new list, dis. 55 to 57½ per cent.
Plate, dis. 52½ to 57½ per cent.

Cattle Leaders.

Nos. 31 and 32, per gross ... 50 ... 9 50

Cement.

Canadian Portland ... 2 50 ... 2 80
English " 3 00
Belgian " ... 2 50 ... 2 75
Canadian hydraulic ... 1 25 ... 1 50

Chalk.

Carpenters Colored, per gross ... 0 45 ... 0 75
White lump, per cwt ... 0 60 ... 0 65
Red ... 0 05 ... 0 06
Crayon, per gross ... 0 14 ... 0 18

Chisels.

Socket, Framing and Firmer.

Broad's, dis. 70 per cent.
Warnock's, dis. 70 per cent.
P. S. & W. Extra 60, 10 and 5 p.c.

Churns.

Revolving Churns, metal frames—No. 0, $8—No. 1, $8.50—No. 2, $9.00—No. 3, $10.00—No. 4, $12.00—No. 5, $16.00 each. Ditto, wood frames—20c. each less than above. Discounts: Delivered from factories, 58 p.c.; from stock in Montreal, 58 p.c. Terms, 4 months or 3 p.c. cash in 30 days.

Clips.

Axle dis. 65 per cent.

Closets.

Plain Ontario Syphon Jet ... $16 00
Emb. Ontario Syphon Jet ... 17 00
Fittings net ... 1 00
Plain Teutonic Syphon Washout ... 10 00
Emb. Teutonic Syphon Washout ... 11 00
Fittings net ... 1 25
Low Down Teutonic, plain ... 16 00
" " " embossed ... 17 00
Plain Richelieu net ... 3 75
Emb. Richelieu net ... 4 00
Fittings net ... 1 25
Low Down Ont. Sy. Jet, plain net ... 19 50
" " " " " emb'd. net ... 20 50
Closet connection net ... 1 25
Basins, round, 14 in ... 1 00
" oval, 17 x 14 in ... 2 50
" " 19 x 15 in ... 3 75
Discount 40 p.c., except on net figures.

Compasses, Dividers, Etc.

American, dis. 62½ to 65 per cent.

Cradles, Grain.

Canadian, dis. 25 to 33⅓ per cent.

Crosscut Saw Handles.

S. & D., No. 3, per pair ... .17½
" " 5, " ... .22½
" " 6, " ... .15
Boynton pattern " ... .20

Door Springs.

Torrey's Rod, per doz ... (15 p.c.) ... 2 00
Coil, per doz ... 0 88 ... 1 60
English, per doz ... 2 00 ... 4 00

Draw Knives.

Coach and Wagon, dis. 50 and 10 per cent.
Carpenters, dis. 70 per cent.

Drills.

Hand and Breast.

Millar's Falls, per doz. net list.

DRILL BITS.

Morse, dis., 37½ to 40 per cent.
Standard dis. 50 and 5 to 55 per cent

Faucets.

Common, cork-lined, dis. 35 per cent.

ELBOWS. (Stovepipe.)

No. 1, per doz ... 1 40
No. 2, per doz ... 1 20
Bright, 20c. per doz. extra.

ESCUTCHEONS.

Discount, 45 per cent.

ESCUTCHEON PINS.

Iron, discount 40 per cent.

FACTORY MILK CANS.

Discount off revised list, 40 per cent.

FILES.

Black Diamond, 70 and 5 per cent.
Kearney & Foote, 60 and 10 to 70 per cent.
Nicholson File Co., 60 to 60 and 10 per cent.
Jowitt's, English list, 25 to 27½ per cent.

FORKS.

Hay, manure, etc., dis., 50 and 10 per cent, revised list.

GLASS—Window—Box Price.

Size United Inches.	Star Per 50 ft.	Star Per 100 ft	D. Diamond Per 50 ft.	D. Diamond Per 100 ft.
Under 26	2 15	4 15	....	6 00
26 to 40	2 30	4 45	....	6 65
41 to 50	....	4 85	....	7 50
51 to 60	....	5 15	....	8 50
61 to 70	....	5 50	....	9 50
71 to 80	....	6 00	....	10 50
81 to 85	....	6 50	....	11 75
86 to 90	....	....	..	14 00
91 to 95	..	....	...	15 50
96 to 100	.. .	....	....	18 00

GAUGES

Marking, Mortise, Etc.

Stanley's dis. 50 to 55 per cent.

Wire Gauges.

Winn's, Nos. 26 to 33, each ... 1 65 ... 2 40

HALTERS.

Rope, ⅜ per gross ... .
" ½ 9 00
" ⅝ to ¾ 14 00
Leather, 1 in., per doz ... 3 87½ ... 4 00
" 1¼ in., " ... 5 15 ... 5 20
Web, — per doz ... 1 87 ... 2 45

HAMMERS.

Nail

Maydole's, dis. 5 to 10 per cent. Can. dis. 25 to 27½ per cent.

Tack.

Magnetic, per doz ... 1 10 ... 1 20

Sledge.

Canadian, per lb ... 0 07½ ... 0 08½

Ball Pean.

English and Can., per lb ... 0 22 ... 0 25

HANDLES.

Axe, per doz. net ... 1 50 ... 2 00
Store door, per doz ... 1 00 ... 1 50

Fork.

C. & B., dis. 40 per cent. rev. list.

Hoe.

C. & B., dis. 40 per cent. rev. list.

Saw.

American, per doz ... 1 00 ... 1 25

Plane.

American, per gross ... 3 15 ... 3 75

Hammer and Hatchet.

Canadian, 40 per cent.

Cross-Cut Saws.

Canadian, per pair ... 0 13½

HANGERS. doz. pairs.

Steel barn door ... 5 85 ... 6 00
Stearns, 4 inch 5 00
" 5 inch 6 50
Lane's covered—
No. 11, 5-ft. run 8 40
No. 11½, 10-ft. run 10 80
No. 12, 10-ft. run 12 60
No. 14, 15-ft. run 21 00
Lane's O.N.T. track, per foot 4½

HARVEST TOOLS.

Discount, 50 and 10 per cent.

HATCHETS.

Canadian, dis. 40 to 42½ per cent.

HINGES.

Blind, Parker's, dis. 50 and 10 to 60 per cent.
Heavy T and strap, 4-in., per lb ... 06½
" " 5-in., " ... 06¼
" " 6-in., " ...
" " 8-in., " ... 06¾
" " 10-in., " ... 0 06¾
Light T and strap, dis. 65 and 5 per cent.
Screw hook and hinge—
6 to 12 in., per 100 lbs 4 50
14 in. up, per 100 lbs 3 50
Per gro. pairs.
Spring 12 00

HOES.

Garden, Mortar, etc., dis. 50 and 10 p.c.
Planter, per doz ... 4 00 ... 4 50

HOLLOW WARE

Discount ... 45 and 5 per cent.

HOOKS.

Cast Iron.

Bird Cage, per doz ... 0 [illegible] ... 1 [illegible]
Clothes Line, per doz ... 0 [illegible] ... 0 [illegible]
Harness, per doz ... 0 50 ... 0 [illegible]
Hat and Coat, per gross ... 1 [illegible] ... 3 10
Chandelier, per doz ... 0 50 ... 1 [illegible]

Wrought Iron.

Wrought Hooks and Staples, Can., dis. 47½ per cent.

Wire.

Hat and Coat, discount 45 per cent.
Belt, per 1,000 ... 0 60
Screw, bright, dis. 55 per cent.

HORSE NAILS.

"C" brand 50 and 7½ p.c. off new list } Oval-
"M" brand 60 per cent. } head
Countersunk, 60 and 10 per cent.

HORSESHOES

F.O.B. Montreal.

Iron Shoes.	No. 2 and larger.	No. 1. and smaller.
Light, medium, and heavy.	3 50	3 75
Snow shoes	3 75	4 00
Steel Shoes.		
Light	3 60	3 85
Featherweight (al lsizes)	4 85	4 85

F.O.B. Toronto, Hamilton, London and Guelph, 10c. per keg additional.

Toe weight steel shoes...... 6 70

JAPANNED WARE.

Discount, 45 and 5 p.c. off list, June 1899

ICE PICKS.

Star per doz.......... 3 00 3 25

KETTLES.

Brass spun, 7½ p.c. dis. off new list.

Copper, per lb.......... 0 30 0 50

American, 60 and 10 to 65 and 5 p.c.

KEYS.

Lock, Can., dis., 45 p.c.

Cabinet, trunk, and padlock, Am. per gross.......... 60

KNOBS.

Door, japanned and N.P., per doz.	1 50	2 50
Bronze, Berlin, per doz.	2 75	3 25
Bronze Genuine, per doz.	6 00	9 00
Shutter, porcelain, F. & L. screw, per gross	1 30	4 00
White door knobs—per doz.		1 25

HAY KNIVES.

Discount, 50 and 10 per cent.

LAMP WICKS.

Discount, 60 per cent.

LANTERNS.

Cold Blast, per doz.	7 00
No. 3 "Wright's"	8 50
Ordinary, with O burner	4 00
Dashboard, cold blast	9 00
No. 0.	5 75

Japanning, 50c. per doz. extra.

LEMON SQUEEZERS.

Porcelain lined, per doz.	2 20	5 60
Galvanized	1 87	3 85
King, wood	2 75	2 90
King, glass	4 00	4 50
All glass	1 20	1 30

LINES.

Fish, per gross	1 05	2 50
Chalk "	1 90	7 40

LOCKS.

Canadian, dis. 45 p.c.

Russell & Erwin, per doz.... 3 00 3 25

Cabinet.

Eagle, dis. 30 p.c.

Padlock.

English and Am., per doz.	50	6 00
Scandinavian, "	1 00	2 40

Eagle, dis. 20 to 25 p.c.

MACHINE SCREWS. Iron and Brass.

Flat head discount 25 p.c.

Round Head, discount 20 p.c.

MALLETS.

Tinsmiths', per doz.	1 25	1 50
Carpenters', hickory, per doz.	1 25	3 75
Lignum Vitae, per doz.	3 85	5 00
Caulking, each	60	2 00

MATTOCKS.

Canadian, per doz.......... 5 50 6 50

MEAT CUTTERS.

American, dis. 25 to 30 p.c.

German, 15 per cent.

MILK CAN TRIMMINGS.

Discount, 25 per cent.

NAILS.

Quotations are:	Cut.	Wire.
2d and 3d	$3 35	$3 85
3d	3 00	3 52
4 and 5d	2 75	3 35
6 and 7d	2 65	3 20
8 and 9d	2 50	3 00
10 and 12d	2 45	2 95
16 and 20d	2 40	2 90
30, 40, 50 and 60d. (base)	2 35	2 85

Wire nails in carlots are $2.77½

Galvanizing 2c. per lb. net extra.

Steel Cut Nails 10c. extra.

Miscellaneous wire nails, dis. 70 and 10 p.c.

Coopers' nails, dis. 30 per cent.

Flour barrel nails, dis. 25 per cent.

NAIL PULLERS.

German and American....... 1 85 3 50

NAIL SETS

Square, round, and octagon, per gross	3 38	4 00
Diamond	12 00	15 00

NETTING.

Poultry, 50 and 5 per cent. for McMullen's

OAKUM. Per 100 lb.

Navy	6 00
U. S. Navy	7 25

OIL.

Water White (U.S.)	0 16½
Prime White (U.S.)	0 15½
Water White (Can.)	0 15
Prime White (Can.)	0 14

OILERS.

McClary's Model galvan. oil can, with pump, 5 gal., per doz. 10 00

Zinc and tin, dis. 50, 50 and 10.

Copper, per doz.	1 25	3 50
Brass, "	1 50	3 50

Malleable, dis. 25 per cent.

GALVANIZED PAILS.

Dufferin pattern pails, dis. 45 p.c.

Flaring pattern, discount 45 per cent.

Galvanized washtubs, discount 45 per cent.

PIECED WARE.

Discount 40 per cent. off list, June, 1899.

10-qt. flaring sap buckets, dis. 45 p.c.

6, 10 and 14-qt. flaring pails, dis. 45 p.c.

Creamer cans, dis. 45 p.c.

PICKS.

Per doz.......... 5 00 9

PICTURE NAILS.

Porcelain head, per gross	1 75	3 00
Brass head "	0 40	1 00

PICTURE WIRE.

Tin and gilt, discount 75 p.c.

PLANES.

Wood, bench, Canadian dis. 50 per cent. American dis. 50.

Wood, fancy Canadian or American 7½ o 40 per cent.

PLANE IRONS.

English, per doz.......... 2 00 5 00

PLIERS AND NIPPERS.

Button's Genuine per doz pairs, dis. 37½ 40 p.c.

Button's Imitation, per doz.	5 00	9 00
German, per doz.	0 60	2 60

PLUMBERS' BRASS GOODS.

Compression work, discount, 60 per cent.

Fuller's work, discount 65 per cent.

Rough stops and stop and waste cocks, discount, 60 per cent.

Jenkins disk globe and angle valves, discount, 55 per cent.

Standard valves, discount, 60 per per cent.

Jenkins' radiator valves discount 55 per cent.

" " " standard, dis., 60 p.c.

Quick opening valves discount, 60 p.c.

No. 1 compression bath cock	2 00
No. 4 " " "	2 00
No. 7, Fuller's	2 50
No 4½, "	3 00

POWDER.

Velox Smokeless Shotgun Powder.

100 lb. or less	0 85
1,000 lb. or more	0 80

Net 30 days.

PRESSED SPIKES.

Discount 20 to 25 per cent.

PULLEYS.

Hothouse, per doz.	0 55	1 00
Axle	0 22	0 33
Screw	0 27	1 00
Awning	0 35	2 50

PUMPS.

Canadian cistern	1 80	3 60
Canadian pitcher spout	1 40	2 10

PUNCHES.

Saddlers', per doz.	1 00	1
Conductors', "	9 00	15
Tinners' solid, per set	0 00	0
" hollow, per inch	0 00	1 00

RANGE BOILERS.

Galvanized, 3 gallons	7
" 35 "	8 00
" 40 "	9 00
Copper, 30 "	22 00
" 35 "	26 00
" 40 "	30 00

Discount off Copper Boilers 10 per cent.

RAKES.

Cast steel and malleable, 50, 10 and 5 p.c.

Wood, 25 per cent.

RASPS AND HORSE RASPS.

New Nicholson horse rasp, discount 60 to 60 and 10 p.c.

Globe File Co.'s rasps, 60 and 10 to 70 p.c.

Heller's Horse rasps, 50 to 50 and 5 p.c.

RAZORS.

	per doz.	
Geo. Butler & Co.'s,	8 00	18 00
Boker's	7 50	11 00
" King Cutter	12 50	50 00
Wade & Butcher's	3 60	10 00
Theile & Quack's	7 00	12 00
Elliot's	4 00	18 00

REAPING HOOKS.

Discount, 50 and 10 per cent.

REGISTERS.

Discount.......... 40 per cent.

RIVETS AND BURRS.

Iron Rivets, black and tinned, discount 60 and 10 per cent.

Iron Burrs, discount 55 per cent.

Extras on Iron Rivets in 1-lb. cartons, ½c. per lb.

Extras on Iron Rivets in ⅛-lb. cartons, 1c. per lb.

Copper Rivets & Burrs, 35 and 5 p.c. dis. and cartons, 1c. per lb. extra, net.

Extras on Tinned or Coppered Rivets ⅛-lb. cartons, 1c. per lb.

RIVET SETS

Canadian, dis. 35 to 37½ per cent.

ROPE ETC.

	Sisal.	Manila.
7-16 in. and larger, per lb	10	13½
⅜ in.	11	14½
¼ and 5-16 in.		15½
Cotton, 3-16 inch and larger		16½
" 5-32 inch		21½
" ⅛ inch		22½
Russia Deep Sea		15½
Jute		8
Lath Yarn		9½
New Zealand Rope		10

RULES.

Boxwood, dis. 75 and 10 p.c.

Ivory, dis. 37½ to 40 p.c.

SAD IRONS. per set.

Mrs. Potts, No. 55, polished	62½
" No. 50, nickle-plated	67½

SAND AND EMERY PAPER.

Dominion Flint Paper, 47½ per cent.

B & A. sand, 40 and 5 per cent.

Emery, 40 per cent.

Garnet (Rurton's), 5 to 10 p.c. advance on list.

SAP SPOUTS.

Bronzed iron with hooks, per doz... 9 50

SAWS.

Hand Disston's, dis. 12½ p.c.

S. & D., 40 per cent.

Crosscut, Disston's, per ft.... 35 0 55

S. & D., dis. 35 p.c. on Nos. 2 and 3.

Hack, complete, each	0 75	2 75
" frame only	....	0 75

SASH WEIGHTS.

Sectional, per 100 lbs.	2 75	3 00
Solid, "	2 00	2 25

SASH CORD.

Per lb.......... 0 23 0 30

SAW SETS.

"Lincoln," per doz. 6 50

SCALES.

B. S. & M. Scales, 45 p.c.

Champion, 65 per cent.

Fairbanks Standard, 35 p.c.

" Dominion, 55 p.c.

" Richelieu, 55 p.c.

Chatillon Spring Balances, 10 p.c.

Warren Champion 65 p.c.

" Standard 45 p.c.

SCREW DRIVERS.

Sargent's per doz.......... 0 65 1 00

SCREWS

Wood, F. H., bright and steel, 87½ and 10 p.c.

Wood R. H., " dis. 82½ and 10 p.c.

" F. H., brass, dis. 80 and 10 p.c.

Wood, R. H., " dis. 75 and 10 p.c.

" F.H., bronze, dis. 75 p.c.

" R.H. " 70 p.c.

Drive Screws, 87½ and 10 per cent.

Bench, wood, per doz.	3 25	4 00
" iron. "	4 25	5 75

Set, Case hardened, 60 per cent.

Square Cap, 50 and 5 per cent.

Hexagon Cap, 45 per cent.

SCYTHES.

Per doz, net.......... 9 00

SCYTHE SNATHS.

Canadian, dis. 45 p.c.

SHEARS.

Bailey Cutlery Co., full nickeled, dis. 60 p.c.

Seymour's, dis. 50 and 10 p.c.

SHOVELS AND SPADES.

Canadian, dis. 40 and 5 per cent.

SINKS.

Steel and galvanized, discount 45 per cent.

SNAPS.

Harness, German, dis. 25 p.c.

Lock, Andrews'.......... 4 50 11

SOLDERING IRONS.

1, 1½ lb., per lb.	0 37
2 lb. or over, per lb.	0 34

SQUARES.

Iron, No. 493, per doz	2 40	2 55
" No. 494, "	3 25	3 40

Steel, dis. 50 and 5 to 50 and 10 p.c., rev. list.

Try and bevel, dis. 50 to 52½ p.c.

STAMPED WARE.

Plain, dis., 75 and 12½ p.c. off revised list.

Retinned, dis., 75 p.c. off revised list.

STAPLES.

Galvanized	3 50	4 00
Plain	3 25	3 75

Coopers', discount 45 per cent.

Poultry netting staples, 40 per cent.

STOCKS AND DIES.

American dis. 25 p.c.

STONE. Per lb.

Washita	0 28	0 60
Hindostan	0 06	0 07
" slip,	0 09	0 09
Labrador	....	0 13
" Axe	....	0 15
Turkey	....	0 50
Arkansas	0 00	1 50
Water-of-Ayr	0 00	0 10
Scythe, per gross	3 50	5 00
Grind, per ton	15 00	18 00

STOVE PIPES.

5 and 6 inch Per 100 lengths	7 00
7 inch " "	7 50

ENAMELINE STOVE POLISH.

No. 4—3 dozen in case, net cash	$4 80
No. 6—3 dozen in case, "	8 40

TACKS BRADS, ETC.

	Per cent.
Strawberry box tacks, bulk	75 & 10
Cheese-box tacks, blued	80 & 12½
Trunk tacks, black and tinned	85
Carpet tacks, blued	80 & 15
" " tinned	80 & 20
" " (in kegs)	40
Cut tacks, blued, in dozens only	80
" ¼ weights	60
Swedes, cut tacks, blued and tinned—	
In bulk	80 & 10
In dozens	75
Swedes, upholsterers', bulk	85, 12½ & 12½
" brush, blued & tinned, bulk	70
" gimp, blued, tinned and japanned	75 & 12½
Zinc tacks	35
Leather carpet tacks	55
Copper tacks	50
Copper nails	52½
Trunk nails, black	65 and 5
Trunk nails, tinned	65 and 10
Clout nails, blued	65 and 5
Chair nails	35
Patent brads	40
Fine finishing	40
Picture frame points	10
Lining tacks, in papers	10

Lining tacks, in bulk		15
" " solid heads, in bulk		75
Saddle nails in papers		10
" " in bulk		15
Tufting buttons, 22 line, in dozens only		60
Tin capped trunk nails		15
Zinc glazier's points		5
Double pointed tacks, papers	90 and	10
" " " bulk		40

TAPE LINES.

English, ass skin, per doz	2 75	5 00
English, Patent Leather	5 50	9 75
Chesterman's each	0 90	2 85
" steel, each	0 80	8 00

THERMOMETERS.

Tin case and dairy, dis. 75 to 75 and 10 p.c.

TRAPS. (Steel.)

Game, Newhouse, dis. 2) p.c.
Game, H. & N., P. S. & W., 65 p.c.
Game, steel, 72½, 75 p.c.

TROWELS.

Disston's discount 10 per cent.

German, per doz	4 75	6 0)

S. & D., discount 35 per cent.

TWINES.

Bag, Russian, per lb	....	0 27
Wrapping, cotton, per lb	0 22	0 26
Wrapping, mottled, per pack	0 53	0 60
Wrapping, cotton, 3-ply		0 20
" " 4-ply	....	0 26
Mattress, per lb	0 33	0 45
Staging, "	0 27	0 35

VISES.

Hand, per doz	4 00	6 00
Bench, parallel, each	2 00	4 50
Coach, each	6 00	7 00
Peter Wright's, per lb	0 12	0 13
Pipe, each	5 50	9 00
Saw, per doz	6 50	13 00

ENAMELLED WARE.

White, Princess, Turquoise, Blue and White, discount 50 per cent.
Diamond, Famous, Premier, 50 and 10 p.c.
Granite or Pearl, Imperial, Crescent, 50, 10 and 10 per cent.

WIRE.

Brass wire, 50 to 50 and 2½ per cent. off the list.

Copper wire, 45 and 10 per cent. net cash 30 days, f.o.b. factory.

Smooth Steel Wire, is quoted at the following net selling prices:

No. 6 to 8 gauge		$2 90
" 9 "		2 80
" 10 "		2 87
" 11 "		2 90
" 12 "		2 95
" 13 "		3 15
" 14 "		3 37
" 15 "		3 50
" 16 "		3 65

Other sizes of plain wire outside of Nos. 9, 10, 11, 12 and 13, and other varieties of plain wire remain at $2.80 base with extras as before. The prices for Nos 9 to 13 include the charge of 10c. for oiling. Extras net per 100 lb.: Coppered wire, 60c.—tinned wire, $2—oiling, 10c.—special hay-bailing wire, 30c.—spring wire, $1—best steel wire, 75c.—bright soft drawn, 15c.—in 50 and 100-lb. bundles net, 10c.—in 25-lb. bundles net 15c.—packed in casks or cases, 15c.—bagging or papering, 10c.

Fine Steel Wire, dis. 17½ per cent. List of extras: In 100-lb. lots: No. 17, $5—No. 18, $5.50—No. 19, $6—No. 20, $6.65—No. 21, $7—No. 22, $7.30—No. 23, $7.65—No. 24, $8—No. 25, $9—No. 26, $9.50—No. 27, $10—No. 28, $11—No. 29, $12—No. 30, $13—No. 31, $14—No. 32, $15 No. 33, $16—No. 34, $17. Extras net—tinned wire, Nos. 17-25, $2—Nos. 26-31 $4—Nos. 32-34, $6. Coppered, 5c.—oiling, 10c.—in 25-lb. bundles, 15c.—in 5 and 10-lb. bundles, 25c.—in 1-lb. hanks, 50c.—in ½-lb. hanks, 75c.—in ¼-lb. hanks, $1—packed in casks or cases, 15c.—bagging or papering, 10c.

Galvanized Wire, per 100 lb.—Nos. 6, 7, 8, $3 50 to $3 8‹—No. 9, $2.85 to $3.15—No. 10, $3.60 to $3.95—No. 11, $3.70 to $4.10—No 12, $3 to $3 30—No. 13, $3.10 to $3 40—No. 14, $4.10 to $4.50—No. 15, $4.60 to $5.05—No. 16, $4.85 to $5 35. Base sizes, Nos. 6 to 9, $2.57½ f.o.b. Cleveland.

Clothes Line Wire, solid 7 strand, No. 17, $4.25; No. 18, $2.65; No. 19, $2.35, f.o.b. Hamilton, Toronto, Montreal.

WIRE FENCING. F.O.B. Toronto

Galvanized barb	....	3 05
Galvanized, plain twist	....	3 05

Galvanized barb, f.o.b. Cleveland, $2.82½ in less than carlots, and $2.70 in carlots.

WIRE CLOTH.

Painted Screen, per 100 sq. ft., net.. 1 35

WASTE COTTON. per lb.

Colored	4½ to 5
White, according to quality	6½ to 7¾

500-lb bale lots shaded.

WRENCHES.

Acme, 35 to 37½ per cent.
Agricultural, 60 p.c.
Coe's Genuine, dis. 20 to 25 p.c.

Towers' Engineer, each	2 00	7 00
" S., per doz	5 80	6 00
G. & K.'s Pipe, per doz	....	3 40
Burrell's Pipe, each	....	3 00
Pocket, per doz	0 25	2 90

WRINGERS.

Leader .... per doz.	$30 0	33 00
Royal Canadian.. "	26 00	28 00
Royal American., "	26 00	28 00
Sampson "	....	30 00

Terms 4 months, or 3 p.c. 30 days.

WROUGHT IRON WASHERS.

Canadian make, discount, 40 and 5 per cent.

PERSONS addressing advertisers will kindly mention having seen their advertisement in Canadian Hardware and Metal Merchant.

"THE EMLYN" SAW BENCH

Made in 6 sizes. Best value obtainable. Specially designed for export., With or without "Emlyn" Patent Guard. Sole maker—

CHARLES D. PHILLIPS,

Cables— "Machinery," Newport.

Emlyn Engineering Works, NEWPORT, MON., ENGLAND.

A retail hardware merchant advises us that he finds

THE DUNDAS AXE

the best seller he ever had.

Dundas Axe Works

DUNDAS.

Lockerby & McComb

AGENTS IN CANADA

FOR THE

Celebrated P. & B.

Cold Storage Lining

AND

.. Ruberoid Roofing ..

P. S.--Prices on Application.

65 Shannon Street, MONTREAL.

BUSINESS NEWS

of any kind that is of value to business men supplied by our Bureau. We can give you market quotations from any town in Canada, reports from the city markets, stock quotations, etc. You can get commercial news from any Canadian paper through us.

Write us, giving us particulars of what you want and where you want it from, and we will quote you prices by return.

"Clippings from any Canadian paper on any subject."

CANADIAN PRESS CLIPPING BUREAU,

232 McGill Street, MONTREAL, QUE.

Telephone Main 1255.

10 Front St. East, Toronto. Telephone 2148.

---

75 YEARS. *ESTABLISHED 1825.* *75 YEARS.*

CELEBRATED HEINISCH SHEARS.

Tailors' Shears,
Trimmers, Scissors,
Tinners' Snips, etc.

ACKNOWLEDGED THE BEST.

R. HEINISCH'S SONS CO.

NEW YORK OFFICE, 90 Chambers St.
NEWARK, N.J., U.S.A.

Not connected with any Shear Combination.

CHAS. F. CLARK, President. JARED CHITTENDEN, Treasurer.

...ESTABLISHED 1849...

BRADSTREET'S

Capital and Surplus, $1,500,000. Offices Throughout the Civilized World.

Executive Offices: Nos. 346 and 348 Broadway, New York City, U.S.A.

THE BRADSTREET COMPANY gathers information that reflects the financial condition and the controlling circumstances of every seeker of mercantile credit. Its business may be defined as of the merchants, by the merchants, for the merchants. In procuring, verifying and promulgating information no effort is spared, and no reasonable expense considered too great, that the results may justify its claim as an authority on all matters affecting commercial affairs and mercantile credit. Its offices and connections have been steadily extended, and it furnishes information concerning mercantile persons throughout the civilized world.

Subscriptions are based on the service furnished, and are available only by reputable wholesale, jobbing and manufacturing concerns, and by responsible and worthy financial, fiduciary and business corporations. Specific terms may be obtained by addressing the Company at any of its offices. **Correspondence Invited.**

—OFFICES IN CANADA—

HALIFAX, N.S.	HAMILTON, ONT.	LONDON, ONT.	MONTREAL, QUE.
OTTAWA, ONT.	QUEBEC, QUE.	ST. JOHN, N.B.	TORONTO, ONT.
VANCOUVER, B.C.	VICTORIA, B.C.	WINNIPEG, MAN.	

THOS. C. IRVING, Gen. Man. Western Canada, Toronto. JOHN A. FULTON, Gen. Man. Eastern Canada, Montreal.

Awarded a Gold Medal at **PARIS EXPOSITION** for superiority. That's proof enough of their quality, and clearly shows that they are the best.

Send for Catalogue and . . . . Price List.

The Bailey

Cutlery Co.

BRANTFORD, ONT.

THE Empire Typewriter

Equal to **any** Machine in every way.

Superior to **all** Machines in several **Important Features.**

Canadian Pacific Railway have 175 Empires in daily use!

Only $60

You can save $60 by purchasing an Empire.

THE WILLIAMS MFG. CO., Limited

MONTREAL.

There are other designations more expensive and less worth
Use LANGWELL'S Babbit, Montreal.

CANADIAN HARDWARE AND METAL MERCHANT

The Weekly Organ of the Hardware, Metal, Heating, Plumbing and Contracting Trades in Canada.

VOL. XIII. MONTREAL AND TORONTO, MAY 4, 1901. NO. 18

NEW CYCLE AND SPORTING GOODS FOR 1901

THE TRADE

will be pleased with our new assortment of goods for the coming season. Our travellers are now on the road with samples ; if they have not seen you, a card to us will bring them to you. It will pay to get our new prices.

WRITE FOR CATALOGUE AND PRICES.

RICE LEWIS & SON

LIMITED

Cor. King and Victoria Sts., TORONTO.

THE INCREASING DEMANDS FOR....

The Fairbanks Asbestos Disc Valves

is a proof that it will pay you to carry these goods in stock. Our renewable Asbestos Disc is a money saver to the steam user.

We carry a full line of Pipe and Fittings.

These goods are used by all the leading steam plants in Canada and the United States.

Write for Catalogues and Prices.

THE FAIRBANKS COMPANY

749 Craig Street, *MONTREAL*

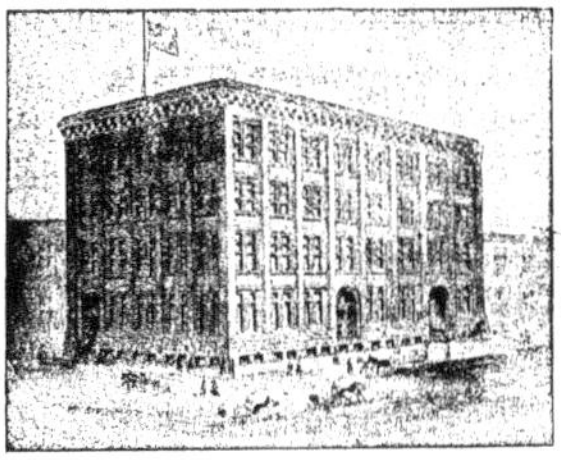

300.20.280
INDELIBLO
WHITE

DAISY

LEWIS BROS. & CO.

Wholesale Hardware

MONTREAL.

FOR YOUR HOT WEATHER TRADE.

The Crown Ice Chipper.

Family Ice Tongs.

Gem Ice Shave.

Peerless Iceland Freezers, all sizes.
REMARKS—Rapid Freezers, Easy to Work, and Clean.

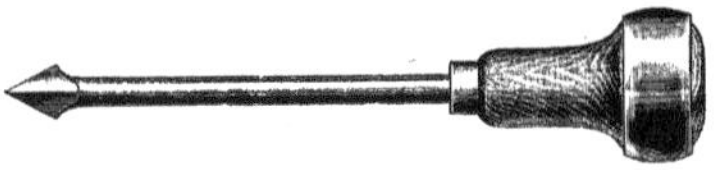

Ice Pick—Arrow Point, Wooden Handle, with heavy Galvanized Malleable Ring.

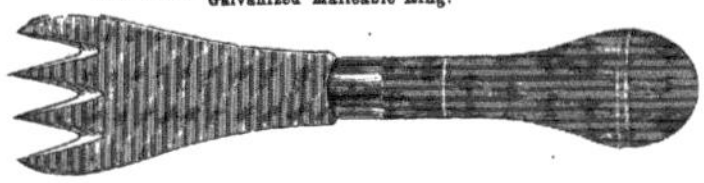

Ice Chisel—Polished Steel Blade, or Galvanized Steel Blade.

Lewis Bros. & Co.,

WRITE FOR PRICES.
MAIL ORDERS SHIPPED SAME DAY AS RECEIVED.

MONTREAL.

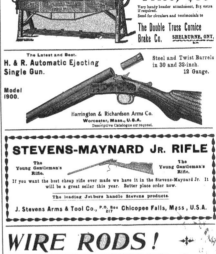

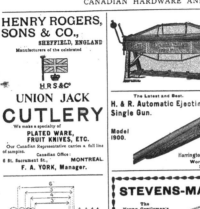

HARDWARE AND METAL

VOL. XIII. MONTREAL AND TORONTO, MAY 4, 1901. NO. 18.

President,
JOHN BAYNE MacLEAN,
Montreal.

THE MacLEAN PUBLISHING CO.
Limited.

Publishers of Trade Newspapers which circulate in the Provinces of British Columbia, North-West Territories, Manitoba, Ontario, Quebec, Nova Scotia, New Brunswick, P.E. Island and Newfoundland.

OFFICES

MONTREAL - - - - - - - 232 McGill Street, Telephone 1255.
TORONTO - - - - - - - 10 Front Street East, Telephone 2148.
LONDON, ENG. - - - - - 109 Fleet Street, E.C., W. H. Miln.
MANCHESTER, ENG. - - - - 18 St Ann Street. H. S. Ashburner,
WINNIPEG - - - - - Western Canada Block. J. J. Roberts,
ST. JOHN, N.B. - - - - No. 3 Market Wharf, J. Hunter White.
NEW YORK. - - - - - - - 176 E. 88th Street.

Subscription, Canada and the United States, $2.00.
Great Britain and elsewhere - - - 12s.

Cable Address { Adscript, London. Adscript, Canada.

Published every Saturday.

MAKING THE METAL MARKET UNIFORM.

ALTHOUGH the market for Canadian manufactured iron and steel, including bar iron and pipe, rules firm, the spot sheet metal market, as governed by English values, has been weakening appreciably during the last few weeks.

The English market has, of course, fallen since there were any large importations made last fall in black sheets, tinplates, Canada plates, galvanized iron and similar goods, but, while our Canadian ports have been shut off from sea commerce by the winter frosts, dealers in Montreal and Toronto have been able to escape the brunt of a falling market and lighten their stocks at fair prices. But now that goods are on the way here prices must come down into touch with the English market. So, during the past week, several reductions in imported sheet metals have been made varying in amount with the different firms.

Black sheets are suffering severely. Last all it cost £12 5s. per ton to lay them down in Montreal; to-day a fair price would be £8 17s. 6d. Montreal houses are, to-day, selling these goods at $2.50 to $2.60 for 28 gauge, according to quantity, a drop of 40 to 50c. per 100 lb.

Tinned plates are also declining. The fall in the English market since the close of navigation last year amounts to about 45c. per box, from 14s. to 12s. for cokes. The wholesale price now ranges about $3.60 to $3.70 for cokes.

Canada plate has been selling at very low figures for import, and there have been some transactions in Montreal this week at $2.40 to $2.50 for 52's. Last fall, the price to import was about £10 per ton or a little lower; now it is about £8. Some dealers have light stocks and are holding their higher prices about $2.60 to $2.70, but others are clearing at about import prices.

In galvanized iron, the drop has amounted to about 35c. per cwt. covering the winter period, while the difference in the import price of terne plates amounts to 3s. or 4s. per box. We have heard that some import orders have been taken here on retail account for terne plates in the neighborhood of $6.75, but prices on spot goods are much above that figure.

There is no doubt these reductions have fallen heavily upon some importers, but they can at least congratulate themselves that the wintry weather has for some months kept out goods purchaseable at these low prices, giving them an opportunity to make the best of an unfortunate turn of the market. At any rate, the condition must be faced in the course of next week.

Although no one is buying heavily in these lines for forward delivery, there is a feeling this week that the English market may take a turn for the better. The American market is very active, and a demand seems to be setting in across the border for English goods. Should this prove true, the depression that has characterized the English market for some time may give way to buoyant tone. The danger of a prolonged coal strike is another circumstance to be considered and duly weighed.

JEALOUSY IN BUSINESS.

JEALOUSY is as much to be discouraged in business as it is in love affairs. The merchant who is possessed of it is laboring under a disadvantage, for it blinds his own judgment and makes his fellow businessmen his enemies.

Those who succeed in business are not the men who are eternally watching with a jealous eye their competitors; they are those who are too busy minding their own affairs to grieve over the success that may be attending the efforts of their fellows.

He is a wise merchant who keeps himself independent of his business confreres. But he is equally wise who keeps on good terms with them.

Jealousy is the root of price-cutting and other trade evils. The spirit of live and let live, on the other hand, is the basis on which legitimate and profitable business is erected.

STEEL RAILS AND THE IRON TRADE.

INTEREST in the iron and steel trades this week centres around the steel rail market, on account of the advance of $2 per ton, which went into effect on Wednesday in the United States. The price of standard steel rails is now $28 per ton.

The manufacture of steel rails in the United States is now practically in the hands of the United States Steel Corporation, or, as it is better known, the Billion Dollar Steel Trust. An advance of $2 per ton so soon after the consummation of the big consolidation, with its $1,100,000,000 of stock and $304,000,000 of bonds, has naturally induced a great deal of comment, the concensus of which is not favorable to the action of the corporation.

The New York Journal of Commerce declares that the profits were already $10 per ton, while "the price of ore for the next year has been reduced." It, therefore, comes to the conclusion that "the increase was justified by nothing."

All that The Journal of Commerce says may be true in regard to the large profit on steel rails, but it must be remembered that the demand is abnormally brisk. Some of the mills in the corporation have practically sold up for the year, while, according to The Iron Trade Review, "probably 90 per cent. of the business the mills can take care of this year has been booked." It is estimated that the mills have in hand orders for over 2,000,000 tons of rails.

In view of such a strong position the corporation has only done what an ordinary manufacturer would have done under the circumstances. Most manufacturers earn as high a profit as they can whether they are acting independently or in unison with others. It may not be right morally, but it is the practice commercially. In saying this we are not endorsing the conditions which give one concern the power to manipulate prices of such an enormous industry as that of steel rails at its own sweet will without any consideration whatever as to the state of the market. That we cannot endorse. The autocrat in commerce is as much to be deplored as the autocrat in politics.

While the demand for steel rails is so active, it is significant that the month of April closed with a quiet trade in pig iron, somewhat contrary to its opening, and The Iron Trade Review points out that "the prominent fact in the situation is the continued unwillingness of the average buyer of pig iron to contract to the end of the year, or, in fact, much beyond July." As is generally known, large buyers on this side of the Atlantic have, as a rule, placed orders for their pig iron requirements for the first half of the year.

Steel continues active and firm, and dealers in both Canada and the United States are still experiencing a great deal of difficulty in getting delivery. The same remarks apply to galvanized sheets. Importers in Canada have been advised by manufacturers in the United States that they will not be able to book orders for delivery before June. As delivery can be obtained sooner than that from England, manufacturers of galvanized sheets there are getting some orders that would otherwise go to the United States.

The demand for wire and wire nails is on the whole good. For barb wire the demand appears to be exceedingly heavy, both in the United States and Canada. Nearly the whole of the demand for the home market is now supplied by United States mills, and, owing to the distance the latter are behind with their orders, there is practically a famine here in barb wire at the moment, although supplies are promised by the end of next week.

Owing to the difficulty of getting barb wire from the United States lately, one of the Canadian factories has recently begun manufacturing again. It will be remembered that shortly after barb wire was placed on the free list the manufacturing of that article ceased in this country.

Now that delivery of barb wire from the United States is so difficult, in quite a few instances consumers are being persuaded to take oiled and annealed wire as a substitute. As this wire is the product of the home factories, sentiment has not a little to do with the efforts that are being made to push its sale at this opportune time.

It is a current fact well worth mentioning that the manufacturers in the United States have never yet been able to make an oiled and annealed wire equal to the Canadian product.

According to our exchanges, the condition of the iron and steel trades in Great Britain and in Germany is a little more satisfactory than it was a short time ago. At the same time, it is in anything but a satisfactory condition. This is evident from the fact that in the Midland districts there has, this week, been made a reduction in the wages of the men employed in the iron trade.

A CANADIAN MINT AT LAST.

IN October last, the Minister of Finance informed a newspaper correspondent that the Canadian Government had obtained the sanction of the Imperial authorities to establish a mint in Canada for the coinage of gold and silver. On Tuesday last, in the House of Commons, Lieut.-Col. Prior, one of the members for Victoria, B.C., reminded the Minister of the statement he made six months ago, and hoped he would implement that promise by bringing in a Bill at this session to establish both a mint and an assay office.

The reply of Mr. Fielding was most satisfactory. He said that notice would be given directly for the introduction of a Bill to provide for the establishment of a Royal mint, and that an assay office would follow later.

At present Canada has no gold coinage of her own. The little gold that is in circulation here is composed of British sovereigns and American eagles. But, while a certain amount of notional pride will be satisfied by the establishment of a mint of our own, probably the most important consideration in connection with a mint or an assay office is the trade as well as the gold which either attracts.

Knowing this, the United States Government took steps to establish an assay office at Seattle as soon as the rich discoveries of gold in the Klondyke were revealed to the outside world, and by the time the output of 1898 was brought out the office was ready to deal with it. There being neither assay office nor mint in British Columbia, the gold went to Seattle, a foreign city, and the great bulk of the trade on Klondyke account as well. According to the United States authorities at Seattle, $14,000,000 worth of Klondyke gold went into the United States last year.

THE BASIS OF PROGRESS.

THE majority of Canadians are ambitious. The educational facilities enjoyed in early life, the civil liberty, the social conditions, and, what may be most influential, the rapid development of our country, present opportunities for progress and create a desire for improvement that does much to stimulate ambition and develop a nation of progressive people.

But here, as elsewhere, the difference between desire and attainment, between ambition and success, is manifest. And observation leads to the conviction that the same factors that have made for progress in other countries are, to the same extent, effectual in Canada, and apply in the mercantile world as in any other condition of life.

The merchant who starts at the bottom—say, with a small business and meagre capital, may hope to attain large success under certain conditions. The primary factor is that he must work. A merchant has no more right to sit down and expect his business to grow and prosper without the expenditure of much thought and labor in its interest than has a clerk or an employe of any kind to sit idle during the week and expect his employer to give him full pay at the week's close. The public demands service for its money just as does a private employer, and it bestows its rewards in the shape of steady patronage.

It is not sufficient to merely do what work is forced upon one. The majority do that, and it is only to the extent that one is superior to his creditors that he will develop beyond them. The progressive merchant is thoughtful, alert. He is ever-ready to take advantage of any conditions, to follow up any line of action, to do any extra work which will extend the scope of his business or increase its net profit. If by working half an hour per day extra he can save a small percentage of his expenses he sacrifices the half hour.

The habit of economy grows when cultivated. The merchant who is bound to succeed soon learns that every dollar saved is a source of strength to his business and jealously guards against unnecessary expenses both in the business and on personal account. This habit of economy is a factor that has meant all the difference between success and failure to many men.

It is easy to live on $1,000 per year, but it is not easy to save money on a net income of $400 or $500 per year. Yet many successful merchants bear witness to the fact that they owe their success to the money they saved when their business yielded even less than either of these amounts. Not only did they watch closely the expense of the business, but maintained the strictest personal economy, and though they saved slowly every dollar made them stronger and helped to extend the scope of their undertaking.

Some merchants lack the necessary courage or boldness to launch out into greater expenditure in order to obtain a larger business and a bigger margin of profit. After working hard and living closely until they have more cash than is necessary to the working of their business, they bank their money rather than invest it in better premises, though they are persuaded that the latter move would not only be safe but profitable. Others are too bold, extending their business before their business has grown sufficiently to warrant it, or before they have accumulated enough capital to insure the success of the extension.

The truly progressive merchant combines courage with caution and extends as his capital increases—and his capital only increases according to the thought and labor expended on it and the proportion of net income set aside from it each year.

ALD. LAMARCHE MAKES A SPEECH.

AT the last meeting of the Montreal Master Plumbers' Association, Ald. Jos. Lamarche gave an interesting talk, a resume of the work he had done in the council in general, and on behalf of the plumbing trade in particular. He gave a brief history of his attempt to reserve the position of sanitary inspector to plumbers alone. In March, 1900, he introduced a motion into the council which had for its object to make all applicants pass an examination before the hygiene committee, only those who had been in the plumbing trade for five years to be eligible for the position. This motion passed successfully, and two plumbers were appointed for the first time. Unhappily the motion was revoked on November 23 last by the hygiene committee ; still, the appointments made since that time have gone to plumbers and his agitation has not been fruitless.

Mr. Lamarche has also been instrumental in having the building by-law passed which had been on the shelf for four years. Much of the credit for this he modestly gives to Ald. Hart. The plumbers' by-laws are Ald. Lamarche's greatest civic progeny. These will involve great improvements for the trade. They have all been passed before the council with the exception of the clause licensing plumbers, the right of the council to pass which has been referred to the city attorneys. However, they have reported in Ald. Lamarche's favor and the whole by-law will probably go into force in a month or so.

Other questions, such as the civic hospital scheme were discussed and Ald. Lamarche's attitude was approved. He has shown himself to be an able and willing champion of the plumber's cause while in the council, and if an alderman ever gets any thanks, which is doubtful, he for one deserves some.

A HANDY MAN TO HAVE.

A HANDY man, a sort of jack-at-all-trades but good at most of them, is a valuable addition to a hardware store, says Stoves and Hardware Reporter. A clerk who can turn his hand to repair work, a shop worker who can go behind the counter and sell goods, is often of more value to his employer in the long run than are two men who occupy different positions but cannot diversify their work. One of the best hardware travellers now going out of St. Louis began to acquire knowledge of the business in a retail store where his first duties were as an errand boy. When not outside, he was either dusting off or cleaning up the stock or was in the tinshop. He wanted to learn, and he succeeded. His first promotion was to a position created for him, that of utility man, although he was nothing but a boy.

He soon learned how to sell a stove and then to set it up, becoming a mechanic-clerk, and even finding time to study the science of accounting and become a walking encyclopædia of hardware. He wanted to go on the road in order to learn more, and that is the reason why he is no longer a utility man, but if there were more like him in the country, or if employers would endeavor to discover such as he or to educate the clerks they now have the result would be an economy in management and a more satisfactory condition of business. The more a clerk knows about the goods he handles, and especially about the way they should be used, the greater is his value to the employer.

THE ANTI-TRADING STAMP ACT.

A MOVEMENT TO MAKE IT EFFECTIVE.

AT the regular monthly meeting of the Toronto branch of the Retail Merchants' Association of Canada, the trading-stamp situation was discussed at some length. The legislation recently passed by the Ontario Legislature empowering municipalities to prohibit merchants from giving trading stamps to be redeemed by a third party or concern was considered so satisfactory that a resolution was unanimously adopted asking the Toronto City Council to pass a by-law to this effect, as they had been empowered.

CONSIDERS THE ACT SOUND.

In speaking to HARDWARE AND METAL regarding the matter, E. M. Trowern, secretary of the association, stated: "The Act is absolutely sound, as the Provincial Legislatures only have control of this class of legislation. The Law Clerk of the Ontario House and several eminent legal authorities examined the Act before it was passed and pronounced it constitutional. Similar legislation was granted by the Nova Scotia Legislature and passed by the city of Halifax. The Montreal Trading Stamp Co. appealed against the Act, but it was upheld in the courts. We are now preparing a by-law which will not only be submitted to the municipal council of Toronto, but that of other municipalities in Ontario."

THE TRADING STAMP CO.'S VIEW.

The representative of HARDWARE AND METAL also saw Mr. Hubbell, secretary of The Dominion Trading Stamp Co., Limited, regarding their proposed course of action in the matter. He stated that his company are confident that the Bill is ultra vires, that this view was taken by one of the leading M.P.P.'s, Mr. Carscallen, of Hamilton, when the Bill was up before the House, who went so far as to say "that the Bill was not worth the paper on which it was printed, and would never stop the trading-stamp business, and that if he was the company's counsel he would advise them to pay no heed to it." The Bill is only a permissive one and does not permit municipalities acting upon it until after January 1. The company in the meantime proposes to contest the validity of the Bill in the highest court if necessary. Mr. Hubbell also states that the company is permanent and intends to continue in business, that, even if the Bill was permitted to stand, it can be evaded in a very simple and effective manner without causing the company any great trouble or expense, that it is the intention of the company to continue in business and meet every difficulty that may arise, that there is nothing illegal in the business he states has been clearly proven, as its entire operation has been reviewed before Chancellor Boyd and Judge Ferguson within the past 18 months, the former having defined the business as "an advertising device whereby local trade was promoted and cash trade stimulated."

"This Bill," said Mr. Hubbell, "as passed is an exact copy of a Bill passed in Rhode Island something more than a year ago, and declared to be unconstitutional by the Supreme Court of that State, Justice Tillinghast rendering the decision, in which he states that the proposition that a merchant can give a stamp or coupon and redeem it himself cannot be prohibited, and that to make a difference between the merchant giving the stamp and redeeming it himself, as against having it redeemed by some other person or company, is not a distinction that requires or will admit of any legislation."

"Has your company experienced legislative opposition in other Provinces?"

"Yes, in British Columbia. The British Columbia Legislature last summer passed an Act compelling the payment of a license to handle stamps in that Province. We are contesting that Act, and are as confident of success there as we are here."

A HARDWARE M. P.

Nicholas Flood Davin, the Ottawa correspondent of The Toronto News gave in a recent issue of that journal a brief and interesting sketch of Col. Prior, one of the hardwaremen who sits in the House of Commons. It read as follows:

"Col. Prior has been in the House of Commons since 1888, in January of which year he succeeded Mr. Noah Shakespeare as one of the representatives for Victoria. He was then 35 years of age, a handsome man, well built, fair, florid, with a military bearing and a bright manner—a great fellow amongst the ladies, admiring and admired. For some years he gave himself more to society than to legislation, but since January, '96, when he entered the Government of Sir Mackenzie Bowell as Comptroller of Inland Revenue, he has been a hard worker and has stuck closely to his place in the House of Commons. He has gained in weight, both in the House and in the estimation of his constituents. Had his party remained in power he would undoubtedly have been Minister of Militia, for which he has special aptitude, having had an excellent military training and possessing good executive power. The son of a Yorkshire clergyman and an engineer by profession, he, after a successful career as mining engineer, went in 1880 into business as an iron and hardware merchant, and the splendid establishments in Victoria and Vancouver which bear his name would seem to attest business success. He is a good fellow, has a dashing way with him and an off-hand manner."

INQUIRIES REGARDING OANADIAN TRADE.

The following were among the recent inquiries relating to Canadian trade received at the High Commissioner's office, in London, England:

1. The names of Canadian egg shippers are asked for by a large firm of importers in the North of England.

2. An Irish correspondent, with a fair capital, inquires through the Dominion Government agency in Dublin for information regarding the tanning and leather trade in Canada, both in respect to the manufacture of sole leather and dressed goods.

3. A Glasgow firm ask for particulars of shipbuilding firms in Canada, being desirous to do business in ship's plates, etc.

4. Further inquiry is made by a Continental house for exporters of seal oil from Canada, liberal advances offered on consignments.

5. A London firm are desirous of importing from Canada small wood discs, such as are used in tops of corks in mineral waters. They are usually packed up in barrels containing 500 gross each. If suitable prices quoted, quantities of 100 to 150 barrels could be taken at a time.

6. A selling agent with a good connection among large wholesale and export firms desires to be placed in touch with Canadian packers of canned meats (especially pigs' tongues).

7. A London agent is inquiring for exporters of good tares from Canada for feeding purposes, there being a demand both in the city and on the Continent.

8. A correspondent asks for names of Canadian exporters of canned yolks of eggs.

Mr. Harrison Watson, curator of the Canadian Section of the Imperial Institute, London, England, is in receipt of the following inquiries regarding Canadian trade:

9. A London firm largely interested in graphite is prepared to hear from Canadian producers of same.

10. A house in Malta desires names of Canadian manufacturers of enamelled ware.

11. A London house, exporting all lines of soft goods, cotton, woollen, linen, etc., would like to arrange to be represented in Canada.

12. A Midlands manufacturer of brooms and brushes desires names of Canadian manufacturers who can supply handles.

[The names of the firms making the above inquiries, can be obtained on application to the editor of HARDWARE AND METAL, Toronto. When asking for names, kindly give number of paragraph and date of issue.]

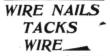

THE COAL MINES OF VANCOUVER ISLAND.

THE most important factors in the prosperity of Vancouver Island are the rich coal mines that have been developed all along its eastern coast, that give employment and livelihood to thousands of people.

Until of late years, the Vancouver Island mines have had no rivals in British Columbia and have practically monopolized the coal trade. But the development of the Crow's Nest coal mines in the southeast portion of the Province, recently purchased by an American syndicate, now in control. has prevented any increase of the consumption of Vancouver Island coal on the mainland, though it is claimed that the Canadian Pacific Railroad, which still uses the island coal, regards it as superior for making steam to the product of the Crow's Nest mines. It is stated that the Crow's Nest mines are the largest on the Coast; capable of producing, when properly worked, 10,000 tons of coal daily, and even at that rate of lasting several hundred years. The present product of these mines amounts to less than 1,000 tons per day and cannot be largely increased until a way is found to reach the United States. A Bill is now before the Legislature and Dominion Parliament asking a charter for a railroad from Michel, British Columbia, to the international boundary, where connection is made with the Great Northern Railroad. The company is ready to build the road without bonus or land grant of any kind, but is, of course, opposed in their application for charter by the Canadian Pacific Railroad, aided by parties who oppose all roads that lead to the United States.

There has been quite a change in the mines of Vancouver Island during the past year. The coal mines at Comox, the most northern, have been worked to their fullest capacity and one or two new shafts dug out. It was in one of these new shafts (No. 6) that a terrible explosion took place on Friday, February 15, 1901, by which 64 miners, all who were working in the shaft, lost their lives. Of the dead miners, 20 were whites, and the remainder Asiatics. Since then work has been resumed, and the mine is again being run to its full capacity.

At Wellington, the mines have been worked out and abandoned—and so has the town itself, practically. Before the work ceased there were frequent "cave-ins," involving loss of life and property. During the past year, the principal buildings there have been moved away to other towns, and, by petition of the remaining inhabitants, the municipal charter has been surrendered and revoked. The repair shops of the Esquimalt and Nanaimo Railway, however, still remain there. It is understood that if the road is continued north to Hardy Bay, the shops will be kept at Wellington. A charter has been applied for to authorize the extension of the railroad, which will probably be granted; but whether the road will be built is yet to be determined.

The mines at Nanaimo, extending from the town under the bay to Protection Island, are worked to their full capacity, and ship as much coal as all the other mines of the island combined. These mines are worked entirely by union men, and no Asiatics are employed underground. The result is that Nanaimo is represented in the Provincial Legislature by two labor men, while the seat in the Dominion Parliament is filled by the secretary of the miners' union. Strikes have so far been avoided, although one was set for March 1, 1901 but the management met the committee representing the miners and showed th contracts of the company, many of which were made before the present Provincial tax of five cents per ton was levied, and the men have so far continued without increase. In these mines, the miners receive 68 cents per ton and make an average of $4 to $4.50 per day.

The Alexandria mine is closed, and has been since last December. The management gives as reason for closing that there is no demand for the coal, which was chiefly sent to San Francisco. Since that time the whole force of 180 men previously employed in the mine has been lying idle, save those who have gone to work in other mines. The miners claim that the real cause for closing the mine is that the management insists that 2,800 pounds should constitute a miner's ton, instead of 2,352 pounds, as had previously been the rule; and, further, that nothing should be allowed for "turning places," which are quite numerous in this mine. The management states that the United States duty of 67 cents per ton, the 5-cent tax, the eight-hour law, and the discovery of oil in California render the market unprofitable; and further intimates that the general output of Vancouver Island coal mines may have to be curtailed in the near future, owing to the competition of the Washington mines, and of Californian oil. A number of the unemployed and their families are now receiving relief from the Nanaimo and other unions.

The Extension mines, owned by the same company (The Wellington Colliery Company), are distant 12 miles only from the Alexandria. It may be stated that all th coal mines of the island are on the east coast, within a radius of less than 40 miles. The Extension mines are running as usual and frequently are unable to supply the demand for coal, vessels having to wait their turn at the wharves before receiving their needed supply. Forty or more of the miners who formerly worked at Alexandria have secured work in the Extension mines, where 72 cents per ton is paid miners, against 66 cents in the Alexandria mine.

The operation of the Extension mines has resulted in the building and incorporation of a new town at Oyster Bay, known as Ladysmith, which already has 1,200 inhabitants, several stores, hotels, etc., and bids fair to become a thriving city; although entirely dependent, as was Wellington and as are Nanaimo and South Wellington, on the mining industry. Here, bunkers capable of holding 20,000 tons of coal have been erected, and tracks run direct to the wharves from the mines at Extension, 14 miles away. Many of the miners reside in small shanties close to the mine, but the management prefers to have them live at Ladysmith. To this end, lots are sold miners at $100 each and neat cottages erected for them, payable on the installment plan. Trains run at convenient hours for the different shifts of miners, and transportation is without cost. All the men are paid at Ladysmith, which is the headquarters of the management of Extension mines.

Victoria, March 16, 1901.

James Shuter, general merchant, Thornhill, Ont., died March 17 last without making his will His wife is asking for letters of administration. The estate is valued at $5 934, $5,050 of which consists of realty at Thornhill and the balance in book debts and stock in trade.

EXIT THE FEMALE DRUMMER.

WOMAN, fair woman, is invading the realm of business, and it really seems as though we should have to look up men slaveys to do our cooking and washing if this thing keeps on. The female drummer has ceased to be a novelty. From all accounts I have not yet heard that she has invaded the shoe business to any serious extent, still it may come, alas, any day when the old war horses, whose faces we have known for years, and who have visited the retail trade spring and fall every year for a decade or two, will be laid upon the shelf and their places taken by more (or less) beautiful females who will attempt to induce shoe retailers to place large orders with them. From what I have heard of the invasion of other trades by these petticoated salesmen, I should judge that the change is not exactly agreeable to many retailers, who, although they may have a naturally high respect for the sex, and who are invariably courteous in their treatment of them when in society, are loth to see them mingling in business matters.

It was only the other day that a merchant in another line of business told me of a case where a firm had failed to sell him a bill of goods, the said retailer being very well satisfied to continue trading with the manufacturers who had served him well for several years. For four successive trips the representative of the rival house called and endeavored to have him place an order, but without success, but last fall a change came over the spirit of his dreams when there marched into his store a trig young woman, dressed in the height of fashion, straight-front corsets, L'Aiglon dingle-dangles, and all the rest of the latest ideas. Following her was a hotel porter with two big grips looking like sample cases. It did not take long to find out that the woman represented the firm whose drummer of the male persuasion had been unable to secure an order during the previous seasons, for the woman, with the volubility for which her sex is noted, opened fire and took the merchant's breath away with the torrent of her praise of the goods.

The dealer was at first courteous and considerate, but firm in his previously announced position that he was well satisfied with the firms with whom he was dealing and did not wish to see the samples. But that made no difference, and for three solid hours that woman talked, wheedled and cajoled in endeavoring to place an order for those goods. The man's patience had deserted him and when the lady commercial ventured to remark: "You don't seem really to like lady drummers," he was hasty enough to give her a most forcible negative, at the same time consigning the entire tribe of female salesmen to the old fashioned orthodox locality of eternal punishment. It was not a gentlemanly remark. He was not in the habit of saying such things, and he was naturally courteous to women, but he claimed afterwards that he lost his temper under the unusual provocation.

The drummer did not sell the goods, and this dealer told me that the change in the policy of the firm in hiring female drummers was abandoned before the saleswoman had reached the further point of her outward trip, and long before the return route was entered upon there was a telegram calling her back to the factory and she was given a liberal cheque for the remainder of the term for which she had been engaged and allowed to seek some other and perhaps more feminine field of industry. — Geo. E. B. Putnam in Boot and Shoe Recorder.

G. M. Farrington is making a two storey brick addition to his store in Picton, Ont.

THE FUTURE OF THE TINSMITH.

A GROWING disposition on the part of old-school tinsmiths to put their sons to some other line of business than that which they have long followed gives rise to the question, if there are, as they contend, fewer opportunities in the trade than there formerly were, says T. Worker in Metal Worker. The fact that the sale of tinware has very largely passed out of the hands of the tinner has made some of the older members of the craft very dubious about the future of the trade, and one frequently hears comparisons made between the trade as it is to day and as is was years ago, when the tinsmith made with his own hands or those of his workmen nearly every article he offered for sale. In those days any man who established himself in a good location, was a fair workman and industrious in his habits, was reasonably sure to succeed. The bulk of his job work came at certain seasons of the year, as it does now, but he could always keep busy, as the dull seasons were employed in making up stock. He had competition, of course, but his competitors were all in the same line as he, the department store not being much in evidence, and barbers, dentists and others did not carry tinware as a side line.

There was, no doubt, less headache connected with doing business in those days than there is now, but whether the opportunities were any greater is extremely doubtful. A great deal of work is now done in the tin shop that was never thought of in years gone by. The job work is more diversified in character, and there is more of it than there was then. In those days the country tinner did no plumbing work to speak of, while now it is considered an important and legitimate branch of his business. New methods of heating have also opened up to him a field for his skill and a demand for his services. In the past a man might succeed by reason of his mechanical skill alone, but under present conditions it is necessary for him to possess business ability as well as technical skill. The changes that have taken place in the trade have come about as a result of improved methods of manufacturing being adopted, and do not necessarily indicate that the trade of the sheet metal worker is on the decline. On the contrary, the demand for his services will probably increase, but along other lines than those followed by his predecessors.

RAILWAYS IN SOUTH AFRICA.

South Africa is a country of immense distances, and it is interesting, for the purpose of realizing its area, to recall the great extent of its railway systems. In the Cape Colony, the open mileage of the Cape Government railways is about 2,000 miles, with 360 miles under construction and 350 miles of privately owned lines. In Natal, there are upwards of 600 miles open, and short extensions, totaling 60 miles, on the north and south coasts, and between Dundee and Vryheid, under construction. In the Transvaal Colony, there are 890 miles open to traffic and 200 miles under construction, exclusive of the Vereeniging-Rand line. The Orange River Colony possesses about 400 miles of open mileage and about 100 miles under construction. Rhodesia, although only a decade old, already possesses about 1,600 miles of open railway, and its three main sections under construction—the Bulawayo-Zambesi, Bulawayo-Salisbury and Bulawayo Tuli—aggregate about 600 miles. Altogether, the open mileage in South Africa amounts to upwards of 5,900 miles, with at least 2,000 miles under construction. These figures are exclusive of the projected extensions for which funds have not yet been provided.—British and South African Export Gazette.

CHANGED HIS ADDRESS.

Knox Henry, Montreal, has moved into a central and commodious office in the Canada Life Building this week. His room number is 32.

TRADE CHAT.

The Dill & Hill Co., general merchants, Wolseley, N.W.T., are opening a branch store at Fernie, B.C.

J. D. Klassen, hardware merchant, Rosthern, N.W.T., has sold his stock to G. E. Knechtel & Co., of that town.

J. W. Husband, hardware dealer, Wallaceburg, Ont., has admitted his son, D. A. Husband, under the style of J. W. Husband & Son.

Two tanks of the Dominion Iron and Steel Co. and their contents, about 300,000 gallons of coal tar, were destroyed by fire April 30. The fire, which is supposed to have caught from a locomotive spark, was an extremely fierce one while it lasted. The loss is fully covered by insurance.

The Sydney Hardware Co., Limited, have been incorporated with $25,000 capital. The new company will be managed by Messrs. Dana & McLennan, who are now engaged in the retail hardware business in Sydney. As soon as possible the company will drop the retail part of the business and devote its energies entirely to jobbing. In May a traveller will be put on the road, and Cape Breton will be thoroughly covered. It is likely a strong effort will be made to cover Prince Edward Island and Newfoundland.

BUSINESS CHANGES.

DIFFICULTIES, ASSIGNMENTS, COMPROMISES.

T. H. Easton, carriagemaker, Minnedosa, Man., has assigned to C. H. Newton, Winnipeg.

A meeting of the creditors of James Grady, carriagemaker, New Glasgow, Que., will be held on May 4.

A meeting of the creditors of J. E. Tremblay, general merchant, St. Anne de Bellevue, Que., has been held.

Colborne & Williamson, stove and tinware dealers, Trenton, Ont., have assigned to George F. Hope, and a meeting of their creditors was held on Tuesday.

PARTNERSHIPS FORMED AND DISSOLVED.

St. Armand Freres, hardware dealers, Montreal, have dissolved, and a new partnership has been registered.

SALES MADE AND PENDING.

Charles Brown, general merchant, Hilton, Man., has sold his hardware stock.

The assets of the estate of the late John Y. Laidlaw, tinsmith, Shubenacadie, N.S., have been sold.

The assets of Miles Birkett, hardware dealer, Ottawa, are to be sold by auction on the 6th inst.

The stock of A. Bertrand, general merchant, Edmunston, N.B., has been seized under bill of sale.

A. Dulmage, general merchant, Lakelet, Ont., has sold out to Wm. Bushfield.

W. R. McCormick, general merchant, Didsbury, N.W.T., has sold out to J. Studer.

The store of the estate of M. Weidman, general merchant, Winnipeg, has been sold at 50c. on the dollar.

CHANGES.

James Ellis & Co. have registered as coal dealers in Quebec.

Hoskins Bros. have registered as tinsmiths, etc., in Brome, Que.

John B. Owens, painter, Montreal, has been succeeded by Peter Houle.

Henri Crevier has registered as Crevier & Fils, hardware dealers, Montreal.

T. Weltin, blacksmith, Dashwood, Ont., has been succeeded by Alex. Zimmer.

Rosina St. Denis has registered as L. B. Pigéon, coal and wood dealer, Lachine, Que.

Henry Redfearn, coal dealer, Brighton, Ont., has been succeeded by D. C. Bullock.

C. M. Allaire has bought out W. R. Walker, tinware dealer, etc., Welland, Ont.

Hypolite Germain has registered as proprietor of H. Germain & Co., blacksmiths, Que.

R. Richardson, general merchant, Bedford, N.S., has registered consent for his wife, Lydia P. Richardson, to do business in her own name.

Adams, Currie & Co., general merchants, etc., Campbellton, N.B., have been succeeded by Wm. Glover.

Albertine Royer has registered as proprietor of Bureau & Cie., hardware dealers, Lake Megantic, Que.

Wilson & Son, blacksmiths, Okotoks, N.W.T., have been succeeded by Wilson Bros.

Cruise & Cummings, electricians, etc., Sydney, N.S., have been succeeded by A. W. Cruise.

Hamilton & Sutton, hardware dealers, Keyes, Man., have been succeeded by Hamilton & Thurston.

Mrs. A. Leroux, who has for some time been proprietress of A. Leroux, hardware dealer, Montreal, has ceased doing business under that style.

FIRES.

The premises of Henry Walters & Sons, axe manufacturers, Hull, Que., have been damaged by fire; insured.

DEATHS.

Gordon Paterson, of Paterson Bros., general merchants, Fisherville, Ont., is dead.

H. B. Elderkin, of H. Elderkin & Co., general merchants, Port Greville, N.S., is dead.

MARKETS AND MARKET NOTES

QUEBEO MARKETS

Montreal, May 3, 1901.

HARDWARE.

THE different wholesale houses and manufacturing establishments are busy making shipments by the first boats and barges, and it is expected that the rush will be maintained till May 15. All spring lines are selling well, with some goods hard to procure. One dealer has told us this week that he could sell 1,000 tons of barb wire if he had it on hand. Nails are now in light stock, and will be till the mills on the canal are in full operation again. Fence wires of all kinds are in heavy demand as well as poultry netting, building paper, harvest tools, ice cream freezers, churns, oil stoves, screens and gas stoves. Plumbers' supplies are also moving well. Shelf goods are in first-class shape, with cutlery and builders' hardware in fair inquiry. One feature of the wholesale trade has been a better demand for white lead; although the outlook in this article is much improved, dealers are still buying from hand to mouth. Early boats are arriving from England, and the pressure that has been felt in some lines will be relieved next week. Iron pipe and bar iron are both strong and advancing.

BARB WIRE — Supplies are urgently needed to supply the demand. Spot stocks are selling at a slight premium. The price is still $3.05 per 100 lb. f.o.b. Montreal.

GALVANIZED WIRE—Trade is exceedingly brisk, and the market presents nothing new. We quote: No. 5, $4.25; Nos. 6, 7 and 8 gauge, $3.55; No. 9, $3.10; No. 10, $3.75; No. 11, $3.85; No. 12, $3.25; No. 13, $3.35; No. 14, $4.25; No. 15, $4.75; No. 16, $5.00.

SMOOTH STEEL WIRE—Oiled and annealed are being shipped in large lots, and the situation seems healthy. We quote oiled and annealed: No. 9, $2.80; No. 10, $2.87; No. 11, $2.90; No. 12, $2.95; No. 13, $3.15 per 100 lb. f.o.b. Montreal, Toronto, Hamilton, London, St. John and Halifax.

FINE STEEL WIRE—Fair amounts are moving at the same discount, 17½ per cent. off the list.

BRASS AND COPPER WIRE — A sma inquiry is reported. The discount on brass is 55 and 2½ per cent., and on copper 50 and 2½ per cent.

FENCE STAPLES—The orders for fence staples are still numerous, and good amounts are being sold. We quote: $3.25 for bright, and $3.75 for galvanized, per keg of 100 lb.

WIRE NAILS — The demand keeps up to the volume of the past two weeks, and supplies are running short on account of the long inactivity of the mills. We quote $2.85 for small lots and $2.77½ for carlots, f.o.b. Montreal, London, Toronto, Hamilton and Gananoque.

CUT NAILS—The demand for cut nails is fair. We quote: $2.35 for small and $2.25 for carlots; flour barrel nails, 25 per cent. discount; coopers' nails, 30 cent. discount.

HORSE NAILS — There is nothing new to report in the market, which varies according to the different transactions. "C" brand is firm at the discount of 50 and 7½ per cent. off new list, while "M" brand is selling in ordinary sized lots at 60 per cent. off old list on oval and city head and 65 per cent. off countersunk head.

HORSESHOES—Although the demand is not very active, there are a few more inquiries this week. We quote as follows: Iron shoes, light and medium pattern, No. 2 and larger, $3.50; No. 1 and smaller, $3.75; snow shoes, No. 2 and larger, $3.75; No. 1 and smaller, $4.00; X L steel shoes, all sizes, 1 to 5, No. 2 and larger, $3.60; No. 1 and smaller, $3.85; feather-weight, all sizes, $4.85; toe weight steel shoes, all sizes, $5.95 f.o.b. Montreal; f.o.b. Hamilton, London and Guelph, 10c. extra.

POULTRY NETTING—Business is reported good this week, some large orders having come to hand. Immediate delivery is now the general request. The discount is still 55 per cent.

GREEN WIRE CLOTH — A fairly heavy trade is being done in green wire cloth at the former quotations, $1.35 per 100 sq. ft.

SCREEN DOORS AND WINDOWS—Business is brisk. We quote: Screen doors, plain cherry finish, $8.25 per doz.; do. fancy, $11.50 per doz.; walnut, $7.40 per doz., and yellow, $7.45; windows, $2.25 to $3.50 per doz.

SCREWS — A good business is doing in screws. Discounts are as follows: Flat head bright, 87½ and 10 per cent. off list; round head bright, 82½ and 10 per cent.; flat head brass, 80 and 10 per cent.; round head brass, 75 and 10 per cent.

BOLTS — Trade is rather brisk in bolts, and fair quantities are moving out. Discounts are as follows: Norway carriage bolts, 65 per cent.; common, 60 per cent.; machine bolts, 60 per cent.; coach screws, 70 per cent.; sleigh shoe bolts, 72½ per cent.; blank bolts, 70 per cent.; bolt ends, 62½ per cent.; plough bolts, 60 per cent.; tire bolts, 67½ per cent.; stove bolts, 67½ per cent. To any retailer an extra discount of 5 per cent. is allowed. Nuts, square, 4c. per lb. off list; hexagon nuts, 4¼c. per lb. off list. To all retailers an extra discount of ¼c. per lb. is allowed.

BUILDING PAPER — The demand continues heavy. We quote as follows: Tarred felt, $1.70 per 100 lb.; 2-ply, ready roofing, 80c. per roll; 3-ply, $1.05 per roll; carpet felt, $2.25 per 100 lb.; dry sheathing, 30c. per roll; tar sheathing, 40c. per roll; dry fibre, 50c. per roll; tarred fibre, 60c. per roll; O.K. and I.X.L., 65c. per roll; heavy straw sheathing, $28 per ton; slaters' felt, 50c. per roll.

RIVETS—A fair number of orders have been filled this week. Discount on best iron rivets, section, carriage, and wagon box, black rivets, tinned do., coopers' rivets and tinned swedes rivets, 60 and 10 per cent.; swedes iron burrs are quoted at 55 per cent. off; copper rivets, 35 and 5 per cent. off; and coppered iron rivets and burrs, in 5-lb. carton boxes, are quoted at 60 and 10 per cent. off list.

BINDER TWINE—The amount of business being done in this line is not large as yet. We quote: Blue Ribbon, 11½c.; Red Cap, 9¾c.; Tiger, 8¾c.; Golden Crown, 8c.; Sisal, 8¼c.

CORDAGE—The demand is hardly as heavy as it was, yet a good business is still being done. Manila is now worth 13½c. per lb. for 7-16 and larger; sisal is selling at 10c., and lathyarn 10c.

SPADES AND SHOVELS—This line has been quite active this week. The discount is 40 and 5 per cent. off the list.

HARVEST TOOLS—The distributing trade is fully up to the usual spring average. The discount is still 50, 10 and 5 per cent.

TACKS—Sorting orders are coming in daily. We quote: Carpet tacks, in dozens and bulk, 80 and 15 per cent.; tinned, 80 and 20 per cent., and cut tacks, blued, in dozens, 80 per cent.

LAWN MOWERS—A good demand has set in for lawn mowers, and stocks are moving out freely. We quote as follows: High wheel, 50 and 5 per cent. f.o.b. Montreal; low wheel, in all sizes, $2.75 each net; high wheel, 11-inch, 30 per cent. off.

FIREBRICKS—The demand for firebricks is fair for small lots, and the movement is about up to the average of the season. Prices are unchanged at $18 to $24 per 1,000 as to brand ex store, and Scotch are quoted at $17.50 to $22 and English at $17 to $21 per 1,000 for arrival ex wharf.

CEMENT—In foreign cement the volume of business is very disappointing for the season of the year, which is entirely due to the fact that, owing to the keen competition on the part of Canadian and American manufacturers, importers are booking few orders, as they find it difficult to compete with the latter on high grades. The indications at present are that the imports of foreign cement this season will show a considerable decrease, as compared with previous years; in fact, importers say that up to the present few orders have been placed with makers. We quote: German cement, $2.30 to $2.50; English, $2.25 to $2.35; American, $2.25 to $2.50, and Belgian, $1.70 to $1.95 for summer delivery, and spot prices are unchanged at $2.45 to $2.55 for German; $2.30 to $2,40 for English, and $1.95 to $2 05 for Belgian, ex store.

METALS.

In bar iron, pipe, chain, and zinc, the market is decidedly firm, but the sheet metals are rather easy in anticipation of arrivals of lower-priced goods from England. The first steamers are now on the river.

PIG IRON—The feeling in pig iron is a little steadier in sympathy with a slight improvement in England. Canadian pig iron is worth about 18 to $19, and Summerlee about $20.50 to $21.

BAR IRON—The tendency of the market here is upward. Dealers are selling merchants' bar at $1.75, and horseshoe at $2.

BLACK SHEETS—A fair amount of business is being done, but the spot market is falling, dealers generally quoting 10 to 15c. lower this week. We quote: 8 to 16 gauge, $2.55; 26 gauge, $2.60, and 28 gauge, $2.65.

GALVANIZED IRON—English stock is now coming in, and deliveries of new goods will soon be made. Spot stocks are rather light on account of the brisk trade that has been doing of late. We quote: No. 28 Queen's Head, $4.65; Apollo, 10¾ oz., $4.50, and Comet, $4.40 to $4.45, with a 15c. reduction for case lots.

INGOT COPPER—There is little demand for copper at present, and business is rather limited. The ruling figure is 18c.

INGOT TIN — Some sales are reported this week at 30c.

LEAD — The price of lead varies. Some quote $3 85, others $3 75 per 100 lb.

LEAD PIPE—Fair quantities are being shipped. We quote: 7c. for ordinary and 7½c. for composition waste, with 25 per cent. off.

IRON PIPE—The predominant tone of the market is strong and the feeling is that further advance will take place in the near future. We quote: Black pipe, ¼, $3 per 100 ft.; ⅜, $3; ½, $3 05; ¾, $3.30; 1-in., $4.70; 1¼, $6.40; 1½, $7.70; 2-in. $10.25. Galvanized, ½, $4.60; ¾, $5.25; 1-in., $7.50; 1¼, $9.80; 1½, $11.75; 2 in., $16.

TINPLATES—The receipt of goods by first steamers weakens the market slightly. We quote: Coke plates, $3.80 to $4; charcoal, $4.15 to $4.25.

CANADA PLATE—Some firms are quoting lower prices this week, on account of expected receipts from England. We quote: 52's, $2.50; 60's, $2.60; 75's, $2.70; full polished, $3.35, and galvanized, $4.20.

STEEL — The steel market is steady and rather brisk. We quote: Sleighshoe, $1.95; tire, $2; bar, $1.95; spring, $2.75; machinery, $2.75 and toe-calk, $2.50.

SHEET STEEL—Business is fairly active aud prices are steady. We quote: Nos. 22 and 24, $3, and Nos. 18 and 20, $2.85.

TOOL STEEL—Black Diamond is worth 8c. and Jessop's, 13c.

TERNE PLATES—The demand for terne plates is fair. The price rules about $7.75.

COIL CHAIN—The market is firm and active. We quote: No. 6, 11½c.; No. 5, 10c.; No. 4, 9½c.; No. 3, 9c.; ¼-inch, 7½c. per lb.; 5-16, $4.85; 5-16 exact, $5.30; ⅜, $4 40; 7-16, $4.20; ½, $3.95; 9-16, $3.85; ⅝, $3.55; ¾, $3.45; ⅞, $3.40; 1-in., $3.35. In carload lots an allowance of 10c. is made.

SHEET ZINC—The English zinc market has advanced £2 on the week. The situation here is firmer in consequence. Spot goods are worth about $5 75 to $6.25.

ANTIMONY—Quiet, at 10c.

ZINC SPELTER—Is worth 5c.

SOLDER—We quote: Bar solder, 18½c.; wire solder, 20c.

GLASS.

The glass market seems to be tending upwards. Trade is rather brisk. We quote as follows: First break, $2; second, $2.10 for 50 feet; first break, 100 feet, $3.80; second, $4; third, $4.50; fourth, $4.75; fifth, $5.25; sixth, $5.75, and seventh, $6.25.

PAINTS AND OILS.

In view of heavy arrivals of red lead, the scarcity which has been apparent will shortly be relieved. Quotations are now reduced 25c. per 100 lb. There is a heavy rush of orders in just now at all the factories, and prices are well maintained all along the line. Whiting keeps firm at present quotations. Kalsomines, and what are termed household colors, such as liquid paints, handy and stains, are being shipped in immense quantities all over the Dominion. The outlook is exceedingly promising. Paris green still keeps sluggish. Turpentine is easier, but linseed oil is experiencing a bright demand, and there is no change in prices either for spot or future delivery. A feature of the past week has been the gratifying inquiry for graphite and oxide of iron for construction purposes, proving that there is a good deal of building going on all through the country. Shingle stains are also being shipped in limited quantities, but a better trade is looked for. We quote:

WHITE LEAD—Best brands, Government standard, $6.25; No. 1, $5 87½; No. 2, $5.50; No. 3, $5.12½, and No. 4, $4.75 all f.o.b. Montreal. Terms, 3 per cent. cash or four months.

DRY WHITE LEAD — $5.50 in casks; kegs, $5.75.

RED LEAD — Casks, $5.00; in kegs, $5.25.

WHITE ZINC PAINT—Pure, dry, 7c.; No. 1, 6c.; in oil, pure, 8c; No. 1, 7c.; No. 2, 6c.

PUTTY—We quote: Bulk, in barrels, $1.90 per 100 lb.; bulk, in less quantity, $2.05; bladders, in barrels, $2.10; bladders, in 100 or 200 lb. kegs or boxes, $2.25; in tins, $2.55 to $2.65; in less than 100-lb. lots, $3 f.o.b. Montreal, Ottawa, Toronto, Hamilton, London and Guelph. Maritime Provinces 10c. higher, f.o.b. St. John and Halifax.

LINSEED OIL—Raw, 72c.; boiled, 75c. in 5 to 9 bbls., 1c. less, 10 to 20 bbl. lots, open, net cash, plus 2c. for 4 months Delivered anywhere in Ontario between Montreal and Oshawa at 2c. per gal. advance and freight allowed.

TURPENTINE—Single bbls., 57c.; 2 to 4 bbls., 56c.; 5 bbls. and over, open terms, the same terms as linseed oil.

MIXED PAINTS—$1.25 to $1.45 per gal.

CASTOR OIL—8¾ to 9¾c. in wholesale lots, and ½c. additional for small lots.

SEAL OIL—47½ to 49c.

COD OIL—32½ to 35c.

NAVAL STORES — We quote: Resins, $2.75 to $4 50, as to brand; coal tar, $3 25 to $3.75; cotton waste, 4½ to 5½c. for colored, and 6 to 7½c. for white; oakum, 5½ to 6½c., and cotton oakum, 10 to 11c.

PARIS GREEN—Petroleum barrels, 16¾c. per lb.; arsenic kegs, 17c.; 50 and 100, lb. drums, 17½c.; 25-lb. drums, 18c.; 1-lb-packages, 18½c.; ½-lb. packages, 20½c.; 1-lb. tins, 19½c.; ½-lb. tins, 21½c. f.o.b. Montreal; terms 3 per cent. 30 days, or four months from date of delivery.

SCRAP METALS.

The scrap metal market is rather easier, in view of arrivals of wrought scrap from England, which will relieve the scarcity that has been felt here. Dealers are paying the following prices in the country: Heavy copper and wire, 13 to 13½c. per lb.; light copper, 12c.; heavy brass, 12c.; heavy yellow, 8½ to 9c.; light brass, 6½ to 7c.; lead, 2½ to 2¾c. per lb.; zinc, 2¼ to 2½c.; iron, No. 1 wrought, $15 to $16 per gross ton f.o.c. Montreal; No. 1 cast, $13 to $14; stove plate, $8 to $9; light iron, No. 2, $4 a ton; malleable and steel, $4; rags, country, 70 to 80c. per 100 lb.; old rubbers, 6¼c. per lb.

HIDES.

The market is steady. We quote: Light hides, 6½c. for No. 1; 5½c. for No. 2, and 4½c. for No. 3. Lambskins, 10c.; sheepskins, 90c.; calfskins, 8c. for No. 1 and 6c. for No. 2.

PETROLEUM.

The demand is falling off. We quote as follows: "Silver Star," 14½ to 15½c.; "Imperial Acme," 16 to 17c.; "S.C. Acme," 18 to 19c., and "Pratt's Astral," 18½ to 19½c.

ONTARIO MARKETS.

TORONTO, May 4, 1901.

HARDWARE.

ALTHOUGH the wholesale hardware trade is not, perhaps, as heavy as it was a week ago, it is still decidedly brisk. Very few changes have taken place in our quotations, but the manufacturers have notified the jobbers of some advances in their figures. For instance, American makers of barn door tracks advise the jobbers here of an advance of 10c. per 100 ft., and sheet zinc is 10s. per ton dearer in the primary market. An advance of 10 per cent. has taken place in asbestos goods. The difficulty noted in previous issues in getting supplies of barb wire still exists, although the manufacturers promise to make some shipments next week. The demand for oiled and annealed wire is fairly good, and the same may be said of galvanized wire. Orders are still fairly good for wire nails, but they are still small for cut nails. Trade in horseshoes and horse nails is fair for the season. Screws, bolts, and rivets and burrs are all meeting with a fairly good demand. A steady trade is being done in rope. In building paper the demand continues active. A fair amount of cutlery is going out, and a number of inquiries are coming in for certain lines of sporting goods. A fair movement is reported in lawn mowers. A good movement is reported in churns, and a little business is to be noted in wringers. Fairly liberal quantities of screen doors and windows and green wire cloth are going out. The demand keeps good for poultry netting. Harvest tools are going out fairly well, and there is quite a little demand for garden tools.

BARB WIRE—The demand for barb wire is quite active, even more so than has been usual for some years. However, a great deal of difficulty is still being experienced in getting delivery from the manufacturers, and this is causing trade to be much more limited than it otherwise would be. The manufacturers in the United States are still from three to four weeks behind with their orders, but they advise jobbers in this country that shipments will be made next week. We quote $3.05 per 100 lb. from stock Toronto; f.o.b. Cleveland $2.82½ per 100 lb. for less than carlots and $2.70 for carlots.

GALVANIZED WIRE—A steady trade is being done in galvanized wire, and stocks are in fairly good shape. The manufacturers, however, are somewhat behind with their orders. Prices are steady and unchanged. We quote as follows: Nos. 6, 7 and 8, $3.50 to $3.85 per 100 lb., according to quantity; No. 9, $2.85 to $3.15; No. 10, $3.60 to $3.95; No. 11, $3.70 to $4 10; No. 12, $3 to $3 30; No. 13, $3 10 to $3 40; No. 14, $4 10 to $4 50; No. 15 $4 60 to $5 05: No. 16, $4.85 to $5.35. Nos. 6 to 9 base f.o.b. Cleveland are quoted at $2.57½ in less than carlots and 12c. less for carlots of 15 tons.

SMOOTH STEEL WIRE—The demand is fairly active for oiled and annealed wire, and, as pointed out last week, the scarcity of barb wire is stimulating business in oiled and annealed. The net selling prices for oiled and annealed are as follows: Nos. 6 to 8, $2 90; 9, $2.80; 10, $2.87; 11, $2.90; 12, $2.95; 13, $3 15; 14, $3.37; 15, $3.50; 16, $3.65. Delivery points, Toronto, Hamilton, London and Montreal, with freights equalized on these points.

WIRE NAILS—Although a very few large orders are coming to hand, there is a fair demand for 5 to 10 keg lots. In the United States, there is an active demand for wire nails, and, in fact, there has been for some weeks past. Prices are unchanged, and we quote less than carlots at $2.85, and carlots at $2.77½. Delivery points: Toronto, Hamilton, London, Gananoque and Montreal.

CUT NAILS—Business in this line still seems to be without any improvement. As noted last week, a few shingle nails are going out. We quote the base price at $2.35 per keg for less than carlots, and $2.25 for carlots. Delivery points: Toronto, Hamilton, London, Montreal and St. John, N.B.

HORSE NAILS—Trade is fairly good, but without any particular feature. Discount on "C" brand, oval head, 50 and 7½ per cent. off new list and on other brands 50, 10 and 5 per cent., off the old list; countersunk, 50, 10 and 10 per cent.

HORSESHOES—A fair trade is being done, but it is confined to small quantities. We quote as follows f.o.b. Toronto: Iron shoes, No. 2 and larger, light, medium and heavy, $3.60; snow shoes, $3.85; light steel shoes, $3.70; featherweight (all sizes), $4.95; iron shoes, No. 1 and smaller, light, medium and heavy (all sizes), $3.85; snow shoes, $4; light steel shoes, $3.95; featherweight (all sizes), $4 95.

SCREWS — A steady trade is still to be reported in screws. We quote discounts: Flat head bright, 87½ and 10 per cent.; round head bright, 82½ and 10 per cent.; flat head brass, 80 and 10 per cent.; round head brass, 75 and 10 per cent.; round head bronze, 65 per cent., and flat head bronze at 70 per cent.

BOLTS AND NUTS—Business in bolts and nuts remains much about the same as it has during the past week, being generally active, and in a satisfactory condition. We quote as follows: Carriage bolts (Norway), full square, 65 per cent.; carriage bolts full square, 65 per cent.; common carriage bolts, all sizes, 60 per cent.; machine bolts, all sizes, 60 per cent.; coach screws, 70 per cent.; sleighshoe bolts, 72½ per cent.; blank bolts, 60 per cent.; bolt ends, 62½ per cent.; nuts, square, 4c. off; nuts, hexagon, 4½c. off; tire bolts, 67½ per cent.; stove bolts, 67½; plough bolts, 60 per cent.; stove rods, 6 to 8c.

RIVETS AND BURRS—A steady trade is being done, and the chief business is still for copper rivets. We quote: Iron rivets, 60 and 10 per cent.; iron burrs, 55 per cent.; copper rivets and burrs, 35 and 5 per cent.

ROPE—A steady trade for small quantities is still reported, and prices are unchanged

and firm. We quote the base price of sisal at 10c. and of manila at 13½c. per lb.

CUTLERY—There is a fair demand this week for table cutlery, and for pocket cutlery. The table cutlery most wanted this week is of the cheaper kind for the summer trade.

SPORTING GOODS—There has been quite an inquiry during the past week for guns and rifles, and quite a little ammunition is going out.

ENAMELLED WARE AND TINWARE—An active business is being done this week in both these lines, and prices rule steady and unchanged. The discounts on enamelled ware are as follows: "Granite," "Pearl," "Crescent" and "Imperial" wares at 50, 10 and 10 per cent.; white, "Princess," "Turquoise," blue and white, 50 per cent.; "Diamond," "Famous" and "Premier," 50 and 10 per cent.

GREEN WIRE CLOTH—Shipments of this are going out fairly well; the price is still $1.35 per 100 sq. ft.

SCREEN DOORS AND WINDOWS—Fairly large quantities of these are also going out this week. We quote as follows: 4 inch styles in doors, $7.20 to $7 80 per doz., and 3-inch styles, 20c. per doz. less; screen windows, $1.60 to $3 60 per doz., according to size and extension.

BUILDING PAPER—Business in this line is active. A number of orders that were booked for shipment early in May are now going forward. We quote: Plain building, 30c.; tarred lining, 40c., and tarred roofing, $1.65.

POULTRY NETTING—Business which has been good all the spring is still active. Discount on Canadian is 55 per cent.

ICE CREAM FREEZERS AND REFRIGERATORS—An active trade is still being done in these lines. Business opened up earlier this spring than usual, and it has been more satisfactory than is even usual at this season.

OIL STOVES—A large trade is being done in oil stoves, but business in gas stoves appears to be but moderate, so far.

CHURNS AND WRINGERS—There is an active movement in churns, and the factories are able to make fairly prompt shipment. Not a great many wringers are going out.

LAWN MOWERS—These are moving fairly well, and quite a number have been shipped during the past week.

HARVEST TOOLS—Although the trade is gradually improving, the movement, generally speaking, is not large. Discount 50, 10 and 5 per cent.

GARDEN TOOLS—There is a fairly good trade being done in such garden tools as spades, rakes, etc.

EAVETROUGH—No doubt stimulated by the low prices, the demand for eavetrough and conductor pipe continues good.

BRIGHT WIRE GOODS—Business continues steady at the discount of 62½ per cent.

BINDER TWINE—Trade is gradually falling off in this line. We quote as follows: American—Sisal and standard, 8½c.; manila, 10½c.; pure manila, 11½ to 12c. per lb. Canadian—Sisal, 8½c.; pure manila, 11½c. per lb.

CHAIN—There is a little demand for trace chain, but otherwise business is quiet.

SLEIGH BELLS—The manufacturers of sleigh bells have issued their prices for the next season. The figures they quote are the same as those which ruled last season.

GARDEN SYRINGES—There is not much business being done yet. A feature of the trade is the fact that both Canadian and English syringes are lower in price than those made in the United States.

BARN DOOR TRACKS—The wholesale dealers here have been advised by the makers in the United States that prices have been advanced 10c. per 100 ft. The cause of the advance is attributed to the high price of steel. The new price took effect on May 1. Local wholesalers report a good trade in this line at present.

HORSE BLANKETS—The makers of horse blankets announce that prices for next season will be 10 per cent. higher.

ASBESTOS GOODS—The manufacturers of such asbestos goods as mill boards, building papers and wicks have advanced their prices 10 per cent.

CEMENT—Prices are easy throughout. Canadian portland is 5 to 10c. lower. We quote barrel lots: Canadian portland $2.25 to $2.80; German, $3 to $3.15; English, $3; Belgian, $2.50 to $2.75; Canadian hydraulic, $1.25 to $1.50.

METALS.

The trade generally, is good this week, and stock shipments are larger than they were a week ago. The demand is chiefly for tinplates, galvanized sheets, and ingot metals.

PIG IRON—The iron market is quiet, with prices steady. No. 2 Canadian pig iron is worth $16.50 in large lots.

BAR IRON—The demand continues active with the mills still behind with their orders. We still quote the base price at $1.85 to $1.90 per 100 lb. from stock.

STEEL—The demand for steel continues active and prices rule firm. We quote tire steel at $2 30 to $2.50; sleighshoe steel at $2.10 to $2.25; reeled machinery steel, $3. Hoop steel, $3.10.

PIG TIN—The demand has been active during the past week and the outside markets are stronger. The monthly trade statistics are more favorable than expected. Locally, quotations are unchanged at 31 to 32c.

TINPLATES—There has been a good demand both from stock and for importation. Prices are unchanged.

TINNED SHEETS—The demand during the past week has been active. We quote 9 to 9½c. per lb.

TERNE PLATES—A little more is being done in this line, although trade is not active.

GALVANIZED SHEETS—Trade is fairly active in galvanized sheets, and shipments are mostly case lots. The market is fairly strong, and the American manufacturers advise that they are unable to book orders for delivery before June. At this rate delivery can be obtained sooner from Great Britain. Ruling prices from stock are still $4 60 for English and $4.50 for American.

BLACK SHEETS—Trade is moderate and prices are unchanged; 28 gauge is still quoted at $2.30.

CANADA PLATES—The shipments from stock are still quiet. Orders continue to be booked for importation. We quote: All dull, $3; half-and half, $3.15, and all bright, $3.65 to $3.75.

COPPER—Trade is fairly active in ingot copper in small lots, and a good business is still being done in sheet copper. Prices rule steady in the outside markets. Locally they are without change. We quote: Ingot, 19c.; bolt or bar, 23½ to 25c.; sheet, 23 to 23½c. per lb.

BRASS—Business is a little better than it

was a week ago, now being reported fair. Discount on rod and sheet is 15 per cent.

SOLDER—Trade is still fair. We quote: Half-and-half, guaranteed, 18½c.; ditto, commercial, 18c.; refined, 18c., and wiping, 17c.

IRON PIPE — This is a quiet season for iron pipe, and very little business is in consequence doing. prices are unchanged at quotations.

LEAD — Trade conrinues fairly active, and prices are steady in the outside markets. We quote 4¼ to 4½c. per lb.

ZINC SPELTER — Trade is rather quiet, and in New York prices are rather easier. Locally, we still quote 5¾ to 6c. per lb.

ZINC SHEET—Importers here have been advised of an advance of 10s. per ton in the price of zinc sheet, but we still quote 6½c. for casks, and 6¾c. for part casks as the ruling prices.

ANTIMONY—Trade is quiet at 10½ to 11c. per lb.

PAINTS AND OILS.

Jobbers report April trade to have reached the largest volume for April in the history of the trade. The movement continues exceedingly brisk. Some jobbers report an improvement in the demand for paris green, but there is not a great deal moving yet. Linseed oil and castor oil are rather scarce. Other lines are in liberal supply and good demand. Turpentine keeps easy at primary points, but linseed oil is firm. We quote:

WHITE LEAD—Ex Toronto, pure white lead, $6.37½; No. 1, $6; No. 2, $5.67½; No. 3, $5.25; No. 4, $4.87½; genuine dry white lead in casks, $5.37½.

RED LEAD—Genuine, in casks of 560 lb., $5.50; ditto, in kegs of 100 lb., $5.75; No. 1, in casks of 560 lb., $5; ditto, kegs of 100 lb., $5.25.

LITHARGE—Genuine, 7 to 7½c.

ORANGE MINERAL—Genuine, 8 to 8½c.

WHITE ZINC—Genuine, French V.M., in casks, $7 to $7.25; Lehigh, in casks, $6.

PARIS WHITE—90c. to $1 per 100 lb.

WHITING — 70c. per 100 lb.; Gilders' whiting, 80c.

GUM SHELLAC — In cases, 22c.; in less than cases, 25c.

PARIS GREEN—Bbls., 16¾c.; kegs, 17c.; 50 and 100 lb. drums, 17½c.; 25-lb. drums, 18c.; 1-lb. papers, 18½c.; 1-lb. tins, 19½c.; ½-lb. papers, 20½c.; ½-lb. tins, 21½c.

PUTTY —Bladders, in bbls., $2.10; bladders, in 100 lb. kegs, $2.25; bulk in bbls., $1.90; bulk, less than bbls. and up to 100 lb., $2.05; bladders, bulk or tins, less than 100 lb., $2.90.

PLASTER PARIS—New Brunswick, $1.90 per bbl.

PUMICE STONE — Powdered, $2.50 per cwt. in bbls., and 4 to 5c. per lb. in less quantity; lump, 10c. in small lots, and 8c. in bbls.

LIQUID PAINTS—Pure, $1.20 to $1.30 per gal.; No. 1 quality, $1 per gal.

CASTOR OIL—East India, in cases, 10 to 10½c. per lb. and 10¾ to 11c. for single tins.

LINSEED OIL—Raw, 1 to 4 barrels, 71c.; boiled, 74c.; 5 to 9 barrels, raw, 70c.; boiled, 73c., delivered. To Toronto, Hamilton, Guelph and London, 1c. less.

TURPENTINE—Single barrels, 54c.; 2 to 4 barrels, 53c., delivered. Toronto, Hamilton and London 1c. less. For less quantities than barrels, 5c. per gallon extra will be added, and for 5-gallon packages, 50c., and 10 gallon packages, 80c. will be charged.

GLASS.

The movement among the jobbers for an advance has been checked by the refusal of one or two houses to make the change. The feeling continues firm, however. We quote as follows: Under 26 in., $4.15 26 to 40 in., $4.45; 41 to 50 in., $4.85; 51 to 60 in., $5.15; 61 to 70 in., $5.50; double diamond, under 26 in., $6; 26 to 40 in., $6.65; 41 to 50 in., $7.50; 51 to 60 in., $8.50; 61 to 70 in., $9.50, Toronto, Hamilton and London. Terms, 4 months or 3 per cent. 30 days.

HIDES, SKINS AND WOOL.

HIDES—The market continues dull, with no change in prices. We quote: Cowhides, No. 1, 6½c.; No. 2, 5½c.; No. 3, 4½c. Steer hides are worth 1c. more. Cured hides are quoted at 7 to 7½c.

SKINS—A fair trade is doing. Prices are unchanged. We quote: No. 1 veal, 8-lb. and up, 8c. per lb.; No. 2, 7c.; dekins, from 40 to 60c.; culls, 20 to 25c. Sheepskins, 90c. to $1.

WOOL—There is nothing doing. We quote: Combing fleece, 14 to 15c., and unwashed, 8 to 9c.

OLD MATERIAL.

There is a free delivery of most lines at steady prices. The demand is moderate. We quote jobbers' prices: Agricultural scrap, 55c. per cwt.; machinery cast, 60c. per cwt.; stove cast, 50c.; No. 1 wrought 50c. per 100 lb.; new light scrap copper, 12c. per lb.; bottoms, 11½c.; heavy copper, 13c.; coil wire scrap, 13c.; light brass, 7c.; heavy yellow brass, 10 to 10½c.; heavy red brass, 10¾ to 11c.; scrap lead, 3c.; zinc, 2c; scrap rubber, 6½c.; good country mixed rags, 65 to 75c.; clean dry bones, 40 to 50c. per 100 lb.

PETROLEUM.

Prices are unchanged. There is not much doing. We quote as follows: Pratt's Astral, 16¾ to 17c. in bulk (barrels, $1 extra); American water white, 16¾ to 17c. in barrels; Photogene, 16 to 16½c.; Sarnia water white, 15¾ to 16c. in barrels; Sarnia prime white, 14¾ to 15c. in barrels.

COAL.

Prices for May delivery are 10c. higher than were noted in April. We quote at international bridges as follows: Grate, $3.75 per gross ton; egg, stove and nut, $5 per gross ton with a rebate of 40c. off for May shipment.

MARKET NOTES.

Horse blankets for next season are 10 per cent. higher.

Coal for May delivery is 10c. per ton higher than for April delivery.

The manufacturers of asbestos goods have advanced their prices 10 per cent.

H. S. Howland, Sons & Co. are in receipt of another carload of screen doors and windows.

The United States manufacturers of barn door tracks have advanced their prices 10c. per 100 feet.

Rice Lewis & Son, Limited, are in receipt of another carload of "Jewel" brand paints, of which they are the agents.

Next season's prices for sleigh bells have been issued by the manufacturers. They are the same as those which ruled last season.

TRADE IN COUNTRIES OTHER THAN OUR OWN.

IRON AND STEEL IN PITTSBURG.

So far as new business is concerned the market in this district has been exceedingly quiet during the past week. There is little buying of pig iron, for the reason that consumers have their requirements well covered until the end of the first half, and the buying of steel billets has been exceedingly limited, for the reason that they are scarcely to be had. In finished lines, with the exception of rails and structural material, little new business has been placed. The heavy tonnage placed during the first three months of the year is the cause of the light buying at the present time but the lull has by no means disconcerted the manufacturers. In fact, they have been wanting a breathing spell in order that they might catch up on deliveries. Specifications on running contracts continue to keep the mills operating at their utmost capacity and immediate shipment is something foreign to the trade at the present time.—Iron Trade Review, April 25.

HARDWARE TRADE IN THE UNITED STATES.

With the progress of the season and the liberal purchasing of the jobbers and retailers there is in many lines a slight relaxing of the demand upon the manufacturers. This occasions, however, no uneasiness, as in most cases their order books are well filled and the difficulty of getting out goods promptly is one of the features of the situation. There is, however, less booking of new business, and manufacturers are looking forward to the opportunity of replenishing stocks. The difficulty of getting material promptly is the cause of some delay in the execution of orders. It is in heavy goods that most complaint is heard of delayed shipments, most lines of shelf and miscellaneous hardware being supplied with a good degree of promptness. The tone of the market remains strong and confident. Some advances are being made from time to time, with a few minor changes in the direction of lower prices. Export business continues in good volume. It feels to some extent the effect of higher prices here as compared with those current abroad, and some trade has been diverted on this account.—Iron Age.

REDUCTION IN BRITISH IRON.

A cable despatch says: "In consequence of the severe depression in the Midland iron trade, a reduction of 20 shillings per ton has been made in Staffordshire marked iron."

THE UNDERSELLING IN WIRE NAILS.

Ironmonger, London, England, says: "After the recent period of underselling in the German wire-nail trade, matters appear to be gliding along a little more smoothly again, although the distinct understanding which used to prevail amongst the five firms concerned is a thing of the past. The persistent underselling which went on in the early part of last year brought about the collapse of the arrangement entered into between the parties concerned, and at the present time there is no fixed and agreed basis of values. The leading firm in the trade, in reply to overtures for the purpose of restoring the status quo ante, appear to have told the firm who approached them that their present price was 7s. 3d. for Nos. 0 to 7, and that they had no present intention of altering it, but they added that if they found others underselling them, they would meet such competition by swift retaliatory measures. For the good of the trade as a whole it is hoped that these measures will not be rendered necessary. There can be no doubt either of the ability of the firm indicated to make matters very unpleasant for 'cutters' or of their intention to fight hard if forced to take action."

THE SHEFFIELD IRON TRADE

Speaking generally, the new quarter has opened unfavorably in the iron and steel trades here, the amount of business done this week being disappointing, especially to those who anticipated a decided revival. Opinions differ as to the position and prospects of local trade, and a few manufacturers think that there are signs of improvement, but the majority say that things are as bad as ever, while some report that the stagnation is more pronounced than it was before the holidays. The present market conditions are certainly such as should stimulate trade if it really possesses any latent life.—Ironmonger, London.

NEW YORK METAL MARKET.

TIN—In anticipation of a statistical showing at the beginning of May that will be more favorable to their interests the bulls in London succeeded in getting price up 17s. 6d. on spot and 10s. on futures, but trading was moderate. The New York market was sympathetically stronger and higher, closing at 25.65c. bid and 25.95c. asked for spot, and 25.50c. bid and 25.80 asked for May.

COPPER—The London cable reported a decline of 1s. 3d. on both spot and futures, the market being quiet, particularly for the latter. The situation here was virtually unaltered. The movement into home consumption is free, though largely on old orders, but there is no new export business. Prices are held steady up to 17c. for Lake Superior and 16 5-8c. for electrolytic and casting.

PIG LEAD—We have still to report a quiet, but steady market, with spot prices based on 4.37 1-2c. for lots of 50 tons or over. The St. Louis and London markets were quiet and unchanged.

SPELTER—The local market was quieter, and if anything not so firm as yesterday. Still there was no quotable change in prices, the range still being 4.02 1-2c. to 4.05c. The St. Louis market was reported firm at 3.85 to 3.87 1-2c. London was unchanged.

ANTIMONY—The demand for regulus is moderate and is supplied at prices within the range of 8 1-2 to 10 1-4c., as to brand.

OLD METALS—Though trade is dull prices are well maintained.

IRON AND STEEL—The quiet conditions in control of the iron and steel markets at present do not detract from the strength of tone which have characterized them for weeks past. There are no indications of decreasing consumption, but the disposition to buy far in advance of requirements has to a great extent disappeared. This lull in trade gives the mills and foundries an opportunity to catch up with orders, and for that reason it does not appear to be unwelcome. There were no developments of fresh interest to-day, though advices from Pittsburg are to the effect that one of the large steel mills has been buying several large blocks of Bessemer pig, and that The Carnegie Steel Company is taking a good deal of basic iron, the latter being drawn chiefly from Eastern furnaces. Since the announcement of the advance in the price of steel rails to take effect May 1 it is reported that little business has been done.

TINPLATE—There is nothing new in the situation, the movement continuing free, through contract deliveries, while a moderate volume of new business is noted.—New York Journal of Commerce, May 1.

HEATING AND PLUMBING

SOME BUILDING NOTES.

COUNCILLOR SIMPSON has begun to erect a new house on North Station street, Weston, Ont.

Davey Bros., Lambeth, Ont., are erecting two new houses.

It is proposed to erect a new Methodist church at Riverview, Man.

N. R. Darrach, architect, is erecting a new Presbyterian church at Cowal, Ont.

Mrs. John McIntyre, Kingsley, Man., is erecting a new house near Miami, Man.

A $25,000 Roman Catholic church will be erected in London, Ont., this summer.

W. A. Pae intends building a cottage at Big Bay Point, Barrie, Ont., this summer.

The Royal Bank of Canada is building a three storey, 100 x 40 ft. building in Sydney, N.S.

Ex Ald. H. M. Douglas is erecting three brick houses on Central avenue, London, Ont.

A new school will be erected by the London, Ont., school board on Park street, London.

A $100,000 Roman Catholic church will be erected in St. Edward Parish, St. Denis and Beaubien streets, Montreal.

Tenders are called before Monday for large additions to the building of the Canadian General Electric Co., at Peterboro', Ont.

Architect A. W. Peene, Hamilton, is going to build 10 brick houses on St. Matthew's avenue, Hamilton, for J. M. Peregrine.

The Muskoka Navigation Company are erecting a summer hotel at Wrenshaw's Point, Muskoka, Ont., which is to be completed by July 1.

THE NATIONAL CONVENTION.

The sub-committee which have charge of the arrangements of the annual convention of the National Plumbers' Association of Canada, consisting of President Meredith, Secretary Mansell and Messrs. Wright and Wilson, have decided on June 26, 27 and 28 as the dates of meeting. This will enable delegates desiring to visit Buffalo, Niagara Falls, or any place outside of Toronto, an opportunity of doing so at "First of July" rates.

The Toronto branch of the association have appointed a committee, consisting of H. Hogarth (chairman), W. J. McGuire, K. J. Allison, James Wilson, J. E. Fullerton, James Sherlock, Alex. Purdy and Geo. Clapperton, to make arrangements for the reception and entertainment of delegates. The personnel of the committee is sufficient guarantee that everything will be done to give the visiting delegates a good time. In addition to this, however, it is rumored that the manufacturers and supply houses may "take a hand" in entertaining the visitors. The meeting this year should be a big one.

BUILDING PERMITS ISSUED.

The following building permits have been issued in Toronto: to the Toronto Railway Co., for a motor factory and blacksmith shop at the corner of Esplanade and Sherbourne streets, to cost $22,000; to Margaret Hayes, for a $1,200 addition to the rear of 50 Jarvis street; to Dale & Harley, for a bakery at the corner of Wolseley and Hackney streets to cost $1,000; to Dr. Davidson, for a $1,500 residence at the corner of Beverley and College street; to Mrs. Teeter, for a pair of houses near College street, on Dovercourt road, to cost $3 500; to the T. Eaton Co., Limited, for a $4,000 addition to 19 and 21 Albert street; to H. J. Finch, for a $2,500 residence at 479 Dovercourt road; to Thomas Lewis, for a $1,000 residence at 329 Davenport road. A total of 101 building permits were issued during April representing a cost of $346,812, as compared with $198,624 for April of last year. The figures for the first four months of 1901 are $606,167, as compared with $504,904 for the corresponding period last year.

Building permits have been issued in Ottawa as follows: James Mather on behalf of the trustees of the French Presbyterian church for a brick church, Wellington street, south side, $4,000; Mrs. Jane McKechnie, brick veneered dwelling, Preston street, $1,200; Thomas Mitchell, brick veneered dwelling, Concession street, $1,800; F. S. Warwicker, double brick veneered dwelling, Nepean street, $3,000.

CHANGING MEETING PLACE.

The Montreal Master Plumbers' Association has decided hereafter to meet in the Liberal Contractors' Hall, 90 St. James street. The first meeting was held in the new quarters on Thursday evening. Since its organization the association has held all its meetings in St. Joseph's Hall, St. Catherine street.

PLUMBING AND HEATING NOTES.

Daze Freres, plumbers, St. Louis de Mile End, Que., have dissolved.

Adelard Binette, plumber, Lachine, Que., has assigned, and a meeting of his creditors will be held on May 6.

Hypolite Gougeon, contractor, St. Henri de Montreal, has filed consent of assignment.

Mrs. Thos. Forest has registered as proprietress of Thos. Forest & Co., plumbers, Montreal.

GETTING BUSINESS.

MOST business men, no matter in what line they may be engaged, prefer to have trade come to them unsolicited, rather than be put to the trouble and expense of going about hunting it up, writes "Solicitor" in Metal Worker. But in these days of sharp competition there are few so situated as to be able to hold a satisfactory share of patronage without some effort on their part. Even the mechanic of to-day can no longer depend on his mechanical skill alone to bring custom to him; he has got to do more or less soliciting, or his more hustling competitors will get a share of patronage away from him, no matter how much he may excel them in ability to do the work. People are getting accustomed to being solicited for their trade and seem to rather like it; so that however disagreeable this feature of the times may be to a mechanic, it is a condition which he has got to face, and he may as well do it with the best grace possible.

There are some men so totally devoid of the book agent instinct that they can never make a success at soliciting trade. When such is the case it is far better for them to have someone in their employ to whom this kind of work is not so repugnant and whom they can trust to do it for them. When one stops to consider how much is required of a mechanic of to-day, it is a wonder that so many succeed.

Experience teaches in connection with soliciting work that it is seldom, if ever, good policy to plead the need of money, or any other personal reason, to induce people to award one their work. For some reason most people prefer to trade with the man who does the most business, which goes to prove the truth of the saying that "Nothing succeeds like success." Neither is it good policy to give a prospective customer the impression that he is solicited simply because other business in the neighborhood called the solicitor that way. Some men are too insistent, others are not insistent enough; the successful man is he who can strike a happy medium and adjust it to different individuals.

MANITOBA MARKETS.

WINNIPEG, April 29, 1901.

BUSINESS is moving quite briskly, especially in such lines as screen doors and windows, wire screening, garden tools, poultry netting and the like. Wire fencings of all kinds are showing good demand. The opening of navigation will mean increased activity also, as there are a good many summer lines waiting for open water to come forward.

Prices of iron and steel are firm, in sympathy with Eastern markets. The discount on bolts has been increased though regular quotations are not yet given at a lower figure. As this is the time of spring cleaning there is a corresponding demand in paints, kalsomining materials and the like. Linseed oil remains at last week's figures but turpentine has dropped 2c. per gallon making the present quotation 63c. Coal oil is firm but is showing the usual spring decrease in consumption.

Quotations for the week are as follows:

Barbed wire, 100 lb.	$3 45
Plain twist	3 45
Staples	3 95
Oiled annealed wire ... 10	3 95
" 11	4 00
" 12	4 05
" 13	4 20
" 14	4 35
" 15	4 45
Wire nails, 30 to 60 dy, keg	3 45
" 16 and 20	3 50
" 10	3 55
" 8	3 65
" 6	3 70
" 4	3 85
" 3	4 10
Cut nails, 30 to 60 dy.	3 00
" 20 to 40	3 05
" 10 to 16	3 10
" 8	3 15
" 6	3 20
" 4	3 30
" 3	3 65
Horsenails, 45 per cent. discount.	
Horseshoes, iron, No. 0 to No 1	4 65
No. 2 and larger	4 40
Snow shoes, No. 0 to No. 1	4 90
No. 2 and larger	4 40
Steel, No. 0 to No. 1	4 95
No. 2 and larger	4 70
Bar iron, $2.50 basis.	
Swedish iron, $4.50 basis.	
Sleigh shoe steel	3 00
Spring steel	3 25
Machinery steel	3 75
Tool steel, Black Diamond, 100 lb	8 50
Jessop	13 00
Sheet iron, black, 10 to 20 gauge, 100 lb.	3 50
20 to 26 gauge	3 75
28 gauge	4 00
Galvanized American, 16 gauge	2 54
18 to 22 gauge	4 50
24 gauge	4 75
26 gauge	5 00
28 gauge	5 25
Genuine Russian, lb	12
Imitation " "	8
Tinned, 24 gauge, 100 lb	7 55
26 gauge	8 80
28 gauge	8 00
Tinplate, IC charcoal, 20 x 28, box	10 75
" IX "	12 75
" IXX "	14 75
Ingot tin	35
Canada plate, 18 x 21 and 18 x 24	3 75
Sheet zinc, cask lots, 100 lb	7 50
Broken lots	8 00
Pig lead, 100 lb	6 00
Wrought pipe, black up to 2 inch ... 50 an 10 p.c.	
" Over 2 inch	50 p.c.
Rope, sisal, 7-16 and larger	$10 00
" ⅜	10 50
" ¼ and 5-16	11 00
Manila, 7-16 and larger	13 50
" ⅜	14 00
" ¼ and 5-16	14 50
Solder	21½
Cotton Rope, all sizes, lb.	16
Axes, chopping	$7 50 to 12 00
" double bitts	12 00 to 18 00
Screws, flat head, iron, bright	87½
Round " "	82½
Flat " brass	80
Round " "	75
Coach	57½ p.c.
Bolts, carriage	55 p.c.
Machine	55 p.c.
Tire	60 p.c
Sleigh shoe	65 p.c.
Plough	40 p.c.
Rivets, iron	50 p.c.
Copper, No. 8	35
Spades and shovels	40 p.c.
Harvest tools	50, and 10 p.c.
Axe handles, turned, s. g. hickory, doz.	$2 50
No. 1	1 50
No. 2	1 25
Octagon extra	1 75
No. 1	1 25
Files common	70, and 10 p.c.
Diamond	60
Ammunition, cartridges, Dominion R.F.	50 p.c.
Dominion, C.F., pistol	30 p.c.
" military	15 p.c.
American R.F.	30 p.c.
C.F. pistol	5 p.c.
C.F. military	10 p.c. advance.
Loaded shells:	
Eley's soft, 12 gauge black	16 50
chilled, 12 guage	18 00
soft, 10 guage	21 00
chilled, 10 guage	23 00
Shot; Ordinary, per 100 lb	6 75
Chilled	7 50
Powder, F.F., keg	4 75
F.F.G.	5 00
Tinware, pressed, retinned	75 and 2½ p.c.
" " plain	70 and 15 p.c.
Graniteware, according to quality	50 p.c.

PETROLEUM.

Water white American	25½c.
Prime white American	24c.
Water white Canadian	22c.
Prime white Canadian	21c.

PAINTS, OILS AND GLASS

Turpentine, pure, in barrels	$ 63
Less than barrel lots	71
Linseed oil, raw	80
Boiled	83
Lubricating oils, Eldorado castor	25½
Eldorado engine	24½
Atlantic red	27½
Renown engine	41
Black oil	23½ to 25
Cylinder oil (according to grade)	55 to 74
Harness oil	61
Neatsfoot oil	$ 1 00
Steam refined oil	85
Sperm oil	1 50
Castor oil ... per lb.	11½
Glass, single glass, first break, 16 to 25 united inches	2 25
26 to 40 ... per 50 ft.	2 50
41 to 50 ... " 100 ft.	5 50
51 to 60 ... " " "	6 00
61 to 70 ... per 100-ft. boxes	6 50
Putty, in bladders, barrel lots ... per lb,	2½
kegs ... "	2¾
White lead, pure ... per cwt.	7 25
No 1 ... "	7 00
Prepared paints, pure liquid colors, according to shade and color, per gal.	$1.30 to $1.90

INTERCOLONIAL ROLLING STOCK

In the House of Commons, on Wednesday, Hon. Mr. Blair, the Minister of Railways, in explanation of the item of $2,000,000 for rolling stock, said that $380,000 of the amount was for the locomotives being constructed at Kingston, and the balance was for freight rolling stock and for passenger and dining cars, including a car for the Royal party. The cars were being built by contract at the lowest tender price. The work was divided between The Rathbun Co., The Crossen Co. and The Rhodes & Curry Co.

INDUSTRIAL GOSSIP.

Those having any items of news suitable for this column will confer a favor by forwarding them to this office addressed the Editor.

The Cleveland Sarnia Saw Mills Co., Limited, Sarnia, Ont., are seeking incorporation.

The John Tetrault Steel and Malleable Iron Works have registered in St. Henri de Montreal.

Hespeler, Ont., passed without a dissenting vote, last Monday, a by-law to provide the Hespeler Furniture Co. with a free site.

The Toronto Street Railway Co. intend erecting a $22,000 motor factory and blacksmith shop at the northwest corner of Sherbourne and Esplanade streets, Toronto.

On Monday, Hanover, Ont., passed a by-law to grant the Knechtel Furniture Co. a bonus of $10,000 and another by-law to spend $25,000 on a system of waterworks.

On Monday the steamship Marian carried from St. John, N.B., 2 700 tons of Kootenay, B.C., lead to Antwerp. This is the largest shipment of Canadian lead ever made.

The Pressed Steel Car & Wheel Co., with a capital of $700,000, head office Perth, has been incorporated, with the following provisional directors: James H. Mitchell, John A. Currie, Neil McLean, Alexander McL. Macdonell and Arthur C. McMaster, all of Toronto. The company is empowered to manufacture pressed steel cars, car wheels, railway equipments, etc. and as a contractor to construct cars and other equipments, build ships, bridges, elevators, etc.

LOOKING TO CANADA FOR HELP.

We have so often referred in this journal to the prospects of the Canadian iron and steel trades, and the great developments in special at Sydney (Cape Breton), that we do not wish to-day to deal again with this subject. But it is a significant fact that from a mere question of trade interest and discussion in trade centres and periodicals, it is now being brought before the general public by daily journals. On Tuesday a long article from its own correspondent at Toronto appeared in The Times, while on Wednesday an interesting leader in The Daily Telegraph referred to the same subject. We have always agreed with the general opinion in Canada of the great future in store for the iron and steel industries of the Dominion, and it appears to us that it may look an appropriate coincidence to the help the colonial children of the Empire have given in the South-African troubles that in the industrial struggle we are promised with the United States Steel Corporation it might be a daughter of the Empire, who will help to maintain its traditional supremacy in the iron and steel trades.—Iron and Steel Trades' Journal.

THE AUTOMATIC CREAM SEPARATOR.

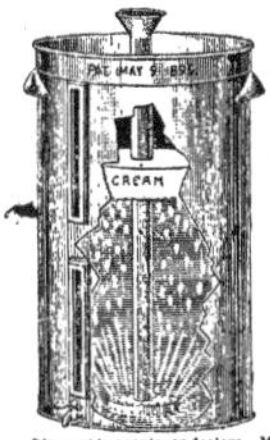
PAT. MAY 9 1899
CREAM

Newest and best thing yet for separating cream from milk. Will do the same work with better results than a $1.0 power separator, as it separates it in the natural way, making nice, sweet, coarse-grained butter. Not so with a power separator, as the rate of speed it goes breaks the particles of cream all to pieces, consequently, making salvy or greasy butter. With this machine the separating is done in the natural way in a very few moments—no expense—taking out about 98 per cent of the cream. Only one or two dishes to wash.

Price, only $2.50

Discount to agents, or dealers. Manufactured by

THE COLLINS MANUFACTURING CO.,
34 Adelaide Street West, TORONTO

KNOX HENRY

Heavy Hardware and Metal Broker
No. 1 Place Royale, MONTREAL.

"SECCOTINE"
FOR STICKING EVERYTHING.

Samples sent free on application.

HORSE NAILS—"C" Brand Horse Nails—Canada Horse Nail Co.

"BRASSITE" GOODS — Gunn Castor Co. Limited, Birmingham, Eng.

The Robin Hood Powder Company

If you want the best Trap or Game load in the world, buy "Robin Hood Smokeless," in "Robin Hood" Shells. It is quick, safe, and reliable. Try it for pattern and penetration from forty to seventy yards against any powder on the market. We make the powder, we make the shells, and we load them. Write for our booklet, "Powder Facts."

The Robin Hood Powder Company
SWANTON, VT.

REPEAT ORDERS FOR

BENNETT'S PATENT SHELF BOX

mean that they are O.K.

836 Yonge St., Toronto, April 9, 1901.

DEAR SIR, We wish that you would make for us another lot of your shelf boxes, per specification attached. We are delighted with the last equipment you made for us. The boxes are all that you claim for them, being light, strong, and exceedingly neat. They fit well too, showing care in turning them out. Altogether we consider the satisfaction we have had from putting in your system to be infinitely beyond comparison with the monetary outlay. Yours, very truly,

J. W. ORAM.

For particulars apply to the patentee and manufacturer.

J. S. BENNETT, 20 Sheridan Ave., TORONTO

CANADIAN ADVERTISING is best done by THE E. DESBARATS ADVERTISING AGENCY, MONTREAL.

BROOM AND MATTRESS WIRE

High Grade, Double Tinned

Fine Annealed Brush Wire. Soft Coppered Wire. Tinned Wire of all kinds.

THE PEERLESS WIRE CO.,
Hamilton, Ont.

Patented April 16, 1889.

BRADLEY STEEL SHELF BRACKETS

Greatly improved recently.

For sale by principal Canadian Jobbers.

Write to factory for free sample

ATLAS MFG. CO.
New Haven, Conn., U.S.A.

Oneida quality Cow Ties

may now be had in either

"American"
(Flat Link)
—OR—
"Niagara"
(Wire Link)

All Patterns.

Open and Closed Ring
Three Chain, and Dominion.

Your jobber can supply you.

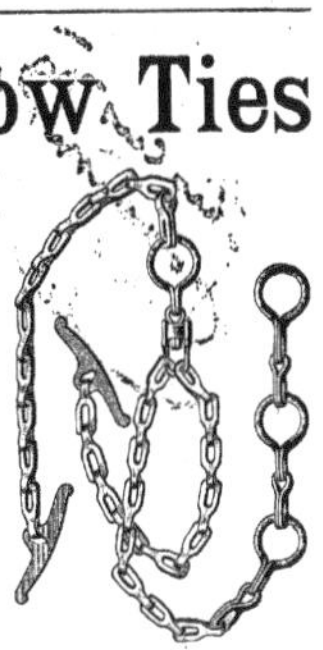

Manufactured by

Oneida Community, Limited Niagara Falls, Ont.

"I CAN'T MANAGE THESE OLD ELBOWS"

WHY DON'T YOU GET ALLEN'S FLAT CRIMPED ELBOW NO LEAK"

FLAT CRIMP

2, 3, and 4-in. Tin and Galvanized Conductor Elbows, for sale by all leading jobbing houses.

Made by GEO. ALLEN
BURLINGTON, ONT.

WHERE VARNISHES ARE MADE.

SINCE The Sherwin-Williams Co. erected their Canadian varnish factory in Montreal, some five years ago, they have been trying to gain for the products of this branch of the business a reputation equal to that held by The Sherwin-Williams' paints. Naturally, this was an object difficult of attainment, but Mr. Ballantyne, the manager of the Canadian division, is now proud in the confidence that the efforts of

Tester's Laboratory.

the firm have not been in vain, but that they are to-day shipping varnish that ranks as high in the varnish field as The Sherwin-Williams' paints do in their own sphere. One could hardly wish for greater recommendation.

The firm has recognized that the excellence of a varnish depends directly upon the excellence af the factory in which it is made, and the efficiency and carefulness of the staff that makes it. Consequently, the company has brought to its aid the best that it can get in both of these directions. The factory at Point St. Charles is a revelation of system and cleanliness. One is impressed with this idea immediately he enters the storage-room. We are apt to associate the idea of sticky filth with varnish, but not so here. All the woodwork is highly polished, the tanks and piping system painted, uniformity and regularity are evident in the rows of tanks, and everywhere are signs significant of a permeating system.

STORAGE ROOMS.

The total storage capacity is 40,000 gallons, comprising eleven 1,000-gal. tanks, thirty-five 3,500-gal. tanks, and the rest made up of innumerable smaller ones. There are three storage rooms, one above the other. A perfect record of the age of each varnish is kept, for the quality of the product depends as much upon the age as upon the propriety of the ingredients and cooking. On each tank is a card showing when the goods were put into storage and bearing the results of the tester's examination, giving the specific gravity, viscosity, working qualities and age at which it can be used. When it is matured, it is again examined by the superintendent, who writes "O.K. for use." Then, and only then, can it be sold.

In the main storage-room the bulk of the shipping is done. Each can is stamped on the bottom showing the date on which the varnish was made, and just what quality it contains. Other measures are also taken to keep track of any varnish that is shipped. A sample is taken out of each shipment, and kept in a separate vial for reference if need be. The Sherwin-Williams Co. can never be blamed for the faults of other people's varnishes, for they can always test their own shipments for years after the shipments are made.

TESTER'S LABORATORY.

Up-stairs, to the rear of the top storage-room, is the tester's laboratory, a neatly arranged room. All the ingredients that enter into the manufacture of the varnishes are tested by a thoroughly practical man. He also tests the finished goods, taking their specific gravity, viscosity and other properties. This is the scientific part of the establishment. He has some very fine instruments here, including a Westphal

Press and Pumping Room.

balance and a viscosity meter. Off this room is the varnishmaker's headquarters, where the veteran varnishmaker, the chief of the staff, does his headwork. Their varnishmaker was previously engaged in many of the large American centres, including New York, Chicago and San Francisco, and is reputed to be at the head of his business, having fortified himself with an experience of 30 years.

PRESS AND PUMPING ROOM.

Down stairs, below this room, one comes to the press and pumping-room. Here is a filtering press, the only one of its kind in Canada, through which all the varnishes are squeezed after manufacture, clearing them of all impurities. From here they are pumped up to the storage rooms. A large weighing kettle also figures here, allowing the varnishmaker to test his results by weight, as the practical tester does by much smaller machines. A washing machine completes the equipment of this room.

FIRE ROOM.

To the rear of the establishment is the fire room, containing five fires, and affording a daily capacity of 1,500 gallons. The fire apparatus includes all the most modern conveniences in the way of furnaces, thermometers and kettles, and even fire appliances. An aluminum kettle is an object of special pride.

After being cooked in the fire-room, the varnishes are taken into the cooking shed, where they are allowed to gradually cool off. From there they go to the thinning-room, where they are reduced down to proper viscosity with turpentine. Various machinery is employed in this room for other purposes. The varnish is then taken from the kettle, and, thereafter, handled by pipes.

The engine-house is separated entirely from the rest of the works, just as is the fire room. Here the power is generated by an electric gasoline engine of 7-horse-power. It is a neat, modern machine. Outside is the gum house, where the gums are kept and sorted, and where the oils are admitted to the factory. It might be mentioned in passing that all oil is kept in storage for one year to have everything capable of settlement turn itself into sediment. Another sign of care.

HOUSE AND FACTORY COMBINED.

To keep the temperature of the storage-rooms always at the same level, a trusted employe lives and sleeps in the building. He has a neat room fitted up off the storage-room at the front of the factory, and, throughout, the place is so clean that he could not have a more comfortable place of residence. Good varnish is essentially a production involving much care in its manufacture, and this seems to be a final proof that The Sherwin-Williams' varnishes could hardly be improved upon.

Thinning Room.

THE MISPLACED SYMPATHIES OF CLERKS.

JEROME K. JEROME.

NOW and again, mingled with other sounds of distress, comes to our ears the cry of the clerk. Between the huge forces of labor and capital, girding their loins for the coming struggle—beside which the feeble contests of the past will, when the entire history of the world comes to be written, appear insignificant by comparison—the clerkly band hovers undecided. With a snobbery which is partly humorous, but more pathetic, it tries to persuade itself that it belongs to the aristocratic faction.

Of course, it pays the classes to encourage the poor fellow in this more or less harmless folly of his. His vote is useful at election times, when capital—an utterly insignificant force in itself—finds it necessary to collect around it all those whom it can cajole or bribe into assisting it. The clerk, poor fellow, only gets the cajolery. Call him "Esquire" and talk to him of "our party" and the poor simple fellow asks no more.

The hidden lives of many of these poor fellows are real tragedies—the semi starvation of the whole family, the overworking of the poor young wife, who from a bright girl in a year or two is turned into a draggle-tailed, worn, and fretful old woman; the cup of coffee and the bun, or the little packet of dry sandwiches for dinner, eaten covertly while walking in solitude the dreary city streets. But then John is a gentleman, with a silk hat and a frock coat, and, maybe, by still more pinching, some little overworked servant is kept, and gentility is the reward.

Until he grasps the fact that he is a laborer, dependent upon his labor and not the labor of others, for his well-being, his position will remain unchanged, may possibly—though it is inconceivable—grow worse. Women are being dragged into the ranks now to compete with him, to still further lower his scanty wage. Had he not better dismiss this idea of his that his interests are those of the railway director and the rich shareholder and grasp the simple fact that his interests are opposed to theirs; that individually he is helpless in the hands of his employers; that only by combination can he hope to gain a living wage?

Numbers is the only weapon in the hands of the laborer. If he voluntarily lay that aside he hands himself over bound hand and foot into the hands of the slave-driver. Even a just employer, an employer wishful to do the right thing, is powerless under the circumstances. For him to attempt by himself to institute better wages, easier conditions of life, would be for him to ruin himself in competition with those whose instincts are to grind down and oppress. Capital is a fixed quantity; labor is worth—what you can get for it. The contest under such circumstances is not a fair one. Until labor by combination fixes itself, good employers and bad employers alike can only take advantage of its weakness.

Fire Room.

Until labor says to capital: "The world is mine as much as it is yours, it is for all of us alike; I have my value and you yours; the respective amounts shall be fixed between us; it shall not be left merely to you to decide," capital will naturally make terms entirely to its own interest. It is not just, and it is not honest. The world could not succeed without the exploiter—the man who thinks and plans and organizes. He on his side can not exist without the laborer. The terms between them are not for either of them to decide without reference to the other. To say that the matter should be left to the individual employer and the individual employed is mere jugglery.

A man for his business needs money, just as he needs labor; he buys it at a price fixed by the money market. If money were not organized for its own protection the same state of things would exist to-day that existed a thousand to five hundred years ago, when the robber barons held their castles, and every man with a groat in his pocket was liable to become their victim. Money organized itself against the robber. It secured its police, its soldiery; it made its laws, and it fashioned its governments.

Unionism is not a new thing. In the Middle Ages, the trade unions were huge forces—the only things, indeed, powerful enough to keep despotism in check. They dictated terms to kings and emperors. The liberties of modern Europe have sprung from them.

The laborer, uneducated, uninformed, untaught to think for himself, has hitherto listened sheepishly. Of late, some glimmering of his own rights, of his own power to enforce them, has come to him—much to the indignation and disgust of those whose interests, viewed from a narrow standpoint, are diametrically opposed to his. But the world cannot exist on injustice—at least, not for long. It is to the interests of the generations to come that a fairer arrangement should be arrived at, lest worse things befall.

INQUIRIES AND ANSWERS.

IDEAS ON NAIL COUNTERS.

Editor HARDWARE AND METAL,—I am going to build a nail counter, and would like to ask my brother hardwaremen, through your inquiry column, to give me their ideas on the construction of same.

SUBSCRIBER.

TAKEN LARGER WAREHOUSE.

Mr. Hector M. MacKenzie, wholesale dealer in heavy metals and plumbers' supplies, Montreal, finds that his business has out-grown his premises on St. Paul St., and has moved this week to more commodious quarters at 745 Craig St. His brother, manager of the Beaver Oil Co., lately at the same address, is moving with him to the new address.

Eliza Brown, general merchant, Bracebridge, Ont., is dead.

THE VALUE OF CHARACTER.

IT has been repeatedly remarked by those who have had large opportunity for observation that there are not enough honest, capable, reliable men to do the world's work, says Michigan Tradesman.

The saying is a true one and its truth exemplified every day. How common it is that men in important positions of trust and responsibility, whether public or private, are found to be defaulters or otherwise false and untrustworthy. How much more common is it that men in public or private service have no other care than to draw their salary or wages and to do as little for it as possible.

Employers or persons in authority always know the degree of reliability and usefulness of the men under them and, except in cases where political or other influences which create discriminations are concerned, the persons whose services are most willingly dispensed with are those who are least desirable and least valuable. Of course, there are exceptions to this rule in times of great industrial depression, when many establishments are closed or are working on short time; but, as an ordinary thing, the really valuable and faithful workers are seldom out of employment for any length of time. There is always something against a man who is unable to hold a place.

This subject comes up in an article on the causes of poverty in The Journal of Ethics for April, by J. G. Phelps Stokes. He holds that, while poverty is usually attributed to lack of employment, vice and crime, it will be found that lack of employment is ordinarily due to some defect of character and qualities in the individual. If persons are given to crime and vice, it is also because of defects in moral nature and disposition.

Of course, poverty is often due to misfortune or to circumstances beyond human control. Undoubtedly much poverty is due to sickness and death; but sickness is most often due to impairment of tissue vitality, to defective physical personality, which results either from unhelathy occupations or environment, or from violation (conscious or unconscious) of the recognized laws of health.

The poverty that is ascribed to drunkenness and to various forms of vice and crime can similarly be traced to defective personality as its cause and fountain head. For drunkenness and wrong-doing are but evidences of moral weakness; are but manifestations of defective personality. The shiftless, idle, drunken father of a family consigns his wife and children to misery and want, and they are the innocent and helpless victims of his misconduct and worthlessness. Any charity that enables such a creature to live without rendering any compensation to society is on a wrong basis. It ought to be so arranged as to help the innocent while excluding from all benefits the cause of their trouble.

If any system of socialism can ever be made practicable, it must be so organized as to punish the persistently idle and to provide that no man shall be allowed to eat who does not render some compensatory service. As matters now stand, a vast burden is placed upon the honest, industrious classes by compelling them to support criminals and the habitually idle and vicious classes.

Mr. Phelps thinks that the outcry of an excess of honest, efficient laborers unable to secure employment is seldom based on fact. The exception is in times of extraordinary commercial and industrial depression. Then great numbers who would otherwise be at work are, from no fault of their own, condemned to enforced idleness. Except under such conditions, he holds that there is everywhere an oversupply of shiftless or inefficient people in whom defective or undeveloped personality, is a conspicuous characteristic.

For the services of people of this latter class there are comparatively few demands, other than of temporary nature. Such people are replaced as speedily as circumstances allow, by workers of more efficient personality. Under ordinary circumstances it is chiefly persons of inefficient or undeveloped personalities who swell so largely the ranks of the unemployed.

The greatest evil of poverty is that it places so many women and children, by no fault of their own, but through the pitiful condition forced upon them by worthless heads of families amid surroundings that familiarize them with vice and are likely to drive them to crime. If they could be rescued from such associations great good would be accomplished, and it is to this that philanthropy should especially address itself.

Charity should be so organized that it would devote itself to rescuing the young of both sexes from vicious surroundings, so that they might be brought up in virtue, honesty and industrious habits. As for habitual adult male idlers, they should all be put in a workhouse and condemned to hard labor.

It is a conspicuous fact that many men who hold foremost places in commercial, industrial and financial affairs in this country started out as poor boys, often with but scanty education. But they had all the elements of character that make men valuable to society. They were honest, industrious, faithful to every duty and responsibility committed to them.

The fact that they had been faithful in humble situations warranted the belief that they would be faithful in still more important positions. Combined with their honesty and reliability, they were industrious, they were intelligent, they were alert to improve themselves and to increase their usefulness and to promote the interests of their employers. Their good qualities and faithful services met due recognition, not probably because of any gratitude on the part of their employers, but because they had urgent need of such men in their business.

And the need for honest, faithful and able men in every department of business is greater than ever before, because business combinations and operations are on a vaster scale than ever before. The stockholders in the great trusts and corporations engaged in carrying on the industries, the commerce, the transportation and general business interests of the country must trust the management and the special details of their enormous concerns to others. These others are required to have all the high qualities necessary for such great responsibilities. The men who start in the lowest places have every opportunity to rise to the highest, and it rests upon them to do so. In attaining success they will only be doing what others like them have done before. Character is one of the most important qualities required. Let that fact be taken to heart by every boy who has a worthy ambition to rise in the world.

J. S. BENNETT INJURED.

J. S. Bennett, manufacturer of Bennett's Patent Shelf Boxes, Toronto, has unfortunately met with an accident that will prevent him giving the personal supervision to the manufacture of his boxes for a few days which is absolutely necessary to insure the satisfaction that has so far attended his business. Mr. Bennett has many orders in hand, but these will not be delayed many days.

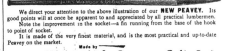

CURRENT MARKET QUOTATIONS.

May 3, 1901.

These prices are for such qualities and quantities as are usually ordered by retail dealers on the usual terms of credit, the lowest figures being for larger quantities and prompt pay. Large cash buyers can frequently make purchases at better prices. The Editor is anxious to be informed at once of any apparent errors in this list as the desire is to make it perfectly accurate.

METALS.

Tin.

Lamb and Flag and Straits—56 and 28 lb. ingots, per lb.	0 31	0 32

Tinplates.

Charcoal Plates—Bright

		Per box
M.L.S., equal to Bradley.		
I.C., usual sizes	....	$6 75
I.X., "	....	8 25
I.X.X., "	....	9 75
Famous—		
I.C.	....	6 75
I.X.	....	8 25
I.X.X.	....	9 50
Raven & Vulture Grades—		
I.C., usual sizes	....	5 00
I.X., "	....	6 00
I.X.X. "	....	7 00
I.XXX., "	....	8 00
D.C., 12½x17	....	4 75
D.X.	....	5 50
D.X.X	....	7 50

Coke Plates—Bright

Bessemer Steel—		
I.C., usual sizes	....	4 50
I.C., special sizes, base	....	4 50
20 x 28	....	8 50

Charcoal Plates—Terne

Dean or J. G. Grade—		
I.C., 20x28, 112 sheets	....	8 25
I.X., Terne Tin	....	10 25

Charcoal Tin Boiler Plates.

		Per lb.
Cookley Grade—		
X X., 14x56, 50 sheet bxs; " 14x60, "; " 14x65, "	0 07	0 07½

Tinned Sheets

72x30 up to 24 gauge	0 08	0 08½
" 26 "	0 08½	0 09
" 28 "	0 09	0 09½

Iron and Steel.

Common Bar, per 100 lbs.	1 85	1 90
Refined " "	....	2 35
Horse Shoe Iron "	....	2 25
Hoop steel, 1½ to 3 in. base, extras for smaller sizes	3 10	3 25
Sleigh Shoe Steel " base	....	2 10
Tire Steel	2 30	2 50
Reeled Machinery	....	3 00
Cast Steel, per lb	0 00	0 00
Toe Calk Steel	....	
T. Firth & Co's special cast steel, per lb.	0 13	
Jessop's Tool Steel	....	0 13

Boiler Tubes.

1½-inch	....	0 12½
2 "	....	0 13
2½ "	....	0 15
3 "	....	0 16
3½ "	....	0 20
4 "	....	0 25

Steel Boiler Plate.

¼ inch	2 50	2 60
3-16 inch	2 60	2 70
⅜ inch and thicker	2 50	2 60

Black Sheets.

18 gauge	....	3 00
20 gauge	....	3 00
22 to 24 "	....	3 10
26 "	....	3 20
28 "	....	3 30

Canada Plates.

Al dull, 52 sheets	....	3 00
Half polished	....	3 15
All bright	3 65	3 75

Iron Pipe.

Black pipe—		
⅛ "	....	4 45
¼ inch	....	3 25
⅜ "	....	3 30
½ "	....	3 50
¾ "	....	3 65
1 "	....	5 15
1¼ "	....	7 00
1½ "	....	8 40
2 "	....	11 25
2½ "	....	20 95
3 "	....	24 55
3½ "	....	31 75
4 "	....	39 00
5 "	....	47 35
6 "	....	62 10
Galvanized pipe—		
½ inch	....	4 90
¾ "	....	5 35
1 "	....	7 55
1¼ "	....	10 30
1½ "	....	12 35
2 "	....	16 50

Galvanized Sheets.

	G C. Comet.		Amer.	Queen's Head.
16 gauge		....	4 5	3 90
18 to 24 gauge	4 15	3 85	4 35	4 15
26 "	4 40	4 10	4 35	4 40
28 "	4 65	4 35	4 60	4 65

28 gauge American equals 26 gauge English.

Chain.

Proof Coil, 3-16 in., per 100 lb	....	....
" ¼ " "	8 00	8 50
" 5-16 " "	4 70	5 00
" ⅜ " "	4 05	4 50
" 7-16 " "	3 90	4 25
" ½ " "	3 70	4 10
" 9-16 " "	3 65	4 05
" ⅝ " "	3 35	3 90
" ¾ " "	3 60	4 00

Halter, kennel and post chains,	40 and 50 p.c.
Cow ties	40 p.c.
Tie-out chains	65 p.c.
Stall fixtures	35 p.c.
Trace chain	45 p.c.

Jack chain, iron, single and double, discount 35 p.c.

Jack chain, brass, single and double, discount 40 p.c.

Copper.

Ingot

English B. S., ton lots	....	0 19
Lake Superior	....	....

Bolt or Bar.

Cut lengths round, ½ to ⅞ in.	0 23½	0 25
" round and square 1 to 2 inches	0 23½	0 25

Sheet.

Plain, 14 oz., and light, 16 oz., 14x48 and 14x60	0 23	0 23
Plain, 14 oz., and light, 16 oz., irregular sizes	0 23	0 23½
Tinned copper sheets	....	0 26
Planished	....	0 32

Braziers (In sheets.)

4x6ft. 25 to 30 lbs. ea., per lb.	....	0 25½
" 35 to 45 " "	....	0 24½
" 50-lb. and above, "	....	0 23½

Boiler and T. K. Pitts

Plain Tinned, per lb	....	0 28
Spun, per lb	....	0 32

Brass.

Rod and Sheet, 14 to 30 gauge, 15 per cent.

Sheets, hard-rolled, 2x4	0 24	0 25
Tubing, base, per lb.	0 24	0 25

Zinc Spelter

Foreign, per lb	0 05½	0 06
Domestic "	....	..

Zinc Sheet.

5 cwt. casks	....	00 6½
Part casks	....	00 6¾

Lead.

Imported Pig, per lb	0 04½	0 4½
Bar, 1 lb.	0 05½	0 05¾
Sheets, 2½ lbs. sq. ft., by	....	0 06¾
Sheets, 3 to 6 lbs., "	....	0 06

NOTE.—Cut sheets ½ cent per lb. extra. Pipe, by the roll, usual weights per yard, lists at 7c. per lb. and 25 p.c. dis. f.o.b. Toronto.

NOTE.—Cut lengths, net price, waste pipe 8-ft. lengths lists at 7½ cents.

Shot.

Common, $6.50 per 100 lb.; chilled, $7.00 per 100 lb.; buck, seal and ball, $7.50. Discount, 25 p.c. Prices are f.o.b. Toronto, Hamilton, Montreal, St. John and Halifax. Terms 3 per cent. cash, freights equalized on Montreal.

Soil Pipe and Fittings.

Discount, 60 and 10 per cent. on medium and extra heavy, and 60 per cent. on light.

Solder.

	Per lb.	Per lb.
Bar half-and-half, guarant'd	....	0 18½
Bar half-and-half, commer'l	....	0 18
Refined	....	0 17½
Wiping	....	0 17

NOTE.—Prices of this graded according to quantity. The prices of other qualities of solder in the market indicated by private brands vary according to composition.

Antimony.

Cookson's, per lb.	0 10½	0 11

White Lead.

	Per 100 lb.
Pure	6 37
No. 1 do	6 00
No. 2 do	5 62½
No. 3 do	5 15
No. 4 do	4 87½
Munro's Select Flake White	7 37½
Elephant and Decorators' Pure	7 12½
Brandram's B B. Genuine	8 00
" " Decorative	7 55
" " No. 1	6 85
" " No. 2	6 00

Red Lead.

Genuine, 560 lb. casks, per cwt.	....	$5 50
Genuine, 100 lb. kegs, per cwt.	....	5 75
No. 1, 560 lb. casks, per cwt.	....	5 25
No. 1, 100 lb. kegs, per cwt	....	5 00

White Zinc Paint.

Elephant Snow White	0 08	0 09
Pure White Zinc	0 08	0 10
No. 1	0 06	0 07½
No. 2	0 05	0 06½

Dry White Lead.

Pure, casks	5 75
Pure, kegs	6 25
No. 1, casks	5 50
No. 1, kegs	5 00

Prepared Paints.

In ¼, ½ and 1 gallon tins.

Pure, per gallon	....	1 25
Second qualities, per gallon	....	1 10
Barn (in bbls.)	0 75	0 85
The Sherwin-Williams Paints	....	1 45
Canada Paint Co's Pure	....	1 25
Toronto Lead & Color Co's Pure	....	1 25
Sanderson Pearcy's Pure	....	1 20
Stewart & Wood's Champion Pure.		1 20

Colors in Oil.

25 lb. tins, Standard Quality.

Venetian Red, per lb.	0 05
Chrome Yellow	0 11
Golden Ochre	0 06
French "	0 05
Marine Black	0 09
" Green	0 09
Chrome "	0 08
French Imperial Green	0 09

Colors, Dry.

Yellow Ochre (J.C.) bbls.	1 35	1
Yellow Ochre (J.F.L.S.), bbls		2
Yellow Ochre (Royal)	1 10	1
Brussels Ochre		2
Venetian Red (best), per cwt.	1 8	1
English Oxides, per cwt.	3	3
American Oxides, per cwt.	1	2
Canadian Oxides, per cwt.	1	2
Super Magnetic Oxides, 93p.c.	2 00	2
Burnt Sienna, pure, per lb.	....	0
" Umber, " "	....	0
do Raw	....	0
Drop Black, pure	....	0
Chrome Yellows, pure	....	0
Chrome Greens, pure, per lb.	....	0 40
Golden Ochre	....	0 03½
Ultramarine Blue in 28-lb. boxes, per lb.	0 08	0 24
Fire Proof Mineral, per 100 lb	....	1 00
Genuine Eng. Litharge, per lb	....	0 07
Mortar Color, per 100 lb.	....	1 25
English Vermillion	....	0 80
Pure Indian Red, No. 45 lb.	....	0 80
Whiting, per 100 lb.	....	0 55

Blue Stone.

Casks, for spraying, per lb.	0 07
100-lb. lots, do. per lb.	0 08

Putty.

Bulk in bbls.	1 90
Bulk in less quantity	2 05
Bladders in bbls.	2 10
Bladders in kegs, boxes or loose	2 25
Bladders in 25-lb. tins	2 35
Bladders in 12½-lb. tins	2 65
Bladders in bulk or tins less than 100 lb.	2 90

Varnishes.

	In 5-gal. lots.).	Per gal.
Carriage, No. 1	2 90	3 30
" body	8 00	9 00
" rubbing	4 00	5 00
Gold Size, Japan	3 00	3 40
Brown Japan	2 40	2 80
Elastic Oak	2 90	3 30
Furniture, extra	2 40	2 80
" No. 1	1 60	2 00
Hard Oil Finish	2 70	3 10
Light Oil Finish	3 20	3 60
Demar	3 30	3 70
Shellac, white	4 40	4 80
" orange	4 00	4 40
Furniture Brown Japan	1 60	2 00
Black Japan	2 40	2 80
" " No. 1	1 60	2 00

The Imperial Varnish & Color Co's, Limited Elastilite Varnish 1 gal. can, each, $3 00.

Granatine Floor Finish, per gal., $2.50.

Maple Leaf Coach Enamels; Size 1, 60c.; Size 2, 35c.; Size 3, 20c. each.

Castor Oil.

East India, in cases, per lb.	0 10	0 10½
" " small lots	0 10½	0 11

Cod Oil, Etc.

Cod Oil per gal	0 50	0 55
Pure Olive	....	1 20
" Neatsfoot	....	90

Glue.

Common	0 08½	0 09
French Medal	0 14	0 14½
Cabinet sheet	0 12	0 13
White, extra	0 18	0 20
Gelatine	0 22	0 30
Strip	0 18	0 20
Coopers	0 19	0 20
Huttner	....	0 18

HARDWARE.

Ammunition.

Cartridges.

B. B Cap Dom. 50 and 5 per cent.
Rim Fire Pistol, dis. 40 p. c., Amer.
Rim Fire Cartridges, Dom., 50 and 5 p. c.
Central Fire Pistol and Rifle, 10 p.c. Amer.
Central Fire Cartridges, pistol sizes, Dom. 30 per cent.
Central Fire Cartridges, Sporting and Military, Dom., 15 and 5 per cent.
Central Fire, Military and Sporting, Amer, add 5 p.c. to list. B.B. Caps, discount 40 per cent, Amer.
Loaded and empty Shells, "Trap" and "Dominion" grades, 25 per cent. Rival and Nitro, net list.
Brass Shot Shells, 55 per cent.
Primers, Dom., 30 per cent.

Wads	per lb.
Best thick white felt wadding, in ½-lb bags	1 00
Best thick brown or grey felt wads, in ½-lb. bags	70
Best thick white card wads, in boxes of 500 each, 12 and smaller gauges	0 99
Best thick white card wads, in boxes of 500 each, 10 gauge	0 35
Best thick white card wads, in boxes of 500 each, 8 gauge	0 55
Thin card wads, in boxes of 1,000 each, 12 and smaller gauges	0 20
Thin card wads, in boxes of 1,000 each, 10 gauge	0 25
Thin card wads in boxes of 1,000 each, 8 gauge	
Chemically prepared black edge grey cloth wads, in boxes of 250 each—	Per M.
11 and smaller gauge	0 60
9 and 10 gauges	0 70
7 and 8 gauges	0 90
5 and 6 gauges	1 10
Superior chemically prepared pink edge, best white cloth wads, in boxes of 250 each—	
11 and smaller gauge	1 15
9 and 10 gauges	1 40
7 and 8 gauges	1 65
5 and 6 gauges	1 90

Adzes.

Discount, 20 per cent.

Anvils.

Per lb	0 10	(12½
Anvil and Vise combined	....	4 50
Wilkinson & Co.'s Anvils..lb.	0 09	0 09½
Wilkinson & Co.'s Vices..lb.	0 09¾	0 10

Augers.

Gilmour's, discount 65 and 5 p.c. off list.

Axes.

Chopping Axes		
Single bit, per doz	5 50	10 00
Double bit, "	11 00	18 00
Bench Axes, 40 p.c.		
Broad Axes, 33⅓ per cent.		
Hunters' Axes	5 50	6 00
Boy's Axes	5 75	6 75
Splitting Axes	6 50	12 00
Handled Axes	7 00	10 00

Axle Grease.

Ordinary, per gross	5 75	6 00
Best quality	13 00	15 00

Bath Tubs.

Zinc 6 00
Copper, discount 15 p.c. off revised list

Baths.

Standard Enameled.

5½-inch rolled rim, 1st quality	25 00
" " " 2nd "	21 00

Anti-Friction Metal.

"Tandem" A	per lb.	0 27
" B	"	0 21
" C	"	0 11½
Magnolia Anti-Friction Metal, per lb.		0 25
Frictionless metal, per lb		0 23

SYRACUSE SMELTING WORKS.

Aluminum, genuine	0 45
Dynamo	0 29
Special	0 25
Aluminum, 99 p.c. pure "Syracuse"	0 50

Bells.

Hand.

Brass, 60 per cent.
Nickel, 55 per cent.

Cow.

American make, discount 66⅔ per cent.
Canadian, discount 45 and 50 per cent.

Door.

Gongs, Sargant's 5 50 8 00
" Peterboro', discount 45 per cent.

Farm.

American, each 1 25 3 00

House.

American, per lb 0 35 0 40

Bellows.

Hand, per doz	3 35	4 75
Moulders', per doz	7 50	10 00
Blacksmiths', discount 40 per cent.		

Belting.

Extra, 60 per cent.
Standard, 60 and 10 per cent.

Bits.

Auger.

Gilmour's, discount 60 and 5 per cent.
Rockford, 50 and 10 per cent.
Jennings' Gen., net list.

Car.

Gilmour's, 47½ to 50 per cent.

Expansive.

Clark's, 40 per cent

Gimlet.

Clark's, per doz	0 65	0 90
Damond, Shell, per doz	1 00	1 50
Nail and Spike, per gross	2 25	5 20

Blind and Bed Staples.

All sizes, per lb 0 07¾ 0 12

Bolts and Nuts. Per cent.

Carriage Bolts, full square, Norway	65
" " full square	65
Common Carriage Bolts, all sizes	60
Machine Bolts, all sizes	60
Coach Screws	70
Sleigh Shoe Bolts	72½
Blank Bolts	60
Bolt Ends	62½
Plough Bolts	60
Nuts, square	4 c. off
Nuts, hexagon	4¼c. off
Tire Bolts	67½
Stove Bolts	67½
Stove rods, per lb	5½ to 6c.

Boot Calks.

Small and medium, ball, per M	4 25
Small heel, per M	4 50

Bright Wire Goods.

Discount .... 62½ per cent.

Broilers.

Light, dis., 65 to 67½ per cent.
Reversible, dis., 65 to 67½ per cent.
Vegetable, per doz., dis. 37½ per cent.

Henis, No. 8, "	....	6 00
Henis, No. 9, "	....	7 00
Queen City "	7 50	0 00

Butchers' Cleavers.

German, per doz	6 00	11 00
American, per doz	12 00	20 00

Building Paper, Etc.

Plain building, per roll	0 30
Tarred lining, per roll	0 40
Tarred roofing, per 100 lb	1 65
Coal Tar, per barrel	3 50
Pitch, per 100-lb	0 85
Carpet felt, per ton	45 00

Bull Rings.

Copper, $2.00 for 2½ in. and $1.90 for 2 in.

Butts.

Wrought Brass, net revised list

Cast Iron.

Loose Pin, dis., 60 per cent.

Wrought Steel.

Fast Joint, dis. 65, 10 and 2½ per cent.
Loose Pin, dis. 65, 10 and 2½ per cent.
Berlin Bronzed, dis. 70, 70 and 5 per cent.
Gen. Bronzed, per pair 0 40 0 65

Carpet Stretchers.

American, per doz	1 00	1 50
Bullard's, per doz	6 50	....

Castors.

Bed, new list, dis. 55 to 57½ per cent.
Plate, dis. 52½ to 57½ per cent.

Cattle Leaders.

Nos. 31 and 32, per gross 50 9 50

Cement.

Canadian Portland	2 50	2 80
English "	....	3 00
Belgian "	2 50	2 75
Canadian hydraulic	1 25	1 50

Chalk.

Carpenters Colored, per gross	0 45	0 75
White lump, per cwt	0 60	0 65
Red	0 05	0 06
Crayon, per gross	0 14	0 18

Chisels.

Socket, Framing and Firmer.

Broad's, dis. 70 per cent.
Warnock's, dis. 70 per cent.
P. S. & W. Extra 60, 10 and 5 p.c.

Churns.

Revolving Churns, metal frames—No. 0, $8—No. 1, $8.50—No. 2, $9.00—No. 3, $10.00—No. 4, $12.00—No. 5, $16.00 each. Ditto, wood frames—20c. each less than above. Discounts: Delivered from factories, 58 p.c.; from stock in Montreal, 58 p.c. Terms, 4 months or 3 p.c. cash in 30 days.

Clips.

Axle dis. 65 per cent.

Closets.

Plain Ontario Syphon Jet	$16 00
Emb. Ontario Syphon Jet	17 00
Fittings net	1 00
Plain Teutonic Syphon Washout	10 00
Emb. Teutonic Syphon Washout	11 00
Fittings net	1 25
Low Down Teutonic, plain	16 00
" " " embossed	17 00
Plain Richelieu net	3 75
Emb. Richelieu net	4 00
Fittings net	1 25
Low Down Ont. Sy. Jet, plain net	19 50
" " " " " emb'd. net	20 50
Closet connection net	1 25
Basins, round, 14 in	1 00
" oval, 17 x 14 in	2 50
" " 19 x 15 in	3 75

Discount 40 p.c., except on net figures.

Compasses, Dividers, Etc.

American, dis. 62½ to 65 per cent.

Cradles, Grain.

Canadian, dis. 25 to 33⅓ per cent.

Crosscut Saw Handles.

S. & D., No. 3, per pair	17½
" " 5, "	22½
" " 6, "	15
Boynton pattern "	20

Door Springs.

Torrey's Rod, per doz (15 p.c.)		2 00
Coil, per doz	0 88	1 60
English, per doz	2 00	4 00

Draw Knives.

Coach and Wagon, dis. 50 and 10 per cent.
Carpenters, dis. 70 per cent.

Drills.

Hand and Breast.

Millar's Falls, per doz. net list.

DRILL BITS.

Morse, dis. 37½ to 40 per cent.
Standard dis. 50 and 5 to 55 per cent.

Faucets.

Common, cork-lined, dis. 35 per cent.

ELBOWS. (Stovepipe.)

No. 1, per doz	1 40
No. 2, per doz	1 20

Bright, 20c. per doz. extra.

ESCUTCHEONS.

Discount, 45 per cent.

ESCUTCHEON PINS.

Iron, discount 40 per cent.

FACTORY MILK CANS.

Discount off revised list, 40 per cent.

FILES.

Black Diamond, 70 and 5 per cent.
Kearney & Foote, 60 and 10 to 70 per cent.
Nicholson File Co., 60 to 60 and 10 per cent.
Jowitt's, English list, 25 to 27½ per cent.

FORKS.

Hay, manure, etc., dis., 50 and 10 per cent. revised list.

GLASS—Window—Box Price.

Size United Inches.	Star Per 50 ft.	Star Per 100 ft.	D. Diamond Per 50 ft.	D. Diamond Per 100 ft.
Under 26	2 15	4 15	....	6 00
26 to 40	2 30	4 45	....	6 65
41 to 50	....	4 85	....	7 50
51 to 60	....	5 15	....	8 50
61 to 70	....	5 50	....	9 50
71 to 80	....	6 00	....	10 50
81 to 85	....	6 50	....	11 75
86 to 90	....	....	..	14 00
91 to 95	..	....	....	15 50
96 to 100	....	....	....	18 00

GAUGES.

Marking, Mortise, Etc.

Stanley's dis. 50 to 55 per cent.

Wire Gauges.

Winn's, Nos. 26 to 33, each 1 65 2 40

HALTERS.

Rope, ⅜ per gross	....	....
" ½ "	....	9 00
" ⅝ to ¾	....	14 00
Leather, 1 in., per doz	3 87½	4 00
" 1¼ in., "	5 15	5 20
Web, — per doz	1 87	2 45

HAMMERS.

Nail

Maydole's, dis. 5 to 10 per cent. Can. dis. 25 to 27½ per cent.

Tack.

Magnetic, per doz 1 10 1 20

Sledge.

Canadian, per lb 0 07½ 0 08½

Ball Pean.

English and Can., per lb 0 22 0 2

HANDLES.

Axe, per doz. net	1 50	2 00
Store door, per doz	1 00	1 50

Fork.

C. & B., dis. 40 per cent. rev. list.

Hoe.

C. & B., dis. 40 per cent. rev. list.

Saw.

American, per doz 1 00 1 25

Plane.

American, per gross 3 15 3 75

Hammer and Hatchet.

Canadian, 40 per cent.

Cross-Out Saws.

Canadian, per pair 0 13½

HANGERS.	doz.	pairs.
Steel barn door	5 85	6 00
Stearns, 4 inch	....	5 00
" 5 inch	....	6 50
Lane's covered—		
No. 11, 5-ft. run	....	8 40
No. 11½, 10-ft. run	....	10 80
No. 12, 10-ft. run	....	12 60
No. 14, 15-ft. run	....	21 00
Lane's O.N.T. track, per foot	....	4½

HARVEST TOOLS.

Discount, 50 and 10 per cent.

HATCHETS.

Canadian, dis. 40 to 42½ per cent.

HINGES.

Blind, Parker's, dis. 50 and 10 to 60 per cent.

Heavy T and strap, 4-in., per lb	....	0 06½
" " 5-in., "	....	0 06¼
" " 6-in., "	....	0 06
" " 8-in., "	....	0 05¾
" " 10-in., "	....	0 05½
Light T and strap, dis. 65 and 5 per cent.		
Screw hook and hinge—		
6 to 12 in., per 100 lbs	....	4 50
14 in. up, per 100 lbs	....	3 50
		Per gro. pairs.
Spring	....	12 00

HOES.

Garden, Mortar, etc., dis. 50 and 10 p.c.
Planter, per doz 4 00 4 50

HOLLOW WARE

Discount .... 45 and 5 per cent.

HOOKS.

Cast Iron.

Bird Cage, per doz	0 50	1 10
Clothes Line, per doz	0 27	0 63
Harness, per doz	0 72	0 88
Hat and Coat, per gross	1 00	3 00
Chandelier, per doz	0 50	1 00

Wrought Iron.

Wrought Hooks and Staples, Can., dis. 47½ per cent.

Wire.

Hat and Coat, discount 45 per cent.
Belt, per 1,000 .... 0 60
Screw, bright, dis. 55 per cent.

HORSE NAILS.

"C" brand 50 and 7½ p.c. off new list } Oval-
"M" brand 60 per cent. } head
Countersunk, 60 and 10 percent.

HORSESHOES

F.O.B. Montreal.

Iron Shoes.	No. 2 and larger.	No. 1. and smaller.
Light, medium, and heavy.	3 50	3 75
Snow shoes	3 75	4 00
Steel Shoes.		
Light	3 60	3 85
Featherweight (all sizes)	4 85	4 85

F.O.B. Toronto, Hamilton, London and Guelph, 10c. per keg additional.

Toe weight steel shoes...... 6 70

JAPANNED WARE.

Discount, 45 and 5 p.c. off list, June 1899

ICE PICKS.

Star per doz.......... 3 00 3 25

KETTLES.

Brass spun, 7½ p.c. dis. off new list.

Copper, per lb.......... 0 30 0 50

American, 60 and 10 to 65 and 5 p.c.

KEYS.

Lock, Can., dis., 45 p.c.

Cabinet, trunk, and padlock, Am. per gross.......... 60

KNOBS.

Door, japanned and N.P., per doz.	1 50	2 50
Bronze, Berlin, per doz.	2 75	3 25
Bronze Genuine, per doz.	6 00	9 00
Shutter, porcelain, F. & L. screw, per gross	1 30	4 00
White door knobs—per doz.		1 25

HAY KNIVES.

Discount, 50 and 10 per cent.

LAMP WICKS.

Discount, 60 per cent.

LANTERNS.

Cold Blast, per doz.	7 00
No. 3 "Wright's"	8 50
Ordinary, with O burner	4 00
Dashboard, cold blast	9 00
No. 0	5 75

Japanning, 50c. per doz. extra.

LEMON SQUEEZERS.

Porcelain lined, per doz.	2 20	5 60
Galvanized	1 87	3 85
King, wood	2 75	2 90
King, glass	4 00	4 50
All glass	1 20	1 30

LINES.

Fish, per gross	1 05	2 50
Chalk "	1 90	7 40

LOCKS.

Canadian, dis. 45 p.c.

Russell & Erwin, per doz.... 3 00 3 25

Cabinet.

Eagle, dis. 30 p.c.

Padlock.

English and Am., per doz.	50	6 00
Scandinavian, "	1 00	2 40

Eagle, dis. 20 to 25 p.c.

MACHINE SCREWS. Iron and Brass.

Flat head discount 25 p.c.

Round Head, discount 20 p.c.

MALLETS.

Tinsmiths', per doz.	1 25	1 50
Carpenters', hickory, per doz.	1 25	3 75
Lignum Vitae, per doz.	3 85	5 00
Caulking each	60	2 00

MATTOCKS.

Canadian, per doz.......... 5 50 6 50

MEAT CUTTERS.

American, dis. 25 to 30 p.c.

German, 15 per cent.

MILK CAN TRIMMINGS.

Discount, 25 per cent.

NAILS.

Quotations are:	Cut.	Wire.
2d and 3d	$3 35	$3 85
3d	3 00	3 52
4 and 5d	2 75	3 35
6 and 7d	2 65	3 20
8 and 9d	2 50	3 00
10 and 12d	2 45	2 95
16 and 20d	2 40	2 90
30, 40, 50 and 60d. (base)	2 35	2 85

Wire nails in carlots are $2.77½

Galvanizing 2c. per lb. net extra.

Steel Cut Nails 10c. extra.

Miscellaneous wire nails, dis. 70 and 10 p. c.

Coopers' nails, dis. 30 per cent.

Flour barrel nails, dis. 25 per cent.

NAIL PULLERS.

German and American....... 1 85 3 50

NAIL SETS

Square, round, and octagon, per gross	3 38	4 00
Diamond	12 00	15 00

NETTING.

Poultry, 50 and 5 per cent. for McMullen's

OAKUM. Per 100 lb.

Navy	6 00
U. S. Navy	7 25

OIL.

Water White (U.S.)	0 16½
Prime White (U.S.)	0 15½
Water White (Can.)	0 15
Prime White (Can.)	0 14

OILERS.

McClary's Model galvan. oil can, with pump, 5 gal., per doz. 10 00

Zinc and tin, dis. 50, 50 and 10.

Copper, per doz.	1 25	3 50
Brass, "	1 50	3 50

Malleable, dis. 25 per cent.

GALVANIZED PAILS.

Dufferin pattern pails, dis. 45 p.c.

Flaring pattern, discount 45 per cent.

Galvanized washtubs, discount 45 per cent.

PIECED WARE.

Discount 40 per cent. off list, June, 1899.

10-qt. flaring sap buckets, dis. 45 p.c.

6, 10 and 14-qt. flaring pails, dis. 45 p.c.

Creamer cans, dis. 45 p.c.

PICKS.

Per doz.......... 6 00 9 00

PICTURE NAILS.

Porcelain head, per gross	1 75	3 00
Brass head "	0 40	1 00

PICTURE WIRE.

Tin and gilt, discount 75 p.c.

PLANES.

Wood, bench, Canadian dis. 50 per cent. American dis. 50.

Wood, fancy Canadian or American 7½ to 40 per cent.

PLANE IRONS.

English, per doz.......... 2 00 5 00

PLIERS AND NIPPERS.

Button's Genuine per doz pairs, dis. 37½ 40 p.c.

Button's Imitation, per doz.	5 00	9 00
German, per doz.	0 60	2 60

PLUMBERS' BRASS GOODS.

Compression work, discount, 60 per cent.

Fuller's work, discount 65 per cent.

Rough stops and stop and waste cocks, discount, 60 per cent.

Jenkins disk globe and angle valves, discount, 55 per cent.

Standard valves, discount, 60 per per cent.

Jenkins' radiator valves, discount 55 per cent.

" " " standard, dis., 60 p.c.

Quick opening valves discount, 60 p.c.

No. 1 compression bath cock	2 00
No. 4 " " "	2 00
No. 7, Fuller's	2 50
No 4½, "	3 00

POWDER.

Velox Smokeless Shotgun Powder.

100 lb. or less	0 85
1,000 lb. or more	0 80

Net 30 days.

PRESSED SPIKES.

Discount 20 to 25 per cent.

PULLEYS.

Hothouse, per doz.	0 55	1 00
Axle	0 22	0 33
Screw	0 27	1 00
Awning	0 35	2 50

PUMPS.

Canadian cistern	1 80	3 60
Canadian pitcher spout	1 40	2 10

PUNCHES.

Saddlers', per doz.	1 00	1 85
Conductors', "	9 00	15 00
Tinners' solid, per set	0 00	0 72
" hollow, per inch	0 00	1 00

RANGE BOILERS.

Galvanized, 3 gallons	7 00
" 35 "	8 25
" 40 "	9 50
Copper, 30 "	22 00
" 35 "	26 00
" 40 "	30 00

Discount off Copper Boilers 10 per cent.

RAKES.

Cast steel and malleable, 50, 10 and 5 p.c.

Wood, 25 per cent.

RASPS AND HORSE RASPS.

New Nicholson horse rasp, discount 60 to 60 and 10 p.c.

Globe File Co.'s rasps, 60 and 10 to 70 p.c.

Heller's Horse rasps, 50 to 50 and 5 p.c.

RAZORS.

	per doz.	
Geo. Butler & Co.'s	8 00	18 00
Boker's	7 50	11 00
" King Cutter	12 50	50 00
Wade & Butcher's	3 60	10 00
Theile & Quack's	7 00	12 00
Elliot's	4 00	18 00

REAPING HOOKS.

Discount, 50 and 10 per cent.

REGISTERS.

Discount.......... 40 per cent.

RIVETS AND BURRS.

Iron Rivets, black and tinned, discount 60 and 10 per cent.

Iron Burrs, discount 55 per cent.

Extras on Iron Rivets in 1-lb. cartons, ½c. per lb.

Extras on Iron Rivets in ½-lb. cartons, 1c. per lb.

Copper Rivets & Burrs, 35 and 5 p.c. dis. and cartons, 1c. per lb. extra, net.

Extras on Tinned or Coppered Rivets ½-lb. cartons, 1c. per lb.

RIVET SETS

Canadian, dis. 35 to 37½ per cent.

ROPE ETC.

	Sisal.	Manila.
7-16 in. and larger, per lb	10	13½
⅜ in.	11	14½
¼ and 5-16 in.		15½
Cotton, 3-16 inch and larger		16½
" 5-32 inch		21½
" ⅛ inch		22½
Russia Deep Sea		15½
Jute		8
Lath Yarn		9½
New Zealand Rope		10

RULES.

Boxwood, dis. 75 and 10 p.c.

Ivory, dis. 37½ to 40 p.c.

SAD IRONS. per set.

Mrs. Potts, No. 55, polished	62½
" No. 50, nickle-plated	67½

SAND AND EMERY PAPER.

Dominion Flint Paper, 47½ per cent.

B & A. sand, 40 and 5 per cent.

Emery, 40 per cent.

Garnet (Rurton's), 5 to 10 p.c. advance on list.

SAP SPOUTS.

Bronzed iron with hooks, per doz... 9 50

SAWS.

Hand Disston's, dis. 12½ p.c.

S. & D., 40 per cent.

Crosscut, Disston's, per ft.... 35 0 55

S. & D., dis. 35 p.c. on Nos. 2 and 3.

Hack, complete, each	0 75	2 75
" frame only	....	0 75

SASH WEIGHTS.

Sectional, per 100 lbs.	2 75	3 00
Solid, "	2 00	2 25

SASH CORD.

Per lb.......... 0 23 0 30

SAW SETS.

"Lincoln," per doz.......... .... 6 50

SCALES.

B. S. & M. Scales, 45 p.c.

Champion, 65 per cent.

Fairbanks Standard, 35 p.c.

" Dominion, 55 p.c.

" Richelieu, 55 p.c.

Chatillon Spring Balances, 10 p.c.

Warren-Champion 65 p.c.

" Standard 45 p.c.

SCREW DRIVERS.

Sargent's per doz.......... 0 65 1 00

SCREWS

Wood, F. H., bright and steel, 87½ and 10 p.c.

Wood R. H., " dis. 82½ and 10 p.c.

" F. H., brass, dis. 80 and 10 p.c.

Wood, R. H., " dis. 75 and 10 p.c.

" F.H., bronze, dis. 75 p.c.

" R.H. " 70 p.c.

Drive Screws, 87½ and 10 per cent.

Bench, wood, per doz.	3 25	4 00
" iron, "	4 25	5 75

Set, Case hardened, 60 per cent.

Square Cap, 50 and 5 per cent.

Hexagon Cap, 45 per cent.

SCYTHES.

Per doz, net.......... 9 00

SCYTHE SNATHS.

Canadian, dis. 45 p.c.

SHEARS.

Bailey Cutlery Co., full nickeled, dis. 60 p.c.

Seymour's, dis. 50 and 10 p.c.

SHOVELS AND SPADES.

Canadian, dis. 40 and 5 per cent.

SINKS.

Steel and galvanized, discount 45 per cent.

SNAPS.

Harness, German, dis. 25 p.c.

Lock, Andrews'.......... 4 50 11

SOLDERING IRONS.

1, 1½ lb., per lb.	0 37
2 lb. or over, per lb.	0 34

SQUARES.

Iron, No. 493, per doz	2 40	2 55
" No. 494, "	3 25	3 40

Steel, dis. 50 and 5 to 50 and 10 p.c., rev. list

Try and bevel, dis. 50 to 52½ p.c.

STAMPED WARE.

Plain, dis. 75 and 12½ p.c. off revised list.

Retinned, dis., 75 p.c. off revised list.

STAPLES.

Galvanized	3 50	4 00
Plain	3 15	3 75

Coopers', discount 45 per cent.

Poultry netting staples, 40 per cent.

STOCKS AND DIES.

American dis. 25 p.c.

STONE. Per lb.

Washita	0 28	0 60
Hindostan	0 06	0 07
" slip,	0 09	0 09
Labrador	....	0 13
" Axe	....	0 15
Turkey	....	0 50
Arkansas	0 00	1 50
Water-of-Ayr	0 00	0 10
Scythe, per gross	3 50	5 00
Grind, per ton	15 00	18 00

STOVE PIPES.

5 and 6 inch Per 100 lengths	7 00
7 inch " "	7 50

ENAMELINE STOVE POLISH.

No. 4—3 dozen in case, net cash	$4 80
No. 6—3 dozen in case, "	8 40

TACKS BRADS, ETC.

	Per cent.
Strawberry box tacks, bulk	75 & 10
Cheese-box tacks, blued	80 & 12½
Trunk tacks, black and tinned	85
Carpet tacks, blued	80 & 15
" " tinned	80 & 20
" " (in kegs)	40
Cut tacks, blued, in dozens only	80
" ¼ weights	60
Swedes, cut tacks, blued and tinned—	
In bulk	80 & 10
In dozens	75
Swedes, upholsterers', bulk	85, 12½ & 12½
" brush, blued & tinned, bulk	70
" gimp, blued, tinned and japanned	75 & 12½
Zinc tacks	35
Leather carpet tacks	55
Copper tacks	50
Copper nails	52½
Trunk nails, black	65 and 5
Trunk nails, tinned	65 and 10
Clout nails, blued	65 and 5
Chair nails	35
Patent brads	40
Fine finishing	40
Picture frame points	10
Lining tacks, in papers	10

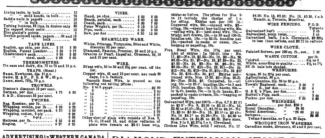

Lining tacks, in bulk	15
" " solid heads, in bulk	75
Saddle nails in papers	10
" " in bulk	15
Tufting buttons, 22 line, in dozens only	60
Tin capped trunk nails	15
Zinc glazier's points	5
Double pointed tacks, papers	90 and 10
" " " bulk	40

TAPE LINES.

English, ass skin, per doz.	2 75	5 00
English, Patent Leather	5 50	9 75
Chesterman's each	0 90	2 85
" steel, each	0 80	8 00

THERMOMETERS.

Tin case and dairy, dis. 75 to 75 and 10 p.c.

TRAPS. (Steel.)

Game, Newhouse, dis. 2½ p.c.
Game, H. & N., P. S. & W., 65 p.c.
Game, steel, 72½, 75 p.c.

TROWELS.

Disston's discount 10 per cent.

German, per doz.	4 75	6 00

S. & D., discount 35 per cent.

TWINES.

Bag, Russian, per lb.	....	0 27
Wrapping, cotton, per lb.	0 22	0 26
Wrapping, mottled, per pack.	0 50	0 60
Wrapping, cotton, 3-ply		0 20
" " 4-ply	....	0 26
Mattress, per lb.	0 33	0 45
Staging, "	0 27	0 35

VISES.

Hand, per doz.	4 00	6 00
Bench, parallel, each	2 00	4 50
Coach, each	6 00	7 00
Peter Wright's, per lb.	0 12	0 13
Pipe, each	5 50	9 00
Saw, per doz	6 50	12 00

ENAMELLED WARE.

White, Princess, Turquoise, Blue and White, discount 50 per cent.
Diamond, Famous, Premier, 50 and 10 p.c.
Granite or Pearl, Imperial, Crescent, 50, 10 and 10 per cent.

WIRE.

Brass wire, 50 to 50 and 2½ per cent. off the list.

Copper wire, 45 and 10 per cent. net cash 30 days, f.o.b. factory.

Smooth Steel Wire, is quoted at the following net selling prices:

No. 6 to 8 gauge		$2 90
" 9	"	2 80
" 10	"	2 87
" 11	"	2 90
" 12	"	2 95
" 13	"	3 15
" 14	"	3 37
" 15	"	3 50
" 16	"	3 65

Other sizes of plain wire outside of Nos. 9, 10, 11, 12 and 13, and other varieties of plain wire remain at $2.81 base with extras as before. The prices for Nos 9 to 13 include the charge of 1 c. for oiling. Extras net per 100 lb.: Coppered wire, 60c.—tinned wire, $2—oiling, 10c.—special hay-bailing wire, 30c.—spring wire, $1—best steel wire, 75c.—bright soft drawn, 15c.—in 50 and 100-lb. bundles net, 10c.—in 25-lb. bundles net 15c.—packed in casks or cases, 15c.—bagging or papering, 10c.

Fine Steel Wire, dis. 17½ per cent. List of extras: In 100-lb. lots: No. 17, $5—No. 18, $5.50—No. 19, $6—No. 20, $6.65—No. 21, $7—No. 22, $7.30—No. 23, $7.65—No. 24, $8—No. 25, $9—No. 26, $9.50—No. 27, $10—No. 28, $11—No. 29, $12—No. 30, $13—No. 31, $14—No. 32, $15 No. 33, $16—No. 34, $17. Extras net—tinned wire, Nos. 17-25, $2—Nos. 26-31 $4—Nos. 32-34, $6. Coppered, 5c.—oiling, 10c.—in 25-lb. bundles, 15c.—in 5 and 10-lb. bundles, 25c.—in 1-lb. hanks, 50c.—in ½-lb. hanks, 75c.—in ¼-lb. hanks, $1—packed in casks or cases, 15c.—bagging or papering, 10c.

Galvanized Wire, per 100 lb.—Nos. 6, 7, 8, $3 50 to $3 85—No. 9, $2.85 to $3.15—No. 10, $3.60 to $3.95—No. 11, $3.70 to $4.10—No. 12, $3 to $3 30—No. 13, $3.10 to $3 40—No. 14 $4.10 to $4.50—No. 15, $4.60 to $5.05—No. 16, $4.85 to $5 35. Base sizes, Nos. 6 to 9, $2.57½ f.o.b. Cleveland.

Clothes Line Wire, solid 7 strand, No. 17 $4.25; No. 18, $2.65; No. 19, $2.35, f.o.b Hamilton, Toronto, Montreal.

WIRE FENCING. F.O.B. Toronto

Galvanized barb	....	3 05
Galvanized, plain twist	....	3 05

Galvanized barb, f.o.b. Cleveland, $2.82½ in less than carlots, and $2.70 in carlots.

WIRE CLOTH.

Painted Screen, per 100 sq. ft., net.	1 35

WASTE COTTON. per lb.

Colored	....	4½ to 5
White, according to quality	...	6½ to 7½

500-lb bale lots shaded.

WRENCHES.

Acme, 35 to 37½ per cent.
Agricultural, 60 p.c.
Coe's Genuine, dis. 20 to 25 p.c.

Towers' Engineer, each	2 00	7
" S., per doz.	5 80	6
G. & K.'s Pipe, per doz.	....	3
Burrell's Pipe, each	....	3 00
Pocket, per doz.	0 25	2 90

WRINGERS.

Leader	per doz.	$30 0	
Royal Canadian.	"	26 00	
Royal American.	"	26 00	33 00
Sampson	"	....	36 00

Terms 4 months, or 3 p.c. 30 days.

WROUGHT IRON WASHERS.

Canadian make, discount, 40 and 5 per cent.

CANADIAN HARDWARE AND METAL MERCHANT

The Weekly Organ of the Hardware, Metal, Heating, Plumbing and Contracting Trades in Canada.

VOL. XIII. MONTREAL AND TORONTO, MAY 11, 1901. NO. 19

To Dominion Radiator Co
10-9-01 see this book
3 page 109

Garden Hose

Discounts for the new season now out.

WRITE FOR PRICES.

* * *

CANADIAN RUBBER CO.,

Montreal. Toronto. Winnipeg.

The Mower

THAT WILL KILL ALL THE WEEDS IN YOUR LAWNS

If you keep the weeds cut so they do not go to seed, and cut your grass without breaking the small feeders of roots, the grass will become thick and weeds will disappear. **The Clipper will do it.**

CANADIAN PATENT FOR SALE.
SEND FOR CATALOGUE AND PRICES.

Clipper Lawn Mower Co.

NORRISTOWN, Pa.

Lightning, Gem Blizzard . . . FREEZERS

ARE
- Well Advertised
- In Demand
- Easily Sold
- Satisfactory in Use
- Of Known Reputation

HAVE
- Cedar Pails with Electric Welded Wire Hoops
- Cans of Heavy Tin with Drawn Steel Bottoms
- AUTOMATIC Twin Scrapers
- "The Ice Cream Freezer Book" tells all about these and our other Freezers, mailed free

EXCEL IN
- Easy Running
- Quick Freezing
- Economy
- Convenience
- Practical Results

North Bros. Mfg. Co., Philadelphia, Pa., U.S.A.

Oxford Gas Ranges

EMBODY ALL THE BEST IDEAS YET CONTRIVED.

They are made in a full line of sizes and styles to meet all demands.

Have large ovens, a special improved oven burner lighter, and the most perfect valves and burners known.

The intense heat furnished by them from a most economical supply of gas, delights every customer. It is a talking point of most convincing worth in making sales.

This year we emphasize two new styles with 16 and 18 inch square ovens, remarkably fine lines that satisfy the popular call for a standard quality Gas Range at a very moderate price.

Correspondence invited. Full Particulars and Price Lists at your service for the asking.

THE GURNEY FOUNDRY CO., LIMITED

TORONTO WINNIPEG VANCOUVER

THE GURNEY-MASSEY CO., Limited, - MONTREAL.

For the first season since 1897, we are in the market for the supply of

Galvanized Wire, Plain Twist AND Barb Wire Fencing

OF OUR OWN MANUFACTURE.

4 Barb, 6" apart "A" — 4 Barb, 4" apart "B"

2 Wires "C" — 3 Wires "D"

2 Barb, 5" apart—Long Barb "G" — 2 Barb, 5" apart—Short Barb "I"

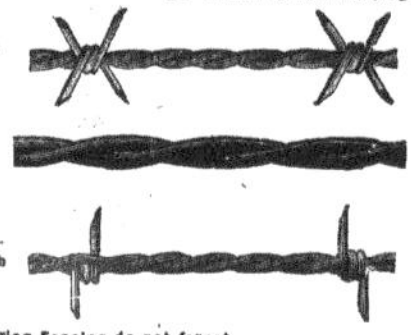

When ordering Fencing do not forget

Bright and Galvanized Staples.

We solicit your orders also for

Wire of all kinds and for all purposes, of Steel, Brass or Copper.

Wire Nails and Wood Screws

Bright Wire Goods — Cotter Pins
Jack Chain — Door Pulls
and "Crescent" Coat and Hat Hooks
Bed, Blind, and Netting Staples.

Prices quoted upon application to

Dominion Wire Manufacturing Co.

MONTREAL and TORONTO.

THE NEW BALDWIN

DRY AIR CLEANABLE

REFRIGERATOR.

133 Modern Varieties. Ash, Oak and Soft-wood Finishes

METAL, PORCELAIN, SPRUCE LININGS.

BALDWIN

Positive Circulation—Sanitary—Odorless.

Latest Cleanable Features—The Strongest and Best System of Patent Removable Metal Air-Flues.

Air-Tight Lever Locks. Ball-Bearing Casters. Swing Base—in and out.

Rubber around Doors and Lids, making them doubly air-tight.

Handsome Designs. Moderate Prices.

Built in the newest, largest and best equipped refrigerator plant in the East; run all the year round on refrigerators exclusively; stock goods; special refrigerators and coolers in sections.

Handsome Trade Catalogue Ready.

Baldwin Refrigerator Co.,

BURLINGTON, VERMONT.

ESTABLISHED 1860. INCORPORATED 1895.

Milk Can Trimmings

Iron Clad Pattern.

In all our Milk Can Bottoms sufficient space is left between bottom proper and rim to allow body of can to be inserted ¾ of an inch, making permanent joint.

Bottoms are thus sweated in with half the solder.

Top Hoops are "**Shouldered**" and have cut-out at joint, making neater and cleaner job in half time than the old style hoop.

All Hoops have retinned edges.

For durability, finish, and economy in making up our Trimmings are unequalled.

Broad Hoop Pattern.

THE THOS. DAVIDSON MFG. CO., Limited, MONTREAL, QUE.

Galvanized Sheets

From stock or for Importation.

SAMUEL, SONS & BENJAMIN, - - LONDON AND LIVERPOOL, ENGLAND

M. & L. Samuel, Benjamin & Co.

General Importers and Exporters and Metal Merchants

27 Wellington Street West, - - TORONTO, ONT.

Henry Disston & Sons (Incorporated)

Philadelphia, Pa.

KEYSTONE FILES.

Will wear longer than any other File made.

Disston's Square Files—All Sizes.

Disston's Round Files—All Sizes.

Disston's Great American Cross Cut Saw Files.

Disston's Mill Saw Files—All Sizes.

Disston's Flat Bastard Files—All Sizes.

Disston's Half-Round Bastard Files—All Sizes.

Disston's Horse Rasps.

Lewis Bros. & Co.
MONTREAL
AGENTS

Manufactured by
Henry Disston & Sons
Philadelphia, Pa.

Henry Rogers, Sons & Co.

Wolverhampton, England.

Manufacturers of

"Union Jack" Galvanized Sheets
Canada and Tin Plates
Black Sheets
Sleigh Shoes and Tyre Steel
Coil Chain, Hoop Iron
Sheet and Pig Lead
Sheet Zinc

Quotations can be had from Canadian Office:
8 St. Sacrament St., - MONTREAL

F. A. YORK, Manager.

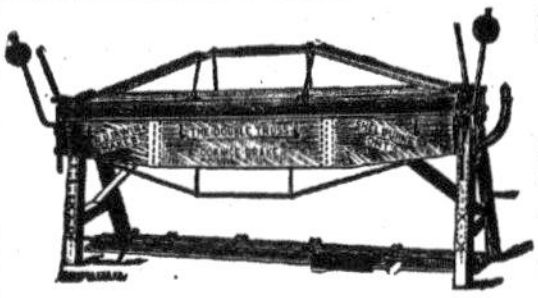

This eight-foot Brake bends 22-gauge iron and lighter, straight and true.

Price, $60

Very handy beader attachment, $15 extra if required.
Send for circulars and testimonials to

The Double Truss Cornice Brake Co. SHELBURNE, ONT.

The Latest and Best.

H. & R. Automatic Ejecting Single Gun.

Steel and Twist Barrels in 30 and 32-inch.
12 Gauge.

Model 1900.

Harrington & Richardson Arms Co.
Worcester, Mass., U.S.A.
Descriptive Catalogue on request.

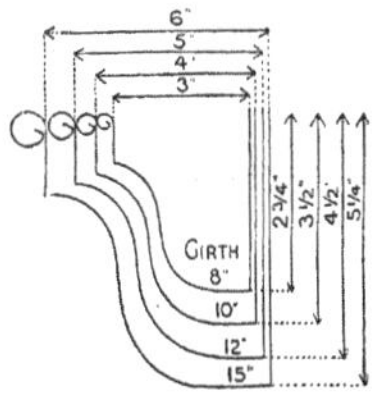

EAVETROUGH AND CONDUCTOR PIPE

Write for prices.

E. T. WRIGHT & CO., Mfrs.
HAMILTON, CANADA.

Oneida quality Cow Ties

may now be had in either

"American"
(Flat Link)
—OR—
"Niagara"
(Wire Link)

All Patterns.

Open and Closed Ring
Three Chain, and Dominion.

Your jobber can supply you.

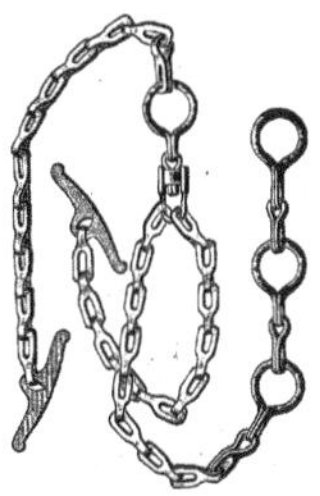

Manufactured by

Oneida Community, Limited Niagara Falls, Ont.

TRADE MARK J REGISTERED.

"JARDINE" HAND DRILLS

Five different sizes to suit all requirements. It pays to sell the best tools.

A. B. JARDINE & CO.
HESPELER, ONT.

A LONDON FENCE MACHINE

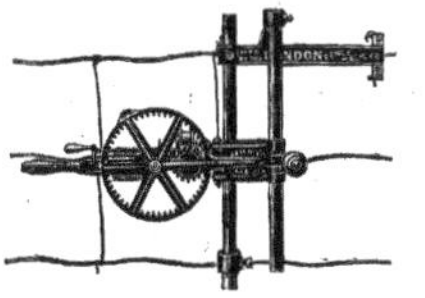

ON EXHIBITION IN YOUR STORE

will create more curiosity and bring in more customers than any other implement. There's money in handling the "London," as other merchants have found. You get a snap if you get the "London" agency for your town.

WRITE FOR TERMS BEFORE YOU MISS IT.

Coiled Spring Wire at right prices to the trade.

The London Fence Machine Co., Manufacturers and Wholesale Dealers, **London, Can.**

Mrs. Potts Sad Irons

No. 55—Plain Polished
No. 50—Nickel Plated

IF YOU ARE BUYING

COMPARE WEIGHT AND FINISH

and we are sure you will insist upon having our make.
Sold by all the leading Jobbers.

A. R. WOODYATT & CO. - GUELPH, ONT.

SEASONABLE SUGGESTIONS.

WATER COOLERS.

Our Water Coolers are well made.
They have Nickel-Plated Faucets.
They are finished in attractive designs and colors.

WATERING POTS.

Six sizes. Plain, Japanned and Galvanized.
All Cans are made with Patent Rose, which can be easily taken apart for cleaning.
Well finished, strong and durable.
That is why we sell so many.

KEMP MANUFACTURING CO.,

TORONTO, Canada

HARDWARE AND METAL

VOL. XIII. MONTREAL AND TORONTO, MAY 11, 1901. NO. 19.

President,
JOHN BAYNE MacLEAN,
Montreal.

THE MacLEAN PUBLISHING CO.
Limited.

Publishers of Trade Newspapers which circulate in the Provinces of British Columbia, North-West Territories, Manitoba, Ontario, Quebec, Nova Scotia, New Brunswick, P.E. Island and Newfoundland.

OFFICES

MONTREAL . . . . 232 McGill Street, Telephone 1255.
TORONTO . . . . 10 Front Street East, Telephone 2148,
LONDON, ENG. . . . . . 109 Fleet Street, E.C., W. H. Miln.
MANCHESTER, ENG. . . . 18 St Ann Street. H. S. Ashburner,
WINNIPEG . . . . . Western Canada Block. J. J. Roberts,
ST. JOHN, N.B. . . . . No. 3 Market Wharf, J. Hunter White.
NEW YORK. . . . . 176 E. 88th Street,

Subscription, Canada and the United States, $2.00.
Great Britain and elsewhere . . . 12s.

Published every Saturday.

Cable Address { Adscript, London. Adscript, Canada.

THE PRICES OF READY-MIXED PAINTS.

SINCE linseed oil made such a rapid decline some months ago, the retail trade has been quite solicitous over the future of the mixed-paint prices. Not a few inquiries have been received by the manufacturers asking when the drop was to take place.

The idea received its quietus this week when the English oil market, nearly bare of seed, advanced more than £2 per ton, putting oil values back to almost what they were last January, and causing an advance in spot values in Montreal of 5c. per gal.

The market is growing in strength, as the Argentine seed seems no longer to be forthcoming and the Indian, Canadian and United States crops will not be procurable till September. Seed will be badly needed before then.

There is but little doubt that, if oil had remained at the low quotations made for May, June and July import some weeks ago, the prices of mixed paints would have been reduced in June, at the latest. The drop could not, perhaps, have come before that time, for the manufacturers have been working on high-priced oil, and no cheaper goods were procurable until the first steamers from England were in port. But, just as the market was shaping itself for a reduction, the oil market suddenly reacted, and a steady bull movement set in, until now oil is quoted above £30 in London. Raw could have been imported some time ago at 46c.; now, it would cost 68c., at least.

Of course, most of the importers of oil "got in" at the low prices, but, just as the spot market of oil has advanced with the latest English quotations, the mixed-paint values will also be maintained in harmony.

Other constituents of paints, such as chromes and ochres, are all in good demand and firm, and, if it were not that zinc and lead were somewhat lower, the manufacturers would be compelled to consider an advance rather than a decline.

MANITOBA'S NEXT CROP.

The C.P.R. and N.P.R. have prepared their May reports on crop indications in Manitoba. The outlook is promising. In nearly every district the reports show that the acreage is greater than last year, the increase ranging from a few acres to 40 per cent., the average increase being placed at 10 per cent.

Even more satisfactory, however, are the reports as to the condition of the crops. In many districts from 75 to 80 per cent. of the crop is in the ground; in others, seeding is from one to two weeks later than a year ago. But the weather has been favorable in practically every section, and, as there is abundance of moisture in the ground, the reports unite in the confidence that next year Manitoba will produce a harvest much in excess of that of last year.

He who cuts prices usually cuts more into his own profits than he does into his competitors' trade.

HAS BOTTOM BEEN TOUCHED?

THE latest news from Wales indicates that the improved tone in sheet metals, mentioned in last week's issue, continues to manifest itself. H. I. Russel & Co.'s (New York) weekly letter from South Wales says that "considerable more activity is discernible in the tinplate trade. There is a good inquiry for plates, and the receipts for the week amounted to 55,500 boxes. The shipments exceeded 90,400 boxes, so that stocks have been denuded to the extent of nearly 35,000 boxes. They are now exceptionally low, standing at 64,220 boxes, and prices have a distinctly hardening tendency. For some sorts they have gone up 1½d. per box, best brands of C 14 x 20 Bessemer cokes selling at 12s. 1½d. and Siemens ditto at 12s. 3d. The demand for steel bars is greater than the supply."

It would seem that an American activity is being slowly imparted to the English market. English pig iron is also much stronger and business is on the increase.

INERTIA OF THE DEPARTMENT OF COMMERCE.

ONE does not hear much these days of the Department of Trade and Commerce. By the monthly report of the Department, which is issued from two to three months after the month whose trade it deals with is passed, we are reminded that it is not dead, and by the Auditor-General's report we are reminded that it is still a charge upon the revenue of the country. Were it not for these two reminders we certainly would never know that the Department had an existence.

Judging the Department by its fruits, it is simply a sinecure, as far as its fulfilling the purposes for which it was created are concerned.

It was called into being, as its name indicates, for the avowed purpose of devising ways and means of developing the trade and commerce of the country; but it has scarcely yet done anything to warrant its existence. It has certainly done something; but it has not yet done anything that could not have been done just as well under the supervision of one of the other Departments, say, for instance, the Department of Customs.

This unsatisfactory state of affairs is not because the Department has no reason for its existence; it is simply and solely because the Department has been badly administered.

The present head of the Department, Sir Richard Cartwright, is a man whom we have always rated above the average politician. His experience in Canadian politics is wide and his attainments are many. And in their attacks upon him his enemies have never called into question his ability. The point of attack has been his economical doctrines or his alleged inconsistencies at times in regard to them. But as Minister of Trade and Commerce he has been a failure, a decided failure.

We refuse to believe that Sir Richard is a failure because he lacked the inherent qualifications to administer the Department. With his intelligence, his experience and his scholastic attainments we believe that there are few in this country who could have done better than he had he applied himself to the task. But therein lies the trouble. He has not applied himself. His administration is, therefore, all the more worthy of condemnation.

It is well known that the office was not one of his own choosing. He was merely assigned it because he could not be ignored in the formation of the Laurier Cabinet in 1896. Sir Richard was unpopular with the business men, and it was considered unwise to place him in charge of the Finance Department. For political exigencies it might have been a good move, but for the business necessities of the country it has been a decidedly bad one.

Whether it be due to unsatisfied political ambition or not we cannot say, but one thing is certain, his administration of the Department of Trade and Commerce has been characterized by ennui of a most positive type.

It must be acknowledged that such members of the Government as Messrs. Paterson, Fielding, Mulock and Tarte have exhibited most commendable energy in the administration of their respective departments, whatever may be our views as to the general policy of the Government of which they form a part. And it is only to be regretted, for the good of the country at any rate, that the same cannot be said of Sir Richard Cartwright.

He has been tried and found wanting. And the necessities of the trade and commerce of the country demand that he give place to someone of business experience and ambition to make the Department of Trade and Commerce what it really should be—one of the most important and aggressive parts of the Government machinery.

We cannot afford to wink at drones in the administration of the Department of Trade and Commerce whatever we may allow in any other.

FREIGHT RATES AND FACTORY LOCATION.

AT a meeting of the British Iron and Steel Institute, London, England, on Thursday, Mr. William Garrett, of Cleveland, urged upon the manufacturers of Great Britain the necessity of locating factories at points most advantageous for shipment, intimating that this was the policy followed in the United States.

While intended for the British manufacturer the suggestion is not without its lesson to the manufacturer in Canada, and particularly so at present.

The prosperity which Canada has enjoyed during the last few years has given a decided stimulus to the manufacturing industry; and to the iron industry probably more so than to any other. This is to be commended, for the manufacturing possibilities of this country are enormous. But there is a grave danger of our ambition getting the better of our discretion in the location of new industries.

Unless the factory is situated where it can command advantageous shipping facilities there is little possibility of its being a permanent success. When trade is abnormally active it may be carried along with the current. But when normal conditions obtain, much less dull trade intervene, nine chances to one the concern will drop out of sight in the commercial stream.

The railways charge as high rates of freight as they possibly can. In getting as much as they can, they are probably no worse than those engaged in any other commercial enterprise. But that does not alter the fact that those contemplating the starting of a factory of almost any kind should carefully ascertain how they will be situated as compared with their competitors in regard to freight rates. There are factories in Canada to-day in which no hum of machinery is heard, chiefly because they are located where they could not command freight rates as advantageous as those of their competitors.

Manufacturers, generally, in this country are laboring under a great disadvantage in regard to both the home and the export trade as compared with their competitors in the United States. In many instances is this so great that it greatly or altogether nullifies the protection which the manufacturers concerned enjoy under the Customs tariff. If, in addition to discriminating railway rates in favor of foreign competitors, a manufacturer is situated where he is at a disadvantage compared with his home competitors as well, his position is indeed grave.

ARE RAILWAYS RESPONSIBLE TO SHIPPERS FOR LOSSES?

A MATTER of the utmost importance to every man who ships merchandise has just been brought to the attention of the Toronto Board of Trade by Mr. P. C. Larkin, of The "Salada" Tea Co.

In the conflagration that visited Ottawa over a year ago, it will be remembered that the freight sheds of the Canadian Pacific Railway were destroyed. Those who had goods in the freight sheds at the time made, in due course, a claim upon the railway company for compensation. The railway company requested them not to press their claims, as it had entered suit against the insurance companies.

The shippers accordingly complied with this request. And this week they were astonished to receive a letter from the Canadian Pacific Railway notifying them that, as it (the railway company) had lost its suit against the insurance companies, it was not responsible to the shippers whose goods were destroyed by fire in the railway freight sheds. The railway, however, signified its willingness to pay an amount equal to 50 per cent. of the loss. Attached to each letter was a slip of paper on which was printed an extract from the judgment of the court. It read as follows:

That fire was an overwhelming catastrophe, not arising through any negligence in any sense attributable to the defendants, but arising on the property of others, and sweeping down upon their property with such irresistible suddenness and force as no reasonable human forethought could have guarded against; nor could any reasonable efforts have prevented the great destruction and loss it caused them (the C.P.R. Co.) among many.

As manufacturers and merchants had, heretofore, considered that transportation companies were responsible for the delivery of the goods placed in their hands, the announcement of the Canadian Pacific Railway came to them like a bolt out of a clear sky.

The matter is of such manifest importance that it should not be allowed to rest, for it opens up a vista of several important questions.

Nothing is said in the judgment in regard to the responsibility of the transportation company to the shipper. It deals only with the responsibility of the insurance companies to the transportation company. But even in this particular it is not without interest to every man that insures his property. Certainly the fire was not due to any carelessness of the railway company. But is payment from an insurance company only to be obtained when carelessness can be traced to the insured? It is absurd to conceive of any such law. And yet, on the face of the judgment quoted, it was upon that plea that the insurance companies are exempted from payment of losses on the goods destroyed in the fire at the railway company's freight shed in Montreal.

It is possible, however, that the insurance policy was a special one. Else how could ever such a judgment be given?

But if the policy is a special one, and such as would relieve the insurance companies, that does not to our mind relieve the railway company from responsibility unless the ideas of shippers and others as to the responsibility of transportation companies have been resting on a foundation of sand.

If the railway companies are not responsible for loss by fire, why should they be responsible for losses created by other means not attributable to their carelessness, say, for instance, the misplacing of a rail by some miscreant which results in the destruction of merchandise as well as rolling stock?

We are pleased to see that the matter has been submitted to the Toronto Board of Trade, and it is to be hoped that it will probe to the very bottom the questions entailed.

STEAMSHIP DEAL AND CANADA.

THE big steamship deal appears to have excited an interest scarcely less general than the famous steel and iron consolidation which Mr. J. Pierpont Morgan brought about a short time ago.

While it is doubtless true, as some allege, that the importance of the deal is exaggerated, one cannot ignore the fact that it is important and that its very uniqueness naturally creates in it a greater interest than would otherwise be attached to it. At the same time, however, the deal is not as comprehensive as a good many people seem to think. The Leyland Line has not been bought out and out by the United States capitalists whom Mr. Morgan represents. What they have done is to secure the controlling interest. According to a statement of Chairman Ellerman at the annual meeting of the Leyland Steamship Co., held a few days ago, the deal left the company English, the boats will be worked by Englishmen, and would fly the English flag. Seeing that Mr. Morgan and his fellow capitalists have the controlling interest, no one can say, however, what might eventually be done.

It is considered certain that the deal also includes the Atlantic Transport Line. If it is true, it includes 76 steamers, of an aggregate of 353,465 tons gross, not to mention eight new steamers building for the Atlantic Transport Line.

Canada is not without direct interest in the deal, as five steamers of the Leyland Line last summer plied regularly between Montreal and Antwerp. These steamers are the Albanian, tonnage 2,930; Almerian, tonnage 2,984; Assyrian, tonnage 2,899; Belgian, tonnage 3,740, and Mexican, tonnage 4,202. The Belgian, which is a new boat, and the Mexican were only put on the route toward the end of the season. The Leyland Line last season also maintained a new service from Quebec to Liverpool, the steamer Albanian taking the first cargo of grain from the former port shortly before the close of navigation.

Some concern has been manifested lest the deal would lead to the withdrawal of the line of steamers from the Montreal route, but the following paragraph from The New York Journal of Commerce, of May 4, would indicate that the contrary is the intention:

The Morgan purchase includes the various Leyland services, except that between Liverpool and the St. Lawrence. A private cable yesterday stated that for five days the Leyland Line held off for the Canadian business, insisting that they should be permitted to run their steamers to Portland, Me., if not to Boston, during the time that navigation on the St. Lawrence is closed by ice. A compromise was effected, whereby Portland was omitted and the demand for a Canadian route granted. This means that for eight months of the year the Leylands will maintain a passenger and freight service between Liverpool and Montreal, with a call at Quebec.

A paragraph which appeared in the pamphlet on the export trade of the port of Montreal for 1900, compiled by the commercial department of The Montreal Gazette and issued shortly after the close of navigation last year, may be opportunely quoted here. In part, it reads as follows:

The Leyland Line have maintained a regular line of steamers from Montreal to Antwerp direct * * * and as these boats were despatched at regular intervals, without calling at intermediate ports, they have been fully appreciated by importers and exporters, and it is expected that next year the fleet will be still further increased with sailings almost every week.

What effect the deal will have on the expected increase in the number of steamers on the Montreal-Antwerp route this season remains to be seen. Fortunately, the outlook at present is not for a decrease.

TRADE IN COUNTRIES OTHER THAN OUR OWN.

ENORMOUS DEMAND FOR SOIL PIPE.

NEVER in the history of the cast iron soil pipe business has there been such an enormous demand as there is at the present time. All the foundries are behind in their orders, and, while there is no immediate prospect of a change, it would not be a surprise to the trade if an advance did take place, on account of the inability of the foundries to handle all the business that is coming to them. The demand for goods is always a factor in the price, and the present continued extraordinary demand for this line should necessarily have a corresponding effect on prices. Stocks with the jobbers are exceedingly low, and it is very hard for the manufacturers to fill a complete specification for stock orders. It is expected it will take two months before the manufacturers of cast iron soil pipe and fittings see their way clear to fill orders at sight. The associated manufacturers of cast iron soil pipe on April 23 made an advance of 5 per cent. in the price of cast iron soil pipe and fittings for all territory west of the Alleghany Mountains.—Metal Worker, New York, May 4.

STEEL CLAD BATH TUBS SCARCE.

There is a scarcity of steel clad and all steel enameled bathtubs reported by the jobbing trade in the vicinity of Greater New York. This is due to the scarcity of the galvanized steel sheets from which the tubs are made. It is said that some manufacturers are resorting to the use of ordinary black sheets, or common Russia iron sheets, and substituting them for the galvanized sheets. Unless a person was apprised that this was so he could never detect the substitution, as the sheets are afterward painted over.—Metal Worker.

THE STEEL TRADE IN SCOTLAND.

The dullness is very acute in the steel trade, and that fact must have been borne very clearly home to the men, for they have just agreed to another reduction of 10 per cent. in wages, after protesting that such a step was not justified. Had they not acquiesced, however, 20 per cent. would certainly have been enforced at the cost of a strike. Some steelmakers advise that their latest advices from Belgium and Germany report a stiffening of prices. During the last day or two it has transpired that North of England makers of steel have been selling largely in the Clyde district, and that of course tells against the local makers. Their southern brethern must have "cut" in stiffly.—Ironmonger, London.

TINPLATE TRADE IN ENGLAND.

The market has shown increased firmness during the past week, and a fair amount of business has been done. It cannot be said that the prices offered for May-June are higher than have been obtainable lately, but there is more inclination to buy for extended delivery, and the fact of so many works being fully booked for some time ahead naturally gives a good backbone to the market. Prompt oilsizes are scarce, and for early delivery a small premium is obtainable.—Iron and Steel Trades' Journal, April 27.

The burning question of the hour has been the probable effect of the proposed export duty of the 1s. a ton on the coal trade, and it is thought by some that this may lead to a downward movement, but that it must take time to allow of its influence being felt.

As regards the tinplate trade, there are now two-thirds of the mills running full time—leaving one-third of the latent power still in suspense—and it is this Sword of Damocles that hangs threateningly over the market, and stifles any attempt at an advance in values; and it is this that leads also to more forward sales, under the ancient wisdom of the strategic policy of the "bird-in-hand" principle.

The demand during the week has been of a general character coming in from all sides, and has been productive of considerable sales in full weights and light plates.

The inquiry has been running to a large extent on prompt shipments, and for such business some advance has been gained in prices where special shipments had to be made.—Iron and Coal Trades' Review, Apr. 26.

NEW YORK METAL MARKET.

The upward movement in the London market continued under a fairly active demand, chiefly in forward deliveries. The close there was firm at an advance of 5s. on spot and of 10s. on futures. Prices here also advanced, but it was said that the higher figures were due to bidding on behalf of the bull interests. There was very little actual buying and the market closed quiet at 26.12 1-2c. bid and 26.25c. bid for spot and 26 to 26.25c. bid and asked respectively. The stock afloat was increased 400 tons to-day, making the total to date 5,089 tons, of which about 2,350 tons will be due within the next ten days

COPPER—The shrinkage in values in the London market continues, though the cable reports indicate that there is still a considerable business in progress there for both prompt and future delivery. The close was quiet at 2s. 6d. under last night's figures on spot and 1s. 3d. on futures. Conditions here were practically the same as for weeks past. With export trade light and diminishing the steady tone of the market is maintained as the result of a large home consumption. The quotations remain at 17c. for Lake Superior and 16 5-8c. for electrolytic and casting.

PIG LEAD—There was a moderate demand for prompt deliveries at prices based on 4.37 1-2c. for lots of 50 tons or over. St. Louis was quiet at 4.20 to 4.22 1-2c. In London soft Spanish appeared to be steady at the previous quotation.

SPELTER—We find little business to report and the tone of the market seems to be rather easy. Shipments are freely offered at 4c., while for spot 3.95c. is bid and 4c. asked. At St. Louis 3.87 1-2c. was nominally quoted. London was unchanged.

ANTIMONY—Regulus finds a moderate jobbing outlet at prices within the quoted range of 8 1-2 to 10 1-4c., as to brand and quality.

OLD METALS—Though the market remains quiet prices are steadily maintained at the quotations.

IRON AND STEEL—Except for deliveries on existing contracts there is not much business reported in any department at present. There is, however, no abatement of the feeling of confidence in the situation, and while there may be a disposition here and there to shade prices on pig iron in order to interest buyers in futures the general tone of the market is one of firmness. In fact, reports from Chicago are to the effect that even in pig iron the feeling there is strong, and that after prolonging negotiations which resulted in no concessions on the part of sellers, several large lots of pig iron have been taken for delivery over the last half of the year.

TINPLATE—The movement of stock into consumption continues on a liberal scale, though chiefly through the medium of deliveries on existing contracts. Prices are maintained and apply to all deliveries up to the end of September.—N. Y. Journal of Commerce, May 9.

WARRANTING TOOLS.

ONE of the vexed questions of the tool trade is that relating to the exchange of implements alleged to be defective. If the tool will not stand the usage to which it is subjected the workman generally complains that the material is bad, and demands another in exchange. In this he resembles Old Nick, who, unable to swim, found fault with the water. But, like most others, this question has a converse side, which a provincial firm of manufacturers plead with no little force. They show that a large proportion of the "faulty" tools returned to them stand the severest tests. The reason of the failure in the workman's hands is due to the unskilful regrinding either by a dry stone or else by unduly reducing the bevel required to give the needful support to the cutting-edge. This contention is undoubtedly valid, and retailers would do well to make a note of it.

In this connection the retailer's position is not altogether a happy one. Often he finds himself between the manufacturer's anvil and the customer's hammer. To tell a mechanic that he does not know how to sharpen his own tools is to offer an affront which the British workman is not inclined to brook. But the remedy is not at this stage of the transaction. If the retailer wishes to avoid both monetary loss and unpleasantness, he will be chary of selling goods on warrant. He should remember that, though he may have confidence in them, he can have but little in the users, and he need be sure of both before giving an unqualified warrantry. Speaking generally, salesmen are much too ready to give guarantees. They would find it far better invariably to decline to be answerable for results.—Ironmonger, London.

The shareholders of The Consolidated Lake Superior Co. and of The Ontario Lake Superior Co., both of which companies were promoted by F. H. Clergue, will meet in New Haven, Conn., to consider a proposal to amalgamate and increase their capital stock to $117,000,000. These two companies will control the Michipicoten iron mines, certain nickel mines in Sudbury, the various manufacturing enterprises at Sault Ste. Marie, the water-power in St. Mary's river and the railway and boat lines promoted by Mr. Clergue.

BUSINESS CHANGES.

DIFFICULTIES, ASSIGNMENTS, COMPROMISES.

George Palmer, general merchant, North Bay, Ont., is offering to compromise.

J. E. Tremblay, general merchant, Ste. Anne de Bellevue, Que., is offering 30c. on the dollar.

J. A. Boyd & Son, tinsmiths and stove dealers, St. Stephen, N.B., have assigned to Geo. J. Clark.

H. Roberts & Co., general merchants, Strathclair, Man., have assigned to C. H. Newton, Winnipeg.

Gagnon & Caron have been appointed curators of James Grady, carriagemaker, New Glasgow, Que.

Assignment has been demanded of Pierre Dauplaise, sash and door manufacturer, St. Cyrille de Wendover, Que.

PARTNERSHIPS FORMED AND DISSOLVED.

Johnson & Thompson, tinsmiths, Danville, Que., have dissolved.

Cameron & Palmer, blacksmiths, Greenwood, B.C., have dissolved.

S. Courser & Co., general merchants, Glen Sutton, Que., have dissolved.

H. Brookbank, blacksmith, Hartney, Man., has admitted W. B. Brookbank as partner.

Pepper & Toole, dealers in agricultural implements, etc., Stonewall, Man., have dissolved.

Sanders & McCann, dealers in agricultural implements, Killarney, Man., have dissolved.

Morley Carscallen, dealer in agricultural implements, Dresden, Ont., has admitted James Anderson into partnership.

W. Wood, H. W. Stevens and [illegible] H. Douglas have formed a partnership under the style of Douglas & Co., hardware dealers, Amherst, N.S.

A. D. Chisholm and K. Sweet have registered a partnership under the style of Chisholm, Sweet & Co., and have bought out McCurdy & Co., general merchants, Antigonish, N.S.

SALES MADE AND PENDING.

The stock of Joseph Quinlan, general merchant, Manotick, Ont., has been sold.

The assets of G. A. Manning, general merchant, Johnville, Que., are to be sold.

The executors of the estate of T. Ross, general merchant, Amqui, Que., have sold the real estate.

Wm. Laidlaw, general merchant, bicycle dealer, etc., Durham, Ont., is advertising his business for sale.

The stock of Miles Birkett, hardware dealer, etc., Ottawa, has been sold at 30 and 3-4c. on the dollar.

The assets of Alp. Letourneau, general merchant, Petit Mechins, Que., are to be sold on the 17th inst.

CHANGES.

Frignon & Jourdain have registered as painters in Montreal.

Lafontaine & Bastien have registered as painters, etc., in Montreal.

Giasson & Frères have registered as carriagemakers, Farnham, Que.

W. E. Murray, harness dealer, Hamilton, Ont., is adding boots and shoes.

E. Prefontaine & Fils have registered as coal and wood dealers at Longueuil, Que.

Richard McDonald, blacksmith, Springfield, Ont., has sold out to Homer Lyons.

W. J. Storey, general merchant, Wendover, Ont., has given up business.

John Kerr, general merchant, Wingham, Ont., has been succeeded by J. and J. H. Kerr.

J. Eilbe[illegible] & Son, general merchants, Crediton, Ont., have sold out to W. W. Kerr.

Gibson & Doner, hardware dealers, Stayner, Ont., have been succeeded by Doner Bros.

W. H. Milburn, general merchant, Tamworth, Ont., has sold out to Wagar & Carscallan.

J. A. Turnbull, carriagemaker, Lawrence Station, Ont., has sold out to David Beattie.

P. Templeman, general merchant, Bonavista, Nfld., has opened a branch at Catalina, Nfld.

C. H. Egan, general merchant, Blind River, Ont., has been succeeded by Dobie & Muncaster.

Geo. T. Leddingham, blacksmith, Nanaimo, B.C., have been succeeded by Leddingham & Ross.

Murphy & Morgan, general merchants, Head of Millstream, N.B., have been succeeded by W. S. Mason.

Starr, Son & Franklyn, hardware dealers, etc., Wolfville, N.S., have been succeeded by C. E. Starr & Son.

FIRES.

I. E. Shantz & Co., founders, Berlin, Ont., have been burned out ; insured.

The pattern shop of J. Matheson & Co., machinists, New Glasgow, N.S., has been destroyed by fire.

Hayes Bros., and Murphy & Morgan, general merchants ; W. S. Mason, sawmiller, and Geo. Stewart, blacksmith, Head of Millstream, N.B., have suffered loss by fire.

DEATHS.

D. Rainville, sawmiller, Coaticook, Que., is dead.

W. T. Beadles, of W. T. Beadles & Co., general merchants, Salmo, B.C., is dead.

Thomas Kerr, of Kerr Bros., general merchants, etc., Farran's Point, Ont., is dead.

WORLD'S COPPER OUTPUT.

According to a German authority on the metal trade the world's production of copper in 1900 aggregated about 479,000 metric tons (2,204 pounds, English weight.) This shows an increase of 8,200 tons as compared with the output of 1899, which is rather small, considering the high price now ruling for the metal, which price is nearly 50 per cent. more than was realized in 1897.

Toward the last year's copper production the United States contributed 268,800 tons and Germany only 32,000 tons. Germany in 1900 imported 83,500 tons of copper and exported 5,500 tons. Her consumption of this metal amounted to 118,500 tons, and her exports of articles composed of copper reached 46,900 tons. Germany's consumption of copper has increased fivefold since 1880, exceeding the consumption of every other country ; the United States alone excepted. Of Germany's total consumption it is estimated that 43,000 tons are used for electrical purposes, 18,000 tons in rolled plates, sheets and bars ; in brass foundries and wire works, 35,000 tons ; in chemical factories (including the manufacture of vitriol), 2,000 tons, and in shipyards, railroad shops, for castings, alloys, etc., about 20,000 tons.

INDUSTRIAL GOSSIP.

Those having any items of news suitable for this column will confer a favor by forwarding them to this office addressed the Editor.

THE Dominion Iron and Steel Co., Limited, expect to "blow in" another furnace in a few days. Two more furnaces are also nearing completion.

Drayton, Ont., passed a by-law on Monday to loan $10,000 to a boot and shoe factory there.

Thos. L. Kay, W. E. Mulholland and Orlando Kelland have been incorporated under the style of The Kay Electric Dynamo and Motor Co., Toronto.

Shantz & Son's foundry, Berlin, Ont., was damaged to the extent of about $4 000 by fire on Friday night, last week. The loss is covered by insurance.

The Canadian Steam Carriage Co. propose erecting in London, Ont., a $25,000 factory to employ 50 hands and turn out seven single carriages per week.

During the past four years 137 factories have been erected or extended in Toronto at a cost of about $1,500,000. In the same time warehouses have been built or enlarged at a cost of about $700,000.

The Northey Manufacturing Co., Toronto, are building for the Montreal Water Power Co. a pumping engine which will weigh about 200 tons, and will probably be the largest single electrical pumping engine in Canada. The contract price is $27,000.

Charles J. McLennan, Henry Mackey, and F. A. Lane, manufacturers, and W. M. Kestin, accountant, Buffalo, and Andrew Dods, solicitor, Toronto, have been incorporated under the style of The McLennan Paint Co., Toronto, to acquire the business of The McLennan-French Paint Co.

The Dominion Coal Co. are sinking a large new shaft which will probably increase the output to 20,000 tons per day. Mr. C. Shields, formerly vice-president and general manager of the Virginia Iron, Coal and Coke Co., and also of the Virginia South-Western Railway, had been appointed vice-president and general manager of the company, and will reside at Sydney.

PERSONAL MENTION.

Mr. W. A. Drummond, who until recently has been the manager of the hardware department of the Robert Simpson Co., Limited, Toronto, left on Monday for London, Ont., where he is to fill a responsible position at Wood's Fair.

MAY ERECT ANOTHER STEEL PLANT.

A Halifax despatch of Wednesday stated that at their meeting in Truro last Monday the Nova Scotia Steel Company bonded all their property to an English and American syndicate for 14 days, dating from May 6, for an exceedingly large figure. The deal includes the iron and steel works at New Glasgow and Trenton, the coal mines and works at Sydney Mines and elsewhere, the iron mines at Bell Island, Newfoundland, the limestone and dolomite deposits in Cape Breton, and all their privileges and franchises.

There is the greatest excitement amongst those already in the secret, as it is understood that the outcome of the deal will undoubtedly be the erection of iron and steel works at North Sydney, rivalling in extent those on the other side of the harbor. The price is not yet known here.

W. T. Andrews is starting a paint and varnish works in Victoria.

H. S. HOWLAND, SONS & CO.

WHOLESALE ONLY — 37-39 Front Street West, **Toronto.** — ONLY WHOLESALE

CORN PLANTERS.

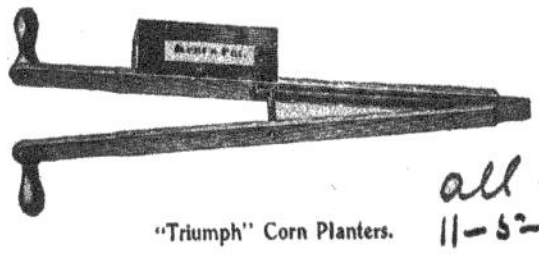

"Triumph" Corn Planters.

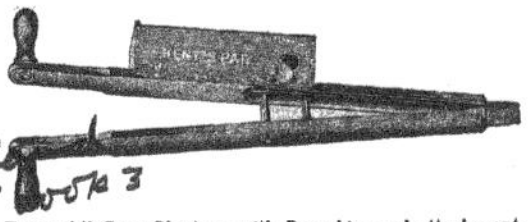

"Triumph" Corn Planters, with Pumpkin seed attachment.

SHEEP SHEARS.

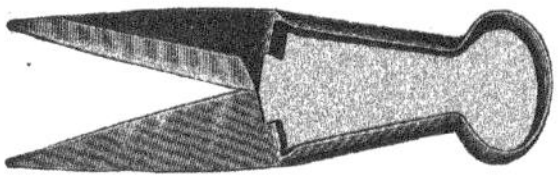

Boker's,

No. 1500.	11	Inch Blade, Bent.
" 1501	12	" " "
" 6654B	11	" " "

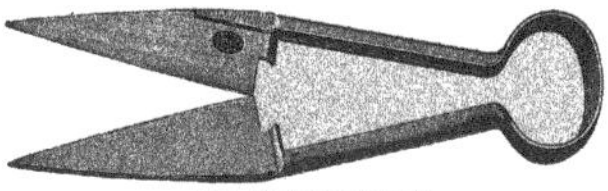

Wm. Wilkinson & Sons.

No. 5G.	6	Inch Blade, Polished, Bent	
" K50.	5½	" " " "	
" "	6	" " " "	
" "	7	" " " "	
" 4	5	" " " "	Trowel Handle.

LAWN MOWERS.

"Star."

Length of Knives.	No. of Knives.	Diameter of Wheel.
12 Inches.	3	9 Inch.
14 "	3	9 "
16 "	3	9 "

"Woodyatt."

Length of Knives	No. of Knives.	Diameter of Wheel.
12 Inches.	4	10½ Inch.
14 "	4	10½ "
16 "	4	10½ "
18 "	4	10½ "
20 "	4	10½ "

H. S. HOWLAND, SONS & CO., Toronto.

OUR PRICES ARE RIGHT — Graham Wire and Cut Nails are the Best. — WE SHIP PROMPTLY.

A FINE NEW BUSINESS BLOCK.

NELSON, B.C., is still to the front in the Kootenays, as instanced by its multiplicity of new and handsome private and business erections, last but not the least being the very imposing solid stone and brick building just erected by Kilpatrick & Wilson, the pioneer grocers of Nelson, a cut of which is herewith given. Built in the most modern style of architecture, their new store is fitted up internally with the newest, up-to-date fittings of cedar and Coast fir woods, making the general appearance look cool, clean and comfortable—in fact, inviting—customers having the comfort of patent piano seats along the counter. There is ample cellar accommodation and every modern convenience for running a large business.

The office and partner's private room are elevated at the back end of the store 12 ft. from the shop level, and are approached by a neat staircase, giving the "tout ensemble" a very decided business appearance. The floor space is 135 x 30 ft. One plate-glass window, measures 18 ft by 15 ft. in height, with a side window, 12 ft. broad by 15 ft. high.

Having erected this—shall we say—monument to their commercial integrity and acumen, they have retired on their well-earned laurels, and have been succeeded by two of the smartest young grocers in Nelson, viz., Mr. T. S. McPherson and M. T. I. McCammon, the first having been for some years manager of one of the largest grocery establishments in the city, the latter (Mr. McCammon) being the well-known prize grocery window dresser, having carried off favors from Nelson and the diploma for "Essay on Window Dressing" lately offered by a Chicago paper. I. M.

A Fine New Business Block in Nelson, B.C.

INTRODUCTION OF MATCHES.

The Atlas, a London newspaper, published on January 10, 1830, the following paragraph under the head of "Instantaneous Light"; "Amongst the different methods invented for obtaining a light instantaneously ought certainly to be recorded that of Mr. Walker, chemist, Stockton-on-Tees. He supplies the purchaser with prepared matches, which are put into tin boxes, but are not liable to change in the atmosphere, and also with a piece of fine glass paper folded in two. Even a strong blow will not inflame the matches, because of the softness of the wood underneath, nor does rubbing upon wood or any common substance produce any effect except that of spoiling the match; but when one is pinched between the folds of the glass paper and suddenly drawn out it is instantly inflamed. Mr. Walker does not make them for extensive sale, but only to supply the small demand in his own neighborhood."

INTO NEW PREMISES.

Leeder & Parsons, general merchants, Quyon, Que., are moving into new premises which have been made thoroughly up-to-date. A plate-glass front, new shelving, counters, etc., have been installed, and a circular glass office is being built in the centre of the store. As the store is much larger than their former one, the firm have put in about $10,000 worth of new stock.

PETROLEUM IN CANADA.

THE United States Consul-General at Montreal sends, under date of April 15, the following interesting account of the petroleum industry in Canada:

"The production of petroleum in Canada is practically confined at present to the counties of Lambton, Kent and Bothwell, in the southwestern part of the Province of Ontario. The town of Petrolia, in Lambton county, is the centre of the principal district of production, while the work of refining is carried on in Sarnia, about 14 miles distant, the crude oil being pumped through pipes to the refineries. In this Ontario oil district, there are about 9,000 oil wells in operation, and the average monthly yield is nearly 60,000 bbls. The ratio of crude petroleum to refined oil is about 100 to 40, so that more than two barrels of petroleum are required to make one barrel of refined oil ready for household use. The process of manufacture is complicated, and an expensive plant is required. The oil is first distilled from the petroleum, the latter being dark in color. The distilled oil is then refined, and, lastly, it is bleached or clarified. From the mud-colored petroleum, an illuminating oil as clear as water is obtained. It is then ready for shipment in barrels or tanks.

"The oil is in every case found in the corniferous limestone, and the different producing areas present local dome structures on the main anticlines, which afford good reservoirs for the accumulation of oil.

"The oil is pumped from an average depth of 465 feet.

"In 1899, there were produced in Canada 11,883,627 gallons of illuminating oil of a value of $1,197,870. The production of that year was equivalent to 748,667 barrels.

"During 1899, there were imported into Canada oils to the value of $1,408,000, upon which duty was paid to the amount of $589,000. This included illuminating oils, animal oils, and vegetable oils. The larger portion of the total importation was made up of coal and kerosene oils, several grades used in Canada coming from the United States. The importations of linseed oil were also very large, running up to $332,000 in value, exclusive of the duties, which amounted to $64,000 upon this one article. Lubricating oils are also very largely imported, the value of the imports in 1899, including duty, being upward of $100,000."

MARKETS AND MARKET NOTES

QUEBEO MARKETS

Montreal, May 10, 1901.

HARDWARE.

THE activity that we have noticed for some weeks continues even in stronger form this week. The first boats left for the west during the last 10 days, and they have carried full cargoes. The market shows few new features from last week. Wire is still scarce, and cut nails are none too plentiful. Horse nails and shoes are selling fairly well. Screen wire cloth, screen doors and windows, hose and lawn mowers are active lines, while cutlery and builders' hardware are moving out freely. Building paper is being shipped in fair quantities, and the wholesale houses are handling immense quantities of paris green and paints.

BARB WIRE — It seems to be absolutely impossible to obtain goods for immediate delivery without paying a premium. Supplies are expected in the course of a week or so. The price is unchanged at $3.05 per 100 lb. f.o.b. Montreal.

GALVANIZED WIRE—This line is very active. We quote: No. 5, $4.25; Nos. 6, 7 and 8 gauge, $3.55; No. 9, $3.10; No. 10, $3.75; No. 11, $3 85; No. 12, $3.25; No. 13' $3.35; No. 14, $4.25; No. 15, $4.75; No. 16, $5.00.

SMOOTH STEEL WIRE — The Canadian mills are even with orders, but some dealers are waiting for goods to come with other varieties ordered. The demand continues quite brisk. We quote oiled and annealed as follows: No. 9, $2.80; No. 10, $2.87; No. 11, $2.90; No. 12, $2.95; No. 13, $3 15 per 100 lb. f.o.b. Montreal, Toronto, Hamilton, London, St. John and Halifax.

FINE STEEL WIRE — The market is steady under a moderate demand. The discount remains at 17½ per cent. off the list.

BRASS AND COPPER WIRE—The demand is limited. The discount on brass is 55 and 2½ per cent., and on copper 50 and 2½ per cent.

FENCE STAPLES—Quite large large quantities of staples have been sold this week. We quote: $3.25 for bright, and $3.75 for galvanized, per keg of 100 lb.

WIRE NAILS — The demand is hardly as brisk as it was last week, yet a good trade continues to be done, and large amounts are moving out. We quote $2.8[illegible] for small lots and $2.77½ for carlots, f.o.b. Montreal, London, Toronto, Hamilton and Gananoque.

CUT NAILS—The cut nail demand shows a little improvement. Stocks are very light. We quote: $2.35 for small and $2.25 for carlots; flour barrel nails, 25 per cent discount; coopers' nails, 30 cent. discount.

HORSE NAILS—The market continues as it has been for some time, discounts varying on most brands with the size of the order. "C" brand is held firmly at discount of 50 and 7½ per cent. off the new list. "M" brand is quoted at 60 per cent. off old list on oval and city head and 65 per cent. off countersunk head.

HORSESHOES—A few more inquiries have been received this week, but the demand is not exceedingly brisk. We quote as follows: Iron shoes, light and medium pattern, No.

2 and larger, $3.50; No. 1 and smaller, $3.75; snow shoes, No. 2 and larger, $3.75; No. 1 and smaller, $4.00; X L steel shoes, all sizes, 1 to 5, No. 2 and larger, $3.60; No. 1 and smaller, $3.85; feather-weight, all sizes, $4.85; toe weight steel shoes, all sizes, $5.95 f.o.b. Montreal; f.o.b. Hamilton, London and Guelph, 10c. extra.

POULTRY NETTING—A moderate trade is being done at the former discount, 55 per cent.

GREEN WIRE CLOTH — This is one of the most active articles on the market. The price is still $1.35 per 100 sq. ft.

SCREEN DOORS AND WINDOWS — The demand is good. We quote: Screen doors, plain cherry finish, $8.25 per doz.; do. fancy, $11.50 per doz; walnut, $7.40 per doz., and yellow, $7.45; windows, $2.25 to $3.50 per doz.

SCREWS—A moderate business is being done. The market is featureless. Discounts are as follows: Flat head bright, 87½ and 10 per cent. off list; round head bright, 82½ and 10 per cent.; flat head brass, 80 and 10 per cent. round head brass, 75 and 10 per cent.

BOLTS—A heavy demand has been experienced from the large manufacturing industries. Discounts are: Norway carriage bolts, 65 per cent.; common, 60 per cent.; machine bolts, 60 per cent.; coach screws, 70 per cent.; sleigh shoe bolts, 72½ per cent.; blank bolts, 70 per cent.; bolt ends, 62½ per cent.; plough bolts, 60 per cent.; tire bolts, 67½ per cent.; stove bolts, 67½ per cent. To any retailer an extra discount of 5 per cent. is allowed. Nuts, square, 4c. per lb. off list; hexagon nuts, 4¼c. per lb. off list. To all retailers an extra discount of ¼c. per lb. is allowed.

BUILDING PAPER — A very satisfactory trade is being done in all kinds of building paper. We quote as follows: Tarred felt, $1.70 per 100 lb.; 2 ply, ready roofing, 80c. per roll; 3-ply, $1.05 per roll; carpet felt, $2.25 per 100 lb.; dry sheathing, 30c. per roll; tar sheathing, 40c. per roll; dry fibre, 50c. per roll; tarred fibre, 60c. per roll; O.K. and I.X.L., 65c. per roll; heavy straw sheathing, $28 per ton; slaters' felt, 50c. per roll.

RIVETS AND BURRS—There is nothing new to report. Discounts on best iron rivets, section, carriage, and wagon box, black rivets, tinned do., coopers' rivets and tinned swedes rivets, 60 and 10 per cent.; swedes iron burrs are quoted at 55 per cent. off; copper rivets, 35 and 5 per cent. off; and coppered iron rivets and burrs, in 5-lb. carton boxes, are quoted at 60 and 10 per cent. off list.

BINDER TWINE—A few more inquiries have been received this week. We quote: Blue Ribbon, 11½c.; Red Cap, 9¼c.; Tiger, 8¾c.; Golden Crown, 8c.; Sisal, 8¼c.

CORDAGE—The demand keeps up exceptionally well. Manila is worth 13½c. per lb. for 7-16 and larger; sisal is selling at 10c., and lathyarn 10c.

SPADES AND SHOVELS—The activity noticed last week continues undiminished. The discount is unchanged at 40 and 5 per cent. off the list.

HARVEST TOOLS—Wholesalers report a continued good trade in harvest tools and manufacturers are quite busy. The discount is as before, 50, 10 and 5 per cent.

TACKS—Unchanged. We quote: Carpet tacks, in dozens and bulk, 80 and 15 per cent.; tinned, 80 and 20 per cent., and cut tacks, blued, in dozens, 80 per cent.

LAWN MOWERS—Quite a number of sales have been effected during the week. We quote as follows: High wheel, 50 and 5 per cent. f.o.b. Montreal; low wheel, in all sizes, $2.75 each net; high wheel, 11-inch, 30 per cent. off.

FIREBRICKS—A fairly good trade is being done at $18 to $24 per 1,000 as to brand ex store, and Scotch are quoted at $17.50 to $22 and English at $17 to $21 per 1,000 ex wharf.

CEMENT—The demand is still first class. We quote: German cement, $2.30 to $2.50; English, $2.25 to $2.35; American, $2.25 to $2.50, and Belgian, $1.70 to $1.95 for summer delivery, and spot prices are unchanged at $2.45 to $2.55 for German; $2.30 to $2.40 for English, and $1.95 to $2 05 for Belgian, ex store.

METALS.

The metal market is in somewhat better shape than it was last week. The tone of the market is more generally firm. The English market is now beginning to feel the effect of the boom on this side of the water and prices seems to have touched bottom. English pig iron is also reported more active and stronger.

PIG IRON—In the English market, which has of late been influencing the Canadian prices, the movements in prices have been irregular, but, on the whole, the tendency has been toward a higher level. Canadian pig is worth $18 to $19, and Summerlee about $20.50 to $21, per ton.

BAR IRON—The market is strong and fairly active. Merchants' bar is worth $1.75, and horseshoe $2.

HOOP IRON — Is rather lower this week. One inch, 19 gauge is worth $2.85, and No. 17, $3 10.

BLACK SHEETS — A fair demand is being experienced. The market is steady. We quote: 8 to 16 gauge, $2.55; 26 gauge, $2.60, and 28 gauge, $2.65.

GALVANIZED IRON — Dealers are buying rather heavily. We quote: No. 28 Queen's Head, $4.65; Apollo, 10¾ oz., $4.50, and Comet, $4 40 to $4 45, with a 15c. reduction for case lots.

INGOT COPPER — The market continues firm and quiet at 18c.

INGOT TIN—In London ingot tin has advanced to £119. Each day the market there shows a slight rise. Sales have been made at 31c. during the week.

LEAD—Unchanged at $3 75 to $3 85.

LEAD PIPE—A fair business is passing. We quote: 7c. for ordinary and 7½c. for composition waste, with 25 per cent. off.

IRON PIPE—The demand is satisfactory. We quote: Black pipe, ¼, $3 per 100 ft.; ⅜, $3; ½, $3 05; ¾, $3.30; 1-in., $4.70; 1¼, $6.40; 1½, $7.70; 2-in. $10 25. Galvanized, ½, $4.60; ¾, $5.25; 1-in., $7.50; 1¼, $9.80; 1½, $11.75; 2 in., $16.

TINPLATES — The feeling is somewha better this week. A fair distributing trade is being done. We quote: Coke plates, $3.80 to $4; charcoal, $4.15 to $4.25.

CANADA PLATE—There has been a fair inquiry from the country this week. We quote: 52's, $2.50; 60's, $2.60; 75's, $2.70; full polished, $3.35, and galvanized, $4.20.

STEEL —The feeling is steady and firm. We quote: Sleighshoe, $1.95; tire, $2;

bar, $1.95; spring, $2.75; machinery, $2.75 and toe-calk, $2.50.

Sheet Steel—There is nothing new to report. We quote: Nos. 22 and 24, $3, and Nos. 18 and 20, $2.85.

Tool Steel—Black Diamond is worth 8c. and Jessop's, 13c.

Terne Plates—Small lots are moving. The general price is about $7.75.

Coil Chain—There is no change. We quote as follows: No. 6, 11½c.; No. 5, 10c.; No. 4, 9½c.; No. 3, 9c.; ¼-inch, 7½c. per lb.; 5-16, $4.85; 5-16 exact, $5.30; ⅜, $4.40; 7-16, $4.20; ½, $3.95; 9-16, $3.85; ⅝, $3.55; ¾, $3.45; ⅞, $3.40; 1-in., $3.35. In carload lots an allowance of 10c. is made.

Sheet Zinc—There is nothing new to report. Spot goods are worth $5 75 to $6.25.

Antimony—Quiet, at 10c.

Zinc Spelter—Is worth 5c.

Solder—We quote: Bar solder, 18½c.; wire solder, 20c.

GLASS.

There is a brisk trade being done in glass. The market is firm. We quote as follows: First break, $2; second, $2.10 for 50 feet; first break, 100 feet, $3.80; second, $4; third, $4.50; fourth, $4.75; fifth, $5.25; sixth, $5.75, and seventh, $6.25.

PAINTS AND OILS.

Linseed oil is 5c. per gal. higher. The general paint and oil market may be described as in the full swing of a very heavy business with the exception of white lead, upon which, it is reported, there has been some cutting. Every other article is extremely firm, and values have a strong undertone. The demand for white lead is, however, improving, and standard makers are selling all they can turn out at our regular scale. Ample supplies of dry white lead and zinc are coming forward. This has had the effect of reducing quotations for the dry article somewhat. Paris green, with the advent of warm weather, is moving out more freely. Prices, if anything, will have a tend upwards from this until the end of the season. Liquid paints have received a tremendous spurt, and the manufacturers have scarcely been able to keep pace with the demand. Kalsomines have been asked for, and carriagemakers throughout the country, having more orders than they can attend to, are liberal buyers of coach-builders' colors and varnishes generally. We quote:

White Lead—Best brands, Government standard, $6.25; No. 1, $5.87½; No. 2, $5.50; No. 3, $5.12½, and No. 4, $4.75 all f.o.b. Montreal. Terms, 3 per cent. cash or four months.

Dry White Lead—$5.25 in casks; kegs, $5.50.

Red Lead—Casks, $5.00; in kegs, $5.25.

Dry White Zinc—Pure, dry, 6¼c.; No. 1, 5¼c.; in oil, pure, 7¼c.; No. 1, 6¼c.; No. 2, 5¼c.

Putty—We quote: Bulk, in barrels, $1.90 per 100 lb.; bulk, in less quantity, $2.05; bladders, in barrels, $2.10; bladders, in 100 or 200-lb. kegs or boxes, $2.25; in tins, $2.55 to $2.65; in less than 100-lb. lots, $3 f.o.b. Montreal, Ottawa, Toronto, Hamilton, London and Guelph. Maritime Provinces 10c. higher, f.o.b. St. John and Halifax.

Linseed Oil—Raw, 77c.; boiled, 80c. in 5 to 9 bbls., 1c. less, 10 to 20 bbl. lots, open, net cash, plus 2c. for 4 months Delivered anywhere in Ontario between Montreal and Oshawa at 2c. per gal. advance and freight allowed.

Turpentine—Single bbls., 57c.; 2 to 4 bbls., 56c.; 5 bbls. and over, open terms, the same terms as linseed oil.

Mixed Paints—$1.25 to $1.45 per gal.

Castor Oil—8¾ to 9¼c. in wholesale lots, and ½c. additional for small lots.

Seal Oil—47½ to 49c.

Cod Oil—32½ to 35c.

Naval Stores—We quote: Resins, $2.75 to $4.50, as to brand; coal tar, $3.25 to $3.75; cotton waste, 4½ to 5½c. for colored, and 6 to 7½c. for white; oakum, 5½ to 6½c., and cotton oakum, 10 to 11c.

Paris Green—Petroleum barrels, 16¾c. per lb.; arsenic kegs, 17c.; 50 and 100-lb. drums, 17½c.; 25-lb. drums, 18c.; 1-lb. packages, 18½c.; ½-lb. packages, 20½c.; 1-lb. tins, 19½c.; ½-lb. tins, 21½c. f.o.b. Montreal; terms 3 per cent. 30 days, or four months from date of delivery.

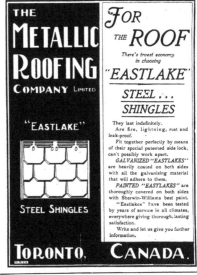

SCRAP METALS.

There has been no striking feature in the market this week. Supplies are somewhat more liberal. Dealers are now paying the following prices in the country: Heavy copper and wire, 13 to 13½c. per lb.; light copper, 12c.; heavy brass, 12c.; heavy yellow, 8½ to 9c.; light brass, 6½ to 7c.; lead, 2½ to 2¾c. per lb.; zinc, 2¼ to 2½c.; iron, No. 1 wrought, $15 to $16 per gross ton f.o.c. Montreal; No. 1 cast, $13 to $14; stove plate, $8 to $9; light iron, No. 2, $4 a ton; malleable and steel, $4; rags, country, 70 to 80c. per 100 lb.; old rubbers, 6¼c. per lb.

HIDES.

Trade is rather quiet. We quote: Light hides, 6½c. for No. 1; 5½c. for No. 2, and 4½c. for No. 3. Lambskins, 10c.; sheepskins, 90c.; calfskins, 8c. for No. 1 and 6c. for No. 2.

PETROLEUM.

Trade is only moderate. We quote as follows: "Silver Star," 14½ to 15½c.; "Imperial Acme," 16 to 17c.; "S.C. Acme," 18 to 19c., and "Pratt's Astral," 18½ to 19½c.

MONTREAL NOTES.

The Canada Hardware Company,

Limited, received quite a large importation of cutlery by an early steamer.

Linseed oil is 5c. higher.

Hoop iron has been marked down this week.

ONTARIO MARKETS.

TORONTO, May 11, 1901.

HARDWARE.

TRADE is still fairly brisk, although, generally speaking, it is perhaps not as pronounced as it was a week ago. Prices are, on the whole, steady. Not many changes have taken place during the week. In such lines as oil stoves, ice cream freezers and refrigerators the demand is in excess of the supply. The same can be said of some lines of garden tools. In fact, in most lines appertaining to the hardware trade the manufacturers appear to be pretty busy. The difficulty noted in previous issues in regard to the scarcity of barb wire is even more marked than it was a week ago, for, while a great many orders have come in, no supplies have come to hand. Much the same condition of affairs also exists in regard to plain galvanized wire and plain twist. Orders for wire nails are numerous, and prices keep steady. Business in cut nails is still nearly altogether confined to shingle nails. A moderate business is being done in horse nails and horseshoes. Both screws and rivets are in active demand. Trade is active in eavetrough and conductor pipe and prices continue unsatisfactory. The manufacturers of stoveboards announce a reduction of about 10 per cent. in prices.

BARB WIRE — Fresh deliveries of barb wire have not yet come to hand, and, as the demand is active and quite a number of orders have been received, the condition of affairs is even worse than it was a week ago. As far as we can learn the jobbers have not yet received even the invoices of goods which the manufacturers promised to ship this week. Nearly all the orders coming in for the wholesale trade are small, and for immediate shipment. We quote $3.05 per 100 lb. from stock Toronto; f.o.b. Cleveland $2.82½ per 100 lb. for less than carlots and $2.70 for carlots.

GALVANIZED WIRE — There is also a scarcity of galvanized wire, while orders are coming every day for small lots. We quote as follows: Nos. 6, 7 and 8, $3.50 to $3.85 per 100 lb., according to quantity; No. 9, $2.85 to $3.15; No. 10, $3.60 to $3.95; No. 11, $3.70 to $4.10; No. 12, $3 to $3.30; No. 13, $3.10 to $3.40; No. 14, $4.10 to $4.50; No. 15, $4.60 to $5.05; No. 16, $4.85 to $5.35. Nos. 6 to 9 base f.o.b. Cleveland are quoted at $2.57½ in less than carlots and 12c. less for carlots of 15 tons.

SMOOTH STEEL WIRE—The scarcity of barb and plain galvanized wire is still stimulating trade in oiled and annealed, business in which continues active. The net selling prices for oiled and annealed are as follows: Nos. 6 to 8, $2.90; 9, $2.80; 10, $2.87; 11, $2.90; 12, $2.95; 13, $3.15; 14, $3.37; 15, $3.50; 16, $3.65. Delivery points, Toronto, Hamilton, London and Montreal, with freights equalized on these points.

WIRE NAILS — There is a brisk movement in wire nails, although the quantities wanted are not large. Prices are unchanged, and we quote less than carlots at $2.85, and carlots at $2.77½. Delivery points: Toronto, Hamilton, London, Gananoque and Montreal.

CUT NAILS—Business in cut nails continues dull, and the kind chiefly wanted is shingle nails. We quote the base price at $2.35 per keg for less than carlots, and $2.25 for carlots. Delivery points: Toronto, Hamilton, London, Montreal and St. John, N.B.

HORSE NAILS—There is a fair quantity moving for this time of the year. Discount on "C" brand, oval head, 50 and 7½ per cent. off new list, and on "M" and other brands, 50, 10 and 5 per cent., off the old list. The discount on countersunk head is 60 per cent.

HORSESHOES—In this line, there is a steady trade being done for this time of the year. We quote as follows f.o.b. Toronto: Iron shoes, No. 2 and larger, light, medium and heavy, $3.60; snow shoes, $3.85; light steel shoes, $3.70; featherweight (all sizes), $4.95; iron shoes, No. 1 and smaller, light, medium and heavy (all sizes), $3.85; snow shoes, $4; light steel shoes, $3.95; featherweight (all sizes), $4.95.

SCREWS — An active sorting-up trade is to be noted in this line. We quote discounts: Flat head bright, 87½ and 10 per cent.; round head bright, 82½ and 10 per cent.; flat head brass, 80 and 10 per cent.; round head brass, 75 and 10 per cent.; round head bronze, 65 per cent., and flat head bronze at 70 per cent.

BOLTS AND NUTS—The bolt trade continues active and the manufacturers report that they are so pushed with orders that they cannot fill them promptly. We quote as follows: Carriage bolts (Norway), full square, 65 per cent.; carriage bolts full square, 65 per cent.; common carriage bolts, all sizes, 60 per cent.; machine bolts, all sizes, 60 per cent.; coach screws, 70 per cent.; sleighshoe bolts, 72½ per cent.; blank bolts, 60 per cent.; bolt ends, 62½ per cent.; nuts, square, 4c. off; nuts, hexagon, 4½c. off; tire bolts, 67½ per cent.; stove bolts, 67½; plough bolts, 60 per cent.; stove rods, 6 to 8c.

RIVETS AND BURRS — A fair trade is being done. We quote: Iron rivets, 60 and 10 per cent.; iron burrs, 55 per cent.; copper rivets and burrs, 35 and 5 per cent.

ROPE—There is quite a little business being done, especially for the kinds wanted for hay fork work and scaffolds. The base price is unchanged at 10c. for sisal, 13¼c. for manila.

COTTON WRAPPING TWINE—The price of cotton wrapping twine has been reduced; also that of bed cord.

CUTLERY—The improvement noted in the cutlery trade last week has been maintained, and quite a nice business is being done for this time of the year.

SPORTING GOODS—Quite a little powder, shot, loaded cartridges and metallic cartridges are going out. There is also quite a demand for double and singled barrel shot guns.

ENAMELLED WARE — The demand for enamelled ware continues brisk. The discounts on enamelled ware are: "Granite," "Pearl," "Crescent" and "Imperial" wares at 50, 10 and 10 per cent.; white, "Princess," "Turquoise." blue and white, 50 per cent.; "Diamond," "Famous" and "Premier," 50 and 10 per cent.

TINWARE—A satisfactory trade is still being done in tinware, especially in such lines as milk pans, milk pails, water pails, kettles and boilers.

STOVE BOARDS — The manufacturers of stove boards have issued a circular to the trade announcing a reduction of 10 per cant. in the price of stove boards for the coming season.

GREEN WIRE CLOTH — Trade is fairly good in this line at $1.35 per 100 sq. ft.

SCREEN DOORS AND WINDOWS—Quite a few of these are being shipped this week.

BUILDING PAPER—The demand for building paper is active, and quite a number of shipments are being made. We quote: Plain building, 30c.; tarred lining, 40c., and tarred roofing, $1.65.

POULTRY NETTING—Business continues fairly good with the discount on Canadian unchanged at 55 per cent.

ICE CREAM FREEZERS AND REFRIGERATORS — Trade in this line is still more than usually active, and in some lines of refrigerators the manufacturers are away behind in their orders.

OIL STOVES — A large trade is still being done in oil stoves.

HARVEST TOOLS—The demand is good, and some lines are hard to get in great enough supply for the demand. This is particularly true of such lines as garden hose. Discount 50, 10 and 5 per cent.

SPADES AND SHOVELS—A fair quantity of these are being shipped. Discount 40 and 5 per cent.

EAVETROUGH, ETC. — The demand for eavetrough and conductor pipe continues brisk, while the manufacturers are still doing business on an unsatisfactory margin.

BINDER TWINE — Trade is quiet. We quote: American—Sisal and standard, 8¼c.; manila, 10¼c.; pure manila, 11¼ to 12c. per lb. Canadian—Sisal, 8¼c.; pure manila, 11¼c. per lb.

CEMENT— A big trade is doing. Another decline of 5c. is noted in Canadian portland. We quote barrel lots: Canadian portland $2.25 to $2.75; German, $3 to $3.15; English, $3; Belgian, $2.50 to $2.75; Canadian hydraulic, $1.25 to $1.50.

METALS.

Business in metals is most active this week. Since our last report trade has improved quite materially. Prices are on the whole firm, and the general tendency is toward a higher rather than toward a lower range of values. The strength of the pig tin market has been one of the features of the trade during the past week.

PIG IRON—Most of the large buyers in Canada seem to have placed their orders for the whole of the year's supply, and we hear of little or no business on that account. As far as we can learn the idea as to price for Canadian iron is $16.50 for No. 2 in large lots, and $16 per ton for No. 3.

BAR IRON — The demand continues brisk and prices firm. Some of the mills are refusing to entertain orders for prompt delivery, so much business have they on their books. The ruling price for bar iron is $1.85 to $1.90 per 100 lb. according to quantity.

STEEL—The steel market continues fairly active with prices unchanged. We quote tire steel at $2.30 to $2.50; sleighshoe steel at $2.10 to $2.25; reeled machinery steel, $3. Hoop steel, $3.10.

PIG TIN—The position of the tin market is a decidedly strong one, particularly in London, Eng., where there have been several advances of late. On Wednesday prices there advanced 12s. 6d. per ton, and in New York the market advanced 25 points during the same day. In both cities the market closed strong. In Canada the demand is fairly good, although it cannot be called active, while prices are unchanged, the idea as to price for Lamb and Flag still being 31 to 32c. per lb.

TINPLATES — Prices are advancing in the British market, but locally quotations are unchanged. The demand locally for tinplates is good.

TERNE PLATES — While the demand is improving, the volume of business is still light.

TINNED SHEETS — Trade is active, but most of the orders are for small lots, large lots being the exception. We quote 9 to 9½c. per lb.

GALVANIZED SHEETS — Business continues good, with stocks light. Some of the orders call for large lots. Spring importations have not yet arrived, but they are expected almost any day, as the invoices and bills of lading are to hand. We quote $4.60 from stock for English and $4.50 for American.

BLACK SHEETS—A good demand is being experienced in this line. We still quote 28 gauge at $2.30.

CANADA PLATES — Although business is still light somewhat better demand is being experienced. Import orders are still being booked. We quote: All dull, $3; half-and-half, $3.15, and all bright, $3.65 to $3.75.

COPPER—The demand this week is good both for ingot and sheet copper, and prices are unchanged locally, and in the outside markets they are fairly steady. We quote:

Ingot, 19c.; bolt or bar, 23½ to 25c.; sheet, 23 to 23½c. per lb.

BRASS—The demand for brass is fair. Discount on rod and sheet is 15 per cent.

SOLDER—This article is going out freely and prices are firm at quotations. We quote: Half-and-half, guaranteed, 18½c.; ditto, commercial, 18c.; refined, 18c., and wiping, 17c.

IRON PIPE—There is nothing new in the situation. Prices are, on the whole, firm, but there is not a great deal of business being done.

LEAD—There is a fairly good demand for lead, and prices rule steady in the outside markets. Locally, we quote 4¼ to 4½c. per lb.

ZINC SPELTER—Locally, trade is quiet, and New York advices report trade dull with prices fairly steady. We quote 6½c. for casks, and 6¾c. for part casks.

ANTIMONY—There is a little better inquiry and some business has been done as the result. We quote 10½ to 11c. per lb.

PAINTS AND OILS.

The feature of the week is the shortage of linseed oil on spot. This, combined with the high basis of prices in England, has resulted in an advance of 4c. Deliveries of turpentine have been delayed by washouts in the Southern States, but, as the trade is not anxious for immediate delivery, prices are unchanged. White lead, prepared paints, varnishes, etc., are in excellent demand at firm prices. We quote:

WHITE LEAD—Ex Toronto, pure white lead, $6.37½; No. 1, $6; No. 2, $5.67½; No. 3, $5.25; No. 4, $4.87½; genuine dry white lead in casks, $5.37½.

RED LEAD—Genuine, in casks of 560 lb., $5.50; ditto, in kegs of 100 lb., $5.75; No. 1, in casks of 560 lb., $5; ditto kegs of 100 lb., $5.25.

LITHARGE—Genuine, 7 to 7½c.

ORANGE MINERAL—Genuine, 8 to 8½c.

WHITE ZINC—Genuine, French V.M., in casks, $7 to $7.25; Lehigh, in casks, $6.

PARIS WHITE—90c. to $1 per 100 lb.

WHITING—70c. per 100 lb.; Gilders' whiting, 80c.

GUM SHELLAC—In cases, 22c.; in less than cases, 25c.

PARIS GREEN—Bbls., 16¾c.; kegs, 17c.; 50 and 100-lb. drums, 17½c.; 25-lb. drums, 18c.; 1-lb. papers, 18½c.; 1-lb. tins, 19½c.; ½-lb. papers, 20½c.; ½-lb. tins, 21½c.

PUTTY—Bladders, in bbls., $2.10; bladders, in 100 lb. kegs, $2.25; bulk in bbls., $1.90; bulk, less than bbls. and up to 100 lb., $2.05; bladders, bulk or tins, less than 100 lb., $2.90.

PLASTER PARIS—New Brunswick, $1.90 per bbl.

PUMICE STONE—Powdered, $2.50 per cwt. in bbls., and 4 to 5c. per lb. in less quantity; lump, 10c. in small lots, and 8c. in bbls.

LIQUID PAINTS—Pure, $1.20 to $1.30 per gal.

CASTOR OIL—East India, in cases, 10 to 10½c. per lb. and 10½ to 11c. for single tins.

LINSEED OIL—Raw, 1 to 4 barrels, 75c.; boiled, 78c.; 5 to 9 barrels, raw, 74c.; boiled, 77c., delivered. To Toronto, Hamilton, Guelph and London, 1c. less.

TURPENTINE—Single barrels, 54c.; 2 to 4 barrels, 53c., delivered. Toronto, Hamilton and London 1c. less. For less quantities than barrels, 5c. per gallon extra will be added, and for 5-gallon packages, 50c., and 10-gallon packages, 80c. will be charged.

GLASS.

There is an excellent demand, and some lines are beginning to give evidence of a shortage. We quote: Under 26 in., $4.15 26 to 40 in., $4.45; 41 to 50 in., $4.85; 51 to 60 in., $5.15; 61 to 70 in., $5.50; double diamond, under 26 in., $6; 26 to 40 in., $6.65; 41 to 50 in., $7.50; 51 to 60 in., $8.50; 61 to 70 in., $9.50, Toronto, Hamilton and London. Terms, 4 months or 3 per cent. 30 days.

OLD MATERIAL.

The movement is more liberal, but there is no change in price. We quote jobbers' prices as follows: Agricultural scrap, 55c. per cwt.; machinery cast, 60c. per cwt.; stove cast, 50c.; No. 1 wrought 50c. per 100 lb.; new light scrap copper, 12c. per lb.; bottoms, 11½c.; heavy copper, 13c.; coil wire scrap, 13c.; light brass, 7c.; heavy yellow brass, 10 to 10½c.; heavy red brass, 10½ to 11c.; scrap lead, 3c.; zinc, 2c; scrap rubber, 6½c.; good country mixed rags, 65 to 75c.; clean dry bones, 40 to 50c. per 100 lb.

PETROLEUM.

There has been a considerable reduction in the price of crude oil, but the refined product is unchanged. There is not much doing, however. We quote: Pratt's Astral, 16½ to 17c. in bulk (barrels, $1 extra); American water white, 16½ to 17c. in barrels; Photogene, 16 to 16½c.; Sarnia water white, 15½ to 16c. in barrels; Sarnia prime white, 14½ to 15c. in barrels.

COAL.

A fairly good movement is noted. We quote at international bridges as follows: Grate, $3.75 per gross ton; egg, stove and nut, $5 per gross ton with a rebate of 40c. off for May shipment.

MARKET NOTES.

Cotton wrapping twine and cotton bed cord are lower.

A reduction of 10 per cent. has been made in the price of stove boards for the coming season.

H. S. Howland, Sons and Co. have placed in stock a shipment of "3 in 1" bicycle oil; also a line of rust remover.

A shipment of clarified leather cement is to hand with H. S. Howland, Sons & Co.

Linseed oil has advanced 4c. per gallon.

PERSONAL MENTION.

Mr. Thomas Ellin, of T. Ellin & Co., Sylvester works, Sheffield, England, is in Toronto this week. It is two years since he last visited Canada.

Mr. A. A. McMichael, of Toronto, the vice-president and general manager of James Robertson & Co., Limited, and also president of the Robertson-Godson Co., of this city, with Mrs. McMichael, is paying a visit to Vancouver. They are the guests of Mr. and Mrs. C. A. Godson.—The Province, Vancouver.

As their premises are too small Boutilier & Morehouse, general merchants, Centreville, N.S., are erecting a new 25 x 50 ft. store.

MALLEABLE IRON RANGES.

MALLEABLE iron is apparently coming to the front and perhaps it is destined to play a more important part than is anticipated by many cast iron manufacturers, particularly in the production of steel ranges and cook stoves. Of its advantages and disadvantages I have no opinion to express, but the fact is that range peddling concerns are working and have been working this thing for years. They make a deep impression on the farmer and his family when they drive up in the farm yard and give the range a fling out of their wagon and it hits the ground with that dull thud, but with a "never-touch-me" expression. This aerial flight, if attempted with a cast iron range, would be very apt to jar it somewhat seriously; but the trick rouses the wonder of the farmer's family when gaping at the caravan, and afterward, when they fire up the range, right out in the open, and coax a batch of dough out of the housewife, and in a jiffy turn out a lot of fine biscuit such as perhaps the rustics never saw before, it is very apt to fetch the wife and daughter, and the sale is made. Now, was it the fine castings, exquisite mountings, etc., of the stove that sold it, or the ocular demonstrations of its strength?

A number of concerns are making malleable iron ranges, and others are considering the matter. It would rather astonish stove dealers in the East to see stoves rolled over like barrels; yet this is done at one of the shipping stations of a maker of malleable iron ranges, as the railroad hands never bother to truck them, but roll them right along in any old way. If fine castings and superior workmanship are a positive necessity, as is so generally claimed, for the sale of a cast iron stove, would its ability, in addition to these qualifications, to stand a flight through the air as a malleable article does add to its salable qualities?—Metal Worker.

ALUMINUM PLATING.

A GERMAN engineer, Herr Sichelstiel, has reported to the Nuernberg section of the Verein Deutscher Ingenieure about the Warkwitz process of aluminum plating. If plates of aluminum, which melts at 700 degrees C., and of copper, melting point about 1,100 degrees C., are placed upon one another and heated, the aluminum will unite with the copper. But it will form a hard, brittle alloy, rich in aluminum. At lower temperatures the two metals will not unite. Yet Warkwitz proceeds somewhat as described, but his aluminum sheet is exceedingly thin, forming, when pressed on hard to avoid all oxidation of the copper, a copper alloy which contains very little of the other metal. This alloy welds under the rolls with other sheet aluminum, and thus the plating is accomplished. In the case of iron and steel the process is less simple; but, on the whole, the same method is applied for uniting any metals of widely differing melting points; for instance, for coppering zinc. Aluminum plated-copper and iron can be worked like white metal, and are, therefore, recommended for kitchen utensils. The lecturer also said that there would be no difficulty in coating aluminum wire 0.8 inch in diameter with 0.08 inch of copper, in order to get good conductors for alternating currents, for which we need only a well conducting outer copper shell. The following results were obtained with sheet iron, aluminum foil, and iron-plated on both sides with aluminum, the combined metals consisting of 70 parts by volume of iron with 15 parts on each side of aluminum. The figures are the average of six tests; the two strengths are expressed in tons per square inch, the elongations in percentages. The tests were conducted at the Bavarian Technical Museum:

	Iron.	Aluminum.	Plated iron.
Tensile strength....	21.14	9.14	11.87
Shearing strength...	17.21	5.14	7.36
Elongation.........	28.5	5.70	16.7

The test pieces were further passed through presses, and hollow semicylinders were shaped of them; cracks and peeling were not observed. It is claimed that aluminum plated zinc, prepared in this way, would be cheaper than when obtained galvanically.

TRANSIT IN BOND.

A United States Treasury Department circular, dated Washington, D.C., April 26, 1901, says:

Article 717 of the Customs regulation of 1899 provides that all merchandise entered and exported to British North America, when the transportation is made through the United States, whether by land or by water, unless conveyed in sealed cars, shall be corded and sealed. The requirement, heretofore enforced, that such merchandise, when destined to other portions of the Dominion of Canada than the Northwest possessions, shall be forwarded in cars specially appropriate for the purpose and secured with Customs fastenings, is hereby modified, and collectors are instructed, in instances where a sufficient number of packages of such merchandise to fill an entire car is not available, to allow such packages to be forwarded, by bonded routes, when properly corded and sealed, in cars not secured by the prescribed Customs fastenings. This ruling will apply to merchandise passing through the United States from places in the Dominion of Canada, for exportation via the seaboard, as well as to goods arriving at the seaboard and destined for places in said Dominion. In all other respects the existing regulations will continue to be observed.

HEATING AND PLUMBING

THE TORONTO PALACE HOTEL CONTRACT.

The contract for plumbing, heating, ventilating and lighting Toronto's palace hotel, which was awarded some time ago to The Bennett & Wright Co., Limited, is undoubtedly the largest contract of the kind ever given out in Canada. The plumbing alone will call for an immense amount of stock and take much time and thought to install. There will be 200 ordinary bathrooms and 50 shower-bath rooms. These will all be fitted in the most modern style, the former averaging about $100 each and the latter $150 each. They will be so placed that almost every suite of rooms will include either a shower or an ordinary bathroom. There will be 250 closets and 50 urinals; 340 washstands and 30 sinks. Another feature will be the Turkish bathroom, which will be most elaborately finished, comprising a complete plant with shower baths, urinals, etc. A complete laundry plant will also be installed, including four sets of laundry tubs.

Both electricity and gas will be used for lighting. There will be over 5,000 incandescent lights and 3,000 gas jets. Between 50 and 60 chandeliers are called for, these at a total cost of $15,000. Some of these, especially those in the rotunda and reception rooms, will be magnificent affairs. The electricity will also be used as motive power for the laundry.

The building will be heated by low-pressure steam heating, and will be ventilated by the Stutvesant system.

It will be readily seen that this hotel will not only be a large one, but will be one of the most magnificent institutions on the continent. A representative of one of the largest supply houses of the United States, who has examined the plans, informed "Hardware and Metal" this week that, outside of New York, Chicago and Boston, there is not a hotel that will surpass it, and that while the finest hotels in these places are both larger and more elaborate, none of them are better planned to insure not only comfort, but the most luxurious living to their guests than the building now being prepared for in Toronto.

A REMEDY FOR ROACHES.

From A. E. Y., Toronto, Ont.—I would thank you to publish a remedy that will banish water bugs or cockroaches from a kitchen sink and around hot water pipes. I have tried liquid poisons, etc., but they do not seriously reduce their numbers.

Answer.—If, after the places where the pests congregated have been washed with hot water in which considerable borax has been used, powdered borax is sprinkled freely over the haunts of the insects, they will seek new quarters. If one of the little blow guns used for distributing insect powder is used to blow borax powder into the cracks and crevices and around boiler pipes it will not take long to cause an entire evacuation.—Metal Worker.

PLUMBING AND HEATING CONTRACTS.

Mashinter & Co., Toronto, have the contract for heating and ventilating the new Manning Chambers, Queen street west; E. J. Lennox, architect.

Lessard & Harris, Montreal, are at work upon the plumbing of The Great Northwestern Telegraph Co.'s building, for which they have secured the contract.

The Bennett & Wright Co., Limited, have the contract for the plumbing, gas-fitting and draining an addition to a factory at 20 Temperance street for Dr. Ritchie; for plumbing, gas-fitting and draining a house for J. M. Bond, Glen road; a house for G. G. Roche, 167 John street; five houses on Cumberland street and three houses on Franklyn Place, and for plumbing, heating and gas-fitting in the Bank of Hamilton building, Orangeville, Ont..

BUILDING PERMITS ISSUED.

Building permits have been issued in Toronto to H. W. Wade for two three-storey brick stores, corner of Queen and Morse streets, to cost $8,000; to John Dill, for alterations to offices at 158 Bay street, $1,900; to Thomas Davies, for a two-storey brick factory, near Queen and River streets, $3,000; to R. Brown, for two two-storey dwellings, College street, near Brock avenue, $2,400; to J. M. Henderson, for a brick residence, at 296 Avenue road, $2,500; to Thos. Vallentyne for a pair of houses, 105 and 107 Delaware avenue, to cost $7,000; The T. Eaton Co., Limited, for a $3,500 residence at 340 Brunswick avenue; to Samuel Crane, for a $5,000 dwelling at the corner of Lowther avenue and Walmer road; to John Malloy for two houses at 76 and 78 Markham street, to cost $4,000; to J. M. Bond for a $5,000 residence near Dale avenue, on Glen road; to Geo. Roche, for a $3,000 residence at 167 John street; to Dr. Ritchie for an addition to factory at 20 Temperance street, to cost $2,500.

During April building permits to the value of $18,282 were taken out in Brantford. Among the permits were the following:

City of Brantford, Marlborough street, erection of Technical School, to cost $2,240; W. H. Inglis, Chatham street, erection of brick residence, to cost $1,800; J. F. Simmons, Colborne street, erection of three brick stores to cost $1,500; Lloyd Harris, Brant avenue, erection of stone stable, to cost $1,400; Lloyd Harris, Brant avenue, erection of a stone cottage, to cost $900; Evangelist Booth, Darling street, the erection of Salvation Army Barracks, to cost $3,000; Miss Sadie Vansickle, erection of brick residence on Chatham street, to cost $1,300.

Building permits have been issued in Ottawa to Shirley Ogilvy, solid brick veneered residence, Somerset street, $8,000; William Farmer, brick veneered store and dwelling, Wellington street, $1,600; Ellen Joanies, brick veneered house, Brond street, $1,100.

PLUMBING AND HEATING NOTES.

A by-law was carried in Parry Sound on Monday in favor of municipal purchase of the local electric light plant.

An expert is valuing the Almonte electric light plant for the council of that town, which proposes to purchase the plant.

Wm. Delahey is erecting a three-storey business block on Dominion avenue, Phoenix, B.C. New buildings are also being erected for The Phoenix Pioneer and The Hunter-Kendrick Company.

SOME BUILDING NOTES.

Wm. Webster is building a house in Melcombe, Ont..

A. Barker intends erecting a house in Glencairn, Man.

A $16,000 opera house is to be erected in Kingston, Ont..

James Collins intends erecting a house in Hintonburg, Ont.

A new Presbyterian church is being built in Manitou, Man.

John Mills is erecting a brick building on Crooks street, Hamilton, Ont.

Contractor Purcell is erecting a house for F. Friday, in Port Arthur, Ont.

S. E. Burton intends erecting a three-storey frame hotel at Kamloops, B.C.

Moses Edey, architect, Ottawa, is designing a new school house for Quyon, Que.

Sterns Bros., Souris, P.E.I., are asking tenders for a new brick store in that place.

John Vance and Isaac Rink intend erecting residences in Tavistock, Ont., this summer.

Wm. Warcup is erecting a two-storey residence on his farm, near Granton Corner, Sherbrooke county, Que.

Architect Graham, Sault Ste. Marie, has prepared plans for a $15,000 residence for D. M. Gregor; and for a $2,000 residence for Mr. McAllister, of that place.

Among the buildings to be erected in Petrolea, Ont., this summer will be a hotel for Ed. Fletcher, a station for the G.T.R., and a factory for The Petrolea Packing Co.

Architect Russell, Winnipeg, is calling for tenders for new dwellings for J. A. M. Aikins, Longside avenue; for Henry Landerkin, Ellen avenue; for Lisgar Laing, on Roslyn road, and for James McKay, on Colony street.

A BIG CONTRACT LET.

The Bank of Montreal has awarded the contract for erecting its new building at Montreal, which is to cost nearly a million dollars and will be one of the finest buildings of the kind, to Norcross Bros., Worcester, Mass., whose tender was considerably below that of the Canadian contractors. The contract comprises the stone, brick, marble, foundations, roofing, painting, glazing, elevators, and plastering, while the contracts for such work as plumbing, heating, electric wiring, fittings, etc., have yet to be awarded. McKim, Mead & White, of New York, and Mr. Andrew T. Taylor, of Montreal, are jointly associated as the architects of the new bank building.

CANADIAN PIG IRON IN ENGLAND.

We mentioned in these columns last week that Canadian pig iron was about to be shipped to this country, and are now able to give a little information on the subject. The brand will be known by the letters "DISC," which plainly indicate the name of the manufacturing company. The iron, which, we understand, is identical with that of Cleveland, will be in Nos. 1, 2, and 3 foundry and basic, and shipments to this country are likely to be large; in fact, six figures have been suggested as a minimum. The effect of these importations on the value of our manufactures here will be watched with interest.—Iron and Steel Trades Journal, London.

MANITOBA MARKETS.

WINNIPEG, May 4' 1901.

THE market here is active in all lines, but without special features of any kind. Among the changes of the week is a decline in stovepipes of 50c. per 100 lengths, and a drop in shot amounting to 40c. There are likely to be more changes during the coming week.

Quotations for the week are as follows :

Barhed wire, 100 lb	$3 45
Plain twist	3 45
Staples	3 95
Oiled annealed wire 10	3 95
" 11	4 00
" 12	4 05
" 13	4 20
" 14	4 35
" 15	4 45
Wire nails, 30 to 60 dy, keg	3 45
" 16 and 20	3 50
" 10	3 55
" 8	3 65
" 6	3 70
" 4	3 85
" 3	4 10
Cut nails, 30 to 60 dy.	3 00
" 20 to 40	3 05
" 10 to 16	3 10
" 8	3 15
" 6	3 20
" 4	3 30
" 3	3 65
Horsenails, 45 per cent. discount.	
Horseshoes, iron, No. 0 to No 1	4 65
No. 2 and larger	4 40
Snow shoes, No. 0 to No. 1	4 90
No. 2 and larger	4 40
Steel, No. 0 to No. 1	4 95
No. 2 and larger	4 70
Bar iron, $2.50 basis.	
Swedish iron, $4.50 basis.	
Sleigh shoe steel	3 00
Spring steel	3 25
Machinery steel	3 75
Tool steel, Black Diamond, 100 lb	8 50
Jessop	13 00
Sheet iron, black, 10 to 20 gauge, 100 lb.	3 50
20 to 26 gauge	3 75
28 gauge	4 00
Galvanized American, 16 gauge	2 54
18 to 22 gauge	4 50
24 gauge	4 75
26 gauge	5 00
28 gauge	5 25
Genuine Russian, lb	12
Imitation " "	8
Tinned, 24 gauge, 100 lb	7 55
26 gauge	8 80
28 gauge	8 00
Tinplate, IC charcoal, 20 x 28, box	10 75
" IX "	12 75
" IXX "	14 75
Ingot tin	33
Canada plate, 18 x 21 and 18 x 24	3 75
Sheet zinc, cask lots, 100 lb	7 50
Broken lots	8 00
Pig lead, 100 lb	6 00
Wrought pipe, black up to 2 inch	50 an 10 p.c.
" Over 2 inch	50 p.c.
Rope, sisal, 7-16 and larger	$10 00
" ⅜	10 50
" ¼ and 5-16	11 00
Manila, 7-16 and larger	13 50
" ⅜	14 00
" ¼ and 5-16	14 50
Solder	21½
Cotton Rope, all sizes, lb	16
Axes, chopping	$ 7 50 to 12 00
" double bits	12 00 to 18 00
Screws, flat head, iron, bright	87½
Round " "	82½
Flat " brass	80
Round " "	75
Coach	57½ p.c.
Bolts, carriage	55 p.c.
Machine	55 p.c.
Tire	60 p.c
Sleigh shoe	65 p.c.
Plough	40 p.c.
Rivets, iron	50 p.c.
Copper, No. 8	35
Spades and shovels	40 p.c.
Harvest tools	50, and 10 p.c.
Axe handles, turned, s. g. hickory, doz.	$2 50
No. 1	1 50
No. 2	1 25
Octagon extra	1 75
No. 1	1 25
Files common	70, and 10 p.c.
Diamond	60
Ammunition, cartridges, Dominion R.F.	50 p.c.
Dominion, C.F., pistol	30 p.c.
" military	15 p.c.
American R.F.	30 p.c.
C.F. pistol	5 p.c.
C.F. military	10 p.c. advance.
Loaded shells :	
Eley's soft, 12 gauge black	16 50
" chilled, 12 guage	18 00
soft, 10 guage	21 00
chilled, 10 guage	23 00
Shot, Ordinary, per 100 lb	6 25
Chilled	7 15
Powder, F.F., keg	4 75
F.F.G.	5 00
Tinware, pressed, retinned	75 and 2½ p.c
" " plain	70 and 15 p.c.
Graniteware, according to quality	50 p.c.

PETROLEUM.

Water white American	25½c.
Prime white American	24c.
Water white Canadian	22c.
Prime white Canadian	21c

PAINTS, OILS AND GLASS

Turpentine, pure, in barrels	$ 63
Less than barrel lots	71
Linseed oil, raw	70
Boiled	73
Lubricating oils, Eldorado castor	25½
Eldorado engine	24½
Atlantic red	27½
Renown engine	41
Black oil	23½ to 25
Cylinder oil (according to grade)	55 to 74
Harness oil	61
Neatsfoot oil	$ 1 00
Steam refined oil	85
Sperm oil	1 50
Castor oil per lb.	11½
Glass, single glass, first break, 16 to 25 united inches	2 25
26 to 40 per 50 ft.	2 50
41 to 50 " 100 ft.	5 50
51 to 60 " " "	6 00
61 to 70 per 100-ft. boxes	6 50
Putty, in bladders, barrel lots per lb.	2½
kegs "	2¾
White lead, pure per cwt.	7 25
No 1 "	7 00
Prepared paints, pure liquid colors, according to shade and color. per gal.	$1.30 to $1.90

BUSINESS AND PLEASURE.

Mr. Alexander Gardner, of the firm of Gardner & Co. the big hardware dealers of Woodstock, Ont., was in Montreal this week for a few days. He was accompanied by his wife. On Wednesday evening they left on a pleasure trip to Quebec and points in the east. He reports business in Western Ontario to be very good. Mr. Gardner is a brother of Mr. James Gardner, so long and favorably connected with Wood, Vallance & Co., of Hamilton.

FIRST DELIVERY.

The Canada Hardware Co., Montreal, have this week made their first delivery of supplies on account of the 52-car Government contract for the Maissonneuve wharf awarded them last fall. The first shipment consisted of six carloads.

THE NEW WORKS IN OPERATION.

The Acme Can Works, Montreal, are now comfortably settled in their new quarters on Ontario street and in full operation. The management would ask their customers to exercise a little forbearance toward them on account of the slow delivery of goods during the removal period of the last month. They now guarantee to fill all orders entrusted to them satisfactorily and promptly.

RULES FOR THE HARDWARE STORE.

THE rules which are given below have been in force for several years in a Boston hardware store, and were printed in last week's issue of The Iron Age:

REMINDER TO OUR CLERKS.

The condition of trade shows us that we are making no money, and that our expenses are nearly as much as when trade was good and a better margin of profits was to be obtained. This being the case, if we are to continue in business and keep our present force of help, it will be absolutely necessary that each one see to it how much can be gained.

There are different ways of doing this, but some of the most important which will have to be kept in mind are, to remove dead or unsalable stock, and that goods may not be in this condition, sell the worst of any particular kind of goods first.

Bronze goods may be unpapered; use these, adding screws if necessary, and in this way much of what otherwise would be dead stock will be sold.

Another way is to get as much profit as possible from each sale. Do not discount 2 or 3c. on each sale to make "even money," as you can readily see that if each employe does this on two or three sales a day a considerable sum is lost.

Do not charge goods at dozen prices except when necessary to regular customers, and then in not less than half-dozens. See Rule No. 2. In getting the price for each, a cent or two can be added to the dozen price.

Another good and more important way is to make a permanent customer out of each one to whom a sale, no matter how small, is made. This can easily be done by politeness and care for the customer's interest, and inquiring as to what may be his possible needs. It should be the ambition of each not simply to sell the articles called for, but something in addition, not, of course, by any "Jew" method.

Other ways are to be constantly on the lookout against waste or extravagance in the little details of business. Promptness in attending to a customers, in filling orders, and in answering any demand made upon you, is more than ever necessary at this time, and you will observe that this is one of the essential qualifications for successful business.

A considerable proportion of loss sustained has been through bad accounts. If each salesman will feel a responsibility for his customers, and watch the accounts, he can very often collect them better than any-one else. Sharp collections will help out on the profits of the future.

We hope each one will use every effort to excel in these respects, and in others which have not been mentioned. Rules have been changed, and we wish each employe to read them until familiar with them.

STORE RULES.

Take every order on order book, or, if not at hand, on pocket order book. First put down complete address, including name of "job." Carry out numbers and prices. If by written order, enter before filling; mark with initials and place on file. In no case enter on pass book or give bill or memo. before charging. Keep pocket order book in store coat pocket and ready for instant use.

Price up charges at time of sale. This must be done before charges are copied. Extend prices by dozens, if half dozens or more. If less, at so much each. See that prices correspond with quantity—i. e., do not put down gross prices where charged by dozen. When more than one item is sold, either cash or charge, always make memo. as goods are laid out; then compare item with memo.

Have each lot of goods, including nails and paper, looked over by one competent. Fill out slips properly and inclose and make note on order book of person calling back.

All goods leaving the store must be receipted for or name of person to whom delivered put with entry of goods.

Make no private memo. charges on pocket order book, or anywhere, except in regular order of day's business. If for good reasons it does not need to be entered on journal, mark "don't copy," and carry any such memo. and balance of orders to first working day of each month. Then going backward, scanning each page, mark top "C" when clear. Nothing but balances of orders to be carried over to the following month without being charged.

Make no new accounts until blank is filled out and authorized by one of the firm, and when done so state with charge, with number of application for account. All accounts made in any other manner will be at the risk of salesman who made the sale and who will be expected to settle for same.

All credits of cash must be made in office by bookkeeper, cashier, or one of the firm, and credits of merchandise must be examined or rechecked by one of the firm. All goods returned must be entered on the pocket order book slip with name of persons returning the same.

Call back each morning, or when requested to do so, and make a "C" around journal folio when called, or put down folio page.

Examine invoices each morning, and compare prices and quantity with sale. If correct, check and put down initials on invoice.

No new tool can be taken from stock to be used except by permission of one of the firm. Those in use about store must be marked and kept in their proper places. Any store tool loaned must be charged to parties taking them and credited if returned in good order.

Each salesman can have for store use one rule, No. 36½, which must be marked in the slide with salesman's name.

Pencils hereafter will be kept in the office and given out by the stenographer. No other article can be taken for store use without permission of one of the firm.

RULES FOR MANAGEMENT OF STOCK IN INDIVIDUAL SECTIONS.

Each employe shall be responsible for the condition of his section, as by observing care and uniformity the stock can be kept clean, in good order and well fronted up.

All goods received should be properly marked and sampled. Sample box should be marked with section and shelf number and the location of overstock, if any.

All goods must be evenly arranged in order from left to right by number, size or grade. Largest packages should be placed underneath, broken packages on top; nothing put behind but overstock of what is directly in front.

Fill orders from overstock in the back, using broken packages and shop-worn goods first.

Unsalable stock must be put in good order to push sales. Memoranda must be taken of shortages in stock and properly reported.

These instructions are to be faithfully followed.

CORROSION OF STEEL AND IRON.

F. H. Williams, of Pennsylvania, has recently, according to an exchange, published the results of some corrosion experiments made with pieces of soft steel containing respectively 0.078, 0.145 and 0.263 per cent. of copper. The steel without copper was found to lose 1.85 per cent. of its weight under the conditions of the experiment, while the pieces containing copper lost only 0.89, 0.75 and 0.74 per cent. in weight. Some experiments with wrought iron showed that the addition of 0.393 per cent. of copper to the iron materially reduced its rate of corrosion.

THE LITTLE MACHINE SHOP.

IF anyone has the idea that this is exclusively the age for vast enterprises and that there is no longer any chance for the small man, a talk with some machinery men will at least partially eradicate that impression. Thus does Iron and Steel corroborate the statistics on accretions to the business world which we publish from time to time. In support of its statement our technical contemporary cites the sales of equipment by machine-tool men to a number of new little shops throughout the country. This growth is particularly true of the west. Here is a description of the genesis of the little machine shop: "At country crossroads enterprising young business men start a general store or a blacksmith shop, a cluster of houses appear, and the hamlet is created. The hamlet becomes a village and the village a town. At thousands of points in territory tributary to Chicago that process of growth is in operation. Finally comes the time in the stage of growth when a machine shop becomes necessary; most likely repair work is the greatest need. In small, as in large affairs, when the man is needed he usually appears. An ambitious young machinist hears of a good opening. He has saved from his wages a few hundred dollars. Perhaps he has a chum or two who have been equally prudent. They pool their issues and start a little repair shop, confident they can make wages, hopeful that in time they can do even a little better. A new industry is thus created. What it is to become depends upon nothing else than the personalities behind it. To occupy time when repair work is slack the manufacture of some specialty is begun. Expenses are light, and the article can be made at little cost with modern facilities. The second step on the road to success has been taken. Thus have been humbly born and self-nurtured many of the industrial plants of to-day. Thus are there now in the earliest stages of apparent weakness and doubt many enterprises that cannot be prevented from developing into strength and magnitude. Location cuts no figure. Industries grow in the byways of trade, then make lasting connection with the world of commerce. The country has lately been putting much currency into circulation. It travels fast, and is generally useful wherever it goes. The modest machine shop, potentially great, is largely the product of these benign conditions. There is a broadcast industrial sowing to-day, so say some of the machine men, and to-morrow, or a week hence, the crop may be plenteous, a delight for impartial contemplation."

BENDING EDGES OF SHEETS.

A simple little device for bending the edges of sheet metal has been the subject of a recent patent issued to two Frenchmen, Messrs. Gaston and Clovis Delachartre, of Paris.

It consists of three strips of rules, hinged to each other, one of the outer rules forming a base whereby it may be secured to a bench or table, the other having its outer edge sharpened and protected by a strip of metal. The intermediate section is provided with upstanding pins, whith are arranged in longitudinal alignment, and fit in sockets arranged in the co-acting face of the outer hinge section or rule. These two sections carry operating handles.

The sheet metal is placed across the base section with its edge resting against the pins of the intermediate section, which, therefore, forms stops for the purpose. The outer hinge section is brought down upon the intermediate section, whereby the edge of the metal is clamped between the two. The two sections are then folded simultaneously upon the base, crimping the edge of the metal. To complete the operation the edge is released from between the two movable sections, and the outer hinge section is brought down upon the base, bending the edge of the metal and forcing it tightly down upon the body of the same.

DO NOT PATRONIZE CUSTOMERS.

There is nothing so ridiculous and so easily discernable in a merchant as an air of patronage towards customers, says a Western exchange. It is a species of vanity which is as ludicrous as it is disagreeable. This weakness is called bumptiousness, and is repelling to the general run of customers, who are quick to distinguish it from a pleasant presence. A pleasant presence in the store is one of the chief essentials towards the success of a merchant. This essential of a pleasant presence is made up of simplicity. Just that and nothing else. Simplicity is the most charming of all qualities and is and always has been possessed by the men and women that the world deems great. The simple man is natural and is possessed of a suavity which is real. Assumed suavity is generally made up of bumptiousness and is as different from the inherent quality as the sweet violet is different from the violet of the millinery counter. The bumptious merchant in his vanity reckons himself somewhat of a philanthropist. His behaviour towards his customers is an offence to those amongst them who are of a keen or sensitive nature. In his overweening vanity he reckons himself the patron and the customer the beneficiary. Every action of his conveys that expression. When he is sympathetic he is condescendingly so; his heartiness is luring and often vulgar; his insincerity is apparant

COST OF DOING BUSINESS.

Estimates of the proper cost of transacting a retail general business are apt to range between 15 and 25 per cent. of the gross turnover, although it is considered by some persons that the higher figures are exaggerated, says Merchants' Review, New York. From 15 to 20 per cent. is the usual estimate. It would seem that when a general merchant finds his expenses running above the 20 per cent. mark it is time for him to look for the leaks. The "expenses" are supposed to include rent, clerk hire, light, heating and motive power, advertising, delivery service and general store expenses.

Most merchants try and keep expenses down to the lowest notch, and when they find the cost is rising, when compared with the gross turnover, they push sales with greater vigor than usual. That is the best way of keeping the cost down, and better results can be obtained in that way than by going over the clerk's salary account or advertising expenses and clipping off a little here and a little there.

A fine roomy store, a good and efficient staff of assistants, and all the other facilities in modern storekeeping should be employed, wherever they can be afforded, the market prices being paid, and then the merchant's calibre will be shown by the amount of business he is able to do with the facilities employed. The better the merchant, the more skilful and energetic the more business he will do with a specified staff, storeroom, etc., and the lower will be his perchntage of cost.

In old business, where dry rot has begun to appear, searching for opportunities to cut down expenses will often reveal plenty of them, but it is questionable whether it is not largely a waste of time to attempt that sort of thing when the business is young and so much remains to be done in respect of developing the selling end of the business.

CURRENT MARKET QUOTATIONS.

May 10, 1901.

These prices are for such qualities and quantities as are usually ordered by retail dealers on the usual terms of credit, the lowest figures being for larger quantities and prompt pay. Large cash buyers can frequently make purchases at better prices. The Editor is anxious to be informed at once of any apparent errors in this list as the desire is to make it perfectly accurate.

METALS.

Tin.

Lamb and Flag and Straits—56 and 28 lb. ingots, per lb.	0 31	0 32

Tinplates.

Charcoal Plates—Bright

M.L.S., equal to Bradley.		Per box
I.C., usual sizes	....	$6 75
I.X., "	....	8 25
I.X.X., "	....	9 75
Famous—		
I.C.	....	6 75
I.X.	....	8 25
I.X.X.	....	9 50
Raven & Vulture Grades—		
I.C., usual sizes	....	5 00
I.X., "	....	6 00
I.X.X. "	....	7 00
I.X.X.X., "	....	8 00
D.C., 12½x17	....	4 75
D.X.	....	5 80
D.X.X	....	7 50

Coke Plates—Bright

Bessemer Steel—		
I.C., usual sizes	....	4 00
I.C., special sizes, base	....	4 50
20 x 28	....	8 60

Charcoal Plates—Terne

Dean or J. G. Grade—		
I.C., 20x28, 112 sheets	....	8 25
I.X., Terne Tin	....	10 25

Charcoal Tin Boiler Plates.

Cookley Grade—		Per lb.
X X., 14x56, 50 sheet bxs; " 14x60, "; " 14x65, "	0 07	0 07½

Tinned Sheets

72x30 up to 24 gauge	0 08	0 08½
" 26 "	0 08½	0 09
" 28 "	0 09	0 09½

Iron and Steel.

Common Bar, per 100 lbs.	1 85	1 90
Refined " "	....	2 35
Horse Shoe Iron "	....	2 25
Hoop steel, 1½ to 3 in. base, extras for smaller sizes	3 10	3 25
Sleigh Shoe Steel " base	....	2 10
Tire Steel	2 30	2 50
Reeled Machinery	....	3 00
Cast Steel, per lb	0 00	0 00
Toe Calk Steel	....	....
Firth & Co's special cast steel, per lb.		0 13
Jessop's Tool Steel	....	0 13

Boiler Tubes.

1½-inch	....	0 12½
2 "	....	0 13
2½ "	....	0 15
3 "	....	0 16
3½ "	....	0 20
4 "	....	0 25

Steel Boiler Plate.

¼ inch	2 50	2 60
3-16 inch	2 60	2 70
⅜ inch and thicker	2 50	2 60

Black Sheets.

18 gauge	....	3 00
20 gauge	....	3 00
22 to 24 "	....	3 10
26 "	....	3 20
28 "	....	3 30

Canada Plates.

Al dull, 52 sheets	....	3 00
Half polished	....	3 15
All bright	3 65	3 75

Iron Pipe.

Black pipe—		
⅛ "	....	4 45
¼ inch	....	3 25
⅜ "	....	3 30
½ "	....	3 50
¾ "	....	3 65
1 "	....	5 15
1¼ "	....	7 00
1½ "	....	8 40
2 "	....	11 95
2½ "	....	20 95
3 "	....	21 55
3½ "	....	31 75
4 "	....	39 00
5 "	....	47 35
6 "	....	63 10
Galvanized pipe—		
½ inch	....	4 90
¾ "	....	6 25
1 "	....	7 85
1¼ "	....	10 30
1½ "	....	12 35
2 "	....	16 50

Galvanized Sheets.

	G C.	Comet.	Amer.	Queen's Head.
16 gauge		...	4 5	3 90
18 to 24 gauge	4 15	3 85	4 35	4 15
26 "	4 40	4 10	4 35	4 40
28 "	4 65	4 55	4 60	4 65

28 gauge American equals 26 gauge English.

Chain.

Proof Coil, 3-16 in., per 100 lb	....	....
" ¼ " "	8 30	8 50
" 5-16 " "	4 70	5 00
" ⅜ " "	4 05	4 20
" 7-16 " "	3 90	4 25
" ½ " "	3 70	4 10
" 9-16 " "	3 65	4 15
" ⅝ " "	3 35	3 90
" ¾ " "	3 60	4 10

Halter, kennel and post chains, 40 to 40 and 5 p.c.

Cow ties	40 p.c.
Tie-out chains	65 p.c.
Stall fixtures	35 p.c.
Trace chain	45 p.c.

Jack chain, iron, single, and double, discount 35 p.c.

Jack chain, brass, single and double, discount 40 p.c.

Copper.

Ingot

English B. S., ton lots	....	0 19
Lake Superior	....	....

Bolt or Bar.

Cut lengths round, ½ to ⅞ in.	0 23½	0 25
" round and square 1 to 2 inches	0 23½	0 25

Sheet.

Plain, 14 oz., and light, 16 oz., 14x48 and 14x60	0 23	0 23
Plain, 14 oz., and light, 16 oz., irregular sizes	0 23	0 23½
Tinned copper sheets	....	0 26
Planished	....	0 32

Braziers (in sheets.)

4x6ft. 25 to 30 lbs. ea., per lb.	....	0 25½
" 35 to 45 " "	....	0 24½
" 50-lb. and above, "	....	0 23½

Boiler and T. K. Pitts

Plain Tinned, per lb	....	0 28
Spun, per lb.	....	0 32

Brass.

Rod and Sheet, 14 to 30 gauge, 15 per cent.

Sheets, hard-rolled, 2x4	0 24	0 25
Tubing, base, per lb.	0 24	0 25

Zinc Spelter

Foreign, per lb	0 05½	0 06
Domestic "	....	..

Zinc Sheet.

5 cwt. casks	....	00 6¼
Part casks	....	00 6¾

Lead.

Imported Pig, per lb	0 04½	0 04½
Bar, 1 lb.	0 05½	0 05¾
Sheets, 2½ lbs. sq. ft., by	....	0 06¾
Sheets, 3 to 6 lbs., "	....	0 06

NOTE.—Cut sheets ½ cent per lb. extra. Pipe, by the roll, usual weights per yard, lists at 7c. per lb. and 25 p.c. dis. f.o.b. Toronto.

NOTE.—Cut lengths, net price, waste pipe 8-ft. lengths lists at 7½ cents.

Shot.

Common, $6.50 per 100 lb.; chilled, $7.00 per 100 lb.; buck, seal and ball, $7.50. Discount, 15 p.c. Prices are f.o.b. Toronto, Hamilton, Montreal, St. John and Halifax. Terms 3 per cent. cash, freights equalized.

Soil Pipe and Fittings.

Discount, 60 and 10 per cent. on medium and extra heavy, and 60 per cent. on light.

Solder.

	Per lb.	Per lb.
Bar half-and-half, guarant'd	....	0 18½
Bar half-and-half, commer'l	....	0 18
Refined	....	0 17½
Wiping	....	0 17

NOTE.—Prices of this graded according to quantity. The prices of other qualities of solder in the market indicated by private brands vary according to composition.

Antimony.

Cookson's, per lb.	0 10½	0 11

White Lead.

	Per 100 lb.
Pure	6 37
No. 1 do	6 00
No. 2 do	5 62½
No. 3 do	5 25
No. 4 do	4 87½
Munro's Select Flake White	7 37½
Elephant and Decorators' Pure	7 12½
Brandram's B B. Genuine	8 00
" " Decorative	7 55
" " No. 1	6 85
" " No. 2	6 00

Red Lead.

Genuine, 560 lb. casks, per cwt.	....	$5 50
Genuine, 100 lb. kegs, per cwt.	....	5 75
No. 1, 560 lb. casks, per cwt.	....	5 25
No. 1, 100 lb. kegs, per cwt	....	5 00

White Zinc Paint.

Elephant Snow White	0 08	0 09
Pure White Zinc	0 08	0 09
No. 1	0 06	0 07½
No. 2	0 05	0 06½

Dry White Lead.

Pure, casks	5 75
Pure, kegs	6 25
No. 1, casks	5 50
No. 1, kegs	5 00

Prepared Paints.

In ¼, ½ and 1 gallon tins.

Pure, per gallon	....	1 25
Second qualities, per gallon	....	1 10
Barn (in bbls.)	0 75	0 85
The Sherwin-Williams Paints	....	1 45
Canada Paint Co's Pure	....	1 25
Toronto Lead & Color Co's Pure	....	1 25
Sanderson Pearcy's Pure	....	1 20
Stewart & Wood's Champion Pure		1 20

Colors in Oil.

25 lb. tins, Standard Quality.

Venetian Red, per lb.	0 05
Chrome Yellow	0 11
Golden Ochre	0 06
French "	0 05
Marine Black	0 09
" Green	0 09
Chrome "	0 08
French Imperial Green	0 09

Colors, Dry.

Yellow Ochre (J.C.) bbls.	1 35	1 40
Yellow Ochre (J.F.L.S.), bbls	...	2 75
Yellow Ochre (Royal)	1 10	1 15
Brussels Ochre	....	2 00
Venetian Red (best), per cwt.	1 80	1 90
English Oxides, per cwt.	3 00	3 25
American Oxides, per cwt.	1 75	2 00
Canadian Oxides, per cwt.	1 75	2 00
Super Magnetic Oxides, 93p.c.	2 00	2 25
Burnt Sienna, pure, per lb.	....	0 10
" Umber, " "	....	0 10
do Raw	....	0 09
Drop Black, pure	....	0 09
Chrome Yellows, pure	....	0 18
Chrome Greens, pure, per lb.	....	0 12
Golden Ochre	....	0 03½
Ultramarine Blue in 28-lb. boxes, per lb.	0 08	0 24
Fire Proof Mineral, per 100 lb	....	1 00
Genuine Eng. Litharge, per lb	....	0 07
Mortar Color, per 100 lb.	....	1 25
English Vermillion	....	0 80
Pure Indian Red, No. 45 lb.	....	0 80
Whiting, per 100 lb.	....	0 55

Blue Stone.

Casks, for spraying, per lb.	0 07
100-lb. lots, do. per lb.	0 08

Putty.

Bulk in bbls.	1 90
Bulk in less quantity	2 05
Bladders in bbls.	2 10
Bladders in kegs, boxes or loose	2 25
Bladders in 25-lb. tins	2 35
Bladders in 12½-lb. tins	2 65
Bladders in bulk or tins less than 100 lb	3 00

Varnishes.

	In 5-gal. lots.)	Per gal.
Carriage, No. 1	2 90	3 30
" body	8 00	9 00
" rubbing	4 00	5 00
Gold Size, Japan	3 00	3 40
Brown Japan	2 40	2 80
Elastic Oak	2 90	3 30
Furniture, extra	2 40	2 80
" No. 1	1 60	2 00
Hard Oil Finish	2 70	3 10
Light Oil Finish	3 20	3 60
Demar	3 30	3 70
Shellac, white	4 40	4 80
" orange	4 00	4 40
Furniture Brown Japan	1 60	2 00
Black Japan	2 40	2 80
" " No. 1	1 60	2 00

The Imperial Varnish & Color Co's, Limited Elastilite Varnish 1 gal. can, each, $3.00.

Granatine Floor Finish, per gal., $2.75.

Maple Leaf Coach Enamels; Size 1, $1 20; Size 2, 70c.; Size 3, 40c. each.

Castor Oil.

East India, in cases, per lb.	0 10	0 10½
" " small lots	0 10½	0 11

Cod Oil, Etc.

Cod Oil per gal.	0 50	0 55
Pure Olive	....	1 20
" Neatsfoot	....	90

Glue.

Common	0 08½	0 09
French Medal	0 14	0 14½
Cabinet sheet	0 12	0 13
White, extra	0 18	0 20
Gelatine	0 22	0 30
Strip	0 18	0 20
Coopers	0 19	0 20
Huttner	....	0 18

HARDWARE.

Ammunition.

Cartridges.

B. B Cap Dom. 50 and 5 per cent.
Rim Fire Pistol, dis. 40 p. c., Amer.
Rim Fire Cartridges, Dom., 50 and 5 p. c.
Central Fire Pistol and Rifle, 10 p.c. Amer.
Central Fire Cartridges, pistol sizes, Dom. 30 per cent.
Central Fire Cartridges, Sporting and Military, Dom., 15 and 5 per cent.
Central Fire, Military and Sporting, Amer, add 5 p.c. to list. B.B. Caps, discount 40 per cent, Amer.
Loaded and empty Shells, "Trap" and "Dominion" grades, 25 per cent. Rival and Nitro, net list.
Brass Shot Shells, 55 per cent.
Primers, Dom., 30 per cent.

Wads — per lb.

Best thick white felt wadding, in ¾-lb bags 1 00
Best thick brown or grey felt wads, in ¼-lb. bags 70
Best thick white card wads, in boxes of 500 each, 12 and smaller gauges 0 99
Best thick white card wads, in boxes of 500 each, 10 gauge 0 35
Best thick white card wads, in boxes of 500 each, 8 gauge 0 55
Thin card wads, in boxes of 1,000 each, 12 and smaller gauges 0 20
Thin card wads, in boxes of 1,000 each, 10 gauge 0 25
Thin card wads in boxes of 1,000 each, 8 gauge ....
Chemically prepared black edge grey cloth wads, in boxes of 250 each— Per M.
11 and smaller gauge 0 60
9 and 10 gauges 0 70
7 and 8 gauges 0 90
5 and 6 gauges 1 10
Superior chemically prepared pink edge, best white cloth wads, in boxes of 250 each—
11 and smaller gauge 1 15
9 and 10 gauges 1 40
7 and 8 gauges 1 65
5 and 6 gauges 1 90

Adzes.

Discount, 20 per cent.

Anvils.

Per lb 0 10 0 12½
Anvil and Vise combined 4 50
Wilkinson & Co.'s Anvils..lb. 0 09 0 09½
Wilkinson & Co.'s Vices..lb. 0 09¾ 0 10

Augers.

Gilmour's, discount 65 and 5 p.c. off list.

Axes.

Chopping Axes
Single bit, per doz 5 50 10 00
Double bit, " 11 00 18 00
Bench Axes, 40 p.c.
Broad Axes, 33⅓ per cent.
Hunters' Axes 5 50 6 00
Boy's Axes 5 75 6 75
Splitting Axes 6 50 12 00
Handled Axes 7 00 10 00

Axle Grease.

Ordinary, per gross 5 75 6 00
Best quality 13 00 15 00

Bath Tubs.

Zinc 6 00
Copper, discount 15 p.c. off revised list

Baths.

Standard Enameled.

5½-inch rolled rim, 1st quality 25 00
" " " 2nd " 21 00

Anti-Friction Metal.

"Tandem" A .... per lb. 0 27
" B .... " 0 21
" C .... " 0 11½
Magnolia Anti-Friction Metal, per lb. 0 25
Frictionless metal, per lb 0 23

SYRACUSE SMELTING WORKS.

Aluminum, genuine 0 45
Dynamo 0 29
Special 0 25
Aluminum, 99 p.c. pure "Syracuse" .. 0 50

Bells.

Hand.

Brass, 60 per cent.
Nickel, 55 per cent.

Cow.

American make, discount 66⅔ per cent.
Canadian, discount 45 and 50 per cent.

Door.

Gongs, Sargant's 5 50 8 00
" Peterboro', discount 45 per cent.

Farm.

American, each 1 25 3 00

House.

American, per lb 0 35 0 40

Bellows.

Hand, per doz 3 35 4 75
Moulders', per doz 7 50 10 00
Blacksmiths', discount 40 per cent.

Belting.

Extra, 60 per cent.
Standard, 60 and 10 per cent.

Bits.

Auger.

Gilmour's, discount 60 and 5 per cent.
Rockford, 50 and 10 per cent.
Jennings' Gen., net list.

Car.

Gilmour's, 47½ to 50 per cent.

Expansive.

Clark's, 40 per cent

Gimlet.

Clark's, per doz 0 65 0 90
Damond, Shell, per doz 1 00 1 50
Nail and Spike, per gross 2 25 5 20

Blind and Bed Staples.

All sizes, per lb 0 07½ 0 12

Bolts and Nuts. Per cent.

Carriage Bolts, full square, Norway 65
" " full square 65
Common Carriage Bolts, all sizes 60
Machine Bolts, all sizes 60
Coach Screws 70
Sleigh Shoe Bolts 72½
Blank Bolts 60
Bolt Ends 62½
Plough Bolts 60
Nuts, square .... 4 c. off
Nuts, hexagon .... 4¼c. off
Tire Bolts 67½
Stove Bolts 67½
Stove rods, per lb .... 5½ to 6c.

Boot Calks.

Small and medium, ball, per M 4 25
Small heel, per M 4 50

Bright Wire Goods.

Discount .... 62½ per cent.

Broilers.

Light, dis., 65 to 67½ per cent.
Reversible, dis., 65 to 67½ per cent.
Vegetable, per doz., dis. 37½ per cent.
Henis, No. 8, " 6
Henis, No. 9, " 7 00
Queen City " 7 50 0 00

Butchers' Cleavers.

German, per doz 6 00 11 00
American, per doz 12 00 20 00

Building Paper, Etc.

Plain building, per roll 0
Tarred lining, per roll 0
Tarred roofing, per 100 lb 1
Coal Tar, per barrel 3
Pitch, per 100-lb 0 30
Carpet felt, per ton 45 00

Bull Rings.

Copper, $2.00 for 2½ in. and $1.90 for 2 in.

Butts.

Wrought Brass, net revised list

Cast Iron.

Loose Pin, dis., 60 per cent.

Wrought Steel.

Fast Joint, dis. 65, 10 and 2½ per cent.
Loose Pin, dis. 65, 10 and 2½ per cent.
Berlin Bronzed, dis. 70, 70 and 5 per cent.
Gen. Bronzed, per pair 0 40 0 65

Carpet Stretchers.

American, per doz 1 00 1 50
Bullard's, per doz .... 6 50 ....

Castors.

Bed, new list, dis. 55 to 57½ per cent.
Plate, dis. 52½ to 57½ per cent.

Cattle Leaders.

Nos. 31 and 32, per gross 50 9 50

Cement.

Canadian Portland 2 50 2 80
English " 3 00
Belgian " 2 50 2 75
Canadian hydraulic 1 25 1 50

Chalk.

Carpenters Colored, per gross 0 45 0 75
White lump, per cwt 0 60 0 65
Red 0 05 0 06
Crayon, per gross 0 14 0 18

Chisels.

Socket, Framing and Firmer.

Broad's, dis. 70 per cent.
Warnock's, dis. 70 per cent.
P. S. & W. Extra 60, 10 and 5 p.c.

Churns.

Revolving Churns, metal frames—No. 0, $8—No. 1, $8.50—No. 2, $9.00—No. 3, $10.00—No. 4, $12.00—No. 5, $16.00 each. Ditto, wood frames—20c. each less than above. Discounts: Delivered from factories, 58 p.c.; from stock in Montreal, 58 p.c. Terms, 4 months or 3 p.c. cash in 30 days.

Clips.

Axle dis. 65 per cent.

Closets

Plain Ontario Syphon Jet $16 00
Emb. Ontario Syphon Jet 17 00
Fittings net 1 00
Plain Teutonic Syphon Washout 10 00
Emb. Teutonic Syphon Washout 11 00
Fittings net 1 25
Low Down Teutonic, plain 16 00
" " " embossed 17 00
Plain Richelieu net 3 75
Emb. Richelieu net 4 00
Fittings net 1 25
Low Down Ont. Sy. Jet, plain net .. 19 50
" " " " " emb'd. net 20 50
Closet connection net 1 25
Basins, round, 14 in 1 00
" oval, 17 x 14 in 2 50
" " 19 x 15 in 3 75
Discount 40 p.c., except on net figures.

Compasses, Dividers, Etc.

American, dis. 62½ to 65 per cent.

Cradles, Grain.

Canadian, dis. 25 to 33⅓ per cent.

Crosscut Saw Handles.

S. & D., No. 3, per pair 17½
" " 5, " 22½
" " 6, " 15
Boynton pattern " 20

Door Springs.

Torrey's Rod, per doz .... (15 p.c.) 2 00
Coil, per doz 0 88 1 60
English, per doz 2 00 4 00

Draw Knives.

Coach and Wagon, dis. 50 and 10 per cent.
Carpenters, dis. 70 per cent.

Drills.

Hand and Breast.

Millar's Falls, per doz. net list.

DRILL BITS.

Morse, dis., 37½ to 40 per cent.
Standard dis. 50 and 5 to 55 per cent

Faucets.

Common, cork-lined, dis. 35 per cent.

ELBOWS. (Stovepipe.)

No. 1, per doz 1 40
No. 2, per doz 1 20
Bright, 20c. per doz. extra.

ESCUTCHEONS.

Discount, 45 per cent.

ESCUTCHEON PINS.

Iron, discount 40 per cent.

FACTORY MILK CANS.

Discount off revised list, 40 per cent.

FILES.

Black Diamond, 70 and 5 per cent.
Kearney & Foote, 60 and 10 to 70 per cent.
Nicholson File Co., 60 to 60 and 10 per cent.
Jowitt's, English list, 25 to 27½ per cent.

FORKS.

Hay, manure, etc., dis., 50 and 10 per cent. revised list.

GLASS—Window—Box Price.

Size United Inches.	Star Per 50 ft.	Star Per 100 ft.	D. Diamond Per 50 ft.	D. Diamond Per 100 ft.
Under 26	2 15	4 15	....	6 00
26 to 40	2 30	4 45	....	6 65
41 to 50	....	4 85	....	7 50
51 to 60	....	5 15	....	8 50
61 to 70	....	5 50	....	9 50
71 to 80	....	6 00	....	10 50
81 to 85	....	6 50	....	11 75
86 to 90	....	....	..	14 00
91 to 95	..	....	...	15 50
96 to 100	.. .	....	....	18

GAUGES

Marking, Mortise, Etc.

Stanley's dis. 50 to 55 per cent.

Wire Gauges.

Winn's, Nos. 26 to 33, each 1 65 2 40

HALTERS.

Rope, ⅜ per gross ....
" ½ " 9 00
" ⅝ to ¾ 14 00
Leather, 1 in., per doz 3 87½ 4 00
" 1¼ in., " 5 15 5 20
Web, — per doz 1 87 2 45

HAMMERS.

Nail

Maydole's, dis. 5 to 10 per cent. Can. dis. 25 to 27½ per cent.

Tack.

Magnetic, per doz 1 10 1 20

Sledge.

Canadian, per lb 0 07½ 0 08½

Ball Pean.

English and Can., per lb 0 22 0 2

HANDLES.

Axe, per doz. net 1 50 2 00
Store door, per doz 1 00 1 50

Fork.

C. & B., dis. 40 per cent. rev. list.

Hoe.

C. & B., dis. 40 per cent. rev. list.

Saw.

American, per doz 1 00 1 25

Plane.

American, per gross 3 15 3 75

Hammer and Hatchet.

Canadian, 40 per cent.

Cross-Cut Saws.

Canadian, per pair 0 13½

HANGERS. doz. pairs.

Steel barn door 5 85 6 00
Stearns, 4 inch 5 00
" 5 inch 6 50
Lane's covered—
No. 11, 5-ft. run 8 40
No. 11½, 10-ft. run 10 80
No. 12, 10-ft. run 12 60
No. 14, 15-ft. run 21 00
Lane's O.N.T. track, per foot 4½

HARVEST TOOLS.

Discount, 50 and 10 per cent.

HATCHETS.

Canadian, dis. 40 to 42½ per cent.

HINGES.

Blind, Parker's, dis. 50 and 10 to 60 per cent.
Heavy T and strap, 4-in., per lb 0 06½
" " 5-in., " 0 06¼
" " 6-in., " 0 06
" " 8-in., " 0 05¾
" " 10-in., " 0 05½
Light T and strap, dis. 65 and 5 per cent.
Screw hook and hinge—
6 to 12 in., per 100 lbs 4 50
14 in. up, per 100 lbs 3 50
Per gro. pairs.
Spring 12 00

HOES.

Garden, Mortar, etc., dis. 50 and 10 p.c.
Planter, per doz .... 4 00 4 5[illegible]

HOLLOW WARE

Discount .... 45 and 5 per cent

HOOKS.

Cast Iron.

Bird Cage, per doz 0 50 1 10
Clothes Line, per doz 0 27 0 63
Harness, per doz 0 72 0 88
Hat and Coat, per gross 1 00 3 00
Chandelier, per doz 0 50 1 00

Wrought Iron.

Wrought Hooks and Staples, Can., dis. 47½ per cent.

Wire.

Hat and Coat, discount 45 per cent.
Belt, per 1,000 .... 0 60
Screw, bright, dis. 55 per cent.

HORSE NAILS.

"C" brand 50 and 7½ p.c. off new list } Oval-
"M" brand 50, 10 and 5 per cent. } head
Countersunk, 60 per cent.

HORSESHOES

F.O.B. Montreal.

Iron Shoes.	No. 2 and larger.	No. 1. and smaller.
Light, medium, and heavy...	3 50	3 75
Snow shoes	3 75	4 00
Steel Shoes.		
Light	3 60	3 85
Featherweight (all sizes)	4 85	4 85

F.O.B. Toronto, Hamilton, London and Guelph, 10c. per keg additional.
Toe weight steel shoes...... 6 70.

JAPANNED WARE.
Discount, 45 and 5 p.c. off list, June 1899

ICE PICKS.
Star per doz.... 3 00 3 25

KETTLES.
Brass spun, 7½ p.c. dis. off new list.
Copper, per lb.... 0 30 0 50
American, 60 and 10 to 65 and 5 p.c.

KEYS.
Lock, Can., dis., 45 p.c.
Cabinet, trunk, and padlock, Am. per gross.... 60

KNOBS.
Door, japanned and N.P., per doz.... 1 50 50
Bronze, Berlin, per doz.... 2 75 25
Bronze Genuine, per doz.... 6 00 2 00
Shutter, porcelain, F. & L. screw, per gross.... 1 30 4 00
White door knobs—per doz. 1 25

HAY KNIVES.
Discount, 50 and 10 per cent.

LAMP WICKS.
Discount, 60 per cent.

LANTERNS.
Cold Blast, per doz.... 7 00
No. 3 "Wright's".... 8 50
Ordinary, with O burner.... 4 60
Dashboard, cold blast.... 9 00
No. 0.... 5 75
Japanning, 50c. per doz. extra.

LEMON SQUEEZERS.
Porcelain lined,......per doz. 2 20 5 60
Galvanized.... 1 87 3 85
King, wood.... 2 75 2 90
King, glass.... 4 00 4 50
All glass.... 1 20 1 30

LINES.
Fish, per gross.... 1 05 2 50
Chalk " 1 90 7 40

LOCKS.
Canadian, dis. 45 p.c.
Russell & Erwin, per doz.... 3 00 3 25

Cabinet.
Eagle, dis. 30 p.c.

Padlock.
English and Am., per doz.... 50 6 00
Scandinavian, " 1 00 2 40
Eagle, dis. 20 to 25 p.c.

MACHINE SCREWS. Iron and Brass.
Flat head discount 25 p.c
Round Head discount 20 p.c.

MALLETS.
Tinsmiths' per doz.... 1 25 1
Carpenters', hickory, per doz. 1 25 3
Lignum Vitae, per doz.... 3 85 5 50
Caulking each.... 60 2 00

MATTOCKS.
Canadian, per doz.... 5 50 6 50

MEAT CUTTERS.
American, dis. 25 to 30 p.c.
German, 15 per cent.

MILK CAN TRIMMINGS.
Discount, 25 per cent.

NAILS.

Quotations are :	Cut.	Wire.
2d and 3d	$3 35	$3 85
3d	3 00	3 52
4 and 5d	2 75	3 35
6 and 7d	2 65	3 20
8 and 9d	2 50	3 00
10 and 12d	2 45	2 95
16 and 20d	2 40	2 90
30, 40, 50 and 60d. (base)	2 35	2 85

Wire nails in carlots are $2.77½
Galvanizing 2c. per lb. net extra.
Steel Cut Nails 10c. extra.
Miscellaneous wire nails, dis. 70 and 10 p.c.
Coopers' nails, dis. 30 per cent.
Flour barrel nails, dis. 25 per cent.

NAIL PULLERS.
German and American.... 1 85 3 50

NAIL SETS
Square, round, and octagon, per gross.... 3 38 4 00
Diamond.... 12 00 15 00

NETTING.
Poultry, 50 and 5 per cent. for McMullen's

OAKUM. Per 100 lb.
Navy.... 6 00
U. S. Navy.... 7 25

OIL.
Water White (U.S.).... 0 16½
Prime White (U.S.).... 0 15½
Water White (Can.).... 0 15
Prime White (Can.).... 0 14

OILERS.
McClary's Model galvan. oil can, with pump, 5 gal., per doz.... 10 00
Zinc and tin, dis. 50, 50 and 10.
Copper, per doz.... 1 25 3 50
Brass, " 1 50 3 50
Malleable, dis. 25 per cent.

GALVANIZED PAILS.
Dufferin pattern pails, dis. 45 p.c.
Flaring pattern, discount 45 per cent.
Galvanized washtubs, discount 45 per cent.

PIECED WARE.
Discount 40 per cent. off list, June, 1899.
10-qt. flaring sap buckets, dis. 45 p.c.
6, 10 and 14-qt. flaring pails, dis. 45 p.c.
Creamer cans, dis 45 p.c.

PICKS.
Per doz.... 6 00 9 00

PICTURE NAILS.
Porcelain head, per gross.... 1 75 3 00
Brass head " 0 40 1 00

PICTURE WIRE.
Tin and gilt, discount 75 p.c.

PLANES.
Wood, bench, Canadian dis. 50 per cent, American dis. 50.
Wood, fancy Canadian or American 7½ to 40 per cent.

PLANE IRONS.
English, per doz.... 2 00 5 00

PLIERS AND NIPPERS.
Button's Genuine per doz pairs, dis. 37½ 40 p.c.
Button's Imitation, per doz.. 5 00 9 00
German, per doz.... 0 60 2 60

PLUMBERS' BRASS GOODS.
Compression work, discount, 60 per cent.
Fuller's work, discount 65 per cent.
Rough stops and stop and waste cocks, discount, 60 per cent.
Jenkins disk globe and angle valves, discount, 55 per cent.
Standard valves, discount, 60 per per cent.
Jenkins' radiator valves discount 55 per cent.
" " " standard, dis., 60 p.c.
Quick opening valves discount, 60 p.c.
No. 1 compression bath cock.... 2 00
No. 4 " " " 2 00
No. 7, Fuller's.... 2 50
No 4½, " 3 00

POWDER.
Velox Smokeless Shotgun Powder.
100 lb. or less.... 0 85
1,000 lb. or more.... 0 80
Net 30 days.

PRESSED SPIKES.
Discount 20 to 25 per cent.

PULLEYS.
Hothouse, per doz.... 0 55 1 00
Axle.... 0 22 0 33
Screw.... 0 27 1 00
Awning.... 0 35 2 50

PUMPS.
Canadian cistern.... 1 80 3 60
Canadian pitcher spout.... 1 40 2 10

PUNCHES.
Saddlers', per doz.... 1 0 1 85
Conductors', " 9 0 15 00
Tinners' solid, per set.... 0 0 72
" hollow, per inch.... 0 0 1 00

RANGE BOILERS.
Galvanized, 3 gallons.... 7 00
" 35 " 8 25
" 40 " 9 50
Copper, 30 " 22 00
" 35 " 26 00
" 40 " 30 00
Discount off Copper Boilers 10 per cent.

RAKES.
Cast steel and malleable, 50, 10 and 5 p.c.
Wood, 25 per cent.

RASPS AND HORSE RASPS.
New Nicholson horse rasp, discount 60 to 60 and 10 p.c.
Globe File Co.'s rasps, 60 and 10 to 70 p.c.
Heller's Horse rasps, 50 to 50 and 5 p.c.

RAZORS.
per doz.
Elliot's.... 4 00 18 00
Geo. Butler & Co.'s.... 8 00 18 00
Boker's.... 7 50 11 00
" King Cutter.... 12 50 50 00
Wade & Butcher's.... 3 60 10 00
Theile & Quack's.... 7 00 12 00

REAPING HOOKS.
Discount, 50 and 10 per cent.

REGISTERS.
Discount,.... 40 per cent.

RIVETS AND BURRS.
Iron Rivets, black and tinned, discount 60 and 10 per cent.
Iron Burrs, discount 55 per cent.
Extras on Iron Rivets in 1-lb. cartons, ½c. per lb.
Extras on Iron Rivets in ¼-lb. cartons, 1c. per lb.
Copper Rivets & Burrs, 35 and 5 p.c. dis. and cartons, 1c. per lb. extra, net.
Extras on Tinned or Coppered Rivets ¼-lb. cartons, 1c. per lb.

RIVET SETS
Canadian, dis. 35 to 37½ per cent.

ROPE ETC.

	Sisal.	Manila.
7-16 in. and larger, per lb	10	13½
⅜ in.	11	14½
¼ and 5-16 in.		15½
Cotton, 3-16 inch and larger		16½
" 5-32 inch		21½
" ⅛ inch		22½
Russia Deep Sea		15½
Jute		8
Lath Yarn		9¼
New Zealand Rope		10

RULES.
Boxwood, dis. 75 and 10 p.c.
Ivory, dis. 37½ to 40 p.c.

SAD IRONS. per set.
Mrs. Potts, No. 55, polished.... 62½
" No. 50, nickle-plated.... 67½

SAND AND EMERY PAPER.
Dominion Flint Paper, 47½ per cent.
B & A. sand, 40 and 5 per cent.
Emery, 40 per cent.
Garnet (Rurton's), 5 to 10 p.c. advance on list.

SAP SPOUTS.
Bronzed iron with hooks, per doz.... 9 50

SAWS.
Hand Disston's, dis. 12½ p.c.
S. & D., 40 per cent.
Crosscut, Disston's, per ft.... 35 0 55
S. & D., dis. 35 p.c. on Nos. 2 and 3.
Hack, complete, each.... 0 75 2 75
" frame only.... 0 75

SASH WEIGHTS.
Sectional, per 100 lbs.... 2 75 3 00
Solid, " 2 00 2 25

SASH CORD.
Per lb.... 0 23 0 30

SAW SETS.
"Lincoln," per doz.... 6 50

SCALES.
B. S. & M. Scales, 45 p.c.
Champion, 65 per cent.
Fairbanks Standard, 35 p.c.
" Dominion, 55 p.c.
" Richelieu, 55 p.c.
Chatillon Spring Balances, 10 p.c.
Warren Champion 65 p.c.
" Standard 45 p.c.

SCREW DRIVERS.
Sargent's per doz.... 0 65 1 00

SCREWS
Wood, F. H., bright and steel, 87½ and 10 p.c.
Wood R. H., " dis. 82½ and 10 p.c.
" F. H., brass, dis. 80 and 10 p.c.
Wood, R. H., " dis. 75 and 10 p.c.
" F.H., bronze, dis. 75 p.c.
" R.H. " 70 p.c.
Drive Screws, 87½ and 10 per cent.
Bench, wood, per doz.... 3 25 4 00
" iron, " 4 25 5 75
Set, Case hardened, 60 per cent.
Square Cap, 50 and 5 per cent.
Hexagon Cap, 45 per cent.

SCYTHES.
Per doz. net.... 9 00

SCYTHE SNATHS.
Canadian, dis. 45 p.c.

SHEARS.
Bailey Cutlery Co., full nickeled, dis. 60 p.c.
Seymour's, dis. 50 and 10 p.c.

SHOVELS AND SPADES.
Canadian, dis. 40 and 5 per cent.

SINKS.
Steel and galvanized, discount 45 per cent.

SNAPS.
Harness, German, dis. 25 p.c.
Lock, Andrews'.... 4 50 11 00

SOLDERING IRONS.
1, 1½ lb., per lb.... 0 37
2 lb. or over, per lb.... 0 34

SQUARES.
Iron, No. 493, per doz.... 2 40 2 55
" No. 494, " 3 25 3 40
Steel, dis. 50 and 5 to 50 and 10 p.c., rev. list.
Try and bevel, dis. 50 to 52½ p.c.

STAMPED WARE.
Plain, dis. 75 and 12½ p.c. off revised list.
Retinned, dis., 75 p.c. off revised list.

STAPLES.
Galvanized.... 3 50 4 00
Plain.... 3 15 3 75
Coopers', discount 45 per cent.
Poultry netting staples, 40 per cent.

STOCKS AND DIES.
American dis. 25 p.c.

STONE. Per lb.
Washita.... 0 28 0 60
Hindostan.... 0 06 0 07
" slip,.... 0 09 0 09
Labrador.... 0 13
" Axe.... 0 15
Turkey.... 0 50
Arkansas.... 0 00 1 50
Water-of-Ayr.... 0 00 0 10
Scythe, per gross 3 50 5 00
Grind, per ton 15 00 18 00

STOVE PIPES.
5 and 6 inch Per 100 lengths.... 7 00
7 inch " " 7 50

ENAMELINE STOVE POLISH.
No. 4—3 dozen in case, net cash.... $4 80
No. 6—3 dozen in case, " 8 40

TACKS BRADS, ETC.
Per cent.
Strawberry box tacks, bulk.... 75 & 10
Cheese-box tacks, blued.... 80 & 12½
Trunk tacks, black and tinned.... 85
Carpet tacks, blued.... 80 & 15
" " tinned.... 80 & 20
" " (in kegs).... 40
Cut tacks, blued, in dozens only.. 80
" ½ weights.... 60
Swedes, cut tacks, blued and tinned—
In bulk.... 80 & 10
In dozens.... 75
Swedes, upholsterers', bulk.... 85, 12½ & 12½
" brush, blued & tinned, bulk.. 70
" gimp, blued, tinned and japanned.... 75 & 12½
Zinc tacks.... 35
Leather carpet tacks.... 55
Copper tacks.... 50
Copper nails.... 52½
Trunk nails, black.... 65 and 5
Trunk nails, tinned.... 65 and 10
Clout nails, blued.... 65 and 5
Chair nails.... 35
Patent brads.... 40
Fine finishing.... 40
Picture frame points.... 10
Lining tacks, in papers.... 10

Lining tacks, in bulk		15
" " solid heads, in bulk		75
Saddle nails in papers		10
" " in bulk		15
Tufting buttons, 22 line, in dozens only		60
Tin capped trunk nails		15
Zinc glazier's points		5
Double pointed tacks, papers		90 and 10
" " " bulk		40

TAPE LINES.

English, ass skin, per doz	2 75	5 00
English, Patent Leather	5 50	9 75
Chesterman's each	0 90	2 85
" steel, each	0 80	8 00

THERMOMETERS.

Tin case and dairy, dis. 75 to 75 and 10 p.c.

TRAPS. (Steel.)

Game, Newhouse, dis. 25 p.c.
Game, H. & N., P. S. & W., 65 p.c.
Game, steel, 72½, 75 p.c.

TROWELS.

Disston's discount 10 per cent.

German, per doz	4 75	6 00

S. & D., discount 35 per cent.

TWINES.

Bag, Russian, per lb	....	0 27
Wrapping, cotton, 3-ply		0 20
" " 4-ply	....	0 26
Mattress, per lb	0 33	0 45
Staging, "	0 27	0 35

VISES.

Hand, pe doz	4 00	6 00
Bench, parallel, each	2 00	4 50
Coach, each	6 00	7 00
Peter Wright's, per lb	0 12	0 13
Pipe, each	5 50	9 00
Saw, per doz	6 50	13 00

ENAMELLED WARE.

White, Princess, Turquoise, Blue and White, discount 50 per cent.
Diamond, Famous, Premier, 50 and 10 p.c.
Granite or Pearl, Imperial, Crescent, 50, 10 and 10 per cent.

WIRE.

Brass wire, 50 to 50 and 2½ per cent. off the list.

Copper wire, 45 and 10 per cent. net cash 30 days, f.o.b. factory.

Smooth Steel Wire, is quoted at the following net selling prices:

No. 6 to 8 gauge	$2 90
" 9 "	2 80
" 10 "	2 87
" 11 "	2 90
" 12 "	2 95
" 13 "	3 15
" 14 "	3 37
" 15 "	3 50
" 16 "	3 65

Other sizes of plain wire outside of Nos. 9, 10, 11, 12 and 13, and other varieties of plain wire remain at $2 81 base with extras as before. The prices for Nos 9 to 13 include the charge of 1 c for oiling. Extras net per 100 lb.: Coppered wire, 60c.—tinned wire, $2—oiling, 10c.—special hay-bailing wire, 30c.—spring wire, $1—best steel wire, 75c.—bright soft drawn, 15c.—in 50 and 100-lb. bundles net, 10c.—in 25-lb. bundles net 15c.—packed in casks or cases, 15c.—bagging or papering, 10c.

Fine Steel Wire, dis. 17½ per cent. List of extras: In 100-lb. lots: No. 17, $5—No. 18, $5.50—No. 19, $6—No. 20, $6.65—No. 21, $7—No. 22, $7.30—No. 23, $7.65—No. 24, $8—No. 25, $9—No. 26, $9.50—No. 27, $10—No. 28, $11—No. 29, $12—No. 30, $13—No. 31, $14—No. 32, $15 No. 33, $16—No. 34, $17. Extras net—tinned wire, Nos. 17-25, $2—Nos. 26-31 $4—Nos. 32-34, $6. Coppered, 5c.—oiling, 10c.—in 25-lb. bundles, 15c.—in 5 and 10-lb. bundles, 25c.—in 1-lb. hanks, 50c.—in ½-lb. hanks, 75c.—in ¼-lb. hanks, $1—packed in casks or cases, 15c.—bagging or papering, 10c.

Galvanized Wire, per 100 lb.—Nos. 6,7 8, $3 51 to $3 8'—No. 9, $2.85 to $3.15—No. 10, $3.60 to $3.95—No. 11, $3 70 to $4.10—No 12, $3 to $3 30—No. 13, $3.10 to $3 4?—No. 14 $4.1? to $1.50—No. 15, $4.60 to $5.05—No 16, $4.85 to $5 35. Base sizes, Nos. 6 to 9, $2 57½ f.o.b. Cleveland.

Clothes Line Wire, solid 7 strand, No. 17 $4.25; No. 18, $2.65; No. 19, $2.35, f.o.b Hamilton, Toronto, Montreal.

WIRE FENCING. F.O.B. Toronto

Galvanized barb	....	3 05
Galvanized, plain twist	....	3 05

Galvanized barb, f.o.b. Cleveland, $2.82½ in less than carlots, and $2.70 in carlots.

WIRE CLOTH.

Painted Screen, per 100 sq. ft., net	1 35

WASTE COTTON. per lb.

Colored	4½ to 5
White, according to quality	6½ to 7½

500-lb bale lots shaded.

WRENCHES.

Acme, 35 to 37½ per cent.
Agricultural, 60 p.c.
Coe's Genuine, dis. 20 to 25 p.c.

Towers' Engineer, each	2 00	7 00
" S., per doz	5 80	6 00
G. & K.'s Pipe, per doz	....	3 40
Burrell's Pipe, each	....	3 00
Pocket, per doz	0 25	2 90

WRINGERS.

Leader	per doz.	$30 0	3? ?0
Royal Canadian	"	76 00	78 00
Royal American	"	56 00	78 ?0
Sampson	"	....	30 00

Terms 4 months, or 3 p.c. 30 days.

WROUGHT IRON WASHERS.

Canadian make, discount, 40 and 5 per cent.

CANADIAN HARDWARE AND METAL MERCHANT

The Weekly Organ of the Hardware, Metal, Heating, Plumbing and Contracting Trades in Canada.

VOL. XIII. MONTREAL AND TORONTO, MAY 18, 1901. NO. 20

The "IDEAL" Ice Cream Freezer

Made only in one size, but will freeze from half pint to the capacity of the freezer (about one gallon).

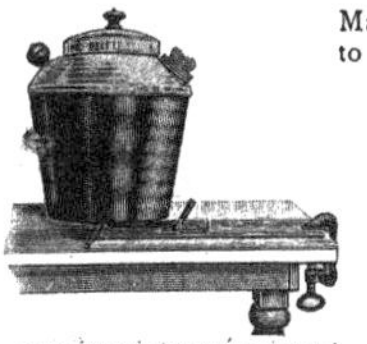

IN POSITION TO RECEIVE CREAM

PRACTICALLY ONLY FOUR PARTS

IN POSITION FOR FREEZING

Will Freeze Cream in from Two to Five Minutes According to Quantity.

Indurated Fibre Tub.
No Hoops to Fall Off.
No Staves to Shrink or Swell.
No Repacking Necessary to Keep Cream 5 Hours.
Saves 50 per cent. in Ice.

Simple in Construction.
Easy in Operation.
Rapid in Results.
No Wooden Tub with Leaky Seams.
No Danger of Spoiling the Cream with Salt.

Wood, Vallance & Co., Hamilton

Toronto Office: 32 Front Street West—H. T. Eager.
British Columbia Representative: W. G. Mackenzie, P.O. Box 460, Vancouver, B.C.

Geo. D. Wood & Co.,
Winnipeg, Man.

COLD WATER PAINT

Did you ever know a cold water paint that was weatherproof, sanitary, washable and fireproof all in one? Not unless you know

INDELIBLO

because it's the only water paint that comes up to this standard. Indeliblo comes in the form of dry powder. The powder is not dear and cold water is free. There is money in it for all. It will clean, freshen and brighten the dirtest hole, shaft, alley-way, elevator, factory or any other that ever existed. Want to learn about it?

WRITE TO

A. RAMSAY & SON, - - - - - - - MONTREAL.
J. H. ASHDOWN, - - - - - - - - WINNIPEG.
McLENNAN, McFEELY & CO., - - - VANCOUVER.

— AGENTS —

"DAISY" CHURN

Has tempered steel cased bicycle ball bearings, strongest, neatest and most convenient frame. Only two bolts to adjust in setting up. Steel Bow Levers, suitable for either a standing or sitting posture. Has four wheels and adjustable feet to hold stand steady while churning. When churn is locked to stand the bow can be used as handles to move it about on the front wheels as handy as a baby carriage. Open on both sides to centre, giving free space for pail. Made with wood or steel stands, with Cranks only, or Bow Levers as desired.

Vollmar Perfect Washer

Has a most enviable record. A perfection of its kind—will wash more clothes in less time, do it better and easier, with less wear and tear, than any other machine.

THE Wortman & Ward Mfg. Co., Limited
LONDON, ONT.
Eastern Branch, 60 McGill Street, Montreal, Que.

IMPROVED STEEL WIRE TRACE CHAINS.

Every chain guaranteed. Most profitable and satisfactory chain to handle.

THE B. GREENING WIRE CO., LIMITED
HAMILTON, ONT., AND MONTREAL, QUE.

ESTABLISHED 1860. INCORPORATED 1895.

Milk Can Trimmings

Iron Clad Pattern.

In all our Milk Can Bottoms sufficient space is left between bottom proper and rim to allow body of can to be inserted ¾ of an inch, making permanent joint.

Bottoms are thus sweated in with half the solder.

Top Hoops are "**Shouldered**" and have cut-out at joint, making neater and cleaner job in half time than the old style hoop.

All Hoops have retinned edges.

For durability, finish, and economy in making up our Trimmings are unequalled.

Broad Hoop Pattern.

THE THOS. DAVIDSON MFG. CO., Limited, MONTREAL, QUE.

IRON PIPE

All Sizes.

OUR STOCK IS NOW COMPLETE.

SAMUEL, SONS & BENJAMIN, - - LONDON AND LIVERPOOL, ENGLAND

M. & L. Samuel, Benjamin & Co.

General Importers and Exporters and Metal Merchants

27 Wellington Street West, - - TORONTO, ONT.

Lewis Bros. & Co.

MONTREAL.

We are **Headquarters for Tools of All Kinds.**

No. 50—Improved Mitre Box.

On Dividers.

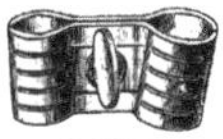

Off Dividers.
Stanley's Patent Pencil Clasp.

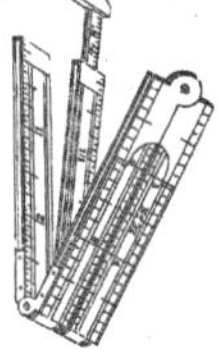

A Complete Line of Rules.

Stanley's No. 2 Plumb Bob.

IRON FORE PLANE.
Style of Nos. 5, 5½, 6, 7 and 8.

No. 77—Pat. Marking and Mortise Gauge.

No. 14—Round Mallet, with Iron Rings

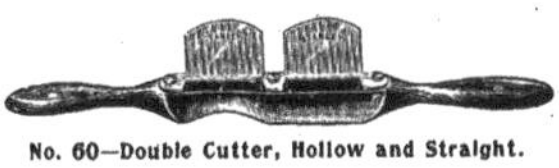

No. 60—Double Cutter, Hollow and Straight.

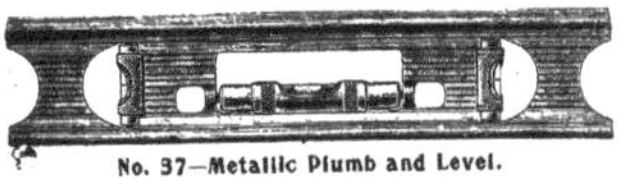

No. 37—Metallic Plumb and Level.

PLUMBS AND LEVELS.
Style of Nos. 0, 00, 104, 11-2

Our Stock is Complete **WRITE FOR PRICES.**

MAIL ORDERS
Shipped Same Day as Received.

LEWIS BROS. & CO.
MONTREAL

CAN'T BE BEAT

HIGHEST
IN
QUALITY.

EASIEST
RUNNING.

BEST
IN
FINISH.

EASIEST
ADJUSTED.

A CHILD CAN USE IT.

MANUFACTURED AND PATENTED BY

A. R. Woodyatt & Co.,

GUELPH, ONT.

Sold only through the wholesale trade.

A MAN WILL APPRECIATE IT.

SEASONABLE SUGGESTIONS.

WATER COOLERS.

Our Water Coolers are well made.
They have Nickel-Plated Faucets.
They are finished in attractive designs and colors.

WATERING POTS.

Six sizes. Plain, Japanned and Galvanized.
All Cans are made with Patent Rose, which can be easily taken apart for cleaning.
Well finished, strong and durable.
That is why we sell so many.

KEMP MANUFACTURING CO.,

TORONTO, Canada

HARDWARE AND METAL

VOL. XIII. MONTREAL AND TORONTO, MAY 18, 1901. NO. 20.

President,
JOHN BAYNE MacLEAN,
Montreal.

THE MacLEAN PUBLISHING CO.
Limited.

Publishers of Trade Newspapers which circulate in the Provinces of British Columbia, North-West Territories, Manitoba, Ontario, Quebec, Nova Scotia, New Brunswick, P.E. Island and Newfoundland.

OFFICES

MONTREAL - - - - - - - 232 McGill Street, Telephone 1255.
TORONTO - - - - - - - 10 Front Street East, Telephone 2148.
LONDON, ENG. - - - - - 109 Fleet Street, E.C., W. H. Miln.
MANCHESTER, ENG. - - - - 18 St Ann Street. H. S. Ashburner,
WINNIPEG - - - - - Western Canada Block. J. J. Roberts.
ST. JOHN, N. B. - - - - No. 3 Market Wharf, J. Hunter White.
NEW YORK. - - - - - - - 176 E. 88th Street,

Subscription, Canada and the United States, $2.00.
Great Britain and elsewhere - - - - 12s.

Published every Saturday.

Cable Address { Adscript, London. Adscript, Canada.

PIG IRON OUTPUT IN ONTARIO.

A REPORT dealing with the mineral output in Ontario for the first three months of the year has just been issued by the Bureau of Mines in that Province. A quarterly report is a new departure, and a departure which is quite acceptable. Hitherto, the annual reports of the Ontario Bureau of Mines, like nearly all reports issued by Governments, both Provincial and Federal, have been issued so long after the time with which they dealt had passed as to have lost much of their value.

The fact that the report is a good one makes its early appearance all the more gratifying.

The total value of the minerals produced in Ontario for the three months was $827,860. As the total for last year was $2,541,191, the output for the quarter under review shows a proportional increase of about 30 per cent., compared with the same period in 1900.

What is particularly gratifying to note is that pig iron has become, in point of value, the most important of the mineral products of Ontario, the output during the three months being valued at $438,659. The quantity was 28,694 tons. There are now three blast furnaces in Ontario, the furnace at Midland having been, it will be remembered, started up in December last.

The two furnaces during the whole of last year produced 54,269 tons of pig iron. This was at the rate of 13,567 tons quarterly. The three furnaces now in operation are producing at more than double that ratio.

Counting the second furnace of The Dominion Iron and Steel Co., which was blown in at Sydney this week, there are now in Canada 10 blast furnaces in operation.

The quantity of iron ore melted in Ontario during the first three months of the year was 48,663 tons, of which 21,083 tons were Provincial ores and 27,580 tons imported. The quantity of iron ore mined in the Province was 36,503 tons, valued at $44,100.

The quantity and value of various metals produced in the Province were as follows:

	Quantity.	Value.
Iron ore, tons	36,503	$ 44,106
Pig iron, tons	28,694	438,659
Nickel, lb	1,805,691	190,858
Copper, lb	1,680,301	75,625
Arsenic, lb	236,054	12,046
Gold, oz	3,150	54,520
Silver, oz	20,077	12,046
Total		$827,860

STEEL KEEPS ACTIVE AND STRONG.

THE strength and activity of the steel market continues to be the chief feature of trade.

In both the United States and Canada, the demand exceeds the supply. In Great Britain, too, there is a little more disposition on the part of consumers to do business, but prices there do not show any improvement, there having been actual declines in some localities. In Chicago, steel billets are nominally $2 per ton dearer, and wire rods are quoted $1 higher at $39 per ton delivered. Firm and all as steel billets are, they are about $5 per ton lower than they were a year ago.

Steel is being produced in enormous quantities in the United States, the different plants of the United States Steel Corporation turning out nearly 800,000 tons of steel ingots during April. The production of pig iron in the United States last month was also large, being 293 915 tons, against 388,766 tons in March.

ENGLISH METAL MARKET BUOYANT

The English market has been a rather interesting one during the past week to importers of metals. One of the features of the market is the strength of black sheets, brokers having been advised of an advance of 7s. 6d., making the price now for 20 gauge base £8 5s.

As we pointed out last week, tinplates are on the upward road, the rise on the week amounting to 4½d., the base price for coke plates being given at 12s. 3d. per box. Canada plates are steady at £5, but they are expected to soar with the other lines, as are also terne plates. Galvanized sheets are firmer and advancing, and Canadian importers have entered into quite a few contracts during the last two weeks.

A STRONG LINSEED OIL MARKET.

THE marked weakness which characterized the linseed oil market a short time ago has given place to one of marked strength.

The centre of the market's strength is Great Britain, the price in London having advanced about £2 during the past week. This makes the price there £33, which would mean a laid-down cost here of about 76c. per gallon for raw oil.

The Argentine crop has faded away and no one in England seems to have secured the needed supplies. Even Russia, that generally sells seed, has been a purchaser of the Argentine crop and the United States is also said to have bought rather more heavily than usual. Altogether, the English crushers find their present stocks decidedly short and the strong feeling is little short of a panic.

Strong as the market is in Great Britain, it is not one whit less so in Canada. The crushers here are completely sold out and they will not, it is understood, accept any more orders until fresh supplies of seed are secured. Stocks, too, in jobbers' hands are light. In Toronto it would be practically impossible to get a carload, as far as we can gather, and it is not likely that much in the way of replenishing stocks can be done for another couple of months.

As a natural outcome of the condition of affairs in this country, as well as in Great Britain, prices here are very strong. In Toronto on Monday prices were advanced 3c. per gal., and it is quite probable that before this issue is in the hands of our readers quotations may be again marked up.

Besides the influence of the conditions already noted, there is another one that has has a tendency to appreciate values in this country. And that is the increase in the primage rates charged by the regular steamship companies. Hitherto, this primage has been 10 per cent. Importers have, however, been notified that it will be 20 per cent. this season, although a rebate will be given, provided they (the importers) do not bring in oil by tramp steamers. Some of the lines who are taking this arbitrary course are doubtless receiving subventions from the treasury of this country.

UNIVERSITY COMMERCIAL EDUCATION.

THE proposal to establish a course of commercial instruction in connection with the University of Toronto is a matter that merits more than passing attention.

Hitherto we have been accustomed to think, as President Loudon has said, that a university course leads almost invariably to theology, law, medicine or pedagogy. It has been held to unfit a man for business, converting his practical endowments into a theoretic unbusinesslike nature. That idea is passing away, as we find not only that men of education make a success of business, but that they can, in many cases, make the greatest success. At times they are even essential to success.

Business is not what it was. In these days of rapid and cheap transportation and communication, the sphere of action of the manager of a large concern is wide—very wide. He must have a brain that measures just as wide. He must have a broad knowledge, and his mind must be capable of quick and accurate thinking. He must be accurate and be possessed of a fund o confidence. In other words, he must be trained. Most business men are trained in the pursuit itself. But why not train them in a specially adapted place, as we train our teachers, our doctors, or our lawyers? The idea seems feasible enough.

Yet, at present in Canada, we have but little of what might be called university commercial education. The nearest approaches to it we have in Toronto University is her political science course, and in McGill which has lately been endowed with a similar chair. And these courses do not furnish, do not pretend to furnish, a complete university commercial education. Additional subjects that should come on it are touched upon but very unsatisfactorily in business colleges and such commercial departments as we have at the Montreal Y.M.C.A. Many of our ambitious young men are trying to get a commercial education through American correspondence schools.

But none of these existing arrangements furnish just what we want. We should probably copy the Germans, the leaders in the present university movement. A commercial course was founded at the University of Leipzig two years ago, and similar courses are being established at Hamburg, Frankfort and Magdeburg. Some of the special subjects treated are: Commercial law, economic theory, economic history, economic geography, public finance, insurance, banking, foreign exchanges and transportation. A beginning in the same direction has also been made on this continent, viz., at the Universities of New York, Pennsylvania, California, Chicago and Wisconsin.

At the University of Toronto, it is proposed to establish at first a two-years' course, looking to a diploma in commerce, not a degree. It will require an entrance standard somewhat similar to the Ontario matriculation, but without Latin or Greek as essential subjects. The special feature of the course will be the study of at least two modern languages during both years, English literature, economics, and chemistry and physics in the first year, and, in the second year, specialization of any of the various subjects to which the students wish to devote themselves, such as applied chemistry, mineralogy, architecture, mechanical drawing or electricity. Regular students will be allowed to combine these with their other subjects.

There is much about this course that is to be commended, but, at the same time, it does not appeal to us as being as practical, on the whole, as that laid down in the curriculum of the German universities. The subjects are not, in a word, hardly what a business man would prescribe. And therein is the test of their fitness.

One of the most significant results we can see for a commercial course in the university is that it will give us a more educated class of business men and all that goes with that qualification. Furnished and fortified with a university culture and knowledge, we may, among other things, expect to see our business men take the places of many of our professional politicians in our legislative chambers.

We hope not only that the movement on foot in Toronto will result in the establishment of the proposed commercial course of instruction, but also that the other Canadian universities will see their way clear to "follow on."

COST OF THE TRADE AND COMMERCE DEPARTMENT.

IN last week's issue we dealt with the inertia of the Canadian Department of Commerce. In this issue we propose to deal with the expenditure of the Department.

The total expenditure of the Deparment last year was $676,542, but, in estimating the cost of the Department, it would be manifestly unfair to include the whole of that sum, for $599,831 was paid out in mail subsidies and steamship subventions, payment of which was authorized by statutory enactment. Then, there is the sum of $43,335 as a bounty on silver ore, and $3,195 on Chinese immigration, which would obviously have to be paid through some Department if the portfolio of Trade and Commerce had no existence.

But omitting these, we think that most people will agree with us that the country is paying out a great deal of money for a Department that has become noted for its inertness rather than for its activity.

Sir Richard Cartwright, the Minister of Trade and Commerce, is, as pointed out last week, responsible for this. There is no man in the Government, or even in the House, who, in a speech, can be more masterful, or, in marshalling facts, can be more effective. But, while in these and many other parts he stands high, his standing as administrator of his Department is anything but high.

The salary bill at Ottawa for the Department of Trade and Commerce last year was $20,712, over one-third of which is for the Minister's salary alone. Printing and stationery cost $941.73. Under the classification of sundries the sum of $2,052.57 was paid. Sir Richard's cab hire at Ottawa, $282.75, is included in this amount. To commercial agencies $8,460.67 is charged. We have here only referred to items of expendituse incurred in carrying on the work for which the Trade and Commerce Department was specifically called into existence to perform and the sum total of these items is $33,166 97. Viewed in the light of Departmental expenditures the amount is small, but when one considers the insignificant work which the Department is doing along the lines of foreign trade development it is a high price indeed.

As far as a factor in developing the foreign trade of this country is concerned, the Department, over which Sir Richard is the head, has become little short of a sinecure, for which the country pays $33,000 per year, or, capitalized at 3 per cent., equal to an addition of $1,100,000 to the public debt; and, as already pointed out, we are only including such sums as are expended purely in the trade and commerce branch of the Department's work. With the High Commissioner's Office in London, on which, last year, there was an expenditure of $32,647, the Trade and Commerce Department has nothing to do.

The article which appeared in last week's issue in regard to the Trade and Commerce Department attracted quite a little attention. The Hamilton Herald, in commenting upon it, urges that the better plan would be to abolish the portfolio of Trade and Commerce or combine it with that of Customs. "The Department," it says, "was established for the purpose of directing the development of trade and commerce; but surely all that is to be done in that direction can be done by the Customs Department."

We cannot agree with our contemporary. As The Herald points out the Department was created for the purpose of developing the trade and commerce of the country. It was therefore conceived that there was a necessity for it. The question is: Does that necessity still exist? Undoubtedly, it does. And if it does there is certainly work for the Trade and Commerce Department to do.

When a manufacturer or a wholesale merchant discovers that one of his departments has become unprofitable because of the inertia or incapacity of the head of that department, he merely removes the head of the department and appoints as a successor one who is deemed to be strong where the other is weak.

It is this business principle that should be applied to the Trade and Commerce Department. The portfolio should be one of the most important and useful in the Government system. And it is not the portfolio, but its head, that needs removal.

THE WHITE LEAD ASSOCIATION.

THE trade will be pleased to learn that at a meeting of the White Lead Association, held in Montreal on May 9, the troubles that have been hampering the organization in its actions during the last few months were amicably settled.

It was reported at one time that there was nothing but dissolution in store for the association. We are glad to know that such an eventuality has been avoided, for this association has amply shown in its past history that it lives to exert a wholesome influence upon the white lead business.

Its success of some years ago in securing legislation which protects the public against adulterated lead is not yet forgotten, while at all times when it has set prices it has guarded against extortion of the consumer, charging only a fair and living profit for the manufacturer and the jobber.

It was expected that if the association were reestablished on a firm footing, as it has been, prices would be advanced, for manufacturers during the last few months have been working at a loss. However, in view of the arrivals of low priced lead, the old prices have been confirmed, and the members seem to be content with a small margin. The advance in linseed oil may justify higher prices later on, particularly as the demand for white lead has materially improved during the past month.

FIRMER GLASS MARKET.

The upward tendency in the price of glass, which we have noted from time to time of late, has resulted in an advance of 15 per cent. this week, according to cable reports just to hand.

DECLINE IN PETROLEUM.

During the week the price of all lines of Canadian petroleum has been lowered ½c. per gallon throughout Ontario, both for city and country delivery. The decline is principally due to the reduction in the price of Canadian crude oil, which has fallen several cents in the past two or three weeks. American crude oil has declined to an even greater extent, but the price of American petroleum is yet unchanged.

TRADE IN COUNTRIES OTHER THAN OUR OWN.

UNFILLED orders for barb wire are occupying the attention of manufacturers to a large exent, although new business is continually being received. Some jobbers are advising customers to order sparingly, as considerable time must elapse before shipments can be made, by which time consumers may not want to take the wire.—Iron Age.

UNITED STATES PIG IRON OUTPUT.

Our monthly blast furnace statistics show that there has been a further expansion in the output during April, the coke and anthracite furnaces entering May with the enormous total capacity of 299,915 tons weekly, as compared with 288,766 tons on April 1. In spite of this increase, the furnace stocks declined 31,000 tons in April, thus showing that consumption was still in excess of the current make.—Iron Age.

DULL ENGINEERING TRADE.

The condition of the engineering trade shows very little change since our last report. There is a slow but steady decline in the industrial barometer. Here and there firms may be found who report that orders are coming in rather more freely than of late, but they are exceptions to the general rule. For the most part orders are being executed more rapidly than they are being replaced, while the steady decline in prices, as a result of the increasing severity in competition, shows no sign of abatement. In the machine tool trade, especially, the accounts to hand are discouraging, a decline in business being notified from most of the principal centres. On the other hand, the electric industry maintains a high level of activity, mainly due to the rapid extension of electric traction and motive power in all parts of the country.—Iron and Coal Trades Review, May 3.

HARDWARE TRADE IN THE UNITED STATES.

Notwithstanding the large movement of hardware from manufacturers and jobbers to the retail trade, there does not appear to be any special increase in retail stocks, and those in jobbers' hands are in many cases broken. The goods sold are apparently going rapidly into consumption in response to demands to meet the requirements of the industries and activities of the country, which are in an exceptionally prosperous condition. In nearly every direction there is evidence of enterprise in commercial and manufacturing lines and the hardware trade, including both the manufacturers and distributers of goods, are reaping the benefit. In building operations there is especial activity and the year promises to be marked by a very satisfactory trade in the branches immediately concerned.—Iron Age.

PIG IRON IN ENGLAND.

There has been a somewhat better feeling in the pig iron market during the past week, and, although the improvement has not developed to any great extent as yet, prices are stiffer in some quarters. In the Cleveland district, for instance, there is a decidedly better feeling. All qualities of pig iron have advanced in price, and makers are apparently in a strong position. Last week the deliveries to local consumers and to others inland by rail assumed almost unprecedented dimensions, and the shipments from Middlesbrough to foreign ports during April have been very large.

Victoria Day.

As Friday next will be kept as a public holiday, HARDWARE AND METAL will be issued on May 23. Consequently it will be necessary for us to have all advertising copy, changes, etc., in hand Monday evening, May 20. The insertion of matter after that day cannot be guaranteed

The Publishers.

The stocks of makers in this district are now very much reduced, and average price of No. 3 during the week has been about 45s. 9d., and little has been procurable at less. Hematite pig has been in somewhat brisker demand this week, owing in some measure to an improvement in the Sheffield steel trade. In Barrow, the market for hematite has undergone no change. It continues quiet all around. From Lancashire there is reported to be more inquiry; but the actual business passing is still somewhat small.—Iron and Coal Trades Review, May 3.

THE BRITISH TINPLATE TRADE.

The depletion of stocks in tinplates at Swansea has given fresh impetus to the buying for prompt shipments, and some considerable business has resulted at enhanced values. The demand for the 'Frisco market has again come to the front—with inquiries running up to November, and with the promise of some further good orders from this desirable quarter.

The "call" for plates in other directions has also been well sustained, and makers are fully supplied with orders for the balance of this half-year, while more indents have been placed for the following half, and at an advance on previous sale prices. It is evident that this continued hand to mouth process of buying has aided the advance in values, as it has produced a continuous stream of orders, which has had to be satisfied with quick shipments, and, consequently, makers have been enabled to hold for and obtain better value.

As regards the bete noir—the wages question—this has been shunted for the next week or so, to allow of the new wages list being promulgated; but it remains to be seen how far the men are willing to fall in with the new regime, and whether a lock-out can be avoided by concessions on both sides.—Iron and Coal Trades Review, May 3.

NEW YORK METAL MARKETS.

TIN—There was a further advance of 5s. in the London market, which was cabled strong and active. The bulk of the business, as heretofore, was in futures, but the latter showed no further advance. The upward movement in the English market was without influence here. On the contrary, the tone was weaker and sellers were seeking business at 25 points under the figures quoted last night. There was scarcely any demand, and this, in the face of large and increasing supplies, has a depressing influence which the improvement in London is not sufficient to overcome. To day the Minneapolis arrived with 425 tons, making the total arrivals for the month to date nearly 1,900 tons, while 525 tons more are due this week. The market closed dull at 26.50c. bid and 26.75c. asked for spot and May; 26.40c. bid and 26.75c. asked for June.

COPPER—Part of yesterday's improvement in the London market was lost, the close being easy at 2s. 6d. below last night's quotation on spot, and 3s. 9d. on futures. In New York, conditions were the same as previously noted, the market being quiet, with prices steady at 17c. for Lake Superior and 16⅞c. on electrolytic and casting.

PIG LEAD—Consumers continue to buy only when the stock is needed, and the market has a tame appearance. Prices are maintained on the basis of 4.37¾c. for lots of 50 tons or over. St. Louis was reported firm at 4.22½ to 4.25c. The advance of 1s. 3d. in the price of soft Spanish noted yesterday was lost to day.

SPELTER—Nothing new was presented in this market, which remained quiet but steady at 3 95 to 4c. The St. Louis market was dull at 3.85c. The London cable reported an advance of 2s. 6d.

ANTIMONY—Regulus remains quiet at the range of 8¾ to 10¾c., as to brand.

OLD METALS—Prices are steady, though there is little demand.

IRON — Buyers of pig iron show little or no disposition to anticipate their needs, and, while current consumption continues on a heavy scale, there is scarcely any business for future delivery reported. Prices are nominal. The movement in finished material continues free, but consists chiefly of deliveries on existing contracts. There is a good demand for bar iron, and sheet steel continues very active, but plates receive less attention. Old iron and steel are dull, but holders seem unwilling to make concessions.

TINPLATE—The distribution through all channels of consumption continues on a liberal scale, and the firm tone of the market is retained.—New York Journal of Commerce, May 15.

A PROPOSAL TO BRITISH IRONMAKERS.

Hon. J. W. Longley, Attorney General for Nova Scotia, who recently returned from a honeymoon trip throughout England, was interviewed in Halifax last week. He was in London while the large dailies there were discussing the great American steel trust. He endeavored to set forth through the press and in other ways the great field which Nova Scotia presented for the development of a great iron industry, and asked the iron men of England to consider the propriety of transferring their works to Nova Scotia where the conditions would enable them to compete with the trust on British soil on equal terms. He found, however, that the English manufacturers were slow to move, and he was afraid they would wake up too late to find that they have neglected their opportunities. They seemed much surprised to learn that iron is being shipped from Sydney, C.B., to the Clyde. The facts he presented regarding the iron production of Nova Scotia excited considerable attention.

ST. JOHN HARDWARE STORES.

At a meeting of the Iron and Hardware Association, St. John, N.B., last week, it was unanimously agreed to close all the hardware stores at 1 o'clock on Saturdays during the months of June, July and August. The association adjourned until the second week in September.

A committee was appointed, consisting of W. H. Thorne, Thomas McAvity, John P. Macintyre, A. M. Rowan and John J. Barry, to consider date and nature of an excursion on the river during the summer. The present idea of the association is—the excursion so be confined to the merchants, employes and families. The McClary Manufacturing Co., of London, Ont., were elected members.

Those signing the agreement to observe the Saturday half-holiday are W. H. Thorne & Co.; T. McAvity & Sons; James Robertson Co., Limited; H. H. Horton & Son; Kerr & Robertson; S. Hayward & Co., Limited; Emerson & Fisher; I. & E. R. Burpee; M. E. Ager; James Addison.

Promoters of the steel shipbuilding enterprise are getting surveys made of two sites in the Dartmouth cove, near Halifax.

The Waterous Co., Brantford, Ont., has just delivered to Ottawa a fire engine which has a guaranteed capacity for pumping 750 gallons per minute, but at the recent test it pumped 1,079 gallons in 1 min. 15 sec., which is about 20 per cent. over the guarantee.

INDUSTRIAL GOSSIP.

Those having any items of news suitable for this column will confer a favor by forwarding them to this office addressed the Editor.

THE Chatham Binder Twine Co., Limited, have let the contract for their new factory. It is to be ready by June 15.

Angrove's foundry, Kingston, was badly damaged by fire last week. It will be rebuilt.

The second blast furnace of the Dominion Iron and Steel Co., Limited, Sydney, N.S., was "blown in" this week.

J. L. Board's proposal to locate a machine tool and metal works in Point Edward, Ont., will likely be carried out.

It is reported that Quebec capitalists have offered to buy the Kingston, Ont., Locomotive Works and remove them to Quebec.

The Peterboro' Shovel Co., composed almost entirely of local capitalists, propose erecting a 50 x 125-ft. factory, with a $25,000 plant to manufacture shovels, etc., in Peterboro', Ont.

An Ottawa despatch states that Hon. J. I. Tarte has intimated that important developments were on foot in regard to the transportation problem. In addition to the employment of grain carriers of the largest size on the upper lakes, a fleet of steel barges, each costing $100,000, will before long be placed on the route between Port Colborne and Montreal.

CATALOGUES, BOOKLETS, ETC.

A SPORTING GOODS CATALOGUE.

The special sporting goods catalogue which Wood, Vallance & Co., Hamilton and Toronto, have just issued, is one of the best, most comprehensive works of the kind ever sent out in Canada. Its illustrations, descriptions and price lists include scores of different styles of guns, rifles, air-guns, carbines, revolvers, sights, reloading implements, cartridges, primers, caps, shells, gun wads, powder, shot, gun oil, blue rocks and traps, bench closers, cleaning rods, cartridge bags, belts, covers, canvas coats and vests, hunting knives, traps, dog collars, fish hooks, flies, minnows, traces, spoon baits, sinkers, floats, fishing lines, rods and bells, safety razors and nut picks and cracks. At the back of the book is an excellent resume of the Ontario Game Laws, which completes a most creditable sporting goods catalogue.

A SEASONABLE CATALOGUE.

Lewis Bros. & Co., Montreal, have issued, for the benefit of their customers, a seasonable little catalogue of lawn requisites. This illustrates one of the most complete lines of lawnkeepers' instruments ever shown in Canada. Of late years, manufacturers have displayed a great deal of ingenuity in providing tools for this purpose, and the range has grown to be a wide one. For that reason hardware dealers throughout the country will find this publication of Lewis Bros. a decided convenience. The book is neatly arranged and well illustrated. Copies may be had on application. The firm will be pleased to quote discounts on the articles shown to intending purchasers.

SALESMAN WANTED.

The Fairbanks Co., 749 Craig street, Montreal, are advertising for a traveller familiar with the hardware trade between Toronto and Sault Ste Marie, to sell pipes, valves and fittings, and mill and mining supplies. This is a good opening for a bright salesman. Applications will be received all next week.

PERSONAL MENTION.

Mr. Geo. Stephens, M.P., hardware merchant, Chatham, was registered at the Windsor Hotel, Montreal, this week.

TRADE CHAT.

A VICTORIA despatch says that two groups of copper mines on Mount Sicker, B.C., were sold to a New York syndicate represented by W. A. Dier, of Victoria, and E. E. Smith, of New York, for $336,000. The purchasers say that in about a month they will have machinery on the ground and put a big force to work, and expect before the close of summer to be shipping.

H. E. Reid has opened as tinsmith in Fort William, Ont.

The Lay Whip Co., Rock Island, Que., have registered as incorporated.

W. H. Johns, hardware dealer, Southampton, Ont., proposes to put a plate-glass front in his store.

The Pennsylvania Coal Co., Limited, have obtained a Quebec charter to do business in Montreal.

The hardware merchants of New Westminster, B.C., have agreed to close their respective places of business on Saturday afternoons from 2 to 7 o'clock to give their employes a half-holiday. This agreement takes effect to-day.

A new store is being built for E. J. Boucher, general merchant, Boucherville, Ont.

Johnson & Thompson, tinsmiths, etc., Danville, Que., have dissolved, Chas. Thompson continuing.

The crockery and hardware store on the corner of Mill and Main streets, St. John, N.B., was sold on Saturday to Messrs. Linton & Sinclair, the former proprietress, Mrs. E. F. Copp, retiring.

H. S. HOWLAND, SONS & CO.

WHOLESALE ONLY — 37-39 Front Street West, **Toronto.** — ONLY WHOLESALE

Garden Syringes.

No. 600—18 x 1½ in. dia., Brass, 2 Nipples.

Insect Exterminators.

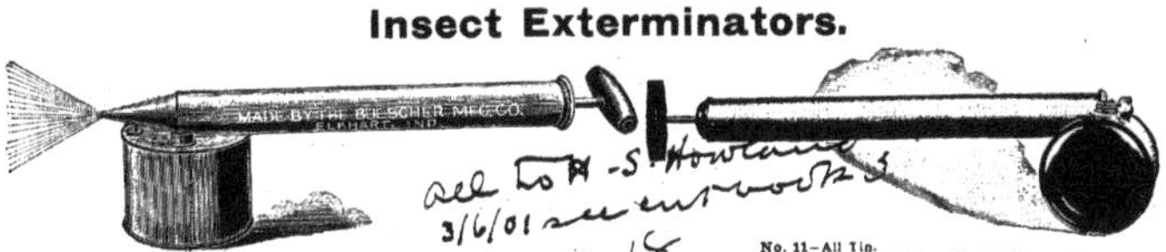

"Cataract," Tin, Lacquered
" All Brass, cannot Rust.

No. 11—All Tin.
14—Tin Barrel, Brass Reservoir.
16—All Brass, Lacquered.

Improved Brass Spray Pumps.

WITH PATENT AGITATORS.

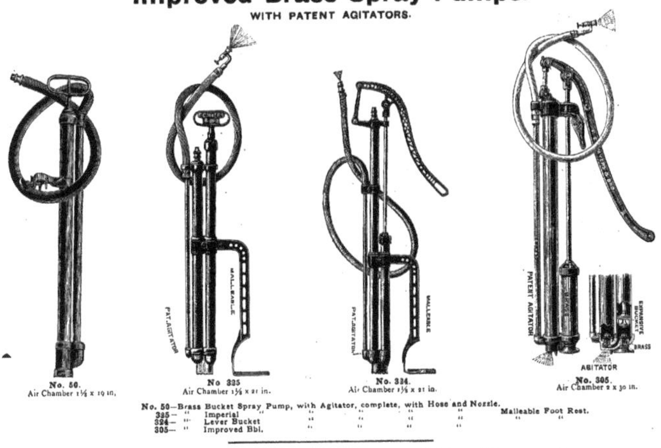

No. 50. Air Chamber 1⅛ x 19 in.

No 335 Air Chamber 1⅜ x 21 in.

No. 334. Air Chamber 1⅜ x 21 in.

No. 305. Air Chamber 2 x 30 in.

No. 50—Brass Bucket Spray Pump, with Agitator, complete, with Hose and Nozzle.
335— " Imperial " " " " " Malleable Foot Rest.
324— " Lever Bucket " " " " " "
305— " Improved Bbl. " " " "

H. S. HOWLAND, SONS & CO., Toronto.

WE SHIP PROMPTLY — Graham Wire and Cut Nails are the Best. — OUR PRICES ARE RIGHT

The Art of Window Dressing.

HARDWARE TRIMS.

HARDWARE is admitted to be the most difficult line of merchandise to display, owing to its lack of colorings, its shapes and weights. To make an attractive display of hardware the

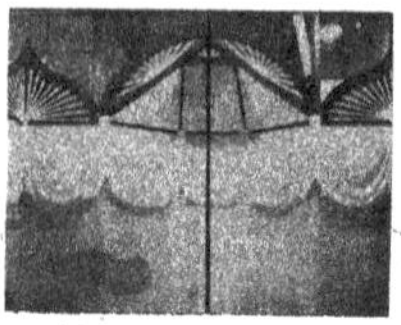

No. 1.

trimmer has to resort to a fancy design in a background or centre piece on which different articles of hardware are arranged. A look at the different hardware windows soon convinces one that there is not much interest taken in displaying the goods properly.

If the man who has the window trimming in charge will only give it a little thought and study he will soon be convinced that beautiful trims can be made with articles from the hardware stock. If the following suggestions were tried some good trims (paying trims) could be carried out. A small outlay of a few dollars would furnish the trimmer with a few wooden frames, such as circles, ovals, arches, half circles, etc. These can be covered with any dark material (black print preferred), and on them could be arranged cutlery, tools, etc.

A hardware display requires a background to it, the same as any other trim. It is no use without it. It is like a photograph; it needs a backing to show it off. The illustration No. 1 shows a pretty background. With a couple of half-circles and white tape, the top part is easily constructed. The back is pleated in white cheesecloth over white paper, which you can get at any local printing office (newspaper stuff). Arrange a nice design in front of this background, with oval frames, circles or arches, cover them with white canton flannel and show nothing but nickel goods. Around the circles, ovals, etc., you can put scissors or any line of nickel-plated ware. They can be kept solid in position with wire brads. Drapery effects can be made with chain, under the arches. Cover the floor with white canton flannel and show, nicely arranged, here and there a piece of nickelware. There are dozens of lines of nickel goods, such as nickel plated teapots, kettles, knives, forks, spoons, soap holders, coffee pots, trays, match boxes, etc. A large card calling attention to your line of nickel goods should be placed in the centre. A neat price card should be placed on each article or each line of goods. These are the kind of trims that are sure to make people stop and admire your goods, and are something out of the every-day, common, monotonous hardware trims as shown in the average window. Many hardwaremen may say on reading this: "Oh, these are all very well for dry goods and other window trims, but don't suit hardware

No. 3.

trims!" I say, by way of experiment, try a hardware trim with a fancy backing as suggested and be convinced that they are what is required to draw attention to your wares.

Illustration No. 2 is a good stand for showing hardware on. Have it made in two sections. When you put it in the window puff some colored stuff over it, and around the top edge of each step puff a piece of colored crepe paper or other material.

Illustration No. 3 is a novel idea and one that is easily carried out. It was the work of a prominent trimmer in one of the big hardware stores in Chicago. A solid circle 4 ft. in diameter was covered with black cloth. On this was drawn the face of one of Chicago's prominent citizens. This was done in chalk and the outlining was filled in with chain, tools, etc. Calipers, dividers, and auger bits were selected as best suited for the purpose, while a jack-chain artistically festooned formed the whiskers.

In order to create a good ad. for your store, you may arouse the interest of the public, by reproducing the picture of some person in your town or city. Have an artist or some good sketcher outline the face (of some prominent person) with crayon and then fill in with hardware. Introduce a guessing scheme, that you will present some article to the first person guessing whose face it is.

For a scissor exhibit, have a solid circle made of lumber as large as your window will permit, cover this over with black cloth. Start in the centre by placing your smallest scissors in the form of a circle and increase size of scissors until the outside edge of circle is reached. Display on the floor everything pertaining to this article.

It is always best to keep exclusively to one line of goods at a time. In this way a better impression is made on the public than if you jumble all kinds of goods in at once.

Now that the warm season will soon be on, the man in charge of the windows can be thinking out a good design on which to show goods required for warm weather. During this month housekeepers will be looking for all kinds of housecleaning articles. There are a "hundred-and-one" lines of goods carried by the hardware merchant that the housewife will be looking for. Here's a chance to make a good window trim of housecleaning articles, such as stepladders, mops, brushes, paints, dustpans, dusters, etc. Then this is also a good time to show all kinds of gardening tools. A good idea is to cover the bottom of your window with sod, make a little flower bed in the centre, arrange a nice background in green, and on the sod place a lawn mower, rake, hoe, spade, grass shears, watering can, garden hose, etc.

Then, in June, make a good display of, as I said before, "Hot-weather goods," viz.: Ice cream freezers, refrigerators, lemon squeezers, corkscrews and dozens of other lines of the same nature. During June and September some very good catchy trims could be made by the hardwareman with sporting goods. These furnish great suggestions for novel trims. H. H.

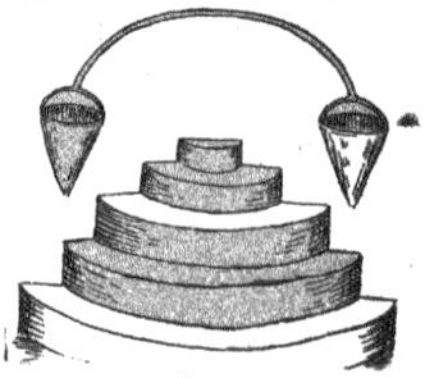

No. 2.

SOME ENLIGHTENING POINTERS.

The Ontario Lantern Co., Hamilton, Ont., recently put on the market a small brass lantern which burns kerosene, which they have named the "Little Bobs," and which is particularly adapted for ladies or where a small light only is required. The lantern can be supplied in either brass, nickel plated or tin. They are also offering a line of colored bulbs with the Radiant Shelby incandescent lamp for illuminating purposes required during the summer holidays. The Radiant Shelby is finding a sale throughout the Dominion where efficiency and long life is required. They expect before the fall trade opens up to have a line of aluminum reflectors of all sizes and shapes, adapted for the Radiant Shelby incandescent lamp. Orders are solicited from the large lighting stations.

MR. JOHN NEIL INJURED.

Many will regret to hear that John Neil, the veteran traveller for Edwin Chown & Son, Kingston, met with an accident on Monday morning last. He had driven up to the office door of the firm to get his samples, etc. In getting out of the buggy he tripped and fell heavily on his face, cutting his nose, lip and hands, and bruising his left arm and had to be helped into the office, whence he was driven to his home shortly after. He will be laid up for a short time. His ground in the meantime is being covered by Edwin Chown, jr., of the office staff. Mr. Neil has been in the employ of the senior member of the firm for 40 years.

The White Pass Railway Co., whose road runs into the Klondyke from Skagway, have decided to purchase their supplies in Vancouver and Victoria. G. C. Glyn has been appointed purchasing agent.

MARKETS AND MARKET NOTES

QUEBEC MARKETS

Montreal, May 17, 1901.

HARDWARE.

BUSINESS maintains its volume remarkably well and all seasonable lines are selling well. There is only one check to trade and that is the difficulty found in getting quick deliveries from manufacturers. Barb wire is cleaned out and orders are some weeks behind. There have been some arrivals of wire from England this week, but these have not begun to satiate the demand. Canadian factories are working overtime. Other kinds of wire are also in short supply and it takes a few days to fill an order in galvanized or plain. Neither are cut nails in the small sizes any too plentiful. The irregularity of the horse nail market continues; 60 and 10 and 10 per cent. is frequently quoted on countersunk heads in fairly large lots, while 66⅔ is also mentioned. Screens are active this week, as also is screen wire cloth. Harvest tools, building paper and other spring goods are moving out freely, and stocks are well reduced. Time is even being asked by the wholesalers on several lines. Gas and coal oil stoves are still meeting with a good inquiry. The discount on steel squares has been raised to 60, 10 and 5 per cent. The English market is reported stronger all around.

BARB WIRE—The difficulty found for some weeks in securing sufficient supplies to supply the trade has not yet been alleviated. Some barb wire has been unloaded from English ships this week, but it was only a drop in the bucket that needs filling. Immediate deliveries cannot be made by most houses. The price is unchanged at $3 05 per 100 lb. f.o.b. Montreal.

GALVANIZED WIRE—Orders for galvanized wire are coming in very freely. Stocks are light. We quote: No. 5, $4.25; Nos. 6, 7 and 8 gauge, $3.55; No. 9, $3.10; No. 10, $3.75; No. 11, $3.85; No. 12, $3.25; No. 13, $3.35; No. 14, $4.25; No. 15, $4.75; No. 16, $5.00.

SMOOTH STEEL WIRE—A good inquiry is still being experienced for all fence wires. We quote oiled and annealed as follows: No. 9, $2.80; No. 10, $2.87; No. 11, $2.90; No. 12, $2.95; No. 13, $3.15 per 100 lb. f.o.b. Montreal, Toronto, Hamilton, London, St. John and Halifax.

FINE STEEL WIRE—This has been a quiet week in this line. The discount is still 17½ per cent. off the list.

BRASS AND COPPER WIRE—There is nothing new to report. The discount on brass is 55 and 2½ per cent., and on copper 50 and 2½ per cent.

FENCE STAPLES—Quite a number of orders have been filled this week. We quote: $3.25 for bright, and $3.75 for galvanized, per keg of 100 lb.

WIRE NAILS—The shipments during the week have been large and a good business is again reported. We quote $2.85 for small lots and $2.77½ for carlots, f.o.b. Montreal, London, Toronto, Hamilton and Gananoque.

CUT NAILS—The demand for cut nails has been rather heavier than usual this spring. Orders still continue to come in for fair-sized lots. We quote as follows: $2.35 for small and $2.25 for carlots; flour barrel nails, 25 per cent. discount; coopers' nails, 30 cent. discount.

HORSE NAILS—The situation is unchanged. "C" brand is held firmly at discount of 50 and 7½ per cent. off the new list. "M" brand is quoted at 60 per cent. off old list on oval and city head and 65 per cent. off countersunk head.

HORSESHOES—The demand is fair and the market steady. We quote as follows: Iron shoes, light and medium pattern, No. 2 and larger, $3.50; No. 1 and smaller, $3.75; snow shoes, No. 2 and larger, $3.75; No. 1 and smaller, $4.00; X L steel shoes, all sizes, 1 to 5, No. 2 and larger, $3.60; No. 1 and smaller, $3.85; feather-weight, all sizes, $4.85; toe weight steel shoes, all sizes, $5.95 f.o.b. Montreal; f.o.b. Hamilton, London and Guelph, 10c. extra.

POULTRY NETTING—This is one of the spring articles that is selling fairly well in small lots. The discount is 55 per cent.

GREEN WIRE CLOTH—A good business is again to be reported. The price is steady at $1.35 per 100 sq. ft.

SCREEN DOORS AND WINDOWS—The demand from the country has been quite heavy this week. We quote: Screen doors, plain cherry finish, $8.25 per doz.; do. fancy, $11.50 per doz.; walnut, $7.40 per doz., and yellow, $7.45; windows, $2.25 to $3.50 per doz.

SCREWS—The wholesale houses uniformly report a brisk trade. Discounts are as follows: Flat head bright, 87½ and 10 per cent. off list; round head bright, 82½ and 10 per cent.; flat head brass, 80 and 10 per cent. round head brass, 75 and 10 per cent.

BOLTS—The active inquiry we have noted for the past few weeks is being maintained. Discounts are: Norway carriage bolts, 65 per cent.; common, 60 per cent.; machine bolts, 60 per cent.; coach screws, 70 per cent.; sleigh shoe bolts, 72½ per cent.; blank bolts, 70 per cent.; bolt ends, 62½ per cent.; plough bolts, 60 per cent.; tire bolts, 67½ per cent.; stove bolts, 67½ per cent. To any retailer an extra discount of 5 per cent. is allowed. Nuts, square, 4c. per lb. off list; hexagon nuts, 4½c. per lb. off list. To all retailers an extra discount of ½c. per lb. is allowed.

BUILDING PAPER—The market remains as reported last week, steady and active. We quote: Tarred felt, $1.70 per 100 lb.; 2-ply, ready roofing, 80c. per roll; 3-ply, $1.05 per roll; carpet felt, $2.25 per 100 lb.; dry sheathing, 30c. per roll; tar sheathing, 40c. per roll; dry fibre, 50c. per roll; tarred fibre, 60c. per roll; O.K. and I.X.L., 65c. per roll; heavy straw sheathing, $28 per ton; slaters' felt, 50c. per roll.

RIVETS AND BURRS — Business is confined to sorting up proportions. Discounts on best iron rivets, section, carriage, and wagon box, black rivets, tinned do., coopers' rivets and tinned swedes rivets, 60 and 10 per cent.; swedes iron burrs are quoted at 55 per cent. off; copper rivets, 35 and 5 per cent. off; and coppered iron rivets and burrs, in 5-lb. carton boxes, are quoted at 60 and 10 per cent. off list.

BINDER TWINE — There is not a great deal doing. We quote: Blue Ribbon, 11½c.; Red Cap, 9¾c.; Tiger, 8¾c.; Golden Crown, 8c.; Sisal, 8¾c.

CORDAGE—Supplies of cordage still continue to be wanted, and a fair business is reported. Manila is worth 13½c. per lb. for 7-16 and larger; sisal is selling at 10c., and lathyarn 10c.

HARVEST TOOLS — Business is good at the former discount, 50, 10 and 5 per cent.

SPADES AND SHOVELS — The orders are diminishing in size. The discount is still 40 and 5 per cent. off the list.

LAWN MOWERS — The inquiry continues to be maintained, and quite a number of lawn mowers have been shipped this week. We quote as follows: High wheel, 50 and 5 per cent. f.o.b. Montreal; low wheel, in all sizes, $2.75 each net; high wheel, 11-inch, 30 per cent. off.

FIREBRICKS—The inquiry for firebricks during the past week has been very encouraging. Ex store they are quoted at $18 to $24, while ex wharf Scotch may be had for $17.50 to $22 and English at $17 to $21 per 1,000.

CEMENT — Trade is good in Canadian and American varieties. We quote as follows: German cement, $2.30 to $2.50; English, $2.25 to $2.35; American, $2.25 to $2:50, and Belgian, $1.70 to $1.95 for summer delivery, and spot prices are unchanged at $2.45 to $2.55 for German; $2.30 to $2.40 for English, and $1.95 to $2.05 for Belgian, ex store.

METALS.

The metal market shows much improvement this week. The market for pig iron is everywhere firm, and Canadian furnaces are filled with orders for some weeks to come. Brighter news also comes from England in regard to sheet metals. Black sheets are cabled 7s. 6d. higber and firm at the rise. Postal advices from England say that the tinplate market is firm, and, although the fact that one-third of the mills are yet closed down makes the market bearish, yet advances are being paid. Hoop iron is a little easier, and terne plates are rather lower. There has been some heavy dealing in galvanized iron. The London ingot tin market has advanced a number of points.

PIG IRON—The pig iron market is firm and the furnaces are filled up with orders for three or four weeks. The price of pig iron here is $18.50 for No. 1. Orders are said to have been taken at Hamilton for $17.50. No. 1 Summerlee is selling at the wharf at $20.50 to $21.

BAR IRON—Prices are steady and business quite active. Merchants' bar is worth $1.75, and horseshoe $2.

HOOP IRON — Hoop iron is rather easier this week from stock. One-inch, 19 gauge is worth $2.75, and No. 17, $3.

BLACK SHEETS — Prices vary with the different firms. The English market has sharply reacted and one broker here quotes an advance of 7s. 6d. We quote: 8 to 16 gauge, $2.50; 26 gauge, $2.55, and 28 gauge, $2.60.

GALVANIZED IRON — There have been some heavy import orders given for galvanized iron during the past week. The market is steady to firm. We quote: No. 28 Queen's Head, $4.65; Apollo, 10¾ oz., $4.50, and Comet, $4.40 to $4.45, with a 15c. reduction for case lots.

COPPER—The market is steady with values unchanged. The selling price now is 18c.

INGOT TIN—The London market is quoted £3 10s. higher on the week, at £122 10s. In New York, however, there is a lack of confidence in the stability of the London market and prices there are easy. Locally, quotations remain at 30 to 31c.

LEAD—The general quotation is $3.75, although we have heard of transactions taking place at lower figures.

LEAD PIPE—A fair trade is reported. We quote: 7c. for ordinary and 7½c. for composition waste, with 25 per cent. off.

IRON PIPE — A good trade is being done at rising figures. We quote as follows: Black pipe, ¼. $3 per 100 ft.; ⅜. $3; ½. $3 05; ¾. $3.30; 1-in., $4.70; 1¼, $6.40; 1½, $7.70; 2-in. $10.25. Galvanized, ½, $4.60; ¾, $5.25; 1-in., $7.50; 1¼, $9.80; 1½, $11.75; 2-in., $16.

TINPLATES—The tinplate market is firm in England. Stocks continue to decline. About one-third of the mills are, however, still quiet and only this prevents a rise. We quote: Coke plates, $3.75 to $4; charcoal, $4 to $4.50.

CANADA PLATE—The feeling in Canada plate is better. We quote: 52's, $2.50 to $2.60; 60's, $2.60 to $2.70; 75's, $2.70 to $2.80; full polished, $3.10, and galvanized, $4.

STEEL—Unchanged. We quote: Sleigh-shoe, $1.95; tire, $2; bar, $1.95; spring, $2.75; machinery, $2 75, and toe-calk, $2.50.

SHEET STEEL—The market is quiet and steady. We quote: Nos. 22 and 24, $3, and Nos. 18 and 20, $2.85.

TOOL STEEL—Black Diamond is worth 8c. and Jessop's, 13c.

TERNE PLATES—The English market is steady. From stock, plates are worth $7.50 and $7.75.

COIL CHAIN—The market is steady. We quote as follows: No. 6, 11½c.; No. 5, 10c.; No. 4, 9½c.; No. 3, 9c.; ¼-inch, 7½c. per lb.; 5-16, $4.85; 5-16 exact, $5.30; ⅜, $4.40; 7-16, $4.20; ½, $3.95; 9-16, $3.85; ⅝, $3.55; ¾, $3.45; ⅞, $3.40; 1-in., $3.35. In carload lots an allowance of 10c. is made.

SHEET ZINC — Business is confined to small lots, which are worth $5.75 to $6.25.

ANTIMONY—Quiet, at 10c.

ZINC SPELTER—Is worth 5c.

SOLDER—We quote: Bar solder, 18½c.; wire solder, 20c.

GLASS.

Cable advices from Belgium indicate that the glass market there has advanced 15 per cent. during the past week. This makes the market here much firmer. We quote as follows: First break, $2; second, $2.10 for 50 feet; first break, 100 feet, $3.80; second, $4; third, $4.50; fourth, $4.75; fifth, $5.25; sixth, $5.75, and seventh, $6.25.

PAINTS AND OILS.

Turpentine is coming forward a little more freely and quotations have been somewhat relaxed, 53c. now being the figure for single barrels. The meeting of the White Lead Association being over and no change in prices having been made the market is now on a much more satisfactory footing as the feeling of uneasiness has been removed. It is anticipated that lead prices, barring any unforseen accident, will remain unchanged for some time. There is no diminution in the immense turnover in liquid paints, and, as we hinted in our last issue, with the continued stiffening on the price of linseed oil, enhanced figures for paints may be looked for in the near future. It is well to point out to prospective buyers of oil that, although the present figures seem high, prices in the Old Country give grounds for further advances, and everything points to an immediate rise, as the laid down cost to import is above our schedule to the retailers. We quote:

WHITE LEAD—Best brands, Government standard, $6.25; No. 1, $5.87½; No. 2, $5.50; No. 3, $5.12½, and No. 4, $4.75 all f.o.b. Montreal. Terms, 3 per cent. cash or four months.

DRY WHITE LEAD—$5.25 in casks; kegs, $5.50.

RED LEAD—Casks, $5.00; in kegs, $5.25.

DRY WHITE ZINC—Pure, dry, 6¾c.; No. 1, 5¾c.; in oil, pure, 7¾c.; No. 1, 6¾c.; No. 2, 5¾c.

PUTTY—We quote: Bulk, in barrels, $1.90 per 100 lb.; bulk, in less quantity, $2 05; bladders, in barrels, $2.10; bladders, in 100 or 200 lb. kegs or boxes, $2.25; in tins, $2.55 to $2.65; in less than 100-lb. lots, $3 f.o.b. Montreal, Ottawa, Toronto, Hamilton, London and Guelph. Maritime Provinces 10c. higher, f.o.b. St. John and Halifax.

LINSEED OIL—Raw, 77c.; boiled, 80c. in 5 to 9 bbls., 1c. less, 10 to 20 bbl. lots, open, net cash, plus 2c. for 4 months Delivered anywhere in Ontario between Montreal and Oshawa at 2c. per gal. advance and freight allowed.

TURPENTINE—Single bbls., 53c.; 2 to 4 bbls., 52c.; 5 bbls. and over, open terms, the same terms as linseed oil.

MIXED PAINTS—$1.25 to $1.45 per gal.

CASTOR OIL—8¾ to 9¾c. in wholesale lots, and ½c. additional for small lots.

SEAL OIL—47½ to 49c.

COD OIL—32½ to 35c.

NAVAL STORES — We quote: Resins, $2.75 to $4.50, as to brand; coal tar, $3.25 to $3.75; cotton waste, 4½ to 5½c. for colored, and 6 to 7½c. for white; oakum, 5½ to 6½c., and cotton oakum, 10 to 11c.

PARIS GREEN—Petroleum barrels, 16¾c. per lb.; arsenic kegs, 17c.; 50 and 100-lb. drums, 17½c.; 25-lb. drums, 18c.; 1-lb. packages, 18½c.; ½-lb. packages, 20½c.; 1-lb. tins, 19½c.; ½-lb. tins, 21½c. f.o.b. Montreal; terms 3 per cent. 30 days, or four months from date of delivery.

SCRAP METALS.

The market is active and firm. Higher prices are looked for on scrap iron. The arrivals from England rather weakened the market, but a reaction is confidently expected. Old rubbers are ½c. per lb. higher. Dealers are now paying the following prices in the country: Heavy copper and wire, 13 to 13½c. per lb.; light copper, 12c.; heavy brass, 12c.; heavy yellow, 8½ to 9c.; light brass, 6½ to 7c.; lead, 2½ to 2¾c. per lb.; zinc, 2¼ to 2½c.; iron, No. 1 wrought, $15 to $16 per gross ton f.o.c. Montreal; No. 1 cast, $13 to $14; stove plate, $8 to $9; light iron, No. 2, $4 a ton; malleable and steel, $4; rags, country, 70 to 80c. per 100 lb.; old rubbers, 6¼c. per lb.

HIDES.

There is no change to report in the hide market as dealers are still paying 6½c. for No. 1 light and tanners 7½c. The receipts are improving in quality, and the demand from tanners prevents any accumulation. Dry hides in New York are still firm and in limited supply. We quote: Light hides, 6½c. for No. 1; 5½c. for No. 2, and 4½c. for No. 3. Lambskins, 10c.; sheepskins, 90c.; calfskins, 8c. for No. 1 and 6c. for No. 2.

PETROLEUM.

There is no change. We quote: "Silver Star," 14½ to 15½c.; "Imperial Acme," 16 to 17c.; "S C. Acme," 18 to 19c., and "Pratt's Astral," 18½ to 19½c.

MONTREAL NOTES.

Old rubbers are ½c. per lb. higher.

Glass has advanced 15 per cent. in Belgium.

The discount on steel squares has been raised to 60, 10 and 5 per cent.

Black sheets have advanced 7s. 6d. per ton in England and tinplates 3d. per box.

ONTARIO MARKETS.

TORONTO, May 18, 1901.

HARDWARE.

WHILE a good trade is being done, it is hardly expected that the volume of business will be as large for May as it was for April. Refrigerators, ice cream freezers, oil stoves, snaths, scythes, churns and enamelled ware are among the most active lines. On Thursday some small shipments of barb wire arrived, but the quantity is in no way sufficient to fill the orders that are on the books, and there is still in consequence a great deal of scarcity. The same remarks apply to galvanized wire. Oiled and annealed wire is still in fairly good demand. Not much is being done in wire nails and the few orders that are coming in are for small quantities. The feature in cut nails is the demand for the shingling description, which somewhat at the moment exceeds the supply. Just a light business is being done in horse nails and horseshoes. A fairly good trade is being experienced in sporting goods. An active trade is being done in screen doors and windows and poultry netting. Trade in building paper continues fairly active. In rivets and burrs and bolts business is still good. Rope is in fairly good demand. Payments are fairly satisfactory.

BARB WIRE — Some shipments were received in Toronto within the last couple of days which were promised early in the month. They are, however, quite insufficient to fill the orders that are on the books, and there is still in consequence quite a scarcity. Prices are steady and unchanged. We quote $3.05 per 100 lb. from stock Toronto; f.o.b. Cleveland $2.82½ per 100 lb. for less than carlots, and $2.70 for carlots.

GALVANIZED WIRE — There is still a scarcity of galvanized wire and the demand is only fair. The scarcity is most marked in Nos. 12 and 13, which are used for fencing purposes. We quote as follows: Nos. 6, 7 and 8, $3.50 to $3.85 per 100 lb., according to quantity; No. 9, $2.85 to $3.15; No. 10, $3.60 to $3.95; No. 11, $3.70 to $4.10; No. 12, $3 to $3.30; No. 13, $3.10 to $3.40; No. 14, $4.10 to $4.50; No. 15, $4.60 to $5.05; No. 16, $4.85 to $5.35. Nos. 6 to 9 base f.o.b. Cleveland are quoted at $2.57½ in less than carlots and 12c. less for carlots of 15 tons.

SMOOTH STEEL WIRE—The scarcity of barb and plain galvanized wire is still stimulating the demand for oiled and annealed wire. The net selling prices for oiled and annealed are as follows: Nos. 6 to 8, $2.90; 9, $2.80; 10, $2.87; 11, $2.90; 12, $2.95; 13, $3.15; 14, $3.37; 15, $3.50; 16, $3.65. Delivery points, Toronto, Hamilton, London and Montreal, with freights equalized on these points.

WIRE NAILS —The demand is light and is only for small quantities. Prices are steady and unchanged at $2.85 for less than carlots and $2.77½ for carlots. Delivery points: Toronto, Hamilton, London, Gananoque and Montreal.

CUT NAILS—There is a good demand for 4d. nails, the shingling description, which are scarce. There is practically none to be had in Toronto, Hamilton or Montreal. Trade in cut nails generally is, however, dull. We quote the base price at $2.35 per keg for less than carlots, and $2.25 for carlots. Delivery points: Toronto, Hamilton, London, Montreal and St. John, N.B.

HORSESHOES—There is only a light demand for horseshoes, but prices are steady. We quote as follows f. o. b. Toronto: Iron shoes, No. 2 and larger, light, medium and heavy, $3.60; snow shoes, $3.85; light steel shoes, $3.70; featherweight (all sizes), $4.95; iron shoes, No. 1 and smaller, light, medium and heavy (all sizes), $3.85; snow shoes, $4; light steel shoes, $3.95; featherweight (all sizes), $4.95.

HORSE NAILS — There is a little being done, but the demand is never brisk at this time of the year. Discount on "C" brand, oval head, 50 and 7½ per cent. off new list, and on "M" and other brands, 50, 10 and 5 per cent., off the old list. The discount on countersunk head is 60 per cent.

SCREWS — A good trade continues to be done in screws at steady prices. We quote discounts: Flat head bright, 87½ and 10 per cent.; round head bright, 82½ and 10 per cent.; flat head brass, 80 and 10 per cent.; round head brass, 75 and 10 per cent.; round head bronze, 65 per cent., and flat head bronze at 70 per cent.

RIVETS AND BURRS — There is quite an active demand for rivets. The boilermakers have been particularly free purchasers lately, as they are all busily employed. The strike which has taken place among the boilermakers, may, however, interfere with business. We quote: Iron rivets, 60 and 10 per cent.; iron burrs, 55 per cent.; copper rivets and burrs, 35 and 5 per cent.

BOLTS AND NUTS — There is a large demand for bolts, notwithstanding that purchases were pretty heavy before the recent advance. We quote: Carriage bolts (Norway), full square, 65 per cent.; carriage bolts full square, 65 per cent.; common carriage bolts, all sizes, 60 per cent.; machine bolts, all sizes, 60 per cent.; coach screws, 70 per cent.; sleighshoe bolts, 72½ per cent.; blank bolts, 60 per cent.; bolt ends, 62½ per cent.; nuts, square, 4c. off; nuts, hexagon, 4½c. off; tire bolts, 67½ per cent.; stove bolts, 67½; plough bolts, 60 per cent.; stove rods, 6 to 8c.

ROPE—There is a good demand for rope of 1 and 1¼-inch sizes. The base price is steady and unchanged at 10c. for sisal, and 13½c. for manila.

BINDER TWINE—It is yet a little early for the sorting up trade, and business in consequence is still quiet.

CUTLERY—Business in cutlery during the past week has been quiet and featureless.

SPORTING GOODS—There has been a good business doing in gunpowder and shot, but very little in loaded shells. A fairly-steady trade is reported for this time of the year in firearms.

ENAMELLED WARE—A large trade has been doing, and prices are steady and unchanged.

TINWARE—Sales in tinware, as far as we can learn, are larger than they were a year ago. The demand for kitchen utensils is particularly good just now.

OIL STOVES—The activity noted in previous issues has been more than maintained, and the outlook is for a good season's trade.

ICE CREAM FREEZERS AND REFRIGERATORS—Trade in both these lines continues active.

CHURNS, ETC.—In churns, washers and wringers the demand is exceedingly brisk, and stocks are short, so that difficulty is being experienced in filling orders.

GREEN WIRE CLOTH—A good deal of green wire cloth has been sold ahead, but the sorting up business has not yet begun. We still quote $1.35 per 100 sq. ft.

SCREEN DOORS AND WINDOWS—There is a good demand for screen doors and windows, and some difficulty is also being experienced in this line in getting delivery from the factories.

BUILDING PAPER—Trade is fairly good and manufacturers claim that this is one of the best seasons they have had. We quote: Plain building, 30c.; tarred lining, 40c., and tarred roofing, $1.65.

POULTRY NETTING—An active demand is being experienced for poultry netting, and some difficulty is being experienced in getting supplies from the manufacturers, who in turn are finding difficulty in getting their raw material from the factories in the United States. Discount on Canadian is still 55 per cent.

HARVEST TOOLS—There is a good movement in snaths, scythes and forks. Discount, 50, 10 and 5 per cent.

SPADES AND SHOVELS—These are going out fairly well. Discount, 40 and 5 per cent.

EAVETROUGH—The demand is not quite as brisk as it was, the trade now being fairly well supplied. There is still, however, a very good movement.

METALS.

The metal trade continues fairly active. Prices on Canada plate are a little easier, and pig tin during the last few days has shown a little weakness. The demand is particularly active in galvanized sheets.

PIG IRON—Business is rather quiet, but prices are steady and nominally unchanged at $16.50 per ton for No. 2 and $16 for No 3.

BAR IRON—There does not appear to be any falling off in the demand for bar iron, and some, at least, of the mills are still refusing to take orders for prompt shipment. The ruling base price is $1.85 to $1.90 per 100 lb., according to quantity.

STEEL—The steel trade continues one of the active features of the metal business, and prices rule firm. We quote as follows: Merchantable cast steel, 9 to 15c. per lb.; drill steel, 7 to 8c. per lb.; "Black Diamond" tool steel, 10 to 11c.; Jessop's and Firth's tool steel, 13c.; toe calk steel, $2.85 to $3; tire steel, $2.30 to $2.50; sleighshoe steel, $2.10 to $2.25; reeled machinery steel, $3; hoop steel, $3.10.

PIG TIN—Up to Wednesday there were some sharp advances in pig tin on the London market, but, as New York failed to respond, prices in the former city on that day dropped 15s. per ton for spot tin and £1 per ton for futures. Locally, the demand is rather light, and 31c. is the ruling price for Lamb and Flag.

TINPLATES—There is a fairly good movement in tinplates and prices are steady, the English market having much improved during the last week or two.

TINNED SHEETS—The demand is pretty well over, and there is not, in consequence, much movement. We still quote 9¼ to 9½c. per lb.

GALVANIZED SHEETS—This is about the most active line in the metal trade, and stocks are light. An occasional order for importation is being placed. We quote $4.60 from stock for English and $4.50 for American.

BLACK SHEETS—The English market is much stronger on black sheets, but, locally, there is no change in quotations. We still quote 28 gauge at $2.30.

CANADA PLATES—These are a little easier and we now quote: All dull, $2.90; half polished, $3, and all bright, $3.50.

COPPER—The copper market has been fairly steady, although prices are a little lower in London, England. Locally there is no change. We quote: Ingot, 19c.; bolt or bar, 23½ to 25c.; sheet, 23 to 23½c. per lb.

SOLDER—The demand is good, and prices unchanged. We quote: Half-and-half, guaranteed, 18½c.; ditto, commercial, 18c.; refined, 18c., and wiping, 17c.

IRON PIPE—Prices continue steady, but there is not much business being done. One-inch black pipe is quoted at $5.15, and 1-inch galvanized at $7.55 per 100 ft.

LEAD—The market is quiet, and prices steady at 4¼ to 4½c. per lb.

ZINC SPELTER—Trade is still only light, but prices are steady. We quote 5½ to 6c.

ZINC SHEET—While the trade does not expect much business at this time of the year, some fairly good quantities have been shipped during the past week. We still quote 6¼c. for casks and 6¾c. for part casks.

ANTIMONY—Trade is quiet and prices steady at 10 to 11½c. per lb.

PAINTS AND OILS.

A big movement continues. The demand for prepared paints, varnishes, white lead, sundries, etc., is excellent. The demand for turpentine is fair. Linseed oil is in good request, but stocks, both in manufacturers' and jobbers' hands are extremely light, and, as the British market is steadily advancing, the feeling here stiffens daily. An advance of 3c. has been made this week, and a further rise is looked for. Other lines are unchanged. Turpentine is easy in the primary market. We quote:

WHITE LEAD—Ex Toronto, pure white lead, $6.37½; No. 1, $6; No. 2, $5.67½; No. 3, $5.25; No. 4, $4.87½; genuine dry white lead in casks, $5.37½.

RED LEAD—Genuine, in casks of 560 lb., $5.50; ditto, in kegs of 100 lb., $5.75; No. 1, in casks of 560 lb., $5; ditto kegs of 100 lb., $5 25.

LITHARGE—Genuine, 7 to 7½c.

ORANGE MINERAL—Genuine, 8 to 8½c.

WHITE ZINC—Genuine, French V.M., in casks, $7 to $7.25; Lehigh, in casks, $6.

PARIS WHITE—90c. to $1 per 100 lb.

WHITING — 70c. per 100 lb.; Gilders' whiting, 80c.

GUM SHELLAC — In cases, 22c.; in less than cases, 25c.

PARIS GREEN—Bbls., 16¾c.; kegs, 17c.; 50 and 100 lb. drums, 17½c.; 25-lb. drums, 18c.; 1-lb. papers, 18½c.; 1-lb. tins, 19½c.; ½-lb. papers, 20½c.; ½-lb. tins, 21½c.

PUTTY — Bladders, in bbls., $2.10; bladders, in 100 lb. kegs, $2.25; bulk in bbls., $1.90; bulk, less than bbls. and up to 100 lb., $2.05; bladders, bulk or tins, less than 100 lb., $2 90.

PLASTER PARIS—New Brunswick, $1.90 per bbl.

PUMICE STONE — Powdered, $2.50 per cwt. in bbls., and 4 to 5c. per lb. in less quantity; lump, 10c. in small lots, and 8c. in bbls.

LIQUID PAINTS—Pure, $1.20 to $1.30 per gal.

CASTOR OIL—East India, in cases, 10 to 10½c. per lb. and 10½ to 11c. for single tins.

LINSEED OIL—Raw, 1 to 4 barrels, 78c.; boiled, 81c.; 5 to 9 barrels, raw, 77c.; boiled, 80c., delivered. To Toronto, Hamilton, Guelph and London, 1c. less.

TURPENTINE—Single barrels, 54c.; 2 to 4 barrels, 53c., delivered. Toronto, Hamilton and London 1c. less. For less quantities than barrels, 5c. per gallon extra will be added, and for 5-gallon packages, 50c., and 10 gallon packages, 80c. will be charged.

GLASS.

There is no change. The feeling keeps firm and the demand steady. We quote as follows: Under 26 in., $4.15 26 to 40 in., $4 45; 41 to 50 in., $4 85; 51 to 60 in., $5 15; 61 to 70 in., $5.50; double diamond, under 26 in., $6; 26 to 40 in., $6.65; 41 to 50 in., $7.50; 51 to 60 in., $8.50; 61 to 70 in., $9.50, Toronto, Hamilton and London. Terms, 4 months or 3 per cent. 30 days.

OLD MATERIAL.

Prices are unchanged. The demand keeps firm. We quote jobbers' prices: Agricultural scrap, 55c. per cwt.; machinery cast, 60c. per cwt.; stove cast, 50c.; No. 1 wrought 50c. per 100 lb.; new light scrap copper, 12c. per lb.; bottoms, 11½c.; heavy copper, 13c.; coil wire scrap, 13c.; light brass, 7c.; heavy yellow brass, 10 to 10½c.; heavy red brass, 10½ to 11c.; scrap lead, 3c.; zinc, 2c; scrap rubber, 6½c.; good country mixed rags, 65 to 75c.; clean dry bones, 40 to 50c. per 100 lb.

PETROLEUM.

A decline of ½c. has been made in all lines but American water white. This is due to the reduced price of crude oil noted last week. We quote: Pratt's Astral, 16 to 16½c. in bulk (barrels, $1 extra); American water white, 16½ to 17c. in barrels; Photogene, 15½ to 16c.; Sarnia water white, 15 to 15½c. in barrels; Sarnia prime white, 14 to 14½c. in barrels.

COAL.

A good demand is reported, owing to the low prices for May delivery. We quote at international bridges as follows: Grate, $3.75 per gross ton; egg, stove and nut, $5 per gross ton with a rebate of 40c. off for May shipment.

MARKET NOTES.

Linseed oil is 3c. per gal. dearer in Toronto.

There has been a decline of ½c. per gal. in all lines of Canadian petroleum.

A little barb wire arrived this week from the manufacturers, but it was soon all shipped out again.

THE BRITISH INCOME TAX.

The British citizen is wont to look upon the income tax as the most objectionable of the various duties levied upon him by the State because it is a direct tax; and though we have now grown accustomed to it as one of the necessary evils to which flesh is heir, time was, and that not so many years ago, when John Bull regarded this imposition with anything but the equanimity of to-day," remarks The London Express. "It is 102 years since an income tax, as we understand it, was first imposed, and from 1792 to 1901 there have been no fewer than thirty-nine changes in the amount of the income tax, which has varied from 2 pence to 2 shillings in the pound. A penny in the pound nowadays realizes £2,312,000. It was in order to furnish the means for defraying the expenses of the war begun with France in 1793 that Pitt proposed, amid great opposition, a tax on incomes. This tax, which produced an average of £5,500,000 annually, was repealed after the peace of Amiens in 1802, but in the following year the income tax was revised under the name of the property tax, though it was finally repealed on Brougham's motion in 1816. The next appearance of the income tax was in 1842, when Sir Robert Peel's Bill imposing a tax at the rate of 7 pence in the pound on incomes of £150 and upward was passed, the measure being limited in its operation to three years, and was renewed for similar periods in 1845 and 1848. Such is the story of the income tax in brief. In 1842 it yielded £5,710,550; in 1852, £5,509,637; in 1862, £10,300,000; in 1872, £9,000,000; in 1882, £9,950,000; in 1892, £13,810,000, and last year, £25,300,000."

There is a movement on foot to have all stores in St. Thomas, Ont., close at 9 p.m. on Saturday evening instead of about midnight. In order to facilitate this, manufacturers in that city have been asked to pay their employes on Friday night, or earlier in the week, and most of them have expressed readiness to do so.

MANITOBA MARKETS.

WINNIPEG, May 13, 1901.

MARKET is steady. With fair business in all lines, there have been some revisions of price lists, notably in linseed oil, nails, Swedish iron, Russia iron, Canada plates, and rope. The bulk of trade at present is coming from the city rather than the country, although a good country trade is anticipated later.

In paints and oils there has been some rather unexpected movements during the past week. Linseed oil has advanced 5c. per gallon, making present quotations 80c. for raw and 85c. for boiled. This advance is in sympathy with the rise in the English market.

Glass is very firm and all advices reaching this point from Belgium indicate that, as the strike troubles are still on, there is little likelihood that orders from this side can be filled. There is equally no possibility of getting the needed supply from the American side. All these things tend to make prices very firm here.

Quotations for the week are as follows:

Barbed wire, 100 lb.	$3 45
Plain twist	3 45
Staples	3 95
Oiled annealed wire 10	3 95
" 11	4 00
" 12	4 05
" 13	4 20
" 14	4 35
" 15	4 45
Wire nails, 30 to 60 dy, keg	3 40
" 16 and 20	3 45
" 10	3 50
" 8	3 60
" 6	3 65
" 4	3 80
" 3	4 05
Cut nails, 30 to 60 dy.	2 90
" 20 to 40	2 95
" 10 to 16	3 00
" 8	3 05
" 6	3 10
" 4	3 20
" 3	3 55
Horsenails, 45 per cent. discount.	
Horseshoes, iron, No. 0 to No 1	4 65
No. 2 and larger	4 40
Snow shoes, No. 0 to No. 1	4 90
No. 2 and larger	4 40
Steel, No. 0 to No. 1	4 95
No. 2 and larger	4 70
Bar iron, $2.50 basis.	
Swedish iron, $5.00 basis.	
Sleigh shoe steel	3 00
Spring steel	3 25
Machinery steel	3 75
Tool steel, Black Diamond, 100 lb	8 50
Jessop	13 00
Sheet iron, black, 10 to 20 gauge, 100 lb.	3 50
20 to 26 gauge	3 75
28 gauge	4 00
Galvanized American, 16 gauge	2 54
18 to 22 gauge	4 50
24 gauge	4 75
26 gauge	5 00
28 gauge	5 25
Genuine Russian, lb	12
Imitation " "	8
Tinned, 24 gauge, 100 lb.	7 75
26 gauge	8 00
28 gauge	8 50
Tinplate, IC charcoal, 20 x 28, box	10 75
" IX "	12 75
" IXX "	14 75
Ingot tin	33
Canada plate, 18 x 21 and 18 x 24	3 25
Sheet zinc, cask lots, 100 lb	7 50
Broken lots	8 00
Pig lead, 100 lb	6 00
Wrought pipe, black up to 2 inch	50 an 10 p.c.
" Over 2 inch	50 p.c.
Rope, sisal, 7-16 and larger	$11 00
" 3/8	11 50
" 1/4 and 5-16	12 00
Manila, 7-16 and larger	14 00
" 3/8	14 50
" 1/4 and 5-16	15 00

Solder	20
Cotton Rope, all sizes, lb.	16
Axes, chopping	$ 7 50 to 12 00
" double bitts	12 00 to 18 00
Screws, flat head, iron, bright	87½
Round " "	82½
Flat " brass	80
Round " "	75
Coach	57½ p.c.
Bolts, carriage	55 p.c.
Machine	55 p.c.
Tire	60 p.c
Sleigh shoe	65 p.c.
Plough	40 p.c.
Rivets, iron	50 p.c.
Copper, No. 8	35
Spades and shovels	40 p.c.
Harvest tools	50, and 10 p.c.
Axe handles, turned, s. g. hickory, doz.	$2 50
No. 1	1 50
No. 2	1 25
Octagon extra	1 75
No. 1	1 25
Files common	70, and 10 p.c.
Diamond	60
Ammunition, cartridges, Dominion R.F.	50 p.c.
Dominion, C.F., pistol	30 p.c.
" military	15 p.c.
American R.F.	30 p.c.
C.F. pistol	5 p.c.
C.F. military	10 p.c. advance.
Loaded shells:	
Eley's soft, 12 gauge black	16 50
chilled, 12 guage	18 00
soft, 10 guage	21 00
chilled, 10 guage	23 00
Shot, Ordinary, per 100 lb	6 25
Chilled	7 15
Powder, F.F., keg	4 75
F.F.G.	5 00
Tinware, pressed, retinned	75 and 2½ p.c.
" " plain	70 and 15 p.c.
Graniteware, according to quality	50 p.c.

PETROLEUM.

Water white American	25½c.
Prime white American	24c.
Water white Canadian	22c.
Prime white Canadian	21c.

PAINTS, OILS AND GLASS.

Turpentine, pure, in barrels	$ 63
Less than barrel lots	71
Linseed oil, raw	70
Boiled	73
Lubricating oils, Eldorado castor	25½
Eldorado engine	24½
Atlantic red	27½
Renown engine	41
Black oil	23½ to 25
Cylinder oil (according to grade)	55 to 74
Harness oil	61
Neatsfoot oil	$ 1 00
Steam refined oil	85
Sperm oil	1 50
Castor oil per lb.	11½
Glass, single glass, first break, 16 to 25 united inches	2 25
26 to 40 per 50 ft.	2 50
41 to 50 " 100 ft.	5 50
51 to 60 " " "	6 00
61 to 70 per 100-ft. boxes	6 50
Putty, in bladders, barrel lots per lb.	2½
kegs "	2¾
White lead, pure per cwt.	7 00
No 1 "	6 75
Prepared paints, pure liquid colors, according to shade and color. per gal.	$1.30 to $1.90

BRITISH BUSINESS CHANCES.

Firms desirous of getting into communication with British manufacturers or merchants, or who wish to buy British goods on the best possible terms, or who are willing to become agents for British manufacturers, are invited to send particulars of their requirements for

FREE INSERTION

in "Commercial Intelligence," to the Editor

'SELL'S COMMERCIAL INTELLIGENCE,' 168 Fleet Street, London, England.

"Commercial Intelligence" circulates all over the United Kingdom amongst the best firms. Firms communicating should give reference as to bona fides.

N.B.—A free specimen copy will be sent on receipt of a post card.

NOTES.

Among the hardware travellers in town this week were Geo. A. Boomer representing The Plume & Atwood Company, manufacturers of brass and copper goods, Connecticut. This is Mr. Boomer's first trip here and he was both surprised and pleased with the city; Edward M. Levy, representing The R. Hoehm Company of New York, thermometers, barometers, etc.; Mr. Griffin, representing The J. Stevens Arms & Tool Company, of Chicopee Falls, Mass.

HEATING AND PLUMBING

AN INTERESTING LAW SUIT.

THERE is now before the courts of Sherbrooke, Que., a lawsuit, Blondin vs. Whiting, that intimately concerns the plumbing trade. A. Blondin & Co., plumbers, of St. Hyacinthe, are suing Madame Whiting, of Sherbrooke, for $5,500, the stipulation of a contract for the heating of a house for the respondent.

It seems that A. Blondin & Co. accepted a contract in which they were obliged to put in 4,400 sq. ft. of radiation. They further agreed to instal more if such was needed to produce a temperature of 70 deg. when the thermometer registered 10 deg. below zero. One special clause of the contract mentioned that they were to get no money till entire satisfaction was given the proprietor.

Mr. Blondin now claims to have put in 600 sq. ft. of radiation over his obligations, and to have put himself to nearly $1,200 expense over his contract in the attempt to please the proprietor. It seems that the building was erected by an American contractor, who was supervised by an American architect, and Mr. Blondin's witnesses held the dwelling was not built for our climate. The argument is then advanced that heating contracts are always taken subject to the normal condition that the building is built so as to be habitable in our climate.

Since the difficulties commenced, the proprietor has taken another plumber into her employ, changed the position of the radiators, nearly doubling their quantity, changed the patterns, put the radiators into the bay windows, and, when the new system was tested, last March, it is alleged that it did not seem, even on a fairly warm day and with special fires, to give any more heat than is ordinarily necessary in a building of the kind.

The judgment shall tell us when a proprietor shall declare himself satisfied. On the other hand, it shows us how careless many contractors are in agreeing to unbusinesslike stipulations.

Mr. Blondin has called in several plumbing experts to his aid, including Messrs. Lamarche, Ballantyne and John Garth, of Montreal, and several architects, including Messrs. Wright and Gauthier, of the same city. These gentlemen have all testified in his favor, the architects saying that the building is fit only for Southern habitation, and the plumbers that Mr. Blondin's work is fully up to the standard.

PLUMBING AND HEATING NOTES.

The Windsor Electric Co. have registered in Windsor Mills, Que.

Couiller & Gauthier, contractors, Boucherville, Que., have dissolved.

The Iroquois Electric Light & Power Co., Limited, Iroquois, Ont., have been incorporated.

J. C. Orr & Co., have opened a plumbing shop at 350 Williams avenue, Winnipeg.

Assignment has been demanded of Mrs. Maxime Deslauriers, contractor, Montreal.

Thorold, Ont., has voted $6,000 for extending the electric light system in that town.

St. Armour & Doucet have been appointed curators of Adelard Binette, plumber, Lachine, Que.

Hormisdas Lewis has registered under the style of J. E. Lewis & Co., plumbers, etc., Nicolet, Que.

The stock of A. Binette, plumber, Lachine, Que., is offered for sale on the 22nd instant.

Gagnon & Caron have been appointed curators of Hypolite Gougeon, contractor, St. Henri de Wendover, Que.

J. A. Chesterfield, plumber, etc., St. Marys, Ont., has moved into the store in the Opera House block, in that place.

BUILDING PERMITS ISSUED.

The following building permits have been issued in Toronto: G. E. Dobson, two-storey $1,000 dwelling near Howland ave. on Simpson ave.; Mr. Prowser, two houses on Woodbine ave., to cost $1,100; to Thos. W. Murray, for a dwelling near Howland ave., on Callendar street, to cost $5,500; to T. K. Haffey, for a house at 216 Wilton ave., and to W. T. Kernahan, for two houses at 31 and 33 Wellesley, to cost $5,000.

Building permits have been taken out in Ottawa by A. Paquette, lot 11, Queen street west, for brick veneered shop and dwelling, $4,000; Hon. E. H. Bronson, lots 1 and 2, Nepean street, solid brick dwelling, $10,000; William H. Munson, lot 5, Lett street, brick veneered dwelling, $1,000; Theodore St. Germain, lot 61, Gilmour street, brick veneered dwelling, $2,000; Wm. Thackray, lot 32, James street, brick veneered dwelling, $3,500; Dr. J. G. Scott, lot 58, Elgin street, solid brick veneered dwelling and office, $6,000; R. J. Cowie, lot 21, James street, double brick house, $4,000; S. J. Davis, lots 41, 42 and 43, Jane street, three frame dwellings, $1,800; Holtby & Shearer, lot 4, Hickey street, brick veneered dwelling, $1,200; E. L. Horwood, on behalf of Sacred Heart Convent, lots 3 and 4, Rideau street, stone building, $8,000; Alex. Garvock, lot 46, Lewis street, solid brick dwelling, $2,500; Thomas Saunders, lot 41, Second ave., brick veneered dwelling, $1,400.

Building permits have been issued in Hamilton to Peter Erskine for the erection of two dwellings and store at the corner of Tom and Dundurn streets, to cost $3,500; W. P. Witton, for alterations to the dwelling owned by W. M. Watson, 176 Hughson street south, to cost $450; W. P. Witton, for the erection of nine brick dwellings for J. M. Peregrine, on St. Matthew ave., between Barton and Birge streets, to cost $9,000.

The following permits have been taken out in Quebec: J. B. Jinchereau, for a two-storey brick dwelling on Julia street, to cost $3,000; F. K. Maheux, for a wood and brick building on d'Aiguillon street; J. Berube, for repairing of property on Madelaine street to extent of $1,400; F. Parent, for building on Fabrique street; Metropolitan Insurance Company, for building on St. John street, to cost $21,000; P. Godbout, for property on Notre-Dame street, to cost $2,500; Ls. Boivin, for repairing property on St. John street to extent of $3,850, and F. U. Leveille, for repairing of property on Desfosses street to extent of $175.

SOME BUILDING NOTES.

A new school is being built at Grenfell, Man.

A. Cottle is building a residence in Exeter, Ont.

L. McCallum is building a new residence in Hubrey, Ont.

O. B. Henry is erecting a hardware store in Drayton, Ont.

Moses Knight will build a house in Bradford, Ont., this summer.

David Kitchen is building a new blacksmith shop in Waterford, Ont.

McKay Bros. are building a new house for Wm. Baker, Bennington, Ont.

Erskine Church, Ottawa, propose erecting a $9,000 church on Concession street.

Agnew & Co. are erecting a two-storey 40 x 36 ft. building in Rossland, B.C.

The Salvation Army, at Brantford, Ont., have asked for tenders for new barracks.

The Bell Telephone Co. have received tenders for new offices in St. John, N.B.

E. W. Stone expects to start the erection of a new hotel in Carstairs, Ont., in a few days.

Joseph Taylor will erect a brick business block in Portage la Prairie, Man., this summer.

Pybus Bros., Napanee, have the contract for a brick residence for H. A. Baker, Colebrook, Ont.

Arthur Davis is building a new harness shop and D. Laport a new blacksmith shop in Warren, Ont.

Benj. Miller, John Miller and Robert McConachie are erecting new houses in Maple Grove, Ont.

About $1,000 will be spent on improvements to St. Alban's church, Grand Valley, Ont., this summer.

J. A. Russell, architect, Stratford, Ont., is calling tenders for a new Methodist church in Midland, Ont.

Mr. Wilson, of the G. T. R., Ed. Burling and O. R. Ferguson are building new residences in Cookstown, Ont.

Tenders are asked by J. W. Brown for alterations and additions to the Hanover, Ont., Presbyterian church before June 1.

Architect Robert Fawcett is preparing plans for a new residence on Vidal street, Sarnia, Ont., for Capt. A. E. McGregor, to cost $2,500.

The Department of Public Works is putting up a modern 150 x 75 ft. building in Quebec, for the use of the U. S. immigration officials.

Contractor Horribin, Vancouver, is building a large new factory and warehouse on the C. P. R. track, south of the drill hall, for The Thorpe Soda Water Works, of that city.

PLUMBING AND HEATING CONTRACTS.

The Bennett & Wright Co., Toronto, have the contract for plumbing and gas-fitting in houses at 93, 95 and 97 Alexander street, for R. Jaffray.

J. Ballantyne & Co., Montreal, have the contract for plumbing, heating and ventilating; The Fensom Elevator Co., Toronto, for the elevators, and The Dominion Bridge Co., Montreal, for the steel work in the Royal Insurance Co.'s new building, Montreal.

Purdy, Mansell & Co., Toronto, have the contract for steam-fitting and plumbing in The Royal Muskoka Hotel, Muskoka, for The Muskoka Navigation Co., and for plumbing and gas-fitting in a house for R. E. Patterson, Wellington street west, Toronto.

BUSINESS CHANGES.

DIFFICULTIES, ASSIGNMENTS, COMPROMISES.

POIRIER, Dorion & Co., general merchants, Shediac, N.B., are offering 65 cents on the dollar, Their liabilities are estimated at about $19,000; their assets at $16,000.

D. Jobin, general merchant, Sacré-Cœur de Marie, Que., has assigned.

V. Leblanc & Co., general merchants, Hull, Que., are asking an extension.

A meeting of the creditors of Roberts Plaining Mill, Renfrew, Ont., has been held.

J. L. Gallagher, general merchant, Frankville, Ont., has assigned to A. E. Baker.

W. G. Armour, general merchant, Myrtle Station, Ont., has assigned to J. F. Paxton.

Morrow Bros., general merchants, Portage la Prairie, Man., have assigned to C. H. Newton.

C. Lindsay, general merchant, Roberval, Que., has suspended and is offering 40c. on the dollar.

Samuel Bricker, hardware and coal dealer, Listowel, Ont., has assigned to Wm. R. Hobbs, jr.

Lefaivre & Taschereau, Quebec, are preparing a statement for the creditors of Lafontaine & Lavoie, general merchants, St. Cyrille de Wendover, Que.

PARTNERSHIPS FORMED AMD DISSOLVED.

Duval & Favreau, machinists, Montreal, have dissolved.

Favreau & Monty, painters, Montreal, have dissolved.

The St. Maurice Foundry and Machine Co., Three Rivers, Que., have dissolved.

Mills, McKenzie & Ross have formed partnership, and have bought out J. C. Mills, general merchant, Sydney, N.S.

Charles E. Starr and George L. Starr have registered partnership under the style of C. E. Starr & Son, hardware dealers, Wolfville, N.S.

SALES MADE AND PENDING.

The business of J. Mills & Son, general merchants, Granville Ferry, N.S., is advertised for sale,

The general store stock of the estate of H. Roberts & Co., general merchants, Strathclair Station, Man., is advertised for sale by auction on the 15th inst.

CHANGES.

F. C. Boles, general merchant, Beaverton, B.C., is removing to Carmi, B.C.

The Empire Oil Co., Fulford, Que., is retiring from business.

Deslaurier & Ouellet have registered as painters, in Levis, Que.

The National Coal Tar Co. have registered to do business in Montreal.

A. M. Ballak, general merchant, Port Alma, Ont., is removing to Merlin.

W. A. Hayward, general merchant, Coldstream, N.B., has closed his business.

M. J. Johnson, Son & Co. have registered as painters, etc., at Magog, Que.

C. W. McLeod & Co., general merchants, Springhill, N.S., have sold out to H. S. Terris.

Rendell & Co., general merchants, Greenwood, B.C., are opening a branch at Rendell.

P. McCaughan, general merchant, etc., St. Francois Xavier, Man., has sold out his cheese factory.

Thos. A. Cannvan has registered under the style of N. Dussault & Co., machinists, etc., Montreal.

M. C. Buckberrough, blacksmith, Bookton, Ont., has given up business there and has gone to Sault Ste. Marie.

Mrs. Joseph Brodeur has registered as proprietress of Joseph Brodeur & Co., general merchants, St. Hyacinthe, Que.

G. B. McDermot & Co., general merchants, Golden, B.C., have been succeeded by The Golden & East Kootenay Trading Co., Limited.

FIRES.

Thos. B. Angrove, founder, Kingston, Ont., has suffered loss by fire.

Chalmers Bros. & Bethune, hardware and lumber dealers, etc., Pilot Mound, Man., have suffered loss by fire.

Louis D'Amour, blacksmith, Ottawa, has been damaged by fire and water; insured.

Mallman Bros.' sawmill at Milford, N.S., has been burned; no insurance.

DEATHS.

James E. Sprague, general merchant, Regina, N.W.T., is dead.

W. S. Smith, of W. S. Smith & Son, blacksmiths, Bath, N.B., is dead.

Joseph Lafontaine, of Lafontaine & Lavoie, general merchants, St. Cyrille de Wendover, Que., is dead.

THEY ARE DOING WELL.

Mr. Adolphe Henry, of J. Henry & Son, hardware merchants, Orono, Ont., was in Toronto on Tuesday, and in company with Mr. H. T. Eager, manager of the Toronto branch of Wood, Vallance & Co., Hamilton, paid a flying visit to the office of "Hardware and Metal." Henry & Son began business after the New Year, and they appropriately named their store the "New Century Store."

"Business is going very well indeed, so far," said Mr. Henry in reply to a question, "and general trade in our part of the country is opening up well."

LOOKING FOR CLOSE BUYERS

One of the most important industries in the Dominion of Canada is that of the Canada Paint Company, of Toronto and Montreal. Their varnish plant is very extensive and it is equipped with the very latest appliances. All dry colors are manufactured by this company in a factory separate and distinct from their grinding works, where all classes of painting material are ground either in paste form or in liquid ready for use. Street car, railway and coach builders' supplies receive special attention. The company are prepared to tender for high-class paint, varnishes and colors for home and foreign trades. We are told that this company own very valuable Canadian deposits of oxide and graphite, and they make the statement that for paint, colors and varnishes, The Canada Paint Company is in the best position to quote to large companies and close buyers in all parts of the world.

GUELPH CLERKS ORGANIZE.

The Guelph, Ont., Retail Clerks and Salesmen's Association met at the Wellington Hotel, Guelph, on Thursday of last week to complete organization. The following officers were elected:

Honorary Presidents—W. A. Knowles, R. J. Stewart, Hugh Macdonald, Gilbert Jackson, Wm. Lillie, Josiah Gould.

President—A. H. Wallace.
1st Vice-President—John Lundy.
2nd " —Jos. Clark.
Secretary—R. A. McGillivray.
Treasurer—Jas. Benson.

Executive Committee—Fred McPherson, Jos. Foitz, Chas. Groom, F. Kickley, Wm. Hood, Geo. McLeod, R. H. McLeod, Evan Macdonald, H. Cull, R. Millar, Ed. Sloan, Alex. Shields, Jas. Ryan, Alex. Henderson, Robt. Dowler and Wm. Raddington.

MATCH FACTORY BURNED.

The plant of The Walkerville Match Co. was totally destroyed by fire on Tuesday morning. It is believed that an explosion of natural gas was responsible for the conflagration. Fifteen minutes from the time the first alarm was turned in the entire plant was practically destroyed.

President Anderson, of the match company, believes the loss will not exceed $115,000. There was only $75,000 insurance on the plant, although a policy for $25,000 more had been applied for. The work of rebuilding will be commenced shortly.

INQUIRIES AND ANSWERS.

EARLY CLOSING BY LAW.

"Subscriber," St. Thomas, writes: Will you please inform me if any municipality can pass a by-law compelling merchants to close their places of business at 9 o'clock on Saturday evening, providing two-thirds of the merchants sign a petition to that effect?

[Remarks: An Act was placed on the Statute Book by the Ontario Legislature about 12 years ago empowering municipalities, on a petition being presented containing the signatures of two-thirds of the merchants interested, to pass a by-law compelling the closing of stores at a certain hour. The hour of closing is not stipulated by the Act; that is fixed by the municipality.—THE EDITOR]

Hirtle, Refuse & Co., general merchants, etc., Lunenburg, N.S., have been incorporated.

J. W. Hughes & Co., Montreal, have the contract for the heating and plumbing of a factory for Hughs, Cooks & Co., Cote St. Paul. They have also been awarded the contract for the heating of Gault Bros.' new factory on Inspector street, Montreal.

METHOD OF DEALING WITH SLOW ACCOUNTS.

THE following, which appeared in a recent issue of Ironmonger, London, may prove of interest to our readers:

Bad debts cannot be avoided altogether, but to make them as few as possible should be the aim of every firm. No single system will suit all businesses, but, after trying many, the writer some years ago adopted one which has proved very successful, reducing losses from 5s. to about 2s. per £100 of turnover, a difference which is of considerable importance in a large concern.

In the writer's firm the value of the sales is about £30,000 yearly to ironmongers, colliery and quarry owners, and contractors, large and small. The business is both wholesale and retail; the accounts number about 1,000, varying in amount from a few shillings to hundreds of pounds; the credit allowed ranges from a few days to twelve months, and the fact that the risk of bad debts is not trifling has been proved by the serious losses met with by other firms appealing to the same circle of customers.

On the 15th of each month statements of accounts due in the same month have always been sent to each customer since the business was begun 25 years ago, the names, addresses, and amounts having been first entered in alphabetical order in an accounts-rendered book. Formerly, if an account so rendered included items entered in that book in a previous month, such entries were marked "carried forward," and when the accounts-rendered book was examined occasionally and irregularly so that overdue accounts might be written for, many of these entries were certain to escape notice. The plan was therefore adopted of going systematically through the ledgers about the 20th of each month, and entering on one or two sheets of foolscap all overdue accounts, with details as follows, carrying forward from the sheets written out the previous month all accounts still unpaid, with notes regarding applications for payment of same:—

Account Dates from	Ledger-folio	Name and Address*	Dates of Application for Payment.
July	32	S. Willcox, Leeds	Dec. 14, Dec 28 and Jan. 12.
April	45	N. Stevens, Oban	Nov. 27 and Dec. 28.
Sept.	71	S. Benson, Edinburgh	
July	103	J. Hudson, Liverpool	Jan. 15. Reply promised end of month.
July	105	M. Strong, Kendal	
Feb.	264	P. Crone, Norwich	Dec. 28.
July	321	J. Mittel, London	Dec. 28.
Sept.	450	N. Crosbie, Cardiff	
Sept.	505	W. Mallis, Cork	Nov. 27 and Dec. 28. Reply promised in 14 days.
Feb.	652	M. Pride, Dublin	Dec. 28.

* The names in this table are all fictitious.

When the list is made up it is carefully checked, and then a letter is sent to each debtor who is not likely to be called on personally for several months, pointing out that the account is overdue, and requesting payment, the date of each letter being marked on the sheet, as shown. Two, three, four, or more letters follow if no attention has been paid to the earlier application, each letter being more curt than the previous one, and the last of all intimating that the account will, without further intimation, be handed to a lawyer for collection unless paid within one week. In the fictitious list given above (the real one generally contains from 40 to 50 names), two applications have been marked as "replied to," promising remittances by certain dates. Such a reply is always a hopeful sign; but it is quite a common experience to write half-a-dozen times without eliciting any answer, and after legal aid has been enlisted, to receive an indignant letter complaining about the account being placed in a lawyer's hands for collection, notwithstanding that it seemed hopeless to recover payment by any other means.

New accounts require careful watching until the financial standing of the customer has been ascertained by experience; but the most worrying accounts to recover are those of buyers who intend to do well but are unable to meet their obligations promptly owing to want of capital, carelessness, or giving too long credit to their own customers. Such men, having paid within a reasonable time for the small purchases made in the past, and finding their credit curtailed in other quarters, begin to send all their orders to the one firm which seems willing to supply them. Pleasure is felt at the increase of business until difficulty is experienced in obtaining payment, when it is found that legal proceedings will only lead to bankruptcy and loss. In such cases the writer has found it an excellent plan to stop further supplies, unless for cash with order, and to insist on the payment of small amounts on account at regular intervals, so that the whole amount owing shall be paid up in six to 12 months.

Debtors of this kind are generally being dunned on all sides, and whenever they can collect £5 it goes to the most pressing creditor; but it is comparatively easy to make them remit £1 weekly or fortnightly if they are regularly reminded of their obligation a day or two before the amount is due.

The tone of letters to customers who are known to be of undoubtedly good financial standing, but dilatory in their payments, must, of course, be very different from that to those who may reasonably be considered as "doubtful." The former will not take the trouble to reply, and may readily take offence if it is suggested that their credit is doubted, but the latter hope to gain time by silence. Such debtors should be given no rest until their debt is greatly reduced, and the account ought to be kept within small limits thereafter. To decide in which category to place an overdue account is very difficult at times, but the study of previous settlements and information gathered from travellers or personal visits may prevent unfortunate errors.

To sum up the three important points—examine the ledgers at frequent and regular intervals, make frequent and regular applications for payment of all overdue accounts, and lose no time in endeavoring to compel a settlement.

THE QUESTION OF CANADIAN LEAD.

THE VIEWS OF AN AUTHORITY.

THE production of lead in Canada is daily becoming a more momentous question, not only to the public in British Columbia but also to the paint and lead product manufacturers throughout Canada. A few days since, "Hardware and Metal" was favored with an interview by Mr. Robert Munro, Managing Director of The Canada Paint Co., Limited, known as one of our keenest students of Canadian minerals in the sense of their adaptation to practical purposes, and we feel sure that his remarks and conclusions on the subject cannot but be instructive and interesting to our readers.

CANADA'S PRODUCTION OF LEAD.

"Do you consider Canada's production of lead as a matter of much significance?."

"Most certainly. The enormous expansion of the production of lead in Canada entirely alters our relation to the rest of the world in the matter of lead supply. In 1894 we produced only 400 tons; in 1898 our output had risen to 20,000 tons, while the probable output for 1901 is being estimated at over 30,000 tons with still further expansion in sight."

"Lead mining is then a comparatively new industry in Canada?"

"Yes; although we have long had a knowledge of our great possession in lead ores. Indeed, there has long existed an import duty on pig lead which was doubtless intended to stimulate the mining and melting of lead in the Dominion, but it has remained a dead letter, and all the lead consumed in the Dominion has been imported from other countries. These imports take the form of pig lead, red lead, dry white lead, litharge, etc., and while we were not producing lead within ourselves. the process of importing seemed quite natural. Now, however, the anomaly exists of our being importers of our manufactured leads while we are actually exporting crude lead in quantities much greater than our imports."

ANOMALOUS CONDITIONS.

"It is this anomaly that leads to the present agitation, is it not?"

"Exactly. Canada has now become a producer of lead far in excess of her own requirements and the present agitation, emanating mainly from British Columbia, is intended to so adjust our tariff that we may manufacture from our own ores all the lead pipe, lead shot, white lead, etc.; and also to export to the best advantage our surplus lead, which is most readily marketable in the form of pig. We are already at the stage of having not only mining of ore in active operation, but have also smelters which reduce these ores to the form of concentrates. A further link is needed to handle these concentrates, hence the call for a refinery in order to separate the precious metals contained in the ores. The refinery once established would from the concentrates produce gold and silver ready for the mint, and pure pig lead ready for the manufacture of the various products named."

AN ATTEMPTED REMEDY.

"Has the Government never offered any encouragement to the establishment of such a refinery?"

"No; but it did try to remedy matters. Some 18 months ago, the Government (evidently seeing the anomaly that had begun to exist), passed an Order-in-Council re-admitting free of duty the pig lead produced from concentrates of our own smelters sent to the United States to be refined. Under this Order-in-Council, pig lead of Canadian origin is being used in the manufacture of our lead pipe, lead shot, etc., which gives us a market for 3,000 or 4,000 tons of our lead. But it is not reasonable or businesslike that our concentrates should be sent abroad for the finishing process, even if The American Smelting Trust operated them at a moderate rate, which they do not. In proof of this we have just shipped 8,700 tons to Antwerp, hauling it by rail from the Kootenays to St. John, N.B., thence by steamer to Antwerp, and it is more than likely that much of this identical lead will find its way back to Canada in some manufactured form.

BOUNTY ASKED FOR.

"The British Columbian delegation have applied for a bounty on pig lead produced in Canada, or, optionally for assistance in establishing a refinery. Looked at dispassionately it appears to me much more important, as well as more reasonable, that our British Columbian friends should be assured of a market for their lead up to the entire requirements of the Dominion for all lead products, and this being assured would probably render any further impetus unnecessary.

THE TARIFF.

"A glance at the Customs tariff makes it plain that it was drafted under conditions entirely different from those now prevailing. For example, pig lead is rated at 15 per cent. while litharge (oxide of lead) is free, while white and red leads pay only 5 per cent. On the other hand, pipe and shot pay 35 per cent.

"There is surely great room for adjustment here, and it appears most reasonable that such arrangement be made as will put a stop to importation of lead in any form, in order to secure our own markets to our miners and smelters to the extent of our domestic requirements. This reservation of good business would enable the smelters to compete with England and the United States in the sale of pig lead, especially for the supply of China and Japan, for which we are favorably located.

"It may be objected that the change of duties suggested would raise the price [illegible] lead products to the consumer."

"I do not recommend protection so much as adjustment. The prices of some products would be increased, but others would be reduced and the average would not show at all unfavorably. I believe it will be cheaper and it is certainly more businesslike to adapt our market to the advantage of our mining industry rather than supplement their operations by payment of a bonus, which itself would not be final, seeing that tariff adjustment is required in any event."

INCREASED USE OF ALUMINUM.

In 1889 the production of aluminum in the United States was about 22 tons, and in all other countries 71 tons. In the calendar year 1900 the United States produced about 4,000 tons of this metal, against 7,500 tons in all other countries. It is believed that in the near future copper telegraph and telephone wires will be replaced to a great extent by those made of aluminum, it having been demonstrated that in order to do the same work copper wires must be twice as heavy as aluminum ones, and it is estimated that 6,000 tons of aluminum used for sheathing for roofs will replace 20,000 tons of copper. Aluminum wires are now being used in many localities, but the scarcity of the metal has until now prevented its general use.

CURRENT MARKET QUOTATIONS.

May 17, 1901.

These prices are for such qualities and quantities as are usually ordered by retail dealers on the usual terms of credit, the lowest figures being for larger quantities and prompt pay. Large cash buyers can frequently make purchases at better prices. The Editor is anxious to be informed at once of any apparent errors in this list as the desire is to make it perfectly accurate.

METALS.

Tin.

Lamb and Flag and Straits—56 and 28 lb. ingots, per lb.	0 00	0 31

Tinplates.

Charcoal Plates—Bright

M.L.S., equal to Bradley.		Per box
I.C., usual sizes	....	$6 75
I.X., "	....	8 25
I.X.X., "	....	9 75
Famous—		
I.C.	....	6 75
I.X	....	8 25
I.X.X	....	9 50
Raven & Vulture Grades—		
I.C., usual sizes	....	5 00
I.X., "	....	6 00
I.X.X "	....	7 00
I.XXX., "	....	8 00
D.C., 12½x17	....	4 75
D.X.	....	5 50
D.X.X	....	7 50

Coke Plates—Bright

Bessemer Steel—		
I.C., usual sizes	....	4 00
I.C., special sizes, base	....	4 50
20 x 28	....	8 50

Charcoal Plates—Terne

Dean or J. G. Grade—		
I.C., 20x28, 112 sheets	....	8 25
I.X., Terne Tin	....	10 25

Charcoal Tin Boiler Plates.

Cookley Grade—		Per lb.
X.X., 14x56, 50 sheet bxs; " 14x60, "; " 14x65, "	0 07	0 07½

Tinned Sheets

72x30 up to 24 gauge	0 08	0 08½
" 26 "	0 08½	0 09
" 28 "	0 09	0 09½

Iron and Steel.

Common Bar, per 100 lbs.	1 85	1 90
Refined " "	....	2 35
Horse Shoe Iron "	....	2 25
Hoop steel, 1½ to 3 in. base, extras for smaller sizes	3 10	3 25
Sleigh Shoe Steel " base	....	2 10
Tire Steel	2 30	2 50
Reeled Machinery	....	3 00
Cast Steel, per lb	0 00	0 00
Toe Calk Steel	....	
T. Firth & Co's special cast steel, per lb.		0 13
Jessop's Tool Steel	....	0 13

Boiler Tubes.

1½-inch	....	0 12½
2 "	....	0 13
2½ "	....	0 15
3 "	....	0 16
3½ "	....	0 20
4 "	....	0 25

Steel Boiler Plate.

¼ inch	2 50	2 60
3-16 inch	2 60	2 70
⅜ inch and thicker	2 50	2 60

Black Sheets.

18 gauge	....	3 00
20 gauge	....	3 00
22 to 24 "	....	3 10
26 "	....	3 20
28 "	....	3 30

Canada Plates.

All dull, 52 sheets	....	2 90
Half polished	....	3 00
All bright	....	3 10

Iron Pipe.

Black pipe—		
⅛ "	....	4 45
¼ inch	....	3 25
⅜ "	....	3 30
½ "	....	3 50
¾ "	....	3 65
1 "	....	5 15
1¼ "	....	7 10
1½ "	....	8 40
2 "	....	11 25
2½ "	....	20 25
3 "	....	24 55
3½ "	....	30 75
4 "	....	39 00
5 "	....	47 35
6 "	....	62 10
Galvanized pipe—		
½ inch	....	4 90
¾ "	....	5 25
1 "	....	7 55
1¼ "	....	10 30
1½ "	....	12 35
2 "	....	16 50

Galvanized Sheets.

	G C.	Comet.	Amer.	Queen's Head.
16 gauge		....	4 .5	3 90
18 to 24 gauge	4 15	3 85	4 35	4 15
26 "	4 40	4 10	4 35	4 40
28 "	4 65	4 25	4 70	4 65

28 gauge American equals 26 gauge English.

Chain.

Proof Coil, 3-16 in., per 100 lb	....	....
" ¼ " "	8 20	8 50
" 5-16 " "	4 70	5 00
" ⅜ " "	4 05	4 20
" 7-16 " "	3 90	4 25
" ½ " "	3 70	4 10
" 9-16 " "	3 65	4 15
" ⅝ " "	3 25	3 90
" ¾ " "	3 60	4 10

Halter, kennel and post chains, 40 to 40 and 5 p.c.

Cow ties	40 p.c.
Tie-out chains	65 p.c.
Stall fixtures	35 p.c.
Trace chain	45 p.c.

Jack chain, iron, single and double, discount 35 p.c.

Jack chain, brass, single and double, discount 40 p.c.

Copper.

Ingot

English B. S., ton lots	....	0 19
Lake Superior	....	....

Bolt or Bar.

Cut lengths round, ¼ to ⅞ in.	0 23½	0 25
" round and square 1 to 2 inches	0 23½	0 25

Sheet.

Plain, 14 oz., and light, 16 oz., 14x48 and 14x60	0 23	0 23
Plain, 14 oz., and light, 16 oz., irregular sizes	0 23	0 23½
Tinned copper sheets	....	0 26
Planished	....	0 32

Braziers (In sheets.)

4x6ft. 25 to 30 lbs. ea., per lb.	....	0 25½
" 35 to 45 " "	....	0 24½
" 50-lb. and above, "	....	0 23½

Boiler and T. K. Pitts

Plain Tinned, per lb	....	0 28
Spun, per lb	....	0 32

Brass.

Rod and Sheet, 14 to 30 gauge, 15 per cent.

Sheets, hard-rolled, 2x4	0 24	0 25
Tubing, base, per lb.	0 24	0 25

Zinc Spelter

Foreign, per lb	0 05½	0 06
Domestic "	....	..

Zinc Sheet.

5 cwt. casks	....	00 6¾
Part casks	....	00 6¾

Lead.

Imported Pig, per lb	0 04½	0 04½
Bar, 1 lb.	0 05½	0 05¾
Sheets, 2½ lbs. sq. ft., by	....	0 06¼
Sheets, 3 to 6 lbs., "	....	0 06

NOTE.—Cut sheets ½ cent per lb. extra. Pipe, by the roll, usual weights per yard, lists at 7c. per lb. and 25 p.c. dis. f.o.b. Toronto.

NOTE.—Cut lengths, net price, waste pipe 8-ft. lengths lists at 7½ cents.

Shot.

Common, $6.50 per 100 lb.; chilled, $7.00 per 100 lb.; buck, seal and ball, $7.50. Discount, 15 p.c. Prices are f.o.b. Toronto, Hamilton, Montreal, St. John and Halifax. Terms 3 per cent. cash, freights equalized.

Soil Pipe and Fittings.

Discount, 60 and 10 per cent. on medium and extra heavy, and 60 per cent. on light.

Solder.

	Per lb.	Per lb.
Bar half-and-half, guarant'd	....	0 18½
Bar half-and-half, commer'l	....	0 18
Refined	....	0 17½
Wiping	....	0 17

NOTE.—Prices of this graded according to quantity. The prices of other qualities of solder in the market indicated by private brands vary according to composition.

Antimony.

Cookson's, per lb.	0 10½	0 11

White Lead.

	Per 100 lb.
Pure	6 37
No. 1 do	6 00
No. 2 do	5 62½
No. 3 do	5 55
No. 4 do	4 87½
Munro's Select Flake White	7 37½
Elephant and Decorators' Pure	7 12½
Brandram's B B. Genuine	8 00
" " Decorative	7 55
" " No. 1	6 85
" " No. 2	6 00

Red Lead.

Genuine, 560 lb. casks, per cwt.	....	$5 50
Genuine, 100 lb. kegs, per cwt.	....	5 75
No. 1, 560 lb. casks, per cwt.	....	5 25
No. 1, 100 lb. kegs, per cwt	....	5 50

White Zinc Paint.

Elephant Snow White	0 08	0 09
Pure White Zinc	0 08	0 09
No. 1	0 06	0 07½
No. 2	0 05	0 06½

Dry White Lead.

Pure, casks	5 75
Pure, kegs	6 25
No. 1, casks	5 50
No. 1, kegs	5 00

Prepared Paints.

In ¼, ½ and 1 gallon tins.

Pure, per gallon	....	1
Second qualities, per gallon	....	1
Barn (in bbls.)	0 75	0
The Sherwin-Williams Paints.	....	1
Canada Paint Co's Pure	....	1
Toronto Lead & Color Co's Pure	....	1
Sanderson Pearcy's Pure	....	1 25
Stewart & Wood's Champion Pure.		1 [illegible]

Colors in Oil.

25 lb. tins, Standard Quality.

Venetian Red, per lb.	0 05
Chrome Yellow	0 11
Golden Ochre	0 06
French "	0 05
Marine Black	0 06
" Green	0 09
Chrome "	0 08
French Imperia Green	0 09

Colors, Dry.

Yellow Ochre (J.C.) bbls.	1 35	1
Yellow Ochre (J.F.L.S.), bbls	....	2
Yellow Ochre (Royal)	1 10	1
Brussels Ochre	....	2
Venetian Red (best), per cwt.	1 80	1
English Oxides, per cwt.	3 00	3
American Oxides, per cwt.	1 75	2
Canadian Oxides, per cwt.	1 75	2
Super Magnetic Oxides, 93p.c.	2 00	2
Burnt Sienna, pure, per lb.	....	0
" Umber, " "	....	0
do Raw	....	0
Drop Black, pure	....	0
Chrome Yellows, pure	....	0
Chrome Greens, pure, per lb.	....	0 10
Golden Ochre	....	0 [illegible]
Ultramarine Blue in 28-lb. boxes, per lb.	0 08	0 24
Fire Proof Mineral, per 100 lb	....	1 00
Genuine Eng. Litharge, per lb	....	0 07
Mortar Color, per 100 lb.	....	1 25
English Vermillion	....	0 80
Pure Indian Red, No. 45 lb.	....	0 80
Whiting, per 100 lb.	....	0 55

Blue Stone.

Casks, for spraying, per lb.	0 07
100-lb. lots, do. per lb.	0 08

Putty.

Bulk in bbls.	1 90
Bulk in less quantity	2 05
Bladders in bbls.	2 10
Bladders in kegs, boxes or loose.	2 25
Bladders in 25-lb. tins	2 35
Bladders in 12½-lb. tins	2 65
Bladders in bulk or tins less than 100lb	2 90

Varnishes.

	In 5-gal. lots.).	Per gal.
Carriage, No. 1	2 90	3 30
" body	8 00	9 00
" rubbing	4 00	5 00
Gold Size, Japan	3 00	3 40
Brown Japan	2 40	2 80
Elastic Oak	2 90	3 30
Furniture, extra	2 40	2 80
" No. 1	1 60	2 00
Hard Oil Finish	2 70	3 10
Light Oil Finish	3 20	3 60
Demar	3 30	3 70
Shellac, white	4 40	4 80
" orange	4 00	4 40
Furniture Brown Japan	1 60	2 00
Black Japan	2 40	2 80
" " No. 1	1 60	2 00

The Imperial Varnish & Color Co's., Limited Elastilite Varnish 1 gal. can, each. $3.00.

Granatine Floor Finish, per gal., $2.75.

Maple Leaf Coach Enamels; Size 1, $1 20; Size 2, 70c.; Size 3, 40c. each.

Castor Oil.

East India, in cases, per lb.	0 10	0 10½
" " small lots	0 10½	0 11

Cod Oil, Etc.

Cod Oil per gal.	0 50	0 55
Pure Olive	....	1 20
" Neatsfoot	....	90

Glue.

Common	0 08½	0 09
French Medal	0 14	0 14½
Cabinet sheet	0 12	0 13
White, extra	0 18	0 20
Gelatine	0 22	0 30
Strip	0 18	0 20
Coopers	0 19	0 20
Huttner	....	0 18

HARDWARE.

Ammunition.

Cartridges.

- B. B Cap Dom. 50 and 5 per cent.
- Rim Fire Pistol, dis. 40 p. c., Amer.
- Rim Fire Cartridges, Dom., 50 and 5 p. c.
- Central Fire Pistol and Rifle, 10 p.c. Amer.
- Central Fire Cartridges, pistol sizes, Dom, 30 per cent.
- Central Fire Cartridges, Sporting and Military, Dom., 15 and 5 per cent.
- Central Fire, Military and Sporting, Amer, add 5 p.c. to list. B.B. Caps, discount 40 per cent, Amer.
- Loaded and empty Shells, "Trap" and "Dominion" grades, 25 per cent. Rival and Nitro, net list.
- Brass Shot Shells, 55 per cent.
- Primers, Dom., 30 per cent.

Wads — per lb.

- Best thick white felt wadding, in ½-lb bags 1 00
- Best thick brown or grey felt wads, in ½-lb. bags 70
- Best thick white card wads, in boxes of 500 each, 12 and smaller gauges 0 99
- Best thick white card wads, in boxes of 500 each, 10 gauge 0 35
- Best thick white card wads, in boxes of 500 each, 8 gauge 0 55
- Thin card wads, in boxes of 1,000 each, 12 and smaller gauges 0 10
- Thin card wads, in boxes of 1,000 each, 10 gauge 0 25
- Thin card wads in boxes of 1,000 each, 8 gauge ....
- Chemically prepared black edge grey cloth wads, in boxes of 250 each— Per M.
  - 11 and smaller gauge 0 60
  - 9 and 10 gauges .... 0 [illegible]
  - 7 and 8 gauges .... 0 7[illegible]
  - 5 and 6 gauges 1 90
- Superior chemically prepared pink edge, best white cloth wads, in boxes of 250 each—
  - 11 and smaller gauge 1 15
  - 9 and 10 gauges 1 40
  - 7 and 8 gauges 1 65
  - 5 and 6 gauges 1 90

Adzes.

Discount, 20 per cent.

Anvils.

- Per lb. 0 10 0 12½
- Anvil and Vise combined 4 50
- Wilkinson & Co.'s Anvils..lb. 0 09 0 09½
- Wilkinson & Co.'s Vices..lb. 0 09¾ 0 10

Augers.

Gilmour's, discount 65 and 5 p.c. off list.

Axes.

- Chopping Axes
  - Single bit, per doz 5 50 10 00
  - Double bit, " 11 00 18 00
- Bench Axes, 40 p.c.
- Broad Axes, 33⅓ per cent.
- Hunters' Axes 5 50 6 00
- Boy's Axes 5 75 6 75
- Splitting Axes 6 50 12 00
- Handled Axes 7 00 10 00

Axle Grease.

- Ordinary, per gross 5 75 6 00
- Best quality 13 00 15 00

Bath Tubs.

- Zinc 6 00
- Copper, discount 15 p.c. off revised list

Baths.

Standard Enameled.

- 5½-inch rolled rim, 1st quality 25 00
- " " " 2nd " 21 00

Anti-Friction Metal.

- "Tandem" A .... per lb. 0 27
- " B .... " 0 21
- " C .... " 0 11½
- Magnolia Anti-Friction Metal, per lb. 0 25
- Frictionless metal, per lb 0 23

SYRACUSE SMELTING WORKS.

- Aluminum, genuine 0 45
- Dynamo 0 29
- Special 0 25
- Aluminum, 99 p.c. pure "Syracuse" .. 0 50

Bells.

Hand.

- Brass, 60 per cent.
- Nickel, 55 per cent.

Cow.

- American make, discount 66⅔ per cent.
- Canadian, discount 45 and 50 per cent.

Door.

- Gongs, Sargant's 5 50 8 00
- " Peterboro', discount 45 per cent.

Farm.

- American, each 1 25 3 00

House.

- American, per lb 0 35 0 40

Bellows.

- Hand, per doz 3 35 4 75
- Moulders', per doz 7 50 10 00
- Blacksmiths', discount 40 per cent.

Belting.

- Extra, 60 per cent.
- Standard, 60 and 10 per cent.

Bits.

Auger.

- Gilmour's, discount 60 and 5 per cent.
- Rockford, 50 and 10 per cent.
- Jennings' Gen., net list.

Car.

- Gilmour's, 47½ to 50 per cent.

Expansive.

- Clark's, 40 per cent

Gimlet.

- Clark's, per doz 0 65 0 90
- Damond, Shell, per doz 1 00 1 50
- Nail and Spike, per gross 2 25 5 20

Blind and Bed Staples.

- All sizes, per lb 0 07½ 0 12

Bolts and Nuts. Per cent.

- Carriage Bolts, full square, Norway 65
- " " full square 65
- Common Carriage Bolts, all sizes 60
- Machine Bolts, all sizes 60
- Coach Screws 70
- Sleigh shoe Bolts 72½
- Blank Bolts 60
- Bolt Ends 62½
- Plough Bolts 60
- Nuts, square .... 4 c. off
- Nuts, hexagon .... 4½c. off
- Tire Bolts 67½
- Stove Bolts 67½
- Stove rods, per lb .... 5½ to 6c.

Boot Calks.

- Small and medium, ball, per M 4 25
- Small heel, per M 4 50

Bright Wire Goods.

- Discount .... 6 ½ per cent.

Broilers.

- Light, dis., 65 to 67½ per cent.
- Reversible, dis., 65 to 67½ per cent.
- Vegetable, per doz., dis. 37½ per cent.
- Henis, No. 8, " 6
- Henis, No. 9, " 7 00
- Queen City " 7 50 0 00

Butchers' Cleavers.

- German, per doz 6 00 11 00
- American, per doz 12 00 20 00

Building Paper, Etc.

- Plain building, per roll 0 30
- Tarred lining, per roll 0 40
- Tarred roofing, per 100 lb 1 65
- Coal Tar, per barrel 3 50
- Pitch, per 100-lb 0 85
- Carpet felt, per ton 45 00

Bull Rings.

- Copper, $2.00 for 2½ in. and $1.90 for 2 in.

Butts.

- Wrought Brass, net revised list

Cast Iron.

- Loose Pin, dis., 60 per cent.

Wrought Steel.

- Fast Joint, dis. 65, 10 and 2½ per cent.
- Loose Pin, dis. 65, 10 and 2½ per cent.
- Berlin Bronzed, dis. 70, 70 and 5 per cent.
- Gen. Bronzed, per pair 0 40 0 65

Carpet Stretchers.

- American, per doz 1 00 1 50
- Bullard's, per doz .... 6 50 ....

Castors.

- Bed, new list, dis. 55 to 57½ per cent.
- Plate, dis. 52½ to 57½ per cent.

Cattle Leaders.

- Nos. 31 and 32, per gross 50 9 50

Cement.

- Canadian Portland 2 50 2 80
- English " 3 00
- Belgian " 2 50 2 75
- Canadian hydraulic 1 25 1 50

Chalk.

- Carpenters Colored, per gross 0 45 0 75
- White lump, per cwt 0 60 0 65
- Red 0 05 0 06
- Crayon, per gross 0 14 0 18

Chisels.

Socket, Framing and Firmer.

- Broad's, dis. 70 per cent.
- Warnock's, dis. 70 per cent.
- P. S. & W. Extra 60, 10 and 5 p.c.

Churns.

Revolving Churns, metal frames—No. 0, $8—No. 1, $8.50—No. 2, $9.00—No. 3, $10.00—No. 4, $12.00—No. 5, $16.00 each. Ditto, wood frames—20c. each less than above. Discounts: Delivered from factories, 58 p.c.; from stock in Montreal, 58 p.c. Terms, 4 months or 3 p.c. cash in 30 days.

Clips.

- Axle dis. 65 per cent.

Closets

- Plain Ontario Syphon Jet $16 00
- Emb. Ontario Syphon Jet 17 00
- Fittings net 1 00
- Plain Teutonic Syphon Washout 10 00
- Emb. Teutonic Syphon Washout 11 00
- Fittings net 1 25
- Low Down Teutonic, plain 16 00
- " " " embossed 17 00
- Plain Richelieu net 3 75
- Emb. Richelieu net 4 00
- Fittings net 1 25
- Low Down Oct. Sy. Jet, plain net .. 19 50
- " " " " " emb'd. net 20 50
- Closet connection net 1 25
- Basins, round, 14 in 1 00
- " oval, 17 x 14 in 2 51
- " " 19 x 15 in 3 75
- Discount 40 p.c., except on net figures.

Compasses, Dividers, Etc.

- American, dis. 62½ to 65 per cent.

Cradles, Grain.

- Canadian, dis. 25 to 33½ per cent.

Crosscut Saw Handles.

- S. & D., No. 3, per pair 17½
- " " 5, " 22½
- " " 6, " 15
- Boynton pattern " 20

Door Springs.

- Torrey's Rod, per doz .... (15 p.c.) 2 00
- Coil, per doz 0 88 1 60
- English, per doz 2 00 4 00

Draw Knives.

- Coach and Wagon, dis. 50 and 10 per cent.
- Carpenters, dis. 70 per cent.

Drills.

Hand and Breast.

- Millar's Falls, per doz. net list.

DRILL BITS.

- Morse, dis., 37½ to 40 per cent.
- Standard dis. 50 and 5 to 55 per cent

Faucets.

- Common, cork-lined, dis 35 per cent.

ELBOWS. (Stovepipe.)

- No. 1, per doz 1 40
- No. 2, per doz 1 20
- Bright, 20c. per doz. extra.

ESCUTCHEONS.

- Discount, 45 per cent.

ESCUTCHEON PINS.

- Iron, discount 40 per cent.

FACTORY MILK CANS.

- Discount off revised list, 40 per cent.

FILES.

- Black Diamond, 70 and 5 per cent.
- Kearney & Foote, 60 and 10 to 70 per cent.
- Nicholson File Co., 60 to 60 and 10 per cent.
- Jowitt's, English list, 25 to 27½ per cent.

FORKS.

- Hay, manure, etc., dis., 50 and 10 per cent. revised list.

GLASS—Window—Box Price.

Size United Inches.	Star Per 50 ft.	Star Per 100 ft.	D. Diamond Per 50 ft.	D. Diamond Per 100 ft.
Under 26	2 15	4 15	....	6 00
26 to 40	2 30	4 45	....	6 85
41 to 50	....	4 85	....	7 50
51 to 60	....	5 15	....	8 50
61 to 70	....	5 50	....	9 50
71 to 80	....	6 00	....	10 50
81 to 85	....	6 50	....	11 70
86 to 90	....	....	..	14 00
91 to 95	..	....	...	15 50
96 to 100	.. .	....	....	18

GAUGES

Marking, Mortise, Etc.

- Stanley's dis. 50 to 55 per cent.

Wire Gauges.

- Winn's, Nos. 26 to 33, each ... 1 65 2 40

HALTERS.

- Rope, ⅝ per gross ....
- " ½ 9 00
- " ⅝ to ¾ 14 00
- Leather, 1 in., per doz 3 87½ 4 00
- " 1¼ in., " 5 15 5 20
- Web, — per doz 1 87 2 45

HAMMERS.

Nail

- Maydole's, dis. 5 to 10 per cent. Can. dis. 25 to 27½ per cent.

Tack.

- Magnetic, per doz 1 10 1 20

Sledge.

- Canadian, per lb 0 07½ 0 08½

Ball Pean.

- English and Can., per lb 0 22 0 25

HANDLES.

- Axe, per doz. net 1 50 2 00
- Store door, per doz 1 00 1 50

Fork.

- C. & B., dis. 40 per cent. rev. list.

Hoe.

- C. & B., dis. 40 per cent. rev. list.

Saw.

- American, per doz 1 00 1 25

Plane.

- American, per gross 3 15 3 75

Hammer and Hatchet.

- Canadian, 40 per cent.

Cross-Cut Saws.

- Canadian, per pair 0 13½

HANGERS. doz. pairs.

- Steel barn door 5 85 6 00
- Stearns, 4 inch 5 00
- " 5 inch 6 50
- Lane's covered—
  - No. 11, 5-ft. run 8 40
  - No. 11½, 10-ft. run 10 80
  - No. 12, 10-ft. run 12 60
  - No. 14, 15-ft. run 21 00
- Lane's O.N.T. track, per foot 4½

HARVEST TOOLS.

- Discount, 50 and 10 per cent.

HATCHETS.

- Canadian, dis. 40 to 42½ per cent.

HINGES.

- Blind, Parker's, dis. 50 and 10 to 60 per cent.
- Heavy T and strap, 4-in., per lb 0 06½
- " " 5-in., " 0 06¼
- " " 6-in., " 0 06
- " " 8-in., " 0 05¾
- " " 10-in., " 0 05½
- Light T and strap, dis. 65 and 5 per cent.
- Screw hook and hinge—
  - 6 to 12 in., per 100 lbs 4 50
  - 14 in. up, per 100 lbs 3 50
- Per gro. pairs.
- Spring 12 00

HOES.

- Garden, Mortar, etc., dis. 50 and 10 p.c.
- Planter, per doz 4 00 4 50

HOLLOW WARE

- Discount, ... 45 and 5 per cent

HOOKS.

Cast Iron.

- Bird Cage, per doz 0 50 1 10
- Clothes Line, per doz 0 27 0 63
- Harness, per doz 0 72 0 88
- Hat and Coat, per gross 1 00 3 00
- Chandelier, per doz 0 50 1 00

Wrought Iron.

- Wrought Hooks and Staples, Can., dis. 47½ per cent.

Wire.

- Hat and Coat, discount 45 per cent.
- Belt, per 1,000 .... 0 60
- Screw, bright, dis. 55 per cent.

HORSE NAILS.

- "C" brand 50 and 7½ p.c. off new list } Oval-
- "M" brand 50, 10 and 5 per cent. } head
- Countersunk, 60 per cent.

HORSESHOES

Iron Shoes.	F.O.B. Montreal. No. 2 and larger.	No. 1. and smaller.
Light, medium, and heavy.	3 50	3 75
Snow shoes	3 75	4 00
Steel Shoes.		
Light	3 60	3 85
Featherweight (all sizes)	4 85	4 85

F.O.B. Toronto, Hamilton, London and Guelph, 10c. per keg additional.

Toe weight steel shoes 6 70

JAPANNED WARE.

Discount, 45 and 5 p.c. off list, June 1899

ICE PICKS.

Star per doz. 3 00 3 25

KETTLES.

Brass spun, 7½ p.c. dis. off new list.
Copper, per lb. 0 30 0 50
American, 60 and 10 to 65 and 5 p.c.

KEYS.

Lock, Can., dis., 45 p.c.
Cabinet, trunk, and padlock, Am. per gross 60

KNOBS.

Door, japanned and N.P., per doz. 1 50 2 50
Bronze, Berlin, per doz. 2 75 3 25
Bronze Genuine, per doz. 6 00 9 00
Shutter, porcelain, F. & L. screw, per gross 1 30 4 00
White door knobs—per doz. 1 25

HAY KNIVES.

Discount, 50 and 10 per cent.

LAMP WICKS.

Discount, 60 per cent.

LANTERNS.

Cold Blast, per doz. 7 00
No. 3 "Wright's" 8 0
Ordinary, with O burner 4 00
Dashboard, cold blast 9 00
No. 0. ...... 5 75
Japanning, 50c. per doz. extra.

LEMON SQUEEZERS.

Porcelain lined, ...per doz. 2 20 5
Galvanized 1 87 3 8
King, wood 2 75 2 9
King, glass 4 00 4 50
All glass 1 20 1 30

LINES.

Fish, per gross 1 05 2 0
Chalk " 1 90 7 40

LOCKS.

Canadian, dis. 45 p.c.
Russell & Erwin, per doz. 3 00 3 25

Cabinet.

Eagle, dis. 30 p.c.

Padlock.

English and Am., per doz. 50 6 00
Scandinavian, " 1 00 2 40
Eagle, dis. 20 to 25 p.c.

MACHINE SCREWS. Iron and Brass.

Flat head discount 25 p.c.
Round Head discount 20 p.c.

MALLETS.

Tinsmiths' per doz. 1 25 1 50
Carpenters', hickory, per doz. 1 25 3 75
Lignum Vitae, per doz. 3 85 5 00
Caulking each 60 2 00

MATTOCKS.

Canadian, per doz. 5 50 6 50

MEAT CUTTERS.

American, dis. 25 to 30 p.c.
German, 15 per cent.

MILK CAN TRIMMINGS.

Discount, 25 per cent.

NAILS.

Quotations are:	Cut.	Wire.
2d and 3d	$3 35	$3 85
3d	3 00	3 52
4 and 5d	2 75	3 35
6 and 7d	2 65	3 20
8 and 9d	2 50	3 00
10 and 12d	2 45	2 95
16 and 20d	2 40	2 90
30, 40, 50 and 60d. (base)	2 35	2 85

Wire nails in carlots are $2.77½
Galvanizing 2c. per lb. net extra.
Steel Cut Nails 10c. extra.
Miscellaneous wire nails, dis. 70 and 10 p.c.
Coopers' nails, dis. 30 per cent.
Flour barrel nails, dis. 25 per cent.

NAIL PULLERS.

German and American 1 85 3 50

NAIL SETS

Square, round, and octagon, per gross 3 38 4 00
Diamond 12 00 15 00

NETTING.

Poultry, 50 and 5 per cent for McMullen's

OAKUM. Per 100 lb.

Navy 6 00
U. S. Navy 7 25

OIL.

Water White (U.S.) 0 16½
Prime White (U.S.) 0 15½
Water White (Can.) 0 15
Prime White (Can.) 0 14

OILERS.

McClary's Model galvan. oil can, with pump, 5 gal., per doz. 10 00
Zinc and tin, dis. 50, 50 and 10.
Copper, per doz. 1 25 3 50
Brass, " 1 50 3 50
Malleable, dis. 25 per cent.

GALVANIZED PAILS.

Dufferin pattern pails, dis. 45 p.c.
Flaring pattern, discount 45 per cent.
Galvanized washtubs, discount 45 per cent.

PIECED WARE.

Discount 40 per cent. off list, June, 1899.
10-qt. flaring sap buckets, dis. 45 p.c.
6, 10 and 14-qt. flaring pails, dis. 45 p.c.
Creamer cans, dis. 45 p.c.

PICKS.

Per doz. 6 00 9 00

PICTURE NAILS.

Porcelain head, per gross 1 75 3 00
Brass head " 0 40 1 00

PICTURE WIRE.

Tin and gilt, discount 75 p.c.

PLANES.

Wood, bench, Canadian dis. 50 per cent. American dis. 50.
Wood, fancy Canadian or American 7½ to 40 per cent.

PLANE IRONS.

English, per doz. 2 00 5 00

PLIERS AND NIPPERS.

Button's Genuine per doz pairs, dis. 37½ 40 p.c.
Button's Imitation, per doz. 5 00 9 00
German, per doz. 0 60 2 60

PLUMBERS' BRASS GOODS.

Compression work, discount, 60 per cent.
Fuller's work, discount 65 per cent.
Rough stops and stop and waste cocks, discount, 60 per cent.
Jenkins disk globe and angle valves, discount, 55 per cent.
Standard valves, discount, 60 per per cent.
Jenkins radiator valves discount 55 per cent.
" " " standard, dis., 60 p.c.
Quick opening valves discount, 60 p.c.
No. 1 compression bath cock 2 00
No. 4 " " " 2 00
No. 7, Fuller's 2 50
No 4½, " 3 00

POWDER.

VeloX Smokeless Shotgun Powder.
100 lb. or less 0 85
1,000 lb. or more 0 80
Net 30 days.

PRESSED SPIKES.

Discount 20 to 25 per cent.

PULLEYS.

Hothouse, per doz. 0 55 1 00
Axle 0 22 0 33
Screw 0 27 1 00
Awning 0 35 2 50

PUMPS.

Canadian cistern 1 80 3 60
Canadian pitcher spout 1 40 2 10

PUNCHES.

Saddlers', per doz. 1 00 1 85
Conductors', " 9 00 15 00
Tinners' solid, per set 0 00 0 72
" hollow, per inch 0 00 1 00

RANGE BOILERS.

Galvanized, 3 gallons 7 00
" 35 " 8 25
" 40 " 9 50

Copper, 30 " 22 00
" 35 " 26 00
" 40 " 30 00
Discount off Copper Boilers 10 per cent.

RAKES.

Cast steel and malleable, 50, 10 and 5 p.c.
Wood, 25 per cent.

RASPS AND HORSE RASPS.

New Nicholson horse rasp, discount 60 to 60 and 10 p.c.
Globe File Co.'s rasps, 60 and 10 to 70 p.c.
Heller's Horse rasps, 50 to 50 and 5 p.c.

RAZORS.

	per doz.	
Elliot's	4 00	18 00
Geo. Butler & Co.'s	8 00	18 00
Boker's	7 50	11 00
" King Cutter	12 50	50 00
Wade & Butcher's	3 60	10 00
Theile & Quack's	7 00	12 00

REAPING HOOKS.

Discount, 50 and 10 per cent.

REGISTERS.

Discount ...... 40 per cent.

RIVETS AND BURRS.

Iron Rivets, black and tinned, discount 60 and 10 per cent.
Iron Burrs, discount 55 per cent.
Extras on Iron Rivets in 1-lb. cartons, ½c. per lb.
Extras on Iron Rivets in ½-lb. cartons, 1c. per lb.
Copper Rivets & Burrs, 35 and 5 p.c. dis. and cartons, 1c. per lb. extra, net.
Extras on Tinned or Coppered Rivets ½-lb. cartons, 1c. per lb.

RIVET SETS

Canadian, dis. 35 to 37½ per cent.

ROPE ETC.

	Sisal.	Manila.
7-16 in. and larger, per lb	10	13½
⅜ in.	11	14½
¼ and 5-16 in.		16½
Cotton, 3-16 inch and larger		16
" 5-32 inch		21
" ⅛ inch		22½
Russia Deep Sea		15½
Jute		8
Lath Yarn		9¼
New Zealand Rope		10

RULES.

Boxwood, dis. 75 and 10 p.c.
Ivory, dis. 37½ to 40 p.c.

SAD IRONS. per set

Mrs. Potts, No. 55, polished 62½
" No. 50, nickle-plated 67½

SAND AND EMERY PAPER.

Dominion Flint Paper, 47½ per cent.
B & A. sand, 40 and 5 per cent.
Emery, 40 per cent.
Garnet (Rurton's), 5 to 10 p.c. advance on list.

SAP SPOUTS.

Bronzed iron with hooks, per doz. 9 50

SAWS.

Hand Disston's, dis. 12½ p.c.
S. & D., 40 per cent.
Crosscut, Disston's, per ft. 35 0 55
S. & D., dis. 35 p.c. on Nos. 2 and 3.
Hack, complete, each 0 75 2 75
" frame only 0 75

SASH WEIGHTS.

Sectional, per 100 lbs. 2 75 3 00
Solid, " 2 00 2 25

SASH CORD.

Per lb. 0 23 0 30

SAW SETS.

"Lincoln," per doz. 6 50

SCALES.

B. S. & M. Scales, 45 p.c.
Champion, 65 per cent.
Fairbanks Standard, 35 p.c.
" Dominion, 55 p.c.
" Richelieu, 55 p.c.
Chatillon Spring Balances, 10 p.c.
Warren Champion 65 p.c.
" Standard 45 p.c.

SCREW DRIVERS.

Sargent's per doz. 0 65 1 00

SCREWS

Wood, F. H., bright and steel, 87½ and 10 p.c.
Wood, R. H., " dis. 82½ and 10 p.c.
" F. H., brass, dis. 80 and 10 p.c.
Wood, R. H., " dis. 75 and 10 p.c.
" F.H., bronze, dis. 75 p.c.
" R.H. " 70 p.c.
Drive Screws, 87½ and 10 per cent.
Bench, wood, per doz. 3 25 4 00
" iron. " 4 25 5 75
Set, Case hardened, 60 per cent.
Square Cap, 50 and 5 per cent.
Hexagon Cap, 45 per cent.

SCYTHES.

Per doz. net 9 00

SCYTHE SNATHS.

Canadian, dis. 45 p.c.

SHEARS.

Bailey Cutlery Co., full nickeled, dis. 60 p.c.
Seymour's, dis. 50 and 10 p.c.

SHOVELS AND SPADES.

Canadian, dis. 40 and 5 per cent.

SINKS.

Steel and galvanized, discount 45 per cent.

SNAPS.

Harness, German, dis. 25 p.c.
Lock, Andrews' 4 50 11 50

SOLDERING IRONS.

1, 1½ lb., per lb. 0 37
2 lb. or over, per lb. 0 34

SQUARES.

Iron, No. 493, per doz 2 40 2 55
" No. 494, " 3 25 3 40
Steel, dis. 60, 10 and 5 p.c.
Try and bevel, dis. 50 to 52½ p.c.

STAMPED WARE.

Plain, dis. 75 and 12½ p.c. off revised list.
Retinned, dis., 75 p.c. off revised list.

STAPLES.

Galvanized 3 50 4 00
Plain 3 25 3 75
Coopers', discount 45 per cent.
Poultry netting staples, 40 per cent.

STOCKS AND DIES.

American dis. 25 p.c.

STONE.	Per lb.	
Washita	0 28	0 60
Hindostan	0 06	0 07
" slip	0 09	0 09
Labrador	....	0 13
" Axe	....	0 15
Turkey	....	0 50
Arkansas	0 00	1 50
Water-of-Ayr	0 00	0 10
Scythe, per gross	3 50	5 00
Grind, per ton	15 00	18 00

STOVE PIPES.

5 and 6 inch Per 100 lengths 7 00
7 inch " " 7 50

ENAMELINE STOVE POLISH.

No. 4—3 dozen in case, net cash $4 80
No. 6—3 dozen in case, " 8 40

TACKS BRADS, ETC.

	Per cent.
Strawberry box tacks, bulk	75 & 10
Cheese-box tacks, blued	80 & 12½
Trunk tacks, black and tinned	85
Carpet tacks, blued	80 & 15
" " tinned	80 & 20
" " (in kegs)	40
Cut tacks, blued, in dozens only	80
" ¼ weights	60
Swedes, cut tacks, blued and tinned—	
In bulk	80 & 10
In dozens	75
Swedes, upholsterers', bulk	85, 12½ & 12½
" brush, blued & tinned, bulk	70
" gimp, blued, tinned and japanned	75 & 12½
Zinc tacks	35
Leather carpet tacks	55
Copper tacks	50
Copper nails	52½
Trunk nails, black	65 and 5
Trunk nails, tinned	65 and 10
Clout nails, blued	65 and 5
Chair nails	35
Patent brads	40
Fine finishing	40
Picture frame points	10
Lining tacks, in papers	10

Lining tacks, in bulk		15
" " solid heads, in bulk		75
Saddle nails in papers		10
" " in bulk		15
Tufting buttons, 22 line, in dozens only		60
Tin capped trunk nails		15
Zinc glazier's points		5
Double pointed tacks, papers	90 and	10
" " " bulk		40

TAPE LINES.

English, ass skin, per doz	2 75	5 00
English, Patent Leather	5 50	9 75
Chesterman's each	0 90	2 85
" steel, each	0 80	8 00

THERMOMETERS.

Tin case and dairy, dis. 75 to 75 and 10 p.c.

TRAPS. (Steel.)

Game, Newhouse, dis. 25 p.c.
Game, H. & N., P. S. & W., 65 p.c.
Game, steel, 72½, 75 p.c.

TROWELS.

Disston's discount 10 per cent.

German, per doz	4 75	6 00

S. & D., discount 35 per cent.

TWINES.

Bag, Russian, per lb	....	0 27
Wrapping, cotton, 3-ply		0 20
" " 4-ply	....	0 26
Mattress, per lb	0 33	0 45
Staging, "	0 27	0 35

VISES.

Hand, pe doz	4 00	6 00
Bench, parallel, each	2 00	4 50
Coach, each	6 00	7 00
Peter Wright's, per lb	0 12	0 13
Pipe, each	5 50	9 00
Saw, per doz	6 50	13 00

ENAMELLED WARE.

White, Princess, Turquoise, Blue and White, discount 50 per cent.
Diamond, Famous, Premier, 50 and 10 p.c.
Granite or Pearl, Imperial, Crescent, 50, 10 and 10 per cent.

WIRE.

Brass wire, 50 to 50 and 2½ per cent. off the list.

Copper wire, 45 and 10 per cent. net cash 30 days, f.o.b. factory.

Smooth Steel Wire, is quoted at the following net selling prices:

No. 6 to 8 gauge	$2 90
" 9 "	2 80
" 10 "	2 87
" 11 "	2 90
" 12 "	2 95
" 13 "	3 15
" 14 "	3 37
" 15 "	3 50
" 16 "	3 65

Other sizes of plain wire outside of Nos. 9, 10, 11, 12 and 13, and other varieties of plain wire remain at $2.80 base with extras as before. The prices for Nos 9 to 13 include the charge of 1 c. for oiling. Extras net per 100 lb.: Coppered wire, 60c.—tinned wire, $2—oiling, 10c.—special hay-balling wire, 30c.—spring wire, $1—best steel wire, 75c.—bright soft drawn, 15c.—in 50 and 100-lb. bundles net, 10c.—in 25-lb. bundles net 15c.—packed in casks or cases, 15c.—bagging or papering, 10c.

Fine Steel Wire, dis. 17½ per cent. List of extras: In 100-lb. lots: No. 17, $5—No. 18, $5.50—No. 19, $6—No. 20, $6.65—No. 21, $7—No. 22, $7.30—No. 23, $7.65—No. 24, $8—No. 25, $9—No. 26, $9.50—No. 27, $10—No. 28, $11—No. 29, $12—No. 30, $13—No. 31, $14—No. 32, $15 No. 33, $16—No. 34, $17. Extras net—tinned wire, Nos. 17-25, $2—Nos. 26-31 $4—Nos. 32-34, $6. Coppered, 5c.—oiling, 10c.—in 25-lb. bundles, 15c.—in 5 and 10-lb. bundles, 25c.—in 1-lb. hanks, 50c.—in ½-lb. hanks, 75c.—in ¼-lb. hanks, $1—packed in casks or cases, 15c.—bagging or papering, 10c.

Galvanized Wire, per 100 lb.—Nos. 6,7,8, $3 50 to $3 8?—No. 9, $2.85 to $3.15—No. 10, $3.60 to $3.95—No. 11, $3.70 to $4.10—No 12, $3 to $3 30—No. 13, $3.10 to $3.40—No. 14, $4.10 to $4.50—No. 15, $4.60 to $5.05—No. 16, $4.85 to $5.35. Base sizes, Nos. 5 to 9, $2.57½ f.o.b. Cleveland.

Clothes Line Wire, solid 7 strand, No. 17. $4.25; No. 18, $2.65; No. 19, $2.35, f.o.b Hamilton, Toronto, Montreal.

WIRE FENCING. F.O.B. Toronto

Galvanized barb	....	3 05
Galvanized, plain twist	....	3 05

Galvanized barb, f.o.b. Cleveland, $2.82½ in less than carlots, and $2.70 in carlots.

WIRE CLOTH.

Painted Screen, per 100 sq. ft., net	1 35

WASTE COTTON.

		per lb.
Colored	....	4½ to 5
White, according to quality	....	6½ to 7½

50-lb bale lots shaded.

WRENCHES.

Acme, 35 to 37½ per cent.
Agricultural, 60 p.c.
Coe's Genuine, dis. 20 to 25 p.c.

Towers' Engineer, each	2 00	7 00
" S., per doz	5 80	6 00
G. & K.'s Pipe, per doz	....	3 40
Burrell's Pipe, each	....	3 00
Pocket, per doz	0 25	2 90

WRINGERS.

Leader per doz.	$30 0	33 00
Royal Canadian "	26 00	28 00
Royal American "	26 00	28 00
Sampson "	....	30 00

Terms 4 months, or 3 p.c. 30 days.

WROUGHT IRON WASHERS.

Canadian make, discount, 40 and 5 per cent.

PERSONS addressing advertisers will kindly mention having seen their advertisement in Canadian Hardware and Metal Merchant.

"THE EMLYN" SAW BENCH

Made in 6 sizes. Best value obtainable. Specially designed for export. With or without "Emlyn" Patent Guard. Sole maker—

CHARLES D. PHILLIPS,

Cables—"Machinery," Newport.

Emlyn Engineering Works, NEWPORT, MON., ENGLAND.

Mr. Retail Hardwareman,

You have had trouble with Axes. Some were poor sellers, others did not give your customers satisfaction.

You can avoid having all this trouble in future. We can tell you how. Write us.

Dundas Axe Works

DUNDAS, CANADA.

"Build to-day then, strong and sure, with a firm and ample base."
—Longfellow.

DO YOU?

Wish thus to build an advertisement in the CONTRACT-RECORD, TORONTO will bring you tenders from the best contractors

---

BUSINESS NEWS

of any kind that is of value to business men supplied by our Bureau. We can give you market quotations from any town in Canada, reports from the city markets, stock quotations, etc. You can get commercial news from any Canadian paper through us.

Write us, giving us particulars of what you want and where you want it from, and we will quote you prices by return.

"Clippings from any Canadian paper on any subject."

CANADIAN PRESS CLIPPING BUREAU,

232 McGill Street, MONTREAL, QUE.

Telephone Main 1255.

10 Front St. East, Toronto. Telephone 2148.

---

75 YEARS. *ESTABLISHED 1825.* *75 YEARS.*

CELEBRATED HEINISCH SHEARS.

Tailors' Shears,
Trimmers, Scissors,
Tinners' Snips, etc.

ACKNOWLEDGED THE BEST.

R. HEINISCH'S SONS CO. NEW YORK OFFICE, 90 Chambers St. NEWARK, N.J., U.S.A.

Not connected with any Shear Combination.

CHAS. F. CLARK, President. JARED CHITTENDEN, Treasurer.

...ESTABLISHED 1849...

BRADSTREET'S

Capital and Surplus, $1,500,000. Offices Throughout the Civilized World.

Executive Offices: Nos. 346 and 348 Broadway, New York City, U.S.A.

THE BRADSTREET COMPANY gathers information that reflects the financial condition and the controlling circumstances of every seeker of mercantile credit. Its business may be defined as of the merchants, by the merchants, for the merchants. In procuring, verifying and promulgating information no effort is spared, and no reasonable expense considered too great, that the results may justify its claim as an authority on all matters affecting commercial affairs and mercantile credit. Its offices and connections have been steadily extended, and it furnishes information concerning mercantile persons throughout the civilized world.

Subscriptions are based on the service furnished, and are available only by reputable wholesale, jobbing and manufacturing concerns, and by responsible and worthy financial, fiduciary and business corporations. Specific terms may be obtained by addressing the Company at any of its offices. **Correspondence Invited.**

—OFFICES IN CANADA—

HALIFAX, N.S.	HAMILTON, ONT.	LONDON, ONT.	MONTREAL, QUE.
OTTAWA, ONT.	QUEBEC, QUE.	ST. JOHN, N.B.	TORONTO, ONT.
VANCOUVER, B.C.	VICTORIA, B.C.	WINNIPEG, MAN.	

THOS. C. IRVING, Gen. Man. Western Canada, Toronto. JOHN A. FULTON, Gen. Man. Eastern Canada, Montreal.

Awarded a Gold Medal at **PARIS EXPOSITION** for superiority. That's proof enough of their quality, and clearly shows that they are the best.

Send for Catalogue and . . . . Price List.

The Bailey Cutlery Co.

BRANTFORD, ONT.

THE MOWER

THAT WILL KILL ALL THE WEEDS IN YOUR LAWNS.

If you keep the weeds cut so they do not go to seed, and cut your grass without breaking the small feeders of roots, the grass will become thick and weeds will disappear. **The Clipper will do it.**

CANADIAN PATENT FOR SALE.
SEND FOR CATALOGUE AND PRICES.

Clipper Lawn Mower Co., NORRISTOWN, PA.

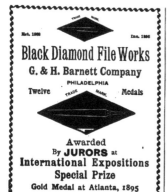

CANADIAN HARDWARE AND METAL MERCHANT

The Weekly Organ of the Hardware, Metal, Heating, Plumbing and Contracting Trades in Canada.

VOL. XIII. MONTREAL AND TORONTO, MAY 25, 1901. NO. 21

TRUCKS.

For Warehouse uses, Hotels, Factory and Office, Railroad and Barrel Trucks.

We have a large and varied assortment in stock, and can ship promptly.

We Make a Specialty of...

Rubber Wheeled Trucks

FOR HOTELS AND WAREHOUSES.

Send for our Truck Catalogue and Price Lists.

THE FAIRBANKS COMPANY

749 Craig Street, *MONTREAL*

Meat & Food Chopper

No. 5, - $2.00

Rapid Grinding & Pulverizing Mills

No. 2½, - $4.50

Fruit, Wine & Jelly Press

No. 34, - $3.00

"ENTERPRISE"

PATENTED

HARDWARE SPECIALTIES

Comprising

Meat and Food Choppers, *35 sizes aud styles for hand and power;* Rapid Grinding and Pulverizing Mills, *32 sizes and styles for hand and power;* Fruit, Wine and Jelly Presses; Meat Juice Extractors; Cherry Stoners; Raisin Seeders, *for hand and power;* Ice Shredders; Vegetable Slicers; Mrs. Potts' Cold Handle Irons; Sausage Stuffers and Lard Presses, etc., etc.

ARE THE BEST

Write for Descriptive Catalogue — *Sold by the Leading Jobbers of the Dominion*

The Enterprise Mg. Co. of Pa.

Philadelphia, Pa., U. S. A.

Raisin Seeder

No. 36, - .85

Meat Juice Extractor

No. 21, - $2.50

Meat & Food Chopper

No. 12, - $2.50

Woman is Responsible

for the sale of ice cream freezers.

Fifteen million women have considerable influence.

Ten thousand of them wrote to us in three months.

We are talking to them all the time—winter and summer.

They are all familiar with the Peerless Iceland Freezer.

Reports from them prove that we are right, and that one motion is enough.

We want your help in distributing Peerless Iceland Freezer.

DANA & CO.,
Cincinnati.

New York Branch, 10 Warren St.
San Francisco Branch, 105 Front St.

FLY TIME will soon be here. YOU will require a stock of DOOR PULLS for Screen Doors, etc.

We Manufacture among our many other lines both Japanned and Copper Coated Wire Door Pulls (5 inches).

"CRESCENT"

HAT and COAT HOOKS

2½ in. 3 in. 3½ in.

In Japanned and Coppered.

—Wood Screws and Wire Nails.—

BRIGHT GOODS

(A full line of these goods always in stock.)

WIRE

Bright, Annealed, Oiled and Annealed.

Coppered, Coppered Spring, Tinned, Brass, Staples, Galvanized and Barb Wire.

COPPER WIRE

For Merchant and Electrical Purposes.

For Prices, etc., apply to

Dominion Wire Manufacturing Co.

MONTREAL and TORONTO.

THE NEW BALDWIN

DRY AIR CLEANABLE

REFRIGERATOR.

135 Modern Varieties. Ash, Oak and Soft-wood Finishes

METAL, PORCELAIN, SPRUCE LININGS.

BALDWIN

Positive Circulation—Sanitary—Odorless.

Latest Cleanable Features—The Strongest and Best System of Patent Removable Metal Air-Flues.

Air-Tight Lever Locks Ball-Bearing Casters. Swing Base—in and out.

Rubber around Doors and Lids, making them doubly air-tight.

Handsome Designs. Moderate Prices.

Built in the newest, largest and best equipped refrigerator plant in the East run all the year round on refrigerators exclusively; stock goods; special refrigerators and coolers in sections.

Handsome Trade Catalogue Ready.

Baldwin Refrigerator Co.,

BURLINGTON, VERMONT.

Lewis Bros. & Co.

WHOLESALE HARDWARE, MONTREAL.

For Your Hot Weather Trade.

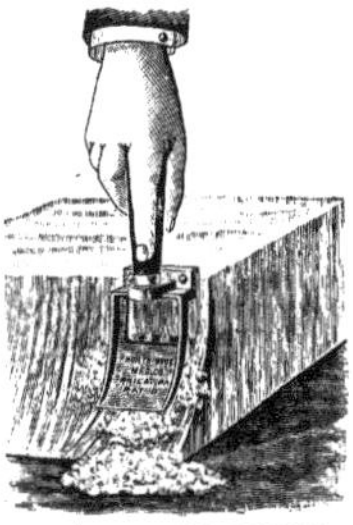

THE CROWN ICE CHIPPER.

PEERLESS ICELAND RAPID ICE CREAM FREEZER

THE GEM ICE SHAVE.

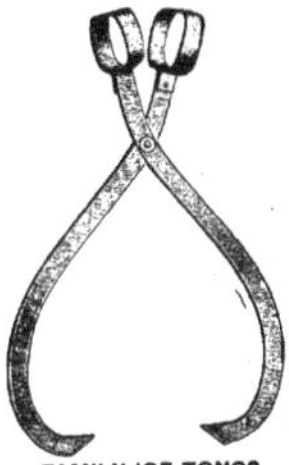

FAMILY ICE TONGS.

NO. 1 ICE TONGS.

ICE PICK—Arrow Point, Wooden Handle, with Heavy Galvanized Malleable Ring.

ICE CHISEL—Polished Steel or Galvanized Blade.

ICE CHISEL—Polished Steel Blade, with Heavy Malleable Galvanized Ring on Handle.

Lewis Bros. & Co.

MAIL ORDERS SHIPPED SAME DAY AS RECEIVED. MONTREAL.

Wadsworth-Howland Co. CHICAGO.

JEWEL

BEST PAINT ON EARTH

BEST PAINT ON EARTH

MIXED PAINTS

BANNER BUGGY PAINT prepared expressly for painting old or new work where a handsome, lasting and glossy surface is required. Dries quickly and will not crack or chip off.

ADAMANTINE FLOOR PAINT AND COTTAGE SHINGLE STAINS

RICE LEWIS & SON

LIMITED.

Canadian Distributors TORONTO.

HARDWARE AND METAL

VOL. XIII. MONTREAL AND TORONTO, MAY 25, 1901. NO. 21.

President,
JOHN BAYNE MacLEAN,
Montreal.

THE MacLEAN PUBLISHING CO.
Limited.

Publishers of Trade Newspapers which circulate in the Provinces of British Columbia, North-West Territories, Manitoba, Ontario, Quebec, Nova Scotia, New Brunswick, P.E. Island and Newfoundland.

OFFICES

MONTREAL - - - - - - - 232 McGill Street, Telephone 1255.
TORONTO - - - - - - - 10 Front Street East, Telephone 2148.
LONDON, ENG. - - - - - 109 Fleet Street, E.C., W. H. Miln.
MANCHESTER, ENG. - - - - 18 St Ann Street. H. S. Ashburner,
WINNIPEG - - - - - Western Canada Block. J. J. Roberts.
ST. JOHN, N.B. - - - - No. 3 Market Wharf, J. Hunter White.
NEW YORK. - - - - - - - 176 E. 88th Street,

Subscription, Canada and the United States, $2.00.
Great Britain and elsewhere - - - 12s.

Published every Saturday.

Cable Address { Adscript, London. Adscript, Canada.

BUSINESS MEN AND TOURIST TRAVEL.

CANADA as a summer resort for tourists from the United States and Great Britain has come to the front in a striking manner during the last few years. But, obviously, we have been handicapped in attracting tourists because of insufficient hotel accommodation at many points.

Most of the people who travel during the summer on pleasure bent do not demand elaborate hotel accommodation, but they do demand clean, wholesome and comfortably equipped hotels and homes. Where these cannot be had they do not resort.

There are many places in Canada rich in natural beauties and in attractions to sportsmen which might have thousands of tourists annually in their midst where they now have scarcely hundreds, and largely on account of the absence of proper accommodation.

As we have repeatedly pointed out, a large share of the money which is spent in a place where tourists resort finds its way into the stores of the merchants there. It is obviously, therefore, to their interests that every effort should be made by business men to improve the accommodation in their respective localities.

Every town that possesses natural attractions for tourists should have its organization, call it tourist association or what you like, whose office it should be to work up enthusiasm among the people of the locality in regard to the tourist question ; to induce the hotelkeepers to improve their premises and to secure lists of private homes where accommodation can be secured.

One of the most energetic tourist associations is that at St. John, N.B., the president of which is Mr. W. S. Fisher, of the wholesale hardware firm of Emmerson & Fisher, and day in and day out that association is preaching the gospel of New Brunswick as a tourist resort. The illustrated booklets it sends out annually are models of what such booklets should be, and the one which has recently been issued is the best of all. Business men in the western part of Canada who want a model upon which to work, or who want to get an inspiration where to go for their holidays, should send to Mr. Charles D. Shaw, St. John, N.B., for a copy of "St. John River, the Rhine of America." But this association is not content with merely issuing a booklet and interesting the newspapers in the Maritime Provinces, but it has a lady employed to visit the different points of interest in New Brunswick, and particularly those on the charming St. John river, which has well been named the "Rhine of America," and interview the hotelkeepers and private housekeepers in regard to accommodation. This will no doubt all be tabulated, and the information thus obtained disseminated for the guidance of tourists.

What the New Brunswick Tourist Association has done, similar organization can do elsewhere.

THE QUALITY OF MANILA ROPE.

A GOOD many complaints are heard this spring about the color of the manila rope that is on the market. As a rule, as the trade well know, the color is bright and clean-looking, but a great deal of that which is now being delivered is dark and dirty-looking. Some retailers have refused to receive their shipments on that account.

The makers claim that it is impossible this season to procure hemp of the usual bright color, but they say that, although defective in color, in strength it is unimpaired. It is understood that the manufacturers of rope in the United States have an experience similar to that of their confreres in Canada.

Sisal rope is as usual bright and clean, but, of course, sisal does not possess the same strength as manila rope.

It is alleged that mixed rope is being palmed off upon the trade as pure manila. The hint to dealers is obvious.

If there is any friction in the attempt to form the proposed powder combination in the United States we may expect to see the scheme explode.

A DELIBERATE "HOLD-UP."

THE decision of the Members of Parliament to increase their sessional indemnity by 50 per cent. is a deliberate and preconcerted plan on the part of Liberals and Conservatives alike to raid the public treasury.

The Premier, in introducing the resolution to increase the indemnity to $1,500, said he believed the "increase would strike the sense of fairness and justice of every member of the House." What hypocrisy! Were it not for the seriousness of the matter, we might term it an unique sense of humor. Sense of justice, to be sure! What other term might those engaged in the "hold-up" be expected to give it?

Fairness and justice, indeed! No doubt the employes of a large mercantile concern would so term their action if it was possible for them to get their heads together and decide, with the connivance of the men appointed to guard the treasury of the firm, to increase their own salaries by 50 per cent.

And then, Mr. R. L. Borden, the leader of the Opposition, who seconded the resolution. His line of argument was different from that of the Premier, but it was none the less weak and illogical. The burden of his plea was that, instead of being an additional incentive to the professional politician to aspire to a seat in the House, it would, on the contrary, induce men of "business interests to offer themselves for election."

What a species of fallacious argument! Mr. Borden, like Sir Wilfrid Laurier, is a man of parts, and much superior in intellect to the average politician, but, like the leader of the Government, he evidently believes that an untenable argument is better than none at all.

The idea that an extra $500 indemnity would be more likely to induce a business man than a professional politician to aspire for a seat in the House of Commons is so absurd that one can scarcely believe that a man of Mr. Borden's intellect would advance it. And the very fact that such silly arguments are being used is one of the best proofs we could have of the weakness of the case for the increase in the indemnity.

At any rate if business men would aspire to a seat in the House of Commons on account of the indemnity that was to be obtained, wherein would they differ from the professional politician who is now so much the bane of our public life? In no other way than in name. In practice they would both be the same.

Of the few men in the House, who, up to the time of writing, have opposed the increase in the indemnity, one of them is a business man, namely, Mr. Blain, who sits for Peel and carries on a hardware store in Brampton. The stand that Mr. Blain has taken is a rebuke to the fallacious argument of the leader of the Opposition.

As the Senators are also to join in the increased indemnity there are 285 members whose allowance will be $500 more than formerly. That number multiplied by 500 means an increase of $142,500 in the annual indemnity, which, capitalized at 3 per cent., equals an addition of $4,750,000 to the national debt. And yet none of these things move the indignation of the leaders on either side of the House.

Then, it must be remembered, there is the mileage allowance. This means another $10,000, which practically all goes into the pockets of the members of the two Houses, for it is notorious that, while the members charge mileage, there is scarcely one of them but carries a pass in his pocket.

While the Government is to be blamed for allowing this raid upon the treasury, the Opposition is equally to blame. If the office of an Opposition is anything, it is to act as a brake on Government expenditure, but in this instance we find it assisting in the authorization of an expenditure that is tantamount to increasing the public debt by nearly $5,000,000.

There is no question about it, our public expenditure is increasing at a rate too rapid. But how can the Opposition hereafter have the temerity to raise even its little finger against extravagant expenditure when it is consenting to an expenditure a part of which goes into the pockets of the members on its side of the House?

The plea that the length of the sessions warrants the increase in the indemnity is the principal stock-in-trade argument. But they are long, not because of the actual business transacted, but because of the extravagant and useless speeches with which the time of the House is taken up. It is not the practice of business men to increase the salaries of their employes who waste their time in wrangling and in delivering long speeches traducing each other. And what is the practice in business should be the practice in the House.

THE MACHINISTS' STRIKE.

THERE are probably few people outside the employers who are opposing the innovation who do not sympathize with the machinists who are now on strike in the United States.

What they are contending for is a reduction of the hours of labor to nine hours a day, the rate of wages to be the same as for the present ten-hour day.

There are few of the skilled trades where the day's labor exceeds nine hours, and in many eight hours is the maximum. Under such conditions the machinists must have, no doubt, for some time felt that their condition was anomalous, as far as hours of labor were concerned, compared with mechanics in other branches of trade. Skilled workmen in some branches of trade have enjoyed the nine-hour day for 30 years or more.

A great many of the employers appear to be only offering a half-hearted resistance to the demands of the machinists, for a number of them have already surrendered and the 50,000 men that quit their work a few days ago have been reduced by probably one-third, and it is estimated that by the beginning of next week fully 90 per cent. of the men will again be at their accustomed places.

The pity is that arbitration had not taken the place of the strike.

FIRES IN MATCH FACTORIES.

Match factories seem to be having their full share of fires in Canada. Little more than a year ago the plant of The E. B. Eddy Co., Limited, was completely destroyed in the big Hull-Ottawa fire. Last January The Walkerville Match Co., Limited, suffered about $25,000 loss, and last week their premises were destroyed, causing a loss of $100,000.

It is estimated that new works can be in operation in three months, but in the meantime there will be a big reduction in the output of matches in Canada.

LINSEED OIL STILL ADVANCING.

THERE seems to be no let up in the stiffening feeling re linseed oil. On Monday, an advance of 3c. was made by the Ontario jobbers, which makes the basis in Toronto, Hamilton, London and Guelph 80c. for raw and 83c. for boiled in single barrels, with an advance of 1c. for delivery to outside points throughout Western Ontario. This rise of 3c. makes a total advance of 12c. in the last month.

Needless to say the present basis of prices is a high one. Last year the range was even higher, but during the ten years previous to 1900, the market during May never got above 18c. for raw and 71c. for boiled. The following prices have ruled on the fourth week in May during the decades for single barrel lots :

	Raw.	Boiled.		Raw.	B'l'd.
1890..........	68c	71c	1896..........	53c	56c
1891..........	64	67	1897..........	41	44
1892..........	56	59	1898..........	49	52
1893..........	55	58	1899..........	52	55
1894..........	53¼	56¼	1900..........	84	87
1895..........	53	56	1901..........	80	83

It will be remembered that the exceptionally high price of seed last season was due to the big demand for, and consequent high price of, flaxseed in America, and the shortage of flaxseed in England, owing to there being few, if any, arrivals from the Argentine Republic, where there was a good crop, which, however, could not be exported, owing to bubonic plague at Buenos Ayres, the port of that country. The fact that the greater portion of the whole season's crop of flaxseed had been kept off the market naturally caused a scarcity of oil, and boosted prices up to exceptionally high figures.

It was just as naturally anticipated, however, that when this seed began to arrive on the British market prices would at least resume their usual level. The result was that, on the British market, during the early months of the present year, prices were so low that Canadian manufacturers and jobbers were forced to reduce prices, though, owing to the manipulation of American speculators, the price of flaxseed was kept at a stiff figure. But, as the season progressed, it became manifest that British refiners could not depend on the great quantity of flaxseed from the Argentine Republic that had been anticipated, and prices at once began to advance there.

In the meantime, American flaxseed, being controlled by speculators, was held at firm prices, so the Canadian refiners have not been working at their full capacity, and have practically no surplus stock on hand. As stocks in jobbers' hands are scarce, advantage of the advancing British markets has been taken to put up figures here.

The indications seem to be that for at least three weeks or a month prices will continue at the present high basis. It would be difficult to prophesy further than that, as by that time British oil, bought during the depression, will be on the market, and holders of this may force a break in prices.

IMPORTANT LEGISLATION.

THE past week has seen some important legislation of interest to business men introduced in the House of Commons.

One was a Bill to establish a branch of the Royal Mint in Ottawa where bronze, silver and gold coins are to be made, and an assay office in either British Columbia or the Yukon.

Another important Bill was one to provide for the giving of a bounty to the lead refining industry of Canada. The bounty is to be $5 per ton, but the amount expended in any one year for this purpose is not to exceed $100,000. The bounty is to decrease by $1 per ton each year, and at the end of five years will cease altogether.

The bounties are to be paid half yearly, but the total sum so payable is not to exceed $100,000 in any year. If the sum earned during any half year exceeds $50,000 the bounties are to be reduced as regards that six months, so that the maximum payable shall not be more than the $50,000. If the sum earned is less than $50,000 the balance is to be carried to the credit of the next half year, and may be paid out in addition to the amount falling due for the latter period.

This form of aid, as our readers are already aware, has been made necessary on account of the smelter combination in the United States, through whose action the closing up of every silver and lead mine in the United States was threatened.

GLASS STRIKE IS OVER.

Cable advices from Belgium received this week state that the great strike of glassblowers in that country, which was declared last August and maintained firmly since by the operatives, had been settled. This strike, as is generally known, has been responsible for the high prices ruling, for, as fully two-thirds of the glass-workers were out, the production of glass was seriously handicapped.

It is likely that the settlement of the strike will give a big impetus to the production of glass, and that Belgian prices will be materially lowered. Import orders, which were refused some months ago, will now be placed on more favorable terms, but these will not likely be delivered until about the close of navigation on the St. Lawrence.

The effect this settlement will have on prices will be to prevent the advance which was threatened by Canadian jobbers on account of the local scarcity. It is not likely, however, that there will be a decline, at least for some months.

REDUCTION IN LEAD PIPE.

One of the effects of the weakness in the lead market to be noted this week is a decline in the price of lead pipe, the discount now being 30 per cent. instead of 25 per cent. as formerly.

This means a reduction of 6⅔ per cent. in the actual selling price.

HARDWAREMEN ORGANIZE.

On Thursday evening last week a good representation of the retail hardware, paint and oil, stove, tin and metal dealers of Toronto gathered together to form the Toronto Retail Hardware Merchants' Association.

The following officers were elected :

President—E. R. Rogers.
1st Vice-President—F. J. Russill.
2nd Vice-President—Albert Welch.
Treasurer—John Caslor.
Secretary—Fred. W. Unitt.
Executive Committee—R. Fletcher, W. Emery, C. Dale, J. W. Peacock, C. Watkins and J. T. Wilson.

The objects of the association are to look after the interests of the retail trade in all matters. Among the matters which will receive early consideration are early-closing, buying in the cheapest market, and selling, as far as possible, at a uniform rate.

TRADE IN COUNTRIES OTHER THAN OUR OWN.

BRITISH ENGINEERING TRADE.

THE position of the engineering trade continues far from satisfactory. In some branches a state of absolute depression prevails, while there is as yet no modification of the general complaint among those firms who are fortunate enough to be working full time, that the orders they have in hand are being executed more rapidly than others are coming in. The recent cheapening of certain items of raw material seems to have had no other effect than to increase the severity of competition and to bring prices down still lower. The electrical and locomotive industries are about the only branches that can be excluded from this somewhat gloomy summary; both of them report crowded order books and full pressure.

The general situation is not improved by the spirit of unrest that seems to prevail among the workmen. There appears to be a disposition among them to magnify the importance and urgency of various industrial questions that have been simmering for some time past, and to bring matters to a more acute stage of discussion. One of the most prominent of these questions is that of the payment of skilled labor wages for the working of automatic machine tools. It is to be hoped that the wiser counsels will prevail, for it is hardly likely that under the existing unsettled conditions employers will be willing, or, indeed, able to make many concessions.—Iron and Coal Trades' Review, May 10.

TRADE IN SHEFFIELD.

Although prices are decidedly firm there have been very few transactions in raw material during the week, and sellers find things flatter than at any time since the Easter holidays. There is also increased difficulty in disposing of scrap metal. Undoubtedly many firms in the bar-iron and steel trades are experiencing depression, and no general improvement appears to be in operation. The export returns of steel for April afford but little encouragement. Canada and the United States only taking about a third of the quantity credited to them in April of last year. Fortunately, Australasia remains a good customer for steel, and the decrease of business with Germany is less than was the case in the earlier months of the present year. The figures relating to Russia, Denmark, Holland and France are bad.—Ironmonger.

THE BRITISH TINPLATE TRADE.

During the past week the demand for tinplates has been well sustained both for prompt and for forward deliveries, and some extensive orders have been placed out for both these positions at advanced values. The "coal scare" has led to many makers withdrawing from the market until the result of this proposed and preposterous disturbance has further developed. In the meantime the colliery proprietors have been quick to advance prices on the chance of a strike, and this has caused considerable excitement among the manufacturers throughout the country, as a stoppage of machinery at this juncture would be very undesirable. Tinplate makers are all well supplied with orders, and are, therefore, inclined to hold off from further sales unless they can secure full prices. On the other hand, merchants are disposed to buy with caution, in face of an expected coal collapse after the present excitement has blown over: they do not consider the present position of affairs as calculated to "rush" the market, and, therefore, decline to pay any fancy figures, in the face of what may yet come about to swing the pendulum the other way. The inquiry has been considerable from all quarters, running largely upon special sizes, light plates, black plates, and Canadas, while wasters have also met with a large amount of favor.—Iron and Coal Trades' Review, May 10.

THE BRITISH PIG IRON TRADE.

The position of the pig-iron market is steadily improving, especially in the Cleveland district, where, to all appearances, the worst has been experienced. Consumers seem to be convinced of this by the way they are coming into the market and by the pressure to buy. In other districts the demand is only moderate, but prices are being well maintained, and in some brands show a tendency to rather more firmness. Producers at Middlesbrough have advanced the price of No. 3 to 46s. 6d. per ton, and have done a large business at 46s. From Lancashire only a slow business is reported, and orders are still restricted to hand-to mouth requirements.—Iron and Coal Trades' Review, May 10.

THE BRITISH GALVANIZED IRON TRADE.

During the last twelve months the galvanized iron trade of South Staffordshire has undergone some very remarkable experiences. Early in the year almost unparalleled prices obtained, but as the year advanced the decline in value was large and continuous until the record in low prices was reached. It is doubtful whether the trade in this district has ever experienced a reaction so remarkable in its severity. The intense competition which set in with the slump in prices resulted in the failure of several important firms of sheet iron makers. The outlook for the galvanized iron trade of the district is at the present moment by no means encouraging, handicapped as the local industry is by almost prohibitive railway rates of freightage. The exports of galvanized iron from England for the past year are the highest on record, but the share that Wolverhampton and South Staffordshire have had in the trade during the last twelve months is distinctly less than for some time past. Wolverhampton is not losing so heavily as the district around, but Messrs. Lysaght continue to transfer more and more of their work to Newport (Mon.), and at the present time the well-known Osier Bed Works are practically closed.—Hardwareman.

UNITED STATES LINSEED OIL MARKET.

We have a firm market to report, the conditions being much the same as outlined in our last review. Inside carload quotations remain 60c. per 7 1-2 pounds (8c. per pound) for raw and 61c. (8 2-15c.) for boiled. Some mills want more money for what oil they have, naming figures 1 to 2c. above prices already quoted. Outside markets generally are strong in sympathy with Chicago, and nowhere except at New York do values appear to be out of harmony with our market here. It is stated that offerings of oil at 59c. for May and June are being made in the eastern market mentioned, but they come from but one crushing interest. Our information is to the effect that it is doubtful if round lots can be purchased there any cheaper than in the west. The developments within the week have for the most part been of a bullish character. Seed in the home market has made an 8 1-2c. advance and oil in the foreign market has gone up sharply. On the other hand, receipts have been unexpectedly large, allowing for such seed as may have come from other receiving points. The demand for oil is quite large, more brisk perhaps in small lots, single barrels selling at 62c. for raw, and 63c. for boiled, and 5-barrel lots 1c. less than last named figures. The export cake market shows increased firmness, Europe being short on her feed supply. Bids this week range at $24 to $24.50 for prompt and nearby forward shipments, an advance of 50c. in some instances. Trading is light, limited by the very small stocks available in crushers' hands.—Paint, Oil & Drug Review, Chicago.

IRON TRADE IN THE UNITED STATES.

Taken as a whole the iron markets have quieted down considerably and it seems likely that some time may elapse before buyers and sellers get together on the question of the basis for the second half of the year. In some sections and in some branches sellers show some uneasiness, which points to concessions in order to start the buying movement for the third and fourth quarters. The effect of the flurry in stocks has probably not been measured yet. It will cut off many wildcat enterprises, which is an advantage rather than a drawback. On the other hand, some buyers show a disposition to hold off in order to watch developments.

It is a significant fact that the agricultural implement makers continue in the market for supplies for the future, notably charcoal pig iron and steel bars, thus showing that the one industry which is nearest to the farming interests enjoys unshaken confidence.—Iron Age, May 16.

UNEXPECTED DROP IN BRITISH BAR IRON.

Some surprise has been occasioned in Midland iron-trade circles by an announcement of the leading marked-bar firms reducing prices £1 per ton as from May 1. This step has apparently been arrived at rather in a hurry, inasmuch as special inquiries made last Thursday elicited the reply that "nothing of the kind was contemplated, at any rate for the present." Of course, it may be that the intention was purposely kept a secret; but, if that is so, the object of such a manœuvre is scarcely obvious, especially when the event itself was so close at hand. The unexpected drop shows the weakness of the commercial position. For some time past producers of the leading brands of iron, although they have been in a better position than the makers of other descriptions of iron, have been doing a hand-to-mouth trade, and there is no doubt that the alteration just declared has been made with the view of attracting orders, both from home and foreign customers. All classes of Midland iron are very weak, and the action of the marked-bar makers is likely further to reduce the value of the commoner qualities of iron.—Ironmonger, London, May 4.

HARDWARE TRADE IN THE UNITED STATES

The past week has developed little change in the business situation. Prices remain very steady, with less than the usual revisions on the part of manufacturers. Manufacturers complain of the difficulty of getting raw material in certain lines, and the trade, on the other hand, are obliged to get along with a tardy execution of their orders for a few kinds of goods. There is perhaps a little let up in the demand made by the trade upon manufacturers, but the volume of business is excellent. There is relatively greater activity in the West than in the East. Reports of the general conditions of Southern trade are very favorable. Export trade continues large, but feels the effect of having prices higher on certain leading lines than are now prevailing abroad. Apart from the large foreign business in staple lines, there is a rapidly extending trade in specialties. Throughout the country hardware merchants are generally doing a good business, and report pretty full employment of labor and an increase in building operations and other enterprises.—Iron Age, May 16.

NEW YORK METAL MARKETS.

TIN—The London cables reported a continuance of the excitement in tin, with the market strong and still advancing. The movement over there, as previously stated, appears to be of a strictly speculative character, but is evidently exerting more influence here than it had at the start. The advance recorded in the English market to-day amounted to £2 5s. on spot and £3 15s. on futures, the difference between the quotations on prompt and forward deliveries being reduced to 10s. Under the stimulus of the news from London and some speculative bidding the New York market went up 60 points, sales of both spot and futures being made at 27.50c. There was less buying interest manifested than toward the end of last week, consumers seeming to have little confidence in the situation and keeping out of the market. At the close 27.50c. was bid and 28c. asked on both spot and futures.

COPPER—Under the stimulus of an increased demand the London market advanced 5s. on spot and 3s. 9d. on futures this morning, but reacted later, closing easy at £69 18s. 9d. for spot and £70 8s. 9d. on futures, which was 2s. 6d. higher than Friday's close on spot and but 1s. 3d. better on futures. Nothing new was presented in the New York market, which was quiet but steady at 17c. for Lake Superior and 16 5-8c. for electrolytic and casting.

PIG LEAD—The market remains steady though there is not much doing. Quotations are maintained on the basis of 4.37 1-2c. St. Louis was firm and unchanged at 4.22 to 4.25c. There was no change in the London market.

SPELTER—The New York market was quiet at 3.95 to 5c., and St. Louis was reported to be in a similar position, the quotation there being 3.77c. London continues to advance, and at the close was 2s. 6d. higher than on Friday last.

ANTIMONY—A moderate demand for regulus is supplied at prices within the range of 8 1-2 to 10 1-4c. as to brand.

OLD METALS—Continued dullness characterizes this market, which, however, is steady at the quotations.

IRON—The conditions governing this market are much the same as for a couple of weeks past. Business in pig iron has narrowed down to small dimensions outside of deliveries on contracts made some time since, and while it is intimated that some concessions would be made to secure business for delivery in the third quarter and beyond sellers do not seem very anxious and are not inclined to force business. In finished products of all kinds the movement continues on a very large scale, but in this department also business in deliveries after July 1 seems to be light, though the outlook is held to be excellent for continued activity for some time to come.

TINPLATE—Nothing new was presented in this market. The consumption continues on a large scale, but the demand for forward delivery does not seem to be urgent. Prices are firm and unchanged.

BUSINESS CHANGES.

DIFFICULTIES, ASSIGNMENTS, COMPROMISES.

Archibald McCormack, general merchant, Sydney, N.S., has assigned.

The creditors of M. Simon, general merchant, Alexandria, Ont., are to meet.

H. Hudon & Co., general merchants, Ste. Angele, Que., are offering 40c. on the dollar

Lafontaine & Lavoie, general merchants, St. Cyrille de Wendover, Que., have suspended.

J. A. Russell, harness dealer, Fort William, Ont., has assigned to Charles W. Jarvis.

Gilbert Lewis, dealer in agricultural implements, Clifford, Ont., has assigned to Anson Spotton.

PARTNERSHIPS FORMED AND DISSOLVED.

Brouillet & Frere, general agents, etc., Montreal, have dissolved.

Thos. H. Norman, general merchant, Ruthven, Ont., has admitted Thomas Dawson.

J. S. Frost & Co., coal dealers, St. John, N.B., have dissolved. Alfred Mills continues.

W. L. Belyea & Co., general merchants, Brown's Flats, N.B., have dissolved. W. L. Belyea continues.

Pierce & Howey, stove and tinware dealers, etc., Port Rowan, Ont., have dissolved. J. R. Pierce continues.

Reeves Bros., general merchants, Port Hawkesbury, N.S., have dissolved; succeeded by J. H. Reeves & Co.

CHANGES.

Thomas Kelly has registered as sawmiller, etc., at Pléssisville, Que.

J. A. Savoie, saddler, Wotton, Que., has removed to St. Paul de Chester.

S. J. Brown, harness dealer, Campbellton, N.B., is returning to Hartland.

Thompson & Co., general merchants, Harrow, Ont., have retired from business.

J. S. McCracken & Co., harness dealers, Brandon, Man., have retired from business.

Burch & Co., general merchants, Red Deer, N.W.T., have sold out to W. Phillips.

Henry Rogers, Sons & Co., have registered as hardware dealers, etc., in Montreal.

Keith & Plummer, general merchants, Hartland, N.B., are building a cheese factory.

Graves & Maguire, general merchants, Vienna, Ont., have sold out to C. M. Wilson.

H. H. Haliburton, general merchant, Port au Port, N.S., has been succeeded by Haliburton & Teroux.

M. Campbell & Co., general merchants, Glace Bay, N.S., are about opening a branch at International Pier.

DEATHS.

G. E. Nugent, of G. E. Nugent, & Co., general merchants, Wapella, Man., is dead.

CANADA'S TRADE STILL GROWING.

THE export trade of the Dominion of Canada continues to expand. The country's aggregate trade on the basis of goods entered for consumption and export exclusive of corn and bullion for the ten months ended April 30 exceeded that for the same period of the previous year by nearly twenty millions. The actual figures were $302,567,352, against $283 517,239, showing an advance of $19,050 113. The imports totalled $142,942,420 for the ten months, as against $143,303 759 in 1900, a falling off of $361,339. The exports, reckoning both foreign and domestic produce, amounted to $159 624,932, as compared pared with $140,213,480 for the preceding ten months.

The following are the exports in detail for the two periods:

Ten months ending April, 1900.

	Domestic.	Foreign.
Mines	$10,831,379	$ 168,866
Fisheries	9,048,910	50,086
Forest	24,577,226	286,149
Animals	48,187,911	960,685
Agriculture	23,410,785	9,991,330
Mnfrs	11,118,976	1,014,448
Misc	284,258	280,471
	$127,461,445	$12,752,035

Ten months ending April, 1901.

	Domestic.	Foreign.
Mines	$30,163,967	$ 121,593
Fisheries	8,665,271	9,759
Forest	23,065,604	263,720
Animals	48,299,828	2,060,058
Agriculture	21,448,758	10,981,329
Mnfrs	12,834,520	1,428,957
Misc	42,955	238,613
Total	$144,520,903	$15,104,029

For the month of April last the exports, including foreign and domestic, show a betterment of $1,622,414, as compared with the same month of last year.

HOG BRISTLES.

A manufacturer of brushes says: "Hogs are fattened and killed young in the United States, and with the constant and widespread improvement of breeds here the hogs have run less to bristles. Some extensive packers collect bristles, but the American supply is probably less than 1 per cent. of the consumption. The longest American bristles are about 4¾ inches in length. Imported bristles come from various parts of Russia, but most largely from Poland and North Germany. Some finer, soft bristles come from France. The bulk of the supply comes from cold countries where the hogs are well protected by thick coats, and many of the bristles are those of wild hogs. The hogs shed their coats as many other animals do, and there are men who gather the bristles of wild hogs, knowing their haunts and where the bristles are to be found. Other bristles are collected in the usual way when hogs are killed. The bristles are subjected to various processes of curing and preparation before they become commercial bristles and ready for the market.

"Imported bristles range from 3¾ to 7¾ inches in length. Various vegetable fibres are now used extensively in the manufacture of cheap brushes, but for the best kinds of brushes only the finest foreign bristles are used, and their importation continues steadily."—National Provisioner, New York.

THE CRAMP ONTARIO STEEL WORKS.

C. D. CRAMP, president of the Cramp Ontario Steel Co., who was in Toronto on Wednesday to attend a meeting of the directors of the company, stated that during this meeting the purchase of 10,000 acres of coking coal in Wise County, Western Virginia, had been ratified. This coal had been reported on as of exceptionally high quality, not only for steam coal, but for coking purposes, by both the company's engineer and the engineers of the United States Geological Department. In two of the five coal seams which occur above water level there can be mined without any sinking 140,000,000 tons of good coal for coking purposes. The company expect to be a large producer of coke and coal, so that as a matter of fact they will have their own coke for their own plant free from cost.

It had generally been understood that the Cramp company would purchase its ore from the Clergue concern, but Mr. Cramp states that the intention of the company is to own its own coal and ore lands, and thus to be independent of other enterprises. They have secured several iron properties in Northern Ontario. One of these properties their engineer has computed to contain 100,000 tons of Bessemer iron ore in sight, and another area has a showing of iron extending over 300 x 1,200 ft., and it is intended to secure other iron property and be in a position to ship iron ore this fall.

"From this date," concluded Mr. Cramp, "We are satisfied we have enough iron ore to do us for years to come, and we will be in a better position than any steel company in Canada by owning our own iron and coal with the exception of the two companies doing business in Cape Breton. Our aim is to be independent of every other ore and coal producer, and I am pleased to say we feel ourselves now in that position. We are now completing arrangements to let the contract for the plant at Collingwood, and expect to commence work shortly."

WORKS OF THE J. STEVENS COMPANY.

Under date of May 16, the J. Stevens Arms and Tool Co., Chicopee Falls, Mass., write: "Mr. H. M. Pope moved his special machinery for the manufacture and boring of his high-grade rifle barrels from Hartford, Conn., to our plant yesterday. That this is now being installed and Mr. Pope is now permanently located here.

"Our force has increased from 44 six years ago to 700 to-day; that a year ago, shortly after acquiring the large Overman plant, which is one of the finest in the East, and gives us an additional 220,000 square feet of floor space, we tried to rent a part of it. To-day we are occupying nearly all of it ourselves and are contemplating an addition to our river plant."

HILL PLANT.

The accompanying cut shows the two plants, which doubtless will interest our readers, few of whom probably realize the present capacity of the works of the Stevens Company.

AN APPRECIATIVE HARDWAREMAN.

"I would not think of being without HARDWARE AND METAL for 10 times its cost," remarked Mr. S. D. Ross, a Brighton hardwareman, a few days ago, to a representative of this journal who was in his place of business. "The quotations are reliable and the hints and pointers it gives are valuable."

PERSONAL MENTION.

Mr. A. Burdette Lee, vice president of Rice Lewis & Son, Limited, sailed on Saturday from Montreal on the ss. Tunisian for Great Britain.

INDUSTRIAL GOSSIP.

Those having any items of news suitable for this column will confer a favor by forwarding them to this office addressed the Editor.

V. E. TANNER, Mount Forest, Ont., proposes to organize a $100,000 cooperative company and build a binder twine factory in Brandon, Man., to have a capacity of three tons per day.

The E. Long Manufacturing Co., Orillia, Ont., are enlarging their works.

Another $10,000 building is to be added to the nickle-copper works at Hamilton.

Henry Walters & Sons, axe manufacturers, Hull, Que., have started the erection of a new factory to cost between $8,000 and $10,000.

It is reported from Oshawa, Ont., that The Detroit Wire Fence Co. have accepted the inducements from that town to establish a wire fence factory to employ 25 hands from the start.

It is reported from Germany that the British Admiralty have ordered eight tubular boilers from the Durr Boilerworks at Ratingen, near Dusseldorf, for the cruiser Medusa. The value of the order amounts to $100,000.

EARLY CLOSING IN ST. THOMAS.

There is a strong movement in favor of early closing in St. Thomas, Ont., but, like the majority of agitations to this end, it has not originated with the merchants of the town, and proposes early closing on Saturday evenings, rather than on the other evenings of the week.

On Thursday evening, last week, a conference was held between members of the Lord's Day Alliance and the Trades and Labor Council, of St. Thomas. The Alliance proposed securing by petitions of the merchants a by-law compelling all stores to close at 10 on Saturday evening. The Trades Council fully endorsed the movement, providing the Alliance would change the hour proposed in the petition from 10 to 9 o'clock, and appointed a committee of three to cooperate with the Alliance committe in circulating the necessary petition among the shopkeepers.

GUARANTEED SILVERWARE.

The S. L. & G. H. Rogers Co., manufacturers of sterling silver and silver-plated ware, Hartford, Con., guarantee their extra plate teaspoons to strip 48 dwts. per gross; dessert spoons and forks, 72 dwts. per gross, and tablespoons and medium forks, 96 dwts. per gross. Their medium knives are guaranteed to strip 12 dwts. per doz. on No. 12; 14 dwt. per doz. on No. 14, and 16 dwts. per doz. on No. 16. All their other extra-plate goods are guaranteed to have the same proportionate plate, which is 20 per cent. above standard plate.

THE COKE INDUSTRY.

A CENSUS bulletin on the coke industry of the United States, based on its condition in the year 1899, shows that a rapid increase in production has occurred since 1880, the first year that the business received attention from the census officials, says Bradstreets. Thus, in 1880 the value of the country's output of coke amounted to $15,250, and in 1890 the production was worth $16,498,345, while in 1899 the total of $35,585,445 had been attained.

There were twenty-two States in which coke was produced in 1899, as against eighteen States in 1889 and nine in 1880. At each census Pennsylvania has stood at the head of the coke-producing States, more than two-thirds of the total coke production of the United States being made in that State. But the proportion of her product to the total has decreased from 84 per cent. in 1880 to 73 per cent. in 1890, and 67 per cent. in 1889. West Virginia, which was third in rank at the two preceding censuses, became second in importance in 1899; and Alabama, which was sixth in 1880 and second in 1899, now stands third. Ohio, which was second in 1880, had fallen to the eighth place in 1889 and held the same relative position in 1899. Virginia, which reported no coke product in 1880, was sixth in 1889 and fourth in 1899. Colorado, the only important coke-producing State west of the Mississippi river, now ranks fifth among the total number, and Tennessee, which stood fourth in 1880 and 1889, was sixth in 1899. Massachusetts, which had no coke product in 1880 or 1889, was seventh in importance in 1899.

The bulletin says that the present tendency toward large industries under one management is illustrated in the statistics of coke production in 1899. The total number of tons of coke produced has increased 96 per cent. and the value of all products has increased 115 per cent., while the number of active establishments reporting for 1899 was only twenty-three, or 10 per cent. more than the number reporting for 1889. The amount of capital invested in the industry in 1899 was $36,502,679, as compared with $17,462,729 in 1889 and $4,769,858 in 1880. The increase of capital in 1899 over 1889 was $19,039,950, or 109 per cent. As compared with 1880, the capital invested in 1899 increased 665 per cent.

The report shows that there is an average of 16,999 persons employed in the business, aside from officers, clerks, etc., the aggregate salary roll amounting to $7,085,736. There were 915 managers, clerks, etc., to whom was paid in the way of salaries $797,296.

Boutilier & Morehouse, general merchants, etc., Centreville, N.S., are building a new, up-to-date two-storey 50 x 25 ft. store

D. & H. Becker, general merchants, New Hamburg, Ont., have built an addition to their storó and are now rebuilding their old premises.

The Classic City Mills, Stratford, Ont., are to be rebuilt and modernized. Their capacity is to be 150 bbls. per day at first, but provision will be made in the new premises for increasing the output above that figure.

MARKETS AND MARKET NOTES

QUEBEO MARKETS

Montreal, May 23, 1901.

HARDWARE.

BUSINESS is fairly active and as brisk as can be expected just at this season of the year, nevertheless, the various wholesale houses are not as busy as they were last week. The spring rush is now over, and so-called spring shipments are all made. Yet seasonable goods continue to be inquired for, and ice cream freezers, lawn requisites, screens, refrigerators, wires, etc., are all moving out in regular order. Barb wire is as scarce as ever, and supplies will not be forthcoming to meet the demand for weeks yet. The only change in prices this week is the raising of the discount on lead pipe to 30 per cent. Sheet metals are firmer this week, and are rather scarce on this market. Payments are first class.

BARB WIRE—There have been but few deliveries of barb wire during the last few weeks, and relief does not seem to be in sight, as the mills are several weeks behind in their orders. Prices are unchanged at $3.05 per 100 lb. f.o.b. Montreal.

GALVANIZED WIRE—There is still a scarcity of galvanized wire. New orders are not as numerous as they were. We quote as follows: No. 5, $4.25; Nos. 6, 7 and 8 gauge, $3.55; No. 9, $3.10; No. 10, $3.75; No. 11, $3.85; No. 12, $3.25; No. 13, $3.35; No. 14, $4.25; No. 15, $4.75; No. 16, $5.00.

SMOOTH STEEL WIRE—Those that have smooth steel wire coming from the other side with their barb wire are waiting for supplies. But the stringency is hardly as great in this line, as our Canadian mills have been able to make shipments when reqnired. We quote oiled and annealed as follows: No. 9, $2.80; No. 10, $2.87; No. 11, $2.90; No. 12, $2.95; No. 13, $3.15 per 100 lb. f.o.b. Montreal, Toronto, Hamilton, London, St. John and Halifax.

FINE STEEL WIRE—A fair trade is passing. The discount is 17½ per cent. off the list.

BRASS AND COPPER WIRE—The usual demand has been experienced. The discount on brass is 55 and 2½ per cent., and on copper 50 and 2½ per cent.

FENCE STAPLES—Business is good, but hardly as brisk as last week. We quote: $3.25 for bright, and $3.75 for galvanized, per keg of 100 lb.

WIRE NAILS—Trade in wire nails has somewhat improved upon that of last week. We quote $2.85 for small lots and $2.77½ for carlots, f.o.b. Montreal, London, Toronto, Hamilton and Gananoque.

CUT NAILS—Business is picking up in cut nails. Prices are unchanged. We quote as follows: $2.35 for small and $2.25 for carlots; flour barrel nails, 25 per cent. discount; coopers' nails, 30 cent. discount.

HORSE NAILS—The demand has slackened during the last few days, as most dealers have laid in stocks since the present discounts went into force. "C" brand is held firmly at discount of 50 and 7½ per cent. off the new list. "M" brand is quoted at 60 per cent. off old list on oval and city head and 65 per cent. off countersunk head.

HORSESHOES—A fair trade is being done in horseshoes. We quote as follows:

Iron shoes, light and medium pattern, No. 2 and larger, $3.50; No. 1 and smaller, $3.75; snow shoes, No. 2 and larger, $3.75; No. 1 and smaller, $4.00; X L steel shoes, all sizes, 1 to 5, No. 2 and larger, $3.60; No. 1 and smaller, $3.85; feather-weight, all sizes, $4.85; toe weight steel shoes, all sizes, $5.95 f.o.b. Montreal; f.o.b. Hamilton, London and Guelph, 10c. extra.

POULTRY NETTING—There is still some inquiry for poultry netting, but shipments are only of a fair size. The discount is 55 per cent.

GREEN WIRE CLOTH—A fairly active demand is to be reported in green wire cloth. The price is steady at $1.35 per 100 sq. ft.

SCREEN DOORS AND WINDOWS—Quite a number of orders have been filled this week. On the whole the screen trade has been a satisfactory one this spring. We quote: Screen doors, plain cherry finish, $8.25 per doz.; do. fancy, $11.50 per doz.; walnut, $7.40 per doz., and yellow, $7.45; windows, $2.25 to $3.50 per doz.

SCREWS—Sorting orders in fair nnmbers and of good size have come to hand this week. Discounts are as follows: Flat head bright, 87½ and 10 per cent. off list; round head bright, 82½ and 10 per cent.; flat head brass, 80 and 10 per cent.; round head brass, 75 and 10 per cent.

BOLTS—The manufacturing trade throughout the country is flourishing and the demand for bolts is quite brisk. Discounts are as follows: Norway carriage bolts, 65 per cent.; common, 60 per cent.; machine bolts, 60 per cent.; coach screws, 70 per cent.; sleigh shoe bolts, 72½ per cent.; blank bolts, 70 per cent.; bolt ends, 62½ per cent.; plough bolts, 60 per cent.; tire bolts, 67½ per cent.; stove bolts, 67½ per cent. To any retailer an extra discount of 5 per cent. is allowed. Nuts, square, 4c. per lb. off list; hexagon nuts, 4½c. per lb. off list. To all retailers an extra discount of ½c. per lb. is allowed.

BUILDING PAPER—A good trade is being done at former quotations. We quote: Tarred felt, $1.70 per 100 lb.; 2-ply, ready roofing, 80c. per roll; 3-ply, $1.05 per roll; carpet felt, $2.25 per 100 lb.; dry sheathing, 30c. per roll; tar sheathing, 40c. per roll; dry fibre, 50c. per roll; tarred fibre, 60c. per roll; O.K. and I.X.L., 65c. per roll; heavy straw sheathing, $28 per ton; slaters' felt, 50c. per roll.

RIVETS AND BURRS—There is nothing new to note. Discounts on best iron rivets, section, carriage, and wagon box, black rivets, tinned do., coopers' rivets and tinned swedes rivets, 60 and 10 per cent.; swedes iron burrs are quoted at 55 per cent. off; copper rivets, 35 and 5 per cent. off; and coppered iron rivets and burrs, in 5-lb. carton boxes, are quoted at 60 and 10 per cent. off list.

BINDER TWINE — There is a little more inquiry for binder twine this week, but trade is not active. We quote: Blue Ribbon, 11½c.; Red Cap, 9¼c.; Tiger, 8¼c.; Golden Crown, 8c.; Sisal, 8¼c.

CORDAGE—Quite large quantities continue to be shipped. Prices are steady. Manila is worth 13½c. per lb. for 7-16 and larger; sisal is selling at 10c., and lathyarn 10c.

HARVEST TOOLS — All lines of harvest tools are selling well. The discount is unchanged at 50, 10 and 5 per cent.

SPADES AND SHOVELS — The inquiry has been well maintained during the week. The discount is 40 and 5 per cent. off the list.

LAWN MOWERS — A moderate demand has been experienced this week. We quote as follows: High wheel, 50 and 5 per cent. f.o.b. Montreal; low wheel, in all sizes, $2.75 each net; high wheel, 11-inch, 30 per cent. off.

FIREBRICKS — The trade in firebricks has only been fair, the demand being chiefly for small lots at $17.50 to $22 for Scotch, and at $17 to $21 for English per 1,000 ex wharf.

CEMENT — The season so far has been very unsatisfactory in foreign brands, but a large trade has been done in American and Canadian makes. A fair jobbing trade is reported. We quote: German cement, $2.35 to $2.50; English, $2.25 to $2.35; Belgian, $1.70 to $1.95 per bbl. ex wharf, and American, $2.30 to $2.45, ex cars.

METALS.

The metal situation is fairly strong, particularly so far as the English market is concerned. Galvanized iron has advanced this week, on some brands as much as 10c. per ton. Orders offered at old prices have been turned down this week. The other metals are also strong. The cable advance we noted last week in black sheets has been confirmed by post, which also says that further advances are anticipated. Canada plate, tinplate and terne plates are all firm and scarce on this market.

PIG IRON—The pig iron market is in a healthy condition, and prices are well maintained. Sales are being made here at $18.50 for Canadian brands. Summerlee No. 1 is quoted on the wharf at $20.50 to $21.

BAR IRON—The feeling is steady and prices are unchanged. Merchants' bar is worth $1.75, and horseshoe $2.

HOOP IRON—Fair quantities of hoop iron have been sold this week at $2.75 for No. 19 gauge and $3 for No. 17.

BLACK SHEETS — The English market is reported still firmer this week, and dealers here show no desire to make concessions. We quote: 8 to 16 gauge, $2.50; 26 gauge, $2.55, and 28 gauge, $2.60.

GALVANIZED IRON — Some brands of English galvanized iron are quoted at an advance of 10s. per ton. Further advances are expected. We quote as follows: No. 28 Queen's Head, $4.65; Apollo, 10¾ oz., $4.50, and Comet, $4.40 to $4.45, with a 15c. reduction for case lots.

COPPER — The market is firm. The selling price now is 18c.

INGOT TIN — The market in London is somewhat firmer, but New York is still below the importation cost. Ingot tin here is worth 31 to 32c.

LEAD — The market at the present moment is steady. The general quotation is $3.75.

LEAD PIPE — The discount has been raised from 25 to 30 per cent. We quote: 7c. for ordinary and 7½c. for composition waste, with 30 per cent. off.

IRON PIPE—The market is strong and active. We quote as follows: Black pipe, ¼, $3 per 100 ft.; ⅜, $3; ½, $3.05; ¾, $3.30; 1-in., $4.70; 1¼, $6.40; 1½, $7.70; 2-in., $10.25. Galvanized, ½, $4.60; ¾,

$5.25; 1-in., $7.50; 1¼, $9.80; 1½, $11.75; 2-in., $16.

TINPLATES—Tinplates are rather scarce on the local market, and dealers have had to scurry around for goods to fill orders. Supplies are expected next week. We quote: Coke plates, $3.75 to $4; charcoal, $4 to $4.50.

CANADA PLATE—There is also a scarcity of some gauges of Canada plate, due to light arrivals from England this spring. More goods are expected next week. We quote: 52's, $2.55 to $2.60; 60's, $2.65 to $2.70; 75's, $2.70 to $2.80; full polished, $3.10, and galvanized, $4.

STEEL—Unchanged. We quote: Sleigh-shoe, $1.95; tire, $2; bar, $1.95; spring, $2.75; machinery, $2 75, and toe-calk, $2.50.

SHEET STEEL — Some good business is being done. We quote: Nos. 22 and 24, $3, and Nos. 18 and 20, $2.85.

TOOL STEEL—Black Diamond is worth 8c. and Jessop's, 13c.

TERNE PLATES—Terne plates are very scarce in Montreal, and spot stock are demanding full prices. The price for import is now said to be $6.90. From stock, goods are being sold at $7.35 to $7.50.

COIL CHAIN—Business is rather quiet. We quote as follows: No. 6, 11½c.; No. 5, 10c.; No. 4, 9½c.; No. 3, 9c.; ¼-inch, 7½c. per lb.; 5-16, $4.85; 5-16 exact, $5.30; ⅜, $4.40; 7-16, $4.20; ½, $3.95; 9-16, $3.85; ⅝, $3.55; ¾, $3.45; ⅞, $3.40; 1-in., $3.35. In carload lots an allowance of 10c. is made.

SHEET ZINC — Business is confined to small lots, which are worth $5 75 to $6.25.

ANTIMONY—Quiet, at 10c.

ZINC SPELTER—Is worth 5c.

SOLDER—We quote: Bar solder, 18½c.; wire solder, 20c.

GLASS.

Although the market is firm, prices here are unchanged. We quote as follows: First break, $2; second, $2.10 for 50 feet; first break, 100 feet, $3.80; second, $4; third, $4.50; fourth, $4.75; fifth, $5.25; sixth, $5.75, and seventh, $6.25.

PAINTS AND OILS.

Linseed oil is 3c. higher. White lead keeps very fair; the demand is brisk. Building repairs are active and this tends to make the paint and oil department in the hardware business bright and active, a large quantity of roofing and structural paints are being shipped and the turnover is exceedingly gratifying. Linseed oil continues to be scarce and its firmness is affecting other lines, such as paints and white lead. It is said to be very difficult to pick up a carload of oil anywhere, except at the top figure. Turpentine is coming forward freely and cables say there is no change in prices. Ready-mixed paints are in great demand and the manufacturers of them are running overtime. Whiting, kalsomine, glue and painters' sundries are all feeling the buoyancy in the market. Red lead, litharge, orange mineral, and dry white lead are coming in by almost every steamer and the manufacturers report a sufficiency of these materials. We quote:

WHITE LEAD—Best brands, Government standard, $6.25; No. 1, $5 87½; No. 2, $5.50; No. 3, $5.12½, and No. 4, $4.75 all f.o.b. Montreal. Terms, 3 per cent. cash or four months.

DRY WHITE LEAD — $5.25 in casks; kegs, $5.50.

RED LEAD — Casks, $5.00; in kegs, $5.25.

DRY WHITE ZINC—Pure, dry, 6¼c.; No. 1, 5¼c.; in oil, pure, 7¼c.; No. 1, 6¼c.; No. 2, 5¼c.

PUTTY—We quote: Bulk, in barrels, $1.90 per 100 lb.; bulk, in less quantity, $2 05; bladders, in barrels, $2.10; bladders, in 100 or 200 lb. kegs or boxes, $2.25; in tins, $2.55 to $2.65; in less than 100-lb. lots, $3 f.o.b. Montreal, Ottawa, Toronto, Hamilton, London and Guelph. Maritime Provinces 10c. higher, f.o.b. St. John and Halifax.

LINSEED OIL—Raw, 77c.; boiled, 80c. in 5 to 9 bbls., 1c. less, 10 to 20 bbl. lots, open, net cash, plus 2c. for 4 months. Delivered anywhere in Ontario between Montreal and Oshawa at 2c. per gal. advance and freight allowed.

TURPENTINE—Single bbls., 53c.; 2 to 4 bbls., 52c.; 5 bbls. and over, open terms, the same terms as linseed oil.

MIXED PAINTS—$1.25 to $1.45 per gal.

CASTOR OIL—8¾ to 9¼c. in wholesale lots, and ½c. additional for small lots.

SEAL OIL—47½ to 49c.

COD OIL—32½ to 35c.

NAVAL STORES — We quote: Resins, $2.75 to $4.50, as to brand; coal tar, $3.25 to $3.75; cotton waste, 4½ to 5½c. for colored, and 6 to 7½c. for white; oakum, 5½ to 6½c., and cotton oakum, 10 to 11c.

PARIS GREEN—Petroleum barrels, 16¼c. per lb.; arsenic kegs, 17c.; 50 and 100 lb. drums, 17½c.; 25-lb. drums, 18c.; 1-lb. packages, 18½c.; ½-lb. packages, 20½c.; 1-lb. tins, 19½c.; ½-lb. tins, 21½c. f.o.b. Montreal; terms 3 per cent. 30 days, or four months from date of delivery.

SCRAP METALS.

The activity we have noticed in the scrap metal market continues. Dealers are now paying the following prices in the country: Heavy copper and wire, 13 to 13½c. per lb.; light copper, 12c.; heavy brass, 12c.; heavy yellow, 8½ to 9c.; light brass, 6½ to 7c.; lead, 2½ to 2¾c. per lb.; zinc, 2¼ to 2½c.; iron, No. 1 wrought, $11 to $16 per gross ton f.o.c. Montreal; No. 5 cast, $13 to $14; stove plate, $8 to $9; light iron, No. 2, $4 a ton; malleable and steel, $4; rags, country, 70 to 80c. per 100 lb.; old rubbers, 6¼c. per lb.

HIDES.

The hide market is without change. We quote: Light hides, 6½c. for No. 1; 5½c. for No. 2, and 4½c. for No. 3. Lambskins, 10c.; sheepskins, 90c.; calfskins, 8c. for No. 1 and 6c. for No. 2.

PETROLEUM.

There is no change. We quote: "Silver Star," 14½ to 15½c.; "Imperial Acme," 16 to 17c.; "S C. Acme," 18 to 19c., and "Pratt's Astral," 18½ to 19½c.

ONTARIO MARKETS.

TORONTO, May 23, 1901.

HARDWARE.

BUSINESS is keeping up fairly well in the wholesale hardware trade, shipments being made pretty freely to all parts of the country. The most unsatisfactory feature of the trade is still the difficulty which is being experienced in getting delivery of plain, galvanized and barb wires from the manufacturers in the United States. In oiled and annealed wire there is still a good supply, and business in it continues active. In wire nails business appears to be a little more satisfactory, although not a large trade is being done. In cut nails there is scarcely any demand for anything outside of shingle nails, which continue scarce. The cooler weather has affected the sale of refrigerators and oil stoves, but it does not seem to have the same effect on ice cream freezers, which continue in good request. The bolt trade is still fairly good, especially for special kinds. Screws continue in good demand and the same can be said of rivets and burrs. Poultry netting is a little firmer and a reduction has taken place in the price of lead pipe. Sporting goods are beginning to move a little better, the demand for oiled and annealed wire has fallen off a little during the week, but in tinware business is being well maintained. A fair demand is to be noted in spades and shovels and harvest tools.

BARB WIRE—The small shipments which arrived as we were going to press last week do not appear to have reached the warehouses, having been re-shipped to customers and still the scarcity is as great as ever, and the manufacturers advise jobbers that they have enough orders on their books to keep them going for the next six weeks or two months. Prices are firm and unchanged. We quote $3.05 per 100 lb. from stock Toronto; f.o.b. Cleveland $2.82½ for less than carlots, and $2.70 for carlots.

GALVANIZED WIRE—None of the wire which came to hand last week went into stock, all being shipped out to customers, and still there are a great many orders that have not been filled, and the manufacturers advise the jobbing trade that they have enough orders to keep them going for several weeks yet. We quote as follows: Nos. 6, 7 and 8, $3.50 to $3.85 per 100 lb., according to quantity; No. 9, $2.85 to $3.15; No. 10, $3.60 to $3.95; No. 11, $3.70 to $4.10; No. 12, $3 to $3.30; No. 13, $3.10 to $3.40; No. 14, $4.10 to $4.50; No. 15, $4.60 to $5.05; No. 16, $4.85 to $5.35. Nos. 6 to 9 base f.o.b. Cleveland are quoted at $2.57½ in less than carlots and 12c. less for carlots of 15 tons.

SMOOTH STEEL WIRE—The scarcity of plain galvanized and barb wire still stimulates the demand for oiled and annealed wire, and, as the supply of this is sufficient for requirements, a large trade is being done. Both jobbers and retailers are endeavoring to get their customers to take oiled and annealed rather than take the risk of waiting several weeks for galvanized and barb wire. The net selling prices for oiled and annealed are as follows: Nos. 6 to 8, $2.90; 9, $2.80; 10, $2.87; 11, $2.90; 12, $2.95; 13, $3.15; 14, $3.37; 15, $3.50; 16, $3.65. Delivery points, Toronto, Hamilton, London and Montreal, with freights equalized on these points.

WIRE NAILS—Business appears to be a little better in wire nails and the manufacturers are fairly well employed. At the same time, the orders are not for large quantities. We still quote $2.85 base for less than carlots, and $2.77½ for carlots. Delivery points: Toronto, Hamilton, London, Gananoque and Montreal.

CUT NAILS—Shingle nails continue in good demand and the scarcity noted last week still exists. Outside this particular kind the demand for cut nails is as dull as ever. The base price $2.35 per keg for less than carlots, and $2.25 for carlots. Delivery points: Toronto, Hamilton, London, Montreal and St. John, N.B.

HORSESHOES—There is the usual steady trade being done, with prices unchanged. We quote as follows f. o. b. Toronto: Iron shoes, No. 2 and larger, light, medium and heavy, $3.60; snow shoes, $3.85; light steel shoes, $3.70; featherweight (all sizes), $4.95; iron shoes, No. 1 and smaller, light, medium and heavy (all sizes), $3.85; snow shoes, $4; light steel shoes, $3.95; featherweight (all sizes), $4.95.

HORSE NAILS—There is the usual quiet seasonable trade being done in this line. Discount on "C" brand, oval head, 50 and 7½ per cent. off new list, and on "M" and other brands, 50, 10 and 5 per cent. off the old list. Countersunk head 60 per cent.

SCREWS—Trade is good and has kept so all the month. "Screws are cheap now, and people are realizing this," remarked a dealer. We quote discounts: Flat head bright, 87½ and 10 per cent.; round head bright, 82½ and 10 per cent.; flat head brass, 80 and 10 per cent.; round head brass, 75 and 10 per cent.; round head bronze, 65 per cent., and flat head bronze at 70 per cent.

RIVETS AND BURRS — Trade in rivets and burrs, both copper and iron, keeps good, with prices unchanged. We quote: Iron rivets, 60 and 10 per cent.; iron burrs, 55 per cent.; copper rivets and burrs, 35 and 5 per cent.

BOLTS AND NUTS—Trade is fairly good in the regular standard bolts, and particularly good just now for special makes, which are rather difficult to get prompt delivery of, as the manufacturers are pretty fully employed on the standard descriptions. We quote: Carriage bolts (Norway), full square, 65 per cent.; carriage bolts full square, 65 per cent.; common carriage bolts, all sizes, 60 per cent.; machine bolts, all sizes, 60 per cent.; coach screws, 70 per cent.; sleighshoe bolts, 72½ per cent.; blank bolts, 60 per cent.; bolt ends, 62½ per cent.; nuts, square, 4c. off; nuts, hexagon, 4½c. off; tire bolts, 67½ per cent.; stove bolts, 67½; plough bolts, 60 per cent.; stove rods, 6 to 8c.

ROPE—Quite a few complaints are heard in regard to the dark color of manila rope, but the manufacturers state that they can get nothing but dark hemp, and that the darkness does not impair the strength of the article. Prices are steady and unchanged, sisal being quoted at 10:. and manila at 13½c. per lb.

CUTLERY—Trade is fairly good. Quite a number of pocket knives are going out, and also a fair quantity of barbers' scissors and barbers' shears, and knives and forks.

ENAMELLED WARE—Trade has fallen a little during the past week, and there is not a great deal now being done.

TINWARE—The demand for this continues active, and prices unchanged.

OIL STOVES — The demand has fallen off during the week, and is generally attributed to the cooler weather.

ICE CREAM FREEZERS AND REFRIGERATORS—The demand keeps up well for ice cream freezers, but in refrigerators the cooler weather seems to have materially affected trade.

GREEN WIRE CLOTH —This is going out pretty freely, but jobbers' stocks are not yet broken. The price is unchanged at $1.35 per 100 sq. ft.

SCREEN DOORS AND WINDOWS—Shipments are still going out and as stocks in the hands of the manufacturers are small, some sizes are rather scarce.

SPORTING GOODS — The American Association manufacturing metallic ammunition has issued a new list of prices. The trade here is beginning to improve, there having been received quite a few orders for guns and ammunition for later shipment.

BUILDING PAPER — There is a good demand for building paper and stocks are in fairly good condition. We quote: Building paper, 30c.; tarred lining, 40c., and tarred roofing, $1.65.

POULTRY NETTING—The price of poultry netting has been advanced by the manufacturers, but the jobbers are still quoting 55 per cent. to the retail trade. The manufacturers have been compelled to advance their prices on account of the high price they are now compelled to pay for wire. The demand for poultry netting is not by any means as heavy as it was.

POULTRY NETTING STAPLES.—The manufacturers in the United States advise the jobbing trade here of a slight advance in the price of poultry netting staples.

HARVEST TOOLS—All kinds of harvest tools are moving fairly well. Discount, 50, 10 and 5 per cent.

SPADES AND SHOVELS—A good business is being done in spades and shovels, and, in fact, in all kinds of goods of this description. Discount, 40 and 5 per cent.

EAVETROUGH — The demand for eavetrough is fair, although, as noted last week, it is not as heavy as it was a short time ago. Prices are still low.

LAWN MOWERS—The demand for lawn mowers keeps quite active.

METALS.

Trade locally is fairly good and quotations rule as before. Tin, which has been very strong has reacted a little during the last day or so, but in other kinds of metals prices are on the whole steady.

PIG IRON—The market is dull, and advices from United States say that holders are willing to make slight concessions in order to secure business. As pointed out in a previous issue, the large buyers in Canada have pretty well anticipated their wants for sometime to come.

BAR IRON—The demand for bar iron continues good and somewhat in excess of the supply, and at least some of the mills are a little firmer in their quotations, the ruling base price is, however, still $1.85 to $1.90 per 100 lb.

STEEL—Steel continues in good demand with prices firm. We quote as follows Merchantable cast steel, 9 to 15c. per lb.; drill steel, 7 to 8c. per lb.; "Black Diamond" tool steel, 10 to 11c.; Jessop's and Firth's tool steel, 13c.; toe calk steel, $2.85 to $3; tire steel, $2.30 to $2.50; sleighshoe steel, $2.10 to $2.25; reeled machinery steel, $3; hoop steel, $3.10.

PIG TIN—The demand has taken quite a spurt during the past week, and quite a nice business has been done. Until within the last day or so the outside markets, and particularly London, have been quite strong. At the time of writing, however, the market has taken a weaker turn, and has declined somewhat in London, although prices appear to be fairly steady in New York. Locally, 31c. is the ruling price, although some round lots have changed hands at 30c.

TINPLATES — The improvement noted last week has been maintained, there having been quite a fair movement during the week.

TINNED SHEETS — There is not a great deal doing, as is, of course, to be expected at this time of the year. We still quote 9¼ to 9½c. per lb.

GALVANIZED SHEETS — The demand during the past week has been quite active, and prices are firm. Advices from the manufacturers in the United States state, that as the demand in the home market is so brisk they are unable to quote for the Canadian market. At any rate on account of the price and the difficulty of getting delivery, the importers in Canada have been devoting their attention to English galvanized sheets for sometime. We quote $4.60

from stock for English, and $4.50 for American.

BLACK SHEETS—The firmness in the English market noted last week has been maintained, and the demand is fair. We quote 28 gauge at $2.30.

CANADA PLATES—Trade continues dull. We quote: All dull, $2.90; half polished, $3, and all bright, $3.50.

COPPER—The English market is weak and lower, but we have no change to note here. We quote: Ingot, 19c.; bolt or bar, 23½ to 25c.; sheet, 23 to 23½c. per lb.

SOLDER—The movement in solder keeps up fairly well. We quote: Half-and-half, guaranteed, 18½c.; ditto, commercial, 18c.; refined, 18c., and wiping, 17c.

IRON PIPE—Just a moderate trade is being done. Prices are steady and unchanged at $5.15 for 1-inch black pipe and $7.55 for 1-inch galvanized.

LEAD—The outside markets are rather easy, and locally there is not much business being done. We quote 4¼ to 4½c. per lb.

ZINC SPELTER—Prices rule steady in the outside markets. Locally, trade is quiet at 5½ to 6c.

ZINC SHEET—Although there is fairly little doing in small lots stocks are running low. We quote casks at 6½c. and part casks at 6¾c. per lb.

ANTIMONY—Trade is just moderate, and prices steady and unchanged at 10 to 11c. per lb.

LEAD PIPE—In sympathy with the outside market prices are lower. The discount now being 30 per cent., instead of 25 per cent.

PAINTS AND OILS.

The feature of the week is the advance of 3c. in linseed oil, which is due to stiffening prices in Great Britain and scarcity here. Turpentine is unchanged at primary points and consequently is steady on the local market. White lead keeps firm as do also prepared paints and varnishes. We quote:

WHITE LEAD—Ex Toronto, pure white lead, $6.37½; No. 1, $6; No. 2, $5.67½; No. 3, $5.25; No. 4, $4.87½; genuine dry white lead in casks, $5.37½.

RED LEAD—Genuine, in casks of 560 lb., $5.50; ditto, in kegs of 100 lb., $5.75; No. 1, in casks of 560 lb., $5; ditto kegs of 100 lb., $5.25.

LITHARGE—Genuine, 7 to 7½c.

ORANGE MINERAL—Genuine, 8 to 8¾c.

WHITE ZINC—Genuine, French V.M., in casks, $7 to $7.25; Lehigh, in casks, $6.

PARIS WHITE—90c. to $1 per 100 lb.

WHITING—70c. per 100 lb.; Gilders' whiting, 80c.

GUM SHELLAC—In cases, 22c.; in less than cases, 25c.

PARIS GREEN—Bbls., 16¾c.; kegs, 17c.; 50 and 100 lb. drums, 17½c.; 25-lb. drums, 18c.; 1-lb. papers, 18½c.; 1-lb. tins, 19½c.; ½-lb. papers, 20½c.; ½-lb. tins, 21½c.

PUTTY—Bladders, in bbls., $2.10; bladders, in 100 lb. kegs, $2.25; bulk in bbls., $1.90; bulk, less than bbls. and up to 100 lb., $2.05; bladders, bulk or tins, less than 100 lb., $2.90.

PLASTER PARIS—New Brunswick, $1.90 per bbl.

PUMICE STONE—Powdered, $2.50 per cwt. in bbls., and 4 to 5c. per lb. in less quantity; lump, 10c. in small lots, and 8c. in bbls.

LIQUID PAINTS—Pure, $1.20 to $1.30 per gal.

CASTOR OIL—East India, in cases, 10 to 10½c. per lb. and 10½ to 11c. for single tins.

LINSEED OIL—Raw, 1 to 4 barrels, 81c.; boiled, 84c.; 5 to 9 barrels, raw, 80c.; boiled, 83c., delivered. To Toronto, Hamilton, Guelph and London, 1c. less.

TURPENTINE—Single barrels, 54c.; 2 to 4 barrels, 53c., delivered. Toronto, Hamilton and London 1c. less. For less quantities than barrels, 5c. per gallon extra will be added, and for 5-gallon packages, 50c., and 10-gallon packages, 80c. will be charged.

GLASS.

News has been received from Belgium to the effect that the strike of glass-blowers in that country has been settled. This will relieve the shortage, and manufacturers will be able to accept orders received to date. This will prevent the threatened advance. We quote as follows: Under 26 in., $4.15 26 to 40 in., $4.45; 41 to 50 in., $4.85; 51 to 60 in., $5.15; 61 to 70 in., $5.50; double diamond, under 26 in., $6; 26 to 40 in., $6.65; 41 to 50 in., $7.50; 51 to 60 in., $8.50; 61 to 70 in., $9.50, Toronto, Hamilton and London. Terms, 4 months or 3 per cent. 30 days.

OLD MATERIAL.

There is a fair movement. Prices are steady. We quote jobbers' prices: Agricultural scrap, 55c. per cwt.; machinery cast, 60c. per cwt.; stove cast, 50c.; No. 1 wrought 50c. per 100 lb.; new light scrap copper, 12c. per lb.; bottoms, 11½c.; heavy copper, 13c.; coil wire scrap, 13c.; light brass, 7c.; heavy yellow brass, 10 to 10½c.; heavy red brass, 10½ to 11c.; scrap lead, 3c.; zinc, 2c; scrap rubber, 6½c.; good country mixed rags, 65 to 75c.; clean dry bones, 40 to 50c. per 100 lb.

HIDES, SKINS AND WOOL.

HIDES—There is little doing, and the market is weak. We quote: Cowhides, No. 1, 6½c.; No. 2, 5½c.; No. 3, 4½c. Steer hides are worth 1c. more. Cured hides are quoted at 7 to 7½c.

SKINS—The market is steady since last week's advance. We quote: No. 1 veal, 8-lb. and up, 9c. per lb.; No. 2, 8c.; dekins, from 60 to 70c.; culls, 20 to 25c. Sheepskins, 90c. to $1.

WOOL—The feeling keeps weak. Prices are unchanged. We quote: Combing fleece, 13 to 14c., and unwashed, 8 to 9c.

PETROLEUM.

The market is steady since the decline. We quote as follows: Pratt's Astral, 16 to 16½c. in bulk (barrels, $1 extra); American water white, 16½ to 17c. in barrels; Photogene, 15½ to 16c.; Sarnia water white, 15 to 15½c. in barrels; Sarnia prime white, 14 to 14½c. in barrels.

COAL.

There is a good movement, owing to the low prices for May delivery. We quote at international bridges as follows: Grate, $3.75 per gross ton; egg, stove and nut, $5 per gross ton with a rebate of 40c. off for May shipment.

MARKET NOTES.

Linseed oil has been advanced another 3c. per gallon.

The discount on lead pipe has been raised to 30 per cent.

The Canadian manufacturer of poultry netting has advanced prices.

Higher prices are being quoted for poultry netting staples by the manufacturers in the United States.

The Gurney Foundry Co., Limited, is in receipt of another shipment of "Quick-meal" wickless and gasoline stoves.

The discount on "Black Diamond" files has for some time been given as 70 and 5 per cent. This was an error, the correct discount being 60 to 60 and 10 per cent.

ATTRACTING TOURIST TRAVEL.

THE following from a St. John, N.B., paper, while designed for the people of New Brunswick, contains some hints which will apply to other parts of the country as well:

"If the Province does not get a larger share of tourist travel than usual this season it will certainly not be the fault of The Tourist Association. The members of that organization are leaving nothing undone that will tend to attract travellers to New Brunswick, and are specially interesting themselves in ways and means to favorably impress the visitors both with the natural beauties of the country, and the hospitality of its people.

"The association naturally feels that, while it has made rapid progress during the past four years, a little more interest and co-operation on the part of the people generally would greatly facilitate matters, and would bring about the desired object more rapidly than, perhaps, any other means.

"In speaking of the prospects of tourist travel this year and the preparations which are being made to divert it in this direction, Mr. W. S. Fisher, president of the association, said this week:

"'There is a strong possibility that the Pan-American Exposition will somewhat interfere with the season's travel, therefore a greater effort on our part is absolutely necessary to offset that influence. The association is making elaborate efforts along various lines, and taking prompt advantage of the knowledge gained in previous years. We have very direct evidence that work in the past has been effective, from the assurance of hotels and transportation companies, of largely increased business; and many subscribers, feeling the direct gain which has come to them in this way, have voluntarily increased their subscriptions to the association.

"'In the way of this season's work we have now ready for distribution 25,000 handsome booklets, entitled: 'St. John River—The Rhine of America.' It contains a comprehensive description of the fairest parts of the Province, written in a bright, attractive manner; and such other matters as experience has enabled us to embody in the work.

"'There is also in preparation a booklet dealing particularly with this city and the unusual attractions which are to be found in its vicinity. The book will be given to visitors upon their arrival, and will tell them where to go, what to see and how to see it. Last season's work in this line was very effective, but we may claim an improvement this year. There are, for instance, separate paragraphs with illustrations devoted to each subject, including the falls, park, surf bathing, trout fishing, harbor excursions, bicycle trips, drives, river and rail excursions, cemetery, Martello Tower, golf links, etc., together with a wealth of other information, compiled specially to help in the pleasure of tourists.

"'One of the most important of this season's moves is the decision to establish a tourist bureau, which will be opened the first of June in the new Board of Trade rooms, on Prince William street.

"'The bureau will be in charge of Mrs. Rupert G. Olive, who, in the meantime, is engaged in collecting information for the benefit of visitors. She will visit the St. John river resorts, and all places between here and Fredericton, Halifax and the Annapolis Valley will also be visited with the same object.

"'The association is preparing and distributing large pictures, extreme care being taken to place them to the best advantage. We are issuing a further supply of illustrated post cards and adding to their variety.

"'Our work heretofore has been in advertising by means of booklets and other literature, and, while we are continuing this, we are also going to try to impress upon the people of this city the necessity of making this particular spot, so pleasant and attractive for travellers that once they reach here they will want to stay right in St. John. We want to enlarge upon the attractions at our doors, for the purpose of holding visitors here.

"'We want to get the people to feel that they have a responsibility in this matter; and attractive surroundings and clean streets will accomplish what no amount of advertising can do. Surely we have hitherto failed to realize that right here in New Brunswick we have the grandest scenery in the world. Men and women who have travelled extensively abroad, visited Italy, Switzerland, the Highlands of Scotland and other lands famed for beauty and grandeur of their scenery, acknowledge that in none of them is found anything more magnificent than our river scenery. The opportunities which we, as a people, possess will yet be recognized at their full value; in the meantime it is the duty of every person in this city and Province to help along the good work of making them known and rendering the country a pleasant place for the traveller and tourist of other lands.'"

A HARDY RAZOR GRINDER.

AN acquaintance of mine attending the recent football cup final at the Crystal Palace, discovered among the Sheffield portion of the crowd a razor-grinder who must surely be the oldest member of his craft. Owing to the deleterious character of the work and its surroundings, the life of the grinder is a short without being a merry one. Many die before reaching the prime of manhood, and those who attain three score years and ten are so rare as to be objects of interest and curiosity. But Mr. Jas. Sadler, who braved the discomforts of two long midnight journeys between Sheffield and London in order to see his favorite football team perform, is only a year short of four score, and he has been engaged in grinding razors ever since he was 13 years old.

At that age he was apprenticed to the well-known firm of George Butler & Co., of Trinity Works, Sheffield, where he has ever since been employed. The virility of this veteran of the grindstone must be phenomenal. He spent 12 sleepless hours in the railway train, and, besides watching the game with close interest from beginning to end, he filled up his spare time in "doing" the sights of London, and turned up at his post punctually on the following morning. Far from being knocked up by his outing, Mr. Sadler was unable to resist the attraction of the replayed tie at Bolton on the following Saturday. What a remarkable constitution he must possess! I wonder how many millions of razors have passed through his hands during the 56 years he has spent at his occupation?—"Vulcan," in Ironmonger.

ANOTHER LINSEED OIL MILL.

The R. W. English Refining Co., of Montreal, has placed with The Steel Storage and Elevator Co., of Buffalo, an order for the erection of a linseed oil mill of 20 press capacity. The contract also provides for the erection of a steel elevator, capacity 400,000 bbls., with warehouses for cake and feed. George M. Metzger, who recently resigned the vice presidency of The American Linseed Co., has joined The R. W. English Co. in the enterprise.

HAS DISPOSED OF HIS INTEREST.

J P. Noonan, of Mount Forest, Ont., has disposed of his interest in the Union Manufacturing Co. of Buffalo, N.Y., to the Crosby Co., who will enlarge the business and continue under the same name.

HEATING AND PLUMBING

THE OUTLOOK FOR THE MASTER PLUMBERS.

THE convention of master plumbers for 1901 will be here and away with those that have been. Just as quick as all great events come and go.

Preparation then should be going on. Every master plumber should sniff with pleasure the breeze that once more tells him another revolution of the wheel has been made, another milestone towards his destiny has been reached. The opening of the twentieth century has been grasped by leading industrial critics to survey the wonders of past achievements and to picture in luminious language the marvellous possibilities of the future. Just pause a while and consider whether the plumber has been unconscious or inactive in the progression of things !

He was no idle theorist who waged war against the reeking pan closet, the unventilated drain, the closed pestiferous sink, and all the incompetence that characterized it. He was simply an honest, practical and progressive plumber.

I am not particularizing, but the individual formed with others and the results of to-day are the outcome of associated effort. It is not alone for the present masters of the trade that associations are formed and conventions held. It is, and I trust I will be above suspicion in so saying, as much for our posterity.

In general our objects, efforts and benefits towards our fellow citizens are strictly mutual and relative. We follow a trade that follows civilization. We outcast plague by installing preventatives. Yet the fact is not recognized as it should be. For instance, how many municipalities are there in the Dominion that have the same concern for bad plumbing as they have for a delinquent taxpayer ? And yet it is a cardinal principle that the health of a community is the basis of its wealth. Then why allow any unit that forms the asset to be discounted by neglecting to enforce the conditions that a clean and healthy life depend upon ?

Business may be business, but the plumber may hoist his pennant stil higher and on it inscribe : "Value for value is business only." He can truthfully say, what doth it profit any community if health is poisoned and the spectre of mortality in epidemical form introduced ? Knowing therefore, the important relation that his work bears on humanity and remembering that all his improvements are due to co-operative effort, he enrolls in the association and secures an identity that he would not otherwise obtain. Any plan he may explain to extend the benefits of the association or the abolition of any evils always receives active and sympathetic support. So he plods along not unmindful of his importance as a man nor of his duty to society. He may be killed by typhoid or wrecked by rheumatism in its performance, in the words of Kipling, though, for the sake of his ideal he is ready to pay-pay-pay.

H. A. KNOX.

Ottawa.

SUMMER ACETYLENE BUSINESS.

OF course as a plumber you know that the traditional time to sell lighting apparatus is in the fall, "as the days begin to lengthen." But has it ever occurred to you that there are chances for profitable contracts to which most men are blind ? And at a season when things are supposed to be dead ? The temporary summer resident in all his varying influences upon local commercial conditions, presents a field for legitimate cultivation which, properly worked, will yield large returns.

THE OPPORTUNITY.

The ideal conditions for placing acetylene are realized in summer residents. They are isolated, accustomed to the comforts and luxuries of life, are frequently without the usual household help, hoping to avoid annoyances, seeking abnormal convenience, avoiding all useless effort, entertaining friends. Generally they have the ready cash to pay the bills. It is your business to give them something for their money.

Who and where are they ? Near all the cities and large towns are the golf and boat clubs and the summer gardens which, if not provided with electric lights, must finally have acetylene. Many of them will use it in any event.

Farther away are the smaller lakes and the great lake, ocean and river fronts lined with resting places of those who have left other homes for these months, during which energy is stored up by coming a little closer to nature's heart. The great mansion, the landed estate, the pretentious villa and the typical resort hotel are principally in the hands of men who believe that all good things come from the city, knowing which the city plumber should undertake to convince them that he has the particular "good thing" in acetylene. Next come the moderate summer homes, neat cottages, bungalows, house boats and camps, perhaps with their wharves, piers or driveways, all presenting shining marks for acetylene.

After these are the houses in the woods and the farms, so many of which attract city people, who come year after year with pale peevish children, soon to go back to the grind of modern schools with stored up health and ideas.

Then there are the village houses, which are rented to deluded persons who imagine they are in the country, and the stores. Have you ever attempted to show the small storekeeper of the towns where summer visitors are so easily induced to spend money, how much of it he can secure if he makes his place, light, bright and attractive ? So much for the opportunity, now for

THE MAN.

Sooner or later, if the local man fails to grasp it, some outsider will go in and take it, the traveling man or the manufacturer, the metropolitan plumber, or the faker, who represents nobody and nothing, but his own irresponsibility. Woe to your acetylene business for a time if the faker strikes your town first.

Owing to the conditions which surround the average acetylene installation, the most logical medium through which generator sales are to be made would seem to be a local man, preferably the plumber. A job of piping must be done, simple it may be, but absolutely tight if it is to be of use when seeking more business.

If the plumber is not wideawake and ready in the small town to do gas, water or heating jobs, he will find part of his field taken up by the tinner, the well driller, the wind-mill man, the general machinist who repairs the threshing engines, the blacksmith, the jeweler, and clock man, or, worst of all, by the country schemer and all round inventor (?) beside whom the traveling faker sinks into nothingness.

HIS WEAK POINTS.

All of us have our faults, but keen observers seem to agree that the average plumber's crowning fault is diffidence, he is too shy. He does not go out and try to make business, but waits for it to grow, this habit develops a certain kind of indolence, not laziness exactly, but the receptive lethargy of Mr. Micawber, expecting something to turn up. The architect's notice, the builder's request to figure, or the unfortunate householder's "hurry up" call are commonly his only gateways to business. Few, if any plumbers try to make prosperous farmers see how much they need bath rooms. A plumber rarely follows up a sewer or water pipe extension, seeking work. It is thought more easy to have a health inspector or physician order the fixing of sanitary appliances. How many plumbers go to their regular trade of the best class to present the merits of the independent domestic water heater ? Did you ever discuss acetylene light ?

Change all this, overcome your inexperience or lack of assurance as a salesman, try to talk and say just what you mean, learn all you can about this branch of your business. Command confidence through your ability to tell your "prospective victim" all about acetylene gas for lighting.

Talk light first and acetylene afterward.

SELECT INTELLIGENTLY.

First look up a number of reputable builders of acetylene apparatus which is approved by underwriters and past the experimental stage. If you have not the means at hand to learn about their commercial standing, ask some friend in the establishment of a city jobber to do so. Look over the files of your trade papers when making up your list. Send for a sample copy of The Acetylene Gas Journal, Chicago. Write for printed matter, tell what you are at, and when [illegible] comes read with care and thought, use ordinary prudence, and judge whether it is accurate and credible. Ask for the terms of the maker's guaranty. If extravagant, vague or unsupported by reasonable responsibility reject it. Demand that the generator shall be guaranteed to be precisely as represented in the printed matter.

In short, apply correct business rules ; buy tried goods from honorable houses prepared to ship promptly and close the best arrangements you can to represent them. Buy no "rights" or "blue sky," pay nothing for the privilege of becoming any man's customer, and do not load up with a

PORTLAND CEMENTS

Best German, Belgian and English Brands.

Fire Bricks, Fire Clay,
Flue Linings,
Drain Pipes,
Calcined Plaster,
Granite Hard Wall Plaster,
Wheelbarrows,
Mortar Stains.

A full stock of Builders' and Contractors' Supplies. Write for Quotations.

W. McNally & Co.,
MONTREAL.

RAMSAYS PAINTS

pay the dealer because, made upon scientific principles from the best pigments to produce the best paint, they never fail the consumer and consequently are always the best friend of the dealer. You can make money out of paints if they are good paints. You can't make it out of poor stuff. Cheap stuff is poor stuff; it can't be anything else. You can make more money out of Ramsays Paints, because they are the best paints that can be made, because they are well known, because they are well advertised, and because we charge them at a price that leaves you a good margin of profit.

A. RAMSAY & SON,
PAINTMAKERS,
Est'd 1842. MONTREAL.

DAVID PHILIP

MANUFACTURERS' AGENT

362½ Main St., WINNIPEG.

Correspondence invited from manufacturers of Staple or Heavy Hardware, Iron or Steel Bolts and Nuts, etc., either by carrying stock in Winnipeg or by selling direct from factory.

GOOD REFERENCES.

Patented April 16, 1889.

BRADLEY STEEL SHELF BRACKETS

Greatly improved recently.
For sale by principal Canadian Jobbers.
Write to factory for free sample

ATLAS MFG. CO.
New Haven, Conn., U.S.A.

CANADIAN ADVERTISING is best done by THE E. DESBARATS ADVERTISING AGENCY, MONTREAL.

ALEXANDER GIBB

Manufacturers' Agent and Metal Broker,
13 St. John Street, Montreal.

Representing British and American Manufacturers. Correspondence invited from firms wishing to be represented in Canada.

LOW TANK WATER CLOSET COMBINATIONS
THE MOST PERFECT ON THE MARKET
NOISELESS IN ACTION
BEAUTIFUL DESIGNS.

Write for Catalogue.
The James Morrison Brass Mfg. Co. Limited
TORONTO, ONT.

STEVENS-MAYNARD JR. RIFLE

The Young Gentleman's Rifle.

The Young Gentleman's Rifle.

If you want the best cheap rifle ever made we have it in the Stevens-Maynard Jr. It will be a great seller this year. Better place order now.

The leading Jobbers handle Stevens products.

J. Stevens Arms & Tool Co., P.O. Box 217 Chicopee Falls, Mass., U.S.A.

HOW TO SAVE GAS

..USE..
Peebles' Automatic Gas Governors

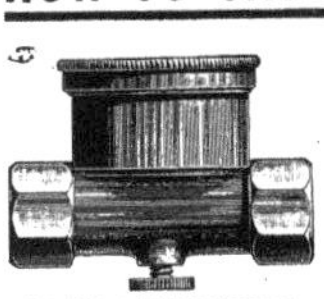

Gas, Fire and Stove Governor.

House Governor Burner.

Governor for Incandescents.

Mercurial Governor for Fixing at Meter

Sole Manufacturers, D. BRUCE PEEBLES & CO., Tay Works, Edinburgh, Scotland.

stock of generators; you have no way to tell what assortment you will need. Count on having one or two samples to show in

MARKET VIGOROUSLY.

The old receipt for rabbit stew commenced." First catch your rabbit." Make a carefully prepared list of all the possible opportunities to place acetylene light that are in your range of travel. It will be found most convenient to use uniform cards or slips of stiff paper for this purpose, a card for each or cheap envelopes, all arranged in a box. Note all the facts you can about each place.

Local owners are easily found, but it will require hard work to learn the permanent addresses of summer visitors. Many of them are likely to subscribe for the local paper, and your small ad. warrants you in asking the publisher for information of this class. The station agent probably will know where many live and a few will be favorably and definitely remembered by the pastors of the local churches. A wide-awake postmaster should be able to fill in vacant places, as well as the bankers, doctors, meat man and grocer, but the great mine of information is frequently found around the livery stable if the "stage" is regularly used.

As fast as you make your list, urge your generator builder to mail to your candidates good acetylene matter, not too much at once, but single pieces at intervals. Make them conscious of your existence, and gain their attention by writing short personal letters which state how well you are situated to do that job of piping and acetylene lighting and how well it will pay them to let you do it. Too busy? Nonsense, you have time enough if you use it right.

Arrange with the acetylene man to call on any owners who may be in cities within his reach, and finally follow up your prospective customers with personal solicitation as soon as they land in town, and do not quit a case till it is hopeless.

Present to the women the cleanliness and sweetness of premises lighted by acetylene, remind the old people how excellently it helps out impaired eyesight, tell the young ladies how beautifully it sets off gowns and complexions, and impress the men with its absolute safety.

Learn about acetylene yourself and your enthusiasm will make you eloquent.

"Be wise to-day; 'tis madness to defer." —Plumbers' Trade Journal.

REDUCTION BY FRICTION.

Editor HARDWARE AND METAL,—If the pump at a waterworks power house registers 125-lb. pressure on a 6-inch service pipe, what will it reduce by friction in two or three miles of service pipe of the same size? In answering question please define amount of water to be forced in order to obtain registered pressure and state what power is lost by friction.

Innerkip, Ont., May 9.

Ans.—This question can only be answered in part. The loss by friction in pounds per square inch, no matter what the pressure at the power house, can always be calculated from the following schedule, which shows the friction loss per mile through 6-in. pipe in proportion to the amount of water discharged from the pipes:

Discharge.	Friction Loss.
100 gal. per minute.	2.65 lb. pressure per mile.
200 " "	9.01 " "
300 " "	19.60 " "
400 " "	34.45 " "
500 " "	50.80 " "

It will be, therefore, readily understood that without knowledge of the discharge (or in other words the amount of water taken from the pipe) one cannot estimate from the pressure at the pumping station what the loss in two miles from friction would be. If the pipes were full and no water was being taken out the pressure at the far end would be as great as at the pumping house. To find the necessary pressure at the pumping station, to secure a given pressure at a certain distance, one must know the head (or the difference in level between the pumping station and the point of discharge), and the rate of discharge.

SOME BUILDING NOTES.

Fred. Fulcher is building a house at Mosa, Ont.

F. Gleb will erect a house in Acton, Ont., this summer.

Several summer residences are being erected at Orilla, Ont.

Herbert Matthews and Dr. Abbott are building residences in London, Ont.

Martyn & Hammett are building a residence of Wm. Chapman, in London, Ont.

M. Mitchell is builning for John A. Campbell two cottages in Fredericton, N.B.

Plans are being prepared by Architect Talbot, Quebec, for a new convent at St. Malo, near Quebec.

J. J. Hall and J. Spence, Portage la Prairie, Man., have gone to Delta, Man., to build a storehouse for fish.

A Presbyterian church will be erected at Minnedosa, Man. H. A. Griffith is asking for building tenders before Monday next.

Oakley & Holmes, Toronto, have the contract for erecting new works in Peterboro', Ont., for The Canadian General Electric Co.

Schultz Bros., Guelph, Ont., have been awarded the contract for building the Massey Hall and Library at the Ontario Agricultural College, near Guelph. The contract price is $40,000.

PLUMBING AND HEATING NOTES.

Dulong & Grignon have registered as plumbers, in Montreal.

Deschatelet & Bertrand have registered as plumbers, in Montreal.

Sussex, N.B., has voted to install waterworks for fire protection.

McCullough & Hawley, plumbers, Rat Portage, Ont., have dissolved.

Anthime Carriere has registered under the style of J. Carriere & Co., contractors, Montreal.

BUILDING PERMITS ISSUED.

Building permits have been issued in Toronto to R. Manning, for two brick dwellings at 734 and 736 Gerrard street east, to cost $3,500; to The Cosgrave Brewing Co., for a $4,000 addition to rear of 293 Niagara street; to Inglis & Son, for a $1,200 addition to their foundry at 14 Strachan avenue; to ——. Babster, for two dwellings at 121 and 121 1-2 Palmerston avenue, to cost $3,200; to Robt. McKee, for a $2,600 dwelling on Lansdowne avenue, near Union street; to Mrs. Frank Fleming, for a $3,500 dwelling at 58 Bernard avenue; to R. J. Copeland, for a $12,000 dwelling at 6 Walmer road; to W. A. Young, for a two-storey brick stable at 145 College street, to cost $2,000; to Wm. Donohue, for a pair of dwellings at 29 and 31 Robert street, to cost $2,000; to J. D. Henderson, for three residences on Spadina avenue, near Wellington Place, to cost $7,000.

The following building permits have been issued in Ottawa: R. B. Whyte, lot 36 Rideau street, wholesale warehouse, 40 x 80, to cost $12,000; Mrs. Law, lot 8 Sherwood street, row of four solid brick houses, $4,000; Eli Rivens, lot 11 Kenny street, frame dwelling, $1,000; Andrew Ker, lot 166 Arthur street, frame dwelling, $1,000.

THE STORE SIGN.

THE sign over a store door is an invitation to enter, remarks Stoves and Hardware Reporter. It designates the nature of the business, tells in a general way what is on sale and to all practical purposes is a standing advertisement. The sign should be no better and no worse than the goods. If it is a glittering array of gilt letters arranged in a fetching manner, the casual observer has a right to believe that the goods are after the same order. But when, on entering the store, he finds the goods of an inferior quality and sees that the management is lax and inefficient, he realizes that the sign is out of place and that he has been caught by a trick, whether intentional or not. But if the sign is a mere daub of painted letters, the same casual observer may be prevented from entering the store simply because he will take it for granted that the goods on sale are in keeping with the sign, which is supposed to represent the business as it is and not to belittle or enlarge it. The same supposition attaches to business stationery. Tidiness and neatness, symmetry of arrangement and the use of good paper make a combination of advantages in the eyes of those who read business letters and commend the writer to those with whom he wishes to be on good terms.

HINTS TO BEGINNERS IN ADVERTISING.

By Charles Austin Bates.

THE best thing for the beginner in advertising to do do is to study the successes in advertising. Don't study them with the idea of finding fault with them. Try to find out what it is about them that has made them successful. If you can't find out the particular point or points that you believe has made them successful, make it the effort of your life to copy the entire plan just as closely as your ability will let you.

A success is a fact. There is no getting around it or getting over it. It is right there in the road, and when a man bumps his head against it, it's time to quit theorizing.

I take off my hat to a success, whether it be a success in advertising or bridge-building or selling peanuts. Just yesterday I heard of a man who is actually getting rich running a press-room in New York. He is cleaning up, beyond his business expenses and living expenses, $12,000 or $13,000 a year in net, cold cash. This man can neither read nor write. He can just manage to sign his own name, but after he has signed it he can't tell which letter in the signature is an "r" and which is a "p."

You would say that it was absurd that such a man should succeed in the printing business, but he has succeeded. And strange as it may seem, the very fact that he cannot read nor write has been one of the elements of his success. His ignorance has made him afraid of business papers of any sort, so that he has never taken a note from any of his customers. The result has been that as he would do work for nobody who did not pay cash he has gathered around him a lot of spot-cash paying customers that the most highly educated printer of them all would be most happy to have.

Of course, you may say that this printer would have succeeded better if he had been properly educated. I am not going to follow my own reasoning far enough to make the assertion that because this illiterate man has succeeded all printers should cultivate illiteracy. The point I make is that it is bad business, and a waste of time, to hunt out the faults in a success. The wise man profits by the successes of others, and he profits by them by finding out the "why" of them.

It is the positive facts that are profitable to know. Negative facts are much less valuable. It is sometimes desirable to know why a man did not succeed, but it is much more valuable to know why some other man did succeed.

Just why an advertiser should be willing to spend his money in buying space in which to put the efforts of a "new beginner" is something that I never should understand.

Why a man of no experience in advertising should consider himself competent to manage and place a large appropriation is also something that I never could understand.

Business men seem to think that advertising ability is a matter of inspiration, and that each man is born with the ability to advertise, just as he is born with the ability to breathe and eat.

A tailor wouldn't employ a "new beginner" to cut clothes for him. He would laugh at a "new beginner" who proposed such a thing. He wouldn't allow him to cut into his cloth in his efforts to learn the tailoring business. The most costly cloth that ever was woven doesn't run into money nearly as fast as advertising space in the newspapers.

If space didn't cost anything, it might be well enough for people to allow the inexperienced artist and writer to try their apprentice hands in it, but it passes my comprehension why they should let an artist learn his trade at their expense, when good, artistic ability is so cheap.

The more I see of advertising the more I am convinced that it is wise to use only the strong papers in any given class, unless you have so much money to spend that you feel like going into absolutely everything that is printed.

There is no use in buying space at retail when you can buy it at wholesale. You can buy 300,000 circulation from one publication cheaper than you can buy the same quantity from 10 publications, and the circulation is likely to be better.

A publication with 300,000 circulation has a sufficient income and does a sufficiently profitable business to make the cost of reading matter, engravings and composition comparatively unimportant. It can afford to pay better prices for its reading matter than can a less prosperous paper.

If a paper with 50,000 publication furnishes the quality and quantity of reading matter and pictures as does the publication of 300,000 circulation, it must of necessity get a higher price per copy and a higher proportionate price for its advertising.

THE PREFERENCE.

"THE COUNTRY"—He's half apologisin', that thar M.P. is, for askin' $500 more for the extry time he talks. I'd willin'ly give him $1,000 if he didn't talk half as much and only sot six weeks.—From The Toronto Daily Star, May 21.

Camlachie, Ont., merchants have agreed to close every evening except Saturdays and nights preceding holidays, at 7 o'clock sharp.

MANITOBA MARKETS.

WINNIPEG, May 20, 1901.

ALL the wholesale houses are busy and lake shipments are arriving rapidly. The only change of importance to note is an advance of 10c. in nails. This is not a surprise, as it was not expected that the recent drop of that amount would hold. Prices are again at $3.45 and $3 base.

Quotations for the week are as follows:

Barbed wire, 100 lb.	$3 45
Plain twist	3 45
Staples	3 95
Oiled annealed wire 10	3 95
" 11	4 00
" 12	4 05
" 13	4 20
" 14	4 35
" 15	4 45
Wire nails, 30 to 60 dy, keg	3 40
" 16 and 20	3 45
" 10	3 50
" 8	3 60
" 6	3 65
" 4	3 80
" 3	4 05
Cut nails, 30 to 60 dy.	2 90
" 20 to 40	2 95
" 10 to 16	3 00
" 8	3 05
" 6	3 10
" 4	3 20
" 3	3 55
Horsenails, 45 per cent. discount.	
Horseshoes, iron, No. 0 to No 1	4 65
No. 2 and larger	4 40
Snow shoes, No. 0 to No. 1	4 90
No. 2 and larger	4 40
Steel, No. 0 to No. 1	4 95
No. 2 and larger	4 70
Bar iron, $2.50 basis.	
Swedish iron, $5.00 basis.	
Sleigh shoe steel	3 00
Spring steel	3 25
Machinery steel	3 75
Tool steel, Black Diamond, 100 lb	8 50
Jessop	13 00
Sheet iron, black, 10 to 20 gauge, 100 lb.	3 50
20 to 26 gauge	3 75
28 gauge	4 00
Galvanized American, 16 gauge	2 54
18 to 22 gauge	4 50
24 gauge	4 75
26 gauge	5 00
28 gauge	5 25
Genuine Russian, lb	12
Imitation " "	8
Tinned, 24 gauge, 100 lb.	7 75
26 gauge	8 00
28 gauge	8 50
Tinplate, IC charcoal, 20 x 28, box	10 75
" IX "	12 75
" IXX "	14 75
Ingot tin	33
Canada plate, 18 x 21 and 18 x 24	3 25
Sheet zinc, cask lots, 100 lb	7 50
Broken lots	8 00
Pig lead, 100 lb	6 00
Wrought pipe, black up to 2 inch	50 an 10 p.c.
" Over 2 inch	50 p.c.
Rope, sisal, 7-16 and larger	$11 00
" ⅜	11 50
" ¼ and 5-16	12 00
Manila, 7-16 and larger	14 00
" ⅜	14 50
" ¼ and 5-16	15 00
Solder	20
Cotton Rope, all sizes, lb	17
Axes, chopping	$ 7 50 to 12 00
" double bitts	12 00 to 18 00
Screws, flat head, iron, bright	87½
Round " "	82½
Flat " brass	80
Round " "	75
Coach	57½ p.c.
Bolts, carriage	55 p.c.
Machine	55 p.c.
Tire	60 p.c
Sleigh shoe	65 p.c.
Plough	40 p.c.
Rivets, iron	50 p.c.
Copper, No. 8	35
Spades and shovels	40 p.c.
Harvest tools	50, and 10 p.c.
Axe handles, turned, s. g. hickory, doz.	$2 50
No. 1	1 50
No. 2	1 25
Octagon extra	1 75
No. 1	1 25
Files common	70, and 10 p.c.
Diamond	60
Ammunition, cartridges, Dominion R.F.	50 p.c.
Dominion, C.F., pistol	30 p.c.
" military	15 p.c.
American R.F.	30 p.c.
C.F. pistol	5 p.c.
C.F. military	10 p.c. advance.
Loaded shells:	
Eley's soft, 12 gauge black	16 50
chilled, 12 guage	18 00
soft, 10 guage	21 00
chilled, 10 guage	23 00
Shot, Ordinary, per 100 lb	6 25
Chilled	6 75
Powder, F.F., keg	4 75
F.F.G.	5 00
Tinware, pressed, retinned	75 and 2½ p.c.
" " plain	70 and 15 p.c.
Graniteware, according to quality	50 p.c.

PETROLEUM.

Water white American	25½c.
Prime white American	24c.
Water white Canadian	22c.
Prime white Canadian	21c.

PAINTS, OILS AND GLASS

Turpentine, pure, in barrels	$ 63
Less than barrel lots	71
Linseed oil, raw	87
Boiled	90
Lubricating oils, Eldorado castor	25½
Eldorado engine	24½
Atlantic red	27½
Renown engine	41
Black oil	23½ to 25
Cylinder oil (according to grade)	55 to 74
Harness oil	61
Neatsfoot oil	$ 1 00
Steam refined oil	85
Sperm oil	1 50
Castor oil per lb.	11½
Glass, single glass, first break, 16 to 25 united inches	2 25
26 to 40 per 50 ft.	2 50
41 to 50 " 100 ft.	5 50
51 to 60 " " "	6 00
61 to 70 per 100-ft. boxes	6 50
Putty, in bladders, barrel lots per lb.	2½
kegs "	2¾
White lead, pure per cwt.	7 00
No 1 "	6 75
Prepared paints, pure liquid colors, according to shade and color per gal.	$1.30 to $1.90

TRADING STAMPS IN QUEBEC.

The city of Quebec, some time ago, placed a tax of $500 on trading-stamp concerns. The Quebec Trading Stamp Co. appealed against the tax. Last week the recorder gave judgment in favor of the city against the company and maintained the special municipal tax as legal. The company, moreover, has been condemned to the payment of the costs of the judicial proceedings which are very heavy. The company, however, will take the case to the Superior Court, and will continue to do business in the meantime. It will continue in business, it declares, even if it has to pay the tax.

RAT PORTAGE CLERKS MEET.

At the annual meeting of the Rat Portage Clerks' Association the following officers were elected:

Honorary President—A. M. Rose.
President—J. N. Murphy.
1st Vice-president—N. J. Cummer.
2nd Vice-president—G. Bolton.
Treasurer—D. D. Stewart.
Guard—F. L. Taylor.
Sentinel—O. A. Mabey.

The treasurer's report showed the association to be in sound financial standing.

MACHINERY AND THE MAN*

THE substitution of automatic or semi-automatic machinery for hand labor in industrial establishments has progressed so rapidly and has attained such large proportions, more especially in this country, during the past few years, that the subject is attracting much attention and a wide diversity of opinion is expressed by students of industrial economics, employers and others, as to the probable influence of this far-reaching evolution upon the future intellectual development and material welfare of the wage-earner.

An address was recently given by a well-known teacher of economics upon the present aspect of labor in this country, and it was an able exposition of the views of one who has apparently studied the subject mainly from a theoretical and scholastic point of view. According to this authority the extensive substitution of automatic machinery for hand labor, now evident in all trades, is of necessity more or less detrimental to the intellectual development of the wage-earner, since the work which he is called upon to perform is reduced to the simplest routine operations involved in feeding a machine with raw material: that the monotony of his task is very depressing, and that the modern system of minute subdivision of labor develops a hopeless feeling in the mind of the operative, because he knows that there is little or no opportunity for him to become a skilled master of any trade through his daily work; that in the old days of the apprenticeship system when boys were indentured to masters and taught the principles and practice of a trade, there was more incentive to ambition, and consequently, a quicker intellectual growth of the young mind, and a keener desire on the part of the youth to become a thorough workman. In a word, we were told that the modern system is injurious to the progress of the wage-earner. This is, perhaps, a natural view for one to take who looks at the subject from a theoretical standpoint only, but daily observation in large industrial works, covering a period of years during which a revolution has occurred in methods of conducting manufacturing industries, has given me a different opinion, based upon a different point of view; moreover, long before the invention of modern automatic machinery, and even before the birth of the factory system, similar views to those which have been given were expressed by the best known writers on economics. In 1776, Adam Smith, in his great work, "The Wealth of Nations," said:

"They (the working people) have little time to spare for education. . . . As soon as they are able to work they must apply to some trade by which they can earn their subsistence. That trade, too, is generally so simple and uniform as to give little exercise to the understanding, while at the same time their labor is both so constant and so severe that it leaves them little leisure and less inclination to apply to or even to think of anything else."

The elimination of exhausting manual labor by the substitution of powerful machinery for puny arms has emancipated labor in our day from its hardest tasks, and has given to the worker both inclination and leisure for the development of his intellect in various ways that were impossible under former conditions.

It is not merely the ability to turn out a maximum amount of work from a modern machine that constitutes a skilled operative. No matter how nearly automatic the machine may be, it is still subject to human guidance, and no matter how nearly perfect its construction, its work is still subject to final correction and control by the hand of the operator. I am satisfied that in all trades where automatic machinery has been extensively introduced for the purpose, it may be, of supplanting hand labor, the ultimate result has proved beneficial to the workers in raising the general average of intelligence, and, furthermore, that it has largely increased the opportunities for labor.

This statement may appear at first sight somewhat paradoxical, but a little examination will, I think, convince you that it follows as a logical sequence. The cheapening of manufactured articles through the aid of machinery enlarges the demand and increases the production to such an extent that those things which were heretofore regarded as luxuries of the rich, soon become ordinary conveniences in the economics of life. This increased production necessitates the employment of a larger number of operatives than were formerly required to make the same articles by hand.

Several years ago a labor-saving machine (an electric travelling crane) was introduced into a certain department of a large manufacturing establishment, and immediately displaced no less than 60 helpers. Since then many other machines of like character have been installed, yet the number of workers in this establishment is more than 50 per cent. greater to-day than before, the total number of wage-earners in these works having risen from a little under 5,000 at the time alluded to, to over 8,000 men at the present time, and the works have grown to be the largest of their kind in the world.

The introduction of labor-saving machinery has proved beneficial to the workers in many other directions. It has shortened the hours of labor; it has improved the sanitary conditions in workshops; it has increased wages; it has increased the purchasing value of wages, and has elevated the social plane of the worker of the present day above that of his predecessors.

Finally, I may say that I believe the opportunities for advancement of the wage-earner in this country are to-day far greater than at any previous time, and that this fortunate condition of affairs is due largely to the educational influence of machinery upon the wage-earner, and to his emancipation from grinding toil by the aid of modern labor-saving machines.

The majority of men holding responsible positions in large industrial establishments to-day have risen from the ranks of the operatives. As a striking illustration I may allude to a remarkable instance, that of a comparatively young man who now stands at the head of the most stupendous industrial corporation the world has ever known, who, 20 years ago, began his work at the bottom of the ladder, and has risen to a position which is entirely unique, being now the central figure in the iron and steel industry of the country, and the president of a corporation with a capital exceeding $1,000,000,000! This is, of course, an extraordinary instance, and is not to be taken as representing an average case, but other illustrations might be given, all tending to show that the substitution of modern labor-saving machinery for hand labor has proved to be one of the greatest of all benefits to the wage-earner. The opportunities for lucrative employment and rapid advancement to young men properly equipped entering the industrial establishments to-day, are greater than at any previous time within my recollection.

* An address to graduating students of the Schools of Drawing Machine Design, and Naval Architecture, of the Franklin Institute, April 26, 1901.

CURRENT MARKET QUOTATIONS.

May 24, 1901.

These prices are for such qualities and quantities as are usually ordered by retail dealers on the usual terms of credit, the lowest figures being for larger quantities and prompt pay. Large cash buyers can frequently make purchases at better prices. The Editor is anxious to be informed at once of any apparent errors in this list as the desire is to make it perfectly accurate.

METALS.

Tin.

Lamb and Flag and Straits—56 and 28 lb. ingots, per lb.	0 00	0 31

Tinplates.

Charcoal Plates—Bright

		Per box
M.L.S., equal to Bradley.		
I.C., usual sizes	....	$6 75
I.X., "	....	8 25
I.X.X., "	....	9 75
Famous—		
I.C.	....	6 75
I.X	....	8 25
I.X.X	....	9 50
Raven & Vulture Grades—		
I.C., usual sizes	....	5 00
I.X., "	....	6 00
I.X.X "	....	7 00
I.XXX., "	....	8 00
D.C., 12½x17	....	4 75
D.X.	....	5 50
D.X.X	....	7 50

Coke Plates—Bright

Bessemer Steel—		
I.C., usual sizes	....	4 00
I.C., special sizes, base	....	4 50
20 x 28	....	8 50

Charcoal Plates—Terne

Dean or J. G. Grade—		
I.C., 20x28, 112 sheets	....	8 25
I.X., Terne Tin	....	10 25

Charcoal Tin Boiler Plates.

		Per lb.
Cookley Grade—X X., 14x56, 50 sheet bxs; " 14x60, "; " 14x65, "	0 07	0 07½

Tinned Sheets

72x30 up to 24 gauge	0 08	0 08½
" 26 "	0 08½	0 09
" 28 "	0 09	0 ½

Iron and Steel.

Common Bar, per 100 lbs.	1 85	1 90
Refined " "	....	[illegible]
Horse Shoe Iron "	....	2 25
Hoop steel, 1½ to 3 in. base, extras for smaller sizes	3 10	
Sleigh Shoe Steel " base	....	
Tire Steel	2 30	
Reeled Machinery	....	[illegible]
Cast Steel, per lb	0 00	[illegible]
Toe Calk Steel	....	
T. Firth & Co's special cast steel, per lb.		0 13
Jessop's Tool Steel	....	0 13

Boiler Tubes.

1½-inch	....	0 12½
2 "	....	0 13
2½ "	....	0 15
3 "	....	0 16
3½ "	....	0 20
4 "	....	0 25

Steel Boiler Plate.

¼ inch	2 50	2 60
3-16 inch	2 60	2 70
⅜ inch and thicker	2 50	2 60

Black Sheets.

18 gauge	....	3 00
20 gauge	....	3 00
22 to 24 "	....	3 10
26 "	....	3 20
28 "	....	3 30

Canada Plates.

All dull, 52 sheets	....	2 90
Half polished	....	3 00
All bright	....	3 50

Iron Pipe.

Black pipe—		
⅛ "	....	4 45
¼ inch	....	3 25
⅜ "	....	3 30
½ "	....	3 50
¾ "	....	3 65
1 "	....	5 15
1¼ "	....	7 40
1½ "	....	8 40
2 "	....	11 25
2½ "	....	20 95
3 "	....	24 55
3½ "	....	30 75
4 "	....	39 00
5 "	....	47 35
6 "	....	62 10
Galvanized pipe—		
½ inch	....	4 90
¾ "	....	5 25
1 "	....	7 55
1¼ "	....	10 30
1½ "	....	12 35
2 "	....	16 50

Galvanized Sheets.

	G. C.	Comet.	Amer.	Queen's Head.
16 gauge		....	4 15	3 90
18 to 24 gauge	4 15	3 85	4 35	4 15
26 "	4 40	4 10	4 35	4 40
28 "	4 65	4 35	4 50	4 65

28 gauge American equals 26 gauge English.

Chain.

Proof Coil, 3-16 in., per 100 lb	....	....
" ¼ " "	8 00	8 50
" 5-16 " "	4 70	5 00
" ⅜ " "	4 05	4 50
" 7-16 " "	3 90	4 25
" ½ " "	3 70	4 10
" 9-16 " "	3 65	4 05
" ⅝ " "	3 55	3 90
" ¾ " "	3 60	4 10

Halter, kennel and post chains, 40 to 40 and 5 p.c.

Cow ties	40 p.c.
Tie-out chains	65 p.c.
Stall fixtures	35 p.c.
Trace chain	45 p.c.

Jack chain, iron, single and double, discount 35 p.c.

Jack chain, brass, single and double, discount 40 p.c.

Copper.

Ingot

English B. S., ton lots	....	0 19
Lake Superior	....	....

Bolt or Bar.

Cut lengths round, ½ to ⅞ in.	0 23½	0 25
" round and square 1 to 2 inches	0 23½	0 25

Sheet.

Plain, 14 oz., and light, 16 oz., 14x48 and 14x60	0 23	0 23
Plain, 14 oz., and light, 16 oz., irregular sizes	0 23	0 23½
Tinned copper sheets	....	0 26
Planished	....	0 32

Braziers (In sheets.)

4x6ft. 25 to 30 lbs. ea., per lb.	....	0 25½
" 35 to 45 " "	....	0 24½
" 50-lb. and above, "	....	0 23½

Boiler and T. K. Pitts

Plain Tinned, per lb	....	0 28
Spun, per lb	....	0 32

Brass.

Rod and Sheet, 14 to 30 gauge, 15 per cent.

Sheets, hard-rolled, 2x4	0 24	0 25
Tubing, base, per lb.	0 24	0 25

Zinc Spelter

Foreign, per lb	0 05½	0 06
Domestic "	....	..

Zinc Sheet.

5 cwt. casks	....	00 6½
Part casks	....	00 6¾

Lead.

Imported Pig, per lb	0 04½	0 04½
Bar, 1 lb.	0 05½	0 06¾
Sheets, 2½ lbs. sq. ft., by	....	0 06½
Sheets, 3 to 6 lbs., "	....	0 06

NOTE.—Cut sheets ½ cent per lb. extra. Pipe, by the roll, usual weights per yard, lists at 7c. per lb. and 30 p.c. dis. f.o.b. Toronto.

NOTE.—Cut lengths, net price, waste pipe 8-ft. lengths lists at 7½ cents.

Shot.

Common, $6.50 per 100 lb.; chilled, $7.00 per 100 lb.; buck, seal and ball, $7.50. Discount, 15 p.c. Prices are f.o.b. Toronto, Hamilton, Montreal, St. John and Halifax. Terms 3 per cent. cash, freights equalized.

Soil Pipe and Fittings.

Discount, 60 and 10 per cent. on medium and extra heavy, and 60 per cent. on light.

Solder.

	Per lb.	Per lb.
Bar half-and-half, guarant'd	....	0 18½
Bar half-and-half, commer'l	....	0 18
Refined	....	0 17½
Wiping	....	0 17

NOTE.—Prices of this graded according to quantity. The prices of other qualities of solder in the market indicated by private brands vary according to composition.

Antimony.

Cookson's, per lb	0 10½	0 11

White Lead.

	Per 100 lb.
Pure	6 37
No. 1 do	6 00
No. 2 do	5 62½
No. 3 do	5 55
No. 4 do	4 87½
Munro's Select Flake White	7 37½
Elephant and Decorators' Pure	7 12½
Brandram's B B. Genuine	8 00
" " Decorative	7 55
" " No. 1	6 85
" " No. 2	6 00

Red Lead.

Genuine, 560 lb. casks, per cwt.	....	$5 50
Genuine, 100 lb. kegs, per cwt.	....	5 75
No. 1, 560 lb. casks, per cwt.	....	5 25
No. 1, 100 lb. kegs, per cwt	....	5 50

White Zinc Paint.

Elephant Snow White	0 08	0 09
Pure White Zinc	0 08	0 09
No. 1	0 06	0 07½
No. 2	0 05	0 06½

Dry White Lead.

Pure, casks	5 75
Pure, kegs	6 25
No. 1, casks	5 50
No. 1, kegs	6 00

Prepared Paints.

In ¼, ½ and 1 gallon tins.

Pure, per gallon	....	1 25
Second qualities, per gallon	....	1 10
Barn (in bbls.)	0 75	0 85
The Sherwin-Williams Paints.	....	1 45
Canada Paint Co's Pure	....	1 25
Toronto Lead & Color Co's Pure	....	1 25
Sanderson Pearcy's Pure	....	1 20
Stewart & Wood's Champion Pure.		1 20

Colors in Oil.

25 lb. tins, Standard Quality.

Venetian Red, per lb.	0 05
Chrome Yellow	0 11
Golden Ochre	0 06
French "	0 05
Marine Black	0 09
" Green	0 09
Chrome "	0 08
French Imperia Green	0 09

Colors, Dry.

Yellow Ochre (J.C.) bbls.	1 35	1 40
Yellow Ochre (J.F.L.S.), bbls	....	2 75
Yellow Ochre (Royal)	1 10	1 15
Brussels Ochre	....	2 00
Venetian Red (best), per cwt.	1 80	1 90
English Oxides, per cwt.	3 00	3 25
American Oxides, per cwt.	1 75	2 00
Canadian Oxides, per cwt.	1 75	2 00
Super Magnetic Oxides, 93p.c.	2 00	2 25
Burnt Sienna, pure, per lb.	....	0 10
" Umber, " "	....	0 10
do Raw	....	0 09
Drop Black, pure	....	0 09
Chrome Yellows, pure	....	0 18
Chrome Greens, pure, per lb.	....	0 12
Golden Ochre	....	0 03½
Ultramarine Blue in 28-lb. boxes, per lb.	0 08	0 24
Fire Proof Mineral, per 100 lb	....	1 00
Genuine Eng. Litharge, per lb	....	0 07
Mortar Color, per 100 lb.	....	1 25
English Vermillion	....	0 80
Pure Indian Red, No. 45 lb.	....	0 80
Whiting, per 100 lb.	....	0 55

Blue Stone.

Casks, for spraying, per lb.	0 07
100-lb. lots, do. per lb.	0 08

Putty.

Bulk in bbls.	1 90
Bulk in less quantity	2 05
Bladders in bbls.	2 10
Bladders in kegs, boxes or loose	2 25
Bladders in 25-lb. tins	2 35
Bladders in 12½-lb. tins	2 65
Bladders in bulk or tins less than 100lb	2 90

Varnishes.

	In 5-gal. lots.).	Per gal.
Carriage, No. 1	2 90	3 30
" body	8 00	9 00
" rubbing	4 00	5 00
Gold Size, Japan	3 00	3 40
Brown Japan	2 40	2 80
Elastic Oak	2 90	3 30
Furniture, extra	2 40	2 80
" No. 1	1 60	2 00
Hard Oil Finish	2 70	3 10
Light Oil Finish	3 20	3 60
Demar	3 30	3 70
Shellac, white	4 40	4 80
" orange	4 00	4 40
Furniture Brown Japan	1 60	2 00
Black Japan	2 40	2 80
" " No. 1	1 60	2 00

The Imperial Varnish & Color Co's., Limited Elastilite Varnish 1 gal. can, each. $3.00.

Granatine Floor Finish per gal., $2.75.

Maple Leaf Coach Enamels; Size 1, $1 20; Size 2, 70c.; Size 3, 40c. each.

Castor Oil.

East India, in cases, per lb.	0 10	0 10½
" " small lots	0 10½	0 11

Cod Oil, Etc.

Cod Oil per gal.	0 50	0 55
Pure Olive	....	1 20
" Neatsfoot	....	90

Glue.

Common	0 08½	0 09
French Medal	0 14	0 14½
Cabinet sheet	0 12	0 13
White, extra	0 18	0 20
Gelatine	0 22	0 30
Strip	0 18	0 20
Coopers	0 19	0 20
Huttner	....	0 18

HARDWARE.

Ammunition.

Cartridges.

B. B Cap Dom. 50 and 5 per cent.
Rim Fire Pistol, dis. 40 p. c., Amer.
Rim Fire Cartridges, Dom., 50 and 5 p. c.
Central Fire Pistol and Rifle, 10 p.c. Amer.
Central Fire Cartridges, pistol sizes, Dom. 30 per cent.
Central Fire Cartr'dges, Sporting and Military, Dom., 15 and 5 per cent.
Central Fire, Military and Sporting, Amer. add 5 p.c. to list. B.B. Caps, discount 40 per cent, Amer.
Loaded and empty Shells, "Trap" and "Dominion" grades, 25 per cent. Rival and Nitro, net list.
Brass Shot Shells, 55 per cent.
Primers, Dom., 30 per cent.

Wads — per lb.

Best thick white felt wadding, in ¼-lb bags 1 00
Best thick brown or grey felt wads, in ½-lb. bags 70
Best thick white card wads, in boxes of 500 each, 12 and smaller gauges 0 99
Best thick white card wads, in boxes of 500 each, 10 gauge 0 35
Best thick white card wads, in boxes of 500 each, 8 gauge 0 55
Thin card wads, in boxes of 1,000 each, 12 and smaller gauges 0 20
Thin card wads, in boxes of 1,000 each, 10 gauge 0 25
Thin card wads in boxes of 1,000 each, 8 gauge ....
Chemically prepared black edge grey cloth wads, in boxes of 250 each— Per M.
11 and smaller gauge 0 60
9 and 10 gauges 0 70
7 and 8 gauges 0 90
5 and 6 gauges 1 10
Superior chemically prepared pink edge, best white cloth wads, in boxes of 250 each—
11 and smaller gauge 1 15
9 and 10 gauges 1 40
7 and 8 gauges 1 65
5 and 6 gauges 1 90

Adzes.

Discount, 20 per cent.

Anvils.

Per lb. 0 10 0 12½
Anvil and Vise combined 4 50
Wilkinson & Co.'s Anvils..lb. 0 09 0 09½
Wilkinson & Co.'s Vices..lb. 0 09¾ 0 10

Augers.

Gilmour's, discount 65 and 5 p.c. off list.

Axes.

Chopping Axes
Single bit, per doz 5 50 10 00
Double bit, " 11 00 18 00
Bench Axes, 40 p.c.
Broad Axes, 33⅓ per cent.
Hunters' Axes 5 50 6 00
Boy's Axes 5 75 6 75
Splitting Axes 6 50 12 00
Handled Axes 7 00 10 00

Axle Grease.

Ordinary, per gross 5 75 6 00
Best quality 13 00 15 00

Bath Tubs.

Zinc 6 00
Copper, discount 15 p.c. off revised list

Baths.

Standard Enameled.

5½-inch rolled rim, 1st quality 25 00
" " " 2nd " 21 00

Anti-Friction Metal.

"Tandem" A .... per lb. 0 27
" B .... " 0 21
" C .... " 0 11½
Magnolia Anti-Friction Metal, per lb. 0 25
Frictionless metal, per lb 0 23

SYRACUSE SMELTING WORKS.

Aluminum, genuine 0 45
Dynamo 0 29
Special 0 25
Aluminum, 99 p.c. pure "Syracuse".. 0 50

Bells.

Hand.

Brass, 60 per cent.
Nickel, 55 per cent.

Cow.

American make, discount 66⅔ per cent.
Canadian, discount 45 and 50 per cent.

Door.

Gongs, Sargant's 5 50 8 00
" Peterboro', discount 45 per cent.

Farm.

American, each 1 25 3 00

House.

American, per lb 0 35 0 40

Bellows.

Hand, per doz 3 35 4 75
Moulders', per doz 7 50 10 00
Blacksmiths', discount 40 per cent.

Belting.

Extra, 60 per cent.
Standard, 60 and 10 per cent.

Bits.

Auger.

Gilmour's, discount 60 and 5 per cent.
Rockford, 50 and 10 per cent.
Jennings' Gen., net list.

Car.

Gilmour's, 47½ to 50 per cent.

Expansive.

Clark's, 40 per cent

Gimlet.

Clark's, per doz 0 65 0 90
Damond, Shell, per doz 1 00 1 50
Nail and Spike, per gross 2 25 5 20

Blind and Bed Staples.

All sizes, per lb 0 07½ 0 12

Bolts and Nuts. Per cent.

Carriage Bolts, full square, Norway 65
" " full square 65
Common Carriage Bolts, all sizes 60
Machine Bolts, all sizes 60
Coach Screws 70
Sleigh Shoe Bolts 72½
Blank Bolts 60
Bolt Ends 62½
Plough Bolts 60
Nuts, square .... 4 c. off
Nuts, hexagon .... 4¼c. off
Tire Bolts 67½
Stove Bolts 67½
Stove rods, per lb .... 5½ to 6c.

Boot Calks.

Small and medium, ball, per M 4 25
Small heel, per M 4 50

Bright Wire Goods.

Discount .... 62½ per cent.

Broilers.

Light, dis., 65 to 67½ per cent.
Reversible, dis., 65 to 67½ per cent.
Vegetable, per doz., dis. 37½ per cent.
Henis, No. 8, " 6
Henis, No. 9, " 7 00
Queen City " 7 50 0 00

Butchers' Cleavers.

German, per doz 6 00 11 00
American, per doz 12 00 20 00

Building Paper, Etc.

Plain building, per roll 0
Tarred lining, per roll 0
Tarred roofing, per 100 lb 1 30
Coal Tar, per barrel 3 50
Pitch, per 100-lb 0 85
Carpet felt, per ton 45 00

Bull Rings.

Copper, $2.00 for 2½ in. and $1.90 for 2 in.

Butts.

Wrought Brass, net revised list

Cast Iron.

Loose Pin, dis., 60 per cent.

Wrought Steel.

Fast Joint, dis. 65, 10 and 2½ per cent.
Loose Pin, dis. 65, 10 and 2½ per cent.
Berlin Bronzed, dis. 70, 70 and 5 per cent.
Gen. Bronzed, per pair 0 40 0 65

Carpet Stretchers.

American, per doz 1 00 1 50
Bullard's, per doz .... 6 50 ....

Castors.

Bed, new list, dis. 55 to 57½ per cent.
Plate, dis. 52½ to 57½ per cent.

Cattle Leaders.

Nos. 31 and 32, per gross 50 9 50

Cement.

Canadian Portland 2 50 2 80
English " 3 00
Belgian " 2 50 2 75
Canadian hydraulic 1 25 1 50

Chalk.

Carpenters Colored, per gross 0 45 0 75
White lump, per cwt 0 60 0 65
Red 0 05 0 06
Crayon, per gross 0 14 0 18

Chisels.

Socket, Framing and Firmer.

Broad's, dis. 70 per cent.
Warnock's, dis. 70 per cent.
P. S. & W. Extra 60, 10 and 5 p.c.

Churns.

Revolving Churns, metal frames—No. 0, $8—No. 1, $8.50—No. 2, $9.00—No. 3, $10.00—No. 4, $12.00—No. 5, $16.00 each. Ditto, wood frames—20c. each less than above. Discounts: Delivered from factories, 58 p.c.; from stock in Montreal, 56 p.c. Terms, 4 months or 3 p.c. cash in 30 days.

Clips.

Axle dis. 65 per cent.

Closets

Plain Ontario Syphon Jet $16 00
Emb. Ontario Syphon Jet 17 00
Fittings net 1 00
Plain Teutonic Syphon Washout 10 00
Emb. Teutonic Syphon Washout 11 00
Fittings net 1 25
Low Down Teutonic, plain 16 00
" " " embossed 17 00
Plain Richelieu net 3 75
Emb. Richelieu net 4 00
Fittings net 1 25
Low Down Ont. Sy. Jet, plain net.. 19 50
" " " " " emb'd. net 20 50
Closet connection net 1 25
Basins, round, 14 in 1 00
" oval, 17 x 14 in 2 50
" " 19 x 15 in 3 75
Discount 40 p.c., except on net figures.

Compasses, Dividers, Etc.

American, dis. 62½ to 65 per cent.

Cradles, Grain.

Canadian, dis. 25 to 33⅓ per cent.

Crosscut Saw Handles.

S. & D., No. 3, per pair 17½
" " 5, " 22½
" " 6, " 15
Boynton pattern " 20

Door Springs.

Torrey's Rod, per doz ...... (15 p.c.) 2 00
Coil, per doz 0 88 1 60
English, per doz. 2 00 4 00

Draw Knives.

Coach and Wagon, dis. 50 and 10 per cent.
Carpenters, dis. 70 per cent.

Drills.

Hand and Breast.

Millar's Falls, per doz. net list.

DRILL BITS.

Morse, dis., 37½ to 40 per cent.
Standard dis. 50 and 5 to 55 per cent

Faucets.

Common, cork-lined, dis. 35 per cent

ELBOWS. (Stovepipe.)

No. 1, per doz 1 40
No. 2, per doz 1 20
Bright, 20c. per doz. extra.

ESCUTCHEONS.

Discount, 45 per cent.

ESCUTCHEON PINS.

Iron, discount 40 per cent.

FACTORY MILK CANS.

Discount off revised list, 40 per cent

FILES.

Black Diamond, 60 to 60 and 10 per cent.
Kearney & Foote, 60 and 10 to 70 per cent.
Nicholson File Co., 60 to 60 and 10 per cent
Jowitt's, English list, 25 to 27½ per cent.

FORKS.

Hay, manure, etc., dis., 50 and 10 per cent. revised list.

GLASS—Window—Box Price.

Size United Inches.	Star Per 50 ft.	Star Per 100 ft.	D. Diamond Per 50 ft.	D. Diamond Per 100 ft.
Under 26	2 15	4 15	....	6 00
26 to 40	2 30	4 45	....	6 65
41 to 50	....	4 85	....	7 50
51 to 60	....	5 15	....	8 50
61 to 70	....	5 50	....	9 50
71 to 80	....	6 00	....	10 50
81 to 85	....	6 50	....	11 70
86 to 90	....	....	..	14 00
91 to 95	..	....	...	15 50
96 to 100	.. .	....	....	18 00

GAUGES

Marking, Mortise, Etc.

Stanley's dis. 50 to 55 per cent.

Wire Gauges.

Winn's, Nos. 26 to 33, each ... 1 65 2 40

HALTERS.

Rope, ⅜ per gross ....
" ½ " 9 00
" ⅝ to ¾ 14 00
Leather, 1 in., per doz 3 87½ 4 00
" 1¼ in., " 5 15 5 20
Web,—per doz 1 87 2 45

HAMMERS.

Nail

Maydole's, dis. 5 to 10 per cent. Can. dis. 25 to 27½ per cent.

Tack.

Magnetic, per doz 1 10 1 20

Sledge.

Canadian, per lb 0 07½ 0 08½

Ball Pean.

English and Can., per lb 0 22 0 25

HANDLES.

Axe, per doz. net 1 50 2 00
Store door, per doz 1 00 1 50

Fork.

C. & B., dis. 40 per cent. rev. list.

Hoe.

C. & B., dis. 40 per cent. rev. list.

Saw.

American, per doz 1 00 1 25

Plane.

American, per gross 3 15 3 75

Hammer and Hatchet.

Canadian, 40 per cent.

Cross-Cut Saws.

Canadian, per pair 0 13½

HANGERS. doz. pairs.

Steel barn door 5 85 6 00
Stearns, 4 inch 5 00
" 5 inch 6 50
Lane's covered—
No. 11, 5-ft. run 8 40
No. 11½, 10-ft. run 10 80
No. 12, 10-ft. run 12 60
No. 14, 15-ft. run 21 00
Lane's O.N.T. track, per foot. 4½

HARVEST TOOLS.

Discount, 50 and 10 per cent.

HATCHETS.

Canadian, dis. 40 to 42½ per cent.

HINGES.

Blind, Parker's, dis. 50 and 10 to 60 per cent.
Heavy T and strap, 4-in., per lb 0 06½
" " 5-in., " 0 06¼
" " 6-in., " 0 06
" " 8-in., " 0 05¾
" " 10-in., " 0 05¾
Light T and strap, dis. 65 and 5 per cent.
Screw hook and hinge—
6 to 12 in., per 100 lbs 4 50
14 in. up, per 100 lbs 3 50
Per gro. pairs.
Spring 12 00

HOES.

Garden, Mortar, etc., dis. 50 and 10 p.c.
Planter, per doz 4 00

HOLLOW WARE

Discount, ... 45 and 5 per cent

HOOKS.

Cast Iron.

Bird Cage, per doz 0 50 1 10
Clothes Line, per doz 0 27 0 63
Harness, per doz 0 72 0 88
Hat and Coat, per gross 1 00 3 00
Chandelier, per doz 0 50 1 00

Wrought Iron.

Wrought Hooks and Staples, Can., dis. 47½ per cent.

Wire.

Hat and Coat, discount 45 per cent.
Belt, per 1,000 .... 0 60
Screw, bright, dis. 55 per cent.

HORSE NAILS.

"C" brand 50 and 7½ p.c. off new list } Oval-
"M" brand 50, 10 and 5 per cent. } head
Countersunk 60 per cent.

HORSESHOES

F.O.B. Montreal.

Iron Shoes.	No. 2 and larger.	No. 1 and smaller.
Light, medium, and heavy	3 50	3 75
Snow shoes	3 75	4 00
Steel Shoes.		
Light	3 60	3 85
Featherweight (all sizes)	4 85	4 85

F.O.B. Toronto, Hamilton, London and Guelph, 10c. per keg additional.
Toe weight steel shoes 6 70

JAPANNED WARE.
Discount, 45 and 5 p.c. off list, June 1899

ICE PICKS.
Star per doz 3 00 3 25

KETTLES.
Brass spun, 7½ p.c. dis. off new list.
Copper, per lb 0 30 0 50
American, 60 and 10 to 65 and 5 p.c.

KEYS.
Lock, Can., dis., 45 p.c.
Cabinet, trunk, and padlock. Am. per gross 60

KNOBS.
Door, japanned and N.P., per doz 1 50 2 50
Bronze, Berlin, per doz 2 75 3 25
Bronze Genuine, per doz 6 00 9 00
Shutter, porcelain, F. & L. screw, per gross 1 30 4 00
White door knobs—per doz. 1 25

HAY KNIVES.
Discount, 50 and 10 per cent.

LAMP WICKS.
Discount, 60 per cent.

LANTERNS.
Cold Blast, per doz 7 00
No. 3 "Wright's" 8 50
Ordinary, with O burner 4 00
Dashboard, cold blast 9 00
No. 0 ...... 5 75
Japanning, 50c. per doz. extra.

LEMON SQUEEZERS.
Porcelain lined, ... per doz. 2 20 5
Galvanized 1 87 3
King, wood 2 75 2
King, glass 4 00 4 60
All glass 1 20 1 30

LINES.
Fish, per gross 1 05 2 50
Chalk " 1 90 7 40

LOCKS.
Canadian, dis. 45 p.c.
Russell & Erwin, per doz 3 00 3 25

Cabinet.
Eagle, dis. 30 p.c.

Padlock
English and Am., per doz 50 6 00
Scandinavian, " 1 00 2 40
Eagle, dis. 20 to 25 p.c.

MACHINE SCREWS. Iron and Brass.
Flat head discount 25 p.c.
Round Head discount 20 p.c.

MALLETS.
Tinsmiths' per doz 1 25 1 50
Carpenters', hickory, per doz. 1 25 3 75
Lignum Vitae, per doz 3 85 5 00
Caulking each 60 2 00

MATTOCKS.
Canadian, per doz 5 50 6 50

MEAT CUTTERS.
[illegible], dis. 25 to 30 p.c.
German, 15 per cent.

MILK CAN TRIMMINGS.
Discount, 25 per cent.

NAILS.

Quotations are:	Cut.	Wire.
2d and 3d	$3 35	$3 55
3d	3 00	3 52
4 and 5d	2 75	3 35
6 and 7d	2 65	3 20
8 and 9d	2 50	3 00
10 and 12d	2 45	2 95
16 and 20d	2 40	2 90
30, 40, 50 and 60d. (base)	2 35	2 85

Wire nails in carlots are $2.77½
Galvanizing 2c. per lb. net extra.
Steel Cut Nails 10c. extra.
Miscellaneous wire nails, dis. 70 and 10 p.c.
Coopers' nails, dis. 30 per cent.
Flour barrel nails, dis. 25 per cent.

NAIL PULLERS.
German and American 1 85 3 50

NAIL SETS
Square, round, and octagon, per gross 3 38 4 00
Diamond 12 00 15 00

NETTING.
Poultry, 55 per cent for McMullen's

OAKUM. Per 100 lb.
Navy 6 00
U. S. Navy 7 25

OIL.
Water White (U.S.) 0 16½
Prime White (U.S.) 0 15½
Water White (Can.) 0 15
Prime White (Can.) 0 14

OILERS.
McClary's Model galvan. oil can, with pump, 5 gal., per doz 10 00
Zinc and tin, dis. 50, 50 and 10.
Copper, per doz 1 25 3 50
Brass, " 1 50 3 50
Malleable, dis. 25 per cent.

GALVANIZED PAILS.
Dufferin pattern pails, dis. 45 p.c.
Flaring pattern, discount 45 per cent.
Galvanized washtubs, discount 45 per cent.

PIECED WARE.
Discount 40 per cent. off list, June, 1899.
10-qt. flaring sap buckets, dis. 45 p.c.
6, 10 and 14-qt. flaring pails, dis. 45 p.c.
Creamer cans, dis. 45 p.c.

PICKS.
Per doz 6 00 9 00

PICTURE NAILS.
Porcelain head, per gross 1 75 3 00
Brass head " 0 40 1 00

PICTURE WIRE.
Tin and gilt, discount 75 p.c.

PLANES.
Wood, bench, Canadian dis. 50 per cent. American dis. 50.
Wood, fancy Canadian or American 7½ to 40 per cent.

PLANE IRONS.
English, per doz 2 00 5 00

PLIERS AND NIPPERS.
Button's Genuine per doz pairs, dis. 37½ 40 p.c.
Button's Imitation, per doz 5 00 9 00
German, per doz 0 60 2 60

PLUMBERS' BRASS GOODS.
Compression work, discount, 60 per cent.
Fuller's work, discount 65 per cent.
Rough stops and stop and waste cocks, discount, 60 per cent.
Jenkins disk globe and angle valves, discount, 55 per cent.
Standard valves, discount, 60 per per cent.
Jenkins radiator valves discount 55 per cent.
" " " standard, dis., 60 p.c.
Quick opening valves discount, 60 p.c.
No. 1 compression bath cock 2 00
No. 4 " " " 2 00
No. 7, Fuller's 2 50
No 4½, " 3 00

POWDER.
VeloX Smokeless Shotgun Powder.
100 lb. or less 0 85
1,000 lb. or more 0 80
Net 30 days.

PRESSED SPIKES.
Discount 20 to 25 per cent.

PULLEYS.
Hothouse, per doz 0 55 1 00
Axle 0 22 0 33
Screw 0 27 1 00
Awning 0 35 2 50

PUMPS.
Canadian cistern 1 80 3 60
Canadian pitcher spout 1 40 2 10

PUNCHES.
Saddlers', per doz 1 00 1 85
Conductors', " 9 00 15 00
Tinners' solid, per set 0 00 0 72
" hollow, per inch 0 00 1 00

RANGE BOILERS.
Galvanized, 3 gallons 7 00
" 35 " 8 25
" 40 " 9 50
Copper, 30 " 22 00
" 35 " 26 00
" 40 " 30 00
Discount off Copper Boilers 10 per cent.

RAKES.
Cast steel and malleable, 50, 10 and 5 p.c.
Wood, 25 per cent.

RASPS AND HORSE RASPS.
New Nicholson horse rasp, discount 60 to 60 and 10 p.c.
Globe File Co.'s rasps, 60 and 10 to 70 p.c.
Heller's Horse rasps, 50 to 50 and 5 p.c.

RAZORS.

per doz.
Elliot's 4 00 18 00
Geo. Butler & Co.'s 8 00 18 00
Boker's 7 50 11 00
" King Cutter 12 50 50 00
Wade & Butcher's 3 60 10 00
Theile & Quack's 7 00 12 00

REAPING HOOKS.
Discount, 50 and 10 per cent.

REGISTERS.
Discount ...... 40 per cent.

RIVETS AND BURRS.
Iron Rivets, black and tinned, discount 60 and 10 per cent.
Iron Burrs, discount 55 per cent.
Extras on Iron Rivets in 1-lb. cartons, ½c. per lb.
Extras on Iron Rivets in ½-lb. cartons, 1c. per lb.
Copper Rivets & Burrs, 35 and 5 p.c. dis. and cartons, 1c. per lb. extra, net.
Extras on Tinned or Coppered Rivets ½-lb. cartons, 1c. per lb.

RIVET SETS
Canadian, dis. 35 to 37½ per cent.

ROPE ETC.

	Sisal.	Manila.
7-16 in. and larger, per lb	10	13½
⅜ in.	11	14½
¼ and 5-16 in.		15½
Cotton, 3-16 inch and larger		16
" 5-32 inch		21
" ⅛ inch		22½
Russia Deep Sea		15½
Jute		8
Lath Yarn		9½
New Zealand Rope		10

RULES.
Boxwood, dis. 75 and 10 p.c.
Ivory, dis. 37½ to 40 p.c.

SAD IRONS. per set.
Mrs. Potts, No. 55, polished 62½
" No. 50, nickle-plated 67½

SAND AND EMERY PAPER.
Dominion Flint Paper, 47½ per cent.
B & A. sand, 40 and 5 per cent.
Emery, 40 per cent.
Garnet (Rurton's), 5 to 10 p.c. advance on list.

SAP SPOUTS.
Bronzed iron with hooks, per doz 9 50

SAWS.
Hand Disston's, dis. 12½ p.c.
S. & D., 40 per cent.
Crosscut, Disston's, per ft 35 0 55
S. & D., dis. 35 p.c. on Nos. 2 and 3.
Hack, complete, each 0 75 2 75
" frame only 0 75

SASH WEIGHTS.
Sectional, per 100 lbs 2 75 3 00
Solid, " 2 00 2 25

SASH CORD.
Per lb 0 23 0 30

SAW SETS.
"Lincoln," per doz 6 50

SCALES.
B. S. & M. Scales, 45 p.c.
Champion, 55 per cent.
Fairbanks Standard, 35 p.c.
" Dominion, 55 p.c.
" Richelieu, 55 p.c.
Chatillon Spring Balances, 10 p.c.
Warren Champion 65 p.c.
" Standard 45 p.c.

SCREW DRIVERS.
Sargent's per doz 0 65 1 00

SCREWS
Wood, F. H., bright and steel, 87½ and 10 p.c.
Wood R. H., " dis. 82½ and 10 p.c.
" F. H., brass, dis. 80 and 10 p.c.
Wood, R. H., " dis. 75 and 10 p.c.
" F.H., bronze, dis. 75 p.c.
" R.H. " 70 p.c.
Drive Screws, 87½ and 10 per cent.
Bench, wood, per doz 3 25 4 00
" iron, " 4 25 5 75
Set, Case hardened, 60 per cent.
Square Cap, 50 and 5 per cent.
Hexagon Cap, 45 per cent.

SCYTHES.
Per doz, net 9 00

SCYTHE SNATHS.
Canadian, dis. 45 p.c.

SHEARS.
Bailey Cutlery Co., full nickeled, dis. 60 p.c.
Seymour's, dis. 50 and 10 p.c.

SHOVELS AND SPADES.
Canadian, dis. 40 and 5 per cent.

SINKS.
Steel and galvanized, discount 45 per cent.

SNAPS.
Harness, German, dis. 25 p.c.
Lock, Andrews' 4 50 11 50

SOLDERING IRONS.
1, 1½ lb., per lb 0 37
2 lb. or over, per lb 0 34

SQUARES.
Iron, No. 493, per doz 2 40 2 55
" No. 494, " 3 25 3 40
Steel, dis. 60, 10 and 5 p.c.
Try and bevel, dis. 50 to 52½ p.c.

STAMPED WARE.
Plain, dis., 75 and 12½ p.c. off revised list.
Retinned, dis., 75 p.c. off revised list.

STAPLES.
Galvanized 3 50 4 00
Plain 3 25 3 75
Coopers', discount 45 per cent.
Poultry netting staples, 40 per cent.

STOCKS AND DIES.
American dis. 25 p.c.

STONE. Per lb.
Washita 0 28 0 60
Hindostan 0 06 0 07
" slip, ... 0 09 0 09
Labrador 0 13
" Axe 0 15
Turkey 0 50
Arkansas 0 00 1 50
Water-of-Ayr 0 00 0 10
Scythe, per gross 3 50 5 00
Grind, per ton 15 00 18 00

STOVE PIPES.
5 and 6 inch Per 100 lengths 7 00
7 inch " " 7 50

ENAMELINE STOVE POLISH.
No. 4—3 dozen in case, net cash $4 80
No. 6—3 dozen in case, " 8 40

TACKS, BRADS, ETC.

Per cent.
Strawberry box tacks, bulk ...... 75 & 10
Cheese-box tacks, blued ...... 80 & 12½
Trunk tacks, black and tinned 85
Carpet tacks, blued ...... 80 & 15
" " tinned ...... 80 & 20
" " (in kegs) 40
Cut tacks, blued, in dozens only 80
" ¼ weights 60
Swedes, cut tacks, blued and tinned—
In bulk ...... 80 & 10
In dozens 75
Swedes, upholsterers', bulk ...... 85, 12½ & 12½
" brush, blued & tinned, bulk 70
" gimp, blued, tinned and japanned ...... 75 & 12½
Zinc tacks 35
Leather carpet tacks 55
Copper tacks 50
Copper nails 52½
Trunk nails, black ...... 65 and 5
Trunk nails, tinned ...... 65 and 10
Clout nails, blued ...... 65 and 5
Chair nails 35
Patent brads 40
Fine finishing 40
Picture frame points 10
Lining tacks, in papers 10

Lining tacks, in bulk		15
" " solid heads, in bulk		75
Saddle nails in papers		10
" " in bulk		15
Tufting buttons, 22 line, in dozens only		60
Tin capped trunk nails		15
Zinc glazier's points		5
Double pointed tacks, papers.....90 and		10
" " " bulk		40

TAPE LINES.

English, ass skin, per doz	2 75	5 00
English, Patent Leather	5 50	9 75
Chesterman's each	0 90	2 85
" steel, each	0 80	8 00

THERMOMETERS.

Tin case and dairy, dis. 75 to 75 and 10 p.c.

TRAPS. (Steel.)

Game, Newhouse, dis. 25 p.c.
Game, H. & N., P. S. & W., 65 p.c.
Game, steel, 72½, 75 p.c.

TROWELS.

Disston's discount 10 per cent.

German, per doz	4 75	6 00

S. & D., discount 35 per cent.

TWINES.

Bag, Russian, per lb	....	0
Wrapping, cotton, 3-ply		0
" " 4-ply	....	0
Mattress, per lb	0 33	0 27
Staging, "	0 27	0 [illegible]

VISES.

Hand, pe doz	4 00	6 00
Bench, parallel, each	2 00	4 50
Coach, each	6 00	7 00
Peter Wright's, per lb	0 12	0 13
Pipe, each	5 50	9 00
Saw, per doz	6 50	13 00

ENAMELLED WARE.

White, Princess, Turquoise, Blue and White, discount 50 per cent.
Diamond, Famous, Premier, 50 and 10 p.c.
Granite or Pearl, Imperial, Crescent, 50, 10 and 10 per cent.

WIRE.

Brass wire, 50 to 50 and 2½ per cent. off the list.

Copper wire, 45 and 10 per cent. net cash 30 days, f.o.b. factory.

Smooth Steel Wire, is quoted at the following net selling prices:

No. 6 to 8 gauge	$2 90
" 9 "	2 80
" 10 "	2 87
" 11 "	2 90
" 12 "	2 91
" 13 "	3 15
" 14 "	3 37
" 15 "	3 50
" 16 "	3 65

Other sizes of plain wire outside of Nos. 9, 10, 11, 12 and 13, and other varieties of plain wire remain at $2.80 base with extras as before. The prices for Nos 9 to 13 include the charge of 10c. for oiling. Extras net per 100 lb.: Coppered wire, 60c.—tinned wire, $2—oiling, 10c.—special hay-bailing wire, 30c.—spring wire, $1—best steel wire, 75c.—bright soft drawn, 15c.—in 50 and 100-lb. bundles net, 10c.—in 25-lb. bundles net 15c.—packed in casks or cases, 15c.—bagging or papering, 10c.

Fine Steel Wire, dis. 17½ per cent. List of extras: In 100-lb. lots: No. 17, $5—No. 18, $5.50—No. 19, $6—No. 20, $6.65—No. 21, $7—No. 22, $7.30—No. 23, $7.65—No. 24, $8—No. 25, $9—No. 26, $9.50—No. 27, $10—No. 28, $11—No. 29, $12—No. 30, $13—No. 31, $14—No. 32, $15 No. 33, $16—No. 34, $17. Extras net—tinned wire, Nos. 17-25, $2—Nos. 26-31 $4—Nos. 32-34, $6. Coppered, 5c.—oiling, 10c.—in 25-lb. bundles, 15c.—in 5 and 10-lb. bundles, 25c.—in 1-lb. hanks, 50c.—in ½-lb. hanks, 75c.—in ¼-lb. hanks, $1—packed in casks or cases, 15c.—bagging or papering, 10c.

Galvanized Wire, per 100 lb.—Nos. 6, 7, 8, $3.50 to $3 85—No. 9, $2.85 to $3.15—No. 10, $3.60 to $3.95—No. 11, $3.70 to $4.10—No 12, $3 to $3 30—No. 13, $3.10 to $3.40—No. 14, $4.10 to $4.50—No. 15, $4.60 to $5.05—No 16, $4.85 to $5.35. Base sizes, Nos. 6 to 9, $2.57½ f.o.b. Cleveland.

Clothes Line Wire, solid 7 strand, No. 17 $4.25; No. 18, $2.65; No. 19, $2.35, f.o.b Hamilton, Toronto, Montreal.

WIRE FENCING. F.O.B. Toronto

Galvanized barb	....	3 45
Galvanized, plain twist	....	3 45

Galvanized barb, f.o.b. Cleveland, $2.82½ in less than carlots, and $2.70 in carlots.

WIRE CLOTH.

Painted Screen, per 100 sq. ft., net	1 35

WASTE COTTON. per lb.

Colored	6½ to 8
White, according to quality	6½ to 7½

500-lb bale lots shaded.

WRENCHES.

Acme, 35 to 37½ per cent.
Agricultural, 60 p.c.
Coe's Genuine, dis. 20 to 25 p.c.

Towers' Engineer, each	2 00	7 00
" S., per doz	5 80	6 00
G. & K.'s Pipe, per doz	....	3 40
Burrell's Pipe, each	....	3 00
Pocket, per doz	0 25	2 90

WRINGERS.

Leader	per doz.	$30 0	33 00
Royal Canadian	"	26 00	28 00
Royal American	"	26 00	28 00
Sampson	"	....	30 00

Terms 4 months, or 3 p.c. 30 days.

WROUGHT IRON WASHERS.

Canadian make, discount, 40 and 5 per cent.

PERSONS addressing advertisers will kindly mention having seen their advertisement in Canadian Hardware and Metal Merchant.

"THE EMLYN" SAW BENCH

Made in 6 sizes. Best value obtainable. Specially designed for export. With or without "Emlyn" Patent Guard. Sole maker—

CHARLES D. PHILLIPS,

[illegible]es— [illegible]chinery," Newport. Emlyn Engineering Works, NEWPORT, MON., ENGLAND.

Mr. Retail Hardwareman,

You have had trouble with Axes. Some were poor sellers, others did not give your customers satisfaction.

You can avoid having all this trouble in future. We can tell you how. Write us.

Dundas Axe Works

DUNDAS, CANADA.

Lockerby & McComb

AGENTS IN CANADA

FOR THE

Celebrated P. & B.

Cold Storage Lining

AND

.. Ruberoid Roofing ..

P. S.—Prices on Application.

65 Shannon Street, MONTREAL.

BUSINESS NEWS

of any kind that is of value to business men supplied by our Bureau. We can give you market quotations from any town in Canada, reports from the city markets, stock quotations, etc. You can get commercial news from any Canadian paper through us.

Write us, giving us particulars of what you want and where you want it from, and we will quote you prices by return.

"Clippings from any Canadian paper on any subject."

CANADIAN PRESS CLIPPING BUREAU,

232 McGill Street, MONTREAL, QUE.

Telephone Main 1255.

10 Front St. East, Toronto. Telephone 2148.

75 YEARS. *ESTABLISHED 1825.* *75 YEARS.*

CELEBRATED HEINISCH SHEARS.

Tailors' Shears,
Trimmers, Scissors,
Tinners' Snips, etc.

ACKNOWLEDGED THE BEST.

R. HEINISCH'S SONS CO. NEWARK, N.J., U.S.A. NEW YORK OFFICE, 90 Chambers St.

Not connected with any Shear Combination.

CHAS. F. CLARK, President. JARED CHITTENDEN, Treasurer.

...ESTABLISHED 1849...

BRADSTREET'S

Capital and Surplus, $1,500,000. Offices Throughout the Civilized World.

Executive Offices: Nos. 346 and 348 Broadway, New York City, U.S.A.

THE BRADSTREET COMPANY gathers information that reflects the financial condition and the controlling circumstances of every seeker of mercantile credit. Its business may be defined as of the merchants, by the merchants, for the merchants. In procuring, verifying and promulgating information no effort is spared, and no reasonable expense considered too great, that the results may justify its claim as an authority on all matters affecting commercial affairs and mercantile credit. Its offices and connections have been steadily extended, and it furnishes information concerning mercantile persons throughout the civilized world.

Subscriptions are based on the service furnished, and are available only by reputable wholesale, jobbing and manufacturing concerns, and by responsible and worthy financial, fiduciary and business corporations. Specific terms may be obtained by addressing the Company at any of its offices. Correspondence Invited.

—OFFICES IN CANADA—

HALIFAX, N.S. OTTAWA, ONT. VANCOUVER, B.C.
HAMILTON, ONT. QUEBEC, QUE. VICTORIA, B.C.
LONDON, ONT. ST. JOHN, N.B. WINNIPEG, MAN.
MONTREAL, QUE. TORONTO, ONT.

THOS. C. IRVING, Gen. Man. Western Canada, Toronto. JOHN A. FULTON, Gen. Man. Eastern Canada, Montreal.

TRADE NOBLES & HOARE NH LONDON MARK

Nobles & Hoare.

CORNWALL ROAD STAMFORD STREET.

LONDON, ENG.

Manufacturers of

HIGH-CLASS VARNISHES ONLY

Which can be obtained direct from the works or from the principal Color Dealers in Canada

THE MOWER THAT WILL KILL ALL THE WEEDS IN YOUR LAWNS.

If you keep the weeds cut so they do not go to seed, and cut your grass without breaking the small feeders of roots, the grass will become thick and weeds will disappear. **The Clipper will do it.**

CANADIAN PATENT FOR SALE.
SEND FOR CATALOGUE AND PRICES.

Clipper Lawn Mower Co., NORRISTOWN, PA.

CANADIAN HARDWARE AND METAL MERCHANT

The Weekly Organ of the Hardware, Metal, Heating, Plumbing and Contracting Trades in Canada.

VOL. XIII. MONTREAL AND TORONTO, JUNE 1, 1901. NO. 22

ISLAND CITY PAINT & VARNISH WORKS

REGISTERED TRADE MARK

ISLAND CITY

Consulting Chemist
Milton L. Hersey, M.Sc.,
(Provincial Analyst)

Manufacturers of

Island City Pure Lead

Island City Pure White Paint

Non-poisonous and covers more than pure White Lead.

Island City Decorators' Pure White Lead

Island City Ready-Mixed Paint

Made by the old process—made with Oil and Turpentine.

Island City Floor Paints

have no equal, dry hard in 8 hours.

P. D. DODS & CO.,

188 and 190 McGill Street, - - MONTREAL.

SUPREMACY IN WARM AIR HEATING.

OUR LITTLE OX Furnaces for Wood and Oxford 400 Series for Coal.

Cover a range of capacities to suit all requirements.

The "Little Ox" Wood Furnaces are the most powerful of their class on the market—they are supplied either portable, as shown, or stationary for brick setting.

Their double, or triple, as shown, return flue radiators afford an immense fire travel—in addition to which they are smoke and dust proof, and will take 3-ft. rough wood.

As they are very easy to manage, and have a capacity of from 15,000 to 50,000 cubic feet, they sell readily in every locality.

Next week we will show our "Oxford 400 Series" furnace for coal.

Full particulars on application.

The Gurney Foundry Co.,

Limited

TORONTO. WINNIPEG. VANCOUVER.

THE GURNEY-MASSEY CO., LIMITED, MONTREAL.

FLY TIME will soon be here. YOU will require a stock of DOOR PULLS for Screen Doors, etc.

We Manufacture among our many other lines both Japanned and Copper Coated Wire Door Pulls (5 inches).

"CRESCENT"

HAT and COAT HOOKS

2½ in. 3 in. 3½ in.

In Japanned and Coppered.

—Wood Screws and Wire Nails.—

BRIGHT GOODS

(A full line of these goods always in stock.)

WIRE

Bright, Annealed, Oiled and Annealed.

Coppered, Coppered Spring, Tinned, Brass, Staples, Galvanized and Barb Wire.

COPPER WIRE

For Merchant and Electrical Purposes.

For Prices, etc., apply to

Dominion Wire Manufacturing Co.

MONTREAL and TORONTO.

THE NEW BALDWIN

DRY AIR CLEANABLE

REFRIGERATOR.

135 Modern Varieties. Ash, Oak and Soft-wood Finishes

METAL, PORCELAIN, SPRUCE LININGS.

BALDWIN

Positive Circulation—Sanitary—Odorless.

Latest Cleanable Features—The Strongest and Best System of Patent Removable Metal Air-Flues.

Air-Tight Lever Locks

Ball-Bearing Casters.

Swing Base—in and out.

Rubber around Doors and Lids, making them doubly air-tight.

Handsome Designs.

Moderate Prices.

Built in the newest, largest and best equipped refrigerator plant in the East run all the year round on refrigerators exclusively; stock goods; special refrigerators and coolers in sections.

Handsome Trade Catalogue Ready.

Baldwin Refrigerator Co.,

BURLINGTON, VERMONT.

The "IDEAL" Ice Cream Freezer

Made only in one size, but will freeze from half pint to the capacity of the freezer (about one gallon)

IN POSITION TO RECEIVE CREAM

PRACTICALLY ONLY FOUR PARTS

IN POSITION FOR FREEZING

Will Freeze Cream in from Two to Five Minutes According to Quantity.

Indurated Fibre Tub.
No Hoops to Fall Off.
No Staves to Shrink or Swell.
No Repacking Necessary to Keep Cream 5 Hours.
Saves 50 per cent. in Ice.

Simple in Construction.
Easy in Operation.
Rapid in Results.
No Wooden Tub with Leaky Seams.
No Danger of Spoiling the Cream with Salt.

Wood, Vallance & Co., Hamilton

Toronto Office: 32 Front Street West—H. T. Eager.
British Columbia Representative: W. G. Mackenzie, P.O. Box 460, Vancouver, B.C.

Geo. D. Wood & Co.,
Winnipeg, Man.

PAINT WITH WATER ONLY.

WHEN YOU BUY

INDELIBLO

you get a dry powder that mixed with cold water makes the only perfect water paint ever known. No mess, no dirt, no smell, clean, healthy, pure white, and clear tones. Washable, weatherproof. A perfect paint for Factories, Breweries, Sawmills, Barns, Fences, anything and everything where cheapness combined with durability and beauty is required.

— AGENTS —

A. RAMSAY & SON, - - - - - - MONTREAL.
J. H. ASHDOWN, - - - - - - - WINNIPEG.
McLENNAN, McFEELY & CO., - - VANCOUVER.

"DAISY" CHURN

Has tempered steel cased bicycle ball bearings, strongest, neatest and most convenient frame. Only two bolts to adjust in setting up. Steel Bow Levers, suitable for either a standing or sitting posture. Has four wheels and adjustable feet to hold stand steady while churning. When churn is locked to stand the bow can be used as handles to move it about on the front wheels as handy as a baby carriage. Open on both sides to centre, giving free space for pail. Made with wood or steel stands, with Cranks only, or Bow Levers as desired.

Vollmar Perfect Washer

Has a most enviable record. A perfection of its kind—will wash more clothes in less time, do it better and easier, with less wear and tear, than any other machine.

THE
Wortman & Ward Mfg. Co., Limited
LONDON, ONT.
Eastern Branch, 60 McGill Street, Montreal, Que.

Special list of low-priced Japanned and Regalvanized Wire Cloth.

24, 30, 36 in. wire, in 50 ft. rolls.

SAMPLES SENT WHEN DESIRED. WRITE FOR PRICES.

The B. GREENING WIRE CO., Limited

Hamilton, Ont., and Montreal, Que.

The
WATSON FOSTER CO.
LIMITED
MONTREAL
SILVER MEDAL
PARIS 1900
CHICAGO 1893
WALL PAPERS
WE WERE AWARDED A SILVER MEDAL AT PARIS LAST SEASON FOR EXCELLENCE OF DESIGN AND COLORING.
THIS SEASON, IT WILL BE OUR AIM TO WIN YOUR APPROVAL OF WHAT WE CLAIM WILL PROVE OUR MOST SUCCESSFUL EFFORT TO PRODUCE IN GREAT VARIETY ALL THAT IS GOOD AND ARTISTIC IN
WALL PAPERS
WHICH YIELD A PROFIT WHEN SOLD, AND CONTINUE AN ADVERTISEMENT WHEN HUNG.

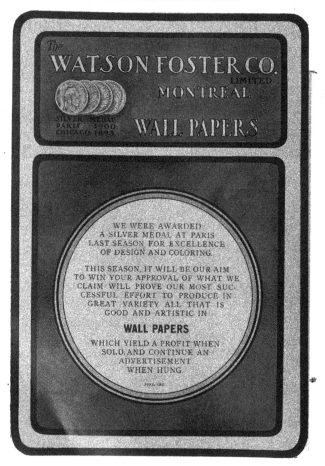

Wadsworth-Howland

Co. CHICAGO.

A MA N

Are Superior to any on the market

And Second to none. One trial will convince you

N

made to paint floors with.

BANNER BUGGY PAINT prepared expressly for painting old or new work where a handsome. lasting and glossy surface is required. Dries quickly and will not crack or chip off.

Jewel Mixed Paints and Roof and Barn Paints.

A single trial will convince you that these paints are all we claim, being far superior in body, more lasting, will retain their color and lustre longer than strictly pure lead and oil mixed by hand. Uniform in shade and consistency, and will not fade unevenly. Desirable paint to use on account of their spreading and covering properties and their great durability.

RICE LEWIS & SON

LIMITED.

Canadian Distributors TORONTO.

HUDSON PATTERN

APPLE PARER

WE GUARANTEE EVERY ONE MADE BY US TO BE EQUAL IN EVERY RESPECT TO THE GENUINE HUDSON, AND THEY WILL COST YOU LESS MONEY.

It will pay you to place your order for CANADIAN MADE.

SOLD ONLY THROUGH THE JOBBING TRADE.

Manufactured by **A. R. WOODYAT & CO., GUELPH, CANADA.**

SEASONABLE SUGGESTIONS.

WATER COOLERS.

Our Water Coolers are well made.
They have Nickel-Plated Faucets.
They are finished in attractive designs and colors.

WATERING POTS.

Six sizes. Plain, Japanned and Galvanized.
All Cans are made with Patent Rose, which can be easily taken apart for cleaning.
Well finished, strong and durable.
That is why we sell so many.

KEMP MANUFACTURING CO.,

TORONTO, Canada

HARDWARE AND METAL

VOL. XIII. MONTREAL AND TORONTO, JUNE 1, 1901. NO. 22.

President,
JOHN BAYNE MacLEAN,
Montreal.

THE MacLEAN PUBLISHING CO.
Limited.

Publishers of Trade Newspapers which circulate in the Provinces of British Columbia, North-West Territories, Manitoba, Ontario, Quebec, Nova Scotia, New Brunswick, P.E. Island and Newfoundland.

OFFICES

MONTREAL - - - - - - 232 McGill Street, Telephone 1255.
TORONTO - - - - - - 10 Front Street East, Telephone 2148,
LONDON, ENG. - - - - 109 Fleet Street, E.C., W. H. Miln.
MANCHESTER, ENG. - - - 18 St Ann Street. H. S. Ashburner,
WINNIPEG - - - - Western Canada Block. J. J. Roberts.
ST. JOHN, N. B. - - - No. 3 Market Wharf, J. Hunter White.
NEW YORK. - - - - - - - 176 E. 88th Street,

Subscription, Canada and the United States, $2.00.
Great Britain and elsewhere - - - 12s.
Published every Saturday.

Cable Address { Adscript, London. Adscript, Canada.

DEFECTS OF THE HIGH COMMISSIONER'S OFFICE.

IT has for a long time been realized that the High Commissioner in London is of little or no use to Canada in a commercial sense.

Business men in Great Britain who have sought information there about Canada, even in simple matters, have time and again told us that the staff seemed to know little or nothing about the country which they are supposed to represent. In fact, outside Lord Strathcona, who cannot be expected to look after details, there is no one in the High Commissioner's office who has any practical knowledge of Canada.

Then, there are manufacturers and other classes of business men in Canada who have visited the High Commissioner's office in quest of local information of a commercial nature; their experience has been similar to that of Englishmen who were in quest of information in regard to Canada.

In fact, to Englishmen on the other side of the Atlantic, and to Canadians on this side, the High Commissioner's office is an absurdity as far as its office as a bureau of information on commercial matters is concerned. And yet, among the list of commercial agents published monthly is given the name of one of the officials of the High Commissioner's office who "will answer inquiries relative to trade matters, and their services are available in furthering the interests of Canadian traders."

We have no complaint to make in regard to the mannerisms of the staff in the High Commissioner's office. As far as we are aware, everyone is courteous and always ready and willing to issue passes for the House of Commons and other interesting places in the British metropolis. But, while this is all right in its place it should at any rate be of secondary importance. The commercial interests of Canada should be of the first importance, which at present they obviously are not.

We have nothing to say against the High Commissioner's office as a diplomatic institution. And as for Lord Strathcona, he is not only an honor to Canada, but he is one which this country should be proud to be represented by in Great Britain. We have no desire to abolish the office of High Commissioner. What we want, and what the business men of this country want, is its reconstruction.

The cost of the office to Canada last year was $30,517. This was made up as follows: Lord Strathcona, salary $10,000; salary of staff, $9,999.98; contingencies, $10,517.11. Capitalized at 3 per cent. the annual cost of the High Commissioner's office represents the sum of $1,017,235. For what Canada receives the cost is altogether too high.

What this country requires in London is a commercial agent, one whose time is devoted to the discovery of new channels for Canada, and one who is a practical business man and a Canadian. Australia and New Zealand have had commercial agents in Great Britain for some time. And, although Canada has recognized the need of one for some time, nothing has yet been done by the Government in the direction of supplying what it has been importuned by different business men's organizations and by the press to supply.

The High Commissioner's office does not come within the purview of the Trade and Commerce Department, but commercial agencies do. Consequently, we must hold Sir Richard Cartwright chiefly responsible for the neglect of the Government to comply with the demand of the business interests of this country in this particular. But Sir Richard is noted for his contempt of the wishes of business men.

By his inertia he has brought the Trade and Commerce Department into disrepute, until, to-day, there is a strong opinion developing in favor of its abolition, a consummation it is to be hoped will never be realized.

But all these things serve to emphasize the necessity of Sir Richard giving place to someone whose sympathy is in accord with the objects of the Trade and Commerce Department and who has ambition to further them.

DISSATISFIED BRITISH COLONIES.

MORE interest has been taken in Canada in the new British Customs tariff than is perhaps apparent on the surface. There are, doubtless, a large number of people in this country who have taken no more than a passing notice, but thinking men who take an interest in public affairs are taking a great deal more than a passing notice. And the result is a deep disappointment: There is a feeling that the British Government has not acted kindly towards the colonies.

Directly, the Dominion is little affected by the tariff. Outside of canned fruits and preserves, we know of no line we export that is likely to bear the burden of taxation under the new order of things, and, while our export trade in that line is a growing one, it is not of sufficient importance to excite general concern.

It is the general belief that the tariff is more likely to increase than diminish the anomalies which the sugars of the British West Indies experience in the United Kingdom in competition with the bounty-fed sugars of Europe that probably excites our displeasure most.

Purely out of sympathy with the deplorable condition of the sugar industry in the British West Indies, the Canadian Government made a preferential reduction of one-third in the duty on British sugars. And what irritates us is the fact that what one of the children of the Empire has done the mother of the Empire has refused to do.

Although one of the political parties in Canada has placed itself on record in the House of Commons as being in favor of asking the British Government to give Canadian products preferential tariff treatment, we believe that the majority of the people in this country consider that, with the British market free to the products of our farms and factories, such a request was hardly fair.

But here is a partial departure from the strictly free trade line of demarcation. We cannot get away from that, while, at the same time, recognizing that revenue and not protection was the motive of the British Government in imposing the tax. In this departure was the Imperial Government's opportunity, not to insert the thin edge of the wedge of protection, but to enunciate the principle of free trade within the Empire, something for which the British people devoutly wish. Right Hon. Joseph Chamberlain, Secretary of State for the Colonies, has placed himself on record in favor of that system of free trade which was to be the basis of closer trade relations within the Empire.

Although there is, of course, no close comparison, we cannot help calling to mind Rehoboam's unwise reply to the people of Israel when they pleaded for an amelioration of their burdens, to the effect that, whereas his father had chastised them with whips he would chastise them with scorpions. As everyone knows, this led to the permanent dismemberment of the kingdom. It is devoutly to be hoped that not an acre of British territory will be alienated by the action of the Chancellor of the Exchequer, but the feeling is very sore in the British West Indies, and the sentiment in favor of annexation to the United States, which has been budding in some of the islands, and notably Jamaica, is not likely to propagate less freely in consequence. That a great deal of discontent has been engendered is evident from the cablegrams which have been received in Great Britain by the West Indian committee from the different colonies in the West Indies. These cablegrams were as follows:

From Antigua:

Meeting protests against opportunity lost countervailing.

From Barbadoes:

Barbadoes protests against continued advantages given bounty-fed sugar under budget; appalled at Chancellor's satisfaction at prospect; flood bounty-fed sugar in face Government's repeated condemnation principle of bounties; implores opportunity be grasped to do justice to British colonies by discriminating against foreign countries granting bounties.

From British Guiana Planters' Association:

Association deeply deplóres neglect of British colonies interested by Mother Country and lost favorable opportunity imposing countervailing duties.

From St. Kitts:

St. Kitts strongly protests failing bounty abolition against duties on colonial sugar.

From Trinidad:

Profound discouragement felt; opportunity lost equalizing colonial and bounty sugars; planters protest.

We will better understand the intense feeling that has been engendered in the West Indies when we remember that the tax will be heavier on the sugars imported from there than on those from Europe.

The tax is not upon the sugar plus the bounty. It is upon the sugar minus the bounty. It is only after the bounty is taxed off that the balance of the tax becomes a tax on sugar. For instance, the tax on refined is barely sufficient to tax off the French bounty. The effect of the duties as set forth in a statement issued by the West Indian committee is as follows: "In the first place, they are necessarily a tax on bounties, and only in the second place, and after the bounty is exhausted do they become a tax upon sugar. These countervailing duties have been established upon sugars coming from the continent, in the receipt of a bounty, but the actual duty on the sugar coming from each of these countries is graduated in inverse ratio to the bounty, i.e., the higher the bounty the less is the duty on the sugar. France will practically pay no duty at all; Germany will pay about £2 18s.; Austria the same; Belgium and Holland about £2 15s., and the British colonies, £4 3s. 4d.

Remember, the sugar product of the British colonies will pay £4 3s. 4d. Is it any wonder that they are irritated? Is it any wonder that the people of Canada sympathize with them? We may have been struck with only a straw, but they have been struck with a club—and a pretty heavy one too.

In order to retain the affection and the fealty of her colonies, the Mother Country must treat them as if they were a part of the Empire, and not as if they were outsiders.

Although we in Canada are not seriously affected by the new tariff, the disapproval of the policy of the British Government is such that its continuance will undoubtedly increase the ranks of those in this country who are opposed to the present preferential tariff on British products.

The world does not travel much faster than does legislation through the House of Commons designed to benefit the members, but a snail travels faster than does legislation designed to benefit the country.

BRITISH HARDWARE EXPORTS TO CANADA.

THE British exports of iron, steel, hardware and kindred lines to Canada cannot be said to have developed under the preferential tariff, as a good many hoped they would. This will be readily gathered from a glance at the following table, compiled from British returns, giving the chief exports to Canada during the nine months ending March 31 and for the corresponding periods in 1900 and 1899:

NINE MONTHS' EXORTS FROM GREAT BRITAIN TO CANADA.

	1899.	1900.	1901.
Pig iron	$ 80,126	$ 204,853	$ 94,608
Bar, angle, bolt and rod	40,726	16 ,001	22,990
Railroad, of all sorts	67,881	845,352	169,146
Hoops, sheet, boiler and armor plates	245,506	884 207	244 788
Galvanized sheets	211,026	241,386	207,797
Tinplates and sheets	584,442	1,134,130	1,136,558
Cast and wrought iron, etc.	110,458	308,047	8,359
Steel, unwrought	146,429	971,666	181,089
Pig lead	182,272	138 667	86,512
Tin, unwrought	60,316	124,398	131,748
Hardware, unenumerated	88,5[illegible]1	96,150	[illegible],923
Cutlery	212,743	197,573	198,954

It is not possible to give figures dealing with the exports to Canada from the United States in order that a comparison may be made, item by item, with the exports to Canada from Great Britain. The following table, however, which is for the eight-month period ending February, will give our readers some idea as to the condition of the trade of both countries with Canada:

EIGHT MONTHS' EXPORTS FROM UNITED STATES TO CANADA.

	1899.	1900.	1901.
Agricultural implements	$ 700,578	$1,083 395	$1,089,762
Carriages, cars, and parts of	184,160	354,685	489,936
Copper, ingot, bars, etc	100,680	112,445	188,740
Cycles, and parts of	241,023	129,447	116,034
Builders' hardware, saws and tools	530,351	505,474	520,802
Steel bars or rails for railways	1,358 666	1,671,447	2,566,832

It is evident from these figures that, while the United States has not made much headway in the articles enumerated outside "steel bars or rails for railways," Great Britain has decreased in her exports to Canada in spite of the favorable influence of the preferential tariff of 33⅓ per cent. Price and other things being equal the people in this country prefer to place their orders with British firms rather than with United States firms. It is because all things are not equal that they do not do so to the extent they otherwise would. Distance has got something to do with it, for, under normal conditions, delivery can often be obtained from points in the United States in as many days as it takes weeks from the United Kingdom. But this is not the most important of the inequalities: The manufacturers pay closer attention to the patterns and styles required by this country and are more ready to make a new line when requested to do so. Then, their travellers make more frequent and more regular visits to the trade centres here. Primage, charges for boxing, casing, binding, transportation charges to the seaboard, dock charges, agents' charges, etc., which the importer usually has to pay on British goods and not on goods imported from the United States, are also sources of irritation which militate against imports from the United Kingdom. Manufacturers in the United States and even those in Germany sell their products delivered at some central point in Canada. This enables the Canadian importer to tell to a cent what his goods will cost him and he can therefore, with confidence, send his travellers out to take orders before the goods are in the warehouse. Under the system by which he buys British goods he naturally has not the same confidence, for he is not positive as to his ultimate cost.

There are now in operation in Canada eight blast furnaces, namely, 3 in Nova Scotia, 3 in Ontario and 2 in Quebec. Three of them have been started up within the last six months, and there are more to follow. It is obvious, therefore, that our importation of pig iron is destined to gradually diminish. Much the same may be expected of steel, although, for some time, it may not be expected to be as marked as in pig iron. Builders' hardware and kindred lines are being supplied in increasing quantities by our own factories, and there is not much hope for the manufacturers of any country in this particular.

In table and pocket cutlery, the British manufacturers still have almost a monopoly in Canada, but both the United States and Germany are gradually coming into greater prominence in regard to these lines.

The United States has been competing on the Canadian market pretty keenly with Great Britain in galvanized sheets, but for some months this competition has been practically nil, for the simple reason that the manufacturers across the border have been so filled with orders that delivery could be obtained much quicker from England. And last week importers in this country were notified that, as the demand in the home market was so heavy, the manufacturers in the United States could not for the present quote for the Canadian market. This condition of affairs is, of course, only temporary.

In tinplates, the British article still possesses the Canadian market, and tinplates, it will be remembered, are on the free list.

While the British manufacturers of iron, steel, and hardware generally are not even holding their own on the Canadian market, and are likely never to occupy the place they once did in it, they can do relatively better in it than they have of late years by paying closer attention to its requirements, and by doing away with some of the methods we have pointed out which are so annoying to importers in this country.

FLAGRANT INCONSISTENCY.

ON the day following the raid on the Dominion Treasury by the members of both sides of the House of Commons, the Conservative newspapers resumed their criticism of the increase of Government expenditure.

We quite believe that the criticism is not without reason. Our public expenditure is increasing too rapidly. But why did not the same newspapers, if they are so concerned about the welfare of the country, oppose the increase in the sessional indemnity, which bled the country for the advantage of the politicians, Conservatives as well as Liberals? Simply because the members of their own party were sharing the spoils.

We have had nothing for many years that has so exposed the inconsistency of the party press as their action in regard to the sessional indemnity.

What is wrong when political opponents alone are enjoying it is all right when friends as well as foes are participants. Is it any wonder we hear so much about the waning influence of the political press?

A DECLINE IN SHOT.

Shot has followed in the wake of the lead pipe market, a reduction in prices having taken place this week. The discount is now 17½ per cent., instead of 15 per cent. as formerly. The reduction is due to the lower cost of raw material.

TRADE IN COUNTRIES OTHER THAN OUR OWN.

THE COPPER MARKET.

THE mid-monthly report of James Lewis & Son, dated May 16, contain the following upon the foreign copper market :

"There is little change in the position of copper since May 1. From £69 8s. 9d. for cash standard advanced on May 3 to £70 10s. In consequence of the purchase of about 1,500 tons of three months' prompt by one firm, 'bears' also rushing to cover; but with the cessation of these purchases, values quickly fell away until £69 7s. 6d. was accepted on May 15. With a reduction in stocks and visible supply, to-day's market closes with buyers of cash at £69 11s. 3d. and of three months' prompt at £70 1s. 3d. per ton.

"High conductivity copper continues very scarce, and wire drawers have difficulty in supplying their immediate requirements, purchases being made for delivery up to September. English refined is in but moderate demand, though 200 tons have been taken by the Admiralty, as Birmingham consumers will only buy from hand to mouth.

"Our statistics show a considerable falling off in European consumption for the past four months, as compared with the same period last year. This reduction is, however, more apparent than real, the stocks in the hands of manufacturers being very much less now than 12 months ago, after a period of heavy arrivals from the United States. Some large orders have been recently placed for American copper for shipment to the Continent, at full prices.

"The Board of Trade returns for the months of January to April, show the following results :

	1898. Tons.	1899. Tons.	1900. Tons.	1901. Tons.
Imports into England	44,745	43,492	54,199	45,863
Exports from England	17,077	21,847	21,722	20,129
Difference	27,668	21,845	32,177	25,754
Difference in stocks Jan. 1 to May 1	*3,529	*1,742	†2,663	†1,389
‡Consumption of England	22,698	17,592	22,778	20,712

* Decrease. † Increase. ‡ Deducting sulphate of copper made and later exported.

PIG IRON IN GREAT BRITAIN.

Iron and Coal Trades' Review, May 17 : "In the majority of cases there appears to be an improved demand for pig iron, and this is especially so in the Cleveland district, where several furnaces that were blown out in the winter months have been restarted, and where it is anticipated others will follow. This result is to some extent a consequence of the increased demands of Scotland, to which the tonnage recently forwarded has been phenomenally large. Foundry iron is reported to be scarce. There is not much probability of prices falling materially below present quotations—the less so, that the price of fuel, so far from having further dropped, shows a tendency to harden. Hematite iron supply is also reported to be rather short of the recent demand. On the whole, this branch of the trade has better prospects. Reference has been made in several journals to the fact that there has been a considerable import of pig iron into Glasgow from Canada within the last few days. This is, no doubt, an interesting fact, but it is not a new one. In April last 50 tons of pig iron were imported into Liverpool from Canada, and small quantities have been imported previous to that date. It is, indeed, extremely unlikely that the pig-iron exports from Canada will affect the British market for many months to come, if at all. It will be at least 12 months before the Dominion would be in a position to export more than 100,000 to 150,000 tons a year to all countries. The movement of stocks during the past week has been distinctly favorable. While the public stocks show an increase of 8,415 on the year up to the present date, the change during the week ended May 13 has been a decrease of 2,080 tons, and the total public stocks on that date amounted to 170,989 tons, the following being the details :

	Tons.	Change during 1901. Decrease.
Connal's at Glasgow	60,410	67
Connal's at Middlesbrough	69,966	638
Railway Stores, Middlesbrough	10,275	
Connal's at Middlesbrough, hematite	6,645	......
Cumberland & Barrow Stores	23,693	1,355

THE BRITISH TINPLATE TRADE.

It is satisfactory to note that the export of tinplates for the month of April keeps up with that of the same month of last year—although the shipments for the first four months of this year fall some 228,240 boxes below the corresponding period of last year; the latter, however, must be remembered as the last "boom" year of the last century, and therefore something out of the common on ordinary average.

Owing to the advance in tin, and the excitement in coal, prices for tinplates have shown a hardening tendency, and better values have consequently been obtained.

Makers generally are comfortably off for orders, and this has produced a scarcity of prompt supplies and helped the upward movement. Should the advance continue there is the probability of the idle works restarting, and already there are rumors of one or two preparing to launch out.

The demand during the week has kept up in full volume, especially in light substances, and squares and odds, while blackplate has also shared in the demand.

The inquiry for Russian oil plates has eased off for the time being, owing to large supplies having been booked in the earlier part of the year. For the States there have been some good inquiries for deliveries over the next half year, and some considerable business resulted. Common, 14 by 20, are quoted at 12s. 4½d. for prompts and 12s. 3d. for forwards, with business at both figures.

Common 20 by 28 Bessemers are quoted at 25s. for 112 sheets, with 56 sheets at 12s. 9d. for prompt delivery.

Common light 14 by 20, 100 lb. are held for 12s., and are in good inquiry ; 95 lb. quoted at 11s. 9d., with buyers at 11s. 7½d.; 90 lb. quoted at 11s. 6d.; 85 lb. at 11s. 3d. to 11s. 4½d., and in good demand ; 80 lb. held for 11s. 1½d. to 11s. 3d. The demand for light plates remains vigorous for sharp shipments, as well as forward deliveries.—Iron and Coal Trades' Review, May 17.

MANUFACTURED IRON AND STEEL IN ENGLAND.

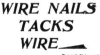

It is impossible to ignore the fact that this branch of the trade is somewhat weaker than it was, and business is not coming forward as it was expected to do. The material reduction last week in marked bars is an evidence of this fact. All manufactured iron prices are naturally and inevitably tending to a lower level, but the rate of movement must largely be affected by the cost of pig iron of fuel, and of labor, all three being at this moment materially above the normal range. The only branches of the trade that are kept really busy are those devoted to munitions of war, which means, of course, mainly the Sheffield trades. The tone of the steel trade is, however, most hopeful. The chief rock ahead is the threatened increase of American competition, when the competitive firms in the United States find home business slacken. There is increasing keenness of competition in shipbuilding material between the North of England and the Scotch manufacturers, and although the tonnage of new work on hand in shipyards is considerably below what it was a year ago, and German manufacturers are now offering plates and angles at low prices, the trade is pretty well monopolized by British firms, who are turning out large quantities of material. — Iron and Coal Trades' Review.

TRADE IN SHEFFIELD.

Nothing very encouraging can be said of the iron and steel branches this week. Although rather more material has changed hands as compared with the two previous weeks, buyers are remarkably timid, and refuse to order forward lest there should be a fall in values. Business being so hard to get, the competition among rolling mills and tilts for that which is to be had is intensely keen. The output of pig iron being now only on a level with the demand, prices are firm, and it is the belief of some that the least spurt in trade would cause an advance. A growing scarcity of hematite irons is being experienced, and it is difficult to give reliable quotations. The market for this material appears to be in a sensitive condition, consumers stating that buyers promptly raise the price should they show a disposition to place a substantial order. Local makers of steel billets consider that the position is slightly stronger. — Ironmonger, May 18.

NEW YORK METAL MARKET.

TIN—The London market opened weak this morning after the holiday, and prices declined £2 5s. during the day, with moderate trading. The decline in futures was even more decided, amounting to £2 12s. 6d. The close there was easy. New York, in sympathy with the news from London, broke about 35 points and closed easy at the decline. At the call on the Exchange, 15 tons spot sold at 28c., 10 tons at 27.95c. and 5 tons at 27.90c., the market closing at 27.90 to 28c. for spot and May. There was no disposition shown to trade in futures. The consuming demand for tin continues very light, buyers seeming to be well supplied. The St. Paul from Southampton brought 10 tons, making the total arrivals since May 1, 2,626 tons.

COPPER—With the opening of the London market after the Whitsuntide holidays, the downward movement of values was resumed, and up to noon the tone was weak. Later, the feeling was steadier under liberal purchases of spot, but prices did not recover, and closed 5s. under last Friday's quotations. Trade here continues dull, and the market is uninteresting. Prices, however, are maintained by the leading interests at 17c. for Lake Superior and 16⅝c. for electrolytic and casting.

PIG LEAD—A fair business is reported, with prices steady on the basis of 4.37½c. for lots of 50 tons or more. Lead is reported to be scarce in St. Louis and that market was strong at 4.27½c. bid for common, 4.30c. for soft Missouri and 4.32c. for chemical. London was firmer and closed at an advance of 2s. 6d. in soft Spanish.

SPELTER—The market is dull and easy at 3.95 to 4c. In St. Louis there was nothing doing, that market closing dull at 3.77½c. bid and 3.80c. asked. A break of 5s. occurred in the London market to-day.

ANTIMONY—Regulus is quiet but steady at the range of 8½ to 10¼c. as to brand.

OLD METALS—There was no change in quotations and business in all kinds was light.

IRON AND STEEL—Nothing of special interest was developed in the iron market to-day. Consumers are drawing freely on existing contracts, but are giving little attention as yet to probable future requirements. There is a fairly steady tone to the market in the absence of any decided pressure to sell. Cables from London report a decline in prices for pig iron after a period of steadiness lasting for a week or more. Scotch warrants at Glasgow were 3d. lower and foundry iron at Middlesboro showed a similar decline.

TINPLATE—The market is quiet so far as future business is concerned, but current consumption is large and prices are firm.— N.Y Journal of Commerce, May 29.

THOMAS PARKINSON DEAD.

Thomas Parkinson, of Parkinson & Co., stove dealers, London, Ont., died at his home in that city on Saturday last, after being confined to his home for the greater part of the past winter with pulmonary trouble. Mr. Parkinson was 28 years of age, and was the eldest son of Mr. Thomas Parkinson, of 192 Richmond street, London. He was born in the Old Country, but had resided in London for 30 years. He leaves four children, the eldest of whom is nine years of age. Mrs. Parkinson died in August, 1899.

Mr. Parkinson was much liked by those who did business with him.

NEW LINSEED OIL MILL.

The Livingston Linseed Oil Co., Baden, Ont., is to start a mill in Montreal. The building formerly occupied by the William Johnston Co. has been secured for the purpose, and is now being fitted up with machinery, which will be of the most modern description. The capacity of the mill will be 100 bbls. per day, and it will be put into operation some time in September.

The linseed oil mill at Baden, will be conducted as heretofore.

PERSONAL MENTION.

Mr. Thos. Robson, hardware merchant, Fenelon Falls, Ont., spent Friday (Victoria Day) and Saturday last week in Toronto.

THE "IDEAL" FREEZER.

ONE of the most unique ice cream freezers that have ever been offered to the Canadian hardware trade is "The Ideal" ice cream freezer, handled in Canada by Wood, Vallance & Co., Hamilton, Toronto and Vancouver, and Geo. D. Wood & Co., Winnipeg. It combines three excellent qualities, in being so simple in construction, easy in operation and rapid in results, that it is confidently claimed to be the most practical and satisfactory freezer ever offered for domestic use. The mechanism of the machine can be readily understood by a study of the three cuts in the advertisement on page 3 of this issue. Other freezers require a rapid moving mechanism inside the can to displace the cream sufficiently to allow the freezing process to take place. "The Ideal" has a breast on the top of the tub which holds the ice and salt securely, so that the freezer can be tilted to an angle of about 45 deg. This secures the displacement of the cream without the necessity of an intricate mechanism, thus, it is claimed, giving a capacity and rapidity fully 25 per cent. greater than other freezers. Only 15 lb. of ice and 5 lb. of granulated rock salt are required to operate the freezer at full capacity. Common barrel salt will do, but rock salt is cheaper and better. As the ice and salt are kept in an air-tight compartment there is no danger of spoiling the cream with salt, as it is impossible for liquefied ice and salt to get into the cream while making or serving, nor is there danger of overflow from the tub to slop the floor.

"The Ideal" will always freeze a quart in three minutes, two quarts in four minutes and three quarts in five minutes, so it is ready to serve. It will freeze any quantity, from one pint up to the capacity of the freezer, with equal perfection, and repeat as often as necessary to serve a reasonable number of people, without renewing ice or salt by simply pouring in more material as fast as the supply of frozen cream is withdrawn, and freeze as before. If the quantity thus served is so great as to actually exhaust the freezing power, it requires less than a minute to empty the ice compartment, and (after the ice is crushed) only four minutes to refill it.

WORKMAN'S TIME SHEETS.

It is essential to the proper and economical conduct of a workshop that the workmen's time sheets should be carefully filled in by themselves and as carefully checked by the employer or overseer. The use of time sheets similar to that illustrated herewith keeps on record the hours worked during the day and week, and also shows the different jobs on which the workman has been engaged. The total column at the right-hand side gives the hours worked per day, and the row at the bottom of the sheet shows the various hours spent on each job. Should the number of jobs handled in any week exceed the number of spaces provided for entries, two or more sheets may be used. The method appears to afford an efficient method of keeping time in small or large shops, and by comparing one week's sheet with another pertaining to similar work it is possible to detect negligence or waste of time. The sheet also acts as a pay voucher at the end of the week. The width of the actual sheet is 9 inches, and 20 spaces across the paper are easily obtained.

WORKMAN'S TIME SHEET.

Workman's Name. No. Week ending.

Number of Job......	106	74	99	Total.
Monday	3	4	2	9
Tuesday		7	2	9
Wednesday	1		8	9
Thursday				
Friday..............				
Saturday				
Totals	4	11	12	
Total for week....	Hours at....£			

—Ironmonger.

WINDOW GLASS TRADE IN THE UNITED STATES.

Never before in the history of the window glass trade have so many new plants been proposed as now, fully 25 tanks, not to mention pot furnaces, now being negotiated for. One fact that came to light last week may cause a decided restriction to the number of these plants materializing. An inventory of all tank-block stock in sight and available for the immediate future shows that practically every tank-block is contracted for until well towards fall, a condition that will make it hard for many to get ready for next fire. And then, again, there are only 2,100 blowers. It takes blowers as well as factories to make glass.—Paint, Oil and Drug Review.

BOOM IN TWEED.

The town of Tweed is one of the smartest small places on the C.P.R. This spring houses are springing up everywhere. Some 10 houses are now in course of erection, including a Bay of Quinte Railway station and a Presbyterian manse and a new store. Business is booming in the two up-to-date hardware stores.

OLD-TIME GUN-PROVERS.

The main argument put forward by the gunmakers in the reign of Charles I. in favor of establishing the Gunmakers' Company was couched in the following terms, the paragraph being an excerpt from the charter granted by the King in 1637:

> Divers blacksmiths and others inexpert in the art of gunmaking had taken upon them to make, try and prove guns after their unskilful way, whereby the trade was not only much damnified, but much harm and danger through such unskilfulness had happened to His Majesty's subjects.

"Damnified," says The Kynoch Journal, is distinctly good. Modern guns made "after an unskilful way" are even nowadays liable to be much "damnified," and deservedly so.—"Vulcan" in Ironmonger.

THE ONTARIO HARDWARE ASSOCIATION.

THE Retail Hardware Merchants' Association, of Ontario, the organization of which was reported in last week's issue of HARDWARE AND METAL, promises to be an exceedingly wide-awake association. An adjourned meeting was held on Thursday night, in Templer's Hall, Bathurst and Queen streets, Toronto, and was attended by a representative gathering of hardwaremen of Toronto and vicinity, including the following: President, E. R. Rogers, who occupied the chair; secretary F. W. Unitt, F. J. Russill, A. Welch, John Castor, J. W. Peacock, J. T. Wilson, C. Dale, R. Fletcher, C. Watkins, W. Emery, S. Hobbs, Wm. Batters, G. Alexander, J. S. Hall and James Ivory, several of whom were added to the membership since the first meeting.

Mr. Fletcher, chairman of the committee appointed to draft a constitution and by-laws, reported that the committee had met and had prepared such constitution and by-laws. These were submitted and adopted by the association.

OBJECTS OF THE ASSOCIATION.

In answer to a question from Mr. Batters, one of the new members, President Rogers stated that the object of the association was, in brief, the mutual benefit of its members. Many matters have to be fully discussed—early closing, for instance, and matters relating to buying and selling. But one of the essential features of the organization would be to draw the members closer together, to bring about a better understanding among the various retailers in the city.

Secretary Unitt followed, by expressing the conviction that the association could do a great work in bringing its members into sympathy with each other, and by getting them to meet together and discuss matters of general interest. By united action they could greatly advance the general good and do much to elevate the retail trade throughout the city.

Mr. Batters believed it would be a great advantage, if it were possible, to adopt means to keep track of credit customers.

EARLY CLOSING.

Mr. Hall thought that early closing was one of the questions that should be considered at once. He had a card in his window that his store was closed at 6.30 p.m., but found difficulty in doing so, as customers insisted on coming later than that hour.

Mr. Fletcher considered that there could be no organization without benefit, and that as union is strength, and as the members are united in the desire to improve the conditions of trade, there was no reason why the association should not be a power for good.

As the election of the third vice-president and of four members of the executive had been deferred at last meeting, the following were added to the list of officers: Third vice-president, S. Hobbs, 1434 Queen street west. Executive—Wm. Batters, 569 Queen street west; G. Alexander, 586 Queen street west; J. S. Hall, 1097 Yonge street and J. Ivory, 682 Queen street west.

It was then moved by Mr. Fletcher, seconded by Mr. Russill, that "The Toronto members of the Hardware Merchant's Association of Ontario do hereby agree to close our places of business at 7.30 p.m. on all days except Saturdays and days before holidays during the months of June, July and August."

After discussion, in which Messrs. Dale, Hobbs, Wilson, Peacock, Ivory, Welch, Fletcher, Unitt and Rogers engaged, it was agreed to close at 7 p.m. instead of 7.30 p.m., and to not make the agreement binding till June 10. In the meantime the members will endeavor to persuade their neighbors to agree to the proposal.

A letter was received from E. M. Trowern, secretary of the Retail Merchants' Association of Canada, stating that that body had for some time been intending to form a hardware section in their body, and asking that a deputation representing their association might wait on the Retail Hardware Association to discuss amalgamation. It was decided to hear the deputation.

President Rogers, Secretary Unitt and A. Welch were appointed a committee to prepare cards to advertise early closing and the association.

On motion of Messrs. Welch and Caslor, the association adjourned to meet on June 14.

HARDWAREMEN WILL CLOSE AT 1 O'CLOCK P.M.

There is a strong movement in favor of closing at 1 o'clock Saturday afternoons during July and August in Ottawa. Last week a petition was circulated among the hardware merchants, they being selected as being known to be probably the most unanimous in favor of the closing, though not more unanimous than some other lines. Practically every hardwareman in the city signed the following petition the first day: "We, the undersigned hardware merchants, agree to close our stores during the months of July and August, each Saturday at 1 o'clock and keep them closed till Monday morning."

The signatures were as follows: Graves Bros., Grant Bros., Butterworth & Co., A. Workman & Co., Wm. Graham, The Ottawa Hardware Co., Blyth & Watt, N. Hay, Thos. Birkett & Son Co., McDougal & Cuzner, McKinley & Northwood, W. G. Charleson, E. G. Laverdure, J. P. & F. W. Esmonde.

INDUSTRIAL GOSSIP.

Those having any items of news suitable for this column will confer a favor by forwarding them to this office addressed the Editor.

A DESPATCH from North Sydney states that the Anglo-American Syndicate have taken up the option, and the properties of The Nova Scotia Steel Co. have passed into the hands of these capitalists. It is understood that the syndicate will push development of these properties, and proceed at no distant date to inaugurate gigantic steel works near Sydney.

A 10-ten drop hammer has been installed in the Galt, Ont., axe factory.

The Algoma Steel Co., Limited, Sault Ste. Marie, Ont., has been incorporated.

The Peat Development Syndicate propose to manufacture gas from peat or some similar material in Toronto Junction, Ont., and have asked that municipality to grant inducements for the erection of a $3,500 factory for that purpose.

The G. L. Cosby Co. of Ontario, Limited, are rebuilding the G.T.R. round-house at Point Edward, Ont., into a large foundry and machine shop. It is stated that, when the plant being installed is completed and being run at its full capacity, it will employ nearly 300 hands.

A despatch from Kingston, dated May 28, says: "At the next meeting of the city council, a local financier will lay a communication before the aldermen asking if the city will fulfil its part of the agreement entered into with Messrs. Meyer, relative to the establishment here of a smelter, providing another company agrees to fulfil the agreement made by the promoters, who did not carry out the enterprise.

Mr. J. H. Roper has been in Toronto this week representing Alexander Gibb, Montreal. He will go on to visit the trade in Hamilton and London.

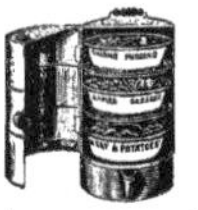

THE

COLLINS STEAM COOKER

Cooks Quickly, Properly, and in Quantities. Everybody likes this Cooker, as it will cook, and cook well, a whole dinner of about 5 or 6 articles at the same time over one hole on the Range, Gas Stove, Coal Oil Stove, or Gasoline Stove. Consequently, it is a great saver of fuel, and time-saver, as you can be doing other housework or go to church as it cannot burn the contents. Tough meat will be made tender and food much more palatable and healthy, as all the flavor and nutrition is retained in the food. Door can be opened and any one article taken out without molesting the whole. Write for catalogue and prices.

Manufactured solely by

THE COLLINS MFG. CO.,

34 Adelaide West TORONTO.

WESTERN ASSURANCE COMPANY

Incorporated 1851.

Fire and Marine

Capital, subscribed	$2,000,000.00
Capital - - -	1,000,000.00
Assets, over - -	2,340,000.00
Annual Income -	2,290,000.00

Head Office: TORONTO, ONT.

Hon. Geo. A. Cox, President. J. J. Kenny, Vice-President. C. C. Foster, Secretary.

Buy the Best.

HERCULES

Sash Cord.

Star Brand Cotton Rope
Star Brand Cotton Clothes Lines
Star Brand Cotton Twine

For Sale by all Wholesale Dealers

McCaskill, Dougall & Co
Fine Varnish & Japan
Manufacturers.
Montreal.
Price Lists on application.

KNOX HENRY

Heavy Hardware and Metal Broker
Room 32, Canada Life Bldg., MONTREAL.

"SECCOTINE"
FOR STICKING EVERYTHING.

Samples sent free on application.

HORSE NAILS—"C" Brand Horse Nails—Canada Horse Nail Co.

"BRASSITE" GOODS—Gunn Castor Co. Limited, Birmingham, Eng.

THE OAKVILLE BASKET CO.,

Manufacturers of

1, 2, 3 Bushel

Grain

AND

Root

BASKETS

THE OAKVILLE BASKET CO.

THE TIME TO INSURE IS

NOW

While you are WELL, STRONG and INSURABLE

THE

Confederation Life

ASSOCIATION issues policies on all approved plans of Insurance, and is a prosperous and progressive Canadian Company.

PROTECTION FOR YOUR FAMILY.
PROFITABLE INVESTMENT FOR YOURSELF.

Pamphlets and full information sent on application.

Hon. Sir W. P. Howland, K.C.M.G., C.B., - - President.
W. H. Beatty, Esq., W. D. Matthews, Esq., - Vice-Presidents.

W. C. MACDONALD, ACTUARY. J. K. MACDONALD, MANAGING DIRECTOR.

HEAD OFFICE, - TORONTO.

BURMAN & SONS' CELEBRATED CLIPPERS

for Horsemen and Barbers.

Established 1871. BIRMINGHAM, ENG.

NO. 3—POWER CLIPPER, with "Wrist Joint."

As supplied to { The Czar of Russia. The King of Denmark. Earl Roberts, Etc., Etc.

NO. 297.

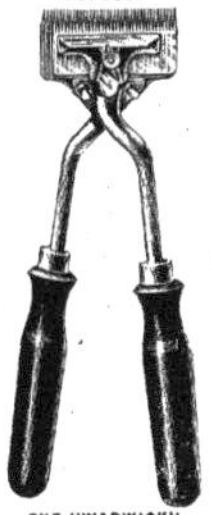

THE "WARWICK" CLIPPER.
Cuts over three teeth.
As supplied to His Majesty's War Department.

THE "LEOPOLD" TOILET.

SEND FOR PRICE LIST AND TERMS.

DELORME BROS., Agents, Debresoles Street, Montreal

A CORNER FOR CLERKS.

By W. T. R.

CHEAP CLERKS.

"A MERCHANT as a rule is unable to estimate the value of a clerk who endeavors to save stock from wasting," said one of our most successful merchants a few days ago. We hear and read these days of "the utilization of the by-production" in the manufacturing line and how large amounts of money are being made from that which was formerly wasted or non-productive. So, in the mercantile line everything must be used to the best advantage in order to make money. "An employer is constantly at the mercy of his clerks" is true as it is simply impossible for him to watch their every transaction in a business of ordinary magnitude. This point is frequently overlooked by the merchant, i.e., his dependance on his clerks No merchant can afford to keep a clerk who has no regard for his employer's interest, or should he be lax in this respect his usefulness to the business may be questioned.

Then again, on his clerks has a merchant to depend to look after the constant leaks which are always going on in every business. Certain goods have not been selling. They are deteriorating in value; it is the duty of the clerk to see that they are brought to the front, cleaned up, and sold. Some men are naturally careless. They never look after anything except their wages. They are constantly complaining because they do not receive more pay, when in reality they don't earn what they now receive, because they do not pay sufficient attention to their employer's interest. except their wages. They are constantly complaining because they do not receive more pay, when in reality they don't earn what they now receive, because they do not pay sufficient attention to their employer's interest.

Stock-keeping is a very important part of a clerk's duties and one that must never be overlooked. "A store is known by its clerks," is an old axiom and one that is true. Chas. F. Jones, in Printer's Ink says: "I doubt if one merchant in ten has a proper appreciation of the influence well-informed, intelligent salespeople exert in the building up of a business. The average merchant in choosing between this class of help and the kind that is neither experienced, intelligent nor well appearing will choose the latter in order to make a fancied saving of a few dollars a week." This, our best and most successful merchants have found out to be false economy. Their experience has proved "the best to be the cheapest" in the matter of this help, and the so-called cheap man may be the most expensive.

There are some young men behind our counters to-day who are unquestionably out of their proper sphere; they lack the essentials required for good salesmen. Tact in dealing with customers, persuasiveness and a keen insight into human nature must be used by the successful salesman. First-class trade demands the most careful attention. Customers in a good store look for, and have a right to expect, the most courteous treatment from the man behind the counter.

"P.W.B." has been brought up in his father's store. He wants to know if it would be beneficial to him to go into a store in another town, to get experience and ideas.

Educationally, it would improve you very much. The only difficulty in a case such as yours, is in getting away; the home ties are usually so strong. However, there can be no question of the benefit. You are sure to learn considerable of value to you in your business. New methods of doing business and frequently new lines of goods are being sold as one goes into new places and one has always much to learn. Let me hear from you again.

"J.R." has had two allowances in one week from the railway company for overcharges on freight. He thinks the boys should weigh their freight when they think they are being overcharged.

I agree with you. They frequently add on more weight than is right and they will always stand watching in this respect. You deserve credit for your carefulness in this matter, "saving is good earning," and I hope this will be appreciated by your employer. Railroad companies have men who sometimes are careless and every clerk has a right to protect his employer's interest.

"W.S." writes: Did you make any money on the recent rise in stocks? A traveller was telling me of one of the boys who made a big haul.

We always hear of the success of others who come out on the right side; and we have very little compassion for those who lose all, who make a failure and wreck their life in speculation. A few men may make a fortune through side deals in wheat and stocks, but they are in a great minority to those who have lost. The man who sticks to his business and lets outside speculation alone is the safe man for his employer and those dependent upon him. He may not get suddenly rich (through money he never earned), but the chances are that he will go through life with an easier conscience than the man who stakes his money on the turn of the market—I am satisfied to let the other fellow make money in this way. If he likes to take the risk, all right; that's his affair, not mine, but just at present I'm not in that business. The worst feature of this to my mind is, that the fellow who cannot afford to lose generally "gets it in the neck" and then hopes to have some one help him out of his difficulty. Reading of Mr. Choate, of Toronto, and his making $300,000 is apt to excite some young men to speculate and to most young men this is bound to prove unprofitable. Let us hope their experience will not cost them too much.

"A.C.": Should a young man work in a store with the idea of starting in opposition to his employer?

It is required of every person to be fair. Take no underhand methods to influence customers in your favor, other than are honorable. The golden rule, "to do unto others as you would have them do unto you," will always hold good. The aspiration of nearly every clerk who stands behind the counter is to own a business; to be his own master—a free man. This is the incentive to work, to save, to learn all about his business, and prompted by this motive when he is able he starts a store and takes no unfair advantage of his former employer, seeking not to injure him, but to gain trade by fair means. There should be no objection to his starting for himself.

PROFESSIONAL STORE ACCOUNTANTS.

A REAL twentieth century departure is announced in the following letter to The Bookkeeper: "I am residing in a town which Uncle Sam recently announced has some 10,000 inhabitants, and am a bookkeeper with some few years of experience, from which I have gained considerable amount of system, and, possessing some hustle, I am establishing the business of thorough, systematic bookkeeping for smaller firms whose business would not justify the salary of a permanent bookkeeper, though the need of such is none the less imperative.

"Upon approaching a client I explain the object of my visit and point out the value and importance of a thorough system of accounts. I soon gain sufficient data upon which to base the remuneration problem. This settled, I make a complete inventory of assets and liabilities, make the proper opening entries upon a new set of books. Then, according to the exigencies of each case, I visit the business at regular periodical times, whether it be half a day, a day or two days a week, and from the entries on their day book I evolve a complete system of bookkeeping and give the firm a statistical comparative balance record of their business at the end of each month. In this way they get the same effective service as the largest firms and only pay for the actual time occupied. I have now on my list a stove and furniture firm, a laundry, a livery stable, a retail grocer and a butcher, with more to come. This gives me a comfortable living. One of the strongest arguments that can be used to induce a merchant or manufacturer to have the work done is to ask the question, 'If a fire overtook you to-night, are you in a position to inform the insurance company what amount of loss you have sustained?'"

J. E. Tremblay, general merchant, Ste. Anne de Bellevue, Que., has compromised at 30c. on the dollar.

BUSINESS CHANGES.

DIFFICULTIES, ASSIGNMENTS, COMPROMISES.

F. PERREAULT & Cie., general merchants, Rimouski, Que., have assigned to V. E. Paradis, provisional guardian.

Bedard, Bertrand & Gauvin, general merchants, Quebec, have assigned.

[illegible] Paquet, general merchant, Roberval, Que., is offering 20c. on the dollar.

Samuel Bricker, hardware dealer, Listowel, Ont., is offering 27 1-2c. on the dollar.

Joseph H. Frigon, general merchant, St. Tite, Que., has assigned to H. Lamarre.

J. J. Boese, general merchant, Rosthern, Man., has assigned to P. Weibe, Rosthern.

M. Simon, general merchant, Alexandria, Ont., has compromised at 75c. on the dollar.

R. Tuplin & Co., general merchants, Kensington, P.E.I., are offering 50c. on the dollar.

Simon Johnson, general merchant, Moose Creek, Ont., has assigned to John C. Milligan.

Herbert A. Bigham, general merchant, Culloden, Ont., has assigned to David T. Cuthbertson.

H. N. Halpenny & Co., hardware dealers, Minnedosa, Man., have assigned to C. H. Newton, Winnipeg.

Lamarche & Benoit have been appointed curators of Pierre Dauplaise, sash and door manufacturer, St. Cyrille de Wendover, Que.

David Jobin, general merchant, Sacré-Cœur de Marie, Que., is offering 25c. on the dollar, and V. E. Paradis has been appointed curator.

PARTNERSHIPS FORMED AND DISSOLVED.

Miller & Sloat, sawmillers, Tracey's Mills, N.B., have dissolved, Mr. Sloat retiring.

Leiser & Hamburger, general merchants, Wellington and Ladysmith, B.C., have dissolved and Simon Leiser continues.

Louis Dana has registered co-partnership under the style of The Sydney Hardware Co., hardware dealers, Sydney, N.S.

Buzzell Bros., grocers, hardware dealers, etc., Cowansville, Que., have dissolved, and a new partnership has been registered.

SALES MADE AND PENDING.

Mrs. Z. Francœur, general merchant, French Village, Que., has sold out.

W. H. Couse, blacksmith, Smithville, Ont., is advertising his business for sale.

The stock, etc., of the estate of W. G. Armour, general merchant, Myrtle Station, Ont., is advertised for sale to-day (Friday).

CHANGES.

George Theriault, general merchant, Bonfield, Ont., is giving up business.

G. R. Vanzant, hardware dealer, Markham, Ont., is closing up his business.

S. J. Brown, harness dealer, Campbellton, N.B., has closed up his business.

Stanislas Lebel, blacksmith, French Village, Que., has removed to Warwick.

Libberman Bros. & Lipson have registered as junk dealers, etc., at Waterloo, Que.

Philippe Marcotte, blacksmith, French Village, Que., has sold out to Benj. Cayer.

T. L. Stetson, harness dealer, Hartland, N.B., has been succeeded by S. J. Brown.

George H. Wallace, tinsmith, Stouffville, Ont., is closing up and removing to Sutton, Ont.

B. L. Bishop, general merchant, Kentville, N.S., has sold out to Spurgeon L. Cross.

Charles E. Darling, blacksmith, Paradise Lane, N.S., has been succeeded by Darling & Burke.

W. S. Santo & Co., general merchants, Peterboro', B.C., have been succeeded by The Peterboro' Trading Co.

Elias Harmer, general merchant, Norton and Mechanics' Settlement, N.B., has sold his branch at the latter place to James Webster.

FIRES.

Nathaniel McNair's sawmill at River Louison, N.B., has been burned; no insurance.

Sylvester Shannahan, general merchant, Sydney, N.S., has been burned out; insured.

DEATHS.

A. S. Pierce, grist and saw miller, Newton, N.B., is dead.

W. K. Secord, general merchant, etc., Winona, Ont., is dead.

Thomas A. Parkinson, of Parkinson & Co., stove dealers, tinsmiths, etc., London, Ont., is dead.

FUMIGATED THE HARDWARE STOCK.

A paper published in an Illinois town, in announcing the purchase of a hardware stock by a new firm, makes this very suggestive remark: "The first thing the boys did was to clean up and fumigate the building, which now presents a neat and clean appearance and shows the work of experienced men." What a comment this language is on the store habits of the original owners! A building that requires fumigation is hardly the place where customers will willingly enter, and it is probably a fact that the business had run down until a sale in bulk was made necessary. There are many such buildings used as retail stores, and every one of them is a blot on the fair name of the business with which it is connected. Dust, dirt and uncleanliness are not proper associates with business unless it is the deliberate intention of the owner to ruin it, yet they are often found and the wonder is that even the few straggling customers continue to patronize a store where they are permitted to remain. No matter how well selected a stock may be, it is sadly out of place in a store that has become a receptacle for dust, dirt and other extraneous matter. Keep the store clean, above all things, so that the business itself may be clean and not suggest the need for fumigation.—Stoves and Hardware Reporter.

A SUGGESTION.

In a Western city a merchant recently displayed in his window the following sign:

Cast Iron Sinks.

After a small boy had discovered the joke the sign attracted considerable attention. Many people who had before passed down the street looking straight ahead paused and glanced in the window at the sign, assumed an amused expression and went on. The sign was talked about and the public was made familiar with the name over the door.—Iron Age.

A CANADIAN CLERK IN ENGLAND.

MR. L. C. A. HOWARD, who was formerly employed with H. S. Howland, Sons & Co., Toronto, and who left for England some time ago, eventually taking a position in a retail hardware store in Southport, a town 18 miles north of Liverpool, was requested by Mr. H. A. Gunn, of H. S. Howland, Sons & Co., to write his opinion of hardware retailing in England. He responded a few weeks ago by an interesting letter to Mr. Gunn, extracts of which are published below :

"I am working in a retail hardware store in Southport, the best store in town for all-round trade, including builders' and general hardware, cutlery, brass goods, electro-plate, grates, mantles, etc. Owing to my bicycle and valise with my papers being stolen, I had difficulty in establishing my identity, which was necessary when I applied for this situation, but is now settled satisfactorily. I have now been here three weeks, and like the place very well.

"The place is too small for the large stock carried, and goods have been shoved and piled into every available corner. On the shelves, wherever a vacant space occurred, something had to be put in, whether it belonged there or not. I have been doing my best to straighten out the mess during the past three weeks.

"I have been greatly surprised at the quantity of American goods in the shop. We have Yale and other locks (all the spring padlocks come from the United States); Whitman & Barnes' goods ; Stanley planes and levels (very few rules, as they think English rules are better marked—caliper rules don't seem to be used, but I am going to get my employer to order some No. 32) ; wrought and brass butts are American, as are also Victor and Torrey door springs ; Dover and express egg-beaters; Irwin auger bits ; Out o' Sight traps (of these traps, mouse size, five gross were sold in about two months at 4d. each, but the price is now 3½d.—they don't know anything about the Gee-Whiz) ; Le Page's glue ; Victor flour sifters ; Edgar's grates ; Eddy's fiber scrubs, and a lot of other stuff. The wire nails we sell are mostly oval, and are of German manufacture. Our chisels are Haworth's, and brass screw hooks, etc., Nettleford's. The spades are all T handled, some of them being Hardy's. I don't care for our cutlery : at least I don't like the patterns of the pocket knives. We handle Clark's razors and Harrison Bros. & Hewson's pocket and table knives, with just a few of Rogers' pocket knives. They don't know anything about I.X L. or Butlers'. We handle a quantity of German goods, such as clamps, small vises, skates and cheaper stuff generally. Some of the skates are 'Starrs.' Those of English manufacture are the old wooden ones. Eclipse and Columbia door springs and Bissell's carpet sweepers have a good sale here.

"A great many articles go by different names here to those usual in Canada. A dishpan is called a washup, an auger bit a twist drill, etc. There is a lake near the town, but a Canadian canoe is unknown. If I stay here any length of time I will certainly send home for one."

CATALOGUES, BOOKLETS, ETC.

CAVERHILL, LEARMONT & CO.

THIS firm have just issued two catalogues which should be secured by all hardwaremen. One of these, their 1901 sporting goods catalogue, describes their comprehensive range of guns, rifles, air rifles, revolvers, cartridges, powder, shells, sights, locks, nipples, implements, tools, oils, lubricants, rifle covers, belts, bags, etc., as well as of skates, hockey sticks, pucks, etc., sleigh bells, shaft bells, saddle gongs and sewing machines. The other catalogue is devoted to bicycles and sundries, and contains, besides the description of the firm's bicycles, illustrations of a large variety of bicycle and driving lamps, bells, cyclometers, wrenches, oilers, whistles, locks, brackets, cement, stands, pumps, clips and pedals. This catalogue also includes sewing machines handled by Caverhill, Learmont & Co.

ELECTRICAL SUPPLIES.

Ness, McLaren & Bate, manufacturers of telephones and electrical supplies, Montreal, have issued a series of booklets that many hardwaremen will be interested in. The series comprise four booklets, devoted to telegraphs, telephones, annunciators and switch boards. The telegraph catalogue contains illustrated descriptions with price lists of the various telegraph instruments and supplies handled by this house, together with detailed instructions for learning telegraphy. The telephone catalogue includes main line, portable and warehouse telephones of various styles made by this firm. The switch-board catalogue gives in addition to illustrations of switch boards, diagrams of working parts and connections which enable one to more readily judge of their value. The most comprehensive of the four, however, is the annunciator catalogue, which contains a large list of hotel, house, steamship, office, railway, stock and special annunciators and annunciator supplies. These catalogues, as well as those previously issued by Ness, McLaren & Bate, should be secured for reference by hardware dealers and electrical contractors.

THE BRADLEY STEEL SHELF BRACKET.

LIKE many articles of more complicated design, the Bradley steel shelf bracket, made by the Atlas Manufacturing Co., New Haven, Conn., is a result of evolution. For over six years, its makers sought to improve their strength, and, in that time, three distinct changes have been made in its construction.

PAT. APR. 5, 1892.
NOV. 8, 1892.
APR. 4, 1893.

First was the changing of the screw holes, which originally were placed one over the other, and not side by side as at present. The original position brought the screw holes in line with the grain of the wood, with the result that very often the wood would split when the screws were driven in. Although it was an item of considerable expense to put the necessary offsets in the wire and to devise the improved form of the bracket at this point, it was done immediately upon attention being called to this defect.

The next thing was the criticism that the wire used in the bracket was too light, and, although the changing of the guage involved making new sets of dies all through, this was also done.

Then it was pointed out that occasionally the lock for fastening the ends of the wire together did not hold. As first made, the ends of the wire were brought together under the metal clasp at the point "A." In a few instances, when subjected to a severe strain, the lock did not hold and the wires pulled apart. This lock was made thus :

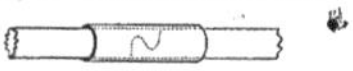

As now made, the wires are brought together at the extreme end of the bracket at the point "B." This secures two solid wires through the whole length of the bracket, eliminating entirely the last fault found with the bracket.

V. Leblanc & Co., general merchants, Hull, Que., have dissolved, and their business will be continued by V. Leblanc under unchanged style.

There is a Difference

In the way ice cream freezers move.
Some go slow, but

White Mountain Freezers

move rapidly and continuously, because there is a demand for them. They always please the purchasers so well that they recommend them. These recommendations make other customers, and so on. The sure and splendid results that can always be depended upon with White Mountain Freezers have been a lasting advertisement that makes them easy to sell, and

ANY DEALER SELLING FREEZERS CAN DOUBLE HIS SALES AND MORE THAN DOUBLE HIS PROFITS BY SELLING THE "WHITE MOUNTAIN."

SEND FOR CATALOGUE.

McCLARY MANUFACTURING CO., Sole Agents for the Dominion.

London, Toronto, Montreal, St. John, N.B., Winnipeg, Vancouver.

MARKETS AND MARKET NOTES

QUEBEC MARKETS

Montreal, May 31, 1901.

HARDWARE.

TRADE continues to show a seasonable volume, the number of orders coming to hand this week being quite up to the market. At present the demand is running along general lines, but spring goods, such as lawn requisites, spray pumps, freezers and screens, are among the most called for goods. Harvest tools have been shipped quite freely during the week. Barb wire continues very scarce, and the various shipments that have come to hand have not begun to fill orders. Wire nails have been exceedingly scarce, and some sizes were almost unobtainable for a few days. There has also been a decided light supply of poultry netting on the market, some sizes being sold out in quite a number of houses. The prices are somewhat higher on this article at the moment. Sheet metals are somewhat firmer again this week, and linseed oil is 3c. higher and in short supply, owing to disappointingly light arrivals from England. Glass has been advanced 10c. per 50 ft. Cutlery has moved out well this week, while the manufacturing demand for bolts, nuts, screws, etc., is quite encouraging. The wholesale houses report a good trade in ready mixed paints and painters' supplies. Collections are improving.

BARB WIRE—The scarcity that has been the important feature of the market for some weeks continues to annoy both wholesaler and retailer. Some shipments have come to hand from the American manufacturers, but they have been quickly reshipped and nothing has gone into stock. The price remains as before, $3.05 per 100 lb. f.o.b. Montreal.

GALVANIZED WIRE—Business is rather quieter than it was. Prices are unchanged. We quote as follows: No. 5, $4.25; Nos. 6, 7 and 8 gauge, $3.55; No. 9, $3.10; No. 10, $3.75; No. 11, $3.85; No. 12, $3.25; No. 13, $3.35; No. 14, $4.25; No. 15, $4.75; No. 16, $5.00.

SMOOTH STEEL WIRE—There is nothing new to note in this article. We quote oiled and annealed as follows: No. 9, $2.80; No. 10, $2.87; No. 11, $2.90; No. 12, $2.95; No. 13, $3.15 per 100 lb. f.o.b. Montreal, Toronto, Hamilton, London, St. John and Halifax.

FINE STEEL WIRE—The demand is moderate. The discount is still 17½ per cent. off the list.

BRASS AND COPPER WIRE—A fair number of orders have been opened this week. The discount on brass is 55 and 2½ per cent., and on copper 50 and 2½ per cent.

FENCE STAPLES—Trade remains much as last reported. We quote: $3.25 for bright, and $3.75 for galvanized, per keg of 100 lb.

WIRE NAILS—During the week there has been quite a scarcity of certain sizes of wire nails, but supplies are now forthcoming. The demand is not brisk. We quote $2.85 for small lots and $2.77½ for carlots, f.o.b. Montreal, London, Toronto, Hamilton and Gananoque.

CUT NAILS—Business is rather slow. There is no change to be reported. We quote as follows: $2.35 for small and $2.25 for carlots; flour barrel nails, 25 per cent. discount; coopers' nails, 30 cent. discount.

HORSE NAILS — The demand at the moment is rather slow. "C" brand is held firmly at discount of 50 and 7½ per cent. off the new list. "M" brand is quoted at 60 per cent. off old list on oval and city head and 65 per cent. off countersunk head.

HORSESHOES—A fairly good business is doing. We quote as follows: Iron shoes, light and medium pattern, No. 2 and larger, $3.50; No. 1 and smaller, $3.75; snow shoes, No. 2 and larger, $3.75; No. 1 and smaller, $4.00; X L steel shoes, all sizes, 1 to 5, No. 2 and larger, $3.60; No. 1 and smaller, $3.85; feather-weight, all sizes, $4.85; toe weight steel shoes, all sizes, $5.95 f.o.b. Montreal; f.o.b. Hamilton, London and Guelph, 10c. extra.

POULTRY NETTING—The market has been very bare of supplies during the past week, particularly in medium sizes. Shipments of new goods are expected next week. The discount is now steady at 55 per cent., the extra 5 per cent. that has been allowed in many cases and the 10 per cent. in 100 roll lots not now being granted.

GREEN WIRE CLOTH — The market is steady and demand quite good. The price is unchanged at $1.35 per 100 sq. ft.

SCREEN DOORS AND WINDOWS—There has been a good trade done in screens this week, the demand keeping up exceptionally well. We quote: Screen doors, plain cherry finish, $8.25 per doz.; do. fancy, $11.50 per doz.; walnut, $7.40 per doz., and yellow, $7.45; windows, $2.25 to $3.50 per doz.

SCREWS—Good sorting orders are to hand this week. Discounts are as follows: Flat head bright, 87½ and 10 per cent. off list; round head bright, 82½ and 10 per cent.; flat head brass, 80 and 10 per cent.; round head brass, 75 and 10 per cent.

BOLTS—There is a good demand coming from manufacturing concerns for bolts, and trade in this line is fairly active. Discounts are as follows: Norway carriage bolts, 65 per cent.; common, 60 per cent.; machine bolts, 60 per cent.; coach screws, 70 per cent.; sleigh shoe bolts, 72½ per cent.; blank bolts, 70 per cent.; bolt ends, 62½ per cent.; plough bolts, 60 per cent.; tire bolts, 67½ per cent.; stove bolts, 67½ per cent. To any retailer an extra discount of 5 per cent. is allowed. Nuts, square, 4c. per lb. off list; hexagon nuts, 4¼c. per lb. off list. To all retailers an extra discount of ¼c. per lb. is allowed.

BUILDING PAPER—Trade is fairly active. We quote: Tarred felt, $1.70 per 100 lb.; 2-ply, ready roofing, 80c. per roll; 3-ply, $1.05 per roll; carpet felt, $2.25 per 100 lb.; dry sheathing, 30c. per roll; tar sheathing,

40c. per roll ; dry fibre, 50c. per roll ; tarred fibre, 60c. per roll ; O.K. and I.X.L., 65c. per roll ; heavy straw sheathing, $28 per ton ; slaters' felt, 50c. per roll.

RIVETS AND BURRS — The demand is moderate. Discounts on best iron rivets, section, carriage, and wagon box, black rivets, tinned do., coopers' rivets and tinned swedes rivets, 60 and 10 per cent.; swedes iron burrs are quoted at 55 per cent. off ; copper rivets, 35 and 5 per cent. off ; and coppered iron rivets and burrs, in 5-lb. carton boxes, are quoted at 60 and 10 per cent. off list.

BINDER TWINE—There has been a little business transacted this week. We quote : Blue Ribbon, 11½c. ; Red Cap, 9¾c. ; Tiger, 8¾c.; Golden Crown, 8c.; Sisal, 8¼c.

CORDAGE — The activity is fairly well maintained, although hardly in so large a volume. Manila is worth 13½c. per lb. for 7-16 and larger ; sisal is selling at 10c., and lathyarn 10c.

HARVEST TOOLS — This is one of the most active lines on the market just now. Some good shipments have been made this week. The discount is unchanged at 50, 10 and 5 per cent.

SPADES AND SHOVELS — A moderate business is to be reported at the former discount, 40 and 5 per cent. off the list.

LAWN MOWERS—Sales have been fairly numerous this week. We quote as follows : High wheel, 50 and 5 per-cent. f.o.b. Montreal ; low wheel, in all sizes, $2.75 each net ; high wheel, 11-inch, 30 per cent. off.

FIREBRICKS — A small distributing trade has been done this week at $17.50 to $22 for Scotch, and at $17 to $21 for English per 1,000 ex wharf.

CEMENT—The sales of foreign cement have been disappointing. The demand continues to run along the Canadian and American varieties. On the whole business is good. We quote : German cement, $2.35 to $2.50; English, $2.25 to $2.35 ; Belgian, $1.70 to $1.95 per bbl. ex wharf, and American, $2.30 to $2.45, ex cars.

METALS.

In Canadian metals the market is steady, but the English sheet metal prices seem to be on the upward move. Tinplates have advanced in sympathy with the advance in pig tin and are quoted 4½d. per box higher by cable. Canada plate is about 3s. 9d. per ton higher, but selling here at ridiculously low figures. Sheet zinc is reported £1 per ton higher this week. Terne plates are also higher at primary points and prices here are hardening. Imports continue to be very light and stocks here are short. In consequence, values are being held quite firmly.

PIG IRON—Business is reported to be rather slow and few contracts are being made. Canadian productions are worth $18.50 here and No. 1 Summerlee, $20.50 to $21.

BAR IRON — The market is unchanged at $1.75 per 100 lb. for Merchants' bar and $2 for horseshoe.

HOOP IRON—Prices are steadier than they were. The demand is moderate. Sales are being made at $2.75 for No. 19 gauge and $3 for No. 17.

BLACK SHEETS — Several brokers have received notification of a further advance this week, but the report is not general. Certainly the market is strong. We quote: 8 to 16 gauge, $2.50 to $2.60 ; 26 gauge, $2.55 to $2 65, and 28 gauge, $2.60 to $2 70.

GALVANIZED IRON — Prices are steady, and, although there have been some advances the market is said to have settled back to pretty much the same figures. The demand from the country is fairly good. We quote as follows : No. 28 Queen's Head, $4.65 ; Apollo, 10¾ oz., $4.50, and Comet, $4.40 to $4.45, with a 15c. reduction for case lots.

COPPER—Values are steady at about 18c.

INGOT TIN — The market is advancing, and quotations here are 1c. higher at 32 to 33c.

LEAD — The lead market is rather firm. The ruling quotation is $3.75 to $3.85.

LEAD PIPE—Fair quantities are selling at the reduced price. We quote: 7c. for ordinary and 7½c. for composition waste, with 30 per cent. off.

IRON PIPE—There have been no new developments in the upward tendency of the market. The demand is good. We quote as follows: Black pipe, ¼, $3 per 100 ft.; ⅜, $3; ½, $3 05; ¾, $3.30; 1-in., $4.70; 1¼, $6.40; 1½, $7.70; 2-in. $10 25. Galvanized, ½, $4.60; ¾, $5.25; 1-in., $7.50; 1¼, $9.80; 1½, $11.75; 2 in., $16.

TINPLATES—Owing to the advance in pig tin, tinplates are cabled 4½d. higher. The incoming supplies are very light. We quote: Coke plates, $3.75 to $4; charcoal, $4 25 to $4.50; extra quality, $5 to $5.10.

CANADA PLATE—Quotations are up 3s.6d. per ton and quotations here are very firm, although some houses are selling below our schedule prices. We quote: 52's, $2.55 to $2.60; 60's, $2.65 to $2.70; 75's, $2.70 to $2.80; full polished, $3.10, and galvanized, $4.

STEEL—Unchanged. We quote: Sleigh-shoe, $1.95; tire, $2; bar, $1.95; spring, $2.75; machinery, $2 75, and toe-calk, $2.50.

SHEET STEEL—We quote: Nos. 22 and 24, $3. and Nos. 18 and 20, $2.85.

TOOL STEEL—Black Diamond, 8c. and Jessop's, 13c.

TERNE PLATES—Supplies are none too plentiful even yet, and, at times, there is difficulty found in filling orders. Business is rather quiet. The English market is decidedly firm. From stock, goods are worth $7.50.

COIL CHAIN—Unchanged. We quote as follows: No. 6, 11½c.; No. 5, 10c.; No. 4, 9½c.; No. 3, 9c.; ¼-inch, 7½c. per lb.; 5-16, $4.85; 5-16 exact, $5.30; ⅜, $4.40; 7-16, $4.20; ½, $3.95; 9-16, $3.85; ⅝, $3.55; ¾, $3.45; ⅞, $3.40; 1-in., $3.35. In carload lots an allowance of 10c. is made.

SHEET ZINC—At primary points sheet zinc is £1 per ton higher, and quotations here have been raised in some instances in sympathy. Small lots are quoted at $6.25.

ANTIMONY—Quiet, at 10c.

ZINC SPELTER—Is worth 5c.

SOLDER—We quote: Bar solder, 18½c.; wire solder, 20c.

GLASS.

In sympathy with the strong foreign markets, glass has been advanced 10c. per 50 ft. on this market. We quote as follows: First break, $2; second, $2.10 for 50 feet; first break, 100 feet, $3.80; second, $4; third, $4.50; fourth, $4.75; fifth, $5.25; sixth, $5.75, and seventh, $6.25.

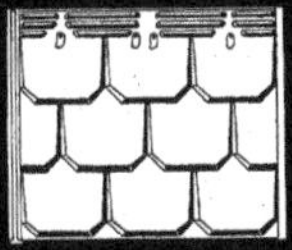

PAINTS AND OILS.

There is not much change to report. A good, healthy feeling prevades the paint and oil trade, and most of the manufacturers report a heavy business. In fact, we hear of complaints from the retail and jobbing trade that they are unable to obtain supplies promptly. The market is utterly bare of linseed oil, but it is hoped that the famine will be relieved next week as oil afloat is expected to hand. Quotations are 3c. higher. Shingle stains and bridge and roof paints are in good request, while ready-mixed colors seem to be still selling well into consumers' hands, as large orders have been received from all parts of the country. Ground white lead is in ample supply, and fair sales are reported. Turpentine is moving out freely and the price is steady. We quote:

WHITE LEAD—Best brands, Government standard, $6.25; No. 1, $5.87½; No. 2, $5.50; No. 3, $5.12½, and No. 4, $4.75 all f.o.b. Montreal. Terms, 3 per cent. cash or four months.

DRY WHITE LEAD—$5.25 in casks; kegs, $5.50.

RED LEAD—Casks, $5.00; in kegs, $5.25.

DRY WHITE ZINC—Pure, dry, 6¾c.; No. 1, 5¾c.; in oil, pure, 7¾c.; No. 1, 6¾c.; No. 2, 5¾c.

PUTTY—We quote: Bulk, in barrels, $1.90 per 100 lb.; bulk, in less quantity, $2.05; bladders, in barrels, $2.10; bladders, in 100 or 200 lb. kegs or boxes, $2.25; in tins, $2.55 to $2.65; in less than 100-lb. lots, $3 f.o.b. Montreal, Ottawa, Toronto, Hamilton, London and Guelph. Maritime Provinces 10c. higher, f.o.b. St. John and Halifax.

LINSEED OIL—Raw, 80c.; boiled, 83c. in 5 to 9 bbls., 1c. less, 10 to 20 bbl. lots, open, net cash, plus 2c. for 4 months. Delivered anywhere in Ontario between Montreal and Oshawa at 2c. per gal. advance and freight allowed.

TURPENTINE—Single bbls., 53c.; 2 to 4

bbls , 52c.; 5 bbls. and over, open terms, the same terms as linseed oil.

Mixed Paints—$1.25 to $1.45 per gal.

Castor Oil—8¼ to 9¼c. in wholesale lots, and ½c. additional for small lots.

Seal Oil—47½ to 49c.

Cod Oil—32½ to 35c.

Naval Stores — We quote : Resins, $2.75 to $4.50, as to brand ; coal tar, $3.25 to $3.75 ; cotton waste, 4½ to 5½c. for colored, and 6 to 7½c. for white ; oakum, 5½ to 6½c., and cotton oakum, 10 to 11c.

Paris Green—Petroleum barrels, 16¼c. per lb.; arsenic kegs, 17c.; 50 and 100. lb. drums, 17½c.; 25-lb. drums, 18c.; 1-lb-packages, 18½c.; ½-lb. packages, 20½c.; 1-lb. tins, 19½c.; ½-lb. tins, 21½c. f.o.b. Montreal; terms 3 per cent. 30 days, or four months from date of delivery.

SCRAP METALS.

The scrap metal market is firm and active. Rags are easier and rubbers somewhat higher. Dealers are now paying the following prices in the country : Heavy copper and wire, 13 to 13½c. per lb.; light copper, 12c.; heavy brass, 12c.; heavy yellow, 8½ to 9c.; light brass, 6½ to 7c.; lead, 2½ to 2¼c. per lb.; zinc, 2¼ to 2½c.; iron, No. 1 wrought, $14 to $16 per gross ton f.o.c. Montreal; No. 5 cast, $13 to $14; stove plate, $8 to $9; light iron, No. 2, $4 a ton; malleable and steel, $4; rags, country, 60 to 70c. per 100 lb.; old rubbers, 7¼c. per lb.

HIDES.

The prices are steady under an easy demand. We quote : Light hides, 6½c. for No. 1; 5½c. for No. 2, and 4½c. for No. 3. Lambskins, 10c.; sheepskins, 90c.; calfskins, 8c. for No. 1 and 6c. for No. 2.

PETROLEUM.

There is no change. We quote : "Silver Star," 14½ to 15½c.; "Imperial Acme," 16 to 17c.; "S.C. Acme," 18 to 19c., and "Pratt's Astral," 18½ to 19½c.

ONTARIO MARKETS.

Toronto, June 1, 1901.

HARDWARE.

THE unfavorable weather is naturally affecting business in a good many lines, particularly in gas and oil stoves, ice cream freezers and refrigerators. It does not follow, however, that even in these lines trade is dull. On the contrary there is quite a nice volume of business passing. This can be said of hardware generally. There is still a scarcity of barb wire and plain galvanized wire. Oiled and annealed wire, as before, is getting the benefit of it. A fair business is being done in wire nails in small lots. The scarcity in shingle nails noted in previous issues has been overcome. Horse nails and horseshoes are seasonably quiet. The demand for harvest tools is good, and a large trade is looked for this season. A moderate quantity of spades and shovels are going out. Trade in screws, bolts and nuts and rivets and burrs continues fairly brisk. Rope is still going out in fair quantities. Just a moderate business is being done in screen doors and windows, and trade is steady in green wire cloth. Business is fair in lawn mowers. A scarcity has already being experienced in some sizes. An active trade is reported in building paper.

Barb Wire—The situation in barb wire is much about the same as a week ago. No fresh shipments have come forward as far as we can learn. The scarcity is becoming even more pronounced, and jobbers report that it will probably be another two or three weeks before they will be able to get their orders filled. Prices are unchanged. We quote $3.05 per 100 lb. from stock Toronto; f.o.b Cleveland $2.82½ for less than carlots, and $2.70 for carlots.

Galvanized Wire—This also continues scarce and the demand good. We quote as follows : Nos. 6, 7 and 8, $3.50 to $3.85 per 100 lb., according to quantity ; No. 9, $2.85 to $3.15 ; No. 10, $3.60 to $3.95 ; No. 11, $3.70 to $4.10 ; No. 12, $3 to $3.30 ; No. 13, $3.10 to $3.40 ; No. 14, $4.10 to $4.50 ; No. 15, $4.60 to $5.05 : No. 16, $4.85 to $5.35. Nos. 6 to 9 base f.o.b. Cleveland are quoted at $2.57½ in less than carlots and 12c. less for carlots of 15 tons.

Smooth Steel Wire — Oiled and annealed wire is still deriving benefit from the scarcity of barb and galvanized wires, and, while the demand continues active, the supply keeps ample. The net selling prices for oiled and annealed are : Nos. 6 to 8, $2.90; 9, $2.80; 10, $2.87; 11, $2.90 ; 12, $2.95 ; 13, $3.15 ; 14, $3.37 ; 15, $3.50 ; 16, $3.65. Delivery points, Toronto, Hamilton, London and Montreal, with freights equalized on these points.

Wire Nails — Trade is fairly good, although the quantities wanted are of a sorting-up nature. We still quote $2.85 base for less than carlots, and $2.77½ for carlots. Delivery points : Toronto, Hamilton, London, Gananoque and Montreal.

Cut Nails—The supply of shingle nails, which has been short during the past couple of weeks, is now sufficient for the demand, and a fair business is being done in this particular kind. Outside shingle nails, the demand for cut nails is still dull. The base price is $2.35 per keg for less than carlots, and $2.25 for carlots. Delivery points : Toronto, Hamilton, London, Montreal and St. John, N.B.

Horse Nails — Business is only small as usual at this time of the year. Discount on "C" brand, oval head, 50

and 7½ per cent. off new list, and on "M" and other brands, 50, 10 and 5 per cent. off the old list. Countersunk head 60 per cent.

HORSESHOES—A small steady trade is being done. We quote f. o. b. Toronto: Iron shoes, No. 2 and larger, light, medium and heavy, $3.60; snow shoes, $3.85; light steel shoes, $3.70; featherweight (all sizes), $4.95; iron shoes, No. 1 and smaller, light, medium and heavy (all sizes), $3.85; snow shoes, $4; light steel shoes, $3.95; featherweight (all sizes), $4.95.

SCREWS — These are moving in good quantities, as they have been for some time. We quote discounts: Flat head bright, 87½ and 10 per cent.; round head bright, 82½ and 10 per cent.; flat head brass, 80 and 10 per cent.; round head brass, 75 and 10 per cent.; round head bronze, 65 per cent., and flat head bronze at 70 per cent.

RIVETS AND BURRS — The demand for rivets keeps good, as the boilermakers are all well employed. We quote: Iron rivets, 60 and 10 per cent.; iron burrs, 55 per cent.; copper rivets and burrs, 35 and 5 per cent.

BOLTS AND NUTS—The demand for bolts keep good, and there is still a particularly good business being done in special bolts. We quote: Carriage bolts (Norway), full square, 65 per cent.; carriage bolts full square, 65 per cent.; common carriage bolts, all sizes, 60 per cent.; machine bolts, all sizes, 60 per cent.; coach screws, 70 per cent.; sleighshoe bolts, 72½ per cent.; blank bolts, 60 per cent.; bolt ends, 62½ per cent.; nuts, square, 4c. off; nuts, hexagon, 4½c. off; tire bolts, 67½ per cent.; stove bolts, 67½; plough bolts, 60 per cent.; stove rods, 6 to 8c.

ROPE—Business continues good in both sisal and manila rope, and prices are steady and unchanged, at 10c. for sisal, and 13½c. for manila.

CUTLERY—A fairly good sorting-up business is reported this week.

SPORTING GOODS — The season is beginging to open up a little better. There is quite a demand for ball and shot shells.

SHOT—On account of the lower prices in raw material a reduction has been made in prices, the discount off the list now being 17½ per cent., instead of 15 per cent.

LAWN MOWERS—A good trade is being done in lawn mowers, and already there is a scarcity in some sizes. We quote: "Star" mowers, $2.35; discount on Woodyatt mowers, 40 and 10 per cent. off list.

ENAMELLED WARE AND TINWARE—The demand in both these lines is just fair, there still being no recovery from the recent quietness. An improvement is looked for during the coming month.

OIL STOVES—Trade is not as brisk as dealers would desire, and no improvement is looked for as long as the present unseasonable weather lasts.

REFRIGERATORS—There have been quite a number of inquiries this week and a fairly good trade is being done, particularly in view of the unseasonable weather.

ICE CREAM FREEZERS—There are a few of these going out, but the demand is not as active as it would be were the weather more seasonable.

GREEN WIRE CLOTH—There is a steady trade being done, although no great rush is being experienced. We quote $1.35 per 100 sq. ft.

SCREEN DOORS AND WINDOWS—Jobbers report that some shipments have come to hand, and a good many orders have been filled, but trade throughout the country has not, so far, been as heavy as the trade would desire. We quote: Screen doors, 4 in. styles, $7.20 to $7.80 per doz.; ditto, 3 in. styles, 20c. per doz. less; screen windows, $1.60 to $3.60 per doz., according to size and extension.

BUILDING PAPER — Trade is active and prices steady and unchanged. We quote: Building paper, 30c.; tarred lining, 40c., and tarred roofing, $1.65.

POULTRY NETTING — Stocks in this line are getting light, but the season is pretty well over as far as the wholesale trade is concerned. Discount is steady at 55 per cent.

HARVEST TOOLS—Stocks in the hands of the wholesale trade are heavy, but they are so on account of the fact that a heavy trade is expected in scythes, snaths, cradles and forks. It is now almost certain that the crop of hay will be a heavy one, and jobbers have supplemented their stocks in anticipation of it. Discount, 50, 10 and 5 per cent.

SPADES AND SHOVELS — A fairly good trade of a sorting-up nature is reported. Discount, 40 and 5 per cent.

EAVETROUGH —Shipments of eavetrough and conductor pipe continue heavy at the low prices which are ruling.

CEMENT—As the building of granolithic walks and cement cellars is brisk, a big movement of cement is recorded. We quote barrel lots: Canadian portland, $2.25 to $2.75; German, $3 to $3.15; English, $3; Belgian, $2.50 to $2.75; Canadian hydraulic, $1.25 to $1.50.

METALS.

The metal trade continues fairly active. About the only change in quotations is an advance of 1c. per lb. in pig tin in sympathy with the outside markets. The demand for ingot metals and galvanized sheets continues active. The feature of the tinplate market is its firmness.

PIG IRON—The market is dull and prices are nominal. One of the features of the pig iron market is the reaction in Great Britain, there having been some declines there during the past week.

BAR IRON—The demand has fallen off somewhat, but the mills have still a large numbers on hand. We still quote the base price at $1.85 to $1.90 per 100 lb.

STEEL—The steel trade is still active and prices firm. We quote as follows: Merchantable cast steel, 9 to 15c. per lb.;

drill steel, 7 to 8c. per lb.; "Black Diamond" tool steel, 10 to 11c.; Jessop's and Firth's tool steel, 13c.; toe calk steel, $2.85 to $3; tire steel, $2 30 to $2.50; sleighshoe steel, $2.10 to $2.25; reeled machinery steel, $3; hoop steel, $3 10.

PIG TIN—The market in the early part of the week showed a great deal of strength both in London and New York, and there were some sharp advances. Towards the middle of the week, however, there was some reaction, but at the time of writing there has been a recovery, and the market is firm at the moment. In sympathy with the outside markets, prices here are higher, Lamb and Flag now being quoted at 32c. per lb.

TINPLATES—The position of the market is a strong one, and, according to advices received this week by importers, it would cost importers 10d. per box more than a week ago to purchase in the British market. The demand here for tinplates has not been as heavy during the past week as it was at the time of our last review.

TINNED SHEETS—Wholesalers have been sending out quite a few shipments of tinned sheets, and also trimmings, during the past week. We still quote 9¼ to 9½c. per lb.

GALVANIZED SHEETS—The demand during the past week has been active and stocks are becoming more reduced. As pointed out last week the import lots have come to hand for customers, but jobbers' stocks have not yet arrived. Case lots of English sheets are quoted at $4 50, and less than case lots at $4 65. American sheets are quoted at $4 40.

BLACK SHEETS—There is a fair business to report. The ruling price for 28 gauge is $3. Through an inadvertence the price last week was given as $2.30.

CANADA PLATES—Very few of these are going out, as is, of course, to be expected at this time of the year. We quote: All dull, $2.90; half polished, $3, and all bright, $3.50.

COPPER—A better tone obtains on the English markets this week according to cable advices, and on Wednesday the market advanced 2s. 6d. and closed firm. The New York market, on the other hand, is dull and uninteresting. Locally, we quote: Ingot, 19c.; bolt or bar, 23½ to 25c.; sheet, 23 to 23½c. per lb.

SOLDER—The demand is fairly good, as is usually the case when galvanized sheets are active. We quote: Half-and-half, guaranteed, 18½c.; ditto, commercial, 18c.; refined, 18c., and wiping, 17c.

IRON PIPE—The situation is much as before. The market is firm as to price, and there is not a great deal of business being done. We quote: 1-inch black pipe, $5.15 per 100 ft., and 1-inch galvanized, $7.55.

LEAD—A rather stronger feeling exists in the outside markets. Locally, there is no change. We quote 4¼ to 4½c. per lb.

ZINC SPELTER—The market is quiet and rather easy. We quote 5¾ to 6c. per lb.

ZINC SHEET—Trade is rather quiet, with prices unchanged at 6½c. for casks and 6¾c. for part casks.

PAINTS AND OILS.

May has been, like April, one of the busiest months in years. "In fact," said one dealer, "April and May have been so busy that, though the early months of the year were quite dull, we have done much more trade in the past half year than during any similar six months for some time." All lines are in good demand. Paris green, which has been the poor seller, is in better request than formerly. Red lead is easier in prices. All other lines are firm, with a possibility of further advances in linseed oil. We quote:

WHITE LEAD—Ex Toronto, pure white lead, $6.37½; No. 1, $6; No. 2, $5.67½; No. 3, $5.25; No. 4, $4.87½; genuine dry white lead in casks, $5.37½.

RED LEAD—Genuine, in casks of 560 lb., $5.50; ditto, in kegs of 100 lb., $5.75; No. 1, in casks of 560 lb., $5; ditto kegs of 100 lb., $5 25.

LITHARGE—Genuine, 7 to 7½c.

ORANGE MINERAL—Genuine, 8 to 8½c.

WHITE ZINC—Genuine, French V.M., in casks, $7 to $7.25; Lehigh, in casks, $6.

PARIS WHITE—90c. to $1 per 100 lb.

WHITING—70c. per 100 lb.; Gilders' whiting, 80c.

GUM SHELLAC—In cases, 22c.; in less than cases, 25c.

PARIS GREEN—Bbls., 16½c.; kegs, 17c.; 50 and 100 lb. drums, 17½c.; 25-lb. drums, 18c.; 1-lb. papers, 18½c.; 1-lb. tins, 19½c.; ½-lb. papers, 20½c.; ½-lb. tins, 21½c.

PUTTY—Bladders, in bbls., $2.10; bladders, in 100 lb. kegs, $2.25; bulk in bbls., $1.90; bulk, less than bbls. and up to 100 lb., $2.05; bladders, bulk or tins, less than 100 lb., $2.90.

PLASTER PARIS—New Brunswick, $1.90 per bbl.

PUMICE STONE—Powdered, $2.50 per cwt. in bbls., and 4 to 5c. per lb. in less quantity; lump, 10c. in small lots, and 8c. in bbls.

LIQUID PAINTS—Pure, $1.20 to $1.30 per gal.

CASTOR OIL—East India, in cases, 10 to 10½c. per lb. and 10½ to 11c. for single tins.

LINSEED OIL—Raw, 1 to 4 barrels, [illegible]; boiled, 84c.; 5 to 9 barrels, raw, 80c.; boiled, 83c., delivered. To Toronto, Hamilton, Guelph and London, 1c. less.

TURPENTINE—Single barrels, 54c.; 2 to 4 barrels, 53c., delivered. Toronto, Hamilton and London 1c. less. For less quantities than barrels, 5c. per gallon extra will be added, and for 5-gallon packages, 50c., and 10 gallon packages, 80c. will be charged.

GLASS.

Notwithstanding the settlement of the strike of the Belgian glassmakers and the consequent increase in supplies for fall delivery, there is a stiff feeling in regard to prices for immediate delivery from stock, and there is still some talk of an advance. We quote as follows: Under 26 in., $4 15; 26 to 40 in., $4.45; 41 to 50 in., $4.85; 51 to 60 in., $5.15; 61 to 70 in., $5.50; double diamond, under 26 in., $6; 26 to 40 in., $6.65; 41 to 50 in., $7.50; 51 to 60 in., $8.50; 61 to 70 in., $9.50, Toronto, Hamilton and London. Terms, 4 months or 3 per cent. 30 days.

OLD MATERIAL.

There is a fair movement. Prices are steady. We quote jobbers' prices: Agricultural scrap, 55c. per cwt.; machinery cast, 60c. per cwt.; stove cast, 50c.; No. 1 wrought 50c. per 100 lb.; new light scrap copper, 12c. per lb.; bottoms, 11½c.; heavy copper, 13c.; coil wire scrap, 13c.; light brass, 7c.; heavy yellow brass, 10 to 10½c.; heavy red brass, 10½ to 11c.; scrap lead, 3c.; zinc, 2c; scrap rubber,

6½c.; good country mixed rags, 65 to 75c.; clean dry bones, 40 to 50c. per 100 lb.

HIDES, SKINS AND WOOL.

HIDES—The market keeps dull, with prices easy at unchanged figures. We quote: Cowhides, No. 1, 6½c.; No. 2, 5½c.; No. 3, 4½c. Steer hides are worth 1c. more. Cured hides are quoted at 7 to 7½c.

SKINS—The market is firm, but there is no great deal doing. We quote: No. 1 veal, 8-lb. and up, 9c. per lb.; No. 2, 8c.; dekins, from 60 to 70c.; culls, 20 to 25c. Sheepskins, 90c. to $1.

WOOL—While this year's wool is being clipped the demand keeps dull, last year's clip still being in Canadian dealers' hands. We quote: Combing fleece, washed, 13c., and unwashed, 8c.

COAL.

A good demand is reported, but, owing to the scarcity of coal and cars, local shippers are behind in their orders. Prices for June shipments are 10c. higher than was noted during May. We quote at international bridges as follows: Grate, $4.75 per gross ton; egg, stove and nut, $5 per gross ton with a rebate of 30c. off for June shipments.

MARKET NOTES.

Pig tin is 1c. higher at 32c.

The discount on shot has been increased to 17½ per cent. It was formerly 15 per cent.

NEW PRODUCT FOR PRESERVING IRON.

New Caledonia, the French convict settlement, has long been known to be richly stored with minerals, but what development has taken place—which is little—in these islands has been effected almost entirely by English capital and English enterprise, says The London Express. The French, however, are stated to have made recently a somewhat singular discovery. Scattered about the Island of Noumeo are deposits of a sort of mineral ochre, which possesses characteristics unlike those of any known substance. The mineral contains some 65 per cent. of proxide of iron, six of alumina, six of sesquioxide of chrome, besides proportions of cobalt, nickel, silica, chalk and magnesia. It is asserted that this substance, applied to the surface in the form of paint, will preserve ironwork from rusting. Should this be the case, Nature will accomplish that which has exercised the fruitless toil of a multitude of chemists, and engineers, henceforward, may be freed from one of their most dangerous and insidious enemies.

Jewel & Co., Toronto, have taken an option on the Grimsby Agricultural Works, Grimsby, Ont.

F. H. Clergue has organized a new company, the British American Express Co., which will do business on the Clergue steamers and on the Algoma Central and the Manitoulin and North Shore railways.

HEATING AND PLUMBING

ALUMINUM-ZINO ALLOYS.

MANY contradictory statements have appeared regarding the properties of aluminum-zinc alloys; and since these statements were all apparently made in good faith, the divergent accounts are probably to be ascribed to the use of impure aluminum or impure zinc, or to a lack of the knowledge of how to properly alloy the pure metals. In fact, the experience gained in the practice of alloying enables one to continually obtain better and better results, such as are wholly unattainable to the unskilled metallurgist, writes Dr. Joseph W. Richards in Aluminum World.

Tissier Brothers, in the first book written about aluminum, describe the alloys of one part of aluminum with 2, 3 and 10 parts of zinc. They found these to be brittle, looking like zinc, fine grained and more fusible than aluminum, but less so than zinc. These alloys were used at first in attempts to solder aluminum, but they would not run liquid, and made poor looking joints. Aside from this, the alloys had no mechanically useful properties. My own experience confirms these statements. Out of eight alloys, which I have made, containing 2, 3, 4, 5, 6, 7, 8 and 9 parts of zinc to 1 part of aluminum, with specific gravity from 4 1-2 to 6, there appears to be none with any specially valuable mechanical properties. Even the alloy of 1 part aluminum to 1 part of zinc, with a specific gravity of 4, may be classed with the above alloys. Alloys containing more aluminum than zinc first commence to possess valuable properties.

The alloy of 2 aluminum to 1 of zinc, containing 33 1-3 per cent. of zinc, is a remarkable alloy. W. F. Durand has described it (Science, 1897, p. 396) as being "equal to cast iron in strength, melts about 800 degrees F. (425 degrees C.), does not readily oxidize, takes a fine finish, perfectly fills the joints of the mold, and is, like cast iron, brittle, but resists corrosion well." After several years' experience with this alloy, commencing in 1895, I can substantiate the general correctness of Durand's statements, and add to them the following observations:

Experience in making this alloy has led to considerable improvement in the results obtained. The first samples made were just about equal to cast iron in stength, tests showing 18,000 to 24,000 pounds breaking strain per square inch, with no perceptible elongation; but at present there is no difficulty in reaching a tensile strength of 40,000 pounds, in castings. It resembles closely in its characteristics a high carbon steel, being extremely rigid, slightly elastic and breaking short with a fine grained fracture. It works well under the tools, in turning or boring, not requiring lubrication. It is the hardest and strongest of the valuable alloys of aluminum and zinc, takes a high polish and keeps its color very well. It is not so resistant to shock as the other alloys containing less zinc. Its specific gravity is 3.8, and calculation shows the remarkable contraction of 17 per cent. taking place during the alloying of its ingredients, which suggests the cause of its great strength. Numerous uses for this strong, rigid alloy will suggest themselves to every mechanical engineer.

The alloy of 3 parts aluminum to 1 part zinc, containing 25 per cent. of zinc, is the most generally useful of all the aluminum-zinc alloys. It is softer than the previous alloy, has a tensile strength of 35,000 pounds per square inch, and elastic limit nearly the same, with a slight elongation before breaking. It is, therefore, not a malleable alloy, but yet it is not brittle, for it bends slightly before breaking. This quality is a valuable one, for it enables one to straighten out a casting to a certain extent under the hammer. Remarkably clean and sharp castings can be made with it, when experience has been obtained with the proper gating of the mold and the exact temperature of casting. As with all these alloys, overheating in the crucible must be scrupulously avoided, as well as the use of iron-stirring implements, since oxide or dross does not separate out of the metals easily, and may thus get poured into the mold and injure the casting. Its specific gravity is 3.4, and the contraction taking place during alloying is, therefore, 14 per cent., which indicates a close and intimate combination. This alloy, when properly made from the pure metals, is all that could be desired in its working qualities, being equal to the finest brass in the lathe, under the drill, and in not clogging the file. It casts sound, takes a high polish, and has as fine a color as the best aluminum. It is not so hard and short as the 33 per cent. zinc alloy, nor quite so strong, but has supplanted it for most purposes because of its better working qualities and greater relinbility under shock. It is at present in use for scale beamd, surveying and astronomical instruments, light machine parts, gear wheels cut from blanks, cash registers, calculating machines, testing machines, indicator drums, surgical appliances, cases and parts of pneumatic tools, &c. It is nonmagnetic, and therefore particularly useful in scientific instruments. The use of this alloy is increasing rapidly, and it bids fair to be the most generally useful of the light aluminum-zinc alloys.

Below 25 per cent. of zinc the strength and hardness of the alloys decrease rapidly, and when the zinc is 15 per cent. or less the alloy can be forged, rolled or drawn. The alloy with 15 per cent. of zinc has in castings an elastic limit 16,000 pounds per square inch, tensile strength 22,330 pounds, elongation in 2 inches 6 per cent., and reduction of area 10 1-2 per cent. It can be rolled and drawn into wire if frequently annealed. Mechanical tests of this alloy after working have not yet been made.

The alloys with less than 15 per cent. of zinc become softer and weaker, roll and draw well, but require lubrication of the tools during working. They may find application for special purposes where the previously mentioned alloys are too hard.

THE ANNUAL CONVENTION.

The Toronto branch of The National Plumbers' and Steamfitters' Association of Canada are making preparations to give the visiting delegates a good time this year. Last week a joint meeting of committees representing the manufacturers and the masters of the city was held to consider plans. On Monday, the Masters' Association met and appointed W. J. McGuire, J. B. Fitzsimons and H. Hogarth representing delegates, with J. K. Allison, George Clapperton and R. Ross as substitutes. They expected a report from the sub-committee appointed to confer with the masters, but these were not ready to report, as another conference is to be held. It is likely, therefore, that the entertainment provided the delegates will possess somewhat of the nature of a surprise. In any case a good programme may be counted on. It has not yet been decided what hall the convention will be held in.

BUILDING PERMITS ISSUED.

Building permits have been issued in Vancouver, B.C., to A, Smith, dwelling, Alexander street, $1,150; W. A. Lightheart, Richards street, dwelling, $1,400; Oscar Bruce Allen, dwelling, Nelson street, $2,000; Joseph Paull, dwelling, Nelson street, $2,000; Mrs. M. Paterson, dwelling, Burrard street, $4,100.

A Winnipeg despatch of May 28 says that permits for new buildings to be erected in Winnipeg to that date number 244, aggregating $753,500, as against 151 for $454,959, for the same period last year.

Permits have been issued in Ottawa to Col. J. M. Wood, New York City, on behalf of The Russell Theatre Co., rebuilding and alterations to old theatre, $45,000; Dowd Milling Co., Limited, Broad street west, frame building, $1,000; E. L. Horwood, lot 15 Waverley street, brick house, $2,500; J. P. MacLaren, on behalf of Erskine Presbyterian church, Concession street, stone church, $8,000.

George Martel, lot 4 Murray street, repairs to hotel, $3,350; The Harris, Campbell & Boyden Company, lot 12 Queen street, addition to factory, $4,000; Bryson, Graham & Co., solid brick building as warehouse, lot 10 Queen street,, $12,000.

E. C. Grant, lot 7 Blackburn avenue, brick residence, $7,000.

Building permits have been issued in Toronto to James Crang, for a $1,700 dwelling on Granby avenue, near Avenue road; to J. F. Drown, for a $3,000 dwelling at 47 Queen street east; to P. Brady, for two houses at 71 and 73 Birch avenue, to cost $2,000; to Frank G. Gushingham, for a $2,300 dwelling on Pearson avenue, near Sorauren avenue; to A. A. Allan, for a $4,500 residence near Dunbar road, on Elm street, and to Wm. Greenway, for a $3,000 dwelling on Brunswick avenue, near Barton avenue.

PLUMBING AND HEATING CONTRACTS.

Purdy, Mansell & Co., Toronto, have the contracts for plumbing in a residence on Dovercourt road for F. G. McCraney, and for plumbing and heating in a residence on Walmer road, for Samuel Crane.

The Bennett & Wright Co., Limited, were awarded this week the contract for plumbing, heating, ventilating and lighting a house in Queen's Park for J. W. Flavelle, This house is to be the largest private residence in Toronto, it having a frontage of 178 feet., and an average depth of 50 ft., so this contract is a big one. Six bath rooms are called for, each to be fitted in the most luxurious style, five of them to include shower baths, and the sixth a fine

needle shower bath. Steam heating and a modern system of ventilation will be installed. The lighting will be by electricity, the contract calling for about 200 incandescent lights.

SOME BUILDING NOTES.

Dr. More, Brandon, Man., intends erecting a new house.

Silas Myers, Coleridge, Ont., is erecting a new brick house.

J. Moulton is building a new house in Prestonvale, Ont.

Nelson Teeter, Maple Grove, Ont., is building a brick house.

A new Episcopal church is to be erected at Millidgeville, N.B.

A new Anglican church will be erected at Cargill, Ont., this summer.

Gray, Cameron & Co. have received tenders for a new house in Wiarton.

Tenders were received on Tuesday for the new court house at Sydney, N.S.

W. Van Allan has started the construction of a new house in Flinton, Ont.

George Whetham has men engaged building a new residence in Sheffield, Ont.

The contract has been let for a new Presbyterian church at Bishop's Mills, Ont.

D. H. Waterbury is asking tenders before June 12, for an armoury at Sussex, N.B.

Thomas Hamilton, Clover Valley, Ont., intends erecting a brick house this summer.

A Roman Catholic church is being erected on the Couchiching Indian Reserve, Fort Frances, Ont.

George Browne, architect, Winnipeg, has received tenders for a bank building in Carberry, Man.

W. M. Dodd, architect, Calgary, N.W.T., is calling for tenders before June 8, for the erection of the Hall block, Calgary.

H. S. Griffith, architect, Winnipeg, has received tenders for a dwelling house for R. H. Myers, M.P.P., at Minnedosa, Man.

John Argue, David Johnston and Dr. Beattie and Mr. Christie will erect buildings at Parry Sound, Ont., this summer.

C. O'Connell, of The Tecumseth House, Winnipeg, has bought a lot on King street, Winnipeg, and intends erecting thereon an up-to-date hotel.

Joseph Bernhardt, of the Cosmopolitan Hotel, Winnipeg, has bought the Franklin house property on the northwest corner of Henry avenue and Main street, Winnipeg, and is going to erect a fine six-storey hotel.

Tenders are asked before June 8 by G. M. Miller & Co., architects, Toronto, for the erection of the large new shops for The Canada Foundry Co., Limited, corner of McKenzie avenue and Davenport road, Toronto.

PLUMBING AND HEATING NOTES.

Mattinson & Pope have registered as plumbers, etc., Montreal.

The Montreal Light, Heat and Power Co. has registered in Montreal.

The Provincial Light, Heat and Power Co. has been incorporated in Montreal.

Tenders are asked before June 10 for heating the Arthur, Ont., public school.

The by-law to raise $30,000 to extend the waterworks system has been carried at Owen Sound.

Mrs. Maxime Deslauriers has registered as proprietress of M. Deslauriers & Cie, contractors, Montreal.

The assets of Adelard Binette, plumber, Lachine, Que., have been sold, and Mrs. Binette has registereed as proprietress of the business.

HONESTY AND ECONOMY.

IT is entirely within the lines of different points of view for the employer to think that he may be paying his help too much and for the employe to think that he is receiving too little for the work he is doing. There can be no dispute that both sides have good points for contention and that both sides are more or less right.

The employer undoubtedly looks at the question from his point of view, and thinks he is always pursuing that course which is best for him and his business, yet that manner of looking at the question is seldom complete, and he is exceedingly apt to overlook the possibility of self injury by the attempt at self protection. The measure of responsibility attached to a position should surely be some sort of gauge for the worth of the work that is being done, and where trust is of necessity a part of the requirements, the deepest incentives to dishonesty should be removed. It is true that all business requires honest men, and no man can succeed if he is dishonest, but there are many places of employment where the temptations to dishonesty are nil, even for the underpaid employe, while other places filled by the equally underpaid employe, offer illimitable temptation for the making up of the deficiency by stealing.

Two cases of this kind have recently attracted public attention. One of them was the case of a bank cashier in an eastern town who defaulted for a considerable sum. The evidence was plainly against him and the judge was compelled to sentence him to a long term in prison, but in pronouncing the sentence this right-minded judge made this forcible statement:

"I only wish that the law permitted me to send along to prison with you every one of the bank directors, who through a long term of years, expected you to do your work, live respectably and becomingly, bring up a large family and be honest—all on a salary of $600."

The other case was in a western city. It was the sentencing, for stealing a tray of diamonds, of a young man who had been employed as salesman in a jewelry store and paid the munificent sum of $6 a week for his work and the temptation to be honest. Before pronouncing the sentence, the judge questioned the young man closely as to his salary and the necessities of his position, then he said in almost as many words, that the jeweler deserved to have his diamonds stolen in punishment for being so miserly and inhuman.

Under our own personal observation came a case of an old and faithful employe, who, for various stated reasons not necessary to tell here, had his weekly pay reduced to $6, and was expected to appear respectably in the store and at the same time support a family. For a year or two this was done through the aid of a little savings, but the time came when he had to steal to keep going. He took small amounts from his sales and falsified the checks sent to the cashier. Detection came, as it always will. There was no prosecution, but he lost his position and with it much of the good name he had previously borne.

In all these cases, how much of the responsibility for crime rested upon the employe and how much upon the employer?

If these employers could honestly enter the plea that their business would not warrant better pay, had they any right, as citizens, to longer engage in that business and submit temptation and aid and abet crime?

We must admit that if the grain of dishonesty is in a man he will steal more or less, whatever his salary, but we must in return contend that no excuse can excuse the responsibility of an employer who will take advantage of an employe by paying a salary that he knows is entirely inadequate to the requirements and responsibilities of the position. Extravagances will not excuse thefts of employes, nor can they be entered as a part of considerations in making engagements, yet the employer knows very well whether he is paying a price that is sufficient to maintain the employe in the manner which is expected of the position.

How many unwilling thieves are made of naturally honest people, and how many business profits are undermined through the false idea that business economy is attained by paying the smallest possible salaries to employes.—Drygoodsman, St. Louis.

FIXING THE PROFITS.

IN estimating the profits that a stock of goods should earn, too little attention is sometimes paid to the cost of carrying on the business, remarks Stoves and Hardware Reporter. This is especially the case with merchants who carry small stocks and are not adepts in the science of figuring. They confine their business management to the details of buying and selling, paying and collecting bills, keeping the store open and the stock in a more or less presentable shape. They have not the time, they think, to waste on figures when their attention is required for more important matters.

No matter how small or large a business may be, its success depends on a thorough understanding of the figures. Invoices must be compared with orders and carefully kept, a cost book be provided and each item be recorded, and every matter of expense be thoroughly guarded against if the business is to be preserved against loss. In figuring cost so as to determine the rate of profit, not a single known or ascertainable item should be omitted. It is generally a good plan for the owner to credit himself with a certain salary each month and to charge the amount against expense or merchandise account. When a credit business is done, a certain percentage should be provided for a loss on bad bills.

In figuring on the rate of profit, it is a common mistake to group all goods under one head and to add a fixed percentage for profit. Under this system, all goods are expected to realize the same percentage even though some are necessarily quick sellers and others comparatively slow. If a certain line is in good demand at a fair price, there is more money in moving it quickly for a reasonable profit than in holding it at a figure which the customers will not feel like paying. Slow selling goods of a staple character can stand a higher percentage of profit and their sale will compensate for the smaller advance over cost made on others. In this way an average rate of profit can be struck and the result will be more satisfactory than if it had been attempted by fixing a uniform rate on each separate line.

Nevertheless, each line should be made to bear its proper burden of expense. This can be determined by the amount of sales for any previous period and by dividing the different lines into departments. Conditions vary from time to time and in this case a rule that was formerly correct may not have a proper or profitable application, but by comparing them together and striking an average it will not be difficult to determine the share of expense which each line should bear and where this has been done the percentage of profit in each case can be easily ascertained.

THE S. L. & G. H. ROGERS CO.

Manufacturers of

Sterling Silver and Silver Plated Ware

P. O. BOX 1205. HARTFORD, CONN.

Factory at Hartford, Conn.

Factory at Wallingford, Conn.

SIMEON L. ROGERS, Son of the late Simeon S. Rogers.

The late SIMEON S. ROGERS, One of the three original Rogers Brothers.

GEORGE H. ROGERS, Son of the late Simeon S. Rogers.

THE ONLY REAL ROGERS BROS. IN THE BUSINESS.

We guarantee our Extra Plate to Strip.

On Tea Spoons, 48 dwts. per gross.
On Dessert Spoons and Forks, 72 dwts. per gross.
On Table Spoons and Medium Forks, 96 dwts. per gross.

We guarantee our Medium Knives to Strip.

On No. 12, 12 dwts. of Silver per dozen.
On No. 14, 14 dwts. of Silver per dozen.
On No. 16, 16 dwts. of Silver per dozen.

All Other Extra-Plate Goods are Guaranteed to have the same proportionate Plate, which is 20 per cent. above Standard Plate.

All our Spoons, Forks, etc., are plated upon 18 per cent. NICKEL SILVER; the Best Base for Plating.

XII OR SECTIONAL PLATE.

☞ All goods plated by our process receive an extra coating of Silver three times the usual thickness upon the parts most exposed to wear, rendering our XII Goods equal to Triple Plate, and adding three times to the durability of goods so plated. The great advantage of our XII Plate will be readily seen, as Spoons and Forks always wear through first on parts exposed to wear, while the plate is yet good upon all other parts of the article. We put our XII or Sectional Plate upon all our Extra, Double and Triple Plates.

Not In the Trust, and no connection with any other Rogers Goods.

See that the Stamp on your Ware is the **Initial ROGERS.**

MANITOBA MARKETS.

WINNIPEG, May 27, 1901.

HARDWARE AND PAINTS, OILS AND GLASS.

THERE are no new features to report this week. Business is brisk, but not out of the ordinary. Of interest to the building trade is the drop of 5c. per roll of both plain and tarred "Anchor" building paper.

Linseed oil has again shown a sharp advance, being now quoted at 90 and 93c., while turpentine shows a steady weakening, the market now being down to 61 and 66c. as against 62 and 71c. last week. Glass is firm, with considerable demand.

Quotations for the week are as follows:

Barbed wire, 100 lb	$3 45
Plain twist	3 45
Staples	3 95
Oiled annealed wire 10	3 95
" 11	4 00
" 12	4 05
" 13	4 20
" 14	4 35
" 15	4 45
Wire nails, 30 to 60 dy, keg	3 40
" 16 and 20	3 45
" 10	3 50
" 8	3 60
" 6	3 65
" 4	3 80
" 3	4 05
Cut nails, 30 to 60 dy.	2 90
" 20 to 40	2 95
" 10 to 16	3 00
" 8	3 05
" 6	3 10
" 4	3 20
" 3	3 55
Horsenails, 45 per cent. discount.	
Horseshoes, iron, No. 0 to No 1	4 65
No. 2 and larger	4 40
Snow shoes, No. 0 to No. 1	4 90
No. 2 and larger	4 40
Steel, No. 0 to No. 1	4 95
No. 2 and larger	4 70
Bar iron, $2.50 basis.	
Swedish iron, $5.00 basis.	
Sleigh shoe steel	3 00
Spring steel	3 25
Machinery steel	3 75
Tool steel, Black Diamond, 100 lb	8 50
Jessop	13 00
Sheet iron, black, 10 to 20 gauge, 100 lb.	3 50
20 to 26 gauge	3 75
28 gauge	4 00
Galvanized American, 16 gauge	2 54
18 to 22 gauge	4 50
24 gauge	4 75
26 gauge	5 00
28 gauge	5 25
Genuine Russian, lb	12
Imitation " "	8
Tinned, 24 gauge, 100 lb	7 75
26 gauge	8 00
28 gauge	8 50
Tinplate, IC charcoal, 20 x 28, box	10 75
" IX "	12 75
" IXX "	14 75
Ingot tin	33
Canada plate, 18 x 21 and 18 x 24	3 25
Sheet zinc, cask lots, 100 lb	7 50
Broken lots	8 00
Pig lead, 100 lb	6 00
Wrought pipe, black up to 2 inch	50 an 10 p.c.
" Over 2 inch	50 p.c.
Rope, sisal, 7-16 and larger	$11 00
" ⅜	11 50
" ¼ and 5-16	12 00
Manila, 7-16 and larger	14 00
" ⅜	14 50
" ¼ and 5-16	15 00
Solder	20
Cotton Rope, all sizes, lb	17
Axes, chopping	$7 50 to 12 00
" double bitts	12 00 to 18 00
Screws, flat head, iron, bright	87½
Round " "	82½
Flat " brass	80
Round " "	75
Coach	57½ p.c.
Bolts, carriage	55 p.c.
Machine	55 p.c.
Tire	60 p.c
Sleigh shoe	65 p.c.
Plough	40 p.c.
Rivets, iron	50 p.c.
Copper, No. 8	35
Spades and shovels	40 p.c.
Harvest tools	50, and 10 p.c.
Axe handles, turned, s. g. hickory, doz.	$2 50
No. 1	1 50
No. 2	1 25
Octagon extra	1 75
No. 1	1 25
Files common	70, and 10 p.c.
Diamond	60
Ammunition, cartridges, Dominion R.F.	50 p.c.
Dominion, C.F., pistol	30 p.c.
" military	15 p.c.
American R.F.	30 p.c.
C.F. pistol	5 p.c.
C.F. military	10 p.c. advance.
Loaded shells:	
Eley's soft, 12 gauge black	16 50
chilled, 12 guage	18 00
soft, 10 guage	21 00
chilled, 10 guage	23 00
Shot, Ordinary, per 100 lb	6 25
Chilled	6 75
Powder, F.F., keg	4 75
F.F.G.	5 00
Tinware, pressed, retinned	75 and 2½ p.c.
" " plain	70 and 15 p.c.
Graniteware, according to quality	50 p.c.

PETROLEUM.

Water white American	25½c.
Prime white American	24c.
Water white Canadian	[illegible]
Prime white Canadian	21c.

PAINTS, OILS AND GLASS.

Turpentine, pure, in barrels	$ 61
Less than barrel lots	66
Linseed oil, raw	90
Boiled	93
Lubricating oils, Eldorado castor	25½
Eldorado engine	24¾
Atlantic red	27½
Renown engine	41
Black oil	23½ to 25
Cylinder oil (according to grade)	55 to 74
Harness oil	61
Neatsfoot oil	$1 00
Steam refined oil	85
Sperm oil	1 50
Castor oil per lb.	11½
Glass, single glass, first break, 16 to 25 united inches	2 25
26 to 40 per 50 ft.	2 50
41 to 50 " 100 ft.	5 50
51 to 60 " " "	6 00
61 to 70 per 100-ft. boxes	6 50
Putty, in bladders, barrel lots per lb.	2½
kegs "	2¾
White lead, pure per cwt.	7 00
No 1 "	6 75
Prepared paints, pure liquid colors, according to shade and color per gal.	$1.30 to $1.90

NOTES.

This week work has again been resumed on the St. Andrew's locks on the Red River.

A new hardware store is reported at Margaret, Manitoba, the proprietor being Mr. J. Meldrum.

The decline in crude Canadian petroleum has not yet affected the price of the refined oil on this market, although a drop of ½c. per gallon has been reported from Ontario.

W. A. Hayward, general merchant, Coldstream, N.B., has closed up his business.

W. D. Yoe, general merchant, Exeter, Ont., will move into larger premises shortly.

McDonald & Hanrahan's general store at Sydney, N.S., was destroyed by fire on Saturday last.

HENRY ROGERS, SONS & CO.,

SHEFFIELD, ENGLAND

Manufacturers of the celebrated

H.R.S.&Co

UNION JACK

CUTLERY

We make a specialty of

PLATED WARE, FRUIT KNIVES, ETC.

Our Canadian Representative carries a full line of samples.

Canadian Office:

6 St. Sacrament St., MONTREAL.

F. A. YORK, Manager.

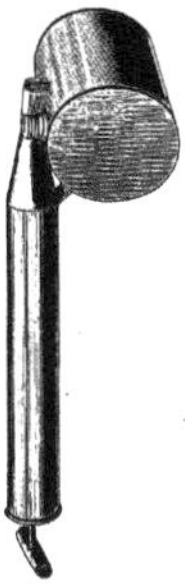

Wright's Insect Sprayers

Plain Tin—Lacquered—all Brass.

"BEST ON EARTH."

Manufactured by

E. T. WRIGHT & CO.,

HAMILTON, ONT.

TRADE MARK J REGISTERED.

"JARDINE" HAND DRILLS

Five different sizes to suit all requirements. It pays to sell the best tools.

A. B. JARDINE & CO.
HESPELER, ONT.

"A WORD TO THE WISE IS SUFFICIENT."

Gold Dollars at 90c. each won't sell until you have convinced the customer that they are genuine. Have you the time to make the argument to do the convincing?

ALABASTINE is the standard wall-coating **"The World Over."** Would any dealer want to spend the time to convince a customer that a substitute, or something else, is "just as good" as **CHURCH'S**

ALABASTINE?

Besides, substitution is suspicious, and often bears earmarks of a swindle.

We do not claim to know all that can be learned about the manufacture of wall coatings, but twenty-five years of experience must count for something, and the fact that **ALABASTINE** is the goods that are in demand speaks volumes in its favor.

Never until **M. B. Church** introduced **ALABASTINE** was it found practicable to make wall-coating adapted to be mixed with cold water.

We own the patents on the cold water process, and this statement is our only argument at this time on that point.

ALABASTINE is healthful and beautiful, has stood the test of time; is manufactured in Canada, and recommended by leading sanitarians throughout the country. There is nothing "just as good." Buy only the genuine.

Manufactured by

THE ALABASTINE CO., LIMITED, Paris, Ont.

We also manufacture **"BUG FINISH,"** the only practical Potato Bug Killer. Farmers will soon be calling for it. Order early and be prepared.

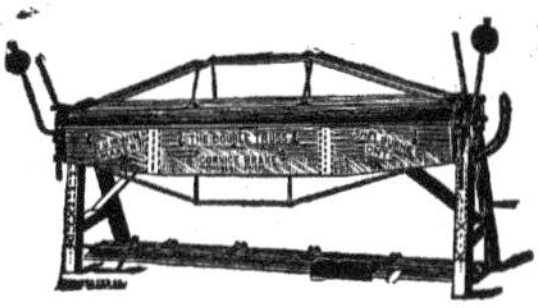

This eight-foot Brake bends 22-gauge iron and lighter, straight and true.

Price, $60

Very handy beader attachment, $15 extra if required.

Send for circulars and testimonials to

The Double Truss Cornice Brake Co. SHELBURNE, ONT.

The Latest and Best.

H. & R. Automatic Ejecting Single Gun.

Steel and Twist Barrels in 30 and 32-inch.
12 Gauge.

Model 1900.

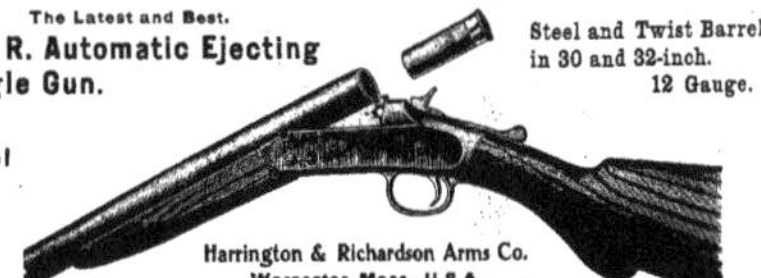

Harrington & Richardson Arms Co.
Worcester, Mass., U.S.A.
Descriptive Catalogue on request.

A LONDON FENCE MACHINE

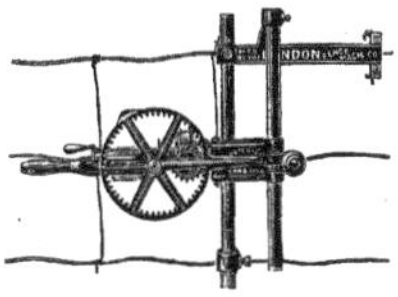

ON EXHIBITION IN YOUR STORE

will create more curiosity and bring in more customers than any other implement. There's money in handling the "London," as other merchants have found. You get a snap if you get the "London" agency for your town.

WRITE FOR TERMS BEFORE YOU MISS IT.

Coiled Spring Wire at right prices to the trade.

The London Fence Machine Co., Manufacturers and Wholesale Dealers, **London, Can.**

POSITIVE KNOWLEDGE.

THERE is no question but what this is a day of positive knowledge; a time when it is one of the first essentials for a man entering into any business enterprise to know positively what he is doing, and know it beforehand, and young America is to-day being duly impressed with that fact and receiving all manner of training in keeping with this spirit, but—and that is the point at issue here—how and what is positive knowledge?

We have long since come to understand that there must be some closer relation between theory and practice than was had in days gone by—that the mainspring of theory must be wound by the key of practice—and have added manual training to our technical schools, have built various and expensive devices for experimenting and demonstrating theories before offering them to the world in a product. All this, too, is as it should be, and there is no criticism to offer on this point, but it will not hurt us to stop and think a little once in a while in the hope that we may see a way to further extend the closer relationship of theory and practice to get positive knowledge of the kind that is in keeping with the spirit of the day. In other words, experiments and manual training schools, while they do much in their way, do not go to the extent that might be desired in surrounding these tests and experiments with some of the various circumstances that arise in the real industrial work of the day. Every once in a while there is something turns up which, while it was entirely unlooked for, and can be easily explained after it is over, might have been expected if not prevented. It is not explanations and excuses that we want in this day, though they are pretty good guides for the future and cannot be entirely eliminated, but it is deeds we want, and as near an exact and positive understanding of what is going to happen as possible beforehand.

For example, there was a recent little tilt between a couple of scientific authorities on riveted joints in a steam boiler as to its efficiency, etc. On the question in controversy here it is not the intention to speak right now, but to point out from this something taken from the impressions of an old-school engineer on the whole question of stress. He looks at a boiler as he sees it at work in the factory, and when he gets to thinking on the stress problem there is present in his mind the strain on the boiler due to its settings. If it simply has a rest under the front end and another under the back end, as is very common, he sees in his mind a strain on the bottom part of the boiler, the part, too, that is subject to the heat, which though rather indistinct in his mind if he tries to reduce it to figures, means to him a representation of the weight of the boiler and the water it contains. From this same source of reasoning he figures it out in the same dim way that there is really a tendency toward compression in the top part of his boiler. In such ideas he has some of the richest soil from which positive knowledge grows, and more of this soil is what we need as a fertilizer for scientific propagation of the branch of knowledge that the spirit of the times demands.

A point that must not be overlooked, however, is the fact that we must have this material of practical experience at first hands, or as nearly so as possible, for like some of the stories of old that were handed down from generation to generation, from mouth to ear, the coloring is likely to change so that ridiculous errors will frequently result. An instance in illustration of this point comes to mind at this writing. An English scientific journal of the very highest class, in treating on the subject of corn as fuel following the time when Kansas set the world talking by growing so much corn that it was used in instances as fuel as being cheaper under the circumstances than coal, had an illustrated headpiece to the article showing men at work in a field of barley corn. This was nothing serious, of course, but it points us to what may happen in handling information for positive knowledge. The puzzling point in the whole problem is then, how are we to combine the two essential features for positive knowledge in the mechanics of the day? How are we to get the unadulterated fertilizer of experience, that which is gained by rugged men who have spent years in harness and have learned, not from schools as we know them now, but from experience unguided except by the brain of the individual, to and mixed with the systematic schooling of to-day? We are progressing, of that there is no doubt, and we are getting nearer each year to the point of positive knowledge in mechanical undertakings, but the soil is losing some of its qualities as we go along and it is in order to look closely after this feature lest a future crop of positive knowledge turns out to be only nubbins at the harvest time.—Zeke in Age of Steel.

CANADIAN ORES IN THE STATES.

IT is quite evident that the product of Canadian iron mines is to play a more important part in the ore business of the lakes than was expected, says The Marine Review. M. A. Hanna & Co., of Cleveland, who have the agency for the Helen mine at Michipicoten, just above Sault Ste. Marie, which is controled by Francis H. Clergue, of Sault water-power fame, have sold ore up to the limit of what can be spared from that mine after the requirements of Canadian furnaces at Midland, Ont., are provided for. A couple of cargoes from Michipicoten were among the first to be delivered at Lake Erie ports this week. Four steamers of The Algoma Steamship Company (a Clergue organization) which went to Europe when navigation closed on the lakes last fall, are again on their way up from Montreal to the Sault with cement and other materials that will be required in the construction of manufacturing plants under way at that point. The vessels are the Monkshaven, Leafield, Theano and Paliki. They will be used during the present season in carrying ore from Michipicoten to Midland and other Canadian ports. The Clergue railway to the Helen mine, twelve miles northeast of Michipicoten, is being extended twelve miles more to reach the newer Josephine mine, which has been opened the past few months. It is a Bessemer ore said to be of better grade than the Helen. Some ore will very probably be shipped from this second mine during the present season. It is not improbable that a second ore dock will have to be built at Michipicoten. At Fort William, Ont. (head of Lake Superior), an ore dock will probably be built next spring, if not during the present season, for the transfer of ore that is to come over The Canadian Pacific Railroad from the Atikokan range, just across the Minnesota border. These mining properties will probably be controled by The United States Steel Corporation, as they were taken under option last fall by The American Mining Company (American Steel and Wire), now a part of the steel corporation, at a price understood to have been $500,000. If the ore is carried by The Canadian Pacific it will be necessary to construct about 45 miles of road in addition to the ore dock at Fort William. It is understood that a rate of less than 75 cents a ton has been guaranteed by the road, and the lake rate will be about equal to that from Duluth.

CURRENT MARKET QUOTATIONS.

May 31, 1901.

These prices are for such qualities and quantities as are usually ordered by retail dealers on the usual terms of credit, the lowest figures being for larger quantities and prompt pay. Large cash buyers can frequently make purchases at better prices. The Editor is anxious to be informed at once of any apparent errors in this list as the desire is to make it perfectly accurate.

METALS.

Tin.

Lamb and Flag and Straits—56 and 28 lb. ingots, per lb.	0 00	0 32

Tinplates.

Charcoal Plates—Bright

	Per box
M.L.S., equal to Bradley.	
I.C., usual sizes	$6 75
I.X., "	8 25
I.X.X., "	9 75
Famous—	
I.C.	6 75
I.X	8 25
I.X.X	9 50
Raven & Vulture Grades—	
I.C., usual sizes	5 00
I.X., "	6 00
I.X.X "	7 00
I.XXX., "	8 00
D.C., 12½x17	4 75
D.X.	5 50
D.X.X	7 50

Coke Plates—Bright

Bessemer Steel—	
I.C., usual sizes	4 00
I.C., special sizes, base	4 50
20 x 28	8 50

Charcoal Plates—Terne

Dean or J. G. Grade—	
I.C., 20x28, 112 sheets	8 25
I.X., Terne Tin	10 25

Charcoal Tin Boiler Plates.

		Per lb.
Cookley Grade—		
X X., 14x56, 50 sheet bxs / " 14x60, " / " 14x65, "	0 07	0 07½

Tinned Sheets

72x30 up to 24 gauge	0 08	0 08½
" 26 "	0 08½	0 09
" 28 "	0 09	0 09½

Iron and Steel.

Common Bar, per 100 lbs.	1 85	1 90
Refined " "	....	2 35
Horse Shoe Iron "	....	2 25
Hoop steel, 1½ to 3 in. base, extras for smaller sizes	3 10	3 25
Sleigh Shoe Steel " base	....	2 10
Tire Steel	2 30	2 50
Reeled Machinery	....	3 00
Cast Steel, per lb	0 00	0 00
Toe Calk Steel	....	
T. Firth & Co's special cast steel, per lb.		0 13
Jessop's Tool Steel	....	0 13

Boiler Tubes.

1½-inch	0 12½
2 "	0 13
2½ "	0 15
3 "	0 16
3½ "	0 20
4 "	0 25

Steel Boiler Plate.

¼ inch	2 50	2 60
3-16 inch	2 60	2 70
⅜ inch and thicker	2 50	2 60

Black Sheets.

18 gauge	3 00
20 gauge	3 00
22 to 24 "	3 10
26 "	3 20
28 "	3 30

Canada Plates.

All dull, 52 sheets	2 90
Half polished	3 00
All bright	3 10

Iron Pipe.

Black pipe—	
⅛ "	4 45
¼ inch	3 25
⅜ "	3 30
½ "	3 50
¾ "	3 65
1 "	5 15
1¼ "	7 10
1½ "	8 40
2 "	11 25
2½ "	20 95
3 "	24 55
3½ "	31 75
4 "	39 60
5 "	47 35
6 "	62 10
Galvanized pipe—	
¼ inch	4 90
⅜ "	5 25
1 "	7 55
1¼ "	10 30
1½ "	12 35
2 "	16 50

Galvanized Sheets.

	G C.	Comet.	Amer.	Queen's Head.
16 gauge		...	4 5	3 90
18 to 24 gauge	4 15	3 85	4 35	4 15
26 "	4 40	4 10	4 35	4 40
28 "	4 65	4 25	4 20	4 65

28 gauge American equals 26 gauge English.

Chain.

Proof Coil, 3-16 in., per 100 lb	....	....
" ¼ " "	8 20	8 50
" 5-16 " "	4 70	5 00
" ⅜ " "	4 05	4 00
" 7-16 " "	3 95	4 25
" ½ " "	3 70	4 10
" 9-16 " "	3 65	4 15
" ⅝ " "	3 25	3 90
" ¾ " "	3 60	4 10

Halter, kennel and post chains, 40 to 40 and 5 p.c.
Cow ties ... 40 p.c.
Tie-out chains ... 65 p.c.
Stall fixtures ... 35 p.c.
Trace chain ... 45 p.c.
Jack chain, iron, single and double, discount 35 p.c.
Jack chain, brass, single and double, discount 40 p.c.

Copper.

Ingot

English B. S., ton lots	....	0 19
Lake Superior	....	....

Bolt or Bar.

Cut lengths round, ½ to ⅞ in.	0 23½	0 25
" round and square 1 to 2 inches	0 23½	0 25

Sheet.

Plain, 14 oz., and light, 16 oz., 14x48 and 14x60	0 23	0 23
Plain, 14 oz., and light, 16 oz., irregular sizes	0 23	0 23½
Tinned copper sheets	....	0 26
Planished	....	0 32

Braziers (in sheets.)

4x6ft. 25 to 30 lbs. ea., per lb.	....	0 25½
" 35 to 45 " "	....	0 24½
" 50-lb. and above, "	....	0 23½

Boiler and T. K. Pitts

Plain Tinned, per lb	....	0 28
Spun, per lb	....	0 32

Brass.

Rod and Sheet, 14 to 30 gauge, 15 per cent.		
Sheets, hard-rolled, 2x4	0 24	0 25
Tubing, base, per lb	0 24	0 25

Zinc Spelter

Foreign, per lb	0 05½	0 06
Domestic "	....	..

Zinc Sheet.

5 cwt. casks	....	00 6¾
Part casks	....	00 6¾

Lead.

Imported Pig, per lb	0 04½	0 04½
Bar, 1 lb.	0 05½	0 06½
Sheets, 2½ lbs. sq. ft., by	....	0 06½
Sheets, 3 to 6 lbs., "	....	0 06

NOTE.—Cut sheets ½ cent per lb. extra. Pipe, by the roll, usual weights per yard, lists at 7c. per lb. and 30 p.c. dis. f.o.b. Toronto.

NOTE.—Cut lengths, net price, waste pipe 8-ft. lengths lists at 7½ cents.

Shot.

Common, $6.50 per 100 lb.; chilled, $7.00 per 100 lb.; buck, seal and ball, $7.50. Discount, 17½ p.c. Prices are f.o.b. Toronto, Hamilton, Montreal, St. John and Halifax. Terms 3 per cent. cash, freights equalized.

Soil Pipe and Fittings.

Discount, 60 and 10 per cent. on medium and extra heavy, and 60 per cent. on light.

Solder.	Per lb.	Per lb.
Bar half-and-half, guarant'd	....	0 18½
Bar half-and-half, commer'l	....	0 18
Refined	....	0 17½
Wiping	....	0 17

NOTE.—Prices of this graded according to quantity. The prices of other qualities of solder in the market indicated by private brands vary according to composition.

Antimony.

Cookson's, per lb.	0 10½	0 11

White Lead.	Per 100 lb.
Pure	6 37
No. 1 do	6 00
No. 2 do	5 62½
No. 3 do	5 25
No. 4 do	4 87½
Munro's Select Flake White	7 37½
Elephant and Decorators' Pure	7 12½
Brandram's B B. Genuine	8 00
" " Decorative	7 55
" " No. 1	6 85
" " No. 2	6 00

Red Lead.

Genuine, 560 lb. casks, per cwt.	....	$5 50
Genuine, 100 lb. kegs, per cwt.	....	5 75
No. 1, 560 lb. casks, per cwt.	....	5 25
No. 1, 100 lb. kegs, per cwt	....	5 50

White Zinc Paint.

Elephant Snow White	0 08	0 09
Pure White Zinc	0 08	0 09
No. 1	0 06	0 07½
No. 2	0 05	0 06½

Dry White Lead.

Pure, casks	5 75
Pure, kegs	6 25
No. 1, casks	5 50
No. 1, kegs	6 00

Prepared Paints.

In ¼, ½ and 1 gallon tins.

Pure, per gallon	....	1 25
Second qualities, per gallon	....	1 10
Barn (in bbls.)	0 75	0 85
The Sherwin-Williams Paints.	....	1 45
Canada Paint Co's Pure	....	1 25
Toronto Lead & Color Co's Pure	....	1 25
Sanderson Pearcy's Pure	....	1 20
Stewart & Wood's Champion Pure,		1 20

Colors in Oil.

25 lb. tins, Standard Quality.

Venetian Red, per lb.	0 05
Chrome Yellow	0 11
Golden Ochre	0 06
French "	0 05
Marine Black	0 09
" Green	0 09
Chrome "	0 08
French Imperial Green	0 09

Colors, Dry.

Yellow Ochre (J.C.) bbls.	1 35	1 40
Yellow Ochre (J.F.L.S.), bbls	...	2 75
Yellow Ochre (Royal)	1 10	1 15
Brussels Ochre	....	2 00
Venetian Red (best), per cwt.	1 80	1 90
English Oxides, per cwt.	3 00	3 25
American Oxides, per cwt.	1 75	2 00
Canadian Oxides, per cwt.	1 75	2 00
Super Magnetic Oxides, 93p.c.	2 00	2 25
Burnt Sienna, pure, per lb.	....	0 10
" Umber, " "	....	0 10
do Raw	....	0 09
Drop Black, pure	....	0 09
Chrome Yellows, pure	....	0 18
Chrome Greens, pure, per lb.	....	0 12
Golden Ochre	....	0 03½
Ultramarine Blue in 28-lb. boxes, per lb.	0 08	0 24
Fire Proof Mineral, per 100 lb	....	1 00
Genuine Eng. Litharge, per lb	....	0 07
Mortar Color, per 100 lb.	....	1 25
English Vermillion	....	0 80
Pure Indian Red, No. 45 lb.	....	0 80
Whiting, per 100 lb.	....	0 55

Blue Stone.

Casks, for spraying, per lb.	0 07
100-lb. lots, do. per lb.	0 08

Putty.

Bulk in bbls.	1 90
Bulk in less quantity	2 05
Bladders in bbls.	2 10
Bladders in kegs, boxes or loose.	2 25
Bladders in 25-lb. tins	2 35
Bladders in 12½-lb. tins	2 65
Bladders in bulk or tins less than 100 lb	2 90

Varnishes.

	In 5-gal. lots.).	Per gal.
Carriage, No. 1	2 90	3 30
" body	8 00	9 00
" rubbing	4 00	5 00
Gold Size, Japan	3 00	3 40
Brown Japan	2 40	2 80
Elastic Oak	2 90	3 30
Furniture, extra	2 40	2 80
" No. 1	1 60	2 00
Hard Oil Finish	2 70	3 10
Light Oil Finish	3 20	3 60
Demar	3 30	3 70
Shellac, white	4 40	4 80
" orange	4 00	4 40
Furniture Brown Japan	1 60	2 00
Black Japan	2 40	2 80
" " No. 1	1 60	2 00

The Imperial Varnish & Color Co's., Limited Elastilite Varnish 1 gal. can, each. $3.00.

Granatine Floor Finish per gal., $2.75.

Maple Leaf Coach Enamels; Size 1, $1 20; Size 2, 70c.; Size 3, 40c. each.

Castor Oil.

East India, in cases, per lb.	0 10	0 10½
" " small lots	0 10½	0 11

Cod Oil, Etc.

Cod Oil per gal	0 50	0 55
Pure Olive	....	1 20
" Neatsfoot	....	90

Glue.

Common	0 08½	0 09
French Medal	0 14	0 14½
Cabinet sheet	0 12	0 13
White, extra	0 18	0 20
Gelatine	0 22	0 30
Strip	0 18	0 20
Coopers	0 19	0 20
Huttner	....	0 18

HARDWARE.

Ammunition.

Cartridges.

B. B Cap Dom. 50 and 5 per cent.
Rim Fire Pistol, dis. 40 p. c., Amer.
Rim Fire Cartridges, Dom., 50 and 5 p. c.
Central Fire Pistol and Rifle, 10 p.c. Amer.
Central Fire Cartridges, pistol sizes, Dom, 30 per cent.
Central Fire Cartr'dges, Sporting and Military, Dom., 15 and 5 per cent.
Central Fire, Military and Sporting, Amer, add 5 p.c. to list. B.B. Caps, discount 40 per cent, Amer.
Loaded and empty Shells, "Trap" and "Dominion" grades, 25 per cent. Rival and Nitro, net list.
Brass shot Shells, 55 per cent.
Primers, Dom., 30 per cent.

Wads — per lb.

Best thick white felt wadding, in ½-lb bags ... 1 00
Best thick brown or grey felt wads, in ½-lb. bags ... 70
Best thick white card wads, in boxes of 500 each, 12 and smaller gauges 0 99
Best thick white card wads, in boxes of 500 each, 10 gauge ... 0 35
Best thick white card wads, in boxes of 500 each, 8 gauge ... 0 55
Thin card wads, in boxes of 1,000 each, 12 and smaller gauges ... 0 20
Thin card wads, in boxes of 1,000 each, 10 gauge ... 0 25
Thin card wads in boxes of 1,000 each, 8 gauge ...
Chemically prepared black edge grey cloth wads, in boxes of 250 each— Per M.
11 and smaller gauge ... 0 60
9 and 10 gauges ... 0 70
7 and 8 gauges ... 0 90
5 and 6 gauges ... 1 10
Superior chemically prepared pink edge, best white cloth wads, in boxes of 250 each—
11 and smaller gauge ... 1 15
9 and 10 gauges ... 1 40
7 and 8 gauges ... 1 65
5 and 6 gauges ... 1 90

Adzes.

Discount, 20 per cent.

Anvils.

Per lb ... 0 10 — 0 12½
Anvil and Vise combined ... — 4 50
Wilkinson & Co.'s Anvils..lb. 0 09 — 0 09½
Wilkinson & Co.'s Vices..lb. 0 09¾ — 0 10

Augers.

Gilmour's, discount 65 and 5 p.c. off list.

Axes.

Chopping Axes
Single bit, per doz ... 5 50 — 10 00
Double bit, " ... 11 00 — 18 00
Bench Axes, 40 p.c.
Broad Axes, 33⅓ per cent.
Hunters' Axes ... 5 50 — 6 00
Boy's Axes ... 5 75 — 6 75
Splitting Axes ... 6 50 — 12 00
Handled Axes ... 7 00 — 10 00

Axle Grease.

Ordinary, per gross ... 5 75 — 6 00
Best quality ... 13 00 — 15 00

Bath Tubs.

Zinc ... — 6 00
Copper, discount 15 p.c. off revised list

Baths.

Standard Enameled.

5½-inch rolled rim, 1st quality ... 25 00
" " " 2nd " ... 21 00

Anti-Friction Metal.

"Tandem" A ... per lb. 0 27
" B ... " 0 21
" C ... " 0 11½
Magnolia Anti-Friction Metal, per lb. 0 25
Frictionless metal, per lb ... 0 23

SYRACUSE SMELTING WORKS.

Aluminum, genuine ... 0 45
Dynamo ... 0 29
Special ... 0 25
Aluminum, 99 p.c. pure "Syracuse" .. 0 50

Bells.

Hand.

Brass, 60 per cent.
Nickel, 55 per cent.

Cow.

American make, discount 66⅔ per cent.
Canadian, discount 45 and 50 per cent.

Door.

Gongs, Sargant's ... 5 50 — 8 00
" Peterboro', discount 45 per cent.

Farm.

American, each ... 1 25 — 3 00

House.

American, per lb ... 0 35 — 0 40

Bellows.

Hand, per doz ... 3 35 — 4 75
Moulders', per doz ... 7 50 — 10 00
Blacksmiths', discount 40 per cent.

Belting.

Extra, 60 per cent.
Standard, 60 and 10 per cent.

Bits.

Auger.

Gilmour's, discount 60 and 5 per cent.
Rockford, 50 and 10 per cent.
Jennings' Gen., net list.

Car.

Gilmour's, 47½ to 50 per cent.

Expansive.

Clark's, 40 per cent

Gimlet.

Clark's, per doz ... 0 65 — 0 90
Damond, Shell, per doz ... 1 00 — 1 50
Nail and Spike, per gross ... 2 25 — 5 20

Blind and Bed Staples.

All sizes, per lb ... 0 07¼ — 0 12

Bolts and Nuts. Per cent.

Carriage Bolts, full square, Norway ... 65
" " full square ... 65
Common Carriage Bolts, all sizes ... 60
Machine Bolts, all sizes ... 60
Coach Screws ... 70
Sleigh Shoe Bolts ... 72½
Blank Bolts ... 60
Bolt Ends ... 62½
Plough Bolts ... 60
Nuts, square ... 4 c. off
Nuts, hexagon ... 4¼c. off
Tire Bolts ... 67½
Stove Bolts ... 67½
Stove rods, per lb ... 5½ to 6c.

Boot Calks.

Small and medium, ball, per M ... 4 25
Small heel, per M ... 4 50

Bright Wire Goods.

Discount ... 62½ per cent.

Broilers.

Light, dis., 65 to 67½ per cent.
Reversible, dis., 65 to 67½ per cent.
Vegetable, per doz., dis. 37½ per cent.
Henis, No. 8, " ... — 6
Henis, No. 9, " ... — 7 00
Queen City " ... 7 50 — 0 00

Butchers' Cleavers.

German, per doz ... 6 00 — 11 00
American, per doz ... 12 00 — 20 00

Building Paper, Etc.

Plain building, per roll ... 0 30
Tarred lining, per roll ... 0 40
Tarred roofing, per 100 lb ... 1 65
Coal Tar, per barrel ... 3 50
Pitch, per 100-lb ... 0 85
Carpet felt, per ton ... 45 00

Bull Rings.

Copper, $2.00 for 2½ in. and $1.90 for 2 in.

Butts.

Wrought Brass, net revised list

Cast Iron.

Loose Pin, dis., 60 per cent.

Wrought Steel.

Fast Joint, dis. 65, 10 and 2½ per cent.
Loose Pin, dis. 65, 10 and 2½ per cent.
Berlin Bronzed, dis. 70, 70 and 5 per cent.
Gen. Bronzed, per pair ... 0 40 — 0 65

Carpet Stretchers.

American, per doz ... 1 00 — 1 50
Bullard's, per doz ... 6 50 — ...

Castors.

Bed, new list, dis. 55 to 57½ per cent.
Plate, dis. 52½ to 57½ per cent.

Cattle Leaders.

Nos. 31 and 32, per gross ... 50 — 9 30

Cement.

Canadian Portland ... 2 50 — 2 80
English " ... — 3 00
Belgian " ... 2 50 — 2 75
Canadian hydraulic ... 1 25 — 1 50

Chalk.

Carpenters Colored, per gross 0 45 — 0 75
White lump, per cwt ... 0 60 — 0 65
Red ... 0 05 — 0 06
Crayon, per gross ... 0 14 — 0 18

Chisels.

Socket, Framing and Firmer.

Broad's, dis. 70 per cent.
Warnock's, dis. 70 per cent.
P. S. & W. Extra 60, 10 and 5 p.c.

Churns.

Revolving Churns, metal frames—No. 0, $8— No. 1, $8.50—No. 2, $9.00—No. 3, $10.00— No. 4, $12.00—No. 5, $16.00 each. Ditto, wood frames—20c. each less than above. Discounts: Delivered from factories, 58 p.c.; from stock in Montreal, 58 p.c. Terms, 4 months or 3 p.c. cash in 30 days.

Clips.

Axle dis. 65 per cent.

Closets

Plain Ontario Syphon Jet ... $16 00
Emb. Ontario Syphon Jet ... 17 00
Fittings net ... 1 00
Plain Teutonic Syphon Washout ... 10 00
Emb. Teutonic Syphon Washout ... 11 00
Fittings net ... 1 25
Low Down Teutonic, plain ... 16 00
" " " embossed ... 17 00
Plain Richelieu net ... 3 75
Emb. Richelieu net ... 4 00
Fittings net ... 1 25
Low Down Ont. Sy. Jet, plain net ... 19 50
" " " " " emb'd. net 20 50
Closet connection net ... 1 25
Basins, round, 14 in ... 1 00
" oval, 17 x 14 in ... 2 50
" " 19 x 15 in ... 3 75
Discount 40 p.c., except on net figures.

Compasses, Dividers, Etc.

American, dis. 62½ to 65 per cent.

Cradles, Grain.

Canadian, dis. 25 to 33⅓ per cent.

Crosscut Saw Handles.

S. & D., No. 3, per pair ... .17½
" " 5, " ... .22½
" " 6, " ... .15
Boynton pattern " ... .20

Door Springs.

Torrey's Rod, per doz ... (15 p.c.) 2 00
Coil, per doz ... 0 88 — 1 60
English, per doz ... 2 00 — 4 00

Draw Knives.

Coach and Wagon, dis. 50 and 10 per cent.
Carpenters, dis. 70 per cent.

Drills.

Hand and Breast.

Millar's Falls, per doz. net list.

DRILL BITS.

Morse, dis., 37½ to 40 per cent.
Standard dis. 50 and 5 to 55 per cent

Faucets.

Common, cork-lined, dis. 35 per cent.

ELBOWS. (Stovepipe.)

No. 1, per doz ... 1 40
No. 2, per doz ... 1 20
Bright, 20c. per doz. extra.

ESCUTCHEONS.

Discount, 45 per cent.

ESCUTCHEON PINS.

Iron, discount 40 per cent.

FACTORY MILK CANS.

Discount off revised list, 40 per cent.

FILES.

Black Diamond, 60 to 60 and 10 per cent.
Kearney & Foote, 60 and 10 to 70 per cent.
Nicholson File Co., 60 to 60 and 10 per cent.
Jowitt's, English list, 25 to 27½ per cent.

FORKS.

Hay, manure, etc., dis., 50 and 10 per cent. revised list.

GLASS—Window—Box Price.

Size United Inches.	Star Per 50 ft.	Star Per 100 ft	D. Diamond Per 50 ft.	D. Diamond Per 100 ft.
Under 26	2 15	4 15	....	6 00
26 to 40	2 30	4 45	....	6 65
41 to 50	....	4 85	....	7 50
51 to 60	....	5 15	....	8 50
61 to 70	....	5 50	....	9 50
71 to 80	....	6 00	....	10 50
81 to 85	....	6 50	....	11 70
86 to 90	....	....	..	14 00
91 to 95	..	....	...	15 50
99 to 100	.. ..	....	....	18 u0

GAUGES

Marking, Mortise, Etc.

Stanley's dis. 50 to 55 per cent.

Wire Gauges.

Winn's, Nos. 26 to 33, each ... 1 65 — 2 40

HALTERS.

Rope, ⅜ per gross ... . — ....
" ½ " ... . — 9 00
" ⅝ to ¾ ... . — 14 00
Leather, 1 in., per doz ... 3 87½ — 4 00
" 1¼ in., " ... 5 15 — 5 20
Web, — per doz ... 1 87 — 2 45

HAMMERS.

Nail

Maydole's, dis. 5 to 10 per cent. Can. dis. 25 to 27½ per cent.

Tack.

Magnetic, per doz ... 1 10 — 1 20

Sledge.

Canadian, per lb ... 0 07½ — 0 08½

Ball Pean.

English and Can., per lb ... 0 22 — 0 25

HANDLES.

Axe, per doz. net ... 1 50 — 2 00
Store door, per doz ... 1 00 — 1 50

Fork.

C. & B., dis. 40 per cent. rev. list.

Hoe.

C. & B., dis. 40 per cent. rev. list.

Saw.

American, per doz ... 1 00 — 1 25

Plane.

American, per gross ... 3 15 — 3 75

Hammer and Hatchet.

Canadian, 40 per cent.

Cross-Cut Saws.

Canadian, per pair ... 0 13½

HANGERS. doz. pairs.

Steel barn door ... 5 85 — 6 00
Stearns, 4 inch ... . — 5 00
" 5 inch ... . — 6 50
Lane's covered—
No. 11, 5-ft. run ... . — 8 40
No. 11½, 10-ft. run ... . — 10 80
No. 12, 10-ft. run ... . — 12 60
No. 14, 15-ft. run ... . — 21 00
Lane's O.N.T. track, per foot ... — 4½

HARVEST TOOLS.

Discount, 50 and 10 per cent.

HATCHETS.

Canadian, dis. 40 to 42½ per cent.

HINGES.

Blind, Parker's, dis. 50 and 10 to 60 per cent.
Heavy T and strap, 4-in., per lb ... 0 06½
" " 5-in., " ... 0 06¼
" " 6-in., " ... 0 06
" " 8-in., " ... 0 05¾
" " 10-in., " ... 0 05½
Light T and strap, dis. 65 and 5 per cent.
Screw hook and hinge—
6 to 12 in., per 100 lbs ... . — 4 50
14 in. up, per 100 lbs ... ... — 3 50
Per gro. pairs.
Spring ... . — 12

HOES.

Garden, Mortar, etc., dis. 50 and 10 p.c.
Planter, per doz ... 4 00 — 4 50

HOLLOW WARE

Discount ... 45 and 5 per cent

HOOKS.

Cast Iron.

Bird Cage, per doz ... 0 50 — 1 10
Clothes Line, per doz ... 0 27 — 0 63
Harness, per doz ... 0 72 — 0 88
Hat and Coat, per gross ... 1 00 — 3 00
Chandelier, per doz ... 0 50 — 1 00

Wrought Iron.

Wrought Hooks and Staples, Can., dis. 47½ per cent.

Wire.

Hat and Coat, discount 45 per cent.
Belt, per 1,000 ... 0 60
Screw, bright, dis. 55 per cent.

HORSE NAILS.

"C" brand 50 and 7½ p.c. off new list } Oval-
"M" brand 50, 10 and 5 per cent. } head
Countersunk, 60 per cent.

HORSESHOES

Iron Shoes.	F.O.B. Montreal. No. 2 and larger.	No. 1. and smaller.
Light, medium, and heavy.	3 50	3 75
Snow shoes	3 75	4 00
Steel Shoes.		
Light	3 60	3 85
Featherweight (all sizes)	4 85	4 85

F.O.B. Toronto, Hamilton, London and Guelph, 10c. per keg additional.
Toe weight steel shoes...... 6 70

JAPANNED WARE.
Discount, 45 and 5 p.c. off list, June 1899

ICE PICKS.
Star per doz........ 3 00 3 25

KETTLES.
Brass spun, 7½ p.c. dis. off new list.
Copper, per lb........ 0 30 0 50
American, 60 and 10 to 65 and 5 p.c.

KEYS.
Lock, Can., dis., 45 p.c.
Cabinet, trunk, and padlock, Am. per gross........ 60

KNOBS.
Door, japanned and N.P., per doz........ 1 50 2 50
Bronze, Berlin, per doz...... 2 75 3 25
Bronze Genuine, per doz.... 6 00 9 00
Shutter, porcelain, F. & L. screw, per gross........ 1 30 4 00
White door knobs—per doz. 1 25

HAY KNIVES.
Discount, 50 and 10 per cent.

LAMP WICKS.
Discount, 60 per cent.

LANTERNS.
Cold Blast, per doz........ 7 00
No. 3 "Wright's"........ 8 50
Ordinary, with O burner........ 4 00
Dashboard, cold blast........ 9 00
No. 0........ 5 75
Japanning, 50c. per doz. extra.

LEMON SQUEEZERS.
Porcelain lined,......per doz. 2 20 5 60
Galvanized........ 1 87 3 85
King, wood........ 2 75 2 90
King, glass........ 4 00 4 50
All glass........ 1 20 1 30

LINES.
Fish, per gross........ 1 05 2 50
Chalk " 1 90 7 40

LOCKS.
Canadian, dis. 45 p.c.
Russel & Erwin, per doz.... 3 00 3 25

Cabinet.
Eagle, dis. 30 p.c.

Padlock
English and Am., per doz.... 50 6 00
Scandinavian, " 1 00 2 40
Eagle, dis. 20 to 25 p.c.

MACHINE SCREWS. Iron and Brass.
Flat head discount 25 p.c
Round Head discount 20 p.c.

MALLETS.
Tinsmiths' per doz........ 1 25 1 50
Carpenters', hickory, per doz. 1 25 3 75
Lignum Vitae, per doz., ... 3 85 5 00
Caulking each........ 60 2 00

MATTOCKS.
Canadian, per doz........ 5 50 6 50

MEAT CUTTERS.
American, dis. 25 to 30 p.c.
German, 15 per cent.

MILK CAN TRIMMINGS.
Discount, 25 per cent.

NAILS.

Quotations are:	Cut.	Wire.
2d and 3d	$3 35	$3 85
3d	3 00	3 52
4 and 5d	2 75	3 35
6 and 7d	2 65	3 20
8 and 9d	2 50	3 00
10 and 12d	2 45	2 95
16 and 20d	2 40	2 90
30, 40, 50 and 60d. (base)	2 35	2 85

Wire nails in carlots are $2.77½
Galvanizing 2c. per lb. net extra.
Steel Cut Nails 10c. extra.
Miscellaneous wire nails, dis. 70 and 10 p.c.
Coopers' nails, dis. 30 per cent.
Flour barrel nails, dis. 25 per cent.

NAIL PULLERS.
German and American....... 1 85 3 50

NAIL SETS
Square, round, and octagon per gross........ 3 38 4 00
Diamond........ 12 00 15 00

NETTING.
Poultry, 55 per cent for McMullen's

OAKUM. Per 100 lb.
Navy........ 6 00
U. S. Navy........ 7 25

OIL.
Water White (U.S.)........ 0 16½
Prime White (U.S.)........ 0 15½
Water White (Can.)........ 0 15
Prime White (Can.)........ 0 14

OILERS.
McClary's Model galvan. oil can, with pump, 5 gal., per doz........ 10 00
Zinc and tin, dis. 50, 50 and 10.
Copper, per doz........ 1 25 3 50
Brass, " 1 50 3 50
Malleable, dis. 25 per cent.

GALVANIZED PAILS.
Dufferin pattern pails, dis. 45 p.c.
Flaring pattern, discount 45 per cent.
Galvanized washtubs, discount 45 per cent.

PIECED WARE.
Discount 40 per cent. off list, June, 1899.
10-qt. flaring sap buckets, dis. 45 p.c.
6, 10 and 14-qt. flaring pails, dis. 45 p.c.
Creamer cans, dis 45 p.c.

PICKS.
Per doz........ 6 00 9 00

PICTURE NAILS.
Porcelain head, per gross.... 1 75 3 00
Brass head " 0 40 1 00

PICTURE WIRE.
Tin and gilt, discount 75 p.c.

PLANES.
Wood, bench, Canadian dis. 50 per cent. American dis. 50.
Wood, fancy Canadian or American 7½ to 40 per cent.

PLANE IRONS.
English, per doz........ 2 00 5 00

PLIERS AND NIPPERS.
Button's Genuine per doz pairs, dis. 37½ 40 p.c.
Button's Imitation, per doz.. 5 00 9 00
German, per doz........ 0 60 2 60

PLUMBERS BRASS GOODS.
Compression work, discount, 60 per cent.
Fuller's work, discount 65 per cent.
Rough stops and stop and waste cocks, discount, 60 per cent.
Jenkins disk globe and angle valves, discount, 55 per cent.
Standard valves, discount, 60 per per cent.
Jenkins' radiator valves discount 55 per cent.
" " " standard, dis., 60 p.c.
Quick opening valves discount, 60 p.c.
No. 1 compression bath cock........ 2 00
No. 4 " " " 2 00
No. 7, Fuller's........ 2 50
No 4½, " 3 00

POWDER.
Velox Smokeless Shotgun Powder.
100 lb. or less........ 0 85
1,000 lb. or more........ 0 80
Net 30 days.

PRESSED SPIKES.
Discount 20 to 25 per cent.

PULLEYS.
Hothouse, per doz........ 0 15 1 00
Axle........ 0 22 0 33
Screw........ 0 27 1 00
Awning........ 0 35 2 50

PUMPS.
Canadian cistern........ 1 80 3 60
Canadian pitcher spout.... 1 40 2 10

PUNCHES.
Saddlers', per doz........ 1 00 1 85
Conductors', " 9 00 15 00
Tinners' solid, per set........ 0 00 0 72
" hollow, per inch.... 0 00 1 00

RANGE BOILERS.
Galvanized, 3 gallons........ 7 00
" 35 " 8 25
" 40 " 9 50
Copper, 30 " 22 00
" 35 " 26 00
" 40 " 30 00
Discount off Copper Boilers 10 per cent.

RAKES.
Cast steel and malleable, 50, 10 and 5 p.c.
Wood, 25 per cent.

RASPS AND HORSE RASPS.
New Nicholson horse rasp, discount 60 to 60 and 10 p.c.
Globe File Co.'s rasps, 60 and 10 to 70 p.c.
Heller's Horse rasps, 50 to 50 and 5 p.c.

RAZORS.

	per doz.	
Elliot's	4 00	18 00
Geo. Butler & Co.'s	8 00	18 00
Boker's	7 50	11 00
" King Cutter	12 50	50 00
Wade & Butcher's	3 60	10 00
Theile & Quack's	7 00	12 00

REAPING HOOKS.
Discount, 50 and 10 per cent.

REGISTERS.
Discount........ 40 per cent.

RIVETS AND BURRS.
Iron Rivets, black and tinned, discount 60 and 10 per cent.
Iron Burrs, discount 55 per cent.
Extras on Iron Rivets in 1-lb. cartons, ½c. per lb.
Extras on Iron Rivets in ½-lb. cartons, 1c. per lb.
Copper Rivets & Burrs, 35 and 5 p.c. dis. and cartons, 1c. per lb. extra, net.
Extras on Tinned or Coppered Rivets ½-lb. cartons, 1c. per lb.

RIVET SETS
Canadian, dis. 35 to 37½ per cent.

ROPE ETC.

	Sisal.	Manila.
7-16 in. and larger, per lb	10	13½
⅜ in.	11	14½
¼ and 5-16 in.		15½
Cotton, 3-16 inch and larger		16
" 5-32 inch		21
" ⅛ inch		22½
Russia Deep Sea		15½
Jute		8
Lath Yarn		9½
New Zealand Rope		10

RULES.
Boxwood, dis. 75 and 10 p.c.
Ivory, dis. 37½ to 40 p.c.

SAD IRONS. per set.
Mrs. Potts, No. 55, polished.. ...... 62½
" No. 50, nickle-plated 67½

SAND AND EMERY PAPER.
Dominion Flint Paper, 47½ per cent.
B & A. sand, 40 and 5 per cent.
Emery, 40 per cent.
Garnet (Rurton's), 5 to 10 p.c. advance on list.

SAP SPOUTS.
Bronzed iron with hooks, per doz... 9 50

SAWS.
Hand Disston's, dis. 12½ p.c.
S. & D., 40 per cent.
Crosscut, Disston's, per ft.... 35 0 55
S. & D., dis. 35 p.c. on Nos. 2 and 3.
Hack, complete, each........ 0 75 2 75
" frame only........ .... 0 75

SASH WEIGHTS.
Sectional, per 100 lbs........ 2 75 3 00
Solid, " 2 00 2 25

SASH CORD.
Per lb........ 0 23 0 30

SAW SETS.
"Lincoln," per doz........ .... 6 50

SCALES.
Standard, 45 p.c.
Champion, 65 p.c.
Spring Balances, 10 p.c.
Fairbanks Standard, 35 p.c.
" Dominion, 55 p.c.
" Richelieu, 55 p.c.

SCREW DRIVERS
Sargent's per doz........ 0 65 1 00

SCREWS
Wood, F. H., bright and steel, 87½ and 10 p.c.
Wood R. H., " dis. 82½ and 10 p.c.
" F. H., brass, dis. 80 and 10 p.c.
Wood, R. H., " dis. 75 and 10 p.c.
" F.H., bronze, dis. 75 p.c.
" R.H. " 70 p.c.
Drive Screws, 87½ and 10 per cent.
Bench, wood, per doz........ 3 25 4 00
" iron. " 4 25 5 75
Set, Case hardered, 60 per cent.
Square Cap, 50 and 5 per cent.
Hexagon Cap, 45 per cent.

SCYTHES.
Per doz. net........ 9 00

SCYTHE SNATHS.
Canadian, dis. 45 p.c.

SHEARS.
Bailey Cutlery Co., full nickeled, dis. 60 p.c.
Seymour's, dis. 50 and 10 p.c.

SHOVELS AND SPADES.
Canadian, dis. 40 and 5 per cent.

SINKS.
Steel and galvanized, discount 45 per cent.

SNAPS.
Harness, German, dis. 25 p.c.
Lock, Andrews'........ 4 50 11 50

SOLDERING IRONS.
1, 1½ lb., per lb........ 0 37
2 lb. or over, per lb........ 0 34

SQUARES.
Iron, No 493, per doz...... 2 40 2 55
" Mo. 494, " 3 25 3 40
Steel, dis. 60 10 and 5 p.c.
Try and bevel, dis. 50 to 52½ p.c.

STAMPED WARE.
Plain, dis.. 75 and 12½ p.c. off revised list.
Retinned, dis., 75 p.c. off revised list.

STAPLES.
Galvanized........ 3 50 4 00
Plain........ 3 15 3 75
Coopers', discount 45 per cent.
Poultry netting staples, 40 per cent.

STOCKS AND DIES.
American dis. 25 p.c.

STONE. Per lb.

Washita		0 28	0 60
Hindostan		0 06	0 07
" slip		0 09	0 09
Labrador		....	0 13
" Axe		....	0 15
Turkey		....	0 50
Arkansas		0 00	1 50
Water-of-Ayr		0 00	0 10
Scythe,	per gross	3 50	5 00
Grind,	per ton	15 00	18 00

STOVE PIPES.
5 and 6 inch Per 100 lengths 7 00
7 inch " " 7 50

ENAMELINE STOVE POLISH.
No. 4—3 dozen in case, net cash $4 80
No. 6—3 dozen in case, " 8 40

TACKS BRADS, ETC.

	Per cent
Strawberry box tacks, bulk	75 & 10
Cheese-box tacks, blued	80 & 12½
Trunk tacks, black and tinned	85
Carpet tacks, blued	80 & 15
" " tinned	80 & 20
" " (in kegs)	40
Cut tacks, blued, in dozens only	80
" ¼ weights	60
Swedes, cut tacks, blued and tinned—	
In bulk	80 & 10
In dozens	75
Swedes, upholsterers', bulk	85, 12½ & 12½
" brush, blued & tinned, bulk	70
" gimp, blued, tinned and japanned	75 & 12½
Zinc tacks	35
Leather carpet tacks	55
Copper tacks	50
Copper nails	52½
Trunk nails, black	65 and 5
Trunk nails, tinned	65 and 10
Clout nails, blued	65 and 5
Chair nails	35
Patent brads	40
Fine finishing	40
Picture frame points	10
Lining tacks, in papers	10

L'n'ug tacks, in bulk		15
" " solid heads, in bulk		75
Saddle nails in papers		10
" " in bulk		15
Tufting buttons, 22 line, in dozens only		60
Tin capped trunk nails		15
Zinc glazier's points		5
Double pointed tacks, papers	90 and	10
" " " bulk		40

TAPE LINES.

English, ass skin, per doz	2 75	5 00
English, Patent Leather	5 50	9 75
Chesterman's each	0 90	2 85
" steel, each	0 80	8 00

THERMOMETERS.

Tin case and dairy, dis. 75 to 75 and 10 p.c.

TRAPS. (Steel.)

Game, Newhouse, dis. 25 p.c.
Game, H. & N., P. S. & W., 65 p.c.
Game, steel, 72½, 75 p.c.

TROWELS.

Disston's discount 10 per cent.

German, per doz	4 75	6 00

S. & D., discount 35 per cent.

TWINES.

Bag, Russian, per lb	....	0 27
Wrapping, cotton, 3-ply		0 20
" " 4-ply	....	0 26
Mattress, per lb	0 33	0 45
Staging, "	0 27	0 35

VISES.

Hand, pe doz	4 00	6 00
Bench, parallel, each	2 00	4 50
Coach, each	6 00	7 00
Peter Wright's, per lb	0 12	0 13
Pipe, each	5 50	9 00
Saw, per doz	6 50	13 00

ENAMELLED WARE.

White, Princess, Turquoise, Blueand White, discount 50 per cent.
Diamond, Famous, Premier, 50 and 10 p.c.
Granite or Pearl, Imperial, Crescent, 50, 10 and 10 per cent.

WIRE.

Brass wire, 50 to 50 and 2½ per cent. off the list.

Copper wire, 45 and 10 per cent. net cash 30 days, f.o.b. factory.

Smooth Steel Wire, is quoted at the following net selling prices:

No. 6 to 8 gauge	$2 90
" 9 "	2 80
" 10 "	2 87
" 11 "	2 90
" 12 "	2 95
" 13 "	3 15
" 14 "	3 37
" 15 "	3 50
" 16 "	3 65

Other sizes of plain wire outside of Nos. 9, 10, 11, 12 and 13, and other varieties of plain wire remain at $2.80 base with extras as before. The prices for Nos 9 to 13 include the charge of 1 c. for oiling. Extras net per 100 lb.: Coppered wire, 60c.—tinned wire, $2—oiling, 10c.—special hay-bailing wire, 30c.—spring wire, $1—best steel wire, 75c.—bright soft drawn, 15c.—in 50 and 100-lb. bundles net, 10c.—in 25-lb. bundles net 15c.—packed in casks or cases, 15c.—bagging or papering, 10c.

Fine Steel Wire, dis. 17½ per cent. List of extras: In 100-lb. lots: No. 17, $5—No. 18, $5.50—No. 19, $6—No. 20, $6.65—No. 21, $7—No. 22, $7.30—No. 23, $7.65—No. 24, $8—No. 25, $9—No. 26, $9.50—No. 27, $10—No. 28, $11—No. 29, $12—No. 30, $13—No. 31, $14—No. 32, $15—No. 33, $16—No. 34, $17. Extras net—tinned wire, Nos. 17-25, $2—Nos. 26-31 $4—Nos. 32-34, $6. Coppered, 5c.—oiling, 10c.—in 25-lb. bundles, 15c.—in 5 and 10-lb. bundles, 25c.—in 1-lb. hanks, 50c.—in ½-lb. hanks, 75c.—in ¼-lb. hanks, $1—packed in casks or cases, 15c.—bagging or papering, 10c.

Galvanized Wire, per 100 lb.—Nos. 6, 7, 8, $3 50 to $3 85—No. 9, $2.85 to $3.15—No. 10, $3.60 to $3.95—No. 11, $3.70 to $4.10—No. 12, $3 to $3.30—No. 13, $3.10 to $3.40—No. 14, $4.10 to $4.50—No. 15, $4.60 to $5.05—No. 16, $4.85 to $5.35. Base sizes, Nos. 6 to 9, $2.57½ f.o.b. Cleveland.

Clothes Line Wire, solid 7 strand, No. 17, $4.25; No. 18, $2.65; No. 19, $2.35, f.o.b. Hamilton, Toronto, Montreal.

WIRE FENCING. F.O.B. Toronto

Galvanized barb	....	3 05
Galvanized, plain twist	....	3 05

Galvanized barb, f.o.b. Cleveland, $2.82½ in less than carlots, and $2.70 in carlots.

WIRE CLOTH.

Painted Screen, per 100 sq. ft., net	1 35

WASTE COTTON. per lb.

Colored	4½ to 5
White, according to quality	6½ to 7¾

500-lb bale lots shaded.

WRENCHES.

Acme, 35 to 37½ per cent.
Agricultural, 60 p.c.
Coe's Genuine, dis. 20 to 25 p.c.

Towers' Engineer, each	2 00	7 00
" S., per doz	5 80	6 00
G. & K.'s Pipe, per doz	....	3 40
Burrell's Pipe, each	....	3 00
Pocket, per doz	0 25	2 90

WRINGERS.

Leader	per doz.	$30 00	33 00
Royal Canadian	"	26 00	28 00
Royal American	"	26 00	28 00
Sampson	"	....	30 00

Terms 4 months, or 3 p.c. 30 days.

WROUGHT IRON WASHERS.

Canadian make, discount, 40 and 5 per cent.

PERSONS addressing advertisers will kindly mention having seen their advertisement in Canadian Hardware and Metal Merchant.

"THE EMLYN" SAW BENCH

Made in 6 sizes. Best value obtainable. Specially designed for export. With or without "Emlyn" Patent Guard. Sole maker—

CHARLES D. PHILLIPS,
Emlyn Engineering Works,
NEWPORT, MON., ENGLAND.

"Machinery," Newport.

Mr. Retail Hardwareman,

You have had trouble with Axes. Some were poor sellers, others did not give your customers satisfaction.

You can avoid having all this trouble in future. We can tell you how. Write us.

Dundas Axe Works

DUNDAS, CANADA.

"BUILD TO-DAY THEN, STRONG AND SURE, WITH A FIRM AND AMPLE BASE." —Longfellow.

DO YOU?

WISH THUS TO BUILD an advertisement in the CONTRACT-RECORD, TORONTO will bring you tenders from the best contractors.

BUSINESS NEWS

of any kind that is of value to business men supplied by our Bureau. We can give you market quotations from any town in Canada, reports from the city markets, stock quotations, etc. You can get commercial news from any Canadian paper through us.

Write us, giving us particulars of what you want and where you want it from, and we will quote you prices by return.

"Clippings from any Canadian paper on any subject."

CANADIAN PRESS CLIPPING BUREAU,

232 McGill Street, MONTREAL, QUE.

Telephone Main 1255.

10 Front St. East, Toronto. Telephone 2148.

75 YEARS. ESTABLISHED 1825. 75 YEARS.

CELEBRATED HEINISCH SHEARS.

Tailors' Shears,
Trimmers, Scissors,
Tinners' Snips, etc.

ACKNOWLEDGED THE BEST.

R. HEINISCH'S SONS CO. NEWARK, N.J., U.S.A.

NEW YORK OFFICE, 90 Chambers St.

Not connected with any Shear Combination.

CHAS. F. CLARK, President. JARED CHITTENDEN, Treasurer.

...ESTABLISHED 1849...

BRADSTREET'S

Capital and Surplus, $1,500,000. Offices Throughout the Civilized World.

Executive Offices: Nos. 346 and 348 Broadway, New York City, U.S.A.

THE BRADSTREET COMPANY gathers information that reflects the financial condition and the controlling circumstances of every seeker of mercantile credit. Its business may be defined as of the merchants, by the merchants, for the merchants. In procuring, verifying and promulgating information no effort is spared, and no reasonable expense considered too great, that the results may justify its claim as an authority on all matters affecting commercial affairs and mercantile credit. Its offices and connections have been steadily extended, and it furnishes information concerning mercantile persons throughout the civilized world.

Subscriptions are based on the service furnished, and are available only by reputable wholesale, jobbing and manufacturing concerns, and by responsible and worthy financial, fiduciary and business corporations. Specific terms may be obtained by addressing the Company at any of its offices. **Correspondence Invited.**

—OFFICES IN CANADA—

HALIFAX, N.S. OTTAWA, ONT. VANCOUVER, B.C.
HAMILTON, ONT. QUEBEC, QUE. VICTORIA, B.C.
LONDON, ONT. ST. JOHN, N.B. WINNIPEG, MAN.
MONTREAL, QUE. TORONTO, ONT.

THOS. C. IRVING, Gen. Man. Western Canada, Toronto. JOHN A. FULTON, Gen. Man. Eastern Canada, Montreal.

THE...
Waggoner Extension Ladder.

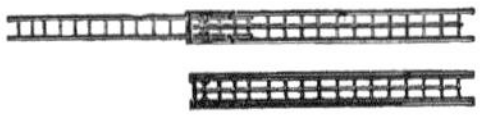

The strongest, lightest and most convenient ladder in the market. The only really satisfactory extension ladder made. Pulls up with a rope. Made in all lengths. Also extension and other step ladders, sawhorses, ironing boards, painters' trestles, etc. All first-class goods. Write for quotations to

The Waggoner Ladder Company, Limited, London, Ont.

THE MOWER

THAT WILL KILL
ALL THE WEEDS
IN YOUR LAWNS.

If you keep the weeds cut so they do not go to seed, and cut your grass without breaking the small feeders of roots, the grass will become thick and weeds will disappear. **The Clipper will do it.**

CANADIAN PATENT FOR SALE.
SEND FOR CATALOGUE AND PRICES.

Clipper Lawn Mower Co., NORRISTOWN, PA.

CANADIAN HARDWARE AND METAL MERCHANT

The Weekly Organ of the Hardware, Metal, Heating, Plumbing and Contracting Trades in Canada.

VOL. XIII. — MONTREAL AND TORONTO, JUNE 8, 1901. NO. 23

ALLOW us to call your attention to our very complete stock of

Incandescent Gas Lamps, Mantles, Burners, Etc.

Sole agents for Canada for the

SOLAR LAMP.

The very best lamp on the market, for natural or artificial gas. Prices on application.

THE ROCHESTER LAMP CO.
OF CANADA

24 Front Street West, TORONTO

HOSE...

WATER	SUCTION
STEAM	ACID
AIR	OIL
FIRE	SODA WATER
BABCOCK	HIGH-PRESSURE

Our **Patent Seamless Tube** is, without doubt, the only perfect construction.

The Canadian Rubber Co.,

CAPITAL - - - $1,500,000.00.

Montreal. Toronto. Winnipeg.

Lightning, Gem Blizzard . . .

FREEZERS

ARE

- Well Advertised
- In Demand
- Easily Sold
- Satisfactory in Use
- Of Known Reputation

HAVE

Cedar Pails with Electric Welded Wire Hoops
Cans of Heavy Tin with Drawn Steel Bottoms
AUTOMATIC Twin Scrapers
"The Ice Cream Freezer Book" tells all about these and our other Freezers, mailed free

EXCEL IN

- Easy Running
- Quick Freezing
- Economy
- Convenience
- Practical Results

North Bros. Mfg. Co., Philadelphia, Pa., U.S.A.

LEWIS BROS. & CO., Montreal.

HARVEST TOOLS.

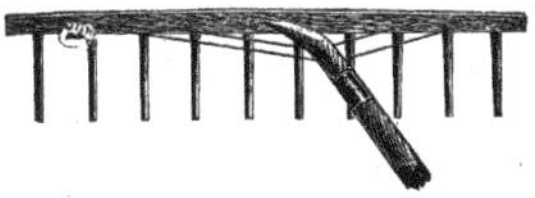

No. 1—Wire Bow, Bent or Straight Handle.
No. 2—Wire Bow, Straight Handle.

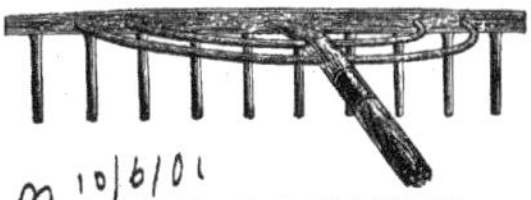

No. 1—Wood Bow, Bent or Straight Handle.
No. 2—Wood Bow, Straight Handle.
No. 1—Oiled Head, Wood Bow, Bent Handle.

Extra and 2nd Growth Bent Fork Handles.

No. 1 and Extra Straight Fork Handles.

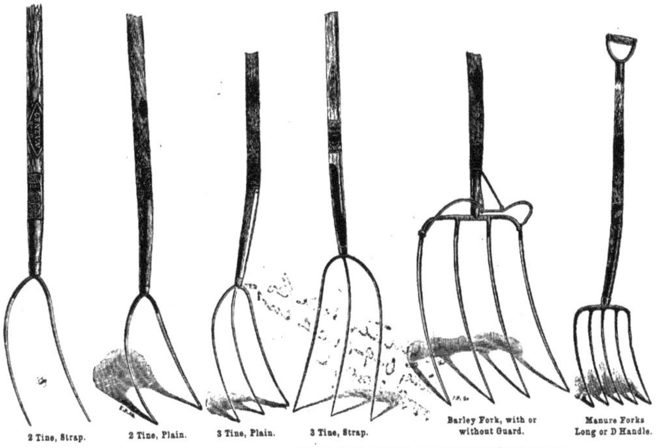

2 Tine, Strap. 2 Tine, Plain. 3 Tine, Plain. 3 Tine, Strap. Barley Fork, with or without Guard. Manure Forks Long or D Handle.

A large Hay Crop is now assured. You will do well to order Harvest Tools at once.

MAIL ORDERS SHIPPED SAME DAY AS RECEIVED. WRITE FOR PRICES.

LEWIS BROS. & CO., Montreal.

Henry Rogers, Sons & Co.

Wolverhampton, England.

Manufacturers of

"Union Jack" Galvanized Sheets
Canada and Tin Plates
Black Sheets
Sleigh Shoes and Tyre Steel
Coil Chain, Hoop Iron
Sheet and Pig Lead
Sheet Zinc

Quotations can be had from Canadian Office:
6 St. Sacrament St., MONTREAL
F. A. YORK, Manager.

Wright's Insect Sprayers
Plain Tin—Lacquered—all Brass.
"BEST ON EARTH."
Manufactured by
E. T. WRIGHT & CO., HAMILTON, ONT.

KNOX HENRY
Heavy Hardware and Metal Broker
Room 32, Canada Life Bldg., MONTREAL.

"SECCOTINE"
FOR STICKING EVERYTHING.

Samples sent free on application.

HORSE NAILS—"C" Brand Horse-Nails
Canada Horse Nail Co.
"BRASSITE" GOODS — Gunn Castor Co Limited, Birmingham, Eng.

McCaskill, Dougall & Co
Fine Varnish & Japan
Manufacturers.
Montreal
Price Lists on application

"A WORD TO THE WISE IS SUFFICIENT."

Gold Dollars at 90c. each won't sell until you have convinced the customer that they are genuine. Have you the time to make the argument to do the convincing?

ALABASTINE is the standard wall-coating **"The World Over."** Would any dealer want to spend the time to convince a customer that a substitute, or something else, is "just as good" as **CHURCH'S**

ALABASTINE?

Besides, substitution is suspicious, and often bears earmarks of a swindle.

We do not claim to know all that can be learned about the manufacture of wall coatings, but twenty-five years of experience must count for something, and the fact that **ALABASTINE** is the goods that are in demand speaks volumes in its favor.

Never until **M. B. Church** introduced **ALABASTINE** was it found practicable to make wall-coating adapted to be mixed with cold water.

We own the patents on the cold water process, and this statement is our only argument at this time on that point.

ALABASTINE is healthful and beautiful, has stood the test of time; is manufactured in Canada, and recommended by leading sanitarians throughout the country. There is nothing "just as good." Buy only the genuine.

Manufactured by

THE ALABASTINE CO., LIMITED, Paris, Ont.

We also manufacture **"BUG FINISH,"** the only practical Potato Bug Killer. Farmers will soon be calling for it. Order early and be prepared.

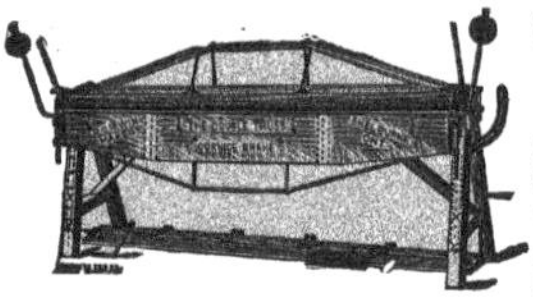

This eight-foot Brake bends 22-gauge iron and lighter, straight and true.

Price, $60

Very handy beader attachment, $15 extra if required.
Send for circulars and testimonials to

The Double Truss Cornice Brake Co. SHELBURNE, ONT.

The Latest and Best.

H. & R. Automatic Ejecting Single Gun.

Model 1900.

Steel and Twist Barrels in 30 and 32-inch.
12 Gauge.

Harrington & Richardson Arms Co.
Worcester, Mass., U.S.A.
Descriptive Catalogue on request.

A LONDON FENCE MACHINE

ON EXHIBITION IN YOUR STORE

will create more curiosity and bring in more customers than any other implement. There's money in handling the "London," as other merchants have found. You get a snap if you get the "London" agency for your town.

WRITE FOR TERMS BEFORE YOU MISS IT.

Coiled Spring Wire at right prices to the trade.

The London Fence Machine Co., Manufacturers and Wholesale Dealers, **London, Can.**

Wadsworth-Howland

Co. CHICAGO.

A M E

Are Superior to any on the market

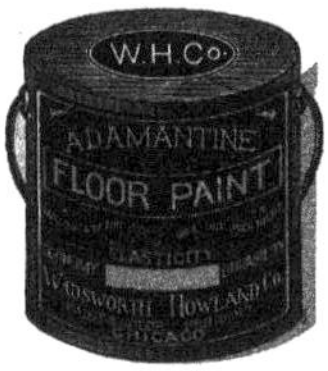

And Second to none. One trial will convince you

F

made to paint floors with.

BANNER BUGGY PAINT prepared expressly for painting old or new work where a handsome. lasting and glossy surface is required. Dries quickly and will not crack or chip off.

Jewel Mixed Paints and Roof and Barn Paints.

A single trial will convince you that these paints are all we claim, being far superior in body, more lasting, will retain their color and lustre longer than strictly pure lead and oil mixed by hand. Uniform in shade and consistency, and will not fade unevenly. Desirable paint to use on account of their spreading and covering properties and their great durability.

RICE LEWIS & SON

LIMITED.

Canadian Distributors TORONTO.

HARDWARE AND METAL

VOL. XIII. MONTREAL AND TORONTO, JUNE 8, 1901. NO. 23.

President,
JOHN BAYNE MacLEAN,
Montreal.

THE MacLEAN PUBLISHING CO.

Limited.

Publishers of Trade Newspapers which circulate in the Provinces of British Columbia, North-West Territories, Manitoba, Ontario, Quebec, Nova Scotia, New Brunswick, P.E. Island and Newfoundland.

OFFICES

MONTREAL - - - - - - - 232 McGill Street, Telephone 1255.
TORONTO - - - - - - - 10 Front Street East, Telephone 2148,
LONDON, ENG. - - - - - 109 Fleet Street, E.C., W. H. Miln.
MANCHESTER, ENG. - - - - 18 St Ann Street. H. S. Ashburner,
WINNIPEG - - - - - Western Canada Block. J. J. Roberts.
ST. JOHN, N.B. - - - - No. 3 Market Wharf, J. Hunter White.
NEW YORK. - - - - - - - 176 E. 88th Street,

Subscription, Canada and the United States, $2.00.
Great Britain and elsewhere - - - 12s.

Published every Saturday.

Cable Address { Adscript, London.
Adscript, Canada.

HEAVY SALES OF MIXED PAINTS.

IT is a point worthy of the attention of the retail trade that manufacturers of mixed paints report they have just closed their banner month's business. Perhaps every firm cannot boast thus, but we have heard enough such pleasant statements to assure ourselves that more mixed paints will be used—are being used—in Canada this season than ever before.

The mixed paint can has now come to occupy even a more important place in the hardware store than the white lead pot. In other words, people are gradually coming to be content to leave the paint-mixing business to experts, securing their paints in a convenient, ready-to-use form.

It is estimated that last year the sales of white lead decreased from those of the year previous by 40 per cent. This was not entirely due to any suddenly-boomed popularity of the paint can, although the splendid advertisements of our wide-awake paint firms must be held partially accountable. But the chief factor was the high price set on white lead.

White lead fairly went beyond consumption value, and many a person refused to paint his house, while others, obedient to the laws of substitution, sought the cheaper mixed paints. And now lead is still high, and oil still at the top notch. Needless to say the consumption of white lead will continue to be restricted. The trade recognize this fact, and again the mixed paint manufacturers are reaping a harvest.

Whether the advent of Canadian refined lead into the market will reduce prices in time and restore the popularity of white lead remains to be seen, but it would not seem foolish prophecy to say that good mixed paints ought to hold the market they have lately won, for use will wear many a prejudice away.

ADVANCE IN MATCHES.

When referring to the fire at the Walkerville match factory, we expressed the opinion that an advance in the price of matches might be looked for. The expected has happened, prices this week having been marked up 10 to 30 per cent.

Prices are now as follows: Telegraph, $4 per case; Telephone, $3.90 per case; Tiger, $3.80 per case; Eagle parlor matches, 200's, $1.70 per case; do., 10's, $1.90 per case; Victoria parlor matches, $3 per case.

On Telephone, Telegraph and Tiger matches the price is 20c. less in 5-case lots, and on Eagle and Victoria matches the price is 10c. less for 5-case lots.

MANITOBA'S CRITICAL WEEK.

THIS is generally looked upon as being what might be termed the pivotal week in the crop experience of Manitoba and the Northwest Territories, particularly as far as danger of damage by frost is concerned, and the pronounced cool wave which is at the moment spreading over that part of the country is naturally creating much concern among those interested.

At the time of writing, the thermometer, according to the report of the Weather Bureau, has not yet touched freezing point, but it is hovering dangerously near it.

It is to be hoped, after the beneficent rains which fell in Manitoba last week and dissipated the threatened drought, that the bright expectations thereby created will not be destroyed by serious damage by frost.

The business men of this country, who now have so much interest centred in Manitoba, are scarcely less solicitous for the welfare of the crop than the farmers themselves.

Advertising is just as essential to the export as to the home trade; and he is a wise business man who recognizes and acts upon it.

LINSEED OIL PRODUCTION IN CANADA.

WITH the starting up of the Livingstone linseed oil mill in Montreal, referred to last week, that city will have two mills in operation, while the total number in Canada will be increased to six, Winnipeg, Guelph, Elora and Baden each having one.

It is estimated that the quantity of linseed oil produced in Canada is about equal to the quantity that has been imported during the last year or two. The quantity imported last year was a little over 1,000,000 gals. From a glance at the following table it will be noticed that the importation of linseed oil during the last five years, with one exception, was near the million-gallon mark. In fact, in 1888 the quantity was slightly in excess of that of 1900, being 1,052,858 gals., valued at $393,894.

IMPORTATION OF LINSEED OIL, RAW OR BOILED.

	Gallons.	Value.
1896	958,226	$336,114
1897	961,075	303,890
1898	539,676	175,316
1899	1,038,308	328,487
1900	1,044,972	479,277

Nearly all the oil imported comes from Great Britain, the quantities from that country being 1,022,235 gals. in 1900 and 1,032,354 gals. in 1899.

A great deal of the seed crushed by the Canadian mills is imported, and it is interesting to note in this connection that the imports of seed during the last two years were the largest on record. The quantities and values during the last five years were as follows:

IMPORTS OF FLAXSEED.

	Quantity in lb.	Value.
1896	1,621,312	$ 30,440
1897	12,992	375
1898	3,349,892	67,432
1899	45,708,682	1,119,539
1900	51,184,541	1,366,436

What proportion of the quantity imported is used by the linseed oil mills we cannot say.

We understand more attention is being given to the production of flax in Canada, particularly in Manitoba. The Dominion experimental farms are engaged in doing a good deal of experimental work in flax cultivation, and we may in time expect good results therefrom. The best result last year, as far as yield of seed per acre is concerned, was at the experimental farm at Indian Head, Northwest Territories, the number of bushels being 10 bushels and 44 lb. to 15 bushels and 18 lb. per acre. At the Brandon, Man., farm the yield per acre was from 5 bushels and 40 lb. to 7 bushels and 8 lb. Experiments were also carried on in Nova Scotia at the Nappan farm, where as high as 15 bushels per acre was obtained in one instance, although in another it was as low as 10 bushels. In the one instance, however, 80 lb. of seed was sown to the acre and in the other 40 lb. At the other experimental farms mentioned the quantity of seed produced from 40 lb. of seed sown was almost invariably within a few pounds of being as large as from 80 lb. of seed.

With the additional mill that is being got ready in Montreal for next season's operations, Canada will be in a position to satisfy more than ever the demand of the home market. The output of the new mill will probably increase the production of linseed oil by a quantity equal to that imported last year.

WASTED BUSINESS ENERGY.

IT is not always the most energetic man who accomplishes most, who makes the greatest progress. Some men seem to be tireless in their vigor, ceaseless in their activity, yet at the end of 10, 20, or 50 years they are not as far advanced, nor have they attained as great results as others who have taken life, as a rule, more leisurely. The result is that the opinion is often expressed: "The harder a man works the less he gets for it."

This conclusion is, however, not only hasty, but erroneous. Hard work, of itself, will not put a man in front of his fellows, for the simple reason that a man's physical powers are so limited that he can excel his fellows but little. The real reason for the power of some men over others is that their energy is wisely directed, being exerted freely when some end is to be reached, but never being expended with the mere purpose of keeping busy.

The cut price that is thrown upon the market is often a boomerang with a sharp edge.

PUBLIC SPIRIT IN MERCHANTS.

IN most of the larger towns and smaller cities of Canada are to be found merchants who have come to the conclusion that they have reached the limit in the extent of their business, basing this belief on the facts that local competition is just as keen and that the population of their district is not materially greater than has been the case for several years. They are generally men of enterprise and ability. Many of them have proved this in the years during which they fought their way from the position of parcel boy to that of proprietor. They have secured the honor and respect of their customers, and, in fact, of their district generally, but have reached the conclusion that it is better to accept the limitations that local conditions put upon their business rather than disturb the harmony of business by price-cutting, or any such means.

There is another class in the same towns or cities—those eager, restless, ambitious men who accept no limitations as natural and effectual, and who are constantly devising methods of extending their trade and of increasing their net profits.

To both of these classes, as indeed to every business man, the present condition of affairs in Canada is full of promise and opportunity. The past quarter of a century, and particularly the past decade, has been an era of discovery and development throughout Canada, from the coal areas of Cape Breton to the gold regions of the Klondike. Thousands of acres of new lands has been taken up for farming purposes in the great West and in New Ontario; our spruce and pine limits are being made use of as never before; the value of our mineral deposits has been recognized and they are now being developed by men of ability and large wealth, which is a guarantee of permanence for the industry. In fact, every possible condition is combining to make Canada ready for a great influx of population.

On the other hand, the development of our iron and coal deposits has led to the erection of smelters in Canada which will insure pig iron at prices on a par with those in the great industrial centres of the United States or Great Britain.

The practical result of these conditions will be to build up, not only the factories and mills throughout the country, but also the various municipalities where these are situated. Here is where the opportunity is offered to business men in all sections of the country to build up their town or city by enterprise and energetic public spirit. If a local industry is in sound condition and has a growing trade, but lacks capital for extensions which are not only desirable but wise, there should be enough of local enterprise to provide the necessary capital.

In another direction business men have the power, by a display of public spirit, to develop their municipality and thus extend the limits of their business. An attractive town or city always attracts to itself a desirable class of people. The number of people who have a competence and are content to live on it is steadily growing as is also the number of pleasure seekers who annually visit the most picturesque or attractive parts of the Dominion. Business men should, therefore, take a keen interest in the appearance of their stores and streets, in the accommodations offered by their hotels and should ever be awake to advertise the attractions of any neighboring resorts and to further any schemes proposed to attract residents, either permanent or transient to their town or district.

WHAT TO DO IN HARDWARE IN JUNE.

WHAT is there to do but keep on gathering the results of your spring preparation? It's the semi-mid-month of the year, with all the getting ready of the months in front of it, and the so called dull months to follow it. Then June should be the summing up of all the half-year's work—the bringing in at the end of it the welcome knowledge of money made, or the disappointing fact of long weeks of work for naught.

JUNE FOR RESULTS

is one of the best months of the year. It is the season of haymaking and of harvest, the time of year in which perhaps more matters of moment are brought together and consummated than at any other.

It is the time for house building, for barn finishing, for sales of hay riggings, mowers and reapers, steel hay rakes and tedders, for the all-important distribution of the tons and tons of binder twine, for cultivator trade, for all that goes to make the farmer and the citizen prosperous, and through them the hardware merchant. Then almost as soon as sales are made settlements of some sort should follow—the sooner the better. For all this class of goods it would seem a capital plan to have at hand and ready a short form of due bill or acceptance, closing every account of moment when made or when full accounts are decided on, with definite time of final settlement thus named, avoiding the loss of time and expense in getting at the same results in the weeks and months to come.

JUNE FOR PLANNING

is not just the best month, because there is so much else to occupy it, but for many of us there must be borne in mind the old-time hoodoo of dull July and August to come. It is the writer's belief that with properly selected stocks, with the right preparation and looking forward to it, there is no excuse for an actual dull month during the entire year. Those of us who are content to go along in the old ruts will, of course, have them. There are instances where location and circumstance may account for them, but the fault is more often our own than that of any combination of place or circumstance. If time can be found at all it is well to make use of it in the effort at finding—if for the first time, then in a small way—stocks that will occupy us during July and August and into September. Other merchants do it; it has been demonstrated that every month can be a busy one; it is worth the trying.

JUNE FOR ADVERTISING.

In the height of the busy and prosperous months of the year we are all prone to neglect our advertising. It is miserable, losing neglect, but occurs all along the line, unless your store be large enough to have it in the hands of one man whose entire time is largely given to it.

At the season of the year when people are all buying, they watch more closely than at any other time. They want to know where to buy the goods without hunting for them. They largely get the prices from the fireside, and the prices should always be given. There has never yet been found a medium to satisfy the general public equaling the daily paper. The public is only satisfied with plain facts and plain prices well placed in a first-class daily.

JUNE FOR COLLECTIONS.

The end of the month should find everything in readiness for extra work at collections. There are times in the year when people expect to pay, that's a good time to ask for money, and next to the first day of the year in importance is the middle or first of July, and, of course, the work of getting ready must be June work. Perhaps you have all noticed the fact in making collections, particularly those called seasonable, that the merchant who is able to get his bills and statements out on the first day of the month receives by far the best results as to payments. There is a reason for it. Your customers' funds will often reach just so far, and, as occasion offers, note the difference where your statements reach the customer on the first, and again where they have gone in on the fifth or sixth. First bills to reach him are always the ones paid, while later ones often go over. It is an important and sometimes a hard matter to impress this fact on a bookkeeper, but it is a fact, nevertheless.

JUNE TO THINK OF A VACATION.

Until of late years very few hardwaremen ever gave themselves the time or considered the benefits of a vacation. No machine can go on forever without resting, overhauling, and oiling. No machine but will last longer and do better work for the resting and reconstruction. If there is a class of men on earth who need vacation it is the hardware merchant, with his never-ending worry over detail, discounts, payments, purchases, etc. It is gratifying to note that some of us are coming to our senses and giving ourselves each year a few weeks of the oil of recreation. We will last the longer and be the better for it. Our work is too arduous to do without it, and the months of June and July should see our work done as nearly as possible, with some sort of real vacation in view later on.

EARLY JUNE AND PERSONAL LETTERS.

The valued country trade—the large buyers for the summer months—are, as a rule, not quickly or always reached through the daily papers, and it is here the value of the stenographer for personal letters comes in. The farming community as no other thoroughly appreciate a personal plea for their custom, and any slack time of the typewriter cannot be used to better advantage than in this way. The letters should go in 2c. or regular postage cover, not under any circumstances as a circular, and should be as personal as it is possible to make them.

In the absence of a stenographer nearly every good printing office is now fitted with the new process of typewriter printing, showing the copy effect, and in reality answering every purpose at very small cost.

A MONTH FOR FARMERS

and their cultivation is the month of June, and there is not another like it in the calendar. Particular attention should be given them on every hand. Show windows should be gotten ready for them and attention called to them through the weekly or country newspaper. They are the most appreciative class in the world, because as a rule so little attention is given them.

Again, when you have made the farmer your friend he can do you untold service among his neighbors, and is more than likely to do it on all occasions, and all out of good feeling for you or your firm. The best investment, the best advertising among country trade, is a half dozen or so sterling friends who are always at hand to say a good word for you.

A SATISFACTORY MONTH

in every sense of the word is June, both in the looking backward and over the work gone through with and in the looking forward and planning for the other half of the year, which begins with the ending of the month. As it is the month of roses socially, so should it be with the merchant in his business, if for no other reason than that it is the busiest one in the whole year, and busy men make happy men.—"H.C.W." in Iron Age.

LAMPLOUGH-STARKE.

A marriage of some importance in Montreal society was solemnized on Wednesday afternoon at the residence of the bride's mother, Simpson street, when Miss Isabella Starke, daughter of Mrs. Geo. K. Starke, and sister of Messrs. Starke, of Howden, Starke & Co., was married to Mr. Frank Westrope Lamplough, of Lamplough & McNaughton. The wedding was a quiet one, only the members of the families, and a few friends being present. The bride was given away by her brother, Lieut.-Col. Starke. The honeymoon is being spent in Boston, New York, Philadelphia and other American centres.

To Mr. Lamplough, HARDWARE AND METAL extends its congratulations, and with his many friends in the trade we hope that the life partnership may be a happy one.

TRADE IN COUNTRIES OTHER THAN OUR OWN.

TRADE in Great Britain in galvanized sheets has improved on foreign and colonial account, and the makers of black sheets are standing out for better terms.

BRITISH MACHINERY TRADE.

Throughout the engineering trade business remains very much as last reported. There is no material change in the general situation, and a feeling of depression is steadily growing. The only possible exception may be found in a report from the Midlands which speaks hopefully of the outlook. An increase in the volume of business received by local engineers has been noted during the past week in that district. The opinion prevails in many quarters that there is plenty of work ready to come forward when buyers are persuaded that the present downward tend in the prices of raw materials has reached its lowest point.

Meanwhile the makers of machine-tool specialties and the electrical industry continue to be practically the sole monopolists of what prosperity exists. Shipbuilders and marine engineers are running short of orders, but in the ordnance factories reports indicate the probability of great pressure for some time to come. Since last writing a Government order for the armor for the new battleship Queen has been divided between the three leading Sheffield firms. Perhaps the greatest depression of all is being experienced by the textile machinery branch, which continues in an extremely stagnant condition.—Iron and Coal Trades' Review.

TINPLATE TRADE IN GREAT BRITAIN.

The market for tinplates has been somewhat excited owing to the continued advance in tin, and the higher values asked for tinplate bars, the combination resulting in increased prices being paid for the manufactured article.

The demand has been well sustained, both for prompt and forward shipments, and has met with considerable sales for both positions. The inquiry has been running on ordinary and special sizes—light plates of all kinds, and with a strong "call" for black plates.

This "betterment" has induced several of the standing works to get ready to come into action, and it is estimated that some five or six works will restart next month with an additional make of half a million boxes to add to the supply. In the present temper of the market this increased make may be absorbed without much detriment to prices, always provided that the demand keeps up in the future as vigorously as in the past.

As regards the oil sizes, buyers for the Russian and American markets are inclining to "take things easy," as both markets have ample supplies to last them for months to come, and seem therefore inclined to go slow. There is some inquiry in for oil sizes for the east, but as delivery is far ahead, makers are not very keen for the business.—Iron and Coal Trades' Review.

BRITISH IRON-MAKING COMPANIES.

The prospects of iron and steel concerns are probably somewhat better than they were a few weeks ago. The reduction of prices all round has helped the trade a good deal, and the outlook is not so depressing as in January, when there was a more complete cessation of inquiries in all branches of the iron trade than has probably ever before occurred, attributable to the fact that British makers maintained a high level of prices long after Americans had begun to undersell them, thus enabling the United States to secure the then pending orders. As things now are prices are materially lower here than in the States. Whether this can be maintained consistently with the payment of good dividends remains to be seen, but there is certainly a more assured tone with reference to iron prospects and properties.—Iron and Coal Trades' Review.

PIG IRON IN GREAT BRITAIN.

On the Glasgow pig iron warrant market business has been very slack, and finally the tone was dull. Once or twice not more than a couple of warrants changed hands, and how a matter of thirty brokers can make ends meet on these conditions it would be hard to explain. Owing to New York trade advices indicating an easier tone in America, sellers have been in evidence, and prices for all kinds of material have rather lost ground. At the same time a fairly good undertone has prevailed, and at the declines there have been several reliable buyers to the fore. The market closed on Wednesday for the Whitsun holidays, and will not reopen until Tuesday next. Connal's Glasgow stocks have decreased 510 tons, to 59,898 tons, but the stocks of the U. K. have increased by 3,104 tons, to 174,773 tons.—Ironmonger, May 25.

BRITISH TRADE IN LEAD.

As far as red lead and foreign white leads are concerned, the position is not different from that of last month, but some British manufacturers of white lead are inclined to take advantage of the spring trade to raise their prices a little. There has been but little change in the metal itself, and what there has been is against rather than in favor of an advance in the value of carbonate; so that the action of these makers does not seem to be justified, and will only lead to an extended use of foreign white lead. The policy of keeping the price of English lead above that of foreign—when there is plenty of margin for profit—appears to us altogether unwise.—Ironmonger.

NEW YORK METAL MARKET.

TIN—There was a further break in the London tin market to-day, and the close was weak at a decline of £1 2s. 6d. on spot and £2 2s. 6d. on futures, according to cables received here. The New York market felt the depressing influence of the London break, and although holders kept the asking price up the bids of buyers indicated a decided lack of confidence on their part. The close was dull and unsettled, with 27.55c. the best bid for spot, while 28.20c. was asked. Futures were not mentioned.

COPPER—There was no fresh developments in this market. The demand continues slow, but prices are held steadily up to the quotation of 17c. for Lake Superior and 16 5-8c. for electrolytic and casting. London cables reported a further advance of 2s. 6d. there, the close being steady, though quiet.

PIG LEAD—A fair demand was reported, with the market steady on the basis of 4.37 1-2c. for lots of 50 tons or more. In St. Louis the tone was easier, the quotations by wire being 4.27 1-2c. to 4.30c. London was unchanged.

SPELTER—We have to report a steady though quiet market, spot stock being quoted at 3.95 to 4.05c. The St. Louis market was steady at 3.80 to 3.85c. In London the quotation was the same as for several days past.

ANTIMONY—Regulus remains steady at the range of 8 1-2 to 10 1-4c., though the demand is moderate.

OLD METALS—Trade is slow and prices are nominal.

IRON—There were few developments in the iron market. Interest in foundry pig iron is lacking, but advices from Pittsburg are to the effect that there is inquiry for a round lot of Bessemer and that The United States Steel Corporation is a prospective buyer. There is little new business reported in steel or finished iron. Old iron and steel are dull and the feeling does not seem to be so firm. Still there has been no quotable change in prices.

TINPLATE—Nothing new has come to the surface in this market. The distribution on old orders is of liberal dimensions, but not much new business is reported. Prices are firm.—New York Journal of Commerce, June 5.

DIVIDENDS OF CANADIAN BANKS

The money disbursed in Canadian bank dividends on June 1 was more than at any period in the history of the institutions. This is due to the fact that in some instances the dividends have been increased, while in others the banks have increased their capital. The total paid-up capital of the banks in the two Provinces of Quebec and Ontario, which pay dividends in June, is now $48,030,000, against $43,785,000, and the aggregate sum to be distributed in dividends amounts to $1,918,000, against $1,724,325 last year.

CUTTING GLAZED TILES.

It is probably no exaggeration to say that in cutting ordinary glazed tiles (as used for grates, hearths, etc.), one out of every three breaks in the wrong place and is more or less useless, whilst another one out of three is cut less truly and neatly than hoped for. This is when a hammer and chisel are used. It is only known to a few persons that a pair of 7-in. carpenter's pincers will do the work without breakages and in much better style. If it is desired to cut a tile in two pieces, and use both parts, then the chisel must be relied on, but when a tile has to be reduced in size or cut to an irregular shape it can be best done by pinching pieces off with pincers. Only about 1-4-in. pieces are taken off at a time, but the end is attained more quickly, and a straighter and sharper line, or a better curve, can be obtained than a chisel-cut will give.—Ironmonger.

BUSINESS CHANGES.

DIFFICULTIES, ASSIGNMENTS, COMPROMISES.

THOMPSON & LAHEY, general merchants, Penetanguishene, Ont., have assigned to Henry Barber & Co., Toronto.

C. Belanger, general merchant, Portneuf, Que., has suspended.

Joseph Parent, general merchant, Rimouski, Que., has assigned.

J. B. Lafrance, general merchant, Crysler, Ont., has assigned to J. G. Hay, Toronto.

V. Leblanc & Co., general merchants, Hull, Que., have obtained extension of time.

Annie L. Graham, general merchant, Owry, Ont., has assigned to Richard L. Gosnell.

Adelard Many, general merchant, St. Sebastien, Que., has assigned to Gagnon & Caron.

J. B. Donville, general merchant, St. Stanislas, Que., has assigned to Lamarche & Benoit.

A meeting of the creditors of S. Bricker, hardware and coal dealer, Listowel, Ont., has been held.

PARTNERSHIPS FORMED AND DISSOLVED.

Wells & Frary, general merchants, Frelighsburg, Que., have dissolved.

The St. Maurice Tool and Axe Works, Three Rivers, Que., have dissolved.

Wm. Dickson, general merchant, Alexander, Man., has admitted a partner.

Larochelle & Co., general merchants, Levis, Que., have registered dissolution.

Godbout & Rathier, general merchants, St. George de Windsor, Que., have dissolved.

R. H. Appleyard & Co., hardware and stove dealers, etc., Humberstone, Ont., have dissolved. R. H. Appleyard continues.

Co-partnership has been registered by John White and William O. Brandis, under the style of John White & Co, stove dealers, etc., Halifax.

Samuel Melanson, general merchant, Bathurst, N.B., has admitted H. A. Melanson, under the style of S. Melanson & Son.

Charles and Harry S. Reeves have registered co-partnership under the style of Reeves Bros., general merchants, Port Hawkesbury, N.S.

SALES MADE AND PENDING.

The stock of R. A. Copeland & Co., general merchants, Grenfell, Man., has been sold.

The business of W. & J. Armstrong, blacksmiths, Kerrwood, Ont., is advertised for sale.

The assets of Z. Paquet, general merchant, Roberval, Que., are to be sold on the 8th inst.

The stock of J. H. Frignon, general merchant, St. Tite, Que., is to be sold to-day (Friday).

The sawmill of the estate of J. R. Owen, Ridgetown, Ont., is advertised for sale by auction on the 11th inst.

CHANGES.

H. M. Keddy, general merchant, Berwick, N.S., has been succeeded by N. W. Keddy.

Boulanger & Co., general merchants, Chaudiere Junction, Que., have removed to Sillery, Que.

Graves & McGuire, general merchants, Vienna, Ont., have been succeeded by C. M. Wilson & Co.

P. A. Dickie has started as harness dealer in Kentville, N.S.

J. W. Stout, sawmiller, Summit City, B.C., has given up business.

The Pennsylvania Coal Co., Limited, have registered in Montreal.

Delorme & Charbonneau have registered as painters, etc., in Montreal.

J. H. Lambert & Cie., have registered as as painters, etc., in Montreal.

Martineau & Duval have registered as painters, in Three Rivers, Que.

Turner Bailey, blacksmith, Crumlin, Ont., has sold out to John Stevenson.

The National Tool and Axe Works have registered in Three Rivers, Que.

Wm. McNair, harnessmaker, Russell, Ont., has been succeeded by Joseph Lascelles.

R. S. Hannah, hardware dealer, Mitchell, Ont., has sold out to Robert Campbell.

Thomas Moxworthy, blacksmith, Thorndale, Ont., has sold out to L. Eberhardt.

Fraser Bros., dealers in agricultural implements, Miniota, Man., have been succeeded by Fraser & Stewart.

The stock of the estate of H. N. Halpenny & Co., hardware dealers, Montreal, has been sold to J. H. Ashdown.

FIRES.

Alphonse Melini, coal oil dealer, Montreal, has been burned out; partially insured.

The Spicer Shingle Mill Co., Limited, Vancouver, B.C., have suffered loss by fire; insured.

A. L. Stickney, general merchant, Stickney, N.B., has been burned out; insurance, $1,300.

DEATHS.

K. J. Scully, painter, Hamilton, Ont., is dead.

Allan McMillan, general merchant, Mabou, N.S., is dead.

WIRELESS TELEPHONY.

SINCE Dr. Simon discovered that the electric arc acts as a telephone, various improvements and new applications of his methods have been made by Duddell, Ruhmer and others. Dr. Simon has himself recently made a remarkable application of his discovery to the transmission of speech to considerable distances by means of the beam of light emitted from the reflector of an ordinary searchlight. The arc of the searchlight is converted into a "speaking arc" by any of the known arrangements such as Duddell's or Ruhmer's. The variations of current produced by the influence of the telephone on the arc, naturally give rise to corresponding variations in the intensity of light emitted by the arc. The receiver is the same as that used by Graham Bell in his photophone, namely, a selenium cell placed at the focus of the parabolic reflector which receives the pulsating beam of light. A telephone in the circuit of the selenium cell translates the pulsations of the light into speech. Experiments carried out by Dr. Simon at Frankfort within the last few months have proved that such a method of "wireless telephony" is quite practicable. It is necessary, however, that the selenium cell should be very sensitive, and that its resistance should be as low as possible. The loudness of the speech emitted was remarkable; and the variations of current were sufficiently strong to impress a record on Poulsen's telegraphone. This invention of Simon's is likely to be of great value in the navy, since the ordinary searchlight may be used to transmit speech to any distance within its range.—Elektrotechnische Zeitschrift, Vol. 22, p. 196.

TORONTO RETAIL HARDWARE ASSOCIATION.

CONSTITUTION.

ARTICLE I.

NAME AND OBJECT.

Sec. 1. The name of the Association shall be The Retail Hardware Mrechants' Association of Ontario.

Sec. 2. The object of the Association shall be to promote the interests of and secure the friendly cooperation of its members.

ARTICLE II

MEMBERSHIP.

Sec. 1. Any person, firm or corporation in the Province of Ontario engaged in the retail business of selling hardware, stoves, tinware, paints, oils and glass, and who carries a general assortment of stock in any of the above lines known and recognized as a regular retail dealer, may become a member of the Association.

ARTICLE III.

OFFICERS.

Sec. 1. The officers of the Association shall consist of a president, three vice-presidents, a secretary and a treasurer and an executive committee consisting of 10 members.

ELECTIONS.

Sec. 2. The officers shall be elected annually by ballot, and shall hold office until their successors are elected.

ARTICLE IV.

MEETINGS.

Sec. 1. The annual meeting for the election of officers and the reception of officers' reports shall be held on the second Thursday in May in each year at 8 o'clock p.m.

Sec. 2. The regular meetings shall be held on the third Thursday in each month at 8 o'clock p.m.

ARTICLE V.

Sec. 1. Amendments to the constitution may be made at any regular meeting by a two-thirds vote of the members present, provided a copy of such proposed amendment be filed with the president of the Association at least one month prior to next regular meeting.

BY-LAWS.

ARTICLE I.

FEES.

Sec. 1. The annual membership fee shall be $2, payable in advance, and shall become due and payable at the annual meeting in each year.

Sec. 2. All fees must be paid before a person can be recognized as a member and entitled to vote in the Association, and any member in default for three months after reasonable notiee may be dropped from the roll of membership by a vote of the majority of the members present at any regular meeting.

Sec. 3. All applications for membership shall be made on the blank form provided, and shall be accompanied by the membership fee.

ARTICLE II.

VACANCIES.

Sec. 1. Vacancies in any office may be filled at any regular meeting of the Association.

ARTICLE III.

DUTIES OF OFFICERS.

Sec. 1. The president shall preside at all meetings of the Association, and in his absence the vice-presidents in their order.

Sec. 2. The secretary shall keep a record of the proceedings of all meetings of the Association and its committees, conduct all correspondence, collect all dues from the members and pay the same to the treasurer, notify all committees of their appointments, new members of their election, notify members of all meetings, and perform such other duties as prescribed by the constitution and by-laws.

Sec. 3. The treasurer shall receive all money due to the Association and keep an account of the same, pay all bills when certified as correct by the secretary and approved by the president of the Association, and make an annual report, and at such other times as the executive committee may require.

ARTICLE IV.

DUTIES OF COMMITTEES.

Sec. 1. The executive committee shall have general charge of the affairs and funds of the Association. They shall have full power, and it shall be their duty to carry out the purposes of the Association according to its constitution and by-laws.

Sec. 2. They shall present at each annual meeting a general report of the affairs of the Association.

Sec. 3. They shall appoint all standing committees for the current official year, not otherwise provided for, and shall have power to make rules for the government of such committees.

Sec. 4. An investigating committee of three members shall be elected annually. It shall be the duty of this committee to secure and investigate all complaints, provided such complaints are made in writing and signed by the party or parties aggrieved, and that such be accompanied by affidavits or other evidence to form a proper basis of complaint or source of investigation. They shall endeavor to adjust amicably all such grievances or complaints, and, if unsuccessful, shall bring the matter before the executive committee for action.

ARTICLE V.

QUORUM.

Sec. 1. Seven members shall constitute a quorum at any regular meeting, and 20 members at any annual meeting of the Association.

ARTICLE VI.

Sec. 1. Each member shall be entitled to one vote (or may in writing appoint a proxy) at any meeting of the Association.

ARTICLE VII.

AMENDMENTS.

Sec. 1. These by-laws may be amended by a two-thirds vote of those present and voting at any regular meeting of the Association, providing notice of such changes have been given at the previous regular meeting.

Adopted at Toronto, Thursday, May 30, 1901.

FRED. W. UNITT, Secretary. E. R. ROGERS, President.

C. A. Anderson, vice-president ; Charles Baxter and W. H. Bullen, of the Walkerville, Ont., Match Co., were in Ottawa this week to consider the advisability of moving their business from Walkerville to Ottawa. It will be remembered that the company's factory at Walkerville, Ont., was destroyed by fire some time ago. If they move to Ottawa, they purpose to manufacture woodenware as well as matches.

CATALOGUES, BOOKLETS, ETC.

A CANADIAN TRADE INDEX.

THE Canadian Manufacturers' Association have taken another step which will strengthen their position as leaders in industrial activity aad Canadian trade expansion. They have recently issued a classified membership directory, which, by virtue of the large proportion of the principal manufacturers of Canada represented, makes the book an excellent Canadian trade index. This index will be distributed free of charge to all foreign merchants interested in Canadian manufactures. It is, in fact, proposed to issue similar editions in different foreign languages, so that trade may be stimulated with the countries where these languages prevail.

The book will be even more useful for domestic trade, as the interchange of commodities among manufacturers is so great that an authoritative and not too comprehensive classified list should prove of inestimable value.

The Index is divided into two parts. Part I. contains an alphabetical list of members of the association. Part II. is an alphabetical list of articles produced, with under each item the names of the firms making it.

A VALUABLE BOOK FOR TINSMITHS.

The imprint of the David Williams Co., New York City, on a book dealing with architecture, mechanics, engineering, metallurgy, electricity or commercial subjects is a stamp of authority when on works dealing with such subjects. From their press has just been issued "The Tinsmiths' Helper and Pattern Book." The first edition of this work was published in 1879, and since then it has had a steady sale, as it was written by a practical tinner in simple, clear style. The new edition is printed from new cuts and type, and the appendix and tables have been brought thoroughly up-to-date. In the work 66 pages are devoted to rules and diagrams, from the simplest to the most complex. Almost as many pages are in the appendix, which contains an epitome of mensuration and of the various tables, rules and recipes which are of value to practical tinners. The book is neat, compact and of a shape to fit in the coat pocket. The price is $1, and should be abundantly worth the money to all tinsmiths.

THE JOSEPH DIXON CRUCIBLE CO.

A catalogue that every merchant should have on his file is one on Graphite Productions, just issued by The Joseph Dixon Crucible Co. It is well printed, excellently illustrated and contains a lot of information that will be of interest to dealers, quite regardless of the fact that it is a catalogue of a particular manufacturer. Any dealer sending a post card asking for one, mentioning HARDWARE AND METAL, will receive a copy without charge, by addressing the offices of the company at Jersey City, U.S.

This is a window display of paints in the store of Hunt & Collister, hardware and paint dealers, Cleveland. It attracted a great deal of attention, and our readers will, no doubt, be able to gather some ideas from it that will be helpful to them in making similar displays.

A DAINTY FIRE ARMS CATALOGUE.

One of the daintiest, yet most practical, catalogues that HARDWARE AND METAL has received for some time is the fire arms catalogue which the Harrington & Richardson Arms Co., Worcester, Mass., have recently issued. The booklet fully describes and illustrates the complete line of fire arms manufactured by this house, and is made doubly interesting by a series of colored plates depicting hunting scenes. This catalogue can be had upon application.

Carpenter: "Well, boy, have you ground all the tools, as I told you, while I've been out?" Boy (newly apprenticed): "Yes, master, all but this 'ere 'andsaw. An' I can't quite get the gaps out of it!"

CODFISH CATCH AND COD LIVER OIL.

With the close of the Lofoden codfishing season come the usual statistics as to the size of the catch and the amount of oil rendered. In the following table, taken from The Oil, Paint and Drug Reporter, we give the total catch of fish and the amount of oil, in hectoliters, rendered each year during the decade from 1892 to 1901, inclusive:

	Fish.	Oil.
1892	30,100,000	8,100
1893	26,700,000	18,[illegible]
1894	28,000,000	12,[illegible]
1895	32,600,000	12,300
1896	32,300,000	8,900
1897	31 300,000	18,300
1898	28,800,000	11,200
1899	24,500,000	18,500
1900	22,700,000	10,800
1901	18,000,000	15,700

It will be seen from the foregoing that, comparatively speaking, the amount of oil rendered this year is out of all proportion to the small catch of fish. The reason for this is that the cod livers this season have been used almost entirely for medicinal oil, and not for tanners' oil, for which purpose a large part of the catch is ordinarily used. It will be further noted that this year's rendering of oil is considerably in excess of that of last year, but a careful observation of the figures will show that on alternate years, in almost every case, there is a large production of oil, and the crop this year is not up to the figures of 1899 by upward of 3,000 hectoliters. This fact will, in a great measure, prevent any material reduction in prices. Below is a table of the high and low prices each year from 1891 to 1900, both inclusive:

	High.	Low.
1891	$23 00	$13.50
1892	23.00	21.00
1893	22.00	19.00
1894	28.00	19.50
1895	49.00	27.00
1896	60 00	43.00
1897	43.00	21.00
1898	25.00	20 00
1899	26.00	19.50
1900	26.00	22.00

PERSONAL MENTION.

Mr. C. S. Archibald, of St. John, N.B., representing the Portland Rolling Mills, has been in Montreal for a week, visiting the different wholesale houses in his firm's interests. They certainly ought to be safe in his hands for he is an exceedingly affable gentleman as well as shrewd and business-like.

HOW TO WIN SUCCESS.

CHAS. M. SCHWAB, President of The United States Steel Corporation, in an address to the graduating class of St. George's Evening Trade School at the commencement exercises in the Memorial Building, New York, a few days ago, declared a college education usually a handicap to one who would succeed in business. He held the boy who got an early start the one more certain to succeed. Chairman Frank E. Havemeyer, introduced Mr. Schwab as a man who had fought battles and won victories in the struggle of life and therefore was well qualified to give boys advice founded on practical experience.

"I will speak to you," began Mr. Schwab, "just as if you had come to my office asking for advice, and the first thing I will say to you is to come alone. Don't come with somebody's backing. Learn to rely upon yourself. That is the first lesson. If you come endorsed by somebody of influence it always will leave room for others to say that whatever position you may get you got by influence and not because of your individual merit. No true success is built on influence. You must win your positions for yourself.

MUST DO MORE THAN HIS DUTY.

"Then, here is another thing that is essential—you must do what you are employed to do a little better than anybody else does it. Everybody is expected to do his duty, but the boy who does his duty, and a little more than his duty, is the boy who is going to succeed in the world. You must take an interest in what you are doing, and it must be a genuine interest."

Here Mr. Schwab told a story which everybody understood referred to himself. Afterward he told another story which it was equally well understood referred to H. C. Frick. The story follows:

"There were ten boys employed by a concern once, and one night the manager said to his subordinate: 'Tell the boys they are to stay a little longer to-night—tell them that they are to stay until six o'clock. Don't tell them why. Just tell them that and watch them.' So this was done, and when six o'clock came around there was just one boy who was wholly interested in his work, and was not watching the clock to see what time it was. That boy was the one the manager wanted, and he was taken into the office, and as he continued to manifest the same interest in his work he was promoted until at last he got a quite responsible place.

ALWAYS ON TIME.

"Then, there was another boy. He began carrying water, and he did it so much better than any other boy, seeing to it always that the men had good water, cool water, and plenty of it, that he attracted attention to himself. He was taken into the office, where he became in time superintendent and then general manager, and he is now the man that is at the head of the great Carnegie Company with thousands of men under him. As a boy he did more than the ordinary run of boys did, and so attracted attention, and that was the secret of his first step upward.

"I was in a bank down town the other day when a newsboy came in and sold the banker a paper. After he had gone out the banker said to me: 'For two years now that boy has been coming in here at the time I told him to come—two o'clock. He does not come before two nor after two, but at two precisely. He has sold me a paper every week-day in that way when I have been here, without a break. He sells it just for one cent—its price. He neither asks more nor seems to expect more.

NEWSBOY WHO HAS A FUTURE.

"'It is a cold commercial transaction. Now, a boy that will attend to business in that way has got stuff in him. He does not know it yet, but I am going to put him in my bank, and you will see that he will be heard from.'

MANITOBA MARKETS.

WINNIPEG, June 3, 1901.

HARDWARE AND PAINTS, OILS AND GLASS.

BUSINESS continues steady in all lines and with few changes in price, excepting nails, which have advanced 10c. per keg, and linseed oil, which has advanced 2c. per gal.

Quotations for the week are as follows:

Barbed wire, 100 lb.	$3 45
Plain twist	3 45
Staples	3 95
Oiled annealed wire 10	3 95
" 11	4 00
" 12	4 05
" 13	4 20
" 14	4 35
" 15	4 45
Wire nails, 30 to 60 dy, keg	3 50
" 16 and 20	3 60
" 10	3 60
" 8	3 70
" 6	3 75
" 4	3 90
" 3	4 15
Cut nails, 30 to 60 dy.	3 00
" 20 to 40	3 05
" 10 to 16	3 10
" 8	3 1
" 6	3 26
" 4	3 30
" 3	3 65
Horsenails, 45 per cent. discount.	
Horseshoes, iron, No. 0 to No 1	4 65
No. 2 and larger	4 40
Snow shoes, No. 0 to No. 1	4 90
No. 2 and larger	4 40
Steel, No. 0 to No. 1	4 95
No. 2 and larger	4 70
Bar iron, $2.50 basis.	
Swedish iron, $5.00 basis.	
Sleigh shoe steel	3 00
Spring steel	3 25
Machinery steel	3 75
Tool steel, Black Diamond, 100 lb	8 50
Jessop	13 00
Sheet iron, black, 10 to 20 gauge, 100 lb.	3 50
20 to 26 gauge	3 75
28 gauge	4 00
Galvanized American, 16 gauge	2 54
18 to 22 gauge	4 50
24 gauge	4 75
26 gauge	5 00
28 gauge	5 25
Genuine Russian, lb	12
Imitation " "	8
Tinned, 24 gauge, 100 lb.	7 75
26 gauge	8 00
28 gauge	8 50
Tinplate, IC charcoal, 20 x 28, box	10 75
" IX "	12 75
" IXX "	14 75
Ingot tin	33
Canada plate, 18 x 21 and 18 x 24	3 25
Sheet zinc, cask lots, 100 lb	7 50
Broken lots	8 00
Pig lead, 100 lb	6 00
Wrought pipe, black up to 2 inch	50 an 10 p.c.
" Over 2 inch	50 p.c.
Rope, sisal, 7-16 and larger	$11 00
" ⅜	11 50
" ¼ and 5-16	12 00
Manila, 7-16 and larger	14 00
" ⅜	14 50
" ¼ and 5-16	15 00
Solder	20
Cotton Rope, all sizes, lb.	17
Axes, chopping	$ 7 50 to 12 00
" double bitts	12 00 to 18 00
Screws, flat head, iron, bright	87½
Round " "	82½
Flat " brass	80
Round " "	75
Coach	57½ p.c.
Bolts, carriage	55 p.c.
Machine	55 p.c.
Tire	60 p.c
Sleigh shoe	65 p.c.
Plough	40 p.c.
Rivets, iron	50 p.c.
Copper, No. 8	35
Spades and shovels	40 p.c.
Harvest tools	50, and 10 p.c.
Axe handles, turned, s. g. hickory, doz.	$2 50
No. 1	1 50
No. 2	1 25
Octagon extra	1 75
No. 1	1 25
Files common	70, and 10 p.c.
Diamond	60
Ammunition, cartridges, Dominion R.F.	50 p.c.
Dominion, C.F., pistol	30 p.c.
" military	15 p.c.
American R.F.	30 p.c.
C.F. pistol	5 p.c.
C.F. military	10 p.c. advance.
Loaded shells:	
Eley's soft, 12 gauge black	16 50
chilled, 12 guage	18 00
soft, 10 guage	21 00
chilled, 10 guage	23 00
Shot, Ordinary, per 100 lb	6 25
Chilled	6 75
Powder, F.F., keg	4 75
F.F.G.	5 00
Tinware, pressed, retinned	75 and 2½ p.c.
" " plain	70 and 15 p.c.
Graniteware, according to quality	50 p.c.

PETROLEUM.

Water white American	25½c.
Prime white American	24c.
Water white Canadian	22c.
Prime white Canadian	21c.

PAINTS, OILS AND GLASS.

Turpentine, pure, in barrels	$ 61
Less than barrel lots	66
Linseed oil, raw	92
Boiled	95
Lubricating oils, Eldorado castor	25½
Eldorado engine	24½
Atlantic red	27½
Renown engine	41
Black oil	23½ to 25
Cylinder oil (according to grade)	55 to 74
Harness oil	61
Neatsfoot oil	$ 1 00
Steam refined oil	85
Sperm oil	1 50
Castor oil per lb.	11½
Glass, single glass, first break, 16 to 25 united inches	2 25
26 to 40 per 50 ft.	2 50
41 to 50 " 100 ft.	5 50
51 to 60 " " "	6 00
61 to 70 per 100-ft. boxes	6 50
Putty, in bladders, barrel lots per lb.	2½
kegs "	2¾
White lead, pure per cwt.	7 00
No 1 "	6 75
Prepared paints, pure liquid colors, according to shade and color. per gal.	$1.30 to $1.90

NOTES.

The community of Winnipeg have been very much shocked over the death, by his own hand, on Friday last at Rat Portage, of Mr. Wm. Hargreaves, one of the best known travellers in the West. Mr. Hargreaves was a man of the most sterling character, and his daily life commanded as much respect as his unquestionable business ability. The reason of the rash act was, without doubt, temporary insanity, induced be severe and prolonged insomina following upon an attack of la grippe last winter. Mr. Hargreaves was last year the president of the N.W.C.T.A., having always been prominently identified with the Masonic body. He was an active and valued member of the First Baptist Church. He leaves a wife and two bright, manly lads, who have the deep sympathy of the community in their terrible bereavement.

MARKETS AND MARKET NOTES

QUEBEC MARKETS

Montreal, June 7, 1901.

HARDWARE.

THE night work has ceased in most of the houses, and, while a good trade is still reported, it can be accommodated in the day time now. In prices there is no change to note this week. There is reported to have been a little cutting going on in screens, but we believe this has stopped. Metals are as firm as ever. Barb wire continues quite scarce, and every mail brings inquiries. A goodly number of orders booked some time ago are not yet filled. Goods that came to hand during the week were not brought near the warehouse. Shipments are expected from Cleveland and England within the next few days. Galvanized wire orders are also in many cases unfilled, and Canadian mills say they cannot catch up for some weeks yet. Oiled and annealed wire are in supply. Wire nails are none too plentiful, and both wire and cut nails have been in exceptionally good request during the week. Poultry netting is in light supply, but the demand is not heavy. Builders' tools are not selling as well as they were, but binder twine and harvest tools are moving in fair quantities this week. Some refrigerators are still wanted, as also are lawn requisites, oil stoves, pipes and elbows. Payments are first class; in fact, cash purchases are the order of the day.

BARB WIRE—There is still a scarcity of barb wire and the wholesale houses are besieged with inquiries for spot goods. Both Canadian and American mills are behind with their shipments. The price is unchanged at $3.05 per 100 lb. f.o.b. Montreal.

GALVANIZED WIRE—Deliveries are still very slow, but the mills are catching up with orders. Prices are unchanged. We quote as follows: No. 5, $4.25; Nos. 6, 7 and 8 gauge, $3.55; No. 9, $3.10; No. 10, $3.75; No. 11, $3.85; No. 12, $3.25; No. 13, $3.35; No. 14, $4.25; No. 15, $4.75; No. 16, $5.00.

SMOOTH STEEL WIRE — The demand for oiled and annealed wire has been very good. We quote oiled and annealed as follows: No. 9, $2.80; No. 10, $2.87; No. 11, $2.90; No. 12, $2.95; No. 13, $3.15 per 100 lb. f.o.b. Montreal, Toronto, Hamilton, London, St. John and Halifax.

FINE STEEL WIRE—A fair trade is reported in this line. The discount is 17½ per cent. off the list.

BRASS AND COPPER WIRE—The demand is moderate. The discount on brass is 55 and 2½ per cent., and on copper 50 and 2½ per cent.

FENCE STAPLES—The demand for small lots is good. We quote: $3.25 for bright, and $3.75 for galvanized, per keg of 100 lb.

WIRE NAILS—There is still a good demand for wire nails, and trade in this line is rather active. The market is firm and steady. Supplies are none too plentiful, and it is difficult at times to have large orders filled promptly. We quote $2.85 for small lots and $2.77½ for carlots, f.o.b. Montreal, London, Toronto, Hamilton and Gananoque.

CUT NAILS—Inquiries for cut nails are more numerous this week, and some 50-keg orders have been filled. Shingle nails are in particularly good request. We quote as follows: $2.35 for small and $2.25 for carlots; flour barrel nails, 25 per cent. discount; coopers' nails, 30 cent. discount.

HORSE NAILS—The feature of the horse nail market is the demand for the higher grade of nails. Those manufacturers who are manufacturing two grades of nails, find that the higher-priced goods are selling better than their inferior brand. This should be a standing illustration of the fact that it is quality the public want. "C" brand is held firmly at discount of 50 and 7½ per cent. off the new list. "M" brand is quoted at 60 per cent. off old list on oval and city head and 66⅔ per cent. off countersunk head.

HORSESHOES—Business is rather slow. We quote: Iron shoes, light and medium pattern, No. 2 and larger, $3.50; No. 1 and smaller $3.75; snow shoes, No. 2 and larger, $3.75; No. 1 and smaller, $4.00; X L steel shoes, all sizes, 1 to 5, No. 2 and larger, $3.60; No. 1 and smaller, $3.85; feather-weight, all sizes, $4.85; toe weight steel shoes, all sizes, $5.95 f.o.b. Montreal; f.o.b. Hamilton, London and Guelph, 10c. extra.

POULTRY NETTING—There is not much demand for poultry netting just now. Stocks are light. The discount is 55 per cent.

GREEN WIRE CLOTH—The demand continues fairly brisk. The price is still $1.35 per 100 sq. ft.

SCREEN DOORS AND WINDOWS—There is reported to have been some slight cutting going on in this line, but prices are now being adhered to. We quote as follows: Screen doors, plain cherry finish, $7.30 per doz.; do. fancy, $11.50 per doz.; walnut, $7.30 per doz., and yellow, $7 45; windows, $2.25 to $3.50 per doz.

SCREWS—A fairly good business is being done in screws Discounts are: Flat head bright, 87½ and 10 per cent. off list; round head bright, 82½ and 10 per cent.; flat head brass, 80 and 10 per cent.; round head brass, 75 and 10 per cent.

BOLTS—The demand for bolts is fair, but hardly as brisk as last week. Discounts are as follows: Norway carriage bolts, 65 per cent.; common, 60 per cent.; machine bolts, 60 per cent.; coach screws, 70 per cent.; sleigh shoe bolts, 72½ per cent.; blank bolts, 70 per cent.; bolt ends, 62½ per cent.; plough bolts, 60 per cent.; tire bolts, 67½ per cent.; stove bolts, 67½ per cent. To any retailer an extra discount of 5 per cent. is allowed. Nuts, square, 4c. per lb. off list; hexagon nuts, 4½c. per lb. off list. To all retailers an extra discount of ½c. per lb. is allowed.

BUILDING PAPER—There is a continued

demand for building paper at unchanged prices. We quote: Tarred felt, $1.70 per 100 lb.; 2-ply, ready roofing, 80c. per roll; 3-ply, $1.05 per roll; carpet felt, $2.25 per 100 lb.; dry sheathing, 30c. per roll; tar sheathing, 40c. per roll; dry fibre, 50c. per roll tarred fibre, 60c. per roll; O.K. and I.X.L., 65c. per roll; heavy straw sheathing, $28 per ton; slaters' felt, 50c. per roll.

RIVETS AND BURRS—A few orders have been filled this week. Discounts on best iron rivets, section, carriage, and wagon box, black rivets, tinned do., coopers' rivets and tinned swedes rivets, 60 and 10 per cent.; swedes iron burrs are quoted at 55 per cent. off; copper rivets, 35 and 5 per cent. off; and coppered iron rivets and burrs, in 5-lb. carton boxes, are quoted at 60 and 10 per cent. off list.

BINDER TWINE—The demand is improving and business in this article will soon be in full swing. Lots of several tons have been sold this week. We quote as follows: Blue Ribbon, 11½c.; Red Cap, 9¾c.; Tiger, 8¾c.; Golden Crown, 8c.; Sisal, 8¼c.

CORDAGE—Some good sales are reported again this week. Manila is worth 13½c. per lb. for 7-16 and larger; sisal is selling at 10c., and lathyarn 10c.

HARVEST TOOLS—Business is rather brisk in all the lines coming under this head. The discount is 50, 10 and 5 per cent.

SPADES AND SHOVELS—A small business has been done in spades and shovels during the week. The discount is 40 and 5 per cent. off the list.

LAWN MOWERS—A number of mail orders have been filled during the past few days and trade is, on the whole, very satisfactory. We quote: High wheel, 50 and 5 per cent. f.o.b. Montreal; low wheel, in all sizes, $2.75 each net; high wheel, 11-inch, 30 per cent. off.

FIREBRICKS—There is a fair business doing in firebricks at $17.50 to $22 for Scotch, and at $17 to $21 for English per 1,000 ex wharf.

CEMENT — Canadian and American cements continue to sell well. We quote: German cement, $2.35 to $2.50; English, $2.25 to $2.35; Belgian, $1.70 to $1.95 per bbl. ex wharf, and American, $2.30 to $2.45, ex cars.

METALS.

The firm feeling that we have reported in English sheet metals has encouraged importers here to operate and a little more business has been done this week. Dealers in the country are also more inclined to do some purchasing. Stocks of all metals are extremely light and on that account values are being well maintained. Ingot tin is higher, while galvanized iron is reported to be advancing. The manufacturers of iron pipe have advanced quotations 15c. this week, but as there has been some cutting among jobbers our schedule is unchanged.

PIG IRON—Business is fair while the market has a firm undertone. Canadian iron is worth about $18.50 per ton and Summerlee, $20.50 to $21 ex wharf.

BAR IRON — A fair demand is reported at unchanged prices. The market is firm at $1.75 for merchants' bar and $2 for horseshoe.

HOOP IRON—There is no change to report in this line. The demand is only fair. We quote $2.75 for No. 19 gauge and $3 for No. 17.

BLACK SHEETS — The market is firm and dealers are more inclined to operate. We quote: 8 to 16 gauge, $2.50 to $2 60; 26 gauge, $2.55 to $2 65, and 28 gauge, $2.60 to $2.70.

GALVANIZED IRON — Cable advices to hand this week report prices somewhat higher on certain brands of galvanized iron. The market is strong on account of the advance in black sheets. We quote as follows: No. 28 Queen's Head, $4.65; Apollo, 10¾ oz., $4.50, and Comet, $4.40 to $4.45, with a 15c. reduction for case lots.

COPPER—Fair sales have been made at 18c.

INGOT TIN—The market is somewhat higher, and prices have been advanced here in sympathy. Lamb and Flag is worth 35 to 36c. in a small way.

LEAD—Unchanged at $3.75 to $3.85.

LEAD PIPE—A fair trade has been done at reduced prices. We quote: 7c. for ordinary and 7½c. for composition waste, with 30 per cent. off.

IRON PIPE—The rolling mills are 15c. higher this week, but, as some of the jobbing houses have been cutting, our schedule will stand still. We quote: Black pipe, ¼, $3 per 100 ft.; ⅜, $3; ½, $3.05; ¾, $3.30; 1-in., $4.75; 1¼, $6.45; 1½, $7.75; 2-in. $10.35. Galvanized, ½, $4.60; ¾, $5.25; 1-in., $7.50; 1¼, $9.80; 1½, $11.75; 2-in., $16.

TINPLATES—The market continues firm, further advances are expected in consequence of the upward trend of the tin market. We quote: Coke plates, $3.75 to $4; charcoal, $4.25 to $4.50; extra quality, $5 to $5.10.

CANADA PLATE—Canada plates are also very firm, and the $2.35 and $2.40 at the wharf offers that were heard of last week have been withdrawn. We quote: 52's, $2.55 to $2.60; 60's, $2.65 to $2.70; 75's, $2.70 to $2.80; full polished, $3.10, and galvanized, $4.

STEEL—Unchanged. We quote: Sleighshoe, $1.95; tire, $2; bar, $1.95; spring, $2.75; machinery, $2.75, and toe-calk, $2.50.

SHEET STEEL—We quote: Nos. 22 and 24, $3. and Nos. 18 and 20, $2.85.

TOOL STEEL—Black Diamond, 8c. and Jessop's, 13c.

TERNE PLATES—On this market terne plates are scarce, the arrivals of new goods having been very light this year. Locally, $7.50 seems to be the prevailing value.

COIL CHAIN—The market is steady. We quote as follows: No. 6, 11½c.; No. 5, 10c.; No. 4, 9½c.; No. 3, 9c.; ¼-inch, 7½c. per lb.; 5-16, $4.85; 5-16 exact, $5.30; ⅜, $4.40; 7-16, $4.20; ½, $3.95; 9-16, $3.85; ⅝, $3.55; ¾, $3.45; ⅞, $3.40; 1-in., $3.35. In carload lots an allowance of 10c. is made.

SHEET ZINC — Fair quantities of zinc are selling at last week's advance. The ruling quotation is $6.25.

ANTIMONY—Quiet, at 10c.

ZINC SPELTER—Is worth 5c.

SOLDER—We quote: Bar solder, 18½c.; wire solder, 20c.

GLASS.

The advance is well maintained. We quote: First break, $2.10; second, $2.20 for 50 feet; first break, 100 feet, $3.90; second, $4.10; third, $4.60; fourth, $4.85; fifth, $5.35; sixth, $5.85, and seventh, $6.35.

PAINTS AND OILS.

Business continues decidedly brisk in mixed paints, and May has been a banner month with more than one manufacturing concern. Linseed oil is still scarce, and large lots of raw have changed hands at 76c. Turpentine is steady and unchanged. We quote:

WHITE LEAD—Best brands, Government standard, $6.25; No. 1, $5.87½; No. 2, $5.50; No. 3, $5.12½, and No. 4, $4.75 all f.o.b. Montreal. Terms, 3 per cent. cash or four months.

DRY WHITE LEAD — $5.25 in casks; kegs, $5.50.

RED LEAD — Casks, $5.00; in kegs, $5.25.

DRY WHITE ZINC—Pure, dry, 6¼c.; No. 1, 5¼c.; in oil, pure, 7¼c.; No. 1, 6¼c.; No. 2, 5¼c.

PUTTY—We quote: Bulk, in barrels, $1.90 per 100 lb.; bulk, in less quantity, $2.05; bladders, in barrels, $2.10; bladders, in 100 or 200 lb. kegs or boxes, $2.25; in tins, $2.55 to $2.65; in less than 100-lb. lots, $3 f.o.b. Montreal, Ottawa, Toronto, Hamilton, London and Guelph. Maritime Provinces 10c. higher, f.o.b. St. John and Halifax.

LINSEED OIL—Raw, 80c.; boiled, 83c. in 5 to 9 bbls., 1c. less, 10 to 20 bbl. lots, open, net cash, plus 2c. for 4 months. Delivered anywhere in Ontario between Montreal and Oshawa at 2c. per gal. advance and freight allowed.

TURPENTINE—Single bbls., 53c.; 2 to 4 bbls., 52c.; 5 bbls. and over, open terms, the same terms as linseed oil.

MIXED PAINTS—$1.25 to $1.45 per gal.

CASTOR OIL—8¾ to 9¼c. in wholesale lots, and ½c. additional for small lots.

SEAL OIL—47½ to 49c.

COD OIL—32½ to 35c.

NAVAL STORES — We quote: Resins, $2.75 to $4.50, as to brand; coal tar, $3.25 to $3.75; cotton waste, 4½ to 5½c. for colored, and 6 to 7½c. for white; oakum, 5½ to 6½c., and cotton oakum, 10 to 11c.

PARIS GREEN—Petroleum barrels, 16¾c. per lb.; arsenic kegs, 17c.; 50 and 100-lb. drums, 17½c.; 25-lb. drums, 18c.; 1-lb-packages, 18½c.; ½-lb. packages, 20½c.; 1-lb. tins, 19½c.; ½-lb. tins, 21½c. f.o.b. Montreal; terms 3 per cent. 30 days, or four months from date of delivery.

SCRAP METALS.

Prices for wrought scrap are decidedly firm, and dealers expect higher prices. Dealers are now paying the following prices in the country: Heavy copper and wire, 13 to 13½c. per lb.; light copper, 12c.; heavy brass, 12c.; heavy yellow, 8½ to 9c.; light brass, 6½ to 7c.; lead, 2½ to 2¾c. per lb.; zinc, 2¼ to 2½c.; iron, No. 1 wrought, $14 to $16 per gross ton f.o.c. Montreal; No. 5 cast, $13 to $14; stove plate, $8 to $9; light iron, No. 2, $4 a ton; malleable and steel, $4; rags, country, 60 to 70c. per 100 lb.; old rubbers, 7¼c. per lb.

HIDES.

The demand is rather light, and prices remain unchanged. We quote: Light hides, for No. 1; 5½c. for No. 2, and 4½c. for 6½c. No. 3. Lambskins, 10c.; sheepskins, 90c.; calfskins, 8c. for No. 1 and 6c. for No. 2.

PETROLEUM.

Trade has now been reduced to small proportion. We quote as follows: "Silver Star," 14½ to 15½c.; "Imperial Acme," 16 to 17c.; "S C. Acme," 18 to 19c., and "Pratt's Astral," 18½ to 19½c.

TRADING STAMPS IN TORONTO.

The trading stamp question is nearing a crisis in Toronto. At their last meeting, the Toronto Retail Grocers' Association passed a resolution asking the city council to pass legislation abolishing trading stamps, as the municipality has been empowered to do by the Ontario Legislature during the last session. This resolution was sent to the city council a fortnight ago, and was passed over by the board of control to the legislative committee.

On Monday, at the meeting of council, Ald. Richardson moved that this committee be instructed to frame and introduce a by-law to abolish stamps. After a brief discussion the resolution was passed.

ONTARIO MARKETS.

TORONTO, June 8, 1901.

HARDWARE.

BUSINESS in the wholesale hardware and kindred trades is keeping up well. The demand is particularly good for scythes, snaths and goods of that kind. Lawn mowers are also in active request, but scarce in some sizes. Business continues active in ice cream freezers, also in oil stoves. In tinware and enamelled ware only a fair trade is reported. The scarcity in barb wire and plain galvanized wire appears to be even more pronounced than before, and oiled and annealed wire, which is in good supply, is still reaping the benefit of the scarcity in the lines mentioned. Business has improved in wire nails, although the quantities wanted are not large. A feature of the trade in cut nails is some good inquiries that have been received from the Coast. Horse nails continue quiet, and in horseshoes trade is fair for this time of the year. Screws, rivets and burrs, and bolts and nuts are all meeting with a good demand. Rope continues active in small quantities. A fair trade is being done in sporting goods. Cutlery is only moving slowly. A good trade is reported in spades and shovels. Building paper is going out freely. Payments are fair.

BARB WIRE—It is still impossible to get sufficient wire to supply the demand and the scarcity, if anything, is even more pronounced than it was a week ago. We quote $3 05 per 100 lb. from stock Toronto; f.o.b. Cleveland $2 82½ for less than carlots, and $2.70 for carlots.

GALVANIZED WIRE—Exactly the same remarks apply to this line as to barb wire. We quote: Nos. 6, 7 and 8, $3.50 to $3.85 per 100 lb., according to quantity; No. 9, $2.85 to $3 15; No. 10, $3.60 to $3.95; No. 11, $3.70 to $4 10; No. 12, $3 to $3 30; No. 13, $3 10 to $3 40; No. 14, $4.10 to $4 50; No. 15, $4.60 to $5.05: No. 16, $4.85 to $5.35. Nos. 6 to 9 base f.o.b. Cleveland are quoted at $2.57½ in less than carlots and 12c. less for carlots of 15 tons.

SMOOTH STEEL WIRE—A good demand is still being experienced for this line, and the same causes, namely, the scarcity in galvanized plain twist and barb wire, are contributing to it. A little hay-baling wire is also going out this week. The net selling prices for oiled and annealed are: Nos. 6 to 8, $2 90; 9. $2.80; 10, $2.87; 11, $2.90; 12, $2.95; 13, $3.15; 14, $3.37; 15. $3.50; 16, $3 65. Delivery points, Toronto, Hamilton, London and Montreal, with freights equalized on these points.

WIRE NAILS—The demand is rather brisk this week, although the quantities wanted are not individually large. The makers report that, whereas a week or two ago stocks were accumulating, they are now decreasing slightly. We still quote $2.85 base for less than carlots, and $2.77½ for carlots. Delivery points: Toronto, Hamilton, London, Gananoque and Montreal.

CUT NAILS—A slight stimulus to business has been given to this line by orders which have been received from the Pacific Coast. The demand, however, in this part of the country, except for shingle nails, is still decidedly dull. The base price is $2.35 per keg for less than carlots, and $2.25 for carlots. Delivery points: Toronto, Hamilton, London, Montreal and St. John, N.B.

HORSE NAILS—Trade is slow, with prices as before. Discount on "C" brand, oval head, 50 and 7½ per cent. off new list, and on "M" and other brands, 50, 10 and 5 per cent. off the old list. Countersunk head 60 per cent.

HORSESHOES—Although not a large trade is being done, business in this line is proportionately better than that in horse nails. We quote f. o. b. Toronto as follows: Iron shoes, No. 2 and larger, light, medium and heavy, $3.60; snow shoes, $3.85; light steel shoes, $3.70; featherweight (all sizes), $4.95; iron shoes, No. 1 and smaller, light, medium and heavy (all sizes), $3.85; snow shoes, $4; light steel shoes, $3.95; featherweight (all sizes), $4.95.

SCREWS—Business continues fairly good, although no new features have developed during the past week. Discounts are: Flat head bright, 87½ and 10 per cent.; round head bright, 82½ and 10 per cent.; flat head brass, 80 and 10 per cent.; round head brass, 75 and 10 per cent.; round head bronze, 65 per cent., and flat head bronze at 70 per cent.

RIVETS AND BURRS — A brisk trade is still being done in rivets and burrs, and prices are steady and unchanged. We quote: Iron rivets, 60 and 10 per cent.; iron burrs, 55 per cent.; copper rivets and burrs, 35 and 5 per cent.

BOLTS AND NUTS—A brisk trade is still the feature of this line, and prices rule steady and unchanged. We quote: Carriage bolts (Norway), full square, 65 per cent.; carriage bolts full square, 65 per cent.; common carriage bolts, all sizes, 60 per cent.; machine bolts, all sizes, 60 per cent.; coach screws, 70 per cent.; sleighshoe bolts, 72½ per cent.; blank bolts, 60 per cent.; bolt ends, 62½ per cent.; nuts, square, 4c. off; nuts, hexagon, 4½c. off; tire bolts, 67½ per cent.; stove bolts, 67½; plough bolts, 60 per cent.; stove rods, 6 to 8c.

ROPE—A good demand is being experienced for sisal and manila rope. One of the features of the trade is the fact that dealers are beginning to place orders for hay-fork rope. The demand for cotton rope is fair. The base price for sisal is still 10c., and that of manila 13½c.

BINDER TWINE—The sorting-up demand has not yet set in, and very little business is in consequence doing. The ruling quotations are: Pure manila, 650 ft., 12c.; manila, 600 ft., 9½c.; mixed, 550 ft., 8½c.; mixed, 500 ft., 8 to 8¼c.

CUTLERY—The movement in this line is light at present, and there are no particular features.

SPORTING GOODS — There is quite a little gunpowder, shot, ammunition and small rifles going out.

LAWN MOWERS — A good many late orders are coming to hand, and, as the Canadian manufacturers have been doing a large business this season, a scarcity is reported in some sizes. We quote: "Star" mowers, $2.35, and discount on Woodyatt mowers, 40 and 10 per cent. off the list.

ENAMELLED WARE AND TINWARE — Trade in both these lines is only fair.

GRINDSTONES — Trade is beginning to open up in grindstones and grindstone fixtures. As a result of the recent combination among the manufacturers, prices are much higher than they were last year. We quote as follows: 2 inch thick, 40 to 20 lb., $25 per ton; 2 inch thick, under 40 lb., $28; under 2 inch thick, $29.

OIL AND GAS STOVES—Business has been good during the past week in oil stoves, but not much is being experienced in gas stoves.

COOKING AND HEATING STOVES — A good many orders are being booked for fall delivery. Shipments will begin about July or August.

REFRIGERATORS—There is a little doing, but business is beginning to taper off. Wholesale dealers, however, report that in some lines they are getting short.

ICE CREAM FREEZERS—These continue to be one of the most active lines in the wholesale hardware trade.

GREEN WIRE CLOTH—Trade is steady at $1.35 per 100 sq. ft.

SCREEN DOORS AND WINDOWS—Trade is only moderate. We quote: Screen doors, 4-in. styles, $7.20 to $7.80 per doz.; ditto, 3-in. styles, 20c. per doz. less; screen windows, $1.60 to $3.60 per doz., according to size and extension.

BUILDING PAPER—Trade continues active and prices unchanged. We quote: Building paper, 30c.; tarred lining, 40c., and tarred roofing, $1.65.

POULTRY NETTING — Not much is being done, as is to be expected at this season of the year. Discount 55 per cent.

HARVEST TOOLS—A good many of these are going out, and prices are steady. Discount, 50, 10 and 5 per cent.

SPADES AND SHOVELS—There is just the ordinary demand to be noted this week. Discount, 40 and 5 per cent.

EAVETROUGH — The movement in eavetrough and conductor pipe continues good. The ruling price of eavetrough is $3.25 per 100 ft. for 10 inch.

CEMENT—There is a big trade doing, and prices are firm, but without change. We quote barrel lots: Canadian portland, $2.25 to $2.75; German, $3 to $3.15; English, $3; Belgian, $2.50 to $2.75; Canadian hydraulic, $1.25 to $1.50.

METALS.

The most active line in metals is still galvanized sheets, which are held firm at quotations. Pig tin rules firm, although local business appears to have been checked somewhat by the higher prices. The steel market is still firm, although the demand is not as active as it was. Pig iron is quiet, but fairly steady as to price. In other lines of metals trade is just moderate.

PIG IRON—We hear of a little inquiry from local consumers of pig iron, and the price quoted for round lots is still about $16 for No. 3 iron. Buyers, however, think they will be able to do a little better than this. As the large buyers in Canada have placed their orders for the bulk of their supplies for the present year, we naturally do not look for much business in pig iron for some time, as far as this country is concerned. In the United States the pig iron market is quiet with prices fairly steady.

BAR IRON—A good trade is being done, although the orders are not, perhaps, as numerous as they were. The mills, however, are still a good deal behind with their orders. The ruling base price for ordinary quantities is $1.85 per 100 lb.

STEEL—A good demand continues to be experienced, although the boom appears to have reached its height for the time being. This is evident from the fact that the makers of steel plates are scarcely as independent in regard to orders as they were a short time ago. We quote as follows: Merchantable cast steel, 9 to 15c. per lb.; drill steel, 7 to 8c. per lb.; "Black Diamond" tool steel, 10 to 11c.; Jessop's and Firth's tool steel, 12½ to 13c.; toe calk steel, $2.85 to $3; tire steel, $2.30 to $2.50; sleighshoe steel, $2.10 to $2.25; reeled machinery steel, $3; hoop steel, $3.

GALVANIZED SHEETS—The market is even more stronger than it was a week ago. Importers here have been advised this week of a further advance of 10s. in the price in England. Quotations which importers are receiving here now are about £1 19s. higher than in March last. The makers in the United States are still withholding prices from the Canadian market. We quote 28 gauge English at $4.50, and American at $4.40. The demand for galvanized sheets is still active, and even more so than in any other line of metals.

BLACK SHEETS—These are also firmer in the British market. A good many orders for importation are being booked by local wholesalers. We quote: 28 gauge, common sheets, at $3, and dead flat at $3.50.

CANADA PLATES—Quotations from England show that prices there are firmer. Some good-sized orders have been placed on importation account, but very little is being done from stock. We quote: All dull, $2.90; half polished, $3, and all bright, $3.50.

PIG TIN—The outside markets ruled strong and higher, particularly in London up till Tuesday last, when prices declined both in London and New York. Since then, however, the market has again taken an upward turn, and at the time of writing prices are firm. Locally, the higher prices have checked consumption, but stocks are rather light. Ruling quotations for Lamb and Flag are 31½ to 32c. per lb.

TINPLATES — The tinplate market still rules firm, but there is not a great deal of business being done. Dealers have had to pay 9d. per box higher this week for importation. The stock of tinplates on the local market is light.

TINNED SHEETS—There is not a large business doing, but the small orders that are coming in are wanted quickly. What tinned sheets are being used for just now is cheese vat purposes.

COPPER — Although some fairly good sized lots have changed hands during the past week in ingot copper, the demand is not what might be termed active. We quote: Ingot, 17¾c., and bar, 23 to 25c.

SOLDER — The demand for solder keeps good. We quote: Half-and-half, guaranteed, 18½c.; ditto, commercial, 18c.; refined, 18c., and wiping, 17c.

IRON PIPE — Trade in iron pipe is fairly good, and prices are steady and unchanged. We quote: 1-inch black pipe,

$5.15 per 100 ft., and 1-inch galvanized, $7.55.

LEAD—The market is fairly steady as to price, and trade is moderate. We quote 4¼ to 4½c.

SPELTER—The outside markets are a little easier, but locally we still quote 5½ to 6c.

ZINC SHEETS—Business continues quiet. We quote 6½c. for casks, 6¾c. for part casks.

ANTIMONY—Prices are steady and trade quiet at 10½ to 11c. per lb.

PAINTS AND OILS.

Owing to the high price of seed the cost of producing linseed oil is claimed to be higher than the jobbing prices ruling, so, as very little is arriving from England, there is a steady stiffening of the market. An advance of 1c. was made early this week, and a further rise is looked for. Turpentine is steady at unchanged figures. Other lines are also steady. There is an excellent trade doing. We quote:

WHITE LEAD—Ex Toronto, pure white lead, $6.37½; No. 1, $6; No. 2. $5.67½; No. 3. $5.25; No. 4. $4 87½; genuine dry white lead in casks, $5.37½.

RED LEAD—Genuine, in casks of 560 lb., $5.50; ditto, in kegs of 100 lb., $5.75; No. 1, in casks of 560 lb., $5; ditto kegs of 100 lb., $5 25.

LITHARGE—Genuine, 7 to 7½c.

ORANGE MINERAL—Genuine, 8 to 8½c.

WHITE ZINC—Genuine, French V.M., in casks, $7 to $7.25; Lehigh, in casks, $6.

PARIS WHITE—90c. to $1 per 100 lb.

WHITING — 70c. per 100 lb.; Gilders' whiting, 80c.

GUM SHELLAC — In cases, 22c.; in less than cases, 25c.

PARIS GREEN—Bbls., 16¾c.; kegs, 17c.; 50 and 100-lb. drums, 17½c.; 25-lb. drums, 18c.; 1-lb. papers, 18½c.; 1-lb. tins, 19½c.; ½-lb. papers, 20½c.; ½-lb. tins, 21½c.

PUTTY—Bladders, in bbls., $2.10; bladders, in 100 lb. kegs, $2.25; bulk in bbls., $1.90; bulk, less than bbls. and up to 100 lb., $2.05; bladders, bulk or tins, less than 100 lb., $2.90.

PLASTER PARIS—New Brunswick, $1.90 per bbl.

PUMICE STONE — Powdered, $2.50 per cwt. in bbls., and 4 to 5c. per lb. in less quantity; lump, 10c. in small lots, and 8c. in bbls.

LIQUID PAINTS—Pure, $1.20 to $1.30 per gal.

CASTOR OIL—East India, in cases, 10 to 10½c. per lb. and 10½ to 11c. for single tins.

LINSEED OIL—Raw, 1 to 4 barrels, 82c.; boiled, 85c.; 5 to 9 barrels, raw, 81c.; boiled, 84c., delivered. To Toronto, Hamilton, Guelph and London, 1c. less.

TURPENTINE—Single barrels, 54c.; 2 to 4 barrels, 53c., delivered. Toronto, Hamilton and London 1c. less. For less quantities than barrels, 5c. per gallon extra will be added, and for 5-gallon packages, 50c., and 10-gallon packages, 80c. will be charged.

GLASS.

A meeting was held this week to consider the advisability of making an advance, but as prices are now at a high basis, it was decided not to raise them further, as the condition of the market warrants. We quote as follows: Under 26 in., $4 15; 26 to 40 in., $4.45; 41 to 50 in., $4.85; 51 to 60 in., $5.15; 61 to 70 in., $5.50; double diamond, under 26 in., $6; 26 to 40 in., $6.65; 41 to 50 in., $7.50; 51 to 60 in., $8.50; 61 to 70 in., $9.50, Toronto, Hamilton and London. Terms, 4 months or 3 per cent. 30 days.

HIDES, SKINS AND WOOL.

HIDES—The market is dull. Prices are unchanged. We quote: Cowhides, No. 1, 6½c.; No. 2, 5½c.; No. 3, 4½c. Steer hides are worth 1c. more. Cured hides are quoted at 7 to 7½c.

SKINS—There is a fair movement at steady prices. We quote: No. 1 veal, 8-lb. and up, 9c. per lb.; No. 2, 8c.; dekins, from 60 to 70c.; culls, 20 to 25c. Sheepskins, 90c. to $1.

WOOL—There is not yet much wool coming in, but dealers have large stocks of last year's wool on hand. We quote: Combing fleece, washed, 13c., and unwashed, 8c.

OLD MATERIAL.

A good trade is noted. Prices are steady. We quote jobbers' prices: Agricultural scrap, 55c. per cwt.; machinery cast, 60c. per cwt.; stove cast, 50c.; No. 1 wrought 50c. per 100 lb.; new light scrap copper, 12c. per lb.; bottoms, 11½c.; heavy copper, 13c.; coil wire scrap, 13c.; light brass, 7c.; heavy yellow brass, 10 to 10½c.; heavy red brass, 10½ to 11c.; scrap lead, 3c.; zinc, 2c; scrap rubber, 6½c.; good country mixed rags, 65 to 75c.; clean dry bones, 40 to 50c. per 100 lb.

COAL.

There is a good demand, but dealers find difficulty in filling orders because of the scarcity of both coal and cars. Prices are unchanged. We quote at international bridges as follows: Grate, $4.75 per gross ton; egg, stove and nut, $5 per gross ton with a rebate of 30c. off for June shipments.

PETROLEUM.

There is a fair movement, with prices unchanged. We quote: Pratt's Astral, 16 to 16½c. in bulk (barrels, $1 extra); American water white, 16½ to 17c. in barrels; Photogene, 15½ to 16c.; Sarnia water white, 15 to 15½c. in barrels; Sarnia prime white, 14 to 14½c. in barrels.

MARKET NOTES.

Linseed oil has advanced 1c., and a further rise is expected.

In a recent issue it was said that cotton twine was lower. It should have read cotton rope.

UNSIGHTLY VIEWS FROM CAR.

A TRAVELLER rapidly touring the country has to size things up from the car window, writes G. M. L. B. in Montreal Herald. He judges our agricultural resources, the prosperity of the towns and cities and our character in general from what he sees in this way. It isn't fair, but how can he help it? He only stops at a few cities for extended sightseeing—the rest of the country has to remain in his memory as a series of snap shots from his seat in the train or from the stations along the route.

This fact people in most sections seem very careful to ignore. The station is often in the dirtiest section of the town. "Very true," you say, "but that is unavoidable!" The line runs through two rows of back yards. "Also unavoidable!" The streets around the station are often the most ill kept in the place. "Can't be helped!" Hold on now—can't it? Shade trees are conspicuously absent—is this irremediable? No policeman is in sight to clear out that batch of dirty children who are crowding the platform; loungers are spitting tobacco juice on the walks; soap advertisers have daubed all the fences in the neighborhood; the river, which the townspeople advertise as possessing great secnic attractions, here presents a vista of half-sunken scow, tumble-down wharf, a garbage pile on its banks, and, possibly, a couple of youngsters preparing for a plunge. Can none of these abuses be rectified? Alas for your claims to handsome, well-kept streets, fine public buildings, beautiful drives, delightful boating, etc. The traveler considers it a squalid, unbusinesslike place to which he never desires an invitation, and to which he certainly will not hie him when on pleasure bent.

If Canada would better advertise her attractions for summer tourists, the first thing to be done is, clearly, to clean up. In this there are few railroads unwilling to co-operate; often, indeed, the latter lead the way. A town which wishes to present a decent appearance to the travelling public could act in its municipal capacity, its board of trade could undertake the task, or a mass meeting of the citizens could be called and a committee formed to act through the strength of public influence, the question being taken up in the local papers and those who hinder the movement or refuse to do their part being shamed into tidiness.

Let the railroad authorities be taken into consultation, and if they have not already done so, they can be easily persuaded to lay out a little garden, keep the buildings and platforms in good condition, and, if the station is old and the town shows signs of progress, may be induced to rebuild. Next approach those owning property along the railroad and ask them what they are willing to do for the town's sake. Urge them to have the rubbish cleared away, the sheds painted, a row of trees or vines planted, to hide unsightly buildings, and appeal to those intending to build that they have some regard for the rear view of the premises. Then urge the municipality to action. Possibly there is a vacant lot in front of the station. Agitate having it taken over and being made a public square. Perhaps the old wharf and rubbish-lined banks of a river in this vicinity are town property. Recommend a clearing up.

If the town proposes erecting a new hall, and the station is central enough, why not build in that locality having grounds in front, with fountain and well-kept flower beds? Plant trees along the streets; send the watering cart there occasionally; keep a policeman on duty when important trains come in; try to strangle the fence and roof advertiser. "Why bless me!" a passer-by will say, "is this Xville?" Then the Canadian just returned from a trip to Europe will not be ashamed of his own land, and our American visitors will have something to praise besides the regions where Nature is still unmolested."

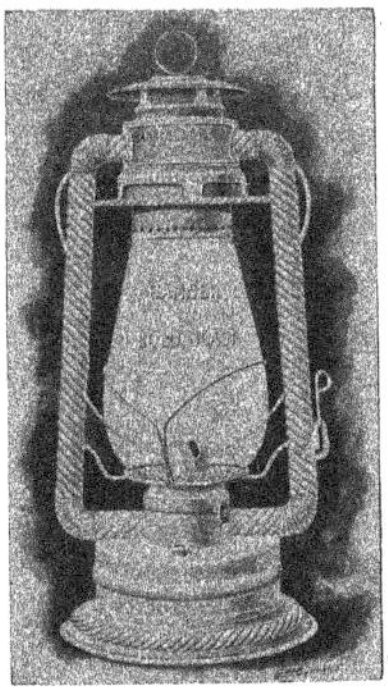

HEATING AND PLUMBING

THE MERITS OF ACETYLENE.

THE aim of the editor in this article is to give convincing proofs to the plumber that the merits of acetylene are such that the most bias minded or prejudiced person is soon convinced of the merits of the new illuminant, and that really all it needs is an introduction to the public, when the real worth of the light speaks for itself. By an introduction I mean show the person interested first the light and then the absolute simplicity of its production, viz., water and calcium carbide brought together.

It cannot be mistaken that the subject of acetylene is one which appeals to the masses, and that they have but to know the true value of the light to have awakened the deepest interest regarding it, for what is more zealously guarded by every human being than the eyesight? And when the public understand but one of the many virtues of the light, viz., its true imitation of the sun's rays, and the consequent restful effects upon the eye, as well as the entire nervous system, it goes without saying that the public are going to demand the use of the light.

The foregoing only accentuates what the editor of this department has endeavored to impress upon the plumber in preceding publications of this journal, i.e., that now is the time for the plumber to get into the business, for, if he does not, someone else is going to have the money for the acetylene installations which are bound to be installed, for have them, the people will, and why not you, Mr. Plumber, be the man to step in and take the opportunity by the forelock, naturally equipped as you are with nearly all that is necessary in this line of work?

Undoubtedly a great deal of harm has been done by the so-called "tin can generator" in the past few years, manufactured out of material both too light and too poor to any more than hold together until installed, but now that the generators have been so thoroughly studied, and constructed upon such scientific lines, it is no longer an experiment when one of these generators is installed, but is rather a charmingly interesting little gas work, compact and pleasing to the eye, which is a thing of admiration to the purchaser as well as the plumbers who install it.

The editor of this department takes this occasion to say that it will be a matter of genuine regret to him as well as surprise, if, from now on, there so much as exists a reputable sanitary and plumbing establishment without its line of generators; in fact, it would seem to the editor that it behooves all first-class plumbing concerns to at least represent some one of the many excellent generators in order that it may be said that this establishment is not lacking in up-to-date appliances and the true progressive spirit of the 20th century.

That the tradesmen may feel the pulse of the public and be convinced of the enthusiasm and general interest displayed by the usually indifferent press, we quote as follows, from one of the leading papers of the country:

"A great many people building fine houses in the country, at some distance from towns or villages, take the precaution to have gas pipes put in during the course of construction on the theory that some time there may be a gas plant near enough to provide them with light. Sometimes gas works are erected in the vicinity and quite as often not. Since the introduction of acetylene gas, the generator for which can be carried about and planted anywhere, many of the owners of such houses are availing themselves of the new gas.

"Experiments by eminent European physicians prove that acetylene does not form any combination with the blood. That it has no specific poisonous action and is much less dangerous in every way than ordinary illuminating gas.

"The rays of acetylene gas being more diffusive than those of any other illuminant, and being in quality equivalent to sunlight, it is not necessary to use as much as ordinary gas. It adds less carbonic acid to the atmosphere. An ordinary gas burner produces carbonic acid equal to that in the exhalations of 18 adults, while acetylene gives off but one-sixth that amount, leaving no injurious effects, and it heats the atmosphere of a room much less.

"Comparisons of deadly and explosive materials and their likelihood to cause fire show that kerosene, gasoline, benzine, city gas indoors and in street mains, electricity, steam boilers, gunpowder and thousands of chemicals and substances in everyday use, and which the world could not conveniently get along without, are far more dangerous than acetylene gas."

The foregoing is but a fair sample of the articles appearing in the press every day all over the country, and necessarily arouse the interest of the reader and which serve to prove that now is the time for the plumber to make his arrangements to represent some good generator and go after business.—Plumbers' Trade Journal.

SOME BUILDING NOTES.

ARCHITECT Mills has prepared plans for a fine house for James Wilson at the corner of Macnab and Hannah streets, Hamilton.

Mr. Gray intends building a residence in Port Credit, Ont.

Burchell & Howe are erecting a new store on Rosser avenue, Brandon, Man.

Mrs. Swail, Mr. Ross and E. Z. Labrosse are erecting new houses at Vankleek Hill, Ont.

The Commercial Hotel, Wiarton, Ont., is undergoing extensive repairs and improvements.

R. Tivey is erecting a new hotel for Commodore Calcutt at Idyl Wyld, near Peterboro', Ont.

Mr. Coatsworth is erecting a two-storey brick building with a 49 ft. front on Talbot street, Leamington, Ont.

Improvements to cost $25,000 are to be made to the Russell House, Ottawa. Many new bathrooms will be added.

Tenders are called by to-day (Saturday), for the erection of the new buildings for the American Cereal Co. at Peterboro', Ont.

A new schoolhouse is to be erected in Port Arthur, Ont., to be finished by September 1. It will be heated by steam.

A third storey is to be added to the Peeter House, Waterford, Ont. D. P. Carey, of the same town, will build a residence.

The Imperial Cotton Co., Hamilton, Ont., intend building 50 houses for their operatives near their factory. C. T. Grantham is manager of the company.

Contracts have been let for a brick warehouse for Martin O'Meara on Clarence street, near the Grand Trunk track, London, Ont. Its dimensions will be 40x75 feet, and it will cost $10,000.

THE NATIONAL CONVENTION.

It has been definitely settled that the business meetings of the annual convention of the National Plumbers and Steam Fitters' Association of Canada will be held in the Temple building, corner Bay and Richmond streets, Toronto.

President Meredith informed HARDWARE AND METAL this week that the preparations

PORTLAND CEMENTS

Best German, Belgian and English Brands.

Fire Bricks, Fire Clay,
Flue Linings,
Drain Pipes,
Calcined Plaster,
Granite Hard Wall Plaster,
Wheelbarrows,
Mortar Stains.

A full stock of Builders' and Contractors' Supplies. Write for Quotations.

W. McNally & Co.,
MONTREAL.

DAVID PHILIP
MANUFACTURERS' AGENT

362½ Main St., - WINNIPEG.

Correspondence invited from manufacturers of Staple or Heavy Hardware, Iron or Steel Bolts and Nuts, etc., either by carrying stock in Winnipeg or by selling direct from factory.

GOOD REFERENCES.

LOW TANK WATER CLOSET COMBINATIONS
THE MOST PERFECT ON THE MARKET
NOISELESS IN ACTION
BEAUTIFUL DESIGNS.

Write for Catalogue.
The James Morrison Brass Mfg. Co. Limited
TORONTO, ONT.

L. G. Fournier & Co., stove and furniture dealers, Ottawa, have moved into larger premises, at 300 Wellington street.

YOUR PAINT TRADE PLEASED

It's well to please your paint trade, because there is money in it. There's more money in paint to-day than there ever was before. Because it's easy to handle, nicely put up, quickly sold and no waste, and it won't do to fool with this trade. A man who offers poor paint will be found out almost as quick as the paint touches a building. Now with

RAMSAYS PAINT

you take no risk. Here's a paint the painter knows and asks for. It's the best that can be made, and that's all he wants or anybody else. It's easy to sell because we help with booklets, cards, fence boards and many other things, and the price is just right too for pure paint.

A. RAMSAY & SON,
PAINTMAKERS,
Est'd 1842. MONTREAL

Patented April 16, 1889.

BRADLEY STEEL SHELF BRACKETS

Greatly improved recently.
For sale by principal Canadian Jobbers.
Write to factory for free sample

ATLAS MFG. CO.
New Haven, Conn., U.S.A.

CANADIAN ADVERTISING is best done by THE E. DESBARATS ADVERTISING AGENCY, MONTREAL.

ALEXANDER GIBB

Manufacturers' Agent and Metal Broker,
13 St. John Street, Montreal.

Representing British and American Manufacturers. Correspondence invited from firms wishing to be represented in Canada.

STEVENS SINGLE BARREL SHOT GUN.

LATEST AND BEST SINGLE GUN ON THE MARKET.

MADE IN THREE STYLES THREE GAUGES UP-TO-DATE.

FINEST SINGLE GUN YET PRODUCED.

Our Goods are Handled by the Leading Jobbers.

J. Stevens Arms & Tool Co., P.O. Box 217 Chicopee Falls, Mass., U.S.A.

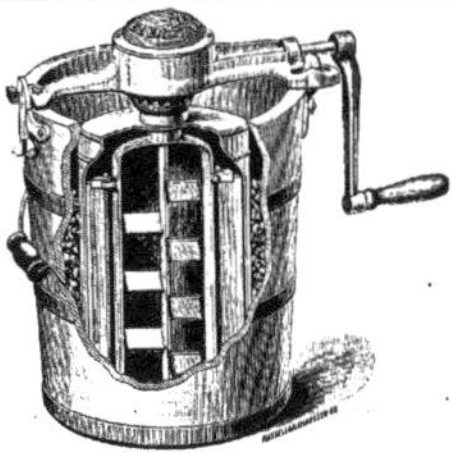

White Mountain Freezers

are made in every size and for all purposes, from the tiny one quart hand freezer for the small family, to the large 40 quart power freezer for hotels and restaurants.

These freezers are invaluable to the home, and, if well pushed, one could be sold to nearly every housekeeper in your town.

Other dealers are sending in orders for them every day. Why shouldn't you?

No delay whatever in shipping.

We are the sole agents in Canada for these perfect freezers.

Write for descriptive pamphlets and prices to

THE McCLARY MANUFACTURING CO.

London, Toronto, Montreal, Winnipeg, Vancouver, and St. John, N.B.

for the convention and for the entertainment of visiting delegates are going ahead satisfactorily, and that both good meetings and a good time generally should be the result.

A committee representing the reception committee of the Toronto association waited on the legislation and reception committee of the Toronto City Council asking for a grant toward the entertainment of visiting delegates. They were promised that tally-hos should be provided to give all delegates a ride around the city, visiting the principal points of interest.

The joint committee representing the local branch of the association and the supply houses of Toronto are quietly preparing to give their brethren a big time. They are not making public their intentions, but it has been whispered that a sail across the lake is almost a certainty, the probable objective point being Niagara Falls.

BUILDING PERMITS ISSUED.

BUILDING permits have been issued in Toronto to James Walsh, for four dwellings at 155 to 161 Park road, to cost $25,000; to Geo. Gooderham, for three dwellings near Huron street, on Prince Arthur avenue, to cost $14,000; to the Confederation Life Association, for $7,600 worth of alterations to their hotel property on Yonge street, near Richmond; to The Walker, Parker Co., for a $10,000 factory at the corner of Wellington and Emily streets; to Chas. Parker, for a $3 000 dwelling on Bernard avenue, near Avenue road; to The Bredin Bread Co., for a $12,000 factory and stable on the corner of Pears avenue and Avenue road; to P. Lzeliski, for a $2,500 dwelling at the corner of Indian road and Radford street; to Robt. McGill, for a $4,900 residence on Tyndall avenue, near King street; to Frank McMahon, for a $3,500 dwelling at 345 Brunswick avenue; to Benj. Brick, for six dwellings on Paul street, near Broadview avenue, to cost $6,000, and to J. M. Downer, for a two storey factory near Simcoe street, on Richmond, to cost $2,500.

The following permits have been issued in Ottawa: H. Moreland, lot 19 Bank street, four stores, $2,500; James Sommers, lots 6 and 7 Second avenue, four tenement houses, $3,000; M. Landreville, lot 54 Albert street, stables and dwelling, $4,000; G. M. Holbrook, lot 6, Cliff street, additions and alterations to present dwelling, $5,500; Joseph Wilkins, lots 11 and 12 Elgin street, dwelling, $2,500; Joseph Burnette, lot 17, Henderson avenue, two dwellings, $3 000; Albert Dunn, lot 16, Henderson avenue, two dwellings, $3,000; Hon. A. G. Bair, Gladstone avenue, stable and coach house, $2,000; Mrs. O'Donnell, lot 8, King street, dwelling, $1,800; F. Peter, lots 23 and 25 Clarence street, dwelling, $2,000.

PLUMBING AND HEATING NOTES.

G. F. McDonald, plumber, Dundas, Ont., is giving up business.

The Canada Brass and Electrical Co., Limited, have been incorporated.

Napoleon Des Chenes & Co., have registered as plumbers in Three Rivers, Que.

Arthur A. Burns, of Boyd, Burns & Co., wholesale plumbers' goods, Vancouver, is dead.

Victor Laramee has registered as protor of Eugene Laramee, plumber, Montreal.

The journeymen plumbers of Ottawa, who are now getting 22½c. an hour, are asking 25c. per hour. They work 9 hours per day.

EARLY GUN-MAKING IN GREAT BRITAIN.

Since the war broke out in South Africa, the Government has placed with The Carron Company, of Scotland, a considerable order for shells, which is interesting as recalling the conditions under which munitions of war were supplied in the eighteenth century. Carron works, which were founded in 1760, are now probably the oldest iron works in this country that have had a continuously active existence for nearly a century and a half. In 1779, they undertook the manufacture of the so-called "Carronades" or "Smashers," the origin of which has been variously attributed to General Robert Melville, to Miller, of Dalswinton, and to a well-known genius of that time named Gascoigne. In the same year the carronades were used against the French fleet. This form of gun carried a relatively large ball, and varied in calibre from 6-pounders to 68-pounders. So rapidly were they applied that by 1781 there were 429 ships in the British navy that mounted them. The "Carronade" was discarded by the British navy in 1852. The modern gun is in every way typical of the vast change in matters metallurgical and mechanical that has occurred in the interval.—Iron and Coal Trades' Review.

THE SPIRIT OF INDEPENDENCE.

"I am a firm believer in the principle that a mutual dependence is a necessity and a blessing," writes the Hustler, in Stoves and Hardware Reporter. "Did it ever occur to you that we all make our living out of each others' wants? I want your hardware and you want my groceries. We supply each others' necessities and establish a mutual relationship which is helpful to us both. We have certain needs in common and go to some one else to fill them and he has needs which we supply. So on it goes in a sort of endless chain until we are bound together by a succession of links which are practically unbreakable unless one of them is given too much strain and breaks, when another is immediately forged into place and the chain becomes as strong as ever.

"When I learn of a business man who talks and acts as if he didn't give a rap for the other fellow, who is so deeply dyed in the sense of his own importance that he is the only color in the spectrum, I at once set him down as a man who is too good for this earth but who ought to gather in a little human intelligence before he leaves it. 'I don't care for his trade' is a not infrequent remark made by some merchant who believes he can get along without it. Perhaps he can, for a time, but doesn't it occur to him that business is the supplying of individual wants, that everything counts and that when one trade is lost others may be lost also? A man has a right to be as independent as his conditions permit, but there are very few of us who realize how actually helpless we are without the aid and support of others and it is dangerous to break any one of the links because our own turn may come next and we'll snap in two on account of the strain that was of our own making.

"The most independent business man that I know of is one who has a good trade which he endeavors to keep, who pays his bills promptly or discounts them, who sells for cash and so has no bad accounts to worry him, who keeps his store on dress parade and his stock insured, who has earned and maintained the respect of his customers and is yet willing to acknowledge that he is dependent upon them for the support of his business. Such a man comes as near realizing the ideality of independence as is possible to human nature. He is close to it but he can't quite touch it, and when he begins to think that he really has it in his grasp he will be promptly reminded of the Irishman's description of a flea: 'Shure, he's wan of them birds that when yez puts your finger on him he ain't there.'"

THE FOREIGN OILSEED REVIEW.

THE confidence prevailing among holders, and referred to a week ago, has borne fruit the last few days and, under an active and healthy consumptive inquiry, prices have daily advanced, finally leaving off with a net gain of 1s. 6d. to 2s. 3d. on the se'nnight. It only needed the rekindling of demand to demonstrate how very bare were the spot supplies and the little stuff likely to be available when the new Calcutta seed began to arrive. The enhanced values have induced America to offer back for sale some of the La Plata recently arrived there, and business to a fair extent is already reported for immediate loading to Hull at 49s. to 49s. 3d. Considering that current spot prices here, compared with the 1st of April, are 7s. 6d. to 8s. 6d. higher, as against those in America are only equal to 3s. to 3s. 4d. per qr., it is but natural to find the latter anxious to resell at such a margin, the freight being but 1s. 8d. per qr. Just now the principal strength of the market is derived from the active buying for Continental account, where there seem to be large "bear" contracts still uncovered. A few weeks back London operators were unsupported in their "bullish" views and the Continent refused to trade, but this week has witnessed a complete change of front and all offers of seed are eagerly sought after by crushers in Germany, France and Holland (mostly the first-named) at a parity of fully 1s. over United Kingdom values.—Dornbusch, May 10.

TOBIN BRONZE.

ONE of the most marvellous metals that has come into prominence during the last few years is Tobin bronze, the invention of a United States naval officer whose name it bears. It is a combination of copper with other metals. When rolled and hot it is remarkable for its high elastic limit, tensile strength, hardness, toughness and uniform texture. It is as strong as ordinary steel rods or plates, and is being used largely for a variety of purposes where a strong, non-corrosive metal is required. When finished it has a bright golden color. One of the most celebrated uses to which this Tobin bronze has been put has been in connection with the construction of the American yachts defending the America Cup, Tobin bronze plates having been used in the underbody of the hull of the Vigilant in 1893, and later of the Columbia. The hulls of both the Constitution and Independence are also constructed of Tobin bronze plates. Many manufacturers also use it for valve stems, piston rods, shafting and air-pump linings, bolt forgings, in fact in any place where a superior article is wanted to give strength or prevent corroding.

Tobin bronze is made by The Ansonia Brass and Copper Company, Ansonia, Conn., and Mr. Alexander Gibb, Montreal, is the Canadian agent. A neat little descriptive pamphlet has been issued on the metal, copies of which may be obtained on application to Mr. Gibb.

DEPARTMENT STORE INSURANCE.

The Colonial Assurance Company, of New York, has issued the following circular to its agents:

"For some time past the losses on retail dry goods and department stores have been very frequent and disastrous. Many causes and hazards combine to produce this result. Defective electric wiring has been the cause of many fires, although we think that this feature has been exaggerated. The modern department store combines the hazard of a large area of sensitive stock with that of touching up and upholstering of furniture, various manufacturing, packing, restaurants, exhibiting of gasoline stoves and numerous other processes more or less dangerous.

"As it is unlikely that we can ever make up what we have lost on this class at the prevailing rates, and as there seems no probability of the rates being placed on a paying basis, we feel that we must either decline to write such risks or obtain premiums therefor which will more nearly pay the losses. We have therefore fixed the following minimum rates, and desire our agents to decline any and all lines on retail dry goods and department stores when the same cannot be obtained.

"On brick buildings or contents having a ground floor area of 10,000 square feet, charge \$2.25; on brick buildings or contents having a ground floor area in excess thereof, for each 1,000 square feet in excess, charge 25 cents.

"We trust that our agents will appreciate our efforts to place this very unprofitable class on a paying basis. These rates do not apply to risks equipped with automatic sprinklers."

BANK OF MONTREAL.

Proceedings at the 83rd Annual Meeting.

A SATISFACTORY REPORT.

Mr. Clouston's Remarks on the General Commercial Situation in Canada.

THE eighty-third annual meeting of the shareholders of the Bank of Montreal was held in the Board Room of the institution, at one o'clock p.m. on Monday, June 3.

There were present: Hon. George A. Drummond, Vice-President; Sir William Macdonald, Hon. James O'Brien, Capt. Benyon, Messrs. R. B. Angus, A. W. Hooper, Hector Mackenzie, David Morris, F. S. Lyman, K.C.; F. T. Judah, K.C.; R. A. Boas, J. G. Snetsinger, W. H. Evans, W. J. Buchanan, E. B. Greenshields, Richard White, A. T. Taylor, J. Try-Davies, Henry Dobell, Hugh Cameron, M. S. Foley, Henry Mason, H. Drummond, A. Walmsley, Nicholas Murphy, John Morrison.

On the motion of Mr. R. B. Angus, Hon. George A. Drummond, Vice-President, was unanimously voted to the chair, in the absence of the President, the Right Hon. Lord Strathcona and Mount Royal.

On the motion of Mr. F. T. Judah, seconded by Mr. Henry Dobell, it was agreed: "That the following gentlemen be appointed to act as scrutineers: Messrs. F. S. Lyman, K.C., and W. J. Buchanan; and that Mr. James Aird be the secretary of the meeting."

DIRECTORS' REPORT.

The report of the directors to the shareholders at their eighty-third annual meeting was then read by Mr. E. S. Clouston, General Manager, as follows:

The directors have pleasure in presenting the eighty-third annual report, showing the result of the Bank's business of the year ended April 30, 1901.

Balance of Profit and Loss Account, 30th April, 1900....		$ 427,180.89
Profits for the year ended 30th April, 1901, after deducting charges of management and making full provision for all bad and doubtful debts		1,537,522.30
		$1,964,703.19
Dividend 5 per cent paid 1st December, 1900— — — —	$600,000.00	
Dividend 5 per cent., payable 1st June, 1901....	600,000.00	
		1,200,000.00
Balance of Profit and Loss carried forward...... ....$		764,703.19

As shareholders are aware, the present bank charters would have expired on the 1st July next. Instead of introducing an entire new Bank Act, the Government proceeded to continue the charters of the banks, and has provided for the changes, which, in its opinion, were advisable by amendments to the Bank Act of 1890.

The accommodation in the bank's building at headquarters having become very inadequate for the proper conduct of the business, it has been found necessary to erect suitable premises on the site recently acquired on Craig street, and the work is now in progress. The new premises are to be connected with the present building by a bridge over Fortification lane.

Premises are also being erected at the corner of Wellington and Magdalen streets, for the use of the Point St. Charles sub-agency, and since the last annual meeting the bank's building at Sydney, N.S., has been completed and occupied by that branch.

It has been decided to open a branch of the Bank at Glace Bay, N.S., at once.

The head office and all the branches have passed through the usual inspection during the year.

STRATHCONA AND MOUNT ROYAL,
President.

Bank of Montreal,
Head Office,
3rd June, 1901.

You will notice that our profits are a little in excess of those of last year, and the statement is one of the strongest we have had the pleasure of laying before you.

As the charters of all the banks would have expired in July of this year, a further extension of ten years was granted, and certain amendments to the Bank Act were enacted.

The chief changes were:

The rate of interest on the notes of suspended banks was reduced from 6 per cent. to 5 per cent.

Power has been given to enable a bank to purchase the assets of another, thus overcoming the barrier which formerly existed to the amalgamation of banks.

In addition to the annual return of unclaimed dividends and balances, we are also required to furnish a statement of all drafts and bills of exchange issued and remaining unpaid,

In the case of a suspended bank, The Canadian Bankers' Association has been given power to appoint a curator. The association has also been entrusted with the work of inspecting and supervising the note circulating accounts of all the banks in the Dominion, an added safeguard, if any were needed, to the circulating currency of the country. In this way the association has practically become an agent of the Government in the administration of the Act.

The form of our statement to the Government has been changed, and fuller details are now required. It was this that rendered advisable the new form of statement now laid before you. Other changes were more of interest to bankers themselves than the public. Generally speaking, the alterations were in the direction of strengthening

THE GENERAL STATEMENT.

The general statement of assets and liabilities of the Bank, 30th April, 1901, was read as follows:

LIABILITIES.

Capital Stock			$12,000,000 00
Rest		$ 7,000,000 00	
Balance of Profits carried forward		764,708 19	
		$ 7,764 708 19	
Unclaimed dividends		2,432 01	
Half-yearly Dividend, payable 1st June, 1901		600,000 00	
			8,367 135 20
			$20,367,135 20
Notes of the Bank in circulation		$ 6,482,214 00	
Deposits not bearing interest		18,184,774 47	
Deposits bearing interest		54,501,858 13	
Balances due to other Banks in Canada		46,082 93	
			70,214,924 53
			$99,582,059 73

ASSETS.

Gold and Silver coin current		$ 2,564,358 30	
Government demand notes		3,472,440 25	
Deposit with Dominion Government required by Act of Parliament for security of general bank note circulation		310,000 00	
Due by agencies of this bank and other banks in Great Britain	$ 2,536,166 61		
Due by agencies of this bank and other banks in Foreign countries	2,264,257 63		
Call and short Loans in Great Britain and United States.	23,536,628 00		
		28,337,052 24	
Dominion and Provincial Government Securities		617,930 93	
Railway and other Bonds, debentures and stocks		2 889,973 17	
Notes and cheques of other Banks		1,690,470 10	
			$30,882,225 05
Bank Premises at Montreal and Branches			600,000 00
Current Loans and discounts in Canada and elsewhere (rebate interest reserved) and other assets		$58,850,449 34	
Debts secured by mortgage or otherwise		131,135 27	
Overdue debts not specially secured (loss provided for)		118,250 07	
			59,099,834 68
			$99,582,059 73

Bank of Montreal,
Montreal, 30th April, 1901.

E. S. CLOUSTON,
General Manager.

THE GENERAL MANAGER.

Mr. Clouston then said:—

The statement before you requires a little explanation, as it is made up to conform to the Amended Bank Act of last session, and now embraces our foreign business as well as our Canadian. Previous statements showed only the balances which would be due us from other countries after our business there had been liquidated. Consequently, our statement now includes all deposits and loans elsewhere than in Canada. This makes a comparison with former statements an impossibility, but for the information of the shareholders, I may say that the principal changes in our Canadian business are as follows:—

Circulation, increase...... ....$	321,000
Deposits not bearing interest, increase......	1,963,000
Deposits bearing interest, increase.	5,422,000
Current loans and discounts, decrease..	360,000

and improving the Act under which we have worked for the last ten years.

At the last session of Parliament the Finance Minister took power to establish a mint. The opinion of the bankers, not from any selfish point of view, but from what we believed to be in the best interest of the country at large, were set forth at the last annual meeting of The Bankers' Association, and I do not propose to say anything more on the subject here. The Act was only permissive, and it may be that on looking more closely into the matter, the Government may decide not to incur considerable expense in order to deteriorate the value of one of our products, as the gold is more available to pay our foreign indebtedness and more valuable for export, as bullion, than if it were minted into coin. As a circulating medium, it will not displace the paper currency here, any more than it does in the United States, while the miners to-day can obtain from the banks the same value for their gold as they would if the mint were established even in British Columbia.

Business during the last year has been generally good, notwithstanding a short crop in the Northwest, and in spite of the unfortunate condition of affairs in the mining districts. In other sections of Canada, even the most pessimistic of farmers should have been satisfied with the results of the last two years. If, from a sentimental point of view, we were eager and willing to aid the Mother Country by the despatch of troops, as will always be the case, the practical result is a magnificent advertisement to Canada, and an additional market established for our products, which will probably recoup the outlay of this country. In the last year there has been an increased demand for its products, in consequence of the Boer War in South Africa.

On the other hand, the woollen manufacturing industry has not been prosperous, and I am sorry to say the outlook for the lumber trade is not of the best, prices ruling low and the markets being congested, and we can only hope for an improvement before the season finishes.

There are also signs of overproduction in textile goods, and in the manufacture of pulp, which only need judicious restraint to be put on a good basis. We must not forget the return of the wave and get so far beyond our depth as to lose our footing.

It is too early to speak of the future crops, though up to the present the reports are good, and if they turn out according to promise, we ought to have another good year, and if that comes you can see that this bank is in a position to take advantage of it.

ADOPTION OF REPORT.

Hon. George A. Drummond said :—

You have heard the statement of the General Manager and the report of the directors, and the statements placed before you appear to me to be so full and complete that I do not consider it necessary to make any further amplification of them. I will content myself, therefore, with moving: "That the report of the Directors now read, be adopted and printed for distribution among the shareholders."

The motion was seconded by Mr. E. B. Greenshields, and after a few remarks by Mr. John Morrison, who thought that the General Manager had taken the correct view with regard to the proposed establishment of a Canadian mint, it was carried unanimously.

Senator O'Brien moved :—

"That the thanks of the meeting be presented to the President, Vice-President, and Directors for their attention to the interests of the bank."

This was seconded by Mr. David Morrice, and was unanimously agreed to.

Sir William Macdonald moved :—

"That the thanks of the meeting be given to the General Manager, the Inspector, the Managers, and other officers of the bank for their services during the past year."

The motion was seconded by Mr. R. B. Angus, and having been unanimously concurred in, was acknowledged by the General Manager.

Mr. B. A. Boas moved :—

"That the ballot now open for the election of directors be kept open until three o'clock unless fifteen minutes elapse without a vote being cast, when it shall be closed, and until that time, and for that purpose only, this meeting be continued."

This was seconded by Mr. Hector Mackenzie, and unanimously agreed to.

On the motion of Mr. John Morrison, seconded by Hon. James O'Brien, a hearty vote of thanks was accorded the chairman for his conduct of the business of the meeting ; and he acknowledged the same.

THE DIRECTORS.

The ballot resulted in the election of the following directors :—

R. B. ANGUS, ESQ.
HON. GEORGE A. DRUMMOND.
A. F. GAULT, ESQ.
E. B. GREENSHIELDS, ESQ.
SIR WILLIAM C. MACDONALD.
A. T. PATERSON, ESQ.
R. G. REID, ESQ.
JAMES ROSS, ESQ.
RIGHT HON. LORD STRATHCONA AND MOUNT ROYAL, G.C.M.G.

PIPE-LINE TRANSPORTATION FOR SOLIDS.

According to Cassier's Magazine, one of the developments of the coming century, worthy of at least passing thought, is the extent to which the pneumatic-tube principle will be employed to expedite transportation which is now entirely dependent on steam locomotives. Consider, for example, the long lines of loaded coal cars on their way from the mines to the seaboard, and back again empty to the mines. If the weight of a car is 25 per cent. of the gross load, there is in this instance more than 50 per cent. loss or non-paying freight, the empty train requiring about as much power to haul it to the interior as was expended in taking it to the shipping port. It does not seem altogether unreasonable, therefore, to think that just as the miles of tank cars loaded with oil, which were seen in former years, have disappeared, and that commodity is now sent hundreds of miles through pipe lines, so may coal, grain and ore be sent speeding through tubes to central stations for local distribution. In the matter of coal transportation, in fact, just such pipe-line conveyance was tried experimentally something like ten or twelve years ago, the coal for that purpose being ground into powder, mixed with water in sufficiently large proportion, and carried through the pipes in semi-liquid form. At the delivery end of the pipe line there were to be settling chambers for the mixture, enabling the water to be drained off and the coal paste to be pressed into cakes and dried for consumption. The project, however, did not extend beyond a brief experimental career. In woollen mills it is a common thing to blow wool from one building to another through pipes by means of fans.

ESSENTIALS TO SUCCESS.

YOUNG man, a college education is something to be desired ; is a help in the struggle of life, but is not at all indispensable to the greatest business success. If your schooling is a limited one, attention to business ; courtesy to everybody ; seizure of chances as they turn up ; a truthful and honest record, will land you at the top, or so near it, you will be in "comf'r'ble" circumstances all the later years of your life.

Read what Charles M. Schwab, President of the billion-dollar United States steel corporation, had to say about it to the boys of the St. George's Evening Trades School, at their commencement exercises. Here is what he said :

"From my long experience I am led to believe that many boys make the mistake of depending upon influence to obtain for them positions of profit. Go yourself to seek work in life and depend upon your own exertions and merits.

"No matter what business you enter, the essential feature to success is that you perform your tasks better than anybody else. This alone will command attention. Everybody is expected to do his duty, but the boy or man who does a little more is certain of promotion.

"Success is not money-making alone. And I want to state that of the truly great men I know in industrial and manufacturing lines, none are college-bred men, but men who received an industrial or mechanical education, and who worked up by perseverance and application.

"Let me advise you all to make an early start in life. The boy with the manual training and the commonfschool education who can start in life at 16 or 17 can leave the boy who goes to college till he is 20 or more so far behind in the race that he can never catch up."

STEEL-MAKING IN GERMANY.

Figures which have been collected and published by The Association of Iron and Steel Manufacturers of Germany show that the total production of steel in that country in 1900 was 6,645,869 tons, of which 6,223,417 tons were produced by the basic process, a system that was not applied until 1879, so that this remarkable advance has been achieved within 20 years. So recently as 1894, the total make of basic steel in Germany was only 3,241,000 tons, so that the output has practically doubled within the last six years. Of the total output of basic steel in 1900, 4,141,587 tons were produced by the Bessemer process, and 1,997,765 tons by the open-hearth process. It may be added that while in Great Britain four-fifths of our total steel output is by the older acid process, in Germany that process yields only one-sixteenth of the whole.

THE WINDOW GLASS STRIKE.

The strike of the window-glass workers in Belgium and France was declared off on Tuesday, the 21st inst., as announced in last week's Review, and the strikers have returned to work at the manufacturers' terms. The men have been on strike for eleven months for recognition of the union and advance in wages, and during that time the glass trade of this country sent them $35,000. The strike affected 8,000 men in Belgium and 2,000 in France. The settlement will have no immediate effect on glass values in this country.

It will be some time before operations will have proceeded to an extent that imports of foreign glass into this market will be possible. The European markets next door to the factories are bare of stocks and this demand will, of necessity, be met first. The trade in the far East have suffered from diminished supplies because of the strike and are clamoring for early shipments to that quarter. So while our home prices are above what under ordinary conditions the foreign glass can be imported at, it is not likely that prices will be forced down to a tariff-wall basis until a surplus available for American importation is produced at the Belgian factories, and this is not likely to occur until after our present stocks are practically exhausted. Hence, we look for a continued steady market on the high basis now ruling.—Paint, Oil and Drug Review, Chicago.

HOW STEEL PENS ARE MADE.

Few people know, remarks an exchange, what a heap of bother and expense it is to make a pen. For instance, the steel is first rolled into big sheets and then cut into strips about three inches wide. These strips are annealed. In other words they are softly heated to a red heat and permitted to cool very gradually, so that the brittleness is all removed and the steel is soft enough to be easily worked. Then the strips are again rolled to the required thickness, or rather thinness, for, as you know, the average steel pen is not thicker than a piece of letter paper. The blank pen is next cut out of the flat strip and the name of the maker stamped upon it. Then comes the molding process. The pen is put in a mold which gives it grace and strength. The rounding enables the pen to hold the requisite ink and to distribute the ink gradually. That little hole which is cut near the end of the slit also helps to make the ink run properly and regulates the elasticity of the pen. Up to this time the metal is soft and lead-like. To make it brittle and springy it is tempered by being heated to a cherry color and then suddenly plunged into cold water. But it is then too brittle for use, so the temper of the steel must be drawn. The elasticity varies with the color, and each color is obtained by suddenly plunging the pen into cold water. Then follow the slitting, polishing, pointing and finishing, all of which is done by expert workmen.

CANADIAN PIG IRON ON THE CLYDE.

The Iron and Coal Trade Review, London, has the following under date of M[illegible] 17 :

A new feature has been the importation of the first half of a consignment of 7,000 tons of Canadian iron to the Clyde. The iron is being imported at the instance of a Glasgow house interested in a Dominion iron company. The Canadian brand will compete in point of quality with Cleveland, but, in the meantime, the Scotch founders and malleable ironmakers have not shown any eagerness to experiment with the iron, as their wants are being fully met in Middlesbrough. The dumping down of Canadian iron on the Clyde is brought about by a bounty of 12s. per ton, but as the freight is 10s. per ton and the cost for reweighing, cartage, and storage amounts to 5s., the bounty is more than absorbed. The Canadian exporters hope to ship the iron at a lower rate."

H. Armstrong intends building a fish-packing house and a general store at Delta, Man.

The stock of the estate of Morrow Bros., general merchants, Portage la Prairie, Man., has been sold to T. Finklestein, of Moosomin, at 50c. on the dollar. Mr. Finklestein has sold his Moosomin stock to Samuel Coppleman.

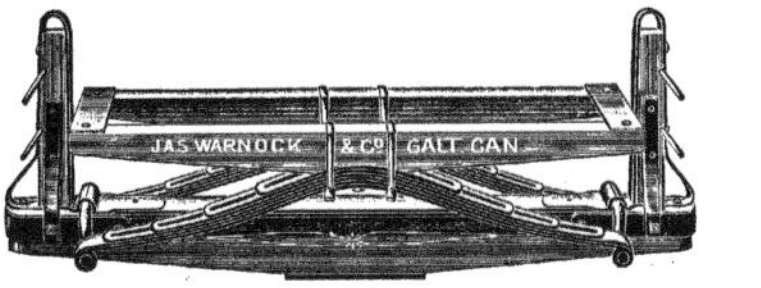

CURRENT MARKET QUOTATIONS

June 7, 1901.

These prices are for such qualities and quantities as are usually ordered by retail dealers on the usual terms of credit, the lowest figures being for larger quantities and prompt pay. Large cash buyers can frequently make purchases at better prices. The Editor is anxious to be informed at once of any apparent errors in this list as the desire is to make it perfectly accurate.

METALS.

Tin.

Lamb and Flag and Straits—56 and 28 lb. ingots, per lb.	0 31½	0 32

Tinplates.

Charcoal Plates—Bright

		Per box
M.L.S., equal to Bradley.		
I.C., usual sizes	....	$6 50
I.X., "	....	8 00
I.X.X., "	....	9 50
Famous—		
I.C.	....	6 50
I.X	....	8 00
I.X.X	....	9 50
Raven & Vulture Grades—		
I.C., usual sizes	....	4 50
I.X., "	....	5 25
I.X.X "	....	6 0)
I.XXX., "	....	6 75
D.C., 12½x17	....	4 00
D.X.	....	4 75
D.X.X	....	5 00

Coke Plates—Bright

Bessemer Steel—		
I.C., usual sizes	....	3 75
I.C., special sizes, base	....	4 00
20 x 28	....	7 75

Charcoal Plates—Terne

Dean or J. G. Grade—		
I.C., 20x28, 112 sheets	....	8 00
I.X., Terne Tin	....	10 (0

Charcoal Tin Boiler Plates.

		Per lb.
Cookley Grade—		
X X., 14x56, 50 sheet bxs / " 14x60, " / " 14x65, "	....	0 06½

Tinned Sheets

72x30 up to 24 gauge	....	0 07½
" 26 "	....	0 08
" 28 "	....	0 08½

Iron and Steel.

Common Bar, per 100 lbs.	1 85	1 90
Refined " "	....	2 35
Horse Shoe Iron "	....	2 35
Hoop steel, 1½ to 3 in. base, extras for smaller sizes	....	3 00
Sleigh Shoe Steel " base	....	2 10
Tire Steel	2 30	2 50
Reeled Machinery	....	3 00
Cast Steel, per lb	0 00	0 00
Toe Calk Steel	....	
T. Firth & Co's cast steel, per lb	0 12½	0 13
Jessop's Tool Steel	0 12½	0 13

Boiler Tubes.

1½-inch	....	0 12½
2 "	....	0 13
2½ "	....	0 15
3 "	....	0 16
3½ "	....	0 20
4 "	....	0 25

Steel Boiler Plate.

¼ inch	2 50	2 60
3-16 inch	2 60	2 70
⅜ inch and thicker	2 50	2 60

Black Sheets.

	Com.	D.Fl.
18 gauge	2 75	3 00
20 gauge	2 75	3 00
22 to 24 "	2 75	3 25
26 "	2 85	....
28 "	3 00	3 50

Canada Plates.

All dull, 52 sheets	....	2 90
Half polished	....	3 00
All bright	....	3 (0

Iron Pipe.

Black pipe—		
⅛ "	....	4 45
¼ inch	....	3 25
⅜ "	....	3 30
½ "	....	3 50
¾ "	....	3 65
1 "	....	5 15
1¼ "	....	7 (0
1½ "	....	8 40
2 "	....	11 25
2½ "	....	20 95
3 "	....	24 55
3½ "	....	30 75
4 "	....	39 00
5 "	....	47 35
6 "	....	62 10
Galvanized pipe—		
½ inch	....	4 90
¾ "	....	5 25
1 "	....	7 55
1¼ "	....	10 30
1½ "	....	12 35
2 "	....	16 50

Galvanized Sheets.

	G C.	Comet.	Amer.	Queen's Head.
16 gauge		...	4 (0	3 75
18 to 24 gauge	4 00	3 85	4 25	4 00
26 "	4 25	4 10	4 25	4 25
28 "	4 50	4 35	4 40	4 50

Case lots 10 to 15c. less.
28 gauge American equals 26 gauge English.

Chain.

Proof Coil, 3-16 in., per 100 lb	....	....
" ¼ " "	8 00	8 50
" 5-16 " "	4 70	5 00
" ⅜ " "	4 05	4 50
" 7-16 " "	3 9)	4 :5
" ½ " "	3 70	4 10
" 9-16 " "	3 75	4 :5
" ⅝ " "	3 :5	3 90
" ¾ " "	3 (0	4 (0

Halter, kennel and post chains, 40 to 40 and 5 p.c.

Cow ties	40 p.c.
Tie-out chains	65 p.c.
Stall fixtures	35 p.c.
Trace chain	45 p.c.

Jack chain, iron, single and double, discount 35 p.c.
Jack chain, brass, single and double, discount 40 p.c.

Copper.

Ingot

English B. S., ton lots	....	0 17¾
Lake Superior	....	....

Bars.

Cut lengths round, ½ to ⅞ in.	0 23	0 25
" round and square 1 to 2 inches	0 23	0 25

Sheet.

Plain, 14 oz., and light, 16 oz., 14x48 and 14x60	0 24	0 14½
Plain, 14 oz., and light, 16 oz., irregular sizes	0 24½	0 25
Tinned copper sheets	....	0 26
Planished	....	0 32

Braziers (In sheets.)

4x6ft. 25 to 30 lbs. ea., per lb.	....	0 25
" 35 to 45 " "	....	0 24
" 50-lb. and above, "	....	0 23

Boiler and T. K. Pitts

Plain Tinned, per lb	....	0 28
Spun, per lb	....	0 32

Brass.

Rod and Sheet, 14 to 30 gauge, 15 per cent.

Sheets, hard-rolled, 2x4	....	0 23
Tubing, base, per lb	....	0 23

Zinc Spelter

Foreign, per lb	0 05½	0 06
Domestic "	....	...

Zinc Sheet.

5 cwt. casks	....	00 6¼
Part casks	....	(0 6¾

Lead.

Imported Pig, per lb	0 04½	0 (4½
Bar, 1 lb.	0 05½	0 06
Sheets, 2½ lbs. sq. ft., by	....	0 06¾
Sheets, 3 to 6 lbs., "	....	0 06

NOTE.—Cut sheets ½ cent per lb. extra. Pipe, by the roll, usual weights per yard, lists at 7c. per lb. and 30 p.c. dis. f.o.b. Toronto.

NOTE.—Cut lengths, net price, waste pipe 8-ft. lengths lists at 7½ cents.

Shot.

Common, $6.50 per 100 lb.; chilled, $7.00 per 100 lb.; buck, seal and ball, $7.50. Discount, 17½ p.c. Prices are f.o.b. Toronto, Hamilton, Montreal, St. John and Halifax. Terms 3 per cent. cash, freights equalized.

Soil Pipe and Fittings.

Discount, 60 and 10 per cent. on medium and extra heavy, and 60 per cent. on light.

Solder.

	Per lb.	Per lb.
Bar half-and-half, guarant'd	....	0 18½
Bar half-and-half, commer'l	....	0 18
Refined	....	0 17½
Wiping	....	0 17

NOTE.—Prices of this graded according to quantity. The prices of other qualities of solder in the market indicated by private brands vary according to composition.

Antimony.

Cookson's, per lb.	0 10½	0 11

White Lead.

	Per 100 lb.
Pure	6 37
No. 1 do	6 00
No. 2 do	5 62½
No. 3 do	5 25
No. 4 do	4 8(½
Munro's Select Flake White	7 37½
Elephant and Decorators' Pure	7 12½
Brandram's B B. Genuine	8 00
" " Decorative	7 55
" " No. 1	6 85
" " No. 2	6 00

Red Lead.

Genuine, 560 lb. casks, per cwt.	....	$5 50
Genuine, 100 lb. kegs, per cwt.	....	5 75
No. 1, 560 lb. casks, per cwt.	....	5 25
No. 1, 100 lb. kegs, per cwt.	....	5 00

White Zinc Paint.

Elephant Snow White	0 08	0 09
Pure White Zinc	0 08	0 (9
No. 1	0 06	0 07½
No. 2	0 05	0 (6½

Dry White Lead.

Pure, casks	5 75
Pure, kegs	6 25
No. 1, casks	5 50
No. 1, kegs	5 00

Prepared Paints.

In ¼, ½ and 1 gallon tins.

Pure, per gallon	....	1
Second qualities, per gallon	....	1
Barn (in bbls.)	0 75	0
The Sherwin-Williams Paints	....	1
Canada Paint Co's Pure	....	1
Toronto Lead & Color Co's Pure	....	1
Sanderson Pearcy's Pure	....	1 25
Stewart & Wood's Champion Pure		1 [illegible]

Colors in Oil.

25 lb. tins, Standard Quality.

Venetian Red, per lb.	0 05
Chrome Yellow	0 11
Golden Ochre	0 06
French "	0 05
Marine Black	0 09
" Green	0 09
Chrome "	0 08
French Imperia Green	0 09

Colors, Dry.

Yellow Ochre (J.C.) bbls.	1 35	1 40
Yellow Ochre (J.F.L.S.), bbls	...	2 75
Yellow Ochre (Royal)	1 10	1 15
Brussels Ochre	....	2 00
Venetian Red (best), per cwt.	1 80	1 90
English Oxides, per cwt.	3 00	3 25
American Oxides, per cwt.	1 75	2 00
Canadian Oxides, per cwt.	1 75	2 00
Super Magnetic Oxides, 93p.c.	2 00	2 25
Burnt Sienna, pure, per lb.	....	0 10
" Umber, " "	....	0 10
do Raw	....	0 09
Drop Black, pure	....	0 09
Chrome Yellows, pure	....	0 18
Chrome Greens, pure, per lb.	....	0 12
Golden Ochre	....	0 03½
Ultramarine Blue in 28-lb. boxes, per lb.	0 08	0 24
Fire Proof Mineral, per 100 lb	....	1 00
Genuine Eng. Litharge, per lb	....	0 07
Mortar Color, per 100 lb.	....	1 25
English Vermillion	....	0 80
Pure Indian Red, No. 45 lb.	....	0 80
Whiting, per 100 lb.	....	0 55

Blue Stone.

Casks, for spraying, per lb.	0 07
100-lb. lots, do. per lb.	0 08

Putty.

Bulk in bbls.	1 9)
Bulk in less quantity	2 05
Bladders in bbls.	2 1)
Bladders in kegs, boxes or loose	2 25
Bladders in 25-lb. tins	2 35
Bladders in 12½-lb. tins	2 6)
Bladders in tu k or tins less than 100 lb	2 9)

Varnishes.

	In 5-gal. lots.).	Per gal.
Carriage, No. 1	2 90	3 30
" body	8 00	9 00
" rubbing	4 00	5 00
Gold Size, Japan	3 00	3 40
Brown Japan	2 40	2 80
Elastic Oak	2 90	3 30
Furniture, extra	2 40	2 80
" No. 1	1 60	2 00
Hard Oil Finish	2 70	3 10
Light Oil Finish	3 20	3 60
Demar	3 30	3 70
Shellac, white	4 40	4 80
" orange	4 00	4 40
Furniture Brown Japan	1 60	2 00
Black Japan	2 40	2 80
" " No. 1	1 60	2 00

The Imperial Varnish & Color Co's., Limited Elastilite Varnish 1 gal. can, each. $3.00.

Granatine Floor Finish per gal., $2.75.

Maple Leaf Coach Enamels; Size 1, $1 20; Size 2, 70c.; Size 3, 40c. each.

Castor Oil.

East India, in cases, per lb.	0 10	0 10½
" " small lots	0 10½	0 11

Cod Oil, Etc.

Cod Oil per gal	0 50	0 55
Pure Olive	....	1 20
" Neatsfoot	....	90

Glue.

Common	0 08½	0 09
French Medal	0 14	0 14½
Cabinet sheet	0 12	0 13
White, extra	0 18	0 20
Gelatine	0 22	0 30
Strip	0 18	0 20
Coopers	0 19	0 20
Huttner	....	0 18

HARDWARE.

Ammunition.

Cartridges.

B. B Cap Dom. 50 and 5 per cent.
Rim Fire Pistol, dis. 40 p. c., Amer.
Rim Fire Cartridges, Dom., 50 and 5 p. c.
Central Fire Pistol and Rifle, 10 p.c. Amer.
Central Fire Cartridges, pistol sizes, Dom. 30 per cent.
Central Fire Cartr'dges, Sporting and Military, Dom., 15 and 5 per cent.
Central Fire, Military and Sporting, Amer. add 5 p.c. to list. B.B. Caps, discount 40 per cent, Amer.
Loaded and empty Shells, "Trap" and "Dominion" grades, 25 per cent. Rival and Nitro, net list.
Brass Shot Shells, 55 per cent.
Primers, Dom., 30 per cent.

Wads	per lb.
Best thick white felt wadding, in ½-lb bags	1 00
Best thick brown or grey felt wads, in ½-lb. bags	70
Best thick white card wads, in boxes of 500 each, 12 and smaller gauges	0 99
Best thick white card wads, in boxes of 500 each, 10 gauge	0 35
Best thick white card wads, in boxes of 500 each, 8 gauge	0 55
Thin card wads, in boxes of 1,000 each, 12 and smaller gauges	0 20
Thin card wads, in boxes of 1,000 each, 10 gauge	0 25
Thin card wads in boxes of 1,000 each, 8 gauge	
Chemically prepared black edge grey cloth wads, in boxes of 250 each—	Per M.
11 and smaller gauge	0 60
9 and 10 gauges	0 70
7 and 8 gauges	0 90
5 and 6 gauges	1 10
Superior chemically prepared pink edge, best white cloth wads, in boxes of 250 each—	
11 and smaller gauge	1 15
9 and 10 gauges	1 40
7 and 8 gauges	1 65
5 and 6 gauges	1 90

Adzes.

Discount, 20 per cent.

Anvils.

Per lb	0 10	0 12½
Anvil and Vise combined	....	4 50
Wilkinson & Co.'s Anvils..lb.	0 09	0 09½
Wilkinson & Co.'s Vices..lb.	0 09¾	0 10

Augers.

Gilmour's, discount 65 and 5 p.c. off list.

Axes.

Chopping Axes		
Single bit, per doz	5 50	10 00
Double bit, "	11 00	18 00
Bench Axes, 40 p.c.		
Broad Axes, 33⅓ per cent.		
Hunters' Axes	5 50	6 00
Boy's Axes	5 75	6 75
Splitting Axes	6 50	12 00
Handled Axes	7 00	10 00

Axle Grease.

Ordinary, per gross	5 75	6 00
Best quality	13 00	15 00

Bath Tubs.

Zinc 6 00
Copper, discount 15 p.c. off revised list

Baths.

Standard Enameled.

5½-inch rolled rim, 1st quality	25 00
" " " 2nd "	21 00

Anti-Friction Metal.

"Tandem" A	per lb.	0 27
" B	"	0 21
" C	"	0 11½
Magnolia Anti-Friction Metal, per lb.		0 25
Frictionless metal, per lb		0 23
SYRACUSE SMELTING WORKS.		
Aluminum, genuine		0 45
Dynamo		0 29
Special		0 25
Aluminum, 99 p.c. pure "Syracuse"		0 50

Bells.

Hand.

Brass, 60 per cent.
Nickel, 55 per cent.

Cow.

American make, discount 66⅔ per cent.
Canadian, discount 45 and 50 per cent.

Door.

Gongs, Sargant's	5 50	8 00
" Peterboro', discount 45 per cent.		
Farm.		
American, each	1 25	3 00
House.		
American, per lb	0 35	0 40

Bellows.

Hand, per doz	3 35	4 75
Moulders', per doz	7 50	10 00
Blacksmiths', discount 40 per cent.		

Belting.

Extra, 60 per cent.
Standard, 60 and 10 per cent.

Bits.

Auger.

Gilmour's, discount 60 and 5 per cent.
Rockford, 50 and 10 per cent.
Jennings' Gen., net list.

Car.

Gilmour's, 47½ to 50 per cent.

Expansive.

Clark's, 40 per cent

Gimlet.

Clark's, per doz	0 65	0 90
Damond, Shell, per doz	1 00	1 50
Nail and Spike, per gross	2 25	5 20

Blind and Bed Staples.

All sizes, per lb	0 07½	0 12

Bolts and Nuts.

	Per cent.
Carriage Bolts, full square, Norway	65
" " full square	65
Common Carriage Bolts, all sizes	60
Machine Bolts, all sizes	60
Coach Screws	70
Sleigh Shoe Bolts	72½
Blank Bolts	60
Bolt Ends	62½
Plough Bolts	60
Nuts, square	4 c. off
Nuts, hexagon	4½c. off
Tire Bolts	67½
Stove Bolts	67½
Stove rods, per lb.	5½ to 6c.

Boot Calks.

Small and medium, ball, per M	4 25
Small heel, per M	4 50

Bright Wire Goods.

Discount .... 6½ per cent.

Broilers.

Light, dis., 65 to 67½ per cent.
Reversible, dis., 65 to 67½ per cent.
Vegetable, per doz., dis. 37½ per cent.

Henis, No. 8, "	....	6 00
Henis, No. 9, "	....	7 00
Queen City "	7 50	0 00

Butchers' Cleavers.

German, per doz	6 00	11 00
American, per doz	12 00	20 00

Building Paper, Etc.

Plain building, per roll	0 30
Tarred lining, per roll	0 40
Tarred roofing, per 100 lb.	1 65
Coal Tar, per barrel	3 50
Pitch, per 100-lb.	0 85
Carpet felt, per ton	45 00

Bull Rings.

Copper, $2.00 for 2½ in. and $1.90 for 2 in.

Butts.

Wrought Brass, net revised list

Cast Iron.

Loose Pin, dis., 60 per cent.

Wrought Steel.

Fast Joint, dis. 65, 10 and 2½ per cent.
Loose Pin, dis. 65, 10 and 2½ per cent.
Berlin Bronzed, dis. 70, 70 and 5 per cent.

Gen. Bronzed, per pair	0 40	0 65

Carpet Stretchers.

American, per doz	1 00	1 50
Bullard's, per doz	6 50	....

Castors.

Bed, new list, dis. 55 to 57½ per cent.
Plate, dis. 52½ to 57½ per cent.

Cattle Leaders.

Nos. 31 and 32, per gross	50	9 50

Cement.

Canadian Portland	2 50	2 80
English "	....	3 00
Belgian "	2 50	2 75
Canadian hydraulic	1 25	1 50

Chalk.

Carpenters Colored, per gross	0 45	0 75
White lump, per cwt	0 60	0 65
Red	0 05	0 06
Crayon, per gross	0 14	0 18

Chisels.

Socket, Framing and Firmer.

Broad's, dis. 70 per cent.
Warnock's, dis. 70 per cent.
P. S. & W. Extra 60, 10 and 5 p.c.

Churns.

Revolving Churns, metal frames—No. 0, $8—No. 1, $8.50—No. 2, $9.00—No. 3, $10.00—No. 4, $12.00—No. 5, $16.00 each. Ditto, wood frames—20c. each less than above. Discounts: Delivered from factories, 58 p.c.; from stock in Montreal, 58 p.c. Terms, 4 months or 3 p.c. cash in 30 days.

Clips.

Axle dis. 65 per cent.

Closets

Plain Ontario Syphon Jet	$16 00
Emb. Ontario Syphon Jet	17 00
Fittings net	1 00
Plain Teutonic Syphon Washout	10 00
Emb. Teutonic Syphon Washout	11 00
Fittings net	1 25
Low Down Teutonic, plain	16 00
" " " embossed	17 00
Plain Richelieu net	3 75
Emb. Richelieu net	4 00
Fittings net	1 25
Low Down Ont. Sy. Jet, plain net	19 50
" " " " emb'd. net	20 50
Closet connection net	1 25
Basins, round, 14 in	1 00
" oval, 17 x 14 in	2 50
" " 19 x 15 in	3 75

Discount 40 p.c., except on net figures.

Compasses, Dividers, Etc.

American, dis. 62½ to 65 per cent.

Cradles, Grain.

Canadian, dis. 25 to 33⅓ per cent.

Crosscut Saw Handles.

S. & D., No. 3, per pair	.17½
" " 5, "	.22½
" " 6, "	.15
Boynton pattern "	.20

Door Springs.

Torrey's Rod, per doz......(15 p.c.)		2 00
Coil, per doz	0 88	1 60
English, per doz	2 00	4 00

Draw Knives.

Coach and Wagon, dis. 50 and 10 per cent.
Carpenters, dis. 70 per cent.

Drills.

Hand and Breast.

Millar's Falls, per doz. net list.

DRILL BITS.

Morse, dis., 37½ to 40 per cent.
Standard dis. 50 and 5 to 55 per cent

Faucets.

Common, cork-lined, dis. 35 per cent.

ELBOWS. (Stovepipe.)

No. 1, per doz	1 40
No. 2, per doz	1 20
Bright, 20c. per doz. extra.	

ESCUTCHEONS.

Discount, 45 per cent.

ESCUTCHEON PINS.

Iron, discount 40 per cent.

FACTORY MILK CANS.

Discount off revised list, 40 per cent.

FILES.

Black Diamond, 60 to 60 and 10 per cent.
Kearney & Foote, 60 and 10 to 70 per cent.
Nicholson File Co., 60 to 60 and 10 per cent.
Jowitt's, English list, 25 to 27½ per cent.

FORKS.

Hay, manure, etc., dis., 50 and 10 per cent. revised list.

GLASS—Window—Box Price.

Size United Inches.	Star Per 50 ft.	Star Per 100 ft.	D. Diamond Per 50 ft.	D. Diamond Per 100 ft.
Under 26	2 15	4 15	....	6 00
26 to 40	2 30	4 45	....	6 65
41 to 50	....	4 85	....	7 50
51 to 60	....	5 15	....	8 50
61 to 70	....	5 50	....	9 50
71 to 80	....	6 00	....	10 50
81 to 85	....	6 50	....	11 70
86 to 90	....	....	..	14 00
91 to 95	..	....	...	15 50
96 to 100	.. ..	....	....	18 00

GAUGES

Marking, Mortise, Etc.

Stanley's dis. 50 to 55 per cent.

Wire Gauges.

Winn's, Nos. 26 to 33, each	1 65	2 40

HALTERS.

Rope, ⅜ per gross	....	....
" ½ "	....	9 00
" ⅝ to ¾	....	14 00
Leather, 1 in., per doz	3 87½	4 00
" 1¼ in., "	5 15	5 20
Web, — per doz	1 87	2 45

HAMMERS.

Nail

Maydole's, dis. 5 to 10 per cent. Can. dis. 25 to 27½ per cent.

Tack.		
Magnetic, per doz	1 10	1 20
Sledge.		
Canadian, per lb	0 07½	0 08½
Ball Pean.		
English and Can., per lb	0 22	0 25

HANDLES.

Axe, per doz. net	1 50	2 00
Store door, per doz	1 00	1 50

Fork.

C. & B., dis. 40 per cent. rev. list.

Hoe.

C. & B., dis. 40 per cent. rev. list.

Saw.		
American, per doz	1 00	1 25
Plane.		
American, per gross	3 15	3 75

Hammer and Hatchet.

Canadian, 40 per cent.

Cross-Cut Saws.

Canadian, per pair 0 13½

HANGERS.	doz.	pairs.
Steel barn door	5 85	6 00
Stearns, 4 inch	....	5 00
" 5 inch	....	6 50
Lane's covered—		
No. 11, 5-ft. run	....	8 40
No. 11½, 10-ft. run	....	10 80
No. 12, 10-ft. run	....	12 60
No. 14, 15-ft. run	....	21 00
Lane's O.N.T. track, per foot	....	4½

HARVEST TOOLS.

Discount, 50 and 10 per cent.

HATCHETS.

Canadian, dis. 40 to 42½ per cent.

HINGES.

Blind, Parker's, dis. 50 and 10 to 60 per cent.

Heavy T and strap, 4-in., per lb	....	06½
" " 5-in., "	....	06¼
" " 6-in., "	....	06
" " 8-in., "	....	0 05¾
" " 10-in., "	....	0 05½

Light T and strap, dis. 65 and 5 per cent.

Screw hook and hinge—

6 to 12 in., per 100 lbs	....	4
14 in. up, per 100 lbs	....	3 50
		Per gro. pairs.
Spring	....	12 00

HOES.

Garden, Mortar, etc., dis. 50 and 10 p.c.

Planter, per doz	4 00	4 50

HOLLOW WARE

Discount .... 45 and 5 per cent

HOOKS.

Cast Iron.

Bird Cage, per doz	0 50	1
Clothes Line, per doz	0 27	0
Harness, per doz	0 72	0
Hat and Coat, per gross	1 00	3 10
Chandelier, per doz	0 50	1 00

Wrought Iron.

Wrought Hooks and Staples, Can., dis. 47½ per cent.

Wire.

Hat and Coat, discount 45 per cent.
Belt, per 1,000 .... 0 60
Screw, bright, dis. 55 per cent.

HORSE NAILS.

"C" brand 50 and 7½ p.c. off new list } Oval
"M" brand 50, 10 and 5 per cent. } head
Countersunk, 60 per cent.

HORSESHOES

F.O.B. Montreal.

Iron Shoes.	No. 2 and larger.	No. 1. and smaller.
Light, medium, and heavy.	3 50	3 75
Snow shoes	3 75	4 00
Steel Shoes.		
Light	3 60	3 85
Featherweight (all sizes)	4 85	4 85

F.O.B. Toronto, Hamilton, London and Guelph, 10c. per keg additional.

Toe weight steel shoes 6 70

JAPANNED WARE.

Discount, 45 and 5 p.c. off list, June 1899

ICE PICKS.

Star per doz 3 00 3 25

KETTLES.

Brass spun, 7½ p.c. dis. off new list.

Copper, per lb 0 30 0 50

American, 60 and 10 to 65 and 5 p.c.

KEYS.

Lock, Can., dis., 45 p.c.

Cabinet, trunk, and padlock, Am. per gross 60

KNOBS.

Door, japanned and N.P., per doz 1 50 2 50

Bronze, Berlin, per doz 2 75 3 25

Bronze Genuine, per doz 6 00 9 00

Shutter, porcelain, F. & L. screw, per gross 1 30 4 00

White door knobs—per doz. 1 25

HAY KNIVES.

Discount, 50 and 10 per cent.

LAMP WICKS.

Discount, 60 per cent.

LANTERNS.

Cold Blast, per doz 7 00

No. 3 "Wright's" 8 50

Ordinary, with O burner 4 00

Dashboard, cold blast 9 00

No. 0 .... 5 75

Japanning, 50c. per doz. extra.

LEMON SQUEEZERS.

Porcelain lined, ... per doz. 2 20 5 60

Galvanized 1 87 3 85

King, wood 2 75 2 90

King, glass 4 00 4 50

All glass 1 20 1 30

LINES.

Fish, per gross 1 05 2 50

Chalk " 1 90 7 40

LOCKS.

Canadian, dis. 45 p.c.

Russel & Erwin, per doz 3 00 3 25

Cabinet.

Eagle, dis. 30 p.c.

Padlock

English and Am., per doz 50 6 00

Scandinavian, " 1 00 2 40

Eagle, dis. 20 to 25 p.c

MACHINE SCREWS. Iron and Brass.

Flat head discount 25 p.c

Round Head discount 20 p.c.

MALLETS.

Tinsmiths' per doz 1 25 1 50

Carpenters', hickory, per doz. 1 25 3 75

Lignum Vitae, per doz 3 85 5 00

Caulking each 60 2 00

MATTOCKS.

Canadian, per doz 5 50 6 50

MEAT CUTTERS.

American, dis. 25 to 30 p.c.

German, 15 per cent.

MILK CAN TRIMMINGS.

Discount, 25 per cent.

NAILS.

Quotations are:	Cut.	Wire.
2d and 3d	$3 35	$3 85
3d	3 00	3 52
4 and 5d	2 75	3 35
6 and 7d	2 65	3 20
8 and 9d	2 50	3 00
10 and 12d	2 45	2 95
16 and 20d	2 40	2 90
30, 40, 50 and 60d. (base)	2 35	2 85

Wire nails in carlots are $2.77½

Galvanizing 2c. per lb. net extra.

Steel Cut Nails 10c. extra.

Miscellaneous wire nails, dis. 70 and 10 p. c.

Coopers' nails, dis. 30 per cent.

Flour barrel nails, dis. 25 per cent.

NAIL PULLERS.

German and American 1 85 3 50

NAIL SETS

Square, round, and octagon per gross 3 38 4 00

Diamond 12 00 15 00

NETTING.

Poultry, 55 per cent for McMullen's

OAKUM. Per 100 lb.

Navy 6 00

U. S. Navy 7 25

OIL.

Water White (U.S.) 0 16½

Prime White (U.S.) 0 15½

Water White (Can.) 0 15

Prime White (Can.) 0 14

OILERS.

McClary's Model galvan. oil can, with pump, 5 gal., per doz 10 00

Zinc and tin, dis. 50, 50 and 10.

Copper, per doz 1 25 3 50

Brass, " 1 50 3 50

Malleable, dis. 25 per cent.

GALVANIZED PAILS.

Dufferin pattern pails, dis. 45 p.c.

Flaring pattern, discount 45 per cent.

Galvanized washtubs, discount 45 per cent.

PIECED WARE.

Discount 40 per cent. off list, June, 1899.

10-qt. flaring sap buckets, dis. 45 p.c.

6, 10 and 14-qt. flaring pails, dis. 45 p.c.

Creamer cans, dis. 45 p c.

PICKS.

Per doz 6 00 9 00

PICTURE NAILS.

Porcelain head, per gross 1 75 3 00

Brass head " 0 40 1 00

PICTURE WIRE.

Tin and gilt, discount 75 p c.

PLANES.

Wood, bench, Canadian dis. 50 per cent. American dis. 50.

Wood, fancy Canadian or American 7½ to 40 per cent.

PLANE IRONS.

English, per doz 2 00 5 00

PLIERS AND NIPPERS.

Button's Genuine per doz pairs, dis. 37½ 40 p.c.

Button's Imitation, per doz .. 5 00 9 00

German, per doz 0 60 2 60

PLUMBERS BRASS GOODS.

Compression work, discount, 60 per cent.

Fuller's work, discount 65 per cent.

Rough stops and stop and waste cocks, discount, 60 per cent.

Jenkins disk globe and angle valves, discount, 55 per cent.

Standard valves, discount, 60 per per cent.

Jenkins' radiator valves discount 55 per cent.

" " " standard, dis., 60 p.c.

Quick opening valves discount, 60 p.c.

No. 1 compression bath cock 2 00

No. 4 " " " 2 00

No. 7, Fuller's 2 50

No 4½, " 3 00

POWDER.

Velox Smokeless Shotgun Powder.

100 lb. or less 0 85

1,000 lb. or more 0 80

Net 30 days.

PRESSED SPIKES.

Discount 20 to 25 per cent.

PULLEYS.

Hothouse, per doz 0 55 1 00

Axle 0 22 0 33

Screw 0 27 1 00

Awning 0 35 2 50

PUMPS.

Canadian cistern 1 80 3 60

Canadian pitcher spout 1 40 2 10

PUNCHES.

Saddlers', per doz 1 00 1 85

Conductors', " 9 00 15 00

Tinners' solid, per set 0 00 0 72

" hollow, per inch 0 00 1 00

RANGE BOILERS.

Galvanized, 3 gallons 7 00

" 35 " 8 25

" 40 " 9 50

Copper, 30 " 22 00

" 35 " 26 00

" 40 " 30 00

Discount off Copper Boilers 10 per cent.

RAKES.

Cast steel and malleable, 50, 10 and 5 p.c.

Wood, 25 per cent.

RASPS AND HORSE RASPS.

New Nicholson horse rasp, discount 60 to 60 and 10 p.c.

Globe File Co.'s rasps, 60 and 10 to 70 p.c.

Heller's Horse rasps, 50 to 50 and 5 p.c

RAZORS.

per doz.

Elliot's 4 00 18 00

Geo. Butler & Co.'s 8 00 18 00

Boker's 7 50 11 00

" King Cutter 12 50 50 00

Wade & Butcher's 3 60 10 00

Theile & Quack's 7 00 12 00

REAPING HOOKS.

Discount, 50 and 10 per cent.

REGISTERS.

Discount .... 40 per cent.

RIVETS AND BURRS.

Iron Rivets, black and tinned, discount 60 and 10 per cent.

Iron Burrs, discount 55 per cent.

Extras on Iron Rivets in 1-lb. cartons, ½c. per lb.

Extras on Iron Rivets in ½-lb. cartons, 1c. per lb.

Copper Rivets & Burrs, 35 and 5 p.c. dis. and cartons, 1c. per lb. extra, net.

Extras on Tinned or Coppered Rivets ½-lb. cartons, 1c. per lb.

RIVET SETS

Canadian, dis. 35 to 37½ per cent.

ROPE ETC.

	Sisal.	Manila.
7-16 in. and larger, per lb	10	13½
⅜ in.	11	14½
¼ and 5-16 in.		15½
Cotton, 3-16 inch and larger		16
" 5-32 inch		21
" ⅛ inch		22½
Russia Deep Sea		15½
Jute		8
Lath Yarn		9½
New Zealand Rope		10

RULES.

Boxwood, dis. 75 and 10 p.c.

Ivory, dis. 37½ to 40 p.c.

SAD IRONS. per set.

Mrs. Potts, No. 55, polished 62½

" No. 50, nickle-plated 67½

SAND AND EMERY PAPER.

Dominion Flint Paper, 47½ per cent.

B & A. sand, 40 and 5 per cent.

Emery, 40 per cent.

Garnet (Rurton's), 5 to 10 p.c. advance on list.

SAP SPOUTS.

Bronzed iron with hooks, per doz ... 9 50

SAWS.

Hand Disston's, dis. 12½ p.c.

S. & D., 40 per cent.

Crosscut, Disston's, per ft 35 0 55

S. & D., dis. 35 p.c. on Nos. 2 and 3.

Hack, complete, each 0 75 2 75

" frame only 0 75

SASH WEIGHTS.

Sectional, per 100 lbs 2 75 3 00

Solid, " 2 00 2 25

SASH CORD.

Per lb 0 23 0 30

SAW SETS.

"Lincoln," per doz 6 50

SCALES.

Standard, 45 p.c.

Champion, 65 p.c.

Spring Balances, 10 p.c.

Fairbanks Standard, 35 p.c.

" Dominion, 55 p.c.

" Richelieu, 55 p.c.

SCREW DRIVERS.

Sargent's per doz 0 65 1 00

SCREWS

Wood, F. H., bright and steel, 87½ and 10 p.c.

Wood R. H., " dis. 82½ and 10 p.c.

" F. H., brass, dis. 80 and 10 p.c.

Wood, R. H., " dis. 75 and 10 p.c.

" F.H., bronze, dis. 75 p.c.

" R.H. " 70 p.c.

Drive Screws, 87½ and 10 per cent.

Bench, wood, per doz 3 25 4 00

" iron. " 4 25 5 75

Set, Case hardened, 60 per cent.

Square Cap, 50 and 5 per cent.

Hexagon Cap, 45 per cent.

SCYTHES.

Per doz. net 9 00

SCYTHE SNATHS.

Canadian, dis. 45 p.c.

SHEARS.

Bailey Cutlery Co., full nickeled, dis. 60 p.c.

Seymour's, dis. 50 and 10 p.c.

SHOVELS AND SPADES.

Canadian, dis. 40 and 5 per cent.

SINKS.

Steel and galvanized, discount 45 per cent.

SNAPS.

Harness, German, dis. 25 p.c.

Lock, Andrews' 4 50 11 50

SOLDERING IRONS.

1, 1½ lb., per lb 0 37

2 lb. or over, per lb 0 34

SQUARES.

Iron, No. 493, per doz 2 40 2 55

" No. 494, " 3 25 3 40

Steel, dis. 60, 10 and 5 p.c.

Try and bevel, dis. 50 to 52½ p.c.

STAMPED WARE.

Plain, dis. 75 and 12½ p.c. off revised list.

Retinned, dis., 75 p.c. off revised list.

STAPLES.

Galvanized 3 50 4 00

Plain 3 25 3 75

Coopers', discount 45 per cent.

Poultry netting staples, 40 per cent.

STOCKS AND DIES.

American dis. 25 p.c.

STONE. Per lb.

Washita	0 28	0 60
Hindostan	0 06	0 07
" slip,	0 09	0 09
Labrador	....	0 13
" Axe	....	0 15
Turkey	....	0 50
Arkansas	0 00	1 50
Water-of-Ayr	0 00	0 10
Scythe, per gross	3 50	5 00
Grind. 2 in. 40 to 200 lb., per ton	....	25 00
" under 40 lb. "	....	28 00
Grind. under 2 in. thick "	....	29 00

STOVE PIPES.

5 and 6 inch Per 100 lengths 7 00

7 inch " " 7 50

ENAMELINE STOVE POLISH.

No. 4—3 dozen in case, net cash $4 80

No. 6—3 dozen in case, " 8 40

TACKS BRADS, ETC.

Per cent

Strawberry box tacks, bulk ...... 75 & 10

Cheese-box tacks, blued .... 80 & 12½

Trunk tacks, black and tinned 85

Carpet tacks, blued .... 80 & 15

" " tinned .... 80 & 20

" " (in kegs) 40

Cut tacks, blued, in dozens only . 80

" ¼ weights 60

Swedes, cut tacks, blued and tinned—

In bulk .... 80 & 10

In dozens 75

Swedes, upholsterers', bulk ... 85, 12½ & 12½

" brush, blued & tinned, bulk .. 70

" gimp, blued, tinned and japanned .... 75 & 12½

Zinc tacks 35

Leather carpet tacks 55

Copper tacks 50

Copper nails 52½

Trunk nails, black .... 65 and 5

Trunk nails, tinned .... 65 and 10

Clout nails, blued .... 65 and 5

Chair nails 35

Patent brads 40

Fine finishing 40

Picture frame points 10

Lining tacks, in papers 10

STANDARD CHAIN CO., PITTSBURGH, U. S. A.

MANUFACTURERS OF

CHAIN

OF ALL KINDS.

Proof Coil, B.B., B.B.B., Crane, Dredge Chain, Trace Chains, Cow Ties etc.

ALEXANDER GIBB, Montreal. For Provinces of Ontario and Quebec. —Canadian Representatives— A. C. LESLIE & CO., Montreal. For other Provinces.

Lining tacks, in bulk	15
" " solid heads, in bulk	75
Saddle nails in papers	10
" " in bulk	15
Tufting buttons, 22 line, in dozens only	60
Tin capped trunk nails	15
Zinc glazier's points	5
Double pointed tacks, papers....90 and	10
" " " bulk	40

TAPE LINES.

English, ass skin, per doz	2 75	5 00
English, Patent Leather	5 50	9 75
Chesterman's each	0 90	2 85
" steel, each	0 80	8 00

THERMOMETERS.

Tin case and dairy, dis. 75 to 75 and 10 p.c.

TRAPS. (Steel.)

Game, Newhouse, dis. 25 p.c.
Game, H. & N., P. S. & W., 65 p.c.
Game, steel, 72½, 75 p.c.

TROWELS.

Disston's discount 10 per cent.

German, per doz	4 75	6 00

S. & D., discount 35 per cent.

TWINES.

Bag, Russian, per lb	....	0 27
Wrapping, cotton, 3-ply		0 20
" " 4-ply	....	0 26
Mattress, per lb	0 33	0 45
Staging, "	0 27	0 35

VISES.

Hand, pe doz	4 00	6 00
Bench, parallel, each	2 00	4 50
Coach, each	6 00	7 00
Peter Wright's, per lb	0 12	0 13
Pipe, each	5 50	9 00
Saw, per doz	6 50	13 00

ENAMELLED WARE.

White, Princess, Turquoise, Blueand White, discount 50 per cent.
Diamond, Famous, Premier, 50 and 10 p.c.
Granite or Pearl, Imperial, Crescent, 50, 10 and 10 per cent.

WIRE.

Brass wire, 50 to 50 and 2½ per cent. off the list.

Copper wire, 45 and 10 per cent. net cash 30 days, f.o.b. factory.

Smooth Steel Wire, is quoted at the following net selling prices:

No. 6 to 8 gauge	$2 90
" 9 "	2 80
" 10 "	2 87
" 11 "	2 90
" 12 "	2 95
" 13 "	3 15
" 14 "	3 37
" 15 "	3 50
" 16 "	3 65

Other sizes of plain wire outside of Nos. 9, 10, 11, 12 and 13, and other varieties of plain wire remain at $2.80 base with extras as before. The prices for Nos 9 to 13 include the charge of 1 c. for oiling. Extras net per 100 lb.: Coppered wire, 60c.—tinned wire, $2—oiling, 10c.—special hay-bailing wire, 30c.—spring wire, $1—best steel wire, 75c.—bright soft drawn, 15c.—in 50 and 100-lb. bundles net, 10c.—in 25-lb. bundles net 15c.—packed in casks or cases, 15c.—bagging or papering, 10c.

Fine Steel Wire, dis. 17½ per cent. List of extras: In 100-lb. lots: No. 17, $5—No. 18, $5.50—No. 19, $6—No. 20, $6.65—No. 21, $7—No. 22, $7.30—No. 23, $7.65—No. 24, $8—No. 25, $9—No. 26, $9.50—No. 27, $10—No. 28, $11—No. 29, $12—No. 30, $13—No. 31, $14—No. 32, $15 No. 33, $16—No. 34, $17. Extras net—tinned wire, Nos. 17-25, $2—Nos. 26-31 $4—Nos. 32-34, $6. Coppered, 5c.—oiling, 10c.—in 25-lb. bundles, 15c.—in 5 and 10-lb. bundles, 25c.—in 1-lb. hanks, 50c.—in ½-lb. hanks, 75c.—in ¼-lb. hanks, $1—packed in casks or cases, 15c.—bagging or papering, 10c.

Galvanized Wire, per 100 lb.—Nos. 6,7,8, $3 50 to $3 85—No. 9, $2.85 to $3.15—No. 10, $3.60 to $3.95—No. 11, $3 70 to $4.10—No 12, $3 to $3.30—No. 13, $3.10 to $3.40—No. 14, $4.10 to $4.50—No. 15, $4.60 to $5.05—No. 16, $4.85 to $5.35. Base sizes, Nos. 6 to 9, $2.57½ f.o.b. Cleveland.

Clothes Line Wire, solid 7 strand, No. 17, $4.25; No. 18, $2.65; No. 19, $2.35, f.o.b Hamilton, Toronto, Montreal.

WIRE FENCING. F.O.B. Toronto

Galvanized barb	....	3 05
Galvanized, plain twist	....	3 05

Galvanized barb, f.o.b. Cleveland, $2.82½ in less than carlots, and $2.70 in carlots.

WIRE CLOTH.

Painted Screen, per 100 sq. ft., net	1 35

WASTE COTTON.

	per lb.
Colored	4½ to 5
White, according to quality	6½ to 7½

500-lb bale lots shaded.

WRENCHES.

Acme, 35 to 37½ per cent.
Agricultural, 60 p.c.
Coe's Genuine, dis. 20 to 25 p.c.

Towers' Engineer, each	2 00	7 00
" S., per doz	5 80	6 00
G. & K.'s Pipe, per doz	....	3 40
Burrell's Pipe, each	....	3 00
Pocket, per doz	0 25	2 90

WRINGERS.

Leader per doz.	$30 0	35 00
Royal Canadian "	26 00	30 00
Royal American "	26 00	28 00
Sampson "	....	30 00

Terms 4 months, or 3 p.c. 30 days.

WROUGHT IRON WASHERS.

Canadian make, discount, 40 and 5 per cent

ADVERTISING in WESTERN CANADA

will be Carefully, Efficiently, and Promptly attended to, by

The Roberts Advertising Agency,
WINNIPEG, CANADA

Standard Paint & Varnish Works
Limited
Makers of High Grade
Varnishes, Japans,
Paints Colors & Enamels
Windsor, Ont.

SHELF BOXES SCREW CASES SAMPLE HOLDERS

For particulars apply to the patentee and manufacturer.

J. S. BENNETT, 20 Sheridan Ave., TORONTO

DIAMOND STOVE PIPE DAMPER AND CLIP.

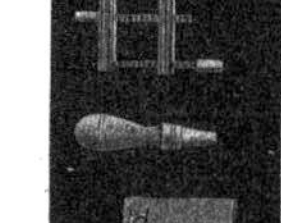

U. S. Patent June 25th, 1895.
Canadian Pat. Dec. 13th, 1894.

Sold by Jobbers of - - -
HARDWARE
TINWARE
and STOVES,
for furnace pipe, to support the sheet steel blade

Manufactured by THE ADAMS COMPANY, Dubuque, Iowa, U.S.A.
A. R. WOODYATT & CO., Guelph, Ontario.

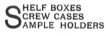

"THE EMLYN" SAW BENCH

Made in 6 sizes. Best value obtainable. Specially designed for export. With or without "Emlyn" Patent Guard. Sole maker—

CHARLES D. PHILLIPS,

Cables—"Machinery," Newport.

Emlyn Engineering Works, NEWPORT, MON., ENGLAND.

SEALED TENDERS addressed to the undersigned, and endorsed "Tender for Iron Superstructure, Des Joachims Interprovincial Bridge, across the Northern Channel," will be received at this office until Wednesday, Jun inclusively, for the construction of an iron superstructure for the Interprovincial Bridge over the Northern Channel of the Ottawa River at Des Joachims, County of Pontiac, P.Q., according to a plan and a specification to be seen at the office of F. S. Rees, Esq., Dockmaster, Dry Dock, Kingston, Ont., at the Public Works Office, Merchants Bank Building, Montreal, Que., on application to the Postmaster at Hamilton, Ont., and at the Department of Public Works, Ottawa.

Tenders will not be considered unless made on the form supplied, and signed with the actual signatures of tenderers.

An accepted cheque on a chartered bank, payable to the order of the Minister of Public Works, for three thousand dollars ($3,000.00), must accompany each tender. The cheque will be forfeited if the party decline the contract or fail to complete the work contracted for, and will be returned in case of non-acceptance of tender.

The Department does not bind itself to accept the lowest or any tender.

By order,

JOS. R. ROY,

Acting Secretary.

Department of Public Works,

Ottawa, May 23, 1901.

Newspapers inserting this advertisement without authority from the Department will not be paid for it. (23)

Lockerby & McComb

AGENTS IN CANADA

FOR THE

Celebrated P. & B.

Cold Storage Lining

AND

. . Ruberoid Roofing . .

P. S.--Prices on Application.

65 Shannon Street, MONTREAL.

BUSINESS NEWS

of any kind that is of value to business men supplied by our Bureau. We can give you market quotations from any town in Canada, reports from the city markets, stock quotations, etc. You can get commercial news from any Canadian paper through us.

Write us, giving us particulars of what you want and where you want it from, and we will quote you prices by return.

"Clippings from any Canadian paper on any subject."

CANADIAN PRESS CLIPPING BUREAU,

232 McGill Street, MONTREAL, QUE.

Telephone Main 1255.

19 Front St. East, Toronto. Telephone 2148.

---

75 YEARS. *ESTABLISHED 1825.* *75 YEARS.*

CELEBRATED HEINISCH SHEARS.

Tailors' Shears,
Trimmers, Scissors,
Tinners' Snips, etc.

ACKNOWLEDGED THE BEST.

R. HEINISCH'S SONS CO. NEW YORK OFFICE, 90 Chambers St. NEWARK, N.J., U.S.A.

Not connected with any Shear Combination.

CHAS. P. CLARK, President. JARED CHITTENDEN, Treasurer.

...ESTABLISHED 1849...

BRADSTREET'S

Capital and Surplus, $1,500,000. **Offices Throughout the Civilized World.**

Executive Offices: Nos. 346 and 348 Broadway, New York City, U.S.A.

THE BRADSTREET COMPANY gathers information that reflects the financial condition and the controlling circumstances of every seeker of mercantile credit. Its business may be defined as of the merchants, by the merchants, for the merchants. In procuring, verifying and promulgating information no effort is spared, and no reasonable expense considered too great, that the results may justify its claim as an authority on all matters affecting commercial affairs and mercantile credit. Its offices and connections have been steadily extended, and it furnishes information concerning mercantile persons throughout the civilized world.

Subscriptions are based on the service furnished, and are available only by reputable wholesale, jobbing and manufacturing concerns, and by responsible and worthy financial, fiduciary and business corporations. Specific terms may be obtained by addressing the Company at any of its offices. **Correspondence Invited.**

—OFFICES IN CANADA—

HALIFAX, N.S.	HAMILTON, ONT.	LONDON, ONT.	MONTREAL, QUE.
OTTAWA, ONT.	QUEBEC, QUE.	ST. JOHN, N.B.	TORONTO, ONT.
VANCOUVER, B.C.	VICTORIA, B.C.	WINNIPEG, MAN.	

THOS. C. IRVING, Gen. Man. Western Canada, Toronto. JOHN A. FULTON, Gen. Man. Eastern Canada, Montreal.

TRADE NOBLES & HOARE NH LONDON MARK

Nobles & Hoare.

CORNWALL ROAD STAMFORD STREET.

LONDON, ENG.

Manufacturers of

HIGH-CLASS VARNISHES ONLY

Which can be obtained direct from the works or from the principal Color Dealers in Canada.

"BRASSITE"

REGISTERED TRADE MARK.

None genuine without the above **"Trade Mark."**

"Gunn's" Patent "Grassite" Goods.

Equal to **Solid Brass** in every particular. Cost less money—look and wear as well. Our sales are increasing all the time. Why not increase your sales?

THE GUNN CASTOR CO.

Limited.

KNOX HENRY, Canadian Agent, Room 32, Canada Life Bldg., **MONTREAL.**

Est. 1863 — Inc. 1895

TRADE MARK

Black Diamond File Works

G. & H. Barnett Company

PHILADELPHIA

Twelve TRADE MARK Medals

Awarded
By **JURORS** at
International Expositions
Special Prize
Gold Medal at Atlanta, 1895

TRADE MARK

1901 GARDEN HOSE. 1901

We manufacture the most popular and best selling lines of Garden Hose in Canada. Our "Maltese Cross" brand we confidently recommend as the finest Garden Hose on the market.

We have other grades not quite so expensive, but good reliable brands, viz.: "Lion" (the popular medium-priced hose), "King" "Sun" and "Leader."

Our "Kinkproof" (wire wound) hose is wired by a special process controlled solely by ourselves, and is so constructed that it may be cut at any wind of the wire without loosening or uncoiling, the wire being self-gripping throughout each length.

THE GUTTA PERCHA AND RUBBER MFG. CO.

OF TORONTO, LIMITED.

Head Office and Warerooms—
49-61-63 West Front St.,

TORONTO, CANADA.

Factories—115-165 West Lodge Ave.
124-148 O'Hara Ave.

Galvanized Steel Sheets

"LEVEL BEST" BRAND.

Black Steel Sheets

"GRIFFIN" BRAND.

PRICES ON APPLICATION TO

B. & S. H. THOMPSON & CO'Y

26 St. Sulpice St., MONTREAL,

CORDAGE

ALL KINDS AND FOR ALL PURPOSES.

Manila Rope	Lathyarn
Sisal Rope	Shingleyarn
Jute Rope	Bale Rope
Russian Rope	Lariat Rope
Marline	Hemp Packing
Houseline	Italian Packing
Hambroline	Jute Packing
Clotheslines	Drilling Cables
Tarred Hemp Rope	Spunyarn
White Hemp Rope	Pulp Cord
Bolt Rope	Lobster Marlin
Hide Rope	Paper Cord
Halyards	Cheese Cord
Deep Sealine	Hay Rope
Ratline	Fish Cord
Plow Lines	Sand Lines

"RED THREAD" Transmission Rope from the finest quality Manila hemp obtainable, laid in tallow.

CONSUMERS CORDAGE COMPANY,

Limited

Western Ontario Representative—
WM. B. STEWART,
Tel 94. 27 Front St. West, TORONTO.

MONTREAL, QUE.

CANADIAN
HARDWARE
AND METAL MERCHANT

The Weekly Organ of the Hardware, Metal, Heating, Plumbing and Contracting Trades in Canada.

VOL. XIII. MONTREAL AND TORONTO, JUNE 15, 1901. NO. 24

FISHING
TACKLE
STEEL
POLES
Including Lines, Reels, Hooks, Spoons, Bait Pails, Poles, Baskets and Minnow Traps.
RICE LEWIS & SON, LIMITED
GUNS
AND
AMMUNITION
Ask for
Hazard Powder.
Toronto.

LEWIS BROS. & CO.

Wholesale Hardware MONTREAL.

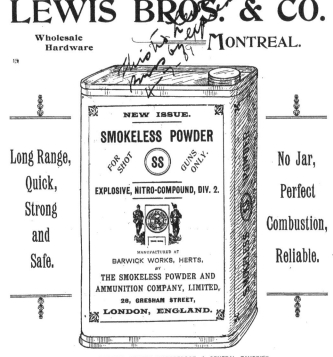

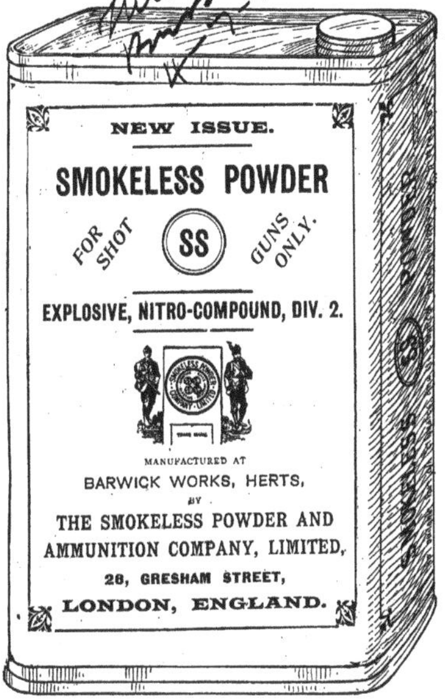
NEW ISSUE.
SMOKELESS POWDER
FOR SHOT
SS
GUNS ONLY.
EXPLOSIVE, NITRO-COMPOUND, DIV. 2.
MANUFACTURED AT
BARWICK WORKS, HERTS,
BY
THE SMOKELESS POWDER AND
AMMUNITION COMPANY, LIMITED,
26, GRESHAM STREET,
LONDON, ENGLAND.
SMOKELESS POWDER

Long Range,
Quick,
Strong
and
Safe.

No Jar,
Perfect
Combustion,
Reliable.

EXTRA HARDENED, DOUBLE WATERPROOF, A GENERAL FAVORITE.

MADE BY

The Smokeless Powder & Ammunition Co., Limited

LONDON, ENGLAND.

We are sole Agents for this Celebrated Powder in Canada and can recommend it as the best Smokeless Powder on the market.

WRITE FOR PRICES. MAIL ORDERS SHIPPED SAME DAY AS RECEIVED.

LEWIS BROS. & CO., Montreal.

Wadsworth-Howland Co. CHICAGO.

JEWEL

BEST PAINT ON EARTH.

ONE PRICE. BEST QUALITY.

. . . . WE PUT OUR NAME ON ONLY ONE BRAND OF . . . .

MIXED PAINTS

The Best Wall Paper Cleaner—The only one offered for sale in the Canadian market.

Jewel Wall Finish—A perfect sanitary Kalsomine. Will not set hard in the pail overnight. Can be tinted with any dry color. Dyes not necessary. Will not crack, peel or rub off the wall.

Terebine Shellac—A substitute for Grain Alcohol Shellac. The most satisfactory first-coat Varnish or Surfacer offered for sale anywhere at any price.

Adamantine Floor Paints are the most reliable goods for the purpose yet produced. The only paint on the American continent that will dry between supper and breakfast.

Write our customers regarding their success in these lines.

RICE LEWIS & SON

LIMITED.

Canadian Distributors TORONTO.

HARDWARE AND METAL

VOL. XIII. MONTREAL AND TORONTO, JUNE 15, 1901. NO. 24.

President,
JOHN BAYNE MacLEAN,
Montreal.

THE MacLEAN PUBLISHING CO.
Limited.

Publishers of Trade Newspapers which circulate in the Provinces of British Columbia, North-West Territories, Manitoba, Ontario, Quebec, Nova Scotia, New Brunswick, P.E. Island and Newfoundland.

OFFICES

MONTREAL - - - - - - - 232 McGill Street, Telephone 1255.
TORONTO - - - - - - - 10 Front Street East, Telephone 2148,
LONDON, ENG. - - - - - 109 Fleet Street, E.C., W. H. Miln.
MANCHESTER, ENG. - - - 18 St Ann Street. H. S. Ashburner,
WINNIPEG - - - - - Western Canada Block, J. J. Roberts.
ST. JOHN, N.B. - - - - No. 3 Market Wharf, J. Hunter White,
NEW YORK. - - - - - - - 176 E. 88th Street.

Subscription, Canada and the United States, $2.00.
Great Britain and elsewhere - - - 12s.

Published every Saturday.

Cable Address { Adscript, London. Adscript, Canada.

ENGLISH METAL MARKET BULLISH.

THE bullish tendency that we have noticed in the English metal markets continues to manifest itself in pronounced form.

The cable quotations received in Montreal this week indicate advances all along the line.

Ingot tin is one of the first on the list, the New York values having reached nearly to 30c. The London speculative movement in this metal is strong and one of those continued upward trends of this market is expected.

The sheet metals are all moving in the same direction. Tinplates, in sympathy with pig tin, have scored an advance of 1s. for coke plates and 1s. 3d. for charcoals, which means a total movement of nearly 2s. per box. Terne plates are quoted at 24s. 3d., an advance of 1s. 9d. to 2s., or about 3s. higher as compared with the price six weeks ago. Canada plates are up 5s. per ton, or a total advance of 7s. 6d. since the beginning of the movement. Galvanized iron is also reported 10s. per ton higher, with every probability of further advances. Sheet zinc is quoted £1 10s. higher, while steel sheets, dead flat, cannot be procured at lower quotations than 10s. per ton above those of some weeks ago.

All the mills are said to be filled with orders, and it is difficult to place contracts for July or August delivery. Most of the mills refuse to guarantee delivery before September.

What aggravates the situation in Canada is the fact that the early shipments contracted for last February and March have, in many instances, not come to hand, and the Canadian metal market is rather bare of supplies. The ss. Assyrian, that went on the rocks a few days ago, had a shipment of sheet zinc on board consigned to Montreal, and there is now a genuine dearth of this article.

As yet Canadian importers are said to be ordering fall importations very lightly.

The Ironmonger, published in London, England, comes to hand this week, ponderous in size, and arranged in a prettily designed cover of five different colored printings. It is the annual spring issue, and the size may be gathered from the fact that it contains about 465 pages, 370 of which is advertising matter.

UNITED STATES TRADE WITH CANADA.

UNITED STATES returns showing the trade of that country with British North America during the nine months, ending March 31, have just been issued. They show a decided gain compared with the same period of 1900, the total trade being $109,529,618, of which $77,891,138 were exports to and $31,638,480 imports from British North America. The total for the nine months of 1900 was $99,353,506, of which $69,780,474 were exports and $29,572,932 imports.

It will be noticed that the increase is proportionately greater in exports to than in the imports from British North America. In the one it is 11.62 per cent. and in the other scarcely 7 per cent.

The increase in the exports to this country is largely due to the increases in the number of cattle and in the quantity of breadstuffs, provisions, steel, carriages, copper, raw cotton, tobacco leaf. The following table shows the exports to British North America from the United States of articles appertaining to the iron and hardware trade during the nine months for the past three years :

Agricultural implements....	$842,449	$1,295,587	$1,368,535
Carriages, cars, and parts of	239 293	378,468	341,938
Copper of different kinds...	119,517	132,211	214,557
Cycles, and parts of.	338,412	205,3.8	173,725
Builders' hardware and saws and tools	62',157	588,495	601,'36
Sewing machines and parts	107,578	145,617	138,993
Typewriting machines.	42,125	38,424	43,953
Steel bars or rails for railways	1,362,199	1,722,826	2,671,432
Turpentine...................... ..	146,160	247,979	208,402
Oils, mineral, refined.........	629,549	889,147	889,530

Among the articles which chiefly contributed to the increase in the imports from British North America were : Coal, tinplates, lead, sugar, hides and skins. Lumber imports declined nearly $1,500,000. In the increases the most important was in lead and manufacturers of, which jumped from $103 814 in 1900 to $1,509,777 in 1901.

BECOMING A SERIOUS MATTER.

RECENT developments in the strike of the machinists in the United States instead of tending toward a shortening of the period of hostilities threaten to extend it. It is not that more men have quit work; it is due to the fact that it has become more than a fight for shorter hours of labor. The employes struck for a nine-hour day at a ten-hour pay; but the employers have now thrown another bone of contention into the fray. They have decided that they cannot allow the union to interfere in the administration of their shops.

The strike is now practically on all fours with that of the famous engineers' strike of a few years ago in Great Britain. The principles involved are the same. And it is this, together with the fact that the hands of both strikers and employers have been strengthened during the past week, that tends to prolong the strike and increase its seriousness.

The machinery manufacturers of the United States have an organization known as the National Metal Trades' Association. Until its members, a few days ago, decided to take an aggressive stand, it had lacked the support of a good many machine manufacturers. Numbers are now, however, flocking to the association's banner.

The platform of principles which the employer's association drew up the other day is both clear and comprehensive. Its principal clauses read as follows:

Since we, as employers, are responsible for the work turned out by our workmen, we must, therefore, have full discretion to designate the men we consider competent to perform the work and to determine the conditions under which that work shall be prosecuted. The question of competency of the men being determined solely by us, and, while disavowing any intention to interfere with the proper functions of labor organizations, we will not admit of any interference with the management of our business.

Disapproving absolutely of strikes and lockouts, the members of this association will not arbitrate any question with men on strike. Neither will this association countenance a lockout on any arbitrable question unless arbitration has failed.

No discrimination will be made against any member of any society or organization. Every workman who elects to work in a shop will be required to work peaceably and harmoniously with all his fellow employes.

The number of apprentices, helpers and handy men to be employed will be determined solely by the employer.

We will not permit employes to place any restriction on the management, methods or production of our shops, and will require a fair day's work for a fair day's pay.

It is the privilege of the employe to leave our employ whenever he sees fit, and it is the privilege of the employer to discharge any workman when he sees fit.

The above principles being absolutely essential to the successful conduct of our business, they are not subject to arbitration.

In case of disagreement concerning matters not covered by the foregoing declaration, we advise our members to meet their employes either individually or collectively, and endeavor to adjust the difficulty on a fair and equitable basis. In case of inability to reach a satisfactory adjustment, we advise that they submit the question to arbitration by a board composed of six persons, three to be chosen by the employe or employes. In order to receive the benefits of arbitration the employe or employes must continue in the service and under the orders of the employer, pending a decision. In case any member refuses to comply with this recommendation he shall be denied the support of this association unless it shall approve the action of said member.

As we stated, there has been a great deal of sympathy with the employes in their effort to secure a nine-hour day; but one thing is liable to be overlooked, and that is that the strike is a direct violation of an agreement entered into between masters and men 13 months ago. This agreement, which is dated May 18, 1900, specifically declared that pending arbitration there should be no strikes or lockouts, but notwithstanding this the machinists refused arbibration and struck.

While we in Canada are not directly interested in the strike, we cannot view it with equanimity, for a prolonged strike would not fail to seriously affect the iron and steel trades. Already its influence is being felt to some extent, and it is estimated that, influenced by the backward season and the strike, the Lake Superior iron ore trade has been shortened by 15 per cent.

WHY NOT, INDEED?

THE American Grocer asks: "Why should we persist in saying 'English' when we speak of American speech?" That is so, although we would prefer it being called "United States" speech, as far as its use in the neighboring Republic is concerned, because Canada is a part of the North American continent, and rather the larger half of it, and in this movement for speech reform we want to take a place.

When referring to the predominating speech in this country, why should we persist in saying "English," when we speak of Canadian speech? Then, there is the Commonwealth of Australia. The people of that country should certainly fall into line. Ceylon, too, cannot afford to ignore the movement. Nor can the islands of the sea.

Of course, this movement is not to be confined to the language that is spoken in the United Kingdom. Whatever the native tongue of a man may be he must not call it by its original name; it must be "United States," "Canadian," "Australian," according to the country in which he happens to be residing.

The politicians in Canada to whom the French language is unsavory have in this movement a panacea for all their woes, for the people in the Province of Quebec who speak the language of France (we must not say French language) will certainly not object to calling it "Canadian."

And then why stop at languages? Why not apply the principle to—well, architecture? When we put up a building in what has heretofore been called the Gothic, Roman, and so forth style, why should we persist in calling it by such names? Let us call it Canadian. And every other nation can do so after its kind. Of course, pillars will have to fall into line with buildings. Why should we persist in saying "Corinthian," "Ionic," and "Doric," when they are made in this country? Let us call them Canadian. The United States can call them "United States."

Antiquated people will probably object. But let them. Forward movements have always met with opposition. Educationalists may also object. But let them. They do not know everything. We are a self-contained people. And so are our cousins to the south.

SUMMER TEMPERATURE OF ST. JOHN.

Arrangements have been made by the tourist association of St. John, N.B., for publishing in one of the New York papers daily reading of the temperature in the former city.

The summer temperature in St. John is delightful, as those who have been fortunate enough to visit there well remember. The ocean breezes always keep the temperature down to a comfortable point, no matter how people may be sweltering elsewhere.

The business men of St. John should lend all the aid they can in disseminating information to the outside world regarding the temperature of their pretty city.

INDUSTRIAL DEVELOPMENT, AND NOT THE CENSUS, THE TEST OF CANADA'S GROWTH.

IF we are to judge by the hints we receive through the Ottawa correspondents of the daily papers, quite a little concern is developing in official circles in regard to the recently-taken census: The figures are apparently not totalling up as well as expected. We are all sorry for this, for we have all been pluming ourselves upon the sure and certain hope of a large increase in the population of the country. And now that we have been weighed by the census we are beginning to fear that we have been overestimating ourselves, although the official figures have yet to be issued. The figures of 1891 were a bitter disappointment. It is only to be hoped that those of 1901 will be a little more to the liking of our palate.

But, after all, are we not disposed to lay too much stress on the importance of population as a factor of national greatness? We believe we are. In saying this we are by no means blind to the need of population in this country. The greater the number of people we have in Canada the greater will be the consumptive demand of the home market.

If size of population was the most important factor China would be in the lead and we should all be trying to emulate her. But Canada has no desire to emulate her, for in every phase of national activity this country far transcends the Celestial Empire with its 400,000,000 people. At any rate, we, as Canadians, are so persuaded in our own minds.

It is evident, therefore, that there are things that should concern us more than population. One of these is the character of the people in whose hands the weal or woe of the country rests. Another is the development of our natural resources. It is our purpose here to deal only with the latter phase of our national development.

In the development of her natural resources, in the expansion of her industrial life, Canada has made strides during the past decade which far exceed the anticipations current at the beginning of that period. Ten years ago there were the disappointments of the census to displease us, and the high rate of the McKinley tariff to worry us. Pessimism so abounded that a great many people seemed to be almost afraid to take a peep into the future. The McKinley tariff was aimed against the agricultural industry of this country. But, instead of of crippling it, as nearly everybody feared, it seemed to put new life and energy into it. At anyrate it is no more the agricultural industry of 10 years ago than is the average stalwart Canadian youth of 20 the child of 10 years ago.

The exports of agricultural products, the produce of our own fields, were over 100 per cent. larger in 1900 than in 1891, when the previous census was taken, the figures being $27,516,609 and $13,666,858 respectively. And it must be remembered that during that period our exports of agricultural products to the United States declined from $7,291,246 in 1891 to $2,041,110 in 1900, but in the meantime our sales to Great Britain have swelled from $5,254,028 to $21,674,965. The following table shows the course of trade in the chief items of export from 1891 to 1900:

CHIEF AGRICULTURAL PRODUCTS EXPORTED.

	1891.	1900.
Fruits	$1,567,137	$3,305,662
Barley	2,929,873	1,010,425
Oats	129,917	2,143,179
Peas	2,032,601	2,145,471
Wheat	1,583,084	11,995,488
Flour	1,388,578	2,791,885
Oatmeal	45,195	474,991
Hay	559,489	1,414,109

The exports of all kinds of grain last year aggregated 30,055,000 bushels, against 10,760,018 in 1891. The decline in the exports of barley is, of course, due to the McKinley tariff, but even in this grain our export trade is again expanding, for, whereas in 1899 the quantity shipped was only 238,948 bushels, last year it was 2,156,282 bushels. When the McKinley tariff went into operation our exports of barley were nearly 10,000,000 bushels.

Turning to animals and their produce, we find that here again is a more than doubling up, for in 1891 our exports under that classification were $25,967,741, while last year they were $56,148,807, largely due to our increased trade with Great Britain, although even with the United States we did a little more in 1900 than 10 years before.

CHIEF EXPORTS OF ANIMALS AND THEIR PRODUCE

	1891	1900.
Horses	$1,417,244	$1,166,981
Horned cattle	8,772,499	9,080,776
Sheep	1,146,465	1,894,012
Furs	1,429 229	1,806 966
Hides and skins	508,925	1,406,339
Butter	602,175	5,122,156
Cheese	9,508,800	19,856,324
Eggs	1,160,359	1,457,902
Bacon and hams	628,469	12,758,025
Beef and mutton	40,044	223,424
Wool	245,503	418,119

The increase in the exports of bacon and hams is one of the most remarkable in the trade experience of the country. It is the reward of strict attention to quality, and conveys a lesson which, in these degenerate days, when price and not quality is so great a factor, business men cannot afford to ignore. That the trade in horned cattle has increaseed at all is rather remarkable, when one considers the high tariff that obtains in the United States and the disabilities under which it labors in Great Britain.

A feature of the cheese industry worthy of note is the fact that the percentage of increase in the number of factories in Canada and the increase in the export trade from 1891 to 1900 was almost the same, it being 110 in the one and 111 in the other.

Our exports of forest products have not shown the same extraordinary expansion as those appertaining to the farm, yet, last year they were, with one exception, the largest on record. The largest on record was in 1897, when $31,258,729 worth was shipped. Last year the figures were $29,663,668; but even that is an increase of over 22 per cent. in the 10 year period.

In all the experience of this country during the past ten years, nothing has probably been more striking or attracted more world-wide attention than the development of our mining industry. The Yukon and the Kootenay gold fields have both come into prominence during that period. Our production of gold last year was about

$28,000,000, whereas ten years ago it was less than $1,000,000. Next to gold, in order of importance, comes coal, and its output last year was valued at $12,668,475, whereas in 1891 it was but $7,000,000. Of lead we only produced 88,665 tons, valued at $3,857 in 1891, but last year the quantity was over 63,000,000 tons and the value $2,760,521. The quantity of nickle produced in Canada last year was the largest on record, being 7,080,227 lb. valued at $3,327,707, against 4,626,627 lb. valued at $2,775,976 in 1891. Copper production increased from 8,928,921 lb. valued at $1,149,598 in 1891 to 18,910,820 lb. valued at $3,063,119 in 1900.

The total value of the minerals produced in Canada last year was $63,775,090, an increase of 236 per cent. in 10 years. And the exports were a reflex of the production, for they grew from $5,784,143 in 1891 to $24,580,266 in 1900, an increase of about 380 per cent.

Among our natural industries, that of fishing occupies an important position. And, although one does not look for a large expansion in this particular industry, it requires no stretch of the imagination to say that the conditions are, on the whole, more satisfactory than they were 10 years ago. We have not yet the figures for 1900, but those of 1899 show the value of the catch in that year to have been $21,891,706, an increase of nearly $3,000,000 compared with that of 10 years ago. The capital invested in the industry increased by a similar approximate amount, the sums being $7,376,186 and $10,149,840 respectively.

In spite of the fact that our shipments of fish to the United States do not grow, and that what was once the Spanish West Indies are not as accessible since falling into the hands of the United States, it is gratifying to note that our export trade last year was the most valuable on record, being $11,169,083, compared with $9,715,401 ten years ago.

Until the census is completed figures cannot be adduced in regard to the manufacturing industry of this country, but everyone who is at all of an observing mind cannot fail to have noticed the development there has been in this particular during the past decade.

The manufacture of iron and steel ten years ago was in a languid condition, and a good many people who were solicitous for its welfare began to question whether it ever would amount to much. To-day everything is changed. We have at Sydney iron and steel works which, in extent and possibilities, were not dreamed of ten years ago. At Sault Ste. Marie we have nickel-steel works in course of construction, which promise to equal anything on this continent. Then there are the iron and steel works in Hamilton, the blast furnace at Midland and another at Deseronto, none of which were even in contemplation when the last census was taken.

Wood pulp making is another industry which has jumped into prominence during the last decade. In different parts of Ontario, Quebec, and the Maritime Provinces pulp mills of enormous size have been constructed within the last few years. With her enormous supplies of spruce and her wealth of water power, Canada's possibilities in the manufacturing of pulp are almost unlimited.

Still another staple industry which has made much headway since 1891 is that of cotton-making. Within the last two or three years some of the larger mills have doubled their capacity and otherwise improved their facilities.

With an increase in the manufacturing capacity of this country we naturally look for an increase in the export trade of manufactured products. And we look not in vain, for, from 1891 to 1900, the increase was 126 per cent., the figures for the two years being $6,296,249 and $14,224,287 respectively. Some of the principal articles of export during the two years in question were :

	1891.	1900.
Agricultural implements......	$252,620	$1,692,155
Cottons......................	159,954	414,259
Iron and steel of various kinds	257,471	1,425,163
Leather......................	950,456	1,871,630
Liquors......................	62,021	406,156
Musical instruments..........	401,553	507,983
Household furniture..........	138,705	380,029
Wood pulp....................	280,619	1,816,016

Now, as to the transportation phase of our industrial condition. The miles of railway in operation in 1891 were 14,009 ; last year they were 17,656. The freight carried increased from 21,753,021 tons in 1891 to 35,946,183 in 1900, and the number of passengers from 13,222,568 to 21,500,175.

During the same period the freight carried through the canals of Canada increased from 2,902,526 tons to 6,225,924 tons, a gain of more than 114 per cent.

In regard to the sea-going vessels, inward and outward, the following table tells the tale.

SEA GOING VESSELS.

	1891	1900
Inward—		
Number of vessels.......	15,548	14,607
Tons register...........	5,273,935	7,262,721
Freight, tons...........	1,028,736	1,587,762
Outward—		
Number of vessels.......	15,773	13,939
Tons register...........	5,421,261	6,912,400
Freight, tons...........	2,100,987	4,163,404

In the coasting trade in 1891 125,564 vessels of all descriptions, with an aggregate tonnage of 24,986,130, were employed and in 1900 there were 143,229 vessels with an aggregate tonnage of 33,631,730.

The returns in regard to navigation on the inland waters of the Dominion are also favorable. In 1891 the number of vessels engaged was 19,008, with a registered tonnage of 4,009,018, and the freight carried was 715,861 tons. In 1900 the number of vessels was 21,195, the tons registered 6,300,050, and the freight carried 817,971 tons.

A large increase in the population of Canada is much to be desired ; but the development of our natural resources and the expansion of our trade are more important. And in the essentials most important Canada has certainly made a tremendous stride during the past decade.

VISIT FROM A PAINTMAN.

Mr. W. A. Campbell, general representative of the Wadsworth-Howland Co., Chicago, was in Toronto last week. He has been calling on the trade in Ontario and was introduced by Rice Lewis & Son, Limited. Business in the east is particularly good and customers are saying some nice things about the "W.-H." paints. Mr. Campbell left for St. John, N.B., on Sunday morning, having a special inquiry there which he desired to answer in person.

The Canadian Wholesale Hardware Association meet at the Windsor Hotel, Montreal, on Friday June 21.

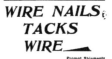

BUSINESS CHANGES.

DIFFICULTIES, ASSIGNMENTS, COMPROMISES.

J. E. SANDERS & CO., general merchants, Fort Frances, Ont., have compromised at 75c. on the dollar.

G. Rioux, general merchant, Trois Pistoles, Que., has assigned.

D. Gauthier, general merchant, St. Felecien, Que., is seeking an extension.

Judson Settle, blacksmith, Cole Harbor, N.S., has assigned to Colin McNab.

N. A. Bigham, general merchant, Culloden, Ont., is offering to compromise.

D. Jobin, general merchant, Sacré-Cœur de Marie, Ont., is offering to compromise.

Adelard Many, general merchant, St. Sebastien, Que., is offering 50 cents on the dollar.

John B. Lafrance, general merchant, Crysler, Ont., has assigned to John G. Hay, Toronto.

The stock of R. W. Richardson, general merchant, Hartland, N.B., has been seized under execution.

A meeting of the creditors of Annie L. Graham, general merchant, Ouvry, Ont., was held on Wednesday.

A meeting of creditors of George D. D'Entremont, general merchant, Middle East Pubnico, N.S., has been held.

Alick E. Chandler, general merchant, Plumas, Man., has assigned to C. H. Newton, and a meeting of creditors has been called for the 14th inst.

Fred. A. Taylor, harness dealer, Fullarton, Ont., has assigned to George Kastner, Sebringville, and a meeting of his creditors will be held on the 18th inst.

PARTNERSHIPS FORMED AND DISSOLVED.

Magnan Freres, hardware dealers, etc., Montreal, have dissolved.

Henry Short & Sons, sporting goods dealers, Victoria, B.C., have dissolved.

Graham & Adams, contractors, Vancouver, B.C., have dissolved partnership.

Benson & Borland, coal dealers, Quebec, have dissolved, and Andrew Borland has registered as sole proprietor of the company.

SALES MADE AND PENDING.

Wm. Hope & Co., tinsmiths, Perth, Ont., have sold out.

The assets of J. B. Frigon, general merchant, St. Tite, Que., have been sold.

Morse & Jack, general merchants, Blenheim, Ont., are advertising to sell out.

The assets of F. Perreault & Cie., general merchants, Rimouski, Que., are to be sold.

The stock of Z. Paquet, general merchant, Roberval, Que., has been sold at 51 1-4c. on the dollar.

The stock of the estate of R. A. Copeland & Co., general merchants, Grenfell, Man., has been sold.

The stock of the estate of Samuel Bricker, hardware and coal dealer, Listowel, Ont., is advertised for sale by auction on the 18th inst.

CHANGES.

Daniel McDonald, blacksmith, Collins' Bay, Ont., has been succeeded by ——. Garret.

A. Brandenburger, dealer in stoves and tinware, Stratford, Ont., has sold out to J. Read & Co.

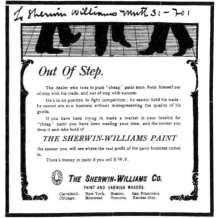

W. C. Brine has registered as sole proprietor of H. H. Fuller & Co., wholesale hardware dealers, Halifax, N.S.

H. S. Law, general merchant, Alberni, B.C., is giving up business.

H. P. Reed, general merchant, Bear River, N.S., has given up business.

N. Piche & Fils have registered as general merchants in St. Raymond, Que.

Frank Schurman, general merchant, etc., Collingwood Corner, N.S., has sold out to James Higgs and John W. Schurman.

FIRES.

James Brown, sawmiller, Acton, Ont., has been burned out; partially insured.

Gross & Granger, wholesale and retail hardware dealers, Whitby, Ont., have been burned out; partially insured.

DEATHS.

J. Pierre Michuad, general merchant, New Brunswick, is dead.

FLAT AND STALE STORE NAMES.

WILL somebody with the inventive faculty please arise in his might and plan out a set of names for the retail clothing stores. We get tired of such overlasting titles as "Globe," "Model," "Boston," "One Price," "Excelsior" and the others which are repeated over and over again. It would seem that when it comes to names the proprietors are as destitute of originality as many of their advertisement writers, which is, we admit, a pretty severe criticism. If the names describe anything, we might pardon the labored staleness, but they don't.

Thus "Model" may be the name of a cheap and trashy little affair holding the same relation to what a really model clothing store is, that an apple stand does to an exposition. "The Excelsior" is applied with flaunting insolence to an E flat concern whose proprietor's sole ambition is to sell the lowest quality of "duds" the times can stand. We have known One Price stores where any old price was accepted rather than let the customers leave the place alive.

Why call a store the "Boston" store when there is absolutely nothing Boston about it, simply because the manager can't think of any other name, and once saw a very good looking store so christened. We have seen "Boston" stores that might better have been called "Gowanus," "Stagnant Pool," "Staring Elk," or "Frozen Mitt," so far as descriptiveness of title was concerned.

When a name indicates a chain of stores, the indefinite signboard business is not so bad, but when the title is simply there because the manager's name can't go up, why not plan out something refreshing? Take the old English tavern signs, for instance; see how effective an advertisement would be if it opened with:

"At the sign of the Copper Pants," etc., or

"Buy Woollen Mitts from the Bone Button Toggery," etc., or

"Great mark down, fire and blood sale at the Square and Compass," etc., or

"Awful slaughter, fierce mangle of prices, dull trade runs amuck and slashes with a two-edged sword, right and left. Ridiculous reductions race red-handed and all our profits perish at the Clothing Cozy Corner," or

In fact, anything, so long as you startle the community by not doing what everybody else does.—Northwestern Clerk.

A NATIONAL COMPANY.

WITH a staff of over 400 employes The Canada Paint Co., of Montreal, is entitled to rank among the largest and most important color and varnish companies in the Dominion of Canada. The Canada Paint Co. is, furthermore, a national Canadian institution par excellence, owned and managed by men of experience who employ the highest skill obtainable to perfect their products. The Magnetic Oxide and Indian Red Works at Malo, Que., have, by the distribution of liberal wages and expenditure for fuel, etc., wrought a complete change, and in no small measure contributed to the advancement and prosperity of the district around Three Rivers, and the products of the mines there are scientifically treated and shipped all over the Dominion, and to the United States and the Old Country.

The company also controls a graphite property in New Brunswick, and the works in Montreal and Toronto are equipped with the finest up-to-date machinery for the purpose of manufacturing white lead, zinc, paints, colors and varnishes by the most approved methods.

Mr. Robert Munro, a prominent member of the Montreal Board of Trade and a gentleman of wide experience and technical knowledge, is the managing director, and he is assisted by a well-trained staff who are acknowledged experts in their various departments.

AN ATTRACTIVE STORE.

Blyth & Watt, of Ottawa, are to be congratulated on the improved appearance of their premises. The store has been remodelled and enlarged and new offices have been added. It is now one of the most attractive hardware stores in the east.

CATALOGUES, BOOKLETS, ETC.

GURNEY'S CATALOGUE, NO. 58.

The Gurney Foundry Co., Limited, Toronto, Winnipeg and Vancouver, seem to have adopted the same rule regarding the production of their catalogues that they observe in the manufacture of their stoves, ranges, etc., namely, to make each edition superior in every respect to those preceding it. The catalogue they are just issuing, "Stove Catalogue, No. 58," is undoubtedly one of the finest publications of the kind that has ever been printed in Canada. The cover is exceptionally artistic, and the typography throughout excellent. In addition to over 100 pages of illustrations, descriptions, and price lists of the great variety of Oxford stoves and ranges, hot water heaters, steam boilers, warm air furnaces, radiators, hollow-ware, general castings, etc., this catalogue comprises many pages of useful information regarding terms, directions for operating stoves, ranges, furnaces, etc. The back cover is a fac-simile in colors of the guarantee bond given by the company with Oxford stoves, ranges and furnaces. This catalogue is being sent out now, and, as it is an exceedingly valuable one, every stove dealer should make sure of securing one.

C. A. Dickie, of Shediac, N.B., has sold out his general store to S. D. White and bought the business of C. C. Hamilton, general merchant, Shediac.

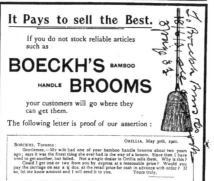

H. S. HOWLAND, SONS & CO.

WHOLESALE ONLY — 37-39 Front Street West, TORONTO. — ONLY WHOLESALE

SCREEN DOORS.

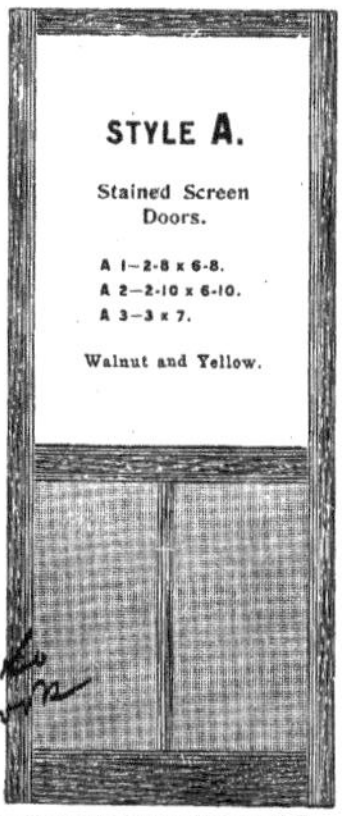

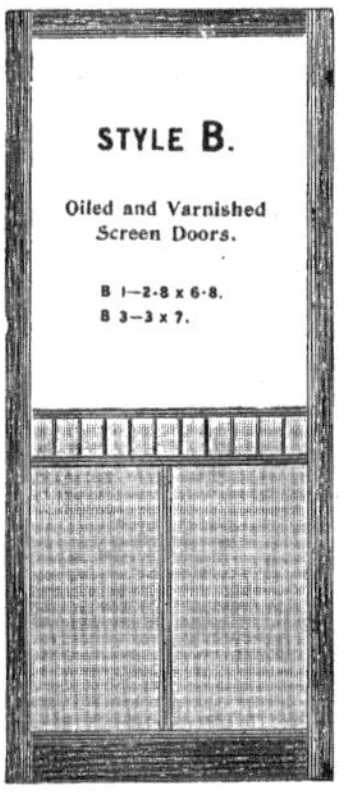

SCREEN WINDOWS.

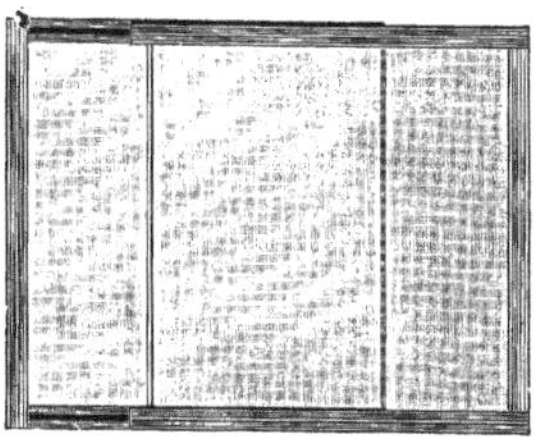

"PERFECTION" SCREENS.

No. 1—18 in. high,	extends	14 x 22 in.,	for Bathrooms.
" 2—18 "	"	20 x 33 in.,	for Windows.
" 3—18 "	"	24 x 40 "	"
" 4—22 "	"	24 x 40 "	"
" 5—24 "	"	24 x 40 "	"

PROMPT SHIPMENT.

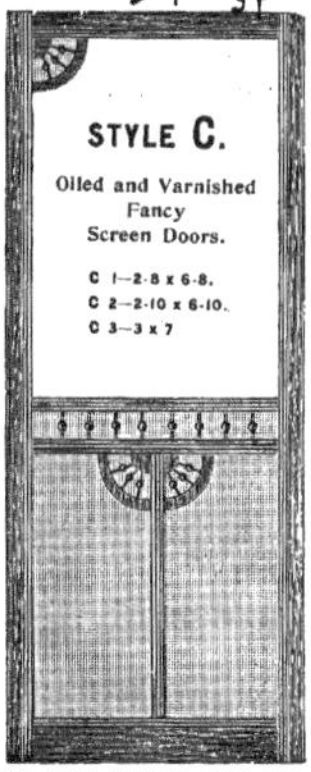

STYLE C.
Oiled and Varnished Fancy Screen Doors.
C 1—2-8 x 6-8.
C 2—2-10 x 6-10.
C 3—3 x 7

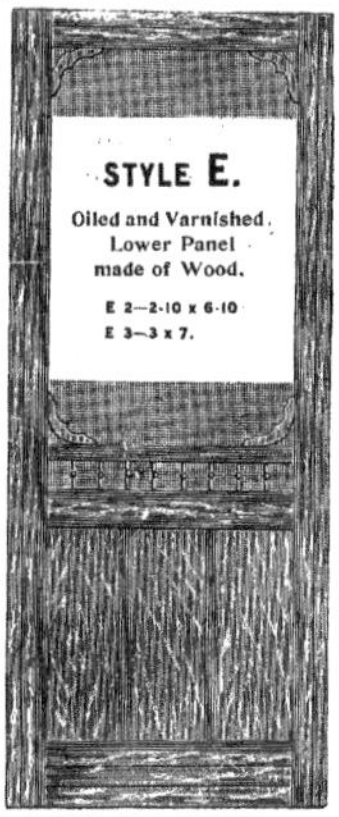

NOTE—OUR DOORS ARE ALL 4-IN. STYLES, AND BOTTOM PART 7-IN. TO ALLOW FOR FITTING.

MARKETS AND MARKET NOTES

QUEBEC MARKETS

Montreal, June 14, 1901.

HARDWARE.

THIS week we have heard several loud complaints about the state of trade, referring both to new business and to payments. Certain lines are not as brisk as they might be and there is not that regular flow that one expects at this time of the year. The Province of Quebec account is probably the least pleasing. Yet, the country is prospering, the markets have rounded into good shape and the outlook must be regarded as hopeful. Probably it is the bad weather of the past month that has restrained business. Wires are not sharing in this dullness, for orders are not nearly filled yet. Nails, too, are among the active lines, certain sizes of wire nails being scarce, with 3-in. hardly obtainable. Poultry netting is scarce and firm, 50 and 5 per cent. being now the discount quoted on the B, C and D lists. American loaded shells are higher this week; the discount has been lowered from 33 1-3 per cent. to 25 per cent. Tin and enamelled ware is selling quite freely. The English sheet metal market shows still further strength this week; supplies are arriving but slowly and goods are scarce.

BARB WIRE—There is nothing new to note. Many retailers will not have their orders filled till the first week in July. The price is $3.05 per 100-lb. f.o.b., Montreal.

GALVANIZED WIRE—The situation is unchanged. We quote as follows: No. 5, $4.25; Nos. 6, 7 and 8 guage, $3.55; No. 9, $3.10; No. 10, $3.75; No. 11, $3.85; No. 12, $3.25; No. 13, $3.35; No. 14, $4.25; No. 15, $4.75; No. 16, $5.

SMOOTH STEEL WIRE—Shipments of fair quantities have been made this week at former quotations. We quote oiled and annealed as follows: No. 9, $2.80; No. 10, $2.87; No. 11, $2.90; No. 12, $2.95; No. 13, $3.15 per 100 lb. f.o.b. Montreal, Toronto, Hamilton, London, St. John and Halifax.

FINE STEEL WIRE—A moderate amount of business is being done. The discount is 17 1-2 per cent. off the list.

BRASS AND COPPER WIRE—The market is stationary. The discount on brass is 55 and 2 1-2 per cent., and on copper, 50 and 2 1-2 per cent.

FENCE STAPLES—Business is good. We quote: $3.25 for bright, and $3.75 for galvanized per keg of 100 lb.

WIRE NAILS—The demand for wire nails is fairly good and supplies are not coming forward any too freely. In fact, 3-inch nails have been hardly obtainable. We quote $2.85 for small lots and $2.77 1-2 for carlots, f.o.b. Montreal, London, Toronto, Hamilton and Gananoque.

CUT NAILS—There is a moderate business being done in cut nails. Shingle nails are rather active. We quote as follows: $2.35 for small and $2.25 for carlots; flour barrel nails, 25 per cent. discount; coopers' nails, 30 per cent. discount,

HORSE NAILS—Dealers seem to have taken sufficient to fill their requirements for some time. "C" brand is held at a discount of 50 and 7 1-2 per cent. off the new list. "M" brand is quoted at 60 per cent. off old list on oval and city head, and 66 2-3 per cent. off countersunk head. Monarch's discount is 66 2-3 per cent.

HORSESHOES—A small business is passing. We quote: Iron shoes, light and medium pattern, No. 2 and larger, $3.50; No. 1 and smaller, $3.75; snow shoes, No. 2 and larger, $3.75; No. 1 and smaller, $4; X L steel shoes, all sizes, 1 to 5, No. 2 and larger, $3.60; No. 1 and smaller, $3.85; feather-weight, all sizes, $4.85; toe weight, steel shoes, all sizes, $5.95 f.o.b. Montreal. f.o.b. Hamilton, London and Guelph, 10c. extra.

POULTRY NETTING—Stocks of poultry netting are very low on this market and prices have been generally advanced to a discount of 50 and 10 per cent. off list A, and 50 and 5 per cent. off lists B, C and D. Supplies are not forthcoming from the manufacturers owing to the difficulty of procuring wire.

GREEN WIRE CLOTH—Unchanged at $1.35. Business is active.

SCREEN DOORS AND WINDOWS—The advent of warm weather has induced a good demand for screens. We quote as follows: Screen doors, plain cherry finish, $7.30 per doz.; do. fancy, $11.50 per doz.; walnut, $7.30 per doz., and yellow, $7.45; windows, $2.25 to $3.50 per doz.

SCREWS—The market is firm and business is good. Discounts are: Flat head, bright, 87 1-2 and 10 per cent. off list; round head, bright, 82 1-2 and 10 per cent.; flat head, brass, 80 and 10 per cent.; round head, brass, 75 and 10 per cent.

BOLTS—Some good orders are to hand this week. Discounts are as follows: Norway carriage bolts, 65 per cent.; common, 60 per cent.; machine bolts, 60 per cent.; coach screws, 70 per cent.; sleigh shoe bolts, 72 1-2 per cent.; blank bolts, 70 per cent.; bolt ends, 62 1-2 per cent.; plough bolts, 60 per cent.; tire bolts, 67 1-2 per cent.; stove bolts, 67 1-2 per cent. To any retailer an extra discount of 5 per cent. is allowed. Nuts, square, 4c. per lb. off list; hexagon nuts, 4 1-4c. per lb. off list. To all retailers an extra discount of 1-4c. per lb. is allowed.

BUILDING PAPER—There is nothing new to note. There is a moderate demand. We quote: Tarred felt, $1.70 per 100 lb.; 2-ply, ready roofing, 80c. per roll; 3-ply, $1.05 per roll; carpet felt, $2.25 per 100 lb.; dry sheathing, 30c. per roll; tar sheathing, 40c. per roll; dry fibre, 50c. per roll; tarred fibre, 60c. per roll; O.K. and I.X.L., 65c. per roll; heavy straw sheathing, $28 per ton; slaters' felt, 50c. per roll.

RIVETS AND BURRS—The demand is fair. Discounts on best iron rivets, section, carriage, and wagon box, black rivets, tinned do., coopers' rivets and tinned swede's rivets, 60 and 10 per cent.; swedes iron burrs are quoted at 55 per cent. off; copper rivets, 35 and 5 per cent. off; and coppered iron rivets and burrs, in 5 lb. carton boxes, are quoted at 60 and 10 per cent. off list.

BINDER TWINE—A fairly good trade is being done in binder twine at unchanged quotations. Some firms are charged higher prices than those we quote but, as a general rule, these prevail: Blue Ribbon, 11 1-2c.; Red Cap, 9 3-4c.; Tiger, 8 3-4c.: Golden Crown, 8c.; Sisal, 8 1-4c.

CORDAGE—Some good orders have come to hand again this week. Manila is worth 13 1-2c. per lb. for 7-16 and larger; sisal is selling at 10c., and lathyarn, 10c.

HARVEST TOOLS—Are very active. The discount is still 50, 10 and 5 per cent.

SPADES AND SHOVELS—A small business is being done. The discount is 40 and 5 per cent.

LAWN MOWERS—Like all lawn requisites lawn mowers are in good request. We quote: High wheel, 50 and 5 per cent. f.o.b. Montreal; low wheel, in all sizes, $2.75 each net; high wheel, 11-inch, 30 per cent. off.

FIREBRICKS—The demand is quiet, as is usual at this season. Prices are steady. We quote: Scotch at $17.50 to $22, and English at $17 to $21 per 1,000, ex-wharf.

CEMENT—The turnover for the season is not large. There are no new features to note. We quote: German cement, $2.35 to $2.50; English, $2.25 to $2.35, Belgian, $1.70 to $1.95 per bbl. ex wharf, and American, $2.30 to $2.45, ex cars.

METALS.

The feature of the market is the all round advance in England, Coke plates are quoted 1s. higher, charcoals are up 1s. 3d.; terne plates, 1s. 9d. to 2s. per box, galvanized iron, 10s. per ton, Canada plates, 5s. per ton and pig tin about 1c. per lb.

Sheet zinc is also scarce and advancing. The early orders placed by the Canadian firms for delivery by early steamers are not coming to hand as expected and stocks are quite light. The English mills are now crowded with orders and will not guarantee shipment before September.

PIG IRON—There is not a great deal of business being done as foundrymen are supplied for some time to come. Canadian iron is worth $18.50 per ton, and Summerlee, $20.50 to $21 ex wharf.

BAR IRON—The market is stationary and active. Merchants' bar is worth $1.75 and horseshoe $2.

HOOP IRON—The demand is moderate. We quote $2.75 for No. 19 guage and $3 for No. 17.

BLACK SHEETS—The market is steady all round. At the moment there is not a heavy demand. We quote: 8 to 16 guage, $2.50 to $2.60; 26 guage, $2.55 to $2.65, and 28 guage, $2.60 to $2.70.

GALVANIZED IRON—Latest advices from England report an advance of 10s. per ton on previous quotations. The market there is stiff and under a fairly good demand. We quote as follows: No. 28 Queen's Head, $4.50; Apollo, 10 3-4 oz., $4.50, and Comet, $4.30 to $4.45, with a 10c. reduction for case lots.

COPPER—Prices are steady at 18c.

INGOT TIN—The market is booming under a heavy speculative movement, till the cost price in New York is nearly 30c. Here 34 and 35c. is asked for Lamb and Flag.

LEAD—There is a fair demand at $3.70 to $3.80.

LEAD PIPE—Quotations are unchanged. We quote: 7c. for ordinary and 7 1-2c. for composition waste, with 20 per cent. off.

IRON PIPE—Trade is good at the advance. We quote: Black pipe, 1-4, $3 per 100 ft.: 3-8, $3; 1-2, $3.05; 3-4, $3.30; 1-in., $4.75; 1 1-4, $6.45; 1 1-2, $7.75; 2-in., $10.35. Galvanized, 1-2, $4.60; 3-4, $5.25; 1-in., $7.50; 1 1-4, $9.80; 1 1-2, $11.75; 2-in., $16.

TINPLATES—Cable advices from England denote a rising market, specifying a rise of 1s. in coke plates and another of 1s. 3d. in charcoals. The demand on spot is fairly good. We quote: 52's, $2.55 to $2.60; 60's, $2.65 to $2.70; 75's, $2.70 to $2.80; full polished, $3.10, and galvanized, $4.

STEEL—Unchanged. We quote: Sleighshoe, $1.95; tire, $2; bar, $1.95; spring, $2.75; machinery, $2.75, and toe-calk, $2.50.

SHEET STEEL—We quote: Nos. 22 and 24, $3, and Nos. 18 and 20, $2.85.

TOOL STEEL—Black Diamond, 8c. and Jessop's, 18c.

TERNE PLATES—Terne plates are also reported 1s. 9d. to 2s. higher in England, the price now being 24s. 3d. The demand here is fair and stocks light. The price is $7.50.

COIL CHAIN—There is nothing new to report. We quote as follows: No. 6, 11 1-2c.; No. 5, 10c.; No. 4, 9 1-2c.; No. 3, 9c.; 1-4-in., 7 1-2c. per lb.; 5-16, $4.85; 5-16 exact, $5.30; 3-8, $4.40; 7-16, $4.20; 1-2, $3.95; 9-16, $3.85; 5-8, $3.55; 3-4, $3.45; 7-8, $3.40; 1-in., $3.35. In carload lots an allowance of 10c. is made.

SHEET ZINC—Sheet zinc is very scarce and the market is firm. The prevailing idea as to price is $6 to $6.25.

ANTIMONY—Quiet, at 10c.

ZINC SPELTER—Is worth 5c.

SOLDER—We quote: Bar solder, 18 1-2c.; wire solder, 20c.

GLASS.

Prices are firm under a fairly good demand. We quote: First break, $2.10; second, $2.20 for 50 ft.; first break, 100 ft., $3.90; second, $4.10; third, $4.60; fourth, $4.85; fifth, $5.35; sixth, $5.85, and seventh, $6.35.

PAINTS AND OILS.

There is a continued brisk demand for mixed paints and manufacturers are still exceedingly busy. Linseed oil is scarce and the arrivals are disappointingly slow. Turpentine is steady. We quote:

WHITE LEAD—Best brands, Government standard, $6.25; No. 1, $5.87 1-2; No. 2, $5.50; No. 3, $5.12 1-2, and No. 4, $4.75 all f.o.b. Montreal. Terms, 3 per cent. cash or four months.

DRY WHITE LEAD—$5.25 in casks; kegs, $5.50.

RED LEAD—Casks, $5; in kegs, $5.25c.

DRY WHITE ZINC—Pure, dry, 6 1-4c.; No. 1, 5 1-4c.; in oil, pure, 7 1-4c.; No. 1, 6 1-4c.; No. 2, 5 1-4c.

PUTTY—We quote: Bulk, in barrels, $1.90 per 100 lb.; bulk, in less quantity, $2.05; bladders, in barrels, $2.10; bladders, in 100 or 200 lb. kegs or boxes, $2.25; in tins, $2.55 to $2.65; in less than 100-lb. lots, $3 f.o.b. Montreal, Ottawa, Toronto, Hamilton, London and Guelph. Maritime Provinces, 10c. higher, f.o.b. St. John and Halifax.

LINSEED OIL—Raw, 80c.; boiled, 83c. in 5 to 9 bbls., 1c. less, 10 to 20 bbl. lots, open, net cash, plus 2c. for 4 months. Delivered anywhere in Ontario between Montreal and Oshawa at 2c. per gal. advance and freight allowed.

TURPENTINE—Single bbls., 53c. ; 2 to 4 bbls., 52c. ; 5 bbls. and over, open terms, the same terms as linseed oil.

MIXED PAINTS—$1.25 to $1.45 per gal.

CASTOR OIL—8 3-4 to 9 1-4c. in wholesale lots, and 1-2c. additional for small lots.

SEAL OIL—47 1-2 to 49c.

COD OIL—32 1-2 to 35c.

NAVAL STORES—We quote: Resins, $2.75 to $4.50, as to brand ; coal tar, $3.25 to $3.75 ; cotton waste, 4 1-2 to 5 1-2c. for colored, and 6 to 7 1-2c. for white ; oakum, 5 1-2 to 6 1-2c., and cotton oakum, 10 to 11c.

PARIS GREEN—Petroleum barrels, 16 3-4c. per lb. ; arsenic kegs, 17c. ; 50 and 100-lb. drums, 17 1-2c. ; 25-lb. drums, 18c. ; 1-lb. packages, 18 1-2c. ; 1-2-lb. packages, 20 1-2c. ; 1-lb. tins, 19 1-2c. ; 1-2-lb. tins, 21 1-2c. f.o.b. Montreal ; terms, 3 per cent. 30 days, or four months from date of delivery.

SCRAP METALS.

The market is rather active and firm. Dealers are now paying the following prices in the country : Heavy copper and wire, 13 to 13 1-2c. per lb. ; light copper, 12c. ; heavy brass, 12c. ; heavy yellow, 8 1-2 to 9c. ; light brass, 6 1-2 to 7c. ; lead, 2 1-2 to 2 3-4c. per lb. ; zinc, 2 1-4 to 2 1-2c. ; iron, No. 1, wrought, $14 to $16 per gross ton f.o.c. Montreal ; No. 5 cast, $13 to $14 ; stove plate, $8 to $9 ; light iron, No. 2, $4 a ton ; malleable and steel, $4 ; rags, country, 60 to 70c. per 100 lb. ; old rubbers, 7 1-4c. per lb.

HIDES.

The demand for green hides is sufficient to prevent any accumulation. We quote : Light hides, 6 1-2c. for No. 1 ; 5 1-2c. for No. 2, and 4 1-2c. for No. 3. Lambskins, 15c. ; sheepskins, 90c. to $1 ; calfskins, 8c. for No. 1 and 6c. for No. 2.

PETROLEUM.

The demand is slow and quotations lower. We quote : "Silver Star," 14 to 15c. ; "Imperial Acme," 15 to 16c. ; "S C. Acme," 17 to 18c., and "Pratt's Astral," 17 to 18 1-2c.

ONTARIO MARKETS.

TORONTO, June 15, 1901.

HARDWARE.

BUSINESS is fairly satisfactory, although nothing particularly new has developed during the past week. Such seasonable goods as snaths, scythes, lawn mowers, screen doors and windows, ice cream freezers and refrigerators, are all going out fairly well. The scarcity of barb wire and galvanized plain wire is as pronounced as ever, and in the meantime oiled and annealed wire continue to derive benefit of this condition of affairs. Quite a nice business is being done in wire nails, bolts, rivets and burrs, and screws. In cut nails there is practically nothing doing outside shingle nails. Rope is still meeting with a fair demand, and the same may be said of building paper. Enamelled ware and tinware is in moderate request only. In bright wire goods a steady trade is reported. Further improvement is to be noted in sporting goods. In horseshoes and horse nails trade is quiet.

BARB WIRE—There is still a scarcity of supplies in barb wire. The manufacturers seem to be about as far as ever behind with their orders. We quote $3 05 per 100 lb. from stock Toronto ; and $2.82½ f.o.b. Cleveland for less than carlots, and $2.70 for carlots.

GALVANIZED WIRE—This too is still scarce and wanted, with prices unchanged. We quote : Nos. 6, 7 and 8, $3.50 to $3.85 per 100 lb., according to quantity ; No. 9. $2.85 to $3 15 ; No. 10, $3.60 to $3.95 ; No. 11, $3.70 to $4 10 ; No. 12, $3 to $3 30 ; No. 13, $3 10 to $3 40 ; No. 14, $4.10 to $4 50 ; No. 15, $4 60 to $5 05 ; No. 16, $4.85 to $5.35. Nos. 6 to 9 base f.o.b. Cleveland are quoted at $2.57½ in less than carlots and 12c. less for carlots of 15 tons.

SMOOTH STEEL WIRE—The demand for oiled and annealed wire is still active and business is larger than is usual. As already pointed out this is to some extent due to the scarcity of barb wire and plain galvanized. On account of the large trade that has been done, some of the factories are this week reported to be running short of stock. Very little is being done in hay-baling wire. The net selling prices for oiled and annealed are: Nos. 6 to 8, $2.90; 9, $2.80; 10, $2.87; 11, $2.90 ; 12, $2.95 ; 13, $3 15 ; 14, $3.37 ; 15, $3.50 ; 16, $3 65. Delivery points, Toronto, Hamilton, London and Montreal, with freights equalized on those points.

WIRE NAILS—The demand is good, and, if anything, rather better than it was a week ago. As far as we can learn the factories have got about all they can do to meet the demand. We still quote $2.85 base for less than carlots, and $2.77½ for carlots. Delivery points : Toronto, Hamilton, London, Gananoque and Montreal.

CUT NAILS—There is a fair demand for shingle nails, but otherwise there is practically nothing doing in cut nails. The base price is $2.35 per keg for less than carlots, and $2.25 for carlots. Delivery points : Toronto, Hamilton, London, Montreal and St. John, N.B.

HORSE NAILS — Very little is being done. Discount on "C" brand, oval head, 50 and 7½ per cent. off new list, and on "M" and other brands, 50, 10 and 5 per cent. off the old list. Countersunk head 60 per cent.

HORSESHOES — These are seasonably quiet with quotations as before. We quote f.o.b. Toronto as follows : Iron shoes, No. 2 and larger, light, medium and heavy, $3.60; snow shoes, $3.85 ; light steel shoes, $3.70; featherweight (all sizes), $4.95; iron shoes, No. 1 and smaller, light, medium and heavy (all sizes), $3.85 ; snow shoes, $4 ; light steel shoes, $3.95; featherweight (all sizes), $4 95.

SCREWS—There does not appear to be any falling off in the demand, the trade still being reported brisk. Discounts are : Flat head bright, 87½ and 10 per cent. ; round head bright, 82½ and 10 per cent. ; flat head brass, 80 and 10 per cent. ; round head brass, 75 and 10 per cent. ; round head bronze, 65 per cent., and flat head bronze at 70 per cent.

RIVETS AND BURRS—A good trade continues to be done in rivets and burrs, although no new features have arisen during the week. We quote : Iron rivets, 60 and 10 per cent.; iron burrs, 55 per cent.; copper rivets and burrs, 25 and 5 per cent.

BOLTS AND NUTS—The situation remains much as before, an active demand still being reported. We quote: Carriage bolts (Norway), full square, 65 per cent.; carriage bolts full square, 65 per cent.; common carriage bolts, all sizes, 60 per cent.; machine bolts, all sizes, 60 per cent.; coach screws, 70 per cent.; sleighshoe bolts, 72½ per cent.; blank bolts, 60 per cent.; bolt ends, 62½ per cent.; nuts, square, 4c. off; nuts, hexagon, 4½c. off; tire bolts, 67½ per cent.; stove bolts, 67½; plough bolts, 60 per cent.; stove rods, 6 to 8c.

ROPE—There is a good steady demand for small quantities and the result is a fairly large trade in the aggregate. Prices are unchanged at 10c. for sisal and 13½c. for manila.

BINDER TWINE—There is a quiet sorting-up demand. We quote: Pure manila, 650 ft., 12c.; manila, 600 ft., 9½c.; mixed, 550 ft., 8½c.; mixed, 500 ft., 8 to 8¼c.

CUTLERY—There is not a great deal being done this week in cutlery, although trade is not quieter than is usual at this time of the year.

SPORTING GOODS — Trade shows further improvement, the demand being specially good for Nos. 6, 7 and 8 loaded shells for target practice.

ENAMELLED WARE AND TINWARE — The situation in both these lines is much the same as a week ago, namely, rather quiet.

GRINDSTONES—A nice steady trade is being done. We quote: 2 inch thick, 40 to 200 lb., $25 per ton; 2 inch thick, under 40 lb., $28; under 2 inch thick, $29.

ICE CREAM FREEZERS AND REFRIGERATORS—Trade in these lines continues fairly good.

GREEN WIRE CLOTH—There is still some movement at $1.35 per 100 ft.

SCREEN DOORS AND WINDOWS—Jobbers have received shipments during the week, and they are now in a better position to fill orders than they were. Quite a large number are going out this week to the retail trade. We quote: Screen doors, 4 in. styles, $7.20 to $7.80 per doz.; ditto, 3-in. styles, 20c. per doz. less; screen windows, $1.60 to $3.60 per doz., according to size and extension.

BUILDING PAPER—A good steady trade is still to be reported in building paper. We quote: Building paper, 30c.; tarred lining, 40c., and tarred roofing, $1.65.

POULTRY NETTING —Stocks have been light for the last 10 days, the makers claiming that they are unable to get a sufficent quantity of wire to fill their orders. They are now, however, better supplied in this respect and are running night and day in order to catch up to their orders. Ruling discount 55 per cent.

HARVEST TOOLS — There has been a fairly good movement during the week, the demand has been particularly brisk for scythes and snaths. Discount 50, 10 and 5 per cent.

SPADES AND SHOVELS — The movement in this line continues only light. Discount 40 and 5 per cent.

EAVETROUGH—There is still a fairly good movement in eavetrough and conductor pipe, the ruling price still being $3.25 per 100 ft. for 10 inch.

CEMENT—The demand continues excellent. We quote barrel lots: Canadian portland, $2.25 to $2.75; German, $3 to $3.15; English, $3; Belgian, $2.50 to $2.75; Canadian hydraulic, $1.25 to $1.50.

METALS.

The metal market continues on the whole in a fairly satisfactory condition, the demand being fairly good for seasonable lines, while prices are, on the whole, fairly steady. The demand is still chiefly for galvanized sheets.

PIG IRON—The pig iron market in the United States is decidedly dull, and prices have an easier tendency in sympathy. In Canada, there is not much being done, as the large users of pig iron have made contracts for some time ahead. The furnaces, however, all appear to be pretty busy supplying these orders, and it is reported that some of them are not delivering fast enough to please purchasers. The idea as to price on track Toronto is $18 per ton for No. 1, $17.50 for No. 2, and $17 for No. 3, for Canadian iron.

BAR IRON—The demand continues good for bar iron, and prices are firm at $1.85 per 100 lb.

STEEL—The demand is not as active as it was, as is to be expected at this season. Prices, however, rule fairly steady. We quote: Merchantable cast steel, 9 to 15c. per lb.; drill steel, 8 to 10c. per lb.; "B C" and "Black Diamond" tool steel, 10 to 11c.; Jessop's, Morton's and Firth's tool steel, 12½ to 13c.; toe calk steel, $2.85 to $3; tire steel, $2 30 to $2.50; seighshoe steel, $2.10 to $2.25; reeled slachinery steel, $3; hoop steel, $3.

GALVANIZED SHEETS—The demand for galvanized sheets continues the feature of the metal trade. Prices are firm, and it is still practically impossible to get delivery from the United States. We quote 28 gauge English at $4 50, and American at $4.40.

BLACK SHEETS — Business is not very active, but prices rule firm. We quote: 28 gauge, common sheets, at $3, and dead flat at $3.50.

CANADA PLATES—Business in this line

continues quiet. We quote: All dull, $2.90; half polished, $3, and all bright, $3.50.

Pig Tin—The outside markets are somewhat irregular as to price at the moment, but since our last report there have been some material advances in London, and, locally, dealers are naturally firmer in their views in sympathy. We quote 32 to 32½c.

Tinplates — This market continues to advance. A cable despatch received on Wednesday announced a further advance of 9d. per box in England. Locally, prices are firmer, and some jobbers are quoting 10c. higher than they were a week ago.

Tinned Sheets—There is just a small business being done, and we quote 28 gauge at 8½c.

Copper—The outside markets are strong, and in London prices advanced 2s. 6d. on Wednesday. Quotations are firm in New York in sympaihy. We quote: Ingot, 17¾c., and bar, 23 to 25c.

Solder — A good demand is still the feature of trade in this line. We quote: Half-and-half, guaranteed, 18½c.; ditto, commercial, 18c.; refined, 18c., and wiping, 17c.

Iron Pipe — A good demand is reported for iron pipe, and, although local quotations have not yet been advanced, the views of the manufacturers are firmer than they were a week ago. A good demand is being experienced. We quote: 1-inch black pipe, $5.15 per 100 ft., and 1-inch galvanized, $7.55.

Lead — The market is rather firm and business moderate. Locally, we still quote at 4¼ to 4½c. per lb.

Spelter—The spelter market is still easy, although local quotations are unchanged at 5½ to 6c. per lb.

Zinc Sheets—Business is only moderate at 6½c. for casks, and 6¾c. for part casks.

Antimony — Trade continues quiet at 10½ to 11c. per lb.

PAINTS AND OILS.

There is a good movement of all paint materials. The warm weather has started much painting that was delayed during May. There is also an improvement in the inquiry for paris green, which is now moving briskly. Prices are steady and unchanged throughout. We quote:

White Lead—Ex Toronto, pure white lead, $6.37½; No. 1, $6; No. 2, $5.67½; No. 3, $5.25; No. 4, $4 87½; genuine dry white lead in casks, $5.37½.

Red Lead—Genuine, in casks of 560 lb., $5.50; ditto, in kegs of 100 lb., $5.75; No. 1, in casks of 560 lb., $5; ditto kegs of 100 lb., $5.25.

Litharge—Genuine, 7 to 7½c.

Orange Mineral—Genuine, 8 to 8½c.

White Zinc—Genuine, French V.M., in casks, $7 to $7.25; Lehigh, in casks, $6.

Paris White—90c. to $1 per 100 lb.

Whiting — 70c. per 100 lb.; Gilders' whiting, 80c.

Gum Shellac — In cases, 22c.; in less than cases, 25c.

Paris Green—Bbls., 16¾c.; kegs, 17c.; 50 and 100 lb. drums, 17½c.; 25-lb. drums, 18c.; 1-lb. papers, 18½c.; 1-lb. tins, 19½c.; ½-lb. papers, 20½c.; ½-lb. tins, 21½c.

Putty — Bladders, in bbls., $2.10; bladders, in 100 lb. kegs, $2.25; bulk in bbls., $1.90; bulk, less than bbls. and up to 100 lb., $2.05; bladders, bulk or tins, less than 100 lb., $2.90.

Plaster Paris—New Brunswick, $1.90 per bbl.

Pumice Stone — Powdered, $2.50 per cwt. in bbls., and 4 to 5c. per lb. in less quantity; lump, 10c. in small lots, and 8c. in bbls.

Liquid Paints—Pure, $1.20 to $1.30 per gal.

Castor Oil—East India, in cases, 10 to 10½c. per lb. and 10½ to 11c. for single tins.

Linseed Oil—Raw, 1 to 4 barrels, 82c.; boiled, 85c.; 5 to 9 barrels, raw, 81c.; boiled, 84c., delivered. To Toronto, Hamilton, Guelph and London, 1c. less.

Turpentine—Single barrels, 54c.; 2 to 4 barrels, 53c., delivered. Toronto, Hamilton and London 1c. less. For less quantities than barrels, 5c. per gallon extra will be added, and for 5-gallon packages, 50c., and 10 gallon packages, 80c. will be charged.

GLASS.

Stocks are running low. It is reported that several hundred boxes were lost on the Assyrian, which went ashore at Newfoundland 10 days or so ago. Prices are stiff, but still remain unchanged. We quote as follows: Under 26 in., $4.15; 26 to 40 in., $4.45; 41 to 50 in., $4.85; 51 to 60 in., $5.15; 61 to 70 in., $5.50; double diamond, under 26 in., $6; 26 to 40 in., $6.65; 41 to 50 in., $7.50; 51 to 60 in., $8.50; 61 to 70 in., $9.50, Toronto, Hamilton and London. Terms, 4 months or 3 per cent. 30 days.

OLD MATERIAL.

There is a fair movement at steady prices. We quote jobbers' prices: Agricultural scrap, 55c. per cwt.; machinery cast, 60c. per cwt.; stove cast, 50c.; No. 1 wrought 50c. per 100 lb.; new light scrap copper, 12c. per lb.; bottoms, 11½c.; heavy copper, 13c.; coil wire scrap, 13c.; light brass, 7c.; heavy yellow brass, 10 to 10½c.; heavy red brass, 10½ to 11c.; scrap lead, 3c.; zinc, 2c; scrap rubber, 6½c.; good country mixed rags, 65 to 75c.; clean dry bones, 40 to 50c. per 100 lb.

HIDES, SKINS AND WOOL.

Hides — There is little doing. Prices are unchanged. We quote: Cowhides, No. 1, 6½c.; No. 2, 5½c.; No. 3, 4½c. Steer hides are worth 1c. more. Cured hides are quoted at 7 to 7½c.

Skins—There is a fair movement at steady prices. We quote: No. 1 veal, 8-lb. and up, 9c. per lb.; No. 2, 8c.; dekins, from 60 to 70c.; culls, 20 to 25c. Sheepskins, 90c. to $1.

Wool—Offerings of wool are liberal, but, as stocks in dealers' hands are heavy, buyers are cautious. Present prices are well maintained. We quote: Combing fleece, washed, 13c., and unwashed, 8c.

COAL.

There is a steady demand, but deliveries are delayed somewhat by scarcity of both coal and cars. Prices are unchanged. We quote at international bridges as follows: Grate, $4.75 per gross ton ; egg, stove and nut, $5 per gross ton with a rebate of 30c. off for June shipments.

PETROLEUM.

A fair movement is reported, with prices steady. We quote : Pratt's Astral, 16 to 16½c. in bulk (barrels, $1 extra) ; American water white, 16½ to 17c. in barrels; Photogene, 15½ to 16c.; Sarnia water white, 15 to 15½c. in barrels; Sarnia prime white, 14 to 14½c. in barrels.

MARKET NOTES.

Manufacturers of iron pipe are holding for higher prices.

A shipment of black rubber snaps with brass springs has been taken into stock by H. S. Howland, Sons & Co.

H. S. Howland, Sons & Co. are in receipt of a shipment of screen doors and windows and are now able to fill orders promptly.

Rice Lewis & Son, Limited, have a display of sporting goods in one of their windows that is attracting a great deal of attention from passers-by. They carry a large stock of these goods.

THEIR STEEL TRADE WITH CANADA IS GROWING.

Mr. A. K. Rhoden, representing B. K. Morton & Co., Sheffield, England, manufacturers of the "B. C." among other brands of steel, is in Toronto this week in the interest of his firm.

This is Mr. Rhoden's fourth annual visit to Canada, and he is making one of his regular business trips around the world. Wood, Vallance & Co., Hamilton, are Morton & Co.'s agents in Ontario, and Mr. Rhoden, in his visits to the trade in Toronto, is accompanied by Mr. H. T. Eager, the manager of Wood, Vallance & Co.'s Toronto branch.

"I have met with quite a deal of success in Canada," said Mr. Rhoden to "Hardware and Metal." "Our trade here is becoming more encouraging from year to year. The special object of my present visit is to work up trade in tool steel with engineers and manufacturers. We have already a good connection in Australia, China and Japan, and we are now trying to work Canada with the same results."

"Do you find the preferential tariff of any assistance to you ?"

"Yes, we do ; but more particularly in the low-priced steel. As far as the general tariff on high grades of steel is concerned it would suit us if it was increased. The general duty on high-grade steel is so low that the preferential tariff makes very little difference."

Mr. Rhoden leaves on Monday for British Columbia en route for Australia, for which country he will sail about the end of July. Mr. B. K. Morton, the head of the firm, is at present in Melbourne on a visit to the branch business of the firm that is situated there.

SS. SMOKELESS POWDER.

Several years ago Lewis Bros., Montreal, accepted the Canadian agency for the S.S. smokeless powder, made by The Smokeless Powder and Ammunition Co., Limited, Barwick, Herts, Eng. This they have made one of their leaders in sporting goods and their efforts to introduce it into Canada have met with such success that last year there was, it is claimed, more S. S. powder consumed in Canada than all other powders of the same class.

S.S. powder is adapted to all uses and has most reliable testimonials in its favor. For clay bird shooting it has the largest number of wins to its credit of all sporting powders. It is now sold by the leading hardware and sporting goods houses in Canada. Lewis Bros. are just receiving importations for the fall trade, and no dealer should be without a supply.

WAR OFFICE PURCHASES IN CANADA.

Lord Strathcona's annual report to the Trade and Commerce Department as High Commissioner for Canada was made public on Tuesday. His Lordship states that both the import and export trade of Canada from and to the United Kingdom during last year seem to have expanded in a satisfactory manner. If the correspondence received at the High Commissioner's Office is any criterion, Canadian trade is attracting more attention than ever in Great Britain. Correspondence on trade matters is considerable and continually growing, and the personal inquiries are also exceedingly numerous. The number of callers at the office during the year was nearly 14,000, of which 2,700 represented travelling Canadians who registered their names. Lord Strathcona gave a list of orders which he induced the War Office and Indian Office to give for the supply from Canada of articles required by the Imperial troops in South Africa and China. The orders aggregated several millions of dollars in value, and included the following : Hay, 53,700 tons ; 1,073 tons of corned beef, one hundred tons of oats, sixteen hundred tons of flour, 280,000 lbs. (10-lb. tins) jam, thirty thousand greatcoats, fifty thousand serge suits, one thousand cases containing 2-lb. cans of baked beans, one thousand cases containing 12,000 10-lb. tins of boneless chicken ; also over 8,000 sets of saddlery. For the China expedition of 1900 the following orders were given : Greatcoats, 33,670 ; 43,300 pairs thick stockings, 29,000 pairs moccasins, 1,500 pairs fur-lined gloves, and 2,320 fur caps.

APPOINTED APPRAISER.

Mr. A. Magnan, of Magnan Freres, St. Lawrence street, Montreal, and Secretary of The Montreal Retail Hardware Association, has been appointed to fill the vacancy in the Appraiser's office, caused by the death of Mr. G. Piché, some months ago. Mr. Magnan is a young man of ability and is certainly well fitted to discharge h[illegible]new duties capably. He has sold out his interest in the retail business to his brother and his departure from the trade has led him to hand in his resignation from the Secretaryship of The Hardware Association.

TRADE CHAT.

F. S. Malcolm, general merchant, Lakeside, Ont., is putting a new floor in his store.

B. R. Bishop, grocer and hardware dealer, Kentville, N.S., has sold his grocery business to Spurgeon L. Cross.

W. H. Miller, general merchant, Severn Bridge, Ont., is erecting a new store and expects to occupy it about July 1.

Charles Labreton, general merchant, Tracadie, N.B., has notified his creditors that he is in difficulties and will have to assign. His liabilities are $3,500 and his assets $2,300.

PERSONAL MENTION.

Mr. W. L. Allen, of Cobourg, was in Toronto last week.

Mr. R. S. Davidson, secretary-treasurer of The Peterboro' Hardware Co., Peterboro', has been confined to his home the last two or three weeks through illness.

Mr. Rupert M. Watson, representing The Dominion Cartridge Co., Montreal, was in Toronto on Thursday. He was showing the new lines manufactured by his company, which he claims will be equal to the very best on the market.

HARDWARE STORE WINDOW DISPLAY.*

ALMOST any kind of advertising effort will show results. Window advertising is the least expensive, and the results are most immediate. Two per cent. of gross sales would be a conservative estimate for newspaper and circular advertising; which means $500 yearly on a $25,000 business. The window advertising will cost practically nothing, as at least cost can be realized out of any of the goods which might become shop worn, and no goods at all should be damaged if windows are properly secured against flies and the trimming changed every week.

THE NEWSPAPER ADVERTISEMENT.

Newspaper, circular and window advertising should be worked in conjunction. The combined result is best. We are unconsciously directed by impressions. The saying that "We are creatures of habit," is simply in line with the psychological fact that impressions once formed in our minds are constantly recurring when anything kindred is under consideration. What we wish to do is to place psychological sign boards of our business in the minds of the public. The newspaper may make a faint impression, and the window display clinch it, or vice versa. They help each other.

If you will write a newspaper advertisement each week and trim your windows each week, and never fail, doing the one will make the other easier. The advertisement suggests the window trim and the window trim the advertisement. This may be hard work for the first six months, but it will gradually become easier, and soon your material for advertising, both newspaper and window, will exceed your space. The passing public will come to recognize the regular changes and look for them.

MAKE GOODS WINDOWS.

As far as possible make "goods windows." Freak windows may be all right occasionally, as during carnival or fair time, but they take a great deal of time, destroy goods and bring no immediate results.

Use neat display cards. They help rivet the impressions you are striving to make. Do not put prices on trade-mark goods, or standard brands, which are carried elsewhere in your city. Your competitors will study your windows, and if your prices are high they will use them against you. If the prices are low, they may go still lower, and the tendency will be to reduce the profits on good staple lines that you all carry. Prices are, of course, always attractive, and may be put on lines of which you control the sale or on any line where qualities vary and the make is not known. Besides prices, display cards might describe new goods or make pointed suggestions. In a builders' hardware window a card might read, "Let us figure on your building bill."

CONSTRUCTION OF THE WINDOW.

As to the windows themselves the window seat should be quite deep, and not more than 15 or 18 inches high inside; the glass not being over two feet from the sidewalk. The entire window should be inclosed with wire cloth screens, made in sections, and held in place by buttons, one section being in a door on loose pin hinges. This will keep insects out in summer, and prevent pilfering of small articles. Sections can be removed at any time to admit articles too large to be taken through the door. If the frames are made light and oil finished, they will obstruct the light very little, are easily cleaned, and will serve as a background for the trimming; though background trimming shuts off the light, and the effect of the window from the customer who has entered the store.

For the bottom of the window a frame, made in sections, raised about six inches at the back and slanting to the front, covered with black cloth, is very serviceable in displaying tools, builders' hardware and small articles.

SHOW ONE LINE OF GOODS.

One line of goods at a time in a window is generally better unless the windows are very large. Large quantities of one article always attract attention. Few people would notice one only of a common, everyday article, like a 10-gallon carrying can or milk cooler, but a window full of either of them in a graduated pile extending to the ceiling would cause any number of people to stop. My neighbor, the grocer, tells me that when he puts on the walk one or two watermelons, very few ask for them, but when he stacks up half a carload, every other man buys one. Thus a great many attractive window displays can be made by using quantities of common articles.

WINDOW DISPLAY BENEFICIAL TO WHOLE STORE.

A study and faithful practice of window trimming will lead to better store service in every way. You become more critical of each individual line as you take it up for display. You ask yourself if you are carrying the right quality of goods, in proper quantity at the right price? In establishing your sign boards in the public mind you will see the more clearly how necessary also is intelligent, courteous and prompt service. This study will shake you out of the rut in which you may be working, you will read the trade journals, scan The Iron Age more eagerly for ideas, and come to realize that the prosperous merchant has no time for kicking against the inevitable; that the retailer cannot look to legislation for success, and if he is making a failure the cause of it is in his own methods.

* A Prize Essay in The Iron Age.

A CHANGE IN MANAGEMENT.

Col. Massey, who has long been identified with the Montreal branch of The Canada Screw Co., has resigned his position and has been succeeded in the management by Mr. James S. Parkes, for 15 years his assistant. Mr. Parkes is well known to the trade and the change will not necessitate the building of new connections.

At present The Canada Screw Company are only occupying two storeys of their large warehouse at 446 St. Paul street, but in the near future Mr. Parkes intends to have the five storeys utilized as sample storerooms, carrying a full line of screws, stove and tire bolts, iron and copper rivets and bright wire goods.

PRICE PRACTICALLY THE SAME.

"I have just returned from a trip through the States of New York and Pennsylvania," said a traveller, "and I found that the price of bar iron and bolts, although not exactly, is practically the same as in Canada. The steel trust, I heard very few people talking about, except theorists. Business men do not seem to be bothering their heads about it.

PLAN OF A NAIL COUNTER.

"Wanderer" writes as follows: "In a recent issue of HARDWARE AND METAL I noticed a request for description or plan for a nail counter. Last week on my route I found one of my customers building one, a plan of which I enclose. The floor of the counter was built on 3 x 4 scantling laid flat with tongued and grooved 1-inch boards on top. The spaces were then made 14 x 14 and 24 inches deep. Then another bottom laid and another row of spaces the same size made on top and the top of the counter then put on, making height when finished 34 inches. He then set up a 2 inch piece in front of each row of spaces and sloped a board back to each floor, allowing it to project enough over the strip in front to permit a scoop to be held under the edge of it and the nails pulled with a claw made for the purpose. If this is any use it can be extended to any number of spaces."

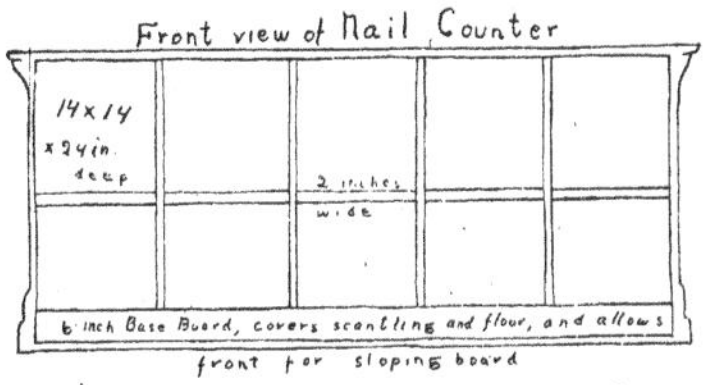

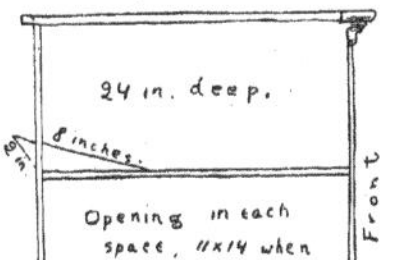

HEATING AND PLUMBING

SOME BUILDING NOTES.

THOS. POCHLMAN & CO., grocers, etc., Hanover, Ont., have bought one of the best business sites in Hanover, and will build a two-storey 80 x 21 block thereon.

The Carleton Baptist Church, St. John, N.B., is to be extended.

Capt. J. J. Grafton intends erecting a new house in Dundas, Ont.

About $3,000 will be spent in improving Queen street school, Chatham, Ont.

The Commercial Bank intend erecting a modern business block in Nanaimo, B.C.

Jones & Hancock, Merlin, Ont., have the contract for erecting a new hotel in Wheatley, Ont.

A new immigration building is to be erected at St. John, N.B. It will cost about $22,000.

Downing & Co. are building a 60 x 26 ft. block at the corner of Pearl and South Water streets, Port Arthur, Ont.

H. R. Halton, architect, Sault Ste. Marie, Ont., has received tenders for a house for H. A. Moore, of the same place.

G. M. Bayley is asking for tenders before June 17, for the erection of a building on Sparks street, Ottawa, for Mayor Morris.

The J. Mickleborough Co., Limited, St. Thomas, Ont., are having plans for a new business block prepared by Architect N. R. Darrach, St. Thomas.

The Parkhill Basket Co., Limited, are erecting in Owen Sound, Ont., a two-storey 60 x 50 ft. warehouse. The exterior of the warehouse will be covered with corrugated iron.

WINNIPEG BUILDING OPERATIONS.

An unusually large amount of building is being done in Winnipeg this year both in new residences up town and new business blocks down town. Among the principal buildings going up are the new blocks for the Bank of Hamilton and the Merchants Bank ; for The Lake of the Woods Milling Co., Limited, for the new Baker, Alexandria, and the Alloway and Pulford blocks are well under way. The Winnipeg Hotel will be extended to have a frontage of 60 feet. The Massey-Harris Company and The Toronto Type Foundry Company intend enlarging their premises. About $35,000 will be spent in converting the Young block into a modern hotel. Residences are being erected by R. Jackson, A. F. Martin, Prof. Osborne, Mrs. Galtic, Welfey & McLean, E. Cass and others.

PLUMBERS EN TOUR.

On Thursday afternoon the steamer Corona brought over from Niagara a party of Washington, U.S.A. plumbers, consisting of Warren W. Briggs, of The Briggs Heating and Ventilating Co. ; G. H. Zellars, of Zellars & Co., and E. J. Febrey, of E. J. Febrey & Co., each of whom were accompanied by their families. The party were met at the boat by W. L. Heliwell, of The Gurney Foundry Co., Limited, and taken for a tally-ho ride through the city, the principal public buildings and points of interest being visited. After the drive, which was thoroughly enjoyed by the entire party, a stop was made at McConkey's, for lunch, which proved to be not the least enjoyable experience of the day. The party returned to Buffalo on the evening boat.

BUILDING PERMITS ISSUED.

Building permits have been issued in Hamilton to Alex. Mercer for seven brick houses at the corner of Mary and Murray streets, which are to cost $7,000, and to W. P. Witton, for a new frame ward at the hospital to cost $1,200.

Building permits have been issued in Ottawa to A. Anderson, lot 7 Wellington street, brick veneered dwelling, $4,800 ; Edward Clarke, lot 12 Frank street, solid brick dwelling, $6,000 ; Michael O'Leary, lot 13 Nicholas street, brick veneered store and dwellings, $5,000.

Dr. Greene, lot 49 Gloucester street, brick dwelling, $4,000 ; Miss Thistle, brick dwelling, Somerset street north, to cost $7,500; W. G. Bronson, stone and brick dwelling Concession street, to cost $13,600, and Geo. Bannister, lots 11 and 12 Elgin street, brick veneered dwelling, $2,500.

PLUMBING AND HEATING NOTES.

S. Rochon & Fils, contractors, Montreal, have dissolved.

Keegan & Hefferman, contractors, Montreal, have dissolved.

Peter Conroy & Bros. have registered as plumbers, etc., in Montreal.

Graham & Pickles, electricians, Halifax, N.S., have sold out to John A. Dunn.

Wm. H. Creed has registered as proprietor of W. H. Creed & Co., plumbers, Montreal.

Thomas Watson, dealer in electrical supplies, etc., Victoria, B.C., has sold his electrical department to C. C. Mackenzie.

PLUMBING AND HEATING CONTRACTS.

The Bennett & Wright Co., Limited, have secured another big contract—that of heating, lighting and plumbing the large residence to be erected for A. E. Kemp, M.P., at Castle Frank, Rosedale. This will be almost as big a job as that in the Flavelle house, the contract calling for six bathrooms, etc.

Joseph Lamarche, Montréal, has secured the contract for the plumbing, heating and roofing of St. Edouard Church, being erected in St. Denis ward. He also has charge of the overhauling of the heating apparatus in the Montreal post office, where three new boilers are being installed.

DELEGATES APPOINTED

The Montreal Master Plumbers' Association held its regular meeting last Thursday evening with Mr. James A. Sadler, President, in the chair and a goodly attendance of members present.

The chief item of business was the election of representatives to the National Convention to be held in Toronto. There were five chosen to go with the three who are connected with the National Executive, Messrs. Harris, Lamarche and Joseph Thibault. The five appointed are President Sadler, J. A. Giroux, John Watson, P. J. Carroll and E. C. Mount. Alternatives are P. C. Ogilvie, J. W. Hughes, G. C. Denman, T. Christie and F. Bonhomme.

ENGLISH COOPERATIVE STORES.

THE great co-operative stores of England report in the aggregate sales last year amounting to £6,968,921, a substantial increase over the sales of any previous year as far back as 1891. The gross profit, says The New York Journal of Commerce, is not only larger in amount than in any previous year, but its percentage upon the sales is larger than in any previous year. The increase over last year is extremely small, but there has been a pretty steady increase in the ratio of gross profit to sales from 11.42 in 1891 to 13.76 in 1900. The salaries and wages have increased a little both absolutely and in their ratio to sales. The percentage of the directors' remuneration remains unchanged. Rent, gas, taxes and insurance have been increasing pretty steadily from .76 per cent. in 1891 to 1 per cent. in 1900. The total expenses have also increased pretty constantly from 8.5 per cent. in 1891 to 9.5 per cent. in 1900.

The net revenue last year was not quite so large as the year before, but with very slight movements in the opposite direction the net revenue has increased from £210,646, or 3.21 per cent. of sales in 1891, to £319,559, or 4.59 per cent. of sales in 1900. Not only does this gross and net profit show a gain during these nine years, but the stock on hand at the close of the year shows a gradual increase, amounting to about £200,000 during the nine years.

Eighteen years ago the gross profit was less than 10 per cent. Viewed from the customer's standpoint, prices are now loaded with a charge of nearly 14 per cent. for costs of management, taxes and interest on capital, as against less than 10 per cent. eighteen years ago. But as the sales have been increasing it is quite evident that the stores have not been driving their customers away by their charges. There has been some increase in the expense of doing the business, but the growing margin between receipts and disbursements is filled to the extent of less than one-half by larger amount of salaries, taxes, etc. The total expense account increased £104,000 in the nine years, and the net revenue increased £115,000.

It is said that the co-operative feature of the business is not the leading one now; originally they sold at cost and expenses ; now they are trying to get as good a return as possible on the capital invested. One of the odd features of the statement is that the larger sales of the Army and Navy stores in 1899 than in 1900 is due to the fact that it had on hand a large quantity of diamonds, which it was enabled to sell at high prices on account of the South-African War.

The sales in the last year of these concerns were as follows :

Army and Navy Co-operative Society	£3,313,995
Civil Service Supply Association.	1,769,656
Army and Navy Auxiliary C. S.	608,009
Junior Army and Navy Stores.	684,712
Civil Service Co-operative Society.	423,610
New Civil Service Co-operation.	109,297

A WEST SUPERIOR HOPPER SCALE TEST.

A VERY instructive hopper scale story comes from West Superior, Wis., U.S.A.

Over a year ago the Great Northern Railway, one of America's greatest transcontinental lines, decided to build a new steel elevator at that point. Every part of its construction is of steel, and it has a storage capacity of 3,200,000 bushels. It stands between the lake and a siding of the Great Northern Railway, and it was necessary to have a track scale on the siding and 18 hopper scales in the elevator to weigh the grain before it was loaded into the boats. The St. Paul house secured the order for a Fairbanks scale for the siding and shipped one in the night to West Superior. It took just 11 hours to set this great scale on the track. The scale is known at the factory as a 42-ft. platform, with a capacity of 60 tons. The Fairbanks people were then asked to bid on 18 hopper scales, each with a capacity of 100,000 lb. They were underbid by a rival Pennsylvanian firm, who filled the order in due time. The State inspector of weights and measures then came to test the scales. The Fairbanks scale was accepted, while the other 18 standard scales were condemned as they were not accurate and were too lightly constructed for the place.

A serious problem now confronted the Great Northern people. Navigation was soon to open on the great lakes, carloads of grain were being loaded into the new elevator; but no wheat could be loaded into the boats without first being weighed on the hopper scales. The Fairbanks, St. Paul house, then secured an order for the 18 scales at the price of the original bid. The manager and mechanical suprintendent went to the factory at St. Johnsbury, Vt., and 10 days later went back with the first scale, sent by express at a cost of about $600. Just 18 working days after the order was received, the last installment of five scales left the St. Johnsbury depot for the Northwest. It was a supreme test of what skilled workmen could do. The State inspector tested them with standard test weights up to their fullest capacity, and they had the most severe test ever given a scale. They weighed perfectly and showed no sign of weaknesss at any point.

This, the Fairbanks Scale Co. regard as quite a triumph. In Canada these scales are well known. They are used in all the C.P.R. elevators, at the Canada Atlantic depot, Harbor Coteau, by the Montreal Transfer Co., as well as by many small elevators. Quite recently the Canadian branch has taken orders from Messrs. Mackenzie & Mann and from the G.T.R. Co.

for scales for the large elevator which they are building at Portland, Me. The latter two will be furnished with the Fairbanks Typer Registering Beam.

CANADA AT GLASGOW.

From Russia to Canada is a far cry, but you may do it with ease in the Exposition. You hardly pass from the echo of the strange Russian tongue till you are greeted with the homely English, with that crisp, sharp ring in it that is peculiarly trans-Atlantic. Recently I met Mr. J. D. Stewart, who is looking after the agricultural products of that great and aspiring colony. Canada is everywhere in the Exposition. She has a magnificent building all to herself, and besides has one of the best sites in the Industrial Hall. "We are bent on being to the front everywhere," observed Mr. Stewart, and they have certainly got there. Among the exhibits of the section are a great quantity of Klondike nuggets. These look very inviting, and excite the cupidity of the onlooker.—Glasgow Evening News.

Charles A. Dickie, general merchant, Shediac, N.B., has sold out his business to Stewart D. White, and has bought out G. C. Hamilton, general merchant, of the same town.

TRADE JOURNALS AS SALESMEN.

THE trade journal is a purveyor of news and information for the business lines in whose interests it is published, writes H. B. Ford in an exchange. It covers its chosen field with an ability that is demonstrated by its success in keeping at it and is very often relied upon to furnish facts and information which are not otherwise obtainable. It goes to the place of business of its subscribers and is read because of the mercantile or trade intelligence that it conveys, being valued in exact accordance with the reader's ideas of what intelligence consists. Even the most unthinking readers, and possibly those who are thoroughly well posted in their particular lines of business, will admit that a very considerable amount of good can be obtained from any trade journal that has facilities for gathering news and is edited by men who understand what they are doing.

The functions of a trade journal are primarily to provide items of interest for its patrons and to consider such questions as concern them in their business, or to publish technical articles for the instruction and benefit of the trade. The news notes are valuable because they keep the business world posted, as do also the market reports. Editorial matter also has its value because of the suggestions therein contained. Most men of business like to know what others are doing and saying in matters either directly or indirectly connected with that business and they are not averse to hearing such opinions as are expressed on subjects that concern their own affairs. These opinions may differ from those entertained by a reader, but if their appearance in print causes him to think and he applies his thought to the betterment of his condition he will derive a benefit from the expressed opinions even if they do not coincide with his own views.

These functions, when well performed, are of great and decided benefit to the business world, but their work is not well done, nor can it be expected to be, without the aid of the advertisements. Practically speaking, a trade journal is a medium for the transmission of trade intelligence and in this sphere of usefulness its value depends as much upon its advertisements as upon any other consideration.

An advertisement is nothing else than a salesman. Instead of going to a few buyers each week, it goes to as many buyers as the journal has subscribers, and in each case makes its offering in a way that is intended to carry conviction. It pays its visits at stated periods, does not encounter any of the difficulties met with by salesmen, is always in a good humor and never aggressive, does not endeavor to run down a rival's goods, even though claiming to offer the very best, and is successful in the same degree that it is convincing. Such is the advertisement as it appears in and becomes a part of the trade journal.

I do not mean to say that the trade journal is a better salesman than the traveller, but I do say most emphatically that in each case and every one of its functions it is an aid to the traveller in the introduction and sale of goods. It goes before him and carries the news. It introduces those whom he represents in cases where they are not known before, or has carried to the buyer's information of what he has to sell. In short, the trade journal, in my opinion, is the advance agent of success for the traveller and gives him an opportunity which he could not find under any other conditions. I say this as a travelling man who reads his trade journals because he finds a profit in so doing, and the only criticism I expect to meet is from other travellers who fear that if they lay too much stress on the value of printed matter as a salesman they may be obliged to seek some other occupation.

ALL THE TIME.

WHY he should advertise all the time is one of the mysteries that bother the new advertiser. "After a time," he thinks to himself, "everyone has seen my advertisement, everyone who will ever be convinced is convinced, so why should I continue to spend money buying advertising space?" When he is told that the man or woman who did not want a thing yesterday may desire it to-day, it does not make much of an impression on him. Let us, therefore, take an instance of how new readers are constantly being secured.

Mr. John Smith is a young man, say 25 years old, who some months ago fell in love with a young lady, who reciprocated his affection, and they engaged themselves to be married in the near future. Previous to this time Mr. Smith had never gazed at any business announcement which related to articles of household use, for, of course, he had no earthly use for such articles. For the past several months, however, Mr. Smith has taken great interest in such advertisements, and has discussed with his fair one the relative merits of the various things in this line that are being brought to public attention. She had also taken but slight interest in such matters. The two are now constant readers of such advertisements.

Not only has this change taken place in Smith's relation to advertisements, but in hundreds, nay, thousands. In this way household advertisers are securing thousands of new readers constantly.

The next year John Smith's wife may have a child. Then Mr. Smith and Mrs. Smith will become interested in advertisements of children's clothing, of baby foods, of toys, and of numerous other things that come in a household with a child. Not only Mr. Smith, of course, but thousands of others, whose trade the advertiser misses if he misses inserting his ad.

Thousands of people to-day apparently in the prime of health, and laughing at all announcements of medicine or treatment, or healthful foods, find themselves in the succeeding year deprived of their vitality and anxiously seeking, through the business announcements in newspapers and magazines, for something from which they may be able to secure again the health that was once their own. The advertiser who advertised only last year is entirely unknown to them; they have probably never heard his name or read a line of his previous announcements. If these advertisements were inserted to-day, what anxious readers and purchasers they would be! And this cumulative effect, this adding of one year's customers to those of previous years, is what makes success in advertising.

It is unnecessary to draw out instances like this. The moral is that new readers for advertising are being made by the conditions of human life; by its constant changes and metamorphoses, its births, its sicknesses and its deaths; that all the people who are interested in one article or one idea can never be reached within a certain limit of time; that to become a successful advertiser, you must advertise day by day and year by year.—Information.

A. D. Cormier's general store and the warehouse of Anthony Grattan, general merchant, Buctouche, N.B., have been burned.

MANITOBA MARKETS.

WINNIPEG, June 10, 1901.

HARDWARE AND PAINTS, OILS AND GLASS.

BUSINESS has materially improved with the coming of an abundant rain, and already orders are coming in more briskly. In paints and oils there are no change of prices to report, but the market is firm.

Quotations for the week are as follows:

Barbed wire, 100 lb	$3 45
Plain twist	3 45
Staples	3 95
Oiled annealed wire 10	3 95
" 11	4 00
" 12	4 05
" 13	4 20
" 14	4 35
" 15	4 45
Wire nails, 30 to 60 dy, keg	3 50
" 16 and 20	3 60
" 10	3 60
" 8	3 70
" 6	3 75
" 4	3 90
" 3	4 15
Cut nails, 30 to 60 dy.	3 00
" 20 to 40	3 05
" 10 to 16	3 10
" 8	3 15
" 6	3 20
" 4	3 30
" 3	3 65
Horsenails, 45 per cent. discount.	
Horseshoes, iron, No. 0 to No 1	4 65
No. 2 and larger	4 40
Snow shoes, No. 0 to No. 1	4 90
No. 2 and larger	4 40
Steel, No. 0 to No. 1	4 95
No. 2 and larger	4 70
Bar iron, $2.50 basis.	
Swedish iron, $5.00 basis.	
Sleigh shoe steel	3 00
Spring steel	3 25
Machinery steel	3 75
Tool steel, Black Diamond, 100 lb	8 50
Jessop	13 00
Sheet iron, black, 10 to 20 gauge, 100 lb.	3 50
20 to 26 gauge	3 75
28 gauge	4 00
Galvanized American, 16 gauge	2 54
18 to 22 gauge	4 50
24 gauge	4 75
26 gauge	5 00
28 gauge	5 25
Genuine Russian, lb	12
Imitation " "	8
Tinned, 24 gauge, 100 lb	7 75
26 gauge	8 00
28 gauge	8 50
Tinplate, IC charcoal, 20 x 28 box	10 75
" IX "	12 75
" IXX "	14 75
Ingot tin	33
Canada plate, 18 x 21 and 18 x 24	3 25
Sheet zinc, cask lots, 100 lb	7 50
Broken lots	8 00
Pig lead, 100 lb	6 00

Wrought pipe, black up to 2 inch	50 an 10 p.c.
" Over 2 inch	50 p.c.
Rope, sisal, 7-16 and larger	$11 00
" ¾	11 50
" ½ and 5-16	12 00
Manila, 7-16 and larger	14 00
" ¾	14 50
" ½ and 5-16	15 00
Solder	20
Cotton Rope, all sizes, lb	17
Axes, chopping	$ 7 50 to 12 00
" double bitts	12 00 to 18 00
Screws, flat head, iron, bright	87½
Round " "	82½
Flat " brass	80
Round " "	75
Coach	57½ p.c.
Bolts, carriage	55 p.c.
Machine	55 p.c.
Tire	60 p.c
Sleigh shoe	65 p.c.
Plough	40 p.c.
Rivets, iron	50 p.c.
Copper, No. 8	35
Spades and shovels	40 p.c.
Harvest tools	50, and 10 p.c.
Axe handles, turned, s. g. hickory, doz.	$2 50
No. 1	1 50
No. 2	1 25
Octagon-extra	1 75
No. 1	1 25
Files common	70, and 10 p.c.
Diamond	60
Ammunition, cartridges, Dominion R.F.	50 p.c.
Dominion, C.F., pistol	30 p.c.
" military	15 p.c.
American R.F.	30 p.c.
C.F. pistol	5 p.c.
C.F. military	10 p.c. advance.
Loaded shells:	
Eley's soft, 12 gauge black	16 50
chilled, 12 guage	18 00
soft, 10 guage	21 00
chilled, 10 guage	23 00
Shot, Ordinary, per 100 lb	6 25
Chilled	6 75
Powder, F.F., keg	4 75
F.F.G.	5 00
Tinware, pressed, retinned	75 and 2½ p.c.
" " plain	70 and 15 p.c.
Graniteware, according to quality	50 p.c.

PETROLEUM.

Water white American	25½c.
Prime white American	24c.
Water white Canadian	22c.
Prime white Canadian	21c.

PAINTS, OILS AND GLASS.

Turpentine, pure, in barrels	$ 61
Less than barrel lots	66
Linseed oil, raw	92
Boiled	95
Lubricating oils, Eldorado castor	25½
Eldorado engine	24½
Atlantic red	27½
Renown engine	41
Black oil	23½ to 25
Cylinder oil (according to grade)	55 to 74
Harness oil	61
Neatsfoot oil	$ 1 00
Steam refined oil	85
Sperm oil	1 50
Castor oil per lb.	11½
Glass, single glass, first break, 16 to 25 united inches	2 25
26 to 40 per 50 ft.	2 50
41 to 50 " 100 ft.	5 50
51 to 60 " " "	6 00
61 to 70 per 100-ft. boxes	6 50
Putty, in bladders, barrel lots per lb.	2½
kegs "	2¾
White lead, pure per cwt.	7 00
No 1 "	6 75
Prepared paints, pure liquid colors, according to shade and color, per gal.	$1.30 to $1.90

LAKE ORE STOCKS.

"The stocks of Lake Superior ore reported on Lake Erie docks on May 1, when account is usually taken," according to The Engineering and Mining Journal, "amounted this year to 3,050,183 long tons, and were the largest reported since 1898. On May 1, 1900, they were only 1,720,655 tons. The accumulation of ore reflects the slackening in furnace activity toward the close of last year, and also the fact that most furnaces using lake ore laid in large stocks during the navigation season. The following statement shows the movement to furnaces through Lake Erie ports during the years ending April 30, 1900 and 1901, in long tons:

	1899-1900.	1900-01.	Changes.
Stocks on Lake Erie docks, May 1	2,073,254	1,720 656	Dec. 352,598
Receipts for the season	15,222,187	15,797,787	Inc. 575,600
Totals	17,295 441	17,518,443	Inc. 223,002
Stocks on docks Apr 30	1,720,656	3,050,183	Inc. 1,329,527
Shipments to furnaces	15,574,785	14,468,260	Dec. 1,106,525

"As the stocks reported on docks on December 1, 1900, at the close of navigation, were 5,904,670 tons, and these had been reduced by May 1 to 3,050,183 tons, the winter movement to furnaces amounted to 2,854,487 tons. The remaining shipments—11,613,773 tons—were made between May 1 and December 1, 1900, much of the ore going almost directly from boats to cars."

CURRENT MARKET QUOTATIONS

June 14, 1901.

These prices are for such qualities and quantities as are usually ordered by retail dealers on the usual terms of credit, the lowest figures being for larger quantities and prompt pay. Large cash buyers can frequently make purchases at better prices. The Editor is anxious to be informed at once of any apparent errors in this list as the desire is to make it perfectly accurate.

METALS.

Tin.

Lamb and Flag and Straits—56 and 28 lb. ingots, per lb.	0 32	0 32½

Tinplates.

Charcoal Plates—Bright

		Per box
M.L.S., equal to Bradley.		
I.C., usual sizes	....	$6 50
I.X., "	....	8 00
I.X.X., "	....	9 50
Famous—		
I.C.	....	6 50
I.X	....	8 00
I.X.X	....	9 50
Raven & Vulture Grades—		
I.C., usual sizes	....	4 50
I.X., "	....	5 25
I.X.X "	....	6 00
I.XXX., "	....	6 75
D.C., 12⅛x17	....	4 00
D.X.	....	4 75
D.X.X	....	5 00

Coke Plates—Bright

Bessemer Steel—		
I.C., usual sizes	....	3 75
I.C., special sizes, base	....	4 00
20 x 28	....	7 75

Charcoal Plates—Terne

Dean or J. G. Grade—		
I.C., 20x28, 112 sheets	....	8 00
I.X., Terne Tin	....	10 00

Charcoal Tin Boiler Plates.

		Per lb.
Cookley Grade—		
X X., 14x56, 50 sheet bxs; " 14x60, "; " 14x65, "	....	0 06¼

Tinned Sheets

72x30 up to 24 gauge	....	0 07½
" 26 "	....	0 08
" 28 "	....	0 08½

Iron and Steel.

Common Bar, per 100 lbs.	1 85	1 90
Refined " "	....	2 35
Horse Shoe Iron "	....	2 35
Hoop steel, 1½ to 3 in. base, extras for smaller sizes	....	3 00
Sleigh Shoe Steel " base	....	2 10
Tire Steel	2 30	2 50
Reeled Machinery	....	3 00
Toe Calk Steel	2 85	3 00
T. Firth & Co's tool steel, per lb	0 12½	0 13
Jessop's tool Steel	0 12½	0 13
Mo[illegible]'s tool steel	0 12½	0 13
Black Diamond and "B.C." tool steel	0 10	0 11
Drill Steel, per lb.	0 08	0 10

Boiler Tubes.

1½-inch	....	0 12½
2 "	....	0 13
2½ "	....	0 15
3 "	....	0 16
3½ "	....	0 20
4 "	....	0 25

Steel Boiler Plate.

¼ inch	2 50	2 60
3-16 inch	2 60	2 70
⅜ inch and thicker	2 50	2 60

Black Sheets.

	Com.	D.Fl.
18 gauge	2 75	3 00
20 gauge	2 75	3 00
22 to 24 "	2 75	3 25
26 "	2 75	....
28 "	3 00	3 50

Canada Plates.

All dull, 52 sheets	...	2 90
Half polished	....	3 00
All bright	....	3 50

Iron Pipe.

Black pipe—		
⅛ "	....	4 45
¼ inch	....	3 25
⅜ "	....	3 30
½ "	....	3 50
¾ "	....	3 65
1 "	....	5 15
1¼ "	....	7 00
1½ "	....	8 40
2 "	....	11 25
2½ "	....	20 95
3 "	....	24 55
3½ "	....	30 75
4 "	....	39 00
5 "	....	47 35
6 "	...	62 10
Galvanized pipe—		
½ inch	....	4 90
¾ "	....	5 25
1 "	....	7 55
1¼ "	....	10 30
1½ "	....	12 35
2 "	....	16 50

Galvanized Sheets.

	G C.	Comet.	Amer.	Queen's Head.
16 gauge		....	4 00	3 75
18 to 24 gauge	4 00	3 85	4 25	4 00
26 "	4 25	4 10	4 25	4 25
28 "	4 50	4 25	4 40	4 50

Case lots 10 to 15c. less.
28 gauge American equals 26 gauge English.

Chain.

Proof Coil, 3-16 in., per 100 lb	....	...
" ¼ " "	6 00	8 50
" 5-16 " "	4 70	5 00
" ⅜ " "	4 05	4 50
" 7-16 " "	3 90	4 75
" ½ " "	3 70	4 10
" 9-16 " "	3 65	4 05
" ⅝ " "	3 25	3 90
" ¾ " "	3 60	4 10

Halter, kennel and post chains, 40 to 40 and 5 p.c.

Cow ties	40 p.c.
Tie-out chains	65 p.c.
Stall fixtures	35 p.c.
Trace chain	45 p.c.

Jack chain, iron, single and double, discount 35 p.c.
Jack chain, brass, single and double, discount 40 p.c.

Copper.

Ingot

English B. S., ton lots	....	0 17¾
Lake Superior	....	....

Bars

Cut lengths round, ½ to ⅞ in.	0 23	0 25
" round and square 1 to 2 inches	0 23	0 25

Sheet.

Plain, 14 oz., and light, 16 oz., 14x48 and 14x60	0 24	0 24½
Plain, 14 oz., and light, 16 oz., irregular sizes	0 24½	0 25
Tinned copper sheets	....	0 26
Planished	....	0 32

Braziers (In sheets.)

4x6ft. 25 to 30 lbs. ea., per lb.	....	0 25
" 35 to 45 " "	....	0 24
" 50-lb. and above, "	....	0 23

Boiler and T. K. Pitts

Plain Tinned, per lb	....	0 28
Spun, per lb.	....	0 32

Brass.

Rod and Sheet, 14 to 30 gauge, 15 per cent.

Sheets, hard-rolled, 2x4	....	0 23
Tubing, base, per lb.	....	0 23

Zinc Spelter

Foreign, per lb	0 06½	0 06
Domestic "	....	...

Zinc Sheet.

5 cwt. casks	....	00 6¼
Part casks	....	00 6¾

Lead.

Imported Pig, per lb	0 04¼	0 04½
Bar, 1 lb.	0 05½	0 06¾
Sheets, 2½ lbs. sq. ft., by	....	0 06¼
Sheets, 3 to 6 lbs., "	....	0 06

NOTE.—Cut sheets ½ cent per lb. extra. Pipe, by the roll, usual weights per yard, lists at 7c. per lb. and 30 p.c. dis. f.o.b. Toronto.

NOTE.—Cut lengths, net price, waste pipe 8-ft. lengths lists at 7½ cents.

Shot.

Common, $6.50 per 100 lb.; chilled, $7.00 per 100 lb.; buck, seal and ball, $7.50. Discount, 17½ p.c. Prices are f.o.b. Toronto, Hamilton, Montreal, St. John and Halifax. Terms 3 per cent. cash, freights equalized.

Soil Pipe and Fittings.

Discount, 60 and 10 per cent. on medium and extra heavy, and 60 per cent. on light.

Solder.

	Per lb.	Per lb.
Bar half-and-half, guarant'd	....	0 18½
Bar half-and-half, commer'l	....	0 18
Refined	....	0 17½
Wiping	....	0 17

NOTE.—Prices of this graded according to quantity. The prices of other qualities of solder in the market indicated by private brands vary according to composition.

Antimony.

Cookson's, per lb.	0 10½	0 11

White Lead.

	Per 100 lb.
Pure	6 37
No. 1 do	6 00
No. 2 do	5 62½
No. 3 do	5 15
No. 4 do	4 87½
Munro's Select Flake White	7 37½
Elephant and Decorators' Pure	7 12½
Brandram's B B. Genuine	8 00
" " Decorative	7 55
" " No. 1	6 85
" " No. 2	6 00

Red Lead.

Genuine, 560 lb. casks, per cwt.	....	$5 50
Genuine, 100 lb. kegs, per cwt.	....	5 75
No. 1, 560 lb. casks, per cwt.	....	5 25
No. 1, 100 lb. kegs, per cwt	....	5 00

White Zinc Paint.

Elephant Snow White	0 08	0 09
Pure White Zinc	0 08	0 09
No. 1	0 06	0 07½
No. 2	0 05	0 06½

Dry White Lead.

Pure, casks	5 75
Pure, kegs	6 25
No. 1, casks	5 60
No. 1, kegs	5 00

Prepared Paints.

In ¼, ½ and 1 gallon tins.

Pure, per gallon	....	1 25
Second qualities, per gallon	....	1 10
Barn (in bbls.)	0 75	0 85
The Sherwin-Williams Paints.	....	1 45
Canada Paint Co's Pure	....	1 25
Toronto Lead & Color Co's Pure	....	1 25
Sanderson Pearcy's Pure	....	1 20
Stewart & Wood's Champion Pure.		1 20

Colors in Oil.

25 lb. tins, Standard Quality.

Venetian Red, per lb.	0 05
Chrome Yellow	0 11
Golden Ochre	0 06
French "	0 05
Marine Black	0 09
" Green	0 09
Chrome "	0 08
French Imperial Green	0 09

Colors, Dry.

Yellow Ochre (J.C.) bbls.	1 35	1 40
Yellow Ochre (J.F.L.S.), bbls	....	2 75
Yellow Ochre (Royal)	1 10	1 15
Brussels Ochre	....	2 00
Venetian Red (best), per cwt.	1 80	1 90
English Oxides, per cwt.	3 00	3 25
American Oxides, per cwt.	1 75	2 00
Canadian Oxides, per cwt.	1 75	2 00
Super Magnetic Oxides, 93p.c.	2 00	2 25
Burnt Sienna, pure, per lb.	....	0 10
" Umber, " "	....	0 10
do Raw	....	0 09
Drop Black, pure	....	0 09
Chrome Yellows, pure	....	0 18
Chrome Greens, pure, per lb.	....	0 12
Golden Ochre	....	0 03½
Ultramarine Blue in 28-lb. boxes, per lb.	0 08	0 24
Fire Proof Mineral, per 100 lb	....	1 00
Genuine Eng. Litharge, per lb	....	0 07
Mortar Color, per 100 lb.	....	1 25
English Vermillion	....	0 80
Pure Indian Red, No. 45 lb.	....	0 80
Whiting, per 100 lb.	....	0 55

Blue Stone.

Casks, for spraying, per lb.	0 07
100-lb. lots, do. per lb.	0 08

Putty.

Bulk in bbls.	1 90
Bulk in less quantity	2 05
Bladders in bbls.	2 10
Bladders in kegs, boxes or loose	2 25
Bladders in 25-lb. tins	2 35
Bladders in 12½-lb. tins	2 65
Bladders in bulk or tins less than 100 lb	2 90

Varnishes.

	In 5-gal. lots.)	Per gal.
Carriage, No. 1	2 90	3 30
" body	8 00	9 00
" rubbing	4 00	5 00
Gold Size, Japan	3 00	3 40
Brown Japan	2 40	2 80
Elastic Oak	2 90	3 30
Furniture, extra	2 40	2 80
" No. 1	1 60	2 00
Hard Oil Finish	2 70	3 10
Light Oil Finish	3 20	3 60
Demar	3 30	3 70
Shellac, white	4 40	4 80
" orange	4 00	4 40
Furniture Brown Japan	1 60	2 00
Black Japan	2 40	2 80
" " No. 1	1 60	2 00

The Imperial Varnish & Color Co's., Limited Elastilite Varnish 1 gal. can, each. $3.00.

Granatine Floor Finish per gal., $2.75.

Maple Leaf Coach Enamels; Size 1, $1 20; Size 2, 70c.; Size 3, 40c. each.

Castor Oil.

East India, in cases, per lb.	0 10	0 10½
" " small lots	0 10½	0 11

Cod Oil, Etc.

Cod Oil per gal.	0 50	0 55
Pure Olive	....	1 20
" Neatsfoot	....	90

Glue.

Common	0 08½	0 09
French Medal	0 14	0 14½
Cabinet sheet	0 12	0 13
White, extra	0 18	0 20
Gelatine	0 22	0 30
Strip	0 18	0 20
Coopers	0 19	0 20
Huttner	....	0 18

HARDWARE.

Ammunition.

Cartridges.

B. B Cap Dom. 50 and 5 per cent.
Rim Fire Pistol, dis. 40 p. c., Amer.
Rim Fire Cartridges, Dom., 50 and 5 p. c.
Central Fire Pistol and Rifle, 10 p.c. Amer.
Central Fire Cartridges, pistol sizes, Dom, 30 per cent.
Central Fire Cartridges, Sporting and Military, Dom., 15 and 5 per cent.
Central Fire, Military and Sporting, Amer, add 5 p.c. to list. B.B. Caps, discount 40 per cent, Amer.
Loaded and empty Shells, "Trap" and "Dominion" grades, 25 per cent. Rival and Nitro, net list.
Brass Shot Shells, 55 per cent.
Primers, Dom., 30 per cent.

Wads — per lb.

Best thick white felt wadding, in ¼-lb bags 1 00
Best thick brown or grey felt wads, in ¼-lb. bags 70
Best thick white card wads, in boxes of 500 each, 12 and smaller gauges 0 99
Best thick white card wads, in boxes of 500 each, 10 gauge 0 35
Best thick white card wads, in boxes of 500 each, 8 gauge 0 55
Thin card wads, in boxes of 1,000 each, 12 and smaller gauges 0 20
Thin card wads, in boxes of 1,000 each, 10 gauge 0 25
Thin card wads in boxes of 1,000 each, 8 gauge ....

Chemically prepared black edge grey cloth wads, in boxes of 250 each— Per M.
11 and smaller gauge 0 60
9 and 10 gauges 0
7 and 8 gauges 0 70
5 and 6 gauges 1 90

Superior chemically prepared pink edge, best white cloth wads, in boxes of 250 each—
11 and smaller gauge 1 15
9 and 10 gauges 1 40
7 and 8 gauges 1 65
5 and 6 gauges 1 90

Adzes.

Discount, 20 per cent.

Anvils.

Per lb. .... 0 10 — 0 12½
Anvil and Vise combined 4 50
Wilkinson & Co.'s Anvils..lb. 0 09 — 0 09½
Wilkinson & Co.'s Vices..lb. 0 09¾ — 0 10

Augers.

Gilmour's, discount 65 and 5 p.c. off list.

Axes.

Chopping Axes
Single bit, per doz .... 5 50 — 10 00
Double bit, " .... 11 00 — 18 00
Bench Axes, 40 p.c.
Broad Axes, 33⅓ per cent.
Hunters' Axes .... 5 50 — 6 00
Boy's Axes .... 5 75 — 6 75
Splitting Axes .... 6 50 — 12 00
Handled Axes .... 7 00 — 10 00

Axle Grease.

Ordinary, per gross .... 5 75 — 6 00
Best quality .... 13 00 — 15 00

Bath Tubs.

Zinc 6 00
Copper, discount 15 p.c. off revised list

Baths.

Standard Enameled.

5½-inch rolled rim, 1st quality 25 00
" " " 2nd " 21 00

Anti-Friction Metal.

"Tandem" A .... per lb. 0 27
" B .... " 0 21
" C .... " 0 11½
Magnolia Anti-Friction Metal, per lb. 0 25
Frictionless metal, per lb 0 23

SYRACUSE SMELTING WORKS.

Aluminum, genuine 0 45
Dynamo 0 29
Special 0 25
Aluminum, 99 p.c. pure "Syracuse" .. 0 50

Bells.

Hand.

Brass, 60 per cent.
Nickel, 55 per cent.

Cow.

American make, discount 66⅔ per cent.
Canadian, discount 45 and 50 per cent.

Door.

Gongs, Sargent's .... 5 50 — 8 00
" Peterboro', discount 45 per cent.

Farm.

American, each .... 1 25 — 3 00

House.

American, per lb .... 0 35 — 0 40

Bellows.

Hand, per doz .... 3 35 — 4 75
Moulders', per doz .... 7 50 — 10 00
Blacksmiths', discount 40 per cent.

Belting.

Extra, 60 per cent.
Standard, 60 and 10 per cent.

Bits.

Auger.

Gilmour's, discount 60 and 5 per cent.
Rockford, 50 and 10 per cent.
Jennings' Gen., net list.

Car.

Gilmour's, 47½ to 50 per cent.

Expansive.

Clark's, 40 per cent

Gimlet.

Clark's, per doz .... 0 65 — 0 90
Diamond, Shell, per doz .... 1 00 — 1 50
Nail and Spike, per gross .... 2 25 — 5 20

Blind and Bed Staples.

All sizes, per lb .... 0 07½ — 0 12

Bolts and Nuts. Per cent.

Carriage Bolts, full square, Norway 65
" " full square 65
Common Carriage Bolts, all sizes 60
Machine Bolts, all sizes 60
Coach Screws 70
Sleigh Shoe Bolts 72½
Blank Bolts 60
Bolt Ends 62½
Plough Bolts 60
Nuts, square .... 4 c. off
Nuts, hexagon .... 4¼c. off
Tire Bolts 67½
Stove Bolts 67½
Stove rods, per lb .... 5½ to 6c.

Boot Calks.

Small and medium, ball, per M 4 25
Small heel, per M 4 50

Bright Wire Goods.

Discount .... 62½ per cent.

Broilers.

Light, dis., 65 to 67½ per cent.
Reversible, dis., 65 to 67½ per cent.
Vegetable, per doz., dis. 37½ per cent.
Henis, No. 8, " 6 00
Henis, No. 9, " 7 00
Queen City " .... 7 50 — 0 00

Butchers' Cleavers.

German, per doz .... 6 00 — 11 00
American, per doz .... 12 00 — 20 00

Building Paper, Etc.

Plain building, per roll 0 30
Tarred lining, per roll 0 40
Tarred roofing, per 100 lb 1 65
Coal Tar, per barrel 3 50
Pitch, per 100-lb 0 85
Carpet felt, per ton 45 00

Bull Rings.

Copper, $2.00 for 2½ in. and $1.90 for 2 in.

Butts.

Wrought Brass, net revised list

Cast Iron.

Loose Pin, dis., 60 per cent.

Wrought Steel.

Fast Joint, dis. 65, 10 and 2½ per cent.
Loose Pin, dis. 65, 10 and 2½ per cent.
Berlin Bronzed, dis. 70, 70 and 5 per cent.
Gen. Bronzed, per pair .... 0 40 — 0 65

Carpet Stretchers.

American, per doz .... 1 00 — 1 50
Bullard's, per doz .... 6 50 — ....

Castors.

Bed, new list, dis. 55 to 57½ per cent.
Plate, dis. 52½ to 57½ per cent.

Cattle Leaders.

Nos. 31 and 32, per gross .... 50 — 9 50

Cement.

Canadian Portland .... 2 50 — 2 80
English " .... — 3 00
Belgian " .... 2 50 — 2 75
Canadian hydraulic .... 1 25 — 1 50

Chalk.

Carpenters Colored, per gross 0 45 — 0 75
White lump, per cwt .... 0 60 — 0 65
Red .... 0 05 — 0 06
Crayon, per gross .... 0 14 — 0 18

Chisels.

Socket, Framing and Firmer.

Broad's, dis. 70 per cent.
Warnock's, dis. 70 per cent.
P. S. & W. Extra 60, 10 and 5 p.c.

Churns.

Revolving Churns, metal frames—No. 0, $8—No. 1, $8.50—No. 2, $9.00—No. 3, $10.00—No. 4, $12.00—No. 5, $16.00 each. Ditto, wood frames—20c. each less than above. Discounts: Delivered from factories, 58 p.c.; from stock in Montreal, 56 p.c. Terms, 4 months or 3 p.c. cash in 30 days.

Clips.

Axle dis. 65 per cent.

Closets

Plain Ontario Syphon Jet $16 00
Emb. Ontario Syphon Jet 17 00
Fittings net 1 00
Plain Teutonic Syphon Washout 10 00
Emb. Teutonic Syphon Washout 11 00
Fittings net 1 25
Low Down Teutonic, plain 16 00
" " " embossed 17 00
Plain Richelieu net 3 75
Emb. Richelieu net 4 00
Fittings net 1 25
Low Down Ont. Sy. Jet, plain net .. 19 50
" " " " " emb'd. net 20 50
Closet connection net 1 25
Basins, round, 14 in 1 00
" oval, 17 x 14 in 2 50
" " 19 x 15 in 3 75
Discount 40 p.c., except on net figures.

Compasses, Dividers, Etc.

American, dis. 62½ to 65 per cent.

Cradles, Grain.

Canadian, dis. 25 to 33⅓ per cent.

Crosscut Saw Handles.

S. & D., No. 3, per pair .... .17½
" " 5, " .... .22½
" " 6, " .... .15
Boynton pattern " .... .20

Door Springs.

Torrey's Rod, per doz .... (15 p.c.) — 2 00
Coil, per doz .... 0 88 — 1 60
English, per doz .... 2 00 — 4 00

Draw Knives.

Coach and Wagon, dis. 50 and 10 per cent.
Carpenters, dis. 70 per cent.

Drills.

Hand and Breast.

Millar's Falls, per doz. net list.

DRILL BITS.

Morse, dis., 37½ to 40 per cent.
Standard dis. 50 and 5 to 55 per cent

Faucets.

Common, cork-lined, dis. 35 per cent.

ELBOWS. (Stovepipe.)

No. 1, per doz 1 40
No. 2, per doz 1 20
Bright, 20c. per doz. extra.

ESCUTCHEONS.

Discount, 45 per cent.

ESCUTCHEON PINS.

Iron, discount 40 per cent.

FACTORY MILK CANS.

Discount off revised list, 40 per cent.

FILES AND RASPS.

Great Western .... 75 and 5 per cent.
Disston .... 75 " 5 "
Arcade .... 75 " 5 "
Kearney & Foot .... 75 " 5 "
American .... 75 " 5 "
McClellan .... 70 " 5 "
Eagle .... 70 10 and 5 "
Nicholson .... 70 " 10 "
Heller .... 70 " 10 "
Royal & Keystone 80 p.c. and 80 and 10 p.c.
Black Diamond, 60 to 60 and 10 per cent.
Jowitt's, English list, 25 to 27½ per cent.

FORKS.

Hay, manure, etc., dis., 50 and 10 per cent. revised list.

GLASS—Window—Box Price.

Size United Inches.	Star Per 50 ft.	Star Per 100 ft	D. Diamond Per 50 ft.	D. Diamond Per 100 ft
Under 26	2 15	4 15	....	5 00
26 to 40	2 30	4 45	....	6 65
41 to 50	....	4 85	....	7 50
51 to 60	....	5 15	....	8 50
61 to 70	....	5 50	....	9 50
71 to 80	....	6 00	....	10 50
81 to 85	....	6 50	....	11 70
86 to 90	....	....	..	14 00
91 to 95	..	....	....	15 50
99 to 100	.. .	....	....	18 00

GAUGES

Marking, Mortise, Etc.

Stanley's dis. 50 to 55 per cent.

Wire Gauges.

Winn's, Nos. 26 to 33, each .... 1 65 — 2 40

HALTERS.

Rope, ⅜ per gross .... — ....
" ½ " .... — 9 00
" ⅝ to ¾ .... — 14 00
Leather, 1 in., per doz .... 3 87½ — 4 00
" 1¼ in., " .... 5 15 — 5 20
Web, — per doz .... 1 87 — 2 45

HAMMERS.

Nail

Maydole's, dis. 5 to 10 per cent. Can. dis. 25 to 27½ per cent.

Tack.

Magnetic, per doz .... 1 10 — 1 20

Sledge.

Canadian, per lb .... 0 07½ — 0 08½

Ball Pean.

English and Can., per lb .... 0 22 — 0 25

HANDLES.

Axe, per doz. net .... 1 50 — 2 00
Store door, per doz .... 1 00 — 1 50

Fork.

C. & B., dis. 40 per cent. rev. list.

Hoe.

C. & B., dis. 40 per cent. rev. list.

Saw.

American, per doz .... 1 00 — 1 25

Plane.

American, per gross .... 3 15 — 3 75

Hammer and Hatchet.

Canadian, 40 per cent.

Cross-Cut Saws.

Canadian, per pair 0 13½

HANGERS. doz. pairs.

Steel barn door .... 5 85 — 6 00
Stearns, 4 inch .... — 5 00
" 5 inch .... — 6 50
Lane's covered—
No. 11, 5-ft. run .... — 8 40
No. 11½, 10-ft. run .... — 10 80
No. 12, 10-ft. run .... — 12 60
No. 14, 15-ft. run .... — 21 00
Lane's O.N.T. track, per foot. .... — 4½

HARVEST TOOLS.

Discount, 50 and 10 per cent.

HATCHETS.

Canadian, dis. 40 to 42½ per cent.

HINGES.

Blind, Parker's, dis. 50 and 10 to 60 per cent.
Heavy T and strap, 4-in., per lb 0 06½
" " 5-in., " 0 06¼
" " 6-in., " 0 06
" " 8-in., " 0 05¾
" " 10-in., " 0 05½
Light T and strap, dis. 65 and 5 per cent.
Screw hook and hinge—
6 to 12 in., per 100 lbs .... — 4 50
14 in. up, per 100 lbs .... — 3 50
Per gro. pairs.
Spring .... — 12 00

HOES.

Garden, Mortar, etc., dis. 50 and 10 p.c.
Planter, per doz .... 4 00 — 4 50

HOLLOW WARE

Discount .... 45 and 5 per cent

HOOKS.

Cast Iron.

Bird Cage, per doz .... 0 50 — 1 10
Clothes Line, per doz .... 0 27 — 0 63
Harness, per doz .... 0 72 — 0 88
Hat and Coat, per gross .... 1 00 — 3 00
Chandelier, per doz .... 0 50 — 1 00

Wrought Iron.

Wrought Hooks and Staples, Can., dis. 47½ per cent.

Wire.

Hat and Coat, discount 45 per cent.
Belt, per 1,000 .... 0 60
Screw, bright, dis. 55 per cent.

HORSE NAILS.

"C" brand 50 and 7½ p.c. off new list } Oval-
"M" brand 50, 10 and 5 per cent. } head
Countersunk, 60 per cent

HORSESHOES

Iron Shoes.	F.O.B. Montreal. No. 2 and larger.	No. 1. and smaller.
Light, medium, and heavy...	3 50	3 75
Snow shoes...	3 75	4 00
Steel Shoes.		
Light...	3 60	3 85
Featherweight (all sizes)....	4 85	4 85

F.O.B. Toronto, Hamilton, London and Guelph, 10c. per keg additional.
Toe weight steel shoes...... 6 70

JAPANNED WARE.

Discount, 45 and 5 p.c. off list, June 1899

ICE PICKS.

Star per doz... 3 00 3 25

KETTLES.

Brass spun, 7½ p.c. dis. off new list.
Copper, per lb... 0 30 0 50
American, 60 and 10 to 65 and 5 p.c.

KEYS.

Lock, Can., dis., 45 p.c.
Cabinet, trunk, and padlock, Am. per gross... 60

KNOBS.

Door, japanned and N.P., per doz... 1 50 2 50
Bronze, Berlin, per doz...... 2 75 3 25
Bronze Genuine, per doz.... 6 00 9 00
Shutter, porcelain, F. & L. screw, per gross... 1 30 4 00
White door knobs—per doz. 1 25

HAY KNIVES.

Discount, 50 and 10 per cent.

LAMP WICKS.

Discount, 60 per cent.

LANTERNS.

Cold Blast, per doz... 7 00
No. 3 "Wright's"... 8 50
Ordinary, with O burner... 4 00
Dashboard, cold blast... 9 00
No. 0... 5 75
Japanning, 50c. per doz. extra.

LEMON SQUEEZERS.

Porcelain lined,......per doz.. 2 20 5
Galvanized... 1 87 3
King, wood... 2 9
King, glass... 4 75 4 60
All glass... 1 90 1 85

LINES.

Fish, per gross... 1 05 2 50
Chalk " ... 1 90 7 40

LOCKS.

Canadian, dis. 45 p.c.
Russel & Erwin, per doz.... 3 00 3 25

Cabinet.

Eagle, dis. 30 p.c.

Padlock

English and Am., per doz.... 50 6 00
Scandinavian, " 1 00 2 40
Eagle, dis. 20 to 25 p.c

MACHINE SCREWS. Iron and Brass.

Flat head discount 25 p.c
Round Head discount 20 p.c.

MALLETS.

Tinsmiths' per doz... 1 25 1 50
Carpenters', hickory, per doz 1 25 3 75
Lignum Vitae, per doz.. .... 3 85 5 00
Caulking each... 60 2 00

MATTOCKS.

Canadian, per doz... 5 50 6 50

MEAT CUTTERS.

American, dis. 25 to 30 p.c.
German, 15 per cent.

MILK CAN TRIMMINGS.

Discount, 25 per cent.

NAILS.

Quotations are:	Cut.	Wire.
2d and 3d...	$3 35	$3 85
3d...	3 00	3 52
4 and 5d...	2 75	3 35
6 and 7d...	2 65	3 20
8 and 9d...	2 50	3 00
10 and 12d...	2 45	2 95
16 and 20d...	2 40	2 90
30, 40, 50 and 60d. (base)...	2 35	2 85

Wire nails in carlots are $2.77½
Galvanizing 2c. per lb. net extra.
Steel Cut Nails 10c. extra.
Miscellaneous wire nails, dis. 70 and 10 p.c.
Coopers' nails, dis. 30 per cent.
Flour barrel nails, dis. 25 per cent

NAIL PULLERS.

German and American....... 1 85 3 50

NAIL SETS

Square, round, and octagon per gross... 3 38 4 00
Diamond... 12 00 15 00

NETTING.

Poultry, 55 per cent for McMullen's

OAKUM. Per 100 lb.

Navy... 6 00
U. S. Navy... 7 25

OIL.

Water White (U.S.)... 0 16½
Prime White (U.S.)... 0 15½
Water White (Can.)... 0 15
Prime White (Can.)... 0 14

OILERS.

McClary's Model galvan. oil can, with pump, 5 gal., per doz... 10 00
Zinc and tin, dis. 50, 50 and 10.
Copper, per doz... 1 25 3 50
Brass, " ... 1 50 3 50
Malleable, dis. 25 per cent.

GALVANIZED PAILS.

Dufferin pattern pails, dis. 45 p.c.
Flaring pattern, discount 45 per cent.
Galvanized washtubs, discount 45 per cent.

PIECED WARE.

Discount 40 per cent. off list, June, 1899.
10-qt. flaring sap buckets, dis. 45 p.c.
6, 10 and 14-qt. flaring pails, dis. 45 p.c.
Creamer cans, dis. 45 p.c.

PICKS.

Per doz... 6 00 9 00

PICTURE NAILS.

Porcelain head, per gross.... 1 75 3 00
Brass head " 0 40 1 00

PICTURE WIRE.

Tin and gilt, discount 75 p.c.

PLANES.

Wood, bench, Canadian dis. 50 per cent, American dis. 50.
Wood, fancy Canadian or American 7½ to 40 per cent.

PLANE IRONS.

English, per doz... 2 00 5 00

PLIERS AND NIPPERS.

Button's Genuine per doz pairs, dis. 37½ 40 p.c.
Button's Imitation, per doz.. 5 00 9 00
German, per doz... 0 60 2 60

PLUMBERS BRASS GOODS.

Compression work, discount, 60 per cent.
Fuller's work, discount 65 per cent.
Rough stops and stop and waste cocks, discount, 60 per cent.
Jenkins disk globe and angle valves, discount, 55 per cent.
Standard valves, discount, 60 per per cent.
Jenkins' radiator valves, discount 55 per cent.
" " " standard, dis., 60 p.c.
Quick opening valves discount, 60 p.c.
No. 1 compression bath cock... 2 00
No. 4 " " " ... 2 00
No. 7, Fuller's... 2 50
No 4½, " ... 3 00

POWDER.

Velox Smokeless Shotgun Powder.
100 lb. or less... 0 85
1,000 lb. or more... 0 80
Net 30 days.

PRESSED SPIKES.

Discount 20 to 25 per cent.

PULLEYS.

Hothouse, per doz... 0 55 1 00
Axle... 0 22 0 33
Screw... 0 27 1 00
Awning... 0 35 2 50

PUMPS.

Canadian cistern... 1 80 3 60
Canadian pitcher spout.... 1 40 2 10

PUNCHES.

Saddlers', per doz... 1 00 1 85
Conductors', " ... 9 00 15 00
Tinners' solid, per set... 0 00 0 72
" hollow, per inch.... 0 00 1 00

RANGE BOILERS.

Galvanized, 3 gallons... 7 00
" 35 " ... 8 25
" 40 " ... 9 50
Copper, 30 " ... 22 00
" 35 " ... 26 00
" 40 " ... 30 00
Discount off Copper Boilers 10 per cent.

RAKES.

Cast steel and malleable, 50, 10 and 5 p.c.
Wood, 25 per cent.

RAZORS.

	per doz.	
Elliot's...	4 00	18 00
Geo. Butler & Co.'s...	8 00	18 00
Boker's...	7 50	11 00
" King Cutter...	12 50	50 00
Wade & Butcher's...	3 60	10 00
Theile & Quack's...	7 00	12 00

REAPING HOOKS.

Discount, 50 and 10 per cent.

REGISTERS.

Discount... 40 per cent.

RIVETS AND BURRS.

Iron Rivets, black and tinned, discount 60 and 10 per cent.
Iron Burrs, discount 55 per cent.
Extras on Iron Rivets in 1-lb. cartons, ½c. per lb.
Extras on Iron Rivets in ½-lb. cartons, 1c. per lb.
Copper Rivets & Burrs, 35 and 5 p.c. dis. and cartons, 1c. per lb. extra, net.
Extras on Tinned or Coppered Rivets ½-lb. cartons, 1c. per lb.

RIVET SETS

Canadian, dis. 35 to 37½ per cent.

ROPE ETC.

	Sisal.	Manila.
7-16 in. and larger, per lb	10	13½
⅜ in...	11	14½
¼ and 5-16 in...		15½
Cotton, 3-16 inch and larger		16
" 5-32 inch...		21
" ⅛ inch...		22½
Russia Deep Sea...		15½
Jute...		8
Lath Yarn...		9½
New Zealand Rope...		10

RULES.

Boxwood, dis. 75 and 10 p.c.
Ivory, dis. 37½ to 40 p.c.

SAD IRONS. per set.

Mrs. Potts, No. 55, polished... 62½
" No. 50, nickle-plated ... 67½

SAND AND EMERY PAPER.

Dominion Flint Paper, 47½ per cent.
B & A. sand, 40 and 5 per cent.
Emery, 40 per cent.
Garnet (Rurton's), 5 to 10 p.c. advance on list.

SAP SPOUTS.

Bronzed iron with hooks, per doz... 9 50

SAWS.

Hand Disston's, dis. 12½ p.c.
S. & D., 40 per cent.
Crosscut, Disston's, per ft.... 35 0 55
S. & D., dis. 35 p.c. on Nos. 2 and 3.
Hack, complete, each... 0 75 2 75
" frame only... .... 0 75

SASH WEIGHTS.

Sectional, per 100 lbs... 2 75 3 00
Solid, " ... 2 00 2 25

SASH CORD.

Per lb... 0 23 0 30

SAW SETS.

"Lincoln," per doz... .... 6 50

SCALES.

Standard, 45 p.c.
Champion, 65 p.c.
Spring Balances, 10 p.c.
Fairbanks Standard, 35 p.c.
" Dominion, 55 p.c.
" Richelieu, 55 p.c.

SCREW DRIVERS.

Sargent's per doz... 0 65 1 00

SCREWS

Wood, F. H., bright and steel, 87½ and 10 p.c.
Wood R. H., " dis. 82½ and 10 p.c.
" F. H., brass, dis. 80 and 10 p.c.
Wood, R. H., " dis. 75 and 10 p.c.
" F.H., bronze, dis. 75 p.c.
" R.H. " 70 p.c.
Drive Screws, 87½ and 10 per cent.
Bench, wood, per doz... 3 25 4 00
" iron, " ... 4 25 5 75
Set, Case hardened, 60 per cent.
Square Cap, 50 and 5 per cent.
Hexagon Cap, 45 per cent.

SCYTHES.

Per doz, net... 9 00

SCYTHE SNATHS.

Canadian, dis. 45 p.c.

SHEARS.

Bailey Cutlery Co., full nickeled, dis. 60 p.c.
Seymour's, dis. 50 and 10 p.c.

SHOVELS AND SPADES.

Canadian, dis. 40 and 5 per cent.

SINKS.

Steel and galvanized, discount 45 per cent.

SNAPS.

Harness, German, dis. 25 p.c.
Lock, Andrews'... 4 50 11 50

SOLDERING IRONS.

1, 1½ lb., per lb... 0 37
2 lb. or over, per lb... 0 34

SQUARES.

Iron, No. 493, per doz... 2 40 2 55
" No. 494, " ... 3 25 3 40
Steel, dis. 60, 10 and 5 p.c.
Try and bevel, dis. 50 to 52½ p.c.

STAMPED WARE.

Plain, dis., 75 and 12½ p.c. off revised list.
Retinned, dis., 75 p.c. off revised list.

STAPLES.

Galvanized... 3 50 4 00
Plain... 3 25 3 75
Coopers', discount 45 per cent.
Poultry netting staples, 40 per cent.

STOCKS AND DIES.

American dis. 25 p.c.

STONE. Per lb.

Washita...	0 28	0 60
Hindostan...	0 06	0 07
" slip,...	0 09	0 09
Labrador...	....	0 13
" Axe...	....	0 15
Turkey...	....	0 50
Arkansas...	0 00	1 50
Water-of-Ayr...	0 00	0 10
Scythe, per gross	3 50	5 00
Grind, 2 in, 40 to 200 lb, per ton	....	25 00
" under 40 lb. "	....	28 00
Grind. under 2 in. thick "	....	29 00

STOVE PIPES.

5 and 6 inch Per 100 lengths 7 00
7 inch " " 7 50

ENAMELINE STOVE POLISH.

No. 4—3 dozen in case, net cash $4 80
No. 6—3 dozen in case, " 8 40

TACKS BRADS, ETC.

Per cent
Strawberry box tacks, bulk... 75 & 10
Cheese-box tacks, blued... 80 & 12½
Trunk tacks, black and tinned... 85
Carpet tacks, blued... 80 & 15
" " tinned... 80 & 20
" " (in kegs)... 40
Cut tacks, blued, in dozens only... 80
" ½ weights... 60
Swedes, cut tacks, blued and tinned—
In bulk... 80 & 10
In dozens... 75
Swedes, upholsterers', bulk... 85, 12½ & 12½
" brush, blued & tinned, bulk.. 70
" gimp, blued, tinned and japanned... 75 & 12½
Zinc tacks... 35
Leather carpet tacks... 55
Copper tacks... 50
Copper nails... 52½
Trunk nails, black... 65 and 5
Trunk nails, tinned... 65 and 10
Clout nails, blued... 65 and 5
Chair nails... 35
Patent brads... 40
Fine finishing... 40
Picture frame points... 10
Lining tacks, in papers... 10

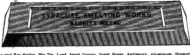

Lining tacks, in bulk		15
" " solid heads, in bulk		75
Saddle nails in papers		10
" " in bulk		15
Tufting buttons, 22 line, in dozens only		60
Tin capped trunk nails		15
Zinc glazier's points		5
Double pointed tacks, papers		90 and 10
" " " bulk		40

TAPE LINES.

English, ass skin, per doz	2 75	5 00
English, Patent Leather	5 50	9 75
Chesterman's each	0 90	2 85
" steel, each	0 80	8 00

THERMOMETERS.

Tin case and dairy, dis. 75 to 75 and 10 p.c.

TRAPS. (Steel.)

Game, Newhouse, dis. 25 p.c.
Game, H. & N., P. S. & W., 65 p.c.
Game, steel, 72½, 75 p.c.

TROWELS.

Disston's discount 10 per cent.

German, per doz	4 75	6 00

S. & D., discount 35 per cent.

TWINES.

Bag, Russian, per lb	....	0 27
Wrapping, cotton, 3-ply		0 20
" " 4-ply	....	0 26
Mattress, per lb	0 33	0 45
Staging, "	0 27	0 35

VISES.

Hand, pe doz	4 00	6 00
Bench, parallel, each	2 00	4 50
Coach, each	6 00	7 00
Peter Wright's, per lb	0 12	0 13
Pipe, each	5 50	9 00
Saw, per doz	6 50	13 00

ENAMELLED WARE.

White, Princess, Turquoise, Blueand White, discount 50 per cent.
Diamond, Famous, Premier, 50 and 10 p.c.
Granite or Pearl, Imperial, Crescent, 50, 10 and 10 per cent.

WIRE.

Brass wire, 50 to 50 and 2½ per cent. off the list.

Copper wire, 45 and 10 per cent. net cash 30 days, f.o.b. factory.

Smooth Steel Wire, is quoted at the following net selling prices:

No. 6 to 8 gauge	$2 90
" 9 "	2 80
" 10 "	2 87
" 11 "	2 90
" 12 "	2 95
" 13 "	3 15
" 14 "	3 35
" 15 "	3 50
" 16 "	3 65

Other sizes of plain wire outside of Nos. 9, 10, 11, 12 and 13, and other varieties of plain wire remain at $2.80 base with extras as before. The prices for Nos 9. to 13 include the charge of 1 c. for oiling. Extras net per 100 lb.: Coppered wire, 60c.—tinned wire, $2—oiling, 10c.—special hay-bailing wire, 30c. —spring wire, $1—best steel wire, 75c.—bright soft drawn, 15c.—in 50 and 100-lb. bundles net, 10c.—in 25-lb. bundles net 15c.—packed in casks or cases, 15c.—bagging or papering, 10c.

Fine Steel Wire, dis. 17½ per cent. List of extras: In 100-lb. lots: No. 17, $5—No. 18, $5.50—No. 19, $6—No. 20, $6.65—No. 21, $7—No. 22, $7.30—No. 23, $7.65—No. 24, $8—No. 25, $9—No. 26, $9.50—No. 27, $10—No. 28, $11—No. 29, $12—No. 30, $13—No. 31, $14—No. 32, $15 No. 33, $16—No. 34, $17. Extras net—tinned wire, Nos. 17-25, $2—Nos. 26-31 $4—Nos. 32-34, $6. Coppered, 5c.—oiling, 10c.—in 25-lb. bundles, 15c.—in 5 and 10-lb. bundles. 25c.—in 1-lb. hanks, 50c.—in ½-lb. hanks, 75c.—in ¼-lb. hanks, $1—packed in casks or cases, 15c.—bagging or papering, 10c.

Galvanized Wire, per 100 lb.—Nos. 6, 7, 8, $3.50 to $3 85—No. 9, $2.85 to $3.15—No. 10 $3.60 to $3.95—No. 11, $3.70 to $4.10—No 12, $3 to $3.30—No. 13, $3.10 to $3.40—No. 14, $4.10 to $4.50—No. 15, $4.60 to $5.05—No. 16, $4.85 to $5 35. Base sizes, Nos. 6 to 9, $2.57½ f.o.b. Cleveland.

Clothes Line Wire, solid 7 strand, No. 17, $4.25; No. 18, $2.65; No. 19, $2.35, f.o.b Hamilton, Toronto, Montreal.

WIRE FENCING. F.O.B. Toronto

Galvanized barb	3 15
Galvanized, plain twist	3 15

Galvanized barb, f.o.b. Cleveland, $2.82½ in less than carlots, and $2.70 in carlots.

WIRE CLOTH.

Painted Screen, per 100 sq. ft., net	1 35

WASTE COTTON. per lb.

Colored	4½ to 5
White, according to quality	6½ to 7½

50-lb bale lots shaded.

WRENCHES.

Acme, 35 to 37½ per cent.
Agricultural, 60 p.c.
Coe's Genuine, dis. 20 to 25 p.c.

Towers' Engineer, each	2 00	7 00
" S., per doz	5 80	6 00
G. & K.'s Pipe, per doz	....	3 40
Burrell's Pipe, each	....	3 00
Pocket, per doz	0 25	2 90

WRINGERS.

Leader	per doz.	$30 0	33 00
Royal Canadian	"	26 00	28 00
Royal American	"	26 00	28 00
Sampson	"	....	30 00

Terms 4 months, or 3 p.c. 30 days.

WROUGHT IRON WASHERS.

Canadian make, discount, 40 and 5 per cent

CANADIAN
HARDWARE AND METAL MERCHANT

The Weekly Organ of the Hardware, Metal, Heating, Plumbing and Contracting Trades in Canada.

VOL. XIII. MONTREAL AND TORONTO, JUNE 22, 1901. NO. 25

"TANDEM" ANTI-FRICTION METAL.

"Tandem" Metals are better than any other for their purpose, and are, therefore:

The Most Economical.
The Least Wearing.
The Most Durable.
Friction Preventing.

Resistance Reducing.
Journal Preserving.
Power Increasing.
Lubricant Saving.

A QUALITY
For Heaviest Pressure and Medium Speed or Heavy Pressure and High Speed.

B QUALITY
For Heavy Pressure and Medium Speed or Medium Pressure and High Speed.

C QUALITY
For Medium Pressure and High Speed or Low Pressure and Highest Speed.

Sole Agents:
LAMPLOUGH & McNAUGHTON, 59 St. Sulpice Street, MONTREAL.

THE TANDEM SMELTING SYNDICATE, LIMITED
The largest smelters of Anti-Friction Metals in Europe.
Queen Victoria St., London, E.C.

Corrugated Iron.

Lysaght's "Orb" and "Redcliffe" brands are the standard—good iron, uniform weights, best workmanship, no risk to the buyer.

A. C. LESLIE & CO., Managers Canadian Branch, MONTREAL.

JOHN LYSAGHT, LIMITED.

Around the Empire

There is an unbroken chain of

"Safford Radiators"

Girdling the British Empire—Girdling the world, voluminous testimony of the solid worth of the "Safford." In thousands of quiet homes and great buildings, here and over the seas, are the links of this great chain, each a recommendation that you disregard to your cost. The world has experimented for you and the result is "SAFFORD" the world over. There's all the advice you need about Radiators. We will be pleased to give you any information you desire. Send for free booklet about heating. Remember we are the largest Radiator Manufacturers under the British Flag—that's the recommendation the British Empire gives the "Safford."

THE DOMINION RADIATOR COMPANY, Limited,
—Toronto.

BUTLER'S

FAMOUS

Sheffield Cutlery.

Fish and Dessert Knives; Spoons and Forks; Cabinets and Cases of Cutlery and Plate.

"BUTLER" was registered as a Trade Mark, A.D. 1768.

Sole Makers of the celebrated "KEEN" Razors, "CAVENDISH" brand of Table Knives and Carvers.

HIGHEST AWARDS. SPECIAL MENTION.

—Full Line of Samples and stock at—

GEORGE BUTLER & CO.'S

LONDON SHOWROOM: 62 HOLBORN VIADUCT, E.C. (Over Snow Hill Station.)

MANUFACTORY: Trinity Works, SHEFFIELD, ENG.

HOSE...

WATER	SUCTION
STEAM	ACID
AIR	OIL
FIRE	SODA WATER
BABCOCK	HIGH-PRESSURE

Our **Patent Seamless Tube** is, without doubt, the only perfect construction.

The Canadian Rubber Co.,

CAPITAL - - - $1,500,000 00.

Montreal. Toronto. Winnipeg.

Lightning, Gem Blizzard . . .

FREEZERS

ARE

- Well Advertised
- In Demand
- Easily Sold
- Satisfactory in Use
- Of Known Reputation

HAVE

Cedar Pails with Electric Welded Wire Hoops
Cans of Heavy Tin with Drawn Steel Bottoms
AUTOMATIC Twin Scrapers
"The Ice Cream Freezer Book" tells all about these and our other Freezers, mailed free

EXCEL IN

- Easy Running
- Quick Freezing
- Economy
- Convenience
- Practical Results

North Bros. Mfg. Co., Philadelphia, Pa., U.S.A.

HENRY DISSTON & SONS (Incorporated)

PHILADELPHIA Pa., U.S.A.

TRADE MARK

Disston's No. 4 Back Saw.

Handles for Hand Saws.

Disston's No. 4 Duplex Saw.

Hand Saw Handle Screws

No 5 No. 4.

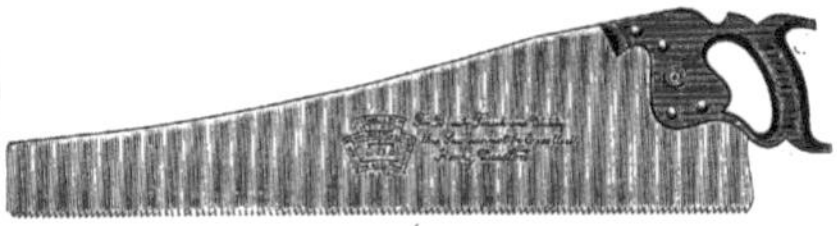

Disston's Hand Saw No. D 8.

No. 1. No. 2.

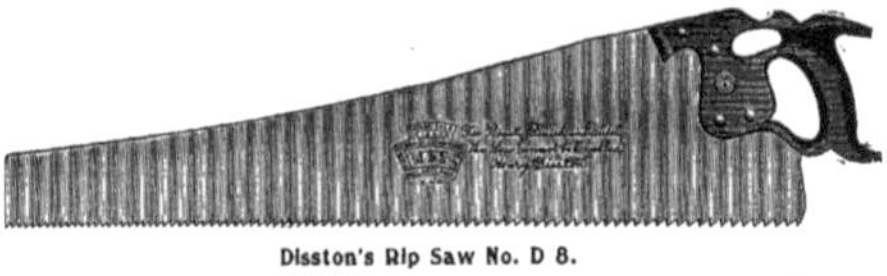

Disston's Rip Saw No. D 8.

Disston's Special Narrow X Cut Saw FOR CANADIAN LUMBER TRADE.

The New St. Lawrence X Cut Saw.

Lewis Bros. & Co., Agents,
MONTREAL.

Made by Henry Disston & Sons,
PHILADELPHIA, Pa., U.S.A. (INCORPORATED)

WRITE FOR PRICES.

MAIL ORDERS SHIPPED SAME DAY AS RECEIVED.

Henry Rogers, Sons & Co.

Wolverhampton, England.

Manufacturers of

"Union Jack" Galvanized Sheets
Canada and Tin Plates
Black Sheets
Sleigh Shoes and Tyre Steel
Coil Chain, Hoop Iron
Sheet and Pig Lead
Sheet Zinc

Quotations can be had from Canadian Office:
6 St. Sacrament St., - MONTREAL

F. A. YORK, Manager.

Wright's Insect Sprayers
Plain Tin—Lacquered—all Brass.
"BEST ON EARTH."
Manufactured by
E. T. WRIGHT & CO.,
HAMILTON, ONT.

KNOX HENRY

Heavy Hardware and Metal Broker
Room 32, Canada Life Bldg., MONTREAL.

"SECCOTINE"
FOR STICKING EVERYTHING.

Samples sent free on application.
HORSE NAILS – "C" Brand Horse - Nails
Canada Horse Nail Co.
"BRASSITE" GOODS – Gunn Castor Co.
Limited, Birmingham, Eng.

McCaskill, Dougall & Co
Fine Varnish & Japan
Manufacturers,
Montreal
Price Lists on application

"A WORD TO THE WISE IS SUFFICIENT."

Gold Dollars at 90c. each won't sell until you have convinced the customer that they are genuine. Have you the time to make the argument to do the convincing?

ALABASTINE is the standard wall-coating "**The World Over.**" Would any dealer want to spend the time to convince a customer that a substitute, or something else, is "just as good" as **CHURCH'S**

ALABASTINE?

Besides, substitution is suspicious, and often bears earmarks of a swindle.

We do not claim to know all that can be learned about the manufacture of wall coatings, but twenty-five years of experience must count for something, and the fact that **ALABASTINE** is the goods that are in demand speaks volumes in its favor.

Never until **M. B. Church** introduced **ALABASTINE** was it found practicable to make wall-coating adapted to be mixed with cold water.

We own the patents on the cold water process, and this statement is our only argument at this time on that point.

ALABASTINE is healthful and beautiful, has stood the test of time; is manufactured in Canada, and recommended by leading sanitarians throughout the country. There is nothing "just as good." Buy only the genuine.

Manufactured by

THE ALABASTINE CO., LIMITED, Paris, Ont.

We also manufacture **"BUG FINISH,"** the only practical Potato Bug Killer. Farmers will soon be calling for it. Order early and be prepared.

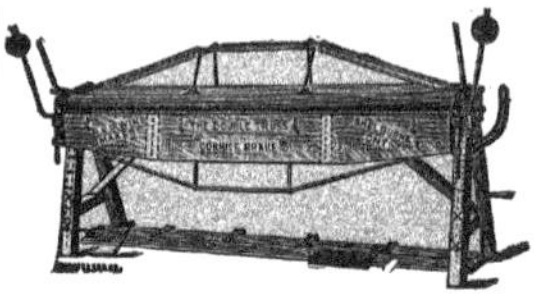

This eight-foot Brake bends 22-gauge iron and lighter, straight and true.

Price, $60

Very handy bender attachment, $15 extra if required.
Send for circulars and testimonials to

The Double Truss Cornice Brake Co. SHELBURNE, ONT.

The Latest and Best.

H. & R. Automatic Ejecting Single Gun.

Model 1900.

Steel and Twist Barrels in 30 and 32-inch.
12 Gauge.

Harrington & Richardson Arms Co.
Worcester, Mass., U.S.A.
Descriptive Catalogue on request.

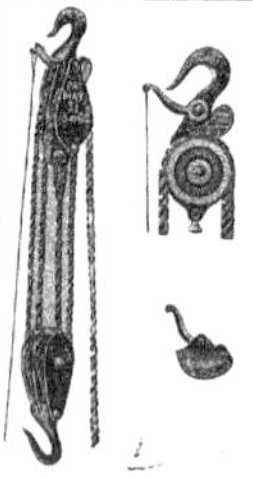

LONDON FENCE MACHINES

Lead on Every Point.

London Safety Tackle Blocks are equally efficient for stretching Coiled Spring Wire and for use as a Hoisting Block. They are A1 and rapid sellers.

TOWNSEND (Lever) STRETCHERS
BERNARD CUTTING PLIERS

Only one agency for our machines in each town. Get our prices, terms and discounts.

Coiled Spring and other Fence Wire at right prices to the trade.

London Fence Machine Co., London, Can.

HARDWARE AND METAL

VOL. XIII. MONTREAL AND TORONTO, JUNE 22, 1901. NO. 25.

President,
JOHN BAYNE MacLEAN,
Montreal.

THE MacLEAN PUBLISHING CO.
Limited.

Publishers of Trade Newspapers which circulate in the Provinces of British Columbia, North-West Territories, Manitoba, Ontario, Quebec, Nova Scotia, New Brunswick, P.E. Island and Newfoundland.

OFFICES

MONTREAL - - - - - - - 232 McGill Street, Telephone 1255.
TORONTO - - - - - - - 10 Front Street East, Telephone 2148,
LONDON, ENG. - - - - 109 Fleet Street, E.C.. W. H. Miln.
MANCHESTER, ENG. - - - 18 St Ann Street. H. S. Ashburner,
WINNIPEG - - - - Western Canada Block. J. J. Roberts.
ST. JOHN, N. B. - - - - No. 3 Market Wharf, J. Hunter White,
NEW YORK. - - - - - - - 176 E. 88th Street.

Subscription, Canada and the United States, $2.00.
Great Britain and elsewhere - - - 12s.

Published every Saturday.

Cable Address { Adscript, London. Adscript, Canada.

AN IRREGULAR TIN MARKET.

MARKED irregularity has characterized the pig tin market during the past week. In New York, up to the close on Saturday last, there was a net decline of 32½ points in five days, while in London, during the week, there was a net decline of £1 15s. The opening of the market this week witnessed a still sharper decline, for on Monday spot tin declined £2 5s., and in New York, on the same day, spot and June tin dropped 60 points. On Tuesday there were signs of a reaction, for, although New York opened at a decline of 2s. 6d. on spot and 10s. on futures, there was subsequently an advance of 5s. on the former, and of 10s. on the latter. New York, although quiet, closed rather firmer. The better tone which developed on Tuesday has been maintained up to the time of writing.

It is the opinion that the advances of a few weeks ago were largely the result of speculative influence, and were unwarranted by the statistical position. Up to the end of May the supplies for the year, according to European compilation, were 68,163 tons, an increase of 6,676 tons compared with the preceding year, while the deliveries were 65,782 tons, against 67,984 tons, a decrease of 2,202 tons. In other words, the statistical position is worse by nearly 9,000 tons than it was a year ago.

THE ADVANCE IN MATCHES.

The item which appeared in our issue of two weeks ago in regard to the advance in matches was somewhat ambiguous. It was stated that the advance was from 10 to 30c. per case. It should have read 10 to 15c. per case on parlor matches, and 30c. per case on sulphur matches.

It is of very little use wanting to be successful in life unless something definite in life is aimed at and persistently striven after.

Philanthropically-inclined people in England are establishing stores for the poor. It will now be in order to inaugurate a movement to provide customers for poor shopkeepers.

THE DEMAND FOR CUT SHINGLE NAILS.

AS will have been noticed by our market reports lately, a brisk trade has been doing in cut shingle nails. We are told by those in the trade that this is not the result of any impulsive and sudden demand ; it is due to a reaction which has been gradually developing in favor of cut nails for shingling purposes.

It is said that after long experience it has been found that the cut shingle nail is less subject to impairment by rust than the wire shingle nail. In the United States the wire nail manufacturers have been trying to overcome this objection by galvanizing their shingle nails. But there arises another difficulty, for galvanizing, while it increases the wearing property of the nails, also increases their cost, and this enhances their price to such an extent as to put them out of serious competition with the cut article.

Owing to the increasing demand, together with the brief temporary closing down of one of the makers, the supply of cut shingle nails in Canada fell below the demand, but this difficulty no longer exists.

OUTLOOK FOR BUILDERS' SUPPLIES

Judging from the large quantity of coarser building materials, such as nails, building paper, roofing materials, cements, etc., which have been sold this spring, a good trade may be expected this fall throughout the country in finishing lines of builders' hardware.

Hardwaremen who realize this are, no doubt, considering ways and means of taking care of this trade.

THE MISMANAGED TRADE DEPARTMENT.

THE articles which have appeared in "Hardware and Metal" in regard to the Trade and Commerce Department have come in for some criticism from The Free Press, of London, Ont. It agrees with us that the Department is doing little or nothing to advance the trade interests of the Dominion, but it at the same time does not agree with the position which we have taken. It does not blame the head of the Department, Sir Richard Cartwright. "It is the office rather that the man that is inert and superfluous," declares our contemporary. In a word, The Free Press would have the Department abolished. "The Department of Trade and Commerce," it holds, "will ever be a sinecure, or fifth wheel to the Government coach, for the good reason that any furtherance of trade by the Government must be by tariff arrangement, and this belongs to the Finance Department." And again: "The only object to be served by the article under notice is to create the semblance of an outcry against Sir Richard himself, and to favor the Government's suspected desire and intention to dethrone him. . . . . It is well-meaning, but it does not go far enough. It should give some indication of the direction in which a more energetic Minister of Trade and Commerce could make himself important and aggressive to the country's advantage."

The Free Press is to be congratulated upon the magnanimity it shows toward Sir Richard. It blames the office, not the man, yet the office was the creation of its own party while Sir Richard is an exponent of the principles of a party it opposes. It is something we do not see every day in papers devoted to the cause of party politics. But mananimity toward Sir Richard does not make untenable the position "The Canadian Grocer" has taken in regard to the Department over which he presides.

We cannot agree with our contemporary that it is the "office rather than the man that is inert and superfluous."

We believe that The Free Press will readily agree with us that the prosperity of this country depends upon its trade. We have no design on other countries than to get into their markets. This recognized, it is obvious that it is the duty of the Government to provide machinery whereby the expansion of trade may be facilitated. We know our contemporary will not dispute that, for it suggests that the Finance Department take the matter in hand. Here are two points upon which we agree.

Now, then, seeing that trade is so important to the country, and that it is the office of the Government to aid in its development, it follows that the Government should delegate the duty of doing so to the Department which is best adapted therefor. It is on this point that we and our contemporary disagree. It asserts that the Finance Department is the proper portfolio. We, on the other hand, claim that the Trade and Commerce Department is the proper portfolio. And we do not think we shall have much difficulty in proving our case.

If there is one phase in the industrial world that stands out prominently to-day it is the decided tendency towards specialization. The individual who would make the best of his vocation must specialize. It is the same with the manufacturer. It is the same with the merchant.

Does it not, therefore, stand to reason that what is most to be desired and what is most aimed at in the industrial world should also be desired and aimed at by the Government in its co-operative duties of developing the trade of the country. This is, in fact, the principle upon which the Government of the country is carried on. Finance, Marine, Railways and Canals, Inland Revenue, Trade and Commerce, etc., have each their several specified Departments. And yet, The Free Press, because, forsooth, Sir Richard Cartwright has inefficiently administered his Department, would do with it what it would not dream of doing in regard to any other should inefficiency be brought home to it as it has against the Trade and Commerce Department.

"If thy right hand offend thee, cut it off," is what we are enjoined to do in Holy Writ. What The Free Press proposes is a reversal of this order. Instead of having the offending head of the Department cut off, it would have the Department destroyed. The illogical character of the proposition is too obvious to be seriously entertained. The business men of this country certainly do not favor such a proposition. At this very moment, for example, The Canadian Manufacturers' Association is trying to endue the Trade and Commerce Department with new life, and not, as The Free Press would have done, take away the little life that it has.

But, suppose, for example, that the duties now appertaining to the Trade and Commerce Department were delegated to the Finance Department, what can we expect to gain thereby? It is generally acknowledged that the Trade and Commerce Department is doing very little toward fulfilling its office. And if it cannot do what it is specially designed to do, how can we expect another Department, designed for another specific purpose, to succeed where the other has failed? Clearly, we could not expect it. The shoemaker must stick to his last.

The premise of The Free Press that "any furtherance of trade by Government must be by tariff arrangement, and this belongs to the Finance Department," is again scarcely logical. The office of the Trade and Commerce Department is to ascertain the requirements of foreign markets, to facilitate transportation, and to hunt up new avenues of trade. In a word, it should be a sort of bureau of commercial intelligence for the business men of the country.

Now, in regard to the suggestion of The Free Press that "'The Canadian Grocer' should give some indication of the direction in which a more energetic Minister of Trade and Commerce could make himself important and aggressive to the country's advantage."

We have already indicated in previous issues the direction in which this should be done; but it doubtless escaped the eye of our contemporary. We will first take Great Britain, with which we have not forgotten, as our contemporary appears to think we have, the bulk of our trade is done. But, while the bulk of our trade is done with the Mother Country, that bulk, measured by the total imports of Great Britain, is almost insignificant.

Great Britain imports over $2,250,000,000 worth of merchandise annually. As our exports to that country last year were $96,500,000, it is obvious there is a great deal of room for expansion notwithstanding the expansion that has been experienced during the last few years. But what is the Trade and Commerce Department doing in the matter? Very little. For two or three years the need of a commercial agent in London has been strongly felt, and although such organizations as boards of trade and The Canadian Manufacturers' Association have repeatedly urged the appointment of such an official Sir Richard has not yet moved in the matter.

A couple of weeks ago "The Canadian Grocer" announced that the British Government had decided to use Manitoba flour in the British navy. As far as we are aware we have nothing to thank the Trade and Commerce Department for in this particular. And yet it could do a great deal of good in little matters of that kind. Then there is South Africa. We have not had any evidence that the Trade and Commerce Department has lost any sleep over getting Canadian products into that part of the world.

Just now the eyes of the Governments of Great Britain, the United States, Germany and Russia are turned toward China and the East where, for the opening up of new markets the possibilities are greater than any other part of the world. But what is the Canadian Trade and Commerce Department doing? Nothing. Our export trade to China and Japan combined last year was the smallest since 1893, and there has been a gradual decline since 1894. The conditions with regard to the West Indian trade are not much better.

As we have said in previous issues, Sir Richard Cartwright is one of the most able men in the House of Commons. In his knowledge of financial matters he stands without a peer in the House. But he has proved his utter unfitness for the portfolio of Trade and Commerce and it is he that should be removed and not the Department abolished.

TRADE IN COUNTRIES OTHER THAN OUR OWN.

HARDWARE TRADE IN THE UNITED STATES.

Business with the jobbing trade continues active, but naturally with some diminution in volume as the season advances. It is, however, on the whole remarkably good. Manufacturers, as usual at this time of the year, find the demands of the trade relaxing somewhat, giving them an opportunity in most lines to accumulate something of a stock and to get factories into shape. It is noticeable that many improvements or enlargements are contemplated or begun, as there is a felling that business is to continue good, and manufacturers desire to be in a position to get their full share of it. The form labor matters are assuming induces an element of uncertainty. The tone of the market in prices may be described as steady rather than strong, with a confidence that no radical reductions are to be anticipated. Some lines are characterized by a decided firmness on account in some cases of difficulty in obtaining raw material. Export business shows a good deal of fluctuation, according to the special conditions affecting the different classes of goods, some kinds of hardware going out in increased volume and others finding it difficult to hold some markets which within a year or two it has been feasible to enter. On the whole the variety of goods sent abroad is larger than ever and the volume and the general conditions are such as to make the outlook for business promising.—Iron Age, June 13.

BRITISH PIG IRON MARKET.

The pig iron market is keeping comparatively steady on the whole, and the weight of orders offering, although not heavy, is not altogether unsatisfactory, and in many instances makers are getting better prices, buyers finding themselves unable to place their contracts at the very low rates which have occasionally been accepted during the past few weeks. These remarks do not apply to the Cleveland district, however, where the week has been a disappointing one and prices have receded, No. 3, which last week was quoted at 46s. per ton, having dropped to 45s. 6d. The hematite market in this district, however, is very strong and active, and producers are holding their own well, notwithstanding the extra production following upon the recent restarting of several furnaces, and the tendency of prices is decidedly upwards. It is noteworthy that the largest shipment of spiegeleisen ever sent from the Tees was sent last week, amounting to 3,500 tons. In. Barrow, too, hematite is steady, and stores of iron are very low.—Iron and Coal Trades' Review, June 7.

THE BRITISH TINPLATE TRADE.

The tone of the market continues as strong as ever under the ample demand, and prompt shipments are scarce and consequently held for a premium. Makers are all well booked and somewhat indifferent about adding to their heavy engagements over the next quarter. The inquiries continue in full force, and although the volume of business passing is not so large as it was, still sellers are contented to wait and take their chances of better things in the future.

Meanwhile, materials are all in a strong position, without any appearance of giving way, the trend being rather in the other direction. Nothing further has been settled with regard to the wages question, and it is thought that the brighter prospects of the market may probably convert this movement into a "back number."

The demand for light plates has been prominent, while squares and odds have come in for their full share of attention, and charcoals also have gained ground again.—Iron and Coal Trades' Review, June 7.

WIRE NAILS IN THE UNITED STATES.

There is a noticeable falling off in the demand for wire nails throughout eastern portions of the country. In the West requirements appear to be about the same as for some time past. Quotations remain unchanged as follows, f.o.b. Pittsburgh, terms, 60 days, or 2 per cent. discount for cash in 10 days:

To jobbers in carlots, $2.30; to jobbers in less than carlots, $1.35; to retailers in carload lots, $2.40; to retailers in less than carlots, $2.50.—Iron Age, June 13.

NEW YORK METAL MARKETS.

TIN — The London pig tin market took an upward turn this morning and by noon showed an advance of £1 5s. on spot and of 10s. on futures over last night's close. Part of the improvement in spot and all of the gain made by futures was lost during the afternoon and the close was easy, spot quotations standing 7s. 6d. below the highest for the day, but 7s. 6d. above last night's price. Futures were quoted at yesterday's closing figures. The early news of a substantial rise in London appeared to give courage to the bulls in this market, as they began bidding up prices and eventually forced them 65 points above the figures buyers were ready to pay for immediate delivery yesterday. The close was quiet at 28.15c. bid and 28.35c. asked for spot and June. The only forward delivery mentioned was September, for which 27.25c. was bid and 27.40c. asked. The Mackinaw landed five tons at Philadelphia to-day, making the total arrivals at Atlantic ports since June 1, 1,585. The Amsterdam has sailed from Rotterdam for New York with 55 tons. The Mesaba is due here from London with 100 tons.

COPPER — The improvement in London prices made further progress to-day, though trading was moderate in spot and very light in futures. The London market closed at 2s. 6d. above last night's quotations. In the New York market conditions were much the same as for some time past. Little new business was reported, but the steady tone of the market was maintained, the quotations remaining 17c. for Lake Superior and 16 5-8c. for electrolytic and casting.

PIG LEAD — The market remains quiet but steady on the basis of 4.37 1-2c. for lots of 50 tons or more. St. Louis was steady at 4.30 to 4.32 1-2c. London closed 1s. 3d. higher.

SPELTER — Trade continues light and prices are nominal at 3.95 to 4c. The St. Louis market also was quiet, the quotation remaining 3.77 1-2 to 3.80c. In London there was a further decline of 2s. 6d.

IRON AND STEEL — There is reported to be rather more interest shown in pig iron, to the extent that buyers are beginning to inquire for prices on deliveries in the last half of the year, but no actual business of consequence is reported. Steel billets and finished products generally are firm, with no surplus of stocks for immediate delivery. Old iron and steel are dull and nominal, with old steel rails 50c. per ton lower.

TINPLATE — Nothing of fresh interest transpired here, but cables from London reported a further advance of 3d., the f.o.b. Swansea price now being 13s. 6d.—New York Journal of Commerce, June 20.

ADVANCE IN TURPENTINE.

The past week has witnessed what seems to be an upward movement in turpentine. It will be remembered that all last year prices were higher than had been the case for some seasons and that during the two months preceding the opening of this season prices fell steadily until the opening price was 19 1-2c. lower than in March, 1900. A further decline of 2c. was reported at Savannah during April, since which time there has been practically no change there until this week, when it became evident that, with a steady demand and with the supply somewhat below the average, the present basis of prices is low.

The result has been an advance of 1 1-2c. during the week at Savannah and other southern points, while in both Montreal and Toronto prices are 2c. higher in sympathy.

A PUSHING HARDWARE FIRM.

One of the enterprising retail concerns of Manitoba is E. Williams & Co., Stonewall, Man. This company have so extended their business that in addition to a general blacksmithing, hardware and tinsmithing business, they manufacture lime and handle stone, etc. This year they have built a 20 x 40 ft. addition to their store, which makes it now 20 x 84 ft. They report business good in their district and state that on account of the favorable weather for grain prospects are still better.

AN AMUSING INCIDENT.

A man recently came into a hardware store in Woodstock with a most extraordinary "tale of woe." It seems he bought a tin of enamel which was guaranteed to dry "over-night." He applied the enamel to his bath-tub during the Saturday half holiday and, on Sunday, essayed to take his weekly "dip." Alas! the enamel came off and various and sundry parts of his anatomy are covered with a beautiful coat of sky-blue paint with an egg-shell gloss which "sticketh closer than a brother" and defies removal. Seeking a remedy, he applied in despair to the hardware shop and eventually purchased a quart tin of Paint and Varnish Remover, which, according to the label, is warranted to remove anything and everything without raising the grain of the wood!

THE TARIFF ON SHIPS.

A curious question has arisen with reference to the importation, or admission to the Canadian register, of foreign-built ships. Our tariff law imposes a ten per cent. duty on such ships. But it is claimed that a vessel built in the United States can register in England, and then, seeing that it has become a British vessel, enter Canadian waters, and participate in the coasting trade free of any charge. The point is to be tested. If the position taken should turn out to be correct, our tariff is practically a dead letter. Any United States vessel can be transferred to Canada via the British register. But we ought to have the privilege of building and selling to our neighbors if this view prevails.—Mail and Empire.

PRICE GUARANTEES IN THE HARDWARE TRADE.

WE have had more or less experience with guarantees in their different forms. Principal among these are:

Price Guaranteed.

Price Guaranteed against decline.

Price Guaranteed against Manufacturer's own decline.

"Price Guaranteed." —

"Price Guaranteed against decline"—expressed in a few words, but of great significance.

I venture to say that in the vocabulary in common use between manufacturer and jobber, there are no other sentences that can be placed on an order or on a contract that mean more. A stipulated price on such an order or contract often means nothing; it is a secondary consideration, being something nominal, and not the invoicing or the settling price. From

THE MANUFACTURERS' POINT OF VIEW.

this broad form of guaranteeing, as it might be called, has its drawbacks, and these are serious. A jobber will make out his order to a certain maker. He may even specify a preferred make or brand of goods, and with "price guaranteed," or "price guaranteed against decline" stated thereon, he is in a position where he can apply a lower price received by him, perhaps under peculiar conditions. He may have received a quotation from another manufacturer on a more favorable specification, or it may be the result of irresponsible competition.

GOODS OF CHEAPER QUALITY.

or of untried make may be offered, a new Richmond in the manufacturing field (and this is a crop that never fails), or, perhaps, in ignorance of what his goods may cost him may offer an enticing price for a first order. He will go to the jobber, who, of course, has no compunction in beating him down: and the price so made, the manufacturer of already established goods is asked to meet, nor is he always asked. In a number of cases he first learns that he must meet, or is expected to meet, by having a deduction made on the remittance sheet. It can readily be seen that a number of such deductions, if allowed, will

PLAY HAVOC

with the profits of the business. It may not occur to the jobber that should he and his brother jobbers place their orders with such parties, there might be difficulty in getting prompt shipment or standard goods, proper credit not being given the manufacturer, who, in good faith, manufactures promptly, or ships from his stock on hand a standard article. This I maintain, gentlemen, is not legitimate trading.

These are only some of the instances that could be cited. It means the returning of checks, the carrying of unpaid balances, and correspondence on both sides, which

TAKES TIME AND CREATES WORRY.

Now and then a jobber will be of the opinion that he is being discriminated against, and that his is an isolated case: but not so: where he has a few manufacturers from whom he buys a certain line, each manufacturer will number his customers by the thousands, and this returning of checks, or carrying of outstanding balances is, under such circumstances, a matter of almost daily occurrence.

Now, while speaking of these broader forms of guarantees, it is but proper to say that there are some jobbers who will always consider a guarantee of price against the manufacturer's own decline, even if not so specifically stated. But unfortunately these gentlemen are few, and, of course, a jobber should not be blamed for getting all he can.

We now come to

ANOTHER FORM OF GUARANTEE,

which is generally expressed: "Price Guaranteed against Manufacturer's own decline." The words "up to date of shipment" are sometimes added to this form, and this is where it becomes susceptible to various meanings. The manufacturer will contend that his lowest price on or at date of shipment should rule. Some jobbers, however, go further and will ask for the benefit of the lowest price that ruled between the date of the placing of the order and the date of shipment. As a consequence, we find guarantees worded: "Price guaranteed at (instead of up to) date of shipment." A change in prepositions making them different propositions.

Broadly guaranteeing up to date of shipment on a contract or order contemplating, say, six months' delivery is, as can readily be seen, a serious matter to the manufacturer, especially on a fluctuating market. He is then in a position, according to some authorities, where on such contracts all goods are sold at lowest price ruling during the period of contract, while, as a matter of fact, the price at time contract was placed and when contract was completed might be very much higher. The fact is, that generally speaking, this matter of guaranteeing prices

OPENS A WIDE DOOR

and if it were done away with entirely, our trade relations would be smoother, more straightforward, and more business like. The manufacturer of hardware staples only hears of price guarantees when he sells the jobber. In other words, he himself finds it impossible to make his purchases in this manner. In England every order is a contract just as "every tub stands on its own bottom." The British manufacturer would not dream of guaranteeing prices any more than he would allow an order once placed to be cancelled. With our natural resources, improved machinery, and up-to-date facilities, we can outdistance England in the manufacturing field, yet in some points of trading she is still our teacher, and her manufacturers would hold up their hands in holy horror if it was even intimated that they should guarantee their prices. This, it seems to me, is an instance where some of us have been progressive in the wrong direction, having been overcome by the blandishments of our friend, the jobber.

THE BUYER OF RAW MATERIAL.

If the manufacturer of staple goods could on his part purchase his billets, rods or wire, as the case may be, with price guaranteed, it would then be an easy matter to make guaranteeing general. But such is not the case. The buyer of raw material is not privileged to deduct a dollar a ton off the face of an invoice by claiming that "another manufacturer has offered this price," or, "this is made to meet one of your competitors." These phases will appeal to you as being inventions of the jobber. They do not emanate from the manufacturer, but come to him. No, the buyer of raw material is not even allowed to cancel an order on a declining market being held strictly to his contract to take out the last pound, no matter how much the market is off. If you gentlemen, did not have the cancelling privilege you would consider yourselves much abused, and here comes the question, "Do you on your part give your customers in the retail trade the benefit of a guarantee against your brother jobbers?" "What is sauce for the goose is sauce for the gander." When it comes to price guarantees, the manufacturers allowing same have evidently been the geese. Why should you not purchase your goods at a fixed price, and sell at a fixed price? We purchase our raw material, our fuel and our labor, all at fixed prices, and expect to make a fair margin of profit thereon. Just as the price on our raw material, fuel and labor is fixed, so should our

SELLING PRICE BE FIXED,

and this rule should obtain in all trading. Like all questions, however, this has its two sides. On one side you will find arrayed the manufacturers or the sellers; on the other, the jobbers, or the buyers, each looking at it from his own point of view, one considering guaranteeing as altogether obnoxious, and the other as something necessary. We should, therefore, strike a happy medium, and reach such a compromise as will be mutually satisfactory. It seems to me, therefore, that the manufacturer, if he give anything, should not give more than a guarantee that at date of shipment, if his price be lower, the jobber will get the benefit of such lower price, also giving him the privilege of timely cancellation of his order, if such order be for staple goods. If he has been quoted a lower price than the manufacturer is willing to give, he has the privilege of cancellation. More than this should not be given on one side, nor should more be expected on the other, and if questions such as these, which might be called "points of variance" between the manufacturer and jobber, were solved in a mutually satisfactory way, through the medium of our Hardware Jobbers' Associations, and were then observed to the letter, much will be gained in the promotion of harmony in our business of trading one with the other.

TORONTO TRAVELERS' MOONLIGHT

The big topic of interest at the meeting of The Toronto City Travellers' Association on Monday evening was the annual moonlight excursion of the Association, which is to be held on Wednesday evening, July 3.

The Excursion Committee, Messrs. W. Anderson, Chairman; W. F. Daniels, Secretary; J. M. Wright, C. H. Wilson, James Scott and Jerry Burns, reported that they had secured the steamer Chippewa, the 48th Highlanders' Band and Glionna's orchestra, and that in every way the excursion this year should be fully up to the standard of those of past years. A big crowd is expected. The double tickets will be 75c.; the single tickets, 50c.

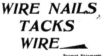

BUSINESS CHANGES.

DIFFICULTIES, ASSIGNMENTS, COMPROMISES.

A MEETING of the creditors of Brown Bros. (W. G. H. Brown, proprietor), dealers in tinware and fancy goods, Campbellford, Ont., has been held.

J. Daly, general merchant, Strathcona, N.W.T., has assigned.

J. B. Douville & Co., general merchants, St. Stanislas, Que., have compromised at 35 cents on the dollar.

PARTNERSHIPS FORMED AND DISSOLVED.

Lamothe & Lemay, sawmillers, St. Sylvere, Que., have dissolved.

Theoret & Frere have registered partnership as grocers, in Montreal.

W. A. Peckham, bicycle dealer, Aylmer, Ont., has admitted E. Millard.

SALES MADE AND PENDING.

Wm. Orr, general merchant, Breckin, Ont., is advertising his business for sale.

O. B. Hulin, dealer in lime, etc., Chatham, Ont., is advertising his business for sale.

The stock of Thompson & Lahey, general merchants, Penetanguishene, Ont., is advertized for sale by auction.

F. Perreault & Cie. general merchants, Rimouski, Que., have sold their stock at 42 cents on the dollar.

The stock, etc., of the estate of J. J. Bosse, general merchant, Rosthern, N.W.T., is advertised for sale by tender on the 21st inst.

The stock of the estate of Samuel Bricker, hardware and coal dealer, Listowel, Ont., has been sold at 47 1-2c. on the dollar to Grant & Co.

CHANGES.

T. S. Hamilton, painter, etc., Winnipeg, has been sold out by bailiff.

Howard P. Jones & Co., have registered as general merchants, at Sabrevois, Que.

S. B. Mitchell, general merchant, Grand View, Man., has sold out to W. J. Swain.

Moses Landry, general merchant, Mariapolis, Man., has sold out to C. Landry.

Jaynes & Co., general merchants, Alvinston, Ont., have sold out to A. S. Harkness & Son.

MacKay & Brooks, hardware dealers, Indian Head, N.W.T., have been succeeded by F. L. McKay & Co.

Charles LeBreton, general merchant, Tracadie, N.B., has sold his general store to Holdengraber & Rosenberg.

J. B. Johnson and J. G. Thomson have registered as proprietors of The Acme Bicycle Agency, Sydney, N.S.

Abraham Silverman has registered under the style of The International Iron and Metal Co., Three Rivers, Que.

C. & X. Kennedy, general merchants, Douglastown and Fox River, Que., have disposed of their Fox River branch.

DEATHS.

N. Lemieux, general merchant, Lake Megantic, Que., is dead.

Wm. Brandis, of John White & Co., stove and tinware dealers, Halifax, N.S., is dead.

FIRES.

C. H. Stiver, general merchant, Unionville, Ont., has suffered loss by fire.

Joseph Morrison, harness dealer, Harrow, Ont., has suffered loss by fire; insured.

The Sumner Co., hardware and lumber dealers, Moncton and Bathurst, N.B., have suffered loss by fire in Bathurst, N.B.

OFF FOR EUROPE.

A merry party left Montreal by the Grand Trunk Railway, from Bonaventure station on Tuesday for Portland, en route by the Dominion Line's palatial steamer New England, for Liverpool. Mr. Andrew A. Brown, the Montreal representative of The McClary Manufacturing Company, of London, Ont., with branches at a number of important points, accompanied by Mrs. Brown, were amongst the number, as also were Mrs. David Milne, and her daughter, from Sarnia. Mr. Davis Milne (of MacKenzie, Milne & Company), came as far as Toronto, but press of business forbade a longer journey at present.

Mr. Brown is on a business and pleasure trip combined. He will visit the manufacturing centres of Great Britain to make purchases of raw material, and will, in turn, make a vigorous push to introduce the numerous specialties of The McClary Manufacturing Company, who, it may be remarked, have an extensive exhibit at Glasgow, which has evoked numerous inquiries from the trade in the Old Land. Having unlimited capital, The McClary Manufacturing Company are naturally looking for a wider field, and their Montreal manager, who has by his energy and personal magnetism built up a magnificent business in Eastern Canada for the "Famous" brand of stoves and tinware, is the one chosen to make their wares known in new territory.

TORONTO SILVER-PLATE WON.

The games scheduled in the Silverware Manufacturing League for Saturday were played in Bayside Park. Much interest was attached to them, on account of a tie for first place between the Toronto Silver Plate Company and Standard Silver Plate Company, which game was played at 2 o'clock, resulting in a clean victory for The Toronto S. P. Co., who completely outclassed their opponents, winning with comparative ease, thus taking the lead. Joe Jordan, the clever twirler of The Toronto S. P. Co., was invincible, only three hits being obtained from his delivery. Fred Wilkes did some clever backstop work for the winners, and the general all-round stick work of the winners was very commendable. The score :—

		R.	H.	E.
Toronto S. P. Co	6 0 0 2 2 2 0 0 0	12	11	2
Standard S. P. Co	0 1 2 1 0 0 0 0 0	4	3	8

Batteries—Jordan and Wilkes; Storey, Brown, Beatty. Umpire—Beard.

The second game at 4 o'clock was a contest between Eckardt Casket Company and Roden Bros., resulting in a defeat for the Eckardts, the game being featureless, with one exception, that of Charlie Leandon, the veteran, who scampered about in left garden like a youngster, doing the best work of the day. The score :—

		R.	H.	E.
Eckardts	3 0 1 2 0 0 0 3 0	9	13	8
Rodens	1 4 0 3 0 2 1 3 *	14	8	6

Batteries—Dine, Hussey, Jobin; Johns and Graham. Umpire—Jim Beard.

The standing of the league :

	Won.	Lost.
Toronto S. P. Co	3	0
Standard S. P. Co	2	1
Rodens	1	2
Eckardt Casket Co	0	3

NATIONAL WEIGHTS AND MEASURES.

THE United States Treasury Department has just issued an immensely valuable list of the weights and measures of all the nations of the globe, with their American equivalents. The names on the list, so odd to American ears, the strange and arbitrary weights and measures for which the foreign denominations stand, will interest many besides the Americans engaged in foreign trade.

The list of weights and measures, the names of the countries in which they are used and their American equivalents follow :

Almude (Portugal), 4.422 gallons.
Ardeb (Egypt), 7.6907 bushels.
Arobe (Paraguay), 25 pounds.
Arroba, liquid, (Cuba, Spain, Venezuela), 4.263 gallons.
Arshine (Russia), 28 inches.
Artel (Morocco), 1 12 pounds.
Baril (Argentine Republic and Mexico), 20.0787 gallons.
Barrel (Spain, raisins), 100 pounds.
Berkovets (Russia), 361.12 pounds.
Bongkal (India) 832 grains.
Bouw (Sumatra), 7,096.5 square metres.
Bu (Japan), 0.1 inch.
Caffiso (Malta), 5.4 gallons.
Candy (India, Bombay), 529 pounds.
Cantar (Morocco), 113 pounds.
Catty (China), 1.333⅓ (1⅓) pounds.
Centaro (Central America), 4 2631 gallons.
Chih (China), 14 inches.
Joch (Austria-Hungary), 1.422 acres.
Ken (Japan), 6 feet.
Klafter (Russia), 216 cubic feet.
Koku (Japan), 4.9629 bushels.
Korree (Russia), 3.5 bushels.
Kwan (Japan), 8.28 pounds.
Last (Belgium and Holland), 85.134 pounds.
Last (Germany), 2 metric tons (4,480 pounds).
Last (Prussia), 112 29 bushels.
Last (Russian Poland), 11¾ bushels.
Last (Spain, salt), 4,760 pounds.
Li (China), 2,115 feet.
Load (England, timber), square, 50 cubic feet; unhewn, 40 cubic feet; inch planks, 600 superficial feet.
Manzana (Costa Rica), 1¾ acres.
Marc (Bolivia), 0.507 pound.
Maund (India), 827 pounds.
Mil (Denmark), 4.68 miles.
Milla (Nicaragua and Honduras), 1.1493 miles.
Morgen (Prussia), 0.63 acre.
Oke (Egypt), 2 7225 pounds.
Oke (Hungary and Wallachia), 2.5 pints.
Pic (Egypt), 41¾ inches.
Picul (Borneo and Celebes), 135.64 pounds.
Pie (Argentine Republic), 0.9478 foot.
Pie (Spain), 0.91407 foot.
Pik (Turkey), 27 9 inches.
Pood (Russia), 36.112 pounds.
Quarter (Great Britain), 8.252 bushels.
Quarter (London, coal), 36 bushels.
Quintal (Argentine Republic), 101.42 pounds.
Sun (Japan), 1.193 inches.
Tael (Cochin China), 590.75 grains. Troy.
Tan (Japan), 0.25 acre.
To (Japan), 2 pecks.
Tonde, cereals (Denmark), 2.84783 bushels.
Tondeland (Denmark), 1.36 acres.
Tsubo (Japan), 6 feet square.
Tsun (China), 1.41 inches.
Tunna (Sweden), 4.5 bushels.
Tunnland (Sweden), 0.22 acres.
Vara (Argentine Republic), 34.1208 inches.
Vedro (Russia), 2.707 gallons.
Vergees (Isle of Jersey), 71.1 square rods.
Vlocka (Russian Poland), 41.98 acres.

HAVE ADOPTED THE METRIC SYSTEM.

The Wadsworth, Howland Co., of Chicago, have adopted the Metric System of liquid measure and hereafter the products will be labelled accordingly. Wadsworth, Howland & Co. are the first people doing business in Canada to adopt this system, which is a legal measure in the Dominion and has the support of the Dominion Government. There are efforts now being made to adopt the system throughout the world.

H. S. HOWLAND, SONS & CO.

WHOLESALE ONLY — 37-39 Front Street West, **Toronto.** — ONLY WHOLESALE

FRUIT and LARD PRESSES.

No. 2, Two Quarts.

A well made and handsome press for making Jellies, Wines, Syrups, etc.

LEMON JUICE EXTRACTORS.

All Glass, Inch Handle.

ICE PICKS.

6-inch long, Polished Handle.

"UNIVERSAL" FOOD CHOPPERS.

Nos. 1, 2, 3.

STEEL ICE TONGS.

12, 16, 24-inch.

ICE CHISEL.

5-inch, Polished Handle.

IRONING WAX.

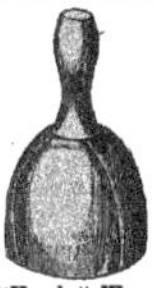

"Handy" Wax.

SWINGS.

Lawn Swing.

FRUIT, WINE and JELLY PRESSES.

"Enterprise" Fruit Presses.

CHERRY STONER.

"Enterprise" Stoner.

all no H. S. Howland Sons Co 27/6/01 see 3 page 41

H. S. HOWLAND, SONS & CO., Toronto.

WE SHIP PROMPTLY — **Graham Wire and Cut Nails are the Best.** — OUR PRICES ARE RIGHT

MARKETS AND MARKET NOTES

QUEBEC MARKETS

Montreal, June 21, 1901.

HARDWARE.

MOST of the wholesale houses continue to give gloomy reports of business. Some lines are fairly active, such as railway supplies and paints and paris green, but, on the whole, the hardware business cannot be said to be what it should be. Even the most optimistic business men confess to that this week. What reason there is for the quiet seems difficult to determine. Manufacturers of wire are beginning to catch up with their orders this week, but it will be some time yet before there will be much barb wire offering. Nails are still scarce, and 3-in. wire nails are by no means plentiful. The demand for lawn requisites is only fair, as is also the call for harvest tools. Cordage is moving in fair-sized quantities. In metals the English market again shows advance, while the scarcity of metals on spot is accentuated. Payments are only fair.

BARB WIRE—The situation remains much about the same as last week. The mills are now catching up with orders, but they are not yet able to ship all that is required. The selling price is $3.05 per 100 lb. f.o.b. Montreal.

GALVANIZED WIRE—There is no new feature to report. We quote: No. 5, $4.25; Nos. 6, 7 and 8 gauge, $3.55; No. 9, $3.10; No. 10, $3.75; No. 11, $3.85; No. 12, $3.25; No. 13, $3.35; No. 14, $4.25; No. 15, $4.75; No. 16, $5.

SMOOTH STEEL WIRE—There has been quite a brisk demand for smooth wire, while the other varieties have been scarce, and the inquiries are still fairly numerous. We quote oiled and annealed as follows: No. 9, $2.80; No. 10, $2.87; No. 11, $2.90; No. 12, $2.95; No. 13, $3.15 per 100 lb. f.o.b. Montreal, Toronto, Hamilton, London, St. John and Halifax.

FINE STEEL WIRE—Moderate amounts are selling at 17½ per cent. off the list.

BRASS AND COPPER WIRE—The market is featureless. The discount on brass is 55 and 2½ per cent., and on copper 50 and 2½ per cent.

FENCE STAPLES—There are not as many orders coming in as there were. We quote: $3.25 for bright, and $3.75 for galvanized, per keg of 100 lb.

WIRE NAILS—There is still a fairly good demand for wire nails, and certain sizes are none too plentiful. We quote: $2.85 for small lots and $2.77½ for carlots, f.o.b. Montreal, London, Toronto, Hamilton and Gananoque.

CUT NAILS—Trade is rather dull in cut nails, only shingle nails moving in quantities. We quote as follows: $2.35 for small and $2.25 for carlots; flour barrel nails, 25 per cent. discount; coopers' nails, 30 per cent. discount.

HORSE NAILS—At present business is quiet, although the better grades are selling in moderate quantities. "C" brand is held at a discount of 50 and 7½ per cent. off the new list. "M" brand is

quoted at 60 per cent. off old list on oval and city head and 66⅔ per cent. off countersunk head. Monarch's discount is 66⅔ per cent., and 70 per cent. in 25 box lots.

HORSESHOES — There is no change to report. We quote as follows: Iron shoes, light and medium pattern, No. 2 and larger, $3.50; No. 1 and smaller, $3.75; snow shoes, No. 2 and larger, $3.75; No. 1 and smaller, $4.00; X L steel shoes, all sizes, 1 to 5, No. 2 and larger, $3.60; No. 1 and smaller, $3.85; feather-weight, all sizes, $4.85; toe weight steel shoes, all sizes, $5.95 f.o.b. Montreal; f.o.b. Hamilton, London and Guelph, 10c. extra.

POULTRY NETTING—The market is unchanged from last week. Goods are scarce but the demand is limited. We quote 50 and 10 per cent. off list A and 50 and 5 per cent. off lists B, C and D.

GREEN WIRE CLOTH—A fair business is passing at $1.35.

SCREEN DOORS AND WINDOWS — Fair amounts are selling, and the demand shows some improvement. We quote as follows: Screen doors, plain cherry finish, $7.30 per doz.; do. fancy, $11.50 per doz.; walnut, $7.30 per doz., and yellow, $7 45; windows, $2.25 to $3.50 per doz.

SCREWS—The market is firm and dealers are ordering freely. Discounts are: Flat head bright, 87½ and 10 per cent. off list; round head bright, 82½ and 10 per cent.; flat head brass, 80 and 10 per cent.; round head brass, 75 and 10 per cent.

BOLTS—The manufacturers throughout the country are good buyers of tire, carriage and stove bolts. Discounts are as follows: Norway carriage bolts, 65 per cent.; common, 60 per cent.; machine bolts, 60 per cent.; coach screws, 70 per cent.; sleigh shoe bolts, 72½ per cent.; blank bolts, 70 per cent.; bolt ends, 62½ per cent.; plough bolts, 60 per cent.; tire bolts, 67½ per cent.; stove bolts, 67½ per cent. To any retailer an extra discount of 5 per cent. is allowed. Nuts, square, 4c. per lb. off list; hexagon nuts, 4¼c. per lb. off list. To all retailers an extra discount of ¼c. per lb. is allowed.

BUILDING PAPER—The ordinary demand is reported. Prices are unchanged. We quote as follows: Tarred felt, $1.70 per 100 lb.; 2-ply, ready roofing, 80c. per roll; 3-ply, $1.05 per roll; carpet felt, $2.25 per 100 lb.; dry sheathing, 30c. per roll; tar sheathing, 40c. per roll; dry fibre, 50c. per oll tarred fibre, 60c. per roll; O.K. and I X.L., 65c. per roll; heavy straw sheathing, $28 per ton; slaters' felt, 50c. per roll.

RIVETS AND BURRS — Trade is moderate. Discounts on best iron rivets, section, carriage, and wagon box, black

rivets, tinned do., coopers' rivets and tinned swedes rivets, 60 and 10 per cent.; swedes iron burrs are quoted at 55 per cent. off; copper rivets, 35 and 5 per cent. off; and coppered iron rivets and burrs, in 5-lb. carton boxes, are quoted at 60 and 10 per cent. off list.

BINDER TWINE—A fairly good trade has been done in binder twine during the week. We quote: Blue Ribbon, 11½c.; Red Cap, 9¾c.; Tiger, 8¾c.; Golden Crown, 8c.; Sisal, 8¾c.

CORDAGE—Some transactions involving fair quantities have been made this week, and altogether cordage is rather active. Manila is worth 13½c. per lb. for 7 16 and larger; sisal is selling at 10c., and lath-yarn, 10c.

HARVEST TOOLS—A moderate business is being done, with the demand not what it should be. The discount is 50, 10 and 5 per cent.

SPADES AND SHOVELS—Inquiries are few. The discount is still 40 and 5 per cent.

LAWN MOWERS—A number of mowers have been sold this week, but business is not exceedingly brisk. We quote: High wheel, 50 and 5 per cent. f.o.b. Montreal; low wheel, in all sizes, $2.75 each net; high wheel, 11-inch, 30 per cent. off.

FIREBRICKS—A small local trade is being done at former prices. We quote: Scotch at $17.50 to $22 and English at $17 to $21 per 1,000 ex wharf.

CEMENT—The importation of foreign cement continues to be limited. We quote: German cement, $2.35 to $2.50; English, $2.25 to $2.35; Belgian, $1.70 to $1.95 per bbl. ex wharf, and American, $2.30 to $2.45, ex cars.

METALS.

The metal market maintains a firm tone, and advances are still reported in England. Stocks here are light and goods are not arriving as expected. Canada plates, 60-sheet, are particularly scarce, as are also terne plates.

PIG IRON—Just now there is not much business being done, as buyers are well filled up. No. 1 Canadian is worth about $18.50 per ton on the track and Summerlee, $20.50 to $21 ex wharf.

BAR IRON—The market is steady, though firm, at $1.75 for merchants' bar and $2 for horseshoe.

BLACK SHEETS—The English market is reported steady to firm. Stocks on the Montreal market are light. We quote: 8 to 16 gauge, $2.50 to $2 60; 26 gauge, $2.55 to $2 65, and 28 gauge, $2.60 to $2 70.

GALVANIZED IRON—Prices are well maintained at the advance quoted last week. We quote: No. 28 Queen's Head, $4.50; Apollo, 10¾ oz., $4.50, and Comet, $4 30, with a 10c. reduction for case lots.

COPPER—The copper market is decidedly firm at 18 to 18½c.

INGOT TIN—Lamb and Flag is from 34 to 35c. The market is strong.

LEAD—The demand for lead is reported rather brisk at $3.70 to $3.80.

LEAD PIPE—A fair business is reported. We quote: 7c. for ordinary and 7½c. for composition waste, with 30 per cent. off.

IRON PIPE—The market is decidedly firm, and the demand is active. We quote as follows: Black pipe, ¼, $3 per 100 ft.; ⅜, $3; ½, $3 05; ¾, $3.30; 1-in., $4.75; 1¼, $6.45; 1½, $7.75; 2-in. $10.35. Galvanized, ½, $4.60; ¾, $5.25; 1 in., $7.50; 1¼, $9.80; 1½, $11.75; 2-in., $16.

TINPLATES—Coke plates are again higher this week, an advance of 3d. to 13s. 6d. being the last news. Dealers here are firm in their ideas, particularly as shipments are not coming to hand as was expected. We quote: Coke plates, $3.75 to $4; char-coal, $4.25 to $4.50; extra quality, $5 to $5.10.

CANADA PLATE—Although Canada plates are quite scarce, dealers have not advanced quotations perceptibly. 60's are hardly to obtained. We quote: 52's, $2.55

to $2.60; 60's, $2.65 to $2.70; 75's, $2.70 to $2.80; full polished, $3.10, and galvanized, $4.

STEEL—Unchanged. We quote: Sleigh-shoe, $1.95; tire, $2; bar, $1.95; spring, $2.75; machinery, $2 75, and toe-calk, $2.50.

SHEET STEEL—We quote: Nos. 22 and 24, $3. and Nos. 18 and 20, $2.85.

TOOL STEEL—Black Diamond, 8c. and Jessop's, 13c.

TERNE PLATES—There is a fair demand for terne plates and supplies are rather light. In a jobbing way the price is $7 50.

COIL CHAIN—The demand at present is slow and prices are unchanged We quote as follows: No. 6, 11½c.; No. 5, 10c.; No. 4, 9½c.; No. 3, 9c.; ¼-inch, 7½c. per lb.; 5-16, $4.85; 5-16 exact, $5.30; ⅜, $4.40; 7-16, $4.20; ½, $3.95; 9-16, $3.85; ⅝, $3.55; ¾, $3.45; ⅞, $3.40; 1-in., $3.35. In carload lots an allowance of 10c. is made.

SHEET ZINC—Although sheet zinc has advanced £1 per ton, and is scarce in this market, holders continue to sell at low figures. Some large lots have changed hands this week at $5.25. The regular jobbing quotation should be $6 to $6 25, while, as a general rule, it is $5 75.

ANTIMONY—Quiet, at 10c.

ZINC SPELTER—Is worth 5c.

SOLDER—We quote: Bar solder, 18½c.; wire solder, 20c.

GLASS.

There is no change to report. The market is rather quiet. We quote as follows: First break, $2.10; second, $2.20 for 50 feet; first break, 100 feet, $3.90; second, $4.10; third, $4.60; fourth, $4.85; fifth, $5.35; sixth, $5.85, and seventh, $6.35.

PAINTS AND OILS.

Turpentine has advanced 2c. per gal. owing to stiffening in the Southern markets. It is now 55c. in single barrels. A change has come over the spirit of the paris green dream. As we mentioned some weeks ago, stocks have been very much attenuated, and a rapid demand has sprung up from all parts of the Dominion. The holders of any fair amount of stocks are seriously thinking of advancing their prices, as it is not at all likely that any more will be made this year. Linseed oil is still not over-plentiful. It is going into consumption the moment it arrives either from England or the Canadian crushers. The air is full of business, and good business at that. White lead is in fair demand on a steady market and dry colors are being called for, while decorators' and painters' supplies are in an exceedingly healthy condition. We quote:

WHITE LEAD—Best brands, Government standard, $6.25; No. 1, $5 87½; No. 2, $5.50; No. 3, $5.12½, and No. 4, $4.75 all f.o.b. Montreal. Terms, 3 per cent. cash or four months.

DRY WHITE LEAD—$5.25 in casks; kegs, $5.50.

RED LEAD—Casks, $5.00; in kegs, $5.25.

DRY WHITE ZINC—Pure, dry, 6¼c.; No. 1, 5¼c.; in oil, pure, 7¼c.; No. 1, 6¼c.; No. 2, 5¼c.

PUTTY—We quote: Bulk, in barrels, $1.90 per 100 lb.; bulk, in less quantity, $2 05; bladders, in barrels, $2.10; bladders, in 100 or 200 lb. kegs or boxes, $2.25; in tins, $2.55 to $2.65; in less than 100-lb. lots, $3 f.o.b. Montreal, Ottawa, Toronto, Hamilton, London and Guelph. Maritime Provinces 10c. higher, f.o.b. St. John and Halifax.

LINSEED OIL—Raw, 80c.; boiled, 83c. in 5 to 9 bbls., 1c. less, 10 to 20 bbl. lots, open, net cash, plus 2c. for 4 months. Delivered anywhere in Ontario between Montreal and Oshawa at 2c. per gal. advance and freight allowed.

TURPENTINE—Single bbls., 55c.; 2 to 4 bbls., 54c.; 5 bbls. and over, open terms, the same terms as linseed oil.

MIXED PAINTS—$1.20 to $1.45 per gal.

CASTOR OIL—8¾ to 9¼c. in wholesale lots, and ½c. additional for small lots.

SEAL OIL—47½ to 49c.

COD OIL—32½ to 35c.

NAVAL STORES—We quote: Resins, $2.75 to $4 50, as to brand; coal tar, $3 25 to $3.75; cotton waste, 4½ to 5½c. for colored, and 6 to 7½c. for white; oakum, 5½ to 6½c., and cotton oakum, 10 to 11c.

PARIS GREEN—Petroleum barrels, 16¼c. per lb.; arsenic kegs, 17c.; 50 and 100-lb. drums, 17½c.; 25-lb. drums, 18c.; 1-lb. packages, 18½c.; ½-lb. packages, 20½c.; 1-lb. tins, 19½c.; ½-lb. tins, 21½c. f.o.b. Montreal; terms 3 per cent. 30 days, or four months from date of delivery.

SCRAP METALS.

The scrap metal market is rather active under a brisk demand. Prices are unchanged. One of the features is a keen demand for old rubbers. Dealers are now paying the following prices in the country: Heavy copper and wire, 13 to 13½c. per lb.; light copper, 12c.; heavy brass, 12c.; heavy yellow, 8½ to 9c.; light brass, 6½ to 7c.; lead, 2½ to 2¾c. per lb.; zinc, 2¼ to 2½c.; iron, No. 1 wrought, $14 to $16 per gross ton f.o.c. Montreal; No. 5 cast, $13 to $14; stove plate, $8 to $9; light iron, No. 2, $4 a ton; malleable and steel, $4; rags, country, 60 to 70c. per 100 lb.; old rubbers, 7¼c. per lb.

HIDES.

Green hides are steady at an advance of ½c., and dealers are now paying 7c. for No. 1 light, with round lots selling to tanners at 7½ to 8c. The American markets are also firmer. We quote: Light hides, 7c. for No. 1; 6c. for No. 2, and 5c. for No. 3. Lambskins, 15c.; sheepskins, 90c. to $1; calfskins, 10c. for No. 1 and 8c. for No. 2.

ONTARIO MARKETS.

TORONTO, June 21, 1901.

HARDWARE.

ALTHOUGH the activity is not as pronounced, generally speaking, as it was a week ago, a good business is still being done in the wholesale hardware trade. The business, however, is largely of a sorting-up character. If anything, business on British Columbian account is rather better than it was, while local business is, perhaps, not as good. A good many screen doors and windows and harvest tools are going out. Quite a good trade is still being done in wire nails, although some of the makers report that new orders are beginning to fall off a little. The demand is still good for cut shingle nails. In screws, bolts and rivets and burrs, a good movement is still being experienced. Building paper continues in good demand. Business is moderate in tinware and enamelled ware. Oil stoves are in active demand, but gas stoves are only in moderate request. A good business is still being done in eavetrough. Horseshoes and horse nails continue quiet. Business is keeping up well in rope.

The crop conditions in both Ontario and Manitoba are more than usually promising. A business man who has been travelling in Ontario for 30 years says he never saw vegetation in such a uniformly healthy condition.

BARB WIRE—The local barb wire market, as far as stocks are concerned, is again in a fairly normal condition. Shipments have been more freely received, and the wholesale trade have not only been able to fill back orders, but now have a little barb wire in stock. The demand continues fairly good and prices steady and unchanged. We quote $3 05 per 100 lb. from stock Toronto; and $2 82½ f.o.b. Cleveland for less than carlots, and $2.70 for carlots.

GALVANIZED WIRE—Quite the same remarks apply to this line as to barb wire, the shipments which have been received having enabled the trade to fill back orders and keep a little on hand for stock. We quote: Nos. 6, 7 and 8, $3.50 to $3.85 per 100 lb., according to quantity; No. 9, $2.85 to $3.15; No. 10, $3.60 to $3.95; No. 11, $3.70 to $4 10; No. 12, $3 to $3.30; No. 13, $3.10 to $3.40; No. 14, $4.10 to $4.50; No. 15, $4.60 to $5.05: No. 16, $4.85 to $5.35. Nos. 6 to 9 base f.o.b. Cleveland are quoted at $2.57½ in less than carlots and 12c. less for carlots of 15 tons.

SMOOTH STEEL WIRE—Business is keeping up well in oiled and annealed wire. Very little is yet being done in hay-baling wire, but, of course, the season is early to expect much business in this line. Net selling prices for oiled and annealed are: Nos. 6 to 8, $2.90; 9, $2.80; 10, $2.87; 11, $2.90; 12, $2.95; 13, $3 15; 14, $3.37; 15, $3.50; 16, $3.65. Delivery points, Toronto, Hamilton, London and Montreal, with freights equalized on those points.

WIRE NAILS — Although some of the makers report that new business is not as brisk as it was, a good trade is still being done, and orders are being filled none too promptly. We quote as follows: $2 85 base for less than carlots, and $2.72½ for carlots. Delivery points: Toronto, Hamilton, London, Gananoque and Montreal.

CUT NAILS — Business has improved a little, but it is more largely confined to shingle nails. Very few cut nails of other descriptions are going out. The base price is $2.35 per keg for less than carlots, and $2.25 for carlots. Delivery points: Toronto, Hamilton, London, Montreal and St. John, N.B.

HORSENAILS—Business in this line is confined to a few sorting-up orders. Discount on "C" brand, oval head, 50 and 7½ per cent. off new list, and on "M" and other brands, 50, 10 and 5 per cent. off the old list. Countersunk head 60 per cent.

HORSESHOES —There is only a small sorting-up trade being done in this line also. We quote f.o.b. Toronto as follows: Iron shoes, No. 2 and larger, light, medium and heavy, $3.60; snow shoes, $3.85; light steel shoes, $3.70; featherweight (all sizes), $4.95; iron shoes, No. 1 and smaller, light, medium and heavy (all sizes), $3.85; snow shoes, $4; light steel shoes, $3.95; feather weight (all sizes), $4.95.

SCREWS—Business continues fairly good and prices steady and unchanged. Discounts are as follows: Flat head bright, 87½ and 10 per cent.; round head bright, 82½ and 10 per cent.; flat head brass, 80 and 10 per cent.; round head brass, 75 and 10 per cent.; round head bronze, 65 per cent., and flat head bronze at 70 per cent.

RIVETS AND BURRS—Business in this line continues fairly good. We quote: Iron rivets, 60 and 10 per cent.; iron burrs, 55 per cent.; copper rivets and burrs, 25 and 5 per cent.

BOLTS AND NUTS—An active business is still the feature of the bolt, nut and washer trade. In some sizes of bolts it is difficult to get a sufficient quantity to satisfy the demand. We quote: Carriage bolts (Norway), full square, 65 per cent.; carriage bolts full square, 65 per cent.; common carriage bolts, all sizes, 60 per cent.; machine bolts, all sizes, 60 per cent.; coach screws, 70 per cent.; sleighshoe bolts, 72½

per cent.; blank bolts, 60 per cent.; bolt ends, 62½ per cent.; nuts, square, 4c. off; nuts, hexagon, 4½c. off; tire bolts, 67½ per cent.; stove bolts, 67½; plough bolts, 60 per cent.; stove rods, 6 to 8c.

ROPE—Business continues good in both sisal and manila rope, and trade continues steady in cotton rope. The base price is unchanged at 10c. for sisal and 13½c. for manila.

BINDER TWINE—Business is still confined to a sorting-up demand. We quote: Pure manila, 650 ft., 12c.; manila, 600 ft., 9½c.; mixed, 550 ft., 8½c.; mixed, 500 ft., 8 to 8½c.

CUTLERY—The cutlery trade is still quiet and without any particularly notable features.

SPORTING GOODS—Business in this line continues to show a slight improvement. During the past week inquiries for guns and ammunition have been more frequent than for some time.

ENAMELLED WARE AND TINWARE—The demand for both these lines is rather better than it was, although it is still only fair.

GAS AND OIL STOVES—There is not a great deal being done in gas stoves, but quite a number of oil stoves are going out.

COOKING AND HEATING STOVES—The manufacturers report that they are booking a good many orders for cooking and heating stoves for later delivery.

ICE CREAM FREEZERS—Some business is still being done, but the season is naturally drawing to a close, as far as the wholesale trade is concerned. The season, taking it all the way round, has been an exceedingly good one in the ice cream freezer trade.

GREEN WIRE CLOTH—Wholesale jobbers report that trade is just fair in this line. We still quote $1.35 per 100 sq. ft.

SCREEN DOORS AND WINDOWS—Shipments in this line continue brisk. We quote: Screen doors, 4 in. styles, $7.20 to $7.80 per doz.; ditto, 3-in. styles, 20c. per doz. less; screen windows, $1.60 to $3.60 per doz., according to size and extension.

BUILDING PAPER—A large business is being done this week in building paper. We quote: Building paper, 30c.; tarred lining, 40c., and tarred roofing, $1.65.

POULTRY NETTING—There is not much doing as the season's trade is about over. Discount 55 per cent.

HARVEST TOOLS—Trade conditions are satisfactory, and the demand is particularly heavy for scythes and snaths. Discount 50, 10 and 5 per cent.

SPADES AND SHOVELS—A nice sorting-up trade is reported in spades and shovels. Discount 40 and 5 per cent.

EAVETROUGH—Business is keeping up well in eavetrough. The low price at which this article is selling this season is evidently inducing dealers to buy the manufacturers' article instead of making it up themselves. The ruling price is $3.25 per 100 ft. for 10 inch.

LEATHER BELTING—A few large contracts have been let recently, but, as a general rule, trade is quiet in this line. Prices are being cut a great deal, but the ruling discounts are as follows: Extra, 60, 10 and 5 per cent.; standard, 70 per cent.; No. 1, 70 and 10 per cent.

CEMENT—There is an excellent demand and prices are firm throughout. We quote barrel lots: Canadian portland, $2.25 to $2.75; German, $3 to $3.15; English, $3; Belgian, $2.50 to $2.75; Canadian hydraulic, $1.25 to $1.50.

METALS.

A feature of the local metal market is the lowness in stocks of tinplates, tinned sheets, and galvanized sheets. Business is fairly good. Pig tin has been very irregular since our last issue, but has again taken a steadier turn during the last few days. There is also a little better feeling in regard to pig iron. Bar iron is steady and in fair demand. Iron pipe has been advanced 5 per cent. by the jobbers in sympathy with a similar advance by the manufacturers.

PIG IRON—While nothing new has developed on the Canadian market, as far as we can learn, there is a somewhat better feeling in the United States on account of the heavy purchases which have been made of Bessemer pig. In foundry iron, however, business does not appear to have shown any appreciable improvement. Canadian iron on track Toronto we quote at $18 per ton for No. 1, $17.50 for No. 2, and $17 for No. 3.

BAR IRON—The demand on new business account is about moderate, but the mills are still decidedly busy filling orders taken some time ago. The ruling base price is still $1.85 to $1.90.

STEEL—There is not a great deal of new business being transacted, but the mills are all busily employed on orders booked some time ago. Prices steady and unchanged. We quote: Merchantable cast steel, 9 to 15c. per lb.; drill steel 8 to 10c. per lb.; "B C" and "Black Diamond" tool steel, 10 to 11c.; Jessop's, Morton's and Firth's tool steel, 12½ to 13c.; toe calk steel, $2.85 to $3; tire steel, $2.30 to $2.50; seighshoe steel, $2.10 to $2.25; reeled slachinery steel, $3; hoop steel, $3.

GALVANIZED SHEETS—Stocks of galvanized sheets are still much reduced, and a great deal of difficulty is being experienced in filling orders. A fairly good business is being done on importation account. The ruling quotation on 28 gauge English is $4.50, and on American $4.40.

BLACK SHEETS—The market continues firm as to price, but there is not a great deal of business being done. We quote: 28 gauge, common sheets, at $3, and dead flat at $3.50.

CANADA PLATES — Prices continue firm, and some wholesalers are holding their import figures a little higher than they did a month ago. Very few shipments are being made from stock, but a few orders continue to be booked for import, although the bulk of the orders on this account were placed some time ago. We quote: All dull, $2.90; half polished, $3, and all bright, $3.50.

PIG TIN—The pig tin market has shown a marked irregularity during the past week, and some sharp breaks in prices took place in both London and New York. A couple of days ago, however, the market took a more favorable turn, and at the time of writing prices are firm. Locally, quotations are slightly lower than they were a week ago, and we quote 31½ to 32c. There is not a great deal of business being done.

TINPLATES —Firmness is still the feature of this market, and within the last few days there has been a further advance of 10d. in the British market. Business here is fair, but stocks are decidedly low, and wholesalers report that they will be placed in a rather uncomfortable position shortly unless they are able to get deliveries more promptly than they have of late.

TINNED SHEETS—There has been quite a little demand during the past week, and stocks in this line are also being reduced. It is the opinion that business is rather better than is usual at this time of the year. We quote 28 gauge at 8½c.

COPPER — Business is moderate. We quote: Ingot, 17¾c., and bars, 23 to 25c. In both New York and London prices have advanced during the past week.

SOLDER—A reasonable quantity of solder has been moving during the past week. Prices rule as before. We quote as follows: Half-and-half, guaranteed, 18½c.; ditto, commercial, 18c.; refined, 18c., and wiping, 17c.

IRON PIPE—In sympathy with the advance in manufacturers' prices noted last week, the local jobbing trade has advanced its prices 5 per cent. One inch black is now quoted at $5.40, and one inch galvanized at $7.95 per 100 ft.

LEAD —Trade is quiet at 4¼ to 4½c.

SPELTER—Business is just moderate at 5½ to 6c. per lb.

ZINC SHEETS—There is a little doing, but business is confined to part casks. We quote 6½c. for casks, and 6¾c. for part casks.

PAINTS AND OILS.

The demand has moderated since the end of May and now there is only a fair sorting trade doing. The movement of paris green is large, and the feeling is decidedly firm as bugs are reported numerous, and the frequent rains necessitate frequent spraying. The sale of both oil and turpentine is light, and it is believed that some houses are cutting prices on oil to force sales. The primary markets in both oil and turpentine are decidedly stiff, however, and the general desire seems to be to maintain local prices firmly. Turpentine is 2c. higher. We quote:

WHITE LEAD—Ex Toronto, pure white lead, $6.37½; No. 1, $6; No. 2, $5.67½; No. 3, $5.25; No. 4, $4.87½; genuine dry white lead in casks, $5.37½.

RED LEAD—Genuine, in casks of 560 lb., $5.50; ditto, in kegs of 100 lb., $5.75; No. 1, in casks of 560 lb., $5; ditto kegs of 100 lb., $5.25.

LITHARGE—Genuine, 7 to 7½c.

ORANGE MINERAL—Genuine, 8 to 8½c.

WHITE ZINC—Genuine, French V.M., in casks, $7 to $7.25; Lehigh, in casks, $6.

PARIS WHITE—90c. to $1 per 100 lb.

WHITING — 70c. per 100 lb.; Gilders' whiting, 80c.

GUM SHELLAC — In cases, 22c.; in less than cases, 25c.

PARIS GREEN—Bbls., 16¾c.; kegs, 17c.; 50 and 100 lb. drums, 17¼c.; 25-lb. drums, 18c.; 1-lb. papers, 18½c.; 1-lb. tins, 19½c.; ½ lb. papers, 20½c.; ½-lb. tins, 21½c.

PUTTY — Bladders, in bbls., $2.10; bladders, in 100 lb. kegs, $2.25; bulk in bbls., $1.90; bulk, less than bbls. and up to 100 lb., $2.05; bladders, bulk or tins, less than 100 lb., $2.90.

PLASTER PARIS—New Brunswick, $1.90 per bbl.

PUMICE STONE — Powdered, $2.50 per cwt. in bbls., and 4 to 5c. per lb. in less quantity; lump, 10c. in small lots, and 8c. in bbls.

LIQUID PAINTS—Pure, $1.20 to $1.30 per gal.

CASTOR OIL—East India, in cases, 10 to 10½c. per lb. and 10½ to 11c. for single tins.

LINSEED OIL—Raw, 1 to 4 barrels, 82c.; boiled, 85c.; 5 to 9 barrels, raw, 81c.; boiled, 84c., delivered. To Toronto, Hamilton, Guelph and London, 1c. less.

TURPENTINE—Single barrels, 56c.; 2 to 4 barrels, 55c., delivered. Toronto, Hamilton and London 1c. less. For less quantities than barrels, 5c. per gallon extra will be added, and for 5-gallon packages, 50c., and 10 gallon packages, 80c. will be charged.

GLASS.

The stocks of some lines are decidedly small, but as some stocks have been received from Belgium, and further stocks are due to arrive during the next few weeks, it is likely that prices will not be changed. We quote as follows: Under 26 in., $4.15; 26 to 40 in., $4.45; 41 to 50 in., $4.85; 51 to 60 in., $5.15; 61 to 70 in., $5.50; double diamond, under 26 in., $6; 26 to 40 in., $6.65; 41 to 50 in., $7.50; 51 to 60 in., $8.50; 61 to 70 in., $9.50, Toronto, Hamilton and London. Terms, 4 months or 3 per cent. 30 days.

OLD MATERIAL.

There is little demand, and jobbers have lowered their prices 5 to 10c. per cwt. on scrap iron, ½c. per lb. on copper, and ¼c. per lb. on lead. We quote jobbers' prices as follows: Agricultural scrap, 50c. per cwt.; machinery cast, 50c. per cwt.; stove cast, 45c.; No. 1 wrought 40c. per 100 lb.; new light scrap copper, 12c. per lb.; bottoms, 11c.; heavy copper, 12½c.; coil wire scrap, 12½c.; light brass, 7c.; heavy yellow brass, 10c.; heavy red brass, 10½c.; scrap lead, 2¼c.; zinc, 2c.; scrap rubber, 6½c.; good country mixed rags, 65 to 75c.; clean dry bones, 40 to 50c. per 100 lb.

HIDES, SKINS AND WOOL.

HIDES—There is a fair trade. Prices are unchanged. We quote: Cowhides, No. 1, 6½c.; No. 2, 5½c.; No. 3, 4½c. Steer hides are worth 1c. more. Cured hides are quoted at 7 to 7½c.

SKINS—There is a good demand with prices unchanged throughout. We quote:

No. 1 veal, 8-lb. and up, 9c. per lb.; No. 2, 8c.; dekins, from 60 to 70c.; culls, 20 to 25c; sheepskins, 90c. to $1.

WOOL—The low prices have caused farmers to hold their wool, but as dealers have large stocks, and the market is not strong, there seems to be no disposition to offer higher prices. We quote: Combing fleece, washed, 13c., and unwashed, 8c.

PETROLEUM.

The movement is light. Prices are unchanged. We quote: Pratt's Astral, 16 to 16½c. in bulk (barrels, $1 extra); American water white, 16½ to 17c. in barrels; Photogene, 15½ to 16c.; Sarnia water white, 15 to 15½c. in barrels; Sarnia prime white, 14 to 14½c. in barrels.

COAL.

There is a good demand. Prices are unchanged. We quote at international bridges as follows: Grate, $4.75 per gross ton; egg, stove and nut, $5 per gross ton with a rebate of 30c. off for June shipments.

MARKET NOTES.

Local stocks of tinplates, turned sheets, and galvanized sheets are getting much reduced.

Iron pipe is quoted 5 per cent. higher by local jobbers. One-inch black pipe is now $5.40 and one-inch galvanized, $7.95.

A NEW MANGLE.

A. R. Woodyatt & Co. Guelph, Ont., have just placed in the hands of the jobbers a new 3-roller mangle. It is called the "Victor," is a handsome machine, and is much appreciated by the trade.

PERSONAL MENTION.

Mr. W. A. Campbell and family, of The Wadsworth, Howland Co., of Chicago, were in Toronto this week on their way to their summer island on the St. Lawrence river.

DEVELOPMENT OF THE HAMMER.

Tracing the development of the hammer, The International Monthly says: "Man's first tool was the uplifted hand grasping a stone, and from this came, after many years, the hammer. As heavier blows became necessary the hammer grew in size, until it was operated by machinery in the form of the tilt or helve hammer. When steam succeeded water as a motive power, a steam cylinder replaced the tripping cam, but the first half of the past century had nearly expired before the original form of this tool was at all changed by James Nasmyth's invention of the upright steam hammer. Since then the falling weight of this design of tool has gradually been increased from a few hundred pounds up to 100 and even 125 tons; but excepting the smaller sizes, up to 25 tons, it has since 1890 been superseded by the hydraulic press, which by its own slow motion produces a more thorough working of the metal. Presses have grown until the capacity of 14,000 tons was reached in 1893, requiring a 15,000 horse-power engine to drive it. Such a tool, with its accompaniment of 200-ton electric cranes for handling the work underneath, is capable of forging ingots over 75 inches in diameter and weighing more than 250,000 pounds. This whole plant, costing over $250,000, was not projected without an adequate understanding that it was to meet the commercial demands of many years to come, and industrial developments, great as they have been, have not as yet called for anything that has tasked its full capacity."

HEATING AND PLUMBING

THE PLUMBERS' CONVENTION PROGRAMME.

THE arrangements for the annual convention of The National Association of Plumbers and Steamfitters of Canada, which have been about completed by the Executive Committee, show that the delegates have reason to expect an exceedingly enjoyable as well as profitable time.

The first session of the convention is called for 9.30 a.m. on Wednesday, when the executive will meet to receive and consider reports and map out the business for the succeeding sessions. Two other sessions on Wednesday, at 2 p.m. and 8 p.m. respectively, will be strictly business meetings. At the first Thursday session at 9.30 a.m., business arising out of the officers' and committees' reports will be considered. Thursday afternoon will be devoted to a drive around the city, during which the principal points of interest will be visited. It is expected that a group photo of the delegates will be taken at Queen's Park.

The annual banquet will be held at McConkey's on Thursday evening. The list of speakers has not yet been made up, but it is intended to invite several prominent representatives of the manufacturing industries.

On Friday morning at 9.30 o'clock the closing business session of the convention will be held. At this meeting officers for 1901-1902, and the next place of meeting, will be chosen and all business not previously settled will be completed. Friday afternoon will be devoted to a trolley ride around the city, and eventually going by The Mimico and Lake Shore Electric Road to Long Branch, where a supper and a good programme of music will be provided.

Delegates desiring to visit Buffalo and the Pan-American Exposition can do so at July 1 rates, leaving Toronto Saturday and returning Monday or Tuesday.

BUILDING PERMITS ISSUED.

The following permits have been issued in Ottawa: Andrew Kerr, for a $1,000 house, Division street; Public School Board, $3,000 school house, Wellington street; W. N. Edge, tenement, Bell street, $1,200; Margaret Brennan, residence and hotel, Clarence street, $1,500; T. W. Alexander, dwelling, Daly avenue, $2,500; L. J. Brown, dwelling, Tackaberry avenue, $1,000; Richard Hare, three dwellings, Maria street, $5,000.

Building permits have been issued in Montreal for the following: Carsley's extension, $70,000; Silverman-Boulter, warehouse on St. Paul street, $20,000; Canada Switch Company, extension to their factory in St. Gabriel Ward, $10,000; Gault Bros., shirt and collar factory, corner Inspector and William streets, $42,000; Mrs. Phillips, widow of the late Harry Phillips, residences on Dorchester street, $8,000; Bellevue Flats, corner of St. Catherine and Metcalfe streets (21 dwellings), $45,000.

Building permits have been issued in Toronto to Wm. Middleton, for a $3,500 residence on McMaster avenue, near Avenue Road; to J. and H. C. Cooper, for a pair of semi-detached houses on Bloor street, near Albany, to cost $8,500; to A. Nicholson, for a $6,000 house at 89 Walmer Road; to M. A. Marshall, for a pair of $2,500 residences, near Bloor on Albany; to R. Bigley, for a $4.500 factory at 3 Mutual street; to The T. Eaton Co., Limited, for stables near James street, on Louisa, to cost $20,000; to T. T. Clark, for a $2,600 factory at 46 Harvard; to J. Agnew, for a $2,500 house at the corner of Jones and Hunter; to Chas. Stark, for two $4,000 houses on Crescent Road, near South Drive, Rosedale.

HOT-WATER FITTERS vs. PLUMBERS

During the discussion on the paper relating to the Plumbers' Registration Bill, read at the last meeting of the Institution of Heating and ventilating Engineers, some members gave their experience of what they considered arbitrary and unfair acts on the part of plumbers with whom they had worked, but probably no one could record such an unique experience as the following: The contract for a large hot-water heating plant was secured and put in hand. In due course the cold service to the boiler had to be run, and then the plumbers threatened trouble. The architect was appealed to, and he pointed out to them that it was an iron pipe and not a lead one, but without avail. The consequence was that a plumber had to be hired. He came, but could not run iron pipe, did not know how to cut it, screw it, or handle it anyhow. The outcome was that the plumber was allowed to have one of the hot-water fitters as a mate, and this extemporized mate did the work while the plumber stood by and handed things to earn his money.—Ironmonger.

PLUMBING AND HEATING CONTRACTS.

Ballantyne & Co., Montreal, have been given the plumbing and heating contract for the Royal Insurance Co.

W. J. McGuire & Co., Montreal, were the successful tenderers for the plumbing and heating of Silverman & Boulter's new warehouse.

Lessard & Harris have been awarded the contract for the roofing of the new Royal Insurance Co., and they will also do the roofing of Silverman & Boulter's new warehouse. Their tender for the plumbing, heating and roofing of a house for Mr. Lavendeau, in Westmount, has recently been accepted.

The John Ritchie Plumbing and Heating Co., Toronto, have contracts for plumbing and heating residences for J. J. Jackson, Oakville, Ont.; F. X. Cousineau, 4 Orde street, Toronto; for steam-heating The Benson House, Oakville; for alterations to the plumbing and heating of the Equitable Chambers, Adelaide street west, and for remodelling the plumbing in Wm. Scott's residence, Church street.

PLUMBING AND HEATING NOTES.

Hugh C. Cann and Wallace E. Wetmore have registered partnership under the style of Cann & Wetmore.

The Chatham, Ont., Electric Light Company, Limited, have sold out.

An electric pump is to be placed in the basement of the Victoria Hospital, London, Ont., to supply the buildings with water. An electric plant will also be installed in the laundry.

WANTED AN EXPLANATION.

Owing to an unfortunate error, The A. Ramsay & Son's advertisement in last week's issue, could hardly be understood by our readers. Two cuts were received at our office and the wrong one was inserted in the advertisement. The reading matter spoke of Ramsay's Barn Paints while the cut used was made to show the great superiority of Ramsay's Exterior Lead. The experiment leading to the making of this cut has been very interesting. A. Ramsay & Son are producing a lead which they believe superior to all others; and have adopted the novel idea of substantiating their contention by a photographic test. Three boards were painted, one with No. 1 white lead, one with absolute pure white lead and one with Ramsay's Exterior. These boards were then exposed to the weather under exactly similar conditions for six months with the result that Ramsay's Exterior lead remained the whitest. The firm has issued a handsome little booklet showing the result of this test in which appears the cut of the boards photographed. This cut appeared without any suitable comment in last week's "ad."

An artist when he wishes to see the strength of light and shade in any object he is sketching, half closes the eyes, to get the full strength of the shadows. So if any one wishes to see the full effect of the superiority of Ramsay's Exterior Lead, they will set up this cut of the three panels about a couple of yards distant and look at it through half-closed eyes. "Any one can see it with half an eye." Or a good way to see the force of the cut is to use the hands as a shade for the eyes in the shape of a telescope, as artists sometimes do. This week appears the rearranged "ad." of Ramsay's Barn Paints.

USING LAMPBLACK.

It sometimes happens that in using lampblack considerable difficulty is experienced in inducing the pigment to mix properly, says Ironmonger. This is generally proof that the black is pure, as in its natural state it is the lightest of all pigments, but when it is adulterated with barytes or with some other black it is, of course, much heavier and mixes with greater ease. If the black used is ground in oil it will mix more readily when thinned with a little turpentine. As a rule, in mixing paint for use, the turpentine must always be added last, but in this case it is permissible to employ a part at first. Lampblack is one of those pigments which require a very large proportion of oil, fully eighteen times as much as does white lead, and four or five times as much as the earth colors, such as yellow ochre, etc.

The Dominion Iron & Steel Company, of Sydney, Cape Breton, have awarded a contract to Thayer & Co., of New York city, for a 5,000 horse-power Cahall boiler equipment, which will be installed in the rolling mill of the Canadian plant.

ALEXANDER GIBB

Manufacturers' Agent and Metal Broker,
13 St. John Street, Montreal.

Representing British and American Manufacturers. Correspondence invited from firms wishing to be represented in Canada.

Will Hold Up a Shelf!

That's what a shelf bracket is for.
For this purpose there can be
NOTHING BETTER
NOTHING CHEAPER
than the ....
BRADLEY STEEL SHELF BRACKET

It is well **Japanned, Strong** and **Light.** The saving in freight is a good profit, aside from the **lower price** at which the goods are sold. Order direct or through your jobber.

ATLAS MFG. CO., New Haven, Conn., U.S.A

LOW TANK WATER CLOSET COMBINATIONS
THE MOST PERFECT ON THE MARKET
NOISELESS IN ACTION
BEAUTIFUL DESIGNS.

Write for Catalogue.
The James Morrison Brass Mfg. Co. Limited
TORONTO, ONT.

FENCE PAINTS BARN PAINTS

In the summer time and between seasons, the farmer thinks of his barns, outhouses, fences, and every wooden thing standing. They require paint! Don't ask him to use our high-grade house paints on rough board barns and fences. He can't afford it, and they don't need it. Now,

RAMSAYS OUTSIDE PAINTS

are made just for rough work on barns, roofs, fences, and such like. They are made of Oxides, Ochres, etc., and heavy Linseed Oil, and there is nothing better in the world than just such ingredients, properly mixed, to withstand the weather and add a double length of years to the farmer's barn and roof. We have shade cards, and there is money in this.

A. Ramsay & Son

PAINTMAKERS, Est'd 1842. MONTREAL.

ONTARIO SILVER CO.,

Limited,
NIAGARA FALLS, CANADA.
Manufacturers of **FLATWARE, CUTLERY and ELECTRO PLATE. . . .**
Ask for our Catalogue and Quotations.

Ontario Nut Works, Paris

BROWN & CO.
Manufacturers of
All sizes of Hot Pressed Nuts, Square and Hexagon.

Good Valves—

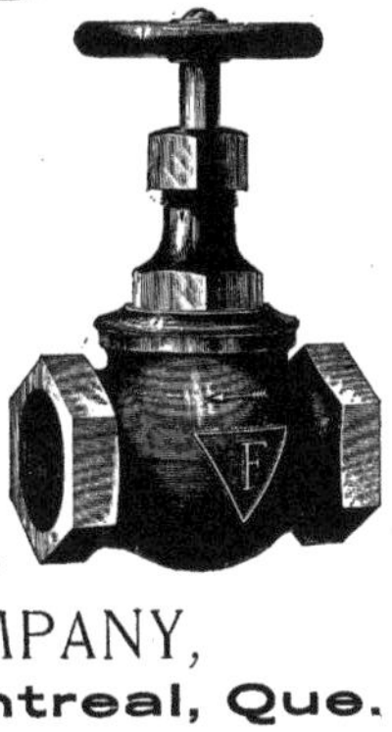

are absolutely essential to the steam-user.

Then it is quite as necessary you should sell an article to fill the bill.

This is what fills it—

The Fairbanks Renewable Disc Valves.

THE FAIRBANKS COMPANY,

749 CRAIG STREET, **Montreal, Que.**

THE SUPERIORITY OF CANADIAN PAINTS.

MR. ROBERT MUNRO, managing-director of The Canada Paint Co., Limited, Montreal, has been in Boston during the past two weeks as a delegate to the Y.M.C.A. Jubilee Convention held in that city. He took advantage of his trip south to visit other American coast cities, among them New York.

Mr. Munro was through quite a number of factories when away, but he claims that none of them compare with the one he manages. According to his ideas, the American manufacturers are not called upon to exert themselves as are the Canadians. They have a large field to cater to and they can easily work up a satisfactory business. They are not called upon to study and scheme as the manufacturers are on this side of the line.

As a result, not only are the factories here better plants, but the formulæ are more scientific and the paints and all finishing materials infinitely better. They do not produce paints that give the gloss of the Canadian paints, at least in those coast cities, and where Canadian paints are used over there, they are easily distinguishable from the dull New York productions. Mr. Munro found the distinction very noticeable in the beautifully finished Casino of The Gorham Silver Plate Co., Providence, R.I., where the white gloss finish of The Canada Paint Co. has been used.

WILL PICNIC AT OSHAWA.

The sixth annual picnic of the employes, office staff and friends of The Gurney Foundry Co., Limited, Toronto, will be held at Prospect Park, Oshawa, on Saturday by steamer Argyle and electric cars. A big programme of sports and games has been arranged for, including a baseball match between the mounters and moulders. The picnics of the Gurney company in past years have always been a success, and the preparations indicate that this year's outing will be right up to the mark.

HEMP FROM THE PHILIPPINES.

The following comparative statement showing the exportation of hemp from the Philippines during the calendar years 1900 and 1899 respectively has been prepared in the Division of Insular Affairs of the United States War Department.

The statement shows an increase in the quantity exported in 1900 of 20,390 tons, or 29.05 per cent., and an increase in the value of $5,296,8:6 over the 1899 figures. The average price per ton for 1899 was $115.77 and for 1900 $148 60. Manila hemp being the leading article of export from the Philippines, it is particularly gratifying to note this increase, which indicates that as peace and order are being restored in the archipelago, agriculture is rapidly returning to its normal condition.

The exports by countries during the respective periods were as follows:

Countries	190[0] Tons.	190[0] Value	1899 Tons.	1899 Value.
United States	20,3[illegible]	$,796,068	26,713	$1,015 7:6
United Kingdom	46,419	7,102,711	21,511	2 574 76)
France	109	1',100	42	3 8 0
Spain	544	116,'54	6(9	53,46)
Canada	10	3 45J	....	....
Egypt	3,1(6	41°,830	....	....
China	13,557	1,930,584	17,6 9	1,974,175
Japan	1,157	215,931	103	22,(93
British East Indies	1,793	2.9,6.3	820	127,310
Australasia	2,631	4 8,8 8	1,485	224,418
Total	89,438	$.3 290,4.0	69,(43	$7,993,574

RUST AND "BUST."

Here is one of the latest salutes to callers which business men are hanging on their walls:

If I Rest
I Rust
If I Trust
I Bust.
Therefore—
No Rest
No Rust
No Trust
No Bust.

It comes in red and blue printing.

A TRAVELER'S DEATH.

The Rat Portage News has the following respecting the recent death of a city traveller: Mr. Wm. Hargraves, the representative of G. F. Stephens & Co., Winnipeg, one of the most popular travellers in Western Canada, died last night in the Hilliard House. The attending physician certified that the immediate cause of death was heart failure. The deceased was last year president of the Western Traveller's Association, was highly respected, and his death cast a gloom over a large circle.

EXPORTING REFRIGERATORS.

Though The Baldwin Refrigerator Co., Burlington, Vt., have an extensive domestic trade, the demand from various foreign countries, including Australia, New South Wales, India, France, England, etc., has become so great that it is now an important part of their business. To secure this export trade has not been an easy matter as the demand from the different countries is quite diversified. The English public, for instance, not only buy larger goods than the American trade usually calls for, but they want good, plain, solidly built goods, and for such goods cheerfully pay a reasonable advance on prices charged in the United States for the same sizes. The Baldwin Company, have, however, adopted their product to the market and have been successful in building up a fine trade. As a result of their doing business on different markets the variety of their output, both of refrigerators and cooling rooms, is particularly large and they are able to meet the wishes of every buyer. Any dealer mentioning "Hardware and Metal" can secure literature describing these refrigerators by writing for them. They are worth having for reference.

ESTABLISHED 1860. INCORPORATED 1895

STOVE PIPE DAMPERS

Steel—"Empire."

Made in

5, 6, 7-Inch.

Prices on Application.

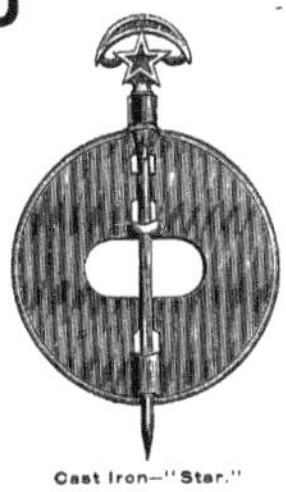

Cast Iron—"Star."

Stove Pipes, Elbows, Stove Boards, Coal Hods, Fire Shovels, Lanterns, etc.

THE THOS. DAVIDSON MFG. CO., Limited, MONTREAL, QUE.

Meat & Food Chopper

No. 5, - $2.00

Rapid Grinding & Pulverizing Mills

No. 2½, - $4.50

Fruit, Wine & Jelly Press

No. 34, - $3.00

"ENTERPRISE"

PATENTED

HARDWARE SPECIALTIES

Comprising

Meat and Food Choppers, *35 sizes aud styles for hand and power;* Rapid Grinding and Pulverizing Mills, *32 sizes and styles for hand and power;* Fruit, Wine and Jelly Presses; Meat Juice Extractors; Cherry Stoners; Raisin Seeders, *for hand and power;* Ice Shredders; Vegetable Slicers; Mrs. Potts' Cold Handle Irons; Sausage Stuffers and Lard Presses, etc., etc.

ARE THE BEST

Write for Descriptive Catalogue — *Sold by the Leading Jobbers of the Dominion*

The Enterprise Mg. Co. of Pa.

Philadelphia, Pa., U. S. A.

Raisin Seeder

No. 36, - .85

Meat Juice Extractor

No. 21, - $2.50

Meat & Food Chopper

No. 12, - $2.50

MANITOBA MARKETS.

WINNIPEG, June 17, 1901.

HARDWARE AND PAINTS, OILS AND GLASS.

THE trade here is feeling the benefit of the generally improved crop conditions and business is good. No change of price in any line is to be recorded, and there is a general absence of "news."

Quotations for the week are as follows:

Barbed wire, 100 lb.	$3 45
Plain twist	3 45
Staples	3 95
Oiled annealed wire 10	3 95
" 11	4 00
" 12	4 05
" 13	4 20
" 14	4 35
" 15	4 45
Wire nails, 30 to 60 dy, keg	3 50
" 16 and 20	3 60
" 10	3 60
" 8	3 70
" 6	3 75
" 4	3 90
" 3	4 15
Cut nails, 30 to 60 dy.	3 00
" 20 to 40	3 05
" 10 to 16	3 10
" 8	3 15
" 6	3 20
" 4	3 30
" 3	3 65
Horsenails, 45 per cent. discount.	
Horseshoes, iron, No. 0 to No 1	4 65
No. 2 and larger	4 40
Snow shoes, No. 0 to No. 1	4 90
No. 2 and larger	4 40
Steel, No. 0 to No. 1	4 95
No. 2 and larger	4 70
Bar iron, $2.50 basis.	
Swedish iron, $5.00 basis.	
Sleigh shoe steel	3 00
Spring steel	3 25
Machinery steel	3 75
Tool steel, Black Diamond, 100 lb	8 50
Jessop	13 00
Sheet iron, black, 10 to 20 gauge, 100 lb.	3 50
20 to 26 gauge	3 75
28 gauge	4 00
Galvanized American, 16 gauge	2 54
18 to 22 gauge	4 50
24 gauge	4 75
26 gauge	5 00
28 gauge	5 25
Genuine Russian, lb	12
Imitation " "	8
Tinned, 24 gauge, 100 lb.	7 75
26 gauge	8 00
28 gauge	8 50
Tinplate, IC charcoal, 20 x 28, box	10 75
" IX "	12 75
" IXX "	14 75
Ingot tin	33
Canada plate, 18 x 21 and 18 x 24	3 25
Sheet zinc, cask lots, 100 lb	7 50
Broken lots	8 00
Pig lead, 100 lb	6 00
Wrought pipe, black up to 2 inch....50 an 10 p.c.	
" Over 2 inch	50 p.c.
Rope, sisal, 7-16 and larger	$11 00
" ⅜	11 50
" ¼ and 5-16	12 00
Manila, 7-16 and larger	14 00
" ⅜	14 50
" ¼ and 5-16	15 00
Solder	20
Cotton Rope, all sizes, lb	17
Axes, chopping	$7 50 to 12 00
" double bitts	12 00 to 18 00
Screws, flat head, iron, bright	87½
Round " "	82½
Flat " brass	80
Round " "	75
Coach	57½ p.c.
Bolts, carriage	55 p.c.
Machine	55 p.c.
Tire	60 p.c
Sleigh shoe	65 p.c.
Plough	40 p.c.
Rivets, iron	50 p.c.
Copper, No. 8	35
Spades and shovels	40 p.c.
Harvest tools	50, and 10 p.c.
Axe handles, turned, s. g. hickory, doz.	$2 50
No. 1	1 50
No. 2	1 25
Octagon extra	1 75
No. 1	1 25
Files common	70, and 10 p.c.
Diamond	60
Ammunition, cartridges, Dominion R.F.	50 p.c.
Dominion, C.F., pistol	30 p.c.
" military	15 p.c.
American R.F	30 p.c.
C.F. pistol	5 p.c.
C.F. military	10 p.c. advance.
Loaded shells:	
Eley's soft, 12 gauge black	16 50
chilled, 12 guage	18 00
soft, 10 guage	21 00
chilled, 10 guage	23 00
Shot, Ordinary, per 100 lb	6 25
Chilled	6 75
Powder, F.F., keg	4 75
F.F.G.	5 00
Tinware, pressed, retinned	75 and 2½ p.c.
" " plain	70 and 15 p.c.
Graniteware, according to quality	50 p.c.

PETROLEUM.

Water white American	25½c.
Prime white American	24c.
Water white Canadian	22c.
Prime white Canadian	21c.

PAINTS, OILS AND GLASS.

Turpentine, pure, in barrels	$ 61
Less than barrel lots	66
Linseed oil, raw	92
Boiled	95
Lubricating oils, Eldorado castor	25½
Eldorado engine	24¾
Atlantic red	27½
Renown engine	41
Black oil	23½ to 25
Cylinder oil (according to grade)	55 to 74
Harness oil	61
Neatsfoot oil	$ 1 00
Steam refined oil	85
Sperm oil	1 50
Castor oil per lb.	11½
Glass, single glass, first break, 16 to 25 united inches	2 25
26 to 40 per 50 ft.	2 50
41 to 50 " 100 ft.	5 50
51 to 60 " " "	6 00
61 to 70 per 100-ft. boxes	6 50
Putty, in bladders, barrel lots per lb.	2¼
kegs "	2½
White lead, pure per cwt.	7 00
No 1 "	6 75
Prepared paints, pure liquid colors, according to shade and color, per gal.	$1.30 to $1.90

NOTES.

The Lac du Bonnet Brick Co. have opened comfortable offices in The Tribune block, and Mr. Crispin, late of Kelly Bros., will act as manager.

A FOUR-DAY ATLANTIC SERVICE.

The first steps for the inauguration of a fast steamship service between Sydney, C.B., and Southampton, Eng., have been taken. H. M. Whitney, of Boston, who is so largely interested in Sydney, has given the contract for three very fast steamships, which will be put on the service as soon as ready. The vessels are intended to make the passage from Southampton to Sydney in four days, and it is claimed that by the new service it will be possible to land passengers in Chicago in the same time as it now takes to get to New York. It is also the intention of the capitalists who are interested with Mr. Whitney to establish an extensive steel shipbuilding plant at Sydney.

BUSINESS GOOD.

Mr. E. G. E. Ffolks, manager of the Wilkinson Plough Co., Toronto, was in Montreal this week on special business, looking exceedingly well. He reports that trade with his firm is exceedingly brisk and that they have orders in sight for ploughs, land-rollers and ensilage cutters that will keep them busy for some time to come.

ORDERS FOR SPECIAL GOODS.

NO wholesaler, however large and assorted his stock may be, keeps everything in his warehouse that appertains to his particular line of business. There are specialties which he must get as they are wanted, sometimes being compelled to get them manufactured. To procure these special goods sometimes requires a good deal of time, and occasionally the customer has refused, upon some ground or other, to accept them. This frequently means that the goods are left on the wholesaler's hands or that they have to be sold at a loss before a customer can be found. Wood, Vallance & Co., the wholesale hardware merchants, Hamilton and Toronto, have been casting about for a remedy, with the result that they have devised a form of agreement which they ask customers to sign when ordering special goods which are not carried in regular stock. This agreement is herewith given. The original form is printed on paper letterhead size.

WOOD, VALLANCE & CO.,
Wholesale Hardware Merchants.

HAMILTON, ONT.,
Canada.

M..

..........................190...

....

We will be pleased to procure for you, as quickly as possible, the special goods which you require. These goods are not regular stock lines with us, and we cannot accept the order unless you will guarantee to accept the goods upon arrival, nor can we be held responsible for delay through causes beyond our control, but we will do all that we can to give you every satisfaction. If you desire us to order the following lines kindly sign and return the attached special order to us.

Yours truly,
WOOD, VALLANCE & CO.

WOOD, VALLANCE & CO.,
Hamilton, Ont. ....190..

Dear Sir:
Kindly procure for us, as quickly as possible, the following goods, which we agree to accept upon arrival:

....

Shipping Instructions..

Yours truly,
....

QUANTITY.			

WOLFVILLE, NOVA SCOTIA.

A booklet which attempts to illustrate or describe the beauty or attractiveness of the Grand Pre, of the "Home of Evangeline," must needs be in itself a work of art and beauty. This standard is reached, however, in the booklet which the Wolfville Board of Trade have gotten up this year to advertise the district, and a copy of which has been sent to "Hardware and Metal" by C. F. Starr & Son. The typography and illustrations are of the highest standard, taken from superb photographs, while the attractiveness of Wolfville and the country adjacent, on account of its historic associations, its romantic interest and its natural beauty, are entrancingly told. Everything considered the booklet is one of the daintiest and most attractive productions of the year in Canada. Persons wishing to enjoy a quiet holiday in the midst of romantic, peaceful surroundings and great natural beauty, or persons desiring to get some pointers in tourist advertising should secure a copy of this booklet by writing to the Wolfville Board of Trade.

SPORTING GOODS FOR HARDWAREMEN.

The J. Stevens Arms & Tool Company, Chicopee Falls, Mass., has recently placed on the market a full line of high-grade golf clubs at popular prices. The company is also notifying the trade that it is prepared to furnish a full line of rifle-cleaning rods—from the plain slotted coppered to the fancy jointed with cocobolo swivel handles—and is making attractive prices for quantity orders.

E. Forrest & Co., general merchants, Ste. Anne de Beaupré, Que., have suffered loss by fire; partially insured.

CURRENT MARKET QUOTATIONS

June 21, 1901.

These prices are for such qualities and quantities as are usually ordered by retail dealers on the usual terms of credit, the lowest figures being for larger quantities and prompt pay. Large cash buyers can frequently make purchases at better prices. The Editor is anxious to be informed at once of any apparent errors in this list as the desire is to make it perfectly accurate.

METALS.

Tin.

Lamb and Flag and Straits—56 and 28 lb. ingots, per lb.	0 31½	0 32

Tinplates.

Charcoal Plates—Bright

		Per box
M.L.S., equal to Bradley.		
I.C., usual sizes	....	$6 50
I.X., "	....	8 00
I.X.X., "	....	9 50
Famous—		
I.C.	....	6 50
I.X	....	8 00
I.X.X	....	9 50
Raven & Vulture Grades—		
I.C., usual sizes	....	4 50
I.X., "	....	5 25
I.X.X. "	....	6 00
I.X.X.X., "	....	6 75
D.C., 12½x17	....	4 40
D.X.	....	4 75
D.X.X	....	5 00

Coke Plates—Bright

Bessemer Steel—		
I.C., usual sizes	....	3 75
I.C., special sizes, base	....	4 00
20 x 28	....	7 75

Charcoal Plates—Terne

Dean or J. G. Grade—		
I.C., 20x28, 112 sheets	....	8 00
I.X., Terne Tin	....	10 (0

Charcoal Tin Boiler Plates.

		Per lb.
Cookley Grade—		
X X., 14x56, 50 sheet bxs } " 14x60, " } " 14x65, " }	....	0 (6½

Tinned Sheets

72x30 up to 24 gauge	....	0 07½
" 26 "	....	0 08
" 28 "	....	0 08½

Iron and Steel.

Common Bar, per 100 lbs.	1 85	1 90
Refined " "	....	2 35
Horse Shoe Iron "	....	2 35
Hoop steel, 1½ to 3 in. base, extras for smaller sizes	....	3 00
Sleigh Shoe Steel " base	...	2 10
Tire Steel	2 3)	2 50
Reeled Machinery	....	3 00
Toe Calk Steel	2 85	3 00
T. Firth & Co's tool steel, per lb	0 12½	0 13
Jessop's tool Steel	0 12½	0 13
Morton's tool steel	0 12½	0 13
Black Diamond and "B.C." tool steel	0 10	0 11
Drill Steel, per lb.	0 (8	0 10

Boiler Tubes.

1½-inch	....	0 12½
2 "	....	0 13
2½ "	....	0 15
3 "	....	0 16
3½ "	....	0 20
4 "	....	0 25

Steel Boiler Plate.

¼ inch	2 50	2 60
3-16 inch	2 60	2 70
⅜ inch and thicker	2 50	2 60

Black Sheets.

	Com.	D.Fl.
18 gauge	2 75	3 00
20 gauge	2 75	3 00
22 to 24 "	2 75	3 25
26 "	2 75	....
28 "	3 00	3 50

Canada Plates.

All dull, 52 sheets	....	2 90
Half polished	....	3 00
All bright	....	3 10

Iron Pipe.

Black pipe—		
⅛ "	....	4 65
¼ inch	....	3 40
⅜ "	....	3 45
½ "	....	3 70
¾ "	....	3 85
1 "	....	5 40
1¼ "	....	7 35
1½ "	....	8 80
2 "	....	11 80
2½ "	....	21 00
3 "	....	2 80
3½ "	....	31 80
4 "	....	40 95
5 "	....	49 70
6 "	....	65 20
Galvanized pipe—		
¼ inch	....	5 15
⅜ "	....	5 50
1 "	....	7 93
1¼ "	....	10 80
1½ "	....	12 95
2 "	....	17 35

Galvanized Sheets.

	G C.	Comet.	Amer.	Queen's Head.
16 gauge		....	4 00	3 75
18 to 24 gauge	4 00	3 85	4 25	4 00
26 "	4 25	4 10	4 25	4 25
28 "	4 50	4 35	4 40	4 50

Case lots 10 to 15c. less.
28 gauge American equals 26 gauge English.

Chain.

Proof Coil, 3-16 in., per 100 lb	....	....
" ¼ " "	6 30	8 50
" 5-16 " "	4 70	5 00
" ⅜ " "	4 05	4 (0
" 7-16 " "	3 9)	4 75
" ½ " "	3 70	4 10
" 9-16 " "	3 65	4 15
" ⅝ " "	3 35	3 90
" ¾ " "	3 60	4 10

Halter, kennel and post chains, 40 to 40 and 5 p.c.

Cow ties	40 p.c.
Tie-out chains	65 p.c.
Stall fixtures	35 p.c.
Trace chain	45 p.c.

Jack chain, iron, single and double, discount 35 p c.

Jack chain, brass, single and double, discount 40 p.c.

Copper.

Ingot

English B. S., ton lots	....	0 17¾
Lake Superior	....	....

Bars.

Cut lengths round, ½ to ⅞ in.	0 23	0 25
" round and square 1 to 2 inches	0 23	0 25

Sheet.

Plain, 14 oz., and light, 16 oz., 14x48 and 14x60	0 24	0 24½
Plain, 14 oz., and light, 16 oz., irregular sizes	0 24½	0 25
Tinned copper sheets	....	0 26
Planished	....	0 32

Braziers (In sheets.)

4x6ft. 25 to 30 lbs. ea., per lb.	....	0 25
" 35 to 45 " "	....	0 24
" 50-lb. and above, "	....	0 23

Boiler and T. K. Pitts

Plain Tinned, per lb	....	0 28
Spun, per lb.	....	0 32

Brass.

Rod and Sheet, 14 to 30 gauge, 15 per cent.

Sheets, hard-rolled, 2x4	....	0 23
Tubing, base, per lb.	....	0 23

Zinc Spelter

Foreign, per lb	0 05½	0 06
Domestic "	....	..

Zinc Sheet.

5 cwt. casks	....	00 6½
Part casks	....	00 6¾

Lead.

Imported Pig, per lb	0 04½	0 04½
Bar, 1 lb.	0 05½	0 06¼
Sheets, 2½ lbs. sq. ft., by	....	0 06¼
Sheets, 3 to 6 lbs., "	....	0 06

NOTE.—Cut sheets ½ cent per lb. extra. Pipe, by the roll, usual weights per yard, lists at 7c. per lb. and 30 p.c. dis. f.o.b. Toronto.

NOTE.—Cut lengths, net price, waste pipe 8-ft. lengths lists at 7½ cents.

Shot.

Common, $6.50 per 100 lb.; chilled, $7.00 per 100 lb.; buck, seal and ball, $7.50. Discount, 17½ p.c. Prices are f o.b. Toronto, Hamilton, Montreal, St. John and Halifax. Terms 3 per cent. cash, freights equalized.

Soil Pipe and Fittings.

Discount, 60 and 10 per cent. on medium and extra heavy, and 60 per cent. on light.

Solder.

	Per lb.	Per lb.
Bar half-and-half, guarant'd	....	0 18¾
Bar half-and-half, commer'l	....	0 18
Refined	....	0 17½
Wiping	....	0 17

NOTE.—Prices of this graded according to quantity. The prices of other qualities of solder in the market indicated by private brands vary according to composition.

Antimony.

Cookson's, per lb.	0 10½	0 11

White Lead.

	Per 100 lb.
Pure	6 37
No. 1 do	6 00
No. 2 do	5 62½
No. 3 do	5 15
No. 4 do	4 87½
Munro's Select Flake White	7 37½
Elephant and Decorators' Pure	7 12½
Brandram's B B. Genuine	8 00
" " Decorative	7 55
" " No. 1	6 85
" " No. 2	6 00

Red Lead.

Genuine, 560 lb. casks, per cwt.	....	$5 50
Genuine, 100 lb. kegs, per cwt.	....	5 75
No. 1, 560 lb. casks, per cwt.	....	5 25
No. 1, 100 lb. kegs, per cwt.	....	5 50

White Zinc Paint.

Elephant Snow White	0 08	0 09
Pure White Zinc	0 08	0 (9
No. 1	0 06	0 07½
No. 2	0 05	0 (6½

Dry White Lead.

Pure, casks	5 75
Pure, kegs	6 25
No. 1, casks	5 50
No. 1, kegs	5 00

Prepared Paints.

In ¼, ½ and 1 gallon tins.

Pure, per gallon	....	1
Second qualities, per gallon	....	1
Barn (in bbls.)	0 75	0
The Sherwin-Williams Paints.	....	1
Canada Paint Co's Pure	....	1
Toronto Lead & Color Co's Pure	....	1
Sanderson Pearcy's Pure	....	1
Stewart & Wood's Champion Pure.		1 25

Colors in Oil.

25 lb. tins, Standard Quality.

Venetian Red, per lb.	0 05
Chrome Yellow	0 11
Golden Ochre	0 06
French "	0 05
Marine Black	0 09
" Green	0 09
Chrome "	0 08
French Imperial Green	0 09

Colors, Dry.

Yellow Ochre (J.C.) bbls.	1 35	1
Yellow Ochre (J.F.L.S.), bbls	....	2
Yellow Ochre (Royal)	1 10	1
Brussels Ochre	....	2
Venetian Red (best), per cwt.	1 80	1
English Oxides, per cwt.	3 00	3
American Oxides, per cwt.	1 75	2
Canadian Oxides, per cwt.	1 75	2
Super Magnetic Oxides, 93p c.	2 00	2
Burnt Sienna, pure, per lb.	....	0
" Umber, " "	....	0
do Raw	....	0
Drop Black, pure	....	0
Chrome Yellows, pure	....	0
Chrome Greens, pure, per lb.	....	0 10
Golden Ochre	....	0 03½
Ultramarine Blue in 28-lb. boxes, per lb.	0 08	0
Fire Proof Mineral, per 100 lb	....	1
Genuine Eng. Litharge, per lb	....	0
Mortar Color, per 100 lb.	....	1
English Vermillion	....	0
Pure Indian Red, No. 45 lb.	....	0
Whiting, per 100 lb.	....	0

Blue Stone.

Casks, for spraying, per lb.	0 07
100-lb. lots, do. per lb.	0 08

Putty.

Bulk in bbls.	1 9)
Bulk in less quantity	2 05
Bladders in bbls.	2 10
Bladders in kegs, boxes or loose.	2 25
Bladders in 25-lb. tins	2 35
Bladders in 12½-lb. tins	2 65
Bladders in bulk or tins less than 100 lb.	2 90

Varnishes.

	In 5-gal. lots.).	Per gal.
Carriage, No. 1	2 90	3 30
" body	8 00	9 00
" rubbing	4 00	5 00
Gold Size, Japan	3 00	3 40
Brown Japan	2 40	2 80
Elastic Oak	2 90	3 30
Furniture, extra	2 40	2 80
" No. 1	1 60	2 00
Hard Oil Finish	2 70	3 10
Light Oil Finish	3 20	3 60
Demar	3 3)	3 70
Shellac, white	4 40	4 80
" orange	4 00	4 40
Furniture Brown Japan	1 60	2 00
Black Japan	2 40	2 8)
" " No. 1	1 60	2 00

The Imperial Varnish & Color Co's., Limited Elastilite Varnish 1 gal. can, each. $3.00.

Granatine Floor Finish per gal., $2.75.

Maple Leaf Coach Enamels; Size 1, $1 2); Size 2, 70c.; Size 3, 40c. each.

Castor Oil.

East India, in cases, per lb.	0 10	0 10½
" " small lots	0 10½	0 11

Cod Oil, Etc.

Cod Oil per gal.	0 50	0 55
Pure Olive	....	1 20
" Neatsfoot	....	90

Glue.

Common	0 08½	0 09
French Medal	0 14	0 14½
Cabinet sheet	0 12	0 13
White, extra	0 18	0 20
Gelatine	0 22	0 30
Strip	0 18	0 20
Coopers	0 19	0 20
Huttner	....	0 18

HARDWARE.

Ammunition.

Cartridges.

B B Cap Dom. 50 and 5 per cent.
Rim Fire Pistol, dis. 40 p. c., Amer.
Rim Fire Cartridges, Dom., 50 and 5 p. c.
Central Fire Pistol and Rifle, 10 p.c. Amer.
Central Fire Cartridges, pistol sizes, Dom. 30 per cent.
Central Fire Cartr'dges, Sporting and Military, Dom., 15 and 5 per cent.
Central Fire, Military and Sporting, Amer, add 5 p.c. to list. B.B. Caps, discount 40 per cent, Amer.
Loaded and empty Shells, "Trap" and "Dominion" grades, 25 per cent. Rival and Nitro, net list.
Brass shot Shells, 55 per cent.
Primers, Dom., 30 per cent.

Wads	per lb.
Best thick white felt wadding, in ¼-lb bags	1 00
Best thick brown or grey felt wads, in ¼-lb. bags	70
Best thick white card wads, in boxes of 500 each, 12 and smaller gauges	0 99
Best thick white card wads, in boxes of 500 each, 10 gauge	0 35
Best thick white card wads, in boxes of 500 each, 8 gauge	0 55
Thin card wads, in boxes of 1,000 each, 12 and smaller gauges	0 20
Thin card wads, in boxes of 1,000 each, 10 gauge	0 25
Thin card wads in boxes of 1,000 each, 8 gauge	
Chemically prepared black edge grey cloth wads, in boxes of 250 each—	Per M.
11 and smaller gauge	0 60
9 and 10 gauges	0 70
7 and 8 gauges	0 90
5 and 6 gauges	1 10
Superior chemically prepared pink edge, best white cloth wads, in boxes of 250 each—	
11 and smaller gauge	1 15
9 and 10 gauges	1 40
7 and 8 gauges	1 65
5 and 6 gauges	1 90

Adzes.

Discount, 20 per cent.

Anvils.

Wright's, 80-lb. and over	0 10½
Hay Budden, 80-lb. and over	0 09½
Brook's, " " "	0 11¼

Augers.

Gilmour's, discount 65 and 5 p.c. off list.

Axes.

Chopping Axes		
Single bit, per doz	5 50	10 00
Double bit, "	11 00	18 00
Bench Axes, 40 p.c.		
Broad Axes, 33⅓ per cent.		
Hunters' Axes	5 50	6 00
Boy's Axes	5 75	6 75
Splitting Axes	6 50	12 00
Handled Axes	7 00	10 00

Axle Grease.

Ordinary, per gross	5 75	6 00
Best quality	13 00	15 00

Bath Tubs.

Zinc 6 00
Copper, discount 15 p.c. off revised list

Baths.

Standard Enameled.

5½-inch rolled rim, 1st quality	25 00
" " " 2nd "	21 00

Anti-Friction Metal.

"Tandem" A	per lb.	0 27
" B	"	0 21
" C	"	0 11½
Magnolia Anti-Friction Metal, per lb.		0 25
Frictionless metal, per lb		0 23

SYRACUSE SMELTING WORKS.

Aluminum, genuine	0 45
Dynamo	0 29
Special	0 25
Aluminum, 99 p.c. pure "Syracuse"	0 50

Bells.

Hand.

Brass, 60 per cent.
Nickel, 55 per cent.

Cow.

American make, discount 66⅔ per cent.
Canadian, discount 45 and 50 per cent.

Door.

Gongs, Sargent's 5 50 8 00
" Peterboro', discount 45 per cent.

Farm.

American, each 1 25 3 00

House.

American, per lb 0 35 0 40

Bellows.

Hand, per doz	3 35	4 75
Moulders', per doz	7 50	10 00

Blacksmiths', discount 40 per cent.

Belting.

Extra, 60 per cent.
Standard, 60 and 10 per cent.

Bits.

Auger.

Gilmour's, discount 60 and 5 per cent.
Rockford, 50 and 10 per cent.
Jennings' Gen., net list.

Car.

Gilmour's, 47½ to 50 per cent.

Expansive.

Clark's, 40 per cent

Gimlet.

Clark's, per doz	0 65	0 90
Damond, Shell, per doz	1 00	1 50
Nail and Spike, per gross	2 25	5 20

Blind and Bed Staples.

All sizes, per lb 0 07½ 0 12

Bolts and Nuts.

	Per cent.
Carriage Bolts, full square, Norway	65
" " full square	65
Common Carriage Bolts, all sizes	60
Machine Bolts, all sizes	60
Coach Screws	70
Sleigh Shoe Bolts	72½
Blank Bolts	60
Bolt Ends	62½
Plough Bolts	60
Nuts, square	4 c. off
Nuts, hexagon	4¼c. off
Tire Bolts	67½
Stove Bolts	67½
Stove rods, per lb	5½ to 6c.

Boot Calks.

Small and medium, ball, per M	4 25
Small heel, per M	4 50

Bright Wire Goods.

Discount .... 62½ per cent.

Broilers.

Light, dis., 65 to 67½ per cent.
Reversible, dis., 65 to 67½ per cent.
Vegetable, per doz., dis. 37½ per cent.

Henis, No. 8, "	....	6 00
Henis, No. 9, "	....	7 60
Queen City "	7 50	0 00

Butchers' Cleavers.

German, per doz	6 00	11 00
American, per doz	12 00	20 00

Building Paper, Etc.

Plain building, per roll	0 30
Tarred lining, per roll	0 40
Tarred roofing, per 100 lb	1 65
Coal Tar, per barrel	3 50
Pitch, per 100-lb	0 85
Carpet felt, per ton	45 00

Bull Rings.

Copper, $2.00 for 2½ in. and $1.90 for 2 in.

Butts.

Wrought Brass, net revised list

Cast Iron.

Loose Pin, dis., 60 per cent.

Wrought Steel.

Fast Joint, dis. 65, 10 and 2½ per cent.
Loose Pin, dis. 65, 10 and 2½ per cent.
Berlin Bronzed, dis. 70, 70 and 5 per cent.
Gen. Bronzed, per pair 0 40 0 65

Carpet Stretchers.

American, per doz	1 00	1 50
Bullard's, per doz	6 50	....

Castors.

Bed, new list, dis. 55 to 57½ per cent.
Plate, dis. 52½ to 57½ per cent.

Cattle Leaders.

Nos. 31 and 32, per gross 50 9 50

Cement.

Canadian Portland	2 50	2 80
English "	....	3 00
Belgian "	2 50	2 75
Canadian hydraulic	1 25	1 50

Chalk.

Carpenters Colored, per gross	0 45	0 75
White lump, per cwt	0 60	0 65
Red	0 05	0 06
Crayon, per gross	0 14	0 18

Chisels.

Socket, Framing and Firmer.

Broad's, dis. 70 per cent.
Warnock's, dis. 70 per cent.
P. S. & W. Extra 60, 10 and 5 p.c.

Churns.

Revolving Churns, metal frames—No. 0, $8—No. 1, $8.50—No. 2, $9.00—No. 3, $10.00—No. 4, $12.00—No. 5, $16.00 each. Ditto, wood frames—20c. each less than above. Discounts: Delivered from factories, 58 p.c.; from stock in Montreal, 58 p.c. Terms, 4 months or 3 p.c. cash in 30 days.

Clips.

Axle dis. 65 per cent.

Closets

Plain Ontario Syphon Jet	$16 00
Emb. Ontario Syphon Jet	17 00
Fittings net	1 00
Plain Teutonic Syphon Washout	10 00
Emb. Teutonic Syphon Washout	11 00
Fittings net	1 25
Low Down Teutonic, plain	16 00
" " " embossed	17 00
Plain Richelieu net	3 75
Emb. Richelieu net	4 00
Fittings net	1 25
Low Down Ont. Sy. Jet, plain net	19 50
" " " " " emb'd. net	20 50
Closet connection net	1 25
Basins, round, 14 in	1 00
" oval, 17 x 14 in	2 50
" " 19 x 15 in	3 75

Discount 40 p.c., except on net figures.

Compasses, Dividers, Etc.

American, dis. 62½ to 65 per cent.

Cradles, Grain.

Canadian, dis. 25 to 33⅓ per cent.

Crosscut Saw Handles.

S. & D., No. 3, per pair	½
" " 5, "	⅞
" " 6, "	17
Boynton pattern "	20

Door Springs.

Torrey's Rod, per doz	(15 p.c.)	2
Coil, per doz	0 88	1 00
English, per doz	2 00	4 00

Draw Knives.

Coach and Wagon, dis. 50 and 10 per cent.
Carpenters, dis. 70 per cent.

Drills.

Hand and Breast.

Millar's Falls, per doz. net list.

DRILL BITS.

Morse, dis., 37½ to 40 per cent.
Standard dis. 50 and 5 to 55 per cent

Faucets.

Common, cork-lined, dis. 35 per cent.

ELBOWS. (Stovepipe.)

No. 1, per doz	1 40
No. 2, per doz	1 20

Bright, 20c. per doz. extra.

ESCUTCHEONS.

Discount, 45 per cent.

ESCUTCHEON PINS.

Iron, discount 40 per cent.

FACTORY MILK CANS.

Discount off revised list, 40 per cent.

FILES AND RASPS.

Great Western	75 and 5 per cent.
Disston	70 " 10 "
Arcade	75 " 5 "
Kearney & Foot	70 " 10 "
American	75 " 5 "
McClellan	70 " 5 "
Eagle	70 10 and 5 "
Nicholson	70 " 10 "
Heller	60 " 10 "

Royal & Keystone 80 p.c. and 80 and 10 p.c.
Black Diamond, 60 to 60 and 10 per cent.
Jowitt's, English list, 25 to 27½ per cent.

FORKS.

Hay, manure, etc., dis., 50 and 10 per cent. revised list.

GLASS—Window—Box Price.

Size United Inches.	Star Per 50 ft.	Star Per 100 ft	D. Diamond Per 50 ft.	D. Diamond Per 100 ft.
Under 26	2 15	4 15	....	6 00
26 to 40	2 30	4 45	....	6 65
41 to 50	4 85	....	....	7 50
51 to 60	5 15	....	....	8 50
61 to 70	5 50	....	....	9 50
71 to 80	6 00	....	....	10 50
81 to 85	6 50	....	....	11 70
86 to 90	....	....	..	14 00
91 to 95	....	....	...	15 50
99 to 100	....	....	....	18 00

GAUGES

Marking, Mortise, Etc.

Stanley's dis. 50 to 55 per cent.

Wire Gauges.

Winn's, Nos. 26 to 33, each 1 65 2 40

HALTERS.

Rope, ⅜ per gross	....	....
" ½ "	....	9 00
" ⅝ to ¾	....	14 00
Leather, 1 in., per doz	3 87½	4 00
" 1¼ in., "	5 15	5 20
Web, — per doz	1 87	2 45

HAMMERS.

Nail

Maydole's, dis. 5 to 10 per cent. Can. dis. 25 to 27½ per cent.

Tack.

Magnetic, per doz 1 10 1 20

Sledge.

Canadian, per lb 0 07½ 0 08½

Ball Pean.

English and Can., per lb 0 22 0 25

HANDLES.

Axe, per doz. net	1 50	2 00
Store door, per doz	1 00	1 50

Fork.

C. & B., dis. 40 per cent. rev. list.

Hoe.

C. & B., dis. 40 per cent. rev. list.

Saw.

American, per doz 1 00 1 25

Plane.

American, per gross 3 15 3 75

Hammer and Hatchet.

Canadian, 40 per cent.

Cross-Cut Saws.

Canadian, per pair 0 13½

HANGERS.

	doz.	pairs.
Steel barn door	5 85	6 00
Stearns, 4 inch	....	5 00
" 5 inch	....	6 50
Lane's covered—		
No. 11, 5-ft. run	....	8 40
No. 11½, 10-ft. run	....	10 80
No. 12, 10-ft. run	....	12 60
No. 14, 15-ft. run	....	21 00
Lane's O.N.T. track, per foot.	....	4½

HARVEST TOOLS.

Discount, 50 and 10 per cent.

HATCHETS.

Canadian, dis. 40 to 42½ per cent.

HINGES.

Blind, Parker's, dis. 50 and 10 to 60 per cent.

Heavy T and strap, 4-in., per lb	0 06½
" " 5-in., "	0 06¼
" " 6-in., "	0 06
" " 8-in., "	0 05¾
" " 10-in., "	0 05½

Light T and strap, dis. 65 and 5 per cent.
Screw hook and hinge—

6 to 12 in., per 100 lbs	4 50
14 in. up, per 100 lbs	3 50
	Per gro. pairs.
Spring	12 00

HOES.

Garden, Mortar, etc., dis. 50 and 10 p.c.
Planter, per doz 4 00 4 50

HOLLOW WARE

Discount .... 45 and 5 per cent

HOOKS.

Cast Iron.

Bird Cage, per doz	0 50	1 10
Clothes Line, per doz	0 27	0 63
Harness, per doz	0 72	0 88
Hat and Coat, per gross	1 00	3 00
Chandelier, per doz	0 50	1 00

Wrought Iron.

Wrought Hooks and Staples, Can., dis. 47½ per cent.

Wire.

Hat and Coat, discount 45 per cent.
Belt, per 1,000 .... 0 60
Screw, bright, dis. 55 per cent.

HORSE NAILS.
"C" brand 50 and 7½ p.c. off new list } Oval-
"M" brand 50, 10 and 5 per cent. } head
Countersunk, 60 per cent.

HORSESHOES
F.O.B. Montreal.

Iron Shoes.	No. 2 and larger.	No. 1. and smaller.
Light, medium, and heavy...	3 50	3 75
Snow shoes	3 75	4 00
Steel Shoes.		
Light	3 60	3 85
Featherweight (all sizes)....	4 85	4 85

F.O.B. Toronto, Hamilton, London and Guelph, 10c. per keg additional.
Toe weight steel shoes...... 6 70

JAPANNED WARE.
Discount, 45 and 5 p.c. off list, June 1899

ICE PICKS.
Star per doz. ... 3 00 3 25

KETTLES.
Brass spun, 7½ p.c. dis. off new list.
Copper, per lb. ... 0 30 0 50
American, 60 and 10 to 65 and 5 p.c.

KEYS.
Lock, Can., dis., 45 p.c.
Cabinet, trunk, and padlock, Am. per gross ... 60

KNOBS.
Door, japanned and N.P., per doz. ... 1 50 2 50
Bronze, Berlin, per doz. ... 2 75 3 25
Bronze Genuine, per doz. ... 6 00 9 00
Shutter, porcelain, F. & L. screw, per gross ... 1 30 4 00
White door knobs—per doz. ... 1 25

HAY KNIVES.
Discount, 50 and 10 per cent.

LAMP WICKS.
Discount, 60 per cent.

LANTERNS.
Cold Blast, per doz. ... 7 00
No. 3 "Wright's" ... 8 50
Ordinary, with O burner ... 4 00
Dashboard, cold blast ... 9 00
No. 0. ... 5 75
Japanning, 50c. per doz. extra.

LEMON SQUEEZERS.
Porcelain lined, ... per doz. 2 20 5 60
Galvanized ... 1 87 3 85
King, wood ... 2 75 2 90
King, glass ... 4 00 4 50
All glass ... 1 20 1 30

LINES.
Fish, per gross ... 1 05 2 50
Chalk " ... 1 90 7 40

LOCKS.
Canadian, dis. 45 p.c.
Russel & Erwin, per doz. ... 3 00 3 25

Cabinet.
Eagle, dis. 30 p.c.

Padlock
English and Am., per doz. ... 50 6 00
Scandinavian, " ... 1 00 2 40
Eagle, dis. 20 to 25 p.c.

MACHINE SCREWS. Iron and Brass.
Flat head discount 25 p.c.
Round Head discount 20 p.c.

MALLETS.
Tinsmiths' per doz. ... 1 25 1 50
Carpenters', hickory, per doz ... 1 25 3 75
Lignum Vitae, per doz. ... 3 85 5 00
Caulking each ... 60 2 00

MATTOCKS.
Canadian, per doz. ... 5 50 6 50

MEAT CUTTERS.
American, dis. 25 to 30 p.c.
German, 15 per cent.

MILK CAN TRIMMINGS.
Discount, 25 per cent.

NAILS.

Quotations are:	Cut.	Wire.
2d and 3d	$3 35	$3 85
3d	3 00	3 52
4 and 5d	2 75	3 35
6 and 7d	2 65	3 20
8 and 9d	2 50	3 00
10 and 12d	2 45	2 95
16 and 20d	2 40	2 90
30, 40, 50 and 60d. (base)	2 35	2 85

Wire nails in carlots are $2.77½
Galvanizing 2c. per lb. net extra.
Steel Cut Nails 10c. extra.
Miscellaneous wire nails, dis. 70 and 10 p.c.
Coopers' nails, dis. 30 per cent.
Flour barrel nails, dis. 25 per cent

NAIL PULLERS.
German and American ... 1 85 3 50

NAIL SETS
Square, round, and octagon per gross ... 3 38 4 00
Diamond ... 12 00 15 00

NETTING.
Poultry, 55 per cent for McMullen's

OAKUM. Per 100 lb.
Navy ... 6 00
U. S. Navy ... 7 25

OIL.
Water White (U.S.) ... 0 16½
Prime White (U.S.) ... 0 15½
Water White (Can.) ... 0 15
Prime White (Can.) ... 0 14

OILERS.
McClary's Model galvan. oil can, with pump, 5 gal., per doz. ... 10 00
Zinc and tin, dis. 50, 50 and 10.
Copper, per doz. ... 1 25 3 50
Brass, " ... 1 50 3 50
Malleable, dis. 25 per cent.

GALVANIZED PAILS.
Dufferin pattern pails, dis. 45 p.c.
Flaring pattern, discount 45 per cent.
Galvanized washtubs, discount 45 per cent.

PIECED WARE.
Discount 40 per cent. off list, June, 1899.
10-qt. flaring sap buckets, dis. 45 p.c.
6, 10 and 14-qt. flaring pails, dis. 45 p.c.
Creamer cans, dis. 45 p.c.

PICKS.
Per doz ... 6 00 9 00

PICTURE NAILS.
Porcelain head, per gross ... 1 75 3 00
Brass head " ... 0 40 1 00

PICTURE WIRE.
Tin and gilt, discount 75 p.c.

PLANES.
Wood, bench, Canadian dis. 50 per cent. American dis. 50.
Wood, fancy Canadian or American 7½ to 40 per cent.

PLANE IRONS.
English, per doz. ... 2 00 5 00

PLIERS AND NIPPERS.
Button's Genuine per doz pairs, dis. 37½ 40 p.c.
Button's Imitation, per doz. ... 5 00 9 00
German, per doz. ... 0 60 2 60

PLUMBERS BRASS GOODS.
Compression work, discount, 60 per cent.
Fuller's work, discount 65 per cent.
Rough stops and stop and waste cocks, discount, 60 per cent.
Jenkins disk globe and angle valves, discount, 55 per cent.
Standard valves, discount, 60 per per cent.
Jenkins radiator valves discount 55 per cent.
" " " standard, dis., 60 p.c.
Quick opening valves discount, 60 p.c.
No. 1 compression bath cock ... 2 00
No. 4 " " " ... 2 00
No. 7, Fuller's ... 2 50
No 4½, " ... 3 00

POWDER.
Velox Smokeless Shotgun Powder.
100 lb. or less ... 0 85
1,000 lb. or more ... 0 80
Net 30 days.

PRESSED SPIKES.
Discount 20 to 25 per cent.

PULLEYS.
Hothouse, per doz. ... 0 55 1 00
Axle ... 0 22 0 33
Screw ... 0 27 1 00
Awning ... 0 35 2 50

PUMPS.
Canadian cistern ... 1 80 3 60
Canadian pitcher spout ... 1 40 2 10

PUNCHES.
Saddlers', per doz. ... 1 00 1 85
Conductors', " ... 9 00 15 00
Tinners' solid, per set ... 0 00 0 72
" hollow, per inch ... 0 00 1 00

RANGE BOILERS.
Galvanized, 3 gallons ... 7 00
" 35 " ... 8 25
" 40 " ... 9 50
Copper, 30 " ... 22 00
" 35 " ... 26 00
" 40 " ... 30 00
Discount off Copper Boilers 10 per cent.

RAKES.
Cast steel and malleable, 50, 10 and 5 p.c.
Wood, 25 per cent.

RAZORS.

	per doz.	
Elliot's	4 00	18 00
Geo. Butler & Co.'s	8 00	14 00
Boker's	7 50	11 00
" King Cutter	12 50	50 00
Wade & Butcher's	3 60	10 00
Theile & Quack's	7 00	12 00

REAPING HOOKS.
Discount, 50 and 10 per cent.

REGISTERS.
Discount, ... 40 per cent.

RIVETS AND BURRS.
Iron Rivets, black and tinned, discount 60 and 10 per cent.
Iron Burrs, discount 55 per cent.
Extras on Iron Rivets in 1-lb. cartons, ½c. per lb.
Extras on Iron Rivets in ½-lb. cartons, 1c. per lb.
Copper Rivets & Burrs, 35 and 5 p.c. dis. and cartons, 1c. per lb. extra, net.
Extras on Tinned or Coppered Rivets ½-lb. cartons, 1c. per lb.

RIVET SETS
Canadian, dis. 35 to 37½ per cent.

ROPE ETC.

	Sisal.	Manila.
7-16 in. and larger, per lb	10	13½
⅜ in.	11	14½
¼ and 5-16 in.		15½
Cotton, 3-16 inch and larger		16
" 5-32 inch		21
" ⅛ inch		22½
Russia Deep Sea		15½
Jute		8
Lath Yarn		9½
New Zealand Rope		10

RULES.
Boxwood, dis. 75 and 10 p.c.
Ivory, dis. 37½ to 40 p.c.

SAD IRONS. per set.
Mrs. Potts, No. 55, polished ... 82½
" No. 50, nickle-plated ... 87½

SAND AND EMERY PAPER.
Dominion Flint Paper, 47½ pc cent.
B & A. sand, 40 and 5 per cent.
Emery, 40 per cent.
Garnet (Rurton's), 5 to 10 p.c. advance on list.

SAP SPOUTS.
Bronzed iron with hooks, per doz. ... 9 50

SAWS.
Hand Disston's, dis. 12½ p.c.
S. & D., 40 per cent.
Crosscut, Disston's, per ft. ... 35 0 55
S. & D., dis. 35 p.c. on Nos. 2 and 3.
Hack, complete, each ... 0 75 2 75
" frame only ... 0 75

SASH WEIGHTS.
Sectional, per 100 lbs. ... 2 75 3 00
Solid, " ... 2 00 2 25

SASH CORD.
Per lb. ... 0 23 0 30

SAW SETS.
"Lincoln," per doz. ... 6 50

SCALES.
Standard, 45 p.c.
Champion, 65 p.c.
Spring Balances, 10 p.c.
Fairbanks Standard, 35 p.c.
" Dominion, 55 p.c.
" Richelieu, 55 p.c.

SCREW DRIVERS.
Sargent's per doz. ... 0 65 1 00

SCREWS
Wood, F. H., bright and steel, 87½ and 10 p.c.
Wood R. H., " dis. 82½ and 10 p.c.
" F. H., brass, dis. 80 and 10 p.c.
Wood, R. H., " dis. 75 and 10 p.c.
" F.H., bronze, dis. 75 p.c.
" R.H. " 70 p.c.
Drive Screws, 87½ and 10 per cent.
Bench, wood, per doz. ... 3 25 4 00
" iron, " ... 4 25 5 75
Set, Case hardened, 60 per cent.
Square Cap, 50 and 5 per cent.
Hexagon Cap, 45 per cent.

SCYTHES.
Per doz, net ... 9 00

SCYTHE SNATHS.
Canadian, dis. 45 p.c.

SHEARS.
Bailey Cutlery Co., full nickeled, dis. 60 p.c.
Seymour's, dis. 50 and 10 p.c.

SHOVELS AND SPADES.
Canadian, dis. 40 and 5 per cent.

SINKS.
Steel and galvanized, discount 45 per cent.

SNAPS.
Harness, German, dis. 25 p.c.
Lock, Andrews' ... 4 50 11 50

SOLDERING IRONS.
1, 1½ lb., per lb. ... 0 37
2 lb. or over, per lb. ... 0 34

SQUARES.
Iron, No. 493, per doz ... 2 40 2 55
" No. 494, " ... 3 25 3 40
Steel, dis. 60 10 and 5 p.c.
Try and bevel, dis. 50 to 52½ p.c.

STAMPED WARE.
Plain, dis., 75 and 12½ p.c. off revised list.
Retinned, dis., 75 p.c. off revised list.

STAPLES.
Galvanized ... 3 50 4.00
Plain ... 3 15 3 75
Coopers', discount 45 per cent.
Poultry netting staples, 40 per cent.

STOCKS AND DIES.
American dis. 25 p.c.

STONE.	Per lb.	
Washita	0 28	0 60
Hindostan	0 06	0 07
" slip,	0 09	0 09
Labrador	....	0 13
" Axe	....	0 15
Turkey	....	0 50
Arkansas	0 00	1 50
Water-of-Ayr	0 00	0 10
Scythe, per gross	3 50	5 00
Grind, 2 in., 40 to 200 lb., per ton	....	25 00
" under 40 lb. "	....	28 00
Grind. under 2 in. thick "	....	29 00

STOVE PIPES.
5 and 6 inch Per 100 lengths ... 7 00
7 inch " " ... 7 50

ENAMELINE STOVE POLISH.
No. 4—3 dozen in case, net cash ... $4 80
No. 6—3 dozen in case, " ... 8 40

TACKS BRADS, ETC.

	Per cent
Strawberry box tacks, bulk	75 & 10
Cheese-box tacks, blued	80 & 12½
Trunk tacks, black and tinned	85
Carpet tacks, blued	80 & 15
" " tinned	80 & 20
" " (in kegs)	40
Cut tacks, blued, in dozens only	80
" ½ weights	60
Swedes, cut tacks, blued and tinned—	
In bulk	80 & 10
In dozens	75
Swedes, upholsterers', bulk	85, 12½ & 12½
" brush, blued & tinned, bulk	70
" gimp, blued, tinned and japanned	75 & 12½
Zinc tacks	35
Leather carpet tacks	55
Copper tacks	50
Copper nails	52½
Trunk nails, black	65 and 5
Trunk nails, tinned	65 and 10
Clout nails, blued	65 and
Chair nails	35
Patent brads	40
Fine finishing	40
Picture frame points	10
Lining tacks, in papers	10

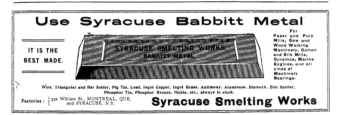

Lining tacks, in bulk		15
" " solid heads, in bulk		75
Saddle nails in papers		10
" " in bulk		15
Tufting buttons, 22 line, in dozens only		60
Tin capped trunk nails		15
Zinc glazier's points		5
Double pointed tacks, papers	90 and	10
" " " bulk		40

TAPE LINES.

English, ass skin, per doz.	2 75	5 00
English, Patent Leather	5 50	9 75
Chesterman's each	0 90	2 85
" steel, each	0 80	8 00

THERMOMETERS.

Tin case and dairy, dis. 75 to 75 and 10 p.c.

TRAPS. (Steel.)

Game, Newhouse, dis. 25 p.c.
Game, H. & N., P. S. & W., 65 p.c.
Game, steel, 72½, 75 p.c.

TROWELS.

Disston's discount 10 per cent.
German, per doz. 4 75 6 00
S. & D., discount 35 per cent.

TWINES.

Bag, Russian, per lb.	....	0 27
Wrapping, cotton, 3-ply		0 20
" " 4-ply	....	0 26
Mattress, per lb.	0 33	0 45
Staging, "	0 27	0 35

VISES.

Wright's	....	0 13¼
Brook's	....	0 12½
Pipe Vise, Hinge, No. 1	....	3 5)
" " " No 2	....	5 50
Saw Vise	4 5)	9 00

ENAMELLED WARE.

White, Princess, Turquoise, Blue and White, discount 50 per cent.
Diamond, Famous, Premier, 50 and 10 p.c.
Granite or Pearl, Imperial, Crescent, 50, 10 and 10 per cent.

WIRE.

Brass wire, 50 to 50 and 2½ per cent. off the list.
Copper wire, 45 and 10 per cent. net cash 30 days, f.o.b. factory.
Smooth Steel Wire, is quoted at the following net selling prices:

No. 6 to 8 gauge	$2 90
" 9 "	2 80
" 10 "	2 87
" 11 "	2 90
" 12 "	2 95
" 13 "	3 15
" 14 "	3 37
" 15 "	3 50
" 16 "	3 65

Other sizes of plain wire outside of Nos. 9, 10, 11, 12 and 13, and other varieties of plain wire remain at $2.8) base with extras as before. The prices for Nos. 9. to 13 include the charge of 1 c. for oiling. Extras net per 100 lb.: Coppered wire, 60c.—tinned wire, $2—oiling, 10c.—special hay-bailing wire, 30c.—spring wire, $1— best steel wire, 75c.—bright soft drawn, 15c.—in 50 and 100-lb. bundles net, 10c.—in 25-lb. bundles net 15c.—packed in casks or cases, 15c.—bagging or papering, 10c.

Fine Steel Wire, dis. 17½ per cent. List of extras: In 100-lb. lots: No. 17, $5—No. 18, $5.50—No. 19, $6—No. 20, $6.65—No. 21, $7—No. 22, $7.30—No. 23, $7.65—No. 24, $8—No. 25, $9—No. 26, $9.50—No. 27, $10—No. 28, $11—No. 29, $12—No. 30, $13—No. 31, $14—No. 32, $15 No. 33, $16—No. 34, $17. Extras net—tinned wire, Nos. 17-25, $2—Nos. 26-31 $4—Nos. 32-34, $6. Coppered, 5c.—oiling, 10c.—in 25-lb. bundles, 15c.—in 5 and 10-lb. bundles. 25c.—in 1-lb. hanks, 50c.—in ½-lb. hanks, 75c.—in ¼-lb. hanks, $1—packed in casks or cases, 15c.—bagging or papering, 10c.

Galvanized Wire, per 100 lb.—Nos. 6,7,8, $3.50 to $3 8?—No. 9, $2.85 to $3.15—No. 10 $3.60 to $3.95—No. 11, $3.70 to $4.10—No 12, $3 to $3 30—No. 13, $3.10 to $3 4?—No. 14, $4.10 to $4.50—No. 15, $4.60 to $5.05—No. 16, $4.85 to $5 35. Base sizes, Nos. 6 to 9, $2.57½ f.o.b. Cleveland.

Clothes Line Wire, solid 7 strand, No. 17, $4.25; No. 18, $2.65; No. 19, $2.35, f.o.b Hamilton, Toronto, Montreal.

WIRE FENCING. F.O.B. Toronto

Galvanized barb	3 05
Galvanized, plain twist	3 05

Galvanized barb, f.o.b. Cleveland, $2.82½ in less than carlots, and $2.70 in carlots.

WIRE CLOTH.

Painted Screen, per 100 sq. ft., net.. 1 35

WASTE COTTON. per lb.

Colored	4½ to 5
White, according to quality	6½ to 7½

500-lb. bale lots shaded.

WRENCHES.

Acme, 35 to 37½ per cent,
Agricultural, 60 p.c.
Coe's Genuine, dis. 20 to 25 p.c.

Towers' Engineer, each	2 00	7 00
" S., per doz.	5 80	6 00
G. & K.'s Pipe, per doz.	....	3 40
Burrell's Pipe, each	....	3 00
Pocket, per doz.	0 25	2 90

WRINGERS.

Leader	per doz.	$30 0	33 00
Royal Canadian..	"	26 00	28 00
Royal American.,	"	26 00	28 00
Sampson	"	....	30 00

Terms 4 months, or 3 p.c. 30 days.

WROUGHT IRON WASHERS.

Canadian make, discount, 40 and 5 per cent

"THE EMLYN" SAW BENCH

Made in 6 sizes. Best value obtainable. Specially designed for export. With or without "Emlyn" Patent Guard. Sole maker—

CHARLES D. PHILLIPS,

Cables—"Machinery," Newport.
Emlyn Engineering Works, NEWPORT, MON., ENGLAND.

PERSONS addressing advertisers will kindly mention having seen their advertisement in Canadian Hardware and Metal Merchant.

IF THE WORDS

"Dundas Axe"

are stamped on an Axe, you can rely on its being the best that can be made.

DUNDAS AXE WORKS

Dundas, Ont.

Lockerby & McComb

AGENTS IN CANADA

FOR THE

Celebrated P. & B.

Cold Storage Lining

AND

.. Ruberoid Roofing ..

P. S.—Prices on Application.

65 Shannon Street, MONTREAL.

BUSINESS NEWS

of any kind that is of value to business men supplied by our Bureau. We can give you market quotations from any town in Canada, reports from the city markets, stock quotations, etc. You can get commercial news from any Canadian paper through us.

Write us, giving us particulars of what you want and where you want it from, and we will quote you prices by return.

"Clippings from any Canadian paper on any subject."

CANADIAN PRESS CLIPPING BUREAU,

232 McGill Street, MONTREAL, QUE.

Telephone Main 1255.

10 Front St. East, Toronto. Telephone 2148.

---

75 YEARS. *ESTABLISHED 1825.* *75 YEARS.*

CELEBRATED HEINISCH SHEARS.

Tailors' Shears,
Trimmers, Scissors,
Tinners' Snips, etc.

ACKNOWLEDGED THE BEST.

R. HEINISCH'S SONS CO. NEW YORK OFFICE, 90 Chambers St. NEWARK, N.J., U.S.A.

Not connected with any Shear Combination.

CHAS. F. CLARK, President. JARED CHITTENDEN, Treasurer.

...ESTABLISHED 1849...

BRADSTREET'S

Capital and Surplus, $1,500,000. Offices Throughout the Civilized World.

Executive Offices: Nos. 346 and 348 Broadway, New York City, U.S.A.

THE BRADSTREET COMPANY gathers information that reflects the financial condition and the controlling circumstances of every seeker of mercantile credit. Its business may be defined as of the merchants, by the merchants, for the merchants. In procuring, verifying and promulgating information no effort is spared, and no reasonable expense considered too great, that the results may justify its claim as an authority on all matters affecting commercial affairs and mercantile credit. Its offices and connections have been steadily extended, and it furnishes information concerning mercantile persons throughout the civilized world.

Subscriptions are based on the service furnished, and are available only by reputable wholesale, jobbing and manufacturing concerns, and by responsible and worthy financial, fiduciary and business corporations. Specific terms may be obtained by addressing the Company at any of its offices. **Correspondence Invited.**

—OFFICES IN CANADA—

HALIFAX, N.S. OTTAWA, ONT. VANCOUVER, B.C.
HAMILTON, ONT. QUEBEC, QUE. VICTORIA, B.C.
LONDON, ONT. ST. JOHN, N.B. WINNIPEG, MAN.
MONTREAL, QUE. TORONTO, ONT.

THOS. C. IRVING, Gen. Man. Western Canada, Toronto. JOHN A. FULTON, Gen. Man. Eastern Canada, Montreal.

TRADE NOBLES & HOARE NH LONDON MARK

Nobles & Hoare,

CORNWALL ROAD STAMFORD STREET.

LONDON, ENG.

Manufacturers of

HIGH-CLASS VARNISHES ONLY

Which can be obtained direct from the works or from the principal Color Dealers in Canada.

"BRASSITE"

REGISTERED TRADE MARK.

None genuine without the above "Trade Mark."

"Gunn's" Patent "Grassite" Goods.

Equal to **Solid Brass** in every particular. Cost less money—look and wear as well. Our sales are increasing all the time. Why not increase your sales?

THE GUNN CASTOR CO.

Limited.

KNOX HENRY, Canadian Agent, Room 32, Canada Life Bldg., **MONTREAL.**

Est. 1808 | Inc. 1895

Black Diamond File Works

G. & H. Barnett Company

PHILADELPHIA

Twelve Medals

Awarded
By **JURORS** at
International Expositions
Special Prize
Gold Medal at Atlanta, 1895

1901 GARDEN HOSE. 1901

We manufacture the most popular and best selling lines of Garden Hose in Canada. Our "Maltese Cross" brand we confidently recommend as the finest Garden Hose on the market.

We have other grades not quite so expensive, but good reliable brands, viz.: "Lion" (the popular medium-priced hose), "King" "Sun" and "Leader."

Our "Kinkproof" (wire wound) hose is wired by a special process controlled solely by ourselves, and is so constructed that it may be cut at any wind of the wire without loosening or uncoiling, the wire being self-gripping throughout each length.

THE GUTTA PERCHA AND RUBBER MFG. CO.

OF TORONTO, LIMITED.

Head Office and Warerooms—
49-61-63 West Front St.,

TORONTO, CANADA.

Factories—115-165 West Lodge Ave.
124-148 O'Hara Ave.

Galvanized Steel Sheets

"LEVEL BEST" BRAND.

Black Steel Sheets

"GRIFFIN" BRAND.

PRICES ON APPLICATION TO

B. & S. H. THOMPSON & CO'Y

26 St. Sulpice St., MONTREAL,

CORDAGE

ALL KINDS AND FOR ALL PURPOSES.

Manila Rope	Lathyarn
Sisal Rope	Shingleyarn
Jute Rope	Bale Rope
Russian Rope	Lariat Rope
Marline	Hemp Packing
Houseline	Italian Packing
Hambroline	Jute Packing
Clotheslines	Drilling Cables
Tarred Hemp Rope	Spunyarn
White Hemp Rope	Pulp Cord
Bolt Rope	Lobster Marlin
Hide Rope	Paper Cord
Halyards	Cheese Cord
Deep Sealine	Hay Rope
Ratline	Fish Cord
Plow Lines	Sand Lines

"RED THREAD" Transmission Rope from the finest quality Manila hemp obtainable, laid in tallow.

CONSUMERS CORDAGE COMPANY, Limited

Western Ontario Representative—
WM. B. STEWART.
Tel 94. 27 Front St. West, TORONTO.

MONTREAL, QUE.

Copper, Tin, Antimony, Etc.
LANGWELL'S BABBITT
Montreal.

CANADIAN HARDWARE AND METAL MERCHANT

The Weekly Organ of the Hardware, Metal, Heating, Plumbing and Contracting Trades in Canada.

VOL. XIII. MONTREAL AND TORONTO, JUNE 29, 1901. NO. 26

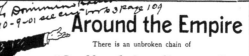

James Cartland & Son

Manufacturers of every description of **Limited**

CABINET, BUILDERS', FURNISHING AND NAVAL BRASSFOUNDRY

BIRMINGHAM, ENGLAND.

London Showrooms: 57 Holborn Viaduct, E.C.

HOSE...

WATER	SUCTION
STEAM	ACID
AIR	OIL
FIRE	SODA WATER
BABCOCK	HIGH-PRESSURE

Our **Patent Seamless Tube** is, without doubt, the only perfect construction.

The Canadian Rubber Co.,

CAPITAL - - - $1,500,000.00.

Montreal. Toronto. Winnipeg.

Other Tools are very good Tools, but

"YANKEE TOOLS" ARE BETTER

No. 15. "Yankee" Ratchet Screw Driver, with Finger Turn on Blade.

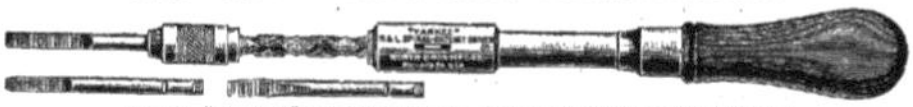

No. 30. "Yankee" Spiral-Ratchet Screw Driver, Right and Left Hand.

No. 41. 'Yankee" Automatic Drill, Eight Drill Points in Handle.

No. 50. "Yankee" Reciprocating Drill, for Iron, Steel, Brass, Wood, etc.

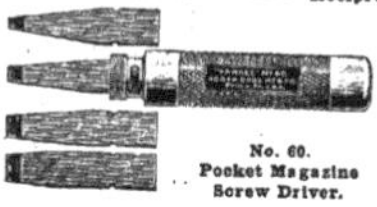

No. 60. Pocket Magazine Screw Driver.

Our "YANKEE" Tool Book tells all about them. Mailed free on application.

Manufacturers also of

LIGHTNING, GEM and BLIZZARD Ice Cream Freezers.

Toy Freezers, Ice Shaves and Ice Chippers, Fluting Machines, Hand Fluters.

Sold by Leading Jobbers in Canada.

NORTH BROS. MFG. CO.,

Philadelphia, Pa., U.S.A.

Oxford Gas Ranges

EMBODY ALL THE BEST IDEAS YET CONTRIVED.

They are made in a full line of sizes and styles to meet all demands.

Have large ovens, a special improved oven burner lighter, and the most perfect valves and burners known.

The intense heat furnished by them from a most economical supply of gas, delights every customer. It is a talking point of most convincing worth in making sales.

This year we emphasize two new styles with 16 and 18 inch square ovens, remarkably fine lines that satisfy the popular call for a standard quality Gas Range at a very moderate price.

Correspondence invited. Full Particulars and Price List at your service for the asking.

THE GURNEY FOUNDRY CO., Limited

TORONTO. WINNIPEG. VANCOUVER.

THE GURNEY-MASSEY CO., LIMITED, MONTREAL.

WE HAVE A COMPLETE STOCK.

Bright Goods, Door Pulls and Hat and Coat Hooks.

ALL ORDERS PROMPTLY ATTENDED TO.

Our Mills are now in full operation, and we are in position to handle any requirements the trade may have.

YOUR ORDERS SOLICITED FOR

Plain, Galvanized and Barb Wire, Wire Nails, Wood Screws, Copper and Brass Wire, Bright and Galvanized Fence Staples, Netting, Blind and Bed Staples, Jack Chain, Cotter Pins.

Prices quoted on application.

Dominion Wire Manufacturing Co.

MONTREAL and TORONTO.

THE NEW BALDWIN

DRY AIR CLEANABLE

REFRIGERATOR.

135 Modern Varieties. Ash, Oak and Soft-wood Finishes

METAL, PORCELAIN, SPRUCE LININGS.

BALDWIN

Positive Circulation—Sanitary—Odorless.

Latest Cleanable Features—The Strongest and Best System of Patent Removable Metal Air-Flues.

Air-Tight Lever Locks Ball-Bearing Casters. Swing Base—in and out.

Rubber around Doors and Lids, making them doubly air-tight.

Handsome Designs. Moderate Prices.

Built in the newest, largest and best equipped refrigerator plant in the East run all the year round on refrigerators exclusively; stock goods; special refrigerators and coolers in sections.

Handsome Trade Catalogue Ready.

Baldwin Refrigerator Co.,

BURLINGTON, VERMONT.

ESTABLISHED 1860. INCORPORATED 1895

STOVE PIPE DAMPERS

Steel—"Empire."

Made in

5, 6, 7-Inch.

Prices on Application.

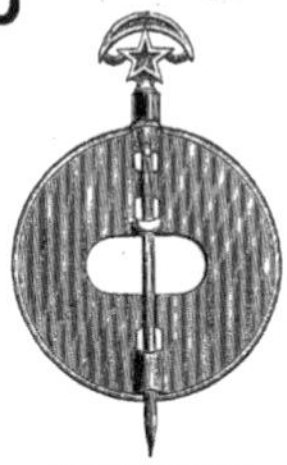

Cast Iron—"Star."

Stove Pipes, Elbows, Stove Boards, Coal Hods, Fire Shovels, Lanterns, etc.

THE THOS. DAVIDSON MFG. CO., Limited, MONTREAL, QUE.

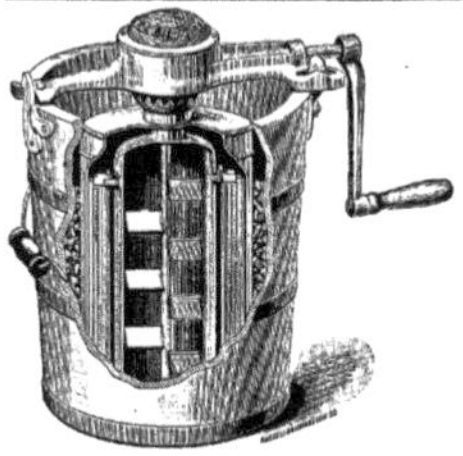

Just the Weather

The past week is the kind of weather that hustles people after Ice Cream Freezers.

And when they are in a hurry they don't want to be bothered examining half a dozen makes.

So they depend on their dealer's word, afterwards blaming him for a poor one or thanking him for a good one.

Sell your customers the

TRIPLE MOTION

WHITE MOUNTAIN FREEZERS,

and you are sure to win their thanks and future custom.

Sold only in Canada by the

THE McCLARY MANUFACTURING CO.

London, Toronto, Montreal, Winnipeg, Vancouver, and St. John, N.B.

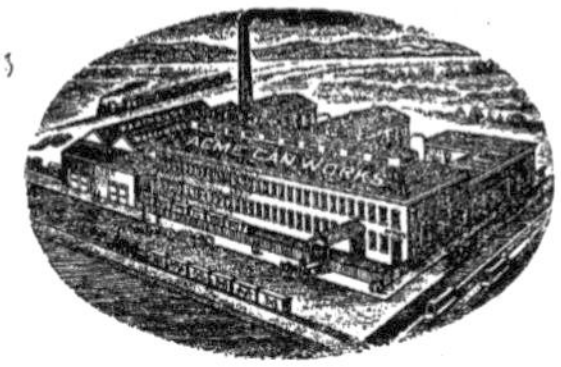

JAS. B. CAMPBELL. WILLIAM PRATT.

THE ACME CAN WORKS

Manufacturers of

Paint and Color Cans, round and square, Varnish and Oil Cans, Paint Irons, Paint Packages, Lye Tins,

and every description of Tin or can required by the trade. Write us for anything you need in our line. We will be pleased to quote you prices.

OFFICE AND FACTORY

Ontario St. and Jeanne D'Arc Avenue, ... MONTREAL

HARDWARE AND METAL

VOL. XIII. MONTREAL AND TORONTO, JUNE 29, 1901. NO. 26.

President,
JOHN BAYNE MacLEAN,
Montreal.

THE MacLEAN PUBLISHING CO.
Limited.

Publishers of Trade Newspapers which circulate in the Provinces of British Columbia, North-West Territories, Manitoba, Ontario, Quebec, Nova Scotia, New Brunswick, P.E. Island and Newfoundland.

OFFICES

MONTREAL - - - - - - - 232 McGill Street, Telephone 1255.
TORONTO - - - - - - - - 10 Front Street East, Telephone 2148,
LONDON, ENG. - - - - 109 Fleet Street, E.C., W. H. Miln.
MANCHESTER, ENG. - - - 18 St Ann Street. H. S. Ashburner.
WINNIPEG - - - - Western Canada Block, J. J. Roberts.
ST. JOHN, N.B. - - - No. 3 Market Wharf, J. Hunter White,
NEW YORK. - - - - - - - 176 E. 88th Street.

Subscription, Canada and the United States, $2.00.
Great Britain and elsewhere - - - 12s.

Published every Saturday.

Cable Address { Adscript, London. Adscript, Canada.

THE HOLIDAY SEASON.

WHAT use should be made of the holiday season? Weather conditions and the slackening of business activity during July and August have combined to make these months the natural holiday season for practically all Canadian business men except the fruiterer and refreshment vendors.

The wise merchant or clerk does not treat his holidays merely as a time of relaxation from business, but recognizes in them an opportunity for recuperation and possibly of enlargement.

The human body is like a large electrical storage battery, and the summer holidays should serve to charge the battery with renewed power, which should energize and invigorate him for another year's work. Therefore the holidays, to be beneficial, should be restful.

Primarily, holidays should be spent in such a way as to be entirely outside the ordinary groove of life. There is nothing which so effectually prevents a merchant from "falling into a rut" as to get completely away from his store and his work for a fortnight or a month, and to spend his time in such a way that business matters shall occupy but little of his time, and that little of a nature to make it possible for him to criticize his own methods and compare them with others.

Fortunately, every section of Canada is blessed with quiet retreats beside a lake or stream where a short season can be spent enjoyably and peacefully. Close to almost every town or city in the country is a resort where many of the business men are able to send their families for the summer and where they can spend their evenings themselves. This is good; the early retiring and equally early rising constitute an excellent antidote for nervous exhaustion. But it is not enough. For at least two weeks every merchant and clerk should be away from the store—away in mind and body.

A COMMERCIAL CURRICULUM.

IN the awakening that is taking place in Canada in regard to the importance of commercial education, it may be profitable for us to keep our eye on the methods that are being employed by European countries with a similar end in view, for we, after all, are only following where they have led in this important movement.

Germany is the centre of this movement; and from time to time we can gain a great deal of information in regard to what is going on there. Cologne has lately come into prominence in this respect, through the opening there on April 1 of a commercial school which aims to be the best of its kind in the world. It is the first of its kind in Germany that has been started as a perfectly independent institution. Every kind or sort of knowledge that can in any way be used in commercial, banking or counting-house life is to be taught in precise detail and from a practical standpoint. The school has attracted the attention of the United States consuls in Germany, and it has been the subject of some of the reports which have been sent to the Government at Washington.

The studies comprise: Science of commerce, knowledge of wares, chemical and mechanical technology, commercial arithmetic, bookkeeping and correspondence, and exercises in foreign languages. The order of study is:

First term—General political economy; commercial geography of the countries outside of Europe; civil law and colonial politics.

Second term—Commercial history up to 1800; civil law; tariff and transportation.

Third term—Agrarian and trade politics; commercial geography of Europe (including statistics); commercial exchange and martime law; trade and social legislation.

Fourth term—Finances; commercial history of the 19th century; international private law; State and Government law; banking; exchange, money and credit.

A young man leaving the college well grounded in these subjects would evidently, provided he also possessed adaptation, be well fortified for the struggle which a commercial career entails. It seems to us that there are at least parts of the course which might be taken up by those who either have not the time or the money to take a commercial course in a university or college.

AN INTERESTING TARIFF REVELATION.

MANY interesting things might be written in regard to the influences unseen and unknown to the public which are at the back of Parliamentary measures if those who write could get behind the scenes more frequently than they do.

One of the most famous Acts of the Dominion Parliament was that of 1879 which created a protective tariff on home industries. Everyone, of course, knows that it was the Government of Sir John Macdonald that introduced and carried the tariff through the House, but very few people know the circumstances surrounding the tariff when it was in actual course of preparation.

Some time after the return of Sir John Macdonald to power on the platform of protection to native industries, The Canadian Manufacturers' Association took a great deal of interest in the tariff question, and in pursuance of this interest a meeting of the Association was held in Ottawa. The President of the Association, a well-known manufacturer in Toronto, had given the question a good deal of consideration and by the time the Association had convened he had his plans ready. His suggestion was to the effect that the members of each branch of manufacturing industry should retire to a separate room and there draft a tariff consonant with its requirements. This was done.

When those interested in the iron duty submitted their draft to the meeting it was found that pig iron had been left on the free list. The President, although a large user of pig iron and consequently interested in keeping its price down, said in effect: "Gentlemen, if we want to build up an iron industry in this country, pig iron must be protected. We must be fair. What we want ourselves we must allow others."

At his suggestion those having in charge the iron tariff retired and when they returned pig iron was on the dutiable list.

When the tariff, as drafted by the different interests was drawn up the President was deputed to lay it before Sir Leonard Tilley, the Finance Minister. "Mr. Tilley," he said, "there is a tariff which has been carefully drawn up by men representing the different manufacturing interests of this country. I would advise you to accept it as it stands."

And the tariff that was introduced in Sir Leonard Tilley's budget speech of that year was, with a few alterations, the same as the President of The Canadian Manufacturers' Association had submitted to him. One of the alterations was the elimination, however, of the pig iron from the dutiable list, to which it was not restored until seven years later.

MERCHANTS AND THE MINT.

CANADA'S having a mint of her own forces into notice the question of what changes it may bring in the use of money.

The new mint will coin Canadian gold coins—$2.50, $5 and $10 pieces. These will be new additions to our money. We imagine they will become very popular, and that the demand for them will exceed expectations. Paper money will continue, as before, to be legal tender. But when people ask for gold, and get to like it, the merchants will have to ask the banks to give them a supply of gold for change. To refuse it might offend good customers. To have gold with which to make change will become a mark of a store doing a high-class trade.

Now, if this taste for gold money becomes general, it will displace paper bills. Who will be to blame? The Canadian Department of Finance which issues the $2 and $1 notes, and the chartered banks which issue the $5, $10 and other notes. These are frequently dirty, offensive-looking, and by no means a credit to the country. What must be the opinion of a tourist from abroad on receiving in change one of these filthy bills? What kind of advertisement of Canada are they?

Our paper is as good as gold any day. Behind it is the credit of Canada which is above par in the markets of the world. Bills are for most people a more convenient form of money than gold coin. But if allowed to deteriorate in appearance, as it has in the last few years, the result will be to deprive it of its old popularity.

We have frequently called attention to this long before a Canadian mint was spoken of, and the Canadian banking authorities have received the same advice from other quarters, but clean bills are still the exception in this country. The authorities will have themselves to thank if our new gold coin displaces bills.

Years ago, when the smallpox scare raged in Central Canada, the dirty bills were soaked and sprinkled with disinfectants. They reeked of carbolic and other ill-smelling stuff. Had the bills being called in and clean ones issued this would not have been necessary.

Canada does not keep its silver money in as good a condition as it should be. Many of the coins are defaced and mutilated. This should not be.

GALVANIZED SHEETS.

The steady demand for galvanized sheets has practically exhausted all stocks of American sheets, and as the demand is equal to the supply in the United States where prices are fully 1c. per lb. higher than in this market, no shipments are being received here from that country. United States manufacturers state that they have domestic orders so far ahead that they will guarantee no deliveries here before October.

In the meantime, stocks of English sheets, which had been greatly depleted during the past few weeks, have been increased by the arrival of several shipments during the past week. As further arrivals are expected, it is not likely there will be a shortage of this stock, the price of which remains firm at $4.50 for case lots and $4.65 to $4.75 for less than case lots. American galvanized sheets are not quoted.

EARLY CLOSING.

From one end of Canada to the other, and in almost every kind of community, from the small village to the principal cities, there has been a general movement in favor of uniform early closing during the summer months. In most cases, the hardwaremen have been party to the movement; in some instances, they have taken the prominent part in locally agitating the movement.

This movement should find favor with hardwaremen everywhere, especially during the hot summer months. There is virtually no necessity for keeping hardware stores open after 6 p.m., except possibly on Saturday evenings. Granting the lack of necessity, there should be enough of mutual good-will among the hardwaremen in any town to make workable an agreement to close at a given hour.

PARIS GREEN ACTIVE.

The movement of paris green during June has been so large and steady that the surplus stocks, which at the beginning of the season were extensive, have been rapidly depleted, and both jobbers and manufacturers are talking of possible shortage.

The demand seems general, no section being blessed (?) with more bugs than others, yet all providing in the aggregate a big consuming demand for paris green. Prices are steady, and are not likely to be changed.

THE GURNEY COMPANY'S PICNIC.

NEARLY everything that successful firms undertake appears to be crowned with success. This is true, at any rate, as far as the Gurney Foundry Co., Limited, of Toronto, is concerned. In its business everyone knows it is successful; in the annual excursion, which it gives its employees, success is again the characteristic feature. This was demonstrated on Saturday last, when the firm's employes had an excursion to Prospect Park, Oshawa.

It was the original intention to go to Oshawa by the Argyle, and that steamer was chartered for the occasion, but, while commercial concerns may propose, the military authorities dispose, for about 4 o'clock on the afternoon preceding the day of the excursion the firm was notified that, as the Argyle had been ordered to report at the wharf at Niagara at 9 o'clock on Saturday morning to convey a portion of the troops home from camp there, she would be unable to fill her contract with the Gurney firm.

Nothing daunted, recourse was at once had to Mr. M. C. Dickson, district passenger agent of the Grand Trunk Railway, and that gentleman immediately relieved them of their dilemma. He promised to have a special train at the foot of Yonge street at 8.15 next morning, and, furthermore, to have excursion tickets printed and in the hands of the firm by the same time. All his promises were complied with, and at 8.25 on Saturday morning a train of six cars containing 400 passengers was speeding towards Oshawa, which was reached a little over an hour afterwards.

Mr. Edward Gurney, the president, and Mr. T. B. Alcock, the secretary of the company, were on board, and during the whole day were as solicitious for the welfare of the excursionists as if they had been their own children. In addition to the employes and their friends there were a number of the firm's customers on board the train. Messrs. W. H. Carrick, vice-president and manager, and W. C. Gurney were in Great Britain, and could not, of course, be present. And they were missed.

The weather was at times rather warm, but, on the whole, was rather delightful, and on arrival at the park each excursionist sought the amusement or recreation which suited him or her the best. And there was such a variety that none needed to lack means of entertainment. There was baseball and various forms of athletic sports; dancing in the pavilion and music, and, for those who preferred to lounge, there was an abundance of shade trees and seats for their comfort.

Shortly after 7 o'clock the excursionists again entrained, and, by 8.15, had reached the city in safety, after having spent an exceedingly pleasant day.

NOTES.

"There," said C. S. Williamson, one of the firm's travellers, as the train swept past Pickering station, "the train has not stopped, and I sold a furnace here the other day."

The 100 yards' race, open to employes who had been with the firm 18 years or over, had a number of competitors, but the winner was old Ambrose Turner, a man of 76 years of age. It was the event of the day.

CATALOGUES, BOOKLETS, ETC.

MARLIN FIREARMS.

ONE of the most striking and, at the same time, thoroughly practical catalogues that has been received by HARDWARE AND METAL for some time has just come to hand from the Marlin Firearms Co. The striking feature of the work is an exceedingly artistic reproduction on the front cover of a full-blooded Indian in native costume examining with evident admiration one of the 1901 Marlin repeaters. Shown in the background is the Indians' wigwam and campfire. This cover is printed in nine colors. The catalogue is fully as useful inside as it is attractive outside. It is, like the 1900 catalogue, arranged in three sections. Part I. is designed for the quick reference of dealers and experienced buyers who desire briefly detail re Marlin arms. Part II. gives more detailed information for general buyers. Part III. comprises hints calculated to interest and help shooters generally. There are altogether over 120 pages in the book. It will be sent to anyone mentioning HARDWARE AND METAL and sending 3c. in stamps to The Marlin Firearms Co., New Haven, Conn., U.S.A.

HEATING BY WARM AIR.

Such is the title of the latest advertising production of the Gurney Foundry Co., Limited, Toronto, an illustrated catalogue and price list of "Oxford" warm air and combination furnaces. In addition to the various lines of furnaces which they have made for some time, this catalogue includes a new line of warm air heaters, the "Oxford 400 series," which is claimed to be superior in both design and construction, and exceptionally efficient and economical in operation. Like most of the Gurney advertising literature this catalogue is not merely an illustrated price list, but is full of detailed descriptions, practical hints and general directions about setting furnaces, etc. The illustrations of the various furnaces are thoroughly practical, sectional views being presented in every case, showing clearly the points of superiority claimed for the "Oxford." In every respect this catalogue is up-to-date and should be secured by any of the trade who are interested in heating. It can be had upon application.

A SHOP BETTERMENT PLAN.

A CONSPICUOUS example of shop betterment institutions in an industry ordinarily considered irremediably given over to grime and discomfort and the severest tax upon the endurance of operatives is, according to The Iron Trade Review, the drop forging plant of a Brooklyn, N.Y., firm. One of the best of the institutions started by this company is the mutual aid society contributed to by the firm and the members—and every employe may be a member—providing weekly cash payments for the sick and immediate cash payments in case of death. A physician under salary attends sick members and provides medicine without further charge. None of the men who are members of this society need pay doctors' bills, medicine bills, or go without income in time of sickness. Should any employe leave who has been a member of the mutual aid a year without receiving sick benefit, he is refunded one-half of all his payments. The men make an annual contribution to local hospitals and the firm gives an equal amount, the total being the largest from any plant in Brooklyn, for hospital work. The men own their own tools, but the company insures these, asking in return the service of the men in the works fire department.

The plan of giving prizes for suggestions of improvement in methods is adopted, and in a given six months prizes of $50, $25, $15, and $10 respectively were given. The company some time ago, without any motion by the men, put its works on a nine-hour day, with 10 hours' pay. The output has been rather increased than diminished. Piece-work is the rule in some departments. The company gives the piece-workers the benefit of rapid work. The reason is that the well-paid man cares for machinery, since he loses by the time required for repairs; nor does he waste material. It is found, also, since it is the rule of the shop that the men replace bad work and pay for the materials, that well-paid men save in the item of imperfect work, as well as in repairs, interest, fuel and other ways.

The comment of the treasurer of the company, upon the work it has done in shop betterments is that it pays. "It pays because a man is more than a machine. It pays because the rate of wages is not the chief factor in cost, but rather the rate of production. A clean man produces more in the long run than a dirty man. A well-informed man produces more than an ignorant man. A justly treated man produces more than one who is unjustly treated. A contented man is a better and cheaper producer than a discontented man. A well paid man is a more economical producer than an ill-paid man."

CANADIAN VS. UNITED STATES PAINTS.

EDITOR "Hardware and Metal"—I am a regular subscriber to your very excellent publication and read it with interest every week.

I have just finished looking over your issue of June 22, on page 24 of which I have read with much astonishment the article headed "The superiority of Canadian Paints."

It so glaringly misrepresents the facts that I feel forced to an endeavor to correct the false impression it is meant to convey.

I want to say at the start that I am still a good Canadian, proud of the country of my birth, and a loyal British subject. Therefore, I greatly dislike making any disparaging remarks regarding Canada or Canadian industries. But as a man with business experience in both countries, and interests still in both, I want to be fair and see fair play all around.

I do not know what paint factories Mr. Munro visited on this side, but I do know he could not have seen the principal ones and truthfully say "that none of them compare with the one he manages" in the sense he meant to convey. Neither could he truthfully say "not only are the factories here (in Canada) better plants, but the formulae are more scientific and the paints and all finished materials infinitely better."

It is a fact known world-wide that the United States is by far the greatest paint market in the universe. It is not strange, therefore, that paint-making should reach the highest perfection here. It would be strange were it not so.

The very argument that Mr. Munro uses to try and establish a reason for American paints being inferior is the one that naturally and satisfactorily explains the cause of their acknowledged superiority. The fact that the field is large is what excites and creates competition. It is so in every line the world over. Where the market is largest there will always be found the greatest competition. Competition enforces study and compels every exertion towards excellence. A manufacturer cannot hope to succeed under such conditions except on merit.

If Mr. Munro had even a slight knowledge of the real conditions over here he would not be foolish enough to say "American manufacturers are not called upon to exert themselves as are the Canadians. They have a large field to cater to and they can easily work up a satisfactory business. They are not called upon to study and scheme as the manufacturers are on this side of the line."

My company are on both sides of the line, and I want to say to Canadian manufacturers they don't know what paint competition such as is engaged in over here is. There is simply no comparison. The Canadian market, to use a forcible slang expression, is a "cinch," compared with the American. Canadian competition is keen. I know that well enough. Over here, it's intense; it's fierce. Where our Canadian salesmen meet eight or ten competitors our salesmen on this side have to meet 40 to 50 on a territory of the same size. The comparison is similar to a race where 10 compete, against one where 100 compete. To win against 99 means something. Against nine it's easy.

The great difference in my mind between Canadian and American paint competition, and I have thoughtfully observed the conditions, is this. In Canada the manufacturers try to see how "cheap" they can make their paints; in America they endeavor to see how good they can make them. To say that Canadian paint plants are better than American is rank absurdity, and must be attributed either to ignorance or prejudice.

With such a tremendous output as the large American factories enjoy they are able to introduce machinery, equipment, method and system that would be impossible in such small plants as Canada affords. I have been through the principal Canadian paint factories and through the leading British factories, and am familiar with the important factories on this side. There is simply no fair comparison between them. It's like comparing a river steamer with an ocean greyhound.

If the Canadian paint makers would give up all this silly, useless "guff" about the inferiority of American paints which I frequently see in certain Canadian papers, and get down to good business sense and make a paint equal to the American, and such as the people want, they would make some substantial headway.

Canadians are not fools. They are no more easily fooled than Americans. They know as quickly as any people on earth when they get a dollar's worth and when they don't. It's an insult to their intelligence to suppose they will buy American paints at an advance of 25 per cent if they are not worth it.

Canada wants good paint, just as good as this country demands. It was a firm belief in this fact that influenced me to induce The Sherwin-Williams Company to join me in supplying such a demand. The fact that every day since we began making paint in Canada our trade has increased, and increased largely, proves we were right. We have succeeded beyond my most sanguine hopes. We have succeeded because we deserved to—because we give the greatest value for a dollar of any one in the business. We solicit trade only on these grounds.

WALTER H. COTTINGHAM,
A Canadian Paint-Maker.

Cleveland, O., June 25, 1901.

THE EARLY-CLOSING MOVEMENT.

The early-closing by-law recently passed by New Westminster, B.C., and which went into force a few days ago, enacts that all stores and business places throughout the city must be closed in the evening as follows: Men's furnishings, clothing, boots, and shoes, dry goods, hardware, groceries, furniture or jewelery stores, not later than 6 p.m.; butcher shops, 6.30 p.m.; booksellers or stationers, 7.30 p.m.; and barber shops, 8 p.m. On Saturdays and other days preceding holidays, 10.30 p.m. will be the closing hour for all except barber shops, which will have till midnight. Exceptions are also provided for exhibition and Christmas weeks.

At a largely attended meeting of The Retail Merchants' Association, of St. Thomas, Ont., a resolution was passed unanimously that if a by-law be passed by the city council regulating the hour for closing on Saturday nights, that the hour be 10 o'clock. A committee was also appointed to secure signatures to an agreement to close stores Wednesday afternoon during July and August, commencing the second week in July.

The merchants in Winchester, Ont. have agreed to close their places of business on Tuesday and Friday evenings of each week at 7 o'clock. A by-law to enforce the agreement will be introduced at the next meeting of the council. In the meantime the stores and shops will commence their early closing to-morrow (Friday) evening.

The merchants of Springfield, N.S., have agreed to close their stores on Wednesday afternoons during July, August and September.

TROUSER OR SKIRT HOLDER.

Rice Lewis & Son, Limited, are showing a cheap and effective trousers or skirt holder, which can be retailed, if necessary, as low

as 10c. each, although usually such goods command a much higher price. Cuts of same are herewith shown.

BACK FROM THE COAST.

Mr. H. Sapery, manager of the Syracuse Smelting Works, Montreal, has just arrived from an extended trip of three months to the Pacific Coast and San Francisco, and reports the condition of affairs good.

BUSINESS CHANGES.

DIFFICULTIES, ASSIGNMENTS, COMPROMISES.

E. FORREST & CO., general merchants, Ste. Anne de Beaupré, Que., have assigned, and V. E. Paradis has been appointed provisional guardian.

Hector Leblanc, hardware dealer, Hull, Que., has assigned to Lamarche & Benoit.

Joseph Morneault, sawmiller, Notre Dame du Lac, Que., is offering 40c. on the dollar.

The creditors of R. W. Richardson, general merchant, Hartland, N.B., meet to-day.

Geo. M. Dalglish, match manufacturer, Hull, Que., has assigned to A. P. Mutchmore.

Charles Lindsay, general merchant, Roberval, Que., has compromised at 50c. on the dollar.

E. J. Belanger, general merchant, Portneuf (Saguenay), Que., is offering to compromise.

Mrs. Octave Beaudet, general merchant and grocer, St. Pierre les Becquets, Que., has assigned.

PARTNERSHIPS FORMED AND DISSOLVED.

Johnson & McPhail, hardware dealers, Vancouver, B.C., have dissolved.

A. Durocher & Sancartier, carriage makers, etc., Montreal, have dissolved.

Joseph, James and Herbert Robb have registered partnership, dating from June 3, under the style of Robb & Sons, machinists, Pictou, N.S.

SALES MADE AND PENDING.

The assets of Wilfrid Renouf, foundryman, Trois Pistoles, Que., have been sold.

Moore & Davis, general merchants, Prince Albert, N.W.T., are reported to be selling out.

James Hyslop, general merchant, Cromarty, Ont., is advertising his business for sale.

The assets of A. Cote & Fils, general merchants, St. Fabien, Que., have been sold.

The stock of H. A. Bigham, general merchant, Culloden, Ont., is advertised for sale by auction.

The stock, etc., of the estate of A. E. Chandler, general merchant, Plumas, Man., has been sold by auction.

The stock of the estate of Annie L. Graham, general merchant, Ouvry, Ont., is advertised for sale by tenuer.

The stock of Thompson & Lahey, general merchants, Penetanguishene, Ont., which was valued at $4,300 has been sold at 76c. on the dollar.

CHANGES.

Alph. Audet has opened a general store at Shawenegan Falls, Que.

John W. Peck, general merchant, Karsdale, N.S., has given up business and gone to Boston, Mass.

Mrs. M. Shields, general merchant, Lyndhurst, Ont., has removed to Osgoode Station.

A. D. McLean, general merchant, Sydney, N.S., has been succeeded by Neil H. McLean.

C. H. Smith, hardware and stove dealer, etc., Listowel, Ont., has sold out to Geo. Zilliax, jr.

Crysler & Stratton, general merchants, Delhi, Ont., have been succeeded by E. D. Heath & Co.

Isaac Villeneuve, carriage aker, Fallowfield, Ont., is about giving up business.

Murphy, Brown & Co., hardware dealers, Carberry, Man., have opened a branch at Wellwood, Man.

Mrs. M. A. Hilton has retired from The Boyce Carriage Company, Winnipeg, and Edward Boyce continues.

Mrs. W. R. Lefebvre has registered as proprietress of W. R. Lefebvre & Co., lumber dealers, etc., Waterloo, Ont.

F. X. Charbonneau, general merchant, etc., Notre Dame de la Sallette, Que., is removing to Ferme Neuve.

Mrs. S. Cartier has registered as proprietress of S. Cartier & Co., coal and wood dealers, St. Henri de Montreal, Que.

DEATHS.

Moses E. Rice, general merchant, Bear River, N.S., is dead.

J. L. Pruneau, general merchant, St. George east, Que., is dead.

BRITISH VS. UNITED STATES LOCOMOTIVES.

Gradually the truth is coming out about the relative merits of British and American locomotives. The Midland Railway saved £400 apiece on the initial price of the American engines ordered about the time of the engineering strike at home; but these locomotives cost 20 1-4 per cent. more for fuel, 50 per cent. more for oil, and 60 per cent. more for repairs than similar engines built in this country. That seems pretty well to settle the question.—Financial News, London.

PHOENIX, B.C., TRADE NOTES.

Phœnix, June 14, 1901.

Morrin, Thompson & Co., general merchants have completed the moving of their old store to the rear end of the lot, where it will be used as a warehouse, connecting with the second storey of the new building which they purpose erecting in the near future. As soon as completed the stock now in the building on Ironside avenue will be placed on the shelves and the new building will be connected with the store they are now doing business in on the corner of Dominion avenue and Phoenix street.

Mr. A. H. Rumberger (late of Houtzdale, Penn.), who has had wide experience in the business has taken charge of the hardware department, and Mr. I. Crawford, late with Cholditch & Co., has taken charge of the office.

When all is completed Morrin, Thompson & Co. will have a neat and well-equipped grocery and hardware store, for which they deserve credit.

Mr. Thos. Hardy, proprietor of "The People's Cash Store," purposes erecting a new building on Dominion avenue almost immediately. If we can judge by appearances and Mr. Hardy's increasing business, the cash system succeeds as well in a mining camp as anywhere else.

Mr. L. Y. Birnie, late with Morrin, Thompson & Co., has accepted a similar position, though somewhat better, with The Hunter-Kendrick Co., Limited, of this place. He has charge of the hardware department.

SUBSCRIBER.

STEEL RANGE PRICES.

THE impression appears to prevail in at least a portion of the trade that prices of steel ranges should be reduced, because the tendency of some items of cost has recently been downward. Those who take this view of steel range prices are not thoroughly conversant with the general situation. Very little change has been made in the price of any material entering into the construction of steel ranges for some time, with the exception of steel sheets and asbestos. Prices of these, however, afford no encouragement to those who are looking for lower prices of ranges. On the contrary, if the sheet steel and asbestos situation should be taken as governing market conditions, the prices of steel ranges should be advanced. Sheets are very much higher than last fall, and fully as high as at the corresponding time last year. The advance over last fall's prices is at least 87 per ton, and in some gauges it is still higher. The percentage of advance in this item of cost is about 12 1-2. Asbestos, also, is much dearer than it was. It will therefore be seen that the cost of the leading items in the production of steel ranges would justify a decided advance in price over those of last fall.

This, however, does not present the entire situation. Sheets have for some time been in such short supply that many manufacturers have been put to serious inconvenience. They have frequently been unable to run their factories to full capacity, simply because it was not possible to secure deliveries according to the terms of the contracts. This necessarily interfered with the smooth running of factories and added to the cost of construction. At present it seems out of the question to expect lower prices for sheets, as the mills are not only crowded with work, but the trade is faced with the strong probability of a general shutdown after July 1, to await the adjustment of wages for the coming year. It is to be hoped that the sheet manufacturers will be able to secure a settlement with their workmen in shorter time than last summer, when, it will be remembered, nearly three months elapsed from the time the mills closed down until the scale for the new year was signed and the mills could be started. If by any disagreement a long stoppage should take place this summer, sheets will be still scarcer in the fall and prices will certainly be prevented from going any lower.—Metal Worker.

THE NEW PARTNER.

There is an old gentleman in Detroit who might be called Peace, so gentle is his nature and so deficient is he in combativeness. He has been greatly imposed upon be some mean enough to take advantage of his disposition. Recently he took a partner in, his son, a strapping young fellow, who pulled in the "varsity" crew and was the all round athelete of his class.

"Is this Peace?" he was asked the other day when he answered the telephone. It was. "Well sir, I got your impudent letter. That stuff I sold you was just what I said it was and just what you agreed to pay for it. What do you mean by telling me I sent you an inferior grade of goods?"

"I think you did." And the lusty youth purposely weakened his voice.

"What, you old shrimp! You dare accuse me of rascality! I'll be down there inside of half an hour, and if you don't apologize and then settle, I'll use you for a club to wreck the office. I'll show you!"

And the roar would have terrified the timid old gentleman.

Young Peace got his father to go and see about some stationery for the new firm, set a table and a few chairs against the walls and then waited. The bully came—big, red, noisy and abusive. For about a minute he had vague thoughts of a windmill, a freight train and an earthquake in conjunction. When the old gentleman returned, everything looked natural except the grinning office boy and the trembling creditor.

"Mr. Peace," said the terror tamely. "I'm sorry I did not send the goods you ordered. I'll take them off your hands, or you can pay me what you think is right."
—Detroit Free Press.

MARKETS AND MARKET NOTES

QUEBEC MARKETS

Montreal, June 28, 1901.

HARDWARE.

IN spite of the intense heat of the past week, business has not suffered appreciably. The mills have now about caught up with their barb wire orders, and next week the wholesalers will be able to ship from stock. Previous to the meeting of the wire nail manufacturers in St. John on Tuesday, there were some rumors that the price would be raised. They proved unfounded, as the price remains unaltered. There is still reported a scarcity of wire nails, the mills not having caught up with the brisk demand that set in this spring. The long shut-down when the water was out of the canal, coupled with the inability to secure quick shipments of wire rods, have been sources of annoyance to the rolling mills. The scarcity of wringers and washing machines is still felt, and the manufacturers say that they are weeks behind. Outside of these facts the market presents little that is new. Although business is not rushing just now, dealers expect to see affairs take a better turn when the bountiful harvest promised us has been husbanded.

BARB WIRE — The manufacturers have pretty well come up with their orders, and wholesalers expect to have stocks next week from which they will be able to fill new orders. The price is $3.05 per 100 lb. f.o.b. Montreal.

GALVANIZED WIRE — The demand is hardly as brisk as it was, yet there is still a fair inquiry reported. Goods are now more plentiful. We quote as follows: No. 5, $4.25; Nos. 6, 7 and 8 gauge, $3.55; No. 9, $3.10; No. 10, $3.75; No. 11, $3.85; No. 12, $3.25; No. 13, $3.35; No. 14, $4.25; No. 15, $4.75; No. 16, $5.

SMOOTH STEEL WIRE—A steady trade is being done in oiled and annealed. We quote oiled and annealed as follows: No. 9, $2.80; No. 10, $2.87; No. 11, $2.90; No. 12, $2.95; No. 13, $3.15 per 100 lb. f.o.b. Montreal, Toronto, Hamilton, London, St. John and Halifax.

FINE STEEL WIRE—The regular trade is passing at 17½ per cent. off the list.

BRASS AND COPPER WIRE — Prices are steady at 55 and 2½ per cent. on brass, and 50 and 2½ per cent. on copper.

FENCE STAPLES—These goods keep going out. We quote: $3.25 for bright, and $3.75 for galvanized, per keg of 100 lb.

WIRE NAILS — The scarcity of wire nails continues to be felt quite keenly. The manufacturers held a meeting at St. John, N.B., on Tuesday, but no change was made in the price. The situation, however, is firm. We quote as follows: $2.85 for small lots and $2.77½ for carlots, f.o.b. Montreal, London, Toronto, Hamilton and Gananoque.

CUT NAILS — The demand is rather slow, although shingle nails have been in good request. We quote: $2.35 for small and $2.25 for carlots; flour barrel nails, 25 per cent. discount; coopers' nails, 30 per cent. discount.

HORSE NAILS—There is not a great deal doing in this line just now. Discounts are unchanged. "C" brand is held at a discount of 50 and 7½ per cent. off the new list. "M" brand quoted at 60 per cent. off old list on oval and city head and 66⅔ per cent. off countersunk head. Monarch's discount is 66⅔ per cent., and 70 per cent. in 25 box lots.

HORSESHOES — The horseshoe market is featureless. We quote as follows: Iron shoes, light and medium pattern, No. 2 and larger, $3.50; No. 1 and smaller, $3.75; snow shoes, No. 2 and larger, $3.75; No. 1 and smaller, $4.00; X L steel shoes, all sizes, 1 to 5, No. 2 and larger, $3.60; No. 1 and smaller, $3.85; feather-weight, all sizes, $4.85; toe weight steel shoes, all sizes, $5.95 f.o.b. Montreal; f.o.b. Hamilton, London and Guelph, 10c. extra.

POULTRY NETTING—The demand is slow; if it were brisk there would not be enough goods to supply it. Stocks are very light. We quote 50 and 10 per cent. off list A and 50 and 5 per cent. off lists B, C and D.

GREEN WIRE CLOTH—There is a continued active demand for screen cloths, intensified by this hot weather. The price is $1.35.

SCREEN DOORS AND WINDOWS — A good business is also being done in screens. We quote as follows: Screen doors, plain cherry finish, $7.30 per doz.; do. fancy, $11.50 per doz.; walnut, $7.30 per doz., and yellow, $7.45; windows, $2.25 to $3.50 per doz.

SCREWS—Prices are steady as also is the demand. Discounts are: Flat head bright, 87½ and 10 per cent. off list; round head bright, 82½ and 10 per cent.; flat head brass, 80 and 10 per cent.; round head brass, 75 and 10 per cent.

BOLTS — Dealers report that the demand for bolts has been good this season and is well maintained. Discounts are as follows: Norway carriage bolts, 65 per cent.; common, 60 per cent.; machine bolts, 60 per cent.; coach screws, 70 per cent.; sleigh shoe bolts, 72½ per cent.; blank bolts, 70 per cent.; bolt ends, 62½ per cent.; plough bolts, 60 per cent.; tire bolts, 67½ per cent.; stove bolts, 67½ per cent. To any retailer an extra discount of 5 per cent. is allowed. Nuts, square, 4c. per lb. off list; hexagon nuts, 4½c. per lb. off list. To all retailers an extra discount of ½c. per lb. is allowed.

BUILDING PAPER—A good inquiry has been felt this week. We quote as follows: Tarred felt, $1.70 per 100 lb.; 2-ply ready roofing, 80c. per roll; 3-ply, $1.05 per roll; carpet felt, $2.25 per 100 lb.; dry sheathing, 30c. per roll; tar sheathing, 40c. per roll; dry fibre, 50c. per oll tarred fibre, 60c. per roll; O.K. and I.X.L., 65c. per roll ; heavy straw sheathing, $28 per ton ; slaters' felt, 50c. per roll.

RIVETS AND BURRS—There is nothing new to report. Discounts on best iron rivets, section, carriage, and wagon box, black rivets, tinned do., coopers' rivets and tinned swedes rivets, 60 and 10 per cent.; swedes iron burrs are quoted at 55 per cent. off ; copper rivets, 35 and 5 per cent. off ; and coppered iron rivets and burrs, in 5-lb. carton boxes, are quoted at 60 and 10 per cent. off list.

BINDER TWINE—There is fairly good demand for binder twine at the moment. Prices are firm. We quote : Blue Ribbon, 11½c. ; Red Cap, 9¾c. ; Tiger, 8¾c.; Golden Crown, 8c.; Sisal, 8¼c.

CORDAGE—Dealers find a ready sale for all kinds of cordage. Manila is worth 13½c. per lb. for 7-16 and larger ; sisal brings 10c. and lath-yarn, 10c.

HARVEST TOOLS—The demand is only fair. The discount is 50, 10 and 5 per cent.

SPADES AND SHOVELS—A small business is passing. The discount is 40 and 5 per cent.

LAWN MOWERS—The trade in this line has rather slackened. We quote : High wheel, 50 and 5 per cent. f.o.b. Montreal ; low wheel, in all sizes, $2.75 each net ; high wheel, 11-inch, 30 per cent. off.

FIREBRICKS—Business is of small proportions. We quote : Scotch at $17.50 to $22 and English at $17 to $21 per 1,000 ex wharf.

CEMENT—The bulk of the business is confined to foreign cement. We quote : German cement, $2.35 to $2.50; English, $2.25 to $2.35 ; Belgian, $1.70 to $1.95 per bbl. ex wharf, and American, $2.30 to $2.45, ex cars.

METALS.

The metal market is steady and firm. The arrivals coming to this market are rather small, and the market continues to be lightly supplied. The demand from the country is moderate, and prices are steady.

PIG IRON—The market for this article is rather dull at the moment. No. 1 Canadian is worth about $18 per ton, and Summerlee $20.50 to $21 ex wharf.

BAR IRON—Bar iron is steady to firm at $1.75 to $1.80 for merchants' bar and $2 for horseshoe.

BLACK SHEETS—Some firms are cleaned out. Prices are steady. We quote : 8 to 16 gauge, $2.50 to $2.60; 26 gauge, $2.55 to $2.65, and 28 gauge, $2.60 to $2.70.

GALVANIZED IRON — Good sales are reported at firm prices. We quote as follows : No. 28 Queen's Head, $4.50 ; Apollo, 10¾ oz., $4.50, and Comet, $4.30, with a 10c. reduction for case lots.

COPPER—There has been a small demand at 18 to 18½c.

INGOT TIN—Lamb and Flag is worth 33 to 34c.

LEAD—Steady at $3.70 to $3.80.

LEAD PIPE—There is no change. Business is fairly brisk. We quote : 7c. for ordinary and 7½c. for composition waste, with 30 per cent. off.

IRON PIPE — The market maintains its firmness. We quote : Black pipe, ¼, $3 per 100 ft.; ⅜, $3 ; ½, $3.05 ; ¾, $3.30 ; 1-in., $4.75 ; 1¼, $6.45 ; 1½, $7.75 ; 2-in. $10.35. Galvanized, ½, $4.60; ¾, $5.25; 1-in., $7.50 ; 1¼, $9.80 ; 1½, $11.75 ; 2-in., $16.

TINPLATES—The prices of tinplates are on the upward trend. The demand is at present quite strong. We quote : Coke plates, $3.75 to $4 ; charcoal, $4.25 to $4.50 ; extra quality, $5 to $5.10.

CANADA PLATE—Canada plate is still rather scarce, and the demand for small lots is active. We quote : 52's, $2.55 to $2.60 ; 60's, $2.65 to $2.70; 75's, $2.70 to $2.80; full polished, $3.10, and galvanized, $4.

STEEL—Unchanged. We quote : Sleighshoe, $1.95 ; tire, $2 ; bar, $1.95; spring, $2.75 ; machinery, $2.75, and toe-calk, $2.50.

SHEET STEEL—We quote: Nos. 22 and 24, $3, and Nos. 18 and 20, $2.85.

TOOL STEEL—Black Diamond, 8c. and Jessop's, 13c.

TERNE PLATES—Supplies are none too plentiful. The jobbing price is $7.50.

COIL CHAIN—There is no change to report. Trade is quiet. We quote as follows: No. 6, 11½c.; No. 5, 10c.; No. 4, 9½c.; No. 3, 9c.; ¼-inch, [illegible] per lb.; 5-16, $4.85; 5-16 exact, $5.30; ⅜, $4.40; 7-16, $4.20; ½, $3.95; 9-16, $3.85; ⅝, $3.55; ¾, $3.45; ⅞, $3.40; 1-in., $3.35. In carload lots an allowance of 10c. is made.

SHEET ZINC—The market is firm since the recent advance in England, but dealers have not taken full advantage of the rise. The selling figure is $5 75 to $6 25.

ANTIMONY—Quiet, at 10c.

ZINC SPELTER—Is worth 5c.

SOLDER—We quote: Bar solder, 18½c.; wire solder, 20c.

GLASS.

Trade is fairly brisk. We quote as follows: First break, $2.10; second, $2.20 for 50 feet; first break, 100 feet, $3.90; second, $4.10; third, $4.60; fourth, $4.85; fifth, $5.35; sixth, $5.85, and seventh, $6.35.

PAINTS AND OILS.

June business is decidedly brisk, and no a complaint from the manufacturers do we hear. Turpentine is steady at the rise. Linseed oil is firming up on forward quotations, which was not expected by the trade. Paris green is in excellent demand, and stocks are running low. We quote:

WHITE LEAD—Best brands, Government standard, $6.25; No. 1, $5.87½; No. 2, $5.50; No. 3, $5.12½, and No. 4, $4.75 all f.o.b. Montreal. Terms, 3 per cent. cash or four months.

DRY WHITE LEAD—$5.25 in casks; kegs, $5.50.

RED LEAD—Casks, $5.00; in kegs, $5.25.

DRY WHITE ZINC—Pure, dry, 6¼c.; No. 1, 5¼c.; in oil, pure, 7¼c.; No. 1, 6¼c.; No. 2, 5¼c.

PUTTY—We quote: Bulk, in barrels, $1.90 per 100 lb.; bulk, in less quantity, $2.05; bladders, in barrels, $2.10; bladders, in 100 or 200 lb. kegs or boxes, $2.25; in tins, $2.55 to $2.65; in less than 100-lb. lots, $3 f.o.b. Montreal, Ottawa, Toronto, Hamilton, London and Guelph. Maritime Provinces 10c. higher, f.o.b. St. John and Halifax.

LINSEED OIL—Raw, 80c.; boiled, 83c. in 5 to 9 bbls., 1c. less, 10 to 20 bbl. lots, open, net cash, plus 2c. for 4 months. Delivered anywhere in Ontario between Montreal and Oshawa at 2c. per gal. advance and freight allowed.

TURPENTINE—Single bbls., 55c.; 2 to 4 bbls., 54c.; 5 bbls. and over, open terms, the same terms as linseed oil.

MIXED PAINTS—$1.20 to $1.45 per gal.

CASTOR OIL—8¾ to 9¼c. in wholesale lots, and ½c. additional for small lots.

SEAL OIL—47½ to 49c.

COD OIL—32½ to 35c.

NAVAL STORES—We quote: Resins, $2.75 to $4 50, as to brand; coal tar, $3.25 to $3.75; cotton waste, 4½ to 5½c. for colored, and 6 to 7½c. for white; oakum, 5½ to 6½c., and cotton oakum, 10 to 11c.

PARIS GREEN—Petroleum barrels, 16¾c. per lb.; arsenic kegs, 17c.; 50 and 100-lb. drums, 17½c.; 25-lb. drums, 18c.; 1-lb. packages, 18½c.; ½-lb. packages, 20½c.; 1-lb. tins, 19½c.; ½-lb. tins, 21½c. f.o.b. Montreal; terms 3 per cent. 30 days, or four months from date of delivery.

SCRAP METALS.

There is no important change to report in this market, which continues active. Dealers are now paying the following prices in the country: Heavy copper and wire, 13 to 13½c. per lb.; light copper, 12c.; heavy brass, 12c.; heavy yellow, 8½ to 9c.; light brass, 6½ to 7c.; lead, 2½ to 2¾c. per lb.; zinc, 2¼ to 2½c.; iron, No. 1 wrought, $14 to $16 per gross ton f.o.c. Montreal; No. 5 cast, $13 to $14; stove plate, $8 to $9; light iron, No. 2, $4 a ton; malleable and steel, $4; rags, country, 60 to 70c. per 100 lb.; old rubbers, 7¼c. per lb.

HIDES.

Hides are steady at the advance quoted last week. The demand from tanners is active and prices are firm. We quote: Light hides, 7c. for No. 1; 6c. for No. 2, and 5c. for No. 3. Lambskins, 15c.; sheepskins, 90c. to $1; calfskins, 10c. for No. 1 and 8c. for No. 2.

ONTARIO MARKETS.

TORONTO, June 28, 1901.

HARDWARE.

THERE has been a great deal of shipping from Toronto during the past week, business being largely of a sorting-up nature, principally in harvest tools, screen doors and windows, poultry netting and fence wires. There has been an excellent demand for cut shingle nails and wire nails of all sizes. Jobbers, as well as manufacturers, are experiencing difficulty in filling orders in several sizes of nails. There is a good movement in building paper, cement, tar, pitch, etc., as building operations seem to be active in all parts of the country. The demand for rope in sizes suitable for hay-fork purposes is excellent. There is a good trade in ice cream freezers, also in oil stoves. The only change in prices during the week is an advance of 10c. per keg in cut nails.

BARB WIRE—The local market is bare of stock, but some shipments are expected to arrive during the next few days. There is a steady demand, with prices unchanged. We quote $3 05 per 100 lb. from stock Toronto; and $2 82½ f.o.b. Cleveland for less than carlots, and $2.70 for carlots.

GALVANIZED WIRE—There is an excellent trade doing, as some arrivals of stock have been noted recently. There is no change in prices. We quote as follows: Nos. 6, 7 and 8, $3.50 to $3.85 per 100 lb., according to quantity; No. 9, $2.85 to $3.15; No. 10, $3.60 to $3.95; No. 11, $3.70 to $4.10; No. 12, $3 to $3 30; No. 13, $3 10 to $3 40; No. 14, $4.10 to $4 50; No. 15, $4.60 to $5.05; No. 16, $4.85 to $5.35. Nos. 6 to 9 base f.o.b. Cleveland are quoted at $2.57½ in less than carlots and 12c. less for carlots of 15 tons.

SMOOTH STEEL WIRE—The demand for both oiled and annealed wire continues excellent, with prices firm. Net selling prices for oiled and annealed are: Nos. 6 to 8, $2.90; 9, $2.80; 10, $2.87; 11, $2.90; 12, $2.95; 13, $3.15; 14, $3.37; 15, $3.50; 16, $3 65. Delivery points, Toronto, Hamilton, London and Montreal, with freights equalized on those points.

WIRE NAILS—There is an excellent demand for all sizes, and though prices are firm, there is no change. We quote: $2.85 base for less than carlots, and $2.72½ for carlots. Delivery points: Toronto, Hamilton, London, Gananoque and Montreal.

CUT NAILS—The feature of this market is the advance of 10c. per keg. The only line moving well is cut shingle nails. The base price is $2.45 per keg for less than carlots, and $2.35 for carlots. Delivery points: Toronto, Hamilton, London, Montreal and St. John, N.B.

HORSENAILS—There is still a fair sorting-up trade doing. Discount on "C" brand, oval head, 50 and 7½ per cent. off new list, and on "M" and other brands, 50, 10 and 5 per cent. off the old list. Countersunk head 60 per cent.

HORSESHOES — A moderate demand is reported. Prices are steady. We quote f.o.b. Toronto as follows: Iron shoes, No. 2 and larger, light, medium and heavy, $3.60; snow shoes, $3.85; light steel shoes, $3.70; featherweight (all sizes), $4.95; iron shoes, No. 1 and smaller, light, medium and heavy (all sizes), $3.85; snow shoes, $4; light steel shoes, $3.95; featherweight (all sizes), $4.95.

SCREWS — Business continues fair, with prices firm and unchanged. Discounts are as follows: Flat head bright, 87½ and 10 per cent.; round head bright, 82½ and 10 per cent.; flat head brass, 80 and 10 per cent.; round head brass, 75 and 10 per cent.; round head bronze, 65 per cent., and flat head bronze at 70 per cent.

RIVETS AND BURRS — A fairly good trade is doing. We quote as follows: Iron rivets, 60 and 10 per cent.; iron burrs, 55 per cent.; copper rivets and burrs, 25 and 5 per cent.

BOLTS AND NUTS—There is a brisk demand. In fact, it is so large that dealers have difficulty in securing supplies to meet it. We quote: Carriage bolts (Norway), full square, 65 per cent.; carriage bolts full square, 65 per cent.; common carriage bolts, all sizes, 60 per cent.; machine bolts, all sizes, 60 per cent.; coach screws, 70 per cent.; sleighshoe bolts, 72½ per cent.; blank bolts, 60 per cent.; bolt ends, 62½ per cent.; nuts, square, 4c. off; nuts, hexagon, 4½c. off; tire bolts, 67½ per cent.; stove bolts, 67½; plough bolts, 60 per cent.; stove rods, 6 to 8c.

ROPE—The principal demand is for sizes suitable for hay-fork purposes. The base price is unchanged at 10c. for sisal and 13½c. for manila.

BINDER TWINE—A fair sorting up trade is doing. We quote: Pure manila, 650 ft., 12c.; manila, 600 ft., 9½c.; mixed, 550 ft., 8½c.; mixed, 500 ft., 8 to 8½c.

SPORTING GOODS—The inqury for shot-guns, rifles and ammunition shows considerable improvement, and it is likely to be brisk in a week or two.

ENAMELLED WARE AND TINWARE — There is a fair demand for most lines of both enamelled ware and tinware.

GAS AND OIL STOVES — The sale of oil stoves show that these stoves are steadily coming into favor for summer purposes, especially in the older towns and cities. There is not much doing in gas stoves.

COOKING AND HEATING STOVES—Orders for later delivery are good, up to the average at this season, though, of course, no deliveries will be made for some time.

ICE CREAM FREEZERS—The hot weather has given a great impetus to the demand for freezers, and a good trade has been done in them all week.

REFRIGERATORS—The demand has kept up rather longer than anticipated, and stocks in jobbers' hands have, as a consequence, been reduced to a very small compass. One house was practically cleaned out early this week.

GREEN WIRE CLOTH—Quite a large trade has been done during the past week, the trade being of a sorting-up nature. While jobbers have not yet experienced difficulty in getting the sizes required, it is predicted that they will have trouble before long. Prices are still at $1.35 per 100 sq. ft.

SCREEN DOORS AND WINDOWS — The demand continues brisk. There is difficulty in securing some sizes and styles. There is no change in prices. We quote: Screen doors, 4-in. styles, $7.20 to $7.80 per doz.; ditto, 3-in. styles, 20c. per doz. less; screen windows, $1.60 to $3.60 per doz., according to size and extension.

BUILDING PAPER—A brisk trade is doing with prices steady. We quote: Building paper, 30c.; tarred paper, 40c., and tarred roofing, $1.65.

POULTRY NETTING—There is a fair trade doing. Discount is still 55 per cent.

HARVEST TOOLS—The demand, particularly for scythes and snaths, and generally for all lines keeps good. Discount 50, 10 and 5 per cent.

SPADES AND SHOVELS—A good sorting-up trade is reported. Discount 40 and 5 per cent.

EAVETROUGH—There is a good trade doing as has been the case for most of the season. The ruling price is $3.25 per 100 ft. for 10 inch.

LEATHER BELTING—There is a moderate trade doing. Prices are being cut somewhat, but the ruling discounts are as follows: Extra, 60, 10 and 5 per cent.; standard, 70 per cent.; No. 1, 70 and 10 per cent.

CEMENT — There is a steady demand with prices firm and unchanged. We quote barrel lots as follows: Canadian portland, $2.25 to $2.75; German, $3 to $3.15; English, $3; Belgian, $2.50 to $2.75; Canadian hydraulic, $1.25 to $1.50.

METALS.

The feature of the market is the firm feeling which is manifested in practically every line, though there is no change whatever in prices. There are no American galvanized sheets on the market. Tinplates are scarce, but some stocks are due to arrive in a few days. Last week's advance in iron pipe was well maintained.

PIG IRON—The market is in a sound condition and prices are firm throughout. Canadian iron on track Toronto we quote at $18 per ton for No. 1, $17.50 for No. 2, and $17 for No. 3.

BAR IRON—There is a moderate demand, but mills are filling old orders. The ruling price is still $1.85 to $1.90.

STEEL—There is considerable movement of old orders, but little new business is reported. Price is steady and unchanged. We quote: Merchantable cast steel, 9 to 15c. per lb.; drill steel, 8 to 10c. per lb.; "B C" and "Black Diamond" tool steel, 10 to 11c.; Jessop's, Morton's and Firth's tool steel, 12½ to 13c.; toe calk steel, $2.85 to $3; tire steel, $2.30 to $2.50; sleighshoe steel, $2.10 to $2.25; reeled machinery steel, $3; hoop steel, $3.

GALVANIZED SHEETS — The market is cleared up of American sheets, but there is a sufficiency of English material to supply the trade. The ruling quotation on 28 gauge English is $4.50 in cask lots.

BLACK SHEETS — There is a moderate trade, and prices are firm. We quote: 28 gauge, common sheets at $3, and dead flat at $3.50.

CANADA PLATES — There is very little doing. Prices are stiff. We quote: All dull, $2.90; half polished, $3, and all bright, $3.50.

PIG TIN—The market is steady this week, but there is little doing. We quote 31½ to 32c.

TINPLATES—There has been considerable scarcity, but stocks are beginning to arrive.

TINNED SHEETS—The season is about over. We quote 28 gauge at 8½c.

COPPER — Business is moderate. We quote: Ingot, 17¾c., and bars, 23 to 25c. In both New York and London prices have advanced during the past week.

SOLDER—There is an excellent demand for solder.

IRON PIPE—The market is firm, since the advance of 5 per cent. noted last week, and a fair trade is doing. One-inch black is now quoted at $5.40, and one inch galvanized at $7.95 per 100 ft.

LEAD—Trade is quiet at 4¼ to 4½c.

SPELTER—There is a fair demand at 5½ to 6c. per lb.

ZINC SHEETS—A fairly steady trade is noted, but generally in small lots. We quote 6½c. for casks, and 6¾c. for part casks.

PAINTS AND OILS.

The feature of the market is the continued big demand for paris green, which has been so great that a scarcity is predicted. The movement is larger than usual during June, but small in comparison with May or April. Prices are firm throughout. We quote:

WHITE LEAD—Ex Toronto, pure white lead, $6.37½; No. 1, $6; No. 2, $5.67½; No. 3, $5.25; No. 4, $4.87½; genuine dry white lead in casks, $5.37½.

RED LEAD—Genuine, in casks of 560 lb., $5.50; ditto, in kegs of 100 lb., $5.75; No. 1, in casks of 560 lb., $5; ditto kegs of 100 lb., $5.25.

LITHARGE—Genuine, 7 to 7½c.

ORANGE MINERAL—Genuine, 8 to 8½c.

WHITE ZINC—Genuine, French V.M., in casks, $7 to $7.25; Lehigh, in casks, $6.

PARIS WHITE—90c. to $1 per 100 lb.

WHITING — 70c. per 100 lb.; Gilders' whiting, 80c.

GUM SHELLAC — In cases, 22c.; in less than cases, 25c.

PARIS GREEN—Bbls., 16¾c.; kegs, 17c.; 50 and 100-lb. drums, 17½c.; 25-lb. drums, 18c.; 1-lb. papers, 18½c.; 1-lb. tins, 19½c.; ½-lb. papers, 20½c.; ½-lb. tins, 21½c.

PUTTY — Bladders, in bbls., $2.10; bladders, in 100 lb. kegs, $2.25; bulk in bbls., $1.90; bulk, less than bbls. and up to 100 lb., $2.05; bladders, bulk or tins, less than 100 lb., $2.90.

PLASTER PARIS—New Brunswick, $1.90 per bbl.

PUMICE STONE — Powdered, $2.50 per cwt. in bbls., and 4 to 5c. per lb. in less quantity; lump, 10c. in small lots, and 8c. in bbls.

LIQUID PAINTS—Pure, $1.20 to $1.30 per gal.

CASTOR OIL—East India, in cases, 10 to 10½c. per lb. and 10¾ to 11c. for single tins.

LINSEED OIL—Raw, 1 to 4 barrels, 82c.; boiled, 85c.; 5 to 9 barrels, raw, 81c.; boiled, 84c., delivered. To Toronto, Hamilton, Guelph and London, 1c. less.

TURPENTINE—Single barrels, 56c.; 2 to 4 barrels, 55c., delivered. Toronto, Hamilton and London 1c. less. For less quantities than barrels, 5c. per gallon extra will be added, and for 5-gallon packages, 50c., and 10-gallon packages, 80c. will be charged.

GLASS.

Prices are firm. New goods are arriving, but the total stocks of some sizes held here are light. We quote: Under 26 in., $4 15; 26 to 40 in., $4.45; 41 to 50 in., $4 85; 51 to 60 in., $5 15; 61 to 70 in., $5.50; double diamond, under 26 in., $6; 26 to 40 in., $6.65; 41 to 50 in., $7 50; 51 to 60 in., $8.50; 61 to 70 in., $9.50. Toronto, Hamilton and London. Terms, 4 months or 3 per cent. 30 days.

OLD MATERIAL.

Prices are firm and unchanged throughout, with a light trade doing. We quote jobbers' prices as follows: Agricultural scrap, 50c. per cwt.; machinery cast, 50c. per cwt.; stove cast, 45c.; No. 1 wrought 40c. per 100 lb.; new light scrap copper, 12c. per lb.; bottoms, 11c.; heavy copper, 12½c.; coil wire scrap, 12½c.; light brass, 7c.; heavy yellow brass, 10c.; heavy red brass, 10½c.; scrap lead, 2¾c.; zinc, 2c.; scrap rubber, 6½c.; good country mixed rags, 65 to 75c.; clean dry bones, 40 to 50c. per 100 lb.

HIDES, SKINS AND WOOL.

HIDES—A fair demand is reported. Prices are unchanged. We quote: Cowhides, No. 1, 6½c.; No. 2, 5½c.; No. 3, 4½c. Steer hides are worth 1c. more. Cured hides are quoted at 7 to 7½c.

SKINS—Prices are firm, but unchanged throughout. We quote: No. 1 veal, 8-lb. and up, 9c. per lb.; No. 2, 8c.; dekins, from 60 to 70c.; culls, 20 to 25c; sheepskins, 90c. to $1.

WOOL—The market is dull and holders are loth to accept present prices and buyers are not disposed to make an advance. We quote: Combing fleece, washed, 13c., and unwashed, 8c.

PETROLEUM.

There is very little doing. Prices are unchanged. We quote: Pratt's Astral, 16 to 16½c. in bulk (barrels, $1 extra); American water white, 16½ to 17c. in barrels; Photogene, 15½ to 16c.; Sarnia water white, 15 to 15½c. in barrels; Sarnia prime white, 14 to 14½c. in barrels.

COAL.

An advance of 10c. is noted for July shipments over June prices. We quote at international bridges: Grate, $4 75 per gross ton; egg, stove and nut, $5 per gross ton with a rebate of 20c. off for June shipments.

MARKET NOTES.

Cut nails are 10c. per keg. dearer.

American galvanized sheets are not offering, and present stocks are practically exhausted.

H. S. Howland, Sons & Co. are in receipt of a shipment of "Robin Hood" smokeless loaded shells, 12 gauge, and 6, 7 and 8 shot. These sizes are particularly well adapted for trap shooting.

CANADIAN SCALES.

C. Wilson & Son, scale manufacturers, Toronto, have just completed their 50th year in the manufacture of scales. Their scales are in use in different parts of the world, but their main business is in Canada. They make over a hundred different kinds of scales, from the 100 ton railroad track scale for weighing a loaded train down to the finest tea scale that will turn with a postage stamp. They make a specialty of fine grocers' computing scales, with agate and ball bearings. They were awarded highest medals at Paris Exposition, France.

PERSONAL MENTION.

Mr. David Parks, who formerly represented The Canada Paint Co. on the road, and who has for some time been foreman in the McLaughlin Carriage Works, Oshawa, is severely ill. Owing to a serious attack of pneumonia his life was despaired of, but late reports give hope of his recovery.

A ROLLING MILL BOILER.

A Sydney, N.S., despatch says that the Dominion Iron and Steel Co. have awarded a contract to Thaher & Co., New York, for a 5,000 horse-power boiler and equipments, which will be installed in the rolling mill of the company. The boiler will be built by the Aultman & Taylor Machinery Co., of Mansfield, Ohio.

A RECOMMENDABLE AIR RIFLE.

A new air rifle that has much to recommend it is being shown to the trade by Mr. Alexander Gibb, of Montreal, The Canadian agent for The Rapid Rifle Company, Limited, of Grand Rapids, Mich. The "Rapid" Rifle is all metal full nickled and looks like a fine high-grade hammerless sporting rifle. It is light and of a size that will please every boy who once sees and shoots it.

The manufacturers of the "Rapid" claim that it has all the good points of old air rifles and is wanting the weak; this new gun is the result of many years of experience. They guarantee no more loose stocks, no worn out plungers, no weak reports, and no broken draw wires. It is sold under an absolute guarantee and, moreover, it is simple and adjustable. In fact, any boy with a screwdriver and a little common sense can take down and put up a "Rapid." It retails in the States for $1.25. BB shot is used as ammunition.

Mr. Gibb reports that the gun has already met with a good reception. Hardware dealers all over the country ought to find it a saleable article next fall and during the Christmas season.

SILVERWARE MANUFACTURING LEAGUE.

The game scheduled for this league for Saturday between the Toronto Silver Plate Company and Roden Bros. was to have been played on Stanley Park at 3.30 p.m. Rodens failed to appear up to 4.15, and Umpire Uniac requested Captain Carpenter of the Toronto Silver Plate Company to place his men on the field, and when Rodens did not appear awarded the game to the T. S. P. Co. by default. Score, 9 to 0.

The Rodens subsequently appeared having obtained the battery and shortstop of the St. Mary's Senior League team, and tried to arrange a game, which, according to the league constitution, was illegal. The Toronto Silver Plate Company rightly refused to participate, judging it a small piece of business on the part of Rodens being parties to a scheme which they consider is on foot to try to besmirch the reputation of the Toronto Silver Plate Company, who have won games purely on their merits. The Toronto Silver Plate Company enjoy the reputation of leading the league by a percentage of 1,000 points.

THE MAKING OF TINPLATES.

THE condition of the Welsh tinplate trade, which is the chief industry of this district, is the source of great anxiety to local manufacturers, according to United States Consul Prees, at Swansea. The cause of this depression is attributed to the keen competition of the United States. Since the boom preceding the adoption of the tariff of 1890 by the United States, the Welsh tinplate trade has been very irregular. All works have been periodically idle, although the reasons given by the masters for closing down are "restriction of make," "remodeling of finishing department," etc.

During the past five years, methods of work in the tin house or finishing department have undergone a complete change, and the coating of the black-plate sheet is now done in an entirely different manner. Formerly, a tinman, wash man, greaser, cleaner, and duster were employed to coat and finish the black-plate; the work is now effectively done by a tinning machine, manipulated by a tinman and a boy.

Following is a description of the method of operating these tinning, cleaning, and dusting machines:

The tinman places a wet black-plate sheet in the tinning pot, which contains a chemical compound called "flux"; from this receptacle the sheet, by means of guides and rolls, which I shall call rolls No. 1, passes through a bed of molten tin. Then, by means of guides and rolls No. 2, the sheet is conducted to a grease pot, which is a bath of hot palm oil. From this bath the sheet, now thoroughly coated, emerges between three pairs of rolls No. 3, and is transferred by the boy to the cleaning machine. In the cleaning machine, the tin sheet is conveyed by small rollers through layers of bran, and all grease or oil spots are removed. The sheet is finally placed in a dusting machine. This last is a very simple affair, consisting of three or more pairs of rollers covered with sheepskin, between which the sheet passes. The latter is then ready for the assorting room.

This method of coating and finishing tinplates is a great improvement on the old style. In addition to saving time—the whole process occupying but a few seconds—labor, and expense, the finished article is better. It bears a clearer and more uniform surface; the "yield," also, is greater, an average of about two pounds of tin being used per standard box of plates.

Notwithstanding the extensive improvements, Welsh manufacturers still find competition keen, and a new workmen's wage list is under consideration, to replace that of 1874. This means a reduction of wages of from 19 to 70 per cent.

The present rate of freight on tinplates to New York is 9s. ($2.19) per ton, and the bulk of plates shipped to that port consists of 18 3-4 by 14 oil plates.

Every effort of shippers is now concentrated on opening up new markets for this trade in the Far East, and frequent shipments are made to Singapore, China and Japan. Extensive exports are also made to France, Germany, Austria, Italy and Russia.

MARITIME BOARD OF TRADE.

The executive officers of the Maritime Board of Trade are making arrangements for the annual convention to be held in Chatham, N.B., in August next. Since the last convention two boards of trade have been organized, one at Digby, N.S., and the other at Woodstock, N.B.

TEMPERATURE OF HALIFAX.

At a meeting of the Nova Scotian Tourist Association in Halifax, a few days ago, it was decided to arrange for the publication in a New York paper of the temperature daily in Halifax during the hot season. The paragraph would read like this:

"Halifax, June 21, ther. 60
New York. 95
Go to a cool place for vacation."

HEATING AND PLUMBING

SANITARY EARTHENWARE.

We understand that the demand for sanitary earthenware at the present time is so great that the majority of the potters are about two months behind on their orders. This is due to the fact that the consumption of all kinds of plumbing material is extraordinarly heavy throughout the entire Western and Middle States, and as there has been considerable labor trouble in the East Liverpool potteries, orders which heretofore went to the Western potteries have been pouring into the Trenton manufacturers in such volume that the regular Eastern trade is put to considerable trouble in getting goods from there. The business in this vicinity is opening up in good volume at the present time. This is particularly true of the summer resorts on the seashore and in the mountains, as all work which is being done in these places has to be done promptly or not at all, in order to insure comfort to the guests who visit them. The plumbers in these localities are being given the preference on shipments over the local trade in Greater New York. The jobbers can always get a higher price for their wares from the out-of-town plumber than they can for those nearby, owing to the fact that the greater part of the out-of-town orders come through the mail, and a man situated at a distance from his point of purchase is unable to shop around and get the lowest prices which are being quoted.—Metal Worker, June 22.

VENTILATING A SCHOOLROOM.

From R. R. M. C., Wilkesbarre, Pa.—In reply to the inquiry of "W. P. G." in The Metal Worker of June 8 in reference to heating and ventilating a schoolroom, 32 x 24 feet in size, I would say that neither of his methods is correct. Allowing 15 square feet of floor surface for each pupil, the room will accommodate 50 scholars. In accordance with the general rule of allowing 30 cubic feet of air per scholar per minute, 1,500 cubic feet of air per minute will be required in the room, or 90,000 cubic feet must be warmed per hour from zero to 100 degrees. This would require 360 square feet of indirect radiating surface, and it would be placed in the basement at the base of the warm air flue. The flue should have an area of four square feet, and be smooth on the inside. There should be an opening from the flue into the room at a point about eight feet from the floor, and the opening should have an area of 20 per cent. larger than that of the flue. The ventilation flue should be on the same side of the room, with an opening near the floor and having an area of about five square feet, this flue to terminate above the highest point of the roof and be warmed with the coils or a stack heater.—Metal Worker.

GAS WATER HEATER IN FRANCE.

The United States Consul at Nice, under date of May 17, writes: "I have had an opportunity to examine a new gas water-heating machine, which it is claimed furnishes hot water to a number of rooms, and the gas in which can be lighted or extinguished at any distance from the heater. The apparatus may be placed in any part of the house. In the case of buildings furnished with water by a reservoir in the garret or mansard, the only requisite is that the apparatus shall be placed not less than five feet below the reservoir. A second contrivance closes that portion of the mechanism used to provide hot water, so that the supply may be received cold, as desired. The apparatus contains a device whereby gas not consumed is prevented from accumulating in the apparatus, and explosions are obviated.

"In the experiment witnessed by me, one faucet was placed immediately at the apparatus and another some 50 feet away. The diminutive gas-jet was lighted and turned into the apparatus, and the water feed-pipe faucet turned. Upon opening the discharge faucet, the gas was instantly lighted, and 10 seconds later about 12 quarts of water at 35 degrees C. was issuing from the faucet pipe per minute. On opening the faucet half way, the supply came cold; on opening it a little more, hot water came; and on closing it, the gas was instantly extinguished. The inventor asks from $60 to $70 each for the apparatuses."

SOME BUILDING NOTES.

A Roman Catholic church is being erected in Yarmouth, N.S.

The Acton, Ont., Flour Mills are being enlarged and improved.

Metcalfe & Son, Portage la Prairie, Man., intend erecting a new warehouse.

F. Gleb intends erecting a new brick house in Acton, Ont., this summer.

Tenders have been received by D. M. Curry, for a new court house in Sydney, N.S.

G. F. & J. Galt, Winnipeg, are to build a new factory for their package tea business in Winnipeg.

Tenders are asked by Alf. Young, Allandale, Ont., before July 26, for a new Presbyterian church in Allandale.

The Sussex Mercantile Co., Sussex, N.B., are building a warehouse for C. W. Upham. Edwin Fairweather has the contract.

An $80,000 wing is to be added to the Laval Normal School in Quebec. The contract has been given to Joseph Gosselin.

THE NATIONAL CONVENTION,

The annual convention of The National Association of Master Plumbers of Canada is being held in the Temple building, Toronto, as we go to press. A report of the proceedings will be given in next week's issue.

PLUMBING AND HEATING NOTES.

Moffatt & Doering, contractors, Sydney, N.S., have dissolved.

Rheume & Gratton have registered as contractors, in Maisonneuve, Que.

Tenders are asked by W. H. Langworthy before July 2, for steamheating the Port Arthur, Ont., public school.

R. H. O'Brien, who has carried on the plumbing business on Regent street, Fredericton, N.B., for a couple of years, formerly McGoldrick & O'Brien, has made an assignment to Sheriff Sterling. The principal creditors are: James S. Neill, Geo. S. Stanger, R. Chestnut & Sons, F. B. Edgecombe, Bank of Nova Scotia, Bank of British North America, Royal Bank of Canada, Jas. P. McManus, R. H. Simonds, Dougald McCatherin, M. A. Tweedale, John Macpherson, Daniel Lucy.

ST. JOHN PLUMBERS MAY ORGANIZE.

Frank Power, of Lunenburg, was at The Royal yesterday en route to Toronto to attend the convention of The Master Plumbers' Association, which will open there on Wednesday. While here he arranged for a meeting of the master plumbers of this city to consider the question of forming a local branch of The Master Plumbers' Association.

The meeting was held at The Royal yesterday afternoon, but the attendance was so small that nothing definite was decided upon, the whole matter being laid over until Mr. Power's return.

The association is not a new one to St. John, as it was formerly in operation here. Interest in it waned, however, with the result that its existence ceased. It is now hoped that Mr. Power will be successful in its reorganization, as the master plumbers feel that such a society is a necessity to them and will do much to justify its existence.—St. John Sun, June 25.

A NEW METAL.

A NEW metal which promises to be of much use to the trade has lately been introduced into Canada by the inventor of the process, Mr. R. R. Ditzel, of New York City.

This product has the color of silver, but it does not tarnish like that metal; it takes a high polish and is claimed to be vastly superior to plated ware as there is nothing in it to show use; it is absolutely impervious to the action of organic acid or salt water and cannot be affected, corroded or otherwise injured by any atmospheric conditions. It is especially adapted to washstands, water tanks, ice coolers, hot water and coffee urns, alcoves, trays, spittoons, bath tubs, butlers' pantries, liquor shakers, spoons, forks, piping and trimming for open plumbing; in fact, for all purposes where the color of silver is required. Practically speaking it can be used in a thousand different ways, being solid, it is susceptible of carrying engraving as well as solid silver and can be worked into any shape desired, spun or stamped.

Another metal manufactured by the same process has the color of nickel; it takes a high polish and is impervious to organic acid or atmospheric conditions; it will hold an edge almost equal to steel and is especially adapted for the manufacture of fine cutlery, carving, table and fruit knives and forks, surgical instruments, springs, optical and electrical purposes.

A company is being organized and has already been incorporated with a capital stock of half a million dollars to manufacture the metal in Toronto for the Canadian trade. A quantity of it has been used for fittings on steamships, etc., and is giving general satisfaction.

HARDWARE WINDOW DRESSING.

TICKET WRITING.

NOTHING is more important in the grocery window trim than the price and show cards. In fact, a window trim is almost useless as a means of inducing the customer to purchase, unless the goods are ticketed. When an article is ticketed it always appears a bargain. For those who cannot or who have never tried to make their own show cards, the following remarks may be of some value: First of all, "practice" is the watchword. Nothing succeeds without constant study and practice. One will be surprised at the advance that can be made if you "stick to it." Naturally, most of those who try it once are not satisfied with their attempt and pitch it up as a bad job.

When I first started ticket-writing I could not form a letter or a figure decently and became discouraged, as I never was a good sketcher; in fact, I could not draw anything, but I made up my mind that it was necessary for me to make my own cards. After a few weeks' practice I was surprised to find I was improving splendidly and found it more interesting each time I tried. By watching the different styles of lettering on bill-boards, illustrated magazines, etc., a person soon gets acquainted with the formation of letters and figures.

Card-writing can be well executed by almost any young man of intelligence with a few weeks' practice, and in most cases it is not only an interesting and instructive pastime, but a trimmer who is able to write his own cards can generally have a dollar or two added to his salary. Every merchant who has a man about the establishment who can make good show cards has an assistant whose services he ought to value, for nothing brings the money in any better than good window and interior displays "well ticketed."

To begin, purchase a couple of sable brushes. Nos. 5 or 6 are good practical sizes for general use. The best paints for general show-card work are water colors. These are to be preferred for the beginner's use, as they are easily made, and do not spoil brushes and soil the beginner's hands like the paints with shellac and turpentine ingredients. After the beginner has had a few practices with the water colors the black lettering should be tried as it makes the neatest card and is the easiest read of any colors, but it is harder to work with than the water colors and should not be attempted by the beginner until he has the lettering with the water colors down fairly well.

The water colors can be made in the following way: Buy a package of "Diamond Dye," the shade you want to use. Put about a half a teaspoonfull of the dye in enough mucilage or liquid glue to form a thick paste, then add about a half a cup of water and stir same until the dye has dissolved. Then it is ready for use. By adding or diminishing the quantity of water the different shades of your coloring can be made. For instance, say you desire a dark red. Put very little water on the red dye, but if a light red add more water. Any color can be made with the dyes except the black. The best black to use is ca[illegible]o paint or a little lampblack dissolved in turpentine, in which drop a few drops of alcohol. The brushes should be kept thoroughly clean after using the water colors. Dip the brush in water and dry it. Never leave the brushes standing on end as it will turn the hair up and your brush is useless after. Alter using the brushes in the black, or any color that has shellac or varnish in it, dip your brush in turpentine several times so as to keep it soft for future use.

By steady practice with the water colors, the beginner will find it does not take long to acquire a marked degree of proficiency and will also find it profitable and interesting work. Don't get discouraged at your first attempt. Nothing can be accomplished by becoming discouraged the first time one tries to make a ticket. Keep your eye on the style of lettering seen on trade newspapers and colored advertisements, and after a while you will be able to form your letters in any style of lettering. Draft out your letters on cardboard with your lead-pencil first. These letters can be erased.

A GRANITEWARE DISPLAY.

Illustration No. 2 is a display of graniteware. The idea is a good one but the arrangement of the pails and dishes under the archway should have been more uni-

No. 1—Sporting Goods Window.

form. The two piles of pails seen on the left of picture should have been divided, one pile to each side of arch and the others arranged at a uniform height in the centre. The idea though, is a neat one. The goods should have been better ticketed. There is not enough of them. Every article, in fact, should have been ticketed in order to make this a selling window.

A SPORTING GOODS DISPLAY.

The illustration No. 1 is a splendid display for a hardware trim to put in during the hunting season. It is very easily made in - sized window and would be sure to draw a crowd, and the results in sales from such a display would more than repay with good interest the time spent in its arrangement. All that is required to carry out this display is about six scantlings 2 x 2-inches, an oval or circular frame, some white cheesecloth for a background and for covering the framework, and a few long brads. The figure of a man or boy can be borrowed or rented from some house having them in stock. The swamp reeds could be substituted by any shrubbery found in the woods. The rifles are supported on revolvers which are kept in place by a long nail driven into the framework. The floor could be covered with excelsior dyed grass green. Fishing tackle could be shown along with this trim. The rifles ought to be price-ticketed. A large card in the centre should read similar to the following:

A COMPLETE RANGE
OF
SPORTING GOODS,
Note the prices. AT RIGHT PRICES.

DISPLAYS FOR DULL SEASONS.

A good idea during the dull season coming on and one that may result in a good many sales is to make a good display of household and other articles ranging in price from 5 to 25c.—not higher. Take a lot of medium sized baskets and fill each with one line of goods. Put a good sized price card on each basket and arrange it to the best advantage in your window, placing the articles that are most likely to attract attention (being the most needed) in the prominent positions. For instance, fill one basket with egg beaters and put a good sized card in this basket reading similar to this:

EGG -:- BEATERS
JUST 10c.

Fill another with, say, corkscrews, another with tin cups, or can openers, tin spoons, forks, salt and pepper shakers, mouse traps, match safes, scissors, tack hammers, screwdrivers; in fact, there are hundreds of lines of goods (small stuffs) that can be shown. People will come in for some of these little articles that they saw ticketed in the window and while in may see something else of much greater value and possibly purchase it. The main thing is to get them in anyway. Have the displays inside ticketed with price and explanatory show-cards, and you'll find hundreds of sales can be made that otherwise would not be effected were it not for the displays and the tickets. The warm season is at hand and this is the time to make a good catchy display of warm-weather goods. Get out your ice cream freezers, your refrigerators, ice picks, lemon squeezers, and everything pertaining to hot-weather necessities. Make a design of some kind and show these articles well ticketed.

"The Hardware and Metal" will be pleased to give its readers any information it possesses or to render them any assistance in getting up good window trims. We request hardware trimmers to send in photos of any hardware window trims that have received special mention. They will be reproduced in these columns if possible. The time to have a trim photographed so as to get the best results is late at night when the window is lit up. A long exposure with a sharp diaphragm will bring the desired results if the lights in the window are all right. Any trimmer desiring any assistance may write "H. H.," care of The Window Department.

No. 2— A Graniteware Window.

MANITOBA MARKETS.

WINNIPEG, June 24, 1901.

HARDWARE AND PAINTS, OILS AND GLASS.

THE hardware market is steady and fairly active in all lines. There has been some slight shading in prices during the week, but nothing of moment. The weather throughout the week has been perfect for crops as it has been turning gradually warmer, until to-day the ther. mometers have again reached 90 in the shade. All reports from country points go to show that present conditions are about as favorable as they can be. In consequence business is firm in tone and increasing in volume, although the country roads have been in too bad condition after the rain to admit of many farmers getting into the towns.

Quotations for the week are as follows:

Barbed wire, 100 lb.		$3 45
Plain twist		3 45
Staples		3 95
Oiled annealed wire	10	3 95
"	11	4 00
"	12	4 05
"	13	4 20
"	14	4 35
"	15	4 45
Wire nails, 30 to 60 dy, keg		3 50
" 16 and 20		3 60
" 10		3 60
" 8		3 70
" 6		3 75
" 4		3 90
" 3		4 15
Cut nails, 30 to 60 dy.		3 00
" 20 to 40		3 05
" 10 to 16		3 10
" 8		3 15
" 6		3 20
" 4		3 30
" 3		3 65
Horsenails, 45 per cent. discount.		
Horseshoes, iron, No. 0 to No 1		4 65
No. 2 and larger		4 40
Snow shoes, No. 0 to No. 1		4 90
No. 2 and larger		4 40
Steel, No. 0 to No. 1		4 95
No. 2 and larger		4 70
Bar iron, $2.50 basis.		
Swedish iron, $5.00 basis.		
Sleigh shoe steel		3 00
Spring steel		3 25
Machinery steel		3 75
Tool steel, Black Diamond, 100 lb		8 50
Jessop		13 00
Sheet iron, black, 10 to 20 gauge, 100 lb.		3 50
20 to 26 gauge		3 75
28 gauge		4 00
Galvanized American, 16 gauge		2 54
18 to 22 gauge		4 50
24 gauge		4 75
26 gauge		5 00
28 gauge		5 25
Genuine Russian, lb		12
Imitation " "		8
Tinned, 24 gauge, 100 lb.		7 75
26 gauge		8 00
28 gauge		8 50
Tinplate, IC charcoal, 20 x 28, box		10 75
" IX "		12 75
" IXX "		14 75
Ingot tin		33
Canada plate, 18 x 21 and 18 x 24		3 25
Sheet zinc, cask lots, 100 lb		7 50
Broken lots		8 00
Pig lead, 100 lb		6 00
Wrought pipe, black up to 2 inch	50 an	10 p.c.
" Over 2 inch		50 p.c.
Rope, sisal, 7-16 and larger		$11 00
" ⅜		11 50
" ¼ and 5-16		12 00
Manila, 7-16 and larger		14 00
" ⅜		14 50
" ¼ and 5-16		15 00
Solder		20
Cotton Rope, all sizes, lb.		17

Axes, chopping	$ 7 50 to 12 00
" double bitts	12 00 to 18 00
Screws, flat head, iron, bright	87½
Round " "	82½
Flat " brass	80
Round " "	75
Coach	57½ p.c.
Bolts, carriage	55 p.c.
Machine	55 p.c.
Tire	60 p.c
Sleigh shoe	65 p.c.
Plough	40 p.c.
Rivets, iron	50 p.c.
Copper, No. 8	35
Spades and shovels	40 p.c.
Harvest tools	50, and 10 p.c.
Axe handles, turned, s. g. hickory, doz.	$2 50
No. 1	1 50
No. 2	1 25
Octagon extra	1 75
No. 1	1 25
Files common	70, and 10 p.c.
Diamond	60
Ammunition, cartridges, Dominion R.F.	50 p.c.
Dominion, C.F., pistol	30 p.c.
" military	15 p.c.
American R.F.	30 p.c.
C.F. pistol	5 p.c.
C.F. military	10 p.c. advance.
Loaded shells:	
Eley's soft, 12 gauge black	16 50
chilled, 12 guage	18 00
soft, 10 guage	21 00
chilled, 10 guage	23 00
Shot, Ordinary, per 100 lb	6 25
Chilled	6 75
Powder, F.F., keg	4 75
F.F.G.	5 00
Tinware, pressed, retinned	75 and 2½ p.c.
" " plain	70 and 15 p.c.
Graniteware, according to quality	50 p.c.

PETROLEUM.

Water white American	25½c.
Prime white American	24c.
Water white Canadian	22c.
Prime white Canadian	21c.

PAINTS, OILS AND GLASS.

Turpentine, pure, in barrels	$ 61
Less than barrel lots	66
Linseed oil, raw	92
Boiled	95
Lubricating oils, Eldorado castor	25½
Eldorado engine	24½
Atlantic red	27½
Renown engine	41
Black oil	23½ to 25
Cylinder oil (according to grade)	55 to 74
Harness oil	61
Neatsfoot oil	$ 1 00
Steam refined oil	85
Sperm oil	1 50
Castor oil per lb.	11½
Glass, single glass, first break, 16 to 25 united inches	2 25
26 to 40 per 50 ft.	2 50
41 to 50 " 100 ft.	5 50
51 to 60 " " "	6 00
61 to 70 per 100-ft. boxes	[illegible]
Putty, in bladders, barrel lots per lb.	2½
kegs "	2¾
White lead, pure per cwt.	7 00
No 1 "	6 75
Prepared paints, pure liquid colors, according to shade and color, per gal.	$1.30 to $1.90

NOTES.

Miller, Morse & Co. announce their intention of erecting a large new warehous this season.

EXPORT REBATE ON GERMAN STEEL.

Under date of May 25, 1901, Consul-General Guenther, of Frankfort, quotes from The Cologne Gazette that the German iron trust, between May 1 and December 31, 1901, will pay a rebate of 10 marks ($2.40) per ton on all exported puddled steel that is bought from the trust. Proof of export must be produced, however, and the rebate will be paid on the basis of the actual raw iron contained in the article exported. It is added that sales to shipyards enjoying the benefit of free importations will be regarded as exports.

L. O. Pepin, of L. O. Pepin & Fils, general merchants, Arthabaskaville, Que., is dead.

ENCOURAGE SUMMER VISITORS.

Fishermen from the United States spend $5,000,000 a year in Canada--Merchants Get Nearly All of It.

THE following from E. T. D. Chambers, Quebec, written June 22, is a practical proof of the actual cash value of visitors to Canada. A great part of this money is spent among the merchants. There are hundreds of waters in every Province where the fishing can be developed under the direction of the local business men and to which every year an increasing number of wealthy men can be brought. They not only spend money on supplies for a couple of weeks' fishing, but many of them will buy land and erect houses. A number who have come to Canada this way have become permanent summer residents. Summer residents are good customers for the average merchant.

Even if they are not the familiar millionaires they spend more money on their vacations than they do at home. Some of them spend large sums. We know of one American who came to this country for a week's fishing about five years ago. Since then he has acquired the land about a lake and has expended over $500,000 in buildings and improvements besides his current living expenses, which must now amount to many thousands a year:

The annual spring migration of American anglers to Canada is at its height just now, and promises to exceed that of any preceding year. So rapid has been the growth of the sporting and tourist travel to the Dominion that new hotels have sprung up like mushrooms in many newly opened up sections of the country, and even Quebec, with its stationary population, has 50 per cent. more hotels running than it had five years ago.

American visitors probably

SPEND UPWARD OF $5,000,000

a year in Canada. The Ontario Commissioner of Fisheries says in his last report: "Rare now is the locality one may visit during the summer months where he will not find the summer visitor with creel slung over his shoulder and rod in hand, meandering along some chattering brook or rushing river, or seated in his buoyant canoe in search of the speckled beauties of their golden and green robed rival, the black bass. Scarcely a day passes from the beginning to the end of these months that whole trainloads of tourists are not carried to our holiday districts; and each year brings an increasing number, all in search of health, rest or recreation, to which the pastime of fishing is so valuable an adjunct. Some of these have beautiful cottages at various points, while others are accommodated at the numerous hotels and boarding houses or among the farmers."

RIGHTS IN QUEBEC.

In the Province of Quebec, where the system of leasing out the fishing rights of various lakes and rivers belonging to the Government is followed, nearly $40,000 is annually collected from this source alone. There are small rights for which not more than $5 or $10 a year are paid, while a club formed by H. W. de Forrest, of New York, pays $7,500 a year for a part of the salmon fishing in the Grand Cascapedia, and J. J. Hill, of St. Paul, pays $3,000 a year for that of the St. John river on the north shore of the Gulf of St. Lawrence. Half a million dollars would not be a very high estimate of the value of the salmon fisheries of the Restigouche, so beautifully described and illustrated in the rare folio of Dean Sage. The shares of this club are worth $7,500 each and the membership is fairly large, including W. K. Vanderbilt, Dr. Webb, the Rev. Dr. Rainsford and several of their friends. The club does not control nearly all the fishing of the Restigouche, yet it pays nearly $1,500 a year rental to the Government of Quebec for the smaller portion of its rights, and rents others from the Government of New Brunswick, besides considerable property which it has purchased outright from former riparian proprietors. And it is not so very long ago that the fishing of the entire river and its tributaries was let for $200 a year.

This is an earlier season for fishing than usual, and many salmon fishermen are already on the banks of their rivers, though it is seldom in other years that anglers go down to their preserves before June 8 or 10. Some of them went down this year to the Restigouche on June 1. Net fishermen report that the salmon are running into the rivers in large quantities, and there is every prospect of a most successful season. This is the more gratifying that last year was a kind of off year for angling of all kinds in Canada.

OUANANICHE AND SALMON.

Canadian anglers are watching with deep interest for the result of a remarkable experiment that has been under way for some time at Lake St. John, looking to the introduction of salmon into "the chosen waters where the ouananiche is waiting." Every fisherman who keeps abreast of the angling literature of the day, even if he has not formed the personal acquaintance of the leaping finny warrior of Lake St. John, knows that the ouananiche is a pure salmon of northern Quebec, dwarfed, of course, by the love of home and consequent abandonment of the ancestral habit of anadromy, but every whit as gamy, in proportion to his size, as the salmon of the sea that may boast descent from the same common stock. For the last two years thousands of young salmon have been hatched at Lake St. John, and planted, at a certain age, in some of its feeders, with the expectation that they will run out to the sea, after the manner of their kind, returning to the river whence they came, to reproduce their species.

A CHIEF ATTRACTION.

Ouananiche fishing has been for some years past one of the chief Canadian attractions for anglers. There are more reasons for this than the rare sport afforded by the fish itself. The Canadian environment of the ouananiche is of scarcely less interest to lovers of the contemplative man's recreation than the angling for the fish. It reaches away far into the uninhabited northern forests, where the fish ascends the feeders of the great lake for hundreds of miles. In his charming introduction to Walton's "Complete Angler," James Russell Lowell asks: "Where now would the fugitive from the espials of our modern life find a sanctuary which telegraph or telephone has not deflowered?" The answer is known to those who have fished the "chosen waters" in the northern part of the Canadian environment of the ouananiche, and explains one of their chieftest charms.

The best fishing of the season for ouananiche is now about opening in the waters of the Grand Discharge, where Lake St. John pours its surplus waters over a series of rocky obstructions down into the deep and dismal Saguenay. Here, in the eddying and foam covered pools, lying amid the roaring rapids, the ouananiche are now lying in wait for the insect life upon which they feed, and greedily take the angler's artificial lures. For the next few weeks there will be good fishing in these waters, and after July anglers will find the best sport with the ouananiche in that far northern forest sanctuary which neither telegraph or telephone has yet deflowered. In this paradise of the sportsman there are many good trout waters, which would be famous if their possibilities were more generally known to anglers. They are reached by several days' canoe journey from Lake St. John. Many American anglers are already in this territory for spring fishing and a few have returned home after making good catches, and will revisit Quebec later in the summer. Most visiting trout fishermen are now at the headquarters of the different American fish and game clubs along the line of railway that runs from Quebec to Lake St. John. Here the sport is at present all that can be desired, and will likely continue so until very hot weather sets in. Among a number of pretty specimens of brook trout brought down from some of these waters, and notably from Lake Edward, were several four and five-pound fish. In fact, the reports from all the trout waters in this Province indicate that the fishing is exceptionally good.

Several parties of New York anglers, principally members of the Laurentian Fish and Game Club, one of which was headed by Mr. and Mrs. J. Grant Lafarge, have returned home after enjoying excellent trout fishing upon the club preserves in the St. Maurice district of the Province of Quebec.

IN THE NEPIGON.

The far-famed Nepigon will next month attract, as usual, those ambitious disciples of the gentle Izaak who are anxious to feel a heavy trout at the end of their line and can spare the time to go to the north of Lake Superior to fish what is probably the most noted stream in the world for large specimens of Salvelinus fontinalis, fish of 11 pounds having been taken here on the fly. The river falls 313 feet in its course of 31 miles, and there is a railway station of the Canadian Pacific at its mouth. There are many beautiful sites for camps all along the river, trout from two to five pounds each are readily taken on any of the best pools, and white fish are plentiful and afford fine sport, rising eagerly at gnat flies.

LAKE TEMISCAMING.

Among the newest and most attractive of the many new northern canoe routes opened up for anglers and tourists in Canada perhaps the most delightful is that to Lake Temagaming, which is a large "haunted" lake, containing 1,400 islands and limpid waters teeming with game fish. Much of the time and trouble hitherto necessary to reach Temiscaming—on the way to beautiful Temiscaming—may now be saved by taking the new branch railway line running from Mattawa, on the upper Ottawa, to the foot of the lake, which thereafter continues on to the famous Kippewa country, where a bewildering variety of canoe routes is at the angler's choice. Most of the smaller streams which flow into the Mattawa are well stocked with brook trout, while almost all the larger waters passed on the way to Temagaming contain small mouth black bass up to five and even seven pounds in weight, lake trout, maskinonge, pike, doré or pickerel, sturgeon and white fish. In Temiscaming salmon trout have been taken up to 50 pounds in weight. From Temiscaming to Temagaming the route is by portage from Haileybury and by canoe up the Montreal river, and thence through the charming Lady Evelyn and Diamond Lakes. The return to Temiscaming can be varied by canoeing through the northeast arm and the Rabbit Lakes.

If a longer and still more northern trip be desired, the angler can reach James Bay in about three weeks from Temiscaming, and can vary the return trip by ascending the Moose river from Moose Fort to Missanabie, north of Lake Superior, on the line of the Canadian Pacific Railroad—a canoe journey of 15 to 18 days.

TO DEVELOP GRAPHITE AREAS.

A despatch from Ottawa says that a large company of American and Canadian capitalists, with Mr. H. P. H. Brumell, of Buckingham, as Canadian manager, has been organized to develop the graphite properties in Labelle county. Mr. Brumell, speaking to a reporter, said his company controls eight thousand acres of graphite lands in Labelle and five water-powers, capable of developing eight thousand horsepower. Graphites is used in crucibles, lubricants, paint, pencils, electrotyping, and in many other processes. Canadian graphite is said to be of superior quality for crucible purposes.

Miller, Morse & Co., hardware dealers, Winnipeg, contemplate erecting a handsome new warehouse on the corner of Adelaide street and McDermot avenue. Plans are to be prepared and the work proceeded with before long.

CURRENT MARKET QUOTATIONS.

June 28, 1901.

These prices are for such qualities and quantities as are usually ordered by retail dealers on the usual terms of credit, the lowest figures being for larger quantities and prompt pay. Large cash buyers can frequently make purchases at better prices. The Editor is anxious to be informed at once of any apparent errors in this list as the desire is to make it perfectly accurate.

METALS.

Tin.

Lamb and Flag and Straits—56 and 28 lb. ingots, per lb.	0 31½	0 32

Tinplates.

Charcoal Plates—Bright

		Per box
M.L.S., equal to Bradley.		
I.C., usual sizes	....	$6 50
I.X., "	....	8 00
I.X.X., "	....	9 50
Famous—		
I.C.	....	6 50
I.X.	....	8 00
I.X.X.	....	9 50
Raven & Vulture Grades—		
I.C., usual sizes	....	4 50
I.X., "	....	5 25
I.X.X "	....	6 00
I.X.X.X., "	....	6 75
D.C., 12½x17	....	4 00
D.X.	....	4 75
D.X.X	....	5 00

Coke Plates—Bright

Bessemer Steel—		
I.C., usual sizes	....	3 75
I.C., special sizes, base	....	4 00
20 x 28	....	7 75

Charcoal Plates—Terne

Dean or J. G. Grade—		
I.C., 20x28, 112 sheets	....	8 00
I.X., Terne Tin	....	10 00

Charcoal Tin Boiler Plates.

		Per lb.
Cookley Grade—		
X X., 14x56, 50 sheet bxs; " 14x60, "; " 14x65, "	....	0 06½

Tinned Sheets

72x30 up to 24 gauge	....	0 07½
" 26 "	....	0 08
" 28 "	....	0 08½

Iron and Steel.

Common Bar, per 100 lbs.	1 85	1 90
Refined " "	....	2 35
Horse Shoe Iron "	....	2 35
Hoop steel, 1½ to 3 in. base, extras for smaller sizes	....	3 00
Sleigh Shoe Steel " base	....	2 10
Tire Steel	2 30	2 50
Reeled Machinery	....	3 00
Toe Calk Steel	2 85	3 00
T. Firth & Co's tool steel, per lb	0 12½	0 13
Jessop's tool Steel	0 12½	0 13
Mo[illegible]'s tool steel	0 12½	0 13
Black Diamond and "B.C." tool steel	0 10	0 11
Drill Steel, per lb.	0 08	0 10

Boiler Tubes.

1½-inch	....	0 12½
2 "	....	0 13
2½ "	....	0 15
3 "	....	0 16
3½ "	....	0 20
4 "	....	0 25

Steel Boiler Plate.

¼ inch	2 50	2 60
3-16 inch	2 60	2 70
⅜ inch and thicker	2 50	2 60

Black Sheets.

	Com.	D.Fl.
18 gauge	2 75	3 00
20 gauge	2 75	3 00
22 to 24 "	2 75	3 25
26 "	2 85	....
28 "	3.00	3 50

Canada Plates.

All dull, 52 sheets	....	2 90
Half polished	....	3 00
All bright	....	3 50

Iron Pipe.

Black pipe—		
⅛ "	....	4 65
¼ inch	....	3 40
⅜ "	....	3 45
½ "	....	3 70
¾ "	....	3 85
1 "	....	5 40
1¼ "	....	7 35
1½ "	....	8 80
2 "	....	11 80
2½ "	....	22 00
3 "	....	27 60
3½ "	....	34 50
4 "	....	40 95
5 "	....	49 70
6 "	....	65 20
Galvanized pipe—		
½ inch	....	5 15
¾ "	....	5 50
1 "	....	7 93
1¼ "	....	10 80
1½ "	....	12 95
2 "	....	17 35

Galvanized Sheets.

	G C. Comet.		Amer.	Queen's Head.
16 gauge		....	4 00	3 75
18 to 24 gauge	4 00	3 85	4 25	4 00
26 "	4 25	4 10	4 25	4 25
28 "	4 50	4 35	4 40	4 50

Case lots 10 to 15c. less.
28 gauge American equals 26 gauge English.

Chain.

Proof Coil, 3-16 in., per 100 lb	....	....
" ¼ " "	8 00	8 50
" 5-16 " "	4 70	5 00
" ⅜ " "	4 05	4 50
" 7-16 " "	3 90	4 15
" ½ " "	3 70	4 10
" 9-16 " "	3 65	4 05
" ⅝ " "	3 55	3 90
" ¾ " "	3 60	4 00

Halter, kennel and post chains, 40 to 40 and 5 p.c.

Cow ties	40 p.c.
Tie-out chains	65 p.c.
Stall fixtures	35 p.c.
Trace chain	45 p.c.

Jack chain, iron, single and double, discount 35 p.c.
Jack chain, brass, single and double, discount 40 p.c.

Copper.

Ingot

English B. S., ton lots	....	0 17¾
Lake Superior	....	....

Bars.

Cut lengths round, ½ to ⅞ in.	0 23	0 25
" round and square 1 to 2 inches	0 23	0 25

Sheet.

Plain, 14 oz., and light, 16 oz., 14x48 and 14x60	0 24	0 24½
Plain, 14 oz., and light, 16 oz., irregular sizes	0 24½	0 25
Tinned copper sheets	....	0 26
Planished	....	0 32

Braziers (In sheets.)

4x6ft. 25 to 30 lbs. ea., per lb.	....	0 25
" 35 to 45 " "	....	0 24
" 50-lb. and above, "	....	0 23

Boiler and T. K. Pitts

Plain Tinned, per lb	....	0 28
Spun, per lb.	....	0 32

Brass.

Rod and Sheet, 14 to 30 gauge, 15 per cent.

Sheets, hard-rolled, 2x4	....	0 23
Tubing, base, per lb.	....	0 23

Zinc Spelter

Foreign, per lb	0 05½	0 06
Domestic "	...	...

Zinc Sheet.

5 cwt. casks	....	00 6½
Part casks	....	00 6¾

Lead.

Imported Pig, per lb	0 04¼	0 04½
Bar, 1 lb.	0 05½	0 05¾
Sheets, 2½ lbs. sq. ft., by	....	0 06¾
Sheets, 3 to 6 lbs., "	....	0 06

NOTE.—Cut sheets ½ cent per lb. extra. Pipe, by the roll, usual weights per yard, lists at 7c. per lb. and 30 p.c. dis. f.o.b. Toronto.

NOTE.—Cut lengths, net price, waste pipe 8-ft. lengths lists at 7½ cents.

Shot.

Common, $6.50 per 100 lb.; chilled, $7.00 per 100 lb.; buck, seal and ball, $7.50. Discount, 17½ p.c. Prices are f.o.b. Toronto, Hamilton, Montreal, St. John and Halifax. Terms 3 per cent. cash, freights equalized.

Soil Pipe and Fittings.

Discount, 60 and 10 per cent. on medium and extra heavy, and 60 per cent. on light.

Solder.

	Per lb.	Per lb.
Bar half-and-half, guarant'd	....	0 18¾
Bar half-and-half, commer'l	....	0 18
Refined	....	0 17½
Wiping	....	0 17

NOTE.—Prices of this graded according to quantity. The prices of other qualities of solder in the market indicated by private brands vary according to composition.

Antimony.

Cookson's, per lb.	0 10¼	0 11

White Lead.

	Per 100 lb.
Pure	6 37
No. 1 do	6 00
No. 2 do	5 62½
No. 3 do	5 15
No. 4 do	4 87½
Munro's Select Flake White	7 37½
Elephant and Decorators' Pure	7 12½
Brandram's B B. Genuine	8 00
" " Decorative	7 55
" " No. 1	6 85
" " No. 2	6 00

Red Lead.

Genuine, 560 lb. casks, per cwt.	....	$5 50
Genuine, 100 lb. kegs, per cwt.	....	5 75
No. 1, 560 lb. casks, per cwt.	....	5 25
No. 1, 100 lb. kegs, per cwt	....	5 00

White Zinc Paint.

Elephant Snow White	0 08	0 09
Pure White Zinc	0 08	0 09
No. 1	0 06	0 07½
No. 2	0 05	0 06½

Dry White Lead.

Pure, casks	5 75
Pure, kegs	6 25
No. 1, casks	5 50
No. 1, kegs	5 00

Prepared Paints.

In ¼, ½ and 1 gallon tins.

Pure, per gallon	....	1 25
Second qualities, per gallon	....	1 10
Barn (in bbls.)	0 75	0 85
The Sherwin-Williams Paints.	....	1 45
Canada Paint Co's Pure	....	1 25
Toronto Lead & Color Co's Pure	....	1 25
Sanderson Pearcy's Pure	....	1 20
Stewart & Wood's Champion Pure.		1 20

Colors in Oil.

25 lb. tins, Standard Quality.

Venetian Red, per lb.	0 05
Chrome Yellow	0 11
Golden Ochre	0 05
French "	0 05
Marine Black	0 09
" Green	0 09
Chrome "	0 08
French Imperial Green	0 09

Colors, Dry.

Yellow Ochre (J.C.) bbls.	1 35	1 40
Yellow Ochre (J.F.L.S.), bbls	....	2 75
Yellow Ochre (Royal)	1 10	1 15
Brussels Ochre	....	2 00
Venetian Red (best), per cwt.	1 80	1 90
English Oxides, per cwt.	3 00	3 25
American Oxides, per cwt.	1 75	2 00
Canadian Oxides, per cwt.	1 75	2 00
Super Magnetic Oxides, 93p.c.	2 00	2 25
Burnt Sienna, pure, per lb.	....	0 10
" Umber, " "	....	0 10
do Raw	....	0 09
Drop Black, pure	....	0 09
Chrome Yellows, pure	....	0 18
Chrome Greens, pure, per lb.	....	0 12
Golden Ochre	....	0 03½
Ultramarine Blue in 28-lb. boxes, per lb.	0 08	0 24
Fire Proof Mineral, per 100 lb	....	1 00
Genuine Eng. Litharge, per lb	....	0 07
Mortar Color, per 100 lb.	....	1 25
English Vermillion	....	0 80
Pure Indian Red, No. 45 lb.	....	0 80
Whiting, per 100 lb.	....	0 55

Blue Stone.

Casks, for spraying, per lb.	0 07
100-lb. lots, do. per lb.	0 08

Putty.

Bulk in bbls.	1 90
Bulk in less quantity	2 05
Bladders in bbls.	2 10
Bladders in kegs, boxes or loose	2 25
Bladders in 25-lb. tins	2 35
Bladders in 12½-lb. tins	2 65
Bladders in bulk or tins less than 100lb	2 90

Varnishes.

	In 5-gal. lots.)	Per gal.
Carriage, No. 1	2 90	3 30
" body	8 00	9 00
" rubbing	4 00	5 00
Gold Size, Japan	3 00	3 40
Brown Japan	2 40	2 80
Elastic Oak	2 90	3 30
Furniture, extra	2 40	2 80
" No. 1	1 60	2 00
Hard Oil Finish	2 70	3 10
Light Oil Finish	3 20	3 60
Demar	3 30	3 70
Shellac, white	4 40	4 80
" orange	4 00	4 40
Furniture Brown Japan	1 60	2 00
Black Japan	2 40	2 80
" " No. 1	1 60	2 00

The Imperial Varnish & Color Co's., Limited Elastilite Varnish 1 gal. can, each. $3.00.

Granatine Floor Finish per gal., $2.75.

Maple Leaf Coach Enamels; Size 1, $1 20; Size 2, 70c.; Size 3, 40c. each.

Castor Oil.

East India, in cases, per lb.	0 10	0 10½
" " small lots	0 10½	0 11

Cod Oil, Etc.

Cod Oil per gal	0 50	0 55
Pure Olive	....	1 20
" Neatsfoot	....	90

Glue.

Common	0 08½	0 09
French Medal	0 14	0 14½
Cabinet sheet	0 12	0 13
White, extra	0 18	0 20
Gelatine	0 22	0 30
Strip	0 18	0 20
Coopers	0 19	0 20
Huttner	....	0 18

HARDWARE.

Ammunition.

Cartridges.

B. B Cap Dom. 50 and 5 per cent.
Rim Fire Pistol, dis. 40 p. c., Amer.
Rim Fire Cartridges, Dom., 50 and 5 p. c.
Central Fire Pistol and Rifle, 10 p.c. Amer.
Central Fire Cartridges, pistol sizes, Dom. 30 per cent.
Central Fire Cartridges, Sporting and Military, Dom., 15 and 5 per cent.
Central Fire, Military and Sporting, Amer. add 5 p.c. to list. B.B. Caps, discount 40 per cent, Amer.
Loaded and empty Shells, "Trap" and "Dominion" grades, 25 per cent. Rival and Nitro, net list.
Brass Shot Shells, 55 per cent.
Primers, Dom., 30 per cent.

Wads	per lb.
Best thick white felt wadding, in ½-lb bags	1 00
Best thick brown or grey felt wads, in ½-lb. bags	70
Best thick white card wads, in boxes of 500 each, 12 and smaller gauges	0 99
Best thick white card wads, in boxes of 500 each, 10 gauge	0 35
Best thick white card wads, in boxes of 500 each, 8 gauge	0 55
Thin card wads, in boxes of 1,000 each, 12 and smaller gauges	0 10
Thin card wads, in boxes of 1,000 each, 10 gauge	0 25
Thin card wads in boxes of 1,000 each, 8 gauge	
Chemically prepared black edge grey cloth wads, in boxes of 250 each—	Per M.
11 and smaller gauge	0 60
9 and 10 gauges	0 70
7 and 8 gauges	0 90
5 and 6 gauges	1 10
Superior chemically prepared pink edge, best white cloth wads, in boxes of 250 each—	
11 and smaller gauge	1 15
9 and 10 gauges	1 40
7 and 8 gauges	1 65
5 and 6 gauges	1 90

Adzes.

Discount, 20 per cent.

Anvils.

Wright's, 80-lb. and over	0 10½
Hay Budden, 80-lb. and over	0 09½
Brook's, " " "	0 11½

Augers.

Gilmour's, discount 65 and 5 p.c. off list.

Axes.

Chopping Axes		
Single bit, per doz	5 50	10 00
Double bit, "	11 00	18 00
Bench Axes, 40 p.c.		
Broad Axes, 33⅓ per cent.		
Hunters' Axes	5 50	6 00
Boy's Axes	5 75	6 75
Splitting Axes	6 50	12 00
Handled Axes	7 00	10 00

Axle Grease.

Ordinary, per gross	5 75	6 00
Best quality	13 00	15 00

Bath Tubs.

Zinc 6 00
Copper, discount 15 p.c. off revised list

Baths.

Standard Enameled.

5½-inch rolled rim, 1st quality	25 00
" " " 2nd "	21 00

Anti-Friction Metal.

"Tandem" A, per lb.	0 27
" B, "	0 21
" C, "	0 11½
Magnolia Anti-Friction Metal, per lb.	0 25
Frictionless metal, per lb	0 23
SYRACUSE SMELTING WORKS.	
Aluminum, genuine	0 45
Dynamo	0 29
Special	0 25
Aluminum, 99 p.c. pure "Syracuse"	0 50

Bells.

Hand.

Brass, 60 per cent.
Nickel, 55 per cent.

Cow.

American make, discount 66⅔ per cent.
Canadian, discount 45 and 50 per cent.

Door.

Gongs, Sargant's	5 50	8 00
" Peterboro', discount 45 per cent.		
Farm.		
American, each	1 25	3 00
House.		
American, per lb.	0 35	0 40

Bellows.

Hand, per doz.	3 35	4 75
Moulders', per doz.	7 50	10 00

Blacksmiths', discount 40 per cent.

Belting.

Extra, 60 per cent.
Standard, 60 and 10 per cent.

Bits.

Auger.

Gilmour's, discount 60 and 5 per cent.
Rockford, 50 and 10 per cent.
Jennings' Gen., net list.

Car.

Gilmour's, 47½ to 50 per cent.

Expansive.

Clark's, 40 per cent

Gimlet.

Clark's, per doz.	0 65	0 90
Damond, Shell, per doz.	1 00	1 50
Nail and Spike, per gross	2 25	5 20

Blind and Bed Staples.

All sizes, per lb.	0 07½	0 12

Bolts and Nuts.

	Per cent.
Carriage Bolts, full square, Norway	65
" " full square	65
Common Carriage Bolts, all sizes	60
Machine Bolts, all sizes	60
Coach Screws	70
Sleigh Shoe Bolts	72½
Blank Bolts	60
Bolt Ends	62½
Plough Bolts	60
Nuts, square	4 c. off
Nuts, hexagon	4¼c. off
Tire Bolts	67½
Stove Bolts	67½
Stove rods, per lb.	5½ to 8c.

Boot Calks.

Small and medium, ball, per M.	4 25
Small heel, per M.	4 50

Bright Wire Goods.

Discount .... 62½ per cent.

Broilers.

Light, dis., 65 to 67½ per cent.
Reversible, dis., 65 to 67½ per cent.
Vegetable, per doz., dis. 37½ per cent.

Henis, No. 8, "		6 00
Henis, No. 9, "		7 00
Queen City "	7 50	0 00

Butchers' Cleavers.

German, per doz.	6 00	11 00
American, per doz.	12 00	20 00

Building Paper, Etc.

Plain building, per roll	0 30
Tarred lining, per roll	0 40
Tarred roofing, per 100 lb.	1 65
Coal Tar, per barrel	3 50
Pitch, per 100-lb.	0 85
Carpet felt, per ton	45 00

Bull Rings.

Copper, $2.00 for 2½ in. and $1.90 for 2 in.

Butts.

Wrought Brass, net revised list

Cast Iron.

Loose Pin, dis., 60 per cent.

Wrought Steel.

Fast Joint, dis. 65, 10 and 2½ per cent.
Loose Pin, dis. 65, 10 and 2½ per cent.
Berlin Bronzed, dis. 70, 70 and 5 per cent.

Gen. Bronzed, per pair	0 40	0 65

Carpet Stretchers.

American, per doz.	1 00	1 50
Bullard's, per doz.	6 50	....

Castors.

Bed, new list, dis. 55 to 57½ per cent.
Plate, dis. 52½ to 57½ per cent.

Cattle Leaders.

Nos. 31 and 32, per gross	50	9 50

Cement.

Canadian Portland	2 50	2 80
English "	....	3 00
Belgian "	2 50	2 75
Canadian hydraulic	1 25	1 50

Chalk.

Carpenters Colored, per gross	0 45	0 75
White lump, per cwt.	0 60	0 65
Red	0 05	0 06
Crayon, per gross	0 14	0 18

Chisels.

Socket, Framing and Firmer.

Broad's, dis. 70 per cent.
Warnock's, dis. 70 per cent.
P. S. & W. Extra 60, 10 and 5 p.c.

Churns.

Revolving Churns, metal frames—No. 0, $8—No. 1, $8.50—No. 2, $9.00—No. 3, $10.00—No. 4, $12.00—No. 5, $16.00 each. Ditto, wood frames—20c. each less than above. Discounts: Delivered from factories, 58 p.c.; from stock in Montreal, 58 p.c. Terms, 4 months or 3 p.c. cash in 30 days.

Clips.

Axle dis. 65 per cent.

Closets

Plain Ontario Syphon Jet	$16 00
Emb. Ontario Syphon Jet	17 00
Fittings net	1 00
Plain Teutonic Syphon Washout	10 00
Emb. Teutonic Syphon Washout	11 00
Fittings net	1 25
Low Down Teutonic, plain	16 00
" " " embossed	17 00
Plain Richelieu net	3 75
Emb. Richelieu net	4 00
Fittings net	1 25
Low Down Ont. Sy. Jet, plain net	19 50
" " " " " emb'd. net	20 50
Closet connection net	1 25
Basins, round, 14 in.	1 00
" oval, 17 x 14 in.	2 50
" " 19 x 15 in.	3 75

Discount 40 p.c., except on net figures.

Compasses, Dividers, Etc.

American, dis. 62½ to 65 per cent.

Cradles, Grain.

Canadian, dis. 25 to 33⅓ per cent.

Crosscut Saw Handles.

S. & D., No. 3, per pair	17½
" " 5, "	22½
" " 6, "	15
Boynton pattern "	20

Door Springs.

Torrey's Rod, per doz. (15 p.c.)		2 00
Coil, per doz.	0 88	1 60
English, per doz.	2 00	4 00

Draw Knives.

Coach and Wagon, dis. 50 and 10 per cent.
Carpenters, dis. 70 per cent.

Drills.

Hand and Breast.

Millar's Falls, per doz. net list.

DRILL BITS.

Morse, dis., 37½ to 40 per cent.
Standard dis. 50 and 5 to 55 per cent

Faucets.

Common, cork-lined, dis. 35 per cent.

ELBOWS. (Stovepipe.)

No. 1, per doz.	1 40
No. 2, per doz.	1 20

Bright, 20c. per doz. extra.

ESCUTCHEONS.

Discount, 45 per cent.

ESCUTCHEON PINS.

Iron, discount 40 per cent.

FACTORY MILK CANS.

Discount off revised list, 40 per cent.

FILES AND RASPS.

Great Western	75 and 5 per cent.
Disston	70 " 10 "
Arcade	75 " 5 "
Kearney & Foot	70 " 10 "
American	75 " 5 "
McClellan	70 " 5 "
Eagle	70 10 and 5 "
Nicholson	70 " 10 "
Heller	60 " 10 "

Royal & Keystone 80 p.c. and 80 and 10 p.c.
Black Diamond, 60 to 60 and 10 per cent.
Jowitt's, English list, 25 to 27½ per cent.

FORKS.

Hay, manure, etc., dis., 50 and 10 per cent. revised list.

GLASS—Window—Box Price.

Size United Inches.	Star Per 50 ft.	Star Per 100 ft	D. Diamond Per 50 ft.	D. Diamond Per 100 ft
Under 26	2 15	4 15	....	6 00
26 to 40	2 30	4 45	....	6 65
41 to 50	4 85	....		7 50
51 to 60	5 15	....		8 50
61 to 70	5 50	....		9 50
71 to 80	6 00	....		10 50
81 to 85	6 50	....		11 70
86 to 90	....	..		14 00
91 to 95	....	...		15 50
99 to 100	....	....		18 00

GAUGES

Marking, Mortise, Etc.

Stanley's dis. 50 to 55 per cent.

Wire Gauges.

Winn's, Nos. 26 to 33, each	1 65	2 40

HALTERS.

Rope, ⅜ per gross	....	....
" ½ "	....	9 00
" ⅝ to ¾	....	14 00
Leather, 1 in., per doz.	3 87½	4 00
" 1¼ in., "	5 15	5 20
Web, — per doz.	1 87	2 45

HAMMERS.

Nail

Maydole's, dis. 5 to 10 per cent. Can. dis. 25 to 27½ per cent.

Tack.		
Magnetic, per doz.	1 10	1 20
Sledge.		
Canadian, per lb.	0 07½	0 08½
Ball Pean.		
English and Can., per lb.	0 22	0 25

HANDLES.

Axe, per doz. net	1 50	2 00
Store door, per doz.	1 00	1 50

Fork.

C. & B., dis. 40 per cent. rev. list.

Hoe.

C. & B., dis. 40 per cent. rev. list.

Saw.		
American, per doz.	1 00	1 25
Plane.		
American, per gross	3 15	3 75

Hammer and Hatchet.

Canadian, 40 per cent.

Cross-Out Saws.

Canadian, per pair 0 13½

HANGERS.	doz. pairs.	
Steel barn door	5 85	6 00
Stearns, 4 inch	....	5 00
" 5 inch	....	6 50
Lane's covered—		
No. 11, 5-ft. run	....	8 40
No. 11½, 10-ft. run	....	10 80
No. 12, 10-ft. run	....	12 60
No. 14, 15-ft. run	....	21 00
Lane's O.N.T. track, per foot.	....	4½

HARVEST TOOLS.

Discount, 50 and 10 per cent.

HATCHETS.

Canadian, dis. 40 to 42½ per cent.

HINGES.

Blind, Parker's, dis. 50 and 10 to 60 per cent.

Heavy T and strap, 4-in., per lb.	0 06½
" " 5-in., "	0 06¼
" " 6-in., "	0 06
" " 8-in., "	0 05¾
" " 10-in., "	0 05½

Light T and strap, dis. 65 and 5 per cent.

Screw hook and hinge—	
6 to 12 in., per 100 lbs.	4 50
14 in. up, per 100 lbs.	[illegible]
	Per gro. pairs.
Spring	12 [illegible]

HOES.

Garden, Mortar, etc., dis. 50 and 10 p.c.

Planter, per doz.	4 00	4 50

HOLLOW WARE

Discount, ... 45 and 5 per cent

HOOKS.

Cast Iron.

Bird Cage, per doz.	0 50	1 10
Clothes Line, per doz.	0 27	0 63
Harness, per doz.	0 72	0 88
Hat and Coat, per gross	1 00	3 00
Chandelier, per doz.	0 50	1 00

Wrought Iron.

Wrought Hooks and Staples, Can., dis. 47½ per cent.

Wire.

Hat and Coat, discount 45 per cent.
Belt, per 1,000 .... 0 60
Screw, bright, dis. 55 per cent.

HORSE NAILS.

"C" brand 50 and 7½ p.c. off new list } Oval-
"M" brand 50, 10 and 5 per cent. } head
Countersunk. 60 per cent

HORSESHOES.

Iron Shoes.	F.O.B. Montreal. No. 2 and larger.	No. 1. and smaller.
Light, medium, and heavy.	3 50	3 75
Snow shoes	3 75	4 00
Steel Shoes.		
Light	3 60	3 85
Featherweight (all sizes)	4 85	4 85

F.O.B. Toronto, Hamilton, London and Guelph, 10c. per keg additional.
Toe weight steel shoes...... 6 70

JAPANNED WARE.

Discount, 45 and 5 p.c. off list, June 1899

ICE PICKS.

Star per doz. 3 00 3 25

KETTLES.

Brass spun, 7½ p.c. dis. off new list.
Copper, per lb. 0 30 0 50
American, 60 and 10 to 65 and 5 p.c.

KEYS.

Lock, Can., dis., 45 p.c.
Cabinet, trunk, and padlock, Am. per gross 60

KNOBS.

Door, japanned and N.P., per doz.	1 50	2 50
Bronze, Berlin, per doz.	2 75	3 25
Bronze Genuine, per doz.	6 00	9 00
Shutter, porcelain, F. & L. screw, per gross	1 30	4 00
White door knobs—per doz.		1 25

HAY KNIVES.

Discount, 50 and 10 per cent.

LAMP WICKS.

Discount, 60 per cent.

LANTERNS.

Cold Blast, per doz.	7 00
No. 3 "Wright's"	8 50
Ordinary, with O burner	4 00
Dashboard, cold blast	9 00
No. 0.	5 75

Japanning, 50c. per doz. extra.

LEMON SQUEEZERS.

Porcelain lined, ...per doz.	2 20	
Galvanized	1 87	5
King, wood	2 75	0
King, glass	4 00	[illegible]
All glass	1 20	[illegible]

LINES.

Fish, per gross	1 05	2 50
Chalk "	1 90	7 40

LOCKS.

Canadian, dis. 45 p.c.
Russel & Erwin, per doz. 3 00 3 25

Cabinet.

Eagle, dis. 30 p.c.

Padlock

English and Am., per doz.	50	6 00
Scandinavian, "	1 00	2 40

Eagle, dis. 20 to 25 p.c

MACHINE SCREWS. Iron and Brass.

Flat head discount 25 p.c
Round Head discount 20 p.c.

MALLETS.

Tinsmiths' per doz.	1 25	1
Carpenters', hickory, per doz	1 25	3
Lignum Vitae, per doz.	3 85	5 50
Caulking each	60	2 00

MATTOCKS.

Canadian, per doz. 5 60 6 50

MEAT CUTTERS.

American, dis. 25 to 30 p.c.
German, 15 per cent.

MILK CAN TRIMMINGS.

Discount, 25 per cent.

NAILS.

Quotations are:	Cut.	Wire.
2d and 3d	$3 35	$3 85
3d	3 00	3 52
4 and 5d	2 75	3 35
6 and 7d	2 65	3 20
8 and 9d	2 50	3 00
10 and 12d	2 45	2 95
16 and 20d	2 40	2 90
30, 40, 50 and 60d. (base)	2 35	2 85

Wire nails in carlots are $2.77½
Galvanizing 2c. per lb. net extra.
Steel Cut Nails 10c. extra.
Miscellaneous wire nails, dis. 70 and 10 p. c.
Coopers' nails, dis. 30 per cent.
Flour barrel nails, dis. 25 per cent

NAIL PULLERS.

German and American 1 85 3 50

NAIL SETS

Square, round, and octagon per gross 3 38 4 00
Diamond 12 00 15 00

NETTING.

Poultry, 55 per cent for McMullen's

OAKUM.

	Per 100 lb.
Navy	6 00
U. S. Navy	7 25

OIL.

Water White (U.S.)	0 16½
Prime White (U.S.)	0 15½
Water White (Can.)	0 15
Prime White (Can.)	0 14

OILERS.

McClary's Model galvan. oil can, with pump, 5 gal., per doz. 10 00
Zinc and tin, dis. 50, 50 and 10.
Copper, per doz. 1 25 3 50
Brass, " 1 50 3 50
Malleable, dis. 25 per cent.

GALVANIZED PAILS.

Dufferin pattern pails, dis. 45 p.c.
Flaring pattern, discount 45 per cent.
Galvanized washtubs, discount 45 per cent.

PIECED WARE.

Discount 40 per cent. off list, June, 1899.
10-qt. flaring sap buckets, dis. 45 p.c.
6, 10 and 14-qt. flaring pails, dis. 45 p.c.
Creamer cans, dis. 45 p.c.

PICKS.

Per doz 6 00 9 00

PICTURE NAILS.

Porcelain head, per gross.... 1 75 3 00
Brass head " 0 40 1 00

PICTURE WIRE.

Tin and gilt, discount 75 p c.

PLANES.

Wood, bench, Canadian dis. 50 per cent. American dis. 50.
Wood, fancy Canadian or American 7½ to 40 per cent.

PLANE IRONS.

English, per doz. 2 00 5 00

PLIERS AND NIPPERS.

Button's Genuine per doz pairs, dis. 37½ 40 p.c.
Button's Imitation, per doz.. 5 00 9 00
German, per doz. 0 60 2 60

PLUMBERS BRASS GOODS.

Compression work, discount, 60 per cent.
Fuller's work, discount 65 per cent.
Rough stops and stop and waste cocks, discount, 60 per cent.
Jenkins disk globe and angle valves, discount, 55 per cent.
Standard valves, discount, 60 per per cent.
Jenkins' radiator valves discount 55 per cent.
" " " standard, dis., 60 p.c.
Quick opening valves discount, 60 p.c.

No. 1 compression bath cock	2 00
No. 4 " " "	2 00
No. 7, Fuller's	2 50
No 4½, "	3 00

POWDER.

Velox Smokeless Shotgun Powder.
100 lb. or less 0 85
1,000 lb. or more 0 80
Net 30 days.

PRESSED SPIKES.

Discount 20 to 25 per cent.

PULLEYS.

Hothouse, per doz.	0 55	1
Axle	0 22	0
Screw	0 27	1 00
Awning	0 35	2 50

PUMPS.

Canadian cistern 1 80 3 60
Canadian pitcher spout 1 40 2 10

PUNCHES.

Saddlers', per doz.	1	1
Conductors', "	9	15
Tinners' solid, per set	0 00	0 85
" hollow, per inch	0 00	1 00

RANGE BOILERS.

Galvanized,	3 gallons		7 60
"	35 "		8 25
"	40 "		9 50
Copper,	30 "		22 00
"	35 "		26 00
"	40 "		30 00

Discount off Copper Boilers 10 per cent.

RAKES.

Cast steel and malleable, 50, 10 and 5 p.c.
Wood, 25 per cent.

RAZORS.

	per doz.	
Elliot's	4 00	18 00
Geo. Butler & Co.'s,	4 00	18 00
Boker's	7 50	11 00
" King Cutter	12 50	50 00
Wade & Butcher's	3 60	10 00
Theile & Quack's	7 00	12 00

REAPING HOOKS.

Discount, 50 and 10 per cent.

REGISTERS.

Discount, ... 40 per cent.

RIVETS AND BURRS.

Iron Rivets, black and tinned, discount 60 and 10 per cent.
Iron Burrs, discount 55 per cent.
Extras on Iron Rivets in 1-lb. cartons, ½c. per lb.
Extras on Iron Rivets in ½-lb. cartons, 1c. per lb.
Copper Rivets & Burrs, 35 and 5 p.c. dis. and cartons, 1c. per lb. extra, net.
Extras on Tinned or Coppered Rivets ½-lb. cartons, 1c. per lb.

RIVET SETS

Canadian, dis. 35 to 37½ per cent.

ROPE ETC.

	Sisal.	Manila.
7-16 in. and larger, per lb	10	13½
⅜ in.	11	14½
¼ and 5-16 in.		15½
Cotton, 3-16 inch and larger		16
" 5-32 inch		21
" ⅛ inch		22½
Russia Deep Sea		15½
Jute		8
Lath Yarn		9½
New Zealand Rope		10

RULES.

Boxwood, dis. 75 and 10 p.c.
Ivory, dis. 37½ to 40 p.c.

SAD IRONS. per set.

Mrs. Potts, No. 55, polished.. ...... 62½
" No. 50, nickle-plated 67½

SAND AND EMERY PAPER.

Dominion Flint Paper, 47½ per cent.
B & A. sand, 40 and 5 per cent.
Emery, 40 per cent.
Garnet (Rurton's), 5 to 10 p.c. advance on list.

SAP SPOUTS.

Bronzed iron with hooks, per doz... 9 50

SAWS.

Hand Disston's, dis. 12½ p.c.
S. & D., 40 per cent.
Crosscut, Disston's, per ft.... 35 0 55
S. & D., dis. 35 p.c. on Nos. 2 and 3.
Hack, complete, each 0 75 2 75
" frame only 0 75

SASH WEIGHTS.

Sectional, per 100 lbs. 2 75 3 00
Solid, " 2 00 2 25

SASH CORD.

Per lb. 0 23 0 30

SAW SETS.

"Lincoln," per doz. 6 50

SCALES.

Standard, 45 p.c.
Champion, 65 p.c.
Spring Balances, 10 p.c.
Fairbanks Standard, 35 p.c.
" Dominion, 55 p.c.
" Richelieu, 55 p.c.

SCREW DRIVERS.

Sargent's per doz. 0 65 1 00

SCREWS

Wood, F. H., bright and steel, 87½ and 10 p.c.
Wood R. H., " dis. 82½ and 10 p.c.
" F. H., brass, dis. 80 and 10 p.c.
Wood, R. H., " dis. 75 and 10 p.c.
" F.H., bronze, dis. 75 p.c.
" R.H. " 70 p.c.
Drive Screws, 87½ and 10 per cent.
Bench, wood, per doz. 3 25 4 00
" iron. " 4 25 5 75
Set, Case hardened, 60 per cent.
Square Cap, 50 and 5 per cent.
Hexagon Cap, 45 per cent.

SCYTHES.

Per doz, net 9 00

SCYTHE SNATHS.

Canadian, dis. 45 p.c.

SHEARS.

Bailey Cutlery Co., full nickeled, dis. 60 p.c.
Seymour's, dis. 50 and 10 p.c.

SHOVELS AND SPADES.

Canadian, dis. 40 and 5 per cent.

SINKS.

Steel and galvanized, discount 45 per cent.

SNAPS.

Harness, German, dis. 25 p.c.
Lock, Andrews' 4 50 11 50

SOLDERING IRONS.

1, 1½ lb., per lb. 0 37
2 lb. or over, per lb. 0 34

SQUARES.

Iron, No. 493, per doz 2 40 2 55
" No. 494, " 3 25 3 40
Steel, dis. 60, 10 and 5 p.c.
Try and bevel, dis. 50 to 52½ p.c.

STAMPED WARE.

Plain, dis., 75 and 12½ p.c. off revised list
Retinned, dis., 75 p.c. off revised list.

STAPLES.

Galvanized 3 50 4 00
Plain 3 15 3 75
Coopers', discount 45 per cent.
Poultry netting staples, 40 per cent.

STOCKS AND DIES.

American dis. 25 p.c.

STONE.

	Per lb.	
Washita	0 28	0 60
Hindostan	0 06	0 07
" slip,	0 09	0 09
Labrador	....	0 13
" Axe	....	0 15
Turkey	....	0 50
Arkansas	0 00	1 50
Water-of-Ayr	0 00	0 10
Scythe, per gross	3 50	5 00
Grind 2 in. 40 to 200 lb. per ton	....	25 00
" under 40 lb. "	....	28 00
Grind. under 2 in. thick "	....	20 00

STOVE PIPES.

5 and 6 inch Per 100 lengths 7 00
7 inch " " 7 50

ENAMELINE STOVE POLISH.

No. 4—3 dozen in case, net cash $4 80
No. 6—3 dozen in case, " 8 40

TACKS BRADS, ETC.

	Per cent
Strawberry box tacks, bulk	75 & 10
Cheese-box tacks, blued	80 & 12½
Trunk tacks, black and tinned	85
Carpet tacks, blued	80 & 15
" " tinned	80 & 20
" " (in kegs)	40
Cut tacks, blued, in dozens only	80
" ½ weights	60
Swedes, cut tacks, blued and tinned—	
In bulk	80 & 10
In dozens	75
Swedes, upholsterers', bulk	85, 12½ & 12½
" brush, blued & tinned, bulk	70
" gimp, blued, tinned and japanned	75 & 12½
Zinc tacks	35
Leather carpet tacks	55
Copper tacks	50
Copper nails	52½
Trunk nails, black	65 and 5
Trunk nails, tinned	65 and 10
Clout nails, blued	65 and 5
Chair nails	35
Patent brads	40
Fine finishing	40
Picture frame points	10
Lining tacks, in papers	10

Lining tacks, in bulk		15
" " solid heads, in bulk		75
Saddle nails in papers		10
" " in bulk		15
Tufting buttons, 22 line, in dozens only		60
Tin capped trunk nails		15
Zinc glazier's points		5
Double pointed tacks, papers		90 and 10
" " " bulk		40

TAPE LINES.

English, ass skin, per doz	2 75	5 00
English, Patent Leather	5 50	9 75
Chesterman's each	0 90	2 85
" steel, each	0 80	8 00

THERMOMETERS.

Tin case and dairy, dis. 75 to 75 and 10 p.c.

TRAPS. (Steel.)

Game, Newhouse, dis. 21 p.c.
Game, H. & N., P. S. & W., 65 p.c.
Game, steel, 72½, 75 p.c.

TROWELS.

Disston's discount 10 per cent.		
German, per doz	4 75	6 00
S. & D., discount 35 per cent.		

TWINES.

Bag, Russian, per lb	....	0
Wrapping, cotton, 3-ply		0
" " 4-ply	....	0
Mattress, per lb	0 33	0 27
Staging, "	0 27	0 [illegible]

VISES.

Wright's	....	0 13½
Brook's	....	0 12½
Pipe Vise, Hinge, No. 1	....	3 57
" " " No 2	....	5 50
Saw Vise	4 50	9 00

ENAMELLED WARE.

White, Princess, Turquoise, Blue and White, discount 50 per cent.
Diamond, Famous, Premier, 50 and 10 p.c.
Granite or Pearl, Imperial, Crescent, 50, 10 and 10 per cent.

WIRE.

Brass wire, 50 to 50 and 2½ per cent. off the list.
Copper wire, 45 and 10 per cent. net cash 30 days, f.o.b. factory.
Smooth Steel Wire, is quoted at the following net selling prices:

No. 6 to 8 gauge	$2 90
" 9 "	2 80
" 10 "	2 87
" 11 "	2 90
" 12 "	2 93
" 13 "	3 15
" 14 "	3 37
" 15 "	3 50
" 16 "	3 65

Other sizes of plain wire outside of Nos. 9, 10, 11, 12 and 13, and other varieties of plain wire remain at $2.80 base with extras as before. The prices for Nos. 9 to 13 include the charge of 10c. for oiling. Extras net per 100 lb.: Coppered wire, 60c.—tinned wire, $2—oiling, 10c.—special hay-bailing wire, 30c.—spring wire, $1—best steel wire, 75c.—bright soft drawn, 15c.—in 50 and 100-lb. bundles net, 10c.—in 25-lb. bundles net 15c.—packed in casks or cases, 15c.—bagging or papering, 10c.

Fine Steel Wire, dis. 17½ per cent. List of extras: In 100-lb. lots: No. 17, $5—No. 18, $5.50—No. 19, $6—No. 20, $6.65—No. 21, $7—No. 22, $7.30—No. 23, 7.65—No. 24, $8—No. 25, $9—No. 26, $9.50—No. 27, $10—No. 28, $11 No. 29, $12—No. 30, $13—No. 31, $14—No. 32, $15 No. 33, $16—No. 34, $17. Extras net—tinned wire, Nos. 17-25, $2—Nos. 26-31 $4—Nos. 32-34, $6. Coppered, 5c.—oiling, 10c.—in 25-lb. bundles, 15c.—in 5 and 10-lb. bundles, 25c.—in 1-lb. hanks, 50c.—in ½-lb. hanks, 75c.—in ¼-lb. hanks, $1—packed in casks or cases, 15c.—bagging or papering, 10c.

Galvanized Wire, per 100 lb.—Nos. 6, 7, 8, $3 50 to $3 85—No. 9, $2.85 to $3.15—No. 10 $3.60 to $3.95—No. 11, $3.70 to $4.10—No 12, $3 to $3.30—No. 13, $3.10 to $3 40—No. 14, $4.10 to $4.50—No. 15, $4.60 to $5.05—No. 16, $4.85 to $5 35. Base sizes, Nos. 5 to 9, $2.57½ f.o.b. Cleveland.

Clothes Line Wire, solid 7 strand, No. 17, $4.25; No. 18, $2.65; No. 19, $2.35, f.o.b Hamilton, Toronto, Montreal.

WIRE FENCING.

	F.O.B. Toronto
Galvanized barb	3 05
Galvanized, plain twist	3 05

Galvanized barb, f.o.b. Cleveland, $2.82½ in less than carlots, and $2.70 in carlots.

WIRE CLOTH.

Painted Screen, per 100 sq. ft., net.. 1 35

WASTE COTTON.

	per lb.
Colored	4½ to 5
White, according to quality	6½ to 7½

500-lb bale lots shaded.

WRENCHES.

Acme, 35 to 37½ per cent.
Agricultural, 60 p.c.
Coe's Genuine, dis. 20 to 25 p.c.

Towers' Engineer, each	2 00	7 00
" S., per doz	5 80	6 00
G. & K.'s Pipe, per doz	....	3 40
Burrell's Pipe, each	....	3 00
Pocket, per doz	0 25	2 90

WRINGERS.

Leader	per doz.	$30 0	31 00
Royal Canadian	"	26 00	28 00
Royal American	"	26 00	28 00
Sampson	"	....	30 00

Terms 4 months, or 3 p.c. 30 days.

WROUGHT IRON WASHERS.

Canadian make, discount, 40 and 5 per cent